BirdLife Conservation Series No. 8

IMPORTANT BIRD AREAS IN EUROPE

Priority sites for conservation

Edited by

Melanie F. Heath and Michael I. Evans

with D. G. Hoccom, A. J. Payne and N. B. Peet

Volume 2: Southern Europe

with major sponsorship from

landbouw, natuurbeheer en visserij

Vogelbescherming NEDERLAND

Royal Society for the Protection of Birds
The BirdLife Partner in the United Kingdom and the main sponsor.

Ministry of Agriculture, Nature Management and Fisheries, Directorate for Nature Management, through the PIN/MATRA Funds of the Ministry of Foreign Affairs, The Netherlands.

Vogelbescherming Nederland
The BirdLife Partner in the Netherlands.

BirdLife INTERNATIONAL

Together for birds and people

Recommended citation
Heath, M. F. and Evans, M. I., eds. (2000) *Important Bird Areas in Europe: Priority sites for conservation.* 2 vols. Cambridge, UK: BirdLife International (BirdLife Conservation Series No. 8).

Where one volume used, indicate title as:
Heath, M. F. and Evans, M. I., eds. (2000) *Important Bird Areas in Europe: Priority sites for conservation.* 2: Southern Europe. Cambridge, UK: BirdLife International (BirdLife Conservation Series No. 8).

For a country chapter, cite as in this example:
Casale, F. and Bino, T. (2000) Albania. Pp. 67–76 in M. F. Heath and M. I. Evans, eds. *Important Bird Areas in Europe: Priority sites for conservation.* 2: Southern Europe. Cambridge, UK: BirdLife International (BirdLife Conservation Series No. 8).

Wellbrook Court, Girton Road, Cambridge, CB3 0NA, United Kingdom
Tel: +44 1223 277318 Fax: +44 1223 277200 email: birdlife@birdlife.org.uk Internet: www.birdlife.net

BirdLife International is a UK-registered charity

ISBN 0 946888 34 5 (Volume 1, softback)
0 946888 35 3 (Volume 2, softback)
0 946888 36 1 (2 Volume set, softback)
0 946888 37 X (2 Volume set, hardback)

British Library-in-Publication Data
A catalogue record for this book is available from the British Library

First published 2000 by BirdLife International

Designed and produced by the Nature Conservation Bureau Limited,
36 Kingfisher Court, Hambridge Road, Newbury, Berkshire RG14 5SJ, United Kingdom

Printed and bound in Great Britain by Butler and Tanner Ltd., Frome and London

Available from: Natural History Book Service, 2–3 Wills Road, Totnes, Devon, TQ9 5XN, United Kingdom
Tel: +44 1803 865913 Fax: +44 1803 865280 email: sales@nhbs.co.uk Internet: www.nhbs.com

B

VOLUME 2: SOUTHERN EUROPE

AUTHORS OF COUNTRY CHAPTERS

ALBANIA

Fabio Casale and Taulant Bino
Lega Italiana Protezione Uccelli (LIPU)*;
Albanian Society for the Protection of Birds and Mammals (ASPBM)*

ANDORRA

Marie-Jo Dubourg-Savage
Associació per a la Defensa de la Natura (ADN)*

ARMENIA

Michael Wilson

AZERBAIJAN

Michael Patrikeev and Michael Wilson

BOSNIA AND HERZEGOVINA

Based on data from Grimmett and Jones (1989)

BULGARIA

Irina Kostadinova
Bulgarian Society for the Protection of Birds (BSPB)*

CROATIA

Based on data from Grimmett and Jones (1989)

CYPRUS

Melis Charalambides

FRANCE

Bernard Deceuninck, Xavier Rufray and Philippe Jourde
Ligue Pour La Protection des Oiseaux (LPO)*

GEORGIA

Michael Wilson

GIBRALTAR

John Cortes
Gibraltar Ornithological and Natural History Society (GONHS)*

GREECE

Stratis Bourdakis and Stella Vareltzidou
Hellenic Ornithological Society (HOS)*

HUNGARY

Szabolcs Nagy
MME/BirdLife Hungary*

ITALY

Fabio Casale, Umberto Gallo-Orsi and Vincenzo Rizzi
Lega Italiana Protezione Uccelli (LIPU)*

FORMER YUGOSLAV REPUBLIC OF MACEDONIA

Based on data from Grimmett and Jones (1989)

MALTA

Ian Mifsud
BirdLife Malta*

MOLDOVA

A. Munteanu and N. Zubcov
Academy of Sciences of Moldova

PORTUGAL

Rui Rufino and Renato Neves
Instituto da Conservação da Natureza

AZORES

Luís R. Monteiro
University of Azores

MADEIRA

Francis Zino, Manuel José Biscoito and Paulo Oliveira

ROMANIA

Dan Munteanu
Romanian Ornithological Society (SOR)*

SLOVENIA

Slavko Polak
Bird Watching and Bird Study Association of Slovenia (DOPPS)*

SPAIN (INCLUDING THE CANARY ISLANDS)

Carlota Viada
Sociedad Española de Ornitología (SEO/BirdLife)*

TURKEY

Gernant Magnin, Güven Eken and Murat Yarar
Doğal Hayati Koruma Derneği (DHKD)*

UKRAINE

Alexander Mikityuk
Ukrainian Union for Bird Conservation (UTOP)*

FEDERAL REPUBLIC OF YUGOSLAVIA

Slobodan Puzović and Bratislav Grubač
Institute for Protection of Nature of Serbia

* Organization is part of the BirdLife International European Partnership (as of December 1999)

CONTENTS

vii **Foreword s**

ix **Acknowledgements**

xi **Summary**

1 **Introduction**
1 Europe's environment
1 The need for bird conservation in Europe
2 Prioritization of conservation action
2 The conservation of sites for birds—Important Bird Areas
3 Important Bird Areas in Europe

5 **Data collection**
5 Geographical definition of Europe
5 The BirdLife International Partnership
6 Sources of data
8 Types of data collected
9 Tools for data collection
9 Data-collection procedure

11 **Identifying Important Bird Areas**
11 Why apply IBA criteria?
11 The categories of IBA criteria
12 The application of IBA criteria
13 Defining the boundaries of an IBA
13 Detailed definitions of IBA criteria
18 How do IBA criteria relate to the identification of Special Protection Areas under the EC Birds Directive?
19 How do IBA criteria relate to the identification of Ramsar Sites under the Ramsar Convention?

21 **Data presentation**
21 National overview
26 Site accounts

29 **Overview of results**
29 Why are IBAs important and has a complete set of sites been found?
52 Land-uses and threats at IBAs
59 How well protected are IBAs?
62 What should be done next?

Country chapters: inventory of Important Bird Areas by country

67 **Albania**

77 **Andorra**

81 **Armenia**

87 **Azerbaijan**

109 **Bosnia and Herzegovina**

111 **Bulgaria**

137 **Croatia**

147 **Cyprus**

157 **France**

247 **Georgia**

257 **Gibraltar**

261 **Greece**

335 **Hungary**

357 **Italy**

431 **Former Yugoslav Republic of Macedonia**

435 **Malta**

439 **Moldova**

445 **Portugal**

463 **Azores**

473 **Madeira**

481 **Romania**

503 **Slovenia**

515 **Spain (including the Canary Islands)**

651 **Turkey**

691 **Ukraine**

725 **Federal Republic of Yugoslavia**

747 **Appendices**
747 **Appendix 1**. Descriptions of international agreements relevant to the protection and management of sites for nature conservation in Europe.
751 **Appendix 2a.** The 515 bird species occurring regularly in Europe with the Important Bird Area criteria (at the global and European levels) that they can potentially fulfil and the numerical threshold(s) if applicable.
765 **Appendix 2b.** Bird species occurring regularly in the European Union that can fulfil Important Bird Area criteria (at the European Union level), and the numerical threshold(s) if applicable.
770 **Appendix 3**. The classification of habitats, land-uses and threats at Important Bird Areas in Europe.
776 **Appendix 4.** A comparison of Important Bird Area criteria categories applied in this publication with those used in the first pan-European inventory (Grimmett and Jones 1989) and European Union study (Grimmett and Gammell 1989).
778 **Appendix 5.** The regions of EU countries used in the application of Important Bird Area criteria for species threatened at the European Union level (criterion C6).
779 **Appendix 6.** The total number and total area of Important Bird Areas by country and the reasons for importance of IBAs by country, indicated by the number of IBAs fulfilling each criterion.
780 **Appendix 7.** Transboundary Important Bird Areas.
782 **Appendix 8.** The number of Important Bird Areas in each country that overlap with Ramsar Sites, World Heritage Sites, Biosphere Reserves and Special Protection Areas.

783 **Index to sites**

FOREWORDS

ONE of the greatest strengths of the Birdlife International Partnership is its ability to focus on priorities for species, sites, habitats and people. There is no better example of putting this focus into practice than BirdLife's Important Bird Area (IBA) programme, now well on its way to identifying and documenting over 20,000 IBAs worldwide.

The beauty of the IBA programme is its simplicity. It identifies important sites for the conservation of the world's birds by applying locally a set of objective criteria to an internationally agreed global standard. This provides the basis for local, national, regional and global action and advocacy. It also generates a conservation agenda from which local and national institutions can strengthen their own capacity and be linked, through IBA programmes across the world, to like-minded institutions.

When the first pan-European inventory of IBAs was launched ten years ago, the designation Important Bird Areas was unknown. Now, IBAs are recognised worldwide and they are rapidly becoming a common and increasingly valuable currency of site conservation. In some places IBAs now have legal status. Both the World Bank and the Global Environment Facility recognise IBAs in their own strategies and action plans, while the European Union has used them as a basis for legal judgements. Communities adjacent to IBAs and dependent on them for their own livelihoods value these sites and play a key role in their conservation. I can think of few concepts in conservation and sustainable development that now have such widespread support and ownership. IBAs unite local people in Burkina Faso, the Philippines, Hungary, Jordan, New York State, Kenya, Spain, Palestine, Panama and many more. They are recognised by inter-governmental bodies and international treaties, such as the Biodiversity and Ramsar Conventions.

This remarkable volume provides detailed information on 3,619 IBAs spread across every country in Europe. It is a unique and powerful source of information that profoundly improves our ability to enable people to conserve birds, biodiversity and the wider environment throughout the region.

Her Majesty Queen Noor of Jordan,
Honorary President of BirdLife International

SINCE its adoption in 1979 the Wild Birds Directive has provided a strong legal basis for the protection of habitats of wild bird species throughout the European Union, especially through the designation of Special Protection Areas (SPAs). The objective is to create a coherent network of protected areas which meets the protection requirements of endangered and migratory bird species and to preserve our common heritage.

The identification and delimitation of sites under this directive is exclusively a scientific exercise. With this in mind, ornithological criteria were elaborated as far back as 1981 when the first inventory of important bird areas of the then European Community was prepared. During the 1980s the European Commission set up a working group, which led to the further identification of Community-wide criteria for the selection of SPAs. This resulted in the preparation of the 1989 inventory of Important Bird Areas (IBAs) in Europe by the forerunner to BirdLife International, the International Council for Bird Preservation (ICBP), with contributions by experts from the Member States.

The 1989 IBA review has proven to be a key scientific reference for the selection of sites to be protected under the Birds Directive. The European Union Court of Justice has concluded that it represents a list of sites of great conservation importance for the conservation of wild birds.

However, more than 10 years have passed since its publication and there have been considerable advances in knowledge on the numbers and distribution of wild birds. The publication of a new inventory of Important Bird Areas is therefore most welcome. I would like to congratulate BirdLife International for the clarity and quality of this work. I am convinced that it will be a standard reference for many years to come.

As most EU Member States have still to actively complete their networks of SPAs it should be a key tool in helping them to fulfil this key requirement of the Birds Directive. Given that it covers all European countries it will also be a valuable reference for those applicants to the European Union as they prepare to take on the nature conservation obligations of membership.

Margot Wallström,
European Commissioner for the Environment

FOLLOWING the publication of the first edition of *Important Bird Areas in Europe* in 1989, the year the iron curtain fell, pan-European cooperation in nature conservation issues has improved enormously. The impressive quantity of new data in this second inventory of *Important Bird Areas of Europe* illustrates just how effectively BirdLife International and its network of national bird conservation societies responded to the new social and political situations. More than 1,000 new IBAs have been designated, mostly in Eastern Europe, and there is valuable new information about existing IBAs. All this was made possible by extensive networks of highly dedicated and scientifically skilled volunteers who undertook field research for this book.

Since 1989, many new NGOs have been established in Europe as a result of BirdLife International programmes. The Dutch Government has supported bird conservation organizations in Central and Eastern Europe—the Russian Federation and the Ukraine in particular—by supporting the collection of field data for this edition of *Important Bird Areas of Europe* and contributed towards its publication. High standards have been applied to submitted data so that the designation Important Bird Area (IBA) is given only to those areas which truly have great ornithological value. In other words, those places which deserve to be protected by existing international conservation frameworks such as Ramsar, the Bonn Convention, the Bern Convention, the EC Birds Directive and the Africa Eurasia Waterbird Agreement.

The work of BirdLife International was of crucial importance in the European Court of Justice case concerning the designation of Special Protection Areas in the Netherlands. Vogelbescherming Nederland, the BirdLife Partner in the Netherlands, had drafted a list of IBAs in the Netherlands, which enabled a quick, efficient designation of a satisfactory number of SPAs, which will be followed by more designations in the future.

The Dutch Government congratulates BirdLife International on the publication of this second edition of *Important Bird Areas of Europe*; without doubt it is a major achievement. We look forward to continuing close cooperation with BirdLife International in conservation projects around the world, particularly with the European Division that is now situated in Wageningen.

G.H. Faber, State Secretary for Agriculture,
Nature Management and Fisheries,
Dutch Government

IT is an honour for me to greet the dear Reader on behalf of the comparatively large family of conservationists in our small country, Hungary.

Biogeographically, Hungary ranks among the richest countries in Europe. Currently, approximately 42,000 animal and 3,000 plant species are known from Hungary, but it is an alarming fact that more than half (62%) of the natural vegetation associations are threatened to some degree and need protection. It was a major step forward in the 1980s and 1990s when it was recognised that it is hopeless to try to protect species without the protection of their habitats. In Hungary, 855 animal and 515 plant species are protected by law. Among them, birds enjoy a particularly favourable status with 278 protected species.

The majority of Important Bird Areas (IBAs) are already protected in Hungary, but in a country that has long been inhabited, urbanised and agriculturally cultivated, isolated protected areas are in a precarious situation and may easily be doomed without the implementation of special protection measures. The total area of Ramsar sites, constituting a major part of IBAs, has increased by 30% during the 1990s and today 19 sites with a total area of 149,000 ha are protected under the Ramsar Convention. The Hungarian BirdLife Partner, the Hungarian Ornithological and Nature Conservation Society, played a key role in the designation of IBAs in Hungary and has established excellent relations with the Hungarian Authority for Nature Conservation, Ministry of the Environment. It is hoped that the future of Hungarian IBAs will be ensured, not only by the declaration of protected status, but also by new measures to support nature-friendly farming.

Many thanks to BirdLife International for their excellent, inspirational and far-reaching programmes, and for creating this remarkable manual. It is an honour that the 106-year-old Hungarian Ornithological Institute and the Hungarian Authority for Nature Conservation could, even if modestly, contribute to this work.

Dr János Tardy PhD, Deputy Secretary of State,
Head of the Authority for Nature Conservation, Hungary

BEING a mayor of a small town like Třeboň, located at the very heart of Třeboňsko Important Bird Area (IBA) in the unique wetland habitat of Třeboň Basin Biosphere Reserve and Protected Landscape Area, is both a professional task and a public service. It allows intimate daily contact with the surrounding landscape. For example, each time I walk to my office or show guests around my home town or around our ancient fish ponds that give the town its specific character.

Třeboň is the kind of place where one can experience the harmony that still exists between people and nature. It is a town where a glance out of the office window will often provide the magnificent spectacle of migrating flocks of Greylag Geese, or Grey Herons heading towards their nearby colony or even the majestic silhouette of a White-tailed Eagle soaring high in the clouds. I am proud of my town and of all the natural treasures in its vicinity and I am aware of my responsibility for preserving them for future generations. I also appreciate the economic importance of international designations, such as the recognition of Třeboň as an IBA. This attracts thousands of national and international visitors to Třeboň – naturalists, field scientists, participants in ornithological conferences, birdwatchers and nature lovers. All these activities help to make the local economy more viable. In this sense, the IBA status helps me in two ways – to help secure better protection of our birds and habitats and to promote environmentally sound local development.

It is my great pleasure and honour to express my personal feelings on the occasion of the publication of this book. I hope that the updated list of European IBAs will efficiently help to protect birds and nature in the whole of Europe.

Jiri Houdek, Mayor of Třeboň, Czech Republic

FOLLOWING the implementation of the Convention on Biological Diversity (CBD) at the end of 1993, over 170 governments have been working to conserve biodiversity and ensure that any use of biological resources is sustainable. Two articles in the CBD set the stage for all others: Article 6 calls for national strategies and action plans; and Article 7 calls for identifying important biodiversity and monitoring its status. Solid science is the foundation upon which both of these critical articles depend, and birds offer a particularly useful group for collecting and presenting information in a way that is useful to politicians, resource managers and the general public. BirdLife International has been working for many years to build a database on birds and their habitats, and now, with the publication of *Important Bird Areas in Europe*, they have produced a valuable source of knowledge that these various interested parties will be able to use to implement the CBD.

BirdLife International, as a member of IUCN, has been a global leader in collecting and presenting essential information on birds and their habitats, and therefore providing solid leadership for implementing the CBD—and other biodiversity-related conventions and agreements—at local, national and regional levels. Their comprehensive approach is particularly useful in enabling countries to coordinate their respective conservation efforts and ensure that transboundary issues are fully addressed. On the basis of this important book, we can confidently expect much more effective conservation action throughout Europe. The challenge now is to extend this innovative and useful approach to the rest of the world and to other taxa. IUCN is pleased to join other partners to help address this challenge.

Jeff McNeely, IUCN Biodiversity Policy Coordination Division

SITE-BASED management is a cornerstone of biodiversity conservation, and fundamental to this is readily available knowledge of the location and importance of such sites. BirdLife International's Important Bird Areas programme draws together a wealth of this vital information. The publication in 1989 by ICBP (BirdLife's forerunner) and the International Waterfowl and Wetlands Research Bureau (forerunner of Wetlands International) of Important Bird Areas in Europe was a milestone in providing accessible information on Europe's many and varied places vital for birds. Many of these areas are wetlands. The IBA programme has greatly contributed to identifying the whole network of key sites which are vital for the continued survival of migratory waterbirds, and has provided the foundation for developing an effective bird conservation strategy for Europe.

Wetlands International is pleased to have contributed its knowledge and information on waterbirds to this impressively expanded and updated edition of Important Bird Areas in Europe, both through our volunteer census networks in individual countries and from our population- and flyway-scale information analyses for waterbirds. The new IBA Europe book provides an essential tool for the 21st century in realising the sustainable conservation of waterbirds in Europe and its delivery through the mechanisms of intergovernmental conventions and agreements, notably the Ramsar Convention and Bonn Convention's African-Eurasian Migratory Waterbird Agreement, and in the designation of Special Protection Areas under the EC Birds Directive.

We warmly congratulate BirdLife International on bringing to fruition the huge task of drawing together this wealth of essential information needed for implementing effective bird conservation in Europe. We look forward to continuing to work closely with BirdLife International in progressing this implementation, and in supporting the continuing IBA programme worldwide with our waterbird knowledge and information.

Robin Schaap, International Director,
Wetlands International

ACKOWLEDGEMENTS

The work of the staff and volunteer networks of the BirdLife International Partnership are at the core of this publication. National Important Bird Areas (IBA) coordinators have worked tirelessly to collate and assess the data in each country through these national networks and have authored individual country chapters. There are hundreds more individuals who have been closely involved in the collation of national data for this project and they are acknowledged in the individual country chapters. For the project team at the BirdLife International Secretariat it has been a great pleasure and honour to collaborate with such a fantastic group of people across Europe. The last five years of working together, often corresponding and exchanging views from a distance, but also the times shared at workshops and meetings, have been of immense importance.

The data, which are the foundations of this inventory, have been collected by many thousands of amateur and professional ornithologists and birdwatchers. They cannot be thanked individually but we owe them a great debt and wish to recognize the important contribution each has made to the identification of Important Bird Areas across Europe.

We are very grateful to the funders of the project and the many BirdLife Partners who have invested substantial resources into the collation of data nationally. We particularly wish to acknowledge the commitment from the Royal Society for the Protection of Birds (RSPB), the BirdLife Partner in the United Kingdom, who have funded the project manager's post and project-coordination activities over the last five years. Mark Avery and David Gibbons (RSPB) arranged this financing and provided advice and words of motivation throughout the course of the project. The Ministry of Agriculture, Nature Management and Fisheries, Directorate for Nature Management, through the PIN/MATRA Funds of the Ministry of Foreign Affairs, The Netherlands, substantially financed this publication. Also Vogelbescherming Nederland, the BirdLife Partner in the Netherlands, have helped in funding overall coordination costs. We thank Mike Rands, Johanna Winkelman and Marco Lambertini for their help in securing these funds.

Many people have provided helpful advice over the course of the project including the BirdLife European Team, which comprised Johanna Winkelman, Neil Burgess, Alistair Gammell, Fritz Hirt, Peter Iankov, Alejandro Sánchez, Joe Sultana, Zoltan Waliczky, David Chandler, Nelly Paleologou and Christel Davidson. We also thank particularly Marco Lambertini and Colin Bibby of the BirdLife Secretariat in Cambridge, for their advice and scientific guidance at all stages of the work and their constructive and helpful comments on draft manuscripts. Our colleague Zoltan Waliczky, the European Conservation Coordinator at the BirdLife Secretariat from 1993–1998, provided substantial help throughout the project including assisting with the development of the IBA database, criteria and thresholds, co-facilitating workshops associated with the project and providing comments on earlier drafts of the manuscript.

Establishing scientifically credible criteria for selecting IBAs was vital to the project. In this regard we thank Lincoln Fishpool and Zoltan Waliczky, who together with MH developed the global categories with input from David Wege and Mike Crosby of the BirdLife International Secretariat. The European criteria, added a further 13 categories to the set of global criteria, the development of which was coordinated by MH in conjunction with the national IBA coordinators in Europe. We also thank Paul Rose, formerly of Wetlands International, for making data available and for generously giving his time to discuss and advise on their application in the context of IBA identification. Mike Brooke and Mark Tasker advised and provided welcome expertise on seabird population estimates. The European Union criteria development was led by Eduard Osieck, together with MH and also David Pritchard, Nelly Paleologou and members of the Birds and Habitats Directive Task Force of BirdLife International. Eduard Osieck is a key figure concerning IBA work in Europe, beginning his involvement in 1981 with the publication of the first work on IBAs, and since then for many years carrying out pioneering work on IBAs in the Netherlands and as Chair of the Birds and Habitats Directive Task Force. He has contributed greatly to this publication and has always being available to offer his expert advice.

Many parts of this publication have been generated from the BirdLife IBA database, an essential tool for storing, managing and analysing the huge quantity of data presented in the country chapters and national and pan-European analysis. We thank Henk van Dijkhuizen for all his work on the original specification of the IBA database, and Mike Crosby and Miriam Langeveld for their work on earlier databases. We recognize the many helpful comments from the BirdLife Partners across Europe during the course of the IBA database development, especially Michael Dvorak, Ian Fisher, Eduard Osieck, Szabolcs Nagy, Tanya Sviridova and Carlota Viada, and also in Asia, particularly Rudyanto (BirdLife Indonesia Programme) and Koyama (Wild Bird Society of Japan, the BirdLife Partner in Japan). Research staff at the BirdLife Secretariat have provided regional input into the development of standardized terms and structure of the database, especially Zoltan Waliczky, Lincoln Fishpool, David Wege, Adrian Long and Mike Crosby. For the programming we are indebted to Peter Bishop of Integra whose skills and patience were fundamental to the success of the computer database we have in use today across more than 35 European countries. We also give special thanks to Andrew Rayner for his skills and always gracious and helpful technical support to the Secretariat and Partners in using and distributing the database.

Very helpful comments were received on drafts of the introductory and analysis chapters and also earlier designs of the site accounts and national overviews from Colin Bibby, Nigel Collar, John Fanshawe, David Gibbons, Marco Lambertini, Adrian Long, Szabolcs Nagy, Eduard Osieck, David Pritchard, Alison Stattersfield, John O'Sullivan, Richard Thomas, Zoltán Waliczky and Johanna Winkelman.

Many colleagues, past and present, at the Secretariat of BirdLife International in Cambridge have given assistance to this project, many of whom have already been mentioned above. We would like to add particular thanks to Mike Rands, Carlos Martín Novella, Laurence Rose, Terry Parker, Chris Spreadbury and Kelly Hague for fundraising, finance and marketing support to the project, and to Mike Rands and Kathleen Rosewarne for negotiations with production companies and designers at earlier stages of the project. Adrian Long oversees the production of the BirdLife Conservation Series and has contributed greatly to this publication through the innovative and constructive input on design and liasing with printers and designers. Margaret Parnwell helped by cataloguing the photographs, and we are very grateful to the photographers who provided them, for the most part, free of charge. Beverly Childs and Anne Collins provided highly valuable secretarial assistance.

Several workshops were held throughout the course of the project and for the generous and excellent logistical help in their organization we would like to thank Dorothy Bashford, Beverley Childs, Anne Collins, Johanna Winkelman and Marianne de Rijk. We also thank the officers working for BirdLife Partners responsible for cross-Partner support for the European Programme for the help they have given in this regard to this project: Johanna Winkelman, Kevin Standring, Norbert Schäffer, Neil Burgess and Fritz Hirt.

Collaboration and support from other organizations was very gratefully received throughout the course of the project. Paul Rose of Wetlands International provided help and data on waterbird populations, Jim Paine and Michael Green of the World Conservation Monitoring Centre (WCMC) provided access to the WCMC Protected Areas Database, Henrik Skov of Ornis Consult advised on marine issues, Dorian Moss from the Institute of Terrestrial Ecology and Mark Roekaerts advised on database systems and Ward Hagermeijer from the European Bird Census Council for advice on atlas data.

For the design, layout, map and graphic production, copy-editing and print management we thank Peter Creed, Rebecca Miles and Tony Vetta of the Nature Conservation Bureau Ltd., whose dedication, professionalism and enthusiasm have added much to the presentation of information in this book.

Finally, this publication builds upon and updates the first pan-European IBA inventory, the authors of which, Richard Grimmett and Tim Jones, are warmly thanked and congratulated for setting such sound foundations upon which the European IBA Programme has evolved to benefit the conservation of IBAs, and from which this new publication has evolved 10 years on.

Project Team

Melanie Heath – Project Manager, 1994–1999 Coordinating the production of this publication as part of her job as European Research Manager, including project development, scientific direction and analysis, editing and the development of the European IBA Database.

Michael Evans – Senior Editor, 1997–1999. Responsible for overall editing standards, standardization of terms and nomenclature, and final proof-reading.

Tony Payne – Research Assistant, 1996–1998. Central to the development, functionality and management of the IBA database and the validation of data within it, and technical support to Partners using the database.

David Hoccom – Research Assistant, 1998–1999. Editor of country chapters and producer of national overview graphics.

Nic Peet – Research Assistant, 1999. Editor of country chapters.

Tim Morrissey – Map production for country chapters and pan-European overview.

Regina Pfaff – Editorial support and incorporating editorial changes into the manuscript.

Paulina Phillips – Data-entry and checking.

SUMMARY

WHAT IS THE BIRDLIFE IBA PROGRAMME?

The Important Bird Areas (IBA) Programme of BirdLife International is a worldwide initiative aimed at identifying and protecting a network of critical sites for the conservation of the world's birds. The European IBA Programme is the longest running of a number of regional IBA programmes and for the last decade has addressed site-oriented research and action, encompassing habitat management, monitoring, education, advocacy, and national and international legal protection. This publication presents essential information on all known Important Bird Areas in Europe and builds on and updates the first pan-European IBA inventory, published in 1989.

DATA FOR PRIORITY-SETTING AND DECISION MAKING

Data collection and analysis play a highly influential role in underpinning the conservation and management of IBAs. Useful for conservationists, ornithologists, governmental and non-governmental agencies, policy-makers, researchers, consultants and planners, the data presented here are intended to guide practical management and actions at IBAs and to target political and legal mechanisms to achieve the adequate protection of IBAs.

AN INTERNATIONALLY IMPORTANT NETWORK OF SITES

Through the rigorous application of standard and quantitative ornithological criteria, the international importance of the network of sites identified is assured. The criteria applied justify the importance of each site, and their global standardization facilitates comparisons between sites at local, national, regional and global levels. These criteria are compatible with those used to designate wetlands of international importance under the Ramsar Convention and Special Protection Areas under the EC Birds Directive.

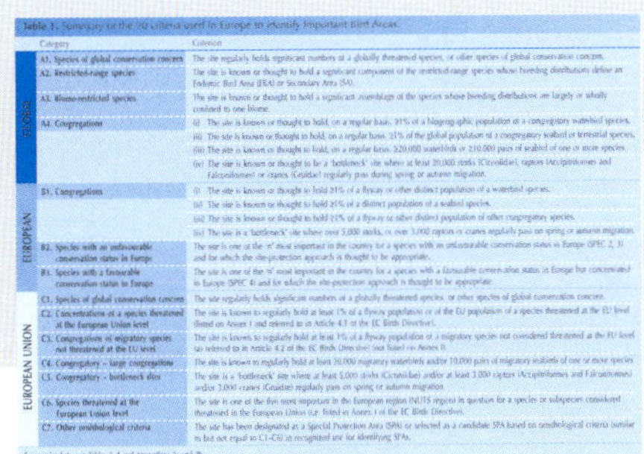

Table 1. Summary of the 20 criteria used in Europe to identify Important Bird Areas.

	Category	Criterion
GLOBAL	A1. Species of global conservation concern	The site regularly holds significant numbers of a globally threatened species, or other species of global conservation concern.
	A2. Restricted-range species	The site is known or thought to hold a significant component of the restricted-range species whose breeding distributions define an Endemic Bird Area (EBA) or Secondary Area (SA).
	A3. Biome-restricted species	The site is known or thought to hold a significant assemblage of the species whose breeding distributions are largely or wholly confined to one biome.
	A4. Congregations	(i) The site is known or thought to hold, on a regular basis, ≥1% of a biogeographic population of a congregatory waterbird species.
		(ii) The site is known or thought to hold, on a regular basis, ≥1% of the global population of a congregatory seabird or terrestrial species.
		(iii) The site is known or thought to hold, on a regular basis, ≥20,000 waterbirds or ≥10,000 pairs of seabird of one or more species.
		(iv) The site is known or thought to be a 'bottleneck' site where at least 20,000 storks (Ciconiidae), raptors (Accipitriformes and Falconiformes) or cranes (Gruidae) regularly pass during spring or autumn migration.
EUROPEAN	B1. Congregations	(i) The site is known or thought to hold ≥1% of a flyway or other distinct population of a waterbird species.
		(ii) The site is known or thought to hold ≥1% of a distinct population of a seabird species.
		(iii) The site is known or thought to hold ≥1% of a flyway or other distinct population of other congregatory species.
		(iv) The site is a 'bottleneck' site where over 5,000 storks, or over 3,000 raptors or cranes regularly pass on spring or autumn migration.
	B2. Species with an unfavourable conservation status in Europe	The site is one of the 'n' most important in the country for a species with an unfavourable conservation status in Europe (SPEC 2, 3) and for which the site-protection approach is thought to be appropriate.
	B3. Species with a favourable conservation status in Europe	The site is one of the 'n' most important in the country for a species with a favourable conservation status in Europe but concentrated in Europe (SPEC 4) and for which the site-protection approach is thought to be appropriate.
EUROPEAN UNION	C1. Species of global conservation concern	The site regularly holds significant numbers of a globally threatened species, or other species of global conservation concern.
	C2. Concentrations of a species threatened at the European Union level	The site is known to regularly hold at least 1% of a flyway population or of the EU population of a species threatened at the EU level (listed on Annex I and referred to in Article 4.1 of the EC Birds Directive).
	C3. Congregations of migratory species not threatened at the EU level	The site is known to regularly hold at least 1% of a flyway population of a migratory species not considered threatened at the EU level (as referred to in Article 4.2 of the EC Birds Directive) (not listed on Annex I).
	C4. Congregatory – large congregations	The site is known to regularly hold at least 20,000 migratory waterbirds and/or 10,000 pairs of migratory seabirds of one or more species.
	C5. Congregatory – bottleneck sites	The site is a 'bottleneck' site where at least 5,000 storks (Ciconiidae) and/or at least 3,000 raptors (Accipitriformes and Falconiformes) and/or 3,000 cranes (Gruidae) regularly pass on spring or autumn migration.
	C6. Species threatened at the European Union level	The site is one of the five most important in the European region (NUTS region) in question for a species or subspecies considered threatened in the European Union (i.e. listed in Annex I of the EC Birds Directive).
	C7. Other ornithological criteria	The site has been designated as a Special Protection Area (SPA) or selected as a candidate SPA based on ornithological criteria (similar to but not equal to C1–C6) in recognized use for identifying SPAs.

For species lists, see Tables 2–4 and Appendices 2a and 2b.

Page 11

THOUSANDS CONTRIBUTE – THE LARGEST NETWORK IN EUROPE

Contributions have been made by a vast network of ornithologists, birdwatchers and conservation experts across Europe. Many hundreds of individuals have been directly involved in the collation of data and many thousands more have laid the foundations for the project, through carrying out field surveys of bird numbers and distributions during the last three decades. In most countries the relevant BirdLife International Partner has coordinated the work nationally, and this has generally involved substantial collaboration with governmental and non-governmental organizations.

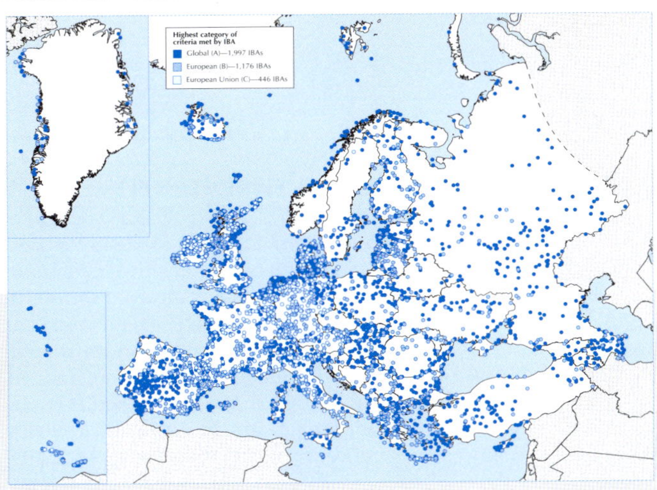

3,619 IBAS COVER 7% OF EUROPE

A comprehensive network of IBAs has been identified across Europe. The effective protection and management of these sites, equivalent to a relatively low percentage of each country's area, is a realistic goal that would make a significant contribution to conserving many bird species in Europe.

OVER 1,000 NEW IBAS IDENTIFIED

A large amount of new data has been gathered as part of this inventory, resulting in improved coverage across the region compared to the previous pan-European IBA inventory in 1989. Since then, no fewer than 1,175 new IBAs have been identified, representing a 48% increase in the number of sites, and a 130% increase in the total surface area covered, with improved coverage particularly in the east of the region.

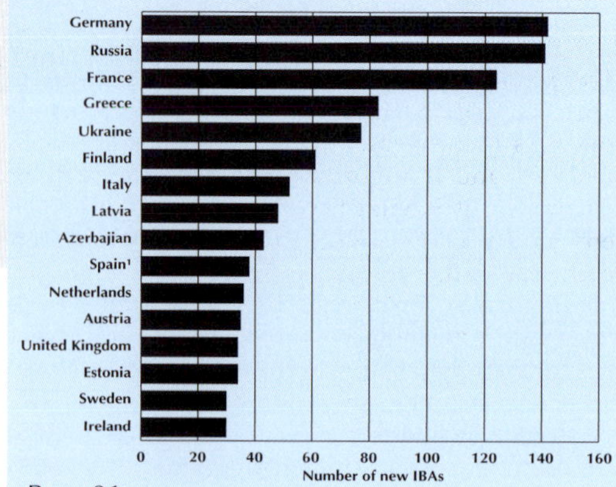

Page 31

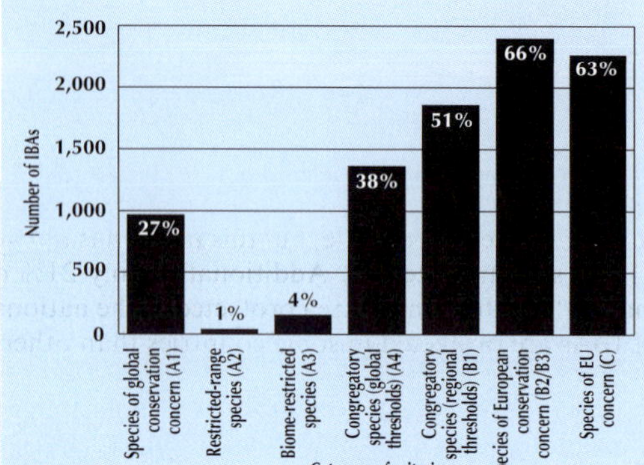

Page 31

SITES IDENTIFIED FOR 378 SPECIES (73% OF EUROPE'S BIRDS)

IBAs are particularly important for species that congregate in large numbers such as wintering and passage waterbirds and breeding seabirds (51% of IBAs in Europe are identified for these species). Many sites have also been identified for species of global conservation concern (27% of IBAs are important for one or more of the 35 such species in Europe), for species of European conservation concern (66% of all IBAs) and for species of concern within the European Union (63% of all IBAs).

ALL MAJOR HABITAT-TYPES COVERED

IBAs encompass all major habitat-types in Europe, with wetlands, forests, grasslands and cultivated/grazed habitats being the most frequent. Each of these habitat-types holds a distinctive but diminishing avifauna which urgently needs recognition and protection against ill-conceived development.

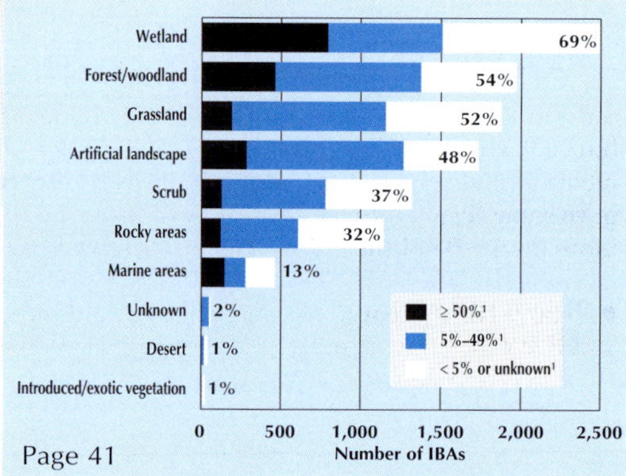

Page 41

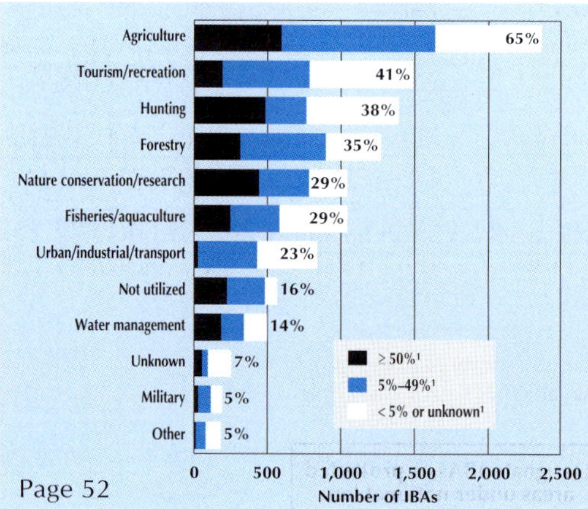

Page 52

IBAS HEAVILY INFLUENCED BY DIVERSE LAND-USE PRACTICES

Land-use practices have been recorded within 95% of IBAs and often cover the entire IBA area, reflecting the high human use of Europe's habitats in general. As a result, the conservation of habitats and birds within IBAs is very much dependent on the land-use practices, policies and programmes affecting these sites and their surroundings. Of particular importance is agriculture, the dominant land-use in Europe, and this is reflected within IBAs, with agricultural activities taking place in 65% of IBAs.

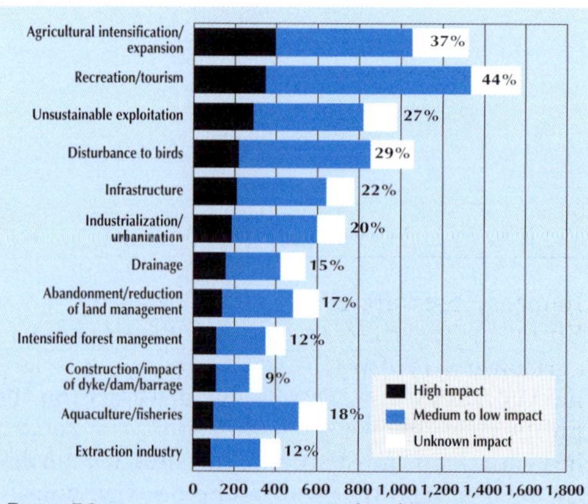

Page 52

93% OF IBAS CONSIDERED THREATENED

Nearly all IBAs are threatened to some degree by at least one factor, and 42% of IBAs are affected by one or more high-impact threats (actual or potential). Agricultural intensification and/or expansion is the most serious threat affecting IBAs, being cited at more than 1,300 sites and with a high impact at nearly 400 of these. Recreation and tourism, unsustainable exploitation, infrastructure, industrialization and urbanization, drainage, abandonment and reduction of land management, intensification of forestry practices, aquaculture and fisheries, the construction of dams, dykes or barrages and the extraction industry each threaten more (and in many cases significantly more) than 300 IBAs (c.10% of European IBAs).

IBAS IN THE CONTEXT OF THE WIDER ENVIRONMENT

Increasingly, many environmental problems and potential threats to IBAs are regional or global in scope or origin, and cannot be solely addressed by good protection and management of IBAs themselves. Furthermore, many species breed in a dispersed, non-congregatory fashion, and therefore cannot be conserved successfully by the IBA programme alone. Therefore, it is vital that conservation of the wider environment and the integration of environmental objectives into all policy sectors are also pursued in addition to, and in the context of, IBA conservation.

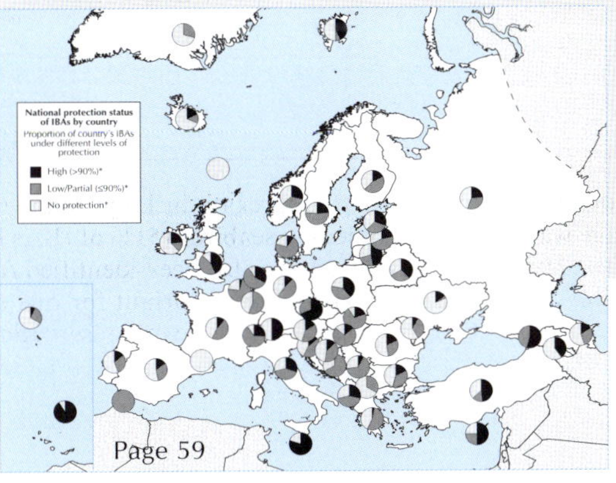

Page 59

40% OF IBAS UNPROTECTED BY NATIONAL LAWS

All European countries have legal and institutional frameworks for the designation and management of protected areas. IBAs should be designated as protected areas under national law where appropriate, but this target has not yet been reached, with 40% of IBAs still unprotected. Additionally, only 21% of European IBAs have more than 90% of their entire area protected at the national level. A higher proportion of IBAs are protected in some countries than others.

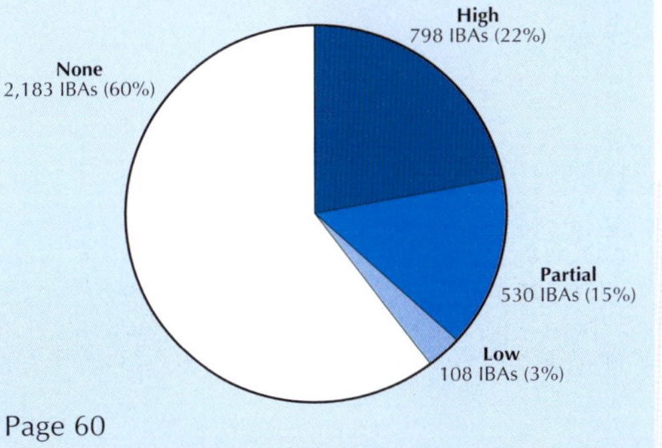

Page 60

60% OF IBAS UNPROTECTED AT THE INTERNATIONAL LEVEL

Substantial international recognition of IBAs has been achieved in the last decade but many IBAs remain inadequately protected or recognized under relevant international agreements, and further designations are required. Only 40% of IBAs have some form of international protection, and in only 22% of IBAs in Europe is more than 90% of the individual IBA area protected at the international level.

- Many IBAs are wetlands of international importance yet only 30% of these have been designated under the Ramsar Convention. Given the compatibility of IBA criteria with Ramsar criteria, it is possible to estimate that a further 1,000 IBAs in Europe should be considered for designation as Ramsar Sites because of the internationally important numbers of waterbirds that they hold.

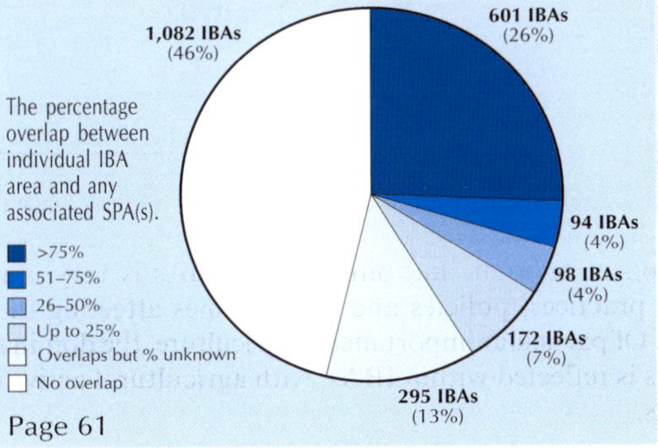

Page 61

- Many IBAs in the European Union are internationally important for migratory and threatened birds but only 54% have been designated as Special Protection Areas (SPAs) under the EC Birds Directive. A further 1,000 IBAs in the European Union should therefore be designated as SPAs. Additionally, only one quarter of IBAs in the EU have more than 75% of their individual area designated as SPAs.

WHAT ACTIONS NEED TO BE TAKEN?

Immense pressure continues to be placed on the European environment through developments in agriculture, forestry, fisheries, transport, energy, industry, tourism/recreation and urbanization. This publication shows such pressures are severely impacting the most important sites for birds in Europe. Much still needs to be done to conserve IBAs adequately.

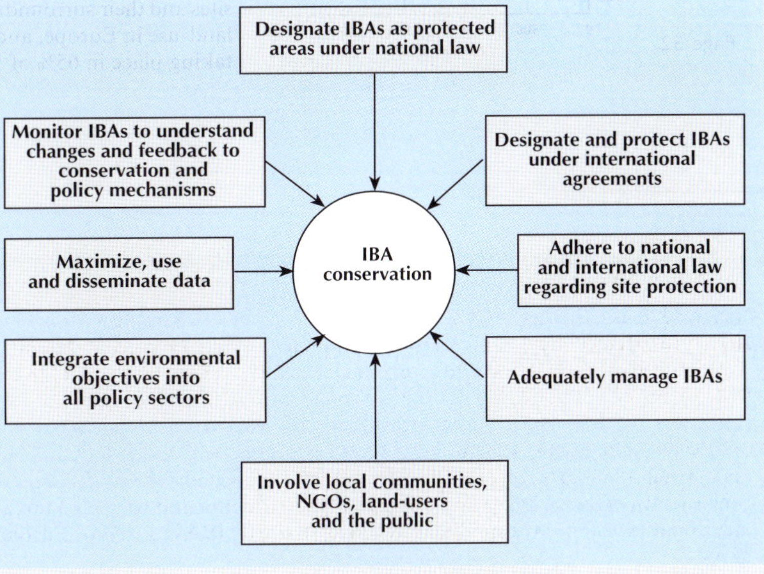

INTRODUCTION

EUROPE'S ENVIRONMENT

Pressure

The 20th century has seen profound changes in the European environment. Across most of the continent, man's use of the land and sea continues to intensify, driven by technological innovations, market forces, and political and socio-economic changes. Very few areas have escaped the impact of recent developments in agriculture, forestry, fisheries, transport, energy, industry and urbanization, and associated pollution.

In particular, the intensification since 1950 of the two most widespread land-uses in Europe, agriculture and forestry, has led to the 'simplification' of ecosystems across vast areas, giving man greater control over natural variations in the productivity of farmland and forests, but also markedly reducing the biological diversity and complexity of these habitats. Increasingly efficient but indiscriminate harvesting methods are reducing the populations of many marine and freshwater organisms in Europe's rivers and seas, altering food-webs in the process. Under these pressures, less intensive forms of natural-resource use have not been able to compete economically, and have been widely abandoned (Pain and Pienkowski 1997, EEA 1998). Many other semi-natural habitats, such as inland wetlands and dry grasslands, have also been greatly reduced in extent and diversity by man's activities, with the remaining fragments increasingly isolated from each other (Tucker and Evans 1997).

Status

These changes in land-use have had a highly detrimental impact on Europe's biodiversity (EEA 1998). About 12% of vertebrate species in Europe are threatened with global extinction (WCMC 1997) and, at least for birds (the only group for which there is adequate data), a much larger proportion of the total (38% of regularly occurring bird species) have an unfavourable conservation status on the continent, most of these species having undergone significant reductions in abundance or range since 1970 (Box 1). Man's activities are implicated as the driving factor in most of these declines. The intensifying use of the environment has also had damaging effects on society within and beyond Europe, e.g. through massive loss of rural livelihoods, over-dependence on subsidies, and the socio-economic impacts of food surpluses and food-safety issues (EEA 1998).

Response

Future directions and policies for land-use in Europe are still decided by the orthodox application of economic theory. As yet, mainstream economics do not take into account the full benefits of biodiversity, nor the long-term costs to society of intensification, in terms of biodiversity loss (UNEP 1995). As a result, decisions relevant to future land-use still ignore many of the projected impacts on biodiversity and society, and the phenomenon of biodiversity loss still receives insufficient attention from most sectors of the economy and society.

Although biodiversity is not yet valued in strict economic terms, it is now widely recognized that these negative environmental trends cannot be allowed to continue indefinitely, since the direct and indirect costs to society will be too great (UNEP 1995). The challenge is to find more sustainable ways of using the environment, so-called 'sustainable development'.

At the governmental level, the world community has responded with initiatives such as *Agenda 21* and the Convention on Biological Diversity (Biodiversity Convention), both products of the 'Earth Summit'—the United Nations Conference on Environment and Development—at Rio de Janeiro in 1992. At the European level, the *Pan-European Biological and Landscape Diversity Strategy* has received unanimous endorsement from Europe's environment ministers (McCloskey 1997), and the European Community has responded with *Towards Sustainability*, the fifth environmental action programme of the European Commission (CEC 1993), followed more recently by the *EC Biodiversity Strategy* (CEC 1998).

These international initiatives, whether master-plans, conventions or policy documents, all share similar objectives, that is to promote, within and between nations, the conservation of biodiversity and the sustainable use of natural resources, through the integration of environmental issues and concerns into all sectors of society (including the economy), encouraging closer cooperation between sectors, and advocating a more participatory form of environmental policy-making.

At the local level, one of the ways that individual citizens can participate in policy-making, and help to solve problems in society, is through creating or supporting non-governmental organizations (NGOs). The number and size of NGOs that address environmental problems have increased enormously in recent years, in Europe and elsewhere, indicating the rising level of concern in society world-wide. The combined membership of such NGOs in Europe alone runs into many millions, and gives these bodies authority and democratic basis when seeking to influence governments, businesses and other powerful sectors of society. Indeed, through their research, publicity and action, these groups of like-minded people are a crucial partner to governments in the search for more sustainable forms of development.

THE NEED FOR BIRD CONSERVATION IN EUROPE

Birds are one of the best-researched and most reliable indicators of biodiversity loss in Europe (Tucker and Heath 1994, Tucker and Evans 1997). Nearly 40% of bird species in Europe currently have an unfavourable conservation status (Box 1), mainly as a result of man's intensifying use of the environment and because of other, more direct threats from people, such as disturbance, persecution and excessive hunting (Tucker and Heath 1994). Faced with these rare or declining populations, the conservation of Europe's birds and other biodiversity has become an increasingly important and necessary activity, requiring coordination on an international scale.

In response, BirdLife International, the world-wide partnership of national NGOs devoted to the conservation of birds and their habitats (Box 2), pursues a programme of research, publicity and action in Europe, designed to:

- Prevent the extinction of any bird species.
- Reduce the number of bird species that are globally threatened.
- Maintain and where possible enhance the conservation status of all bird species.
- Conserve sites and habitats important for birds.

Box 1. The status of birds in Europe (Tucker and Heath 1994).

A total of 515 bird species occur regularly in Europe, comprising thousands of millions of individuals which depend on the continent as a breeding, staging or wintering ground. On a world scale, Europe is particularly important for 136 species, supporting the majority of their global population at some stage of their annual cycle.

Nearly 40% of the continent's bird species have an unfavourable conservation status in Europe, based on their rarity or their rate of population decline since 1970. Of these 195 species, 20 are globally threatened with extinction (Collar *et al.* 1994) and a further 15 are also of global conservation concern.

Most bird species in Europe occur widely across the continent, with populations in many different countries. Additionally, the great majority of bird species in Europe are migratory in at least part of their range, crossing many national frontiers, with some species regularly reaching not only southern Africa but also as far as Canada, east Siberia, India and the sub-Antarctic. An international perspective is therefore all-important for most bird-conservation initiatives in Europe.

Box 2. BirdLife International.

BirdLife International consists, in Europe, of 32 Partner organizations with a combined membership of over 2 million people and staff of more than 1,700. In addition, 22 other national organizations in Europe are affiliated to the Partnership, as Representatives or Associates.

BirdLife International, through seeking to conserve all bird species and their habitats on earth, works for the world's biodiversity and for the sustainability of human use of natural resources.

Why birds?

Birds play a special and unique role in connecting humanity to biodiversity and the environment. They occupy almost all habitats on the earth's surface, and are widely dispersed in all countries and regions. They are the best known and documented major taxonomic group of organisms in the world.

Birds are important to society because:

- They are valuable in their own right
- They are sensitive indicators of biological richness and environmental condition
- They are vital for ecological functions in the natural environment
- They have a direct and indirect economic and cultural value to people
- They provide a useful means to improve our scientific knowledge and understanding of the environment
- They are beautiful and inspirational, and a source of happiness and pleasure for many people
- They are very useful for promoting conservation awareness
- Like all species, they have a moral right to a continued existence on Earth.

PRIORITIZATION OF CONSERVATION ACTION

BirdLife International approaches the conservation of birds, in Europe as elsewhere, in three different ways:

1. Conservation of species
2. Conservation of sites
3. Conservation of habitats

Each approach is essential for the effective conservation of a large number of species and, while different, each is complementary and interdependent. Box 3 summarizes the rationale behind each approach.

BirdLife attaches great importance to the identification of priorities, to set targets and guide conservation action on species, sites and habitats. The BirdLife International Partnership has therefore conducted baseline studies across Europe to identify the conservation priorities for species, sites and habitats (Box 3). This publication defines the priority sites for bird conservation in Europe.

THE CONSERVATION OF SITES FOR BIRDS—IMPORTANT BIRD AREAS

Some areas in Europe are particularly important for birds:

- Sites for globally threatened species (Collar *et al.* 1994) and other species of European conservation concern and of European Union concern (Tucker and Heath 1994, CEC 1994).
- Sites for migratory species which congregate in high numbers (Rose and Scott 1997).
- Sites for species unique to a small region (Stattersfield *et al.* 1998).
- Sites that support a species assemblage that is highly representative of a distinct biome.

BirdLife International classifies such sites as Important Bird Areas (IBAs), where a significant part of these species' populations can be found on a regular basis. The protection and management of these high-priority sites, in a network across all relevant countries in the world, could contribute greatly to the effective conservation of these species. One of the main aims of BirdLife International is to build such a global network, through its IBA Programme (Box 4).

Legal mechanisms exist within Europe to protect the ecological integrity of these sites, at a variety of geographical scales. All European countries have legal and institutional frameworks for the designation and management of protected areas, and the perspective beyond national borders (so important for highly mobile organisms such as birds) is provided by a number of international conventions and initiatives that explicitly encourage the designation of international networks of sites that serve a shared conservation purpose (see in Box 2, 'Overview of results' chapter). However, despite this relatively well-developed statutory 'infrastructure', implementation has often been relatively slow, and many IBAs remain without official protection or recognition.

Box 3. Three ways of conserving biodiversity.

1. Conservation of species

This is the protection of species through conservation action, which may take many forms, such as legislation, monitoring, research, prioritization, management of populations, and the acquisition and management of land. Such an approach is often taken for species of economic or cultural importance, and increasingly for species that are threatened with extinction at a local, national or global level.

BirdLife International has identified those species that are threatened with global extinction (Collar *et al.* 1994) or that are otherwise of conservation concern in Europe (Tucker and Heath 1994), and has recently coordinated the production of up-to-date, international action plans for the conservation of the most threatened species in Europe (Heredia *et al.* 1996, CoE in prep.).

2. Conservation of sites

This is the protection of sites to conserve species and habitats and to maintain the integrity of ecosystem processes, by designating areas for the conservation of natural resources, and regulating and managing them according to the needs of the biodiversity which they contain. Apart from the long-term benefits gained by conserving biodiversity, sites also have other important functions for society, including education and research, and have non-material and non-monetary values as well as providing the focus for local or national pride. In moving towards a more sustainable society, all these functions will be increasingly highly valued. The protection and conservation of sites is an integral part of any attempt to achieve sustainability.

The BirdLife International Important Bird Area (IBA) Programme (see Box 4), a world-wide initiative aimed at identifying and protecting a network of critical sites for the world's birds, aims to facilitate the conservation of birds via this approach. This book, which builds upon the first pan-European IBA inventory (Grimmett and Jones 1989), is intended to guide practical on-the-ground conservation management and also to target political and legal mechanisms to achieve adequate protection of Europe's IBAs.

3. Conservation of habitats

A great threat to birds in Europe, and to biodiversity in general, lies in the continuing erosion of the quality and extent of habitats across the entire landscape ('the wider environment'). The loss and degradation is driven by the increasing intensity of human uses of the environment. The conservation of habitat extent and quality across the landscape cannot be achieved solely by the protection of representative areas: a wider approach is needed.

National governments and regional or local authorities can favour and encourage the conservation of the wider environment in the way that they formulate and use their laws, policies, plans, programmes, initiatives, subsidies, taxes, funds, inter-governmental relations and other broad measures. Ideally, these should be fully integrated into land-use policies, regulations and plans, across all sectors of the economy and at all scales. BirdLife International has recently outlined a strategy for conserving birds and their habitats in the wider European environment over the next 20 years (Tucker and Evans 1997).

Box 4. The Important Bird Area Programme of BirdLife International.

- The function of the Important Bird Area (IBA) Programme is to identify, protect and manage a network of sites that are important for the long-term viability of naturally occurring bird populations, across the geographical range of those bird species for which a site-based approach is appropriate.
- The continued ecological integrity of these sites will be decisive in maintaining and conserving such birds. Legal protection, management and monitoring of these crucial sites will all be important targets for action, and many (but not all) bird species may be effectively conserved by these means. Patterns of bird distribution are such that, in most cases, it is possible to select sites that support many species.
- The IBA Programme is global in scale, and it is anticipated that up to 20,000 IBAs will be identified world-wide, using standard, internationally recognized criteria for selection.
- The sites are identified on the basis of the bird numbers and species' complements that they hold, and are selected such that, taken together, they form a network throughout the species' biogeographic distributions.
- This network may be considered as a minimum essential to ensure the survival of these species across their ranges, should there occur a net loss of remaining habitat elsewhere through human, or other, modification. Therefore the consequences of the loss of any one of these sites may be disproportionately large.
- The programme aims to guide the implementation of national conservation strategies, through the promotion and development of national protected-area programmes. It is also intended to assist the conservation activities of international organisations and to promote the implementation of global agreements and regional measures.

There are also, in Europe, many non-statutory protected areas, often smaller, which contribute greatly to biodiversity conservation, such as those owned and/or managed by non-governmental organizations, private bodies, villages, farmers, and urban authorities.

IMPORTANT BIRD AREAS IN EUROPE

The first IBA inventory to cover the whole of Europe was published in 1989 (Grimmett and Jones 1989). *Important Bird Areas in Europe* provided key information on 2,444 high-priority sites in 39 countries or autonomous regions. It represented a major step towards realizing a bird-conservation strategy for Europe, and it accelerated progress in maintaining and enhancing the conservation value of all IBAs. Over the last ten years, BirdLife Partners across 32 countries have increasingly focused their site-conservation activities towards the IBA network, and local volunteers have mobilized at numerous IBAs to help protect, manage or monitor 'their' sites, on the ground. Facilitated since 1990 by a coordinator at the BirdLife International Secretariat and, increasingly, by national IBA coordinators in individual countries, the actions of many individuals have coalesced into a formal IBA programme (Box 4).

As part of this programme, 12 international workshops have been held across Europe, to develop the skills of national IBA coordinators (and their networks of local volunteers and experts) in fund-raising, awareness-raising, monitoring, site management, lobbying, advocacy, dealing with the media, and so on. In addition, the IBA programme has generated hundreds of 'interventions' over the past decade, coordinated actions involving Partners and other NGOs, their members and the general public in concerted lobbying against unsustainable developments at particular sites.

As well as this influence at the grassroots, the inventory has also stimulated advances in statutory or 'top-down' conservation of sites, notably through two judgements of the European Court of Justice, one giving precedent for the legal protection of an IBA (Santoña marshes, Spain, 1993), the other requiring an EU member state to designate a more complete network of Special Protection Areas under the EC Birds Directive, citing the IBA inventory as an example of such a network (case lost by the Netherlands government, 1998). In the same vein, data and data analyses have been used to influence and have strong input into several international legal frameworks for site conservation, notably the EC Birds Directive (leading to Natura 2000), the Ramsar Convention and HELCOM (further details of these instruments are given in Appendix 1).

The new IBA inventory

This book represents an update of the first IBA inventory, ten years on. It presents essential information on all known sites of international importance for the conservation of birds in Europe, targeted at a number of audiences:

- decision-makers and policy-makers
- land-use planners and regulators
- funders
- land managers
- conservationists
- birdwatchers and ornithologists
- environmental consultants
- academic and research bodies.

Since 1989 a considerable amount of new data on birds and sites has become available (Box 5), and as much as possible has been included in this book. Thus 3,619 IBAs are listed and described for Europe, a net increase of 50% since 1989. This book, and the computer database from which it was produced, are two products of five years' work by the BirdLife International Partnership in Europe, together with thousands of other ornithologists and birdwatchers, in 51 countries or autonomous regions.

Objectives of the inventory

The principal objectives of this book are:

- to identify and promote awareness of the most important sites in Europe for the conservation of birds
- to help direct conservation activity and available funding towards these sites
- to present the ornithological value of each site in a standardized but also reliable way, using numerical criteria
- to provide a tool for planning and management, at practical and political levels, through the presentation of key information on birds, habitats, land-uses, threats, legal protection, and conservation status
- to develop networks of local experts, fieldworkers and volunteers, and motivate them to monitor and protect IBAs
- to stimulate national and international cooperation and co-ordination in conserving Europe's most important sites for birds

Box 5. Progress in identifying and documenting Important Bird Areas in Europe.

1981: The first ever inventory of IBAs is prepared by the International Council for Bird Preservation (ICBP) (now BirdLife International) for the European Commission (Osieck and Mörzer Bruyns 1981), covering 694 sites in the (then) nine member states of the European Community (EC).

1983–1986: Further IBA inventories are prepared by ICBP for the European Commission, covering new member states of the EC, as well as threatened IBAs in the expanded EC.

1989: The first pan-European IBA inventory is published, covering 2,444 sites in 39 countries or autonomous regions (Grimmett and Jones 1989). In the same year, a separate inventory, commissioned by the European Commission, is produced by ICBP on the IBAs in the (then) 12 member states of the EC (Grimmett and Gammell 1989).

1989–2000: Many new IBAs are identified as a result of field surveys, literature review and expert knowledge, especially in the east and south of Europe. National IBA inventories are published in 16 countries, with substantially revised and enhanced data in many cases (listed in Box 1, 'Data collection' chapter).

2000 – Second pan-European IBA inventory (this book) published, covering 3,619 sites in 51 countries or autonomous regions.

- to establish a more rigorous baseline for measuring Europe's success or failure in conserving its most important sites for birds
- to facilitate the comparison of information at local, national and international scales
- to promote awareness of the value of a site-based approach for the conservation of birds and biodiversity

Components of the inventory

This publication is divided into four main sections:

1. Introductory chapters

- Data collection: information on standardized methods and classification schemes.
- Site selection: ascertaining the importance of sites for birds against agreed scientific criteria.
- Data presentation: how to use the book.

2. 'Overview of results' chapter

- Pan-European summary analysis of the data gathered, with conclusions and recommendations.

3. Country chapters: inventory of Important Bird Areas by country

For each country, there is:

- A national overview of the IBAs .
- A detailed site-account for each IBA.

Due to the large number of countries and sites involved, the inventory has been divided into two volumes: each starts with the Introductory chapters and 'Overview of results' chapter (sections 1 and 2, above), followed by a group of country chapters, as follows.

Volume 1: **Northern Europe**—Austria, Belarus, Belgium, Czech Republic, Denmark, Faroe Islands, Greenland, Estonia, Finland, Germany, Iceland, Republic of Ireland, Latvia, Liechtenstein, Lithuania, Luxembourg, Netherlands, Norway, Svalbard and Jan Mayen, Poland, Russia, Slovakia, Sweden, Switzerland, United Kingdom (including Channel Islands and Isle of Man).

Volume 2: **Southern Europe**—Albania, Andorra, Armenia, Azerbaijan, Bosnia and Herzegovina, Bulgaria, Croatia, Cyprus, France, Georgia, Gibraltar, Greece, Hungary, Italy, Former Yugoslav Republic of Macedonia, Malta, Moldova, Portugal, Azores, Madeira, Romania, Slovenia, Spain (including Canary Islands), Turkey, Ukraine, Yugoslavia.

4. Appendices

A series of Appendices give more detailed information, descriptions and analyses, in support of the introductory and overview chapters (the Appendices are provided in both volumes of the publication).

REFERENCES

CEC [Commission of the European Communities] (1993) *Towards sustainability. The European Community programme of policy and action in relation to the environment and sustainable development*. Luxembourg: Official Journal of the EC (OJ C138, 17 May 1993).

CEC [Commission of the European Communities] (1994) *Special Protection Areas*. Luxembourg: Official Journal of the EC.

CEC [Commission of the European Communities] (1998) *The European Community Biodiversity Strategy*. Luxembourg: Official Journal of the EC.

Collar, N. J., Crosby, M. J. and Stattersfield, A. J. (1994) *Birds to watch 2: the world list of threatened birds*. Cambridge, UK: BirdLife International (Conservation Series No. 4.).

CoE [Council of Europe] (in prep.) Second tranche of international action plans for high-priority bird species in Europe. Strasbourg, France: CoE.

EEA [European Environmental Agency] (1998) *Europe's environment: the second assessment*. Luxembourg: Office of Official Publications of the European Communities/Oxford, UK: Elsevier.

Grimmett R. F. A. and Gammell, A. B. (1989) Inventory of Important Bird Areas in the European Community. (Unpublished report prepared for the Directorate-General for the Environment, Consumer Protection and Nuclear Safety of the European Community, Study contract B6610-54-88.) Cambridge, UK: International Council for Bird Preservation.

Grimmett, R. F. A. and Jones, T. A. (1989) *Important Bird Areas in Europe*. Cambridge, U.K.: International Council for Bird Preservation (Techn. Publ. 9).

Heredia, B., Rose, L. and Painter, M., eds. (1996) *Globally threatened birds in Europe: action plans*. Strasbourg, France: Council of Europe.

McCloskey, C. (1997) *P.E.B.L.D.S. explained*. Tilburg, Netherlands: IUCN European Programme.

Osieck, E. R. and Mörzer Bruyns, M. F. (1981) *Important bird areas in the European community*. Cambridge, UK: International Council for Bird Preservation.

Pain, D. J. and Pienkowski, M. W., eds. (1997) *Farming and birds in Europe: the Common Agricultural Policy and its implications for bird conservation*. London: Academic Press.

Rose, P. M. and Scott, D. A. (1997) *Waterfowl population estimates*. Second edition. Wageningen, Netherlands: Wetlands International (Publication 44).

Stattersfield, A. J., Crosby, M. J., Long, A. J. and Wege, D. C. (1998) *Endemic bird areas of the world: priorities for bird conservation*. Cambridge, UK: BirdLife International (Conservation Series No. 7).

Tucker, G. M. and Evans, M. I., eds. (1997) *Habitats for birds in Europe: a conservation strategy for the wider environment*. Cambridge, UK: BirdLife International (Conservation Series No. 6).

Tucker, G. M. and Heath, M. F. (1994) *Birds in Europe: their conservation status*. Cambridge, UK: BirdLife International (Conservation Series No. 3).

UNEP [United Nations Environment Programme] (1995) *Global Biodiversity Assessment*. Nairobi: UNEP.

WCMC [World Conservation Monitoring Centre] (1997) Red Data Book of European vertebrates. [Final draft for review at the 17th meeting of the Standing Committee of the Convention on the Conservation of European Wildlife and Natural Habitats, Strasbourg, 1–5 December 1997.] Strasbourg, France: Council of Europe (internal document T-PVS (97) 61 Eng. orig.).

DATA COLLECTION

GEOGRAPHICAL DEFINITION OF EUROPE

The geographical area covered by this publication is defined in Figure 1, and is the same as that covered by the previous pan-European Important Bird Area (IBA) inventory 10 years ago (Grimmett and Jones 1989). Thus, Europe here includes the Atlantic archipelagos of the Canary Islands, the Azores and Madeira, as well as western Russia (all administrative regions lying mainly west of the Ural mountains and Ural river), Armenia, Azerbaijan, Georgia, the Faroe Islands, Iceland, Svalbard and Jan Mayen, and the whole of Greenland, Turkey and Cyprus.

In total, 51 countries or autonomous regions are covered in this publication, compared to 39 such entities in the previous pan-European IBA inventory. This difference is due to the large political changes that have taken place during the past decade. No IBAs were identified in Monaco, San Marino or Vatican City (Holy See).

THE BIRDLIFE INTERNATIONAL PARTNERSHIP

A vast network of ornithologists, birdwatchers and conservation experts across Europe have contributed to this inventory (Figure 2). Many hundreds have been involved in the collation of data specifically for this project, and are acknowledged in each country chapter. Indirectly, many thousands more have laid the foundation for the inventory, through carrying out surveys of bird numbers and distributions during the last three decades.

The existence of the BirdLife International Partnership (see Box 2, 'Introduction' chapter) has greatly facilitated the assembly of this huge and diverse network. The 32 Partners of BirdLife International in Europe have endorsed and are implementing the BirdLife IBA Programme, initiated in 1990. Most have an IBA coordinator (or team), responsible for delivering this programme within the country concerned.

In most countries, therefore, it has been the IBA coordinators/teams within Partners who have organized the collation of new and revised IBA information, and who have been responsible for a large network of volunteer contributors to the project (Figure 2). The networks of local IBA experts are usually those who are active in the protection, management and monitoring of IBAs in their area (Figure 2). In a few countries this network has a formal structure (e.g. 'IBA caretaker network') and procedures, e.g. regular reporting to the national coordinator on the status of sites, or distribution of a national IBA newsletter. In most such cases, the IBA coordinator has also been the principal author of the relevant country overview in this publication. In countries without a BirdLife Partner or IBA coordinator, an expert was commissioned to coordinate the collection of data nationally.

In addition, much valuable information has come from beyond the BirdLife Partnership. At local and national levels there has been substantial collaboration with other bodies that collect and hold relevant data, both governmental and non-governmental, and these are duly acknowledged in each country chapter. Special mention should be made of Wetlands International, the leading non-profit

Figure 1. The geographical definition of 'Europe' as used in this book.

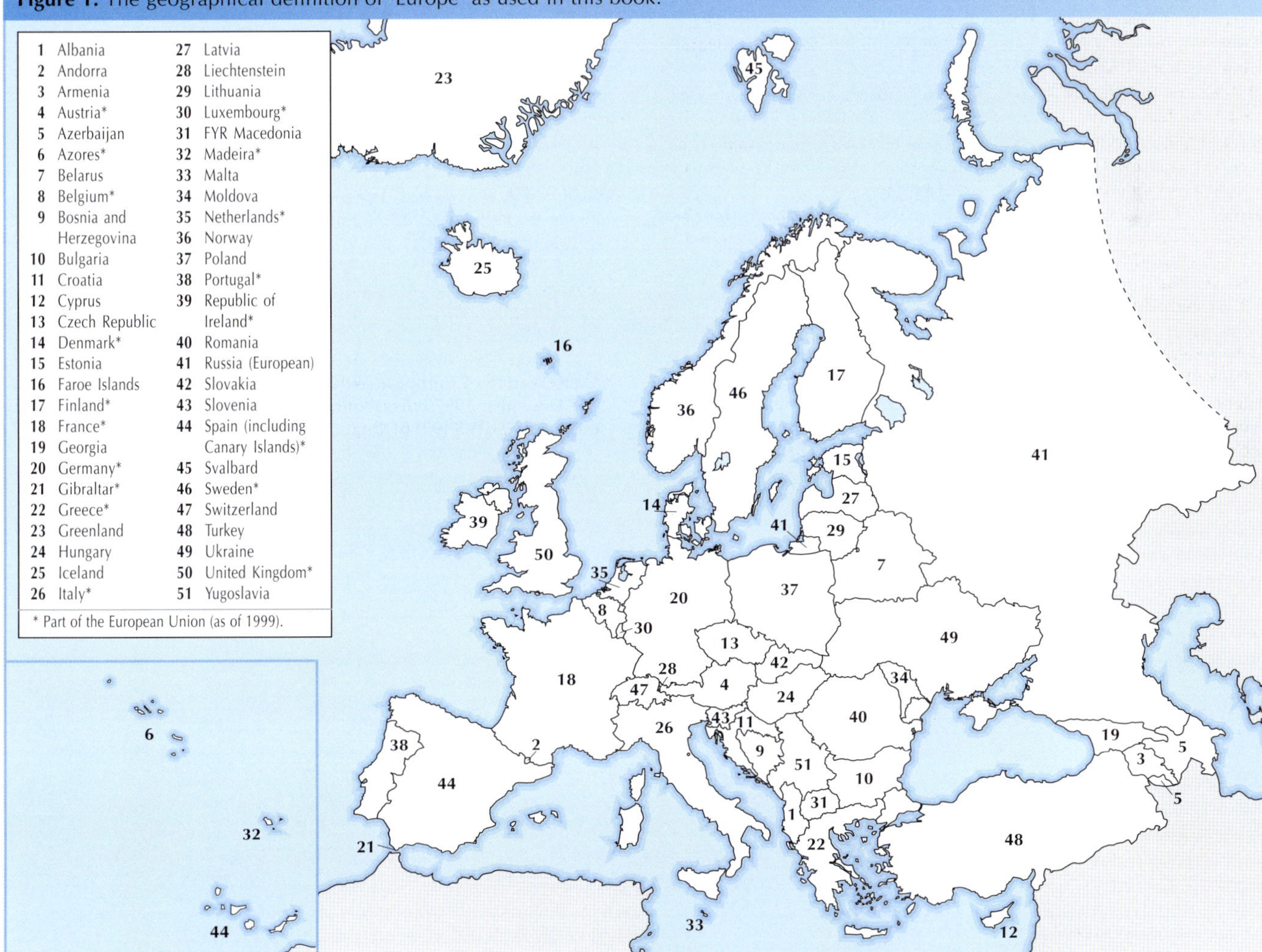

Figure 2: The Important Bird Area data-collection network.

organization devoted to the conservation of wetlands, which co-produced the first pan-European IBA inventory (Grimmett and Jones 1989) and which has helped in the production of this revised inventory, through the provision of data on population thresholds of waterbirds and, at the national level, on particular wetlands.

SOURCES OF DATA

■ National IBA inventories

The previous pan-European IBA inventory (Grimmett and Jones 1989) stimulated, during the past decade, the production of national IBA inventories for 16 European countries (Box 1). In a high proportion of these publications, the national IBA inventory was a completely revised version, or otherwise contained much new, updated or corrected data. These inventories were a primary source of information for this new pan-European inventory.

■ Other publications

As well as publications focusing on IBAs, a large number of other relevant studies have been published over the last 10 years that have stimulated or facilitated data-collection for IBAs:

- *Birds in Europe: their conservation status* (Tucker and Heath 1994): this BirdLife publication was a comprehensive assessment of the conservation status of all bird species in Europe, based on population sizes and trends, and range trends. Through identifying which species are of European conservation concern, and determining their national population numbers, the results have been central to the identification of IBAs for these species, based on numerical criteria.
- *Birds to Watch 2* (Collar *et al.* 1994): on behalf of IUCN–The World Conservation Union, the universally recognized authority on globally threatened species, BirdLife identified in this publication those species of bird that are of global conservation concern, in Europe and elsewhere. This update of previous such

Box 1: National Important Bird Area inventories that have been produced in Europe since 1989.

All inventories are in the relevant national language; those marked with an asterisk (*) are also available in English. Apart from those indicated below, national IBA inventories are currently also in preparation for Estonia, Finland, Germany, Romania, Russia, Slovenia and Ukraine.

Country	Year	Citation
Austria	1995	Dvorak, M. and Karner, E. *Important Bird Areas in Österreich.* Wien: Bundesministerium für Umwelt (Monographien Bund 71). 454 pp.
Bulgaria	1997	Kostadinova, I. [*Important Bird Areas in Bulgaria.*] Sofia: Bulgarian Society for the Protection of Birds (BSPB Conservation Series, Book 1). 176 pp. (In Bulgarian.)
Czech Republic	1992*	Hora, J. and Kanuch, P. *Vyznamna ptaci uzemi v Evrope* [*Important Bird Areas in Europe—Czechoslovakia*]. Prague: Czechoslovakia Section of the International Council for Bird Preservation. 124 pp.
France	1994	Rocamora, G. *Les Zones Importantes pour la Conservation des Oiseaux en France.* Angoulême, France: Ligue Pour La Protection des Oiseaux/Ministère de l'Environment. 339 pp.
(East) Germany	1991	Naturschutzbund Deutschland (NABU). *Die Europäischen Vogelschutzgebiete (IBA) in den fünf neuen Bundesländern* [*IBAs in the five new Bundesländer of the Federal Republic of Germany*]. Bonn: NABU.
Greece	1994	Hellenic Ornithological Society (HOS). [*IBAs for the birds of Greece*]. Athens: HOS. 271 pp. (In Greek.)
Hungary	1992	Waliczky, Z. *Európai Jelentóségú Madárélóhelyek Magyarországon.* Budapest: Magyar Madártani és Természetvédelmi Egyesület/International Council for Bird Preservation. 118 pp.
	1998	Nagy, S. *Fontos Madárélóhelyek Magyarországon.* Budapest: Magyar Madártani és Természetvédelmi Egyesület. 138 pp.
Ireland	1995	Magee, E. and Coveney, J. *Important Bird Areas (IBAs): threats and protection status.* Monkstown, Ireland: Irish Wildbird Conservancy.
Italy (+ in prep.)	1991	Lambertini, M., Gustin, M., Faralli, U. and Tallone, G. *IBA—Italia. Aree di Importanza Europea per gli Uccelli Selvatici in Italia.* Parma, Italy: Lega Italiana Protezione Uccelli. 263 pp.
Latvia (+ in prep.)	1994*	Viksne, J. *Putniem Nozimigas Vietas Latvija* [*Important Bird Areas in Latvia*]. Riga: Latvijas Ornitologijas Biedriba. 45 pp.
Netherlands	1994*	van den Tempel, R. and Osieck, E. R. [*Areas important for birds in the Netherlands: wetlands and other areas of international or European importance for birds.*] Zeist, Netherlands: Vogelbescherming Nederland (Techn. Rep. 13E). (In Dutch.) 126 pp.
	1996	Eggenhuizen, T. and van den Tempel, R. *Belangrijke Vogelgebieden.* Zeist, Netherlands: Vogelbescherming Nederland. 160 pp.
Poland	1994	Gromadzki, M., Dyrcz, A., Glowacinski, Z. and Wieloch, M. *Ostoje Ptaków W Polsce.* Gdansk: Ogólnopolskie Towarzystwo Ochrony Ptaków. 403 pp.
Slovakia	1992*	Hora, J. and Kanuch, P. [*Important Bird Areas in Europe—Czechoslovakia.*] Prague: Czechoslovakia Section of the International Council for Bird Preservation. 124 pp.
Spain	1992	de Juana, E., ed. *Áreas Importantes para las Aves en España.* Madrid: Sociedad Española de Ornitología (Monografía 3). 183 pp.
	1998	Viada, C. *Áreas Importantes para las Aves en España.* Madrid: Sociedad Española de Ornitología/BirdLife (Monografía 5). 398 pp.
Turkey	1989	Ertan, A., Kiliç A. and Kasparek, M. *Türkiye'nin Önemli Kus Alanlari.* Istanbul: Doğal Hayatı Koruma Derneği. 155 pp.
	1997*	Magnin, G. and Yarar, M. [*Important Bird Areas in Turkey.*] Istanbul: Doğal Hayatı Koruma Derneği. 313 pp.
United Kingdom	1992	Pritchard, D. E., Housden, S. D., Mudge, G. P., Galbraith, C. A. and Pienkowski, M. W., eds. *Important Bird Areas in the UK including the Channel Islands and the Isle of Man.* Sandy, UK: Royal Society for the Protection of Birds. 540 pp.

Box 2. The different types of data collected on Important Bird Areas, and available in the IBA database.

Area accuracy		
Reliable	A	accurate to within 10%
Incomplete	B	accurate to within 50%
Poor	C	definitely not accurate to within 50%
Unknown	U	

Land ownership
Private
State
Communal
Religious group
International waters
Mixed
Other
Unknown

Season types	Code	Description
Breeding resident	R	Species breeds in IBA and remains throughout the year.
Breeding visitor	B	Species breeds in IBA but is not present for parts of the year.
Winter visitor	W	Species spends a substantial part of the winter in IBA.
Passage visitor	P	Species stages in IBA during migration.
Non-breeding visitor	N	Species occurs in IBA but does not breed (usually over-summering immature birds or post-breeding moult-gatherings).
Unknown	U	Breeding or seasonal status of species in IBA is unknown or uncertain.

Population abundance	
Abundant	Encountered in large numbers in preferred habitat.
Common	Encountered singly or in small numbers in preferred habitat.
Frequent	Often, but not always, met with in preferred habitat.
Uncommon	Encountered sporadically in preferred habitat.
Rare	Rarely seen, often implying less than 10 or so records.
Unknown	Not possible to assess abundance on available information.

Accuracy of Population size/Trend		
Reliable	A	accurate to within 10%
Incomplete	B	accurate to within 50%
Poor	C	definitely not accurate to within 50%
Unknown	U	

Estimate of trend (1985–1995)	
+2	Large increase
+1	Small increase
0	Stable
-1	Small decrease
-2	Large decrease
F	Fluctuating
N	New breeder
X	Extinct
U	Unknown

Impact of threat
High
Medium
Low
Unknown

Relationship of protected area/IBA
Protected area is contained by IBA
Protected area contains IBA
Protected area overlaps with IBA
Protected area is adjacent to IBA
Relationship unknown

Geographical data	
Compiler	Person responsible for the IBA data provided.
Date	Date of completion of data compilation.
IBA codes	Current site-code; national IBA code; previous pan-European inventory (1989) code.
Site names	International name in English; national name in national language.
Country	Country in which IBA is located.
Administrative regions	Administrative region(s) in which IBA is located (at primary and secondary levels).
Area of IBA	Area of IBA in hectares (ha); 100 hectares = 1 km^2.
Area accuracy	Accuracy to which area of IBA is known.
Central coordinates	Central coordinates of IBA, in degrees and minutes (latitude/longitude; Greenwich).
Altitude	Altitudinal range spanned by IBA (in metres above/below sea level).
Map	Whether a map showing IBA boundaries (in paper or digitized form) is available.
General description	A general description of the IBA, its location and general appearance.
Ownership	An indication of the dominant type of ownership of the land within the IBA (should cover >50% of the IBA area).
Management plan	An indication of whether the IBA is covered (partly or wholly) by any existing management plan(s).
Criteria	
Endemic Bird Areas	Name of Endemic Bird Area, if A2 criterion is met; see the following chapter for further explanation.
Biomes	Name of biome(s), if A3 criterion is met; see the following chapter for further explanation.
Criteria	The reasons why the site is considered ornithologically important (summary list of IBA criteria fulfilled at the site). See the following chapter for an explanation of the criteria.
Ornithological data	
Species name	See Appendix 2a for a list of scientific and English names of birds used in this book. Taxonomy and nomenclature follow Cramp *et al.* (1977–1994).
Season	The season in which the species occurs in the IBA.
Year	The year of the latest data on which the population estimate is based.
Population size (minimum and maximum)	An estimate of minimum and maximum population size of the species at the IBA.
Population size accuracy	Accuracy of population-size estimate.
Population abundance	Qualitative estimate of population size. Only completed if minimum and maximum values not available.
Trend	An indication of the population-size trend at the site over the last 10 years.
Trend accuracy	Accuracy of indicated trend.
Criteria	IBA criteria fulfilled by species's population at site (see the following chapter for an explanation of the criteria).
Habitat data	
Habitat type	Habitat types covering >5% of the IBA area. Two levels of habitat data can be provided (see Box 3 for classification).
Percentage cover	The percentage of the IBA covered by the habitat type.
Land-use data	
Land-use type	Land-uses covering >5% of the IBA area. (See Box 4 for classification.)
Percentage cover	The percentage of the IBA covered by the land-use.
Threat data	
Threat type	Key threats impacting on the IBA. (See Box 5 for classification.)
Impact of threat	The seriousness of the threat (the assessment of impact is explained in Appendix 3).
Protection status	
Name	The name of the protected area.
Designation	The national or international designation-type (e.g. National Park).
Year	The year of initial designation.
IUCN category	The IUCN protected-area management category (I–VI) (IUCN 1994).
Area	The area, in hectares (ha; 100 ha = 1 km^2), of the protected area.
Central coordinates	The central coordinates of the protected area (in degrees/minutes of latitude/longitude; Greenwich).
Relationship to IBA	The spatial relationship between the IBA and the protected area.
Overlap	The extent of overlap in hectares (ha) between the IBA and protected area.
Other data	
General ornithological description	A general description of the ornithological importance of the IBA.
Other flora/fauna	Other significant flora and fauna present in the IBA.
Habitats/Land-uses/Threats	Additional text on habitats, land-uses or threats.
Research/conservation projects	Information on research, conservation or management activities at the IBA. Further details on protection, including any proposed protection measures.

publications, together with data in *Birds in Europe*, has allowed more authoritative identification of IBAs for globally threatened species, based on standard, numerical criteria.

- Wetlands International has published new and more accurate information on the sizes and geographical ranges of waterbird populations in Europe (Rose and Scott 1994, 1997)—analysed and mapped in more detail for swans, geese and ducks (Scott and Rose 1996)—and has also stimulated the production of national wetland inventories in Europe, which contain much new information on wetland sites. Together, these data-sources have allowed more reliable and comprehensive identification of IBAs for waterbird species.
- The process of drawing up the most recent international action plans for bird species of global conservation concern in Europe, which was coordinated by BirdLife International, and the subsequent publication of the plans (Heredia *et al.* 1996), have together stimulated many national-scale field surveys of these species. Again, this has led to the discovery of a large number of new IBAs for these key species, and has improved or updated information on known IBAs.
- *United Nations List of Protected Areas* (WCMC/WCPA 1998): this publication has provided much useful information on the location and extent of statutory protected areas in each country, making it much easier to categorize and analyse the legal protection status of IBAs.

■ Field surveys of IBAs

The national monitoring and research programmes of BirdLife Partners have updated and filled gaps in the original 1989 data on IBAs and have improved monitoring of these key sites. In other European countries as well, much new fieldwork has been conducted on IBAs over the last decade, resulting in more detailed and up-to-date information about each site, as well as the identification of many new IBAs. The collation of more detailed information on IBAs has also depended heavily on local IBA experts or caretaker groups (Figure 2). Apart from contributing their own knowledge of the current status of IBAs locally, they have often succeeded in involving many of the other key organizations, individuals and bird-counting networks that study or use IBAs in their area, in the provision of data (often unpublished or otherwise difficult of access).

TYPES OF DATA COLLECTED

For each IBA across Europe, key data have been collected on:

- Location
- Bird species
- Reasons for importance
- Habitats and land-uses
- Threats
- Protection status
- Conservation action

Box 2 presents a fuller explanation of the types of data collected.

For many countries in Europe, especially in the centre and east, this is the first time that such detailed data on sites have been collated. In other countries, such information already existed, at least in part, but was spread across a broad range of sources.

Compared to the previous pan-European IBA inventory (Grimmett and Jones 1989), the data collected for this inventory have been more quantitative, which has allowed a more rigorous analysis. The methods used for compiling and classifying this information have been standardized as much as possible. Standard lists were developed for several of the data-types, to simplify data-collection, and to facilitate any subsequent comparison and analysis of data between sites at local, national, continental and global levels. Thus, for habitats, land-uses and threats, a standard classification was drawn up for each, covering all the possible options that can be encountered at IBAs in Europe, and these are listed in Boxes 3, 4 and 5.

The importance of an individual site for bird conservation was also categorized in a standard way, such that a site may qualify as an IBA on the basis of one or more of 20 ornithological criteria

Box 3. Classification of habitats at Important Bird Areas in Europe. Full definitions are given in Appendix 3. Primary habitats are comparable with other IBAs globally.

Primary	Secondary
Forest/woodland	broadleaved deciduous forest native coniferous forest mixed forest alluvial/very wet forest broadleaved evergreen forest wooded steppe wooded tundra treeline ecotone wooded desert/semi-desert
Scrub	scrub heathland sclerophyllous scrub/garrigue/maquis
Grassland	steppe/dry calcareous grassland dry siliceous grassland alpine/subalpine/boreal grassland humid grassland mesophile grassland tundra machair
Desert	desert semi-desert
Wetland	tidal river/enclosed tidal water mudflat/sandflat saltmarsh sand-dunes/sand beach shingle/stony beach coastal lagoon standing fresh water standing brackish and salt water river/stream raised bog blanket bog water-fringe vegetation fen/transition mire/spring
Marine areas	open sea sea inlet/coastal features
Rocky areas	sea cliff/rocky shore rock stacks/islets scree/boulders inland sand-dunes caves
Artificial landscape	highly improved reseeded grassland arable land perennial crops/orchards/groves forestry plantation urban parks/gardens other urban/industrial areas ruderal land
Introduced/exotic vegetation	
Unknown	

Based on: *A classification of Palaearctic habitats* (Devillers and Devillers-Terschuren 1996).

Box 4. Classification of land-uses at Important Bird Areas in Europe. Full definitions are given in Appendix 3.

Agriculture
Fisheries/aquaculture
Forestry
Hunting
Military
Nature conservation/research
Tourism/recreation
Urban/industrial/transport
Water management
Not utilized
Other
Unknown

Box 5. Classification of threats at Important Bird Areas in Europe. Full definitions are given in Appendix 3.

Abandonment/reduction of land management
Afforestation
Agricultural intensification/expansion
Aquaculture/fisheries
Burning of vegetation
Consequences of animal/plant introductions
Construction/impact of dyke/dam/barrage
Deforestation (commercial)
Disturbance to birds
Drainage
Dredging/canalization
Extraction industry
Filling-in of wetlands
Firewood collection
Forest grazing
Groundwater abstraction
Industrialization/urbanization
Infrastructure
Intensified forest management
Natural events
Recreation/tourism
Selective logging/cutting
Shifting agriculture
Unsustainable exploitation
Other
Unknown

(see the following chapter). The ornithological data provided for each site were analysed systematically against these 20 numerical criteria (BirdLife International 1995), to ensure that all sites accepted as IBAs were truly of international importance and that the reasons for qualification were clearly documented. These standard, numerical criteria are fundamental to the IBA concept, and are fully explained in the following chapter.

TOOLS FOR DATA COLLECTION

Two main tools were developed to gather, store and manipulate the data on IBAs:

- Data-form (paper questionnaire)
- Computer database

Important qualities that were required of these tools were compatibility, uniformity and flexibility.

Compatibility

The tools were developed as part of the regional IBA programmes of Europe and other continents (particularly Africa), so as to ensure that all user-requirements were fully considered. In addition, elements of the data-collection mechanisms and classifications used by other international organizations, or by directly relevant pan-European initiatives, which were collecting information on sites in Europe—e.g. the Protected Areas Database of the World Conservation Monitoring Centre, the Ramsar Sites Database of the Ramsar Convention Bureau, the Natura 2000 network of the European Commission, and the CORINE Biotopes database of the European Environmental Agency—were taken into account in designing the tools, so as to maximize compatibility (European Commission 1995, WCMC 1996).

Uniformity

Both tools have a very similar structure, the IBA Database being a computerized version of the paper data-form, although it must be stressed that the IBA Database forms part of a larger World Bird Database and has many other valuable functions apart from data-collection, explained more fully in Box 6.

Since the tools were designed to collect standardized data, it is possible to compare directly any attribute of an IBA, such as location, ornithological importance or habitats, with that at any other IBA or IBAs, within and between countries or even continents, without encountering methodological problems.

Flexibility

The creation of two tools, one paper-based and one computer-based, allowed for the breadth in user-requirements when collecting data. In general, local experts found it most convenient to compile the original IBA data onto paper data-forms (often translated into the local language by the national coordinator), and then sent these in to the national coordinator, who organized the transfer of data onto the IBA computer database. Nearly all national (and sub-national) IBA coordinators used the database in this way. The tools were also designed so as to allow different quantities and detail of data to be collected per site, depending on the level of knowledge, and also for more data to be added at a later date without difficulty.

DATA-COLLECTION PROCEDURE

The general procedure for data-collection is illustrated in Figure 3.

In 1995, 1996 and 1998, international workshops were held involving national IBA coordinators from across the region, in order to share and pass on useful skills and experiences in planning, training and communication. Similarly, many countries held national workshops involving many local coordinators, contributors and experts, to publicize this project and to involve, train and enthuse participants. Most countries translated the data-form into their local language, so as to enable the widest possible dissemination and use of this data-gathering tool. However, the

Box 6. The Important Bird Area Database—a part of the World Bird Database

The IBA Database forms the basis of this book, and contains much additional information that could not be published here due to space constraints. Currently used in more than 35 countries across Europe, the IBA Database is a key part of the larger World Bird Database, which is a tool used by BirdLife International for managing, analysing and reporting on the full spectrum of its data on sites and species, for the purpose of conserving birds and their habitats. The World Bird Database is a distribution network, enabling data to flow between the people who collect the data or update it, those who collate and verify it, and those who make the analyses to turn data into information and targets, in order to influence policy and decision-making—moving from science to conservation action.

Key benefits of the World Bird Database are the ability to:

- Manage and validate a large volume of information on IBAs and globally threatened species
- Analyse trends in data and monitor changes
- Link site (IBA) data with species data, e.g. for international conventions, directives and other initiatives or programmes
- Determine the conservation status of sites or species internationally, regionally and nationally, with background information about the birds
- Produce focused, targeted reports for specific purposes
- Improve the sharing of information and expertise between Partners
- Improve electronic links to non-BirdLife data and information, such as socio-economic and non-bird species data, for use in analyses
- Link to geographic information systems (GIS) for presentation purposes

An overall principle of BirdLife International is to maximize the availability and use of its data for conservation purposes. Information on IBAs in Europe will therefore be made available beyond the BirdLife International Partnership, following the guidelines on use and dissemination of data that are stipulated in the *IBA Data Access Policy* (BirdLife International 1996). This agreement was drawn up by the European Partnership in February 1996 and applies to the IBA Database as it currently stands, plus subsequent updates.

The World Bird Database runs in single-user and multi-user environments and can be made available over the Internet. The development tools are industrial-strength products (Microsoft Visual Basic and Microsoft Access), which offer suitable upgrade paths when new technologies become available. The World Bird Database is simple to use and intuitive in operation, adopting the same look and feel as other widely used Windows products.

Figure 3. The procedure for collecting Important Bird Area data.

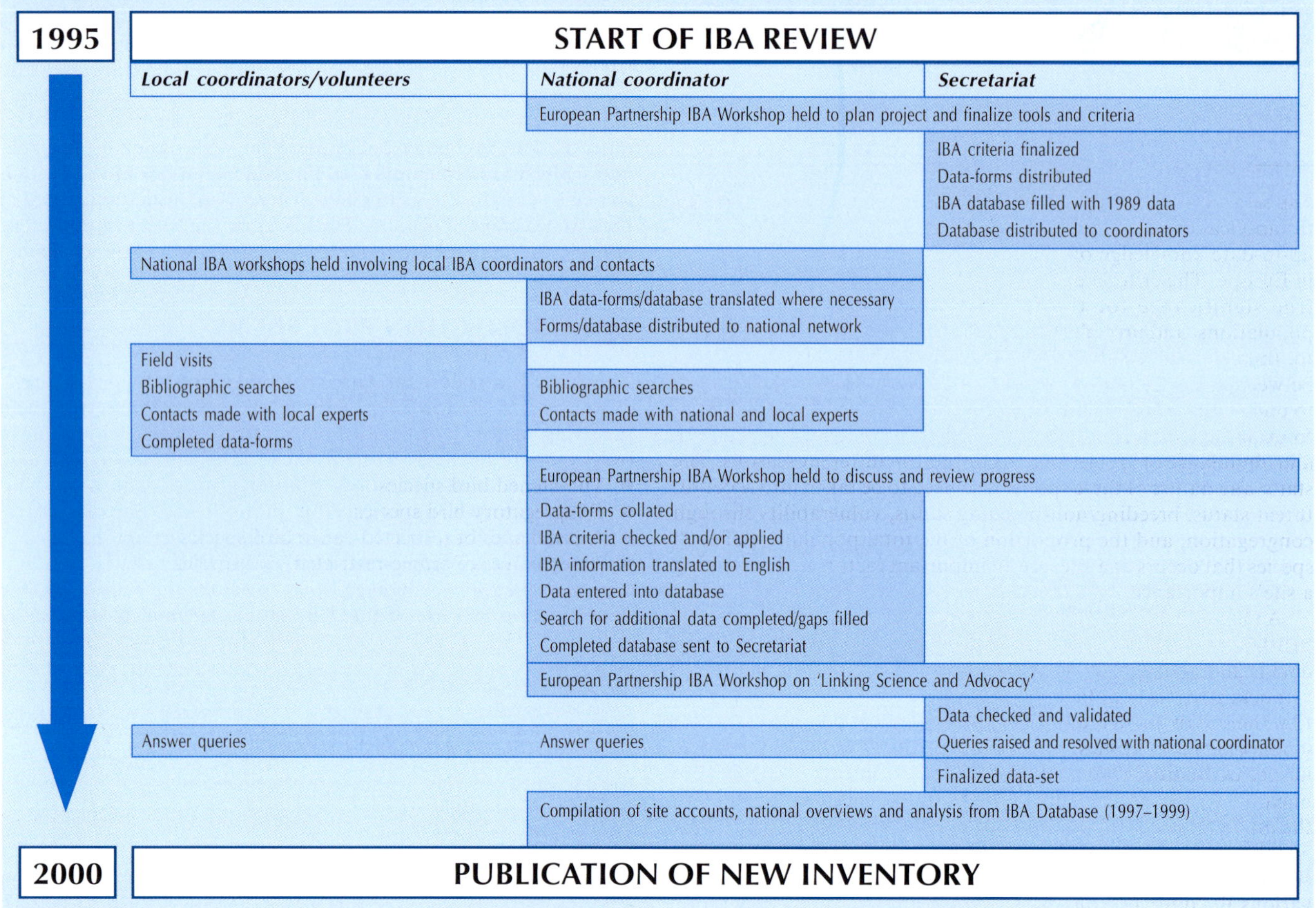

exact mechanism for collecting data within a country varied according to the resources available (in terms of time, people-power, expertise, money, and so on) and according to the existing methods for the collection of national bird data. Therefore Figure 3 only shows the general procedure followed.

■ Data checking/validation

Once the national network had submitted a suite of sites as potential IBAs, the data provided were then checked for errors and inconsistencies by national IBA coordinators and the BirdLife Secretariat, before the importance of each site was evaluated against the 20 ornithological criteria mentioned earlier. This evaluation process is explained more fully in the following chapter.

The use of pre-defined selection lists and standardized terms during the data-collection phase helped to reduce the amount of checking needed, and the computer database also allowed the checking and validation of huge quantities of data to be partially automated, at both international and national levels (see Box 6).

REFERENCES

BirdLife International (1995) IBA criteria: categories and thresholds. Cambridge, UK: BirdLife International (internal document).

BirdLife International (1996) IBA data access policy. Cambridge, UK: BirdLife International (internal document).

Collar, N. J., Crosby, M. J. and Stattersfield, A. J. (1994) *Birds to watch 2: the world list of threatened birds*. Cambridge, UK: BirdLife International (Conservation Series No. 4.).

Cramp, S. *et al.* (1977–1994) *The birds of the western Palearctic*. Vols 1–9. Oxford, UK: Oxford University Press.

Devillers, P. and Devillers-Terschuren (1996) *A classification of Palaearctic habitats*. Strasbourg: Council of Europe (Nature and environment, No. 78).

European Commission (1995) *Natura 2000 Network: Standard Data Form*. March 1995 Version.

Grimmett, R. F. A. and Jones, T. A. (1989) *Important Bird Areas in Europe*. Cambridge, UK: International Council for Bird Preservation (Techn. Publ. 9).

Heredia, B., Rose, L. and Painter, M., eds. (1996) *Globally threatened birds in Europe: action plans*. Strasbourg, France: Council of Europe.

IUCN (1994) *Guidelines for protected area management categories*. Gland, Switzerland: IUCN—The World Conservation Union.

Rose, P. M. and Scott, D. A. (1994) *Waterfowl population estimates*. Slimbridge, UK: Wetlands International.

Rose, P. M. and Scott, D. A. (1997) *Waterfowl population estimates*. Second edition. Wageningen, Netherlands: Wetlands International (Publication 44).

Scott, D. A. and Rose, P. M. (1996) *Atlas of Anatidae populations in Africa and western Eurasia*. Wageningen, Netherlands: Wetlands International (Publication 41).

Tucker, G. M. and Heath, M. F. (1994) *Birds in Europe: their conservation status*. Cambridge, UK: BirdLife International (BirdLife Conservation Series no. 3).

WCMC/WCPA [World Conservation Monitoring Centre/World Commission on Protected Areas] (1998) *1997 United Nations List of Protected Areas*. Gland, Switzerland and Cambridge, UK: IUCN.

WCMC [World Conservation Monitoring Centre] (1996) *Guide to information management in the Context of the Convention of Biological Diversity*. Nairobi, Kenya: United Nations Environment Programme.

IDENTIFYING IMPORTANT BIRD AREAS

WHY APPLY IBA CRITERIA?

The selection of Important Bird Areas (IBAs) is achieved through the application of quantitative ornithological criteria, grounded in up-to-date knowledge of the sizes and trends of bird populations in Europe. The criteria ensure that the sites selected as IBAs have true significance for the international conservation of bird populations, and provide a common currency that all IBAs adhere to, thus creating consistency among, and enabling comparability between, sites at national, continental and global levels. It is crucial to understand why a site is important, and to do this it is necessary to examine its international significance in terms of the presence and abundance of species that occur there in different seasons. The status and nature of these species also need to be taken into account: threat status, breeding/non-breeding status, vulnerability through congregation, and the proportion of the total population of each species that occurs at a site, are all important factors in determining a site's importance.

A main aim of the IBA Programme of BirdLife International is to attain protection for IBAs, and the provision of convincing bird data is an essential part of any argument for statutory protection. Importantly, the application of criteria to significant species, together with future data-gathering and the development of monitoring programmes, permit not only the assessment of changes in species' numbers but also an examination of how these changes impact on the overall importance of the site, thus helping to guide the management and conservation of the area. The more specific, quantitative and comprehensive is the information available on IBAs, with links showing the fulfilment of obligations laid out in various EC directives and international conventions, the stronger is the case for protection. To this end, the criteria build upon existing international legal instruments such as the EC Birds Directive which obliges the designation of Special Protection Areas in the European Community, and the Ramsar Convention under which contracting parties must designate at least one Ramsar Site.

THE CATEGORIES OF IBA CRITERIA

Twenty IBA criteria have been developed for the selection of IBAs in Europe. These allow the identification of IBAs, based on a site's **international** importance for:

- Threatened bird species
- Congregatory bird species
- Assemblages of restricted-range bird species
- Assemblages of biome-restricted bird species

Criteria have been developed such that, by applying different ('staggered') numerical thresholds, the international importance of a site for a species may be categorized at three distinct geographical levels:

- Global ('A' criteria)
- European ('B' criteria)
- European Union ('C' criteria)

A summary of each of the 20 criteria is given in Table 1. These 20 criteria are based on the criteria used in the first pan-European IBA inventory (Grimmett and Jones 1989), which in turn took

Table 1. Summary of the 20 criteria used in Europe to identify Important Bird Areas.

	Category	Criterion
GLOBAL	**A1. Species of global conservation concern**	The site regularly holds significant numbers of a globally threatened species, or other species of global conservation concern.
	A2. Restricted-range species	The site is known or thought to hold a significant component of the restricted-range species whose breeding distributions define an Endemic Bird Area (EBA) or Secondary Area (SA).
	A3. Biome-restricted species	The site is known or thought to hold a significant assemblage of the species whose breeding distributions are largely or wholly confined to one biome.
	A4. Congregations	(i) The site is known or thought to hold, on a regular basis, ≥1% of a biogeographic population of a congregatory waterbird species.
		(ii) The site is known or thought to hold, on a regular basis, ≥1% of the global population of a congregatory seabird or terrestrial species.
		(iii) The site is known or thought to hold, on a regular basis, ≥20,000 waterbirds or ≥10,000 pairs of seabird of one or more species.
		(iv) The site is known or thought to be a 'bottleneck' site where at least 20,000 storks (Ciconiidae), raptors (Accipitriformes and Falconiformes) or cranes (Gruidae) regularly pass during spring or autumn migration.
EUROPEAN	**B1. Congregations**	(i) The site is known or thought to hold ≥1% of a flyway or other distinct population of a waterbird species.
		(ii) The site is known or thought to hold ≥1% of a distinct population of a seabird species.
		(iii) The site is known or thought to hold ≥1% of a flyway or other distinct population of other congregatory species.
		(iv) The site is a 'bottleneck' site where over 5,000 storks, or over 3,000 raptors or cranes regularly pass on spring or autumn migration.
	B2. Species with an unfavourable conservation status in Europe	The site is one of the 'n' most important in the country for a species with an unfavourable conservation status in Europe (SPEC 2, 3) and for which the site-protection approach is thought to be appropriate.
	B3. Species with a favourable conservation status in Europe	The site is one of the 'n' most important in the country for a species with a favourable conservation status in Europe but concentrated in Europe (SPEC 4) and for which the site-protection approach is thought to be appropriate.
EUROPEAN UNION	**C1. Species of global conservation concern**	The site regularly holds significant numbers of a globally threatened species, or other species of global conservation concern.
	C2. Concentrations of a species threatened at the European Union level	The site is known to regularly hold at least 1% of a flyway population or of the EU population of a species threatened at the EU level (listed on Annex I and referred to in Article 4.1 of the EC Birds Directive).
	C3. Congregations of migratory species not threatened at the EU level	The site is known to regularly hold at least 1% of a flyway population of a migratory species not considered threatened at the EU level (as referred to in Article 4.2 of the EC Birds Directive) (not listed on Annex I).
	C4. Congregatory – large congregations	The site is known to regularly hold at least 20,000 migratory waterbirds and/or 10,000 pairs of migratory seabirds of one or more species.
	C5. Congregatory – bottleneck sites	The site is a 'bottleneck' site where at least 5,000 storks (Ciconiidae) and/or at least 3,000 raptors (Accipitriformes and Falconiformes) and/or 3,000 cranes (Gruidae) regularly pass on spring or autumn migration.
	C6. Species threatened at the European Union level	The site is one of the five most important in the European region (NUTS region) in question for a species or subspecies considered threatened in the European Union (i.e. listed in Annex I of the EC Birds Directive).
	C7. Other ornithological criteria	The site has been designated as a Special Protection Area (SPA) or selected as a candidate SPA based on ornithological criteria (similar to but not equal to C1–C6) in recognized use for identifying SPAs.

For species lists, see Tables 2–4 and Appendices 2a and 2b.

account of several previous studies of IBA criteria at the level of the European Community (Osieck and Mörzer Bruyns 1981, Grimmett and Gammell 1989). Appendix 4 gives a full comparison between the current criteria and those used in 1989. Adaptation of the 1989 criteria has been driven particularly by the globalization of the IBA programme, with IBAs being identified in the Middle East (Evans 1994), Africa (Fishpool in prep.), Asia and the Americas. This has resulted in:

- The inclusion of criteria to identify sites within areas of high avian endemism, termed Endemic Bird Areas (criterion A2) and listed in Stattersfield *et al.* (1998), and sites within certain biomes (criterion A3).
- The creation of staggered numerical thresholds (A, B and C levels) to allow meaningful comparison between sites across regions of the world. This is particularly significant in that many countries outside Europe, which are just beginning their IBA programmes, may not have completed an assessment of the threat status of species at the continental or sub-regional level, yet can still proceed in identifying IBAs (of global importance, i.e. using 'A' criteria) at this stage.
- The identification of sites for species of European conservation concern (Box 1), under the B2/B3 criteria.
- The strengthening of links to the EC Birds Directive, through the creation of criteria categories C1–C7 with associated numerical thresholds.

The IBA criteria have been developed by BirdLife International, involving all BirdLife European Partners in several stages of consultation and in workshops held between 1993 and 1996 (BirdLife International 1995, Heath 1995, 1996). The 'Birds and Habitats Directives' Task Force of BirdLife International played a leading role in developing the EU-specific categories and thresholds (C1–C7), to maximize their utility in guiding the selection of Special Protection Areas in EU countries (Osieck 1998).

A total of 515 bird species occur regularly in Europe. IBAs have been identified on the basis of the occurrence of many of these species, provided that they occur in sufficient numbers to meet numerical population thresholds. All 515 bird species are listed in Appendix 2a of this publication, together with the criteria which they can potentially fulfil and the numerical threshold(s) if applicable.

THE APPLICATION OF IBA CRITERIA

For most categories of criteria, application of a criterion involved comparing the data provided for each relevant bird species at the site—usually in the form of an estimate of the number of individuals or pairs of the species using the site—against a numerical threshold for the species concerned (normally representing 1% of the species' population in question, e.g. its global population if applying 'A1' or 'C1' criteria). For only two criteria, A2 and A3, was the application

Box 1. Defining species of European conservation concern (SPECs).

Species of conservation concern on a European scale have been identified by BirdLife International (Tucker and Heath 1994), based on their European Threat Status and the proportion of their world population in Europe.

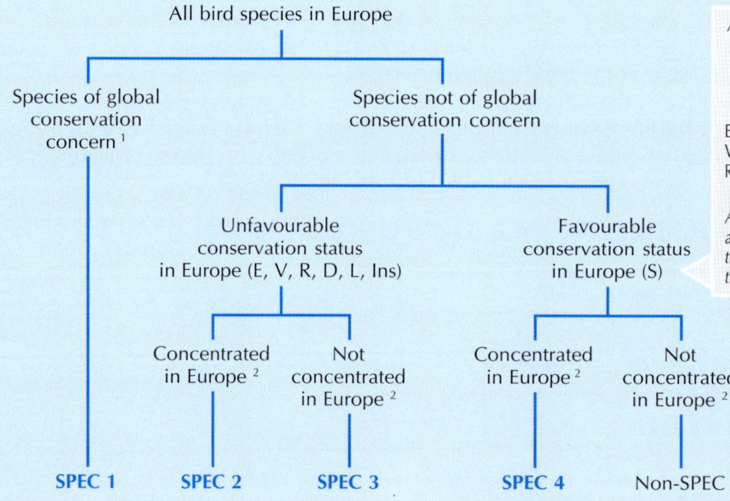

Allocation of conservation status as Unfavourable or Favourable is determined by **European Threat Status** *(see below)*

E Endangered D Declining
V Vulnerable L Localized
R Rare S Secure
Ins Insufficiently Known

A European Threat Status category is also assigned to SPEC 1 species, but these species are not dependent on this for their SPEC classification.

1. Species listed as Globally Threatened, Conservation Dependent or Data Deficient by Collar *et al.* (1994).
2. Concentrated in Europe: species with more than 50% of their global population or range lying within Europe.

European Threat Status: summary of criteria and categories

Criteria: European population size/trend	<250 pairs	<2,500 pairs	<10,000 pairs	>10,000 pairs
Large decline [3]	ENDANGERED	ENDANGERED	ENDANGERED	VULNERABLE
Moderate decline [4]	ENDANGERED	ENDANGERED	VULNERABLE	DECLINING
No decline	ENDANGERED	VULNERABLE	RARE	SECURE

In addition, species that have more than 10,000 pairs in Europe are categorized as LOCALIZED if more than 90% of the European population occurs at 10 sites or fewer in Europe. See Tucker and Heath (1994) for full details of criteria.

3. Large decline: applied to a breeding or wintering population which has declined in size or range by at least 20% in at least 66% of the population or by at least 50% in at least 25% of the population between 1970 and 1990, and where the total size of populations that declined is greater than the total size of populations that increased.
4. Moderate decline: applied to a breeding or wintering population which has declined in size or range by at least 20% in 33–65% of the population or by at least 50% in 12–24% of the population between 1970 and 1990, and where the total size of populations that declined is greater than the total size of populations that increased.

Note that:

- Winter population criteria use flyway population levels of less than 1,000, 10,000 and 40,000 individuals as respective equivalents to the figures of 250, 2,500 and 10,000 pairs used above for breeding populations.
- Due to inadequate data for most species, declines in winter populations are only considered for Anatidae, Haematopodidae, Charadriidae and Scolopacidae.

process more qualitative, requiring only that particular assemblages of bird species be present.

Each criterion is associated with a list of relevant species, with each species being accompanied, where necessary, by a numerical population threshold which must be matched or exceeded in order to allow the site to qualify under that criterion. These population thresholds were derived, wherever possible, from internationally recognised sources of bird population data.

In some countries it has not been possible to apply the criteria fully to all relevant species, due to a lack of data on some species. This is known to be the case for a number of bird taxa that are considered threatened at the scale of the European Union, being listed in Annex I of the EC Birds Directive. Therefore, the identification of sites for some of these species, or the data presented on qualifying species at these sites, may be incomplete.

The definitions of the criteria given in this chapter are guidelines for the identification of IBAs. They have been followed as far as possible but, since definitions of this sort cannot cover all possibilities, they are not inflexible rules. The need for scientific objectivity and standardization has had to be balanced by common sense and the practical objectives of the exercise.

DEFINING THE BOUNDARIES OF AN IBA

- A site is defined so that, as far as possible, it:
  - i) is different in character or habitat or ornithological importance from the surrounding area;
  - ii) exists as an actual or potential protected area, with or without buffer zones, or is an area which can be managed in some way for nature conservation;
  - iii) is, alone or with other sites, a self-sufficient area which provides all the requirements of the birds (that it is important for) which use it during the time that they are present.
- Where extensive tracts of continuous habitat occur which are important for birds, only characteristics ii) and iii) apply. This definition is not applicable to migratory bottleneck sites.
- Practical considerations of how best the site may be conserved are the foremost consideration.
- Simple, conspicuous boundaries such as roads or rivers can often be used to delimit site margins, while features such as watersheds, ridge-lines and hilltops can help in places where there are no obvious discontinuities in habitat (transitions of vegetation or substrate). Boundaries of ownership are also relevant.
- There is no fixed maximum or minimum size for IBAs—the biologically sensible should be tempered with the practical. Neither is there a definitive answer on how to treat cases where a number of small sites lie near each other. Whether these are best considered as a series of separate IBAs, or as one larger site containing areas lacking ornithological significance, depends upon the local situation with regard to conservation and management.

DETAILED DEFINITIONS OF OF IBA CRITERIA

A: Important Bird Areas – global importance

Globally threatened species – Category A1

The site regularly holds significant numbers of a globally threatened species, or other species of global conservation concern.

Under this criterion, sites are identified for those species most threatened with extinction at a global level. This includes species classified as 'Critical', 'Endangered' and 'Vulnerable', according to the most recent, universally recognized criteria for global threat status (IUCN 1994), as well as those species classified as 'Conservation Dependent', 'Data Deficient' or 'Near-threatened'. These latter types of species, although not strictly globally threatened, are considered here to be of sufficient global conservation concern to merit the identification of Important Bird Areas at the global level. All of these types of species are listed in Table 2 and in *Birds to Watch 2* (Collar *et al.* 1994).

This category thus allows the identification of IBAs for 35 species in Europe (Table 2). All of these species are also species of European conservation concern (see Box 1). The regular presence of a Critical or Endangered species at a site, irrespective of its abundance at the site, is considered sufficient to propose the site as an IBA. The only such species in Europe are *Pterodroma madeira* and *Numenius tenuirostris*. The remaining 33 species have to be present at a site in 'significant' numbers for a site to qualify under this criterion. The relevant numerical threshold for each species is calculated from the size of the species's global population and also depends on whether the species has a relatively large or small body-size, and whether it has primarily dispersed or colonial nesting habits (see footnote to Table 2).

The words 'regular' and 'significant' in these definitions are intended to exclude instances of vagrancy, marginal occurrence, and ancient or historical records. 'Regularly' includes seasonal presence (and presence at longer intervals, if suitable conditions themselves only occur at extended intervals, e.g. at temporary wetlands).

Table 2. Numerical thresholds for species of global conservation concern in Europe.

Species	Global threat status	Threshold (pairs)
Pterodroma feae Fea's Petrel	VU	5
Pterodroma madeira Zino's Petrel	CR	0
Phalacrocorax pygmeus Pygmy Cormorant	NT	10
Pelecanus crispus Dalmatian Pelican	VU	10
Anser erythropus Lesser White-fronted Goose	VU	5
Branta ruficollis Red-breasted Goose	VU	60 ind
Marmaronetta angustirostris Marbled Teal	VU	5
Aythya nyroca Ferruginous Duck	VU	20
Polysticta stelleri Steller's Eider	VU	30 ind
Oxyura leucocephala White-headed Duck	VU	5
Haliaeetus albicilla White-tailed Eagle	NT	5
Aegypius monachus Cinereous Vulture	NT	5
Circus macrourus Pallid Harrier	NT	10
Aquila clanga Greater Spotted Eagle	VU	2
Aquila heliaca Imperial Eagle	VU	2
Aquila adalberti Spanish Imperial Eagle	VU	2
Falco naumanni Lesser Kestrel	VU	10
Tetrao mlokosiewiczi Caucasian Black Grouse	NT	20
Crex crex Corncrake	VU	20
Tetrax tetrax Little Bustard	NT	60 ind
Otis tarda Great Bustard	VU	30 ind
Glareola nordmanni Black-winged Pratincole	NT	10
Chettusia gregaria Sociable Plover	VU	10
Gallinago media Great Snipe	NT	20
Numenius tenuirostris Slender-billed Curlew	CR	0
Larus audouinii Audouin's Gull	CD	20
Columba trocaz Long-toed Pigeon	CD	10
Columba bollii Dark-tailed Laurel Pigeon	VU	5
Columba junoniae White-tailed Laurel Pigeon	VU	5
Saxicola dacotiae Fuerteventura Chat	NT	5
Acrocephalus paludicola Aquatic Warbler	VU	10
Sitta whiteheadi Corsican Nuthatch	NT	10
Fringilla teydea Blue Chaffinch	CD	10
Loxia scotica Scottish Crossbill	DD	5
Emberiza cineracea Cinereous Bunting	NT	5

All figures are in pairs unless "ind" is given, indicating individuals. To convert between individuals and pairs, a multiplying/dividing factor of 3 was used.

Globally threat status follows Collar *et al.* (1994):
CR Critical
VU Vulnerable
CD Conservation Dependent
DD Data Deficient
NT Near Threatened

Threshold calculations: For Vulnerable, Conservation Dependent, Data Deficient and Near-threatened species the following threshold levels apply at any site:

Threshold	European population (pairs) <1,000	1,000–10,000	>10,000
large sized and/or fairly dispersed species	2	5	10
small sized and/or colonial nesting species	5	10	20

Restricted-range species – Category A2

The site is known or thought to hold a significant component of a group of species whose breeding distributions define an Endemic Bird Area or a Secondary Area.

Under this criterion, the most important sites within Endemic Bird Areas (EBAs) and Secondary Areas are identified.

An EBA is defined as a region to which two or more restricted-range bird species are confined, with 'restricted range' defined as a world distribution of less than 50,000 km^2 (Stattersfield *et al.* 1998). There are 218 EBAs globally, three of which are in Europe (Table 3): Madeira and the Canary Islands (with nine restricted-range species), the Caucasus (three restricted-range species) and Cyprus (two restricted-range species). Two of these species are also globally threatened, and nearly all are species of European conservation concern (Table 3), thus also qualifying under other criteria.

For many EBAs, which hold a large number of restricted-range species, it is necessary that a network of sites be chosen, using complementarity analysis, to protect adequately all relevant species. The term 'significant component' in the definition of the criterion is intended to avoid the selection of sites solely on the presence of one or more restricted-range species that are common and adaptable within the EBA and which may, therefore, occur at other (many other) chosen sites. Additional sites can, however, be chosen for one or a few species that would otherwise be under-represented or not represented at all.

Also included within this category are species of Secondary Areas. A Secondary Area supports one or more restricted-range species, but does not qualify as an EBA because only a single species is entirely confined to it (Stattersfield *et al.* 1998). There are three Secondary Areas in Europe: 'Corsican mountains' (France) for *Sitta whiteheadi*, 'Caledonian pine forest' (United Kingdom) for *Loxia scotica*, and the Azores (Portugal) for *Serinus canaria* and *Pyrrhula murina*.

Table 3. Restricted-range bird species in Europe.

Endemic Bird Area	Global threat status	SPEC category/European threat status
Madeira and the Canary Islands		
Columba trocaz Long-toed Pigeon	CD	1 Vulnerable
Columba bollii Dark-tailed Laurel Pigeon	VU	1 Vulnerable
Columba junoniae White-tailed Laurel Pigeon	VU	1 Vulnerable
Apus unicolor Plain Swift	—	4 Secure
Anthus berthelotii Berthelot's Pipit	—	4 Secure
Saxicola dacotiae Fuerteventura Chat	NT	2 Vulnerable
Regulus teneriffae Tenerife Goldcrest	—	4 Secure
Fringilla teydea Blue Chaffinch	CD	1 Vulnerable
Serinus canaria Canary	—	4 Secure
Caucasus		
Tetrao mlokosiewiczi Caucasian Black Grouse	NT	2 Insufficiently known
Tetraogallus caucasicus Caucasian Snowcock	—	4 Secure
Phylloscopus lorenzii Caucasian Chiffchaff	—	— (Secure)
Cyprus		
Oenanthe cypriaca Cyprus Pied Wheatear	—	2 Rare
Sylvia melanothorax Cyprus Warbler	—	2 Rare
Secondary Area		
Corsican mountains		
Sitta whiteheadi Corsican Nuthatch	NT	2 Vulnerable
Azores		
Serinus canaria Canary	—	4 Secure
Pyrrhula murina Azores Bullfinch	[not assessed]	
Caledonian pine forest		
Loxia scotica Scottish Crossbill	DD	1 Insufficiently known

Global threat status follows Collar *et al.* (1994):
VU Vulnerable
CD Conservation Dependent
DD Data Deficient
NT Near Threatened

For key to SPEC category and European threat status, see Box 1.

Biome-restricted assemblages – Category A3

The site is known or thought to hold a significant component of the group of species whose distributions are largely or wholly confined to one biome.

This category applies to groups of bird species with largely shared distributions (each individual species's distribution being greater than 50,000 km^2), which occur mostly or wholly within a particular biome and which are, therefore, of global importance. Many of these assemblages occur in large areas of relatively intact and continuous habitat where delimiting IBAs is particularly difficult.

Biome-restricted species are those species whose entire (global) breeding distribution lies within the defined boundaries of the biome (with a few exceptions where a small part of the distribution may extend to another biome). Seabirds are excluded from these lists because their distributions are thought to be influenced by different factors to those affecting terrestrial species, and their conservation

Table 4. Biome-restricted bird species in Europe.

Arctic/tundra biome
Gavia adamsii White-billed Diver
Cygnus columbianus Bewick's Swan
Anser brachyrhynchus Pink-footed Goose
Anser albifrons White-fronted Goose
Anser erythropus Lesser White-fronted Goose
Branta leucopsis Barnacle Goose
Branta bernicla Brent Goose
Aythya marila Scaup
Somateria spectabilis King Eider
Clangula hyemalis Long-tailed Duck
Melanitta nigra Common Scoter
Buteo lagopus Rough-legged Buzzard
Falco rusticolus Gyrfalcon
Pluvialis squatarola Grey Plover
Calidris canutus Knot
Calidris alba Sanderling
Calidris minuta Little Stint
Calidris temminckii Temminck's Stint
Calidris maritima Purple Sandpiper
Limosa lapponica Bar-tailed Godwit
Tringa erythropus Spotted Redshank
Phalaropus lobatus Red-necked Phalarope
Phalaropus fulicarius Grey Phalarope
Stercorarius pomarinus Pomarine Skua
Stercorarius longicaudus Long-tailed Skua
Larus hyperboreus Glaucous Gull
Pagophila eburnea Ivory Gull
Nyctea scandiaca Snowy Owl
Anthus cervinus Red-throated Pipit
Carduelis hornemanni Arctic Redpoll
Calcarius lapponicus Lapland Bunting
Plectrophenax nivalis Snow Bunting

Boreal biome
Podiceps auritus Slavonian Grebe
Mergus albellus Smew
Lymnocryptes minimus Jack Snipe
Tringa nebularia Greenshank
Surnia ulula Hawk Owl
Strix nebulosa Great Grey Owl
Bombycilla garrulus Waxwing
Phylloscopus borealis Arctic Warbler
Parus cinctus Siberian Tit
Perisoreus infaustus Siberian Jay
Fringilla montifringilla Brambling
Loxia leucoptera Two-barred Crossbill
Loxia pytyopsittacus Parrot Crossbill
Pinicola enucleator Pine Grosbeak
Emberiza rustica Rustic Bunting

Mediterranean biome
Falco eleonorae Eleonora's Falcon
Alectoris graeca Rock Partridge
Caprimulgus ruficollis Red-necked Nightjar
Oenanthe cypriaca Cyprus Pied Wheatear
Oenanthe hispanica Black-eared Wheatear
Oenanthe leucura Black Wheatear
Hippolais olivetorum Olive-tree Warbler
Sylvia sarda Marmora's Warbler
Sylvia conspicillata Spectacled Warbler
Sylvia cantillans Subalpine Warbler
Sylvia melanocephala Sardinian Warbler
Sylvia melanothorax Cyprus Warbler
Sylvia rueppelli Rüppell's Warbler
Sitta krueperi Krüper's Nuthatch
Sitta whiteheadi Corsican Nuthatch
Sitta neumayer Rock Nuthatch
Lanius nubicus Masked Shrike
Sturnus unicolor Spotless Starling
Emberiza cineracea Cinereous Bunting
Emberiza caesia Cretzschmar's Bunting
Emberiza melanocephala Black-headed Bunting

Eurasian high-montane (alpine) biome
Tetrao mlokosiewiczi Caucasian Black Grouse
Tetraogallus caucasicus Caucasian Snowcock
Tetraogallus caspius Caspian Snowcock
Prunella collaris Alpine Accentor
Sitta tephronota Eastern Rock Nuthatch
Tichodroma muraria Wallcreeper
Pyrrhocorax graculus Alpine Chough
Montifringilla nivalis Snowfinch
Serinus citrinella Citril Finch
Carpodacus rubicilla Great Rosefinch

Eurasian steppe biome
Circus macrourus Pallid Harrier
Accipiter brevipes Levant Sparrowhawk
Aquila heliaca Imperial Eagle
Anthropoides virgo Demoiselle Crane
Glareola nordmanni Black-winged Pratincole
Chettusia gregaria Sociable Plover
Larus ichthyaetus Great Black-headed Gull
Melanocephala leucoptera White-winged Lark
Melanocephala yeltoniensis Black Lark

is covered through the application of other criteria categories (see criterion A4).

A biome is defined as a major regional ecological community characterized by distinctive life forms and principal plant species. No global classification of biomes has been found which is suitable for generating bird-species lists for BirdLife's IBA Programme. This has necessitated a regional approach to the identification of biomes and has resulted in inter-regional differences between the biome classifications used but, as far as possible, the overall scale at which biome divisions are recognised—the 'depth' of treatment—is comparable.

Five biomes have been treated under this criterion in Europe: the Arctic/tundra biome (with 32 characteristic bird species in Europe), the boreal biome (15 species), the Mediterranean biome (21 species), the Eurasian high-montane biome (10 species) and the Eurasian steppe biome (nine species). The geographical extent of these biomes in Europe is shown in Box 1of the 'Overview of results' chapter, and Table 4 lists the bird species characteristic of each biome.

In applying the criterion, there were some important considerations, as follows:

- ***Number and area of sites:*** The number of sites selected per country under this category took into account both the size of the country and the relative amount of a given biome within it. The size of the site is also relevant; it is preferable to select a few, large sites that reflect the distribution of biome across the country rather than many small ones confined to only a part of it. This ensures that a greater number of species are represented per site and takes account of their geographical distribution. Sites should not, however, be so large that they are not amenable to conservation and, in some cases, small sites with high population densities may be preferable to large ones with lower densities.
- ***Coverage of all biome species:*** Common sense was used to ensure that sites chosen were rich in biome-restricted species with each in high numbers wherever possible. Some sites, however, were chosen for one or a few species which would otherwise be under-represented, such as those species confined to a relatively small part of the biome.
- ***Geographical spread of sites throughout biome:*** All of the biomes in Europe (that are treated under this criterion) cross political boundaries, and most have a wide geographical extent. In identifying a network of sites under these criteria, the geographical spread of the biome across political boundaries was taken into account, to try to ensure that the network of IBAs identified covered much of the biological and political extent of the biome.

Globally important congregations – Category A4

The site may qualify on any one of the four criteria listed below:

i) The site is known or thought to hold, on a regular basis, 1% or more of a biogeographic population of a congregatory waterbird species.

ii) The site is known or thought to hold, on a regular basis, 1% or more of the global population of a congregatory seabird or terrestrial species.

iii) The site is known or thought to hold, on a regular basis, at least 20,000 waterbirds, or at least 10,000 pairs of seabird, of one or more species.

iv) The site is known or thought to be a 'bottleneck site' where at least 20,000 storks (Ciconiidae), raptors (Accipitriformes and Falconiformes) or cranes (Gruidae) pass regularly during spring or autumn migration.

This category was applied to those species that are vulnerable, at the population level, to the destruction or degradation of sites, by virtue of their congregatory behaviour when breeding, wintering or on passage. A total of 160 species were treated when applying these criteria in Europe, and 1% thresholds for all of these species are given in Appendix 2a. A few species of waterbird and raptor that are not considered to be congregatory (in Europe), or which have small, marginal populations in Europe, are not treated under this criteria category.

Criteria A4i and A4iii identify wetlands of international importance (Ramsar Sites), being similar to Ramsar criteria 6 and 5 respectively (see Box 4).

Definition of 'waterbird' and 'seabird'

The term 'waterbird' is used in the same sense as that used for 'waterfowl' under the Ramsar Convention, and covers (in Europe) all bird species in the following families (Rose and Scott 1997): Gaviidae (divers), Podicipedidae (grebes), Pelecanidae (pelecans), Phalacrocoracidae (cormorants), Ardeidae (herons), Ciconiidae (storks), Threskiornithidae (ibises), Phoenicopteridae (flamingos), Anatidae (wildfowl), Gruidae (cranes), Rallidae (rails), Haematopodidae (oystercatchers), Recurvirostridae (stilts, avocets), Burhinidae (stone-curlews), Glareolidae (pratincoles), Charadriidae (plovers), Scolopacidae (sandpipers and allies) and Laridae (gulls and terns). By this definition waterbirds include, for example, cormorants, gulls and terns, which some authors have more traditionally considered as seabirds. The term 'seabird' covers, in Europe, all bird species in the following families: Procellaridae (fulmars, petrels, shearwaters), Hydrobatidae (storm-petrels), Sulidae (gannets), Stercoraridae (skuas) and Alcidae (auks). A list of the criteria categories within which these species are considered is presented in Appendix 2a.

Definition of 'biogeographic population'

'Biogeographic' is used in the sense of a zoogeographic realm, e.g. the Palearctic or Afrotropical realms, which are large geographical regions in which the organisms present tend to be different from those of other realms. Thus such regions are characterized largely through the shared distribution patterns of many species. For European IBAs, the 'biogeographic' region approximates to the western Palearctic, but excludes the Middle East (for which, see Evans 1994). All 'populations' of a given species that are resident or migratory through this region are combined to form the 'biogeographic population'. For most waterbird species, the biogeographic population is taken here to equal the European breeding population. For a few species, part or all of the total passage/wintering population in Europe breeds outside the continent (e.g. in Siberia), therefore the biogeographic populations are calculated by summing the respective national wintering/passage population estimates across Europe. Feral populations of all qualifying species have, as far as possible, been excluded when applying these criteria.

Setting 1% thresholds and applying the criteria

One-percent threshold figures have been defined for all congregatory waterbird species (listed in Appendix 2a), including species for which no thresholds are currently recognised under the Ramsar Convention. Wetlands International has collaborated in generating numeric thresholds from range estimates and from unpublished population data.

There is a logical inconsistency between criterion A4i for waterbirds (1% or more of the **biogeographic** population) and criterion A4ii for seabirds (1% or more of **global** population of seabirds). It was felt, however, that the alternative of using 1% of the global population for waterbirds would, as well as departing from the criteria used under the Ramsar Convention, have insufficient biological justification, since relatively well-defined, discrete flyway populations can be distinguished within Europe for many migratory waterbird species. Taking 1% of global population would over-emphasise waterbirds endemic to Europe, since many widely distributed species may rarely occur at congregations exceeding 1% of the global population, over much of their range. For the same reason, in Europe the biogeographic and global populations are considered the same for regionally endemic waterbird species.

The A4iii and A4iv criteria are applied at the site level only, not to individual species. The most common species under these criteria are recorded in the text of the site-account where possible. The use of the A4iii criterion alone for a site has been discouraged in Europe when bird data for the site exceed 1% threshold levels for one or more individual species (cf. Atkinson-Willes 1976).

The A1iv criterion embraces sites over which flying migrants concentrate, e.g. at narrow sea-crossings, along mountain ranges or through mountain passes. Although it is the airspace here that is important, conservation of the land beneath may be necessary to protect the site and its birds from threats such as shooting and the construction of lethal obstacles such as power-lines and high radio-masts. Also included under A1iv are migratory stop-over sites and nocturnal roosts which may not hold 20,000 or more storks, raptors or cranes at any one time but which, nevertheless, do hold such numbers over a relatively short period due to the rapid turnover of birds on passage.

B: Important Bird Areas – European importance

Regionally important congregations – Category B1

The site may qualify on <u>any one</u> of the four criteria listed below:

i) The site is known or thought to hold 1% or more of a flyway population or other distinct population of a waterbird species.

ii) The site is known or thought to hold 1% or more of a distinct population of a seabird species.

iii) The site is known or thought to hold 1% or more of a flyway population or other distinct population of a congregatory species other than a waterbird or seabird.

iv) The site is a 'bottleneck site' where 5,000 or more storks (Ciconiidae), or 3,000 or more raptors (Accipitriformes and Falconiformes) or cranes (Gruidae), pass regularly on spring or autumn migration.

The aim of this category of criteria is the same as that for the A4 category, that is to identify important sites for species which are vulnerable at sites because of their congregatory nature. However, the B1 category sets lower numerical thresholds, based largely on 1% values of the flyway population or other distinct regional population of congregatory species, and the thresholds for 'bottleneck' sites are also lower.

For species without distinct populations in Europe, as is the case for many of the seabirds, the global and regional thresholds are the same. Appendix 2a lists, for each relevant species, the geographical limits of each flyway population (if any) and the corresponding 1% threshold used, following Rose and Scott (1994, 1997).

Criterion B1i covers wetlands of international importance (Ramsar Sites) identified under Ramsar criterion 6 (see Box 4).

Definition of 'flyway population'

For waterbird species, flyway or other distinct populations in Europe have been identified by Wetlands International (Rose and Scott 1997, Scott and Rose 1996), who refer to these units interchangeably as 'geographic regions', 'biogeographic regions', 'biogeographic populations' or 'geographic limits of every known distinct population of a species or subspecies'. These biogeographic areas vary from species to species, and the resulting 1% thresholds can be applied in different seasons.

Species with an unfavourable conservation status in Europe – Category B2

The site is one of the 'n' most important sites in a country for a species with an unfavourable conservation status in Europe (endangered, vulnerable, rare, declining, localized or insufficiently known in Europe), and for which the site-protection approach is thought to be appropriate.

Under this criterion, sites are identified for those species of European conservation concern (SPECs) of categories 1, 2 and 3 (see Box 1 and Appendix 2a) for which the site-protection approach is thought to be appropriate.

Setting thresholds and applying the criteria

In order to identify a network of IBAs covering a substantial proportion of the European population of each relevant species, throughout its European range, numerical thresholds were defined, as follows.

- For each country holding 1% or more of the minimum European breeding population of a given species, those sites which support 1% or more of the minimum national breeding population should be selected.

Assuming that these thresholds are met for a particular species in a particular country, there is also an upper limit—'n'—to the number of sites allowed to be identified in that country for that species, ranging from five to 100 depending on the circumstances. The procedure for determining this value is explained further in Box 2.

This criterion addresses the problem of identifying IBAs for species that are widely dispersed across the landscape but which are amenable to conservation through site protection, and is framed so as to limit the maximum number of qualifying sites in countries with large total populations of any species (Box 2). The criterion should, however, be used with caution, for example in countries where absolute populations of a species are low (e.g. 100 pairs or less), since use of the 1% level loses meaning if a site qualifies on the basis of a single pair. Also, for countries which hold less than 1% of the European population of a given species, or for countries which comprise less than 1% of the land area of Europe (i.e. less than c.100,000 km^2), sites may still be selected under this criterion if they support similar numbers of the species as sites in other countries which meet this criterion in a standard fashion.

For many widely dispersed SPEC2 and SPEC3 species, the site-protection approach may not be appropriate over large parts of their ranges in Europe. Yet, because they are dispersed, many IBAs identified for other species are likely to hold, overall, a sizeable population of these species. In addition, towards the edge of their range they may occur in well-defined sites, which could be considered important for maintaining the overall range of the species even if the absolute numbers occurring at any one of these 'edge' sites are low.

The B2 criterion is applied to bird data for the season in which the species qualifies as a SPEC. As a result, the great majority of applications of this criterion concern the breeding season (and are computed using breeding-population data), since only a few species in Europe have been identified as SPECs on the basis of their non-breeding populations (see Box 1).

European and national population data

European and national population data are taken from the European Bird Database (1994 version) of BirdLife International/ European Bird Census Council. This database was updated in 1998 but as the majority of the IBAs had been proposed by this time, the new figures could not be used for most countries. However, where changes are substantial for a country, these new figures have been taken into account when finalizing the IBA identification.

The proportion of the national population of each SPEC falling within IBAs has been calculated for each country and these figures are presented in the respective national overviews. The figures should be interpreted with care as the national population estimate may not have been updated as recently as the population estimates for each IBA in the country. In addition, the stated count for a species at a site may be the maximum or average over recent years, and summing these may record more birds than are present nationally in any single year. Full documentation of the national population estimates used for each species, the proportions of the European total and calculation ratios cannot be presented in this publication because of space constraints. However, much of the relevant data is published in Tucker and Heath (1994) and Heath and Borggreve (2000).

Species with a favourable conservation status but concentrated in Europe – Category B3

The site is one of the 'n' most important sites in a country for a species with a favourable conservation status in Europe but with its global range concentrated in Europe, and for which the site-protection approach is thought to be appropriate.

This criterion applies to those species of European conservation concern (SPECs) of category 4 (see Box 1)—i.e. those with a favourable conservation status in Europe but with more than 50% of their global range lying within Europe (Appendix 2a)—for which the site-protection approach is thought to be appropriate. In effect, the criterion identifies sites for species with ranges greater than 50,000 km^2 and which are not restricted to a biome treated under the A3 criterion. The principles and methods used for setting thresholds, calculating the maximum number of sites per species in each country (see Box 2), and applying the criteria, are the same as for the B2 criterion.

Box 2. An example of the application of the B2/B3 criteria.

Identifying IBAs for Common Gull *Larus canus* in Finland

Minimum European breeding population estimate = 420,000 pairs
Minimum national breeding population estimate = 50,000 pairs

1. The country should hold at least 1% of the total European population.
Finland holds 12% of the European breeding population of Larus canus.
2. The proposed site should hold at least 1% of the national population.
A total of 15 sites in Finland each hold at least 500 pairs of Larus canus.
3. Calculate the proportion of the national population relative to the European one. This % value corresponds to a guideline figure for the maximum number of sites ('n' max) that can be proposed for the given species in a country.
Finland holds 12% of the European breeding population of Larus canus.

The 'n' max figure is determined by the minimum size of the national population relative to the minimum estimate of the total European population. The table provides guidelines for calculating the **maximum** number of sites ('n' max) for any given species. It is emphasized that these are guidelines for the **maximum** number of sites that can be proposed and, where it is appropriate to identify a lower number of sites for a given species in a country, this is entirely acceptable.

Proportion (%) of total European population or range held by the country in question	Maximum number of sites that may be identified in the country in question ('n' max)
1–5	5
10	10
20	16–25
30	26–35
40	36–45
50	46–55
60	56–65
70	66–75
80	76–85
90	86–95
100	96–100

Therefore up to 10 IBAs may be identified for Larus canus *in Finland, with each having at least 500 pairs of the species. From the 15 sites that actually hold at least 500 pairs, a maximum of 10 (usually those with the highest population estimate) were selected as IBAs.*

Note: For countries which hold less than 1% of the total European population of a given species, it is still possible to propose sites which are thought to hold internationally important numbers of the species: population levels at these sites were then compared to sites identified in other countries using the above procedure. Countries with a surface area less than 1% of the total European territory (see Step 1) were particularly encouraged to use this method (with caution) to identify potential IBAs for selected species.

C: Important Bird Areas – European Union importance

The 'C' categories of criteria are used for selecting sites in the European Union which qualify, under the EC Birds Directive, as Special Protection Areas (SPAs). For details of the EC Birds Directive, see Appendix 1. These ornithological criteria represent a consolidation of the criteria which have been used, to date, by the different member states of the EU. Full details of the development of the C criteria are presented in Osieck (1998).

Importantly, the C criteria categories are based on those used in the most recent inventory of IBAs in the European Community (Grimmett and Gammell 1989), which received legal recognition as a scientific reference in a ruling by the European Court of Justice against the government of the Netherlands in 1998 (see Box 3). The criteria presented here are a logical progression from those of Grimmett and Gammell (1989), introducing some additional quantitative thresholds (Appendix 4). The criteria take into account the conservation requirements of species within the EU territory, with the geographical spread of sites representing the full extent of each species' range in the EU (irrespective of the pattern of regional abundance) as well as sites selected on a basis of relative abundance. The criteria do not produce a more restricted selection of sites than those criteria applied in the past.

Seven categories of criteria have been applied (C1–C7), several of which emulate the higher categories under the global (A) and European (B) level criteria. It should be noted that Annex I of the EC Birds Directive lists a number of subspecies that should be

Box 3. Legal recognition of the value of the European Important Bird Area inventory by the European Court of Justice.

The 1989 IBA inventory for the European Community (Grimmett and Gammell 1989) received legal recognition in 1998 in the European Court of Justice ruling against the government of the Netherlands. On 19 May 1998 the Court of Justice delivered a significant judgment against the Netherlands in an infringement case (Case C-3/96 Commission v Netherlands [1998] ECR 1-3031, paragraphs 60–63). The Court reasoned as follows:

"...while the Member States have a certain margin of discretion in the choice of SPAs, the classification of those areas is nevertheless subject to certain ornithological criteria determined by the Directive [...]. It follows that the Member States' margin of discretion in choosing the most suitable territories for classification as SPAs does not concern the appropriateness of classifying as SPAs the territories which appear the most suitable according to ornithological criteria, but only the application of those criteria for identifying the most suitable territories for conservation of the species listed in Annex I to the Directive. Consequently, Member States are obliged to classify as SPAs all the sites which, applying ornithological criteria, appear to be the most suitable for conservation of the species in question. Thus where it appears that a Member State has classified as SPAs sites the number and total area of which are manifestly less than the number and total area of the sites considered to be the most suitable for conservation of the species in question, it will be possible to find that the Member State has failed to fulfil its obligation under Article 4(1) of the Directive".

The Court accordingly dismisses the Netherlands Government's argument that the Commission must establish, territory by territory, specific infringements of that provision.

[1] "The Court went on to acknowledge the relevance of the *Inventory of Important Bird Areas in the European Community* [Grimmett and Gammell 1989] prepared for the competent Directorate-General of the Commission by the 'Eurogroup for the Conservation of Birds and Habitats' in conjunction with the International Council for Bird Preservation [now BirdLife International] and in cooperation with Commission experts. [Grimmett and Gammell (1989) lists, with one exception, the same sites as Grimmett and Jones (1989).] That inventory, although not legally binding on the Member States concerned, could, by reason of its acknowledged scientific value in the present case, be used by the Court as a basis of reference for assessing the extent to which the Kingdom of the Netherlands had complied with its obligation to classify SPAs. In the circumstances, IBA89 [Grimmett and Gammell 1989] had proved to be the only document containing scientific evidence making it possible to assess whether the defendant State had fulfilled its obligation to classify as SPAs the most suitable territories in number and area for conservation of the protected species. The situation would have been different if the Kingdom of the Netherlands had produced scientific evidence in particular to show that the obligation in question could be fulfilled by classifying as SPAs territories whose number and total area were less than those resulting from IBA89."

"The Commission is continuing Article 171 proceedings against the Netherlands, to obtain implementation of the judgments." [This means a subsequent judgment, in which the Court is asked to confirm the non-implementation of the first judgement and to impose a fine of up to 220,000 Euro per day in the case of the Netherlands.]

1. Source: Page 78 in *XVIth Report on monitoring the application of Community law*. Brussels: European Commission (COM(1999)301 final, adopted on 09/07/99).

treated independently of their 'mother' species when applying C criteria. These subspecies are listed in Appendix 2b.

Species of global conservation concern – C1

The site regularly holds significant numbers of a globally threatened species, or other species of global conservation concern.

This criterion is identical to the A1 criterion, and the same thresholds apply (see Appendices 2a and 2b).

Concentrations of species threatened at the European Union level – C2

The site is known to regularly hold at least 1% of the flyway or EU population of a species considered to be threatened in the EU.

'Threatened species' refers to species, subspecies and populations listed in Annex I of the EC Birds Directive, for which Special Protection Areas (SPAs) are designated under Article 4.1 of the Directive. The definition of 'flyway population' is the same as that given for the B1 criteria. However, for a small number of species where the European breeding population is significantly larger than the EU breeding population, lower numerical thresholds have been set (see Appendix 2b). This criterion has also been applied for a number of dispersed species on the basis that the site holds more than 1% of the European population of the species.

Migratory non-threatened species – C3

The site is known to regularly hold at least 1% of a flyway population of a migratory species that is not considered to be threatened in the EU.

'Migratory species not considered threatened' refer to species considered under Article 4.2 of the Birds Directive (i.e. regularly occurring migratory species not listed in Annex I). 'Migration' is defined as seasonal long-distance movements from and to breeding areas. The word 'migratory' therefore excludes populations which are largely sedentary or short-distance dispersive (e.g. *Larus argentatus* and *Cepphus grylle* in western Europe). This criterion covers wetlands of international importance (Ramsar Sites) identified under Ramsar criteria category 6, to which reference is made in Article 4.2 of the Birds Directive. Wetlands of international importance uniquely qualifying for waterbirds listed in Annex I of the Birds Directive are covered by IBA criterion C2.

The definition of 'flyway population' is the same as that given for the B1 criteria. Lower numerical thresholds (than those used for B1 criteria) have not been set.

Large congregations – C4

The site is known to regularly hold at least 20,000 migratory waterbirds, or at least 10,000 pairs of migratory seabird, of one or more species.

This criterion is the same as the A4iii criterion and covers all wetlands of international importance identified under Ramsar criteria category 5.

Large congregations – 'bottleneck' sites – C5

The site is a 'bottleneck' site where at least 5,000 storks (Ciconiidae), or at least 3,000 migratory raptors (Accipitriformes and Falconiformes) or cranes (Gruidae), regularly pass on spring or autumn migration.

This criterion is the same as the B1iv criterion. As most of the species concerned are listed in Annex I of the Birds Directive, this criterion refers mainly to sites important in the context of Article 4.1.

Species threatened at the European Union level – C6

The site is one of the five most important in the European region in question for a species or subspecies considered threatened in the European Union.

'Threatened species' refers to species, subspecies and populations listed in Annex I of the EC Birds Directive. 'European region' refers to what are known as NUTS regions (Appendix 5). The Nomenclature of Territorial Units for Statistics (NUTS) was established by Eurostat—the EC Statistical Office—to provide a single uniform breakdown of approximately equal territorial units for the production of regional statistics for the European Union. Although the NUTS has no legal value *per se*, it has been used since 1988 in Community legislation. NUTS regions are equalized on the basis of human population density: regions are larger where population density is lower. The NUTS approach is not ideal for birds, because many species of birds occur preferentially in remote, sparsely populated areas. Different levels of NUTS region have therefore been selected for the purpose of IBA identification, such that the geographical size of the NUTS region used is roughly the same across the European Union (see Appendix 2b).

In general, up to five sites per NUTS region may be identified for a species—however, in exceptional cases, there may be grounds for increasing the number of sites per NUTS region to slightly more than five. In the previous pan-European IBA inventory of 1989, a total of ten sites could be selected per country for those countries which were not, then, part of the EU but which now are (Austria, Finland, German Democratic Republic, Sweden); this option has now been deleted, as NUTS levels have now been defined for these countries or territories as well.

If two or more sites in a given region hold the same number of pairs or individuals of a particular taxon, the relative priority of the sites for selection as IBAs is ranked according to the overall number of threatened (Annex I) species that occurs at each site. The C6 criterion has generally been applied to breeding populations, but may also be applied for non-breeding occurrences if these are not covered well by other criteria in the country concerned. The rationale of the criterion, overall, is to achieve a wide geographical coverage of sites throughout the species' range in the European Union.

Sites meeting C6 should hold appreciable numbers (appreciable at the EU level) of the species or subspecies concerned. This additional condition is necessary to exclude irregular occurrences and sites holding a low number of birds (1% of the regional breeding population or 0.1% of the biogeographical population are suggested as minimum levels), although different countries have adopted different approaches in their definition of 'appreciable'.

Other ornithological criteria – C7

A site which has been designated as a Special Protection Area (SPA), or has been selected as a candidate SPA, based on ornithological criteria (similar to, but not equal to, C1–C6) in recognized use for identifying SPAs.

Application of this criterion is confined to designated SPAs, and to sites which have been selected as SPAs in the framework of a national inventory which has been used by government agencies as such (although not necessarily officially accepted). This criterion should be applied only to a minority of exceptional cases where it would be inadvisable to exclude the sites concerned from the IBA inventory.

HOW DO THE IBA CRITERIA RELATE TO THE IDENTIFICATION OF SPECIAL PROTECTION AREAS UNDER THE EC BIRDS DIRECTIVE?

As the Birds and Habitats Directives are the most important international legal instruments for site protection in the European Union, all sites which qualify under Article 4 of the Birds Directive should be listed as IBAs. Keeping the aim of SPA designation in mind, when selecting IBAs, serves a practical conservation purpose but also helps to keep the process within sensible bounds. All potential Special Protection Areas (SPAs) should meet at least one of the criteria of the IBA criteria category C.

HOW DO THE IBA CRITERIA RELATE TO THE IDENTIFICATION OF RAMSAR SITES UNDER THE RAMSAR CONVENTION?

The Ramsar (or Wetlands) Convention defines a wetland as "an area of marsh, fen, peatland or water, whether natural or artificial, permanent or temporary, with water that is static or flowing, fresh, brackish or salt, including areas of marine water the depth of which at low tide does not exceed 6 m" (Article 1). Article 2.1 of the Convention also states that "the boundaries of each wetland [...] may incorporate riparian and coastal zones adjacent to the wetlands, and islands or bodies of marine water deeper than 6 m at low tide lying within the wetlands, especially where these have importance as waterfowl habitat".

The criteria for identifying wetlands of international importance under the Ramsar Convention, as adopted at the Conference of the Parties on 7 May 1999, fall into eight categories (Box 4). There is a strong relationship between the Ramsar categories for waterbirds and the IBA criteria. Ramsar criterion category 6 was one of the main criteria used in identifying IBAs. It has been used for both non-breeding populations of waterbirds and for breeding concentrations of some congregatory species. Averages (preferably five-year) of seasonal peak numbers have been used to assess whether the 1% threshold is met, although this was not possible for every country due to a lack of five-year data-sets. Ramsar criterion category 5 has also been applied to many IBAs. However, it has not always been applied as an independent criterion, because criterion 6 was given preference where suitable data were available. Many IBAs also comply with Ramsar criteria categories 2, 3 and 4, particularly wetland sites (such as mires, peat-bogs, etc.) that are important for birds other than waterbirds.

Thus, overall, IBA criteria comply with the Ramsar criteria for birds. However, one divergence is that IBA criteria categories A4i, B1i, C2 and C3 may be applied to congregations of waterbirds in grassland and marine habitats (not classifiable as wetland habitat under the Ramsar definition). For example, agricultural grassland IBAs in the Netherlands regularly meet the 1% threshold for *Anser albifrons*, and *Cygnus columbianus* and *Branta leucopsis* also use non-intensive grassland areas for feeding during the winter in the Netherlands. Similarly, many IBAs in Europe contain both coastal wetlands and some marine habitat deeper than 6 m. Thus, even though the 1% thresholds for some waterbirds may be met in these grassland and marine areas (Ramsar criterion 6), the Ramsar wetland definition excludes these sites from consideration under the Convention, therefore their eligibility for designation as Ramsar Sites has to be considered on a case-by-case basis.

Box 4. Categories of criteria for site selection under the Ramsar Convention (adopted at the Conference of the Parties, 7 May 1999).

1. Representative, rare, or unique example of a natural or near-natural wetland type found within the appropriate biogeographic region.
2. Supports vulnerable, endangered, or critically endangered species or threatened ecological communities.
3. Supports populations of plant and/or animal species important for maintaining the biological diversity of a particular biogeographic region.
4. Supports plant and/or animal species at a critical stage in their life cycles, or provides refuge during adverse conditions.
5. Regularly supports 20,000 or more waterbirds.
6. Regularly supports 1% of the individuals in a population of one species or subspecies of waterbird.
7. Supports a significant proportion of indigenous fish subspecies, species or families, life-history stages, species interactions and/or populations that are representative of wetland benefits and/or values and thereby contributes to global biological diversity.
8. An important source of food for fishes, spawning ground, nursery and/or migration path on which fish stocks, either within the wetland or elsewhere, depend.

REFERENCES

ATKINSON-WILLES, G. L. (1976) The numerical distribution of ducks, swans and coots as a guide in assessing the importance of wetlands in midwinter. Pp. 199–255 in M. Smart, ed. *Proc. Int. Conf. on the Conservation of Wetlands and Waterfowl, Heiligenhafen*. Slimbridge, UK: International Waterfowl and Wetlands Research Bureau.

BIRDLIFE INTERNATIONAL (1995) IBA criteria. Categories and thresholds. Cambridge, UK: BirdLife International (unpublished report).

COLLAR, N. J., CROSBY, M. J. AND STATTERSFIELD, A. J. (1994) *Birds to watch 2: the world list of threatened birds*. Cambridge, UK: BirdLife International (Conservation Series 4).

EVANS, M. I., ED. (1994) *Important Bird Areas in the Middle East*. Cambridge, UK: BirdLife International (BirdLife Conservation Series 2).

FISHPOOL, L. F. A. (in prep.) *Important Bird Areas in Africa*. Cambridge, UK: BirdLife International.

GRIMMETT R. F. A. AND GAMMELL, A. B. (1989) Inventory of Important Bird Areas in the European Community. (Unpublished report prepared for the Directorate-General for the Environment, Consumer Protection and Nuclear Safety of the European Community, study contract B6610-54-88.) Cambridge, UK: International Council for Bird Preservation.

GRIMMETT, R. F. A. AND JONES, T. A. (1989) *Important Bird Areas in Europe*. Cambridge, UK: International Council for Bird Preservation (ICBP Technical Publication 9).

HEATH, M. F. (1995) Important Bird Areas inventory review project. First workshop, Arnhem 1995. Proceedings. Cambridge, UK: BirdLife International (unpublished report).

HEATH, M. F. (1996) Important Bird Areas inventory review project. Second workshop, Finland 1996. Proceedings. Cambridge, UK: BirdLife International (unpublished report).

HEATH, M. F. AND BORGGREVE, C. (2000) *BirdLife International/EBCC European Bird Database 1998*. Cambridge, UK: BirdLife International.

IUCN (1994) *IUCN Red List categories*. Gland, Switzerland: IUCN—The World Conservation Union.

OSIECK, E.R. (1999) IBA review and the EU Birds Directive. Cambridge, UK: BirdLife International (Internal report; 6th and final draft. 18 April 1998).

OSIECK, E. R. AND MÖRZER BRUYNS, M. F. (1981) *Important bird areas in the European Community*. Cambridge, UK: International Council for Bird Preservation.

ROSE, P. M. AND SCOTT, D. A. (1994) *Waterfowl population estimates*. UK: International Waterfowl and Wetlands Research Bureau (IWRB Publ. 29).

ROSE P. M. AND SCOTT D. A. (1997) *Waterfowl population estimates*. Second edition. Wageningen, Netherlands: Wetlands International (Publication 44).

SCOTT, D. A. AND ROSE, P. M. (1996) *Atlas of Anatidae populations in Africa and Western Europe*. Wageningen, Netherlands: Wetlands International (Wetlands International Publ. No. 41).

STATTERSFIELD, A. J., CROSBY, M. J., LONG, A. J. AND WEGE, D. C. (1998) *Endemic bird areas of the world: priorities for bird conservation*. Cambridge, UK: BirdLife International (BirdLife Conservation Series no. 7).

TUCKER, G. M. AND HEATH, M. F. (1994) *Birds in Europe: their conservation status*. Cambridge, UK: BirdLife International (BirdLife Conservation Series no. 3).

DATA PRESENTATION

Each chapter contains a national overview of the status of Important Bird Areas (IBAs) and their conservation, followed by a series of site accounts describing the IBAs in the country.

The national overviews have been written in a standard way, summarizing key information on IBAs in the country. The tables, maps and figures in each national overview summarize the data within the site accounts which have, in turn, been derived from the BirdLife International IBA Database. Space for each site account is limited, given the inclusion of 3,619 sites in this publication, and the text has been restricted to key facts. More detailed information is contained within the IBA Database and, for several countries, has also been published in national IBA inventories (see Box 1, 'Data collection' chapter).

NATIONAL OVERVIEW

General introduction

This includes basic geographical information about the country, information on the number and distribution of IBAs, the percentage of the country's surface area covered by IBAs, and a comparison of the current IBA inventory with the one presented 10 years previously (Grimmett and Jones 1989) and with any national IBA publications. This section is accompanied by a table listing all IBAs and the reasons for their importance, and a map displaying their location.

International and national names of sites.

Primary administrative region(s) in which IBA is located.

Area of site in hectares (100 ha = 1 km²).

Number of IBAs in the country and total area covered by them.

EXAMPLE TABLE 1

Table 1. Summary of Important Bird Areas in Austria. 55 IBAs covering 12,388 km²

IBA code	1989 code	International name	National name	Administrative region	Area (ha)	Criteria (see p. 11)
001	AT003	Austrian part of Hanság	Österreichischer Teil des Hanság	Burgenland	7,000	B2, C6
002	AT004	Parndorfer Platte	Parndorfer Platte	Burgenland	18,000	B3, C6
003	AT002	Southern Seewinkel and	Südlicher Seewinkel und Zitzmannsdorfer Wiesen	Burgenland	14,000	A4i, B1i, B2, C3, C6
				Burgenland	23,272	A1, A4i, B1i, B2, B3, C2, C3, C6
053	—	Slope fores...				
054	—	Meadows on the lower Ill	Streuwiesen an der Unteren Ill	Vorarlberg	279	A1, C1, C6
055	AT020	Delta of the Rhine and Lauteracher Ried	Rheindelta und Lauteracher Ried	Vorarlberg	2,760	C6

Sites identified in the previous inventory of IBAs in Europe (Grimmett and Jones 1989) but no longer considered to be IBAs
AT006 Braunsberg-Hundsheimer Berg; AT014 Seetaler Alpen; AT019 Karwendel.

Cross-reference to number(s) used to identify this site in previous pan-European inventory (Grimmett and Jones 1989).
Cross-reference to number(s) used to identify this site in national IBA inventory (where one has been produced) are shown in a separate column.

For definitions of criteria, see the previous chapter.

Number used to refer to the site in this inventory.

Sites included in the previous pan-European IBA inventory (Grimmett and Jones 1989) that are no longer IBAs are listed at the bottom of the table.

IBA 031 is of global importance (i.e. it meets one or more 'A' criteria) and has an area in the range 25,000–178,700 hectares.

EXAMPLE MAP 1

Map 1. Location, area and criteria category of Important Bird Areas in Austria.
The location of IBA 005 is confidential and so is ommitted from the map.

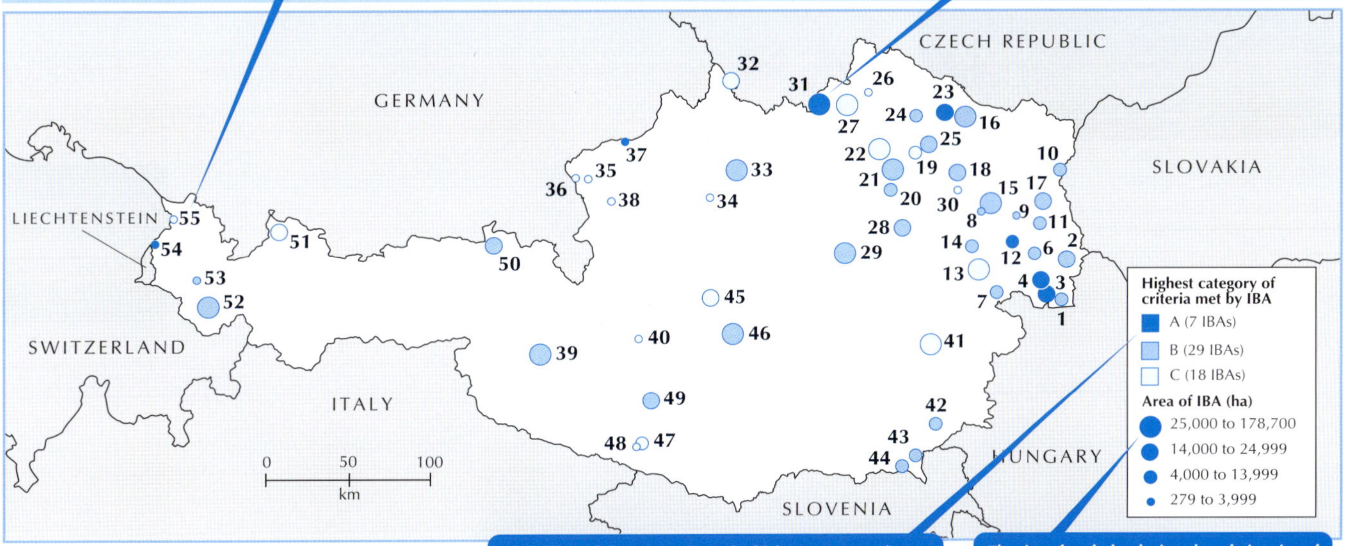

The colour of each dot reflects the highest category of criteria applied to the IBA: global importance (A), European importance (B) or European Union importance (C).

The size of each dot depicts the relative size of the IBA: four categories are displayed, each corresponding to an area range (quartile).

■ Ornithological importance

This section summarizes the key facts about birds and IBAs in the country, giving special attention to species meeting IBA criteria and emphasizing those species for which the country is important in an international context, especially species of global or European conservation concern (Collar *et al.* 1994, Tucker and Heath 1994). This section includes up to three tables illustrating the importance of the IBA network for:

- species of global conservation concern
- congregatory species
- species of European conservation concern and species threatened in the European Union (listed on Annex I of the EC Birds Directive).

SPECIES OF GLOBAL CONSERVATION CONCERN

All species of global conservation concern in Europe are listed in Table 2 of the previous chapter.

For definition of A1 criterion, see the previous chapter.

Table 2. Important Bird Areas in Hungary that are important for species of global conservation concern (meeting criterion A1).

Species	IBA code
Phalacrocorax pygmeus Pygmy Cormorant	032
Anser erythropus Lesser White-fronted Goose	032
Aythya nyroca Ferruginous Duck	004, 006, 007, 008, 011, 013, 026, 030, 032
Haliaeetus albicilla White-tailed Eagle	007, 010, 026, 030, 032
Aquila heliaca Imperial Eagle	013, 015, 018, 032, 036, 040, 041, 042, 043
Crex crex Corncrake	019, 035, 039, 043
Otis tarda Great Bustard	001, 022, 029, 031, 032, 036, 037
Numenius tenuirostris Slender-billed Curlew	032
Acrocephalus paludicola Aquatic Warbler	032

Sites where species of global conservation concern meet the A1 criterion in Hungary (at some stage in their annual cycle).

CONGREGATORY SPECIES

Table 3. Important Bird Areas in Estonia that support important numbers of one or more congregatory species (i.e. meeting criteria A4 and/or B1). IBAs meeting both criteria A4 and B1 for the species are shown in **bold**. IBAs meeting only criterion B1 for the species concerned, and not A4, are shown in normal type. For key to 'Season', see p. 7.

Species	Season	IBA code
Gavia stellata Red-throated Diver	W	**017**
	P	**027**, **029**
Gavia arctica Black-throated Diver	W	**017**
	P	**027**, **029**
Phalacrocorax carbo Cormorant	B	**010**, 019
	P	007, 027
Cygnus olor Mute Swan	P	007, 013
Cygnus columbianus Bewick's Swan	P	**003**, **006**, **007**, **008**, **010**, **023**, **024**, **026**, **027**, **028**, **029**, **030**, **032**, **036**, **037**, **038**, **040**
Cygnus cygnus Whooper Swan	P	003, **007**, **008**, **010**, **023**, **024**, 027, 028
Anser fabalis Bean Goose	P	**010**, **031**, **040**
Anser albifrons White-fronted Goose	P	**031**
Anser anser Greylag Goose	B	010
	P	002, 003, **005**, 006, 007, **010**, 012, 013, 016, 017, 018, 020, **023**, 028

At least 1% of the biogeographic population of Red-throated Diver overwinters at site 017.

At least 1% of the flyway population of Whooper Swan stages at site 024 while on passage.

SPECIES OF EUROPEAN CONSERVATION CONCERN AND SPECIES THREATENED IN THE EUROPEAN UNION

Red-throated Diver is a species:
- which is of European conservation concern and /or is listed in Annex I of the EC Birds Directive
- for which one or more IBAs have been identified in Finland.

Table 4. Species of European conservation concern and species listed on Annex I of the EC Birds Directive with significant breeding populations at IBAs in Finland (meeting any IBA criteria).

Species [1]	Minimum national breeding population (pairs) [2]	Proportion (%) of national population breeding at all IBAs in Finland
Gavia stellata Red-throated Diver	900	11
Gavia arctica Black-throated Diver	7,000	5
Podiceps auritus Slavonian Grebe	3,000	5
[illegible] Bittern	100	84
Perisoreus infaustus Siberian Jay	30,000	11
Loxia pytyopsittacus Parrot Crossbill	10,000	9

1. Only those species of European conservation concern (see Box 1, p. 12) that meet IBA criteria in Finland are listed, together with those species listed on Annex I of the EC Birds Directive that fulfil criterion C6 in IBAs in Finland.
2. Data are taken from the BirdLife/EBCC European Bird Database 1998 (Heath and Borggreve 2000).
3. The percentage of the national population in IBAs exceeds 100%. Usually this is because the national population estimate has not been updated recently whilst the IBA population estimate has been recently updated with new data as a result of comprehensive surveys of IBAs themselves. Also, the individual site count for a species may be the maximum or average over recent years, and summing these may record more birds than are present nationally in any single year.

It is estimated that at least 11% of the breeding population of Red-throated Diver in Finland breeds within IBAs.

Habitats

This section gives a brief outline of the predominant habitat-types in the country. It examines the most widespread and common habitat-types found within IBAs and discusses the representativeness of IBA coverage from a habitat perspective in a national context.

A bar chart illustrates the occurrence and coverage of different habitat-types within IBAs. For a fuller explanation of the habitat classification used, see Appendix 3.

For some sites, and some countries, the percentage cover of habitat-types has not been estimated; in these cases the bar chart indicates only the frequency of occurrence of habitats at IBAs.

Habitat-types are listed and described in more detail in Appendix 3.

HABITATS AT IBAs

At 35% of IBAs in Slovenia, forest/woodland covers at least 50% of the area of the individual IBA.

At 85% of IBAs in Slovenia, forest/woodland covers at least 5% of the area of the individual IBA.

Figure 1. Habitats at Important Bird Areas in Slovenia (see Appendix 3 for definitions of habitats).

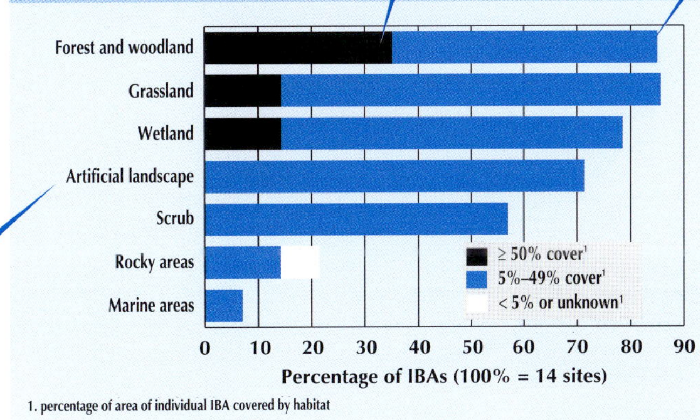

Impacts on IBAs—land-use and threats

The most widespread and common land-uses within IBAs are described and presented in a bar chart. For a fuller explanation of the land-use classification, see Appendix 3.

For some sites, and some countries, the percentage cover of land-use has not been estimated; in these cases the bar chart indicates only the frequency of occurrence of land-uses at IBAs.

Key factors threatening the sites, and their level of impact, are described and summarized in a bar chart. For a fuller explanation of the classification of threats, see Appendix 3.

For some sites, and some countries, the impact of threats has not been estimated; in these cases, the bar chart indicates only the frequency of occurrence of threats at IBAs.

Any links between land-uses, threats and land-ownership are addressed and the key species or habitats affected are described. Those IBAs considered most threatened are highlighted and any relevant national trends in land-uses and threats are outlined and contrasted with trends at IBAs.

Land-uses are listed and defined in Appendix 3.

LAND-USES AT IBAs

At 20% of IBAs in Russia, at least 50% of the area of the individual IBA is used for forestry.

At 40% of IBAs in Russia, between 5% and 49% of the area of the individual IBA is used for forestry.

Figure 2. Land-uses at Important Bird Areas in Russia (see Appendix 3 for definitions of land-uses).

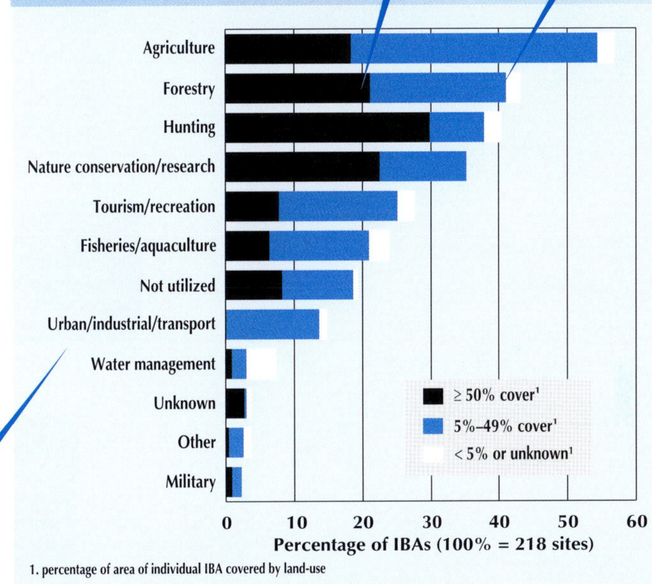

THREATS AT IBAs

Ongoing drainage is having (or planned drainage will have) a high impact at about 5% of IBAs in Belarus.

Agricultural intensification/expansion poses a threat at 48% of IBAs in Belarus: at 42% of IBAs in Belarus, the impact is considered to be medium or low, while at 6% of IBAs in Belarus, the degree of impact is not known.

Over 70% of IBAs in Belarus are threatened, to some extent, by drainage.

Figure 3. Threats at Important Bird Areas in Belarus (see Appendix 3 for definitions of threat types and impact categories).

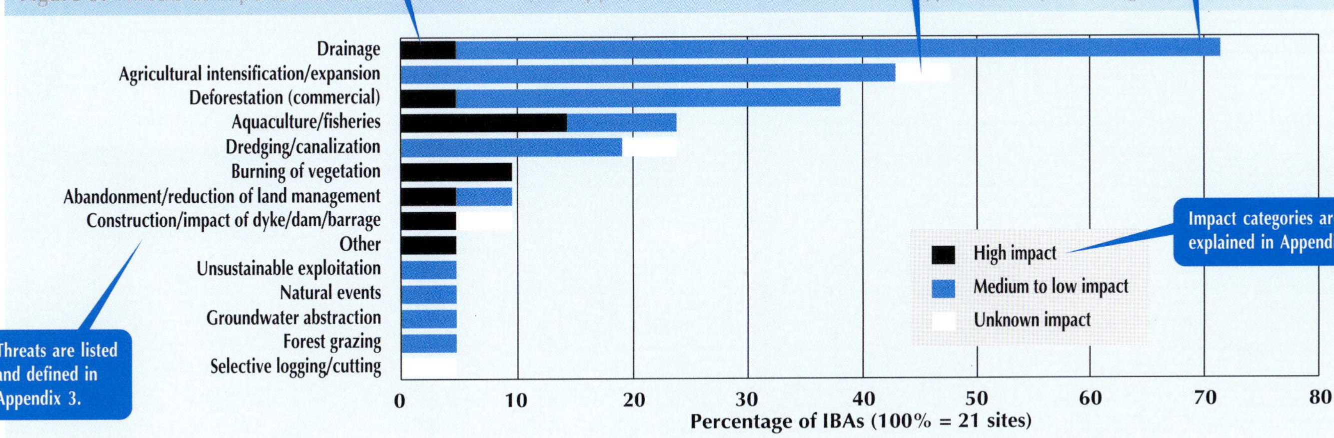

Impact categories are explained in Appendix 3.

Threats are listed and defined in Appendix 3.

PROTECTION STATUS OF IBAs

Table 5. Protection status of Important Bird Areas in the Czech Republic. A tick (✓) indicates that an IBA overlaps with a protected area (to any extent).

IBA code	International name	National: National Park	National: National Nature Reserve	National: Nature Reserve	National: National Natural Monument	National: Natural Monument	National: Protected Landscape Area	National: Temporary Protected Area	International: Ramsar Site	International: Biosphere Reserve
001	Krkonoše mountains (Giant mountains)	✓							✓	✓
002	Žehuňský rybník pond		✓			✓				
003	Křivoklátsko (Křivoklátsko region)		✓	✓		✓	✓			✓
004	Doupov hills		✓	✓						
005	Šumava mountains (Bohemian forest)	✓	✓	✓	✓	✓	✓		✓	✓
006	Řežabinec pond		✓							
007	Dehtář pond							✓		
008	Třeboňsko (Třeboň region)		✓	✓	✓	✓	✓		✓	✓
009	Nové Mlýny middle reservoir			✓					✓	
010	Pálava		✓	✓		✓	✓			✓
011	Lednické rybníky ponds (Lednice fish-ponds)		✓						✓	
012	Confluence of the Morava (March) and Dyje (Thaya) rivers		✓						✓	
013	Beskydy mountains		✓	✓	✓	✓	✓			
014	Poodří		✓	✓			✓		✓	
015	Heřmanský stav pond–Stružka wetlands			✓						
016	Jeseníky mountains		✓	✓		✓	✓			
	Total number of IBAs	2	12	10	3	7	7	1	7	5

'Pálava' IBA overlaps partly or wholly with one or more National Nature Reserves (see individual site-account for more details).

The types of protected area are further described in the 'Overview' text. In this table, as far as possible, protected-area types are listed from left to right in order of IUCN management category, from I to VI (IUCN 1994).

12 IBAs in the Czech Republic overlap to some extent with National Nature Reserves.

16% of IBAs in Spain (62 IBAs) have more than 90% of their individual area covered by the national system of protected areas.

NATIONAL PROTECTION STATUS

Figure 4. The national protection status of Important Bird Areas in Spain.

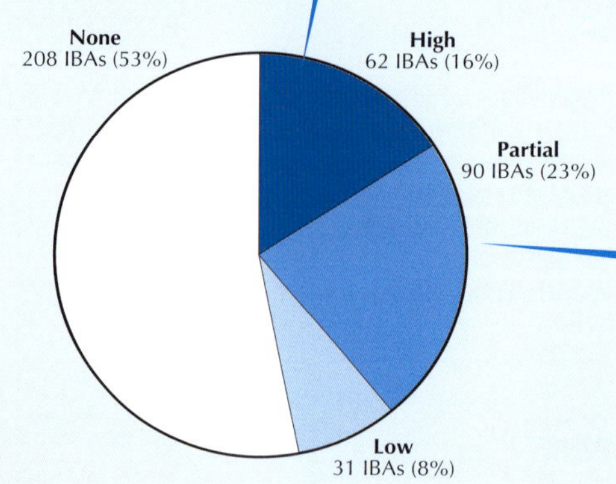

% of area of individual IBA overlaps with national protected areas:
High >90%
Partial 10–90%
Low <10%
None

Total area of overlap between IBA network in Spain and national protected-area system (see Table 5 for categories) = 29,141–34,266 km² (18–22% of total IBA area).

47% of IBAs in Spain overlap to some extent with the national system of protected areas.

INTERNATIONAL LEGISLATION AND INITIATIVES

Box 1. International legislation and initiatives that are relevant to site conservation in Greece (see Appendix 1 for a general description of these agreements).

Global	
Biodiversity Convention	✔
Ramsar Convention	✔
Bonn Convention	✔
World Heritage Convention	✔
MAB Programme	✔
Pan-European	
Bern Convention	✔
Regional	
EC Birds Directive	✔
EC Habitats Directive	✔
Barcelona Convention	✔

✔ Convention ratified/initiative supported
(✔) Convention signed

INTERNATIONAL PROTECTION STATUS

Figure 5. The international protection status of Important Bird Areas in the United Kingdom.

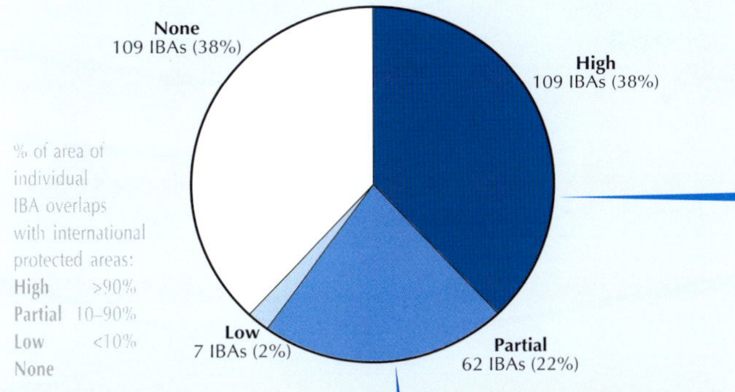

Total area of overlap between IBA network in the United Kingdom and international protected-area system (see Table 5 for categories) = 7,832–7,882 km² (25% of total IBA area).

62% of IBAs in the UK overlap, partly or wholly, with one or more 'international' protected area (areas designated under international conventions, directives or programmes).

22% of IBAs in the UK (62 IBAs) have 10–90% of their individual area covered by 'international' protected areas.

■ Protection status

This section analyses the extent to which IBAs are legally protected, through their overlap with areas protected by national law and by international agreements.

The overlap (or not) of each IBA with national and international protected areas is recorded in a table.

The main protected-area types in the country are described, together with their legal status, type of management and the authority responsible for implementing protection. In many countries, this list of national designation-types has been limited to those of truly national standing with legally binding status. Therefore most sub-national, local or non-statutory categories of protected area have not been discussed, analysed or otherwise covered, in these countries.

The extent of overlap between individual IBAs and national protected areas is summarized in a pie chart, and the total area of IBAs under national protection is stated. It is possible that the total areas of IBAs under protection is underestimated due to incomplete data on area of overlap between protected areas and IBAs.

Regarding international protection, the main international legislation and initiatives relevant to site-based biodiversity conservation in the country are described.

A box summarizes all of the international legislation and initiatives that can give international recognition or protection to sites in the given country, and indicates whether the country has ratified or signed such conventions or takes part in such initiatives.

The extent of overlap between individual IBAs and internationally designated areas is summarized in a pie chart, and the total area of IBAs under international protection is stated.It is possible that the total area of IBAs under protection is underestimated due to incomplete data on area of overlap between protected areas and IBAs.

■ Conservation

This section summarizes key research, monitoring, site-conservation or site-management initiatives that affect IBAs in the country, drawing together any national trends or conservation issues (e.g. the coverage of IBAs by management plans), as well as any relevant information that is otherwise oft-repeated in the individual site-accounts.

■ Analytical methods

This section provides important information on the preparation of the national inventory and the methodology followed in collecting and analysing the data. It includes information on:

- bird counts, especially the conventions used in counting or estimating numbers of birds at sites at different seasons, and the conventions used in presenting such information, as well as the units used;
- important species in the country that are not adequately covered by the IBA network, either because of their dispersed nature or because of inadequacies in data or observer coverage;
- dates and main sources of the information on IBAs;
- any difficulties encountered in obtaining or categorizing data or in assigning criteria, and how these problems were addressed, as well as any standards or conventions that were followed (if different from, or expanding on, the IBA project guidelines).

■ Glossary

This gives explanations of technical or non-standard terms or abbreviations used in the chapter.

■ Acknowledgements

This section acknowledges all contributors to the IBA review in the country, including financial contributors if appropriate.

SITE ACCOUNTS

EXAMPLE SITE ACCOUNT

Internationally recognized site name.

Primary administrative region(s) in which IBA is located.

Central coordinates of the IBA (latitude/longitude; Greenwich).

Criteria under which the site qualifies as an IBA (see the previous chapter for definitions).

Altitudinal range of IBA, in metres above/below sea level (a minus sign indicates that the site is below sea level).

IBA code: number which identifies this site in this publication.

Thumbnail map showing location of the IBA.

Danube plain A1, A4i, B1i, B2 **022**

Admin region Bács-Kiskun, Pest
Coordinates 46°49′N 19°15′E
Altitude 93–103 m **Area** 82,000 ha

Site description

A flood-plain lying between natural levees and overlain by loess soils, situated adjacent to the River Danube and main Duna-völgyi canal between Kiskunlacháza and Szakmár. Human activities include cattle- and sheep-grazing, arable farming, angling and hunting. This area includes two sites that were treated as separate IBAs in the previous international IBA inventory (Grimmett and Jones 1989): 'Kiskunsági szikes-tavak' (former site HU021) and 'Harta-Akasztói puszta' (former site HU042).

Habitats Grassland (40%; steppe/dry calcareous grassland). Wetland (5%; standing brackish and salt water; water-fringe vegetation). Artificial landscape (55%; highly improved reseeded grassland; arable land)

Land-use Agriculture (90%), Nature conservation/research (60%), Tourism/recreation (5%), Water management (5%)

Habitats are defined in Appendix 3; percentages refer to proportion of IBA area that is covered by primary level habitats. Lack of a percentage indicates either (a) lack of relevant knowledge, or (b) cover is less than 5%.

Land-uses are defined in Appendix 3; percentages refer to proportion of IBA area that is utilized. Lack of a percentage indicates either (a) lack of relevant knowledge, or (b) cover is less than 5%.

Area of the IBA in hectares (100 ha = 1 km²)

40% of the area of the IBA comprises steppe/dry calcareous grassland habitat.

90% of the area of the IBA is affected by agricultural activities.

Birds

Species	Season	Year	Pop min	Pop max	Acc	Criteria
[1] *Ardea purpurea* Purple Heron	P	1996	50	50	C	B2
[2] *Anser albifrons* White-fronted Goose	P	1996	2,000	15,000	A	B1i
[2] *Anser anser* Greylag Goose	P	1996	200	5,000	A	A4i, B1i
[1] *Anas acuta* Pintail	P	1996	260	260	A	B1i
[1] *Falco vespertinus* Red-footed Falcon	B	1996	85	85	B	B2
Otis tarda Great Bustard	R	1996	190	290	A	A1, B2
[2] *Limosa limosa* Black-tailed Godwit	B	1996	50	80	B	B2
[3] *Limosa limosa* Black-tailed Godwit	P	1996	150	8,000	A	A4i, B1i
[2] *Numenius arquata* Curlew	P	1996	Common		—	A4i, B1i
[1] *Chlidonias hybridus* Whiskered Tern	B	1996	—	175	A	B2
[1] *Chlidonias niger* Black Tern	B	1996	40	40	B	B2

1. 1992–1996.
2. 1986–1996.
3. Large decrease 1986–1996.

An important area for lowland farmland birds and waterbirds. Species of global conservation concern that do not meet IBA criteria: *Phalacrocorax pygmeus* and *Branta ruficollis* (both on passage), *Aythya nyroca* (6 breeding pairs), and *Haliaeetus albicilla* (2 breeding pairs and 4 wintering birds).

Year of population estimate. In cases where the population estimate is derived from a multi-year series, the year given here is the last in the series, and a footnote may give more details of the period covered, if this aspect has not been dealt with already in the 'Analytical methods' section of the national overview.

Season in which species meets criterion/criteria:
- R Breeding resident
- B Breeding visitor
- W Winter visitor
- P Passage visitor
- N Non-breeding visitor
- U Status uncertain.

Minimum and maximum population estimate (see 'Analytical methods' section of national overview for more details on conventions followed).

Quality of population data (see 'Data collection' chapter for more details):
- A Reliable
- B Incomplete
- C Poor
- — Unknown

Breeding population data (season code 'R' or 'B') are given in pairs unless stated otherwise in a footnote or in the 'Analytical methods' section of the national overview.

If population figures are unavailable, a qualitative assessment of abundance may be given (see 'Data collection' chapter for further explanation).

Species are listed in taxonomic order. Taxonomy and nomenclature follow *The Birds of the Western Palearctic* (Cramp *et al.* 1977–1994).

Protection status

National Partial **International** Partial

21,200 ha of IBA covered by National Park (Kiskunsági, 53,429 ha). 3,903 ha of IBA covered by Ramsar Site (Kiskunság, 3,903 ha). The IBA also overlaps with the Kiskunság Biosphere Reserve (22,095 ha).

Overlap between IBA and protected area(s), measured as a percentage of the IBA's area, is 'Partial' at the national level and 'Partial' at the international level:
- High >90% overlap
- Partial 10%–90% overlap
- Low <10% overlap
- None No overlap

Area of IBA in hectares (100 ha = 1 km²) that overlaps with protected area. If the entire IBA is covered by the protected area this figure will equal the area of the IBA given in the header.

Area of protected area (in hectares; 100 ha = 1 km²).

Official name of protected area.

Type of protected area (further described in national overview).

Conservation issues

Threats Abandonment/reduction of land management (A), Agricultural intensification/expansion (B), Disturbance to birds (C), Drainage (B), Natural events (B), Recreation/tourism (C)

The main threats are changes in habitat composition and quality caused both by the abandonment of grazing and by drought. A *Perdix perdix* recovery project has begun in Apajpuszta, and the National Park Authority is working to increase the breeding success of *Otis tarda*. MME/BirdLife Hungary provides artificial nesting-sites, and a management plan exists for the area. Wildfowl shooting is banned, and the enlargement of the National Park is in process.

Threats are defined in Appendix 3.

Importance score: Calculation of the score is explained in Appendix 3. Lack of a score implies that impact is unknown, and does not imply minimal impact.
- A High impact
- B Medium impact
- C Low impact
- U Unknown impact

Header

Each site account has a boxed header which gives key information about the site, including international site name, site code, administrative region(s), central coordinates, altitudinal range, surface area, a thumb-nail map showing the location of the site within the country, and a list of the criteria under which the site qualifies as an IBA.

Site description

This section includes a short general description of the site covering relevant information on location and topographical features.

The section includes a table listing the habitats and land-uses at the site, and the percentage cover of each where this is known. Two levels of habitat are listed, e.g. 'Forest/woodland' at the primary level, and 'broadleaved deciduous forest', or 'native coniferous forest' at the secondary level. The percentage cover is given for the primary habitats only. Primary habitats which cover less than 5% of the IBA are usually not listed. If habitat-types are rare or otherwise difficult to classify, they may be listed at the primary level only.

Birds

A table presents data on all species meeting IBA criteria at the site. Much more information on these species, and on species not meeting IBA criteria at each site, is contained within the IBA database but, because of space constraints, it has not been possible to publish these extra data.

The table may be supported by additional text summarizing the overall reasons for the site being important, and including non-specific data, for example listing total counts for a group of species such as waterbirds or migrating raptors. If the relevant criteria have been met, reference is also made to the site's occurrence within a biome or an Endemic Bird Area. All species of global conservation concern that occur at the site but which do not meet IBA criteria (sometimes due to lack of data rather than due to insufficient numbers) are also mentioned in the text. Additionally, for some countries, all species that are listed in Annex I of the EC Birds Directive and that breed at the site have also been listed in the text. For some countries, species present at the site in nationally important numbers (e.g. with at least 1% of the national population occurring) are also listed, especially if this factor is relevant to the national application of the EC Birds Directive.

Protection status

This section indicates the national and international protection status (High, Partial, Low, None) of the site. This has been calculated as follows:

High	More than 90% of the area of an individual IBA overlaps with national/international protected area(s)
Partial	10–90% of the area of an individual IBA overlaps with national/international protected area(s)
Low	Less than 10% of the area of an individual IBA overlaps with national/international protected area(s)
None	No overlap with national/international protected area(s)

This provides an indication of the coverage of each IBA by protected areas (in this publication, 'protected area' usually refers only to statutory areas, with official recognition of their purpose) and is summarized from the more detailed information on the degree of overlap between each individual protected area and the IBA. Where details of overlap are unknown, the 'Partial' category has usually been assigned as a default since it presents the broadest range of overlap. Where this applies, it has been explained in the country overview. It is important to note that the terms 'High', 'Partial' or 'Low' protection do not refer to the strictness of protection prescribed by law, only to the spatial extent of protection, nor do they refer to the success or otherwise of management authorities in implementing protection.

This is followed by details of the protected areas and the extent of their overlap with the IBA. The 'Protection status' portion of the national overview gives more details on each protected-area type, its status in national law, management regime, etc.

Conservation issues

This section includes a table summarizing the main threats (actual and potential) impacting on the site and a score of the importance of each (as explained in Appendix 3). Additional text concerning these threats may be given, as may further detail on protection status, including any proposed designations. Key research/monitoring activities and conservation issues may be mentioned, as well as the existence or otherwise of any management plan(s) for the IBA or associated protected areas. This section should be read in conjunction with the 'Conservation' section of the national overview, at the start of the chapter, which summarizes any common conservation issues that would otherwise be repeated across several or many IBAs.

REFERENCES

CRAMP, S. *ET AL.* (1977–1994) *The birds of the western Palearctic.* Vols 1–9. Oxford, UK: Oxford University Press.

GRIMMETT, R. F. A. AND JONES, T. A. (1989) *Important Bird Areas in Europe.* Cambridge, UK: International Council for Bird Preservation (ICBP Techn. Publ. 9).

IUCN (1994) *Guidelines for protected area management categories.* Gland, Switzerland: IUCN—The World Conservation Union.

TUCKER, G. M. AND HEATH, M. F. (1994) *Birds in Europe: their conservation status.* Cambridge, UK: BirdLife International (BirdLife Conservation Series no. 3).

■ OVERVIEW OF RESULTS

This chapter presents a pan-European analysis of the data given in this publication and details some of the key findings of this first review of Important Bird Areas (IBAs) in Europe.

WHY ARE IBAs IMPORTANT AND HAS A COMPLETE SET OF SITES BEEN FOUND?

■ IBA coverage by country

A comprehensive network of internationally important sites for birds has been identified across Europe. A total of 3,619 Important Bird Areas are distributed among 51 countries or autonomous regions in the continent, covering 931,700 km², which equates to 7% of the continental land area (Figure 1). There has been a tremendous amount of new data gathered as part of this inventory, resulting in improved coverage across the region compared to the previous IBA inventory, 10 years ago (Grimmett and Jones 1989).

The number of IBAs identified in each country depends on many different factors, such as the size of the country, its importance for different species, the extent of particular habitats remaining in the country, the national capacity to gather data and, importantly, the types of criteria applied to identify IBAs. An additional category of criteria ('C' criteria) was applied within the European Union (EU), with lower numeric thresholds, which resulted in a greater number of sites being identified in these countries. Indeed, 2,342 IBAs (65% of the European total) fall within the 15 member states of the EU. A full list of the number and area of IBAs within each country is given in Appendix 6.

The countries with more than 90 IBAs and the countries with over 20,000 km² in IBAs are listed in Figures 2 and 3; it should be noted that the two sets of countries are not the same. Spain has the highest number of IBAs (Figure 2), with 391 in total, covering 160,000 km² or 31% of the land area of the country (Figure 3). Russia has 218 IBAs but these tend to be larger sites (average area 800 km²), including many large protected areas, with the total IBA area being 174,500 km², higher than Spain but equal to just 4% of the land area of European Russia. Overall, the size of IBAs in Europe varies greatly, ranging from 1 hectare to 19,000 km², with a mean area of 259 km².

The percentage of the area of each country covered by IBAs is shown in Figure 4. In the majority of countries, IBAs cover between 5% and 15% of the national land area. The adequate protection and management of the 3,619 IBAs in Europe, equivalent to a relatively low percentage of each country's surface area, is therefore a realistic goal that would make a very significant contribution to conserving those bird species in Europe for which a site-protection approach is appropriate. In several countries, particularly in the east of Europe, in parts of Scandinavia and the Balkans, IBAs cover less than 5% of the national area and it is probable that more sites will be identified in the future to complete the network of IBAs.

Figure 1. The location and criteria category of Important Bird Areas in Europe. There are 3,619 IBAs in Europe, covering 931,700 km², equivalent to 7% of the land area of Europe.

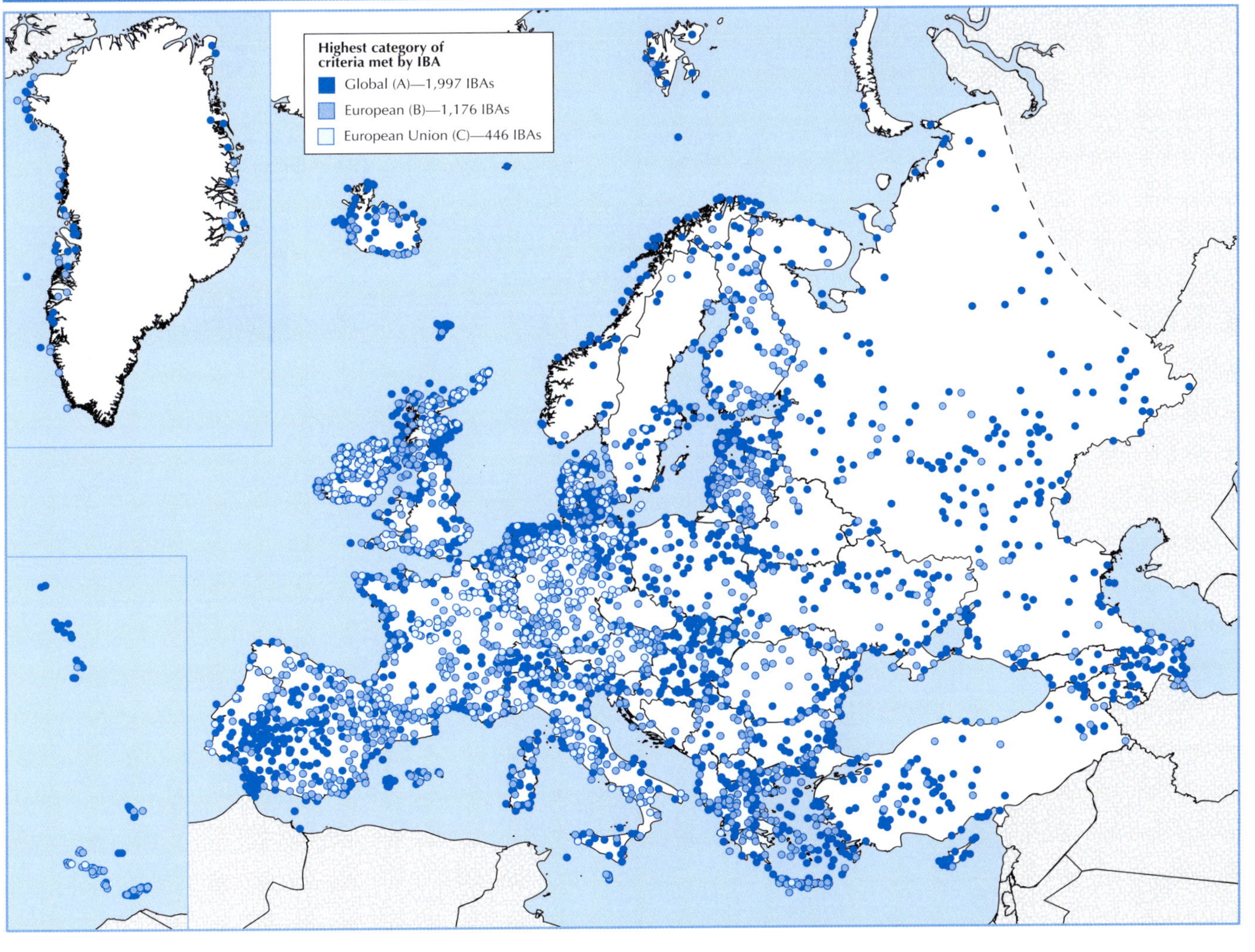

Figure 2. The countries in Europe with more than 90 Important Bird Areas.

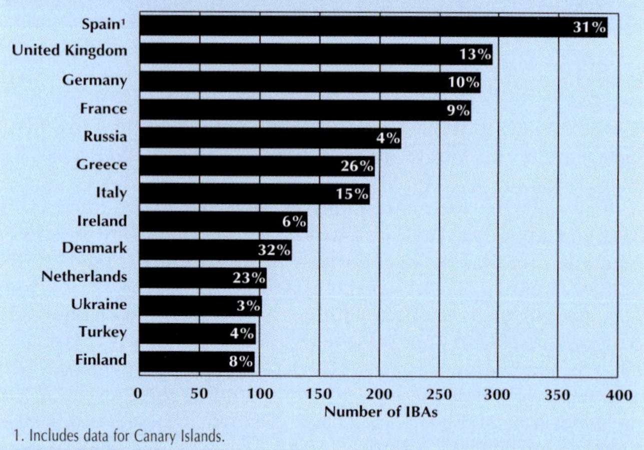

1. Includes data for Canary Islands.

The percentage at the end of each bar indicates the proportion of the country's land area that is equivalent to the total IBA area (for those countries with extensive marine IBAs, e.g. Denmark, this figure exaggerates the actual extent of land covered).

Figure 3. The countries in Europe where the total area of Important Bird Areas exceeds 20,000 km².

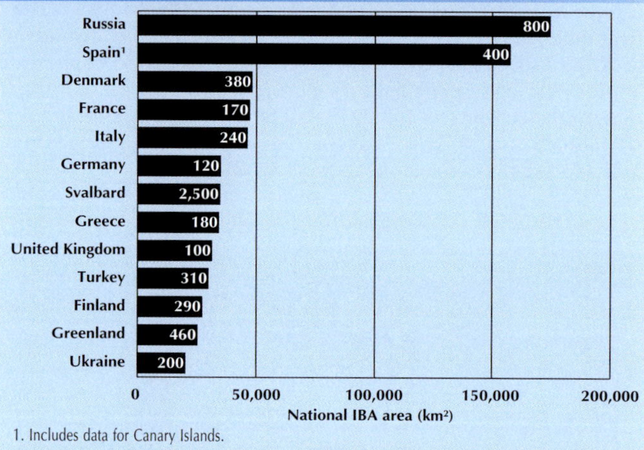

1. Includes data for Canary Islands.

The figure at the end of each bar indicates the mean size (in km²) of IBAs in that country. Across all countries, IBAs range in size from 1 ha to 19,000 km², and their mean size is 259 km².

A total of 1,175 new IBAs have been identified since the previous pan-European IBA inventory was published 10 years ago (Grimmett and Jones 1989), representing a 48% increase in the number of sites, and a 130% increase in the total surface area covered, from 404,600 km² to 931,700 km². Those countries with the greatest change in number of IBAs and area covered since 1989 are shown in Figures 5 and 6. In many countries all over Europe, but particularly in the east, both the number of IBAs and the total area of IBAs have substantially increased. This is particularly evident in Russia where 141 new IBAs have been identified, adding a further 105,000 km² to the total area covered by IBAs. In several smaller countries such as Azerbaijan, Estonia, Georgia, Hungary, Latvia and Ukraine, increases in terms of total number or area of IBAs are also particularly significant. Additionally, many of the original (1989) IBAs have been subdivided into two or more sites, and many others have been amalgamated to form larger ones. With more in-depth knowledge of IBAs, the boundaries of many sites have been refined as well, changing the sites' area.

Overall, 240 sites that were included in the previous (1989) inventory have been excluded from this book. The three main reasons for exclusion were: (1) the availability of improved data on the site showing that it was no longer considered of international importance, although most such sites remain important at the national level; (2) the application of higher population thresholds for some species-based criteria, based on improved knowledge of European population sizes; and (3) the deterioration of the quality of the site, resulting in declines in numbers of birds (however, sites were retained as IBAs if it was considered that the value of the site for birds could be restored).

There are still some countries in Europe where not all IBAs have been identified, and ongoing field and literature studies will identify further sites in the future. Countries where a larger network of IBAs will be identified in the future, either in terms of area and/or number, include:

- Belarus—currently 21 IBAs cover 3% of the country;
- Russia—work is ongoing to identify further IBAs (some regions still have no IBAs whatsoever), e.g. 151 proposed sites were surveyed during 1998–1999 (listed in the national overview of the Russian chapter) and will be described in the forthcoming national inventory for Russia;
- Ukraine—102 IBAs cover 3% of the country; fieldwork during 1998–1999 identified further IBAs which will be included in the forthcoming national IBA publication;

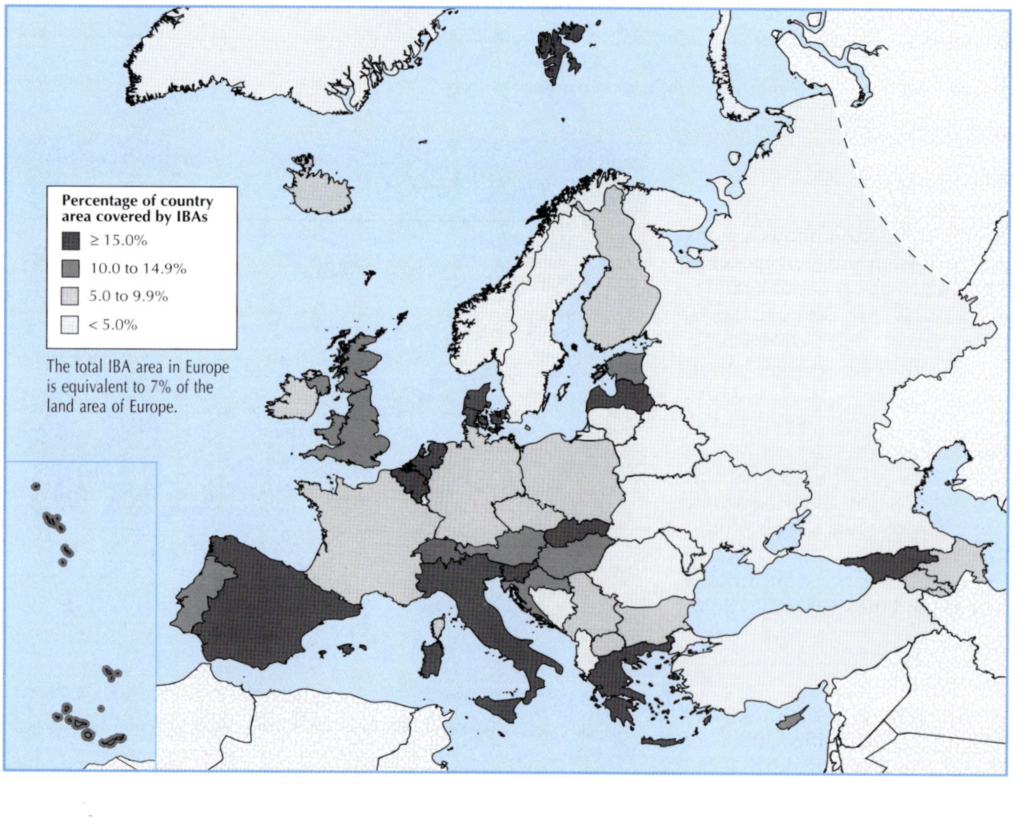

Figure 4. The proportion of the national land area covered by Important Bird Areas in each country in Europe.

Figure 5. The countries in Europe where the total number of new Important Bird Areas, identified since the previous pan-European IBA inventory (Grimmett and Jones 1989), exceeds 30.

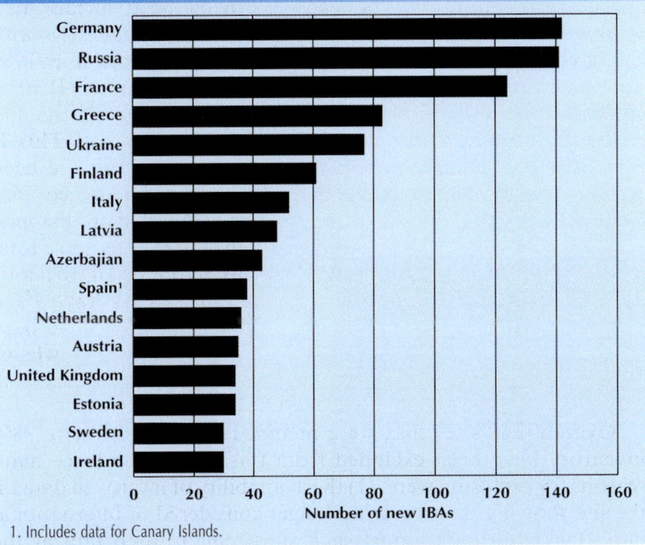

1. Includes data for Canary Islands.

Figure 6. The countries in Europe where the total area of new Important Bird Areas exceeds 10,000 km².

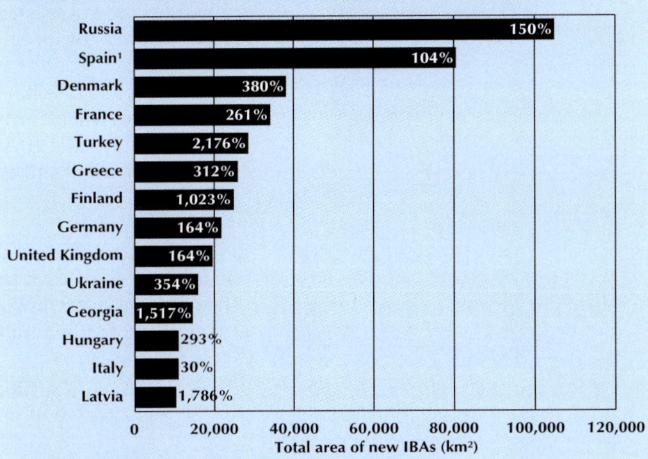

1. Includes data for Canary Islands.

The percentage at the end of each bar indicates the percentage change in the national total IBA area since the previous pan-European IBA inventory (Grimmett and Jones 1989). Such change is mainly attributable to the addition of new IBAs but also involves, in most countries, the enlargement or reduction in size of many original (1989) IBAs as well as the deletion from the current IBA inventory of a small minority of the original sites (rejected as IBAs for various reasons).

- Turkey—97 IBAs cover 4% of the country; many non-wetland IBAs have yet to be identified;
- Sweden—63 IBAs cover 3% of the country.

In addition, for a small number of countries or autonomous regions (Croatia, Macedonia, Faroe Islands, and Bosnia and Herzegovena) fully updated information was not obtained for this publication, thus future revisions will almost certainly result in further changes to the number and area of sites in these countries. With the anticipated accession of several countries to the European Union during the coming decade, the EU-specific criteria for IBAs (category 'C') will be applied, which is likely to result in the identification of additional sites, particularly for species considered threatened in the EU (listed in Annex I of the EC Birds Directive).

A number of IBAs span the borders of two or more countries, often wetlands in river valleys or tracts of forest and alpine habitat in mountainous areas. Details of the these transboundary sites are summarized in Appendix 7. International collaboration for the protection and management of these sites, and the species they hold, is essential.

IBA coverage by species and criteria

Through the rigorous application of quantitative ornithological criteria, it is assured that the network of 3,619 identified sites is of international importance. The criteria permit one to distinguish the importance of sites at three geographical levels: global, European and European Union. Such distinctions are particularly significant when comparing IBAs across regions of the world where, to date, effort has been prioritized on identifying sites of global importance only. A high proportion of Europe's IBAs (55%, 1,997 IBAs) are of global significance (meeting 'A' criteria). A further 1,176 IBAs are important at the European level (and not global) and the remaining 446 IBAs qualify as important at the EU level only (Figure 7). Sites of global importance are distributed across all European countries (Figure 1, Appendix 6).

The reasons for the importance of IBAs, indicated by the criteria that each site fulfils, are shown in Figure 8. A site may be important for many different species and thus may qualify under more than one type of criterion. In general terms, a high percentage of IBAs are important for congregatory species (51% of all IBAs in Europe), species of European conservation concern (66%) and species listed in Annex I of the EC Birds Directive (63%). A lower proportion (27%), though very significant given the small number of species involved, are important for one or more of the 35 species of global conservation concern in Europe. IBAs identified for assemblages of species with small global ranges (A2 criterion), or species restricted to a biome in Europe (A3 criterion), are relatively few but still individually significant at the global level.

Sites are often important for several species, each of which fulfils the same type of criterion, for example, several species of waterbird may congregate at a site in a given season, each in significant numbers that meet the B1i criterion. Overall, over 30,000 applications of criteria have been made to species at the 3,619 IBAs. This is an indication of the strength and substance of the scientific evidence that supports the treatment of each site as worthy of international recognition.

Figure 7. The proportion of Important Bird Areas in Europe that are important at the global, European and EU levels.

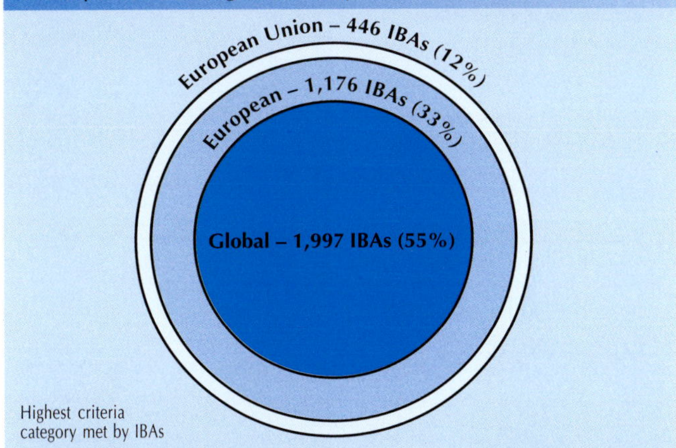

Figure 8. The number of sites meeting different Important Bird Area criteria in Europe.

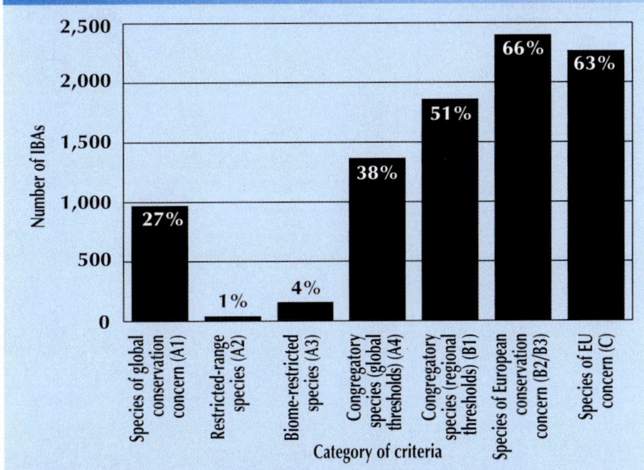

N.B. A site may be important for more than one bird species (or in more than one way for a particular species) and may thus qualify under more than one criterion. In addition, criteria are nested such that, if higher thresholds are met, criteria are also assigned at lower levels. For example, if a site in the European Union is important because it holds more than 1% of the biogeographic population of a migratory waterbird, then not only is criterion A4i applied, but also B4i and C2/C3.

Species of global conservation concern

A network of 938 IBAs (27% of IBAs in Europe) support significant numbers of one or more species of global conservation concern, thus meeting the A1 criterion. Sites have been identified in 41 countries or autonomous regions for these species, and those countries with the highest number of sites identified include Russia (152 sites), Spain (121), Greece (73), Turkey (56) and Italy (51).

The number of sites identified for each species of global conservation concern (under the A1 criterion), and the proportion of the European breeding population within IBAs, are shown in Table 1. (A further 111 IBAs in Europe meet criteria other than A1 (B2, C1, C2, C6) for these species.) Species for which very few sites have been identified include *Pterodroma feae*, *Pterodroma madeira*, *Polysticta stelleri*, *Columba trocaz*, *Saxicola dacotiae*, *Sitta whiteheadi*, *Fringilla teydea* and *Emberiza cineracea*. However, the IBAs for several of these species hold nearly the entire world population of the species concerned. For other species with larger European ranges, more sites have been identified, e.g. *Phalacrocorax pygmeus*, *Haliaeetus albicilla*, *Falco naumanni*, *Crex crex* and *Otis tarda*. For all species, the conservation of IBAs makes a significant contribution to their survival (Table 1) but for those species with highly dispersed distributions, such as *Crex crex*, it is particularly important that conservation of sites is accompanied by conservation of the wider environment (Tucker and Evans 1997).

Dalmatian Pelican *Pelecanus crispus* at Gediz Delta, Turkey (IBA 016). A globally threatened species classified as Vulnerable (Collar *et al.* 1994), its global decline is due to disturbance and degradation of wetlands, hunting and colony destruction by fishermen.
(PHOTO: CÜNEYT OGUZTÜZÜN)

Recent surveys in Belarus and Ukraine have discovered new IBAs for the globally threatened Aquatic Warbler *Acrocephalus paludicola*.
(PHOTO: ALEXANDER KOZULIN)

Lesser Kestrel *Falco naumanni* is a globally threatened species, classified as Vulnerable (Collar *et al.* 1994). It has shown major population declines in large parts of its western Palearctic breeding range.
(PHOTO: CÜNEYT OGUZTÜZÜN)

Table 1. Internationally important sites for species of global conservation concern in Europe (meeting the A1 criterion).

	Number of IBAs qualifying for species: Season – Breeding	Non-breeding	Status unknown	Total[1]	Percentage of the minimum European breeding population occurring at IBAs [2,3]
Pterodroma feae Fea's Petrel	1	—	—	1	68
Pterodroma madeira Zino's Petrel	1	—	—	1	71
Phalacrocorax pygmeus Pygmy Cormorant	34	90	1	115	100[4]
Pelecanus crispus Dalmatian Pelican	22	50	1	69	100[4]
Anser erythropus Lesser White-fronted Goose	2	33	1	36	2
Branta ruficollis Red-breasted Goose	—	33	—	33	—
Marmaronetta angustirostris Marbled Duck	19	9	1	19	96
Aythya nyroca Ferruginous Duck	62	37	1	92	25
Polysticta stelleri Steller's Eider	—	9	—	9	—
Oxyura leucocephala White-headed Duck	21	41	1	52	100[4]
Haliaeetus albicilla White-tailed Eagle	56	46	1	96	25
Aegypius monachus Cinereous Vulture	30	9	1	40	94
Circus macrourus Pallid Harrier	7	3	—	10	13
Aquila clanga Greater Spotted Eagle	33	17	—	50	13
Aquila heliaca Imperial Eagle	62	8	1	71	48
Aquila adalberti Spanish Imperial Eagle	19	6	—	24	88
Falco naumanni Lesser Kestrel	76	18	2	94	73
Tetrao mlokosiewiczi Caucasian Black Grouse	13	—	—	13	100[4]
Crex crex Corncrake	263	1	—	264	20
Tetrax tetrax Little Bustard	61	17	—	75	15
Otis tarda Great Bustard	74	8	—	78	70
Glareola nordmanni Black-winged Pratincole	10	4	1	15	5
Chettusia gregaria Sociable Plover	—	—	—	—	<1
Gallinago media Great Snipe	45	5	—	50	1
Numenius tenuirostris Slender-billed Curlew	—	31	—	31	—
Larus audouinii Audouin's Gull	43	16	—	58	100[4]
Columba trocaz Long-toed Pigeon	5	—	—	5	41
Columba bollii Dark-tailed Laurel Pigeon	11	—	—	11	38
Columba junoniae White-tailed Laurel Pigeon	11	—	—	11	41
Saxicola dacotiae Fuerteventura Chat	9	—	—	9	8
Acrocephalus paludicola Aquatic Warbler	31	5	—	36	100[4]
Sitta whiteheadi Corsican Nuthatch	4	—	—	4	67
Fringilla teydea Blue Chaffinch	4	—	—	4	9
Loxia scotica Scottish Crossbill	12	3	—	14	2
Emberiza cineracea Cinereous Bunting	4	—	—	4	11

1. The total number of sites identified for a species of global conservation concern does not necessarily equal the sum of breeding, non-breeding and unknown-seasonal-status sites, since a species may meet criteria at an IBA in more than one season.
2. Percentage calculated using the geometric mean of the minimum and maximum population size in IBAs and the geometric mean of the European breeding population.
3. All IBAs where the species breeds (regardless of whether the IBA criteria are met for this species) are included in the total population breeding in IBAs.
4. The percentage of the national population in IBAs exceeds 100%. Usually this is because the national population estimate has not been updated recently whilst the IBA population estimate has been recently updated with new data as a result of comprehensive surveys of IBAs themselves. Also, the individual site count for a species may be the maximum or average over recent years, and summing these may record more birds than are present nationally in any single year.

Species with restricted global ranges

A total of 47 IBAs in Europe meet the A2 criterion for restricted-range species, i.e. those species with global ranges of less than 50,000 km² (Table 2). Since the entire world ranges of most of these species are confined to 'Endemic Bird Areas' (Stattersfield *et al.* 1998), it is particularly important to conserve the network of IBAs which have been identified for these species. Their restriction to such a small area and number of sites makes them particularly vulnerable to habitat change and other threats. IBAs identified for restricted-range species in Europe are situated within three Endemic Bird Areas (Madeira and the Canary Islands; Cyprus; Caucasus) and three 'Secondary Areas' (Azores; Corsican mountains; Caledonian pine forest) (see 'Identifying IBAs' chapter for definition of Secondary Area). Overall, the network of IBAs for these species is thought to be complete.

The globally threatened Dark-tailed Laurel Pigeon *Columba bollii* is confined to the Canary Islands (La Palma, La Gomera, Tenerife and El Hierro) where its principal habitat is closed-canopy laurel forest. (PHOTO: N. MARTIN)

Table 2. Internationally important sites for restricted-range bird species in Europe (meeting the A2 criterion).

Restricted-range species	Endemic Bird Area	Range states in Europe	Number	IBA codes
Tetrao mlokosiewiczi Caucasian Black Grouse	Caucasus	Armenia, Azerbaijan, Georgia, Russia, Turkey	9	GE001–GE003, GE005–GE008, RU163, RU167
Tetraogallus caucasicus Caucasian Snowcock	Caucasus	Azerbaijan, Georgia, Russia	7	GE001–GE003, GE006, GE007, RU163, RU167
Columba trocaz Long-toed Pigeon	Madeira and the Canary Islands	Madeira	5	ZZ001–ZZ005
Columba bollii Dark-tailed Laurel Pigeon	Madeira and the Canary Islands	Spain (Canary Islands)	4	ES366, ES376, ES379, ES380
Columba junoniae White-tailed Laurel Pigeon	Madeira and the Canary Islands	Spain (Canary Islands)	4	ES366, ES376, ES379, ES380
Apus unicolor Plain Swift	Madeira and the Canary Islands	Madeira, Spain (Canary Islands)	7	ZZ005, ES340, ES346, ES355, ES366, ES376, ES379
Anthus berthelotii Berthelot's Pipit	Madeira and the Canary Islands	Madeira, Spain (Canary Islands)	13	ZZ005, ZZ008, ZZ009, ZZ010, ES338, ES340, ES346, ES354, ES355, ES366, ES370, ES376, ES379
Saxicola dacotiae Fuerteventura Chat	Madeira and the Canary Islands	Spain (Canary Islands)	3	ES338, ES340, ES346
Oenanthe cypriaca Cyprus Pied Wheatear	Cyprus	Cyprus	13	CY002, CY003, CY004, CY005, CY006, CY007, CY008, CY010, CY012, CY013, CY014, CY015, CY016
Sylvia melanothorax Cyprus Warbler	Cyprus	Cyprus	13	CY002, CY003, CY004, CY005, CY006, CY007, CY008, CY010, CY012, CY013, CY014, CY015, CY016
Phylloscopus lorenzii Caucasian Chiffchaff	Caucasus	Armenia, Azerbaijan, Georgia, Russia, Turkey	2	GE007, GE008
Regulus teneriffae Tenerife Goldcrest	Madeira and the Canary Islands	Spain (Canary Islands)	6	ES366, ES369, ES370, ES376, ES379, ES380
Sitta whiteheadi Corsican Nuthatch	Corsican mountains*	France	2	FR269, FR270
Fringilla teydea Blue Chaffinch	Madeira and the Canary Islands	Spain (Canary Islands)	4	ES354, ES355, ES369, ES370
Serinus canaria Canary	Madeira and the Canary Islands; Azores*	Madeira, Spain (Canary Islands); Azores	14	ZZ008, ZZ009, ES354, ES355, ES366, ES369, ES370, ES376, ES379, ES380, QQ001, QQ003, QQ004, QQ013
Pyrrhula murina Azores Bullfinch	Azores*	Azores	1	QQ013

* Secondary Area. It was considered inappropriate to identify IBAs for Scottish Crossbill *Loxia scotica* (Secondary Area: Caledonian pine forest) on the basis of its restricted range. However, several IBAs have been identified for the taxon under other criteria, as it is also a species of European conservation concern and is listed in Annex I of the EC Birds Directive (see UK chapter).

Box 1. Internationally important sites for assemblages of biome-restricted bird species in Europe (meeting the A3 criterion).

A total of 155 Important Bird Areas have been identified on the basis that they support important assemblages of species that are restricted (when breeding in Europe) to a particular biome (thus meeting the A3 criterion). In applying this criterion in Europe, five biomes have been considered:

- Arctic/tundra biome
- Boreal biome
- Mediterranean biome
- Eurasian high-montane (alpine) biome
- Eurasian steppe biome.

Biomes are the world's major (terrestrial) ecological communities, characterized by the occurrence together of similar vegetation, animals and climate. A 'biome' encompasses all flora, fauna, and soils within the geographical region concerned. Biomes are classified by climax vegetation type (e.g. forests, grasslands) and are characterized by the adaptations of their organisms to the particular conditions in which they occur, for example high temperatures, low rainfall, poor soils or periodic flooding. The distributions of biomes therefore strongly match patterns of regional climate (e.g. temperate, tropical). A given biome (e.g. temperate grassland) may be composed of different taxa on different continents. Continent-specific associations of species within a given biome are known as formations and often are known by different local names. For example, the temperate grassland biome is variously called prairie, steppe, pampa, or veld, depending on where it occurs (North America, Eurasia, South America, and South Africa, respectively).

Number of IBAs meeting the A3 criterion

	Arctic/tundra biome	Boreal biome	Mediterranean biome	Eurasian high-montane biome	Eurasian steppe biome	Total
Armenia	—	—	—	1	—	1
Azerbaijan	—	—	—	9	—	9
Bulgaria	—	—	3	1	—	4
Croatia	—	—	1	—	—	1
Cyprus	—	—	3	—	—	3
Finland	5	11	—	—	—	14[1]
France	—	—	4	1	—	5
Georgia	—	—	—	8	—	8
Germany	—	—	—	1	—	1
Greece	—	—	13	5	—	17[1]
Greenland	3	—	—	—	—	3
Italy	—	—	7	12	—	19
Norway	1	1	—	—	—	2
Poland	—	—	—	1	—	1
Russia	5	12	2	3	7	25[1]
Slovakia	—	—	—	1	—	1
Slovenia	—	—	—	1	—	1
Spain	—	—	7	5	—	12
Svalbard	3	—	—	—	—	3
Sweden	4	1	—	—	—	5
Switzerland	—	—	—	14	—	14
Turkey	—	—	—	1	—	1
Ukraine	—	—	—	—	2	2
Yugoslavia	—	—	1	2	—	3
Total number of IBAs	**21**	**25**	**41**	**66**	**9**	**155[1]**

1. The national total is less than the sum of sites qualifying for individual biomes, since some IBAs qualify for two biomes (usually large sites that span the ecotone between biomes).

ARCTIC/TUNDRA BIOME

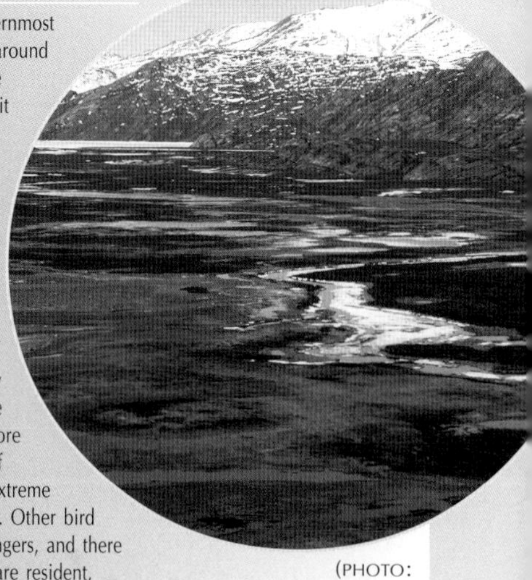

The Arctic/tundra biome is the northernmost terrestrial biome, restricted to a belt around the Arctic Ocean and adjoined by the boreal biome to the south. In Europe it covers about 680,000 km² (c.7% of the continent). Winters are long and cold and the subsoil is frozen year-round (permafrost), as a result of which there are no trees; mosses dominate the vegetation, with lichens, sedges, grasses, and occasional low bushes and dwarf shrubs. Most bird species characteristic of this biome are highly migratory waterbirds which spend the brief summer raising their young, before departing south, ahead of the onset of winter. Some of these species show extreme adaptation to long-distance migration. Other bird species are mainly raptors and scavengers, and there are few passerines; very few species are resident, possessing adaptations to survive the harsh tundra winter.

(PHOTO: DAVID BOERTMANN)

BOREAL BIOME

The boreal biome, also known in Europe as the taiga, is the largest terrestrial biome on earth. Confined to the northern hemisphere, it extends as a broad and near-continuous band of coniferous forest and open mires across Eurasia and North America, lying immediately south of the Arctic/tundra biome. In Europe this belt and covers at least 1,000,000 km² (10% or more of the continent). The forest is a mosaic of different-aged stands, dominated in Europe by two conifer species, Norway spruce *Picea abies* and Scots pine *Pinus sylvestris*. A few deciduous tree species also occur, especially in disturbed or wet areas. Large, tree-less mires form in poorly drained, glacial depressions, and tundra-like habitats occur above the treeline on higher mountains within this zone (here trated as part of the Arctic/tundra biome). Among birds, insectivores and waterbirds tend to be migratory, leaving after the breeding season since freezing conditions eliminate their food-supply in winter. Small-mammal predators (e.g. owls), seed-eaters (e.g. finches) and omnivores (e.g. ravens) tend to be year-round residents.

(PHOTO: MIKA HONKALINNA)

MEDITERRANEAN BIOME

The Mediterranean biome as considered in this publication—the land in Europe, North Africa and the Levant surrounding the Mediterranean Sea—is the largest of several formations of the Mediterranean biome around the world, covering about 500,000 km² in Europe. In the Mediterranean proper, the vegetation and soils have long been impacted by humans, especially through the frequent use of fire and the grazing of livestock, and dry shrublands with thin, rocky soils are characteristic. Shrubs and trees tend to be evergreen with small, leathery leaves. Where fire and grazing are reduced or absent, shrubland is replaced by regenerating woodland and forest, dominated by evergreen oaks *Quercus* and pines *Pinus*. The flora and fauna are characterized by a high degree of endemism. Among birds, most of the characteristic species are found in shrublands and open, rocky areas rather than in forest habitats.

(PHOTO: MARCO LAMBERTINI)

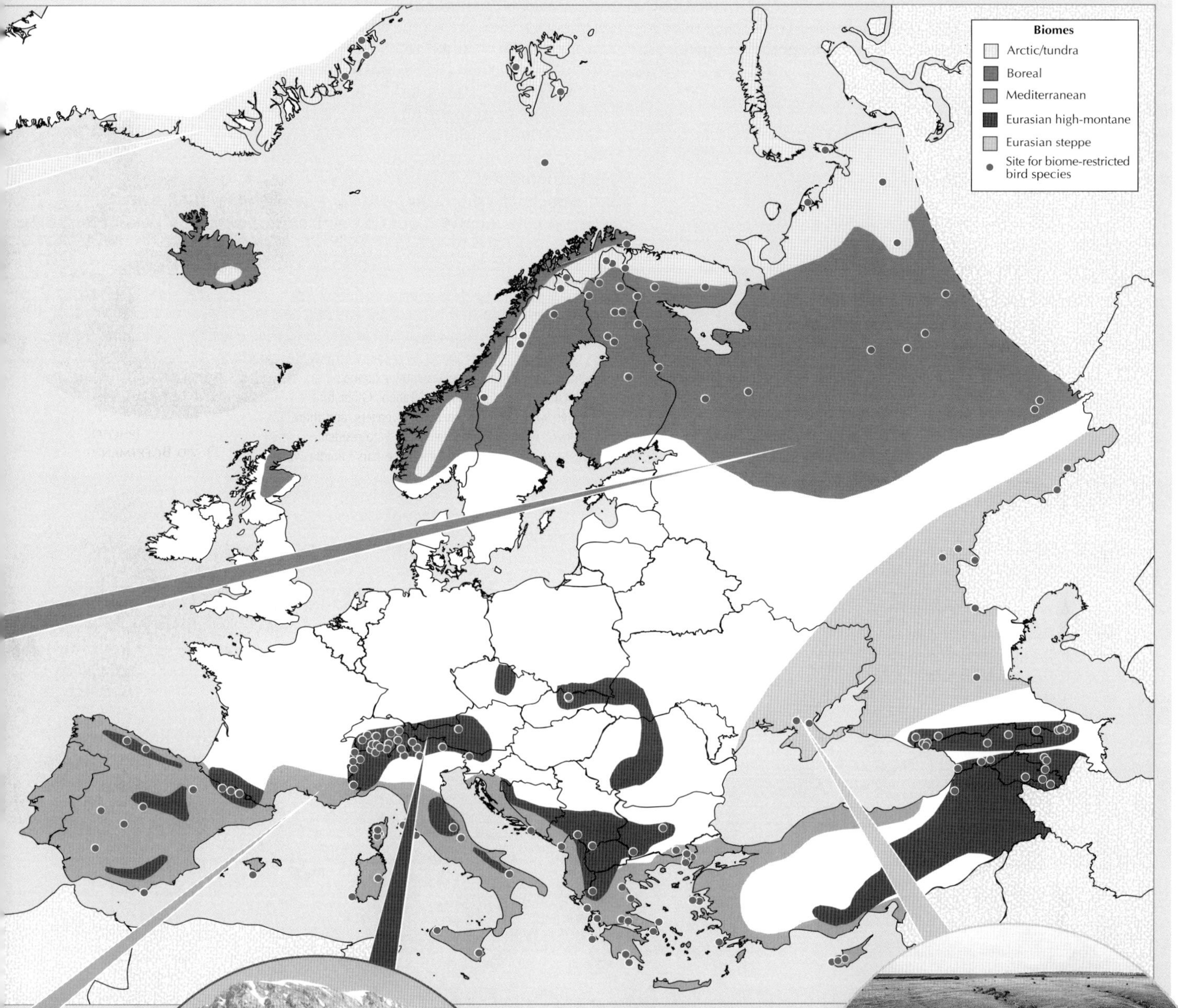

(PHOTO: PETER BALWIN/ SCHWEIZER VOGELSCHUTZ)

RASIAN HIGH-
)NTANE (ALPINE) BIOME

e Eurasian high-montane (or alpine)
me is found high on mountains in
asia. The climate is severe, receiving
isture mostly in the form of snow, with high winds, drastic daily temperature changes
d sometimes heavy snowfall. Permafrost occurs at the higher altitudes and there are
v plants and no trees. Alpine vegetation grows low to the ground above the
mmholz (stunted-tree zone), in alpine meadows and alpine stone-fields. Animals
ernate, migrate, or live under the snow to survive the harsh winter conditions.
atively few birds in Europe are restricted to this biome during the breeding season,
h the majority being passerines, unlike the situation in the Arctic.

(PHOTO: PAUL GORIUP)

EURASIAN STEPPE BIOME

The Eurasian steppes are temperate grasslands, found in mid-latitude regions with relatively cold winter temperatures. In Eurasia and elsewhere, such grasslands occur in the interior of continents and in rain-shadows. The climate is semi-arid and continental, with warm to hot summers and cold winters. In Europe, before the onset of economic development in the 16th century, the steppes covered perhaps 800,000 km^2 or more, extending in a belt 300–1,000 km wide from the lower Danube valley east across the flat lowlands of Ukraine and southern Russia, with temperate broadleaved and boreal forest lying to the north, and semi-desert to the south around the Caspian Sea. Vegetation was dominated by perennial grasses and forbs and there were very few trees except along rivers and streams. However, since the development of the steel plough nearly all temperate grassland in Europe, indeed Eurasia, has been converted to agricultural land, and has become densely populated and industrialized. Among birds restricted to the biome when breeding, open-country species naturally predominate, and some of the species more dependent on original, unploughed steppe, such as Sociable Plover *Chettusia gregaria*, are now globally threatened through lack of habitat.

Species restricted to biomes

A total of 155 IBAs have been identified on the basis that they support important assemblages of species that are restricted (when breeding in Europe) to a particular biome (thus meeting the A3 criterion). In applying this criterion in Europe, five biomes have been considered:

- Arctic/tundra
- Boreal
- Mediterranean
- Eurasian high-montane
- Eurasian steppe.

The IBAs and their distribution are shown in Box 1.

Congregatory species

Individuals of many bird species traverse political boundaries at some stage in their annual cycle, and many also congregate in large numbers at sites across the continent at some time each year. IBAs are particularly suited to conserving these species, with over 51% of IBAs in Europe (1,858 sites) being important for congregations of birds: 1,364 sites qualify because the site supports globally important congregations of at least one species (meeting the A4 criteria), and an additional 494 sites qualify because the site holds regionally important congregations of at least one species (B1 criteria).

Appendix 6 shows the number of sites identified under each IBA criterion, by country. Countries with at least 60 sites important for

Box 2. Internationally important sites for congregations of non-breeding waterbirds or seabirds in Europe (meeting the A4i/ii/iii and/or B1i/ii criteria).

The Wadden Sea (many IBAs) is the largest intertidal area in Europe, spanning the North Sea coasts of Germany, the Netherlands and Denmark. Important habitats include mudflats, sandflats, saltmarshes, shallowly submerged sandbanks, and estuaries. The area is of major international importance for breeding, moulting, passage and wintering waterbirds, particularly wildfowl and waders.

The Oulu region wetlands, Finland (FI028) are a very large and diverse wetland complex, including sea bays, archipelago and farmland, in the vicinity of Oulu City. A uniquely important breeding and staging area for numerous bird species. A very high proportion of the populations of many northern waterbird species stage in the area.

Total number of sites=1,340 IBAs.

The Wash, United Kingdom (UK077) is an extremely important area of estuarine mudflats, sandbanks and saltmarsh, supporting large concentrations of passage and wintering waders and wildfowl. The IBA holds over 300,000 wintering and 250,000 passage waterbirds on a regular basis.

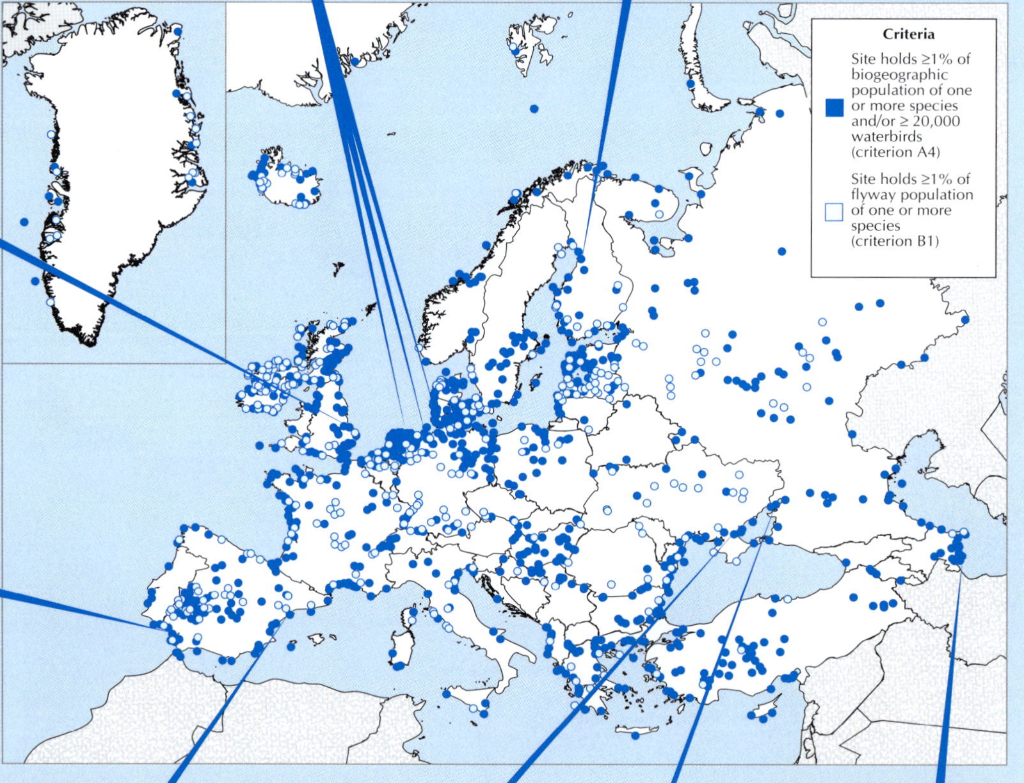

The Guadalquivir marshes (ES259), lying at the mouth of the River Guadalquivir, are one of the largest wetlands in Europe. They are the most important wetland in Spain for breeding, passage and wintering waterbirds and passerines. Over 360 bird species have been recorded. Wintering waterbird numbers reach 400,000 individuals, increasing to over 6 million birds during migration periods.

The Ebro delta, Spain (ES148) is a large delta that protrudes into the Mediterranean Sea, forming a complex of shallow brackish lagoons, saltmarshes, salt-lakes and sand beaches with dunes. One of the most important sites in the Mediterranean for breeding, passage and wintering waterbirds, with 27,000 pairs of breeding waterbird in summer and up to 180,000 waterbirds in winter.

Sivash Bay, Sea of Azov, Ukraine (UA069), is an extensive system of lagoons with sandy islands and shores in the coastal zone of the Black Sea and Sea of Azov. Over 25 species congregate in internationally important numbers on passage.

The Don delta, Russia (RU141) supports hundreds of thousands of migrating waterbirds that use the area as a staging post, with ducks, gulls and terns comprising most of the total.

Kizil Agach Bay, Azerbaijan (AZ048), an open bay connected to the Caspian Sea, was until recently an immensely important stop-over and wintering site, but waterbird numbers (ducks and Coot *Fulica atra*) have fallen dramatically: from 10 million in 1930, to 5–7 million in the late 1950s, to 1,500,000 by the end of the 1960s, to 338,000–715,000 during the 1970s, and down to 200,000–400,000 birds at the beginning of the 1980s.

1,340 IBAs are important for congregations of non-breeding waterbirds or seabirds. Some key sites from across the region are described.

Oystercatcher *Haematopus ostralegus* congregating at Snettisham, The Wash, in the United Kingdom (IBA 077). (PHOTO: MIKE READ)

Whooper Swan *Cygnus cygnus*, a highly congregatory species, wintering at Shabla lake complex, Bulgaria (IBA 049). (PHOTO: WERNER MÜLLER)

Gannet *Sula bassana* colony at Bass Rock, Forth Islands, in the United Kingdom (IBA 148). Over 90% of the European breeding population of *Sula bassana* breed at IBAs. (PHOTO: C. H. GOMERSALL)

congregatory species (qualifying under the A4 and/or B1 criteria) are Denmark, France, Germany, Greece, Ireland, Italy, Netherlands, Norway, Russia, Spain, Turkey, Ukraine and United Kingdom.

Under these criteria, IBAs can be important for different groups of congregatory species: congregations of waterbirds (criteria A4i and B1i), congregations of seabirds (A4ii and B1ii), congregations of more than 20,000 waterbirds or 10,000 pairs of seabirds (A4iii), 'bottleneck' concentrations of migrating storks, raptors and cranes (A4iv and B1iv) and congregations of other species (B1iii). These criteria may be applied in any season and IBAs have been identified for migrating, wintering, non-breeding and breeding congregations of birds. A total of 1,340 sites are important for congregations of non-breeding waterbirds and seabirds (Box 2) and 811 for breeding waterbirds or seabirds. One hundred and seven sites in Europe are important bottleneck sites through which large numbers of storks, raptors or cranes pass on migration (Box 3): at 45 such IBAs, at least 20,000 storks/raptors/cranes pass through per season (criterion A4iv), while at the remaining 62 sites, the seasonal total exceeds 5,000 storks or 3,000 raptors/cranes (criterion B1iv).

Box 3. Internationally important 'bottleneck' sites for migrating storks, raptors and cranes in Europe (meeting the A4iv/B1iv criteria).

A total of 107 sites in Europe meet the B1iv criterion, with at least 5,000 storks or at least 3,000 raptors or cranes passing through the site. Forty-five out of the 107 sites also meet the A4iv criterion, with at least 20,000 storks, raptors or cranes passing through the site.

The Alps form a major barrier to migrating birds in Europe. In France, 'Haute Chaîne du Jura: défilé de l'écluse, Etournel et Mont Vuache' (FR172) is a major migratory bottleneck, where more than 20,000 raptors pass on a regular basis, including 10,000 Honey Buzzard *Pernis apivorus*, 20,000 Black Kite *Milvus migrans* and 20,000 Buzzard *Buteo buteo*.

Large numbers of raptors, mostly Fennoscandian, pass through Denmark, especially the eastern part, during both spring and autumn migration, with the most significant bottlenecks being at Gilleleje (DK126, spring), Hellebæk (DK124, spring and autumn), Stigsnæs (part of DK095, autumn) and Stevns (DK127, autumn). The most numerous species are Honey Buzzard *Pernis apivorus*, Buzzard *Buteo buteo* and Sparrowhawk *Accipiter nisus*.

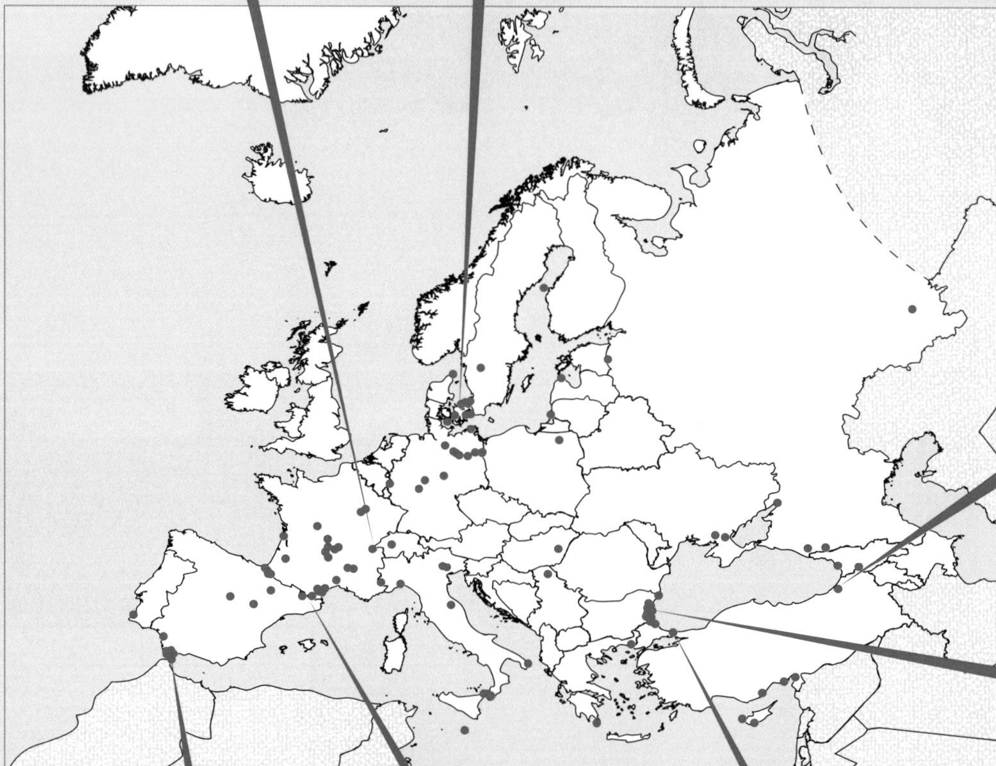

The 'North-east Turkey' IBA (TR060) covers three major mountain ranges (Soganli, Kaçkar and Karçal), and includes a major migratory bottleneck, where large numbers of raptors regularly pass in autumn. A count of 380,220 birds of prey was made in 1976, including 138,000 Honey Buzzard *Pernis apivorus*, 5,775 Black Kite *Milvus migrans* and 205,000 Buzzard *Buteo buteo*. Actual numbers may be higher as important new passage points have been discovered in recent years.

Along the Bulgarian Black Sea coast, nine IBAs are bottleneck sites for migratory birds. These sites are the most important concentrations of large soaring birds along the western Black Sea migration route ('Via Pontica'). It is believed that large proportions of the European populations of White Stork *Ciconia ciconia*, White Pelican *Pelecanus onocrotalus*, Honey Buzzard *Pernis apivorus*, Buzzard *Buteo buteo* and Imperial Eagle *Aquila heliaca* migrate along the western Black Sea coast, especially over Cape Emine (BG043) and Atanasovo lake (BG036).

Gibraltar is a well-known migratory bottleneck site for birds of prey, with an estimated 250,000 raptors crossing the Strait per season. If the winds are right most of these birds pass over Gibraltar (GI001), and numbers may in fact be higher. The species involved include Honey Buzzard *Pernis apivorus*, Black Kite *Milvus migrans*, Egyptian Vulture *Neophron percnopterus*, Short-toed Eagle *Circaetus gallicus*, Montagu's Harrier *Circus pygargus* and Booted Eagle *Hieraaetus pennatus*. Large numbers of birds of prey also cross IBAs surrounding the Strait of Gibraltar (ES244, ES245, ES246, ES248, ES250) before moving to Africa, and some stop over at the Guadalquivir marshes (ES259) before crossing to Africa over Ceuta (ES247).

Birds that breed in western Europe cross the Pyrenees at several sites in Spain and France, amongst the most important being the 'Roncesvalles-Irati-Abodi mountain range' (ES085), Cadí mountain range (ES135), 'Haute Soule: Forêt d'Irraty, Organbidexka et Pic des Escaliers' (FR183; more than 20,000 raptors pass regularly, including 18,000 Honey Buzzard *Pernis apivorus*, 13,000 Black Kite *Milvus migrans*, 4,600 Red Kite *M. milvus* and 15,000 Crane *Grus grus*), and 'Etangs de Leucate et Lapalme' (FR214; including 15,000 *Pernis apivorus* and 3,000 *Milvus migrans*).

The Bogaziçi strait (TR005), in western Turkey, runs north–south for c.33 km, and is c.1.5 km wide. Total counts made in the mid-1960s and early 1970s, covering the entire autumn migration period, include 8,318 Black Stork *Ciconia nigra*, 338,353 White Stork *Ciconia ciconia*, 25,751 Honey Buzzard *Pernis apivorus*, 2,707 Black Kite *Milvus migrans*, 32,895 Buzzard *Buteo buteo* and 18,898 Lesser Spotted Eagle *Aquila pomarina*. More recent autumn records include 11,703 *Aquila pomarina* at Çamlica and 11,379 of the same species at Sariyer, both on the same day in September 1990.

IBAs shown on the map: Belgium 044; Bulgaria 034, 036, 040, 041, 043–045, 047, 049; Cyprus 007, 014; Denmark 072, 095, 102, 124–127; Finland 045; France 068, 070, 123, 143–145, 149, 151, 158, 162, 163, 172, 177, 178, 183, 191, 192, 213–215, 217, 218, 222, 225; Georgia 004, 008; Germany 037, 106, 122, 124, 138, 142, 143, 149, 151, 172, 179; Gibraltar 001; Greece 006, 129; Hungary 029; Italy 035, 036, 054, 055, 085, 147, 150, 151, 153; Latvia 015, 021; Lithuania 004; Malta 004; Poland 006, 022; Portugal 031; Russia 050, 141, 163, 167, 212; Spain 078, 085, 095, 116, 135, 244–248, 250, 259; Sweden 042, 057, 061; Switzerland 017; Turkey 002, 005, 060, 073, 078, 081; Ukraine 068, 069; Yugoslavia 012.

Species of European conservation concern

The IBA programme has an important role to play in conserving populations of many species of European conservation concern (SPECs). These species are in particular need of conservation measures as they either have an unfavourable conservation status in Europe (because they are declining in numbers, are present in small numbers or are highly localized) or they are concentrated in Europe (a substantial proportion of their global population occurs in Europe and thus Europe has a special responsibility for their conservation). A total of 2,399 IBAs (66% of all IBAs in Europe) are important for SPECs, meeting B2/B3 criteria, with sites identified for 269 of the 278 species (97%).

The IBA network supports a significant proportion of the European population of many SPECs. Each country chapter gives a summary of the proportions of the national populations of SPECs covered by IBAs. From the data presented in this publication, it is possible to calculate, for individual species, the proportion of the European population found within IBAs across the region. For example, the European breeding population of *Platalea leucorodia* is well covered by the IBA network, with a total of 81 IBAs (distributed throughout the European breeding range) holding significant breeding numbers of the species, which together represent approximately 80% of the breeding population in Europe (Figure 9).

Barrow's Goldeneye *Bucephala islandica* at Lake Mývatn/River Laxá, Iceland (IBA 046), a site which holds 85–90% of the European population of the species.
(PHOTO: JÓHANN ÓLI HILMARSSON)

The breeding distribution of Broad-billed Sandpiper *Limicola falcinellus*, a species of European conservation concern, is poorly known, but it seems likely that at least 10–20% of the European population breeds within IBAs.
(PHOTO: PEKKA KOMI)

Spoonbill *Platalea leucorodia*, a species of European conservation concern, in Atanasovo lake, Bulgaria (IBA 036). The species's breeding sites are well covered by IBAs (see Figure 9).
(PHOTO: SVETISLAV SPOSOV)

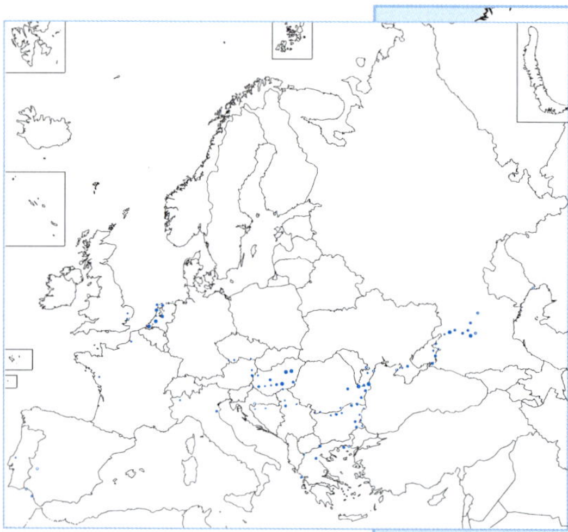

The European breeding distribution of Spoonbill *Platalea leucorodia*. Source: *EBCC Atlas of European breeding birds* (Hagemeijer and Blair 1997).

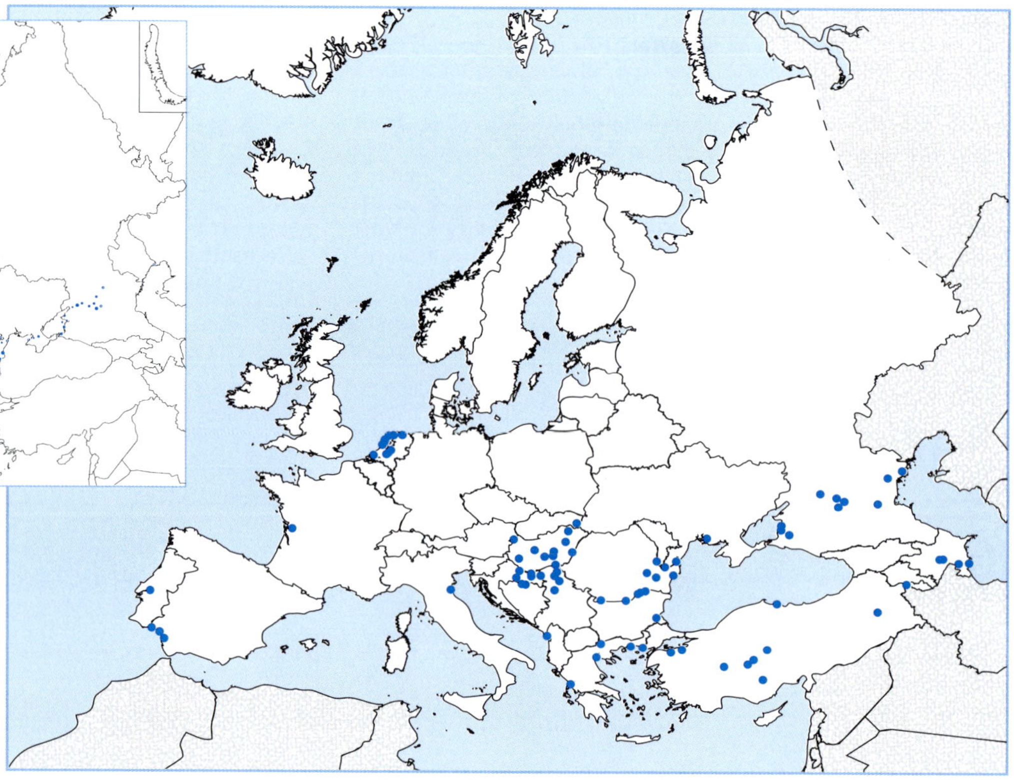

Figure 9. Internationally important breeding sites for Spoonbill *Platalea leucorodia* in Europe (meeting Important Bird Area criteria of any category).

However, some species of European conservation concern are not well covered by IBAs. Reasons for this may include a lack of data on sites and populations—it is recognized that the network of sites is not yet complete for some SPECs. In addition, many species of European conservation concern breed in a dispersed, non-congregatory fashion, and therefore cannot be conserved successfully by the IBA programme alone. For many SPECs, even though the IBA programme makes a significant contribution to their conservation, it is vital that conservation of the wider environment is also pursued in parallel (Tucker and Evans 1997).

Species meeting criteria in the European Union (level C)

In total, 2,342 IBAs have been identified in the 15 member states of the European Union. Most of the criteria for the importance of IBAs at the EU level parallel the criteria at the European and global levels but some are associated with lower population thresholds. Sites have been identified for species of global conservation concern (criterion C1), migratory and congregatory species (criteria C2, C3, C4 and C5), species or subspecies considered threatened in the EU (criterion C6) or for other reasons (C7). The numbers of IBAs meeting the different 'C' criteria are shown in Figure 10 and are summarized by country in Appendix 6.

Figure 10. The number of sites meeting the different 'C' criteria in the European Union.

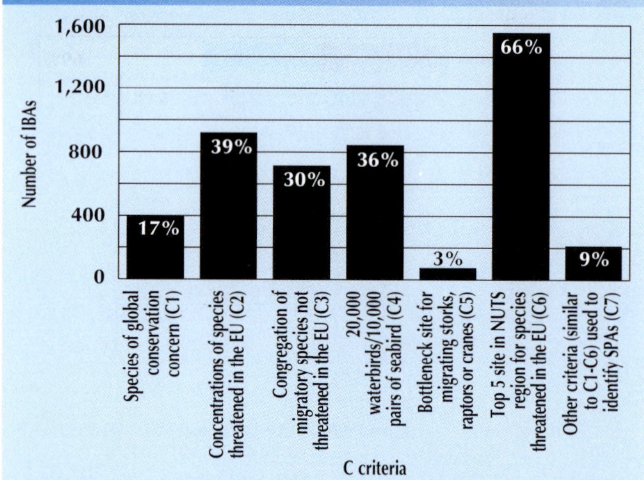

The percentage above each column indicates the proportion of the 2,342 IBAs in the EU that meet the criterion.

N.B. A site may be important for more than one bird species (or in more than one way for a particular species) and may thus qualify under more than one criterion.
For more details of the 'C' criteria, see the 'Identifying IBAs' chapter.

Stone Curlew *Burhinus oedicnemus*, listed in Annex I of the EC Birds Directive and a species of European conservation concern. A total of 94 IBAs across nine countries support significant breeding numbers of this species. (PHOTO: CARLOS SANCHEZ)

Egyptian Vulture *Neophron percnopterus*, listed in Annex I of the EC Birds Directive and a species of European conservation concern. A total of 89 IBAs across 11 countries support significant numbers of this species.
(PHOTO: JUAN M. SIMON)

While 2,264 IBAs in the EU (97% of the EU total) qualify as IBAs of European Union importance (meeting C criteria), the remaining 78 sites in the EU do not meet C criteria. This is because they qualify for SPECs which are not currently listed in Annex I of the EC Birds Directive, or because they do not meet any criteria for migratory species under Article 4.2 of this Directive, or because they have not been designated as Special Protection Areas (SPAs) under similar national criteria. Since the IBA data provided for some Annex I species are incomplete for some EU countries, it is likely that many of these sites will also meet the C6 criterion in the future, following the availability of more data.

Within the EU, 31 species of global conservation concern occur, all of which are listed in Annex I of the EC Birds Directive. Overall, 396 sites (17% of all IBAs in the EU) are important for these species, with Spain and Greece being particularly important, holding 121 and 73 identified sites respectively (see Appendix 6 for details).

Many sites are important at the EU-level for populations of congregatory species (meeting the C2, C3 and C4 criteria), qualifying for concentrations of species that are threatened in the EU (criterion C2; 918 IBAs) or migratory species that are not threatened in the EU (criterion C3; 712 IBAs). EU countries identifying particularly high numbers of IBAs for congregatory species include France, Germany, Greece, Ireland, Italy, Netherlands, Spain and United Kingdom (Appendix 6). A total of 842 sites in the EU support at least 20,000 waterbirds or 10,000 pairs of seabird (C4 criterion) and 71 sites are important bottlenecks for migrating storks, raptors or cranes (C5 criterion).

In the EU, 1,550 IBAs (66% of the EU total) support important numbers of one or more species or subspecies threatened in the European Union, i.e. listed in Annex I of the EC Birds Directive, thus meeting the C6 criterion. IBAs have been identified for 135 of the 181 species or subspecies listed in this Annex. Countries or autonomous regions with a high proportion (more than 85%) of their IBAs qualifying under criterion C6 include Austria, France, Luxembourg, Portugal, Madeira and Spain (Appendix 6). Others, such as the Netherlands and United Kingdom, have a slightly lower application rate for C6, but many of the sites in these countries have been identified for concentrations of species (under criteria C2, C3 and C4). In some countries, including Germany, Denmark and Finland, the C6 criterion has not been fully applied because of a lack of data on the abundance of Annex I species at sites (see relevant country chapters). In other countries, full networks of sites may not yet have been identified for all species listed in Annex I and thus further applications of C6 are likely to be made in the future.

A total of 210 IBAs (9% of the EU total) meet the C7 criterion, having been designated as Special Protection Areas (SPAs), or selected as candidate SPAs, based on ornithological criteria that are similar to, but not exactly the same as, criteria C1–C6. This criterion has been applied in several countries but efforts have been made to minimize its use and, where possible, apply other C criteria (more species-based and quantitative) instead.

IBA coverage by habitat

All of the main habitat-types in Europe occur at IBAs, with wetland, forest and grassland being the most extensive (Figure 11).

- Wetland IBAs are particularly important for large congregations of birds. Many wetland sites have been identified, with 69% of all IBAs having some wetland habitat. At a large number of sites (at least 22%), wetland is predominant, covering at least 50% of the area of the individual IBA.
- 54% of IBAs in Europe hold forest/woodland, this being the dominant habitat (covering more than 50% of the individual IBA area) at 12% of IBAs.
- Grassland habitats are also frequent within IBAs, being recorded at 52% of all IBAs, although covering more than 50% of the individual IBA area at only 5% of the IBAs in Europe.
- Cultivated or grazed habitats cover most of Europe, and artificial landscapes are thus widespread in IBAs, being recorded at 48% of all sites, and being predominant at 8% of sites.
- Many IBAs contain some scrub or rocky areas but this is usually not the predominant vegetation-type.

Most IBAs are composed of mosaics of habitats on which different bird species depend, therefore several habitat-types are usually recorded within the same IBA.

Figure 11. The number of Important Bird Areas in Europe that contain particular habitats.

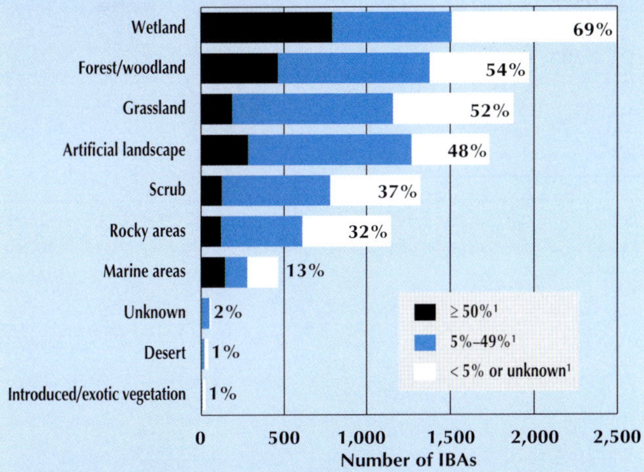

1. Proportion of individual IBA area covered by habitat

The percentage at the end of each bar indicates the proportion of IBAs in Europe that contain the habitat (no habitat data were recorded at five IBAs). See Appendix 3 for a more detailed description of the habitat classification used.

Liepaja lake in Latvia (IBA 004) is a shallow, nutrient-poor coastal lake with extensive areas of water-fringe vegetation. (PHOTO: EDMUNDS RAČINSKIS)

The flood-plains of the Biebrza river valley in Poland (IBA 025) are one of the most important breeding places in central and western Europe for some threatened or localized species such as Great Snipe *Gallinago media*, Corncrake *Crex crex*, Spotted Crake *Porzana porzana*, Greater Spotted Eagle *Aquila clanga* and White-winged Black Tern *Chlidonias leucopterus*, and are also an important site for migrating waders, ducks, geese and Crane *Grus grus*. (PHOTO: PAWEŁ OLAF SIDŁO)

'Kemihaara mires', in Finland (IBA 017), is a very large complex of aapa mires, riverside forests, herb-rich mires and forested hills and ridges along the upper course of the Kemi river, important for its boreal avifauna, especially breeding waders. (PHOTO: MIKA HONKALINNA)

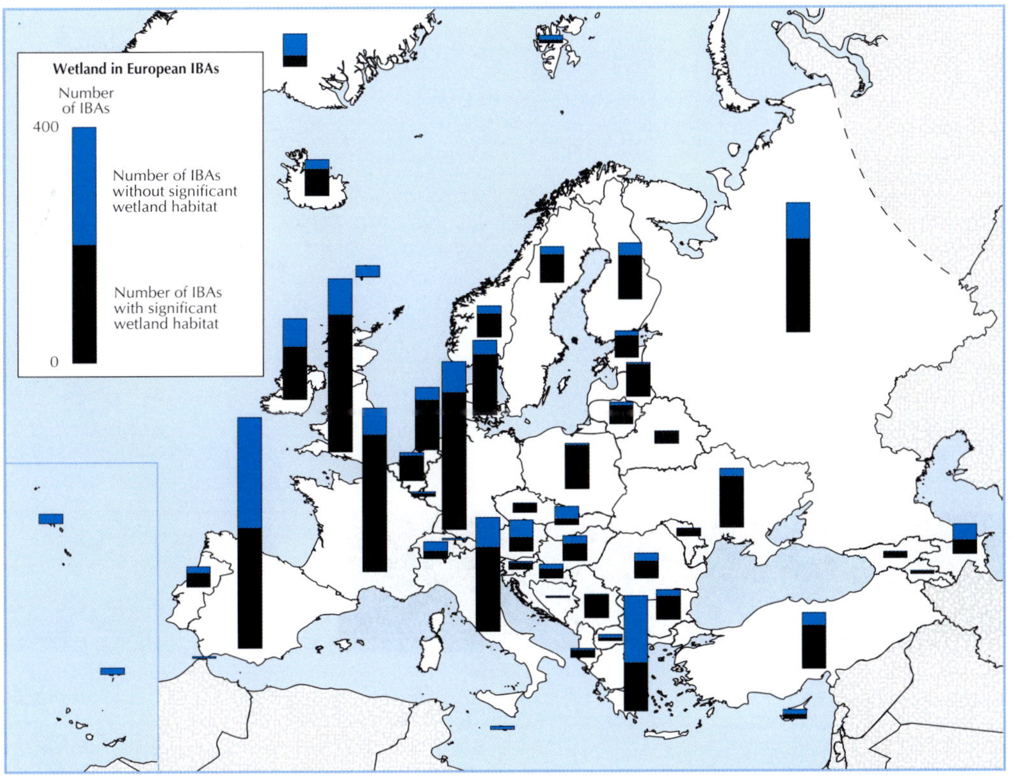

Figure 12. The number of Important Bird Areas in Europe, by country, that contain a significant extent of wetland.

The height of the column is proportional to the total number of IBAs in the country. Overall, at least 2,494 IBAs in Europe contain a significant extent of wetland (i.e. covering at least 5% of the individual IBA area).

Wetland

Sixty-nine percent of all IBAs have at least 5% of their individual area covered by wetland habitats. Figure 12 shows the proportion of IBAs in each country that have a significant extent of wetland habitat. Most countries have a high proportion of wetland IBAs, reflecting the importance of many coastal and inland sites across the region for congregatory waterbirds. There is a broad range of wetland types and their presence in IBAs is shown in Figure 13.

- Of those IBAs with any wetland habitat, 57% include freshwater lakes or ponds (including fish-ponds), many with associated water-fringe vegetation such as reedbeds.
- Rivers and streams occur in 44% of wetland IBAs.
- Concerning mires, 19% of wetland IBAs hold fen or transition mire, 13% hold raised bog and 4% hold blanket bog.
- Many coastal sites include one or more of the following types of wetland habitat: standing brackish/salt water; enclosed tidal water; coastal lagoon; saltmarsh; intertidal mudflat/sandflat; sand-dunes; sand/shingle/stony beach.

Figure 13. The types of wetland present at Important Bird Areas in Europe.

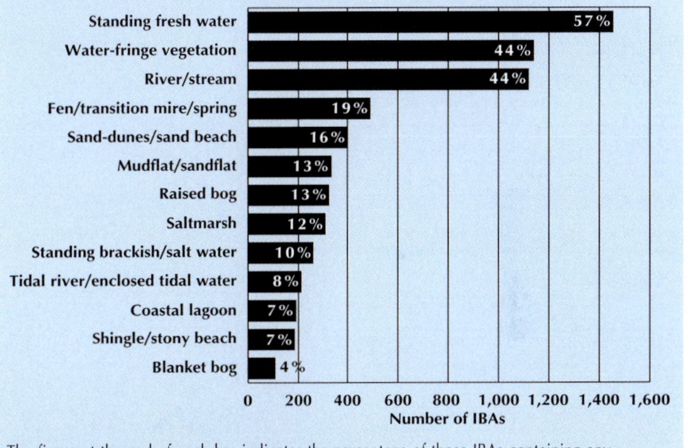

The figure at the end of each bar indicates the percentage of those IBAs containing any wetland (2,563 sites) that actually contain the wetland-type in question. At 2% of the 2,563 wetland IBAs, no types of wetland were distinguished.

Of major international importance for waterbirds all-year-round, the Wadden Sea is the largest intertidal area in Europe, shared between Germany, the Netherlands and Denmark, and comprises many IBAs. Habitats include mudflats, sandflats, saltmarshes, shallow subtidal waters and estuaries. (PHOTO: NATURSCHUTZBUND DEUTSCHLAND)

Broadleaved deciduous forests in central and southern Germany hold rich populations of forest birds such as Red Kite *Milvus milvus*, Grey-headed Woodpecker *Picus canus* and Firecrest *Regulus ignicapillus*. (PHOTO: CHRISTOPH HEINRICH)

Mixed forest occurs at medium altitudes in Yugoslavia, and is an important habitat for breeding raptors at several of the 13 IBAs that are predominantly forested. (PHOTO: BRATISLAV GRUBAČ)

Native coniferous forest and montane meadows at Veľká Fatra, Slovakia (IBA 014), an extensive range of forested mountains in the inner Western Carpathians. (PHOTO: VEĽKÁ FATRA PLA)

Well preserved riverine forests along the River Mura in eastern Slovenia (IBA 013). The area includes the largest alder *Alnus* forest in central Europe, and is dense and uninhabited. (PHOTO: BORUT ŠTUMBERGER)

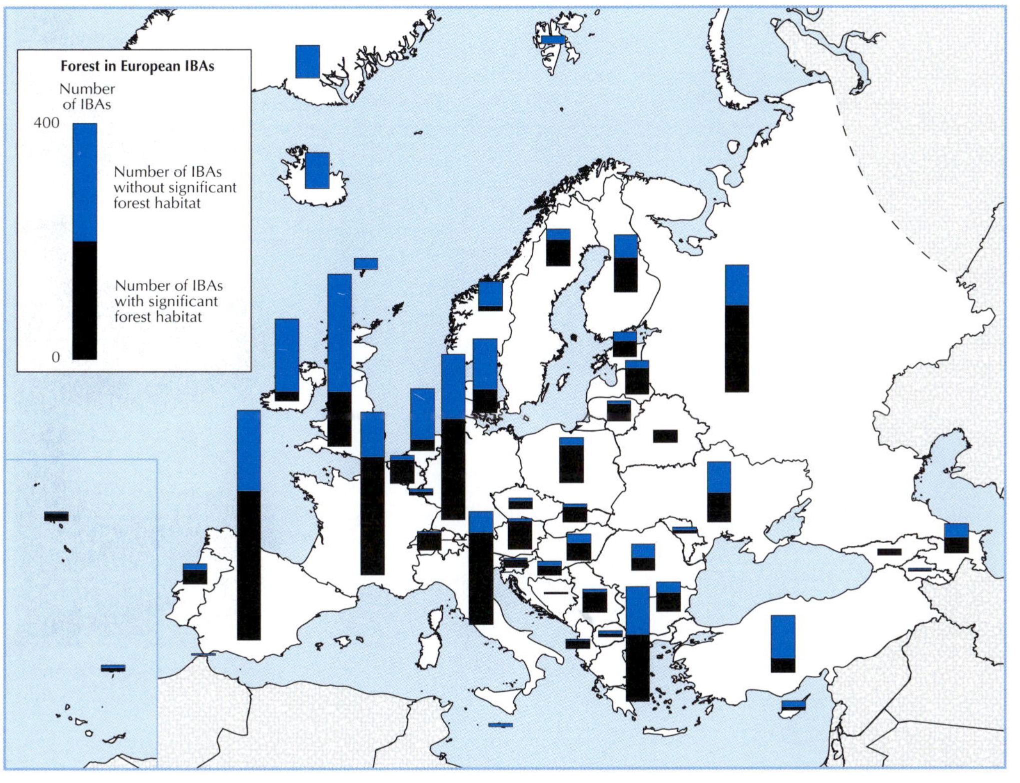

Figure 14. The number of Important Bird Areas in Europe, by country, that contain a significant extent of forest/woodland.

The height of the column is proportional to the total number of IBAs in the country. Overall, at least 1,970 IBAs in Europe contain a significant extent of forest/woodland (i.e. covering at least 5% of the individual IBA area).

Forest and woodland

The total area of forest in Europe has been increasing since the 1920s, due to widespread afforestation of open land and due to forest regeneration following abandonment of agriculture. Forest currently covers about 30% of Europe's land area, but most of this exists as intensively managed plantations, often monocultures of fast-growing conifers or non-native tree species, or as highly fragmented small patches of semi-natural forest. The only large and unbroken tracts of primaeval forest left on the continent are in the north-east of European Russia.

Fifty-four percent of IBAs in Europe hold forest/woodland. Figure 14 shows the proportion of IBAs in each country that have a significant extent of forest/woodland (more than 5% cover of individual IBA). Many countries in the centre, east and south of the continent have a high proportion of IBAs with some forest/woodland cover. The types of forest and woodland found within these IBAs are shown in Figure 15.

- Broadleaved deciduous forest is present in nearly 50% of IBAs with any forest/woodland habitat.
- Native coniferous forest is present in 34% of forest/woodland IBAs.
- Mixed forest (a 50:50 mixture of conifers and broadleaved deciduous trees) is present in 31% of forest/woodland IBAs.
- Alluvial/wet forest is present in 35% of forest/woodland IBAs.
- Broadleaved evergreen forest is present in 14% of forest/woodland IBAs.
- Types of forest/woodland that are less frequent in IBAs include wooded tundra, wooded steppe and treeline ecotone.

Figure 15. The types of forest/woodland present at Important Bird Areas in Europe.

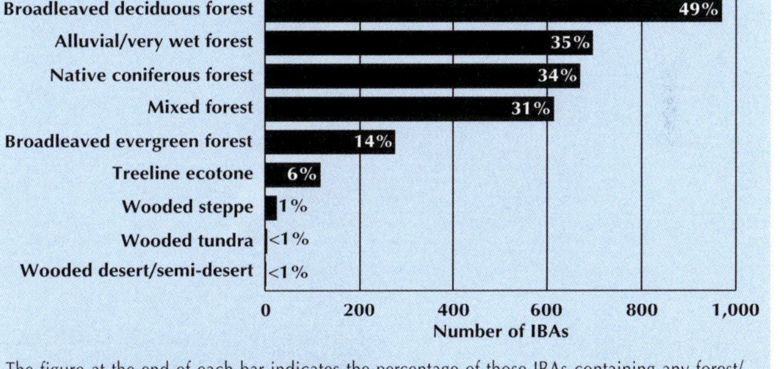

The figure at the end of each bar indicates the percentage of those IBAs containing any forest/woodland (1,996 sites) that actually contain the forest-type in question. At 8% of the 1,996 forest/woodland IBAs, no types of forest/woodland were distinguished.

In Spain, the 'dehesas'—open parklands of broadleaved evergreen forest—are important for the conservation of globally threatened or near-threatened species such as Spanish Imperial Eagle *Aquila adalberti* and Cinereous Vulture *Aegypius monachus*. (PHOTO: FERNANDO BARRIO (SEO/BIRDLIFE))

Lake Cerknica (IBA 008) is situated on the largest polje (karst field) in Slovenia. A seasonal, nutrient-rich lake which is usually totally dry in summer and late winter, with huge expanses of wet grassland. Important numbers of the globally threatened Corncrake *Crex crex* breed.
(PHOTO: SLAVKO POLAK)

Natural steppe was formerly the dominant habitat in southern Ukraine, but most has been ploughed for cultivation and only small fragments of this habitat remain, as here at Bagerovo (IBA 102).
(PHOTO: PAUL GORIUP)

Mesophile grassland in the Bukovské hills, Slovakia (IBA 027), important for meadow-breeding species such as Corncrake *Crex crex* and Red-backed Shrike *Lanius collurio*.
(PHOTO: Š. PČOLA)

Augstmatthorn, Switzerland (IBA 018), a mountainous area with alpine meadows at higher elevations. In Switzerland, alpine and subalpine grasslands are found at 19 IBAs and are important for species such as Rock Partridge *Alectoris graeca*, Alpine Accentor *Prunella collaris*, Rock Thrush *Monticola saxatilis*, Wallcreeper *Tichodroma muraria*, Alpine Chough *Pyrrhocorax graculus* and Snowfinch *Montifringilla nivalis*.
(PHOTO: LORENZ HEER)

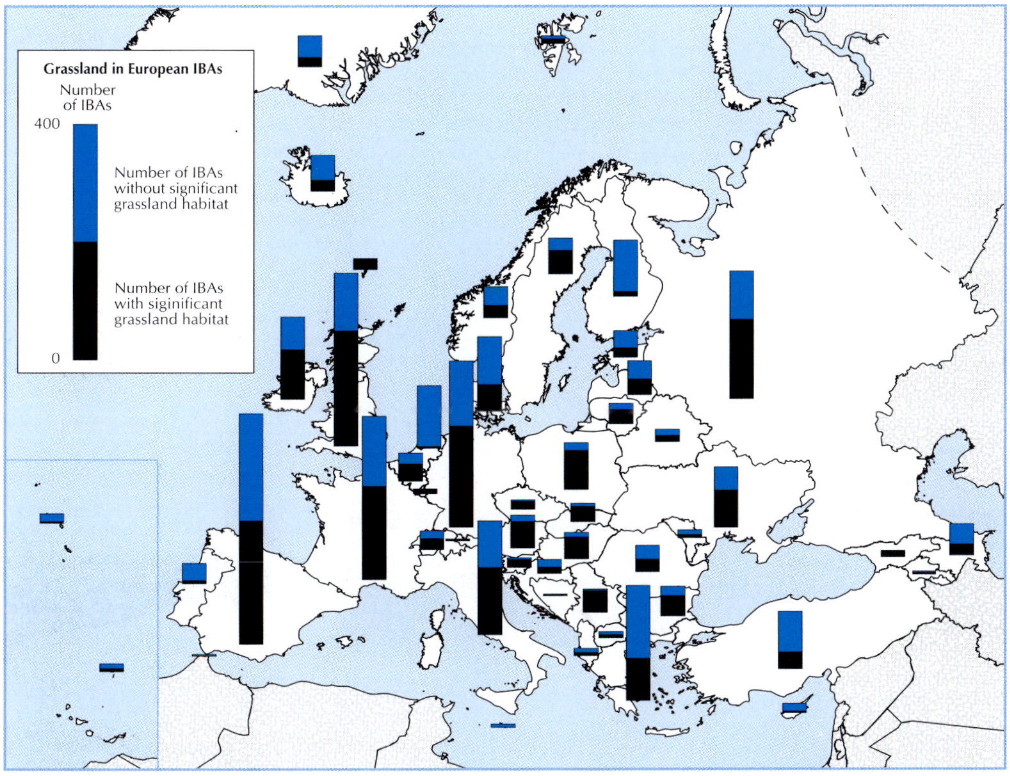

Figure 16. The number of Important Bird Areas in Europe, by country, that contain a significant extent of grassland.

The height of the column is proportional to the total number of IBAs in the country. Overall, at least 1,877 IBAs in Europe contain a significant extent of grassland (i.e. covering at least 5% of the individual IBA area).

Grassland

Grassland habitats are recorded at 52% of all IBAs. Figure 16 shows the proportion of IBAs in each country that have a significant extent of grassland habitat (at least 5% cover of individual IBA). More than half of IBAs in most countries have significant grassland, and in the countries with alpine habitats (and smaller numbers of IBAs) the proportion is higher. In terms of the number of IBAs with grassland habitat, France, Germany, Ireland, Italy, Russia, Spain and United Kingdom rate highly. The types of grassland found within IBAs are shown in Figure 17.

- Humid grassland is the most frequently recorded type of grassland within IBAs, being recorded at 45% of IBAs with any grassland habitat.
- Steppe or dry calcareous grassland is present at 23% of IBAs with any grassland habitat.
- Mesophile grassland (permanent pasture in the temperate zone) is present at 22% of IBAs with any grassland habitat.
- Alpine, subalpine or boreal grassland is present at 14% of grassland IBAs.
- Tundra and machair are grassland types that are much more rarely recorded at IBAs, occurring at 3% and 1% of grassland IBAs respectively.

Figure 17. The types of grassland present at Important Bird Areas in Europe.

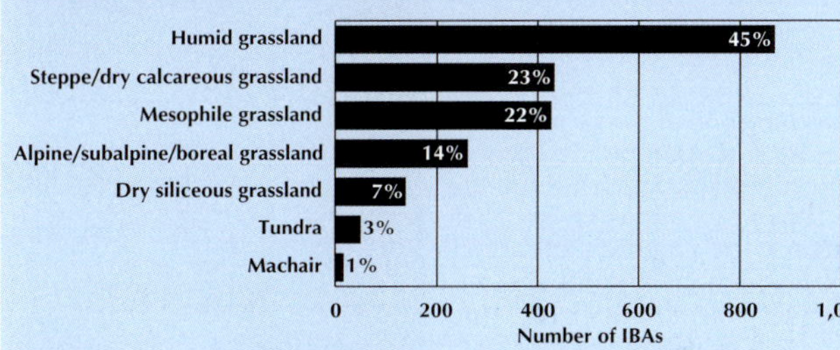

The figure at the end of each bar indicates the percentage of those IBAs containing any grassland (1,918 sites) that actually contain the grassland-type in question. At 16% of the 1,918 grassland IBAs, no types of grassland were distinguished.

Extensive tundra at Itsako, western Greenland (IBA 018), an important site for moulting and breeding White-fronted Goose *Anser albifrons flavirostris*. (PHOTO: DAVID BOERTMANN)

Extensive tracts of artificial landscape occur in 36% of IBAs in Spain. Rain-fed arable cultivation, which occurs at more than 22% of Spanish IBAs, is of great importance in conserving Great Bustard *Otis tarda*, Little Bustard *Tetrax tetrax* and Lesser Kestrel *Falco naumanni*, three species of global conservation concern. (PHOTO: PAUL GORIUP)

Most perennial crops, orchards and groves in Europe are now managed rather intensively, as large commercial operations, with the notable exception of olive production in the Mediterranean, most groves of which remain family-owned. Olive groves consequently display a higher diversity of structure, making them attractive to many woodland species, together with their oil-rich fruit. (PHOTO: MARCO LAMBERTINI)

Much of the forest cover in Europe is now composed of plantations, as here in Germany. If situated and managed appropriately, such plantations can provide valuable habitat for forest birds and other wildlife. (PHOTO: CHRISTOPH HEINRICH)

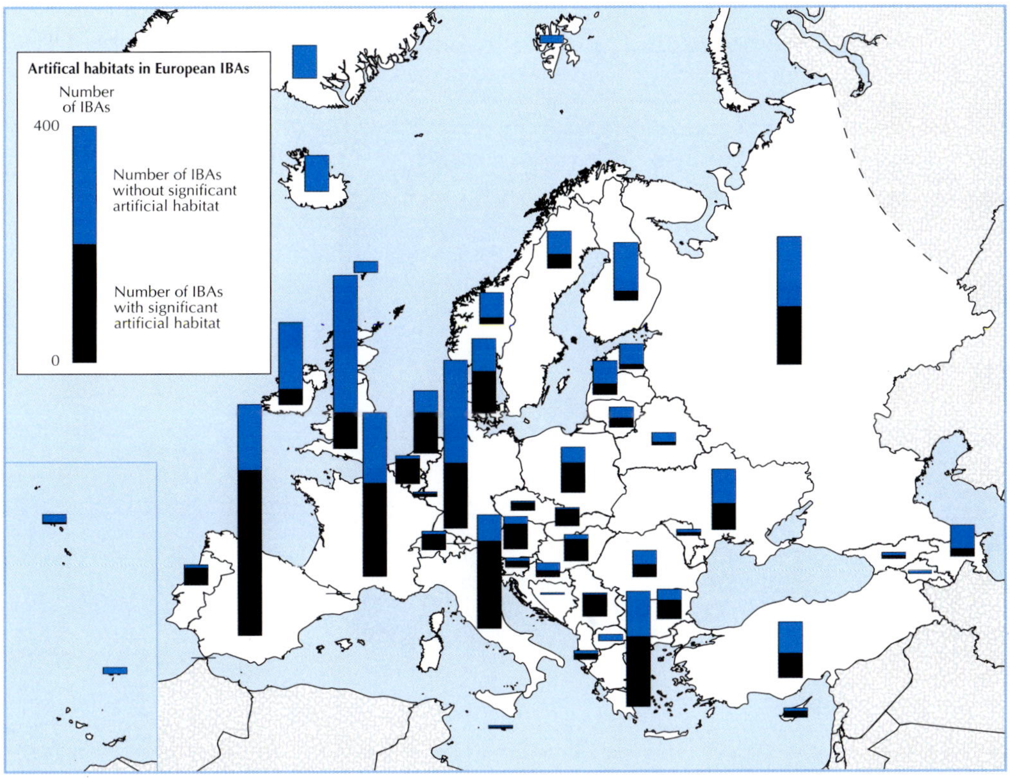

Figure 18. The number of Important Bird Areas in Europe, by country, that contain a significant extent of artificial landscape.

The height of the column is proportional to the total number of IBAs in the country. Overall, at least 1,734 IBAs in Europe contain a significant extent of artificial landscape (i.e. covering at least 5% of the individual IBA area).

Artificial landscapes

Artificial landscapes, especially those that are extensively or traditionally managed for agriculture or forestry, support important populations of many bird species. Artificial landscapes have been recorded at IBAs in most countries (Figure 18), with the exception of Greenland, Iceland, Svalbard/Jan Mayen and the Faroe Islands, being recorded in 52% of all IBAs. In Scandinavia and the Baltic countries, the proportion of IBAs with artificial habitats is also very low. In contrast, Belgium, the Netherlands and many central and southern European countries have a high proportion of IBAs containing a significant extent of artificial landscape. The types of artificial landscape found within IBAs are further described in Figure 19.

- Arable land is present in 69% of IBAs with artificial landscapes.
- Perennial crops, orchards and groves are present in 30% of IBAs with artificial landscapes.
- Forestry plantations are present in 26% of IBAs with artificial landscapes.
- Highly improved reseeded grassland is present in 23% of IBAs with artificial landscapes.
- Urban or industrial areas (other than parks/gardens and ruderal land) are present in 21% of IBAs with artificial landscapes.

Figure 19. The types of artificial landscape present at Important Bird Areas in Europe.

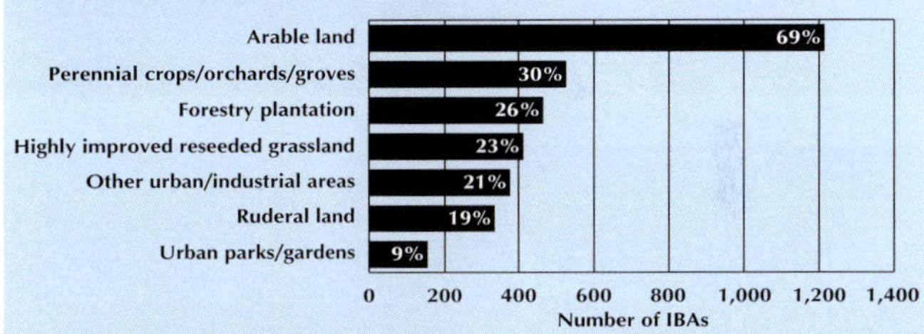

The figure at the end of each bar indicates the percentage of those IBAs containing any artificial landscape (1,768 sites) that actually contain the type of artificial landscape in question. At 5% of the 1,768 IBAs, no types of artificial landscape were distinguished.

Improved reseeded grassland at Wexford Wildfowl Reserve, Ireland (IBA 102). This is one of the two most important sites in the world for wintering White-fronted Goose of the Greenland subspecies *Anser albifrons flavirostris*. (PHOTO: ALYN WALSH)

Scrub is a particularly important habitat in the Mediterranean biome, for such species as warblers *Sylvia*, shrikes *Lanius*, nightjars *Caprimulgus* and many other birds. (PHOTO: ALBERTO NARDI/NHPA)

Semi-desert landscape in Azerbaijan. (PHOTO: MICHAEL PATRIKEEV)

Seabird colony on a coastal cliff face in Germania Land, north-eastern Greenland (IBA 053). (PHOTO: DAVID BOERTMANN)

Other habitats

- In Europe, 36% of IBAs have a significant extent of scrub habitat, with 30% of these having sclerophyllous scrub, garrigue or maquis, and 31% heathland.
- Over 30% of IBAs in Europe have a significant extent of rocky areas. Approximately half of these IBAs are coastal, comprising such rocky areas as sea cliffs, rocky shores, rock stacks and islets, with coastal cliffs being particularly important for breeding seabirds. The remaining half of these IBAs are inland areas, with inland cliffs, scree and boulders usually occurring at IBAs in mountainous regions.
- Over 10% of IBAs in Europe are partly or wholly marine. In about two-thirds of these IBAs, 'marine' areas refer to sea inlets and coastal features, whilst the remaining one-third are open sea. See Box 4 for a more detailed discussion of marine IBAs and the difficulties involved in defining and identifying such sites.
- Located mainly in the south-eastern part of the Iberian peninsula and the Canary Islands, as well as in the lowlands of the Caucasus/Caspian region, desert and semi-desert habitats are present in 43 IBAs in Europe.

Box 4. Important marine areas for birds in Europe.

The importance of marine areas for birds

The marine environment is important for birds in Europe. European seas support 62 species of bird that either have an unfavourable conservation status in Europe (Tucker and Heath 1994) or are marine specialists which are highly dependent upon the continuing integrity and quality of marine ecosystems (Tucker and Evans 1997). These 62 species include four species that are globally threatened or near-threatened with extinction. Many of these 62 species congregate at certain times of the year, such as seabirds at breeding colonies or divers *Gavia* and seaduck (Anatidae) at favourable feeding grounds in winter. Yet despite its importance for bird conservation, the marine environment has, until recently, received relatively little attention in terms of the identification of Important Bird Areas. This is due primarily to the difficulty of defining, identifying and delineating priority sites in areas of homogeneous appearance, coupled with the difficulty of stimulating international co-operation to establish integrated protection and management schemes for important sites outside national territorial waters.

This inventory identifies 470 IBAs in Europe that have some marine habitat, including 147 IBAs that have more than 50% marine habitat cover and at least 25 IBAs are solely marine. However, no specific attempt has been made, within the scope of this IBA review, to provide complete listings of marine areas important for birds in Europe.

Previous inventories of marine areas that are internationally important for birds

Two major studies have defined, identified and delineated important areas for birds in European seas: *Important bird areas for seabirds in the North Sea* (Skov *et al.* 1995) and *Important marine areas for wintering birds in the Baltic Sea* (Durinck *et al.* 1994). Further work on the identification of IBAs is in progress, as part of the BirdLife Baltic Sea Programme, the results of which are expected to be published in 2000.

North Sea The study area covered the entire North Sea, the Channel and the Kattegat, bounded by the coastlines of the United Kingdom, Norway, Sweden, Denmark, Germany, the Netherlands, Belgium and France (Skov *et al.* 1995, Skov *et al.* in prep.). It included all territorial and international waters between the western Channel, the Northern Atlantic and the Kattegat. Data were analyzed from coastal and offshore seabird surveys during 1979–1994. The distribution of 30 species provided the basis for identifying important areas for birds. Twenty areas were identified as internationally important for birds in the area, of which the top six sites comprising less than 5 % of the region hold more than 80% of the cumulated sum of proportions for the species in question.

Baltic Sea The study area encompassed the entire Baltic Sea, the Danish Straits and the Kattegat (Durinck *et al.* 1994, Skov *et al.* in prep). Data were presented for 30 species for which 1% or more of their biogeographic population winters in the Baltic Sea. Thirty-nine areas were identified as internationally important for wintering birds in the region, of which the top 10 sites, which cover less than 5% of the region, hold about 90% of the total estimated number of wintering birds.

The conservation of marine areas

The IVth World Congress on National Parks and Protected Areas (IUCN 1993) highlighted the need for increased establishment of marine protected areas in order to conserve biodiversity. Some site-based agreements that are important for the conservation of such areas in Europe are as follows (further details can be found in Appendix 1).

EC Directives: Special Protection Areas (SPAs) and Special Areas for Conservation (SACs) The Habitats Directive calls for the establishment by 2004 of a coherent network of protected areas for biodiversity conservation in the European Union, the Natura 2000 network. The network will be composed of SPAs (designated under national law in fulfillment of the Birds Directive) and SACs (similarly, in fulfillment of the Habitats Directive). These designations apply within the territorial waters of EU Member States. The application of EU law to the offshore Exclusive Economic Zone (EEZ) of member states has recently been clarified by the EU Commission. Although EU law *per se* applies to the territorial sea of coastal member states, it also applies to their EEZ in cases when a member state considers itself competent to apply national law within the same field to the EEZ.

Ministerial Conference on the North Sea This process, devoted to the protection of the North Sea ecosystem, began in 1984 and involves the ministers of North Sea states, Switzerland and the European Commission. The significance of ministerial conferences is that they may consider subjects outside existing conventions and treaties, such as the protection and management of marine IBAs outside territorial waters.

OSPAR Convention (Oslo and Paris Convention) The focus of this convention has been the reduction of land- and sea-based pollution in the north-east Atlantic, but the scope of OSPAR was broadened in July 1998 when ministers, at the meeting of the OSPAR Commission in Sintra (Portugal), adopted a new Annex V to the OSPAR Convention. This opens the way for the Commission to take measures to protect and conserve ecosystems and biological diversity in the north-east Atlantic. One of the agreed tools for doing this is to work towards the establishment of a network of marine protected areas (MPAs). In protecting species and habitats, the OSPAR Commission can recommend to the competent authorities a wide range of measures affecting all relevant sectors, including shipping and fisheries.

Helsinki Convention This convention brings countries bordering the Baltic Sea together with the European Community under the Helsinki Commission (HELCOM). One of HELCOM's most important decisions has been to recommend the establishment of a network of Baltic Sea Protected Areas (BSPAs). Included in this recommendation was provision for the identification of areas important for concentrations of feeding, wintering or staging seabirds (Hägerhäll and Skov 1998).

Barcelona Convention Since 1994–1995 this convention for the protection of the Mediterranean Sea ecosystem has included a greater emphasis on biodiversity conservation. Contracting parties are now required to draw up lists of Mediterranean Specially Protected Areas (MedSPAs) within their territorial waters (previously, this was optional). Such sites should be important for the protection of ecosystems or of special scientific, aesthetic or cultural value. There is also an Annex to the MedSPA Protocol which lists species considered by the contracting parties to be endangered and threatened in the Mediterranean, including seabirds. Parties are required to take appropriate measures to protect these species and their habitats.

Summary and conclusions

- Many bird species that are dependent on the marine environment congregate at certain times of the year. Several hundred marine IBAs have been identified in this publication for such species, and the protection and suitable management of these areas would be a major step towards conserving these marine species in Europe.
- Despite the inclusion of these marine areas within this pan-European inventory, a comprehensive documentation of marine areas important for birds was not achieved. However, work previously undertaken in the North Sea and Baltic Sea (Durinck *et al.* 1994, Skov *et al.* 1995) shows that entirely marine areas important for birds, including congregations of wintering auks, divers and seaducks, can be defined, identified and delineated.
- More than one conservation mechanism is appropriate to the conservation of the marine environment for birds. Most marine birds are distributed in a highly dispersed fashion at some stage of their annual cycle, and given the size, socio-economic importance, and international nature of marine areas that are likely to be important for birds, site-based measures—e.g. the establishment of networks of strictly protected areas suitable for highly congregatory species—need to balanced with the conservation of the wider marine environment, involving the integration of conservation across all relevant sectors, programmes and policies.
- A site-based approach to seabird conservation would be greatly assisted by marine extensions to terrestrial (coastal) protected areas, in order to protect key areas such as off-colony foraging areas and loafing areas at sea.
- A number of different site-based conservation initiatives are in the early stages of implementation, under a variety of international legal agreements.
- Outside territorial waters, international cooperation and coordination will be vital if the marine environment is to be managed to conserve its biodiversity.

LAND-USES AND THREATS AT IBAs

■ Land-uses at IBAs

The great variety and high intensity of land-uses recorded within IBAs reflect the high human use of Europe's habitats in general. Land-use activities have been recorded in 95% of IBAs, and in only 15% of these IBAs is part or all of the individual IBA area classified as 'Not utilized'. As a result, the conservation of habitats and birds within IBAs is very much dependent on the land-use practices, policies and programmes that affect these sites. Figure 20 shows the types of land-use recorded in IBAs:

- Agriculture is the dominant land-use in Europe and this is reflected within IBAs, with agricultural activities taking place in 65% of IBAs and being predominant at 17% of sites (Figure 20). The conservation of most IBAs is therefore much influenced by agricultural practices in the individual IBA and its surroundings, and suitable management is required to conserve species dependent on these habitats.
- Tourism and recreation are the second-most frequently cited land-use, affecting 41% of IBAs. The very high human population densities over much of Europe, the attractiveness as destinations of natural areas such as IBAs, and the relative ease of access to most parts of the continent, explain much of this statistic.
- Hunting takes place at 38% of sites and may be even more wide-spread at IBAs, since this land-use was probably under-recorded.
- At least 54% of IBAs contain forest/woodland or forestry plantation. Much of the area of these forest habitats in Europe is actively managed, with forestry taking place in at least 35% of IBAs.
- Fisheries and/or aquaculture take place in at least 29% of IBAs, especially in marine and coastal areas and, inland, at fish-ponds.
- At least 29% of IBAs are used for nature conservation and research. Generally, this category was recorded only when land was used primarily for this purpose.
- Urban, industrial and transport activities affect at least 23% of IBAs but, in general, affect only a small proportion of the area of each site.
- Water management affects at least 14% of IBAs and includes the drainage, storage and extraction of water, for such purposes as irrigation, flood control and power generation.

Figure 20. The number of Important Bird Areas in Europe affected by particular land-uses.

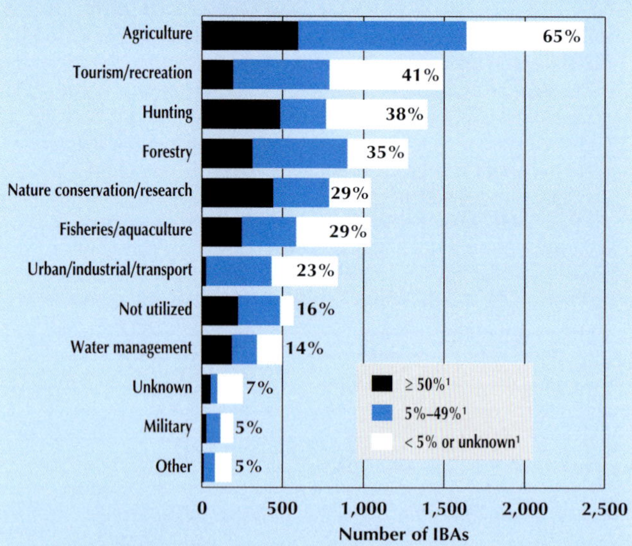

1. Proportion of individual IBA area affected by land-use.

The percentage at the end of each bar indicates the proportion of IBAs in Europe that are affected by the land-use (no land-use data were recorded at 155 IBAs). See Appendix 3 for a more detailed description of the land-use classification used.

■ Threats at IBAs

At least 93% of IBAs are considered to be threatened to some degree by at least one factor, and 42% of IBAs in Europe are affected by one or more high-impact threats (Figure 21). Most IBAs are threatened by more than one factor (Figure 22), with the majority (58%) being threatened by 2–5 different types of threat.

Threats facing IBAs are diverse, resulting from lack of effective management and wardening, illegal land-use, regional, industrial or urban development and impact of land-uses from neighbouring areas (e.g. pollution, drainage). The key types of threats affecting IBAs are shown in Figure 23, with the intensification and expansion of agriculture being the most significant, impacting more than 1,300 IBAs. The following sections present further information, with selected examples, on the 12 most common high-impact threats

Figure 21. The highest level of threat recorded against individual Important Bird Areas in Europe.

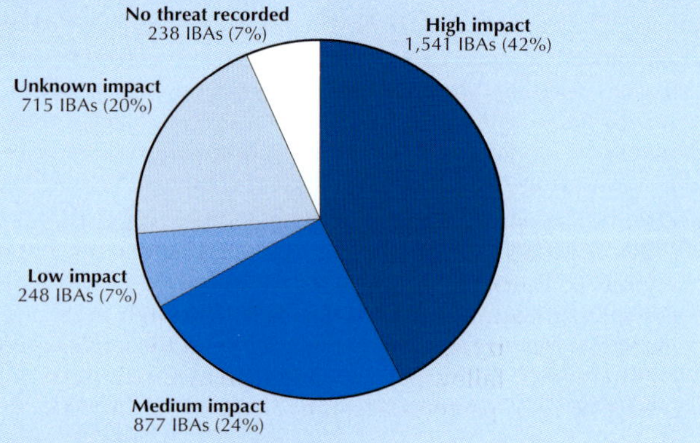

Highest level of impact on individual IBA

The percentage in brackets indicates the proportion of IBAs in Europe that are affected, at this level of impact, by one or more threats. See Appendix 3 for a more detailed description of the system used for estimating the magnitude of impact.

Figure 22. The total number of threats affecting individual Important Bird Areas in Europe.

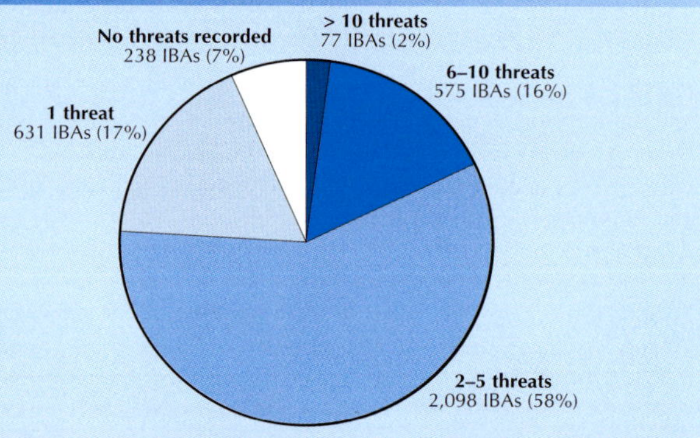

The percentage in brackets indicates the proportion of IBAs in Europe that are affected.

Figure 23. The high-impact threats that are most prevalent at Important Bird Areas in Europe.

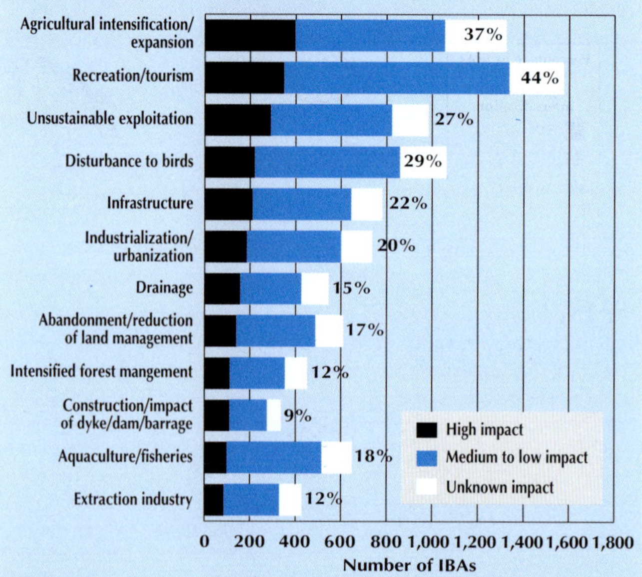

Only those threats impacting more than 10% of IBAs in Europe are listed.

The percentage at the end of each bar indicates the proportion of IBAs in Europe that are impacted by the threat. See Appendix 3 for a more detailed description of the threat classification used.

at IBAs. Other common threats, each impacting more than 250 IBAs, include afforestation (352 IBAs), burning of vegetation (287 IBAs), the consequences of animal or plant introductions (270 IBAs), and commercial deforestation (261 IBAs).

Agricultural intensification and expansion

Agricultural intensification and/or expansion is the most serious threat affecting IBAs across Europe, being cited at more than 1,300 sites and with a high impact at nearly 400 of these. In most countries, with the exception of those with large areas of land at high latitude or altitude, agricultural intensification and expansion threatens more than 20% of the country's IBAs (Figure 24). This threat covers a range of agricultural activities, including:

- the replacement of dry grasslands, rain-fed cereals and steppe with dense, irrigated, fast-growing, heavily fertilized and pesticide-treated crops;
- the lowering of the water-tables of wet grasslands, followed by reseeding, fertilizer applications and high stocking levels; and the conversion to intensive silage crops of mesophile grasslands (hay-meadows and permanent pastures) across the temperate regions of Europe;
- the intensification of arable farmland through high inputs of inorganic fertilizers, crop specialization (at the landscape level), elimination of fallow periods, and increased field-sizes;
- overgrazing of pastures accompanied by excessive soil erosion and compaction; and
- the loss from the farmed landscape of small semi-natural areas and important features, including strips of meadow, hedgerows, copses, small ponds, and tree stands along rivers and lakes.

Some examples of agricultural intensification and/or expansion posing a threat (actual or potential) to IBAs (excerpted from national overviews in this publication)

- In Portugal, agricultural intensification affects 50% of IBAs and is a particular problem for species dependent on non-intensive farming systems (for example Great Bustard *Otis tarda* and Black-bellied Sandgrouse *Pterocles orientalis*), and for farmland species in areas that have already undergone some intensification and are now seeing further measures being taken to increase productivity.
- In Spain, EU livestock headage payments have caused overgrazing problems in steppe areas, increasing erosion and disturbing key breeding birds. Arable production has become more intensive, particularly through the expansion of irrigated crops, resulting in serious loss of steppe habitats and traditional non-irrigated crops, and threatening the most important areas for Great Bustard *Otis tarda*, Little Bustard *Tetrax tetrax* and Lesser Kestrel *Falco naumanni*, including sites 038, 039, 041, 043, 058, 061, 062, 072, 074, 105, 106, 111, 193, 199, 202, 207, 229, 232, 237, 238, 264, 276, 280, 285, 294 and 295.
- In Turkey, agricultural intensification affects 49 out of 97 IBAs. Since the 1950s the land area under cultivation has more than doubled, with the expansion coming partly at the expense of available grazing land. This has led to an increase in grazing pressure in remaining pastures and hence problems of overgrazing and excessive soil erosion. Wetlands have been drained for conversion to arable land and many others have been adversely affected by water-diversion schemes to feed irrigated agriculture. At least 1,300,000 ha of wetland habitat have been drained or otherwise destroyed this century, mainly since 1960. The Konya basin of central Anatolia provides a shocking example of the consequences of large-scale irrigation projects. Two large wetlands with a total area of 30,000 ha have dried up, and three large wetland IBAs totalling 106,000 ha (sites 067–069) have sustained serious damage. A further three wetland IBAs (070–072) now risk long-term damage as a result of pollution from untreated irrigation waste-water.

Steppe land in Spain, which supports most of the European population of the globally threatened Great Bustard *Otis tarda*, is increasingly being converted to irrigated cultivation. (PHOTO: GUY DUKE)

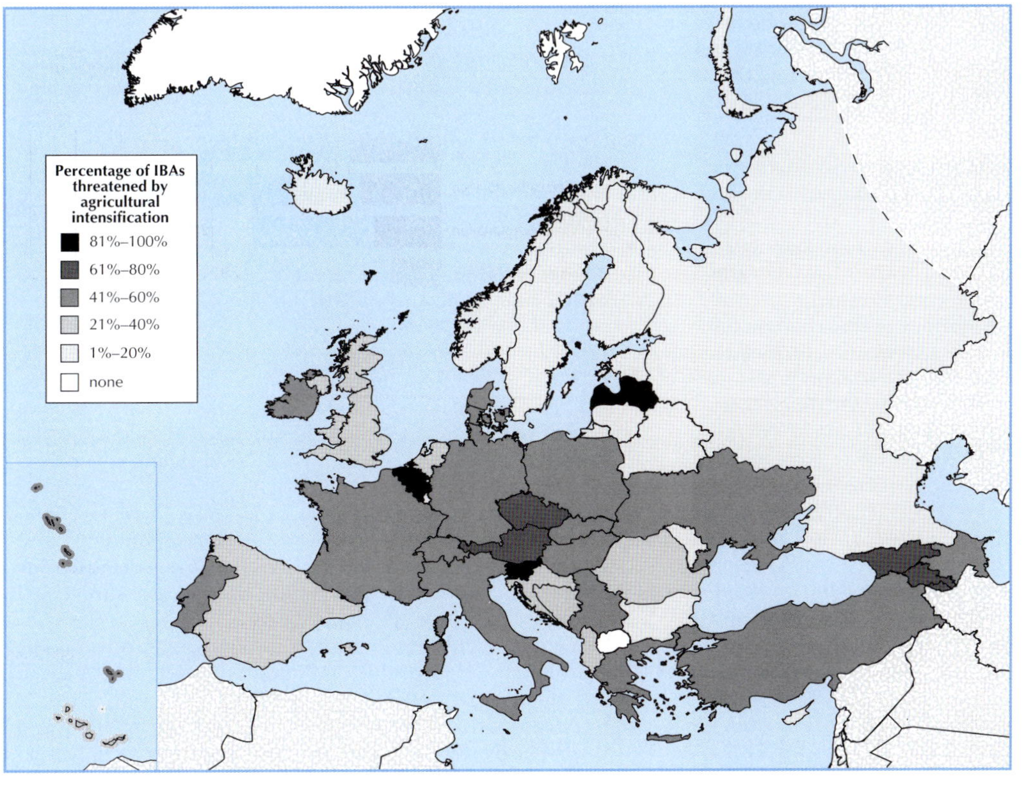

Figure 24. The percentage of Important Bird Areas, per country, that are threatened by agricultural intensification.

- In Ukraine, the intensification of agriculture in recent decades has caused pollution of soil and fresh water with fertilizers and pesticides. Agricultural intensification is recorded as a high-impact threat at sites 025, 032, 037, 047, 071, 094 and 102.
- In Ireland, agricultural intensification is considered to have a high impact at 11 IBAs, primarily areas where the maintenance of non-intensive farming is crucial for the continued breeding of the globally threatened *Crex crex* (sites 002, 007, 012–014, 041 and 131), sites which have suffered severe overgrazing such as at Owenduff–Nephin Beg (site 046) and water-bodies threatened by nutrient pollution as a result of agricultural run-off (e.g. sites 123 and 138).
- In Denmark 22% of sites are seriously affected by agricultural intensification and expansion, mostly through high inputs of fertilisers and pesticides, which has resulted in uniform landscapes that are poor in the habitat features required by many bird species of European conservation concern. Furthermore, nutrient-laden run-off from intensive agricultural land is believed to be the main factor behind the nutrient pollution of many wetland sites in Denmark.
- In Austria, agricultural intensification and expansion have negatively affected 69% of IBAs, and are regarded as a high-impact threat at 29% of sites, including 003, 007, 014, 015, 019, 024, 025, 028, 030, 038, 041, 042, 043, 045, 054 and 055.
- In Poland, as is the case in many countries in the east of the region, at agricultural IBAs the main threat is posed by an anticipated change in the Polish farming system, away from the current predominance of small, non-intensive farms towards a landscape dominated by large, intensive farms. At least 43% of IBAs in Poland are threatened, to some degree, by the intensification or expansion of agricultural practices.

The development of tourism threatens many coastal areas in the Mediterranean and elsewhere. (PHOTO: MARCO LAMBERTINI)

- In Bulgaria, the growth of recreation and urbanization has been very rapid in recent years, especially along the Black Sea coast. Wetlands near large settlements and industrial zones are particularly vulnerable, with the most affected sites including those bordering the towns of Burgas and Varna: the Mandra–Poda complex (site 034), Burgasko lake (035), Atanasovo lake (036) and Varna–Beloslav lakes complex (047).
- Mountain areas are also particularly affected by recreation and tourism, including activities such as skiing, hiking, climbing, off-road driving and biking. For example, in France skiing or climbing are recorded as a significant threat at Forêts d'altitude du Jura (097), Haute Soule: massif de la Pierre-St-Martin (188) and Bois de Palayson, du Rouet et de Malvoisin (265).

Recreation and tourism

Tourism and/or recreation have been cited as a threat at 1,575 IBAs, with impact considered to be high at 348 IBAs. This threat is particularly serious in Mediterranean wetland, forest, shrubland and rocky habitats, where tourism is probably the main cause of habitat destruction and degradation, as well as of disturbance to birds. The Mediterranean countries are the world's most popular tourist destination. Existing tourist developments are most concentrated along the coasts of Spain, France, Portugal and Italy, but other countries such as Greece and Turkey continue to show explosive growth in this sector. From a 1990 level of 157 million visitors per year to these countries, numbers are predicted to increase to 380 million per year by 2050 if overall economic growth is weak, or up to 760 million if growth is strong (Stanners and Bourdeau 1995). It should be noted that in some cases this threat has also been recorded under 'Disturbance to birds'.

Some examples of recreation/tourism posing a threat (actual or potential) to IBAs (excerpted from national overviews in this publication)

- In Spain, recreation/tourism is the most frequently cited threat to IBAs, often in connection with infrastructural developments. Impacts tend to be concentrated in coastal areas, particularly along the Mediterranean coast and on the Balearic and Canary Islands, where urbanization (driven by tourism) has severely affected or destroyed many habitats that were previously of high importance for birds, especially wetlands. Related infrastructural developments such as major roads and airports are often funded by the EU. Sites where recreation or tourism is regarded a high threat include IBAs 307, 308, 310, 311, 315, 319, 321, 323, 328, 330, 332, 337, 340, 343, 345, 356, 367 and 372.
- Recreation and tourism have a high impact at 33% of Azorean IBAs. Tern *Sterna* colonies along the coast of Flores (001), coast of Faial (003), coast and islets of Graciosa (007) and south-east coast of Terceira (009) are most vulnerable to these threats, which have led to colony desertion and low breeding success. Tourists pose a serious problem at Vila Franca islet (012), where they inadvertently trample and destroy nest-sites of Cory's Shearwater *Calonectris diomedea*.

Unsustainable exploitation

Unsustainable hunting (whether legal or illegal) is considered a threat at more than 950 IBAs, and is particularly significant at nearly 300 sites. As well as shooting, other forms of unsustainable bird exploitation which pose a threat at some IBAs include trapping, the use of lime, and the collection of eggs and young.

Some examples of unsustainable exploitation posing a threat (actual or potential) at IBAs (excerpted from national overviews in this publication)

- Particularly badly affected is Cyprus, where the killing of migratory birds is a serious problem and is listed as a threat at nearly 90% of IBAs. Although the island's hunting (shooting) regulations are fairly comprehensive, law enforcement is inadequate and thousands of birds of many species are illegally shot each year.
- Illegal bird-shooting is a serious problem in Malta, e.g. at Buskett and Wied il-Luq (site 003) which is a major passage site for raptors and nationally important for wintering passerines as well as many trans-Saharan autumn migrants.
- Hunting is a widespread activity in Greece, occurring in almost all IBAs, and is a threat at more than 80% of IBAs, with a high impact at 36% of all sites. Illegal hunting is a particularly severe threat at sites 019, 022, 084, 095, 100, 105, 116, 130, 140, 146 and 172.
- In Spain, hunting is considered a threat at over 40% of IBAs. Additionally, at several globally important wetlands such as the Ebro delta (site 148) and El Hondo wetland (165), lead poisoning of waterbirds, as a result of ingesting spent shot, is a problem.
- In Bulgaria, hunting pressure is particularly high in wetland areas such as the Shabla lakes complex (site 049) and Durankulak lake (050) which both hold large concentrations of wintering waterbirds, including the globally threatened Red-breasted Goose *Branta ruficollis*.
- The predominant threat to IBAs in Georgia is unsustainable exploitation (legal and illegal hunting) which affects seven out of the 11 IBAs (63%), with a high impact at five sites.
- In Azerbaijan, the main threat at IBAs is the unsustainable level of bird exploitation, affecting 63% of IBAs (mainly legal

Freshly killed White Pelican *Pelecanus onocrotalus*. Hunting remains a significant threat at many IBAs, in terms of disturbance as well as mortality. (PHOTO: GERNANT MAGNIN)

and illegal hunting at over 30 IBAs). Waterbirds at Azerbaijan's most important wetlands are severely affected by hunting and poaching, also by the accompanying disturbance which, for some sites, may be a more serious threat than the direct mortality.

Infrastructure

A total of 779 IBAs are threatened by planned or ongoing infrastructural development. This problem looks set to increase, with new transportation infrastructure expected to claim 8,500–12,500 km² in the EU from other land-uses during 1990–2010 (EEA 1999). This will increase the fragmentation and isolation of Europe's remaining patches of semi-natural habitat, and is likely to have a major negative impact on biodiversity.

The total length of motorway in the majority of countries in the EU increased by more than 300% between 1970 and 1994 (EEA 1998). The Trans-European Network planned by the EU will involve around 140 projects for further road construction and improvement in the European Union. Approximately 15,000 km of new road are proposed, as well as rail links, combined transport and waterways. These links are likely to be extended to central and eastern Europe as trade increases and as the EU is enlarged (EEA 1998).

A pilot study (Bina *et al.* 1995) assessed the potential impact of the Trans-European Network on IBAs and statutory protected areas, based on the previous pan-European IBA inventory (Grimmett and Jones 1989). It found that more than 12% of IBAs in the EU are within 10 km of planned road and rail developments, with more than 2% of IBAs in the EU being situated within 2 km of planned road and rail developments.

Some examples of infrastructure posing a threat (actual or potential) to IBAs (excerpted from national overviews in this publication)

- Infrastructure and associated industrialization and urbanization are long-term threats to 50–65% of IBAs in Belgium. Schorren en Polder van de Beneden Schelde (site 011), an important wintering site for waders and other waterbirds, will largely be destroyed by the expansion of Antwerp harbour facilities.

The total length of motorway in the majority of European countries increased by more than 300% between 1970 and 1994 (EEA 1998). (PHOTO: BIRDLIFE INTERNATIONAL)

- In France several very significant IBAs, holding rare species and habitats, remain under threat even though they are designated as Special Protection Areas under the EC Birds Directive. The Crau (site 240), the only large semi-arid plain in France, supporting important populations of typical steppe species, is threatened by state development projects (military activities, motorways).
- In Ireland, individual infrastructure development projects may have a marked impact on specific IBAs, such as road developments within Cork Harbour (site 088) and at Malahide/Swords (113), and planned road development within Dublin Bay (109), while port development threatens the Boyne estuary (119).
- Recent government plans to further develop the road network in Austria could pose a major threat to several IBAs (e.g. sites 006, 012, 045). Plans for other infrastructure development, such as golf courses and recreation parks, and the planned construction of dams on rivers are a constant and recurring threat to several sites (e.g. 013). If implemented, the planned Elbe–Oder–Danube canal could alter dramatically the ecosystem of the March river (010).
- Infrastructure development in Switzerland, particularly increasing road density and other transport networks, is a major threat, affecting 55% of IBAs (13% with a high impact), especially as it facilitates access to remote areas, resulting in more intense disturbance of birds such as Black Grouse *Tetrao tetrix*.
- In Spain, infrastructure development is a high-impact threat at 14% of IBAs and also threatens a further 27% of sites. Most of these threats have arisen since Spain joined the European Union in 1986. Since then, transport and water-related infrastructure (reservoirs, hydroelectric power, irrigation) have been improved but much of the investment has affected natural areas detrimentally. Environmental impact assessments have failed to adequately examine alternative scenarios or projects. Many further water-related developments are planned that will affect IBAs, some of them without real justification and potentially with a highly deleterious impact on sites and their species. Hydroelectric power plants are proliferating on rivers in mountainous areas and few rivers remain unregulated. New wind-energy plants are being constructed in mountainous areas where key bird species migrate, breed or feed.

Disturbance to birds

Disturbance to birds is listed as a threat at over 1,000 IBAs, and is a high-impact threat at 22% of sites. With intensifying use and increasing fragmentation of Europe's habitats by many factors, including infrastructure, urbanization, industrialization, recreation/tourism, agriculture and forestry, the level of human activity in remaining patches of habitat increases, as does disturbance to birds' vital activities such as breeding or foraging.

The impact of disturbance at a particular site is often difficult to quantify, as its effects on the reproductive success of individual species, at the level of national or European population, remain unknown.

Breeding colony of Roseate Tern *Sterna dougallii* being disturbed by fishermen at Alagoa complex on the coast of Flores, Azores (IBA 001). (PHOTO: L. R. MONTEIRO/IMAGDOP)

Industrialization and urbanization

Industrialization and urbanization threaten at least 735 IBAs, and the extent of urban areas in the EU is predicted to increase by 5–8% between 1990 and 2010 (EEA 1999).

Some examples of industrialization and urbanization threatening IBAs (taken from national overviews in this publication)

- More than half of the total forest area in the Czech Republic is affected by industrial air-pollution, a proportion higher than that in any other European country. IBAs containing forests that are being badly impacted by acid rain include Šumava mountains (site 005), Jeseníky mountains (016) and Krkonoše mountains (001).
- In Latvia, the possibility of pollution from Butinge oil terminal in Lithuania is thought to pose a major potential threat to two marine/coastal IBAs in Latvia, on the west coast (site 005) and the Irbe strait (014).
- Industrialization, urbanization and infrastructural developments, such as new industrial areas, housing and roads, are a major threat in Germany, affecting over 20% of IBAs.

Industrialization and/or urbanization threaten more than 700 IBAs in Europe. (PHOTO: MARCO LAMBERTINI)

- Industrial, agricultural, infrastructural and tourist development in Italy, often EU-funded, have in the past destroyed large areas of semi-natural habitat, and in some areas this is still occurring. Industrialization/urbanization is recorded as a threat at 67 IBAs and is considered to pose a high-impact threat at 21 sites.

Drainage

For centuries European wetlands have been drained, particularly for agricultural purposes, and very few now remain in a natural condition. The lower degree of industrialization, urbanization and agricultural intensification in central and eastern Europe, and lower population densities in Fennoscandia, mean that in these areas more extensive areas of natural wetland occur, but expected changes in central and eastern European countries, especially the intensification of agricultural practices, pose a severe threat. For example, 70% of wetlands in Lithuania have been drained during the last 30 years (EEA 1999). Overall, 544 IBAs in Europe are threatened by drainage, of which 157 are threatened to a high degree.

Some examples of drainage posing a threat (actual or potential) to IBAs (excerpted from national overviews in this publication)

- Wetland drainage has affected several wetlands of global importance in eastern Europe, including the Danube delta (site 001 in Romania) and the Volga delta (site 179 in Russia).
- In Hungary, drainage is a widespread threat, affecting over 50% of the land area of 19 IBAs and is listed as a high-impact threat at 11 IBAs (002, 003, 006, 007, 020, 025, 027, 028, 035, 037 and 038).
- In Belarus, a national drainage campaign over the last 30 years has resulted in 25,000 km² of mires being drained. The wetlands that remain (including the IBAs) continue to be influenced by the drastically lowered groundwater levels and by the pollution with nutrients and other chemicals that have followed previous drainage. The unique fens of Dikoe (site 010), Sporovo (014) and Zvanets (016) are the most threatened IBAs in Belarus because of the drainage of surrounding areas, together with the cessation of traditional land-uses, in particular manual haymaking.
- Drainage has also been particularly prevalent in the Mediterranean region, where wetland drainage has affected many IBAs such as the Evros delta (site 006 in Greece and 001 Turkey) and the Nestos delta (012) in Greece, both important sites all-year-round for waterbirds, raptors and passerines.
- Wet meadows are the most threatened habitat for birds in Slovenia, 11 IBAs being seriously affected by drainage, dredging and canalization (sites 002, 004–010, 012–014).
- In Italy, large areas of wetland have been drained for land-claim: while there were 7,640 km² of wetland in Italy in 1865, there are now only 1,800 km² (a 75% reduction). This remains a threat at 28 IBAs including River Po wetlands (sites 028 and 068).

Wetland drainage for agriculture in Turkey.
(PHOTO: TOBIAS SALATHÉ)

Abandonment and reduction of land management

One of the indirect effects of agricultural intensification is that remaining marginal land becomes less profitable to farm. Many bird species that have adapted to artificial landscapes such as farmland require methods of non-intensive habitat management for their continued survival in such habitats. The abandonment of such practices is considered a threat at 609 IBAs.

Land abandonment is a significant threat to IBAs in many parts of central and eastern Europe (Figure 25), especially in the 12 countries due to join the European Union, where livestock numbers (especially sheep and cattle) decreased by 30–45% between 1989 and 1997 (Peters 1998). Land abandonment is also an important threat in the Mediterranean region, and in Sweden and other Baltic states.

Some examples of abandonment and reduction of land management posing a threat (actual or potential) to IBAs (taken from national overviews in this publication)

- In Sweden, abandonment and reduction of land management is the single most important threat, affecting 31 sites (49%). Undergrazing is a particular problem at many of these IBAs (for example sites 016, 023, 031, 036, 054, 058 and 059).

Grazed areas such as the coastal meadows on Öland Island, Sweden (IBA 054) hold high densities of breeding waders such as Redshank *Tringa totanus*, Dunlin *Calidris alpina* and Avocet *Recurvirostra avosetta*, and are also important stop-over sites for geese, ducks and waders. (PHOTO: BERTIL BREIFE)

- In Latvia, the state's transition to a market economy has been accompanied by a dramatic decrease in agricultural output in many marginal areas, whilst in more productive areas the intensification of agriculture continues. Large expanses of farmland have been set-aside or abandoned. For the first few years, this land is of increased importance for some birds, for instance the globally threatened *Crex crex*, but it soon loses its suitability for such 'open-country' species as the ground vegetation becomes too dense and the land becomes overgrown with bushes and small trees. This process threatens 30% of IBAs in Latvia.
- In Russia, the management of millions of hectares of agricultural land has been neglected or abandoned, and haymaking and grazing have stopped on many meadows and pastures. Land abandonment is recorded as a threat at 27 sites, with a high impact at nine (sites 027, 042, 098, 109, 119, 161; 206, 214 and 217).
- In Estonia, reduction or abandonment of agricultural land-use activity is having a detrimental effect on key bird species at 29 IBAs (67% of IBAs), largely due to overgrowth of coastal meadows by reed *Phragmites* as a result of undergrazing (e.g. Hiiumaa islands, site 004) and due to the overgrowth of flood-plains by scrub where haymaking has been abandoned (e.g. Soomaa mire complex, site 032).
- In Poland, as in most of the transition countries of central Europe, the main threat to birds at fish-pond IBAs is the likelihood that fishery management will become uneconomic (in the short term) and will be abandoned, leading to the disappearance of these man-maintained wetlands.
- In Slovakia, the abandonment of agricultural land and changes in management practices (for example, reductions in grazing intensity) over the last decade have affected 21 IBAs (66%), impacting three to a high degree (sites 011, 030 and 032).
- The abandonment of traditional land-uses and management practices in Greece is a threat at over 20% of IBAs. In particular, traditional pasture lands and arable cultivation have been abandoned, resulting in natural regeneration of forest and consequent changes in bird communities. The same pattern can be seen in many Mediterranean countries.

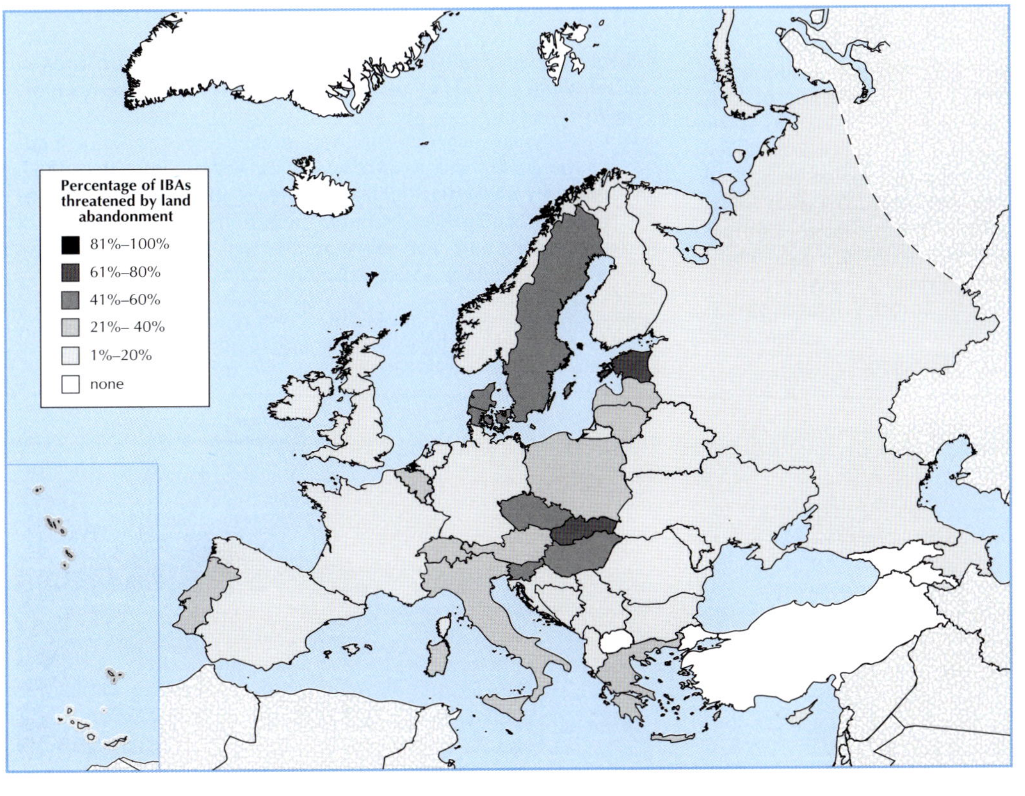

Figure 25. The percentage of Important Bird Areas, per country, that are threatened by land abandonment.

Construction/impact of dyke/dam/barrage

A total of 335 IBAs in Europe are threatened by the construction or downstream impact of dykes, dams or barrages.

Some examples of the construction/impact of dykes, dams or barrages posing a threat (actual or potential) to IBAs (excerpted from national overviews in this publication)

- Srebarna lake (site 033), Shabla lakes complex (049) and Durankulak lake (050) in Bulgaria, all particularly important for wintering geese (including the globally threatened Red-breasted Goose *Branta ruficollis*), are affected by dyke construction which lowers water-levels and destroys habitats such as temporary freshwater bodies and areas of shallow or open water.
- Two IBAs in Iceland, Thjórsárver (site 059) and Eyjabakkar (061), will be submerged if reservoirs, planned for hydroelectric power-stations, are constructed and filled. The proposals involve the complete submersion of Eyjabakkar and the partial submersion of the Ramsar Site at Thjórsárver. Part of the Sog river (site 021) has already been damaged by reservoirs and other infrastructure for hydroelectric power-stations.
- Wetlands in Slovenia are seriously affected by drainage, dredging and canalization. Such water management, together with the construction of dykes and dams and attempts to agriculturally improve arable land, all have an influence on groundwater levels and thus indirectly on wetlands. The riverine forests of the River Drava (site 012), River Mura (013) and Krakovo forest (009) are threatened by such activities.
- The Loire valley (sites 031, 128, 154), containing the last unregulated river in France, is seriously threatened by the possible construction of barrages.

Over 300 IBAs in Europe are threatened by the construction or impact of dykes, dams or barrages. (PHOTO: WWF)

Intensified forest management

A total of 450 IBAs across Europe are threatened by intensified forest management, which can involve the clear-felling of large tracts, re-planting with uniform-aged trees of few species (often non-native), addition of fertilizer in some cases, harvesting of trees long before biological maturity, and extensive use of heavy soil-compacting machinery. Such management is also often accompanied by the construction of new forest roads, leading to increased habitat fragmentation and increased disturbance to birds (through facilitating access to previously remote areas).

In total 924 IBAs are threatened by forestry-related threats encompassing intensified forest management, commercial deforestation, afforestation and selective logging.

Some examples of intensified forest management posing a threat (actual or potential) to IBAs (excerpted from national overviews in this publication)

- In Finland, the pressure to convert old-growth forest to plantations is very high outside of protected areas. The areas of untouched wilderness in northern and eastern Finland have declined over recent decades as a result of active commercial forestry. This has caused population declines in many species dependent on old-growth forest. Fortunately, the most important IBAs for boreal forest, and its associated birds, are nearly all protected from commercial forestry. However, conversion to plantations is still a threat to 10% of IBAs and selective logging/cutting to a further 9%.

Forestry is widespread in Hungary and forest management is generally intensive. (PHOTO: HONCS)

- Forestry in Latvia is intensifying and has grown to be the major state export; felling activity has doubled since 1990/1991. Harvesting during the breeding season is causing critical disturbance to sensitive bird species, such as Black Stork *Ciconia nigra* and Lesser Spotted Eagle *Aquila pomarina*, countrywide. Forest structure is changing towards a much lower percentage of older and less disturbed stands. Moreover, a considerable proportion of forest in Latvia has been privatized and thereafter it has become hard if not impossible to influence management towards conservation objectives. Intensified forest management is recorded as a threat in 30 IBAs (51% of IBAs in Latvia).
- Many lowland forests in Greece are threatened by intensified forest management, in particular through the construction of new forest roads. The latter cause fragmentation and disturbance and greatly facilitate access to previously remote areas. A total of 25 IBAs in Greece are threatened by intensified forest management.
- The intensification of forestry activities in Ukraine, through increased clear-felling and selective logging, are causing a decrease in the national populations of some raptors. IBAs affected include sites 001, 006, 028, 042 and 046.
- Intensive forestry practices (logging, construction of forest roads, use of heavy machinery, the introduction of non-native tree species) pose a significant threat to 19 IBAs (59%) in Slovakia.

Aquaculture/fisheries

A total of 648 IBAs in Europe are threatened by fisheries or aquaculture activities.

Some examples of aquaculture/fisheries posing a threat (actual or potential) to IBAs (excerpted from national overviews in this publication)

- Commercial fisheries are a threat at 15% of the IBAs in the Netherlands, and at nearly 10% the threat is rated as 'high impact'. This specifically refers to shellfish harvesting on intertidal and shallow subtidal mudflats (which causes habitat destruction and food shortage for birds) and to over-fishing on the IJsselmeer and Markermeer (sites 034, 038).

Shellfish are 'mined' from mudflats by suction-dredgers in the Netherlands, as here in the Wadden Sea (IBA 001). (PHOTO: MARTIJN DE JONGE)

- A high proportion of Norwegian IBAs are important for their seabird colonies, and fisheries pose the most serious threat to these birds. In the past, over-exploitation of fish stocks has caused dramatic declines in several seabird populations along the coast of northern Norway, especially Guillemot *Uria aalge* and Puffin *Fratercula arctica*. Sustainable management of the fish stocks is crucial for most of the Norwegian seabird colonies listed in this inventory.
- Clam-dredging is a very localized activity in Greenland, but may occur at the same places as King Eider *Somateria spectabilis* moulting grounds, as at IBA 025 where this activity seems to have contributed to a major decline in numbers of this seaduck at this previously important moulting site.
- In Ireland, intertidal shellfish cultivation is a serious threat at a number of IBAs, including Trawbreaga Bay (site 003), Drumcliff Bay (033) and Dungarvan Harbour (094). Potentially it causes a loss of mudflat habitat and also disturbance to wintering waterbirds.

Extraction industry

A total of 426 IBAs in Europe are threatened by extraction industries, including mining and oil, gas and peat extraction.

Some examples of the extraction industries posing a threat (actual or potential) to IBAs (excerpted from national overviews in this publication)

- Peat mining has remained a constant threat to most peatland IBAs in Latvia. Regeneration can be poor or non-existent.

Peat extraction threatens Cena bog (IBA 030), one of the largest raised bogs in Latvia, important for thousands of geese *Anser* on passage and for breeding and migrating Cranes *Grus grus*. (PHOTO: EDMUNDS RAČINSKIS)

- In the United Kingdom, mineral extraction poses a significant threat to the East Devon Heaths (site 020). These heaths support a range of breeding heathland birds and are reported to be the best lowland heaths for butterflies in England.
- In Belgium, quarrying is destroying valuable habitats at Entre Sambre et Meuse (site 032), Haute Meuse (033), Lesse et Lomme (036) and Sinémurienne (047).
- In Iceland, mining of the bottom sediments of Lake Mývatn, to supply a diatomite factory by the lake, seriously threatens Mývatn–Laxá (site 044), a Ramsar Site. The site holds more than 20,000 waterbirds on a regular basis and several duck species breed at higher densities than anywhere else in Europe.

HOW WELL PROTECTED ARE IBAs?

The legal protection status of IBAs

National protection

A total of 2,191 IBAs in Europe (60% of the total) have at least some protection under national law, through their overlap with (mainly statutory) protected areas (Figure 26). However, in only 21% of IBAs in Europe is the extent of overlap classified as 'high', i.e. where more than 90% of the area of the individual IBA is protected. Indeed, in

Figure 26. The national protection status of Important Bird Areas in Europe.

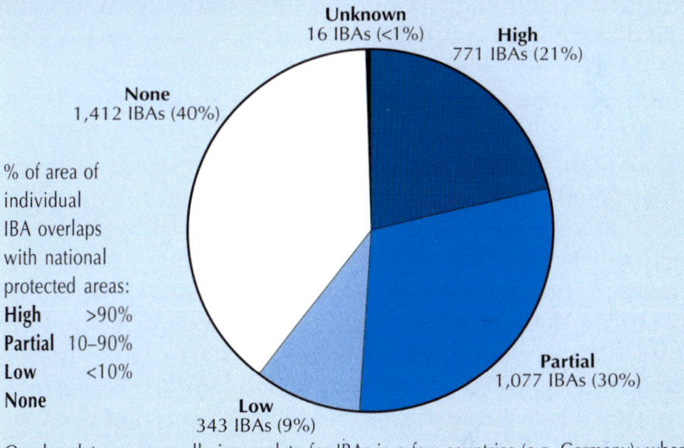

Overlap data are generally incomplete for IBAs in a few countries (e.g. Germany); where an overlap of unknown size exists, the IBA protection status has generally been recorded as 'Partial' by default.

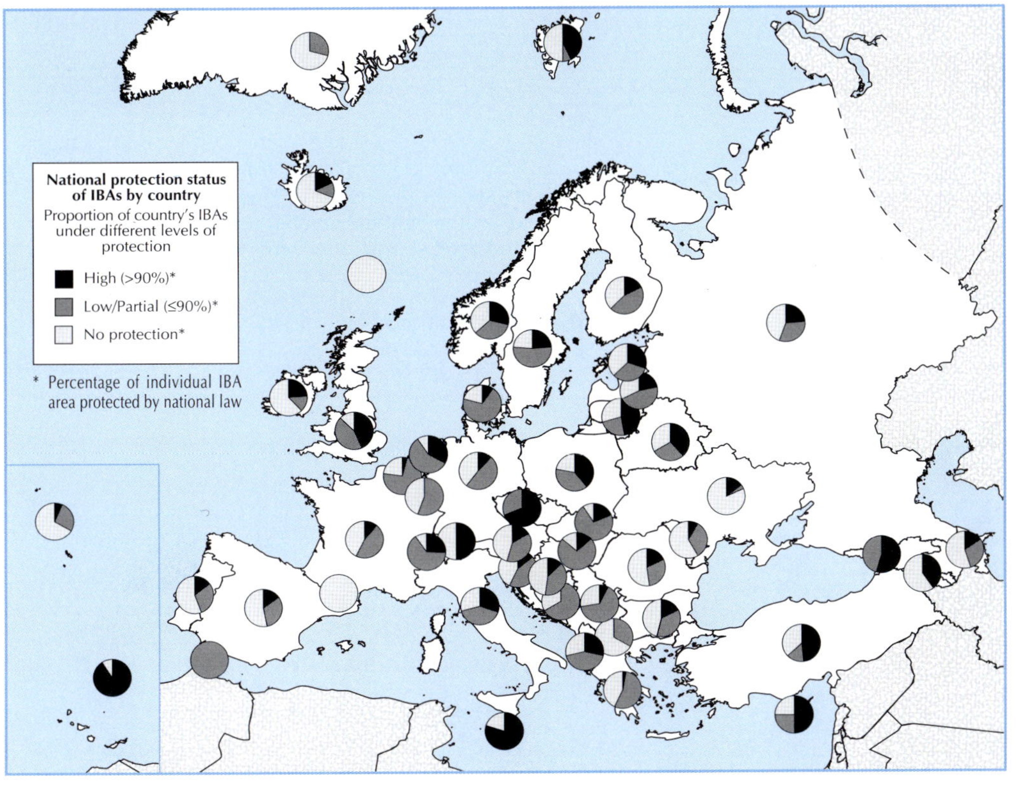

Figure 27. The national protection status of Important Bird Areas in each European country.

343 IBAs less than 10% of the area of the individual IBA overlaps with protected areas, and in 1,412 IBAs (39% of those in Europe) no form of legal protection is in place at the national level.

A higher proportion of IBAs are protected in some countries than in others (Figure 27). Countries with less than a third of their IBAs having any protection at the national level include Greenland, Faroe Islands, Iceland and Ukraine (see national chapters for more information). Those countries with more than a third of their IBAs receiving a high extent of protection, in terms of individual IBA area covered, include several island-states such as Madeira, Malta and Cyprus, as well as the United Kingdom, Belarus, Poland, Lithuania and Turkey. The Czech Republic stands out particularly in having a high proportion of its IBAs protected and most of these having, individually, a high extent of protection. For some countries the data presented on the extent of national protection of individual IBAs is incomplete.

In many countries, especially those preparing for accession to the EU, changes in the national protected-area legislation are ongoing or imminent, and thus the data presented in this publication may change in future. In addition, despite designation of statutory protected areas at the national level, in many countries the designated type of protection and management may be too weak or otherwise inappropriate for the site in question, and the level of enforcement is often under-resourced or poor.

Figure 28. The international protection status of Important Bird Areas in Europe.

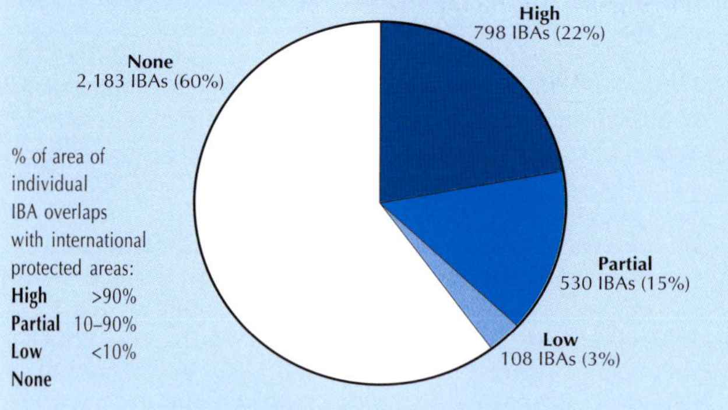

Overlap data are generally incomplete for IBAs in a few countries (e.g. Germany); where an overlap of unknown size exists, the IBA protection status has generally been recorded as 'Partial' by default.

International protection

Substantial international recognition of IBAs has been achieved in the last decade but many IBAs remain inadequately protected or recognized at the international level, given their relatively small overlap with areas that are listed under international conventions or designated under EC Directives (Figure 28). Only 1,436 IBAs in Europe (40% of the total) have some form of international protection, and in only 22% of all IBAs is the extent of protection classified as 'high', i.e. where more than 90% of the area of the individual IBA is protected. Indeed, at 108 IBAs less than 10% of the area of the individual IBA is protected at the international level, and at 2,183 IBAs (60% of the total) no form of international protection is in place.

A higher proportion of IBAs receive international protection in some countries than in others (Figure 29). Appendix 8 shows the number of IBAs in each country that are wholly or partly covered by the main international site-protection designations. Some forms of international legislation are only applicable in certain countries, for instance the EC Birds Directive is only relevant to the 15 member states of the European Union (see Appendix 1 for details). Due mainly to the application of this Directive, the proportion of internationally protected IBAs is generally much lower in non-EU than EU countries. Countries outside of the EU with more than one third of their IBAs designated internationally at least in part are Estonia, Czech Republic, Slovakia and Hungary, largely reflecting the designation of Ramsar Sites.

For further details of the key international site-protection mechanisms, see Appendix 1. Currently, there are 23 natural or joint natural/cultural World Heritage Sites in Europe, 17 of which overlap with IBAs, and there are 126 Biosphere Reserves in Europe, 117 of which overlap with IBAs. Ramsar Sites and Special Protection Areas are assessed in more detail as follows (Appendix 8).

Assessment of Ramsar designations

As of 29 September 1999, 636 Ramsar Sites have been designated in Europe, 594 of which overlap partly or wholly with 556 IBAs (see Appendix 1 for details of the Ramsar Convention). A total of 1,700 IBAs in Europe are important at the European level for congregations of waterbirds because they support at least 1% of a biogeographic population of one or more species (criterion A4i), or at least 1% of the flyway population of one or more species (criterion B1i) or at least 20,000 waterbirds overall (criterion A4iii). These three criteria match criteria 5 and 6 of the Ramsar Convention for the identification of wetlands of international importance (Ramsar Sites). However, criteria A4i, A4iii and B1i can also be applied to congregations of waterbirds in certain grassland and

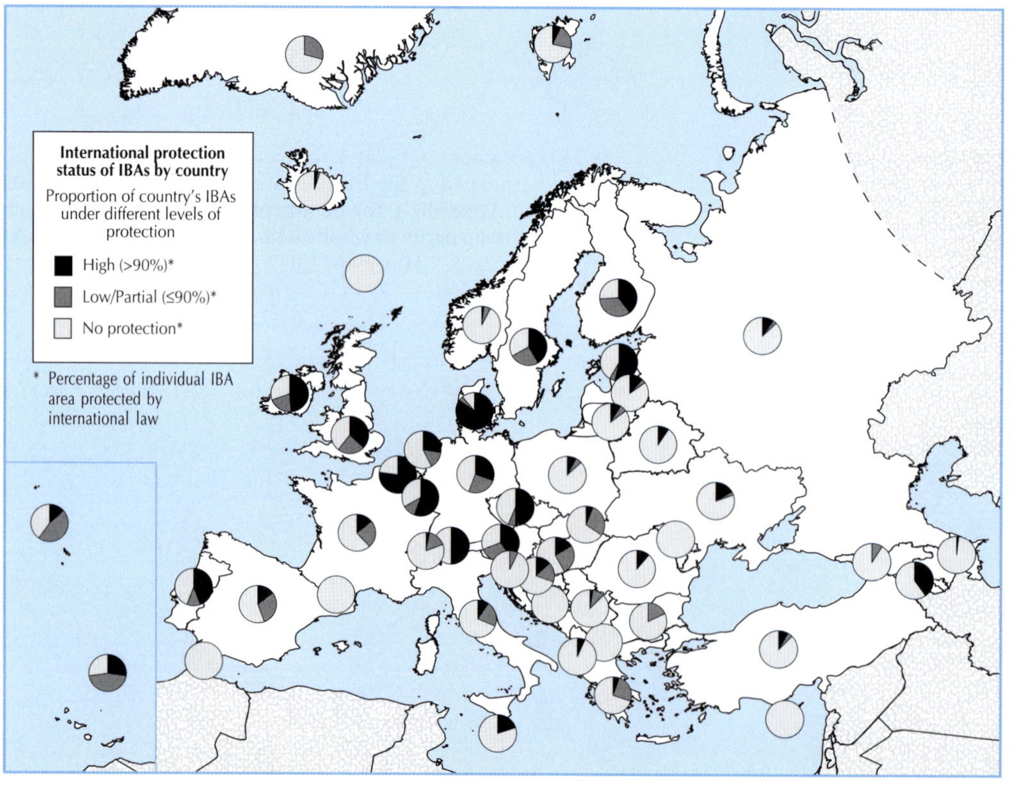

Figure 29. The international protection status of Important Bird Areas in each European country.

Figure 30. The coverage of Important Bird Areas in Europe by Ramsar Sites and by Special Protection Areas: a comparison of the situation in 1989 and 1999.

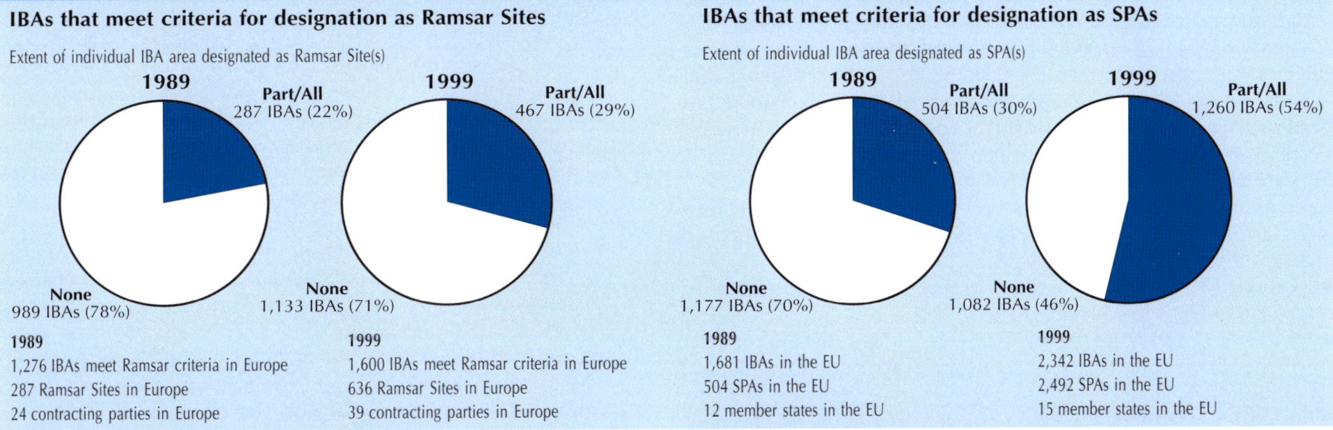

Figure 31. The extent of overlap between individual Important Bird Areas and Ramsar Sites in Europe.

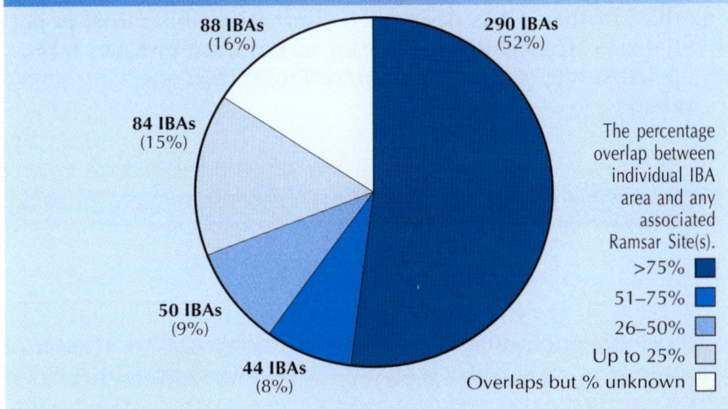

Figure 33. The extent of overlap between individual Important Bird Areas and Special Protection Areas in the European Union.

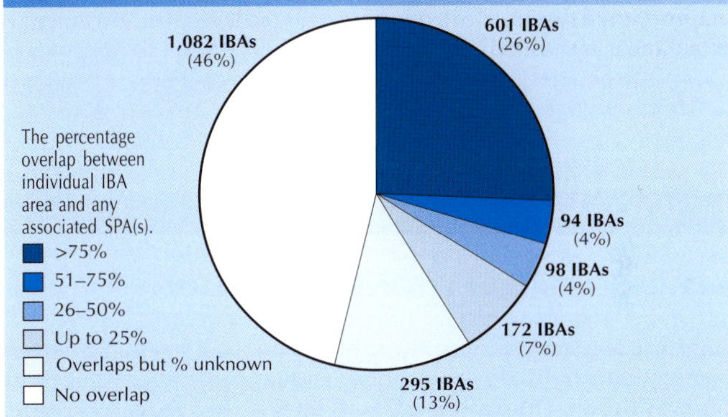

marine habitats that do not qualify as wetland habitat under the Ramsar Convention (see Appendix 1). Approximately 100 of the 1,700 sites are totally marine or grassland. The eligibility of such sites for inclusion in the Ramsar List of wetlands has, therefore, to be considered on a case-by-case basis. Overall it is estimated that 1,600 IBAs in Europe should be considered as potential Ramsar Sites based on concentrations of waterbirds; 467 (29%) of them are actually covered, partly or wholly, by Ramsar Sites (Figure 30).

Many wetland IBAs that are important for non-waterbirds (such as mires) also comply with Ramsar criteria 2, 3 and 4, and thus represent potential Ramsar Sites too. An assessment of this additional number of potential sites has not been made for this publication. However, 90 IBAs that do not qualify as IBAs for congregations of waterbirds are designated as Ramsar Sites; they may have been designated for terrestrial species dependent on wetlands such as raptors, or they may fulfil other non-ornithological Ramsar Site criteria.

It is also important to know the extent to which individual IBAs are covered by Ramsar Sites (Figure 31). In general, there is a high degree of overlap: in 52% of IBAs which overlap with Ramsar Sites, the Ramsar designation covers more than 75% of the area of the individual IBA.

Figure 32. The coverage of Important Bird Areas by Special Protection Areas in the European Union.

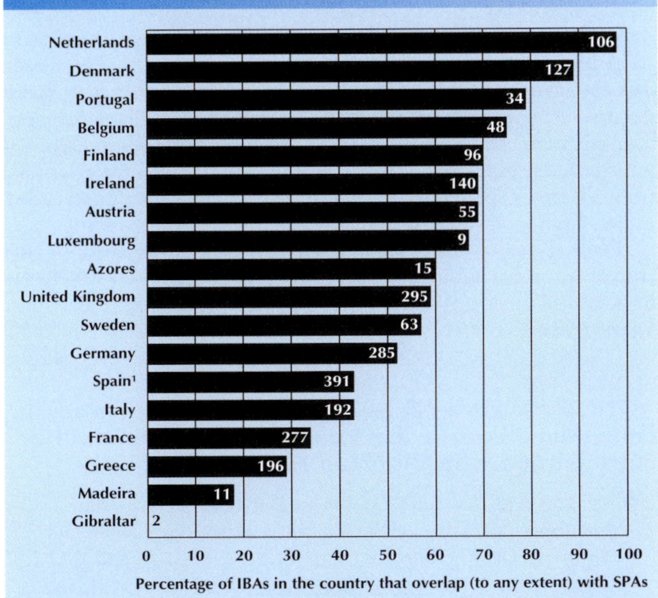

The figure at the end of each bar equals the total number of IBAs in the country concerned.
1. Includes data for Canary Islands.

Assessment of SPA designations

As of 23 September 1999, 2,492 Special Protection Areas (SPAs) have been designated in the European Union under the EC Birds Directive (see Appendix 1 for details of the Directive), of which 1,375 SPAs overlap partly or wholly with 1,260 IBAs—54% of IBAs in the EU (Figure 30). However, 1,082 IBAs (46%) in the EU are not covered at all by SPAs and thus should be considered for designation. Some countries have proceeded further than others in the designation of IBAs as SPAs, with the Netherlands, Denmark, Portugal and Belgium being the only countries that have designated over 75% of their IBAs (partly or wholly) as SPAs (Figure 32). The extent of coverage of individual IBAs is also important to note—in just 26% of IBAs in the European Union do SPAs cover more than 75% of the area of the individual IBA (Figure 33).

Changes in protection status of IBAs between 1989 and 1999

According to available information, the proportion of IBAs under statutory national protection has increased in most countries since 1989, although it is difficult to assess this properly in the absence of comprehensive or detailed data from 1989. Any comparison also needs to compensate for the fact that many new IBAs have been identified since 1989, and a (relatively small) number of 1989 sites have lost their IBA status. Analysis indicates that the proportion of all European IBAs designated (partly or wholly) as Ramsar Sites has increased from 22% in 1989 (Langeveld and Grimmett 1990), to 29% in 1999 (Figure 30). Although there has been a substantial increase in Ramsar designations (Frazier 1996, 1999), 71% of potential sites (identified in this publication) remain unprotected by this Convention. Similarly, regarding SPA designations, 54% of IBAs are partly or wholly covered by SPAs in 1999, compared to 30% in 1989 (Figure 30). However, given that 46% of IBAs in the EU are unprotected by SPAs and considered to be of international importance, further designations are necessary.

Threats and protection status

A large number of IBAs that are protected at national and international levels remain threatened. Analysis of threat and protection status data shows that:

- A total of 12,600 records of threat were recorded against 3,381 of the 3,619 IBAs in Europe.
- A total of 1,114 IBAs are threatened by at least one high-impact threat, yet are partly or wholly covered by some form of legal protection.
- A total of 904 IBAs of global importance (meeting one or more 'A' criteria) are threatened by one or more high-impact threats. Of these, 644 are protected (partly or wholly) at either national or international levels, while 408 of them are protected at the international level.
- Of the 1,260 IBAs in the European Union at least partly covered by Special Protection Areas (SPAs), 559 IBAs (44%) are threatened by at least one high-impact threat. Thus the designation of an IBA as an SPA has not prevented harmful development, or the risk of it in the near future, at these sites.

Development activities in SPAs

At SPAs, member states are required to "avoid [...] deterioration [...] of the habitats of species for which areas have been designated, in so far as such disturbance could be significant in relation to the objectives of this Directive" (as amended by Art. 6(2) of the Habitats Directive 92/43/EEC). The preparation of management plans, although not legally required, would be an appropriate way to achieve the above objective but very few member states have made significant progress in this respect. Contrary to the original provision of the Birds Directive Article 6 of the Habitats Directive contains a derogation of the general protection regime. Any plan or project likely to have a significant effect on the site, shall only be agreed upon "after having ascertained that it will not adversely affect the integrity of the site concerned. If, in spite of a negative assessment of the implications for the site and in the absence of alternative solutions, a plan or project must nevertheless be carried out for imperative reasons of overriding public interest all compensatory measures shall be taken necessary to ensure that the overall coherence of Natura 2000 is protected."

- Of the 557 IBAs that are at least partly covered by Ramsar Sites, 271 IBAs (49%) are also threatened by at least one high-impact threat.

WHAT SHOULD BE DONE NEXT?

KEY RECOMMENDATIONS FOR ACTION

Immense pressure continues to be placed on the European environment through developments in agriculture, forestry, fisheries, transport, energy, industry and urbanization, and, as this publication shows, such pressures are severely impacting the most important sites for birds in Europe. Much still needs to be done to adequately conserve Important Bird Areas.

ACTION: Designate IBAs as protected areas under national law

- **40% of IBAs in Europe are unprotected by national legislation.**
- **39% of IBAs in Europe that have some protection by national law are still highly threatened.**

Most sites in this inventory should be protected by national law and, where appropriate, managed to conserve their ornithological and biological value. The required level of protection and management will vary from site to site, and the options immediately available will be dictated largely by the protected-area legislation already in force in each country. The appropriate level of protection will vary from protected landscapes, where the management objective is to maintain the landscape character whilst sustaining traditional or non-intensive land-uses and ecological diversity, to strictly protected nature reserves where human presence is minimized. In countries with inadequate nature-conservation legislation, improved legislation is essential. Cooperation between governments and non-governmental organizations can be an effective way to achieve these objectives.

There is also a need to implement existing laws better, including improved commitment to adequate staff and resources, as well as policies and funding that prevent destruction, damage and neglect, and that encourage positive action and management of important biodiversity areas.

- **All 31 European national reports to the most recent Conference of the Parties to the Convention on Biological Diversity mention measures of in-situ conservation, including protected areas, but only eight reports refer to IBAs.**

The populations of most bird species in Europe are shared between many countries, and individuals of most species for which IBAs are identified traverse political boundaries at some stage in their annual cycle. Actions to conserve birds and other biodiversity gain in effectiveness if they are framed in an international context.

The international agreements most relevant to the conservation of IBAs are outlined in Box 2. Apart from the EC Directives, official recognition or listing of an IBA under these international agreements does not normally lead immediately to greater legal powers of protection, but does confer a higher profile and status. Site-based aspects of such agreements are key mechanisms for bringing about positive change, through increased political and public awareness, commitment and support. They provide for the creation of international networks of sites on the ground, and international frameworks of people and institutions to facilitate conservation understanding, awareness and cooperation between nations.

Further details about each agreement and its relevance to IBA conservation are outlined in Appendix 1. All possible efforts should be taken to ensure that IBAs are properly protected under these agreements.

ACTION: Protect IBAs under relevant international agreements

- **60% of IBAs in Europe are not protected under any international convention, directive or programme (of these, 49% have some national protection).**
- **At least 1,100 wetland IBAs in Europe meet Ramsar criteria but are not designated as Ramsar Sites.**
- **46% of IBAs in the European Union have no part of their area designated as SPAs.**

ACTION: Countries pursuing accession to the European Union should ensure that the protection of IBAs will be adequate under the Birds and Habitats Directives

- **To date, 433 IBAs have been identified in the 12 accession countries, and more will be identified in the future when the European Union criteria (category 'C' criteria) are applied.**
- **Only 103 (24%) of these IBAs are partly or wholly covered by any form of international protection, while 305 (70%) are covered by national protection.**

Box 2. International agreements relevant to site conservation in Europe (see Appendix 1 for more details).

Convention/initiative	Year of adoption	Year in force	Region covered	Participation in Europe (no. of countries)[1]: Ratified or applicable	Signed	Not yet participating	Site-conservation mechanism	No. of sites designated in Europe	Total area in Europe (km²)	As of:
WITH AN ACTIVE SITE-CONSERVATION MECHANISM										
Ramsar Convention	1971	1975	Worldwide	39	—	6[2]	Ramsar Site	636	112,186	29 Sep 99
Man and Biosphere Programme	1970	n/a	Worldwide	35[3]	—	—	Biosphere Reserve	126	808,370	4 Jan 99
World Heritage Convention	1972	1975	Worldwide	39	—	6[2]	World Heritage Site	23[4]	51,711	2 Dec 98
Bern Convention [5]	1979	1982	Member states of Council of Europe (39 states) and invited non-member states in Europe and N and W Africa	33[6]	3	9	Biogenetic Reserves[5]	340	>30,000	21 Dec 98
EC Birds Directive	1979	1981	European Union	15	—	—	Special Protection Area (SPA) — Natura 2000	2,492	169,823	23 Sep 99
EC Habitats Directive	1992	1994	European Union	15	—	—	Special Area for Conservation (SAC) — Natura 2000	9,942[7]	341,344	23 Sep 99
Barcelona Convention	1976	1978	All states riparian to the Mediterranean Sea	11[6]	—	—	Mediterranean Specially Protected Area (MedSPA)	>52	>5,907	12 Oct 99
[New] Helsinki Convention	1992	—	All states in catchment of the Baltic Sea	12[6]	1	—	Baltic Sea Protected Area (BSPA)	65[9]	—	early 1998
CURRENTLY WITHOUT AN ACTIVE SITE-CONSERVATION MECHANISM										
Convention on Biological Diversity	1992	1994	Worldwide	40[6]	3	2				15 Jan 99
Convention on the Conservation of Migratory Species of Wild Animals (Bonn Convention)	1976	1983	Worldwide	27[6]	—	18				1 Sep 99
African-Eurasian Migratory Waterbirds Agreement (under Bonn Convention)	1995	1999	Africa, Europe and west Asia (c.120 sites)	9	6[6]	30				1 Nov 99
OSPAR Convention	1992	1998	All states in catchment of north-east Atlantic Ocean	15[6]	—	—				2 Dec 98
Black Sea Environmental Programme	1992	1994	All states riparian to the Black Sea	6	—	—				26 Apr 97
Pan-European Biological and Landscape Diversity Strategy [8]	1995	n/a	Europe	45	—	—				n/a

1. Totals exclude autonomous or dependant territories (Azores, Canary Islands, Faroe Islands, Greenland, Gibraltar, Madeira, Svalbard/Jan Mayen) as well as Monaco, Vatican City (Holy See) and San Marino.
2. European Union has not ratified/signed as an entity.
3. Total number of active National Man and Biosphere (MAB) Committees.
4. Total is for 'natural' or 'natural/cultural' sites only.
5. The EMERALD Network [new pan-European site-conservation mechanism] is in process of being conceptualized.
6. Figure does not include the European Union, which has also ratified/signed as an entity.
7. Proposed Sites of Community Interest
8. The Pan-European Ecological Network [new integrative framework for all existing site-conservation mechanisms in Europe] is in process of being conceptualized.
9. Number of proposed BSPAs (none have been designated as such, although some are protected at the national level).

In June 1993 the heads of state and governments of the European Union countries agreed that those countries of central and eastern Europe which so desired would be able to join the EU once they met the agreed political and economic conditions. Accession negotiations were opened on 31 March 1998 with the first six countries recommended by the Commission: Czech Republic, Estonia, Hungary, Poland, Slovenia and Cyprus. The second group of accession countries comprises Bulgaria, Latvia, Lithuania, Malta, Romania and Slovakia.

Before EU membership can be attained, each country needs to adopt the 'acquis communautaire' (the accumulated laws and obligations of EU membership), including the extensive 'environmental acquis'. Adoption of the acquis entails the alignment of the national legislation of candidate countries with EU law, and comprises three distinct activities:

- transposition (adoption or change of national laws, rules and procedures so that the requirements of the relevant EU law are fully incorporated into the national legal order);
- implementation or practical application (providing the institutions and budgets necessary to carry out the laws and regulations); and
- enforcement of EU legislation (providing the necessary controls and penalties to ensure that the law is being complied with fully and properly).

Thus the 'approximation' of national nature-conservation legislation to the Birds and Habitats Directives of the EU, including the identification and designation of Special Protection Areas for inclusion in the EU-wide ecological network of protected areas (entitled 'Natura 2000'), is a key nature-conservation objective of the accession process. It should therefore be ensured that the conservation of IBAs is adequately addressed by this process.

ACTION: Prevent breaches of obligations under national and international law

Using the legal opportunities provided by national laws and the EC Birds and Habitats Directives, it is possible to object to activities that may be harmful to IBAs and to steer development towards sustainability. By registering formal complaints at the European Commission, concerned citizens or organizations can complain against decisions by EU member states which are considered to breach the legal obligations of the Directives. BirdLife Partners have successfully opposed harmful developments at many IBAs, and have highlighted the deficient protective strategies taken by some governments.

ACTION: Plan and carry out adequate management at all IBAs, including habitat or species restoration if necessary, and involve local communities, NGOs and the public in IBA conservation

- **17% of IBAs in Europe (619 sites) have management plans that cover part or all of their area.**

Much of the landscape of Europe has been moulded by human influence over thousands of years, and some management is required at most IBAs to maintain, enhance or restore their value for birds. Such management should take into account all biodiversity, as well as social, economic, political and other relevant factors, including the interests of all stakeholders. The nature of this management may best be decided, implemented and monitored through the development of a management plan for the site or wider region. Management planning should take fully into account existing and potential human-uses, and should ensure, where human-use is truly in conflict with conservation objectives, that such use is effectively controlled (Welch 1994).

Often, IBAs can only be conserved effectively with the help and participation of local communities and through collaboration with NGOs and the public. When conservation actions involve or affect local communities, the building of support and mutual understanding more often leads to a sustainable outcome. Local communities have an important stewardship role, and in the long term can benefit from the conservation of IBAs, although this may sometimes conflict with the short-term interests of some economic sectors.

Building local and national capacity to protect these areas requires coordination. A growing number of BirdLife Partners have established 'caretaker' groups that, at the local level, carry out practical conservation actions to safeguard IBAs. The results of such activities include management plans that have been accepted by the competent authorities, site-use agreements that have ended disturbances to birds, and practical habitat-management that has been implemented on the ground.

ACTION: Integrate environmental objectives, including biodiversity conservation, into all policy sectors, and conserve the wider environment

- **93% of IBAs in Europe are threatened to some extent. Of these, 1,640 IBAs are threatened by agriculture, 1,575 IBAs by recreation/tourism, 1,057 by man-made changes to the hydrological regime, 924 IBAs by forestry and 779 IBAs by infrastructural development.**

Increasingly, many environmental problems and potential threats to IBAs are global or regional in scope or origin, and cannot be effectively addressed or solved solely by good protection and management of the IBAs themselves. Many threats facing birds at sites result from generally unsustainable human development activities. Valuable sites for birds, even when protected, cannot be isolated from surrounding communities or society: IBAs not only receive harmful impacts from human society but also offer a wide range of services from aesthetic and recreation functions to the provision of clean water and air buffering against pollution and protection against erosion. Integrated conservation projects which aim to address broader habitat-conservation issues should be promoted as a tool for conserving those IBAs whose continued importance is dependent upon land-use and development policies in the surrounding countryside.

Financial measures are needed to provide incentives to encourage good management and the protection of wildlife, while perverse incentives, which encourage the destruction and degradation of IBAs, need to be removed and financial penalties and other legal sanctions provided. International funding for development programmes (e.g. through the Structural Funds in the EU) needs to be conditional on the outcomes of environmental impact assessments. Funding should aim to achieve economic development while at the same time being environmentally sustainable, with targeted provisions for biodiversity conservation (BirdLife International 1995).

BirdLife has published habitat conservation strategies for the major habitat-types in Europe (Tucker and Evans 1997), outlining directions for tackling these wider policy issues outside IBAs. It is clear that a broad range of legislation and policies have significant impacts on IBAs, particularly policies on agriculture, forestry, land-use planning, privatization, land ownership, environmental impact assessment and hunting. By ensuring that these laws take into account the principles of nature conservation, damage to important sites can be minimized. Strategic environmental assessments and environmental impact assessments need to be applied to major forestry, agriculture, infrastructural and other programmes and projects in order to avoid detrimental impacts to IBAs and the wider environment.

ACTION: Maximize the use of data on IBAs and disseminate information to promote their conservation

- **BirdLife IBA database constructed, coherent across countries.**

Much data on IBAs is presented in this publication, with the primary purpose of guiding the conservation of these important sites. This information thus needs to be disseminated widely at local, national and international levels, to ensure that decisions affecting Europe's environment adequately take into account the sites, and the site-based issues such as species status, threats, land-use and protection status, that are most important for bird conservation.

The efficient storage and management of data means that it can be analysed, interpreted and communicated effectively to decision makers and those who need it. Therefore, in addition to this publication, the BirdLife International IBA Database (see Box 6, 'Data collection' chapter) has been established, holding additional data on these sites. Public access to this system is possible, subject to the terms agreed in the IBA Data Access Policy agreed by the European Partnership of BirdLife International (BirdLife International 1996). This database will continue to be updated by the BirdLife International Partnership.

The effective conservation of IBAs will also depend heavily on the work and agendas of other national and international governmental and non-governmental organizations. There is therefore great value in mobilizing scientific information within collaborative networks in support of policy development and decision-making at all levels of society. One international mechanism for the public dissemination of biodiversity data, still in the early stages of development, is the Biodiversity Conservation Information System (BCIS). This is to be created by a coalition of 12 organizations which manage biodiversity information at an international scale (BirdLife International, Botanical Gardens Conservation International, Conservation International, International Species Information System, IUCN Commission on Ecosystem Management, IUCN Environmental Law Programme, IUCN Species Survival Commission, IUCN World Commission on Protected Areas, The Nature Conservancy, TRAFFIC International, Wetlands International, World Conservation Monitoring Centre). Their common challenge lies in harmonizing their data and knowledge, and improving public access to it.

ACTION: Monitor IBAs and continue to collect data for defined purposes

- **Scientific approach to design of site-selection criteria. Continuity with previous IBA criteria and in parallel with criteria of other site-based mechanisms.**
- **Data in this publication and the IBA database are standardized, permitting easy analysis for monitoring and trend assessments.**

Data are required not only to set priorities but also to understand the reasons for change in the status of sites and the species dependent on them. This publication and the BirdLife IBA Database (see Box 6, 'Data collection' chapter), together with the previous pan-European IBA inventory published 10 years ago and national IBA publications, set important baselines from which progress in the conservation of IBAs can be monitored. It is important that the status of IBAs continue to be monitored to identify trends and provide early warning. Results can be used to assess priorities for further conservation action and research, to generate policy-relevant arguments for advocacy programmes, and to judge environmental impacts.

Also important is the continuing identification and conservation of IBAs for European migratory birds when outside Europe, predominantly in Africa, the Middle East, and the Indian subcontinent, and Arctic Siberia and Canada. IBA programmes have been established in these regions and approximately 20,000 IBAs are likely to be identified worldwide.

REFERENCES

BINA, O., BRIGGS, B. AND BUNTING, G. (1995) *The impact of the Trans-European Networks on nature conservation: a pilot project*. Sandy, UK: Royal Society for the Protection of Birds.

BIRDLIFE INTERNATIONAL (1995) *The Structural Funds and biodiversity conservation*. Sandy, UK: Royal Society for the Protection of Birds.

BIRDLIFE INTERNATIONAL (1996) IBA criteria: categories and thresholds. Cambridge, UK: BirdLife International (unpublished report).

COLLAR, N. J., CROSBY, M. J. AND STATTERSFIELD, A. J. (1994) *Birds to watch 2: the world list of threatened birds*. Cambridge, UK: BirdLife International (Conservation Series 4).

DURINCK, J., SKOV, H., JENSEN, F. P. AND PIHL, S. (1994) *Important marine areas for wintering birds in the Baltic Sea*. Copenhagen: Ornis Consult (report on European Commission DG XI research contract no. 224/90-09-01).

EEA [EUROPEAN ENVIRONMENTAL AGENCY] (1998) *Europe's environment: the second assessment*. Luxembourg: Office of Official Publications of the European Communities/Oxford, UK: Elsevier.

EEA [EUROPEAN ENVIRONMENT AGENCY] (1999) *Environment in the European Union at the turn of the century*. Luxembourg: Office for Official Publications of the European Communities (Environment Assessment Report No. 2.).

FRAZIER, S. (1996) *An overview of the world's Ramsar Sites*. Slimbridge, UK: Wetlands International (Wetlands International Publ. 39).

FRAZIER, S. (1999) *Ramsar Sites overview: a synopsis of the world's wetlands of international importance*. Wageningen, Netherlands: Wetlands International.

GRIMMETT, R. F. A. AND JONES, T. A. (1989) *Important Bird Areas in Europe*. Cambridge, UK: International Council for Bird Preservation (Techn. Publ. 9).

HAGEMEIJER, W. J. M. AND BLAIR, M., EDS (1997) *The EBCC atlas of European breeding birds: their distribution and abundance*. London: T. & A. D. Poyser.

HÄGERHÄLL, B. AND SKOV, H. (1998) Proposal for marine Baltic Sea Protected Areas (BSPAs). Copenhagen: Ornis Consult (report to HELCOM Environment Committee).

ICBP (1989) Important Bird Areas in the European Community: a shadow list of Special Protection Areas. Cambridge, UK: International Council for Bird Preservation (unpublished report).

IUCN (1993) *Parks for Life: Report of the Ivth World Congress on National Parks and Protected Areas.* Gland, Switzerland: IUCN.

LANGEVELD, M. J. AND GRIMMETT, R. F. A. (1990) Important Bird Areas in Europe: wetlands for the shadow list of Ramsar Sites. Cambridge and Slimbridge, UK: International Council for Bird Preservation/International Waterfowl and Wetlands Research Bureau (unpublished report).

PETERS, R. (1998) *Agricultural situation and prospects in the central and eastern European countries. Summary report* Brussels: European Commission (Directorate General for Agriculture [DG VI]). Working document.

SKOV, H., DURINCK, J., LEOPOLD, M. F. AND TASKER, M. L. (1995) *Important bird areas for seabirds in the North Sea*. Sandy, UK: Royal Society for the Protection of Birds.

SKOV, H., DURINCK, J., LEOPOLD, M. F. AND TASKER, M. L. (in prep.) A quantitative method for evaluating the importance of marine areas for the conservation of birds.

STANNERS, D. AND BOURDEAU, P., EDS (1995) *Europe's environment: the Dobris report*. Copenhagen: European Environmental Agency.

STATTERSFIELD, A. J., CROSBY, M. J., LONG, A. J. AND WEGE, D. C. (1998) *Endemic bird areas of the world: priorities for bird conservation*. Cambridge, UK: BirdLife International (Conservation Series 7).

TUCKER, G. M. AND EVANS, M. I., EDS. (1997) *Habitats for birds in Europe: a conservation strategy for the wider environment*. Cambridge, UK: BirdLife International (Conservation Series 6).

TUCKER, G. M. AND HEATH, M. F. (1994) *Birds in Europe: their conservation status*. Cambridge, UK: BirdLife International (BirdLife Conservation Series no. 3).

WELCH, G. (1994) Site management and planning for conservation. Workshop proceedings, Gdansk, Poland 1994. Cambridge, UK: BirdLife International (unpublished report).

ALBANIA

Fabio Casale and Taulant Bino

Participants on an expedition to monitor IBAs in Albania. (PHOTO: FABIO CASALE/LIPU)

GENERAL INTRODUCTION

Albania is bordered by south-western Yugoslavia (Montenegro and Kosovo), the Former Yugoslav Republic of Macedonia, north-western Greece, and the southern Adriatic and northern Ionian Seas. It is one of the smaller European states, having a surface area of just 28,750 km². About two-thirds of the country's 2.9 million inhabitants (an average population density of 100 persons per km²) live in rural settlements, whilst the largest urban area is the capital city Tiranë, which holds approximately 220,000 people.

Topographically, Albania can be divided into four zones: an alpine zone, north of the River Drin, with peaks over 2,000 m; a mountainous zone between the Drin and Osum valleys with alternating forested mountains and cultivated depressions; a region of limestone mountains and valleys south of the Osum valley; and a coastal plain which is up to 60 km wide and traversed by the Rivers Vijosë, Seman, Shkumbin, Drin, and Erzen.

Fifteen Important Bird Areas (IBAs) have been identified in Albania, covering a total surface area of 903 km² (Table 1, Map 1). This compares to 13 IBAs identified in the previous pan-European IBA inventory (Grimmett and Jones 1989), all of which remain IBAs (1989 site AL011 having been joined with 1989 site AL005 in this inventory), and a further three new sites have been identified (Table 1). Site 011 ('Marash mountains'), listed as AL012 in the previous European inventory, has also been included in this inventory but, despite investigation, it could not be located or, therefore, updated.

Table 1. Summary of Important Bird Areas in Albania. 15 IBAs covering 903 km²

IBA code	1989 code	International name	National name	Administrative region	Area (ha)	Criteria (see p.11)
001	AL001	Lake Shkodra (Lake Scadar)	Liqeni i Shkodres	Shkodër	14,000	A1, A4i, A4iii, B1i
002	AL002	Lake Ohrid	Liqeni i Ohrit	Pogradec	9,700	A1, A4i, A4iii, B1i
003	AL003	Lake Megali Prespa	Liqeni i Prespes se Madhe	Korçë	10,000	A1, A4i, B1i
004	AL004	Lake Mikri Prespa	Liqeni i Prespes se Vogel	Korçë	800	A1
005	AL005, AL011	Narta lagoon	Laguna e Nartes	Vlorë	4,180	A1, A4i, A4iii, B1i
006	AL006	Karavasta lagoon	Laguna e Karavastase	Fier, Lushnjë	5,450	A1, A4i, A4iii, B1i, B2
007	AL007	Drini delta	Delta e Drinit	Lezhë, Shkodër	2,188	A1, A4i, B1i
008	AL008	Thethi	Thethi	Shkodër	2,300	B2
009	AL009	Lura	Lura	Peshkopi	1,280	B2
010	AL005, AL010	Vlora Bay, Karaburun peninsula and Cika mountain	Vlora Bay, Karaburun, Mali i Cikes	Vlorë	35,000	B2
011	AL012	'Marash mountains'	'Mali i Marashit'	—	—	B2
012	AL013	Lake Butrinti	Liqeni i Butrintit	Sarandë	1,900	B2
013	—	Velipoja	Velipoja	Shkodër	1,500	A1
014	—	Patoku lagoon	Laguna e Patokut	Kruje	1,211	A1, A4i, B1i
015	—	Lalzi Bay	Gjiri i Lalzit	Durresi	800	B1i

Map 1. Location, area and criteria category of Important Bird Areas in Albania. Site 011 is not plotted (location unknown).

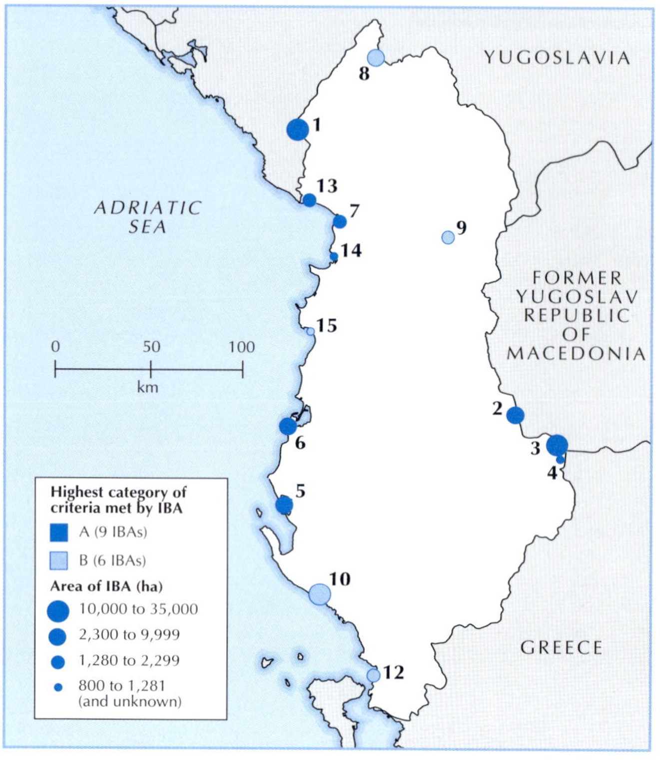

ORNITHOLOGICAL IMPORTANCE

Much of the ornithological information presented in this chapter is taken from Zekhuis and Tempelman (1996) and Kayser *et al.* (1997).

Seven IBAs are important for significant numbers of globally threatened species (Table 2). In the 1950s and 1960s Albania was known to be of world importance for breeding *Phalacrocorax pygmeus* and *Pelecanus crispus*. The latter species was thought to have become extinct as an Albanian breeder by the early 1980s, but a population of at least 11 pairs was located in 1985. The present status of *Phalacrocorax pygmeus* is unclear, with a large colony suspected at Lake Shkodra (001). However, significant declines are thought to have occurred as a result of wetland modification and destruction.

Eight sites are important for internationally important numbers of wintering waterbirds (Table 4). One hundred and thirty-five species of European conservation concern breed in Albania (Tucker and Heath 1994). However, little data are available on these species at IBAs, hence only a small number of species are listed in Table 4.

On current knowledge of the species breeding across the border in Yugoslavia and Greece, it is reasonable to assume that Albania still supports significant populations of *Neophron percnopterus*, *Gyps fulvus*, *Falco naumanni*, *F. biarmicus*, *Tetrao urogallus*, *Alectoris graeca* and *Dendrocopos leucotos*.

Table 2. Important Bird Areas in Albania that are important for species of global conservation concern (meeting criterion A1).

Species	IBA code
Phalacrocorax pygmeus Pygmy Cormorant	001, 002, 003, 004, 006, 007, 013
Pelecanus crispus Dalmatian Pelican	005, 006
Oxyura leucocephala White-headed Duck	005
Aquila clanga Greater Spotted Eagle	006, 013
Numenius tenuirostris Slender-billed Curlew	014

Table 3. Important Bird Areas in Albania that support important numbers of one or more congregatory species (i.e. meeting criteria A4 and/or B1). IBAs meeting both criteria A4 and B1 for the species are shown in **bold**. IBAs meeting only criterion B1 for the species concerned, and not A4, are shown in normal type. For key to 'Season', see p. 7.

Species	Season	IBA code
Podiceps nigricollis Black-necked Grebe	W	**002**
Phalacrocorax carbo Cormorant	W	001
	P	006, 007
Phalacrocorax pygmeus Pygmy Cormorant	W	**001**, **002**, **003**, **007**
	P	**006**, **007**
Pelecanus crispus Dalmatian Pelican	R	**006**
	W	**005**, **006**
Egretta alba Great White Egret	P	**006**
Anas penelope Wigeon	W	**005**, **006**, 015
Anas crecca Teal	W	006
Anas clypeata Shoveler	W	**006**
Netta rufina Red-crested Pochard	W	**002**
Bucephala clangula Goldeneye	W	002, 005, 007
Fulica atra Coot	W	**002**, **005**
Recurvirostra avosetta Avocet	W	**006**, **014**
Glareola pratincola Collared Pratincole	B	006

Table 4. Species of European conservation concern with significant breeding populations at Important Bird Areas in Albania (meeting criteria B2/B3).

Species [1]	Minimum national breeding population (pairs) [2]	Proportion (%) of national population breeding at all IBAs in Albania
Pelecanus crispus Dalmatian Pelican	40	28
Gyps fulvus Griffon Vulture	50	2
Aquila chrysaetos Golden Eagle	50	4
Glareola pratincola Collared Pratincole	100	50
Sterna albifrons Little Tern	500	15

1. Only those species of European conservation concern (see Box 1, p. 12) that meet B2/B3 criteria at IBAs in Albania are listed.
2. Data are taken from the BirdLife/EBCC European Bird Database 1998 (Heath and Borggreve 2000).

HABITATS

Albania holds four very large lakes, all of which are IBAs, shared with the Federal Republic of Yugoslavia (Montenegro), the FYR of Macedonia and/or Greece, namely: Lake Shkodra (001), Lake Ohrid (002), Lake Megali Prespa (003) and Lake Mikri Prespa (004). Several deltas and lagoons along the coast of Albania are IBAs, including Narta lagoon (005), Karavasta lagoon (006), Drini delta (007), Vlora Bay (010), Velipoja (013), Patoku lagoon (014) and Lalzi Bay (015). Overall 87% of IBAs in Albania hold some wetland habitat (Figure 1). The coastal sites also encompass some marine areas, with 6 IBAs containing some marine habitat.

The flora in Albania is rich and diverse, reflecting a varied topography. Despite intensification of agriculture and a rapidly growing human population, there are still some extensive tracts of relatively undisturbed forested/mountainous terrain, and 11 IBAs hold some forest (Figure 1).

IMPACTS ON IBAs – LAND-USE AND THREATS

Hunting, agriculture, recreation, fisheries and forestry are the most common types of land-use within IBAs in Albania, all present in more than 60% of sites (Figure 2).

Figure 1. Habitats at Important Bird Areas in Albania (see Appendix 3 for definitions of habitats).

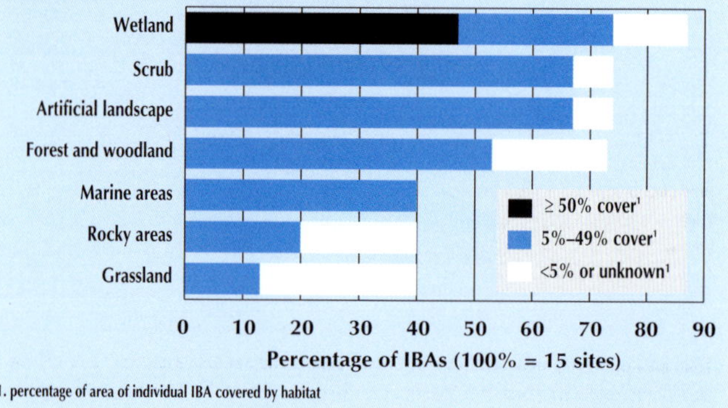

1. percentage of area of individual IBA covered by habitat

Figure 2. Land-uses at Important Bird Areas in Albania (see Appendix 3 for definitions of land-uses).

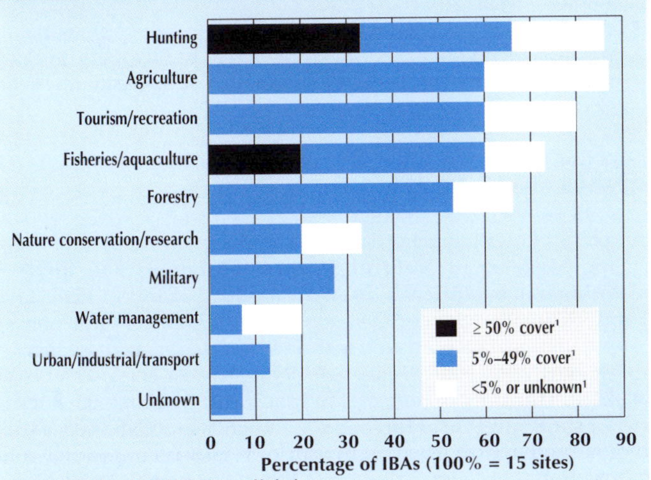

1. percentage of area of individual IBA covered by land-use

These human activities also threaten several sites (Figure 3). Disturbance to birds is a common threat affecting over 85% of IBAs, and caused by a range of factors including farmers, fishermen, gatherers, tourists and grazing animals. Many forests within IBAs are being degraded and destroyed, mainly as a consequence of commercial deforestation, but also for the collection of firewood. The littoral zone formerly held large marshes and lagoons, but these have been drained in a post-war drive towards increased agricultural productivity and the eradication of malaria. Between 1946 and 1974, 60,000 ha were drained and a further 170,000 ha were 'improved' for agriculture. Breeding sites for *Phalacrocorax pygmeus* and *Pelecanus crispus* are known to have been drained and many of these areas are now being used intensively, mainly for cattle-grazing (IBAs 001, 002, 003). However, socio-economic reforms have also resulted in the creation of artificial wetlands such as reservoirs and canals, although these are thought to be not as important for birds. Hunting and poaching are also a threat, affecting over 50% of IBAs.

PROTECTION STATUS

■ National protection

The National Environmental Agency (NEA)—the former Committee of Environmental Protection—operates as the only state body, answerable to the Council of Ministers, dealing with the environment, based on the legal framework approved by the Law on Environmental Protection in 1993. NEA implements and develops governmental policy in the field of environmental protection.

The development of a modern legislative system for environmental protection only started after 1991 and is still far from complete. One aim of the Agency is to complete and improve the legislation concerning nature protection. With the help of legal experts (provided by the PHARE programme of the European Commission) it is working on two principal laws, one for Protected Areas and one for the Protection of Nature and Biodiversity.

It is understood that six protected area categories exist in Albania (Z. Dedej *in litt.*), however only four have been recorded in relation to IBAs: (1) **National Parks** (strictly protected with grazing, forestry and hunting all prohibited), (2) **Natural Reserves** (created for the management of one or more species), (3) **Game Reserves** and, (4) **Protected Landscape Areas**. Overall 10 IBAs are at least partly protected by one or more of these protected-area designations (Table 5, Figure 4).

Figure 3. Threats at Important Bird Areas in Albania (see Appendix 3 for definitions of threat types and impact categories).

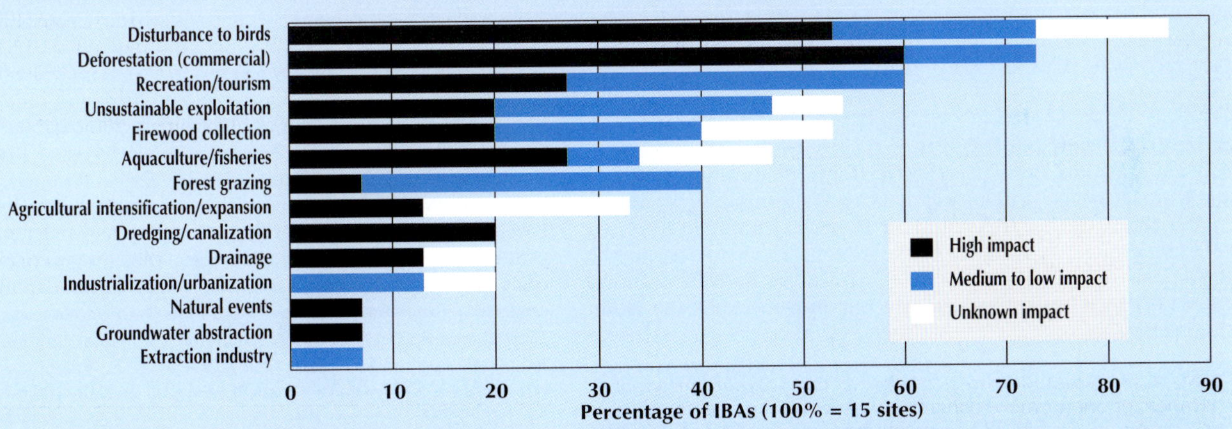

Figure 4. The national protection status of Important Bird Areas in Albania.

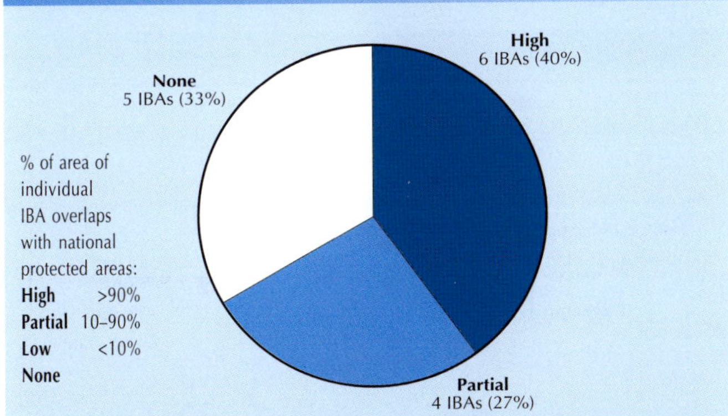

Total area of overlap between IBA network in Albania and national protected-area system (see Table 5 for categories) = 272–284 km² (30–31% of total IBA area).

Figure 5. The international protection status of Important Bird Areas in Albania.

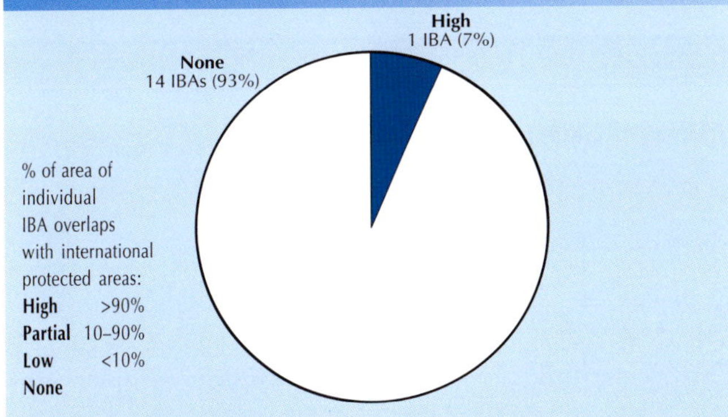

Total area of overlap between IBA network in Albania and international protected-area system (see Table 5 for categories) = 55 km² (6% of total IBA area).

Table 5. Protection status of Important Bird Areas in Albania. A tick (✔) indicates that an IBA overlaps with a protected area (to any extent).

IBA code	International name	National: National Park	National: Natural Reserve	National: Game Reserve	National: Protected Landscape Area	International: Ramsar Site
001	Lake Shkodra (Lake Scadar)					
002	Lake Ohrid				✔	
003	Lake Megali Prespa	✔				
004	Lake Mikri Prespa	✔				
005	Narta lagoon					
006	Karavasta lagoon	✔		✔		✔
007	Drini delta		✔			
008	Thethi	✔				
009	Lura	✔				
010	Vlora Bay, Karaburun peninsula and Cika mountain					
011	'Marash mountains'					
012	Lake Butrinti			✔		
013	Velipoja			✔		
014	Patoku lagoon			✔		
015	Lalzi Bay					
Total number of IBAs		5	1	4	1	1

Box 1. International legislation and initiatives that are relevant to site conservation in Albania (see Appendix 1 for a general description of these agreements).

Global	
Biodiversity Convention	✔
Ramsar Convention	✔
Bonn Convention	
World Heritage Convention	✔
MAB Programme	
Pan-European	
Bern Convention	✔
Regional	
Barcelona Convention	✔

✔ Convention ratified/initiative supported
(✔) Convention signed

International protection

Albania is a contracting party to the Ramsar Convention (Box 1) and Karavasta lagoon (006) is designated as a Ramsar Site (Table 5, Figure 5). Six further IBAs fulfil the Ramsar criteria but have not been designated as Ramsar Sites. Albania is also party to the Barcelona Convention and in 1998 ratified the Bern Convention (Box 1). Procedures are currently ongoing to ratify the Bonn Convention.

CONSERVATION

- A project, financed by the World Bank with $1.7 million, started in 1998 at Lake Prespa (003, 004). This aims to create a laboratory near the lake in order to monitor and further study the lake, and to involve local people in monitoring and managing the watershed.
- At Narta lagoon (005) a management plan will be prepared as part of a Global Environment Facility biodiversity project, coordinated by the Station Biologique de la Tour du Valat (France).

ANALYTICAL METHODS

- The Station Biologique de la Tour du Valat, in cooperation with the Museum of Natural Sciences (Albania) and PPNEA (Albania), organized midwinter waterfowl counts in 1994–1996 in Albanian wetlands (Kayser *et al.* 1997). Much of the wintering population data presented in this inventory is from these surveys (Kayser *et al.* 1995; Bino *et al.* 1996a; Kayser *et al.*1997).
- LIPU (BirdLife Partner in Italy) and ASBPM (Albanian Society for the Protection of Birds and Mammals), in collaboration with the Museum of Natural History of Tirana, organized four ornithological expeditions in April 1995, January 1996, April 1996 and May 1996, with the participation of LIPU, BTO (British Trust for Ornithology) and RSPB (Royal Society for the Protection of Birds, UK) staff members, visiting the coastal wetlands in winter and spring and also some inland mountain areas. Much of the data presented in this inventory on mountain species and migratory waterbirds comes from these surveys (Bashford and Wilson 1996; Bego *et al.* 1995; Bino *et al.* 1996b; Casale 1997).
- In June 1996 an ornithological expedition, under the auspices of the Foundation International Working Group Waterbird and Wetland Research (WIWO, The Netherlands), visited all the main wetlands of Albania (Zekhuis and Tempelman 1996). Much of the waterbird breeding population data presented in this inventory are from this survey.
- IBA 011 (believed to be called 'Marash mountains'), described in the previous pan-European inventory, has also been included in this inventory but its location remains unknown.

ACKNOWLEDGEMENTS

We would like to thank Barbara Lombatti (LIPU staff member), Prof. Ferdinand Bego (Museum of Natural History of Tirana) and the Station Biologique de la Tour du Valat, who participated in the organization of the LIPU-ASPBM ornithological expeditions and their participants: Richard Bashford and Andrew Wilson (BTO, UK), Marcus Brew (UK), Simon Busuttil (RSPB, UK), Jaqueline Crozier and Brian J. Dore (Andorra), Lorenzo Dotti (Italy), Fokko Erhart (The Netherlands), Giorgio Grazzini (Italy), Ilir Kagca (Albania), Ardian Laci (Albania), Piero Papa (Italy), John Richardson (UK), Franco Roscelli (Italy), Ruta Vaiciunaite (Lithuania). We are thankful to Nikolai Petkov and K. Ruskov (BSPB, Bulgaria), F. Willems and E. De Vries (University of Nijmegen, The Netherlands) who gave useful information. Also we thank Tobias Salathé of the Station Biologique de la Tour du Valat (and latterly the Ramsar Bureau) for providing helpful comments on the chapter, and Zamir Dedej (Nature Protection Directory in Albania) and Clare Shine for providing information on protection and conservation issues.

SITE ACCOUNTS

Lake Shkodra (Lake Scadar) A1, A4i, A4iii, B1i **001**

Admin region Shkodër
Coordinates 42°10'N 19°20'E
Altitude 0–10 m **Area** 14,000 ha

Site description

A large freshwater lake in north-east Albania (also known as Lake Scutari), on the border with Yugoslavia (Montenegro). Oligotrophic and generally shallow except for a few very deep parts. Approximately one third of the lake lies within Albania, the remainder lies in Montenegro (Skadarsko Jezero, site 038). The main human activities are fisheries, extensive agriculture, hunting and tourism. Fisheries are the most important activity and in places are very intensive.

Habitats Grassland, Wetland (standing fresh water; river/stream; water-fringe vegetation), Artificial landscape (arable land; ruderal land)
Land-use Agriculture, Fisheries/aquaculture, Hunting, Tourism/recreation

Birds

Species	Season	Year	Pop min	Pop max	Acc	Criteria
Phalacrocorax carbo Cormorant	W	1996	97	1,526	A	B1i
[1] ***Phalacrocorax pygmeus*** Pygmy Cormorant	W	1996	596	1,073	A	A1, A4i, B1i

Species ... continued	Season	Year	Pop min	Pop max	Acc	Criteria
[2] ***Phalacrocorax pygmeus*** Pygmy Cormorant	N	1995	150	150	B	A1

1. Also a potential breeding site. >200 breeding pairs on Montenegran side of the Lake.
2. April 1995 record.

This is an important site for wintering waterfowl, especially *Phalacrocorax pygmeus* and *Phalacrocorax carbo*. Several wintering waterfowl counts have been carried out; about 13,000 wintering waterfowl were counted in January 1995 and 1996, and 30,682 waterbirds (representing 26 species) were counted during 2–19 January 1997. Wetlands International have carried out earlier counts on the Montenegro side of the lake.

Protection status

National None **International** None

Conservation issues

Threats Agricultural intensification/expansion (A), Aquaculture/fisheries (U), Disturbance to birds (U), Drainage (U), Unsustainable exploitation (A)

The main threats are land claim for agricultural purposes, and hunting. The lake shores in the Albanian sector have been heavily claimed and have lost most of their original natural value, now being used intensively for agriculture, mainly cattle-grazing, which is also a cause of disturbance. Hunting is also a serious problem with 150 shooting hides recorded along 3 km of shoreline in 1993, and the number increased in 1995 and 1996.

Lake Ohrid — A1, A4i, A4iii, B1i — 002

Admin region Pogradec
Coordinates 41°00′N 20°45′E
Altitude 750–750 m **Area** 9,700 ha

Site description

A large freshwater lake shared with the Former Yugoslav Republic of Macedonia (site 005); its total area is 34,800 ha. The lake has deep waters and stony shores with no reedbeds or other emergent vegetation. On the Albanian side all the surrounding area is cultivated, mainly with corn and wheat. The lake is used for many purposes, of which fisheries are the most important.

Habitats Forest and woodland, Scrub (5%; heathland), Wetland (85%; standing fresh water), Rocky areas, Artificial landscape (10%; arable land; forestry plantation)
Land-use Agriculture (5%), Fisheries/aquaculture (85%), Forestry (5%), Hunting (10%), Tourism/recreation (5%), Urban/industrial/transport (5%)

Birds

Species	Season	Year	Pop min	Pop max	Acc	Criteria
Podiceps nigricollis Black-necked Grebe	W	1997	991	–	–	A4i, B1i
[1] ***Phalacrocorax pygmeus*** Pygmy Cormorant	W	1996	182	215	A	A1, A4i, B1i
Netta rufina Red-crested Pochard	W	1996	726	1,405	A	A4i, B1i
Bucephala clangula Goldeneye	W	1997	235	–	–	B1i
Fulica atra Coot	W	1996	26,633	40,765	A	A4i, B1i

1. 102 individuals 1997.

This is an important site for wintering waterfowl, the most numerous species being *Fulica atra*. Counts were carried out in 1995 and 1996 (numbers given refer to waterfowl numbers for the Albanian part of the lake only); 31,000 individuals (19 species) in 1995, 46,000 individuals (17 species) in January 1996 and 28,018 individuals (17 species) in 1997.

Protection status

National High **International** None
9,700 ha of IBA covered by Protected Landscape Area.

Conservation issues

Threats Aquaculture/fisheries (A), Deforestation (commercial) (A), Disturbance to birds (B), Extraction industry (C), Industrialization/urbanization (B), Recreation/tourism (B), Unsustainable exploitation (C)

The shores of Lake Ohrid are degraded, mainly as a result of cattle-grazing and the close proximity of a road. These factors mean that breeding possibilities for waterbirds are minimal along the whole shore. The lake is a proposed World Heritage Site. With an age of 2–4 million years, this is one of the most ancient lakes in the world, harbouring many endemic species among its aquatic fauna.

Lake Megali Prespa — A1, A4i, B1i — 003

Admin region Korçë
Coordinates 40°53′N 20°57′E
Altitude 950–950 m **Area** 10,000 ha

Site description

A large freshwater lake shared with the Former Yugoslav Republic of Macedonia (site 006) and Greece (site 047). The total area of the lake is 27,000 ha. Up until the recent past the lake used to form one lake with Lake Mikri Prespa (IBA 004), but is now separated and 3 m lower in altitude. The Albanian sector of the lake has no marshes or reedbeds. All the surrounding areas are cultivated.

Habitats Scrub (10%; scrub; heathland), Wetland (85%; standing fresh water; water-fringe vegetation), Rocky areas (5%; scree/boulders; caves)
Land-use Fisheries/aquaculture (85%), Hunting (5%), Unknown (10%)

Birds

Species	Season	Year	Pop min	Pop max	Acc	Criteria
[1] ***Phalacrocorax pygmeus*** Pygmy Cormorant	W	1996	462	462	A	A1, A4i, B1i

1. January 1996 data only.

The lake is of high importance for wintering waterbirds. A total of 12,000 individuals were counted in January 1996, including *Phalacrocorax pygmeus*, 171 *Podiceps nigricollis*, 1,268 ducks of six species and over 10,000 *Fulica atra*. The importance of the site for migrating birds is not known. Because of lack of shore vegetation, and disturbance, the Albanian part of the lake is of minor importance for breeding waterbirds but it is part of the foraging area of several species breeding in Macedonian and Greek parts of the lake, including *Pelecanus crispus*, *Pelecanus onocrotalus* and *Phalacrocorax pygmeus*.

Protection status

National High **International** None
10,000 ha of IBA covered by National Park.

Conservation issues

Threats Agricultural intensification/expansion, Aquaculture/fisheries (A), Deforestation (commercial) (A), Disturbance to birds (U), Natural events (A), Unsustainable exploitation (U)

As a result of human activities, the area has lost much of its ecological value, especially concerning waterbirds. Many of the slopes surrounding the lake and the greater part of the shores have lost or are losing vegetation, mostly as a result of cattle-grazing; erosion is a significant problem that will increase in the near future. The impact of disturbance by fishermen and hunting is unknown. A project, financed by the World Bank with $1.7 million, started in 1998. This aims to create a laboratory near the lake in order to monitor and further study the lake, and to involve local people in monitoring and managing the watershed.

Lake Mikri Prespa — A1 — 004

Admin region Korçë
Coordinates 40°50′N 21°05′E
Altitude 853–900 m **Area** 800 ha

Site description

The western end of a mesotrophic, shallow, natural freshwater lake, shared with Greece (site 047). It is linked to the north with the much larger Lake Megali Prespa and is fed by seasonal streams. The main surface of the Albanian part of the lake is covered by reedbeds and surrounded by rocky scrub. In winter it freezes almost completely. The area is used for many purposes including grazing by large numbers of cattle.

Habitats Scrub (10%), Wetland (80%; standing fresh water), Rocky areas (caves), Artificial landscape (10%; arable land)
Land-use Agriculture, Fisheries/aquaculture, Hunting, Water management

Birds

Species	Season	Year	Pop min	Pop max	Acc	Criteria
Phalacrocorax pygmeus Pygmy Cormorant	W	1997	—	81	A	A1

The lake supports very important breeding populations of *Pelecanus crispus* and *Phalacrocorax pygmeus*—although the breeding sites are in the Greek section of the lake, the birds also forage within the Albanian section. In winter the Albanian part of the lake is important for *Phalacrocorax pygmeus*, but in general for wintering waterbirds the area is less important as it is covered by ice for several months. Little data are available on breeding birds but *Chlidonias hybridus* (30–50 pairs) and *Ixobrychus minutus* (2–20 pairs) are known to breed.

Protection status

National High **International** None
800 ha of IBA covered by National Park.

Conservation issues

Threats Agricultural intensification/expansion (U), Unsustainable exploitation (A)

The main threat to the site is the unsustainable use of water resources: water is extracted to irrigate the plains between Korça and Pogradeci, causing fluctuations and a general reduction of the water-levels and siltation. The Greek side of the lake is protected as a National Park.

Narta lagoon — A1, A4i, A4iii, B1i — 005

Admin region Vlorë
Coordinates 40°35′N 19°23′E
Altitude 0–5 m **Area** 4,180 ha

Site description

Narta is one of the largest and most important coastal wetlands of Albania. The site comprises a large, shallow and brackish lake (Lake Nartes) and a number of coastal lagoons between the Vjose river delta and the city of Vlore. Large parts of the area have been converted into agricultural land or saltpans. The lake surface decreases by 30% (especially when the connection with the sea is blocked) during the summer and large mudflats are exposed which support saltmarsh. Coastal habitats include dunes, beaches and rocky shores, with *Pinus pinea* and *Pinus halepensis*. Surrounding habitats consist of olive-groves, pine plantations and scattered poplar plantations. The site includes the small Pishe Poro lagoon (former (1989) IBA AL011). Previously, the Narta lagoon IBA also included the Karaburun peninsula and Sazan island, but these are now included in site 010.

Habitats Forest and woodland (10%; broadleaved deciduous forest), Grassland (20%; humid grassland), Wetland (60%; mudflat/sandflat; saltmarsh; sand-dunes/sand beach; coastal lagoon; standing brackish and salt water; river/stream; water-fringe vegetation), Marine areas (3%; sea inlet/coastal features), Rocky areas (2%; sea cliff/rocky shore), Artificial landscape (5%; arable land; forestry plantation; ruderal land)
Land-use Agriculture (10%), Fisheries/aquaculture (40%), Forestry (10%), Hunting (100%), Military (5%), Tourism/recreation (5%), Water management (20%)

Birds

Species	Season	Year	Pop min	Pop max	Acc	Criteria
Pelecanus crispus Dalmatian Pelican	W	1996	8	28	A	A1, A4i, B1i
Anas penelope Wigeon	W	1997	—	22,000	A	A4i, B1i
Bucephala clangula Goldeneye	W	1996	188	382	A	B1i
Oxyura leucocephala White-headed Duck	W	1993	—	4	A	A1
Fulica atra Coot	W	1997	—	30,256	A	A4i, B1i

An important site for wintering and migratory waterfowl, especially *Pelecanus crispus*; 19,663 waterfowl were counted in January 1996, and 81,223 individuals counted in 1997. No waterbird breeding colonies seem to occur around the lagoon (1996). It is considered as a potentially very favourable site for the globally threatened *Numenius tenuirostris*.

Protection status

National None **International** None

Conservation issues

Threats Agricultural intensification/expansion, Aquaculture/fisheries, Deforestation (commercial) (A), Disturbance to birds (A), Dredging/canalization (A), Groundwater abstraction (A), Industrialization/urbanization, Recreation/tourism (B), Unsustainable exploitation (B)

The main threats are water pollution from a hydrochloric acid factory located on the north-east shore, urban expansion, deforestation, overgrazing, disturbance to birds, dynamite fishing, hunting, tourism and land claim for enlarging the saltpans. A management plan will be prepared as part of a GEF Biodiversity project, coordinated by the Station Biologique de la Tour du Valat.

Karavasta lagoon — A1, A4i, A4iii, B1i, B2 — 006

Admin region Fier, Lushnjë
Coordinates 40°53′N 19°25′E
Altitude 0–5 m **Area** 5,450 ha

Site description

Karavasta is the largest wetland in Albania, with the Karavasta lagoon forming the largest part of the complex. The lagoon is shielded from the sea by a sandy bar (Divjaka) covered with pine forest dominated by *Pinus pinaster* and *P. pinea*. Karavasta comprises a shallow inner lagoon, and a smaller outer lagoon. The inner lagoon has many peninsulas and small, low islands with muddy shores and some areas of bare sand. A sandy bar covered by pines separates the two lagoons. In the inner lagoon only artisanal fishery is practised, but in the outer one there are more intensive fisheries.

Habitats Forest and woodland (13%; broadleaved deciduous forest; native coniferous forest; mixed forest; alluvial/very wet forest), Scrub (5%; scrub), Wetland (67%; mudflat/sandflat; saltmarsh; sand-dunes/sand beach; coastal lagoon; standing brackish and salt water; river/stream; water-fringe vegetation), Marine areas (10%; open sea), Artificial landscape (5%; arable land; other urban/industrial areas)
Land-use Agriculture (5%), Fisheries/aquaculture (85%), Forestry (15%), Hunting (35%), Nature conservation/research (15%), Tourism/recreation (10%)

Birds

Species	Season	Year	Pop min	Pop max	Acc	Criteria
Phalacrocorax carbo Cormorant	P	1996	1,800	1,800	A	B1i
Phalacrocorax pygmeus Pygmy Cormorant	P	1996	400	400	A	A1, A4i, B1i
Pelecanus crispus Dalmatian Pelican	R	1996	11	62	A	A1, A4i, B1i
Pelecanus crispus Dalmatian Pelican	W	1996	139	171	A	A1, A4i, B1i
Egretta alba Great White Egret	P	1992	220	220	A	A4i, B1i
Anas penelope Wigeon	W	1997	—	24,643	A	A4i, B1i
Anas crecca Teal	W	1996	5,241	9,720	A	B1i
Anas clypeata Shoveler	W	1997	—	5,162	—	A4i, B1i
Aquila clanga Greater Spotted Eagle	W	1996	—	6	A	A1
Aquila clanga Greater Spotted Eagle	N	1992	3	3	A	A1
Recurvirostra avosetta Avocet	W	1996	1,079	1,763	A	A4i, B1i
Glareola pratincola Collared Pratincole	B	1996	50	130	A	B1i, B2
Sterna albifrons Little Tern	B	1993	75	100	B	B2

The site is important for wintering waterbirds (45,000 individuals in 1996; 68,171 individuals in 1997) and for breeding colonial waterbirds, including *Pelecanus crispus*. The site is also potentially very important for the globally threatened *Numenius tenuirostris*.

Protection status

National Partial **International** High
1,250 ha of IBA covered by National Park (Divjaka). 4,200 ha of IBA covered by Game Reserve. 5,450 ha of IBA covered by Ramsar Site (Karavasta Lagoon; 20,000 ha).

Conservation issues

Threats Deforestation (commercial) (B), Disturbance to birds (B), Firewood collection (C), Forest grazing (C), Recreation/tourism (B)

A management plan has been developed. The area is threatened by plans to build a new recreation complex and by plans to drain parts of the lagoon (1996). Disturbance is high from tourists, local people and illegal fishing.

Drini delta — A1, A4i, B1i — 007

Admin region Lezhë, Shkodër
Coordinates 41°47′N 19°37′E
Altitude 0–5 m **Area** 2,188 ha

Site description

Drini delta is a complex of relatively intact coastal lakes, marshes and forests. The site is divided in two parts: Kune is the northern part of the Drini river and Ceka the southern part. The north and north-east borders of this site are formed by the Barbalush mountains, the south by the Drini river. The main habitats are brackish lagoons, sandbars and beaches, marshes, reedbeds and scattered woodland areas and pine plantations. The surrounding area is cultivated and has a large system of dykes, dams, ditches and channels.

Habitats Forest and woodland (15%; broadleaved deciduous forest; mixed forest), Scrub (5%; scrub; sclerophyllous scrub/garrigue/maquis), Wetland (65%; mudflat/sandflat; saltmarsh; sand-dunes/sand beach; coastal lagoon; standing fresh water; standing brackish and salt water; river/stream), Marine areas (5%; open sea), Artificial landscape (10%; arable land)
Land-use Agriculture (10%), Fisheries/aquaculture (45%), Forestry (15%), Hunting (70%), Nature conservation/research (20%), Tourism/recreation (5%), Water management

Birds

Species	Season	Year	Pop min	Pop max	Acc	Criteria
Phalacrocorax carbo Cormorant	P	1992	1,100	1,100	A	B1i
Phalacrocorax pygmeus Pygmy Cormorant	P	1992	650	650	A	A1, A4i, B1i
Phalacrocorax pygmeus Pygmy Cormorant	W	1996	120	357	A	A1, A4i, B1i
Bucephala clangula Goldeneye	W	1996	124	431	A	B1i

The area is important for several waterbirds; winter waterbird censuses recorded 17,000 individuals in January 1995, and 9,000 individuals in January 1996.

Protection status

National Partial **International** None
The IBA includes parts of four Natural Reserves, totalling 5,500 ha.

Conservation issues

Threats Aquaculture/fisheries (A), Deforestation (commercial) (A), Disturbance to birds (A), Drainage (A), Dredging/canalization (A), Firewood collection (C), Forest grazing (C), Industrialization/urbanization (B), Recreation/tourism (A), Unsustainable exploitation (A)

Harvesting of wood has destroyed a large part of the broadleaved forest and older pine plantations. Birds in parts of the area are disturbed by tourists. There has been an EU-funded project to start management planning, including a workshop.

Thethi — B2 — 008

Admin region Shkodër
Coordinates 42°30′N 19°45′E
Altitude 725–2,570 m **Area** 2,300 ha

Site description

A mountainous area with *Fagus* and *Abies alba* forests, thickets of *Pinus heldreichii* and high-altitude alpine pastures.

Habitats Forest and woodland (broadleaved deciduous forest), Scrub, Grassland (alpine/subalpine/boreal grassland), Rocky areas (scree/boulders; inland cliff)
Land-use Agriculture, Forestry, Nature conservation/research, Tourism/recreation

Birds

Species	Season	Year	Pop min	Pop max	Acc	Criteria
Gyps fulvus Griffon Vulture	R	—	—	—	—	B2

This is an important site for large raptors and *Tetrao urogallus* (1989 data only, no recent data available).

Protection status

National High **International** None
2,300 ha IBA covered by National Park (Thethi, 2,300 ha).

Conservation issues

Threats Deforestation (commercial) (A), Disturbance to birds (A), Firewood collection (U)

Lura — B2 — 009

Admin region Peshkopi
Coordinates 41°45′N 20°15′E
Altitude 1,375–2,121 m **Area** 1,280 ha

Site description

A mountainous area with *Fagus* and *Abies alba* forests, lakes and bogs with *Pinus mugo* and areas of *Taxus baccata*.

Habitats Forest and woodland (broadleaved deciduous forest; native coniferous forest; treeline ecotone), Grassland (alpine/subalpine/boreal grassland), Wetland (standing fresh water; river/stream; fen/transition mire/spring)
Land-use Agriculture, Forestry, Hunting, Nature conservation/research, Tourism/recreation

Birds

Species	Season	Year	Pop min	Pop max	Acc	Criteria
Aquila chrysaetos Golden Eagle	R	—	—	—	—	B2

This is an important site for breeding *Aquila chrysaetos*, and also *Tetrao urogallus* and *Alectoris graeca* (1989 data only, no recent data available).

Protection status

National High **International** None
1,280 ha of IBA covered by National Park (Lura, 1,280 ha).

Conservation issues

Threats Deforestation (commercial) (A), Disturbance to birds (A), Firewood collection (U)

The main threat is deforestation by clear-felling.

Vlora Bay, Karaburun peninsula and Cika mountain — B2 — 010

Admin region Vlorë
Coordinates 40°10′N 19°40′E
Altitude 475–2,018 m **Area** 35,000 ha

Site description

This site includes Vlora Bay, Karaburun peninsula, Orikumi lagoon, Sazan island and Cika mountain. Vlora Bay includes the sea gulf of Vlora and part of the coastal area with rocky shores and cliffs, sandy beaches and a small lagoon (Orikumi). Karaburun peninsula and Sazan island are rocky areas with scattered Mediterranean scrub, and are military areas. Cika mountain (previously Llogara IBA AL010) is a mountain area with forests of *Pinus austriaca* and flora typical of calcareous soil.

Habitats Forest and woodland (20%; broadleaved deciduous forest; native coniferous forest; mixed forest), Scrub (35%; sclerophyllous scrub/garrigue/maquis), Grassland (5%; alpine/subalpine/boreal grassland), Wetland (5%; coastal lagoon), Marine areas (40%; open sea; sea inlet/coastal features), Rocky areas (5%; sea cliff/rocky shore; rock stacks/islets; scree/boulders; inland cliff; caves), Artificial landscape (5%; arable land; urban parks/gardens)
Land-use Agriculture (5%), Forestry (20%), Hunting (30%), Military (35%), Nature conservation/research (5%), Tourism/recreation (5%)

Birds

Species	Season	Year	Pop min	Pop max	Acc	Criteria
Aquila chrysaetos Golden Eagle	R	1989	—	4	—	B2

This site is believed to be important for breeding raptors. *Neophron percnopterus*, *Circaetus gallicus*, *Aquila chrysaetos* and *Hieraaetus fasciatus* were observed in 1991 and 1996, but breeding numbers are unknown.

Protection status

National None **International** None

Conservation issues

Threats Deforestation (commercial) (B), Disturbance to birds (B), Firewood collection (C), Forest grazing (C), Recreation/tourism (B), Unsustainable exploitation (C)

'Marash mountains' B2 011

Admin region —
Coordinates —
Altitude 1,000–1,500 m **Area** —

Site description

This site was included in the previous European inventory (Grimmett and Jones 1989) as site AL012 ('Marshabit'; 42°05′N 20°07′E; c.1,000 ha; Curri district) but despite substantial investigation the site cannot be located on the ground. As a precautionary measure it remains in this inventory. The text from 1989 reads as follows: "a mountainous area with heterogeneous forests containing five major tree species: *Pinus austrica*, *Abies alba*, *Fagus* sp., *Tilia cordata*, and *Castanea sativa*. Although the area is unprotected there is no forestry, and the structure of the woods is typical of old mountainous forests."

Habitats Forest and woodland (broadleaved deciduous forest)
Land-use Unknown

Birds

"In May 1986, at an altitude of 1,100 m, six different species of raptors were observed: *Buteo rufinus*, *Aquila chrysaetos*, *Hieraaetus fasciatus*, *Falco tinnunculus*, *F. subbuteo*, and *F. peregrinus*. Among the birds observed in May 1987, six different species of woodpeckers were noted including *Dryocopus martius* and *Dendrocopos medius*. *Bonasa bonasia* is killed every year by local hunters and *Tetrao urogallus* is present."

Protection status

National None **International** None

Conservation issues

Threats Unknown

Lake Butrinti B2 012

Admin region Sarandë
Coordinates 39°50′N 20°00′E
Altitude 0–0 m **Area** 1,900 ha

Site description

A brackish coastal lake in the southern part of Albania, with an outlet to the sea via a channel of about 4 km. The lake (1,630 ha) is used for extensive mussel-farming and hunting, and is surrounded by hills with orchards and scattered maquis. Water-fringe vegetation is poor, apart from an area (50 ha) on the north side that includes *Phragmites*, *Scirpus* and wet grasslands. The channel to the sea has *Salicornia*-covered islands and small coastal marshes. The nearby small lake of Bufi, included within the IBA, is fringed with reedbeds.

Habitats Forest and woodland (10%; broadleaved deciduous forest), Scrub (10%; sclerophyllous scrub/garrigue/maquis), Wetland (45%; shingle/stony beach; coastal lagoon; standing fresh water; standing brackish and salt water; water-fringe vegetation), Marine areas (10%; open sea; sea inlet/coastal features), Rocky areas (15%; scree/boulders; inland cliff), Artificial landscape (5%; arable land)
Land-use Agriculture (10%), Fisheries/aquaculture (35%), Hunting (30%), Tourism/recreation (10%), Urban/industrial/transport (5%), Unknown (10%)

Birds

Given that there are few large natural tectonic lakes with reedbeds in the Balkans, the site is believed to be of importance, especially for birds breeding in reedbeds, and criterion B2 has been applied. However, data are incomplete and numbers currently fall below the IBA thresholds; disturbance may also contribute to the low counts. The site also supports substantial numbers of wintering waterbirds; censuses recorded 7,700 individuals in January 1995, and 8,300 individuals in January 1996.

Protection status

National Partial **International** None
IBA covered by Game Reserve.

Conservation issues

Threats Disturbance to birds (A), Recreation/tourism (A)

Velipoja A1 013

Admin region Shkodër
Coordinates 41°52′N 19°26′E
Altitude 0–50 m **Area** 1,500 ha

Site description

The site includes two coastal areas: the Viluni (or Velipoja lagoon) (390 ha) and surrounding Velipoja Game Reserve (694 ha); and the inland Dumi wetland (Keneta e Dumit) comprising a large reedbed. Viluni lagoon is a large, shallow coastal lagoon at the feet of a rocky and forested mountain area (Bregulbunes mountains). The surrounding marshes and the Dumi wetland are drained by a channel. Sand-dunes, beaches and small brackish pools are found along the coastline, while along the Buna river and its delta (to the north and within the IBA) there is riparian deciduous woodland. The main land-uses are hunting, fishing, agriculture and tourism.

Habitats Forest and woodland (20%; broadleaved deciduous forest), Scrub (15%; scrub; sclerophyllous scrub/garrigue/maquis), Wetland (45%; sand-dunes/sand beach; coastal lagoon; standing fresh water; standing brackish and salt water; river/stream; water-fringe vegetation), Marine areas (10%; sea inlet/coastal features), Artificial landscape (10%; arable land; forestry plantation)
Land-use Agriculture (5%), Fisheries/aquaculture (25%), Forestry (20%), Hunting (60%), Military (20%), Tourism/recreation (5%)

Birds

Species	Season	Year	Pop min	Pop max	Acc	Criteria
Phalacrocorax pygmeus Pygmy Cormorant	W	1996	33	62	A	A1
Phalacrocorax pygmeus Pygmy Cormorant	P	1996	49	49	A	A1
Aquila clanga Greater Spotted Eagle	W	1996	—	3	A	A1

Herons, *Platalea leucorodia* and *Phalacrocorax pygmeus* used to breed inside the reserve, close to the mouth of the Buna river, but they no longer do so, probably due to increased illegal disturbance and wood-cutting inside the reserve. The birds now breed on Ada island, on the Montenegro side of the river, but still use the habitat on the Albanian side for feeding. The site is also important for wintering waterfowl (max. 8,000 individuals in 1993) and for migratory waterbirds.

Protection status

National Partial **International** None
694 ha of IBA covered by Game Reserve (694 ha).

Conservation issues

Threats Aquaculture/fisheries (B), Deforestation (commercial) (A), Disturbance to birds (A), Firewood collection (A), Forest grazing (A), Recreation/tourism (A)

Drainage, poaching, hunting and disturbance to birds are the main threats. Stricter adherence to the legislation within the reserve may make it possible for the breeding colony to re-establish.

Patoku lagoon A1, A4i, B1i 014

Admin region Kruje
Coordinates 41°38′N 19°36′E
Altitude 0–1 m **Area** 1,211 ha

Site description

The site includes two brackish, shallow, coastal lagoons, separated from the sea by two sandbars, with an outlet to the Adriatic Sea in the centre. At low tide several hectares of mudflats are exposed. The site includes the riparian Fushe-Kushe forest with *Alnus glutinosa* and

Populus alba, extending for approximately 50 ha. Adjacent marshes have reedbeds with *Juncus maritimus* and *Typha angustifolia.* Until recent drainage the site encompassed 4,200 ha of wetland. The main land-uses are fishing, agriculture, grazing, hunting (the area is a Game Reserve) and tourism (along the sandbars).

Habitats Forest and woodland (10%; native coniferous forest; alluvial/very wet forest), Scrub (10%), Wetland (60%; mudflat/sandflat; saltmarsh; sand-dunes/sand beach; coastal lagoon; standing brackish and salt water; river/stream; water-fringe vegetation), Artificial landscape (20%; arable land)
Land-use Agriculture (20%), Fisheries/aquaculture (40%), Forestry (10%), Hunting (80%), Tourism/recreation (10%)

Birds

Species	Season	Year	Pop min	Pop max	Acc	Criteria
Recurvirostra avosetta Avocet	W	1996	87	1,034	A	A4i, B1i
[1] ***Numenius tenuirostris*** Slender-billed Curlew	P	1992	—	5	A	A1

1. Observed November 1992.

The area is particularly important for migratory and wintering waders (c.4,000 waterbirds present each year), including the globally endangered *Numenius tenuirostris.* Formerly a colony of 20–30 pairs of *Pelecanus crispus* bred on Cabok island.

Protection status

National High **International** None
1,211 ha of IBA covered by Game Reserve (1,211 ha).

Conservation issues

Threats Agricultural intensification/expansion (A), Aquaculture/fisheries (A), Deforestation (commercial) (A), Disturbance to birds (A), Drainage (A), Dredging/canalization (A), Firewood collection (A), Forest grazing (B), Recreation/tourism (B), Unsustainable exploitation (B)

The main threats are wetland drainage and human disturbance. Other threats include illegal activities such as dynamite fishing, wood-cutting inside the hunting reserve, and hunting at night.

Lalzi Bay

B1i **015**

Admin region Durresi
Coordinates 41°18′N 19°30′E
Altitude 0–0 m **Area** 800 ha

Site description

A coastal area north of the town of Durresi, including mainly sea inlets and coastal features important for wintering ducks, and a sandy shore with a few small brackish lagoons. The site also includes the outlet of the Erzeni river. The main land-uses are agriculture, grazing, hunting, and fishing.

Habitats Forest and woodland (10%; native coniferous forest; mixed forest), Scrub (10%), Grassland (dry siliceous grassland), Wetland (10%; mudflat/sandflat; saltmarsh; sand-dunes/sand beach; coastal lagoon; standing brackish and salt water; river/stream; water-fringe vegetation), Marine areas (40%; sea inlet/coastal features), Artificial landscape (30%; arable land; forestry plantation)
Land-use Agriculture (20%), Fisheries/aquaculture (10%), Forestry (30%), Hunting (80%), Military (10%), Tourism/recreation (10%)

Birds

Species	Season	Year	Pop min	Pop max	Acc	Criteria
Anas penelope Wigeon	W	1996	643	5,000	B	B1i

This is an important site for wintering and migratory waterbirds, with 10,000 individuals recorded in January 1996.

Protection status

National None **International** None

Conservation issues

Threats Deforestation (commercial) (A), Disturbance to birds (A), Firewood collection (A), Forest grazing (B), Recreation/tourism (A)

As in most Albanian wetlands, disturbance by farmers, fishermen, gatherers, tourists and grazing animals limit the reproductive success of breeding waterbirds.

REFERENCES

Barbieri, F., Bogliani, G. and Prigioni, C. (1986) Note sull'ornitofauna dell'Albania. *Riv. Ital. Orn.* 56(1/2): 53–66.

Bashford, R. and Wilson, A. (1996) Bird survey of mountainous areas of Albania: 18–27 May 1996. Unpublished report. Tring, UK: British Trust for Ornithology.

Bego, F., Casale, F., Dotti, L., Grazzini, G., Papa, P., Roscelli, F. and Lombatti, B. (1995) Ornithological observations in Albania: 13–18 April 1995. Unpublished report. Parma, Italy: Lega Italiana Protezione Uccelli.

Bino, T. (1996) Wintering and breeding waterbirds in Albania. The Karavasta lagoon complex. Implications for the conservation of aquatic avifauna. Unpublished PhD thesis.

Bino, T., Tourenq, C., Kayser, Y., Busuttil, S., Crozier, J., Dore, B. J., and Bego, F. (1996) Recensement des oiseaux d'eau hivernants en Albanie (14–31 Janvier 1996). Unpublished report. Arles, France: Station Biologique de la Tour du Valat/Parma, Italy: Lega Italiana Protezione Uccelli/Tirana: Museum of Natural History of Tirana.

Bino, T., Casale, F., Brew, M., Erhart, F., Richardson, J., and Bego, F. (1996) Ornithological observations in Albania: 7–14 April 1996. Unpublished report. Parma, Italy: Lega Italiana Protezione Uccelli.

Bogliani, G., Barbieri, F. and Prigioni, C. (1987) Ornithological research in Albania: an annotated bibliography. *Avocetta* 11: 63–66.

Bouvier, M. and Kempf, C. (1987) La nature en Albanie. *Le Courier de la Nature* 109.

Casale, F. (1997) Osservazioni ornitologiche in Albania: 1995–1996. *Riv. Ital. Orn.* 67(1): 21–31. Milano.

Grimmett, R. F. A. and Jones T. A. (1989) *Important Bird Areas in Europe.* Cambridge, UK: International Council for Bird Preservation (Tech. Publ. 9).

Heath, M. F. and Borggreve, C. (2000) *BirdLife International/EBCC European Bird Database 1998.* Cambridge, UK: BirdLife International.

Kayser, Y., Bino, T., Bego, F., Fremuth, W. and Jorgo, J. (1997) Recensement des oiseaux d'eau hivernants en Albanie (1997). Unpublished report. Arles, France: Station Biologique de la Tour du Valat/Tirana: Museum of Natural History of Tirana.

Kayser, Y., Bino, T. and Gauthier-Clerc, M. (1995) Recensement des oiseaux d'eau hivernants en Albanie (17 janvier–7 février 1995). Unpublished report. Arles, France: Station Biologique de la Tour du Valat.

Lamani, F. and Zeko, I. (1983) L'Albanie, pays propre à l'hivernage des anseriformes. *Biologia Gallo-Hellenica* 10(1): 333–337.

Shine, C. (1996) *Legal and institutional analysis. European Commission PHARE programme: Karavasta lagoon wetland management project.* Arles, France: Station Biologique de la Tour du Valat.

Tucker, G. M. and Heath, M. F. (1994) *Birds in Europe: their conservation status.* Cambridge, UK: BirdLife International (BirdLife Conservation Series no. 3).

Zekhuis, M. J. and Tempelman, D. (1998) *Breeding birds of the Albanian wetlands, spring 1996.* Zeist, Netherlands: Stichting WIWO (WIWO Rep. No. 64).

ANDORRA

Marie-Jo Dubourg-Savage

Pirineo de Andorra (IBA 001). (Photo: Josep Argelich)

GENERAL INTRODUCTION

The Principality of Andorra is located on the southern slopes of the eastern Pyrenees, on the border between France and Spain. It forms part of the catchment area of the Valira river (Spain) and the Ariège river (France), and contains narrow valleys and gorges surrounded by high mountains.

There is only one Important Bird Area (IBA) in Andorra and it covers much of the country, with the exception of the main urban area in the Valira valley (Table 1). Andorra has an area of 468 km^2 and a population (in 1996) of about 64,000 divided amongst seven parishes (an average population density of 136 people per km^2).

ORNITHOLOGICAL IMPORTANCE

Sixty-six species of European conservation concern (SPECs) breed in Andorra. The majority of these do not meet IBA criteria, due to the small numbers present. Andorra only covers approximately 0.003% of the total land area of Europe and, as such, holds less than 1% of the total European populations of virtually all species occurring there. However, one large site has been identified as an IBA due to the presence of three SPECs, *Pyrrhocorax pyrrhocorax*, *Serinus citrinella* and *Emberiza cia*, which breed in internationally important numbers (Table 2).

Owing to the altitude of the high passes, the extent of visible migration over Andorra depends mainly on the weather. During spring, northerly winds trap birds in the valleys. Passage migrants include *Ciconia ciconia*, *Pernis apivorus* and *Milvus migrans*.

HABITATS

More than a third of the country is covered with broadleaved and coniferous forests, and rich alpine meadows occur at higher altitudes. All of the montane and plateau areas are included within the IBA.

IMPACTS ON IBAs – LAND-USE AND THREATS

Agriculture has declined as a land-use in Andorra, and now occupies only 7% of the land area. Nearly all meadows have either been replaced by urban development or succeeded by scrub and woodland as a result of undergrazing. The main agricultural activities are tobacco cultivation and the grazing of free-range cattle. Tourism is the predominant industry, and many of the mountain slopes have been developed for skiing, causing problems of disturbance and loss of habitat. Hunting and the use of poisoned baits pose additional threats.

PROTECTION STATUS

Andorra has no legally protected areas, and no legislation for the conservation of habitats or for the protection of flora. An act for the conservation of fauna was voted on in July 1998, and should result in a list of protected species being fixed by decree. All birds of prey have been protected by law since February 1988.

Table 1. Summary of Important Bird Areas in Andorra. 1 IBA covering 468 km^2

IBA code	1989 code	International name	National name	Administrative region	Area (ha)	Criteria (see p. 11)
001	AD001	Pirineo de Andorra	IBA D'Andorra	Principat d'Andorra	46,800	B2, B3

Table 2. Species of European conservation concern with significant breeding populations at Important Bird Areas in Andorra.

Species [1]	Minimum national breeding population (pairs) [2]	Proportion (%) of national population breeding at all IBAs in Andorra
Pyrrhocorax pyrrhocorax Chough	300	100
Serinus citrinella Citril Finch	500	100
Emberiza cia Rock Bunting	200	100

1. Only those species of European conservation concern (see Box 1, p. 12) that meet IBA criteria in Andorra are listed.
2. Data are taken from the BirdLife/EBCC European Bird Database 1998 (Heath and Borggreve 2000).

Box 1. International legislation and initiatives that are relevant to site conservation in Andorra (see Appendix 1 for a general description of these agreements).

Global
- Biodiversity Convention
- Ramsar Convention
- Bonn Convention
- World Heritage Convention ✔
- MAB Programme

Pan-European
- Bern Convention

✔ Convention ratified/initiative supported
(✔) Convention signed

Approximately 3,400 ha of the IBA are covered by two hunting reserves (Vedat d'Enclar and Vedat de Xixerella) where hunting is provisionally prohibited but other types of activity are allowed. A further zone exists where flora is protected at the local level.

The Andorran government is unable to make decisions regarding the use of land within the seven parishes. Furthermore, private landowners have complete control over their land, and neither the parish nor the government can intervene in land-use decisions. Expropriation does not exist in Andorra, and there is no official register of land deeds. This consequently makes it very difficult to know whom the land belongs to.

Although Andorra now has full status as a state (it entered the United Nations in 1993), and has been a member of the Council of Europe since 1994, it is not a member of the European Community and has signed only one international conservation treaty relevant to site conservation (Box 1).

CONSERVATION

- Projecte Horitzo: In 1995 two members of l'Associació per a la Defensa de la Natura (ADN) presented a sustainable development project for Andorra. Entitled *Nature and Ecotourism in Andorra: an option for the future*, it included an action plan for National Parks and Nature Reserves in Andorra (Palau Puigvert and Argelich Baró 1996). The document was publicized and copies distributed to the Andorran population and decision-makers.

ANALYTICAL METHODS

- Bird data were collected over the period 1989–1997.
- 1997 was the final year of data collection for the *Atlas of Breeding Birds of Andorra 1992–1997*, which is to be published by ADN in 1999. This project has allowed more accurate estimates to be made of numbers of breeding birds than was possible for the previous pan-European IBA inventory (Grimmett and Jones 1989).
- Breeding data are given as pairs.
- Data on threats were gathered over the period 1973–1998.

GLOSSARY

ADN l'Associació per a la Defensa de la Natura (BirdLife Representative in Andorra).

ACKNOWLEDGEMENTS

The following contributed to the IBA census: Josep Argelich, Wiebke Berger, Alex Clamens, Jacquie Crozier, Brian Dore, Marie-Jo Dubourg-Savage, Antoni Lopez, Ann Matschke, Barbara Melville, Jordi Palau and Sebastià Semene.

SITE ACCOUNT

Pirineo de Andorra — B2, B3 — 001

Admin region Principat d'Andorra
Coordinates 42°35'N 1°35'E
Altitude 838–2,942 m **Area** 46,800 ha

Site description

Forests of *Pinus*, *Abies*, *Quercus* and *Betula* cover 35% of the IBA, and rich alpine meadows are also widespread (37% cover). Tobacco cultivation and the grazing of free-range cattle are the main agricultural activities, with hay pastures and potato fields increasingly being replaced by tobacco fields. The main human activities are tourism, commerce and skiing.

Habitats Forest and woodland (35%; broadleaved deciduous forest; native coniferous forest; treeline ecotone), Scrub (7%), Grassland (44%; alpine/subalpine/boreal grassland), Rocky areas (6%), Artificial landscape (8%; arable land; other urban/industrial areas)
Land-use Agriculture (7%), Tourism/recreation (5%), Urban/industrial/transport

Birds

Species	Season	Year	Pop min	Pop max	Acc	Criteria
[1] ***Pyrrhocorax pyrrhocorax*** Chough	R	1997	300	500	C	B2
Serinus citrinella Citril Finch	R	1997	500	1,000	B	B3
Emberiza cia Rock Bunting	R	1996	200	1,000	B	B2

1. Very approximate estimate.

The IBA is important for forest and high-mountain birds, and holds internationally important numbers of three breeding species. The following SPECs also breed: *Gypaetus barbatus* (1 pair, attempted to breed in 1996), *Circaetus gallicus* (3–5 pairs), *Aquila chrysaetos* (3 pairs), *Falco tinnunculus* (35–90 pairs), *Falco peregrinus* (3–6 pairs), *Alectoris rufa* (7–12 pairs), *Perdix perdix hispaniensis* (50–80 pairs), *Coturnix coturnix* (7–12 pairs), *Otus scops* (2–3 pairs), *Bubo bubo* (2–3 pairs), *Caprimulgus europaeus* (25–30 pairs), *Picus viridis* (50–100 pairs), *Lullula arborea* (50–75 pairs), *Alauda arvensis* (200–750 pairs), *Hirundo rustica* (15–20 pairs), *Saxicola torquata* (40–80 pairs), *Monticola saxatilis* (50–100 pairs), *Monticola solitarius* (2–6 pairs), *Sylvia cantillans* (60–100 pairs), *Sylvia undata* (20–30 pairs), *Lanius collurio* (75–150 pairs) and *Emberiza hortulana* (4–6 pairs). Several non-SPECs breed in important numbers, including: *Lagopus mutus*, *Tetrao urogallus*, *Charadrius morinellus*, *Scolopax rusticola*, *Aegolius funereus* and *Dryocopus martius*.

Protection status

National None **International** None

Conservation issues

Threats Abandonment/reduction of land management (B), Agricultural intensification/expansion (A), Disturbance to birds (A), Industrialization/urbanization (A), Infrastructure (C), Natural events (C), Recreation and tourism (A), Unsustainable exploitation (A)

Poaching and shooting of species protected under the EC Birds Directive (*Gyps fulvus*, *Gypaetus barbatus*, *Aquila chrysaetos*) is a problem. What was once a traditional pastime has intensified, with the use of semi-automatic weapons and night-vision scopes. Controls are inadequate and legal 'seasons' are flouted. Ski resorts currently cover about 5% of the country and new developments are planned, including one in an area where *Charadrius morinellus* sometimes breeds. The use of snowmobiles, 4-wheel drive vehicles, scrambling bikes and mountain bikes is unregulated. Nearly all hay meadows in the south of the IBA have been converted into tobacco fields with uncontrolled use of chemicals (fertilizers and pesticides).

REFERENCES

L'ASSOCIACIÓ PER A LA DEFENSA DE LA NATURA (in prep.) [*Atlas of the breeding birds of Andorra 1992–1997*]. (In Catalan and English.)

CLAMENS, A. (1993) [The habitat of Subalpine Warbler *Sylvia cantillans* in Andorra (Pyrenees): influence of the change in human activities]. *Alauda* 61(2): 100–104. (In Catalan.)

CLAMENS, A. AND CROZIER, J. (1992) [First confirmed breeding of Blue Rock Thrush *Monticola solitarius* in Andorra (Pyrenees)]. *Alauda* 60(2): 116. (In Catalan.)

FOLCH I GUILLÈN, R. *ET AL.* (1984) [*Andorra's natural heritage: Andorran ecosystems and their use*]. Barcelona, Spain: Ketres Editora. (In Catalan.)

GRIMMETT, R. F. A. AND JONES, T. A. (1989) *Important Bird Areas in Europe.* Cambridge, UK: International Council for Bird Preservation (Techn. Publ. 9).

HEATH, M. F. AND BORGGREVE, C. (2000) *BirdLife International/EBCC European Bird Database 1998*. Cambridge, UK: BirdLife International.

PALAU PUIGVERT, J. AND ARGELICH BARÓ, J. (1996) [*Nature and ecotourisn in Andorra: an option for the future*]. Andorra: Credit Andorra. (In Catalan.)

TUCKER, G. M. AND HEATH, M. F. (1994) *Birds in Europe: their conservation status.* Cambridge, UK: BirdLife International (BirdLife Conservation Series no. 3).

VIGO I BONADA, J. (1976) [*Flora and vegetation of the high mountains of Catalunya*]. Barcelona, Spain: Centre Excursionista de Catalunya, Mountblanc-Martin, ed. (In Catalan.)

ARMENIA

Michael Wilson

Vast tablelands surrounded by ridges are characteristic of the Armenian landscape.
(Photo: Shahen Mnatsakanian)

GENERAL INTRODUCTION

Independent since 1991, Armenia is a small land-locked country lying between the Black and Caspian Seas. It occupies an area of some 29,800 km^2 in the south-western Transcaucasian region of the former USSR and is bordered in the east by the main territory of Azerbaijan, in the south-west by Nakhichevan (also part of Azerbaijan), in the north by Georgia, in the south by Iran and in the west by Turkey. The country is characteristically mountainous, consisting of several vast tablelands surrounded by ridges. There are clear landscape zones covering elevations from 400 m to over 4,000 m above sea-level. More than 90% of the country lies above 1,000 m. The major Armenian river systems are the Araks, Hrazdan, Kashakh, Vorotan, Metsamor, Debt, Aghstev and Arpa river systems. The two largest natural lakes are Sevan—the largest inland water-body in Transcaucasia and one of the largest freshwater mountain lakes in Eurasia—and, in the north-west, Lake Arpi.

A total of five Important Bird Areas (IBAs) have been identified in Armenia, although it is likely that several more IBAs will be identified in the future (see also 'Conservation', below). The five

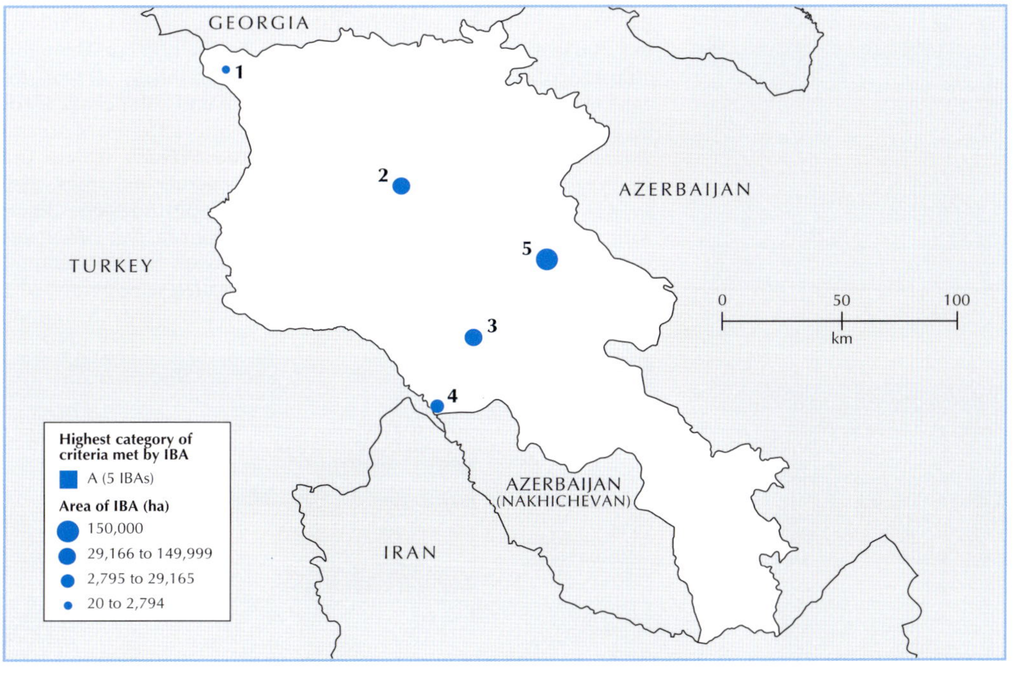

Map 1. Location, area and criteria category of Important Bird Areas in Armenia.

Table 1. Summary of Important Bird Areas in Armenia. 5 IBAs covering 1,820 km^2

IBA code	1989 code	International/national name	Area (ha)	Criteria (see p. 11)
001	—	Lake Arpi	2,000	A4i, B1i
002	—	Pombak mountain chain	—	A1, A3
003	—	Khosrov Reserve	29,166	A1, B2, A3
004	—	Armash fish-farm	2,795	A1, A4i, B1i, B2
005	—	Lake Sevan	150,000	A1, A4i, B1i

cover at least 1,820 km^2, or approximately 6% of the country's surface area (Table 1, Map 1). No Armenian IBAs (as part of the former USSR) were identified in the previous pan-European IBA inventory (Grimmett and Jones 1989).

ORNITHOLOGICAL IMPORTANCE

There are 113 species of European conservation concern (SPECs) breeding regularly in Armenia (Tucker and Heath 1994; Adamian and Klem 1997, 1999). Of these, *Marmaronetta angustirostris*, *Aythya nyroca*, *Oxyura leucocephala*, *Aquila heliaca* and *Falco naumanni* are globally threatened. Two additional species potentially in this category for Armenia are *Pelecanus crispus* (occurs, but does not breed regularly) and *Crex crex* (more information is required to determine breeding status). Globally near-threatened breeding species include *Phalacrocorax pygmeus*, *Aegypius monachus*, *Tetrao mlokosiewiczi* and *Glareola nordmanni*, while a further 69 (perhaps up to 73) breeding SPECs have an unfavourable conservation status in Europe. Forty-five SPECs occur regularly on passage or overwinter, and among these are the globally threatened or near-threatened *Anser erythropus*, *Haliaeetus albicilla*, *Circus macrourus*, *Aquila clanga* and *Gallinago media*.

Table 1 lists the criteria which each of the five sites meets in order to qualify as an IBA. Four sites qualify because they hold (or are believed to hold) significant numbers of globally threatened or near-threatened species (criterion A1; see Table 2). Three out of the five sites meet criteria A4i and B1i by supporting 1% or more of the biogeographic or flyway population of waterbird species (Table 3): the near-threatened *Phalacrocorax pygmeus* at IBA 004 and *Larus armenicus*, restricted as a breeding bird to Transcaucasia, Turkey and Iran, at Lake Arpi (IBA 001) and Lake Sevan (005). Two out of the five IBAs support species of European conservation concern in important numbers, thus meeting criterion B2. Estimated breeding-population figures are not available for the majority of breeding SPECs in Armenia (see Table 4).

Table 2. Important Bird Areas in Armenia that are important for species of global conservation concern (meeting criterion A1).

Species	IBA code
Phalacrocorax pygmeus Pygmy Cormorant	004
Anser erythropus Lesser White-fronted Goose	005
Marmaronetta angustirostris Marbled Duck	004
Aythya nyroca Ferruginous Duck	004, 005
Oxyura leucocephala White-headed Duck	004
Aegypius monachus Cinereous Vulture	003
Tetrao mlokosiewiczi Caucasian Black Grouse	002
Glareola nordmanni Black-winged Pratincole	004

Table 3. Important Bird Areas in Armenia that support important numbers of one or more congregatory species (i.e. meeting criteria A4 and/or B1). IBAs meeting both criteria A4 and B1 for the species are shown in **bold**. IBAs meeting only criterion B1 for the species concerned, and not A4, are shown in normal type. For key to 'Season' see p. 7.

Species	Season	IBA code
Phalacrocorax pygmeus Pygmy Cormorant	B	**004**
Aythya nyroca Ferruginous Duck	N	**004**
Larus armenicus Armenian Gull	B	**001,005**

Table 4. Species of European conservation concern with significant breeding populations at Important Bird Areas in Armenia (meeting any IBA criteria).

Species [1]	Minimum national breeding population (pairs) [2]	Proportion (%) of national population breeding at all IBAs in Armenia
Phalacrocorax pygmeus Pygmy Cormorant	—	—[3]
Marmaronetta angustirostris Marbled Duck	15	53
Aythya nyroca Ferruginous Duck	—	—[3]
Oxyura leucocephala White-headed Duck	—	—[3]
Gypaetus barbatus Lammergeier	—	—[3]
Neophron percnopterus Egyptian Vulture	—	—[3]
Gyps fulvus Griffon Vulture	—	—[3]
Aegypius monachus Cinereous Vulture	15	47
Buteo rufinus Long-legged Buzzard	—	—[3]
Tetrao mlokosiewiczi Caucasian Black Grouse	—	—[3]
Tetraogallus caspius Caspian Snowcock	166	1
Alectoris chukar Chukar	—	—[3]
Glareola pratincola Collared Pratincole	—	—[3]
Otus scops Scops Owl	—	—[3]
Bubo bubo Eagle Owl	—	—[3]
Merops apiaster Bee-eater	—	—[3]
Monticola solitarius Blue Rock Thrush	—	—[3]

1. Only those species of European conservation concern (see Box 1, p. 12) that meet IBA criteria in Armenia are listed.
2. Data are taken from Snow and Perrins (1998).
3. No national population estimate available.

The country's geographical position means that it is a major intersection of migration routes from all directions, with birds moving between Europe and wintering areas lying south of Transcaucasia. Few detailed studies have been undertaken, but there is a prominent raptor passage in the autumn, birds heading from north of Mount Aragats south-west to Sardarapat steppe and the Araks valley or south-east between the Pombak (IBA 002) and Bazoom mountains towards Lake Sevan (005) and on southward, either in unbroken flights or with stopovers. Wetland habitats, among them Lake Arpi (IBA 001) and, more especially, site 004 (the Armash fish-farming ponds) and Lake Sevan (005), are important staging sites for a wide range of migratory waterbirds and others.

HABITATS

Armenia is located in the north-eastern part of the Armenian Upland at the junction of three geological-botanical regions: the mesophilic Caucasian, desert/semi-desert Iranian and subtropical Eastern Mediterranean. Semi-desert, with rock, barren soil and scrub, predominates in the south of the country, covering much of the Araks river depression (Ararat plain) and adjacent foothills up to 1,500 m above sea-level, but primarily below 1,000 m. Drought-adapted plant communities comprise *Artemisia*, *Artemisia*-ephemeral, *Artemisia*-grass, *Achillea*, scrub and solonchak. Birds favouring semi-desert habitats include bee-eaters *Merops*, *Coracias garrulus*, larks (Alaudidae), *Hippolais* and *Sylvia* warblers and *Carpospiza brachydactyla*; where there is scrub and trees, also shrikes *Lanius* and *Irania gutturalis*. Higher-altitude semi-desert with rocky areas supports many raptors, *Apus melba*, *Monticola solitarius*, nuthatches *Sitta*, *Sturnus roseus* and *Bucanetes githagineus*. Agriculture has increasingly invaded semi-desert, with the cultivation of wheat, viticulture and fruit orchards. Mountain

Figure 1. Occurrence of habitats at Important Bird Areas in Armenia (see Appendix 3 for definitions of habitats).

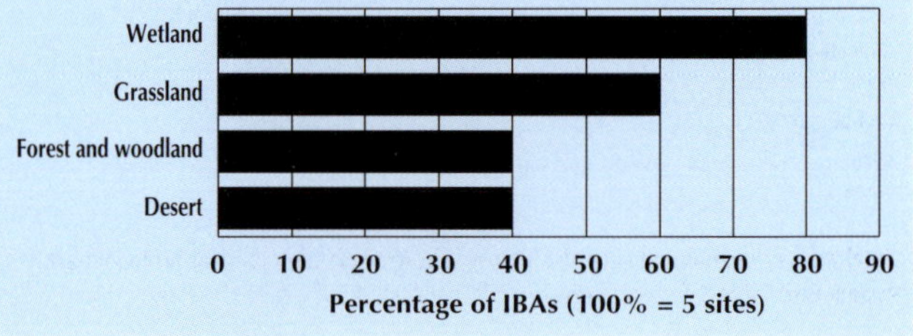

Figure 2. Occurrence of land-uses at Important Bird Areas in Armenia (see Appendix 3 for definitions of land-uses).

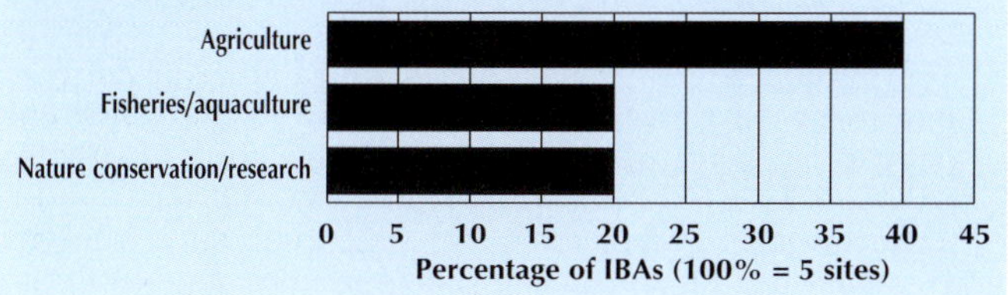

steppe, a characteristically diverse plant assemblage, is the predominant vegetation type in Armenia. It extends up to 500–550 m in the wetter north, up to 2,300–2,400 m in drier areas and there are two main types: arid mountain steppe (close to semi-desert, also ornithologically) and meadow steppe. Most mountain steppes are dominated by *Poa bulbosa* and *Catabrosella humilis*. Among species found in arid habitat of this kind are *Alectoris chukar*, wheatears *Oenanthe*, *Rhodopechys sanguinea*, *Emberiza buchanani* and *E. cia*, while raptors haunt rocky outcrops; it is also important for wintering passerines. The most widely distributed type of meadow steppe is 'multiherbaceous'; there is also grass and sedge *Carex humilis*. Typical birds include *Perdix perdix*, *Coturnix coturnix*, many passerines such as *Carduelis flavirostris* and *Carpodacus erythrinus*, and more variety in stream valleys with rocks. Mountain meadows are alpine and subalpine habitat at 2,400–3,500 m (higher on Mount Aragats). Dwarf vegetation is typical of alpine meadows, with carpets on plateaus and gradual slopes; other features of these altitudes are sedge meadows, dry upland meadows, also boulder fields, talus and scree slopes. This habitat supports *Tetraogallus caspius*, *Eremophila alpestris*, *Anthus spinoletta*, in scrub *Prunella ocularis*, also (in rocky areas) *P. collaris*, *Tichodroma muraria* and choughs *Pyrrhocorax*.

Timber cutting and widespread grazing have reduced the now mostly discontinuous forested areas in Armenia to c.12% of the land cover, at elevations between 550 and 2,600 m. The main forest-types are *Fagus*, mixed *Quercus*, *Carpinus* (particularly widespread), open *Juniperus* woodland and *Pinus*, though stands of native conifers are rare in Armenia. Destruction of *Quercus-Carpinus* forest has given rise to sibljax (steppe scrub) with *Juniperus* and *Paliurus*; also, in rocky, arid hills, *Pistacia*, *Cotinus*, *Rhamnus*, etc. Typical forest birds are woodpeckers, raptors (*Pernis apivorus*, *Accipiter brevipes*, *Aquila pomarina*, *Hieraaetus pennatus*), and many passerines (*Ficedula parva*, *F. semitorquata*); *Tetrao mlokosiewiczi* uses subalpine meadows for breeding, but moves into forest (*Rhododendron*, *Betula*, *Sorbus*) in winter. In the south of the country, dense deciduous forest or scattered *Juniperus* have *Aegypius monachus*, *Aegolius funereus*, *Dendrocopos syriacus*, *Sylvia* warblers, *Parus lugubris* and *Serinus pusillus*.

Armenian wetlands comprise over 400 natural and man-made lakes, reservoirs, fish-ponds and rivers. Of outstanding importance for birds are Lake Sevan (005) and fish-farming ponds in the Araks valley at Armash (IBA 004), Masees and Yeghegnoot. These ponds generally have surrounding emergent vegetation or 'islands' of such growth (*Phragmites*, *Typha*, etc.). Because of the falling water-level at Lake Sevan and the consequent loss of the adjacent Lake Gilli and its marshlands, the fish-ponds have become increasingly important for breeding waterbirds (some globally threatened), including *Phalacrocorax pygmeus*, *Plegadis falcinellus*, *Aythya nyroca*, *Oxyura leucocephala*, *Recurvirostra avosetta*, *Charadrius leschenaultii*, *Chettusia leucura*, *Burhinus oedicnemus* and pratincoles *Glareola*. Pelicaniformes and Ciconiiformes, wildfowl, shorebirds and others occur on passage or as rare visitors.

Figure 1 shows the main habitat types represented at the five IBAs in Armenia.

IMPACTS ON IBAs – LAND-USE AND THREATS

Figure 2 summarizes land-use at the five Armenian IBAs. Little is known of threats at four out of the five sites. Fluctuating water-levels are a threat to nesting birds at Lake Arpi (IBA 001) and many waterbirds, including *Phalacrocorax pygmeus* and *Pelecanus onocrotalus*, are shot (legally or illegally) at Armash and other ponds because of the perceived conflict with fish-farming interests, while the shooting of migrating and other raptors appears to be a widespread threat throughout the country.

PROTECTION STATUS

National protection

In common with other now-independent states of the former USSR, Armenia has a protected-area system comprising **Zapovedniks** (strict nature reserves), **Zakazniks** (temporary reserves with a less strict regime) and **National Parks**.

Only two out of the five Armenian IBAs overlap with a protected area (Table 5, Figure 4).

Figure 3. Occurrence of threats at Important Bird Areas in Armenia (see Appendix 3 for definitions of threat types and impact categories).

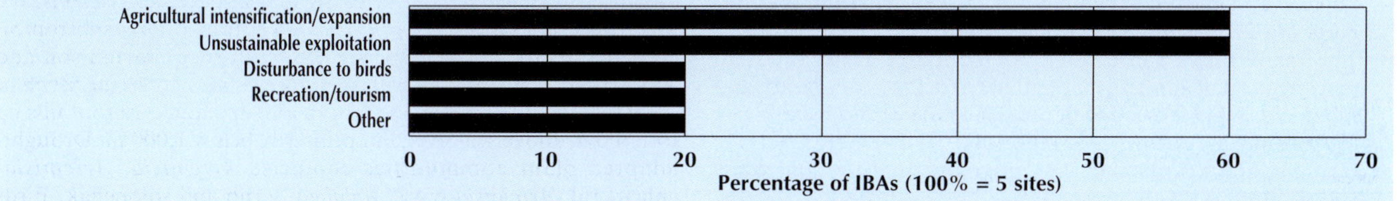

Figure 4. The national protection status of Important Bird Areas in Armenia.

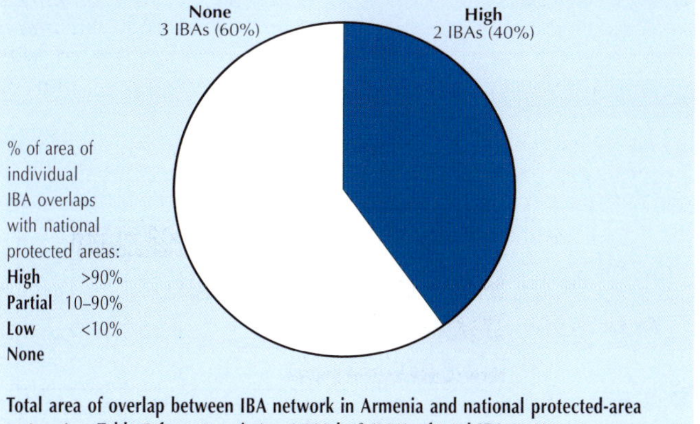

Total area of overlap between IBA network in Armenia and national protected-area system (see Table 5 for categories) = 1,792 km² (98% of total IBA area).

Figure 5. The international protection status of Important Bird Areas in Armenia.

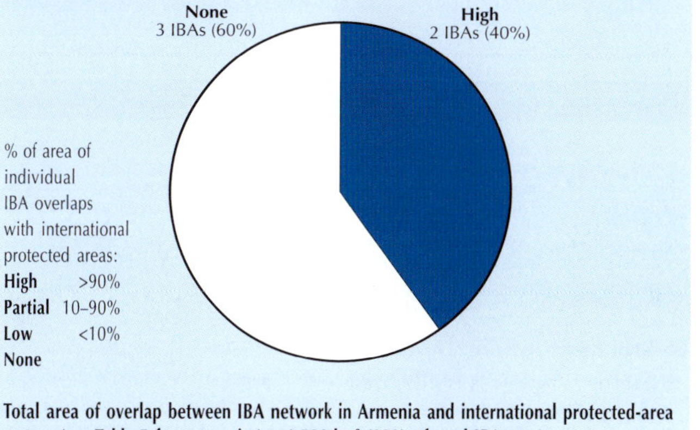

Total area of overlap between IBA network in Armenia and international protected-area system (see Table 5 for categories) = 1,520 km² (83% of total IBA area).

Table 5. Protection status of Important Bird Areas in Armenia. A tick (✔) indicates that an IBA overlaps with a protected area (to any extent).

		National		International
IBA code	International name	Zapovednik	National Park	Ramsar Site
001	Lake Arpi			✔
002	Pombak mountain chain			
003	Khosrov Reserve	✔		
004	Armash fish-farm			
005	Lake Sevan		✔	✔
Total number of IBAs		1	1	2

Box 1. International legislation and initiatives that are relevant to site conservation in Armenia (see Appendix 1 for a general description of the agreements).

Global	
Biodiversity Convention	(✔)
Ramsar Convention	✔
Bonn Convention	
World Heritage Convention	✔
MAB Programme	✔
Pan-European	
Bern Convention	

✔ Convention ratified/initiative supported
(✔) Convention signed

International protection

Armenia is not yet committed to the Bonn Convention, but is a contracting party to the World Heritage Convention and supports the Biodiversity and Ramsar Conventions (Box 1). Two areas are designated Ramsar Sites, Lake Arpi (001) and Lake Sevan (005).

CONSERVATION

- The following four nature reserves (Zapovedniks) are not covered as IBAs in this inventory. 1) Dilijan (established 1958): 27,995 ha of forested mountainous country in north-eastern Armenia. 2) Shikahogh Reserve (1959): c.10,000 ha on the northern slopes of the Meghri mountains in the south-eastern Little Caucasus. 3) Sevleech Reserve (1987): 240 ha in southern Armenia on the border with Azerbaijan. 4) Ereboonee Reserve (1981): 89 ha situated between Yerevan and site 003 (Khosrov).
- Preliminary observations indicate a large and diverse raptor migration through Armenia. Several of these migration watch-points may in the future be identified as IBAs, most notably Gree gorge (40°21′N 45°05′E, 55 km east-north-east of Yerevan).
- The importance for birds of two further areas requires investigation: Khoostop mountain, and the semi-desert and mountain steppe in the Meghri region in the extreme south-east of Armenia.
- Contemporary Armenian laws governing nature (and birds specifically) are as inherited from the Soviet period and the available evidence suggests that the current level of enforcement of laws relating to the protection of wildlife and natural resources is far from satisfactory. New national laws unique to Armenia (including 'The Law of the Republic of Armenia on the Animal Kingdom') and under consideration by the country's legislators were expected to be enacted by the end of 1998 or early 1999. The country's leading ornithologist, M. S. Adamian, was appointed to a government committee dealing with wildlife conservation laws in 1997. Such new legislation, if enforced effectively by the appropriate government agencies, holds great promise for improving bird protection and conservation. Considerable impetus has come from the American University of Armenia (AUA), which was opened simultaneously with the country's declaration of independence in 1991 and which houses a Department of Environmental Management having an influence on the Ministry of Environmental Protection and running courses in environmental awareness which all students, regardless of their field of study, are required to follow.
- The international Birds of Armenia Project involved cooperation between the Armenian National Academy of Sciences and the American University of Armenia in comprehensively reviewing the literature and organizing fieldwork in the period 1992–1995. This project led to the publication of two key references: a field guide (Adamian and Klem 1997) and a more recent handbook (Adamian and Klem 1999). These books carry a strong conservation message, describing in detail protected areas and identifying species of special concern within the country. Of the c.350 species recorded in Armenia, about 30% are deemed to be of special concern (Endangered, Threatened, Undetermined or Extirpated). The Red Data Book of Armenia (1987) contains 67 such species. The handbook (1999) should serve as a basis for a national scheme to protect, conserve and, where possible, to restore the country's birds and to promote and justify relevant laws.
- Against the background of harsh economic restraints and the heavy impact on the environment of human activities, more ornithological studies, especially population censuses and surveys, need to be encouraged, and greater involvement of young people in these and other schemes is felt to be essential.

ANALYTICAL METHODS

- Data for the IBAs in this inventory were compiled from the literature and recent (1993–1997) ornithological research in the country. The lack of abundance studies using census plots and the line-transect method has meant that the population size of most species is unknown.

GLOSSARY

solonchak intrazonal (i.e. affected more by local factors than by climate) saline soil without structure.

Transcaucasia the area, in the former USSR, lying to the south of the Great Caucasus mountain range and comprising the republics of Armenia, Azerbaijan and Georgia.

ACKNOWLEDGEMENTS

This chapter is based almost entirely on published and unpublished information collected by the entire research team of the Birds of Armenia Project, including S. Acopian, M. S. Adamian, D. Klem Jr. and P. Saenger. P. Saenger also kindly provided comments on this chapter.

SITE ACCOUNTS

Lake Arpi — A4i, B1i — 001

Admin region —
Coordinates 41°03′N 43°35′E
Altitude 2,000–2,000 m **Area** 2,000 ha

Site description

Arpi reservoir, located in the very north-west of Armenia, evolved in the 1950s from the small Lake Arpi. Fed by thaws and four streams, it is the source of the Akhurian river. The landscape is mountain steppe with mostly grassy alpine vegetation. The site is adjacent to wetlands. In spring the reservoir floods, and water covers an area of 20 km², and two islands (2 ha and 4 ha) form. By autumn, when the lake waters are used for irrigation, the shores of the two islands merge.

Habitats Grassland (steppe/dry calcareous grassland;alpine/subalpine/boreal grassland), Wetland (standing fresh water)
Land-use Agriculture

Birds

More than 100 species have been observed around Lake Arpi. Of special concern is the large breeding colony of *Larus armenicus*; 7,000–

10,000 pairs are known to breed on the two islands, comprising 60% of the population in Armenia. Species of global conservation concern that do not meet IBA criteria: *Pelecanus crispus* (breeding every 2–5 years, juveniles visiting annually).

Species	Season	Year	Pop min	Pop max	Acc	Criteria
Larus armenicus Armenian Gull	B	—	7,000	10,000	—	A4i, B1i

Protection status

National None **International** High

2,000 ha of IBA covered by Ramsar Site (Lake Arpi, 2,000 ha).

Conservation issues

Threats Agricultural intensification/expansion (U), Disturbance to birds (U)

The colony is extremely vulnerable due to annual water-level fluctuations of the reservoir. Birds are sometimes subject to disturbance when fluctuating water-levels allow access to nesting areas.

Pombak mountain chain — A1, A3 — 002

Admin region —
Coordinates 40°38′N 44°32′E
Altitude 1,000–3,000 m **Area** —

Site description

Pombak chain is one of the largest mountain ranges in the central part of northern Armenia, extending for 65 km from the north-west to south-east and evolving two ridges split by the Marmareek river. The mountain range has clearly defined landscape zones; mountain steppe, and subalpine and alpine meadows interspersed with mosaics of woodlands covering steep, north-facing slopes.

Habitats Forest and woodland, Grassland (steppe; alpine/subalpine/boreal grassland)
Land-use Agriculture

Birds

Species	Season	Year	Pop min	Pop max	Acc	Criteria
Tetrao mlokosiewiczi Caucasian Black Grouse	R	—	—	—	—	A1, A3
Tetraogallus caspius Caspian Snowcock	R	—	—	—	—	A3
Prunella collaris Alpine Accentor	R	—	—	—	—	A3
Tichodroma muraria Wallcreeper	R	—	—	—	—	A3
Pyrrhocorax graculus Alpine Chough	R	—	—	—	—	A3
Montifringilla nivalis Snowfinch	R	—	—	—	—	A3

Tetrao mlokosiewiczi is confined to the edge of the snow-line in the upper edges of the forest and in the subalpine and alpine meadows. The following species of special concern are also present: *Gypaetus barbatus, Gyps fulvus, Accipiter gentilis, Aquila pomarina, A. clanga, A. chrysaetos, Prunella ocularis* and *Phoenicurus erythrogaster* (current status uncertain).

Protection status

National None **International** None

Conservation issues

Threats Agriculture intensification/expansion (U), Unsustainable exploitation (U)

The deeply-incised mountain range and the mosaic of woodland result in a patchy distribution of the *Tetrao mlokosiewiczi* population, thus lowering its viability and productivity. In addition, the species is threatened by heavy livestock-grazing during the breeding season and year-round poaching.

Khosrov Reserve — A1, B2 — 003

Admin region —
Coordinates 40°02′N 44°56′E
Altitude 800–2,800 m **Area** 29,196 ha

Site description

The Khosrov Reserve lies in a central part of southern Armenia. The terrain is rather complicated. The entire area is typically alpine, densely traversed by primary and side-gorges, high plateaus and inter-montane depressions, massive towering mountain ranges, volcanic expanses and isolated mountain peaks. Two alpine rivers, the Azat and Verdi, with tributaries and a small lake (0.3 ha) are within the IBA. The area extends over four landscape zones: desert/semi-desert, mountain steppe, woodland, and alpine and subalpine meadows, though much valuable high-altitude habitat (meadow steppe on plateaus and rocky areas) lies outside the reserve.

Habitats Forest and woodland (wooded steppe), Grassland (alpine/subalpine/boreal grassland), Desert (semi-desert), Wetland (standing fresh water; river/stream)
Land-use Nature conservation/research

Birds

Species	Season	Year	Pop min	Pop max	Acc	Criteria
Gypaetus barbatus Lammergeier	R	—	3	—	—	B2
Neophron percnopterus Egyptian Vulture	B	—	5	6	—	B2
Gyps fulvus Griffon Vulture	R	—	15	20	—	B2
Aegypius monachus Cinereous Vulture	R	—	7	9	—	A1, B2
Buteo rufinus Long-legged Buzzard	R	—	4	5	—	B2
Tetraogallus caspius Caspian Snowcock	R	—	2	3	—	A3, B2
Alectoris chukar Chukar	R	—	300	—	—	B2
Otus scops Scops Owl	B	—	50	60	—	B2
Bubo bubo Eagle Owl	R	—	7	10	—	B2
Merops apiaster Bee-eater	B	—	200	300	—	B2
Prunella collaris Alpine Accentor	R	—	—	—	—	A3
Monticola solitarius Blue Rock Thrush	B	—	30	35	—	B2
Sitta tephronota Great Rock Nuthatch	R	—	60	70	—	A3
Tichodroma muraria Wallcreeper	R	—	—	—	—	A3
Pyrrhocorax graculus Alpine Chough	R	—	—	—	—	A3

At least 156 species have been recorded and 76 of these have been proved to breed. An outstanding site for raptors, with at least 21 species breeding and three possibly breeding, including *Aquila chrysaetos, A. pomarina, Circaetus gallicus, Accipiter brevipes* and, uniquely in Armenia, a small relict population of *Aegolius funereus*. A colony of *Apus melba* in Azat Gorge numbers some 2,000 birds. Among other breeding birds are *Dendrocopos medius, Oenanthe hispanica, Monticola saxatilis, M. solitarius* and *Bucanetes githagineus*.

Protection status

National High **International** None

29,196 ha of IBA covered by Strict Nature Reserve (Zapovednik) (Khosrov, 29,196 ha).

Conservation issues

Threats Agricultural intensification/expansion (U), Other (U), Recreation/tourism (U), Unsustainable exploitation (U)

Fingers of grazing land 200–500 m wide and 1–5 km long penetrate into the reserve territory and herds are regularly driven through the reserve from adjacent farms. Until recently, there was hay-making and the collecting of fruit and seeds of edible, decorative and medicinal plants. Serious damage to the flora and fauna is caused by villagers resident in the reserve. There is also some poaching. The construction of tourist facilities, roads running through the reserve and the lack of a buffer zone are further causes for concern. A hydroelectric power-station, planned for a site on the Azat river within the reserve, would, it is believed, have a very detrimental affect on Khosrov and its wildlife.

Armash fish-farm — A1, A4i, B1i, B2 — 004

Admin region —
Coordinates 37°45′N 44°46′E
Altitude — **Area** 2,795 ha

Site description

Armash area is the largest fishery in the Araks valley of Armenia, with 1,514 ha covered by a total of 29 fish-ponds, each between 11 ha and 95 ha in size. The Araks valley is mainly semi-desert with extremely salty soils. The vegetation is basically halophytic, with mostly goosefoot *Chenopodium*, pepper-grass *Lepidium*, saltwort *Salsola*, camel's thorn *Alhagi* and bean caper *Zygophyllum*. Marshes, wetlands, channel shores and ponds are densely vegetated with reeds and forbs, mixed with *Typha*, *Carex* and sparse tamarisk.

Habitats Desert (semi-desert), Wetland (standing fresh water; water-fringe vegetation)
Land-use Aquaculture/fisheries

Birds

Species	Season	Year	Pop min	Pop max	Acc	Criteria
[1] ***Phalacrocorax pygmeus*** Pygmy Cormorant	B	—	150	200	—	A1, A4i, B1i
Marmaronetta angustirostris Marbled Duck	B	—	8	10	—	A1, B2
Aythya nyroca Ferruginous Duck	B	—	25	30	—	A1
Aythya nyroca Ferruginous Duck	N	1995	—	700	—	A1, A4i, B1i
Oxyura leucocephala White-headed Duck	B	—	4	6	—	A1
Glareola nordmanni Black-winged Pratincole	B	—	15	20	—	A1

1. 80 nests 1994, 27 nests 1995.

The reedbeds cover edges of ponds, or grow in islet-shaped patches taking up approximately 30–50% of the water area, thus providing suitable cover and breeding places for grebes, cormorants, egrets/herons, rails, ducks and other birds.

Protection status

National None **International** None

Conservation issues

Threats Unsustainable exploitation (U)

Many waterbirds, including *Phalacrocorax pygmeus* and *Pelecanus onocrotalus* are shot (legally or illegally) at Armash and other ponds because of the perceived conflict with fish-farming interests.

Lake Sevan — A1, A4i, B1i — 005

Admin region —
Coordinates 40°20′N 45°20′E
Altitude — **Area** 150,000 ha

Site description

A large lake in east-central Armenia, the country's largest protected area (125,300 ha of lake and 24,800 ha of land).

Habitats Wetland (standing fresh water; water-fringe vegetation)
Land-use Unknown

Birds

Species	Season	Year	Pop min	Pop max	Acc	Criteria
[1] ***Anser erythropus*** Lesser White-fronted Goose	U	1995	—	26	—	A1
Aythya nyroca Ferruginous Duck	N	1995	—	60	—	A1
Larus armenicus Armenian Gull	B	—	—	5,000	—	A4i, B1i

1. Three counts in 1995 with the highest of 26 individuals.

This is an important site for breeding *Larus armenicus* and large numbers of wintering waterfowl.

Protection status

National High **International** High
150,000 ha of IBA covered by National Park (Sevan). 150,000 ha of IBA covered by Ramsar Site (Lake Sevan, 489,100 ha).

Conservation issues

Threats Unknown

The surrounding Lake Sevan basin (339,000 ha) is designated as an External Protective Zone (EPZ).

REFERENCES

Adamian, M. S. and Klem, D. (1997) *A field guide to birds of Armenia.* Oakland, USA: American University of Armenia.

Adamian, M. S. and Klem, D. (1999) *Handbook of the birds of Armenia.* Oakland, USA: American University of Armenia.

Anon. (1996) *Whitaker's Almanack 1997.* 129th edition. London: Whitaker and Sons Ltd.

Grimmett, R. F. A. and Jones T. A. (1989) *Important Bird Areas in Europe.* Cambridge, UK: International Council for Bird Preservation (Tech. Publ. 9).

Movsesian, S. O. and Ayrumian, K. A., eds. (1987) [*Red Data Book of the Armenian SSR.*] Yerevan: Hayastan Publishing House. (In Russian.)

Snow, D. W. and Perrins, C. M. (1998) *The birds of the western Palearctic. Concise edition.* Vols 1 and 2. Oxford, UK: Oxford University Press.

Sokolov, V. E. and Syroechkovskiy, E. E. (1990) [*The nature reserves of the USSR: reserves of the Caucasus*]. Moscow. (In Russian.)

Tucker, G. M. and Heath, M. F. (1994) *Birds in Europe: their conservation status.* Cambridge, UK: BirdLife International (BirdLife Conservation Series no. 3).

AZERBAIJAN

MICHAEL PATRIKEEV AND MICHAEL WILSON

Mixed colony of herons, egrets, ibises and cormorants at Lake Mahmud-chala (IBA 045). (PHOTO: MICHAEL PATRIKEEV)

GENERAL INTRODUCTION

Covering an area of 86,600 km², Azerbaijan lies in the eastern part of Transcaucasia, its eastern border being formed by the shore of the Caspian Sea. The country is bounded by Iran in the south, in the west by Armenia, and in the north by Georgia and Russia. To the south-west, but geographically separated from the rest of Azerbaijan by Armenian territory, lies the Autonomous Republic of Nakhichevan, which also borders on Iran and Turkey. The south-eastern part of the Great (Greater) Caucasus extends into and dominates the north-eastern part of the country, while the lower mountains of the Little (Lesser) Caucasus lie in the south-western part and the Talysh mountains are situated on the border with Iran in the south-east. The central part of Azerbaijan is a depression drained by the Kura river and the lower reaches of its tributary, the Aras.

Azerbaijan has 52 Important Bird Areas (IBAs), covering 6,161 km² or 7% of the country's surface area (Table 1). IBAs are relatively evenly distributed over the country, though there are gaps in the Nagorno-Karabakh region (scene of a recent war with

Table 1. Summary of Important Bird Areas in Azerbaijan. 52 IBAs covering 6,161 km²

IBA code	1989 code	International name	National name	Administrative region	Area (ha)	Criteria (see p. 11)
001	SU075	Zakataly	Zakataly (Zakataly goryhy)	Zakataly, Belokany	23,844	A1, A3
002	—	Ilisu area	Ilisu	Kakhi	9,345	A1
003	—	Alazani river valley	Alazani river valley	Zakataly, Belokany, Kakhi	5,000	A1, B2
004	—	Sheki upland	Sheki upland	Sheki	10,400	A1
005	SU073	Karayazi forest	Karayazi	Kazakh, Akstafa	10,000	A1
006	—	Akstafa-chai valley	Akstafa-chai valley	Kazakh	200	B2
007	—	Shamkhor area	Shamkhor area	Shamkhor	10,000	A1, B2
008	—	Korchai area	Korchai	Khanlar, Mingechaur	15,000	A1, B2
009	—	Varvara reservoir	Varvara reservoir	Mingechaur	4,000	A1
010	SU081	Turianchai	Turian-Chai, Evlakh, and Agdash	Agdash, Evlakh	12,600	B2
011	—	Gekchai Bozdag mountains	Gekchai Bozdag	Gabala (Kutgashen), Oguz (Vartashen)	5,000	A1, B2
012	SU074	Lake Gey Gel	Gei Gel (Gei Gel goryhy)	Khanlar	7,131	A1, A3
013	—	Mount Giamysh	Giamyshdag	Khanlar, Karabakh	8,500	A1, A3, B2
014	—	Mount Dalidag	Dalidag	Kelbadjhar, Lachin	10,000	A3, B2
015	—	Lachin area	Lachin	Lachin	20,000	A1
016	—	Shahbuz area	Shahbuz	Nakhichevan, Shahbuz	7,000	A3, B2
017	—	Ordubad area	Ordubad	Nakhichevan, Ordubad	40,000	A3, B2
018	—	Mount Ilandag	Ilandag	Nakhichevan, Babek	4,000	B2
019	—	Negram mountains	Negramdag	Nakhichevan, Babek	3,500	B2

Table 1 ... continued. Summary of Important Bird Areas in Azerbaijan. 52 IBAs covering 6,161 km²

IBA code	1989 code	International name	National name	Administrative region	Area (ha)	Criteria (see p. 11)
020	—	Mount Bazar-Duzu	Bazar-Duzu	Kuba, Gabala (Kutgashen), Kusari	4,000	A1, A3
021	—	Mount Shahdag	Shakhdag	Kusari	3,500	A1, A3
022	—	Kusari (Gusari) area	Kusari (Gusari)	Kusari	15,000	B2
023	—	Samur delta	Samur delta	Khachmas	6,000	B2
024	—	Divichi liman (or Lake Akzibir)	Akzibir	Divichi	7,000	A1, A4i, A4iii, B1i, B2
025	—	Mount Babadag	Babadag	Ismailly, Kuba	9,000	A1, A3
026	—	Ismailly area	Ismailly	Ismailly	5,778	A1
027	—	Alty Agach area	Alty Agach	Khizi	5,500	A1, B2
028	SU079	Pirgulu (Pirkuli)	Pirgulu	Shemakha	1,520	B2
029	—	Barda tugai forest	Barda tugai forest	Agdash, Barda	4,000	A1, B2
030	SU076	Lake Aggel	Aggel, Ahgel	Agjabedy	9,173	A1, A4i, B1i, B2
031	—	Lake Boz-Koba	Boz-Koba	Agjabedy, Beilagan, Imishli	4,000	A4iii
032	—	Lake Sarysu	Sarysu	Imishli	20,000	A1, A4i, B1i, B2
033	—	Yashma island	Yashma island	Absheron	200	B1i
034	—	Absheron archipelago (north) and Artem Bay	Absheron archipelago (north) and Artem Bay	Baku	1,000	B1i
035	—	Shah Cape (Shakhova Kosa)	Shah Cape (Shah gilasy)	Baku	500	A4i, A4iii, B1i
036	—	Lake Krasnoye and other water-bodies of the Absheron peninsula	Kizilgel		—	A1, A4i, B1i
037	—	Mount Kargabazar and Mount Gush-gaya	Kargabazardag and Gush-gaya	Maraza	3,000	A1
038	—	Gobustan area	Gobustan	Karadag, Maraza	5,000	B3
039	—	Sangachal Bay	Sangachal Bay	Karadag	1,000	A4i, A4iii, B1i
040	—	Glynanyi island	Glynanyi island	Karadag	200	A4i, B1i, B2
041	—	Lake Hadjikabul	Lake Hadjikabul		1,500	A1, A4i, B1i
042	—	Pirsagat islands and Los island	Pirsagat islands and Los island	Karadag	250	A4i, B1i, B2
043	SU078/080	Shorgel lakes/Shirvan Reserve	Beuk Shorgel, Kichik Shorgel/Shirvan	Saliany, Neftechala	26,000	A1, A4i, A4iii, B1i, B2
044	—	Mugan steppe	Mugan steppe	Saliany, Imishli, Pushkin, Saatly, Sabirabad	100,000	A1
045	—	Lake Mahmud-chala	Mahmud-chala	Pushkin, Djalilabad, Massally	10,000	A1, A4i, A4iii, B1i
046	—	Kura delta	Kura delta	Neftechala	10,000	A1, A4i, B1i
047	—	Lake Ych-chala (Novogolovka-chala)	Ych Chala (Novogolovka-chala)	Saliany, Massally	2,500	A1, A4i, B1i
048	SU077	Kizil Agach Bay	Kizil Agash, Khazar	Lenkoran	132,500	A1, A4i, B1i
049	—	Vilajchai valley	Vilajchai valley	Lenkoran, Yardymly	1,000	B2
050	—	Zuvand upland	Zuvand upland	Yardymly, Lerik	15,000	B2
051	—	Girkan forest	Girkan Meshasi	Astara, Lenkoran	4,500	B2
052	—	Astara-chai valley	Astara-chai valley	Astara	2,000	B2, B3

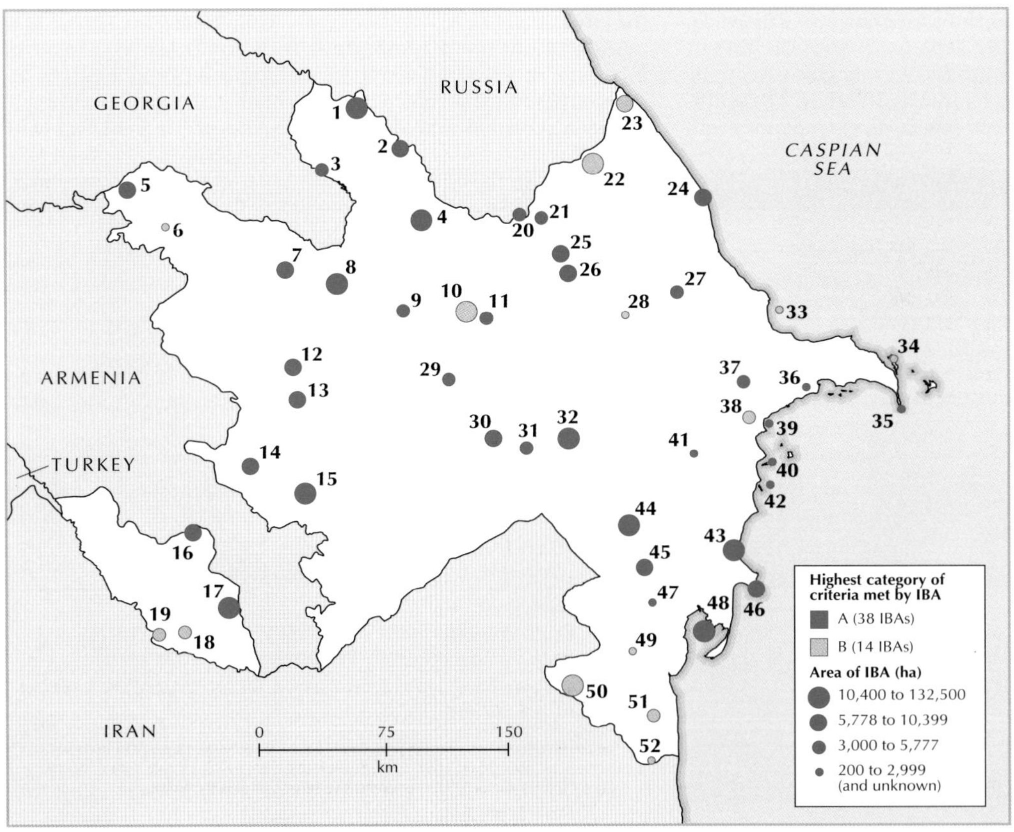

Map 1. Location, area and criteria category of Important Bird Areas in Azerbaijan.

Table 2. Important Bird Areas in Azerbaijan that are important for species of global conservation concern (meeting criterion A1).

Species	IBA code
Phalacrocorax pygmeus Pygmy Cormorant	009, 024, 030, 032, 043, 045, 046, 047, 048
Pelecanus crispus Dalmatian Pelican	024, 030, 032, 043, 046, 048
Anser erythropus Lesser White-fronted Goose	048
Branta ruficollis Red-breasted Goose	048
Marmaronetta angustirostris Marbled Duck	030, 032, 043, 048
Oxyura leucocephala White-headed Duck	024, 030, 032, 036, 041
Aegypius monachus Cinereous Vulture	011, 015
Circus macrourus Pallid Harrier	008
Aquila heliaca Imperial Eagle	003, 005, 007, 011, 025, 026, 027
Falco naumanni Lesser Kestrel	029, 037
Tetrao mlokosiewiczi Caucasian Black Grouse	001, 002, 012, 013, 020, 021, 025, 026
Tetrax tetrax Little Bustard	004, 008, 043, 044, 048

Table 3. Important Bird Areas in Azerbaijan that support important numbers of one or more congregatory species (i.e. meeting criteria A4 and/or B1). IBAs meeting both criteria A4 and B1 for the species are shown in **bold**. IBAs meeting only criterion B1 for the species concerned, and not A4, are shown in normal type. For key to 'Season', see p. 7.

Species	Season	IBA code
Tachybaptus ruficollis Little Grebe	W	**048**
Podiceps nigricollis Black-necked Grebe	W	**048**
Phalacrocorax pygmeus Pygmy Cormorant	R	**032**, **045**, **046**
	B	**030**, **048**
	W	**032**, **043**, **047**, **048**
	P	**024**
Pelecanus onocrotalus White Pelican	W	**048**
Pelecanus crispus Dalmatian Pelican	W	**030**, **043**, **046**, **048**
	P	**024**
Nycticorax nycticorax Night Heron	B	**048**
Ardeola ralloides Squacco Heron	B	**030**, 032, **045**, **046**, **048**
Bubulcus ibis Cattle Egret	B	**045**, **048**
Egretta garzetta Little Egret	R	**030**
	B	**045**, **046**, **048**
Egretta alba Great White Egret	B	**030**
	W	**048**
Plegadis falcinellus Glossy Ibis	B	**030**, **032**, **045**, **048**
	P	035
Platalea leucorodia Spoonbill	B	**030**, **045**
	P	**024**
Phoenicopterus ruber Greater Flamingo	W	**048**
Cygnus olor Mute Swan	W	**048**
Cygnus cygnus Whooper Swan	W	**030**, 034, 039, 040, **042**, **048**
	P	**024**, 035, **039**
Anser albifrons White-fronted Goose	W	**048**
Anser erythropus Lesser White-fronted Goose	W	**048**
Anser anser Greylag Goose	W	**048**
Anas penelope Wigeon	W	**045**, **048**
Anas strepera Gadwall	W	**048**
Anas crecca Teal	W	**032**, **048**
Anas acuta Pintail	W	**032**
Anas clypeata Shoveler	W	**045**, **048**
Marmaronetta angustirostris Marbled Duck	B	**030**, **032**
	W	**043**, **048**
Netta rufina Red-crested Pochard	W	**032**, **039**, **045**, **048**
	P	**039**
Aythya ferina Pochard	W	**032**, **045**, **048**
Aythya fuligula Tufted Duck	W	**048**
	P	033, 035, 039
Aythya marila Scaup	P	033, 035
Oxyura leucocephala White-headed Duck	W	**030**, **036**, **041**
Fulica atra Coot	W	**030**, **032**, **048**
	P	**035**, 039
Himantopus himantopus Black-winged Stilt	P	024
Glareola pratincola Collared Pratincole	B	040, **042**
Tringa stagnatilis Marsh Sandpiper	P	**024**
Xenus cinereus Terek Sandpiper	P	024
Larus genei Slender-billed Gull	B	**042**
Larus cachinnans Yellow-legged Gull	B	**040**
Sterna sandvicensis Sandwich Tern	B	**042**

Armenia) (Map 1). The previous pan-European IBA inventory (Grimmett and Jones 1989) identified nine IBAs in Azerbaijan (as part of the former USSR), all of which are included in the current total of 52, two sites (previously IBAs SU078 and 080) (i.e. the combined water-bodies of Shirvan steppe/Shirvan reserve) having been merged to make current site 043 (Table 1).

ORNITHOLOGICAL IMPORTANCE

There are 119 species of European conservation concern (SPECs) breeding regularly in Azerbaijan (Tucker and Heath 1994; M. Patrikeev *in litt.*). Of these, *Marmaronetta angustirostris, Aythya nyroca, Aquila heliaca* and *Falco naumanni* are globally threatened, *Phalacrocorax pygmeus, Haliaaetus albicilla* (perhaps no longer breeding regularly), *Aegypius monachus* and *Tetrao mlokosiewiczi* are globally near-threatened and a further 77 species have an unfavourable conservation status in Europe. Fifty-four SPECs occur regularly on passage or overwinter, and these include the globally threatened or near-threatened *Pelecanus crispus, Anser erythropus, Branta ruficollis, Oxyura leucocephala, Circus macrourus, Aquila clanga, Crex crex, Tetrax tetrax, Otis tarda, Glareola nordmanni* (has bred), *Chettusia gregaria, Numenius tenuirostris* (possibly regular at IBA 033) and *Gallinago media* (Tucker and Heath 1994; M. Patrikeev *in litt.*)

Table 1 lists the criteria which each site meets in order to qualify as an IBA. Thirty sites qualify because they hold significant numbers of globally threatened or near-threatened species (criterion A1; see Table 2). *Phalacrocorax pygmeus*, of which Azerbaijan supports a substantial proportion of the European population, occurs at nine of the sites, breeding at six (possibly at one more—IBA 047). *Pelecanus crispus* is a former breeder, having last nested at Lake Boz-Koba (IBA 031) in the late 1980s or early 1990s (4–6 pairs), and it may even still breed there; otherwise, it occurs as a passage- or winter-visitor at six sites. *Marmaronetta angustirostris* breeds at two sites and possibly at a third (Kizil Agach Bay IBA 048), *Aegypius monachus* meets the A1 criterion at two sites, *Aquila heliaca*

Table 4. Species of European conservation concern with significant breeding populations at Important Bird Areas in Azerbaijan (meeting any IBA criteria).

Species [1]	Minimum national breeding population (pairs) [2]	Proportion (%) of national population breeding at all IBAs in Azerbaijan
Phalacrocorax pygmeus Pygmy Cormorant	4,800	71
Nycticorax nycticorax Night Heron	7,000	90
Ardeola ralloides Squacco Heron	15,000	95
Plegadis falcinellus Glossy Ibis	12,000	100[3]
Platalea leucorodia Spoonbill	500	81
Marmaronetta angustirostris Marbled Duck	70	100[3]
Neophron percnopterus Egyptian Vulture	80	36
Gyps fulvus Griffon Vulture	100	35
Aegypius monachus Cinereous Vulture	55	33
Accipiter brevipes Levant Sparrowhawk	—	—[4]
Aquila pomarina Lesser Spotted Eagle	50	40
Aquila heliaca Imperial Eagle	40	50
Hieraaetus pennatus Booted Eagle	60	18
Falco naumanni Lesser Kestrel	500	7
Tetrao mlokosiewiczi Caucasian Black Grouse	500	55
Tetraogallus caspius Caspian Snowcock	133	100[3]
Alectoris chukar Chukar	33,000	18
Francolinus francolinus Black Francolin	1,000	43
Porphyrio porphyrio Purple Gallinule	10,000	88
Glareola pratincola Collared Pratincole	500	31
Sterna sandvicensis Sandwich Tern	3,800	100[3]
Dendrocopos medius Middle Spotted Woodpecker	—	—[4]
Sitta neumayer Rock Nuthatch	—	—[4]

1. Only those species of European conservation concern (see Box 1, p. 12) that meet B2/B3 criteria at IBAs in Azerbaijan are listed.
2. Data are taken from the BirdLife/EBCC European Bird Database 1998 (Heath and Borggreve 2000).
3. The percentage of the national population in IBAs exceeds 100%. This is most likely to be because the national population estimate needs to be reassessed (often as a result of new data on species numbers from comprehensive surveys of IBAs themselves).
4. No national population estimate available.

at seven, *Falco naumanni* at two, and *Tetrao mlokosiewiczi*, another key species in Azerbaijan, at seven or eight IBAs.

Table 3 shows that 16 sites qualify as IBAs in holding 1% or more of the biogeographic population of one or more waterbird species (criteria A4i and/or B1i), seven of these for breeding species (sites 030, 032, 040, 042, 045, 046, 048), four sites (IBA 024 Divichi liman the most important, and 033, 035, 039) for staging during migration and thirteen for wintering species, the outstanding site for these being Kizil Agach Bay (048). Fourteen of the 52 sites do not qualify as globally Important Bird Areas (i.e. under any A criteria), but are of regional importance (under B2/B3 criteria) for SPECs (Table 1).

Although some species (e.g. certain raptors, *Grus grus*) migrate on a broad front through Azerbaijan, wildfowl, shorebirds, other waterbirds, and some passerines (about whose movements far less information is available), tend to follow a coastal route, such that IBAs Divichi liman (site 024) in the north, the Kura delta (046), Sangachal Bay (039), Shah Cape (035), Yashma island (033), Kizil Agach Bay (048), and Lake Mahmud-chala (045), further inland but still very much within the Caspian coastal belt, function as key staging posts for these species. Other migration routes are along the Aras river and across the Kura valley, including the nearby lakes (site 030 Lake Aggel, etc.). Large, in some cases spectacular, concentrations of wintering waterbirds are a feature of several sites, the most notable being Kizil Agach Bay (048) where formerly millions of birds occurred, though numbers have since shown a conspicuous decline, the Kura delta (046), and the inland lakes of Mahmud-chala (045), Sarysu (032) and Aggel (030).

IBAs were identified for the following bird species most often (number of sites in brackets): *Phalacrocorax pygmeus* (nine), *Cygnus cygnus* (eight), *Tetrao mlokosiewiczi* (nine), *Aquila heliaca* (seven), *Pelecanus crispus* (six), *Ardeola ralloides* (five), *Plegadis falcinellus* (five), *Fulica atra* (five) and *Tetrax tetrax* (five). Table 4 lists SPECs which meet criteria B2/B3 together with the proportion of their national populations covered by IBAs. The table shows that the IBAs in Azerbaijan support more than 30% of the national breeding populations of several waterbird species, raptors and gamebirds.

The most outstanding IBAs in the country are all wetlands: Divichi liman (site 024), Lake Mahmud-chala (045), Lake Sarysu (032), Lake Aggel (030) and Kizil Agach Bay (048). Between them, these qualify on the basis of global (criteria A1, A4i, A4iii) and regional importance (criteria B1i, B2) for a total of 37 species.

Figure 1. Occurrence of habitats at Important Bird Areas in Azerbaijan (see Appendix 3 for definitions of habitats).

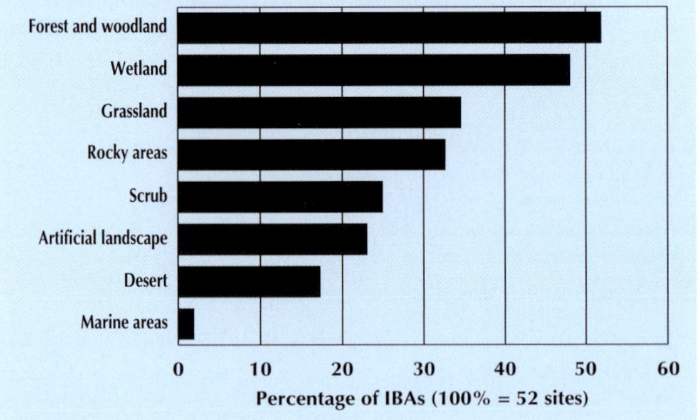

Figure 2. Occurrence of land-uses at Important Bird Areas in Azerbaijan (see Appendix 3 for definitions of land-uses).

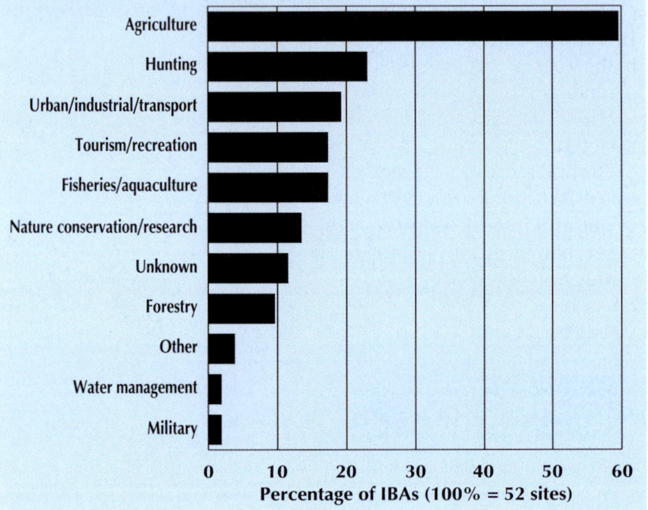

HABITATS

Azerbaijan has an altitudinal range extending from areas below sea-level on the Caspian littoral up to the rocky landscapes, snowfields and glaciers of high peaks in the Great Caucasus at over 4,000 m. The mountains of the Great Caucasus show the typical succession of vegetation and soils from lower to higher slopes related to changes in temperature, precipitation and gradient. In the lower belts, and in plains by the foothills (including in river valleys), mixed broadleaved forests of the warm-temperate type (*Castanea, Fagus, Quercus*) flourish, often with evergreen undergrowth. Higher up, *Fagus* predominates; higher still, conifer forest (*Picea, Abies*) and some broadleaved trees. On the upper treeline, there is dwarf *Fagus* and *Betula*, with dwarf *Juniperus* in drier places and, in the subalpine belt, *Rhododendron* and meadows; in the alpine belt similarly, short-grass meadows. Mountain steppe and mountain meadows prevail in the Little Caucasus, while the Talysh mountains have slopes clothed with broadleaved forests, the summits with meadow steppe. Some lower mountain areas have sparse arid forest chiefly of *Juniperus*. The Kura-Aras lowland is a semi-desert plain with a natural steppe region in the Lower Kura valley, where steppe or desert soils require irrigation for crop production (notably cotton). Wetlands in Azerbaijan include coastal bays, river deltas and marshes along the Caspian, as well as lakes (saline and freshwater, some man-made) there and further inland, river valleys and irrigation channels across plains.

Figure 1 shows that 27 of the 52 IBAs (52%) are predominantly forested, 25 (48%) are predominantly wetlands, 18 (35%) are predominantly grassland (lowland steppe/semi-desert and mountain meadows) and 17 (33%) rocky areas. Based on incomplete data, there are at least nine IBAs where more than 50% of the surface area is wooded.

Key species in the main habitats are as follows: *Ciconia nigra*, raptors, *Coracias garrulus* and woodpeckers are important in wooded river valleys. Raptors also occur higher up in mountains where there is a mixture of forest and rocky areas. *Tetrao mlokosiewiczi, Tetraogallus caucasicus* and *T. caspius*, and passerines such as *Anthus spinoletta, Prunella collaris* (very locally, *P. ocularis*), *Phoenicurus erythrogaster* and *Carpodacus rubicilla* occur in areas of rocks, shrubs and mountain meadows. Wintering geese and *Tetrax tetrax* occur on lowland plains, with resident *Francolinus francolinus* also occurring in thickets along rivers and canals there. Wetlands in Azerbaijan are immensely important, both nationally and internationally, for breeding, wintering and passage waterbirds, such as *Phalacrocorax pygmeus, Marmaronetta angustirostris, Oxyura leucocephala* and *Porphyrio porphyrio*.

IMPACTS ON IBAs – LAND-USE AND THREATS

Figure 2 summarizes land-use at IBAs in Azerbaijan. The dominant land-use in the country is agriculture, this affecting 31 IBAs (60%), while 12 (23%) are used for hunting, ten (19%) for industrial purposes (primarily, the oil industry), nine (17%) for tourism and recreation, nine (17%) for fisheries/aquaculture, seven (13%) for nature conservation/research (this figure is based on incomplete information and may not be accurate; see also 'Protection status' and 'Conservation', below) and five (9%) for forestry.

All IBAs are state-owned. At least eight wetland sites, including some of the most valuable IBAs, are rented or part-rented by the Hunters' Society of Azerbaijan. All nature reserves except Kizil

Figure 3. Occurrence of threats at Important Bird Areas in Azerbaijan (see Appendix 3 for definitions of threat types and impact categories).

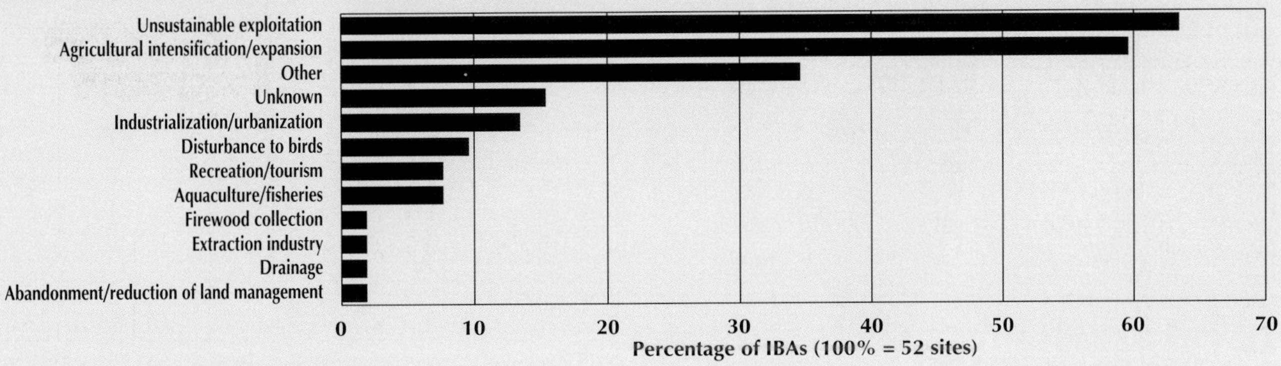

Agach and Gobustan Nature Monument were formerly managed by Azerbaijan's State Committee for Nature Conservation. Kizil Agach Reserve was (and continues to be) managed by the Department of Nature Reserves (Zapovedniks) of the former USSR in Moscow and Gobustan Nature Monument by Azerbaijan's Ministry of Culture. No details were available for this inventory on the role of other ministries (Oil, Forestry) in respect of site ownership and management.

Threats to IBAs are summarized in Figure 3. The level of the impact at individual sites is largely unknown.

The main threats are unsustainable levels of exploitation, affecting 64% of IBAs (mainly legal and illegal hunting at over 30 IBAs), and agricultural intensification/expansion (60%) (Figure 3). Waterbirds at Azerbaijan's most important wetlands are severely affected by hunting and poaching, also by the accompanying disturbance which, as has been suggested for some sites, is the more serious threat. Through habitat loss caused by grazing, agricultural intensification/expansion has an impact on upland species such as *Tetrao mlokosiewiczi, Tetraogallus caucasicus* and *Tetraogallus caspius*, and on geese at IBA 043. Problems are similarly created at wetlands (the most threatened IBAs) through drainage and run-off of pesticides from cotton fields; the hydrology of coastal wetlands has been further altered by the fall and rise of the Caspian Sea over the last 70 years.

PROTECTION STATUS

Table 5 and Figures 4 and 5 summarize the protection status of IBAs in Azerbaijan.

■ National protection

Regarding the overlap of individual IBAs with nationally protected areas, seven are each covered by more than 90%, whilst 28 (54%) are not legally protected at all (Figure 4).

Table 5. Protection status of Important Bird Areas in Azerbaijan.
A tick (✔) indicates that an IBA overlaps with a protected area (to any extent).

		National				International
IBA code	International name	Zapovednik	Zakaznik	Game Reserve	Nature Monument	Ramsar Site
001	Zakataly	✔				
002	Ilisu area	✔				
003	Alazani river valley					
004	Sheki upland			✔		
005	Karayazi forest	✔				
006	Akstafa-chai valley					
007	Shamkhor area			✔		
008	Korchai area			✔		
009	Varvara reservoir					
010	Turianchai	✔				
011	Gekchai Bozdag mountains					
012	Lake Gey Gel	✔				
013	Mount Giamysh					
014	Mount Dalidag					
015	Lachin area			✔		
016	Shahbuz area					
017	Ordubad area			✔		
018	Mount Ilandag					
019	Negram mountains					
020	Mount Bazar-Duzu					
021	Mount Shahdag					
022	Kusari (Gusari) area			✔		
023	Samur delta					
024	Divichi liman (or Lake Akzibir)					
025	Mount Babadag			✔		
026	Ismailly area	✔				
027	Alty Agach area	✔				
Subtotal of IBAs		7	0	7	0	0
028	Pirgulu (Pirkuli)	✔				
029	Barda tugai forest					
030	Lake Aggel	✔	✔			
031	Lake Boz-Koba					
032	Lake Sarysu					
033	Yashma island					
034	Absheron archipelago (north) and Artem Bay					
035	Shah Cape (Shakhova Kosa)			✔		
036	Lake Krasnoie and other water-bodies of the Absheron peninsula					
037	Mount Kargabazar and Mount Gush-gaya					
038	Gobustan area				✔	
039	Sangachal Bay					
040	Glynanyi island			✔		
041	Lake Hadjikabul					
042	Pirsagat islands and Los island					
043	Lakes Shorgel/Shirvan Reserve	✔				
044	Mugan steppe					
045	Lake Mahmud-chala					
046	Kura delta			✔		
047	Lake Ych-chala (Novogolovka-chala)	✔				
048	Kizil Agach Bay	✔	✔			✔
049	Vilajchai (Vilyazhchay) valley					
050	Zuvand upland			✔		
051	Girkan forest	✔				
052	Astara-chai valley					
Total number of IBAs		**13**	**2**	**11**	**1**	**1**

Figure 4. The national protection status of Important Bird Areas in Azerbaijan.

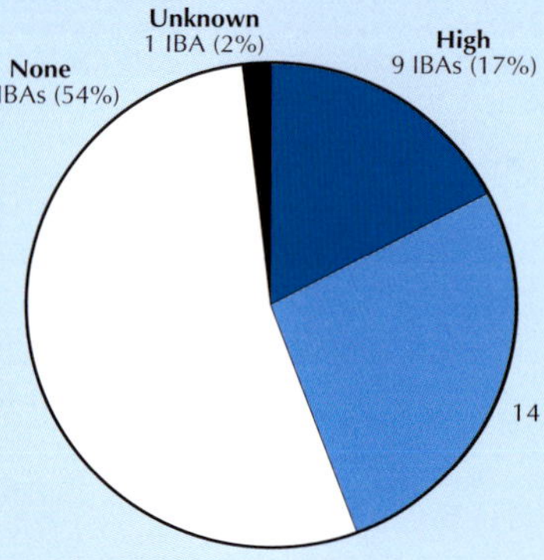

Total area of overlap between IBA network in Azerbaijan and national protected-area system (see Table 5 for categories) = 2,357 km² (38% of total IBA area).

Figure 5. The international protection status of Important Bird Areas in Azerbaijan.

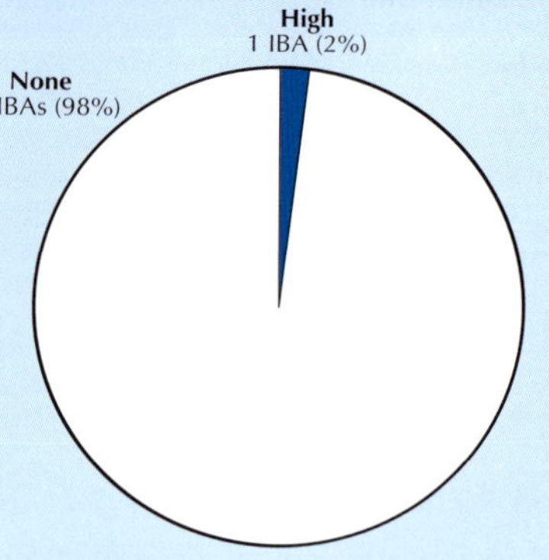

Total area of overlap between IBA network in Azerbaijan and international protected-area system (see Table 5 for categories) = 1,325 km² (22% of total IBA area).

Box 1. International legislation and initiatives that are relevant to site conservation Azerbaijan (see Appendix 1 for a general description of these agreements).

Global	
Biodiversity Convention	(✔)
Ramsar Convention	(✔)
Bonn Convention	
World Heritage Convention	✔
MAB Programme	
Pan-European	
Bern Convention	

✔ Convention ratified/initiative supported
(✔) Convention signed

In common with other constituent republics of the former USSR, Azerbaijan's protected-area system comprised both **Zapovedniks** (strict nature reserves generally set aside solely for wildlife protection and research; **goryh** in Azeri) and **Zakazniks** (temporary reserves with less protection, sometimes acting as a buffer zone to Zapovedniks; **yasaglakh** in Azeri). Further protected-area categories include **Game Reserve**, **Nature Monument** (represented here by IBA 038), and **National Park** (none at present in Azerbaijan). Prior to its secession from the former Soviet Union in August 1991, there were 14 Zapovedniks (15 are treated in the present inventory) and 18 Zakazniks in Azerbaijan, these reserves covering, respectively, 191,200 ha and 266,100 ha, or approximately 2.2% and 3.1% of the national territory. The Azerbaijan Committee for Nature Conservation is in charge of all nature reserves (Zapovedniks) and game reserves in Azerbaijan except Kizil Agach Reserve (managed from Moscow) and Gobustan Nature Monument (Ministry of Culture).

International protection

Azerbaijan is party to some of the international agreements relevant to site conservation (Box 1). Only one IBA (Kizil Agach Bay 048) overlaps with a Ramsar Site (Table 5). Sites 032, 043 and 045 certainly meet Ramsar criteria but have not yet been designated as Ramsar Sites. Sites 024, 030, 031 and 033 probably also meet Ramsar criteria; with the potential for Lake Boz-Koba (IBA 031) and Lake Aggel (IBA 030) to be combined with Lake Sarysu (IBA 032) as a single unit—'Water-bodies of the Mil Steppe'.

CONSERVATION

- No information was available for this inventory on either the more recent development of plans to establish further Zapovedniks at IBAs 014, 016, 019 and 029, the role of Zakazniks in Azerbaijan since independence (for brief mention, see sites 030, 043 and 048), or on what has become of plans to designate as National Parks IBAs 020 and 021.
- Against the background of political and social changes as well as severe economic difficulties in Azerbaijan in recent years, the Azeri authorities have had insufficient funds and manpower to secure and manage reserves, so that the country's protected-area system has largely collapsed. Pressure on reserves has increased markedly from both illegal hunting and incursions by local communities suffering from the economic decline and with little appreciation of the importance of protected areas. Therefore, it may be difficult to justify maintaining state reserves as exclusive areas. Bold, new management practices need to be developed in order to rescue and revive the country's reserve system and, especially, to strengthen wetland protection. Crucially, the extent of hunting and its influence on waterbird populations requires careful investigation in this context. Within the framework of an agreement between Azerbaijan's State Committee for Ecology (SCE) and the British Petroleum/Statoil Alliance (1992), a joint Fauna-and-Flora-International/BP/Statoil team proposed for immediate attention (in 1995): 1) The development of a management plan for the Kizil Agach Reserve (IBA 048), this also to serve as a model for other reserves (the lack of such plans is a serious deficiency); 2) The development of a conservation-education programme at the Shirvan Reserve for local children and improvement of visitor facilities there with the aim of attracting foreign fee-paying visitors from Baku; 3) The assessment of wetland sites (043, 048 and others) for breeding and wintering waterbirds and, especially, to determine the status of the globally threatened *Anser erythropus*, *Branta ruficollis* and *Numenius tenuirostris*.
- Exploration and exploitation of three Caspian oilfields was agreed between Azerbaijan and a consortium of western oil firms in September 1994. A programme subsequently drawn up by E. Sultanov (Academy of Sciences and Ornithological Society of Azerbaijan) for the oil companies and Wetlands International initially to cover the period April 1997 to May 1998 envisaged the following: regular monitoring during the breeding season, migration periods and in winter along pipelines, at the Sangachal oil terminal and in other potentially dangerous zones for birds on the Caspian littoral (thus covering several IBAs described in this inventory), this to involve the setting-up of a network of observers (amateurs, with some professional supervision), each to cover a c.10 km stretch weekly and to submit results to the Azerbaijan Ornithological Society; annual winter counts of waterbirds at 18 sites (including IBAs 009, 024, 030, 031, 032, 033, 043, 046 and 048); investigation of bird-scaring techniques to combat the threat of oil pollution (the acute danger this represents is illustrated by the fact that some 30,000 birds perished in an oil-pollution incident off Gil island during March–June 1998).

ANALYTICAL METHODS

- Most of the bird data (species composition and numbers) in this inventory were collected, and the majority of IBAs identified, by M. Patrikeev (Institute for Conservation of World Biodiversity, Canada; formerly at the Ecological Centre of Azerbaijan) in the period 1988–1991, with additional information (mainly helicopter-based counts, 1993) for IBAs 024, 046 (extending beyond delta to cover entire Kura estuary), 005, 032, 033, 042 and 045 supplied by E. Sultanov (Azerbaijan Ornithological Society). The accounts for four wetland sites (IBAs 030, 032, 045, 048) were expanded on the basis of counts (estimates and more exact counts from boat, helicopter and on

Box 2. Potential IBAs in Azerbaijan. No recent data are available on these sites. With more current information or study these sites may well in the future be considered as IBAs.

1. Coastal waters between towns of Lenkoran and Astara
2. Lenkoran lowland
3. Ah-chala and other wetlands of northern and central Mugan
4. Shilian marsh
5. Karasy marsh
6. Lake Mehman
7. Mingechaur reservoir (east)
8. Baku Bay and Caspian Sea south of Baku

foot) made during the winter of 1996 through a collaborative effort between Azerbaijan (State Committee for Ecology, Academy of Sciences) and international agencies (Wildfowl and Wetlands Trust, NOF/BirdLife Norway) coordinated by Fauna and Flora International, with logistical support coming from BP/Statoil.

- Significant numbers of certain SPECs are considered to breed at a number of sites (018, 019, 022, 023, 028, 049 and 051), although no quantitative data are available, therefore the B2 criterion has been applied at the level of the whole site, rather than to the relevant species.
- Information on land-uses and threats is derived from the above sources and the literature.
- Some sites were surveyed very well. At others, rapid assessment was carried out, with time spent at many sites varying from one day to a week. Only five to six sites were visited for longer periods or repeatedly with over a month of fieldwork per site. Nature reserve staff, hunters, teachers, and game wardens throughout Azerbaijan were interviewed and/or completed questionnaires.
- Some older data were available on eight further sites (Box 2) but, based on these data, it was not possible to assign IBA criteria. In the future, with more information, these sites may be considered as IBAs.

GLOSSARY

chala marsh or other wetland with reeds or other emergent vegetation.

tugai periodically flooded riparian forest in arid, semi-desert or semi-savanna regions; in Azerbaijan, originally with tall *Populus* and *Salix* (outside protected areas, most such trees are now felled) and dense, impenetrable understorey of *Crataegus*, *Rubus*, lianas, etc.

place-names Azeri names are used in the majority of cases, but some are transliterations from Russian. The Aras is the Azeri name for the river otherwise known as the Araks.

ACKNOWLEDGEMENTS

Elchin Sultanov, Chairman of the Ornithological Society of Azerbaijan (Academy of Sciences), supplied valuable information on eight IBAs, winter waterbird counts and other monitoring work particularly at sites vulnerable to the impact of the oil industry. Peter Cranswick (Wildfowl and wetlands Trust) reported on recent sightings of *Oxyura leucocephala* at two IBAs.

SITE ACCOUNTS

Zakataly — A1, A3 — 001

Admin region Belokany, Zakataly
Coordinates 41°44'N 46°40'E
Altitude 630–3,648 m **Area** 23,844 ha

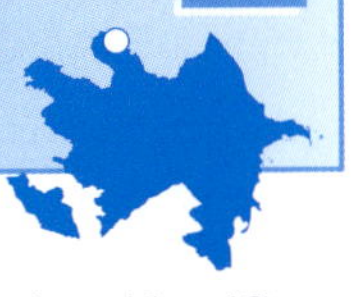

Site description

An area on the southern slope of the main Caucasian ridge (Great Caucasus). All the high-altitude vegetation zones are well defined, from broadleaved forest (mainly of *Fagus*, also *Quercus*, *Carpinus*, *Acer*, etc.) to alpine meadows. The relief is mountainous with deep chasms.

Habitats Forest and woodland (67%; broadleaved deciduous forest), Grassland (28%; alpine/subalpine/boreal grassland), Rocky areas
Land-use Unknown

Birds

Species	Season	Year	Pop min	Pop max	Acc	Criteria
[1] ***Tetrao mlokosiewiczi*** Caucasian Black Grouse	R	—	300	—	—	A1, A3
Tetraogallus caucasicus Caucasian Snowcock	R	—	—	—	—	A3
Montifringilla nivalis Snowfinch	R	—	—	—	—	A3

1. Average in 1980s.

This IBA has a diverse upland avifauna (104 species recorded), including at least three of the 10 species in Europe that are restricted (when breeding) to the Eurasian high-montane biome, as well as (e.g.) several raptors (*Gypaetus barbatus*, *Accipiter brevipes*, *Aquila chrysaetos*, *Otus scops*, *Bubo bubo*, etc.), *Alectoris chukar*, *Coturnix coturnix*, woodpeckers, and many passerines, among which are *Cinclus cinclus*, *Phoenicurus erythrogaster* and *Turdus torquatus*. Species of global conservation concern that do not meet IBA criteria: *Aegypius monachus*.

Protection status

National High **International** None
23,800 ha of IBA covered by Zapovednik (Zakataly goryhy, 23,800 ha).

Conservation issues

Tetrao mlokosiewiczi and *Tetraogallus caucasicus* have suffered from habitat loss since reserve boundaries changed, the reserve was split into two and some meadows were lost to grazing. It is recommended that the meadows (3,660 ha) separating the two sections of the reserve should be acquired for protection and that geological prospecting should cease in the Filizchai basin, returning at least 1,000 ha, taken away in 1978, to the reserve.

Threats Agricultural intensification/expansion (U)

Ilisu area — A1 — 002

Admin region Kakhi (Kah)
Coordinates 41°34'N 47°01'E
Altitude 700–2,000 m **Area** 9,345 ha

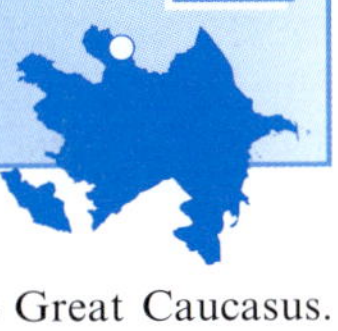

Site description

The site is situated on the southern slopes of the Great Caucasus. Forests are the dominant vegetation, mostly broadleaved, especially *Fagus* and (above 1,800 m) *Betula*. All land-uses are prohibited.

Habitats Forest and woodland (93%; broadleaved deciduous forest)
Land-use Nature conservation/research (100%)

Birds

Species	Season	Year	Pop min	Pop max	Acc	Criteria
[1] ***Tetrao mlokosiewiczi*** Caucasian Black Grouse	R	—	25	—	—	A1

1. 50+ birds.

Over 90 species breed at this IBA, including at least 11 raptors (*Gypaetus barbatus*: status uncertain, probably visits the area for foraging) and several woodpeckers; *Dryocopus martius* being rare and *Dendrocopos medius* common. Breeding species of global conservation concern that do not meet IBA criteria: *Aegypius monachus* (2–3 pairs).

Protection status

National High **International** None
9,345 ha of IBA covered by Zapovednik (Ilisu goryhy, 9,345 ha).

Conservation issues

Threats Unknown

The site was designated as a Zapovednik in 1987 and is under the control of the Azerbaijan Committee for Nature Conservation.

Alazani river valley — A1, B2 — 003

Admin region Kakhi (Kah), Zakataly, Belokany
Coordinates 41°27′N 46°29′E
Altitude 100–500 m **Area** 5,000 ha

Site description

Patches of tugai forests with *Populus*, Salix, etc. in the Alazani river valley adjacent to the Georgian border. Land-uses include forestry and some cattle- and sheep-grazing.

Habitats Forest and woodland (100%; broadleaved deciduous forest)
Land-use Agriculture, Forestry

Birds

Species	Season	Year	Pop min	Pop max	Acc	Criteria
Aquila pomarina Lesser Spotted Eagle	B	—	12	15	—	B2
[1] ***Aquila heliaca*** Imperial Eagle	B	—	8	10	—	A1

1. Probably partly on Georgian side

This is a notable site for forest raptors, especially *Aquila heliaca*; with *Accipiter brevipes* probably breeding and *Pandion haliaetus* a rare visitor throughout the year. Species of global conservation concern that do not meet IBA criteria: *Haliaeetus albicilla* (formerly bred, with probably 1 pair surviving). Significant proportion (≥1%) of national population breeding at site: *Ciconia nigra* (5–6 pairs).

Protection status

National None **International** None

Conservation issues

Threats Other (U)

Threatened by illegal tree-felling ('Other' threat).

Sheki upland — A1 — 004

Admin region Sheki
Coordinates 41°12′N 47°10′E
Altitude 100–1,500 m **Area** 10,400 ha

Site description

The lower foothills of the Great Caucasian mountains, with a few patches of broadleaved forest (mainly in the central part), with bushes and scrub in the river valleys. Land-uses include sheep- and cattle-grazing and some cultivation/arable farming.

Habitats Artificial landscape (arable land), Forest and woodland (broadleaved deciduous forest), Scrub (scrub)
Land-use Agriculture

Birds

Species	Season	Year	Pop min	Pop max	Acc	Criteria
Tetrax tetrax Little Bustard	W	—	200	300	—	A1

A wintering area for *Tetrax tetrax*, most of which occur outside the Game Reserve. Species of global conservation concern that do not meet IBA criteria: *Aegypius monachus* (visitor throughout the year, formerly bred), *Circus macrourus* (non-breeding summer visitor), *Aquila heliaca* (breeds), *Falco naumanni* (status uncertain, probably breeds), *Crex crex* (common on passage in some years), *Otis tarda* (extremely rare on migration and in winter). Among breeding raptors are *Milvus migrans*, *Circaetus gallicus*, *Accipiter brevipes* and *Aquila pomarina*, while *Aquila nipalensis* is common on passage. *Phasianus colchicus* and *Alectoris chukar* are very common; *Francolinus francolinus* is rare.

Protection status

National Partial **International** None
About half of the IBA is covered by a Game Reserve.

Conservation issues

Threats Agricultural intensification/expansion (U), Other (U), Unsustainable exploitation (U)

The main threats are poaching of *Tetrax tetrax* and *Phasianus colchicus*, excessive grazing in some areas and illegal tree-felling ('Other' threat).

Karayazi forest — A1 — 005

Admin region Kazakh, Akstafa
Coordinates 41°19′N 45°10′E
Altitude 0–200 m **Area** 10,000 ha

Site description

The reserve is bounded to the north by the Tbilisi–Baku railway, and to the south by the Kura waterway. Tugai forest in Kura valley, with *Populus*, *Salix*, *Morus*, etc. and, above valley, *Quercus*, *Pistacia* and *Morus*. There are also pastures, arable land, temporary islands.

Habitats Artificial landscape (arable land), Forest and woodland (broadleaved deciduous forest)
Land-use Nature conservation/research

Birds

Species	Season	Year	Pop min	Pop max	Acc	Criteria
[1] ***Aquila heliaca*** Imperial Eagle	B	—	3	4	—	A1

1. Late 1980s/early 1990s.

Over 70 species (c.50 passerines) recorded in reserve. Breeding species include *Ardea cinerea*, *Ciconia nigra*, *Pernis apivorus*, *Buteo buteo*, *Aquila pomarina*, *Phasianus colchicus*, *Sterna albifrons*, *Columba oenas*, *Strix aluco* and *Picus viridis*. *Coturnix coturnix* and *Gallinago gallinago* are numerous on passage.

Protection status

National Partial **International** None
4,900 ha of IBA covered by Zapovednik (Karayazi goryhy, 4,900 ha).

Conservation issues

Threats Unknown

This is the only Zapovednik protecting the threatened tugai forests of Azerbaijan, with the exception of a small patch in Turianchai Zapovednik. The forest has benefited since the establishment of the reserve (in 1978), large mammals and birds less so (there are frequent violations of the Zapovednik regime). However, the best stands of tugai forest are not included in the Zapovednik.

Akstafa-chai valley — B2 — 006

Admin region Kazakh
Coordinates 41°08′N 45°26′E
Altitude 200–500 m **Area** 200 ha

Site description

This site is close to Karayazi forest (IBA 005). The area contains the last remaining tugai forest on the banks of the River Akstafa, a right-bank tributary of the Kura river. Main land-uses are firewood-collection ('Other' land-use) and tourism. There is also some sheep-grazing.

Habitats Forest and woodland (100%; broadleaved deciduous forest)
Land-use Agriculture, Other, Tourism/recreation

Birds

Species	Season	Year	Pop min	Pop max	Acc	Criteria
Aquila pomarina Lesser Spotted Eagle	B	—	1	—	—	B2

Species of global conservation concern that do not meet IBA criteria: *Aquila heliaca* (visits the site throughout the year from IBA 005, but probably does not breed). Significant proportion (≥1%) of national population breeding at site: *Aquila pomarina* (1 pair, possibly more). *Alcedo atthis* and *Coracias garrulus* are very common breeding birds.

Protection status

National None **International** None

Conservation issues

Threats Agricultural intensification/expansion (U), Other (U)

Threats include illegal tree-felling ('Other' threat), which is being phased out, and grazing on the edge of the forest which inhibits regeneration. A game reserve for *Phasianus colchicus* is planned.

Shamkhor area
A1, B2 **007**

Admin region Shamkhor
Coordinates 40°56'N 46°15'E
Altitude 0–200 m **Area** 10,000 ha

Site description
Tugai forests with *Populus*, *Salix* in the Kura valley and remnants of tugai forests in the Iori delta (Mingechaur reservoir). There are also pastures, semi-arid steppe and scrub. Land-uses include grazing (buffalo and sheep) and some cultivation/arable farming.

Habitats Artificial landscape (arable land), Forest and woodland (broadleaved deciduous forest), Grassland (steppe/dry calcareous grassland), Scrub (scrub), Wetland (river/stream; standing fresh water)
Land-use Agriculture

Birds

Species	Season	Year	Pop min	Pop max	Acc	Criteria
Aquila heliaca Imperial Eagle	B	—	1	2	—	A1
[1] ***Francolinus francolinus*** Black Francolin	R	—	50	100	—	B2

1. 150-200 birds.

Species of global conservation concern that do not meet IBA criteria: *Haliaeetus albicilla* (visitor throughout the year, formerly bred). Significant proportion (≥1%) of national population breeding at site: *Phalacrocorax carbo* (15–20 pairs), *Aquila pomarina* (2–3 pairs, possibly more). Other breeding birds include *Alcedo atthis* (very common), *Coracias garrulus* (common) and woodpeckers. *Ciconia nigra* (status uncertain, probably breeds).

Protection status
National Partial **International** None
IBA partly covered by Game Reserve.

Conservation issues

Threats Agricultural intensification/expansion (U), Other (U)

The Game Reserve was established mainly to protect *Phasianus colchicus*, *Francolinus francolinus* and game animals. The Iori delta section is not protected. Grazing inhibits forest regeneration and illegal tree-felling ('Other' threat) also occurs.

Korchai area
A1, B2 **008**

Admin region Khanlar, Mingechaur
Coordinates 40°52'N 46°36'E
Altitude 100–500 m **Area** 15,000 ha

Site description
The area consists of semi-desert plains with low hills, low arid mountains, scrub along rivers, and a reservoir. Land-uses include sheep-grazing, agriculture in river valleys, and oil prospecting.

Habitats Desert (semi-desert), Scrub (scrub), Wetland (standing fresh water)
Land-use Agriculture, Urban/industrial/transport

Birds

Species	Season	Year	Pop min	Pop max	Acc	Criteria
Neophron percnopterus Egyptian Vulture	W	—	15	20	—	B2
Circus macrourus Pallid Harrier	W	—	5	10	—	A1
Francolinus francolinus Black Francolin	R	—	250	—	—	B2
Tetrax tetrax Little Bustard	W	—	300	600	—	A1

Breeding species: *Tadorna ferruginea* (10–15 pairs), *Buteo rufinus* (3 pairs), *Alectoris chukar* (150 pairs), *Pterocles orientalis* (very rare). Species of global conservation concern that do not meet IBA criteria: *Aegypius monachus* (non-breeding visitor throughout the year), *Falco naumanni* (breeds), *Otis tarda* (winter visitor: 2–5, not annual). *Aquila nipalensis* occurs, with 3–4 wintering and some tens passing through; *Anthropoides virgo* is rare on migration.

Protection status
National Partial **International** None
IBA partly covered by Game Reserve.

Conservation issues

Threats Agricultural intensification/expansion (U), Other (U), Unsustainable exploitation (U)

The main threats are overgrazing, poaching for *Tetrax tetrax* and *Francolinus francolinus*, and the presence of insecure stores of pesticides ('Other' threat).

Varvara reservoir
A1 **009**

Admin region Mingechaur
Coordinates 40°44'N 47°05'E
Altitude 0–100 m **Area** 4,000 ha

Site description
The reservoir was created in 1956–1957, thereby flooding tugai forests in the Kura river basin. Reedbeds have grown up along the western and eastern shores and there is also some *Tamarix* on islands. The reservoir water is used for various purposes; there are fish-ponds, and hunting takes place in autumn and winter.

Habitats Wetland (100%; standing fresh water; water-fringe vegetation)
Land-use Fisheries/aquaculture, Hunting, Water management

Birds

Species	Season	Year	Pop min	Pop max	Acc	Criteria
[1] ***Phalacrocorax pygmeus*** Pygmy Cormorant	R	—	40	50	—	A1

1. 50+ birds wintering.

The site supports a range of breeding waterbirds: herons, *Plegadis falcinellus*, *Porphyrio porphyrio*, terns. *Egretta alba* and *Phalacrocorax carbo* bred in dead trees standing in the reservoir until 1962. Species of global conservation concern that do not meet IBA criteria: *Haliaeetus albicilla* (rare winter visitor). Up to 7,000–8,000 waterbirds (ducks and *Fulica atra*) winter, *Larus ichthyaetus* occurs rarely in the same season; *Larus minutus* is a rare, non-breeding summer visitor.

Protection status
National None **International** None

Conservation issues

Threats Unknown

Threats are unknown, but intensive hunting is probably the reason why many waterbirds move to Mingechaur reservoir.

Turianchai
B2 **010**

Admin region Agdash, Evlakh
Coordinates 40°44'N 47°29'E
Altitude 400–650 m **Area** 12,634 ha

Site description
A reserve on the heavily-eroded slopes of the Bozdag ridge, in the southern foothills of the Great Caucasus. The principal habitats are steppe and sparse, arid forest mainly of *Juniperus* with *Pistacia* and, less so, *Quercus*. A small patch of tugai forest along Turianchai river of *Alnus*, *Salix*, *Populus*, etc.

Habitats Forest and woodland (61%; native coniferous forest; broadleaved deciduous forest), Grassland (steppe/dry calcareous grassland), Wetland (river/stream)
Land-use Unknown

Birds

Species	Season	Year	Pop min	Pop max	Acc	Criteria
Alectoris chukar Chukar	R	—	—	6,000	—	B2

At least 112 species recorded, with c.60 passerines. Breeding species of global conservation concern that do not meet IBA criteria: *Aegypius monachus*, *Aquila heliaca*, *Falco naumanni* (status uncertain, perhaps only on passage). Among other raptors in the best area for birds of prey in Azerbaijan are: *Gypaetus barbatus*, *Neophron percnopterus*, *Gyps fulvus*, *Circaetus gallicus*, *Aquila pomarina*, *A. clanga* (passage only), *Hieraaetus fasciatus* (pair in April 1991). *Merops superciliosus*, *Coracias garrulus* and *Emberiza melanocephala* also occur and huge numbers of *Turdus* thrushes winter in the reserve when there is a good *Juniperus* berry crop.

Protection status

National High **International** None

12,600 ha of IBA covered by Zapovednik (Turianchai goryhy, 12,600 ha).

Conservation issues

Threats Unknown

Lack of natural regeneration in the arid, sparse forests results from earlier overgrazing. The situation has improved since the reserve was established in 1958 and erosion has been reduced. Return to the reserve of the 16,000 ha (i.e. part of Gekchai Bozdag, IBA 011) lost in 1961 is considered essential for the better protection of Azerbaijan's arid landscapes.

Gekchai Bozdag mountains — A1, B2 — 011

Admin region Gabala (Kutgashen), Oguz (Vartashen)
Coordinates 40°42′N 47°37′E
Altitude 200–1,000 m **Area** 5,000 ha

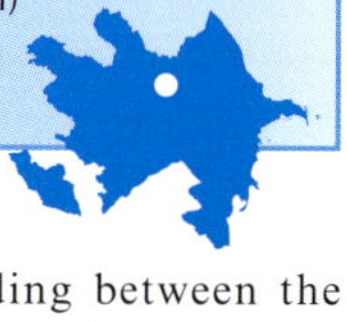

Site description

Low mountains with *Juniperus* forests extending between the Turianchai and Gekchai rivers; there are also cliffs, ravines, etc. This site is situated close to IBA 010 (Turianchai).

Habitats Forest and woodland (mixed forest), Rocky areas (inland cliff)
Land-use Agriculture, Forestry

Birds

Species	Season	Year	Pop min	Pop max	Acc	Criteria
[1] ***Gyps fulvus*** Griffon Vulture	R	—	12	15	—	B2
Aegypius monachus Cinereous Vulture	R	—	6	7	—	A1
Aquila heliaca Imperial Eagle	R	—	2	4	—	A1

1. Two colonies.

Breeding species of global conservation concern that do not meet IBA criteria: *Falco naumanni.* Significant proportion (≥1%) of national population breeding at site: *Neophron percnopterus* (5+ pairs). Other notable breeding birds: *Circaetus gallicus*, *Buteo rufinus*, *Aquila pomarina* and *Bubo bubo*.

Protection status

National None **International** None

Conservation issues

Threats Agricultural intensification/expansion (U), Unsustainable exploitation (U)

The main threats are from illegal tree-felling, and grazing.

Lake Gey Gel — A1, A3 — 012

Admin region Khanlar
Coordinates 40°26′N 46°19′E
Altitude 1,100–3,065 m **Area** 7,131 ha

Site description

An area of forest (in main reserve of 6,739 ha: *Quercus*, *Fagus*, *Carpinus*, etc.), subalpine and alpine meadows, with mountain lakes, rivers and springs, on the deeply incised slopes of Mount Kyapaz (3,065 m) in the north-eastern Murovdag range (Little Caucasus).

Habitats Forest and woodland (52%; broadleaved deciduous forest), Grassland (36%; alpine/subalpine/boreal grassland), Wetland (standing fresh water)
Land-use Unknown

Birds

Species	Season	Year	Pop min	Pop max	Acc	Criteria
Tetrao mlokosiewiczi Caucasian Black Grouse	R	—	—	—	—	A1, A3
[1] ***Tetraogallus caspius*** Caspian Snowcock	R	1990	84	—	—	A3
Prunella collaris Alpine Accentor	U	—	—	—	—	A3
Pyrrhocorax graculus Alpine Chough	U	—	—	—	—	A3

1. Individuals (autumn count).

Over 50 species breed, including 35 in the forest zone. Breeding species of global conservation concern that do not meet IBA criteria: *Aegypius monachus*. Other breeding vulture species are *Gypaetus barbatus*, *Neophron percnopterus* and *Gyps fulvus*. The subalpine and alpine zones support at least four of the 10 species in Europe that are restricted (when breeding) to the Eurasian high-montane biome, as well as *Alectoris chukar*, *Anthus spinoletta*, *Serinus pusillus*, *Carduelis flavirostris*, *Carpodacus erythrinus*, among others.

Protection status

National High **International** None

7,131 ha of IBA covered by Zapovednik (Geygel goryhy, 7,131 ha).

Conservation issues

Threats Agricultural intensification/expansion (U), Other (U), Recreation/tourism (U)

The main threats are the drawing of water from Lake Gey Gel for Gandja (Kirovabad) ('Other' threat), increased access for tourists to the lake environs, and associated development of facilities, also strip farming and cattle-driving. The separate northern section of the reserve exists primarily to protect the rare endemic pine *Pinus eldarica*.

Mount Giamysh — A1, A3, B2 — 013

Admin region Khanlar, Karabakh
Coordinates 40°16′N 46°21′E
Altitude 1,500–3,724 m **Area** 8,500 ha

Site description

The highest peak of the Murovdag ridge (Little Caucasus). A mountainous area with some forests and bushes. The main land-uses are sheep- and cattle-grazing, and tourism.

Habitats Forest and woodland, Rocky areas (scree/boulders; inland cliff), Scrub (scrub)
Land-use Agriculture, Tourism/recreation

Birds

Species	Season	Year	Pop min	Pop max	Acc	Criteria
[1] ***Tetrao mlokosiewiczi*** Caucasian Black Grouse	R	—	75	—	—	A1, A3
Tetraogallus caspius Caspian Snowcock	R	—	25	40	—	A3, B2
Pyrrhocorax graculus Alpine Chough	R	—	—	—	—	A3

1. 150+ birds.

The alpine and subalpine zones support at least three of the 10 species in Europe that are restricted (when breeding) to the Eurasian high-montane biome. This is also a notable site for breeding raptors, including *Gypaetus barbatus*, *Gyps fulvus*, *Neophron percnopterus*, *Aquila chrysaetos*, *Otus scops* and *Bubo bubo*. Species of global conservation concern that do not meet IBA criteria: *Aegypius monachus* (breeds), *Aquila heliaca* (status uncertain; probably does not breed).

Protection status

National None **International** None

Conservation issues

Threats Agricultural intensification/expansion (U), Other (U), Unsustainable exploitation (U)

Threats include intensive grazing and poaching. Some damage may have been caused by the war between Azerbaijan and Armenia ('Other' threat).

Mount Dalidag

A3, B2 **014**

Admin region Kelbadjhar, Lachin
Coordinates 39°55'N 46°03'E
Altitude 1,500–3,616 m **Area** 10,000 ha

Site description

The highest peak of the Karabakh uplands (Little Caucasus). Precipices, ravines, cliffs, rocks and steep slopes, and some forest near the foot of the mountain. The main land-use is (summer) sheep-grazing.

Habitats Forest and woodland, Grassland (steppe/dry calcareous grassland; alpine/subalpine/boreal grassland), Rocky areas (scree/boulders; inland cliff)
Land-use Agriculture

Birds

Species	Season	Year	Pop min	Pop max	Acc	Criteria
[1] ***Tetraogallus caspius*** Caspian Snowcock	R	—	75	—	—	A3, B2

1. 150+ adults.

Probably the most important area in Azerbaijan and the Transcaucasus for *Tetraogallus caspius*, a species restricted to the Eurasian high-montane biome. Other breeding birds include *Gypaetus barbatus*, *Neophron percnopterus*, *Gyps fulvus*, *Circaetus gallicus* (possibly breeds), *Aquila chrysaetos*, *Falco peregrinus*, *Bubo bubo*, *Alectoris chukar* and *Melanocorypha bimaculata*. Species of global conservation concern that do not meet IBA criteria: *Aegypius monachus* (status uncertain).

Protection status

National None **International** None

Conservation issues

Threats Agricultural intensification/expansion (U), Other (U), Unsustainable exploitation (U)

Among threats are intensive grazing and poaching, and the war between Azerbaijan and Armenia may also have had a deleterious impact ('Other' threat). A nature reserve (Zapovednik) was planned.

Lachin area

A1 **015**

Admin region Lachin
Coordinates 39°47'N 46°25'E
Altitude 500–1,800 m **Area** 20,000 ha

Site description

Low mountains of Karabakh ridge and Karabakh upland. More than half the site is covered by *Juniperus* forests, and there is mountain steppe in the north. The main land-use is sheep-grazing.

Habitats Forest and woodland, Grassland (steppe/dry calcareous grassland)
Land-use Agriculture

Birds

Species	Season	Year	Pop min	Pop max	Acc	Criteria
Aegypius monachus Cinereous Vulture	R	—	5	—	—	A1

Breeding species of global conservation concern that do not meet IBA criteria: *Aquila heliaca*, *Falco naumanni*, *Tetrao mlokosiewiczi*. *Ciconia ciconia* and *C. nigra*, at least 14 raptor species and *Alectoris chukar* (very common) also breed. *Milvus milvus* is a very rare summer visitor and *Gypaetus barbatus* possibly breeds.

Protection status

National Partial **International** None
IBA partly covered by Game Reserve.

Conservation issues

Threats Agricultural intensification/expansion (U), Other (U), Unsustainable exploitation (U)

Threats include intensive grazing, poaching and tree-felling. The area was probably damaged as a result of the war with Armenia ('Other' threat). The game reserve was established primarily to protect wild goat *Capra aegagrus*.

Shahbuz area

A3, B2 **016**

Admin region Shahbuz, Nakhichevan
Coordinates 39°34'N 45°41'E
Altitude 1,500–3,120 m **Area** 7,000 ha

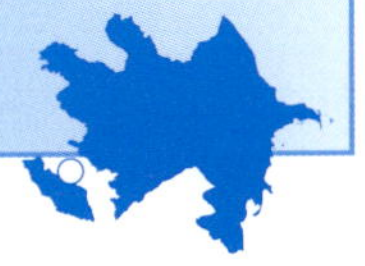

Site description

High mountains of the Little Caucasus (including Mount Kukudag, 3,120 m) with alpine meadows, ravines, precipices, etc. and the Bichenek high-altitude forest. Land-uses include sheep- and cattle-grazing, and mining.

Habitats Forest and woodland (broadleaved deciduous forest; mixed forest), Grassland (alpine/subalpine/boreal grassland), Rocky areas (inland cliff)
Land-use Agriculture, Urban/industrial/transport

Birds

Species	Season	Year	Pop min	Pop max	Acc	Criteria
Tetraogallus caspius Caspian Snowcock	R	—	20	—	—	A3, B2

The alpine and subalpine zones support at least one of the 10 species in Europe that are restricted (when breeding) to the Eurasian high-montane biome. Breeding species of global conservation concern that do not meet IBA criteria: *Aegypius monachus*. Other breeding raptors include *Gypaetus barbatus*, *Gyps fulvus*, *Neophron percnopterus*, *Aquila chrysaetos* (possibly breeds) and *Bubo bubo*.

Protection status

National None **International** None

Conservation issues

Threats Agricultural intensification/expansion (U), Other (U)

Intensive grazing is a threat and the war with Armenia is likely to have caused damage ('Other' threat). This IBA contains the only forest in the Nakhichevan region. A nature reserve is planned.

Ordubad area

A3, B2 **017**

Admin region Ordubad, Nakhichevan
Coordinates 39°11'N 45°56'E
Altitude 1,000–3,904 m **Area** 40,000 ha

Site description

The wildest part of Nakhichevan: high mountains of the Zangezur ridge (including the highest peak, Mount Kapydjik, at 3,904 m) with rocky slopes, cliffs, ravines, alpine meadows. There are also low, semi-arid mountains and some areas of semi-desert. Land-uses include grazing (intensive), some agriculture and mining.

Habitats Desert (semi-desert), Grassland (alpine/subalpine/boreal grassland), Rocky areas (scree/boulders; inland cliff)
Land-use Agriculture, Urban/industrial/transport

Birds

The alpine and subalpine zones support at least two of the 10 species in Europe that are restricted (when breeding) to the Eurasian high-montane biome. Breeding species of global conservation concern that do not meet IBA criteria: *Aegypius monachus*, *Aquila heliaca* (probably breeds), *Falco naumanni*. Raptors also breeding at this site include *Gypaetus barbatus*, *Neophron percnopterus* and *Gyps fulvus*. Among breeding passerines are *Pyrrhocorax pyrrhocorax*, *Irania gutturalis* and *Bucanetes githagineus* (*B. mongolicus* was also locally numerous in 1969–1970).

Species	Season	Year	Pop min	Pop max	Acc	Criteria
Tetrao mlokosiewiczi Caucasian Black Grouse	R	—	—	—	—	A3
[1] ***Tetraogallus caspius*** Caspian Snowcock	R	—	25	—	—	A3, B2

1. 50+ birds.

Protection status

National High **International** None
40,000 ha of IBA covered by Game Reserve.

Conservation issues

Threats Agricultural intensification/expansion (U), Other (U), Unsustainable exploitation (U)

Threats include some poaching, overgrazing and potential damage from the war with Armenia ('Other' threat). The Game Reserve, where only hunting is prohibited, was established primarily for the conservation of *Tetraogallus caspius* and certain mammals. Virtually unexplored by ornithologists.

Mount Ilandag — B2 — 018

Admin region Babek, Nakhichevan
Coordinates 39°03'N 45°39'E
Altitude 500–1,500 m **Area** 4,000 ha

Site description

Low, semi-arid mountain surrounded by undulating semi-arid plain, steep slopes, rocky area, some bushes.

Habitats Desert (semi-desert), Rocky areas (scree/boulders), Scrub (scrub)
Land-use Agriculture

Birds

Breeding species include several raptors, *Alectoris chukar* (very common), *Irania gutturalis*, *Oenanthe finschii*, *Pyrrhocorax pyrrhocorax*, *Petronia petronia*, *Bucanetes githagineus*. Species of global conservation concern that do not meet IBA criteria: *Falco naumanni* (breeds), *Tetrax tetrax* (rare passage and winter visitor).

Protection status

National None **International** None

Conservation issues

Threats Agricultural intensification/expansion (U), Unsustainable exploitation (U)

Main threats come from intensive grazing and poaching.

Negram mountains — B2 — 019

Admin region Babek, Nakhichevan
Coordinates 39°02'N 45°29'E
Altitude 1,000–1,500 m **Area** 3,500 ha

Site description

Semi-arid low mountains with cliffs, ravines, rocky areas, etc. Formerly part of the border zone, with restricted access and land-use. There is now some sheep-grazing.

Habitats Desert (semi-desert), Rocky areas (scree/boulders; inland cliff)
Land-use Agriculture

Birds

The site supports many species typical of semi-arid mountains—raptors, *Alectoris chukar*, *Pterocles orientalis* and passerines such as *Pyrrhocorax pyrrhocorax*, *Petronia petronia* and *Bucanetes githagineus*. Breeding species of global conservation concern that do not meet IBA criteria: *Falco naumanni*. *Aquila nipalensis* also occurs on passage. Status uncertain: *Chlamydotis undulata* (recorded here in early 19th century; unconfirmed reports of irregular occurrence in late 1980s), *Apus affinis* (possibly breeds; occasional records in Nakhichevan, including at this site, following late 19th century observations elsewhere in Azerbaijan).

Protection status

National None **International** None

Conservation issues

Threats Agricultural intensification/expansion (U), Unsustainable exploitation (U)

Threats include overgrazing and poaching. A nature reserve (Zapovednik) was planned.

Mount Bazar-Duzu — A1, A3 — 020

Admin region Kuba, Gabala (Kutgashen), Kusari
Coordinates 41°14'N 47°50'E
Altitude 1,500–4,466 m **Area** 4,000 ha

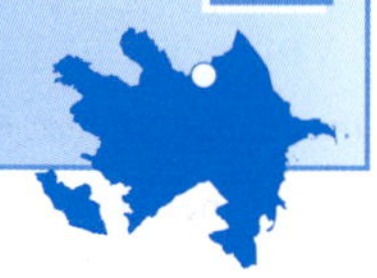

Site description

The highest peak in Azerbaijan and part of the mountain lying in Dagestan (Russia). Habitats range from broadleaved woodland up to the zone of eternal snow.

Habitats Forest and woodland (broadleaved deciduous forest; native coniferous forest; mixed forest; treeline ecotone), Grassland (alpine/subalpine/boreal grassland), Rocky areas (scree/boulders)
Land-use Agriculture, Tourism/recreation

Birds

Species	Season	Year	Pop min	Pop max	Acc	Criteria
Tetrao mlokosiewiczi Caucasian Black Grouse	R	—	—	—	—	A1, A3
Tetraogallus caucasicus Caucasian Snowcock	R	—	—	—	—	A3
Pyrrhocorax graculus Alpine Chough	R	—	—	—	—	A3
Carpodacus rubicilla Great Rosefinch	R	—	—	—	—	A3

This site supports a rich diversity of typical mountain birds, including at least four of the 10 species in Europe that are restricted (when breeding) to the Eurasian high-montane biome, as well as raptors (*Gypaetus barbatus*, *Gyps fulvus*, *Neophron percnopterus*, *Aquila chrysaetos*) and such passerines as *Phoenicurus erythrogaster*. Breeding species of global conservation concern that do not meet IBA criteria: *Aegypius monachus*.

Protection status

National None **International** None

Conservation issues

Threats Agricultural intensification/expansion (U), Unsustainable exploitation (U)

Threats include sheep-grazing and some poaching. A National Park was planned.

Mount Shahdag — A1, A3 — 021

Admin region Kusari
Coordinates 41°13'N 47°59'E
Altitude 3,000–4,243 m **Area** 3,500 ha

Site description

One of the highest mountains in Azerbaijan (4,243 m) with ravines, cliffs, alpine and subalpine meadows, glaciers and eternal snow. Land-uses include sheep-grazing and mountaineering.

Habitats Grassland (alpine/subalpine/boreal grassland), Rocky areas (scree/boulders; inland cliff)
Land-use Agriculture, Tourism/recreation

Birds

Species	Season	Year	Pop min	Pop max	Acc	Criteria
Tetrao mlokosiewiczi Caucasian Black Grouse	R	—	—	—	—	A1, A3
Tetraogallus caucasicus Caucasian Snowcock	R	—	—	—	—	A3
Pyrrhocorax graculus Alpine Chough	R	—	—	—	—	A3
Carpodacus rubicilla Great Rosefinch	R	—	—	—	—	A3

The alpine and subalpine zones support at least four of the 10 species in Europe that are restricted (when breeding) to the Eurasian high-montane biome. The site also supports breeding raptors (*Gypaetus barbatus*, *Gyps fulvus*, etc.) and, among passerines, *Cinclus cinclus*, *Phoenicurus erythrogaster* and *Pyrrhocorax pyrrhocorax*. Species of global conservation concern that do not meet IBA criteria: *Aegypius monachus* (possibly breeds), *Aquila heliaca* (possibly breeds).

Protection status

National None **International** None

Conservation issues

Threats Agricultural intensification/expansion (U), Unsustainable exploitation (U)

Threats include intensive grazing and some poaching. There was a plan to designate the area as a National Park.

Kusari (Gusari) area — B2 — 022

Admin region Kusari
Coordinates 41°29′N 48°23′E
Altitude 200–1,500 m **Area** 15,000 ha

Site description

Forests in the Samur valley, including bushes and scrub, on low hills and extending up to 1,000–1,500 m into the mountains. The main land-uses are grazing, arable farming and forestry.

Habitats Artificial landscape (arable land), Forest and woodland (broadleaved deciduous forest), Grassland (alpine/subalpine/boreal grassland), Scrub (scrub)
Land-use Agriculture, Forestry

Birds

The site supports at least 16 raptor species, most of which are breeders. *Neophron percnopterus* only possibly breeds, while *Pandion haliaetus* is a rare visitor throughout the year. Species of global conservation concern that do not meet IBA criteria: *Aegypius monachus* (status uncertain; common visitor throughout the year), *Aquila heliaca* (breeds), *Crex crex* (occurs on passage).

Protection status

National Partial **International** None
IBA partly covered by Game Reserve.

Conservation issues

Threats Agricultural intensification/expansion (U), Unsustainable exploitation (U)

There is some illegal tree-felling and poaching. Grazing hinders forest regeneration. The game reserve was established mainly for the protection of *Phasianus colchicus* and game mammals (wild boar *Sus scrofa* and roe deer *Capreolus capreolus*, etc.).

Samur delta — B2 — 023

Admin region Khachmas
Coordinates 41°47′N 48°42′E
Altitude (-27)–75 m **Area** 6,000 ha

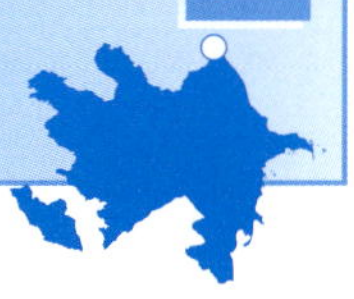

Site description

Remnants of broadleaved forests (*Quercus*, *Tilia*, *Ulmus*, *Fagus*, etc.) in the lowland river delta of north-eastern Azerbaijan; the site also extends into Dagestan (Russia). Sand beaches along Caspian coast have been largely lost (inundated) since mid-1980s—a few pebble beaches remain. Main land-uses around, and to some extent within, forest patches are crop-growing and cattle-grazing, while increasing numbers of tourists and others frequent the Caspian coast.

Habitats Artificial landscape (arable land), Forest and woodland (broadleaved deciduous forest), Wetland (shingle/stony beach; river/stream)
Land-use Agriculture, Tourism/recreation

Birds

Species of global conservation concern that do not meet IBA criteria: *Haliaeetus albicilla* (1–2 breeding pairs still present on the Dagestan side of this site). Other breeding species include *Ciconia nigra*, *Aquila pomarina*, *Alcedo atthis* (30+ pairs), and *Coracias garrulus* (50+ pairs). There is a congregation of seabirds off the delta from July to August (some thousands, but never counted) including *Pelecanus* sp. (some tens).

Protection status

National None **International** None

Conservation issues

Threats Agricultural intensification/expansion (U), Recreation/tourism (U), Unsustainable exploitation (U)

Threats come from fragmentation of the forests for agriculture and disturbance from tourism. There is also some poaching.

Divichi liman (or Lake Akzibir) — A1, A4i, A4iii, B1i, B2 — 024

Admin region Divichi
Coordinates 41°19′N 49°05′E
Altitude (-27)–(-27) m **Area** 7,000 ha

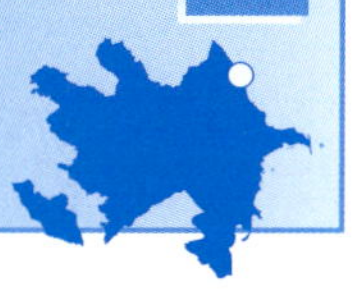

Site description

A shallow (0.5–1.2 m) lake with vast reedbeds on its western side in the Khachmas coastal lowland. It is leased by the Hunters' Society of Azerbaijan. The lake area varies from 1,600 to 7,000 ha depending on irrigation and rainfall.

Habitats Wetland (100%; standing fresh water; water-fringe vegetation)
Land-use Fisheries/aquaculture, Hunting

Birds

Species	Season	Year	Pop min	Pop max	Acc	Criteria
Phalacrocorax pygmeus Pygmy Cormorant	P	—	200	300	—	A1, A4i, B1i
[1] ***Pelecanus crispus*** Dalmatian Pelican	P	—	50	100	—	A1, A4i, B1i
[2] ***Platalea leucorodia*** Spoonbill	P	—	300	400	—	A4i, B1i
[3] ***Cygnus cygnus*** Whooper Swan	P	—	600	1,000	—	A4i, B1i
[4] ***Oxyura leucocephala*** White-headed Duck	P	—	—	—	—	A1
[5] ***Himantopus himantopus*** Black-winged Stilt	P	—	500	600	—	B1i
[5] ***Recurvirostra avosetta*** Avocet	P	—	100	200	—	B2
Glareola pratincola Collared Pratincole	B	—	20	—	—	B2
[6] ***Tringa stagnatilis*** Marsh Sandpiper	P	—	200	300	—	A4i, B1i
[6] ***Xenus cinereus*** Terek Sandpiper	P	—	300	—	—	B1i

1. Early October.
2. August–October.
3. November and March.
4. Some tens in November
5. August.
6. August–September.

Among breeding species are large colonies of herons (including *Botaurus stellaris*, *Ixobrychus minutus* and *Ardea purpurea*). *Porzana parva* and *Porphyrio porphyrio* also breed. The site is especially important for the autumn migration of waterbirds, more than 70,000–80,000 pass each year, with about 5,000 wildfowl wintering (*Anas platyrhynchos*, *A. clypeata*, *A. crecca*, *A. querquedula*, *A. strepera*, *Aythya fuligula*, *Netta rufina*), and up to 5,000 *Fulica atra*. Species of global conservation concern that do not meet IBA criteria: *Marmaronetta angustirostris* (very rare passage visitor), *Aythya nyroca* (two estimates: 5+ breeding pairs; several hundreds breeding).

Protection status

National None **International** None

Conservation issues

Threats Agricultural intensification/expansion (U), Other (U), Unsustainable exploitation (U)

The main threat comes from very intensive hunting, this being one of the most visited sites in Azerbaijan (about 225,000 shots per year). There are frequent changes in the water-level ('Other' threat), and it was probably polluted by pesticides from agricultural run-off.

Mount Babadag A1, A3 025

Admin region Ismailly, Kuba
Coordinates 41°02′N 48°07′E
Altitude 1,000–3,629 m **Area** 9,000 ha

Site description

One of the four highest peaks of the Great Caucasus in Azerbaijan. Habitats include ravines, subalpine meadows, rock, mountain forests (on south-western slopes only). There are some abandoned orchards on the southern slopes. Part of the area is leased by the Hunters' Society of Azerbaijan.

Habitats Artificial landscape (perennial crops/orchards/groves), Forest and woodland (mixed forest; treeline ecotone), Grassland (alpine/subalpine/boreal grassland), Rocky areas (scree/boulders)
Land-use Agriculture, Hunting

Birds

Species	Season	Year	Pop min	Pop max	Acc	Criteria
Aquila heliaca Imperial Eagle	R	—	2	—	—	A1
Tetrao mlokosiewiczi Caucasian Black Grouse	R	—	100	—	—	A1, A3
Tetraogallus caucasicus Caucasian Snowcock	R	—	Common		—	A3

The subalpine and alpine zones support at least two of the 10 species in Europe that are restricted (when breeding) to the Eurasian high-montane biome. Breeding species of global conservation concern that do not meet IBA criteria: *Aegypius monachus* (3–4 pairs). Significant proportion (≥1%) of national population breeding at site: *Gypaetus barbatus* (1+ pair), *Neophron percnopterus* (4–5 pairs), *Gyps fulvus* (8–10 pairs) and *Bubo bubo* (2+ pairs).

Protection status

National Partial **International** None
IBA partly covered by Game Reserve (Ismailly).

Conservation issues

Threats Agricultural intensification/expansion (U), Unsustainable exploitation (U)

Subalpine meadows are damaged by intensive grazing and poaching is a further threat. There is a hunting lodge near the mountain.

Ismailly area A1 026

Admin region Ismailly
Coordinates 40°56′N 48°10′E
Altitude 600–2,250 m **Area** 5,778 ha

Site description

The reserve lies on the southern slope of the Great Caucasus. Broadleaved forest (*Fagus*, *Carpinus*, *Quercus*, etc.) predominates, with only 227 ha of subalpine meadows. All land-uses are prohibited.

Habitats Forest and woodland (87%; broadleaved deciduous forest), Grassland (4%; alpine/subalpine/boreal grassland)
Land-use Nature conservation/research (100%)

Birds

Species	Season	Year	Pop min	Pop max	Acc	Criteria
Aquila heliaca Imperial Eagle	B	—	2	—	—	A1
[1] ***Tetrao mlokosiewiczi*** Caucasian Black Grouse	R	—	15	20	—	A1

1. 30–40 adults.

Over 100 species typical of upland forests have been recorded. Breeding birds include raptors, *Coracias garrulus* and woodpeckers, *Dryocopus martius* being rare.

Protection status

National High **International** None
5,778 ha of IBA covered by Zapovednik (Ismailly goryhy, 5,778 ha).

Conservation issues

Threats Unknown

Secondary meadows above 1,900 m were damaged by grazing before the reserve's establishment in 1981.

Alty Agach area A1, B2 027

Admin region Khizi
Coordinates 40°50′N 48°54′E
Altitude 800–2,205 m **Area** 5,500 ha

Site description

Low mountains on the easternmost part of the Great Caucasus. Most of the site is forested and there are also areas of steppe and numerous small lakes.

Habitats Artificial landscape (arable land), Forest and woodland (broadleaved deciduous forest; mixed forest), Grassland (steppe/dry calcareous grassland), Wetland (standing fresh water)
Land-use Agriculture, Tourism/recreation

Birds

Species	Season	Year	Pop min	Pop max	Acc	Criteria
Aquila heliaca Imperial Eagle	B	—	—	—	—	A1
Hieraaetus pennatus Booted Eagle	B	—	6	—	—	B2

A preliminary survey found more than 65 species, including 20 raptors. Species of global conservation concern that do not meet IBA criteria: *Aegypius monachus* (2+ pairs), *Falco naumanni* (breeds), *Tetrao mlokosiewiczi* (very rare, possibly breeds), *Crex crex* (common on passage). Significant proportion (≥1%) of national population breeding at site: *Hieraaetus pennatus* (6+ pairs). *Aquila nipalensis* is common on migration.

Protection status

National Partial **International** None
4,400 ha of IBA covered by Zapovednik (Alty-Agach goryhy, 4,400 ha).

Conservation issues

Threats Unknown

Threats probably include illegal tree-cutting and firewood-collection. Alty-Agach Zapovednik was established in 1990. Few field studies have been carried out at the site.

Pirgulu (Pirkuli) B2 028

Admin region Shemakha
Coordinates 40°43′N 48°33′E
Altitude 800–2,000 m **Area** 1,520 ha

Site description

A reserve in three sections: the basin of the Kyrkbulak stream; Mount Pirgulu; wooded spurs east of Chagan. Habitats include open areas, ancient pasture, hay-meadows, cliffs, scree, open water, and *Carpinus*, *Fagus*, and *Quercus* forest.

Habitats Forest and woodland (81%; broadleaved deciduous forest), Grassland (humid grassland), Rocky areas (inland cliff), Wetland (river/stream)
Land-use Unknown

Birds

Species of global conservation concern that do not meet IBA criteria: *Aegypius monachus* (status uncertain), *Aquila heliaca* (status uncertain). Breeding birds include: *Pernis apivorus*, *Neophron percnopterus*, *Gyps fulvus* (recorded in breeding season), *Buteo buteo*, *Hieraaetus pennatus*, *Alectoris chukar* and woodpeckers.

■ Protection status

National High **International** None
1,500 ha of IBA covered by Zapovednik (Pirgulu goryhy, 1,500 ha).

■ Conservation issues

Threats Unknown

Prior to the establishment of the reserve (1968), the main threat was destruction of the forest habitat and consequent erosion. The present reserve is considered too small (an increase by c.4,000 ha is recommended) and its division into three separate sections a disadvantage. The fauna, including birds, has been little studied.

Barda tugai forest — A1, B2 — 029

Admin region Agdash, Barda
Coordinates 40°23′N 47°22′E
Altitude 0–100 m **Area** 4,000 ha

■ Site description

The site encompasses tugai forest (mainly *Populus* and *Salix*), along the meandering Kura river. This is one of the two last big areas of this habitat in Azerbaijan. The main land-uses are forestry and agriculture (including grazing).

Habitats Artificial landscape (arable land), Forest and woodland (broadleaved deciduous forest), Scrub (scrub), Wetland (standing fresh water; river/stream)
Land-use Agriculture, Forestry

■ Birds

Species	Season	Year	Pop min	Pop max	Acc	Criteria
Accipiter brevipes Levant Sparrowhawk	B	—	5	—	—	B2
Falco naumanni Lesser Kestrel	B	—	30	—	—	A1
Francolinus francolinus Black Francolin	R	—	100	150	—	B2

Species of global conservation concern that do not meet IBA criteria: *Haliaeetus albicilla* (formerly bred; 3–5 birds winter), *Aquila heliaca* (probably breeds) and *Crex crex* (uncommon on passage). Significant proportion (≥1%) of national population breeding at site: *Ciconia nigra* (2+ pairs). *Coracias garrulus* (50+ pairs) and both *Dendrocopos syriacus* and *D. medius* are common.

■ Protection status

National None **International** None

■ Conservation issues

Threats Agricultural intensification/expansion (U), Firewood collection (U), Other (U), Unsustainable exploitation (U)

The main threats are illegal tree-felling ('Other' threat), firewood-collection, grazing and poaching. The area is under the jurisdiction of the Ministry of Forestry. The area is partly an unprotected game reserve. A nature reserve was planned.

Lake Aggel — A1, A4i, B1i, B2 — 030

Admin region Agjabedy
Coordinates 40°05′N 47°40′E
Altitude — **Area** 9,173 ha

■ Site description

Shallow (0.5–3.0 m) saline lake (freshwater in the past) with vast *Phragmites*, *Typha* and *Scirpus* beds, open-water areas linked by channels through reeds, together with saltmarsh fringes and semi-desert islands with *Salsola*, etc., lying between the Kura and Aras rivers. Surrounding Mil steppe is slightly undulating alluvial plain with semi-desert and desert vegetation. The wetland area has recently decreased in size, having previously increased from 4,400 ha to 10,000 ha as a result of receiving water from irrigation systems. This wetland, together with lakes Boz Koba (IBA 010) and Sarysu (IBA 032), forms the water-body system of the Mil steppe.

Habitats Wetland (standing brackish and salt water; water-fringe vegetation)
Land-use Fisheries/aquaculture

■ Birds

Species	Season	Year	Pop min	Pop max	Acc	Criteria
Phalacrocorax pygmeus Pygmy Cormorant	B	—	—	1,800	—	A1, A4i, B1i
Pelecanus crispus Dalmatian Pelican	W	1991	60	80	—	A1, A4i, B1i
Ardeola ralloides Squacco Heron	B	1988	4,000	6,000	—	A4i, B1i, B2
Egretta garzetta Little Egret	R	1988	4,000	5,000	—	A4i, B1i
Egretta alba Great White Egret	B	1988	50	60	—	A4i, B1i
Plegadis falcinellus Glossy Ibis	B	1988	6,000	8,000	—	A4i, B1i
Platalea leucorodia Spoonbill	B	1988	200	600	—	A4i, B1i
Cygnus cygnus Whooper Swan	W	1991	900	900	—	A4i, B1i
Marmaronetta angustirostris Marbled Duck	B	—	60	80	—	A1, A4i, B1i
Oxyura leucocephala White-headed Duck	W	1991	3,000	3,000	—	A1, A4i, B1i
Porphyrio porphyrio Purple Gallinule	R	—	—	6,000	—	B2
Fulica atra Coot	W	1991	65,000	65,000	—	A4i, B1i

At least 115 species have been recorded and at least 87 breed, including a colony of Pelecaniformes and Ciconiiformes (20,000–25,000 pairs), also *Cygnus olor*, *Anser anser*, *Netta rufina*, *Francolinus francolinus* (a few broods each year), *Himantopus himantopus*, *Recurvirostra avosetta*, *Glareola nordmanni*, and *Chettusia leucura*. Species of global conservation concern that do not meet IBA criteria: *Haliaeetus albicilla* (winter visitor in small numbers). About 300,000 birds (presumably mainly waterbirds—mostly wildfowl) formerly wintered at the site, with a lower count of 110,000 in January 1991 (50,000–60,000 ducks).

■ Protection status

National Partial **International** None
c.7,000 ha of IBA covered by Zapovednik (Ak-Gel goryhy, 7,000 ha). Additional 2,173 ha are Zakaznik, acting as protective buffer zone.

■ Conservation issues

Threats Aquaculture/fisheries, Drainage

The main threats are fluctuations in the water area and salinity, including the artificial lowering of the water-level in a quarter of the lake in order to create sheep pastures, and the drowning of birds in fishing nets. There is some poaching, but this does not constitute a serious threat. A Zakaznik of 9,173 ha was established in 1964, 4,400 ha being designated as a Zapovednik in 1978 (expanded to 7,000 ha in 1988 or 1989). It is recommended that the size of the reserve be increased by 2,500 ha, creating a buffer zone c.500 m wide, that commercial fishing should cease, and that fresh water be fed into the lake and other measures implemented to stabilize the hydrological regime. Lake Aggel is probably the most important wetland in Azerbaijan since Kizil Agach (IBA 048) lost some of its importance as a wintering site.

Lake Boz-Koba — A4iii — 031

Admin region Agjabedy, Beilagan, Imishli
Coordinates 40°02′N 47°53′E
Altitude — **Area** 4,000 ha

■ Site description

A chain of oxbow lakes and channels in the ancient valley of the Kura river uniting to form a single lake in flood and lying between lakes Aggel (IBA 030) and Sarysu (IBA 032), and containing shallows, vast reedbeds and sparse *Tamarix* scrub. The main land-uses are legal hunting, oil extraction, grazing, etc. Part of the area is rented by the Hunters' Society of Azerbaijan, and part is under the jurisdiction of the Ministry of Oil.

Habitats Scrub, Wetland (standing brackish and salt water; river/stream; water-fringe vegetation)
Land-use Agriculture, Hunting, Urban/industrial/transport

Birds

Undoubtedly a site of great significance for waterbirds with 10,000–20,000 birds wintering in the late 1980s (although formerly 40,000–50,000). Breeding waterbirds include *Botaurus stellaris*, *Ixobrychus minutus* and colonies of other Ciconiiformes totalling 1,000+ pairs. Species of global conservation concern that do not meet IBA criteria: *Phalacrocorax pygmeus* (breeds), *Pelecanus crispus* (the last site in Azerbaijan where this species bred; formerly 5–15 pairs, could still survive), *Marmaronetta angustirostris* (breeds), *Oxyura leucocephala* (non-breeding winter visitor), *Haliaeetus albicilla* (non-breeding, 2–3 winter). Other breeding birds include *Tadorna ferruginea*, *Circus pygargus*, *Francolinus francolinus*, *Porphyrio porphyrio*, *Chettusia leucura* and passerines.

Protection status

National None **International** None

Conservation issues

Threats Industrialization/urbanization (U), Unsustainable exploitation (U)

Threats include poaching and pollution.

Lake Sarysu — A1, A4i, B1i, B2 — 032

Admin region Imishli
Coordinates 40°05′N 48°10′E
Altitude — **Area** 20,000 ha

Site description

A system of saline lakes with vast reedbeds, 3 areas of open water (the eastern by far the largest), a few small islands, shallows, etc.; the bottom is covered by *Potamogeton*. Formerly connected to the Kura river during the spring flood, but water-levels are now dependent on irrigation channels. Most of the site is rented by the Hunters' Society of Azerbaijan. There is also fishing (including with nets), grazing and oil extraction.

Habitats Wetland (100%; standing brackish and salt water; water-fringe vegetation)
Land-use Agriculture, Fisheries/aquaculture, Hunting, Urban/industrial/transport

Birds

Species	Season	Year	Pop min	Pop max	Acc	Criteria
Phalacrocorax pygmeus Pygmy Cormorant	R	—	100	—	—	A1, A4i, B1i
Phalacrocorax pygmeus Pygmy Cormorant	W	1996	1,158	—	—	A1, A4i, B1i
Pelecanus crispus Dalmatian Pelican	W	—	20	30	—	A1
Ardeola ralloides Squacco Heron	B	—	300	—	—	B1i
Plegadis falcinellus Glossy Ibis	B	—	100	—	—	A4i, B1i
Anas crecca Teal	W	1996	27,661	—	—	A4i, B1i
Anas acuta Pintail	W	1993	45,000	—	—	A4i, B1i
Marmaronetta angustirostris Marbled Duck	B	—	20	—	—	A1, A4i, B1i
Netta rufina Red-crested Pochard	W	1993	247,500	—	—	A4i, B1i
Aythya ferina Pochard	W	1996	97,250	—	—	A4i, B1i
Oxyura leucocephala White-headed Duck	W	—	—	—	—	A1
Porphyrio porphyrio Purple Gallinule	R	—	800	1,000	—	B2
Fulica atra Coot	W	1996	24,826	—	—	A4i, B1i

One of the best sites for waterbirds in the interior of Azerbaijan with 70,000–80,000 in years with good conditions (usually 40,000–50,000) in the beginning of winter. Species of global conservation concern that do not meet IBA criteria: *Aythya nyroca* (30+ pairs), *Haliaeetus albicilla* (formerly bred, 2–3 winter). Other breeding birds include herons (Ardeidae), shorebirds (*Himantopus himantopus*, *Chettusia leucura*, etc.), *Francolinus francolinus* and *Alcedo atthis*. *Phoenicopterus ruber* is a rare visitor throughout the year and *Asio flammeus* occurs rarely in winter.

Protection status

National None **International** None

Conservation issues

Threats Agricultural intensification/expansion (U), Unsustainable exploitation (U)

Threats include poaching (alongside legal hunting), and pesticide run-off from cotton fields is also a problem. Lake Sarysu undoubtedly merits protection and designation as a Ramsar Site (since it supports at least 20,000 waterbirds and eight species meet the ≥1% population criterion).

Yashma island — B1i — 033

Admin region Absheron
Coordinates 40°44′N 49°35′E
Altitude (-27)–(-27) m **Area** 200 ha

Site description

A low sandy island near the Absheron peninsula. There are shallows with patches of reeds. The island and the adjacent mainland shore are rented by the Hunters' Society of Azerbaijan.

Habitats Wetland (100%; sand-dunes/sand beach; water-fringe vegetation)
Land-use Fisheries/aquaculture, Hunting, Urban/industrial/transport

Birds

Species	Season	Year	Pop min	Pop max	Acc	Criteria
[1] ***Aythya fuligula*** Tufted Duck	P	—	5,000	—	—	B1i
Aythya marila Scaup	P	—	2,000	—	—	B1i

1. 1,000+ wintering.

Breeding birds include herons, gulls and terns. A staging site of great significance for passage waterbirds, with *Phalacrocorax carbo*, herons, wildfowl (*Cygnus cygnus* and ducks, some also wintering), *Fulica atra* (10,000+) and, especially, a wide range of shorebirds, some 20,000–30,000 annually. Species of global conservation concern that do not meet IBA criteria: *Pelecanus crispus* (rare passage visitor), *Numenius tenuirostris* (Yashma Island is the only site in the country where this species occurs at present; 1 shot in 1988 and, according to hunters, an annual passage-visitor, but extremely rare, with flocks of 2–4 birds).

Protection status

National None **International** None

Conservation issues

Threats Aquaculture/fisheries (A), Industrialization/urbanization (U), Unsustainable exploitation (U)

Main threats come from pollution from the oil industry and (especially affecting *Aythya fuligula*) fishing nets. Yashma island has not been properly surveyed by ornithologists.

Absheron archipelago (north) and Artem Bay — B1i — 034

Admin region Baku
Coordinates 40°28′N 50°20′E
Altitude (-27)–(-27) m **Area** 1,000 ha

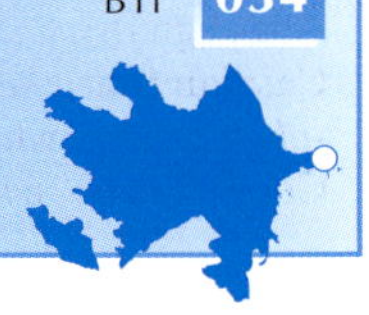

Site description

The partly reed-covered shores of the Artem Bay, near Baku, and small islands near the north-eastern coast of the Absheron peninsula. Part of the bay is leased by the Hunters' Society of Azerbaijan.

Habitats Marine areas (sea inlet/coastal features), Rocky areas (rock stacks/islets), Wetland (sand-dunes/sand beach; water-fringe vegetation)
Land-use Fisheries/aquaculture, Hunting, Tourism/recreation

Birds

Species	Season	Year	Pop min	Pop max	Acc	Criteria
Cygnus cygnus Whooper Swan	W	—	300	500	—	B1i

Breeding birds include *Tadorna ferruginea*, *Porphyrio porphyrio*, *Larus cachinnans*, *Sterna hirundo* and (unconfirmed) *S. sandvicensis*. Wintering area for 2,000–3,000 wildfowl, notably *Cygnus cygnus*, also *Aythya fuligula*, *A. marila*, *A. ferina*, *Bucephala clangula* and *Fulica atra*. Species of global conservation concern that do not meet IBA criteria: *Circus macrourus* (3–6 winter).

Protection status

National None **International** None

Conservation issues

Threats Industrialization/urbanization (U), Recreation/tourism (U), Unsustainable exploitation (U)

The area suffers from pollution by oil and industrial waste, disturbance from recreation and intensive hunting, including the use of power-boats for poaching (especially for *Cygnus*, *Fulica atra*, etc.). Few field investigations have been carried out.

Shah Cape (Shakhova Kosa) A4i, A4iii, B1i 035

Admin region Baku
Coordinates 40°17′N 50°20′E
Altitude (-27)–(-27) m **Area** 500 ha

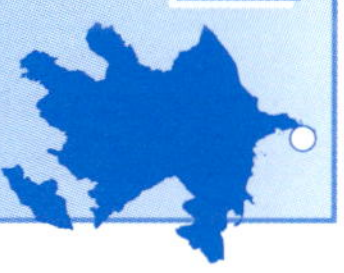

Site description

The site is the tip of the Absheron peninsula near Baku, formerly with sandy beaches, about a third of which was flooded as the level of the Caspian Sea rose during the 1980s. The flooded area is now being colonized by reeds. All land-uses are prohibited.

Habitats Wetland (100%; sand-dunes/sand beach; water-fringe vegetation)
Land-use Nature conservation/research

Birds

Species	Season	Year	Pop min	Pop max	Acc	Criteria
Plegadis falcinellus Glossy Ibis	P	—	100	—	—	B1i
Cygnus cygnus Whooper Swan	P	—	500	600	—	B1i
Aythya fuligula Tufted Duck	P	—	10,000	—	—	B1i
Aythya marila Scaup	P	—	3,000	—	—	B1i
Fulica atra Coot	P	—	40,000	—	—	A4i, B1i

The IBA is notable for passage and wintering Pelecaniformes and Ciconiiformes. At least 20,000 ducks and 40,000 *Fulica atra* migrate through the area. Species of global conservation concern that do not meet IBA criteria: *Phalacrocorax pygmeus* (non-breeding: 2–4 winter in some years), *Circus macrourus* (3–4 winter).

Protection status

National High **International** None
IBA is covered by Game Reserve.

Conservation issues

Threats Unsustainable exploitation (U)

Although the whole site is protected as a Game Reserve (for waterbirds, Caspian Seal *Phoca caspica* and reintroduced Goitred gazelle *Gazella subgutturosa*), which is owned by the Azerbaijan Committee for Nature Conservation, heavy poaching is a threat.

Lake Krasnoye and other water-bodies of the Absheron peninsula A1, A4i, B1i 036

Admin region —
Coordinates 40°20′N 49°45′E
Altitude (-27)–(-27) m **Area** —

Site description

In the 1920s, Puta Bay of the Caspian Sea was dyked and partly land-claimed. The remaining water-body was known thereafter as Lake Krasnoye (Ozero Krasnoye) or Kizilgel.

Habitats Wetland
Land-use Unknown

Birds

In the early 1940s, ducks, *Fulica atra* and *Phoenicopterus ruber* wintered at Lake Krasnoye. Some stands of *Typha* were present there then (Vereshchagin 1946, 1950). In the late 1980s, the lake no longer supported any emergent vegetation. A chemical factory was constructed on its shores, causing severe water pollution. However, 140 *Oxyura leucocephala* were seen at Lake Krasnoye in February 1998 (P. Cranswick *et al.* 1998).

Species	Season	Year	Pop min	Pop max	Acc	Criteria
Oxyura leucocephala White-headed Duck	W	1998	—	140	—	A1, A4i, B1i

Protection status

National Unknown **International** None

Conservation issues

Threats Industrialization/urbanization (A)

Apart from the *Oxyura leucocephala* data in 1998 no up-to-date information is available on the conservation status of the lake.

Mount Kargabazar and Mount Gush-gaya A1 037

Admin region Maraza
Coordinates 40°22′N 49°20′E
Altitude 200–284 m **Area** 3,000 ha

Site description

An area of low, rocky mountains in the semi-desert of Gobustan upland. Land-uses include intensive sheep-grazing and some oil exploration.

Habitats Desert (semi-desert), Rocky areas
Land-use Agriculture, Urban/industrial/transport

Birds

Species	Season	Year	Pop min	Pop max	Acc	Criteria
Falco naumanni Lesser Kestrel	B	—	5	10	—	A1

Notable breeding species: *Tadorna ferruginea* (5+ pairs, 10–12 km from nearest water-body), *Falco biarmicus* (1 pair formerly bred; only site in USSR), *Apus melba* (20–25 pairs, only colony in Gobustan upland), *Merops superciliosus*. Among very common passerines are *Oenanthe isabellina*, *O. finschii* and *Sitta neumayer*.

Protection status

National None **International** None

Conservation issues

Threats Agricultural intensification/expansion (U), Unsustainable exploitation (U)

The main threat is overgrazing. There is also some poaching.

Gobustan area B3 038

Admin region Karadag, Maraza
Coordinates 40°11′N 49°22′E
Altitude 0–200 m **Area** 5,000 ha

Site description

Low, semi-arid mountains, including the flat summits of Beukdash, Kichikdash, Kagnizdag, Djingirdag and the dry stream of Djeirankechmez. Some trees and bushes on the eastern slopes (wild *Pyrus*, *Ficus*, *Punica*, etc.). Quarrying for limestone has taken place in the past.

Habitats Desert (semi-desert), Forest and woodland (wooded desert/semi-desert), Rocky areas (scree/boulders), Scrub (scrub)
Land-use Agriculture, Tourism/recreation

Birds

Species	Season	Year	Pop min	Pop max	Acc	Criteria
Sitta neumayer Rock Nuthatch	R	—	30	40	—	B3

Species of global conservation concern that do not meet IBA criteria: *Falco naumanni* (non-breeding; rare summer visitor). Breeding species include *Buteo rufinus*, *Alectoris chukar* (rare), *Cercotrichas galactotes*, *Oenanthe pleschanka*, *Oenanthe finschii*, *Pyrrhocorax pyrrhocorax*, *Petronia petronia* and *Emberiza melanocephala*.

Protection status

National Partial **International** None

IBA partly covered by Nature Monument (Gobustan Nature Monument, 4,400 ha).

Conservation issues

Threats Agricultural intensification/expansion (U), Disturbance to birds (U), Recreation/tourism (U)

Threats come from sheep- and cattle-grazing, and disturbance from tourists and low-flying military aircraft. The central and southern part of the IBA is covered wholly, and the northern section partly, by the Gobustan Nature Monument. The area is a Nature Monument (virtually unprotected), established to safeguard archaeological artifacts.

Sangachal Bay

A4i, A4iii, B1i **039**

Admin region Karadag
Coordinates 40°09'N 49°30'E
Altitude (-27)–(-27) m **Area** 1,000 ha

Site description

A bay in the Caspian Sea with sandy beaches many of which are becoming overgrown with reeds and typical coastal vegetation. Part of the bay is rented by the Hunters' Society of Azerbaijan. There is a military base and the area was formerly important for oil production, although only half of the derricks are now used.

Habitats Wetland (100%; sand-dunes/sand beach; water-fringe vegetation)
Land-use Hunting, Military, Urban/industrial/transport

Birds

Species	Season	Year	Pop min	Pop max	Acc	Criteria
Cygnus cygnus Whooper Swan	W	—	200	400	—	B1i
Cygnus cygnus Whooper Swan	P	—	1,000	—	—	A4i, B1i
Netta rufina Red-crested Pochard	W	—	1,000	—	—	A4i, B1i
Netta rufina Red-crested Pochard	P	—	5,000	—	—	A4i, B1i
Aythya fuligula Tufted Duck	P	—	10,000	—	—	B1i
Fulica atra Coot	P	—	30,000	—	—	B1i

The site is notable for staging and wintering wildfowl and *Fulica atra*. At least 20,000 diving ducks and 30,000 *Fulica atra* stage here every autumn. Species of global conservation concern that do not meet IBA criteria: *Pelecanus crispus* (very rare migrant), *Branta ruficollis* (extremely rare, not seen every year). Other passage visitors include *Pelecanus onocrotalus* (200–300 per year), *Plegadis falcinellus* (500+), *Himantopus himantopus* (200+), *Sterna caspia* (August).

Protection status

National None **International** None

Conservation issues

Threats Industrialization/urbanization (U), Unsustainable exploitation (U)

The area suffers from pollution (oil and other industrial waste) and some poaching—its proximity to Baku makes it very popular with hunters. It is rarely visited by ornithologists, however, and there have been no systematic studies.

Glynanyi island

A4i, B1i, B2 **040**

Admin region Karadag
Coordinates 39°57'N 49°31'E
Altitude (-27)–(-23) m **Area** 200 ha

Site description

The largest island of the Baku archipelago, Glynanyi is volcanic in origin and almost devoid of vegetation, because of its large rabbit population.

Habitats Rocky areas
Land-use Nature conservation/research

Birds

Species	Season	Year	Pop min	Pop max	Acc	Criteria
Cygnus cygnus Whooper Swan	W	—	500	—	—	B1i
Glareola pratincola Collared Pratincole	B	—	50	—	—	B1i, B2
Larus cachinnans Yellow-legged Gull	B	—	10,000	11,000	—	A4i, B1i

The site of Azerbaijan's largest colony of *Larus cachinnans* and one of the largest in the Caspian Basin. Significant proportion (≥1%) of national population breeding at site: *Tadorna ferruginea* (50 pairs), *Sterna hirundo* (min. 500 pairs), *S. albifrons* (80 pairs).

Protection status

National Partial **International** None

IBA covered by Game Reserve.

Conservation issues

Threats Unsustainable exploitation (U)

The island is a protected Game Reserve, with no access during the breeding season, except to scientists. Hunting is prohibited, but there are frequent violations, and egg-collecting is a major threat.

Lake Hadjikabul

A1, A4i, B1i **041**

Admin region —
Coordinates 40°00'N 49°00'E
Altitude — **Area** 1,500 ha

Site description

Lake Hadjikabul lies in the Kura valley near the town of Ali-Bairamly.

Habitats Wetland (standing fresh water; water-fringe vegetation)
Land-use Fisheries/aquaculture

Birds

Species	Season	Year	Pop min	Pop max	Acc	Criteria
Oxyura leucocephala White-headed Duck	W	1998	—	620	—	A1, A4i, B1i

Although this lake had relatively few reed stands, many waterbirds rest there in winter (e.g. *Anas platyrhynchos*, diving ducks and sometimes *Fulica atra*). In the past, good numbers of *Anser anser*, *A. albifrons*, *A. erythropus* and *Branta ruficollis* were recorded resting during the day after feeding in the Shirvan steppe. The three *Anser* species still occur at Lake Hadjikabul, but their numbers are very low (Vereshchagin 1950; M. Patrikeev, *in litt.*). In February 1998, *Oxyura leucocephala* were observed at Lake Hadjikabul (Cranswick *et al.* 1998).

Protection status

National None **International** None

Conservation issues

Threats Unsustainable exploitation

Pirsagat islands and Los island

A4i, B1i, B2 **042**

Admin region Karadag
Coordinates 39°50'N 49°30'E
Altitude (-27)–(-10) m **Area** 250 ha

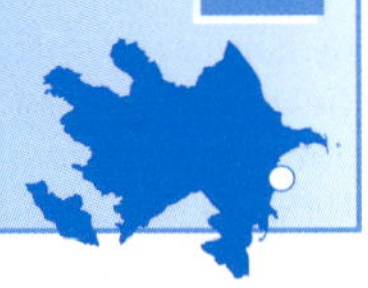

Site description

The Pirsagat islands (six main limestone ones and many smaller rocks) form part of the Baku archipelago (which also includes the volcanic Los island) and extend as a chain from the Cape of Pirsagat into the Caspian Sea. They are rented by the Hunters' Society of Azerbaijan. There was formerly oil-drilling on Los island.

Habitats Rocky areas (rock stacks/islets)
Land-use Hunting, Tourism/recreation, Urban/industrial/transport

Birds

Species	Season	Year	Pop min	Pop max	Acc	Criteria
[1] ***Cygnus cygnus*** Whooper Swan	W	—	500	1,000	—	A4i, B1i
Glareola pratincola Collared Pratincole	B	1989	80	90	—	A4i, B1i, B2
Larus genei Slender-billed Gull	B	1989	550	550	—	A4i, B1i
[2] ***Sterna sandvicensis*** Sandwich Tern	B	1989	4,000	—	—	A4i, B1i

1. 500–600, also on migration.
2. First breeding in Azerbaijan since 1955.

A total of about 5,500–6,000 pairs of seabird breed; 4,000+ on Pelikanyi, 1,000+ on Baburyi, 300 on Los. Gulls in 1989 included 28 pairs of *Larus melanocephalus* (the only colony in the Caspian Sea) and other breeding terns in the same year were *Gelochelidon nilotica* (75 pairs; first recorded breeding in Azerbaijan), *Sterna hirundo* (1,000+ pairs) and *S. albifrons* (25 pairs). Species of global conservation concern that do not meet IBA criteria: *Branta ruficollis* (vagrant, probably a common migrant formerly).

Protection status

National None **International** None

Conservation issues

Threats Disturbance to birds (U), Industrialization/urbanization (U), Unsustainable exploitation (U)

Threats come from egg-collection, disturbance caused to breeding birds by fishermen, hunting (autumn) and pollution from the oil industry.

Shorgel lakes/ Shirvan Reserve — A1, A4i, A4iii, B1i, B2 — 043

Admin region Saliany, Neftechala
Coordinates 39°30′N 49°15′E
Altitude (-25)–(-20) m **Area** 26,000 ha

Site description

An alluvial plain extending inland from the Caspian Sea. Predominantly dry semi-desert steppe with *Artemisia* and a high diversity of grasses. Canals run across the steppe and there are also saline lakes. The Shorgel lakes, with their rich aquatic vegetation, are fed by irrigation water. Part (c.3,500 ha) of this wetland lies within the reserve, the remainder extending into a contiguous Zakaznik.

Habitats Grassland (40%; steppe/dry calcareous grassland), Wetland (standing brackish and salt water; water-fringe vegetation)
Land-use Nature conservation/research

Birds

Species	Season	Year	Pop min	Pop max	Acc	Criteria
Phalacrocorax pygmeus Pygmy Cormorant	W	—	—	1,000	—	A1, A4i, B1i
Pelecanus crispus Dalmatian Pelican	W	—	—	50	—	A1, A4i, B1i
Marmaronetta angustirostris Marbled Duck	W	—	—	240	—	A1, A4i, B1i
Porphyrio porphyrio Purple Gallinule	B	1990	—	4,000	—	B2
Tetrax tetrax Little Bustard	W	1996	—	900	—	A1

Among breeding birds at this IBA are *Ardea purpurea*, *Tadorna ferruginea*, *T. tadorna*, *Netta rufina*, *Circus aeruginosus*, *Francolinus francolinus*, *Burhinus oedicnemus*, *Charadrius alexandrinus* and *Gelochelidon nilotica.* Species of global conservation concern that do not meet IBA criteria (all observed in winter): *Anser erythropus* (27 in 1996), *Aegypius monachus*, *Haliaeetus albicilla*, *Aquila heliaca.* Other wintering birds include *Phoenicopterus ruber* (200–300), also wildfowl (*Cygnus cygnus* and others), *Fulica atra*, etc. Report of 20,000–35,000 wintering wildfowl, while 150,000–200,000 stage at the site in autumn. *Aquila nipalensis* also occurs on passage.

Protection status

National High **International** None

26,000 ha of IBA covered by Zapovednik (Shirvan goryhy, 26,000 ha). Zakaznik of 22,000 ha adjacent.

Conservation issues

Threats Agricultural intensification/expansion (U), Aquaculture/fisheries (U), Disturbance to birds (U), Extraction industry (U)

The chief threats are grazing by sheep (100,000 sheep graze in the reserve legally, a condition under which the reserve was gazetted) and, more recently, cattle, leading to destruction of goose pastures, continuing oil extraction and fishing (fishermen's camps are located within the Zapovednik and the Zakaznik), pollution of some lakes by pesticides and disturbance by shepherds. The oil-drilling operation should be isolated from the reserve. Shirvan Reserve (Zapovednik) was established in 1969 partly within the Byandovan Zakaznik, the aim being to give better protection to the Goitred Gazelle *Gazella subgutturosa*. Some new facilities have been built and the site has considerable education potential.

Mugan steppe — A1 — 044

Admin region Saliany, Imishli, Pushkin, Saatly, Sabirabad
Coordinates 39°38′N 48°34′E
Altitude 0–100 m **Area** 100,000 ha

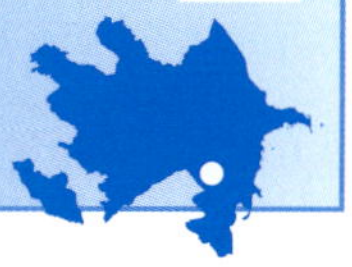

Site description

A semi-desert plain, cut across by many irrigation channels, with pastures and, to a lesser extent, cotton fields.

Habitats Artificial landscape (arable land), Desert (semi-desert)
Land-use Agriculture

Birds

Species	Season	Year	Pop min	Pop max	Acc	Criteria
[1] ***Tetrax tetrax*** Little Bustard	W	—	10,000	30,000	—	A1

1. Peak numbers in October–November, some may disperse to the south.

Certainly the most important area for wintering *Tetrax tetrax* in Azerbaijan, probably also in the former USSR; discovered only in 1989. Breeding species include *Ciconia ciconia* (in some villages), *Francolinus francolinus* (rare, in scrub along channels), *Burhinus oedicnemus*, *Merops superciliosus* (common) and larks (Alaudidae). Species of global conservation concern that do not meet IBA criteria: *Branta ruficollis* (formerly very common winter visitor, now very rare and not annual), *Otis tarda* (very rare winter visitor, not annual). *Aquila nipalensis* is common on migration, when *Grus grus* and *Anthropoides virgo* also occur.

Protection status

National None **International** None

Conservation issues

Threats Disturbance to birds (U), Unsustainable exploitation (U)

Main threats are poaching and disturbance of *Tetrax tetrax* by shepherds and other local people.

Lake Mahmud-chala — A1, A4i, A4iii, B1i — 045

Admin region Massally, Pushkin, Djalilabad
Coordinates 39°25′N 48°40′E
Altitude — **Area** 4,000 ha

Site description

A shallow saline lake (depth 0.5–2.0 m) with a few small islands. Reedbeds cover large parts of the lake, especially in the central and southern parts. There are some *Tamarix* groves (largely flooded) along the southern shore. The lake is surrounded by saltwort *Salsola* semi-desert in the south and cotton fields in the north; there are also some saltmarshes. The lake is rented by the Hunters' Society of Azerbaijan and sheep graze the edges.

Habitats Artificial landscape (arable land), Desert (semi-desert), Scrub (scrub), Wetland (saltmarsh; standing brackish and salt water; water-fringe vegetation)
Land-use Agriculture, Hunting

■ Birds

Species	Season	Year	Pop min	Pop max	Acc	Criteria
Phalacrocorax pygmeus Pygmy Cormorant	R	—	500	800	—	A1, A4i, B1i
Ardeola ralloides Squacco Heron	B	—	1,000	—	—	A4i, B1i
Bubulcus ibis Cattle Egret	B	1990	1,000	1,500	—	A4i, B1i
Egretta garzetta Little Egret	B	—	3,000	3,000	—	A4i, B1i
Plegadis falcinellus Glossy Ibis	B	—	5,500	6,000	—	A4i, B1i
Platalea leucorodia Spoonbill	B	—	200	—	—	A4i, B1i
Anas penelope Wigeon	W	—	—	19,350	—	A4i, B1i
Anas clypeata Shoveler	W	—	—	6,940	—	A4i, B1i
Netta rufina Red-crested Pochard	W	—	—	—	—	A4i, B1i
Aythya ferina Pochard	W	—	—	9,390	—	A4i, B1i

This is an important site for breeding waterbirds, totalling 12,000–15,000 pairs. Species of global conservation concern that do not meet IBA criteria: *Marmaronetta angustirostris* (summer visitor, possibly breeds; 1 bird also winter 1996), *Aythya nyroca* (5+ pairs), *Oxyura leucocephala* (occasional in winter). Terns and shorebirds (including *Charadrius alexandrinus* and *Chettusia leucura*) also occur. Winter waterbird numbers range from 10,000 to 40,000, or more.

■ Protection status

National None **International** None

■ Conservation issues

Threats Agricultural intensification/expansion (U), Other (U), Unsustainable exploitation (U)

Pollution caused by pesticide run-off from cotton fields threatens both breeding and wintering birds. Threats also include intensive hunting, trampling of nests on islands by buffaloes grazing in water and the effects of fluctuating water-levels (dependent on watering of cotton fields further north) on the timing of breeding by Pelecaniformes and Ciconiiformes ('Other' threat, above). In supporting at least 20,000 waterbirds and at least 1% of the regional population of three species (*Anas penelope*, *A. clypeata*, *Aythya ferina*), the IBA qualifies for designation as a Ramsar Site.

Kura delta

A1, A4i, B1i **046**

Admin region Neftechala
Coordinates 39°15'N 49°20'E
Altitude (-27)–(-27) m **Area** 10,000 ha

■ Site description

Delta of the Kura, the largest river in the Transcaucasus, with reedbeds, flooded areas, bushes and shallows. Land-uses include hunting, fishing, grazing and fish-ponds. To the south there are oil fields.

Habitats Wetland (100%; standing fresh water; river/stream; water-fringe vegetation)
Land-use Agriculture, Fisheries/aquaculture, Hunting, Urban/industrial/transport

■ Birds

Species	Season	Year	Pop min	Pop max	Acc	Criteria
Phalacrocorax pygmeus Pygmy Cormorant	R	—	1,300	1,300	—	A1, A4i, B1i
[1] ***Pelecanus crispus*** Dalmatian Pelican	W	1988	60	1,118	—	A1, A4i, B1i
Ardeola ralloides Squacco Heron	B	—	800	1,000	—	A4i, B1i
Egretta garzetta Little Egret	B	—	800	1,000	—	A4i, B1i

1. 60–400 in 1986, up to 1118 in 1988.

One of the two most important sites for wintering *Pelecanus* in Azerbaijan. Other breeding Ciconiiformes include *Botaurus stellaris*, *Ixobrychus minutus*, *Egretta alba*, *Ardea cinerea*, *A. purpurea* and *Platalea leucorodia*, and there are many thousands of other passage and winter visitors.

■ Protection status

National None **International** None

■ Conservation issues

Threats Agricultural intensification/expansion (U), Industrialization/urbanization (U), Unsustainable exploitation (U)

Threats include some poaching and possible pollution from pesticides and oil. Undoubtedly important for wintering waterbirds, but has yet to be surveyed properly.

Lake Ych-chala (Novogolovka-chala)

A1, A4i, B1i **047**

Admin region Saliany, Massally
Coordinates 39°14'N 48°43'E
Altitude — **Area** 2,500 ha

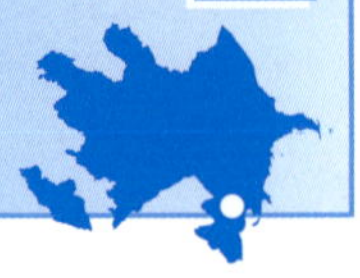

■ Site description

Adjacent to Kizil Agach Zapovednik (IBA 048). A very shallow lake (0.5–1.0 m in depth) which is mainly covered by reeds. It occasionally dries out in very hot years, but can flood the surrounding area during autumn rains. The water-level is also affected by irrigation. Land-use includes grazing domestic cattle and buffaloes.

Habitats Wetland (100%; standing brackish/salt water; water-fringe vegetation)
Land-use Agriculture, Hunting

■ Birds

Species	Season	Year	Pop min	Pop max	Acc	Criteria
[1] ***Phalacrocorax pygmeus*** Pygmy Cormorant	W	—	300	400	—	A1, A4i, B1i

1. Possibly breeds.

Significantly, many wildfowl wintering in Kizil Agach Reserve come to this IBA to feed and about 10,000–15,000 wildfowl and shorebirds winter. Species of global conservation concern that do not meet IBA criteria: *Branta ruficollis* (very rare winter visitor), *Marmaronetta angustirostris* (summer visitor, probably breeds), *Aythya nyroca* (breeds), *Oxyura leucocephala* (very rare on passage), *Haliaeetus albicilla* (5–8 birds, sometimes only 1–2, wintering). The site holds breeding herons, *Plegadis falcinellus*, *Platalea leucorodia* (status uncertain, possibly breeds), *Porphyrio porphyrio*, etc.

■ Protection status

National None **International** None

■ Conservation issues

Threats Agricultural intensification/expansion (U), Unsustainable exploitation (U)

The main threats are intensive poaching and contamination of the lake by pesticide run-off from cotton fields. The IBA is immediately adjacent to a Zapovednik (Kizil Agach, 88,360 ha), but there is no overlap and so the lake is not protected. One of the least-known wetlands in Azerbaijan despite its proximity to Kizil Agach Zapovednik (which has scientific staff).

Kizil Agach Bay

A1, A4i, B1i **048**

Admin region Lenkoran
Coordinates 39°05'N 49°03'E
Altitude (-29)–(-24) m **Area** 132,500 ha

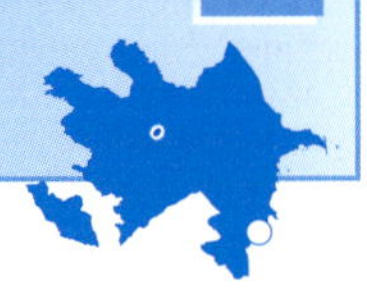

■ Site description

Kizil Agach Bay, an open bay connected to the Caspian, and Kichik (Little) Kizil Agach Bay, which is a freshwater reservoir separated from the sea by an artificial dam and is greatly silted-up and overgrown. The relief is characterized by alternating low ridges, open hollows and ancient, silted-up channels. Emergent vegetation includes beds of *Phragmites*, *Scirpus* and *Juncus*. Adjacent, low-lying semi-desert areas with *Artemisia*, *Salsola* and dry scrub with *Rubus*.

Habitats Artificial landscape (arable land), Desert (semi-desert), Grassland (steppe/dry calcareous grassland), Scrub, Wetland (coastal lagoon; standing fresh water; standing brackish and salt water; water-fringe vegetation)
Land-use Agriculture, Fisheries/aquaculture, Nature conservation/research

Birds

Species	Season	Year	Pop min	Pop max	Acc	Criteria
Tachybaptus ruficollis Little Grebe	W	1996	—	5,640	—	A4i, B1i
Podiceps nigricollis Black-necked Grebe	W	1996	—	1,982	—	A4i, B1i
Phalacrocorax pygmeus Pygmy Cormorant	B	—	1,400	2,000	—	A1, A4i, B1i
Phalacrocorax pygmeus Pygmy Cormorant	W	1996	—	7,054	—	A1, A4i, B1i
Pelecanus onocrotalus White Pelican	W	—	—	—	—	A4i, B1i
Pelecanus crispus Dalmatian Pelican	W	1996	—	134	—	A1, A4i, B1i
Nycticorax nycticorax Night Heron	B	—	6,000	7,200	—	A4i, B1i
Ardeola ralloides Squacco Heron	B	—	8,000	9,600	—	A4i, B1i
Bubulcus ibis Cattle Egret	B	—	1,400	4,800	—	A4i, B1i
Egretta garzetta Little Egret	B	—	3,000	5,000	—	A4i, B1i
Egretta alba Great White Egret	W	1996	—	983	—	A4i, B1i
Plegadis falcinellus Glossy Ibis	B	—	900	3,000	—	A4i, B1i
Phoenicopterus ruber Greater Flamingo	W	1996	—	4,780	—	A4i, B1i
Cygnus olor Mute Swan	W	1985	—	9,200	—	A4i, B1i
Cygnus cygnus Whooper Swan	W	—	—	9,200	—	A4i, B1i
Anser albifrons White-fronted Goose	W	—	—	25,000	—	A4i, B1i
Anser erythropus Lesser White-fronted Goose	W	—	—	1,058	—	A1, A4i, B1i
Anser anser Greylag Goose	W	—	—	40,000	—	A4i, B1i
Branta ruficollis Red-breasted Goose	W	—	3	45	—	A1
Anas penelope Wigeon	W	1996	—	109,895	—	A4i, B1i
Anas strepera Gadwall	W	1996	—	1,300	—	A4i, B1i
Anas crecca Teal	W	1996	—	67,245	—	A4i, B1i
Anas clypeata Shoveler	W	1996	—	35,188	—	A4i, B1i
Marmaronetta angustirostris Marbled Duck	W	1996	—	44	—	A1, A4i, B1i
Netta rufina Red-crested Pochard	W	1996	—	33,631	—	A4i, B1i
Aythya ferina Pochard	W	1996	—	43,591	—	A4i, B1i
Aythya fuligula Tufted Duck	W	1996	—	52,632	—	A4i, B1i
Fulica atra Coot	W	—	35,000	265,000	—	A4i, B1i
Tetrax tetrax Little Bustard	W	1990	10,000	30,000	—	A1

The IBA holds the largest colonies of Pelecaniformes and Ciconiiformes in Europe, though there has been a decline: from 225,000 pairs of 10 species in 1957 to c.60,000 pairs in the 1980s. Other breeding species include *Botaurus stellaris*, *Ardea purpurea*, *Tadorna ferruginea*, *Himantopus himantopus*, *Chlidonias leucopterus*, *Merops superciliosus* (500–700 pairs), and passerines such as *Passer hispaniolensis* (colonies of 500–10,000, once 70,000, nests). Until recently, an immensely important stop-over and wintering site, but waterbird numbers (ducks and *Fulica atra*) have fallen dramatically: from 10 million in 1930, to 5–7 million in the late 1950s, to 1,500,000 by the end of the 1960s, to 338,000–715,000 in the 1970s and to 200,000–400,000 at the beginning of the 1980s. A total of 50,000–170,000 diving ducks were reported during the mid-1980s (see also *F. atra* in Table), but total (ducks and *F. atra*) in 1989 only 40,000. In January 1991 there were 628,370 waterbirds (mostly wildfowl, but also 243,250 *Fulica atra*, 12,970 *Phoenicopterus ruber*, 1,750 *Pelecanus*), then in 1996 there were 180,000–362,000 waterbirds. Species of global conservation concern that do not meet IBA criteria: *Haliaeetus albicilla*, *Aquila clanga* (both winter). At least 19 species of shorebirds winter (e.g. 1,200 *Recurvirostra avosetta* in 1996) and large flocks of *Alauda arvensis* in 1996 contained many *Melanocorypha leucoptera*.

Protection status

National Partial **International** High

88,360 ha of IBA covered by Zapovednik (Kizil Agach, 88,360 ha): includes larger bay and northern section (5,000 ha) of smaller bay. Southern part of smaller bay partly protected as a Zakaznik (10,700 ha). 132,500 ha of IBA covered by Ramsar Site (Kirov Bays, 132,500 ha).

Conservation issues

Threats Abandonment/reduction of land management (U), Agricultural intensification/expansion (U), Aquaculture/fisheries (U), Disturbance to birds (U), Other (U), Unsustainable exploitation (U)

Removal of water for irrigation and regulation of water have created unfavourable conditions for waterbirds, with changes in the aquatic vegetation (submerged vegetation has changed since the 1960s with *Potamogeton* replacing *Zostera*). Outflow of water via many channels to the Caspian has led to the drying-out of habitats, causing birds to leave for Iran or the interior of Azerbaijan, while the rising sea-level may kill off freshwater habitats ('Other' threat). Water contaminated with pesticides flowing through the southern part of the reserve causes *Tamarix* thickets to dry out, thus threatening the colonies of Ciconiiformes, etc. there. Agricultural changes (from rice and grain to vegetables, grapes and cotton) have affected the availability of food for geese, but there are plans to improve the freshwater wintering sites for dabbling ducks, and fields have been sown with barley to attract the geese. The presence of fish-farms has adversely affected the surrounding steppe, which is now criss-crossed by channels, dams, and embankments and is gradually disappearing. In addition, there is disturbance and intensive hunting (several shots per minute in 1989), even involving reserve wardens and including the use of power-boats. This is largely subsistence hunting essential to local people (lots of shot or netted ducks and *Fulica atra* are sold at the roadside), but complete refuges for waterbirds are urgently needed (Lake Kara Gush in the north of the Zapovednik serves to some extent as such a refuge, its boggy shores making access difficult). There was no indication in the early 1990s that any conservation measures were being taken by the Azeri government to save Kizil Agach. Significant spawning grounds of several fish species are situated at this site. The reserve was initially (in 1926) designated as a Zakaznik, then (in 1929) as a Zapovednik (of 180,000 ha, later reduced to 132,500 ha and finally to 88,360 ha); the rest is Zakaznik or unprotected.

Vilajchai valley

B2 **049**

Admin region Lenkoran, Yardymly
Coordinates 38°59′N 48°35′E
Altitude 100–1,000 m **Area** 1,000 ha

Site description

Broadleaved forests with bushes and scrub in the valley of the Vilajchai river, which flows eastwards from the Talysh mountains.

Habitats Forest and woodland (broadleaved deciduous forest), Scrub (scrub)
Land-use Agriculture

Birds

An important area for endemic subspecies of *Phasianus colchicus*—*P. c. talishensis* (25+ pairs). Significant proportion (≥1%) of national population breeding at site: *Ciconia nigra* (2+ pairs). Other breeding species include *Milvus migrans* (common), *Aquila pomarina* and *Hieraeetus pennatus*, *Alcedo atthis* (very common), *Coracias garrulus* (common), woodpeckers and passerines. *Milvus milvus* is extremely rare in winter, *Falco cherrug* a rare winter visitor.

Protection status

National None **International** None

Conservation issues

Threats Agricultural intensification/expansion (U), Other (U), Unsustainable exploitation (U)

The main threats are intensive grazing, which prevents forest regeneration, poaching of *Phasianus colchicus*, and illegal tree-felling ('Other' threat).

Zuvand upland

B2 **050**

Admin region Yardymly, Lerik
Coordinates 38°50′N 48°15′E
Altitude 500–2,500 m **Area** 15,000 ha

Site description

Semi-arid plateaus and mountains near the Iranian border with some broadleaved forests. The area, which has a high population density, was formerly part of the border zone, with all land-uses prohibited, but there is now intensive sheep-grazing and some forestry.

Habitats Artificial landscape (other urban/industrial areas), Forest and woodland (broadleaved deciduous forest), Grassland (steppe/dry calcareous grassland)
Land-use Agriculture, Forestry

Birds

Species	Season	Year	Pop min	Pop max	Acc	Criteria
Neophron percnopterus Egyptian Vulture	B	—	15	—	—	B2
Gyps fulvus Griffon Vulture	R	—	15	—	—	B2

Among other breeding raptors at this site are *Circaetus gallicus*, *Buteo rufinus*, *Aquila chrysaetos* and *Bubo bubo*. *Alectoris chukar* and *Perdix perdix* also breed, as do passerines such as *Phoenicurus erythrogaster*, *Pyrrhocorax pyrrhocorax*, *Carpospiza brachydactyla* and *Emberiza buchanani*.

Protection status

National Partial **International** None
IBA partly covered by a Game Reserve.

Conservation issues

Threats Agricultural intensification/expansion (U), Other (U)

Threats come from grazing and some illegal tree-felling ('Other' threat).

Girkan forest B2 051

Admin region Astara, Lenkoran
Coordinates 38°39'N 48°43'E
Altitude 15–1,000 m **Area** 4,500 ha

Site description

Ancient broadleaved forests, mainly situated on the Ulyasy ridge, part of the Talysh mountains, but one small area (81 ha only) is in the lowlands. There is also a reservoir (400 ha).

Habitats Forest and woodland (broadleaved deciduous forest), Wetland (standing fresh water)
Land-use Unknown

Birds

Breeding birds include *Ciconia nigra* (probably still 1 pair), *Hieraaetus pennatus* (1+ pair), *Otus scops* and four woodpeckers, *Dendrocopos syriacus* and *D. medius* being common. *Milvus milvus*: status uncertain, mainly a very rare winter visitor. The site holds a good representation of restricted-range subspecies (chiefly passerines).

Protection status

National Partial **International** None
2,900 ha of IBA covered by Zapovednik (Girkan goryhy, 2,900 ha).

Conservation issues

Threats Unknown

All land-uses are prohibited. Some illegal tree-cutting and firewood-collection take place, but this was not deemed a threat, nor were other threats evident, in 1991. The Girkan goryhy formerly called Girkanskiy Zapovednik, founded in 1936, is the only real nature reserve in the Talysh mountains.

Astara-chai valley B2, B3 052

Admin region Astara
Coordinates 38°25'N 48°42'E
Altitude 0–500 m **Area** 2,000 ha

Site description

Broadleaved forests and scrub in the Astara-chai valley. When part of the Soviet Union, the area had been within the restricted border zone, requiring a special pass to enter ('Other' land-use).

Habitats Forest and woodland (broadleaved deciduous forest), Scrub (scrub)
Land-use Other

Birds

Species	Season	Year	Pop min	Pop max	Acc	Criteria
Aquila pomarina Lesser Spotted Eagle	B	—	3	—	—	B2
Dendrocopos medius Middle Spotted Woodpecker	R	—	—	—	—	B3

The only area in Azerbaijan where the *P.c. talishensis* subspecies of *Phasianus colchicus* has a good chance of survival (up to 75 breeding pairs). *Accipiter badius* possibly breeds, though there have been no definite records since 1953. There are a number of restricted-range passerine subspecies. Significant proportion (≥1%) of national population breeding at site: *Aquila pomarina* (3+ pairs). *Alcedo atthis*, *Coracias garrulus* and *Dendrocopos medius* are all very common breeding birds. *Milvus milvus* is a rare visitor, especially in the winter, and possibly breeds.

Protection status

National None **International** None

Conservation issues

Threats Other (U), Unsustainable exploitation (U)

Current threats include illegal tree-felling ('Other' threat) and poaching of *Phasianus colchicus*.

REFERENCES

ANON. (1996) *Whitaker's Almanack 1997*. 129th edition. London: J. Whitaker and Sons Ltd.

BORODIN, A. M. AND SYROECHKOVSKIY, E. E. (1983) [*The nature reserves of the USSR.*] Moscow: Lesnaya promyshlennost'. (In Russian.)

BROWN, A., FENNELL, J., KASER, M., AND WILLETS, H. T., EDS. (1982) *The Cambridge encyclopedia of Russia and the Soviet Union.* Cambridge, UK: Cambridge University Press.

CRANSWICK, P. A., PAYNTER, D. B., SULTANOV, E. H., ABULADZE, A., HEARN, R. D. AND QUINN, J. L. (1998) Ornithological surveys of the western route export pipeline in Azerbaijan and Georgia: September/October 1997 and February/March 1998. Unpublished report by the Wetlands Advisory service of the Wildfowl and Wetlands Trust (WWT) to Environment and Resource Technology Ltd of surveys commissioned by Azerbaijan International Operating Company.

GRIMMETT, R. F. A. AND JONES, T. A., EDS. (1989) *Important bird areas in Europe*. Cambridge, UK: International Council for Bird Preservation (Techn. Publ. 3).

HEATH, M. F. AND BORGGREVE, C. (2000) *BirdLife International/EBCC European Bird Database 1998*. Cambridge, UK: BirdLife International.

KNYSTAUTAS, A. (1987) *The natural history of the USSR*. London: Century.

LITVINOVA, N. A. (1989) [On the number of pelicans wintering in Kizil Agach Nature Reserve.] Pp.144–145 in [*Proc. USSR Conf. on Wildlife Cadastre and Census, Part 3.*] Ufa, Russia. (In Russian.)

PATRIKEEV, M. (1999) The birds of Azerbaijan. Unpublished manuscript.

PAYNTER, D., AARVAK, T., AND SULTANOV, E. (1996) Conservation of wetland reserves in Azerbaijan. Unpublished report. Cambridge, UK: Flora and Fauna International (FFI).

SNOW, D. W. AND PERRINS, C. M., EDS. (1998) *The birds of the Western Palearctic: Concise Edition*, vols 1 and 2. Oxford, UK: Oxford University Press.

SOKOLOV, V. E. AND SYROECHKOVSKIY, E. E., EDS. (1990) [*The nature reserves of the USSR: reserves of the Caucasus*]. Moscow: Mysl'. (In Russian.)

TUAEV, D. G. (1975) [On the waterfowl of Kura–Aras lowland.] Pp. 151–188 in [*Data on fauna and ecology of terrestrial vertebrates of Azerbaijan.*] Baku, Azerbaijan. (In Russian.)

TUAEV, D. G. AND VASILIEV, V. I. (1972) [On the species composition and biology of birds at the Mingechaur and Varvara reservoirs.] *Ornitologiya* 10: 260–265. (In Russian.)

TUCKER, G. M. AND HEATH, M. F. (1994) *Birds in Europe: their conservation status*. Cambridge, UK: BirdLife International (BirdLife Conservation Series no. 3).

VERESHCHAGIN, N. K. (1946) [Bird mortality caused by oil pollution in Azerbaijan.] [*Zool. Journal*] 25: 69-80. (In Russian.)

VERESHCHAGIN, N. K (1950) [Wintering wildfowl and the wildfowling trade in Azerbaijan.] [*Proc. Inst. Zool. Acad. Sci. Az. SSR*] 14: 133–214. (In Russian.)

BOSNIA AND HERZEGOVINA

Ferruginous Duck *Aythya nyroca*. (PHOTO: MANFRED DANEGGER/NHPA)

OVERVIEW

Bosnia and Herzegovina occupies 51,130 km^2 with a population of 2,900,000 people. It has not been possible to collect new data for this inventory. Therefore, the three Important Bird Areas (IBAs) presented here (Table 1, Map 1) are identical to those included in the previous pan-European inventory (Grimmett and Jones 1989).

Three globally threatened or near-threatened species occur in significant numbers, meeting the A1 criterion: *Phalacrocorax pygmeus*, *Aythya nyroca* and *Crex crex*. The three IBAs all contain wetland habitats together with alluvial woodlands and humid grasslands.

Table 1. Summary of Important Bird Areas in Bosnia and Herzegovina. 3 IBAs covering 68 km^2

IBA code	1989 code	International/national name	Area (ha)	Criteria (see p. 11)
001	YU026	Hutovo blato	6,144	A1
002	YU027	Boracko jezero	—	B2
003	YU028	Bardaca	700	A1

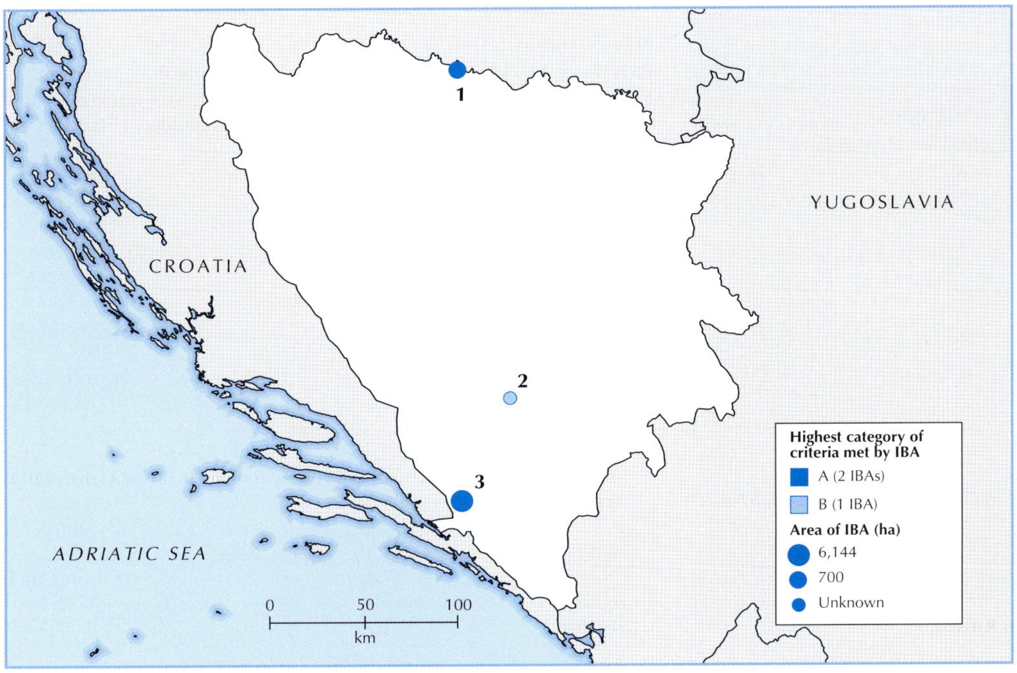

Map 1. Location, area and criteria category of Important Bird Areas in Bosnia and Herzegovina.

SITE ACCOUNTS

Hutovo blato
A1 **001**

Admin region —
Coordinates 43°03′N 17°46′E
Altitude — **Area** 6,144 ha

Site description
Areas of open water with extensive stands of emergent vegetation (*Phragmites*, *Scirpus*), wet meadows, riverine and water-edge forests (*Salix*, *Populus*, *Alnus*), and degraded *Quercus* and *Carpinus* woodland on high ground. Human activities include fishing and hunting.

Habitats Forest and woodland (broadleaved deciduous forest), Wetland (standing fresh water; water-fringe vegetation)
Land-use Fisheries/aquaculture, Hunting

Birds

Species	Season	Year	Pop min	Pop max	Acc	Criteria
Phalacrocorax pygmeus Pygmy Cormorant	B	—	—	—	—	A1
Aythya nyroca Ferruginous Duck	B	—	—	—	—	A1

Two species of global conservation concern breed at the site. Other breeding species include *Phalacrocorax carbo*, *Botaurus stellaris*, *Ardeola ralloides*, *Nycticorax nycticorax*, *Egretta garzetta* and *Ardea purpurea*.

Protection status
National Partial **International** None
IBA covered by Ornithological Reserve.

Conservation issues
Threats Agricultural intensification and expansion (U), Drainage (U), Industrialization/urbanization (U), Recreation/tourism (U)

Drainage, agricultural intensification, urbanization and tourism are problems at the site.

Boracko jezero
B2 **002**

Admin region —
Coordinates 43°34′N 18°01′E
Altitude — **Area** —

Site description
A lake with gravel edges and little or no vegetation, bordered by *Alnus* woodland, meadows, orchards, and surrounded by steep mountains covered in deciduous forest. The valley is used for farming and tourism.

Habitats Forest and woodland (broadleaved deciduous forest), Grassland, Wetland (standing fresh water), Artificial landscape (perennial crops/orchards/groves)
Land-use Agriculture, Tourism/recreation

Birds
Breeding species include *Strix uralensis*, *Dendrocopos medius* and *D. leucotos*. Criterion B2 has been assigned at the site level assuming that the populations of these species are important, but no population data are available.

Protection status
National None **International** None

Conservation issues
Threats Unknown

Bardaca
A1 **003**

Admin region —
Coordinates 45°07′N 17°27′E
Altitude — **Area** 700 ha

Site description
Fish-ponds (with islands and reedbeds), marshes, grazing-meadows, and remnant alluvial forest along the River Sava.

Habitats Forest and woodland (alluvial/very wet forest), Grassland, Wetland (standing fresh water; water-fringe vegetation)
Land-use Unknown

Birds

Species	Season	Year	Pop min	Pop max	Acc	Criteria
Aythya nyroca Ferruginous Duck	B	—	—	—	—	A1
Crex crex Corncrake	B	—	—	—	—	A1

The globally threatened *Aythya nyroca* and *Crex crex* breed at the site. Other breeding species include *Ixobrychus minutus*, *Nycticorax nycticorax*, *Ardeola ralloides*, *Egretta garzetta*, *Ardea purpurea*, *Plegadis falcinellus*, *Platalea leucorodia*, *Sterna hirundo* (30+ pairs), and *Chlidonias hybridus* (40+ pairs). *Ciconia ciconia* and the globally near-threatened *Haliaeetus albicilla* feed, and possibly breed, at the site. The area is also very important for passage migrants.

Protection status
National Partial **International** None
IBA covered by Ornithological Reserve.

Conservation issues
Threats Drainage (U), Other (U), Unsustainable exploitation (U)

Drainage, hunting, and forestry are problems at the site.

REFERENCES

GRIMMETT, R. F. A. AND JONES T. A. (1989) *Important Bird Areas in Europe*. Cambridge, UK: International Council for Bird Preservation (Tech. Publ. 9).

BULGARIA

Irina Kostadinova

White Pelican *Pelecanus onocrotalus* in Atanasovo lake (IBA 036). (Photo: Svetoslav Spasov/BSPB)

GENERAL INTRODUCTION

Bulgaria covers 110,994 km^2 and is bounded to the east by the Black Sea, to the south by Turkey and Greece, to the west by Yugoslavia and the Former Yugoslav Republic of Macedonia, and to the north by Romania. The country can be divided into five main regions: (1) the fertile plain of the River Danube along the border with Romania; (2) the Stara Planina mountains, a 600-km-long chain running east–west across the country, with the highest point (Botev) at 2,376 m; (3) the central plain and valleys of the Maritsa river immediately south of the Stara Planina mountains; (4) the high mountains of Rila, Pirin and Rhodopi in the south-west, with the highest point at 2,925 m; and (5) the Black Sea coast, which is mostly flat but with some sea cliffs in the north.

There are 50 Important Bird Areas (IBAs) in Bulgaria (Table 1), covering a total area of 7,002 km^2, or 6.3% of the land area of the country (Kostadinova 1997). The IBAs lie mainly along the River Danube and the Black Sea coast, as well as in the central and eastern Stara Planina mountains and in the south-west of the country. The first pan-European IBA inventory (Grimmett and Jones 1989)

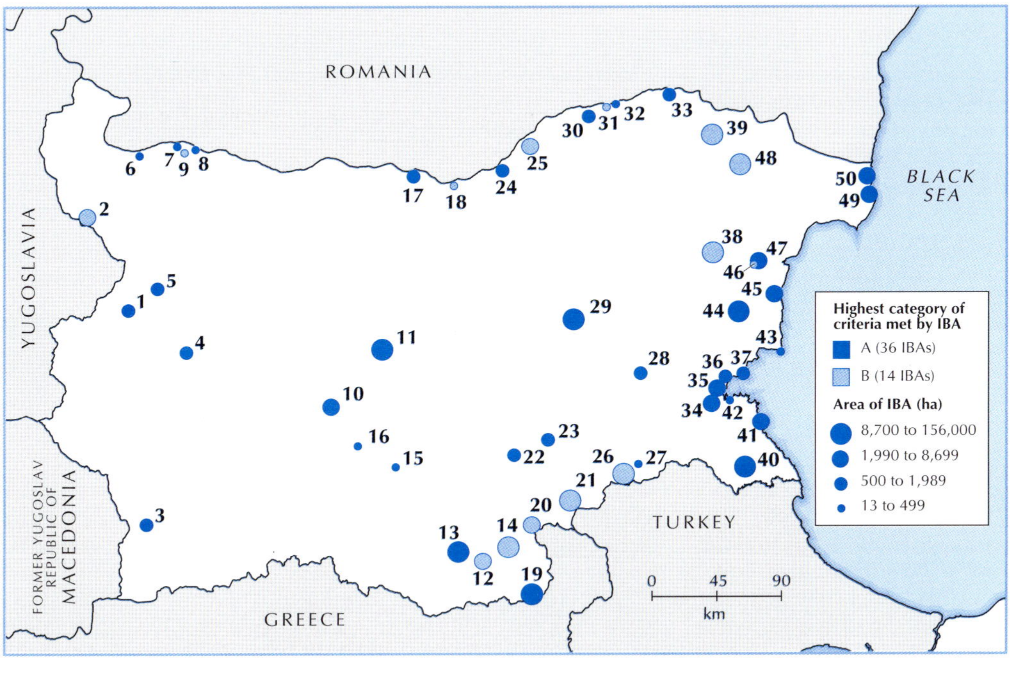

Map 1. Location, area and criteria category of Important Bird Areas in Bulgaria.

Table 1. Summary of Important Bird Areas in Bulgaria. 50 IBAs covering 7,002 km²

IBA code	National code[1]	1989 code	International name	National name	Administrative region	Area (ha)	Criteria (see p. 11)
001	BG001	—	Rayanovci meadows	Vlazni Livadi Rayanovci	Sofia	1,500	A1, B2
002	BG002	—	Tchuprene	Tchuprene	Montana	1,990	B2
003	BG003	—	Tissata	Tissata	Sofia	1,450	A3, B2
004	BG004	—	Dolni Bogrov–Kazichene	Dolni Bogrov–Kazichene	Sofia	1,000	A1
005	BG005	—	Zimevitza meadows	Vlazni Livadi Zimevitza	Sofia	500	A1
006	BG006	—	Orsoya fish-ponds	Ribarnici Orsoya	Montana	360	A1, B2
007	BG007	—	Ibisha island	Ostrov Ibisha	Montana	70	A1, B2
008	BG008	—	Island near Gorni Tzibar	Ostrov do Gorni Tzibar	Montana	60	A1, A4i, B1i
009	BG009	—	Funiyata	Funiyata	Montana	13	B2
010	BG010	—	Pyasachnik reservoir	Yazovir Pyasachnik	Plovdiv	4,000	A4i, A4iii, B1i
011	BG011	BG019–021	Central Balkan	Centralen Balkan	Khaskovo, Lovech, Sofia, Plovdiv	73,262	A1, A3, B2, B3
012	BG012	—	Krumovitza	Krumovitza	Khaskovo	3,500	B2
013	BG013	BG013	Studen Kladenetz	Studen Kladenetz	Khaskovo	18,000	A3, B2
014	BG014	—	Madjarovo	Madjarovo	Khaskovo	8,700	B2
015	BG015	—	Konush reservoir	Yazovir Konush	Plovdiv	30	A1, A4i, B1i, B2
016	BG016	—	Plovdiv fish-ponds	Ribarnici Plovdiv	Plovdiv	98	A1, A4i, B1i
017	BG017	BG001	Belene islands complex	Complex Belenski Ostrovi	Lovech	1,714	A1, B2
018	BG018	BG002	Vardim island	Ostrov Vardim	Lovech	458	B1i, B2
019	BG019	—	Byala Reka	Byala Reka	Khaskovo	18,000	A1, A3, B2
020	BG020	—	Radinchevo	Radinchevo	Khaskovo	5,000	B2
021	BG021	—	Southern Sakar	Yuzen Sakar	Khaskovo	45,000	B2
022	BG022	—	Rozov Kladenetz reservoir	Yazovir Rozov Kladenetz	Khaskovo	600	A1, A4i, B1i
023	BG023	BG003	Ovtcharitza reservoir	Yazovir Ovtcharitza	Khaskovo	630	A1, A4i, A4iii, B1i
024	BG024	—	Mechka fish-ponds	Ribarnici Mechka	Ruse	800	A1, A4i, B1i, B2
025	BG025	BG014	Lomovete	Lomovete	Ruse	4,320	B2
026	BG026	—	Derventski Vuzvishenia	Derventski Vuzvishenia	Burgas	20,000	B2
027	BG027	BG004	Malko Sharkovo reservoir	Yazovir Malko Sharkovo	Burgas	380	A4i, B1i
028	BG028	—	Tzerkovski reservoir	Yazovir Tzerkovski	Burgas	500	A1, B1i
029	BG029	—	Kotlenska mountain	Kotlenska Planina	Burgas	156,000	A1, B2, B3
030	BG030	BG012	Kalimok complex	Complex Kalimok	Ruse	1,000	A1, A4i, B1i, B2, B3
031	BG031	—	Stenata	Stenata	Ruse	15	B2
032	BG032	—	Pozharevo island	Ostrov Pozharevo	Ruse	170	A1, A4i, B1i, B2
033	BG033	BG008	Srebarna lake	Ezero Srebarna	Ruse	1,445	A1, A4i, B1i, B2, B3
034	BG034	BG005	Mandra–Poda complex	Complex Mandra–Poda	Burgas	2,270	A1, A4i, A4iii, A4iv, B1i, B1iv, B2, B3
035	BG035	BG007	Burgasko lake	Burgasko Ezero	Burgas	2,800	A1, A4i, A4iii, B1i
036	BG036	BG010	Atanasovo lake	Atanasovsko Ezero	Burgas	1,950	A1, A4i, A4iv, B1i, B1iv, B2, B3
037	BG037	—	Pomorie lake	Pomorijsko Ezero	Burgas	850	A1, A4i, B1i, B3
038	BG038	—	Provadijsko–Royaksko Plato	Provadijsko–Royaksko Plato	Varna	100,000	B2, B3
039	BG039	—	Hursovska Reka	Hursovska Reka	Varna	20,000	B2
040	BG040	—	Strandja	Strandja	Burgas	116,136	A4iv, B1iv, B2, B3
041	BG041	BG015	Ropotamo complex	Complex Ropotamo	Burgas	4,800	A1, A4i, B1i, B1iv, B2
042	BG042	—	Chengene Skele	Zaliv Tchengene Skele	Burgas	160	A1, B2
043	BG043	BG018	Emine	Emine	Burgas	100	A1, A4iv, B1iv, B2
044	BG044	—	Kamchijska mountain	Kamtchijska Planina	Varna	55,000	A4iv, B1iv, B2, B3
045	BG045	BG022	Kamchia complex	Complex Kamchia	Varna	3,000	A4iv, B1i, B1iv, B2, B3
046	BG046	BG011	Yatata	Yatata	Varna	154	B1i, B2
047	BG047	—	Varna–Beloslav lakes complex	Varnensko–Beloslavsko Ezero	Varna	2,300	A1, A4i, A4iv, B1i, B1iv, B3
048	BG048	—	Suha Reka	Suha Reka	Varna	15,000	B2
049	BG049	BG006	Shabla lake complex	Shablenski Ezeren Complex	Varna	3,100	A1, A4i, A4iii, B1i, B1iv, B2
050	BG050	BG009	Durankulak lake	Durankulak	Varna	2,000	A1, A4i, B1i, B2

1. Kostadinova (1997).

Sites identified in the previous inventory of IBAs in Europe (Grimmett and Jones 1989) but no longer considered to be IBAs
BG016 Gintzi; BG017 Cape Kaliakra.

identified 22 IBAs in Bulgaria, and field surveys during the last eight years, combined with literature research and collaboration with other organisations in 1997, have resulted in a further 32 IBAs being identified in this review (Table 1). Three of the original (1989) IBAs (former sites 019–021) have been merged into one site (Central Balkan, site 011) in this review, and two of the original sites (former sites 016 and 017) no longer qualify as IBAs, after comparison of improved and updated information against the current (revised) IBA criteria (see Table 1).

ORNITHOLOGICAL IMPORTANCE

There are 140 species of European conservation concern (SPECs) which regularly breed in Bulgaria. Five of these are globally threatened (*Pelecanus crispus*, *Aythya nyroca*, *Aquila heliaca*, *Falco naumanni* and *Crex crex*), three of them are near-threatened (*Phalacrocorax pygmeus*, *Haliaeetus albicilla* and *Aegypius monachus*), and a further 81 have an unfavourable conservation status in Europe (Tucker and Heath 1994). Thirteen other globally threatened species occur mainly on migration and in winter, and an additional 14 species, which have an unfavourable conservation status in Europe due to the state of their European wintering populations, occur regularly in winter in Bulgaria.

Twenty-nine sites qualify as IBAs because they support important numbers of species of global conservation concern, i.e. they qualify under criterion A1 (Tables 1 and 2). Of particular note is the occurrence of more that half of the European wintering population (and almost all of the Bulgarian wintering population) of *Branta ruficollis* at two IBAs, Durankulak lake (050) and Shabla lake complex (049).

Twenty-three of the IBAs in Bulgaria support important congregations of waterbirds (Tables 1 and 3), mainly in winter or while staging on migration. Nineteen of the 23 qualify on the basis of their globally important congregations of waterbirds (A4i

Table 2. Important Bird Areas in Bulgaria that are important for species of global conservation concern (meeting criterion A1).

Species	IBA code
Phalacrocorax pygmeus Pygmy Cormorant	006, 007, 015, 016, 017, 022, 023, 024, 030, 032, 033, 034, 035, 036, 042, 047, 049, 050
Pelecanus crispus Dalmatian Pelican	008, 023, 024, 030, 032, 033, 034, 036, 037, 041, 043, 047
Branta ruficollis Red-breasted Goose	028, 030, 032, 034, 035, 036, 049, 050
Aythya nyroca Ferruginous Duck	006, 024, 030, 036, 049
Oxyura leucocephala White-headed Duck	034, 035
Aegypius monachus Cinereous Vulture	019
Aquila clanga Greater Spotted Eagle	036
Aquila heliaca Imperial Eagle	011, 029
Crex crex Corncrake	001, 004, 005, 029
Numenius tenuirostris Slender-billed Curlew	034, 036, 042

Table 3. Important Bird Areas in Bulgaria that support important numbers of one or more congregatory species (i.e. meeting criteria A4 and/or B1). IBAs meeting both criteria A4 and B1 for the species are shown in **bold**. IBAs meeting only criterion B1 for the species concerned, and not A4, are shown in normal type. For key to 'Season', see p. 7.

Species	Season	IBA code
Phalacrocorax carbo Cormorant	R	033
	B	018
	W	**022**, **023**, 032, 034, 035
	P	008, 024
	N	008
Phalacrocorax pygmeus Pygmy Cormorant	W	**015**, **016**, **034**, **035**, **050**
	P	**015**, **016**, **024**, **030**, **036**, **050**
Pelecanus onocrotalus White Pelican	P	**034**
Pelecanus crispus Dalmatian Pelican	B	**033**
	W	**023**, **024**, **032**, **034**, **036**
	P	**008**, **030**, **037**, **041**, **047**
	N	**008**, **030**
Egretta alba Great White Egret	W	**016**, **034**
Ciconia ciconia White Stork	P	**030**
Plegadis falcinellus Glossy Ibis	P	**030**
Platalea leucorodia Spoonbill	P	036
Cygnus olor Mute Swan	W	036, 037, 046, 050
Cygnus cygnus Whooper Swan	W	034, 035, 045, 049
Anser albifrons White-fronted Goose	W	**010**, **023**, **027**, 028, **033**, **034**, **035**, 036, **049**, **050**
Anser anser Greylag Goose	W	030, 032, 033
Branta ruficollis Red-breasted Goose	W	**034**, **035**, **049**, **050**
Tadorna tadorna Shelduck	W	**036**
Anas platyrhynchos Mallard	W	028, 049
Aythya ferina Pochard	W	035, 047
Aythya nyroca Ferruginous Duck	P	**024**
Aythya fuligula Tufted Duck	W	035
Oxyura leucocephala White-headed Duck	W	**034**
Recurvirostra avosetta Avocet	B	**036**
	P	**036**
Sterna sandvicensis Sandwich Tern	B	036

Table 4. Species of European conservation concern with significant breeding populations at Important Bird Areas in Bulgaria (meeting any IBA criteria).

Species [1]	Minimum national breeding population (pairs) [2]	Proportion (%) of national population breeding at all IBAs in Bulgaria
Phalacrocorax pygmeus Pygmy Cormorant	60	78
Pelecanus crispus Dalmatian Pelican	80	100
Ixobrychus minutus Little Bittern	200	94
Nycticorax nycticorax Night Heron	560	98
Ardeola ralloides Squacco Heron	500	29
Ciconia nigra Black Stork	150	53
Plegadis falcinellus Glossy Ibis	200	6
Platalea leucorodia Spoonbill	100	57
Tadorna ferruginea Ruddy Shelduck	50	34
Aythya nyroca Ferruginous Duck	150	29
Neophron percnopterus Egyptian Vulture	100	39
Circaetus gallicus Short-toed Eagle	50	44
Accipiter brevipes Levant Sparrowhawk	20	35
Buteo rufinus Long-legged Buzzard	200	24
Aquila pomarina Lesser Spotted Eagle	60	42
Aquila heliaca Imperial Eagle	20	10
Aquila chrysaetos Golden Eagle	130	21
Hieraaetus pennatus Booted Eagle	30	57
Falco tinnunculus Kestrel	1,500	3
Falco cherrug Saker	30	3
Alectoris graeca Rock Partridge	1,000	7
Coturnix coturnix Quail	7,000	8
Crex crex Corncrake	8,000	1
Recurvirostra avosetta Avocet	300	100[3]
Charadrius alexandrinus Kentish Plover	200	39
Sterna sandvicensis Sandwich Tern	100	10
Otus scops Scops Owl	1,000	9
Bubo bubo Eagle Owl	100	30
Athene noctua Little Owl	4,000	1
Alcedo atthis Kingfisher	1,000	1
Merops apiaster Bee-eater	5,000	5
Coracias garrulus Roller	1,000	1
Picus canus Grey-headed Woodpecker	500	9
Dendrocopos medius Middle Spotted Woodpecker	300	63
Lullula arborea Woodlark	10,000	4
Erithacus rubecula Robin	5,000,000	1
Saxicola torquata Stonechat	10,000	2
Oenanthe hispanica Black-eared Wheatear	1,000	11
Monticola saxatilis Rock Thrush	1,000	3
Monticola solitarius Blue Rock Thrush	300	3
Turdus torquatus Ring Ouzel	5,000	100
Turdus merula Blackbird	1,000,000	2
Turdus philomelos Song Thrush	100,000	14
Turdus viscivorus Mistle Thrush	50,000	7
Hippolais olivetorum Olive-tree Warbler	150	65
Sylvia cantillans Subalpine Warbler	1,000	31
Sylvia melanocephala Sardinian Warbler	500	13
Sylvia atricapilla Blackcap	500,000	5
Regulus regulus Goldcrest	100,000	7
Ficedula semitorquata Semi-collared Flycatcher	1,000	41
Parus caeruleus Blue Tit	500,000	1
Sitta neumayer Rock Nuthatch	100	37
Lanius collurio Red-backed Shrike	100,000	1
Lanius minor Lesser Grey Shrike	1,000	5
Lanius nubicus Masked Shrike	100	10
Fringilla coelebs Chaffinch	1,000,000	9
Carduelis cannabina Linnet	50,000	1
Emberiza melanocephala Black-headed Bunting	5,000	15

1. Only those species of European conservation concern (see Box 1, p. 12) that meet IBA criteria in Bulgaria are listed.
2. Data are taken from the BirdLife/EBCC European Bird Database 1998 (Heath and Borggreve 2000).
3. The percentage of the national population in IBAs exceeds 100%. Usually this is because the national population estimate has not been updated recently whilst the IBA population estimate has been recently updated with new data as a result of comprehensive surveys of IBAs themselves. Also, the individual site count for a species may be the maximum or average over recent years, and summing these may record more birds than are present nationally in any single year.

criterion), and the other four are important at the European flyway level (B1i criterion). Most of these sites are wetlands and support particularly important numbers of wintering *Pelecanus crispus*, *Anser albifrons*, *Branta ruficollis*, *Tadorna tadorna* and *Oxyura leucocephala* (almost the whole Bulgarian wintering population is concentrated within IBAs), as well as of *Phalacrocorax carbo*, *Phalacrocorax pygmeus* and *Egretta alba*. Among the most important sites are Atanasovo lake (036), Durankulak lake (050) and Shabla lake (049) (all are designated Ramsar Sites) as well as Mandra–Poda complex (034), Burgasko lake (035), Ovtcharitza reservoir (023) and Pyasachik reservoir (010).

Along the Black Sea coast, nine IBAs are migratory 'bottleneck' sites, qualifying under criteria A4iv or B1iv (Table 1). These sites are the most important concentrations of large soaring birds along the western Black Sea migration route (Via Pontica). It is believed that the majority of migrating *Ciconia ciconia*, *Pelecanus onocrotalus*, *Pernis apivorus*, *Buteo buteo* and *Aquila heliaca* in Europe pass along the western Black Sea coast, especially over Cape Emine (043) and Atanasovo lake (036), which are among the top sites in Europe for visible migration of large, soaring birds.

Forty IBAs meet the B2/B3 criteria, in that they support important numbers of species of European conservation concern (SPECs) during the breeding season (Table 1). For several such SPECs significant proportions of their national breeding populations breed in IBAs (Table 4). More than 80% of the minimum national breeding population occurs in IBAs for *Pelecanus crispus* (033), *Ixobrychus minutus* (at sites 006, 024, 030, 033 and 050), *Nycticorax nycticorax* (007, 015, 018, 032, 033) and *Recurvirostra avosetta* (034, 036, 037, 047). These are wetland birds with relatively small but important populations in Bulgaria. Good knowledge of these sites makes it comparatively easy to determine their relative importance. Most raptor species in Bulgaria have rather dispersed populations, and consequently a smaller proportion (1–57%) of their minimum national populations are covered by the IBA network (Table 4), nevertheless this proportion is still significant for most species. Available data are not so complete for other SPECs, therefore additional IBAs may be identified for them in future.

Ten key species in Bulgaria are inadequately covered or not covered at all by the IBA inventory (Table 4). These include breeding populations of the globally threatened *Aythya nyroca* and *Aquila heliaca*, as well as populations of *Falco cherrug* (breeding) and the near-threatened *Phalacrocorax pygmeus* (wintering). For these species it is vital that parallel conservation approaches, in addition to the conservation of IBAs, are followed in order to ensure the maintenance of these populations. In addition, wider scale surveys are needed to identify further important sites for several species (Table 4), notably for breeding or wintering populations of *Aquila clanga*, *Falco naumanni* and *Otis tarda* (all three being globally threatened and currently not covered by IBAs through lack of information). In addition, the important feeding areas of two globally threatened species, *Anser erythropus* and *Branta ruficollis*, and those of the near-threatened *Aegypius monachus*, are not covered fully by IBAs.

HABITATS

The lowlands of Bulgaria have been largely converted to agricultural land. In 1995, 61,640 km² (56% of the land surface) was used for agriculture, of which 46,930 km² (42% of the land surface) was cultivated. In 1995, forests covered 38,760 km² (35% of the land surface), of which 34% were conifer-dominated and 26% were oak *Quercus*-dominated. The climax vegetation is dominated by *Quercus* up to an altitude of 1,000 m, by beech *Fagus* between 1,000 m and 1,500 m, and by coniferous woodland up to 2,200 m. Forests of pine *Pinus peuce* are particularly well developed, forming stands above 1,700 m in the Rila, Pirin and Rhodopi mountains. However, vast areas of forest in the uplands have been severely damaged, mainly by sheep-grazing. The main wetlands lie along the River Danube and the Black Sea coast, but many of Bulgaria's marshlands and coastal lakes have been drained or modified since 1944, mainly for agriculture, but also as a result of touristic and industrial development. The north-eastern part of the Black Sea coast has some remnant steppe, similar to the adjacent Dobrudja plain of Romania.

Figure 1. Habitats at Important Bird Areas in Bulgaria (see Appendix 3 for definitions of habitats).

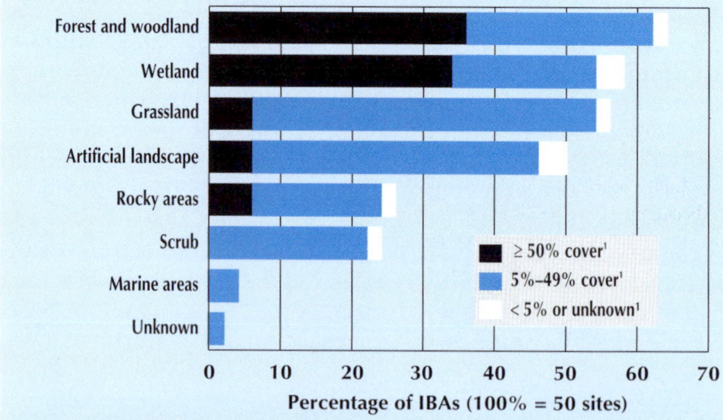

1. percentage of area of individual IBA covered by habitat

Figure 2. Land-uses at Important Bird Areas in Bulgaria (see Appendix 3 for definitions of land-uses).

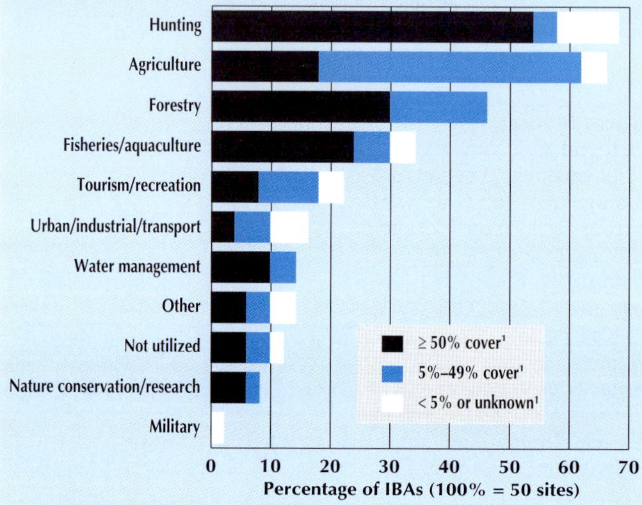

1. percentage of area of individual IBA covered by land-use

Sixty-four percent of IBAs contain some forest or woodland habitat, and in 36% of sites this covers more than half of the IBA's area (Figure 1). The majority of forest sites comprise broadleaved forests of *Quercus* and *Fagus* (e.g. sites 019, 029, 040 and 044), as well as alluvial/very wet forests along the River Danube and Black Sea coast (e.g. 007, 032, 045). Most IBAs contain some wetland, and at 34% of sites such habitat covers more than half of the site's area (Figure 1). In the main, wetland IBAs comprise standing fresh or brackish water (lakes, reservoirs and fish-ponds), often with reedbeds *Phragmites*, and rivers with water-fringe vegetation. Uncommon, but very important for *Numenius tenuirostris*, are mudflats (Chengene Skele, site 042), confined in Bulgaria to the southern Black Sea coast. Grasslands, artificial landscapes, rocky areas and scrub occur within many IBAs but each of these generally covers less than 50% of any particular IBA's area (Figure 2).

Thirty-four of the 50 IBAs (68%) overlap wholly or partly with "sites of major importance of nature conservation" identified under the CORINE Biotopes Programme of the European Commission. Further details for individual sites are stored in the BirdLife International IBA Database.

IMPACTS ON IBAs – LAND-USE AND THREATS

The most common forms of land-use in IBAs in Bulgaria are summarized in Figure 2. Sixty-eight percent of IBAs are used for hunting, a result of most IBAs being unprotected or only partly protected by national law. Forty-six percent of IBAs are used for forestry, and in 30% of IBAs forestry is the major economic activity, impacting more than 50% of the IBA's area. Approximately two-thirds of IBAs in Bulgaria experience some agricultural activity, many being under intensive crop production, especially in north- and south-

Figure 3. Threats at Important Bird Areas in Bulgaria (see Appendix 3 for definitions of threat types and impact categories).

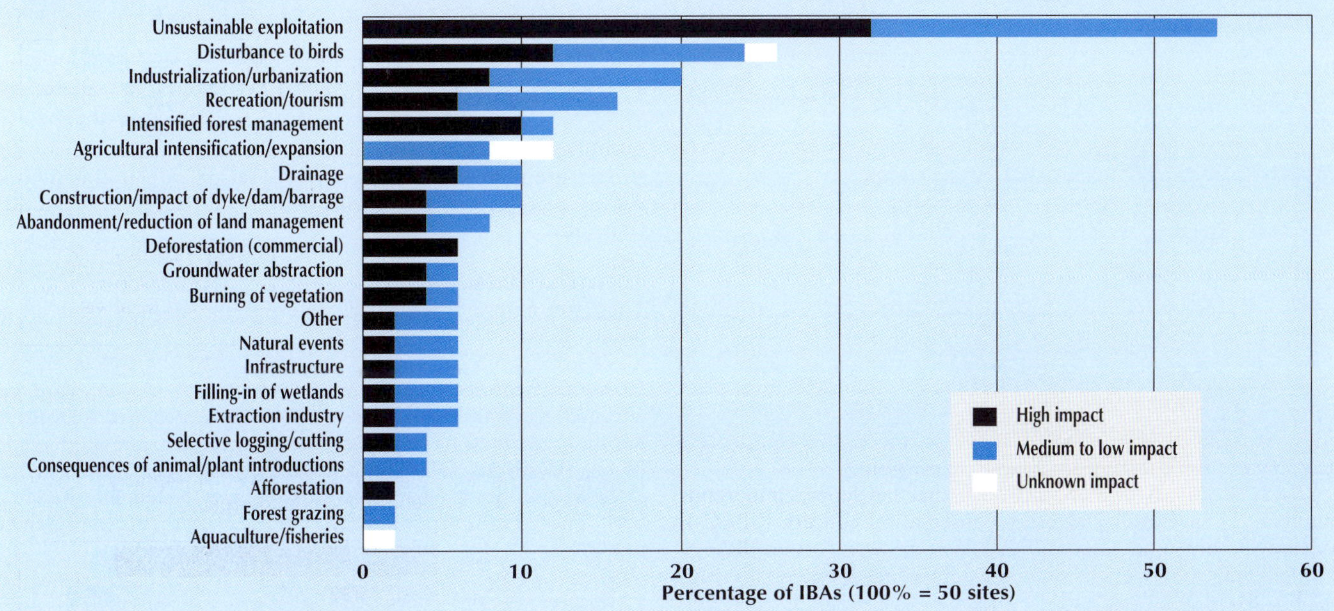

eastern Bulgaria where the largest congregations of wintering geese are found. Grasslands are grazed by cattle, and wet meadows are used for haymaking. Wetlands are used mainly for fisheries and for water management, and the wetlands along the Black Sea coast are surrounded by industrial and recreation areas. Typical and traditional salt-production activities endure at two coastal salt-lakes, Atanasovo (036) and Pomorie (037). The ownership of most IBAs is mixed, which may give rise to difficulties in managing these sites in the future.

Key threats and impacts to IBAs in Bulgaria are presented in Figure 3. Unsustainable exploitation of birds, particularly through illegal or excessive hunting, is the most widespread threat at IBAs, affecting 54% of sites, and in most cases it is believed to have a high impact. Hunting pressure is particularly high in wetland areas such as Shabla lakes complex (site 049) and Durankulak lake (050) which hold large concentrations of wintering waterfowl.

Other serious impacts on IBAs include the development of recreation and urbanization, very rapid in recent years, especially along the Black Sea coast. Wetlands near large settlements and industrial zones are particularly vulnerable, with the most affected sites including those bordering the towns of Burgas and Varna–the Mandra–Poda complex (site 034), Burgasko lake (035), Atanasovo lake (036) and Varna–Beloslav lakes complex (047). Some sites, such as Srebarna lake (033), Shabla lakes complex (049) and Durankulak lake (050), are affected by dyke construction which in turn lowers water-levels and destroys habitats such as temporary freshwater bodies and areas of shallow or open water. Forest habitats are also under threat from intensified forest management in all unprotected IBAs. At the moment, agricultural and grassland areas are relatively little threatened. However, the expected forthcoming membership of Bulgaria in the European Union and the ensuing processes of agricultural intensification pose a potential threat to such habitats.

PROTECTION STATUS

The Ministry of the Environment and Water is the central authority responsible for nature conservation in Bulgaria, and the Central Laboratory for General Ecology coordinates ecological research within the Bulgarian Academy of Science. Table 5 and Figures 4 and 5 summarize the national and international protection status of Bulgarian IBAs.

■ National protection

Until recently, there were four types of protected area relevant to nature conservation: National Parks, Protected Landscapes, Nature Reserves and Natural Monuments. The definitions for these are given below, and the details of protection afforded to IBAs follow this classification (Table 5, Figure 4). These protected areas may also have Buffer Zones, in which land-use is less strictly controlled. The extent of these Buffer Zones is included within the summary statistics on national protection status.

1. National Parks

These are territories distinguished by their outstanding natural beauty. They cover comparatively large areas and are suitable for recreation and tourism.

Figure 4. The national protection status of Important Bird Areas in Bulgaria.

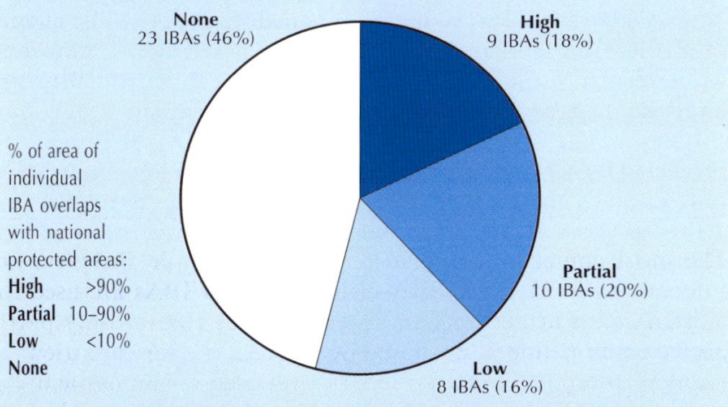

Total area of overlap between IBA network in Bulgaria and national protected-area system (see Table 5 for categories) = 2,028–2,090 km² (29–30% of total IBA area).

Figure 5. The international protection status of Important Bird Areas in Bulgaria.

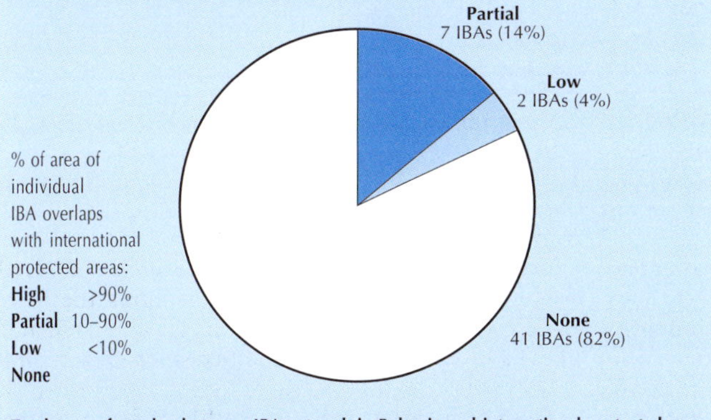

Total area of overlap between IBA network in Bulgaria and international protected-area system (see Table 5 for categories) = 180 km² (3% of total IBA area).

Table 5. Protection status of Important Bird Areas in Bulgaria. A tick (✔) indicates that an IBA overlaps with a protected area (to any extent).

IBA code	International name	National: National Park	National: Protected Landscape	National: Nature Reserve	National: National Monument	National: Buffer Zone	International: Ramsar Site	International: World Heritage Site	International: Biosphere Reserve
001	Rayanovci meadows								
002	Tchuprene			✔		✔			✔
003	Tissata			✔	✔	✔			
004	Dolni Bogrov–Kazichene								
005	Zimevitza meadows								
006	Orsoya fish-ponds								
007	Ibisha island			✔					
008	Island near Gorni Tzibar								
009	Funiyata								
010	Pyasachnik reservoir								
011	Central Balkan	✔		✔		✔			✔
012	Krumovitza								
013	Studen Kladenetz			✔					
014	Madjarovo		✔		✔				
015	Konush reservoir								
016	Plovdiv fish-ponds								
017	Belene islands complex			✔	✔				
018	Vardim island		✔						
019	Byala Reka								
020	Radinchevo								
021	Southern Sakar								
022	Rozov Kladenetz reservoir								
023	Ovtcharitza reservoir								
024	Mechka fish-ponds								
025	Lomovete	✔		✔		✔			
026	Derventski Vuzvishenia								
027	Malko Sharkovo reservoir								
028	Tzerkovski reservoir								
029	Kotlenska mountain			✔	✔	✔			
030	Kalimok complex		✔						
031	Stenata								
032	Pozharevo island		✔						
033	Srebarna lake			✔		✔	✔	✔	✔
034	Mandra–Poda complex		✔						
035	Burgasko lake		✔						
036	Atanasovo lake			✔		✔	✔		
037	Pomorie lake								
038	Provadijsko–Royaksko Plato				✔				
039	Hursovska Reka								
040	Strandja	✔	✔	✔	✔	✔			✔
041	Ropotamo complex		✔		✔	✔	✔		
042	Chengene Skele		✔						
043	Emine				✔				
044	Kamchijska mountain			✔	✔	✔			
045	Kamchia complex		✔	✔		✔			✔
046	Yatata				✔				
047	Varna–Beloslav lakes complex		✔						
048	Suha Reka								
049	Shabla lake complex		✔				✔		
050	Durankulak lake				✔		✔		
Total number of IBAs		3	12	13	11	11	5	1	5

Box 1. International legislation and initiatives that are relevant to site conservation in Bulgaria (see Appendix 1 for a general description of these agreements).

Global	
Biodiversity Convention	✔
Ramsar Convention	✔
Bonn Convention	✔
World Heritage Convention	✔
MAB Programme	✔
Pan-European	
Bern Convention	✔

✔ Convention ratified/initiative supported
(✔) Convention signed

2. Protected Landscapes
These are part of the natural scenery (such as mountain ridges and gorges) and other characteristic landscapes which are suitable for rest and recreation.

3. Nature Reserves
These areas are defined by rare plant or animal communities of scientific value. Activities which may adversely affect the site are strictly prohibited, including agricultural or industrial activities. Admittance to, and movement within, the reserves are strictly controlled.

4. Natural Monuments
These are natural phenomena where preservation is a matter of public interest.

However, according to the new Protected Areas Act (not yet adopted) there are now six categories of protected area relevant to nature conservation in Bulgaria. All protected areas will have to be re-categorized, following this Act, by the year 2000. This re-classification is not yet complete and the new system has therefore been ignored in this publication, but the categories are presented below.

1. Reserve
These areas are model natural ecosystems that include typical or remarkable wild plant and animal species and their habitats.

2. National Park
National Parks should not contain populated areas within their boundaries. They include natural ecosystems with a high flora, fauna and habitat diversity, with typical and remarkable scenery and non-living natural features.

3. Natural Monument
Natural Monuments are typical or remarkable non-living natural features such as rock formations, rock excavations of scientific value, earth pyramids, caves, caverns, waterfalls, fossil and mineral deposits, sand-dunes and other features of exceptional value due to their inherent rarity, representativeness, aesthetic value, or scientific or cultural significance.

4. Maintained Reserve
Maintained Reserves contain ecosystems that include rare and/or endangered wild plant and animal species and their habitats.

5. Natural Park
These areas include diverse ecosystems with multiformity of plant and animal species and habitats thereof, with typical and remarkable scenery and non-living natural features.

6. Protected Locality
Protected Localities include:

i. Areas with typical or remarkable scenery including those that are the result of the harmonious existence of man and nature and which stand out for their significant aesthetic value;
ii. Habitats of endangered, rare or vulnerable plant and animal species and communities.

National Parks, Reserves and Maintained Reserves will be owned, administered and monitored by the state, while Natural Monuments, Natural Parks and Protected Localities will be owned and administered by their owners (not necessarily the state).

Only about 35% of the total area of all IBAs in Bulgaria is protected by national law, with just 5% of the IBA area under strict protection as Nature Reserves (Figure 4). Twenty-seven IBAs are afforded some legal protection through their overlap with the national protected-area network (Figure 4). Three IBAs overlap fully or partly with National Parks, 13 IBAs with Nature Reserves, 12 IBAs with Protected Landscapes, and 11 IBAs with Natural Monuments (Table 5). It should be stressed that large portions of many of these IBAs are not legally protected (Figure 4), and for eight sites (013, 014, 029, 030, 038, 043, 044, 047), this unprotected portion comprises more than 90% of the IBA's area. In addition, 23 IBAs are not afforded any protection by national law.

International protection

Bulgaria is party to all of the international agreements that are relevant to site protection (Box 1). There are five designated Ramsar Sites in Bulgaria, which are covered by five IBAs (033, 036, 049, 050 and part of 041). Given that 15 other IBAs qualify as potential Ramsar Sites, since they meet relevant criteria (A4i, A4iii; see Tables 1 and 3), it should be a priority to obtain official Ramsar status for these very important sites (008, 010, 015, 016, 022, 023, 024, 027, 028, 030, 032, 034, 035, 037 and 047). Srebarna lake (033) is a World Heritage Site. Five IBAs contain Biosphere Reserves, including Central Balkan (011) which holds four Biosphere Reserves (Tsaritchina, Boatin, Steneto and Djendema). Overall, 41 IBAs are unprotected at the international level (Figure 5).

Twenty-three IBAs receive no protection at all through official national or international legislation (Table 5), although ten of these are in the process of being designated. Nine of the unprotected IBAs are considered of global importance for the numbers of waterbirds that they support, and thirteen of them support important numbers of globally threatened species.

CONSERVATION

- Mapping of breeding birds in Bulgaria is conducted throughout the entire IBA network and aims to determine the current status of breeding bird populations in Bulgaria, as well as to provide data for a forthcoming edition of the Bulgarian Red Data Book.
- Monitoring of migrating birds is carried out at IBAs along the Black Sea coast.
- Midwinter counts of waterbirds are carried out each year at all wetland IBAs by the Bulgarian Society for the Protection of Birds (BSPB), together with the Regional Offices of the Ministry of Environment and Water and some other NGOs.
- Management plans have been written for only seven of the IBAs. Plans for a further two IBAs are in preparation.
- The National Parks of Central Balkan (011) and Strandja (040) have administrations under the control of the Ministry of Environment and Water.

ANALYTICAL METHODS

- The quoted size of the breeding population of any particular bird species at an IBA is based on mapping of its breeding populations during the 1990–1997 period. Figures are given as a minimum/ maximum number of pairs (or calling males) on an annual basis.
- Midwinter waterbird counts are of good quality and are based on counts during the 1990–1997 period.
- Counts of passage birds are given as the minimum/maximum number of individuals per season on an annual basis for the 1990–1996 period. For some sites, e.g. Ropotamo complex (041), exact figures for one-off observations are given, because of the lack of monitoring throughout the whole migration period. This means that the real figures may be higher than is presented.
- Sites not meeting criteria are not included in this review. Some sites that meet criteria are very important for certain species which, however, are not mentioned in the text for reasons of confidentiality. Qualifying criteria are not given for these species or for the relevant site.
- Most bird data are based on field research or publications from the 1990–1997 period. Only the data for some strict forest reserves are taken from publications before 1990, but no earlier than 1985.
- Estimations of land-uses and threats at the IBAs are based on information available up to 1996. Excessive or illegal hunting has been included under the threat category 'Unsustainable exploitation' for Bulgaria.
- With regard to land-use, it is difficult to determine the ownership of the sites. The privatization of land in Bulgaria has not yet been completed, and until now there has not been any good information source to check the ownership.

ACKNOWLEDGEMENTS

We thank the BSPB expert and fieldworkers group comprising Antoaneta Petrova, Anton Antonov, Anton Kovachev, Boris Burov, Boris Ivanov, Bozhidar Ivanov, Christo Christov, Dimitar Georgiev, Eberhard Undzian, Emilian Stoinov, Georgi Gruev, Jeko Spiridonov, Iordan Kutzarov, Irina Kostadinova, Ivan Mitev, Kalin Velev, Konstantin Niagolov, Lubomir Profirov, Marin Kurtev, Milen Marinov, Milko Dimitrov, Nikola Mihov, Nikolai Petkov, Pavel Zehtindziev, Petar Iankov, Sergei Dereliev, Snezana Kotzeva, Stoicho Stoichev, Tzeno Petrov, Vencislav Delov, Ursula Georgieva, Valentin Katrandziev and Viktor Vasilev. BSPB also cooperated with the following experts and organizations: Atanas Lalev, Milko Dimitrov, Tanu Michev, Ministry of the Environment and Water, Bulgarian–Swiss Biodiversity Conservation Programme, and the European Commission's CORINE Biotope Programme. Funds have been generously provided by the Royal Society for the Protection of Birds (BirdLife Partner in the UK) and by the Bulgarian–Swiss Biodiversity Conservation Programme.

SITE ACCOUNTS

Rayanovci meadows — A1, B2 — 001

Admin region Sofia
Coordinates 42°56′N 23°02′E
Altitude 800–900 m **Area** 1,500 ha

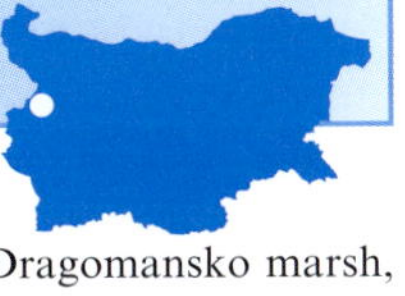

Site description

Rayanovci meadows, together with the former Dragomansko marsh, is in western Bulgaria. Vegetation is dominated by *Carex*, *Poa* and *Alopecurus*, and there are groups of *Salix* and *Populus* trees along drainage canals. Human activities include cattle-grazing and haymaking.

Habitats Grassland (80%; humid grassland), Wetland (20%; standing fresh water; water-fringe vegetation)
Land-use Agriculture (80%), Hunting (100%)

Birds

Species	Season	Year	Pop min	Pop max	Acc	Criteria
Coturnix coturnix Quail	B	1996	500	500	A	B2
Crex crex Corncrake	B	1996	25	40	A	A1
Lullula arborea Woodlark	B	1996	100	100	A	B2
Lanius collurio Red-backed Shrike	B	1996	1,000	1,000	A	B2

The site is important for breeding species such as *Crex crex*, *Coturnix coturnix*, *Lullula arborea* and *Lanius collurio*. Species of global conservation concern that do not meet IBA criteria: *Falco naumanni* (rare, on passage).

Protection status

National None **International** None

Conservation issues

Threats Agricultural intensification/expansion (B)

Conversion of the meadows to arable land is a possible threat. The site is proposed as a Protected Landscape. Investigation of the flora and fauna of the meadows by BSPB is ongoing.

Tchuprene — B2 — 002

Admin region Montana
Coordinates 43°26′N 22°41′E
Altitude 1,350–2,100 m **Area** 1,990 ha

Site description

Old-growth coniferous, broadleaved and mixed forests on the slopes of the western Balkan mountains at 1,400–2,004 m, near the border with Serbia. Nearly 10% of the area is covered by high mountain grasslands, rocky outcrops and pastures, as well as some bogs. The forests are dominated by *Picea*, *Abies*, *Fagus*, *Acer* and *Betula*. Human activities are forbidden within the strict nature reserve.

Habitats Forest and woodland (96%; broadleaved deciduous forest; native coniferous forest), Scrub (3%; scrub), Grassland (1%; dry siliceous grassland)
Land-use Agriculture (27%), Not utilized (73%), Tourism/recreation (27%)

Birds

Species	Season	Year	Pop min	Pop max	Acc	Criteria
[1] ***Alectoris graeca*** Rock Partridge	R	1996	10	—	B	B2

1. Site qualifies according to 1997 national population estimate.

The area is important for *Alectoris graeca*. Species of global conservation concern that do not meet IBA criteria: *Crex crex* (max. 3 pairs breeding). Significant proportion (≥1%) of national population breeding at site: *Hieraaetus pennatus* (1 pair), *Bubo bubo* (1 pair).

Protection status

National High **International** Partial
1,440 ha of IBA covered by Nature Reserve (Tchuprene, 1,440 ha). 542 ha of IBA covered by Buffer Zone (Tchuprene, 542 ha). 1,440 ha of IBA covered by Biosphere Reserve (Tchuprene, 1,440 ha).

Conservation issues

Threats Recreation/tourism (B), Unsustainable exploitation (A)

Illegal recreation and hunting are threats (the Midzhur hunting area is nearby).

Tissata — A3, B2 — 003

Admin region Sofia
Coordinates 41°45′N 23°10′E
Altitude 250–850 m **Area** 1,450 ha

Site description

An area in the southern part of Kresna gorge, along the Struma river, where the slopes are steep and rocky, and covered in Mediterranean forest and shrubland. The forests are dominated by *Juniperus exelsa* (100–200 years old), *Juniperus oxycedrus*, *Phyllirea* and other Mediterranean tree species (*Pyrus*, *Pistacia*, *Jasminum*, etc.). There are also secondary *Quercus* forests, alluvial forests (*Platanus*, *Celtis*, *Tamarix*), and some forestry plantations of *Pinus*. Human activities are forbidden in the strict nature reserve. Grazing, mainly of goats, is allowed in the buffer zone, and forests are used for forestry and cattle-grazing.

Habitats Forest and woodland (62%; broadleaved deciduous forest; native coniferous forest; alluvial/very wet forest), Scrub (20%; scrub; sclerophyllous scrub/garrigue/maquis), Grassland (6%; steppe/dry calcareous grassland), Wetland (2%; river/stream), Rocky areas (2%; inland cliff), Artificial landscape (8%; perennial crops/orchards/groves)
Land-use Agriculture (50%), Forestry (30%), Not utilized (40%)

Birds

Species	Season	Year	Pop min	Pop max	Acc	Criteria
Otus scops Scops Owl	B	1996	10	10	A	B2
Oenanthe hispanica Black-eared Wheatear	B	1996	30	50	A	A3
Monticola saxatilis Rock Thrush	B	1996	10	10	A	B2
[1] ***Monticola solitarius*** Blue Rock Thrush	B	1996	4	5	A	B2
Hippolais olivetorum Olive-tree Warbler	B	1996	50	50	A	A3, B2
Sylvia cantillans Subalpine Warbler	B	1996	40	60	A	A3
Sylvia melanocephala Sardinian Warbler	B	1996	4	5	A	A3
Sitta neumayer Rock Nuthatch	B	1996	12	15	A	A3
Lanius nubicus Masked Shrike	B	1996	1	5	B	A3, B2
Emberiza melanocephala Black-headed Bunting	B	1996	20	20	A	A3

1. Site qualifies according to 1997 national population estimate.

The site is important for its assemblage of Mediterranean-restricted species.

Protection status

National High **International** None
575 ha of IBA covered by Nature Reserve (Tissata, 575 ha). 185 ha of IBA covered by National Monument (Moravska, 185 ha). 625 ha of IBA covered by Buffer Zone (Tissata, 625 ha).

Conservation issues

Threats Forest grazing (B), Recreation/tourism (B)

Building activity is a potential threat. The three types of protected area cover the whole IBA.

Dolni Bogrov–Kazichene — A1 — 004

Admin region Sofia
Coordinates 42°42′N 23°28′E
Altitude 500–500 m **Area** 1,000 ha

Site description

Meadows and marshes, drained by canals, in the Sofia region. The vegetation is dominated by meadows of *Festuca*, *Poa* and *Lolium*, stands of *Phragmites* and *Typha*, and *Salix* thickets. Human activities include cattle-grazing and haymaking.

Habitats Grassland (95%; humid grassland; mesophile grassland), Wetland (5%; water-fringe vegetation)
Land-use Agriculture (100%), Hunting (100%)

Birds

Species	Season	Year	Pop min	Pop max	Acc	Criteria
Crex crex Corncrake	B	1996	21	21	A	A1

The area is important for breeding *Crex crex*. Species of global conservation concern that do not meet IBA criteria: *Aythya nyroca* (one breeding pair), *Falco naumanni* (rare, on passage), *Acrocephalus paludicola* (rare, on passage).

Protection status

National None **International** None

Conservation issues

Threats Agricultural intensification/expansion (B), Unsustainable exploitation (A)

Hunting is a threat, and conversion of the site to arable land is a possible threat. An investigation of the flora and fauna of the meadows by BSPB is ongoing.

Zimevitza meadows — A1 — 005

Admin region Sofia
Coordinates 43°03′N 23°15′E
Altitude 1,000–1,100 m **Area** 500 ha

Site description

Meadows dominated by *Festuca*, *Agrostis* and *Cynosurus* grasses in western Bulgaria, and surrounded by broadleaved deciduous forests of *Fagus* and *Carpinus*, rocky cliffs and arable land. Human activities include grazing and haymaking in the meadows, forestry and hunting.

Habitats Forest and woodland (5%; broadleaved deciduous forest), Grassland (92%; humid grassland; mesophile grassland), Rocky areas (3%; sea cliff/rocky shore)
Land-use Agriculture (92%), Forestry (5%), Hunting (100%)

Birds

Species	Season	Year	Pop min	Pop max	Acc	Criteria
Crex crex Corncrake	B	1996	27	27	A	A1

The area is important for breeding *Crex crex*. Significant proportion (≥1%) of national population breeding at site: *Buteo rufinus* (2 pairs), *Bubo bubo* (1 pair).

Protection status

National None **International** None

Conservation issues

Threats Agricultural intensification/expansion (B)

Conversion of the grasslands to arable land is a possible threat. An investigation of the flora and fauna of the meadows by BSPB is ongoing.

Orsoya fish-ponds A1, B2 006

Admin region Montana
Coordinates 43°47′N 23°07′E
Altitude 100–150 m **Area** 360 ha

Site description

The former Orsoya marshes, on the bank of the River Danube, which have been converted to fish-ponds and overgrown by stands of *Phragmites* and *Typha*. In deep parts of the ponds there are water-lilies *Nymphoides* and *Nymphaea*. The ponds are surrounded by meadows, scrub and trees. Land-uses include fish-farming, haymaking, and hunting.

Habitats Forest and woodland (5%; alluvial/very wet forest), Grassland (20%; humid grassland), Wetland (70%; standing fresh water; water-fringe vegetation), Artificial landscape (5%; perennial crops/orchards/groves)
Land-use Fisheries/aquaculture (60%), Hunting (70%), Other (20%)

Birds

Species	Season	Year	Pop min	Pop max	Acc	Criteria
Phalacrocorax pygmeus Pygmy Cormorant	P	1996	40	180	A	A1
Ixobrychus minutus Little Bittern	B	1996	100	100	A	B2
Aythya nyroca Ferruginous Duck	B	1996	20	20	A	A1
Aythya nyroca Ferruginous Duck	P	1996	48	79	A	A1

This site is important for species of global conservation concern (*Aythya nyroca* and *Phalacrocorax pygmeus*), and supports large numbers of breeding *Ixobrychus minutus*. Breeding species of global conservation concern that do not meet IBA criteria: *Crex crex* (uncommon).

Protection status

National None **International** None

Conservation issues

Threats Burning of vegetation (A), Unsustainable exploitation (A)

The main problem is the cutting and burning of reedbeds during the breeding season.

Ibisha island A1, B2 007

Admin region Montana
Coordinates 43°52′N 23°27′E
Altitude 50–50 m **Area** 70 ha

Site description

An island on kilometre 717 of the River Danube, covered by seasonally flooded forest of *Alnus*, *Salix* and *Populus*. The western part of the island is a sandy clearing, covered by a plantation of *Populus*. Human activities include intensive forest management in the plantation.

Habitats Forest and woodland (50%; alluvial/very wet forest), Artificial landscape (50%; forestry plantation)
Land-use Forestry (50%), Not utilized (50%)

Birds

Species	Season	Year	Pop min	Pop max	Acc	Criteria
[1] ***Phalacrocorax pygmeus*** Pygmy Cormorant	B	1996	12	12	A	A1, B2
Nycticorax nycticorax Night Heron	B	1996	250	250	A	B2
Ardeola ralloides Squacco Heron	B	1996	50	50	A	B2
Platalea leucorodia Spoonbill	B	1996	10	10	A	B2

1. Also occurs on passage.

The island supports the largest nesting colonies of herons and cormorants in the country, being especially important for its large numbers of *Phalacrocorax pygmeus*, *Nycticorax nycticorax*, *Ardeola ralloides* and *Platalea leucorodia*.

Protection status

National Partial **International** None
34 ha of IBA covered by Nature Reserve (Ibisha, 34 ha).

Conservation issues

Threats Disturbance to birds (A), Selective logging/cutting (B)

The main problem is disturbance to nesting birds caused by people visiting the island during the breeding season. There is some illegal cutting of trees in the reserve.

Island near Gorni Tzibar A1, A4i, B1i 008

Admin region Montana
Coordinates 43°49′N 23°32′E
Altitude 100–100 m **Area** 60 ha

Site description

A newly formed island on kilometre 715 of the River Danube. The island is sandy, encloses a small lake, and is covered by forest of *Salix* and *Populus* in the central part. Human activities include firewood-collection during the summer ('Other' land-use, below) and recreation, as well as boat-traffic in surrounding waters.

Habitats Forest and woodland (40%; alluvial/very wet forest), Wetland (1%; river/stream), Rocky areas (59%; inland sand-dunes)
Land-use Tourism/recreation, Other

Birds

Species	Season	Year	Pop min	Pop max	Acc	Criteria
Phalacrocorax carbo Cormorant	P	1996	600	2,300	A	B1i
Phalacrocorax carbo Cormorant	N	1996	184	1,800	A	B1i
Pelecanus crispus Dalmatian Pelican	P	1996	150	252	A	A1, A4i, B1i
Pelecanus crispus Dalmatian Pelican	N	1996	40	210	A	A1, A4i, B1i

The island is an important resting place for *Pelecanus crispus* (which uses sandy areas) and *Phalacrocorax carbo* during the breeding season and migration. Species of global conservation concern that do not meet IBA criteria: *Phalacrocorax pygmeus* (min. 15 in winter), *Aythya nyroca* (min. 3 on passage).

Protection status

National None **International** None

Conservation issues

Threats Afforestation (A), Disturbance to birds

The main problems are afforestation of the sandy areas with *Salix*, and disturbance of nesting birds by human activities during the summer.

Funiyata B2 009

Admin region Montana
Coordinates 43°48′N 23°27′E
Altitude 100–100 m **Area** 13 ha

Site description

Man-made cliffs of loess at the western end of the Zlatia plateau in north-western Bulgaria, surrounded by steppe-like habitats and scrub. *Ailanthus* covers part of the loess walls. Ownership of the IBA may be communal but this is uncertain. Human activities include sheep-grazing and loess excavation in small quantities.

Habitats Rocky areas (100%; inland cliff)
Land-use Agriculture, Urban/Industrial/Transport

Birds

Species	Season	Year	Pop min	Pop max	Acc	Criteria
Merops apiaster Bee-eater	B	—	100	—	A	B2

This is one of the most important nesting colonies of *Merops apiaster* in the country.

■ Protection status

National None **International** None

■ Conservation issues

There are no serious threats to this site.

Pyasachnik reservoir A4i, A4iii, B1i 010

Admin region Plovdiv
Coordinates 42°24′N 24°33′E
Altitude 500–500 m **Area** 4,000 ha

■ Site description

A freshwater reservoir in south-eastern Bulgaria, surrounded by low hills with forests of *Quercus* and forestry plantations of *Pinus*. Land-uses include hunting, fishing and irrigation.

Habitats Forest and woodland (15%; broadleaved deciduous forest), Grassland (10%; steppe/dry calcareous grassland), Wetland (73%; standing fresh water), Artificial landscape (2%; perennial crops/orchards/groves)
Land-use Agriculture (2%), Fisheries/aquaculture (73%), Hunting (60%), Water management (73%)

■ Birds

Species	Season	Year	Pop min	Pop max	Acc	Criteria
Anser albifrons White-fronted Goose	W	1996	9,072	25,388	A	A4i, B1i

Anser albifrons winters in important numbers, and the site holds 20,000 or more waterbirds on a regular basis. Species of global conservation concern that do not meet IBA criteria: *Phalacrocorax pygmeus* (max. 21 in winter), *Anser erythropus* (max. 3 on passage), *Branta ruficollis* (max. 12 in winter), *Aythya nyroca* (max. 10 on passage).

■ Protection status

National None **International** None

■ Conservation issues

Threats Unsustainable exploitation (A)

Illegal and uncontrolled hunting of waterbirds poses a serious threat.

Central Balkan A1, A3, B2, B3 011

Admin region Khaskovo, Lovech, Sofia, Plovdiv
Coordinates 42°43′N 24°56′E
Altitude 500–2,376 m **Area** 73,262 ha

■ Site description

The northern and southern slopes of the highest, central part of the Stara Planina mountains. The limestone karst topography is outstanding, with ancient forests, pastures, subalpine scrub and grasslands. Deciduous forests (up to 250 years old) of *Carpinus*, *Quercus*, *Ostrya* and *Fagus* occur in the higher parts of the area, mixed with *Pinus* and with *Abies*. Coniferous forests are dominated by *Picea* and *Pinus*, and subalpine scrub by *Juniperus* and *Vaccinium*. Human activities (outside the strict reserves) include tourism, and cattle-grazing in the high mountain grasslands.

Habitats Forest and woodland (45%; broadleaved deciduous forest; native coniferous forest), Scrub (21%; scrub; sclerophyllous scrub/garrigue/maquis), Grassland (25%; steppe/dry calcareous grassland; dry siliceous grassland; alpine/subalpine/boreal grassland; humid grassland), Wetland (1%; standing fresh water; river/stream; fen/transition mire/spring), Rocky areas (6%; inland cliff; caves), Artificial landscape (1%; perennial crops/orchards/groves)
Land-use Agriculture (43%), Forestry (20%), Tourism/recreation (72%)

■ Birds

This IBA is important for its alpine-restricted species, including two subspecies that are endemic to the Balkans (*Prunella collaris subalpinum* and *Eremophila alpestris balcanica*), and for its large numbers of a wide

Species	Season	Year	Pop min	Pop max	Acc	Criteria
Buteo rufinus Long-legged Buzzard	B	1996	6	7	A	B2
Aquila heliaca Imperial Eagle	B	1996	1	2	A	A1
Aquila chrysaetos Golden Eagle	B	1996	8	9	A	B2
Hieraaetus pennatus Booted Eagle	B	1996	2	2	A	B2
[1] ***Falco tinnunculus*** Kestrel	B	1996	40	50	A	B2
Falco cherrug Saker	B	1996	1	1	A	B2
Alectoris graeca Rock Partridge	B	1996	60	60	A	B2
Athene noctua Little Owl	R	1996	30	40	A	B2
Picus canus Grey-headed Woodpecker	R	1996	15	20	A	B2
Lullula arborea Woodlark	B	1996	100	100	A	B2
Prunella collaris Alpine Accentor	B	1996	500	1,500	A	A3
Erithacus rubecula Robin	R	1996	55,000	65,000	A	B3
Saxicola torquata Stonechat	B	1996	200	300	A	B2
Monticola saxatilis Rock Thrush	B	1996	20	20	A	B2
Turdus torquatus Ring Ouzel	B	1997	5,000	6,500	A	B3
Turdus merula Blackbird	B	1996	15,000	20,000	A	B3
Turdus pilaris Fieldfare	W	1996	2,000	3,000	A	B3
Turdus philomelos Song Thrush	B	1996	14,000	16,000	A	B3
Turdus viscivorus Mistle Thrush	B	1996	3,400	4,000	A	B3
Sylvia atricapilla Blackcap	B	1996	25,000	40,000	A	B3
Regulus regulus Goldcrest	B	1996	6,500	9,000	B	B3
Ficedula semitorquata Semi-collared Flycatcher	B	1996	100	250	B	B2
Parus caeruleus Blue Tit	B	1996	6,000	7,000	A	B3
Tichodroma muraria Wallcreeper	R	1996	10	15	A	A3
Fringilla coelebs Chaffinch	R	1996	90,000	115,000	A	B3
Carduelis cannabina Linnet	R	1996	500	—	A	B3

1. Site qualifies according to 1997 national population estimate.

variety of other breeding species, especially raptors such as *Aquila heliaca*, *A. chrysaetos* and *Falco cherrug*. Significant proportion (≥1%) of national population breeding at site: *Circaetus gallicus* (1–2 pairs), *Aquila pomarina* (1 pair), *Dendrocopos medius* (7–8 pairs).

■ Protection status

National High **International** Partial

73,262 ha of IBA covered by National Park (Central Balkan, 73,262 ha). 1,597 ha of IBA covered by Nature Reserve (Boatin, 1,597 ha). 4,220 ha of IBA covered by Nature Reserve (Djendema, 4,220 ha). 904 ha of IBA covered by Nature Reserve (Kozya Stena, 904 ha). 1,466 ha of IBA covered by Nature Reserve (Peeshti Skali, 1,466 ha). 1,610 ha of IBA covered by Nature Reserve (Severen Djendem, 1,610 ha). 1,250 ha of IBA covered by Nature Reserve (Sokolna, 1,250 ha). 1,975 ha of IBA covered by Nature Reserve (Stara reka, 1,975 ha). 3,579 ha of IBA covered by Nature Reserve (Steneto, 3,579 ha). 3,419 ha of IBA covered by Nature Reserve (Tsaritchina, 3,419 ha). 942 ha of IBA covered by Buffer Zone (Boatin, 942 ha). 2,117 ha of IBA covered by Buffer Zone (Djendema, 2,117 ha). 114 ha of IBA covered by Buffer Zone (Kozya Stena, 114 ha). 213 ha of IBA covered by Buffer Zone (Severen Djendem, 213 ha). 169 ha of IBA covered by Buffer Zone (Sokolna, 169 ha). 131 ha of IBA covered by Buffer Zone (Stara reka, 131 ha). 233 ha of IBA covered by Buffer Zone (Steneto, 233 ha). 1,062 ha of IBA covered by Buffer Zone (Tsaritchina, 1,062 ha). 1,597 ha of IBA covered by Biosphere Reserve (Boatin, 1,597 ha). 1,775 ha of IBA covered by Biosphere Reserve (Djendema, 1,775 ha). 3,579 ha of IBA covered by Biosphere Reserve (Steneto, 3,579 ha). 3,419 ha of IBA covered by Biosphere Reserve (Tsaritchina, 3,419 ha).

■ Conservation issues

Currently there are no serious problems in the area, although the new Protected Area Law may change the status of some protected areas. The IBA is designated as a National Park, and incorporates nine strict reserves and four Biosphere Reserves. There is a management plan for the area, prepared by the Ministry of Environment under the Bulgarian–Swiss Biodiversity Conservation Programme.

Krumovitza B2 012

Admin region Khaskovo
Coordinates 41°33′N 25°41′E
Altitude 150–300 m **Area** 3,500 ha

■ Site description

The valley of the Krumovitza river and the Diushun river in the eastern Rhodopi mountains, together with the surrounding low mountain slopes. The rivers flow through deep canyons and open valleys, with the vegetation along the riverbanks dominated by *Alnus*, *Salix*, *Populus*, *Rubus*, *Rosa* and *Tamarix*. Dry grassland, scrub and broadleaved forests cover the neighbouring hills. The forests are dominated by *Quercus*, *Carpinus* and Mediterranean tree species. *Juniperus* is predominant in the sclerophyllous scrub. Land-uses are extensive rearing of sheep and cattle, hunting and forestry, but the area is becoming progressively depopulated due to local emigration.

Habitats Forest and woodland (10%; broadleaved deciduous forest), Scrub (45%; scrub; sclerophyllous scrub/garrigue/maquis), Grassland (15%; steppe/dry calcareous grassland), Wetland (5%; river/stream), Rocky areas (5%; inland cliff), Artificial landscape (20%; perennial crops/orchards/groves)
Land-use Agriculture (80%), Fisheries/aquaculture (5%), Forestry (5%), Hunting

Birds

Species	Season	Year	Pop min	Pop max	Acc	Criteria
Ciconia nigra Black Stork	B	1996	10	10	A	B2
Neophron percnopterus Egyptian Vulture	B	1996	7	7	A	B2
Hieraaetus pennatus Booted Eagle	B	1996	2	2	A	B2

The site is one of the most important areas in Bulgaria for breeding *Ciconia nigra* and *Neophron percnopterus*. Significant proportion (≥1%) of national population breeding at site: *Circaetus gallicus* (1 pair), *Bubo bubo* (1–2 pairs), *Hippolais olivetorum* (2–5 pairs).

Protection status

National None **International** None

Conservation issues

Threats Abandonment/reduction of land management (A), Intensified forest management (B), Unsustainable exploitation (B)

The rural depopulation has led to decreases in available food resources (livestock carrion) for vultures in the region. Uncontrolled hunting and fishing are further threats.

Studen Kladenetz — A3, B2 — 013

Admin region Khaskovo
Coordinates 41°36'N 25°30'E
Altitude 300–879 m **Area** 18,000 ha

Site description

A dam in a rocky gorge of the Arda river, east of the town of Kardjali. The numerous rocky cliffs and banks are covered by sub-Mediterranean vegetation, mainly forests of *Quercus* with *Carpinus*, *Fraxinus* and Mediterranean tree species. *Juniperus*, *Jasminum* and *Paliurus* are predominant in the sclerophyllous shrubland. *Dichantium* and *Poa* predominate in the grasslands. Human activities include hunting, forestry, fishing, and extensive rearing of sheep and cattle.

Habitats Forest and woodland (57%; broadleaved deciduous forest), Scrub (12%; heathland; sclerophyllous scrub/garrigue/maquis), Grassland (7%; steppe/dry calcareous grassland), Wetland (14%; standing fresh water; river/stream), Rocky areas (6%; inland cliff), Artificial landscape (4%; perennial crops/orchards/groves)
Land-use Agriculture (11%), Fisheries/aquaculture (14%), Forestry (56%), Hunting (60%), Water management (12%)

Birds

The site is important for its assemblage of Mediterranean-restricted species, and for its high numbers of breeding *Ciconia nigra*, *Neophron percnopterus* (the best site in Bulgaria) and *Bubo bubo*. Breeding species of global conservation concern that do not meet IBA criteria: *Aegypius monachus* (1–2 pairs), *Falco naumanni* (3–5 pairs). Significant proportion (≥1%) of national population breeding at site: *Circaetus gallicus* (1–3 pairs), *Aquila pomarina*, *Hieraaetus pennatus* (1 pair).

Species	Season	Year	Pop min	Pop max	Acc	Criteria
Ciconia nigra Black Stork	B	1996	30	30	A	B2
Neophron percnopterus Egyptian Vulture	B	1996	5	5	A	B2
Bubo bubo Eagle Owl	B	1996	3	3	A	B2
Oenanthe hispanica Black-eared Wheatear	B	1996	10	10	A	A3
[1] ***Monticola solitarius*** Blue Rock Thrush	R	1996	5	6	A	B2
Hippolais olivetorum Olive-tree Warbler	B	1996	3	5	A	A3, B2
Sylvia cantillans Subalpine Warbler	B	1996	30	30	A	A3
Sylvia melanocephala Sardinian Warbler	R	1996	20	20	A	A3
Sitta neumayer Rock Nuthatch	R	1996	10	10	A	A3
Lanius nubicus Masked Shrike	B	1996	5	5	A	A3, B2
Emberiza melanocephala Black-headed Bunting	B	1996	5	15	A	A3

1. Site qualifies according to 1997 national population estimate.

Protection status

National Low **International** None
775 ha of IBA covered by Nature Reserve (Valchi Dol, 775 ha).

Conservation issues

Threats Abandonment/reduction of land management (B), Industrialization/urbanization (A), Unsustainable exploitation (B)

One problem is rural depopulation, which has led to a decrease in food resources (livestock carrion) for vultures in the region. A BSPB project for eastern Rhodopi mountains (the IBA is part of this region) is ongoing under the Bulgarian–Swiss Biodiversity Conservation Programme.

Madjarovo — B2 — 014

Admin region Khaskovo
Coordinates 41°38'N 25°52'E
Altitude 100–1,200 m **Area** 8,700 ha

Site description

An ancient volcano, cut by the gorge of the Arda river. The shingle and sand riverbanks are covered by *Salix* trees, bushes and other vegetation. There are scattered forests of *Quercus* and Mediterranean trees (*Juniperus*, *Colutea*) on the rocky mountain slopes. *Jasminum* and *Paliurus* are predominant in the sclerophyllous shrubland. Land-uses are the extensive rearing of cattle, sheep and goats, forestry and mining. The area is becoming depopulated due to ongoing rural emigration.

Habitats Forest and woodland (45%; broadleaved deciduous forest), Scrub (17%; scrub; sclerophyllous scrub/garrigue/maquis), Grassland (3%; steppe/dry calcareous grassland), Wetland (5%; river/stream), Rocky areas (25%; inland cliff), Artificial landscape (5%; perennial crops/orchards/groves)
Land-use Agriculture (25%), Forestry (45%), Hunting (70%), Tourism/recreation (50%), Urban/industrial/transport (1%)

Birds

Species	Season	Year	Pop min	Pop max	Acc	Criteria
Ciconia nigra Black Stork	B	1996	6	15	A	B2
Otus scops Scops Owl	B	1996	25	70	A	B2
Hippolais olivetorum Olive-tree Warbler	B	1996	5	10	A	B2

The site is one of the most important sites in Bulgaria for breeding *Ciconia nigra* and *Otus scops*. It also supports the only breeding colony of *Gyps fulvus* in Bulgaria. Species of global conservation concern that do not meet IBA criteria: *Aegypius monachus* (common non-breeding resident). Significant proportion (≥1%) of national population breeding at site: *Neophron percnopterus* (3 pairs), *Circaetus gallicus* (1 pair), *Bubo bubo* (3 pairs).

Protection status

National Low **International** None
79 ha of IBA covered by National Monument (Kovan Kaya, 79 ha).
19 ha of IBA covered by Protected Landscape (Momina Skala, 19 ha).
180 ha of IBA covered by Protected Landscape (Patronka, 180 ha).

Conservation issues

Threats Abandonment/reduction of land management (A), Extraction industry (A), Infrastructure (B)

Mining drives away vultures and other large cliff-nesting birds, through open-air and underground explosions, drilling activities, and the disposal of waste material. In addition, rural depopulation has led to a decrease in available food (livestock carrion) for vultures in the region. Only a very small area of the IBA has been protected. A BSPB project for the eastern Rhodopi mountains (of which this area is a part) is ongoing under the Bulgarian–Swiss Biodiversity Conservation Programme.

Konush reservoir — A1, A4i, B1i, B2 — 015

Admin region Plovdiv
Coordinates 42°04'N 25°02'E
Altitude 150–1,500 m **Area** 30 ha

Site description

A small, open reservoir near Konush village, with beds of *Typha*, and a group of trees on its south bank, and surrounded by arable land. Land-uses include fish-farming, irrigation and hunting.

Habitats Wetland (98%; standing fresh water; water-fringe vegetation), Artificial landscape (2%; arable land)
Land-use Agriculture (2%), Fisheries/aquaculture (98%), Hunting (100%), Water management (98%)

Birds

Species	Season	Year	Pop min	Pop max	Acc	Criteria
Phalacrocorax pygmeus Pygmy Cormorant	P	1996	30	800	A	A1, A4i, B1i
Phalacrocorax pygmeus Pygmy Cormorant	W	1997	50	800	A	A1, A4i, B1i
Nycticorax nycticorax Night Heron	B	1996	100	160	A	B2

The site is important for migrating and wintering *Phalacrocorax pygmeus*, and supports one of the most important breeding colonies of *Nycticorax nycticorax* in Bulgaria.

Protection status

National None **International** None

Conservation issues

Threats Disturbance to birds (A), Unsustainable exploitation (A)

The main problems are the cutting-down of nesting trees at the colonies, the burning of reedbeds, hunting, and disturbance of nesting birds by people close to colonies.

Plovdiv fish-ponds — A1, A4i, B1i — 016

Admin region Plovdiv
Coordinates 42°11'N 24°45'E
Altitude 150–150 m **Area** 98 ha

Site description

The fish-ponds are situated north of the town of Plovdiv, and are partially covered by aquatic vegetation of *Ceratophyllum*, *Potamogeton*, *Trapa* and *Nymphoides*, with fringing beds of *Typha*, *Phragmites*, *Carex*, *Juncus* and *Salix* along the banks.

Habitats Wetland (100%; standing fresh water; water-fringe vegetation)
Land-use Fisheries/aquaculture (100%), Hunting (100%)

Birds

Species	Season	Year	Pop min	Pop max	Acc	Criteria
Phalacrocorax pygmeus Pygmy Cormorant	W	1997	800	1,000	A	A1, A4i, B1i
Phalacrocorax pygmeus Pygmy Cormorant	P	1996	20	800	A	A1, A4i, B1i
Egretta alba Great White Egret	W	1997	140	168	A	A4i, B1i

The ponds are important for wintering and migrating *Phalacrocorax pygmeus* and for wintering *Egretta alba*. Species of global conservation concern that do not meet IBA criteria: *Aythya nyroca* (min. 3 on passage). Significant proportion (≥1%) of national population breeding at site: *Ixobrychus minutus* (5–6 pairs).

Protection status

National None **International** None

Conservation issues

Threats Unsustainable exploitation (A), Other (C)

The main problem is reed-cutting during the breeding season ('Other' threat, above).

Belene islands complex — A1, B2 — 017

Admin region Lovech
Coordinates 43°40'N 25°10'E
Altitude 25–26 m **Area** 1,714 ha

Site description

The eastern part of the largest Bulgarian island (Belene) along the River Danube. Three large marshes, with open-water areas partly covered by floating water-plants and fringed by reedbeds of *Phragmites*, *Sparganium* and *Alisma*, and which are surrounded by seasonally flooded forest (*Salix*, *Populus*) and wet meadows (*Cynodon*, *Scirpus*). There are two smaller islands (one called Kitka), also covered with flooded forest. The complex lies between kilometres 576 and 560 of the Danube. Human activities include cattle-grazing in the buffer zone of the Persinski marshes reserve, agriculture in surrounding areas and hunting (including international hunting tourism).

Habitats Forest and woodland (60%; alluvial/very wet forest), Grassland (35%; humid grassland), Wetland (5%; standing fresh water)
Land-use Agriculture (30%), Fisheries/aquaculture (10%), Forestry (55%)

Birds

Species	Season	Year	Pop min	Pop max	Acc	Criteria
Phalacrocorax pygmeus Pygmy Cormorant	P	1992	—	55	C	A1
Lanius minor Lesser Grey Shrike	B	1996	2	10	B	B2

Formerly one of the most important breeding sites in the country for *Phalacrocorax pygmeus* (max. 40 pairs) and *Plegadis falcinellus* (max. 60 pairs), which stopped breeding on the island about 10 years ago. They would probably breed again if water-levels were restored, but such site-management depends on finding the necessary funding. Breeding species of global conservation concern that do not meet IBA criteria: *Haliaetus albicilla* (1 pair). Significant proportion (≥1%) of national population breeding at site: *Platalea leucorodia* (rare).

Protection status

National Partial **International** None
25 ha of IBA covered by Nature Reserve (Kitka Island, 25 ha). 385 ha of IBA covered by Nature Reserve (Persinski marshes, 385 ha). 719 ha of IBA covered by National Monument (Persina Iztok, 719 ha).

Conservation issues

Threats Construction/impact of dyke/dam/barrage (A), Drainage (A), Filling-in of wetlands (B), Industrialization/urbanization (A), Unsustainable exploitation (B)

The main problems are the lowered water-levels in the Danube and the subsequent drying-out of wetlands, resulting from the 'Iron Gate' Dam in Serbia and from dykes built along the banks of Belene. Special measures are needed to restore water-levels. Illegal tree-cutting, grazing and hunting are also problems, as is the potential construction of a nuclear power-station near Belene.

Vardim island — B1i, B2 — 018

Admin region Lovech
Coordinates 43°37'N 25°28'E
Altitude 16–16 m **Area** 458 ha

Site description

An island in the River Danube located between kilometres 576 and 560, lying opposite Vardim village, east of the town of Svistov. Alluvial forest of *Quercus*, *Ulmus* and *Populus* covers the eastern part of the island, while a *Populus* plantation covers the western part. There are also sandbars covered by seasonally flooded forest. Human activities include intensive management of the plantation, and hunting (including international hunting tourism) during the winter.

Habitats Forest and woodland (95%; alluvial/very wet forest), Rocky areas (5%; inland sand-dunes)
Land-use Forestry (78%), Hunting (100%)

Birds

Species	Season	Year	Pop min	Pop max	Acc	Criteria
Phalacrocorax carbo Cormorant	B	1996	350	375	A	B1i
[1] ***Nycticorax nycticorax*** Night Heron	B	1996	40	45	A	B2
[2] ***Platalea leucorodia*** Spoonbill	B	1996	9	20	A	B2

1. 300 pairs before 1989.
2. 10 pairs before 1989.

One of the most important nesting sites in the country for *Phalacrocorax carbo*, *Nycticorax nycticorax* and *Platalea leucorodia*. Breeding species of global conservation concern that do not meet IBA criteria: *Phalacrocorax pygmeus* (c.20 pairs before 1989, but now extinct), *Haliaeetus albicilla* (1 pair before 1989, now extinct).

Protection status

National Partial **International** None
99 ha of IBA covered by Protected Landscape (Stariat Dab, 99 ha).

Conservation issues

Threats Construction/impact of dyke/dam/barrage (B), Deforestation (commercial) (A)

Problems are (1) the disturbance to nesting waterbirds caused by forestry activities near the colonies, (2) the lowered water-levels of the Danube (due to the 'Iron Gate' dam in Serbia) have caused the drying-out of forest, and (3) breeding colonies have shifted outside of the protected area. The eastern part of the IBA is protected.

Byala Reka

A1, A3, B2 **019**

Admin region Khaskovo
Coordinates 41°22'N 26°03'E
Altitude 100–1,226 m **Area** 18,000 ha

Site description

The Byala river valley and neighbouring hills, situated in the eastern Rhodopi mountains in southern Bulgaria. The vegetation is rich, consisting of old broadleaved deciduous forests of *Fagus*, *Quercus* and *Carpinus*, and sclerophyllous scrub (pseudo-maquis) of *Juniperus*, *Phyllirea*, *Paliurus* and *Fraxinus*, with *Salix* trees along the riverbanks. There are also coniferous plantations and rocky outcrops. Land-uses include hunting, forestry, fishing, extensive stock-rearing and cultivation. There is a small abandoned marble quarry, but drilling for minerals and other geological investigations still go on.

Habitats Forest and woodland (60%; broadleaved deciduous forest), Scrub (10%; sclerophyllous scrub/garrigue/maquis), Grassland (2%; steppe/dry calcareous grassland), Wetland (1%; river/stream), Artificial landscape (27%; perennial crops/orchards/groves)
Land-use Agriculture (29%), Forestry (60%), Hunting (70%), Tourism/recreation (2%)

Birds

Species	Season	Year	Pop min	Pop max	Acc	Criteria
[1] ***Aegypius monachus*** Cinereous Vulture	N	1996	15	—	A	A1
Circaetus gallicus Short-toed Eagle	B	1996	2	3	A	B2
Accipiter brevipes Levant Sparrowhawk	B	1996	1	3	A	B2
Hieraaetus pennatus Booted Eagle	B	1996	4	4	A	B2
Otus scops Scops Owl	B	1996	54	90	A	B2
Lullula arborea Woodlark	B	1996	190	500	A	B2
Oenanthe hispanica Black-eared Wheatear	B	1996	Common	—		A3
Hippolais olivetorum Olive-tree Warbler	B	1996	25	45	A	A3, B2
Sylvia cantillans Subalpine Warbler	B	1996	80	165	A	A3
Sylvia melanocephala Sardinian Warbler	B	1996	30	40	A	A3
Lanius nubicus Masked Shrike	B	1996	3	5	A	A3, B2
Emberiza melanocephala Black-headed Bunting	B	1996	475	900	A	A3

1. Site used for resting.

The site is important for its assemblage of Mediterranean-restricted species, and is one of the most important breeding areas in Bulgaria for raptors (*Aegypius monachus*, *Circaetus gallicus*, *Accipiter brevipes*, *Hieraaetus pennatus*, *Otus scops*) and *Lullula arborea*. Significant proportion (≥1%) of national population breeding at site: *Ciconia nigra* (3–4 pairs), *Neophron percnopterus* (2 pairs), *Aquila pomarina* (1–3 pairs), *Dendrocopos medius* (5–10 pairs).

Protection status

National None **International** None

Conservation issues

Threats Extraction industry (B), Intensified forest management (A), Disturbance to birds (C)

Some nesting bird species are disturbed by hunting. A BSPB project for the eastern Rhodopi mountains (of which this area is a part) is ongoing under the Bulgarian–Swiss Biodiversity Conservation Programme.

Radinchevo

B2 **020**

Admin region Khaskovo
Coordinates 41°45'N 26°03'E
Altitude 100–200 m **Area** 5,000 ha

Site description

An open, rocky area on the bank of the Maritza river in the south-west Sakar mountains. There is sclerophyllous scrub (*Paliurus*, *Jasminum*, *Nigella*) and scattered broadleaved deciduous forest of *Quercus* and *Carpinus*. Human activities are mainly extensive stock-rearing and agriculture.

Habitats Forest and woodland (30%; broadleaved deciduous forest), Scrub (20%; scrub; sclerophyllous scrub/garrigue/maquis), Grassland (35%; dry siliceous grassland), Artificial landscape (15%; arable land; perennial crops/orchards/groves)
Land-use Agriculture (70%), Forestry (30%), Hunting

Birds

Species	Season	Year	Pop min	Pop max	Acc	Criteria
Buteo rufinus Long-legged Buzzard	B	1996	5	5	A	B2
Lanius nubicus Masked Shrike	B	1996	1	2	A	B2

One of the most important sites for breeding *Buteo rufinus* in the country. Species of global conservation concern that do not meet IBA criteria: *Phalacrocorax pygmeus* (min. 8 on passage), *Crex crex* (max. 1 on passage), *Otis tarda* (max. 9 on passage). Significant proportion (≥1%) of national population breeding at site: *Ciconia nigra* (4 pairs), *Neophron percnopterus* (1 pair), *Aquila pomarina* (2 pairs).

Protection status

National None **International** None

Conservation issues

Threats Disturbance to birds (B), Natural events (B), Unsustainable exploitation (A)

Problems include the disturbance and persecution of nesting raptors by local people (and treasure-hunters), illegal hunting, and wild-fires during the summer. A part of the site is proposed as a protected area.

Southern Sakar

B2 **021**

Admin region Khaskovo
Coordinates 41°50'N 26°21'E
Altitude 100–650 m **Area** 45,000 ha

Site description

A rugged region of low, rounded hills and relatively open valleys, with few small villages and a poorly developed network of roads, close to the border with Turkey. There are dry grasslands, Mediterranean shrubland of *Paliurus* and *Jasminum*, scattered forests of Mediterranean broadleaved type (mainly *Quercus* and *Carpinus*), and some rocky areas. Land-uses are extensive stock-rearing and cultivation, hunting and forestry.

Habitats Forest and woodland (30%; broadleaved deciduous forest), Scrub (10%; scrub; sclerophyllous scrub/garrigue/maquis), Grassland (15%; dry siliceous grassland), Wetland (2%; river/stream), Artificial landscape (43%; arable land)
Land-use Agriculture (70%), Forestry (30%), Hunting

Birds

Species	Season	Year	Pop min	Pop max	Acc	Criteria
Buteo rufinus Long-legged Buzzard	B	1996	3	3	A	B2
Picus canus Grey-headed Woodpecker	R	1996	6	—	A	B2

One of the most important breeding areas for *Buteo rufinus* and *Picus canus* in Bulgaria. Significant proportion (≥1%) of national population breeding at site: *Neophron percnopterus* (1 pair), *Circaetus gallicus* (1 pair), *Hippolais olivetorum* (min. 3 pairs).

Protection status

National None **International** None

Conservation issues

Threats Selective logging/cutting (A)

The main problem is the cutting of mature trees along the rivers. There are large populations of this rodent and of brown hare *Lepus europaeus* in the area, providing much small-mammal prey for raptors.

Rozov Kladenetz reservoir — A1, A4i, B1i — 022

Admin region Khaskovo
Coordinates 42°08′N 25°55′E
Altitude 100–100 m **Area** 600 ha

Site description

A water-storage reservoir which does not freeze in winter, since an associated thermoelectric power-station uses the water as coolant. The water is also used for irrigation.

Habitats Wetland (99%; standing fresh water), Artificial landscape (1%; perennial crops/ orchards/groves; other urban/industrial areas)
Land-use Fisheries/aquaculture (99%), Urban/industrial/transport (1%), Water management (99%)

Birds

Species	Season	Year	Pop min	Pop max	Acc	Criteria
Phalacrocorax carbo Cormorant	W	1997	116	8,062	A	A4i, B1i
Phalacrocorax pygmeus Pygmy Cormorant	W	1997	16	142	A	A1

The reservoir supports large and important numbers of *Phalacrocorax carbo* and *P. pygmeus* in winter. Species of global conservation concern that do not meet IBA criteria: *Pelecanus crispus* (6–17 in winter), *Aythya nyroca* (max. 2 in winter).

Protection status

National None **International** None

Conservation issues

Threats Industrialization/urbanization (B), Unsustainable exploitation (A)

The main problem is illegal and uncontrolled hunting. The reservoir is proposed as a protected area.

Ovtcharitza reservoir — A1, A4i, A4iii, B1i — 023

Admin region Khaskovo
Coordinates 42°13′N 26°10′E
Altitude 100–100 m **Area** 630 ha

Site description

A dam on the Ovtcharitza river, surrounded by low hills with arable land (mainly winter wheat). The reservoir is used for cooling an electric power-station at the site ('Other' land-use, below), and therefore the water does not freeze in winter. Human activities include hunting, fishing and irrigation.

Habitats Grassland (3%; steppe/dry calcareous grassland), Wetland (81%; standing fresh water; river/stream), Artificial landscape (11%; arable land), Unknown (5%)
Land-use Agriculture (20%), Fisheries/aquaculture (80%), Hunting (100%), Other (80%), Urban/industrial/transport (5%)

Birds

Species	Season	Year	Pop min	Pop max	Acc	Criteria
Phalacrocorax carbo Cormorant	W	1997	6,635	8,020	A	A4i, B1i
Phalacrocorax pygmeus Pygmy Cormorant	W	1997	—	52	A	A1
Pelecanus crispus Dalmatian Pelican	W	1997	8	211	A	A1, A4i, B1i
Anser albifrons White-fronted Goose	W	1997	200	34,755	A	A4i, B1i

Important in winter for fish-eating birds (*Phalacrocorax* spp., *Pelecanus crispus*) and for *Anser albifrons*. The site holds 20,000 or more waterbirds on a regular basis.

Protection status

National None **International** None

Conservation issues

Threats Unsustainable exploitation (A)

Uncontrolled hunting of waterbirds is a serious threat. The reservoir is proposed as a protected area.

Mechka fish-ponds — A1, A4i, B1i, B2 — 024

Admin region Ruse
Coordinates 43°44′N 25°49′E
Altitude 150–150 m **Area** 800 ha

Site description

Fish-ponds overgrown with reeds, situated close to the bank of the River Danube. The aquatic vegetation is dominated by *Phragmites*, *Trapa*, *Nymphaea*, as well as by *Typha* in some places. In the western part of the ponds there is a seasonally flooded forest of *Fraxinus*, *Tilia*, *Ulmus* and *Acer*, as well as a small river fully overgrown with shrubs, acacia and grass. The ponds are surrounded by dry hills overgrown with acacia and steppe-like vegetation. Human activities includes fishery, reed-cutting ('Other' land-use, below), and hunting.

Habitats Forest and woodland (14%; broadleaved deciduous forest; alluvial/very wet forest), Grassland (6%; humid grassland), Wetland (70%; standing fresh water; river/ stream; water-fringe vegetation), Artificial landscape (10%; arable land)
Land-use Agriculture (16%), Fisheries/aquaculture (60%), Hunting (70%), Other (30%)

Birds

Species	Season	Year	Pop min	Pop max	Acc	Criteria
Phalacrocorax carbo Cormorant	P	1996	40	1,400	A	B1i
Phalacrocorax pygmeus Pygmy Cormorant	W	1996	10	43	A	A1
Phalacrocorax pygmeus Pygmy Cormorant	P	1996	—	1,400	B	A1, A4i, B1i
Pelecanus crispus Dalmatian Pelican	W	1997	12	43	A	A1, A4i, B1i
Ixobrychus minutus Little Bittern	B	1996	15	50	A	B2
Aythya nyroca Ferruginous Duck	B	1996	10	30	A	A1
Aythya nyroca Ferruginous Duck	P	1996	100	3,000	A	A1, A4i, B1i

The ponds are of global importance for breeding and migrating *Aythya nyroca*, migrating *Phalacrocorax pygmeus* and wintering *Pelecanus crispus*, and are one of the most important sites in Bulgaria for breeding *Ixobrychus minutus*. Species of global conservation concern that do not meet IBA criteria: *Anser erythropus* (max. 1 in winter), *Branta ruficollis* (max. 20 in winter), *Haliaeetus albicilla* (1 breeding pair; 2–7 birds along river in winter), *Aquila heliaca* (max. 2 on passage), *Acrocephalus paludicola* (max. 1 on passage). Significant proportion (≥1%) of national population breeding at site: *Tadorna ferruginea* (1 pair), *Porzana pusilla* (1 pair).

Protection status
National None **International** None

Conservation issues

Threats Drainage (A), Natural events (C), Unsustainable exploitation (A), Other (A)

The main problems are reed-cutting and the drying-out of ponds during the breeding season.

Lomovete B2 025

Admin region Ruse
Coordinates 43°50′N 26°02′E
Altitude 50–150 m **Area** 4,320 ha

Site description
A complex of river valleys, of the Beli Lom, Cherni Lom, Malki Lom and Rusenski Lom rivers, south-east from the town of Ruse. Limestone cliffs form vertical canyons up to 100 m deep. Forests of *Quercus*, *Carpinus*, *Fraxinus* and *Tilia* predominate, and there are also plantations of *Robinia* and *Pinus*, meadows, and arable land. The riverbanks are overgrown with *Salix* and *Populus*. The main human activities are forestry, cultivation, livestock-grazing and recreation.

Habitats Forest and woodland (70%; broadleaved deciduous forest), Scrub (2%; scrub), Grassland (8%; steppe/dry calcareous grassland), Wetland (3%; river/stream), Rocky areas (7%; inland cliff), Artificial landscape (10%; perennial crops/orchards/groves)
Land-use Agriculture (8%), Fisheries/aquaculture (1%), Forestry (70%), Hunting (50%), Tourism/recreation (75%)

Birds

Species	Season	Year	Pop min	Pop max	Acc	Criteria
Tadorna ferruginea Ruddy Shelduck	B	1996	10	12	A	B2
Neophron percnopterus Egyptian Vulture	B	1996	5	6	A	B2
Circaetus gallicus Short-toed Eagle	B	1996	3	4	A	B2
Buteo rufinus Long-legged Buzzard	B	1996	15	15	A	B2
Aquila pomarina Lesser Spotted Eagle	B	1996	3	6	A	B2
Hieraaetus pennatus Booted Eagle	B	1996	2	2	B	B2
Bubo bubo Eagle Owl	B	1996	5	5	A	B2

Important numbers of a wide variety of large raptors breed here, as well as *Tadorna ferruginea*. Breeding species of global conservation concern that do not meet IBA criteria: *Aquila heliaca* (1 pair), *Falco naumanni* (2–3 pairs). Significant proportion (≥1%) of national population breeding at site: *Ciconia nigra* (2 pairs).

Protection status
National High **International** None
3,259 ha of IBA covered by National Park (Rusenski Lom, 3,259 ha). 773 ha of IBA covered by Nature Reserve (Beli Lom, 773 ha). 278 ha of IBA covered by Buffer Zone (Beli Lom, 278 ha).

Conservation issues

Threats Deforestation (commercial) (A), Recreation/tourism (B), Unsustainable exploitation (A)

Illegal human activities in the protected area, such as building, driving of vehicles, use of pesticides and wood-cutting, are the main problem, and the regulations protecting the area are often violated.

Derventski Vuzvishenia B2 026

Admin region Burgas
Coordinates 42°00′N 26°45′E
Altitude 100–556 m **Area** 20,000 ha

Site description
The Derventski hills are situated between the Tundzha river and the border with Turkey. They are partly covered by broadleaved deciduous forests of *Quercus* and *Carpinus*, with Mediterranean trees and shrubs, and there are also open grassy and bushy areas with scattered trees, and arable land. Shrubland is dominated by *Paliurus*, *Jasminum* and *Nigella*. Land-uses are extensive stock-rearing and cultivation, hunting and forestry.

Habitats Forest and woodland (58%; broadleaved deciduous forest), Scrub (23%; scrub; sclerophyllous scrub/garrigue/maquis), Grassland (7%; steppe/dry calcareous grassland; mesophile grassland), Artificial landscape (12%; arable land)
Land-use Agriculture (42%), Forestry (58%), Hunting

Birds

Species	Season	Year	Pop min	Pop max	Acc	Criteria
Aquila pomarina Lesser Spotted Eagle	B	1996	5	5	A	B2

One of the most important breeding sites for *Aquila pomarina* in Bulgaria. Significant proportion (≥1%) of national population breeding at site: *Circaetus gallicus* (1 pair), *Bubo bubo* (1 pair), *Dendrocopos medius* (min. 3 pairs), *Hippolais olivetorum* (min. 2 pairs).

Protection status
National None **International** None

Conservation issues

Threats Unknown

Malko Sharkovo reservoir A4i, B1i 027

Admin region Burgas
Coordinates 42°05′N 26°50′E
Altitude 100–100 m **Area** 380 ha

Site description
A water-storage reservoir in south-eastern Bulgaria, which does not freeze in winter. It is surrounded by low hills covered by arable land (mainly winter wheat). Human activities include hunting, fishing and irrigation.

Habitats Forest and woodland (broadleaved deciduous forest), Scrub (scrub), Wetland (standing fresh water), Artificial landscape (arable land)
Land-use Agriculture (10%), Hunting (100%), Water management (90%)

Birds

Species	Season	Year	Pop min	Pop max	Acc	Criteria
Anser albifrons White-fronted Goose	W	1997	16,000	16,810	A	A4i, B1i

The site is important for wintering *Anser albifrons*. Species of global conservation concern that do not meet IBA criteria: *Phalacrocorax pygmeus* (wintering).

Protection status
National None **International** None

Conservation issues

Threats Unsustainable exploitation (A)

Uncontrolled hunting of waterbirds is a serious threat.

Tzerkovski reservoir A1, B1i 028

Admin region Burgas
Coordinates 42°35′N 26°51′E
Altitude 120–120 m **Area** 500 ha

Site description
A reservoir which replaced the former Straldza marsh, surrounded by meadows and arable land. It does not freeze in winter. *Typha* is predominant along the banks. Wet meadows are dominated by *Carex*, *Poa* and *Lolium*. There are lines of *Populus* and *Robinia* trees. Land-uses include hunting, fish-farming and haymaking.

Habitats Grassland (10%; humid grassland), Wetland (90%; standing fresh water; water-fringe vegetation), Artificial landscape (perennial crops/orchards/groves)
Land-use Agriculture (8%), Fisheries/aquaculture (90%), Hunting

Birds

Species	Season	Year	Pop min	Pop max	Acc	Criteria
Anser albifrons White-fronted Goose	W	1997	6,178	8,500	A	B1i
Branta ruficollis Red-breasted Goose	W	1997	46	210	A	A1
Anas platyrhynchos Mallard	W	1997	—	22,755	B	B1i

Important for waterbirds in winter, especially *Anser albifrons*, the globally threatened *Branta ruficollis*, and *Phalacrocorax carbo*.

Protection status

National None **International** None

Conservation issues

Threats Unsustainable exploitation (A)

Illegal and uncontrolled hunting of waterbirds poses a serious threat.

Kotlenska mountain A1, B2, B3 029

Admin region Burgas
Coordinates 42°53′N 26°21′E
Altitude 315–1,128 m **Area** 156,000 ha

Site description

The central ridge of the eastern Stara Planina mountains, covered by broadleaved deciduous forests, pasture and arable land. A great number of minor but distinct ridges slope down north and south to the valleys of the Rivers Titcha and Luda Kamtchia. These side-valleys are narrow and deeply incised, with steep slopes. The underlying rock is limestone and sandstone, and there are some sand quarries in the region. The forests are dominated by *Fagus* or *Quercus*, with *Carpinus* in some places, and there are natural mixed forests of *Abies* and *Fagus* in the western part of the area.

Habitats Forest and woodland (67%; broadleaved deciduous forest), Scrub (1%; scrub), Grassland (15%; steppe/dry calcareous grassland; mesophile grassland), Wetland (1%; standing fresh water; river/stream), Rocky areas (1%; inland cliff), Artificial landscape (15%; arable land; forestry plantation; other urban/industrial areas)
Land-use Agriculture (27%), Forestry (66%), Tourism/recreation (5%), Urban/industrial/transport (1%)

Birds

Species	Season	Year	Pop min	Pop max	Acc	Criteria
Ciconia nigra Black Stork	B	1996	11	12	A	B2
Circaetus gallicus Short-toed Eagle	B	1996	2	4	A	B2
Buteo rufinus Long-legged Buzzard	R	1996	3	4	A	B2
Aquila pomarina Lesser Spotted Eagle	B	1996	3	4	A	B2
Aquila heliaca Imperial Eagle	B	1996	1	2	A	A1
Aquila chrysaetos Golden Eagle	R	1996	4	6	A	B2
Crex crex Corncrake	B	1996	28	34	A	A1
Picus canus Grey-headed Woodpecker	R	1996	6	8	B	B2
Dendrocopos medius Middle Spotted Woodpecker	R	1996	20	22	B	B3
Ficedula semitorquata Semi-collared Flycatcher	B	1996	15	—	B	B2

An important breeding area for globally threatened species (*Crex crex*, *Aquila heliaca*) as well as for *Ciconia nigra*, a wide variety of large raptors, and two woodpeckers (*Picus canus* and *Dendrocopos medius*). Species of global conservation concern that do not meet IBA criteria: *Phalacrocorax pygmeus* (3–12 on passage), *Pelecanus crispus* (max. 1 on passage), *Aythya nyroca* (1 breeding pair; 4–16 on passage), *Aquila clanga* (max. 1 on passage), *Falco naumanni* (max. 3 pairs). Significant proportion (≥1%) of national population breeding at site: *Bubo bubo* (2–4 pairs).

Protection status

National Low **International** None
100 ha of IBA covered by Nature Reserve (Kersenlika-Ardachlaka, 100 ha). 567 ha of IBA covered by Nature Reserve (Orlitza, 567 ha). 5 ha of IBA covered by National Monument (Orlovi Pesteri, 5 ha). 10 ha of IBA covered by National Monument (Urushki Skali, 10 ha). 358 ha of IBA covered by National Monument (Zlosten, 358 ha). 90 ha of IBA covered by Buffer Zone (Kersenlika-Ardachlaka, 90 ha).

Conservation issues

Threats Consequences of animal/plant introductions (B), Deforestation (commercial) (A), Disturbance to birds (A), Intensified forest management (A), Recreation/tourism (B)

Problems include the unsustainable exploitation of forests, and disturbance to breeding raptors and *Ciconia nigra* caused by uncontrolled recreation, tourist activity and quarrying near nests.

Kalimok complex A1, A4i, B1i, B2, B3 030

Admin region Ruse
Coordinates 44°00′N 26°28′E
Altitude 45–45 m **Area** 1,000 ha

Site description

A former fish-farm, situated on the bank of the River Danube, north of the village of Nova Cherna. The open water of the ponds is partly covered by rich marsh vegetation of *Hydrocharis*, *Nymphaea*, *Nymphoides* and *Trapa*. The ponds are fringed by dense reedbeds of *Phragmites*, *Typha* and *Schoenoplectus*, with some groups of *Salix* trees, and are separated from the river by a seasonally flooded forest of *Salix* and *Populus*, which also covers a nearby island in the Danube (part of this complex). Plantations of *Robinia* and *Elaeagnus* cover the surrounding hills. Fish-farming ceased in 1993 because of financial problems. 'Other' land-use (below) is reed-cutting.

Habitats Forest and woodland (7%; alluvial/very wet forest), Grassland (29%; steppe/dry calcareous grassland; humid grassland), Wetland (64%; standing fresh water; river/stream; water-fringe vegetation)
Land-use Fisheries/aquaculture, Tourism/recreation, Other

Birds

Species	Season	Year	Pop min	Pop max	Acc	Criteria
Phalacrocorax pygmeus Pygmy Cormorant	P	1996	3	220	B	A1, A4i, B1i
Pelecanus crispus Dalmatian Pelican	W	1997	9	40	A	A1
Pelecanus crispus Dalmatian Pelican	P	1996	20	40	A	A1, A4i, B1i
[1] ***Pelecanus crispus*** Dalmatian Pelican	N	1996	20	40	A	A1, A4i, B1i
Ixobrychus minutus Little Bittern	B	1996	15	30	A	B2
Ardeola ralloides Squacco Heron	B	1996	2	6	A	B2
[2] ***Ciconia ciconia*** White Stork	P	1996	10,000	15,000	A	A4i, B1i
[2] ***Plegadis falcinellus*** Glossy Ibis	P	1996	100	500	B	A4i, B1i
Platalea leucorodia Spoonbill	B	1996	1	6	A	B2
Anser anser Greylag Goose	W	1997	300	500	A	B1i
Branta ruficollis Red-breasted Goose	W	1997	120	200	A	A1
Aythya nyroca Ferruginous Duck	B	1996	6	20	A	A1
Coturnix coturnix Quail	B	1996	40	100	B	B2
Coracias garrulus Roller	B	1996	5	10	B	B2
Turdus pilaris Fieldfare	W	1996	500	2,000	A	B3

1. Birds roost and feed here during the breeding season.
2. Birds roost here during autumn migration.

The site is important during the breeding season for nesting, feeding and resting waterbirds, e.g. the globally threatened *Pelecanus crispus*, *Branta ruficollis* and *Aythya nyroca*, and is also an important staging and roosting area for migrating waterbirds, e.g. *Phalacrocorax pygmeus*, *Ciconia ciconia* and *Plegadis falcinellus*. Breeding species of global conservation concern that do not meet IBA criteria: *Phalacrocorax pygmeus* (5 pairs before 1989, now only 1 pair left). Significant proportion (≥1%) of national population breeding at site: *Nycticorax nycticorax* (30–50 pairs).

Protection status

National Low **International** None
74 ha of IBA covered by Protected Landscape (Bezimenen Island, 74 ha).

Conservation issues

Threats Burning of vegetation (A), Disturbance to birds (B), Drainage (B)

Problems include reductions in water-levels, and disturbance of birds by reed-cutters and collectors of herbs and snails. The procedure for designation of 663 ha of the site as a protected area is not finished.

Stenata — B2 — 031

Admin region Ruse
Coordinates 44°03′N 26°36′E
Altitude 50–62 m **Area** 15 ha

Site description

A 12-m-high, vertical loess cliff, located on a major regional road south-west of the town of Tutrakan. The surrounding open lands are covered by steppe-like vegetation of grasslands with some scrub (*Crataegus*, *Rosa*, *Rubus*) and forests of *Robinia*. Human activities include stock-grazing and loess excavation.

Habitats Rocky areas (100%; inland cliff)
Land-use Agriculture, Urban/Industrial/Transport

Birds

Species	Season	Year	Pop min	Pop max	Acc	Criteria
Merops apiaster Bee-eater	B	1996	150	—	A	B2

The site holds the largest nesting colony of *Merops apiaster* in Bulgaria.

Protection status

National None **International** None

Conservation issues

There are no serious threats to this site. Children sometimes disturb the birds and destroy nests.

Pozharevo island — A1, A4i, B1i, B2 — 032

Admin region Ruse
Coordinates 44°04′N 26°40′E
Altitude 45–45 m **Area** 170 ha

Site description

An island located between kilometres 424–423 in the River Danube, covered by seasonally flooded forest of *Salix* and *Populus*. There are sandbars near the banks. Human activities include hunting during the winter.

Habitats Forest and woodland (98%; alluvial/very wet forest), Wetland (2%; river/stream)
Land-use Fisheries/aquaculture (2%), Hunting (98%)

Birds

Species	Season	Year	Pop min	Pop max	Acc	Criteria
Phalacrocorax carbo Cormorant	W	1997	500	1,200	A	B1i
Phalacrocorax pygmeus Pygmy Cormorant	B	1996	10	30	A	A1, B2
Pelecanus crispus Dalmatian Pelican	W	1997	20	40	A	A1, A4i, B1i
Nycticorax nycticorax Night Heron	B	1997	50	150	A	B2
Ardeola ralloides Squacco Heron	B	1996	5	50	A	B2
Anser anser Greylag Goose	W	1997	100	500	A	B1i
Branta ruficollis Red-breasted Goose	W	1997	100	200	A	A1

The island is important for its breeding colonies of *Phalacrocorax pygmeus*, *Nycticorax nycticorax* and *Ardeola ralloides*, and for the large numbers of some wintering waterbirds, notably of the globally threatened *Pelecanus crispus* and *Branta ruficollis*. Significant proportion (≥1%) of national population breeding at site: *Platalea leucorodia* (1–2 pairs).

Protection status

National Partial **International** None
71 ha of IBA covered by Protected Landscape (Maluk Pozarevo, 71 ha).

Conservation issues

Threats Disturbance to birds (A)

The main problem is disturbance of nesting birds by bird-ringers.

Srebarna lake — A1, A4i, B1i, B2, B3 — 033

Admin region Ruse
Coordinates 44°07′N 27°04′E
Altitude 11–13 m **Area** 1,445 ha

Site description

A hyper-eutrophic freshwater lake directly east of the village of Srebarna, on the bank of the River Danube between kilometres 393 and 391. The open-water surface is partly covered by *Hydrocharis*, *Nymphaea* and *Potamogeton*, and is fringed by reedbeds of *Phragmites* and *Typha*. There are some groups of *Salix* in the reedbeds. Komluka island (in the Danube), which is part of the reserve, is covered by flooded forest of *Salix* and *Populus*. Plantations of *Robinia* and *Eleagnus* cover the surrounding hills.

Habitats Wetland (100%; standing fresh water; water-fringe vegetation)
Land-use Nature conservation/research (100%), Not utilized (63%), Tourism/recreation (37%)

Birds

Species	Season	Year	Pop min	Pop max	Acc	Criteria
[1] ***Phalacrocorax carbo*** Cormorant	R	1996	90	540	A	B1i
[2] ***Phalacrocorax pygmeus*** Pygmy Cormorant	B	1996	15	15	A	A1, B2
Phalacrocorax pygmeus Pygmy Cormorant	W	1997	3	58	A	A1
[3] ***Pelecanus crispus*** Dalmatian Pelican	B	1996	20	74	A	A1, A4i, B1i
Ixobrychus minutus Little Bittern	B	1996	20	20	A	B2
Nycticorax nycticorax Night Heron	B	1996	50	150	A	B2
Ardeola ralloides Squacco Heron	B	1996	60	60	A	B2
Plegadis falcinellus Glossy Ibis	B	1996	5	5	A	B2
Platalea leucorodia Spoonbill	B	1996	6	20	A	B2
Anser albifrons White-fronted Goose	W	1996	1,000	50,550	A	A4i, B1i
Anser anser Greylag Goose	W	1996	17	505	B	B1i
Turdus pilaris Fieldfare	W	1997	5,000	10,000	B	B3

1. Breeding resumed after canal built.
2. 20 pairs before 1989.
3. The only breeding site in Bulgaria. 29–127 pairs before 1989.

The site is important for breeding waterbirds, especially the globally threatened *Pelecanus crispus*, and for wintering geese. Species of global conservation concern that do not meet IBA criteria: *Anser erythropus* (has occurred in the past), *Branta ruficollis* (12–50 in winter), *Haliaeetus albicilla* (1–5 in winter; extinct as a breeding species).

Protection status

National High **International** Partial
902 ha of IBA covered by Nature Reserve (Srebarna, 902 ha). 543 ha of IBA covered by Buffer Zone (Srebarna, 543 ha). 600 ha of IBA covered by Ramsar Site (Srebarna, 600 ha). 902 ha of IBA covered by Biosphere Reserve (Srebarna, 902 ha). 902 ha of IBA covered by World Heritage Site (Srebarna, 902 ha).

Conservation issues

Threats Construction/impact of dyke/dam/barrage (A), Filling-in of wetlands (A), Groundwater abstraction (A), Natural events (A)

Threats include nest-predation by foxes *Vulpes vulpes*, jackals *Canis aureus* and boars *Sus scrofa*, which affect the pelican colony ('Natural events', above). Some restoration measures were taken by the Ministry of Environment in response to the disturbed water-levels in the lake.

Mandra–Poda complex — A1, A4i, A4iii, A4iv, B1i, B1iv, B2, B3 — 034

Admin region Burgas
Coordinates 42°25′N 27°23′E
Altitude 0–3 m **Area** 2,270 ha

Site description

A dammed reservoir on the Black Sea coast, below which lies a brackish lagoon linked by an outlet to the sea, next to the town of Burgas. The open-water surface covers 1,300 ha, and is fringed by *Phragmites*, *Typha*, *Juncus* and *Bolboschoenus*. Along the shore, salt-tolerant vegetation is dominated by *Puccinellia* and *Salicornia*, and sand-dunes are overgrown with *Leymus* grass and *Gypsophila*. The lake is surrounded by wet meadows, marshes, arable land and areas of *Paliurus* scrub, with some forest of *Quercus* and *Acer* on the south bank.

Habitats Grassland (1%; mesophile grassland), Wetland (98%; standing fresh water; standing brackish and salt water; water-fringe vegetation), Artificial landscape (1%; arable land)
Land-use Agriculture (2%), Fisheries/aquaculture (90%), Forestry (1%), Hunting (85%), Nature conservation/research (4%), Urban/industrial/transport (10%)

Birds

Species	Season	Year	Pop min	Pop max	Acc	Criteria
Phalacrocorax carbo Cormorant	W	1997	15	1,873	A	B1i
[1] ***Phalacrocorax pygmeus*** Pygmy Cormorant	W	1997	19	320	A	A1, A4i, B1i
[1,2] ***Pelecanus onocrotalus*** White Pelican	P	1996	203	2,000	B	A4i, B1i
[1] ***Pelecanus crispus*** Dalmatian Pelican	W	1997	11	203	A	A1, A4i, B1i
Ardeola ralloides Squacco Heron	B	1996	30	30	A	B2
[1] ***Egretta alba*** Great White Egret	W	1997	4	244	A	A4i, B1i
Plegadis falcinellus Glossy Ibis	B	1996	—	30	A	B2
Platalea leucorodia Spoonbill	B	1996	30	37	A	B2
Cygnus cygnus Whooper Swan	W	1997	9	599	A	B1i
Anser albifrons White-fronted Goose	W	1997	10,000	30,370	A	A4i, B1i
Branta ruficollis Red-breasted Goose	W	1997	2	1,850	A	A1, A4i, B1i
Oxyura leucocephala White-headed Duck	W	1997	24	202	A	A1, A4i, B1i
Recurvirostra avosetta Avocet	B	1996	5	15	A	B3
Numenius tenuirostris Slender-billed Curlew	P	1996	0	1	A	A1

1. Formerly bred.
2. Migrants rest on lake.

The complex is important for many bird species, notably for globally threatened species (e.g. *Numenius tenuirostris*, *Oxyura leucocephala*), breeding waterbirds (e.g. *Platalea leucorodia*, *Recurvirostra avosetta*), and wintering waterbirds. The site holds 20,000 or more waterbirds on a regular basis, and is also a notable migratory bottleneck site in autumn, where more than 3,000 raptors (notably *Aquila pomarina*) and up to 20,000 *Ciconia ciconia* pass overhead. Species of global conservation concern that do not meet IBA criteria while staging at the site: *Pelecanus crispus*, *Aythya nyroca*. Significant proportion (≥1%) of national population breeding at site: *Nycticorax nycticorax* (28–30 pairs).

Protection status

National Partial **International** None
151 ha of IBA covered by Protected Landscape (Izvorska Mouth, 151 ha). 101 ha of IBA covered by Protected Landscape (Poda Lagoon, 101 ha).

Conservation issues

Threats Burning of vegetation (C), Industrialization/urbanization (B), Infrastructure (B), Unsustainable exploitation (A)

Problems include illegal hunting in the protected area, mining of sand from sand-dunes, and chemical pollution of water. The management plan for Poda Lagoon Protected Landscape was prepared by BSPB under the Bulgarian–Swiss Biodiversity Conservation Programme, and BSPB manages this area.

Burgasko lake — A1, A4i, A4iii, B1i — 035

Admin region Burgas
Coordinates 42°30′N 27°25′E
Altitude 0–10 m **Area** 2,800 ha

Site description

A shallow, brackish, coastal lake just to the west of, and bordering, the town of Burgas, and connected to the Black Sea by a canal with a sluice. Reedbeds of *Phragmites* and *Typha* fringe most of the shoreline, where salt-tolerant vegetation is dominated by *Puccinellia*.

Habitats Wetland (100%; standing fresh water; standing brackish and salt water; water-fringe vegetation)
Land-use Agriculture (5%), Fisheries/aquaculture (80%), Hunting (50%), Urban/industrial/transport (50%)

Birds

Species	Season	Year	Pop min	Pop max	Acc	Criteria
Phalacrocorax carbo Cormorant	W	1997	70	1,046	A	B1i
[1] ***Phalacrocorax pygmeus*** Pygmy Cormorant	W	1997	22	257	A	A1, A4i, B1i
Phalacrocorax pygmeus Pygmy Cormorant	P	1996	9	92	B	A1
Cygnus cygnus Whooper Swan	W	1997	20	174	A	B1i
Anser albifrons White-fronted Goose	W	1997	6,593	23,931	A	A4i, B1i
Branta ruficollis Red-breasted Goose	W	1997	7	1,975	A	A1, A4i, B1i
Aythya ferina Pochard	W	1997	150	10,905	A	B1i
Aythya fuligula Tufted Duck	W	1997	250	15,807	A	B1i
Oxyura leucocephala White-headed Duck	W	1997	5	69	A	A1
Oxyura leucocephala White-headed Duck	P	1997	19	43	A	A1

1. Breeding not confirmed over last 7 years.

The complex is important for wintering and migrating waterbirds, especially for wildfowl such as geese (*Anser albifrons* and the globally threatened *Branta ruficollis*) and ducks (e.g. *Aythya* spp. and the globally threatened *Oxyura leucocephala*). Pelicans *Pelecanus* also rest here in notable numbers. Species of global conservation concern that do not meet IBA criteria: *Pelecanus crispus* (max. 10 in winter and on passage), *Aythya nyroca* (breeding).

Protection status

National Partial **International** None
397 ha of IBA covered by Protected Landscape (Vaya, 397 ha).

Conservation issues

Threats Industrialization/urbanization (B), Unsustainable exploitation (B)

The IBA and surroundings suffer from high levels of industrial and chemical pollution, and the continued construction of industrial and residential districts also poses a threat. The lake is monitored by BSPB under the Bulgarian–Swiss Biodiversity Conservation Programme.

Atanasovo lake — A1, A4i, A4iv, B1i, B1iv, B2, B3 — 036

Admin region Burgas
Coordinates 42°34′N 27°29′E
Altitude 0–2 m **Area** 1,950 ha

Site description

A hyper-saline lake on the Black Sea coast, just north of the suburbs of the town of Burgas. Most of the lake is now a complex of evaporation saltpans, with smaller settling pools for salt-extraction. Smaller freshwater marshes, including a system of canals overgrown with marsh vegetation, surround the lake. The banks of the saltpans are covered by *Salicornia*, while the freshwater areas are fringed by beds of *Typha* and *Phragmites*. Most of the area is used for salt production, but traditional and non-intensive production methods continue to be used ('Other' land-use, below).

Habitats Grassland (mesophile grassland), Wetland (100%; saltmarsh; standing brackish and salt water; water-fringe vegetation)
Land-use Agriculture (1%), Nature conservation/research (100%), Urban/industrial/transport (1%), Other (99%)

Birds

The lake is a very important staging, wintering and breeding area for migratory waterbirds, including globally threatened species such as *Numenius tenuirostris*, *Pelecanus crispus*, *Branta ruficollis*, *Aythya nyroca* and *Aquila clanga*. It is also a major migratory bottleneck site, where huge numbers of soaring raptors and waterbirds pass overhead in autumn. The commonest such species are as follows (minimum and maximum seasonal totals over the 1990–1996 period are given): *Pelecanus onocrotalus* (17,645–37,228), *Ciconia nigra* (2,781–4,574), *Ciconia ciconia* (80,592–204,423), *Pernis apivorus* (13,406–23,052), *Buteo buteo* (14,441–30,662), *Aquila pomarina* (9,009–25,796) and *Falco vespertinus* (438–3,110). Species of global conservation concern that do not meet IBA criteria: *Anser erythropus* (max. 1 in winter), *Marmaronetta angustirostris* (one winter record), *Oxyura leucocephala* (max. 10 in winter), *Haliaeetus albicilla* (max. 2 on passage), *Circus macrourus* (7–20 per season on

Species	Season	Year	Pop min	Pop max	Acc	Criteria
Phalacrocorax pygmeus Pygmy Cormorant	W	1997	1	32	A	A1
Phalacrocorax pygmeus Pygmy Cormorant	P	1996	14	565	B	A1, A4i, B1i
Pelecanus crispus Dalmatian Pelican	W	1997	18	100	A	A1, A4i, B1i
Pelecanus crispus Dalmatian Pelican	P	1996	158	1,695	A	A1
Platalea leucorodia Spoonbill	P	1996	15	130	B	B1i
Cygnus olor Mute Swan	W	1997	34	875	A	B1i
Anser albifrons White-fronted Goose	W	1997	2,250	10,000	A	B1i
Branta ruficollis Red-breasted Goose	W	1997	60	320	A	A1
Tadorna tadorna Shelduck	W	1997	1,102	4,036	A	A4i, B1i
Aythya nyroca Ferruginous Duck	P	1996	2	88	B	A1
Aquila clanga Greater Spotted Eagle	P	1996	6	16	A	A1
Recurvirostra avosetta Avocet	B	1996	557	725	A	A4i, B1i, B3
Recurvirostra avosetta Avocet	P	1997	1,134	7,570	—	A4i, B1i
[1] ***Charadrius alexandrinus*** Kentish Plover	B	1996	70	80	A	B2
Numenius tenuirostris Slender-billed Curlew	P	1996	1	3	A	A1
Numenius tenuirostris Slender-billed Curlew	N	1996	0	1	A	A1
[2] ***Sterna sandvicensis*** Sandwich Tern	B	1996	10	772	A	B1i

1. Site qualifies according to 1997 national population estimate.
2. The only breeding site in Bulgaria.

passage), *Aquila heliaca* (2–4 on passage), *Falco naumanni* (max. 3 on passage), *Crex crex* (max. 3 pairs breeding). Significant proportion (≥1%) of national population breeding at site: *Tadorna ferruginea*.

■ Protection status

National High **International** Partial

1,050 ha of IBA covered by Nature Reserve (Atanasovo lake, 1,050 ha). 900 ha of IBA covered by Buffer Zone (Atanasovo lake, 900 ha). 1,050 ha of IBA covered by Ramsar Site (Atanasovo lake, 1,050 ha).

■ Conservation issues

Threats Infrastructure (A), Disturbance to birds (B), Other (B)

Some problems are bird mortality from collision with powerlines, as well as possible expansion of the airport (which might increase bird-scaring activities, to reduce the risk of collision with aircraft), and intensification of salt-production methods ('Other' threat, above). The management plan for Atanasovo lake was prepared by the Ministry of Environment under the Bulgarian–Swiss Biodiversity Conservation Programme.

Pomorie lake — A1, A4i, B1i, B3 — 037

Admin region Burgas
Coordinates 42°35′N 27°37′E
Altitude 0–5 m **Area** 850 ha

■ Site description

A hyper-saline lake on the Black Sea coast (a former coastal lagoon), most of which is utilized for salt production. There are shallow saltpans of varying salinity, separated by narrow dykes; a freshwater canal between the saltpans; freshwater marshes with rich vegetation; dry, saline areas covered by salt-tolerant vegetation; and a sand beach along the coast. The vegetation around the saltpans is dominated by *Salicornia* and *Puccinellia*, while *Typha*, *Phragmites* and *Schoenoplectus* dominate in the more freshwater areas. Most of the lake is used for artisanal (non-intensive) salt-production ('Other' land-use, below). There is a mud-cure establishment.

Habitats Wetland (95%; sand-dunes/sand beach; coastal lagoon; standing fresh water; standing brackish and salt water; water-fringe vegetation), Artificial landscape (5%; arable land; perennial crops/orchards/groves; other urban/industrial areas)
Land-use Hunting (100%), Urban/industrial/transport (5%), Other (80%)

■ Birds

Species	Season	Year	Pop min	Pop max	Acc	Criteria
Pelecanus crispus Dalmatian Pelican	P	1996	100	—	B	A1, A4i, B1i
Cygnus olor Mute Swan	W	1997	4	635	A	B1i
[1] ***Recurvirostra avosetta*** Avocet	B	1996	20	50	A	B3

1. Max. 30 birds in winter; max. 130 on passage.

The lake is important for breeding *Recurvirostra avosetta*, migrating *Pelecanus crispus* and wintering *Cygnus olor*. Species of global conservation concern that do not meet IBA criteria: *Phalacrocorax pygmeus* (min. 24 in winter; max. 29 on passage), *Aythya nyroca* (max. 15 on passage), *Oxyura leucocephala* (3–5 in winter), *Crex crex* (occurs on passage). Significant proportion (≥1%) of national population breeding at site: *Ixobrychus minutus* (5 pairs).

■ Protection status

National None **International** None

■ Conservation issues

Threats Abandonment/reduction of land management (C), Industrialization/urbanization (B), Unsustainable exploitation (B)

The main problems are caused by intensification of salt-production techniques (including abandonment of some salt-pans), restitution of pans to former owners, and water pollution from the surrounding area (all are classed under 'Industrialization/urbanization' threat, above). Hunting is also a problem. The IBA has been proposed as a Protected Landscape.

Provadijsko–Royaksko Plato — B2, B3 — 038

Admin region Varna
Coordinates 43°15′N 27°23′E
Altitude 100–389 m **Area** 100,000 ha

■ Site description

A low, hilly area between two river valleys, with some rocky massifs. Land cover is dominated by broadleaved forests of *Quercus*, *Carpinus* and *Fraxinus*, cultivated land, and pastures overgrown with bushes.

Habitats Forest and woodland (50%; broadleaved deciduous forest), Scrub (1%; scrub), Grassland (18%; steppe/dry calcareous grassland), Wetland (river/stream), Rocky areas (1%; inland cliff), Artificial landscape (30%; arable land; other urban/industrial areas)
Land-use Agriculture (47%), Forestry (50%), Hunting (100%)

■ Birds

Species	Season	Year	Pop min	Pop max	Acc	Criteria
Neophron percnopterus Egyptian Vulture	B	1996	4	5	A	B2
Circaetus gallicus Short-toed Eagle	B	1996	2	2	A	B2
Buteo rufinus Long-legged Buzzard	B	1996	4	4	A	B2
Aquila chrysaetos Golden Eagle	R	1996	2	3	A	B2
Bubo bubo Eagle Owl	R	1996	3	6	A	B2
Dendrocopos medius Middle Spotted Woodpecker	R	1996	30	30	B	B3
Lanius minor Lesser Grey Shrike	B	1996	25	45	A	B2

This is one of the most important breeding areas in the country for several raptors (*Neophron percnopterus*, *Circaetus gallicus*, *Buteo rufinus*, *Aquila chrysaetos*, *Bubo bubo*), as well as for *Dendrocopos medius* and *Lanius minor*. Significant proportion (≥1%) of national population breeding at site: *Aquila pomarina* (1–2 pairs).

■ Protection status

National Low **International** None

74 ha of IBA covered by National Monument (Slaveikova Gora, 74 ha).

■ Conservation issues

Threats Intensified forest management (A)

Some current problems are excessive exploitation of forests for timber, disturbance of nesting raptors by forestry activities, and illegal hunting.

Hursovska Reka — B2 — 039

Admin region Varna
Coordinates 43°54′N 27°23′E
Altitude 50–300 m **Area** 20,000 ha

■ Site description

An arid river valley with rocky massifs, covered by broadleaved forest and dry grassland on the ridges of the hills. Forests are dominated by *Quercus*, *Carpinus* and *Acer*.

Habitats Forest and woodland (70%; broadleaved deciduous forest), Grassland (15%; dry siliceous grassland), Rocky areas (5%; inland cliff), Artificial landscape (10%; arable land; perennial crops/orchards/groves)
Land-use Agriculture (25%), Forestry (70%), Hunting (60%)

Birds

Species	Season	Year	Pop min	Pop max	Acc	Criteria
Tadorna ferruginea Ruddy Shelduck	B	1994	3	3	A	B2
Buteo rufinus Long-legged Buzzard	B	1994	3	3	A	B2

The area is one of the most important breeding sites for *Tadorna ferruginea* and *Buteo rufinus* in the country. *Gyps fulvus* and *Aquila chrysaetos* used to breed in the area many years ago. Significant proportion (≥1%) of national population breeding at site: *Neophron percnopterus* (1 pair), *Circaetus gallicus* (1 pair), *Aquila pomarina* (1 pair), *Hieraaetus pennatus* (1 pair), *Bubo bubo* (3 pairs).

Protection status

National None **International** None

Conservation issues

Threats Disturbance to birds (A), Extraction industry (B)

The main problem is disturbance of nesting birds (by treasure-hunters).

Strandja

A4iv, B1iv, B2, B3 **040**

Admin region Burgas
Coordinates 42°04'N 27°38'E
Altitude 0–701 m **Area** 116,136 ha

Site description

A rather low mountain in the south-easternmost part of the country, through which the Veleka and Resovska rivers flow. The mountain is covered by mature broadleaved forests and pastures. Natural forest of *Fagus* predominates, although *Quercus* is common in higher areas. Mediterranean grassland and shrubland are widespread. Alluvial forests occur mainly at the river-mouths and are dominated by *Fraxinus*, with dense *Salix* and *Alnus*. Temporarily flooded areas along the rivers are overgrown with aquatic vegetation. Land-uses are forestry, hunting and extensive cultivation and stock-rearing.

Habitats Forest and woodland (59%; broadleaved deciduous forest; alluvial/very wet forest), Scrub (8%; scrub; sclerophyllous scrub/garrigue/maquis), Grassland (10%; steppe/dry calcareous grassland; humid grassland), Wetland (1%; sand-dunes/sand beach; standing fresh water; river/stream; water-fringe vegetation), Rocky areas (sea cliff/rocky shore; inland cliff; caves), Artificial landscape (22%; arable land; perennial crops/orchards/groves; other urban/industrial areas)
Land-use Agriculture (22%), Fisheries/aquaculture (1%), Forestry (57%), Hunting (60%), Tourism/recreation (1%), Urban/industrial/transport (3%)

Birds

Species	Season	Year	Pop min	Pop max	Acc	Criteria
Ciconia nigra Black Stork	B	1996	8	11	A	B2
Neophron percnopterus Egyptian Vulture	B	1996	4	4	A	B2
Circaetus gallicus Short-toed Eagle	B	1996	4	4	A	B2
Aquila pomarina Lesser Spotted Eagle	B	1996	5	9	A	B2
Aquila chrysaetos Golden Eagle	R	1996	5	7	A	B2
Hieraaetus pennatus Booted Eagle	B	1996	3	4	A	B2
Bubo bubo Eagle Owl	R	1996	4	7	A	B2
Alcedo atthis Kingfisher	B	1996	8	10	B	B2
Picus canus Grey-headed Woodpecker	R	1996	20	31	B	B2
Dendrocopos medius Middle Spotted Woodpecker	R	1996	45	46	B	B3
Ficedula semitorquata Semi-collared Flycatcher	B	1996	31	35	B	B2

This is one of the most important breeding areas in the country for *Ciconia nigra*, for raptors such as *Neophron percnopterus*, *Circaetus gallicus*, *Aquila pomarina*, *Aquila chrysaetos*, *Hieraaetus pennatus* and *Bubo bubo*, and for forest species such as *Picus canus*, *Dendrocopos medius* and *Ficedula semitorquata*. The mountain is also a major migratory bottleneck site, where more than 33,000 *Ciconia ciconia* pass overhead each autumn. Species of global conservation concern that do not meet IBA criteria: *Phalacrocorax pygmeus* (common on passage), *Pelecanus crispus* (common on passage), *Crex crex* (occurs on passage).

Protection status

National High **International** Low

116,136 ha of IBA covered by National Park (Strandja, 116,136 ha). 390 ha of IBA covered by Nature Reserve (Silkossia, 390 ha). 608 ha of IBA covered by Nature Reserve (Sredoka, 608 ha). 749 ha of IBA covered by Nature Reserve (Tissovitsa, 749 ha). 2,582 ha of IBA covered by Nature Reserve (Uzunbodzak, 2,582 ha). 1,112 ha of IBA covered by Nature Reserve (Vitanovo, 1,112 ha). 22 ha of IBA covered by National Monument (Bratanova pestera, 22 ha). 5 ha of IBA covered by National Monument (Elenina Dupka, 5 ha). 19 ha of IBA covered by National Monument (Konadzika, 19 ha). 8 ha of IBA covered by National Monument (Pesterata, 8 ha). 15 ha of IBA covered by National Monument (Piren, 15 ha). 52 ha of IBA covered by Protected Landscape (Bosna, 52 ha). 5 ha of IBA covered by Protected Landscape (Dokuzak, 5 ha). 7 ha of IBA covered by Protected Landscape (Dupkata, 7 ha). 19 ha of IBA covered by Protected Landscape (Kalkata, 19 ha). 109 ha of IBA covered by Protected Landscape (Krivinizovo, 109 ha). 47 ha of IBA covered by Protected Landscape (Marina Reka, 47 ha). 103 ha of IBA covered by Protected Landscape (Moryane, 103 ha). 989 ha of IBA covered by Protected Landscape (Paroria, 989 ha). 15 ha of IBA covered by Protected Landscape (Rudenovo, 15 ha). 773 ha of IBA covered by Protected Landscape (Silistar, 773 ha). 1,546 ha of IBA covered by Protected Landscape (Veleka, 1,546 ha). 1,511 ha of IBA covered by Protected Landscape (Veleka Mouth, 1,511 ha). 91 ha of IBA covered by Buffer Zone (Silkossia, 91 ha). 160 ha of IBA covered by Buffer Zone (Sredoka, 160 ha). 502 ha of IBA covered by Buffer Zone (Tissovitsa, 502 ha). 488 ha of IBA covered by Buffer Zone (Uzunbodzak, 488 ha). 185 ha of IBA covered by Buffer Zone (Vitanovo, 185 ha). 2,575 ha of IBA covered by Biosphere Reserve (Uzunbodzak, 2,575 ha).

Conservation issues

Threats Construction/impact of dyke/dam/barrage (C), Industrialization/urbanization (A), Intensified forest management (A), Recreation/tourism (A), Unsustainable exploitation (B)

Intensive tourism along the Black Sea coast during the past seven years has led to destruction of alluvial forests. There are several projects for the construction of reservoirs on the Veleka river, possibly disturbing the water regime. People often destroy nests of raptors, their eggs and chicks when cutting trees. Uncontrolled hunting and fisheries, and unsustainable forest management, are also problems. The IBA is designated as a National Park, comprising five Strict Reserves with Buffer Zones, 12 Protected Landscapes and five Nature Monuments. The management plan for the National Park was prepared by the Ministry of Environment under the Bulgarian–Swiss Biodiversity Conservation Programme.

Ropotamo complex

A1, A4i, B1i, B1iv, B2 **041**

Admin region Burgas
Coordinates 42°19'N 27°45'E
Altitude 0–250 m **Area** 4,800 ha

Site description

A river mouth on the Black Sea coast with diverse habitats, including riparian forest of *Fraxinus*, *Ulmus* and *Crataegus*, nearby mountain ridges with broadleaved deciduous forests of *Quercus*, coastal freshwater marshes, rocky seashore, reedbeds of *Phragmites*, *Typha* and *Schoenoplectus*, sand beaches with dunes and inlets, and a small offshore island. Open fresh water is dominated by *Hydrocharis*, *Nymphaea*, and *Lemna*. Sand-dunes have a cover of grasses and dwarf-shrubs. There is some scrub dominated by *Paliurus*. Land-uses are confined to the buffer zone and unprotected areas.

Habitats Forest and woodland (71%; broadleaved deciduous forest; alluvial/very wet forest), Scrub (4%; sclerophyllous scrub/garrigue/maquis), Grassland (2%; steppe/dry calcareous grassland), Wetland (21%; tidal river/enclosed tidal water; sand-dunes/sand beach; standing fresh water; river/stream; water-fringe vegetation), Rocky areas (2%; sea cliff/rocky shore)

Land-use Forestry (58%), Hunting (40%), Tourism/recreation (7%), Urban/industrial/transport (2%)

Birds

Species	Season	Year	Pop min	Pop max	Acc	Criteria
Pelecanus crispus Dalmatian Pelican	P	1996	100	—	C	A1, A4i, B1i
Ficedula semitorquata Semi-collared Flycatcher	B	1996	10	10	A	B2

The IBA is important as a breeding site for *Ficedula semitorquata*, and is also a notable migratory bottleneck site, where a minimum of 10,000 *Ciconia ciconia* pass overhead each autumn. Breeding species of global conservation concern that do not meet IBA criteria: *Phalacrocorax pygmeus* (4 pairs; max. 54 in winter), *Aythya nyroca* (3 pairs). Significant proportion (≥1%) of national population breeding at site: *Ixobrychus minutus* (10–15 pairs).

Protection status

National Partial **International** Low

1,001 ha of IBA covered by Nature Reserve (Ropotamo, 1,001 ha). 167 ha of IBA covered by National Monument (Alepu lake, 167 ha). 18 ha of IBA covered by National Monument (Maslen Nos, 18 ha). 12 ha of IBA covered by National Monument (Pyassachni Dyuni, 12 ha). 40 ha of IBA covered by Protected Landscape (Stamopolu, 40 ha). 771 ha of IBA covered by Buffer Zone (Ropotamo, 771 ha). 97 ha of IBA covered by Ramsar Site (Arkoutino, 97 ha).

Conservation issues

Threats Construction/impact of dyke/dam/barrage (B), Industrialization/urbanization (B), Recreation/tourism (A), Unsustainable exploitation (B)

The main problems are heavy tourist pressure, building encroachment, and changes to the water regime (caused by upstream water-abstraction, and eutrophication of marshes). The management plan for Ropotamo Complex was prepared by the Ministry of Environment under the Bulgarian–Swiss Biodiversity Conservation Programme.

Chengene Skele — A1, B2 — 042

Admin region Burgas
Coordinates 42°26′N 27°31′E
Altitude 0–30 m **Area** 160 ha

Site description

The mouth of a small river in a small bay on the Black Sea coast. There is a sand beach and a mudflat, a large area of reedbeds of *Phragmites*, *Typha* and *Schoenoplectus*, and secondary broadleaved woodland on the nearby slopes.

Habitats Forest and woodland (20%; broadleaved deciduous forest), Wetland (74%; mudflat/sandflat; sand-dunes/sand beach; river/stream; water-fringe vegetation), Marine areas (5%; sea inlet/coastal features), Rocky areas (1%; sea cliff/rocky shore)
Land-use Fisheries/aquaculture (70%), Urban/industrial/transport (2%)

Birds

Species	Season	Year	Pop min	Pop max	Acc	Criteria
[1] ***Phalacrocorax pygmeus*** Pygmy Cormorant	B	1996	5	10	A	A1, B2
Plegadis falcinellus Glossy Ibis	B	1996	5	10	A	B2
Numenius tenuirostris Slender-billed Curlew	W	1997	1	2	A	A1
Numenius tenuirostris Slender-billed Curlew	P	1996	1	2	A	A1

1. Up to 3 on passage and in winter.

The site is of global importance for *Numenius tenuirostris* on passage and in winter, and is one of the most important sites in the country for breeding *Phalacrocorax pygmeus* and *Plegadis falcinellus*. Species of global conservation concern that do not meet IBA criteria: *Pelecanus crispus* (common on passage). Significant proportion (≥1%) of national population breeding at site: *Ixobrychus minutus* (2–3 pairs).

Protection status

National High **International** None

160 ha of IBA covered by Protected Landscape (Chengene Skele Cove, 160 ha).

Conservation issues

Threats Disturbance to birds (B), Unsustainable exploitation (B)

Illegal hunting occurs in the protected area. People and nearby road traffic disturb birds during the breeding season.

Emine — A1, A4iv, B1iv, B2 — 043

Admin region Burgas
Coordinates 42°42′N 27°54′E
Altitude 0–300 m **Area** 100 ha

Site description

A coastal headland formed where the easternmost point of the Stara Planina mountains meets the Black Sea. The coast is rocky, with vertical cliffs, dry grassland, *Paliurus* scrub, and sparse stands of secondary *Quercus* forest.

Habitats Forest and woodland (60%; broadleaved deciduous forest), Scrub (6%; scrub), Grassland (27%; steppe/dry calcareous grassland), Marine areas (2%; open sea), Rocky areas (5%; sea cliff/rocky shore)
Land-use Agriculture (20%), Forestry (60%), Hunting (60%), Military, Tourism/recreation (50%)

Birds

Species	Season	Year	Pop min	Pop max	Acc	Criteria
Pelecanus crispus Dalmatian Pelican	P	1996	230	—	B	A1
Hippolais olivetorum Olive-tree Warbler	B	1996	8	20	A	B2

The headland is a major migratory bottleneck site, where huge numbers of soaring waterbirds and raptors pass overhead in autumn, especially *Ciconia ciconia* (92,272–172,928 per season during 1990–1996). It is also one of the most important breeding areas for *Hippolais olivetorum* in the country.

Protection status

National Low **International** None

1 ha of IBA covered by National Monument (Cape Emine, 1 ha).

Conservation issues

Threats Industrialization/urbanization (C)

There are no serious problems now, but urbanization of the area is possible in the future.

Kamchijska mountain — A4iv, B1iv, B2, B3 — 044

Admin region Varna
Coordinates 42°55′N 27°35′E
Altitude 0–627 m **Area** 55,000 ha

Site description

This mountainous area forms the north-easternmost part of the Stara Planina mountains. There are rocky massifs, two reservoirs—Eleshnitza (100 ha) and Tzonevo (300 ha)—and three rivers. Broadleaved forests of *Quercus*, *Fagus* and *Tilia* predominate, with cultivated land and pastures scattered between forests and villages. Human activity is only absent from the Strict Reserves.

Habitats Forest and woodland (85%; broadleaved deciduous forest), Scrub (1%; scrub), Grassland (7%; steppe/dry calcareous grassland), Wetland (1%; sand-dunes/sand beach; standing fresh water; river/stream; water-fringe vegetation), Rocky areas (1%; sea cliff/rocky shore; inland cliff), Artificial landscape (5%; arable land; perennial crops/orchards/groves; other urban/industrial areas)
Land-use Agriculture (14%), Fisheries/aquaculture (1%), Forestry (85%), Hunting (60%), Not utilized, Tourism/recreation (2%), Urban/industrial/transport (1%)

Birds

This is one of the most important breeding areas in the country for some raptors (*Neophron percnopterus*, *Aquila chrysaetos*) and forest

Species	Season	Year	Pop min	Pop max	Acc	Criteria
Neophron percnopterus Egyptian Vulture	B	1996	3	4	A	B2
Aquila chrysaetos Golden Eagle	R	1996	2	2	A	B2
Dendrocopos medius Middle Spotted Woodpecker	R	1996	20	50	B	B3
Ficedula semitorquata Semi-collared Flycatcher	B	1996	20	40	B	B2

species (*Dendrocopos medius*, *Ficedula semitorquata*). The eastern part of the mountain is also a major migratory bottleneck site, where huge numbers of large soaring migrants pass overhead in autumn, especially *Ciconia ciconia* (up to 80,000 per season) and *Aquila pomarina*. Significant proportion (≥1%) of national population breeding at site: *Ciconia nigra* (3–4 pairs), *Circaetus gallicus* (1 pair), *Buteo rufinus* (2 pairs), *Aquila pomarina* (1 pair), *Bubo bubo* (2–3 pairs).

Protection status

National Low **International** None

54 ha of IBA covered by Nature Reserve (Kirov Dol, 54 ha). 44 ha of IBA covered by Nature Reserve (Valtchi Prehod, 44 ha). 71 ha of IBA covered by Nature Reserve (Varbov Dol, 71 ha). 60 ha of IBA covered by National Monument (Orlov kamak, 60 ha). 4 ha of IBA covered by National Monument (Siniat Vir, 4 ha). 13 ha of IBA covered by National Monument (Tchudnite Skali, 13 ha). 27 ha of IBA covered by Buffer Zone (Kirov Dol, 27 ha). 87 ha of IBA covered by Buffer Zone (Valtchi Prehod, 87 ha). 61 ha of IBA covered by Buffer Zone (Varbov Dol, 61 ha).

Conservation issues

Threats Intensified forest management (A), Recreation/tourism (B)

Some current problems are caused by unsustainable forest exploitation for timber (forestry activity near raptors' nests also causes disturbance), uncontrolled recreation, and quarries. There are three Strict Reserves with Buffer Zones and three National Monuments, covering less than 1% of the area.

Kamchia complex — A4iv, B1i, B1iv, B2, B3 — 045

Admin region Varna
Coordinates 43°01′N 27°51′E
Altitude 0–10 m **Area** 3,000 ha

Site description

The lower reaches and estuary of the Kamchia river, with remnant riverine forests, small freshwater marshes of *Phragmites* and *Typha* along the riverbank, arable land (the former Staro–Oryachovo marshes), a beach with sand-dunes, and a sea bay. The forests are dominated by *Fraxinus*, *Ulmus*, *Quercus*, *Acer* and *Alnus*, with scrub of *Crataegus*, *Cornus*, *Paliurus* and *Ligustrum*, and lianas of *Clematis*, *Smilax* and *Periploca*.

Habitats Forest and woodland (40%; alluvial/very wet forest), Scrub (1%; scrub), Wetland (19%; tidal river/enclosed tidal water; sand-dunes/sand beach; standing fresh water; river/stream; water-fringe vegetation), Marine areas (4%; open sea), Artificial landscape (36%; arable land; forestry plantation)
Land-use Agriculture (34%), Forestry (40%), Nature conservation/research (50%), Tourism/recreation (5%), Urban/industrial/transport

Birds

Species	Season	Year	Pop min	Pop max	Acc	Criteria
Cygnus cygnus Whooper Swan	W	1997	15	170	A	B1i
Dendrocopos medius Middle Spotted Woodpecker	R	1996	59	59	A	B3
Ficedula semitorquata Semi-collared Flycatcher	B	1996	230	230	A	B2

The site is important for wintering *Cygnus cygnus* and is the most important breeding area in Bulgaria for *Dendrocopos medius* and *Ficedula semitorquata*. The locality is also a major migratory bottleneck site, where 60,000 or more *Ciconia ciconia* pass overhead each autumn.

Protection status

National Partial **International** Partial

842 ha of IBA covered by Nature Reserve (Kamtchia, 842 ha). 373 ha of IBA covered by Protected Landscape (Kamtchiyski Pyassatzi, 373 ha). 230 ha of IBA covered by Buffer Zone (Kamtchia, 230 ha). 842 ha of IBA covered by Biosphere Reserve (Kamtchia, 842 ha).

Conservation issues

Threats Consequences of animal/plant introductions (B), Drainage (A), Recreation/tourism (A), Unsustainable exploitation (B)

The water regime has been altered following the construction of embankments (along the riverside) and of reservoirs (six; throughout the catchment), and the periphery of the dense forest is gradually drying out. Part (30 ha) of the Protected Landscape was afforested with *Pinus*. A management plan has been prepared for Kamchia Biosphere Reserve by the Ministry of Environment under the Bulgarian–Swiss Biodiversity Conservation Programme.

Yatata — B1i, B2 — 046

Admin region Varna
Coordinates 43°11′N 27°42′E
Altitude 2–15 m **Area** 154 ha

Site description

A former sedimentation reservoir, close to the town of Beloslav, south of Varnensko lake. There are reedbeds of *Phragmites* and *Typha*, open-water areas, and low cliffs along the southern margin. The banks are covered by *Salicornia*, with *Poa*/*Lolium* grasslands and forests of *Quercus*, *Ulmus* and *Carpinus* on surrounding hills. The reservoir is not used for human activity, which is prohibited under the protected-area status.

Habitats Grassland (40%; dry siliceous grassland), Wetland (58%; standing fresh water; river/stream; water-fringe vegetation), Rocky areas (2%; inland cliff)
Land-use Agriculture (53%), Not utilized (47%)

Birds

Species	Season	Year	Pop min	Pop max	Acc	Criteria
Cygnus olor Mute Swan	W	1997	143	460	A	B1i
Tadorna ferruginea Ruddy Shelduck	B	1996	1	2	A	B2

The IBA is important for wintering *Cygnus olor* and is one of the most important breeding sites in the country for *Tadorna ferruginea*. Species of global conservation concern that do not meet IBA criteria: *Phalacrocorax pygmeus* (max. 20 on passage, max. 9 in winter), *Pelecanus crispus* (max. 5 on passage), *Aythya nyroca* (1 pair sometimes breeds; max. 18 on passage). Significant proportion (≥1%) of national population breeding at site: *Recurvirostra avosetta* (3 pairs).

Protection status

National High **International** None

154 ha of IBA covered by National Monument (Yatata, 154 ha).

Conservation issues

Threats Disturbance to birds (B), Drainage (B), Unsustainable exploitation (B)

Nesting birds are disturbed by passing people and cars. BSPB monitors the site.

Varna–Beloslav lakes complex — A1, A4i, A4iv, B1i, B1iv, B3 — 047

Admin region Varna
Coordinates 43°12′N 27°44′E
Altitude 0–5 m **Area** 2,300 ha

Site description

A natural coastal lake to the west of Varna, the largest Bulgarian sea-port. The northern shore is fringed by reedbeds of *Phragmites*, *Typha* and *Schoenoplectus*, while other parts are steeply sloping. There are sand-banks, as well as small saltpans in the westernmost part of the complex, whose banks are overgrown with *Salicornia*. The lake was converted in the 1970s into a ship-canal connecting the Black Sea and Lake Beloslav, and there are many industrial installations (factories, ports, etc.), roads and railways along the lake-shore.

Habitats Wetland (95%; standing fresh water; standing brackish and salt water; water-fringe vegetation), Artificial landscape (5%; other urban/industrial areas)
Land-use Fisheries/aquaculture (90%), Hunting (90%), Urban/industrial/transport (100%)

Birds

Species	Season	Year	Pop min	Pop max	Acc	Criteria
Phalacrocorax pygmeus Pygmy Cormorant	W	1996	14	145	A	A1
Phalacrocorax pygmeus Pygmy Cormorant	P	1996	200	—	A	A1
Pelecanus crispus Dalmatian Pelican	P	1996	30	—	C	A1, A4i, B1i
Aythya ferina Pochard	W	1997	1,125	10,240	A	B1i
[1] ***Recurvirostra avosetta*** Avocet	B	1996	13	13	A	B3

1. Max. 25 on passage.

The complex is important for breeding *Recurvirostra avosetta*, which nest at the saltpans. It is the only wetland between Shabla and Durankulak lakes in the north and the Burgas lakes to the south, and thus is an important staging and wintering area for waterbirds, especially *Pelecanus crispus* and *Phalacrocorax pygmeus*. It is also a major migratory bottleneck site, where up to 50,000 *Ciconia ciconia* pass overhead each autumn. Species of global conservation concern that do not meet IBA criteria: *Aythya nyroca* (min. 9 on passage), *Oxyura leucocephala* (max. 3 in winter).

Protection status

National Low **International** None
125 ha of IBA covered by Protected Landscape (Kazashko, 125 ha).

Conservation issues

Threats Industrialization/urbanization (A), Unsustainable exploitation (B)

Problems include water pollution by industry and the overexploitation of fish stocks.

Suha Reka — B2 — 048

Admin region Varna
Coordinates 43°44′N 27°36′E
Altitude 100–300 m **Area** 15,000 ha

Site description

A section of the arid valley between the village of Karapelit and the Romanian border. There are many high, rocky massifs among the large river-meanders between the village of Balik and the village of Brestnitza, but elsewhere massifs are not very high. Forests are dominated by *Quercus*, *Carpinus* and *Fraxinus*. There is a reservoir, nearly 7 km long, whose open water is partly covered by *Lemna* and *Ceratophyllum*.

Habitats Forest and woodland (50%; broadleaved deciduous forest), Grassland (20%; dry siliceous grassland), Wetland (1%; standing fresh water; water-fringe vegetation), Rocky areas (9%; inland cliff), Artificial landscape (20%; arable land; perennial crops/orchards/groves)
Land-use Agriculture (40%), Fisheries/aquaculture (1%), Forestry (50%), Hunting (50%)

Birds

Species	Season	Year	Pop min	Pop max	Acc	Criteria
Tadorna ferruginea Ruddy Shelduck	B	1994	2	2	A	B2
Buteo rufinus Long-legged Buzzard	B	1994	3	6	A	B2
Aquila pomarina Lesser Spotted Eagle	B	1994	3	3	A	B2

The area is one of the most important breeding sites in the country for *Tadorna ferruginea*, *Buteo rufinus* and *Aquila pomarina*. Significant proportion (≥1%) of national population breeding at site: *Neophron percnopterus* (1 pair), *Circaetus gallicus* (1 pair), *Hieraaetus pennatus* (1 pair), *Bubo bubo* (1–3 pairs).

Protection status

National None **International** None

Conservation issues

Threats Agricultural intensification/expansion (U), Disturbance to birds (A)

The main problems are the disturbance of nesting birds (by treasure-hunters), and water pollution by waste-water from farms. The exact sources of the pollution are unknown because of the well-drained karst terrain.

Shabla lake complex — A1, A4i, A4iii, B1i, B1iv, B2 — 049

Admin region Varna
Coordinates 43°34′N 28°34′E
Altitude 0–4 m **Area** 3,100 ha

Site description

A complex of three lakes in north-eastern Bulgaria, 5 km north-east from the village of Shabla. Two lakes are connected by an artificial canal and separated from the Black Sea by a sandbar; both have indented shorelines fringed with reedbeds of *Phragmites*, *Typha* and *Carex*. There is a hyper-saline lake (Shablenska tuzla) just to the south. There are planted trees and bushes of *Elaeagnus*, *Syringa*, *Ligustrum*, *Cotinus* and *Crataegus* in the former state residence, and forestry plantations to the north (*Robinia*, *Fraxinus*) and south (*Populus*) of the lake.

Habitats Grassland (6%; steppe/dry calcareous grassland; mesophile grassland), Wetland (19%; sand-dunes/sand beach; standing brackish and salt water; water-fringe vegetation), Marine areas (13%; open sea), Artificial landscape (62%; arable land; forestry plantation; other urban/industrial areas)
Land-use Agriculture (32%), Fisheries/aquaculture, Hunting (30%), Tourism/recreation (1%), Water management (68%)

Birds

Species	Season	Year	Pop min	Pop max	Acc	Criteria
Phalacrocorax pygmeus Pygmy Cormorant	W	1997	13	162	A	A1
Phalacrocorax pygmeus Pygmy Cormorant	P	1997	5	392	A	A1
Cygnus cygnus Whooper Swan	W	1997	162	261	B	B1i
Anser albifrons White-fronted Goose	W	1997	61,900	177,335	A	A4i, B1i
Branta ruficollis Red-breasted Goose	W	1997	20,000	55,845	A	A1, A4i, B1i
Anas platyrhynchos Mallard	W	1997	3,045	62,210	A	B1i
Aythya nyroca Ferruginous Duck	P	1996	10	88	A	A1
Lanius minor Lesser Grey Shrike	B	1996	8	14	A	B2

The complex is very important in winter for the globally threatened *Branta ruficollis* and for *Anser albifrons*, which both occur in huge numbers, as well as for other wintering and migrating waterbirds such as *Phalacrocorax pygmeus*, *Cygnus cygnus*, *Anas platyrhynchos* and *Aythya nyroca* (one pair sometimes remains to breed). The site is also a notable migratory bottleneck site in autumn, when up to 6,500 *Ciconia ciconia* have been counted passing overhead (the site may qualify under the A4iv criterion but more counts are needed). Species of global conservation concern that do not meet IBA criteria: *Anser erythropus* (1 or 2 on passage and in winter). Significant proportion (≥1%) of national population breeding at site: *Tadorna ferruginea*.

Protection status

National Partial **International** Partial
511 ha of IBA covered by Protected Landscape (Shabla lake, 511 ha).
403 ha of IBA covered by Ramsar Site (Shabla lake, 403 ha).

Conservation issues

Threats Agricultural intensification/expansion (U), Aquaculture/fisheries (U), Groundwater abstraction (A), Unsustainable exploitation (A)

The main problem is illegal hunting, including international hunting tourism. Illegal net-fishing and cattle-grazing cause disturbance to birds using the site. The management plan for Durankulak lake was prepared by Ministry of Environment under the Bulgarian–Swiss Biodiversity Conservation Programme.

Durankulak lake — A1, A4i, B1i, B2 — 050

Admin region Varna
Coordinates 43°40'N 28°33'E
Altitude 0–3 m **Area** 2,000 ha

■ Site description

A natural firth on the Black Sea coast in north-western Bulgaria, near the Romanian border. The water is fringed with reedbeds of *Phragmites*, *Typha* and *Schoenoplectus*, and is surrounded by arable land and plantations of *Populus*, *Fraxinus* and *Prunus*. The water-level varies widely, due to demand for irrigation and household-consumption in summer. There are some groups of *Salix* trees around the lake, and scrub to the south-west is dominated by *Paliurus*, *Crataegus* and *Rosa*.

Habitats Scrub (4%; scrub), Grassland (5%; steppe/dry calcareous grassland; humid grassland; mesophile grassland), Wetland (22%; sand-dunes/sand beach; standing brackish and salt water; water-fringe vegetation), Artificial landscape (69%; arable land; perennial crops/orchards/groves)
Land-use Agriculture (74%), Hunting (75%), Nature conservation/research (22%), Water management (20%)

■ Birds

The site is of global importance for wintering *Branta ruficollis*, *Anser albifrons* and *Phalacrocorax pygmeus*, and also important for some other wintering or migrating waterbirds, e.g. *Cygnus olor*. Species of global conservation concern that do not meet IBA criteria: *Anser erythropus* (max. 5 on passage and in winter).

Species	Season	Year	Pop min	Pop max	Acc	Criteria
Phalacrocorax pygmeus Pygmy Cormorant	W	—	4	356	A	A1, A4i, B1i
Phalacrocorax pygmeus Pygmy Cormorant	P	1996	47	228	A	A1, A4i, B1i
Ixobrychus minutus Little Bittern	B	1996	14	17	A	B2
Cygnus olor Mute Swan	W	1997	22	460	A	B1i
Anser albifrons White-fronted Goose	W	1997	5,657	73,531	A	A4i, B1i
Branta ruficollis Red-breasted Goose	W	1997	3,020	39,233	A	A1, A4i, B1i
[1] ***Charadrius alexandrinus*** Kentish Plover	B	1996	8	17	A	B2
Lanius minor Lesser Grey Shrike	B	1996	18	18	A	B2

1. Site qualifies according to 1997 national population estimate.

■ Protection status

National Partial **International** Partial

350 ha of IBA covered by National Monument (Durankulak lake, 350 ha). 350 ha of IBA covered by Ramsar Site (Durankulak lake, 350 ha).

■ Conservation issues

Threats Agricultural intensification/expansion (B), Filling-in of wetlands (B), Groundwater abstraction (B), Unsustainable exploitation (A)

The main problem is illegal hunting, including international hunting tourism. The management plan for Durankulak lake was prepared by the Ministry of Environment under the Bulgarian–Swiss Biodiversity Conservation Programme.

REFERENCES

ASENOV, L. (1994) Status of the 'Atanasovo Lake' Reserve. Report. Sofia. Unpublished.

BAROV, B. (1995) Report on the field work carried out for the project 'Eastern Rhodopi' Visitors Center. Unpublished.

BAROV, B. (1996) The results of the mapping of the breeding birds in the region of Byala Reka River, eastern Rhodopi mountains. BSPB. Unpublished.

BONDEV, I. (1991) [*The vegetation of Bulgaria: 1:600,000 map with explanatory text*.] Sofia: St. Kliment Ohridski University Press. (In Bulgarian.)

BORISOV, B. (1988) [Ornithological observations in south-west part of Sakar mountains.] *Orn. Inf. Newsl.* 23–24: 24–38. (In Bulgarian.)

BOTEV, B. AND PESHEV, TZ., EDS. (1985) [*Red Data Book of Republic Bulgaria. 2: Animals*.] Sofia: Bulgarian Academy of Science. (In Bulgarian.)

DELOV, V. (1995) [Investigations on Corncrake (*Crex crex*) in the region of Sofia.] *Annuaire de l'Universite de Sofia 'St. Kliment Ohridski'*. 88(4): 25–31. (In Bulgarian.)

DIMITROVA, E. AND PETKOV, P. (1995) [Information on the avifauna of Depo-2b in the area of the Beloslav.] *Orn. Inf. Newsl.* 19–20: 102–106. (In Bulgarian.)

DOIKOV, V. (1990) [Vardim Island.] *Priroda* 2: 43–44. (In Bulgarian.)

FILIPOVA-MARINOVA, M., DIMITROV, D. AND IVANOV, D. (in prep.) [*Flora and vegetation of 'Shabla Lake' Protected Area*.] Varna: Bulgarian-Swiss Biodiversity Conservation Programme. (In Bulgarian.)

GEORGIEV, G. (1993) [*National parks and reserves in Bulgaria*.] Sofia: Prosveta. (In Bulgarian.)

GEORGIEV, G. AND SAKELAROVA, L. (1991) [The reserves in Strandja mountains.] *Priroda* 3–4: 66–71. (In Bulgarian.)

GRIMMETT, R. F. A. AND JONES, T. A. (1989) *Important Bird Areas in Europe*. Cambridge, UK: International Council for Bird Preservation (Techn. Publ. 9).

HAUFF, P. (1981) Beitrag zur Kenntnis der Vogelwelt Thrakiens. *Faun. Abh. Staatl. Mus. fur Tierk. in Dresden*. 8: 67–73.

HEATH, M. F. AND BORGGREVE, C. (2000) *BirdLife International/ EBCC European Bird Database 1998*. Cambridge, UK: BirdLife International.

IANKOV, P. (1986) [Some observations on bird species from the orders Anseriformes and Charadriiformes in eastern Rhodopi mountains.] *Orn. Inf. Newsl.* 19–20: 111–121. (In Bulgarian.)

IANKOV, P. (1991) [The birds of the eastern Rhodopes. I: Time of presence and dynamics of the ornithofauna.] *Ecology* 24: 26–43. (In Bulgarian.)

IANKOV, P. (1995) The complex of Burgas Lakes. Pp. 33–44 in T. Michev, ed. *National action plan for the conservation of the most important wetlands in Bulgaria*. Sofia: Ministry of Environment.

IANKOV, P. (1995) The complex of Kamchia River. Pp. 30–32 in T. Michev, ed. *National action plan for the conservation of the most important wetlands in Bulgaria*. Sofia: Ministry of Environment.

IANKOV, P. AND PROFIROV, L. (1985) [The necessity and prerequistes for the establishment of East Rhodope National Park.] Pp. 50–62 in *Collection of reports. International Conference on MAB-8 project (UNESCO) 'Conservation of the Natural Areas and Their Genetic Fund'*, 2. Sofia: Bulgarian Academy of Science. (In Bulgarian.)

IANKOV, P. AND NYAGOLOV, K. (1987) [Observations on birds in eastern Rhodopi mountains during 1984.] *Orn. Inf. Newsl.* 21–22: 42–54. (In Bulgarian.)

IVANOV, B. (1995) The complex of Belene Islands. Pp. 14–17 in T. Michev, ed. *National action plan for the conservation of the most important wetlands in Bulgaria*. Sofia: Ministry of Environment.

IVANOV, B. (1995) The complex of Dourankoulak-Shabla. Pp. 24–29 in T. Michev, ed. *National action plan for the conservation of the most important wetlands in Bulgaria*. Sofia: Ministry of Environment.

IVANOV, B. (1995) The complex of Srebarna Nature Reserve. Pp. 20–22 in T. Michev, ed. *National action plan for the conservation of the most important wetlands in Bulgaria*. Sofia: Ministry of Environment.

KOSTADINOVA, I. (1997) *Important Bird Areas in Bulgaria*. Sofia: Bulgarian Society for the Protection of Birds. (In Bulgarian with English summary.)

LALEV, L. (1988) [Rare birds in the region of village Pozharevo.] *Orn. Inf. Newsl.* 23–24: 19–21. (In Bulgarian.)

MARINOV, M. (in print) *Management Plan of 'Poda Lagoon' Protected Area, located close to Burgas*. Burgas: Bulgarian Society for the Protection of Birds. (In Bulgarian and English.)

MICHEV, T. AND IANKOV, P. (1993) Ornithofauna. Pp. 585–614 in M. Sakalyan, ed. *The National Biological Diversity Strategy. Main reports*, 1. Sofia: Biodiversity Support Program c/o World Wildlife Fund. (In Bulgarian and English.)

MILCHEV, B. (1994) Breeding Bird Atlas of the Srandja Mountains, South-east Bulgaria. *Sandgrouse* 16: 2–27.

MILCHEV, B. (1994) [Investigations on wintering waterfowl in Strandja mountain and contiguous Black Sea coast.] *Priroda*. 3–4: 66–71. (In Bulgarian.)

MILCHEV, B. (1995) [The animals in 'Veleka Mouth' Protected Area and 'Silistar' Protected Area.] Pp.71–80 in *Management Plan for 'Veleka Mouth' Protected Area and 'Silistar' Protected Area*. Sofia: Ministry of Environment. (In Bulgarian.)

MILCHEV, B. AND KOVACHEV, A. (1995) [On migration of the White Stork along the West Black Sea migration road.] *Annual of the Sofia University*. 86–87: 43–48. (In Bulgarian.)

MILCHEW, B. (1996) Widerheistellung eines Horstes des Schmitzgeiers. *Orn. Mitteil.* 38(5/6): 327–334.

Milchew, B. and Georgiewa, U. (1992) Eine Studie zum Bestand, zur Brutbiologie und Ernahrung des Steinadlers *Aquila chrysaetos* im Strandsha Gebirge. *Beitr. Vogelkd.* 38(5): 124–126.

Milchew, B. and Georgiewa, U. (1994) Verbreitung, Fortoflanzungsverhalten und Ernahrung des Schmitzgeiers *Neophron percnopterus* im Strandsha Gebirge. *Ecology*. 26: 68–77.

Mitchev, T. (in press) *Management Plan of 'Atanasovo Lake' Reserve*. Sofia: Bulgarian-Swiss Biodiversity Conservation Programme. (In Bulgarian and English.)

Nankinov, D. and Djuninski, E. (1985) [On species composition of the 'Tchuprene' Biosphere Reserve, Vidin Region.] Pp. 45–54 in in *Collection of reports. International Conference on MAB-8 project (UNESCO) 'Conservation of the Natural Areas and Their Genetic Fund'*, 3. Sofia: Bulgarian Academy of Science. (In Bulgarian.)

National Bank for Ornithological Information with BSPB. Sofia: Bulgarian Society for the Protection of Birds. Unpublished.

Nedyalkov, C. (1984) Working up theoretically substantiated proposals for developing of network of protected areas in Kotel Municipality.

Petrov, T. (1981) [Birds of the Sredna gora mountain.] *News of the Natural Museums in South Bulgaria* 7: 9–49. (In Bulgarian.)

Petrov, T. (1996) Birds of the Pyasachnik Reservoir and its surrounding area. Unpublished.

Profirov, L. and Undzian, E. (1985) [About the breeding avifauna in 'Rusenski Lom' National Park.] Pp. 218–227 in *Collection of reports. International Conference on MAB-8 project (UNESCO) 'Conservation of the Natural Areas and Their Genetic Fund'*. Sofia: Bulgarian Academy of Science. (In Bulgarian.)

Rose, P. M. (1992) *Western Palearctic and South-West Asia waterfowl census 1992*. Slimbridge, UK: International Waterfowl and Wetlands Research Bureau.

Rose, P. M. and Scott, D. A. (1994) *Waterfowl population estimates*. Slimbridge, UK: International Waterfowl and Wetlands Research Bureau (IWRB Spec. Publ. 29).

Rose, P. M. and Taylor, V. (1993) *Western Palearctic and South-West Asia waterfowl census 1993*. Slimbridge, UK: International Waterfowl and Wetlands Research Bureau.

Simeonov, C. and Milchev, B. (1994) [Investigations on Eagle Owl *Bubo bubo* in Strandja mountain. I: Distribution, habitat and breeding biology.] *Ecology* 26: 78–87. (In Bulgarian.)

Simeonov, P. (1983) [Breeding avifauna in 'Boatin' Biosphere Reserve.] Pp. 83–90 in *Collection of reports. International Conference on MAB-8 project (UNESCO) 'Conservation of the Natural Areas and Their Genetic Fund'*, 2. Sofia: Bulgarian Academy of Science. (In Bulgarian.)

Spiridonov, Z. and Simeonov, P. (1988) [Investigations on breeding avifauna in the 'Tissata' Reserve.] *Fauna of the South-west Bulgaria* 2: 23–29. (In Bulgarian.)

Spiridonov, Z. and Mileva, L. (1988) [Threatened and rare birds in 'Steneto' Reserve.] *Orn. Inf. Newsl.* 23–24: 99–105. (In Bulgarian.)

Spiridonov, Z. Observation on avifauna in Central Balkan 1987–1996. Unpublished data.

Spiridonov, Z., Mileva, L. and Spasov, N. (1987) [Breeding avifauna in 'Tsarichina' Biosphere Reserve.] Pp. 104–121 in *Collection of reports 'Tsarichina' Biosphere Reserve*. Krakov, Poland. (In Russian.)

Stanchev, C. (1988) [Investigations on the avifauna of the 'Rusrenski Lom' Reserve 1985–1987.] *Orn. Inf. Newsl.* 23–24: 140–151. (In Bulgarian.)

Tucker, G. M. and Heath, M. F. (1994) *Birds in Europe: their conservation status*. Cambridge, UK: BirdLife International (BirdLife Conservation Series no. 3).

Undzian, E. (1996) Tourist and recreation resources in the 'Rusenski Lom' National Park. Unpublished report.

Velchev, V., ed. (1984) [*Red Data Book of Republic Bulgaria*. 1: *Plants*.] Sofia: Bulgarian Academy of Science. (In Bulgarian.)

Zehtindziev, P. and Kutzarov, I. (in press) Ornithological investigations of some typical biotopes of Danube Plain in Bulgaria.

CROATIA

Willow forest along the River Drava (IBA 011). (PHOTO: T. MIKUSKA)

OVERVIEW

Croatia occupies 56,540 km² with a population of 4,726,000 people. Twenty-three Important Bird Areas (IBAs) are identified, covering 7,546 km² (approximately 13%) of the country's surface area. This inventory represents a partial update of the previous pan-European inventory (Grimmett and Jones 1989). New data have been compiled for some sites by Tibor Mikuska (see site-account species tables) but were unavailable for many existing sites and potential new IBAs. With the exception of two sites (YU008 and YU018), all sites from the previous inventory are included here, whether or not new data are available (Table 1, Map 1). Sites YU008 and YU018 did not meet criteria in the previous inventory but were judged important for migratory species and therefore included. As these sites do not meet the revised criteria for this inventory either, they have not been confirmed as IBAs (Table 1). In addition to the 17 sites retained from the previous inventory, six new sites have been identified (Table 1, Map 1).

Several globally threatened and near-threatened bird species (Collar *et al.* 1994) occur in IBAs, meeting the A1 criterion. *Aythya nyroca* meets the criterion at 12 sites, with nine sites holding important numbers of breeding birds, two sites significant numbers of passage birds and one site wintering birds. *Haliaeetus albicilla* breeds in important numbers at six sites and also winters in

Table 1. Summary of Important Bird Areas in Croatia. 23 IBAs covering 7,546 km²

IBA code	1989 code	International name	National name	Administrative region	Area (ha)	Criteria (see p. 11)
001	—	Alluvial wetlands of the River Danube	Poplavno podrucje rijeke Dunav	Osjecko-baranjska zupanija	30,000	A1, A4i, A4iii, B1i, B2
002	YU014	Kopacki Rit Special Zoological Reserve	Posebni zooloski rezervat Kopacki rit	Osjecko-baranjska zupanija	7,000	A1, A4i, A4iii, B1i, B2
003	YU013	River Drava	Drava od Donjie Miholjca do Osijeka (ukljucujuci Dravske sume)	Osjecko-baranjska zupanija	25,000	A1, B2
004	—	Nasicka Breznica fish-ponds	Ribnjaci kod Nasicke Breznice	Osjecko-baranjska zupanija	1,345	A1, A4i, B1i, B2
005	—	Alluvial forests from Podravska Slatina to Vuka	Poplavne sume hrasta luznjaka izmedju Podravske Slatine i Vuke	Viroviticko-podravska zupanija, Pozesko-slavonska zupanija, Osjecko-baranjska zupanija	100,000	A1, B2
006	—	Donji Miholjac fish-ponds	Ribnjaci kod Donjeg Miholjca	Osjecko-baranjska zupanija	981	A1, A4i, B1i, B2
007	—	Grudnjak fish-ponds	Ribnjaci Grudnjak	Viroviticko-podravska zupanija	1,020	A1, B1i, B2, B3
008	YU012	Jelas field	Jelas polje	Brodsko-posavska zupanija	10,000	A1, A4i, A4iii, B1i, B2
009	YU010	Alluvial wetlands of the River Sava	Poplavno podrucje rijeke Save	Sisacko-moslavacka zupanija, Brodsko-posavska zupanija, Vukovarsko-srijemska zupanija	210,000	A1, A4i, B1i, B2
010	YU011	Koncanica fish-ponds and surrounding area	Ribnjaci Koncanica s okolicom	Bjelovarsko-bilogorska zupanija	10,000	A1, B1i, B2
011	—	Alluvial wetlands of the River Drava	Poplavno podrucje rijeke Drave	Koprivnicko-krizevacka zupanija, Viroviticko-podravska zupanija, Osjecko-baranjska zupanija	50,000	A1, A4i, B1i, B2

Table 1 ... continued. Summary of Important Bird Areas in Croatia. 23 IBAs covering 7,546 km²

IBA code	1989 code	International name	National name	Administrative region	Area (ha)	Criteria (see p. 11)
012	YU009	Turopolje	Turopolje	Sisacko-moslavacka zupanija, Grad Zagreb	20,000	B2
013	YU007	Pokupsko depression	Pokupska Depresija	Zagrebacka zupanija, Karlovacka zupanija	196,600	A1, A4i, A4iii, B1i, B2
014	YU015	Special Ornithological Reserve on Krk island	Posebni ornitološki rezervat na otoku Krku	Primorsko-goranska zupanija	1,000	B2
015	YU017	Prvic island	Otok Prvic	Primorsko-goranska zupanija	7,000	B2
016	YU016	Special Ornithological Reserves on Cres island	Posebni ornitološki rezervati na otoku Cresu	Primorsko-goranska zupanija	1,450	B2
017	YU019	Paklenica National Park	Nacionalni park Paklenica	Zadarsko-kninska zupanija	3,617	B2
018	YU020	Kornati National Park and Telascica Nature Park	Nacionalni park Kornati i Park prirode Telascica	Zadarsko-kninska zupanija, Sibenska zupanija	30,200	B2
019	YU022	Lake Vransko	Vransko jezero	Zadarsko-kninska zupanija	3,000	A4i, A4iii, B1i
020	YU021	Krka National Park	Nacionalni park Krka	Zadarsko-kninska zupanija, Sibenska zupanija	14,200	B2
021	YU023	Sinjsko field	Sinjsko polje	Splitsko-dalmatinska zupanija	640	B2
022	YU024	Dinara mountain	Dinara planina	Zadarsko-kninska zupanija, Splitsko-dalmatinska zupanija	20,000	B2
023	YU025	Delta of the Neretva river	Delta rijeke Neretve	Dubrovacko-neretvanska zupanija	11,500	A3, B1i

Sites identified in the previous inventory of IBAs in Europe (Grimmett and Jones 1989) but no longer considered to be IBAs
YU008 St Eufemia Bay, Rab; YU018 Bokanjacko blato and Ninske solane (Nin saltpan).

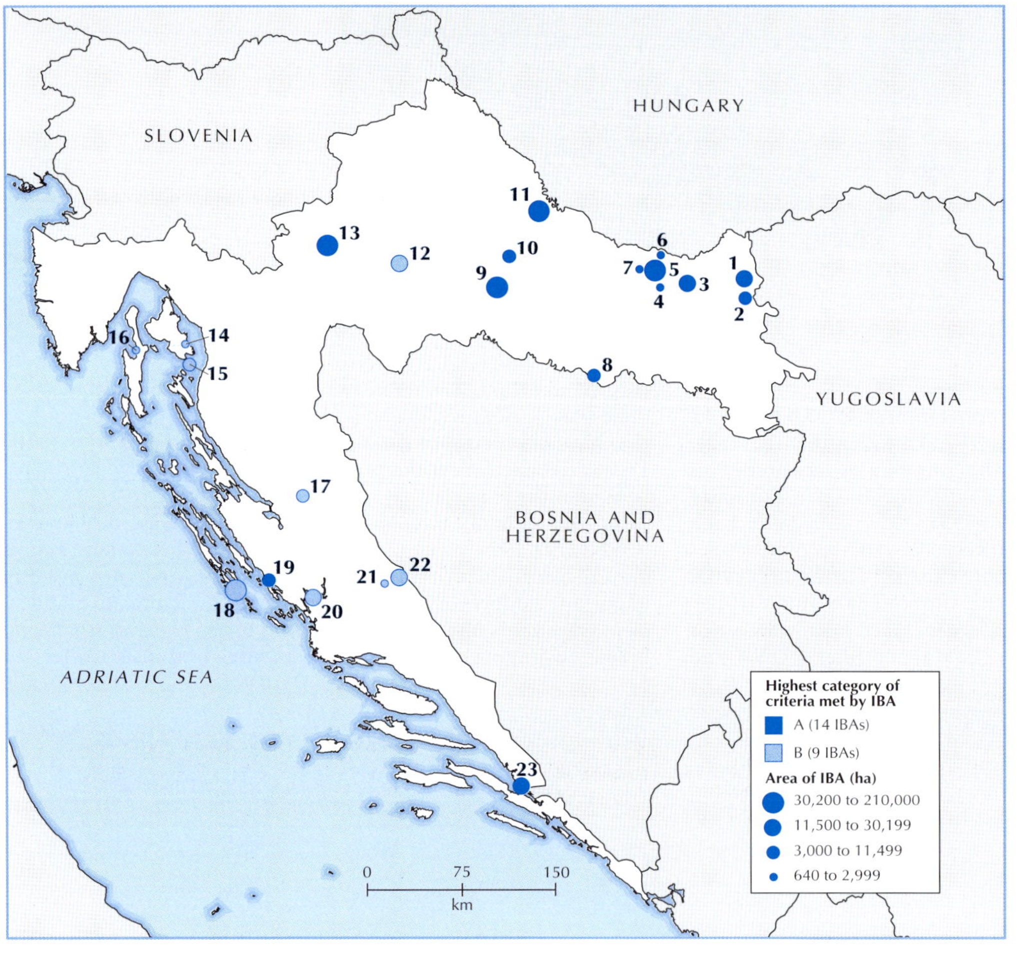

Map 1. Location, area and criteria category of Important Bird Areas in Croatia.

several IBAs. *Crex crex* breeds in significant numbers in a single site and *Aquila clanga* (wintering) meets the A1 criterion at one site. *Falco naumanni* occurs but does not meet criteria. There are two records of *Numenius tenuirostris* from 008 and 023 but these remain unverified and the species is therefore not considered to meet criteria.

The majority of sites selected as IBAs are wetland-forest complexes, with extensive areas of broadleaved alluvial forest in association with natural water-bodies and fish-ponds. These sites hold important breeding populations of species with an unfavourable conservation status in Europe (meeting the B2/B3 criteria), in particular herons and storks, including *Ixobrychus minutus*, *Nycticorax nycticorax*, *Ardeola ralloides*, *Ciconia nigra* and *Ciconia ciconia*. Several of these IBAs are also important for passage and wintering waterbirds, five sites regularly holding more than 20,000 waterbirds (meeting the A4iii criterion).

Alluvial forest and wetlands are threatened by a wide variety of factors. Of particular importance are agricultural expansion, forestry, drainage and river canalization. Canalization prevents natural flood-plain dynamics and causes the drying of forests and surrounding wetlands, as inundation no longer occurs and the water-table is lowered. In IBAs with fish-ponds, hunting is a serious threat as the ponds are often leased to foreign hunting groups. Hunting causes both significant mortality and disturbance.

SITE ACCOUNTS

Alluvial wetlands of the River Danube — A1, A4i, A4iii, B1i, B2 — 001

Admin region Osjecko-baranjska zupanija
Coordinates 45°40′N 18°50′E
Altitude 84–89 m **Area** 30,000 ha

Site description

Alluvial wetlands along the River Danube from the Croatian–Hungarian border south to the mouth of the River Drava. Area includes alluvial *Quercus* forests, extensive freshwater wetlands and reedbeds (*Phragmites*) and commercial fish-ponds. Since 1991 the area has been completely occupied by military forces.

Habitats Forest and woodland (40%; broadleaved deciduous forest; alluvial/very wet forest), Grassland (5%; humid grassland), Wetland (50%; standing fresh water; river/stream; water-fringe vegetation), Artificial landscape (5%; arable land; forestry plantation; other urban/industrial areas)
Land-use Agriculture (10%), Fisheries/aquaculture (5%), Forestry (40%), Hunting (80%), Nature conservation/research (50%), Tourism/recreation (30%)

Birds

Species	Season	Year	Pop min	Pop max	Acc	Criteria
Ixobrychus minutus Little Bittern	B	—	100	100	B	B2
[1] ***Nycticorax nycticorax*** Night Heron	B	—	100	400	B	B2
[2,3] ***Ardeola ralloides*** Squacco Heron	B	—	—	70	B	B1i, B2
[4] ***Anas platyrhynchos*** Mallard	W	—	20,000	20,000	—	B1i
Aythya nyroca Ferruginous Duck	B	1991	50	50	B	A1, B2
Aythya nyroca Ferruginous Duck	P	1994	1,000	1,000	—	A1, A4i
Haliaeetus albicilla White-tailed Eagle	R	—	15	20	A	A1, B2

1. 1980–1990.
2. 1987–1990.
3. Large decrease.
4. 1989.

The site is important for breeding herons. The site holds 20,000 or more waterbirds in winter, on a regular basis. Species of global conservation concern that do not meet IBA criteria: *Aquila heliaca* (passage), *Falco naumanni*, *Crex crex* (breeding).

Protection status

National Partial **International** Partial
7,000 ha of IBA covered by Special Reserve (Kopacki Rit, 7,000 ha). 17,200 ha of IBA covered by Nature Park (Kopacki Rit, 17,200 ha). 17,770 ha of IBA covered by Ramsar Site (Kopacki Rit, 17,770 ha).

Conservation issues

Threats Agricultural intensification/expansion (B), Aquaculture/fisheries (C), Burning of vegetation (C), Deforestation (commercial) (A), Disturbance to birds (A), Dredging/canalization (A), Natural events (B), Other (A), Recreation/tourism (B)

There are plans for river canalization, forestry and use of the site as an exclusive state hunting property. The entire area has been proposed for protection as a National Park.

Kopacki Rit Special Zoological Reserve — A1, A4i, A4iii, B1i, B2 — 002

Admin region Osjecko-baranjska zupanija
Coordinates 45°34′N 18°51′E
Altitude — **Area** 7,000 ha

Site description

An area of flood-plain of the Rivers Danube and Drava, with extensive *Salix*, *Populus* and *Quercus* forests. There are numerous channels and oxbows, extensive reedbeds (*Phragmites*) and a large fish-pond complex.

Habitats Forest and woodland (broadleaved deciduous forest; alluvial/very wet forest), Wetland (standing fresh water; river/stream; water-fringe vegetation)
Land-use Fisheries/aquaculture, Hunting, Tourism/recreation

Birds

Species	Season	Year	Pop min	Pop max	Acc	Criteria
Ixobrychus minutus Little Bittern	B	—	100	—	—	B2
Nycticorax nycticorax Night Heron	B	—	400	400	—	B2
Ardeola ralloides Squacco Heron	B	—	10	100	—	B2
Ardea purpurea Purple Heron	B	—	10	200	—	B2
Ciconia nigra Black Stork	B	—	30	30	—	B2
Ciconia ciconia White Stork	B	—	20	50	—	B2
Platalea leucorodia Spoonbill	P	—	500	10,000	—	A4i, B1i
Anser fabalis Bean Goose	W	—	10,000	50,000	—	A4i, B1i
Anas platyrhynchos Mallard	W	—	50,000	50,000	—	B1i
Aythya nyroca Ferruginous Duck	B	—	100	200	—	A1, A4i, B1i, B2
Milvus migrans Black Kite	B	—	50	50	—	B2
Haliaeetus albicilla White-tailed Eagle	R	—	15	20	—	A1, B2
Aquila clanga Greater Spotted Eagle	W	—	2	5	—	A1
Falco cherrug Saker	B	—	4	5	—	B2
Grus grus Crane	P	—	50	500	—	B1i
Chlidonias hybridus Whiskered Tern	B	—	100	100	—	B1i, B2

This is an important site for breeding herons. The site holds 20,000 or more waterbirds in winter, on a regular basis. Species of global conservation concern that do not meet IBA criteria: *Aquila heliaca, Falco naumanni, Crex crex*.

Protection status

National Partial **International** Partial
6,234 ha of IBA covered by Special Zoological Reserve (Kopacki Rit, 6,234 ha). IBA overlaps with Nature Park (Kopacki Rit, 17,200 ha). IBA overlaps with Ramsar Site (Kopacki Rit, 17,770 ha).

Conservation issues

Threats Aquaculture/fisheries (U), Other (U)

Threats include fisheries, eutrophication, siltation, and frog collection.

River Drava — A1, B2 — 003

Admin region Osjecko-baranjska zupanija
Coordinates 45°37′N 18°25′E
Altitude 85–95 m **Area** 25,000 ha

Site description

A stretch of the River Drava, extending from Donji Miholjac to Osijek (including Dravske forest).

Habitats Forest and woodland (alluvial/very wet forest; broadleaved evergreen forest), Wetland (river/stream)
Land-use Forestry

Birds

Species	Season	Year	Pop min	Pop max	Acc	Criteria
Haliaeetus albicilla White-tailed Eagle	B	—	5	6	—	A1, B2

Protection status

National None **International** None

Conservation issues

Threats Intensified forest management (U)

Nasicka Breznica fish-ponds — A1, A4i, B1i, B2 — 004

Admin region Osjecko-baranjska zupanija
Coordinates 45°35′N 18°13′E
Altitude 95–105 m **Area** 1,345 ha

Site description

A complex of fish-ponds with extensive reedbeds (*Phragmites*) and ponds with emergent vegetation.

Habitats Wetland (90%; standing fresh water; water-fringe vegetation), Artificial landscape (10%; ruderal land)
Land-use Fisheries/aquaculture (100%), Hunting (100%)

Birds

Species	Season	Year	Pop min	Pop max	Acc	Criteria
[1] ***Ardeola ralloides*** Squacco Heron	B	1995	10	15	A	B2
[2] ***Egretta alba*** Great White Egret	W	1996	6	313	A	A4i, B1i
Aythya nyroca Ferruginous Duck	B	1993	20	50	B	A1, B2

1. 1993–1995.
2. 1993–1996.

Species of global conservation concern that do not meet IBA criteria: *Haliaeetus albicilla*.

Protection status

National None **International** None

Conservation issues

Threats Abandonment/reduction of land management (B), Aquaculture/fisheries (B), Burning of vegetation (C), Disturbance to birds (B), Recreation/tourism (B), Unsustainable exploitation (U)

The main threats are fishery management practices, the leasing of the ponds to foreign hunters, and disturbance caused by the shooting of *Phalacrocorax carbo* during the breeding season.

Alluvial forests from Podravska Slatina to Vuka

A1, B2 **005**

Admin region Viroviticko-podravska zupanija, Pozesko-slavonska zupanija, Osjecko-baranjska zupanija
Coordinates 45°40′N 18°10′E
Altitude 90–100 m **Area** 100,000 ha

Site description

Alluvial *Quercus* forests streching from the town of Podravska Slatina south-east to the village of Vuka.

Habitats Forest and woodland (90%; broadleaved deciduous forest; alluvial/very wet forest), Artificial landscape (10%; other urban/industrial areas)
Land-use Forestry (100%), Hunting

Birds

Species	Season	Year	Pop min	Pop max	Acc	Criteria
Ciconia nigra Black Stork	B	—	10	50	B	B2
Milvus migrans Black Kite	B	—	10	20	C	B2
Haliaeetus albicilla White-tailed Eagle	R	—	5	10	B	A1, B2

An important site for breeding storks and raptors. Species of global conservation concern that do not meet IBA criteria: *Falco naumanni*.

Protection status

National None **International** None

Conservation issues

Threats Deforestation (commercial) (B), Infrastructure (B), Natural events (B), Unsustainable exploitation (A)

The main threat is from the intensification of forestry, particularly clear-felling. Falling groundwater levels, as a result of drought and possibly drainage, are also a problem.

Donji Miholjac fish-ponds

A1, A4i, B1i, B2 **006**

Admin region Osjecko-baranjska zupanija
Coordinates 45°45′N 18°12′E
Altitude 90–92 m **Area** 981 ha

Site description

A complex of a fish-ponds along the River Drava. The site includes large ponds with extensive reedbeds (*Phragmites*) and other emergent vegetation. The surrounding area includes abandoned ponds and a former oxbow lake.

Habitats Wetland (90%; standing fresh water; water-fringe vegetation), Artificial landscape (10%; ruderal land)
Land-use Fisheries/aquaculture (100%), Hunting (100%)

Birds

Species	Season	Year	Pop min	Pop max	Acc	Criteria
[1] ***Phalacrocorax carbo*** Cormorant	N	1990	400	400	B	B1i
[2] ***Egretta alba*** Great White Egret	W	1996	1	177	B	A4i, B1i
[3] ***Ardea purpurea*** Purple Heron	B	1995	3	58	A	B2
[4] ***Anser anser*** Greylag Goose	W	1987	500	500	A	B1i
Aythya nyroca Ferruginous Duck	B	1993	10	40	B	A1, B2
[5] ***Mergus merganser*** Goosander	W	1996	33	33	A	B1i

1. Birds breed on Hungarian part of site but feed in Croatia.
2. 1993–1996.
3. 1990–1993.
4. 1987.
5. 1996.

This is an important site for breeding and wintering waterbirds. Species of global conservation concern that do not meet IBA criteria: *Haliaeetus albicilla* (1–6 wintering).

Protection status

National None **International** None

Conservation issues

Threats Aquaculture/fisheries (B), Burning of vegetation (C), Disturbance to birds (B), Unsustainable exploitation (U)

The main threats are from fisheries management, and disturbance and mortality caused by foreign hunters. Shooting of *Phalacrocorax carbo* causes great disturbance to breeding birds including *Aythya nyroca*.

Grudnjak fish-ponds

A1, B1i, B2, B3 **007**

Admin region Viroviticko-podravska zupanija
Coordinates 45°40′N 18°03′E
Altitude 99–102 m **Area** 1,020 ha

Site description

A complex of fish-ponds surrounded by water-fringe vegetation and some completely covered with emergent vegetation. Wet meadows and pastures are adjacent to the fish-ponds.

Habitats Forest and woodland, Grassland (5%; humid grassland), Wetland (90%; standing fresh water; water-fringe vegetation), Artificial landscape (5%)
Land-use Agriculture (5%), Fisheries/aquaculture (95%), Hunting (95%)

Birds

Species	Season	Year	Pop min	Pop max	Acc	Criteria
[1] ***Ixobrychus minutus*** Little Bittern	B	1995	15	20	B	B2
[1] ***Nycticorax nycticorax*** Night Heron	B	1995	23	100	A	B2
[2] ***Ardeola ralloides*** Squacco Heron	B	1995	2	20	A	B2
Aythya ferina Pochard	B	1993	17	17	B	B3
Aythya nyroca Ferruginous Duck	B	1993	20	50	B	A1, B2
Chlidonias hybridus Whiskered Tern	B	1993	200	200	B	B1i, B2

1. 1993–1995.
2. 1994–1995.

This is an important site for breeding herons and terns. Species of global conservation concern that do not meet IBA criteria: *Aythya nyroca*, *Haliaeetus albicilla* (1–9 wintering).

Protection status

National None **International** None

Conservation issues

Threats Aquaculture/fisheries (B), Burning of vegetation (C), Disturbance to birds (A), Recreation/tourism (A), Unsustainable exploitation (U)

The fish-ponds are leased to foreign hunters. Breeding birds are disturbed by fishermen and by the shooting of *Phalacrocorax carbo*.

Jelas field — A1, A4i, A4iii, B1i, B2 — 008

Admin region Brodsko-posavska zupanija
Coordinates 45°05'N 17°45'E
Altitude 87–100 m **Area** 10,000 ha

Site description

Alluvial flood-plain of the River Sava, situated west of the town of Slavonski Brod. The site includes commercial fish-ponds, alluvial *Quercus* forests, wet meadows and pastures, as well as arable land with traditional villages.

Habitats Forest and woodland (20%; broadleaved deciduous forest; alluvial/very wet forest), Grassland (20%; humid grassland), Wetland (30%; standing fresh water; river/stream; water-fringe vegetation), Artificial landscape (30%; highly improved reseeded grassland; arable land; perennial crops/orchards/groves; other urban/industrial areas)
Land-use Agriculture (30%), Fisheries/aquaculture (25%), Forestry (25%), Hunting (25%)

Birds

Species	Season	Year	Pop min	Pop max	Acc	Criteria
[1] ***Phalacrocorax carbo*** Cormorant	W	1996	27	3,500	A	B1i
[2] ***Phalacrocorax pygmeus*** Pygmy Cormorant	W	1996	242	400	A	A1, A4i, B1i
[3] ***Ixobrychus minutus*** Little Bittern	B	1995	25	26	B	B2
[4] ***Nycticorax nycticorax*** Night Heron	B	1995	122	320	A	B2
[4] ***Ardeola ralloides*** Squacco Heron	B	1995	60	180	A	B2
[1] ***Egretta alba*** Great White Egret	W	1996	12	157	A	A4i, B1i
[5] ***Platalea leucorodia*** Spoonbill	B	1996	100	125	A	B1i, B2
[6] ***Anas platyrhynchos*** Mallard	W	1996	2,500	11,000	A	B1i
Aythya nyroca Ferruginous Duck	B	—	120	160	—	A1, B1i, B2
[7] ***Aythya nyroca*** Ferruginous Duck	W	1996	11	52	A	A1
[1] ***Larus ridibundus*** Black-headed Gull	W	1996	42	5,500	A	B1i

1. 1992–1996.
2. 1995–1996.
3. 1993–1996.
4. 1991–1995.
5. 1991–1994.
6. 1991–1997.
7. 1991–1996.

This is an important site for breeding herons. The site holds 20,000 or more waterbirds in winter, on a regular basis. Species of global conservation concern that do not meet IBA criteria: *Haliaeetus albicilla, Falco naumanni, Crex crex.*

Protection status

National None **International** None

Conservation issues

Threats Abandonment/reduction of land management (A), Agricultural intensification/expansion (B), Aquaculture/fisheries (B), Burning of vegetation (C), Disturbance to birds (B), Intensified forest management (B)

Local authorities protect the mixed heron colony as a Special Ornithological Reserve.

Alluvial wetlands of the River Sava — A1, A4i, B1i, B2 — 009

Admin region Sisacko-moslavacka zupanija, Brodsko-posavska zupanija, Vukovarsko-srijemska zupanija
Coordinates 45°30'N 17°00'E
Altitude 90–105 m **Area** 210,000 ha

Site description

A flood-plain area of the Rivers Sava, Lonja and Strug. The site inculdes alluvial *Quercus* and *Populus* forests, marshes, meadows, fish-ponds, oxbow lakes and rivers and streams. Some areas have been drained and cultivated.

Habitats Forest and woodland (40%; broadleaved deciduous forest; alluvial/very wet forest), Grassland (10%; humid grassland), Wetland (20%; standing fresh water; river/stream; water-fringe vegetation), Artificial landscape (40%; arable land; perennial crops/orchards/groves; other urban/industrial areas; ruderal land)

Land-use Agriculture, Forestry

Birds

Species	Season	Year	Pop min	Pop max	Acc	Criteria
Nycticorax nycticorax Night Heron	B	—	100	200	—	B2
Ciconia nigra Black Stork	B	—	50	80	—	B1i, B2
Ciconia ciconia White Stork	B	—	570	570	—	B2
Platalea leucorodia Spoonbill	B	—	150	170	—	A4i, B1i, B2
Milvus migrans Black Kite	B	—	10	—	—	B2
Haliaeetus albicilla White-tailed Eagle	B	—	15	15	—	A1, B2
Aquila pomarina Lesser Spotted Eagle	B	—	25	—	—	B2
Crex crex Corncrake	B	—	120	150	—	A1
Chlidonias hybridus Whiskered Tern	B	—	150	—	—	B1i, B2

This is an important site for breeding storks and raptors. Species of global conservation concern that do not meet IBA criteria: *Aythya nyroca, Aquila heliaca.*

Protection status

National Low **International** None
25 ha of IBA covered by Bird Sanctuary (Krapje Dol, 25 ha). 430 ha of IBA covered by Bird Sanctuary (Rakita, 430 ha). 77 ha of IBA covered by Bird Sanctuary (Vrazje Blato, 77 ha).

Conservation issues

Threats Drainage (U), Filling-in of wetlands (U)

The main threats are drainage and land reclamation.

Koncanica fish-ponds and surrounding area — A1, B1i, B2 — 010

Admin region Bjelovarsko-bilogorska zupanija
Coordinates 45°40'N 17°04'E
Altitude 129–144 m **Area** 10,000 ha

Site description

A complex of fish-ponds along the Ilova river consisting of large ponds with water-fringe vegetation (*Phragmites*) covering 1,475 ha and wet grasslands covering over 700 ha. Alluvial *Qurecus* forests occupy 2,500 ha.

Habitats Forest and woodland (30%; broadleaved deciduous forest; alluvial/very wet forest), Grassland (10%; humid grassland), Wetland (30%; standing fresh water; river/stream; water-fringe vegetation), Artificial landscape (40%; arable land; perennial crops/orchards/groves; other urban/industrial areas)
Land-use Agriculture (40%), Fisheries/aquaculture (30%), Forestry (30%), Hunting (30%)

Birds

Species	Season	Year	Pop min	Pop max	Acc	Criteria
[1] ***Phalacrocorax carbo*** Cormorant	W	1994	3	2,500	A	B1i
Ciconia ciconia White Stork	B	—	20	30	A	B2
[2] ***Anser fabalis*** Bean Goose	W	1987	2,000	2,000	A	B1i
Aythya nyroca Ferruginous Duck	B	—	10	50	B	A1, B2
Chlidonias hybridus Whiskered Tern	B	—	15	15	B	B2

1. 1992–1994.
2. 1987.

Species of global conservation concern that do not meet IBA criteria: *Haliaeetus albicilla, Crex crex.*

Protection status

National None **International** None

Conservation issues

Threats Agricultural intensification/expansion (B), Aquaculture/fisheries (B), Deforestation (commercial) (B), Disturbance to birds (B), Other (B), Unsustainable exploitation

The main threats include conversion of grasslands to arable cultivation, increasingly intensive forestry, hunting, pesticide run-off, and lead poisoning from spent shot ('Other', threats).

Alluvial wetlands of the River Drava

A1, A4i, B1i, B2 **011**

Admin region Koprivnicko-krizevacka zupanija, Viroviticko-podravska zupanija, Osjecko-baranjska zupanija
Coordinates 45°55′N 17°15′E
Altitude 85–130 m **Area** 50,000 ha

Site description

Natural alluvial landscapes along the River Drava from the mouth of the River Mura, east to its mouth with the River Danube. Area includes extensive alluvial *Quercus* forests, wet meadows, pastures, and small freshwater wetlands. The river has created 40-m-high cliffs and numerous meanders, as well as gravel islands.

Habitats Forest and woodland (40%; broadleaved deciduous forest; alluvial/very wet forest), Grassland (10%; humid grassland), Wetland (25%; standing fresh water; river/stream; water-fringe vegetation), Artificial landscape (25%; highly improved reseeded grassland; arable land; perennial crops/orchards/groves; other urban/industrial areas)
Land-use Agriculture (60%), Forestry (40%), Hunting, Water management (10%)

Birds

Species	Season	Year	Pop min	Pop max	Acc	Criteria
Phalacrocorax carbo Cormorant	R	—	2,000	2,000	A	A4i
Ixobrychus minutus Little Bittern	B	—	10	43	B	B2
Nycticorax nycticorax Night Heron	B	—	200	423	A	B2
Ardeola ralloides Squacco Heron	B	—	5	22	A	B2
Ardea purpurea Purple Heron	B	—	144	144	A	B2
Ciconia nigra Black Stork	B	—	10	41	B	B2
Ciconia ciconia White Stork	B	—	100	120	B	B2
Platalea leucorodia Spoonbill	B	—	5	31	A	B2
Aythya nyroca Ferruginous Duck	B	—	100	200	B	A1, A4i, B1i, B2
Haliaeetus albicilla White-tailed Eagle	R	—	10	27	B	A1, B2
Riparia riparia Sand Martin	B	—	4,000	4,700	B	B2

This is an important site for breeding herons and storks.

Protection status

National Low **International** None
3,886 ha of IBA covered by Nature Park (Repas Nature Park). 700 ha of IBA covered by Special Reserve (Veliki Pazut, 700 ha).

Conservation issues

Threats Agricultural intensification/expansion (B), Burning of vegetation (C), Construction/impact of dyke/dam/barrage (A), Deforestation (commercial) (B), Disturbance to birds (B), Dredging/canalization (A), Extraction industry (B), Filling-in of wetlands (B), Industrialization/urbanization (B), Infrastructure (B)

There are plans to build power-stations along the River Drava and to canalize parts of the river. Gravel mining is a further threat. A joint conservation project between Austrian, Slovenian, Hungarian and Croatian NGOs, under the leadership of Euronatur, has proposed the whole area for protection.

Turopolje

B2 **012**

Admin region Sisacko-moslavacka zupanija, Grad Zagreb
Coordinates 45°34′N 16°15′E
Altitude 97–105 m **Area** 20,000 ha

Site description

An area regularly inundated by the River Odra and its tributaries. The site includes extensive pastures, wet grasslands and alluvial *Quercus* and *Fagus* forests.

Habitats Forest and woodland (25%; broadleaved deciduous forest; alluvial/very wet forest), Grassland (65%; humid grassland), Wetland (5%; standing fresh water; river/stream), Artificial landscape (5%; arable land)
Land-use Agriculture (70%), Forestry (30%), Hunting

Birds

This is an important site for breeding storks, herons and raptors. Species of global conservation concern that do not meet IBA criteria: *Aythya nyroca*, *Haliaeetus albicilla* (2–3 breeding pairs), *Crex crex* (10 breeding pairs).

Species	Season	Year	Pop min	Pop max	Acc	Criteria
Ciconia nigra Black Stork	B	—	5	10	B	B2
Ciconia ciconia White Stork	B	—	15	20	B	B2
Milvus migrans Black Kite	B	—	2	5	B	B2
Aquila pomarina Lesser Spotted Eagle	B	—	2	3	B	B2

Protection status

National None **International** None

Conservation issues

Threats Agricultural intensification/expansion (A), Burning of vegetation (C), Construction/impact of dyke/dam/barrage (B), Deforestation (commercial) (B), Industrialization/urbanization (B), Intensified forest management (B), Natural events (B)

There are a number of major threats to the site. A nuclear power plant has been planned for the site and a major highway is likely to run through or close to the area. It has been suggested as a location for an urban waste-disposal site. River canalization, in combination with drought, has already caused a reduction in the groundwater level and drying of the alluvial forests. Further canalization and drainage work is planned.

Pokupsko depression

A1, A4i, A4iii, B1i, B2 **013**

Admin region Zagrebacka zupanija, Karlovacka zupanija
Coordinates 45°37′N 15°42′E
Altitude 116–200 m **Area** 196,600 ha

Site description

The site includes a number of subsites: Crna Mlaka Special Ornithological Reserve; Pisarovina fish-ponds; Draganici fish-ponds; and Lonjsko and Mokro fields. Draganici and Crna Malka are both fish-pond complexes with emergent vegetation and extensive water-fringe vegetation, surrounded by alluvial *Quercus* forests. Lonjsko and Mokro fields include the flood-plains of the Rivers Sava, Lonja and Strug. The area includes extensive alluvial *Quercus* forests (52,000 ha), wet grasslands (12,000 ha), pastures (11,900 ha) and numerous small temporary and permanent wetlands. Other habitats include old settlements, fish-ponds (1,400 ha), arable land and perennial crops.

Habitats Forest and woodland (alluvial/very wet forest), Wetland (standing fresh water; water-fringe vegetation)
Land-use Agriculture, Fisheries/aquaculture, Forestry, Hunting

Birds

Species	Season	Year	Pop min	Pop max	Acc	Criteria
Ixobrychus minutus Little Bittern	B	—	10	20	C	B2
Egretta alba Great White Egret	U	1992	23	120	B	A4i, B1i
Anas platyrhynchos Mallard	W	—	12	10,000	B	B1i
Anas platyrhynchos Mallard	P	—	10,000	10,000	B	B1i
Aythya nyroca Ferruginous Duck	B	—	70	120	B	A1, B2
Aythya nyroca Ferruginous Duck	P	—	3,200	5,200	—	A1, A4i, B1i, B2
Aquila pomarina Lesser Spotted Eagle	B	—	1	2	—	B2

The site holds 20,000 or more waterbirds on passage, on a regular basis. Species of global conservation concern that do not meet IBA criteria: *Haliaeetus albicilla*, *Crex crex*.

Protection status

National Low **International** Partial
625 ha of IBA covered by Special Ornithological Reserve (Crna Mlaka, 625 ha). 625 ha of IBA covered by Ramsar Site (Crna Mlaka, 625 ha). IBA overlaps with Ramsar Site (Lonjsko Polje and Mokro Poljie, 50,560 ha).

Conservation issues

Threats Agricultural intensification/expansion (U), Aquaculture/fisheries (U), Dredging/canalization (U), Infrastructure (U), Intensified forest management (U)

Special Ornithological Reserve on Krk island
B2 **014**

Admin region Primorsko-goranska zupanija
Coordinates 45°01′N 14°45′E
Altitude — **Area** 1,000 ha

Site description
Canyons and sea cliffs on the north-eastern side of Krk island.

Habitats Rocky areas (sea cliff/rocky shore; inland cliff)
Land-use Tourism/recreation

Birds

Species	Season	Year	Pop min	Pop max	Acc	Criteria
Gyps fulvus Griffon Vulture	R	—	10	15	—	B2

Species of global conservation concern that do not meet IBA criteria: *Aythya nyroca, Falco naumanni, Crex crex.*

Protection status
National High **International** None
1,000 ha of IBA covered by Ornithological Reserve (Krk, 1,000 ha).

Conservation issues
Threats Recreation/tourism (U)

Prvic island
B2 **015**

Admin region Primorsko-goranska zupanija
Coordinates 44°55′N 14°48′E
Altitude — **Area** 7,000 ha

Site description
The island has high sea cliffs in the north-east and south-west. The upland plateau is dominated by garrigue with some sheep-grazed grassland.

Habitats Scrub (sclerophyllous scrub/garrigue/maquis), Grassland (steppe/dry calcareous grassland), Rocky areas (sea cliff/rocky shore)
Land-use Tourism/recreation

Birds

Species	Season	Year	Pop min	Pop max	Acc	Criteria
Gyps fulvus Griffon Vulture	R	—	10	15	—	B2

This is an important site for breeding raptors.

Protection status
National Partial **International** None
IBA overlaps with Botanical and Zoological Reserve (Prvic Island, 7,000 ha).

Conservation issues
Threats Recreation/tourism (U)

Special Ornithological Reserves on Cres island
B2 **016**

Admin region Primorsko-goranska zupanija
Coordinates 44°57′N 14°24′E
Altitude — **Area** 1,450 ha

Site description
Quercus, Carpinus, Ostrya and *Castanea* forest and high sea cliffs.

Habitats Forest and woodland (broadleaved deciduous forest), Rocky areas (sea cliff/rocky shore)
Land-use Tourism/recreation

Birds
This is an important site for breeding raptors.

Species	Season	Year	Pop min	Pop max	Acc	Criteria
Gyps fulvus Griffon Vulture	B	—	25	25	—	B2
Circaetus gallicus Short-toed Eagle	B	—	2	4	—	B2

Protection status
National High **International** None
1,450 ha of IBA covered by Ornithological Reserves (Cres, 1,450 ha).

Conservation issues
Threats Recreation/tourism (U)

Paklenica National Park
B2 **017**

Admin region Zadarsko-kninska zupanija
Coordinates 44°19′N 15°45′E
Altitude — **Area** 3,617 ha

Site description
Two large river canyons with high cliffs and forests dominated by *Fagus, Quercus, Carpinus* and *Pinus.*

Habitats Forest and woodland (broadleaved deciduous forest; native coniferous forest; mixed forest), Rocky areas (inland cliff)
Land-use Tourism/recreation

Birds

Species	Season	Year	Pop min	Pop max	Acc	Criteria
Gyps fulvus Griffon Vulture	R	—	20	20	—	B2
Falco peregrinus Peregrine	R	—	2	2	—	B2

This is an important site for breeding raptors.

Protection status
National High **International** High
3,617 ha of IBA covered by National Park (Paklenica, 3,617 ha). 3,617 ha of IBA covered by Specially Protected Area (Paklenica, 3,617 ha).

Conservation issues
Threats Recreation/tourism (U)

Kornati National Park and Telascica Nature Park
B2 **018**

Admin region Zadarsko-kninska zupanija, Sibenska zupanija
Coordinates 43°45′N 15°18′E
Altitude — **Area** 30,200 ha

Site description
An archipelago of several hundred islands in the central part of the Adriatic Sea. The islands are not forested but there are patches of scrub.

Habitats Scrub (sclerophyllous scrub/garrigue/maquis), Rocky areas (sea cliff/rocky shore)
Land-use Tourism/recreation

Birds

Species	Season	Year	Pop min	Pop max	Acc	Criteria
Circaetus gallicus Short-toed Eagle	B	—	10	10	—	B2
Falco eleonorae Eleonora's Falcon	B	—	10	10	—	B2
Falco peregrinus Peregrine	R	—	1	3	—	B2

Species of global conservation concern that do not meet IBA criteria: *Falco naumanni.*

Protection status
National Partial **International** Partial

IBA overlaps with Nature Park (Telascica). IBA overlaps with National Park (Kornati, 30,200 ha). IBA overlaps with Specially Protected Area (Kornati, 30,200 ha).

Conservation issues

Threats Recreation/tourism (U)

Lake Vransko — A4i, A4iii, B1i — 019

Admin region Zadarsko-kninska zupanija
Coordinates 43°54′N 15°35′E
Altitude 0–109 m **Area** 3,000 ha

Site description

A natural lake connected to the sea by a canal, fringed with reedbeds (*Phragmites*) and agricultural land along the western shore. The site is used for recreation, fishing, aquaculture, reed-harvesting, hunting and tourism.

Habitats Wetland (standing fresh water; water-fringe vegetation), Artificial landscape (arable land)
Land-use Agriculture, Fisheries/aquaculture, Hunting, Other, Tourism/recreation

Birds

Species	Season	Year	Pop min	Pop max	Acc	Criteria
[1] ***Fulica atra*** Coot	W	1994	30,000	60,000	A	A4i, B1i

1. 1993–1994.

The site holds 20,000 or more waterbirds in winter on a regular basis. Species of global conservation concern that do not meet IBA criteria: *Aythya nyroca, Crex crex.*

Protection status

National Low **International** None
30 ha of IBA covered by Special Reserve (Vransko, 30 ha).

Conservation issues

Threats Agricultural intensification/expansion (U), Aquaculture/fisheries (U), Other (U), Recreation/tourism (U), Unsustainable exploitation (U)

Krka National Park — B2 — 020

Admin region Zadarsko-kninska zupanija, Sibenska zupanija
Coordinates 43°48′N 15°55′E
Altitude — **Area** 14,200 ha

Site description

Sea coast, canyons, lakes and *Quercus, Carpinus* and *Ostrya* forests.

Habitats Forest and woodland (broadleaved deciduous forest), Wetland (standing freshwater; river/stream), Marine areas (sea inlet/coastal features), Rocky areas (inland cliff)
Land-use Tourism/recreation

Birds

Species	Season	Year	Pop min	Pop max	Acc	Criteria
Circaetus gallicus Short-toed Eagle	B	—	4	5	—	B2
Falco peregrinus Peregrine	R	—	2	3	—	B2

Species of global conservation concern that do not meet IBA criteria: *Aythya nyroca, Aquila heliaca, Falco naumanni, Crex crex.*

Protection status

National High **International** High
14,200 ha of IBA covered by National Park (Krka, 14,200 ha). 14,200 ha of IBA covered by Specially Protected Area (Krka, 14,200 ha).

Conservation issues

Threats Recreation/tourism

Tourism is the main threat.

Sinjsko field — B2 — 021

Admin region Splitsko-dalmatinska zupanija
Coordinates 43°55′N 16°25′E
Altitude 295–301 m **Area** 640 ha

Site description

An area of meadows at the head of the Cetina river.

Habitats Grassland (humid grassland), Wetland (river/stream)

Birds

The B2 criterion has been applied at the site level. The site was in the previous pan-European inventory (Grimmett and Jones 1989) as Pasko field (YU023). However, bird data are lacking. *Caprimulgus europaeus* and *Lullula arborea* are common with 'many' breeding pairs reported.

Protection status

National None **International** None

Conservation issues

Threats Agricultural intensification/expansion (U)

Dinara mountain — B2 — 022

Admin region Zadarsko-kninska zupanija, Splitsko-dalmatinska zupanija
Coordinates 43°58′N 16°32′E
Altitude 0–1,831 m **Area** 20,000 ha

Site description

An area of montane meadows, cliffs, deciduous and conifer woodland. Land-uses include sheep- and cattle-grazing.

Habitats Forest and woodland (broadleaved deciduous forest; native coniferous forest), Grassland (alpine/subalpine/boreal grassland), Rocky areas (inland cliff)
Land-use Agriculture

Birds

Species	Season	Year	Pop min	Pop max	Acc	Criteria
Circaetus gallicus Short-toed Eagle	B	—	5	5	—	B2
Aquila chrysaetos Golden Eagle	R	—	1	2	—	B2
Falco peregrinus Peregrine	R	—	2	3	—	B2

This is an important site for breeding raptors.

Protection status

National None **International** None

Delta of the Neretva river — A3, B1i — 023

Admin region Dubrovacko-neretvanska zupanija
Coordinates 42°57′N 17°33′E
Altitude — **Area** 11,500 ha

Site description

Delta of the Neretva river, the largest wetland on the Croatian coast. The site includes extensive areas of brackish and freshwater marshes, reedbeds (*Phragmites*) and other emergent vegetation, as well as coastal sand beaches.

Habitats Wetland (tidal river/enclosed tidal water; saltmarsh; sand-dunes/sand beach; standing fresh water; standing brackish and salt water; water-fringe vegetation)

Birds

Species	Season	Year	Pop min	Pop max	Acc	Criteria
Alectoris graeca Rock Partridge	R	1994	—	—	—	A3
Oenanthe hispanica Black-eared Wheatear	B	1994	—	—	—	A3

Species ... continued	Season	Year	Pop min	Pop max	Acc	Criteria
Hippolais olivetorum Olive-tree Warbler	B	1994	—	—	—	A3
Sylvia conspicillata Spectacled Warbler	B	1994	—	—	—	A3
Sylvia cantillans Subalpine Warbler	B	1994	—	—	—	A3
Sylvia melanocephala Sardinian Warbler	B	1994	—	—	—	A3
Sitta neumayer Rock Nuthatch	R	1994	—	—	—	A3
Emberiza melanocephala Black-headed Bunting	B	1994	—	—	—	A3

The B1i criterion has been applied at the site level as data are missing but the site is known to be important for spring and autumn passage waterbirds. Breeding birds include eight of the 21 species that are restricted in Europe to the Mediterranean biome (when breeding). Species of global conservation concern that do not meet IBA criteria: *Aythya nyroca*, *Haliaeetus albicilla*, *Aquila heliaca*, *Falco naumanni*, *Crex crex*.

Protection status

National Partial **International** High

1,250 ha of IBA covered by Special Reserve (Pod Gredon / Prud / Orepak / south-east delta, 1,250 ha). 11,500 ha of IBA covered by Ramsar Site (Neretva Delta, 11,500 ha). 11,500 ha of IBA covered by Specially Protected Area (Neretva Delta, 11,500 ha).

Conservation issues

Threats Agricultural intensification/expansion (U), Industrialization/urbanization (U), Infrastructure (U), Unsustainable exploitation (U)

Drainage, agricultural development, urbanization, road building and hunting are all threats to the site.

REFERENCES

Bartovsky, V., Kletecki, E., Radovic, D., Stipcevic, M. and Susic, G. (1987) Breeding waders in Yugoslavia. *Wader Study Group Bull.* 51: 33–37.

Collar, N. J., Crosby, M. J. and Stattersfield, A. J. (1994) *Birds to watch 2: the world list of threatened birds*. Cambridge, UK: BirdLife International (Conservation Series 4).

Delic, A. (1989) [Waterbirds of the Koncanica fishponds.] *Larus* 40: 25–33. (In Croatian.)

Ern, H. (1990) Autumn and winter observations of Great White Herons (*Egretta alba*) on fishponds in Croatia. *Troglodytes* 3: 11–12.

Getz, D. (1994) Cesta kroz Kopacki rit. [Road through Koapcki rit.] *Novosti IBA projekta* 3/4: 1–2.

Grimmett, R. F. A. and Jones T. A. (1989) *Important Bird Areas in Europe*. Cambridge, UK: International Council for Bird Preservation (Tech. Publ. 9).

Jurcevic, I. (1998) [Winter fauna and ecology of waterfowl, Anseriformes, in Croatia.] Zagreb: University of Zagreb (M.Sc. thesis). (In Croatian.)

Lukac, G. (1994) Umjetni mocvarni biotopi—znacajna stanista mocvarnih ptica. [Artificial wetlands—important habitats for waterbirds.] *Novosti IBA projekta* 3/4: 3–5.

Lukaè, G. and Stipèeviæ M. (1997) Birds of National Park Paklenica, Croatia. *Natura Croatica* 6: 60.

Majer, J. (1995) Data to the fish fauna (Pisces) of Drava river (Hungary) and to the quality of backwaters. Pp. 189–202 in A. Uherkovics, ed. *The fauna of the Drava region, Vol. 1., Studia Pannonica, (A) Series Historico-naturalis*. Pecs, Croatia.

Mikuska, J. (1979) Ekoloske osobine i zastita Specijalnog zooloskog rezervata "Kopacki rit" s posebnim osvrtom na ekologiju kraljesnjaka. [Ecological characteristics and protection of the Special zoological reserve "Kopacki rit".] Zagreb: University of Zagreb (Ph.D. thesis).

Mikuska, J. and Mikuska, T. (1994) Ptice Dunava na podrucju Hrvatske. [Birds of the Danube valley in Croatia.] *Anali Zavoda za znanstveni rad u Osijeku* 10: 109–175.

Mikuska, T. (1988/89) Prijeti li ekoloska katastrofa "Kopackom ritu". [Does ecological catastrophe threaten "Kopacki rit"?] *Priroda* 77(9): 6–7.

Mikuska, T. (1990) Veliki vranac (Kormoran). [Great Cormorant.] *Priroda* 79 (9/10): 24–25.

Mikuska, T. (1990) Ornithological report from delta of river Neretva on June 29th 1990. *Troglodytes* 3: 22–23.

Mikuska, T. (1991) [Red-throated Diver near Velika Gorica.] *Troglodytes* 4: 37. (In Croatian.)

Mikuska, T. (1991) Goosander on the Kopacko Lake. *Troglodytes* 4: 38.

Mikuska, T. (1991) Little Egret *Egretta garzetta* flying in mixed flocks. *Troglodytes* 4: 36.

Mikuska, T. (1993) Status of the Black Stork in Croatia. P. 61 in M. Strazds, ed. *1st International Black Stork Conservation and Ecology Symposium, Abstract Book*. Latvia: Jurmala.

Mikuska, T. and Martinovic D. (1993) Medjunarodna Drava Konferencija. [International "Drava" Conference.] *Novosti IBA projekta* 2/3: 1–2.

Mikuska, T. and Mikuska, J. (1992) Baranja i Kopacki rit u ratu. [Baranja and Kopacki rit during war.] Pp. 47–57 in P. Durbesic and M. Kerovec, eds. *Posljedice rata na okolis u Hrvatskoj*. [*Consequences of the war on the Croatian environment*.] Zagreb: Hrvatsko Ekolosko Drustvo (Mala Ekoloska biblioteka Knjiga 6).

Mikuska, T. and Mikuska, J. (1993) Vijesti s IBA podrucja. Kopacki rit—IBA 014. [News from IBAs. Kopacki rit.] *Novosti IBA projekta* 1: 4–6.

Mikuska, T., Setina, M. and Mikuska, J. (1993) Vijesti s IBA podrucja. Jelas polje—IBA 014. [News from IBAs. Jelas field.] *Novosti IBA projekta* 1: 4.

Mikuska, T., Nola, I. A., Kajfez, R. and Kovacic, D. (1991) Otter distribution in view of habitat quality in Turopolje. Pp. 333–334 in D. Serman, ed. *Proceedings of the Fourth Congress of Croatian Biologists. Abstract Book*. Zagreb: Hrvatsko Biolosko Drustvo.

Mrakovcic, M. and Turk. M. (1988) [Application of ornithological research in the protection and management of fisheries in Croatia.] Pp. 175–180 in M. Mestrov and G. Susic, eds. *Ornithology in Croatia*. Zagreb: JAZU.

Rucner, D. (1962) Prilog poznavanju ornitofaune Kopackog rita i okolice u Baranji. [Contribution to the knowledge of ornithofauna of Kopacki rit and surrounding area in the Baranja region.] *Larus* 14: 84–121.

Schneider-Jacoby, M. (1993) Vogel als Indikatoren fur das okologische Potential der Saveaue und Moglischkeiten fur deren erhaltung. Konstanz, Germany: University of Konstanz (Ph.D. thesis).

Schneider-Jacoby, M. (1994) Sava and Drava—Ecological value and future of the two main rivers in Croatia. *Periodicum Biologorum* 4: 348–356.

Schneider-Jacoby, M. (1995) *Drau und Mur—Leben durch Flussdynamik*. Uberlingen, Germany: Naturerbe Verlag Jurgen Resch.

Schneider-Jacoby, M. (1997) Die Drau—ein Stuck europaische Urlandschaft. *Natur und Mensch* 2: 10–15.

Stipcevic, M. (1996) A contribution to the Croatian list of rare and scarce birds recorded from 1985–1995. *Natura Croatica* 5(1): 53–81.

Topic, J. (1989) Vegetation of the special zoological reserve of Kopacki rit. *Hydrobiologia* 182: 149–160.

Toth, S. (1995) Preliminary report on the dragonfly (Odonata) fauna of the Drava region, Hungary. Pp. 41–52 in A. Uherkovics, ed. *The fauna of the Drava region, Vol. 1., Studia Pannonica, (A) Series Historico-naturalis*. Pecs, Croatia.

Tucker, G. M. and Heath, M. F. (1994) *Birds in Europe: their conservation status*. Cambridge, UK: BirdLife International (BirdLife Conservation Series no. 3).

■ CYPRUS

MELIS CHARALAMBIDES

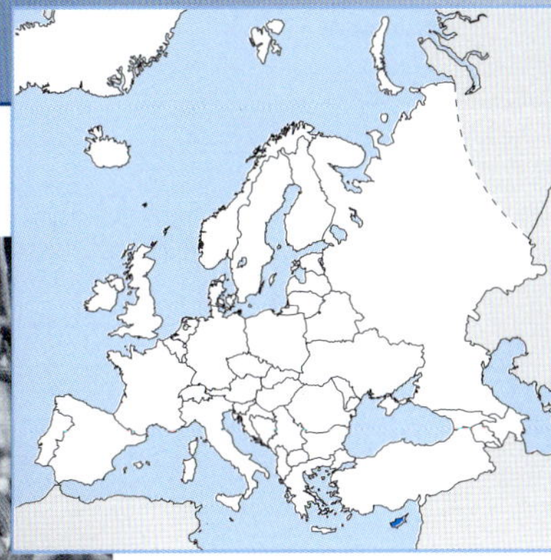

Cyprus Warbler *Sylvia melanothorax*, a species endemic to the island when breeding. (PHOTO: LOUIS KOURTELLARIDES)

GENERAL INTRODUCTION

Cyprus is the third largest island of the Mediterranean, with a total surface area of 9,250 km^2. The island is positioned 70 km south of Turkey, 105 km west of Syria and about 360 km north of Egypt, and has a population of 700,000 (in 1996); an average population density of 75 people per km^2.

The island has two mountain ranges separated by a central plain. The southern Troodos mountain range with its hilly extensions occupies a large part of the west and south of the island. The highest peak is Mount Olympus with an altitude of 1,952 m. The northern Pentadactylos mountain range runs along the north coast from Cape Kormakitis in the west to Cape Andres in the east. The highest peak is Kyparissovouno (altitude 1,024 m) at its western end. The narrow coastal strip along the north coast is fairly intensively cultivated.

Between the two mountain ranges lies the central plain of Mesaoria. This is the 'bread-basket' of Cyprus where the rain-fed crops of barley and wheat are grown. The fertile red-soil areas to the south and south-east corner of the plain are the main potato and vegetable growing areas of Cyprus.

There are no perennial rivers on Cyprus, only a few springs and streams in Troodos. The largest seasonal river is the Pedhieos, with a total length of about 95 km, it originates from the Troodos mountain range, passes through the Mesaoria plain and flows into Famagusta Bay, but only in the winter months.

Cyprus has 16 Important Bird Areas (IBAs) covering a total area of 1,305 km^2, which is 14% of the land area of the island (Map 1, Table 1). Seventeen IBAs were identified in 1989 (Grimmett and Jones 1989), of which two no longer meet the IBA criteria, and in 1996 a new IBA was identified (Table 1, Map 1).

In 1974 the Turkish army invaded Cyprus and occupied almost 40% of the island. Since then a buffer zone has been created called the Green Line and Greek Cypriots live to the south of the line and Turkish Cypriots to the north; access is not allowed to the northern Turkish occupied areas. Three IBAs lie in the northern occupied part: the Klidhes Islands (001), Karpas peninsula (002) and Kyrenia range (003). There are no recent records or surveys available for these IBAs.

ORNITHOLOGICAL IMPORTANCE

Every year millions of migrants stop or pass over Cyprus. Nearly 200 species occur as regular passage migrants and 90 species are regular winter visitors. Of the 114 species that have bred in Cyprus, 80 species breed regularly and 50 are resident.

Map 1. Location, area and criteria category of Important Bird Areas in Cyprus.

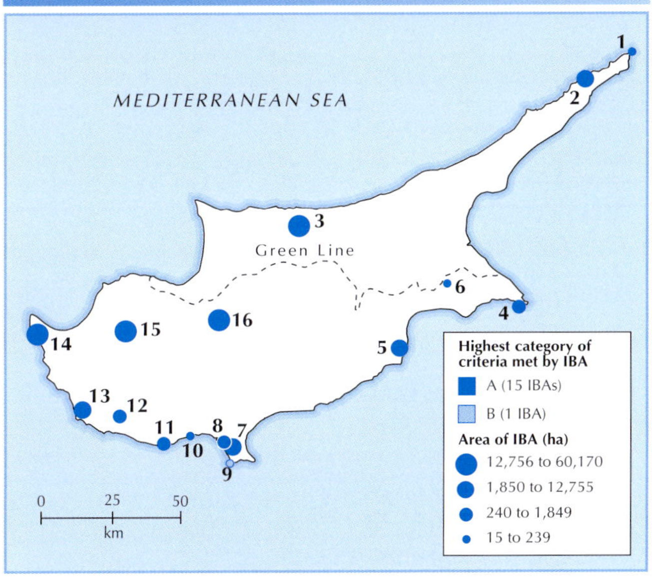

Table 1. Summary of Important Bird Areas in Cyprus. 16 IBAs covering 1,305 km²

IBA code	1989 code	International/national name	Administrative region	Area (ha)	Criteria (see p. 11)
001	CY001	Klidhes Islands	Famagusta	15	A1, B1i
002	CY002	Karpas peninsula	Famagusta	8,000	A2, B2
003	CY003	Kyrenia range between Kornos and Pentadactylos peaks	Kyrenia	16,000	A2
004	CY004	Cape Greco	Famagusta	385	A2
005	CY005	Larnaca salt-lakes	Larnaca	1,850	A1, A2, A4i, B1i, B2
006	—	Akhna Dam	Famagusta	125	A2
007	CY007	Akrotiri salt-lake including Bishop's Pool	Limassol	4,000	A1, A2, A4i, B1i, B1iv, B2
008	CY008	Phasouri reedbeds	Limassol	323	A2, B2
009	CY009	Akrotiri Cliffs	Limassol	200	B2
010	CY010	Episkopi Cliffs	Limassol	175	A2, B2
011	CY011	Cape Aspro	Limassol	240	A4ii, B1i, B1iii, B2
012	CY012	Asprokremmos Dam	Paphos	260	A2, B2
013	CY013	Paphos plain	Paphos	10,500	A2, B2
014	CY014	Akamas peninsula	Paphos	15,500	A2, A3, A4i, B1i, B1iv
015	CY015	Paphos State Forest and Stavros tis Psokas Forest Station	Paphos, Nicosia	60,170	A1, A2, A3, B2
016	CY016	Adelphi Forest	Nicosia	12,756	A1, A2, A3, B2

Sites identified in the previous inventory of IBAs in Europe (Grimmett and Jones 1989) but no longer considered to be IBAs
CY006 Dhypotamos Dam; CY017 Yeri near Nicosia

There are 10 species of European conservation concern (SPECs) which regularly breed in Cyprus (Tucker and Heath 1994), two of which, *Aquila heliaca* and *Larus audouinii*, are of global conservation concern and breed in sufficient numbers to satisfy the A1 criterion at three sites (Klidhes Islands 001, Paphos State Forest 015 and Adelphi Forest 016) (Table 2). Additionally the globally threatened *Oxyura leucocephala* fulfils the A1 criterion in winter at Larnaca salt-lakes (005) and Akrotiri salt-lake (007) (Table 2) and *Aythya nyroca* occurs at both sites. Two species are endemic to the island, *Oenanthe cypriaca* and *Sylvia melanothorax*, and although common and widespread throughout the island, have very small global ranges and are the reasons for Cyprus being an Endemic Bird Area (Stattersfield *et al.* 1998). Three IBAs have also been identified on the basis of assemblages of Mediterranean-biome species (criterion A3) such as *Sylvia conspicillata*, *Sylvia melanothorax*, *Oenanthe cypriaca*, *Lanius nubicus*, *Emberiza caesia* and *Emberiza melanocephala*.

Four sites are IBAs for congregatory species (Table 3), including Cape Aspro (011) which is particularly important for breeding *Falco eleonorae*, and Akrotiri salt-lake (007) and the Akamas peninsula (014) over which fly thousands of *Grus grus*, and also *Anthropoides virgo*, on passage.

Eleven IBAs in Cyprus qualify on the basis of significant breeding populations of species with an unfavorable conservation status in Europe (criterion B2) (Table 1). The proportions of the national populations of the eight SPECs meeting criterion B2 that breed in IBAs are shown in Table 4. Over 50% of the national population of *Phalacrocorax aristotelis*, *Aquila heliaca*, *Hieraaetus fasciatus*, *Falco eleonorae* and *Larus audouinii* breed in IBAs. Most of the European breeding population of *Francolinus francolinus* is restricted to Cyprus and about 50% of the national population breeds within IBAs (Table 4).

Table 2. Important Bird Areas in Cyprus that are important for species of global conservation concern (meeting criterion A1).

Species	IBA code
Oxyura leucocephala White-headed Duck	005, 007
Aquila heliaca Imperial Eagle	015, 016
Larus audouinii Audouin's Gull	001

Table 3. Important Bird Areas in Cyprus that support important numbers of one or more congregatory species (i.e. meeting criteria A4 and/or B1). IBAs meeting both criteria A4 and B1 for the species are shown in **bold**. IBAs meeting only criterion B1 for the species concerned, and not A4, are shown in normal type. For key to 'Season', see p. 7.

Species	Season	IBA code
Phalacrocorax aristotelis Shag	R	001, 011
Phoenicopterus ruber Greater Flamingo	W	**005**, **007**
Tadorna tadorna Shelduck	W	007
Falco eleonorae Eleonora's Falcon	B	**011**
Grus grus Crane	P	**007**, **014**

Table 4. Species of European conservation concern with significant breeding populations at Important Bird Areas in Cyprus (meeting any IBA criteria).

Species [1]	Minimum national breeding population (pairs) [2]	Proportion (%) of national population breeding at all IBAs in Cyprus
Phalacrocorax aristotelis Shag	80	60
Aquila heliaca Imperial Eagle	2	100[3]
Hieraaetus fasciatus Bonelli's Eagle	10	50
Falco eleonorae Eleonora's Falcon	100	100[4]
Francolinus francolinus Black Francolin	1,000	50[4]
Larus audouinii Audouin's Gull	10	100[3]
Oenanthe cypriaca Cyprus Pied Wheatear	3,000	30[4]
Sylvia melanothorax Cyprus Warbler	4,000	20[4]

1. Only those species of European conservation concern (see Box 1, p. 12) that meet IBA criteria in Cyprus are listed.
2. Data are taken from the BirdLife/EBCC European Bird Database 1998 (Heath and Borggreve 2000).
3. The percentage of the national population in IBAs exceeds 100%. Usually this is because the national population estimate has not been updated recently whilst the IBA population estimate has been recently updated with new data as a result of comprehensive surveys of IBAs themselves. Also, the individual site count for a species may be the maximum or average over recent years, and summing these may record more birds than are present nationally in any single year.
4. Data are incomplete, therefore approximate percentages have been added based on knowledge of the populations in Cyprus

HABITATS

Five IBAs are predominantly wetland sites (Figure 1) with a further four holding some wetland habitats. These sites are important for passage and wintering waterbirds. Two of these sites are salt-lakes, Akrotiri (007) and Larnaca (005). Both salt-lakes dry up in the summer and their water-levels depend greatly on the amount of rainfall. Akrotiri salt-lake is less saline than Larnaca salt-lake and for this reason it also differs ecologically. Both sites are particularly important for wintering *Phoenicopterus ruber*. Significant water development projects, primarily in the form of dams constructed in the last few decades, have resulted in the creation of some important bird habitats including the dams of Akhna (006), Asprokremmos (012), Evretou Kourris, Yermasoyia and Kiti.

Four IBAs, including Klidhes Islands (001) at Cape Andreas in the north-east of the island, are either islands or rocky coastal cliffs (Figure 1), important for breeding populations of *Falco eleonorae*, *Gyps fulvus*, *Falco peregrinus*, *Larus audouinii* and *Phalacrocorax aristotelis*,

Figure 1. Habitats at Important Bird Areas in Cyprus (see Appendix 3 for definitions of habitats).

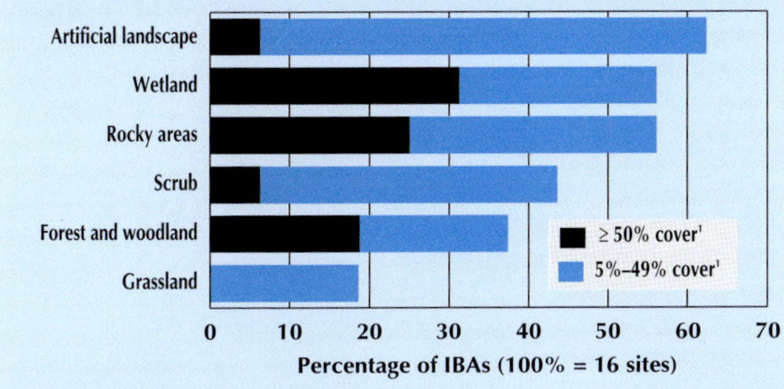

1. percentage of area of individual IBA covered by habitat

Figure 2. Land-uses at Important Bird Areas in Cyprus (see Appendix 3 for definitions of land-uses).

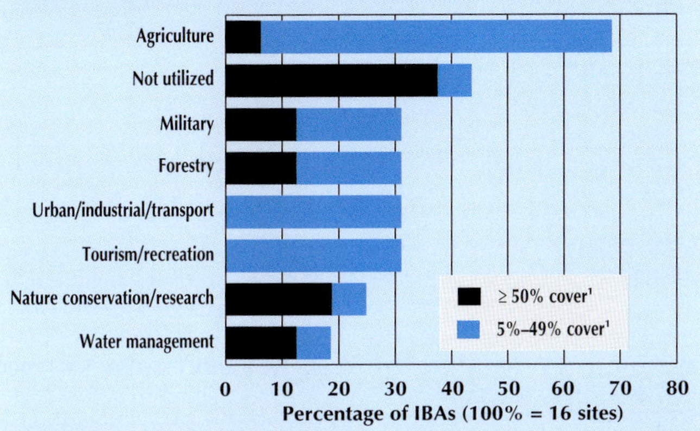

1. percentage of area of individual IBA covered by land-use

Many of the IBAs include areas of agricultural land, scrub (maquis and garrigue) and forests, but this generally constitutes less than 50% of the total area of an IBA. The Karpas peninsula (002) and the Akamas peninsula (014) are mainly covered with *Juniperus phoenicea*. Many of these sites hold breeding birds of prey (*Aquila heliaca*, *Hieraaetus fasciatus*, *Accipiter gentilis*), as well as the endemics *Oenanthe cypriaca* and *Sylvia melanothorax*, *Francolinus francolinus* and many passage migrants.

Clearance over many centuries and the unrestricted grazing by goats have destroyed much of the islands' original forest, which is now confined mainly to the Troodos and Pentadactylos mountains. Five IBAs are predominantly forest sites: 003, 015 and 016, with 002 and 014 having just under 50% forest cover. These IBAs are large and thus the overall forest cover is nearly 75% of the total area of all IBAs in Cyprus. There are two pine zones: the *Pinus brutia* zone occupies the altitudes from sea-level to 1,200 m and covers a total area of 100,000 ha or 11% of the island's total area, and on the higher Troodos, above 1,200 m to 1,950 m, the *Pinus nigra* zone forms pure forest and covers an area of 6,000 ha. Sixty-nine percent of the area of the IBAs (89,500 ha) is state-owned forest.

IMPACTS ON IBAs – LAND-USE AND THREATS

Almost 70% of the IBAs have agriculture as a significant land-use and this includes arable land, mostly cultivated with barley, perennial crops, orchards, groves, and annuals such as vegetables and potatoes (Figure 2).

The killing of migratory birds is a serious problem in Cyprus. This threatens birds across Cyprus, including those within IBAs; unsustainable exploitation is listed as a threat at nearly 90% of IBAs (Figure 3). Although the island's hunting (shooting) regulations are fairly comprehensive, the enforcement of the laws is inadequate and thousands of birds of many species are illegally shot each year. Other problems include liming and netting. Liming (using sticks covered in glue to trap birds) is a traditional method of catching birds for food, but is now strictly illegal. However, it still takes place from August to May on a massive scale, with little or no effort being made to stop it. The illegal use of nets to catch birds is also a problem.

Disturbance to birds is also listed as a threat at nearly 80% of IBAs (Figure 3). The creation of new roads, especially forest roads, and the popularity of four-wheel drive cars has made every corner of the IBAs more accessible to humans, thus increasing disturbance to birds, especially raptors. Also the dramatic increase in use of leisure boats makes it easier for people to reach even the most inaccessible sea cliffs on the island.

PROTECTION STATUS

Table 5 and Figure 4 summarize the national and international protection status and coverage at the 16 IBAs in Cyprus. Three-quarters of IBAs are protected by national legislation to some extent (Figure 4) but none are protected internationally.

■ National protection

The two major categories of protected areas protecting birds from hunting are (1) Permanent Game Reserves and (2) Temporary Game Reserves. Forty-nine per cent (63,680 ha) of the total IBA area is covered by Permanent Game Reserves: 678 ha of Larnaca salt-lakes (005) covering the main lake, 2,000 ha of Akrotiri salt-lake (007) covering the main lake, 200 ha of Akrotiri Cliffs (009), 175 ha of Episkopi Cliffs (010), 10,500 ha of Paphos plain (013) and 50,000 ha of Paphos State Forest (015) (Table 5). Additionally, part (1,172 ha) of Larnaca salt-lakes (005), 125 ha of Akhna Dam (006) and 260 ha of Asprokremmos Dam (012) are designated Temporary Game Reserves.

Temporary Game Reserves cover areas (such as dams) designated from time to time by the government, usually to protect possible breeding areas, heavily hunted areas (mainly to protect *Alectoris chukar* and *Francolinus francolinus*, the two main game

Figure 3. Threats at Important Bird Areas in Cyprus (see Appendix 3 for definitions of threat types and impact categories).

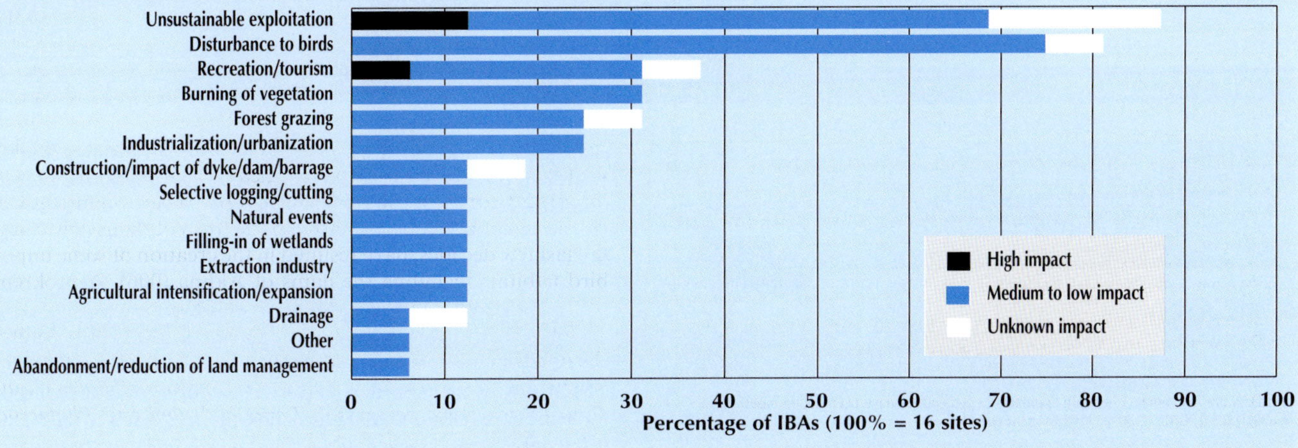

Figure 4. The national protection status of Important Bird Areas in Cyprus.

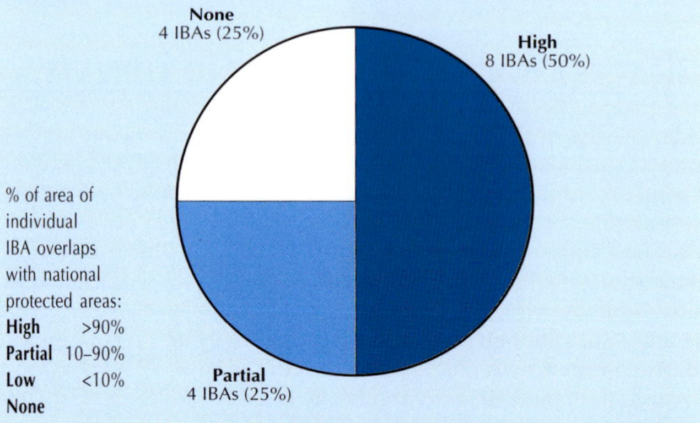

Total area of overlap between IBA network in Cyprus and national protected-area system (see Table 5 for categories) = 648–655 km² (50% of total IBA area).

birds) and also extensive areas of orchards and other cultivation that need protection from hunters. Anyone caught hunting in these areas is subject to fairly strong penalties; liming and netting, although illegal, still go on in some of these areas. In this way, the government aims to maintain some sort of control over hunting. Apart from these restrictions, the licenses issued to cover each open hunting season clearly stipulate which species may be shot. However, the vast majority of the 55,000 licensed hunters (1998) shoot birds indiscriminately, posing a major conservation problem.

There are proposals to establish a National Park at the Akamas peninsula (014) to control the area from further tourist development, but hunting is still allowed and thousands of birds are indiscriminately shot every year. Cape Greco (004) is the only National Forest Park. Klidhes Islands (001) are the only Nature Reserve, and 8,000 ha of Karpas peninsula (002), is a proposed site for a Nature Reserve.

Akrotiri salt-lake (007), Phasouri reedbeds (008), Akrotiri Cliffs (009) and Episkopi Cliffs (010) are within the UK Sovereign Base Area (SBA), and the responsibility of policing the above areas is with the SBA police. However, the SBA Authorities normally agree with any changes proposed by the Cyprus government regarding hunting in the above areas, unless defence considerations dictate otherwise.

A total of 57,000 ha, or 44% of the total area, of the IBAs are totally unprotected, meaning that hunting is allowed during the designated hunting seasons.

International protection

The international legislation which Cyprus is party to is shown in Box 1. Although party to the World Heritage Convention it has not designated any natural World Heritage Sites. Also Cyprus has ratified the Barcelona Convention; only one site has been designated a Mediterranean Specially Protected Area, its importance being for turtles. No IBAs are afforded international protection.

Table 5. Protection status of Important Bird Areas in Cyprus. A tick (✔) indicates that an IBA overlaps with a protected area (to any extent).

		National			
IBA code	International name	National Forest Park	Temporary Game Reserve	Permanent Game Reserve	Nature Reserve
001	Klidhes Islands				✔
002	Karpas peninsula				
003	Kyrenia range between Kornos and Pentadactylos peaks		✔		
004	Cape Greco	✔			
005	Larnaca salt-lakes		✔	✔	
006	Akhna Dam		✔		
007	Akrotiri salt-lake including Bishop's Pool			✔	
008	Phasouri reedbeds		✔		
009	Akrotiri Cliffs			✔	
010	Episkopi Cliffs			✔	
011	Cape Aspro				
012	Asprokremmos Dam		✔		
013	Paphos plain			✔	
014	Akamas peninsula				
015	Paphos State Forest and Stavros tis Psokas Forest Station			✔	
016	Adelphi Forest				
Total number of IBAs		1	5	6	1

Box 1. International legislation and initiatives that are relevant to site conservation in Cyprus (see Appendix 1 for a general description of these agreements).

Global	
Biodiversity Convention	✔
Ramsar Convention	
Bonn Convention	
World Heritage Convention	✔
MAB Programme	
Pan-European	
Bern Convention	✔
Regional	
Barcelona Convention	✔

✔ Convention ratified/initiative supported
(✔) Convention signed

ACKNOWLEDGEMENTS

The site descriptions were compiled from *Important Bird Areas in Europe* (Grimmett and Jones 1989), with the exception of Akhna Dam (006). Updated information was also provided by Louis Kourtellarides of the Forestry Department, Antonis Antoniou of the Environment Department, and the Cyprus Ornithological Society Reports (1988–1991) edited by Melis Charalambides and Pavlos Neophytou.

I would also like to thank Melanie Heath of BirdLife International for her guidance and encouragement, which was extremely helpful.

SITE ACCOUNTS

Klidhes Islands — A1, B1i — 001

Admin region Famagusta
Coordinates 35°42′N 34°36′E
Altitude 0–20 m Area 15 ha

Site description

Habitats Rocky areas (100%; sea cliff/rocky shore; rock stacks/islets)
Land-use Nature conservation/research (100%)

A chain of six small, rocky islands off the north-east point of Cyprus. Low cliffs surround the main island, the vegetation of which is dominated by grasses, with *Salicornia* and *Atriplex* forming short dense swards in places. With the exception of a lighthouse, the islands are uninhabited.

Birds

Species	Season	Year	Pop min	Pop max	Acc	Criteria
Phalacrocorax aristotelis Shag	R	—	25	50	B	B1i
Larus audouinii Audouin's Gull	B	1987	15	15	B	A1

The islands are an important stop-over site during migration, especially for waterbirds. One of the larger islands, the one closest to the mainland, is the only known site in Cyprus where *Larus audouinii* breeds.

Protection status

National High **International** None
15 ha of IBA covered by Nature Reserve (Klidhes Islands, 15 ha).

Conservation issues

Threats Disturbance to birds (C)

Human predation of Laridae eggs formerly occurred, but is no longer thought to be a significant problem.

Karpas peninsula A2, B2 002

Admin region Famagusta
Coordinates 35°38′N 34°25′E
Altitude 0–157 m **Area** 8,000 ha

Site description

A long, narrow peninsula north-east of the village of Rizokarpasso, with small offshore islands. The IBA includes hills covered mainly by low juniper *Juniperus phoenicea* forest, with maquis occurring towards the coast.

Habitats Forest and woodland (45%; native coniferous forest), Scrub (30%; sclerophyllous scrub/garrigue/maquis), Wetland (10%; sand-dunes/sand beach; shingle/stony beach; sea cliff/rocky shore), Artificial landscape (15%; arable land; perennial crops/orchards/groves)
Land-use Agriculture (30%), Military (15%), Not utilized (50%), Tourism/recreation (5%)

Birds

Species	Season	Year	Pop min	Pop max	Acc	Criteria
[1] ***Francolinus francolinus*** Black Francolin	R	—	Abundant		—	B2
[1] ***Oenanthe cypriaca*** Cyprus Pied Wheatear	B	—	Common		—	A2
[1] ***Sylvia melanothorax*** Cyprus Warbler	R	—	Common		—	A2

1. No recent surveys or records.

An important area for *Francolinus francolinus*, which is increasing in numbers following the introduction of a ban on shooting. It has spread throughout the peninsula and south to the Famagusta area, and has been recorded near Avgorou and the Ahna Dam. Many common species are resident or visiting breeders, many more overwinter or occur on passage. It is not uncommon to record 120 different species in a day in April, ranging from *Phoenicopterus ruber* and raptors to the smallest passerines. Cape Andreas is by far the best birdwatching site for spring and autumn migrants in Cyprus.

Protection status

National None **International** None

Conservation issues

Threats Forest grazing (C), Unsustainable exploitation (C)

Threats to the IBA have not been fully assessed. There is a proposal for the designation of a protected area covering 5,635 ha of the Karpas peninsula.

Kyrenia range between Kornos and Pentadactylos peaks A2 003

Admin region Kyrenia
Coordinates 35°17′N 33°18′E
Altitude 600–1,024 m **Area** 16,000 ha

Site description

Habitats Forest and woodland (60%; native coniferous forest; mixed forest), Scrub (30%; sclerophyllous scrub/garrigue/maquis), Rocky areas (15%; inland cliff)
Land-use Agriculture (5%), Forestry (10%), Not utilized (80%), Tourism/recreation (5%)

A long, narrow ridge of rocky mountains forming part of the Pentadactylos range, stretching from Kornos (946 m) to Pentadactylos peak (740 m) in the east. The mountains support a mixture of coniferous woodland, scrub, maquis and grassland. Human activities include forestry, and the grazing of goats and sheep in all suitable areas.

Birds

Species	Season	Year	Pop min	Pop max	Acc	Criteria
[1] ***Oenanthe cypriaca*** Cyprus Pied Wheatear	B	—	Common		—	A2
[1] ***Sylvia melanothorax*** Cyprus Warbler	R	—	Common		—	A2

1. No recent surveys or records.

There are no recent accurate data on bird populations due to inaccessibility.

Protection status

National Partial **International** None
Approximately 15% of the IBA is included within Game Reserves.

Conservation issues

Threats Extraction industry (C), Forest grazing (U), Natural events (B), Unsustainable exploitation (U)

Forest fires and increased access via forestry roads make the area more accessible and thus cause disturbance to birds. 9,600 ha of IBA covered by State Forest land (Kyrenia Range, 50,000 ha).

Cape Greco A2 004

Admin region Famagusta
Coordinates 34°57′N 34°04′E
Altitude 0–100 m **Area** 385 ha

Site description

A rocky escarpment ending in steep cliffs on the southern seaward side. The IBA comprises mainly scrub (with rocky outcrops) and some cultivated areas; the tip of the Cape is closed to the public due to the presence of a radio station.

Habitats Forest and woodland (10%; native coniferous forest), Scrub (50%; scrub), Rocky areas (15%; sea cliff/rocky shore; inland cliff), Artificial landscape (30%; perennial crops/orchards/groves)
Land-use Agriculture (30%), Not utilized (50%), Urban/industrial/transport (20%)

Birds

Species	Season	Year	Pop min	Pop max	Acc	Criteria
Oenanthe cypriaca Cyprus Pied Wheatear	B	—	—	10	B	A2
Sylvia melanothorax Cyprus Warbler	R	—	10	—	B	A2

An important site for migrants (particularly passerines, but also raptors), providing the first arrival point for northbound birds in spring.

Protection status

National High **International** None
385 ha of IBA covered by National Forest Park (Cape Greco, 385 ha).

Conservation issues

Threats Disturbance to birds (C), Recreation/tourism (U), Unsustainable exploitation (C)

Part of the IBA is a hunting-dog training area, which causes disturbance to birds. Also liming and netting is illegally practiced, especially in autumn.

Larnaca salt-lakes A1, A2, A4i, B1i, B2 005

Admin region Larnaca
Coordinates 34°52′N 33°37′E
Altitude 0–20 m **Area** 1,850 ha

Site description

The IBA covers four salt-lakes, which dry out in summer, surrounding agricultural land and *Salicornia* scrub. The north-eastern corner of the site borders the suburbs of Larnaca and several military camps, whilst Larnaca International Airport adjoins on the eastern side (one of the lakes is bisected by a runway).

Habitats Scrub (10%; scrub), Wetland (70%; mudflat/sandflat; shingle/stony beach; standing brackish and salt water), Artificial landscape (20%; arable land; other urban/industrial areas)
Land-use Agriculture (20%), Nature conservation/research (70%), Urban/industrial/transport (10%)

Birds

Species	Season	Year	Pop min	Pop max	Acc	Criteria
[1] ***Phoenicopterus ruber*** Greater Flamingo	W	1995	2,000	12,000	B	A4i, B1i
Oxyura leucocephala White-headed Duck	W	1995	—	14	B	A1
[1] ***Oenanthe cypriaca*** Cyprus Pied Wheatear	B	1995	10	40	B	A2, B2
[1] ***Sylvia melanothorax*** Cyprus Warbler	B	1995	—	10	B	A2

1. 1990–1995.

The site is very important for wintering waterbirds and as a resting-place for many species of waterbirds, waders and passerines during migration. Species of global conservation concern that do not meet IBA criteria: *Branta ruficollis* (wintering), *Circus macrourus* (passage).

Protection status

National High **International** None
678 ha of IBA covered by Permanent Game Reserve (Larnaca Lake, 678 ha). 1,172 ha of IBA covered by Temporary Game Reserve.

Conservation issues

Threats Agricultural intensification/expansion (C), Construction/impact of dyke/dam/barrage (B), Disturbance to birds (B), Filling-in of wetlands (C), Industrialization/urbanization (B), Infrastructure (A), Recreation/tourism (B), Unsustainable exploitation (A)

The area is coming under threat from development for tourism and the encroachment of both the airport and the Larnaca suburbs, and is affected by pollution from the airport. A desalination plant is to be constructed near the Larnaca sewage-treatment plant at the end of 1999. Night hunting is a threat, which, due to its indiscriminate nature, can lead to protected species being shot. The Environment Department has drafted a management plan, which states that the lakes and marginal land within 500 m of the shore should be protected. The lakes to the south of the airport and east of the road from Meneou to Spyros beach are a Permanent Game Reserve, and hold the Larnaca sewage-treatment plant. The main lake to the north and west of the main road is also a Permanent Game Reserve, with the designation only covering the area that is under water.

Akhna Dam — A2 — 006

Admin region Famagusta
Coordinates 35°03'N 33°49'E
Altitude 25–40 m **Area** 125 ha

Site description

The dam was constructed in 1987 to enable the irrigation of the Famagusta area. Water is supplied to the reservoir (capacity 5.8 million m^3) via a pipeline from Kourris Dam near Limasol. After ten years of operation the reservoir is surrounded by well-established trees (*Tamarix*, *Mimosa*) and reedbeds (*Phragmites*), and has developed into a diverse wetland of great importance. Human activities include fishing.

Habitats Wetland (100%; standing fresh water; water-fringe vegetation)
Land-use Tourism/recreation (5%), Water management (95%)

Birds

Species	Season	Year	Pop min	Pop max	Acc	Criteria
Oenanthe cypriaca Cyprus Pied Wheatear	B	—	Common		—	A2
Sylvia melanothorax Cyprus Warbler	R	—	Common		—	A2

The site has attracted a variety of birds since its construction, and today is one of the most important wetlands in Cyprus, both for wintering and migrating waterbirds, raptors and waders. Species of global conservation concern that do not meet IBA criteria: *Aythya nyroca* (wintering, no data), *Oxyura leucocephala* (wintering, no data), *Circus macrourus* (small numbers on passage), *Glareola nordmanni* (passage, no data), *Gallinago media* (passage, no data).

Protection status

National High **International** None
125 ha of IBA overlaps with Temporary Game Reserve.

Conservation issues

Threats Disturbance to birds (C), Recreation/tourism (C), Unsustainable exploitation (C)

Unsustainable exploitation through illegal hunting remains a threat to birds in the area, even though the site is a Temporary Game Reserve. Also, due to the prolonged drought, there is a very big threat of the dam drying out.

Akrotiri salt-lake including Bishop's Pool — A1, A2, A4i, B1i, B1iv, B2 — 007

Admin region Limassol
Coordinates 34°37'N 32°58'E
Altitude 0–64 m **Area** 4,000 ha

Site description

The IBA includes Akrotiri salt-lake, reedbeds and gravel pits, Bishop's Pool, Zakaki pools, and Ladie's Mile beach. The salt-lake dries out in summer, and is surrounded by reedbeds and saltmarsh. There are vineyards and orange groves in the vicinity of Bishop's Pool, which is a man-made reservoir taking surplus treated effluent from the Akrotiri sewage-treatment works.

Habitats Scrub (5%; scrub), Wetland (75%; mudflat/sandflat; saltmarsh; sand-dunes/sand beach; coastal lagoon; standing fresh water; standing brackish and salt water; water-fringe vegetation), Artificial landscape (20%; perennial crops/orchards/groves; forestry plantation)
Land-use Agriculture (10%), Forestry (10%), Military (20%), Nature conservation/research (50%), Water management (10%)

Birds

Species	Season	Year	Pop min	Pop max	Acc	Criteria
[1] ***Phoenicopterus ruber*** Greater Flamingo	W	1995	2,000	20,000	B	A4i, B1i
[1] ***Tadorna tadorna*** Shelduck	W	1995	300	1,000	B	B1i
Oxyura leucocephala White-headed Duck	W	—	5	10	C	A1
[2] ***Grus grus*** Crane	P	—	Common		—	A4i, B1i
[1] ***Oenanthe cypriaca*** Cyprus Pied Wheatear	B	1995	10	30	B	A2, B2
[1] ***Sylvia melanothorax*** Cyprus Warbler	B	1995	10	30	B	A2

1. 1990–1995.
2. Hundreds on spring passage, thousands on autumn passage.

This site is of major importance as a staging area during spring and autumn passage. It is also important for wintering waterbirds, especially *Phoenicopterus ruber* and *Anas* spp. Bird data are averaged for 1990–1995. Many species are recorded regularly every year, ranging from raptors and large waterbirds to small warblers (Sylviidae) and finches (Fringillidae). Species of global conservation concern that do not meet IBA criteria: *Aythya nyroca* (up to 20 wintering birds) and *Circus macrourus* (1–2 passage birds).

Protection status

National Partial **International** None
2,000 ha of IBA covered by Permanent Game Reserve (Akrotiri Lake, 2,000 ha).

Conservation issues

Threats Burning of vegetation (B), Construction of dykes/dams/barrage (C), Disturbance to birds (C), Drainage (U), Extraction industry (C), Filling-in of wetlands (C), Industrialization/urbanization (C), Unsustainable exploitation (B)

The area is threatened with drying-out following the construction of a dam on the River Kouris, which was a source of water during the winter. Shooting from the lake shores, especially at night, indiscriminately kills birds. There is a need for the entire area to be designated as a Permanent Game Reserve.

Phasouri reedbeds — A2, B2 — 008

Admin region Limassol
Coordinates 34°38′N 32°56′E
Altitude 0–5 m **Area** 323 ha

Site description

An extensive area of reedbeds (*Phragmites*) and water-meadows (unique in Cyprus), adjacent to Akrotiri salt-lake.

Habitats Grassland (40%; humid grassland), Wetland (60%; standing brackish and salt water; water-fringe vegetation)
Land-use Agriculture (20%), Not utilized (80%)

Birds

Species	Season	Year	Pop min	Pop max	Acc	Criteria
Francolinus francolinus Black Francolin	R	—	10	20	B	B2
Oenanthe cypriaca Cyprus Pied Wheatear	B	—	2	10	B	A2
Sylvia melanothorax Cyprus Warbler	B	—	3	10	B	A2

A very important stop-over site during spring and autumn migration; also important for winter visitors. Bird data are averaged for 1990–1995, although many more species are recorded every year, ranging from raptors and large waterbirds to small warblers (Sylviidae) and finches (Fringillidae). Species of global conservation concern that do not meet IBA criteria: *Crex crex*, *Gallinago media* (both passage).

Protection status

National Partial **International** None
Part of the IBA is protected by a Temporary Game Reserve.

Conservation issues

Threats Burning of vegetation (B), Construction of dykes/dams/barrage (U), Disturbance to birds (A), Other (B), Unsustainable exploitation (B)

The area is threatened with drying-out following the construction of the dam on the River Kouris, which formerly supplied water to the area. Inappropriate reed-cutting ('Other' threat) and burning and encroachment by grazing animals is resulting in extensive damage to the reedbeds. Indiscriminate shooting is also a major problem. The Environment Department has drafted a management plan.

Akrotiri Cliffs — B2 — 009

Admin region Limassol
Coordinates 34°34′N 32°57′E
Altitude 0–60 m **Area** 200 ha

Site description

The site covers the sea cliffs situated between Cape Gata and Cape Zevgari, and is included within the UK Sovereign Base Area.

Habitats Rocky areas (100%; sea cliff/rocky shore; rock stacks/islets)
Land-use Military (100%)

Birds

Species	Season	Year	Pop min	Pop max	Acc	Criteria
[1] ***Falco eleonorae*** Eleonora's Falcon	B	—	10	25	B	B2

1. No recent surveys.

Large numbers of birds pass through on migration.

Protection status

National High **International** None
200 ha of IBA covered by Permanent Game Reserve.

Conservation issues

Threats Disturbance to birds (C)

The site is located in a military zone and is therefore inaccessible for the purposes of conducting surveys. Low-flying aircraft cause disturbance to birds.

Episkopi Cliffs — A2, B2 — 010

Admin region Limassol
Coordinates 34°40′N 32°49′E
Altitude 0–80 m **Area** 175 ha

Site description

The IBA comprises sea cliffs, rocky shores and occasional sandy beaches, and covers Kensington Cliffs, Tunnel Beach, Quarry Beach, Curium, and Gibraltar Point. The cliffs contain steep valleys, gorges and caves, and are included within the UK Sovereign Base Area.

Habitats Wetland (10%; shingle/stony beach), Rocky areas (85%; sea cliff/rocky shore; inland cliff; caves), Artificial landscape (5%; other urban/industrial areas)
Land-use Military (50%), Not utilized (50%)

Birds

Species	Season	Year	Pop min	Pop max	Acc	Criteria
Falco eleonorae Eleonora's Falcon	B	—	15	25	B	B2
Oenanthe cypriaca Cyprus Pied Wheatear	B	—	20	60	B	A2
Sylvia melanothorax Cyprus Warbler	R	—	20	60	B	A2

Thousands of migrants pass over and frequent the area in spring and autumn.

Protection status

National High **International** None
175 ha of IBA covered by Permanent Game Reserve.

Conservation issues

Threats Disturbance to birds (C), Unsustainable exploitation (U)

The IBA falls within a military zone and is inaccessible for the purposes of conducting surveys. This means that it is also inaccessible to hunters, thus reducing the disturbance to birds.

Cape Aspro — A4ii, B1i, B1iii, B2 — 011

Admin region Limassol
Coordinates 34°39′N 32°43′E
Altitude 0–120 m **Area** 240 ha

Site description

These are the highest sea cliffs on Cyprus, extending for 8 km from the village of Pissouri (Cape Aspro) to the Petra tou Romiou area, and are accessible only by sea.

Habitats Rocky areas (100%; sea cliff/rocky shore; rock stacks/islets)
Land-use Not utilized (100%)

Birds

Species	Season	Year	Pop min	Pop max	Acc	Criteria
Phalacrocorax aristotelis Shag	R	—	10	50	B	B1i
[1] ***Falco eleonorae*** Eleonora's Falcon	B	—	50	60	B	A4ii, B1iii, B

1. No recent surveys.

Surveys are difficult due to the inaccessibility of the cliffs, which support the largest breeding colony of *Falco eleonorae* in Cyprus.

Protection status

National None **International** None

Conservation issues

Threats Disturbance to birds (U), Unsustainable exploitation (U)

The cliffs are inaccessible; the only threats being disturbance from boats and helicopters.

Asprokremmos Dam — A2, B2 — 012

Admin region Paphos
Coordinates 34°45′N 32°34′E
Altitude 0–80 m **Area** 260 ha

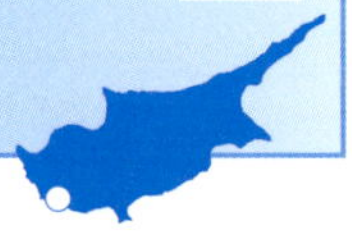

Site description

The reservoir was formed following the damming of the River Xeros, and is surrounded mainly by rough grassland (used for goat- and sheep-grazing) and arable land planted with cash crops. Small pools of water and reeds *Phragmites* are located under the dam wall. Constructed in 1982, this is the second largest reservoir in Cyprus with a capacity of 51 million m^3 of water.

Habitats Grassland (10%; steppe/dry calcareous grassland), Wetland (75%; standing fresh water), Rocky areas (5%; inland cliff), Artificial landscape (10%; arable land; perennial crops/orchards/groves)
Land-use Agriculture (15%), Water management (85%)

Birds

Species	Season	Year	Pop min	Pop max	Acc	Criteria
Francolinus francolinus Black Francolin	R	—	5	15	B	B2
Oenanthe cypriaca Cyprus Pied Wheatear	B	—	—	10	B	A2
Sylvia melanothorax Cyprus Warbler	B	—	5	15	B	A2

The site attracts a variety of passage and overwintering waterbirds, waders, raptors and passerines.

Protection status

National High **International** None
260 ha of IBA covered by Temporary Game Reserve (Asprokremmos Dam, 800 ha).

Conservation issues

Threats Disturbance to birds (B), Drainage (B), Unsustainable exploitation (B)

Drying-out of the pools and dam, mainly due to drought, is a serious threat.

Paphos plain — A2, B2 — 013

Admin region Paphos
Coordinates 34°47′N 32°26′E
Altitude 0–250 m **Area** 10,500 ha

Site description

A narrow coastal plain situated between Ay Georghios Peyias and the mouth of the River Khapotami, and bisected by the Ezousas, Dhiarizos and Xeros rivers. The site comprises cliffs, rocky shores and a few sandy bays on the seaward side, and low rolling hills inland. Cash crops are grown on an area of irrigated land.

Habitats Wetland (40%; sand-dunes/sand beach; shingle/stony beach; river/stream), Rocky areas (5%; sea cliff/rocky shore), Artificial landscape (55%; arable land; perennial crops/orchards/groves; urban parks/gardens; other urban/industrial areas)
Land-use Agriculture (50%), Not utilized (10%), Tourism/recreation (20%), Urban/industrial/transport (20%)

Birds

Species	Season	Year	Pop min	Pop max	Acc	Criteria
Francolinus francolinus Black Francolin	R	—	Common		—	B2
[1] ***Oenanthe cypriaca*** Cyprus Pied Wheatear	B	—	Common		—	A2
[1] ***Sylvia melanothorax*** Cyprus Warbler	R	—	Common		—	A2

1. Tens of pairs.

One of the main breeding areas for *Francolinus francolinus*, and a favoured resting-place for migrants during spring and autumn. Passage species of global conservation concern that do not meet IBA criteria: *Circus macrourus* (2–3 birds), *Falco naumanni* (up to 20 birds), *Glareola nordmanni* (up to 10 birds).

Protection status

National High **International** None
The entire IBA is covered by Permanent Game Reserves.

Conservation issues

Threats Disturbance to birds (B), Industrialization/urbanization (B), Recreation/tourism (A), Unsustainable exploitation (B)

The main problem is posed by development for tourism along the shoreline on either side of Paphos town.

Akamas peninsula — A2, A3, A4i, B1i, B1iv — 014

Admin region Paphos
Coordinates 35°02′N 32°18′E
Altitude 0–500 m **Area** 15,500 ha

Site description

A narrow range of hills covered with scrub, pockets of coniferous woodland, farmland and maquis, and considerable areas of rank grassland. The area extends from the Latchi, Prodromi, Droushia, Inia road down to Cape Drepanum, and is used mainly for grazing goats and sheep, with a small amount of arable farming and fruit-growing.

Habitats Forest and woodland (45%; native coniferous forest), Scrub (20%; scrub; sclerophyllous scrub/garrigue/maquis), Grassland (10%; steppe/dry calcareous grassland), Wetland (5%; sand-dunes/sand beach; shingle/stony beach), Rocky areas (5%; sea cliff/rocky shore), Artificial landscape (40%; arable land; perennial crops/orchards/groves)
Land-use Agriculture (40%), Forestry (25%), Military (20%), Nature conservation/research (5%), Tourism/recreation (10%)

Birds

Species	Season	Year	Pop min	Pop max	Acc	Criteria
[1] ***Grus grus*** Crane	P	—	Common		—	A4i, B1i
[2] ***Oenanthe cypriaca*** Cyprus Pied Wheatear	B	—	Common		—	A2, A3
Sylvia conspicillata Spectacled Warbler	R	—	2	4	B	A3
[2] ***Sylvia melanothorax*** Cyprus Warbler	B	—	Common		—	A2, A3
[2] ***Lanius nubicus*** Masked Shrike	B	—	Common		—	A3
[2] ***Emberiza caesia*** Cretzschmar's Bunting	B	—	Common		—	A3
[2] ***Emberiza melanocephala*** Black-headed Bunting	B	—	Common		—	A3

1. Thousands on autumn passage.
2. Tens of pairs.

One of the most important bird areas in Cyprus with hundreds of species recorded every year, including many which stop over whilst on passage. Species of global conservation concern that do not meet IBA criteria: *Larus audouinii* (single birds on passage).

Protection status

National None **International** None

Conservation issues

Threats Burning of vegetation (C), Disturbance to birds (B), Forest grazing (B), Industrialization/urbanization (B), Recreation/tourism (C), Unsustainable exploitation (B)

Probably the most significant threat is the hunting of birds within the IBA. Also there is considerable pressure to allow the development of coastal areas for tourism. Another threat is posed by the use of the area by British troops for training exercises and as a rifle range; during 1984–1989 about 60 ha were burned. The area is a proposed National Park, and there is an Environment Department management plan. 7,000 ha is State Forest land. Lara Bay within the IBA is the only site in Cyprus designated as a Mediterranean Specially Protected Area under the Barcelona Convention; its importance is for breeding turtles (*Chelonia mydas* and *Caretta caretta*).

Paphos State Forest and Stavros tis Psokas Forest Station

A1, A2, A3, B2 **015**

Admin region Paphos, Nicosia
Coordinates 35°01′N 32°37′E
Altitude 100–1,362 m **Area** 60,170 ha

Site description

Stavros tis Psokos is the best-known forest station in Cyprus, and is situated in the middle of Paphos Forest. The site comprises mainly coniferous woodland with rocky outcrops. A large replanting scheme, covering 8,500 ha out of a total of 21,600 ha, was implemented after this area was destroyed by fire in 1974. 400 ha of the IBA are covered with the endemic Cyprus cedar *Cedrus brevifolia*.

Habitats Forest and woodland (90%; broadleaved deciduous forest; native coniferous forest; mixed forest), Scrub (5%; sclerophyllous scrub/garrigue/maquis), Artificial landscape (15%; arable land; perennial crops/orchards/groves; urban parks/gardens; other urban/industrial areas)
Land-use Agriculture (15%), Forestry (80%), Urban/industrial/transport (5%)

Birds

Species	Season	Year	Pop min	Pop max	Acc	Criteria
Aquila heliaca Imperial Eagle	R	—	2	2	B	A1, B2
Hieraaetus fasciatus Bonnelli's Eagle	R	—	3	3	B	B2
Oenanthe cypriaca Cyprus Pied Wheatear	B	—	—	—	B	A2, A3
Sylvia melanothorax Cyprus Warbler	R	—	—	—	B	A2, A3
Lanius nubicus Masked Shrike	B	—	Common		—	A3
Emberiza caesia Cretzschmar's Bunting	B	—	Common		—	A3

The area is very important for breeding raptors and for birds preferring forest habitat.

Protection status

National Partial **International** None
50,000 ha of IBA covered by Permanent Game Reserve.

Conservation issues

Threats Agricultural intensification/expansion (C), Burning of vegetation (C), Disturbance to birds (C), Forest grazing (C), Recreation/tourism (C), Selective logging/cutting (C), Unsustainable exploitation (C)

Paphos Forest is the largest Permanent Game Reserve in Cyprus. The construction of forestry access roads has made the area more accessible and increased disturbance to birds. 60,170 ha is State Forest land.

Adelphi Forest

A1, A2, A3, B2 **016**

Admin region Nicosia
Coordinates 35°01′N 32°58′E
Altitude 400–1,612 m **Area** 12,756 ha

Site description

One of the largest areas of forest in Cyprus, situated mainly on the north-eastern slopes of the Troodos massif.

Habitats Forest and woodland (70%; broadleaved deciduous forest; native coniferous forest), Rocky areas (2%; inland cliff), Artificial landscape (30%; perennial crops/orchards/groves)
Land-use Agriculture (30%), Forestry (60%), Urban/industrial/transport (10%)

Birds

Species	Season	Year	Pop min	Pop max	Acc	Criteria
Aquila heliaca Imperial Eagle	R	—	1	2	B	A1, B2
Hieraaetus fasciatus Bonnelli's Eagle	R	—	2	4	B	B2
[1] ***Oenanthe cypriaca*** Cyprus Pied Wheatear	B	—	Common		—	A2, A3
[1] ***Sylvia melanothorax*** Cyprus Warbler	R	—	Common		—	A2, A3
[1] ***Lanius nubicus*** Masked Shrike	B	—	Common		—	A3
[1] ***Emberiza caesia*** Cretzschmar's Bunting	B	—	Common		—	A3
[1] ***Emberiza melanocephala*** Black-headed Bunting	B	—	Common		—	A3

1. Tens of pairs.

The site is important for breeding raptors and species of forest and open cultivated land. There have been recent sightings of *Aquila heliaca* in the area, suggesting that between one and three birds are present. In 1988 a nest containing two eaglets was found near Orkontas.

Protection status

National None **International** None

Conservation issues

Threats Abandonment/reduction of land management (C), Burning of vegetation (B), Disturbance to birds (C), Forest grazing (C), Natural events (C), Selective logging/cutting (C), Unsustainable exploitation (B)

Expansion of villages within the IBA and the construction of forest roads have increased accessibility to the area and thus made bird shooting and disturbance easier. The entire area is State Forest land.

REFERENCES

Charalambides, M. and Neophytou, P., Eds. (1988–1991) *Cyprus Ornithological Society Annual Reports.* Nicosia.

Flint, P. and Stewart, P. (1992) *The Birds of Cyprus.* UK: British Ornithologists' Union (BOU Check-list No. 6 Second Edition).

Grimmett, R. F. A. and Jones T. A. (1989) *Important Bird Areas in Europe.* Cambridge, UK: International Council for Bird Preservation (Tech. Publ. 9).

Heath, M. F. and Borggreve, C. (2000) *BirdLife International/EBCC European Bird Database 1998.* Cambridge, UK: BirdLife International.

Kourtellarides, L. (1998) *Breeding Birds of Cyprus.* Nicosia: Bank of Cyprus Group.

Stattersfield, A. J., Crosby, M. J., Long, A. J. and Wege, D. C. (1998) *Endemic Bird Areas of the World: Priorities for Biodiversity Conservation.* Cambridge, UK: BirdLife International (BirdLife Conservation Series no. 7).

Tucker, G. M. and Heath, M. F. (1994) *Birds in Europe: their conservation status.* Cambridge, UK: BirdLife International (BirdLife Conservation Series no. 3).

FRANCE

Bernard Deceuninck, Xavier Rufray and Philippe Jourde

Alpine habitat in the Pyrenees: Vallée de Melles, cols d'Aoueran et d'Artisagou et Mont Vallier (IBA 204). (Photo: Bernard Deceuninck)

GENERAL INTRODUCTION

France covers 551,600 km² and is bordered by the Atlantic Ocean to the west, the English Channel, Belgium, Luxembourg and Germany to the north, Switzerland and Italy to the east and Spain and the Mediterranean Sea to the south. The previous pan-European inventory (Grimmett and Jones 1989) identified 153 Important Bird Areas (IBAs). The current inventory lists 277 IBAs, representing 8.6% or 47,248 km² of the total land area (Table 1, Map 1). A total of 134 new sites have been identified since 1989, six sites have been split into 19 new IBAs, nine sites have been merged to form three IBAs and 17 sites have been dropped. In 1994, a new national inventory of IBAs was compiled (Rocamora 1994). Eight sites from this inventory did not meet criteria for the present IBA inventory.

Geographically, IBAs are distributed throughout the country, however the Mediterranean region, which has the greatest diversity of habitats and species in the country, also has the highest density of IBAs (Map 1). The regions of the Aquitaine basin and the Parisian basin have the lowest numbers of IBAs because of intensive agricultural land-use.

Table 1. Summary of Important Bird Areas in France. 277 IBAs covering 47,248 km²

IBA code	National code	1989 code	International/National name	Administrative region	Area (ha)	Criteria (see p. 11)
001	BT01	FR050	Baie de Saint-Brieuc	Bretagne	3,130	A4i, A4iii, B1i, B2, C3, C4
002	BT02	FR051	Archipel des Sept-Iles	Bretagne	4,300	A4ii, B1ii, B2, B3, C3, C6
003	BT03	FR052	Ile de Goulmedec	Bretagne	600	B2, C6
004	BT04	—	Estuaires du Trieux et du Jaudy	Bretagne	9,900	B2, C6
005	BT05	FR053	Iles de la Colombière, de la Nellière et des Haches	Bretagne	4,400	B2, C6
006	BT06	FR054	Ilôts de Trévorc'h	Bretagne	470	C6
007	BT07	FR055	Archipel de Molène	Bretagne	10,900	A4i, B1i, B3, C3, C6
008	BT08	FR056	Rade de Brest: Baie de Daoulas et Anse du Poulmic	Bretagne	9,000	A4i, A4iii, B1i, C3, C4
009	BT09	FR057	Cap Sizun	Bretagne	590	B3
010	BT10	FR059	Marais de la Baie d'Audierne	Bretagne	3,100	A1, A4i, B1i, C1, C3, C6
011	BT11	FR060	Archipel des Glénan	Bretagne	4,100	A4i, B1i, B2, B3, C3, C6
012	BT12	FR061	Baies de Morlaix et de Carantec	Bretagne	7,900	A4i, A4iii, B1i, B2, B3, C2, C3, C4, C6
013	BT13	FR062	Ile d'Ouessant	Bretagne	5,850	B3, C6
014	BT14	FR063, FR064	Golfe du Morbihan et Etier de Penerf	Bretagne	18,800	A4i, A4iii, B1i, B2, B3, C2, C3, C4, C6
015	BT15	—	Cap Frehel	Bretagne	2,070	B3
016	BT16	FR065	Baie de Vilaine	Bretagne	4,500	A4i, A4iii, B1i, B2, C2, C3, C4
017	BT17	FR058	Presqu'île de Crozon, Tas de Pois et Rochers du Toulinguet	Bretagne	1,200	B3, C6

Table 1 ... continued. Summary of Important Bird Areas in France. 277 IBAs covering 47,248 km²

IBA code	National code	1989 code	International/National name	Administrative region	Area (ha)	Criteria (see p.11)
018	BT18	—	Rade de Lorient	Bretagne	2,800	A4iii, C4
019	BT19	—	Baie de Quiberon	Bretagne	14,750	A4i, B1i, C3
020	BT20	—	Baie de Goulven	Bretagne	2,000	A4i, A4iii, B1i, C2, C3, C4
021	PL01	FR041	Traicts et marais salants de la Presqu'île Guérandaise	Pays de la Loire	4,650	A4i, A4iii, B1i, B2, B3, C2, C3, C4, C6
022	PL02	FR042	Marais de Brière	Pays de la Loire	18,900	B1i, B2, B3, C2, C3, C6
023	PL03	FR043	Estuaire de la Loire	Pays de la Loire	21,400	A1, A4i, A4iii, B1i, B2, B3, C1, C2, C3, C4, C6
024	PL04	FR044	Lac de Grand-Lieu	Pays de la Loire	5,600	A4i, A4iii, B1i, B2, C2, C3, C4, C6
025	PL05	FR045	Baie de Bourgneuf et marais breton	Pays de la Loire	47,000	A4i, A4iii, B1i, B2, B3, C2, C3, C4, C6
026	PL06	FR046	Basses Vallées Angevines	Pays de la Loire	6,450	A1, A4i, A4iii, B1i, B2, B3, C1, C2, C3, C4, C6
027	PL07	FR047	Marais salants de Noirmoutier	Pays de la Loire	1,650	B1i, B3, C2, C6
028	PL08	FR048	Marais et forêt d'Olonne	Pays de la Loire	3,250	A4i, A4iii, B1i, B2, B3, C2, C3, C4, C6
029	PL09	—	Forêt du Gavre	Pays de la Loire	4,900	C6
030	PL10	—	Marais de Mazerolles - Petit Mars	Pays de la Loire	2,700	B1i, C3
031	PL11	FR017	Vallée de la Loire : de Nantes à Montsoreau	Pays de la Loire	12,000	A1, B1i, B2, C1, C2, C6
032	PL12	—	Plaine calcaire du sud Vendée	Pays de la Loire	18,700	B3, C6
033	PL13	FR049	Marais poitevin et baie de l'Aiguillon	Pays de la Loire	77,900	A4i, A4iii, B1i, B2, B3, C2, C3, C4, C6
034	BN01	FR018	Falaise du Bessin	Basse-Normandie	2,300	C7
035	BN02	FR019	Baie des Veys et Marais du Cotentin	Basse-Normandie	37,500	A1, A4i, A4iii, B1i, B2, C1, C2, C3, C4, C6
036	BN03	FR020	Iles Saint Marcouf	Basse-Normandie	1,540	B3
037	BN04	FR022	Iles Chausey	Basse-Normandie	18,400	A4i, B1i, B3, C2, C6
038	BN05	—	Havre de la Sienne	Basse-Normandie	5,150	B1i, C3
039	BN06	—	Estuaire de l'Orne	Basse-Normandie	1,000	B1i, C2
040	BN07	—	Littoral Augeron	Basse-Normandie	23,100	A4i, A4iii, B1i, B2, C2, C3, C4
041	BN08	—	Forêts du Perche	Basse-Normandie	50,850	B2, B3, C6
042	BN09	FR023	Baie du Mont Saint Michel et Ile des Landes	Basse-Normandie	63,800	A4i, A4iii, B1i, B2, B3, C2, C3, C4
043	HN01	—	Cap Fagnet	Haute-Normandie	5,700	C7
044	HN02	—	Boucle de Pose et de Muids	Haute-Normandie	5,200	A4iii, B1i, B2, C3, C4, C6
045	HN03	FR011	Estuaire et embouchure de la Seine	Haute-Normandie	21,100	A1, A4i, A4iii, B1i, B2, C1, C2, C3, C4, C6
046	IF01	FR001	Boucle de Moisson	Ile de France	6,150	C6
047	IF02	—	Massifs de Fontainebleau et zones humides adjacentes	Ile de France	36,600	B2, B3, C2, C6
048	IF03	—	Bassée et plaines adjacentes	Ile de France	39,900	C6
049	IF04	—	Marais de Fontenay-le-Vicomte et d'Itteville	Ile de France	710	C6
050	IF05	—	Etang et Forêt de Villefermoy	Ile de France	4,350	C6
051	PE01	—	Marais arrières littoraux picards	Picardie	5,550	B3, C6
052	PE02	—	Etangs et marais du bassin de la Somme	Picardie	6,900	B2, C6
053	PE03	—	Forêts picardes: Compiègne, Laigue, Ourscamp	Picardie	32,700	C6
054	PE04	—	Forêts picardes: massif de Retz	Picardie	27,650	C6
055	PE05	—	Forêts picardes: massif de Saint-Gobain	Picardie	19,300	C6
056	PE06	FR010	Marais de Sacy	Picardie	2,350	C6
057	PE07	—	Vallée de l'Oise de Thourotte à Vendeuil	Picardie	12,050	A1, C1, C6
058	PE08	—	Marais de la Souche	Picardie	4,950	C6
059	PE09	—	Forêts picardes: massif des Trois Forêts et bois du Roi	Picardie	32,200	C6
060	PE10	FR009	Estuaires picards: baies de Somme et d'Authie	Picardie	18,200	A4i, A4iii, B1i, B2, B3, C2, C3, C4, C6
061	NC01	FR026	Vallées de la Scarpe et de l'Escaut	Nord - Pas-de-Calais	11,150	B1i, C3, C6
062	NC02	FR009	Estuaire de la Canche	Nord - Pas-de-Calais	5,050	A4i, A4iii, B1i, C3, C4, C6
063	NC03	FR027	Marais de Balançon et de Villiers	Nord - Pas-de-Calais	1,500	C6
064	NC04	—	Cap Gris-nez	Nord - Pas-de-Calais	8,950	A4i, A4iii, B1i, B2, C2, C3, C4
065	NC05	—	Plateaux agricoles des environs de Frencq	Nord - Pas-de-Calais	11,450	C6
066	NC06	—	Forêts de Thiérache: Trelon, Fourmies, Hirson et St-Michel	Nord - Pas-de-Calais	22,400	B3, C6
067	CA01	FR003	Plateau Ardennais	Champagne-Ardenne	94,800	B3, C6
068	CA02	FR004	Lacs de la Forêt d'Orient	Champagne-Ardenne	35,800	A4i, A4iii, B1i, B1iv, B2, C2, C3, C4, C5, C6
069	CA04	FR007	Etangs d'Argonne	Champagne-Ardenne	44,100	B2, C6
070	CA05	FR008	Lac du Der-Chantecoq et étangs latéraux	Champagne-Ardenne	56,000	A4i, A4iii, A4iv, B1i, B1iv, B2, C2, C3, C4, C5, C6
071	CA06	—	Barois et Forêt de Clairvaux	Champagne-Ardenne	66,550	B2, C6
072	CA07	—	Vallée de l'Aube, de la Superbe et Marigny	Champagne-Ardenne	21,500	C6
073	CA08	—	Vallée de l'Aisne	Champagne-Ardenne	18,800	A4i, A4iii, B1i, C3, C4, C6
074	CA09	—	Confluent des vallées de la Meuse et de la Chiers	Champagne-Ardenne	4,860	A4iii, B3, C4, C6
075	CA10	—	Bassigny	Champagne-Ardenne	98,400	B3, C6
076	LE01	FR028	Etangs de la Woëvre: lac de Madin	Lorraine	1,730	C6
077	LE02	FR029	Etangs de la Woëvre: la Chaussée	Lorraine	6,200	A4i, B1i, C2, C6
078	LE03	FR030	Etang de Lindre	Lorraine	1,660	B1i, B2, C3, C6
079	LE04	—	Vallée de la Meuse	Lorraine	18,100	A1, B2, C1, C6
080	LE05	—	Plateau de Delme, Val de Petite Seille	Lorraine	11,950	C6
081	LE06	—	Val de Chiers et environs de Spincourt	Lorraine	16,900	B1i, B3, C2, C6

Table 1 ... continued. Summary of Important Bird Areas in France. 277 IBAs covering 47,248 km²

IBA code	National code	1989 code	International/National name	Administrative region	Area (ha)	Criteria (see p. 11)
082	LE07	—	Fresnes-en-Woëvre–Mars-la-Tour	Lorraine	11,700	C6
083	LE08	—	Bazoncourt–Vigy	Lorraine	10,700	C6
084	LE10	—	Forêt de la Reine	Lorraine	16,650	C6
085	AC01	FR031	Ried de Colmar et Sélestat	Alsace	10,100	A4i, A4iii, B1i, B3, C3, C4, C6
086	AC02	—	Forêt domaniale de la Harth	Alsace	15,700	B2, B3, C6
087	AC03	—	Forêt de Haguenau	Alsace	21,350	B3, C2, C6
088	AC04	—	Zones agricoles de la Hardt	Alsace	21,800	B2, C6
089	AC05	—	Massif des Vosges: crêtes de Schneeberg au Donon	Alsace	8,450	C6
090	AC06	FR032–FR036	Vallée du Rhin: Village-neuf à Biescheim	Alsace	5,690	A4iii, B1i, B2, C3, C4, C6
091	AC07	—	Vallée du Rhin: Marckolsheim à Strasbourg	Alsace	10,600	A4i, A4iii, B1i, B2, B3, C3, C4, C6
092	AC08	—	Vallée du Rhin: Strasbourg à Lauterbourg	Alsace	11,000	A4i, A4iii, B1i, B2, C3, C4, C6
093	AC09	—	Massif des Vosge : Hautes Vosges	Alsace	88,200	B2, B3, C6
094	FC01	FR039	Bassin du Drugeon: Pontarlier-Frasne	Franche-Comté	12,500	C6
095	FC02	—	Forêts d'altitude du Jura: forêt du Risol et Mont d'Or	Franche-Comté	12,300	B3, C6
096	FC03	—	Forêts d'altitude du Jura: forêt du Risoux	Franche-Comté	4,400	B3, C6
097	FC04	—	Forêts d'altitude du Jura: forêt du Massacre	Franche-Comté	1,970	B3, C6
098	FC05	—	Forêt de Chaux	Franche-Comté	21,700	B2, C6
099	FC06	—	Vallée de la Saône de Corre à Broye	Franche-Comté	14,050	B2, B3, C6
100	FC07	FR040	Basse vallée du Doubs: Dole Sud	Franche-Comté	11,850	B2, C6
101	PC01	—	Estuaire de la Charente	Poitou-Charentes	5,400	B1i, B2, C3, C6
102	PC02	FR066	Vallée de la Charente et de la Seugne (Cabariot-Pons/St-Sever-de-Saintonge)	Poitou-Charentes	8,300	A1, B2, B3, C1, C2, C6
103	PC03	—	Vallée de la Charente: amont d'Angoulême	Poitou-Charentes	4,650	A1, B2, C1, C6
104	PC05	FR069	Anse de Fouras, baie d'Yves et marais de Rochefort	Poitou-Charentes	17,550	A4iii, B1i, B2, B3, C2, C3, C4, C6
105	PC06	FR070	Ile d'Oléron, marais de Brouage-Saint-Agnant	Poitou-Charentes	26,500	A4i, A4iii, B1i, B2, B3, C2, C3, C4, C6
106	PC07	FR068	Anse du Fiers d'Ars en Ré	Poitou-Charentes	4,350	A4i, A4iii, B1i, B2, C2, C3, C4, C6
107	PC08	FR073	Bonne Anse	Poitou-Charentes	1,060	C6
108	PC09	FR074	Plaine de Niort (sud-est)	Poitou-Charentes	14,300	A1, B2, C1, C6
109	PC10	FR075	Plaine de Niort (nord-ouest)	Poitou-Charentes	12,500	A1, B2, B3, C1, C6
110	PC11	FR076	Plaines de Saint-Jouin de Marne et d'Assais-les-Jumeaux	Poitou-Charentes	12,400	A1, B2, B3, C1, C6
111	PC12	FR078	Le Pinail, Forêt de Moulière et bois du Défens	Poitou-Charentes	6,150	C6
112	PC14	FR080	Plateau de Bellefonds	Poitou-Charentes	2,550	B1i, C2
113	PC16	FR082	Plaines de Mirebeau et de Neuville-du-Poitou	Poitou-Charentes	10,250	C6
114	PC17	—	Region de Pressac, étang de Combourg	Poitou-Charentes	3,500	C6
115	PC18	—	Marais et estuaire de la Seudre	Poitou-Charentes	14,800	A4i, B1i, C2, C6
116	PC19	FR077	Plaine de Saint-Jean-de-Sauves	Poitou-Charentes	6,550	C6
117	PC20	FR067	Plaine de Villefagnan	Poitou-Charentes	9,050	A1, C1, C6
118	PC21	FR083	Estuaire de la Gironde: marais de la rive nord	Poitou-Charentes	2,580	A1, A4i, B1i, B2, C1, C2, C3, C6
119	CE01	FR017	Vallée de la Loire: environs de Blois	Centre	22,200	C6
120	CE02	—	Vallée de la Conie et Beauce centrale	Centre	71,600	B1i, B2, B3, C2, C6
121	CE03	—	Petite Beauce	Centre	52,200	B1i, B2, C2, C6
122	CE04	—	Vallée de l'Yèvre	Centre	2,200	A1, C1, C6
123	CE05	FR012	Brenne centrale	Centre	22,000	A4i, A4iii, B1i, B1iv, B2, C2, C3, C4, C5, C6
124	CE06	FR012	Brenne Sud	Centre	6,660	C6
125	CE07	FR012	Queue de Brenne: Etang des Loges, Grand et Petit Mez	Centre	770	C6
126	CE08	—	Brenne: Forêt de la Petite-Brenne	Centre	12,560	C6
127	CE09	—	Brenne: Forêt de Lancosme	Centre	5,950	C6
128	CE10	FR017	Vallée de la Loire: confluence Loire-Vienne	Centre	5,080	C6
129	CE11	—	Basse vallée de L'Indre	Centre	1,900	A1, C1, C6
130	CE12	FR017	Vallée de la Loire: environs de Montlouis-sur-Loire	Centre	510	C6
131	CE13	FR014	Etangs de Sologne: St Viatre, Marcilly-en-Gault et forêt de Bruadan	Centre	37,500	B1i, B2, C3, C6
132	CE14	FR015	Etangs de Sologn: étang de l'Arche	Centre	165	C6
133	CE15	—	Vallée du Fouzon	Centre	700	A1, C1, C6
134	CE16	FR013	Lac de Rillé et forêts voisines	Centre	18,800	C6
135	CE17	—	Vallée de la Loire: Orléanais	Centre	7,750	B1i, C3, C6
136	CE18	—	Forêt d'Orléans: massifs d'Ingrannes et de Lorris	Centre	40,400	B2, C6
137	CE20	FR024	Mars sur Allier	Centre	770	C6
138	CE21	FR017	Vallée de la Loire: La Charité-sur-Loire	Centre	1,400	C6
139	BE01	—	Forêt de Jugny	Bourgogne	6,800	C6
140	BE02	—	Forêt de Citeaux et environs	Bourgogne	13,100	B3, C2, C6
141	BE03	—	Forêt de Châtillon et environs	Bourgogne	21,900	B2, C6
142	BE04	—	Arrière Côte de Dijon et de Beaune	Bourgogne	63,600	B2, C6
143	LN01	FR102	Etang des Landes	Limousin	1,800	B1i, B1iv, C2, C5, C6

Table 1 ... continued. Summary of Important Bird Areas in France. 277 IBAs covering 47,248 km²

IBA code	National code	1989 code	International/National name	Administrative region	Area (ha)	Criteria (see p. 11)
144	LN02	—	Plateaux de Millevaches et de Gentioux	Limousin	64,900	B1i, B1iv, B2, C2, C5, C6
145	LN03	FR101	Gorges de la Dordogne	Limousin	45,700	B1iv, B2, C5, C6
146	AE01	FR024	Val d'Allier Bourbonnais	Auvergne	17,900	A4i, B1i, B2, C2, C6
147	AE02	—	Haut Val d'Allier	Auvergne	52,700	B2, B3, C6
148	AE03	—	Gorges de la Sioule	Auvergne	25,250	B2, B3, C6
149	AE04	FR114	Monts et Plomb du Cantal	Auvergne	6,450	B1iv, C5
150	AE05	FR115	Planèze de Saint-Flour	Auvergne	24,150	C6
151	AE06	FR113	Montagne de la Serre	Auvergne	2,600	B1i, B1iv, C2, C5, C6
152	AE07	—	Sologne Bourbonnaise	Auvergne	22,000	A4i, B1i, B2, C2, C6
153	AE08	FR017	Vallée de la Loire: Pit majeur d'Iguerande à Décize	Auvergne	29,800	A1, C1, C6
154	AE09	FR017	Vallée de la Loire: gorges de la Loire	Auvergne	63,000	B2, B3, C6
155	AE10	FR095	Gorges de la Truyère	Auvergne	21,500	B2, B3, C6
156	AE11	—	Les Couzes Nord	Auvergne	19,500	B2, C6
157	AE12	—	Les Couzes Sud	Auvergne	16,400	B2, C6
158	AE13	—	Val d'Allier: Saint-Yorre-Joze	Auvergne	5,500	B1i, B1iv, C2, C5, C6
159	RA01	FR103	La Dombes	Rhône-Alpes	79,800	A4i, A4iii, B1i, B2, C2, C3, C4, C6
160	RA02	—	Val de Saône	Rhône-Alpes	12,250	A1, B2, C1, C2, C6
161	RA03	—	Forêt de Saou	Rhône-Alpes	5,050	C6
162	RA04	FR104	Val de Drôme: Les Ramières-printegarde	Rhône-Alpes	2,100	A4i, B1i, B1iv, C3, C5, C6
163	RA05	FR107	Col de l'Escrinet	Rhône-Alpes	4,600	A4iv, B1i, B1iv, C2, C5
164	RA06	—	Basse Ardêche	Rhône-Alpes	46,000	B2
165	RA07	FR106	Hauts Plateaux du Vercors et Forêt des Coulmes	Rhône-Alpes	52,800	B2, C6
166	RA08	—	Iles du Haut-Rhône	Rhône-Alpes	3,400	B1i, C3, C6
167	RA09	FR108	Plaine du Forez	Rhône-Alpes	80,850	A4i, A4iii, B1i, B2, B3, C2, C3, C4, C6
168	RA10	—	Ile de la Platière	Rhône-Alpes	470	B2
169	RA11	—	Parc National de la Vanoise	Rhône-Alpes	67,600	A3, B2, C2, C6
170	RA12	FR111	Lac Léman (partie française)	Rhône-Alpes	24,300	A4i, A4iii, B1i, C3, C4
171	RA13	FR109	Lac et marais du Bourget	Rhône-Alpes	9,350	B1i, C3
172	RA14	—	Haute chaîne du Jura: défilé de l'écluse, Etournel et Mont Vuache	Rhône-Alpes	12,400	A4iv, B1iv, B2, C5, C6
173	RA15	FR110	Montagne des Frêtes et plateau des Glières	Rhône-Alpes	8,200	B2, C6
174	RA16	FR112	Les Bauges	Rhône-Alpes	17,750	B2, C6
175	RA17	—	Haut Giffre	Rhône-Alpes	20,850	B2, C6
176	AN01	FR086	Bassin d'Arcachon et Banc d'Arguin	Aquitaine	20,100	A4i, A4iii, B1i, B2, C2, C3, C4, C6
177	AN02	FR087	Pointe de Grave	Aquitaine	800	B1i, B1iv, C3, C5, C6
178	AN03	FR084	Champ de Tir du Poteau	Aquitaine	12,200	A4i, A4iv, B1i, B1iv, C2, C5, C6
179	AN04	FR088	Barthes de l'Adour	Aquitaine	16,100	B2, C6
180	AN05	FR085	Rochers de Biarritz: le Bouccalot et la Roche Ronde	Aquitaine	250	C6
181	AN06	—	Haute Soule: Massif Forestier, Gorges d'Holzarte et d'Olhadubi	Aquitaine	2,400	C6
182	AN07	FR089	Haute Soule: Forêt des Arbailles	Aquitaine	6,850	B2, C6
183	AN08	FR089	Haute Soule: Forêt d'Irraty, Organbidexka et Pic des Escaliers	Aquitaine	5,500	A4iv, B1iv, C5, C6
184	AN09	FR090	Rive Gauche de la Vallée d'Aspe et Pic d'Anie	Aquitaine	10,750	C6
185	AN10	FR091	Hautes Vallées d'Aspe et d'Ossau	Aquitaine	37,650	B1iii, B2, C2, C6
186	AN11	—	Eth Turoun des Aureys	Aquitaine	2,100	B2, C6
187	AN12	—	Pènes du Moulle de Jaut	Aquitaine	4,300	B2, C6
188	AN13	—	Haute Soule: massif de la Pierre-St-Martin	Aquitaine	18,450	B2, C2, C6
189	AN14	—	Pic de l'Estibet et Mondragon	Aquitaine	4,550	C6
190	AN15	FR092	Barrage d'Artix et Saligue du Gave de Pau	Aquitaine	3,200	C2, C6
191	AN16	—	Col de Lizarrieta	Aquitaine	1,500	A4i, B1i, B1iv, C2, C3, C5, C6
192	AN17	—	Vallée de la Nive des Aldudes–Col de Lindux	Aquitaine	14,500	A4i, A4iv, B1i, B1iv, C2, C5
193	AN18	—	Domaine d'Orx	Aquitaine	4,250	A4iii, B1i, B2, C3, C4, C6
194	AN19	—	Garonne: marais de Bordeaux	Aquitaine	6,100	B2, C6
195	AN20	—	Estuaire de la Bidassoa et Baie de Fontarabie	Aquitaine	9,300	A4i, A4iii, B1i, B2, C2, C3, C4
196	AN21	—	Réserve Naturelle du courant d'Huchet	Aquitaine	820	B2, C6
197	AN22	—	Arjuzanx	Aquitaine	2,430	A4i, B1i, C2, C6
198	AN23	—	Estuaire de la Gironde: marais du Nord Médoc	Aquitaine	22,250	B2, C6
199	AN24	FR072	Estuaire de la Gironde: marais du Blayais	Aquitaine	7,050	C6
200	AN25	—	Haute Cize: Pic d'Errozate et Forêt d'Orion	Aquitaine	6,450	B2, C6
201	AN26	—	Côte Médocaine: dunes boisées et dépressions humides	Aquitaine	3,900	B1i, C2, C6
202	MP01	FR097	Vallées du Lis et de la Pique	Midi-Pyrénées	9,600	B2, C6
203	MP02	FR098	Cirque de Gavarnie	Midi-Pyrénées	7,600	B2, C6
204	MP03	—	Vallée de Melles, cols d'Aoueran et d'Artisagou et Mont Vallier	Midi-Pyrénées	40,100	B2, C6
205	MP04	—	Zones rupestres du Tarasconnais et massif d'Aston	Midi-Pyrénées	23,300	B2, C6
206	MP05	—	Vallée de la Garonne: Boussens à Carbonne	Midi-Pyrénées	2,300	B2, C6
207	MP06	FR099	Vallée de la Garonne: Palayre et environs	Midi-Pyrénées	1,700	A4i, B1i, C2, C6

Table 1 ... continued. Summary of Important Bird Areas in France. 277 IBAs covering 47,248 km²

IBA code	National code	1989 code	International/National name	Administrative region	Area (ha)	Criteria (see p. 11)
208	MP07	FR100	Vallée de la Garonne: Moissac	Midi-Pyrénées	1,300	A4i, B1i, C2, C6
209	MP08	—	Vallée de la Garonne: méandre de Grisolles	Midi-Pyrénées	1,050	C6
210	MP09	—	Gorges de la Frau et Belestat	Midi-Pyrénées	14,000	C6
211	MP10	FR096	Gorges du Tarn et de la Jonte	Midi-Pyrénées	11,100	A1, B2, C1, C6
212	MP11	FR094	Gorges de la Dourbie et causses avoisinants	Midi-Pyrénées	23,000	A1, B2, C1, C6
213	LR02	—	Montagne de la Clape	Languedoc-Roussillon	10,400	B1iv, C5, C6
214	LR03	FR123	Etangs de Leucate et Lapalme	Languedoc-Roussillon	13,650	A4i, A4iv, B1i, B1iv, B2, C2, C3, C5, C6
215	LR04	FR122	Etangs Narbonnais	Languedoc-Roussillon	10,600	A4i, A4iii, B1i, B1iv, B2, C2, C3, C4, C5, C6
216	LR05	FR124	Etang de Capestang	Languedoc-Roussillon	1,360	B2, C2, C6
217	LR06	FR125	Hautes Corbières	Languedoc-Roussillon	74,750	B1iv, B2, C5, C6
218	LR07	—	Basses Corbières	Languedoc-Roussillon	47,400	A4iv, B1iv, B2, B3, C5, C6
219	LR08	FR126	Gorges de la Vis et cirque de Navacelles	Languedoc-Roussillon	24,400	C6
220	LR09	FR128	Etangs Montpellierains	Languedoc-Roussillon	12,700	A1, A4i, A4iii, B1i, B2, B3, C1, C2, C3, C4, C6
221	LR10	FR129	Massif des Albères	Languedoc-Roussillon	17,850	B2, C6
222	LR11	—	Massif du Canigou-Carança	Languedoc-Roussillon	53,550	A4iv, B1iv, B2, B3, C5, C6
223	LR12	—	Puig Carlit et environs	Languedoc-Roussillon	33,850	B2, B3, C2, C6
224	LR13	—	Gorges du Gardon	Languedoc-Roussillon	19,500	A1, B2, B3, C1, C6
225	LR14	—	Hautes garrigues du Montpellierais	Languedoc-Roussillon	90,700	A1, B1iv, B2, C1, C5, C6
226	LR15	—	Etang du Bagnas	Languedoc-Roussillon	561	C2, C6
227	LR16	—	Cordon lagunaire de Sète à Agde	Languedoc-Roussillon	800	A4i, B1i, B2, C2, C6
228	LR17	—	Etang de Thau	Languedoc-Roussillon	1,000	A4i, B1i, C3, C6
229	LR18	FR119, FR120	Etangs de Vendres, Pissevache et Lespignan	Languedoc-Roussillon	4,850	B1i, B2, C2, C6
230	LR19	—	Minervois	Languedoc-Roussillon	19,000	B2, B3, C6
231	LR20	—	Etangs de Canet et de Villeneuve-de-la-Raho et embouchure du Tech	Languedoc-Roussillon	2,500	A4i, B1i, B2, C2, C6
232	LR21	—	Gorges du Rieutord–Fage-Cagnasses	Languedoc-Roussillon	6,200	C6
233	LR22	—	Massif du Madres et Mont Coronat	Languedoc-Roussillon	19,300	B2, C6
234	LR23	—	Petite Camargue fluvio-lacustre	Languedoc-Roussillon	19,300	A1, A4i, B1i, B2, C1, C2, C6
235	LR24	FR127	Petite Camargue laguno-marine	Languedoc-Roussillon	21,000	A4i, B1i, B2, B3, C2, C3, C6
236	LR25	—	Parc National des Cévennes	Languedoc-Roussillon	92,000	B2, B3, C6
237	LR26	—	Montagnes de Marcou, de l'Espinouse et du Caroux	Languedoc-Roussillon	14,500	C6
238	PAC01	FR143	Moyenne Vallée de la Durance	Provence-Alpes-Côte d'Azur	11,300	A4i, B1i, B2, C3, C6
239	PAC02	FR130	Camargue	Provence-Alpes-Côte d'Azur	76,500	A4i, A4iii, B1i, B2, B3, C2, C3, C4, C6
240	PAC03	FR131	Crau	Provence-Alpes-Côte d'Azur	40,100	A1, B2, C1, C2, C6
241	PAC04	FR133	Chaîne des Alpilles	Provence-Alpes-Côte d'Azur	21,800	B2, B3, C6
242	PAC05	FR134	Salines de l'Etang de Berre	Provence-Alpes-Côte d'Azur	530	C6
243	PAC06	FR135	Montagne Sainte-Victoire	Provence-Alpes-Côte d'Azur	15,000	B2, B3, C6
244	PAC07	FR136	Iles Marseillaises: Maire, Jarron, Jarron, Jarre, Riou, Calseraigne, Conglour et Pomègues	Provence-Alpes-Côte d'Azur	2,350	B2, C6
245	PAC08	FR132	Marais entre Crau et Grand Rhône: Meyranne, Chanoine, Plan de Bourg et Salins du Caban	Provence-Alpes-Côte d'Azur	5,650	A4iii, B1i, B2, C2, C3, C4, C6
246	PAC09	FR137	Massif du Petit Lubéron	Provence-Alpes-Côte d'Azur	14,000	B2, B3, C6
247	PAC10	FR138	Salins d'Hyères et de Pesquiers	Provence-Alpes-Côte d'Azur	1,000	A4i, B1i, B2, B3, C2, C6
248	PAC11	FR139	Iles d'Hyères	Provence-Alpes-Côte d'Azur	6,600	A4i, B1i, B2, C3, C6
249	PAC12	—	Plateau de Valensole	Provence-Alpes-Côte d'Azur	22,500	A1, B2, B3, C1, C6
250	PAC13	—	Plateau de l'Arbois, garrigues de Lancon et chaîne des Côtes	Provence-Alpes-Côte d'Azur	34,500	A1, B2, C1, C6
251	PAC14	—	Plaine des Maures	Provence-Alpes-Côte d'Azur	7,500	B2, C2, C6
252	PAC15	—	Etangs de Citis, Lavalduc, Engrenier, Pourra, l'Estomac, Fos et salines de Rassuen et de Fos	Provence-Alpes-Côte d'Azur	1,250	A4i, B1i, C2, C3, C6
253	PAC16	—	Marais de L'Ile Vieille	Provence-Alpes-Côte d'Azur	1,100	B1i, C2, C6
254	PAC17	FR143	Basse Vallée de la Durance	Provence-Alpes-Côte d'Azur	6,000	B1i, C3, C6
255	PAC18	FR143	Vallée de la Durance: de Tallard à Sisteron	Provence-Alpes-Côte d'Azur	1,650	B2, C6
256	PAC19	—	Vallée du Haut-Guil	Provence-Alpes-Côte d'Azur	5,000	B2, C6
257	PAC20	—	Marais de Manteyer	Provence-Alpes-Côte d'Azur	62	B2, B3, C6
258	PAC21	FR141	Bois des Ayes	Provence-Alpes-Côte d'Azur	450	C6
259	PAC22	—	Bois du Chapitre	Provence-Alpes-Côte d'Azur	250	C6
260	PAC23	FR142	Bec de Crigne	Provence-Alpes-Côte d'Azur	400	B2, C6
261	PAC24	—	Parc National du Mercantour	Provence-Alpes-Côte d'Azur	68,500	B2, B3, C2, C6
262	PAC25	—	Basse vallée du Var	Provence-Alpes-Côte d'Azur	1,100	A4i, B1i, C2, C6
263	PAC26	—	Préalpes de Grasse	Provence-Alpes-Côte d'Azur	8,500	B2, C6
264	PAC27	FR140	Parc National des Ecrins	Provence-Alpes-Côte d'Azur	120,000	B2, C2, C6
265	PAC28	—	Bois de Palayson, du Rouet et de Malvoisin	Provence-Alpes-Côte d'Azur	14,150	C7
266	CS01	FR144	Détroit de Bonifaccio et Iles Lavezzi	Corse	16,600	B1i, B1ii, B2, B3, C2, C6
267	CS02	FR145	Iles Cerbicale	Corse	5,040	B1i, B1ii, B2, B3, C2, C6

Table 1 ... continued. Summary of Important Bird Areas in France. 277 IBAs covering 47,248 km²

IBA code	National code	1989 code	International/National name	Administrative region	Area (ha)	Criteria (see p.11)
268	CS03	FR146	Golfe de Porto, presqu'île de Scandola et golfe de Galeria	Corse	28,700	A4i, B1i, B2, B3, C2, C3, C6
269	CS04	FR147	Forêts domaniales de Corse	Corse	50,217	A1, A2, A3, B2, C1, C2, C6
270	CS05	FR148	Vallée d'Asco	Corse	3,800	A1, A2, A3, B2, C1, C2, C6
271	CS06	FR149	Vallée de la Restonica	Corse	6,400	A1, A3, B2, C1, C2, C6
272	CS07	FR150	Etang de Biguglia	Corse	1,980	A4iii, B1i, C3, C4
273	CS08	—	Vallée du Régino	Corse	9,500	B2, B3, C2, C6
274	CS09	—	Etang d'Urbino et zones humides périphériques	Corse	5,200	B1i, C3
275	CS10	FR151	Iles Sanguinaires	Corse	175	B1i, C2, C6
276	CS11	FR153	Iles Finocchiarola et côte de Tamarone è Centuri	Corse	1,140	A1, B2, C1, C6
277	CS12	FR152	Vallée du Verghello	Corse	2,740	A1, A3, B2, C1, C2, C6

Sites identified in the previous inventory of IBAs in Europe (Grimmett and Jones 1989) but no longer considered to be IBAs

FR002 Etangs de St Hubert; FR005 Etangs de la Horre; FR006 Marais de Saint-Gond; FR016 Etangs de Galetas; FR021 Marais de la Sangurière et des Gorges; FR025 Bresse et zones inondables du Val de Saône; FR037 Rhin: Robertsau–Dalhunden; FR038 Rhin: Dalhunden–Munchhouse; FR071 Plaines de Pons–Rouffiac; FR079 Camp de Montmorillon; FR081 Bois de l'Hospice, étang de Beaufour et environs; FR093 Haute Vallée de l'Aude et massif des Pyrénées-Orientales; FR105 Etang de Suze la Rousse; FR116 Site name and location confidential; FR117 Site name and location confidential; FR118 Site name and location confidential; FR121 Aérodrome de Lézignan-Corbières.

Map 1. Location, area and criteria category of Important Bird Areas in France.

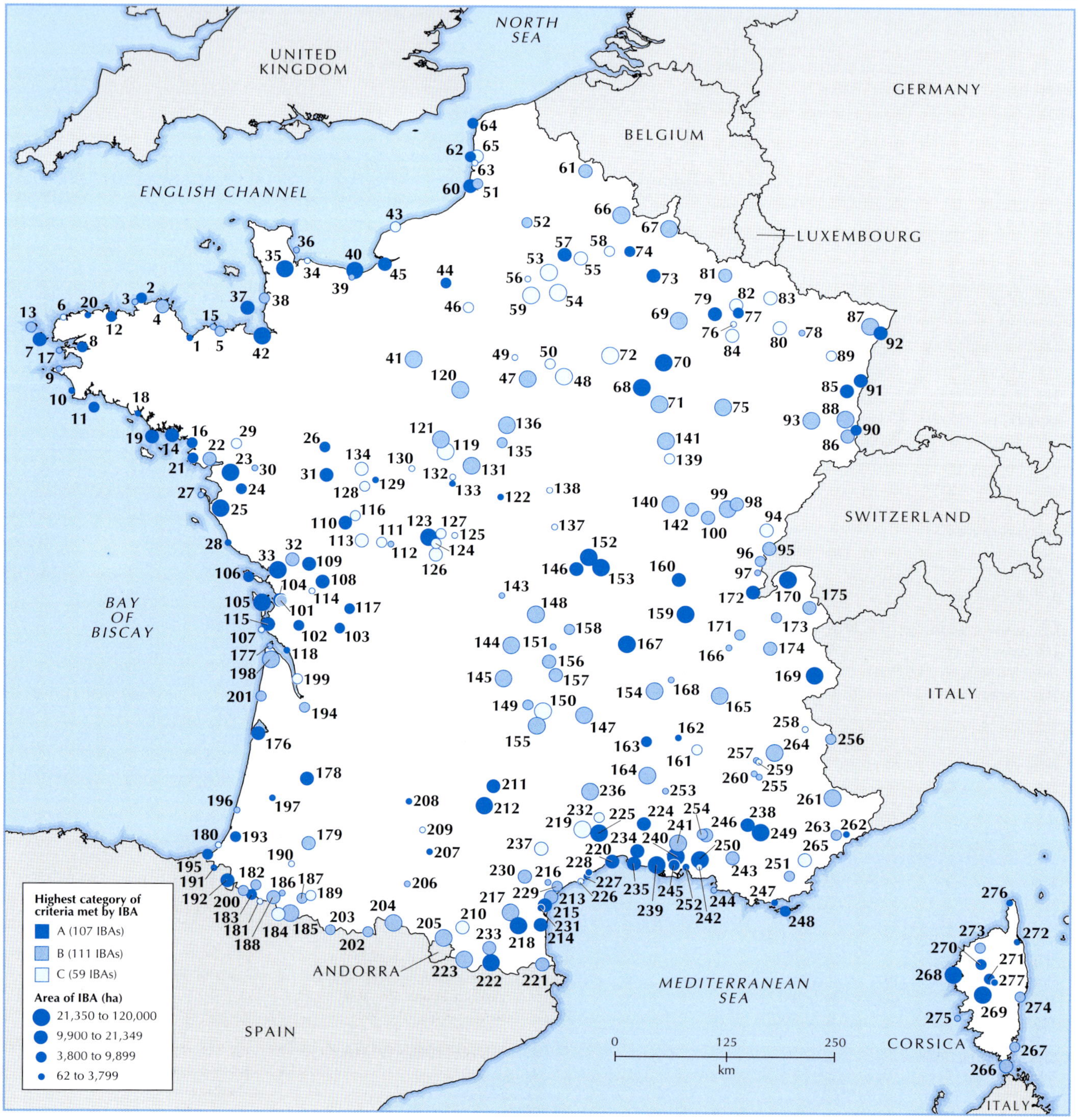

Table 2. Important Bird Areas in France that are important for species of global conservation concern (meeting criterion A1).

Species	IBA code
Aegypius monachus Cinereous Vulture	211
Falco naumanni Lesser Kestrel	240
Crex crex Corncrake	023, 026, 031, 035, 045, 057, 079, 102, 103, 122, 129, 133, 160
Tetrax tetrax Little Bustard	108, 109, 110, 117, 153, 212, 220, 224, 225, 234, 240, 249, 250
Larus audouinii Audouin's Gull	276
Acrocephalus paludicola Aquatic Warbler	010, 118
Sitta whiteheadi Corsican Nuthatch	269, 270, 271, 277

Table 3. Important Bird Areas inFrance that support important numbers of one or more congregatory species (i.e. meeting criteria A4 and/or B1). IBAs meeting both criteria A4 and B1 for the species are shown in **bold**. IBAs meeting only criterion B1 for the species concerned, and not A4, are shown in normal type. For key to 'Season', see p. 7.

Species	Season	IBA code
Tachybaptus ruficollis Little Grebe	P	**064**
Podiceps cristatus Great Crested Grebe	W	170
	P	**170**
Podiceps grisegena Red-necked Grebe	P	064
Podiceps auritus Slavonian Grebe	W	040, 106
Podiceps nigricollis Black-necked Grebe	B	**159**, **167**
	W	**008**, **014**, **220**, **228**, **252**
	P	**220**, **228**, **252**
Calonectris diomedea Cory's Shearwater	B	266
Hydrobates pelagicus Storm Petrel	B	267
Sula bassana Gannet	B	**002**
Phalacrocorax carbo Cormorant	W	135, 162, 228, 235, 238, 239, 252, 254, 274
	P	**035**, 070, **162**, 195, 220, 228, **239**, 272
Phalacrocorax aristotelis Shag	B	266, 267, 268, 275
Nycticorax nycticorax Night Heron	B	**207**, **208**
Egretta garzetta Little Egret	B	025, **105**, **115**, **176**, **234**, **235**, **239**
	W	**021**, 106, **176**, 220, **239**, 253
	P	**220**
Platalea leucorodia Spoonbill	B	024
	W	176
	P	014, 022, **028**, **035**, 039, **045**, 060, **064**, 105, 118, 176
Phoenicopterus ruber Greater Flamingo	B	**239**
	W	**214**, **215**, **220**, **235**, **247**, **252**
	P	**214**, **215**, **220**, **227**, **231**, **234**, **235**, **247**
Anser fabalis Bean Goose	W	**085**
Anser anser Greylag Goose	W	070
	P	033, **176**, **191**, **195**
Branta bernicla Brent Goose	W	**001**, **014**, **016**, **021**, **025**, **033**, 038, 042, **105**, **106**, **176**
	P	**025**, **035**, **064**
Tadorna tadorna Shelduck	W	**014**, 021, **025**, **060**, 235, **239**
	P	**064**
Anas penelope Wigeon	W	**042**, 239
Anas strepera Gadwall	B	**159**
	W	090, **091**, **092**, 123, **239**
	P	068, **070**, 123, 159
Anas crecca Teal	W	023, **042**, 193, **239**
	P	070
Anas platyrhynchos Mallard	W	239
Anas acuta Pintail	W	001, 014, 016, 021, 023, 028, 035, 045, 060, 105, 176, 193
	P	022, 024, 026, 028, 030, 035, 060, 064, 073, 105, 159, 176
Anas clypeata Shoveler	B	025
	W	014, 021, 022, 023, 024, 025, 028, 030, 070, 104, 105, 123, 131, 193, **239**
	P	**024**, 026, **028**, 035, 061, 064, 068, 070, 073, 078, 104, 105, 123, 159, 176
Netta rufina Red-crested Pochard	B	159, **239**
	W	**239**, 245
	P	159
Aythya ferina Pochard	B	159
	W	044, **070**, 092, 123, 166, 171, 252
	P	123
Aythya fuligula Tufted Duck	W	091, **092**, 170, 252
Aythya marila Scaup	W	**170**
Melanitta nigra Common Scoter	W	**042**
	P	**042**
Mergus serrator Red-breasted Merganser	W	**014**
	P	**014**
Gyps fulvus Griffon Vulture	W	185
Fulica atra Coot	W	239
	P	024
Grus grus Crane	W	**070**, **178**, **197**
	P	**068**, **070**, **077**, 081, **123**, 143, 144, **146**, 151, **152**, 158, 163, **178**, **191**, **192**, **195**, 201
Haematopus ostralegus Oystercatcher	B	220, 235, 239
	W	001, **035**, **042**, **060**, **064**
	P	064
Himantopus himantopus Black-winged Stilt	B	**025**, 220, 234, 239
	P	252
Recurvirostra avosetta Avocet	B	**025**, 027, **028**, **235**, **239**
	W	014, **016**, **021**, **023**, **025**, **033**, **045**, **106**, 220, **239**
	P	028, **033**, **045**, **105**, **176**
Charadrius dubius Little Ringed Plover	P	**060**
Charadrius hiaticula Ringed Plover	W	**176**
	P	**020**, **035**, **042**, **064**, **239**
Charadrius alexandrinus Kentish Plover	P	**064**, **239**
Pluvialis apricaria Golden Plover	P	112, 118, 120, 121
Pluvialis squatarola Grey Plover	W	**014**, **025**, **033**, **042**, **060**, 105, **106**
	P	**025**, **033**, **042**, **060**, **064**, 105, **176**
Vanellus vanellus Lapwing	P	**073**
Calidris canutus Knot	W	**033**, **042**, 060, **105**
	P	**033**, 060, 101, 105, **176**, 177
Calidris alba Sanderling	P	**010**, **020**, **060**
Calidris minuta Little Stint	W	**215**
Calidris ferruginea Curlew Sandpiper	P	**176**, **239**
Calidris alpina Dunlin	W	**014**, 025, **042**, 060, 104, **105**, 106, **176**
	P	025, 033, 064, 105, **176**
Limosa limosa Black-tailed Godwit	P	**026**, **028**, **033**, **042**, **239**, 245
Limosa lapponica Bar-tailed Godwit	W	**025**, **042**
	P	**020**, **025**, **033**, **045**, **064**, **176**
Numenius phaeopus Whimbrel	P	**028**, **033**
Numenius arquata Curlew	W	**025**, **035**, **042**, **060**
	P	**064**
Tringa erythropus Spotted Redshank	P	**070**, **176**, **239**
Tringa totanus Redshank	B	025
	P	014, 020, **028**, 033, **045**, **064**, 104, 105, **118**, 176
Arenaria interpres Turnstone	W	**012**, **021**, **106**
	P	**019**, **020**
Larus melanocephalus Mediterranean Gull	P	**262**
Larus minutus Little Gull	P	**033**, **040**
Larus genei Slender-billed Gull	B	**239**
Larus fuscus Lesser Black-backed Gull	B	**007**, **011**
	P	**040**, **062**, **064**
Larus argentatus Herring Gull	B	007, 011
	W	021, 062, **064**
	P	064
Larus cachinnans Yellow-legged Gull	B	**214**, **215**, **235**, **248**, **268**
	W	**195**, **214**, **220**, **238**, **252**
Larus marinus Great Black-backed Gull	W	**062**
	P	**064**
Gelochelidon nilotica Gull-billed Tern	B	234, **235**, **239**
Sterna caspia Caspian Tern	P	239
Sterna sandvicensis Sandwich Tern	B	**012**, **176**, 235, **239**
	P	035, **037**, **040**, 045, **064**
Sterna dougallii Roseate Tern	B	**012**
	P	**064**
Sterna hirundo Common Tern	B	239
	P	**035**, **040**, 045, **064**
Sterna paradisaea Arctic Tern	P	**064**
Sterna albifrons Little Tern	B	031, 227, **235**, **239**
	P	**035**, 040, **064**
Chlidonias hybridus Whiskered Tern	B	**123**, **159**, **167**
	P	**123**, 159, 215, 220, 229, 234, **239**
Chlidonias niger Black Tern	P	**035**, **239**

Table 4. Species of European conservation concern and species listed on Annex I of the EC Birds Directive with significant breeding populations at IBAs in France.

Species [1]	Minimum national breeding population (pairs) [2]	Proportion (%) of national population breeding at all IBAs in France
Calonectris diomedea Cory's Shearwater	810	100[3]
Hydrobates pelagicus Storm Petrel	500	64
Sula bassana Gannet	13,500	100
Phalacrocorax aristotelis Shag	5,000	61
Botaurus stellaris Bittern	300	97
Ixobrychus minutus Little Bittern	242	95
Nycticorax nycticorax Night Heron	4,176	88
Ardeola ralloides Squacco Heron	127	69
Egretta garzetta Little Egret	9,850	61
Ardea purpurea Purple Heron	1,978	87
Ciconia nigra Black Stork	22	27
Ciconia ciconia White Stork	315	29
Platalea leucorodia Spoonbill	53	19
Phoenicopterus ruber Greater Flamingo	13,500	74
Anas strepera Gadwall	1,000	89
Netta rufina Red-crested Pochard	270	100[3]
Aythya ferina Pochard	2,600	91
Pernis apivorus Honey Buzzard	8,000	11
Milvus migrans Black Kite	5,800	25
Milvus milvus Red Kite	2,000	18
Gypaetus barbatus Lammergeier	30	80
Neophron percnopterus Egyptian Vulture	63	59
Gyps fulvus Griffon Vulture	340	45
Circaetus gallicus Short-toed Eagle	770	38
Circus aeruginosus Marsh Harrier	1,000	88
Circus cyaneus Hen Harrier	2,500	14
Circus pygargus Montagu's Harrier	3,000	28
Aquila chrysaetos Golden Eagle	274	57
Hieraaetus pennatus Booted Eagle	250	27
Hieraaetus fasciatus Bonelli's Eagle	25	84
Pandion haliaetus Osprey	42	43
Falco naumanni Lesser Kestrel	48	88
Falco tinnunculus Kestrel	35,000	<1
Falco peregrinus Peregrine	700	22
Bonasa bonasia Hazel Grouse	2,000	49
Tetrao tetrix Black Grouse	8,000	16
Tetrao urogallus Capercaillie	4,000	13
Alectoris graeca Rock Partridge	2,000	49
Perdix perdix Partridge	900,000	<1
Porzana porzana Spotted Crake	100	61
Porzana parva Little Crake	1	100[3]
Crex crex Corncrake	1,249	83
Tetrax tetrax Little Bustard	544	100[3]
Himantopus himantopus Black-winged Stilt	1,850	65
Recurvirostra avosetta Avocet	2,500	94
Burhinus oedicnemus Stone Curlew	5,000	23
Glareola pratincola Collared Pratincole	6	100[3]
Charadrius alexandrinus Kentish Plover	1,500	62
Charadrius morinellus Dotterel	2	100
Philomachus pugnax Ruff	1	100[3]
Tringa totanus Redshank	1,400	71
Larus melanocephalus Mediterranean Gull	871	79
Larus audouinii Audouin's Gull	54	100[3]
Larus fuscus Lesser Black-backed Gull	23,563	47
Larus marinus Great Black-backed Gull	2,310	85
Gelochelidon nilotica Gull-billed Tern	225	100[3]
Sterna sandvicensis Sandwich Tern	7,000	87
Sterna dougallii Roseate Tern	100	100[3]
Sterna hirundo Common Tern	4,461	77
Sterna albifrons Little Tern	1,000	100[3]
Chlidonias hybridus Whiskered Tern	2,117	48
Chlidonias niger Black Tern	237	54
Pterocles alchata Pin-tailed Sandgrouse	115	100[3]
Otus scops Scops Owl	1,000	8
Bubo bubo Eagle Owl	950	40
Glaucidium passerinum Pygmy Owl	50	100[3]
Athene noctua Little Owl	10,000	1
Asio flammeus Short-eared Owl	10	100
Aegolius funereus Tengmalm's Owl	1,000	21
Caprimulgus europaeus Nightjar	20,000	8
Alcedo atthis Kingfisher	2,500	30
Merops apiaster Bee-eater	6,800	3
Coracias garrulus Roller	500	42
Jynx torquilla Wryneck	5,000	<1
Picus canus Grey-headed Woodpecker	1,000	31
Dryocopus martius Black Woodpecker	3,000	32
Dendrocopos medius Middle Spotted Woodpecker	5,000	24
Dendrocopos leucotos White-backed Woodpecker	200	12
Picoides tridactylus Three-toed Woodpecker	10	10
Melanocorypha calandra Calandra Lark	81	43
Calandrella brachydactyla Short-toed Lark	1,000	100[3]
Galerida theklae Thekla Lark	10	100[3]
Lullula arborea Woodlark	10,000	23
Riparia riparia Sand Martin	50,000	3
Anthus campestris Tawny Pipit	20,000	3
Luscinia svecica Bluethroat	3,000	67
Monticola saxatilis Rock Thrush	1,250	42
Monticola solitarius Blue Rock Thrush	500	8
Turdus torquatus Ring Ouzel	3,000	24
Locustella luscinioides Savi's Warbler	1,000	4
Acrocephalus melanopogon Moustached Warbler	1,000	68
Sylvia sarda Marmora's Warbler	30,000	1
Sylvia undata Dartford Warbler	40,000	8
Sylvia cantillans Subalpine Warbler	10,000	19
Sylvia hortensis Orphean Warbler	1,000	<1
Sitta whiteheadi Corsican Nuthatch	2,000	58
Lanius collurio Red-backed Shrike	160,000	2
Lanius minor Lesser Grey Shrike	52	38
Lanius senator Woodchat Shrike	5,900	2
Pyrrhocorax pyrrhocorax Chough	1,000	100[3]
Serinus serinus Serin	100,000	1
Emberiza hortulana Ortolan Bunting	10,000	7

1. Only those species of European conservation concern (see Box 1, p. 12) that meet IBA criteria in France are listed, together with those species listed on Annex I of the EC Birds Directive that fulfil criterion C6 in IBAs in France.
2. Data are taken from the BirdLife/EBCC European Bird Database 1998 (Heath and Borggreve 2000).
3. The percentage of the national population in IBAs exceeds 100%. Usually this is because the national population estimate has not been updated recently whilst the IBA population estimate has been recently updated with new data as a result of comprehensive surveys of IBAs themselves. Also, the individual site count for a species may be the maximum or average over recent years, and summing these may record more birds than are present nationally in any single year.

ORNITHOLOGICAL IMPORTANCE

As a result of France's geographical position and habitat diversity, two-thirds of European bird species regularly occur in the country (Cruon *et al.* 1992). One hundred and sixty-two species of European conservation concern (SPECs1–4; Tucker and Heath 1994) breed regularly, six are of global conservation concern (SPEC 1): *Aegypius monachus*, *Falco naumanni*, *Crex crex*, *Tetrax tetrax*, *Larus audouinii* and *Sitta whiteheadi*. Furthermore, the globally threatened *Acrocephalus paludicola* occurs on passage.

Twenty-six breeding SPECs have an unfavourable conservation status in Europe and have their populations concentrated in Europe (SPEC 2) and 69 have an unfavourable conservation status in Europe (SPEC 3) but their world population is not concentrated in Europe.

Thirty-four sites are selected on the basis of holding internationally important populations of species of global conservation concern, meeting criteria A1 (Table 2). The entire French breeding populations of *Larus audouinii* and 88% of the breeding population of *Falco naumanni* are found in IBAs (Table 4). Eighty-three percent of the meadows in which *Crex crex* breed are in IBAs. France holds the entire world population of the European endemic *Sitta whiteheadi*, with nearly 60% of its global population found within IBAs (Tables 2 and 4).

One hundred and seventy IBAs support important populations of SPECs, meeting criteria B2/B3 (Table 1). Substantial breeding populations of the SPECs listed in Table 4 occur in IBAs. In addition, 252 IBAs support important populations of the species listed on Annex I of the EC Birds Directive, meeting criterion C6 (Tables 1 and 4).

France is particularly important for congregatory waterbirds with 116 sites supporting the species listed in Table 3, meeting criteria A4 and B1. A further 44 sites regularly hold 20,000 or more waterbirds, either breeding, on passage, or in winter. Coastal wetlands are particularly important for passage waders moving to and from their breeding and wintering grounds, as well as wintering waterbirds including *Phoenicopterus ruber* (40% of the European population winters at IBA 239), *Branta bernicla, Anas strepera* and *Recurvirostra avosetta.* Of particularly high importance for waterbirds are the Camargue (239), Vallée du Rhin (090, 091, 092), Bassin d'Arcachon et Banc d'Arguin (176), Baie de l'Aiguillon et Pointe d'Arçay (033) and Golfe du Morbihan et Etier de Penerf (014).

In addition to passage waterbirds, France has a number of important bottleneck sites for diurnal passage raptors, particularly *Pernis apivorus, Milvus migrans* and *Milvus milvus* and also storks, cranes and passerines. These are predominantly in the Massif Central, the Alps and Pyrenees. Some of the most important sites are Haute Soule : Forêt d'Irraty, Organbidexka et Pic des Escaliers (183), Monts et Plomb du Cantal (149), Etangs de Leucate et Lapalme (214), Col de l'Escrinet (163) and Pointe de Grave (177).

Four sites (269, 270, 271, 277) are important for species characteristic of the Mediterranean biome, in particular for the presence of the endemic *Sitta whiteheadi*. A single site meets the A3 criterion for the Eurasian high-montane biome (169), holding characteristic species such as *Tichodroma muraria* and *Montifringilla nivalis*.

Some dispersed breeding species in France are poorly covered by the present inventory, including *Circus cyaneus, Burhinus oedicnemus, Otus scops, Emberiza hortulana, Lanius collurio* and *Jynx torquilla*.

HABITATS

Four main vegetation zones can be identified in France: the alpine zone in the Alps, Corsica and the Pyrenees, which is characterized by formations of *Pinus, Rhododendron, Vaccinium* and alpine grasslands; the Mediterranean zone, which is dominated by formations of *Quercus* spp., *Pinus* and scrub; the Atlantic zone, which stretches from the Basque region in the south to Normandy in the north, has important areas of heathland and on calcareous soils a northern extension of Mediterranean vegetation; and the mid-European zone, which occupies the east of France and the Massif Central.

Within IBAs, wetland habitats are the most widespread, occurring in 84% (Figure 1), reflecting the importance of France for passage and wintering waterbirds such as *Branta bernicla, Anas clypeata*, and *Pluvialis squatarola.* Most wetlands lie along the Atlantic and Mediterranean coasts and in the regions of Dombes, Sologne, Brenne and Alsace. In France as a whole, only 3% of the national land-area is covered by wetlands, with more than a quarter of these in coastal areas. In addition, coastal habitats support important breeding and passage congregations of seabirds, including *Puffinus yelkouan, Calonectris diomedea,* and *Larus audouinii* (sites 244, 266, 276).

Woodland occupies approximately 25% of France's land area and wooded habitats are also widely distributed in IBAs, being found in 72% (Figure 1). They are particularly characteristic of eastern France and the Massif Central region, where most forests are dominated by *Fagus, Quercus* and *Abies* spp., for example in IBAs 086, 087, 095, 140, 169, 188, 269. These forests support species such as *Dryocopus martius, Ficedula albicollis, Pernis apivorus, Bonasa bonasia, Hieraaetus pennatus, Picoides tridactylus* and *Glaucidium passerinum.*

Figure 1. Occurrence of habitats at Important Bird Areas in France (see Appendix 3 for definitions of habitats).

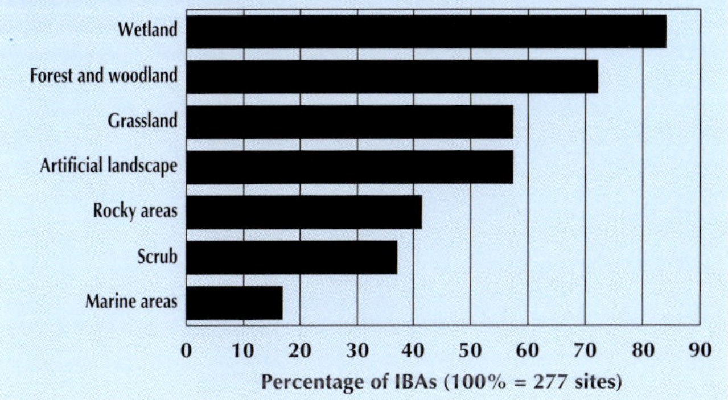

Natural, semi-natural and artificial grasslands cover 28–30% of France and are found in 57% of IBAs (Figure 1). These range from alpine grasslands above the treeline to dry steppe/calcareous grasslands. The latter, such as those in site 240, support species such as *Falco naumanni, Tetrax tetrax*, and *Melanocorypha calandra.* Wet meadows also have a characteristic bird fauna and are particularly important for *Crex crex* (for example sites 026 and 102).

Scrub habitats include heathlands, which occur in the Atlantic zone on acid soils, and garrigue and maquis in the Mediterranean region. The later are important for species such as *Hieraaetus fasciatus, Bubo bubo, Coracias garrulus* and several *Sylvia* warblers, particularly *S. conspicillata, S. cantillans* and *S. melanocephala.* Overall, scrub habitats are found in 37% of IBAs (Figure 1).

IMPACT ON IBAs – LAND-USE AND THREATS

The extent of different land-uses within individual IBAs is generally poorly known. Hunting, urban, industrial and transport, agriculture, tourism and recreation, and fisheries and aquaculture are widespread land-uses, all occurring in over 70% of IBAs (Figure 2). The other significant form of land-use is forestry, which occurs in 50% of sites.

According to Rocamora *et al.* (1995), 78 IBAs identified in the national inventory had already suffered serious damage, and the current inventory identifies damage to 133 sites. The most widespread land-uses are generally the activities which also pose the greatest threat to IBAs and their bird faunas; the most common threats being tourism and recreation activities, agricultural intensification, hunting, infrastructure developments and forestry (Figure 3). In terms of the total land area affected, rather than the number of IBAs, agricultural intensification and expansion

Figure 2. Occurrence of land-uses at Important Bird Areas in France (see Appendix 3 for definitions of land-uses).

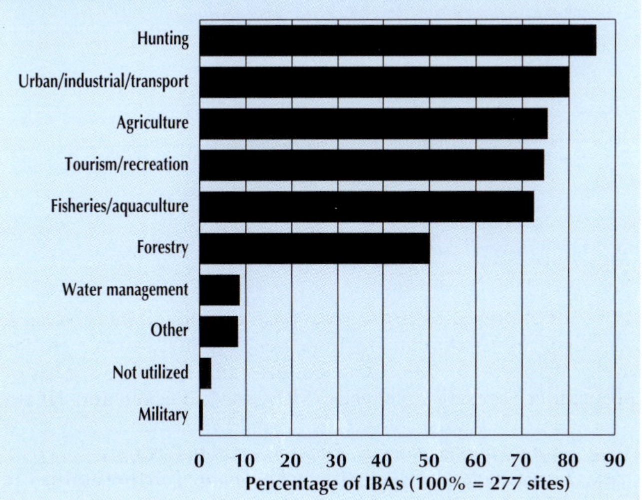

Figure 3. Threats at Important Bird Areas in France (see Appendix 3 for definitions of threat types and impact categories).

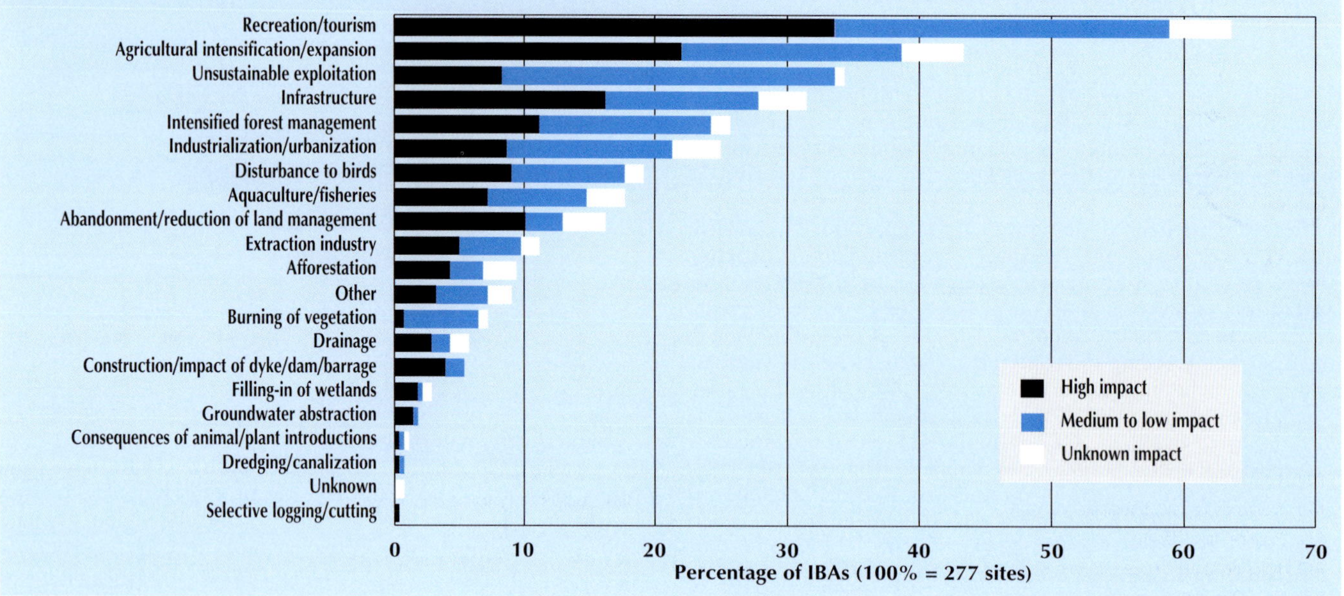

represents the principal cause of habitat degradation in IBAs (Rocomora *et al.* 1995).

Several very significant IBAs, holding rare species and habitats, remain under threat even though they are designated as Special Protection Areas under the EC Birds Directive. The Crau (240) is threatened by state development projects (military activities, motorways, etc.), and the Loire Valley (031, 128, 154), containing the last unregulated river in France, is seriously threatened by the possible construction of barrages.

PROTECTION STATUS

National protection

At the national level only 58% of IBAs are afforded some degree of protection, with only 11% having more than 90% of their total area protected (Figure 4). The two main laws governing the conservation of natural habitats in France are the National Parks Act (1960) and the Nature Conservation Act (1976). A number of different types of protected area can be designated according to these acts, including:

1. National Park
Established by a decree from the Council of State, National Parks include a core zone in which human activities are restricted in order to ensure the integrity of the natural environment. A peripheral zone acts as a buffer between the core and the exterior of the park.

2. Nature Reserve
Reserves are created by ministerial decree and aim to conserve the natural environment, its fauna, flora and geological features. Human activities are restricted and sites are managed by state or private organizations.

3. Biological State Reserve
Sites are owned by the State and are managed by the Office National des Forêts. Human activities are restricted and sites are managed to maintain their biological interest.

4. Private Reserve
Protected areas owned by private individuals or organizations. Human activities are restricted.

5. Hunting Reserve
Created by the Ministry of the Environment, hunting is forbidden in these areas.

International protection

Only 38% of IBAs receive any kind of international protection, with only 14% having 90% or more of their area under protection (Figure 5). Box 1 shows the conventions and EC directives ratified by France. France has designated eight Biosphere Reserves, six of which are IBAs (Table 5). There are currently 15 designated Ramsar Sites, which overlap with 22 IBAs, covering nearly 9% of the total area of IBAs in France (Table 5). One hundred and seven Special

Figure 4. The national protection status of Important Bird Areas in France.

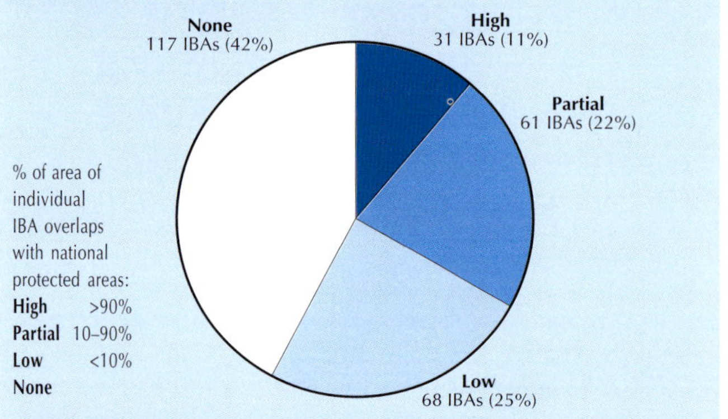

Total area of overlap between IBA network in France and national protected-area system (see Table 5 for categories) = 6,492–7,179 km² (14–15% of total IBA area).

Figure 5. The international protection status of Important Bird Areas in France.

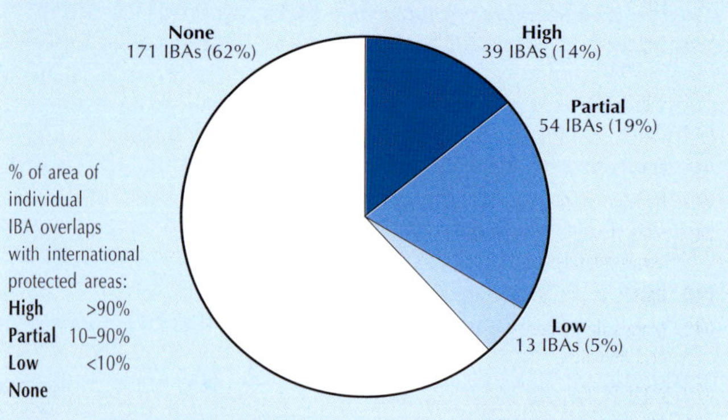

Total area of overlap between IBA network in France and international protected-area system (see Table 5 for categories) = 11,168–11,281 km² (24% of total IBA area).

Table 5. Protection status of Important Bird Areas in France.
A tick (✔) indicates that an IBA overlaps with a protected area (to any extent).

		National					International		
IBA code	International name	National Park	Nature Reserve	Biological State Reserve	Private Reserve	Hunting Reserve	Ramsar Site	Biosphere Reserve	Special Protection Area
001	Baie de Saint-Brieuc		✔			✔			✔
002	Archipel des Sept-Iles		✔			✔			✔
003	Ile de Goulmedec					✔			✔
004	Estuaires du Trieux et du Jaudy					✔			✔
005	Iles de la Colombière, de la Nellière et des Haches				✔	✔			✔
006	Ilôts de Trévorc'h				✔	✔			✔
007	Archipel de Molène		✔		✔	✔		✔	✔
008	Rade de Brest: Baie de Daoulas et Anse du Poulmic					✔			✔
009	Cap Sizun				✔	✔			✔
010	Marais de la Baie d'Audierne				✔	✔			✔
011	Archipel des Glénan		✔		✔	✔			✔
012	Baies de Morlaix et de Carantec				✔	✔			✔
013	Ile d'Ouessant				✔	✔			✔
014	Golfe du Morbihan et Etier de Penerf		✔		✔	✔	✔		✔
015	Cap Frehel					✔			✔
016	Baie de Vilaine					✔			✔
017	Presqu'île de Crozon, Tas de Pois et Rochers du Toulinguet				✔	✔			
018	Rade de Lorient					✔			✔
019	Baie de Quiberon					✔			✔
020	Baie de Goulven					✔			
021	Traicts et marais salants de la Presqu'île Guérandaise					✔	✔		✔
022	Marais de Brière					✔	✔		
023	Estuaire de la Loire					✔			✔
024	Lac de Grand-Lieu		✔				✔		✔
025	Baie de Bourgneuf et marais breton				✔	✔			
026	Basses Vallées Angevines				✔		✔		✔
027	Marais salants de Noirmoutier		✔						
028	Marais et forêt d'Olonne					✔			
029	Forêt du Gavre								
030	Marais de Mazerolles – Petit Mars				✔	✔			
031	Vallée de la Loire: de Nantes à Montsoreau					✔			
032	Plaine calcaire du sud Vendée								
033	Marais poitevin et baie de l'Aiguillon		✔		✔	✔			✔
034	Falaise du Bessin								✔
035	Baie des Veys et Marais du Cotentin		✔			✔	✔		✔
036	Iles Saint Marcouf				✔	✔			✔
037	Iles Chausey				✔	✔			✔
038	Havre de la Sienne								
039	Estuaire de l'Orne					✔			✔
040	Littoral Augeron								
041	Forêts du Perche								
042	Baie du Mont Saint Michel et Ile des Landes				✔	✔	✔		✔
043	Cap Fagnet					✔			✔
044	Boucle de Pose et de Muids				✔				
045	Estuaire et embouchure de la Seine		✔			✔			✔
046	Boucle de Moisson								
047	Massifs de Fontainebleau et zones humides adjacentes				✔			✔	
048	Bassée et plaines adjacentes					✔			
049	Marais de Fontenay-le-Vicomte et d'Itteville								✔
050	Etang et Forêt de Villefermoy								
051	Marais arrières littoraux picards								
052	Etangs et marais du bassin de la Somme		✔			✔			
	Subtotal of IBAs	0	11	0	19	37	7	2	32
053	Forêts picardes: Compiègne, Laigue, Ourscamp								
054	Forêts picardes: massif de Retz								
055	Forêts picardes: massif de Saint-Gobain								
056	Marais de Sacy								
057	Vallée de l'Oise de Thourotte à Vendeuil								✔
058	Marais de la Souche								
059	Forêts picardes: massif des Trois Forêts et bois du Roi								
060	Estuaires picards: baies de Somme et d'Authie		✔		✔	✔	✔		✔
061	Vallées de la Scarpe et de l'Escaut								
062	Estuaire de la Canche		✔			✔			✔
063	Marais de Balançon et de Villiers								✔
064	Cap Gris-nez					✔			✔
065	Plateaux agricoles des environs de Frencq								
066	Forêts de Thiérache: Trelon, Fourmies, Hirson et St-Michel								
067	Plateau Ardennais		✔						
068	Lacs de la Forêt d'Orient					✔	✔		✔
069	Etangs d'Argonne						✔		
070	Lac du Der-Chantecoq et étangs latéraux				✔	✔	✔		✔
071	Barois et Forêt de Clairvaux								
072	Vallée de l'Aube, de la Superbe et Marigny								✔
073	Vallée de l'Aisne								
074	Confluent des vallées de la Meuse et de la Chiers								
075	Bassigny								
076	Etangs de la Woëvre: lac de Madine					✔			✔
077	Etangs de la Woëvre: la Chaussée				✔		✔		✔
078	Etang de Lindre								
079	Vallée de la Meuse								
080	Plateau de Delme, Val de Petite Seille								
081	Val de Chiers et environs de Spincourt								
082	Fresnes-en-Woëvre–Mars-la-Tour								
083	Bazoncourt–Vigy								
084	Forêt de la Reine								
085	Ried de Colmar et Sélestat		✔		✔				
086	Forêt domaniale de la Harth			✔					
087	Forêt de Haguenau								
088	Zones agricoles de la Hardt								
089	Massif des Vosges: crêtes de Schneeberg au Donon			✔					
090	Vallée du Rhin: Village-neuf à Biescheim		✔			✔			
091	Vallée du Rhin: Marckolsheim à Strasbourg		✔			✔			
092	Vallée du Rhin: Strasbourg à Lauterbourg		✔	✔		✔			
093	Massif des Vosges: Hautes Vosges		✔	✔					
094	Bassin du Drugeon: Pontarlier-Frasne				✔				
095	Forêts d'altitude du Jura: forêt du Risol et Mont d'Or					✔			
096	Forêts d'altitude du Jura: forêt du Risoux								
097	Forêts d'altitude du Jura: forêt du Massacre								
098	Forêt de Chaux					✔			
099	Vallée de la Saône de Corre à Broye								
100	Basse vallée du Doubs: Dole Sud		✔						
101	Estuaire de la Charente								✔
	Subtotal of IBAs	0	20	4	24	48	12	2	43

Table 5 ... continued. Protection status of Important Bird Areas in France.
A tick (✔) indicates that an IBA overlaps with a protected area (to any extent).

		National					International		
IBA code	International name	National Park	Nature Reserve	Biological State Reserve	Private Reserve	Hunting Reserve	Ramsar Site	Biosphere Reserve	Special Protection Area
102	Vallée de la Charente et de la Seugne (Cabariot-Pons/St-Sever-de-Saintonge)								✔
103	Vallée de la Charente: amont d'Angoulême								
104	Anse de Fouras, baie d'Yves et marais de Rochefort		✔						✔
105	Ile d'Oléron, marais de Brouage–Saint-Agnant		✔						✔
106	Anse du Fiers d'Ars en Ré		✔			✔			✔
107	Bonne Anse					✔			
108	Plaine de Niort (sud-est)								
109	Plaine de Niort (nord-ouest)								
110	Plaines de Saint-Jouin de Marne et d'Assais-les-Jumeaux								
111	Le Pinail, Forêt de Moulière et bois du Défens		✔						✔
112	Plateau de Bellefonds								
113	Plaines de Mirebeau et de Neuville-du-Poitou								
114	Region de Pressac, étang de Combourg								
115	Marais et estuaire de la Seudre								
116	Plaine de Saint-Jean-de-Sauves								
117	Plaine de Villefagnan					✔			
118	Estuaire de la Gironde: marais de la rive nord					✔			
119	Vallée de la Loire: environs de Blois					✔			
120	Vallée de la Conie et Beauce centrale								
121	Petite Beauce		✔						
122	Vallée de l'Yèvre								
123	Brenne centrale		✔			✔	✔		✔
124	Brenne Sud						✔		
125	Queue de Brenne: Etang des Loges, Grand et Petit Mez						✔		
126	Brenne: Forêt de la Petite-Brenne						✔		
127	Brenne: Forêt de Lancosme						✔		
128	Vallée de la Loire: confluence Loire-Vienne								
129	Basse vallée de L'Indre								
130	Vallée de la Loire: environs de Montlouis-sur-Loire					✔			
131	Etangs de Sologne: St Viatre, Marcilly-en-Gault et forêt de Bruadan					✔			
132	Etangs de Sologne: étang de l'Arche								
133	Vallée du Fouzon								
134	Lac de Rillé et forêts voisines					✔			
135	Vallée de la Loire: Orléanais		✔			✔			
136	Forêt d'Orléans: massifs d'Ingrannes et de Lorris								
137	Mars sur Allier					✔			✔
138	Vallée de la Loire: La Charité-sur-Loire		✔			✔			✔
139	Forêt de Jugny								
140	Forêt de Citeaux et environs			✔					
141	Forêt de Châtillon et environs								
142	Arrière Côte de Dijon et de Beaune								
143	Etang des Landes								
144	Plateaux de Millevaches et de Gentioux								
145	Gorges de la Dordogne								
146	Val d'Allier Bourbonnais		✔						✔
147	Haut Val d'Allier								
148	Gorges de la Sioule								
149	Monts et Plomb du Cantal								✔
150	Planèze de Saint-Flour								
Subtotal of IBAs		0	29	5	24	60	17	2	53

		National					International		
IBA code	International name	National Park	Nature Reserve	Biological State Reserve	Private Reserve	Hunting Reserve	Ramsar Site	Biosphere Reserve	Special Protection Area
151	Montagne de la Serre								
152	Sologne Bourbonnaise								
153	Vallée de la Loire: Pit majeur d'Iguerande à Décize					✔			
154	Vallée de la Loire: gorges de la Loire					✔			
155	Gorges de la Truyère								
156	Les Couzes Nord								
157	Les Couzes Sud								
158	Val d'Allier: Saint-Yorre-Joze				✔	✔			
159	La Dombes								
160	Val de Saône		✔						✔
161	Forêt de Saou								
162	Val de Drôme: Les Ramières-printegarde		✔			✔			✔
163	Col de l'Escrinet								
164	Basse Ardêche		✔						
165	Hauts Plateaux du Vercors et Forêt des Coulmes		✔						✔
166	Iles du Haut-Rhône				✔				✔
167	Plaine du Forez					✔			
168	Ile de la Platière		✔						
169	Parc National de la Vanoise	✔	✔						✔
170	Lac Léman (partie française)		✔			✔	✔		✔
171	Lac et marais du Bourget		✔			✔			✔
172	Haute chaîne du Jura: défilé de l'écluse, Etournel et Mont Vuache		✔			✔			
173	Montagne des Frêtes et plateau des Glières			✔					
174	Les Bauges					✔			
175	Haut Giffre		✔			✔			
176	Bassin d'Arcachon et Banc d'Arguin		✔		✔	✔			✔
177	Pointe de Grave								
178	Champ de Tir du Poteau								✔
179	Barthes de l'Adour					✔			✔
180	Rochers de Biarritz: le Bouccalot et la Roche Ronde				✔				
181	Haute Soule: Massif Forestier, Gorges d'Holzarte et d'Olhadubi								
182	Haute Soule: Forêt des Arbailles								
183	Haute Soule: Forêt d'Irraty, Organbidexka et Pic des Escaliers								
184	Rive Gauche de la Vallée d'Aspe et Pic d'Anie	✔							
185	Hautes Vallées d'Aspe et d'Ossau	✔							✔
186	Eth Turoun des Aureys								
187	Pènes du Moulle de Jaut		✔						✔
188	Haute Soule: massif de la Pierre-St-Martin								
189	Pic de l'Estibet et Mondragon								
190	Barrage d'Artix et Saligue du Gave de Pau					✔			
191	Col de Lizarrieta								
192	Vallée de la Nive des Aldudes–Col de Lindux								
193	Domaine d'Orx		✔						✔
194	Garonne: marais de Bordeaux		✔						✔
195	Estuaire de la Bidassoa et Baie de Fontarabie					✔			
196	Réserve Naturelle du courant d'Huchet		✔			✔			✔
197	Arjuzanx					✔			
198	Estuaire de la Gironde: marais du Nord Médoc								✔
199	Estuaire de la Gironde: marais du Blayais					✔			
200	Haute Cize: Pic d'Errozate et Forêt d'Orion								
Subtotal of IBAs		3	44	6	28	77	18	2	69

Table 5 ... continued. Protection status of Important Bird Areas in France.
A tick (✔) indicates that an IBA overlaps with a protected area (to any extent).

		National					International		
IBA code	International name	National Park	Nature Reserve	Biological State Reserve	Private Reserve	Hunting Reserve	Ramsar Site	Biosphere Reserve	Special Protection Area
201	Côte Médocaine: dunes boisées et dépressions humides		✔						✔
202	Vallées du Lis et de la Pique					✔			
203	Cirque de Gavarnie	✔							✔
204	Vallée de Melles, cols d'Aoueran et d'Artisagou et Mont Vallier					✔			
205	Zones rupestres du Tarasconnais et massif d'Aston								
206	Vallée de la Garonne: Boussens à Carbonne					✔			
207	Vallée de la Garonne: Palayre et environs					✔			
208	Vallée de la Garonne: Moissac								
209	Vallée de la Garonne: méandre de Grisolles								
210	Gorges de la Frau et Belestat								
211	Gorges du Tarn et de la Jonte								✔
212	Gorges de la Dourbie et causses avoisinants	✔						✔	✔
213	Montagne de la Clape								✔
214	Etangs de Leucate et Lapalme					✔			
215	Etangs Narbonnais					✔			
216	Etang de Capestang								
217	Hautes Corbières					✔			
218	Basses Corbières					✔			
219	Gorges de la Vis et cirque de Navacelles								
220	Etangs Montpellierains		✔			✔			✔
221	Massif des Albères		✔						
222	Massif du Canigou-Carança		✔	✔		✔			✔
223	Puig Carlit et environs					✔			
224	Gorges du Gardon								✔
225	Hautes garrigues du Montpellierais								
226	Etang du Bagnas		✔						✔
227	Cordon lagunaire de Sète à Agde								
228	Etang de Thau		✔						
229	Etangs de Vendres, Pissevache et Lespignan					✔			
230	Minervois								
231	Etangs de Canet et de Villeneuve-de-la-Raho et embouchure du Tec		✔			✔			
232	Gorges du Rieutord–Fage-Cagnasses		✔			✔			
233	Massif du Madres et Mont Coronat		✔			✔			
234	Petite Camargue fluvio-lacustre				✔		✔		
235	Petite Camargue laguno-marine					✔	✔		
236	Parc National des Cévennes	✔				✔		✔	✔
237	Montagnes de Marcou, de l'Espinouse et du Caroux			✔		✔			
238	Moyenne Vallée de la Durance								
239	Camargue		✔		✔		✔	✔	✔
	Subtotal of IBAs	6	54	8	30	95	21	5	80
240	Crau		✔		✔				✔
241	Chaîne des Alpilles								
242	Salines de l'Etang de Berre								
243	Montagne Sainte-Victoire		✔						✔
244	Iles Marseillaise: Maire, Jarron, Jarre, Riou, Calseraigne, Congloue et Pomègues								
245	Marais entre Crau et Grand Rhône: Meyranne, Chanoine, Plan de Bourg et Salins du Caban								
246	Massif du Petit Lubéron				✔	✔			✔
247	Salins d'Hyères et de Pesquiers								
248	Iles d'Hyères	✔							✔
249	Plateau de Valensole								
250	Plateau de l'Arbois, garrigues de Lancon et chaîne des Côtes				✔				✔
251	Plaine des Maures								
252	Etangs de Citis, Lavalduc, Engrenier, Pourra, l'Estomac, Fos et salines de Rassuen et de Fos								
253	Marais de L'Ile Vieille								
254	Basse Vallée de la Durance				✔	✔			
255	Vallée de la Durance: de Tallard à Sisteron					✔			
256	Vallée du Haut-Guil								
257	Marais de Manteyer					✔			
258	Bois des Ayes			✔					
259	Bois du Chapitre			✔					
260	Bec de Crigne					✔			
261	Parc National du Mercantour	✔							✔
262	Basse vallée du Var					✔			
263	Préalpes de Grasse					✔			
264	Parc National des Ecrins	✔	✔						✔
265	Bois de Palayson, du Rouet et de Malvoisin								
266	Détroit de Bonifaccio et Iles Lavezzi		✔			✔			✔
267	Iles Cerbicale		✔			✔			✔
268	Golfe de Porto, presqu'île de Scandola et golfe de Galeria		✔			✔			✔
269	Forêts domaniales de Corse							✔	
270	Vallée d'Asco					✔			✔
271	Vallée de la Restonica								✔
272	Etang de Biguglia		✔				✔		✔
273	Vallée du Régino								
274	Etang d'Urbino et zones humides périphériques					✔			✔
275	Iles Sanguinaires								✔
276	Iles Finocchiarola et côte de Tamarone à Centuri		✔						✔
277	Vallée du Verghello								
	Total number of IBAs	**9**	**62**	**10**	**34**	**107**	**22**	**6**	**96**

Box 1. International legislation and initiatives that are relevant to site conservation in France (see Appendix 1 for a general description of these agreements).

Global	
Biodiversity Convention	✔
Ramsar Convention	✔
Bonn Convention	✔
World Heritage Convention	✔
MAB Programme	✔
Pan-European	
Bern Convention	✔
Regional	
EC Birds Directive	✔
EC Habitats Directive	✔
Barcelona Convention	✔

✔ Convention ratified/initiative supported
(✔) Convention signed

Protection Areas have been designated under the EC Birds Directive, 103 of which overlap with 96 IBAs, although this equates to just 16% of the total area of IBAs (Table 5). More than 70 sites, including several IBAs, have been declared as Specially Protected Areas under the Barcelona Convention. France has also declared 35 Biogenetic Reserves, many of which overlap with parts of several IBAs.

CONSERVATION ISSUES

- The European Union has financed the maintenance of environmentally-friendly agricultural practices through the creation of geographically-based agri-environment schemes (Opérations Groupées d'Aménagements Fonciers, Environnement: OGAF-Environment). Between 1989–1997, 260 such schemes were implemented in France (Anon. 1997a), many of which affect IBAs.

- A number of studies aimed at defining the conservation and management measures needed for IBAs have been carried out through projects financed under the EC LIFE-Nature regulation. These have included: a study of *Crex crex* in IBAs (023, 026, 046, 102, 103, 160); the conservation of coastal lagoons in Languedoc, southern France, which encompassed several coastal IBAs (214, 220, 227, 228); and an initiative on *Tetrax tetrax* in IBAs 108, 110, 113, 234, 240.

ANALYTICAL METHODS

- Most breeding bird data and data on migratory birds have been taken from the nationwide IBA surveys of 1990 and 1991 (Rocamora and Thauront 1992; Rocomora 1994). This detailed inventory of IBAs has been carried out by the Ligue pour la Protection des Oiseaux (LPO) and Ecosphère, with the help of a large network of observers, for the French Ministry of the Environment.
- For a few breeding species (waders, herons, storks, *Tetrax tetrax*, *Crex crex* and *Larus melanocephalus*), data are from more recent surveys (Broyer *et al.* 1994; Deceuninck and Mahéo 1998; Duquet and Michel 1994; Duquet 1995; Jolivet 1996; Jolivet 1997; Marion 1997; Yésou 1997).
- Data for 30 rare breeding species are from the rare and threatened species network (Sériot, Trotignon et les coordinateurs espèces 1996; Sériot et les coordinateurs espèces 1997).
- Wintering waterbird data are from the Wetlands International/ LPO database (Deceuninck 1998) and are based on minimum and maximum numbers between 1993 and 1997.
- Threat data are based on Rocamora *et al.* (1995).
- Data concerning the protection of IBAs largely dates from 1990 and 1991 (Rocamora and Thauront 1992; Rocomora 1994), with the exception of SPAs which were updated in 1998 (Deceuninck and Jourde 1998), Nature Reserves (Cans and Reille 1997, Mosse 1996) and Ramsar sites (Barnaud 1995).

GLOSSARY

bocage an agricultural landscape with pockets of land separated by hedgerows.
LPO Ligue pour la Protection des Oiseaux (BirdLife International Partner in France).
garrigue a community of low, scattered, often spiny and aromatic shrubs found in the Mediterranean region.
maquis dense, mostly evergreen shrub community 1–3 m high, characteristic of the Mediterranean region.
LIFE an EU funding line (Regulation 1973/92/EEC) which supports EU environmental legislation, including the Birds and Habitats Directives.
SPA Special Protection Area (designated under Article 4 of the EC Birds Directive.

ACKNOWLEDGEMENTS

We would like to thank all the regional IBA coordinators and the 172 participants from 71 NGOs and organizations (full list published in Rocamora 1994) without whom the site inventory in France could not have been carried out, and all the teams from the Ligue pour la Protection des Oiseaux (LPO) who worked on this project, especially Nadège Maillet for the building of the IBA database, and Alison Duncan for the translation of this text. The 1990–1992 inventory and its publication were funded by the Ministry of the Environment. The Secrétariat Faune Flore (currently Service du Patrimoine Naturel) of the Muséum National d'Histoire Naturelle contributed to the description of sites and mapping.

SITE ACCOUNTS

Baie de Saint-Brieuc — A4i, A4iii, B1i, B2, C3, C4 — 001

Admin region Bretagne
Coordinates 48°32'N 2°40'W
Altitude 0–1 m **Area** 3,130 ha

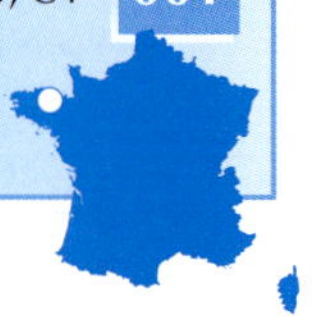

Site description
A large bay with mudflats and islets.

Habitats Wetland (mudflat/sandflat), Marine areas (sea inlet/coastal features)
Land-use Fisheries/aquaculture, Hunting, Tourism/recreation

Birds

Species	Season	Year	Pop min	Pop max	Acc	Criteria
Branta bernicla Brent Goose	W	1991	2,420	4,000	—	A4i, B1i, C3
Anas acuta Pintail	W	1991	220	616	—	B1i, C3
Haematopus ostralegus Oystercatcher	W	1991	1,600	9,750	—	B1i, C3
Calidris alpina Dunlin	W	1991	3,000	4,450	—	B2
Limosa lapponica Bar-tailed Godwit	W	1991	500	980	—	B2

The site holds 20,000 or more waterbirds in winter, on a regular basis.

Protection status
National Partial **International** Partial
1,140 ha of IBA covered by Nature Reserve (Baie de Saint Brieuc, 1,140 ha). 650 ha of IBA covered by Hunting Reserve (Baie de Saint Brieuc, 650 ha). 1,370 ha of IBA covered by Special Protection Area (Baie de Saint-Brieuc, 1,370 ha).

Conservation issues

Threats Agricultural intensification/expansion (B), Aquaculture/fisheries (B), Recreation/ tourism (B), Unsustainable exploitation (C)

Recreation and hunting cause disturbance to birds.

Archipel des Sept-Iles — A4ii, B1ii, B2, B3, C3, C6 — 002

Admin region Bretagne
Coordinates 48°53'N 3°28'W
Altitude 0–56 m **Area** 4,300 ha

Site description
Several rocky islands with sand and shingle-beaches and areas of open sea.

Habitats Wetland (shingle/stony beach), Marine areas (open sea), Rocky areas (rock stacks/islets)
Land-use Not utilized

Birds

Species	Season	Year	Pop min	Pop max	Acc	Criteria
Hydrobates pelagicus Storm Petrel	B	1997	14	17	—	C6
Sula bassana Gannet	B	1997	13,500	13,500	—	A4ii, B1ii, B2, C3
Phalacrocorax aristotelis Shag	B	1997	346	346	—	B3
Larus fuscus Lesser Black-backed Gull	B	1997	821	821	—	B3

This is the most important seabird colony in France.

Protection status
National High **International** High
4,300 ha of IBA covered by Hunting Reserve (Les Sept-Iles, 4,300 ha). 280 ha of IBA covered by Nature Reserve (Sept-Iles, 280 ha). 4,300 ha of IBA covered by Special Protection Area (Archipel des Sept-Iles, 4,300 ha).

Conservation issues

Threats Consequences of animal/plant introductions (B), Recreation/tourism (U), Other (A)

Oil spills pose the biggest risk to the site ('Other' threats). It has already been severely affected by several major incidents including the Torrey

Canyon in 1965 and Amoco Cadiz in 1978. Fishing and recreational activities do not pose direct threats.

Ile de Goulmedec — B2, C6 — 003

Admin region Bretagne
Coordinates 48°50′N 3°34′W
Altitude 0–5 m **Area** 600 ha

Site description

The site is mainly marine, with islets with rocky coastlines and sand and shingle-beaches. Human activities include fishing and sailing.

Habitats Wetland (shingle/stony beach), Marine areas (95%; open sea), Rocky areas (rock stacks/islets)
Land-use Fisheries/aquaculture, Tourism/recreation

Birds

Species	Season	Year	Pop min	Pop max	Acc	Criteria
Sterna sandvicensis Sandwich Tern	B	1991	2	50	—	C6
Sterna dougallii Roseate Tern	B	1996	—	—	—	B2
Sterna albifrons Little Tern	B	1991	10	12	—	C6

This is an important site for tern colonies.

Protection status

National Partial **International** Partial
490 ha of IBA covered by Hunting Reserve (Goulmedec, 490 ha). 490 ha of IBA covered by Special Protection Area (Ile de Goulmedec, 490 ha).

Conservation issues

Threats Aquaculture/fisheries (U)

No serious threats are known at this site.

Estuaires du Trieux et du Jaudy — B2, C6 — 004

Admin region Bretagne
Coordinates 48°50′N 3°07′W
Altitude 0–25 m **Area** 9,900 ha

Site description

A bay with mudflats and rocky islets, sand-dunes, two estuaries and sand and shingle beaches. Fishing and hunting are the main human activities.

Habitats Wetland (tidal river/enclosed tidal water; mudflat/sandflat; sand-dunes/sand beach; shingle/stony beach), Marine areas (sea inlet/coastal features), Rocky areas (rock stacks/islets)
Land-use Fisheries/aquaculture, Hunting, Tourism/recreation, Urban/industrial/transport

Birds

Species	Season	Year	Pop min	Pop max	Acc	Criteria
Melanitta fusca Velvet Scoter	W	1997	40	40	—	B2
Calidris canutus Knot	W	1997	2,300	2,600	—	B2
Limosa lapponica Bar-tailed Godwit	W	1997	400	810	—	B2
Sterna albifrons Little Tern	B	1991	20	25	—	C6

This is an important site for wintering waterbirds and for a tern colony, with nationally important numbers of *Sterna hirundo* (110 pairs).

Protection status

National Partial **International** High
1,450 ha of IBA covered by Hunting Reserve (Embouchure du Jaudy, 1,450 ha). 9,900 ha of IBA covered by Special Protection Area (Estuaires du Trieux et du Jaudy, 11,770 ha).

Conservation issues

Threats Aquaculture/fisheries, Infrastructure, Recreation/tourism

Tern colonies have suffered disturbance, mainly from recreational activities. Local authorities are now aware of the problem and disturbance has been reduced.

Iles de la Colombière, de la Nellière et des Haches — B2, C6 — 005

Admin region Bretagne
Coordinates 48°38′N 2°12′W
Altitude 0–10 m **Area** 4,400 ha

Site description

Small islets and open sea. Human activities include fishing and boating.

Habitats Marine areas (open sea), Rocky areas (rock stacks/islets)
Land-use Fisheries/aquaculture, Not utilized, Tourism/recreation

Birds

Species	Season	Year	Pop min	Pop max	Acc	Criteria
Sterna sandvicensis Sandwich Tern	B	1991	300	380	—	B2, C6
Sterna dougallii Roseate Tern	B	1996	—	—	—	B2
Sterna hirundo Common Tern	B	1991	65	120	—	C6

This is an important site for tern colonies.

Protection status

National High **International** Partial
4,400 ha of IBA covered by Hunting Reserve (4,400 ha). 12 ha of IBA covered by Private Reserve (12 ha). 1,710 ha of IBA covered by Special Protection Area (Iles de la Colombière, de la Nellière et des Hache, 1,710 ha).

Conservation issues

Threats Aquaculture/fisheries (B), Disturbance to birds (A), Recreation/tourism (A)

A well protected site but recreation and tourism disturb the tern colonies.

Ilôts de Trévorc'h — C6 — 006

Admin region Bretagne
Coordinates 48°36′N 4°38′W
Altitude 0–5 m **Area** 470 ha

Site description

Open sea and several rocky islets with beaches. Human activities include fishing and boating.

Habitats Wetland (sand-dunes/sand beach), Marine areas (open sea), Rocky areas (rock stacks/islets)
Land-use Fisheries/aquaculture, Tourism/recreation

Birds

Species	Season	Year	Pop min	Pop max	Acc	Criteria
Sterna sandvicensis Sandwich Tern	B	1991	20	350	—	C6
Sterna hirundo Common Tern	B	1991	50	150	—	C6

This site is important for tern colonies.

Protection status

National Partial **International** Partial
380 ha of IBA covered by Hunting Reserve (Ilôts de Trevors, 380 ha). 1 ha of IBA covered by Private Reserve (Ilôts de Trevors, 1 ha). 381 ha of IBA covered by Special Protection Area (Ilôts de Trevors, 381 ha).

Conservation issues

Threats Recreation/tourism (B)

Recreational activities cause some disturbance to birds.

Archipel de Molène — A4i, B1i, B3, C3, C6 — 007

Admin region Bretagne
Coordinates 48°22′N 4°56′W
Altitude 0–15 m **Area** 10,900 ha

Site description

An archipelago of rocky islands with sand and shingle-beaches, rocky habitats and brackish lagoons.

Habitats Wetland (sand-dunes/sand beach; shingle/stony beach; coastal lagoon), Marine areas (open sea), Rocky areas (rock stacks/islets)
Land-use Fisheries/aquaculture, Hunting, Tourism/recreation

Birds

Species	Season	Year	Pop min	Pop max	Acc	Criteria
Hydrobates pelagicus Storm Petrel	B	1991	150	200	—	C6
Larus fuscus Lesser Black-backed Gull	B	1991	6,000	6,500	—	A4i, B1i, B3, C3
Larus argentatus Herring Gull	B	1991	5,500	5,500	—	B1i
Larus marinus Great Black-backed Gull	B	1991	400	450	—	B3
Sterna albifrons Little Tern	B	1991	10	40	—	C6

The area is important for seabird colonies.

Protection status

National High **International** High
39 ha of IBA covered by Nature Reserve (Mer d'Iroise, 39 ha). 10,000 ha of IBA covered by Hunting Reserve (10,000 ha). 40 ha of IBA covered by Private Reserve (40 ha). 10,000 ha of IBA covered by Special Protection Area (Archipel de Molène, 10,000 ha). 10,900 ha of IBA covered by Biosphere Reserve (21,400 ha).

Conservation issues

No serious threats are known at this site. It is fairly well protected and wardened, although there is possible disturbance from tourism and sailing.

Rade de Brest: Baie de Daoulas et Anse du Poulmic

A4i, A4iii, B1i, C3, C4 **008**

Admin region Bretagne
Coordinates 48°20'N 4°17'W
Altitude 0–3 m **Area** 9,000 ha

Site description

A bay with mudflats, rocky coastlines and beaches. Human activities include fishing, recreation (including sailing) and there is a harbour.

Habitats Wetland (mudflat/sandflat; sand-dunes/sand beach; shingle/stony beach), Marine areas (sea inlet/coastal features)
Land-use Fisheries/aquaculture, Tourism/recreation, Urban/industrial/transport

Birds

Species	Season	Year	Pop min	Pop max	Acc	Criteria
Podiceps nigricollis Black-necked Grebe	W	1997	222	1,900	—	A4i, B1i, C3

The site holds 20,000 or more waterbirds in winter, on a regular basis (including *Podiceps nigricollis* 1,900; *Anas penelope* 2,270; *Calidris alpina* 4,815).

Protection status

National Partial **International** Partial
6,800 ha of IBA covered by Hunting Reserve (Baie de Daoulas et Anse du Poulmic, 6,800 ha). 6,800 ha of IBA covered by Special Protection Area (Rade de Brest: Baie de Daoulas et Anse du Poulmic, 6,800 ha).

Conservation issues

Threats Unknown

Cap Sizun

B3 **009**

Admin region Bretagne
Coordinates 48°05'N 4°36'W
Altitude 0–20 m **Area** 590 ha

Site description

Habitats Scrub (heathland), Wetland (shingle/stony beach), Marine areas (open sea), Rocky areas (sea cliff/rocky shore; rock stacks/islets)
Land-use Agriculture, Fisheries/aquaculture, Tourism/recreation

A stretch of rocky coastline with cliffs, rocky islets, heathland and open sea.

Birds

Species	Season	Year	Pop min	Pop max	Acc	Criteria
Phalacrocorax aristotelis Shag	B	1991	200	200	—	B3

The site supports several important seabird colonies.

Protection status

National High **International** High
500 ha of IBA covered by Hunting Reserve (Cap Sizun, 500 ha). 30 ha of IBA covered by Private Reserve (30 ha). 530 ha of IBA covered by Special Protection Area (Cap Sizun, 530 ha).

Conservation issues

Threats Aquaculture/fisheries (A), Other (A)

Oil pollution ('Other' threats, above) and fishing nets are responsible for declines in populations of auks (Alcidae).

Marais de la Baie d'Audierne

A1, A4i, B1i, C1, C3, C6 **010**

Admin region Bretagne
Coordinates 47°53'N 4°22'W
Altitude 0–5 m **Area** 3,100 ha

Site description

Saltwater and freshwater lagoons with reedbeds (*Phragmites*), sand-dunes, beaches, rocky coast, wet meadows, agricultural and uncultivated land.

Habitats Scrub (heathland), Grassland (humid grassland), Wetland (sand-dunes/sand beach; shingle/stony beach; coastal lagoon; standing brackish and salt water; water-fringe vegetation), Artificial landscape (arable land)
Land-use Agriculture, Hunting, Tourism/recreation, Urban/industrial/transport

Birds

Species	Season	Year	Pop min	Pop max	Acc	Criteria
Ixobrychus minutus Little Bittern	B	1991	1	10	—	C6
Circus aeruginosus Marsh Harrier	B	1991	10	20	—	C6
Calidris alba Sanderling	P	1997	1,000	10,000	—	A4i, B1i, C3
Acrocephalus paludicola Aquatic Warbler	P	1991	250	300	—	A1, C1
Sylvia undata Dartford Warbler	B	1991	10	10	—	C6

The site is important for passage and wintering waterbirds. Species of global conservation concern that do not meet IBA criteria: *Aythya nyroca* (passage).

Protection status

National Low **International** Partial
52 ha of IBA covered by Private Reserve (SEPNB Reserve, 52 ha). 850 ha of IBA covered by Hunting Reserve (Baie d'Audierne, 850 ha). 1,600 ha of IBA covered by Special Protection Area (Baie d'Audierne, 1,600 ha).

Conservation issues

Threats Agricultural intensification/expansion (C), Disturbance to birds (A), Infrastructure (B), Recreation/tourism (B)

The main threats are from recreational development on and around the site, including a golf course and swimming pool.

Archipel des Glénan

A4i, B1i, B2, B3, C3, C6 **011**

Admin region Bretagne
Coordinates 47°45'N 4°00'W
Altitude 0–10 m **Area** 4,100 ha

Site description

Small islands and islets with beaches and rocky habitats, surrounded by open sea.

Habitats Wetland (sand-dunes/sand beach; shingle/stony beach), Marine areas (open sea), Rocky areas (rock stacks/islets)
Land-use Fisheries/aquaculture, Hunting, Tourism/recreation, Urban/industrial/transport

Birds

Species	Season	Year	Pop min	Pop max	Acc	Criteria
Larus fuscus Lesser Black-backed Gull	B	1991	3,000	3,500	—	A4i, B1i, B3, C3
Larus argentatus Herring Gull	B	1991	5,500	6,000	—	B1i
Larus marinus Great Black-backed Gull	B	1991	100	120	—	B3
Sterna sandvicensis Sandwich Tern	B	1991	—	250	—	B2, C6

An important site for breeding seabirds.

Protection status

National High **International** High
3,800 ha of IBA covered by Hunting Reserve (Archipel des Glénan, 3,800 ha). 5 ha of IBA covered by Private Reserve (5 ha). 2 ha of IBA covered by Nature Reserve (Saint-Nicolas-des-Glénan, 2 ha). 3,800 ha of IBA covered by Special Protection Area (Archipel des Glenans, 3,800 ha).

Conservation issues

Threats Aquaculture/fisheries (B), Extraction industry (C), Recreation/tourism (B), Unsustainable exploitation (A)

There is disturbance from sailing and sand extraction.

Baies de Morlaix et de Carantec — A4i, A4iii, B1i, B2, B3, C2, C3, C4, C6 — 012

Admin region Bretagne
Coordinates 48°40′N 3°54′W
Altitude 0–15 m **Area** 7,900 ha

Site description

An estuary and bay with mudflats, sand and shingle-beaches, rocky areas, islets and cliffs. Human activities include fishing, hunting, tourism and sailing.

Habitats Wetland (tidal river/enclosed tidal water; mudflat/sandflat; sand-dunes/sand beach; shingle/stony beach), Marine areas (sea inlet/coastal features), Rocky areas (sea cliff/rocky shore; rock stacks/islets)
Land-use Fisheries/aquaculture, Hunting, Tourism/recreation

Birds

Species	Season	Year	Pop min	Pop max	Acc	Criteria
Hydrobates pelagicus Storm Petrel	B	1991	10	20	—	C6
Calidris alpina Dunlin	W	1997	10,000	12,800	—	B2
Arenaria interpres Turnstone	W	1997	575	1,480	—	A4i, B1i, C3
Larus fuscus Lesser Black-backed Gull	B	1991	350	350	—	B3
Larus marinus Great Black-backed Gull	B	1991	440	440	—	B3
Sterna sandvicensis Sandwich Tern	B	1997	750	1,100	—	A4i, B1i, B2, C2, C6
Sterna dougallii Roseate Tern	B	1996	105	110	—	A4i, B1i, B2, C2
Sterna hirundo Common Tern	B	1991	190	210	—	C6
Pyrrhocorax pyrrhocorax Chough	B	1991	12	13	—	C6

The site supports important tern colonies and is also important for wintering waterbirds. The site holds 20,000 or more waterbirds in winter, on a regular basis (including *Branta bernicla* 2,170; *Haematopus ostralegus* 1,590; *Larus ridibundus* 38,500 individuals).

Protection status

National Partial **International** Partial
5 ha of IBA covered by Private Reserve (5 ha). 2,000 ha of IBA covered by 2 Hunting Reserves (Baie de Morlaix; Baie de Carantec; total area 2,000 ha). 1,800 ha of IBA covered by Special Protection Area (Baie de Morlaix, 1,800 ha).

Conservation issues

Threats Disturbance to birds (B), Recreation/tourism (B)

Recreation activities and tourism are the main threats affecting breeding birds.

Ile d'Ouessant — B3, C6 — 013

Admin region Bretagne
Coordinates 48°28′N 5°06′W
Altitude 0–32 m **Area** 5,850 ha

Site description

Agriculture, cattle-raising and fishing are the most common human activities.

Habitats Forest and woodland (broadleaved deciduous forest), Scrub (heathland), Wetland (shingle/stony beach; standing fresh water; water-fringe vegetation), Marine areas (open sea), Rocky areas (sea cliff/rocky shore; rock stacks/islets)
Land-use Agriculture, Fisheries/aquaculture, Hunting, Tourism/recreation, Urban/industrial/transport

Birds

Species	Season	Year	Pop min	Pop max	Acc	Criteria
Hydrobates pelagicus Storm Petrel	B	1991	10	20	—	C6
Larus fuscus Lesser Black-backed Gull	B	1991	350	350	—	B3
Larus marinus Great Black-backed Gull	B	1991	440	440	—	B3
Pyrrhocorax pyrrhocorax Chough	B	1991	12	13	—	C6

This is an important site for breeding seabirds. The site also supports large numbers of passage waterbirds and passerines.

Protection status

National Partial **International** Partial
7 ha of IBA covered by Private Reserve (7 ha). 2,200 ha of IBA covered by Hunting Reserve (Ouessant, 2,200 ha). 2,200 ha of IBA covered by Special Protection Area (Ile d'Ouessant, 2,200 ha).

Conservation issues

No threats are reported.

Golfe du Morbihan et Etier de Penerf — A4i, A4iii, B1i, B2, B3, C2, C3, C4, C6 — 014

Admin region Bretagne
Coordinates 47°34′N 2°46′W
Altitude 0–5 m **Area** 18,800 ha

Site description

A bay with mudflats, saltwater marshes, saltwater lagoons and raised saltmarsh, islands, rocky coast, beaches, grassland and arable land.

Habitats Scrub (heathland), Wetland (38%; mudflat/sandflat; saltmarsh; shingle/stony beach; coastal lagoon; standing brackish and salt water; water-fringe vegetation), Marine areas (sea inlet/coastal features), Rocky areas (sea cliff/rocky shore; rock stacks/islets)
Land-use Agriculture, Fisheries/aquaculture, Hunting, Tourism/recreation, Urban/industrial/transport

Birds

Species	Season	Year	Pop min	Pop max	Acc	Criteria
Podiceps nigricollis Black-necked Grebe	W	1997	970	1,493	—	A4i, B1i, C3
Egretta garzetta Little Egret	B	1991	50	100	—	C6
Platalea leucorodia Spoonbill	P	1997	20	60	—	B1i, C2
Branta bernicla Brent Goose	W	1997	4,860	8,360	—	A4i, B1i, C3
Tadorna tadorna Shelduck	W	1997	2,450	4,250	—	A4i, B1i, C3
Anas acuta Pintail	W	1997	1,075	2,350	—	B1i, C3
Anas clypeata Shoveler	W	1997	180	652	—	B1i, C3
Mergus serrator Red-breasted Merganser	W	1997	1,530	1,965	—	A4i, B1i, C3
Mergus serrator Red-breasted Merganser	P	1997	2,000	2,000	—	A4i, B1i, C3
Circus aeruginosus Marsh Harrier	B	1991	5	8	—	C6
Himantopus himantopus Black-winged Stilt	B	1995	54	54	—	C6
Recurvirostra avosetta Avocet	B	1997	140	142	—	B3, C6
Recurvirostra avosetta Avocet	W	1997	515	905	—	B1i, B2, C2
Pluvialis squatarola Grey Plover	W	1997	1,720	2,510	—	A4i, B1i, C3
Calidris alpina Dunlin	W	1997	15,200	29,780	—	A4i, B1i, B2, C3
Tringa totanus Redshank	P	1997	1,000	2,000	—	B1i, C3
Larus fuscus Lesser Black-backed Gull	B	1991	100	500	—	B3
Sterna hirundo Common Tern	B	1991	50	200	—	C6

The site holds 20,000 or more waterbirds on passage and in winter, on a regular basis.

■ Protection status

National Low **International** High

1,500 ha of IBA covered by Hunting Reserve (Golfe du Morbihan, 1,500 ha). 30 ha of IBA covered by Private Reserve (30 ha). 410 ha of IBA covered by Nature Reserve (Marais de Séné, 410 ha) 1,490 ha of IBA covered by Special Protection Area (Etier de Penerf, 1,490 ha). 5,830 ha of IBA covered by Special Protection Area (Golfe du Morbihan, 5,830 ha). 18,800 ha of IBA covered by Ramsar Site (Golfe du Morbihan, 23,000 ha).

■ Conservation issues

Threats Aquaculture/fisheries (A), Disturbance to birds (A), Drainage, Recreation/tourism (A), Unsustainable exploitation (B)

Wetland areas are threatened by drainage and the encroachment of vegetation. Tourists cause some disturbance to birds.

Cap Frehel — B3 — 015

Admin region Bretagne
Coordinates 48°40′N 2°19′W
Altitude 0–73 m **Area** 2,070 ha

■ Site description

An area of open sea, rocky islets, cliffs and heathland.

Habitats Scrub (20%; heathland), Wetland (raised bog), Marine areas (80%; open sea), Rocky areas (sea cliff/rocky shore; rock stacks/islets; inland cliff)
Land-use Hunting, Tourism/recreation

■ Birds

Species	Season	Year	Pop min	Pop max	Acc	Criteria
Phalacrocorax aristotelis Shag	B	1992	230	400	—	B3

This is an important area for seabird colonies.

■ Protection status

National High **International** High

1,750 ha of IBA covered by Hunting Reserve (Cap Frehel, 1,750 ha). 2,020 ha of IBA covered by Special Protection Area (Cap Frehel, 2,020 ha).

■ Conservation issues

Threats Aquaculture/fisheries (U), Other (U)

Marine pollution from oil tankers ('Other' threats, above) is the major threat to seabird breeding colonies and wintering populations.

Baie de Vilaine — A4i, A4iii, B1i, B2, C2, C3, C4 — 016

Admin region Bretagne
Coordinates 47°30′N 2°28′W
Altitude 0–2 m **Area** 4,500 ha

■ Site description

A bay and estuary with mudflats, coastal cliffs, sand and shingle-beaches and saltmarshes. Human activities include fishing, hunting, and recreation, including sailing.

Habitats Wetland (tidal river/enclosed tidal water; mudflat/sandflat; saltmarsh; sand-dunes/sand beach; shingle/stony beach), Marine areas (sea inlet/coastal features), Rocky areas (sea cliff/rocky shore)
Land-use Fisheries/aquaculture, Hunting, Tourism/recreation

■ Birds

Species	Season	Year	Pop min	Pop max	Acc	Criteria
Branta bernicla Brent Goose	W	1997	1,420	3,610	—	A4i, B1i, C3
Anas acuta Pintail	W	1997	108	691	—	B1i, C3
Aythya marila Scaup	W	1997	1,950	5,440	—	B2
Recurvirostra avosetta Avocet	W	1997	1,294	2,103	—	A4i, B1i, B2, C2
Numenius arquata Curlew	W	1997	215	1,596	—	B2

The site holds 20,000 or more waterbirds in winter, on a regular basis.

■ Protection status

National Partial **International** Partial

500 ha of IBA covered by Hunting Reserve (500 ha). 1,270 ha of IBA covered by Special Protection Area (Baies de Kervoyal et de Vilaine, 1,270 ha).

■ Conservation issues

Threats Industrialization/urbanization (C), Infrastructure (A)

The planned construction of a marina threatens the site.

Presqu'île de Crozon, Tas de Pois et Rochers du Toulinguet — B3, C6 — 017

Admin region Bretagne
Coordinates 48°16′N 4°38′W
Altitude 0–10 m **Area** 1,200 ha

■ Site description

An area of coastal cliffs, open sea and rocky islets.

Habitats Marine areas (open sea), Rocky areas (sea cliff/rocky shore; rock stacks/islets)
Land-use Fisheries/aquaculture, Urban/industrial/transport

■ Birds

Species	Season	Year	Pop min	Pop max	Acc	Criteria
Hydrobates pelagicus Storm Petrel	B	1991	30	50	—	C6
Phalacrocorax aristotelis Shag	B	1991	410	410	—	B3

The site is important for breeding seabirds.

■ Protection status

National High **International** None

1,200 ha of IBA covered by Hunting Reserve (Presqu'île de Crozon, 1,200 ha). 3 ha of IBA covered by Private Reserve (Presqu'île de Crozon, Ilots des Tas de Pois et Rochers du Toulinguet, 3 ha).

■ Conservation issues

Threats Aquaculture/fisheries (B), Other (B)

Marine pollution ('Other' threats) from oil tankers is a significant threat.

Rade de Lorient — A4iii, C4 — 018

Admin region Bretagne
Coordinates 47°44′N 3°20′W
Altitude 0–30 m **Area** 2,800 ha

■ Site description

A bay with mudflats, saltmarshes, dunes, sand and shingle-beaches and a lagoon.

Habitats Wetland (tidal river/enclosed tidal water; mudflat/sandflat; saltmarsh; sand-dunes/sand beach; shingle/stony beach; coastal lagoon; standing fresh water), Marine areas (open sea; sea inlet/coastal features)
Land-use Agriculture, Fisheries/aquaculture, Hunting, Tourism/recreation, Urban/industrial/transport

■ Birds

The site is of international importance for wintering and migrating waterbirds, particularly waders. The site holds 20,000 or more waterbirds in winter (including *Branta bernicla* 2,550; *Fulica atra* 2,396; *Calidris alpina* 9,180; *Larus ridibundus* 10,000 individuals).

■ Protection status

National Low **International** Partial

130 ha of IBA covered by Hunting Reserve (130 ha). 480 ha of IBA covered by Special Protection Area (Rade de Lorient, 480 ha).

Conservation issues

Threats Agricultural intensification/expansion (B), Aquaculture/fisheries (U), Infrastructure (C)

Marine pollution from agricultural effluent and the construction of a marina are the most significant threats to the site.

Baie de Quiberon — A4i, B1i, C3 — 019

Admin region Bretagne
Coordinates 47°32′N 3°04′W
Altitude 0–30 m **Area** 14,750 ha

Site description

A bay and estuary with rocky islets, mudflats, saltmarshes, sand and shingle-beaches and coastal dunes.

Habitats Wetland (tidal river/enclosed tidal water; mudflat/sandflat; saltmarsh; sand-dunes/sand beach; shingle/stony beach; coastal lagoon), Marine areas (open sea; sea inlet/coastal features), Rocky areas (rock stacks/islets)
Land-use Fisheries/aquaculture, Hunting, Tourism/recreation, Urban/industrial/transport

Birds

Species	Season	Year	Pop min	Pop max	Acc	Criteria
Arenaria interpres Turnstone	P	1997	300	700	—	A4i, B1i, C3

This is an important site for wintering waterbirds, particularly *Branta bernicla* and *Calidris alpina*.

Protection status

National Low **International** Low
1,300 ha of IBA covered by Hunting Reserve (Iles des Théviec et Gevric, 1,300 ha). 860 ha of IBA covered by Special Protection Area (Baie de Quiberon, 860 ha).

Conservation issues

Threats Infrastructure (A), Recreation/tourism

Fishing, hunting and boating are the main human activities, leading to some disturbance within the site.

Baie de Goulven — A4i, A4iii, B1i, C2, C3, C4 — 020

Admin region Bretagne
Coordinates 48°39′N 4°16′W
Altitude 0–5 m **Area** 2,000 ha

Site description

A bay with mudflats, an estuary, saltmarsh, shingle and sandy beaches, rocky islets, coastal dunes and marine cliffs.

Habitats Wetland (mudflat/sandflat; saltmarsh; sand-dunes/sand beach; shingle/stony beach), Marine areas (open sea; sea inlet/coastal features), Rocky areas (sea cliff/rocky shore; rock stacks/islets)
Land-use Fisheries/aquaculture, Tourism/recreation

Birds

Species	Season	Year	Pop min	Pop max	Acc	Criteria
Charadrius hiaticula Ringed Plover	P	1997	2,000	2,500	—	A4i, B1i, C3
Calidris alba Sanderling	P	1997	1,300	2,000	—	A4i, B1i, C3
Limosa lapponica Bar-tailed Godwit	P	1997	600	2,000	—	A4i, B1i, C2
Tringa totanus Redshank	P	1997	500	2,000	—	B1i, C3
Arenaria interpres Turnstone	P	1997	900	2,000	—	A4i, B1i, C3

The site holds 20,000 or more waterbirds on passage and in winter, on a regular basis (including *Charadrius hiaticula* 2,500 (passage); *Vanellus vanellus* 1,500 (passage) 2,660 (winter); *Calidris alpina* 2,500 (passage) 4,060 (winter)).

Protection status

National High **International** None
1,850 ha of IBA covered by Hunting Reserve (1,850 ha).

Conservation issues

No serious threats are known at this site.

Traicts et marais salants de la Presqu'île Guérandaise — A4i, A4iii, B1i, B2, B3, C2, C3, C4, C6 — 021

Admin region Pays de la Loire
Coordinates 47°21′N 2°26′W
Altitude 0–2 m **Area** 4,650 ha

Site description

A brackish bay with associated marshes (including salinas), mudflats, beaches and sand-dunes, woodlands and heathland.

Habitats Forest and woodland (broadleaved deciduous forest), Scrub (heathland), Wetland (mudflat/sandflat; saltmarsh; sand-dunes/sand beach), Marine areas (sea inlet/coastal features)
Land-use Agriculture, Fisheries/aquaculture, Hunting, Tourism/recreation, Urban/industrial/transport

Birds

Species	Season	Year	Pop min	Pop max	Acc	Criteria
Egretta garzetta Little Egret	B	1997	77	77	—	C6
Egretta garzetta Little Egret	W	1997	522	1,902	—	A4i, B1i, C2
Branta bernicla Brent Goose	W	1997	1,758	3,785	—	A4i, B1i, C3
Tadorna tadorna Shelduck	W	1997	2,065	3,748	—	B1i, C3
Anas acuta Pintail	W	1997	201	610	—	B1i, C3
Anas clypeata Shoveler	W	1997	15	2,300	—	B1i, C3
Aythya marila Scaup	W	1997	4	62	—	B2
Himantopus himantopus Black-winged Stilt	B	1996	113	113	—	C6
Recurvirostra avosetta Avocet	B	1997	175	178	—	B3, C6
Recurvirostra avosetta Avocet	W	1997	633	2,000	—	A4i, B1i, B2, C2
Arenaria interpres Turnstone	W	1997	693	1,240	—	A4i, B1i, C3
Larus argentatus Herring Gull	W	1997	520	26,713	—	B1i
Sterna hirundo Common Tern	B	1979	180	210	—	C6
Luscinia svecica Bluethroat	B	1979	300	400	—	C6

The site holds 20,000 or more waterbirds in winter, on a regular basis.

Protection status

National Partial **International** High
550 ha of IBA covered by Hunting Reserve (Traicts du Groisic, 550 ha). 4,650 ha of IBA covered by Ramsar Site (Marais de Guérande, 5,200 ha). 550 ha of IBA covered by Special Protection Area (Traicts et Marais de Guérande, 550 ha).

Conservation issues

Threats Abandonment/reduction of land management (U), Other (B), Recreation/tourism (B)

The abandonment of the management of saltmarshes, recreation and chemical treatment against mosquitoes ('Other' threats, above) are the main threats to the site.

Marais de Brière — B1i, B2, B3, C2, C3, C6 — 022

Admin region Pays de la Loire
Coordinates 47°22′N 2°10′W
Altitude 0–10 m **Area** 18,900 ha

Site description

Marshes and wet meadows, reedbeds (*Phragmites*), ponds and arable fields.

Habitats Grassland (humid grassland; mesophile grassland), Wetland (standing brackish and salt water; water-fringe vegetation)
Land-use Agriculture, Fisheries/aquaculture, Hunting, Tourism/recreation, Urban/industrial/transport

Birds

Species	Season	Year	Pop min	Pop max	Acc	Criteria
Botaurus stellaris Bittern	B	1997	25	45	—	B2, C2, C6
Platalea leucorodia Spoonbill	P	1991	40	50	—	B1i, C2
Anas acuta Pintail	P	1997	100	2,500	—	B1i, C3
Anas clypeata Shoveler	W	1997	106	401	—	B1i, C3

Species ... continued	Season	Year	Pop min	Pop max	Acc	Criteria
Milvus migrans Black Kite	B	1991	30	40	—	C6
Circus aeruginosus Marsh Harrier	B	1997	35	45	—	C6
Circus cyaneus Hen Harrier	B	1991	5	10	—	C6
Porzana porzana Spotted Crake	B	1991	5	10	—	B3, C6
Chlidonias hybridus Whiskered Tern	B	1991	—	70	—	B2, C6
Chlidonias niger Black Tern	B	1991	50	200	—	C2, C6
Luscinia svecica Bluethroat	B	1991	150	200	—	C6

This is an important site for breeding and wintering waterbirds. It is also the best site for *Chlidonias niger* in France. Species of global conservation concern that do not meet IBA criteria: *Haliaeetus albicilla* (passage); *Aquila clanga* (passage); *Crex crex* (passage); *Gallinago media* (passage); *Acrocephalus paludicola* (passage).

Protection status

National Low **International** High

1,075 ha of IBA covered by Hunting Reserve (1,075 ha). 18,900 ha of IBA covered by Ramsar Site (Grande-Brière, 19,000 ha).

Conservation issues

Threats Agricultural intensification/expansion (C), Drainage (B), Filling-in of wetlands (B), Recreation/tourism (B), Unsustainable exploitation (C)

The Grande Brière is insufficiently flooded during summer (due to changes to the Loire estuary, see site 023) and the resulting high salinity is having a serious effect on the vegetation at the site. There is also considerable hunting pressure. Drainage and infilling of the surrounding areas are also reported threats.

Estuaire de la Loire — 023

A1, A4i, A4iii, B1i, B2, B3, C1, C2, C3, C4, C6

Admin region Pays de la Loire
Coordinates 47°16′N 1°49′W
Altitude 0–6 m **Area** 21,400 ha

Site description

An estuary with mudflats, beaches and alluvial islands, marshes, wet meadows and woodland.

Habitats Forest and woodland (broadleaved deciduous forest; alluvial/very wet forest), Grassland (humid grassland), Wetland (tidal river/enclosed tidal water; mudflat/sandflat; saltmarsh; sand-dunes/sand beach; standing brackish and salt water; river/stream; water-fringe vegetation), Rocky areas (rock stacks/islets), Artificial landscape (arable land)
Land-use Agriculture, Fisheries/aquaculture, Hunting, Tourism/recreation, Urban/industrial/transport

Birds

Species	Season	Year	Pop min	Pop max	Acc	Criteria
Botaurus stellaris Bittern	B	1991	5	10	—	C6
Anas crecca Teal	W	1997	4,300	13,049	—	B1i, C3
Anas acuta Pintail	W	1997	65	706	—	B1i, C3
Anas clypeata Shoveler	W	1997	233	2,940	—	B1i, C3
Porzana porzana Spotted Crake	B	1991	10	15	—	B3, C6
Crex crex Corncrake	B	1994	52	53	—	A1,B2,C1,C2,C6
Himantopus himantopus Black-winged Stilt	B	1996	18	19	—	C6
Recurvirostra avosetta Avocet	W	1997	850	2,590	—	A4i, B1i, B2, C2
Chlidonias niger Black Tern	B	1991	10	20	—	C6
Luscinia svecica Bluethroat	B	1991	50	50	—	C6

The site holds 20,000 or more waterbirds in winter, on a regular basis. Species of global conservation concern that do not meet IBA criteria: *Gallinago media* (passage).

Protection status

National Low **International** Partial

3,913 ha of IBA covered by 3 Hunting Reserves (Massereau, 393 ha; Estuaire de la Loire, 3,400 ha; Ile de la Pierre rouge, 120 ha). 17,700 ha of IBA covered by Special Protection Area (Estuaire de la Loire, 17,700 ha).

Conservation issues

Threats Disturbance to birds (A), Filling-in of wetlands (A), Infrastructure (A), Recreation/tourism (A), Unsustainable exploitation (B)

The site is seriously threatened by water-level modifications (which have altered the freshwater-saltwater ratios), infilling of saltmarshes, meadows and mudflats, and embankment of the River Loire.

Lac de Grand-Lieu — 024

A4i, A4iii, B1i, B2, C2, C3, C4, C6

Admin region Pays de la Loire
Coordinates 47°05′N 1°40′W
Altitude 2–5 m **Area** 5,600 ha

Site description

A lake, with associated reedbeds (*Phragmites*) and wet meadows, and woodlands, hedgerows and cultivated land.

Habitats Forest and woodland (broadleaved deciduous forest), Scrub (heathland), Grassland (29%; humid grassland), Wetland (58%; standing fresh water; river/stream; water-fringe vegetation), Artificial landscape (18%; arable land; perennial crops/orchards/groves)
Land-use Agriculture (48%), Fisheries/aquaculture (100%), Forestry (2%), Hunting (94%), Urban/industrial/transport (1%)

Birds

Species	Season	Year	Pop min	Pop max	Acc	Criteria
Ixobrychus minutus Little Bittern	B	1991	1	10	—	C6
Nycticorax nycticorax Night Heron	B	1994	55	55	—	C6
Egretta garzetta Little Egret	B	1991	180	180	—	C6
Ardea purpurea Purple Heron	B	1991	40	80	—	C2, C6
Platalea leucorodia Spoonbill	B	1991	10	15	—	B1i, B2, C2, C6
Anas strepera Gadwall	B	1997	40	50	—	B2
Anas acuta Pintail	P	1997	1,000	1,200	—	B1i, C3
Anas clypeata Shoveler	W	1997	1,800	4,000	—	B1i, C3
Anas clypeata Shoveler	P	1997	3,000	5,000	—	A4i, B1i, C3
Milvus migrans Black Kite	B	1991	15	25	—	C6
Circus aeruginosus Marsh Harrier	B	1997	20	40	—	C6
Fulica atra Coot	P	1997	20,000	20,000	—	B1i, C3

The site holds 20,000 or more waterbirds on passage and in winter, on a regular basis. Species of global conservation concern that do not meet IBA criteria: *Aythya nyroca* (wintering); *Crex crex* (passage).

Protection status

National Partial **International** High

2,694 ha of IBA covered by Nature Reserve (Lac de Grand-Lieu, 2,694 ha). 5,600 ha of IBA covered by Special Protection Area (Lac de Grand-Lieu, 6,000 ha). 5,600 ha of IBA covered by Ramsar Site (Grand-Lieu, 6,300 ha).

Conservation issues

Threats Agricultural intensification/expansion (U), Other (U)

The site is drying out as it no longer receives sufficient water from the River Loire ('Other' threats, above). Problems of eutrophication are common due to agricultural pollution.

Baie de Bourgneuf et marais breton — 025

A4i, A4iii, B1i, B2, B3, C2, C3, C4, C6

Admin region Pays de la Loire
Coordinates 46°54′N 1°55′W
Altitude 0–3 m **Area** 47,000 ha

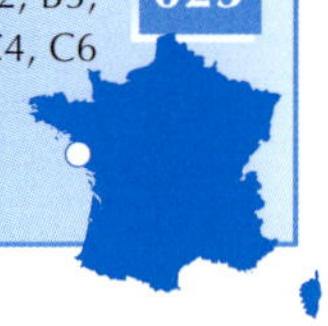

Site description

A bay with saltwater and freshwater lagoons, saltwater marshes, mudflats and wet meadows.

Habitats Grassland (humid grassland), Wetland (saltmarsh; coastal lagoon; standing fresh water; river/stream), Marine areas (sea inlet/coastal features)
Land-use Agriculture, Fisheries/aquaculture, Hunting, Tourism/recreation, Urban/industrial/transport

Birds

The site holds 20,000 or more waterbirds on passage and in winter, on a regular basis. Species of global conservation concern that do not meet IBA criteria: *Haliaeetus albicilla* (passage).

Species	Season	Year	Pop min	Pop max	Acc	Criteria
Egretta garzetta Little Egret	B	1997	310	310	—	B1i, C2, C6
Branta bernicla Brent Goose	W	1997	4,947	8,745	—	A4i, B1i, C3
Branta bernicla Brent Goose	P	1997	1,000	10,000	—	A4i, B1i, C3
Tadorna tadorna Shelduck	W	1997	1,728	4,445	—	A4i, B1i, C3
Anas clypeata Shoveler	B	1997	400	450	—	B1i, C3
Anas clypeata Shoveler	W	1997	32	514	—	B1i, C3
Milvus migrans Black Kite	B	1995	10	30	—	C6
Circus aeruginosus Marsh Harrier	B	1997	80	120	—	C6
Circus pygargus Montagu's Harrier	B	1995	50	100	—	B3, C6
Himantopus himantopus Black-winged Stilt	B	1995	212	249	—	A4i, B1i, C2, C6
Recurvirostra avosetta Avocet	B	1997	284	404	—	A4i, B1i, B3, C2, C6
Recurvirostra avosetta Avocet	W	1997	1,450	1,788	—	A4i, B1i, B2, C2
Pluvialis squatarola Grey Plover	W	1997	1,578	6,110	—	A4i, B1i, C3
Pluvialis squatarola Grey Plover	P	1997	3,000	3,000	—	A4i, B1i, C3
Calidris alpina Dunlin	W	1997	4,853	15,582	—	B1i, B2, C3
Calidris alpina Dunlin	P	1997	10,000	20,000	—	B1i, C3
Limosa lapponica Bar-tailed Godwit	W	1997	42	1,090	—	A4i, B1i, B2, C2
Limosa lapponica Bar-tailed Godwit	P	1997	1,500	1,500	—	A4i, B1i, C2
Numenius arquata Curlew	W	1997	700	3,530	—	A4i, B1i, B2, C3
Tringa totanus Redshank	B	1997	676	816	—	B1i, C3
Sterna hirundo Common Tern	B	1997	42	50	—	C6
Asio flammeus Short-eared Owl	B	1995	2	5	—	C6
Luscinia svecica Bluethroat	B	1995	500	600	—	C6

Protection status

National Low **International** None

4,200 ha of IBA covered by Hunting Reserve (Baie de Bourgneuf, 4,200 ha). 80 ha of IBA covered by Private Reserve (LPO-BirdLife, 80 ha).

Conservation issues

Threats Abandonment/reduction of land management (A), Agricultural intensification/expansion (B), Aquaculture/fisheries (C), Drainage, Industrialization/urbanization (B), Infrastructure (C), Recreation/tourism (A), Unsustainable exploitation (B)

The site is threatened by drainage and cultivation, marine aquaculture and the digging of ponds for aquaculture. Additional threats include increasing hunting pressure.

Basses Vallées Angevines — A1, A4i, A4iii, B1i, B2, B3, C1, C2, C3, C4, C6 — 026

Admin region Pays de la Loire
Coordinates 47°34′N 0°28′W
Altitude 16–16 m **Area** 6,450 ha

Site description

The site encompasses rivers, streams, wet meadows and *Populus* plantations.

Habitats Grassland (95%; humid grassland), Wetland (standing fresh water; river/stream; water-fringe vegetation), Artificial landscape (arable land)
Land-use Agriculture (95%), Fisheries/aquaculture, Forestry (5%), Hunting

Birds

Species	Season	Year	Pop min	Pop max	Acc	Criteria
Anas acuta Pintail	P	1997	300	4,000	—	B1i, C3
Anas clypeata Shoveler	P	1997	100	450	—	B1i, C3
Porzana porzana Spotted Crake	B	1991	5	10	—	B3, C6
Crex crex Corncrake	B	1997	380	411	—	A1, B2, C1, C2, C6
Limosa limosa Black-tailed Godwit	P	1991	4,000	30,000	—	A4i, B1i, C3

This is the best site in France for breeding *Crex crex*. The site holds 20,000 or more waterbirds on passage, on a regular basis.

Protection status

National Low **International** High

361 ha of IBA covered by Private Reserve (LPO-BirdLife, 361 ha). 6,450 ha of IBA covered by Ramsar Site (Basses Vallées Angevines, 6,450 ha).

Conservation issues

The site is threatened by agricultural improvements and commercial *Populus* plantations. Nevertheless, strong conservation measures are applied to this area, including agri-environmental schemes, prohibition of *Populus* plantations and land purchase under LIFE contracts and as a result of public fundraising. 6,450 ha of IBA is planned to be designated as a Special Protection Area (Basses Vallées Angevines) in 1999. More than 300 ha of the site have been purchased by the LPO, to be managed for the conservation of meadows and *Crex crex*.

Threats Abandonment/reduction of land management (C), Afforestation (A), Agricultural intensification/expansion (A), Extraction industry (C), Infrastructure (B), Intensified forest management (B), Recreation/tourism (A), Unsustainable exploitation (B)

Marais salants de Noirmoutier — B1i, B3, C2, C6 — 027

Admin region Pays de la Loire
Coordinates 46°59′N 2°16′W
Altitude 1–3 m **Area** 1,650 ha

Site description

Saltwater marshes, salinas and raised saltmarsh.

Habitats Wetland (saltmarsh)
Land-use Fisheries/aquaculture, Hunting, Tourism/recreation

Birds

Species	Season	Year	Pop min	Pop max	Acc	Criteria
Circus pygargus Montagu's Harrier	B	1991	10	30	—	C6
Himantopus himantopus Black-winged Stilt	B	1991	30	50	—	C6
Recurvirostra avosetta Avocet	B	1991	115	255	—	B1i, B3, C2, C6
Sterna hirundo Common Tern	B	1991	120	120	—	C6

The site supports important colonies of waders and terns.

Protection status

National Low **International** None

48 ha of IBA covered by Nature Reserve (Marais de Müllembourg, 48 ha).

Conservation issues

Threats Abandonment/reduction of land management, Agricultural intensification/expansion (B), Aquaculture/fisheries (C), Industrialization/urbanization (B), Recreation/tourism (A)

Stock-raising is declining which is leading to the abandonment of meadows and their conversion to intensive cultivation.

Marais et forêt d'Olonne — A4i, A4iii, B1i, B2, B3, C2, C3, C4, C6 — 028

Admin region Pays de la Loire
Coordinates 46°32′N 1°48′W
Altitude 1–34 m **Area** 3,250 ha

Site description

The site consists of saltwater lagoons, beaches, saltwater marshes, woodland and arable land.

Habitats Forest and woodland (mixed forest; alluvial/very wet forest; broadleaved evergreen forest), Scrub (heathland), Wetland (56%; sand-dunes/sand beach; coastal lagoon; standing fresh water; standing brackish and salt water; river/stream; water-fringe vegetation), Artificial landscape (arable land; perennial crops/orchards/groves; ruderal land)
Land-use Agriculture, Fisheries/aquaculture, Forestry, Hunting, Tourism/recreation, Urban/industrial/transport

Birds

Species	Season	Year	Pop min	Pop max	Acc	Criteria
Platalea leucorodia Spoonbill	P	1997	100	500	—	A4i, B1i, C2
Anas acuta Pintail	W	1997	387	973	—	B1i, C3
Anas acuta Pintail	P	1997	1,000	10,000	—	B1i, C3
Anas clypeata Shoveler	W	1997	82	784	—	B1i, C3
Anas clypeata Shoveler	P	1997	1,000	10,000	—	A4i, B1i, C3
Milvus migrans Black Kite	B	1991	10	100	—	B2, C6
Recurvirostra avosetta Avocet	B	1997	400	400	—	A4i, B1i, B3, C2, C6
Recurvirostra avosetta Avocet	W	1997	128	375	—	B2
Recurvirostra avosetta Avocet	P	1997	250	750	—	B1i, C2

Species ... continued	Season	Year	Pop min	Pop max	Acc	Criteria
Limosa limosa Black-tailed Godwit	P	1991	1,000	10,000	—	A4i, B1i, C3
Numenius phaeopus Whimbrel	P	1991	10,000	20,000	—	A4i, B1i, C3
Tringa totanus Redshank	P	1997	1,000	10,000	—	A4i, B1i, C3
Caprimulgus europaeus Nightjar	B	1991	10	100	—	C6
Lullula arborea Woodlark	B	1991	10	100	—	C6

The site holds 20,000 or more waterbirds on passage, on a regular basis. Species of global conservation concern that do not meet IBA criteria: *Aythya nyroca* (wintering).

Protection status

National Low **International** None

40 ha of IBA covered by Hunting Reserve (Chanteloup, 40 ha).

Conservation issues

Threats Industrialization/urbanization (C), Intensified forest management (B), Recreation/tourism (B)

This site is threatened by the infilling of the saltmarsh. New forest tracks have increased disturbance.

Forêt du Gavre — C6 — 029

Admin region Pays de la Loire
Coordinates 47°32′N 1°48′W
Altitude 20–20 m **Area** 4,900 ha

Site description

Mixed forests and remnant moorlands.

Habitats Forest and woodland (mixed forest)
Land-use Forestry (100%), Hunting (100%), Tourism/recreation

Birds

Species	Season	Year	Pop min	Pop max	Acc	Criteria
Circus cyaneus Hen Harrier	B	1991	10	11	—	C6
Dendrocopos medius Middle Spotted Woodpecker	B	1991	100	115	—	C6

Protection status

National None **International** None

Conservation issues

Threats Intensified forest management (B), Recreation/tourism (B), Unsustainable exploitation (C)

Intensive forestry, high hunting pressure and increased tourism are the main threats.

Marais de Mazerolles – Petit Mars — B1i, C3 — 030

Admin region Pays de la Loire
Coordinates 47°18′N 1°30′W
Altitude 5–10 m **Area** 2,700 ha

Site description

The site includes marshes, reedbeds (*Phragmites*) and aquatic vegetation.

Habitats Wetland (100%; water-fringe vegetation)
Land-use Agriculture, Fisheries/aquaculture, Hunting, Tourism/recreation, Urban/industrial/transport

Birds

Species	Season	Year	Pop min	Pop max	Acc	Criteria
Anas acuta Pintail	P	1991	500	4,000	—	B1i, C3
Anas clypeata Shoveler	W	1991	500	2,500	—	B1i, C3

Protection status

National Low **International** None

69 ha of IBA covered by 3 Hunting Reserves (total area 69 ha). 150 ha Private Reserve (LPO-BirdLife, 150 ha).

Conservation issues

Threats Disturbance to birds (U), Extraction industry (A), Recreation/tourism (B), Other, Unsustainable exploitation (C)

The main threats are sand and peat extraction and problems with water management during the summer ('Other' threats, above).

Vallée de la Loire: de Nantes à Montsoreau — A1, B1i, B2, C1, C2, C6 — 031

Admin region Pays de la Loire
Coordinates 47°18′N 0°24′W
Altitude 3–30 m **Area** 12,000 ha

Site description

The site includes a river, marshes, aquatic vegetation, wet meadows, riverine forest, *Populus* plantations, bocages and arable fields.

Habitats Forest and woodland (alluvial/very wet forest), Grassland (humid grassland; mesophile grassland), Wetland (river/stream; water-fringe vegetation), Artificial landscape (highly improved reseeded grassland; arable land; perennial crops/orchards/groves)
Land-use Agriculture, Fisheries/aquaculture, Forestry, Hunting, Tourism/recreation, Urban/industrial/transport

Birds

Species	Season	Year	Pop min	Pop max	Acc	Criteria
Crex crex Corncrake	B	1994	224	235	—	A1, B2, C1, C2, C6
Sterna hirundo Common Tern	B	1991	225	400	—	C6
Sterna albifrons Little Tern	B	1991	150	280	—	B1i, B2, C2, C6
Alcedo atthis Kingfisher	B	1991	100	200	—	B2, C6
Riparia riparia Sand Martin	B	1991	1,400	1,520	—	B2

This is an important site for breeding waterbirds, *Crex crex* and terns, and as a wintering site for waterbirds, in particular *Vanellus vanellus* (14,000 individuals).

Protection status

National Low **International** None

175 ha of IBA covered by Hunting Reserve (175 ha).

Conservation issues

Threats Afforestation (A), Agricultural intensification/expansion (A), Extraction industry (B), Intensified forest management (B), Recreation/tourism (A), Unsustainable exploitation (B)

Recreation (boating), conversion of wet meadows into arable fields and afforestation with *Populus* threaten the site.

Plaine calcaire du sud Vendée — B3, C6 — 032

Admin region Pays de la Loire
Coordinates 46°26′N 0°49′W
Altitude 10–60 m **Area** 18,700 ha

Site description

The site contains cereal crops, dry meadows, woodlots, bocages and abandoned fields.

Habitats Grassland (steppe/dry calcareous grassland), Artificial landscape (highly improved reseeded grassland; arable land; perennial crops/orchards/groves; ruderal land)
Land-use Agriculture, Fisheries/aquaculture, Hunting, Tourism/recreation, Urban/industrial/transport

Birds

Species	Season	Year	Pop min	Pop max	Acc	Criteria
Circus pygargus Montagu's Harrier	B	1991	20	40	—	B3, C6
Tetrax tetrax Little Bustard	B	1996	12	12	—	C6
Burhinus oedicnemus Stone Curlew	B	1991	40	50	—	C6

The site is important for species of open agricultural landscapes. Species of global conservation concern that do not meet IBA criteria: *Tetrax tetrax* (passage).

Protection status

National None **International** None

Conservation issues

Threats Agricultural intensification/expansion (B), Groundwater abstraction (A), Infrastructure (A)

Water pollution from agricultural run-off, over-exploitation of water resources as a result of irrigation, and ongoing highway construction, are the main threats.

Marais poitevin et baie de l'Aiguillon

A4i, A4iii, B1i, B2, B3, C2, C3, C4, C6 — **033**

Admin region Pays de la Loire
Coordinates 46°20'N 1°00'W
Altitude 0–25 m **Area** 77,900 ha

Site description

A complex of many habitat-types. Human activities include stock and arable farming, oyster and mussel farming, and grazing on reclaimed saltmarshes.

Habitats Forest and woodland (broadleaved deciduous forest; native coniferous forest; alluvial/very wet forest), Scrub (heathland), Grassland (steppe/dry calcareous grassland; humid grassland), Wetland (tidal river/enclosed tidal water; mudflat/sandflat; saltmarsh; sand-dunes/sand beach; shingle/stony beach; coastal lagoon; standing fresh water; standing brackish and salt water; river/stream; water-fringe vegetation; fen/transition mire/spring), Marine areas (open sea; sea inlet/coastal features), Rocky areas (inland cliff), Artificial landscape (highly improved reseeded grassland; arable land; perennial crops/orchards/groves; urban parks/gardens; ruderal land)

Land-use Agriculture, Fisheries/aquaculture, Forestry, Hunting, Tourism/recreation, Urban/industrial/transport

Birds

Species	Season	Year	Pop min	Pop max	Acc	Criteria
Ixobrychus minutus Little Bittern	B	1991	5	10	—	C6
Nycticorax nycticorax Night Heron	B	1997	52	52	—	C6
Ardea purpurea Purple Heron	B	1997	159	159	—	B2, C2, C6
Ciconia ciconia White Stork	B	1997	10	10	—	C6
Anser anser Greylag Goose	P	1997	2,000	2,000	—	B1i, C3
Branta bernicla Brent Goose	W	1997	745	4,110	—	A4i, B1i, C3
Milvus migrans Black Kite	B	1991	10	100	—	B2, C6
Circus aeruginosus Marsh Harrier	B	1997	10	100	—	C6
Circus pygargus Montagu's Harrier	B	1991	10	100	—	B3, C6
Himantopus himantopus Black-winged Stilt	B	1995	65	68	—	C6
Recurvirostra avosetta Avocet	W	1997	2,075	7,973	—	A4i, B1i, B2, C2
Recurvirostra avosetta Avocet	P	1997	1,000	1,500	—	A4i, B1i, C2
Pluvialis squatarola Grey Plover	W	1997	983	1,975	—	A4i, B1i, C3
Pluvialis squatarola Grey Plover	P	1997	3,400	4,500	—	A4i, B1i, C3
Calidris canutus Knot	W	1997	810	10,360	—	A4i, B1i, B2, C3
Calidris canutus Knot	P	1997	20,000	20,000	—	A4i, B1i, C3
Calidris alpina Dunlin	P	1997	20,000	20,000	—	B1i, C3
Limosa limosa Black-tailed Godwit	P	1997	40,000	80,000	—	A4i, B1i, C3
Limosa lapponica Bar-tailed Godwit	P	1997	350	1,500	—	A4i, B1i, C2
Numenius phaeopus Whimbrel	P	1991	9,000	17,000	—	A4i, B1i, C3
Tringa totanus Redshank	P	1997	500	2,000	—	B1i, C3
Larus minutus Little Gull	P	1991	1,500	2,000	—	A4i, B1i, C3
Chlidonias niger Black Tern	B	1996	20	30	—	C6
Alcedo atthis Kingfisher	B	1991	10	50	—	C6
Luscinia svecica Bluethroat	B	1991	300	300	—	C6

The site holds 20,000 or more waterbirds, on passage and in winter on a regular basis. Species of global conservation concern that do not meet IBA criteria: *Crex crex* (5–6 breeding pairs); *Tetrax tetrax* (20–50 passage individuals); *Acrocephalus paludicola* (passage).

Protection status

National Low **International** Partial

1,328 ha of IBA covered by 4 Hunting Reserves. 207 ha of IBA covered by Nature Reserve (Saint-Denis-du-Payré, 207 ha). 2,300 ha of IBA covered by Nature Reserve (Baie de l'Aiguillon, 2,300 ha). 255 ha of IBA covered by Private Reserve (LPO-BirdLife parcels of Champagné, 255 ha). 5,290 ha of IBA covered by Special Protection Area (Marais Poitevin Baie de l'Aiguillon, 5,290 ha). 19,960 ha of IBA covered by Special Protection Area (Marais Poitevin Intérieur, 19,960 ha). 1,000 ha of IBA covered by Special Protection Area (Marais Poitevin Pointe d'Arçay, 1,000 ha). 3,540 ha of IBA covered by Special Protection Area (Marais doux charentais, 3,540 ha).

Conservation issues

Threats Abandonment/reduction of land management (B), Agricultural intensification/expansion (A), Drainage (A), Infrastructure (A), Recreation/tourism (A), Unsustainable exploitation (A)

The site is seriously threatened by drainage of the wet meadow areas. Intensive agricultural practices, channelization of rivers and streams, removal of hedges, scrub and tree-lined channels, drainage of saltmarshes and salinas, and extension of oyster and mussel farming on the mudflats also pose significant problems. In addition, communal meadows of the Vendeé marshes are affected by the creation of lakes for wind-surfing. Approximately 10,000 ha of the IBA is proposed as a Special Protection Area (13,000 ha).

Falaise du Bessin

C7 — **034**

Admin region Basse-Normandie
Coordinates 49°24'N 0°57'W
Altitude 0–35 m **Area** 2,300 ha

Site description

A stretch of rocky coastline and open sea with cliffs, moorland, and arable land.

Habitats Scrub (heathland), Marine areas (99%; open sea), Rocky areas (sea cliff/rocky shore), Artificial landscape (arable land)

Land-use Agriculture, Not utilized (90%), Tourism/recreation

Birds

The most important site in France for breeding *Fulmarus glacialis* and *Rissa tridactyla*.

Protection status

National High **International** Partial

1,200 ha of IBA covered by Special Protection Area (Falaise du Bessin occidental, 1,200 ha).

Conservation issues

Threats Disturbance to birds (C), Recreation/tourism (C)

There is considerable pressure from tourism leading to increasing disturbance.

Baie des Veys et Marais du Cotentin

A1, A4i, A4iii, B1i, B2, C1, C2, C3, C4, C6 — **035**

Admin region Basse-Normandie
Coordinates 49°20'N 1°15'W
Altitude 0–7 m **Area** 37,500 ha

Site description

A bay with mudflats, saltmarshes, wet meadows, aquatic vegetation, and peatbogs.

Habitats Grassland (humid grassland), Wetland (tidal river/enclosed tidal water; mudflat/sandflat; saltmarsh; standing fresh water; river/stream; raised bog; water-fringe vegetation; fen/transition mire/spring), Marine areas (open sea; sea inlet/coastal features), Artificial landscape (highly improved reseeded grassland; perennial crops/orchards/groves)

Land-use Agriculture, Fisheries/aquaculture, Hunting, Tourism/recreation, Urban/industrial/transport

Birds

Species	Season	Year	Pop min	Pop max	Acc	Criteria
Phalacrocorax carbo Cormorant	P	1997	750	5,000	—	A4i, B1i, C3
Ciconia ciconia White Stork	B	1991	1	5	—	C6
Platalea leucorodia Spoonbill	P	1991	50	250	—	A4i, B1i, C2
[1] ***Branta bernicla*** Brent Goose	P	1997	10,000	20,000	—	A4i, B1i, C3
Anas acuta Pintail	W	1997	236	1,215	—	B1i, C3
Anas acuta Pintail	P	1997	750	5,000	—	B1i, C3
Anas clypeata Shoveler	P	1997	500	1,000	—	B1i, C3
Melanitta fusca Velvet Scoter	W	1997	2	250	—	B2
Circus pygargus Montagu's Harrier	B	1991	3	5	—	C6

Species ... continued	Season	Year	Pop min	Pop max	Acc	Criteria
Crex crex Corncrake	B	1992	26	31	—	A1, C1, C6
Haematopus ostralegus Oystercatcher	W	1997	1,500	11,500	—	A4i, B1i, C3
Charadrius hiaticula Ringed Plover	P	1997	1,000	10,000	—	A4i, B1i, C3
Calidris alpina Dunlin	W	1997	6,900	12,520	—	B2
Numenius arquata Curlew	W	1997	793	13,298	—	A4i, B1i, B2, C3
Sterna sandvicensis Sandwich Tern	P	1997	1,000	10,000	—	A4i, B1i, C2
Sterna hirundo Common Tern	P	1991	1,000	10,000	—	A4i, B1i, C2
Sterna albifrons Little Tern	P	1991	500	1,000	—	A4i, B1i, C2
Chlidonias niger Black Tern	P	1991	1,000	10,000	—	A4i, B1i, C2
Sylvia undata Dartford Warbler	B	1991	1	10	—	C6

1. *B. b. hrota*

The site holds 20,000 or more waterbirds on passage and in winter, on a regular basis. Species of global conservation concern that do not meet IBA criteria: *Aythya nyroca* (wintering); *Crex crex* (passage).

Protection status

National Low **International** High

1,907 ha of IBA covered by Hunting Reserve (1,907 ha). 600 ha of IBA covered by Nature Reserve (Domaine de Beauguillot, 600 ha). 400 ha of IBA covered by Nature Reserve (Marais de la Sangsurière et de l'Adriennerie, 400 ha). 20,715 ha of IBA covered by Special Protection Area (Basses vallées du Cotentin et baie des Veys, 20,715 ha). 32,500 ha of IBA covered by Ramsar Site (Marais du Cotentin et du Bessin, Baie de Veys, 32,500 ha).

Conservation issues

Threats Agricultural intensification/expansion (B), Groundwater abstraction (A), Recreation/tourism (A), Unsustainable exploitation (A)

The high hunting pressure and the management of water-levels are not compatible with species and habitat conservation.

Iles Saint Marcouf — B3 — 036

Admin region Basse-Normandie
Coordinates 49°30'N 1°08'W
Altitude 0–5 m **Area** 1,540 ha

Site description

A group of small islands with rocky coastlines in the English Channel.

Habitats Marine areas (99%; open sea), Rocky areas (rock stacks/islets)
Land-use Fisheries/aquaculture, Tourism/recreation

Birds

Species	Season	Year	Pop min	Pop max	Acc	Criteria
Larus fuscus Lesser Black-backed Gull	B	1991	300	300	—	B3

This is an important site for breeding seabirds.

Protection status

National High **International** High

1,400 ha of IBA covered by Hunting Reserve (Iles Saint Marcouf, 1,400 ha). 3 ha of IBA covered by Private Reserve (3 ha). 1,400 ha of IBA covered by Special Protection Area (Iles Saint Marcouf, 1,400 ha).

Conservation issues

Threats Aquaculture/fisheries (A), Recreation/tourism (A)

There is considerable disturbance from tourism (boats); further tourism development and possible aquaculture projects pose additional threats to the site.

Iles Chausey — A4i, B1i, B3, C2, C6 — 037

Admin region Basse-Normandie
Coordinates 48°54'N 1°48'W
Altitude 5–15 m **Area** 18,400 ha

Site description

Islands in the Baie du Mont-St-Michel with rocky coasts, sand and shingle-beaches, scrub, grassland and arable land.

Habitats Scrub (heathland), Wetland (mudflat/sandflat; sand-dunes/sand beach; shingle/stony beach), Marine areas (open sea), Rocky areas (rock stacks/islets), Artificial landscape (arable land)
Land-use Agriculture, Fisheries/aquaculture, Tourism/recreation, Urban/industrial/transport

Birds

Species	Season	Year	Pop min	Pop max	Acc	Criteria
Hydrobates pelagicus Storm Petrel	B	1991	1	10	—	C6
Phalacrocorax aristotelis Shag	B	1991	525	525	—	B3
Larus marinus Great Black-backed Gull	B	1991	350	350	—	B3
Sterna sandvicensis Sandwich Tern	P	1991	750	5,000	—	A4i, B1i, C2
Sterna hirundo Common Tern	B	1991	20	25	—	C6
Sylvia undata Dartford Warbler	B	1991	1	10	—	C6

The site is important for breeding seabirds.

Protection status

National High **International** High

18,054 ha of IBA covered by Hunting Reserve (Iles Chausey Maritime, 18,000 ha). 54 ha of IBA covered by Hunting Reserve (Iles Chausey terrestre, 54 ha). 90 ha of IBA covered by Private Reserve (90 ha). 18,000 ha of IBA covered by Special Protection Area (Iles Chausey, 18,000 ha).

Conservation issues

Threats Aquaculture/fisheries (A), Recreation/tourism (A)

There is disturbance from tourism and sailing.

Havre de la Sienne — B1i, C3 — 038

Admin region Basse-Normandie
Coordinates 49°00'N 1°34'W
Altitude 0–0 m **Area** 5,150 ha

Site description

An area of open sea and an estuary with mudflats, saltmarshes and sandy beaches. Agriculture and cattle-raising are the main human activities.

Habitats Wetland (tidal river/enclosed tidal water; mudflat/sandflat; saltmarsh; sand-dunes/sand beach), Marine areas (open sea)
Land-use Agriculture, Fisheries/aquaculture, Hunting, Tourism/recreation, Urban/industrial/transport

Birds

Species	Season	Year	Pop min	Pop max	Acc	Criteria
[1] ***Branta bernicla*** Brent Goose	W	1991	250	750	—	B1i, C3

1. *B.b. hrota.*

This is an important site for wintering seabirds and waders.

Protection status

National None **International** None

Conservation issues

Threats Agricultural intensification/expansion (B), Infrastructure (A), Recreation/tourism (A)

Disturbance is caused by recreational activities, particularly boating.

Estuaire de l'Orne — B1i, C2 — 039

Admin region Basse-Normandie
Coordinates 49°16'N 0°14'W
Altitude 0–2 m **Area** 1,000 ha

Site description

The site includes open sea, mudflats, sandy beaches, saltmarsh and wet meadows.

Habitats Grassland (humid grassland), Wetland (tidal river/enclosed tidal water; mudflat/sandflat; saltmarsh; sand-dunes/sand beach), Marine areas (open sea)

Land-use Agriculture (10%), Fisheries/aquaculture (15%), Hunting (27%), Other, Tourism/recreation (3%), Urban/industrial/transport (1%)

Birds

Species	Season	Year	Pop min	Pop max	Acc	Criteria
Platalea leucorodia Spoonbill	P	1989	10	100	—	B1i, C2

This site is important for passage waterbirds.

Protection status

National Partial **International** High
500 ha of IBA covered by Hunting Reserve (500 ha). 1,000 ha of IBA covered by Special Protection Area (Estuaire de l'Orne, 1,000 ha).

Conservation issues

Threats Disturbance to birds (A), Infrastructure (A), Recreation/tourism (A), Unsustainable exploitation (B)

Hunting pressure is increasing and leads to disturbance.

Littoral Augeron A4i, A4iii, B1i, B2, C2, C3, C4 040

Admin region Basse-Normandie
Coordinates 49°22'N 0°09'W
Altitude 0–0 m **Area** 23,100 ha

Site description

An area of open sea, mudflats, sandy beaches and cliffs.

Habitats Wetland (tidal river/enclosed tidal water; mudflat/sandflat; sand-dunes/sand beach), Marine areas (open sea), Rocky areas (sea cliff/rocky shore)
Land-use Fisheries/aquaculture, Tourism/recreation

Birds

Species	Season	Year	Pop min	Pop max	Acc	Criteria
Podiceps auritus Slavonian Grebe	W	1991	10	100	—	B1i, C2
Aythya marila Scaup	W	1997	35	300	—	B2
Melanitta fusca Velvet Scoter	W	1997	1,074	3,403	—	B2
Larus minutus Little Gull	P	1997	1,000	10,000	—	A4i, B1i, C3
Larus fuscus Lesser Black-backed Gull	P	1997	1,000	10,000	—	A4i, B1i, C3
Sterna sandvicensis Sandwich Tern	P	1991	1,000	10,000	—	A4i, B1i, C2
Sterna hirundo Common Tern	P	1991	1,000	10,000	—	A4i, B1i, C2
Sterna albifrons Little Tern	P	1991	100	500	—	B1i, C2

The site holds 20,000 or more waterbirds on passage and in winter, on a regular basis.

Protection status

National None **International** None

Conservation issues

Threats Agricultural intensification/expansion (B), Aquaculture/fisheries (C), Industrialization/urbanization (A), Infrastructure (A), Other (A), Recreation/tourism (A)

Marine pollution from oil tankers ('Other' threat, above), industry and ports, and also fishing and recreation are major threats.

Forêts du Perche B2, B3, C6 041

Admin region Basse-Normandie
Coordinates 48°31'N 0°51'E
Altitude 200–273 m **Area** 50,850 ha

Site description

An area of deciduous and conifer forests, ponds, rivers, marshes, aquatic vegetation, meadows, and hedgerows.

Habitats Forest and woodland (broadleaved deciduous forest; native coniferous forest; mixed forest), Scrub (heathland), Grassland (humid grassland), Wetland (standing fresh water; river/stream; raised bog; water-fringe vegetation), Artificial landscape (highly improved reseeded grassland; arable land; perennial crops/orchards/groves)
Land-use Agriculture, Fisheries/aquaculture, Forestry, Urban/industrial/transport

Birds

Species	Season	Year	Pop min	Pop max	Acc	Criteria
Pernis apivorus Honey Buzzard	B	1991	50	100	—	B3, C6
Circus cyaneus Hen Harrier	B	1991	1	10	—	C6
Caprimulgus europaeus Nightjar	B	1991	10	100	—	C6
Picus canus Grey-headed Woodpecker	B	1991	10	100	—	B2, C6
Dryocopus martius Black Woodpecker	B	1991	10	100	—	C6
Lullula arborea Woodlark	B	1991	10	10	—	C6

The site is of special interest for forest and wetland species.

Protection status

National None **International** None

Conservation issues

No serious threats are known at this site. Forestry and hunting practices are well managed.

Baie du Mont Saint Michel et Ile des Landes A4i, A4iii, B1i, B2, B3, C2, C3, C4 042

Admin region Basse-Normandie
Coordinates 48°39'N 1°31'W
Altitude 0–70 m **Area** 63,800 ha

Site description

A very large bay with islands, mudflats (22,000 ha), raised saltmarsh (>1,000 ha), grassland and arable land, rocky coast and cliffs. Human activities include cattle-raising.

Habitats Grassland (humid grassland), Wetland (66%; mudflat/sandflat; saltmarsh; shingle/stony beach; standing fresh water; raised bog; water-fringe vegetation), Marine areas (sea inlet/coastal features), Rocky areas (sea cliff/rocky shore; rock stacks/islets), Artificial landscape (highly improved reseeded grassland; arable land)
Land-use Agriculture, Fisheries/aquaculture, Hunting, Tourism/recreation, Urban/industrial/transport

Birds

Species	Season	Year	Pop min	Pop max	Acc	Criteria
Phalacrocorax aristotelis Shag	B	1991	200	200	—	B3
Branta bernicla Brent Goose	W	1991	2,000	3,000	—	B1i, C3
Anas penelope Wigeon	W	1991	50	25,000	—	A4i, B1i, C3
Anas crecca Teal	W	1991	200	1,500	—	A4i, B1i, C3
Melanitta nigra Common Scoter	W	1991	10,000	15,000	—	A4i, B1i, C3
Melanitta nigra Common Scoter	P	1991	10,000	20,000	—	A4i, B1i, C3
Haematopus ostralegus Oystercatcher	W	1996	10,000	20,000	—	A4i, B1i, C3
Charadrius hiaticula Ringed Plover	P	1991	4,000	4,000	—	A4i, B1i, C3
Pluvialis squatarola Grey Plover	W	1991	1,500	3,000	—	A4i, B1i, C3
Pluvialis squatarola Grey Plover	P	1991	1,000	10,000	—	A4i, B1i, C3
Calidris canutus Knot	W	1991	3,000	7,000	—	A4i, B1i, B2, C3
Calidris alpina Dunlin	W	1991	22,000	35,000	—	A4i, B1i, B2, C3
Limosa limosa Black-tailed Godwit	P	1991	1,000	10,000	—	A4i, B1i, C3
Limosa lapponica Bar-tailed Godwit	W	1991	300	1,500	—	A4i, B1i, B2, C2
Numenius arquata Curlew	W	1991	2,000	5,000	—	A4i, B1i, B2, C3

This is an important site, holding 20,000 or more wintering and passage waterbirds on a regular basis. It is also an important area for passage passerines.

Protection status

National Partial **International** High
7,850 ha of IBA covered by Hunting Reserve (Baie du Mont Saint Michel, 7,850 ha). 22 ha of IBA covered by Private Reserve (Ile des Landes, 22 ha). 18,000 ha of IBA covered by Special Protection Area (Baie du Mont Saint Michel, 18,000 ha). 62,000 ha of IBA covered by Ramsar Site (Baie du Mont Saint Michel, 62,000 ha).

Conservation issues

Threats Drainage, Recreation/tourism (A)

The drainage of wet meadows for cultivation is slowly causing the drying out the whole area. Breeding sites for waterbirds are being lost.

Cap Fagnet

C7 **043**

Admin region Haute-Normandie
Coordinates 49°48′N 0°25′E
Altitude 0–95 m **Area** 5,700 ha

Site description
An area of open sea, shingle beach and marine cliffs.

Habitats Wetland (shingle/stony beach), Marine areas (open sea), Rocky areas (sea cliff/rocky shore)
Land-use Fisheries/aquaculture, Hunting, Tourism/recreation

Birds
This is an important site for breeding seabirds.

Protection status
National High **International** High
5,700 ha of IBA covered by Hunting Reserve (Falaise de la Pointe Fagnet, 5,700 ha). 5,700 ha of IBA covered by Special Protection Area (Falaise de la Pointe Fagnet, 5,700 ha).

Conservation issues
No serious threats are known at this site.

Boucle de Pose et de Muids

A4iii, B1i, B2, C3, C4, C6 **044**

Admin region Haute-Normandie
Coordinates 49°16′N 1°16′E
Altitude 0–4 m **Area** 5,200 ha

Site description
The site includes a river, ponds, marshes and aquatic vegetation, meadows, arable fields and abandoned fields.

Habitats Forest and woodland (mixed forest; alluvial/very wet forest), Scrub (heathland), Wetland (standing fresh water; river/stream; water-fringe vegetation), Artificial landscape (highly improved reseeded grassland; arable land; urban parks/gardens; other urban/industrial areas; ruderal land)
Land-use Agriculture, Fisheries/aquaculture, Hunting, Tourism/recreation, Urban/industrial/transport

Birds

Species	Season	Year	Pop min	Pop max	Acc	Criteria
Aythya ferina Pochard	W	1991	4,500	7,000	—	B1i, C3
Aythya marila Scaup	W	1991	10	100	—	B2
Pernis apivorus Honey Buzzard	B	1991	1	10	—	C6
Burhinus oedicnemus Stone Curlew	B	1991	1	10	—	C6
Dendrocopos medius Middle Spotted Woodpecker	B	1991	1	10	—	C6

The site holds 20,000 or more waterbirds in winter, on a regular basis (including *Aythya fuligula* 2,000; *Fulica atra* 4,000 and *Vanellus vanellus* 5,000). Species of global conservation concern that do not meet IBA criteria: *Aythya nyroca* (wintering).

Protection status
National Low **International** None
70 ha of IBA covered by Private Reserve (70 ha).

Conservation issues
Threats Extraction industry (A), Recreation/tourism (B)

Illegal hunting and recreation are the main threats; both cause disturbance to birds.

Estuaire et embouchure de la Seine

A1, A4i, A4iii, B1i, B2, C1, C2, C3, C4, C6 **045**

Admin region Haute-Normandie
Coordinates 49°26′N 0°18′E
Altitude 0–5 m **Area** 21,100 ha

Site description
An estuary with mudflats, saltmarshes, sand and shingle-beaches, reedbeds (*Phragmites*), wet meadows, urban and industrial areas. The area includes the port of Le Havre.

Habitats Grassland (15%; humid grassland), Wetland (62%; tidal river/enclosed tidal water; mudflat/sandflat; saltmarsh; sand-dunes/sand beach; shingle/stony beach; coastal lagoon; standing fresh water; standing brackish and salt water; river/stream; water-fringe vegetation), Artificial landscape (20%; other urban/industrial areas; ruderal land)
Land-use Agriculture (10%), Fisheries/aquaculture, Hunting (40%), Tourism/recreation, Urban/industrial/transport (15%)

Birds

Species	Season	Year	Pop min	Pop max	Acc	Criteria
Platalea leucorodia Spoonbill	P	1991	140	140	—	A4i, B1i, C2
Anas acuta Pintail	W	1997	277	1,631	—	B1i, C3
Aythya marila Scaup	W	1997	1	160	—	B2
Melanitta fusca Velvet Scoter	W	1997	1,500	1,500	—	B2
Circus aeruginosus Marsh Harrier	B	1997	3	5	—	C6
Porzana porzana Spotted Crake	B	1991	1	10	—	C6
Crex crex Corncrake	B	1996	26	31	—	A1, C1, C6
Himantopus himantopus Black-winged Stilt	B	1995	4	7	—	C6
Recurvirostra avosetta Avocet	B	1997	10	10	—	C6
Recurvirostra avosetta Avocet	W	1997	437	1,360	—	A4i, B1i, B2, C2
Recurvirostra avosetta Avocet	P	1997	5,000	5,000	—	A4i, B1i, C2
Limosa lapponica Bar-tailed Godwit	P	1997	500	1,000	—	A4i, B1i, C2
Tringa totanus Redshank	P	1997	1,000	10,000	—	A4i, B1i, C3
Sterna sandvicensis Sandwich Tern	P	1991	1,500	2,000	—	B1i, C2
Sterna hirundo Common Tern	P	1991	1,500	2,000	—	B1i, C2
Luscinia svecica Bluethroat	B	1991	1	10	—	C6

The site holds 20,000 or more waterbirds on passage and in winter, on a regular basis (including *Podiceps cristatus* 4,706 (wintering); *Melanitta nigra* 4,006 (wintering); *Haematopus ostralegus* 4,950 (wintering)). Species of global conservation concern that do not meet IBA criteria: *Acrocephalus paludicola* (passage).

Protection status
National Partial **International** Low
1,000 ha of IBA covered by Nature Reserve (Estuaire de la Seine, 3,770 ha). 93 ha of IBA covered by Nature Reserve (Les Mannevilles, 93 ha). 9,000 ha of IBA covered by 2 Hunting Reserves (Baie de Seine; Banc herbeux; total area 9,000 ha). 1,300 ha of IBA covered by Special Protection Area (Estuaire de la Seine, 2,750 ha).

Conservation issues
Threats Filling-in of wetlands (A), Industrialization/urbanization (A), Infrastructure (A), Recreation/tourism (A), Unsustainable exploitation (A)

The expansion of the harbour is the major threat to the site. Other threats include sand and gravel extraction and considerable hunting pressure at night.

Boucle de Moisson

C6 **046**

Admin region Ile de France
Coordinates 49°02′N 1°38′E
Altitude 10–150 m **Area** 6,150 ha

Site description
Human activities include arable farming, tourism and recreation (including boating) and hunting.

Habitats Forest and woodland (broadleaved deciduous forest; mixed forest), Scrub (heathland), Grassland (dry siliceous grassland), Wetland (standing fresh water; river/stream; water-fringe vegetation), Rocky areas (inland cliff), Artificial landscape (arable land; perennial crops/orchards/groves; urban parks/gardens; other urban/industrial areas; ruderal land)
Land-use Agriculture, Forestry, Hunting, Tourism/recreation, Urban/industrial/transport

Birds

Species	Season	Year	Pop min	Pop max	Acc	Criteria
Burhinus oedicnemus Stone Curlew	B	1991	7	7	—	C6

An important site for passage and wintering waterbirds. Species of global conservation concern that do not meet IBA criteria: *Crex crex* (breeding).

■ **Protection status**

National None **International** None

■ **Conservation issues**

Threats Afforestation (A), Extraction industry (A), Intensified forest management (A), Recreation/tourism (A)

The main threats are the expansion of gravel-pits and further development of recreation activities.

Massifs de Fontainebleau et zones humides adjacentes — B2, B3, C2, C6 — 047

Admin region Ile de France
Coordinates 48°22′N 2°38′E
Altitude 40–145 m **Area** 36,600 ha

■ **Site description**

An area of mixed woodland, lakes, marshes, aquatic vegetation, meadows, heathland and residential areas.

Habitats Forest and woodland (mixed forest), Scrub (heathland), Grassland (steppe/dry calcareous grassland; dry siliceous grassland; humid grassland; mesophile grassland), Wetland (standing fresh water; river/stream; water-fringe vegetation; fen/transition mire/spring), Rocky areas (scree/boulders; inland cliff; caves), Artificial landscape (perennial crops/orchards/groves; urban parks/gardens; ruderal land)
Land-use Agriculture, Forestry, Hunting, Tourism/recreation, Urban/industrial/transport

■ **Birds**

Species	Season	Year	Pop min	Pop max	Acc	Criteria
Nycticorax nycticorax Night Heron	B	1991	5	15	—	C6
Pernis apivorus Honey Buzzard	B	1991	30	40	—	C6
Caprimulgus europaeus Nightjar	B	1991	100	200	—	B2, C6
Picus canus Grey-headed Woodpecker	B	1991	50	250	—	B2, C6
Dryocopus martius Black Woodpecker	B	1991	50	50	—	C6
Dendrocopos medius Middle Spotted Woodpecker	B	1991	50	250	—	B3, C2, C6
Sylvia undata Dartford Warbler	B	1991	100	100	—	C6
Lanius collurio Red-backed Shrike	B	1991	10	10	—	C6

■ **Protection status**

National Low **International** Partial

100 ha of IBA covered by Private Reserve (100 ha). Part of IBA covered by Biosphere Reserve (Forêt de Fontainebleau).

■ **Conservation issues**

Threats Afforestation (A), Intensified forest management (A), Recreation/tourism (A)

Recreation, the intensification of forestry, and urbanization around the site boundaries are the main threats.

Bassée et plaines adjacentes — C6 — 048

Admin region Ile de France
Coordinates 48°24′N 3°11′E
Altitude 55–172 m **Area** 39,900 ha

■ **Site description**

The site contains lakes, rivers, alluvial forests, marshes and meadows.

Habitats Forest and woodland (alluvial/very wet forest), Scrub (heathland), Grassland (steppe/dry calcareous grassland; humid grassland; mesophile grassland), Wetland (standing fresh water; river/stream; water-fringe vegetation), Artificial landscape (highly improved reseeded grassland; arable land; perennial crops/orchards/groves; urban parks/gardens; other urban/industrial areas; ruderal land)
Land-use Agriculture, Fisheries/aquaculture, Forestry, Hunting, Tourism/recreation, Urban/industrial/transport

■ **Birds**

Species	Season	Year	Pop min	Pop max	Acc	Criteria
Sterna hirundo Common Tern	B	1991	90	150	—	C6
Alcedo atthis Kingfisher	B	1991	20	20	—	C6

Species of global conservation concern that do not meet IBA criteria: *Aythya nyroca* (passage).

■ **Protection status**

National Low **International** None

2,000 ha of IBA covered by Hunting Reserve (2,000 ha).

■ **Conservation issues**

Threats Agricultural intensification/expansion (A), Dredging/canalization (A), Extraction industry (B), Industrialization/urbanization (A), Infrastructure (A), Intensified forest management (B), Recreation/tourism (B)

Afforestation with *Populus*, agricultural intensification and recreation are the main threats.

Marais de Fontenay-le-Vicomte et d'Itteville — C6 — 049

Admin region Ile de France
Coordinates 48°33′N 2°23′E
Altitude 44–54 m **Area** 710 ha

■ **Site description**

The site includes lakes, marshes, rivers, deciduous forests, bogs, *Populus* plantations and suburban parks.

Habitats Forest and woodland (broadleaved deciduous forest; alluvial/very wet forest), Wetland (standing fresh water; river/stream; water-fringe vegetation; fen/transition mire/spring), Artificial landscape (perennial crops/orchards/groves; urban parks/gardens; ruderal land)
Land-use Fisheries/aquaculture, Forestry, Hunting, Tourism/recreation, Urban/industrial/transport

■ **Birds**

Species	Season	Year	Pop min	Pop max	Acc	Criteria
Ixobrychus minutus Little Bittern	B	1991	10	10	—	C6

An important site for species associated with reedbeds (*Phragmites*).

■ **Protection status**

National None **International** Partial

480 ha of IBA covered by Special Protection Area (Marais de Fontenay-le-Vicomte et d'Itteville, 480 ha).

■ **Conservation issues**

Threats Agricultural intensification/expansion (B), Industrialization/urbanization (B), Intensified forest management (A)

Changes in the water regime, declining water quality and afforestation threaten the site.

Etang et Forêt de Villefermoy — C6 — 050

Admin region Ile de France
Coordinates 48°30′N 2°56′E
Altitude 100–128 m **Area** 4,350 ha

■ **Site description**

The site has deciduous forest with lakes, a river, mesophile meadows, bocage and arable fields.

Habitats Forest and woodland (broadleaved deciduous forest), Grassland (mesophile grassland), Wetland (standing fresh water; river/stream), Artificial landscape (arable land; perennial crops/orchards/groves; ruderal land)
Land-use Fisheries/aquaculture, Forestry, Hunting, Tourism/recreation

■ **Birds**

Species	Season	Year	Pop min	Pop max	Acc	Criteria
Pernis apivorus Honey Buzzard	B	1991	10	10	—	C6

Species of global conservation concern that do not meet IBA criteria: *Haliaeetus albicilla* (wintering).

■ **Protection status**

National None **International** None

Conservation issues

Threats Intensified forest management (A)

The intensification of forestry and high hunting pressure are threats.

Marais arrières littoraux picards

B3, C6 **051**

Admin region Picardie
Coordinates 50°16′N 1°42′E
Altitude 5–5 m **Area** 5,550 ha

Site description

The site includes marshes, reedbeds (*Phragmites*), ponds, riverine forest, wet meadows and bogs.

Habitats Forest and woodland (alluvial/very wet forest), Grassland (humid grassland), Wetland (standing fresh water; river/stream; water-fringe vegetation; fen/transition mire/spring), Artificial landscape (perennial crops/orchards/groves)
Land-use Agriculture, Fisheries/aquaculture, Hunting, Tourism/recreation, Urban/industrial/transport

Birds

Species	Season	Year	Pop min	Pop max	Acc	Criteria
Botaurus stellaris Bittern	B	1991	10	12	—	C6
Circus aeruginosus Marsh Harrier	B	1991	10	13	—	C6
Porzana porzana Spotted Crake	B	1991	2	12	—	B3, C6
Luscinia svecica Bluethroat	B	1991	25	50	—	C6

This is an important site for breeding waterbirds.

Protection status

National None **International** None

Conservation issues

Threats Afforestation (A), Agricultural intensification/expansion (A), Infrastructure (A), Other (A), Recreation/tourism (A), Unsustainable exploitation (A)

The abandonment of grazing, water pollution from agricultural run-off, a planned road-construction project and high hunting pressure are the main threats.

Etangs et marais du bassin de la Somme

B2, C6 **052**

Admin region Picardie
Coordinates 49°54′N 2°31′E
Altitude 5–5 m **Area** 6,900 ha

Site description

The area contains ponds, marshes, rivers with riverine forest, *Populus* plantations, wet meadows, bogs, arable fields and bocages.

Habitats Forest and woodland (alluvial/very wet forest), Grassland (humid grassland), Wetland (standing fresh water; river/stream; raised bog; water-fringe vegetation; fen/transition mire/spring), Artificial landscape (highly improved reseeded grassland; arable land; perennial crops/orchards/groves)
Land-use Agriculture, Fisheries/aquaculture, Forestry, Hunting, Tourism/recreation, Urban/industrial/transport

Birds

Species	Season	Year	Pop min	Pop max	Acc	Criteria
Botaurus stellaris Bittern	B	1997	10	10	—	C6
Ixobrychus minutus Little Bittern	B	1991	20	25	—	B2, C6
Circus aeruginosus Marsh Harrier	B	1991	20	23	—	C6
Luscinia svecica Bluethroat	B	1991	20	25	—	C6

The site is important for species that breed in reedbeds (*Phragmites*).

Protection status

National Low **International** None
13 ha of IBA covered by Nature Reserve (Boves, 13 ha). Part of IBA covered by Hunting Reserve (canal de la Somme).

Conservation issues

Threats Abandonment/reduction of land management (A), Agricultural intensification/expansion (A), Industrialization/urbanization (B), Recreation/tourism (A), Unsustainable exploitation (A)

Industrial and agricultural pollution, high hunting pressure and the development of urbanization for tourism are the main threats.

Forêts picardes: Compiègne, Laigue, Ourscamp

C6 **053**

Admin region Picardie
Coordinates 49°26′N 2°55′E
Altitude 30–145 m **Area** 32,700 ha

Site description

An area of deciduous woodlands, conifer plantations, ponds, marshes and riverine forest.

Habitats Forest and woodland (97%; broadleaved deciduous forest; native coniferous forest; alluvial/very wet forest), Wetland (standing brackish and salt water; water-fringe vegetation)
Land-use Forestry (100%), Hunting (100%), Tourism/recreation (100%), Urban/industrial/transport

Birds

Species	Season	Year	Pop min	Pop max	Acc	Criteria
Pernis apivorus Honey Buzzard	B	1991	35	35	—	C6
Alcedo atthis Kingfisher	B	1991	10	10	—	C6
Dryocopus martius Black Woodpecker	B	1991	100	100	—	C6
Dendrocopos medius Middle Spotted Woodpecker	B	1991	100	100	—	C6

Protection status

National None **International** None

Conservation issues

Threats Intensified forest management (A), Recreation/tourism (B)

Exploitation of very old trees and the development of sporting activities (including motor-cross, mountain-biking and off-road vehicles) are the main threats.

Forêts picardes: massif de Retz

C6 **054**

Admin region Picardie
Coordinates 49°14′N 3°04′E
Altitude 70–240 m **Area** 27,650 ha

Site description

The site includes deciduous and conifer forests, rivers, ponds, marshes, riverine forest and wet meadows.

Habitats Forest and woodland (91%; broadleaved deciduous forest; native coniferous forest; alluvial/very wet forest), Grassland (steppe/dry calcareous grassland; humid grassland), Wetland (5%; standing fresh water; river/stream; water-fringe vegetation)
Land-use Agriculture (25%), Fisheries/aquaculture, Forestry (80%), Hunting (100%), Tourism/recreation (100%), Urban/industrial/transport

Birds

Species	Season	Year	Pop min	Pop max	Acc	Criteria
Pernis apivorus Honey Buzzard	B	1991	20	20	—	C6
Dryocopus martius Black Woodpecker	B	1991	40	40	—	C6
Dendrocopos medius Middle Spotted Woodpecker	B	1991	20	20	—	C6

The site is important for forest birds, particularly woodpeckers and raptors.

Protection status

National None **International** None

Conservation issues

Threats Industrialization/urbanization (C), Infrastructure (C), Recreation/tourism

The development of recreational activities, including off-road vehicles and mountain-biking, causes disturbance throughout the year.

Forêts picardes: massif de Saint-Gobain C6 055

Admin region Picardie
Coordinates 49°34′N 3°23′E
Altitude 207–207 m **Area** 19,300 ha

■ Site description
An area of deciduous forests, rivers and ponds.

Habitats Forest and woodland (100%; broadleaved deciduous forest)
Land-use Forestry, Hunting, Tourism/recreation, Urban/industrial/transport

■ Birds

Species	Season	Year	Pop min	Pop max	Acc	Criteria
Dryocopus martius Black Woodpecker	B	1991	10	40	—	C6
Dendrocopos medius Middle Spotted Woodpecker	B	1991	10	20	—	C6

This site is important for breeding forest species.

■ Protection status
National None **International** None

■ Conservation issues

Threats Recreation/tourism (C)

Increased tourism and afforestation with conifers are the main threats.

Marais de Sacy C6 056

Admin region Picardie
Coordinates 49°20′N 2°33′E
Altitude 32–62 m **Area** 2,350 ha

■ Site description
The site includes marshes, bogs, deciduous forests, reedbeds (*Phragmites*) and moorlands.

Habitats Forest and woodland (30%; broadleaved deciduous forest; alluvial/very wet forest), Wetland (70%; standing brackish and salt water; raised bog; water-fringe vegetation; fen/transition mire/spring)
Land-use Hunting (100%), Urban/industrial/transport

■ Birds

Species	Season	Year	Pop min	Pop max	Acc	Criteria
Botaurus stellaris Bittern	B	1991	6	9	—	C6
Ixobrychus minutus Little Bittern	B	1991	8	8	—	C6
Luscinia svecica Bluethroat	B	1991	7	10	—	C6

This is an important breeding and wintering site for waterbirds.

■ Protection status
National None **International** None

■ Conservation issues

Threats Unsustainable exploitation (B)

High hunting pressure is the only reported threat to this site.

Vallée de l'Oise de Thourotte à Vendeuil A1, C1, C6 057

Admin region Picardie
Coordinates 49°36′N 3°08′E
Altitude 40–40 m **Area** 12,050 ha

■ Site description
The site contains marshes, wet meadows, alluvial forests, *Populus* plantations, the River Oise, arable fields and hedgerows.

Habitats Forest and woodland (alluvial/very wet forest), Grassland (humid grassland), Wetland (standing brackish and salt water; water-fringe vegetation), Artificial landscape (highly improved reseeded grassland; arable land; perennial crops/orchards/groves)
Land-use Agriculture, Fisheries/aquaculture, Forestry, Hunting, Tourism/recreation, Urban/industrial/transport

■ Birds

Species	Season	Year	Pop min	Pop max	Acc	Criteria
Crex crex Corncrake	B	1994	22	24	—	A1, C1, C6

The most important breeding site for *Crex crex* in northern France.

■ Protection status
National None **International** Partial
5,400 ha of IBA covered by Special Protection Area (Moyenne Vallée de l'Oise, 5,400 ha).

■ Conservation issues

Threats Afforestation (B), Agricultural intensification/expansion (A), Extraction industry (B), Intensified forest management (C)

The main threats are gravel extraction, the loss of wet meadows as a result of ploughing for conversion to arable cultivation, and afforestation with *Populus*.

Marais de la Souche C6 058

Admin region Picardie
Coordinates 49°38′N 3°49′E
Altitude 66–75 m **Area** 4,950 ha

■ Site description
The site includes marshes, ponds, wet meadows, alluvial forests, bogs, arable fields and hedgerows.

Habitats Forest and woodland (alluvial/very wet forest), Grassland (humid grassland), Wetland (standing fresh water; water-fringe vegetation; fen/transition mire/spring), Artificial landscape (highly improved reseeded grassland; arable land; perennial crops/orchards/groves)
Land-use Agriculture, Fisheries/aquaculture, Forestry, Hunting, Urban/industrial/transport

■ Birds

Species	Season	Year	Pop min	Pop max	Acc	Criteria
Botaurus stellaris Bittern	B	1991	8	10	—	C6
Ixobrychus minutus Little Bittern	B	1991	6	10	—	C6
Circus aeruginosus Marsh Harrier	B	1991	12	12	—	C6

■ Protection status
National None **International** None

■ Conservation issues

Threats Afforestation (B), Agricultural intensification/expansion (A), Aquaculture/fisheries (A), Disturbance to birds (A), Industrialization/urbanization (B), Intensified forest management (C), Recreation/tourism (A), Unsustainable exploitation (B)

Afforestation with *Populus*, conversion of wet meadows into arable fields and high hunting pressure (with late closure dates) are the main threats.

Forêts picardes: massif des Trois Forêts et bois du Roi C6 059

Admin region Picardie
Coordinates 49°12′N 2°39′E
Altitude 25–211 m **Area** 32,200 ha

■ Site description
An area of deciduous and conifer forests, ponds, marshes, river and aquatic vegetation.

Habitats Forest and woodland (95%; broadleaved deciduous forest; native coniferous forest), Wetland (5%; standing fresh water; river/stream; water-fringe vegetation)

Land-use Forestry (95%), Hunting, Tourism/recreation (100%), Urban/industrial/transport

Birds

Species	Season	Year	Pop min	Pop max	Acc	Criteria
Pernis apivorus Honey Buzzard	B	1991	30	30	—	C6
Caprimulgus europaeus Nightjar	B	1991	25	25	—	C6
Alcedo atthis Kingfisher	B	1991	10	10	—	C6
Dryocopus martius Black Woodpecker	B	1991	30	30	—	C6
Dendrocopos medius Middle Spotted Woodpecker	B	1991	20	20	—	C6

Protection status

National None **International** None

Conservation issues

Threats Disturbance to birds (A), Industrialization/urbanization (C), Recreation/tourism (B)

Reported threats to the site are urbanization in areas surrounding the site, and an increased use of the area by tourists.

Estuaires picards: baies de Somme et d'Authie — A4i, A4iii, B1i, B2, B3, C2, C3, C4, C6 — 060

Admin region Picardie
Coordinates 50°15'N 1°36'E
Altitude 0–32 m Area 18,200 ha

Site description

A large estuary and bay with mudflats and wet meadows, sand and rocky beaches, woodland, marshes, and agricultural land. The site also includes a harbour.

Habitats Forest and woodland (8%; native coniferous forest; mixed forest; alluvial/very wet forest), Grassland (5%; humid grassland), Wetland (94%; tidal river/enclosed tidal water; mudflat/sandflat; saltmarsh; sand-dunes/sand beach; shingle/stony beach; standing brackish and salt water; river/stream; water-fringe vegetation), Artificial landscape (arable land)
Land-use Agriculture, Fisheries/aquaculture, Hunting, Tourism/recreation, Urban/industrial/transport

Birds

Species	Season	Year	Pop min	Pop max	Acc	Criteria
Botaurus stellaris Bittern	B	1997	3	5	—	C6
Egretta garzetta Little Egret	B	1997	10	12	—	C6
Platalea leucorodia Spoonbill	P	1997	10	100	—	B1i, C2
Tadorna tadorna Shelduck	W	1997	8,000	16,800	—	A4i, B1i, C3
Anas acuta Pintail	W	1997	438	1,229	—	B1i, C3
Anas acuta Pintail	P	1997	500	1,500	—	B1i, C3
Melanitta fusca Velvet Scoter	W	1997	100	100	—	B2
Porzana porzana Spotted Crake	B	1991	1	5	—	C6
Porzana parva Little Crake	B	1991	1	4	—	C6
Haematopus ostralegus Oystercatcher	W	1997	5,426	11,400	—	A4i, B1i, C3
Recurvirostra avosetta Avocet	B	1997	89	94	—	B3, C6
Charadrius dubius Little Ringed Plover	P	1996	1,000	3,500	—	A4i, B1i, C3
Charadrius alexandrinus Kentish Plover	B	1997	48	51	—	B2
Pluvialis squatarola Grey Plover	W	1997	305	5,053	—	A4i, B1i, C3
Pluvialis squatarola Grey Plover	P	1997	500	6,000	—	A4i, B1i, C3
Calidris canutus Knot	W	1997	55	8,070	—	B1i, B2, C3
Calidris canutus Knot	P	1997	500	6,000	—	B1i, C3
Calidris alba Sanderling	P	1997	1,000	1,500	—	A4i, B1i, C3
Calidris alpina Dunlin	W	1997	6,300	21,851	—	B1i, B2, C3
Limosa lapponica Bar-tailed Godwit	W	1997	15	800	—	B2
Numenius arquata Curlew	W	1997	551	5,302	—	A4i, B1i, B2, C3
Asio flammeus Short-eared Owl	B	1991	2	3	—	C6
Caprimulgus europaeus Nightjar	B	1991	25	25	—	C6
Luscinia svecica Bluethroat	B	1991	35	40	—	C6

The site holds 20,000 or more waterbirds on passage and in winter, on a regular basis.

Protection status

National Partial **International** High
3,146 ha of IBA covered by Nature Reserve (Baie de Somme, 3,146 ha). 2,000 ha of IBA covered by Hunting Reserve (Baie de Somme Maritime, 2,000 ha). 13 ha of IBA covered by Hunting Reserve (Baie de Somme Terrestre, 13 ha). 13,000 ha of IBA covered by Hunting Reserve (Authie-Somme, 13,000 ha). 188 ha of IBA covered by Private Reserve (Parc Ornithologique du Marqueterre, 188 ha). 17,000 ha of IBA covered by Ramsar Site (Baie de Somme, 17,000 ha). 15,000 ha covered by Special Protection Area (Estuaires Picards: baies de Somme et d'Authie, 15,000 ha).

Conservation issues

Threats Abandonment/reduction of land management (B), Agricultural intensification/expansion (B), Infrastructure (U), Other (C), Recreation/tourism (A), Unsustainable exploitation (C)

The main threats are road construction, the abandonment of grazing, the conversion of meadows into arable fields and inappropriate management of water-levels.

Vallées de la Scarpe et de l'Escaut — B1i, C3, C6 — 061

Admin region Nord - Pas-de-Calais
Coordinates 50°26'N 3°26'E
Altitude 15–35 m Area 11,150 ha

Site description

The site includes the Escaut and Scarpe Rivers and valleys where human activities include arable farming, cattle-farming, fishing, tourism, recreation and hunting.

Habitats Forest and woodland (broadleaved deciduous forest; alluvial/very wet forest), Grassland (humid grassland), Wetland (standing fresh water; river/stream; raised bog; water-fringe vegetation), Artificial landscape (arable land; perennial crops/orchards/groves)
Land-use Agriculture, Fisheries/aquaculture, Forestry, Hunting, Tourism/recreation, Urban/industrial/transport

Birds

Species	Season	Year	Pop min	Pop max	Acc	Criteria
Ixobrychus minutus Little Bittern	B	1991	4	10	—	C6
Anas clypeata Shoveler	P	1997	50	500	—	B1i, C3
Pernis apivorus Honey Buzzard	B	1991	5	10	—	C6
Circus aeruginosus Marsh Harrier	B	1991	3	8	—	C6
Caprimulgus europaeus Nightjar	B	1991	10	10	—	C6
Alcedo atthis Kingfisher	B	1991	10	10	—	C6
Dryocopus martius Black Woodpecker	B	1991	5	5	—	C6
Luscinia svecica Bluethroat	B	1991	100	100	—	C6

This is an important site for wintering waterbirds.

Protection status

National None **International** None

Conservation issues

Threats Construction/impact of dyke/dam/barrage (A), Industrialization/urbanization (B), Infrastructure (A), Other (A)

Many factors threaten the site including industrial development of the surrounding area, highway construction, *Populus* plantations and conversion of wet meadows into arable fields. Contaminated mud is also stored on the site ('Other' threats, above).

Estuaire de la Canche — A4i, A4iii, B1i, C3, C4, C6 — 062

Admin region Nord - Pas-de-Calais
Coordinates 50°32'N 1°34'E
Altitude 0–43 m Area 5,050 ha

Site description

The River Canche estuary, with mudflats, raised saltmarsh, beaches, sand-dunes, freshwater marshes and wet meadows.

Habitats Forest and woodland (6%; native coniferous forest; mixed forest; alluvial/very wet forest), Grassland (humid grassland), Wetland (61%; tidal river/enclosed tidal water; mudflat/sandflat; saltmarsh; sand-dunes/sand beach; standing fresh water; standing brackish and salt water; river/stream; water-fringe vegetation), Marine areas (40%; open sea), Artificial landscape (urban parks/gardens)

Land-use Fisheries/aquaculture (2%), Forestry (2%), Hunting (30%), Not utilized (40%), Tourism/recreation (30%), Urban/industrial/transport

Birds

Species	Season	Year	Pop min	Pop max	Acc	Criteria
Larus fuscus Lesser Black-backed Gull	P	1991	1,000	10,000	—	A4i, B1i, C3
Larus argentatus Herring Gull	W	1991	10,000	20,000	—	B1i
Larus marinus Great Black-backed Gull	W	1991	2,000	10,000	—	A4i, B1i, C3
Caprimulgus europaeus Nightjar	B	1991	10	10	—	C6
Lullula arborea Woodlark	B	1991	1	10	—	C6

The site is important for passage and wintering gulls. The site holds 20,000 or more waterbirds in winter, on a regular basis.

Protection status

National High **International** High

505 ha of IBA covered by Nature Reserve (Baie de Canche, 505 ha). 4,000 ha of IBA covered by Hunting Reserve (Estuaire de la Canche, 4,000 ha). 4,465 ha of IBA covered by Special Protection Area (Estuaire de la Canche, 4,465 ha).

Conservation issues

Threats Disturbance to birds (B), Infrastructure (A), Recreation/tourism (B)

No major threats are reported.

Marais de Balançon et de Villiers — C6 — 063

Admin region Nord - Pas-de-Calais
Coordinates 50°28′N 1°38′E
Altitude 0–2 m **Area** 1,500 ha

Site description

An area of marshes with ponds, reedbeds (*Phragmites*), wet meadows and deciduous woodlands.

Habitats Forest and woodland (broadleaved deciduous forest), Grassland (5%; humid grassland), Wetland (93%; standing fresh water; river/stream; water-fringe vegetation)
Land-use Agriculture, Fisheries/aquaculture, Hunting, Urban/industrial/transport

Birds

Species	Season	Year	Pop min	Pop max	Acc	Criteria
Luscinia svecica Bluethroat	B	1991	10	20	—	C6

The area is important for birds breeding in reedbeds and passage waterbirds.

Protection status

National None **International** Partial

1,200 ha of IBA covered by Special Protection Area (Marais de Balançon, 1,200 ha).

Conservation issues

Threats Burning of vegetation (U), Recreation/tourism (A), Unsustainable exploitation

Uncontrolled burning of vegetation and hunting pressure are the main threats to the site, which is not adequately managed.

Cap Gris-nez — A4i, A4iii, B1i, B2, C2, C3, C4 — 064

Admin region Nord - Pas-de-Calais
Coordinates 50°52′N 1°35′E
Altitude 0–50 m **Area** 8,950 ha

Site description

A largely marine area with sand and shingle beaches, cliffs, heathlands, meadows and arable fields.

Habitats Scrub (heathland), Grassland (humid grassland), Wetland (sand-dunes/sand beach; shingle/stony beach; standing brackish and salt water; river/stream), Marine areas (95%; open sea), Rocky areas (sea cliff/rocky shore), Artificial landscape (arable land; ruderal land)
Land-use Agriculture, Fisheries/aquaculture, Hunting, Tourism/recreation, Urban/industrial/transport

Birds

Species	Season	Year	Pop min	Pop max	Acc	Criteria
Tachybaptus ruficollis Little Grebe	P	1997	1,000	10,000	—	A4i, B1i, C3
Podiceps grisegena Red-necked Grebe	P	1997	100	500	—	B1i, C3
Platalea leucorodia Spoonbill	P	1991	100	500	—	A4i, B1i, C2
Branta bernicla Brent Goose	P	1997	1,000	10,000	—	A4i, B1i, C3
Tadorna tadorna Shelduck	P	1997	1,000	10,000	—	A4i, B1i, C3
Anas acuta Pintail	P	1997	500	1,000	—	B1i, C3
Anas clypeata Shoveler	P	1997	500	1,000	—	B1i, C3
Haematopus ostralegus Oystercatcher	W	1997	338	13,044	—	A4i, B1i, C3
Haematopus ostralegus Oystercatcher	P	1997	1,000	10,000	—	B1i, C3
Charadrius hiaticula Ringed Plover	P	1997	1,000	10,000	—	A4i, B1i, C3
Charadrius alexandrinus Kentish Plover	P	1997	500	1,000	—	A4i, B1i, C3
Pluvialis squatarola Grey Plover	P	1997	1,000	10,000	—	A4i, B1i, C3
Calidris alpina Dunlin	P	1997	10,000	20,000	—	B1i, C3
Limosa lapponica Bar-tailed Godwit	P	1997	1,000	10,000	—	A4i, B1i, C2
Numenius arquata Curlew	W	1997	75	1,408	—	B2
Numenius arquata Curlew	P	1997	1,000	10,000	—	A4i, B1i, C3
Tringa totanus Redshank	P	1997	1,000	10,000	—	A4i, B1i, C3
Larus fuscus Lesser Black-backed Gull	P	1997	1,000	10,000	—	A4i, B1i, C3
Larus argentatus Herring Gull	W	1997	80	30,000	—	A4i, B1i, C3
Larus argentatus Herring Gull	P	1997	10,000	20,000	—	B1i
Larus marinus Great Black-backed Gull	P	1997	1,000	10,000	—	A4i, B1i, C3
Sterna sandvicensis Sandwich Tern	P	1991	10,000	20,000	—	A4i, B1i, C2
Sterna dougallii Roseate Tern	P	1991	10	100	—	A4i, B1i, C2
Sterna hirundo Common Tern	P	1991	10,000	20,000	—	A4i, B1i, C2
Sterna paradisaea Arctic Tern	P	1991	1,000	10,000	—	A4i, B1i, C2
Sterna albifrons Little Tern	P	1991	1,000	10,000	—	A4i, B1i, C2

The site holds 20,000 or more waterbirds on passage, on a regular basis. Species of global conservation concern that do not meet criteria: *Aythya nyroca* (passage and wintering).

Protection status

National High **International** High

8,600 ha of IBA covered by Hunting Reserve (Cap Gris-nez, 8,600 ha). 8,600 ha of IBA covered by Special Protection Area (Cap Gris-nez, 8,600 ha).

Conservation issues

Threats Other (U)

Oil pollution of marine areas from shipping ('Other' threats, above) is a major threat and frequently affects seabirds.

Plateaux agricoles des environs de Frencq — C6 — 065

Admin region Nord - Pas-de-Calais
Coordinates 50°33′N 1°41′E
Altitude 0–200 m **Area** 11,450 ha

Site description

The site includes cereal fields, mesophile meadows and deciduous woodlands.

Habitats Forest and woodland (10%; broadleaved deciduous forest), Grassland (10%; mesophile grassland), Artificial landscape (90%; arable land)
Land-use Agriculture (100%), Hunting (100%), Urban/industrial/transport (20%)

Birds

Species	Season	Year	Pop min	Pop max	Acc	Criteria
Circus cyaneus Hen Harrier	B	1991	10	10	—	C6
Circus pygargus Montagu's Harrier	B	1991	9	9	—	C6

This is an important site for breeding species associated with agricultural habitats.

Protection status

National None **International** None

Conservation issues

Threats Infrastructure (A)

The main threat to the site is the ongoing construction of a highway.

Forêts de Thiérache: Trelon, Fourmies, Hirson et St-Michel — B3, C6 — 066

Admin region Nord - Pas-de-Calais
Coordinates 50°01′N 4°02′E
Altitude 50–250 m **Area** 22,400 ha

Site description

A large area of forest that is contiguous with site 067 and the Belgian site 032 'Entre-Sambre-et-Meuse'.

Habitats Forest and woodland (90%; broadleaved deciduous forest; native coniferous forest), Scrub (heathland), Grassland (20%; steppe/dry calcareous grassland; humid grassland), Wetland (20%; standing fresh water; river/stream; water-fringe vegetation), Artificial landscape (10%; perennial crops/orchards/groves)
Land-use Forestry (90%), Hunting (90%), Tourism/recreation (10%), Urban/industrial/transport (10%)

Birds

Species	Season	Year	Pop min	Pop max	Acc	Criteria
Pernis apivorus Honey Buzzard	B	1991	30	40	—	C6
Alcedo atthis Kingfisher	B	1991	10	15	—	C6
Dryocopus martius Black Woodpecker	B	1991	5	8	—	C6
Dendrocopos medius Middle Spotted Woodpecker	B	1991	100	150	—	B3, C6

The site is important for forest birds.

Protection status

National None **International** None

Conservation issues

Threats Recreation/tourism

The only major threat is the planned construction of a tourist centre within the best-conserved area of forest.

Plateau Ardennais — B3, C6 — 067

Admin region Champagne-Ardenne
Coordinates 49°53′N 4°47′E
Altitude 106–504 m **Area** 94,800 ha

Site description

Habitats Forest and woodland (broadleaved deciduous forest; native coniferous forest; alluvial/very wet forest), Scrub (heathland), Wetland (standing fresh water; river/stream; raised bog; water-fringe vegetation), Rocky areas (inland cliff; inland sand-dunes; caves), Artificial landscape (highly improved reseeded grassland; arable land; perennial crops/orchards/groves)
Land-use Agriculture, Fisheries/aquaculture, Forestry, Hunting, Tourism/recreation, Urban/industrial/transport

Birds

Species	Season	Year	Pop min	Pop max	Acc	Criteria
Pernis apivorus Honey Buzzard	B	1991	80	80	—	B3, C6
Milvus migrans Black Kite	B	1991	10	10	—	C6
Milvus milvus Red Kite	B	1991	20	20	—	C6
Circus cyaneus Hen Harrier	B	1991	10	10	—	C6
Bonasa bonasia Hazel Grouse	B	1991	100	200	—	C6
Tetrao tetrix Black Grouse	B	1991	10	15	—	C6
Aegolius funereus Tengmalm's Owl	B	1991	5	10	—	C6
Caprimulgus europaeus Nightjar	B	1991	10	10	—	C6
Alcedo atthis Kingfisher	B	1991	50	50	—	C6
Picus canus Grey-headed Woodpecker	B	1991	20	20	—	C6
Dryocopus martius Black Woodpecker	B	1991	20	50	—	C6
Lullula arborea Woodlark	B	1991	10	30	—	C6
Lanius collurio Red-backed Shrike	B	1991	100	200	—	C6

This is an important site for woodland birds, particularly breeding raptors. It is also one of the most important breeding sites in France for *Bonasa bonasia*.

Protection status

National Low **International** None
135 ha of IBA covered by Nature Reserve (Vireux-Molhain, 135 ha).

Conservation issues

Threats Disturbance to birds (A), Recreation/tourism (B), Unsustainable exploitation (B)

The development of tourism and recreation activities, such as climbing, are significant threats.

Lacs de la Forêt d'Orient — A4i, A4iii, B1i, B1iv, B2, C2, C3, C4, C5, C6 — 068

Admin region Champagne-Ardenne
Coordinates 48°18′N 4°22′E
Altitude 120–200 m **Area** 35,800 ha

Site description

A lake surrounded by wet meadows, woodland, hedges, grassland and arable land. Human activities include agriculture, forestry, cattle-raising, hunting and recreation (including sailing).

Habitats Forest and woodland (35%; broadleaved deciduous forest; native coniferous forest; alluvial/very wet forest), Grassland (humid grassland), Wetland (15%; standing fresh water; river/stream; water-fringe vegetation), Artificial landscape (44%; highly improved reseeded grassland; arable land; perennial crops/orchards/groves; other urban/industrial areas)
Land-use Agriculture (20%), Fisheries/aquaculture (15%), Forestry (34%), Hunting (80%), Tourism/recreation (60%), Urban/industrial/transport (4%), Water management (14%)

Birds

Species	Season	Year	Pop min	Pop max	Acc	Criteria
Ixobrychus minutus Little Bittern	B	1991	4	6	—	C6
Cygnus columbianus Bewick's Swan	W	1997	2	29	—	B2
Anas strepera Gadwall	P	1997	100	300	—	B1i, C3
Anas clypeata Shoveler	P	1997	200	700	—	B1i, C3
Pernis apivorus Honey Buzzard	B	1991	10	25	—	C6
Milvus migrans Black Kite	B	1991	25	40	—	C6
Circus aeruginosus Marsh Harrier	B	1991	3	6	—	C6
Grus grus Crane	P	1997	6,000	13,500	—	A4i, B1i, C2
Dendrocopos medius Middle Spotted Woodpecker	B	1991	40	50	—	C6
Lanius collurio Red-backed Shrike	B	1991	30	50	—	C6

This is an important site for migrating and wintering waterbirds and the most important wintering site for *Grus grus* in France. The site holds 20,000 or more waterbirds on passage and in winter, on a regular basis (including *Podiceps cristatus*, 4,000 (passage); *Anser fabalis* 2,000 (passage); *Anas crecca* 3,000 (passage), 3,309 (winter); *Anas platyrhynchos* 10,000 (passage); *Aythya ferina* 2,500 (passage); *Larus ridibundus* 21,000 (wintering)). Species of global conservation concern that do not meet IBA criteria: *Branta ruficollis* (passage); *Aythya nyroca* (passage); *Oxyura leucocephala* (passage); *Haliaeetus albicilla* (1–3 passage, 3–5 wintering individuals).

Protection status

National High **International** High
2,500 ha of IBA covered by Hunting Reserve (Lac Seine, 2,500 ha). 35,800 ha of IBA covered by Ramsar Site (Etangs de Champagne humide, 135,000 ha). 2,500 ha of IBA covered by Special Protection Area (Lac de la forêt d'Orient, 2,500 ha).

Conservation issues

Threats Aquaculture/fisheries (A), Disturbance to birds (B), Industrialization/urbanization (A), Recreation/tourism (A)

The site is threatened by the creation of an artificial beach and access road, together with the development of tourist facilities and the use of powered hang-gliders.

Etangs d'Argonne B2, C6 069

Admin region Champagne-Ardenne
Coordinates 48°58'N 4°56'E
Altitude 110–250 m **Area** 44,100 ha

Site description

Lakes with marshes, surrounded by woodlands, wet meadows, grassland and arable land.

Habitats Forest and woodland (46%; broadleaved deciduous forest; native coniferous forest; mixed forest), Grassland (humid grassland), Wetland (5%; standing fresh water; river/stream; water-fringe vegetation), Artificial landscape (44%; highly improved reseeded grassland; arable land; perennial crops/orchards/groves)

Land-use Agriculture (42%), Fisheries/aquaculture (4%), Forestry (48%), Hunting (95%), Urban/industrial/transport (2%)

Birds

Species	Season	Year	Pop min	Pop max	Acc	Criteria
Botaurus stellaris Bittern	B	1991	8	12	—	C6
Ixobrychus minutus Little Bittern	B	1991	6	15	—	C6
Pernis apivorus Honey Buzzard	B	1991	20	30	—	C6
Milvus migrans Black Kite	B	1991	12	15	—	C6
Milvus milvus Red Kite	B	1991	6	10	—	C6
Circus aeruginosus Marsh Harrier	B	1991	12	15	—	C6
Circus cyaneus Hen Harrier	B	1991	6	10	—	C6
Hieraaetus pennatus Booted Eagle	B	1991	1	3	—	B2, C6
Alcedo atthis Kingfisher	B	1991	8	15	—	C6
Picus canus Grey-headed Woodpecker	B	1991	4	15	—	C6
Dryocopus martius Black Woodpecker	B	1991	10	15	—	C6
Dendrocopos medius Middle Spotted Woodpecker	B	1991	15	30	—	C6

This is an important site for birds breeding in reedbeds (*Phragmites*) and forests. Species of global conservation concern that do not meet IBA criteria: *Aythya nyroca* (passage); *Haliaeetus albicilla* (passage, wintering); *Crex crex* (breeding).

Protection status

National None **International** High

44,100 ha of IBA covered by Ramsar Site (Etangs de Champagne humide, 135,000 ha).

Conservation issues

Threats Abandonment/reduction of land management (A), Agricultural intensification/expansion (B), Aquaculture/fisheries (B), Construction/impact of dyke/dam/barrage (A), Disturbance to birds (C), Infrastructure (B), Recreation/tourism (A), Unsustainable exploitation (A)

There is disturbance from hunting, fishing and sailing.

Lac du Der-Chantecoq et étangs latéraux A4i, A4iii, A4iv, B1i, B1iv, B2, C2, C3, C4, C5, C6 070

Admin region Champagne-Ardenne
Coordinates 48°33'N 4°42'E
Altitude 110–170 m **Area** 56,000 ha

Site description

A lake with marshes and wet meadows, surrounded by woodland, hedgerows, grassland and arable land.

Habitats Forest and woodland (20%; broadleaved deciduous forest; mixed forest; alluvial/very wet forest), Grassland (humid grassland), Wetland (12%; standing fresh water; river/stream; water-fringe vegetation), Artificial landscape (65%; highly improved reseeded grassland; arable land; perennial crops/orchards/groves; other urban/industrial areas)

Land-use Agriculture (65%), Fisheries/aquaculture (10%), Forestry (21%), Hunting (90%), Tourism/recreation (34%), Urban/industrial/transport, Water management (10%)

Birds

Species	Season	Year	Pop min	Pop max	Acc	Criteria
Phalacrocorax carbo Cormorant	P	1997	800	2,000	—	B1i, C3
Botaurus stellaris Bittern	B	1997	4	5	—	C6
Ixobrychus minutus Little Bittern	B	1991	8	15	—	C6
Ardea purpurea Purple Heron	B	1991	9	10	—	B2, C6
Cygnus columbianus Bewick's Swan	W	1997	3	27	—	B2
Anser anser Greylag Goose	W	1997	856	2,169	—	B1i, C3
Anas strepera Gadwall	B	1997	30	50	—	B2
Anas strepera Gadwall	P	1997	300	1,800	—	A4i, B1i, C3
Anas crecca Teal	P	1997	2,000	5,000	—	B1i, C3
Anas clypeata Shoveler	W	1997	12	449	—	B1i, C3
Anas clypeata Shoveler	P	1997	500	1,300	—	B1i, C3
Aythya ferina Pochard	W	1997	66	1,363	—	A4i, B1i, C3
Pernis apivorus Honey Buzzard	B	1991	15	20	—	C6
Milvus migrans Black Kite	B	1991	40	55	—	C6
Milvus milvus Red Kite	B	1991	10	15	—	C6
Circus aeruginosus Marsh Harrier	B	1991	8	12	—	C6
Hieraaetus pennatus Booted Eagle	B	1991	1	3	—	B2, C6
Grus grus Crane	W	1997	2,582	6,775	—	A4i, B1i, C2
Grus grus Crane	P	1997	15,000	42,000	—	A4i, B1i, C2
Tringa erythropus Spotted Redshank	P	1997	300	800	—	A4i, B1i, C3
Alcedo atthis Kingfisher	B	1991	15	25	—	C6
Dendrocopos medius Middle Spotted Woodpecker	B	1991	30	50	—	C6
Lanius collurio Red-backed Shrike	B	1991	60	100	—	C6

The site holds 20,000 or more waterbirds on passage and in winter, on a regular basis (including *Anas penelope* (4,500 passage); *Anas crecca* (2,572 (wintering); *Anas platyrhynchos* (5,000 passage, 3,537 wintering); *Vanellus vanellus* (30,000 passage); *Larus ridibundus* (20,000 passage, 15,250 wintering)). The site is a major migratory bottleneck, where more than 20,000 *Grus grus* pass, on a regular basis. Species of global conservation concern that do not meet IBA criteria include: *Branta ruficollis* (wintering); *Aythya nyroca* (passage, wintering); *Haliaeetus albicilla* (passage, 2–5 wintering individuals); *Crex crex* (passage, breeding); *Gallinago media* (passage).

Protection status

National Partial **International** High

5,000 ha of IBA covered by Hunting Reserve (Etang du Der, 5,000 ha). 60 ha of IBA covered by Private Reserve (60 ha). 56,000 ha of IBA covered by Ramsar Site (Etangs de Champagne humide, 135,000 ha). 5,000 ha of IBA covered by Special Protection Area (Lac du Der, 5,000 ha).

Conservation issues

Threats Agricultural intensification/expansion (A), Industrialization/urbanization (B), Infrastructure (A), Intensified forest management (B), Recreation/tourism (B), Unsustainable exploitation (C)

The main threat is the development of tourist activities.

Barois et Forêt de Clairvaux B2, C6 071

Admin region Champagne-Ardenne
Coordinates 48°08'N 4°38'E
Altitude 150–350 m **Area** 66,550 ha

Site description

A large forested area with deciduous and mixed woodland with streams, cultivated land, meadows, orchards and hedgerows.

Habitats Forest and woodland (broadleaved deciduous forest; mixed forest), Scrub (heathland), Grassland (steppe/dry calcareous grassland; mesophile grassland), Wetland (river/stream), Artificial landscape (highly improved reseeded grassland; arable land; perennial crops/orchards/groves)

Land-use Agriculture, Forestry, Hunting, Urban/industrial/transport

Birds

Species	Season	Year	Pop min	Pop max	Acc	Criteria
Circus cyaneus Hen Harrier	B	1991	20	25	—	B2, C6
Circus pygargus Montagu's Harrier	B	1991	15	20	—	C6
Bonasa bonasia Hazel Grouse	B	1991	10	10	—	C6
Aegolius funereus Tengmalm's Owl	B	1991	1	10	—	C6
Caprimulgus europaeus Nightjar	B	1991	15	25	—	C6
Picus canus Grey-headed Woodpecker	B	1991	10	20	—	C6
Dryocopus martius Black Woodpecker	B	1991	20	50	—	C6
Lullula arborea Woodlark	B	1991	100	100	—	C6

This is an important area for breeding forest birds.

Protection status
National None **International** None

Conservation issues
Threats Afforestation (A), Agricultural intensification/expansion (A), Disturbance to birds (A), Intensified forest management (B)

Vallée de l'Aube, de la Superbe et Marigny — C6 — 072
Admin region Champagne-Ardenne
Coordinates 48°37'N 3°53'E
Altitude 100–100 m **Area** 21,500 ha

Site description
An area of conifer woodland with heathland, marshes, bogs, wet meadows, arable land, orchards and hedges.

Habitats Forest and woodland (native coniferous forest; alluvial/very wet forest), Scrub (heathland), Grassland (steppe/dry calcareous grassland; humid grassland; mesophile grassland), Wetland (river/stream; water-fringe vegetation; fen/transition mire/spring), Artificial landscape (arable land; perennial crops/orchards/groves)
Land-use Agriculture, Fisheries/aquaculture, Forestry, Hunting, Urban/industrial/transport

Birds

Species	Season	Year	Pop min	Pop max	Acc	Criteria
Burhinus oedicnemus Stone Curlew	B	1991	3	6	—	C6
Caprimulgus europaeus Nightjar	B	1991	1	10	—	C6

Species of global conservation concern that do not meet IBA criteria: *Crex crex* (breeding); *Tetrax tetrax* (breeding).

Protection status
National None **International** Low
1,424 ha of IBA covered by Special Protection Area (Etang de la Horre, 1,424 ha).

Conservation issues
Threats Agricultural intensification/expansion (A), Aquaculture/fisheries (A), Disturbance to birds (A), Drainage (A), Infrastructure (B), Recreation/tourism (B)

Agricultural intensification is a major threat for breeding *Tetrax tetrax*, and drainage threatens waterbirds.

Vallée de l'Aisne — A4i, A4iii, B1i, C3, C4, C6 — 073
Admin region Champagne-Ardenne
Coordinates 49°24'N 4°31'E
Altitude 60–130 m **Area** 18,800 ha

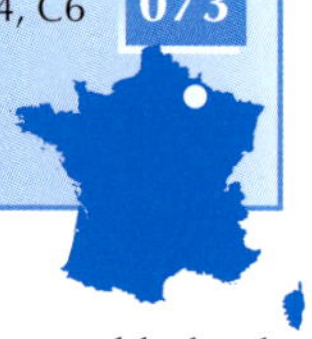

Site description
A valley with deciduous woodland, rivers, meadows, arable land, hedges, ponds and gravel-pits.

Habitats Forest and woodland (broadleaved deciduous forest), Grassland (17%; humid grassland), Wetland (6%; standing fresh water; river/stream; water-fringe vegetation), Artificial landscape (74%; highly improved reseeded grassland; arable land; perennial crops/orchards/groves)
Land-use Agriculture (85%), Fisheries/aquaculture, Forestry (5%), Hunting (90%), Urban/industrial/transport

Birds

Species	Season	Year	Pop min	Pop max	Acc	Criteria
Anas acuta Pintail	P	1991	300	1,200	—	B1i, C3
Anas clypeata Shoveler	P	1991	150	800	—	B1i, C3
Milvus milvus Red Kite	B	1991	6	10	—	C6
Circus cyaneus Hen Harrier	B	1991	6	8	—	C6
Crex crex Corncrake	B	1994	9	9	—	C6
Vanellus vanellus Lapwing	P	1991	20,000	80,000	—	A4i, B1i, C3
Alcedo atthis Kingfisher	B	1991	20	50	—	C6
Luscinia svecica Bluethroat	B	1991	2	6	—	C6
Lanius collurio Red-backed Shrike	B	1991	50	150	—	C6

The site holds 20,000 or more waterbirds on passage, on a regular basis.

Protection status
National None **International** None

Conservation issues
Threats Afforestation (B), Agricultural intensification/expansion (A), Aquaculture/fisheries (C), Drainage (A), Intensified forest management (C), Recreation/tourism (B), Unsustainable exploitation (C)

The drainage of wet meadows is destroying *Crex crex* habitat.

Confluent des vallées de la Meuse et de la Chiers — A4iii, B3, C4, C6 — 074
Admin region Champagne-Ardenne
Coordinates 49°38'N 4°08'E
Altitude 160–170 m **Area** 4,860 ha

Site description
An area of meadows, rivers, ponds and gravel-pits, arable land and *Populus* groves.

Habitats Grassland (12%; humid grassland), Wetland (7%; standing fresh water; river/stream; water-fringe vegetation), Artificial landscape (81%; highly improved reseeded grassland; arable land; perennial crops/orchards/groves)
Land-use Agriculture (90%), Fisheries/aquaculture (5%), Forestry (12%), Hunting (90%), Urban/industrial/transport

Birds

Species	Season	Year	Pop min	Pop max	Acc	Criteria
Crex crex Corncrake	B	1991	5	10	—	C6
Turdus pilaris Fieldfare	W	1991	—	1,000	—	B3

This is an important site for passage waterbirds and thrushes (Turdidae), and wintering waterbirds. The site holds 20,000 or more waterbirds on passage, on a regular basis (including *Vanellus vanellus* 20,000–50,000 individuals). Species of global conservation concern that do not meet IBA criteria: *Crex crex* (10–15 passage individuals).

Protection status
National None **International** None

Conservation issues
Threats Afforestation (A), Agricultural intensification/expansion (A), Infrastructure (B), Intensified forest management (B)

The main threats are the cultivation of wet meadows and afforestation with *Populus*.

Bassigny — B3, C6 — 075
Admin region Champagne-Ardenne
Coordinates 48°06'N 5°36'E
Altitude 315–500 m **Area** 98,400 ha

Site description
The site includes deciduous woodland, meadows, arable land, rivers, orchards, and bocages.

Habitats Forest and woodland (49%; broadleaved deciduous forest; native coniferous forest; mixed forest), Grassland (humid grassland), Wetland (river/stream; water-fringe vegetation), Artificial landscape (36%; highly improved reseeded grassland; arable land; perennial crops/orchards/groves)
Land-use Agriculture (35%), Fisheries/aquaculture (2%), Forestry (50%), Hunting (95%), Urban/industrial/transport (9%)

Birds

Species	Season	Year	Pop min	Pop max	Acc	Criteria
Pernis apivorus Honey Buzzard	B	1991	15	25	—	C6
Milvus migrans Black Kite	B	1991	15	20	—	C6

Species ... continued	Season	Year	Pop min	Pop max	Acc	Criteria
Milvus milvus Red Kite	B	1991	25	30	—	B3, C6
Circus cyaneus Hen Harrier	B	1991	10	15	—	C6
Circus pygargus Montagu's Harrier	B	1991	27	30	—	C6
Crex crex Corncrake	B	1991	5	10	—	C6
Alcedo atthis Kingfisher	B	1991	25	50	—	C6
Picus canus Grey-headed Woodpecker	B	1991	5	10	—	C6
Dryocopus martius Black Woodpecker	B	1991	10	20	—	C6
Dendrocopos medius Middle Spotted Woodpecker	B	1991	20	30	—	C6
Lanius collurio Red-backed Shrike	B	1991	50	100	—	C6

The site is important for breeding raptors, *Crex crex* and woodpeckers. Species of global conservation concern that do not meet IBA criteria: *Crex crex* (passage).

Protection status

National None **International** None

Conservation issues

Threats Agricultural intensification/expansion (A), Construction/impact of dyke/dam/barrage (A), Disturbance to birds (B), Intensified forest management (B)

The cultivation of meadows and significant river pollution are the main threats.

Etangs de la Woëvre: lac de Madine — C6 — 076

Admin region Lorraine
Coordinates 48°54′N 5°44′E
Altitude 225–350 m **Area** 1,730 ha

Site description

A lake with reedbeds (*Phragmites*), wet meadows, and alluvial forests.

Habitats Forest and woodland (12%; broadleaved deciduous forest), Grassland (5%; humid grassland), Wetland (82%; standing fresh water; standing brackish and salt water; water-fringe vegetation), Artificial landscape (perennial crops/orchards/groves)
Land-use Agriculture, Fisheries/aquaculture, Forestry, Tourism/recreation

Birds

Species	Season	Year	Pop min	Pop max	Acc	Criteria
Milvus migrans Black Kite	B	1991	15	30	—	C6
Milvus milvus Red Kite	B	1991	3	10	—	C6

The site is important for breeding and wintering waterbirds. Species of global conservation concern that do not meet IBA criteria: *Aythya nyroca* (passage); *Haliaeetus albicilla* (wintering).

Protection status

National Partial **International** Partial
1,100 ha of IBA covered by Special Protection Area (Etangs de la Woëvre: Etang de Madine, 1,100 ha). 1,100 ha of IBA covered by Hunting Reserve (1,100 ha).

Conservation issues

Threats Agricultural intensification/expansion (A), Recreation/tourism (A)

Agriculture intensification, including the conversion of meadows to arable fields, and the development of recreation activities threaten the site.

Etangs de la Woëvre: la Chaussée — A4i, B1i, C2, C6 — 077

Admin region Lorraine
Coordinates 49°01′N 5°49′E
Altitude 227–350 m **Area** 6,200 ha

Site description

Habitats Forest and woodland (12%; broadleaved deciduous forest), Grassland (5%; humid grassland), Wetland (82%; standing fresh water; standing brackish and salt water; water-fringe vegetation), Artificial landscape (perennial crops/orchards/groves)
Land-use Agriculture, Fisheries/aquaculture, Forestry, Tourism/recreation, Urban/industrial/transport

A lake with reedbeds (*Phragmites*), marshes, wet meadows and woodland.

Birds

Species	Season	Year	Pop min	Pop max	Acc	Criteria
Botaurus stellaris Bittern	B	1991	4	6	—	C6
Ardea purpurea Purple Heron	B	1991	4	8	—	C6
Milvus migrans Black Kite	B	1991	8	30	—	C6
Circus aeruginosus Marsh Harrier	B	1991	10	15	—	C6
Grus grus Crane	P	1991	5,000	10,000	—	A4i, B1i, C2

Species of global conservation concern that do not meet IBA criteria: *Haliaeetus albicilla* (wintering).

Protection status

National Low **International** Partial
400 ha of IBA covered by Private Reserve (400 ha). 5,300 ha of IBA covered by Ramsar Site (Etangs de la Petite Woëvre, 5,300 ha). 368 ha of IBA covered by Special Protection Area (Etangs de la Woëvre: Etang de la Chaussée, 368 ha). 38 ha of IBA covered by Special Protection Area (Etangs de la Woëvre: Marais de Pagny, 38 ha).

Conservation issues

Threats Agricultural intensification/expansion (A), Infrastructure (B)

The main threats are the conversion of grazing meadows to arable land and the planned construction of a railway line close to the site.

Etang de Lindre — B1i, B2, C3, C6 — 078

Admin region Lorraine
Coordinates 48°48′N 6°47′E
Altitude 210–232 m **Area** 1,660 ha

Site description

Lake and marshes, surrounded by woodland, grassland and arable land.

Habitats Forest and woodland (broadleaved deciduous forest), Wetland (standing fresh water; water-fringe vegetation), Artificial landscape (arable land)
Land-use Agriculture, Forestry, Hunting, Military, Tourism/recreation, Urban/industrial/transport

Birds

Species	Season	Year	Pop min	Pop max	Acc	Criteria
Botaurus stellaris Bittern	B	1991	4	4	—	C6
Ixobrychus minutus Little Bittern	B	1991	8	8	—	C6
Ardea purpurea Purple Heron	B	1991	7	7	—	C6
Cygnus columbianus Bewick's Swan	W	1997	9	9	—	B2
Anas clypeata Shoveler	P	1997	500	500	—	B1i, C3

The area is important for breeding and wintering waterbirds.

Protection status

National None **International** None

Conservation issues

Threats Agricultural intensification/expansion (C), Disturbance to birds (A), Infrastructure (A), Unsustainable exploitation (A), Aquaculture/fisheries

The expansion and intensification of fish-farming is resulting in the construction of new dykes. Agricultural pollution also impacts on the site. Local authorities own the site and are trying to purchase the surrounding areas for conservation purposes.

Vallée de la Meuse — A1, B2, C1, C6 — 079

Admin region Lorraine
Coordinates 49°01′N 5°28′E
Altitude 165–256 m **Area** 18,100 ha

Site description

The site includes wet meadows, a river, alluvial forest, marshes and aquatic vegetation.

Habitats Forest and woodland, Grassland (humid grassland), Wetland (5%; river/stream; water-fringe vegetation)
Land-use Agriculture, Fisheries/aquaculture, Hunting, Tourism/recreation, Urban/industrial/transport

Birds

Species	Season	Year	Pop min	Pop max	Acc	Criteria
Circus aeruginosus Marsh Harrier	B	1991	5	10	—	C6
Crex crex Corncrake	B	1993	25	25	—	A1, C1, C6
Alcedo atthis Kingfisher	B	1991	10	100	—	B2, C6
Picus canus Grey-headed Woodpecker	B	1991	1	10	—	C6

This is an important site for species associated with wet meadows.

Protection status

National None **International** None

Conservation issues

Threats Agricultural intensification/expansion (A), Construction/impact of dyke/dam/barrage (A), Dredging/canalization (B), Extraction industry (A), Infrastructure (C)

The drainage and ploughing of the wet meadows is destroying the site.

Plateau de Delme, Val de Petite Seille

C6 **080**

Admin region Lorraine
Coordinates 48°52′N 6°28′E
Altitude 188–307 m **Area** 11,950 ha

Site description

The site includes cereal fields, meadows, bocages, deciduous woodlands, rivers and *Populus* plantations.

Habitats Forest and woodland (5%; broadleaved deciduous forest), Wetland (river/stream), Artificial landscape (85%; highly improved reseeded grassland; arable land; perennial crops/orchards/groves; other urban/industrial areas)
Land-use Agriculture (80%), Fisheries/aquaculture, Forestry, Hunting, Other, Urban/industrial/transport

Birds

Species	Season	Year	Pop min	Pop max	Acc	Criteria
Circus pygargus Montagu's Harrier	B	1991	31	31	—	C6

This is an important site for species associated with open habitats such as *Circus pygargus*.

Protection status

National None **International** None

Conservation issues

Threats Agricultural intensification/expansion (B), Infrastructure (C)

The ploughing of meadows is the major threat.

Val de Chiers et environs de Spincourt

B1i, B3, C2, C6 **081**

Admin region Lorraine
Coordinates 49°24′N 5°38′E
Altitude 205–396 m **Area** 16,900 ha

Site description

The site includes cereal fields, woodlots, deciduous woodlands, rivers, bocage and abandoned fields.

Habitats Forest and woodland (10%; broadleaved deciduous forest), Wetland (river/stream), Artificial landscape (87%; arable land; perennial crops/orchards/groves; other urban/industrial areas; ruderal land)
Land-use Agriculture (90%), Fisheries/aquaculture, Forestry (10%), Hunting (75%), Urban/industrial/transport

Birds

Species	Season	Year	Pop min	Pop max	Acc	Criteria
Pernis apivorus Honey Buzzard	B	1991	10	100	—	B3, C6
Circus pygargus Montagu's Harrier	B	1991	27	27	—	C6
Grus grus Crane	P	1991	4,000	4,000	—	B1i, C2
Dryocopus martius Black Woodpecker	B	1991	1	10	—	C6
Lanius collurio Red-backed Shrike	B	1991	10	100	—	C6

This site is important for species breeding in forest and agricultural habitats.

Protection status

National None **International** None

Conservation issues

Threats Agricultural intensification/expansion (B)

The ploughing of meadows is the main threat.

Fresnes-en-Woëvre – Mars-la-Tour

C6 **082**

Admin region Lorraine
Coordinates 49°06′N 5°48′E
Altitude 200–310 m **Area** 11,700 ha

Site description

The site includes cereal fields, orchards, deciduous woodlands, abandoned fields, and rivers.

Habitats Forest and woodland (5%; broadleaved deciduous forest), Wetland (river/stream), Artificial landscape (87%; arable land; perennial crops/orchards/groves; other urban/industrial areas; ruderal land)
Land-use Agriculture (80%), Fisheries/aquaculture, Forestry (5%), Hunting, Urban/industrial/transport

Birds

Species	Season	Year	Pop min	Pop max	Acc	Criteria
Circus pygargus Montagu's Harrier	B	1991	28	28	—	C6
Dryocopus martius Black Woodpecker	B	1991	1	10	—	C6

This is an important site for breeding raptors and forest species.

Protection status

National None **International** None

Conservation issues

Threats Agricultural intensification/expansion (B), Industrialization/urbanization (C)

The ploughing of meadows is the major threat.

Bazoncourt–Vigy

C6 **083**

Admin region Lorraine
Coordinates 49°10′N 6°20′E
Altitude 220–316 m **Area** 10,700 ha

Site description

An area of cereal fields, deciduous woodlands, wet meadows, bocages and rivers.

Habitats Forest and woodland (10%; broadleaved deciduous forest), Grassland (5%; humid grassland), Wetland (river/stream), Artificial landscape (77%; arable land; perennial crops/orchards/groves; other urban/industrial areas)
Land-use Agriculture, Fisheries/aquaculture, Forestry, Hunting, Tourism/recreation, Urban/industrial/transport

Birds

Species	Season	Year	Pop min	Pop max	Acc	Criteria
Circus pygargus Montagu's Harrier	B	1991	15	15	—	C6
Dryocopus martius Black Woodpecker	B	1991	1	10	—	C6

This site is important for breeding raptors.

Protection status

National None **International** None

Conservation issues

Threats Agricultural intensification/expansion (B), Industrialization/urbanization

The ploughing of natural meadows and urbanization are the major threats.

Forêt de la Reine C6 084

Admin region Lorraine
Coordinates 48°48′N 5°44′E
Altitude 225–285 m **Area** 16,650 ha

Site description

An area of deciduous woodlands, open fields, ponds, marshes, reedbeds (*Phragmites*), rivers and riverine forest.

Habitats Forest and woodland (55%; broadleaved deciduous forest; alluvial/very wet forest), Wetland (11%; standing brackish and salt water; river/stream; water-fringe vegetation), Artificial landscape (35%; arable land)
Land-use Agriculture, Fisheries/aquaculture, Hunting, Urban/industrial/transport

Birds

Species	Season	Year	Pop min	Pop max	Acc	Criteria
Botaurus stellaris Bittern	B	1991	10	12	—	C6
Circus aeruginosus Marsh Harrier	B	1991	5	10	—	C6
Dryocopus martius Black Woodpecker	B	1991	1	10	—	C6

This is an important site for forest and wetland species.

Protection status

National None **International** None

Conservation issues

Threats Agricultural intensification/expansion (A), Drainage (A), Intensified forest management (A)

Drainage of forest ponds, the intensification of forestry and the ploughing of meadows to convert them to arable land are the main threats.

Ried de Colmar et Sélestat A4i, A4iii, B1i, B3, C3, C4, C6 085

Admin region Alsace
Coordinates 48°13′N 7°28′E
Altitude 170–170 m **Area** 10,100 ha

Site description

This site contains wet grassland, peatbogs, deciduous woodlands, *Populus* grove, rivers, riverine forest, arable land and bocage.

Habitats Forest and woodland (broadleaved deciduous forest; alluvial/very wet forest), Grassland (steppe/dry calcareous grassland; humid grassland; mesophile grassland), Wetland (river/stream; fen/transition mire/spring), Artificial landscape (highly improved reseeded grassland; arable land; perennial crops/orchards/groves)
Land-use Agriculture, Fisheries/aquaculture, Forestry, Hunting, Tourism/recreation, Urban/industrial/transport

Birds

Species	Season	Year	Pop min	Pop max	Acc	Criteria
Anser fabalis Bean Goose	W	1991	750	5,000	—	A4i, B1i, C3
Pernis apivorus Honey Buzzard	B	1991	10	100	—	B3, C6
Milvus migrans Black Kite	B	1991	15	20	—	C6
Dendrocopos medius Middle Spotted Woodpecker	B	1991	40	100	—	C6
Lanius collurio Red-backed Shrike	B	1991	10	100	—	C6

The site holds 20,000 or more waterbirds in winter, on a regular basis.

Protection status

National Partial **International** None

1,799 ha of IBA covered by Voluntary Nature Reserve (l'Ill Wald, 1,799 ha). 23 ha of IBA covered by Private Reserve (23 ha).

Conservation issues

Threats Agricultural intensification/expansion, Infrastructure

The site is threatened by agricultural intensification (ploughing of meadows) and gravel extraction. The construction of a motorway is planned, which will result in the alteration and dredging of rivers.

Forêt domaniale de la Harth B2, B3, C6 086

Admin region Alsace
Coordinates 47°46′N 7°27′E
Altitude 182–240 m **Area** 15,700 ha

Site description

The site encompasses deciduous and mixed woodlands, heathland and clearings.

Habitats Forest and woodland (80%; broadleaved deciduous forest; mixed forest), Scrub (heathland)
Land-use Forestry, Hunting, Urban/industrial/transport

Birds

Species	Season	Year	Pop min	Pop max	Acc	Criteria
Pernis apivorus Honey Buzzard	B	1990	10	100	—	B3, C6
Milvus migrans Black Kite	B	1990	10	100	—	B2, C6
Milvus milvus Red Kite	B	1990	5	50	—	B3, C6
Dryocopus martius Black Woodpecker	B	1990	10	100	—	C6
Dendrocopos medius Middle Spotted Woodpecker	B	1990	100	200	—	B3, C6

The avifauna of lowland forests is well represented, with several species of raptor and woodpecker.

Protection status

National Low **International** None

30 ha of IBA covered by Biological State Reserve (30 ha).

Conservation issues

Threats Infrastructure (B), Intensified forest management

Forestry is the main human activity and major threat as practices are too intensive. Highway and road construction are destroying the site.

Forêt de Haguenau B3, C2, C6 087

Admin region Alsace
Coordinates 48°52′N 7°53′E
Altitude 130–183 m **Area** 21,350 ha

Site description

The site includes deciduous and conifer woodlands, heathland, wet grassland, peatbogs and rivers.

Habitats Forest and woodland (mixed forest), Scrub (heathland), Grassland (humid grassland), Wetland (river/stream; raised bog)
Land-use Forestry, Hunting, Tourism/recreation, Urban/industrial/transport

Birds

Species	Season	Year	Pop min	Pop max	Acc	Criteria
Pernis apivorus Honey Buzzard	B	1991	10	100	—	B3, C6
Milvus milvus Red Kite	B	1991	1	10	—	C6
Caprimulgus europaeus Nightjar	B	1991	10	100	—	C6
Picus canus Grey-headed Woodpecker	B	1991	5	50	—	C6
Dryocopus martius Black Woodpecker	B	1991	10	100	—	C6
Dendrocopos medius Middle Spotted Woodpecker	B	1991	100	500	—	B3, C2, C6
Lullula arborea Woodlark	B	1991	1	10	—	C6
Lanius collurio Red-backed Shrike	B	1991	10	100	—	C6

The site is important for raptors, woodpeckers and heathland-edge species.

Protection status

National None **International** None

Conservation issues

Threats Infrastructure (C), Intensified forest management (A), Recreation/tourism

The main threats are intensive forestry, tourism and ongoing road construction projects.

Zones agricoles de la Hardt — B2, C6 — 088

Admin region Alsace
Coordinates 47°57′N 7°27′E
Altitude 190–212 m **Area** 21,800 ha

Site description

This site includes areas of cereal cultivation, deciduous woodland and rivers.

Habitats Forest and woodland (20%; broadleaved deciduous forest), Wetland (river/stream), Artificial landscape (80%; arable land)
Land-use Agriculture (80%), Forestry (20%), Hunting, Urban/industrial/transport

Birds

Species	Season	Year	Pop min	Pop max	Acc	Criteria
Circus pygargus Montagu's Harrier	B	1991	10	10	—	C6
Burhinus oedicnemus Stone Curlew	B	1991	100	150	—	B2, C6
Anthus campestris Tawny Pipit	B	1991	1	10	—	C6

This area is important for species associated with agricultural habitats.

Protection status

National None **International** None

Conservation issues

Threats Agricultural intensification/expansion (A)

The main threat is the conversion of meadows into maize fields.

Massif des Vosges: crêtes de Schneeberg au Donon — C6 — 089

Admin region Alsace
Coordinates 48°34′N 7°14′E
Altitude 400–1,010 m **Area** 8,450 ha

Site description

An area of mixed woodland, rivers, peatbogs, cliffs and rocky slopes.

Habitats Forest and woodland (95%; mixed forest), Scrub (heathland), Wetland (river/stream; raised bog), Rocky areas (inland cliff)
Land-use Fisheries/aquaculture, Forestry (95%), Hunting, Tourism/recreation

Birds

Species	Season	Year	Pop min	Pop max	Acc	Criteria
Bonasa bonasia Hazel Grouse	B	1991	10	100	—	C6
Tetrao urogallus Capercaillie	B	1991	20	20	—	C6
Aegolius funereus Tengmalm's Owl	B	1991	1	10	—	C6
Picus canus Grey-headed Woodpeck	B	1991	5	50	—	C6
Dryocopus martius Black Woodpecker	B	1991	10	100	—	C6
Lanius collurio Red-backed Shrike	B	1991	10	100	—	C6

This is an important area for forest species.

Protection status

National Low **International** None
456 ha of IBA covered by Biological State Reserve (456 ha).

Conservation issues

Threats Intensified forest management (A), Recreation/tourism (A)

The main threats are intensive forestry and the development of winter tourism.

Vallée du Rhin: Village-neuf à Biescheim — A4iii, B1i, B2, C3, C4, C6 — 090

Admin region Alsace
Coordinates 47°49′N 7°34′E
Altitude 220–220 m **Area** 5,690 ha

Site description

An area of rivers, riverine forest, deciduous woodland, heathland, marsh, reedbed (*Phragmites*) and arable land.

Habitats Forest and woodland (10%; broadleaved deciduous forest; alluvial/very wet forest), Scrub (heathland), Grassland (steppe/dry calcareous grassland), Wetland (river/stream; water-fringe vegetation), Artificial landscape (10%; arable land; other urban/industrial areas; ruderal land)
Land-use Agriculture (10%), Fisheries/aquaculture, Tourism/recreation, Urban/industrial/transport (50%), Water management

Birds

Species	Season	Year	Pop min	Pop max	Acc	Criteria
Anas strepera Gadwall	W	1991	200	500	—	B1i, C3
Milvus migrans Black Kite	B	1991	10	100	—	B2, C6
Picus canus Grey-headed Woodpecker	B	1991	10	100	—	B2, C6
Dendrocopos medius Middle Spotted Woodpecker	B	1991	10	100	—	C6
Lanius collurio Red-backed Shrike	B	1991	10	100	—	C6

The site holds 20,000 or more waterbirds in winter, on a regular basis (including *Anas platyrhynchos* 12,000; *Aythya ferina* 3,000; *Aythya fuligula* 3,500; *Fulica atra* 3,000; *Larus ridibundus* 10,000). Species of global conservation concern that do not meet IBA criteria: *Aythya nyroca* (1–10 passage individuals).

Protection status

National Partial **International** None
120 ha of IBA covered by Nature Reserve (Petite Camargue Alsacienne, 120 ha). 2,000 ha of IBA covered by Hunting Reserve (2,000 ha).

Conservation issues

Threats Agricultural intensification/expansion, Industrialization/urbanization (B), Recreation/tourism (B)

Reported threats are high human pressure on the river (boating and fishing) and on the edges of the site, as well as agriculture, recreation and industry.

Vallée du Rhin: Marckolsheim à Strasbourg — A4i, A4iii, B1i, B2, B3, C3, C4, C6 — 091

Admin region Alsace
Coordinates 48°19′N 7°41′E
Altitude 220–220 m **Area** 10,600 ha

Site description

Habitats Forest and woodland (10%; broadleaved deciduous forest; alluvial/very wet forest), Scrub (heathland), Grassland (steppe/dry calcareous grassland; mesophile grassland), Wetland (1%; standing fresh water; river/stream; water-fringe vegetation), Artificial landscape (perennial crops/orchards/groves; other urban/industrial areas; ruderal land)
Land-use Fisheries/aquaculture, Forestry, Hunting, Tourism/recreation, Urban/industrial/transport

Birds

Species	Season	Year	Pop min	Pop max	Acc	Criteria
Anas strepera Gadwall	W	1991	1,500	2,500	—	A4i, B1i, C3
Aythya fuligula Tufted Duck	W	1991	7,000	14,000	—	B1i, C3
Pernis apivorus Honey Buzzard	B	1991	10	100	—	B3, C6
Milvus migrans Black Kite	B	1991	10	100	—	B2, C6
Circus aeruginosus Marsh Harrier	B	1991	1	10	—	C6
Sterna hirundo Common Tern	B	1991	130	150	—	C6
Alcedo atthis Kingfisher	B	1991	10	100	—	B2, C6
Picus canus Grey-headed Woodpecker	B	1991	10	100	—	B2, C6

Species ... continued	Season	Year	Pop min	Pop max	Acc	Criteria
Dendrocopos medius Middle Spotted Woodpecker	B	1991	100	100	—	C6
Lanius collurio Red-backed Shrike	B	1991	10	100	—	C6

The site holds 20,000 or more waterbirds in winter, on a regular basis (including *Anas platyrhynchos* 25,000; *Fulica atra* 5,000). The site is also of importance for breeding woodpeckers. Species of global conservation concern that do not meet IBA criteria: *Aythya nyroca* (1–10 breeding individuals).

Protection status

National Partial **International** None

180 ha of IBA covered by Nature Reserve (Forêt d'Erstein, 180 ha). 310 ha of IBA covered by Nature Reserve (l'Ile de Rohrschollen, 310 ha). 311 ha of IBA covered by Nature Reserve (Ile de Rhinau, 311 ha). 2,140 ha of IBA covered by Hunting Reserve (2,140 ha).

Conservation issues

Threats Disturbance to birds (B), Industrialization/urbanization (C), Recreation/tourism (B)

Further planned industrial development is a threat.

Vallée du Rhin: Strasbourg à Lauterbourg — A4i, A4iii, B1i, B2, C3, C4, C6 — 092

Admin region Alsace
Coordinates 48°47′N 8°01′E
Altitude 125–125 m **Area** 11,000 ha

Site description

The site includes a river and riverine forest, canals, pools, marshes, meadows, broadleaved forests and heathlands.

Habitats Forest and woodland (broadleaved deciduous forest; alluvial/very wet forest), Scrub (heathland), Grassland (humid grassland; mesophile grassland), Wetland (river/stream; water-fringe vegetation), Artificial landscape (perennial crops/orchards/groves; other urban/industrial areas; ruderal land)

Land-use Fisheries/aquaculture, Forestry, Hunting, Tourism/recreation, Urban/industrial/transport, Water management

Birds

Species	Season	Year	Pop min	Pop max	Acc	Criteria
Anas strepera Gadwall	W	1997	2,936	4,495	—	A4i, B1i, C3
Aythya ferina Pochard	W	1997	2,876	5,823	—	B1i, C3
Aythya fuligula Tufted Duck	W	1997	11,917	18,991	—	A4i, B1i, C3
Milvus migrans Black Kite	B	1991	10	100	—	B2, C6
Circus aeruginosus Marsh Harrier	B	1991	1	10	—	C6
Larus melanocephalus Mediterranean Gull	B	1996	20	20	—	C6
Alcedo atthis Kingfisher	B	1991	10	100	—	B2, C6
Picus canus Grey-headed Woodpecker	B	1991	10	100	—	C6
Dryocopus martius Black Woodpecker	B	1991	10	100	—	C6
Dendrocopos medius Middle Spotted Woodpecker	B	1991	10	100	—	C6
Luscinia svecica Bluethroat	B	1991	1	10	—	C6
Lanius collurio Red-backed Shrike	B	1991	10	100	—	C6

The site holds 20,000 or more waterbirds in winter, on a regular basis (including *Anas platyrhynchos* 25,940). Species of global conservation concern that do not meet IBA criteria: *Branta ruficollis* (wintering); *Aythya nyroca* (1–10 passage individuals); *Haliaeetus albicilla* (wintering).

Protection status

National Partial **International** None

486 ha of IBA covered by Nature Reserve (Delta de la Sauer, 486 ha). 60 ha of IBA covered by Nature Reserve (Forêt d'Offendorf, 60 ha). 147 ha of IBA covered by Biological State Reserve (147 ha). 1,064 ha of IBA covered by Hunting Reserve (1,064 ha).

Conservation issues

Threats Extraction industry (C), Recreation/tourism

Forestry, industry, boating and recreation are the main threats to the site.

Massif des Vosges: Hautes Vosges — B2, B3, C6 — 093

Admin region Alsace
Coordinates 47°57′N 6°56′E
Altitude 200–1,362 m **Area** 88,200 ha

Site description

A large montane area encompassing broadleaved forest, heathlands, meadows, peatbogs, scree, rocky slopes and alpine grasslands.

Habitats Forest and woodland (broadleaved deciduous forest; native coniferous forest; mixed forest), Scrub (heathland), Grassland (dry siliceous grassland; alpine/subalpine/boreal grassland; humid grassland; mesophile grassland), Wetland (standing fresh water; river/stream; raised bog), Rocky areas (scree/boulders; inland cliff), Artificial landscape (highly improved reseeded grassland; arable land; perennial crops/orchards/groves; ruderal land)

Land-use Agriculture, Fisheries/aquaculture, Forestry (80%), Hunting, Other, Tourism/recreation (55%), Urban/industrial/transport (20%), Water management (5%)

Birds

Species	Season	Year	Pop min	Pop max	Acc	Criteria
Pernis apivorus Honey Buzzard	B	1991	10	100	—	B3, C6
Falco peregrinus Peregrine	B	1991	10	12	—	B2, C6
Bonasa bonasia Hazel Grouse	B	1991	50	250	—	C6
Tetrao urogallus Capercaillie	B	1991	80	100	—	C6
Aegolius funereus Tengmalm's Owl	B	1991	10	100	—	C6
Dryocopus martius Black Woodpecker	B	1991	50	250	—	C6
Lanius collurio Red-backed Shrike	B	1991	100	500	—	C6

The site is important for breeding raptors and other species associated with mountain habitats.

Protection status

National Low **International** None

746 ha of IBA covered by Nature Reserve (Frankental-Missheimle, 746 ha). 505 ha of IBA covered by Nature Reserve (Le Tanet–Gazon du Faing, 505 ha). 1,647 ha of IBA covered by Nature Reserve (Massif du Ventron, 1,647 ha). 145 ha of IBA covered by Nature Reserve (Tourbière de Machais, 145 ha). 1,881 ha of IBA covered by Biological State Reserve (1,881 ha).

Conservation issues

Threats Intensified forest management (A), Recreation/tourism (A)

Increasing use of the site for recreational activities creates disturbance. Parts of the forest are now managed to conserve species of Tetraonidae.

Bassin du Drugeon: Pontarlier-Frasne — C6 — 094

Admin region Franche-Comté
Coordinates 46°51′N 6°12′E
Altitude 850–1,100 m **Area** 12,500 ha

Site description

Deciduous and conifer woodlands, peatbogs, ponds, wet grasslands, rivers and riverine vegetation.

Habitats Forest and woodland (broadleaved deciduous forest; native coniferous forest), Grassland (humid grassland), Wetland (standing fresh water; raised bog; water-fringe vegetation)

Land-use Agriculture, Fisheries/aquaculture, Forestry, Hunting, Tourism/recreation, Urban/industrial/transport

Birds

Species	Season	Year	Pop min	Pop max	Acc	Criteria
Pernis apivorus Honey Buzzard	B	1991	1	1	—	C6
Milvus migrans Black Kite	B	1991	40	50	—	C6
Bonasa bonasia Hazel Grouse	B	1991	20	50	—	C6
Porzana porzana Spotted Crake	B	1991	2	5	—	C6
Crex crex Corncrake	B	1992	2	2	—	C6

The site is of special importance for wet meadow and forest birds.

Protection status

National Low **International** None

160 ha of IBA covered by Private Reserve (160 ha).

Conservation issues

Threats Afforestation (A), Agricultural intensification/expansion (C), Drainage (B), Extraction industry (A), Industrialization/urbanization (A), Intensified forest management (C)

Afforestation with spruce *Picea* is degrading the site.

Forêts d'altitude du Jura: forêt du Risol et Mont d'Or — B3, C6 — 095

Admin region Franche-Comté
Coordinates 46°40'N 6°14'E
Altitude 950–1,464 m **Area** 12,300 ha

Site description

Conifer and deciduous woodlands with scree slopes, cliffs, and alpine grassland.

Habitats Forest and woodland (native coniferous forest; mixed forest), Grassland (steppe/dry calcareous grassland), Rocky areas (scree/boulders; inland cliff)
Land-use Agriculture, Forestry, Hunting, Tourism/recreation, Urban/industrial/transport

Birds

Species	Season	Year	Pop min	Pop max	Acc	Criteria
Bonasa bonasia Hazel Grouse	B	1991	250	250	—	C6
Tetrao urogallus Capercaillie	B	1991	40	50	—	C6
Glaucidium passerinum Pygmy Owl	B	1991	10	15	—	C6
Aegolius funereus Tengmalm's Owl	B	1991	10	100	—	C6
Dryocopus martius Black Woodpecker	B	1991	10	40	—	C6
Turdus torquatus Ring Ouzel	B	1991	100	500	—	B3

The site is important for breeding montane species.

Protection status

National Low **International** None
504 ha of IBA covered by Hunting Reserve (504 ha).

Conservation issues

Threats Abandonment/reduction of land management (A), Agricultural intensification/expansion (A), Industrialization/urbanization (A), Infrastructure (A), Recreation/tourism (A)

The increasing number of tourists visiting the area (largely due to the ski-resort facilities) and the abandonment of traditional cattle-raising are the main threats.

Forêts d'altitude du Jura: forêt du Risoux — B3, C6 — 096

Admin region Franche-Comté
Coordinates 46°32'N 6°05'E
Altitude 1,050–1,280 m **Area** 4,400 ha

Site description

An area of conifer and mixed woodland, scree, cliffs and subalpine grasslands.

Habitats Forest and woodland (85%; native coniferous forest), Grassland (steppe/dry calcareous grassland), Rocky areas (scree/boulders; inland cliff)
Land-use Agriculture, Forestry, Tourism/recreation, Urban/industrial/transport

Birds

Species	Season	Year	Pop min	Pop max	Acc	Criteria
Pernis apivorus Honey Buzzard	B	1991	5	10	—	C6
Bonasa bonasia Hazel Grouse	B	1991	100	120	—	C6
Tetrao urogallus Capercaillie	B	1991	20	30	—	C6
Glaucidium passerinum Pygmy Owl	B	1991	10	20	—	C6
Aegolius funereus Tengmalm's Owl	B	1991	10	100	—	C6
Turdus torquatus Ring Ouzel	B	1991	100	500	—	B3

This is an important site for forest and mountain species.

Protection status

National None **International** None

Conservation issues

Threats Disturbance to birds (A), Extraction industry (A), Recreation/tourism (A)

The main threats are the development of recreation and increased access to the forests.

Forêts d'altitude du Jura: forêt du Massacre — B3, C6 — 097

Admin region Franche-Comté
Coordinates 46°25'N 6°02'E
Altitude 1,150–1,465 m **Area** 1,970 ha

Site description

Conifer woodland with screes, cliffs and subalpine grasslands.

Habitats Forest and woodland (native coniferous forest; mixed forest), Grassland (steppe/dry calcareous grassland), Rocky areas (scree/boulders; inland cliff)
Land-use Agriculture, Forestry, Tourism/recreation

Birds

Species	Season	Year	Pop min	Pop max	Acc	Criteria
Bonasa bonasia Hazel Grouse	B	1991	50	100	—	C6
Tetrao urogallus Capercaillie	B	1991	15	20	—	C6
Glaucidium passerinum Pygmy Owl	B	1991	1	10	—	C6
Aegolius funereus Tengmalm's Owl	B	1991	10	100	—	C6
Turdus torquatus Ring Ouzel	B	1991	100	500	—	B3

This is an important site for mountain and forest birds.

Protection status

National None **International** None

Conservation issues

Threats Disturbance to birds (A), Recreation/tourism (A)

The development of tourist activities, particularly skiing, is the main threat.

Forêt de Chaux — B2, C6 — 098

Admin region Franche-Comté
Coordinates 47°05'N 5°39'E
Altitude 220–270 m **Area** 21,700 ha

Site description

Deciduous woodland with ponds, marshes, wet forest and grassland. Forestry is the main human activity.

Habitats Forest and woodland (90%; broadleaved deciduous forest; alluvial/very wet forest), Grassland (mesophile grassland), Wetland (standing fresh water; water-fringe vegetation)
Land-use Forestry, Hunting, Tourism/recreation, Urban/industrial/transport

Birds

Species	Season	Year	Pop min	Pop max	Acc	Criteria
Pernis apivorus Honey Buzzard	B	1991	1	10	—	C6
Milvus migrans Black Kite	B	1991	10	20	—	C6
Milvus milvus Red Kite	B	1991	1	10	—	C6
Hieraaetus pennatus Booted Eagle	B	1991	1	2	—	C6
Picus canus Grey-headed Woodpecker	B	1991	10	100	—	B2, C6
Dendrocopos medius Middle Spotted Woodpecker	B	1991	40	100	—	C6

This site is important for breeding raptors and woodpeckers.

Protection status

National Low **International** None
1,408 ha of IBA covered by Hunting Reserve (1,408 ha).

Conservation issues

Threats Intensified forest management (A)

Clear-felling is considered the main threat to the site.

Vallée de la Saône de Corre à Broye

B2, B3, C6 **099**

Admin region Franche-Comté
Coordinates 47°07′N 5°46′E
Altitude 190–200 m **Area** 14,050 ha

Site description

A river valley with associated riverine forest, wet and mesophile grassland, deciduous woodland, marshes and reedbeds (*Phragmites*).

Habitats Forest and woodland (broadleaved deciduous forest; alluvial/very wet forest), Grassland (humid grassland; mesophile grassland), Wetland (river/stream; water-fringe vegetation), Artificial landscape (highly improved reseeded grassland)
Land-use Agriculture, Fisheries/aquaculture, Forestry, Hunting, Tourism/recreation, Urban/industrial/transport

Birds

Species	Season	Year	Pop min	Pop max	Acc	Criteria
Ixobrychus minutus Little Bittern	B	1991	5	10	—	C6
Pernis apivorus Honey Buzzard	B	1991	1	10	—	C6
Milvus migrans Black Kite	B	1991	10	100	—	B2, C6
Milvus milvus Red Kite	B	1991	25	100	—	B3, C6
Circus aeruginosus Marsh Harrier	B	1991	1	5	—	C6
Circus pygargus Montagu's Harrier	B	1991	1	10	—	C6
Porzana porzana Spotted Crake	B	1991	1	10	—	C6
Crex crex Corncrake	B	1992	2	2	—	C6
Alcedo atthis Kingfisher	B	1991	5	50	—	C6
Picus canus Grey-headed Woodpecker	B	1991	1	10	—	C6
Dendrocopos medius Middle Spotted Woodpecker	B	1991	20	40	—	C6
Lanius collurio Red-backed Shrike	B	1991	10	100	—	C6

The site is important for birds breeding in wetlands and riverine forest.

Protection status

National None **International** None

Conservation issues

Threats Afforestation (B), Agricultural intensification/expansion (A), Drainage (B), Intensified forest management (C)

The intensification of agriculture, particularly the ploughing of grasslands for arable crops, is the main threat. The afforestation of natural wet meadows is also a significant threat. The site is included in an agri-environmental scheme.

Basse vallée du Doubs: Dole Sud

B2, C6 **100**

Admin region Franche-Comté
Coordinates 46°59′N 5°20′E
Altitude 190–250 m **Area** 11,850 ha

Site description

A slow-flowing river, with associated wet meadows, marshes, cultivated land, sandbanks, and woodland.

Habitats Forest and woodland (broadleaved deciduous forest; alluvial/very wet forest), Grassland (humid grassland), Wetland (river/stream; water-fringe vegetation), Artificial landscape (highly improved reseeded grassland; arable land)
Land-use Agriculture, Fisheries/aquaculture, Hunting, Urban/industrial/transport

Birds

Species	Season	Year	Pop min	Pop max	Acc	Criteria
Ixobrychus minutus Little Bittern	B	1991	10	15	—	C6
Nycticorax nycticorax Night Heron	B	1991	40	60	—	C6
Ardea purpurea Purple Heron	B	1991	10	20	—	C6
Milvus migrans Black Kite	B	1991	10	20	—	C6
Circus aeruginosus Marsh Harrier	B	1991	1	5	—	C6
Circus cyaneus Hen Harrier	B	1991	10	20	—	C6
Circus pygargus Montagu's Harrier	B	1991	5	10	—	C6
Burhinus oedicnemus Stone Curlew	B	1995	2	2	—	C6
Alcedo atthis Kingfisher	B	1991	10	100	—	B2, C6
Merops apiaster Bee-eater	B	1991	10	100	—	B2
Luscinia svecica Bluethroat	B	1991	80	80	—	C6

Species of global conservation concern that do not meet IBA criteria: *Aythya nyroca* (wintering); *Crex crex* (breeding).

Protection status

National Low **International** None
95 ha of IBA covered by Nature Reserve (l'Ile du Girard, 95 ha).

Conservation issues

Threats Construction/impact of dyke/dam/barrage (A), Extraction industry (A), Infrastructure (A)

Grasslands are threatened by agriculture intensification and conversion to arable fields.

Estuaire de la Charente

B1i, B2, C3, C6 **101**

Admin region Poitou-Charentes
Coordinates 45°57′N 1°00′W
Altitude 0–5 m **Area** 5,400 ha

Site description

An estuary with mudflats, saltmarshes, reedbeds (*Phragmites*) and aquatic vegetation, freshwater marshes, wet meadows, and arable fields.

Habitats Grassland (humid grassland), Wetland (tidal river/enclosed tidal water; mudflat/sandflat; saltmarsh; standing fresh water; standing brackish and salt water; river/stream; water-fringe vegetation), Artificial landscape (highly improved reseeded grassland; arable land; perennial crops/orchards/groves)
Land-use Agriculture, Fisheries/aquaculture, Hunting, Tourism/recreation, Urban/industrial/transport

Birds

Species	Season	Year	Pop min	Pop max	Acc	Criteria
Ardea purpurea Purple Heron	B	1997	27	27	—	C6
Himantopus himantopus Black-winged Stilt	B	1996	12	20	—	C6
Recurvirostra avosetta Avocet	W	1991	300	300	—	B2
Calidris canutus Knot	P	1991	5,000	5,000	—	B1i, C3
Luscinia svecica Bluethroat	B	1997	25	50	—	C6

The site is important for breeding waterbirds.

Protection status

National None **International** Partial
2,500 ha of IBA covered by Special Protection Area (Marais de Fouras, Baie d'Yves et Marais de Rochefort, 11,679 ha).

Conservation issues

Threats Agricultural intensification/expansion (A), Industrialization/urbanization (A), Recreation/tourism (A), Unsustainable exploitation (B)

Urbanization, high hunting pressure (including illegal night hunting) and the conversion of wet meadows into arable fields are the main threats.

Vallée de la Charente et de la Seugne (Cabariot-Pons/ St-Sever-de-Saintonge)

A1, B2, B3, C1, C2, C6 **102**

Admin region Poitou-Charentes
Coordinates 45°46′N 0°40′W
Altitude 3–50 m **Area** 8,300 ha

Site description

An area of wet meadows, bogs and marshes, with riverine forest and *Populus* plantations.

Habitats Forest and woodland (alluvial/very wet forest), Grassland (humid grassland), Wetland (river/stream; water-fringe vegetation), Artificial landscape (perennial crops/orchards/groves)
Land-use Agriculture, Hunting, Urban/industrial/transport

Birds

Species	Season	Year	Pop min	Pop max	Acc	Criteria
Milvus migrans Black Kite	B	1997	30	40	—	C6
Porzana porzana Spotted Crake	B	1991	5	10	—	B3, C6

Species ... continued	Season	Year	Pop min	Pop max	Acc	Criteria
Crex crex Corncrake	B	1997	60	65	—	A1, B2, C1, C2, C6
Alcedo atthis Kingfisher	B	1996	20	30	—	C6

The site is of national importance for breeding *Crex crex*.

Protection status

National None **International** Partial

250 ha of IBA covered by Special Protection Area (Marais de Fouras, Baie d'Yves et Marais de Rochefort, 11,676 ha).

Conservation issues

Threats Abandonment/reduction in land management (A), Afforestation (A), Agricultural intensification/expansion (A), Industrialization/urbanization (A), Infrastructure (B), Intensified forest management (A), Recreation/tourism (A), Unsustainable exploitation (C)

The main threats are urbanization, afforestation with *Populus*, the abandonment of grazing and the conversion of meadows into arable fields. Conservation efforts for *Crex crex* stand to benefit from a new agri-environmental scheme in the area, which should encourage farmers to maintain wet meadows.

Vallée de la Charente: amont d'Angoulême

A1, B2, C1, C6 **103**

Admin region Poitou-Charentes
Coordinates 45°46'N 0°04'W
Altitude 35–80 m **Area** 4,650 ha

Site description

A river valley with wet meadows, *Populus* plantations and arable fields.

Habitats Grassland (mesophile grassland), Wetland (river/stream), Artificial landscape (highly improved reseeded grassland; arable land; perennial crops/orchards/groves)
Land-use Agriculture, Fisheries/aquaculture, Hunting, Urban/industrial/transport

Birds

Species	Season	Year	Pop min	Pop max	Acc	Criteria
Ixobrychus minutus Little Bittern	B	1991	1	10	—	C6
Crex crex Corncrake	B	1996	27	35	—	A1, C1, C6
Alcedo atthis Kingfisher	B	1991	10	100	—	B2, C6

Protection status

National None **International** None

Conservation issues

Threats Agricultural intensification/expansion (A)

The main threats are the conversion of wet meadows into arable fields and the destruction of nests and young *Crex crex* due to early mowing. An agri-environment scheme is now in operation in the area.

Anse de Fouras, baie d'Yves et marais de Rochefort

A4iii, B1i, B2, B3, C2, C3, C4, C6 **104**

Admin region Poitou-Charentes
Coordinates 46°01'N 0°57'W
Altitude 0–22 m **Area** 17,550 ha

Site description

A bay with sandy beaches, saltwater lagoons and marshes, mudflats, rivers, ponds, wet meadows, small woodlands, grassland, and arable land.

Habitats Forest and woodland (broadleaved deciduous forest), Grassland (humid grassland), Wetland (mudflat/sandflat; saltmarsh; sand-dunes/sand beach; standing fresh water; standing brackish and salt water; river/stream; water-fringe vegetation), Marine areas (sea inlet/coastal features), Artificial landscape (arable land)
Land-use Agriculture, Fisheries/aquaculture, Hunting, Tourism/recreation, Urban/industrial/transport

Birds

Species	Season	Year	Pop min	Pop max	Acc	Criteria
Ardea purpurea Purple Heron	B	1997	148	150	—	B2, C2, C6
Ciconia ciconia White Stork	B	1997	5	5	—	C6
Anas clypeata Shoveler	W	1997	125	624	—	B1i, C3
Anas clypeata Shoveler	P	1997	500	500	—	B1i, C3
Milvus migrans Black Kite	B	1997	20	40	—	C6
Circus aeruginosus Marsh Harrier	B	1997	20	40	—	C6
Circus pygargus Montagu's Harrier	B	1995	35	45	—	B3, C6
Himantopus himantopus Black-winged Stilt	B	1997	30	40	—	C6
Calidris alpina Dunlin	W	1997	10,500	16,517	—	B1i, B2, C3
Tringa totanus Redshank	P	1997	2,000	2,000	—	B1i, C3
Chlidonias niger Black Tern	B	1996	10	20	—	C6

The site holds 20,000 or more waterbirds in winter, on a regular basis.

Protection status

National Low **International** Partial

185 ha of IBA covered by Nature Reserve (Le Marais d'Yves, 185 ha). 800 ha of IBA covered by Hunting Reserve (Anse de Fouras, 1,800 ha) 8,926 ha of IBA covered by Special Protection Area (Marais de Fouras, Baie d'Yves et Marais de Rochefort, 11,676 ha).

Conservation issues

Threats Abandonment/reduction of land management (A), Agricultural intensification/expansion (A), Drainage (A), Recreation/tourism (A), Unsustainable exploitation (B)

The site is threatened by drainage, cultivation, the inadequate management of water-levels and high hunting pressure. The rate of drainage within the SPA area has been reduced by agri-environmental schemes but the threat remains.

Ile d'Oléron, marais de Brouage–Saint-Agnant

A4i, A4iii, B1i, B2, B3, C2, C3, C4, C6 **105**

Admin region Poitou-Charentes
Coordinates 45°54'N 1°10'W
Altitude 0–9 m **Area** 26,500 ha

Site description

A bay with saltwater marshes and mudflats, ponds, wet meadows with ditches, woodland, grassland, and arable land.

Habitats Forest and woodland (broadleaved deciduous forest), Grassland (humid grassland), Wetland (mudflat/sandflat; saltmarsh; standing fresh water; river/stream; water-fringe vegetation), Marine areas (sea inlet/coastal features), Artificial landscape (arable land)
Land-use Agriculture, Fisheries/aquaculture, Hunting, Tourism/recreation, Urban/industrial/transport

Birds

Species	Season	Year	Pop min	Pop max	Acc	Criteria
Egretta garzetta Little Egret	B	1997	570	580	—	A4i, B1i, C2, C6
Ardea purpurea Purple Heron	B	1997	17	17	—	C6
Ciconia ciconia White Stork	B	1997	38	38	—	C6
Platalea leucorodia Spoonbill	P	1997	50	100	—	B1i, C2
Branta bernicla Brent Goose	W	1997	3,515	7,000	—	A4i, B1i, C3
Anas acuta Pintail	W	1997	316	860	—	B1i, C3
Anas acuta Pintail	P	1997	900	900	—	B1i, C3
Anas clypeata Shoveler	W	1997	76	846	—	B1i, C3
Anas clypeata Shoveler	P	1997	600	600	—	B1i, C3
Milvus migrans Black Kite	B	1997	50	80	—	C6
Circus aeruginosus Marsh Harrier	B	1997	100	200	—	C6
Circus pygargus Montagu's Harrier	B	1996	30	60	—	B3, C6
Himantopus himantopus Black-winged Stilt	B	1997	40	70	—	C6
Recurvirostra avosetta Avocet	W	1997	284	1,815	—	B2
Recurvirostra avosetta Avocet	P	1997	2,000	2,000	—	A4i, B1i, C2
Pluvialis squatarola Grey Plover	W	1997	1,755	3,720	—	B1i, C3
Pluvialis squatarola Grey Plover	P	1997	2,000	2,000	—	B1i, C3
Calidris canutus Knot	W	1997	8,860	13,500	—	A4i, B1i, B2, C3
Calidris canutus Knot	P	1997	6,000	6,000	—	B1i, C3
Calidris alpina Dunlin	W	1997	23,060	35,800	—	A4i, B1i, B2, C3
Calidris alpina Dunlin	P	1997	15,000	20,000	—	B1i, C3
Numenius arquata Curlew	W	1997	756	1,849	—	B2
Tringa totanus Redshank	P	1997	1,000	1,500	—	B1i, C3
Chlidonias niger Black Tern	B	1997	10	20	—	C6
Luscinia svecica Bluethroat	B	1997	100	200	—	C6
Lanius collurio Red-backed Shrike	B	1997	20	40	—	C6

One of the most important wintering sites for waterbirds in France. The site holds 20,000 or more waterbirds on passage and in winter, on a regular basis.

Protection status

National Partial **International** Partial

6,714 ha of IBA covered by Nature Reserve (Réserve Naturelle de Moëze, 6,714 ha). 15,233 ha of IBA covered by Special Protection Area (Ile d'Oléron, Marais de Brouage: Saint-Agnant, 22,000 ha).

Conservation issues

Threats Abandonment/reduction of land management (A), Agricultural intensification/expansion (A), Drainage (A), Infrastructure (B), Recreation/tourism (A), Unsustainable exploitation (A)

The site is threatened by drainage, cultivation and pressure from tourism (mostly on Oléron island). Pressure from hunting, including hunting at night, is also high.

Anse du Fiers d'Ars en Ré — A4i, A4iii, B1i, B2, C2, C3, C4, C6 — 106

Admin region Poitou-Charentes
Coordinates 46°13'N 1°27'W
Altitude 0–2 m **Area** 4,350 ha

Site description

A bay with saltwater lagoons, saltwater marshes and mudflats.

Habitats Wetland (mudflat/sandflat; saltmarsh; coastal lagoon), Marine areas (sea inlet/coastal features)
Land-use Fisheries/aquaculture, Hunting, Urban/industrial/transport

Birds

Species	Season	Year	Pop min	Pop max	Acc	Criteria
Podiceps auritus Slavonian Grebe	W	1997	42	119	—	B1i, C2
Egretta garzetta Little Egret	B	1997	220	220	—	C6
Egretta garzetta Little Egret	W	1997	233	830	—	B1i, C2
Branta bernicla Brent Goose	W	1997	8,500	13,420	—	A4i, B1i, C3
Circus aeruginosus Marsh Harrier	B	1991	15	20	—	C6
Himantopus himantopus Black-winged Stilt	B	1997	5	20	—	C6
Recurvirostra avosetta Avocet	B	1997	31	69	—	C6
Recurvirostra avosetta Avocet	W	1997	420	1,835	—	A4i, B1i, B2, C2
Pluvialis squatarola Grey Plover	W	1997	930	2,300	—	A4i, B1i, C3
Calidris alpina Dunlin	W	1997	12,330	17,050	—	B1i, B2, C3
Arenaria interpres Turnstone	W	1997	292	730	—	A4i, B1i, C3

The site holds 20,000 or more waterbirds in winter, on a regular basis. The site is also important for breeding waders.

Protection status

National Partial **International** High

195 ha of IBA covered by Nature Reserve (Lilleau des Niges, 195 ha). 1,300 ha of IBA covered by Hunting Reserve (Il de Ré, 1,300 ha). 4,350 ha of IBA covered by Special Protection Area (Anse du Fiers d'Ars en Ré, 5,080 ha).

Conservation issues

Threats Abandonment/reduction of land management (A), Agricultural intensification/expansion (B), Aquaculture/fisheries (A), Filling-in of wetlands (A), Recreation/tourism (A), Unsustainable exploitation (B)

Disturbance of birds by tourists and aquaculture activities are problems. The possible extension of a golf course in the area also poses a threat.

Bonne Anse — C6 — 107

Admin region Poitou-Charentes
Coordinates 45°41'N 1°12'W
Altitude 0–16 m **Area** 1,060 ha

Site description

A coastal lagoon on the northern side of the Girond estuary. It is surrounded by sand-dunes and sea on the western side and *Pinus* woodland on the eastern edge.

Habitats Wetland (mudflat/sandflat; sand-dunes/sand beach), Marine areas (open sea)
Land-use Fisheries/aquaculture, Other, Tourism/recreation

Birds

Species	Season	Year	Pop min	Pop max	Acc	Criteria
[1] ***Anthus campestris*** Tawny Pipit	B	1998	10	10	—	C6

1. One of the top 5 sites in the Poitou-Charentes region.

A wintering site for 3,000–10,000 waterbirds and an important migration stop-over site (2,000–5,000 individuals).

Protection status

National High **International** None

1,060 ha of IBA covered by Hunting Reserve (Baie de la Bonne Anse, 1,060 ha).

Conservation issues

Threats Other (B), Recreation/tourism (B)

Tourism, including boating, and open access to the site for the public, results in disturbance of wintering and breeding birds. Pollution from heavy metals from the Gironde estuary ('Other' threats) is also a threat. There is a project to create a new nature reserve to protect the lagoon and sand-dunes.

Plaine de Niort (sud-est) — A1, B2, C1, C6 — 108

Admin region Poitou-Charentes
Coordinates 46°14'N 0°21'W
Altitude 60–100 m **Area** 14,300 ha

Site description

Habitats Artificial landscape (arable land)
Land-use Agriculture, Hunting, Other, Urban/industrial/transport

Birds

Species	Season	Year	Pop min	Pop max	Acc	Criteria
Circus pygargus Montagu's Harrier	B	1991	30	30	—	C6
Tetrax tetrax Little Bustard	B	1996	35	35	—	A1, C1, C6
Tetrax tetrax Little Bustard	P	1996	1,000	1,000	—	A1, C1
Burhinus oedicnemus Stone Curlew	B	1991	120	120	—	B2, C6

Protection status

National None **International** None

Conservation issues

Threats Agricultural intensification/expansion (A), Infrastructure (A)

Agricultural intensification, mostly through irrigation and the planting of monocultures, is the main threat to *Tetrax tetrax*. The other major threat is from the erection of powerlines.

Plaine de Niort (nord-ouest) — A1, B2, B3, C1, C6 — 109

Admin region Poitou-Charentes
Coordinates 46°24'N 0°34'W
Altitude 30–70 m **Area** 12,500 ha

Site description

Habitats Artificial landscape (arable land)
Land-use Agriculture, Hunting, Other, Urban/industrial/transport

Birds

Species	Season	Year	Pop min	Pop max	Acc	Criteria
Circus pygargus Montagu's Harrier	B	1991	40	40	—	B3, C6
Tetrax tetrax Little Bustard	B	1996	100	100	—	A1, B2, C1, C6
Burhinus oedicnemus Stone Curlew	B	1991	120	120	—	B2, C6

■ Protection status
National None **International** None

■ Conservation issues

Threats Agricultural intensification/expansion (A), Infrastructure (A)

Agricultural intensification, mostly through irrigation and the planting of monocultures, are the main threats to *Tetrax tetrax*.

Plaines de Saint-Jouin de Marne et d'Assais-les-Jumeaux
A1, B2, B3, C1, C6 **110**

Admin region Poitou-Charentes
Coordinates 46°50′N 0°04′W
Altitude 60–120 m **Area** 12,400 ha

■ Site description
Agricultural land.

Habitats Artificial landscape (highly improved reseeded grassland; arable land)
Land-use Agriculture, Hunting, Other, Urban/industrial/transport

■ Birds

Species	Season	Year	Pop min	Pop max	Acc	Criteria
Circus pygargus Montagu's Harrier	B	1991	50	60	—	B3, C6
Tetrax tetrax Little Bustard	B	1996	50	50	—	A1, B2, C1, C6
Burhinus oedicnemus Stone Curlew	B	1991	60	80	—	C6

■ Protection status
National None **International** None

■ Conservation issues

Threats Agricultural intensification/expansion (A)

Agricultural intensification, especially the use of irrigation, intensive market-gardening and the planting of monocultures, are the main threats to *Tetrax tetrax* and other steppe species.

Le Pinail, Forêt de Moulière et bois du Défens
C6 **111**

Admin region Poitou-Charentes
Coordinates 46°39′N 0°29′E
Altitude 140–140 m **Area** 6,150 ha

■ Site description
Ilex heathlands, with recently afforested areas of *Pinus*, some mixed forests and small ponds. Human activities include forestry and hunting.

Habitats Forest and woodland (mixed forest), Scrub (heathland), Wetland (standing fresh water)
Land-use Forestry, Hunting, Other, Tourism/recreation, Urban/industrial/transport

■ Birds

Species	Season	Year	Pop min	Pop max	Acc	Criteria
Circus cyaneus Hen Harrier	B	1991	17	17	—	C6
Caprimulgus europaeus Nightjar	B	1991	40	80	—	C6
Dendrocopos medius Middle Spotted Woodpecker	B	1991	40	50	—	C6
Sylvia undata Dartford Warbler	B	1991	200	250	—	C6

■ Protection status
National Low **International** Partial
135 ha of IBA covered by Nature Reserve (Le Pinail, 135 ha). 3,797 ha of IBA covered by Special Protection Area (Le Pinial et Forêt de Moulière, 3,797 ha).

■ Conservation issues

Threats Afforestation, Intensified forest management (A), Recreation/tourism (B)

The site is threatened by forestry exploitation: ancient woodland is being felled and replaced with *Pinus* plantations and other exotic species.

Plateau de Bellefonds
B1i, C2 **112**

Admin region Poitou-Charentes
Coordinates 46°38′N 0°37′E
Altitude 70–140 m **Area** 2,550 ha

■ Site description
Bogs and marshes, agricultural land and wet meadows. Human activities include arable farming, tourism and recreation.

Habitats Grassland (humid grassland), Artificial landscape (arable land)
Land-use Agriculture, Hunting, Other, Tourism/recreation, Urban/industrial/transport

■ Birds

Species	Season	Year	Pop min	Pop max	Acc	Criteria
Pluvialis apricaria Golden Plover	P	1991	2,600	10,000	—	B1i, C2

Species of global conservation concern that do not meet criteria: *Tetrax tetrax* (breeding).

■ Protection status
National None **International** None

■ Conservation issues

Threats Agricultural intensification/expansion (A)

The main threat is the conversion of meadows to arable fields.

Plaines de Mirebeau et de Neuville-du-Poitou
C6 **113**

Admin region Poitou-Charentes
Coordinates 46°40′N 0°12′E
Altitude 80–100 m **Area** 10,250 ha

■ Site description
Agricultural land.

Habitats Artificial landscape (arable land)
Land-use Agriculture, Hunting, Urban/industrial/transport

■ Birds

Species	Season	Year	Pop min	Pop max	Acc	Criteria
Tetrax tetrax Little Bustard	B	1996	18	18	—	C6
Burhinus oedicnemus Stone Curlew	B	1991	30	45	—	C6
Emberiza hortulana Ortolan Bunting	B	1991	10	10	—	C6

■ Protection status
National None **International** None

■ Conservation issues

Threats Agricultural intensification/expansion (A), Infrastructure (A)

Agricultural intensification, mostly through irrigation and the production of monocultures, is the main threat to breeding birds in this site.

Region de Pressac, étang de Combourg
C6 **114**

Admin region Poitou-Charentes
Coordinates 46°07′N 0°31′W
Altitude 20–255 m **Area** 3,500 ha

■ Site description
The site includes ponds, moorlands, *Quercus* forests, marshes and aquatic vegetation, cereal fields and bocages.

Habitats Forest and woodland (broadleaved deciduous forest), Scrub (heathland), Grassland (humid grassland; mesophile grassland), Wetland (standing fresh water; water-fringe vegetation), Artificial landscape (arable land; perennial crops/orchards/groves; ruderal land)
Land-use Agriculture, Fisheries/aquaculture, Forestry, Hunting, Urban/industrial/transport

Birds

Species	Season	Year	Pop min	Pop max	Acc	Criteria
Lanius collurio Red-backed Shrike	B	1991	20	30	—	C6

The site is regionally important for breeding and wintering waterbirds. Species of global conservation concern that do not meet IBA criteria: *Aythya nyroca* (passage); *Haliaeetus albicilla* (wintering).

Protection status

National None **International** None

Conservation issues

Threats Consequences of animal/plant introductions (A), Recreation/tourism (B), Unsustainable exploitation (B)

High hunting pressure and the destruction of reedbeds (*Phragmites*) by coypu *Myocaster coypus* are the main threats.

Marais et estuaire de la Seudre — A4i, B1i, C2, C6 — 115

Admin region Poitou-Charentes
Coordinates 45°46′N 1°06′W
Altitude 0–5 m **Area** 14,800 ha

Site description

The site includes an estuary with mudflats and saltmarshes, brackish lagoons, abandoned salinas and wet meadows.

Habitats Wetland (tidal river/enclosed tidal water; mudflat/sandflat; saltmarsh; standing fresh water; standing brackish and salt water)
Land-use Agriculture, Fisheries/aquaculture, Hunting, Tourism/recreation, Urban/industrial/transport

Birds

Species	Season	Year	Pop min	Pop max	Acc	Criteria
Egretta garzetta Little Egret	B	1997	1,522	1,530	—	A4i, B1i, C2, C6
Milvus migrans Black Kite	B	1997	10	20	—	C6
Circus aeruginosus Marsh Harrier	B	1997	20	30	—	C6
Himantopus himantopus Black-winged Stilt	B	1997	20	30	—	C6
Luscinia svecica Bluethroat	B	1997	150	250	—	C6

This is an important site for waterbirds, particularly herons.

Protection status

National None **International** None

Conservation issues

Threats Abandonment/reduction in land management (A), Aquaculture/fisheries (A), Disturbance to birds (A), Infrastructure (A), Intensified forest management (C), Recreation/tourism (A), Unsustainable exploitation (A)

The abandonment of grazing and high hunting pressure, including illegal night hunting, are the main threats. The erection of powerlines and the exploitation of timber in the main heronry are also threats.

Plaine de Saint-Jean-de-Sauves — C6 — 116

Admin region Poitou-Charentes
Coordinates 46°54′N 0°04′E
Altitude 50–90 m **Area** 6,550 ha

Site description

Agricultural land.

Habitats Artificial landscape (highly improved reseeded grassland; arable land)
Land-use Agriculture, Hunting, Other, Urban/industrial/transport

Birds

Species	Season	Year	Pop min	Pop max	Acc	Criteria
Tetrax tetrax Little Bustard	B	1996	12	12	—	C6

Protection status

National None **International** None

Conservation issues

Threats Agricultural intensification/expansion (A)

Agricultural intensification, mostly through irrigation and the production of monocultures, is the main threat to breeding birds in this site.

Plaine de Villefagnan — A1, C1, C6 — 117

Admin region Poitou-Charentes
Coordinates 45°58′N 0°04′E
Altitude 100–120 m **Area** 9,050 ha

Site description

An area of agricultural land.

Habitats Artificial landscape (arable land)
Land-use Agriculture, Hunting, Other, Urban/industrial/transport

Birds

Species	Season	Year	Pop min	Pop max	Acc	Criteria
Tetrax tetrax Little Bustard	B	1996	20	30	—	A1, C1, C6
Burhinus oedicnemus Stone Curlew	B	1991	40	40	—	C6

One of the main breeding sites for *Tetrax tetrax* in western France.

Protection status

National Partial **International** None
1,800 ha of IBA covered by Hunting Reserve (1,800 ha).

Conservation issues

Threats Agricultural intensification/expansion (A)

The intensification of agricultural practices, particularly the use of irrigation and the planting of monocultures, are the main threats to breeding birds at the site.

Estuaire de la Gironde: marais de la rive nord — A1, A4i, B1i, B2, C1, C2, C3, C6 — 118

Admin region Poitou-Charentes
Coordinates 45°26′N 0°48′W
Altitude 0–19 m **Area** 2,580 ha

Site description

An estuary with mudflats, marshes, reedbeds (*Phragmites*), cliffs, wet meadows and arable fields.

Habitats Wetland (tidal river/enclosed tidal water; mudflat/sandflat; saltmarsh; standing brackish and salt water; water-fringe vegetation), Rocky areas (inland cliff), Artificial landscape (arable land)
Land-use Agriculture, Fisheries/aquaculture, Hunting, Tourism/recreation, Urban/industrial/transport

Birds

Species	Season	Year	Pop min	Pop max	Acc	Criteria
Platalea leucorodia Spoonbill	P	1991	10	100	—	B1i, C2
Milvus migrans Black Kite	B	1991	15	15	—	C6
Pluvialis apricaria Golden Plover	P	1991	1,000	10,000	—	B1i, C2
Tringa totanus Redshank	P	1991	1,000	10,000	—	A4i, B1i, C3
Alcedo atthis Kingfisher	B	1991	10	100	—	B2, C6
Luscinia svecica Bluethroat	B	1991	10	100	—	C6
Acrocephalus paludicola Aquatic Warbler	P	1991	100	500	—	A1, C1
Lanius collurio Red-backed Shrike	B	1991	10	100	—	C6

The reedbeds are important for roosting passage and wintering passerines.

■ Protection status

National Partial **International** None
1,290 ha of IBA covered by Hunting Reserve (Estuaire de la Gironde, 1,290 ha).

■ Conservation issues

Threats Agricultural intensification/expansion (A), Recreation/tourism (A), Unsustainable exploitation (A)

Water pollution from agricultural run-off is significant. Hunting pressure, including night hunting, is very high, and leads to disturbance.

Vallée de la Loire: environs de Blois — C6 — 119

Admin region Centre
Coordinates 47°37′N 1°25′E
Altitude 60–78 m **Area** 22,200 ha

■ Site description

The site includes a river, ponds, wet meadows, riverine forest, hedgerows and abandoned fields.

Habitats Forest and woodland (alluvial/very wet forest), Grassland (humid grassland), Wetland (standing fresh water; river/stream), Artificial landscape (perennial crops/orchards/groves; ruderal land)
Land-use Agriculture, Fisheries/aquaculture, Hunting, Tourism/recreation, Urban/industrial/transport, Water management

■ Birds

Species	Season	Year	Pop min	Pop max	Acc	Criteria
Sterna hirundo Common Tern	B	1991	40	55	—	C6
Sterna albifrons Little Tern	B	1991	20	36	—	C6

This site important for the tern colonies which breed on alluvial islands in the river.

■ Protection status

National Low **International** None
1.2 km of river in IBA covered by Hunting Reserve (1.2 km).

■ Conservation issues

No serious threats are known at this site.

Vallée de la Conie et Beauce centrale — B1i, B2, B3, C2, C6 — 120

Admin region Centre
Coordinates 48°14′N 1°36′E
Altitude 120–120 m **Area** 71,600 ha

■ Site description

Arable land with deciduous woodland, rivers, marshes, dry calcareous grassland and scrub.

Habitats Forest and woodland (broadleaved deciduous forest), Grassland (steppe/dry calcareous grassland), Wetland (river/stream; water-fringe vegetation), Artificial landscape (70%; arable land; perennial crops/orchards/groves; ruderal land)
Land-use Agriculture (70%), Fisheries/aquaculture, Hunting, Urban/industrial/transport

■ Birds

Species	Season	Year	Pop min	Pop max	Acc	Criteria
Pernis apivorus Honey Buzzard	B	1991	6	10	—	C6
Circus aeruginosus Marsh Harrier	B	1991	8	10	—	C6
Circus cyaneus Hen Harrier	B	1991	30	45	—	B2, C6
Circus pygargus Montagu's Harrier	B	1991	4	7	—	C6
Burhinus oedicnemus Stone Curlew	B	1991	10	20	—	C6
Pluvialis apricaria Golden Plover	P	1991	1,000	10,000	—	B1i, C2
Calandrella brachydactyla Short-toed Lark	B	1991	5	8	—	C6
Locustella luscinioides Savi's Warbler	B	1991	1	10	—	B3

The site is important for species breeding in agricultural habitats, such as *Calandrella brachydactyla*, and also wetlands.

■ Protection status

National None **International** None

■ Conservation issues

Threats Afforestation (U), Agricultural intensification/expansion (A), Extraction industry (U), Filling-in of wetlands (U), Unsustainable exploitation (A)

The degradation of reedbeds (*Phragmites*), and their conversion to arable land, is a major threat.

Petite Beauce — B1i, B2, C2, C6 — 121

Admin region Centre
Coordinates 47°44′N 1°20′E
Altitude 80–13 m **Area** 52,200 ha

■ Site description

Arable land with dry calcareous grasslands and scrub, deciduous woodlands, ponds and aquatic vegetation.

Habitats Forest and woodland (broadleaved evergreen forest), Grassland (steppe/dry calcareous grassland), Wetland (standing fresh water; river/stream; water-fringe vegetation), Artificial landscape (arable land)
Land-use Agriculture, Hunting, Urban/industrial/transport

■ Birds

Species	Season	Year	Pop min	Pop max	Acc	Criteria
Circus cyaneus Hen Harrier	B	1991	52	52	—	B2, C6
Circus pygargus Montagu's Harrier	B	1991	8	8	—	C6
Burhinus oedicnemus Stone Curlew	B	1991	10	100	—	B2, C6
Pluvialis apricaria Golden Plover	P	1991	1,000	10,000	—	B1i, C2

This is an important site for species associated with open agricultural landscapes and grassland.

■ Protection status

National Low **International** None
296 ha of IBA covered by Nature Reserve (Grand-Pierre et Vitain, 296 ha).

■ Conservation issues

Threats Unknown

Vallée de l'Yèvre — A1, C1, C6 — 122

Admin region Centre
Coordinates 47°09′N 2°14′E
Altitude 115–120 m **Area** 2,200 ha

■ Site description

The site contains rivers associated riverine vegetation, wet grasslands, *Populus* groves, arable land and scrub.

Habitats Grassland (humid grassland; mesophile grassland), Wetland (river/stream; water-fringe vegetation), Artificial landscape (highly improved reseeded grassland; arable land; perennial crops/orchards/groves; ruderal land)
Land-use Agriculture, Fisheries/aquaculture, Forestry, Hunting, Urban/industrial/transport

■ Birds

Species	Season	Year	Pop min	Pop max	Acc	Criteria
Crex crex Corncrake	B	1995	20	20	—	A1, C1, C6

This site is important for species associated with wet meadows.

■ Protection status

National None **International** None

■ Conservation issues

The main threats are the cultivation of meadows and urbanization.

Threats Agricultural intensification/expansion (A), Construction/impact of dyke/dam/barrage (A), Industrialization/urbanization (B), Infrastructure (U)

Brenne centrale

A4i, A4iii, B1i, B1iv, B2, C2, C3, C4, C5, C6 **123**

Admin region Centre
Coordinates 46°45′N 1°13′E
Altitude 100–100 m **Area** 22,000 ha

Site description

The site includes ponds, marshes, reedbeds (*Phragmites*), wet grasslands, deciduous woodland, heathland, hedgerows and bocage.

Habitats Forest and woodland (mixed forest), Scrub (heathland), Grassland (dry siliceous grassland), Wetland (standing fresh water; water-fringe vegetation), Artificial landscape (perennial crops/orchards/groves; ruderal land)
Land-use Agriculture, Fisheries/aquaculture, Hunting, Tourism/recreation, Urban/industrial/transport

Birds

Species	Season	Year	Pop min	Pop max	Acc	Criteria
Botaurus stellaris Bittern	B	1997	30	35	—	B2, C2, C6
Ixobrychus minutus Little Bittern	B	1995	8	12	—	C6
Nycticorax nycticorax Night Heron	B	1994	35	45	—	C6
Egretta garzetta Little Egret	B	1997	15	20	—	C6
Ardea purpurea Purple Heron	B	1994	150	150	—	C2, C6
Anas strepera Gadwall	W	1997	33	818	—	B1i, C3
Anas strepera Gadwall	P	1997	100	300	—	B1i, C3
Anas clypeata Shoveler	W	1997	259	1,427	—	B1i, C3
Anas clypeata Shoveler	P	1997	—	600	—	B1i, C3
Aythya ferina Pochard	W	1997	540	5,137	—	B1i, C3
Aythya ferina Pochard	P	1997	1,000	4,000	—	B1i, C3
Pernis apivorus Honey Buzzard	B	1991	8	15	—	C6
Milvus migrans Black Kite	B	1991	15	20	—	C6
Circus aeruginosus Marsh Harrier	B	1991	55	60	—	C6
Grus grus Crane	P	1997	500	5,000	—	A4i, B1i, C2
Himantopus himantopus Black-winged Stilt	B	1995	6	7	—	C6
Burhinus oedicnemus Stone Curlew	B	1991	10	15	—	C6
Chlidonias hybridus Whiskered Tern	B	1996	650	670	—	A4i, B1i, B2, C2, C6
Chlidonias hybridus Whiskered Tern	P	1996	1,000	1,000	—	A4i, B1i, C2
Chlidonias niger Black Tern	B	1996	7	8	—	C6
Caprimulgus europaeus Nightjar	B	1991	15	25	—	C6
Alcedo atthis Kingfisher	B	1991	50	75	—	C6
Lanius collurio Red-backed Shrike	B	1991	50	75	—	C6

This is an important site for breeding, passage and wintering waterbirds. The site holds 20,000 or more waterbirds on passage and in winter, on a regular basis. The site is a migratory bottleneck, where more than 3,000 *Grus grus* regularly pass. Species of global conservation concern that do not meet IBA criteria: *Aythya nyroca* (wintering); *Tetrax tetrax* (3–4 breeding pairs, passage).

Protection status

National Low **International** High
145 ha of IBA covered by Nature Reserve (Chérine, 145 ha). 112 ha of IBA covered by Hunting Reserve (112 ha). 22,000 ha of IBA covered by Ramsar Site (La Brenne, 140,000 ha). 145 ha of IBA covered by Special Protection Area (Brenne: Marais de Chérine, 145 ha).

Conservation issues

Threats Aquaculture/fisheries (A), Recreation/tourism (A), Unsustainable exploitation (B)

The conversion of wet meadows to hunting ponds and the intensification of fish farming are the main threats.

Brenne Sud

C6 **124**

Admin region Centre
Coordinates 46°40′N 1°18′E
Altitude 100–100 m **Area** 6,660 ha

Site description

An area of ponds and marshes with reedbeds (*Phragmites*), heathland and deciduous woodland.

Habitats Forest and woodland (broadleaved deciduous forest), Scrub (heathland), Grassland (dry siliceous grassland), Wetland (standing fresh water; water-fringe vegetation), Artificial landscape (perennial crops/orchards/groves; ruderal land)
Land-use Agriculture, Fisheries/aquaculture, Hunting, Tourism/recreation, Urban/industrial/transport

Birds

Species	Season	Year	Pop min	Pop max	Acc	Criteria
Ardea purpurea Purple Heron	B	1991	10	10	—	C6
Milvus migrans Black Kite	B	1991	5	10	—	C6
Circus aeruginosus Marsh Harrier	B	1991	5	12	—	C6
Lanius collurio Red-backed Shrike	B	1991	30	40	—	C6

This is an important site for wetland and reedbed breeding species.

Protection status

National None **International** High
6,600 ha of IBA covered by Ramsar Site (La Brenne, 140,000 ha).

Conservation issues

Threats Abandonment/reduction of land management (B), Agricultural intensification/expansion (C), Aquaculture/fisheries (A), Recreation/tourism (B), Unsustainable exploitation (C)

Land abandonment, destruction of aquatic vegetation and pond creation for hunting purposes are the major factors degrading the site.

Queue de Brenne: Etang des Loges, Grand et Petit Mez

C6 **125**

Admin region Centre
Coordinates 46°45′N 1°34′E
Altitude 120–130 m **Area** 770 ha

Site description

An area of ponds, marshes, reedbeds (*Phragmites*), riverine forest and wet grasslands.

Habitats Forest and woodland (alluvial/very wet forest), Wetland (standing fresh water; water-fringe vegetation)
Land-use Fisheries/aquaculture, Hunting

Birds

Species	Season	Year	Pop min	Pop max	Acc	Criteria
Nycticorax nycticorax Night Heron	B	1991	5	10	—	C6
Ardea purpurea Purple Heron	B	1991	5	15	—	C6

The site is important for breeding waterbirds.

Protection status

National None **International** High
770 ha of IBA covered by Ramsar Site (La Brenne, 140,000 ha).

Conservation issues

Threats Aquaculture/fisheries (A), Unsustainable exploitation (A)

The intensification of fisheries is the major threat.

Brenne: Forêt de la Petite-Brenne

C6 **126**

Admin region Centre
Coordinates 46°34′N 1°19′E
Altitude 135–135 m **Area** 12,560 ha

Site description

An area of deciduous woodland, ponds, heathland, wet grassland and bocage.

Habitats Forest and woodland (broadleaved deciduous forest), Scrub (heathland), Grassland (humid grassland), Wetland (standing fresh water), Artificial landscape (perennial crops/orchards/groves)
Land-use Agriculture, Fisheries/aquaculture, Forestry, Hunting, Urban/industrial/transport

Birds

Species	Season	Year	Pop min	Pop max	Acc	Criteria
Nycticorax nycticorax Night Heron	B	1991	5	5	—	C6
Pernis apivorus Honey Buzzard	B	1991	10	10	—	C6
Circus cyaneus Hen Harrier	B	1991	10	10	—	C6
Circus pygargus Montagu's Harrier	B	1991	5	5	—	C6

The site is important for forest, heathland and wetland birds.

Protection status

National None **International** High
12,560 ha of IBA covered by Ramsar Site (La Brenne, 140,000 ha).

Conservation issues

Threats Abandonment/reduction of land management (A), Agricultural intensification/expansion (A), Aquaculture/fisheries (A)

The abandonment of cattle-raising and the intensification of fishing activities are the main threats.

Brenne: Forêt de Lancosme — C6 — 127

Admin region Centre
Coordinates 46°46′N 1°22′E
Altitude 110–110 m **Area** 5,950 ha

Site description

Mixed woodland with heathland, ponds, *Populus* groves and bocage.

Habitats Forest and woodland (60%; mixed forest), Scrub (30%; heathland), Wetland (5%; standing fresh water), Artificial landscape (perennial crops/orchards/groves; urban parks/gardens)
Land-use Forestry, Hunting, Tourism/recreation, Urban/industrial/transport

Birds

Species	Season	Year	Pop min	Pop max	Acc	Criteria
Pernis apivorus Honey Buzzard	B	1991	5	8	—	C6
Caprimulgus europaeus Nightjar	B	1991	30	50	—	C6

This is an important area for forest and heathland birds.

Protection status

National None **International** High
5,950 ha of IBA covered by Ramsar Site (La Brenne, 140,000 ha).

Conservation issues

Threats Intensified forest management (A), Recreation/tourism (C), Disturbance to birds

The main threats are recreation, afforestation with conifers, and stag hunting until the end of March, which disturbs early breeding species.

Vallée de la Loire: confluence Loire-Vienne — C6 — 128

Admin region Centre
Coordinates 47°12′N 0°11′E
Altitude 28–34 m **Area** 5,080 ha

Site description

Rivers, ponds, wet grasslands, riverine forest, *Populus* groves, marshes, arable land and bocage. Agriculture and forestry are the two main activities at the site.

Habitats Forest and woodland (alluvial/very wet forest), Grassland (dry siliceous grassland; humid grassland; mesophile grassland), Wetland (standing fresh water; river/stream; water-fringe vegetation), Artificial landscape (arable land; perennial crops/orchards/groves)
Land-use Agriculture, Fisheries/aquaculture, Forestry, Hunting, Tourism/recreation, Urban/industrial/transport

Birds

Species	Season	Year	Pop min	Pop max	Acc	Criteria
Crex crex Corncrake	B	1992	10	15	—	C6

Crex crex breeds in wet meadows in the valley.

Protection status

National None **International** None

Conservation issues

Threats Abandonment/reduction of land management (A), Afforestation, Agricultural intensification/expansion (B), Extraction industry (A), Groundwater abstraction (A)

The valley is threatened by afforestation with *Populus*.

Basse vallée de L'Indre — A1, C1, C6 — 129

Admin region Centre
Coordinates 47°16′N 0°20′E
Altitude 32–42 m **Area** 1,900 ha

Site description

A river valley with wet grasslands, riverine forest, marshes, *Populus* groves, arable land and bocage.

Habitats Forest and woodland (alluvial/very wet forest), Grassland (88%; humid grassland; mesophile grassland), Wetland (6%; river/stream; water-fringe vegetation), Artificial landscape (36%; arable land; perennial crops/orchards/groves)
Land-use Agriculture (72%), Fisheries/aquaculture, Forestry (20%), Hunting (100%), Tourism/recreation

Birds

Species	Season	Year	Pop min	Pop max	Acc	Criteria
Crex crex Corncrake	B	1992	15	20	—	A1, C1, C6

Species of global conservation concern that do not meet IBA criteria: *Crex crex* (passage).

Protection status

National None **International** None

Conservation issues

Threats Abandonment/reduction of land management (A), Agricultural intensification/expansion (B), Disturbance to birds (A), Extraction industry (A), Groundwater abstraction (A)

Drainage, land abandonment and early mowing are the main threats.

Vallée de la Loire: environs de Montlouis-sur-Loire — C6 — 130

Admin region Centre
Coordinates 47°24′N 0°52′E
Altitude 50–50 m **Area** 510 ha

Site description

An area of rivers, marshes, dry grassland, riverine forest, *Populus* groves and scrubland.

Habitats Forest and woodland (alluvial/very wet forest), Grassland (dry siliceous grassland), Wetland (standing fresh water; river/stream; water-fringe vegetation), Artificial landscape (perennial crops/orchards/groves; ruderal land)
Land-use Agriculture, Fisheries/aquaculture, Forestry, Hunting, Tourism/recreation, Urban/industrial/transport

Birds

Species	Season	Year	Pop min	Pop max	Acc	Criteria
Larus melanocephalus Mediterranean Gull	B	1996	3	9	—	C6
Sterna hirundo Common Tern	B	1992	54	75	—	C6
Sterna albifrons Little Tern	B	1992	20	35	—	C6

The alluvial islands support breeding terns and gulls. Species of global conservation concern that do not meet IBA criteria: *Crex crex* (breeding, passage).

Protection status

National Partial **International** None
300 ha of IBA covered by Hunting Reserve (300 ha).

Conservation issues

Threats Afforestation (B), Intensified forest management (C), Other (C)

The regulation of the river has led to the spread of vegetation to the alluvial islets, making them less attractive for breeding terns ('Other' threats, above).

Etangs de Sologne: St Viatre, Marcilly-en-Gault et forêt de Bruadan B1i, B2, C3, C6 131

Admin region Centre
Coordinates 47°29'N 1°49'E
Altitude 90–110 m **Area** 37,500 ha

Site description

Habitats Forest and woodland (broadleaved deciduous forest), Scrub (heathland), Grassland (humid grassland), Wetland (standing fresh water; water-fringe vegetation), Artificial landscape (arable land; perennial crops/orchards/groves)
Land-use Agriculture, Fisheries/aquaculture, Forestry, Hunting, Tourism/recreation, Urban/industrial/transport

Birds

Species	Season	Year	Pop min	Pop max	Acc	Criteria
Botaurus stellaris Bittern	B	1991	2	10	—	C6
Ixobrychus minutus Little Bittern	B	1991	1	10	—	C6
Nycticorax nycticorax Night Heron	B	1991	15	20	—	C6
Ardea purpurea Purple Heron	B	1991	15	30	—	C6
Anas clypeata Shoveler	W	1997	9	512	—	B1i, C3
Pernis apivorus Honey Buzzard	B	1991	20	30	—	C6
Circaetus gallicus Short-toed Eagle	B	1991	5	7	—	C6
Circus aeruginosus Marsh Harrier	B	1997	60	100	—	C6
Circus cyaneus Hen Harrier	B	1991	10	20	—	C6
Chlidonias hybridus Whiskered Tern	B	1991	20	100	—	B2, C6
Chlidonias niger Black Tern	B	1996	10	15	—	C6
Caprimulgus europaeus Nightjar	B	1991	10	20	—	C6
Alcedo atthis Kingfisher	B	1991	30	50	—	C6
Picus canus Grey-headed Woodpecker	B	1991	10	20	—	C6
Dryocopus martius Black Woodpecker	B	1991	30	40	—	C6
Lullula arborea Woodlark	B	1991	30	40	—	C6

This is an important site for wintering and passage waterbirds.

Protection status

National Low **International** None
77 ha of IBA covered by Hunting Reserve (Etang de Malzone, 77 ha).

Conservation issues

Threats Abandonment/reduction of land management (A), Agricultural intensification/expansion (A), Aquaculture/fisheries (A), Recreation/tourism (A)

Habitats are threatened by the abandonment of stock-raising and the intensification of fishing activities, the latter involving the removal of reedbeds.

Etangs de Sologne: étang de L'Arche C6 132

Admin region Centre
Coordinates 47°20'N 1°30'E
Altitude 105–105 m **Area** 165 ha

Site description

Ponds, marshes and reedbeds (*Phragmites*) with deciduous woodlands.

Habitats Forest and woodland (30%; broadleaved deciduous forest), Wetland (70%; standing fresh water; water-fringe vegetation)
Land-use Fisheries/aquaculture, Hunting

Birds

Species	Season	Year	Pop min	Pop max	Acc	Criteria
Chlidonias hybridus Whiskered Tern	B	1991	5	20	—	C6
Chlidonias niger Black Tern	B	1996	10	15	—	C6

During winter, the site is important for wildfowl (Anatidae). Species of global conservation concern that do not meet IBA criteria: *Aythya nyroca* (passage); *Haliaeetus albicilla* (wintering); *Gallinago media* (1–7 passage).

Protection status

National None **International** None

Conservation issues

Threats Abandonment/reduction of land management (A), Agricultural intensification/expansion (U), Aquaculture/fisheries (U), Infrastructure (U), Recreation/tourism (U)

The main threats are the intensification of fisheries and the abandonment of meadows.

Vallée du Fouzon A1, C1, C6 133

Admin region Centre
Coordinates 47°16'N 1°30'E
Altitude 100–100 m **Area** 700 ha

Site description

Habitats Grassland (humid grassland), Wetland (river/stream), Artificial landscape (perennial crops/orchards/groves; ruderal land)
Land-use Agriculture, Fisheries/aquaculture, Hunting, Tourism/recreation

Birds

Species	Season	Year	Pop min	Pop max	Acc	Criteria
Crex crex Corncrake	B	1991	14	30	—	A1, C1, C6

The area is important for species breeding in wet meadows. Species of global conservation concern that do not meet IBA criteria: *Tetrax tetrax* (passage).

Protection status

National None **International** None

Conservation issues

Threats Abandonment/reduction of land management (U), Agricultural intensification/expansion (U)

The main threat is the abandonment of meadows and their conversion to arable fields.

Lac de Rillé et forêts voisines C6 134

Admin region Centre
Coordinates 47°23'N 0°08'E
Altitude 75–75 m **Area** 18,800 ha

Site description

A lake with reedbeds (*Phragmites*), wet meadows, heathland, woodland, and arable land. Human activities include agriculture and stock-raising.

Habitats Forest and woodland (native coniferous forest), Scrub (heathland), Grassland (humid grassland), Wetland (standing fresh water; water-fringe vegetation), Artificial landscape (highly improved reseeded grassland; arable land; perennial crops/orchards/groves)
Land-use Agriculture, Tourism/recreation, Urban/industrial/transport

Birds

Species	Season	Year	Pop min	Pop max	Acc	Criteria
Ciconia nigra Black Stork	B	1991	3	3	—	C6

This site is important for woodland species, particularly *Accipiter gentilis*, *Milvus migrans* and *Dryocopus martius*.

Protection status

National Low **International** None
250 ha of IBA covered by Hunting Reserve (250 ha).

Conservation issues

Threats Aquaculture/fisheries (A), Construction/impact of dyke/dam/barrage (C), Intensified forest management (A), Recreation/tourism (B), Unsustainable exploitation (C)

Breeding birds are disturbed by tourism and recreation activities, fishing and hunting (including illegal hunting).

Vallée de la Loire: Orléanais

B1i, C3, C6 **135**

Admin region Centre
Coordinates 47°42′N 2°14′E
Altitude 100–110 m **Area** 7,750 ha

Site description

The site includes rivers, ponds, natural sandy islets and riverine forest.

Habitats Wetland (100%; river/stream)
Land-use Agriculture, Fisheries/aquaculture, Hunting, Tourism/recreation, Urban/industrial/transport

Birds

Species	Season	Year	Pop min	Pop max	Acc	Criteria
Phalacrocorax carbo Cormorant	W	1997	500	2,224	—	B1i, C3
Milvus migrans Black Kite	B	1991	8	10	—	C6
Larus melanocephalus Mediterranean Gull	B	1996	14	27	—	C6
Sterna hirundo Common Tern	B	1991	192	204	—	C6
Sterna albifrons Little Tern	B	1991	116	118	—	C6

The site is important for breeding terns.

Protection status

National Partial **International** None

899 ha of IBA covered by 6 Hunting Reserves (total area 899 ha). 20 ha of IBA covered by Nature Reserve (Ile de Saint-Pryvé-Saint-Mesmin, 20 ha).

Conservation issues

Threats Construction/impact of dyke/dam/barrage (C), Disturbance to birds (B), Industrialization/urbanization (B), Infrastructure (A)

Plans to construct a sewage farm and build a bridge threaten parts of the site.

Forêt d'Orléans: massifs d'Ingrannes et de Lorris

B2, C6 **136**

Admin region Centre
Coordinates 47°54′N 2°20′E
Altitude 146–146 m **Area** 40,400 ha

Site description

An area of mixed and coniferous forests, ponds, heathland and streams.

Habitats Forest and woodland (100%; mixed forest)
Land-use Forestry (100%), Urban/industrial/transport

Birds

Species	Season	Year	Pop min	Pop max	Acc	Criteria
Hieraaetus pennatus Booted Eagle	B	1991	10	10	—	B2, C6
Pandion haliaetus Osprey	B	1996	5	5	—	C6
Caprimulgus europaeus Nightjar	B	1991	10	100	—	C6
Dryocopus martius Black Woodpecker	B	1991	10	100	—	C6
Dendrocopos medius Middle Spotted Woodpecker	B	1991	10	100	—	C6

The site is important for breeding raptors, and is the only regular breeding site in continental France for *Pandion haliaetus*. Species of global conservation concern that do not meet IBA criteria: *Haliaeetus albicilla* (wintering).

Protection status

National None **International** None

Conservation issues

Threats Disturbance to birds (B), Intensified forest management (B), Recreation/tourism (B), Unsustainable exploitation (C)

The main threats are recreation and forest management activities, which extend into the breeding season, causing disturbance. However, the forests are managed to protect *Pandion haliaetus* nests.

Mars sur Allier

C6 **137**

Admin region Centre
Coordinates 46°52′N 3°03′E
Altitude 175–175 m **Area** 770 ha

Site description

The site includes rivers, sandy islets, marshes, riverine forest and arable land.

Habitats Forest and woodland (alluvial/very wet forest), Wetland (river/stream; water-fringe vegetation), Artificial landscape (arable land)
Land-use Fisheries/aquaculture, Tourism/recreation

Birds

Species	Season	Year	Pop min	Pop max	Acc	Criteria
Nycticorax nycticorax Night Heron	B	1991	130	130	—	C6
Burhinus oedicnemus Stone Curlew	B	1991	11	11	—	C6

Heron colonies are present along the river. *Burhinus oedicnemus* and terns breed on the natural sandy islands.

Protection status

National Partial **International** Partial

150 ha of IBA covered by Hunting Reserve (150 ha). 413 ha of IBA covered by Special Protection Area (Mars sur Allier, 413 ha).

Conservation issues

Threats Agricultural intensification/expansion (U), Aquaculture/fisheries (C), Construction/impact of dyke/dam/barrage (A), Disturbance to birds (C), Industrialization/urbanization (U), Recreation/tourism (C)

Fishing and recreational activities are threats.

Vallée de la Loire: La Charité-sur-Loire

C6 **138**

Admin region Centre
Coordinates 47°14′N 2°58′E
Altitude 150–150 m **Area** 1,400 ha

Site description

The River Loire is the last unaltered and free-flowing large river in Europe. This site encompasses part of the river, sandy islets, marshes, wet grasslands and riverine forest. Human activities include arable farming and fishing.

Habitats Forest and woodland (alluvial/very wet forest), Grassland (humid grassland), Wetland (river/stream; water-fringe vegetation)
Land-use Agriculture, Fisheries/aquaculture, Tourism/recreation

Birds

Species	Season	Year	Pop min	Pop max	Acc	Criteria
Burhinus oedicnemus Stone Curlew	B	1990	15	15	—	C6
Sterna hirundo Common Tern	B	1990	40	40	—	C6
Sterna albifrons Little Tern	B	1990	30	30	—	C6

The site is important for species which breed on the alluvial islets in the river.

Protection status

National High **International** Partial

925 ha of IBA covered by Hunting Reserve (La Charité sur Loire, 18 km river, 925 ha). 1,400 ha of IBA covered by Nature Reserve (Val de Loire, 1,900 ha). 925 ha of IBA covered by Special Protection Area (La Charité sur Loire, 925 ha).

Conservation issues

Threats Agricultural intensification/expansion (U), Disturbance to birds (U), Industrialization/urbanization (U)

This site is threatened by urban and agricultural pollution.

Forêt de Jugny — C6 — 139

Admin region Bourgogne
Coordinates 47°34′N 4°45′E
Altitude 330–520 m **Area** 6,800 ha

Site description

The site contains deciduous forests, conifer plantations, meadows and cultivated fields, and rivers.

Habitats Forest and woodland (90%; broadleaved deciduous forest), Grassland (5%; steppe/dry calcareous grassland), Wetland (river/stream; fen/transition mire/spring), Artificial landscape (5%; highly improved reseeded grassland; arable land)
Land-use Agriculture, Fisheries/aquaculture, Forestry (90%), Hunting, Urban/industrial/transport

Birds

Species	Season	Year	Pop min	Pop max	Acc	Criteria
Ciconia nigra Black Stork	B	1991	1	10	—	C6
Aegolius funereus Tengmalm's Owl	B	1991	10	100	—	C6
Picus canus Grey-headed Woodpecker	B	1991	10	30	—	C6
Dendrocopos medius Middle Spotted Woodpecker	B	1991	40	50	—	C6

This site holds six species of woodpecker.

Protection status

National None **International** None

Conservation issues

No serious threats are known at this site.

Forêt de Citeaux et environs — B3, C2, C6 — 140

Admin region Bourgogne
Coordinates 47°04′N 5°06′E
Altitude 220–230 m **Area** 13,100 ha

Site description

The site includes deciduous forest, meadows, marshes, and ponds.

Habitats Forest and woodland (95%; broadleaved deciduous forest), Wetland (standing fresh water), Artificial landscape (highly improved reseeded grassland)
Land-use Agriculture, Forestry, Hunting, Tourism/recreation, Urban/industrial/transport

Birds

Species	Season	Year	Pop min	Pop max	Acc	Criteria
Circus cyaneus Hen Harrier	B	1991	1	10	—	C6
Circus pygargus Montagu's Harrier	B	1991	1	10	—	C6
Picus canus Grey-headed Woodpecker	B	1991	40	50	—	C6
Dendrocopos medius Middle Spotted Woodpecker	B	1991	100	500	—	B3, C2, C6

Six species of woodpecker breed.

Protection status

National Partial **International** None
4,200 ha of IBA covered by Biological State Reserve (4,200 ha).

Conservation issues

No serious threats are known at this site. It is well protected.

Forêt de Châtillon et environs — B2, C6 — 141

Admin region Bourgogne
Coordinates 47°46′N 4°44′E
Altitude 270–430 m **Area** 21,900 ha

Site description

A large forest area (deciduous forest and *Pinus* plantations), with dry meadows, peatbogs, crops, rivers and small ponds.

Habitats Forest and woodland (90%; broadleaved deciduous forest), Grassland (steppe/dry calcareous grassland), Wetland (river/stream; fen/transition mire/spring), Artificial landscape (arable land)
Land-use Agriculture, Fisheries/aquaculture, Forestry (90%), Hunting, Urban/industrial/transport

Birds

Species	Season	Year	Pop min	Pop max	Acc	Criteria
Ciconia nigra Black Stork	B	1991	1	10	—	C6
Aegolius funereus Tengmalm's Owl	B	1991	10	100	—	C6
Picus canus Grey-headed Woodpecker	B	1991	40	150	—	B2, C6
Dryocopus martius Black Woodpecker	B	1991	10	40	—	C6
Dendrocopos medius Middle Spotted Woodpecker	B	1991	40	50	—	C6

The site is of special interest for woodpeckers, with six species breeding.

Protection status

National None **International** None

Conservation issues

Threats Afforestation (U), Agricultural intensification/expansion (U), Infrastructure (U)

Disturbance from newly constructed roads is a threat; nesting *Ciconia nigra* recently failed to breed as a result of this.

Arrière Côte de Dijon et de Beaune — B2, C6 — 142

Admin region Bourgogne
Coordinates 47°08′N 4°48′E
Altitude 222–641 m **Area** 63,600 ha

Site description

An area of deciduous, mixed and coniferous forests, heathlands, recent plantations, calcareous cliffs, rivers, vineyards and other crops.

Habitats Forest and woodland (70%; broadleaved deciduous forest; mixed forest), Scrub (10%; heathland; sclerophyllous scrub/garrigue/maquis), Grassland (steppe/dry calcareous grassland), Wetland (river/stream), Rocky areas (scree/boulders; inland cliff; inland sand-dunes; caves), Artificial landscape (10%; arable land; perennial crops/orchards/groves)
Land-use Agriculture, Fisheries/aquaculture (95%), Forestry (80%), Hunting, Tourism/recreation (100%), Urban/industrial/transport

Birds

Species	Season	Year	Pop min	Pop max	Acc	Criteria
Circaetus gallicus Short-toed Eagle	B	1992	1	5	—	C6
Circus cyaneus Hen Harrier	B	1992	10	20	—	C6
Falco peregrinus Peregrine	B	1992	5	7	—	C6
Aegolius funereus Tengmalm's Owl	B	1992	5	50	—	C6
Picus canus Grey-headed Woodpecker	B	1992	40	50	—	C6
Dryocopus martius Black Woodpecker	B	1992	10	40	—	C6
Lullula arborea Woodlark	B	1992	100	500	—	B2, C6
Sylvia hortensis Orphean Warbler	B	1992	1	10	—	B2
Lanius collurio Red-backed Shrike	B	1992	100	500	—	C6

This area is important for forest and heathland species.

Protection status

National None **International** None

Conservation issues

Threats Abandonment/reduction of land management (U)

The abandonment of cattle-grazing has led to a decrease in the biological interest of grassland areas. A nature reserve is under designation.

Etang des Landes — B1i, B1iv, C2, C5, C6 — 143

Admin region Limousin
Coordinates 46°10′N 2°18′E
Altitude 378–380 m **Area** 1,800 ha

Site description

Lakes, marshes, and wet meadows, surrounded by woodland, hedges and cultivated land, and plains. Human activities include agriculture, stock-raising, fishing and hunting.

Habitats Forest and woodland (mixed forest; alluvial/very wet forest), Grassland (humid grassland), Wetland (standing fresh water; water-fringe vegetation), Artificial landscape (arable land)
Land-use Agriculture, Fisheries/aquaculture, Hunting, Tourism/recreation, Urban/industrial/transport

Birds

Species	Season	Year	Pop min	Pop max	Acc	Criteria
Ardea purpurea Purple Heron	B	1989	30	30	—	C6
Grus grus Crane	P	1989	1,000	10,000	—	B1i, C2

This wetland area is important for breeding waterbirds and passage storks, raptors and cranes. The site is a migratory bottleneck, where more than 3,000 *Grus grus* regularly pass.

Protection status

National None **International** None

Conservation issues

Threats Aquaculture/fisheries

Inadequate management is the main threat. Late spring fishing resulted in the failure of breeding waterbirds in 1999.

Plateaux de Millevaches et de Gentioux — B1i, B1iv, B2, C2, C5, C6 — 144

Admin region Limousin
Coordinates 45°42′N 2°30′E
Altitude 650–980 m **Area** 64,900 ha

Site description

The site includes heathlands, meadows, ponds and marshes, deciduous and coniferous woodlands, bogs, hedgerows and bocages.

Habitats Forest and woodland (broadleaved deciduous forest; native coniferous forest; mixed forest; alluvial/very wet forest), Scrub (heathland), Grassland (dry siliceous grassland; humid grassland), Wetland (5%; standing fresh water; standing brackish and salt water; river/stream; raised bog), Artificial landscape (perennial crops/orchards/groves)
Land-use Agriculture, Fisheries/aquaculture, Forestry, Hunting, Tourism/recreation, Urban/industrial/transport, Water management

Birds

Species	Season	Year	Pop min	Pop max	Acc	Criteria
Milvus migrans Black Kite	B	1991	1	10	—	C6
Milvus milvus Red Kite	B	1991	1	10	—	C6
Circaetus gallicus Short-toed Eagle	B	1991	3	10	—	C6
Circus cyaneus Hen Harrier	B	1991	1	10	—	C6
Circus pygargus Montagu's Harrier	B	1991	5	10	—	C6
Grus grus Crane	P	1991	1,000	10,000	—	B1i, C2
Caprimulgus europaeus Nightjar	B	1991	10	100	—	C6
Dryocopus martius Black Woodpecker	B	1991	10	100	—	C6
Lullula arborea Woodlark	B	1991	100	500	—	B2, C6
Lanius collurio Red-backed Shrike	B	1991	100	500	—	C6

The site is important for forest and heathland birds and migrating raptors. The site is a migratory bottleneck where more than 3,000 *Grus grus* regularly pass.

Protection status

National None **International** None

Conservation issues

Threats Abandonment/reduction of land management (A), Intensified forest management (B)

The abandonment of grazing is leading to the expansion of forests. In particular, this threatens *Circus* spp. and *Lullula arborea*.

Gorges de la Dordogne — B1iv, B2, C5, C6 — 145

Admin region Limousin
Coordinates 45°22′N 2°24′E
Altitude 270–660 m **Area** 45,700 ha

Site description

A fast-flowing river with pools, marshes, woodland, cultivated and uncultivated land, heathland, cliffs, and reservoirs. Human activities include mining and quarrying.

Habitats Forest and woodland (85%; broadleaved deciduous forest; native coniferous forest; mixed forest), Scrub (heathland), Wetland (standing fresh water; river/stream; water-fringe vegetation), Rocky areas (inland cliff), Artificial landscape (arable land; perennial crops/orchards/groves)
Land-use Agriculture, Fisheries/aquaculture, Forestry, Hunting, Tourism/recreation, Urban/industrial/transport

Birds

Species	Season	Year	Pop min	Pop max	Acc	Criteria
Pernis apivorus Honey Buzzard	B	1991	50	75	—	C6
Milvus migrans Black Kite	B	1991	50	50	—	C6
Milvus milvus Red Kite	B	1991	20	20	—	C6
Circaetus gallicus Short-toed Eagle	B	1991	7	7	—	C6
Hieraaetus pennatus Booted Eagle	B	1991	14	14	—	B2, C6
Falco peregrinus Peregrine	B	1991	10	10	—	B2, C6
Lanius collurio Red-backed Shrike	B	1991	10	100	—	C6

The site is important for passage and breeding raptors. The site is a migratory bottleneck, where more than 3,000 raptors pass on a regular basis (including *Pernis apivorus* 10,000; *Milvus migrans* 1,000; *Milvus milvus* 1,000).

Protection status

National None **International** None

Conservation issues

Threats Intensified forest management (B), Recreation/tourism (A)

The site is threatened by forestry exploitation and the replacement of natural woodland with exotic species. Sailing causes disturbance to birds.

Val d'Allier Bourbonnais — A4i, B1i, B2, C2, C6 — 146

Admin region Auvergne
Coordinates 46°28′N 3°24′E
Altitude 200–200 m **Area** 17,900 ha

Site description

The River Allier with alluvial sand beaches and islets, marshes, ponds, riverine forest and arable fields.

Habitats Forest and woodland (alluvial/very wet forest), Scrub (heathland), Wetland (standing fresh water; river/stream), Artificial landscape (arable land)
Land-use Agriculture, Fisheries/aquaculture, Hunting, Tourism/recreation, Urban/industrial/transport

Birds

Species	Season	Year	Pop min	Pop max	Acc	Criteria
Nycticorax nycticorax Night Heron	B	1991	210	250	—	C6
Egretta garzetta Little Egret	B	1991	20	35	—	C6
Cygnus columbianus Bewick's Swan	W	1991	—	15	—	B2
Milvus migrans Black Kite	B	1991	20	40	—	C6
Circus cyaneus Hen Harrier	B	1991	5	10	—	C6
Circus pygargus Montagu's Harrier	B	1991	5	10	—	C6
Grus grus Crane	P	1997	1,000	10,000	—	A4i, B1i, C2
Burhinus oedicnemus Stone Curlew	B	1991	50	200	—	B2, C6
Sterna hirundo Common Tern	B	1991	50	100	—	C6
Sterna albifrons Little Tern	B	1991	20	40	—	C6
Alcedo atthis Kingfisher	B	1991	15	30	—	C6

The site is important for wintering waterfowl. *Sterna hirundo* and *S. albifrons* breed on natural gravel islets. Species of global conservation concern that do not meet IBA criteria; *Aythya nyroca* (passage, wintering); *Haliaeetus albicilla* (passage, wintering); *Tetrax tetrax* (passage).

Protection status

National Partial **International** Partial

1,450 ha of IBA covered by Nature Reserve (Val d'Allier, 1,450 ha). 1,300 ha of IBA covered by Special Protection Area (Val d'Allier, 1,300 ha).

Conservation issues

Threats Aquaculture/fisheries (B), Construction/impact of dyke/dam/barrage (A), Extraction industry (A), Infrastructure (A), Recreation/tourism (A)

The area is threatened by proposed dams, the expansion of recreational activities and new gravel-pits.

Haut Val d'Allier — B2, B3, C6 — 147

Admin region Auvergne
Coordinates 45°01'N 3°34'E
Altitude 460–1,200 m **Area** 52,700 ha

Site description

Habitats Forest and woodland (71%; broadleaved deciduous forest; native coniferous forest; mixed forest; alluvial/very wet forest), Scrub (5%; heathland), Grassland (15%; dry siliceous grassland), Wetland (standing fresh water; river/stream), Rocky areas (5%; inland cliff), Artificial landscape (13%; arable land; other urban/industrial areas)

Land-use Agriculture (30%), Fisheries/aquaculture (5%), Forestry, Hunting (90%), Tourism/recreation, Urban/industrial/transport (11%), Water management (1%)

Birds

Species	Season	Year	Pop min	Pop max	Acc	Criteria
Pernis apivorus Honey Buzzard	B	1991	50	100	—	B3, C6
Milvus migrans Black Kite	B	1991	40	100	—	B2, C6
Milvus milvus Red Kite	B	1991	50	100	—	B3, C6
Circaetus gallicus Short-toed Eagle	B	1991	15	50	—	B2, C6
Circus cyaneus Hen Harrier	B	1991	10	20	—	C6
Circus pygargus Montagu's Harrier	B	1991	5	10	—	C6
Hieraaetus pennatus Booted Eagle	B	1991	4	6	—	B2, C6
Bubo bubo Eagle Owl	B	1991	12	15	—	C6
Caprimulgus europaeus Nightjar	B	1991	100	200	—	B2, C6
Alcedo atthis Kingfisher	B	1991	10	20	—	C6
Dryocopus martius Black Woodpecker	B	1991	40	100	—	C6
Lullula arborea Woodlark	B	1991	100	500	—	B2, C6
Lanius collurio Red-backed Shrike	B	1991	100	500	—	C6

The site is mainly important for forest birds, including raptors.

Protection status

National None **International** None

Conservation issues

Water pollution and the development of mass tourism are reported threats to the site.

Threats Agricultural intensification/expansion (B), Construction/impact of dyke/dam/barrage (B), Drainage (B), Extraction industry (C), Industrialization/urbanization (C), Infrastructure (B), Intensified forest management (A), Recreation/tourism (A), Unsustainable exploitation (C)

Gorges de la Sioule — B2, B3, C6 — 148

Admin region Auvergne
Coordinates 46°01'N 2°51'E
Altitude 320–512 m **Area** 25,250 ha

Site description

A mosaic of habitats including a river, cliffs, heathlands, broadleaved and coniferous forests, meadows and hedgerows.

Habitats Forest and woodland (broadleaved deciduous forest), Scrub (heathland), Wetland (river/stream), Rocky areas (inland cliff)

Land-use Agriculture, Fisheries/aquaculture, Hunting, Tourism/recreation, Urban/industrial/transport

Birds

Species	Season	Year	Pop min	Pop max	Acc	Criteria
Pernis apivorus Honey Buzzard	B	1991	10	30	—	C6
Milvus milvus Red Kite	B	1991	20	30	—	B3, C6
Circus cyaneus Hen Harrier	B	1991	8	10	—	C6
Circus pygargus Montagu's Harrier	B	1991	5	10	—	C6
Hieraaetus pennatus Booted Eagle	B	1991	3	3	—	B2, C6

The cliff and forest areas are important for breeding raptors and passage storks.

Protection status

National None **International** None

Conservation issues

Threats Recreation/tourism

Climbers disturb raptors breeding on the cliffs.

Monts et Plomb du Cantal — B1iv, C5 — 149

Admin region Auvergne
Coordinates 45°05'N 2°44'E
Altitude 850–1,855 m **Area** 6,450 ha

Site description

Fast-flowing rivers and streams, peatbogs, deciduous and conifer woodlands, uncultivated land, mountains and cliffs. Human activities include forestry, cattle-grazing, hunting and recreation.

Habitats Forest and woodland (broadleaved deciduous forest; native coniferous forest; mixed forest), Scrub (heathland), Wetland (river/stream; raised bog), Rocky areas (scree/boulders; inland cliff), Artificial landscape)

Land-use Agriculture, Fisheries/aquaculture, Forestry, Hunting, Tourism/recreation, Urban/industrial/transport

Birds

The site is important for passage migrants, particularly raptors, *Ciconia ciconia, C. nigra, Columba palumbus* and *Turdus torquatus.* The site is a migratory bottleneck, where more than 3,000 raptors regularly pass (including *Pernis apivorus* 2,000; *Milvus migrans* 2,000; *Milvus milvus* 1,800).

Protection status

National Low **International** Partial

1,500 ha of IBA covered by Special Protection Area (Monts et Plombs du Cantal, 1,500 ha).

Conservation issues

Threats Disturbance to birds (A), Infrastructure (A), Recreation/tourism (B), Unsustainable exploitation (C)

The site is threatened by plans to expand the existing ski-station and other development projects. Disturbance from trail bikes, four-wheel-drive vehicles and hunting activities are also major threats.

Planèze de Saint-Flour — C6 — 150

Admin region Auvergne
Coordinates 45°03′N 2°59′E
Altitude 930–1,300 m **Area** 24,150 ha

Site description

The site includes marshes, wet meadows, hedges and cultivated land, grasslands and uncultivated land. There are villages and major roads within the area.

Habitats Forest and woodland (broadleaved deciduous forest), Grassland (humid grassland), Wetland (river/stream; water-fringe vegetation), Artificial landscape (highly improved reseeded grassland; arable land; perennial crops/orchards/groves)
Land-use Agriculture, Fisheries/aquaculture, Forestry, Hunting, Tourism/recreation, Urban/industrial/transport

Birds

Species	Season	Year	Pop min	Pop max	Acc	Criteria
Porzana porzana Spotted Crake	B	1991	1	4	—	C6
Asio flammeus Short-eared Owl	B	1991	—	2	—	C6
Lanius collurio Red-backed Shrike	B	1991	100	500	—	C6

Protection status

National None **International** None

Conservation issues

Threats Extraction industry (U), Unsustainable exploitation

The site is threatened by fluctuating water-levels, hedge removal and high hunting pressure. The planned opening of a new quarry threatens breeding *Porzana porzana*.

Montagne de la Serre — B1i, B1iv, C2, C5, C6 — 151

Admin region Auvergne
Coordinates 45°40′N 3°04′E
Altitude 480–990 m **Area** 2,600 ha

Site description

The site has broadleaved and coniferous forests, meadows, heathlands and cliffs.

Habitats Forest and woodland (40%; broadleaved deciduous forest; native coniferous forest; mixed forest), Scrub (20%; heathland), Grassland (15%; steppe/dry calcareous grassland; dry siliceous grassland; mesophile grassland), Wetland (water-fringe vegetation), Rocky areas (inland cliff), Artificial landscape (23%; highly improved reseeded grassland; arable land; perennial crops/orchards/groves; ruderal land)
Land-use Agriculture (20%), Forestry (10%), Hunting (95%), Urban/industrial/transport

Birds

Species	Season	Year	Pop min	Pop max	Acc	Criteria
Grus grus Crane	P	1991	50	750	—	B1i, C2
Lullula arborea Woodlark	B	1991	10	100	—	C6
Lanius collurio Red-backed Shrike	B	1991	10	100	—	C6

This is an important site for migratory species. The site is a migratory bottleneck, where more than 3,000 raptors regularly pass (including *Pernis apivorus* 2,000; *Milvus migrans* 3,000; *Milvus milvus* 2,000).

Protection status

National None **International** None

Conservation issues

Threats Abandonment/reduction of land management (A), Agricultural intensification/expansion (A)

Sologne Bourbonnaise — A4i, B1i, B2, C2, C6 — 152

Admin region Auvergne
Coordinates 46°36′N 3°36′E
Altitude 245–300 m **Area** 22,000 ha

Site description

The area includes broadleaved forests, meadows, hedgerows, bocages, ponds, rivers and marshes.

Habitats Forest and woodland (36%; broadleaved deciduous forest; mixed forest; alluvial/very wet forest), Scrub (5%; heathland), Wetland (6%; standing fresh water; river/stream), Artificial landscape (49%; highly improved reseeded grassland; arable land; perennial crops/orchards/groves; urban parks/gardens; other urban/industrial areas; ruderal land)
Land-use Agriculture (55%), Fisheries/aquaculture (5%), Forestry (30%), Hunting (60%), Not utilized (5%), Tourism/recreation (6%), Urban/industrial/transport (6%)

Birds

Species	Season	Year	Pop min	Pop max	Acc	Criteria
Pernis apivorus Honey Buzzard	B	1991	10	20	—	C6
Milvus milvus Red Kite	B	1991	5	10	—	C6
Circus cyaneus Hen Harrier	B	1991	5	10	—	C6
Circus pygargus Montagu's Harrier	B	1991	5	10	—	C6
Hieraaetus pennatus Booted Eagle	B	1991	1	5	—	B2, C6
Grus grus Crane	P	1991	100	10,000	—	A4i, B1i, C2
Burhinus oedicnemus Stone Curlew	B	1991	20	50	—	C6
Alcedo atthis Kingfisher	B	1991	10	20	—	C6
Lullula arborea Woodlark	B	1991	100	500	—	B2, C6
Lanius collurio Red-backed Shrike	B	1991	100	500	—	C6

The site is important for breeding and wintering waterbirds. Species of global conservation concern that do not meet IBA criteria: *Aythya nyroca* (passage, wintering); *Haliaeetus albicilla* (passage, wintering); *Aquila clanga* (wintering); *Acrocephalus paludicola* (passage).

Protection status

National None **International** None

Conservation issues

Threats Agricultural intensification/expansion (A), Aquaculture/fisheries (A), Disturbance to birds (A), Recreation/tourism (A), Unsustainable exploitation (B)

The major reported threats include agricultural pollution (pesticides and fertilizers) and disturbance caused by hunting and fishing.

Vallée de la Loire: Pit majeur d'Iguerande à Décize — A1, C1, C6 — 153

Admin region Auvergne
Coordinates 46°30′N 3°47′E
Altitude 222–238 m **Area** 29,800 ha

Site description

Habitats Forest and woodland (alluvial/very wet forest), Grassland (humid grassland), Wetland (shingle/stony beach; standing fresh water; river/stream), Artificial landscape (perennial crops/orchards/groves)
Land-use Agriculture, Fisheries/aquaculture, Hunting, Urban/industrial/transport

Birds

Species	Season	Year	Pop min	Pop max	Acc	Criteria
Nycticorax nycticorax Night Heron	B	1991	100	100	—	C6
Egretta garzetta Little Egret	B	1991	1	10	—	C6
Tetrax tetrax Little Bustard	P	1991	25	100	—	A1, C1
Sterna hirundo Common Tern	B	1991	50	100	—	C6
Sterna albifrons Little Tern	B	1991	1	10	—	C6

The site is important for breeding herons and terns.

Protection status

National Partial **International** None
Part of IBA covered by Hunting Reserve (Les Grands Verziaux).

Conservation issues

Threats Extraction industry (A), Drainage

The main threats are the expansion of gravel extraction and embankment projects.

Vallée de la Loire: gorges de la Loire — B2, B3, C6 — 154

Admin region Auvergne
Coordinates 45°16′N 4°34′E
Altitude 430–1,169 m **Area** 63,000 ha

Site description

A rocky area with cliffs, gorges and forests.

Habitats Forest and woodland (broadleaved deciduous forest; native coniferous forest; mixed forest; alluvial/very wet forest), Scrub (heathland), Grassland (dry siliceous grassland; humid grassland), Wetland (standing fresh water; river/stream; water-fringe vegetation), Rocky areas (inland cliff), Artificial landscape (highly improved reseeded grassland; perennial crops/orchards/groves)
Land-use Agriculture (25%), Fisheries/aquaculture, Forestry, Hunting, Tourism/recreation, Urban/industrial/transport, Water management

Birds

Species	Season	Year	Pop min	Pop max	Acc	Criteria
Pernis apivorus Honey Buzzard	B	1991	25	50	—	C6
Milvus migrans Black Kite	B	1991	50	100	—	B2, C6
Milvus milvus Red Kite	B	1991	40	60	—	B3, C6
Circaetus gallicus Short-toed Eagle	B	1991	15	20	—	B2, C6
Circus cyaneus Hen Harrier	B	1991	10	15	—	C6
Circus pygargus Montagu's Harrier	B	1991	5	10	—	C6
Bubo bubo Eagle Owl	B	1991	22	22	—	B2, C6
Caprimulgus europaeus Nightjar	B	1991	100	200	—	B2, C6
Dryocopus martius Black Woodpecker	B	1991	40	50	—	C6
Lullula arborea Woodlark	B	1991	100	500	—	B2, C6
Lanius collurio Red-backed Shrike	B	1991	100	500	—	C6
Emberiza hortulana Ortolan Bunting	B	1991	10	20	—	C6

An important site for breeding raptors. Species of global conservation concern that do not meet IBA criteria: *Crex crex* (breeding).

Protection status

National Low **International** None
108 ha of IBA covered by Hunting Reserve (Bas-en-Basset, 108 ha).

Conservation issues

Threats Agricultural intensification/expansion (B), Aquaculture/fisheries (A), Burning of vegetation (C), Disturbance to birds (B), Industrialization/urbanization (A), Infrastructure (C), Recreation/tourism (A), Selective logging/cutting (A)

Hedgerow destruction, water pollution, urbanization and recreation are the main threats.

Gorges de la Truyère — B2, B3, C6 — 155

Admin region Auvergne
Coordinates 44°54′N 2°54′E
Altitude 235–960 m **Area** 21,500 ha

Site description

Habitats Forest and woodland (broadleaved deciduous forest; native coniferous forest; mixed forest), Scrub (heathland), Wetland (standing fresh water; river/stream), Rocky areas (inland cliff), Artificial landscape (arable land; perennial crops/orchards/groves)
Land-use Agriculture, Fisheries/aquaculture, Forestry, Hunting, Other, Tourism/recreation (67%), Urban/industrial/transport

Birds

Species	Season	Year	Pop min	Pop max	Acc	Criteria
Pernis apivorus Honey Buzzard	B	1991	30	40	—	C6
Milvus migrans Black Kite	B	1991	100	100	—	B2, C6
Milvus milvus Red Kite	B	1991	40	50	—	B3, C6
Circaetus gallicus Short-toed Eagle	B	1991	13	15	—	B2, C6
Circus cyaneus Hen Harrier	B	1991	7	12	—	C6
Hieraaetus pennatus Booted Eagle	B	1991	15	20	—	B2, C6
Falco peregrinus Peregrine	B	1991	3	4	—	C6
Dryocopus martius Black Woodpecker	B	1991	50	50	—	C6

This site is important for breeding raptors.

Protection status

National None **International** None

Conservation issues

Threats Intensified forest management (U), Recreation/tourism (B)

The site is threatened by forestry exploitation and the replacement of natural woodland with exotic species. There is also heavy hunting pressure and disturbance from wind-surfing and sailing.

Les Couzes Nord — B2, C6 — 156

Admin region Auvergne
Coordinates 45°32′N 3°02′E
Altitude 600–1,200 m **Area** 19,500 ha

Site description

Valley and plateaus with mixed forests, cliffs and rocky areas, meadows, crops, hedgerows and bushy heathlands.

Habitats Forest and woodland (mixed forest), Grassland (dry siliceous grassland), Rocky areas (scree/boulders; inland cliff), Artificial landscape (arable land; perennial crops/orchards/groves)
Land-use Agriculture, Fisheries/aquaculture, Hunting

Birds

Species	Season	Year	Pop min	Pop max	Acc	Criteria
Circaetus gallicus Short-toed Eagle	B	1992	3	4	—	C6
Circus pygargus Montagu's Harrier	B	1992	5	10	—	C6
Crex crex Corncrake	B	1992	2	3	—	C6
Bubo bubo Eagle Owl	B	1992	15	20	—	B2, C6
Caprimulgus europaeus Nightjar	B	1992	50	100	—	C6
Emberiza hortulana Ortolan Bunting	B	1992	5	10	—	C6

This is an important area for migrating storks and raptors, but no quantitative data are available.

Protection status

National None **International** None

Conservation issues

Threats Agricultural intensification/expansion (U), Disturbance to birds (U)

The main threats are disturbance, due to hunting, and hedgerow destruction caused by changes in land tenure.

Les Couzes Sud — B2, C6 — 157

Admin region Auvergne
Coordinates 45°24′N 3°08′E
Altitude 400–1,100 m **Area** 16,400 ha

Site description

Habitats Forest and woodland (mixed forest), Grassland (dry siliceous grassland), Rocky areas (scree/boulders; inland cliff)
Land-use Agriculture, Fisheries/aquaculture, Hunting, Urban/industrial/transport, Tourism/recreation

Birds

Species	Season	Year	Pop min	Pop max	Acc	Criteria
Circaetus gallicus Short-toed Eagle	B	1992	5	5	—	C6
Bubo bubo Eagle Owl	B	1992	15	15	—	B2, C6

The area is important for breeding raptors.

Protection status
National None **International** None

Conservation issues

Threats Abandonment/reduction of land management (U)

The abandonment of grazing is the main threat, leading to a decrease in habitat quality.

Val d'Allier: Saint-Yorre-Joze B1i, B1iv, C2, C5, C6 158

Admin region Auvergne
Coordinates 45°51'N 3°18'E
Altitude 260–320 m **Area** 5,500 ha

Site description
A natural river valley encompassing a river with oxbow lakes, alluvial beaches, riverine forest, natural ponds, wet meadows and marshes, dry meadows and bushy heathlands.

Habitats Forest and woodland (alluvial/very wet forest), Grassland (humid grassland), Wetland (standing fresh water; river/stream)
Land-use Agriculture, Fisheries/aquaculture, Hunting, Tourism/recreation, Urban/industrial/transport

Birds

Species	Season	Year	Pop min	Pop max	Acc	Criteria
Nycticorax nycticorax Night Heron	B	1992	100	150	—	C6
Egretta garzetta Little Egret	B	1992	5	6	—	C6
Milvus migrans Black Kite	B	1992	30	50	—	C6
Grus grus Crane	P	1992	500	2,000	—	B1i, C2
Burhinus oedicnemus Stone Curlew	B	1992	20	30	—	C6
Sterna hirundo Common Tern	B	1992	—	5	—	C6
Lullula arborea Woodlark	B	1992	100	100	—	C6
Lanius collurio Red-backed Shrike	B	1992	30	50	—	C6

This is an important site for breeding and wintering waterbirds, passage raptors, and *Grus grus*. The site is a migratory bottleneck, where more than 3,000 raptors regularly pass (including *Milvus migrans* 2,000; *Milvus milvus* 1,000).

Protection status
National Partial **International** None
Part of IBA covered by Hunting Reserve (12 km of the river). Part of IBA covered by Private Reserve (11 ha).

Conservation issues

Threats Aquaculture/fisheries (C), Recreation/tourism (B), Unsustainable exploitation (C)

Disturbance caused by fishing, hunting and recreation, is the main threat.

La Dombes A4i, A4iii, B1i, B2, C2, C3, C4, C6 159

Admin region Rhône-Alpes
Coordinates 46°02'N 5°01'E
Altitude 275–300 m **Area** 79,800 ha

Site description
The area includes ponds, marshes, wet meadows, deciduous forests, moorlands and arable fields.

Habitats Forest and woodland (15%; broadleaved deciduous forest), Scrub (7%; heathland), Grassland (18%; humid grassland), Wetland (11%; standing fresh water; water-fringe vegetation), Artificial landscape (15%; arable land)
Land-use Agriculture (70%), Fisheries/aquaculture (35%), Forestry (20%), Hunting (90%), Urban/industrial/transport (5%)

Birds

Species	Season	Year	Pop min	Pop max	Acc	Criteria
Podiceps nigricollis Black-necked Grebe	B	1997	200	400	—	A4i, B1i, C3
Ixobrychus minutus Little Bittern	B	1991	20	20	—	B2, C6
Nycticorax nycticorax Night Heron	B	1991	70	150	—	C6
Ardeola ralloides Squacco Heron	B	1991	1	5	—	B2
Egretta garzetta Little Egret	B	1997	10	50	—	C6
Ardea purpurea Purple Heron	B	1991	80	100	—	B2, C2, C6
Ciconia ciconia White Stork	B	1997	3	5	—	C6
Anas strepera Gadwall	B	1997	500	600	—	A4i, B1i, B2, C3
Anas strepera Gadwall	P	1997	200	1,100	—	B1i, C3
Anas acuta Pintail	P	1997	250	600	—	B1i, C3
Anas clypeata Shoveler	P	1997	200	800	—	B1i, C3
Netta rufina Red-crested Pochard	B	1997	80	120	—	B1i, C3
Netta rufina Red-crested Pochard	P	1997	—	300	—	B1i, C3
Aythya ferina Pochard	B	1997	1,550	1,550	—	B1i, C3
Circus aeruginosus Marsh Harrier	B	1991	15	20	—	C6
Himantopus himantopus Black-winged Stilt	B	1991	10	10	—	C6
Chlidonias hybridus Whiskered Tern	B	1991	200	1,100	—	A4i, B1i, B2, C2, C6
Chlidonias hybridus Whiskered Tern	P	1991	100	300	—	B1i, C2

The site holds 20,000 or more waterbirds on passage and in winter, on a regular basis (including *Anas platyrhynchos* 2,500 (passage), 10,820 (winter); *Aythya ferina* 6,000 (passage), 9,317 (winter); *Fulica atra* 6,640 (winter); *Vanellus vanellus* 8,000 (passage), 3,444 (winter)). Species of global conservation concern that do not meet IBA criteria: *Aythya nyroca* (passage, wintering); *Haliaeetus albicilla* (passage, wintering); *Aquila clanga* (wintering).

Protection status
National None **International** None

Conservation issues

Threats Agricultural intensification/expansion (A), Aquaculture/fisheries (C), Burning of vegetation (B), Recreation/tourism (A), Unsustainable exploitation (B)

Early mowing and the removal of aquatic vegetation, by burning and cutting in summer, destroys the nests and chicks of rare species. High hunting pressure (including early opening dates) and lead pollution threaten ducks.

Val de Saône A1, B2, C1, C2, C6 160

Admin region Rhône-Alpes
Coordinates 46°22'N 4°54'E
Altitude 150–150 m **Area** 12,250 ha

Site description
A complex of wet meadows, a river, riverine forest and marshes.

Habitats Forest and woodland (alluvial/very wet forest), Grassland (90%; humid grassland), Wetland (river/stream; water-fringe vegetation)
Land-use Agriculture, Hunting, Urban/industrial/transport

Birds

Species	Season	Year	Pop min	Pop max	Acc	Criteria
Egretta garzetta Little Egret	B	1991	5	50	—	C6
Milvus migrans Black Kite	B	1991	5	50	—	C6
Circus pygargus Montagu's Harrier	B	1991	1	10	—	C6
Crex crex Corncrake	B	1992	100	100	—	A1, B2, C1, C2, C6
Lanius collurio Red-backed Shrike	B	1991	100	500	—	C6

This is an important site for species breeding in wet meadows, and wintering waterbirds.

Protection status
National Low **International** Low
93 ha of IBA covered by Nature Reserve (Truchère–Ratenelle, 93 ha). 93 ha of IBA covered by Special Protection Area (Val de Saône: La Truchère, 93 ha).

Conservation issues

Threats Abandonment/reduction of land management (U), Afforestation, Agricultural intensification/expansion (U)

The main reported threats are afforestation with *Populus*, meadow abandonment, conversion of meadows into arable fields and early mowing of remnant meadows.

Forêt de Saou C6 161

Admin region Rhône-Alpes
Coordinates 44°39′N 5°08′E
Altitude 350–1,559 m **Area** 5,050 ha

Site description

This site contains deciduous and coniferous forests, cliffs, rocky slopes and subalpine meadows.

Habitats Forest and woodland (82%; broadleaved deciduous forest; native coniferous forest; alluvial/very wet forest), Grassland (10%; alpine/subalpine/boreal grassland), Wetland (river/stream; water-fringe vegetation), Rocky areas (10%; scree/boulders; inland cliff)
Land-use Fisheries/aquaculture, Forestry (40%), Hunting, Urban/industrial/transport

Birds

Species	Season	Year	Pop min	Pop max	Acc	Criteria
Pernis apivorus Honey Buzzard	B	1991	4	4	—	C6
Milvus migrans Black Kite	B	1991	10	10	—	C6
Circaetus gallicus Short-toed Eagle	B	1991	4	4	—	C6
Lanius collurio Red-backed Shrike	B	1991	10	100	—	C6
Emberiza hortulana Ortolan Bunting	B	1991	10	100	—	C6

This site is important for breeding raptors and forest birds.

Protection status

National None **International** None

Conservation issues

No serious threats are known at this site.

Val de Drôme: Les Ramières-Printegarde A4i, B1i, B1iv, C3, C5, C6 162

Admin region Rhône-Alpes
Coordinates 44°46′N 4°52′E
Altitude 86–160 m **Area** 2,100 ha

Site description

The site comprises both slow and fast-flowing rivers, marshes, woodland and dry calcareous grassland. Human activities include gravel exploitation, agriculture, fishing, hunting and tourism, including boating.

Habitats Forest and woodland (25%; alluvial/very wet forest), Grassland (steppe/dry calcareous grassland), Wetland (69%; standing fresh water; river/stream; water-fringe vegetation), Rocky areas (inland cliff), Artificial landscape (arable land)
Land-use Agriculture (4%), Fisheries/aquaculture (90%), Forestry (1%), Hunting (60%), Tourism/recreation, Urban/industrial/transport

Birds

Species	Season	Year	Pop min	Pop max	Acc	Criteria
Phalacrocorax carbo Cormorant	W	1997	14	3,012	—	B1i, C3
Phalacrocorax carbo Cormorant	P	1997	1,000	10,000	—	A4i, B1i, C3
Nycticorax nycticorax Night Heron	B	1991	30	40	—	C6
Egretta garzetta Little Egret	B	1997	30	40	—	C6
Milvus migrans Black Kite	B	1991	30	40	—	C6

The site is a migratory bottleneck, where more than 3,000 raptors regularly pass. Species of global conservation concern that do not meet criteria: *Aythya nyroca* (wintering).

Protection status

National Partial **International** Partial
346 ha of IBA covered by Nature Reserve (Les Ramières, 346 ha). 200 ha of IBA covered by Hunting Reserve (200 ha). 346 ha of IBA covered by Special Protection Area (Val de Drôme: Les Ramières, 346 ha).

Conservation issues

Threats Recreation/tourism (U)

No major threats are known at this site.

Col de l'Escrinet A4iv, B1i, B1iv, C2, C5 163

Admin region Rhône-Alpes
Coordinates 44°44′N 4°25′E
Altitude 790–790 m **Area** 4,600 ha

Site description

A mountain pass with moorland and woodlots.

Habitats Scrub (heathland)
Land-use Agriculture (5%), Forestry, Hunting (95%), Urban/industrial/transport

Birds

Species	Season	Year	Pop min	Pop max	Acc	Criteria
Grus grus Crane	P	1991	500	1,000	—	B1i, C2

The site is a major migratory bottleneck, where more than 20,000 raptors regularly pass (including *Pernis apivorus* 10,000; *Milvus migrans* 10,000).

Protection status

National None **International** None

Conservation issues

Threats Afforestation (A), Burning of vegetation (C), Intensified forest management (B), Recreation/tourism (A), Unsustainable exploitation (A)

The main threats are afforestation with conifers and high hunting pressure on migratory birds, including shooting during prohibited periods.

Basse Ardêche B2 164

Admin region Rhône-Alpes
Coordinates 44°25′N 4°28′E
Altitude 45–430 m **Area** 46,000 ha

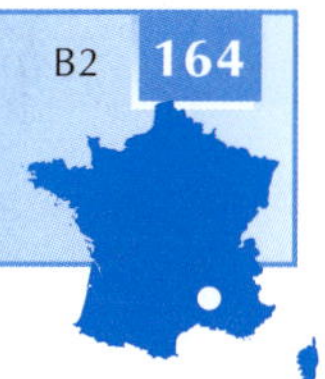

Site description

An area of garrigue and *Quercus ilex* scrub, with a river, cliffs, rocky slopes, arable fields and orchards.

Habitats Forest and woodland (30%; broadleaved evergreen forest), Scrub (70%; sclerophyllous scrub/garrigue/maquis)
Land-use Fisheries/aquaculture, Forestry, Hunting, Tourism/recreation, Urban/industrial/transport

Birds

Species	Season	Year	Pop min	Pop max	Acc	Criteria
Hieraaetus fasciatus Bonnelli's Eagle	B	1996	2	2	—	B2

The site is important for breeding raptors.

Protection status

National Low **International** None
1,570 ha of IBA covered by Nature Reserve (Gorges de l'Ardèche, 1,570 ha).

Conservation issues

Threats Disturbance to birds (A), Industrialization/urbanization (B), Infrastructure (A), Intensified forest management (A), Recreation/tourism (A), Unsustainable exploitation (A)

Recreation (including boating), and the construction of new tracks and tourist facilities are the main threats.

Hauts Plateaux du Vercors et Forêt des Coulmes B2, C6 165

Admin region Rhône-Alpes
Coordinates 45°13′N 5°30′E
Altitude 450–2,120 m **Area** 52,800 ha

Site description

Habitats Forest and woodland (broadleaved deciduous forest), Scrub (heathland), Wetland (river/stream), Rocky areas (inland cliff)
Land-use Agriculture, Fisheries/aquaculture, Forestry, Hunting, Other, Urban/industrial/transport

Mountains and cliffs with fast-flowing rivers, woodlands, moorlands and arable fields.

■ Birds

Species	Season	Year	Pop min	Pop max	Acc	Criteria
Pernis apivorus Honey Buzzard	B	1991	1	10	—	C6
Circaetus gallicus Short-toed Eagle	B	1991	1	10	—	C6
Falco peregrinus Peregrine	B	1991	10	15	—	B2, C6
Tetrao tetrix Black Grouse	B	1991	100	200	—	C6
Bubo bubo Eagle Owl	B	1991	5	10	—	C6
Aegolius funereus Tengmalm's Owl	B	1991	30	40	—	C6
Dryocopus martius Black Woodpecker	B	1991	10	100	—	C6

The site is important for forest and mountain species.

■ Protection status

National Partial **International** Partial

16,662 ha of IBA covered by Nature Reserve (Hauts Plateaux du Vercors, 16,662 ha). 16,662 ha of IBA covered by Special Protection Area (Hauts Plateaux du Vercors, 16,662 ha)

■ Conservation issues

Threats Extraction industry (U), Infrastructure (U), Intensified forest management (U), Recreation/tourism (U)

The development of tourism is considered a threat to the site.

Iles du Haut-Rhône

B1i, C3, C6 **166**

Admin region Rhône-Alpes
Coordinates 45°40′N 5°36′E
Altitude 208–208 m **Area** 3,400 ha

■ Site description

A large river bordered by riverine forest, ponds, heathland, orchards and arable fields.

Habitats Forest and woodland (broadleaved deciduous forest; alluvial/very wet forest), Scrub (heathland), Wetland (standing fresh water; river/stream; water-fringe vegetation), Artificial landscape (highly improved reseeded grassland; arable land; perennial crops/ orchards/groves; ruderal land)
Land-use Agriculture, Fisheries/aquaculture, Forestry, Hunting, Tourism/recreation, Urban/industrial/transport

■ Birds

Species	Season	Year	Pop min	Pop max	Acc	Criteria
Nycticorax nycticorax Night Heron	B	1991	40	50	—	C6
Aythya ferina Pochard	W	1997	1,028	6,024	—	B1i, C3

The site is important for breeding herons and wintering waterbirds. Species of global conservation concern that do not meet IBA criteria: *Aythya nyroca* (wintering).

■ Protection status

National Low **International** Partial

310 ha of IBA covered by Private Reserve (310 ha). 273 ha of IBA covered by Special Protection Area (Iles du haut Rhone, 273 ha).

■ Conservation issues

Threats Recreation/tourism (C)

The site has undergone much degradation, due to dam construction and afforestation with *Populus* plantations. The current threat is tourism development.

Plaine du Forez

A4i, A4iii, B1i, B2, B3, C2, C3, C4, C6 **167**

Admin region Rhône-Alpes
Coordinates 45°44′N 4°10′E
Altitude 350–700 m **Area** 80,850 ha

■ Site description

An area of ponds, marshes, hedges and cultivated land, grassland and arable land.

Habitats Forest and woodland (broadleaved deciduous forest; alluvial/very wet forest), Scrub (heathland), Grassland (dry siliceous grassland; humid grassland), Wetland (standing fresh water; river/stream; water-fringe vegetation), Artificial landscape (arable land)
Land-use Agriculture, Fisheries/aquaculture, Forestry, Hunting, Tourism/recreation, Urban/industrial/transport, Water management

■ Birds

Species	Season	Year	Pop min	Pop max	Acc	Criteria
Podiceps nigricollis Black-necked Grebe	B	1991	100	500	—	A4i, B1i, C3
Botaurus stellaris Bittern	B	1997	1	10	—	C6
Egretta garzetta Little Egret	B	1991	10	100	—	C6
Ardea purpurea Purple Heron	B	1991	25	40	—	C6
Anas strepera Gadwall	B	1997	200	200	—	B2
Milvus migrans Black Kite	B	1991	10	100	—	B2, C6
Milvus milvus Red Kite	B	1991	10	100	—	B3, C6
Circaetus gallicus Short-toed Eagle	B	1991	1	10	—	C6
Circus aeruginosus Marsh Harrier	B	1991	10	100	—	C6
Circus pygargus Montagu's Harrier	B	1991	1	10	—	C6
Falco tinnunculus Kestrel	B	1991	100	500	—	B2
Burhinus oedicnemus Stone Curlew	B	1991	10	100	—	B2
Philomachus pugnax Ruff	B	1997	1	10	—	C6
Chlidonias hybridus Whiskered Tern	B	1991	100	500	—	A4i, B1i, B2, C2, C6
Chlidonias niger Black Tern	B	1991	1	10	—	C6
Alcedo atthis Kingfisher	B	1991	10	100	—	C6
Lanius collurio Red-backed Shrike	B	1991	100	500	—	C6

The site is of importance for several species of heron and raptors. The site holds 20,000 or more waterbirds in winter, on a regular basis. Species of global conservation concern that do not meet IBA criteria: *Haliaeetus albicilla* (passage).

■ Protection status

National Low **International** None

50 ha of IBA covered by Hunting Reserve.

■ Conservation issues

Threats Agricultural intensification/expansion (U), Extraction industry (B), Infrastructure (B)

The site is threatened by a plan to build a motorway through the area, agricultural development, and mineral exploitation. Hunting, shooting and fishing cause considerable disturbance.

Ile de la Platière

B2 **168**

Admin region Rhône-Alpes
Coordinates 45°21′N 4°46′E
Altitude 136–136 m **Area** 470 ha

■ Site description

The Rhône river with riverine forest and surrounding arable fields.

Habitats Forest and woodland (21%; alluvial/very wet forest), Grassland (steppe/dry calcareous grassland), Wetland (45%; river/stream), Artificial landscape (7%; arable land)
Land-use Agriculture (7%), Forestry (11%), Tourism/recreation, Urban/industrial/ transport, Water management (29%)

■ Birds

Species	Season	Year	Pop min	Pop max	Acc	Criteria
Merops apiaster Bee-eater	B	1991	10	100	—	B2

This is a nationally important site for passage and wintering waterbirds.

■ Protection status

National High **International** None

470 ha of IBA covered by Nature Reserve (L'Ile de la Platière, 484 ha).

■ Conservation issues

Threats Afforestation (B), Intensified forest management (C)

The main threat to the site is afforestation with *Populus*.

Parc National de la Vanoise A3, B2, C2, C6 169

Admin region Rhône-Alpes
Coordinates 45°24′N 6°52′E
Altitude 1,000–3,855 m **Area** 67,600 ha

Site description

A mid- to high-altitude mountain area with conifer forests, dry meadows, alpine meadows, scree and cliffs, glaciers and permanent lying snow.

Habitats Forest and woodland (native coniferous forest), Grassland (steppe/dry calcareous grassland; dry siliceous grassland; alpine/subalpine/boreal grassland), Rocky areas (scree/boulders; inland cliff)
Land-use Agriculture, Fisheries/aquaculture, Forestry, Hunting, Tourism/recreation, Urban/industrial/transport, Water management

Birds

Species	Season	Year	Pop min	Pop max	Acc	Criteria
Gypaetus barbatus Lammergeier	B	1996	3	3	—	B2, C6
Circaetus gallicus Short-toed Eagle	B	1991	1	10	—	C6
Aquila chrysaetos Golden Eagle	B	1991	17	17	—	B2, C6
Falco peregrinus Peregrine	B	1991	3	10	—	B2, C6
Bonasa bonasia Hazel Grouse	B	1991	100	200	—	C6
[1] ***Lagopus mutus*** Ptarmigan	B	1991	500	1,000	—	C6
Tetrao tetrix Black Grouse	B	1991	120	150	—	C6
Alectoris graeca Rock Partridge	B	1991	500	600	—	B2, C6
Bubo bubo Eagle Owl	B	1991	5	10	—	C6
Glaucidium passerinum Pygmy Owl	B	1991	1	10	—	C6
Aegolius funereus Tengmalm's Owl	B	1991	10	100	—	C6
Caprimulgus europaeus Nightjar	B	1991	5	50	—	C6
Picoides tridactylus Three-toed Woodpecker	B	1991	1	10	—	C6
Monticola saxatilis Rock Thrush	B	1991	500	1,000	—	B2
[2] ***Tichodroma muraria*** Wallcreeper	B	1991	250	750	—	A3
Pyrrhocorax pyrrhocorax Chough	B	1991	100	500	—	B2, C2, C6
[2] ***Montifringilla nivalis*** Snowfinch	B	1991	500	1,000	—	A3

1. *L. m. helveticus*.
2. Eurasian high-montane biome.

The site holds important populations of mountain, forest and raptor species. Species of global conservation concern that do not meet IBA criteria: *Crex crex* (irregular breeder).

Protection status

National Partial **International** Partial
53,000 ha of IBA covered by National Park (Parc national de la Vanoise, 53,900 ha). 1,112 ha of IBA covered by Nature Reserve (Plan de Tueda, 1,112 ha). 2,230 ha of IBA covered by Nature Reserve (Sassières, 2,230 ha). 52,800 ha of IBA covered by Special Protection Area (La Vanoise, 52,800 ha).

Conservation issues

Threats Infrastructure (A), Recreation/tourism (B)

Development of ski-stations in the areas surrounding the IBA is a threat.

Lac Léman (partie française) A4i, A4iii, B1i, C3, C4 170

Admin region Rhône-Alpes
Coordinates 46°22′N 6°31′E
Altitude 372–372 m **Area** 24,300 ha

Site description

French part of Lake Geneva with reedbeds (*Phragmites*), marshes and aquatic vegetation (Swiss part of the lake contains IBAs 007, 008, 009).

Habitats Wetland (standing fresh water; water-fringe vegetation)
Land-use Fisheries/aquaculture, Hunting, Tourism/recreation

Birds

Species	Season	Year	Pop min	Pop max	Acc	Criteria
Podiceps cristatus Great Crested Grebe	W	1997	1,584	9,351	—	B1i, C3
Podiceps cristatus Great Crested Grebe	P	1997	20,000	20,000	—	A4i, B1i, C3
Aythya fuligula Tufted Duck	W	1997	4,671	19,823	—	A4i, B1i, C3
Aythya marila Scaup	W	1997	3	13	—	A4i, B1i, C3

The area is of international importance for wintering and passage waterbirds. The site holds 20,000 or more waterbirds in winter, on a regular basis. Species of global conservation concern that do not meet IBA criteria: *Aythya nyroca* (wintering). All bird data are for the French part of the lake only.

Protection status

National Low **International** Partial
52 ha of IBA covered by Nature Reserve (Delta de la Dranse, 52 ha). 1,440 ha of IBA covered by 2 Hunting Reserves (Baie de Sciez; Hermance; total area 1,440 ha). 45 ha of IBA covered by Special Protection Area (Lac Léman: Delta de la Dranse, 45 ha). 3,335 ha of IBA covered by Ramsar Site (Rives du Lac Léman, 3,335 ha).

Conservation issues

Threats Aquaculture/fisheries (A), Extraction industry (A), Industrialization/urbanization, Infrastructure (C), Recreation/tourism (A), Unsustainable exploitation (A)

The lake is threatened by pollution from urban areas that is causing eutrophication. Fishing causes disturbance to wintering birds.

Lac et marais du Bourget B1i, C3 171

Admin region Rhône-Alpes
Coordinates 45°48′N 5°46′E
Altitude 231–231 m **Area** 9,350 ha

Site description

The site comprises a lake, marshes, wet meadows, and agricultural land. Human activities include gravel extraction.

Habitats Grassland (humid grassland), Wetland (standing fresh water; water-fringe vegetation), Artificial landscape (arable land; perennial crops/orchards/groves)
Land-use Agriculture, Fisheries/aquaculture, Forestry, Hunting, Tourism/recreation, Urban/industrial/transport

Birds

Species	Season	Year	Pop min	Pop max	Acc	Criteria
Aythya ferina Pochard	W	1997	265	4,889	—	B1i, C3

The site is important for wintering waterbirds.

Protection status

National Low **International** Low
474 ha of IBA covered by Nature Reserve (Marais de Lavours, 474 ha). 772 ha of IBA covered by 4 Hunting Reserves (Anse Nord, 200 ha; Memars, 20 ha; Marais du Nord du Lac du Bourget, 500 ha; Canal de Savières, 4.2 km). 473 ha of IBA covered by Special Protection Area (Lac et Marais du Bourget: Marais de Lavours, 473 ha).

Conservation issues

Threats Disturbance to birds (A), Extraction industry (A), Infrastructure (A), Recreation/tourism (A)

Recreational activities cause disturbance to birds.

Haute Chaîne du Jura: défilé de l'écluse, Etournel et Mont Vuache A4iv, B1iv, B2, C5, C6 172

Admin region Rhône-Alpes
Coordinates 46°14′N 5°59′E
Altitude 330–1,621 m **Area** 12,400 ha

Site description

An area of ponds, a river with riverine forest, cliffs and rocky slopes, conifer forests, and alpine meadows.

Habitats Forest and woodland (broadleaved deciduous forest; native coniferous forest; mixed forest; alluvial/very wet forest), Grassland (alpine/subalpine/boreal grassland), Wetland (standing fresh water; river/stream; water-fringe vegetation), Rocky areas (scree/boulders; inland cliff), Artificial landscape (arable land; perennial crops/orchards/groves; other urban/industrial areas)

Land-use Agriculture, Fisheries/aquaculture, Hunting, Tourism/recreation, Urban/industrial/transport, Water management

Birds

Species	Season	Year	Pop min	Pop max	Acc	Criteria
Pernis apivorus Honey Buzzard	B	1991	1	10	—	C6
Falco peregrinus Peregrine	B	1991	3	10	—	B2, C6
Bonasa bonasia Hazel Grouse	B	1991	100	500	—	C6
Tetrao urogallus Capercaillie	B	1991	50	100	—	C6
Glaucidium passerinum Pygmy Owl	B	1991	1	10	—	C6

This is an important site for breeding mountain and forest species. The site is a major migratory bottleneck, where more than 20,000 raptors pass on a regular basis (including *Pernis apivorus* 10,000; *Milvus migrans* 20,000; *Buteo buteo* 20,000).

Protection status

National High **International** None
10,781 ha of IBA covered by Nature Reserve (Haute Chaîne du Jura, 10,781 ha). 300 ha of IBA covered by Hunting Reserve (300 ha).

Conservation issues

Threats Disturbance to birds (A), Intensified forest management (A), Recreation/tourism (A), Unsustainable exploitation (B)

High levels of tourism and hunting cause disturbance to breeding birds.

Montagne des Frêtes et plateau des Glières — B2, C6 — 173

Admin region Rhône-Alpes
Coordinates 45°58′N 6°18′E
Altitude 500–1,700 m **Area** 8,200 ha

Site description

Coniferous forests, alpine meadows, scree and cliffs. The main human activities are forestry and agriculture.

Habitats Forest and woodland (native coniferous forest), Grassland (alpine/subalpine/boreal grassland), Rocky areas (scree/boulders; inland cliff)
Land-use Agriculture, Forestry, Hunting, Tourism/recreation, Urban/industrial/transport

Birds

Species	Season	Year	Pop min	Pop max	Acc	Criteria
Falco peregrinus Peregrine	B	1991	4	5	—	C6
Bonasa bonasia Hazel Grouse	B	1991	10	100	—	C6
Tetrao tetrix Black Grouse	B	1991	250	500	—	B2, C6
Aegolius funereus Tengmalm's Owl	B	1991	10	20	—	C6

Several mountain species breed at this site.

Protection status

National Low **International** None
244 ha of IBA covered by Biological Reserve (244 ha).

Conservation issues

Threats Afforestation (B), Abandonment/reduction of land management

Under-exploitation of the *Picea* forests is the main threat. The forests are becoming too dense, and are less attractive for edge species and breeding *Tetrao urogallus* and *T. tetrix*.

Les Bauges — B2, C6 — 174

Admin region Rhône-Alpes
Coordinates 45°40′N 6°13′E
Altitude 500–2,200 m **Area** 17,750 ha

Site description

Coniferous forests, alpine meadows, scree and cliffs. Human activities include forestry, cattle-grazing and fishing.

Habitats Forest and woodland (native coniferous forest), Grassland (alpine/subalpine/boreal grassland), Rocky areas (scree/boulders; inland cliff)
Land-use Agriculture, Fisheries/aquaculture, Forestry, Hunting, Other, Tourism/recreation, Urban/industrial/transport

Birds

Species	Season	Year	Pop min	Pop max	Acc	Criteria
Falco peregrinus Peregrine	B	1991	5	10	—	B2, C6
Bonasa bonasia Hazel Grouse	B	1991	10	100	—	C6
Tetrao tetrix Black Grouse	B	1991	300	500	—	B2, C6
Alectoris graeca Rock Partridge	B	1991	40	50	—	C6
Aegolius funereus Tengmalm's Owl	B	1991	10	20	—	C6

Protection status

National Low **International** None
7,500 ha of IBA covered by 2 Hunting Reserves (total area 7,500 ha).

Conservation issues

Threats Abandonment/reduction of land management (A)

The abandonment of cattle-grazing and the subsequent succession of grassland to forest poses a threat to *Tetrao tetrix*.

Haut Giffre — B2, C6 — 175

Admin region Rhône-Alpes
Coordinates 46°04′N 6°48′E
Altitude 450–3,099 m **Area** 20,850 ha

Site description

A mountainous area with lakes, broadleaved and coniferous forests, pasture and alpine meadows, moorlands and peatbogs, screes, cliffs, glaciers and permanent lying snow.

Habitats Forest and woodland (broadleaved deciduous forest; native coniferous forest; mixed forest; alluvial/very wet forest), Scrub (heathland), Grassland (steppe/dry calcareous grassland; alpine/subalpine/boreal grassland; humid grassland; mesophile grassland), Wetland (standing fresh water; river/stream; raised bog; fen/transition mire/spring), Rocky areas (scree/boulders; inland cliff; inland sand-dunes; caves)
Land-use Agriculture, Fisheries/aquaculture, Forestry, Hunting, Tourism/recreation, Urban/industrial/transport

Birds

Species	Season	Year	Pop min	Pop max	Acc	Criteria
Gypaetus barbatus Lammergeier	B	1996	3	4	—	B2, C6
Aquila chrysaetos Golden Eagle	B	1991	1	10	—	B2
Falco peregrinus Peregrine	B	1991	1	10	—	B2
Bonasa bonasia Hazel Grouse	B	1991	100	500	—	C6
[1] ***Lagopus mutus*** Ptarmigan	B	1991	100	500	—	C6
Tetrao tetrix Black Grouse	B	1991	100	500	—	B2, C6
Tetrao urogallus Capercaillie	B	1991	10	50	—	C6
Alectoris graeca Rock Partridge	B	1991	10	50	—	C6
Aegolius funereus Tengmalm's Owl	B	1991	10	20	—	C6

1. *L. m. helveticus.*

An important area for breeding gamebirds and other mountain and forest species.

Protection status

National High **International** None
1,717 ha of IBA covered by Nature Reserve (Passy, 1,717 ha). 11,000 ha of IBA covered by Hunting Reserve (Arvegiffre, 11,000 ha). 9,200 ha of IBA covered by Nature Reserve (Sixt, 9,200 ha).

Conservation issues

Threats Intensified forest management (A)

The development of ski-based tourism in the surrounding area is threatening breeding birds, as much as the tourist pressure within the site itself.

Bassin d'Arcachon et Banc d'Arguin — 176

A4i, A4iii, B1i, B2, C2, C3, C4, C6

Admin region Aquitaine
Coordinates 44°40'N 1°07'W
Altitude 0–2 m **Area** 20,100 ha

Site description

A bay and estuary with islands and sand-dunes, *Zostera* beds, saltwater and freshwater lagoons, wet meadows, marshes and mudflats.

Habitats Forest and woodland (broadleaved deciduous forest), Grassland (humid grassland), Wetland (tidal river/enclosed tidal water; mudflat/sandflat; saltmarsh; sand-dunes/sand beach; coastal lagoon; standing brackish and salt water; river/stream; water-fringe vegetation), Marine areas (sea inlet/coastal features)
Land-use Fisheries/aquaculture, Hunting, Tourism/recreation, Urban/industrial/transport

Birds

Species	Season	Year	Pop min	Pop max	Acc	Criteria
Egretta garzetta Little Egret	B	1997	630	630	—	A4i, B1i, C2, C6
Egretta garzetta Little Egret	W	1997	55	1,810	—	A4i, B1i, C2
Ciconia ciconia White Stork	B	1997	3	3	—	C6
Platalea leucorodia Spoonbill	W	1997	20	37	—	B1i, C2
Platalea leucorodia Spoonbill	P	1997	100	100	—	B1i, C2
Anser anser Greylag Goose	P	1997	15,000	20,000	—	A4i, B1i, C3
Branta bernicla Brent Goose	W	1997	31,841	39,701	—	A4i, B1i, C3
Anas acuta Pintail	W	1997	680	2,560	—	B1i, C3
Anas acuta Pintail	P	1997	3,000	5,000	—	B1i, C3
Anas clypeata Shoveler	P	1997	1,000	2,000	—	B1i, C3
Milvus migrans Black Kite	B	1991	30	30	—	C6
Circaetus gallicus Short-toed Eagle	B	1991	5	5	—	C6
Circus cyaneus Hen Harrier	B	1991	3	5	—	C6
Recurvirostra avosetta Avocet	W	1997	8	136	—	B2
Recurvirostra avosetta Avocet	P	1997	2,000	3,000	—	A4i, B1i, C2
Charadrius hiaticula Ringed Plover	W	1997	18	105	—	A4i, B1i, C3
Charadrius alexandrinus Kentish Plover	B	1997	30	40	—	B2
Pluvialis squatarola Grey Plover	P	1997	2,000	3,000	—	A4i, B1i, C3
Calidris canutus Knot	P	1997	3,000	50,000	—	A4i, B1i, C3
Calidris ferruginea Curlew Sandpiper	P	1997	1,000	20,000	—	A4i, B1i, C3
Calidris alpina Dunlin	W	1997	13,843	25,000	—	A4i, B1i, B2, C3
Calidris alpina Dunlin	P	1997	20,000	30,000	—	A4i, B1i, C3
Limosa lapponica Bar-tailed Godwit	P	1997	2,000	3,000	—	A4i, B1i, C2
Numenius arquata Curlew	W	1997	310	2,501	—	B2
Tringa erythropus Spotted Redshank	P	1997	500	1,000	—	A4i, B1i, C3
Tringa totanus Redshank	P	1997	1,000	2,000	—	B1i, C3
Sterna sandvicensis Sandwich Tern	B	1997	4,200	4,200	—	A4i, B1i, B2, C2
Luscinia svecica Bluethroat	B	1991	120	130	—	C6

The site holds 20,000 or more waterbirds on passage and in winter, on a regular basis.

Protection status

National Partial **International** Partial
1,000 ha of IBA covered by Nature Reserve (Banc d'Arguin, 1,000 ha). 495 ha of IBA covered by Nature Reserve (Prés d'Arès et de Lège, 495 ha). 600 ha of IBA covered by Hunting Reserve (Bassin d'Arachon, 600 ha). 120 ha of IBA covered by Private Reserve (Parc ornithologique du Teich, 120 ha). 500 ha of IBA covered by Special Protection Area (Bassin d'Arcachon: Banc d'Arguin, 500 ha). 1,100 ha of IBA covered by Special Protection Area (Emboudure de la Leyre, 1,100 ha). 495 ha of IBA covered by Special Protection Area (Prés salés d'Arès, 495 ha).

Conservation issues

Threats Construction/impact of dyke/dam/barrage (C), Extraction industry (C), Filling-in of wetlands (A), Industrialization/urbanization (C), Recreation/tourism (A), Unsustainable exploitation (B)

Threats include pollution from anti-fouling paint used on boats, intensive hunting and tourism.

Pointe de Grave — 177

B1i, B1iv, C3, C5, C6

Admin region Aquitaine
Coordinates 45°32'N 1°05'W
Altitude 0–35 m **Area** 800 ha

Site description

The area includes woodland, marshes, wet meadows, dunes and beaches, agricultural land and an estuary.

Habitats Forest and woodland (60%; mixed forest), Grassland (humid grassland), Wetland (sand-dunes/sand beach; water-fringe vegetation), Artificial landscape (arable land)
Land-use Agriculture, Fisheries/aquaculture, Forestry, Hunting, Other, Tourism/recreation, Urban/industrial/transport

Birds

Species	Season	Year	Pop min	Pop max	Acc	Criteria
Pandion haliaetus Osprey	P	1991	1	10	—	C6
Calidris canutus Knot	P	1991	5,000	5,000	—	B1i, C3

The site is a migratory bottleneck, where more than 3,000 raptors regularly pass in spring (including *Milvus migrans* 1,600 individuals; *Pernis apivorus* 650 individuals).

Protection status

National None **International** None

Conservation issues

Threats Intensified forest management, Recreation/tourism (A), Unsustainable exploitation (A)

There is considerable pressure from illegal spring hunting (mainly of *Streptopelia turtur*). Areas of old mixed forest have been logged as a result of intensive forest management.

Champ de Tir du Poteau — 178

A4i, A4iv, B1i, B1iv, C2, C5, C6

Admin region Aquitaine
Coordinates 44°15'N 0°23'W
Altitude 103–126 m **Area** 12,200 ha

Site description

The site consists of heathland and grassland, arable land and woodland.

Habitats Forest and woodland (broadleaved deciduous forest), Scrub (heathland; sclerophyllous scrub/garrigue/maquis), Grassland (humid grassland), Wetland (standing fresh water; raised bog; water-fringe vegetation), Artificial landscape (arable land)
Land-use Agriculture, Hunting, Military, Other, Urban/industrial/transport

Birds

Species	Season	Year	Pop min	Pop max	Acc	Criteria
Circaetus gallicus Short-toed Eagle	B	1990	1	10	—	C6
Grus grus Crane	W	1990	500	2,200	—	A4i, B1i, C2
Grus grus Crane	P	1990	20,000	30,000	—	A4i, B1i, C2
Sylvia undata Dartford Warbler	B	1990	10	100	—	C6

This is an important site for passage and wintering *Grus grus* and breeding raptors. The site is a major migratory bottleneck, where more than 20,000 *Grus grus* regularly pass.

Protection status

National None **International** Partial
10,200 ha of IBA covered by Special Protection Area (Champ de tir du Poteau, 10,200 ha).

Conservation issues

Threats Agricultural intensification/expansion (A), Disturbance to birds (C), Drainage (A), Infrastructure (A), Intensified forest management (A), Recreation/tourism (A), Unsustainable exploitation (B)

The site is seriously threatened by drainage and the construction of access roads.

Barthes de l'Adour — 179

B2, C6

Admin region Aquitaine
Coordinates 43°36'N 0°18'W
Altitude 2–110 m **Area** 16,100 ha

Site description

Marshes, wet meadows, and lakes along the River Adour, with woodland, hedgerows and arable land.

Habitats Forest and woodland (55%; broadleaved deciduous forest; mixed forest; alluvial/very wet forest), Grassland (40%; humid grassland), Wetland (35%; standing fresh water; river/stream; raised bog; water-fringe vegetation), Artificial landscape (53%; arable land; perennial crops/orchards/groves)
Land-use Agriculture, Fisheries/aquaculture, Forestry, Hunting, Other, Tourism/recreation, Urban/industrial/transport, Water management

Birds

Species	Season	Year	Pop min	Pop max	Acc	Criteria
Nycticorax nycticorax Night Heron	B	1991	80	120	—	C6
Ciconia ciconia White Stork	B	1991	4	4	—	C6
Pernis apivorus Honey Buzzard	B	1991	5	50	—	C6
Milvus migrans Black Kite	B	1991	100	500	—	B2, C6
Hieraaetus pennatus Booted Eagle	B	1991	5	50	—	B2, C6
Lanius collurio Red-backed Shrike	B	1991	10	100	—	C6

Species of global conservation concern that do not meet IBA criteria: *Aythya nyroca* (wintering); *Haliaeetus albicilla* (wintering); *Crex crex* (passage).

Protection status

National Partial **International** Low
881 ha of IBA covered by Hunting Reserve (Saint-Martin de Seignanx, 881 ha). 93 ha of IBA covered by Special Protection Area (Barthes de l'Adour: Saint-Martin de Seignanx, 93 ha).

Conservation issues

Threats Abandonment/reduction of land management (C), Agricultural intensification/expansion (C), Disturbance to birds (A), Other (B), Recreation/tourism (B), Unsustainable exploitation (C)

The abandonment of grazing and the conversion of meadows to maize plantations are the main threats to the site.

Rochers de Biarritz: le Bouccalot et la Roche Ronde — C6 — 180

Admin region Aquitaine
Coordinates 43°30'N 1°34'W
Altitude 0–20 m **Area** 250 ha

Site description

A small island with cliffs.

Habitats Rocky areas (sea cliff/rocky shore; rock stacks/islets)
Land-use Tourism/recreation

Birds

Species	Season	Year	Pop min	Pop max	Acc	Criteria
Hydrobates pelagicus Storm Petrel	B	1991	22	32	—	C6

Protection status

National Low **International** None
4 ha of IBA covered by Private Reserve (4 ha).

Conservation issues

Threats Disturbance to birds (B), Other (C), Recreation/tourism (B)

There is considerable pressure from tourist activities.

Haute Soule: Massif Forestier, Gorges d'Holzarte et d'Olhadubi — C6 — 181

Admin region Aquitaine
Coordinates 42°58'N 0°56'W
Altitude 950–1,450 m **Area** 2,400 ha

Site description

Quercus forests, cliffs and rocky areas, scree and mountain rivers.

Habitats Forest and woodland (broadleaved deciduous forest), Wetland (river/stream), Rocky areas (scree/boulders; inland cliff; inland sand-dunes; caves)
Land-use Agriculture, Forestry, Hunting, Tourism/recreation

Birds

Species	Season	Year	Pop min	Pop max	Acc	Criteria
Milvus milvus Red Kite	B	1991	1	10	—	C6

The site is important for raptors and forest breeding birds.

Protection status

National None **International** None

Conservation issues

Threats Agricultural intensification/expansion (C), Burning of vegetation (B), Intensified forest management (B), Recreation/tourism (A), Unsustainable exploitation (B)

High hunting pressure, recreation and forestry are the main threats.

Haute Soule: Forêt des Arbailles — B2, C6 — 182

Admin region Aquitaine
Coordinates 43°08'N 1°00'W
Altitude 600–1,220 m **Area** 6,850 ha

Site description

A mountain area holding *Fagus* forest, cliffs, screes and mountain rivers.

Habitats Forest and woodland (broadleaved deciduous forest), Wetland (river/stream), Rocky areas (scree/boulders; inland cliff; inland sand-dunes; caves)
Land-use Agriculture (30%), Fisheries/aquaculture (5%), Forestry (70%), Hunting (90%), Tourism/recreation (95%), Urban/industrial/transport

Birds

Species	Season	Year	Pop min	Pop max	Acc	Criteria
Milvus milvus Red Kite	B	1991	1	10	—	C6
Gyps fulvus Griffon Vulture	B	1991	10	15	—	B2, C6
Hieraaetus pennatus Booted Eagle	B	1991	1	10	—	B2, C6
Dryocopus martius Black Woodpecker	B	1991	5	50	—	C6

The site is important for raptors and forest breeding birds.

Protection status

National None **International** None

Conservation issues

Threats Agricultural intensification/expansion (C), Burning of vegetation (B), Intensified forest management (A), Recreation/tourism (B), Unsustainable exploitation (C)

Forestry activities have lead to increased track construction causing disturbance and habitat loss.

Haute Soule: Forêt d'Irraty, Organbidexka et Pic des Escaliers — A4iv, B1iv, C5, C6 — 183

Admin region Aquitaine
Coordinates 43°02'N 1°03'W
Altitude 530–1,440 m **Area** 5,500 ha

Site description

A large mountain area encompassing mixed forests of *Fagus* and *Picea*, heathland, cliffs, screes and rivers.

Habitats Forest and woodland (mixed forest), Scrub (heathland), Wetland (river/stream), Rocky areas (scree/boulders; inland cliff)
Land-use Agriculture, Forestry (80%), Hunting, Tourism/recreation, Urban/industrial/transport

Birds

Species	Season	Year	Pop min	Pop max	Acc	Criteria
Milvus milvus Red Kite	B	1991	1	10	—	C6

The site is a major migratory bottleneck, where more than 20,000 raptors regularly pass (including *Pernis apivorus* 18,000; *Milvus migrans* 13,000; *Milvus milvus* 4,600) and 15,000 *Grus grus*.

Protection status

National None **International** None

Conservation issues

Threats Agricultural intensification/expansion (C), Burning of vegetation (B), Intensified forest management (A), Industrialization/urbanization (B), Recreation/tourism (B), Unsustainable exploitation (B)

Hunting pressure is high. The construction of small houses is also a threat.

Rive Gauche de la Vallée d'Aspe et Pic d'Anie

C6 **184**

Admin region Aquitaine
Coordinates 42°52′N 0°39′W
Altitude 1,450–2,510 m **Area** 10,750 ha

Site description

A high mountain area with alpine meadows, mixed forests, cliffs, scree and mountain rivers. Hunting and recreational activities take place throughout the site.

Habitats Forest and woodland (native coniferous forest), Grassland (alpine/subalpine/boreal grassland), Wetland (river/stream), Rocky areas (scree/boulders; inland cliff)
Land-use Agriculture, Forestry, Hunting, Tourism/recreation

Birds

Species	Season	Year	Pop min	Pop max	Acc	Criteria
Lagopus mutus Ptarmigan	B	1991	10	100	—	C6

An important site for raptors, gamebirds, and woodpeckers.

Protection status

National Partial **International** None
4,000 ha of IBA covered by National Park (Parc national des Pyrénées, 45,700 ha).

Conservation issues

Threats Agricultural intensification/expansion (C), Burning of vegetation (B), Industrialization/urbanization (B), Infrastructure (U), Intensified forest management (B), Recreation/tourism (B), Unsustainable exploitation (C)

Hunting and recreational activities cause much disturbance, the latter also resulting in habitat destruction, through the construction of roads, ski installations and housing.

Hautes Vallées d'Aspe et d'Ossau

B1iii, B2, C2, C6 **185**

Admin region Aquitaine
Coordinates 42°54′N 0°28′W
Altitude 450–2,974 m **Area** 37,650 ha

Site description

A large mountainous area in the Pyrenees encompassing fast-flowing rivers and streams, lakes, forests, moorlands, grasslands and arable land, mountains and cliffs. Human activities include extensive traditional stock-raising. There are major roads into the area, giving easy access for mass tourism.

Habitats Forest and woodland (32%; broadleaved deciduous forest; native coniferous forest; mixed forest), Scrub (10%; heathland), Grassland (20%; alpine/subalpine/boreal grassland), Wetland (standing fresh water; river/stream; raised bog), Rocky areas (10%; scree/boulders; inland cliff; inland sand-dunes; caves), Artificial landscape (21%; highly improved reseeded grassland; arable land; perennial crops/orchards/groves)
Land-use Agriculture (63%), Fisheries/aquaculture (3%), Forestry (15%), Hunting (90%), Tourism/recreation (100%), Urban/industrial/transport (6%)

Birds

Species	Season	Year	Pop min	Pop max	Acc	Criteria
Neophron percnopterus Egyptian Vulture	B	1991	3	4	—	B2, C6
Gyps fulvus Griffon Vulture	W	1991	100	500	—	B1iii, C2
Aquila chrysaetos Golden Eagle	B	1991	1	2	—	C6
Falco peregrinus Peregrine	B	1991	3	4	—	C6
[1] ***Lagopus mutus*** Ptarmigan	B	1991	150	200	—	C6
Tetrao urogallus Capercaillie	B	1991	100	200	—	C6
[2] ***Perdix perdix*** Partridge	B	1991	100	500	—	C6
Dryocopus martius Black Woodpecker	B	1991	50	100	—	C6
Dendrocopos leucotos White-backed Woodpecker	B	1991	10	50	—	C2, C6
Pyrrhocorax pyrrhocorax Chough	B	1991	10	100	—	C6
Emberiza hortulana Ortolan Bunting	B	1991	10	20	—	C6

1. *L. m. pyrenaicus.*
2. *P. p. hispaniensis.*

Protection status

National Partial **International** Partial
10,000 ha of IBA covered by National Park (Parc national des Pyrénées, 45,700 ha). 11,000 ha of IBA covered by Special Protection Area (Hautes Vallées d'Aspe et d'Ossau, 11,000 ha).

Conservation issues

Threats Agricultural intensification/expansion (C), Burning of vegetation (B), Infrastructure (B), Intensified forest management (B), Recreation/tourism (A), Unsustainable exploitation (B)

The site is threatened by intensive forestry, new track construction, extension of a ski-station, and urban expansion.

Eth Turoun des Aureys

B2, C6 **186**

Admin region Aquitaine
Coordinates 43°04′N 0°38′W
Altitude 600–1,300 m **Area** 2,100 ha

Site description

The site contains scree, cliffs, rocky slopes, heathland, meadows and montane *Fagus* forests.

Habitats Forest and woodland (broadleaved deciduous forest), Scrub (5%; heathland), Grassland (mesophile grassland), Wetland (river/stream), Rocky areas (scree/boulders; inland cliff)
Land-use Agriculture, Forestry, Hunting, Tourism/recreation

Birds

Species	Season	Year	Pop min	Pop max	Acc	Criteria
Gyps fulvus Griffon Vulture	B	1991	10	25	—	B2, C6

Protection status

National None **International** None

Conservation issues

Threats Agricultural intensification/expansion (C), Burning of vegetation (B), Disturbance to birds (B), Intensified forest management (B), Recreation/tourism (A), Unsustainable exploitation (C)

The main threat is easy public access leading to disturbance of *Gyps fulvus* colonies and a consequent reduction in their breeding success.

Pènes du Moulle de Jaut

B2, C6 **187**

Admin region Aquitaine
Coordinates 43°02′N 0°21′W
Altitude 550–1,230 m **Area** 4,300 ha

Site description

The site includes areas of scree, cliffs, rocky slopes, montane *Fagus* forests and alpine habitats.

Habitats Forest and woodland (broadleaved deciduous forest), Grassland (alpine/subalpine/boreal grassland), Rocky areas (scree/boulders; inland cliff)
Land-use Agriculture, Forestry, Hunting, Tourism/recreation, Urban/industrial/transport

Birds

Species	Season	Year	Pop min	Pop max	Acc	Criteria
Neophron percnopterus Egyptian Vulture	B	1991	3	3	—	B2, C6
Gyps fulvus Griffon Vulture	B	1991	70	70	—	B2, C6

This site is important for breeding raptors, particularly *Gyps fulvus*.

Protection status

National Low **International** Low

83 ha of IBA covered by Nature Reserve (Réserve Naturelle d'Ossau, 83 ha). 82 ha of IBA covered by Special Protection Area (Réserve Naturelle de Bielle, 82 ha).

Conservation issues

Threats Agricultural intensification/expansion (B), Burning of vegetation (A), Industrialization/urbanization (C), Infrastructure (B), Intensified forest management (C), Recreation/tourism (C), Unsustainable exploitation (C)

The burning of vegetation and the erection of powerlines are the main threats. Hunting (including the illegal shooting of raptors), and recreation (skiing and climbing) are also threats.

Haute Soule: massif de la Pierre-St-Martin — B2, C2, C6 — 188

Admin region Aquitaine
Coordinates 43°02′N 0°44′W
Altitude 589–2,038 m **Area** 18,450 ha

Site description

A montane area with *Fagus* forests, coniferous forests, scree, cliffs, rocky slopes and mountain rivers.

Habitats Forest and woodland (broadleaved deciduous forest; native coniferous forest; mixed forest), Grassland (alpine/subalpine/boreal grassland), Wetland (river/stream), Rocky areas (scree/boulders; inland cliff; inland sand-dunes; caves)
Land-use Agriculture, Forestry, Hunting, Tourism/recreation, Urban/industrial/transport

Birds

Species	Season	Year	Pop min	Pop max	Acc	Criteria
Milvus milvus Red Kite	B	1991	1	10	—	C6
Neophron percnopterus Egyptian Vulture	B	1991	6	6	—	B2, C6
Gyps fulvus Griffon Vulture	B	1991	10	15	—	B2, C6
Falco peregrinus Peregrine	B	1991	5	5	—	C6
Tetrao urogallus Capercaillie	B	1991	10	100	—	C6
[1] ***Perdix perdix*** Partridge	B	1991	10	100	—	C6
Dendrocopos leucotos White-backed Woodpecker	B	1991	5	50	—	C2, C6

1. *P. p. hispaniensis*.

The site is important for breeding raptors, gamebirds and woodpeckers.

Protection status

National None **International** None

Conservation issues

Threats Agricultural intensification/expansion (C), Burning of vegetation (B), Intensified forest management (B), Recreation/tourism (A), Unsustainable exploitation (A)

High hunting pressure, disturbance from increasing recreational activities (climbing) and intensive forestry are the main threats.

Pic de l'Estibet et Mondragon — C6 — 189

Admin region Aquitaine
Coordinates 43°04′N 0°14′W
Altitude 718–1,609 m **Area** 4,550 ha

Site description

An area of *Fagus* forests, scree, cliffs, rocky slopes and mountain rivers.

Habitats Forest and woodland (broadleaved deciduous forest), Wetland (river/stream), Rocky areas (scree/boulders; inland cliff)
Land-use Agriculture, Forestry, Hunting, Tourism/recreation, Urban/industrial/transport

Birds

Species	Season	Year	Pop min	Pop max	Acc	Criteria
Gyps fulvus Griffon Vulture	B	1991	1	5	—	C6

The area is important for breeding raptors and mountain birds.

Protection status

National None **International** None

Conservation issues

No serious threats are known at this site. Management plans limit forestry exploitation, resulting in good conservation of the old forests.

Barrage d'Artix et Saligue du Gave de Pau — C2, C6 — 190

Admin region Aquitaine
Coordinates 43°22′N 0°32′W
Altitude 100–115 m **Area** 3,200 ha

Site description

A reservoir, marshes, and wet forest along the River Pau, with some woodland and arable land. Human activities include agriculture, fishing, tourism (including boating), and gravel extraction.

Habitats Forest and woodland (broadleaved deciduous forest; alluvial/very wet forest), Wetland (standing fresh water; river/stream; water-fringe vegetation), Artificial landscape (arable land)
Land-use Agriculture, Fisheries/aquaculture, Hunting (20%), Other, Tourism/recreation (20%), Urban/industrial/transport

Birds

Species	Season	Year	Pop min	Pop max	Acc	Criteria
Nycticorax nycticorax Night Heron	B	1991	296	296	—	C2, C6
Egretta garzetta Little Egret	B	1997	38	38	—	C6
Milvus migrans Black Kite	B	1991	20	20	—	C6

Species of global conservation concern that do not meet IBA criteria: *Aythya nyroca* (wintering).

Protection status

National Low **International** None

130 ha of IBA covered by Hunting Reserve (Barrage d'Artix, 130 ha).

Conservation issues

Threats Agricultural intensification/expansion (B), Aquaculture/fisheries (C), Recreation/tourism (B), Unsustainable exploitation (C)

Most land-use activities cause disturbance.

Col de Lizarrieta — A4i, B1i, B1iv, C2, C3, C5, C6 — 191

Admin region Aquitaine
Coordinates 43°16′N 1°36′W
Altitude 100–700 m **Area** 1,500 ha

Site description

A lower montane area with broadleaved forests (*Quercus* and *Fagus*) and heathland.

Habitats Forest and woodland (broadleaved deciduous forest), Scrub (heathland)
Land-use Forestry, Hunting

Birds

Species	Season	Year	Pop min	Pop max	Acc	Criteria
Anser anser Greylag Goose	P	1991	1,000	10,000	—	A4i, B1i, C3
Pandion haliaetus Osprey	P	1991	10	50	—	C6
Grus grus Crane	P	1991	1,000	15,000	—	A4i, B1i, C2

The site is a migratory bottleneck, where more than 3,000 cranes and 3,000 raptors regularly pass (including *Pernis apivorus* 2,500; *Milvus milvus* 3,500; *Grus grus* 15,000).

Protection status

National None **International** None

Conservation issues

Threats Burning of vegetation (B), Infrastructure (B), Recreation/tourism (A), Unsustainable exploitation (B)

The high hunting pressure on all migratory birds is the main threat.

Vallée de la Nive des Aldudes–Col de Lindux A4i, A4iv, B1i, B1iv, C2, C5 192

Admin region Aquitaine
Coordinates 43°10'N 1°23'W
Altitude 1,000–1,000 m **Area** 14,500 ha

Site description

The site has deciduous forests, heathland, meadows, scree, cliffs, hedgerows and rivers.

Habitats Forest and woodland (broadleaved deciduous forest)
Land-use Agriculture, Forestry, Hunting, Urban/industrial/transport

Birds

Species	Season	Year	Pop min	Pop max	Acc	Criteria
Grus grus Crane	P	1991	1,000	15,000	—	A4i, B1i, C2

The site is a major migratory bottleneck, where more than 20,000 raptors and 3,000 cranes regularly pass (including *Pernis apivorus* 10,000; *Milvus migrans* 5,000; *Milvus milvus* 2,500; *Grus grus* 15,000).

Protection status

National None **International** None

Conservation issues

Threats Agricultural intensification/expansion (B), Burning of vegetation (B), Intensified forest management (B), Recreation/tourism (A), Unsustainable exploitation (B)

There is high hunting pressure on all migratory birds, irrespective of their conservation status.

Domaine d'Orx A4iii, B1i, B2, C3, C4, C6 193

Admin region Aquitaine
Coordinates 43°36'N 1°20'W
Altitude 5–7 m **Area** 4,250 ha

Site description

A lake, *Pinus* forests, heathland, marshes, wet meadows and cultivated fields.

Habitats Forest and woodland (25%; native coniferous forest), Scrub (30%; heathland), Grassland (humid grassland), Wetland (26%; standing fresh water; water-fringe vegetation), Artificial landscape (13%; arable land)
Land-use Agriculture (22%), Urban/industrial/transport

Birds

Species	Season	Year	Pop min	Pop max	Acc	Criteria
Nycticorax nycticorax Night Heron	B	1991	20	20	—	C6
Egretta garzetta Little Egret	B	1997	20	20	—	C6
Anas crecca Teal	W	1997	1,770	6,660	—	B1i, C3
Anas acuta Pintail	W	1997	414	859	—	B1i, C3
Anas clypeata Shoveler	W	1997	504	2,054	—	B1i, C3
Himantopus himantopus Black-winged Stilt	B	1995	3	3	—	C6
Numenius arquata Curlew	W	1997	38	2,500	—	B2
Alcedo atthis Kingfisher	B	1991	20	30	—	C6

The site holds 20,000 or more waterbirds in winter, on a regular basis (including *Fulica atra* 2,562; *Vanellus vanellus* 3,500). Species of global conservation concern that do not meet IBA criteria: *Aythya nyroca* (wintering); *Haliaeetus albicilla* (wintering); *Acrocephalus paludicola* (passage).

Protection status

National Partial **International** Partial
775 ha of IBA covered by Nature Reserve (Marais d'Orx, 775 ha). 4,250 ha of IBA covered by Hunting Reserve (4,250 ha). 785 ha of IBA covered by Special Protection Area (Domaine d'Orx, 785 ha).

Conservation issues

Threats Agricultural intensification/expansion (A), Industrialization/urbanization (B), Recreation/tourism (B), Unsustainable exploitation (C)

The site was protected in 1986 and has since become of international importance for waterbirds. Nevertheless, water pollution from agricultural run-off remains a threat.

Garonne: marais de Bordeaux B2, C6 194

Admin region Aquitaine
Coordinates 44°57'N 0°30'W
Altitude 2–5 m **Area** 6,100 ha

Site description

A complex of wet meadows, marshes, ponds, rivers and wet forests.

Habitats Forest and woodland (alluvial/very wet forest), Grassland (60%; humid grassland), Wetland (standing fresh water; river/stream; water-fringe vegetation), Artificial landscape (arable land)
Land-use Agriculture, Fisheries/aquaculture, Hunting, Urban/industrial/transport

Birds

Species	Season	Year	Pop min	Pop max	Acc	Criteria
Ardea purpurea Purple Heron	B	1991	7	20	—	C6
Milvus migrans Black Kite	B	1991	100	200	—	B2, C6
Circus aeruginosus Marsh Harrier	B	1997	1	10	—	C6
Alcedo atthis Kingfisher	B	1991	10	100	—	B2, C6
Lanius collurio Red-backed Shrike	B	1991	10	100	—	C6
Lanius senator Woodchat Shrike	B	1991	10	100	—	B2

Protection status

National Low **International** Low
262 ha of IBA covered by Nature Reserve (Marais de Bruges, 262 ha). 276 ha of IBA covered by Special Protection Area (Marais de Bruges, 276 ha).

Conservation issues

Threats Aquaculture/fisheries (B), Industrialization/urbanization (C), Other (B), Recreation/tourism (A), Unsustainable exploitation (B)

Water pollution ('Other' threats, above), angling and high hunting pressure are the main threats.

Estuaire de la Bidassoa et Baie de Fontarabie A4i, A4iii, B1i, B2, C2, C3, C4 195

Admin region Aquitaine
Coordinates 43°24'N 1°42'W
Altitude 0–0 m **Area** 9,300 ha

Site description

An estuary, bay and open sea, with mudflats, sandy beaches, cliffs and rocky islets.

Habitats Wetland (tidal river/enclosed tidal water; mudflat/sandflat; saltmarsh; sand-dunes/sand beach; shingle/stony beach; standing brackish and salt water; river/stream), Marine areas (open sea; sea inlet/coastal features), Rocky areas (sea cliff/rocky shore; inland cliff)
Land-use Fisheries/aquaculture, Tourism/recreation, Urban/industrial/transport

Birds

This is an important site for wintering and passage waterbirds and passage auks (Alcidae). The site holds 20,000 or more waterbirds on passage and in winter, on a regular basis (including *Phalacrocorax carbo* 4,000 (passage); *Anser anser* 10,000 (passage); *Vanellus vanellus* 6,000 (winter)). Species of global conservation concern that do not meet IBA criteria: *Tetrax tetrax* (passage).

Species	Season	Year	Pop min	Pop max	Acc	Criteria
Phalacrocorax carbo Cormorant	P	1991	4,000	4,000	—	B1i, C3
Anser anser Greylag Goose	P	1991	5,000	10,000	—	A4i, B1i, C3
Grus grus Crane	P	1991	2,000	5,000	—	A4i, B1i, C2
Numenius arquata Curlew	W	1991	500	3,000	—	B2
Larus cachinnans Yellow-legged Gull	W	1991	10,000	10,000	—	A4i, B1i, C3

■ Protection status

National High **International** None
9,300 ha of IBA covered by Hunting Reserve (9,300 ha).

■ Conservation issues

No serious threats are known at this site. Disturbance has declined since entry to the estuary was prohibited.

Réserve Naturelle du courant d'Huchet — B2, C6 — 196

Admin region Aquitaine
Coordinates 43°52′N 1°21′W
Altitude 0–30 m **Area** 820 ha

■ Site description

A lake with surrounding *Pinus* and cork oak *Quercus suber* forests, marshes, heathland, beaches and coastal dunes. Forestry, fishing and hunting are the most frequent human activities.

Habitats Forest and woodland (55%; native coniferous forest; alluvial/very wet forest; broadleaved evergreen forest), Scrub (heathland), Wetland (48%; tidal river/enclosed tidal water; sand-dunes/sand beach; standing fresh water; river/stream; raised bog; water-fringe vegetation)
Land-use Fisheries/aquaculture (25%), Forestry (25%), Hunting, Tourism/recreation (100%), Urban/industrial/transport

■ Birds

Species	Season	Year	Pop min	Pop max	Acc	Criteria
Ixobrychus minutus Little Bittern	B	1991	20	20	—	B2, C6
Alcedo atthis Kingfisher	B	1991	10	20	—	C6

The site is important for migrating raptors and passerines. Species of global conservation concern that do not meet IBA criteria: *Haliaeetus albicilla* (passage); *Acrocephalus paludicola* (passage).

■ Protection status

National Partial **International** Partial
617 ha of IBA covered by Nature Reserve (Courant d'Huchet, 617 ha). 140 ha of IBA covered by Hunting Reserve (140 ha). 656 ha of IBA covered by Special Protection Area (Courant d'Huchet, 656 ha).

■ Conservation issues

No serious threats are known at this site.

Arjuzanx — A4i, B1i, C2, C6 — 197

Admin region Aquitaine
Coordinates 44°01′N 0°52′W
Altitude 7–106 m **Area** 2,430 ha

■ Site description

Lakes, *Pinus* forests, acid bogs, meadows, abandoned fields and industrial areas all occur in the site. The main human activity was mining, abandoned in 1995.

Habitats Forest and woodland (10%; native coniferous forest), Grassland (dry siliceous grassland), Wetland (37%; standing fresh water; raised bog), Artificial landscape (52%; highly improved reseeded grassland; other urban/industrial areas; ruderal land)
Land-use Urban/industrial/transport

■ Birds

Species	Season	Year	Pop min	Pop max	Acc	Criteria
Circaetus gallicus Short-toed Eagle	B	1991	1	10	—	C6
Grus grus Crane	W	1997	14,493	14,493	—	A4i, B1i, C2
Caprimulgus europaeus Nightjar	B	1991	10	100	—	C6
Lanius collurio Red-backed Shrike	B	1991	10	100	—	C6

The site is important for breeding raptors and wintering *Grus grus*. Species of global conservation concern that do not meet IBA criteria: *Aythya nyroca* (passage).

■ Protection status

National High **International** None
2,430 ha of IBA covered by Hunting Reserve (2,430 ha).

■ Conservation issues

Threats Recreation/tourism (B)

A new threat is the development of recreation activities around the hunting reserve.

Estuaire de la Gironde: marais du Nord Médoc — B2, C6 — 198

Admin region Aquitaine
Coordinates 45°26′N 1°00′W
Altitude 0–10 m **Area** 22,250 ha

■ Site description

An estuary with mudflats, marshes, wet meadows, cultivated fields, hedgerows and abandoned fields.

Habitats Grassland (humid grassland), Wetland (tidal river/enclosed tidal water; mudflat/sandflat; saltmarsh; standing fresh water; standing brackish and salt water; river/stream; water-fringe vegetation), Artificial landscape (arable land; perennial crops/orchards/groves; ruderal land)
Land-use Agriculture, Fisheries/aquaculture, Hunting, Tourism/recreation, Urban/industrial/transport

■ Birds

Species	Season	Year	Pop min	Pop max	Acc	Criteria
Milvus migrans Black Kite	B	1991	10	100	—	B2, C6
Circus aeruginosus Marsh Harrier	B	1997	1	10	—	C6
Circus pygargus Montagu's Harrier	B	1991	20	30	—	C6
Recurvirostra avosetta Avocet	W	1997	51	160	—	B2
Alcedo atthis Kingfisher	B	1991	10	100	—	B2, C6
Lanius collurio Red-backed Shrike	B	1991	100	500	—	B2, C6
Lanius senator Woodchat Shrike	B	1991	10	100	—	B2

This site is of regional importance for wintering waterbirds.

■ Protection status

National None **International** Low
1,243 ha of IBA covered by Special Protection Area (Zones humides du Nord Ouest Médoc, 1,243 ha).

■ Conservation issues

Threats Abandonment/reduction of land management (A), Agricultural intensification/ expansion (B), Aquaculture/fisheries, Unsustainable exploitation (B)

The abandonment of cattle-grazing, significant hunting pressure on all species, and the creation of pools for aquaculture are all threats.

Estuaire de la Gironde: marais du Blayais — C6 — 199

Admin region Aquitaine
Coordinates 45°14′N 0°38′W
Altitude 0–2 m **Area** 7,050 ha

■ Site description

Habitats Forest and woodland (broadleaved deciduous forest), Grassland (humid grassland), Wetland (mudflat/sandflat; saltmarsh; standing brackish and salt water; water-fringe vegetation), Artificial landscape (arable land)
Land-use Agriculture, Fisheries/aquaculture, Hunting, Other, Urban/industrial/transport

Estuary with streams, reedbeds (*Phragmites*), marshes and wet meadows, *Quercus* and *Salix* woodlands, mudflats, saltmarshes and peatbogs.

Birds

Species	Season	Year	Pop min	Pop max	Acc	Criteria
Ardea purpurea Purple Heron	B	1991	47	47	—	C6
Circus aeruginosus Marsh Harrier	B	1997	6	6	—	C6
Circus cyaneus Hen Harrier	B	1991	4	4	—	C6
Circus pygargus Montagu's Harrier	B	1991	23	27	—	C6

This site is important for species associated with reedbeds.

Protection status

National Low **International** None
125 ha of IBA covered by Hunting Reserve (125 ha).

Conservation issues

Threats Aquaculture/fisheries (A), Extraction industry (B), Recreation/tourism (A), Unsustainable exploitation (A)

Recreational development in the area is one of the major threats.

Haute Cize: Pic d'Errozate et Forêt d'Orion — B2, C6 — 200

Admin region Aquitaine
Coordinates 43°04′N 1°10′W
Altitude 400–1,345 m **Area** 6,450 ha

Site description

A mountain area with deciduous and conifer forests, cliffs and rocky slopes, meadows, cultivated fields, peatbogs and rivers.

Habitats Forest and woodland (broadleaved deciduous forest; native coniferous forest), Grassland (mesophile grassland), Wetland (river/stream; raised bog), Rocky areas (scree/boulders; inland cliff), Artificial landscape (arable land)
Land-use Agriculture, Fisheries/aquaculture, Hunting, Tourism/recreation, Urban/industrial/transport

Birds

Species	Season	Year	Pop min	Pop max	Acc	Criteria
Neophron percnopterus Egyptian Vulture	B	1991	3	3	—	B2, C6
Falco peregrinus Peregrine	B	1991	3	3	—	C6

The site is important for breeding raptors, including *Gypaetus barbatus*.

Protection status

National None **International** None

Conservation issues

Threats Agricultural intensification/expansion (C), Aquaculture/fisheries (C), Burning of vegetation (B), Intensified forest management (B), Recreation/tourism (A), Unsustainable exploitation (B)

High hunting and fishing pressure, and the development of recreation and sporting activities, such as off-road driving, are the main threats.

Côte Médocaine: dunes boisées et dépressions humides — B1i, C2, C6 — 201

Admin region Aquitaine
Coordinates 45°02′N 1°08′W
Altitude 0–63 m **Area** 3,900 ha

Site description

Wooded sand-dunes (with *Pinus* and *Quercus*), marshes, ponds, wet meadows and abandoned fields.

Habitats Forest and woodland (broadleaved deciduous forest; native coniferous forest; mixed forest), Scrub (heathland), Grassland (dry siliceous grassland; humid grassland; mesophile grassland), Wetland (sand-dunes/sand beach; standing fresh water; raised bog; water-fringe vegetation), Artificial landscape (ruderal land)
Land-use Agriculture, Fisheries/aquaculture, Forestry, Hunting, Tourism/recreation

Birds

Species	Season	Year	Pop min	Pop max	Acc	Criteria
Milvus migrans Black Kite	B	1991	20	25	—	C6
Circaetus gallicus Short-toed Eagle	B	1991	3	3	—	C6
Grus grus Crane	P	1991	800	1,000	—	B1i, C2
Caprimulgus europaeus Nightjar	B	1991	50	50	—	C6
Lullula arborea Woodlark	B	1991	25	50	—	C6
Anthus campestris Tawny Pipit	B	1991	10	25	—	C6

This site is important for breeding raptors as a stop-over area for migratory waterbirds.

Protection status

National Partial **International** Partial
600 ha of IBA covered by Nature Reserve (Etang de Cousseau, 600 ha). 600 ha of IBA covered by Special Protection Area (Etang de Cousseau, 600 ha).

Conservation issues

Threats Unsustainable exploitation (C)

The illegal hunting of breeding and migrating birds of prey is still common at this site.

Vallées du Lis et de la Pique — B2, C6 — 202

Admin region Midi-Pyrénées
Coordinates 42°44′N 0°36′E
Altitude 826–3,116 m **Area** 9,600 ha

Site description

A mountainous area in the Pyrenees with cliffs, alpine grassland, arable land and woodlands.

Habitats Forest and woodland (50%; native coniferous forest; mixed forest), Grassland (40%; alpine/subalpine/boreal grassland), Rocky areas (9%; scree/boulders; inland cliff), Artificial landscape (highly improved reseeded grassland; arable land)
Land-use Agriculture, Fisheries/aquaculture, Forestry, Hunting, Other, Tourism/recreation, Urban/industrial/transport

Birds

Species	Season	Year	Pop min	Pop max	Acc	Criteria
Gypaetus barbatus Lammergeier	B	1991	1	1	—	B2
Tetrao urogallus Capercaillie	B	1991	100	500	—	C6
[1] ***Perdix perdix*** Partridge	B	1991	100	500	—	C6
Tetrax tetrax Little Bustard	B	1996	5	10	—	C6
Dendrocopos leucotos White-backed Woodpecker	B	1991	1	10	—	C6
Pyrrhocorax pyrrhocorax Chough	B	1991	10	100	—	C6

1. *P. p. hispaniensis*.

Protection status

National Partial **International** None
1,000 ha of IBA covered by Hunting Reserve (1,000 ha).

Conservation issues

Threats Disturbance to birds (B), Intensified forest management (B), Recreation/tourism (C), Unsustainable exploitation (C)

The site is threatened by the construction of new roads for forestry and tourism. There are also plans to build a new ski-station.

Cirque de Gavarnie — B2, C6 — 203

Admin region Midi-Pyrénées
Coordinates 42°44′N 0°04′E
Altitude 1,374–3,248 m **Area** 7,600 ha

Site description

Habitats Forest and woodland (10%; native coniferous forest), Grassland (70%; alpine/subalpine/boreal grassland), Wetland (river/stream), Rocky areas (20%; scree/boulders; inland cliff)
Land-use Agriculture, Forestry, Hunting, Tourism/recreation

Birds

Species	Season	Year	Pop min	Pop max	Acc	Criteria
Gypaetus barbatus Lammergeier	B	1991	1	1	—	B2
[1] ***Perdix perdix*** Partridge	B	1991	10	100	—	C6
Pyrrhocorax pyrrhocorax Chough	B	1991	10	100	—	C6

1. *P. p. hispaniensis*.

Protection status

National Partial **International** Partial

3,800 ha of IBA covered by National Park (Parc National des Pyrénées, 45,700 ha). 3,000 ha of IBA covered by Special Protection Area (Cirque de Gavarnie, 3,000 ha).

Conservation issues

Threats Recreation/tourism (A), Infrastructure (U)

The development of tourism, the construction of roads, and the erection of powerlines threaten the site.

Vallée de Melles, cols d'Aoueran et d'Artisagou et Mont Vallier — B2, C6 — 204

Admin region Midi-Pyrénées
Coordinates 42°50'N 0°58'E
Altitude 700–2,838 m **Area** 40,100 ha

Site description

Alpine meadows, with conifer and deciduous forests, scree and rocky slopes, glaciers and permanent snow.

Habitats Forest and woodland (broadleaved deciduous forest; native coniferous forest), Grassland (alpine/subalpine/boreal grassland), Rocky areas (scree/boulders; inland cliff)
Land-use Agriculture, Fisheries/aquaculture, Forestry, Hunting, Tourism/recreation, Urban/industrial/transport

Birds

Species	Season	Year	Pop min	Pop max	Acc	Criteria
Gypaetus barbatus Lammergeier	B	1996	1	2	—	B2
Aquila chrysaetos Golden Eagle	B	1991	4	5	—	C6

The site is important for breeding raptors and gamebirds but population sizes are not well known.

Protection status

National Partial **International** None

8,815 ha of IBA covered by Hunting Reserve (Haut-Vallier, 8,815 ha).

Conservation issues

No serious threats are known at this site.

Zones rupestres du Tarasconnais et massif d'Aston — B2, C6 — 205

Admin region Midi-Pyrénées
Coordinates 42°44'N 1°41'E
Altitude 550–2,730 m **Area** 23,300 ha

Site description

Habitats Forest and woodland (broadleaved deciduous forest; native coniferous forest; mixed forest), Grassland (alpine/subalpine/boreal grassland), Rocky areas (5%; scree/boulders; inland cliff)
Land-use Agriculture, Forestry, Hunting, Other, Tourism/recreation, Urban/industrial/transport, Water management

Birds

Species	Season	Year	Pop min	Pop max	Acc	Criteria
Gypaetus barbatus Lammergeier	B	1991	1	1	—	B2
Falco peregrinus Peregrine	B	1991	1	10	—	B2, C6

The site is important for montane birds.

Protection status

National None **International** None

Conservation issues

Threats Agricultural intensification/expansion (C), Burning of vegetation (B), Recreation/tourism (B), Unsustainable exploitation (C)

The construction of an access road for tourists and the development of a ski-station threaten the site.

Vallée de la Garonne: Boussens à Carbonne — B2, C6 — 206

Admin region Midi-Pyrénées
Coordinates 43°14'N 1°06'E
Altitude 205–238 m **Area** 2,300 ha

Site description

The site includes a river, reservoir, ponds, wet and mesophile meadows, patches of woodland, marshes and aquatic vegetation, arable fields and *Populus* plantations.

Habitats Grassland (humid grassland; mesophile grassland), Wetland (75%; standing fresh water; river/stream; water-fringe vegetation), Artificial landscape (arable land; perennial crops/orchards/groves, forestry plantation)
Land-use Agriculture, Fisheries/aquaculture, Hunting, Tourism/recreation, Urban/industrial/transport, Water management

Birds

Species	Season	Year	Pop min	Pop max	Acc	Criteria
Ixobrychus minutus Little Bittern	B	1991	4	4	—	C6
Nycticorax nycticorax Night Heron	B	1991	210	210	—	C6
Egretta garzetta Little Egret	B	1991	19	19	—	C6
Milvus migrans Black Kite	B	1991	10	100	—	B2

Protection status

National Low **International** None

100 ha of IBA covered by Hunting Reserve (100 ha).

Conservation issues

Threats Disturbance to birds (A), Extraction industry, Industrialization/urbanization, Infrastructure (C), Recreation/tourism (A)

Industrial pollution of the river, destruction of riverine vegetation, the extension of gravel-pits, and recreation (boating) are the main threats.

Vallée de la Garonne: Palayre et environs — A4i, B1i, C2, C6 — 207

Admin region Midi-Pyrénées
Coordinates 43°34'N 1°24'E
Altitude 120–120 m **Area** 1,700 ha

Site description

Habitats Forest and woodland (alluvial/very wet forest), Wetland (standing fresh water; river/stream; water-fringe vegetation), Artificial landscape (arable land; urban parks/gardens; other urban/industrial areas; ruderal land)
Land-use Agriculture, Fisheries/aquaculture, Hunting, Tourism/recreation, Urban/industrial/transport

Birds

Species	Season	Year	Pop min	Pop max	Acc	Criteria
Ixobrychus minutus Little Bittern	B	1991	1	10	—	C6
Nycticorax nycticorax Night Heron	B	1991	530	730	—	A4i, B1i, C2, C6
Egretta garzetta Little Egret	B	1991	7	26	—	C6

The site holds an important heronry.

Protection status

National Low **International** None

Part of IBA covered by Hunting Reserve.

Conservation issues

Threats Aquaculture/fisheries (A), Industrialization/urbanization (B), Recreation/tourism (C), Unsustainable exploitation (A)

Industrial pollution of the river, urbanization, and the development of recreation (boating) are the main threats.

Vallée de la Garonne: Moissac — A4i, B1i, C2, C6 — 208

Admin region Midi-Pyrénées
Coordinates 44°04′N 1°04′E
Altitude 130–130 m **Area** 1,300 ha

Site description

The site contains a river, reservoir, riverine forest, marshes, reedbeds (*Phragmites*), *Populus* plantations, orchards and abandoned fields.

Habitats Forest and woodland (alluvial/very wet forest), Wetland (standing fresh water; river/stream; water-fringe vegetation), Artificial landscape (perennial crops/orchards/groves; ruderal land)
Land-use Fisheries/aquaculture, Hunting, Tourism/recreation, Urban/industrial/transport, Water management

Birds

Species	Season	Year	Pop min	Pop max	Acc	Criteria
Nycticorax nycticorax Night Heron	B	1991	900	900	—	A4i, B1i, C2, C6
Sterna hirundo Common Tern	B	1991	9	9	—	C6

Species of global conservation concern that do not meet IBA criteria: *Aythya nyroca* (wintering).

Protection status

National None **International** None

Conservation issues

Threats Aquaculture/fisheries (B), Recreation/tourism (A), Unsustainable exploitation (B)

Disturbance due to fishing and recreation (boating) threaten the *Sterna hirundo* colony.

Vallée de la Garonne: méandre de Grisolles — C6 — 209

Admin region Midi-Pyrénées
Coordinates 43°47′N 1°17′E
Altitude 100–100 m **Area** 1,050 ha

Site description

The site contains a river with oxbow lakes, a reservoir, riverine forest, marshes, reedbeds (*Phragmites*), *Populus* plantations and abandoned fields.

Habitats Forest and woodland (alluvial/very wet forest), Wetland (river/stream; water-fringe vegetation), Artificial landscape (ruderal land)
Land-use Fisheries/aquaculture, Hunting, Water management

Birds

Species	Season	Year	Pop min	Pop max	Acc	Criteria
Ardea purpurea Purple Heron	B	1991	25	25	—	C6

Protection status

National None **International** None

Conservation issues

Threats Agricultural intensification/expansion (A), Drainage (A), Extraction industry (B)

Oxbow lakes are drying out as a result of being used to irrigate surrounding fields. Agricultural pollution and the extension of gravel-pits are the other main threats.

Gorges de la Frau et Belestat — C6 — 210

Admin region Midi-Pyrénées
Coordinates 42°50′N 1°55′E
Altitude 549–1,868 m **Area** 14,000 ha

Site description

A mountain area with mixed *Fagus–Abies* forests, areas of scree, cliffs and rocky slopes.

Habitats Forest and woodland (broadleaved deciduous forest; native coniferous forest), Rocky areas (scree/boulders; inland cliff)
Land-use Forestry, Hunting, Tourism/recreation, Urban/industrial/transport

Birds

Species	Season	Year	Pop min	Pop max	Acc	Criteria
Falco peregrinus Peregrine	B	1991	4	4	—	C6
[1] ***Perdix perdix*** Partridge	B	1991	10	100	—	C6

1. *P. p. hispaniensis*.

Protection status

National None **International** None

Conservation issues

Threats Intensified forest management (A)

In the future, forest exploitation may threaten *Tetrao urogallus* leks.

Gorges du Tarn et de la Jonte — A1, B2, C1, C6 — 211

Admin region Midi-Pyrénées
Coordinates 44°16′N 2°17′E
Altitude 390–1,000 m **Area** 11,100 ha

Site description

A fast-flowing river with woodland, scrub and cliffs. Major roads run through the area.

Habitats Forest and woodland (native coniferous forest), Scrub (sclerophyllous scrub/garrigue/maquis), Wetland (river/stream), Rocky areas (scree/boulders; inland cliff)
Land-use Agriculture, Fisheries/aquaculture, Forestry, Hunting, Tourism/recreation, Urban/industrial/transport

Birds

Species	Season	Year	Pop min	Pop max	Acc	Criteria
Gyps fulvus Griffon Vulture	B	1997	50	50	—	B2, C6
Aegypius monachus Cinereous Vulture	W	1997	16	16	—	A1, C1
Circaetus gallicus Short-toed Eagle	B	1991	8	12	—	B2, C6
Aquila chrysaetos Golden Eagle	B	1991	4	4	—	C6
Falco peregrinus Peregrine	B	1991	3	4	—	C6
Bubo bubo Eagle Owl	B	1991	5	5	—	C6
Dryocopus martius Black Woodpecker	B	1991	10	20	—	C6
Pyrrhocorax pyrrhocorax Chough	B	1991	30	40	—	C6

This is an important site for breeding raptors. Species of global conservation concern that do not meet IBA criteria: *Aegypius monachus* (breeding).

Protection status

National None **International** Low
130 ha of IBA covered by Special Protection Area (Gorge de la Jonte, 130 ha).

Conservation issues

Threats Disturbance to birds (A), Intensified forest management (A), Recreation/tourism (A)

Mass tourism disturbs breeding species. *Aegypius monachus* and *Gyps fulvus* have been successfully reintroduced.

Gorges de la Dourbie et causses avoisinants — A1, B2, C1, C6 — 212

Admin region Midi-Pyrénées
Coordinates 44°05′N 2°11′E
Altitude 360–1,200 m **Area** 23,000 ha

Site description

This site comprises rocky habitats, a river, agricultural land, woodland and garrigue.

Habitats Forest and woodland (broadleaved deciduous forest; native coniferous forest; alluvial/very wet forest), Scrub (heathland; sclerophyllous scrub/garrigue/maquis), Grassland (steppe/dry calcareous grassland), Wetland (river/stream), Rocky areas (scree/boulders; inland cliff; inland sand-dunes; caves), Artificial landscape (highly improved reseeded grassland; arable land; perennial crops/orchards/groves)
Land-use Agriculture, Fisheries/aquaculture, Forestry, Hunting, Other, Tourism/recreation, Urban/industrial/transport

Birds

Species	Season	Year	Pop min	Pop max	Acc	Criteria
Circaetus gallicus Short-toed Eagle	B	1991	5	10	—	C6
Circus cyaneus Hen Harrier	B	1991	4	8	—	C6
Circus pygargus Montagu's Harrier	B	1991	3	7	—	C6
Falco peregrinus Peregrine	B	1991	3	4	—	C6
Tetrax tetrax Little Bustard	B	1991	10	20	—	A1, C1, C6
Burhinus oedicnemus Stone Curlew	B	1991	10	100	—	B2, C6
Bubo bubo Eagle Owl	B	1991	12	12	—	C6
Caprimulgus europaeus Nightjar	B	1991	10	100	—	C6
Dryocopus martius Black Woodpecker	B	1991	10	100	—	C6
Anthus campestris Tawny Pipit	B	1991	10	100	—	C6
Sylvia undata Dartford Warbler	B	1991	100	500	—	B2, C6
Lanius collurio Red-backed Shrike	B	1991	10	100	—	C6
Pyrrhocorax pyrrhocorax Chough	B	1991	10	100	—	C6

Protection status

National Partial **International** Partial
6,400 ha of IBA covered by National Park (Parc National des Cévennes, 91,500 ha). 6,400 ha of IBA covered by Biosphere Reserve (9,600 ha). 6,400 ha of IBA covered by Special Protection Area (Les Cévennes, 84,000 ha).

Conservation issues

Threats Agricultural intensification/expansion, Intensified forest management (A), Recreation/tourism (A)

Two subsites, Causses Noir and du Larzac, are threatened by agricultural intensification, forestry operations and tourism. Hunting pressure is also high.

Montagne de la Clape — B1iv, C5, C6 — 213

Admin region Languedoc-Roussillon
Coordinates 43°10′N 3°08′E
Altitude 10–200 m **Area** 10,400 ha

Site description

Habitats Forest and woodland (native coniferous forest), Scrub (sclerophyllous scrub/garrigue/maquis), Rocky areas (scree/boulders; inland cliff; inland sand-dunes; caves)
Land-use Agriculture, Hunting, Tourism/recreation, Urban/industrial/transport

Birds

Species	Season	Year	Pop min	Pop max	Acc	Criteria
Bubo bubo Eagle Owl	B	1998	10	15	—	C6
Caprimulgus europaeus Nightjar	B	1997	20	50	—	C6
Calandrella brachydactyla Short-toed Lark	B	1997	10	20	—	C6
Anthus campestris Tawny Pipit	B	1997	20	50	—	C6
Emberiza hortulana Ortolan Bunting	B	1997	20	50	—	C6

The site is important for species typical of Mediterranean habitats, such as *Sylvia conspicillata*, *S. undata* and *Calandrella brachydactyla*. The site is a migratory bottleneck, where 3,000 or more raptors regularly pass in spring (including *Pernis apivorus* 2,000; *Milvus migrans* 3,000). Species of global conservation concern that do not meet IBA criteria: *Circus macrourus* (passage); *Aquila clanga* (passage).

Protection status

National None **International** Partial
7,500 ha of IBA covered by Special Protection Area (Montagne de la Clape, 7,500 ha).

Conservation issues

Threats Agricultural intensification/expansion (B), Industrialization/urbanization (A), Intensified forest management (A), Recreation/tourism (A), Unsustainable exploitation (A)

The main threats are the expansion of intensive viticulture, conifer plantations and recreation during summer. Spring hunting is also a threat.

Etangs de Leucate et Lapalme — A4i, A4iv, B1i, B1iv, B2, C2, C3, C5, C6 — 214

Admin region Languedoc-Roussillon
Coordinates 42°53′N 3°00′E
Altitude 0–69 m **Area** 13,650 ha

Site description

A complex of coastal habitats, including saltmarshes and lagoons.

Habitats Forest and woodland (native coniferous forest; alluvial/very wet forest; broadleaved evergreen forest), Scrub (sclerophyllous scrub/garrigue/maquis), Grassland (steppe/dry calcareous grassland), Wetland (tidal river/enclosed tidal water; mudflat/sandflat; saltmarsh; sand-dunes/sand beach; coastal lagoon; standing brackish and salt water; river/stream; water-fringe vegetation; fen/transition mire/spring), Rocky areas (sea cliff/rocky shore; rock stacks/islets), Artificial landscape (highly improved reseeded grassland; arable land; perennial crops/orchards/groves; urban parks/gardens; ruderal land)
Land-use Agriculture, Fisheries/aquaculture, Hunting, Tourism/recreation, Urban/industrial/transport

Birds

Species	Season	Year	Pop min	Pop max	Acc	Criteria
Phoenicopterus ruber Greater Flamingo	W	1997	529	805	—	A4i, B1i, C2
Phoenicopterus ruber Greater Flamingo	P	1997	500	1,500	—	A4i, B1i, C2
Circus aeruginosus Marsh Harrier	B	1997	10	15	—	C6
Recurvirostra avosetta Avocet	B	1997	50	60	—	C6
Charadrius alexandrinus Kentish Plover	B	1997	100	180	—	B2
Larus genei Slender-billed Gull	B	1997	50	100	—	C2
Larus cachinnans Yellow-legged Gull	B	1997	4,000	4,000	—	A4i, B1i, C3
Larus cachinnans Yellow-legged Gull	W	1997	5,615	8,700	—	A4i, B1i, C3
Sterna albifrons Little Tern	B	1997	50	100	—	C6
Calandrella brachydactyla Short-toed Lark	B	1997	50	250	—	C6
Acrocephalus melanopogon Moustached Warbler	B	1997	50	250	—	C2, C6

The site is very important for breeding and wintering wetland birds and passage raptors and passerines. The site is a major migratory bottleneck, where 20,000 or more raptors regularly pass (including *Pernis apivorus* 15,000; *Milvus migrans* 3,000). Species of global conservation concern that do not meet IBA criteria: *Larus audouinii* (passage).

Protection status

National Low **International** None
1,293 ha of IBA covered by 2 Hunting Reserves (Salses–Leucate, 1,250 ha; La Rigole; total area 1,293 ha).

Conservation issues

Threats Agricultural intensification/expansion (A), Industrialization/urbanization (A), Recreation/tourism (A), Unsustainable exploitation (C)

Urbanization, the consolidation of land into larger more homogeneous units, and the expansion of tourism are the main threats.

Etangs Narbonnais — A4i, A4iii, B1i, B1iv, B2, C2, C3, C4, C5, C6 — 215

Admin region Languedoc-Roussillon
Coordinates 43°05′N 3°03′E
Altitude 0–69 m **Area** 10,600 ha

Site description

The area contains beaches, sand-dunes, islands, saltwater lagoons and marshes, mudflats, woodland, maquis and cliffs.

Habitats Forest and woodland (broadleaved deciduous forest; native coniferous forest), Scrub (heathland; sclerophyllous scrub/garrigue/maquis), Grassland (steppe/dry calcareous grassland), Wetland (mudflat/sandflat; saltmarsh; sand-dunes/sand beach; coastal lagoon; standing brackish and salt water; water-fringe vegetation), Rocky areas (rock stacks/islets; scree/boulders; inland cliff; inland sand-dunes; caves)
Land-use Fisheries/aquaculture, Hunting, Tourism/recreation, Urban/industrial/transport

Birds

Species	Season	Year	Pop min	Pop max	Acc	Criteria
Egretta garzetta Little Egret	B	1997	100	100	—	C6
Phoenicopterus ruber Greater Flamingo	W	1997	—	3,338	—	A4i, B1i, C2
Phoenicopterus ruber Greater Flamingo	P	1997	1,200	1,200	—	A4i, B1i, C2
Charadrius alexandrinus Kentish Plover	B	1997	180	205	—	B2
Calidris minuta Little Stint	W	1997	—	2,100	—	A4i, B1i, C3
Larus cachinnans Yellow-legged Gull	B	1997	3,500	3,500	—	A4i, B1i, C3
Sterna albifrons Little Tern	B	1997	50	200	—	B2, C6
Chlidonias hybridus Whiskered Tern	P	1997	100	300	—	B1i, C2
Calandrella brachydactyla Short-toed Lark	B	1997	20	50	—	C6

The site holds 20,000 or more waterbirds in winter, on a regular basis (including *Phoenicopterus ruber* 3,338; *Anas platyrhynchos* 2,520; *Aythya ferina* 2,200; *Fulica atra* 20,000). The site is a migratory bottleneck, where more than 3,000 raptors regularly pass (including *Pernis apivorus* 2,000; *Milvus migrans* 3,000). Species of global conservation concern that do not meet IBA criteria: *Aquila clanga* (passage, wintering); *Larus audouinii* (passage).

Protection status

National Partial **International** None
2,130 ha of IBA covered by 3 Hunting Reserves (Etangs de Bage et Sigean; Etang de St Paul; Etang de Campignol; total area 2,130 ha).

Conservation issues

Threats Agricultural intensification/expansion (A), Industrialization/urbanization (A), Recreation/tourism (A), Unsustainable exploitation (A)

High hunting pressure, eutrophication and tourism development, including boating, are the main threats to the site.

Etang de Capestang — B2, C2, C6 — 216

Admin region Languedoc-Roussillon
Coordinates 43°18′N 3°04′E
Altitude 0–3 m **Area** 1,360 ha

Site description

A lake and marshes with aquatic vegetation.

Habitats Grassland (humid grassland), Wetland (standing fresh water; water-fringe vegetation)
Land-use Agriculture, Fisheries/aquaculture, Hunting, Tourism/recreation, Urban/industrial/transport

Birds

Species	Season	Year	Pop min	Pop max	Acc	Criteria
Botaurus stellaris Bittern	B	1997	5	10	—	C6
Ixobrychus minutus Little Bittern	B	1997	5	10	—	C6
Ardea purpurea Purple Heron	B	1990	—	150	—	B2, C2, C6
Chlidonias hybridus Whiskered Tern	B	1997	20	30	—	B2, C6
Acrocephalus melanopogon Moustached Warbler	B	1997	40	250	—	C2, C6

The site is particularly important for species found in reedbeds (*Phragmites*).

Protection status

National None **International** None

Conservation issues

Threats Agricultural intensification/expansion (A), Industrialization/urbanization (B), Infrastructure (A), Unsustainable exploitation (B)

Illegal night hunting using decoys, water-level management and industrial and urban pollution, all threaten the site.

Hautes Corbières — B1iv, B2, C5, C6 — 217

Admin region Languedoc-Roussillon
Coordinates 43°01′N 2°36′E
Altitude 89–1,230 m **Area** 74,750 ha

Site description

The IBA includes maquis, garrigue, forest and cliffs.

Habitats Forest and woodland (broadleaved deciduous forest; native coniferous forest; mixed forest; alluvial/very wet forest; broadleaved evergreen forest), Scrub (heathland; sclerophyllous scrub/garrigue/maquis), Grassland (steppe/dry calcareous grassland), Wetland (standing fresh water; river/stream; raised bog; fen/transition mire/spring), Rocky areas (scree/boulders; inland cliff; inland sand-dunes; caves), Artificial landscape (arable land; perennial crops/orchards/groves; ruderal land)
Land-use Agriculture, Fisheries/aquaculture, Forestry, Hunting, Tourism/recreation, Urban/industrial/transport

Birds

Species	Season	Year	Pop min	Pop max	Acc	Criteria
Pernis apivorus Honey Buzzard	B	1997	10	30	—	C6
Circaetus gallicus Short-toed Eagle	B	1997	20	50	—	B2, C6
Falco peregrinus Peregrine	B	1991	5	10	—	B2, C6
Bubo bubo Eagle Owl	B	1998	30	50	—	B2, C6
Caprimulgus europaeus Nightjar	B	1997	50	100	—	C6
Sylvia undata Dartford Warbler	B	1997	100	200	—	C6
Lanius collurio Red-backed Shrike	B	1997	100	200	—	C6
Pyrrhocorax pyrrhocorax Chough	B	1991	10	100	—	C6

The site is a migratory bottleneck, where more than 3,000 raptors regularly pass (including *Pernis apivorus* 3,000; *Milvus migrans* 2,000). Species of global conservation concern that do not meet IBA criteria: *Tetrax tetrax* (breeding).

Protection status

National Low **International** None
Part of IBA covered by Hunting Reserve.

Conservation issues

Threats Abandonment/reduction of land management (B), Agricultural intensification/expansion (C), Recreation/tourism (A)

The abandonment of grazing, and recreation activities are threats.

Basses Corbières — A4iv, B1iv, B2, B3, C5, C6 — 218

Admin region Languedoc-Roussillon
Coordinates 42°53′N 2°43′E
Altitude 34–969 m **Area** 47,400 ha

Site description

The site contains maquis, garrigue, vineyards and rocky slopes.

Habitats Forest and woodland (broadleaved deciduous forest; native coniferous forest; alluvial/very wet forest; broadleaved evergreen forest), Scrub (heathland; sclerophyllous scrub/garrigue/maquis), Grassland (steppe/dry calcareous grassland), Wetland (standing fresh water; river/stream), Rocky areas (scree/boulders; inland cliff; inland sand-dunes; caves), Artificial landscape (perennial crops/orchards/groves; ruderal land)
Land-use Agriculture, Fisheries/aquaculture, Forestry, Hunting, Tourism/recreation, Urban/industrial/transport

Birds

This is an important site for species associated with Mediterranean habitats. It is also a major migratory bottleneck, where more than

20,000 raptors regularly pass (including *Pernis apivorus* 20,000; *Milvus migrans* 5,000). Species of global conservation concern that do not meet IBA criteria: *Tetrax tetrax* (breeding).

Species	Season	Year	Pop min	Pop max	Acc	Criteria
Circaetus gallicus Short-toed Eagle	B	1997	20	30	—	B2, C6
Circus pygargus Montagu's Harrier	B	1997	30	50	—	B3, C6
Hieraaetus fasciatus Bonnelli's Eagle	B	1996	2	2	—	B2
Bubo bubo Eagle Owl	B	1997	30	50	—	B2, C6
Caprimulgus europaeus Nightjar	B	1997	50	100	—	C6
Galerida theklae Thekla Lark	B	1996	50	100	—	C6
Sylvia undata Dartford Warbler	B	1997	200	300	—	C6

■ Protection status

National Low **International** None
285 ha of IBA covered by Hunting Reserve (285 ha).

■ Conservation issues

Threats Abandonment/reduction of land management (A), Agricultural intensification/expansion (A), Industrialization/urbanization (B), Infrastructure (A), Recreation/tourism (A)

Land abandonment and recreational developments are the main threats.

Gorges de la Vis et cirque de Navacelles — C6 — 219

Admin region Languedoc-Roussillon
Coordinates 43°52′N 3°34′E
Altitude 207–862 m **Area** 24,400 ha

■ Site description

A mountainous area with cliffs, maquis, woodland, agricultural land, marshes, fast-flowing rivers and streams.

Habitats Forest and woodland (broadleaved deciduous forest; native coniferous forest; alluvial/very wet forest; broadleaved evergreen forest), Scrub (heathland; sclerophyllous scrub/garrigue/maquis), Grassland (steppe/dry calcareous grassland), Wetland (river/stream; water-fringe vegetation; fen/transition mire/spring), Rocky areas (scree/boulders; inland cliff; inland sand-dunes; caves), Artificial landscape (highly improved reseeded grassland; arable land; perennial crops/orchards/groves; ruderal land)
Land-use Agriculture, Fisheries/aquaculture, Forestry, Hunting, Tourism/recreation, Urban/industrial/transport

■ Birds

Species	Season	Year	Pop min	Pop max	Acc	Criteria
Lullula arborea Woodlark	B	1997	50	100	—	C6
Anthus campestris Tawny Pipit	B	1991	10	100	—	C6
Emberiza hortulana Ortolan Bunting	B	1997	25	100	—	C6

The site is important for raptors, particularly *Aegypius monachus* and *Gyps fulvus*, whose numbers are increasing. Species of global conservation concern that do not meet IBA criteria: *Aegypius monachus* (1–10 wintering individuals); *Tetrax tetrax* (breeding).

■ Protection status

National None **International** None

■ Conservation issues

Threats Abandonment/reduction of land management (A), Agricultural intensification/expansion (A), Disturbance to birds (A), Infrastructure (B), Recreation/tourism (A)

The development of tourism in the area is a major threat. There is a LIFE project on vulture conservation.

Etangs Montpellierains — A1, A4i, A4iii, B1i, B2, B3, C1, C2, C3, C4, C6 — 220

Admin region Languedoc-Roussillon
Coordinates 43°32′N 3°58′E
Altitude 0–5 m **Area** 12,700 ha

■ Site description

Saltwater lagoons and saltmarshes with reedbeds (*Phragmites*), abandoned salinas and orchards.

Habitats Wetland (saltmarsh; sand-dunes/sand beach; shingle/stony beach; coastal lagoon; standing brackish and salt water; water-fringe vegetation), Artificial landscape (arable land)
Land-use Agriculture, Fisheries/aquaculture, Hunting, Tourism/recreation, Urban/industrial/transport

■ Birds

Species	Season	Year	Pop min	Pop max	Acc	Criteria
Podiceps nigricollis Black-necked Grebe	W	1997	184	2,318	—	A4i, B1i, C3
Podiceps nigricollis Black-necked Grebe	P	1997	200	1,500	—	A4i, B1i, C3
Phalacrocorax carbo Cormorant	P	1997	1,000	2,000	—	B1i, C3
Botaurus stellaris Bittern	B	1997	6	7	—	C6
Egretta garzetta Little Egret	W	1997	445	888	—	B1i, C2
Egretta garzetta Little Egret	P	1997	500	1,000	—	A4i, B1i, C2
Ardea purpurea Purple Heron	B	1991	—	50	—	C2
Phoenicopterus ruber Greater Flamingo	B	1997	—	30	—	C6
Phoenicopterus ruber Greater Flamingo	W	1997	3,153	14,600	—	A4i, B1i, C2
Phoenicopterus ruber Greater Flamingo	P	1997	10,000	15,000	—	A4i, B1i, C2
Milvus migrans Black Kite	B	1997	10	15	—	C6
Circus aeruginosus Marsh Harrier	B	1997	15	20	—	C6
Tetrax tetrax Little Bustard	W	1997	100	260	—	A1, C1
Haematopus ostralegus Oystercatcher	B	1997	10	12	—	B1i, C3
Himantopus himantopus Black-winged Stilt	B	1997	50	200	—	B1i, C2, C6
Recurvirostra avosetta Avocet	B	1997	100	200	—	B3, C6
Recurvirostra avosetta Avocet	W	1997	350	786	—	B1i, B2, C2
Charadrius alexandrinus Kentish Plover	B	1997	100	200	—	B2
Larus cachinnans Yellow-legged Gull	W	1997	3,866	13,503	—	A4i, B1i, C3
Sterna hirundo Common Tern	B	1997	200	350	—	C6
Sterna albifrons Little Tern	B	1997	100	200	—	B2, C6
Chlidonias hybridus Whiskered Tern	P	1997	100	300	—	B1i, C2
Alcedo atthis Kingfisher	B	1997	10	50	—	C6
Acrocephalus melanopogon Moustached Warbler	B	1991	20	100	—	C6

The site holds 20,000 or more waterbirds on passage and in winter, on a regular basis. Species of global conservation concern that do not meet IBA criteria: *Tetrax tetrax* (breeding, passage).

■ Protection status

National Low **International** Low
791 ha of IBA covered by 4 Hunting Reserves (Arnel; Vagaran; Méjean; Les Salins; total area 791 ha). 78 ha of IBA covered by Nature Reserve (Etang de l'Estagnol, 78 ha). 78 ha of IBA covered by Special Protection Area (Etang de l'Estagnol, 78 ha).

■ Conservation issues

Threats Construction/impact of dyke/dam/barrage (A), Filling-in of wetlands (A), Industrialization/urbanization (A), Infrastructure (A), Recreation/tourism (A)

The IBA is adversely affected by shellfish gathering at night and by night hunting using live decoys. There is also industrial development at Pérols. Expansion of the airport within the IBA is possible.

Massif des Albères — B2, C6 — 221

Admin region Languedoc-Roussillon
Coordinates 42°29′N 3°02′E
Altitude 100–1,200 m **Area** 17,850 ha

■ Site description

Economic activities include viticulture.

Habitats Forest and woodland (broadleaved evergreen forest), Scrub (sclerophyllous scrub/garrigue/maquis), Rocky areas (scree/boulders; inland cliff)
Land-use Agriculture, Forestry, Hunting, Tourism/recreation, Urban/industrial/transport

■ Birds

Species	Season	Year	Pop min	Pop max	Acc	Criteria
Galerida theklae Thekla Lark	B	1991	100	500	—	C6
Lullula arborea Woodlark	B	1991	100	500	—	B2, C6

This is an important area for species typical of Mediterranean habitats.

Protection status

National Low **International** None

336 ha of IBA covered by Nature Reserve (Massane, 336 ha).

Conservation issues

Threats Agricultural intensification/expansion (A), Industrialization/urbanization (A), Intensified forest management (A), Recreation/tourism (A)

The area is threatened by the development of tourism and the construction of houses.

Massif du Canigou-Carança — A4iv, B1iv, B2, B3, C5, C6 — 222

Admin region Languedoc-Roussillon
Coordinates 42°30′N 2°21′E
Altitude 300–2,784 m **Area** 53,550 ha

Site description

An area of mountains, with wooded valleys, scree, rocky slopes, cliffs, moorlands, alpine meadows and permanent lying snow.

Habitats Forest and woodland (broadleaved deciduous forest; native coniferous forest; mixed forest), Grassland (dry siliceous grassland; alpine/subalpine/boreal grassland), Wetland (standing fresh water; raised bog; fen/transition mire/spring), Rocky areas (scree/boulders; inland cliff)
Land-use Agriculture, Fisheries/aquaculture, Forestry, Hunting, Tourism/recreation, Urban/industrial/transport, Water management

Birds

Species	Season	Year	Pop min	Pop max	Acc	Criteria
Gypaetus barbatus Lammergeier	B	1996	1	1	—	B2
Circaetus gallicus Short-toed Eagle	B	1991	5	10	—	C6
Falco peregrinus Peregrine	B	1991	7	7	—	C6
[1] ***Lagopus mutus*** Ptarmigan	B	1991	100	500	—	C6
Tetrao urogallus Capercaillie	B	1991	50	250	—	C6
[2] ***Perdix perdix*** Partridge	B	1991	100	500	—	C6
Aegolius funereus Tengmalm's Owl	B	1991	15	100	—	C6
Caprimulgus europaeus Nightjar	B	1991	250	400	—	B2, C6
Dryocopus martius Black Woodpecker	B	1991	10	100	—	C6
Monticola saxatilis Rock Thrush	B	1991	10	100	—	B2
Turdus torquatus Ring Ouzel	B	1991	250	750	—	B3
Lanius collurio Red-backed Shrike	B	1991	250	400	—	C6
Pyrrhocorax pyrrhocorax Chough	B	1991	80	100	—	B2, C6

1. *L. m. pyrenaicus.*
2. *P. p. hispaniensis.*

An important site for breeding and passage raptors. The site is a major migratory bottleneck, where more than 20,000 raptors pass on a regular basis (including *Pernis apivorus* 14,000; *Milvus migrans* 4,500).

Protection status

National Partial **International** Partial

3,028 ha of IBA covered by Nature Reserve (Mantet, 3,028 ha). 1,177 ha of IBA covered by Nature Reserve (Vallée d'Eyne, 1,177 ha). 2,185 ha of IBA covered by Nature Reserve (Prats de Mollo La Preste, 2,185 ha). 3,930 ha of IBA covered by Nature Reserve (Py, 3,930 ha). 5,200 ha of IBA covered by Biological State Reserve (5,200 ha). 4,650 ha of IBA covered by 3 Hunting Reserves (St Martin; Py; Hauts de Carança; total area 4,650 ha). 9,350 ha of IBA covered by Special Protection Area (Py, Mantet et Prats de Mollo, 9,350 ha).

Conservation issues

Threats Disturbance to birds (B), Recreation/tourism (B)

The development of open-air sports and other recreational facilities threaten the site.

Puig Carlit et environs — B2, B3, C2, C6 — 223

Admin region Languedoc-Roussillon
Coordinates 42°30′N 1°55′E
Altitude 1,600–2,921 m **Area** 33,850 ha

Site description

The site includes high mountain valleys with large coniferous forests, lakes, scree, rocky slopes, alpine meadows and permanent lying snow.

Habitats Forest and woodland (broadleaved deciduous forest; native coniferous forest; mixed forest), Scrub (heathland), Grassland (dry siliceous grassland; alpine/subalpine/boreal grassland), Wetland (standing fresh water; fen/transition mire/spring), Rocky areas (scree/boulders; inland cliff)
Land-use Agriculture, Fisheries/aquaculture, Forestry, Hunting, Tourism/recreation, Urban/industrial/transport, Water management

Birds

Species	Season	Year	Pop min	Pop max	Acc	Criteria
Gypaetus barbatus Lammergeier	B	1996	1	1	—	B2
[1] ***Lagopus mutus*** Ptarmigan	B	1991	100	500	—	C6
Tetrao urogallus Capercaillie	B	1991	10	100	—	C6
[2] ***Perdix perdix*** Partridge	B	1991	100	500	—	C6
Charadrius morinellus Dotterel	B	1996	2	10	—	C6
Aegolius funereus Tengmalm's Owl	B	1991	10	10	—	C6
Dryocopus martius Black Woodpecker	B	1991	10	100	—	C6
Turdus torquatus Ring Ouzel	B	1991	100	500	—	B3
Pyrrhocorax pyrrhocorax Chough	B	1991	50	250	—	B2, C2, C6

1. *L. m. pyrenaicus.*
2. *P. p. hispaniensis.*

This is an important site for montane species and breeding raptors.

Protection status

National Partial **International** None

6,593 ha of IBA covered by 5 Hunting Reserves (Carlit; Fort-Vites; Soula de Carol; Motte nègre; Camporeils; total area 6,593 ha).

Conservation issues

Threats Groundwater abstraction (B), Intensified forest management (B), Recreation/tourism (C)

High tourist pressure, the construction of new forest tracks and water abstraction from lakes, to supply artificial snow machines, are the main threats to the site.

Gorges du Gardon — A1, B2, B3, C1, C6 — 224

Admin region Languedoc-Roussillon
Coordinates 43°55′N 4°24′E
Altitude 28–227 m **Area** 19,500 ha

Site description

An area of garrigue and forest, traversed by the River Gardon, with associated cliffs and a canyon.

Habitats Forest and woodland (broadleaved deciduous forest; alluvial/very wet forest; broadleaved evergreen forest), Scrub (heathland; sclerophyllous scrub/garrigue/maquis), Grassland (steppe/dry calcareous grassland), Wetland (standing fresh water; river/stream), Rocky areas (scree/boulders; inland cliff; inland sand-dunes; caves), Artificial landscape (highly improved reseeded grassland; arable land; perennial crops/orchards/groves; ruderal land)
Land-use Agriculture, Fisheries/aquaculture, Hunting, Tourism/recreation, Urban/industrial/transport

Birds

Species	Season	Year	Pop min	Pop max	Acc	Criteria
Nycticorax nycticorax Night Heron	B	1991	5	10	—	C6
Milvus migrans Black Kite	B	1991	5	10	—	C6
Neophron percnopterus Egyptian Vulture	B	1997	2	2	—	C6
Circus pygargus Montagu's Harrier	B	1991	10	45	—	B3, C6
Hieraaetus fasciatus Bonnelli's Eagle	B	1996	2	2	—	B2
Tetrax tetrax Little Bustard	B	1996	30	30	—	A1, C1, C6
Coracias garrulus Roller	B	1991	10	15	—	C6

This is an important area for breeding raptors and species associated with Mediterranean habitats. Species of global conservation concern that do not meet IBA criteria: *Crex crex* (passage).

Protection status

National None **International** Low

328 ha of IBA covered by Special Protection Area (Gorges du Gardon, 328 ha).

Conservation issues

Threats Abandonment/reduction of land management (A), Agricultural intensification/expansion (A), Industrialization/urbanization, Recreation/tourism (A)

This area is highly threatened by the development of tourist activities (including boating), water eutrophication, the abandonment of pastoralism and urbanization.

Hautes garrigues du Montpellierais — A1, B1iv, B2, C1, C5, C6 — 225

Admin region Languedoc-Roussillon
Coordinates 43°50′N 3°48′E
Altitude 33–942 m **Area** 90,700 ha

Site description

An area of garrigue with vineyards, *Pinus* woodlands, rivers, cliffs and rocky slopes.

Habitats Forest and woodland (broadleaved deciduous forest; native coniferous forest; alluvial/very wet forest; broadleaved evergreen forest), Scrub (heathland; sclerophyllous scrub/garrigue/maquis), Grassland (steppe/dry calcareous grassland), Wetland (standing fresh water; river/stream), Rocky areas (scree/boulders; inland cliff; inland sand-dunes; caves), Artificial landscape (highly improved reseeded grassland; arable land; perennial crops/orchards/groves; other urban/industrial areas; ruderal land)
Land-use Agriculture, Fisheries/aquaculture, Forestry, Hunting, Tourism/recreation, Urban/industrial/transport, Water management

Birds

Species	Season	Year	Pop min	Pop max	Acc	Criteria
Milvus migrans Black Kite	B	1997	10	20	—	C6
Circaetus gallicus Short-toed Eagle	B	1997	20	25	—	B2, C6
Circus pygargus Montagu's Harrier	B	1997	20	30	—	C6
Hieraaetus fasciatus Bonnelli's Eagle	B	1996	4	4	—	B2, C6
Tetrax tetrax Little Bustard	B	1996	25	30	—	A1, C1, C6
Bubo bubo Eagle Owl	B	1997	30	40	—	B2, C6
Caprimulgus europaeus Nightjar	B	1997	100	250	—	C6
Merops apiaster Bee-eater	B	1997	50	250	—	B2
Coracias garrulus Roller	B	1997	30	50	—	B2, C6
Lullula arborea Woodlark	B	1997	100	300	—	C6
Anthus campestris Tawny Pipit	B	1997	100	500	—	B2, C6
Sylvia undata Dartford Warbler	B	1997	250	750	—	B2, C6
Lanius senator Woodchat Shrike	B	1997	50	250	—	B2
Emberiza hortulana Ortolan Bunting	B	1997	50	250	—	B2, C6

The site is a migratory bottleneck, where 3,000 or more raptors regularly pass (including *Pernis apivorus* 2,000; *Milvus migrans* 1,000).

Protection status

National None **International** None

Conservation issues

Threats Extraction industry (A), Industrialization/urbanization (B), Infrastructure (A), Recreation/tourism (A), Unsustainable exploitation (A)

High hunting pressure, recreation and urbanization are the main threats.

Etang du Bagnas — C2, C6 — 226

Admin region Languedoc-Roussillon
Coordinates 43°19′N 3°31′E
Altitude 0–2 m **Area** 561 ha

Site description

A shallow brackish lagoon with sandy beaches, dunes, marshes and reedbeds (*Phragmites*).

Habitats Wetland (mudflat/sandflat; saltmarsh; sand-dunes/sand beach; coastal lagoon; standing brackish and salt water; water-fringe vegetation)
Land-use Hunting, Tourism/recreation, Urban/industrial/transport

Birds

This is an important site for breeding and wintering waterbirds.

Species	Season	Year	Pop min	Pop max	Acc	Criteria
Ardea purpurea Purple Heron	B	1991	1	100	—	C2, C6

Protection status

National High **International** High
561 ha of IBA covered by Nature Reserve (Bagnas, 561 ha). 561 ha of IBA covered by Special Protection Area (Etang du Bagnas, 561 ha).

Conservation issues

Threats Disturbance to birds (B), Industrialization/urbanization (B), Infrastructure (C)

Water pollution from domestic effluent, urbanization, and disturbance of birds by air traffic are the main threats.

Cordon lagunaire de Sète à Agde — A4i, B1i, B2, C2, C6 — 227

Admin region Languedoc-Roussillon
Coordinates 43°22′N 3°36′E
Altitude 0–2 m **Area** 800 ha

Site description

A brackish lagoon, saltmarshes, dunes and sandy beaches, ponds, ditches, *Pinus* woodlots and abandoned fields.

Habitats Forest and woodland (native coniferous forest), Wetland (saltmarsh; sand-dunes/sand beach; standing brackish and salt water), Artificial landscape (ruderal land)
Land-use Fisheries/aquaculture, Hunting, Tourism/recreation, Urban/industrial/transport

Birds

Species	Season	Year	Pop min	Pop max	Acc	Criteria
Egretta garzetta Little Egret	B	1997	245	245	—	C6
Phoenicopterus ruber Greater Flamingo	P	1991	500	1,000	—	A4i, B1i, C2
Himantopus himantopus Black-winged Stilt	B	1997	20	80	—	C6
Recurvirostra avosetta Avocet	B	1991	50	100	—	C6
Sterna sandvicensis Sandwich Tern	B	1997	100	200	—	B2, C2, C6
Sterna hirundo Common Tern	B	1991	100	200	—	C6
Sterna albifrons Little Tern	B	1991	50	150	—	B1i, C2, C6
Calandrella brachydactyla Short-toed Lark	B	1997	10	15	—	C6

An important site for breeding species associated with reedbeds (*Phragmites*) and saltmarshes.

Protection status

National None **International** None

Conservation issues

Threats Agricultural intensification/expansion, Industrialization/urbanization (A), Recreation/tourism (A)

Water eutrophication, as a result of agricultural run-off, and the development of recreation activities are the main threats.

Etang de Thau — A4i, B1i, C3, C6 — 228

Admin region Languedoc-Roussillon
Coordinates 43°24′N 3°37′E
Altitude 0–2 m **Area** 1,000 ha

Site description

A brackish lagoon with marshes, dunes and sand beaches, reedbeds (*Phragmites*) and a *Pinus* plantation.

Habitats Forest and woodland (alluvial/very wet forest), Wetland (mudflat/sandflat; saltmarsh; coastal lagoon; standing fresh water; water-fringe vegetation)
Land-use Fisheries/aquaculture, Hunting, Tourism/recreation, Urban/industrial/transport

Birds

This is an important site for breeding and passage waterbirds.

Species	Season	Year	Pop min	Pop max	Acc	Criteria
Podiceps nigricollis Black-necked Grebe	W	1997	500	2,000	—	A4i, B1i, C3
Podiceps nigricollis Black-necked Grebe	P	1997	1,000	2,000	—	A4i, B1i, C3
Phalacrocorax carbo Cormorant	W	1997	500	1,000	—	B1i, C3
Phalacrocorax carbo Cormorant	P	1997	1,000	2,000	—	B1i, C3
Himantopus himantopus Black-winged Stilt	B	1997	50	100	—	C6
Recurvirostra avosetta Avocet	B	1997	10	20	—	C6

Protection status

National Low **International** None

10 ha of IBA covered by Voluntary Nature Reserve (Les Prés de Baugé, 10 ha).

Conservation issues

Threats Agricultural intensification/expansion (A), Industrialization/urbanization (A), Recreation/tourism (A), Unsustainable exploitation (B)

Water pollution from agricultural and urban sources, high hunting pressure and tourist activities are considered the main threats.

Etangs de Vendres, Pissevache et Lespignan

B1i, B2, C2, C6 **229**

Admin region Languedoc-Roussillon
Coordinates 43°16'N 3°12'E
Altitude 0–8 m **Area** 4,850 ha

Site description

A Mediterranean coastal site comprising a shallow brackish lagoon, saltmarshes, dunes and sand beaches, ponds, ditches, *Pinus* plantations and abandoned fields.

Habitats Forest and woodland (alluvial/very wet forest), Wetland (tidal river/enclosed tidal water; saltmarsh; sand-dunes/sand beach; coastal lagoon; standing fresh water; standing brackish and salt water; river/stream; water-fringe vegetation), Artificial landscape (highly improved reseeded grassland; arable land; perennial crops/orchards/groves; ruderal land)
Land-use Agriculture, Fisheries/aquaculture, Hunting, Tourism/recreation, Urban/industrial/transport

Birds

Species	Season	Year	Pop min	Pop max	Acc	Criteria
Botaurus stellaris Bittern	B	1997	30	30	—	B2, C2, C6
Ixobrychus minutus Little Bittern	B	1997	5	15	—	C6
Ardea purpurea Purple Heron	B	1994	234	234	—	B2, C2, C6
Circus aeruginosus Marsh Harrier	B	1997	10	15	—	C6
Himantopus himantopus Black-winged Stilt	B	1997	30	80	—	C6
Sterna hirundo Common Tern	B	1997	1	60	—	C6
Sterna albifrons Little Tern	B	1997	1	50	—	C6
Chlidonias hybridus Whiskered Tern	P	1997	100	500	—	B1i, C2
Coracias garrulus Roller	B	1997	10	10	—	C6
Lullula arborea Woodlark	B	1991	10	100	—	C6
Acrocephalus melanopogon Moustached Warbler	B	1997	230	230	—	C2, C6
Lanius minor Lesser Grey Shrike	B	1996	18	22	—	C6

The site is important for species breeding in reedbeds (*Phragmites*) and saltmarshes. Species of global conservation concern that do not meet IBA criteria: *Crex crex* (passage); *Tetrax tetrax* (breeding); *Larus audouinii* (5–10 passage individuals).

Protection status

National Partial **International** None

113 ha of IBA covered by Hunting Reserve (Saint-Louis-de-la-Mer, 113 ha).

Conservation issues

Threats Abandonment/reduction of land management (A), Agricultural intensification/expansion (A), Construction/impact of dyke/dam/barrage (A), Dredging/canalization (A), Industrialization/urbanization (A), Recreation/tourism (A), Unsustainable exploitation (B)

Water pollution from agricultural and urban sources, significant tourist and hunting pressure and the abandonment of traditional grazing are reported threats.

Minervois

B2, B3, C6 **230**

Admin region Languedoc-Roussillon
Coordinates 43°22'N 2°46'E
Altitude 68–696 m **Area** 19,000 ha

Site description

An area of *Quercus* and *Pinus* forests, garrigue and rocky areas, vineyards, orchards, arable fields and rivers. Forestry and agriculture are the main human activities.

Habitats Forest and woodland (native coniferous forest; alluvial/very wet forest; broadleaved evergreen forest), Scrub (heathland; sclerophyllous scrub/garrigue/maquis), Wetland (standing fresh water; river/stream), Rocky areas (scree/boulders; inland cliff; inland sand-dunes; caves), Artificial landscape (arable land; perennial crops/orchards/groves; ruderal land)
Land-use Agriculture, Fisheries/aquaculture, Forestry, Hunting, Tourism/recreation, Urban/industrial/transport

Birds

Species	Season	Year	Pop min	Pop max	Acc	Criteria
Circus pygargus Montagu's Harrier	B	1991	10	100	—	B3, C6
Falco peregrinus Peregrine	B	1997	1	10	—	B2
Bubo bubo Eagle Owl	B	1991	12	20	—	B2, C6
Lullula arborea Woodlark	B	1991	10	100	—	C6
Anthus campestris Tawny Pipit	B	1997	10	100	—	C6
Sylvia undata Dartford Warbler	B	1991	100	500	—	B2, C6
Emberiza hortulana Ortolan Bunting	B	1997	10	100	—	C6

This is an important site for breeding raptors.

Protection status

National None **International** None

Conservation issues

Threats Infrastructure (A), Intensified forest management (A), Recreation/tourism (A), Unsustainable exploitation (B)

The main threats are intensive forestry, river pollution, high hunting pressure (including illegal hunting during spring migration) and recreation.

Etangs de Canet et de Villeneuve-de-la-Raho et embouchure du Tech

A4i, B1i, B2, C2, C6 **231**

Admin region Languedoc-Roussillon
Coordinates 43°02'N 2°58'E
Altitude 0–25 m **Area** 2,500 ha

Site description

An area of shallow brackish water with associated freshwater ponds, marshes, reedbeds (*Phragmites*), rivers, riverine forest, vineyards and arable fields.

Habitats Forest and woodland (alluvial/very wet forest), Wetland (tidal river/enclosed tidal water; mudflat/sandflat; saltmarsh; sand-dunes/sand beach; coastal lagoon; standing fresh water; standing brackish and salt water; river/stream; water-fringe vegetation), Artificial landscape (arable land; perennial crops/orchards/groves; ruderal land)
Land-use Agriculture, Fisheries/aquaculture, Hunting, Tourism/recreation

Birds

Species	Season	Year	Pop min	Pop max	Acc	Criteria
Ixobrychus minutus Little Bittern	B	1997	10	30	—	B2, C6
Ardea purpurea Purple Heron	B	1991	20	30	—	C6
Phoenicopterus ruber Greater Flamingo	P	1997	300	800	—	A4i, B1i, C2
Merops apiaster Bee-eater	B	1991	100	200	—	B2
Calandrella brachydactyla Short-toed Lark	B	1997	10	50	—	C6

This site is important for breeding species of herons. Species of global conservation concern that do not meet IBA criteria: *Aythya nyroca* (wintering).

Protection status

National Low **International** None

145 ha of IBA covered by Nature Reserve (Mas Larrieu, 145 ha). Part of IBA covered by Hunting Reserve.

Conservation issues

Threats Aquaculture/fisheries (B), Other (A), Recreation/tourism (B)

Water-levels are maintained at a level too high for many species ('Other' threat, above) during the breeding season. Disturbance is caused by tourists.

Gorges du Rieutord–Fage-Cagnasses — C6 — 232

Admin region Languedoc-Roussillon
Coordinates 43°58'N 3°46'E
Altitude 180–922 m **Area** 6,200 ha

Site description

A woody valley, with *Quercus* and *Pinus* forests, cliffs, sclerophyllous vegetation and some vineyards.

Habitats Forest and woodland (broadleaved deciduous forest; native coniferous forest; alluvial/very wet forest; broadleaved evergreen forest), Scrub (heathland; sclerophyllous scrub/garrigue/maquis), Grassland (steppe/dry calcareous grassland), Wetland (river/stream), Rocky areas (scree/boulders; inland cliff; inland sand-dunes; caves), Artificial landscape (highly improved reseeded grassland; arable land; ruderal land)
Land-use Agriculture, Fisheries/aquaculture, Forestry, Hunting, Tourism/recreation, Urban/industrial/transport

Birds

Species	Season	Year	Pop min	Pop max	Acc	Criteria
Lullula arborea Woodlark	B	1991	10	100	—	C6
Anthus campestris Tawny Pipit	B	1991	10	100	—	C6
Lanius collurio Red-backed Shrike	B	1991	10	100	—	C6

Protection status

National Low **International** None

150 ha of IBA covered by Hunting Reserve (150 ha). 60 ha of IBA covered by Voluntary Nature Reserve (60 ha).

Conservation issues

Threats Abandonment/reduction of land management (A), Afforestation, Agricultural intensification/expansion (A), Disturbance to birds (A), Intensified forest management (A), Recreation/tourism (A)

Land abandonment, afforestation of garrigue with *Pinus*, and the construction of new forest tracks, making access to the area easier, are the main threats.

Massif du Madres et Mont Coronat — B2, C6 — 233

Admin region Languedoc-Roussillon
Coordinates 42°38'N 2°18'E
Altitude 400–2,469 m **Area** 19,300 ha

Site description

A mountain area, with mixed forests, bogs, scree and rocky slopes, alpine meadows and permanent lying snow.

Habitats Forest and woodland (broadleaved deciduous forest; native coniferous forest; mixed forest), Grassland (dry siliceous grassland; alpine/subalpine/boreal grassland), Wetland (raised bog; fen/transition mire/spring), Rocky areas (scree/boulders; inland cliff)
Land-use Agriculture, Fisheries/aquaculture, Forestry, Hunting, Tourism/recreation, Urban/industrial/transport

Birds

Species	Season	Year	Pop min	Pop max	Acc	Criteria
Falco peregrinus Peregrine	B	1991	1	10	—	B2
Tetrao urogallus Capercaillie	B	1991	10	100	—	C6
[1] ***Perdix perdix*** Partridge	B	1991	100	500	—	C6
Pyrrhocorax pyrrhocorax Chough	B	1991	10	100	—	C6

1. *P. p. hispaniensis.*

An important site for species characteristic of montane and forest habitats.

Protection status

National Partial **International** None

549 ha of IBA covered by Nature Reserve (Conat-Betlans, 549 ha). 472 ha of IBA covered by Nature Reserve (Jujols, 472 ha). 2,137 ha of IBA covered by Nature Reserve (Nohèdes, 2,137 ha). 2,325 ha of IBA covered by 4 Hunting Reserves (Odeillo–Sausa–Réal; Mosset; Olettes; Nohèdes; total area 2,325 ha).

Conservation issues

Threats Burning of vegetation (B), Recreation/tourism (C)

Petite Camargue fluvio-lacustre — A1, A4i, B1i, B2, C1, C2, C6 — 234

Admin region Languedoc-Roussillon
Coordinates 43°38'N 4°18'E
Altitude 0–4 m **Area** 19,300 ha

Site description

A brackish lagoon with associated aquatic vegetation, and woodlands, vineyards and other cultivation.

Habitats Forest and woodland (native coniferous forest; alluvial/very wet forest; broadleaved evergreen forest), Wetland (standing fresh water; standing brackish and salt water; river/stream; water-fringe vegetation), Rocky areas (scree/boulders; inland sand-dunes), Artificial landscape (highly improved reseeded grassland; arable land; perennial crops/orchards/groves)
Land-use Agriculture, Fisheries/aquaculture, Hunting, Tourism/recreation, Urban/industrial/transport

Birds

Species	Season	Year	Pop min	Pop max	Acc	Criteria
Botaurus stellaris Bittern	B	1997	35	40	—	B2, C2, C6
Ixobrychus minutus Little Bittern	B	1997	5	10	—	C6
Nycticorax nycticorax Night Heron	B	1991	50	100	—	C6
Ardeola ralloides Squacco Heron	B	1997	5	10	—	B2, C6
Egretta garzetta Little Egret	B	1991	400	800	—	A4i, B1i, C2, C6
Ardea purpurea Purple Heron	B	1997	165	165	—	B2, C2, C6
Phoenicopterus ruber Greater Flamingo	P	1991	500	1,000	—	A4i, B1i, C2
Milvus migrans Black Kite	B	1997	15	20	—	C6
Circus aeruginosus Marsh Harrier	B	1997	40	60	—	C6
Porzana porzana Spotted Crake	B	1997	1	10	—	C6
Tetrax tetrax Little Bustard	W	1996	100	260	—	A1, C1
Himantopus himantopus Black-winged Stilt	B	1997	100	200	—	B1i, C2, C6
Gelochelidon nilotica Gull-billed Tern	B	1991	20	50	—	B1i, B2, C2
Chlidonias hybridus Whiskered Tern	P	1997	100	500	—	B1i, C2
Chlidonias niger Black Tern	B	1997	—	5	—	C6
Alcedo atthis Kingfisher	B	1997	10	50	—	C6
Coracias garrulus Roller	B	1997	15	20	—	C6
Acrocephalus melanopogon Moustached Warbler	B	1997	200	500	—	C2, C6

This is an important site for waterbirds. Species of global conservation concern that do not meet IBA criteria: *Aythya nyroca* (2–10 passage individuals, 1–10 passage individuals); *Tetrax tetrax* (5 breeding pairs).

Protection status

National Low **International** Partial

Part of IBA covered by Private Reserve. 16,000 ha of IBA covered by Ramsar Site (La Petite Camargue, 37,000 ha).

Conservation issues

Threats Agricultural intensification/expansion (A), Industrialization/urbanization (A), Other (A), Recreation/tourism (A), Unsustainable exploitation (B)

The main threats include the development of rice farming, water pollution from chemical treatment against mosquitoes ('Other' threats, above) and high hunting pressure.

Petite Camargue laguno-marine A4i, B1i, B2, B3, C2, C3, C6 235

Admin region Languedoc-Roussillon
Coordinates 43°31'N 4°16'E
Altitude 0–9 m **Area** 21,000 ha

Site description

The site includes a brackish lagoon, saltmarshes, coastal dunes, riverine forest, *Pinus* woods, vineyards and other cultivation.

Habitats Forest and woodland (native coniferous forest; alluvial/very wet forest), Wetland (tidal river/enclosed tidal water; saltmarsh; sand-dunes/sand beach; coastal lagoon; standing brackish and salt water; river/stream; water-fringe vegetation), Rocky areas (scree/boulders; inland sand-dunes), Artificial landscape (highly improved reseeded grassland; arable land; perennial crops/orchards/groves)
Land-use Agriculture, Fisheries/aquaculture, Hunting, Tourism/recreation, Urban/industrial/transport

Birds

Species	Season	Year	Pop min	Pop max	Acc	Criteria
Phalacrocorax carbo Cormorant	W	1991	1,000	4,000	—	B1i, C3
Botaurus stellaris Bittern	B	1997	10	15	—	C6
Ixobrychus minutus Little Bittern	B	1997	1	10	—	C6
Egretta garzetta Little Egret	B	1991	100	800	—	A4i, B1i, C2, C6
Ardea purpurea Purple Heron	B	1997	1	150	—	B2, C2, C6
Phoenicopterus ruber Greater Flamingo	W	1991	2,000	5,000	—	A4i, B1i, C2
Phoenicopterus ruber Greater Flamingo	P	1991	5,000	10,000	—	A4i, B1i, C2
Tadorna tadorna Shelduck	W	1991	1,000	1,500	—	B1i, C3
Milvus migrans Black Kite	B	1997	10	15	—	C6
Circus aeruginosus Marsh Harrier	B	1997	15	25	—	C6
Haematopus ostralegus Oystercatcher	B	1997	62	65	—	B1i, C3
Recurvirostra avosetta Avocet	B	1991	300	400	—	A4i, B1i, B3, C2, C6
Recurvirostra avosetta Avocet	P	1991	—	—	—	B2
Charadrius alexandrinus Kentish Plover	B	1997	20	50	—	B2
Larus melanocephalus Mediterranean Gull	B	1997	50	120	—	C2
Larus genei Slender-billed Gull	B	1997	100	250	—	C2
Larus cachinnans Yellow-legged Gull	B	1991	1,800	1,800	—	A4i, B1i, C3
Gelochelidon nilotica Gull-billed Tern	B	1991	150	400	—	A4i, B1i, B2, C2
Sterna sandvicensis Sandwich Tern	B	1991	1	800	—	B1i, B2, C2, C6
Sterna hirundo Common Tern	B	1991	600	800	—	C6
Sterna albifrons Little Tern	B	1991	150	400	—	A4i, B1i, B2, C2, C6

The site is particularly important for many wetland breeding and wintering species.

Protection status

National Low **International** High
Part of IBA covered by Hunting Reserve (Petit Rhône). 21,000 ha of IBA covered by Ramsar Site (La Petit Camargue, 37,000 ha).

Conservation issues

Threats Abandonment/reduction of land management (A), Agricultural intensification/expansion (A), Industrialization/urbanization (A), Recreation/tourism (A)

Declining water quality, an increasing number of tourists, abandonment of salt-pans and the development of rice-fields are the main threats.

Parc National des Cévennes B2, B3, C6 236

Admin region Languedoc-Roussillon
Coordinates 44°15'N 3°40'E
Altitude 380–1,700 m **Area** 92,000 ha

Site description

A mountainous area with deciduous and conifer forests, large areas of heathlands, scree and rocky slopes and alpine meadows.

Habitats Forest and woodland (51%; broadleaved deciduous forest; native coniferous forest; mixed forest; alluvial/very wet forest; broadleaved evergreen forest), Scrub (30%; heathland), Grassland (17%; steppe/dry calcareous grassland; alpine/subalpine/boreal grassland; humid grassland; mesophile grassland), Wetland (standing fresh water; river/stream; raised bog), Rocky areas (scree/boulders; inland cliff; inland sand-dunes; caves), Artificial landscape (highly improved reseeded grassland; arable land; perennial crops/orchards/groves)
Land-use Agriculture, Fisheries/aquaculture, Forestry, Hunting, Tourism/recreation, Urban/industrial/transport

Birds

Species	Season	Year	Pop min	Pop max	Acc	Criteria
Pernis apivorus Honey Buzzard	B	1991	30	100	—	B3, C6
Circaetus gallicus Short-toed Eagle	B	1991	20	100	—	B2, C6
Circus cyaneus Hen Harrier	B	1991	10	100	—	B2, C6
Circus pygargus Montagu's Harrier	B	1991	10	100	—	B3, C6
Falco peregrinus Peregrine	B	1991	1	10	—	B2
Tetrao urogallus Capercaillie	B	1991	10	100	—	C6
Burhinus oedicnemus Stone Curlew	B	1991	10	100	—	B2, C6
Dryocopus martius Black Woodpecker	B	1991	10	100	—	C6
Monticola saxatilis Rock Thrush	B	1991	10	100	—	B2
Lanius collurio Red-backed Shrike	B	1991	100	500	—	C6
Pyrrhocorax pyrrhocorax Chough	B	1991	10	100	—	C6

This is an important site for species associated with open upland habitats and forest, particularly breeding raptors. Species of global conservation concern that do not meet IBA criteria: *Tetrax tetrax* (1–10 breeding pairs).

Protection status

National High **International** Partial
85,100 ha of IBA covered by National Park (Les Cévennes, 91,500 ha). 1,000 ha of IBA covered by Hunting Reserve (1,000 ha). 84,000 ha of IBA covered by Biosphere Reserve (84,000 ha). 77,600 ha of IBA covered by Special Protection Area (Les Cévennes, 84,000 ha).

Conservation issues

Threats Afforestation (A), Agricultural intensification/expansion (A), Infrastructure (B), Intensified forest management (A), Recreation/tourism (A), Unsustainable exploitation (C)

Afforestation with conifers, changes in cattle-raising practices and recreation are the main threats to the site.

Montagnes de Marcou, de l'Espinouse et du Caroux C6 237

Admin region Languedoc-Roussillon
Coordinates 43°39'N 2°59'E
Altitude 200–1,124 m **Area** 14,500 ha

Site description

The site includes deciduous montane forests, heathlands, peatbogs, scree and rocky slopes.

Habitats Forest and woodland (broadleaved deciduous forest), Scrub (heathland; sclerophyllous scrub/garrigue/maquis), Wetland (river/stream; raised bog; fen/transition mire/spring), Rocky areas (scree/boulders; inland cliff)
Land-use Agriculture, Forestry, Hunting, Tourism/recreation, Urban/industrial/transport

Birds

Species	Season	Year	Pop min	Pop max	Acc	Criteria
Emberiza hortulana Ortolan Bunting	B	1991	10	100	—	C6

The site is important for montane and Mediterranean species, including *Aquila chrysaetos, Circaetus gallicus, Circus cyaneus, Circus pygargus, Sylvia undata, Emberiza hortulana, Caprimulgus europaeus* and *Lanius collurio*.

Protection status

National Partial **International** None
Part of IBA covered by Biological State Reserve. Part of IBA covered by Hunting Reserve (le Caroux).

Conservation issues

Threats Abandonment/reduction of land management (U), Afforestation (A), Intensified forest management (A), Recreation/tourism (A)

High numbers of tourists, the abandonment of cattle-grazing and intensified forest management are threatening the site.

Moyenne Vallée de la Durance — A4i, B1i, B2, C3, C6 — 238

Admin region Provence-Alpes-Côte d'Azur
Coordinates 43°54′N 5°51′E
Altitude 236–430 m **Area** 11,300 ha

Site description

The site consists of ponds, reedbeds (*Phragmites*), riverine vegetation and cliffs.

Habitats Forest and woodland (alluvial/very wet forest; broadleaved evergreen forest), Grassland (humid grassland), Wetland (standing fresh water; river/stream; water-fringe vegetation), Rocky areas (inland cliff), Artificial landscape (arable land; perennial crops/orchards/groves)
Land-use Agriculture, Fisheries/aquaculture, Hunting, Tourism/recreation, Urban/industrial/transport

Birds

Species	Season	Year	Pop min	Pop max	Acc	Criteria
Phalacrocorax carbo Cormorant	W	1991	500	1,000	—	B1i, C3
Nycticorax nycticorax Night Heron	B	1991	30	60	—	C6
Milvus migrans Black Kite	B	1991	10	100	—	B2, C6
Larus cachinnans Yellow-legged Gull	W	1991	1,000	10,000	—	A4i, B1i, C3
Melanocorypha calandra Calandra Lark	B	1991	10	100	—	C6

The site is important for breeding and wintering waterbirds. Species of global conservation concern that do not meet IBA criteria: *Tetrax tetrax* (1–10 breeding pairs).

Protection status

National None **International** None

Conservation issues

Threats Construction/impact of dyke/dam/barrage (A), Extraction industry (B)

Cultivation and irrigation of the dry steppes, gravel extraction and hunting (including illegal hunting) are the main threats to the site.

Camargue — A4i, A4iii, B1i, B2, B3, C2, C3, C4, C6 — 239

Admin region Provence-Alpes-Côte d'Azur
Coordinates 43°31′N 4°36′E
Altitude 0–2 m **Area** 76,500 ha

Site description

The largest wetland area in France, comprising a large delta with beaches, sand-dunes, islands, saltwater lagoons, marshes, and slow-flowing rivers; woodlands, grasslands and agricultural land.

Habitats Wetland (tidal river/enclosed tidal water; saltmarsh; sand-dunes/sand beach; coastal lagoon; standing fresh water; standing brackish and salt water; river/stream; water-fringe vegetation), Rocky areas (rock stacks/islets)
Land-use Agriculture, Fisheries/aquaculture, Hunting, Tourism/recreation, Urban/industrial/transport

Birds

Species	Season	Year	Pop min	Pop max	Acc	Criteria
Phalacrocorax carbo Cormorant	W	1997	1	2,783	—	B1i, C3
Phalacrocorax carbo Cormorant	P	1997	6,000	6,000	—	A4i, B1i, C3
Botaurus stellaris Bittern	B	1991	25	50	—	B2, C2, C6
Nycticorax nycticorax Night Heron	B	1991	300	300	—	C2, C6
Ardeola ralloides Squacco Heron	B	1991	50	50	—	B2, C2, C6
Egretta garzetta Little Egret	B	1997	1,000	1,000	—	A4i, B1i, C2, C6
Egretta garzetta Little Egret	W	1997	13	1,289	—	A4i, B1i, C2
Ardea purpurea Purple Heron	B	1991	400	400	—	B2, C2, C6
Phoenicopterus ruber Greater Flamingo	B	1997	10,000	18,000	—	A4i, B1i, B2, C2
Cygnus columbianus Bewick's Swan	W	1997	4	120	—	B2
Tadorna tadorna Shelduck	W	1997	1,307	2,781	—	A4i, B1i, C3
Anas penelope Wigeon	W	1997	10,170	15,010	—	B1i, C3
Anas strepera Gadwall	W	1997	6,640	13,570	—	A4i, B1i, C3
Anas crecca Teal	W	1997	16,182	25,300	—	A4i, B1i, C3
Anas platyrhynchos Mallard	W	1997	13,466	30,800	—	B1i, C3
Anas clypeata Shoveler	W	1997	6,220	13,815	—	A4i, B1i, C3
Netta rufina Red-crested Pochard	B	1997	200	300	—	A4i, B1i, C3
Netta rufina Red-crested Pochard	W	1997	500	2,110	—	A4i, B1i, C3
Circus aeruginosus Marsh Harrier	B	1997	50	80	—	C6
Fulica atra Coot	W	1997	16,500	30,779	—	B1i, C3
Haematopus ostralegus Oystercatcher	B	1997	50	97	—	B1i, C3
Himantopus himantopus Black-winged Stilt	B	1996	195	219	—	B1i, C2
Recurvirostra avosetta Avocet	B	1997	439	439	—	A4i, B1i, B3, C2, C6
Recurvirostra avosetta Avocet	W	1997	599	1,944	—	A4i, B1i, B2, C2
Glareola pratincola Collared Pratincole	B	1996	8	20	—	C6
Charadrius hiaticula Ringed Plover	P	1997	1,000	10,000	—	A4i, B1i, C3
Charadrius alexandrinus Kentish Plover	B	1997	145	170	—	B2
Charadrius alexandrinus Kentish Plover	P	1997	1,000	10,000	—	A4i, B1i, C3
Calidris ferruginea Curlew Sandpiper	P	1991	1,000	10,000	—	A4i, B1i, C3
Limosa limosa Black-tailed Godwit	P	1997	1,000	10,000	—	A4i, B1i, C3
Tringa erythropus Spotted Redshank	P	1997	1,000	10,000	—	A4i, B1i, C3
Larus melanocephalus Mediterranean Gull	B	1997	600	700	—	C2
Larus genei Slender-billed Gull	B	1997	580	800	—	A4i, B1i, C2
Gelochelidon nilotica Gull-billed Tern	B	1996	300	300	—	A4i, B1i, B2, C2, C6
Sterna caspia Caspian Tern	P	1991	10	100	—	B1i, C2
Sterna sandvicensis Sandwich Tern	B	1997	670	1,000	—	A4i, B1i, B2, C2
Sterna hirundo Common Tern	B	1991	500	1,500	—	B1i, C2, C6
Sterna albifrons Little Tern	B	1991	70	350	—	A4i, B1i, B2, C2, C6
Chlidonias hybridus Whiskered Tern	B	1996	—	10	—	C6
Chlidonias hybridus Whiskered Tern	P	1996	1,000	10,000	—	A4i, B1i, C2
Chlidonias niger Black Tern	P	1991	1,000	10,000	—	A4i, B1i, C2
Coracias garrulus Roller	B	1991	10	100	—	B2, C2, C6
Sylvia cantillans Subalpine Warbler	B	1991	100	500	—	B3

This is one of the most important sites for birds in France, regularly holding more than 20,000 waterbirds breeding, on passage and in winter. Species of global conservation concern that do not meet criteria: *Falco naumanni* (passage).

Protection status

National Partial **International** High

13,117 ha of IBA covered by Nature Reserve (La Camargue, 13,117 ha). 3,481 ha of IBA covered by Private Reserve (Tour du Valat, 3,481). 16,000 ha of IBA covered by Special Protection Area (Camargue, 16,000 ha). 76,500 ha of IBA covered by Ramsar Site (Camargue, 85,000 ha). 13,117 ha of IBA covered by Biosphere Reserve (13,117 ha).

Conservation issues

Threats Agricultural intensification/expansion, Infrastructure (A), Industrialization/urbanization, Recreation/tourism (A)

The area is threatened by the development of rice fields with associated modifications of water-levels, tourist development, the use of toxic chemicals in agriculture and air pollution from the chemical plants in Fos-sur-Mer.

Crau — A1, B2, C1, C2, C6 — 240

Admin region Provence-Alpes-Côte d'Azur
Coordinates 43°36′N 4°52′E
Altitude 20–100 m **Area** 40,100 ha

Site description

An area of mainly dry calcareous grassland with sclerophyllous scrub, and old orchards. The main land-uses are extensive traditional sheep-farming; and growing almonds and olives.

Habitats Scrub (5%; sclerophyllous scrub/garrigue/maquis), Grassland (90%; steppe/dry calcareous grassland; humid grassland), Wetland (standing fresh water; water-fringe vegetation), Artificial landscape (10%; arable land; perennial crops/orchards/groves)
Land-use Agriculture, Hunting, Tourism/recreation, Urban/industrial/transport

Birds

Species	Season	Year	Pop min	Pop max	Acc	Criteria
Ardea purpurea Purple Heron	B	1991	10	50	—	C2, C6
Milvus migrans Black Kite	B	1991	50	50	—	C6
Circus pygargus Montagu's Harrier	B	1991	10	10	—	C6
Falco naumanni Lesser Kestrel	B	1996	42	50	—	A1, C1, C6
Tetrax tetrax Little Bustard	B	1996	450	500	—	A1, B2, C1, C6
Tetrax tetrax Little Bustard	W	1996	1,600	1,600	—	A1, C1
Burhinus oedicnemus Stone Curlew	B	1991	300	300	—	B2, C6
Pterocles alchata Pin-tailed Sandgrouse	B	1996	150	170	—	B2, C2, C6
Athene noctua Little Owl	B	1991	100	100	—	B2
Coracias garrulus Roller	B	1991	70	70	—	B2, C6
Melanocorypha calandra Calandra Lark	B	1996	10	30	—	C6
Calandrella brachydactyla Short-toed Lark	B	1991	1,000	1,000	—	C6

This is the only large semi-arid plain in France, and supports important populations of typical steppe breeding species.

Protection status

National Low **International** Partial
146 ha of IBA covered by Voluntary Nature Reserve (La Poitevine, 146 ha). 11,500 ha of IBA covered by Special Protection Area (Crau Sèche, 11,500 ha).

Conservation issues

Threats Industrialization/urbanization (U), Infrastructure (U)

The site is seriously threatened by: the cultivation and modification of steppe-meadows; the use of irrigation; the felling of old almond and olive orchards and their replacement with commercial groves; the leasing out of military training areas for agricultural exploitation; disturbance from recreational activities and hunting; plans to build a motorway. These changes seriously threaten the populations of *Tetrax tetrax*, *Burhinus oedicnemus* and *Pterocles alchata*.

Chaîne des Alpilles — B2, B3, C6 — 241

Admin region Provence-Alpes-Côte d'Azur
Coordinates 43°44′N 4°54′E
Altitude 30–481 m **Area** 21,800 ha

Site description

Mediterranean mountains, cliffs, and hills with maquis, woodlands and arable fields.

Habitats Forest and woodland (50%; native coniferous forest; broadleaved evergreen forest), Scrub (80%; sclerophyllous scrub/garrigue/maquis), Rocky areas (5%; scree/boulders; inland cliff; inland sand-dunes; caves), Artificial landscape (10%; arable land; perennial crops/orchards/groves)
Land-use Agriculture (20%), Forestry, Hunting (100%), Tourism/recreation (70%), Urban/industrial/transport (40%)

Birds

Species	Season	Year	Pop min	Pop max	Acc	Criteria
Neophron percnopterus Egyptian Vulture	B	1991	5	5	—	B2, C6
Hieraaetus fasciatus Bonnelli's Eagle	B	1996	2	2	—	B2
Otus scops Scops Owl	B	1991	10	15	—	B2
Bubo bubo Eagle Owl	B	1991	45	45	—	B2, C6
Caprimulgus europaeus Nightjar	B	1991	100	100	—	C6
Lullula arborea Woodlark	B	1991	100	200	—	C6
Monticola solitarius Blue Rock Thrush	B	1991	10	15	—	B2
Sylvia undata Dartford Warbler	B	1991	1,000	1,000	—	B2, C6
Sylvia cantillans Subalpine Warbler	B	1991	1,000	1,000	—	B3

This area is important for breeding raptors and Mediterranean avifauna. Species of global conservation concern that do not meet IBA criteria: *Tetrax tetrax* (1–10 breeding pairs).

Protection status

National None **International** None

Conservation issues

Threats Abandonment/reduction of land management (A), Agricultural intensification/expansion (B), Recreation/tourism (A), Industrialization/urbanization

The site is threatened by urban development and forestry operations. The construction of new roads and tracks, which will lead to increased disturbance from tourism and hiking, are additional threats.

Salines de l'Etang de Berre — C6 — 242

Admin region Provence-Alpes-Côte d'Azur
Coordinates 43°28′N 5°09′E
Altitude 0–1 m **Area** 530 ha

Site description

Saltmarshes and salinas around the Berre lagoon. Human activities include hunting and salt exploitation.

Habitats Wetland (100%; saltmarsh)
Land-use Hunting (100%), Urban/industrial/transport (100%)

Birds

Species	Season	Year	Pop min	Pop max	Acc	Criteria
Himantopus himantopus Black-winged Stilt	B	1995	6	6	—	C6
Recurvirostra avosetta Avocet	B	1995	59	59	—	C6
Sterna hirundo Common Tern	B	1991	76	418	—	C6
Sterna albifrons Little Tern	B	1991	40	40	—	C6

The site is particularly important for breeding waders.

Protection status

National None **International** None

Conservation issues

Threats Industrialization/urbanization (B), Infrastructure (C), Recreation/tourism

The site is threatened by urban and industrial development, tourism and hunting pressure. Water pollution from industry is also a problem.

Montagne Sainte-Victoire — B2, B3, C6 — 243

Admin region Provence-Alpes-Côte d'Azur
Coordinates 43°34′N 5°38′E
Altitude 400–1,000 m **Area** 15,000 ha

Site description

The site includes mountains and hills with cliffs, maquis and garrigue, dry calcareous grassland and agricultural land.

Habitats Forest and woodland (alluvial/very wet forest), Scrub (83%; sclerophyllous scrub/garrigue/maquis), Grassland (steppe/dry calcareous grassland), Wetland (standing fresh water; river/stream), Rocky areas (5%; scree/boulders; inland cliff; inland sand-dunes; caves), Artificial landscape (arable land; perennial crops/orchards/groves; ruderal land)
Land-use Agriculture, Fisheries/aquaculture, Forestry, Hunting, Tourism/recreation, Urban/industrial/transport

Birds

Species	Season	Year	Pop min	Pop max	Acc	Criteria
Circaetus gallicus Short-toed Eagle	B	1991	23	23	—	B2, C6
Bubo bubo Eagle Owl	B	1991	7	10	—	C6
Caprimulgus europaeus Nightjar	B	1991	50	75	—	C6
Lullula arborea Woodlark	B	1991	100	200	—	C6
Monticola solitarius Blue Rock Thrush	B	1991	1	10	—	B2
Sylvia undata Dartford Warbler	B	1991	500	500	—	B2, C6
Sylvia cantillans Subalpine Warbler	B	1991	500	500	—	B3

An important site for breeding raptors and Mediterranean species.

Protection status

National Low **International** Partial

139 ha of IBA covered by Nature Reserve (Reserve Naturelle géologique de la Sainte-Victoire, 139 ha). 5,300 ha of IBA covered by Special Protection Area (Montagne Sainte-Victoire, 5,300 ha).

Conservation issues

Threats Disturbance to birds (B), Infrastructure (B), Intensified forest management (C), Recreation/tourism (A), Unsustainable exploitation (C)

The site is seriously threatened by the construction of new forest roads and tourist trails.

Iles Marseillaises: Maire, Jarron, Jarre, Riou, Calseraigne, Congloue et Pomègues — B2, C6 — 244

Admin region Provence-Alpes-Côte d'Azur
Coordinates 43°14'N 5°21'E
Altitude 0–191 m **Area** 2,350 ha

Site description

An archipelago of islands, covered in maquis, with rocky coastlines and cliffs.

Habitats Scrub (100%; sclerophyllous scrub/garrigue/maquis), Marine areas (open sea), Rocky areas (100%; sea cliff/rocky shore; rock stacks/islets)
Land-use Fisheries/aquaculture, Tourism/recreation

Birds

Species	Season	Year	Pop min	Pop max	Acc	Criteria
Calonectris diomedea Cory's Shearwater	B	1991	265	325	—	B2, C6
Hydrobates pelagicus Storm Petrel	B	1991	10	30	—	C6
Monticola solitarius Blue Rock Thrush	B	1991	10	10	—	B2

The cliffs of the small islands are breeding sites for several pelagic species.

Protection status

National None **International** None

Conservation issues

Threats Recreation/tourism (B), Consequences of animal/plant introductions (U)

There is planned urban development and stray domestic pets are endangering the breeding species.

Marais entre Crau et Grand Rhône: Meyranne, Chanoine, Plan de Bourg et Salins du Caban — A4iii, B1i, B2, C2, C3, C4, C6 — 245

Admin region Provence-Alpes-Côte d'Azur
Coordinates 43°29'N 4°48'E
Altitude 0–10 m **Area** 5,650 ha

Site description

An area of marshland lying between the Crau steppes and the River Rhone, alongside the Camargue.

Habitats Grassland (humid grassland), Wetland (saltmarsh; standing fresh water; standing brackish and salt water; river/stream; water-fringe vegetation)
Land-use Hunting, Urban/industrial/transport

Birds

Species	Season	Year	Pop min	Pop max	Acc	Criteria
Botaurus stellaris Bittern	B	1991	25	25	—	C2, C6
Nycticorax nycticorax Night Heron	B	1991	80	80	—	C6
Ardeola ralloides Squacco Heron	B	1991	30	30	—	B2, C2, C6
Egretta garzetta Little Egret	B	1991	100	100	—	C6
Ardea purpurea Purple Heron	B	1991	30	100	—	B2, C2, C6
Netta rufina Red-crested Pochard	W	1991	250	250	—	B1i, C3

Species ... continued	Season	Year	Pop min	Pop max	Acc	Criteria
Circus aeruginosus Marsh Harrier	B	1991	20	20	—	C6
Himantopus himantopus Black-winged Stilt	B	1991	10	10	—	C6
Recurvirostra avosetta Avocet	B	1991	10	10	—	C6
Charadrius alexandrinus Kentish Plover	B	1991	50	50	—	B2
Limosa limosa Black-tailed Godwit	P	1991	3,000	4,000	—	B1i, C3
Sterna albifrons Little Tern	B	1991	30	30	—	C6
Coracias garrulus Roller	B	1991	20	20	—	C6
Calandrella brachydactyla Short-toed Lark	B	1991	100	500	—	C6
Acrocephalus melanopogon Moustached Warbler	B	1991	100	500	—	C2, C6

The site holds 20,000 or more waterbirds in winter, on a regular basis (including *Anas crecca* 4,500; *Anas platyrhynchos* 3,000; *Anas clypeata* 1,000; *Fulica atra* 7,000).

Protection status

National None **International** None

Conservation issues

Threats Agricultural intensification/expansion (A), Aquaculture/fisheries (B), Filling-in of wetlands (U), Industrialization/urbanization (A), Infrastructure (A), Recreation/tourism (A), Unsustainable exploitation (B)

Water pollution from industry, the expansion of rice cultivation, reedbed exploitation, drying out and embankment of the wetland, and high hunting pressure are the main threats at this site.

Massif du Petit Lubéron — B2, B3, C6 — 246

Admin region Provence-Alpes-Côte d'Azur
Coordinates 43°48'N 5°16'E
Altitude 100–950 m **Area** 14,000 ha

Site description

Mountains, hills and cliffs with woodland, garrigue, grassland, fast-flowing rivers, scree and rocky slopes.

Habitats Forest and woodland (70%; broadleaved deciduous forest; native coniferous forest; broadleaved evergreen forest), Scrub (30%; sclerophyllous scrub/garrigue/maquis), Grassland (10%; steppe/dry calcareous grassland), Wetland (river/stream), Rocky areas (inland cliff; inland sand-dunes; caves), Artificial landscape (arable land; perennial crops/orchards/groves)
Land-use Agriculture (11%), Fisheries/aquaculture, Forestry (40%), Hunting (90%), Tourism/recreation, Urban/industrial/transport

Birds

Species	Season	Year	Pop min	Pop max	Acc	Criteria
Neophron percnopterus Egyptian Vulture	B	1991	4	5	—	B2, C6
Burhinus oedicnemus Stone Curlew	B	1991	10	100	—	B2, C6
Bubo bubo Eagle Owl	B	1991	20	20	—	B2, C6
Caprimulgus europaeus Nightjar	B	1991	100	500	—	B2, C6
Lullula arborea Woodlark	B	1991	250	750	—	B2, C6
Anthus campestris Tawny Pipit	B	1991	100	500	—	B2, C6
Sylvia undata Dartford Warbler	B	1991	250	750	—	B2, C6
Sylvia cantillans Subalpine Warbler	B	1991	250	750	—	B3
Lanius collurio Red-backed Shrike	B	1991	100	500	—	C6
Serinus serinus Serin	B	1991	1,000	10,000	—	B3
Emberiza hortulana Ortolan Bunting	B	1991	100	500	—	B2, C6

This site is of special interest for breeding raptors and Mediterranean avifauna.

Protection status

National Low **International** High

1,100 ha of IBA covered by Hunting Reserve (1,100 ha). 600 ha of IBA covered by Private Reserve (600 ha). 14,000 ha of IBA covered by Special Protection Area (Massif du petit Luberon, 16,679 ha). 14,000 ha of IBA covered by Biosphere Reserve (14,000 ha).

Conservation issues

Threats Abandonment/reduction of land management (B), Infrastructure (B), Intensified forest management (B), Recreation/tourism (B), Unsustainable exploitation (C)

The site is threatened by the replacement of natural forest with non-native tree species and by the abandonment of agricultural activities.

Salins d'Hyères et de Pesquiers — A4i, B1i, B2, B3, C2, C6 — 247

Admin region Provence-Alpes-Côte d'Azur
Coordinates 43°06'N 6°11'E
Altitude 0–3 m **Area** 1,000 ha

Site description

The IBA contains saltmarshes, ponds, brackish marshes, reedbeds (*Phragmites*), coastal dunes and sandy beaches.

Habitats Wetland (100%; saltmarsh; sand-dunes/sand beach; standing brackish and salt water)
Land-use Other (95%), Tourism/recreation (10%)

Birds

Species	Season	Year	Pop min	Pop max	Acc	Criteria
Phoenicopterus ruber Greater Flamingo	W	1997	123	1,281	—	A4i, B1i, C2
Phoenicopterus ruber Greater Flamingo	P	1997	1,000	10,000	—	A4i, B1i, C2
Himantopus himantopus Black-winged Stilt	B	1996	41	46	—	C6
Recurvirostra avosetta Avocet	B	1997	63	73	—	B3, C6
Charadrius alexandrinus Kentish Plover	B	1997	50	55	—	B2
Sterna albifrons Little Tern	B	1996	31	31	—	C6

The site is important for breeding waders and wintering waterbirds.

Protection status

National None **International** None

Conservation issues

Threats Industrialization/urbanization (C), Infrastructure (A), Other (C), Recreation/tourism (A), Unsustainable exploitation (C)

The main threats are high hunting pressure, urbanization, tourism and road construction.

Iles d'Hyères — A4i, B1i, B2, C3, C6 — 248

Admin region Provence-Alpes-Côte d'Azur
Coordinates 43°01'N 6°20'E
Altitude 0–50 m **Area** 6,600 ha

Site description

A small archipelago with garrigue, Mediterranean woodlands, sand and shingle-beaches, cliffs, arable fields and abandoned fields.

Habitats Forest and woodland (native coniferous forest; broadleaved evergreen forest), Scrub (sclerophyllous scrub/garrigue/maquis), Wetland (sand-dunes/sand beach; shingle/stony beach; coastal lagoon), Marine areas (open sea), Rocky areas (sea cliff/rocky shore; rock stacks/islets), Artificial landscape (arable land; perennial crops/orchards/groves; ruderal land)
Land-use Agriculture, Fisheries/aquaculture, Forestry, Urban/industrial/transport

Birds

Species	Season	Year	Pop min	Pop max	Acc	Criteria
Calonectris diomedea Cory's Shearwater	B	1991	195	245	—	B2, C6
Larus cachinnans Yellow-legged Gull	B	1991	5,000	5,000	—	A4i, B1i, C3
Otus scops Scops Owl	B	1991	50	50	—	B2

The coastal cliffs provide good breeding sites for seabirds.

Protection status

National Partial **International** Partial
685 ha of IBA covered by National Park (Parc national de Port-Cros, 690 ha). 2,475 ha of IBA covered by Special Protection Area (Port Cros, 2,475 ha).

Conservation issues

Threats Aquaculture/fisheries (C), Infrastructure (B), Recreation/tourism (B), Unsustainable exploitation (C)

The development of tourism and recreation, including associated urban development, is the main threat to the area.

Plateau de Valensole — A1, B2, B3, C1, C6 — 249

Admin region Provence-Alpes-Côte d'Azur
Coordinates 43°50'N 6°03'E
Altitude 540–820 m **Area** 22,500 ha

Site description

The site contains a cultivated plateau with garrigue, dry meadows, woodlands and abandoned fields.

Habitats Forest and woodland (10%; broadleaved deciduous forest),, Grassland (10%; steppe/dry calcareous grassland), Artificial landscape (85%; arable land; perennial crops/orchards/groves; ruderal land)
Land-use Agriculture (85%), Forestry, Hunting (10%), Urban/industrial/transport

Birds

Species	Season	Year	Pop min	Pop max	Acc	Criteria
Circus pygargus Montagu's Harrier	B	1991	10	100	—	B3, C6
Tetrax tetrax Little Bustard	B	1996	20	20	—	A1, C1, C6
Burhinus oedicnemus Stone Curlew	B	1991	10	100	—	B2, C6
Otus scops Scops Owl	B	1991	1	10	—	B2
Athene noctua Little Owl	B	1991	10	100	—	B2
Melanocorypha calandra Calandra Lark	B	1991	10	100	—	C6
Calandrella brachydactyla Short-toed Lark	B	1991	10	100	—	C6

An important site for breeding species found in Mediterranean habitats.

Protection status

National None **International** None

Conservation issues

Threats Agricultural intensification/expansion (A)

The main threat is expansion of the areas under irrigation into large parts of the site.

Plateau de l'Arbois, garrigues de Lancon et chaîne des Côtes — A1, B2, C1, C6 — 250

Admin region Provence-Alpes-Côte d'Azur
Coordinates 43°34'N 5°12'E
Altitude 135–479 m **Area** 34,500 ha

Site description

The site contains garrigue, cultivated areas, dry meadows, Mediterranean woodlands, cliffs, a river and a reservoir.

Habitats Forest and woodland (broadleaved evergreen forest), Scrub (sclerophyllous scrub/garrigue/maquis), Grassland (steppe/dry calcareous grassland), Wetland (standing fresh water; river/stream), Rocky areas (inland cliff), Artificial landscape (arable land; perennial crops/orchards/groves)
Land-use Agriculture, Forestry, Hunting, Tourism/recreation, Urban/industrial/transport, Water management

Birds

Species	Season	Year	Pop min	Pop max	Acc	Criteria
Milvus migrans Black Kite	B	1991	10	100	—	B2, C6
Circaetus gallicus Short-toed Eagle	B	1991	5	10	—	C6
Hieraaetus fasciatus Bonnelli's Eagle	B	1996	5	5	—	B2, C6
Tetrax tetrax Little Bustard	B	1996	10	20	—	A1, C1, C6
Bubo bubo Eagle Owl	B	1991	10	20	—	B2, C6
Coracias garrulus Roller	B	1991	10	50	—	B2, C6

The site is important for breeding raptors, *Tetrax tetrax*, and species associated with Mediterranean habitats. Species of global conservation concern that do not meet IBA criteria: *Falco naumanni* (irregularly breeds).

Protection status

National Low **International** Low

285 ha of IBA covered by Private Reserve (La Barben, 285 ha). 285 ha of IBA covered by Special Protection Area (Plateau de L'Arbois: la Barben, 285 ha).

Conservation issues

Threats Industrialization/urbanization (A), Infrastructure (A)

The construction of a new railway and railway station within the site are the main threats.

Plaine des Maures — B2, C2, C6 — 251

Admin region Provence-Alpes-Côte d'Azur
Coordinates 43°22'N 6°24'E
Altitude 47–166 m **Area** 7,500 ha

Site description

Habitats Forest and woodland (mixed forest; broadleaved evergreen forest), Scrub (80%; sclerophyllous scrub/garrigue/maquis), Grassland (dry siliceous grassland), Wetland (standing fresh water; river/stream; water-fringe vegetation), Artificial landscape (5%; arable land)
Land-use Agriculture (10%), Hunting (100%), Tourism/recreation, Urban/industrial/transport (5%)

Birds

Species	Season	Year	Pop min	Pop max	Acc	Criteria
Otus scops Scops Owl	B	1991	1	10	—	B2
Merops apiaster Bee-eater	B	1991	10	100	—	B2
Coracias garrulus Roller	B	1991	10	100	—	B2, C2, C6
Sylvia undata Dartford Warbler	B	1991	100	500	—	B2, C6
Emberiza hortulana Ortolan Bunting	B	1991	100	500	—	B2, C6

Protection status

National None **International** None

Conservation issues

Threats Industrialization/urbanization (A), Infrastructure (A), Other (B), Recreation/tourism (B)

Threats include tourism and other recreation activities, urbanization, industry and the planned construction of a new railway that will run through the site.

Etangs de Citis, Lavalduc, Engrenier, Pourra, l'Estomac, Fos et salines de Rassuen et de Fos — A4i, B1i, C2, C3, C6 — 252

Admin region Provence-Alpes-Côte d'Azur
Coordinates 43°28'N 4°58'E
Altitude 0–0 m **Area** 1,250 ha

Site description

Brackish marshes and lagoons with canals, reedbeds (*Phragmites*), dry and wet meadows and garrigue.

Habitats Scrub (sclerophyllous scrub/garrigue/maquis), Grassland (steppe/dry calcareous grassland; mesophile grassland), Wetland (coastal lagoon; standing fresh water; standing brackish and salt water; river/stream; water-fringe vegetation)
Land-use Agriculture (10%), Hunting

Birds

Species	Season	Year	Pop min	Pop max	Acc	Criteria
Podiceps nigricollis Black-necked Grebe	W	1997	5	6,415	—	A4i, B1i, C3
Podiceps nigricollis Black-necked Grebe	P	1997	3,000	4,000	—	A4i, B1i, C3
Phalacrocorax carbo Cormorant	W	1997	4	1,730	—	B1i, C3
Phoenicopterus ruber Greater Flamingo	W	1997	325	1,807	—	A4i, B1i, C2
Aythya ferina Pochard	W	1997	2,232	12,559	—	B1i, C3
Aythya fuligula Tufted Duck	W	1997	2,116	6,125	—	B1i, C3
Himantopus himantopus Black-winged Stilt	B	1991	10	100	—	C6
Himantopus himantopus Black-winged Stilt	P	1991	100	500	—	B1i, C2
Recurvirostra avosetta Avocet	B	1997	10	10	—	C6
Larus cachinnans Yellow-legged Gull	W	1997	453	5,768	—	A4i, B1i, C3

The site is important for breeding and wintering waterbirds. Species of global conservation concern that do not meet IBA criteria: *Aythya nyroca* (wintering).

Protection status

National None **International** None

Conservation issues

Threats Infrastructure (A)

The main threat is the planned construction of a highway that would run through the site.

Marais de L'Ile Vieille — B1i, C2, C6 — 253

Admin region Provence-Alpes-Côte d'Azur
Coordinates 44°14'N 4°41'E
Altitude 0–10 m **Area** 1,100 ha

Site description

The site includes freshwater marshes, reedbeds (*Phragmites*), riverine forest and cultivated and abandoned fields.

Habitats Forest and woodland (alluvial/very wet forest), Wetland (standing fresh water; water-fringe vegetation), Artificial landscape (perennial crops/orchards/groves; ruderal land)
Land-use Agriculture, Fisheries/aquaculture, Hunting (50%), Urban/industrial/transport

Birds

Species	Season	Year	Pop min	Pop max	Acc	Criteria
Nycticorax nycticorax Night Heron	B	1991	55	75	—	C6
Egretta garzetta Little Egret	B	1991	30	30	—	C6
Egretta garzetta Little Egret	W	1991	100	1,000	—	B1i, C2

This site is mainly important for reedbed and marshland birds. Species of global conservation concern that do not meet IBA criteria: *Crex crex* (passage).

Protection status

National None **International** None

Conservation issues

Threats Disturbance to birds (C), Infrastructure (C)

Wood extraction may disturb the heron colonies.

Basse vallée de la Durance — B1i, C3, C6 — 254

Admin region Provence-Alpes-Côte d'Azur
Coordinates 43°48'N 5°12'E
Altitude 20–255 m **Area** 6,000 ha

Site description

The site includes a river, ponds, riverine vegetation, woodlands, dry meadows and abandoned fields.

Habitats Forest and woodland (broadleaved deciduous forest; alluvial/very wet forest), Grassland (steppe/dry calcareous grassland), Wetland (standing fresh water; standing brackish and salt water; river/stream; water-fringe vegetation), Artificial landscape (other urban/industrial areas; ruderal land)
Land-use Agriculture, Fisheries/aquaculture, Hunting, Tourism/recreation, Urban/industrial/transport, Water management

Birds

Species	Season	Year	Pop min	Pop max	Acc	Criteria
Phalacrocorax carbo Cormorant	W	1991	3,000	4,000	—	B1i, C3
Ixobrychus minutus Little Bittern	B	1991	15	15	—	C6
Nycticorax nycticorax Night Heron	B	1991	170	170	—	C6
Egretta garzetta Little Egret	B	1991	50	50	—	C6

Species ... continued	Season	Year	Pop min	Pop max	Acc	Criteria
Milvus migrans Black Kite	B	1991	50	60	—	C6
Sterna hirundo Common Tern	B	1991	50	50	—	C6
Alcedo atthis Kingfisher	B	1991	50	50	—	C6
Calandrella brachydactyla Short-toed Lark	B	1991	10	10	—	C6

Species of global conservation concern that do not meet IBA criteria: *Aythya nyroca* (wintering).

Protection status

National Low **International** None

35 ha of IBA covered by Hunting Reserve (35 ha). 15 ha of IBA covered by Private Reserve (15 ha).

Conservation issues

Threats Extraction industry (C), Industrialization/urbanization (B), Infrastructure (A)

The erection of powerlines has created a permanent hazard for large raptors. Stone quarrying is also a threat.

Vallée de la Durance: de Tallard à Sisteron

B2, C6 **255**

Admin region Provence-Alpes-Côte d'Azur
Coordinates 44°22'N 6°00'E
Altitude 480–615 m **Area** 1,650 ha

Site description

A river with associated riverine vegetation, cliffs, rocky slopes, garrigue, meadows and arable land.

Habitats Forest and woodland (30%; broadleaved deciduous forest; alluvial/very wet forest), Scrub (15%; heathland; sclerophyllous scrub/garrigue/maquis), Grassland (18%; steppe/dry calcareous grassland; mesophile grassland), Wetland (7%; river/stream; water-fringe vegetation), Rocky areas (40%; inland cliff), Artificial landscape (10%; arable land)
Land-use Agriculture, Fisheries/aquaculture, Hunting, Urban/industrial/transport

Birds

Species	Season	Year	Pop min	Pop max	Acc	Criteria
Ixobrychus minutus Little Bittern	B	1991	10	20	—	C6
Milvus migrans Black Kite	B	1991	10	100	—	B2, C6
Caprimulgus europaeus Nightjar	B	1991	50	250	—	B2, C6
Alcedo atthis Kingfisher	B	1991	10	100	—	B2, C6
Lullula arborea Woodlark	B	1991	50	250	—	C6
Anthus campestris Tawny Pipit	B	1991	50	250	—	B2, C6
Lanius collurio Red-backed Shrike	B	1991	50	250	—	C6

Species of global conservation concern that do not meet IBA criteria: *Gallinago media* (1–10 passage individuals).

Protection status

National Partial **International** None

124 ha of IBA covered by Hunting Reserve (124 ha).

Conservation issues

Threats Industrialization/urbanization (C), Infrastructure (A), Other (C)

The main threat is the planned construction of a highway within the IBA. Other reported threats are the dumping of rubbish and the erection of powerlines.

Vallée du Haut-Guil

B2, C6 **256**

Admin region Provence-Alpes-Côte d'Azur
Coordinates 44°44'N 7°02'E
Altitude 1,670–3,300 m **Area** 5,000 ha

Site description

Habitats Forest and woodland (14%; native coniferous forest; alluvial/very wet forest), Scrub (10%; heathland), Grassland (53%; alpine/subalpine/boreal grassland), Wetland (standing fresh water; river/stream; fen/transition mire/spring), Rocky areas (scree/boulders; inland cliff)
Land-use Agriculture, Fisheries/aquaculture, Forestry, Hunting, Tourism/recreation

The site includes alpine meadows, scree and rocky slopes, cliffs, moorlands, conifer forest, rivers and riverine forest.

Birds

Species	Season	Year	Pop min	Pop max	Acc	Criteria
Tetrao tetrix Black Grouse	B	1991	25	100	—	C6
Alectoris graeca Rock Partridge	B	1991	10	100	—	B2, C6
Glaucidium passerinum Pygmy Owl	B	1991	1	10	—	C6
Dryocopus martius Black Woodpecker	B	1991	10	100	—	C6
Pyrrhocorax pyrrhocorax Chough	B	1991	10	100	—	C6

This area is of special interest for mountain and forest breeding species.

Protection status

National None **International** None

Conservation issues

Threats Abandonment/reduction of land management (U)

No major threats are known at this site. Since closure of the access road, tourist pressure has declined. There is a project to establish a nature reserve.

Marais de Manteyer

B2, B3, C6 **257**

Admin region Provence-Alpes-Côte d'Azur
Coordinates 44°32'N 5°58'E
Altitude 935–946 m **Area** 62 ha

Site description

The largest wetland in the high Alps.

Habitats Scrub (10%; heathland), Grassland (20%; humid grassland; mesophile grassland), Wetland (61%; standing fresh water; water-fringe vegetation), Artificial landscape (9%; arable land; perennial crops/orchards/groves)
Land-use Agriculture, Hunting, Urban/industrial/transport

Birds

Species	Season	Year	Pop min	Pop max	Acc	Criteria
Ixobrychus minutus Little Bittern	B	1991	5	50	—	B2, C6
Porzana porzana Spotted Crake	B	1991	5	50	—	B3, C6

The site is important for migratory waterbirds and raptors.

Protection status

National High **International** None

62 ha of IBA covered by Hunting Reserve (62 ha).

Conservation issues

Threats Abandonment/reduction of land management (U)

The main threat is the abandonment of grazing, which is leading to loss of open habitat.

Bois des Ayes

C6 **258**

Admin region Provence-Alpes-Côte d'Azur
Coordinates 44°50'N 6°40'E
Altitude 1,980–2,446 m **Area** 450 ha

Site description

The site includes conifer woodland, moorland and peatbogs.

Habitats Forest and woodland (80%; native coniferous forest), Scrub (18%; heathland), Wetland (fen/transition mire/spring)
Land-use Agriculture (10%), Forestry (100%), Hunting (100%)

Birds

Species	Season	Year	Pop min	Pop max	Acc	Criteria
Tetrao tetrix Black Grouse	B	1991	15	100	—	C6
Glaucidium passerinum Pygmy Owl	B	1991	10	10	—	C6

Protection status
National Partial **International** None
250 ha of IBA covered by Biological State Reserve (250 ha).

Conservation issues
Threats Recreation/tourism (B)

While many threats were resolved by the designation of the site as a Biological State Reserve, recreation and hunting still cause disturbance.

Bois du Chapitre C6 259
Admin region Provence-Alpes-Côte d'Azur
Coordinates 44°31'N 6°00'E
Altitude 1,250–2,062 m **Area** 250 ha

Site description
The site is dominated by mixed forest and alpine meadows.

Habitats Forest and woodland (75%; mixed forest), Scrub (heathland), Grassland (24%; alpine/subalpine/boreal grassland), Wetland (river/stream), Rocky areas (scree/boulders)
Land-use Forestry, Not utilized

Birds

Species	Season	Year	Pop min	Pop max	Acc	Criteria
Bonasa bonasia Hazel Grouse	B	1991	10	20	—	C6
Tetrao tetrix Black Grouse	B	1991	15	100	—	C6
Glaucidium passerinum Pygmy Owl	B	1991	2	10	—	C6

The site is important for alpine and montane forest breeding species.

Protection status
National High **International** None
250 ha of IBA covered by Biological State Reserve (250 ha).

Conservation issues
Threats Recreation/tourism (C)

The only reported threat is tourists visiting the site and causing disturbance.

Bec de Crigne B2, C6 260
Admin region Provence-Alpes-Côte d'Azur
Coordinates 44°24'N 5°56'E
Altitude 680–1,375 m **Area** 400 ha

Site description
The site holds a diverse set of habitats including garrigue, mountains, forests and cliffs.

Habitats Forest and woodland (25%; broadleaved deciduous forest), Scrub (60%; heathland; sclerophyllous scrub/garrigue/maquis), Grassland (5%; steppe/dry calcareous grassland), Rocky areas (10%; scree/boulders; inland cliff)
Land-use Agriculture, Forestry, Tourism/recreation

Birds

Species	Season	Year	Pop min	Pop max	Acc	Criteria
Alectoris graeca Rock Partridge	B	1991	10	100	—	B2, C6
Lullula arborea Woodlark	B	1991	100	500	—	B2, C6

The site is important for breeding raptors and montane species.

Protection status
National High **International** None
400 ha of IBA covered by Hunting Reserve (400 ha).

Conservation issues
Threats Agricultural intensification/expansion (B), Intensified forest management (B), Recreation/tourism (B)

Forestry activities (construction of new tracks), recreation (including climbing), and hunting disturb breeding birds.

Parc National du Mercantour B2, B3, C2, C6 261
Admin region Provence-Alpes-Côte d'Azur
Coordinates 44°10'N 7°04'E
Altitude 490–3,143 m **Area** 68,500 ha

Site description
The site consists of cliffs, rocky slopes and scree, deciduous and conifer forests, alpine meadows, moorlands and garrigue.

Habitats Forest and woodland (20%; native coniferous forest), Scrub (8%; sclerophyllous scrub/garrigue/maquis), Grassland (26%; alpine/subalpine/boreal grassland), Wetland (standing fresh water), Rocky areas (44%; scree/boulders; inland cliff)
Land-use Agriculture, Fisheries/aquaculture, Forestry, Tourism/recreation, Urban/industrial/transport

Birds

Species	Season	Year	Pop min	Pop max	Acc	Criteria
Pernis apivorus Honey Buzzard	B	1991	10	100	—	B3, C6
Gypaetus barbatus Lammergeier	B	1996	1	1	—	B2
Circaetus gallicus Short-toed Eagle	B	1991	10	100	—	B2, C6
Aquila chrysaetos Golden Eagle	B	1991	25	25	—	B2, C6
Falco peregrinus Peregrine	B	1991	1	10	—	B2
[1] ***Lagopus mutus*** Ptarmigan	B	1991	100	500	—	C6
Tetrao tetrix Black Grouse	B	1991	100	500	—	B2, C6
Alectoris graeca Rock Partridge	B	1991	100	500	—	B2, C6
Glaucidium passerinum Pygmy Owl	B	1991	10	100	—	C6
Aegolius funereus Tengmalm's Owl	B	1991	10	100	—	C6
Dryocopus martius Black Woodpecker	B	1991	10	100	—	C6
Pyrrhocorax pyrrhocorax Chough	B	1991	100	500	—	B2, C2, C6

1. *L. m. helveticus.*

This is an important site for waterbirds, particularly during migration.

Protection status
National High **International** High
68,500 ha of IBA covered by National Park (Parc national du Mercantour, 68,500 ha). 68,500 ha of IBA covered by Special Protection Area (Le Mercantour, 68,500 ha).

Conservation issues
Threats Agricultural intensification/expansion (A), Infrastructure (A), Intensified forest management (A), Recreation/tourism (B)

The main threats are forestry, overgrazing, tourism and the planned construction of a tunnel.

Basse vallée du Var A4i, B1i, C2, C6 262
Admin region Provence-Alpes-Côte d'Azur
Coordinates 43°46'N 7°12'E
Altitude 0–50 m **Area** 1,100 ha

Site description
The Var River estuary with ponds, natural islets and riverine vegetation.

Habitats Wetland (tidal river/enclosed tidal water; mudflat/sandflat; shingle/stony beach; coastal lagoon; standing fresh water; standing brackish and salt water; river/stream), Marine areas (open sea)
Land-use Fisheries/aquaculture (80%), Urban/industrial/transport

Birds

Species	Season	Year	Pop min	Pop max	Acc	Criteria
Larus melanocephalus Mediterranean Gull	P	1991	3,000	10,000	—	A4i, B1i, C2
Sterna hirundo Common Tern	B	1991	50	160	—	C6

Protection status
National Partial **International** None
600 ha of IBA covered by Hunting Reserve (600 ha).

Conservation issues
Threats Infrastructure (A)

Fishing and recreation cause high levels of disturbance.

Préalpes de Grasse

B2, C6 **263**

Admin region Provence-Alpes-Côte d'Azur
Coordinates 43°46'N 7°04'E
Altitude 500–1,300 m **Area** 8,500 ha

Site description

The site includes garrigue, dry meadows, cliffs and rocky slopes.

Habitats Scrub (37%; sclerophyllous scrub/garrigue/maquis), Grassland (42%; steppe/dry calcareous grassland; alpine/subalpine/boreal grassland), Wetland (river/stream), Rocky areas (10%; inland cliff), Artificial landscape (10%; perennial crops/orchards/groves)
Land-use Agriculture, Fisheries/aquaculture, Hunting, Not utilized (100%), Tourism/recreation, Urban/industrial/transport

Birds

Species	Season	Year	Pop min	Pop max	Acc	Criteria
Monticola solitarius Blue Rock Thrush	B	1991	1	10	—	B2
Pyrrhocorax pyrrhocorax Chough	B	1991	10	100	—	C6

The site is important for breeding raptors, Mediterranean and mountain species. Species of global conservation concern that do not meet IBA criteria: *Tetrax tetrax* (breeding).

Protection status

National Partial **International** None
2,415 ha of IBA covered by Hunting Reserve (2,415 ha).

Conservation issues

Threats Abandonment/reduction of land management (U), Industrialization/urbanization (A)

Abandonment of stock-raising is the main threat, leading to growth of bushes and trees in alpine habitats. There is also a risk of urbanization. There is a project to designate an SPA in the site.

Parc National des Ecrins

B2, C2, C6 **264**

Admin region Provence-Alpes-Côte d'Azur
Coordinates 44°38'N 6°16'E
Altitude 800–4,100 m **Area** 120,000 ha

Site description

This large site comprises alpine meadows, cliffs and rocky slopes, glaciers and permanent snow, moorland, deciduous and conifer forests and a reservoir. The main human activity is stock-raising.

Habitats Forest and woodland (10%; broadleaved deciduous forest; native coniferous forest; mixed forest; alluvial/very wet forest), Scrub (12%; heathland), Grassland (21%; steppe/dry calcareous grassland; dry siliceous grassland; alpine/subalpine/boreal grassland; humid grassland; mesophile grassland), Wetland (standing fresh water; river/stream; raised bog; water-fringe vegetation; fen/transition mire/spring), Rocky areas (48%; scree/boulders; inland cliff; caves), Artificial landscape (perennial crops/orchards/groves)
Land-use Agriculture (42%), Fisheries/aquaculture (5%), Forestry (5%), Tourism/recreation (1%), Urban/industrial/transport

Birds

Species	Season	Year	Pop min	Pop max	Acc	Criteria
Pernis apivorus Honey Buzzard	B	1991	30	30	—	C6
Milvus migrans Black Kite	B	1991	30	40	—	C6
Circaetus gallicus Short-toed Eagle	B	1991	10	11	—	C6
Aquila chrysaetos Golden Eagle	B	1991	33	33	—	B2, C6
Falco peregrinus Peregrine	B	1991	6	7	—	C6
Bonasa bonasia Hazel Grouse	B	1991	50	75	—	C6
[1] ***Lagopus mutus*** Ptarmigan	B	1991	200	300	—	C6
Tetrao tetrix Black Grouse	B	1991	250	300	—	B2, C6
Alectoris graeca Rock Partridge	B	1991	300	400	—	B2, C6
Glaucidium passerinum Pygmy Owl	B	1991	5	10	—	C6
Aegolius funereus Tengmalm's Owl	B	1991	10	10	—	C6
Caprimulgus europaeus Nightjar	B	1991	100	200	—	B2, C6
Dryocopus martius Black Woodpecker	B	1991	100	200	—	C6
Lullula arborea Woodlark	B	1991	500	1,000	—	B2, C6
Lanius collurio Red-backed Shrike	B	1991	500	1,000	—	C6
Pyrrhocorax pyrrhocorax Chough	B	1991	500	600	—	B2, C2, C6
Emberiza hortulana Ortolan Bunting	B	1991	100	500	—	B2, C6

1. *L. m. helveticus.*

The site is important for alpine birds. Species of global conservation concern that do not meet IBA criteria: *Crex crex* (breeding).

Protection status

National Partial **International** Partial
145 ha of IBA covered by Nature Reserve (Grand Lac des Estaris, 145 ha). 155 ha of IBA covered by Nature Reserve (Haute vallée de la Séveraise, 155 ha). 85 ha of IBA covered by Nature Reserve (Haute vallée du Béranger, 85 ha). 20 ha of IBA covered by Nature Reserve (Vallée du Torrent de Saint-Pierre, 20 ha). 90 ha of IBA covered by Nature Reserve (Vallée du Vénéon, 90 ha). 685 ha of IBA covered by Nature Reserve (Pics de Combeynot, 685 ha). 91,800 ha of IBA covered by National Park (Parc national des Ecrins, 91,800 ha). 91,800 ha of IBA covered by Special Protection Area (Les Ecrins, 91,800 ha).

Conservation issues

Threats Afforestation (U), Industrialization/urbanization (U), Recreation/tourism (U)

The site is being degraded by afforestation with *Pinus* and tourists disturb breeding birds. In addition, nearby urban development threatens the site.

Bois de Palayson, du Rouet et de Malvoisin

C7 **265**

Admin region Provence-Alpes-Côte d'Azur
Coordinates 43°32'N 6°38'E
Altitude 20–560 m **Area** 14,150 ha

Site description

The site includes cork oak *Quercus suber* forest, cliffs, rivers, riverine forest and cultivated areas.

Habitats Forest and woodland (alluvial/very wet forest; broadleaved evergreen forest), Wetland (standing fresh water; river/stream), Rocky areas (inland cliff), Artificial landscape (arable land)
Land-use Agriculture, Fisheries/aquaculture, Hunting, Urban/industrial/transport

Protection status

National None **International** None

Conservation issues

Threats Industrialization/urbanization (A), Infrastructure (A), Recreation/tourism (A), Unsustainable exploitation (C)

The main threats are disturbance from hunting and tourism (including climbing). Urbanization and the construction of a highway are degrading the site.

Détroit de Bonifaccio et Iles Lavezzi

B1i, B1ii, B2, B3, C2, C6 **266**

Admin region Corse
Coordinates 41°22'N 9°14'E
Altitude 0–50 m **Area** 16,600 ha

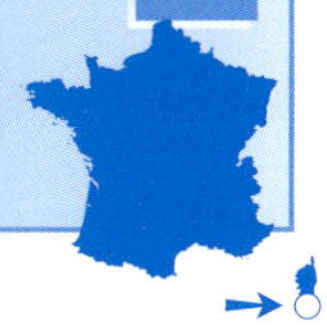

Site description

Open sea with rocky islands and calcareous cliffs.

Habitats Marine areas (open sea; sea inlet/coastal features), Rocky areas (rock stacks/islets)
Land-use Fisheries/aquaculture, Tourism/recreation

Birds

Species	Season	Year	Pop min	Pop max	Acc	Criteria
Calonectris diomedea Cory's Shearwater	B	1991	312	466	—	B1ii, B2, C2, C6
Phalacrocorax aristotelis Shag	B	1991	350	500	—	B1i, B3, C2, C6

This is an important site for breeding seabirds.

Protection status

National High **International** Partial

15,000 ha of IBA covered by Hunting Reserve (Bouches de Bonifaccio, 15,000 ha). 5,130 ha of IBA covered by Nature Reserve (Iles Lavezzi, 5,130 ha). 5,150 ha of IBA covered by Special Protection Area (Iles Lavezzi, 5,150 ha).

Conservation issues

Threats Industrialization/urbanization (B), Other (U), Recreation/tourism (C)

The main threats arise from development for recreation and the risks of marine pollution from oil tankers ('Other' threats, above).

Iles Cerbicale — B1i, B1ii, B2, B3, C2, C6 — 267

Admin region Corse
Coordinates 41°33'N 9°21'E
Altitude 0–35 m **Area** 5,040 ha

Site description

Open sea, rocky islets with coastal cliffs covered by Mediterranean scrub.

Habitats Forest and woodland (broadleaved evergreen forest), Wetland (sand-dunes/sand beach), Marine areas (open sea), Rocky areas (rock stacks/islets)
Land-use Hunting, Tourism/recreation

Birds

Species	Season	Year	Pop min	Pop max	Acc	Criteria
Calonectris diomedea Cory's Shearwater	B	1991	79	115	—	B2, C6
Hydrobates pelagicus Storm Petrel	B	1991	70	75	—	B1ii, C2
Phalacrocorax aristotelis Shag	B	1991	120	185	—	B1i, B3, C2, C6
Larus audouinii Audouin's Gull	B	1995	2	2	—	B2

This is an important area for breeding seabirds. It is also one of the few breeding sites for *Larus audouinii* in France.

Protection status

National High **International** High
60 ha of IBA covered by Nature Reserve (Iles Cerbicale, 60 ha). 5,000 ha of IBA covered by Hunting Reserve (Iles Cerbicale, 5,000 ha). 5,036 ha of IBA covered by Special Protection Area (Iles Cerbicale, 5,036 ha).

Conservation issues

Threats Disturbance to birds (C), Other (A), Recreation/tourism (C)

Marine pollution with hydrocarbons ('Other' threats, above) is the most significant threat to the site.

Golfe de Porto, presqu'île de Scandola et golfe de Galeria — A4i, B1i, B2, B3, C2, C3, C6 — 268

Admin region Corse
Coordinates 42°20'N 8°37'E
Altitude 0–630 m **Area** 28,700 ha

Site description

An area of rocky shore with numerous rocky inlets, cliffs, beaches and islands, as well as Mediterranean scrub and sclerophyllous woodland.

Habitats Forest and woodland (alluvial/very wet forest; broadleaved evergreen forest), Scrub (sclerophyllous scrub/garrigue/maquis), Wetland (tidal river/enclosed tidal water; shingle/stony beach; river/stream), Marine areas (open sea), Rocky areas (sea cliff/rocky shore; rock stacks/islets)
Land-use Agriculture, Fisheries/aquaculture, Hunting, Tourism/recreation, Urban/industrial/transport

Birds

Species	Season	Year	Pop min	Pop max	Acc	Criteria
Phalacrocorax aristotelis Shag	B	1991	50	270	—	B1i, B3, C2, C6
Falco peregrinus Peregrine	B	1991	10	15	—	B2
Larus audouinii Audouin's Gull	B	1995	2	2	—	B2
Larus cachinnans Yellow-legged Gull	B	1991	1,400	2,000	—	A4i, B1i, C3

Protection status

National Partial **International** High
16,500 ha of IBA covered by Hunting Reserve (16,500 ha). 1,919 ha of IBA covered by Nature Reserve (Scandola, 1,919 ha). 26,960 ha of IBA covered by Special Protection Area (Golfe de Porto presqu'ile de Scandola, 26,960 ha).

Conservation issues

Threats Recreation/tourism (C)

The site is threatened by the construction of holiday homes and the associated increase in human disturbance.

Forêts domaniales de Corse — A1, A2, A3, B2, C1, C2, C6 — 269

Admin region Corse
Coordinates 42°07'N 9°00'E
Altitude 100–2,300 m **Area** 50,217 ha

Site description

Old conifer and deciduous forests with streams, scree, cliffs and alpine grasslands.

Habitats Forest and woodland (broadleaved deciduous forest; native coniferous forest; mixed forest; broadleaved evergreen forest), Grassland (alpine/subalpine/boreal grassland), Wetland (river/stream), Rocky areas (scree/boulders; inland cliff)
Land-use Agriculture, Fisheries/aquaculture, Forestry (100%), Hunting, Tourism/recreation

Birds

Species	Season	Year	Pop min	Pop max	Acc	Criteria
Gypaetus barbatus Lammergeier	B	1996	3	5	—	B2
Aquila chrysaetos Golden Eagle	B	1991	10	20	—	B2
[1] ***Sitta whiteheadi*** Corsican Nuthatch	B	1991	1,000	1,500	—	A1, A2, A3, B2, C1, C2, C6

1. Mediterranean biome.

The most important area in Corsica for the endemic *Sitta whiteheadi*.

Protection status

National Low **International** Partial
15 ha of IBA covered by Biological State Reserve (Tavignano, 15 ha). 6,500 ha of IBA covered by Biosphere Reserve (Réserve de la Biosphère du Fango, 6,500 ha).

Conservation issues

Threats Burning of vegetation (A), Intensified forest management (B), Recreation/tourism (C), Unsustainable exploitation (C)

Frequent non-natural fires during the summer months are the most significant threat.

Vallée d'Asco — A1, A2, A3, B2, C1, C2, C6 — 270

Admin region Corse
Coordinates 42°24'N 8°58'E
Altitude 950–2,710 m **Area** 3,800 ha

Site description

Sclerophyllous scrub and *Pinus* forests, mixed woodland, scree, rocky slopes, cliffs, alpine grasslands and permanent snow.

Habitats Forest and woodland (native coniferous forest), Scrub (sclerophyllous scrub/garrigue/maquis), Grassland (alpine/subalpine/boreal grassland), Rocky areas (scree/boulders; inland cliff)
Land-use Agriculture, Fisheries/aquaculture, Forestry, Hunting, Tourism/recreation, Urban/industrial/transport

Birds

Species	Season	Year	Pop min	Pop max	Acc	Criteria
Gypaetus barbatus Lammergeier	B	1996	1	1	—	B2
[1] ***Sitta whiteheadi*** Corsican Nuthatch	B	1991	10	100	—	A1, A2, A3, B2, C1, C2, C6

1. Mediterranean biome.

An important site for the endemic *Sitta whiteheadi* and endemic subspecies such as *Accipiter gentilis arrigonii*, *Buteo buteo arrigonii*, *Troglodytes troglodytes koenigi* and *Muscicapa striata tyrrhenica*.

■ Protection status

National Partial **International** High

1,000 ha of IBA covered by Hunting Reserve (1,000 ha). 3,800 ha of IBA covered by Special Protection Area (Haute Vallée d'Asco, 6,800 ha).

■ Conservation issues

Threats Abandonment/reduction of land management (A), Intensified forest management (B), Recreation/tourism (B), Unsustainable exploitation (C)

The main threats are the abandonment of cattle-grazing and the increasing numbers of tourists visiting the area.

Vallée de la Restonica — A1, A3, B2, C1, C2, C6 — 271

Admin region Corse
Coordinates 42°15'N 9°04'E
Altitude 550–2,620 m **Area** 6,400 ha

■ Site description

Gorges with sclerophyllous scrub and woodland, conifer woodland, dry grasslands, cliffs and mountain lakes.

Habitats Forest and woodland (native coniferous forest), Scrub (sclerophyllous scrub/garrigue/maquis), Wetland (standing fresh water), Rocky areas (scree/boulders; inland cliff)
Land-use Agriculture, Fisheries/aquaculture, Forestry, Hunting, Tourism/recreation

■ Birds

Species	Season	Year	Pop min	Pop max	Acc	Criteria
Gypaetus barbatus Lammergeier	B	1996	1	1	—	B2
[1] ***Sitta whiteheadi*** Corsican Nuthatch	B	1991	100	500	—	A1, A3, B2, C1, C2, C6

1. Mediterranean biome.

An important site for the endemic *Sitta whiteheadi* and endemic subspecies such as *Accipiter gentilis arrigonii*, *Buteo buteo arrigonii*, *Troglodytes troglodytes koenigi* and *Muscicapa striata tyrrhenica*.

■ Protection status

National None **International** High

6,000 ha of IBA covered by Special Protection Area (Vallée de la Restonica, 6,000 ha).

■ Conservation issues

Threats Abandonment/reduction of land management (A), Agricultural intensification/expansion (B), Recreation/tourism (B), Unsustainable exploitation (C)

As livestock farming is abandoned, some habitats are becoming less suitable for some species. Breeding raptors are adversely affected by tourism, recreational activities and hunting, some of which is illegal.

Etang de Biguglia — A4iii, B1i, C3, C4 — 272

Admin region Corse
Coordinates 42°36'N 9°28'E
Altitude 0–3 m **Area** 1,980 ha

■ Site description

A shallow, brackish lagoon with aquatic vegetation, reedbeds (*Phragmites*) and riparian woodlands.

Habitats Scrub (sclerophyllous scrub/garrigue/maquis), Wetland (saltmarsh; sand-dunes/sand beach; shingle/stony beach; coastal lagoon)
Land-use Fisheries/aquaculture

■ Birds

Species	Season	Year	Pop min	Pop max	Acc	Criteria
Phalacrocorax carbo Cormorant	P	1997	500	1,500	—	B1i, C3

The site holds 20,000 or more waterbirds in winter, on a regular basis (including *Aythya ferina* 4,800; *Aythya fuligula* 4,500; *Fulica atra* 9,000 individuals). Species of global conservation concern that do not meet IBA criteria: *Aythya nyroca* (1–10 wintering individuals); *Larus audouinii* (1–10 passage individuals).

■ Protection status

National High **International** High

1,800 ha of IBA covered by Nature Reserve (Etang de Biguglia, 1,800 ha). 1,790 ha of IBA covered by Special Protection Area (Etang de Biguglia, 1,790 ha). 1,980 ha of IBA covered by Ramsar Site (1,980 ha).

■ Conservation issues

Threats Agricultural intensification/expansion (A), Disturbance to birds (C), Industrialization/urbanization (C), Infrastructure (C)

Industrial and agricultural pollution, originating outside the site, are the main threats. Also there is considerable hunting pressure, including illegal hunting in the vicinity. A reintroduction project for *Oxyura leucocephala* is being carried out under a LIFE contract.

Vallée du Régino — B2, B3, C2, C6 — 273

Admin region Corse
Coordinates 42°34'N 8°58'E
Altitude 0–1,680 m **Area** 9,500 ha

■ Site description

A large area of sclerophyllous scrub and Mediterranean woodland lying between the sea and mountains in northern Corsica.

Habitats Forest and woodland (alluvial/very wet forest; broadleaved evergreen forest), Scrub (sclerophyllous scrub/garrigue/maquis), Grassland (dry siliceous grassland), Wetland (5%; tidal river/enclosed tidal water; sand-dunes/sand beach; shingle/stony beach; standing fresh water; river/stream; water-fringe vegetation), Rocky areas (25%; scree/boulders; inland cliff; inland sand-dunes; caves), Artificial landscape (perennial crops/orchards/groves)
Land-use Agriculture (100%), Fisheries/aquaculture, Forestry, Hunting, Tourism/recreation, Urban/industrial/transport, Water management

■ Birds

Species	Season	Year	Pop min	Pop max	Acc	Criteria
Milvus milvus Red Kite	B	1991	25	25	—	B3
Otus scops Scops Owl	B	1991	10	100	—	B2
Caprimulgus europaeus Nightjar	B	1991	10	100	—	C6
Merops apiaster Bee-eater	B	1991	10	100	—	B2
Jynx torquilla Wryneck	B	1991	10	100	—	B2
Lullula arborea Woodlark	B	1991	50	250	—	C6
Anthus campestris Tawny Pipit	B	1991	10	100	—	C6
Monticola solitarius Blue Rock Thrush	B	1991	10	100	—	B2
Sylvia sarda Marmora's Warbler	B	1991	250	750	—	B3, C2, C6
Sylvia undata Dartford Warbler	B	1991	100	500	—	C6
Lanius collurio Red-backed Shrike	B	1991	10	100	—	C6
Lanius senator Woodchat Shrike	B	1991	10	100	—	B2

This site is important for breeding raptors and species associated with Mediterranean habitats.

■ Protection status

National None **International** None

■ Conservation issues

Threats Abandonment/reduction of land management (B), Agricultural intensification/expansion (C), Burning of vegetation, Disturbance to birds (B)

Abandonment of land management and anthropogenic fires threaten the site.

Etang d'Urbino et zones humides périphériques — B1i, C3 — 274

Admin region Corse
Coordinates 42°03'N 9°28'E
Altitude 3–30 m **Area** 5,200 ha

■ Site description

A shallow brackish lagoon and marshes surrounded by arable land and sclerophyllous scrub.

Habitats Grassland (humid grassland), Wetland (tidal river/enclosed tidal water; saltmarsh; sand-dunes/sand beach; standing brackish and salt water; river/stream), Artificial landscape (arable land)
Land-use Agriculture (25%), Fisheries/aquaculture (50%), Hunting, Tourism/recreation

■ Birds

Species	Season	Year	Pop min	Pop max	Acc	Criteria
Phalacrocorax carbo Cormorant	W	1997	400	1,830	—	B1i, C3

The site is important for wintering waterbirds.

■ Protection status

National Partial **International** Partial
1,300 ha of IBA covered by Hunting Reserve (Casabianda, 1,760 ha). 2,300 ha of IBA covered by Special Protection Area (Urbino, 2,300 ha).

■ Conservation issues

Threats Aquaculture/fisheries (B), Recreation/tourism (B), Unsustainable exploitation (B)

Water pollution due to fisheries, and hunting are threats.

Iles Sanguinaires B1i, C2, C6 275

Admin region Corse
Coordinates 41°52′N 8°36′E
Altitude 0–40 m **Area** 175 ha

■ Site description

Rocky islets and open sea.

Habitats Marine areas (open sea), Rocky areas (rock stacks/islets)
Land-use Fisheries/aquaculture, Tourism/recreation, Urban/industrial/transport

■ Birds

Species	Season	Year	Pop min	Pop max	Acc	Criteria
Phalacrocorax aristotelis Shag	B	1991	80	110	—	B1i, C2, C6

■ Protection status

National None **International** Partial
33 ha of IBA covered by Special Protection Area (Iles Sanguinaires, 33 ha).

■ Conservation issues

Threats Disturbance to birds (B), Recreation/tourism (B)

Recreation, mostly boating, disturbs the seabird colonies.

Iles Finocchiarola et côte de Tamarone à Centuri A1, B2, C1, C6 276

Admin region Corse
Coordinates 43°00′N 9°24′E
Altitude 0–350 m **Area** 1,140 ha

■ Site description

An area of rocky islets, rocky shorelines and open sea.

Habitats Forest and woodland (alluvial/very wet forest), Scrub (sclerophyllous scrub/garrigue/maquis), Wetland (saltmarsh; sand-dunes/sand beach; coastal lagoon), Marine areas (open sea), Rocky areas (sea cliff/rocky shore; rock stacks/islets)
Land-use Agriculture, Fisheries/aquaculture, Hunting, Tourism/recreation

■ Birds

Species	Season	Year	Pop min	Pop max	Acc	Criteria
Calonectris diomedea Cory's Shearwater	B	1991	30	30	—	B2, C6
Larus audouinii Audouin's Gull	B	1995	85	85	—	A1, B2, C1, C6

This is an important site for breeding seabirds, particularly *Larus audouinii* for which this is the main site in France.

■ Protection status

National Low **International** Partial
3 ha of IBA covered by Nature Reserve (Iles Finocchiarola, 3 ha). 840 ha of IBA covered by Special Protection Area (Iles Finocchiarola et Côte Nord, 840 ha).

■ Conservation issues

Threats Disturbance to birds (B), Other (A), Recreation/tourism (C)

Larus cachinnans predates other seabirds, and is considered a threat ('Other', above). Recreation, mostly boating, disturbs breeding colonies.

Vallée du Verghello A1, A3, B2, C1, C2, C6 277

Admin region Corse
Coordinates 42°12′N 9°07′E
Altitude 435–2,450 m **Area** 2,740 ha

■ Site description

The site includes mountain streams, sclerophyllous scrub, conifer woodland and rocky areas.

Habitats Forest and woodland (native coniferous forest; broadleaved evergreen forest), Scrub (sclerophyllous scrub/garrigue/maquis), Grassland (alpine/subalpine/boreal grassland), Wetland (standing fresh water; river/stream), Rocky areas (scree/boulders; inland cliff)
Land-use Agriculture, Hunting, Tourism/recreation

■ Birds

Species	Season	Year	Pop min	Pop max	Acc	Criteria
Gypaetus barbatus Lammergeier	B	1996	1	1	—	B2
[1] ***Sitta whiteheadi*** Corsican Nuthatch	B	1991	50	250	—	A1, A3, B2, C1, C2, C6

1. Mediterranean biome

This is an important site for species and subspecies endemic to Corsica.

■ Protection status

National None **International** None

■ Conservation issues

Threats Abandonment/reduction of land management (A), Agricultural intensification/expansion (B), Burning of vegetation (B), Recreation/tourism (B), Unsustainable exploitation (C)

Land abandonment and the construction of forest tracks are the main threats to the site. A project exists to create a nature reserve on part of the site.

REFERENCES

ANON. (1997a) Les mesures agri-environnementales. Premiers bilans des expériences européennes. Actes du Colloque du 3–4 novembre 1997 à Paris. Société Française d'Economie Rurale et Centre National pour l'Aménagement des Structures des Exploitations Agricoles.

BARNAUD, G. (1995) *Quinze Sites Ramsar*. Paris: Rapport Ministère de l' Environnement.

BEZZEL, E. (1981) *Evolution des menaces existantes pour le maintien des espèces d'oiseaux nichant en Europe et de l'importance de leur biotope en tant qu'élément de base pour les mesures de conservation*. Bruxelles: Commission des Communautés Européennes (Rapport final).

BROYER, J., ROCAMORA, G., LANG, B. AND MÉTAIS, M. (1994) Enquête Râles des genêts 1991–1992. Synthèse nationale. Rochefort, France: Rapport LPO/Ministère de l'Environnement.

CANS, CH., AND REILLE, A. (1997) *Guide Delachaux et Niestlé des 134 Réserves Naturelles de France*. Lausanne, Switzerland: Delachaux et Niestlé.

CRUON, R., ERARD, C., LEBRETON, J. D. AND NICOLAU-GUILLAUMET, P. (1992) Liste des oiseaux de France. *L'Oiseau et RFO* 62: 97–115.

DECEUNINCK, B. (1998) Base de données des oiseaux d'eau hivernant en France à la mi-janvier. Rochefort, France: LPO/Wetlands International.

DECEUNINCK, B. AND BLANCHON, J. J. (1996) *Programme LIFE Râle des genêts. Bilan études et conservation 1996*. Synthèse nationale. Rochefort, France: Rapport LPO/Ministère de l'Environnement/Union Européenne.

DECEUNINCK, B. AND JOURDE, P. (1998) *Inventaire avifaunistique provisoire des ZPS de France*. Rochefort, France: Rapport LPO/MATE.

DECEUNINCK, B. AND MAHÉO, R. (1998) *Limicoles nicheurs de France. Synthèse de l'enquête nationale 1995–1996*. Rochefort, France: Rapport LPO/ Ministère de l'Environnement.

DUQUET, M. (1995) La Cigogne blanche *Ciconia ciconia* en France en 1995. Résultats du 5° recensement international. *Ornithos* 3(2): 58–62.

DUQUET, M. AND MICHEL, H. (1994) La nidification de la Cigogne noire *Ciconia nigra* en France: historique et statut actuel. *Ornithos* 1(2): 67–71.

GRIMMETT, R. F. A. AND GAMMELL, A. (1987) Preliminary inventory of Important Bird Areas in Spain and Portugal. International Council for Bird Preservation European Community Working Group. Unpublished report to the Commission of the European Community.

GRIMMETT, R. F. A. AND GAMMELL, A. (1989) *Inventory of Important Bird Areas in the European Community*. Rapport Commission des Communautés/Eurogroup/International Council for Bird Preservation.

GRIMMETT, R. F. A. AND JONES, T. A. (1989) *Important Bird Areas in Europe*. Cambridge, UK: International Council for Bird Preservation (Techn. Publ. 9).

HEATH, M. F. AND BORGGREVE, C. (2000) *BirdLife International/EBCC European Bird Database 1998*. Cambridge, UK: BirdLife International.

JOLIVET, C. (1996) L'Outarde canepetière *Tetrax tetrax* en déclin en France. Situation en 1995. *Ornithos* 3(2): 73–77.

JOLIVET, C. (1997) L'Outarde canepetière *Tetrax tetrax* en France: le déclin s'accentue. *Ornithos* 4(2): 73–77.

JOURDE, P. AND ROCAMORA, G. (1996) La conservation des habitats de l'avifaune en France. Evolution et perspectives. *Alauda* 64(1): 58–60.

JUIN, D. (1992) *L'application de la législation communautaire environnement en France*. Paris: Institut pour une Politique Européenne de l'Environnement (Tome II).

MARION, L. (1997) *Inventaire national des Héronnières de France*. Paris: Muséum National d'Histoire Naturelle.

MOSSE, F. (1996). *A la découverte des Réserves Naturelles de France*. Paris: RNF. Edition Nathan.

RICHARD, D. (1999) Changes and loss of biodiversity. Pp. 285–310 in EEA (coordinators) *Environment in the European Union at the Turn of the Century*. Copenhagen: Environmental assessment report no.2.

ROCAMORA, G. (1994a) *Les zones importantes pour la conservation des oiseaux en France*. Rochefort, France: LPO/BirdLife International/ Ministère de l'Environnement.

ROCAMORA, G. (1994b) L'évolution de l'avifaune nicheuse en France au cours des vingt dernières années. Pp. 32–46 in D. Yeatman-Berthelot and G. Jarry (Editors). *Nouvel Atlas des oiseaux nicheurs de France*. Paris: Société Ornithologique de France.

ROCAMORA, G. AND THAURONT, M. (1992) *Inventaire français des Zones de Grand Intérêt pour la conservation des oiseaux sauvages dans la Communauté Europénne*. Rochefort, France: Rapport Ministère de l'Environnement/ LPO/Ecosphère.

ROCAMORA, G., HOTTE, J. F. AND MAILLET, N. (1995) La conservation des ZICO en France: recherche de priorités en fonction de l'intérêt ornithologique et des niveaux de menaces. Rochefort, France: Rapport LPO/Ministère de l'Environnement.

SÉRIOT, J. ET LES COORDINATEURS ESPÈCES (1997) Les oiseaux rares et menacés en France en 1996. *Ornithos* 4(3): 95–115.

SÉRIOT, J., TROTIGNON, J. ET LES COORDINATEURS ESPÈCES (1996) Les oiseaux rares et menacés en France en 1994 et 1995. *Ornithos* 3(3): 97–117.

THAURONT, M. (1990) *Réactualisation de l'inventaire des Zones de Grand Intérêt pour la conservation des oiseaux sauvages dans la Communauté Européenne: méthodologie et analyse*. Rochefort, France: Rapport LPO/CIPO.

TUCKER, G. AND HEATH, M. F. (1994) *Birds in Europe: their conservation status*. Cambridge, UK: BirdLife International (BirdLife Conservation Series no. 3).

YEATMAN-BERTHELOT, D. (1991) *Atlas des oiseaux de France en hiver*. Paris: Société Ornithologique de France.

YEATMAN-BERTHELOT, D. AND JARRY, G. (1994) *Nouvel Atlas des oiseaux nicheurs de France* (1985–1989). Paris: Société Ornithologique de France.

YÉSOU, P. (1997) Nidification de la Mouette mélanocéphale *Larus melanocephalus* en France, 1965–1996. *Ornithos* 4(2): 54–62.

GEORGIA

MICHAEL WILSON

Mount Mkinvartsven (5,033 m), Kazbegi (IBA 007). (PHOTO: ALEXANDER GAVASHELISHVILI)

GENERAL INTRODUCTION

Georgia, one of three Transcaucasian republics which were formerly part of the USSR, occupies an area of some 69,700 km² and lies in the extreme east of the Mediterranean region. In the west, its coastline is washed by the waters of the Black Sea. It shares a common border with Russia in the north, with Azerbaijan in the south-east, Armenia in the south and Turkey in the south-west. The country contains the autonomous republics of Abkhazeti (Abkhazia) and Adjara (Adjaria), also the disputed region of South Ossetia (Tskhinvali region). Georgia is a predominantly mountainous country (54% of the area is higher than 1,000 m above sea-level), with the Great (Greater) Caucasus range in the north, containing many peaks over 3,000 m and three exceeding 5,000 m, and the lower mountains of the Little or Lesser Caucasus in the south. The other main relief features are the Transcaucasian volcanic plateau and the lowlands of Kolkhida and Iberia lying between the two main mountain ranges; this low-lying area is further subdivided into an eastern and western section by the Surami (or Likhi) ridge.

There are 11 Important Bird Areas (IBAs) in Georgia, these covering 15,463 km² or 22% of the country's surface area (Table 1) and with notable gaps in the north-west and in the west-to-east central belt (Map 1). The previous pan-European IBA inventory (Grimmett and Jones 1989) identified eight IBAs in Georgia ("Gruziya", as part of the former USSR), two of which have been merged with new IBAs here (see site accounts for details).

ORNITHOLOGICAL IMPORTANCE

Some doubt remains about the status of certain bird species in Georgia, but there are at least 127 species of European conservation concern (SPECs) breeding regularly in the country (Boehme *et al.* 1987; Tucker and Heath 1994; A. Gavashelishvili). Of these, *Pelecanus crispus*, *Aythya nyroca*, *Aquila heliaca*, *Falco naumanni*

Table 1. Summary of Important Bird Areas in Georgia. 11 IBAs covering 15,463 km²

IBA code	1989 code	International name	National name	Administrative region	Area (ha)	Criteria (see p. 11)
001	SU067	Ritsa	Ritsa	Abkhazeti, Gudauta	16,289	A2, A3
002	SU065	Pskhu	Pskhu	Abkhazeti	27,333	A2, A3
003	SU066	Gumista	Gumista	Abkhazeti, Sukhumi	13,400	A2, A3
004	SU064	Kolkheti	Kolkheti	Samegrelo-Svaneti, Guria, Adjara	150,000	A1, A4i, A4iv, B1i, B1iv, B2
005	SU068	Kintrishi	Kintrishi	Adjara, Kobuleti	13,893	A2, A3
006	SU070	Liakhvi	Liakhvi	Tskhinvali, South Ossetia	6,388	A2, A3
007	SU069	Eastern Caucasus	Agmosavlet Kavkasioni	Kakheti, Mtskheta-Mtianeti	678,976	A1, A2, A3, B2, B3
008	—	Meskheti	Meskheti	Samtskhe-Javakheti	160,000	A1, A2, A3, A4iv, B1iv, B2
009	—	Javakheti plateau	Javakhetis Plato	Samtskhe-Javakheti, Kvemo Kartli	200,000	A1, A3, B2
010	SU071	Algeti	Algeti	Tetri-Tskaro	6,000	A1
011	—	Iori region	Iovris Zegani	Kakheti	273,996	A1, B2, B3

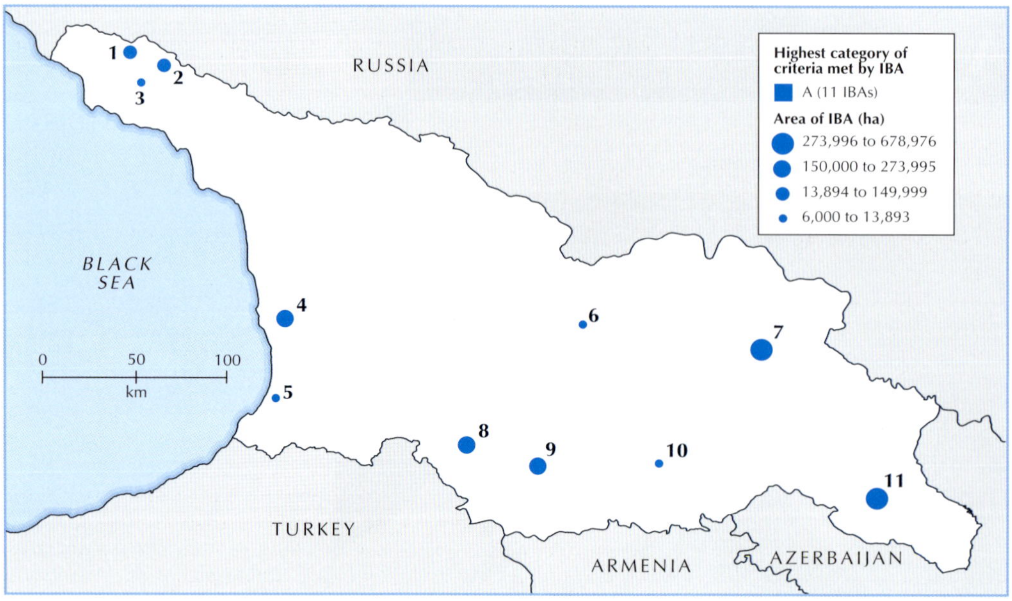

Map 1. Location, area and criteria category of Important Bird Areas in Georgia.

and *Crex crex* are globally threatened, and *Phalacrocorax pygmeus, Haliaeetus albicilla, Aegypius monachus, Circus macrourus* and *Tetrao mlokosiewiczi* are globally near-threatened. The current status of *Marmaronetta angustirostris* (recorded at site 009 in 1996) has yet to be determined. A further 75 SPECs have an unfavourable conservation status in Europe. Thirty-five SPECs are regular on migration or in winter and among them are the globally threatened or near-threatened *Anser erythropus*, *Aquila clanga* (non-breeding visitor, present throughout the year), *Tetrax tetrax*, *Glareola nordmanni*, *Chettusia gregaria* and *Gallinago media*. The following species in these categories occur (or have occurred) in Georgia, but their current status is uncertain: *Oxyura leucocephala, Otis tarda* and *Numenius tenuirostris*.

Six sites were selected as IBAs because they hold significant numbers of globally threatened or near-threatened species (meeting criterion A1: see Table 2). Seven sites also qualify on the basis of supporting a significant component of both restricted-range species (Caucasus Endemic Bird Area) and of the Eurasian high-montane (alpine) biome species (meeting criteria A2 and A3), with one further site—IBA 009—meeting criterion A3 but not A2. Table 3 shows that just one IBA out of 11 meets criteria A4i and B1i by supporting 1% or more of the biogeographic population of one waterbird species; lack of information on numbers prevented these criteria being met at other sites (notably Javakheti plateau, IBA 009) and for other species. Five IBAs (004, 007, 008, 009, 011) are of regional importance (under B2/B3 criteria) for SPECs (Table 4).

Georgia is of great significance for migrating birds, lying as it does at the intersection of several migration routes. Millions of birds are believed to pass through annually, especially on the north-south route linking breeding areas in eastern Europe, Fennoscandia, western Russia (west of the Urals) and wintering grounds in Africa and the Middle East. Kolkheti (site 004) serves as a staging site for birds before they cross the Caucasus, the steppes north of the mountains and the semi-deserts of the Middle East. It is of outstanding importance for migrating raptors, cranes (Gruidae) and others, is also used for staging and wintering by a wide range of

Table 2. Important Bird Areas in Georgia that are important for species of global conservation concern (meeting criterion A1).

Species	IBA code
Phalacrocorax pygmeus Pygmy Cormorant	007, 009
Pelecanus crispus Dalmatian Pelican	009
Anser erythropus Lesser White-fronted Goose	004, 009
Branta ruficollis Red-breasted Goose	004
Marmaronetta angustirostris Marbled Duck	009
Aythya nyroca Ferruginous Duck	004, 009
Oxyura leucocephala White-headed Duck	004, 009
Aegypius monachus Cinereous Vulture	008, 011
Aquila clanga Greater Spotted Eagle	008
Aquila heliaca Imperial Eagle	007, 008, 010, 011
Falco naumanni Lesser Kestrel	008
Crex crex Corncrake	004, 007, 009, 011
Otis tarda Great Bustard	009

Table 3. Important Bird Areas in Georgia that support important numbers of one or more congregatory species (i.e. meeting criteria A4 and/or B1). IBAs meeting both criteria A4 and B1 for the species are shown in **bold**. IBAs meeting only criterion B1 for the species concerned, and not A4, are shown in normal type. For key to 'Season', see p. 7.

Species	Season	IBA code
Podiceps cristatus Great Crested Grebe	W	**004**

Table 4. Species of European conservation concern with significant breeding populations at Important Bird Areas in Georgia (meeting criteria B2/B3).

Species [1]	Minimum national breeding population (pairs) [2]	Proportion (%) of national population breeding at all IBAs in Georgia
Pelecanus crispus Dalmatian Pelican	–	–[4]
Ixobrychus minutus Little Bittern	–	–[4]
Ciconia nigra Black Stork	14	100[3]
Aythya nyroca Ferruginous Duck	–	–[4]
Gypaetus barbatus Lammergeier	22	45
Aegypius monachus Cinereous Vulture	17	100[3]
Circaetus gallicus Short-toed Eagle	4	100[3]
Buteo rufinus Long-legged Buzzard	40	50
Aquila pomarina Lesser Spotted Eagle	80	38
Aquila heliaca Imperial Eagle	10	80
Aquila chrysaetos Golden Eagle	25	80
Tetrao mlokosiewiczi Caucasian Black Grouse	–	–[4]
Tetraogallus caucasicus Caucasian Snowcock	–	–[4]
Tetraogallus caspius Caspian Snowcock	–	–[4]
Crex crex Corncrake	–	–[4]
Charadrius alexandrinus Kentish Plover	–	–[4]
Dendrocopos medius Middle Spotted Woodpecker	–	–[4]
Prunella ocularis Radde's Accentor	–	–[4]
Phoenicurus erythrogaster Güldenstädt's Redstart	–	–[4]
Ficedula semitorquata Semi-collared Flycatcher	–	–[4]
Lanius minor Lesser Grey Shrike	–	–[4]
Carpodacus rubicilla Great Rosefinch	500[5]	–[4]

1. Only those species of European conservation concern (see Box 1, p. 12) that meet B2/B3 criteria at IBAs in Georgia are listed.
2. Data are taken from Snow and Perrins (1998).
3. The percentage of the national population in IBAs exceeds 100%. This is most likely to be because the national population estimate needs to be reassessed (often as a result of new data on species numbers from comprehensive surveys of IBAs themselves).
4. No national population estimate available.
5. Data for Russia and Georgia combined.

Figure 1. Occurence of habitats at Important Bird Areas in Georgia (see Appendix 3 for definitions of habitats).

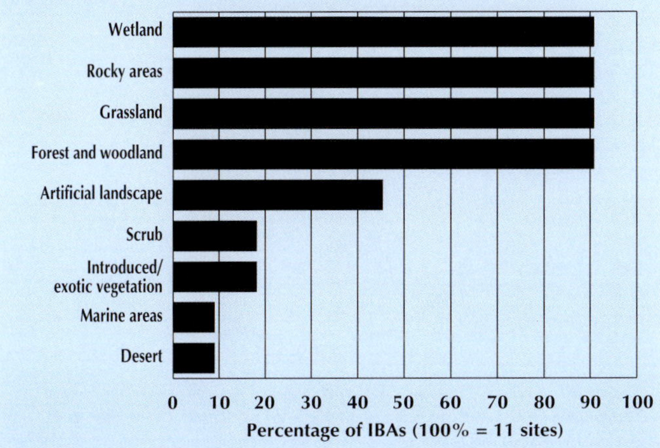

Figure 2. Occurrence of land-uses at Important Bird Areas in Georgia (see Appendix 3 for definitions of land-uses).

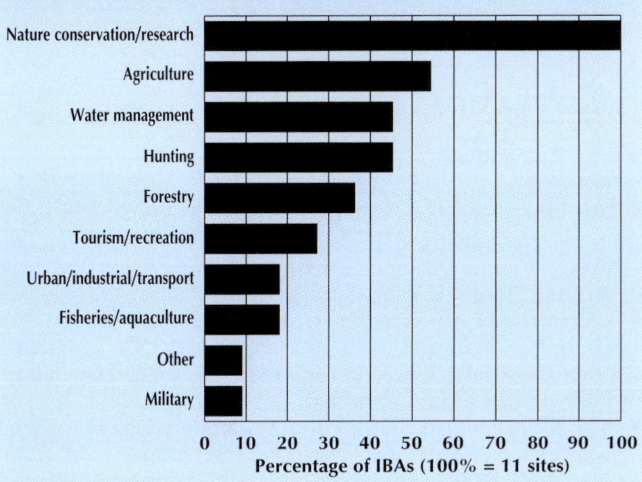

waterbirds and passerines and qualifies, as does site 008 (Meskheti), as a 'bottleneck' site (meeting criteria A4iv and B1iv).

HABITATS

Varied and contrasting natural conditions, climate, the conspicuously fractured relief and the geographical location of Georgia at the junction of landscapes of different genesis have given rise to a rich and diverse vegetation. Habitats range from the luxuriant subtropical forests of the Black Sea coastlands and Kolkhida lowland in western Georgia to semi-deserts in the foothills of the east, from warm-temperate forests on the western slopes of the Great Caucasus through various vegetation belts up to specific cold-resistant flora in the subalpine and alpine zones of the mountains.

Figure 1 shows that the dominant habitats at ten out of 11 sites (91%) are wetland (comprising the seacoast, lowland and plateau lakes, rivers and marshes), rocky areas, grassland (plains with steppe and semi-desert, also mountain meadows) and forest and woodland (broadleaved, mixed and native coniferous), also artificial landscape at five (46%).

Wetlands support notable breeding species such as *Pelecanus onocrotalus* and *P. crispus*, *Ixobrychus minutus*, *Aythya nyroca*, *Melanitta fusca*, *Haliaeetus albicilla*, *Grus grus*, *Himantopus himantopus*, *Charadrius alexandrinus*, *Larus armenicus* and *Chlidonias hybridus*, also an impressive range of wildfowl and others on passage and in winter. Forests, in river valleys and higher up on mountain slopes, have breeding raptors (*Pernis apivorus*, *Milvus migrans*, *Circaetus gallicus*, *Aquila pomarina*, *Hieraaetus pennatus*), woodpeckers and the passerines *Phylloscopus nitidus*, *P. lorenzii* and *Ficedula semitorquata*. Rocky areas with mountain meadows and some scrub hold more raptors, *Tetraogallus caucasicus* and *T. caspius*, *Alectoris chukar*, *Prunella ocularis*, *Phoenicurus erythrogaster* and *Carpodacus rubicilla*. Plains with steppe-type grasslands or semi-desert are visited by staging *Anthropoides virgo* and *Tetrax tetrax*, while *Circus macrourus* winters in such habitat and *Francolinus francolinus* breeds where there is scrub.

IMPACTS ON IBAs – LAND-USE AND THREATS

Nature conservation/research is a land-use at all 11 IBAs in Georgia (Figure 2), other significant land-uses being agriculture (affecting six IBAs), water management and hunting (five each) and forestry (four sites).

More information on the level of impact is available for threats (Figure 3), the predominant ones being unsustainable exploitation (legal and illegal hunting) which affects seven out of the 11 IBAs (63%, five sites at a high impact-level) and agricultural intensification/expansion (seven sites, albeit with medium to low impact at four sites, while the impact is unknown at four). Firewood collection and commercial deforestation both rank as serious threats at four sites and the same is true of selective logging/cutting and drainage at three. According to current knowledge, the wetland site 004 (Kolkheti) has been and apparently still is subject to the widest range of threats and this has resulted in the severe depletion of its breeding bird populations.

Figure 3. Threats at Important Bird Areas in Georgia (see Appendix 3 for definitions of threat types and impact categories).

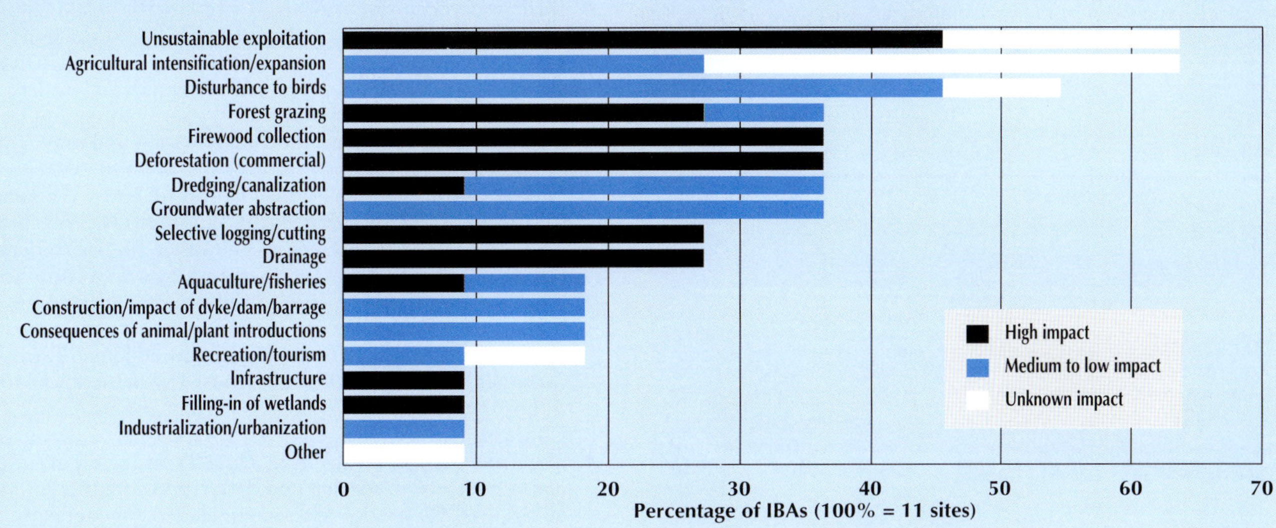

Figure 4. The national protection status of Important Bird Areas in Georgia.

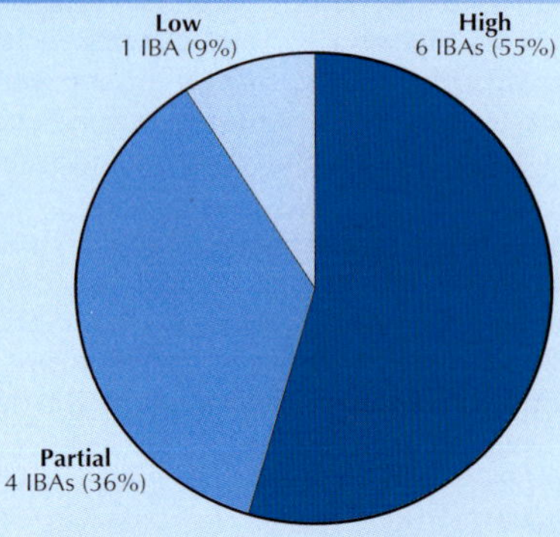

Not possible to calculate total area of overlap between IBA network in Georgia and national protected-area system (see Table 5 for categories) due to incomplete overlap-area data.

Figure 5. The international protection status of Important Bird Areas in Georgia.

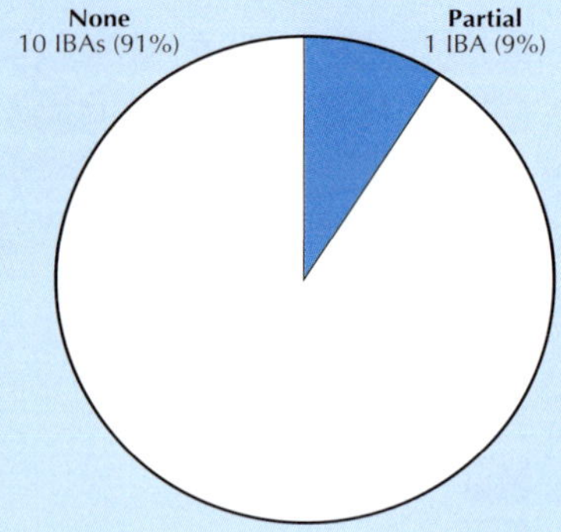

Not possible to calculate total area of overlap between IBA network in Georgia and international protected-area system (see Table 5 for categories) due to incomplete overlap-area data.

PROTECTION STATUS

Table 5 and Figures 4 and 5 summarize the national and international protection status of IBAs in Georgia.

National protection

Individual IBAs overlap with nationally protected areas as follows: six are each covered by more than 90%, four are covered by 10%–90% and one IBA is covered by under 10%.

Georgia inherited the protected-area system of the former USSR of which it was a part, with **Zapovedniks** (strict nature reserves; Nakrdzali in Georgian), **Zakazniks** (temporary reserves with lower protection status; Agkvetili in Georgian) and **National Parks**. However, a new protected-area system has been developed and is currently being established in Georgia.

International protection

Georgia is party to the World Heritage Convention and takes part in the UNESCO Man and the Biosphere Programme (Box 1). It also supports the Ramsar Convention, part of one IBA (Kolkheti, IBA 004) having been designated as a Ramsar Site, while a second Ramsar Site (Javakheti plateau, IBA 009) is in the process of being established (Box 1, Table 5, Figure 5).

Table 5. Protection status of Important Bird Areas in Georgia. A tick (✔) indicates that an IBA overlaps with a protected area (to any extent).

		National				International
IBA code	International name	Managed Nature Reserve	National Park	Nature Reserve	Zapovednik	Ramsar Site
001	Ritsa				✔	
002	Pskhu				✔	
003	Gumista				✔	
004	Kolkheti				✔	✔
005	Kintrishi				✔	
006	Liakhvi				✔	
007	Eastern Caucasus			✔		
008	Meskheti		✔	✔		
009	Javakheti plateau		✔			
010	Algeti				✔	
011	Iori region	✔			✔	
Total number of IBAs		1	2	2	8	1

Box 1. International legislation and initiatives that are relevant to site conservation in Georgia (see Appendix 1 for a general description of these agreements).

Global	
Biodiversity Convention	✔
Ramsar Convention	✔
Bonn Convention	
World Heritage Convention	✔
MAB Programme	✔

✔ Convention ratified/initiative supported
(✔) Convention signed

CONSERVATION

- The new protected-area system in Georgia, which is currently being set up, envisages a comprehensive network of conservation areas based on several different national categories of protected area corresponding to IUCN categories I–VI and including legal possibilities for the designation of global categories (Biosphere Reserves, World Heritage Sites and Ramsar Sites). Provisionally, there are ten protected areas, including one National Park (Borjomi-Kharagauli), some overlapping with IBAs as described here. The strategy aims eventually to have 8–10% of the country's land and sea surface protected.
- The most detailed information available relates to IBA 004 where management guidelines for the establishment of Kolkheti National Park were prepared with the methodological assistance of WWF in cooperation with the World Bank/GEF and within the framework of the protected-area system of Georgia. This was preceded by Government of Georgia Decree No. 11 "On establishment of national parks on the territory of Georgia" (Order No. 562 of 12 May, 1992). The next stage envisages a full-scale management plan designed to create favourable conditions for nature protection and the socio-economic development of the region. Also associated with this site is the project "Establishment of the Black Sea Youth EcoAcademy" initiated by the Georgian Youth EcoMovement (SAEM) and implemented within the framework of the Integrated Coastal Zone Management Programme (ICZM).
- Bird research in Georgia is at present the responsibility of various government institutions (Tbilisi State University, Institute of Zoology, etc.) and NGOs. Comprehensive field surveys are needed to update information, and existing studies require better management, coordination and funding. Based on a desire to see a thorough revision and reinvigoration of Georgian ornithology, and in the hope of securing more effective protection and conservation of birds, the Georgian Centre for the Conservation of Wildlife (GCCW) in Tbilisi plans to undertake taxonomic and other studies to determine the exact composition of the country's avifauna and the distribution of each species; to establish the position of the main Trans-Caucasus migration routes; to investigate the biology and ecology of the Caucasian endemic *Tetraogallus caucasicus* and near-endemic *Tetrao mlokosiewiczi*, and to make a thorough study of *Gypaetus barbatus* and its ecology in the Caucasus.
- The first edition of the Red Data Book of Georgia (1982) contains 33 bird species and, after revisions (deletions and additions), 27 species are recommended for inclusion in the second edition.

ANALYTICAL METHODS

- Data for IBAs 004, 007–009 were compiled from a variety of published and unpublished sources by A. Gavashelishvili (Georgian Centre for the Conservation of Wildlife).
- Lack of relevant information has made it impossible to include percentage figures for 'Habitats' and 'Land-uses'.
- Data are generally of good quality, at least for IBAs 004, 007–009.
- As part of the 1998 Winter Ornithological Monitoring Survey administered by the Georgian Pipeline Company (GPC), midwinter (February) bird counts were carried out at nine sites in coastal Kolkheti by A. Gavashelishvili, Z. Javakhishvili and G. Darchiashvili. Among the 80 species recorded, the most numerous were *Podiceps cristatus* and dabbling ducks (*Anas platyrhynchos*, *A. crecca*).

GLOSSARY

Kolkheti (Kolkhida) historical and geographical region, particularly rich in wetlands, occupying most of lowland Georgia west of the Surami (Likhi) ridge.
tugai riparian forest in arid, semi-desert and semi-savanna regions; in Georgia, characteristically includes *Tamarix*.

ACKNOWLEDGEMENTS

Alexander Gavashelishvili, Deputy Director of the Georgian Centre for the Conservation of Wildlife in Tbilisi, reported on the Centre's activities and future projects, searched the literature and completed data-sheets for five IBAs (004, 007–009). Paata Shanshiashvili, Conservation Programme Coordinator (WWF Office, Tbilisi), provided information on the development of Georgia's new protected-area system.

SITE ACCOUNTS

Ritsa — A2, A3 — 001

Admin region Gudauta, Abkhazeti
Coordinates 43°28′N 40°34′E
Altitude 300–2,500 m **Area** 16,289 ha

Site description
Lake Ritsa and the wooded slopes (forest predominantly of *Abies* and *Fagus*) surrounding it on the southern spurs of the Great Caucasus range. There are also small areas of meadows, and deep river gorges.

Habitats Forest and woodland (92%; broadleaved deciduous forest; native coniferous forest; mixed forest), Grassland (alpine/subalpine/boreal grassland), Rocky areas (scree/boulders; inland cliff)
Land-use Nature conservation/research, Tourism/recreation

Birds

Species	Season	Year	Pop min	Pop max	Acc	Criteria
Tetrao mlokosiewiczi Caucasian Black Grouse	R	—	—	—	—	A2, A3
Tetraogallus caucasicus Caucasian Snowcock	R	—	—	—	—	A2, A3

The avifauna includes at least two of the 10 species in Europe that are restricted (when breeding) to the Eurasian high-montane biome. A total of 110 species was recorded at the site in the early 1980s; at least 69 species breed, including *Gypaetus barbatus*, *Gyps fulvus*, *Aquila chrysaetos*, *Falco peregrinus*, *Bubo bubo* and other owls, as well as woodpeckers. Predominant in the subalpine zone are *Anthus spinoletta*, *Carduelis flavirostris* and *Carpodacus erythrinus*. Species of global conservation concern that do not meet IBA criteria: *Aegypius monachus* (occurs).

Protection status
National High **International** None
16,289 ha of IBA covered by Nature Reserve (Zapovednik) (Ritsa, 16,289 ha).

Conservation issues

Threats Recreation/tourism (U)

The main threat is the impact of tourists on Lake Ritsa and its environs. The site was first designated as a Zapovednik in 1930. It lies within the proposed Abkhazeti Protected Area, part of the Georgian protected-area system whose establishment is now under way.

Pskhu — A2, A3 — 002

Admin region Abkhazeti
Coordinates 43°25′N 40°49′E
Altitude 500–2,400 m **Area** 27,333 ha

Site description
The wooded basin of the upper and middle Bzyb river and its tributaries, all typical fast-flowing mountain rivers with many rapids. The valley lies between the Great Caucasus in the north, the Bzyb range in the south and the Gagra range in the west, towards which the valley widens. There are steep, in places precipitous, slopes on the left bank; the right bank is highly fractured by tributaries of the Bzyb. The forest is mainly of *Fagus*, *Abies*, *Carpinus*, *Quercus*, *Alnus*, with *Rhododendron*, etc. There is a long and continuing tradition of bee-keeping at this site ('Other' land-use).

Habitats Forest and woodland (93%; broadleaved deciduous forest; native coniferous forest; mixed forest; broadleaved evergreen forest), Grassland (alpine/subalpine/boreal grassland), Rocky areas (scree/boulders; inland cliff), Wetland (river/stream)
Land-use Nature conservation/research, Other

Birds

Species	Season	Year	Pop min	Pop max	Acc	Criteria
Tetrao mlokosiewiczi Caucasian Black Grouse	R	—	—	—	—	A2, A3
Tetraogallus caucasicus Caucasian Snowcock	R	—	—	—	—	A2, A3

The following refers to Pskhu–Gumista, i.e. sites 002 and 003 combined. The avifauna includes at least two of the 10 species in Europe that are restricted (when breeding) to the Eurasian high-montane biome. A total of 142 species has been recorded at this site, including 86 passerines and at least 24 raptors. Species of global conservation concern that do not meet IBA criteria: *Aegypius monachus* (non-breeding visitor), *Falco naumanni* (occurs on passage). Among breeding raptors are *Gypaetus barbatus* (2–3 pairs) and *Aquila chrysaetos* (up to 2 pairs), while *Pernis apivorus*, *Neophron percnopterus*, *Circaetus gallicus* (very rare), *Circus cyaneus*, *C. pygargus*, *Aquila pomarina*, *Hieraaetus pennatus*, *Pandion haliaetus* and *Falco vespertinus* pass through on migration. Other breeding species include six woodpeckers, *Phylloscopus nitidus* and *Certhia brachydactyla*.

Protection status
National High **International** None
27,333 ha of IBA covered by Nature Reserve (Zapovednik) (Pskhu–Gumista, 40,818 ha).

Conservation issues

Threats Agricultural intensification/expansion (U), Unsustainable exploitation (U)

The main threat comes from the use for grazing of the unprotected flower-rich alpine and subalpine meadows lying between the Pskhu and Gumista Reserves: the grazing animals have to be driven through the protected forest areas to reach the meadows; there is pressure from grazing on the forest edge; poaching is a problem in an area above the tree-line which is the habitat of such rare species as *Tetrao mlokosiewiczi*. This IBA, together with Gumista (site 003), lies within the Pskhu–Gumista Nature Reserve, which falls within the proposed Abkhazeti Protected Area, part of the Georgian protected-area system whose establishment is now under way.

Gumista

A2, A3 **003**

Admin region Sukhumi, Abkhazeti
Coordinates 43°19′N 40°40′E
Altitude 250–2,200 m **Area** 13,400 ha

Site description

The gorge of the Eastern Gumista river and densely-wooded slopes of the Bzyb range, beyond which to the north lies Pskhu (site 002), between the Black Sea and the Great Caucasus. The principal tree species are *Fagus*, *Castanea*, *Carpinus*, *Quercus*, *Abies*, and *Rhododendron*.

Habitats Forest and woodland (94%; broadleaved deciduous forest; native coniferous forest; mixed forest; broadleaved evergreen forest), Rocky areas (scree/boulders; inland cliff), Wetland (river/stream)
Land-use Nature conservation/research

Birds

Species	Season	Year	Pop min	Pop max	Acc	Criteria
Tetrao mlokosiewiczi Caucasian Black Grouse	R	—	—	—	—	A2, A3
Tetraogallus caucasicus Caucasian Snowcock	R	—	—	—	—	A2, A3

These records are for the whole Pskhu–Gumista Nature Reserve; see the site-account for Pskhu (site 002), which describes the avifauna of both sites (002 and 003).

Protection status

National High **International** None
13,400 ha of IBA covered by Nature Reserve (Zapovednik) (Pskhu–Gumista, 40,818 ha).

Conservation issues

Threats Agricultural intensification/expansion (U), Unsustainable exploitation (U)

The main threat comes from the use for grazing of the unprotected flower-rich alpine and subalpine meadows lying between the Pskhu and Gumista Reserves: the grazing animals have to be driven through the protected forest areas to reach the meadows; there is pressure from grazing on the forest edge; poaching is a problem in an area above the tree-line which is the habitat of such rare species as *Tetrao mlokosiewiczi*. This IBA, together with Pskhu (site 002), lies within the Pskhu-Gumista Nature Reserve, which falls within the proposed Abkhazeti Protected Area, part of the Georgian protected-area system whose establishment is now under way.

Kolkheti

A1, A4i, A4iv, B1i, B1iv, B2 **004**

Admin region Samegrelo-Svaneti, Guria, Adjara
Coordinates 42°10′N 41°50′E
Altitude 0–200 m **Area** 150,000 ha

Site description

An eastern coastal region of the Black Sea consisting of wetlands and damp woodlands. Evergreen vegetation covers much of the coastline within the site. The IBA encompasses the Lower Rioni river and Lake Paliastomi, listed as site SU064 in the previous pan-European IBA inventory (Grimmett and Jones 1989). The water surface of the Black Sea in the west and the pattern of the Little Caucasus mountain range in the east come together to form a migration bottleneck in the southern part of the site.

Habitats Artificial landscape (arable land; perennial crops/orchards/groves; urban parks/gardens), Forest and woodland (broadleaved deciduous forest; alluvial/very wet forest; broadleaved evergreen forest), Grassland (humid grassland), Introduced/exotic vegetation, Marine areas (open sea; sea inlet/coastal features), Rocky areas (inland cliff), Wetland (saltmarsh; sand-dunes/sand beach; shingle/stony beach; standing fresh water; standing brackish and salt water; river/stream; water-fringe vegetation; blanket bog)
Land-use Agriculture, Fisheries/aquaculture, Forestry, Hunting, Nature conservation/research, Tourism/recreation, Urban/industrial/transport, Water management

Birds

Species	Season	Year	Pop min	Pop max	Acc	Criteria
Podiceps cristatus Great Crested Grebe	W	1998	9,164	—	A	A4i, B1i
Ixobrychus minutus Little Bittern	B	1996	Frequent		—	B2
Ciconia nigra Black Stork	B	1996	2	3	B	B2
Anser erythropus Lesser White-fronted Goose	W	1998	Uncommon		—	A1
Branta ruficollis Red-breasted Goose	W	1996	Rare		—	A1
Aythya nyroca Ferruginous Duck	P	1996	Frequent		—	A1
Oxyura leucocephala White-headed Duck	W	1998	Uncommon		—	A1
[1] ***Crex crex*** Corncrake	B	1998	Frequent		—	A1
Charadrius alexandrinus Kentish Plover	B	1996	Uncommon		—	B2

1. Also frequent on passage.

Migrating raptors (at least 25 species) and other large, soaring birds are concentrated here in spring and autumn, due to their aversion to flying over the sea or high mountains. Counts have covered only small periods of the spring and autumn migration seasons, but include (seasonal total in brackets) *Ciconia nigra* (350–500, autumn), *Pernis apivorus* (8,420, autumn), *Milvus migrans* (2,300, spring), *Accipiter brevipes*/*A. nisus* (1,500, spring), *Aquila pomarina* (268, autumn), *Aquila nipalensis* (127, autumn), *Buteo buteo* (44,900, spring), *Hieraaetus pennatus* (156, spring), *Circus aeruginosus* (181, autumn), and several species of falcon *Falco*. The site is also an important wintering ground and staging place for numerous waterbirds as well as passerines. Wetlands at the site attract a wide variety of passage and wintering birds such as *Gavia arctica*, grebes , *Puffinus yelkouan*, Ciconiiformes, wildfowl (several dabbling duck species, also *Clangula hyemalis*, *Melanitta fusca*, *Mergus albellus*), shorebirds, gulls and terns. Species of global conservation concern that do not meet IBA criteria: *Pelecanus crispus* (uncommon in winter), *Haliaeetus albicilla* (at least two breeding pairs and 13 birds in winter), *Circus macrourus* and *Aquila clanga* (both frequent on passage).

Protection status

National Low **International** Partial
554 ha of IBA covered by Nature Reserve (Zapovednik) (Kolkheti, 554 ha). Part of IBA covered by Ramsar Site encompassing Kolkheti Nature Reserve (Central Kolkheti wetlands of Churia, Nabada, Pichora and Ispani II).

Conservation issues

Threats Agricultural intensification/expansion (B), Aquaculture/fisheries (A), Consequences of animal/plant introductions (C), Construction/impact of dyke/dam/barrage (B), Deforestation (commercial) (A), Disturbance to birds (B), Drainage (A), Dredging/canalization (A), Filling-in of wetlands (A), Firewood collection (A), Forest grazing (C), Groundwater abstraction (C), Industrialization/urbanization (B), Infrastructure (A), Recreation/tourism (B), Selective logging/cutting (A), Unsustainable exploitation (A)

The site was formerly rich in wildlife, but is now impoverished. Breeding birds in particular have been severely affected by the impact of human activities in what is a densely populated region, and have declined dramatically since the beginning of the 20th century. The main threats are pollution, poaching, deforestation and drainage. The area of the IBA includes the Kolkheti Ramsar Site and the proposed Kolkheti Protected Area (National Park), part of the Georgian protected-area system whose establishment is now under way. A management plan exists for this site.

Kintrishi

A2, A3 **005**

Admin region Kobuleti, Adjara
Coordinates 41°45′N 41°49′E
Altitude 450–2,600 m **Area** 13,893 ha

Site description

An area of the west and north-west slopes of the Meskheti range, with the densely-wooded gorge of the River Kintrishi and many tributaries with waterfalls and torrents, also subalpine lakes. Main tree species are *Fagus*, *Castanea*, also *Carpinus*, *Alnus*, *Abies*, *Picea*, etc.; evergreen shrubs such as *Laurus*, *Rhododendron*.

Habitats Forest and woodland (95%; broadleaved deciduous forest; native coniferous forest; broadleaved evergreen forest), Grassland (alpine/subalpine/boreal grassland), Rocky areas (inland cliff), Wetland (standing fresh water; river/stream)
Land-use Agriculture, Nature conservation/research

Birds

Species	Season	Year	Pop min	Pop max	Acc	Criteria
Tetrao mlokosiewiczi Caucasian Black Grouse	R	—	Rare		—	A2, A3
Tetraogallus caspius Caspian Snowcock	R	—	—	—	—	A3

The avifauna of this site is little-studied, with a preliminary list comprising only 56 species and virtually no information on status and population. However, the site supports at least two of the 10 species in Europe that are restricted (when breeding) to the Eurasian high-montane biome. Species of global conservation concern that do not meet IBA criteria: *Falco naumanni* (occurs). Also rare at this site are *Gypaetus barbatus*, *Aquila chrysaetos* and *Falco peregrinus*. Other breeding birds include *Hieraaetus pennatus*, *Alectoris chukar*, *Bubo bubo*, woodpeckers, *Oriolus oriolus*, *Corvus corax*, etc.

Protection status

National High **International** None
13,893 ha of IBA covered by Nature Reserve (Zapovednik) (Kintrishi, 13,893 ha).

Conservation issues

Threats Unknown

There are some violations of the nature reserve (Zapovednik) regime, with temporary or permanent residence by humans who practise forest grazing, growing maize, melons, etc. The site lies within the Adjara-Imeretian Protected Area, part of the protected-area system which is now being established in Georgia.

Liakhvi — A2, A3 — 006

Admin region Tskhinvali, South Ossetia
Coordinates 42°15′N 43°55′E
Altitude 1,200–2,300 m **Area** 6,388 ha

Site description

An area on the southern slopes of the main Caucasus range, including the upper valley of the River Patara-Liakhvi, which divides the site in two. The relief is complex, with numerous gorges, etc., and the principal habitat is forest, predominantly of *Fagus* (between 1,000 and 1,700 m), also *Betula*, *Quercus*, *Carpinus* and some *Alnus*, *Fraxinus* and *Corylus*.

Habitats Forest and woodland (87%; broadleaved deciduous forest), Grassland (alpine/subalpine/boreal grassland), Rocky areas (scree/boulders; inland cliff), Wetland (river/stream)
Land-use Nature conservation/research

Birds

Species	Season	Year	Pop min	Pop max	Acc	Criteria
Tetrao mlokosiewiczi Caucasian Black Grouse	R	—	—	—	—	A2, A3
Tetraogallus caucasicus Caucasian Snowcock	R	—	—	—	—	A2, A3
Carpodacus rubicilla Great Rosefinch	R	—	—	—	—	A3

The avifauna includes at least three of the 10 species in Europe that are restricted (when breeding) to the Eurasian high-montane biome. At least 71 species have been recorded at this site, including *Aquila chrysaetos* (possibly 1–2 breeding pairs) and, among other raptors, 3–4 owl species; *Actitis hypoleucos* and up to five species of woodpeckers also breed. Passerines are well represented, with *Phylloscopus nitidus* in forests and *Eremophila alpestris* and *Anthus spinoletta* higher up in meadows.

Protection status

National High **International** None
6,388 covered by Nature Reserve (Zapovednik) (Liakhvi Nature Reserve, 6,388 ha).

Conservation issues

Threats Agricultural intensification/expansion, Disturbance to birds

Only small fragments of alpine and subalpine meadows lie within the site, while grazing on unprotected meadows adjoining it causes disturbance; nests, including of rare species such as *Tetrao mlokosiewiczi* and *Tetraogallus caucasicus*, are trampled by grazing herds which also make incursions into the reserve area. A buffer zone for better protection is urgently needed. The site was designated as a Zapovednik in 1977 and is situated just outside the Eastern Caucasus Protected Area (IBA 007). The fauna of the site, including birds, has not been well studied.

Eastern Caucasus — A1, A2, A3, B2, B3 — 007

Admin region Kakheti, Mtskheta-Mtianeti
Coordinates 42°10′N 45°10′E
Altitude 200–5,047 m **Area** 678,976 ha

Site description

The site covers the entire eastern part of the Greater Caucasus in Georgia. Natural landscapes vary from riparian forests to high mountain peaks with glaciers. Lakes are found in the alpine zone of Lagodekhi Nature Reserve. The forests are predominantly deciduous, with tree species such as *Fagus*, *Betula*, *Quercus*, *Carpinus*, *Castanea*, also *Tilia*, *Populus* and some *Pinus*; *Hippophae* occurs along streams. Geologically, the site is mainly made up of Jurassic slates, sandstone, conglomerates and clay with Cretaceous sediments. The eastern part of the site is well cultivated. The site encompasses Lagodekhi (site SU069 in the previous pan-European IBA inventory: Grimmett and Jones 1989).

Habitats Artificial landscape (arable land; perennial crops/orchards/groves), Forest and woodland (broadleaved deciduous forest; mixed forest), Grassland (alpine/subalpine/boreal grassland), Rocky areas (scree/boulders; inland cliff; caves), Wetland (standing fresh water; river/stream; water-fringe vegetation)
Land-use Agriculture, Forestry, Hunting, Nature conservation/research, Tourism/recreation, Urban/industrial/transport, Water management

Birds

Species	Season	Year	Pop min	Pop max	Acc	Criteria
Phalacrocorax pygmeus Pygmy Cormorant	W	1998	Uncommon		—	A1
Ixobrychus minutus Little Bittern	B	1998	Common		—	B2
Ciconia nigra Black Stork	B	1998	6	—	A	B2
Gypaetus barbatus Lammergeier	R	1998	10	—	A	B2
Aquila pomarina Lesser Spotted Eagle	B	1998	18	—	A	B2
Aquila heliaca Imperial Eagle	R	1998	2	—	A	A1
Aquila chrysaetos Golden Eagle	R	1998	14	—	A	B2
Tetrao mlokosiewiczi Caucasian Black Grouse	R	1998	Common		—	A2, A3
Tetraogallus caucasicus Caucasian Snowcock	R	1998	Common		—	A2, A3
Crex crex Corncrake	B	1998	Frequent		—	A1
Dendrocopos medius Middle Spotted Woodpecker	R	1998	Frequent		—	B3
Prunella collaris Alpine Accentor	R	1998	Common		—	A3
Phoenicurus erythrogaster Güldenstädt's Redstart	R	1998	Uncommon		—	B2
Ficedula semitorquata Semi-collared Flycatcher	B	1998	Frequent		—	B2
Tichodroma muraria Wallcreeper	R	1998	Uncommon		—	A3
Lanius minor Lesser Grey Shrike	B	1998	Frequent		—	B2
Pyrrhocorax graculus Alpine Chough	R	1998	Abundant		—	A3
Montifringilla nivalis Snowfinch	R	1998	Frequent		—	A3
Carpodacus rubicilla Great Rosefinch	R	1998	Uncommon		—	A3, B2
Phylloscopus lorenzii Caucasian Chiffchaff	B	1998	Frequent		—	A2

The site is of global importance for Caucasian endemic birds such as *Tetraogallus caucasicus*, whose entire population is restricted to the Great Caucasus range, and for near-endemics such as *Tetrao mlokosiewiczi* and *Phylloscopus lorenzii*, both of which occur in the Little and Great Caucasus ranges. The site supports seven out of the

10 species in Europe that are restricted (when breeding) to the Eurasian high-montane biome—a higher species complement than any other site on the continent. The site is also of European importance for its numbers of *Gypaetus barbatus* and *Phoenicurus erythrogaster*, and is nationally important as a wintering ground for various forest birds, as well as for *Phalacrocorax pygmeus* and *Egretta alba*, being also one of the most important sites in the country for *Aquila chrysaetos* and *Dendrocopos medius*. Species of global conservation concern that do not meet IBA criteria: *Branta ruficollis* (occurs on passage), *Haliaeetus albicilla* (possibly 1 breeding pair), *Aegypius monachus* (status uncertain; possibly breeds), *Circus macrourus* (winter). Other breeding raptors include *Neophron percnopterus* (16 pairs), *Gyps fulvus* (35 pairs) and seven species of owl; the status of *Accipiter brevipes* (reported from Lagodekhi) is uncertain. *Grus grus* and *Anthropoides virgo* pass through on migration and among typical mountain passerines at the site are *Eremophila alpestris*, *Ptyonoprogne rupestris*, *Anthus spinoletta*, *Monticola saxatilis* and *Serinus pusillus*.

Protection status

National Partial **International** None

16,297 ha of IBA covered by Nature Reserve (Zapovednik) (Akhmeta, 16,297 ha). 8,707 ha of IBA covered by Nature Reserve (Zapovednik) (Kazbegi, 8,707 ha). 17,818 ha of IBA covered by Nature Reserve (Zapovednik) (Lagodekhi, 17,818 ha).

Conservation issues

Threats Agricultural intensification/expansion (B), Consequences of animal/plant introductions (C), Deforestation (commercial) (A), Disturbance to birds (B), Drainage (A), Dredging/canalization (B), Firewood collection (A), Forest grazing (A), Groundwater abstraction (C), Selective logging/cutting (A), Unsustainable exploitation (A)

The main threats are habitat destruction: lowland forests, including the Alazani riparian forest, are threatened by logging and firewood collection, while mountain forests are being overgrazed by domestic sheep. Racoon-dog *Procyon lotor*, introduced into Azerbaijan, has spread into the Alazani forests of Georgia and has begun to prey on *Phasianus colchicus* and their eggs. The outline of the site is about the same as that of the proposed East Caucasus Protected Area, part of the protected-area system which is now being established in Georgia. A management plan exists for this site.

Meskheti — A1, A2, A3, A4iv, B1iv, B2 — 008

Admin region Samtskhe-Javakheti
Coordinates 41°35′N 43°10′E
Altitude 700–2,963 m **Area** 160,000 ha

Site description

Lying on Tertiary sediments and volcanic bedrock, this site represents a intermediate zone between the humid 'kolkheti' of western Georgia and the dry mountain ranges of Anatolia (Turkey) and northern Iran. It abounds in relict and endemic species. The site includes the gorges of the Mtkvari (Kura) and Paravani rivers in southern Georgia.

Habitats Artificial landscape (arable land; perennial crops/orchards/groves), Forest and woodland (broadleaved deciduous forest; native coniferous forest; mixed forest), Grassland (alpine/subalpine/boreal grassland), Rocky areas (scree/boulders; inland cliff; caves), Scrub (scrub), Wetland (river/stream; water-fringe vegetation)
Land-use Agriculture, Forestry, Hunting, Nature conservation/research, Water management

Birds

Species	Season	Year	Pop min	Pop max	Acc	Criteria
Gypaetus barbatus Lammergeier	U	1996	Uncommon		—	B2
Aegypius monachus Cinereous Vulture	P	1998	Uncommon		—	A1
Circaetus gallicus Short-toed Eagle	B	1998	Uncommon		—	B2
Buteo rufinus Long-legged Buzzard	R	1998	Frequent		—	B2
Aquila clanga Greater Spotted Eagle	P	1996	Frequent		—	A1
Aquila heliaca Imperial Eagle	P	1996	Uncommon		—	A1
Aquila chrysaetos Golden Eagle	R	1998	4	—	A	B2
Falco naumanni Lesser Kestrel	U	1998	Frequent		—	A1
Tetrao mlokosiewiczi Caucasian Black Grouse	R	1996	Frequent		—	A2, A3

Species ... continued	Season	Year	Pop min	Pop max	Acc	Criteria
Tetraogallus caspius Caspian Snowcock	R	1996	Uncommon		—	A3, B2
Prunella ocularis Radde's Accentor	B	1996	Frequent		—	B2
Lanius minor Lesser Grey Shrike	B	1998	Frequent		—	B2
Phylloscopus lorenzii Caucasian Chiffchaff	B	1996	Frequent		—	A2

The avifauna includes at least two of the 10 species in Europe that are restricted (when breeding) to the Eurasian high-montane biome. The site is of great significance for raptors, with at least 32 species recorded, and is one of the most important in the country for breeding species, notably *Hieraaetus pennatus*, *Gyps fulvus* (46 pairs in 1996) and others. There have been two recent records of *Hieraaetus fasciatus* (status uncertain). The site is an important migration bottleneck for raptors, some of which actually stage at the site rather than just passing through; species include *Circaetus gallicus* (common), *Pernis apivorus*, *Milvus migrans*, harriers *Circus*, *Accipiter gentilis*, *Accipiter nisus*, *Accipiter brevipes*, *Aquila pomarina* (frequent), *Aquila nipalensis*, *Falco columbarius* and *Falco cherrug*. Species of global conservation concern that do not meet IBA criteria: *Circus macrourus* (frequent on passage).

Protection status

National Partial **International** None

17,948 ha of IBA covered by Nature Reserve (Zapovednik) (Borjomi, 17,948 ha). 50,000 ha of IBA covered by National Park (Borjomi-Kharagauli, 50,400 ha).

Conservation issues

Threats Deforestation (commercial) (A), Disturbance to birds (B), Firewood collection (A), Forest grazing (A), Groundwater abstraction (C), Selective logging/cutting (A), Unsustainable exploitation (A)

The main threat is unsustainable hunting. A large part of the site's area is covered by the Borjomi-Kharagauli National Park and the proposed Erusheti Protected Area, part of Georgia's protected-area system which is now being established, as is Javakheti Ramsar Site, Georgia's second site with this protection status. A management plan exists for this site.

Javakheti plateau — A1, A3, B2 — 009

Admin region Samtskhe-Javakheti, Kvemo Kartli
Coordinates 41°30′N 43°40′E
Altitude 1,500–3,300 m **Area** 200,000 ha

Site description

The site comprises a plateau mainly at altitudes of 1,900–2,100 m and almost completely devoid of forest cover. Comparatively dry subalpine meadows and mountain steppes are the dominant landscapes. Scattered over the high plateau are about 60 natural lakes, six of them large, creating a unique type of high-altitude, treeless wetlands. The human population density is low. The inner part, especially the Samaras ridge and its surroundings, are practically virgin lands. Rising up from the plateau like huge cones are some dormant volcanoes, a vast area around each strewn with boulders and scree.

Habitats Artificial landscape (arable land; forestry plantation), Grassland (steppe/dry calcareous grassland; alpine/subalpine/boreal grassland), Rocky areas (scree/boulders; inland cliff), Wetland (standing fresh water; river/stream; water-fringe vegetation)
Land-use Agriculture, Fisheries/aquaculture, Hunting, Nature conservation/research, Water management

Birds

Species	Season	Year	Pop min	Pop max	Acc	Criteria
Phalacrocorax pygmeus Pygmy Cormorant	W	1998	Uncommon		—	A1
Pelecanus crispus Dalmatian Pelican	B	1996	Frequent		—	A1
Ixobrychus minutus Little Bittern	B	1998	Common		—	B2
Anser erythropus Lesser White-fronted Goose	P	1996	Uncommon		—	A1
Marmaronetta angustirostris Marbled Duck	U	1996	Rare		—	A1
Aythya nyroca Ferruginous Duck	B	1996	Frequent		—	A1

Species ... continued	Season	Year	Pop min	Pop max	Acc	Criteria
Oxyura leucocephala White-headed Duck	U	1996	Rare		—	A1
Aquila chrysaetos Golden Eagle	R	1998	2	—	A	B2
Tetraogallus caspius Caspian Snowcock	R	1996	Uncommon		—	A3, B2
Crex crex Corncrake	B	1998	Frequent		—	A1
[1] ***Otis tarda*** Great Bustard	P	1996	Uncommon		—	A1
Charadrius alexandrinus Kentish Plover	B	1996	Uncommon		—	B2
Prunella ocularis Radde's Accentor	B	1996	Uncommon		—	B2

1. Large decline in last 20 years.

Two shallow lakes at this site, Madatapa and Khanchali, are apparently the most important in the country for breeding and staging waterbirds, including the following species of global conservation concern: *Pelecanus crispus*, *Phalacrocorax pygmeus*, *Marmaronetta angustirostris*, *Aythya nyroca*, *Oxyura leucocephala*, *Anser erythropus* and *Crex crex*. Breeding non-waterbirds include *Tetraogallus caspius*, one of only 10 species in Europe that are restricted (when breeding) to the Eurasian high-montane biome. Species of global conservation concern that do not meet IBA criteria: *Tetrax tetrax* (uncommon passage visitor). Virtually the entire Georgian population of *Ciconia ciconia* (53–61 pairs in 1996) and *Grus grus* (at least 10 pairs) is confined to the site. Both *Pelecanus crispus* and *P. onocrotalus* breed only at Kartsakhi lake on the border with Turkey, but occur commonly in other parts of the site. The site holds one of the largest populations of *Bubo bubo* in the country. Among other notable breeding species are *Ardeola ralloides*, *Melanitta fusca* (common), *Circus pygargus* (9 pairs), *Larus armenicus* (abundant) and *Motacilla citreola*. The status of *Rhodopechys sanguinea* is uncertain.

Protection status

National Partial **International** None

Small part of IBA covered by National Park (Borjomi-Kharagauli, 50,400 ha).

Conservation issues

Threats Aquaculture/fisheries (B), Construction/impact of dyke/dam/barrage (B), Disturbance to birds (C), Drainage (A), Dredging/canalization (C), Unsustainable exploitation (A)

The main threats are drainage, mowing and disturbance to birds. Very little research and conservation have so far been carried out at the site.

Algeti — A1 — 010

Admin region Tetri–Tskaro
Coordinates 41°33'N 44°30'E
Altitude 500–2,000 m **Area** 6,000 ha

Site description

The site lies in the basin of the upper Algeti river, on the wooded southern slopes of the eastern Trialeti range west of Tbilisi. The relief is rugged, with gorges of many rivers and cliffs. Forests are composed mainly of *Picea*, *Abies*, *Quercus*, *Carpinus*, *Fraxinus* and *Betula*. Meadows are also present.

Habitats Forest and woodland (84%; broadleaved deciduous forest; native coniferous forest; mixed forest), Grassland (alpine/subalpine/boreal grassland), Rocky areas (inland cliff), Wetland (river/stream)
Land-use Nature conservation/research

Birds

Species	Season	Year	Pop min	Pop max	Acc	Criteria
[1] ***Aquila heliaca*** Imperial Eagle	U	—	—	—	—	A1

1. Present, possibly breeds.

The site supports over 80 species, but a complete list has yet to be compiled. Among those known to occur are *Accipiter gentilis*, *Otus scops*, *Coracias garrulus*, *Upupa epops*, *Dryocopus martius*, *Ptyonoprogne rupestris*, *Monticola saxatilis*, *Sitta krueperi* and *Lanius minor*.

Protection status

National High **International** None

6,000 ha of IBA covered by Nature Reserve (Zapovednik) (Algeti, 6,000 ha).

Conservation issues

Threats Agricultural intensification/expansion (U), Other (U)

Main threats are the beetle *Dendroctonus micans* infesting *Picea* forests, and the chemical and biological control being used to combat this problem ('Other' threat); also the presence of human settlements and agricultural land breaking up the reserve area. The site was designated as a Zapovednik in 1965 with the primary aim of protecting forests of *Picea orientalis* and *Abies nordmanniana* on the eastern limit of their range in the Little Caucasus.

Iori region — A1, B2, B3 — 011

Admin region Kakheti
Coordinates 41°25'N 46°00'E
Altitude 100–900 m **Area** 273,996 ha

Site description

A characteristically hilly region, with low ridges whose slopes are naturally eroded, and many gorges that accumulate to form valleys or taper off into plains. This site belongs to an area with the driest climate in Georgia; there are steppes and savanna-like landscapes, arid woodlands and semi-deserts, with an abundance of relict and endemic plant species. Vegetation at the site is diverse: relict *Bothriochloa* and typical *Stipa* steppes predominate, frequently represented by *Paliurus* vegetation; there are scattered savanna-like woodlands of *Celtis* and xerophytic species of wild *Pyrus*; arid woodlands composed of relict and rare species of *Juniperus* clothe the slopes of ridges, while semi-deserts are represented by *Artemisia*, *Salsola*, etc.; riparian forests along the Iori and Alazani rivers include some of the tugai-type, dominated by *Salix* and *Populus*.

Habitats Artificial landscape (arable land; perennial crops/orchards/groves), Desert (semi-desert), Forest and woodland (broadleaved deciduous forest; mixed forest; wooded desert/semi-desert), Grassland (steppe/dry calcareous grassland), Introduced/exotic vegetation, Rocky areas (inland cliff; caves), Scrub (scrub), Wetland (saltmarsh; river/stream; water-fringe vegetation)
Land-use Agriculture, Forestry, Hunting, Military, Nature conservation/research, Water management

Birds

Species	Season	Year	Pop min	Pop max	Acc	Criteria
Ciconia nigra Black Stork	B	1998	8	—	A	B2
Gypaetus barbatus Lammergeier	U	1998	5	—	A	B2
Aegypius monachus Cinereous Vulture	R	1998	18	—	A	A1
Circaetus gallicus Short-toed Eagle	B	1998	8	—	A	B2
Buteo rufinus Long-legged Buzzard	R	1998	20	—	A	B2
Aquila pomarina Lesser Spotted Eagle	B	1998	12	—	A	B2
Aquila heliaca Imperial Eagle	R	1998	6	—	A	A1
Falco cherrug Saker	R	1999	2	—	A	B2
Crex crex Corncrake	B	1998	Frequent		—	A1
Dendrocopos medius Middle Spotted Woodpecker	R	1998	Uncommon		—	B3
Lanius minor Lesser Grey Shrike	B	1998	Common		—	B2

This is an outstanding site for raptors, with at least 25 diurnal raptors and six owls recorded: *Neophron percnopterus* (12 pairs in 1998), *Gyps fulvus* (39 pairs in 1998), *Accipiter brevipes* (occurs on passage), *Aquila nipalensis* (passage and winter visitor), *Falco biarmicus* (two breeding pairs in 1988), *F. cherrug* (occurs in winter, but breeding recently confirmed for one pair and suspected for a second). Species of global conservation concern that do not meet IBA criteria: *Haliaeetus albicilla* (three resident pairs; marked decline), *Circus macrourus* (common in winter), *Falco naumanni* (uncommon breeder; strong decline), *Tetrax tetrax* (passage visitor) and *Otis tarda* (formerly a breeder in small numbers, now only a rare passage migrant and winter visitor). This is the only site in Georgia supporting *Francolinus francolinus* (common) and also has the country's largest population of *Alectoris chukar*. Counts of staging *Anthropoides virgo* in 1986 totalled 2,500–3,500 birds and *Grus grus* also passes through on migration. Among other breeding species are *Delichon urbica* (large cliff colonies), *Cercotrichas galactotes* (rare and irregular), *Sitta tephronota*, *S. neumayer* and *Sturnus roseus* (abundant), while *Tichodroma muraria* visits the area in winter.

Protection status

National Partial **International** None

12,451 ha of the IBA covered by managed nature reserves (Zakazniks): Chachuna (5,200 ha), Gardabani (3,315 ha), Iori (1,336 ha) and Korugi (2,600 ha). 8,034 ha of IBA covered by Nature Reserve (Zapovednik) (Vashlovani, 8,034 ha).

Conservation issues

Threats Agricultural intensification/expansion (B), Deforestation (commercial) (A), Disturbance to birds (B), Dredging/canalization (C), Firewood collection (A), Forest grazing (A), Groundwater abstraction (C), Unsustainable exploitation (A)

The main threats are overgrazing, unsustainable levels of hunting, and firewood collection. The limits of the site correspond roughly to those of the proposed Iori Protected Area, part of the protected-area system which is now being established in Georgia. A management plan exists for this site.

REFERENCES

Abashidze, I. et al., eds. (1975–1987) [*Georgian Soviet Encyclopedia.*] Vols 1–11. Tbilisi: Special Scientific Editorial Office. (In Georgian.)

Abuladze, A. (1993) Seasonal migration of the Black Stork in East Pontica. P. 21 in *Abstr. 1st internat. Black Stork Conserv. and Ecol. Symp.* Jurmala, Latvia.

Abuladze, A. (1994) Birds of prey in Georgia in the 20th century. Pp. 23–28 in B. U. Meyburg and R. D. Chancellor, eds. *Raptor conservation today.* Mountfield, UK: Pica Press (World Working Group on Birds of Prey).

Abuladze, A. (1995) Seasonal migration of Demoiselle Cranes in Georgia. Pp. 302–303 in H. Prange, ed. *Crane research and protection in Europe.* Halle-Wittenberg, Germany: Martin-Luther-Universität.

Abuladze, A. V., Eligulashvili, V. and Rostiashvili, G. (1991) [On the status of the Lanner in the USSR.] Pp. 26–28 in A. M. Dorofeev, ed. [*Proc. 10th All-Union Orn. Conf.*] Part 1. Minsk: Navuka i tekhnika. (In Russian.)

Anon. (1995) *Protected area system of Georgia.* Tbilisi: World Wide Fund for Nature.

Anon. (1996) *Whitaker's Almanack 1997.* 129th edition. London: J. Whitaker and Sons Ltd.

Boehme, R. L., Zhordaniya, R. G. and Kuznetsov, A. A. (1987) [*Birds of Georgia.*] Tbilisi: Sabchota Sakartvelo. (In Russian.)

Borodin, A. M. and Syroechkovskiy, E. E., eds. (1983) [*The nature reserves of the USSR.*] Moscow: Lesnaya promyshlennost'. (In Russian.)

Brown, A., Fennell, J., Kaser, M. and Willets, H. T., eds. (1982) *The Cambridge encyclopedia of Russia and the Soviet Union.* Cambridge, UK: Cambridge University Press.

Celmins, A. and Bergmanis, U. (1990) [Observations in the Kolkhida lowland, Georgia, in September 1985.] *Putni daba* 3: 192–193. (In Latvian.)

Chatwin, M., Kikodze, D. et al. (1996) Georgian Biodiversity Country Study Report. NACRES, The Ministry of Environment of Georgia. Report to UNEP (unpublished).

Gavashelishvili, A. (1997) Tracking the Bearded Vulture. *Russian Conservation News* 13: 24–25.

Gavashelishvili, A. (1999) The 1996 census of the White Stork population in Georgia. Pp. 323–325 in H. Schulz, ed. *Weißstorch im Aufwind? White Storks on the up? Proceedings of the International Symposium on the White Stork. Hamburg, Germany September 1996.* Bonn: NABU (Naturschutzbund Deutschland). (In German and English.)

Gavashelishvili, A., Javakhishvili, Z. and Darchiashvili, G. (1998) Midwinter bird counts in Kolkheti. In unpublished report by A. Abuladze to the Georgian Pipeline Company.

Gokhelashvili, R., Gavashelishvili, A. et al. (1998) Faunistic inventory of some regions in Georgia. GCCW's report to the World Bank Small Grants Program (unpublished).

Grimmett, R. F. A. and Jones, T. A., eds. (1989) *Important bird areas in Europe.* Cambridge, UK: International Council for Bird Preservation (Techn. Publ. 3).

Hottola, P. (1987) Bird records in Georgian SSR. *Dutch Birding* 9(3): 123.

Javakhishvili, Z. (1998) [Vultures of the Iori region.] Tbilisi: Tbilisi State University (M.Sc. thesis). (In Georgian.)

Kacharava, W., ed. (1982) [*Red Data Book of the Georgian SSR: rare and endangered species of animals and plants, and some monuments of inorganic nature.*] Tbilisi: Sabchota Sakartvelo. (In Georgian.)

Knystautas, A. (1987) *The natural history of the USSR.* London: Century.

Kurashvili, B. E. (1985) [*Conservation and wise use of the animals of Georgia.*] Moscow: Nauka. (In Russian.)

Lofgren, S. (1982) [Observing raptor migration in Batumi in the autumn of 1981.] *Ornis Karelika* 8: 54–57. (In Finnish.)

Snow, D. W. and Perrins, C. M., eds. (1998) *The birds of the western Palearctic: concise edition,* vols. 1 and 2. Oxford: Oxford University Press.

Sojamo, E. (1982) [Observing raptor migration in Batumi, Georgia, in the spring of 1982.] *Ornis Karelika* 8: 58–62. (In Finnish.)

Sojamo, E. (1983) [Observations of raptor migration in Batumi, Georgia, in the spring of 1983.] *Ornis Karelika* 9. (In Finnish.)

Sokolov, V. E. and Syroechkovskiy, E. E., eds. (1990) [*The nature reserves of the USSR: reserves of the Caucasus.*] Moscow: Mysl'. (In Russian.)

Tucker, G. M. and Heath, M. F. (1994) *Birds in Europe: their conservation status.* Cambridge, UK: BirdLife International (BirdLife Conservation Series no. 3).

GIBRALTAR

JOHN CORTES

Windmill Hill Flats, Rock of Gibraltar (IBA 001), an important stop-over site for migrants. (PHOTO: JOHN CORTES, GONHS)

GENERAL INTRODUCTION

Gibraltar is a British overseas territory situated at the southern tip of the Iberian peninsula. It has its own legislative assembly, although the United Kingdom is responsible for Gibraltar's external affairs.

Gibraltar has two Important Bird Areas or IBAs (Table 1, Map 1), one of which (001) covers all 600 ha of the Rock of Gibraltar. The second IBA, the Strait of Gibraltar (002), includes 30% of the United Kingdom territorial waters adjacent to Gibraltar, in addition to Spanish, Moroccan and international waters.

The north and east sides of Gibraltar consist of high sea cliffs and sandy slopes, whilst the lower western slopes hold the main built-up areas.

The IBAs have not altered in extent since the first inventory of European IBAs (Grimmett and Jones 1989).

ORNITHOLOGICAL IMPORTANCE

Gibraltar is particularly important as a bottleneck site for migratory species (criteria A4iv, B1iv). Given the right wind conditons, the majority of raptors that migrate to Africa across the Strait of Gibraltar (an estimated 250,000 birds) could pass over the Rock in a season. These include *Pernis apivorus*, *Milvus migrans*, *Neophron percnopterus*, *Circaetus gallicus*, *Circus pygargus* and *Hieraeetus pennatus*.

The slopes of the Rock also serve as a staging site for large numbers of passerine and near-passerine migrants. The following species occur regularly on migration: *Falco eleonorae*, *Caprimulgus ruficollis*, *Oenanthe hispanica*, *Phylloscopus collybita brehmii*, *Sylvia conspicillata* and *Sylvia cantillans*. Most migratory western European species can occur at Gibraltar during the northward or southward migration periods (Cortes 1996). Scrub habitat on the Rock provides shelter for many of these birds and for substantial numbers of wintering *Sylvia atricapilla*, *Phylloscopus collybita*, *Erithacus rubecula* and *Phoenicurus ochruros*.

Map 1. Location, area and criteria category of Important Bird Areas in Gibraltar.

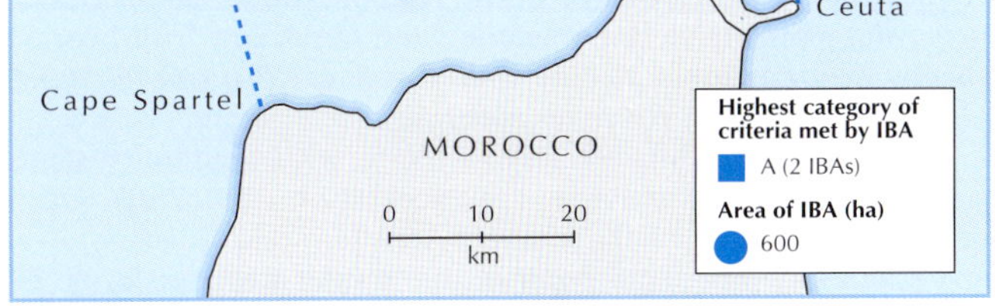

Table 1. Summary of Important Bird Areas in Gibraltar.

IBA code	1989 code	International/National name	Administrative region	Area (ha)	Criteria (see p. 11)
001	GI001	Rock of Gibraltar	Gibraltar	600	A1, A4iv, B1iv, B2, C1, C5
002	GI002	Strait of Gibraltar	Gibraltar	—	A1, A4i, A4ii, A4iii, B1i, B1ii, C1, C2, C3

Table 2. Important Bird Areas in Gibraltar that are important for species of global conservation concern (meeting criterion A1).

Species	IBA code
Falco naumanni Lesser Kestrel	001
Larus audouinii Audouin's Gull	002

Table 3. Important Bird Areas in Gibraltar that support important numbers of one or more congregatory species (i.e. meeting criteria A4 and/or B1). IBAs meeting both criteria A4 and B1 for the species are shown in **bold**. IBAs meeting only criterion B1 for the species concerned, and not A4, are shown in normal type. For key to 'Season', see p. 7.

Species	Season	IBA code
Calonectris diomedea Cory's Shearwater	P	**002**
Puffinus puffinus Manx Shearwater	P	**002**
Larus audouinii Audouin's Gull	P	**002**
Larus fuscus Lesser Black-backed Gull	P	**002**
Larus cachinnans Yellow-legged Gull	B	**002**
	W	**002**
Alca torda Razorbill	P	**002**
Fratercula arctica Puffin	P	002

Table 4. Species of European conservation concern and species listed on Annex I of the EC Birds Directive with significant breeding populations at IBAs in Gibraltar.

Species [1]	Minimum national breeding population (pairs) [2]	Proportion (%) of national population breeding at all IBAs in Gibraltar
Falco naumanni Lesser Kestrel	4	100
Alectoris barbara Barbary Partridge	50	100

1. Only those species of European conservation concern (see Box 1, p. 12) that meet IBA criteria in Gibraltar are listed, together with those species listed on Annex I of the EC Birds Directive that fulfil criterion C6 in IBAs in Gibraltar.
2. Data are taken from the BirdLife/EBCC European Bird Database 1998 (Heath and Borggreve 2000).

Twenty-three species of European conservation concern (SPECs) breed regularly in Gibraltar (Tucker and Heath 1994). Two species of global conservation concern occur regularly: *Falco naumanni* nests on the North Face of the Rock and occurs on passage (Table 2, Table 4), and *Larus audouinii* occurs during the winter and on passage (Table 2). The Rock is the only breeding site of *Alectoris barbara* in mainland Europe (Table 4).

Up to 200,000 seabirds migrate between the Mediterranean and the Atlantic, passing through the Strait (criteria A4ii, B1ii). Table 3 provides a summary of those that meet IBA criteria. Of those that do not, the following pass through on a regular basis: *Puffinus griseus*, *P. gravis* and *Sula bassana* (up to 10,000 birds in a season; Finlayson and Cortes 1984). These waters also support several thousand seabirds in winter, including *Sula bassana*, *Larus ridibundus*, *Larus cachinnans* and *Alca torda*.

HABITATS

The woodland which once covered the limestone of the Rock has been progressively removed, with regeneration being prevented by grazing goats. Following removal of the goats, maquis has become the dominant vegetation type on Gibraltar, and its species composition remains unique in the region. About 500 species of flowering plants have been recorded and important fruit-bearing shrubs of the maquis include *Olea europaea, Pistacia lentiscus, Osyris quadripartita, Rhamnus alaternus* and *Asparagus albus*.

The eastern sea cliffs ascend to c.250 m, with windblown sands supporting an open garrigue-type vegetation (matorral). Open vegetation also exists on the southern stony terraces, whilst matorral and patches of woodland occur on the gentler western slopes. Of those species that meet IBA criteria only *Alectoris barbara* relies solely on these terrestrial habitats. Other species of passerines and near-passerines rely on the matorral when wintering or stopping-over whilst on migration. The majority of raptors and marine species overfly or by-pass the Rock itself, with little feeding effort. Wintering *Larus audouinii*, *L. fuscus* and *L. cachinnans* do feed in the territorial waters, with the last-mentioned also feeding in terrestrial habitats.

IMPACT ON IBAs – LAND-USE AND THREATS

The lower slopes of the Rock of Gibraltar are largely built-up, with approximately 40% of the land area in urban use. Of the remainder, 30% is military land held by the Ministry of Defence, and used as aerial farms for telecommunications purposes and/or for military training, and 8% is designated as a nature reserve. Sections of the remaining areas of vegetated land are under threat from commercial and industrial development, although this pressure has been reduced over the last decade as a result of reclamation of land from the sea. No agricultural activities take place on the Rock.

The Strait of Gibraltar is a major route for passenger, military and commercial vessels, including oil tankers. Inshore waters are used by recreational boats, and support extensive Spanish and Moroccan fisheries. Activities include conch raking and the use of seine and drift nets, which carry with them a threat to seabirds. Such methods are banned from the part of the Strait under Gibraltar's jurisdiction, although political pressure is threatening the continuing protected status of these waters. Recreational boats can cause disturbance to birds inshore, especially in winter, but there is no evidence that migrating birds are adversely affected.

PROTECTION STATUS

Table 5 summarizes the national and international protection status of the two IBAs in Gibraltar.

National protection

The Nature Protection Ordinance (1991) effectively protects all wildlife throughout the territory and waters of Gibraltar, and includes all species of bird. The Ordinance also provides for the declaration of nature reserves, and this led to the Upper Rock being declared a Nature Reserve in 1993. The Marine Reserve, which extends to all British territorial waters around Gibraltar, was created in 1996. This confers legal protection, although full implementation of the regulation is still pending. The protected areas, both marine and terrestrial, belong to the British Crown, but are under the jurisdiction either of the Government of Gibraltar or the Ministry of Defence.

International protection

Several international conventions have been ratified by the United Kingdom Government on behalf of Gibraltar (Box 1). European Union Directives and Regulations, including the EC Birds Directive, also apply to the territory. Parts of both IBAs (001 and 002) have been proposed as SPAs under the Birds Directive and as SACs under the EC Habitats Directive.

Table 5. Protection status of Important Bird Areas in Gibraltar. A tick (✔) indicates that an IBA overlaps with a protected area (to any extent).

IBA code	International name	National: Nature Reserve	National: Marine Reserve
001	Rock of Gibraltar	✔	
002	Strait of Gibraltar		✔
Total number of IBAs		1	1

Box 1. International legislation and initiatives that are relevant to site conservation in Gibraltar (see Appendix 1 for a general description of these agreements).

Global	
Biodiversity Convention	✔
Ramsar Convention	✔
Bonn Convention	✔
World Heritage Convention	✔
MAB Programme	
Pan-European	
Bern Convention	
Regional	
EC Birds Directive	✔
EC Habitats Directive	✔
Barcelona Convention	

✔ Convention ratified/initiative supported
(✔) Convention signed

CONSERVATION

- Bird migration has been monitored in Gibraltar since the 1960s (Cortes 1996). Monitoring work has included direct observation of raptor and seabird migration, and both direct observation and ringing of passerines and near-passerines (20,321 birds were ringed between 1991 and 1997).
- Inadequate management of the habitats on the Upper Rock has led to vegetation succession, resulting in a progressive decline in species of more open ground (for example *Alectoris barbara* and *Sylvia melanocephala*). Management initiatives are consequently being planned in order to resolve the problem.

ANALYTICAL METHODS

- Difficulties in deciding which figures to present for migratory species arose due to the large yearly fluctuations in numbers caused by varying wind conditions. Minimum and maximum figures given in the species tables are therefore based on averages of seasonal minima and maxima recorded between 1967 and 1997 by observers at the Strait of Gibraltar Bird Observatory (SGBO). Some of these records have been published (Bernis 1980, Cortes *et al.* 1980, Finlayson 1991, Finlayson & Cortes 1984 and 1987, Garcia 1973, SEO 1998, *SGBO Reports 1987–1995*, Telleria 1981), whilst other Gibraltar Ornithological and Natural History Society (GONHS)/SGBO records are as yet unpublished.
- Observations for the Strait as a whole are not possible due to its width. Many more birds may occur than can be estimated from observations made purely from Gibraltar's territorial waters, which extend for three miles from Europa Point into the north-east corner of the Strait. Comparable data for seabirds from Spain and Morocco are not available, whilst recent records for soaring birds are available only from Spain for 1997.
- The figures given in the footnotes to the species tables for the Strait as a whole are therefore estimates (based upon Bernis 1980, GONHS records (unpubl.), SEO 1998, *SGBO Reports 1987–1995*). In exceptional weather conditions it would be theoretically possible for the bulk of these birds to cross the Strait at Gibraltar.
- All the information on threats and land-use are current at the time of writing (September 1998).
- The European Union has not assigned a NUTS code for Gibraltar.

GLOSSARY

GONHS Gibraltar Ornithological and Natural History Society (BirdLife International Partner Designate in Gibraltar).
NUTS Nomenclature des Unités Territoriales Statistiques – the administrative regions of the European Union are called NUTS regions, and are designated by the EC Statistical Office.
SAC Special Area for Conservation.
SGBO Strait of Gibraltar Bird Observatory.
SPA Special Protection Area (designated under Article 4 of the EC Birds Directive).

ACKNOWLEDGEMENTS

The following observers contributed records used in this report: Paul Acolina, Richard Banham, Steven Barea, Jouaquin Bensusan, Keith Bensusan, John Cortes, Francis Cosquieri, W. F. Curtis, Ian Farrell, Dennis Federico, Clive Finlayson, Geraldine Finlayson, Andrew Fortuna, Ernest F. J. Garcia, Joakim Hagstrom, John Henwood, Yvonne Henwood, Edmund Hoare, Steve Holliday, Terry Jesty, Niclas Jansen, John Licudi, Edgar Lopez, Tommy Matos, Mario Mosquera, John Murphy, Stanley Olivero, Cristina Parkes, Charles E. Perez, David Price, Darrien Ramos, Nigel Ramos, Fernando Ribeiro, Vincent Robba, Paul Rocca, Andrew Romero, Christopher Romero, Roger Rutherford, Jeffrey Saez, Tony Santana, Giovanni Santini, Daniel Sciacaluga, Eric Shaw, A. B. Sheldon, Michael Short, Brian Thomas, Magnus Ullman, Harry Vangils, W. A. Venables, Simon Waite, Tom Walsh, Stephen Warr and Albert Yome.

SITE ACCOUNTS

Rock of Gibraltar — A1, A4iv, B1iv, B2, C1, C5 — 001

Admin region Gibraltar
Coordinates 36°17′N 5°21′W
Altitude 0–396 m **Area** 600 ha

Site description

The Rock of Gibraltar is dominated by a dense cover of maquis and garrigue, including many important fruit-bearing shrubs that support large passerine populations on passage and during the winter.

Habitats Scrub (40%; sclerophyllous scrub/garrigue/maquis), Rocky areas (20%; sea cliff/rocky shore; rock stacks/islets), Artificial landscape (40%; other urban/industrial areas)
Land-use Military (30%), Nature conservation/research (8%), Tourism/recreation (30%), Urban/industrial/transport (40%)

Birds

Species	Season	Year	Pop min	Pop max	Acc	Criteria
[1] ***Pernis apivorus*** Honey Buzzard	P	—	5,000	50,000	—	A4iv, B1iv, C5
[2] ***Milvus migrans*** Black Kite	P	—	5,000	12,000	—	A4iv, B1iv, C5
[3] ***Neophron percnopterus*** Egyptian Vulture	P	—	25	250	—	A4iv, B1iv, C5
[4] ***Circaetus gallicus*** Short-toed Eagle	P	—	101	250	—	A4iv, B1iv, C5
[5] ***Circus pygargus*** Montagu's Harrier	P	—	101	500	—	A4iv, B1iv, C5
[6] ***Hieraaetus pennatus*** Booted Eagle	P	—	251	2,500	—	A4iv, B1iv, C5
Falco naumanni Lesser Kestrel	B	—	4	10	—	A1, C1
Falco naumanni Lesser Kestrel	P	—	11	100	—	A1, C1
[7] ***Alectoris barbara*** Barbary Partridge	B	—	50	75	—	B2

Data cover period 1967–1998.
The following footnotes are estimates for the Strait of Gibraltar as a whole.
1. Up to 100,000 individuals.
2. Up to 140,000 individuals.
3. Up to 3,500 individuals.
4. Up to 20,000 individuals.
5. Up to 1,500 individuals.
6. Up to 20,000 individuals.
7. Only breeding site in mainland Europe.

Gibraltar is a well-known migratory bottleneck site for birds of prey, with an estimated 250,000 raptors crossing the Strait in a season. If the winds are right then most of these pass over Gibraltar, and numbers may in fact be higher. It is also a stop-over site for passerines and near-passerines, with a rapid turnover of birds on passage during both northward and southward migration.

Protection status

National Low **International** None
45 ha of IBA covered by Nature Reserve (Upper Rock, 45 ha).

Conservation issues

Threats Consequences of animal/plant introductions (U), Industrialization/urbanization (U), Natural events (U)

A number of military sites are designated as conservation areas. 197 ha of the Rock of Gibraltar are proposed as a SPA under the EC Birds Directive, and as a SAC under the EC Habitats Directive.

Strait of Gibraltar — A1, A4i, A4ii, A4iii, B1i, B1ii, C1, C2, C3 — 002

Admin region Gibraltar
Coordinates 36°02′N 5°40′W
Altitude 0–0 m **Area** —

Site description

Habitats Marine areas (100%; open sea; sea inlet/coastal features)
Land-use Fisheries/aquaculture, Tourism/recreation, Urban/industrial/transport

The Strait of Gibraltar is defined here as the sea area demarcated in the west by a straight line joining Cape Trafalgar on the European coast with Cape Spartel on the North African coast, and in the east by a straight line joining the Rock of Gibraltar on the European coast to Ceuta (enclave of Spain) on the North African coast. It includes international waters and waters under the jurisdiction of the United Kingdom, Spain and Morocco, and is 15 km wide at its narrowest point.

Birds

Species	Season	Year	Pop min	Pop max	Acc	Criteria
Calonectris diomedea Cory's Shearwater	P	—	10,000	100,000	—	A4ii, B1ii, C2
Puffinus puffinus Manx Shearwater	P	—	1,001	10,000	—	A4ii, B1ii, C3
Larus audouinii Audouin's Gull	W	—	51	100	—	A1, C1
[1] **Larus audouinii** Audouin's Gull	P	—	1,001	10,000	—	A1,A4i,B1i,C1,C2
Larus fuscus Lesser Black-backed Gull	P	—	1,001	10,000	—	A4i, B1i, C3
Larus cachinnans Yellow-legged Gull	B	—	12,000	—	—	A4i, B1i, C3
Larus cachinnans Yellow-legged Gull	W	—	1,001	10,000	—	A4i, B1i, C3
Alca torda Razorbill	P	—	1,001	10,000	—	A4ii, B1ii, C3
[2] **Fratercula arctica** Puffin	P	—	1,001	10,000	—	B1ii, C3

Data cover period 1967–1998.
The following footnotes are estimates for the Strait of Gibraltar as a whole.
1. Estimated up to 12,000 individuals, July to September.
2. Estimated up to 40,000 individuals.

Hundreds of thousands of birds that migrate between the Mediterranean and the Atlantic over the sea pass through the Strait of Gibraltar, including highly significant numbers of *Calonectris diomedea*, *Larus audouinii* and *Fratercula arctica*.

Protection status

National Partial **International** None
4,900 ha of IBA covered by Marine Reserve (4,900 ha).

Conservation issues

Threats Aquaculture/fisheries (U), Other (U)

Heavy commercial shipping traffic carries with it the threat of major oil spills ('Other threat', above). 4,900 ha of the territorial waters of the United Kingdom are proposed as a SPA under the EC Birds Directive, and as a SAC under the EC Habitats Directive. Commercial fishing is prohibited in these territorial waters, and protection is afforded to cetaceans, turtles, birds and other marine life.

REFERENCES

Bernis, F. (1980) *La migración de las aves en el estrecho de Gibraltar. Volumen I: Aves planeadoras.* Madrid: Universidad Complutense.

Cortes, J. E. (1996) Windmill Hill Flats: a good view of migration across the Straits of Gibraltar. *Almoraima* 15: 163–184.

Cortes, J. E., Finlayson, J. C., Garcia, E. F. J. and Mosquera, M. A. J. (1980) *The birds of Gibraltar.* Gibraltar: Gibraltar Books.

Elkins, N. and Etheridge, B. (1974) The Crag Martin in winter quarters at Gibraltar. *British Birds* 67: 376–389.

Finlayson, J. C. (1980) The recurrence in winter quarters at Gibraltar of some scrub passerines. *Ringing and Migration* 3: 32–34.

Finlayson, J. C. (1991) *Birds of the Strait of Gibraltar.* London: T. & A.D. Poyser.

Finlayson, J. C. and Cortes, J. E. (1984) The migration of Gannets *Sula bassana* past Gibraltar in spring. *Seabird* 7: 19–22.

Finlayson, J. C. and Cortes, J. E. (1987) The Birds of the Strait of Gibraltar—its waters and northern shore. *Alectoris* 6: 1–74.

Finlayson, J. C., Garcia, E. F. J., Mosquera, M. A. J. and Bourne, W. R. P. (1976) Raptor migration across the Strait of Gibraltar. *British Birds* 69: 77–87.

Garcia, E. F. J. (1973) Seabird activity in the Strait of Gibraltar. *Seabird Report* 3: 30–36.

Grimmett, R. F. A. and Jones T. A. (1989) *Important Bird Areas in Europe.* Cambridge, UK: International Council for Bird Preservation (Tech. Publ. 9).

Heath, M. F. and Borggreve, C. (2000) *BirdLife International/EBCC European Bird Database 1998.* Cambridge, UK: BirdLife International.

Linares, L. L. (1983) A checklist of the Gibraltar flora. *Alectoris* 5: 24–39.

Pineau, J. and Giraud-Audine, M. (1979) *Les oiseaux de la péninsule tingitane.* Rabat, Morocco: Institut Scientifique.

Sociedad Española de Ornitología (1998) *Programa MIGRES, Seguimiento de la Migracion en el Estrecho. Proyecto Piloto 1997.* Sevilla, Spain: Junta de Andalucia.

Strait of Gibraltar Bird Observatory (1988–1996) *Bird Reports 1987–1995.* Gibraltar: The Gibraltar Ornithological & Natural History Society.

Telleria, J. L. (1981) *La migración de las aves en el estrecho de Gibraltar. Volumen II: Aves no planeadoras.* Madrid: Universidad Complutense.

Tucker, G. M. and Heath, M. F. (1994) *Birds in Europe: their conservation status.* Cambridge, UK: BirdLife International (BirdLife Conservation Series no. 3).

GREECE

STRATIS BOURDAKIS AND STELLA VARELTZIDOU

Megali Dragonera islet, Kithira island (IBA 129). (PHOTO: A. BONETTI)

GENERAL INTRODUCTION

Greece lies on the southern edge of the Balkan peninsula, in south-eastern Europe. Its land-borders are with Albania, the Former Yugoslav Republic of Macedonia and Bulgaria in the north and Turkey in the east. The remaining eastern border and the southern and western borders are coastal, adjoining the Aegean, Mediterranean and Ionian Seas respectively. The total surface area of Greece is approximately 132,000 km^2 with one fifth of this being islands. Greece has extensive areas of island archipelagos, with more than 2,000 islands, and a long (c.15,000 km), highly indented coastline characterized by numerous peninsulas.

Topographically, Greece is characterized by a mountainous interior (average altitude is nearly 500 m), intersected by valleys and mountain-basin lakes, and fringed by coastal lowlands of varying extent. In the west, the mountains are close to the sea so that the lowland fringe is extremely narrow. In the east and north, however, there are several areas of lowland associated with

Table 1. Summary of Important Bird Areas in Greece. 196 IBAs covering 34,332 km^2

IBA code	1989 code	International name	National name	Administrative region	Area (ha)	Criteria (see p. 11)
001	—	Northern Evros and Ardas riparian forest	Parapotamio Dasos Voriou Evrou kai Arda	East Macedonia and Thrace	20,000	A1, B2, B3, C1, C6
002	—	Asvestades–Vrisika area	Periochi Asvestadon–Vrisikon	East Macedonia and Thrace	9,500	B2, C6
003	GR002	Dadias–Dereiou–Aisimis forest	Dasos Dadias–Dereiou–Aisimis	East Macedonia and Thrace	50,629	A1, A3, B2, C1, C6
004	GR002	Protected area of Dadia–Lefkimi–Soufli forest	Prostatevomeni periochi Dasous Dadias–Leukimis–Soufliou	East Macedonia and Thrace	42,370	A1, A3, B1iii, B2, C1, C2, C6
005	GR002	South forest complex of Evros prefecture	Notio Dasiko Symplegma Nomou Evrou	East Macedonia and Thrace	71,800	A1, A3, B1iii, B2, B3, C1, C2, C6
006	GR001	Evros delta	Delta Evrou	East Macedonia and Thrace	19,000	A1, A4i, A4iii, B1i, B1iv, B2, B3, C1, C2, C3, C4, C5, C6
007	—	Samothraki island	Nissos Samothraki	East Macedonia and Thrace	17,000	B1i, C2
008	GR005	Filiouri river valley and east Rodopi mountains	Kilada Filiouri kai Anatoliki Rodopi	East Macedonia and Thrace	77,000	A1, B2, B3, C1, C6
009	GR006	Kompsatos valley	Kilada Kompsatou	East Macedonia and Thrace	16,000	A1, B2, B3, C1, C6
010	GR003	Lake Mitrikou (Ismarida)	Limni Mitrikou (Ismarida)	East Macedonia and Thrace	6,500	A1, A4i, A4iii, B1i, B2, B3, C1, C2, C3, C4, C6
011	GR004	Porto Lagos, Lake Vistonis, and coastal lagoons (Lakes of Thrace)	Porto Lagos, Limni Vistonida, and parakties Limnothalasses	East Macedonia and Thrace	15,300	A1, A4i, A4iii, B1i, B1ii, B2, B3, C1, C2, C3, C4, C6
012	GR007	Nestou delta and coastal lagoons	Delta Nestou and parakties Limnothalasses	East Macedonia and Thrace	22,000	A1, A4i, B1i, B1ii, B2, B3, C1, C2, C3, C6
013	GR009	Nestos gorge	Stena Nestou	East Macedonia and Thrace	14,500	A1, B2, C1, C6
014	GR010	Central Rodopi mountains and Nestos valley	Kentriki Rodopi and Kilada Nestou	East Macedonia and Thrace	150,000	B2, B3, C6
015	—	Mount Falakro	Oros Falakro	East Macedonia and Thrace	20,000	B2, C6

Table 1 ... continued. Summary of Important Bird Areas in Greece. 196 IBAs covering 34,332 km²

IBA code	1989 code	International name	National name	Administrative region	Area (ha)	Criteria (see p. 11)
016	—	North and north-east Thasos island	Voria kai Vorioanatoliki Nissos Thasos	East Macedonia and Thrace	15,000	B1ii, C2
017	GR014	Mount Pangaio	Oros Pangaio	East Macedonia and Thrace	20,000	B2, B3
018	GR016	Strymon delta	Ekvoles Strymona	East Macedonia and Thrace, Central Macedonia	1,300	A1, A4i, B1i, C1, C2
019	GR011	Mount Kouskouras	Kouskouras Oros	Central Macedonia	27,543	B3, C6
020	GR012	Lake Kerkini	Limni Kerkini	Central Macedonia	12,000	A1, A4i, A4iii, B1i, B2, C1, C2, C3, C4, C6
021	GR017	Mount Kroussia (or Mavrovouni)	Kroussia Ori (or Mavrovouni)	Central Macedonia	18,000	A1, B2, C1, C6
022	GR013	Mount Beles (Kerkini)	Oros Beles (Kerkini)	Central Macedonia	34,000	B2, C6
023	—	Lake Doirani	Limni Doirani	Central Macedonia	2,000	A1, A4i, B1i, C1, C2
024	GR015	Artzan marshes	Elos Artzan	Central Macedonia	1,700	A1, A4i, B1i, C1, C2
025	GR018	River Axios	Axios Potamos	Central Macedonia	12,400	A1, A4i, B1i, C1, C2
026	—	Area of Anthofito	Periochi Anthofitou	Central Macedonia	1,200	A1, B2, C1, C6
027	—	Lake Pikrolimni	Limni Pikrolimni	Central Macedonia	2,000	A1, C1
028	GR019	Axios, Loudias, and Aliakmon estuaries	Delta Axiou, Loudia, and Aliakmona	Central Macedonia	20,000	A1, A4i, A4iii, B1i, B2, B3, C1, C2, C3, C4, C6
029	—	Gallikos estuary and Kalohori lagoon	Ekvoles Gallikou kai Limnothalassa Kalohoriou	Central Macedonia	1,800	A1, A4i, B1i, B3, C1, C2, C6
030	GR020	Alyki Angelochoriou lagoon (Megalou Emvolou)	Limnothalassa Alykis Angelochoriou (Megalou Emvolou)	Central Macedonia	357	A1, A4i, B1i, C1, C2, C3
031	GR021	Epanomi lagoon	Limnothalassa Epanomis	Central Macedonia	450	B1i, B2, C2, C3
032	GR022	Lakes Volvi–Langada (or Koronia–Ag. Vasiliou) and Rentina gorge (or Makedonika Tempi)	Limnes Volvi–Langada (or Koronia–Ag. Vasiliou) kai Stena Rentinas (or Makedonika Tempi)	Central Macedonia	43,000	A1, A4i, A4iii, B1i, B2, C1, C2, C3, C4, C6
033	GR024	Mount Cholomontas	Ori Cholomontas	Central Macedonia	27,500	C7
034	GR025	Agios Mamas marsh	Elos Agiou Mama	Central Macedonia	633	A1, B1i, B2, C1, C2
035	—	Nea Fokea marshes	Eli Neas Fokeas	Central Macedonia	340	B1i, C2
036	GR026	Mount Itamos	Oros Itamos	Central Macedonia	27,000	B2, C6
037	GR027	Mount Athos	Oros Athos	Agion Oros	12,000	B1ii, C3
038	GR028	Mounts Tzena and Pinovo	Ori Tzena kai Pinovo	Central Macedonia	13,154	B3, C6
039	GR029	Mount Voras (Kaimaktchalan)	Oros Voras (Kaimaktsalan)	Central Macedonia, Western Macedonia	52,286	B2, B3, C6
040	GR030	Apsalos and Moglenitsa gorges	Stena Apsalou and Moglenitsas	Central Macedonia	6,110	B2, C6
041	GR031	Agra artificial lake	Techniti Limni Agra	Central Macedonia	1,386	B2, C6
042	GR038	Aliakmon gorge	Stena Aliakmona	Central Macedonia	4,400	B2
043	GR039	Alyki Kitros lagoon	Limnothalassa Alyki Kitros	Central Macedonia	1,500	A4i, B1i, B2, B3, C2, C3, C6
044	GR040	Mount Olympos	Oros Olympos	Central Macedonia, Thessalia	42,000	A3, B2, B3, C6
045	GR033	Lake Vegoritis and Lake Petron	Limni Vegoritida and Limni Petron	Central Macedonia, Western Macedonia	17,000	A1, B2, C1, C6
046	GR034	Lake Chimaditis and Lake Zazaris	Limni Chimaditida and Limni Zazari	Western Macedonia	5,390	A1, A4i, B1i, B2, C1, C2, C6
047	GR032	Lake Mikri Prespa and Lake Megali Prespa	Limni Mikri Prespa and Limni Megali Prespa	Western Macedonia	33,000	A1, A4i, B1i, B2, C1, C2, C3, C6
048	GR035	Lake Kastoria (Orestiada)	Limni Kastorias (Orestiada)	Western Macedonia	3,400	A1, A4i, B1i, B2, C1, C2, C3, C6
049	GR036	Mount Gramos	Oros Gramos	Western Macedonia, Ipiros	45,000	B2, C6
050	—	Mounts Orliakas and Tsourgiakas	Ori Orliakas kai Tsourgiakas	Western Macedonia	10,500	B2, C6
051	GR037	Mount Vourinos	Oros Vourinos	Western Macedonia	31,000	B2, C6
052	GR046	Kerketion mountain (Koziakas)	Oros Kerketio (Koziakas)	Thessalia	23,000	B2, B3, C6
053	GR045	Antikhassia mountains and Meteora	Antikhassia Ori and Meteora	Thessalia	65,000	B2, B3, C2, C6
054	GR047	Kalamaki gorge–Mount Zarkou	Stena Kalamakiou–Ori Zarkou	Thessalia	8,650	B2, C6
055	—	Elassona area	Periochi Elassonas	Thessalia	7,500	A1, B2, C1, C6
056	—	Tyrnavos area	Periochi Tyrnavou	Thessalia	8,500	A1, B2, C1, C6
057	—	Mati of Tirnavos	Mati Tirnavou	Thessalia	500	A1, C1
058	GR048	Mounts Kato Olympos and Ossa, and Tembi ravine	Kato Olympos, Tembi, Ossa	Thessalia	69,000	B2, C6
059	GR048	Pinios delta	Delta Piniou	Central Macedonia, Thessalia	2,600	A1, C1
060	GR049	Mavrovouni mountain, Larissa	Oros Mavrovouni, Larissa	Thessalia	17,000	A1, A3, B2, B3, C1, C2, C6
061	—	Reservoirs of former Lake Karla	Tamieftires teos Limnis Karlas	Thessalia	1,200	A1, A4i, B1i, C1, C2, C3
062	—	Thessaly plain	Thessalikos Kampos	Thessalia	180,000	A1, A4ii, B1iii, B2, B3, C1, C2, C6
063	—	Farsala area	Periochi Farsalon	Thessalia	4,500	A1, B1iii, B2, C1, C2, C6
064	GR068	Mount Pilion	Pilion Oros	Thessalia	31,751	B2, C6
065	GR107	National Marine Park of Alonissos (North Sporades)	Nissi Kyra-Panaghia/Ghioura/Piperi/ Skantzoura (Vorii Sporades)	Thessalia	5,800	A1, A4ii, B1ii, B1iii, B2, B3, C1, C2, C3, C6
066	GR042	Timfi (Gamila) and Smolikas mountains	Oros Timfi (Gamila) and Oros Smolikas	Ipiros	55,000	A3, B2, C6
067	GR043	Central part of Zagori and Mount Mitsikeli	Kentriko Zagori kai Anatoliko Oros Mitsikeli	Ipiros	46,500	B2, B3, C6
068	GR044	Valia Kalda (Pindos National Park)	Valia Kalda (Ethnikos Drymos Pindou)	Western Macedonia, Ipiros	8,000	B2, C7
069	—	Peristeri mountain	Oros Peristeri	Ipiros, Thessalia	21,300	B2, C6
070	—	Lake Pamvotida (Ioanninon)	Limni Pamvotida (Ioanninon)	Ipiros	2,707	A1, B1i, C1, C3
071	—	Ioannina city and neighbouring area	Periochi Polis Ioanninon	Ipiros	8,000	A1, B2, C1, C6

Table 1 ... continued. Summary of Important Bird Areas in Greece. 196 IBAs covering 34,332 km²

IBA code	1989 code	International name	National name	Administrative region	Area (ha)	Criteria (see p. 11)
072	GR041	Area of Oriokastro, Lake Delvinaki, Meropis forest, Gormou valley and Mounts Kasidiaris and Douskon	Oriokastro, Limni Delvinakiou, Dasos Meropis, Kilada Gormou kai Oros Kasidiaris	Ipiros	36,000	B2, B3, C6
073	—	Mounts Tsamanta, Filiaton, Pharmakovouni, and Megali Rahi	Ori Tsamanta, Ori Filiaton, Pharmakouni, Megali Rahi	Ipiros	21,000	A1, B2, C1, C6
074	GR051	Kalamas estuary	Ekvoles Kalama	Ipiros	8,531	A1, A4i, B1i, C1, C2, C3, C6
075	GR050	Kalamas gorge	Stena Kalama	Ipiros	4,000	A1, B2, C1, C6
076	—	Mount Paramithias	Ori Paramithias	Ipiros	20,000	B3, C6
077	GR053	Acheron delta and gorge	Ekvoli and Stena Acherondos	Ipiros	6,000	A3, A4i, B1i, C2, C6
078	—	Lake Kalodiki, Margariti and Karteri marshes	Limni Kalodiki, Eli Margaritiou kai Karteriou	Ipiros	1,650	B2, C6
079	GR056	Athamanika mountains (Tzoumerka)	Ori Athamanika (Tzoumerka)	Ipiros, Thessalia	55,000	B2, C6
080	GR057	Acheloos valley	Kilada Acheloou	Ipiros, Thessalia	10,600	B2, C7
081	GR055	Amvrakikos gulf	Amvrakikos Kolpos	Ipiros, Western Greece	25,000	A1, A4i, A4iii, B1i, B2, C1, C2, C3, C4, C6
082	—	Diapontia islands (Othoni, Erikoussa, Mathraki islands and rocky islets	Diapontia Nissia (Othoni, Erikousa, Mathraki kai vrachonisides	Ionian Islands	1,500	B2, C6
083	—	Lagoons of Kerkyra	Limnothalasses Kerkyras	Ionian Islands	1,100	A1, A4i, B1i, C1, C2, C3
084	—	Echinades islands, Kalamos and Atokos	Nissoi Echinades, Kalamos and Atokos	Ionian Islands	4,700	C7
085	—	The Agia Dinati and Kokkini Rachi mountains of Kefallinia	Ori Agia Dynati, Kokkini Rachi, Kefallinia	Ionian Islands	20,000	C7
086	—	West and north Zakynthos	Dytiki kai Voreia Zakynthos	Ionian Islands	13,600	A3, B1i, C2
087	GR080	Strofades islands	Nissi Strofades	Ionian Islands	220	B1ii, C2, C6
088	—	Valtou mountains	Ori Valtou	Western Greece	43,500	B2, B3, C6
089	—	Lake Amvrakia	Limni Amvrakia	Western Greece	2,205	B1i, C2, C3
090	GR058	Akarnanika mountains	Ori Akarnanika	Western Greece	39,000	A3, B2, B3, C6
091	—	Lakes Trichonida and Lysimachia	Limnes Trichonida and Lysimachia	Western Greece	14,279	A1, C1
092	GR060	Mesolongi and Aetoliko lagoons, and Acheloos and Evinos estuaries	Limnothalassa Mesolongiou and Aetolikou, Ekvoles Acheloou and Evinou	Western Greece	50,000	A1, A4i, A4iii, B1i, B2, C1, C2, C3, C4, C6
093	GR076	Alyki lagoon, Aegion	Limnothalassa Alyki Aegion	Western Greece	32	B1i, C2
094	GR077	Barbas and Klokos mountains and Selinous gorge	Ori Barbas, Klokos, Farangi Selinounta	Western Greece	11,570	B2, C6
095	GR078	Vouraikos gorge including Kalavrita	Faranghi Vouraikou kai Kalavrita	Western Greece	5,600	C7
096	—	Mount Helmos (Aroania)	Oros Helmos (Aroania)	Western Greece, Peloponnissos	17,500	B2
097	—	Mount Erimanthos	Oros Erimanthos	Western Greece	36,900	B2, B3
098	GR075	Kalogria lagoon, Strofilia forest, and Lamia marshes	Limnothalassa Kalogrias, Dasos Strofilias kai Eli Lamias	Western Greece	6,000	A4i, B1i, B2, C2, C3, C6
099	GR074	Kotychi lagoon	Limnothalassa Kotychi	Western Greece	2,700	A1, A4i, B1i, C1, C2
100	GR079	Mount Foloi	Oros Foloi	Western Greece	16,500	B3, C6
101	—	Delimidi and Fteri mountains, Agrafa	Ori Delimidi kai Fteri, Agrafa	Thessalia, Sterea Ellada	21,200	B2, C6
102	GR066	Mount Othrys	Oros Othrys	Thessalia, Sterea Ellada	84,000	B2, B3, C6
103	GR065	Sperchios valley and delta–Maliakos gulf	Koilada kai Ekvoles Sperchiou–Maliakos Kolpos	Sterea Ellada	34,000	A1, A4i, A4iii, B1i, B2, C1, C2, C4, C6
104	GR064	Mount Iti	Oros Iti	Sterea Ellada	18,800	B2, B3, C6
105	GR061	Mount Vardousia	Oros Vardousia	Sterea Ellada	21,300	B2, B3, C6
106	GR062	Mount Ghiona, Reka ravine, Lazorema, and Vathia Lakka	Oros Ghiona, Kharadra Rekas, Lazorema, and Vathia Lakka	Sterea Ellada	25,000	A3, B2, B3, C6
107	GR063	Southern and eastern part of Mount Parnassos	Notio kai Anatoliko Tmima tou Orous Parnassos	Sterea Ellada	19,400	A3, B2, B3, C6
108	—	Galaxidi	Galaxidi	Sterea Ellada	2,000	A1, B1iii, B2, C1, C2, C6
109	GR069	Megalo and Mikro Livari lagoons, Istieas	Limnothalasses Megalo kai Mikro Livari Istieas	Sterea Ellada	1,008	B1i, C3
110	GR070	Kandili mountain	Oros Kandili	Sterea Ellada	5,700	B3, C6
111	GR071	Dirfi, Xirovouni, Skotini, Mavrovouni, Alokteri and Ortari mountains and the vicinity of Kimi	Ori Dirfis, Xirovouni, Skotini, Mavrovouni, Alokteri, Ortari, and vicinity of Kimi	Sterea Ellada	45,000	B3, C6
112	GR072	Lake Distos	Limni Distos	Sterea Ellada	2,600	A3, B2, C2, C6
113	GR073	Mount Ochi and surrounding area	Oros Ochi and surrounding area	Sterea Ellada	18,000	A4ii, B1ii, B2, B3, C2, C3, C6
114	—	North-west Skyros and Ag. Fokas islets	Vorioditiki Skyros kai nisides Ag. Foka	Sterea Ellada	2,700	A1, C1, C6
115	—	Mount Kohilas, Skyros island	Oros Kohilas, Nissos Skyros	Sterea Ellada	5,500	A1, B2, B3, C1, C6
116	—	Mount Ziria (Killini)	Oros Ziria (Killini)	Peloponnissos	19,400	B2
117	GR081	Lake Stymphalia	Limni Stymphalia	Peloponnissos	1,309	B2, C2, C6
118	—	Mounts Artemisio and Lirkio	Ori Artemisio kai Lirkio	Peloponnissos	8,900	B3, C6
119	GR084	Divari Pilou lagoon (Gialova)	Limnothalassa Divari Pilou (Gialova)	Peloponnissos	600	A4i, B1i, C2, C3
120	GR083	Mount Taigetos	Oros Taigetos	Peloponnissos	70,000	A1, A3, B2, B3, C1, C2, C6
121	GR085	Southern Mani, Mount Sanghias and Cape Tenaro	Notia Mani, Oros Sanghias kai Akrotirio Tenaro	Peloponnissos	25,000	A3, B2, B3
122	GR086	Evrota delta	Ekvoles Evrota	Peloponnissos	3,000	A4i, B1i, C2

Table 1 ... continued. Summary of Important Bird Areas in Greece. 196 IBAs covering 34,332 km²

IBA code	1989 code	International name	National name	Administrative region	Area (ha)	Criteria (see p. 11)
123	GR087	East Lakonia mountains	Ori Anatolikis Lakonias	Peloponnissos	80,000	B2, B3, C6
124	GR088	Mount Parnitha	Oros Parnitha	Attiki	19,400	B3, C6
125	—	Schinias marsh	Schinias marsh	Attiki	2,100	A4i, B1i, C2
126	—	Mount Ymittos	Oros Ymittos	Attiki	8,000	A3, B3, C6
127	—	Sounio area	Periochi Souniou	Attiki	7,000	B3, C6
128	—	Falkonera islet	Nisida Falkonera	Attiki	140	B1iii, B2, C2, C6
129	—	North, east and south Kithira island	Voria, anatolika kai notia Kithira	Attiki	18,000	A1, A4ii, B1i, B1ii, B1iv, B2, B3, C1, C2, C3, C5, C6
130	—	Antikithira island, Prasouda and Lagouvardos islets	Nisos Antikithira kai Nisides Prasouda kai Lagouvardos	Attiki	2,000	A4ii, B1iii, B2, C2, C6
131	—	Rocky islets of Limnos island	Vrachonisida Nissou Limnou	North Aegean	4	A1, A4ii, B2, C1, C2, C6
132	GR108	Lakes Khortaro and Alyki, Moudros gulf, Diapori fen, and Fakos peninsula	Limnes Khortaro and Alyki, Kolpos Moudrou, Elos Diapori, and Hersonissos Fakos	North Aegean	14,000	A1, A4i, A4ii, B1i, B1iii, B2, C1, C2, C3, C6
133	—	Agios Efstratios island	Nissos Agios Efstratios	North Aegean	4,600	A4ii, B1iii, B2, C2
134	—	South-west peninsula–petrified forest, Lesvos	ND Hersonisos-Apolithomeno Dasos, Lesvos	North Aegean	34,000	A1, A3, B2, B3, C1, C2, C6
135	—	Mount Lepetimnos, Lesvos	Oros Lepetimnos, Lesvos	North Aegean	9,000	A1, B1ii, B2, B3, C1, C2, C3, C6
136	—	Tokmakia islets	Tokmakia	North Aegean	150	A1, C1
137	GR109	Kallonis gulf	Kolpos Kallonis	North Aegean	7,800	A4i, B1i, B2, C2, C6
138	—	Gera Gulf–Dipi and Haramida marshes, Lesvos	Kolpos Geras-Eli Dipi kai Haramida, Lesvos	North Aegean	5,000	A4i, B1i, B2, C2
139	—	Mount Olympos, Lesvos	Oros Olympos, Lesvos	North Aegean	15,000	A3, B3, C2, C6
140	—	Psara and Antipsara islands	Nissi Psara and Antipsara	North Aegean	3,900	A4ii, B1i, B1ii, B1iii, B2, B3, C2, C3, C6
141	—	North Chios	Voreia Chios	North Aegean	38,500	B2, C6
142	—	Venetiko islet	Nisida Venetiko	North Aegean	9	B2
143	—	Ikaria island	Nisos Ikaria	North Aegean	26,700	A1, B2, C1, C6
144	GR111	Fourni islands	Nissi Fourni	North Aegean	5,000	A1, A4ii, B1ii, B1iii, B2, B3, C1, C2, C3, C6
145	GR110	Mount Kerkis	Oros Kerkis	North Aegean	7,000	B3, C6
146	—	Aliki Psilis Ammou lagoon, Samos	Aliki Psilis Ammou, Samos	North Aegean	180	B2, C6
147	—	Central Andros island	Kentriki Andros	South Aegean	22,750	B3, C6
148	—	North-east Tinos island	Vorianatoliki Tinos	South Aegean	12,000	B2
149	GR105	Northern Syros	Voria Syros	South Aegean	4,000	B2
150	—	Serifos island	Nissos Serifos	South Aegean	7,400	B2
151	GR104	Northern Sifnos	Northern Sifnos	South Aegean	3,000	B2
152	—	Western Milos, Antimilos and Polyegos islands	Ditiki Milos, Antimilos and Polyegos	South Aegean	11,000	B2
153	—	Paros, Antiparos and adjacent islands	Paros, Antiparos and giro Nisia	South Aegean	10,000	A1, A4i, B1i, B2, C1, C3, C6
154	GR106	Mounts Dias, Mavrovouni and Koronos	Oros Dias, Oros Mavrovouni, Oros Koronos	South Aegean	27,000	B2, B3, C6
155	—	Mikres Kyklades	Mikres Kyklades	South Aegean	5,700	A4ii, B1ii, B1iii, B2, B3, C2, C3
156	—	Amorgos island	Nisos Amorgos	South Aegean	12,100	A1, B2, C1, C6
157	—	Ios, Sikinos, Folegandros island group Folegandros	Nisiotiko Symplegma Iou–Sikinou–	South Aegean	19,000	A4ii, B1ii, B2, B3, C3, C6
158	—	Christiana islets	Nisides Christiana	South Aegean	200	A4ii, B1iii, B2, C2
159	—	Anafi island	Nissos Anafi	South Aegean	3,900	A4ii, B1ii, B1iii, B2, B3, C2, C3, C6
160	—	Islets of North Dodekanisa	Nisides and Vrachonisides Vorion Dodekanison	South Aegean	4,275	A1, A4ii, B1i, B1ii, B1iii, B2, B3, C1, C2, C6
161	—	North and south Patmos island and islet	Voria kai notia Patmos kai Nisides	South Aegean	3,200	A4ii, B1iii, B2, C2
162	—	North and east Kalimnos, Telendos, and Kalolimnos islands and islets	Voria kai Anatoliki Kalimnos, Telendos, Kalolimnos kai Nisides	South Aegean	4,700	A1, A4ii, B1iii, B2, C1, C2, C6
163	—	Kinaros and Levitha islands and rocky islets	Kinaros, Levitha kai Vrachonisides	South Aegean	1,725	A1, A4ii, B1iii, B2, C1, C2, C6
164	—	Astypalea island and nearby islets	Nissos Astypalea kai giro Nissides	South Aegean	11,500	A1, A4ii, B1ii, B1iii, B2, B3, C1, C2, C3, C6
165	—	Sirna island and nearby islets	Nissos Sirna kai giro Nissides	South Aegean	1,150	A4ii, B1ii, B1iii, B2, B3, C2, C3, C6
166	GR112	Mount Dikios, Cape Louros, Lake Psalidi, and Alyki	Vouno Dikios, Akrotiri Louros, Limni Psalidi, and Alyki	South Aegean	9,916	A1, B2, C1, C6
167	—	Nisiros island and nearby islets	Nisos Nisiros and nearby islets	South Aegean	5,325	A4ii, B1iii, B2, B3, C2, C6
168	GR113	Tilos island	Nissos Tilos	South Aegean	6,300	A1, A4ii, B1iii, B2, C1, C2
169	—	Simi island and adjacent islets	Nissos Simi kai gitonikes nissides	South Aegean	6,500	A1, B2, C1
170	—	Chalki island and nearby islets	Nisos Chalki and nearby islets	South Aegean	3,900	A1, B2, B3, C1, C6
171	—	Central Rodos and the areas of Lindos, Mesanagros and Katavia-Prassonisi	Kentriki Rodos kai Perioches Lindou, Mesanagrou kai Katavias-Prasonisiou	South Aegean	50,000	B2, B3, C6
172	GR103	Saria island and northern Karpathos	Nissos Saria kai Voria Karpathos	South Aegean	8,791	B2, C6
173	—	Islets of Karpathian Sea	Nisides Karpathiou Pelagous	South Aegean	400	A4ii, B1iii, B2, C2, C6
174	GR102	Kasos island	Nissos Kasos	South Aegean	8,000	A1, A4ii, B1iii, B2, C1, C2, C6
175	—	Tiganis peninsula and the islands of Gramvouses and Pontikonisi	Chersonisos Tiganis kai Nissi Gramvouses and Pontikonisi	Crete	5,544	A4ii, B1iii, B2, C2
176	GR090	Rodopos peninsula	Chersonissos Rodopou	Crete	7,000	B2, C6
177	—	Agioi Theodori islet	Nisida Agioi Theodori	Crete	80	A4ii, B1iii, B2, C2

Table 1 ... continued. Summary of Important Bird Areas in Greece. 196 IBAs covering 34,332 km²

IBA code	1989 code	International name	National name	Administrative region	Area (ha)	Criteria (see p. 11)
178	GR089	Mount Koutroulis, Mount Ag. Dikaios and Modia plateau	Oros Koutroulis, Oros Ag. Dikaios and Oropedio Modia	Crete	20,000	B2, C6
179	GR091	Lefka mountains	Lefka Ori	Crete	54,000	B2, C6
180	GR092	Lake Kourna, Almyrou delta, and Georgioupolis beach	Limni Kourna, Delta Almyrou, and Paralia Georgioupolis	Crete	4,430	C7
181	—	Gavdos and Gavdopoula islands	Nissi Gavdos kai Gavdopoula	Crete	3,600	B1i, B2, C2
182	—	Krioneritis mountain	Oros Krioneritis	Crete	15,000	B2, B3, C6
183	GR093	Mount Kedros and Kourtaliotiko gorge	Oros Kedros and Kourtaliotiko Farangi	Crete	12,000	B2
184	GR094	Prassiano gorge	Faranghi Prassion	Crete	975	B2, C6
185	GR095	Mount Idi	Oros Idi	Crete	44,000	B2, B3, C6
186	—	Geropotamos estuary	Ekvoli Geropotamou	Crete	450	A4i, B1i, C2
187	GR096	Mount Asterousia (Kofinas)	Asterousia Ori (Kofinas)	Crete	29,000	B2, B3, C6
188	GR097	Mount Jouchtas	Jouchtas Oros	Crete	600	B2, C6
189	GR098	Dia island	Nissos Dia	Crete	1,250	A4ii, B1iii, B2, C2, C6
190	GR099	Mount Dikti	Oros Dikti	Crete	54,000	B2, C6
191	GR100	Mount Thryptis and Mount Ornon	Ori Thryptis and Ornon	Crete	15,000	B2, C6
192	GR101	Dionisiades islands	Nissi Dionisiades	Crete	506	A1, A4ii, B1ii, B1iii, B2, C1, C2, C6
193	—	North-eastern edge of Crete	Vorioanatoliko Akro Kritis	Crete	7,500	A1, A4ii, B1iii, B2, C1, C2
194	—	Mount Zakrou	Ori Zakrou	Crete	13,000	B2, B3, C6
195	—	Kavalli islets	Nisides Kavalli	Crete	10	B1iii, B2, C2
196	—	Koufonisi island	Nisos Koufonisi	Crete	525	B2

Sites identified in the previous inventory of IBAs in Europe (Grimmett and Jones 1989) but no longer considered to be IBAs
GR008 Vouna Limnia (Limnia Hills); GR052 Limni Limnopoula (Lake Limnopoula); GR054 Vouna Zalongou (Zalongos Mountains); GR059 Kalon Oros (Mount Kalon); GR067 Ormos Sourpis (Sourpis Bay); GR082 Limni Taka (Lake Taka).

Map 1. Location, area and criteria category of Important Bird Areas in Greece.

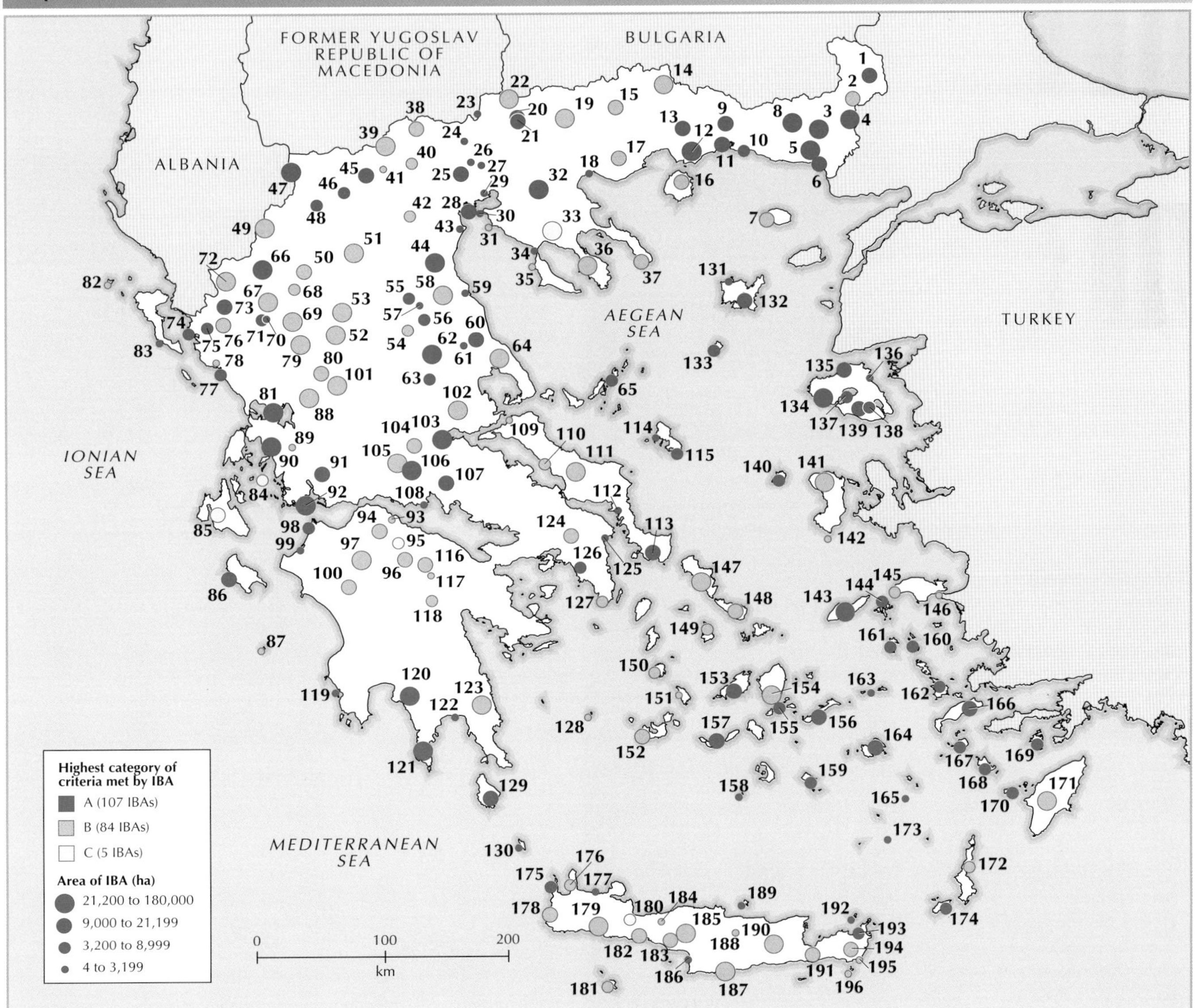

Table 2. Important Bird Areas in Greece that are important for species of global conservation concern (meeting criterion A1).

Species	IBA code
Phalacrocorax pygmeus Pygmy Cormorant	001, 006, 009, 010, 011, 012, 013, 018, 020, 023, 025, 028, 029, 032, 045, 046, 047, 048, 059, 074
Pelecanus crispus Dalmatian Pelican	006, 010, 011, 012, 018, 020, 023, 024, 028, 032, 046, 047, 048, 074, 081
Anser erythropus Lesser White-fronted Goose	006, 010, 011, 012, 020
Branta ruficollis Red-breasted Goose	006
Aythya nyroca Ferruginous Duck	012, 046, 048, 061, 081, 091, 103, 129
Oxyura leucocephala White-headed Duck	011, 020
Aegypius monachus Cinereous Vulture	003, 004, 005, 008
Aquila clanga Greater Spotted Eagle	004, 005, 006, 012, 020, 021, 074, 081
Aquila heliaca Imperial Eagle	005
Falco naumanni Lesser Kestrel	012, 026, 027, 046, 055, 056, 057, 060, 061, 062, 063, 070, 071, 073, 075, 092, 099, 108, 120, 132, 134, 144, 162, 166, 192, 193
Numenius tenuirostris Slender-billed Curlew	006, 010, 011, 012, 028, 029, 030, 034, 074, 081, 083, 092, 103, 166
Larus audouinii Audouin's Gull	065, 114, 129, 131, 136, 144, 153, 156, 160, 162, 163, 164, 168, 169, 170, 174
Emberiza cineracea Cinereous Bunting	115, 134, 135, 143

Table 3. Important Bird Areas in Greece that support important numbers of one or more congregatory species (i.e. meeting criteria A4 and/or B1). IBAs meeting both criteria A4 and B1 for the species are shown in **bold**. IBAs meeting only criterion B1 for the species concerned, and not A4, are shown in normal type. For key to 'Season', see p. 7.

Species	Season	IBA code
Tachybaptus ruficollis Little Grebe	W	**011**
Podiceps cristatus Great Crested Grebe	W	032
Podiceps nigricollis Black-necked Grebe	W	**010**, **011**, **081**
Calonectris diomedea Cory's Shearwater	R	129
	B	159, 192
	N	087
Puffinus yelkouan Yelkouan Shearwater	R	016, **129**, 135
	B	065, 140, **144**, **155**, **157**, 159, **160**, 164, 165
	W	011, 012
	N	**113**
	U	037
Phalacrocorax carbo Cormorant	R	020
	B	047
	W	**006**, 010, 011, 012, 020, 031, 070, 074, 081, 083, 089, 092, 119
Phalacrocorax aristotelis Shag	R	007, 012, 086, 129, 140, 160, 181
Phalacrocorax pygmeus Pygmy Cormorant	R	**020**, **047**
	W	**006**, **010**, **011**, **012**, **020**, **025**, **028**, **029**, **048**
Pelecanus onocrotalus White Pelican	P	**006**, **020**
Pelecanus crispus Dalmatian Pelican	R	**081**
	B	**047**
	W	**006**, **010**, **011**, **012**, **020**, **023**, **032**
	P	**011**, **018**, **020**, **024**, **028**, **032**, **048**
	N	**046**
Nycticorax nycticorax Night Heron	B	**020**, **028**
Ardeola ralloides Squacco Heron	B	**020**, **028**, **081**
Egretta garzetta Little Egret	B	011, 012, **020**, **028**, 081, 092
	W	119
	P	029, 048, 077, **081**, 093, 098, **099**, 109, 117, **119**, 180
	N	010
Egretta alba Great White Egret	B	**081**
	W	**010**, **011**, **012**, **020**, **032**, **061**, **074**, **083**, **099**, **119**
Ciconia ciconia White Stork	P	**006**
Plegadis falcinellus Glossy Ibis	P	**006**, **010**, **011**, **020**, **074**, **077**, **081**, **092**, **099**, **119**, **122**, **125**, **186**
Platalea leucorodia Spoonbill	R	**010**, **020**
	B	028
	P	**006**
Phoenicopterus ruber Greater Flamingo	W	**006**, **011**, **132**, **137**
	P	**010**
	N	**032**
Cygnus olor Mute Swan	W	006, 010, 011, 012, 109
Anser erythropus Lesser White-fronted Goose	W	006
Anser anser Greylag Goose	W	020
Branta ruficollis Red-breasted Goose	W	**006**
Tadorna tadorna Shelduck	W	006, 010, 011, 020, 043, 132
Anas penelope Wigeon	W	006, 010, **081**, 092
Anas strepera Gadwall	W	006
Anas crecca Teal	W	**006**, **020**, 061, **081**
Anas platyrhynchos Mallard	W	020
Anas acuta Pintail	W	**081**
Anas clypeata Shoveler	W	**081**
Aythya ferina Pochard	W	**011**, **020**, 032, **081**, 092
Aythya fuligula Tufted Duck	W	047
Mergus merganser Goosander	W	048
Oxyura leucocephala White-headed Duck	W	**011**
Aegypius monachus Cinereous Vulture	R	004
	P	005
Falco naumanni Lesser Kestrel	B	**062**, 063, 108, **132**
Falco eleonorae Eleonora's Falcon	B	**065**, 128, **130**, 131, **133**, **140**, **144**, **155**, **158**, **159**, **160**, **161**, **162**, **163**, **164**, **165**, **167**, **168**, **173**, **174**, **175**, **177**, **189**, **192**, **193**, 195
Fulica atra Coot	W	011, **081**
Haematopus ostralegus Oystercatcher	R	006, 092
	B	028, 081
	W	006
Himantopus himantopus Black-winged Stilt	B	028, 029, 034, 035, **081**, **098**
	P	032
Recurvirostra avosetta Avocet	W	006, **010**, 043, **103**
	P	**020**, **032**, **043**
Glareola pratincola Collared Pratincole	B	**006**, **010**, **012**, **028**, 029, 031, 034, 043, **081**, **092**
	P	119
Charadrius alexandrinus Kentish Plover	R	**081**, 092
	B	028
	W	**081**
	P	030, **098**
Calidris minuta Little Stint	W	**006**
	P	**081**, **098**
Calidris ferruginea Curlew Sandpiper	P	**081**
Limosa limosa Black-tailed Godwit	P	028
Tringa totanus Redshank	W	081
Tringa stagnatilis Marsh Sandpiper	P	**006**, **098**, **119**
Tringa glareola Wood Sandpiper	P	**119**
Larus melanocephalus Mediterranean Gull	B	**028**
	P	**138**
	N	**030**
Larus minutus Little Gull	P	**010**
Larus genei Slender-billed Gull	W	**103**
Larus cachinnans Yellow-legged Gull	R	**153**
	W	**028**
Gelochelidon nilotica Gull-billed Tern	B	**081**, **092**
	P	093, 119
Sterna caspia Caspian Tern	P	089
Sterna albifrons Little Tern	B	006, **012**, **081**, **092**
Chlidonias hybridus Whiskered Tern	B	020

Table 4. Species of European conservation concern and species listed on Annex I of the EC Birds Directive with significant breeding populations at IBAs in Greece (meeting any IBA criteria).

Species [1]	Minimum national breeding population (pairs) [2]	Proportion (%) of national population breeding at all IBAs in Greece
Calonectris diomedea Cory's Shearwater	5,000	32
Puffinus yelkouan Yelkouan Shearwater	8,000	22
Phalacrocorax aristotelis Shag	1,000	81
Phalacrocorax pygmeus Pygmy Cormorant	1,250	60
Pelecanus onocrotalus White Pelican	100	100
Pelecanus crispus Dalmatian Pelican	545	100
Ixobrychus minutus Little Bittern	600	57
Nycticorax nycticorax Night Heron	500	100[3]
Ardeola ralloides Squacco Heron	200	100[3]
Egretta garzetta Little Egret	1,055	100[3]
Egretta alba Great White Egret	10	100[3]
Ciconia ciconia White Stork	2,000	30
Platalea leucorodia Spoonbill	113	100[3]
Tadorna ferruginea Ruddy Shelduck	35	100[3]
Aythya nyroca Ferruginous Duck	200	54
Gypaetus barbatus Lammergeier	12	100[3]
Neophron percnopterus Egyptian Vulture	150	39
Gyps fulvus Griffon Vulture	400	99
Aegypius monachus Cinereous Vulture	90 individuals	100
Circaetus gallicus Short-toed Eagle	300	48
Accipiter brevipes Levant Sparrowhawk	1,000	14
Buteo rufinus Long-legged Buzzard	150	66
Aquila pomarina Lesser Spotted Eagle	60	100[3]
Aquila chrysaetos Golden Eagle	140	51
Hieraaetus pennatus Booted Eagle	100	31
Hieraaetus fasciatus Bonnelli's Eagle	85	81
Falco naumanni Lesser Kestrel	2,950	100[3]
Falco tinnunculus Kestrel	5,000	8
Falco eleonorae Eleonora's Falcon	2,500	100[3]
Falco biarmicus Lanner	30	43
Falco peregrinus Peregrine	100	100[3]
Alectoris chukar Chukar	1,000	6
Alectoris graeca Rock Partridge	2,000	15
Himantopus himantopus Black-winged Stilt	500	100[3]
Recurvirostra avosetta Avocet	300	100[3]
Burhinus oedicnemus Stone Curlew	300	49
Glareola pratincola Collared Pratincole	500	100[3]
Charadrius alexandrinus Kentish Plover	500	100[3]
Hoplopterus spinosus Spur-winged Plover	32	100[3]
Larus melanocephalus Mediterranean Gull	2,000	100[3]
Larus audouinii Audouin's Gull	750	87
Gelochelidon nilotica Gull-billed Tern	50	100[3]
Sterna sandvicensis Sandwich Tern	50	100[3]
Sterna albifrons Little Tern	1,000	100[3]
Chlidonias hybridus Whiskered Tern	250	100[3]
Tyto alba Barn Owl	2,000	2
Otus scops Scops Owl	5,000	5
Bubo bubo Eagle Owl	200	13
Athene noctua Little Owl	5,000	6
Strix aluco Tawny Owl	10,000	4
Merops apiaster Bee-eater	2,000	9
Picus viridis Green Woodpecker	5,000	5
Dendrocopos syriacus Syrian Woodpecker	5,000	6
Dendrocopos medius Middle Spotted Woodpecker	2,000	8
Dendrocopos leucotos White-backed Woodpecker	200	8
Calandrella brachydactyla Short-toed Lark	20,000	10
Galerida cristata Crested Lark	40,000	2
Anthus campestris Tawny Pipit	10,000	1
Oenanthe hispanica Black-eared Wheatear	30,000	3
Monticola saxatilis Rock Thrush	10,000	8
Monticola solitarius Blue Rock Thrush	5,000	12
Hippolais pallida Olivaceous Warbler	100,000	1
Hippolais olivetorum Olive-tree Warbler	5,000	7
Sylvia cantillans Subalpine Warbler	100,000	<1
Sylvia melanocephala Sardinian Warbler	500,000	<1
Sylvia rueppelli Rüppell's Warbler	3,000	11
Sylvia hortensis Orphean Warbler	3,000	6
Phylloscopus bonelli Bonelli's Warbler	10,000	5
Ficedula semitorquata Semi-collared Flycatcher	1,000	5
Parus lugubris Sombre Tit	10,000	4
Sitta krueperi Krüper's Nuthatch	200	16
Sitta neumayer Rock Nuthatch	10,000	11
Lanius collurio Red-backed Shrike	20,000	2
Lanius minor Lesser Grey Shrike	2,000	9
Lanius senator Woodchat Shrike	5,000	10
Lanius nubicus Masked Shrike	600	10
Pyrrhocorax pyrrhocorax Chough	1,000	28
Corvus monedula Jackdaw	100,000	2
Emberiza cirlus Cirl Bunting	50,000	1
Emberiza cia Rock Bunting	10,000	3
Emberiza cineracea Cinereous Bunting	100	100[3]
Emberiza hortulana Ortolan Bunting	20,000	4
Emberiza caesia Cretzschmar's Bunting	5,000	22
Emberiza melanocephala Black-headed Bunting	100,000	1
Miliaria calandra Corn Bunting	300,000	1

1. Only those species of European conservation concern (see Box 1, p. 12) that meet IBA criteria in Greece are listed, together with those species listed on Annex I of the EC Birds Directive that fulfil criterion C6 in IBAs in Greece.
2. Data are taken from the BirdLife/EBCC European Bird Database 1998 (Heath and Borggreve 2000).
3. The percentage of the national population in IBAs exceeds 100%. Usually this is because the national population estimate has not been updated recently whilst the IBA population estimate has been recently updated with new data as a result of comprehensive surveys of IBAs themselves. Also, the individual site count for a species may be the maximum or average over recent years, and summing these may record more birds than are present nationally in any single year.

valleys, flood-plains and the deltas of major rivers draining the mountains.

Greece has 196 Important Bird Areas (IBAs) covering 34,332 km² (approximately 26%) of its surface area (Table 1). One hundred and twelve of these are on mainland Greece and 84 are on the islands (Map 1). The previous pan-European inventory (Grimmett and Jones 1989) identified 113 IBAs in Greece, six of which do not meet current IBA criteria and have been dropped (Table 1). Two sites from the previous inventory have been merged to form a single IBA whilst one site has been split into three, and another site split into two separate IBAs (Table 1).

ORNITHOLOGICAL IMPORTANCE

One hundred and forty-one species of European conservation concern (SPECs) regularly breed in Greece (Tucker and Heath 1994). Among these, four are globally threatened species (*Pelecanus crispus*, *Aythya nyroca*, *Falco naumanni* and *Larus audouinii*), and four are near-threatened (*Phalacrocorax pygmeus*, *Haliaeetus albicilla*, *Aegypius monachus* and *Emberiza cineracea*) (Collar *et al.* 1994). Ninety-three SPECs have an unfavourable conservation status in Europe (Tucker and Heath 1994). Fifty-five SPECs regularly occur on passage or overwinter and include the globally threatened *Anser erythropus*, *Branta ruficollis*, *Oxyura leucocephala*, *Aquila clanga*, *Aquila heliaca*, and *Numenius tenuirostris* and the near-threatened *Gallinago media*.

A total of 107 IBAs in Greece were selected under one or more A criteria; a further 84 meet B (but not A) criteria, and five C (but not A or B) criteria (Table 1). Nineteen IBAs hold more than ten species meeting criteria: 003, 004, 005, 006, 008, 010, 011, 012, 020, 028, 032, 060, 062, 081, 090, 092, 112, 120, 134. Seventy-three sites are selected as IBAs on the basis of holding internationally important populations of globally threatened or near-threatened species, meeting criterion A1 (Table 2). No sites are selected for *Gallinago media* as passage and wintering numbers are very low.

Thirty-two wetland sites are selected as IBAs as they hold more than 1% of the biogeographic population of those species listed in Table 3 and therefore meet criterion A4i. A further 49 sites hold greater than 1% of the flyway populations of one or more species of waterbird, meeting the criterion B1i (Table 3). A number of sites are also important for breeding seabirds, holding either more than 1% of the global population of a seabird species (meeting criterion A4ii) or more than 1% of a distinct population of a seabird species (B1ii) (Table 3).

Thirteen IBAs are important for their assemblages of breeding species characteristic of the Mediterranean biome, meeting criterion A3 (003, 004, 005, 060, 077, 086, 090, 112, 120, 121, 126, 134, 139). Four IBAs are important for breeding species characteristic of the Eurasian high montane biome (044, 066, 106, 107).

The majority of Greek IBAs (158) are important because they support significant breeding populations of the SPECs listed in Table 4, meeting criteria B2 and B3. Greece holds more than 10% of the European breeding populations of the following SPECs, for which several IBAs have been identified: *Pelecanus crispus* (two IBAs), *Accipiter brevipes* (11 IBAs), *Falco naumanni* (10 IBAs), *Falco eleonorae* (40 IBAs), *Falco biarmicus* (10 IBAs), *Monticola saxatilis* (nine IBAs), *Monticola solitarius* (10 IBAs), *Hippolais olivetorum* (six IBAs), *Sylvia rueppelli* (11 IBAs), *Ficedula semitorquata* (five IBAs), *Sitta neumayer* (10 IBAs), *Emberiza cineracea* (five IBAs) and *Emberiza caesia* (19 IBAs). Five IBAs (006, 010, 043, 092, 103) meet the B3 criterion for wintering SPECs, all for *Recurvirostra avosetta*.

Those species present in Greece that are not adequately covered by the IBA inventory, due to the dispersed nature of their distribution, include *Accipiter brevipes, Monticola saxatilis, Hippolais pallida, Hippolais olivetorum, Sylvia melanocephala, Sylvia rueppelli* and *Ficedula semitorquata* as well as other passerines. The coverage of *Calonectris diomedea, Puffinus yelkouan* and many passerines, by the IBA network, cannot be precisely estimated as a result of a lack of quantitative data.

HABITATS

Arable agricultural habitats are the most widespread in Greece, covering 30% of the land area; forested habitats occupy 25%; partly forested land 24%; pasture/grazing land 13% and 8% of the land is covered by urban and industrial areas, wetlands, bare ground and rocky areas (Handrinos and Akriotis 1997). Greek territory includes large marine areas.

Scrub is the most abundant habitat in Greek IBAs, being found in 81% of sites, with total cover exceeding 50% in 31% of IBAs (Figure 1). Scrub habitats are often associated with extensive grazing systems. Communities include typical Mediterranean assemblages such as phrygana and maquis which support birds such as *Alectoris graeca, Sylvia rueppelli* and *Emberiza caesia.*

Sixty-four percent of IBAs contain rocky habitats although in only 5% of IBAs do these exceed 50% of the total area (Figure 1). Inland rocky habitats include cliffs important for breeding raptors such as *Gyps fulvus*, *Hieraaetus fasciatus* and *Falco peregrinus*, and alpine areas important for species such as *Monticola saxatilis* and *Pyrrhocorax graculus*. Coastal cliffs support breeding species such as *Puffinus yelkouan, Phalacrocorax aristotelis, Falco eleonorae* and *Larus audouinii*.

Agricultural habitats are also widespread in IBAs, including arable farmland, perennial crops and olive groves. On the Thessaly Plain (062) arable land is particularly important for *Falco naumanni*.

Forested habitats, predominantly *Quercus*, *Pinus* and *Abies* forests, are found in 58% of IBAs, in 16% occupying over 50% of the total area (Figure 1). Typical species include *Ciconia nigra, Pernis apivorus, Phylloscopus bonelli* and woodpecker species.

IMPACTS ON IBAs – LAND-USE AND THREATS

The dominant land-use in Greek IBAs is agriculture which occurs in 90% of IBAs (Figure 2). This includes both arable areas, perennial crops and large areas that are used for livestock-grazing. Agricultural intensification is a serious threat to many IBAs, occurring in over 50% and regarded as a high-impact threat in 27% (Figure 3). Land redistribution schemes (amalgamation of fields and the consequent loss of hedges and isolated trees) are taking place, or are planned by the Ministry of Agriculture, all over Greece, even in the close vicinity of Ramsar Sites. The effects of the intensive use of agri-chemicals are a problem for many birds, however, the impacts of their widespread use have not yet been adequately studied and evaluated. In some areas, particularly in Mediterranean scrub communities, overgrazing is a threat.

Hunting is a widespread activity in Greece occurring in almost all IBAs, although as relatively little land is specifically set aside for hunting this is not indicated by Figure 2. It is a threat in over 80% of IBAs and is a high-impact threat in 36% (Figure 3). Hunting is not permitted in Wildlife Refuges, the core areas of the ten National Parks, or parts of Ramsar Sites. Hunting is indiscriminate of species and in addition to direct mortality effects, causes serious disturbance to birds, particularly in wetlands. The construction of roads to remote areas has dramatically increased the negative effects of indiscriminate shooting for many species.

In addition to hunting, the widespread illegal use of poisoned baits, in order to control populations of carnivorous mammals, has led to the local extirpation of many raptors, particularly vultures. Seven *Aegypius monachus* were poisoned in 1995 at Dadia (003) and a small population of the same species at Mount Olympos (044) was also poisoned.

Recreation and tourism are both widespread land-uses and threats in IBAs, in particular around coastal areas (Figures 2 and 3). Associated developments and activities also often occur in and around wetlands, the most threatened habitat in Greece. Additionally, wetlands are threatened by expansion and intensification of agriculture, mismanagement of freshwater sources, heavy hunting pressure, pollution, drainage, infilling, housing development, infrastructure developments and the creation of large fish-ponds.

Figure 1. Habitats at Important Bird Areas in Greece (see Appendix 3 for definitions of habitats).

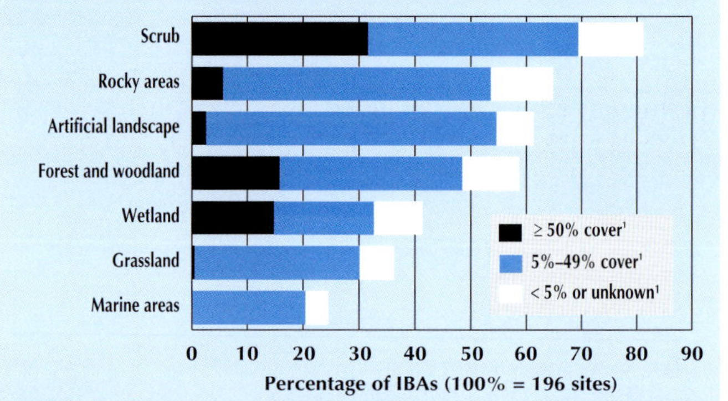

1. percentage of area of individual IBA covered by habitat

Figure 2. Land-uses at Important Bird Areas in Greece (see Appendix 3 for definitions of land-uses).

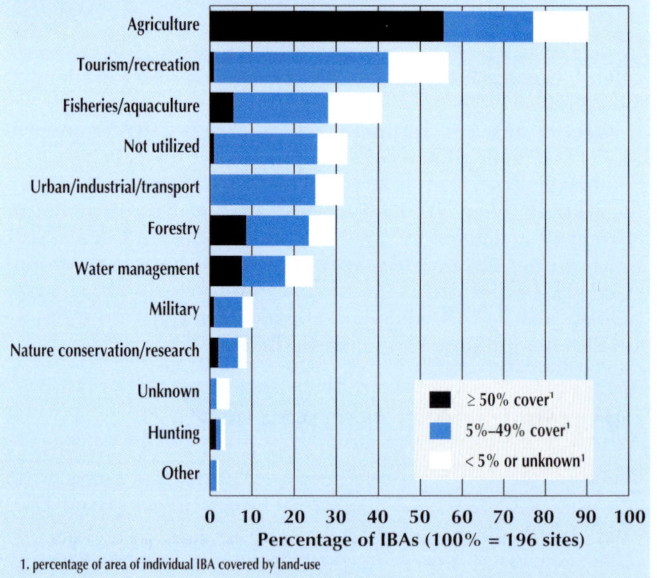

1. percentage of area of individual IBA covered by land-use

Figure 3. Threats at Important Bird Areas in Greece (see Appendix 3 for definitions of threat types and impact categories).

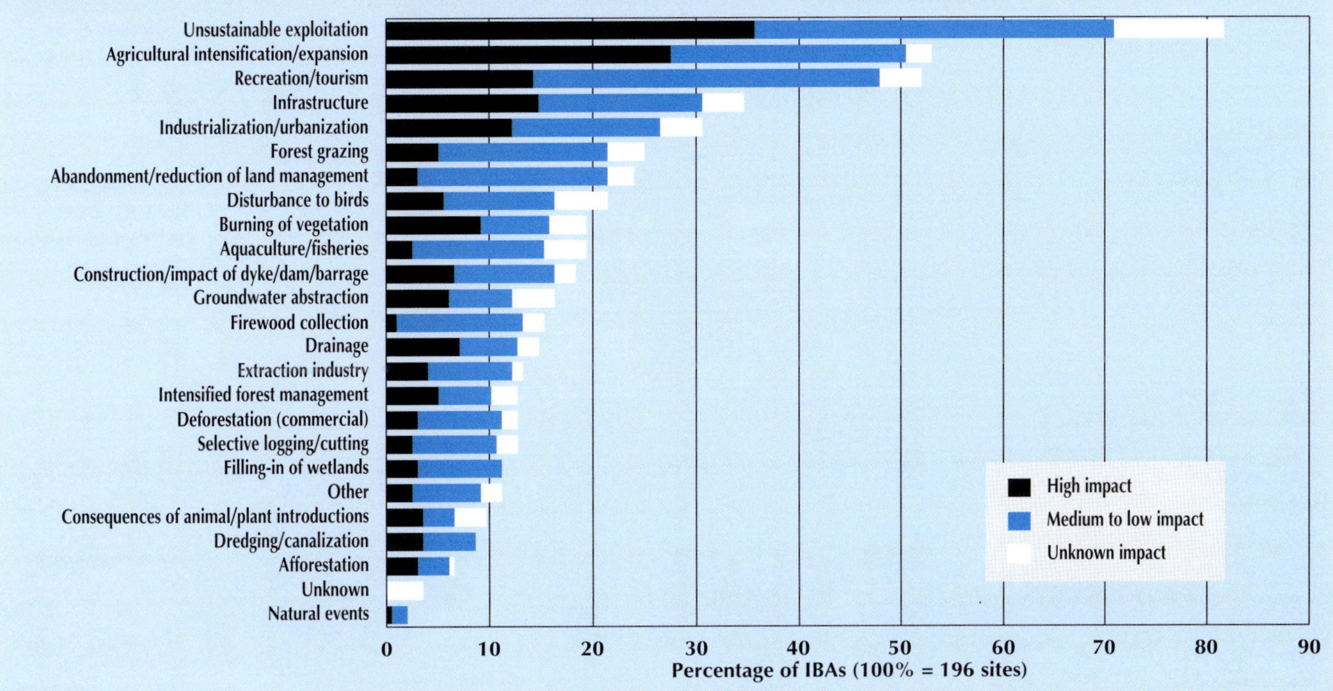

Lowland forests are threatened by fire and in many areas by increasingly intensified forest management, in particular the construction of new forest roads. The latter causes fragmentation, disturbance and provides access to hunters. Forestry practices are not restricted in order to benefit bird conservation.

The abandonment of traditional land-uses and management practices is a threat in over 20% of IBAs (Figure 3). In particular, traditional pasture lands and arable cultivation have been abandoned resulting in afforestation and consequent changes in bird communities.

PROTECTION STATUS

National protection

Fifty-nine percent of IBAs are, at least in part, under some sort of national protection (Figure 4). However, only 3% have over 90% of their area under protection. Caution should be used in interpreting these data as the area of overlap between IBAs and protected areas is unknown for some sites in which case a 'partial' overlap has been applied as a default option. Table 5 details overlaps between IBAs and protected areas.

The Ministry of Environment, Physical Planning and Public Works is the coordinating body for all environmental activities, whilst the Greek Forestry Service of the Ministry of Agriculture is responsible for forested areas and the protected areas established under the Forest Code. The classification of protected areas in Greece is complicated. Protected Areas have been established under the Presidential Decree 996/1971, the Legislative Decree 86/1969, the Public Law 177/1975 and the Public Law 2637/1988, which enable the Ministry of Agriculture to establish National Parks, Aesthetic Forests, Protected Natural Monuments, Wildlife Refuges, Breeding Stations, Controlled Hunting Areas and Protected Forests.

However, since 1986 Public Law 1650/1986 has provided a more complete classification of protected areas, identifying Areas of Strict Protection, Areas of Protection of Nature, National Parks (including Marine Parks and existing National Parks), Protected Natural Formations, Protected Landscapes and Features of the Landscape, and Areas of Eco-development. However, this law is yet to be fully implemented as a result of disagreements between the Ministry of Environment, Physical Planning and Public Works and the Ministry of Agriculture on the legislation used and the responsibilities of each Ministry. Therefore, for the purposes of this inventory we primarily follow protected-area designations relating to the earlier set of laws (see list below and Table 5):

1. **National Park**
2. **Marine Park**

Figure 4. The national protection status of Important Bird Areas in Greece.

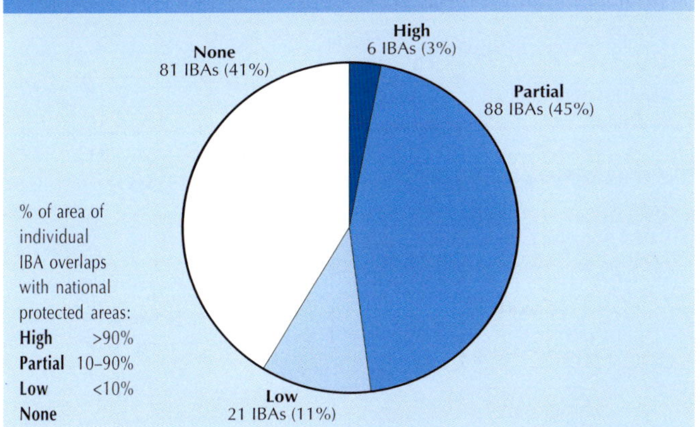

Total area of overlap between IBA network in Greece and national protected-area system (see Table 5 for categories) = 2,667–3,723 km² (8–11% of total IBA area).

Figure 5. The international protection status of Important Bird Areas in Greece.

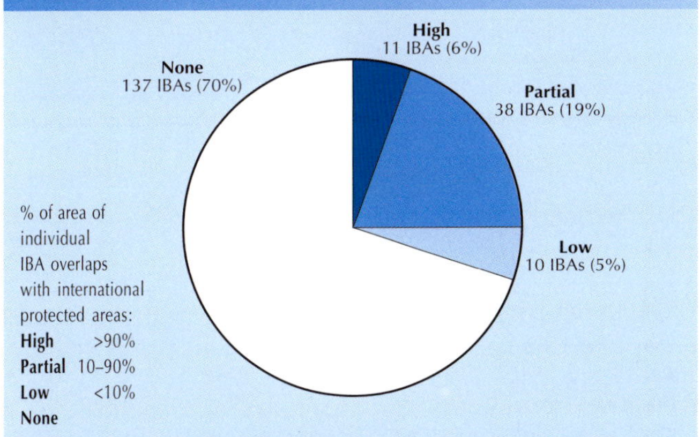

Total area of overlap between IBA network in Greece and international protected-area system (see Table 5 for categories) = 4,632–4,638 km² (13–14% of total IBA area).

Table 5. Protection status of Important Bird Areas in Greece.
A tick (✔) indicates that an IBA overlaps with a protected area (to any extent).

IBA code	International name	National: National Park	National: Marine Park	National: Protected Area	National: Natural Monument	National: Wildlife Refuge	National: Aesthetic Forest	National: Controlled Hunting Area	National: Breeding Station	International: Ramsar Site	International: World Heritage Site	International: Biosphere Reserve	International: Biogenetic Reserve	International: Special Protection Area
001	Northern Evros and Ardas riparian forest					✔								
002	Asvestades–Vrisika area					✔								
003	Dadias–Dereiou–Aisimis forest					✔								
004	Protected area of Dadia–Lefkimi–Soufli forest					✔								✔
005	South forest complex of Evros prefecture					✔								
006	Evros delta			✔		✔				✔				✔
007	Samothraki island					✔								
008	Filiouri river valley and east Rodopi mountains					✔								
009	Kompsatos valley													
010	Lake Mitrikou (Ismarida)			✔		✔				✔				✔
011	Porto Lagos, Lake Vistonis, and coastal lagoons (Lakes of Thrace)			✔		✔				✔				✔
012	Nestou delta and coastal lagoons			✔		✔				✔				✔
013	Nestos gorge					✔	✔							✔
014	Central Rodopi mountains and Nestos valley				✔	✔							✔	✔
015	Mount Falakro													
016	North and north-east Thasos island					✔								
017	Mount Pangaio					✔								
018	Strymon delta													
019	Mount Kouskouras					✔								✔
020	Lake Kerkini									✔				✔
021	Mount Kroussia (or Mavrovouni)													
022	Mount Beles (Kerkini)													
023	Lake Doirani				✔									
024	Artzan marshes													
025	River Axios					✔								
026	Area of Anthofito													
027	Lake Pikrolimni													
028	Axios, Loudias, and Aliakmon estuaries			✔		✔				✔				✔
029	Gallikos estuary and Kalohori lagoon													
030	Alyki Angelochoriou lagoon (Megalou Emvolou)													
031	Epanomi lagoon					✔								
032	Lakes Volvi–Langada (or Koronia–Ag. Vasiliou) and Rentina gorge (or Makenonika Tempi)					✔				✔				✔
033	Mount Cholomontas					✔		✔						✔
034	Agios Mamas marsh													
035	Nea Fokea marshes													
036	Mount Itamos					✔								
037	Mount Athos										✔			
038	Mounts Tzena and Pinovo					✔								✔
039	Mount Voras (Kaimaktchalan)				✔	✔							✔	✔
040	Apsalos and Moglenitsa gorges													
041	Agra artificial lake					✔								✔
042	Aliakmon gorge													
043	Alyki Kitros lagoon					✔								✔
044	Mount Olympos	✔				✔						✔	✔	✔
045	Lake Vegoritis and Lake Petron													
046	Lake Chimaditis and Lake Zazaris					✔								✔
047	Lake Mikri Prespa and Lake Megali Prespa	✔				✔				✔			✔	✔
048	Lake Kastoria (Orestiada)					✔								
049	Mount Gramos				✔	✔							✔	
050	Mounts Orliakas and Tsourgiakas													
051	Mount Vourinos					✔								
052	Kerketion mountain (Koziakas)							✔						
053	Antikhassia mountains and Meteora					✔					✔			✔
054	Kalamaki gorge–Mount Zarkou					✔								
055	Elassona area													
056	Tyrnavos area					✔								
057	Mati of Tirnavos													
058	Mounts Kato Olympos and Ossa, and Tembi ravine					✔	✔	✔						✔
059	Pinios delta					✔								
Subtotal of IBAs		2	0	5	4	37	2	3	0	8	2	1	5	21

Table 5 ... continued. Protection status of Important Bird Areas in Greece.
A tick (✔) indicates that an IBA overlaps with a protected area (to any extent).

		National								International				
IBA code	International name	National Park	Marine Park	Protected Area	Natural Monument	Wildlife Refuge	Aesthetic Forest	Controlled Hunting Area	Breeding Station	Ramsar Site	World Heritage Site	Biosphere Reserve	Biogenetic Reserve	Special Protection Area
060	Mavrovouni mountain, Larissa					✔								✔
061	Reservoirs of former Lake Karla													
062	Thessaly plain					✔								
063	Farsala area					✔								
064	Mount Pilion					✔								
065	National Marine Park of Alonissos (North Sporades)		✔		✔	✔								✔
066	Timfi (Gamila) and Smolikas mountains	✔				✔								✔
067	Central part of Zagori and Mount Mitsikeli					✔								
068	Valia Kalda (Pindos National Park)	✔				✔							✔	✔
069	Peristeri mountain					✔								
070	Lake Pamvotida (Ioanninon)													
071	Ioannina city and neighbouring area													
072	Area of Oriokastro, Lake Delvinaki, Meropis forest, Gormou valley and Mounts Kasidiaris and Douskon					✔								
073	Mounts Tsamanta, Filiaton, Pharmakovouni, and Megali Rahi					✔								
074	Kalamas estuary					✔								✔
075	Kalamas gorge					✔								
076	Mount Paramithias					✔								
077	Acheron delta and gorge													
078	Lake Kalodiki, Margariti and Karteri marshes					✔								
079	Athamanika mountains (Tzoumerka)					✔								
080	Acheloos valley					✔								✔
081	Amvrakikos gulf					✔				✔				✔
082	Diapontia islands (Othoni, Erikoussa, Mathraki islands and rocky islets)													
083	Lagoons of Kerkyra													
084	Echinades islands, Kalamos and Atokos													✔
085	The Agia Dinati and Kokkini Rachi mountains of Kefallinia	✔				✔								✔
086	West and north Zakynthos													
087	Strofades islands													
088	Valtou mountains					✔								
089	Lake Amvrakia					✔								
090	Akarnanika mountains					✔								
091	Lakes Trichonida and Lysimachia													
092	Mesolongi and Aetoliko lagoons, and Acheloos and Evinos estuaries				✔	✔				✔			✔	✔
093	Alyki lagoon, Aegion					✔								
094	Barbas and Klokos mountains and Selinous gorge													✔
095	Vouraikos gorge including Kalavrita					✔								✔
096	Mount Helmos (Aroania)					✔								
097	Mount Erimanthos					✔								
098	Kalogria lagoon, Strofilia forest, and Lamia marshes									✔				✔
099	Kotychi lagoon									✔				✔
100	Mount Foloi													
101	Delimidi and Fteri mountains, Agrafa					✔								
102	Mount Othrys					✔								
103	Sperchios valley and delta–Maliakos gulf					✔								✔
104	Mount Iti	✔				✔							✔	✔
105	Mount Vardousia					✔		✔						
106	Mount Ghiona, Reka ravine, Lazorema, and Vathia Lakka					✔		✔						
107	Southern and eastern part of Mount Parnassos	✔				✔								✔
108	Galaxidi					✔								
109	Megalo and Mikro Livari lagoons, Istieas				✔									
110	Kandili mountain					✔								
111	Dirfi, Xirovouni, Skotini, Mavrovouni, Alokteri and Ortari mountains and the vicinity of Kimi					✔	✔							
112	Lake Distos					✔								
113	Mount Ochi and surrounding area													
114	North-west Skyros and Ag. Fokas islets													
115	Mount Kohilas, Skyros island													
116	Mount Ziria (Killini)					✔								
117	Lake Stymphalia					✔								
	Subtotal of IBAs	7	1	5	7	77	3	5	0	12	2	1	8	38

Table 5 ... continued. Protection status of Important Bird Areas in Greece. A tick (✔) indicates that an IBA overlaps with a protected area (to any extent).

IBA code	International name	National: National Park	National: Marine Park	National: Protected Area	National: Natural Monument	National: Wildlife Refuge	National: Aesthetic Forest	National: Controlled Hunting Area	National: Breeding Station	International: Ramsar Site	International: World Heritage Site	International: Biosphere Reserve	International: Biogenetic Reserve	International: Special Protection Area
118	Mounts Artemisio and Lirkio													
119	Divari Pilou lagoon (Gialova)					✔								
120	Mount Taigetos					✔								
121	Southern Mani, Mount Sanghias and Cape Tenaro													
122	Evrota delta													
123	East Lakonia mountains					✔								
124	Mount Parnitha	✔				✔								✔
125	Schinias marsh													
126	Mount Ymittos					✔								
127	Sounio area	✔												
128	Falkonera islet													
129	North, east and south Kithira island					✔								✔
130	Antikithira island, Prasouda and Lagouvardos islets													
131	Rocky islets of Limnos island													
132	Lakes Khortaro and Alyki, Moudros gulf, Diapori fen, and Fakos peninsula					✔								✔
133	Agios Efstratios island													
134	South-west peninsula–petrified forest, Lesvos				✔	✔								
135	Mount Lepetimnos, Lesvos													
136	Tokmakia islets													
137	Kallonis Gulf					✔								✔
138	Gera gulf–Dipi and Haramida marshes, Lesvos					✔								
139	Mount Olympos, Lesvos					✔								
140	Psara and Antipsara islands					✔								
141	North Chios													
142	Venetiko islet													
143	Ikaria island													
144	Fourni islands													
145	Mount Kerkis													✔
146	Aliki Psilis Ammou lagoon, Samos													
147	Central Andros island					✔								
148	North-east Tinos island													
149	Northern Syros					✔								
150	Serifos island					✔								
151	Northern Sifnos					✔								
152	Western Milos, Antimilos and Polyegos islands								✔					
153	Paros, Antiparos and adjacent islands					✔								
154	Mounts Dias, Mavrovouni and Koronos					✔								
155	Mikres Kyklades													
156	Amorgos island													
157	Ios, Sikinos, Folegandros island group					✔								
158	Christiana islets													
159	Anafi island													
160	Islets of North Dodekanisa													✔
161	North and south Patmos island and islets													
162	North and east Kalimnos, Telendos, and Kalolimnos islands and islets													✔
163	Kinaros and Levitha islands and rocky islets													✔
164	Astypalea island and nearby islets													✔
165	Sirna island and nearby islets													✔
166	Mount Dikios, Cape Louros, Lake Psalidi, and Alyki					✔								
167	Nisiros island and nearby islets													
168	Tilos island													
169	Simi island and adjacent islets													
170	Chalki island and nearby islets													
171	Central Rodos and the areas of Lindos, Mesanagros and Katavia-Prassoni					✔							✔	
172	Saria island and northern Karpathos					✔								
173	Islets of Karpathian Sea													
174	Kasos island													
175	Tiganis peninsula and the islands of Gramvouses and Pontikonisi													
176	Rodopos peninsula					✔								
177	Agioi Theodori islet													
	Subtotal of IBAs	9	1	5	8	100	3	5	1	12	2	1	9	48

Table 5 ... continued. Protection status of Important Bird Areas in Greece.
A tick (✔) indicates that an IBA overlaps with a protected area (to any extent).

IBA code	International name	National: National Park	Marine Park	Protected Area	Natural Monument	Wildlife Refuge	Aesthetic Forest	Controlled Hunting Area	Breeding Station	International: Ramsar Site	World Heritage Site	Biosphere Reserve	Biogenetic Reserve	Special Protection Area
178	Mount Koutroulis, Mount Ag. Dikaios and Modia plateau													✔
179	Lefka mountains	✔				✔						✔	✔	✔
180	Lake Kourna, Almyrou delta, and Georgioupolis beach													
181	Gavdos and Gavdopoula islands													
182	Krioneritis mountain													
183	Mount Kedros and Kourtaliotiko gorge													✔
184	Prassiano gorge					✔								
185	Mount Idi				✔	✔								✔
186	Geropotamos estuary													
187	Mount Asterousia (Kofinas)													✔
188	Mount Jouchtas													
189	Dia island							✔						
190	Mount Dikti					✔								✔
191	Mount Thryptis and Mount Ornon					✔								
192	Dionisiades islands													✔
193	North-eastern edge of Crete						✔							✔
194	Mount Zakrou													
195	Kavalli islets													
196	Koufonisi island													
Total number of IBAs		**10**	**1**	**5**	**9**	**105**	**4**	**6**	**1**	**12**	**2**	**2**	**10**	**56**

3. **Protected Area**
4. **Natural Monument**
5. **Wildlife Refuge**
6. **Aesthetic Forest**
7. **Controlled Hunting Area**
8. **Breeding Station**

During the last two years there have been important changes to environmental protection legislation with a series of Ministerial Decisions and Decrees. These include the decentralization of various responsibilities from the relevant Ministries to regional authorities. Ministries have only kept responsibilities for national protected areas and the Natura 2000 network. There are ongoing plans to establish legal protection for 30 sites, the majority of which are IBAs. Also the Ministry of Environment, Physical Planning and Public Works is planning to address the protection of more IBAs over the next five years. However, legislation allowing the formation of management bodies for protected areas is lacking which is a serious problem preventing their effective management.

International protection

Only 30% of IBAs have any part of their area under international protection, with only 6% having more than 90% of their area protected (Figure 5). Caution should be used in interpreting the data as the area of overlap between IBAs and protected areas is unknown for a few sites, in which cases a 'partial' overlap has been applied as a default option. Table 5 details overlaps between IBAs and protected areas. Twelve IBAs overlap with Ramsar Sites and 56 IBAs overlap with Special Protection Areas. Greece has signed and ratified various international conventions and initiatives (Box 1). Ministerial Decisions have been issued for the provisional protection of most Ramsar Sites and Presidential Decrees are under preparation for their permanent protection. Two hundred and thirty sites have been proposed as candidate Special Areas for Conservation (SACs) by the Greek government, under the EC Habitats Directive, many of which overlap with IBAs.

Box 1. International legislation and initiatives that are relevant to site conservation in Greece (see Appendix 1 for a general description of these agreements).

Global	
Biodiversity Convention	✔
Ramsar Convention	✔
Bonn Convention	✔
World Heritage Convention	✔
MAB Programme	✔
Pan-European	
Bern Convention	✔
Regional	
EC Birds Directive	✔
EC Habitats Directive	✔
Barcelona Convention	✔

✔ Convention ratified/initiative supported
(✔) Convention signed

CONSERVATION

- In general IBAs, even those with some national protection status, are not adequately protected, and in some cases they are destroyed by works funded by the European Union and the Greek government, including the Ministry of Agriculture and the Ministry of Environment, Physical Planning and Public Works.
- Seabird surveys have been carried out by HOS since 1995 and are ongoing, funded by the European Union.
- LIFE-Nature projects on threatened species are ongoing for *Phalacrocorax pygmaeus, Anser erythropus, Gyps barbatus, Numenius tenuirostris* and *Larus audouinii.*

ANALYTICAL METHODS

- Population estimates for most species are based on data obtained during 1995–1997.
- Data for wintering waterbirds are of good quality and are based on annual mid-winter counts. Breeding data for waterbirds are mostly of good quality, particularly for the less common species.
- Quantitative data for migrating birds are in general under-estimated or missing due to a lack of relevant research in Greece. This has led to the exclusion of several small wetlands from this inventory. These wetlands possibly meet criteria set for migratory species but no quantitative data are available.
- Population data on common species are usually estimates or missing. Also the importance of remote IBAs (mountains and rocky islets) is usually underestimated.
- The measure of overlap between IBAs and Wildlife Refuges in the north-eastern Aegean islands, the Dodecanese and Crete may be incomplete due to lack of relevant maps. The overlap between Special Protection Areas and IBAs was the subject of a report produced by HOS. Where the degree of overlap between an IBA and a protected area is unknown, the protection status has been set to 'partial' (10–90% area of overlap).

- There is another category of protection in Greece, 'Landscape of Special Natural (Scenic) Beauty'. However, because no maps exist for these areas, they were not used in the IBA review.
- The lack of good habitat and land-use maps has resulted in most data for these variables being estimated.

GLOSSARY

HOS Hellenic Ornithological Society, BirdLife International Partner in Greece.
LIFE an EU funding line (Regulation 1973/92/EEC) which supports EU environmental legislation including the Birds and Habitats Directives.
maquis dense, mostly evergreen shrub communities 1–3 m high, characteristic of the Mediterranean region.
Natura 2000 the name given by the European Commission to the 'coherent ecological network' of Special Protection Areas (SPAs) and Special Areas for Conservation (SACs) that is due to be established within the European Union member states by June 2004, under the Birds and Habitats Directives.
phrygana a Greek term denoting low scrub developed over dry stony soils in the Mediterranean (Balkan) region, equivalent to the term garrigue which is used in the western Mediterranean.
SPA Special Protection Area (designated under Article 4 of the EC Birds Directive).

ACKNOWLEDGEMENTS

Kostas Papakonstantinou was reponsible for organizing the large scale research that took part all over Greece in 1996. Haralambos Alivizatos and Claire Papazoglou provided valuable assistance. Special thanks to the volunteers who provided the data for the site accounts: Stella Adamantopoulou, Haralambos Alivizatos, Andrea Bonneti, Stratis Bourdakis, Dimitris Bousbouras, Martin Gaetlich, Antonia Galanaki, Vassilis Goutner, Maria Dimaki, Anastasios Dimalexis, Phillipos Dragoumis, Michalis Dretakis, Ben Hallman, Patric Harris, Vassilis Hatzirvasanis, Hans Jerrentrup, Stavros Kalpakis, Niki Kardakari, George Katsadorakis, Savas Kazantzidis, Stella Kladara, Theodoros Kominos, George Korakis, Yannis Ioannidis, Panagiotis Latsoudis, George Mertzanis, Kostas Misiakos, Theodoros Naziridis, Kyriakos Oikonomidis, Maria Panayotopoulou, Haritakis Papaioannou, Kosta Papakostantinou, Photis Pergantis, Kostas Poirazidis, Nikos Probonas, Yannis Rousopoulos, Anastasios Sakoulis, Anastasios Sfougaris, Theodora Skartsi, Evangelos Spinthakis, Rigas Tsiakiris, Yannis Tsougrakis, Stavros Xirouhakis, Stamatis Zogaris. Also many thanks to a large number of members of Hellenic Ornithological Society and other ornithologists and birdwatchers that provided valuable data for the completion of this review. Funds were provided by the Royal Society for the Protection of Birds, the Ministry of Environment, Physical Planning and Public Works, and The Dutch Embassy in Athens (Department of Agriculture Policy).

SITE ACCOUNTS

Northern Evros and Ardas riparian forest — A1, B2, B3, C1, C6 — 001

Admin region East Macedonia and Thrace
Coordinates 41°30'N 26°30'E
Altitude 30–60 m **Area** 20,000 ha

Site description

This site stretches along the River Evros, from Pithio to Ormenio, and part of the River Ardas. Riverine forest is dominated by *Populus* and *Salix* species. The forest exists in a mosaic of grassland, scrub, freshwater ponds with reedbeds (*Phragmites*) and some arable land. A large part of the area is a military zone.

Habitats Forest and woodland (60%; alluvial/very wet forest), Scrub (5%; scrub), Grassland (5%; humid grassland), Wetland (25%; mudflat/sandflat; standing fresh water; river/stream; water-fringe vegetation), Artificial landscape (10%; arable land; other urban/industrial areas)
Land-use Agriculture (20%), Fisheries/aquaculture (10%), Military (80%), Urban/industrial/transport (10%), Water management (30%)

Birds

Species	Season	Year	Pop min	Pop max	Acc	Criteria
Phalacrocorax pygmeus Pygmy Cormorant	N	1997	40	—	—	A1, C1
Nycticorax nycticorax Night Heron	B	1997	120	—	A	B2, C6
Accipiter brevipes Levant Sparrowhawk	B	1997	20	—	A	B2
Dendrocopos syriacus Syrian Woodpecker	R	1997	Common		—	B3, C6
Lanius minor Lesser Grey Shrike	B	1997	Common		—	B2, C6

The site is very important for breeding, migrating and wintering waterbirds, raptors and passerines. Two heronries with more than 250 pairs of *Nycticorax nycticorax*, *Egretta garzetta* and *Ardea cinerea* lie within the IBA.

Protection status

National Low **International** None
400 ha of IBA covered by Wildlife Refuge (Kalos Gialas/Plati–Arzou–Rizion–Filakiou, 400 ha).

Conservation issues

Threats Agricultural intensification/expansion (B), Dredging/canalization (C), Selective logging/cutting (A), Groundwater abstraction (B), Other (C), Unsustainable exploitation (B)

The heronry at Dikaia is threatened by logging of the trees. There is a rubbish tip in the forest of the River Arda ('Other' threats). A further threat is from agricultural intensification, particularly increased use of fertilizers and pesticides, irrigation and drainage. Hunting is a threat in winter.

Asvestades–Vrisika area — B2, C6 — 002

Admin region East Macedonia and Thrace
Coordinates 41°20'N 26°20'E
Altitude 60–195 m **Area** 9,500 ha

Site description

A hilly area with oak *Quercus* forest, riverine vegetation and agricultural land. Human activities include forestry and agriculture.

Habitats Forest and woodland (60%; broadleaved deciduous forest; alluvial/very wet forest), Scrub (5%; scrub), Artificial landscape (35%; arable land; perennial crops/orchards/groves; ruderal land)
Land-use Agriculture (30%), Forestry (60%), Military, Tourism/recreation

Birds

Species	Season	Year	Pop min	Pop max	Acc	Criteria
Hippolais olivetorum Olive-tree Warbler	B	—	—	—	—	B2, C6
Lanius nubicus Masked Shrike	B	1996	Common		—	B2

This is an important area for species typical of a mosaic of forest, non-intensive agriculture and rivers.

Protection status

National None **International** None
Part of IBA covered by Wildlife Refuge (Votsis/Asvestadon, 1,330 ha).

Conservation issues

Threats Agricultural intensification/expansion (U)

No serious threats known at this site.

Dadias–Dereiou–Aisimis forest — A1, A3, B2, C1, C6 — 003

Admin region East Macedonia and Thrace
Coordinates 41°07'N 26°00'E
Altitude 0–1,044 m **Area** 50,629 ha

Site description

The site is dominated by mixed *Quercus* and *Fagus* forest with *Pinus* plantations.

Habitats Forest and woodland (51%; broadleaved deciduous forest), Scrub (14%; sclerophyllous scrub/garrigue/maquis), Grassland (6%; steppe/dry calcareous grassland), Wetland (7%; river/stream), Artificial landscape (28%; arable land; forestry plantation)
Land-use Agriculture (14%), Forestry (85%)

Birds

Species	Season	Year	Pop min	Pop max	Acc	Criteria
Neophron percnopterus Egyptian Vulture	B	—	3	4	B	B2, C6
Aegypius monachus Cinereous Vulture	N	—	Frequent		—	A1, C1
Aquila pomarina Lesser Spotted Eagle	B	—	7	8	B	B2, C6
Picus viridis Green Woodpecker	R	—	Abundant		—	B2
Oenanthe hispanica Black-eared Wheatear	B	—	Common		—	A3
Hippolais olivetorum Olive-tree Warbler	B	—	Uncommon		—	A3
Sylvia cantillans Subalpine Warbler	B	—	Common		—	A3
Sylvia melanocephala Sardinian Warbler	R	—	Frequent		—	A3
Ficedula semitorquata Semi-collared Flycatcher	B	—	—	—	—	B2, C6
Sitta neumayer Rock Nuthatch	R	—	—	—	—	A3
Emberiza caesia Cretzschmar's Bunting	R	—	Frequent		—	A3
Emberiza melanocephala Black-headed Bunting	B	—	Frequent		—	A3

The area is important for breeding and passage raptors and resident species of forest, scrub and arable land. Breeding birds include seven of the 21 species that are restricted in Europe to the Mediterranean biome (when breeding). Species of global conservation concern that do not meet IBA criteria: *Aquila clanga* (passage), *Aquila heliaca* (passage), *Falco naumanni*.

Protection status

National Partial **International** None
1,930 ha of IBA covered by Wildlife Refuge (Poulia/Mikrou Dereiou–Soufliou, 1,930 ha). 9,100 ha of IBA covered by Wildlife Refuge (Trion Vryseon/Dadias–Mikrou Dereiou, 9,100 ha).

Conservation issues

Threats Abandonment/reduction of land management (C), Forest grazing (B), Intensified forest management (B)

Part of the area is a candidate SAC.

Protected area of Dadia–Lefkimi–Soufli forest — 004

A1, A3, B1iii, B2, C1, C2, C6

Admin region East Macedonia and Thrace
Coordinates 41°11'N 26°18'E
Altitude 10–620 m **Area** 42,370 ha

Site description

The forest lies at the south-eastern end of Mount Rodopi. The site is a mosaic of mature forest, meadows, small areas of cultivation and rivers.

Habitats Forest and woodland (60%; broadleaved deciduous forest; native coniferous forest; mixed forest), Scrub (5%; sclerophyllous scrub/garrigue/maquis), Grassland (5%; steppe/dry calcareous grassland), Wetland (5%; river/stream; water-fringe vegetation), Artificial landscape (25%; arable land; forestry plantation; other urban/industrial areas; ruderal land)
Land-use Agriculture (20%), Forestry (62%), Military (6%), Nature conservation/research (20%), Tourism/recreation (5%)

Birds

Species	Season	Year	Pop min	Pop max	Acc	Criteria
Nycticorax nycticorax Night Heron	B	—	70	100	A	B2, C6
Neophron percnopterus Egyptian Vulture	B	—	9	11	A	B2, C6
[1] ***Gyps fulvus*** Griffon Vulture	B	—	40	60	A	C6
[2] ***Aegypius monachus*** Cinereous Vulture	R	—	90	100	A	A1, B1iii, B2, C1, C2, C6
Circaetus gallicus Short-toed Eagle	B	—	15	20	A	B2, C6
Accipiter brevipes Levant Sparrowhawk	B	—	8	12	C	B2
Aquila pomarina Lesser Spotted Eagle	B	—	15	17	A	B2, C6
Aquila clanga Greater Spotted Eagle	W	—	6	15	A	A1, C1
Aquila chrysaetos Golden Eagle	R	—	4	6	A	B2, C6
Hieraaetus pennatus Booted Eagle	B	—	7	9	A	B2, C6
Bubo bubo Eagle Owl	R	—	5	6	A	B2, C6
Oenanthe hispanica Black-eared Wheatear	B	—	Common		—	A3
Hippolais olivetorum Olive-tree Warbler	B	—	Frequent		—	A3, B2, C6
Sylvia cantillans Subalpine Warbler	B	—	Common		—	A3
Sylvia melanocephala Sardinian Warbler	R	—	Common		—	A3
Sylvia hortensis Orphean Warbler	B	—	Frequent		—	B2
Sitta neumayer Rock Nuthatch	R	—	—	—	—	A3
Lanius nubicus Masked Shrike	B	—	Common		—	A3, B2
Emberiza caesia Cretzschmar's Bunting	B	—	Frequent		—	A3
Emberiza melanocephala Black-headed Bunting	B	—	Abundant		—	A3

1. Non-breeding individuals.
2. Individuals.

This is a very important site for breeding and wintering raptors with 36 species recorded. Breeding birds include eight of the 21 species that are restricted in Europe to the Mediterranean biome. Species of global conservation concern that do not meet IBA criteria: *Haliaeetus albicilla* (3–6 wintering individuals), *Aquila heliaca* (breeding, 2–4 wintering individuals), *Falco naumanni*.

Protection status

National Partial **International** Partial
7,200 ha of IBA covered by Wildlife Refuge (Dadia–Lefkimi–Lykofis Forest, 7,200 ha). 1,186 ha of IBA covered by Wildlife Refuge (Zioga–Kalivia/Dadias, Soufliou, 1,187 ha). 7,465 ha of IBA covered by Special Protection Area (Dasos Dadias– Soufli, 7,465 ha).

Conservation issues

Threats Abandonment/reduction of land management (C), Afforestation (B), Agricultural intensification/expansion (B), Burning of vegetation (A), Construction/impact of dyke/dam/barrage (C), Firewood collection (B), Forest grazing (C), Groundwater abstraction (B), Intensified forest management (B), Unsustainable exploitation (B)

The main threats are from intensive forest management, afforestation with conifer plantations and fire. The use of poisoned baits to kill raptors and the recent building of a military camp are also threats. A core area of 7,258 ha is strictly protected. The area is a candidate SAC.

South forest complex of Evros prefecture — 005

A1, A3, B1iii, B2, B3, C1, C2, C6

Admin region East Macedonia and Thrace
Coordinates 40°58'N 25°55'E
Altitude 20–950 m **Area** 71,800 ha

Site description

A forest complex at the south-eastern end of the Rodopi mountain. It is characterized by low hills covered with *Quercus*, *Caprinus*, *Fraxinus* and *Acer* woodland and scrub. *Fagus* forest dominates the northern area; *Pinus* forest and plantations occur in the south. Small wetlands are also found in the area.

Habitats Forest and woodland (62%; broadleaved deciduous forest; native coniferous forest; mixed forest), Scrub (7%; sclerophyllous scrub/garrigue/maquis), Wetland (14%; river/stream; water-fringe vegetation), Rocky areas (7%; inland cliff), Artificial landscape (25%; arable land; forestry plantation)
Land-use Agriculture (30%), Forestry (70%), Military (10%), Tourism/recreation (5%), Urban/industrial/transport (5%)

Birds

Species	Season	Year	Pop min	Pop max	Acc	Criteria
Neophron percnopterus Egyptian Vulture	B	—	3	4	A	C6
Aegypius monachus Cinereous Vulture	P	—	Common		—	A1, B1iii, C1, C2
Circaetus gallicus Short-toed Eagle	B	—	8	10	A	B2, C6

Species ... continued	Season	Year	Pop min	Pop max	Acc	Criteria
Buteo rufinus Long-legged Buzzard	R	—	6	9	A	B2, C6
Aquila pomarina Lesser Spotted Eagle	B	—	7	8	A	B2, C6
Aquila clanga Greater Spotted Eagle	W	—	15	20	A	A1, C1
Aquila heliaca Imperial Eagle	W	—	4	6	—	A1, C1
Aquila chrysaetos Golden Eagle	R	—	3	4	A	B2, C6
Hieraaetus pennatus Booted Eagle	B	—	6	7	A	B2, C6
Bubo bubo Eagle Owl	R	—	3	4	A	B2, C6
Oenanthe hispanica Black-eared Wheatear	B	—	Common		—	A3
Hippolais olivetorum Olive-tree Warbler	B	—	Abundant		—	A3, B2, C6
Sylvia cantillans Subalpine Warbler	B	—	Abundant		—	A3
Sylvia melanocephala Sardinian Warbler	R	—	Abundant		—	A3
Phylloscopus bonelli Bonelli's Warbler	B	—	Abundant		—	B3
Sitta neumayer Rock Nuthatch	R	—	—	—	—	A3
Lanius nubicus Masked Shrike	B	—	Abundant		—	A3, B2
Emberiza caesia Cretzschmar's Bunting	B	—	Frequent		—	A3
Emberiza melanocephala Black-headed Bunting	B	—	Abundant		—	A3

This is an important site for breeding and wintering raptors and other resident species associated with forest. Breeding birds include eight of the 21 species that are restricted in Europe to the Mediterranean biome (when breeding). Species of global conservation concern that do not meet IBA criteria: *Haliaeetus albicilla* (breeding), *Falco naumanni*.

Protection status

National Low **International** None

534 ha of IBA covered by Wildlife Refuge (Drimou/Nipsas, 534 ha). Part of IBA covered by Wildlife Refuge (Pylaia/Kavisou–Ferron, 650 ha).

Conservation issues

Threats Abandonment/reduction of land management (C), Afforestation (A), Agricultural intensification/expansion (B), Consequences of animal/plant introductions (C), Forest grazing (B), Intensified forest management (B)

The main threats are from afforestation with plantations, forest grazing and increasingly intensive forest management. The site is a candidate SAC.

Evros delta — A1, A4i, A4iii, B1i, B1iv, B2, B3, C1, C2, C3, C4, C5, C6 — 006

Admin region East Macedonia and Thrace
Coordinates 40°52′N 26°00′E
Altitude 0–0 m **Area** 19,000 ha

Site description

A coastal wetland complex formed by the delta of the River Evros on the Greece/Turkey border (adjacent to Turkish IBA 001).

Habitats Forest and woodland (6%; alluvial/very wet forest), Grassland (5%; humid grassland), Wetland (63%; tidal river/enclosed tidal water; saltmarsh; coastal lagoon; standing fresh water; standing brackish and salt water; water-fringe vegetation), Artificial landscape (23%; arable land)

Land-use Agriculture (80%), Fisheries/aquaculture (20%), Military (50%), Nature conservation/research, Tourism/recreation (30%), Water management (100%)

Birds

Species	Season	Year	Pop min	Pop max	Acc	Criteria
Phalacrocorax carbo Cormorant	W	1997	7,000	—	A	A4i, B1i, C3
Phalacrocorax pygmeus Pygmy Cormorant	W	1997	26,000	—	A	A1, A4i, B1i, C1, C2
Pelecanus onocrotalus White Pelican	P	1997	2,600	—	A	A4i, B1i, C2
[1] ***Pelecanus crispus*** Dalmatian Pelican	W	1995	—	164	—	A1, A4i, B1i, C1, C2
Ciconia ciconia White Stork	P	—	—	—	—	A4i, B1i, C2
Plegadis falcinellus Glossy Ibis	P	1997	Abundant		—	A4i, B1i, C2
Platalea leucorodia Spoonbill	P	1997	500	—	A	A4i, B1i, C2
Phoenicopterus ruber Greater Flamingo	W	1997	5,000	—	A	A4i, B1i, C2
Cygnus olor Mute Swan	W	1993	744	—	—	B1i, C3
Anser erythropus Lesser White-fronted Goose	W	1988	—	116	—	A1, B1i, C1, C2

Species ... continued	Season	Year	Pop min	Pop max	Acc	Criteria
Branta ruficollis Red-breasted Goose	W	—	—	—	—	A1, A4i, B1i, C1, C2
Tadorna tadorna Shelduck	W	1995	1,630	—	—	B1i, C3
Anas penelope Wigeon	W	1995	—	10,850	—	B1i, C3
Anas strepera Gadwall	W	1995	—	380	—	B1i, C3
Anas crecca Teal	W	1995	—	13,400	—	A4i, B1i, C3
Aquila clanga Greater Spotted Eagle	W	—	—	—	—	A1, C1
Haematopus ostralegus Oystercatcher	R	—	—	—	—	B1i, C3
Haematopus ostralegus Oystercatcher	W	1995	212	—	—	B1i, C3
Recurvirostra avosetta Avocet	B	1997	100	—	A	B3, C6
Recurvirostra avosetta Avocet	W	1995	—	460	—	B1i, B2, C2, C6
[2] ***Glareola pratincola*** Collared Pratincole	B	1997	150	250	B	A4i, B1i, B2, C2
Hoplopterus spinosus Spur-winged Plover	B	1997	10	—	A	B2, C6
Calidris minuta Little Stint	W	1995	2,300	—	—	A4i, B1i, C3
Numenius tenuirostris Slender-billed Curlew	P	—	—	—	—	A1, C1
Tringa stagnatilis Marsh Sandpiper	P	1992	380	—	A	A4i, B1i, C3
Larus melanocephalus Mediterranean Gull	B	1997	250	—	A	B3, C2, C6
[2] ***Gelochelidon nilotica*** Gull-billed Tern	B	1997	60	—	B	B2, C6
Sterna albifrons Little Tern	B	1997	200	—	A	B1i, B2, C2, C6
Merops apiaster Bee-eater	B	—	—	—	—	B2
Calandrella brachydactyla Short-toed Lark	B	—	—	—	—	B2, C6

1. Large increase.
2. Individuals.

A very important site for breeding, passage and wintering waterbirds, raptors and passerines. Species of global conservation concern that do not meet IBA criteria: *Aythya nyroca* (breeding, passage), *Oxyura leucocephala* (wintering), *Haliaeetus albicilla* (breeding, wintering), *Aquila heliaca* (non-breeding), *Falco naumanni* (passage), *Larus audouinii* (non-breeding). The site is a major migratory bottleneck, where more than 5,000 *Ciconia ciconia* regularly pass.

Protection status

National Partial **International** Partial

4,600 ha of IBA covered by Wildlife Refuge (Delta Evrou, 4,600 ha). Part of IBA covered by Protected Area. 9,267 ha of IBA covered by Ramsar Site (Evros Delta, 9,267 ha). 9,218 ha of IBA covered by Special Protection Area (Delta Evrou, 9,218 ha).

Conservation issues

Threats Agricultural intensification/expansion (A), Construction/impact of dyke/dam/barrage (C), Drainage (A), Dredging/canalization (B), Firewood collection (B), Forest grazing (B), Groundwater abstraction (A), Industrialization/urbanization (C), Recreation/tourism (B), Unsustainable exploitation (A)

There are many threats to the site including the effects of intensive agriculture, drainage, groundwater abstraction and hunting. There have been two recent LIFE projects on *Numenius tenuirostris* and on *Phalacrocorax pygmaeus* and *Anser erythropus*. Part of the area is a candidate SAC.

Samothraki island — B1i, C2 — 007

Admin region East Macedonia and Thrace
Coordinates 40°28′N 25°30′E
Altitude 0–1,611 m **Area** 17,000 ha

Site description

A mountainous island with oak *Quercus* forest in the north and extensive olive-groves and scrub in the south. The south-east is largely inaccessible.

Habitats Forest and woodland (33%; broadleaved deciduous forest; alluvial/very wet forest; broadleaved evergreen forest; treeline ecotone), Scrub (20%; sclerophyllous scrub/garrigue/maquis), Grassland (10%; alpine/subalpine/boreal grassland), Rocky areas (15%; sea cliff/rocky shore; scree/boulders; inland cliff), Artificial landscape (20%; arable land; perennial crops/orchards/groves)

Land-use Agriculture (17%), Fisheries/aquaculture, Military (10%), Not utilized (15%), Other (10%), Tourism/recreation (20%)

Birds

The site is important for migratory species.

Species	Season	Year	Pop min	Pop max	Acc	Criteria
[1] ***Phalacrocorax aristotelis*** Shag	R	—	110	—	A	B1i, C2

1. Individuals.

Protection status

National Partial **International** None

500 ha of IBA covered by Wildlife Refuge (Alonoudia/Alonia–Samothrakis, 500 ha). 1,900 ha of IBA covered by Wildlife Refuge (Katsampas/Samothraki, 1,900 ha).

Conservation issues

Threats Aquaculture/fisheries (B), Firewood collection (C), Forest grazing (A), Infrastructure (C), Unsustainable exploitation (U)

The main threat is from forest grazing. Part of the area is a candidate SAC.

Filiouri river valley and east Rodopi mountains — A1, B2, B3, C1, C6 — 008

Admin region East Macedonia and Thrace
Coordinates 41°10′N 25°45′E
Altitude 50–745 m **Area** 77,000 ha

Site description

The site encompasses hills to the south-east of Mount Rodopi and the catchment of the Filiouri river. Vegetation cover is predominantly scrub with some grazed oak *Quercus* forest. Land-use is traditional and non-intensive.

Habitats Forest and woodland (10%; broadleaved deciduous forest), Scrub (65%; sclerophyllous scrub/garrigue/maquis), Grassland (10%; steppe/dry calcareous grassland), Rocky areas (5%; inland cliff), Artificial landscape (10%; arable land; perennial crops/orchards/groves)
Land-use Agriculture (95%)

Birds

Species	Season	Year	Pop min	Pop max	Acc	Criteria
[1] ***Neophron percnopterus*** Egyptian Vulture	B	1995	5	—	B	B2, C6
Gyps fulvus Griffon Vulture	R	1995	2	5	B	C6
Aegypius monachus Cinereous Vulture	N	1996	Frequent		—	A1, C1
Circaetus gallicus Short-toed Eagle	B	1995	10	—	B	B2, C6
[1] ***Aquila chrysaetos*** Golden Eagle	R	1995	2	4	B	C6
Strix aluco Tawny Owl	R	1992	Frequent		—	B3
Dendrocopos syriacus Syrian Woodpecker	R	1995	Common		—	B3, C6
Dendrocopos medius Middle Spotted Woodpecker	R	1995	Common		—	C6
Sylvia hortensis Orphean Warbler	B	1993	Uncommon		—	B2
Phylloscopus bonelli Bonelli's Warbler	B	1995	Common		—	B3
Ficedula semitorquata Semi-collared Flycatcher	B	1990	Rare		—	B2, C6
Lanius collurio Red-backed Shrike	B	1995	Common		—	B2, C6
Emberiza hortulana Ortolan Bunting	B	1995	Common		—	B2, C6

1. Large decrease.

This is an important site for breeding and passage raptors and species associated with forest and scrub. Species of global conservation concern that do not meet IBA criteria: *Aquila heliaca* (breeding), *Falco naumanni* (passage).

Protection status

National Partial **International** None

Part of IBA covered by Wildlife Refuge (Pefkodasas Nimfeas, 400 ha).

Conservation issues

Threats Deforestation (commercial) (B), Firewood collection (C), Forest grazing (C), Infrastructure (B), Intensified forest management (A), Unsustainable exploitation (B)

The main threats are from road construction, deforestation and intensified forest management. Part of area is a candidate SAC.

Kompsatos valley — A1, B2, B3, C1, C6 — 009

Admin region East Macedonia and Thrace
Coordinates 41°10′N 25°07′E
Altitude 20–800 m **Area** 16,000 ha

Site description

A deep river valley surrounded by hills with deciduous forest, scrub and grassland. The main human activity is livestock-farming.

Habitats Forest and woodland (75%; broadleaved deciduous forest), Scrub (10%; sclerophyllous scrub/garrigue/maquis), Grassland (10%; dry siliceous grassland), Artificial landscape (5%; arable land)
Land-use Agriculture (50%), Forestry (50%)

Birds

Species	Season	Year	Pop min	Pop max	Acc	Criteria
Phalacrocorax pygmeus Pygmy Cormorant	W	—	20	40	B	A1, C1
Dendrocopos syriacus Syrian Woodpecker	R	1996	Common		—	B3, C6
Ficedula semitorquata Semi-collared Flycatcher	B	1990	Rare		—	B2, C6
Emberiza hortulana Ortolan Bunting	B	1995	Common		—	B2, C6

This is a very important area for breeding and passage raptors. Species of global conservation concern that do not meet IBA criteria: *Haliaeetus albicilla* (resident), *Aegypius monachus* (passage), *Aquila heliaca* (passage).

Protection status

National None **International** None

Conservation issues

Threats Construction/impact of dyke/dam/barrage (A), Deforestation (commercial) (A), Firewood collection (A), Forest grazing (U), Infrastructure (B), Intensified forest management (A), Recreation/tourism (C), Unsustainable exploitation (B)

The main threats to the site are from deforestation and intensive forest management. Long-term monitoring of raptors is carried out. Part of the area is a candidate SAC.

Lake Mitrikou (Ismarida) — A1, A4i, A4iii, B1i, B2, B3, C1, C2, C3, C4, C6 — 010

Admin region East Macedonia and Thrace
Coordinates 40°58′N 25°17′E
Altitude 0–54 m **Area** 6,500 ha

Site description

A coastal wetland complex. It consists of the freshwater Lake Mitrikou, surrounded by extensive reedbeds (*Phragmites*), coastal lagoons and saltmarsh.

Habitats Wetland (90%; tidal river/enclosed tidal water; mudflat/sandflat; saltmarsh; sand-dunes/sand beach; coastal lagoon; standing fresh water; standing brackish and salt water; river/stream; water-fringe vegetation), Artificial landscape (10%; arable land)
Land-use Agriculture (10%), Fisheries/aquaculture (85%), Tourism/recreation (5%), Water management (85%)

Birds

Species	Season	Year	Pop min	Pop max	Acc	Criteria
Podiceps nigricollis Black-necked Grebe	W	—	—	1,000	—	A4i, B1i, C3
Phalacrocorax carbo Cormorant	W	1996	700	2,500	—	B1i, C3
Phalacrocorax pygmeus Pygmy Cormorant	R	1996	Frequent		—	B2, C6
Phalacrocorax pygmeus Pygmy Cormorant	W	1997	110	350	—	A1, A4i, B1i, C1, C2
Pelecanus crispus Dalmatian Pelican	W	1996	50	600	—	A1, A4i, B1i, C1, C2
Pelecanus crispus Dalmatian Pelican	P	1996	Abundant		—	A1, C1
Ixobrychus minutus Little Bittern	B	—	20	50	—	C6
Ardeola ralloides Squacco Heron	B	—	—	80	—	B2, C2, C6
Egretta alba Great White Egret	W	—	—	280	—	A4i, B1i, C2

Species ... continued	Season	Year	Pop min	Pop max	Acc	Criteria
Plegadis falcinellus Glossy Ibis	P	1996	50	2,500	—	A4i, B1i, C2
[1] ***Platalea leucorodia*** Spoonbill	R	—	—	110	—	A4i, B1i, B2, C2, C6
Phoenicopterus ruber Greater Flamingo	P	1996	1,500	5,500	—	A4i, B1i, C2
Cygnus olor Mute Swan	W	1996	500	1,400	—	B1i, C3
Anser erythropus Lesser White-fronted Goose	W	1997	20	—	—	A1, C1
Tadorna ferruginea Ruddy Shelduck	R	—	Uncommon		—	B2, C6
Tadorna tadorna Shelduck	W	1996	800	2,500	—	B1i, C3
Anas penelope Wigeon	W	1996	1,500	6,000	—	B1i, C3
Aythya nyroca Ferruginous Duck	R	—	Common		—	B2, C6
Recurvirostra avosetta Avocet	W	1996	700	2,500	—	A4i, B1i, B2, C6
Burhinus oedicnemus Stone Curlew	B	—	Common		—	C6
Glareola pratincola Collared Pratincole	B	—	20	100	—	A4i, B1i, B2, C2
Hoplopterus spinosus Spur-winged Plover	B	1993	—	4	—	B2, C6
Numenius tenuirostris Slender-billed Curlew	P	1997	0	1	—	A1, C1
Larus melanocephalus Mediterranean Gull	B	—	—	600	—	B3, C2, C6
Larus minutus Little Gull	P	—	—	1,200	—	A4i, B1i, C3
Chlidonias hybridus Whiskered Tern	B	—	30	110	—	B2, C6
Merops apiaster Bee-eater	B	—	Abundant		—	B2

1. Individuals.

The area is important for breeding, passage and wintering waterbirds, and breeding species associated with reedbeds. Species of global conservation concern that do not meet IBA criteria: *Branta ruficollis* (0–3 wintering individuals), *Oxyura leucocephala* (wintering), *Haliaeetus albicilla* (wintering), *Aquila heliaca* (passage), *Falco naumanni* (passage), *Larus audouinii*. The site holds 20,000 waterbirds in winter, on a regular basis (including *Tadorna tadorna* 2,500; *Anas penelope* 6,000; *Anas crecca* 3,500; *Anas platyrhynchos* 10,000; *Recurvirostra avosetta* 2,500; *Larus ridibundus* 3,000).

Protection status

National Partial **International** Partial

Part of IBA covered by Wildlife Refuge (Mavromatiou–Limni Mitrikou/Imerou–Mavromatiou, 1,900 ha). Part of IBA covered by Protected Area. Part of IBA covered by Ramsar Site (Lake Vistonis, Porto Lagos, Lake Ismaris and adjoining lagoons, 24,396 ha). 4,900 ha of IBA covered by Special Protection Area (Limnes Vistonis–Ismaris–Limnothalasses Porto Lagos, Alyki, Ptelea, Xirolimni, Karatza, 14,647 ha).

Conservation issues

Threats Agricultural intensification/expansion (A), Aquaculture/fisheries (B), Drainage (U), Groundwater abstraction (B), Industrialization/urbanization (A), Infrastructure (A), Recreation/tourism (B), Unsustainable exploitation (A)

Chemical and sewerage pollution, water abstraction, hunting, the expansion of agricultural land and infrastructure developments are the main threats. The area is a candidate SAC.

Porto Lagos, Lake Vistonis, and coastal lagoons (Lakes of Thrace)

A1, A4i, A4iii, B1i, B1ii, B2, B3,C1, C2, C3, C4, C6

011

Admin region East Macedonia and Thrace
Coordinates 41°01′N 25°05′E
Altitude 0–99 m **Area** 15,300 ha

Site description

A wetland complex of freshwater lakes and coastal lagoons, with reedbeds (*Phragmites*), wet meadows and saltmarsh.

Habitats Grassland (10%; humid grassland), Wetland (75%; tidal river/enclosed tidal water; mudflat/sandflat; saltmarsh; sand-dunes/sand beach; coastal lagoon; standing fresh water; standing brackish and salt water; river/stream; water-fringe vegetation), Marine areas (5%; sea inlet/coastal features), Artificial landscape (10%; arable land)

Land-use Agriculture (20%), Fisheries/aquaculture (70%), Nature conservation/research (5%), Tourism/recreation (5%), Water management (20%)

Birds

Species	Season	Year	Pop min	Pop max	Acc	Criteria
Tachybaptus ruficollis Little Grebe	W	—	—	2,000	—	A4i, B1i, C3
Podiceps nigricollis Black-necked Grebe	W	1995	1,150	6,000	—	A4i, B1i, C3
Puffinus yelkouan Yelkouan Shearwater	W	—	—	500	—	B1ii, C3
Phalacrocorax carbo Cormorant	W	—	—	2,200	—	B1i, C3
Phalacrocorax pygmeus Pygmy Cormorant	W	1995	2,350	3,300	—	A1, A4i, B1i, C1, C2
Pelecanus crispus Dalmatian Pelican	W	1995	120	250	—	A1, A4i, B1i, C1, C2
Pelecanus crispus Dalmatian Pelican	P	1995	350	700	—	A1, A4i, B1i, C1, C2
Ixobrychus minutus Little Bittern	B	—	50	150	—	B2, C2, C6
Egretta alba Great White Egret	W	1996	45	550	—	A4i, B1i, C2
Ciconia ciconia White Stork	B	—	80	130	—	B2, C6
Plegadis falcinellus Glossy Ibis	P	—	—	400	—	A4i, B1i, C2
Phoenicopterus ruber Greater Flamingo	W	1996	1,500	5,000	—	A4i, B1i, C2
Cygnus olor Mute Swan	W	—	—	800	—	B1i, C3
Anser erythropus Lesser White-fronted Goose	W	1990	—	40	—	A1, C1
Tadorna tadorna Shelduck	W	1996	250	1,200	—	B1i, C3
Aythya ferina Pochard	W	1996	3,000	14,000	—	A4i, B1i, C3
Aythya nyroca Ferruginous Duck	R	—	Frequent		—	C6
Oxyura leucocephala White-headed Duck	W	1997	—	2,300	—	A1, A4i, B1i, C1, C2
Fulica atra Coot	W	1996	2,000	29,000	—	B1i, C3
Recurvirostra avosetta Avocet	R	—	Abundant		—	B3, C6
Burhinus oedicnemus Stone Curlew	B	—	Abundant		—	B2, C6
Glareola pratincola Collared Pratincole	B	—	Abundant		—	B2
Charadrius alexandrinus Kentish Plover	R	—	Abundant		—	B2
[1] ***Hoplopterus spinosus*** Spur-winged Plover	B	1995	5	15	—	B2, C6
Numenius tenuirostris Slender-billed Curlew	P	—	0	1	—	A1, C1
Larus melanocephalus Mediterranean Gull	B	1997	—	400	B	B3, C2, C6
Sterna albifrons Little Tern	B	—	Abundant		—	C6
Merops apiaster Bee-eater	B	—	Abundant		—	B2
Calandrella brachydactyla Short-toed Lark	B	—	Abundant		—	B2, C6
Lanius minor Lesser Grey Shrike	B	—	Abundant		—	B2, C6
Lanius nubicus Masked Shrike	B	—	Common		—	B2

1. Large decrease.

The area is important for breeding, passage and wintering waterbirds and raptors, and breeding species associated with reedbeds. Species of global conservation concern that do not meet IBA criteria: *Branta ruficollis* (wintering), *Haliaeetus albicilla* (2–4 non-breeding individuals), *Aquila clanga* (5 wintering individuals), *Aquila heliaca* (wintering), *Falco naumanni* (passage).

Protection status

National High **International** High

400 ha of IBA covered by Wildlife Refuge (Fanariou/Porto Lagos, 400 ha). 930 ha of IBA covered by Wildlife Refuge (Limni Vistonida/Lagou, Kessanis, 930 ha). Part of IBA covered by Protected Area. Part of IBA covered by Ramsar Site (Lake Vistonis, Porto Lagos, Lake Ismaris and adjoining lagoons, 24,396 ha). 14,500 ha of IBA covered by Special Protection Area (Lake Vistonis, Porto Lagos, Lake Ismaris and adjoining lagoons, 24,396 ha).

Conservation issues

Threats Agricultural intensification/expansion (A), Aquaculture/fisheries (U), Drainage (A), Groundwater abstraction (A), Infrastructure (A), Industrialization/urbanization (B), Recreation/tourism (C), Unsustainable exploitation (B)

The site is threatened by drainage in the eastern part of Vistonis, and water pollution from intensive agricultural, industrial and domestic sources. The wetland is also threatened by agricultural and urban expansion, excessive water abstraction and hunting. A land-use zonation system exists but has yet to be established legally. Part of the area is a candidate SAC.

Nestou delta and coastal lagoons — A1, A4i, B1i, B1ii, B2, B3, C1, C2, C3, C6 — 012

Admin region East Macedonia and Thrace
Coordinates 40°58′N 24°48′E
Altitude 0–10 m **Area** 22,000 ha

Site description

The site consists of a series of brackish lagoons and arable fields with fragmented patches of freshwater marsh, grassland, saltmarsh and forest. Human activities include agriculture, fishing, forestry and tourism.

Habitats Forest and woodland (10%; broadleaved deciduous forest), Grassland (10%; humid grassland), Wetland (58%; tidal river/enclosed tidal water; mudflat/sandflat; sand-dunes/sand beach; coastal lagoon; standing fresh water), Marine areas (7%; sea inlet/coastal features), Artificial landscape (20%; arable land; forestry plantation)
Land-use Agriculture (20%), Fisheries/aquaculture (30%), Forestry (10%), Tourism/recreation (5%), Water management (50%)

Birds

Species	Season	Year	Pop min	Pop max	Acc	Criteria
Puffinus yelkouan Yelkouan Shearwater	W	1996	—	400	—	B1ii, C3
Phalacrocorax carbo Cormorant	W	1996	150	3,300	—	B1i, C3
Phalacrocorax aristotelis Shag	R	—	—	50	—	B1i, C2
Phalacrocorax pygmeus Pygmy Cormorant	R	1996	Common		—	B2, C6
Phalacrocorax pygmeus Pygmy Cormorant	W	1996	250	1,800	—	A1, A4i, B1i, C1, C2
Pelecanus crispus Dalmatian Pelican	W	1996	10	35	—	A1, A4i, B1i, C1, C2
Pelecanus crispus Dalmatian Pelican	P	1996	Abundant		—	A1, C1
Ixobrychus minutus Little Bittern	B	—	Abundant		—	C6
Egretta alba Great White Egret	W	1996	120	500	—	A4i, B1i, C2
Ciconia ciconia White Stork	B	—	Abundant		—	B2, C6
Cygnus olor Mute Swan	W	1996	200	1,300	—	B1i, C3
Anser erythropus Lesser White-fronted Goose	W	—	—	26	—	A1, C1
Aythya nyroca Ferruginous Duck	P	—	Abundant		—	A1, C1
Accipiter brevipes Levant Sparrowhawk	B	—	10	—	—	B2
Aquila clanga Greater Spotted Eagle	W	1997	4	16	—	A1, C1
Falco naumanni Lesser Kestrel	P	—	Abundant		—	A1, C1
Burhinus oedicnemus Stone Curlew	B	1996	20	40	—	C6
Glareola pratincola Collared Pratincole	B	1996	20	100	—	A4i, B1i, B2, C2
Charadrius alexandrinus Kentish Plover	R	—	40	70	—	B2
[1] ***Hoplopterus spinosus*** Spur-winged Plover	B	1993	25	33	—	B2, C6
Numenius tenuirostris Slender-billed Curlew	P	—	Rare		—	A1, C1
Larus melanocephalus Mediterranean Gull	B	—	300	700	—	B3, C2, C6
Sterna albifrons Little Tern	B	1996	80	450	—	A4i, B1i, B2, C2, C6
Dendrocopos syriacus Syrian Woodpecker	R	—	Abundant		—	B3, C6
Calandrella brachydactyla Short-toed Lark	B	—	Abundant		—	C6
Lanius minor Lesser Grey Shrike	B	—	Abundant		—	B2, C6
Lanius nubicus Masked Shrike	B	—	Common		—	B2

1. Large decrease.

This is an important site for breeding, passage and wintering waterbirds, raptors and passerines associated with reedbeds. Species of global conservation concern that do not meet IBA criteria: *Branta ruficollis* (wintering), *Haliaeetus albicilla* (5 resident individuals), *Aquila heliaca* (wintering), *Gallinago media* (25 wintering individuals), *Larus audouinii* (passage).

Protection status

National Partial **International** Partial
930 ha of IBA covered by Wildlife Refuge (Dasohoriou (Foni tis Amerikis)/Erasmiou, 930 ha). 7,800 ha of IBA covered by Wildlife Refuge (Dasos Nestou (Kotza Orman), 7,800 ha). Part of IBA overlaps with Protected Area. Part of IBA covered by Ramsar Site (Nestos delta, 21,930 ha). 9,459 ha of IBA covered by Special Protection Area (Delta Nestou, 9,459 ha).

Conservation issues

Threats Agricultural intensification/expansion (A), Aquaculture/fisheries (B), Construction/impact of dyke/dam/barrage (A), Deforestation (commercial) (C), Drainage (B), Industrialization/urbanization (B), Infrastructure (B), Unsustainable exploitation (B)

The main threats are from the intensification of agriculture, the construction of hydroelectric dams upriver of the IBA, overgrazing and hunting. Part of the area is a candidate SAC.

Nestos gorge — A1, B2, C1, C6 — 013

Admin region East Macedonia and Thrace
Coordinates 41°08′N 24°43′E
Altitude 10–1,268 m **Area** 14,500 ha

Site description

The site has a large gorge with steep rocky slopes. Forest occupies the bottom of the gorge whilst the surrounding mountain slopes are covered with a mixture of forest and maquis. Human activities include agriculture, mining, tourism and hunting.

Habitats Forest and woodland (40%; broadleaved deciduous forest), Scrub (45%; sclerophyllous scrub/garrigue/maquis), Wetland (5%; river/stream), Rocky areas (5%; inland cliff), Artificial landscape (5%; arable land)
Land-use Agriculture (45%), Fisheries/aquaculture (5%), Forestry (35%), Nature conservation/research (20%), Tourism/recreation (5%), Urban/industrial/transport (5%), Water management (5%)

Birds

Species	Season	Year	Pop min	Pop max	Acc	Criteria
Phalacrocorax pygmeus Pygmy Cormorant	W	—	60	—	—	A1, C1
[1] ***Gyps fulvus*** Griffon Vulture	B	1996	16	—	—	C6
Bubo bubo Eagle Owl	B	—	2	3	—	B2, C6
Merops apiaster Bee-eater	B	—	50	100	—	B2

1. Individuals.

This is an important site for breeding, passage and wintering raptors. Species of global conservation concern that do not meet IBA criteria: *Haliaeetus albicilla* (wintering), *Aegypius monachus* (passage), *Aquila clanga* (wintering), *Aquila heliaca* (wintering), *Falco naumanni* (passage), *Gallinago media* (passage).

Protection status

National Partial **International** Partial
2,380 ha of IBA covered by Aesthetic Forest (Stena Nestou, 2,380 ha). 1,125 ha of IBA covered by Wildlife Refuge (Stena Nestou/Kehrokambos, 1,125 ha). 3,000 ha of IBA covered by Wildlife Refuge (Stena Nestou/Stavroupolis, 3,000 ha). 2,336 ha of IBA covered by Special Protection Area (Stena Nestou, 2,336 ha).

Conservation issues

Threats Aquaculture/fisheries (U), Construction/impact of dyke/dam/barrage (A), Disturbance to birds (U), Extraction industry (A), Infrastructure (A), Intensified forest management (B), Recreation/tourism (B), Unsustainable exploitation (B)

The main threats include the expansion of marble quarries, road building, hunting and uncontrolled tourism development. Part of the area is a candidate SAC.

Central Rodopi mountains and Nestos valley — B2, B3, C6 — 014

Admin region East Macedonia and Thrace
Coordinates 41°27′N 24°32′E
Altitude 500–2,000 m **Area** 150,000 ha

Site description

An extensive area of forest between the River Nestos and the border with Bulgaria. The site includes the primary *Fagus* and *Pinus* forest of Paranestion. Human activities include forestry, stock-raising, and hunting.

Habitats Forest and woodland (broadleaved deciduous forest; native coniferous forest), Grassland (alpine/subalpine/boreal grassland)
Land-use Agriculture, Forestry (90%), Water management

■ Birds

Species	Season	Year	Pop min	Pop max	Acc	Criteria
Picus viridis Green Woodpecker	R	1989	Common		—	B2
Dendrocopos medius Middle Spotted Woodpecker	R	1989	Frequent		—	C6
Phylloscopus bonelli Bonelli's Warbler	B	1992	Frequent		—	B3

The site is important for species associated with upland forests.

■ Protection status

National Partial **International** Low
3,900 ha of IBA covered by Wildlife Refuge (Aetorahis/Paranestiou, 3,900 ha). 2,200 ha of IBA covered by Dasiko Symplegma Kariofytou, 2,200 ha). 903 ha of IBA covered by Wildlife Refuge (Tsanaktere/Kariofytou, 903 ha). Part of IBA covered by Wildlife Refuge (D. Symplegma Kentaurou-Pahnis/Mykis, Ehinou, Kotil, 4,000 ha). Part of IBA covered by Wildlife Refuge, Geraka–Oraiou/Stavroupolis, 3,270 ha). 5,312 ha of IBA covered by Wildlife Refuge (Parthenou Dasous, Paranesti–Sellis, 5,312 ha). 550 ha of IBA covered by Natural Monument (Virgin Forest of Central Rodopi, 550 ha). 550 ha of IBA covered by Biogenetic Reserve (Council of Europe) (Partheno Dasos Kentrikis Rodopis, 550 ha). 569 ha of IBA covered by Special Protection Area (Partheno Dasos Kentrikis Rodopis, 569 ha).

■ Conservation issues

Threats Afforestation (A), Construction/impact of dyke/dam/barrage (A), Deforestation (commercial) (A), Infrastructure (A), Intensified forest management (A), Unsustainable exploitation (B)

The forest is heavily exploited and the main threats come from deforestation and increasingly intensive forest management. Further threats come from dam construction along the Nestos, road building and hunting. Part of the area is a candidate SAC.

Mount Falakro — B2, C6 — 015

Admin region East Macedonia and Thrace
Coordinates 41°17′N 24°05′E
Altitude 70–2,232 m **Area** 20,000 ha

■ Site description

A mountainous area dissected by deep valleys with extensive alpine grassland, areas of scrub and some forest.

Habitats Forest and woodland (20%; broadleaved deciduous forest; native coniferous forest; mixed forest; broadleaved evergreen forest; treeline ecotone), Scrub (40%; sclerophyllous scrub/garrigue/maquis), Grassland (30%; steppe/dry calcareous grassland; alpine/subalpine/boreal grassland), Rocky areas (5%; inland cliff)
Land-use Agriculture (80%), Forestry (20%)

■ Birds

Species	Season	Year	Pop min	Pop max	Acc	Criteria
Circaetus gallicus Short-toed Eagle	B	1995	6	10	C	B2, C6
Aquila chrysaetos Golden Eagle	R	1995	2	4	B	C6
Burhinus oedicnemus Stone Curlew	B	1990	10	25	C	B2, C6

This is an important area for breeding raptors, forest birds and species characteristic of open grassland.

■ Protection status

National Partial **International** None
Part of IBA covered by Wildlife Refuge (Korilovou–Kallifytou/Dramas, 2,480 ha). Part of IBA covered by Wildlife Reserve (Xiropotanou/ Xiropotanou–Petrousas, 2,310 ha).

■ Conservation issues

Threats Deforestation (commercial) (B), Industrialization/urbanization (A), Recreation/tourism (C), Unsustainable exploitation (B)

The main threat is from industrial expansion around the lower slopes. Part of the area is a candidate SAC.

North and north-east Thasos island — B1ii, C2 — 016

Admin region East Macedonia and Thrace
Coordinates 40°40′N 24°40′E
Altitude 0–1,213 m **Area** 15,000 ha

■ Site description

A coastal mountain with steep rocky slopes, large subalpine meadows, pine *Pinus* forest, and broadleaved forest. There is also a wetland with reedbeds (*Phragmites*).

Habitats Forest and woodland (20%; broadleaved deciduous forest; native coniferous forest; mixed forest), Grassland (5%; alpine/subalpine/boreal grassland), Scrub (40%; sclerophyllous scrub/garrigue/maquis), Rocky areas (5%; sea cliff/rocky shore; inland cliff), Wetland (10%; sand-dunes/sand beach; standing fresh water; river/stream; water-fringe vegetation), Artificial landscape (15%; arable land; perennial crops/orchards/groves; urban parks/gardens)
Land-use Agriculture (60%), Fisheries/aquaculture (5%), Forestry (20%), Tourism/recreation (30%), Urban/industrial/transport (10%)

■ Birds

Species	Season	Year	Pop min	Pop max	Acc	Criteria
Puffinus yelkouan Yelkouan Shearwater	R	1996	Common		—	B1ii, C2

The site is important for raptors and migratory species.

■ Protection status

National Partial **International** None
Part of IBA covered by Wildlife Refuge (Portes Pyrgoudia/Limenarion–Theologou–Potamias–Prinou–Marion, 3,174 ha).

■ Conservation issues

Threats Aquaculture/fisheries (B), Burning of vegetation (A), Deforestation (commercial) (A), Extraction industry (A), Forest grazing (A), Recreation/tourism (B), Selective logging/cutting (B), Unsustainable exploitation (A)

Seventy-five percent of the forested area has been burned during the last 10 years. Other threats are overgrazing by cattle, quarrying, intensive hunting and logging in the remnant *Pinus* forest.

Mount Pangaio — B2, B3 — 017

Admin region East Macedonia and Thrace
Coordinates 40°55′N 24°07′E
Altitude 0–1,956 m **Area** 20,000 ha

■ Site description

A mountain dominated by *Fagus* and *Quercus* forest with extensive alpine pastures.

Habitats Forest and woodland (70%; broadleaved deciduous forest; native coniferous forest; mixed forest; broadleaved evergreen forest), Scrub (20%; sclerophyllous scrub/garrigue/maquis), Grassland (10%; alpine/subalpine/boreal grassland), Rocky areas (5%; inland cliff)
Land-use Agriculture (60%), Forestry (40%), Military (5%)

■ Birds

Species	Season	Year	Pop min	Pop max	Acc	Criteria
Alectoris graeca Rock Partridge	R	—	Common		—	B2
Phylloscopus bonelli Bonelli's Warbler	B	1996	Frequent		—	B3

This is an important site for resident forest species.

■ Protection status

National Partial **International** None
4,500 ha of IBA covered by Wildlife Refuge (Periochi Pangaiou Orous, 4,500 ha). Part of IBA covered by Wildlife Refuge (Kalias-Iliokomis/Protis–Iliokomis, 800 ha). Part of IBA covered by Wildlife Refuge (Kormitsas–Iliokomis, 1,560 ha).

Conservation issues

Threats Afforestation (B), Firewood collection (B), Forest grazing (B), Selective logging/cutting (C), Unsustainable exploitation (A)

The main threats are from hunting and afforestation with *Pinus* species. Part of the area is a candidate SAC.

Strymon delta — A1, A4i, B1i, C1, C2 — 018

Admin region East Macedonia and Thrace, Central Macedonia
Coordinates 40°48′N 23°50′E
Altitude 0–2 m **Area** 1,300 ha

Site description

A coastal river delta with saltmarsh, brackish coastal lagoons, wet grassland and reedbeds (*Phragmites*).

Habitats Grassland (5%; humid grassland), Wetland (55%; coastal lagoon; tidal river/enclosed tidal water; saltmarsh; sand-dunes/sand beach; standing fresh water; water-fringe vegetation), Artificial landscape (40%; arable land; other urban/industrial areas; perennial crops/orchards/groves)
Land-use Agriculture (50%), Fisheries/aquaculture (30%), Urban/industrial/transport (10%), Water management (30%)

Birds

Species	Season	Year	Pop min	Pop max	Acc	Criteria
Phalacrocorax pygmeus Pygmy Cormorant	W	—	Common		—	A1, C1
Pelecanus crispus Dalmatian Pelican	P	—	—	250	—	A1, A4i, B1i, C1, C2

This is an important site for breeding, passage and wintering waterbirds.

Protection status

National Partial **International** None
220 ha of IBA covered by Wildlife Reserve (Tsourakia–Gefyra Amfipolis/Amfipolis–Neon, 220 ha).

Conservation issues

Threats Agricultural intensification/expansion (A), Aquaculture/fisheries (B), Disturbance to birds (U), Industrialization/urbanization (A), Infrastructure (B), Recreation/tourism (B), Unsustainable exploitation (B)

The main threats are from the expansion of agriculture and increased industrial and tourism developments. The area is a candidate SAC.

Mount Kouskouras — B3, C6 — 019

Admin region Central Macedonia
Coordinates 41°12′N 23°36′E
Altitude 400–1,800 m **Area** 27,543 ha

Site description

A sparsely forested area of ravines and valleys. Livestock-farming is widespread.

Habitats Forest and woodland (50%; broadleaved deciduous forest; native coniferous forest; broadleaved evergreen forest), Scrub (10%; scrub), Grassland (30%; alpine/subalpine/boreal grassland), Rocky areas (10%; inland cliff)
Land-use Agriculture (60%), Forestry (30%), Not utilized (10%)

Birds

Species	Season	Year	Pop min	Pop max	Acc	Criteria
Dendrocopos syriacus Syrian Woodpecker	R	—	Common		—	B3, C6

This is an important site for species breeding in a mosaic of open forest, scrub and grassland.

Protection status

National None **International** High
Part of IBA covered by Wildlife Refuge (Viras/Eptamilon, Dinousas, Serron, 870 ha). 26,512 ha of IBA covered by Special Protection Area (Koilada Timiou Prodromou–Menoikiou Koryfes Orous Menoiko–Oros, 26,512 ha).

Conservation issues

Threats Deforestation (commercial) (B), Forest grazing (U), Infrastructure (A), Intensified forest management (A), Selective logging/cutting (B), Unsustainable exploitation (A)

The main threats are from intensified forest management and illegal hunting. Part of the area is a candidate SAC.

Lake Kerkini — A1, A4i, A4iii, B1i, B2, C1, C2, C3, C4, C6 — 020

Admin region Central Macedonia
Coordinates 41°12′N 23°09′E
Altitude 32–50 m **Area** 12,000 ha

Site description

Kerkini is a freshwater reservoir with reedbeds (*Phragmites*), some wet forest, and arable land.

Habitats Forest and woodland (15%; alluvial/very wet forest), Wetland (80%; standing fresh water; river/stream; water-fringe vegetation), Artificial landscape (5%; arable land)
Land-use Agriculture (5%), Fisheries/aquaculture (80%), Forestry (10%), Tourism/recreation, Water management (80%)

Birds

Species	Season	Year	Pop min	Pop max	Acc	Criteria
Phalacrocorax carbo Cormorant	R	1993	1,125	—	—	B1i, C3
Phalacrocorax carbo Cormorant	W	1993	2,500	—	—	B1i, C3
Phalacrocorax pygmeus Pygmy Cormorant	R	1991	430	—	—	A1, A4i, B1i, B2, C1, C2, C6
Phalacrocorax pygmeus Pygmy Cormorant	W	1991	500	—	—	A1, A4i, B1i, C1, C2
Pelecanus onocrotalus White Pelican	P	1996	—	900	—	A4i, B1i, C2
Pelecanus crispus Dalmatian Pelican	W	1996	—	450	—	A1, A4i, B1i, C1, C2
Pelecanus crispus Dalmatian Pelican	P	1996	—	1,200	—	A1, A4i, B1i, C1, C2
Nycticorax nycticorax Night Heron	B	1996	530	900	—	A4i, B1i, B2, C2, C6
Ardeola ralloides Squacco Heron	B	1996	170	260	—	A4i, B1i, B2, C2, C6
Egretta garzetta Little Egret	B	1996	445	640	—	A4i, B1i, C2
Egretta alba Great White Egret	W	1996	70	360	—	A4i, B1i, C2
Ciconia ciconia White Stork	B	1996	110	150	—	B2, C6
Plegadis falcinellus Glossy Ibis	P	1996	120	600	—	A4i, B1i, C2
Platalea leucorodia Spoonbill	R	1996	55	120	—	A4i, B1i, B2, C2, C6
Anser erythropus Lesser White-fronted Goose	P	1996	12	110	—	A1, C1
Anser anser Greylag Goose	W	1996	85	550	—	B1i, C3
Tadorna tadorna Shelduck	W	1996	130	1,200	—	B1i, C3
Anas crecca Teal	W	1995	31,500	—	—	A4i, B1i, C3
Anas platyrhynchos Mallard	W	1995	22,000	—	—	B1i, C3
Aythya ferina Pochard	W	1995	15,000	—	—	A4i, B1i, C3
Aythya nyroca Ferruginous Duck	R	1995	Frequent		—	C6
Oxyura leucocephala White-headed Duck	W	1993	3	100	—	A1, C1
Accipiter brevipes Levant Sparrowhawk	B	—	—	10	—	B2
Aquila clanga Greater Spotted Eagle	W	1996	8	13	—	A1, C1
Recurvirostra avosetta Avocet	P	—	—	1,200	—	A4i, B1i, C2
Chlidonias hybridus Whiskered Tern	B	1993	250	—	—	B1i, B2, C2, C6
Chlidonias niger Black Tern	B	1993	50	—	—	C2

A very important area for breeding, passage and wintering waterbirds and raptors and breeding and migratory passerines. The site is valuable as a foraging area for raptors nesting elsewhere. Species of global conservation concern that do not meet IBA criteria: *Haliaeetus albicilla* (non-breeding, 2–4 wintering individuals), *Aquila heliaca* (1–3 wintering individuals), *Falco naumanni* (passage), *Gallinago media* (wintering).

Protection status

National None **International** High
10,996 ha of IBA covered by Ramsar Site (Kerkini reservoir, 10,996 ha). 10,995 ha of IBA covered by Special Protection Area (Techniti Limni Kerkinis, 10,995 ha).

Conservation issues

Threats Agricultural intensification/expansion (B), Aquaculture/fisheries (B), Construction/impact of dyke/dam/barrage (A), Deforestation (commercial) (U), Disturbance to birds (C), Filling-in of wetlands (B), Industrialization/urbanization (U), Infrastructure (B), Other (B), Selective logging/cutting (U), Unsustainable exploitation (B)

The major threat is alteration of the dyke system to raise water-levels. Illegal land-reclamation, road construction, rubbish tipping ('Other' threats) and fishing are also threats. Part of the area is a candidate SAC.

Mount Kroussia (or Mavrovouni) — A1, B2, C1, C6 — 021

Admin region Central Macedonia
Coordinates 41°10'N 23°09'E
Altitude 100–1,150 m **Area** 18,000 ha

Site description

Forested hills south of Lake Kerkini. *Fagus* forest dominates at higher altitudes and *Quercus* on the lower slopes.

Habitats Forest and woodland (80%; broadleaved deciduous forest), Scrub (20%; scrub)
Land-use Agriculture (20%), Forestry (80%)

Birds

Species	Season	Year	Pop min	Pop max	Acc	Criteria
Aquila clanga Greater Spotted Eagle	W	—	Common		—	A1, C1
Hieraaetus pennatus Booted Eagle	B	1996	1	5	—	B2, C6

This is an important IBA for breeding and passage raptors and breeding forest passerines. Species of global conservation concern that do not meet IBA criteria: *Haliaeetus albicilla* (2–4 resident individuals), *Aquila heliaca* (passage), *Falco naumanni* (passage).

Protection status

National None **International** None

Conservation issues

Threats Deforestation (commercial) (A), Disturbance to birds (U), Firewood collection (U), Infrastructure (U), Intensified forest management (C), Selective logging/cutting (U), Unsustainable exploitation (A)

The main threats are logging and hunting. The area is a candidate SAC.

Mount Beles (Kerkini) — B2, C6 — 022

Admin region Central Macedonia
Coordinates 41°20'N 23°04'E
Altitude 100–2,031 m **Area** 34,000 ha

Site description

A remote mountain range characterized by bare summits, wooded slopes, gorges and scree slopes.

Habitats Forest and woodland (90%; broadleaved deciduous forest; native coniferous forest), Grassland (10%; alpine/subalpine/boreal grassland)
Land-use Agriculture (10%), Forestry (85%), Military (5%)

Birds

Species	Season	Year	Pop min	Pop max	Acc	Criteria
Accipiter brevipes Levant Sparrowhawk	B	1996	5	10	—	B2
Aquila pomarina Lesser Spotted Eagle	B	1996	4	8	—	B2, C6

This is an important area for breeding and passage raptors and alpine and forest species. Species of global conservation concern that do not meet IBA criteria: *Haliaeetus albicilla* (non-breeding), *Aquila clanga* (wintering), *Aquila heliaca* (wintering), *Falco naumanni* (passage).

Protection status

National None **International** None

Conservation issues

Threats Afforestation (A), Deforestation (commercial) (U), Infrastructure (U), Intensified forest management (U), Selective logging/cutting (C), Unsustainable exploitation (A)

The main threats are the replacement of native forest with *Pinus* plantations and illegal hunting. The area is a candidate SAC.

Lake Doirani — A1, A4i, B1i, C1, C2 — 023

Admin region Central Macedonia
Coordinates 41°13'N 22°46'E
Altitude 145–150 m **Area** 2,000 ha

Site description

Lake Doirani is fringed with reedbeds (*Phragmites*) and riverine forest. The site lies adjacent to site 010 in the FYR Macedonia.

Habitats Forest and woodland (alluvial/very wet forest), Scrub (scrub), Wetland (sand-dunes/sand beach; standing fresh water; water-fringe vegetation), Artificial landscape (arable land)
Land-use Agriculture, Fisheries/aquaculture, Urban/industrial/transport, Water management

Birds

Species	Season	Year	Pop min	Pop max	Acc	Criteria
Phalacrocorax pygmeus Pygmy Cormorant	W	1988	—	112	A	A1, C1
Pelecanus crispus Dalmatian Pelican	W	1994	—	76	A	A1, A4i, B1i, C1, C2

The site is important for wintering waterbirds.

Protection status

National Low **International** None
9 ha of IBA covered by Natural Monument (Mouries Oak and Ash Forest, 9 ha).

Conservation issues

Threats Agricultural intensification/expansion (A), Drainage (A), Groundwater abstraction (A), Industrialization/urbanization (B), Other (A), Unsustainable exploitation (U)

The main threats to the site are from the intensification of agriculture and the resulting over-exploitation of water, pollution and hunting. The water-level has dropped in recent years as a result of ground- and surface-water abstraction for irrigation ('Other' threats). Part of the area is a candidate SAC.

Artzan marshes — A1, A4i, B1i, C1, C2 — 024

Admin region Central Macedonia
Coordinates 41°01'N 22°39'E
Altitude 25–100 m **Area** 1,700 ha

Site description

A freshwater lake, marshes and wet meadows 2 km north-west of Vafiohori village. The area is divided by a canal. Human activities include arable agriculture, stock-raising, and hunting.

Habitats Forest and woodland (alluvial/very wet forest), Grassland (humid grassland), Wetland (standing fresh water; water-fringe vegetation; fen/transition mire/spring), Rocky areas (inland cliff), Artificial landscape (arable land)
Land-use Agriculture (90%), Water management (90%)

Birds

Species	Season	Year	Pop min	Pop max	Acc	Criteria
Pelecanus crispus Dalmatian Pelican	P	—	—	—	—	A1, A4i, B1i, C1, C2

This is an important site for passage and wintering waterbirds. Species of global conservation concern that do not meet IBA criteria: *Haliaeetus albicilla* (passage).

Protection status

National None **International** None

Conservation issues

Threats Agricultural intensification/expansion (A), Burning of vegetation (U), Construction/impact of dyke/dam/barrage (A), Disturbance to birds (A), Drainage (A), Dredging/canalization (A), Infrastructure (A), Unsustainable exploitation (A)

The site is seriously threatened by construction of a large dam in the marshland which began in 1996. This will flood areas of marshland and wet meadows. Other threats include hunting and the expansion of agriculture. A quarry is to reopen at the southern end of the site.

River Axios — A1, A4i, B1i, C1, C2 — 025

Admin region Central Macedonia
Coordinates 40°47′N 22°38′E
Altitude 0–70 m **Area** 12,400 ha

Site description

This site contains the River Axios together with patches and islands of riverine forest and pasture.

Habitats Forest and woodland (10%; alluvial/very wet forest), Wetland (70%; river/stream), Artificial landscape (20%; arable land)
Land-use Agriculture (20%), Water management (30%)

Birds

Species	Season	Year	Pop min	Pop max	Acc	Criteria
Phalacrocorax pygmeus Pygmy Cormorant	W	1997	5,000	—	A	A1, A4i, B1i, C1, C2

The site is an important corridor for migratory species. Species of global conservation concern that do not meet IBA criteria: *Aquila clanga* (1–3 wintering individuals).

Protection status

National Partial **International** None
1,310 ha of IBA covered by Wildlife Reserve (Palmitsi–Kampas/Evzonan–M. Dasous, 1,310 ha). 800 ha of IBA covered by (Altsaki/Idomenis, 800 ha).

Conservation issues

Threats Agricultural intensification/expansion (A), Construction/impact of dyke/dam/barrage (B), Extraction industry (A), Industrialization/urbanization (A), Unsustainable exploitation (B)

A major threat, particularly to waterbirds, is the extraction of sand from the river-bed, sandbanks and forested areas. Chemical industries deposit solid toxic waste along the river. Urban and agricultural pollution are also problems. The river is dammed and water is extracted for irrigation.

Area of Anthofito — A1, B2, C1, C6 — 026

Admin region Central Macedonia
Coordinates 40°52′N 22°43′E
Altitude 40–70 m **Area** 1,200 ha

Site description

A village 8 km west of Lake Pikrolimni. It is surrounded by cereal farming.

Habitats Artificial landscape (100%; arable land; other urban/industrial areas)
Land-use Agriculture (95%), Urban/industrial/transport (5%)

Birds

Species	Season	Year	Pop min	Pop max	Acc	Criteria
[1] ***Falco naumanni*** Lesser Kestrel	B	1996	80	—	A	A1, B2, C1, C6

1. Individuals.

The site is important for species associated with arable agriculture.

Protection status

National None **International** None

Conservation issues

Threats Agricultural intensification/expansion (A), Disturbance to birds (U), Other (B)

The main threats include intensification of agriculture, disturbance and the demolition of the old buildings in which there is a colony of *Falco naumanni* ('Other' threats).

Lake Pikrolimni — A1, C1 — 027

Admin region Central Macedonia
Coordinates 40°50′N 22°49′E
Altitude 47–53 m **Area** 2,000 ha

Site description

A salt lake with large seasonal variations in water level. The lake is surrounded by cereal crops and marsh.

Habitats Wetland (80%; mudflat/sandflat; saltmarsh; standing brackish and salt water; water-fringe vegetation; fen/transition mire/spring), Artificial landscape (20%; arable land)
Land-use Agriculture (55%), Tourism/recreation (10%), Not utilized (35%)

Birds

Species	Season	Year	Pop min	Pop max	Acc	Criteria
Falco naumanni Lesser Kestrel	N	1997	40	—	B	A1, C1

This is an important site for breeding, passage and wintering waterbirds and also for raptors. Species of global conservation concern that do not meet IBA criteria: *Pelecanus crispus* (non-breeding).

Protection status

National None **International** None

Conservation issues

Threats Agricultural intensification/expansion (B), Disturbance to birds (U), Other (A), Recreation/tourism (C), Unsustainable exploitation (U)

Threats come from new tourist developments, intensified agriculture and the cutting of reedbeds to provide fodder for livestock ('Other' threats). *Circus pygargus* nests in the cereal fields but these are usually harvested before fledging. Part of the area is a candidate SAC.

Axios, Loudias, and Aliakmon estuaries — A1, A4i, A4iii, B1i, B2, B3, C1, C2, C3, C4, C6 — 028

Admin region Central Macedonia
Coordinates 40°30′N 22°43′E
Altitude 0–5 m **Area** 20,000 ha

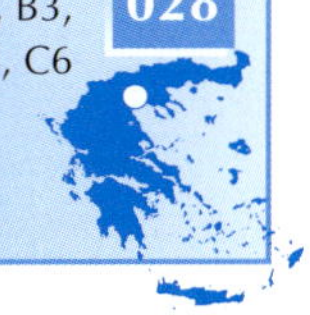

Site description

A coastal wetland of small lagoons, sandflats, and alluvial forests dominated by *Tamarix*, *Alnus*, and *Salix*. Human activities include rice production, livestock farming, fishing and mussel culture.

Habitats Forest and woodland (10%; alluvial/very wet forest), Wetland (60%; tidal river/enclosed tidal water; saltmarsh; coastal lagoon; river/stream; water-fringe vegetation), Artificial landscape (30%; arable land)
Land-use Agriculture (80%), Fisheries/aquaculture (30%), Hunting (50%), Nature conservation/research (30%), Tourism/recreation (10%)

Birds

Species	Season	Year	Pop min	Pop max	Acc	Criteria
Phalacrocorax pygmeus Pygmy Cormorant	W	1997	5,000	—	A	A1, A4i, B1i, C1, C2
Pelecanus crispus Dalmatian Pelican	P	1990	5	50	B	A1, A4i, B1i, C1, C2
Ixobrychus minutus Little Bittern	B	1990	30	80	C	B2, C2, C6
Nycticorax nycticorax Night Heron	B	1990	420	800	A	A4i, B1i, B2, C2, C6
Ardeola ralloides Squacco Heron	B	1990	150	300	A	A4i, B1i, B2, C2, C6
Egretta garzetta Little Egret	B	1990	274	700	A	A4i, B1i, C2
Platalea leucorodia Spoonbill	B	1995	16	50	A	B1i, B2, C2, C6
Haematopus ostralegus Oystercatcher	B	1996	5	15	B	B1i, C3
Himantopus himantopus Black-winged Stilt	B	1996	100	150	A	B1i, C2
Recurvirostra avosetta Avocet	B	1996	50	70	A	B3, C6

Species ... continued	Season	Year	Pop min	Pop max	Acc	Criteria
Glareola pratincola Collared Pratincole	B	1996	50	100	B	A4i, B1i, B2, C2
Charadrius alexandrinus Kentish Plover	B	1990	50	100	B	B1i, B2, C3
Limosa limosa Black-tailed Godwit	P	1990	100	4,000	B	B1i, C3
Numenius tenuirostris Slender-billed Curlew	P	—	1	—	B	A1, C1
[1] ***Larus melanocephalus*** Mediterranean Gull	B	1996	800	1,400	B	A4i, B1i, B3, C2, C6
Larus cachinnans Yellow-legged Gull	W	1990	—	5,000	B	A4i, B1i
Sterna albifrons Little Tern	B	1996	70	130	B	B2, C6
Calandrella brachydactyla Short-toed Lark	B	1990	Frequent		—	C6

1. Large increase.

This is a very important site for breeding, passage and wintering waterbirds. Species of global conservation concern that do not meet IBA criteria: *Aythya nyroca* (passage), *Haliaeetus albicilla* (wintering), *Aquila clanga* (1–3 wintering individuals), *Gallinago media* (passage). The site holds 20,000 or more waterbirds in winter, on a regular basis (including *Phalacrocorax pygmeus* 5,000; *Anas penelope* 1,000; *Calidris alpina* 2,500; *Larus ridibundus* 7,000; *Larus cachinnans* 5,000).

Protection status

National Partial **International** Partial

2,381 ha of IBA covered by Wildlife Refuge (Stergiou–Aliakmon/Aiginiou, 2,381 ha). 1,200 ha of IBA covered by Wildlife Refuge (Delta Aliakmona/Kleidiou, 1,200 ha). 2,150 ha of IBA covered by Wildlife Refuge (Delta Axiou–Chalastras, 2,150 ha). Part of IBA overlaps with Protected Area. 11,808 ha of IBA covered by Ramsar Site (Delta Axiou, Loudia and Aliakmon–Alyki Kitrous, 11,808 ha). 13,763 ha of IBA covered by Special Protection Area (Delta Axiou–Loudia–Aliakmona–Alyki Kitrous, 13,763 ha).

Conservation issues

Threats Agricultural intensification/expansion (A), Construction/impact of dyke/dam/barrage (C), Dredging/canalization (A), Industrialization/urbanization (A), Unsustainable exploitation (B)

Threats include illegal housing, intensive livestock farming, and water pollution from industry and intensive agriculture. Conservation projects have included LIFE projects on *Numenius tenuirostris*, *Phalacrocorax pygmeus*, and *Anser erythropus*. The area is a candidate SAC.

Gallikos estuary and Kalohori lagoon — A1, A4i, B1i, B3, C1, C2, C6 — 029

Admin region Central Macedonia
Coordinates 40°38′N 22°51′E
Altitude 0–5 m **Area** 1,800 ha

Site description

A coastal delta, with salt and freshwater marshes and lagoons.

Habitats Scrub (5%; scrub), Wetland (70%; mudflat/sandflat; saltmarsh; coastal lagoon; standing fresh water; standing brackish and salt water; water-fringe vegetation), Marine areas (20%; sea inlet/coastal features), Artificial landscape (5%; arable land; other urban/industrial areas)
Land-use Agriculture, Fisheries/aquaculture, Urban/industrial/transport

Birds

Species	Season	Year	Pop min	Pop max	Acc	Criteria
[1] ***Phalacrocorax pygmeus*** Pygmy Cormorant	W	1989	10	400	A	A1, A4i, B1i, C1, C2
Himantopus himantopus Black-winged Stilt	B	1996	100	—	A	B1i, C2
Recurvirostra avosetta Avocet	B	1996	—	80	A	B3, C6
Glareola pratincola Collared Pratincole	B	1996	10	40	B	B1i, C2
Numenius tenuirostris Slender-billed Curlew	P	1988	1	—	A	A1, C1
Larus melanocephalus Mediterranean Gull	B	1991	20	50	B	C2

1. Large decrease.

This is an important area for breeding, passage and wintering waterbirds and other passage migrants. Species of global conservation concern that do not meet IBA criteria: *Pelecanus crispus* (passage), *Aythya nyroca* (passage), *Aquila clanga* (wintering).

Protection status

National Partial **International** None

Part of IBA covered by protected area (details unknown).

Conservation issues

Threats Construction/impact of dyke/dam/barrage (A), Disturbance to birds (U), Filling-in of wetlands (A), Industrialization/urbanization (A), Infrastructure (U), Unsustainable exploitation (A), Other (A)

Rubbish dumping ('Other' threats, above) and infilling threaten the important Kalohori lagoon which is also threatened by effluent discharges from nearby industries. The centre of the wetland is also threatened by the expansion of livestock farms. There are plans for the construction of a racecourse in the area.

Alyki Angelochoriou lagoon (Megalou Emvolou) — A1, A4i, B1i, C1, C2, C3 — 030

Admin region Central Macedonia
Coordinates 40°29′N 22°49′E
Altitude 0–2 m **Area** 357 ha

Site description

The site includes a coastal lagoon, a saltpan, saltmarshes and a marine zone. Human activities include salt production ('Other' land-use, below), agriculture, hunting and tourism.

Habitats Wetland (80%; saltmarsh; sand-dunes/sand beach; coastal lagoon; standing brackish and salt water; water-fringe vegetation), Marine areas (10%; sea inlet/coastal features)
Land-use Agriculture (30%), Fisheries/aquaculture (10%), Tourism/recreation, Water management (45%), Other (10%)

Birds

Species	Season	Year	Pop min	Pop max	Acc	Criteria
Charadrius alexandrinus Kentish Plover	P	—	—	384	A	B1i, C3
Numenius tenuirostris Slender-billed Curlew	P	—	0	1	A	A1, C1
Larus melanocephalus Mediterranean Gull	N	—	—	5,500	A	A4i, B1i, C2

This is an important site for passage and wintering waders. Species of global conservation concern that do not meet IBA criteria: *Phalacrocorax pygmeus* (up to 18 wintering individuals).

Protection status

National None **International** None

Conservation issues

Threats Aquaculture/fisheries (B), Recreation/tourism (C), Unsustainable exploitation (A)

The main threat is from hunting. Part of the area is a candidate SAC.

Epanomi lagoon — B1i, B2, C2, C3 — 031

Admin region Central Macedonia
Coordinates 40°23′N 22°54′E
Altitude 0–2 m **Area** 450 ha

Site description

An almost flat peninsula of low sand-dunes with a lagoon, saltmarsh and *Tamarix* scrub.

Habitats Scrub (20%; scrub), Wetland (75%; saltmarsh; sand-dunes/sand beach; coastal lagoon)
Land-use Agriculture (60%), Fisheries/aquaculture, Tourism/recreation

Birds

Species	Season	Year	Pop min	Pop max	Acc	Criteria
Phalacrocorax carbo Cormorant	W	—	—	2,350	A	B1i, C3
Glareola pratincola Collared Pratincole	B	—	10	45	B	B1i, B2, C2

This is an important site for wintering ducks and breeding, passage and wintering waders. Species of global conservation concern that do not meet IBA criteria: *Branta ruficollis* (up to five wintering individuals), *Aythya nyroca* (passage).

Protection status

National High **International** None
450 ha of IBA covered by Wildlife Refuge (Tsairi/Epanomis, 500 ha).

Conservation issues

Threats Dredging/canalization (B), Recreation/tourism (A)

The main threats are from dredging and increased tourism. The area is a candidate SAC.

Lakes Volvi–Langada (or Koronia–Ag. Vasiliou) and Rentina gorge (or Makedonika Tempi)

A1, A4i, A4iii, B1i, B2, C1, C2, C3, C4, C6 **032**

Admin region Central Macedonia
Coordinates 40°41′N 23°22′E
Altitude 0–700 m **Area** 43,000 ha

Site description

The site comprises two freshwater lakes and a gorge. Lake Koronia is shallow, eutrophic and polluted. Lake Volvi is deeper and overflows through the Rentina gorge. The gorge is dominated by maquis and riverine scrub.

Habitats Scrub (20%; sclerophyllous scrub/garrigue/maquis), Wetland (27%; standing fresh water), Artificial landscape (47%; arable land)
Land-use Agriculture (55%), Fisheries/aquaculture (27%)

Birds

Species	Season	Year	Pop min	Pop max	Acc	Criteria
Podiceps cristatus Great Crested Grebe	W	1997	6,729	6,729	A	B1i, C3
Phalacrocorax pygmeus Pygmy Cormorant	W	1997	43	43	A	A1, C1
Pelecanus crispus Dalmatian Pelican	W	1995	249	249	A	A1, A4i, B1i, C1, C2
Pelecanus crispus Dalmatian Pelican	P	1995	—	473	A	A1, A4i, B1i, C1, C2
Egretta alba Great White Egret	W	1997	35	35	A	A4i, B1i, C2
Ciconia ciconia White Stork	B	1993	80	100	A	B2, C6
[1] ***Phoenicopterus ruber*** Greater Flamingo	N	1995	—	1,410	A	A4i, B1i, C2
Aythya ferina Pochard	W	1997	8,029	8,029	A	B1i, C3
Accipiter brevipes Levant Sparrowhawk	B	1992	Uncommon		—	B2
Himantopus himantopus Black-winged Stilt	P	1995	—	600	B	B1i, C2
Recurvirostra avosetta Avocet	P	1995	—	1,150	B	A4i, B1i, C2

1. Large increase.

This site is important for passage and wintering waterbirds. Species of global conservation concern that do not meet IBA criteria: *Aythya nyroca* (minimum 43 wintering individuals), *Oxyura leucocephala* (wintering), *Haliaeetus albicilla* (wintering), *Falco naumanni* (passage). The site holds 20,000 or more waterbirds, on a regular basis (including *Podiceps cristatus* 6,729; *Aythya ferina* 8,029; *Aythya fuligula* 1,361; *Fulica atra* 12,736).

Protection status

National Partial **International** Partial
2,100 ha of IBA covered by Wildlife Refuge (Limni Lagada, 2,100 ha). 500 ha of IBA covered by Wildlife Refuge (Paralimnio Dasos Apollonias, 500 ha). 669 ha of IBA covered by Wildlife Refuge (Richios Potamos, 669 ha). 16,388 ha of IBA covered by Ramsar Site (Lakes Volvi and Langada (Koronia), 16,388 ha). 15,671 ha of IBA covered by Special Protection Area (Limnes Volvi and Lagkada and Stena Rentinas, 15,671 ha).

Conservation issues

Threats Agricultural intensification/expansion (A), Burning of vegetation (A), Forest grazing (B), Groundwater abstraction (A), Industrialization/urbanization (A), Infrastructure (B), Natural events (B)

Declining water-levels and water pollution, from agricultural and industrial sources, are a severe threat. A pollution incident in 1995 left thousands of waterbirds dead. In the gorge, fire is a serious problem and has previously caused severe damage. Part of the area is a candidate SAC.

Mount Cholomontas

C7 **033**

Admin region Central Macedonia
Coordinates 40°23′N 23°30′E
Altitude 50–1,000 m **Area** 27,500 ha

Site description

A mountainous area covered by *Fagus*, *Quercus*, and *Castanea* forest and *Pinus* plantations. Evergreen broadleaved forests also occur. Human activities include agriculture, forestry, and hunting.

Habitats Forest and woodland (70%; broadleaved deciduous forest; broadleaved evergreen forest), Scrub (15%; sclerophyllous scrub/garrigue/maquis), Artificial landscape (15%; arable land; forestry plantation)
Land-use Agriculture (20%), Forestry (20%), Hunting (7%), Nature conservation/ research (50%)

Birds

This is an important site for breeding forest species.

Protection status

National Partial **International** Partial
2,100 ha of IBA covered by Wildlife Refuge (Pr. Ilius/Poligirou, 2,100 ha). 1,650 ha of IBA covered by Wildlife Refuge (Cholomonta–Arneas, 1,650 ha). 1,446 ha of IBA covered by Controlled Hunting Area (Pravitas, 1,446 ha). 21,248 ha of IBA covered by Special Protection Area (Taxiarchis–Polygyros, 21,248 ha)

Conservation issues

Threats Afforestation (A), Firewood collection (C), Selective logging/cutting (C)

Replacement of native forest with plantations is the most serious threat to the site. Part of the area is owned by and used for ecological research by the Aristotelian University of Thessaloniki. Part of the area is a candidate SAC.

Agios Mamas marsh

A1, B1i, B2, C1, C2 **034**

Admin region Central Macedonia
Coordinates 40°14′N 23°20′E
Altitude 0–4 m **Area** 633 ha

Site description

A small coastal wetland with standing water, sand-dunes, and reedbeds (*Phragmites*).

Habitats Wetland (100%; sand-dunes/sand beach; standing brackish and salt water; water-fringe vegetation)
Land-use Tourism/recreation (40%), Not utilized, Military

Birds

Species	Season	Year	Pop min	Pop max	Acc	Criteria
Himantopus himantopus Black-winged Stilt	B	1996	70	300	A	B1i, C2
[1] ***Glareola pratincola*** Collared Pratincole	B	1996	20	50	B	B1i, B2, C2
Numenius tenuirostris Slender-billed Curlew	P	1987	1	—	A	A1, C1

1. Large decrease.

This is an important location for breeding and passage waders.

■ **Protection status**

National None **International** None

■ **Conservation issues**

Threats Dredging/canalization (C), Filling-in of wetlands (B), Industrialization/urbanization (B), Recreation/tourism (B), Other (C), Unsustainable exploitation (A)

The main threats are from illegal house construction, filling-in of the wetland and rubbish tipping ('Other' threats). A small part of the wetland is used by the military. The area is a candidate SAC.

Nea Fokea marshes

B1i, C2 **035**

Admin region Central Macedonia
Coordinates 40°07'N 23°19'E
Altitude 0–20 m **Area** 340 ha

■ **Site description**

A coastal wetland consisting of two periodically inundated brackish marshes, sand-dunes and a 90 ha pine *Pinus* woodland. The existing drainage works in both marshes drain the major part of the wetland in the summer.

Habitats Forest and woodland (25%; native coniferous forest), Wetland (75%; sand-dunes/sand beach; standing brackish and salt water)
Land-use Tourism/recreation (50%), Water management (70%)

■ **Birds**

Species	Season	Year	Pop min	Pop max	Acc	Criteria
Himantopus himantopus Black-winged Stilt	B	1996	90	100	A	B1i, C2

This is an important site for breeding and passage waterbirds. Species of global conservation concern that do not meet IBA criteria: *Aythya nyroca* (4 passage individuals), *Haliaeetus albicilla*.

■ **Protection status**

National None **International** None

■ **Conservation issues**

Threats Drainage (A), Dredging/canalization (A), Recreation/tourism (A), Unsustainable exploitation (A), Other (C)

Drainage is a major threat to the area, with a pumping station and canals within the site. Garbage dumping is leading to a gradual infilling of part of the site ('Other' threats, above). The beach and sand-dune area are used for recreation in the summer. The numbers of wintering birds are thought to be small as a result of hunting.

Mount Itamos

B2, C6 **036**

Admin region Central Macedonia
Coordinates 40°08'N 23°50'E
Altitude 0–823 m **Area** 27,000 ha

■ **Site description**

A hilly coastal region dominated at higher altitudes by conifer forest. At lower altitudes there is a mosaic of maquis, small wetlands and olive-groves.

Habitats Forest and woodland (70%; native coniferous forest), Scrub (12%; sclerophyllous scrub/garrigue/maquis), Wetland (7%), Marine areas (5%; sea inlet/coastal features)
Land-use Agriculture (30%), Forestry (70%), Tourism/recreation (10%)

■ **Birds**

Species	Season	Year	Pop min	Pop max	Acc	Criteria
Falco biarmicus Lanner	R	1996	Rare		—	B2, C6

This is an important area for breeding and passage raptors and resident forest species.

■ **Protection status**

National Partial **International** None

8,500 ha of IBA covered by Wildlife Refuge (Dragounteli N. Marmara/Sithonias Sikeas–Sartis, 8,500 ha). Part of IBA covered by Wildlife Refuge (Ktima Karra/N. Marmara, 1,700 ha).

■ **Conservation issues**

Threats Agricultural intensification/expansion (C), Filling-in of wetlands (B), Industrialization/urbanization (B), Recreation/tourism (B), Selective logging/cutting (A), Unsustainable exploitation (B)

Logging of the *Pinus nigra* forest is a major threat, particularly to large raptors. Other threats include tourism developments along the coast, spreading urbanization, and hunting. A management plan has existed since 1996. Part of the area is a candidate SAC.

Mount Athos

B1ii, C3 **037**

Admin region Agion Oros
Coordinates 40°10'N 24°20'E
Altitude 0–2,033 m **Area** 12,000 ha

■ **Site description**

A coastal area of rugged hills and mountains. The site covers the southern part of the peninsula of Agion Oros.

Habitats Forest and woodland (60%; broadleaved deciduous forest; native coniferous forest; mixed forest; broadleaved evergreen forest; treeline ecotone), Scrub (40%; scrub; sclerophyllous scrub/garrigue/maquis), Rocky areas (10%; sea cliff/rocky shore; scree/boulders; inland cliff)
Land-use Forestry (50%), Nature conservation/research (20%), Not utilized (50%)

■ **Birds**

Species	Season	Year	Pop min	Pop max	Acc	Criteria
Puffinus yelkouan Yelkouan Shearwater	U	1987	Abundant		—	B1ii, C3

This is a nationally important area for breeding and passage raptors. Shearwaters occur offshore. Species of global conservation concern that do not meet IBA criteria: *Falco naumanni*, *Larus audouinii*.

■ **Protection status**

National None **International** High

12,000 ha of IBA covered by World Heritage Site (Mount Athos, 31,000 ha).

■ **Conservation issues**

Threats Burning of vegetation (U), Deforestation (commercial) (C), Infrastructure (U), Intensified forest management (B), Other (B), Selective logging/cutting (A)

The main threats are from logging, intensive forest management and a disease that kills *Castanea sativa* ('Other' threats).

Mounts Tzena and Pinovo

B3, C6 **038**

Admin region Central Macedonia
Coordinates 41°06'N 22°12'E
Altitude 400–2,182 m **Area** 13,154 ha

■ **Site description**

Mounts Tzena and Pinovo form the central part of the Aridaia mountain range. The vegetation comprises extensive broadleaved forests and alpine and subalpine grassland. The site lies adjacent to site 009 in the FYR Macedonia.

Habitats Forest and woodland (48%; broadleaved deciduous forest; treeline ecotone), Scrub (12%; sclerophyllous scrub/garrigue/maquis), Grassland (32%; alpine/subalpine/boreal grassland), Rocky areas (8%; inland cliff)
Land-use Agriculture (20%), Forestry (30%), Military (10%), Not utilized (20%), Tourism/recreation (10%), Water management (5%)

Birds

Species	Season	Year	Pop min	Pop max	Acc	Criteria
Dendrocopos medius Middle Spotted Woodpecker	R	1996	Common		—	B3, C6

This is an important breeding and foraging site for raptors. Species of global conservation concern that do not meet IBA criteria: *Aquila heliaca* (non-breeding).

Protection status

National Partial **International** High

5,000 ha of IBA covered by Wildlife Refuge (Pinovo–Tzena/Notias, 5,199 ha). 5,000 ha of IBA covered by Special Protection Area (Ori Tzena, 12,579 ha).

Conservation issues

Threats Abandonment/reduction of land management (A), Afforestation (A), Firewood collection (C), Infrastructure (A), Intensified forest management (A), Unsustainable exploitation (A)

The main threats are from hunting, the afforestation of abandoned cultivation and grassland with plantations of *Robinia pseudacacia*, and intensified forest management. Part of the area is a candidate SAC.

Mount Voras (Kaimaktchalan)

B2, B3, C6 **039**

Admin region Central Macedonia, Western Macedonia
Coordinates 40°58′N 21°55′E
Altitude 500–2,524 m **Area** 52,286 ha

Site description

A tall mountain with densely forested valleys. The main human activities include forestry, livestock-farming and winter sports.

Habitats Forest and woodland (63%; broadleaved deciduous forest; native coniferous forest; mixed forest; treeline ecotone), Scrub (12%; scrub), Grassland (15%; alpine/subalpine/boreal grassland), Artificial landscape (8%; arable land; perennial crops/orchards/groves; ruderal land)
Land-use Agriculture (30%), Forestry (60%), Tourism/recreation (20%)

Birds

Species	Season	Year	Pop min	Pop max	Acc	Criteria
Hieraaetus fasciatus Bonnelli's Eagle	R	1996	2	—	—	C6
Strix aluco Tawny Owl	R	1996	Common		—	B3
Emberiza cia Rock Bunting	R	1996	Frequent		—	B2

This is an important area for resident raptors, forest species and species associated with open montane habitats. Species of global conservation concern that do not meet IBA criteria: *Aquila heliaca*, *Falco naumanni* (non-breeding).

Protection status

National Partial **International** Partial

3 ha of IBA covered by Natural Monument (Pefkoto Beech Forest, 3 ha). 192 ha of IBA covered by Natural Monument (Almopia Forest, 192 ha). 3,775 ha of IBA covered by Wildlife Refuge (Ano Koryfis–Dragos–Ormas/Ormas–Sarakinon, 3,775 ha). 4,800 ha of IBA covered by Wildlife Refuge (Skopou, 4,800 ha). 192 ha of IBA covered by Biogenetic Reserve (Council of Europe) (Natural Monument of Mixed Forest Almopias Aridaias, 192 ha). 40,326 ha of IBA covered by Special Protection Area (Koryfes Orous Voras, 40,326 ha).

Conservation issues

Threats Abandonment/reduction of land management (C), Afforestation (C), Construction/impact of dyke/dam/barrage (C), Extraction industry (A), Firewood collection (B), Infrastructure (A), Intensified forest management (A), Recreation/tourism (A), Unsustainable exploitation (B)

Threats include the construction of a ski centre and its associated infrastructure and more widespread tourism-based development. A dam is planned for the Papadia river. Part of the area is a candidate SAC.

Apsalos and Moglenitsa gorges

B2, C6 **040**

Admin region Central Macedonia
Coordinates 40°50′N 22°10′E
Altitude 60–500 m **Area** 6,110 ha

Site description

Two small river valleys and surrounding hills dominated by scrub and pastures. Human activities include livestock-farming, hunting and mining.

Habitats Forest and woodland (5%; alluvial/very wet forest), Scrub (60%; scrub; sclerophyllous scrub/garrigue/maquis), Grassland (20%; steppe/dry calcareous grassland), Rocky areas (5%; inland cliff)
Land-use Agriculture (80%), Not utilized, Urban/industrial/transport (5%)

Birds

Species	Season	Year	Pop min	Pop max	Acc	Criteria
Hippolais olivetorum Olive-tree Warbler	B	1996	Frequent		—	B2, C6

The area is important for breeding raptors and passerines characteristic of mixed pasture and forest.

Protection status

National None **International** None

Conservation issues

Threats Agricultural intensification/expansion (C), Extraction industry (B), Industrialization/urbanization (B), Unsustainable exploitation (B)

Threats come from a cement factory in the Apsalos valley and plans to regulate the Moglenista river. The area is a candidate SAC.

Agra artificial lake

B2, C6 **041**

Admin region Central Macedonia
Coordinates 40°47′N 21°54′E
Altitude 470–700 m **Area** 1,386 ha

Site description

An artificial lake with extensive reedbeds (*Phragmites*). The adjacent hills are dominated by scrub, pastures and orchards with small areas of forest.

Habitats Grassland (20%; steppe/dry calcareous grassland; humid grassland), Wetland (70%; standing fresh water; water-fringe vegetation; fen/transition mire/spring), Artificial landscape (10%; perennial crops/orchards/groves)
Land-use Agriculture (10%), Fisheries/aquaculture (30%), Water management (70%)

Birds

Species	Season	Year	Pop min	Pop max	Acc	Criteria
Chlidonias hybridus Whiskered Tern	B	1996	5	15	B	B2, C6

This is an important site for species associated with reedbeds. Species of global conservation concern that do not meet IBA criteria: *Phalacrocorax pygmeus* (passage), *Pelecanus crispus* (passage), *Aythya nyroca* (4 breeding pairs).

Protection status

National Partial **International** High

500 ha of IBA covered by Wildlife Refuge (Elos–Limni Agra/Agra–Nisiou-Bryton, 500 ha). 1,386 ha of IBA covered by Special Protection Area (Limni kai Fragna Agra, 1,386 ha).

Conservation issues

Threats Agricultural intensification/expansion (B), Disturbance to birds (C), Intensified forest management (A), Unsustainable exploitation (C)

The main threats are from agricultural intensification and clear-cutting of the remaining forests. Part of the area is a candidate SAC.

Aliakmon gorge — B2 — 042

Admin region Central Macedonia
Coordinates 40°27'N 22°10'E
Altitude 100–600 m **Area** 4,400 ha

Site description

The site encompasses the lower Aliakmon valley close to Veria. The area is dominated by scrub and forest but also includes an artificial lake.

Habitats Forest and woodland (30%; broadleaved deciduous forest; mixed forest), Scrub (50%; sclerophyllous scrub/garrigue/maquis), Wetland (10%; standing fresh water), Rocky areas (10%; inland cliff)
Land-use Forestry (30%), Not utilized (30%), Tourism/recreation (20%), Urban/industrial/transport (5%), Water management (20%)

Birds

Species	Season	Year	Pop min	Pop max	Acc	Criteria
Emberiza cia Rock Bunting	R	1995	Frequent		—	B2

This is an important area for resident raptors.

Protection status

National None **International** None

Conservation issues

Threats Construction/impact of dyke/dam/barrage (B), Deforestation (commercial) (B), Forest grazing (C), Industrialization/urbanization (B), Infrastructure (A), Intensified forest management (B), Recreation/tourism (C), Unsustainable exploitation (C)

The main threats are a planned road through the site, dam projects and more intensive use of the forests. Part of the area is a candidate SAC.

Alyki Kitros lagoon — A4i, B1i, B2, B3, C2, C3, C6 — 043

Admin region Central Macedonia
Coordinates 40°22'N 22°38'E
Altitude 0–20 m **Area** 1,500 ha

Site description

A coastal lagoon surrounded by dunes, saltpans, saltmarsh, holiday homes and residential areas.

Habitats Wetland (89%; saltmarsh; sand-dunes/sand beach; coastal lagoon; standing brackish and salt water), Marine areas (5%; open sea), Artificial landscape (6%; arable land)
Land-use Agriculture (35%), Tourism/recreation (5%), Urban/industrial/transport (41%), Water management (20%)

Birds

Species	Season	Year	Pop min	Pop max	Acc	Criteria
Tadorna tadorna Shelduck	W	1996	—	1,000	—	B1i, C3
Recurvirostra avosetta Avocet	B	1989	128	—	A	B3, C6
Recurvirostra avosetta Avocet	W	1991	486	—	A	B1i, B2, C2, C6
Recurvirostra avosetta Avocet	P	1989	—	1,240	A	A4i, B1i, C2
Glareola pratincola Collared Pratincole	B	1996	60	—	—	B1i, B2, C2
Larus genei Slender-billed Gull	B	1992	45	—	—	C2
Gelochelidon nilotica Gull-billed Tern	B	1996	—	48	—	B2, C6
Sterna albifrons Little Tern	B	1996	—	60	B	C6

The site is important for passage and wintering waterbirds. Species of global conservation concern that do not meet IBA criteria: *Phalacrocorax pygmeus* (1–5 wintering individuals), *Branta ruficollis* (wintering), *Oxyura leucocephala* (wintering), *Haliaeetus albicilla* (wintering), *Aquila clanga* (wintering).

Protection status

National High **International** High
1,500 ha of IBA covered by Wildlife Refuge (Alyki Kitrous, 1,784 ha). 1,200 ha of IBA covered by Special Protection Area (Delta Axiou, Loudia, Aliakmona–Alyki Kitrous, 13,763 ha).

Conservation issues

Threats Industrialization/urbanization (A), Recreation/tourism (U), Unsustainable exploitation (U)

The major threat is from housing development in the areas of dunes and saltmarsh. *Larus melanocephalus* no longer occurs as a breeding species as a result of disturbance. Part of the area is a candidate SAC.

Mount Olympos — A3, B2, B3, C6 — 044

Admin region Central Macedonia, Thessalia
Coordinates 40°08'N 22°25'E
Altitude 200–2,917 m **Area** 42,000 ha

Site description

The highest mountain in Greece with extensive subalpine meadows, cliffs and *Pinus* woodland. At lower altitudes there are areas of broadleaved evergreen woodland.

Habitats Forest and woodland (broadleaved deciduous forest; native coniferous forest; broadleaved evergreen forest), Scrub (sclerophyllous scrub/garrigue/maquis), Grassland (alpine/subalpine/boreal grassland), Rocky areas (scree/boulders; inland cliff)
Land-use Agriculture, Forestry, Nature conservation/research, Tourism/recreation

Birds

Species	Season	Year	Pop min	Pop max	Acc	Criteria
[1] ***Gypaetus barbatus*** Lammergeier	R	1995	—	2	—	C6
Alectoris graeca Rock Partridge	R	1996	—	—	—	B2
Strix aluco Tawny Owl	R	1996	—	—	—	B3
Prunella collaris Alpine Accentor	R	1996	—	—	—	A3
Monticola saxatilis Rock Thrush	B	1996	—	—	—	B2
Tichodroma muraria Wallcreeper	B	1988	—	—	—	A3
Pyrrhocorax graculus Alpine Chough	R	1996	—	100	—	A3
Montifringilla nivalis Snowfinch	B	1988	—	—	—	A3

1. Individuals.

This is an important area for resident raptors, forest species and species associated with open alpine areas. Breeding birds include four of the 10 species in Europe that are restricted (when breeding) to the Eurasian high-montane biome. Species of global conservation concern that do not meet IBA criteria: *Aegypius monachus*.

Protection status

National Partial **International** Partial
6,000 ha of IBA covered by Wildlife Refuge (Siopoto–Paliabela/Kokkinoplou, 6,000 ha). 1,525 ha of IBA covered by Wildlife Reserve (Magouta–Alonaki–Karamana/Elassonas, 1,525 ha). 4,000 ha of IBA covered by Wildlife Reserve (Karias, 4,000 ha). 3,998 ha of IBA covered by National Park (Mount Olympos, 3,998 ha). 1,226 ha of IBA covered by Wildlife Refuge (Ag. Ioanni–Trohalo/Litohoriou, 1,226 ha). 4,000 ha of IBA covered by Biogenetic Reserve (Council of Europe) (Etnikos Drymos Olympou (Core area), 4,000 ha). 4,000 ha of IBA covered by Biosphere Reserve (Mount Olympus National Park, 4,000 ha). 4,595 ha of IBA covered by Special Protection Area (Ethnikos Drymos Olympou, 4,595 ha).

Conservation issues

Threats Infrastructure (C), Recreation/tourism (B), Unsustainable exploitation (U)

Threats include disorganized tourism development and the construction of numerous communication antennae. Part of the area is a candidate SAC.

Lake Vegoritis and Lake Petron — A1, B2, C1, C6 — 045

Admin region Central Macedonia, Western Macedonia
Coordinates 40°45'N 21°45'E
Altitude 523–560 m **Area** 17,000 ha

Site description

Two freshwater lakes in an area of hills and mountains.

Habitats Grassland (10%; humid grassland), Wetland (80%; standing fresh water), Artificial landscape (10%; arable land)
Land-use Agriculture (50%), Fisheries/aquaculture (50%), Water management (100%)

Birds

Species	Season	Year	Pop min	Pop max	Acc	Criteria
Phalacrocorax pygmeus Pygmy Cormorant	B	1993	3	10	—	A1, B2, C1, C6

The site is important for wintering ducks and geese. Species of global conservation concern that do not meet IBA criteria: *Aythya nyroca* (wintering), *Haliaeetus albicilla* (non-breeding).

Protection status

National Partial **International** None
Part of IBA covered by Wildlife Refuge (Ampelia/Amyntaiou, 2,500 ha). 440 ha of IBA covered by Wildlife Refuge (Maniakiou, 440 ha).

Conservation issues

Threats Agricultural intensification/expansion (B), Construction/impact of dyke/dam/barrage (B), Dredging/canalization (C), Other (A)

The main threats are water abstraction from Lake Vegoritida ('Other' threats) and pollution of Lake Petron ('Other' threats). Part of the area is a candidate SAC.

Lake Chimaditis and Lake Zazaris — A1, A4i, B1i, B2, C1, C2, C6 — 046

Admin region Western Macedonia
Coordinates 40°37'N 21°33'E
Altitude 590–1,100 m **Area** 5,390 ha

Site description

Two eutrophic lakes at the foot of Mount Verno. They are surrounded by extensive reedbeds (*Phragmites*). Beyond the lakes is a mosaic of woodland, pastures and arable fields.

Habitats Forest and woodland (20%; broadleaved deciduous forest), Grassland (13%; mesophile grassland), Wetland (24%; standing fresh water; water-fringe vegetation), Artificial landscape (40%; arable land)
Land-use Agriculture (54%), Fisheries/aquaculture (24%), Forestry (7%)

Birds

Species	Season	Year	Pop min	Pop max	Acc	Criteria
Phalacrocorax pygmeus Pygmy Cormorant	W	1996	—	27	A	A1, C1
Pelecanus crispus Dalmatian Pelican	N	1989	—	63	A	A1, A4i, B1i, C1, C2
Aythya nyroca Ferruginous Duck	B	1996	30	60	B	A1, B2, C1, C6
Falco naumanni Lesser Kestrel	B	1996	5	15	B	A1, C1
Falco naumanni Lesser Kestrel	N	1996	Common		—	A1, C1
Chlidonias hybridus Whiskered Tern	B	1996	30	50	C	B2, C6

This is an important site for species that breed in reedbeds. Species of global conservation concern that do not meet IBA criteria: *Haliaeetus albicilla* (non-breeding), *Gallinago media* (passage).

Protection status

National Partial **International** Partial
1,750 ha of IBA covered by Wildlife Refuge (Hintsko–Heimadi/Limnohoriou, 3,240 ha). 4,064 ha of IBA covered by Special Protection Area (Limnes Cheimaditida–Zazari, 4,064 ha).

Conservation issues

Threats Abandonment/reduction of land management (C), Agricultural intensification/expansion (A), Aquaculture/fisheries (C), Construction/impact of dyke/dam/barrage (B), Drainage (B), Extraction industry (B), Forest grazing (A), Groundwater abstraction (B), Infrastructure (B), Recreation/tourism (A), Unsustainable exploitation (B)

The main threat is increased eutrophication from agricultural run-off as water-levels fall. Part of the area is a candidate SAC.

Lake Mikri Prespa and Lake Megali Prespa — A1, A4i, B1i, B2, C1, C2, C3, C6 — 047

Admin region Western Macedonia
Coordinates 40°45'N 21°03'E
Altitude 849–2,334 m **Area** 33,000 ha

Site description

Two freshwater lakes, Mikri and Megali Prespa, in a mountainous area. The lakes have extensive reedbeds (*Phragmites*) and small areas of wet meadow. The site lies adjacent to Albanian IBAs 003 and 004 and to site 006 in the FYR Macedonia.

Habitats Forest and woodland (40%; broadleaved deciduous forest; native coniferous forest; mixed forest; treeline ecotone), Grassland (alpine/subalpine/boreal grassland; humid grassland), Wetland (30%; standing fresh water), Rocky areas (sea cliff/rocky shore; rock stacks/islets), Artificial landscape (10%; arable land)
Land-use Agriculture (15%), Fisheries/aquaculture (27%), Forestry (40%), Nature conservation/research (5%), Water management

Birds

Species	Season	Year	Pop min	Pop max	Acc	Criteria
Phalacrocorax carbo Cormorant	B	1996	733	733	A	B1i, C3
Phalacrocorax pygmeus Pygmy Cormorant	R	1996	314	400	A	A1, A4i, B1i, B2, C1, C2, C6
Pelecanus onocrotalus White Pelican	B	1996	100	100	A	B2, C6
[1] ***Pelecanus crispus*** Dalmatian Pelican	B	1996	500	500	A	A1, A4i, B1i, B2, C1, C2, C6
[2] ***Ardeola ralloides*** Squacco Heron	B	1996	10	20	B	C2, C6
Aythya fuligula Tufted Duck	W	1994	150	8,850	A	B1i, C3

1. Large increase.
2. Large decrease.

This is an important site for breeding, passage and wintering waterbirds. Species of global conservation concern that do not meet IBA criteria: *Aythya nyroca* (3 breeding pairs).

Protection status

National Partial **International** Partial
900 ha of IBA covered by Wildlife Refuge (Vronterou, 900 ha). Part of IBA covered by Wildlife Refuge (Sfina, Prespon, 6,700 ha). 19,470 ha of IBA covered by National Park (Prespes, 19,470 ha). 12 ha of IBA covered by Biogenetic Reserve (Council of Europe) (Ethnikos Drymos Prespon (Dasos Kedron), 12 ha). 8,000 ha of IBA covered by Ramsar Site (Lakes Mikri Prespa and Megali Prespa, 8,000 ha). 26,620 ha of IBA covered by Special Protection Area (Ethnikos Drymos Prespon, 26,620 ha).

Conservation issues

Threats Abandonment/reduction of land management (B), Agricultural intensification/expansion (B), Consequences of animal/plant introductions (B), Disturbance to birds (B), Forest grazing (C), Industrialization/urbanization (C), Infrastructure (C), Intensified forest management (C), Recreation/tourism (C), Selective logging/cutting (C), Unsustainable exploitation (B)

The main threats are from hunting, disturbance and agricultural intensification. The area is a candidate SAC.

Lake Kastoria (Orestiada) — A1, A4i, B1i, B2, C1, C2, C3, C6 — 048

Admin region Western Macedonia
Coordinates 40°31'N 21°18'E
Altitude 620–620 m **Area** 3,400 ha

Site description

An inland freshwater lake with the town of Kastoria on its shore. Around the north-western part of the lake an area of riverine forest remains.

Habitats Forest and woodland (10%; alluvial/very wet forest), Grassland (5%; humid grassland), Wetland (60%; standing fresh water; water-fringe vegetation), Artificial landscape (30%; arable land; other urban/industrial areas)
Land-use Agriculture (20%), Fisheries/aquaculture, Tourism/recreation (30%), Urban/industrial/transport (10%)

Birds

An important foraging area for *Pelecanus crispus*. Species of global conservation concern that do not meet IBA criteria: *Haliaeetus albicilla* (passage).

Species	Season	Year	Pop min	Pop max	Acc	Criteria
Phalacrocorax pygmeus Pygmy Cormorant	W	1993	—	500	B	A1, A4i, B1i, C1, C2
Pelecanus crispus Dalmatian Pelican	P	1993	50	100	B	A1, A4i, B1i, C1, C2
Ardeola ralloides Squacco Heron	B	1993	10	50	B	B2, C2, C6
Aythya nyroca Ferruginous Duck	B	1993	—	30	—	A1, B2, C1, C6
Mergus merganser Goosander	W	1993	5	10	B	B1i, C3

■ Protection status

National Partial **International** None

3,400 ha of IBA covered by Wildlife Refuge (Limni, Vouno Kastorias, 7,050 ha).

■ Conservation issues

Threats Agricultural intensification/expansion (A), Firewood collection (B), Industrialization/urbanization (B), Recreation/tourism (B)

The remaining riverine forest is threatened by agricultural expansion and road building. The lake water is polluted by agricultural run-off and sewerage from Kastoria. Tourist facilities are likely to expand in the future. The site has an Environmental Education Centre. A management plan is under preparation. The area is a candidate SAC.

Mount Gramos — B2, C6 — 049

Admin region Western Macedonia, Ipiros
Coordinates 40°20′N 20°50′E
Altitude 800–2,520 m **Area** 45,000 ha

■ Site description

A mountain range on the Albanian border, with extensive pasture land, forests and cliffs. Human activities include forestry, livestock-farming, and hunting.

Habitats Forest and woodland (broadleaved deciduous forest; native coniferous forest; mixed forest; treeline ecotone), Grassland (alpine/subalpine/boreal grassland), Wetland (river/stream), Rocky areas (inland cliff)
Land-use Agriculture, Forestry

■ Birds

Species	Season	Year	Pop min	Pop max	Acc	Criteria
Aquila chrysaetos Golden Eagle	R	1997	3	5	A	B2, C6
Anthus campestris Tawny Pipit	B	1996	Frequent		—	B2, C6
Monticola saxatilis Rock Thrush	B	1997	Frequent		—	B2

This is an important area for resident raptors and species associated with alpine habitats.

■ Protection status

National Partial **International** Partial

2,769 ha of IBA covered by Wildlife Refuge (Patoma, Mavri Petra, Limes, Arena, Lioukou, Stavro, 2,769 ha). 1,100 ha of IBA covered by Wildlife Refuge (Katafiki/Gramou, 1,100 ha). 3,200 ha of IBA covered by Wildlife Refuge (Arrenon Gramou/Gramou, 3,200 ha). 4,100 ha of IBA covered by Wildlife Reserve (Psoriarika–Tsarno, 4,100 ha). 130 ha of IBA covered by Natural Monument (Grammos Forest, 130 ha). Part of IBA covered by Biogenetic Reserve.

■ Conservation issues

Threats Abandonment/reduction of land management (C), Disturbance to birds (U), Infrastructure (U), Unsustainable exploitation (B)

Part of the area is a candidate SAC.

Mounts Orliakas and Tsourgiakas — B2, C6 — 050

Admin region Western Macedonia
Coordinates 40°02′N 21°13′E
Altitude 800–1,400 m **Area** 10,500 ha

■ Site description

An area of low limestone mountains and gorges around the River Venetikos. The river has large areas of shingle along its course. The main habitats are oak *Quercus* forest, species-rich mixed forest and open wooded pasture. Human activities include forestry, recreation, livestock-farming and hunting.

Habitats Forest and woodland (broadleaved deciduous forest; native coniferous forest; mixed forest), Scrub (sclerophyllous scrub/garrigue/maquis), Grassland (dry siliceous grassland), Wetland (river/stream), Rocky areas (scree/boulders; inland cliff), Artificial landscape (arable land)
Land-use Agriculture, Forestry, Tourism/recreation, Urban/industrial/transport, Water management

■ Birds

Species	Season	Year	Pop min	Pop max	Acc	Criteria
Falco biarmicus Lanner	R	1996	1	1	—	B2, C6

This is an important site for a range of species associated with forest and pasture.

■ Protection status

National None **International** None

■ Conservation issues

Threats Abandonment/reduction of land management (C), Dredging/canalization (A), Infrastructure (B), Recreation/tourism (B), Unsustainable exploitation (A)

The main threats are from water abstraction from the river, unsustainable hunting, road building, recreational activities and undergrazing of the pastures.

Mount Vourinos — B2, C6 — 051

Admin region Western Macedonia
Coordinates 40°11′N 21°40′E
Altitude 450–1,866 m **Area** 31,000 ha

■ Site description

Most slopes of Mount Vourinos are covered by grassland or scrub, except for the central valley which has *Pinus*, *Abies* and *Fagus* forests.

Habitats Forest and woodland (10%; broadleaved deciduous forest; native coniferous forest; mixed forest; alluvial/very wet forest), Scrub (40%; scrub; sclerophyllous scrub/garrigue/maquis), Grassland (30%; steppe/dry calcareous grassland; dry siliceous grassland; alpine/subalpine/boreal grassland), Rocky areas (10%; inland cliff), Artificial landscape (5%; arable land)
Land-use Agriculture (70%), Forestry (10%), Nature conservation/research (10%), Tourism/recreation (5%)

■ Birds

Species	Season	Year	Pop min	Pop max	Acc	Criteria
Falco biarmicus Lanner	B	1994	Rare		—	B2, C6
Lanius minor Lesser Grey Shrike	B	1995	Frequent		—	B2, C6

The site is important for raptors and species associated with Mediterranean scrub and grassland.

■ Protection status

National Low **International** None

1,425 ha of IBA covered by Wildlife Refuge (Tservena–Bourinos/Dim. Siatistas–Palaiokastrou, 1,425 ha).

■ Conservation issues

Threats Deforestation (commercial) (B), Forest grazing (C), Infrastructure (B), Unsustainable exploitation (C)

Hunting is prohibited on part of the site although poisoning occurs. Part of the area is a candidate SAC.

Kerketion mountain (Koziakas) — B2, B3, C6 — 052

Admin region Thessalia
Coordinates 39°35′N 21°32′E
Altitude 300–1,901 m **Area** 23,000 ha

■ Site description

A partly forested mountain area. Human activities include forestry, stock-raising, hunting, arable agriculture, and tourism.

Habitats Forest and woodland (broadleaved deciduous forest; native coniferous forest), Scrub (sclerophyllous scrub/garrigue/maquis), Grassland (alpine/subalpine/boreal grassland), Rocky areas (inland cliff), Artificial landscape (arable land)
Land-use Agriculture (90%), Forestry (65%), Hunting (95%), Nature conservation/research, Tourism/recreation (20%)

Birds

Species	Season	Year	Pop min	Pop max	Acc	Criteria
Falco biarmicus Lanner	R	1996	1	1	A	B2, C6
Parus lugubris Sombre Tit	R	1996	Frequent		—	B3

This is an important area for breeding raptors. Species of global conservation concern that do not meet IBA criteria: *Aquila heliaca*.

Protection status

National Partial **International** None
Part of IBA covered by Controlled Hunting Area (Koziakas, 48,360 ha)

Conservation issues

Threats Construction/impact of dyke/dam/barrage (A), Infrastructure (A), Intensified forest management (U), Recreation/tourism (C), Selective logging/cutting (U), Unsustainable exploitation (B)

Major threats come from the construction of the Pili dam and the creation of a reservoir. The construction of two roads threatens a *Gyps fulvus* colony and breeding *Neophron percnopterus*. Part of the area is a candidate SAC.

Antikhassia mountains and Meteora B2, B3, C2, C6 053

Admin region Thessalia
Coordinates 39°45′N 21°35′E
Altitude 200–1,416 m **Area** 65,000 ha

Site description

A mountainous site with broadleaved deciduous woodland, cliffs, and riverine forest in the valleys. Human activities include forestry, stock-raising, hunting, and recreation.

Habitats Forest and woodland (30%; broadleaved deciduous forest; alluvial/very wet forest), Scrub (30%; sclerophyllous scrub/garrigue/maquis), Grassland (15%; steppe/dry calcareous grassland), Artificial landscape (25%; arable land), Rocky areas (5%; inland cliff)
Land-use Agriculture (80%), Forestry (30%), Tourism/recreation

Birds

Species	Season	Year	Pop min	Pop max	Acc	Criteria
Neophron percnopterus Egyptian Vulture	B	1996	—	10	—	B2, C6
Circaetus gallicus Short-toed Eagle	B	1994	7	10	—	B2, C6
Accipiter brevipes Levant Sparrowhawk	B	1994	14	30	—	B2
Aquila pomarina Lesser Spotted Eagle	B	1994	6	8	—	B2, C6
Athene noctua Little Owl	R	1994	—	100	—	B2
Picus viridis Green Woodpecker	R	1994	50	100	—	B2
Dendrocopos medius Middle Spotted Woodpecker	R	1994	50	100	—	B3, C6
Dendrocopos leucotos White-backed Woodpecker	R	1994	5	20	—	C2
Lanius collurio Red-backed Shrike	B	1994	100	300	—	B2, C6
Emberiza caesia Cretzschmar's Bunting	B	1994	20	100	—	B3, C6

This is an important area for breeding raptors, forest species and species associated with open montane habitats. Species of global conservation concern that do not meet IBA criteria: *Aquila heliaca* (non-breeding), *Falco naumanni* (passage).

Protection status

National Partial **International** Partial
855 ha of IBA covered by Wildlife Refuge (Arkoudorema–Plastira/Spathadon, 855 ha). 2,000 ha of IBA covered by Wildlife Refuge (Gerakariou, 2,000 ha). 700 ha of IBA covered by Wildlife Refuge (Kanalaki/Kastrakiou–Vlachavas, 700 ha). Part of IBA covered by Controlled Hunting Area (Koziakas, 48,360 ha). 375 ha of IBA covered by World Heritage Site (Meteora Group of Monasteries, 375 ha). 55,220 ha of IBA covered by Special Protection Area (Potamos Pinios–Antichasia Ori, 55,220 ha).

Conservation issues

Threats Agricultural intensification/expansion (U), Extraction industry (U), Intensified forest management (U), Recreation/tourism (B), Unsustainable exploitation (B)

The main threats come from the extraction industry and recreational activities. Part of the area is a candidate SAC.

Kalamaki gorge–Mount Zarkou B2, C6 054

Admin region Thessalia
Coordinates 39°38′N 22°11′E
Altitude 70–734 m **Area** 8,650 ha

Site description

A rugged mountain dominated by scrub. The adjacent gorge and valley contains some riverine forest. Human activities include stock-raising and hunting.

Habitats Forest and woodland (5%; alluvial/very wet forest), Scrub (70%), Grassland (5%; steppe/dry calcareous grassland), Rocky areas (5%; inland cliff), Artificial landscape (15%; arable land)
Land-use Agriculture (60%), Water management (5%)

Birds

Species	Season	Year	Pop min	Pop max	Acc	Criteria
Falco biarmicus Lanner	R	1994	1	2	B	B2, C6

This is an important area for resident raptors and species characteristic of open habitats.

Protection status

National Partial **International** None
1,750 ha of IBA covered by Wildlife Refuge (Pinindos, 1,750 ha).

Conservation issues

Threats Agricultural intensification/expansion (C), Construction/impact of dyke/dam/barrage (C), Disturbance to birds (C), Forest grazing (C), Unsustainable exploitation (A)

The main threat is from hunting. Part of the area is a candidate SAC.

Elassona area A1, B2, C1, C6 055

Admin region Thessalia
Coordinates 39°52′N 22°11′E
Altitude 400–564 m **Area** 7,500 ha

Site description

An area of plain south of Elassona. It is dominated by arable agriculture.

Habitats Scrub (5%; sclerophyllous scrub/garrigue/maquis), Artificial landscape (95%; arable land; other urban/industrial areas)
Land-use Agriculture (95%), Urban/industrial/transport

Birds

Species	Season	Year	Pop min	Pop max	Acc	Criteria
[1] ***Falco naumanni*** Lesser Kestrel	B	1995	49	49	A	A1, B2, C1, C6

1. Large decrease.

The site is important for species associated with arable agriculture.

Protection status

National None **International** None

Conservation issues

Threats Abandonment/reduction of land management (B), Agricultural intensification/expansion (A), Industrialization/urbanization (U), Other (B), Unsustainable exploitation (U)

The main threat is from the intensification of agriculture. The demolition of old buildings is a particular threat to *Falco naumanni* ('Other' threats).

Tyrnavos area — A1, B2, C1, C6 — 056

Admin region Thessalia
Coordinates 39°43′N 22°20′E
Altitude 70–300 m **Area** 8,500 ha

Site description

An area of arable agriculture and villages around the town of Tyrnavos.

Habitats Scrub (5%; sclerophyllous scrub/garrigue/maquis), Artificial landscape (95%; arable land; other urban/industrial areas)
Land-use Agriculture (95%), Urban/industrial/transport (5%)

Birds

Species	Season	Year	Pop min	Pop max	Acc	Criteria
[1] ***Falco naumanni*** Lesser Kestrel	B	1995	95	95	A	A1, B2, C1, C6

1. Large decrease.

The site is important for species associated with arable agriculture.

Protection status

National Partial **International** None
Part of IBA covered by Wildlife Refuge (Kladaries–Tripimeno–Magoula Samari, 1,800 ha).

Conservation issues

Threats Abandonment/reduction of land management (B), Agricultural intensification/expansion (A), Industrialization/urbanization (U), Other (B), Unsustainable exploitation (U)

The main threat is from the intensification of agriculture. The demolition of old buildings is a particular threat to *Falco naumanni* ('Other' threats).

Mati of Tirnavos — A1, C1 — 057

Admin region Thessalia
Coordinates 39°49′N 22°17′E
Altitude 74–100 m **Area** 500 ha

Site description

A spring-fed wetland dominated by reedbeds (*Phragmites*) and surrounded by arable agriculture.

Habitats Forest and woodland (5%; alluvial/very wet forest), Scrub (5%; sclerophyllous scrub/garrigue/maquis), Grassland (20%; humid grassland), Wetland (65%; standing fresh water; river/stream; water-fringe vegetation), Artificial landscape (10%; arable land; forestry plantation)
Land-use Agriculture (20%), Fisheries/aquaculture, Tourism/recreation (5%), Water management (90%)

Birds

Species	Season	Year	Pop min	Pop max	Acc	Criteria
Falco naumanni Lesser Kestrel	N	1996	Frequent		—	A1, C1

The site is important for species breeding in reedbeds and passage waterbirds. Species of global conservation concern that do not meet IBA criteria: *Aythya nyroca*.

Protection status

National None **International** None

Conservation issues

Threats Agricultural intensification/expansion (A), Aquaculture/fisheries (C), Construction/impact of dyke/dam/barrage (U), Drainage (A), Groundwater abstraction (C), Recreation/tourism (C), Unsustainable exploitation (A)

Water abstraction, both groundwater and surface water, for irrigation of crops is a serious threat.

Mounts Kato Olympos and Ossa, and Tembi ravine — B2, C6 — 058

Admin region Thessalia
Coordinates 39°54′N 22°30′E
Altitude 10–1,978 m **Area** 69,000 ha

Site description

The site includes two mountains separated by the Tembi valley. The western part of Kato Olympos is predominantly grassland and scrub, whilst eastern areas are well forested.

Habitats Forest and woodland (35%; broadleaved deciduous forest; native coniferous forest; mixed forest; broadleaved evergreen forest), Scrub (30%; sclerophyllous scrub/garrigue/maquis), Grassland (20%; steppe/dry calcareous grassland; humid grassland), Rocky areas (5%; inland cliff), Artificial landscape (10%; arable land; perennial crops/orchards/groves)
Land-use Agriculture (55%), Forestry (35%), Hunting (40%)

Birds

Species	Season	Year	Pop min	Pop max	Acc	Criteria
Gyps fulvus Griffon Vulture	B	1996	—	2	A	C6
Accipiter brevipes Levant Sparrowhawk	B	1996	6	12	B	B2
Falco biarmicus Lanner	B	1993	Uncommon		—	B2, C6
Merops apiaster Bee-eater	B	1996	Frequent		—	B2
Ficedula semitorquata Semi-collared Flycatcher	B	1996	Uncommon		—	B2, C6
Emberiza hortulana Ortolan Bunting	B	1996	Frequent		—	B2, C6

The site is important for breeding and passage raptors and as a migration corridor. Species of global conservation concern that do not meet IBA criteria: *Aegypius monachus* (non-breeding), *Aquila heliaca* (non-breeding), *Falco naumanni* (passage).

Protection status

National Partial **International** Partial
2,250 ha of IBA covered by Wildlife Refuge (Aspropetra–Ailia/Kalipefkis, 2,250 ha). 470 ha of IBA covered by Wildlife Refuge (Tsaira–Goliani/Rapsanis, 470 ha). Part of IBA covered by Controlled Hunting Area (Ossa, 26,377 ha). 1,690 ha of IBA covered by Aesthetic Forest (Ossa Aesthetic Forest, 1,690 ha). Part of IBA covered by Aesthetic Forest (Tembi Valley, 1,762 ha). 1,336 ha of IBA covered by Special Protection Area (Aisthitiko Dasos Koiladas Tempon, 1,336 ha). 19,519 ha of IBA covered by Special Protection Area (Aisthitiko Dasos Ossas, 19,519 ha).

Conservation issues

Threats Afforestation (C), Burning of vegetation (C), Construction/impact of dyke/dam/barrage (B), Deforestation (commercial) (B), Firewood collection (C), Forest grazing (C), Groundwater abstraction (C), Infrastructure (B), Recreation/tourism (C), Selective logging/cutting (B), Unsustainable exploitation (B)

Logging and scrub burning are leading to gradual degradation of the habitat. Plans for a dam and reservoir in the gorge are a serious threat. The laying of poison baits continues and has already caused the loss of *Gypaetus barbatus* and *Aegypius monachus* from the site. Part of the area is a candidate SAC.

Pinios delta — A1, C1 — 059

Admin region Central Macedonia, Thessalia
Coordinates 39°55′N 22°42′E
Altitude 0–3 m **Area** 2,600 ha

Site description

The site contains a variety of wetland and drier habitats including riverine and coastal woodland, sand-dunes, lagoons, saltmarsh and freshwater marsh.

Habitats Forest and woodland (10%; alluvial/very wet forest), Scrub (10%; sclerophyllous scrub/garrigue/maquis), Grassland (10%; steppe/dry calcareous grassland), Wetland (45%; tidal river/enclosed tidal water; sand-dunes/sand beach; coastal lagoon; standing fresh water; standing brackish and salt water; water-fringe vegetation), Artificial landscape (20%; arable land; perennial crops/orchards/groves)

Land-use Agriculture (80%), Fisheries/aquaculture (20%), Tourism/recreation (20%), Urban/industrial/transport (5%)

■ Birds

Species	Season	Year	Pop min	Pop max	Acc	Criteria
Phalacrocorax pygmeus Pygmy Cormorant	W	1996	—	80	B	A1, C1

An important site for wintering and passage waterbirds. Species of global conservation concern that do not meet IBA criteria: *Aythya nyroca, Larus audouinii* (passage).

■ Protection status

National Partial **International** None

370 ha of IBA covered by Wildlife Refuge (Delta Piniou/Stomiou, 370 ha).

■ Conservation issues

Threats Agricultural intensification/expansion (A), Aquaculture/fisheries (C), Drainage (A), Filling-in of wetlands (A), Firewood collection (B), Forest grazing (C), Groundwater abstraction (A), Recreation/tourism (B), Unsustainable exploitation (B)

The site is seriously threatened by uncontrolled house construction, drainage and infilling of the wetland, and the destruction of woodland.

Mavrovouni mountain, Larissa A1, A3, B2, B3, C1, C2, C6 060

Admin region Thessalia
Coordinates 39°35′N 22°49′E
Altitude 100–1,054 m **Area** 17,000 ha

■ Site description

A mountain between the Thessalian plain and Aegean Sea. The north-east facing slopes are well forested, whilst the south-western slopes are dominated by scrub.

Habitats Forest and woodland (50%; broadleaved deciduous forest; mixed forest), Scrub (35%; sclerophyllous scrub/garrigue/maquis), Grassland (5%; steppe/dry calcareous grassland), Rocky areas (5%; inland cliff)
Land-use Agriculture (40%), Forestry (50%), Tourism/recreation

■ Birds

Species	Season	Year	Pop min	Pop max	Acc	Criteria
Accipiter brevipes Levant Sparrowhawk	B	1995	10	20	B	B2
Buteo rufinus Long-legged Buzzard	R	1995	6	8	B	B2, C6
Aquila pomarina Lesser Spotted Eagle	B	1995	4	5	B	C6
Hieraaetus pennatus Booted Eagle	B	1994	4	6	C	B2, C6
Falco naumanni Lesser Kestrel	P	1995	Abundant		—	A1, C1
Falco biarmicus Lanner	R	1995	1	3	B	B2, C6
Alectoris graeca Rock Partridge	R	1995	20	50	C	A3
Dendrocopos medius Middle Spotted Woodpecker	R	1995	50	100	B	B3, C6
Dendrocopos leucotos White-backed Woodpecker	R	1995	10	30	C	C2
Oenanthe hispanica Black-eared Wheatear	B	1995	Common		—	A3
Hippolais olivetorum Olive-tree Warbler	B	1995	20	50	B	A3, C6
Sylvia cantillans Subalpine Warbler	B	1995	Abundant		—	A3
Sylvia melanocephala Sardinian Warbler	R	1995	Common		—	A3
Sylvia hortensis Orphean Warbler	B	1995	50	100	B	B2
Sitta neumayer Rock Nuthatch	R	1995	Frequent		—	A3
Lanius collurio Red-backed Shrike	B	1995	100	300	B	B2, C6
Lanius senator Woodchat Shrike	B	1995	100	300	B	B2
Lanius nubicus Masked Shrike	B	1995	5	10	B	A3, B2
Emberiza caesia Cretzschmar's Bunting	B	1995	10	30	C	A3
Emberiza melanocephala Black-headed Bunting	B	1995	Abundant		—	A3

The site is important for its resident raptors and forest species. Breeding birds include nine of the 21 species that are restricted in Europe to the Mediterranean biome (when breeding). Species of global conservation concern that do not meet IBA criteria: *Aquila clanga* (wintering), *Aquila heliaca* (non-breeding).

■ Protection status

National Partial **International** High

Part of IBA covered by Wildlife Refuge (Dasia Platania/Skitis–Amigdalis–Potamias, 804 ha). Part of IBA covered by Wildlife Refuge (Mavrovouni/Sklithrou, 1,091 ha). 0 ha of IBA covered by Wildlife Refuge (Polidendriou, 3,530 ha). 17,000 ha of IBA covered by Special Protection Area (Oros Mavrovouni, 36,454 ha).

■ Conservation issues

Threats Agricultural intensification/expansion (C), Extraction industry (B), Firewood collection (U), Forest grazing (C), Infrastructure (A), Intensified forest management (A), Recreation/tourism (A), Selective logging/cutting (B), Unsustainable exploitation (B)

The main threats are from infrastructure developments and more intensive forest management. The area is a candidate SAC.

Reservoirs of former Lake Karla A1, A4i, B1i, C1, C2, C3 061

Admin region Thessalia
Coordinates 39°32′N 22°42′E
Altitude 40–60 m **Area** 1,200 ha

■ Site description

A complex of seven shallow artificial lakes constructed at the bottom of the former Lake Karla.

Habitats Wetland (100%; standing fresh water; water-fringe vegetation)
Land-use Fisheries/aquaculture (50%), Tourism/recreation, Water management

■ Birds

Species	Season	Year	Pop min	Pop max	Acc	Criteria
Egretta alba Great White Egret	W	1996	—	220	B	A4i, B1i, C2
Anas crecca Teal	W	1996	—	11,000	B	B1i, C3
Aythya nyroca Ferruginous Duck	P	1996	40	—	B	A1, C1
Falco naumanni Lesser Kestrel	N	1996	Abundant		—	A1, C1

A site of great importance for migratory and overwintering waterbirds and foraging raptors. Species of global conservation concern that do not meet IBA criteria: *Phalacrocorax pygmeus* (wintering), *Aquila clanga* (wintering), *Gallinago media* (passage).

■ Protection status

National None **International** None

■ Conservation issues

Threats Agricultural intensification/expansion (A), Aquaculture/fisheries (C), Construction/impact of dyke/dam/barrage (B), Disturbance to birds (A), Drainage (A), Groundwater abstraction (A), Other

Threats are from agricultural intensification around the IBA, water-level fluctuations, drainage, groundwater abstraction and surface water abstraction ('Other' threats). There is a plan to build a large reservoir. Part of the area is a candidate SAC.

Thessaly plain A1, A4ii, B1iii, B2, B3, C1, C2, C6 062

Admin region Thessalia
Coordinates 39°28′N 22°25′E
Altitude 70–729 m **Area** 180,000 ha

■ Site description

The largest plain in Greece. The site is predominantly arable land and villages. Hills covered with phrygana are included.

Habitats Scrub (5%; sclerophyllous scrub/garrigue/maquis), Artificial landscape (95%; arable land; perennial crops/orchards/groves; other urban/industrial areas)

Land-use Agriculture (95%), Urban/industrial/transport (5%), Water management (95%)

Birds

Species	Season	Year	Pop min	Pop max	Acc	Criteria
Ciconia ciconia White Stork	B	1996	Common		—	B2, C6
Accipiter brevipes Levant Sparrowhawk	B	1996	Frequent		—	B2
[1] ***Falco naumanni*** Lesser Kestrel	B	1995	2,342	2,342	A	A1, A4ii, B1iii, B2, C1, C2, C6
Tyto alba Barn Owl	R	1996	Frequent		—	B2
Otus scops Scops Owl	B	1996	Frequent		—	B2
Athene noctua Little Owl	R	1996	Common		—	B2
Calandrella brachydactyla Short-toed Lark	B	1996	Abundant		—	B2, C6
Galerida cristata Crested Lark	R	1996	Abundant		—	B2
Hippolais pallida Olivaceous Warbler	B	1996	Abundant		—	B2
Lanius minor Lesser Grey Shrike	B	1996	Frequent		—	B2, C6
Corvus monedula Jackdaw	R	1996	Abundant		—	B3
Emberiza melanocephala Black-headed Bunting	B	1996	Abundant		—	B2
Miliaria calandra Corn Bunting	B	1996	Abundant		—	B3

1. Large decrease.

The site is important for species found in arable land and holds the largest number of breeding *Falco naumanni* in Greece.

Protection status

National Low **International** None

2,300 ha of IBA covered by Wildlife Refuge (Mavrovouni/Agnanteris, 2,300 ha).

Conservation issues

Threats Abandonment/reduction of land management (B), Agricultural intensification/expansion (A), Construction/impact of dyke/dam/barrage (U), Industrialization/urbanization (U), Other (B), Unsustainable exploitation (U)

The main threat is from agricultural intensification. The demolition of old buildings is a particular threat to *Falco naumanni* ('Other' threats, above).

Farsala area — A1, B1iii, B2, C1, C2, C6 — 063

Admin region Thessalia
Coordinates 39°17'N 22°24'E
Altitude 100–524 m **Area** 4,500 ha

Site description

Hills on the southern edge of the Thessalian plain covered with phrygana and arable cultivation.

Habitats Scrub (65%; sclerophyllous scrub/garrigue/maquis), Rocky areas (10%; inland cliff), Artificial landscape (25%; arable land; other urban/industrial areas)
Land-use Agriculture (85%), Not utilized (10%), Urban/industrial/transport (5%)

Birds

Species	Season	Year	Pop min	Pop max	Acc	Criteria
[1] ***Falco naumanni*** Lesser Kestrel	B	1995	100	100	A	A1, B1iii, B2, C1, C2, C6

1. Large decrease.

This is an important site for species associated with scrub and arable land.

Protection status

National Partial **International** None

Part of IBA covered by Wildlife Refuge (Patima–Kyra Aposkia/Farsalon, 210 ha)

Conservation issues

Threats Abandonment/reduction of land management (B), Agricultural intensification/expansion (A), Industrialization/urbanization (U), Other (B), Unsustainable exploitation (U)

The main threats are from the intensification of agriculture and hunting. The demolition of old buildings is a particular threat to *Falco naumanni* ('Other' threats, above).

Mount Pilion — B2, C6 — 064

Admin region Thessalia
Coordinates 39°27'N 23°02'E
Altitude 0–1,624 m **Area** 31,751 ha

Site description

A forested mountain, dominated by *Fagus* and *Quercus*, with broadleaved evergreen forest in the eastern part of the site and evergreen scrub at lower altitudes.

Habitats Forest and woodland (54%; broadleaved deciduous forest; broadleaved evergreen forest), Scrub (27%; sclerophyllous scrub/garrigue/maquis), Artificial landscape (18%; arable land; perennial crops/orchards/groves; other urban/industrial areas)
Land-use Agriculture (25%), Forestry (54%), Tourism/recreation (10%), Urban/industrial/transport (5%)

Birds

Species	Season	Year	Pop min	Pop max	Acc	Criteria
Ficedula semitorquata Semi-collared Flycatcher	B	1996	—	—	—	B2, C6

This is an important site for breeding forest species.

Protection status

National Partial **International** None

1,100 ha of IBA covered by Wildlife Refuge (Ag. Dimitrios–Itamo/Mileon–Xourichtiou, 1,100 ha). 1,800 ha of IBA covered by Wildlife Refuge (Dasos I.M. Flamouriou/Keramidiou, 1,800 ha). 2,300 ha of IBA covered by Wildlife Refuge (Dasos Palias Mitzelas/Pouriou, 2,300 ha).

Conservation issues

Threats Agricultural intensification/expansion (B), Extraction industry (B), Infrastructure (B), Recreation/tourism (C), Unsustainable exploitation (A)

The main threat is from intense hunting pressure outside the Wildlife Refuges. The road network is increasing. The area is a candidate SAC.

National Marine Park of Alonissos (North Sporades) — A1, A4ii, B1ii, B1iii, B2, B3, C1, C2, C3, C6 — 065

Admin region Thessalia
Coordinates 39°18'N 24°04'E
Altitude 0–570 m **Area** 5,800 ha

Site description

A group of small, uninhabited rocky islands in the northern Sporades, covered by phrygana, except Piperi island which is wooded. Human activities include stock-raising and fishing.

Habitats Forest and woodland (5%; native coniferous forest), Scrub (83%; sclerophyllous scrub/garrigue/maquis), Marine areas (9%; sea inlet/coastal features)
Land-use Agriculture, Fisheries/aquaculture, Nature conservation/research (100%)

Birds

Species	Season	Year	Pop min	Pop max	Acc	Criteria
Calonectris diomedea Cory's Shearwater	B	1995	100	—	B	C6
Puffinus yelkouan Yelkouan Shearwater	B	1995	Common		—	B1ii, B3, C3
Falco eleonorae Eleonora's Falcon	B	1996	400	—	—	A4ii, B1iii, B2, C2, C6
Larus audouinii Audouin's Gull	B	1996	51	100	B	A1, C1

This is an important site for species breeding on coastal cliffs, for migratory waders, raptors and passerines.

Protection status

National High **International** High

5,800 ha of IBA covered by Marine Park (Northern Sporades, 100,000 ha). 1,226 ha of IBA covered by Wildlife Refuge (Ag. Ioannis–Trohalo/Litohoriou, 1,226 ha). 438 ha of IBA covered by Natural Monument (Piperi Island, 438 ha). 5,400 ha of IBA covered by Special Protection Area (Nisia Kyra Panagia, Piperi, Psathoura kai gyro Nis, 12,701 ha).

Conservation issues

Threats Agricultural intensification/expansion (C), Disturbance to birds

All the islands are overgrazed. Disturbance from military activities caused the abandonment of a *Larus audouinii* colony in 1996 and 1997. Part of the area is a candidate SAC.

Timfi (Gamila) and Smolikas mountains — A3, B2, C6 — 066

Admin region Ipiros
Coordinates 40°02′N 20°50′E
Altitude 500–2,637 m **Area** 55,000 ha

Site description

A remote mountain range crossed by the Rivers Aoos and Voidomatis. The site is dominated by broadleaved, conifer and mixed forests and extensive subalpine grasslands at higher altitudes. Human activities include forestry, stock-raising, hunting and recreation.

Habitats Forest and woodland (60%; broadleaved deciduous forest; native coniferous forest; mixed forest), Scrub (5%; scrub), Grassland (25%; alpine/subalpine/boreal grassland; mesophile grassland), Rocky areas (10%; scree/boulders; inland cliff)
Land-use Agriculture (80%), Forestry (40%), Tourism/recreation (20%), Water management (5%)

Birds

Species	Season	Year	Pop min	Pop max	Acc	Criteria
Neophron percnopterus Egyptian Vulture	B	1997	5	—	—	B2, C6
Circaetus gallicus Short-toed Eagle	B	—	—	—	—	C6
Aquila chrysaetos Golden Eagle	R	—	2	3	—	C6
Alectoris graeca Rock Partridge	R	1996	Frequent		—	B2
Prunella collaris Alpine Accentor	R	—	—	—	—	A3
Monticola saxatilis Rock Thrush	B	1997	Frequent		—	B2
Tichodroma muraria Wallcreeper	R	—	—	—	—	A3
Pyrrhocorax graculus Alpine Chough	R	1997	700	—	B	A3
Pyrrhocorax pyrrhocorax Chough	R	1997	Uncommon		—	B2, C6
Montifringilla nivalis Snowfinch	R	—	—	—	—	A3

This is an important site for breeding raptors, alpine and forest species. Breeding birds include four of the 10 species in Europe that are restricted (when breeding) to the Eurasian high-montane biome.

Protection status

National Partial **International** Partial

12,600 ha of IBA covered by National Park (Vikos–Aoos, 12,600 ha). 2,080 ha of IBA covered by Wildlife Refuge (Padon, 2,080 ha). 1,017 ha of IBA covered by Wildlife Refuge (Kedras–Papigou, 1,017 ha). 3,900 ha of IBA covered by Wildlife Refuge (Parletzi–Avalos/Vrisochoriou–Hliochoriou, 3,900 ha). 3,600 ha of IBA covered by Wildlife Refuge (Charadras Aoou/Konitsas–Eleftherou-Papigou, 3,600 ha). 12,793 ha of IBA covered by Special Protection Area (Ethnikos Drymos Vikou–Aoou, 12,793 ha).

Conservation issues

Threats Construction/impact of dyke/dam/barrage (A), Deforestation (commercial) (B), Extraction industry (C), Forest grazing (B), Intensified forest management (A), Recreation/tourism (A), Selective logging/cutting (A), Unsustainable exploitation (A)

The main threats are from plans for large-scale tourism development, existing water management schemes and hunting. A number of conservation and development studies have been conducted. A management plan exists. Part of the area is a candidate SAC.

Central part of Zagori and Mount Mitsikeli — B2, B3, C6 — 067

Admin region Ipiros
Coordinates 39°48′N 20°54′E
Altitude 500–1,810 m **Area** 46,500 ha

Site description

The site is dominated by hills covered with *Quercus* forest.

Habitats Forest and woodland (85%; broadleaved deciduous forest; mixed forest), Scrub (10%; sclerophyllous scrub/garrigue/maquis), Grassland (5%; alpine/subalpine/boreal grassland)
Land-use Agriculture (70%), Forestry (70%), Tourism/recreation (5%)

Birds

Species	Season	Year	Pop min	Pop max	Acc	Criteria
Neophron percnopterus Egyptian Vulture	B	1996	2	4	B	C6
Circaetus gallicus Short-toed Eagle	B	1997	5	—	B	C6
Strix aluco Tawny Owl	R	—	—	—	—	B3
Picus viridis Green Woodpecker	R	1997	Common		—	B2
Dendrocopos medius Middle Spotted Woodpecker	R	1997	Common		—	C6
Phylloscopus bonelli Bonelli's Warbler	B	1997	Frequent		—	B3

An important site for forest species.

Protection status

National Low **International** None

710 ha of IBA covered by Wildlife Refuge (Ag. Athanasios–Sipitoura/Aspraggelon–Elatis, 710 ha).

Conservation issues

Threats Abandonment/reduction of land management (U), Deforestation (commercial) (B), Extraction industry (C), Intensified forest management (B), Recreation/tourism (C), Selective logging/cutting (A), Unsustainable exploitation (A)

The site is threatened by intensive forestry, deforestation and hunting. Part of the area is a candidate SAC.

Valia Kalda (Pindos National Park) — B2, C7 — 068

Admin region Western Macedonia, Ipiros
Coordinates 39°54′N 21°08′E
Altitude 1,100–2,177 m **Area** 8,000 ha

Site description

A mountainous valley dominated by *Pinus* and *Fagus* forests. Human activities include livestock farming and tourism.

Habitats Forest and woodland (70%; broadleaved deciduous forest; native coniferous forest; mixed forest), Grassland (20%; alpine/subalpine/boreal grassland; mesophile grassland), Rocky areas (10%; scree/boulders)
Land-use Agriculture (90%), Forestry (35%), Tourism/recreation (5%)

Birds

This is an important area for breeding raptors and forest species. Species of global conservation concern that do not meet IBA criteria: *Aquila heliaca* (non-breeding).

Protection status

National Partial **International** Partial

Part of IBA covered by Wildlife Refuge (Dimasio Dasos Flabourarious and Vovousas, 861 ha). 6,927 ha of IBA covered by National Park (Pindos, 6,927 ha). 3,360 ha of IBA covered by Biogenetic Reserve (Council of Europe) (Pindos National Park (core area), 3,360 ha). 3,294 ha of IBA covered by Special Protection Area (Ethnikos Drymos Pindou, 3,294 ha).

Conservation issues

Threats Construction/impact of dyke/dam/barrage (A), Forest grazing (U), Recreation/tourism (B), Selective logging/cutting (B), Unsustainable exploitation (B)

The most important threat is a plan to divert the Arkoudorema river. There is high grazing pressure in the core area of the National Park. A number of conservation and development studies have been conducted. Part of the area is a candidate SAC.

Peristeri mountain — B2, C6 — 069

Admin region Ipiros, Thessalia
Coordinates 39°40′N 21°08′E
Altitude 800–2,125 m **Area** 21,300 ha

Site description

A large limestone massif, south of the town of Metsovo. The site has an extensive area of alpine grassland, cliffs, scree and gorges. Part of the site is covered by fir *Abies* and beech *Fagus* forest. The main human activity is livestock-farming.

Habitats Forest and woodland (broadleaved deciduous forest; native coniferous forest; mixed forest; treeline ecotone), Grassland (alpine/subalpine/boreal grassland), Rocky areas (scree/boulders; inland cliff)
Land-use Agriculture, Forestry

Birds

Species	Season	Year	Pop min	Pop max	Acc	Criteria
Gyps fulvus Griffon Vulture	R	1993	Common		—	C6
Monticola saxatilis Rock Thrush	B	1993	Abundant		—	B2

The site is important for species associated with alpine grassland, cliffs and scree.

Protection status

National Partial **International** None
1,867 ha of IBA covered by Wildlife Refuge (Koutsouyo–Tzakouta–Mnimata/Katafigiou–Plakas–Anthousas, 1,867 ha). 1,550 ha of IBA covered by Wildlife Refuge (Kalouta–Nikoulitsa/Kalariton, 1,550 ha).

Conservation issues

Threats Unsustainable exploitation (B)

The main threat is from illegal hunting, particularly of raptors. Part of the area is a candidate SAC.

Lake Pamvotida (Ioanninon) — A1, B1i, C1, C3 — 070

Admin region Ipiros
Coordinates 39°40′N 20°53′E
Altitude 470–700 m **Area** 2,707 ha

Site description

A large freshwater lake with extensive reedbeds (*Phragmites*).

Habitats Wetland (90%; standing fresh water; water-fringe vegetation), Artificial landscape (10%; arable land; other urban/industrial areas)
Land-use Agriculture (5%), Fisheries/aquaculture (90%), Urban/industrial/transport (5%), Water management (90%)

Birds

Species	Season	Year	Pop min	Pop max	Acc	Criteria
Phalacrocorax carbo Cormorant	W	1989	—	1,315	A	B1i, C3
Falco naumanni Lesser Kestrel	N	—	—	—	—	A1, C1

This is an important site for wintering ducks (Anatidae) and breeding species associated with reedbeds. Species of global conservation concern that do not meet IBA criteria: *Phalacrocorax pygmeus* (wintering), *Aythya nyroca* (breeding).

Protection status

National None **International** None

Conservation issues

Threats Agricultural intensification/expansion (A), Groundwater abstraction (U), Industrialization/urbanization (A), Unsustainable exploitation (C)

The main threats come from agricultural intensification and building around the lake.

Ioannina city and neighbouring area — A1, B2, C1, C6 — 071

Admin region Ipiros
Coordinates 39°40′N 20°51′E
Altitude 0–470 m **Area** 8,000 ha

Site description

The IBA includes the city of Ioannina and the surrounding arable land and hills.

Habitats Scrub (10%; sclerophyllous scrub/garrigue/maquis), Artificial landscape (90%; arable land; forestry plantation; other urban/industrial areas)
Land-use Agriculture (85%), Unknown (10%), Urban/industrial/transport (5%)

Birds

Species	Season	Year	Pop min	Pop max	Acc	Criteria
Falco naumanni Lesser Kestrel	B	1996	50	150	C	A1, B2, C1, C6

The site is important for species associated with arable agriculture.

Protection status

National None **International** None

Conservation issues

Threats Abandonment/reduction of land management (B), Agricultural intensification/expansion (A), Industrialization/urbanization (U), Other (B), Unsustainable exploitation (U)

The main threat is the intensification of agriculture. The demolition of old buildings is a particular threat to *Falco naumanni* ('Other' threats).

Area of Oriokastro, Lake Delvinaki, Meropis forest, Gormou valley and Mounts Kasidiaris and Douskon — B2, B3, C6 — 072

Admin region Ipiros
Coordinates 39°56′N 20°30′E
Altitude 400–2,200 m **Area** 36,000 ha

Site description

A hilly area dominated by *Quercus* forest. Human activities include agriculture, forestry, livestock-farming, and hunting.

Habitats Forest and woodland (65%; broadleaved deciduous forest), Scrub (10%; scrub; sclerophyllous scrub/garrigue/maquis), Grassland (10%; steppe/dry calcareous grassland), Rocky areas (5%; inland cliff), Artificial landscape (10%; arable land)
Land-use Agriculture (70%), Forestry (50%), Water management (10%)

Birds

Species	Season	Year	Pop min	Pop max	Acc	Criteria
[1] ***Neophron percnopterus*** Egyptian Vulture	B	1996	2	4	B	C6
Accipiter brevipes Levant Sparrowhawk	B	1996	Frequent		—	B2, C6
Hieraaetus pennatus Booted Eagle	B	1996	2	4	B	C6
Dendrocopos medius Middle Spotted Woodpecker	R	1996	Common		—	B3, C6

1. Large decrease.

This is an important site for forest species and resident raptors. Species of global conservation concern that do not meet IBA criteria: *Phalacrocorax pygmeus*.

Protection status

National Partial **International** None
Part of IBA covered by Wildlife Refuge (Prosilio–Ronitsa/Delvinakiou, 1,525 ha).

Conservation issues

Threats Drainage (A), Firewood collection (U), Forest grazing (U), Selective logging/cutting (B), Unsustainable exploitation (U)

Threats include the logging of mature trees, a drainage plan for the Gormos valley and illegal hunting.

Mounts Tsamanta, Filiaton, Pharmakovouni, and Megali Rahi — A1, B2, C1, C6 — 073

Admin region Ipiros
Coordinates 39°45′N 20°30′E
Altitude 200–1,806 m **Area** 21,000 ha

Site description

A mountainous area on the border with Albania. At higher altitudes there are extensive areas of alpine pasture and some conifer forests. Lower down, deciduous forest and scrub dominates. Human activities include stock-raising, hunting, and forestry.

Habitats Forest and woodland (60%; broadleaved deciduous forest; native coniferous forest), Scrub (20%; sclerophyllous scrub/garrigue/maquis), Grassland (20%; alpine/subalpine/boreal grassland), Rocky areas (5%; inland cliff), Artificial landscape (5%; ruderal land)
Land-use Agriculture (50%), Forestry, Not utilized

Birds

Species	Season	Year	Pop min	Pop max	Acc	Criteria
Neophron percnopterus Egyptian Vulture	B	1997	3	—	A	C6
Aquila chrysaetos Golden Eagle	R	1997	2	3	A	C6
Falco naumanni Lesser Kestrel	P	1997	Common		—	A1, C1
Monticola saxatilis Rock Thrush	B	1997	Common		—	B2
Pyrrhocorax pyrrhocorax Chough	R	1997	10	20	B	B2, C6

This is an important site for raptors. Species of global conservation concern that do not meet IBA criteria: *Aquila heliaca*.

Protection status

National Low **International** None
1,350 ha of IBA covered by Wildlife Refuge (Panagia–Kourouza–Makralexi/Lavdani–Ag. Marina Pogo, 1,350 ha).

Conservation issues

Threats Abandonment/reduction of land management (B), Agricultural intensification/expansion (C), Infrastructure (B), Unknown (U), Unsustainable exploitation (A)

The main threats are from shooting, the use of poisoned baits, and road construction. In 1997 HOS carried out a survey of *Aquila heliaca*.

Kalamas estuary — A1, A4i, B1i, C1, C2, C3, C6 — 074

Admin region Ipiros
Coordinates 39°32′N 20°11′E
Altitude 0–509 m **Area** 8,531 ha

Site description

The site comprises a canal with a delta of mudflats, saltmarshes, and lagoons. Two low hills are covered with dense scrub. Much of the area has been reclaimed for arable farming.

Habitats Scrub (18%; sclerophyllous scrub/garrigue/maquis), Wetland (26%; mudflat/sandflat; saltmarsh; coastal lagoon; river/stream), Marine areas (4%; sea inlet/coastal features), Artificial landscape (48%; arable land; perennial crops/orchards/groves; other urban/industrial areas)
Land-use Agriculture (75%), Fisheries/aquaculture (10%), Tourism/recreation (5%), Urban/industrial/transport (5%), Water management

Birds

Species	Season	Year	Pop min	Pop max	Acc	Criteria
Phalacrocorax carbo Cormorant	W	1995	58	1,017	A	B1i, C3
Phalacrocorax pygmeus Pygmy Cormorant	W	1988	—	65	A	A1, C1
Pelecanus crispus Dalmatian Pelican	W	1996	1	22	—	A1, C1
Egretta alba Great White Egret	W	1993	—	219	A	A4i, B1i, C2
Plegadis falcinellus Glossy Ibis	P	1996	445	—	A	A4i, B1i, C2
Aquila clanga Greater Spotted Eagle	W	1996	—	7	A	A1, C1
Numenius tenuirostris Slender-billed Curlew	P	1997	2	—	A	A1, C1
Sterna albifrons Little Tern	B	1992	40	—	—	C6
Calandrella brachydactyla Short-toed Lark	B	—	—	—	—	C6

This site is important for breeding, passage and overwintering waterbirds. Species of global conservation concern that do not meet IBA criteria: *Aythya nyroca* (passage, wintering), *Haliaeetus albicilla* (wintering), *Aquila heliaca* (wintering), *Gallinago media* (passage).

Protection status

National Partial **International** Partial
1,780 ha of IBA covered by Wildlife Refuge (Vatatsa–Divari/Igoumenitsas, 1,780 ha). 5,521 ha of IBA covered by Special Protection Area (Ygrotopos Ekvolon (Delta) Kalama, 5,521 ha).

Conservation issues

Threats Agricultural intensification/expansion (A), Burning of vegetation (B), Drainage (B), Infrastructure (B), Recreation/tourism (C), Unsustainable exploitation (A)

The main threats come from habitat destruction and water pollution as a result of expanding intensive agriculture, road construction, over-grazing and hunting. The area is a candidate SAC.

Kalamas gorge — A1, B2, C1, C6 — 075

Admin region Ipiros
Coordinates 39°35′N 20°21′E
Altitude 100–1,200 m **Area** 4,000 ha

Site description

A gorge with alluvial forest and surrounding hills covered by phrygana. The site includes a small wetland. Human activities include stock-raising, arable agriculture, quarrying and hunting.

Habitats Forest and woodland (alluvial/very wet forest), Scrub (sclerophyllous scrub/garrigue/maquis), Wetland (river/stream), Rocky areas (inland cliff), Artificial landscape (arable land)
Land-use Agriculture (80%), Unknown (10%), Not utilized (10%)

Birds

Species	Season	Year	Pop min	Pop max	Acc	Criteria
Falco naumanni Lesser Kestrel	B	1996	Common		—	A1, B2, C1, C6

This is an important area for raptors and species found in phrygana. Species of global conservation concern that do not meet IBA criteria: *Aquila heliaca* (non-breeding).

Protection status

National Partial **International** None
Part of the IBA is covered by Wildlife Refuge (Palaikastra/Elaias–Arahovas–Sideris, 2,350 ha).

Conservation issues

Threats Agricultural intensification/expansion (B), Drainage (B), Extraction industry (A), Unsustainable exploitation (B)

Threats include the expansion of arable agriculture, drainage and quarrying.

Mount Paramithias — B3, C6 — 076

Admin region Ipiros
Coordinates 39°37′N 20°30′E
Altitude 400–1,658 m **Area** 20,000 ha

Site description

An inaccessible rugged mountain chain with high cliffs in the western part.

Habitats Forest and woodland (broadleaved deciduous forest; native coniferous forest), Scrub (sclerophyllous scrub/garrigue/maquis), Rocky areas (scree/boulders; inland cliff)
Land-use Agriculture (90%), Not utilized (10%)

Birds

Species	Season	Year	Pop min	Pop max	Acc	Criteria
Neophron percnopterus Egyptian Vulture	B	1996	3	—	B	C6
[1] ***Gyps fulvus*** Griffon Vulture	R	1996	10	15	B	C6
Aquila chrysaetos Golden Eagle	R	1996	3	—	B	C6
Hieraaetus fasciatus Bonnelli's Eagle	R	1996	2	—	B	C6
Sitta neumayer Rock Nuthatch	R	1996	Common		—	B3

1. Individuals.

The site is of great importance for raptors. Species of global conservation concern that do not meet IBA criteria: *Falco naumanni* (breeding).

Protection status

National Partial **International** None
Part of IBA is covered by Wildlife Refuge (Ravenis–Malouniou, 2,500 ha).

Conservation issues

Threats Unknown (U), Unsustainable exploitation (U)

Hunting is a potential threat. Part of the area is a candidate SAC.

Acheron delta and gorge — A3, A4i, B1i, C2, C6 — 077

Admin region Ipiros
Coordinates 39°15′N 20°30′E
Altitude 0–350 m **Area** 6,000 ha

Site description

The inland part of the site consists of arable cultivation and a gorge with cliffs, deciduous forest and scrub. The coastal delta includes brackish marshes and low, scrub-dominated hills. Human activities include arable agriculture, livestock-farming, recreation and hunting.

Habitats Forest and woodland (20%; broadleaved deciduous forest; broadleaved evergreen forest), Scrub (15%; sclerophyllous scrub/garrigue/maquis), Wetland (40%; sand-dunes/sand beach; standing fresh water; river/stream; water-fringe vegetation), Rocky areas (5%; scree/boulders; inland cliff), Artificial landscape (20%; arable land)
Land-use Agriculture (60%), Tourism/recreation (20%)

Birds

Species	Season	Year	Pop min	Pop max	Acc	Criteria
Plegadis falcinellus Glossy Ibis	P	1996	250	—	—	A4i, B1i, C2
Hieraaetus fasciatus Bonnelli's Eagle	R	1994	2	2	—	C6
Alectoris graeca Rock Partridge	R	1994	Common		—	A3
Oenanthe hispanica Black-eared Wheatear	B	1995	Common		—	A3
Hippolais olivetorum Olive-tree Warbler	B	1994	Uncommon		—	A3
Sylvia cantillans Subalpine Warbler	B	1995	Common		—	A3
Sylvia melanocephala Sardinian Warbler	R	1995	Common		—	A3
Sitta neumayer Rock Nuthatch	R	1995	Frequent		—	A3
Emberiza caesia Cretzschmar's Bunting	B	1994	Uncommon		—	A3
Emberiza melanocephala Black-headed Bunting	B	1994	Frequent		—	A3

The site is important for breeding, passage and wintering waterbirds and raptors. Breeding birds include eight of the 21 species in Europe that are restricted (when breeding) to the Mediterranean biome. Species of global conservation concern that do not meet IBA criteria: *Phalacrocorax pygmeus* (wintering), *Pelecanus crispus* (non-breeding), *Aythya nyroca* (non-breeding), *Gallinago media* (passage).

Protection status

National None **International** None

Conservation issues

Threats Agricultural intensification/expansion (B), Disturbance to birds (B), Drainage (A), Forest grazing (C), Recreation/tourism (A), Unsustainable exploitation (A)

Major threats are from drainage, agricultural expansion and hunting. The area is a candidate SAC.

Lake Kalodiki, Margariti and Karteri marshes — B2, C6 — 078

Admin region Ipiros
Coordinates 39°20′N 20°27′E
Altitude 96–220 m **Area** 1,650 ha

Site description

An area of freshwater lakes, fen and wet meadows. An area of forest exists on an island in Lake Kalodiki. The surrounding slopes are covered with maquis.

Habitats Forest and woodland (native coniferous forest), Scrub (sclerophyllous scrub/garrigue/maquis), Wetland (standing fresh water; water-fringe vegetation; fen/transition mire/spring)
Land-use Agriculture, Tourism/recreation

Birds

Species	Season	Year	Pop min	Pop max	Acc	Criteria
Aythya nyroca Ferruginous Duck	B	1997	10	15	B	B2, C6

Species of global conservation concern that do not meet IBA criteria: *Falco naumanni* (breeding).

Protection status

National Partial **International** None
300 ha of IBA covered by Wildlife Refuge (Valtos Kalodikiou/Margaritiou–Eleftheriou–Spatharaion, 300 ha).

Conservation issues

Threats Agricultural intensification/expansion (U), Drainage (U), Groundwater abstraction (U), Other (U), Unsustainable exploitation (A)

Threats include the dumping of waste ('Other' threats, above), the expansion of agriculture, uncontrolled water extraction, water pollution, and hunting. There are plans to develop tourism. Part of the area is a candidate SAC.

Athamanika mountains (Tzoumerka) — B2, C6 — 079

Admin region Ipiros, Thessalia
Coordinates 39°30′N 21°13′E
Altitude 600–2,470 m **Area** 55,000 ha

Site description

A mountain in southern Pindos with well-forested slopes, alpine grasslands and cliffs.

Habitats Forest and woodland (60%; broadleaved deciduous forest; native coniferous forest; mixed forest), Scrub (10%; sclerophyllous scrub/garrigue/maquis), Grassland (10%; alpine/subalpine/boreal grassland), Rocky areas (10%; inland cliff)
Land-use Agriculture (50%), Forestry (30%), Tourism/recreation (5%), Unknown (15%), Water management

Birds

Species	Season	Year	Pop min	Pop max	Acc	Criteria
Aquila chrysaetos Golden Eagle	R	1996	—	2	A	C6
Alectoris graeca Rock Partridge	R	1996	Frequent		—	B2

This is an important area for raptors and forest and scrub species.

Protection status

National Low **International** None
1,440 ha of IBA covered by Wildlife Refuge (Ag. Paraskeui–Kranies/Pramanton, 1,440 ha). 1,572 ha of IBA covered by Wildlife Refuge (Melissourgon, 1,572 ha). 1,000 ha of IBA covered by Wildlife Refuge (Vourgareliou, 1,000 ha).

Conservation issues

Threats Deforestation (commercial) (A), Firewood collection (B), Forest grazing (B), Infrastructure (B), Intensified forest management (U), Recreation/tourism (C), Unsustainable exploitation (C)

The main threats are from deforestation and increasing holiday home construction. Part of the area is a candidate SAC.

Acheloos valley

B2, C7 **080**

Admin region Ipiros, Thessalia
Coordinates 39°18′N 21°25′E
Altitude 400–1,750 m **Area** 10,600 ha

Site description

A mountainous area with scrub and deciduous forests and conifer forest at higher altitudes. A gorge with steep cliffs runs through the centre of the site. Human activities include livestock-farming, hunting, and forestry.

Habitats Forest and woodland (40%; broadleaved deciduous forest; native coniferous forest; broadleaved evergreen forest; treeline ecotone), Scrub (40%; sclerophyllous scrub/garrigue/maquis), Rocky areas (20%; scree/boulders; inland cliff)
Land-use Agriculture (60%), Forestry (10%), Not utilized (30%)

Birds

Species	Season	Year	Pop min	Pop max	Acc	Criteria
Lanius senator Woodchat Shrike	B	1994	Abundant		—	B2

This is an important site for species characteristic of forest, scrub and open montane habitats.

Protection status

National Partial **International** Partial
2,275 ha of IBA covered by Wildlife Refuge (Mesountas, 2,275 ha). 9,082 ha of IBA covered by Special Protection Area (Koilada Acheloou, 9,082 ha).

Conservation issues

Threats Abandonment/reduction of land management (B), Afforestation (U), Construction/impact of dyke/dam/barrage (A), Infrastructure (B), Forest grazing (B), Selective logging/cutting (C), Unsustainable exploitation (U)

A dense road network causes problems of soil erosion and also provides access to the forest for loggers and illegal hunters. Afforestation with *Pinus* species is a potential threat to the deciduous forest. Part of the area is a candidate SAC.

Amvrakikos Gulf

A1, A4i, A4iii, B1i, B2, C1, C2, C3, C4, C6 **081**

Admin region Ipiros, Western Greece
Coordinates 39°00′N 21°00′E
Altitude 0–450 m **Area** 25,000 ha

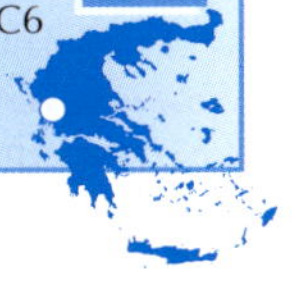

Site description

The northern section of the site is formed by the deltas of the Louros and Arachtos rivers. It consists of brackish lagoons, a sandy coastal strip, saltmarsh, reedbeds (*Phragmites*), wet meadows, mudflats and remnant patches of alluvial forest. The lagoons are all used as fisheries.

Habitats Scrub (5%; sclerophyllous scrub/garrigue/maquis), Grassland (5%; humid grassland), Wetland (70%; saltmarsh; sand-dunes/sand beach; coastal lagoon; standing fresh water; standing brackish and salt water; river/stream; water-fringe vegetation; fen/transition mire/spring), Marine areas (5%; sea inlet/coastal features), Artificial landscape (20%; arable land; perennial crops/orchards/groves; other urban/industrial areas)
Land-use Agriculture (25%), Fisheries/aquaculture (70%), Nature conservation/research, Tourism/recreation (5%), Urban/industrial/transport (5%), Water management (70%)

Birds

Species	Season	Year	Pop min	Pop max	Acc	Criteria
Podiceps nigricollis Black-necked Grebe	W	1996	900	2,200	—	A4i, B1i, C3
Phalacrocorax carbo Cormorant	W	1996	1,200	1,800	—	B1i, C3
Pelecanus crispus Dalmatian Pelican	R	1996	45	50	—	A1, A4i, B1i, B2, C1, C2, C6
Ixobrychus minutus Little Bittern	B	1996	30	60	—	C2, C6
Nycticorax nycticorax Night Heron	B	1996	50	100	—	B2, C6
Ardeola ralloides Squacco Heron	B	1996	80	150	—	A4i, B1i, B2, C2, C6
Egretta garzetta Little Egret	P	1996	800	1,500	—	A4i, B1i, C2
Egretta alba Great White Egret	B	1996	130	200	—	A4i, B1i, C2
Ciconia ciconia White Stork	B	1996	70	80	—	C6
Plegadis falcinellus Glossy Ibis	P	1996	400	1,200	—	A4i, B1i, C2
Platalea leucorodia Spoonbill	B	1996	6	10	—	B2, C6
Anas penelope Wigeon	W	1996	20,000	50,000	—	A4i, B1i, C3
Anas crecca Teal	W	1996	8,000	30,000	—	A4i, B1i, C3
Anas acuta Pintail	W	1996	8,000	15,000	—	A4i, B1i, C3
Anas clypeata Shoveler	W	1996	7,000	26,000	—	A4i, B1i, C3
Aythya ferina Pochard	W	1996	5,000	22,000	—	A4i, B1i, C3
Aythya nyroca Ferruginous Duck	B	1996	25	100	—	A1, B2, C1, C6
Aquila clanga Greater Spotted Eagle	W	1996	3	4	—	A1, C1
Fulica atra Coot	W	1996	20,000	47,000	—	A4i, B1i, C3
Haematopus ostralegus Oystercatcher	B	1996	8	12	—	B1i, C3
Himantopus himantopus Black-winged Stilt	B	1996	220	360	—	A4i, B1i, C2
Burhinus oedicnemus Stone Curlew	B	1996	8	30	—	B2, C6
Glareola pratincola Collared Pratincole	B	1996	100	160	—	A4i, B1i, B2, C2, C6
Charadrius alexandrinus Kentish Plover	R	1996	300	400	—	A4i, B1i, B2, C3
Charadrius alexandrinus Kentish Plover	W	1996	600	2,500	—	A4i, B1i, C3
Calidris minuta Little Stint	P	1996	3,000	8,000	—	A4i, B1i, C3
Calidris ferruginea Curlew Sandpiper	P	1996	2,000	6,000	—	A4i, B1i, C3
Numenius tenuirostris Slender-billed Curlew	P	—	Rare		—	A1, C1
Tringa totanus Redshank	W	1996	800	2,300	—	B1i, C3
Gelochelidon nilotica Gull-billed Tern	B	1996	80	120	—	A4i, B1i, B2, C2, C6
Sterna sandvicensis Sandwich Tern	R	1996	45	90	—	C2
Sterna albifrons Little Tern	B	1996	400	550	—	A4i, B1i, B2, C2, C6
Chlidonias hybridus Whiskered Tern	B	1997	5	6	A	B2, C6
Calandrella brachydactyla Short-toed Lark	B	1996	Abundant		—	B2, C6

This is an important site for wintering ducks and geese, breeding, passage and wintering raptors and migratory waders. Species of global conservation concern that do not meet IBA criteria: *Phalacrocorax pygmeus* (wintering), *Haliaeetus albicilla* (wintering), *Aquila heliaca* (non-breeding), *Falco naumanni* (passage), *Gallinago media* (passage).

Protection status

National Partial **International** Partial
10,500 ha of IBA covered by Wildlife Refuge (Limnothalassa Tsoukalio kai Valtos Rodias, 10,500 ha). 23,649 ha of IBA covered by Ramsar Site (Amvrakikos Gulf, 23,649 ha). 20,900 ha of IBA covered by Special Protection Area (Amvrakikos Kolpos, 21,239 ha).

Conservation issues

Threats Agricultural intensification/expansion (A), Aquaculture/fisheries (A), Construction/impact of dyke/dam/barrage (U), Disturbance to birds (A), Drainage (B), Dredging/canalization (A), Extraction industry (B), Groundwater abstraction (A), Industrialization/urbanization (A), Infrastructure (U), Recreation/tourism (A), Unsustainable exploitation (A)

The site is threatened by water-management schemes, large-scale work to improve fish production in the lagoons, intensive agriculture, hunting, and urban and industrial expansion. Part of the area is a candidate SAC.

Diapontia islands (Othoni, Erikoussa, Mathraki islands and rocky islets)

B2, C6 **082**

Admin region Ionian islands
Coordinates 39°51′N 19°24′E
Altitude 0–393 m **Area** 1,500 ha

Site description

A group of small islands and rocky islets north-west of the island of Corfu. Human activities include agriculture, fishing and tourism.

Habitats Scrub (sclerophyllous scrub/garrigue/maquis), Rocky areas (sea cliff/rocky shore), Artificial landscape (arable land)
Land-use Agriculture, Fisheries/aquaculture, Tourism/recreation

Birds

Species	Season	Year	Pop min	Pop max	Acc	Criteria
Falco eleonorae Eleonora's Falcon	B	—	20	40	B	B2, C6

This is an important site for breeding seabirds and raptors

Protection status

National None **International** None

Conservation issues

Threats Unknown

More research is needed in order to evaluate the importance of the site for migratory species and seabirds.

Lagoons of Kerkyra — A1, A4i, B1i, C1, C2, C3 — 083

Admin region Ionian Islands
Coordinates 39°27′N 19°55′E
Altitude 0–496 m **Area** 1,100 ha

Site description

The site covers four separate coastal wetlands in Corfu. These include brackish lagoons (Korission, Antinioti, and Chalikopoulou), saltpans (Lefkimis) and freshwater marshes and sand-dunes (Korission–Antinioti).

Habitats Scrub (scrub), Wetland (mudflat/sandflat; saltmarsh; sand-dunes/sand beach; coastal lagoon; standing brackish and salt water; fen/transition mire/spring), Artificial landscape (other urban/industrial areas)
Land-use Agriculture, Fisheries/aquaculture, Tourism/recreation, Urban/industrial/transport

Birds

Species	Season	Year	Pop min	Pop max	Acc	Criteria
Phalacrocorax carbo Cormorant	W	—	400	1,000	—	B1i, C3
Egretta alba Great White Egret	W	—	100	300	—	A4i, B1i, C2
Numenius tenuirostris Slender-billed Curlew	P	1974	1	—	A	A1, C1

Korission and Antinioti lagoons are important for cormorants, herons and ducks, while Lefkimi lagoon is important for shorebirds. Species of global conservation concern that do not meet IBA criteria: *Phalacrocorax pygmeus* (passage), *Aythya nyroca* (passage), *Aquila clanga* (wintering).

Protection status

National None **International** None

Conservation issues

Threats Abandonment/reduction of land management (B), Agricultural intensification/expansion (B), Construction/impact of dyke/dam/barrage (C), Drainage (B), Dredging/canalization (C), Filling-in of wetlands (B), Groundwater abstraction (B), Industrialization/urbanization (A), Recreation/tourism (B), Unsustainable exploitation (B)

The main threats are from water management and drainage and intensive tourism development. Chalikopoulou lagoon is seriously threatened from infilling, urbanization and the expansion of an airport. Part of the area is a candidate SAC.

Echinades islands, Kalamos and Atokos — C7 — 084

Admin region Ionian Islands
Coordinates 38°30′N 20°56′E
Altitude 0–745 m **Area** 4,700 ha

Site description

A group of 25 small islands close to the mainland Akarnanika mountains and the Acheloos estuary–Mesolongi wetland area. The islands are predominantly covered with scrub or bare rock. However, some are wooded including Kalamos and Petala. The main human activities are livestock-farming and tourism.

Habitats Forest and woodland (10%; broadleaved deciduous forest; native coniferous forest), Scrub (40%; scrub; sclerophyllous scrub/garrigue/maquis), Wetland (shingle/stony beach), Marine areas (open sea; sea inlet/coastal features), Rocky areas (20%; sea cliff/rocky shore; rock stacks/islets; scree/boulders)
Land-use Agriculture (60%), Not utilized (20%), Urban/industrial/transport (5%)

Birds

The site is particularly important for breeding and wintering raptors. Species of global conservation concern that do not meet IBA criteria: *Aegypius monachus* (non-breeding), *Aquila clanga* (1–4 wintering individuals), *Aquila heliaca* (1–2 wintering individuals).

Protection status

National None **International** Partial
2,000 ha of IBA covered by Special Protection Area (Delta Acheloou, Limnothalassa Mesolongiou–Aitoliko kai Ekvoles Evinou, 34,941 ha).

Conservation issues

Threats Abandonment/reduction of land management (B), Agricultural intensification/expansion (B), Consequences of animal/plant introductions (U), Deforestation (commercial) (U), Disturbance to birds (B), Firewood collection (B), Forest grazing (B), Industrialization/urbanization (B), Recreation/tourism (A), Unsustainable exploitation (A)

Illegal hunting and tourist developments are the major threats. Part of the area is a candidate SAC.

The Agia Dinati and Kokkini Rachi mountains of Kefallinia — C7 — 085

Admin region Ionian Islands
Coordinates 38°14′N 20°33′E
Altitude 0–1,131 m **Area** 20,000 ha

Site description

An area of low mountains predominantly covered by maquis and phrygana with olive groves at lower altitudes. Human activities include livestock-farming and hunting.

Habitats Forest and woodland (10%; native coniferous forest; broadleaved evergreen forest), Scrub (70%; sclerophyllous scrub/garrigue/maquis), Rocky areas (5%; scree/boulders), Artificial landscape (15%; perennial crops/orchards/groves)
Land-use Agriculture (80%), Not utilized (5%), Tourism/recreation (5%), Urban/industrial/transport (5%)

Birds

The site is important for raptors and a range of passerines associated with a mosaic of forest, scrub and agricultural land.

Protection status

National Low **International** Low
575 ha of IBA covered by National Park (Ainos, 2,862 ha). 1,787 ha of IBA covered by Wildlife Refuge (Vouno Atros/Ag. Nikolaou–Porou, 1,787 ha). 575 ha of IBA covered by Special Protection Area (Ethnikos Drymos Ainou, 2,779 ha).

Conservation issues

Threats Agricultural intensification/expansion (U), Forest grazing (B), Industrialization/urbanization (B), Infrastructure (A), Recreation/tourism (U), Unsustainable exploitation (A)

The main threats are from infrastructural developments, hunting and agricultural intensification. Part of the area is a candidate SAC.

West and north Zakynthos — A3, B1i, C2 — 086

Admin region Ionian Islands
Coordinates 37°46′N 20°41′E
Altitude 0–431 m **Area** 13,600 ha

Site description

The area is dominated by maquis, *Pinus* forests and agricultural land.

Habitats Forest and woodland (15%; native coniferous forest), Scrub (60%; sclerophyllous scrub/garrigue/maquis), Marine areas (5%; sea inlet/coastal features), Rocky areas (5%; sea cliff/rocky shore), Artificial landscape (20%; perennial crops/orchards/groves)
Land-use Agriculture (80%), Fisheries/aquaculture (5%), Tourism/recreation (10%), Urban/industrial/transport (5%)

Birds

Species	Season	Year	Pop min	Pop max	Acc	Criteria
[1] ***Phalacrocorax aristotelis*** Shag	R	1996	70	150	B	B1i, C2
Falco eleonorae Eleonora's Falcon	B	1996	2	6	B	A3
Alectoris graeca Rock Partridge	R	—	—	—	—	A3
Oenanthe hispanica Black-eared Wheatear	B	—	—	—	—	A3
Hippolais olivetorum Olive-tree Warbler	B	—	—	—	—	A3
Sylvia cantillans Subalpine Warbler	B	—	—	—	—	A3
Sylvia melanocephala Sardinian Warbler	R	—	—	—	—	A3
Emberiza caesia Cretzschmar's Bunting	B	—	—	—	—	A3
Emberiza melanocephala Black-headed Bunting	B	—	—	—	—	A3

1. Individuals.

The site is important for species associated with scrub and coastal areas. Breeding birds include eight of the 21 species in Europe that are restricted (when breeding) to the Mediterranean biome.

Protection status

National None **International** None

Conservation issues

Threats Disturbance to birds (B), Infrastructure (B), Recreation/tourism (B), Unsustainable exploitation (A)

The main threats are from shooting and excessive tourism development. Part of the area is a candidate SAC.

Strofades islands — B1ii, C2, C6 — 087

Admin region Ionian Islands
Coordinates 37°15'N 21°00'E
Altitude 0–22 m **Area** 220 ha

Site description

Two small rocky islands vegetated with sparse scrub.

Habitats Scrub (70%; scrub; sclerophyllous scrub/garrigue/maquis), Marine areas (10%; sea inlet/coastal features), Rocky areas (15%; sea cliff/rocky shore), Artificial landscape (5%; arable land)
Land-use Agriculture (50%)

Birds

Species	Season	Year	Pop min	Pop max	Acc	Criteria
Calonectris diomedea Cory's Shearwater	B	1995	50	100	A	C6
Calonectris diomedea Cory's Shearwater	N	1995	1,400	—	—	B1ii, C2

This is an important area for breeding seabirds and passage migrants. Species of global conservation concern that do not meet IBA criteria: *Aythya nyroca* (passage), *Aquila clanga* (passage), *Falco naumanni* (passage), *Larus audouinii* (passage).

Protection status

National None **International** None

Conservation issues

Threats Burning of vegetation (B), Consequences of animal/plant introductions (A), Disturbance to birds (B), Unsustainable exploitation (A)

The main threats include hunting, despite its being prohibited, introduced rats *Rattus*, disturbance and fires. The area is a candidate SAC.

Valtou mountains — B2, B3, C6 — 088

Admin region Western Greece
Coordinates 39°07'N 21°19'E
Altitude 500–1,852 m **Area** 43,500 ha

Site description

A long ridge west of the River Acheloos. The site is dominated by maquis, oak *Quercus* woodland, and at higher altitudes fir *Abies* forests. Human activities include stock-raising and arable agriculture.

Habitats Forest and woodland (50%; broadleaved deciduous forest; native coniferous forest; broadleaved evergreen forest; treeline ecotone), Scrub (40%; scrub; sclerophyllous scrub/garrigue/maquis), Wetland (5%), Rocky areas (8%; inland cliff), Artificial landscape (10%; arable land)
Land-use Agriculture (70%), Not utilized (30%)

Birds

Species	Season	Year	Pop min	Pop max	Acc	Criteria
[1,2] ***Gyps fulvus*** Griffon Vulture	R	1996	6	20	A	C6
Dendrocopos medius Middle Spotted Woodpecker	R	1996	Common		—	C6
Sitta neumayer Rock Nuthatch	R	1996	Abundant		—	B3
Lanius senator Woodchat Shrike	B	1996	Abundant		—	B2

1. Individuals.
2. Large decrease.

The site is important for species typical of Mediterranean woodland and scrub. Species of global conservation concern that do not meet IBA criteria: *Falco naumanni* (4–8 breeding pairs).

Protection status

National Partial **International** None
3,600 ha of IBA covered by Wildlife Refuge (Kanala Valtou, 3,600 ha). 1,560 ha of IBA covered by Wildlife Refuge (Mesonyhi–Fountes/Klidi–Skoulikaria, 1,560 ha).

Conservation issues

Threats Abandonment/reduction of land management (B), Afforestation (C), Agricultural intensification/expansion (A), Construction/impact of dyke/dam/barrage (A), Forest grazing (A), Unsustainable exploitation (B)

The site is threatened by agricultural intensification, grazing preventing forest regeneration and a planned construction of a new dam on the River Acheloos.

Lake Amvrakia — B1i, C2, C3 — 089

Admin region Western Greece
Coordinates 38°45'N 21°11'E
Altitude 16–16 m **Area** 2,205 ha

Site description

A deep oligotrophic lake with steep western shores. The northern part is shallow with small reedbeds (*Phragmites*).

Habitats Wetland (100%; standing fresh water; water-fringe vegetation)
Land-use Agriculture, Fisheries/aquaculture, Water management

Birds

Species	Season	Year	Pop min	Pop max	Acc	Criteria
Phalacrocorax carbo Cormorant	W	1997	50	1,200	B	B1i, C3
Sterna caspia Caspian Tern	P	—	20	50	C	B1i, C2

The site is important for wintering ducks and geese.

Protection status

National Partial **International** None
Part of IBA covered by Wildlife Refuge (Petalas/Amfilochias–Kechrinias–Papadatou–Stanou, 13,300 ha).

Conservation issues

The main threat is drainage and the over-abstraction of surface water ('Other' threats). There are plans to construct a road on the lake shore. The area is a candidate SAC.

Threats Agricultural intensification/expansion (A), Drainage (A), Dredging/canalization (C), Filling-in of wetlands (B), Industrialization/urbanization (B), Other (A), Unsustainable exploitation (C)

Akarnanika mountains — A3, B2, B3, C6 — 090

Admin region Western Greece
Coordinates 38°45′N 21°00′E
Altitude 0–1,589 m **Area** 39,000 ha

Site description

A mountain range extending from Amvrakikos gulf to the north of the Acheloos plain. The site is dominated by open *Quercus* forests, maquis and phrygana. Human activities include livestock-farming, arable agriculture, and aquaculture/fishing.

Habitats Forest and woodland (25%; broadleaved deciduous forest; native coniferous forest), Scrub (50%; sclerophyllous scrub/garrigue/maquis), Marine areas (5%; sea inlet/coastal features), Rocky areas (5%; sea cliff/rocky shore; inland cliff), Artificial landscape (15%; arable land)
Land-use Agriculture (85%), Fisheries/aquaculture (5%)

Birds

Species	Season	Year	Pop min	Pop max	Acc	Criteria
[1] ***Gyps fulvus*** Griffon Vulture	R	1996	7	12	A	C6
Circaetus gallicus Short-toed Eagle	B	1996	4	6	A	C6
Hieraaetus fasciatus Bonnelli's Eagle	R	1996	2	4	A	B2, C6
[1] ***Alectoris graeca*** Rock Partridge	R	1996	Common		—	A3
Dendrocopos medius Middle Spotted Woodpecker	R	1996	Frequent		—	C6
Oenanthe hispanica Black-eared Wheatear	R	1996	Abundant		—	A3
Monticola solitarius Blue Rock Thrush	R	1996	Abundant		—	B2
Hippolais olivetorum Olive-tree Warbler	B	1996	Rare		—	A3
Sylvia cantillans Subalpine Warbler	B	1996	Frequent		—	A3
Sylvia melanocephala Sardinian Warbler	R	1996	Abundant		—	A3
Sitta neumayer Rock Nuthatch	R	1996	Abundant		—	A3, B3
Corvus monedula Jackdaw	R	1996	1,200	2,000	A	B3
Emberiza caesia Cretzschmar's Bunting	B	1996	Rare		—	A3
Emberiza melanocephala Black-headed Bunting	B	1996	Common		—	A3

1. Large decrease.

This is an important site for species characteristic of open montane habitats. Breeding birds include eight of the 21 species in Europe that are restricted (when breeding) to the Mediterranean biome. Species of global conservation concern that do not meet IBA criteria: *Falco naumanni* (1–4 breeding pairs).

Protection status

National Partial **International** None
3,960 ha of IBA covered by Wildlife Refuge (Perganti–Xiromerou/Monastiraki–Vatou–Thiriou, 3,960 ha).

Conservation issues

Threats Abandonment/reduction of land management (C), Agricultural intensification/expansion (B), Aquaculture/fisheries (C), Forest grazing (B), Groundwater abstraction (U), Industrialization/urbanization (A), Recreation/tourism (C), Selective logging/cutting (U)

The *Quercus* forest is threatened by heavy grazing pressure which is preventing regeneration. Harbour facilities in the south-western section of the IBA may be developed into an oil refinery.

Lakes Trichonida and Lysimachia — A1, C1 — 091

Admin region Western Greece
Coordinates 38°34′N 21°28′E
Altitude 18–300 m **Area** 14,279 ha

Site description

The site has two large freshwater lakes with reedbeds (*Phragmites*) and surrounding arable land.

Habitats Wetland (69%; standing fresh water; water-fringe vegetation), Artificial landscape (22%; arable land)
Land-use Agriculture (25%), Fisheries/aquaculture (70%), Tourism/recreation (10%), Water management (70%)

Birds

Species	Season	Year	Pop min	Pop max	Acc	Criteria
Aythya nyroca Ferruginous Duck	W	1989	—	225	A	A1, C1

This is an important site for wintering ducks.

Protection status

National None **International** None

Conservation issues

Threats Agricultural intensification/expansion (A), Aquaculture/fisheries (U), Groundwater abstraction (A), Industrialization/urbanization (A), Unsustainable exploitation (B)

The main threats are from agricultural intensification and surface and groundwater abstraction. The area is a candidate SAC.

Mesolongi and Aetoliko lagoons, and Acheloos and Evinos estuaries — A1, A4i, A4iii, B1i, B2, C1, C2, C3, C4, C6 — 092

Admin region Western Greece
Coordinates 38°20′N 21°20′E
Altitude 0–450 m **Area** 50,000 ha

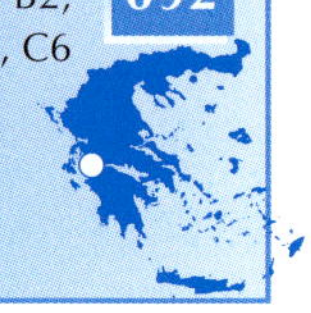

Site description

A wetland complex with lagoons, saltpans ('Other' land-use, below), extensive saltmarshes, freshwater marshes, sandbanks, and a remnant area of alluvial forest.

Habitats Forest and woodland (7%; broadleaved deciduous forest; alluvial/very wet forest), Scrub (8%; sclerophyllous scrub/garrigue/maquis), Wetland (80%; mudflat/sandflat; saltmarsh; sand-dunes/sand beach; coastal lagoon; river/stream)
Land-use Agriculture, Fisheries/aquaculture (50%), Forestry, Other (5%), Tourism/recreation

Birds

Species	Season	Year	Pop min	Pop max	Acc	Criteria
[1] ***Phalacrocorax carbo*** Cormorant	W	1988	1,550	1,550	A	B1i, C3
[1] ***Plegadis falcinellus*** Glossy Ibis	P	1990	500	1,800	A	A4i, B1i, C2
Anas penelope Wigeon	W	1988	10,000	10,000	A	B1i, C3
Aythya ferina Pochard	W	1988	11,100	11,100	A	B1i, C3
Gyps fulvus Griffon Vulture	R	—	Common		—	C6
Falco naumanni Lesser Kestrel	B	1990	12	15	A	A1, C1
Haematopus ostralegus Oystercatcher	R	1990	25	25	A	B1i, C3
Recurvirostra avosetta Avocet	B	1990	23	23	A	C6
Recurvirostra avosetta Avocet	W	1988	300	300	A	B2, C6
Burhinus oedicnemus Stone Curlew	B	1990	39	39	A	B2, C6
[2] ***Glareola pratincola*** Collared Pratincole	B	1990	66	77	A	A4i, B1i, B2, C2, C6
Charadrius alexandrinus Kentish Plover	R	1990	200	200	A	B1i, C3
Numenius tenuirostris Slender-billed Curlew	P	—	—	—	—	A1, C1
Gelochelidon nilotica Gull-billed Tern	B	1990	110	110	A	A4i, B1i, B2, C2, C6
[3] ***Sterna caspia*** Caspian Tern	R	1996	10	20	B	C2
Sterna albifrons Little Tern	B	1990	489	500	A	A4i, B1i, B2, C2, C6
Calandrella brachydactyla Short-toed Lark	B	1990	1,200	1,250	A	B2, C6

1. Large increase.
2. Large decrease.
3. Individuals.

This is an important site for breeding, passage and wintering waterbirds and raptors. Species of global conservation concern that do not meet IBA criteria: *Phalacrocorax pygmeus* (wintering), *Pelecanus crispus* (10 wintering individuals), *Aythya nyroca* (passage), *Haliaeetus albicilla* (wintering), *Aegypius monachus* (wintering), *Aquila clanga* (wintering), *Aquila heliaca* (2 resident individuals), *Gallinago media* (passage). The site holds 20,000 or more waterbirds in winter, on a regular basis (including *Anas penelope* 10,000; *Anas crecca* 3,000; *Anas clypeata* 3,000; *Fulica atra* 15,500).

Protection status

National Partial **International** Partial

1,000 ha of IBA covered by Wildlife Refuge (Varasova/Galata–Perithoriou, 1,000 ha). 550 ha of IBA covered by Wildlife Refuge (Pouri–Asprolith/A. Kerasovou, 550 ha). Part of IBA covered by Wildlife Refuge (Kantilia–Valti/Astakou, 3,500 ha). 46 ha of IBA covered by Natural Monument (Lesini Ash Forest, 46 ha). Part of IBA covered by Ramsar Site (Messolonghi Lagoons, 33,687 ha). 33,000 ha of IBA covered by Special Protection Area (Delta Acheloou, Limnothalassa Mesolongiou–Aitoliko kai Ekvoles Evinou, 34,941 ha). Part of IBA covered by Biogenetic Reserve.

Conservation issues

Threats Agricultural intensification/expansion (A), Aquaculture/fisheries (C), Construction/impact of dyke/dam/barrage (A), Drainage (B), Filling-in of wetlands (B), Industrialization/urbanization (A), Unsustainable exploitation (B)

The main threat to the site is from intensive agriculture and associated drainage and water management schemes. The completed diversion of the Evinas river and the current project to divert the Acheloos river are also major threats. Part of the area is a candidate SAC.

Alyki lagoon, Aegion — B1i, C2 — 093

Admin region Western Greece
Coordinates 38°15′N 22°06′E
Altitude 0–0 m **Area** 32 ha

Site description

A small coastal lagoon east of the town of Aegio, with reedbeds (*Phragmites*), saltmarsh and shingle-beach. The site is popular for recreation.

Habitats Wetland (100%; saltmarsh; sand-dunes/sand beach; shingle/stony beach; coastal lagoon; standing brackish and salt water; water-fringe vegetation; fen/transition mire/spring)
Land-use Nature conservation/research (100%), Tourism/recreation (60%)

Birds

Species	Season	Year	Pop min	Pop max	Acc	Criteria
Gelochelidon nilotica Gull-billed Tern	P	1995	100	300	A	B1i, C2

The site is important for overwintering ducks, migrating waders and other passage migrants. Species of global conservation concern that do not meet IBA criteria: *Aythya nyroca* (5–25 passage individuals), *Larus audouinii* (5–35 passage individuals).

Protection status

National Partial **International** None

20 ha of IBA covered by Wildlife Refuge (Limnothalassa Alykis Aigiou, 100 ha).

Conservation issues

Threats Groundwater abstraction (U), Filling-in of wetlands (B), Industrialization/urbanization (B), Natural events (C), Recreation/tourism (A)

The main threats are disturbance from recreation and limited infilling. The area is a candidate SAC.

Barbas and Klokos mountains and Selinous gorge — B2, C6 — 094

Admin region Western Greece
Coordinates 38°10′N 22°00′E
Altitude 100–1,779 m **Area** 11,570 ha

Site description

A mountainous area including the gorge of the Selinous river. The lower slopes are covered with *Pinus* forests and maquis. At high altitudes *Abies* forest and subalpine grasslands dominate. Human activities include forestry, stock-raising, and hunting.

Habitats Forest and woodland (20%; native coniferous forest), Scrub (20%; sclerophyllous scrub/garrigue/maquis), Grassland (20%; alpine/subalpine/boreal grassland), Wetland (5%; river/stream), Rocky areas (20%; inland cliff), Artificial landscape (15%; arable land)
Land-use Agriculture, Forestry, Tourism/recreation, Urban/industrial/transport

Birds

Species	Season	Year	Pop min	Pop max	Acc	Criteria
Falco biarmicus Lanner	B	1996	1	1	A	B2, C6
Bubo bubo Eagle Owl	R	1996	2	4	B	B2, C6

This is an important area for breeding raptors. Species of global conservation concern that do not meet IBA criteria: *Falco naumanni* (5 breeding pairs).

Protection status

National None **International** High

10,762 ha of IBA covered by Special Protection Area (Zoumpata–Agios Charalampos–Marmpas–Klokos, 10,762 ha).

Conservation issues

Threats Agricultural intensification/expansion (B), Burning of vegetation (B), Deforestation (commercial) (B), Infrastructure (A), Other (U), Unsustainable exploitation (B)

The main threats include road construction, illegal logging, pollution of the river ('Other' threats), and illegal hunting. A LIFE-funded project is producing a management plan. Part of the area is a candidate SAC.

Vouraikos gorge including Kalavrita — C7 — 095

Admin region Western Greece
Coordinates 38°05′N 22°10′E
Altitude 100–1,500 m **Area** 5,600 ha

Site description

A steep ravine with sheer cliffs, *Pinus* forest and olive-groves. At higher altitudes pasture land and *Abies* forests dominate.

Habitats Forest and woodland (broadleaved deciduous forest; native coniferous forest; alluvial/very wet forest; treeline ecotone), Scrub (sclerophyllous scrub/garrigue/maquis), Wetland (river/stream), Rocky areas (scree/boulders; inland cliff; caves), Artificial landscape (perennial crops/orchards/groves; forestry plantation)
Land-use Agriculture, Forestry, Tourism/recreation

Birds

This is an important site for raptors. Species of global conservation concern that do not meet IBA criteria: *Falco naumanni* (passage).

Protection status

National Partial **International** Partial

Part of IBA covered by Wildlife Refuge (Krastoni-Sokara/Vrachniou-Kalavriton, 950 ha). Part of IBA covered by Wildlife Refuge (Trapezas–Diakoftou, 2,000 ha). 1,683 ha of IBA covered by Special Protection Area (Aisthitiko Dasos Kalavryton, 1,683 ha).

Conservation issues

Threats Abandonment/reduction of land management (U), Agricultural intensification/expansion (U), Firewood collection (U), Forest grazing (U), Industrialization/urbanization (U), Recreation/tourism (B), Unsustainable exploitation (A)

Threats include deforestation and erosion, illegal hunting and disturbance from recreation and tourism. Part of the area is a candidate SAC.

Mount Helmos (Aroania) — B2 — 096

Admin region Western Greece, Peloponnissos
Coordinates 37°58′N 22°14′E
Altitude 700–2,355 m **Area** 17,500 ha

Site description

A rugged limestone mountain. The northern part is covered by extensive *Pinus* forest, other areas with fir *Abies* forest, and the higher parts with alpine grassland. The Kalavrita side of the mountain is used for winter tourism. Elsewhere the main activity is livestock-farming.

Habitats Forest and woodland (native coniferous forest; treeline ecotone), Scrub (sclerophyllous scrub/garrigue/maquis), Grassland (alpine/subalpine/boreal grassland), Wetland (river/stream), Rocky areas (scree/boulders; inland cliff)
Land-use Forestry, Tourism/recreation

Birds

Species	Season	Year	Pop min	Pop max	Acc	Criteria
Alectoris graeca Rock Partridge	R	1987	Common		—	B2
Monticola saxatilis Rock Thrush	B	1987	Common		—	B2

This is an important site for species associated with conifer forests, alpine pasture and upland rivers.

Protection status

National Low **International** None
520 ha of IBA covered by Wildlife Refuge (Lagada–Fteres–Vageni–Kastraki/Kastrion, 520 ha).

Conservation issues

Threats Recreation/tourism (B), Unsustainable exploitation (B)

The main threats are from continued tourism developments and illegal hunting. Part of the area is a candidate SAC.

Mount Erimanthos — B2, B3 — 097

Admin region Western Greece
Coordinates 37°57'N 21°51'E
Altitude 500–2,221 m **Area** 36,900 ha

Site description

A rugged limestone mountain with many inaccessible rocky slopes. Limited fir *Abies* forests cover the northern and southern parts of the site. The human presence is low and the main activity is stock-raising.

Habitats Forest and woodland (native coniferous forest; treeline ecotone), Scrub (sclerophyllous scrub/garrigue/maquis), Rocky areas (scree/boulders; inland cliff)
Land-use Agriculture, Not utilized

Birds

Species	Season	Year	Pop min	Pop max	Acc	Criteria
Sitta neumayer Rock Nuthatch	R	1996	Abundant		—	B3
Emberiza cia Rock Bunting	R	1996	Common		—	B2

This is an important area for raptors and alpine species.

Protection status

National Low **International** None
950 ha of IBA covered by Wildlife Refuge (Doxara–Koriza/Kalentzi–Drosia–Spartia, 950 ha).

Conservation issues

Threats Unsustainable exploitation (B)

The main threat is illegal hunting. Part of the area is a candidate SAC.

Kalogria lagoon, Strofilia forest, and Lamia marshes — A4i, B1i, B2, C2, C3, C6 — 098

Admin region Western Greece
Coordinates 38°10'N 21°22'E
Altitude 0–250 m **Area** 6,000 ha

Site description

A coastal wetland with saltmarshes, brackish lagoons, extensive reedbeds (*Phragmites*), sand-dunes, and *Pinus* forest. Human activities include arable agriculture, stock-raising, fishing, hunting, and recreation.

Habitats Forest and woodland (17%; native coniferous forest), Scrub (18%; sclerophyllous scrub/garrigue/maquis), Wetland (43%; saltmarsh; sand-dunes/sand beach; coastal lagoon; standing fresh water; fen/transition mire/spring), Artificial landscape (15%; arable land)
Land-use Agriculture (37%), Fisheries/aquaculture (7%), Tourism/recreation (20%), Urban/industrial/transport (5%), Water management (40%)

Birds

Species	Season	Year	Pop min	Pop max	Acc	Criteria
Ixobrychus minutus Little Bittern	B	1994	50	60	A	B2, C2, C6
Himantopus himantopus Black-winged Stilt	B	1994	270	270	A	A4i, B1i, C2
Charadrius alexandrinus Kentish Plover	P	1994	775	—	A	A4i, B1i, C3
Calidris minuta Little Stint	P	1994	—	3,845	A	A4i, B1i, C3
Tringa stagnatilis Marsh Sandpiper	P	1994	250	—	A	A4i, B1i, C3
Hippolais olivetorum Olive-tree Warbler	B	1994	70	—	A	B2, C6

This is an important IBA for passage migrants and breeding and wintering waterbirds. Species of global conservation concern that do not meet IBA criteria: *Phalacrocorax pygmeus* (wintering), *Aythya nyroca* (minimum 30 passage individuals), *Aquila clanga* (wintering), *Falco naumanni* (passage).

Protection status

National None **International** Partial
3,523 ha of IBA covered by Special Protection Area (Limnothalassa Kalogrias, Dasos Strofilias kai Elos, 3,523 ha). Part of IBA covered by Ramsar Site (Kotychi lagoon, 6309 ha).

Conservation issues

Threats Agricultural intensification/expansion (B), Aquaculture/fisheries (U), Burning of vegetation (U), Forest grazing (B), Infrastructure (U), Recreation/tourism (A), Unsustainable exploitation (A)

The main threats are from intensive tourism development, hunting and forest grazing. The shallow lagoons are sensitive to manipulation of the water-levels. Part of the area is a candidate SAC.

Kotychi lagoon — A1, A4i, B1i, C1, C2 — 099

Admin region Western Greece
Coordinates 38°00'N 21°18'E
Altitude 0–2 m **Area** 2,700 ha

Site description

A small brackish lagoon separated from the sea by dunes and surrounded by patches of saltmarsh and agricultural land. Human activities include aquaculture, agriculture, and tourism.

Habitats Forest and woodland (21%; native coniferous forest), Scrub (20%; sclerophyllous scrub/garrigue/maquis), Wetland (45%; saltmarsh; sand-dunes/sand beach; coastal lagoon; water-fringe vegetation; fen/transition mire/spring), Artificial landscape (14%; arable land; other urban/industrial areas)
Land-use Agriculture (40%), Fisheries/aquaculture (20%), Tourism/recreation (15%), Urban/industrial/transport (5%), Water management (20%)

Birds

Species	Season	Year	Pop min	Pop max	Acc	Criteria
Egretta garzetta Little Egret	P	—	500	1,000	—	A4i, B1i, C2
Egretta alba Great White Egret	W	1996	40	120	A	A4i, B1i, C2
Plegadis falcinellus Glossy Ibis	P	—	200	500	—	A4i, B1i, C2
Falco naumanni Lesser Kestrel	B	—	6	10	—	A1, C1

The site is important for breeding, passage and wintering waterbirds. Species of global conservation concern that do not meet IBA criteria: *Pelecanus crispus* (5–11 non-breeding individuals), *Aythya nyroca* (passage).

Protection status

National None **International** Partial
Part of IBA covered by Ramsar Site (Kotychi Lagoon, 3,700 ha). 1,647 ha of IBA covered by Special Protection Area (Limnothalassa Kotychi, 1,647 ha)

■ **Conservation issues**

Threats Agricultural intensification/expansion (B), Aquaculture/fisheries (B), Forest grazing (U), Infrastructure (B), Recreation/tourism (B), Unsustainable exploitation (A)

Hunting is a serious problem, despite a partial prohibition. Additional threats are from aquaculture, road construction and tourism development. Part of the area is a candidate SAC.

Mount Foloi — B3, C6 — 100

Admin region Western Greece
Coordinates 37°45′N 21°45′E
Altitude 200–650 m **Area** 16,500 ha

■ **Site description**

An area of old olive-groves, *Quercus* woodland and *Pinus* forests, interspersed with scrub and some pasture and cultivation. Human activities include agriculture, forestry, stock-raising, and hunting.

Habitats Forest and woodland (60%; broadleaved deciduous forest; native coniferous forest), Scrub (40%; sclerophyllous scrub/garrigue/maquis)
Land-use Agriculture (50%), Forestry (40%), Tourism/recreation (10%)

■ **Birds**

Species	Season	Year	Pop min	Pop max	Acc	Criteria
Dendrocopos medius Middle Spotted Woodpecker	R	1996	Common		—	B3, C6
Emberiza caesia Cretzschmar's Bunting	B	1997	—	—	—	B3, C6

This is an important site for species characteristic of a mosaic of forest, pasture and olive-groves.

■ **Protection status**

National None **International** None

■ **Conservation issues**

Threats Agricultural intensification/expansion (C), Burning of vegetation (B), Deforestation (commercial) (B), Recreation/tourism (C), Unsustainable exploitation (A)

The main threats include illegal hunting, illegal logging and the burning of vegetation. Part of the area is a candidate SAC.

Delimidi and Fteri mountains, Agrafa — B2, C6 — 101

Admin region Thessalia, Sterea Ellada
Coordinates 39°13′N 21°34′E
Altitude 700–2,163 m **Area** 21,200 ha

■ **Site description**

An isolated, mountainous area with ravines and gorges and limestone scree in the alpine zone. Forest cover is dominated by fir *Abies* and pine *Pinus* with small patches of deciduous woodland. The main human activity is livestock-farming.

Habitats Forest and woodland (55%; native coniferous forest; treeline ecotone), Grassland (5%; mesophile grassland), Rocky areas (35%; scree/boulders; inland cliff)
Land-use Agriculture (60%), Forestry (30%)

■ **Birds**

Species	Season	Year	Pop min	Pop max	Acc	Criteria
Alectoris graeca Rock Partridge	R	1995	Abundant		—	B2
Monticola saxatilis Rock Thrush	B	1995	Common		—	B2
Monticola solitarius Blue Rock Thrush	R	1995	Common		—	B2
Pyrrhocorax pyrrhocorax Chough	R	1995	Common		—	B2, C6

The area is important for alpine species and raptors.

■ **Protection status**

National Partial **International** None
1,420 ha of IBA covered by Wildlife Refuge (Tsouka Sakka–Asprorema/Epinianon, Trovatou, 1,420 ha). Part of IBA covered by Wildlife Refuge (Ag. Nikalaou/Monastirakiou, 815 ha).

■ **Conservation issues**

Threats Unsustainable exploitation (B)

The main threat is from hunting. Part of the area is a candidate SAC.

Mount Othrys — B2, B3, C6 — 102

Admin region Thessalia, Sterea Ellada
Coordinates 39°04′N 22°40′E
Altitude 400–1,726 m **Area** 84,000 ha

■ **Site description**

A mountainous site with grazed plateaus and some, mainly *Abies*-dominated, forest. Human activities include livestock-farming, forestry, and hunting.

Habitats Forest and woodland (15%; broadleaved deciduous forest; native coniferous forest), Scrub (40%; scrub; sclerophyllous scrub/garrigue/maquis), Grassland (40%; steppe/dry calcareous grassland; dry siliceous grassland; alpine/subalpine/boreal grassland), Rocky areas (5%; inland cliff)
Land-use Agriculture (90%), Forestry (10%), Tourism/recreation (5%)

■ **Birds**

Species	Season	Year	Pop min	Pop max	Acc	Criteria
Circaetus gallicus Short-toed Eagle	B	1996	4	6	B	C6
Falco biarmicus Lanner	R	1996	Uncommon		—	B2, C6
Monticola saxatilis Rock Thrush	B	1996	Frequent		—	B2
Emberiza cirlus Cirl Bunting	R	1996	Abundant		—	B3

The site is important for breeding raptors, forest and upland species.

■ **Protection status**

National Partial **International** None
1,310 ha of IBA covered by Wildlife Refuge (Bathi Greki/Kokoton–I.M. Xenias, 1,310 ha). 1,470 ha of IBA covered by Wildlife Refuge (Platanos, 1,470 ha). Part of IBA covered by Wildlife Refuge (Pr. Ilias-Afrati/Lamias-Divris, 3,550 ha). 1,700 ha of IBA covered by Wildlife Refuge (Sapounas/Neraida, 1,700 ha). 1,230 ha of IBA covered by Wildlife Refuge (Neraida–Logitsi, 1,230 ha). 1,450 ha of IBA covered by Wildlife Refuge (Taratsa/Pelasgias-Milon, 1,450 ha). 540 ha of IBA covered by Wildlife Refuge (Fardikoli-Peria-Ag. Vlasias/Vrinainas-Ag. Ioannou, 540 ha).

■ **Conservation issues**

Threats Agricultural intensification/expansion (B), Deforestation (commercial) (C), Firewood collection (C), Forest grazing (C), Infrastructure (B), Unsustainable exploitation (A)

The main threats are from hunting and the use of poison baits, and infrastructure developments.

Sperchios valley and delta–Maliakos gulf — A1, A4i, A4iii, B1i, B2, C1, C2, C4, C6 — 103

Admin region Sterea Ellada
Coordinates 38°51′N 22°32′E
Altitude 0–1,000 m **Area** 34,000 ha

■ **Site description**

The site contains the estuary of the Sperchios river, an area of extensive saltmarshes and reedbeds (*Phragmites*). Along the river is a mosaic of riverine forest, wet grassland and freshwater marsh.

Habitats Forest and woodland (5%; alluvial/very wet forest), Scrub (10%; sclerophyllous scrub/garrigue/maquis; humid grassland), Wetland (50%; tidal river/enclosed tidal water; mudflat/sandflat; saltmarsh; sand-dunes/sand beach; river/stream; water-fringe vegetation), Marine areas (30%; sea inlet/coastal features), Artificial landscape (arable land)
Land-use Agriculture, Fisheries/aquaculture, Water management

■ **Birds**

Species	Season	Year	Pop min	Pop max	Acc	Criteria
Aythya nyroca Ferruginous Duck	P	—	—	—	—	A1, C1

Species ... continued	Season	Year	Pop min	Pop max	Acc	Criteria
Recurvirostra avosetta Avocet	W	1990	1,310	1,310	B	A4i, B1i, B2, C6
Numenius tenuirostris Slender-billed Curlew	N	1963	1	—	A	A1, C1
Larus genei Slender-billed Gull	W	1994	Frequent		—	A4i, B1i, C2
Sterna albifrons Little Tern	B	1996	Frequent		—	C6

This is an important area for species characteristic of riverine forest, reedbed and open water. Species of global conservation concern that do not meet IBA criteria: *Phalacrocorax pygmeus* (passage, wintering), *Aquila clanga* (wintering), *Falco naumanni* (breeding). The site regularly holds 20,000 or more waterbirds in winter (including *Anas crecca* 4,250; *Recurvirostra avosetta* 1,310).

Protection status

National Low **International** Partial

1,225 ha of IBA covered by Wildlife Refuge (Ekvoles Sperchiou, 1,225 ha). 3,800 ha of IBA covered by Special Protection Area (Ygrotopos Ekvolon (Delta) Sperchiou, 3,966 ha).

Conservation issues

Threats Agricultural intensification/expansion (A), Construction/impact of dyke/dam/barrage (B), Drainage (A), Filling-in of wetlands (B), Firewood collection (C), Forest grazing (C), Infrastructure (A), Unsustainable exploitation (B)

The main threats are from agricultural expansion, drainage and infrastructure developments. Part of the riverine forest within the SPA has been cleared as a result of European Union funded flood-control work. The area is a candidate SAC.

Mount Iti — B2, B3, C6 — 104

Admin region Sterea Ellada
Coordinates 38°48′N 22°17′E
Altitude 100–2,152 m **Area** 18,800 ha

Site description

A mountain with four deep gorges. The slopes are dominated by *Abies* forest, with many open meadows. Forestry is confined to the area outside the National Park.

Habitats Forest and woodland (70%; native coniferous forest), Scrub (12%; sclerophyllous scrub/garrigue/maquis), Grassland (23%; alpine/subalpine/boreal grassland; mesophile grassland), Rocky areas (7%; inland cliff)
Land-use Agriculture (70%), Forestry (20%)

Birds

Species	Season	Year	Pop min	Pop max	Acc	Criteria
Alectoris graeca Rock Partridge	R	1996	Common		—	B2
Picus viridis Green Woodpecker	R	1996	Abundant		—	B2
Monticola solitarius Blue Rock Thrush	R	1985	Common		—	B2
Parus lugubris Sombre Tit	R	1988	Common		—	B3
Emberiza hortulana Ortolan Bunting	B	1996	Abundant		—	B2, C6

The site is important for forest species, and species associated with open montane habitat.

Protection status

National Partial **International** Partial

7,210 ha of IBA covered by National Park (Mount Oiti, 7,210 ha). Part of IBA covered by Wildlife Refuge (Oitis–Paulianis, 2,950 ha). 960 ha of IBA covered by Wildlife Refuge (Frantzi–Dio Vounon, 960 ha). 3,010 ha of IBA covered by Biogenetic Reserve (Council of Europe) (Ethnikos Drymos Oitis (Core area), 3,010 ha). 9,114 ha of IBA covered by Special Protection Area (Ethnikos Drymos Oitis, 9,114 ha).

Conservation issues

Threats Recreation/tourism (C), Unsustainable exploitation (B)

The main threats are from tourism and hunting. A management plan for the National Park is under preparation. Part of the area is a candidate SAC.

Mount Vardousia — B2, B3, C6 — 105

Admin region Sterea Ellada
Coordinates 38°40′N 22°08′E
Altitude 500–2,437 m **Area** 21,300 ha

Site description

A rugged mountain with an extensive alpine zone, limited *Abies* woodland, and extensive grasslands. The main human activity is livestock-farming.

Habitats Forest and woodland (43%; native coniferous forest; treeline ecotone), Scrub (10%; sclerophyllous scrub/garrigue/maquis), Grassland (34%; alpine/subalpine/boreal grassland; mesophile grassland), Wetland (5%; river/stream), Rocky areas (20%; scree/boulders; inland cliff)
Land-use Agriculture (70%), Hunting, Not utilized

Birds

Species	Season	Year	Pop min	Pop max	Acc	Criteria
Gypaetus barbatus Lammergeier	R	1995	1	—	—	B2, C6
Alectoris graeca Rock Partridge	R	1990	Common		—	B2
Parus lugubris Sombre Tit	R	1996	Common		—	B3

An important site for resident raptors, forest and alpine grassland.

Protection status

National Partial **International** None

1,653 ha of IBA covered by Wildlife Refuge (Makria Rachi, Dryodasos Pentagion/Kokkinou, 1,653 ha). Part of IBA covered by Controlled Hunting Area (Orini Parnassidus, 24,0341 ha). Part of IBA covered by Controlled Hunting Area (Makrin Rachi, Dryodaxsos Pentagion, 1,767 ha).

Conservation issues

Threats Disturbance to birds (B), Unsustainable exploitation (A)

The main threats are illegal hunting and the use of poison baits. Part of the area is a candidate SAC.

Mount Ghiona, Reka ravine, Lazorema, and Vathia Lakka — A3, B2, B3, C6 — 106

Admin region Sterea Ellada
Coordinates 38°37′N 22°16′E
Altitude 600–2,510 m **Area** 25,000 ha

Site description

A mountain with extensive rocky areas and alpine grasslands. Human activities include livestock-farming, hunting, mining and recreation. The Reka ravine is used for the water-supply to the town of Amfissa.

Habitats Forest and woodland (40%; native coniferous forest; treeline ecotone), Scrub (5%; sclerophyllous scrub/garrigue/maquis), Grassland (43%; alpine/subalpine/boreal grassland; mesophile grassland), Rocky areas (15%; scree/boulders; inland cliff)
Land-use Agriculture (90%), Forestry (30%), Hunting

Birds

Species	Season	Year	Pop min	Pop max	Acc	Criteria
[1] ***Gypaetus barbatus*** Lammergeier	R	1995	1	2	B	B2, C6
Prunella collaris Alpine Accentor	R	1996	50	200	C	A3
Monticola solitarius Blue Rock Thrush	R	1993	Common		—	B2
Parus lugubris Sombre Tit	R	1996	Common		—	B3
Sitta neumayer Rock Nuthatch	R	1996	Abundant		—	B3
Tichodroma muraria Wallcreeper	R	1987	Common		—	A3
Pyrrhocorax graculus Alpine Chough	R	1996	200	600	C	A3
[1] ***Pyrrhocorax pyrrhocorax*** Chough	R	1996	10	50	C	B2, C6
Montifringilla nivalis Snowfinch	R	1987	Rare		—	A3

1. Individuals.

This is an important area for raptors and alpine species. Breeding birds include four of the 10 species in Europe that are restricted (when breeding) to the Eurasian high-montane biome.

Protection status

National Partial **International** None

2,894 ha of IBA covered by Wildlife Refuge (Leukaditiou/Sykeas, 2,894 ha). 3,300 ha of IBA covered by Wildlife Refuge (Rekas/Prosiliou, 3,300 ha). Part of IBA covered by Controlled Hunting Area (Orini Parnassidas, 24,034 ha).

Conservation issues

Threats Extraction industry (A), Unsustainable exploitation (A)

The main threats are from open-cast bauxite mining and the use of poisoned baits. Part of the area is a candidate SAC.

Southern and eastern part of Mount Parnassos — A3, B2, B3, C6 — 107

Admin region Sterea Ellada
Coordinates 38°32′N 22°35′E
Altitude 200–2,457 m **Area** 19,400 ha

Site description

A rugged area of tall cliffs, grassland and *Abies* and *Pinus* forests. The main human activities are tourism and livestock-farming.

Habitats Forest and woodland (25%; native coniferous forest), Scrub (20%; sclerophyllous scrub/garrigue/maquis), Grassland (40%; alpine/subalpine/boreal grassland; mesophile grassland), Rocky areas (20%; scree/boulders; inland cliff)
Land-use Agriculture (50%), Tourism/recreation (20%), Not utilized

Birds

Species	Season	Year	Pop min	Pop max	Acc	Criteria
Gypaetus barbatus Lammergeier	R	1997	1	1	—	B2, C6
Aquila chrysaetos Golden Eagle	R	1990	Common		—	C6
Falco peregrinus Peregrine	R	1986	Common		—	C6
Alectoris graeca Rock Partridge	R	1990	Common		—	B2
Prunella collaris Alpine Accentor	R	1993	Common		—	A3
Monticola solitarius Blue Rock Thrush	R	1988	Common		—	B2
Sylvia rueppelli Rüppell's Warbler	B	1988	Common		—	B3, C6
Sitta neumayer Rock Nuthatch	R	1993	Abundant		—	B3
Tichodroma muraria Wallcreeper	R	1986	Common		—	A3
Pyrrhocorax graculus Alpine Chough	R	1993	Abundant		—	A3
Montifringilla nivalis Snowfinch	R	1985	Uncommon		—	A3
Emberiza caesia Cretzschmar's Bunting	B	1985	Frequent		—	B3, C6

An important area for raptors and alpine species. Breeding birds include four of the 10 species in Europe that are restricted (when breeding) to the Eurasian high-montane biome.

Protection status

National Partial **International** Low
Part of IBA covered by National Park (Parnassos, 3,513 ha). 2,650 ha of IBA covered by Wildlife Refuge (Asprochoma–Psilo–Prontoli–Kelari/Arachovas, 2,650 ha). 625 ha of IBA covered by Special Protection Area (Ethnikos Drymos Parnassou, 3,695 ha).

Conservation issues

Threats Recreation/tourism (B), Unsustainable exploitation (A)

The main threats are from tourism development and hunting. A management plan for the Parnassos National Park is under preparation. Part of the area is a candidate SAC.

Galaxidi — A1, B1iii, B2, C1, C2, C6 — 108

Admin region Sterea Ellada
Coordinates 38°22′N 22°23′E
Altitude 0–750 m **Area** 2,000 ha

Site description

Habitats Scrub (70%; sclerophyllous scrub/garrigue/maquis), Artificial landscape (28%; arable land; perennial crops/orchards/groves; other urban/industrial areas)
Land-use Agriculture (20%), Tourism/recreation, Urban/industrial/transport

This coastal site is dominated by phrygana and perennial crops, particularly olive-groves. The village of Galaxidi lies within the IBA. Human activities include livestock-farming, arable agriculture and hunting.

Birds

Species	Season	Year	Pop min	Pop max	Acc	Criteria
Falco naumanni Lesser Kestrel	B	1996	100	120	A	A1, B1iii, B2, C1, C2, C6

The site is important for species associated with a mixture of scrub and agricultural land.

Protection status

National Partial **International** None
Part of IBA covered by Wildlife Refuge (Xirotri/Galaxidiou–Penteorioan, 1,300 ha).

Conservation issues

Threats Other (U)

A change in the methods of building roofs and walls in the village would be a potential threat to *Falco naumanni* ('Other' threats, above).

Megalo and Mikro Livari lagoons, Istieas — B1i, C3 — 109

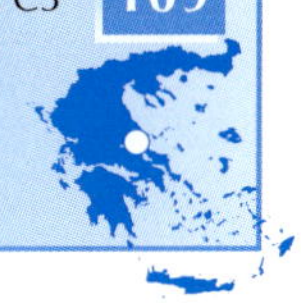

Admin region Sterea Ellada
Coordinates 39°00′N 23°08′E
Altitude 0–3 m **Area** 1,008 ha

Site description

A small coastal wetland near Istiea, fed by freshwater springs and streams, with two lagoons, and riverine forest.

Habitats Forest and woodland (alluvial/very wet forest), Grassland (9%; humid grassland), Wetland (45%; sand-dunes/sand beach; coastal lagoon; water-fringe vegetation; fen/transition mire/spring), Marine areas (44%; open sea), Artificial landscape (arable land; ruderal land)
Land-use Agriculture (22%), Fisheries/aquaculture (62%), Tourism/recreation (11%), Urban/industrial/transport (5%)

Birds

Species	Season	Year	Pop min	Pop max	Acc	Criteria
Cygnus olor Mute Swan	W	1993	10	1,600	A	B1i, C3

This is an important site for passage and wintering waterbirds. Species of global conservation concern that do not meet IBA criteria: *Aythya nyroca* (6–10 passage individuals), *Falco naumanni* (passage).

Protection status

National Low **International** None
1 ha of IBA covered by Natural Monument (Istiaia Remnant of Swamp-forest, 1 ha).

Conservation issues

Threats Agricultural intensification/expansion (C), Aquaculture/fisheries (B), Construction/impact of dyke/dam/barrage (C), Deforestation (commercial) (B), Drainage (A), Dredging/canalization (A), Filling-in of wetlands (A), Firewood collection (B), Groundwater abstraction (B), Industrialization/urbanization (A), Recreation/tourism (A), Unsustainable exploitation (A)

The main threats are illegal building of holiday houses, dredging and infilling at Livari, expansion of arable land, logging of the alluvial forest, hunting, and grazing pressure. The area is a candidate SAC.

Kandili mountain — B3, C6 — 110

Admin region Sterea Ellada
Coordinates 38°41′N 23°28′E
Altitude 0–1,246 m **Area** 5,700 ha

Site description

A low mountain in northern Evia, with extensive conifer forests, mixed forests and scrub. Human activities include forestry, livestock-farming, and hunting.

Habitats Forest and woodland (79%; broadleaved deciduous forest; native coniferous forest; broadleaved evergreen forest), Scrub (15%; sclerophyllous scrub/garrigue/maquis), Artificial landscape (10%; other urban/industrial areas)
Land-use Agriculture (30%), Forestry (70%), Urban/industrial/transport (5%)

Birds

Species	Season	Year	Pop min	Pop max	Acc	Criteria
Emberiza caesia Cretzschmar's Bunting	B	1996	Frequent		—	B3, C6

This is an important site for breeding raptors and for passage migrants.

Protection status

National Partial **International** None
Part of IBA covered by Wildlife Refuge (Iera Moni Galtaki, 900 ha). Part of IBA covered by Wildlife Refuge (Palioplatana/Dafnontos–Pagontos–Kamaritsas, 2,500 ha).

Conservation issues

Threats Abandonment/reduction of land management (U), Agricultural intensification/expansion (C), Extraction industry (C), Forest grazing (C), Industrialization/urbanization (C), Infrastructure (B), Intensified forest management (C), Recreation/tourism (C), Unsustainable exploitation (B)

The main threats are from road construction and tourism developments. Other threats are from overgrazing, quarrying and agricultural and urban expansion.

Dirfi, Xirovouni, Skotini, Mavrovouni, Alokteri and Ortari mountains and the vicinity of Kimi

B3, C6 **111**

Admin region Sterea Ellada
Coordinates 38°38′N 23°45′E
Altitude 0–1,743 m **Area** 45,000 ha

Site description

Steep mountains with numerous cliffs and *Castanea* and *Abies* forest. Human activities include forestry, stock-raising, and hunting.

Habitats Forest and woodland (broadleaved deciduous forest; native coniferous forest), Scrub (sclerophyllous scrub/garrigue/maquis), Grassland (steppe/dry calcareous grassland), Rocky areas (sea cliff/rocky shore; inland cliff)
Land-use Agriculture, Forestry, Tourism/recreation

Birds

Species	Season	Year	Pop min	Pop max	Acc	Criteria
Emberiza caesia Cretzschmar's Bunting	B	1996	Frequent		—	B3, C6

This is an important area for breeding raptors and species characteristic of forest and scrub. Species of global conservation concern that do not meet IBA criteria: *Falco naumanni* (non-breeding).

Protection status

National Partial **International** None
674 ha of IBA covered by Aesthetic Forest (Steni Aesthetic Forest, 674 ha). 677 ha of IBA covered by Wildlife Refuge (Stenis Dirfys Chalkidas, 677 ha). 212 ha of IBA covered by Wildlife Refuge (Tsirloneri–Ortari/Kymis–Vitalou, 212 ha). 2,500 ha of IBA covered by Wildlife Refuge (Makiolivado–Virgini–Koutas/Pagontas, Markaton, Ag. Sofiar, Attalis, 2,500 ha). 500 ha of IBA covered by Wildlife Refuge (Klimataria–Koutoupas/Glifades, 500 ha). 3,400 ha of IBA covered by Wildlife Refuge (Kaditiko–Pasiotioka/Kadiou–Ag.Vlassiou–Manikion, 3,400 ha).

Conservation issues

Threats Burning of vegetation (A), Deforestation (commercial) (B), Recreation/tourism (U), Unsustainable exploitation (B)

The main threats are from fires, deforestation and hunting. Part of the area is a candidate SAC.

Lake Distos

A3, B2, C2, C6 **112**

Admin region Sterea Ellada
Coordinates 38°21′N 24°08′E
Altitude 16–590 m **Area** 2,600 ha

Site description

An eutrophic freshwater lake largely covered by reed *Phragmites*. In recent years the lake has been almost completely drained.

Habitats Scrub (60%; sclerophyllous scrub/garrigue/maquis), Wetland (20%; standing fresh water; water-fringe vegetation), Artificial landscape (20%; arable land)
Land-use Agriculture (80%), Water management (20%)

Birds

Species	Season	Year	Pop min	Pop max	Acc	Criteria
Ixobrychus minutus Little Bittern	B	1991	20	100	B	B2, C2, C6
Aythya nyroca Ferruginous Duck	B	1988	10	—	—	C6
Alectoris graeca Rock Partridge	R	1986	—	—	—	A3
Oenanthe hispanica Black-eared Wheatear	B	1991	—	—	—	A3
Hippolais olivetorum Olive-tree Warbler	B	—	—	—	—	A3
Sylvia cantillans Subalpine Warbler	B	1984	—	—	—	A3
Sylvia melanocephala Sardinian Warbler	R	1984	—	—	—	A3
Sylvia rueppelli Rüppell's Warbler	B	1986	—	—	—	A3
Sitta neumayer Rock Nuthatch	R	1991	Frequent		—	A3
Emberiza caesia Cretzschmar's Bunting	B	1991	Common		A	A3
Emberiza melanocephala Black-headed Bunting	B	1991	—	—	—	A3

This is an important area for breeding herons. Breeding birds include nine of the 21 species in Europe that are restricted (when breeding) to the Mediterranean biome.

Protection status

National Partial **International** None
Part of IBA covered by Wildlife Refuge (Argriro–Pr. Hlias–Panagia/Aliveriou–Argirou–Pras, 5,440 ha).

Conservation issues

Threats Agricultural intensification/expansion (A), Burning of vegetation (B), Drainage (A), Groundwater abstraction (A), Industrialization/urbanization (B), Unsustainable exploitation (A)

The main threat is continued abstraction of water for industry and agriculture that threatens the lake with complete drainage. Another serious threat is the planned construction of two wind-farms in the hills surrounding the site.

Mount Ochi and surrounding area

A4ii, B1ii, B2, B3, C2, C3, C6 **113**

Admin region Sterea Ellada
Coordinates 38°03′N 24°27′E
Altitude 0–1,398 m **Area** 18,000 ha

Site description

A coastal mountain dominated by grazed scrub with numerous springs and streams. Livestock-farming is the main activity.

Habitats Forest and woodland (21%; broadleaved deciduous forest; alluvial/very wet forest; broadleaved evergreen forest), Scrub (55%; scrub; sclerophyllous scrub/garrigue/maquis), Grassland (5%; steppe/dry calcareous grassland; humid grassland), Wetland (5%; river/stream; water-fringe vegetation; fen/transition mire/spring), Marine areas (5%; sea inlet/coastal features), Rocky areas (6%; sea cliff/rocky shore; scree/boulders; rock stacks/islets; inland cliff), Artificial landscape (13%; arable land; perennial crops/orchards/groves)
Land-use Agriculture (75%), Fisheries/aquaculture (10%), Forestry (10%), Tourism/recreation (5%)

Birds

The site is particularly important for passage migrants. Species of global conservation concern that do not meet IBA criteria: *Larus audouinii* (non-breeding).

Species	Season	Year	Pop min	Pop max	Acc	Criteria
Puffinus yelkouan Yelkouan Shearwater	N	1996	—	900	A	A4ii, B1ii, C3
Bubo bubo Eagle Owl	R	1989	5	—	C	B2, C6
Sitta neumayer Rock Nuthatch	R	1996	200	—	—	B3
Emberiza caesia Cretzschmar's Bunting	B	1996	150	—	—	B3, C2, C6

Protection status

National None **International** None

Conservation issues

Threats Abandonment/reduction of land management (C), Agricultural intensification/expansion (B), Deforestation (commercial) (A), Extraction industry (A), Filling-in of wetlands (B), Forest grazing (A), Industrialization/urbanization (A), Infrastructure (A), Intensified forest management (A), Recreation/tourism (B), Unsustainable exploitation (A)

Expansion of the road network and the ensuing development of tourism pose a serious threat to the area. Other threats are from mining in the western and northern parts of the site, abandonment of land, overgrazing, the use of poisoned baits, and the planned construction of wind-farms within the area. Part of the area is a candidate SAC.

North-west Skyros and Ag. Fokas islets — A1, C1, C6 — 114

Admin region Sterea Ellada
Coordinates 38°53′N 24°28′E
Altitude 0–376 m **Area** 2,700 ha

Site description

An area of low hills reaching the coast. *Pinus* forests occupy the majority of the area.

Habitats Forest and woodland (55%; native coniferous forest), Scrub (40%; scrub; sclerophyllous scrub/garrigue/maquis), Rocky areas (7%; sea cliff/rocky shore)
Land-use Agriculture (5%), Not utilized, Tourism/recreation

Birds

Species	Season	Year	Pop min	Pop max	Acc	Criteria
Larus audouinii Audouin's Gull	B	1996	55	55	A	A1, C1, C6

The site is important for breeding raptors, seabirds and species found in coastal scrub.

Protection status

National None **International** None

Conservation issues

Threats Abandonment/reduction of land management (B), Extraction industry (B) Recreation/tourism (B), Unsustainable exploitation (B)

The main threat is from tourism developments and disturbance of the *Larus audouinii* colony by tourists.

Mount Kohilas, Skyros island — A1, B2, B3, C1, C6 — 115

Admin region Sterea Ellada
Coordinates 38°46′N 24°40′E
Altitude 0–792 m **Area** 5,500 ha

Site description

A rugged mountain chain which forms high cliffs along the south-east coast of Skyros. The dominant vegetation is phrygana with small patches of grassland and *Quercus* and *Acer* woodland.

Habitats Scrub (70%; scrub; sclerophyllous scrub/garrigue/maquis), Grassland (5%; steppe/dry calcareous grassland), Rocky areas (25%; sea cliff/rocky shore; scree/boulders; inland cliff)
Land-use Agriculture (95%), Not utilized (5%)

Birds

This is an important site for seabirds, raptors and species associated with scrub.

Species	Season	Year	Pop min	Pop max	Acc	Criteria
[1] ***Falco eleonorae*** Eleonora's Falcon	B	1996	—	60	A	B2, C6
Emberiza cineracea Cinereous Bunting	B	1994	—	—	—	A1, B2, C1, C6
Emberiza caesia Cretzschmar's Bunting	B	1996	Common		—	B3, C6

1. Individuals.

Protection status

National None **International** None

Conservation issues

Threats Firewood collection (A), Forest grazing (A)

The most important threat is overgrazing of the phrygana. This is also preventing regeneration of the remnant patches of woodland. Part of the area is a candidate SAC.

Mount Ziria (Killini) — B2 — 116

Admin region Peloponnissos
Coordinates 37°56′N 22°25′E
Altitude 700–2,374 m **Area** 19,400 ha

Site description

A limestone mountain with areas of scrub, *Abies* and *Pinus* forest. A plateau area is covered by scrub and temporary ponds. There is a limited human presence and the main activities are stock-raising, forestry, tourism, and hunting.

Habitats Forest and woodland (native coniferous forest), Scrub (sclerophyllous scrub/garrigue/maquis), Rocky areas (scree/boulders; inland cliff)
Land-use Agriculture, Forestry, Tourism/recreation

Birds

Species	Season	Year	Pop min	Pop max	Acc	Criteria
Alectoris graeca Rock Partridge	R	1987	Abundant		—	B2

This is an important site for species found in conifer forests and scrub.

Protection status

National Low **International** None
660 ha of IBA covered by Wildlife Refuge (Boutsiou/Karias, 660 ha).

Conservation issues

Threats Unsustainable exploitation (A)

The main threat is from illegal hunting. Part of the area is a candidate SAC.

Lake Stymphalia — B2, C2, C6 — 117

Admin region Peloponnissos
Coordinates 37°51′N 22°28′E
Altitude 600–800 m **Area** 1,309 ha

Site description

A lake partly fed by springs and drained by underground streams and by a canal. In some years the lake dries up and is used for cultivation. The surrounding area is a mixture of scrub and agricultural areas. Human activities include arable agriculture, stock-raising and hunting.

Habitats Forest and woodland (11%; native coniferous forest), Scrub (18%; scrub; sclerophyllous scrub/garrigue/maquis), Wetland (40%; standing fresh water; fen/transition mire/spring), Artificial landscape (38%; arable land; perennial crops/orchards/groves)
Land-use Fisheries/aquaculture, Agriculture (60%), Water management (100%)

Birds

Species	Season	Year	Pop min	Pop max	Acc	Criteria
[1] ***Ixobrychus minutus*** Little Bittern	B	1997	30	70	B	B2, C2, C6
[1] ***Aythya nyroca*** Ferruginous Duck	B	1996	2	10	A	C6

1. Individuals.

The site is important for breeding, passage and wintering waterbirds. Species of global conservation concern that do not meet IBA criteria: *Haliaeetus albicilla*.

Protection status

National Partial **International** None

200 ha of IBA covered by Wildlife Refuge (Gidomandra/Laukas, 1,400 ha).

Conservation issues

Threats Abandonment/reduction of land management (C), Agricultural intensification/expansion (A), Construction/impact of dyke/dam/barrage (C), Dredging/canalization (C), Forest grazing (B), Groundwater abstraction (C), Natural events (A), Other (B), Recreation/tourism (C), Unsustainable exploitation (A)

The lake is threatened by siltation ('Other' threats). There are plans to further manage the surface and groundwater for irrigation and water supply. The area is a candidate SAC.

Mounts Artemisio and Lirkio B3, C6 118

Admin region Peloponnissos
Coordinates 37°40'N 22°29'E
Altitude 700–1,771 m **Area** 8,900 ha

Site description

An area of rocky slopes, scrub, inland cliffs and a small gorge. There is also a terraced plateau with old hedges. The main human activity is livestock-farming.

Habitats Forest and woodland (10%; native coniferous forest), Scrub (50%; sclerophyllous scrub/garrigue/maquis), Grassland (5%; mesophile grassland), Rocky areas (40%; scree/boulders; inland cliff)
Land-use Agriculture (80%), Not utilized (20%)

Birds

Species	Season	Year	Pop min	Pop max	Acc	Criteria
Sylvia rueppelli Rüppell's Warbler	B	1987	Abundant		—	B3, C6

This is an important area for species typical of Mediterranean scrub and rocky habitats.

Protection status

National None **International** None

Conservation issues

Threats Unsustainable exploitation (B)

The main threat is from illegal hunting.

Divari Pilou lagoon (Gialova) A4i, B1i, C2, C3 119

Admin region Peloponnissos
Coordinates 36°58'N 21°40'E
Altitude 0–0 m **Area** 600 ha

Site description

The wetland can be divided into brackish lagoons and freshwater marshes. Beyond these is a buffer zone of cultivated fields and maquis. The main human activities are fishing, livestock- and arable-farming, tourism and hunting.

Habitats Scrub (10%; sclerophyllous scrub/garrigue/maquis), Wetland (70%; mudflat/sandflat; saltmarsh; sand-dunes/sand beach; coastal lagoon; standing fresh water; standing brackish and salt water; fen/transition mire/spring), Rocky areas (5%; sea cliff/rocky shore), Artificial landscape (20%; arable land; forestry plantation)
Land-use Agriculture (14%), Fisheries/aquaculture (65%), Tourism/recreation (30%)

Birds

Species	Season	Year	Pop min	Pop max	Acc	Criteria
[1] ***Phalacrocorax carbo*** Cormorant	W	1996	1,000	1,000	A	B1i, C3
Egretta garzetta Little Egret	P	1995	1,500	2,000	A	A4i, B1i, C2
[1] ***Egretta alba*** Great White Egret	W	1996	280	360	A	A4i, B1i, C2
Plegadis falcinellus Glossy Ibis	P	1996	1,600	1,800	A	A4i, B1i, C2
Glareola pratincola Collared Pratincole	P	1996	117	117	A	B1i, C2
Tringa stagnatilis Marsh Sandpiper	P	1996	300	500	A	A4i, B1i, C3
Tringa glareola Wood Sandpiper	P	1996	7,000	11,000	A	A4i, B1i, C2
Gelochelidon nilotica Gull-billed Tern	P	1996	210	220	A	B1i, C2

1. Large increase.

This is an important site for wintering and passage waterbirds. Species of global conservation concern that do not meet IBA criteria: *Pelecanus crispus* (passage), *Aythya nyroca* (10–20 passage individuals), *Falco naumanni* (10–20 passage individuals), *Gallinago media* (10–20 passage individuals).

Protection status

National High **International** None

600 ha of IBA covered by Wildlife Refuge (Limni Divari–Valtos/Pylou–Romanou, 700 ha).

Conservation issues

Threats Agricultural intensification/expansion (C), Aquaculture/fisheries (A), Groundwater abstraction (A), Infrastructure (A), Recreation/tourism (B), Unsustainable exploitation (A)

The main threats are fishing, increasing salinity in the freshwater marsh as a result of ground and surface water abstraction for agriculture, infilling of the wetland and hunting. The area is a candidate SAC.

Mount Taigetos A1, A3, B2, B3, C1, C2, C6 120

Admin region Peloponnissos
Coordinates 36°58'N 22°19'E
Altitude 150–2,407 m **Area** 70,000 ha

Site description

A mountainous area with *Pinus* and *Abies* forests, treeline ecotone habitat and maquis.

Habitats Forest and woodland (55%; native coniferous forest; treeline ecotone), Scrub (30%; sclerophyllous scrub/garrigue/maquis), Rocky areas (15%; scree/boulders; inland cliff), Artificial landscape (5%; perennial crops/orchards/groves)
Land-use Forestry, Other, Tourism/recreation, Water management

Birds

Species	Season	Year	Pop min	Pop max	Acc	Criteria
Hieraaetus fasciatus Bonnelli's Eagle	R	1996	3	4	—	B2, C6
Falco naumanni Lesser Kestrel	B	1996	3	10	—	A1, C1
Alectoris graeca Rock Partridge	R	1996	Uncommon		—	A3
Oenanthe hispanica Black-eared Wheatear	B	1996	Abundant		—	A3
Monticola solitarius Blue Rock Thrush	R	1996	Common		—	B2
Sylvia cantillans Subalpine Warbler	B	1996	Common		—	A3
Sylvia melanocephala Sardinian Warbler	R	1996	Abundant		—	A3
Sylvia rueppelli Rüppell's Warbler	B	1996	Uncommon		—	A3, B3
Sylvia hortensis Orphean Warbler	B	1996	Uncommon		—	B2
Sitta neumayer Rock Nuthatch	R	1996	Common		—	A3
Lanius senator Woodchat Shrike	B	1996	Common		—	B2
Pyrrhocorax graculus Alpine Chough	R	1996	Uncommon		—	A3
Pyrrhocorax pyrrhocorax Chough	R	1997	200	—	A	C2
Emberiza caesia Cretzschmar's Bunting	B	1996	Uncommon		—	A3

This is an important site for breeding raptors and species associated with open montane habitats. Breeding birds include seven of the 21 species in Europe that are restricted (when breeding) to the Mediterranean biome.

Protection status

National Low **International** None

1,000 ha of IBA covered by Wildlife Refuge (Doubitsa–Kentreas–Vribitsa/Exochoriou, 1,000 ha). 1,300 ha of IBA covered by Wildlife Refuge (Lada/Karveliou, 1,300 ha).

Conservation issues

Threats Burning of vegetation (A), Consequences of animal/plant introductions (U), Disturbance to birds (B), Extraction industry (B), Forest grazing (A), Infrastructure (B), Selective logging/cutting (C), Unsustainable exploitation (B)

The major threats are from forest fires and the subsequent erosion and overgrazing. Expansion of the road network and quarrying are additional threats. The introduction of *Alectoris chukar* is a potential threat to the *Alectoris graeca* population. Part of the area is a candidate SAC.

Southern Mani, Mount Sanghias and Cape Tenaro — A3, B2, B3 — 121

Admin region Peloponnissos
Coordinates 36°34′N 22°27′E
Altitude 0–1,214 m **Area** 25,000 ha

Site description

A rocky, mountainous peninsula in the southern Peloponissos. The area is dominated by phrygana and oak *Quercus* forests.

Habitats Forest and woodland (10%), Scrub (40%; sclerophyllous scrub/garrigue/maquis), Rocky areas (40%; sea cliff/rocky shore; inland cliff), Artificial landscape (10%; arable land; perennial crops/orchards/groves)
Land-use Agriculture (70%), Tourism/recreation (30%)

Birds

Species	Season	Year	Pop min	Pop max	Acc	Criteria
Alectoris graeca Rock Partridge	R	1996	Common		—	A3
Athene noctua Little Owl	R	1996	Common		—	B2
Oenanthe hispanica Black-eared Wheatear	B	1996	Abundant		—	A3, B2
Monticola solitarius Blue Rock Thrush	R	1996	Common		—	B2
Sylvia cantillans Subalpine Warbler	B	1996	Common		—	A3
Sylvia melanocephala Sardinian Warbler	R	1996	Abundant		—	A3
Sylvia rueppelli Rüppell's Warbler	B	1996	Uncommon		—	A3
Sitta neumayer Rock Nuthatch	R	1996	Abundant		—	A3, B3
Emberiza caesia Cretzschmar's Bunting	B	1996	Rare		—	A3

This is an important site for passage raptors and passerines. Breeding birds include seven of the 21 species in Europe that are restricted (when breeding) to the Mediterranean biome. Species of global conservation concern that do not meet IBA criteria: *Aquila heliaca* (non-breeding).

Protection status

National None **International** None

Conservation issues

Threats Abandonment/reduction of land management (B), Agricultural intensification/expansion (B), Burning of vegetation (A), Consequences of animal/plant introductions (A), Disturbance to birds (A), Forest grazing (B), Recreation/tourism (A), Unsustainable exploitation (A)

Major threats are from hunting, land abandonment in the remoter areas, infilling of small wetlands for tourism, fires and forest grazing.

Evrota delta — A4i, B1i, C2 — 122

Admin region Peloponnissos
Coordinates 36°49′N 22°43′E
Altitude 0–100 m **Area** 3,000 ha

Site description

A delta with dunes, saltmarshes, riverine vegetation, meadows and reedbeds (*Phragmites*).

Habitats Scrub (sclerophyllous scrub/garrigue/maquis), Grassland (dry siliceous grassland), Wetland (tidal river/enclosed tidal water; mudflat/sandflat; saltmarsh; sand-dunes/sand beach; coastal lagoon; river/stream; water-fringe vegetation; fen/transition mire/spring), Marine areas (sea inlet/coastal features), Rocky areas (sea cliff/rocky shore), Artificial landscape (arable land; perennial crops/orchards/groves; other urban/industrial areas)
Land-use Agriculture (50%), Fisheries/aquaculture (5%), Nature conservation/research (10%), Tourism/recreation (10%), Urban/industrial/transport (5%), Water management (50%)

Birds

Species	Season	Year	Pop min	Pop max	Acc	Criteria
Plegadis falcinellus Glossy Ibis	P	1997	500	—	B	A4i, B1i, C2

The area is important for passage waterbirds and passerines, and for breeding and passage raptors. Species of global conservation concern that do not meet IBA criteria: *Aquila clanga* (wintering), *Aquila heliaca* (wintering), *Falco naumanni* (minimum 3 breeding pairs).

Protection status

National None **International** None

Conservation issues

Threats Abandonment/reduction of land management (C), Agricultural intensification/expansion (A), Aquaculture/fisheries (A), Construction/impact of dyke/dam/barrage (B), Drainage (U), Dredging/canalization (B), Filling-in of wetlands (B), Groundwater abstraction (A), Industrialization/urbanization (C), Infrastructure (B), Recreation/tourism (C), Unsustainable exploitation (A)

The major threats are the expansion of agriculture into wetland areas and tourism-related development. The area is a candidate SAC.

East Lakonia mountains — B2, B3, C6 — 123

Admin region Peloponnissos
Coordinates 36°55′N 22°57′E
Altitude 0–1,329 m **Area** 80,000 ha

Site description

An area of mountains, sea cliffs and gorges. Phrygana, maquis and sclerophyllous scrub dominate, with areas of *Abies* forest. The main human activities are livestock-farming and hunting.

Habitats Forest and woodland (8%; native coniferous forest), Scrub (80%; sclerophyllous scrub/garrigue/maquis), Marine areas (5%; sea inlet/coastal features), Rocky areas (10%; sea cliff/rocky shore; inland cliff), Artificial landscape (5%; perennial crops/orchards/groves)
Land-use Agriculture (80%), Fisheries/aquaculture (5%), Forestry (8%), Not utilized (5%)

Birds

Species	Season	Year	Pop min	Pop max	Acc	Criteria
Circaetus gallicus Short-toed Eagle	B	1996	3	20	B	C6
[1] ***Hieraaetus fasciatus*** Bonnelli's Eagle	R	1995	2	6	A	B2, C6
Oenanthe hispanica Black-eared Wheatear	B	1996	Common		—	B2
Monticola solitarius Blue Rock Thrush	R	1996	Frequent		—	B2
Hippolais olivetorum Olive-tree Warbler	B	1996	Uncommon		—	B2, C6
Sylvia rueppelli Rüppell's Warbler	B	1996	Uncommon		—	B3, C6
Sylvia hortensis Orphean Warbler	B	1996	Uncommon		—	B2
Sitta neumayer Rock Nuthatch	R	1996	Common		—	B3
Lanius senator Woodchat Shrike	B	1996	Frequent		—	B2
Emberiza caesia Cretzschmar's Bunting	B	1989	Rare		—	B3, C6

1. Large decrease.

The site is an important corridor for migratory passerines and breeding and migrant raptors.

Protection status

National Low **International** None
800 ha of IBA covered by Wildlife Refuge (Kourkoula/Molaon, 800 ha). 3,000 ha of IBA covered by Wildlife Refuge (Zarakos/Koupion–Kremasti–Richeas–Lambokampou, 3,000 ha).

Conservation issues

Threats Abandonment/reduction of land management (A), Unsustainable exploitation (A)

The main threats are land abandonment and hunting. Part of the area is a candidate SAC.

Mount Parnitha

B3, C6 **124**

Admin region Attiki
Coordinates 38°10′N 23°43′E
Altitude 200–1,413 m **Area** 19,400 ha

Site description

The higher parts of the mountain are dominated by *Abies* forest and lower areas by *Pinus* forest, maquis and phrygana. Part of the site is a military area. The main human activity is recreation and tourism.

Habitats Forest and woodland (55%; native coniferous forest), Scrub (41%; sclerophyllous scrub/garrigue/maquis), Grassland (5%; steppe/dry calcareous grassland), Rocky areas (5%; inland cliff)
Land-use Military (5%), Tourism/recreation (10%), Not utilized

Birds

Species	Season	Year	Pop min	Pop max	Acc	Criteria
Emberiza caesia Cretzschmar's Bunting	B	1997	Common		—	B3, C6

This is an important area for species associated with conifer forests and upland scrub.

Protection status

National Partial **International** Partial
3,812 ha of IBA covered by National Park (Parnitha, 3,812 ha). Part of IBA covered by Wildlife Refuge (Voutima/Avlonas, 1,188 ha). 10,070 ha of IBA covered by Wildlife Refuge (Dasoktima Tatoiou–Salonikis–Loimikou–Sinidioktito, 10,070 ha). 4,012 ha of IBA covered by Special Protection Area (Ethnikos Drymos Parnithas, 4,012 ha).

Conservation issues

Threats Burning of vegetation (B), Disturbance to birds (B), Recreation/tourism (B)

The main threats are from disturbance and tourism. A National Park management plan is currently being prepared. Part of the area is a candidate SAC.

Schinias marsh

A4i, B1i, C2 **125**

Admin region Attiki
Coordinates 38°09′N 24°01′E
Altitude 0–250 m **Area** 2,100 ha

Site description

A coastal marsh enclosed by sand-dunes, *Pinus* forest and maquis covered hills.

Habitats Forest and woodland (native coniferous forest), Scrub (sclerophyllous scrub/garrigue/maquis), Wetland (saltmarsh; sand-dunes/sand beach; standing brackish and salt water; water-fringe vegetation; fen/transition mire/spring), Marine areas (sea inlet/coastal features), Rocky areas (sea cliff/rocky shore; inland cliff), Artificial landscape (other urban/industrial areas)
Land-use Not utilized, Tourism/recreation, Urban/industrial/transport, Water management

Birds

Species	Season	Year	Pop min	Pop max	Acc	Criteria
Plegadis falcinellus Glossy Ibis	P	1988	325	—	—	A4i, B1i, C2

A total of 175 species have been recorded at this site.

Protection status

National None **International** None

Conservation issues

Threats Construction/impact of dyke/dam/barrage (B), Disturbance to birds (B), Drainage (B), Dredging/canalization (B), Filling-in of wetlands (C), Groundwater abstraction (B), Infrastructure (A), Recreation/tourism (C)

There are plans to establish an Olympic rowing and canoeing centre and golf course on this site, which would destroy most of the area. Other threats include urban development and expansion, both legal and illegal.

Mount Ymittos

A3, B3, C6 **126**

Admin region Attiki
Coordinates 37°56′N 23°48′E
Altitude 200–1,100 m **Area** 8,000 ha

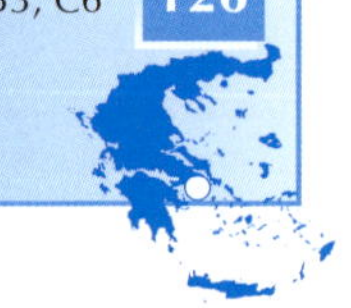

Site description

A mountainous massif between Athens and the Mesogia plain. Much of the area consists of rocky slopes with phrygana and other scrub. The western area of the site is a *Pinus* plantation, now regenerating naturally.

Habitats Forest and woodland (native coniferous forest), Scrub (sclerophyllous scrub/garrigue/maquis), Rocky areas (scree/boulders; inland cliff)
Land-use Agriculture, Not utilized, Tourism/recreation, Urban/industrial/transport

Birds

Species	Season	Year	Pop min	Pop max	Acc	Criteria
Oenanthe hispanica Black-eared Wheatear	B	1996	Common		—	A3
Sylvia cantillans Subalpine Warbler	B	1996	Abundant		—	A3
Sylvia melanocephala Sardinian Warbler	R	1996	Abundant		—	A3
Sylvia rueppelli Rüppell's Warbler	B	1996	Common		—	A3, B3
Sitta neumayer Rock Nuthatch	R	1996	Abundant		—	A3, B3
Emberiza caesia Cretzschmar's Bunting	B	1996	Frequent		—	A3, B3, C6
Emberiza melanocephala Black-headed Bunting	B	—	—	—	—	A3

The site is particularly important for species associated with phrygana. Breeding birds include seven of the 21 species in Europe that are restricted (when breeding) to the Mediterranean biome. Species of global conservation concern that do not meet IBA criteria: *Falco naumanni* (passage).

Protection status

National Partial **International** None
2,000 ha of IBA covered by Wildlife Refuge (Dim. Dasos Ymittou, 2,000 ha).

Conservation issues

Threats Burning of vegetation (A), Industrialization/urbanization (A), Infrastructure (B)

The major threat to the area is the expansion of urban and industrial developments and their associated infrastructure. Fires are also a frequent problem. The area is a candidate SAC.

Sounio area

B3, C6 **127**

Admin region Attiki
Coordinates 37°41′N 24°00′E
Altitude 0–356 m **Area** 7,000 ha

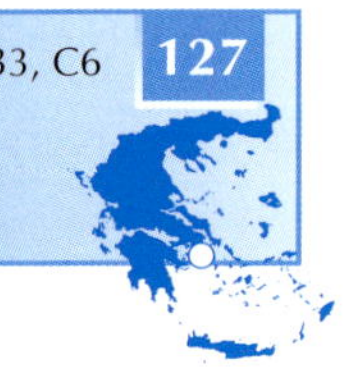

Site description

A coastal site including *Pinus* forest, maquis, phrygana and cliffs. It includes the islet of Patroklou.

Habitats Forest and woodland (20%; native coniferous forest), Scrub (60%; sclerophyllous scrub/garrigue/maquis), Rocky areas (5%; sea cliff/rocky shore; rock stacks/islets; inland cliff), Artificial landscape (15%; perennial crops/orchards/groves; other urban/industrial areas)
Land-use Agriculture (60%), Tourism/recreation (10%), Urban/industrial/transport (5%)

Birds

Species	Season	Year	Pop min	Pop max	Acc	Criteria
Emberiza caesia Cretzschmar's Bunting	B	1996	Common		—	B3, C6

The site is important for species that occur in coastal scrub and forest.

Protection status

National Partial **International** None
3,500 ha of IBA covered by National Park (Sounio, 3,500 ha).

Conservation issues

Threats Burning of vegetation (A), Disturbance to birds (C), Industrialization/urbanization (A), Recreation/tourism (C), Unsustainable exploitation (A)

The main threats are from burning, urbanization and hunting. Part of the area is a candidate SAC.

Falkonera islet — B1iii, B2, C2, C6 — 128

Admin region Attiki
Coordinates 36°50′N 23°53′E
Altitude 0–218 m **Area** 140 ha

Site description

A rocky islet east of Peloponnissos.

Habitats Scrub (sclerophyllous scrub/garrigue/maquis), Rocky areas (rock stacks/islets)
Land-use Fisheries/aquaculture

Birds

Species	Season	Year	Pop min	Pop max	Acc	Criteria
[1] ***Falco eleonorae*** Eleonora's Falcon	B	1994	100	—	A	B1iii, B2, C2, C6

1. Individuals.

This is an important site for species associated with cliffs.

Protection status

National None **International** None

Conservation issues

Threats Unknown

The area is a candidate SAC

North, east and south Kithira island — A1, A4ii, B1i, B1ii, B1iv, B2, B3, C1, C2, C3, C5, C6 — 129

Admin region Attiki
Coordinates 36°14′N 23°03′E
Altitude 0–507 m **Area** 18,000 ha

Site description

The site includes parts of Kithira and the adjacent islets. Human activities include arable agriculture, livestock-farming, tourism, hunting, and fishing. There is also some military use of the site.

Habitats Scrub (60%; sclerophyllous scrub/garrigue/maquis), Wetland (sand-dunes/sand beach; shingle/stony beach), Marine areas (10%; open sea; sea inlet/coastal features), Rocky areas (10%; sea cliff/rocky shore; rock stacks/islets; inland cliff), Artificial landscape (30%; arable land; perennial crops/orchards/groves; forestry plantation; ruderal land)
Land-use Agriculture (50%), Fisheries/aquaculture (10%), Military (5%), Tourism/recreation (35%)

Birds

Species	Season	Year	Pop min	Pop max	Acc	Criteria
[1] ***Calonectris diomedea*** Cory's Shearwater	R	1996	300	600	B	B1ii, B2, C2, C6
Puffinus yelkouan Yelkouan Shearwater	R	1996	600	650	B	A4ii, B1ii, B3, C3
Phalacrocorax aristotelis Shag	R	1996	45	70	B	B1i, C2
Aythya nyroca Ferruginous Duck	P	1996	5	70	C	A1, C1
Falco tinnunculus Kestrel	R	1996	60	70	A	B2
Falco eleonorae Eleonora's Falcon	B	1996	Abundant	—		B2, C6
Falco peregrinus Peregrine	R	1996	13	15	A	B2, C6
Larus audouinii Audouin's Gull	R	1996	30	—	A	A1, B2, C1, C6
Emberiza caesia Cretzschmar's Bunting	B	1996	100	150	B	B3, C2, C6

1. Individuals.

Two hundred and thirty-six species have been recorded. The site is a migratory bottleneck, where more than 3,000 raptors regularly pass in autumn. Species of global conservation concern that do not meet IBA criteria: *Aquila heliaca* (up to 2 wintering individuals), *Falco naumanni* (passage).

Protection status

National Partial **International** Low

3,000 ha of IBA covered by Wildlife Refuge (Thalaria Kythiron, 3,000 ha). 75 ha of IBA covered by Special Protection Area (Nisides Kythiron: Prasonisi, Dragonera, Antidragonera, 989 ha).

Conservation issues

Threats Afforestation (A), Agricultural intensification/expansion (B), Burning of vegetation (B), Industrialization/urbanization (B), Recreation/tourism (B), Unsustainable exploitation (A)

Threats come from uncontrolled grazing, fire, and land reclamation for building. Hunting is also a threat, particularly during autumn. Part of the area is a candidate SAC.

Antikithira island, Prasouda and Lagouvardos islets — A4ii, B1iii, B2, C2, C6 — 130

Admin region Attiki
Coordinates 35°52′N 23°18′E
Altitude 0–378 m **Area** 2,000 ha

Site description

A small island, and its associated islets, between Kithira and Crete. Antikithira is covered mainly by low maquis and phrygana with a few arable areas and olive-groves. Prasouda is covered by low maquis and phrygana. Human activities include stock-raising, fishing, agriculture, and hunting.

Habitats Scrub (80%; sclerophyllous scrub/garrigue/maquis), Marine areas (5%; open sea; sea inlet/coastal features), Rocky areas (5%; sea cliff/rocky shore; rock stacks/islets; inland cliff), Artificial landscape (5%; arable land; perennial crops/orchards/groves; ruderal land)
Land-use Agriculture, Fisheries/aquaculture

Birds

Species	Season	Year	Pop min	Pop max	Acc	Criteria
Falco eleonorae Eleonora's Falcon	B	1996	300	—	B	A4ii,B1iii,B2,C2,C6

This is an important area for breeding and passage raptors and seabirds.

Protection status

National None **International** None

Conservation issues

Threats Agricultural intensification/expansion (A), Aquaculture/fisheries (A), Natural events (B), Unsustainable exploitation (A)

The major threats are from uncontrolled grazing and illegal hunting. The area is used as a swordfish fishery. The area is a candidate SAC.

Rocky islets of Limnos island — A1, A4ii, B2, C1, C2, C6 — 131

Admin region North Aegean
Coordinates 40°01′N 25°08′E
Altitude 0–50 m **Area** 4 ha

Site description

The islets are approximately 300 m from the shore and are vegetated with phrygana. The main human activities are water-sports and angling.

Habitats Scrub (70%; sclerophyllous scrub/garrigue/maquis), Rocky areas (100%; rock stacks/islets)
Land-use Not utilized, Water management, Tourism/recreation

Birds

Species	Season	Year	Pop min	Pop max	Acc	Criteria
Falco eleonorae Eleonora's Falcon	B	1998	120	—	A	A4ii, B2, C2
Larus audouinii Audouin's Gull	B	1998	50	—	A	A1, C1, C6

The site is important for species associated with coastal cliffs.

Protection status

National None **International** None

Conservation issues

Threats Recreation/tourism (U), Disturbance to birds (U)

Potential threats to the site are disturbance from tourists and fishermen.

Lakes Khortaro and Alyki, Moudros gulf, Diapori fen, and Fakos peninsula

A1, A4i, A4ii, B1i, B1iii, B2, C1, C2, C3, C6 **132**

Admin region North Aegean
Coordinates 39°53′N 25°17′E
Altitude 0–319 m **Area** 14,000 ha

Site description

A wetland complex including coastal lagoons, a seasonal brackish marsh, extensive saltmarshes and sandflats and a freshwater marsh with reedbeds (*Phragmites*). Both lakes dry out in late summer. Human activities include arable cultivation, fishing, hunting, and aquaculture.

Habitats Grassland (10%; humid grassland), Wetland (40%; mudflat/sandflat; sand-dunes/sand beach; coastal lagoon; standing fresh water; standing brackish and salt water; fen/transition mire/spring; water-fringe vegetation), Marine areas (20%; sea inlet/coastal features), Artificial landscape (30%; arable land)
Land-use Agriculture (50%), Fisheries/aquaculture (15%), Military (8%), Not utilized (10%), Urban/industrial/transport (5%), Water management (15%)

Birds

Species	Season	Year	Pop min	Pop max	Acc	Criteria
Phoenicopterus ruber Greater Flamingo	W	1996	—	5,000	A	A4i, B1i, C2
Tadorna ferruginea Ruddy Shelduck	R	1996	—	8	A	B2, C6
Tadorna tadorna Shelduck	W	1996	—	850	A	B1i, C3
Falco naumanni Lesser Kestrel	B	1996	—	220	A	A1, A4ii, B1iii, B2, C1, C2, C6
Recurvirostra avosetta Avocet	B	1996	—	45	A	C6
Burhinus oedicnemus Stone Curlew	B	1996	20	—	A	B2, C6

The site is important for migratory waders and raptors and wintering wildfowl. Species of global conservation concern that do not meet IBA criteria: *Aythya nyroca* (passage).

Protection status

National Partial **International** Low

Part of IBA covered by Wildlife Refuge (Komi–Tourles–Chortarolimni/Romanou–Roussopouliou, 950 ha). Part of IBA covered by Wildlife Refuge (Palaiokastro–Alepotripes–Voutsadika/Panagias–Plaka, 1,200 ha). 1,302 ha of IBA covered by Special Protection Area (Ygrotopoi Kolpou Kallonis, 1,302 ha).

Conservation issues

Threats Aquaculture/fisheries (C), Disturbance to birds (B), Dredging/canalization (A), Filling-in of wetlands (A), Industrialization/urbanization (A), Other (B), Unsustainable exploitation (B)

Lake Aliki is threatened by plans to convert it to an industrialized saltpan. Other threats include the filling-in of wetland, rubbish dumping ('Other' threats, above) and hunting. Part of the area is a candidate SAC.

Agios Efstratios island

A4ii, B1iii, B2, C2 **133**

Admin region North Aegean
Coordinates 39°31′N 25°00′E
Altitude 0–303 m **Area** 4,600 ha

Site description

Habitats Forest and woodland (26%; broadleaved deciduous forest), Scrub (60%; sclerophyllous scrub/garrigue/maquis), Wetland (5%; sand-dunes/sand beach), Rocky areas (20%; sea cliff/rocky shore; rock stacks/islets), Artificial landscape (5%; perennial crops/orchards/groves; other urban/industrial areas)
Land-use Agriculture (85%), Not utilized (15%)

A remote island and the surrounding rocky islets in the northern Aegean Sea. The dominant vegetation is phrygana.

Birds

Species	Season	Year	Pop min	Pop max	Acc	Criteria
Falco eleonorae Eleonora's Falcon	B	1995	70	—	B	A4ii, B1iii, B2, C2

An important site for breeding seabirds, raptors and for passage migrants. Species of global conservation concern that do not meet IBA criteria: *Falco naumanni* (passage).

Protection status

National None **International** None

Conservation issues

Threats Agricultural intensification/expansion (A), Burning of vegetation (U), Forest grazing (A), Recreation/tourism (C), Unsustainable exploitation (C)

The main threats come from grazing of scrub and forest and from deliberate burning of the phrygana. The area is a candidate SAC.

South-west peninsula– petrified forest, Lesvos

A1, A3, B2, B3, C1, C2, C6 **134**

Admin region North Aegean
Coordinates 39°10′N 26°00′E
Altitude 0–799 m **Area** 34,000 ha

Site description

A rugged area dominated by phrygana with dispersed *Quercus macrolepis*. The inaccessible northern areas of the site are dominated by maquis. Human activities include livestock-farming, tourism and quarrying.

Habitats Forest and woodland (10%; broadleaved deciduous forest), Scrub (70%; sclerophyllous scrub/garrigue/maquis), Wetland (5%; water-fringe vegetation), Rocky areas (10%; rock stacks/islets; scree/boulders; inland cliff), Artificial landscape (10%; perennial crops/orchards/groves; other urban/industrial areas)
Land-use Agriculture (70%), Nature conservation/research (10%), Tourism/recreation (10%), Not utilized (10%), Urban/industrial/transport (5%)

Birds

Species	Season	Year	Pop min	Pop max	Acc	Criteria
Circaetus gallicus Short-toed Eagle	B	1996	Frequent		—	C6
Buteo rufinus Long-legged Buzzard	R	1996	—	5	A	B2
Falco naumanni Lesser Kestrel	B	1996	5	10	A	A1, C1
Falco eleonorae Eleonora's Falcon	B	1996	Frequent		—	A3
Dendrocopos medius Middle Spotted Woodpecker	R	1996	Frequent		—	C6
Oenanthe hispanica Black-eared Wheatear	B	1996	Abundant		—	A3
Sylvia cantillans Subalpine Warbler	B	1996	Common		—	A3
Sylvia melanocephala Sardinian Warbler	R	1996	Common		—	A3
Sitta neumayer Rock Nuthatch	R	1996	Common		—	A3
Lanius nubicus Masked Shrike	B	1996	Uncommon		—	A3, B2
Emberiza cineracea Cinereous Bunting	B	1996	100	250	A	A1, A3, B2, C1, C2, C6
Emberiza caesia Cretzschmar's Bunting	B	1996	Common		—	A3, B3, C6
Emberiza melanocephala Black-headed Bunting	B	1996	Common		—	A3

The site is important for breeding and passage raptors and also other migrants. Breeding birds include nine of the 21 species in Europe that are restricted (when breeding) to the Mediterranean biome. Species of global conservation concern that do not meet IBA criteria: *Haliaeetus albicilla* (non-breeding), *Aquila clanga* (non-breeding), *Larus audouinii* (non-breeding).

Protection status

National Partial **International** None

Part of IBA covered by Wildlife Refuge (Mesoros/Agra Lesvou, 600 ha). Part of IBA covered by Wildlife Refuge (Apolithomena/Sigriou, 800 ha). Part of IBA covered by Wildlife Refuge (Skamnioudа/Antissas, 600 ha). Part of IBA covered by Wildlife Refuge (Vathia Koilada–Kelemia

Antissas, 400 ha). 15,000 ha of IBA covered by Natural Monument (Petrified Forest of Lesbos Island, 15,000 ha).

Conservation issues

Threats Agricultural intensification/expansion (B), Drainage (C), Filling-in of wetlands (C), Groundwater abstraction (U), Industrialization/urbanization (C), Recreation/tourism (B), Unsustainable exploitation (C)

The main threats are overgrazing and tourist developments. A large dam and associated infrastructure are currently being constructed. Part of the area is a candidate SAC.

Mount Lepetimnos, Lesvos — A1, B1ii, B2, B3, C1, C2, C3, C6 — 135

Admin region North Aegean
Coordinates 39°22′N 26°12′E
Altitude 0–968 m **Area** 9,000 ha

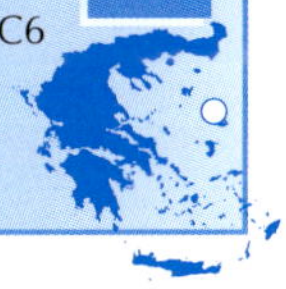

Site description

A mountain in the north of Lesvos. The vegetation is a mix of dense maquis, oak *Quercus* woodland, phrygana and olive-groves.

Habitats Forest and woodland (10%), Scrub (45%; sclerophyllous scrub/garrigue/maquis), Wetland (5%; shingle/stony beach), Rocky areas (10%; rock stacks/islets; inland cliff), Artificial landscape (30%; perennial crops/orchards/groves)
Land-use Agriculture (40%), Tourism/recreation (15%), Urban/industrial/transport (10%)

Birds

Species	Season	Year	Pop min	Pop max	Acc	Criteria
Puffinus yelkouan Yelkouan Shearwater	R	1996	—	200	A	B1ii, C2
Dendrocopos medius Middle Spotted Woodpecker	R	1996	Common		—	C6
Sylvia rueppelli Rüppell's Warbler	B	1996	Frequent		—	B3
Lanius nubicus Masked Shrike	B	1996	Uncommon		—	B2
Emberiza cineracea Cinereous Bunting	B	1996	10	20	A	A1, B2, C1, C2, C6

The site is of importance for breeding and passage raptors. Species of global conservation concern that do not meet IBA criteria: *Larus audouinii* (non-breeding).

Protection status

National None **International** None

Conservation issues

Threats Construction/impact of dyke/dam/barrage (U), Groundwater abstraction (U), Industrialization/urbanization (B), Recreation/tourism (B), Unsustainable exploitation (B)

Threats include a developing road network, tourism development and illegal housing development.

Tokmakia islets — A1, C1 — 136

Admin region North Aegean
Coordinates 39°18′N 26°26′E
Altitude 0–58 m **Area** 150 ha

Site description

A group of small islands east of Lesvos in the north-east Aegean. Human activities include stock-raising, fishing and tourism.

Habitats Scrub (sclerophyllous scrub/garrigue/maquis), Rocky areas (sea cliff/rocky shore)
Land-use Agriculture, Fisheries/aquaculture

Birds

Species	Season	Year	Pop min	Pop max	Acc	Criteria
Larus audouinii Audouin's Gull	B	1998	28	—	A	A1, C1

The site is important for breeding coastal species.

Protection status

National None **International** None

Conservation issues

Threats Disturbance to birds (A)

Hunting is prohibited in part of the area.

Kallonis gulf — A4i, B1i, B2, C2, C6 — 137

Admin region North Aegean
Coordinates 39°10′N 26°13′E
Altitude 0–100 m **Area** 7,800 ha

Site description

A large bay with a shallow brackish zones, small freshwater marshes, saltmarshes, and saltpans. Human activities include agriculture, salt production, stock-raising, fishing, hunting, and recreation.

Habitats Forest and woodland (native coniferous forest), Scrub (sclerophyllous scrub/garrigue/maquis), Wetland (tidal river/enclosed tidal water; saltmarsh; sand-dunes/sand beach; coastal lagoon; standing brackish and salt water), Marine areas (sea inlet/coastal features), Artificial landscape (perennial crops/orchards/groves; other urban/industrial areas)
Land-use Agriculture, Fisheries/aquaculture, Tourism/recreation, Urban/industrial/transport, Water management

Birds

Species	Season	Year	Pop min	Pop max	Acc	Criteria
[1] ***Phoenicopterus ruber*** Greater Flamingo	W	1997	—	1,120	A	A4i, B1i, C2
[2] ***Tadorna ferruginea*** Ruddy Shelduck	B	1995	—	50	—	B2, C6
Recurvirostra avosetta Avocet	B	1993	34	39	A	C6
Lanius nubicus Masked Shrike	B	1995	Frequent		—	B2

1. Large increase.
2. Individuals.

This is an important site for migratory and wintering waterbirds. Species of global conservation concern that do not meet IBA criteria: *Phalacrocorax pygmeus* (maximum 15 wintering individuals), *Aythya nyroca* (passage), *Aquila clanga* (wintering), *Falco naumanni* (breeding), *Gallinago media* (passage), *Larus audouinii*.

Protection status

National Partial **International** Partial
Part of IBA covered by Wildlife Refuge (Prophitis Ilias/Polychnitou, 430 ha). 5,950 ha of IBA covered by Special Protection Area (Paraktioi Ygrotopoi Kolpou Kallonis, 6,461 ha).

Conservation issues

Threats Agricultural intensification/expansion (B), Aquaculture/fisheries (U), Burning of vegetation (C), Construction/impact of dyke/dam/barrage (C), Filling-in of wetlands (B), Industrialization/urbanization (A), Recreation/tourism (B), Unsustainable exploitation (A)

Threats to the site include hunting, water pollution from industry and agriculture, filling-in of wetlands, the creation of embankments, overgrazing, fires, the expansion of cultivation, and new tourism development. Part of the area is a candidate SAC.

Gera Gulf–Dipi and Haramida marshes, Lesvos — A4i, B1i, B2, C2 — 138

Admin region North Aegean
Coordinates 39°05′N 26°25′E
Altitude 0–30 m **Area** 5,000 ha

Site description

A shallow gulf which includes a river delta, coastal lagoons, freshwater marshes, rivers, streams, wet meadows, and reedbeds (*Phragmites*). Human activities include fishing, agriculture, water management, tourism, quarrying, and hunting.

Habitats Wetland (60%; tidal river/enclosed tidal water; mudflat/sandflat; standing fresh water; river/stream; water-fringe vegetation; fen/transition mire/spring), Marine areas (35%; sea inlet/coastal features), Artificial landscape (5%; arable land; perennial crops/orchards/groves)
Land-use Agriculture (40%), Fisheries/aquaculture (5%), Not utilized (30%), Urban/industrial/transport (10%), Water management (10%)

Birds

Species	Season	Year	Pop min	Pop max	Acc	Criteria
Larus melanocephalus Mediterranean Gull	P	1997	5,000	—	—	A4i, B1i, C2
Lanius nubicus Masked Shrike	B	1996	Frequent		—	B2

The site is important for passage migrants and as a foraging area for raptors. Species of global conservation concern that do not meet IBA criteria: *Phalacrocorax pygmeus* (up to 10 wintering individuals).

Protection status

National Low **International** None

Part of IBA covered by Wildlife Refuge (Divolo–Akothi/Loutron Lesvou, 300 ha).

Conservation issues

Threats Agricultural intensification/expansion (B), Aquaculture/fisheries (C), Disturbance to birds (A), Drainage (B), Extraction industry (B), Filling-in of wetlands (B), Groundwater abstraction (B), Industrialization/urbanization (A), Recreation/tourism (A), Unsustainable exploitation (B)

The site is seriously threatened. There is intense pressure from housing, tourist and industrial developments. Wetland areas are subject to infilling and rubbish dumping. Water quality is declining as a result of agricultural intensification and sewerage, and salinity is increasing as a result of groundwater abstraction. Quarrying produces large amounts of solid waste. The fishery is being over-exploited. Part of the area is a candidate SAC.

Mount Olympos, Lesvos — A3, B3, C2, C6 — 139

Admin region North Aegean
Coordinates 39°05′N 26°20′E
Altitude 50–968 m **Area** 15,000 ha

Site description

The top of the mountain is unvegetated. Below this is a band of mixed *Pinus* and *Castanea* woodland, with olive-groves in the lowland area. Human activities include livestock-farming, forestry and quarrying.

Habitats Forest and woodland (80%; broadleaved deciduous forest; native coniferous forest; mixed forest), Rocky areas (5%; inland cliff), Artificial landscape (20%; perennial crops/orchards/groves; forestry plantation)
Land-use Agriculture (40%), Forestry (10%), Not utilized (35%), Tourism/recreation (5%), Urban/industrial/transport (10%)

Birds

Species	Season	Year	Pop min	Pop max	Acc	Criteria
Dendrocopos medius Middle Spotted Woodpecker	R	1996	Frequent		—	C6
Oenanthe hispanica Black-eared Wheatear	B	1996	Frequent		—	A3
Sylvia cantillans Subalpine Warbler	B	1996	Common		—	A3
Sylvia melanocephala Sardinian Warbler	R	1996	Common		—	A3
Sitta krueperi Krüper's Nuthatch	R	1996	30	50	A	A3, B3, C2, C6
Sitta neumayer Rock Nuthatch	R	1996	Uncommon		—	A3
Emberiza melanocephala Black-headed Bunting	B	1996	Frequent		—	A3

This is an important site for species typical of *Pinus* forest and non-intensive olive groves. Breeding birds include six of the 21 species in Europe that are restricted (when breeding) to the Mediterranean biome.

Protection status

National Partial **International** None

Part of IBA covered by Wildlife Refuge (Koutra–Chorafi Agrias–Chorafelia/Agiasou Lesv, 700 ha). Part of IBA covered by Wildlife Refuge (Plati/Asomaton–Sikounta–Mychou).

Conservation issues

Threats Burning of vegetation (A), Disturbance to birds (C), Firewood collection (B), Selective logging/cutting (B), Unsustainable exploitation (C)

Threats include fires, forestry operations, overgrazing and road building. Part of the area is a candidate SAC.

Psara and Antipsara islands — A4ii, B1i, B1ii, B1iii, B2, B3, C2, C3, C6 — 140

Admin region North Aegean
Coordinates 38°34′N 25°35′E
Altitude 0–564 m **Area** 3,900 ha

Site description

A group of islands covered with maquis and phrygana and few trees. The main activities include fishing and stock-raising and some tourism.

Habitats Scrub (90%; sclerophyllous scrub/garrigue/maquis), Rocky areas (10%; sea cliff/rocky shore; rock stacks/islets)
Land-use Agriculture (90%), Fisheries/aquaculture (10%)

Birds

Species	Season	Year	Pop min	Pop max	Acc	Criteria
Calonectris diomedea Cory's Shearwater	B	1993	100	200	C	C6
Puffinus yelkouan Yelkouan Shearwater	B	1993	100	200	C	B1ii, B3, C3
Phalacrocorax aristotelis Shag	R	1993	100	180	B	B1i, C2
Falco eleonorae Eleonora's Falcon	B	1993	40	70	C	A4ii, B1iii, B2, C2

The site is important for breeding seabirds, and breeding and passage raptors. Species of global conservation concern that do not meet IBA criteria: *Falco naumanni* (passage).

Protection status

National Partial **International** None

1,230 ha of IBA covered by Wildlife Refuge (Xirokampos–Ftelia/Psaron, 1,230 ha).

Conservation issues

Threats Aquaculture/fisheries (B), Disturbance to birds (B), Unsustainable exploitation (A), Consequences of animal/plant introductions (C)

The main threat is from intensive illegal hunting. Overfishing, disturbance and introduced cats and rats are also problems.

North Chios — B2, C6 — 141

Admin region North Aegean
Coordinates 38°33′N 26°00′E
Altitude 0–1,297 m **Area** 38,500 ha

Site description

The site includes Mount Pelineo. It is dominated by scrub and contains four small wetlands.

Habitats Forest and woodland (15%; native coniferous forest), Scrub (90%; sclerophyllous scrub/garrigue/maquis), Artificial landscape (5%; arable land; other urban/industrial areas)
Land-use Agriculture (80%), Not utilized (5%), Tourism/recreation (5%), Urban/industrial/transport (5%)

Birds

Species	Season	Year	Pop min	Pop max	Acc	Criteria
Emberiza cineracea Cinereous Bunting	B	1991	4	4	A	B2, C6

This is an important area for species typical of Mediterranean scrub.

Protection status

National None **International** None

Conservation issues

Threats Agricultural intensification/expansion (B), Infrastructure (A), Recreation/tourism (C), Unsustainable exploitation (A)

The trapping of passerines during autumn migration, particularly with lime, is a serious threat. Part of the area is a candidate SAC.

Venetiko islet — B2 — 142

Admin region North Aegean
Coordinates 38°08′N 26°01′E
Altitude 0–50 m **Area** 9 ha

Site description

A rocky islet south of Chios.

Habitats Rocky areas (100%; rock stacks/islets)
Land-use Unknown

Birds

Species	Season	Year	Pop min	Pop max	Acc	Criteria
Falco eleonorae Eleonora's Falcon	B	1992	25	35	—	B2

This is an important site for species of coastal cliffs.

Protection status

National None **International** None

Conservation issues

Threats Disturbance to birds (B), Unsustainable exploitation (A)

The main threat is hunting and nest destruction of *Falco eleonorae*. Visitors also cause some disturbance.

Ikaria island — A1, B2, C1, C6 — 143

Admin region North Aegean
Coordinates 37°36′N 26°10′E
Altitude 0–1,037 m **Area** 26,700 ha

Site description

A mountainous island forested in the centre with pine *Pinus* and oak *Quercus*. The western side of the island is dominated by maquis, the east by phrygana. There are a number of small coastal wetlands and artificial wetlands. Human activities include livestock-farming, angling, hunting, and tourism.

Habitats Forest and woodland (25%; broadleaved deciduous forest; native coniferous forest), Scrub (20%; sclerophyllous scrub/garrigue/maquis), Wetland (sand-dunes/sand beach; standing brackish and salt water; river/stream), Marine areas (sea inlet/coastal features), Rocky areas (30%; sea cliff/rocky shore; inland cliff), Artificial landscape (20%; arable land)
Land-use Agriculture (50%), Fisheries/aquaculture (5%), Not utilized (10%), Tourism/recreation (20%), Urban/industrial/transport (5%), Water management (10%)

Birds

Species	Season	Year	Pop min	Pop max	Acc	Criteria
Buteo rufinus Long-legged Buzzard	R	1996	5	5	B	B2
Hieraaetus fasciatus Bonnelli's Eagle	R	1996	3	3	B	B2
Falco eleonorae Eleonora's Falcon	B	1996	Abundant		—	B2
Emberiza cineracea Cinereous Bunting	B	—	—	—	—	A1, B2, C1, C6

The site is important for breeding raptors and passage migrants. Species of global conservation concern that do not meet IBA criteria: *Haliaeetus albicilla*, *Falco naumanni* (3–5 breeding pairs).

Protection status

National None **International** None

Conservation issues

Threats Abandonment/reduction of land management (C), Agricultural intensification/expansion (A), Deforestation (commercial) (B), Firewood collection (B), Forest grazing (A), Infrastructure (B), Selective logging/cutting (B), Unsustainable exploitation (C)

The main threats are the grazing of livestock and the building of new roads. Hunting and tourism are limited. Part of the area is a candidate SAC.

Fourni islands — A1, A4ii, B1ii, B1iii, B2, B3, C1, C2, C3, C6 — 144

Admin region North Aegean
Coordinates 37°40′N 26°30′E
Altitude 0–514 m **Area** 5,000 ha

Site description

A complex of 16 islands characterized by cliffs and rocky shores. The dominant vegetation is phrygana and maquis with some olive-groves and cereal fields. Human activities include livestock-farming, fishing, hunting and recreation.

Habitats Scrub (65%; sclerophyllous scrub/garrigue/maquis), Marine areas (10%; sea inlet/coastal features), Rocky areas (20%; sea cliff/rocky shore; rock stacks/islets), Artificial landscape (5%; arable land; perennial crops/orchards/groves)
Land-use Agriculture (65%), Fisheries/aquaculture (10%), Not utilized (25%), Tourism/recreation (5%)

Birds

Species	Season	Year	Pop min	Pop max	Acc	Criteria
Calonectris diomedea Cory's Shearwater	B	—	—	—	—	B2, C6
Puffinus yelkouan Yelkouan Shearwater	B	1996	Abundant		—	A4ii, B1ii, B3, C3
Falco naumanni Lesser Kestrel	B	1996	18	20	A	A1, C1
Falco eleonorae Eleonora's Falcon	B	1997	100	—	B	A4ii, B1iii, B2, C2
Larus audouinii Audouin's Gull	R	1996	40	55	A	A1, B2, C1, C6

This is an important area for breeding seabirds and species typical of Mediterranean scrub. Species of global conservation concern that do not meet IBA criteria: *Haliaeetus albicilla*.

Protection status

National None **International** None

Conservation issues

Threats Abandonment/reduction of land management (C), Aquaculture/fisheries (U), Forest grazing (C), Recreation/tourism (A), Unsustainable exploitation (A)

Increasing tourism is the most important threat. The area is a candidate SAC.

Mount Kerkis — B3, C6 — 145

Admin region North Aegean
Coordinates 37°44′N 26°37′E
Altitude 0–1,434 m **Area** 7,000 ha

Site description

A mountain with a variety of habitats including conifer forests, subalpine grassland, scrub, cliffs and arable land.

Habitats Forest and woodland (35%; native coniferous forest), Scrub (25%; sclerophyllous scrub/garrigue/maquis), Grassland (20%; alpine/subalpine/boreal grassland), Rocky areas (10%; sea cliff/rocky shore; scree/boulders), Artificial landscape (20%; arable land)
Land-use Agriculture (70%), Forestry (25%), Not utilized (5%)

Birds

Species	Season	Year	Pop min	Pop max	Acc	Criteria
Sylvia rueppelli Rüppell's Warbler	B	1995	20	40	—	B3, C6

This is an important site for raptors and passerines typical of Mediterranean scrub and forest.

Protection status

National None **International** Partial
5,300 ha of IBA covered by Special Protection Area (Samos: Oros Kerketefs–Mikro kai Megalo Seitani–Dasos Kastanias kai Lekkas, 6,400 ha).

Conservation issues

Threats Agricultural intensification/expansion (B), Burning of vegetation (A), Infrastructure (B), Recreation/tourism (U), Unsustainable exploitation (B)

The main threats are the burning of vegetation, expansion of the road network, and hunting. Part of the area is a candidate SAC.

Aliki Psilis Ammou lagoon, Samos — B2, C6 — 146

Admin region North Aegean
Coordinates 37°42′N 27°01′E
Altitude 0–10 m **Area** 180 ha

Site description

A small coastal lagoon with a reedbed (*Phragmites*). In the summer the lagoon almost dries up.

Habitats Scrub (sclerophyllous scrub/garrigue/maquis), Wetland (saltmarsh; sand-dunes/sand beach; coastal lagoon; water-fringe vegetation)
Land-use Agriculture, Tourism/recreation

Birds

Species	Season	Year	Pop min	Pop max	Acc	Criteria
[1] ***Tadorna ferruginea*** Ruddy Shelduck	R	1996	10	12	A	B2, C6

1. Individuals.

This is an important site for wintering and passage waterbirds.

Protection status

National None **International** None

Conservation issues

Threats Disturbance to birds (B), Unsustainable exploitation (A)

The main threat is from illegal hunting. The area is a candidate SAC.

Central Andros island — B3, C6 — 147

Admin region South Aegean
Coordinates 37°50′N 24°53′E
Altitude 0–990 m **Area** 22,750 ha

Site description

An inaccessible area of gorges and sea cliffs, dominated by scrub. Human activities include livestock-farming and limited arable agriculture.

Habitats Forest and woodland (5%), Scrub (80%; sclerophyllous scrub/garrigue/maquis), Rocky areas (10%; sea cliff/rocky shore; inland cliff)
Land-use Agriculture (80%), Not utilized (10%), Tourism/recreation (5%), Urban/industrial/transport (5%)

Birds

Species	Season	Year	Pop min	Pop max	Acc	Criteria
Emberiza caesia Cretzschmar's Bunting	B	1997	Common		—	B3, C6

The site is important for raptors and species associated with cliffs and scrub.

Protection status

National Partial **International** None

650 ha of IBA covered by Wildlife Refuge (Stenies–M. Ag. Nikolaou/Androu, 650 ha). 1,700 ha of IBA covered by Wildlife Refuge (Steno Kothiou, 1,700 ha). 470 ha of IBA covered by Wildlife Refuge (Kolympar–Zagarniari/Palaiopoleas–Pitrofou–Messarias, 470 ha). 250 ha of IBA covered by Wildlife Reserve (Synetiou/Synetiou-Androu, 250 ha).

Conservation issues

Threats Recreation/tourism (C), Unsustainable exploitation (U)

The scale of the threat posed by hunting is currently unknown. Part of the area is a candidate SAC.

North-east Tinos island — B2 — 148

Admin region South Aegean
Coordinates 37°37′N 25°11′E
Altitude 0–600 m **Area** 12,000 ha

Site description

Main cover is phrygana scrub with a small area of *Quercus* woodland. Main human activities are livestock-farming and limited tourism.

Habitats Forest and woodland (5%; broadleaved deciduous forest), Scrub (80%; sclerophyllous scrub/garrigue/maquis), Rocky areas (10%; sea cliff/rocky shore)
Land-use Agriculture (90%), Not utilized (10%), Tourism/recreation (5%)

Birds

Species	Season	Year	Pop min	Pop max	Acc	Criteria
Falco eleonorae Eleonora's Falcon	B	1996	—	—	—	B2

The site is important for species associated with coastal cliffs and scrub.

Protection status

National None **International** None

Conservation issues

Threats Unsustainable exploitation (U)

Part of the area is a candidate SAC.

Northern Syros — B2 — 149

Admin region South Aegean
Coordinates 37°29′N 24°56′E
Altitude 0–442 m **Area** 4,000 ha

Site description

The northern part of Syros and the surrounding rocky islets, characterized by coastal and inland cliffs, and dominated by phrygana. Human activities include livestock-farming, fishing and hunting.

Habitats Scrub (75%; sclerophyllous scrub/garrigue/maquis), Rocky areas (20%; sea cliff/rocky shore; rock stacks/islets; inland cliff), Artificial landscape (5%; arable land)
Land-use Agriculture (80%), Fisheries/aquaculture, Not utilized

Birds

Species	Season	Year	Pop min	Pop max	Acc	Criteria
Falco eleonorae Eleonora's Falcon	B	1997	Common		—	B2

An important site for passage raptors. Species of global conservation concern that do not meet IBA criteria: *Pelecanus crispus* (passage), *Haliaeetus albicilla* (passage), *Larus audouinii* (20 wintering individuals).

Protection status

National Low **International** None

300 ha of IBA covered by Wildlife Refuge (Akro BD Pleura/Ano Syrou, 300 ha).

Conservation issues

Threats Agricultural intensification/expansion (B), Unsustainable exploitation (A)

The main threats to are from high hunting and grazing pressure. Part of the area is a candidate SAC.

Serifos island — B2 — 150

Admin region South Aegean
Coordinates 37°10′N 24°28′E
Altitude 0–502 m **Area** 7,400 ha

Site description

A typical Aegean island covered with phrygana and with a small coastal wetland. The site includes the adjacent rocky islets of Vous, Serfopoula and Piperi. The main human activities are arable agriculture, livestock-farming, fishing, tourism, and hunting.

Habitats Scrub (75%; sclerophyllous scrub/garrigue/maquis), Marine areas (5%; sea inlet/coastal features), Rocky areas (10%; sea cliff/rocky shore; rock stacks/islets; inland cliff), Artificial landscape (15%; arable land; other urban/industrial areas)
Land-use Agriculture (60%), Fisheries/aquaculture (5%), Not utilized (25%), Tourism/recreation (5%), Urban/industrial/transport (5%)

■ Birds

Species	Season	Year	Pop min	Pop max	Acc	Criteria
Falco eleonorae Eleonora's Falcon	B	1997	Common		—	B2

This is an important area for raptors. Species of global conservation concern that do not meet IBA criteria: *Larus audouinii*.

■ Protection status

National Partial **International** None
200 ha of IBA covered by Wildlife Refuge (Chora–Ag. Kiriaki/Serifos, 200 ha). 800 ha of IBA covered by Wildlife Refuge (Kentarchos–Galini/Serifos, 800 ha).

■ Conservation issues

Threats Aquaculture/fisheries (U), Infrastructure (B), Recreation/tourism (B), Unsustainable exploitation (A)

The main threats are from hunting, tourist and infrastructure developments. Part of the area is a candidate SAC.

Northern Sifnos — B2 — 151

Admin region South Aegean
Coordinates 37°00'N 24°41'E
Altitude 0–530 m **Area** 3,000 ha

■ Site description

An area of cliffs covered by maquis, phrygana, and olive-groves. Human activities include agriculture, tourism, fishing and hunting.

Habitats Scrub (55%; sclerophyllous scrub/garrigue/maquis), Wetland (5%; sand-dunes/sand beach), Marine areas (5%; sea inlet/coastal features), Rocky areas (10%; sea cliff/rocky shore; inland cliff), Artificial landscape (25%; perennial crops/orchards/groves; other urban/industrial areas)
Land-use Agriculture (80%), Fisheries/aquaculture (5%), Not utilized (5%), Tourism/recreation (5%), Urban/industrial/transport (5%)

■ Birds

Species	Season	Year	Pop min	Pop max	Acc	Criteria
Falco eleonorae Eleonora's Falcon	B	1996	5	50	C	B2

This is an important site for breeding raptors and for passage migrants.

■ Protection status

National Partial **International** None
150 ha of IBA covered by Wildlife Refuge (Profitis Ilias/Apollonia Sifnou, 150 ha).

■ Conservation issues

Threats Drainage (B), Industrialization/urbanization (B), Infrastructure (B), Recreation/tourism (B), Unsustainable exploitation (A)

The main threats are from hunting, infrastructure and tourism developments. Part of the area is a candidate SAC.

Western Milos, Antimilos and Polyegos islands — B2 — 152

Admin region South Aegean
Coordinates 36°42'N 24°22'E
Altitude 0–751 m **Area** 11,000 ha

■ Site description

The site is dominated by maquis, phrygana and limited arable land. A small lake and some saltpans are included. There are many mines and quarries.

Habitats Scrub (70%; sclerophyllous scrub/garrigue/maquis), Marine areas (sea inlet/coastal features), Rocky areas (10%; sea cliff/rocky shore; rock stacks/islets
Land-use Agriculture (90%), Urban/industrial/transport

■ Birds

Species	Season	Year	Pop min	Pop max	Acc	Criteria
Falco eleonorae Eleonora's Falcon	B	1996	Abundant		—	B2

This is an important site for migrating waterbirds and species associated with coastal cliffs and scrub. Species of global conservation concern that do not meet IBA criteria: *Aythya nyroca*, *Falco naumanni* (5 breeding pairs), *Larus audouinii*.

■ Protection status

National Low **International** None
795 ha of IBA covered by Breeding Station (Antimilos Island, 795 ha).

■ Conservation issues

Threats Disturbance to birds (C), Extraction industry (A), Infrastructure (A), Unsustainable exploitation (B)

The most serious threat is from the existence and expansion of the mines and quarries. Tourism is limited to the beaches at Achivadolimni. Part of the area is a candidate SAC.

Paros, Antiparos and adjacent islands — A1, A4i, B1i, B2, C1, C3, C6 — 153

Admin region South Aegean
Coordinates 37°02'N 25°10'E
Altitude 0–771 m **Area** 10,000 ha

■ Site description

A group of islands and islets, excluding the main settlements. Paros and Antiparos both have wetlands and arable land. The centre of Paros and the surrounding islets are covered with scrub and phrygana.

Habitats Scrub (50%; sclerophyllous scrub/garrigue/maquis), Marine areas (5%), Rocky areas (30%; sea cliff/rocky shore; rock stacks/islets; inland cliff), Artificial landscape (20%; arable land; perennial crops/orchards/groves; ruderal land)
Land-use Agriculture (80%), Fisheries/aquaculture (5%), Tourism/recreation (15%)

■ Birds

Species	Season	Year	Pop min	Pop max	Acc	Criteria
Falco eleonorae Eleonora's Falcon	B	1996	30	—	C	B2
Falco peregrinus Peregrine	R	1996	4	5	A	B2, C6
Larus audouinii Audouin's Gull	R	1997	—	22	A	A1, C1
Larus cachinnans Yellow-legged Gull	R	1996	2,000	3,000	A	A4i, B1i, C3

The site is important for migratory waterbirds and breeding raptors.

■ Protection status

National Partial **International** None
Part of IBA covered by Wildlife Refuge (Stroudoular–Ag. Pantes/Lefkon–Kostou, 600 ha).

■ Conservation issues

Threats Abandonment/reduction of land management (C), Agricultural intensification/expansion (C), Extraction industry (B), Filling-in of wetlands (B), Groundwater abstraction (B), Industrialization/urbanization (A), Recreation/tourism (A), Unsustainable exploitation (B)

Part of the area is a candidate SAC.

Mounts Dias, Mavrovouni and Koronos — B2, B3, C6 — 154

Admin region South Aegean
Coordinates 37°00'N 25°30'E
Altitude 0–1,004 m **Area** 27,000 ha

■ Site description

Mountains covered with phrygana and to a lesser degree with maquis and forest. Human activities include livestock-farming, fishing and hunting. The pastures are managed with fire.

Habitats Forest and woodland (10%; broadleaved deciduous forest; native coniferous forest; broadleaved evergreen forest), Scrub (45%; scrub; sclerophyllous scrub/garrigue/maquis), Marine areas (5%; open sea; sea inlet/coastal features), Rocky areas (15%; sea cliff/rocky shore; inland cliff), Artificial landscape (25%; perennial crops/orchards/groves; ruderal land)
Land-use Agriculture (80%), Fisheries/aquaculture (5%), Not utilized (10%), Water management (5%)

Birds

Species	Season	Year	Pop min	Pop max	Acc	Criteria
[1] ***Buteo rufinus*** Long-legged Buzzard	R	1997	5	6	B	B2
Hieraaetus fasciatus Bonnelli's Eagle	R	1996	2	3	A	B2
[1] ***Athene noctua*** Little Owl	R	1996	Common		—	B2
Emberiza caesia Cretzschmar's Bunting	B	1995	Common		—	B3, C6

1. Large increase.

This is an important site for passage and breeding raptors and species associated with scrub.

Protection status

National Partial **International** None
346 ha of IBA covered by Wildlife Refuge (Fanari/Filotiou Naxou, 346 ha). Part of IBA covered by Wildlife Refuge (Troulos/Filotiou Naxou, 1,200 ha). 480 ha of IBA covered by Wildlife Refuge (Za/Koronou Naxou, 480 ha).

Conservation issues

Threats Abandonment/reduction of land management (B), Agricultural intensification/expansion (A), Aquaculture/fisheries (C), Burning of vegetation (A), Disturbance to birds (A), Firewood collection (C), Forest grazing (C), Unsustainable exploitation (A)

The main threats are from overgrazing, fire and the use of poisoned baits. Part of the area is a candidate SAC.

Mikres Kyklades — A4ii, B1ii, B1iii, B2, B3, C3 — 155

Admin region South Aegean
Coordinates 36°54′N 25°34′E
Altitude 0–418 m **Area** 5,700 ha

Site description

A complex of small islands and islets east of Naxos. The majority are covered with maquis and phrygana. Human activities include fishing, stock-raising, tourism and hunting.

Habitats Scrub (70%; sclerophyllous scrub/garrigue/maquis), Marine areas (5%; open sea; sea inlet/coastal features), Rocky areas (15%; sea cliff/rocky shore; rock stacks/islets; inland cliff), Artificial landscape (10%; arable land; other urban/industrial areas)
Land-use Agriculture (60%), Fisheries/aquaculture (5%), Not utilized (20%), Tourism/recreation (10%)

Birds

Species	Season	Year	Pop min	Pop max	Acc	Criteria
Puffinus yelkouan Yelkouan Shearwater	B	1996	300	—	—	A4ii, B1ii, B3, C3
Falco eleonorae Eleonora's Falcon	B	1997	80	—	B	A4ii, B1iii, B2, C2

This is an important site for breeding seabirds and raptors associated with coastal cliffs. Species of global conservation concern that do not meet IBA criteria: *Larus audouinii*.

Protection status

National None **International** None

Conservation issues

Threats Aquaculture/fisheries (B), Consequences of animal/plant introductions (A), Recreation/tourism (A), Unsustainable exploitation (U)

The main threats are from tourism developments and introduced rats and cats. Part of the area is a candidate SAC.

Amorgos island — A1, B2, C1, C6 — 156

Admin region South Aegean
Coordinates 36°50′N 25°55′E
Altitude 0–821 m **Area** 12,100 ha

Site description

The easternmost island of the Cyclades. A rugged, rocky island, including Mount Krikelos. The south-east of the island has steep cliffs. Vegetation cover includes scrub, maquis, phrygana and remnant *Quercus* forest. A number of rocky islets fall within the site. Human activities include livestock-farming, tourism, fishing, and hunting.

Habitats Scrub (70%; scrub; sclerophyllous scrub/garrigue/maquis), Rocky areas (20%; sea cliff/rocky shore; rock stacks/islets; scree/boulders; inland cliff), Artificial landscape (10%; arable land)
Land-use Agriculture (70%), Fisheries/aquaculture, Not utilized (20%), Tourism/recreation (10%)

Birds

Species	Season	Year	Pop min	Pop max	Acc	Criteria
Hieraaetus fasciatus Bonnelli's Eagle	R	1996	2	2	A	B2
Falco eleonorae Eleonora's Falcon	B	1996	Abundant		—	B2
Larus audouinii Audouin's Gull	B	1997	51	51	A	A1, B2, C1, C6

The site is important for breeding and passage raptors, and species associated with cliffs and scrub.

Protection status

National None **International** None

Conservation issues

Threats Abandonment/reduction of land management (C), Agricultural intensification/expansion (B), Burning of vegetation (A), Unsustainable exploitation (B)

Threats include burning of vegetation for grazing purposes, overgrazing by goats and hunting. Part of the area is a candidate SAC.

Ios, Sikinos, Folegandros island group — A4ii, B1ii, B2, B3, C3, C6 — 157

Admin region South Aegean
Coordinates 36°40′N 25°01′E
Altitude 0–713 m **Area** 19,000 ha

Site description

The scrub-dominated islands in the South Cyclades and the adjacent rocky islets. Human activities include tourism, fishing, arable agriculture and livestock-farming.

Habitats Scrub (70%; sclerophyllous scrub/garrigue/maquis), Marine areas (5%; open sea; sea inlet/coastal features), Rocky areas (15%; sea cliff/rocky shore; rock stacks/islets; scree/boulders; inland cliff), Artificial landscape (10%; arable land; other urban/industrial areas)
Land-use Agriculture (70%), Fisheries/aquaculture (5%), Not utilized (20%), Tourism/recreation (5%), Urban/industrial/transport (5%)

Birds

Species	Season	Year	Pop min	Pop max	Acc	Criteria
Puffinus yelkouan Yelkouan Shearwater	B	1996	100	1,000	C	A4ii, B1ii, B3, C3
Buteo rufinus Long-legged Buzzard	R	1996	2	7	B	B2
Hieraaetus fasciatus Bonnelli's Eagle	R	1996	2	4	A	B2
Falco peregrinus Peregrine	R	1996	3	7	B	B2, C6
Emberiza caesia Cretzschmar's Bunting	B	1997	Frequent		—	B3, C6

The area is important for breeding seabirds and raptors. Species of global conservation concern that do not meet IBA criteria: *Larus audouinii*.

Protection status

National Partial **International** None

3,700 ha of IBA covered by Wildlife Refuge (Anemomyloi–Psathokalamos–Pr. Ilias Iou, 3,700 ha). 400 ha of IBA covered by Wildlife Refuge (Rachidia–Trypales/ Folegandros, 400 ha).

Conservation issues

Threats Aquaculture/fisheries (B), Disturbance to birds (U), Extraction industry (C), Infrastructure (A), Recreation/tourism (B), Unsustainable exploitation (A)

The main threats are tourism and infrastructure developments and hunting. Part of the area is a candidate SAC.

Christiana islets A4ii, B1iii, B2, C2 158

Admin region South Aegean
Coordinates 36°15′N 25°12′E
Altitude 0–285 m **Area** 200 ha

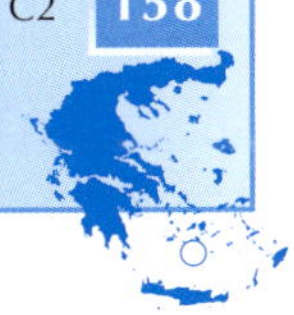

Site description

A group of rocky islets 20 km south-west of Thira island.

Habitats Scrub (sclerophyllous scrub/garrigue/maquis), Rocky areas (sea cliff/rocky shore; rock stacks/islets)
Land-use Fisheries/aquaculture, Unknown

Birds

Species	Season	Year	Pop min	Pop max	Acc	Criteria
Falco eleonorae Eleonora's Falcon	B	1997	100	—	C	A4ii, B1iii, B2, C2

This is an important site for species associated with cliffs and islands.

Protection status

National None **International** None

Conservation issues

Threats Unknown

The area is a candidate SAC.

Anafi island A4ii, B1ii, B1iii, B2, B3, C2, C3, C6 159

Admin region South Aegean
Coordinates 36°21′N 25°50′E
Altitude 0–473 m **Area** 3,900 ha

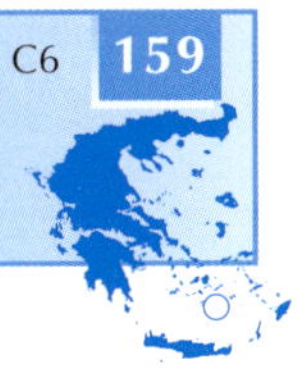

Site description

A remote island and associated islets at the south-eastern edge of Cyclades. The island is dominated by phrygana.

Habitats Scrub (60%; sclerophyllous scrub/garrigue/maquis), Marine areas (20%; open sea; sea inlet/coastal features), Rocky areas (20%; sea cliff/rocky shore; rock stacks/islets)
Land-use Agriculture (65%), Fisheries/aquaculture (20%), Not utilized (5%), Tourism/recreation (10%)

Birds

Species	Season	Year	Pop min	Pop max	Acc	Criteria
Calonectris diomedea Cory's Shearwater	B	1997	—	—	—	B1ii, B2, C2, C6
Puffinus yelkouan Yelkouan Shearwater	B	1997	—	—	—	B1ii, B3, C3
Falco eleonorae Eleonora's Falcon	B	1997	50	—	B	A4ii, B1iii, B2, C2

This is an important site for breeding seabirds. Species of global conservation concern that do not meet IBA criteria: *Larus audouinii* (7 breeding pairs).

Protection status

National None **International** None

Conservation issues

Threats Burning of vegetation (B), Consequences of animal/plant introductions (U), Recreation/tourism (A), Unsustainable exploitation (U)

The main threat is the development of tourism. Rats have been introduced. Part of the area is a candidate SAC.

Islets of North Dodekanisa A1, A4ii, B1i, B1ii, B1iii, B2, B3, C1, C2, C3, C6 160

Admin region South Aegean
Coordinates 37°20′N 26°46′E
Altitude 0–277 m **Area** 4,275 ha

Site description

A complex of 42 islets and the solitary Pharmakonisi islet. The area is generally overgrazed.

Habitats Scrub (70%; sclerophyllous scrub/garrigue/maquis), Rocky areas (20%; sea cliff/rocky shore; rock stacks/islets), Artificial landscape (25%; arable land; perennial crops/orchards/groves; urban parks/gardens)
Land-use Agriculture (85%), Fisheries/aquaculture, Military (15%), Urban/industrial/transport (10%)

Birds

Species	Season	Year	Pop min	Pop max	Acc	Criteria
Puffinus yelkouan Yelkouan Shearwater	B	1996	200	500	C	A4ii, B1ii, B3, C3
Phalacrocorax aristotelis Shag	R˙	1995	50	80	B	B1i, C2
Buteo rufinus Long-legged Buzzard	R	1995	5	7	B	B2
Falco eleonorae Eleonora's Falcon	B	1995	80	150	C	A4ii, B1iii, B2, C2
[1] *Larus audouinii* Audouin's Gull	B	1997	65	65	A	A1, B2, C1, C6

1. Large increase.

The site is important for breeding raptors and seabirds, and for passage migrants. Species of global conservation concern that do not meet IBA criteria: *Falco naumanni* (2–5 breeding pairs).

Protection status

National None **International** Low
50 ha of IBA covered by Special Protection Area (Vrachonisides Voreiou Dodekanisou, 374 ha).

Conservation issues

Threats Abandonment/reduction of land management (A), Agricultural intensification/expansion (A), Aquaculture/fisheries (B), Consequences of animal/plant introductions (A), Disturbance to birds (A), Firewood collection (C), Industrialization/urbanization (C), Recreation/tourism (B), Unsustainable exploitation (A)

The main threats are from overgrazing and hunting. The site was included in a LIFE project on the conservation of *Larus audouinii*. Part of the area is a candidate SAC.

North and south Patmos island and islets A4ii, B1iii, B2, C2 161

Admin region South Aegean
Coordinates 37°20′N 26°34′E
Altitude 0–243 m **Area** 3,200 ha

Site description

The site consists of rocky hills, overgrazed scrub, steep rocky beaches, several islets and two temporary wetlands. The major part of the area belongs to the Monastery of Patmos.

Habitats Scrub (70%; scrub; sclerophyllous scrub/garrigue/maquis), Wetland (5%), Rocky areas (30%; sea cliff/rocky shore; rock stacks/islets), Artificial landscape (15%; arable land; urban parks/gardens)
Land-use Agriculture (50%), Military (10%), Not utilized (10%), Tourism/recreation (10%), Urban/industrial/transport (10%)

Birds

Species	Season	Year	Pop min	Pop max	Acc	Criteria
Falco eleonorae Eleonora's Falcon	B	1995	50	60	C	A4ii, B1iii, B2, C2

This is an important site for breeding raptors. Species of global conservation concern that do not meet IBA criteria: *Larus audouinii* (up to 5 breeding pairs).

Protection status

National None **International** None

Conservation issues

Threats Abandonment/reduction of land management (B), Agricultural intensification/expansion (A), Consequences of animal/plant introductions (A), Extraction industry (C), Filling-in of wetlands (A), Recreation/tourism (A), Unsustainable exploitation (B)

The main threats to the site are from tourism developments and disturbance from tourism. The wetlands are threatened by infilling. Introduced rats and cats are a threat. A LIFE project on 'The conservation of *Larus audouinii*' has been carried out.

North and east Kalimnos, Telendos, and Kalolimnos islands and islets

A1, A4ii, B1iii, B2, C1, C2, C6 — **162**

Admin region South Aegean
Coordinates 37°02′N 27°00′E
Altitude 0–590 m **Area** 4,700 ha

Site description

This site includes the north and east part of Kalimnos and islets from Telendos to Kalolimnos and Limnia. Kalimnos is heavily grazed and rocky.

Habitats Scrub (40%; sclerophyllous scrub/garrigue/maquis), Grassland (steppe/dry calcareous grassland), Wetland (5%; shingle/stony beach), Marine areas (open sea; sea inlet/coastal features), Rocky areas (40%; sea cliff/rocky shore; rock stacks/islets; scree/boulders; inland cliff; caves), Artificial landscape (10%; arable land; urban parks/gardens)
Land-use Agriculture (60%), Military (15%), Not utilized (10%), Tourism/recreation (10%), Urban/industrial/transport (5%)

Birds

Species	Season	Year	Pop min	Pop max	Acc	Criteria
Buteo rufinus Long-legged Buzzard	R	1995	3	6	B	B2
Falco naumanni Lesser Kestrel	B	1995	20	25	C	A1, B2, C1, C6
Falco eleonorae Eleonora's Falcon	B	1995	30	100	C	A4ii, B1iii, B2, C2
Falco peregrinus Peregrine	R	1995	3	6	B	B2, C6
Larus audouinii Audouin's Gull	B	1996	36	36	A	A1, C1

An important site for breeding raptors and species associated with cliffs and scrub. Species of global conservation concern that do not meet IBA criteria: *Haliaeetus albicilla* (non-breeding).

Protection status

National None **International** Low
175 ha of IBA covered by Special Protection Area (Vrachonisides Voreiou Dodekanisou, 374 ha).

Conservation issues

Threats Abandonment/reduction of land management (B), Agricultural intensification/expansion (A), Consequences of animal/plant introductions (B), Disturbance to birds (A), Extraction industry (U), Industrialization/urbanization (U), Infrastructure (U), Recreation/tourism (A), Unsustainable exploitation (A)

The main threats are hunting, overgrazing and tourist developments. A management plan was produced as a result of a LIFE project. Part of the area is a candidate SAC.

Kinaros and Levitha islands and rocky islets

A1, A4ii, B1iii, B2, C1, C2, C6 — **163**

Admin region South Aegean
Coordinates 37°00′N 26°23′E
Altitude 0–296 m **Area** 1,725 ha

Site description

The islands lie to the north-east of the island of Amorgos. Kinaros is characterized by coastal cliffs. Vegetation is generally overgrazed scrub. Only Levitha is inhabited.

Habitats Scrub (60%; sclerophyllous scrub/garrigue/maquis), Rocky areas (55%; sea cliff/rocky shore; rock stacks/islets; scree/boulders), Artificial landscape (5%; arable land)
Land-use Agriculture (75%), Fisheries/aquaculture, Not utilized (20%)

Birds

Species	Season	Year	Pop min	Pop max	Acc	Criteria
Falco eleonorae Eleonora's Falcon	B	1995	150	300	—	A4ii,B1iii,B2,C2,C6
Larus audouinii Audouin's Gull	B	1995	32	38	A	A1, C1

This is an important area for seabirds and raptors.

Protection status

National None **International** Low
75 ha of IBA covered by Special Protection Area (Vrachonisides Voreiou Dodekanisou: Liadia and Glaros Lerou, Pitta, Agathonisi and Fragkos Leipson, Piganousa Lerou, Telendos and Epano Kalymnou, 374 ha).

Conservation issues

Threats Abandonment/reduction of land management (B), Agricultural intensification/expansion (A), Consequences of animal/plant introductions (U), Disturbance to birds (A), Unsustainable exploitation (U)

Overgrazing is the main threat to this site. Disturbance by fishermen caused the abandonment of the *Larus audouinii* in 1997. Rats and cats have been introduced. A management plan was produced in 1997–1999 as part of a LIFE project. The IBA is situated within a candidate SAC.

Astypalea island and nearby islets

A1, A4ii, B1ii, B1iii, B2, B3, C1, C2, C3, C6 — **164**

Admin region South Aegean
Coordinates 36°36′N 26°25′E
Altitude 0–250 m **Area** 11,500 ha

Site description

An island of sea cliffs and rocky shores, dominated by phrygana and other scrub.

Habitats Scrub (60%; sclerophyllous scrub/garrigue/maquis), Grassland (20%; steppe/dry calcareous grassland), Marine areas (5%; sea inlet/coastal features), Rocky areas (10%; sea cliff/rocky shore; rock stacks/islets), Artificial landscape (15%; arable land; perennial crops/orchards/groves; other urban/industrial areas; ruderal land)
Land-use Agriculture (80%), Fisheries/aquaculture (5%), Tourism/recreation (10%), Urban/industrial/transport (5%)

Birds

Species	Season	Year	Pop min	Pop max	Acc	Criteria
Puffinus yelkouan Yelkouan Shearwater	B	1997	50	—	A	B1ii, B3, C3
Falco eleonorae Eleonora's Falcon	B	1997	200	—	A	A4ii,B1iii,B2,C2,C6
Larus audouinii Audouin's Gull	B	1996	18	23	A	A1, C1

The site is important for breeding seabirds.

Protection status

National None **International** Low
380 ha of IBA covered by Special Protection Area (Vrachonisides Astypalaias (Kounoupoi, Fteno, Katsagreli) kai tria nisia (Mesonisi, Plakidi, Stefania), 2,144 ha)

Conservation issues

Threats Abandonment/reduction of land management (A), Agricultural intensification/expansion (A), Aquaculture/fisheries (A), Burning of vegetation (A), Filling-in of wetlands (C), Industrialization/urbanization (B), Recreation/tourism (A), Unsustainable exploitation (A)

Threats include abandonment of agriculture, overgrazing, burning of vegetation, intense shooting, overfishing, urbanization, and potential disturbance to *Larus audouinii* nests by tourists. Road construction in remote areas is a threat to raptors. Part of the area is a candidate SAC.

Sirna island and nearby islets — A4ii, B1ii, B1iii, B2, B3, C2, C3, C6 — 165

Admin region South Aegean
Coordinates 36°13′N 26°40′E
Altitude 0–322 m **Area** 1,150 ha

Site description

The site lies to the south-east of the island of Astypalea. The islets are dominated by *Juniperus* scrub and phrygana. Human activities include stock-raising, fishing, and arable agriculture.

Habitats Scrub (70%; sclerophyllous scrub/garrigue/maquis), Grassland (8%; steppe/dry calcareous grassland), Marine areas (5%; open sea; sea inlet/coastal features), Rocky areas (5%; sea cliff/rocky shore; rock stacks/islets), Artificial landscape (10%; arable land)
Land-use Agriculture (90%), Fisheries/aquaculture (10%)

Birds

Species	Season	Year	Pop min	Pop max	Acc	Criteria
Calonectris diomedea Cory's Shearwater	B	1997	100	—	C	B2, C6
Puffinus yelkouan Yelkouan Shearwater	B	1997	50	—	B	B1ii, B3, C3
Falco eleonorae Eleonora's Falcon	B	1997	95	100	B	A4ii, B1iii, B2, C2

The area is important for passage migrants and breeding seabirds and raptors. Species of global conservation concern that do not meet IBA criteria: *Larus audouinii* (11 breeding pairs).

Protection status

National None **International** Partial
200 ha of IBA covered by Special Protection Area (Vrachonisides Astypalaias kai tria nisia (Mesonisis, Plakidi, Stefania), 2,144 ha).

Conservation issues

Threats Abandonment/reduction of land management (A), Agricultural intensification/expansion (A), Aquaculture/fisheries (B), Consequences of animal/plant introductions (A), Unsustainable exploitation (A)

Threats include overgrazing, introduced rats and rabbits, abandonment of agriculture, illegal fishing, and hunting. Part of the area is a candidate SAC.

Mount Dikios, Cape Louros, Lake Psalidi, and Alyki — A1, B2, C1, C6 — 166

Admin region South Aegean
Coordinates 36°52′N 27°16′E
Altitude 0–843 m **Area** 9,916 ha

Site description

A mountain dominated by maquis and *Pinus* and *Cupressus* forest. The IBA also includes two coastal wetlands with reedbeds (*Phragmites*), a lagoon and saltmarsh.

Habitats Forest and woodland (32%; native coniferous forest), Scrub (50%; sclerophyllous scrub/garrigue/maquis), Wetland (5%; coastal lagoon; saltmarsh; standing brackish and salt water; water-fringe vegetation), Artificial landscape (9%; arable land; other urban/industrial areas)
Land-use Agriculture (60%), Forestry (20%), Military, Tourism/recreation (5%), Urban/industrial/transport (5%)

Birds

Species	Season	Year	Pop min	Pop max	Acc	Criteria
[1] ***Tadorna ferruginea*** Ruddy Shelduck	B	1996	—	35	A	B2, C6
Buteo rufinus Long-legged Buzzard	R	1996	4	7	B	B2
Falco naumanni Lesser Kestrel	B	1996	10	15	B	A1, C1
Numenius tenuirostris Slender-billed Curlew	P	1988	2	—	A	A1, C1

1. Individuals.

This is an important site for breeding and passage raptors. Species of global conservation concern that do not meet IBA criteria: *Branta ruficollis* (passage), *Larus audouinii*.

Protection status

National Partial **International** None
1,901 ha of IBA covered by Wildlife Refuge (Ag. Fokas–Psalidi/Kos, 1,901 ha). Part of IBA covered by Wildlife Refuge (Agginaropi–Galomandra–Pythos–Alyki–Vai/Pyliou–Asfe, 437 ha). 1,369 ha of IBA covered by Wildlife Refuge (Lagoudi–Konario/Asfendiou–Pyliou–Ko, 1,369 ha).

Conservation issues

Threats Agricultural intensification/expansion (C), Drainage (U), Industrialization/urbanization (A), Infrastructure (A), Intensified forest management (U), Recreation/tourism (B), Unsustainable exploitation (B)

The main threats include urban development and expansion, road construction, tourism development and hunting. The area is a candidate SAC.

Nisiros island and nearby islets — A4ii, B1iii, B2, B3, C2, C6 — 167

Admin region South Aegean
Coordinates 36°35′N 27°10′E
Altitude 0–698 m **Area** 5,325 ha

Site description

A volcanic island south of the island of Kos. The island is covered by a mixture of phrygana, maquis and areas of *Quercus* and *Pinus* woodland. Half of the islet of Giali is *Pinus* forest the other half a pumice-stone mine. Human activities include livestock-farming, tourism, and hunting.

Habitats Forest and woodland (5%; broadleaved deciduous forest; native coniferous forest; broadleaved evergreen forest), Scrub (70%; sclerophyllous scrub/garrigue/maquis), Marine areas (5%; open sea; sea inlet/coastal features), Rocky areas (15%; sea cliff/rocky shore; scree/boulders), Artificial landscape (15%; arable land; other urban/industrial areas)
Land-use Agriculture (80%), Fisheries/aquaculture, Not utilized (10%), Urban/industrial/transport (10%)

Birds

Species	Season	Year	Pop min	Pop max	Acc	Criteria
Hieraaetus fasciatus Bonnelli's Eagle	R	1996	2	—	A	B2
Falco eleonorae Eleonora's Falcon	B	1997	60	—	A	A4ii, B1iii, B2, C2
[1] ***Alectoris chukar*** Chukar	R	1996	Abundant		—	B2
Sylvia rueppelli Rüppell's Warbler	B	1996	Frequent		—	B3
Emberiza caesia Cretzschmar's Bunting	B	1996	Common		—	B3, C6

1. Large decrease.

The area is important for migratory species.

Protection status

National None **International** None

Conservation issues

Threats Abandonment/reduction of land management (C), Extraction industry (B), Firewood collection (B), Forest grazing (B), Recreation/tourism (C), Unsustainable exploitation (B)

The main threats to the site are from forest destruction for firewood, pumice-stone mining on Giali, overgrazing, fires set by shepherds, increasing tourism and seasonal hunting. Part of the area is a candidate SAC.

Tilos island — A1, A4ii, B1iii, B2, C1, C2 — 168

Admin region South Aegean
Coordinates 36°25′N 27°23′E
Altitude 0–654 m **Area** 6,300 ha

Site description

A group of rocky islands dominated by phrygana. Human activities include agriculture, livestock-farming, and fishing.

Habitats Scrub (70%; sclerophyllous scrub/garrigue/maquis), Grassland (5%; dry siliceous grassland), Rocky areas (10%; sea cliff/rocky shore), Artificial landscape (10%; perennial crops/orchards/groves)
Land-use Agriculture (90%), Tourism/recreation (5%)

Birds

Species	Season	Year	Pop min	Pop max	Acc	Criteria
Falco biarmicus Lanner	R	1989	—	—	C	C6
Monticola solitarius Blue Rock Thrush	R	1989	Frequent	—		B2

The area is important for passage migrants and breeding raptors. Species of global conservation concern that do not meet IBA criteria: *Haliaeetus albicilla* (wintering).

Protection status

National Partial **International** None
2,300 ha of IBA covered by Wildlife Refuge (Riaka–Piaki–Ormos Menies/Afraton–Rodopou, 2,300 ha).

Conservation issues

Threats Agricultural intensification/expansion (C), Recreation/tourism (C), Unsustainable exploitation (A)

The main threat is from hunting. Part of the area is a candidate SAC.

Agioi Theodori islet — A4ii, B1iii, B2, C2 — 177

Admin region Crete
Coordinates 35°33′N 23°58′E
Altitude 0–156 m **Area** 80 ha

Site description

A rocky island covered with phrygana.

Habitats Scrub (90%; sclerophyllous scrub/garrigue/maquis), Marine areas (5%; sea inlet/coastal features), Rocky areas (5%; sea cliff/rocky shore)
Land-use Fisheries/aquaculture (5%), Nature conservation/research (95%)

Birds

Species	Season	Year	Pop min	Pop max	Acc	Criteria
Falco eleonorae Eleonora's Falcon	B	1982	55	65	A	A4ii, B1iii, B2, C2

An important site for species associated with coastal cliffs and scrub.

Protection status

National Partial **International** None
80 ha of IBA covered by Breeding Station (Island of Thodorou, 80 ha).

Conservation issues

Threats Consequences of animal/plant introductions (A), Disturbance to birds (B)

Threats include grazing by introduced goats and brown hare *Lepus europaeus*, the impacts of rats *Rattus norvegicus* on nesting birds, and disturbance.

Mount Koutroulis, Mount Ag. Dikaios and Modia plateau — B2, C6 — 178

Admin region Crete
Coordinates 35°23′N 23°35′E
Altitude 200–1,100 m **Area** 20,000 ha

Site description

This site is dominated by maquis and phrygana. Fires have created large barren areas. The main human activity is livestock-farming.

Habitats Scrub (40%; sclerophyllous scrub/garrigue/maquis), Rocky areas (80%; rock stacks/islets; scree/boulders; inland cliff)
Land-use Agriculture

Birds

The site is important for species characteristic of Mediterranean scrub.

Species	Season	Year	Pop min	Pop max	Acc	Criteria
Gypaetus barbatus Lammergeier	R	1996	1	1	A	B2, C6

Protection status

National None **International** Partial
2,210 ha of IBA covered by Special Protection Area (Meterizia–Agios Dikaios–Tsounara–Vitsilia Lefkon Oren, 2,210 ha).

Conservation issues

Threats Agricultural intensification/expansion (A), Burning of vegetation (A), Infrastructure (A), Unsustainable exploitation (B)

The main threats are from fire and infrastructural developments. Part of the area is a candidate SAC.

Lefka mountains — B2, C6 — 179

Admin region Crete
Coordinates 35°18′N 24°00′E
Altitude 600–2,450 m **Area** 54,000 ha

Site description

A rugged mountainous area of Crete. The northern area of the site is dominated by maquis, the south by conifer forests. The main human activities are hunting, livestock-farming and tourism.

Habitats Forest and woodland (15%; native coniferous forest), Scrub (20%; sclerophyllous scrub/garrigue/maquis), Grassland (20%; alpine/subalpine/boreal grassland), Rocky areas (60%; sea cliff/rocky shore; scree/boulders; inland cliff)
Land-use Agriculture (70%), Tourism/recreation (30%), Water management (5%)

Birds

Species	Season	Year	Pop min	Pop max	Acc	Criteria
Gypaetus barbatus Lammergeier	R	1996	—	2	A	B2
Aquila chrysaetos Golden Eagle	R	1996	—	4	A	B2, C6
Falco peregrinus Peregrine	R	1996	Frequent	—		C6
Alectoris chukar Chukar	R	1996	Abundant	—		B2
Pyrrhocorax pyrrhocorax Chough	R	1996	Common	—		B2, C6

This is an important site for breeding and passage raptors, and for species characteristic of scrub and montane habitats.

Protection status

National Partial **International** Partial
4,850 ha of IBA covered by National Park (Samaria, 4,850 ha). Part of IBA covered by Wildlife Refuge (Stylou–Katochoriou, 1,050 ha). 4,850 ha of IBA covered by Biogenetic Reserve (Council of Europe) (Ethnikos Drymos Samarias (Core area), 4,850 ha). 4,840 ha of IBA covered by Biosphere Reserve (Gorge of Samaria National Park, 4,840 ha). 5,068 ha of IBA covered by Special Protection Area (Ethnikos Drymos Samarias Lefka Ori, 5,068 ha).

Conservation issues

Threats Forest grazing (A), Infrastructure (A), Recreation/tourism (A)

The construction of new roads is a major threat. Part of the area is a candidate SAC.

Lake Kourna, Almyrou delta, and Georgioupolis beach — C7 — 180

Admin region Crete
Coordinates 35°21′N 24°16′E
Altitude 0–527 m **Area** 4,430 ha

Site description

Habitats Forest and woodland (5%; alluvial/very wet forest), Scrub (30%; sclerophyllous scrub/garrigue/maquis), Grassland (5%; humid grassland), Wetland (35%; tidal river/enclosed tidal water; saltmarsh; standing fresh water; river/stream; fen/transition mire/spring), Rocky areas (10%; sea cliff/rocky shore; inland cliff), Artificial landscape (20%; arable land; other urban/industrial areas)

Land-use Agriculture (60%), Tourism/recreation (10%), Urban/industrial/transport (10%)

The site contains one of the few freshwater lakes in Crete, as well as coastal marshes, a brackish lake, sand-dunes, cliffs and a marine zone.

■ Birds

This is an important site for breeding, migrant and wintering waterbirds and passerines. Species of global conservation concern that do not meet IBA criteria: *Phalacrocorax pygmeus* (wintering), *Aythya nyroca* (wintering), *Aquila clanga* (passage), *Gallinago media* (passage, wintering), *Larus audouinii* (wintering).

■ Protection status

National None **International** None

■ Conservation issues

Threats Agricultural intensification/expansion (B), Filling-in of wetlands (A), Groundwater abstraction (A), Recreation/tourism (A), Unsustainable exploitation (A)

The main threats are from the building of hotels, disturbance from tourists, groundwater abstraction and hunting. There has been a LIFE project investigating protection and management of the site. The site is a candidate SAC.

Gavdos and Gavdopoula islands

B1i, B2, C2 **181**

Admin region Crete
Coordinates 34°51′N 24°05′E
Altitude 0–386 m **Area** 3,600 ha

■ Site description

Two islands south of Crete. Gavdos is covered with maquis, phrygana, and *Pinus* and *Juniperus* forest. Gavdopoula is covered with phrygana.

Habitats Forest and woodland (30%; native coniferous forest), Scrub (50%; sclerophyllous scrub/garrigue/maquis), Marine areas (10%; open sea), Rocky areas (10%; sea cliff/rocky shore), Artificial landscape (10%; perennial crops/orchards/groves)
Land-use Agriculture (10%), Fisheries/aquaculture (10%), Tourism/recreation (10%), Not utilized

■ Birds

Species	Season	Year	Pop min	Pop max	Acc	Criteria
[1] ***Phalacrocorax aristotelis*** Shag	R	1996	150	—	B	B1i, C2
Otus scops Scops Owl	B	1996	Abundant	—		B2

1. Individuals.

This is an important area for passage migrants.

■ Protection status

National None **International** None

■ Conservation issues

Threats Abandonment/reduction of land management (A), Afforestation (B), Burning of vegetation (B), Industrialization/urbanization (A), Infrastructure (A), Unsustainable exploitation (U)

The main threats include uncontrolled tourism development and plans to construct an industrial unit on Gavdopoula islet. The area is a candidate SAC.

Krioneritis mountain

B2, B3, C6 **182**

Admin region Crete
Coordinates 35°14′N 24°21′E
Altitude 300–1,300 m **Area** 15,000 ha

■ Site description

A rugged mountain covered with phrygana and rocky sites. The main human activity is livestock-farming.

Habitats Scrub (70%; sclerophyllous scrub/garrigue/maquis), Rocky areas (40%; scree/boulders; inland cliff)
Land-use Agriculture (70%), Not utilized (30%)

■ Birds

Species	Season	Year	Pop min	Pop max	Acc	Criteria
Gypaetus barbatus Lammergeier	R	1996	1	1	A	B2
Sylvia rueppelli Rüppell's Warbler	B	1996	Frequent	—		B3, C6

This is an important site for raptors and species occurring in scrub.

■ Protection status

National None **International** None

■ Conservation issues

Threats Agricultural intensification/expansion (A), Burning of vegetation (B), Recreation/tourism (C)

The main threat is from the high grazing pressure and anthropogenic fires. Part of the area is a candidate SAC.

Mount Kedros and Kourtaliotiko gorge

B2 **183**

Admin region Crete
Coordinates 35°12′N 24°37′E
Altitude 100–1,776 m **Area** 12,000 ha

■ Site description

A steep mountain and gorge dominated by phrygana. The main human activity is livestock-farming.

Habitats Scrub (50%; scrub; sclerophyllous scrub/garrigue/maquis), Rocky areas (60%; scree/boulders; inland cliff)
Land-use Agriculture (60%), Not utilized (40%)

■ Birds

Species	Season	Year	Pop min	Pop max	Acc	Criteria
Gypaetus barbatus Lammergeier	R	1996	Frequent	—		B2

An important site for breeding raptors, montane and scrub species.

■ Protection status

National None **International** Partial

6,021 ha of IBA covered by Special Protection Area (Zoros–Agathi–Kedros, 6,021 ha).

■ Conservation issues

Threats Agricultural intensification/expansion (A), Firewood collection (B), Industrialization/urbanization (B)

Part of the area is a candidate SAC.

Prassiano gorge

B2, C6 **184**

Admin region Crete
Coordinates 35°20′N 24°32′E
Altitude 100–300 m **Area** 975 ha

■ Site description

A steep rocky gorge with slopes covered with phrygana. The main human activity is stock-raising.

Habitats Scrub (30%; sclerophyllous scrub/garrigue/maquis), Rocky areas (90%; inland cliff)
Land-use Agriculture (20%), Not utilized (80%), Tourism/recreation (10%)

■ Birds

Species	Season	Year	Pop min	Pop max	Acc	Criteria
[1] ***Gyps fulvus*** Griffon Vulture	R	1996	40	45	A	B2, C6

1. Individuals.

This site is important for breeding raptors.

■ Protection status

National Partial **International** None

Part of IBA covered by Wildlife Refuge (Vrysinas/Armenon–Kare–Seliou–Prasson, 1,700 ha).

Conservation issues

Threats Agricultural intensification/expansion (C)

The area is a candidate SAC.

Mount Idi — B2, B3, C6 — 185

Admin region Crete
Coordinates 35°15′N 24°45′E
Altitude 400–2,457 m **Area** 44,000 ha

Site description

A mountain dominated by heavily grazed phrygana and sclerophyllous scrub. The main human activities are livestock-farming, tourism and recreation and hunting.

Habitats Scrub (60%; sclerophyllous scrub/garrigue/maquis), Rocky areas (60%; scree/boulders; inland cliff)
Land-use Agriculture (70%), Tourism/recreation (30%)

Birds

Species	Season	Year	Pop min	Pop max	Acc	Criteria
Gypaetus barbatus Lammergeier	R	1996	2	2	A	B2
[1] ***Gyps fulvus*** Griffon Vulture	R	1996	30	40	A	B2, C6
Aquila chrysaetos Golden Eagle	R	1996	3	3	A	C6
Alectoris chukar Chukar	R	1996	Common		—	B2
Sylvia rueppelli Rüppell's Warbler	B	1996	Frequent		—	B3
Pyrrhocorax pyrrhocorax Chough	R	1996	Frequent		—	B2, C6

1. Individuals.

This is an important site for breeding and passage raptors.

Protection status

National Low **International** Partial
1,500 ha of IBA covered by Wildlife Refuge (Nida/Anogion, 1,500 ha). 1 ha of IBA covered by Natural Monument (Endangered Cretan Cephalantera, 1 ha). 7,876 ha of IBA covered by Special Protection Area (Krousonas–Vromonero Idis, 7,876 ha)

Conservation issues

Threats Agricultural intensification/expansion (A)

The main threat is from overgrazing which is preventing scrub regeneration. Part of the area is a candidate SAC.

Geropotamos estuary — A4i, B1i, C2 — 186

Admin region Crete
Coordinates 35°03′N 24°46′E
Altitude 0–5 m **Area** 450 ha

Site description

A river delta on Crete with a mix of grassland, sand-dunes and scrub.

Habitats Grassland (40%; steppe/dry calcareous grassland; mesophile grassland), Wetland (45%; tidal river/enclosed tidal water; saltmarsh; sand-dunes/sand beach; river/stream; water-fringe vegetation), Marine areas (5%; sea inlet/coastal features), Artificial landscape (10%; other urban/industrial areas)
Land-use Agriculture (70%), Urban/industrial/transport (10%)

Birds

Species	Season	Year	Pop min	Pop max	Acc	Criteria
Plegadis falcinellus Glossy Ibis	P	1994	300	—	—	A4i, B1i, C2

The site is important for passage waterbirds.

Protection status

National None **International** None

Conservation issues

Threats Aquaculture/fisheries (U), Industrialization/urbanization (B), Infrastructure (B), Recreation/tourism (A), Unsustainable exploitation (B)

The main threats are from tourism, housing developments and hunting. Part of the area is a candidate SAC.

Mount Asterousia (Kofinas) — B2, B3, C6 — 187

Admin region Crete
Coordinates 34°58′N 25°05′E
Altitude 600–1,230 m **Area** 29,000 ha

Site description

A coastal mountain dominated by phrygana, with a small *Pinus* woodland. Human activities include arable agriculture, livestock-farming and hunting.

Habitats Scrub (40%; sclerophyllous scrub/garrigue/maquis), Rocky areas (50%; sea cliff/rocky shore; scree/boulders; inland cliff), Artificial landscape (20%; arable land; perennial crops/orchards/groves)
Land-use Agriculture (80%), Tourism/recreation (20%)

Birds

Species	Season	Year	Pop min	Pop max	Acc	Criteria
Gypaetus barbatus Lammergeier	R	1996	—	—	—	B2
[1] ***Gyps fulvus*** Griffon Vulture	R	1996	50	60	A	B2, C6
Hieraaetus fasciatus Bonnelli's Eagle	R	1995	3	4	B	B2
Falco biarmicus Lanner	R	1996	—	—	—	C6
Alectoris chukar Chukar	R	1996	Common		—	B2
Sylvia rueppelli Rüppell's Warbler	B	1996	Frequent		—	B3
Pyrrhocorax pyrrhocorax Chough	R	1996	Frequent		—	B2, C6

1. Individuals.

This is an important site for breeding raptors and species associated with open, rocky upland areas.

Protection status

National None **International** Partial
3,403 ha of IBA covered by Special Protection Area (Dytika Asterousia, 3,403 ha). 21,450 ha of IBA covered by Special Protection Area (Anatolika Asterousia, 25,074 ha).

Conservation issues

Threats Agricultural intensification/expansion (A), Burning of vegetation (A), Infrastructure (A), Recreation/tourism (B)

The main threats are from the construction of roads and illegal housing. Part of the area is a candidate SAC.

Mount Jouchtas — B2, C6 — 188

Admin region Crete
Coordinates 35°15′N 25°10′E
Altitude 130–809 m **Area** 600 ha

Site description

A rocky mountain with very steep slopes covered with phrygana, maquis, and olive-groves. Human activities include stock-raising, agriculture and hunting.

Habitats Scrub (75%; sclerophyllous scrub/garrigue/maquis), Rocky areas (5%; inland cliff), Artificial landscape (20%; perennial crops/orchards/groves)
Land-use Agriculture (95%), Not utilized (5%)

Birds

Species	Season	Year	Pop min	Pop max	Acc	Criteria
[1] ***Gyps fulvus*** Griffon Vulture	R	1996	25	30	A	B2, C6

1. Individuals.

This is an important site for species associated with Mediterranean scrub and olive-groves.

Protection status

National None **International** None

Conservation issues

Threats Burning of vegetation (U), Disturbance to birds (C), Infrastructure (A), Recreation/tourism (U), Unsustainable exploitation (A)

The main threats are from hunting and infrastructure development. The area is protected as an important archaeological site. Part of the area is a candidate SAC.

Dia island A4ii, B1iii, B2, C2, C6 189

Admin region Crete
Coordinates 35°27′N 25°13′E
Altitude 0–268 m **Area** 1,250 ha

Site description

An island in the Gulf of Iraklio surrounded by small rocky islets. The island has steep cliffs covered with phrygana and other scrub.

Habitats Scrub (90%; sclerophyllous scrub/garrigue/maquis), Marine areas (5%; sea inlet/coastal features), Rocky areas (25%; sea cliff/rocky shore; rock stacks/islets)
Land-use Fisheries/aquaculture (5%), Hunting (90%), Tourism/recreation (5%)

Birds

Species	Season	Year	Pop min	Pop max	Acc	Criteria
Falco eleonorae Eleonora's Falcon	B	1982	300	380	A	A4ii, B1iii, B2, C2, C6

This is an important site for breeding seabirds and species associated with coastal cliffs.

Protection status

National High **International** None
1,250 ha of IBA covered by Controlled Hunting Area (Dia, 1,250 ha).

Conservation issues

Threats Agricultural intensification/expansion (A), Recreation/tourism (A)

Introduced goats have caused serious overgrazing. Tourism development is a serious threat for the near future. The area is a candidate SAC.

Mount Dikti B2, C6 190

Admin region Crete
Coordinates 35°10′N 25°30′E
Altitude 300–2,150 m **Area** 54,000 ha

Site description

A wooded mountain in Crete dominated by *Pinus* and *Quercus*. The woodland cover has been reduced by recent fires. Human activities include livestock-farming and tourism.

Habitats Forest and woodland (20%; native coniferous forest; broadleaved evergreen forest), Scrub (60%; sclerophyllous scrub/garrigue/maquis), Rocky areas (50%; scree/boulders; inland cliff)
Land-use Agriculture (60%), Not utilized (20%), Tourism/recreation (20%)

Birds

Species	Season	Year	Pop min	Pop max	Acc	Criteria
Gypaetus barbatus Lammergeier	R	1996	2	2	A	B2
[1] ***Gyps fulvus*** Griffon Vulture	R	1996	65	80	A	B2, C6
Aquila chrysaetos Golden Eagle	R	1996	4	4	A	B2, C6
Alectoris chukar Chukar	R	1996	Common		—	B2

1. Individuals.

An important site for passage and breeding raptors and alpine species.

Protection status

National Partial **International** Partial
Part of IBA covered by Wildlife Refuge (Chalasia/Zenion–Exo Potamon–Mesa Lasithiou, 1,100 ha). Part of IBA covered by Wildlife Refuge (Dasos Viannou–Amiron/Viannou–Amiron–Kefalovryso, 1,300 ha). 10,896 ha of IBA covered by Special Protection Area (Lazaros Koryfi Madara Diktis, 10,896 ha)

Conservation issues

Threats Agricultural intensification/expansion (A), Burning of vegetation (A), Firewood collection (B), Recreation/tourism (U)

The main threats are from fire and agricultural expansion. Part of the area is a candidate SAC.

Mount Thryptis and Mount Ornon B2, C6 191

Admin region Crete
Coordinates 35°05′N 25°50′E
Altitude 400–1,470 m **Area** 15,000 ha

Site description

Two mountains dominated by scrub with some *Pinus* woodland, recently reduced in extent by fire. Human activities include livestock-farming and hunting.

Habitats Forest and woodland (10%; native coniferous forest), Scrub (60%; sclerophyllous scrub/garrigue/maquis), Rocky areas (30%; scree/boulders; inland cliff)
Land-use Agriculture (80%), Not utilized (20%)

Birds

Species	Season	Year	Pop min	Pop max	Acc	Criteria
Gypaetus barbatus Lammergeier	R	1996	1	1	A	B2, C6

This is an important site for resident raptors and species associated with upland scrub.

Protection status

National Partial **International** None
Part of IBA covered by Wildlife Refuge (Monastiraki/Pachias Ammou–Kato Choriou, 700 ha).

Conservation issues

Threats Abandonment/reduction of land management (C), Agricultural intensification/expansion (A), Burning of vegetation (U), Firewood collection (B), Forest grazing (C), Selective logging/cutting (B)

The main threat is from agricultural expansion. Part of the area is a candidate SAC.

Dionisiades islands A1, A4ii, B1ii, B1iii, B2, C1, C2, C6 192

Admin region Crete
Coordinates 35°20′N 26°10′E
Altitude 0–150 m **Area** 506 ha

Site description

A complex of steep, rocky islands vegetated with low scrub.

Habitats Scrub (80%; scrub; sclerophyllous scrub/garrigue/maquis), Rocky areas (20%; sea cliff/rocky shore)
Land-use Agriculture, Fisheries/aquaculture, Not utilized

Birds

Species	Season	Year	Pop min	Pop max	Acc	Criteria
Calonectris diomedea Cory's Shearwater	B	1996	—	900	A	B1ii, B2, C2, C6
Falco naumanni Lesser Kestrel	P	—	Frequent		—	A1, C1
Falco eleonorae Eleonora's Falcon	B	1994	—	750	A	A4ii, B1iii, B2, C2

This is an important site for breeding seabirds and passage migrants. Species of global conservation concern that do not meet IBA criteria: *Gallinago media* (passage), *Larus audouinii* (up to 6 resident pairs).

Protection status

National None **International** High
506 ha of IBA covered by Special Protection Area (Dionysades Nisoi, 506 ha)

Conservation issues

Introduced rats are a threat. The area is a candidate SAC.

Threats Consequences of animal/plant introductions (B), Recreation/tourism (B), Unsustainable exploitation (U)

North-eastern edge of Crete A1, A4ii, B1iii, B2, C1, C2 193

Admin region Crete
Coordinates 35°14′N 26°14′E
Altitude 0–215 m **Area** 7,500 ha

Site description

The site includes part of the north-eastern coastline of Crete and the island of Elasa. The area is dominated by phrygana and maquis scrub with some olive-groves and vineyards.

Habitats Scrub (80%; sclerophyllous scrub/garrigue/maquis), Rocky areas (5%; sea cliff/rocky shore), Artificial landscape (20%; perennial crops/orchards/groves)
Land-use Agriculture (70%), Military (5%), Tourism/recreation (20%)

Birds

Species	Season	Year	Pop min	Pop max	Acc	Criteria
Falco naumanni Lesser Kestrel	P	—	Common		—	A1, C1
Falco eleonorae Eleonora's Falcon	B	1982	40	60	A	A4ii, B1iii, B2, C2

This is an important site for species associated with coastal cliffs and scrub and for passage migrants.

Protection status

National Low **International** Low
20 ha of IBA covered by Aesthetic Forest (Vai, 20 ha). 19 ha of IBA covered by Special Protection Area (Aisthitiko Dasos Vai, 19 ha).

Conservation issues

Threats Agricultural intensification/expansion (A), Recreation/tourism (C), Unsustainable exploitation (A)

The main threats are from agricultural intensification and hunting. The area is a candidate SAC.

Mount Zakrou B2, B3, C6 194

Admin region Crete
Coordinates 35°07′N 26°10′E
Altitude 100–650 m **Area** 13,000 ha

Site description

An area of hills and gorges dominated by phrygana vegetation. The main human activities are agriculture, particularly livestock-farming, hunting and tourism.

Habitats Scrub (40%; sclerophyllous scrub/garrigue/maquis), Rocky areas (50%; inland cliff), Artificial landscape (10%; arable land)
Land-use Agriculture (40%), Not utilized (40%), Tourism/recreation (30%)

Birds

Species	Season	Year	Pop min	Pop max	Acc	Criteria
Falco peregrinus Peregrine	R	1996	6	—	B	B2, C6
Emberiza caesia Cretzschmar's Bunting	R	1997	Common		—	B3, C6

This is an important site for passage migrants and raptors and species associated with cliffs and hills.

Protection status

National None **International** None

Conservation issues

Threats Agricultural intensification/expansion (A), Industrialization/urbanization (B), Unsustainable exploitation (B)

The main threats are from agricultural intensification and hunting.

Kavalli islets B1iii, B2, C2 195

Admin region Crete
Coordinates 35°02′N 26°14′E
Altitude 0–59 m **Area** 10 ha

Site description

Rocky islets off the south-eastern coast of Crete.

Habitats Scrub (90%; sclerophyllous scrub/garrigue/maquis), Rocky areas (10%; sea cliff/rocky shore)
Land-use Unknown

Birds

Species	Season	Year	Pop min	Pop max	Acc	Criteria
Falco eleonorae Eleonora's Falcon	B	1989	30	50	—	B1iii, B2, C2

This is an important site for species restricted to coastal cliffs and scrub.

Protection status

National None **International** None

Conservation issues

Threats Consequences of animal/plant introductions (U), Disturbance to birds (B), Unknown (U), Unsustainable exploitation (C)

Use of dynamite for fishing causes disturbance. Illegal hunting is a threat.

Koufonisi island B2 196

Admin region Crete
Coordinates 34°56′N 26°08′E
Altitude 0–64 m **Area** 525 ha

Site description

A small island and adjacent islets off south-eastern Crete. Dry grassland is the dominant vegetation cover. Human activities include fishing and tourism, although there are no permanent inhabitants on the island.

Habitats Scrub (15%; sclerophyllous scrub/garrigue/maquis), Grassland (70%; steppe/dry calcareous grassland), Wetland (5%; sand-dunes/sand beach), Marine areas (5%; sea inlet/coastal features), Rocky areas (5%; sea cliff/rocky shore)
Land-use Fisheries/aquaculture (5%), Tourism/recreation, Unknown

Birds

Species	Season	Year	Pop min	Pop max	Acc	Criteria
Falco eleonorae Eleonora's Falcon	B	1978	10	40	B	B2

This is an important site for breeding raptors associated with coastal areas and seabirds.

Protection status

National None **International** None

Conservation issues

Threats Aquaculture/fisheries (B), Recreation/tourism (A)

The main threats are from increasing tourism and more intensive fishing. The area is a candidate SAC.

REFERENCES

Adamakopoulos, T., Gatzoyiannis, S. and Poirazidis, K., eds. (1995) [*Special environmental study of Dadia Forest.*] Athens: WWF-Greece—Ministry of Environment—Ministry of Agriculture (In Greek.)

Adamakopoulos, T., Hatzirvassanis, V. and Matsouka, P. (1986) [*The mountains of Roumeli.*]. Athens: Pitsilos. (In Greek.)

Adamakopoulos, T., Matsouka, P. and Hatzirvassanis, V. (1988) [*The mountains of Moria.*]: Athens: Pitsilos. (In Greek.)

Adamakopoulos, T., Sfikas, G. and Hatzirvassanis, V. (1988) [*Evaluation and management of Giona ecosystem.*] Athens: Kinisi Prostasias Ethnikon Drimon. (In Greek.)

ADAMANTOPOULOU, S. AND ANDROUKAKI, E. (1987) [Study on the birds of the National Park of Northern Sporades.] Athens: University of Athens (thesis). (In Greek.)

ADAMANTOPOULOU, S. AND ANDROUKAKI, E. (1989) [*A management proposal of Dadia-Soufli Forest as a habitat of birds of prey and population monitoring for their protection.*] Athens: Ministry of Environment, Physical Planning and Public Works—University of Athens, Department of Biology. (In Greek.)

AKRIOTIS, T., HANDRINOS, G. AND KONDYLIS, G. (1994) Avifauna and conservation of the Spercheios Delta. *Biologia Gallo-hellenica* 22: 123–141.

AKRIOTIS, T., PREUSS, N. O., JOENSEN, A. AND MADSEN, J. (1982) Bird observations at Antikithira island, 7–17/9/82. Unpublished report.

ALIVIZATOS, H. (1996) [Biology and breeding ecology of Long-legged Buzzard (*Buteo rufinus*) in Evros Prefecture.] Athens: Agriculture University of Athens (PhD thesis). (In Greek.)

ANON. (1983) [*Report on the evaluation and conservation of the valley Mesiano Nero Mt. Vourinos.*] Greece: National Council for Planning and Environment, Ministry of Co-ordination. (In Greek.)

ANON. (1986) [*Delineation program for the wetlands of Ramsar Convention.*] Athens: Ministry of Environment Physical Planning and Public Works. (In Greek.)

ANON. (1996) Midwinter Waterfowl Census 1995. Hellenic Ornithological Society—HSPN—Greek Biotope/Wetland Centre—HBRC. Unpublished.

ANON. (1997) Midwinter Waterfowl Census 1996. Hellenic Ornithological Society—HSPN—Greek Biotope/Wetland Centre—HBRC. Unpublished.

ATHANASIOU, C. AND ANAGNOSTOPOULOU, M. (1993) [*Report about the wetlands of village N. Fokaias, Chalkidikis.*] Thessaloniki: National Centre of Wetlands/Biotopes. (In Greek.)

BAUER, W. AND BOHR, H. J. (1987) Zur Kennthis der Budliche Areal-grenzen einiger Vogelarten in der griechischen Rhodopen. *Die Vogelwelt* 108(1): 1–13.

BONAZOUNTAS, M., KALIDROMITOU, D., PERGANTIS, P. AND COLLABORATORS (1993) [Integrated management of Mesolongi–Aetoliko wetland complex of Aitoloakarnania Prefecture]. Athens. C.E.C. Ministry of Environment. (In Greek.)

BOURDAKIS, S. (1994) [*Brief activities report. Wetlands monitoring project 1994, Lake Stimfalia, Lake Taka and Moustos Lagoon.*] Greece: Hellenic Ornithological Society. (In Greek.)

BOUSBOURAS, D. AND BOURDAKIS, S. (1992) *Ymittos, Enas apo tous teleutaious pneumones prasinou tis Attikis.* Greece: Hellenic Ornithological Society. (In Greek.)

BRITTON, R. H., *ET AL.* (1978) Proposals for Nature Conservation in the Evros Delta. Athens. Ministry of Coordination (unpublished report).

BROGGI, M. AND WILLI, G. (1986) Ornithologische Beobachtungen auf Naxos. *Ber. Bot. Zool. Ges. Liecht.* 15: 91–102.

BROOKS, R. (1995) *Birding in Lesvos.* Norfolk, UK: Brookside Publishing.

CEBRIAN, D. AND ROUSSOPOULOS, G. (1996) *Ornithological survey to the islands and islets among Antimilos and Amorgos, South Cyclades.* Athens: Hellenic Ornithological Society.

COLLAR, N. J., CROSBY, M. J. AND STATTERSFIELD, A. J. (1994) *Birds to watch 2: the world list of threatened birds.* Cambridge, UK: BirdLife International (Conservation Series 4).

COMMISSION OF EUROPEAN COMMUNITIES (CEC) (1991) Corine Biotopes Manual. Habitats of the European Community: Greece. Brussels.

CORINE BIOTOPES DATABASE (1989) Data submitted to the Commission of European Communities, Brussels.

COSSON, F. (1985) Some recent data of the raptors of Rhodes (Greece). *Bull. World Working Group on Birds of Prey* 2: 57–60.

DAFIS, S., PAPASTERGIADOU, E., GEORGHIOU, K., BABALONAS, D., GEORGIADIS, T., PAPAGEORGIOU, M. AND LAZARIDOU, T. (1996)[The Greek 'Habitat' Project (Directive 92/43/EEC) NATURA 2000: An Overview.] Final Project Report. Life contract B4-3200/94/756. Commission of the European Communities DGXI, The Goulandris Natural History Museum – Greek Biotope/Wetland Centre. 1000p

DIMAKI, M., SCHOGOLEV, I. AND IOANNIDIS, J. (1996) [Systematic bird observations in Samos island.] *Samiakes Meletes* 2: 398–442. (In Greek.)

DIMITROPOULOS, A. (1989) [*The avifauna of Syros.*] Ermoupolis, Greece: Syriana Grammata 7. (In Greek.)

DRAGOUMIS, P. (1984) [Observations in Sifnos.] *Hellenic Orn. Soc. Bull.* 1: 5–10 (In Greek.)

DRETAKIS, M. (1996) [The waterfowl and the shorebirds at four wetlands of Crete]. Iraklio, Crete: Department of Biology, University of Crete (MSc thesis). (In Greek.)

ECONOMIDOU, E. AND VASILAKIS, K. (1995) [Assesment of the ecological value of the Kalamas river delta and preliminary management proposals (Phase 2: 1994).] Greece: Greek Biotope/Wetland Centre-The Goulandris Natural History Museum. (In Greek.)

ECONOMIDOU, E. *ET AL.* (1981) [National Programme of Coast Management, Coastal Wetlands, Secretariat of National Council of House Planning, Ministry of Co-ordination.] (In Greek.)

ECONOMIDOU, E. *ET AL.* (1988) [*Location and study of wetlands and other biotopes important for the avifauna of Crete.*] Patra, Greece: Patras University, M.E.P.P.P.W. (Final Report, Vol. 1). (In Greek.)

ECONOMIDOU, E., VASILAKIS, K., BOUSBOURAS, D. AND WITTE, R. (1993) [Assessment of the ecological value of the Kalamas river delta and preliminary management proposals (Phase 1: 1992).] Greece: Greek Biotope/Wetland Centre and Univ. of Patras, Dept. of Biology. (In Greek.)

FOPPEN, R., PERGANTIS, P., ROUSSOPOULOS, Y. AND SARIS, F. (1995) Breeding birds in the Messolongi wetlands, Spring 1990. in W. T. De Nobel, ed. *Birds of the Messolongi wetlands.* International Wader and Waterfowl Research: WIWO report no. 53.

GEORGHIOU, K. *ET AL.* (1996) [Specific Management Plan for the site: Koilada kai Ekvoles Spercheiou–Maliakos Kolpos (GR 2440002).] Thermi, Greece: Goulandris Natural History Museum—Greek Biotope/Wetland Centre—University of Athens. (In Greek.)

GEORGIADIS, T. *ET AL.* (1994) [Ecological impacts and management of the wetlands of Kalamas, Acherontas and Kalodiki.] Patras, Greece: Envireg—Ministry of National Economy—University of Patras. 463 pp. (In Greek.)

GILLINGS, S., HALLMANN, B. AND HATZIRVASANIS, V. (1996) Details of an expedition to Peloponissos. Unpublished.

GOUTNER, V. AND HANDRINOS, G. (1990) The occurrence of Slender-billed Curlews *Numenius tenuirostris* in Greece. *Biol. Conserv.* 53: 47–60.

GOUTNER, V., KAZANTZIDIS, S. AND LATSOUDIS, P. (1992) [Wetlands of Gallikos river delta–Kalochori marsh, M. Emvolo salina and Epanomi's area biotopes: Evaluation of ornithological importance and delineation proposal.] Thessaloniki, Greece: Society for the Study and Protection of Wetlands. (In Greek.). Unpublished.

GOUTNER, V., KAZANTZIDIS, S., CHARALABIDOU, I. AND PAPOULIA, S. (1996) [Importance of artificial and natural biotopes for the conservation of threatened shorebirds populations in Axios delta.] Thessaloniki , Greece: Aristotelian University of Thessaloniki (Final report 1571 PENED). (In Greek.)

GOUTNER, V. AND PAPAKOSTAS, G. (1992) Evaluation of the ornithological importance of the Alyki Kitrous wetland, Macedonia, Greece: a priority for conservation. *Biol. Conserv.* 62: 131–138.

GRETTON, A. (1991) *The ecology and conservation of the Slender-billed Curlew (*Numenius tenuirostris*).* Cambridge, UK: International Council for Bird Preservation (Monogr. 6).

GRIMMETT, R. F. A. AND JONES, T. A. (1989) *Important Bird Areas in Europe.* Cambridge, UK: International Council for Bird Preservation (Techn. Publ. 9).

HAENSEL, J. (1992) Zur Vogelwelt der griechiscen Insel Kos (Dodekanes/Agais). *Kartierung mediterr. Brutvögel* 8: 21–28.

HALLMANN, B. (1982) Important Bird Areas in the European Community: Preliminary list of Important Bird Areas in Greece. Internal report to ICBP EC Working Group, Brussels.

HALLMANN, B. (1985) The fauna of Mt. Oiti National Park. Report to the Ministry of Agriculture. Unpublished.

HALLMANN, B. (1985) The wildlife of Mt. Parnassos. Report to Ministry of Agriculture. Unpublished.

HALLMANN, B. (1986) Raptor surveys Greece 1980–85. IUCN—WWF Project 1921 (internal report).

HALLMANN, B. (1998) Preliminary study on the birds of Tembi valley and Mount Ossa Aesthetic forests, Ministry of Agriculture, report.

HALLMANN, B. (1989) Conservation management for vultures in the Kato Olympos area. Unpublished.

HALLMANN, B. *ET AL.* (1995) Lesser Kestrel survey, Thessaly 1995. Hellenic Ornithological Society, RSPB, Birdlife International.

HANDRINOS, G. (1989) The IWRB Midwinter waterfowl counts in Greece 1967–1987: a preliminary analysis of the populations of Anatidae. *Biologia Gallo-hellenica* 15: 219–236.

HANDRINOS, G. AND AKRIOTIS, T. (1997) *The Birds of Greece.* London, UK: Helm Publ.

HATZILAKOU, D. (1993) The distribution of the globally threatened Dalmatian Pelican (*Pelecanus crispus*, Bruch) in Greece. Threats pertaining to its habitats and recommendations for protection. Sandy, UK: The Royal Society for the Protection of Birds. Unpublished report.

HATZILAKOU, D. AND RIZOU, E. (1997) Conservation and wardening of the Dalmatian Pelican colony in Amvrakikos 1996 breeding season. Athens: Hellenic Ornithological Society, Vogelbescherming. Unpublished report.

HATZIRVASSANIS, V. (1995) [Mountainous ecosystem of Giona: study of development with aim the protection.] Hellenic Mountaineering Club of Amfissa. Unpublished. (In Greek.)

HATZIRVASSANIS, V. (1996) Bird list of Pindos. Athens. Unpublished report.

HEATH, M. F. AND BORGGREVE, C. (2000) *BirdLife International/EBCC European Bird Database 1998.* Cambridge, UK: BirdLife International.

HOS [HELLENIC ORNITHOLOGICAL SOCIETY] (1985) [Alyki Aegiou lagoon.] Athens: Sub-Ministry of New Generation and Sports—HOS. (In Greek.)

HOS [Hellenic Ornithological Society] (1990) [*Report on the illegal housing activities on the Alyki Kitrous wetland.*] Thessaloniki: HOS. (In Greek.)

HOS [Hellenic Ornithological Society] (1990) [*Proposal for the protection of Epanomi cape biotope, Thessaloniki prefecture.*] Thessaloniki, Greece: HOS. (In Greek.)

HOS [Hellenic Ornithological Society] (1992) [*Proposal for the establishment of a model ornithological park in Epanomi cape, Thessaloniki prefecture.*] Thessaloniki, Greece: HOS. (In Greek.)

HOS [Hellenic Ornithological Society] (1994) [*IBA for the birds of Greece.*] Athens: HOS. (In Greek.)

HOS [Hellenic Ornithological Society] (1987–1994) Midwinter Waterfowl Census 1987–1994. Unpublished.

Holzinger, J. (1991) Wetlands of the Greek island Limnos. A bird habitat of international significance. *Kartierung Mediterr. Brutvögel* 6: 19–25.

Holzinger, J. (1995) The Cinereous Bunting *Emberiza cineracea* breeding on Skyros (Greece). *Zool. Middle East* 11: 31–36.

Jerrentrup, H., Gaethlich, M., Joensen, A., Nohr, H. and Brogger-Jensen, S. (1988) Urgent action plan to safeguard three endangered bird species in Greece and EC: Pygmy Cormorant (*Phalacrocorax pygmaeus*), Great White Egret (*Egretta alba*), White tailed Eagle (*Haliaeetus albicilla*). Ornis Consult [Denmark]—Natural History Museum Århus [Denmark]—Hellenic Ornithological Society. Report to the Commission of European Communities. 153 pp.

Joensen, A. H. and Jerrentrup, H. (1988) The Agios Mamas lagoon, Halkidiki Greece, an area of international importance for breeding waders. *Natura Jutlandica* 22 (11): 185–188.

Joensen, A. H. and Mansen, J. (1985) Waterfowl and raptors wintering in wetlands of western Greece, 1983–85. *Natura Jutlandica.* 21(11): 169–200.

Kalliris, P. (1993) [*The natural ecosystem of Lake Stimfalia.*] Korinthos, Greece: Society for the Environmental Education and Awareness. (In Greek.)

Katsadorakis, G. (1985) [Preliminary ecological study of Lake Kourna.] Forestry Service Chania. (In Greek.)

Katsadorakis, G. (1985) [Preliminary report on the vertebrates of Vikos/ Aoos National Park], Forestry Department of Ioannina. Unpublished. (In Greek.)

Katsadorakis, G., Adamantopoulou, S., Androukaki, S. and Bousbouras, D. (1985) [Report of a visit at Northern Sporades.] Unpublished. (In Greek.)

Kazantzidis, S. (1994) Bird counts during spring migration in lakes Chimaditis and Zazari. Greek Biotope/Wetland Centre. Unpublished report.

Kazantzidis, S. and Athanasiou, H. (1990) Wetland habitat loss in Thessaloniki plain, Greece. London, UK: University College London (MSc dissertation).

Kazantzidis, S. and Athanasiou, H. (1992) [Axios-Loudias-Aliakmonas and Alyki Kitrous wetland. Evaluation of ornithological importance.] Topiotechniki EPE. Ministry of Environment, Physical Planning and Public Works, Athens. Society for the study and protection of wetlands. Thessaloniki. (In Greek.)

Kazantzidis, S. and Goutner, V. (1996) Foraging ecology and conservation of feeding habitats of Little Egrets (*Egretta garzetta*) in the Axios river delta, Macedonia, Greece. *Col. Waterb.* 19 (Sp. Publ. 1): 115–121.

Kazantzidis, S., Anagnostopoulou, M. and Gerakis, P. (1995) Problems of 35 Greek wetlands and confrontation actions. Wetlands Monitoring Programme 1992–1994. Goulandris Natural History Museum—Greek Biotope/Wetland Centre. 249 pp.

Kilikidis, S. *et al.* (1992) [Artzan marsh, Lakes Cheimaditis and Zazari, Lake Agras, wetland of Kalama estuary: Assessment of their eligibility for inclusion in the Community network of Special Protection Areas by implementing Article 4 of Directive 79/409/EEC on the conservation of wild birds] Aristotelian University of Thessaloniki, Ministry of Environment, Physical Planning and Public Works, Directorate of Environmental Planning. (In Greek.)

Knijff, P. and Alivizatos, H. (1997) Cinereous Bunting. P. 750 in E. J. M. Hagemeijer and M. J. Blair, eds. *The EBCC Atlas of European Breeding Birds: their distribution and abundance.* London: T. and A. D. Poyser.

Kominos, T. (1995) [Kithira, on the migratory path of birds.] Thessaloniki, Greece: Etairia Kythiraikon Spoudon. (In Greek.)

Kominos, T., Bourdakis, S., Roussopoulos, Y. and Sakoulis, A. (1996) [Mission to western Greece, June 1996.] Greece: Hellenic Ornithological Society. Unpublished. (In Greek.)

Koroni, M. and Adamogiannis, E. (1993) [Integrated management study of Pinios river Delta. Ktimatiki Etaireia Dimosiou.] 367 pp. (In Greek.)

Kretzschmar, E. (1992) Die Brutvogel von Alykes Kitros, (Pieria, Griechenland), einem Feuchtgebiet von internationaler bedeutung. Kartierung mediterr. *Brutvögel* 7: 27–38.

Latsoudis, P. (1992) [Cape Epanomi: A next coast biotope.] Thessaloniki. Oikotopia 19: 20–24. (In Greek.)

Latsoudis, P. (1992) The basins of M. Emvolo salina as a bird breeding ground. *Avocetta* 16: 86–88.

Latsoudis, K. P. (1993) The natural and artificial dunes of cape Epanomi. In A. H. P. M. Salman and M. Bonazountas, eds. *Proceedings of the 4th EUCC Conference in Marathon, Greece. Congress of the European Union for Coastal Conservation.* Leiden, Netherlands: European Union for Coastal Conservation.

Legakis, A. (1990) [*Research programme: study of the coasts of Crete that present ecological disturbances.*] Greece: Department of Biology, University of Crete. (In Greek.)

Magioris, S. N. (1994) The avifauna of the Cyclades. *Hellenic Zoological Archives* 2: 1–16.

Magioris, S. N. and Sfenthourakis, S. (1994) Some aspects on the avifauna on the uninhabited islets of Greece. *Scientific Annals of the School of Biology, Aristotle University. Thessaloniki. Bios* 2: 289–294.

Makatsch, W. (1969) Ornithologishe beobachtungen in Griechenland II. *Ber. Abh. Mus. Tierkunde. Dresden* 30(1): 1–55.

Malakou, M. (1983) [Study of the avifauna of Naxos island, east area.] Athens: Department of Biology, University of Athens (BA thesis). (In Greek.)

Malakou, M. and Katsadorakis, G. (1992) Past and present situation of the wetlands of Crete with special reference to their birds. *Biologia Gallo-hellenica* 19(2): 59–72, 143–156.

Malamidis, G. Andreadakis, P., Malakou, M., Pitsokas, L. and Romanos, L. (1988) [Olympos National Park – Master Plan.] Prefecture of Pieria—Forest Service. (In Greek.)

Malson, R. and Ross, B. (1995) Wetland avifauna in Kos. ASKPE—Royal Society for the Protection of Birds [UK]—Hellenic Ornithological Society. Unpublished report.

Mayol, J. (1996) El Halcón de eleonora (*Falco eleonorae*): situacion de la especie y de su conocimiento. Pp. 117–125 in J. Muntaner and J. Mayol, eds. *Biologia y conservacion de las Rapaces Mediterraneas, 1994.* Madrid: Sociedad Ornitología de Española (Monografias No. 4).

Meininger, P., ed. (1990) *Birds of the wetlands in north-eastern Greece, spring 1987.* Zeist, Netherlands: Foundation Working Group International Wader and Waterfowl Research (WIWO)—Hellenic Ornithological Society (WIWO report No. 20). 317 pp.

Millet, T. J. (1995) *A preliminary study of the birds of Kalogria, Achaia Region, Greece.* Athens: Hellenic Ornithological Society.

National Foundation for Agricultural Research—Hellenic Ornithological Society (1995) [Recognition and evaluation of biotopes for avifauna for inclusion in the Community Network of Special Protection Areas according to Article 4 of Directive 79/409/EEC.] Rapid ornithological assessment of 16 biotopes. ENVIREG Project. Athens: Ministry of Environment, Physical Planning and Public Works. (In Greek.)

Papagrigoriou, S. *et al.* (1994) [Specific environmental study of Agios Mamas wetland.] Chalkidiki, Greece: ENVECO A.E., Ag. Mamas Community. (In Greek.)

Papakonstantinou, K. (1992) [Aegio lagoon.] *Physis* 80: 50–53. (In Greek.)

Papakonstantinou, K., Economidou, E., Bousbouras, D., Kardakari, N., Zogaris, S., Tsiakiris, R. and Roussopoulos, Y. (1995) [*Ecological survey in northern Dodecanese.*] Athens: Hellenic Ornithological Society. (In Greek.)

Papakostas, G. (1990) [Evaluation of the ornithological importance of the Alyki Kitrous wetland, Pieria.] Thessaloniki, Greece: Department of Biology, Aristotelian University of Thessaloniki (BSc thesis) (In Greek.)

Papayannis, T., Pergantis, P. and Papayannakis, L. (1986) [*Amvrakikos Gulf: Sources, development and environmental protection.*] Athens: MEPPPW and Brussels: Commission of the European Communities. (In Greek.)

Paraskeuopoulos-Georgiadis-Kriaris EPE (1993) [*Management plan of Evros Delta.*] State Revenue Office of Evros Prefecture. (In Greek.)

Pergantis, P. (1988) *Study on the integrated management of the coastal wetlands in western Greece.* Brussels: Commission of the European Communities (DG XI).

Pergantis, P. (1989) Compiling small-scale ornithogeographical data to delineate important bird habitats in the Amvrakikos area. *Biologia Gallo-hellenica* 15: 201–218.

Pergantis, P. (1993) Study for the conservation of birds and the managenent of the peripheral reed bed and the riparian forest of Kastoria lake. In A. Economopoulos *et al. Study for the restoration and development of Kastoria lake.* Greece: Democrition University of Thrace.

Pergantis, P. and Handrinos, G. (1992) A preliminary check-list of the birds of the Messolongi Wetlands (1860–1986) with seasonal occurrence and population data. *Biologia Gallo-hellenica* 19: 5–27.

Pergantis, P. and Roussopoulos, Y. (1994) Effects of pisciculture development on the breeding populations of *Himantopus himantopus* and *Glareola pratincola* in the Amvrakikos wetland area, west Greece. *Biologia Gallo-hellenica* 22: 151–159.

Pergantis, P., Goutner, V., Handrinos, G. and Akriotis, T. (1994) The breeding status of the Gull-billed Tern (*Gelochelidon nilotica*) in western Greece. *Biologia Gallo-hellenica* 22: 171–179.

POIRAZIDIS, K. (1989) [Study of the distribution and ecology of the Capercaillie (*Tetrao urogallus*) in Greece.] Thessaloniki, Greece: School of Geotechnic Studies, University of Thessaloniki (dissert.). (In Greek.)

RISTOW, D., WINK, C. AND WINK, M. (1986) Assessment on Mediterranean autumn migration by prey analysis of Eleonora's Falcon. *Suppl. Ric. Biol. Selvaggina* 10: 285–295.

RISTOW, D., FELDMANN, F., SCHARLAU, W., WINK, C. AND WINK, M. (1991) Population dynamics of Cory's Shearwater (*Calonectris diomedea*) and Eleonora's Falcon in the Eastern Mediterranean region. Pp. 192–212 in A. Seitz and V. Loeschke, eds. *Species conservation: a population-biological approach*. Basel: Birkhauser Verlag.

ROUND, P. D. AND SWANN, R. I. (1977) Aspects of the breeding of Cory's Shearwater (*Calonectris diomedea*) in Crete. *Ibis* 119: 350–353

ROUSSOPOULOS, Y. (1996) [Spring and autumn report for Amvrakikos Gulf.] Conservation project of Slender-Billed Curlew (*Numenius tenuirostris*) by Institute Royal des Sciences Naturelles de Belgique—Ministry of Agriculture—Greek Biotope/Wetland Centre—Hellenic Ornithological Society. Unpublished report.

ROUSSOPOULOS, Y. AND PERGANTIS, P. (1990) The decline of the Lesser Kestrel (*Falco naumanni*) in Aetolia, Western Greece. *Biologia Gallo-hellenica* 22: 161–171.

ROY, D. (1989) The conservation and management of birds of prey and their habitats in Evros, Greece. Sandy, UK: Royal Society for the Protection of Birds. Unpublished report.

SAKOULIS, A., BOURDAKIS, S., HALLMANN, B. AND ALIVIZATOS, H. (1997): The status of the Imperial Eagle (*Aquila heliaca*) in Greece. Athens: Hellenic Ornithological Society (internal report).

SCHARLAU, A. AND SCHARLAU, W. (1990) Die Vogelwelt der Agais, Teil 5: Die Brutvogel von Naxos und Einigen Benachbarten Inseln. *Kartierung mediterr. Brutvögel*.

SCHARLAU, V. W. (1989) Die Vogelwelt der Agais. Teil 1: Die Brutvogel von Rodos. *Kartierung mediterr. Brutvögel* 3: 3–23.

SCHARLAU, W. (1989) Die Vogelwelt der Agais. Teil 2: Die Brutvogel von Ostkreta. *Kartierung mediterr. Brutvögel* 3: 24–40.

SCHOGOLEV, I. (1996–97) Interim reports of the project 'Conservation measures for the Slender-billed Curlew in Greece'. Athens: Ministry of Agriculture—Hellenic Ornithological Society. Unpublished.

SCHOGOLEV, I. AND DIMAKI, M. (1995) [Bird migration in Strofades.] Unpublished report for WWF-Greece Ionian Project. (In Greek.)

SECLIZIOTIS, S. AND PAPAKONSTANTINOU, A. (1992) Study of the border designation and management of the lakes Volvi-Koronia. Athens: Ministry of Environment Final Report.

SPINTHAKIS, E. (1995–1996) A marine and island park on Psara archipelago. In A. H. P. M. Salman and M. Bonazountas, eds. *Proceedings of the 4th EUCC Conference in Marathon, Greece. Congress of the European Union for Coastal Conservation*. Leiden, Netherlands: European Union for Coastal Conservation.

SPINTHAKIS, E., CHOULIS, D. AND CHOREMIS, J. (1993) [*The birds of Chios Island*.] Municipality of Homeroupolis, Greece. (In Greek.)

SPIROPOULOU, S. (1991) Black Vulture conservation and forest management in Evros, Greece. MSc dissertation. London: University College London.

STUDY GROUP (1985) [Report for the protection and management of the National Park of Parnitha.] Protovoulia gia ti sotiria tis Parnithas. Unpublished report. (In Greek.)

SZIJJ, J. (1983) [Okologische Vertanalyse des Acheloos Deltas, Westgriechenland.] Essen, Germany: Universitat Essen. (In German.)

TEWES, E. AND HALLMANN, B. (1991) Black Vulture Recovery Programme in Greece. European Black Vulture Conservation Foundation. Unpublished report.

TRIANTAFILLAKOS, P. (1998) [*Hunting guides, hunting regions of Macedonia—Thraki, Ipiros, Thessaly, Sterea Ellada—Kyklades and Peloponnisos*.] Athens. (In Greek.)

TROUMBIS, A. (1992) [*Ecological and management study of the Dipi-Larsos wetland*.] Mytilini, Greece: Department of Environment of the University of Aegean. (In Greek.)

TSIOURIS, S. *ET AL.* (1996) *Specific management plan for the area: Limnes Cheimaditida-Zazari*. Thessaloniki, Greece: Goulandris Natural History Museum—Greek Biotope/Wetland Centre. (In Greek.)

TSOUNIS, G. AND DIMITROPOULOS, A. (1992) The avifauna of the north-eastern Aegean islands Lesvos and Chios: differences and similarities. *Biologia Gallo-hellenica* 19(2): 73–102.

TSOUNIS, G., MALAKOU, M. AND RAVASINI, M. (1985) Study of the fauna of the National Park of Pindos, Hellenic Society for the Protection of Nature. Unpublished report.

TUCKER, G. M. AND HEATH, M. F. (1994) *Birds in Europe: their conservation status*. Cambridge, UK: BirdLife International (BirdLife Conservation Series no. 3).

VITTERY, A. *ET AL.* (1996) Recent observations on the avifauna of Kefalonia. Poster presentation at the 7th Int. Congr. Zoogeography and Ecology of Greece and the Adjacent Regions.

VLAHOS, C. (1989) [The ecology of lesser spotted eagle (*Aquila pomarina*) in Dadia forest, Evros, Thrace, Greece.] Thessaloniki, Greece: Aristotle University of Thessaloniki (PhD thesis). (In Greek.)

WALMSLEY, J. AND PARADISI, G. (1993) A geo-marine park for the Psara archipelago: a Medmaravis scientific mission to Psara, Greece, 6–25 May 1993. Unpublished report.

WALTER, H. (1979) *Eleonora's Falcon: adaptation to prey and habitat in a social raptor*. Chicago, USA: Chicago University Press.

WATSON, G. E. (1964) Ecology and evolution of passerine birds on the islands of the Aegean Sea. New Haven, USA: Yale University (PhD thesis).

XIROUCHAKIS, S. (1995) [The avifauna of Strofades.] WWF-Greece, Ionian Project. Unpublished report. (In Greek.)

ZALIDIS, C. G. AND MANTZAVELAS, A. L., EDS. (1994) *Inventory of Greek wetlands as natural resources (first approximation)*. Thermi, Greece: Greek Biotope/Wetland Centre.

ZANAKI, K., XANTHOPOULOU, P. AND HABERIDIS, D. (1992) [Delineation study of the wetland Mati Tyrnavou.] Greece: Prefecture of Larisa. (In Greek.)

ZOGARIS, S. (1992) [List of the birds of the area of Southern Karystia.] Unpublished report. (In Greek.)

ZOGARIS, S. (1995) Birds and nature conservation in southern Karystia, Evvia, Greece. Unpublished thesis. Simon Fraser University, B.C., Canada.

ZOGARIS, S. (1997) The avifauna of Northern Pindos. Preliminary study for the protection of the natural environment. Unpublished report.

ZOGARIS, S., VLAMI, V. AND PROBONAS, N. (1996) *Alyki Naxos lagoon, preliminary study of environmental assesment and management*. Naxos, Greece: Hellenic Ornithological Society.

HUNGARY

Szabolcs Nagy

White-tailed Eagles *Haliaeetus albicilla* on the Biharugra fish-ponds (IBA 030). (PHOTO: MIHÁLY BODNÁR)

GENERAL INTRODUCTION

Hungary lies in the Carpathian Basin of central Europe, and is split roughly in half by the River Danube. The country covers 93,032 km² and comprises six geographical regions: the Great Plain, the Little Plain, the Alp Foothills, the Transdanubian Hills, the Transdanubian Uplands and the Northern Hills. Hungary has 43 Important Bird Areas (IBAs) covering 14,662 km², equalling 15.8% of the country's surface area (Table 1, Map 1).

The previous pan-European IBA inventory identified 46 IBAs (Grimmett and Jones 1989), 13 of which no longer meet the qualifying criteria and were not retained during the review process (Table 1). A further five IBAs have been merged with three other IBAs for this inventory (see Table 1 and site accounts for details).

ORNITHOLOGICAL IMPORTANCE

There are 125 species of European conservation concern (SPECs) breeding regularly in Hungary, of which five are globally threatened (*Aythya nyroca*, *Aquila heliaca*, *Otis tarda*, *Crex crex*, *Acrocephalus paludicola*), and a further 68 have an unfavourable conservation

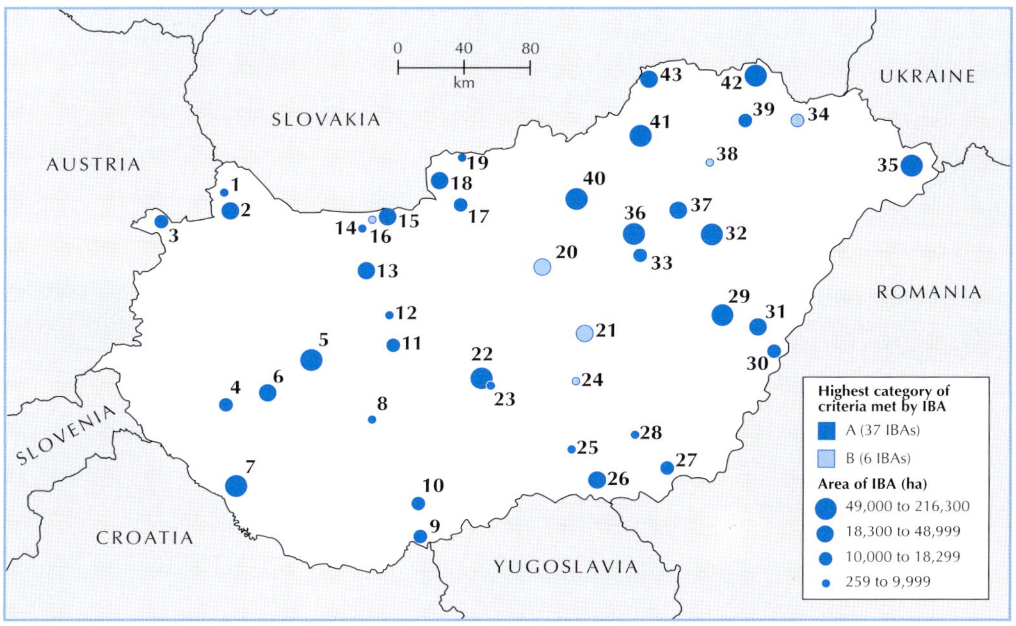

Map 1. Location, area and criteria category of Important Bird Areas in Hungary.

Table 1. Summary of Important Bird Areas in Hungary. 43 IBAs covering 14,662 km²

IBA code	1989 code	International name	National name	Administrative region	Area (ha)	Criteria (see p. 11)
001	—	Moson plain	Mosoni-sík	Győr-Sopron-Moson	4,310	A1, B2
002	HU003	Hanság	Hanság	Győr-Sopron-Moson	27,700	A4i, B1i, B2
003	HU001	Lake Fertő	Fertő-tó	Győr-Sopron-Moson	12,542	A4i, B1i, B2, B3
004	HU009	Kis-balaton	Kis-balaton	Somogy, Zala	14,745	A1, A4i, B1i, B2, B3
005	—	Lake Balaton	Balaton	Somogy, Zala, Veszprém	59,800	A4i, B1i
006	—	Nagyberek	Nagyberek	Somogy	19,400	A1, A4i, B1i, B2
007	HU010	Inner Somogy	Belső-Somogy	Somogy	216,300	A1, A4i, B1i, B2
008	—	Pacsmag fish-ponds	Pacsmagi-halastavak	Tolna	487	A1, A4i, B1i, B2, B3
009	—	Béda-Karapancsa	Béda-Karapancsa	Baranya	11,900	A4i, B1i, B2
010	HU031	Gemenc	Gemenc	Bács-Kiskun, Tolna	17,779	A1, A4i, B1i, B2
011	—	Sárvíz valley	Sárvíz-völgye	Fejér	14,700	A1, A4i, B1i, B2
012	HU008	Lake Velence and Dinnyés marshes	Velencei-tó és a Dinnyési Fertő	Fejér	4,800	A4i, B1i
013	HU006	Vértes mountains and Zámoly basin	Vértes-hegység a Zámolyi medencével	Fejér, Komárom-Esztergom	29,400	A1, B2
014	HU002	Lake Öreg at Tata	Tatai Öreg-tó	Komárom-Esztergom	259	A4i, B1i
015	HU029	Gerecse mountains	Gerecse-hegység	Komárom-Esztergom	20,300	A1, B2
016	—	Danube reach between Gönyű and Szob	Duna Gönyű-Szob közötti szakasza	Győr-Sopron-Moson, Komárom-Esztergom	4,840	B1i
017	—	Danube bend	Dunakanyar	Pest	15,200	A4i, B1i
018	HU004	Börzsöny mountains	Börzsöny-hegység	Nógrád, Pest	46,400	A1, B2
019	—	Middle reach of River Ipoly	Ipoly középső szakasza	Nógrád	2,700	A1
020	—	Tápió valley	Tápió-vidék	Pest, Jász-Nagykun-Szolnok	18,300	B2
021	—	Puszta at Jászkarajenő	Jászkarajenői puszták	Bács-Kiskun, Pest	18,900	B2
022	HU021, HU042	Danube plain	Duna-menti síkság	Bács-Kiskun, Pest	82,000	A1, A4i, B1i, B2
023	HU022	Lake Kolon	Kolon-tó	Bács-Kiskun	2,962	A4i, B1i
024	HU035	Flood-plain at Tiszaalpár	Tiszaalpári rét	Bács-Kiskun, Jász-Nagykun-Szolnok	5,000	B2
025	—	Lake Fehér at Gátér	Gátéri Fehér-tó	Bács-Kiskun, Csongrád	1,400	A4i, B1i, B2
026	HU027	Pusztaszer Landscape Protection Area	Pusztaszeri Tájvédelmi Körzet	Csongrád	22,320	A1, A4i, B1i, B2
027	HU045	Puszta at Pitvaros	Pitvarosi puszták	Békés, Csongrád	13,100	A4i, B1i
028	HU025	Lake Fehér at Kardoskút	Kardoskúti Fehér-tó	Békés, Csongrád	8,800	A4i, B1i
029	HU023	Dévaványa plain	Dévaványai-sík	Békés, Jász-Nagykun-Szolnok	85,000	A1, A4i, B1i, B1iv, B2
030	HU036	Biharugra fish-ponds	Biharugrai-halastavak	Békés	16,000	A1, A4i, B1i, B2
031	—	Bihar plain	Bihari puszták	Hajdú-Bihar	48,000	A1, B2
032	HU016, HU018, HU020, HU043	Hortobágy	Hortobágy	Hajdú-Bihar, Jász-Nagykun-Szolnok	136,300	A1, A4i, A4ii, B1i, B1iii, B2
033	HU017	Kisköre reservoir	Kiskörei-víztározó	Borsod-Abaúj-Zemplén, Heves, Jász-Nagykun-Szolnok	12,700	A4i, B1i, B2
034	HU014	Upper Tisza	Tisza Tuzsér és Rakamaz között	Borsod-Abaúj-Zemplén, Szabolcs-Szatmár-Bereg	10,000	B2
035	—	Szatmár-Bereg plains	Szatmár-Beregi-sík	Szabolcs-Szatmár-Bereg	117,300	A1, B2
036	—	Heves plains	Hevesi-sík	Heves, Jász-Nagykun-Szolnok	49,000	A1, A4ii, B1iii, B2
037	HU039	Borsodi-Mezőség	Borsodi-Mezőség	Borsod-Abaúj-Zemplén	28,900	A1, A4i, B1i, B2
038	HU034, HU037	Kesznyéten	Tiszaluc-Kesznyéteni puszta	Borsod-Abaúj-Zemplén, Szabolcs-Szatmár-Bereg	7,000	B2
039	HU032	Bodrog flood-plain	Bodrogzug	Borsod-Abaúj-Zemplén, Szabolcs-Szatmár-Bereg	10,000	A1, A4i, B1i, B2
040	HU030	Mátra mountains	Mátra-hegység	Heves, Nógrád	58,100	A1, B2
041	HU013	Bükk mountains	Bükk-hegység	Borsod-Abaúj-Zemplén, Heves	61,000	A1, B2
042	HU012	Zemplén mountains	Zempléni-hegység	Borsod-Abaúj-Zemplén	102,000	A1, B2
043	HU028	Aggtelek karst and Bódva valley	Aggteleki-karszt és a Bódva-völgy	Borsod-Abaúj-Zemplén	28,600	A1, B2

Sites identified in the first inventory of IBAs in Europe (Grimmett and Jones 1989) but no longer considered to be IBAs
HU005 Pilis; HU007 Ócsa; HU011 Szaporca Ó-Dráva; HU015 Közép-Tisza; HU019 Áároktő-tiszacsegei hullámtér; HU024 Péteri-tó; HU026 Mártély; HU033 Sárrét; HU038 Müller-szék; HU040 Hajdúsági erdőspuszta; HU041 Nagykunsági puszta; HU044 Bélmegyeri puszta; HU046 Szabadkígyósi puszta.

status in Europe (Tucker and Heath 1994). Thirty-one SPECs regularly occur on passage or overwinter, and include the globally threatened *Anser erythropus*, *Branta ruficollis* and *Numenius tenuirostris*.

Twenty-five sites were selected as IBAs on the basis of internationally important populations of globally threatened or near-threatened species, meeting criterion A1 (Table 2). The majority of IBAs in Hungary are important as staging areas for migrating wildfowl. Twenty-five wetland sites qualify as IBAs as they hold more than 1% of the biogeographic populations of those species listed in Table 3, and therefore meet criterion A4i. A further site (016) meets criterion B1i only (Table 3).

Thirty-four IBAs support important populations of SPECs, meeting criteria B2 and B3 (Table 4). Also of significance is that Hungary holds a large proportion (c.10%) of European populations of the following SPECs, for which several IBAs have been identified: *Ixobrychus minutus* (IBAs 003, 004, 008, 030, 032), *Nycticorax nycticorax* (IBAs 002, 004, 008, 009, 024, 026, 029, 032, 033, 038, 039), *Platalea leucorodia* (IBAs 023, 026, 032), *Falco vespertinus* (IBAs 020, 021, 022, 029, 031, 032, 035, 036, 037) and *Falco cherrug*.

Table 2. Important Bird Areas in Hungary that are important for species of global conservation concern (meeting criterion A1).

Species	IBA code
Phalacrocorax pygmeus Pygmy Cormorant	032
Anser erythropus Lesser White-fronted Goose	032
Aythya nyroca Ferruginous Duck	004, 006, 007, 008, 011, 013, 026, 030, 032
Haliaeetus albicilla White-tailed Eagle	007, 010, 026, 030, 032
Aquila heliaca Imperial Eagle	013, 015, 018, 032, 036, 040, 041, 042, 043
Crex crex Corncrake	019, 035, 039, 043
Otis tarda Great Bustard	001, 022, 029, 031, 032, 036, 037
Numenius tenuirostris Slender-billed Curlew	032
Acrocephalus paludicola Aquatic Warbler	032

Table 3. Important Bird Areas in Hungary that support important numbers of one or more congregatory species (i.e. meeting criteria A4 and/or B1). IBAs meeting both criteria A4 and B1 for the species are shown in **bold**. IBAs meeting only criterion B1 for the species concerned, and not A4, are shown in normal type. For key to 'Season', see p. 7.

Species	Season	IBA code
Phalacrocorax carbo Cormorant	W	**017**
	P	026, 030
Egretta alba Great White Egret	R	**002, 012, 033**
	B	**004, 008, 023, 025, 026, 033**
	P	**003, 006, 007, 010, 011**
Ciconia nigra Black Stork	P	**010, 039**
Platalea leucorodia Spoonbill	B	004, **012, 023, 026, 032**
	P	**003, 030**, 037
Anser fabalis Bean Goose	W	**003, 004, 007, 009, 010, 012, 014**, 016, **017**
	P	**005, 006, 011, 032**
Anser albifrons White-fronted Goose	W	003, 004, 007, 009, 010, 012, 014, 026, 033
	P	011, 022, 027, **028**, 030, **032**
Anser erythropus Lesser White-fronted Goose	P	**032**
Anser anser Greylag Goose	B	003, 004, 006, 032
	W	**004**, 007
	P	**003**, 005, 006, 009, 011, 012, **022**, 026, 028, 030, 032, 033
Anas strepera Gadwall	P	003
Anas crecca Teal	P	003, **032**
Anas platyrhynchos Mallard	W	011, 016
	P	005, 026, 030
Anas acuta Pintail	P	022
Anas clypeata Shoveler	P	**032**
Netta rufina Red-crested Pochard	P	**003**
Aythya ferina Pochard	P	**005**
Bucephala clangula Goldeneye	W	**005**, 010, 016, 017
	P	032
Falco vespertinus Red-footed Falcon	B	**032**
	P	**036**
Fulica atra Coot	W	**004**
Grus grus Crane	W	**028**
	P	**026, 027, 032, 037**, 039
Recurvirostra avosetta Avocet	P	026
Philomachus pugnax Ruff	P	**032**
Limosa limosa Black-tailed Godwit	P	**022, 026**, 028, **029, 030, 032**
Numenius arquata Curlew	P	**022, 032**
Tringa erythropus Spotted Redshank	P	**026, 032**
Larus cachinnans Yellow-legged Gull	P	026
Chlidonias hybridus Whiskered Tern	B	**025, 032**

Table 4. Species of European conservation concern with significant breeding populations at Important Bird Areas in Hungary (meeting any IBA criteria).

Species [1]	Minimum national breeding population (pairs) [2]	Proportion (%) of national population breeding at all IBAs in Hungary
Phalacrocorax pygmeus Pygmy Cormorant	10	100[3]
Botaurus stellaris Bittern	400	28
Ixobrychus minutus Little Bittern	3,500	13
Nycticorax nycticorax Night Heron	1,800	93
Ardeola ralloides Squacco Heron	200	55
Ardea purpurea Purple Heron	1,600	28
Ciconia nigra Black Stork	150	57
Ciconia ciconia White Stork	4,500	29
Platalea leucorodia Spoonbill	600	100[3]
Aythya nyroca Ferruginous Duck	500	89
Haliaeetus albicilla White-tailed Eagle	55	51
Circaetus gallicus Short-toed Eagle	40	45
Aquila pomarina Lesser Spotted Eagle	100	20
Aquila heliaca Imperial Eagle	45	64
Falco vespertinus Red-footed Falcon	2,000	28
Perdix perdix Partridge	15,000	3
Porzana porzana Spotted Crake	500	34
Porzana parva Little Crake	3,000	8
Porzana pusilla Baillon's Crake	30	100[3]
Crex crex Corncrake	300	100
Otis tarda Great Bustard	1,000 individuals	100[3]
Limosa limosa Black-tailed Godwit	1,000	27
Chlidonias hybridus Whiskered Tern	1,700	57
Chlidonias niger Black Tern	500	42
Alcedo atthis Kingfisher	800	17
Coracias garrulus Roller	300	62
Riparia riparia Sand Martin	60,000	26
Acrocephalus paludicola Aquatic Warbler	610	85
Lanius minor Lesser Grey Shrike	5,000	8

1. Only those species of European conservation concern (see Box 1, p. 12) that meet IBA criteria in Hungary are listed.
2. Data are taken from the BirdLife/EBCC European Bird Database 1998 (Heath and Borggreve 2000).
3. The percentage of the national population in IBAs exceeds 100%. Usually this is because the national population estimate has not been updated recently whilst the IBA population estimate has been recently updated with new data as a result of comprehensive surveys of IBAs themselves. Also, the individual site count for a species may be the maximum or average over recent years, and summing these may record more birds than are present nationally in any single year.

Those species present in Hungary that are not adequately covered by the IBA inventory, due to the dispersed nature of their distribution, include *Ciconia ciconia*, *Falco tinnunculus*, *Athene noctua*, *Coracias garrulus*, *Caprimulgus europaeus*, as well as doves, woodpeckers and passerines.

HABITATS

Hungary is dominated by artificial landscapes, with arable land covering 50% of its land area. Grasslands, which occupy 12%, are predominantly semi-natural in origin and are found mainly on saline (clay), sandy or peaty soils in the lowlands, as well as in some hilly areas. 19% of the country is forested (mainly *Quercus*, *Carpinus*, *Fagus*), with plantations of non-native *Robinia pseudacacia*, *Populus* and coniferous species making up half of the total. Fourteen percent of the land area remains uncultivated, and includes urban areas, infrastructure, wetlands and military areas.

Thirty-six Hungarian IBAs (84%) support grasslands (Figure 1), with cover exceeding 50% at 10 IBAs. Most of these are either wet grassland flood-plains or areas of steppe. Steppic habitats are often associated with non-intensive methods of arable farming, and have a fragmented distribution following the expansion of intensive agricultural practices. They are important for *Otis tarda*, *Falco vespertinus*, *Buteo rufinus*, *Burhinus oedicnemus*, *Coracias garrulus* and *Lanius minor*. Steppic habitats have been successfully restored on the Moson plain (001) by setting aside arable land, as evidenced by a substantial increase in *Otis tarda* numbers.

Thirty IBAs (70%) contain areas of wetland. River corridors support *Nycticorax nycticorax*, *Ardeola ralloides*, and *Riparia riparia* colonies; lakes hold *Anser albifrons* and *A. fabalis* on passage; saline lakes hold breeding *Recurvirostra avosetta*, *Charadrius*

Figure 1. Habitats at Important Bird Areas in Hungary (see Appendix 3 for definitions of habitats).

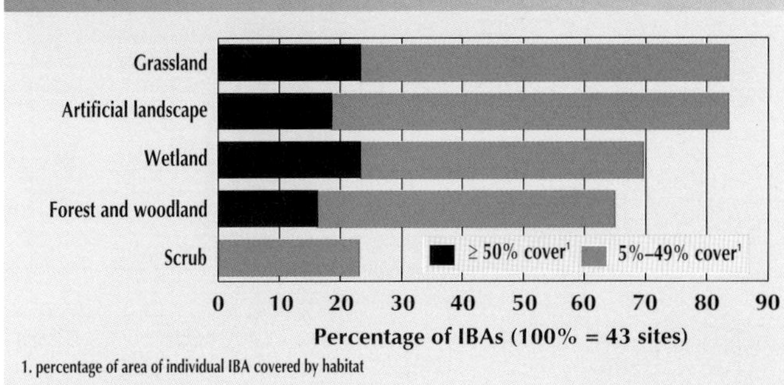

Figure 2. Land-uses at Important Bird Areas in Hungary (see Appendix 3 for definitions of land-uses).

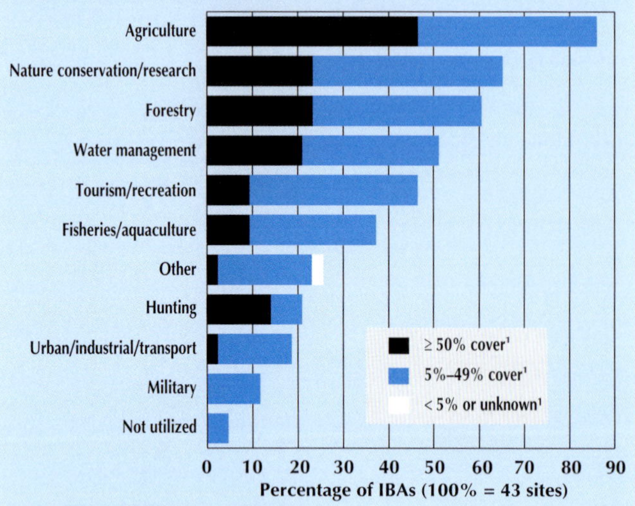

1. percentage of area of individual IBA covered by land-use

alexandrinus, *Tringa totanus*, *Limosa limosa* and migrating waders and Anatidae; and fish-ponds support colonies of *Platalea leucorodia*, *Egretta alba* and *Chlidonias hybridus*, breeding *Aythya nyroca* and wintering *Haliaeetus albicilla*.

Forest occurs in 28 IBAs (65%), and there are seven IBAs where more than 50% of the surface area is wooded. Low mountains are predominantly covered by broadleaved deciduous forest, whilst flood-plains are characterized by both softwood and hardwood alluvial forest. Forested areas are important for *Ciconia nigra*, raptors (*Haliaeetus albicilla*, *Aquila heliaca*, *Aquila pomarina*, *Circaetus gallicus*) and woodpeckers. Ten IBAs (23%) contain some scrub habitat, which usually takes the form of scrub encroaching onto meadows, but also includes karst-rock forests.

IMPACTS ON IBAs – LAND-USE AND THREATS

The dominant land-uses in Hungarian IBAs are agriculture and forestry (Figure 2). Forestry is widespread and management is generally intensive, although restrictions on forestry practices were recently introduced by the Conservation and Forestry Acts. The share of nature conservation is also high, although it is a priority only in strictly protected areas such as National Parks and Ramsar Sites.

Hunting is a more widespread activity than is inferred by the site accounts, as areas are not usually designated exclusively for this purpose. Hunting does, in fact, take place in every IBA, although wildfowl shooting is prohibited on some Ramsar Sites (Ministry of Agriculture Decree No. 8/1993). Fisheries exist in 16 Hungarian IBAs (37%), with four sites supporting fisheries over 50% or more of their area. Conflicts are most obvious on fish-ponds where the persecution of birds and the burning of vegetation are very common.

The abandonment of land-management practices and the intensification of agriculture threaten a large proportion of Hungarian IBAs (Figure 3). Both are likely to become more of a problem in the future, although the former poses a more immediate threat than the latter.

Undergrazing is a problem on saline grasslands, affecting species such as *Glareola pratincola* and *Charadrius alexandrinus* that are associated with bare soil and require high densities of insects. *Aquila heliaca* and *Falco cherrug* are also affected due to a decrease in the area of habitat available to their prey *Citellus citellus*. Agricultural intensification is evidenced by increasing applications of fertiliser and pesticides, increased production of sunflower and maize replacing crops such as wheat and rape, silage-making, irrigation, and the conversion of grasslands into arable land. The latter practice directly affects farmland birds such as *Otis tarda* and *Perdix perdix*.

Disturbance to birds affects 31 IBAs (72%), and is of medium to low impact at 24 sites. The number of people collecting mushrooms and herbs has increased following substantial job losses, with a consequent increase in disturbance to birds such as geese, raptors, *Grus grus* and *Otis tarda*.

Drainage is another widespread threat, affecting over 50% of the land area of 19 IBAs. Tourism affects 18 IBAs (42%), mainly those comprising forested hills and wetlands. Natural events (17 IBAs (39%)) include droughts, which have been very severe over the last 10 years, and have badly affected many saline lakes and humid grasslands.

PROTECTION STATUS

Table 5 and Figures 4 and 5 summarize the national and international protection status of all Hungarian IBAs.

National protection

The Ministry of the Environment (KTM) is responsible for nature conservation at a national level. The regional authorities are called National Park Directorates, and are responsible for all protected areas within their region. All protected areas in Hungary fall into one of the following categories.

Figure 3. Threats at Important Bird Areas in Hungary (see Appendix 3 for definitions of threat types and impact categories).

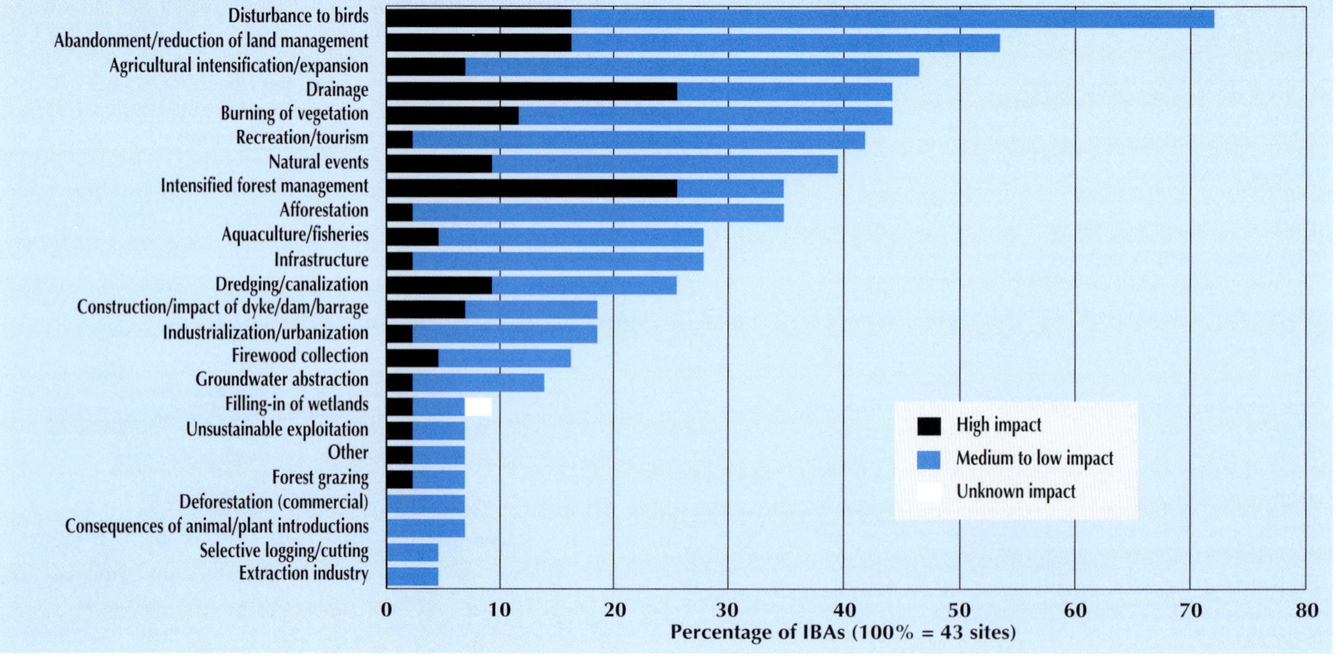

Figure 4. The national protection status of Important Bird Areas in Hungary.

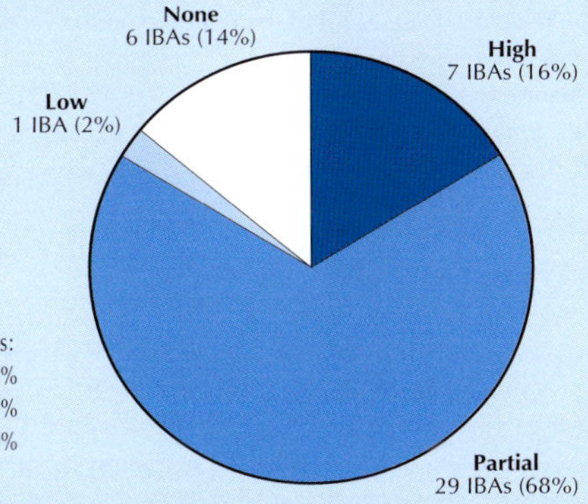

% of area of individual IBA overlaps with national protected areas:
High >90%
Partial 10–90%
Low <10%
None

Total area of overlap between IBA network in Hungary and national protected-area system (see Table 5 for categories) = 4,525–5,039 km² (31–34% of total IBA area).

Figure 5. The international protection status of Important Bird Areas in Hungary.

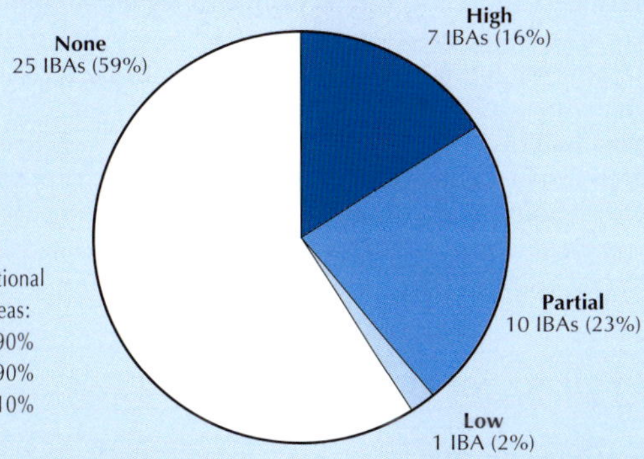

% of area of individual IBA overlaps with international protected areas:
High >90%
Partial 10–90%
Low <10%
None

Total area of overlap between IBA network in Hungary and international protected-area system (see Table 5 for categories) = 2,055–2,216 km² (14–15% of total IBA area).

Table 5. Protection status of Important Bird Areas in Hungary. A tick (✔) indicates that an IBA overlaps with a protected area (to any extent).

		National			International		
IBA code	International name	National Park	Nature Conservation Area	Landscape Protected Area	Ramsar Site	World Heritage Site	Biosphere Reserve
001	Moson plain						
002	Hanság	✔					
003	Lake Fertő	✔			✔		✔
004	Kis-balaton	✔			✔		
005	Lake Balaton				✔		
006	Nagyberek		✔				
007	Inner Somogy	✔	✔	✔			
008	Pacsmag fish-ponds		✔		✔		
009	Béda-Karapancsa	✔			✔		
010	Gemenc	✔			✔		
011	Sárvíz valley		✔	✔	✔		
012	Lake Velence and Dinnyés marshes		✔		✔		
013	Vértes mountains and Zámoly basin			✔			
014	Lake Öreg at Tata				✔		
015	Gerecse mountains			✔			
016	Danube reach between Gönyű and Szob						
017	Danube bend	✔					
018	Börzsöny mountains	✔		✔			
019	Middle reach of River Ipoly	✔					
020	Tápió valley			✔			
021	Puszta at Jászkarajenő						
022	Danube plain	✔			✔		✔
023	Lake Kolon	✔			✔		
024	Flood-plain at Tiszaalpá	✔					
025	Lake Fehér at Gátér						
026	Pusztaszer Landscape Protection Area			✔	✔		
027	Puszta at Pitvaros	✔					
028	Lake Fehér at Kardosk	✔			✔		
029	Dévaványa plain	✔					
030	Biharugra fish-ponds	✔			✔		
031	Bihar plain		✔	✔			
032	Hortobágy	✔			✔		✔
033	Kisköre reservoir	✔			✔		
034	Upper Tisza		✔				
035	Szatmár-Bereg plains			✔			
036	Heves plains			✔			
037	Borsodi-Mezőség			✔			
038	Kesznyéten			✔			
039	Bodrog flood-plain		✔	✔	✔		
040	Mátra mountains		✔	✔			
041	Bükk mountains	✔					
042	Zemplén mountains			✔			
043	Aggtelek karst and Bódva valley	✔				✔	✔
Total number of IBAs		**20**	**9**	**15**	**17**	**1**	**4**

Box 1. International legislation and initiatives that are relevant to site conservation in Hungary (see Appendix 1 for a general description of these agreements).

Global	
Biodiversity Convention	✔
Ramsar Convention	✔
Bonn Convention	✔
World Heritage Convention	✔
MAB Programme	✔
Pan-European	
Bern Convention	✔

✔ Convention ratified/initiative supported
(✔) Convention signed

1. National Parks
Extensive areas of outstanding natural value that serve nature conservation, educational and recreational purposes, with activities arranged according to management zones in line with IUCN standards (14/1997 Order of KTM).

2. Nature Conservation Areas
These are usually small areas (under 1,000 ha) of important natural value, and can be designated at a national and/or local level. They are established by the Minister of the Environment and regional policy initiatives (National Nature Conservation Areas), or by local municipalities (Local Nature Conservation Areas). Bird sanctuaries are usually protected in this way.

3. Landscape Protected Areas
These provide protection for medium- to large-sized areas (over 1,000 ha) with characteristic natural and scenic value.

Hungary has nine National Parks, 138 Nature Conservation Areas, 35 Landscape Protected Areas, and more than 1,500 Local Nature Conservation Areas (E. Lakics pers. comm. 1998).

■ International protection

Hungary is party to various international conventions and initiatives (Box 1). Seventeen IBAs overlap with Ramsar Sites (Table 5), including Lake Balaton (005), which has a temporary Ramsar designation that does not confer legal protection according to Hungarian legislation. Six IBAs meet the Ramsar criteria but have not, as yet, been designated as Ramsar Sites (002, 006, 017, 025, 027, 029). Sections of four IBAs are designated as Biosphere Reserves, and one IBA is designated as a World Heritage Site (Table 5).

CONSERVATION

- Statutory nature conservation activity is organized according to the Nature Conservation Act (53/1996) and subsequent legislation.
- A six-year National Environmental Programme was approved by the Hungarian Parliament in September 1997. The main elements that affect IBAs are:

- a basic ecological survey and management plan to be prepared for each protected area;
- environmentally sensitive areas to be designated, where farmers can receive annual payments for maintaining extensive forms of land-use;
- a national biomonitoring programme to be implemented, to include monitoring of bird populations.

- Almost half of the IBAs in Hungary now have management plans written for them, and the number is increasing rapidly.
- The designation of zonal programmes according to European Union Regulation 2078/92 will follow accession negotiations that began in March 1998.

ANALYTICAL METHODS

- Population estimates for rare and colonial birds are based mainly on good-quality data from 1992–1996.
- Data for waterbirds are of good quality and are based on regular (monthly) synchronized counts from the period 1985–1996. Where data allowed, figures are given as average annual peaks, but in some cases (where fluctuation is high) minimum and maximum counts are given instead.
- Population data on common passerines, woodpeckers and crakes (except *Crex crex*) are usually rough estimates.
- Threats and land-use data are from the period 1993–1996.
- Site boundaries were digitized using official topographic maps (scale 1:100,000) and stored in a geographical information system held by the University of Gödöllő, and are available from MME/BirdLife Hungary.

GLOSSARY

holm an islet in a river.
karst a limestone region with underground drainage and many cavities and passages caused by the dissolution of the rock.
loess a deposit of fine, light-coloured, wind-blown dust, found especially in the basins of large rivers.
LPA Landscape Protected Area.
MME/BirdLife Hungary Magyar Madártani és Természetvédelmi Egyesület (BirdLife International Partner in Hungary).
PHARE Programme an EC initiative to provide grants in support of economic and social transformation in the countries of central and eastern Europe.
puszta the flat, almost treeless steppe-grassland of the Hungarian plains.

ACKNOWLEDGEMENTS

János Mille provided valuable assistance by inputting and checking data. The following also gave valuable input into the IBA inventory: Ferenc Márkus, Zoltán Waliczky (former co-ordinator of the Hungarian IBA Project), György Szimuly (in charge of the Monitoring Centre at MME), Dr. Sándor Faragó (leader of the Hungarian Goose Research Group at the University of Sopron) and E. Lakics. Special thanks go to the volunteers who provided the data for the site accounts: Béla Ambrus, János Bagyura, László Bank, József Berdó, István Béres, Péter Bod, Mihály Bodnár, Péter Borbáth, Emil Boros, László Csihar, Péter Csonka, Tamás Deme, László Demeter, Szilárd Drexler, Zoltán Ecsedi, Tibor Ezsől, Imre Fatér, László Fenyősi, Zoltán Filotás, István Fintha, Gábor Firmánszky, Balázs Forgách, Elemér Futó, Tibor Fülöp, József Gyurácz, Nándor Hivatal, Jenő Horváth, Róbert Horváth, Zoltán Horváth, Zoltán Ilonczay, Ferenc Jánoska, Ádám Kaczkó, Béla Kalivoda, Béla Kalocsa, Zsolt Kalotás, Roland Kern, Gábor Kovács, István Kurpé, András Lelkes, Csaba Megyer, József László Mercsák, Sándor Mogyorósi, Gyula Molnár, Zoltán Molnár, László Musicz, Tamás Nagy, Sándor Palkó, Attila Pellinger, László Puskás, István Sándor, Ádám Selmeczi Kovács, István Staudinger, Balázs Szabó, Sándor Szász, Zsolt Szegedi, Antal Széll, Tibor Szép, Tamás Szitta, László Tajti, János Tar, László Tóth, Tamás Tóth, János Tőgye, Tibor Tömösváry, Sándor Urbán, Róbert Ványi, András Vasas, Zsolt Végvári, Róbert Veprik, Tibor Vincze, Levente Viszló, Walter Wettstein, György Zsigrai.

Funds have been generously provided by the Ministry of the Environment, the Central Environmental Fund of Hungary, the Know How Fund, the Royal Society for the Protection of Birds, Vogelbescherming, and the Ministry of Agriculture, Forestry and Nature Management of the Netherlands.

SITE ACCOUNTS

Moson plain — A1, B2 — 001

Admin region Győr-Sopron-Moson
Coordinates 47°51′N 17°12′E
Altitude 119–128 m **Area** 4,310 ha

Site description

Lowland underlain by alluvial deposits (gravel from the River Danube) and silty loess, which results in thin, poor soils. The town of Mosonmagyaróvár and the borders with Austria and Slovakia surround the site, which is in a predominantly agricultural area. Set-aside is covered by permanent grass with narrow strips of oil-seed rape.

Habitats Grassland (5%; mesophile grassland), Artificial landscape (95%; highly improved reseeded grassland; arable land; forestry plantation)
Land-use Agriculture (75%), Forestry (5%), Hunting (100%), Nature conservation/research (25%)

Birds

Species	Season	Year	Pop min	Pop max	Acc	Criteria
Perdix perdix Partridge	R	1996	180	180	A	B2
Otis tarda Great Bustard	R	1996	82	82	A	A1, B2

An important area for farmland birds, especially for *Perdix perdix* (the largest local population in Hungary) and *Otis tarda*.

Protection status

National None **International** None

Conservation issues

Threats Agricultural intensification/expansion (B), Deforestation (commercial) (B), Groundwater abstraction (C), Other (B)

Privatization ('Other' threat) of land is reducing field sizes, threatening *Otis tarda*. Habitat improvement projects have been implemented by a collaboration between the Lajta-Hanság Joint Stock Company, the Wildlife Management Department, the University of Sopron, Fertő-Hanság National Park and the World Wide Fund for Nature – Austria. The Lajta Project promotes the creation of chemical-free headlands, delayed harvesting of field edges, the creation of harrowed fallow strips, and the control of corvids. The Moson Project, a set-aside programme designed to benefit *Otis tarda*, covers 1,232 ha.

Hanság — A4i, B1i, B2 — 002

Admin region Győr-Sopron-Moson
Coordinates 47°45′E 17°15′E
Altitude 110–120 m **Area** 27,700 ha

Site description

A poorly-drained flood-plain, situated between Kapuvár and Csorna. Human activities are partially controlled, and include mowing and grazing of grasslands, peat extraction, reed-harvesting and hunting.

Habitats Forest and woodland (11%; alluvial/very wet forest), Grassland (30%; humid grassland), Wetland (5%; standing fresh water; raised bog; water-fringe vegetation), Artificial landscape (54%; forestry plantation)
Land-use Agriculture (35%), Forestry (60%), Nature conservation/research (70%)

Birds

An important area for marshland birds. Numbers of breeding *Egretta alba* have increased markedly due to artificial flooding on c.1,000 ha. Species of global conservation concern that do not meet IBA criteria: *Aythya nyroca* (2–3 breeding pairs), *Haliaeetus albicilla* (wintering), *Crex crex* (4–5 breeding pairs), *Otis tarda* (1–2 breeding individuals).

Species	Season	Year	Pop min	Pop max	Acc	Criteria
Nycticorax nycticorax Night Heron	B	1996	200	200	A	B2
Egretta alba Great White Egret	R	1996	110	160	A	A4i, B1i
Ciconia ciconia White Stork	B	1996	250	300	A	B2

Protection status

National Partial **International** None
7,086 ha of IBA covered by National Park (Fertő-Hanság, 12,543 ha).

Conservation issues

Threats Abandonment/reduction of land management (A), Afforestation (C), Burning of vegetation (C), Drainage (A), Natural events (B)

The main threat is the abandonment of grassland management practices, leading to loss of habitat. A further 3,500 ha, the majority of which is already owned by the Fertő-Hanság National Park Directorate, is being designated as a protected area. A management plan exists, and the IBA is a proposed Biosphere Reserve.

Lake Fertő — A4i, B1i, B2, B3 — 003

Admin region Győr-Sopron-Moson
Coordinates 47°36′N 16°38′E
Altitude 115–261 m **Area** 12,542 ha

Site description

A large saline lake situated on the Austrian-Hungarian border. The IBA includes the Hungarian part of the lake (an area of 75 km²), of which 88% is covered by reedbeds (*Phragmites*), and adjacent mesophile and wet saline grasslands. Human activities include reed-harvesting ('Other' land-use, below), a small-scale fishery and recreation.

Habitats Grassland (5%; humid grassland; mesophile grassland), Wetland (65%; standing brackish and salt water; water-fringe vegetation), Artificial landscape (30%; highly improved reseeded grassland; arable land)
Land-use Agriculture (15%), Fisheries/aquaculture (65%), Hunting (98%), Nature conservation/research (100%), Other, Water management (65%)

Birds

Species	Season	Year	Pop min	Pop max	Acc	Criteria
Ixobrychus minutus Little Bittern	B	1995	35	40	C	B2
Egretta alba Great White Egret	P	1996	2,600	—	A	A4i, B1i
Platalea leucorodia Spoonbill	P	1996	200	—	C	A4i, B1i
[1] ***Anser fabalis*** Bean Goose	W	1995	4,000	16,000	A	A4i, B1i
Anser albifrons White-fronted Goose	W	1995	100	3,600	A	B1i
Anser anser Greylag Goose	B	1995	150	—	A	B1i
[2] ***Anser anser*** Greylag Goose	P	1996	100	11,000	A	A4i, B1i
Anas strepera Gadwall	P	1996	1,000	—	B	B1i
Anas crecca Teal	P	1995	10,000	—	C	B1i
Netta rufina Red-crested Pochard	P	1995	800	—	A	A4i, B1i
Porzana parva Little Crake	B	1996	100	—	C	B3

1. Minimum/maximum annual peaks.
2. Minimum/maximum passage birds 1986–1991.

An important breeding and staging area for waterbirds. More than 255 bird species have been recorded.

Protection status

National High **International** High
12,542 ha of IBA covered by National Park (Fertő-Hanság, 12,543 ha). 2,870 ha of IBA covered by Ramsar Site (Fertő-tó, 2,870 ha). 12,542 ha of IBA covered by Biosphere Reserve (Lake Fertő, 12,542 ha).

Conservation issues

Threats Abandonment/reduction of land management (C), Agricultural intensification/expansion (C), Aquaculture/fisheries (B), Construction/impact of dyke/dam/barrage (A), Disturbance to birds (B), Drainage (A), Dredging/canalization (C), Industrialization/urbanization (C), Infrastructure (C), Natural events (B), Recreation/tourism (B)

Wildfowl shooting is banned, although shooting of red fox *Vulpes vulpes* and wild boar *Sus scrofa* is undertaken for conservation purposes. A wetland restoration project at Mexikópuszta has led to the site becoming extremely important for waterbirds. PHARE has funded an eco-tourism project within the transboundary National Park, and a management plan exists for the area.

Kis-balaton — A1, A4i, B1i, B2, B3 — 004

Admin region Somogy, Zala
Coordinates 46°40′N 17°13′E
Altitude 105–110 m **Area** 14,745 ha

Site description

Situated south-west of Lake Balaton at the mouth of the Zala river. Originally part of the lake, the site has silted up and reverted to mire. In 1992 51 km² of adjacent agricultural lands were flooded, and open water and extensive reedbeds (*Phragmites*) and sedgebeds (*Carex*) are now the main habitat types. Human activities include construction works, angling, fishing and reed-harvesting ('Other' land-use, below).

Habitats Wetland (50%), Artificial landscape (45%)
Land-use Agriculture (20%), Fisheries/aquaculture (7%), Forestry (10%), Nature conservation/research (50%), Not utilized (5%), Other (40%), Water management (60%)

Birds

Species	Season	Year	Pop min	Pop max	Acc	Criteria
Ixobrychus minutus Little Bittern	B	1996	50	100	C	B2
Nycticorax nycticorax Night Heron	B	1996	100	150	A	B2
Ardeola ralloides Squacco Heron	B	1996	20	50	—	B2
Egretta alba Great White Egret	B	1996	40	50	A	A4i, B1i
Ardea purpurea Purple Heron	B	1996	40	50	A	B2
Platalea leucorodia Spoonbill	B	1996	10	50	A	B1i, B2
Anser fabalis Bean Goose	W	1991	10,000	30,000	A	A4i, B1i
Anser albifrons White-fronted Goose	W	1995	3,000	8,000	A	B1i
Anser anser Greylag Goose	B	1996	200	250	B	B1i
[1] ***Anser anser*** Greylag Goose	W	1995	800	10,500	A	A4i, B1i
Aythya nyroca Ferruginous Duck	B	1996	100	200	C	A1, B2
Porzana porzana Spotted Crake	B	1996	100	200	C	B3
Porzana parva Little Crake	B	1996	50	50	C	B3
Fulica atra Coot	W	1996	30,000	40,000	A	A4i, B1i
[2] ***Chlidonias hybridus*** Whiskered Tern	B	1996	10	400	A	B2

1. Large increase since 1991.
2. Fluctuates according to status of floating vegetation.

This site is important for breeding herons (Ardeidae) and migratory wildfowl, especially geese. Species of global conservation concern that do not meet IBA criteria: *Haliaeetus albicilla* (1–2 breeding pairs; 10–15 wintering birds), *Crex crex* (10–15 breeding pairs). Nationally important for breeding *Phalacrocorax carbo* (minimum 950 pairs).

Protection status

National High **International** High
14,745 ha of IBA covered by National Park (Balaton-felvidéki, 56,696 ha). 14,745 ha of IBA covered by Ramsar Site (Kis-Balaton, 14,745 ha).

Conservation issues

Threats Abandonment/reduction of land management (C), Afforestation (C), Aquaculture/fisheries (C), Dredging/canalization (C), Natural events (B), Other (A)

A change in the ecological character of the area has been noted following recent land-management work. Spring floods destroy the clutches of reedbed-nesting birds, and fish-kills and cases of botulism occur regularly due to pollution from sewage outfalls and contaminated sediments ('Other' threat, above). An environmental impact-monitoring programme was initiated in 1992, and wildfowl hunting has been banned.

Lake Balaton — A4i, B1i — 005

Admin region Somogy, Zala, Veszprém
Coordinates 46°55′N 17°54′E
Altitude 105–105 m **Area** 59,800 ha

Site description

A large (77 km long), shallow (1–14 m deep), highly regulated lake, the shore of which is densely built-up. Submerged vegetation covers c.5% of the total surface area of the lake, and the water is moderately rich in nutrients. Human activities include fishing, reed- and ice-cutting, sailing and coarse angling.

Habitats Wetland (100%; standing fresh water)
Land-use Fisheries/aquaculture (100%), Tourism/recreation (100%), Urban/industrial/transport (20%), Water management (100%)

Birds

Species	Season	Year	Pop min	Pop max	Acc	Criteria
Anser fabalis Bean Goose	P	1995	5,000	8,000	A	A4i, B1i
Anser anser Greylag Goose	P	1995	1,000	1,500	A	B1i
Anas platyrhynchos Mallard	P	1995	20,000	25,000	A	B1i
Aythya ferina Pochard	P	1995	15,000	17,000	A	A4i, B1i
Bucephala clangula Goldeneye	W	1995	13,000	15,000	A	A4i, B1i

This site is an important area for wintering wildfowl, especially diving ducks.

Protection status

National None **International** High
59,800 ha of IBA covered by Ramsar Site (Balaton, 59,800 ha).

Conservation issues

Threats Burning of vegetation (C), Disturbance to birds (B), Filling-in of wetlands (C), Industrialization/urbanization (A), Recreation/tourism (B)

The lake is affected by run-off from surrounding agricultural land, development of the lakeshore, and fragmentation of reedbeds (*Phragmites*). The Hungarian government is making efforts to revert the lake to its original, nutrient-poor state through sediment removal and sewage treatment. Wildfowl hunting is prohibited, and a long-term research programme is underway. A section of the north shore is part of the proposed Balaton-felvidéki National Park.

Nagyberek — A1, A4i, B1i, B2 — 006

Admin region Somogy
Coordinates 46°44′N 17°33′E
Altitude 103–120 m **Area** 19,400 ha

Site description

A former marshland on the southern side of Lake Balaton, where only small fragments of original habitat remain following land reclamation. The IBA comprises hay meadows and arable land with scattered *Alnus* and *Salix* woodlands, wet meadows, reedbeds (*Phragmites*), an extensive ditch network and several fish-ponds. Human activities include arable farming, cattle-grazing, haymaking and reed-harvesting ('Other' land-use, below).

Habitats Forest and woodland (10%; alluvial/very wet forest), Scrub (10%; scrub), Grassland (50%; humid grassland; mesophile grassland), Wetland (14%; standing fresh water; water-fringe vegetation), Artificial landscape (16%; highly improved reseeded grassland; arable land; forestry plantation)
Land-use Agriculture (50%), Fisheries/aquaculture (5%), Hunting (35%), Other (5%), Tourism/recreation (10%)

Birds

Species	Season	Year	Pop min	Pop max	Acc	Criteria
Egretta alba Great White Egret	P	1995	300	400	A	A4i, B1i
Anser fabalis Bean Goose	P	1995	4,000	4,500	A	A4i, B1i
Anser anser Greylag Goose	B	1995	150	200	A	B1i
Anser anser Greylag Goose	P	1995	1,000	1,500	A	B1i
Aythya nyroca Ferruginous Duck	B	1995	27	30	A	A1, B2

The IBA is important for both breeding and migratory wildfowl. Species of global conservation concern that do not meet IBA criteria: *Haliaeetus albicilla* (breeding and wintering), *Crex crex* (breeding, no data). 206 species have been recorded either as having bred or as having occurred whilst on migration.

Protection status

National Low **International** None
1,537 ha of IBA covered by Nature Conservation Area (Nagybereki Fehérvíz, 1,537 ha).

Conservation issues

Threats Aquaculture/fisheries (B), Construction/impact of dyke/dam/barrage (B), Disturbance to birds (A), Drainage (A), Infrastructure (B), Recreation/tourism (C)

Grasslands and reedbeds are threatened by uncontrolled burning, whilst the development of the M7 motorway will probably destroy the most valuable habitats. The designation of several smaller protected areas is in process.

Inner Somogy — A1, A4i, B1i, B2 — 007

Admin region Somogy
Coordinates 46°13′N 17°18′E
Altitude 110–180 m **Area** 216,300 ha

Site description

A forested landscape situated between the River Dráva and Marcali, and interspersed with smaller blocks of a variety of habitats, including an old fish-pond system, orchards, and wooded pasture. Human activities include arable farming and angling. This area includes one site from the previous international IBA inventory (Grimmett and Jones 1989): 'Barcs' (former site HU010).

Habitats Forest and woodland (35%; broadleaved deciduous forest; mixed forest; alluvial/very wet forest), Grassland (5%; steppe/dry calcareous grassland; humid grassland), Wetland (10%; standing fresh water; river/stream; water-fringe vegetation), Artificial landscape (50%; arable land; perennial crops/orchards/groves; forestry plantation)
Land-use Agriculture (45%), Fisheries/aquaculture (5%), Forestry (40%), Hunting (20%), Nature conservation/research (5%), Tourism/recreation (5%), Water management (5%)

Birds

Species	Season	Year	Pop min	Pop max	Acc	Criteria
Egretta alba Great White Egret	P	1996	200	300	—	A4i, B1i
Ciconia nigra Black Stork	B	1996	25	30	—	B2
Ciconia ciconia White Stork	B	1996	150	200	A	B2
Anser fabalis Bean Goose	W	1996	8,000	10,000	A	A4i, B1i
Anser albifrons White-fronted Goose	W	1996	500	1,000	A	B1i
Anser anser Greylag Goose	W	1996	500	1,000	—	B1i
Aythya nyroca Ferruginous Duck	B	1996	50	—	C	A1
Haliaeetus albicilla White-tailed Eagle	R	1996	14	14	A	A1

The IBA holds significant numbers of breeding *Ciconia nigra*, *Aythya nyroca* and *Haliaeetus albicilla*. Species of global conservation concern that do not meet IBA criteria: *Crex crex* (breeding).

Protection status

National Partial **International** None
174 ha of IBA covered by Nature Conservation Area (Balata-to, 174 ha). 7,833 ha of IBA covered by Landscape Protected Area (Boronka melléki, 7,833 ha). 424 ha of IBA covered by Nature Conservation Area (Csokonyavisontai fás legelő, 424 ha). 83 ha of IBA covered by Nature Conservation Area (Rinyaszentkirályi erdo, 83 ha). 16,090 ha of IBA covered by National Park (Duna-Dráva, 49,479 ha).

Conservation issues

Threats Abandonment/reduction of land management (A), Afforestation (C), Agricultural intensification/expansion (B), Aquaculture/fisheries (C), Burning of vegetation (C), Construction/impact of dyke/dam/barrage (B), Disturbance to birds (B), Drainage (A), Groundwater abstraction (B), Intensified forest management (A), Natural events (B), Recreation/tourism (B)

The main threats are posed by intensified forest management, severe drought leading to the drying out of wetlands, and the potential damming of the River Dráva in Croatia. A management plan is being prepared for Duna-Dráva National Park, and wetland restoration work has begun in Barcs LPA (now part of the National Park). Fish-ponds are managed according to the requirements of *Aythya nyroca*, *Ciconia nigra* and *Haliaeetus albicilla*.

Pacsmag fish-ponds — A1, A4i, B1i, B2, B3 — 008

Admin region Tolna
Coordinates 46°35′N 18°23′E
Altitude 122–128 m **Area** 487 ha

Site description

A system of fish-ponds and adjacent marshes in the valley of the River Koppány, 10 km to the south-east of the town of Tamási. The fish-ponds are overgrown with beds of *Phragmites* and *Typha*, whilst the adjacent marshlands are dominated by sedge *Carex*. Human activities include cattle-grazing and reed-harvesting ('Other' land-use, below).

Habitats Forest and woodland (20%; alluvial/very wet forest), Grassland (15%; humid grassland), Wetland (60%; standing fresh water; water-fringe vegetation), Artificial landscape (5%; arable land)
Land-use Agriculture (25%), Fisheries/aquaculture (20%), Hunting (80%), Not utilized (30%), Other (5%)

Birds

Species	Season	Year	Pop min	Pop max	Acc	Criteria
[1] ***Ixobrychus minutus*** Little Bittern	B	1996	40	45	A	B2
[1] ***Nycticorax nycticorax*** Night Heron	B	1996	50	55	A	B2
[1] ***Egretta alba*** Great White Egret	B	1996	40	45	A	A4i, B1i
Aythya nyroca Ferruginous Duck	B	1996	50	60	A	A1, B2
[1] ***Porzana parva*** Little Crake	B	1996	20	—	A	B3

1. 1991–1996.

An important breeding site for herons (Ardeidae) and other waterbirds, and a feeding area for raptors. Breeding species of global conservation concern that do not meet IBA criteria: *Haliaeetus albicilla*, *Crex crex* (no data).

Protection status

National High **International** High
487 ha of IBA covered by Nature Conservation Area (Pacsmagi-tavak, 487 ha). 485 ha of IBA covered by Ramsar Site (Pacsmag fish-ponds, 485 ha).

Conservation issues

Threats Afforestation (C), Agricultural intensification/expansion (B), Aquaculture/fisheries (C), Disturbance to birds (A), Drainage (C), Unsustainable exploitation (A)

Wildfowl hunting and associated disturbance to birds on the fish-ponds have resulted in the colony of *Egretta alba* moving elsewhere. A management plan exists for the site.

Béda-Karapancsa — A4i, B1i, B2 — 009

Admin region Baranya
Coordinates 45°56'N 18°46'E
Altitude 85–88 m **Area** 11,900 ha

Site description

A 45-km reach of the River Danube between Dunaszekcső and the Serbian border, including Béda-Karapancsa and Riha lake. Both permanently flooded and well-drained flood-plains support ancient forests, plantations, backwaters and reedbeds (*Phragmites*). Human activities include fishing, angling, navigation and hunting.

Habitats Forest and woodland (25%; broadleaved deciduous forest; alluvial/very wet forest), Wetland (40%; standing fresh water; river/stream; water-fringe vegetation), Artificial landscape (35%; arable land; forestry plantation)
Land-use Agriculture (20%), Forestry (50%), Military (5%), Nature conservation/research (10%), Tourism/recreation (10%), Urban/industrial/transport (20%), Water management (30%)

Birds

Species	Season	Year	Pop min	Pop max	Acc	Criteria
Nycticorax nycticorax Night Heron	B	1992	150	300	A	B2
Ardea purpurea Purple Heron	B	1996	36	42	—	B2
[1] ***Anser fabalis*** Bean Goose	W	1995	8,333	8,333	B	A4i, B1i
[1] ***Anser albifrons*** White-fronted Goose	W	1995	2,100	2,100	A	B1i
[1] ***Anser anser*** Greylag Goose	P	1996	744	744	A	B1i

1. Average of annual peaks 1986–1995.

An important area for wintering geese and diving ducks (Anatidae). Breeding species of global conservation concern that do not meet IBA criteria: *Haliaeetus albicilla* (2–4 pairs).

Protection status

National Partial **International** Partial
6,949 ha of IBA covered by National Park (Duna-Dráva, 49,479 ha). 1,149 ha of IBA covered by Ramsar Site (Béda-karapansca, 1,149 ha).

Conservation issues

Threats Aquaculture/fisheries (C), Disturbance to birds (C), Dredging/canalization (A), Infrastructure (B), Intensified forest management (B), Recreation/tourism (C)

River traffic is heavy on this reach of the Danube, and causes problems of disturbance to birds, and damage to the river channel itself through dredging and canalization.

Gemenc — A1, A4i, B1i, B2 — 010

Admin region Bács-Kiskun, Tolna
Coordinates 46°07'N 18°45'E
Altitude 88–92 m **Area** 17,779 ha

Site description

A large stretch of the Danube flood-plain, north of the town of Baja. The IBA covers a complex of habitats, including backwaters, marshes, meadows and both softwood and hardwood flood-plain forest, although approximately half of the forested area consists of plantations of non-native poplar *Populus*.

Habitats Forest and woodland (40%; broadleaved deciduous forest; alluvial/very wet forest), Grassland (10%; humid grassland), Wetland (10%; standing fresh water; river/stream; water-fringe vegetation), Artificial landscape (40%; arable land; forestry plantation)
Land-use Agriculture (10%), Fisheries/aquaculture (10%), Forestry (80%), Hunting (90%), Nature conservation/research (100%), Tourism/recreation (15%), Water management (15%)

Birds

Species	Season	Year	Pop min	Pop max	Acc	Criteria
[1] ***Egretta alba*** Great White Egret	P	1996	160	200	B	A4i, B1i
Ciconia nigra Black Stork	B	1996	30	35	A	B2
[2] ***Ciconia nigra*** Black Stork	P	1996	300	400	B	A4i, B1i
[3] ***Anser fabalis*** Bean Goose	W	1995	8,000	10,000	B	A4i, B1i
[3] ***Anser albifrons*** White-fronted Goose	W	1995	2,000	4,000	B	B1i
Bucephala clangula Goldeneye	W	1995	—	1,000	A	B1i
[3] ***Haliaeetus albicilla*** White-tailed Eagle	R	1996	5	6	A	A1, B2
Haliaeetus albicilla White-tailed Eagle	W	1995	20	30	A	A1
Falco cherrug Saker	B	1995	3	5	A	B2

1. Peak of daily maximum counts 1994–1996.
2. Average of daily maximum counts.
3. 1991–1996.

An important area for wintering wildfowl and local breeding populations of lowland forest birds.

Protection status

National High **International** High
17,779 ha of IBA covered by National Park (Duna-Dráva, 49,479 ha). 16,873 ha of IBA covered by Ramsar Site (Gemenc, 16,873 ha).

Conservation issues

Threats Disturbance to birds (B), Intensified forest management (A), Recreation/tourism (C)

The main threat is posed by forestry and hunting, both of which lead to the degradation of ancient woodlands (in the case of the latter due to overstocking with game species). These activities also cause disturbance to breeding and wintering birds. A long-term research and nest protection effort is ongoing for *Ciconia nigra* and *Haliaeetus albicilla*.

Sárvíz valley — A1, A4i, B1i, B2 — 011

Admin region Fejér
Coordinates 47°00'N 18°33'E
Altitude 100–110 m **Area** 14,700 ha

Site description

Habitats Forest and woodland (15%; broadleaved deciduous forest; alluvial/very wet forest), Grassland (20%; steppe/dry calcareous grassland; humid grassland; mesophile grassland), Wetland (25%; standing fresh water; water-fringe vegetation), Artificial landscape (40%; arable land; forestry plantation; other urban/industrial areas)
Land-use Agriculture (40%), Fisheries/aquaculture (20%), Forestry (15%), Hunting (60%), Water management (25%)

A well-drained low-lying flood-plain situated along the Sárvíz canal, between the two loess plateaus of Mezőföld. The IBA includes fish-ponds with fringing reedbeds (*Phragmites*), and surrounding *Populus* and *Robinia* woodlands, pastures and saline lakes. Human activities include grazing, arable farming and reed-harvesting.

■ Birds

Species	Season	Year	Pop min	Pop max	Acc	Criteria
Ixobrychus minutus Little Bittern	B	1996	30	40	A	B2
Ardeola ralloides Squacco Heron	B	1996	10	15	A	B2
Egretta alba Great White Egret	P	1996	200	250	—	A4i, B1i
[1] ***Anser fabalis*** Bean Goose	P	1993	10,000	20,000	A	A4i, B1i
[1] ***Anser albifrons*** White-fronted Goose	P	1994	1,000	3,000	A	B1i
[1] ***Anser anser*** Greylag Goose	P	1994	1,500	3,000	A	B1i
Anas platyrhynchos Mallard	W	1993	30,000	30,000	—	B1i
Aythya nyroca Ferruginous Duck	B	1993	18	18	—	B2
Aythya nyroca Ferruginous Duck	P	1993	80	100	B	A1

1. 1984–1994.

An important area for wintering geese. Species of global conservation concern that do not meet IBA criteria: *Aythya nyroca* (18 breeding pairs), *Haliaeetus albicilla* (3–5 wintering birds).

■ Protection status

National Partial **International** Partial

1,496 ha of IBA covered by Nature Conservation Area (Rétszilasi-tavak, 1,496 ha). 3,616 ha of IBA covered by Landscape Protected Area (Sárvíz-völgy, 3,616 ha). 1,496 ha of IBA covered by Ramsar Site (Rétszilasi, 1,496 ha).

■ Conservation issues

Threats Abandonment/reduction of land management (B), Aquaculture/fisheries (C), Burning of vegetation (B), Disturbance to birds (B)

The main threats are inappropriate water management in the fish-ponds, disturbance to birds caused by both hunting and attempts to protect fish stocks, and the burning of reeds *Phragmites*. Wildfowl shooting on Rétszilas fish-ponds is banned, and management planning is underway.

Lake Velence and Dinnyés marshes — A4i, B1i — 012

Admin region Fejér
Coordinates 47°10'N 18°31'E
Altitude 105–110 m **Area** 4,800 ha

■ Site description

Lake Velence is the third-largest lake in the country, measuring 10.5 km in length and 3.3 km in breadth, with an average depth of 1–2 m. There are extensive reed-beds with *Phragmites* and *Typha*, and adjacent wet meadows. Human activities include fishing, angling, sailing, boating and reed-harvesting ('Other' land-use, below).

Habitats Grassland (10%; humid grassland), Wetland (70%; standing fresh water; water-fringe vegetation), Artificial landscape (20%; arable land; forestry plantation; other urban/industrial areas)
Land-use Agriculture (15%), Nature conservation/research (30%), Other (25%), Tourism/recreation (40%), Water management (70%)

■ Birds

Species	Season	Year	Pop min	Pop max	Acc	Criteria
Egretta alba Great White Egret	R	1996	30	40	A	A4i, B1i
Platalea leucorodia Spoonbill	B	1996	50	70	A	A4i, B1i
[1] ***Anser fabalis*** Bean Goose	W	1996	20,000	35,000	A	A4i, B1i
[2] ***Anser albifrons*** White-fronted Goose	W	1995	3,487	3,487	—	B1i
Anser anser Greylag Goose	P	1995	500	1,000	—	B1i

1. 1984–1996.
2. Average of annual peaks 1986–1996.

An important site for breeding herons (Ardeidae) and ducks, and as a roosting site for geese in winter. Breeding species of global conservation concern that do not meet IBA criteria: *Aythya nyroca* (4–5 pairs).

■ Protection status

National Partial **International** Partial

539 ha of IBA covered by Nature Conservation Area (Dinnyési Fertő, 539 ha). 420 ha of IBA covered by Nature Conservation Area (Velencei madarrezervatum, 420 ha). 965 ha of IBA covered by Ramsar Site (Velence–Dinnyés, 965 ha).

■ Conservation issues

Threats Burning of vegetation (C), Disturbance to birds (A), Natural events (A), Recreation/tourism (B)

Main threats are disturbance to birds, and over-abstraction. Water-levels are artificially maintained in order to protect the ecological integrity of the site, and to prolong opportunities for recreational use. Wildfowl shooting is banned, and a management plan for Fertő was prepared in 1996. Expansion of the protected area at Dinnyés by 600 ha is underway.

Vértes mountains and Zámoly basin — A1, B2 — 013

Admin region Fejér, Komárom-Esztergom
Coordinates 47°25'N 18°20'E
Altitude 130–480 m **Area** 29,400 ha

■ Site description

The Vértes mountains are part of the Transdanubian mountain chain, and are dolomitic in composition with deep ravines. The southern slopes experience a sub-Mediterranean climate, whilst the northern slopes are subalpine. The Zámoly basin is situated in the foothills to the south-east of Vértes. Human activities include grazing, haymaking, hunting, tourism and mining.

Habitats Forest and woodland (60%; broadleaved deciduous forest; wooded steppe), Scrub (10%), Grassland (22%; steppe/dry calcareous grassland; humid grassland; mesophile grassland)
Land-use Agriculture (30%), Forestry (50%), Nature conservation/research (10%), Water management (5%)

■ Birds

Species	Season	Year	Pop min	Pop max	Acc	Criteria
Aythya nyroca Ferruginous Duck	B	1996	10	20	C	A1, B2
Aquila heliaca Imperial Eagle	R	1996	2	2	A	A1, B2
Falco cherrug Saker	R	1996	3	5	B	B2

An important area for raptors and waterbirds. *Otis tarda* and *Burhinus oedicnemus* bred in the Zámoly basin until the mid-1970s. Breeding species of global conservation concern that do not meet IBA criteria: *Aythya nyroca* (10–20 pairs).

■ Protection status

National Partial **International** None

15,035 ha of IBA covered by Landscape Protected Area (Vertesi, 15,035 ha).

■ Conservation issues

Threats Abandonment/reduction of land management (C), Burning of vegetation (C), Disturbance to birds (C), Drainage (C), Intensified forest management (A)

The Pro Vértes Land Trust has acquired c.1,000 ha of grassland, and management planning and the restoration of wet meadows has begun. An expansion of the existing Vértes LPA is in progress. An additional 6,700 ha in the northern Vértes mountains (not included in the LPA) is now protected and is being managed by Pro Vértes.

Lake Öreg at Tata — A4i, B1i — 014

Admin region Komárom-Esztergom
Coordinates 47°39'N 18°18'E
Altitude 129–140 m **Area** 259 ha

■ Site description

The lake is situated along the Által stream, and is bordered by the town of Tata. It is the largest area of standing water in the district, and has an average depth of 2.3 m when full. Following temporary drainage for fish-farming, shallow water pools and inlets cover large areas. Human activities include boating and angling.

Habitats Forest and woodland (11%; broadleaved deciduous forest; mixed forest; alluvial/very wet forest), Wetland (88%; standing fresh water; river/stream; water-fringe vegetation)

Land-use Fisheries/aquaculture (85%), Forestry (11%), Tourism/recreation (4%), Water management (86%)

Birds

Species	Season	Year	Pop min	Pop max	Acc	Criteria
[1] ***Anser fabalis*** Bean Goose	W	1996	16,000	16,000	A	A4i, B1i
Anser albifrons White-fronted Goose	W	1996	1,800	1,800	A	B1i

1. Average of annual peaks 1991–1996.

An important roosting site for wintering wildfowl, especially for *Anser fabalis*. A total of 220 bird species have been recorded at the site.

Protection status

National None **International** High

259 ha of IBA covered by Ramsar Site (Tata, Old Lake, 269 ha).

Conservation issues

Threats Aquaculture/fisheries (B), Construction/impact of dyke/dam/barrage (B), Disturbance to birds (A), Dredging/canalization (C), Filling-in of wetlands (A), Groundwater abstraction (C), Industrialization/urbanization (B), Infrastructure (B), Natural events (B)

Threats include recreational disturbance and nutrient pollution from inappropriate development and fish-farming. The site is protected by a municipal alliance and, between 1 October and 30 April, by a temporary Ramsar designation. Wildfowl shooting is banned, and a management plan exists for the site. Research into, and measures for, the rehabilitation of the lake and its catchment have been implemented.

Gerecse mountains

A1, B2 **015**

Admin region Komárom-Esztergom
Coordinates 47°43′N 18°30′E
Altitude 200–634 m **Area** 20,300 ha

Site description

A low limestone mountain situated between the town of Tatabánya and the River Danube. Deciduous oak *Quercus* forests cover the lower slopes, with submontane forests of *Quercus*, *Carpinus*, *Fagus* and karst scrub at higher altitudes.

Habitats Forest and woodland (70%; broadleaved deciduous forest), Scrub (5%; scrub), Grassland (10%; steppe/dry calcareous grassland), Artificial landscape (15%; arable land; forestry plantation; other urban/industrial areas)
Land-use Agriculture (20%), Forestry (70%), Military (30%), Tourism/recreation (10%)

Birds

Species	Season	Year	Pop min	Pop max	Acc	Criteria
Aquila heliaca Imperial Eagle	B	1996	2	3	—	A1, B2
Falco cherrug Saker	B	1996	3	3	A	B2

An important area for raptors and other forest birds.

Protection status

National Partial **International** None

8,617 ha of IBA covered by Landscape Protected Area (Gerecsei, 8,617 ha).

Conservation issues

Threats Disturbance to birds (B), Intensified forest management (A)

The main threat to the site is posed by intensive forestry practices. MME's local chapter is actively involved in the protection of raptors.

Danube reach between Gönyű and Szob

B1i **016**

Admin region Győr-Sopron-Moson, Komárom-Esztergom
Coordinates 47°46′N 18°25′E
Altitude 109–115 m **Area** 4,840 ha

Site description

An 83-km long reach of the River Danube forming the border with Slovakia, and characterized by a large industrial agglomeration comprising several important towns. The river section itself is braided into a relatively small number of branches, and includes shallows, holms and masonry dams. Native *Salix* and *Populus* riverine forests occur on the holms. Human activities include shipping and dredging.

Habitats Wetland (100%; river/stream)
Land-use Fisheries/aquaculture (20%), Hunting (50%), Water management (100%)

Birds

Species	Season	Year	Pop min	Pop max	Acc	Criteria
[1] ***Anser fabalis*** Bean Goose	W	1995	3,700	3,700	A	B1i
[2] ***Anas platyrhynchos*** Mallard	W	1991	15,000	25,000	A	B1i
[3] ***Bucephala clangula*** Goldeneye	W	1992	2,997	2,997	A	B1i

1. 1986–1995.
2. 1982–1991.
3. Average January count (1982–1992).

An important inland site for wintering wildfowl, especially for diving ducks. Species of global conservation concern that do not meet IBA criteria: *Haliaeetus albicilla* (wintering).

Protection status

National None **International** None

Conservation issues

Threats Disturbance to birds (B), Dredging/canalization (A), Industrialization/urbanization (C), Natural events (B)

A long-term monitoring programme has been underway since 1982.

Danube bend

A4i, B1i **017**

Admin region Pest
Coordinates 47°47′N 19°05′E
Altitude 86–244 m **Area** 15,200 ha

Site description

The site covers a reach of the River Danube between Szob and Budapest, and includes Szentendre island and adjacent flood-plains. The area is densely populated and one of the most popular areas for recreation in Hungary.

Habitats Forest and woodland (5%; alluvial/very wet forest), Scrub (5%; scrub), Grassland (5%; steppe/dry calcareous grassland), Wetland (50%; river/stream), Artificial landscape (35%; arable land; perennial crops/orchards/groves; forestry plantation; other urban/industrial areas)
Land-use Agriculture (40%), Nature conservation/research (5%), Tourism/recreation (60%), Urban/industrial/transport (50%), Water management (100%)

Birds

Species	Season	Year	Pop min	Pop max	Acc	Criteria
Phalacrocorax carbo Cormorant	W	1995	2,500	4,629	A	A4i, B1i
Anser fabalis Bean Goose	W	1995	600	5,000	B	A4i, B1i
[1] ***Bucephala clangula*** Goldeneye	W	1996	1,500	1,763	A	B1i

1. 1995–1996.

The site is especially important for wintering diving ducks and geese (Anatidae). A total of 210 bird species have been recorded in the area, with the majority occurring on passage.

Protection status

National Partial **International** None

The IBA overlaps with Duna-Ipoly National Park.

Conservation issues

Threats Afforestation (C), Agricultural intensification/expansion (B), Construction/impact of dyke/dam/barrage (A), Disturbance to birds (B), Dredging/canalization (C), Extraction industry (C), Firewood collection (B), Infrastructure (B), Recreation/tourism (A)

The site is threatened by the construction of the Nagymaros Dam following a decision by the Hague Court. Recreational activities destroy riverbank habitats, and are an important threat. A management plan exists for the area.

Börzsöny mountains A1, B2 018

Admin region Nógrád, Pest
Coordinates 47°55'N 18°55'E
Altitude 120–938 m **Area** 46,400 ha

Site description

A range of low mountains situated between the Danube bend and Ipoly river. The lower slopes are covered by oak *Quercus* forest, the higher slopes by submontane *Quercus*, *Carpinus* and *Fagus* forest. Montane *Fagus* forest extends to near the peaks. Human activities include horticulture, arable farming and skiing; pastures are mostly abandoned.

Habitats Forest and woodland (70%; broadleaved deciduous forest), Scrub (5%; scrub), Grassland (5%; steppe/dry calcareous grassland), Artificial landscape (20%; arable land; perennial crops/orchards/groves; forestry plantation)
Land-use Agriculture (10%), Forestry (90%), Nature conservation/research (20%), Tourism/recreation (30%)

Birds

Species	Season	Year	Pop min	Pop max	Acc	Criteria
Aquila heliaca Imperial Eagle	B	1993	3	3	A	A1, B2
Falco cherrug Saker	R	1996	3	3	A	B2

An important area for raptors and forest birds.

Protection status

National Partial **International** None
17,897 ha of IBA covered by Landscape Protected Area (Börzsönyi, 17,897 ha). 18,028 ha of IBA covered by National Park (Duna-Ipoly, 60,314 ha).

Conservation issues

Threats Abandonment/reduction of land management (A), Agricultural intensification/expansion (C), Disturbance to birds (B), Infrastructure (B), Intensified forest management (A), Recreation/tourism (B)

The abandonment of pasture management may lead to scrub encroachment, with a consequent loss of feeding areas for raptors (the nest sites of which are protected). Intensified forest management causes problems of changing age structure and disturbance to birds. Large-scale tourism results in disturbance to birds and habitat degradation, and necessitates road construction.

Middle reach of River Ipoly A1 019

Admin region Nógrád
Coordinates 48°05'N 19°05'E
Altitude 96–180 m **Area** 2,700 ha

Site description

A meandering river on the Slovakian border, situated between the villages of Hont and Ipolyszög. Wet grasslands and marshes dominate the IBA, with smaller areas of *Alnus* woodland and sandy grassland. The area is relatively unaffected by human activity.

Habitats Forest and woodland (10%; alluvial/very wet forest), Grassland (60%; humid grassland), Wetland (20%; river/stream; water-fringe vegetation), Artificial landscape (10%; forestry plantation)
Land-use Agriculture (70%), Forestry (20%), Urban/industrial/transport (5%), Water management (5%)

Birds

Species	Season	Year	Pop min	Pop max	Acc	Criteria
[1] ***Crex crex*** Corncrake	B	1994	25	—	A	A1

1. 1993–1994.

The area is important for breeding and migratory waterbirds, but only *Crex crex* meets IBA criteria.

Protection status

National Partial **International** None
The IBA overlaps with Duna-Ipoly National Park.

Conservation issues

Threats Abandonment/reduction of land management (B), Agricultural intensification/expansion (C), Burning of vegetation (B), Construction/impact of dyke/dam/barrage (A), Dredging/canalization (A), Forest grazing (C)

The main threat to the site is river regulation and engineering. A management plan exists for the area, which is also especially important as a breeding ground for amphibians.

Tápió valley B2 020

Admin region Pest, Jász-Nagykun-Szolnok
Coordinates 47°26'N 19°44'E
Altitude 90–130 m **Area** 18,300 ha

Site description

A variety of lowland habitats are present, including the River Tápió and its tributaries, wet meadows, reedbeds (*Phragmites*), fish-ponds and alluvial *Salix* and *Populus* woodlands. These form a mosaic with dry grassland, arable land and forest plantations of non-native *Robinia*, *Pinus* and *Populus*. Human activities include arable farming, haymaking, grazing, goose-farming, reed-harvesting ('Other' land-use, below), hunting and angling.

Habitats Forest and woodland (5%; broadleaved deciduous forest; alluvial/very wet forest), Grassland (60%; steppe/dry calcareous grassland; humid grassland), Wetland (25%; standing fresh water; standing brackish and salt water; river/stream; water-fringe vegetation), Artificial landscape (10%; arable land; forestry plantation; ruderal land)
Land-use Agriculture (65%), Forestry (10%), Other (10%), Water management (25%)

Birds

Species	Season	Year	Pop min	Pop max	Acc	Criteria
Botaurus stellaris Bittern	B	1996	30	60	B	B2
Ixobrychus minutus Little Bittern	B	1991	30	40	—	B2
Falco vespertinus Red-footed Falcon	B	1993	40	45	—	B2
Limosa limosa Black-tailed Godwit	B	1991	30	60	—	B2

An important area for birds of lowland grassland. Breeding species of global conservation concern that do not meet IBA criteria: *Crex crex* (no data).

Protection status

National Partial **International** None
4,516 ha of IBA covered by Landscape Protected Area (Tápió-Hajta vidéke, 4,516 ha).

Conservation issues

Threats Abandonment/reduction of land management (B), Afforestation (B), Agricultural intensification/expansion (C), Burning of vegetation (A), Disturbance to birds (C), Drainage (A), Dredging/canalization (B), Natural events (B)

The main threat is the drainage of wetlands, especially during periods of drought. Both afforestation and conversion to arable land threaten grassland, and the practice of early cutting can destroy the nests of waders. Recreation also causes some disturbance to birds. Municipalities protect about half of the area.

Puszta at Jászkarajenő B2 021

Admin region Bács-Kiskun, Pest
Coordinates 47°04'N 20°04'E
Altitude 87–97 m **Area** 18,900 ha

Site description

Habitats Grassland (90%; steppe/dry calcareous grassland), Artificial landscape (10%; arable land; perennial crops/orchards/groves; forestry plantation)
Land-use Agriculture (100%)

An area of well-drained lowland overlain by meadow, saline and sandy soils, situated between the Gerje and Körös streams south-west of the town of Szolnok. Saline grassland covers the majority of the IBA, interspersed by arable land, orchards, small saline lakes and abandoned rice-fields. Human activities include grazing, haymaking, arable farming, horticulture, fish-farming and hunting.

Birds

Species	Season	Year	Pop min	Pop max	Acc	Criteria
Falco vespertinus Red-footed Falcon	B	1996	30	35	A	B2
Limosa limosa Black-tailed Godwit	B	1996	25	50	A	B2
Coracias garrulus Roller	B	1996	30	35	A	B2
Lanius minor Lesser Grey Shrike	B	1996	50	—	B	B2

An important area for lowland farmland birds. Species of global conservation concern that do not meet IBA criteria: *Haliaeetus albicilla* (4–5 wintering birds), *Otis tarda* (1–12 breeding individuals).

Protection status

National None **International** None

Conservation issues

Threats Afforestation (C), Burning of vegetation (B), Disturbance to birds (B), Drainage (B), Natural events (B), Unsustainable exploitation (C)

The main threats are the drainage of wetlands, and disturbance to birds caused by hunting and other human activities. Wetland restoration works have been implemented, and the site is a proposed LPA.

Danube plain — A1, A4i, B1i, B2 — 022

Admin region Bács-Kiskun, Pest
Coordinates 46°49′N 19°15′E
Altitude 93–103 m **Area** 82,000 ha

Site description

A flood-plain lying between natural levees and overlain by loess soils, situated adjacent to the River Danube and main Duna-völgyi canal between Kiskunlacháza and Szakmár. Human activities include cattle- and sheep-grazing, arable farming, angling and hunting. This area includes two sites that were treated as separate IBAs in the previous international IBA inventory (Grimmett and Jones 1989): 'Kiskunsági szikes-tavak' (former site HU021) and 'Harta-Akasztói puszta' (former site HU042).

Habitats Grassland (40%; steppe/dry calcareous grassland), Wetland (5%; standing brackish and salt water; water-fringe vegetation), Artificial landscape (55%; highly improved reseeded grassland; arable land)
Land-use Agriculture (90%), Nature conservation/research (60%), Tourism/recreation (5%), Water management (5%)

Birds

Species	Season	Year	Pop min	Pop max	Acc	Criteria
[1] **Ardea purpurea** Purple Heron	P	1996	50	50	C	B2
[2] **Anser albifrons** White-fronted Goose	P	1996	2,000	15,000	A	B1i
[2] **Anser anser** Greylag Goose	P	1996	200	5,000	A	A4i, B1i
[1] **Anas acuta** Pintail	P	1996	260	260	A	B1i
[1] **Falco vespertinus** Red-footed Falcon	B	1996	85	85	B	B2
Falco cherrug Saker	R	1996	2	4	B	B2
Otis tarda Great Bustard	R	1996	190	290	A	A1, B2
[2] **Limosa limosa** Black-tailed Godwit	B	1996	50	80	B	B2
[3] **Limosa limosa** Black-tailed Godwit	P	1996	150	8,000	A	A4i, B1i
[2] **Numenius arquata** Curlew	P	1996	400	12,000	—	A4i, B1i
[1] **Chlidonias hybridus** Whiskered Tern	B	1996	—	175	A	B2
[1] **Chlidonias niger** Black Tern	B	1996	40	40	B	B2

1. 1992–1996.
2. 1986–1996.
3. Large decrease 1986–1996.

An important area for lowland farmland birds and waterbirds. Species of global conservation concern that do not meet IBA criteria: *Phalacrocorax pygmeus* and *Branta ruficollis* (both on passage), *Aythya nyroca* (6 breeding pairs), and *Haliaeetus albicilla* (2 breeding pairs and 4 wintering birds).

Protection status

National Partial **International** Partial

21,200 ha of IBA covered by National Park (Kiskunsági, 53,429 ha). 3,903 ha of IBA covered by Ramsar Site (Kiskunság, 3,903 ha). The IBA also overlaps with the Kiskunság Biosphere Reserve (22,095 ha).

Conservation issues

Threats Abandonment/reduction of land management (A), Agricultural intensification/expansion (B), Disturbance to birds (C), Drainage (B), Natural events (B), Recreation/tourism (C)

The main threats are changes in habitat composition and quality caused both by the abandonment of grazing and by drought. A *Perdix perdix* recovery project has begun in Apajpuszta, and the National Park Authority is working to increase the breeding success of *Otis tarda*. MME/BirdLife Hungary provides artificial nesting-sites, and a management plan exists for the area. Wildfowl shooting is banned, and the enlargement of the National Park is in process.

Lake Kolon — A4i, B1i — 023

Admin region Bács-Kiskun
Coordinates 46°46′N 19°19′E
Altitude 95–110 m **Area** 2,962 ha

Site description

The IBA is situated on the sandy alluvial fan of Kiskunság, near to the village of Izsák. It consists of a large lake almost entirely covered by reeds *Phragmites*, and surrounded by wet grassland and sand-dunes. Human activities include grazing, haymaking and reed-harvesting ('Other' land-use, below).

Habitats Forest and woodland (15%; alluvial/very wet forest; wooded steppe), Grassland (15%; steppe/dry calcareous grassland; humid grassland), Wetland (40%; standing fresh water; water-fringe vegetation), Artificial landscape (30%; highly improved reseeded grassland; forestry plantation)
Land-use Agriculture (30%), Forestry (30%), Nature conservation/research (50%), Other (50%)

Birds

Species	Season	Year	Pop min	Pop max	Acc	Criteria
Egretta alba Great White Egret	B	1995	80	150	B	A4i, B1i
Platalea leucorodia Spoonbill	B	1995	100	—	B	A4i, B1i

An important site for breeding herons (Ardeidae). Breeding species of global conservation concern that do not meet IBA criteria: *Haliaeetus albicilla*, *Crex crex* (no data), *Otis tarda* (2–3 individuals).

Protection status

National High **International** High

2,962 ha of IBA covered by National Park (Kiskunsági, 53,429 ha). 2,962 ha of IBA covered by Ramsar Site (Lake Kolon, 2,962 ha).

Conservation issues

Threats Agricultural intensification/expansion (C), Burning of vegetation (A), Groundwater abstraction (A), Natural events (A)

Threats to the IBA include natural succession, inappropriate reedbed and grassland management, and the early cutting of hay. Restoration work to create more open water is being funded through PHARE, and a management plan exists for the site.

Flood-plain at Tiszaalpár — B2 — 024

Admin region Bács-Kiskun, Jász-Nagykun-Szolnok
Coordinates 46°48′N 20°00′E
Altitude 82–83 m **Area** 5,000 ha

Site description

The IBA covers an area of flood-plain of the River Tisza, which is subject to occasional inundation, and includes areas of softwood flood-plain forest, meadows and oxbow lakes.

Habitats Forest and woodland (15%; alluvial/very wet forest), Grassland (65%; humid grassland), Wetland (20%; river/stream; water-fringe vegetation)
Land-use Agriculture (65%), Forestry (15%), Tourism/recreation (10%)

Birds

An important site for birds associated with flood-plain forest and wetland complexes. Breeding species of global conservation concern that do not meet IBA criteria: *Aythya nyroca* (3–5 pairs).

Species	Season	Year	Pop min	Pop max	Acc	Criteria
Nycticorax nycticorax Night Heron	B	1995	60	180	A	B2
Platalea leucorodia Spoonbill	B	1995	12	40	—	B2

Protection status

National Partial **International** None

600 ha of IBA covered by National Park (Kiskunsági, 53,429 ha).

Conservation issues

Threats Disturbance to birds (B), Dredging/canalization (B), Natural events (A)

The main threats to the IBA are drought and disturbance to birds.

Lake Fehér at Gátér — A4i, B1i, B2 — 025

Admin region Bács-Kiskun, Csongrád
Coordinates 46°25′N 19°58′E
Altitude 83–83 m **Area** 1,400 ha

Site description

The IBA is situated in a depression between the villages of Gátér, Pálmonostora and Tömörkény, and contains large expanses of reedbed (*Phragmites*) along with surrounding saline grasslands and arable fields. Human activities include grazing, haymaking and hunting.

Habitats Grassland (40%; mesophile grassland), Wetland (60%; standing brackish and salt water; water-fringe vegetation)
Land-use Agriculture (50%), Water management (100%)

Birds

Species	Season	Year	Pop min	Pop max	Acc	Criteria
Egretta alba Great White Egret	B	1996	100	100	A	A4i, B1i
Chlidonias hybridus Whiskered Tern	B	1996	450	450	A	A4i, B1i, B2

Protection status

National None **International** None

Conservation issues

Threats Abandonment/reduction of land management (B), Drainage (A), Dredging/canalization (A), Unsustainable exploitation (C)

The main threats are a reduction in grassland management, and excessive drainage during the breeding season. A protected area is in the process of being designated.

Pusztaszer Landscape Protection Area — A1, A4i, B1i, B2 — 026

Admin region Csongrád
Coordinates 46°15′N 20°10′E
Altitude 81–88 m **Area** 22,320 ha

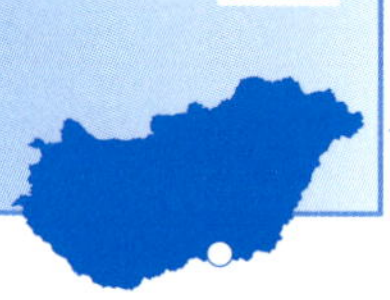

Site description

The IBA is situated between the River Tisza and the sandy highlands of Kiskunság, to the north of the town of Szeged. Fish-ponds and saline lakes are surrounded by saline grassland, arable land, orchards and forestry plantations. Some of the fish-ponds contain large reedbeds (*Phragmites*), and native forest exists along the River Tisza. Human activities include arable farming, horticulture, reed-harvesting ('Other' land-use, below) and hunting.

Habitats Forest and woodland (8%; broadleaved deciduous forest; alluvial/very wet forest), Grassland (20%; steppe/dry calcareous grassland; humid grassland; mesophile grassland), Wetland (22%; standing fresh water; standing brackish and salt water; water-fringe vegetation), Artificial landscape (50%; highly improved reseeded grassland; arable land; perennial crops/orchards/groves; forestry plantation; other urban/industrial areas)
Land-use Agriculture (60%), Fisheries/aquaculture (12%), Forestry (13%), Nature conservation/research (5%), Other (5%)

Birds

An important area for both breeding and passage waterbirds.

Species	Season	Year	Pop min	Pop max	Acc	Criteria
[1] ***Phalacrocorax carbo*** Cormorant	P	1996	2,000	2,500	A	B1i
[1] ***Ixobrychus minutus*** Little Bittern	B	1996	35	—	C	B2
[1] ***Nycticorax nycticorax*** Night Heron	B	1996	300	300	A	B2
[1] ***Ardeola ralloides*** Squacco Heron	B	1996	50	50	A	B2
[1] ***Egretta alba*** Great White Egret	B	1996	150	150	A	A4i, B1i
[1] ***Ardea purpurea*** Purple Heron	B	1996	80	80	A	B2
[1] ***Platalea leucorodia*** Spoonbill	B	1996	250	250	A	A4i, B1i, B2
[1] ***Anser albifrons*** White-fronted Goose	W	1996	5,000	5,000	A	B1i
[1] ***Anser anser*** Greylag Goose	P	1996	700	700	A	B1i
[1] ***Anas platyrhynchos*** Mallard	P	1996	25,000	25,000	A	B1i
Aythya nyroca Ferruginous Duck	B	1996	60	—	A	A1, B2
[1] ***Haliaeetus albicilla*** White-tailed Eagle	W	1996	15	20	A	A1
Grus grus Crane	P	1996	6,000	6,000	A	A4i, B1i
[1] ***Recurvirostra avosetta*** Avocet	P	1996	450	450	A	B1i
[1] ***Limosa limosa*** Black-tailed Godwit	P	1996	7,000	7,000	A	A4i, B1i
[1] ***Tringa erythropus*** Spotted Redshank	P	1996	1,500	1,500	A	A4i, B1i
[1] ***Larus cachinnans*** Yellow-legged Gull	P	1996	3,500	3,500	A	B1i
[1] ***Lanius minor*** Lesser Grey Shrike	B	1996	50	50	B	B2

1. 1986–1996.

Protection status

National High **International** Partial

22,320 ha of IBA covered by Landscape Protected Area (Pusztaszeri, 22,320 ha). 5,000 ha of IBA covered by Ramsar Site (Pusztaszer, 5,000 ha).

Conservation issues

Threats Abandonment/reduction of land management (B), Aquaculture/fisheries (A), Burning of vegetation (A), Deforestation (commercial) (B), Disturbance to birds (B), Forest grazing (C), Infrastructure (C), Intensified forest management (A)

The main threats to the site arise from conflicts of interest with fish farmers (inappropriate management of water and reedbeds, disturbance and persecution of birds). Wildfowl shooting is banned, and a management plan exists for the area.

Puszta at Pitvaros — A4i, B1i — 027

Admin region Békés, Csongrád
Coordinates 46°19′N 20°43′E
Altitude 83–86 m **Area** 13,100 ha

Site description

Low-lying flood-plain surrounded by alluvial fans in the vicinity of the villages of Pitvaros, Tótkomlós and Csanádalberti. Wet saline grasslands cover the majority of the site, with loess grasslands and arable fields on the higher ground. Human activities include grazing, arable farming and hunting.

Habitats Grassland (95%; humid grassland), Artificial landscape (5%; arable land; forestry plantation)
Land-use Agriculture (80%), Nature conservation/research (5%)

Birds

Species	Season	Year	Pop min	Pop max	Acc	Criteria
Anser albifrons White-fronted Goose	P	1996	1,000	3,000	C	B1i
Grus grus Crane	P	1996	3,000	4,000	C	A4i, B1i

An important area for steppic birds. Species of global conservation concern that do not meet IBA criteria: *Haliaeetus albicilla* (2–6 wintering birds), *Otis tarda* (15–25 resident individuals).

Protection status

National Partial **International** None

3,156 ha of IBA covered by National Park (Körös-Maros, 42,636 ha).

Conservation issues

Threats Agricultural intensification/expansion (B), Drainage (A), Filling-in of wetlands (U), Groundwater abstraction (B), Natural events (C)

The main threats to the IBA are the drainage of seasonal ponds and the mechanization of agriculture. A management plan exists for the area.

Lake Fehér at Kardoskút A4i, B1i 028

Admin region Békés, Csongrád
Coordinates 46°30'N 20°28'E
Altitude 85–86 m **Area** 8,800 ha

Site description

Situated on a poorly-drained flood-plain, and underlain by wind-blown soils, the site consists of a salt-lake and surrounding arable fields and saline grassland. Human activities include grazing and arable farming.

Habitats Grassland (40%; steppe/dry calcareous grassland), Wetland (10%; standing brackish and salt water), Artificial landscape (40%; arable land)
Land-use Agriculture (90%), Nature conservation/research (20%)

Birds

Species	Season	Year	Pop min	Pop max	Acc	Criteria
[1] ***Anser albifrons*** White-fronted Goose	P	1995	11,900	11,900	A	A4i, B1i
Anser anser Greylag Goose	P	1996	200	400	C	B1i
Grus grus Crane	W	1994	500	11,000	A	A4i, B1i
Limosa limosa Black-tailed Godwit	P	1996	3,000	4,000	C	B1i

1. Average of annual peaks 1986–1995.

An important staging area for migrating waterbirds, especially for *Grus grus*. Passage species of global conservation concern that do not meet IBA criteria: *Anser erythropus* (no data), *Branta ruficollis* (2–11 birds), *Numenius tenuirostris* (has not been seen for 20 years).

Protection status

National High **International** Low
8,800 ha of IBA covered by National Park (Körös-Maros, 42,636 ha). 488 ha of IBA covered by Ramsar Site (Kardoskút, 488 ha).

Conservation issues

Threats Abandonment/reduction of land management (B), Afforestation (C), Agricultural intensification/expansion (B), Burning of vegetation (C), Disturbance to birds (A), Drainage (A), Natural events (A)

The main threat to the site is a lack of water, caused by inappropriate land-reclamation work and severe drought. As a result artificial water supplies have been provided since 1992. Wildfowl shooting is prohibited, and the enlargement of the existing protected area is underway. A management plan exists for the area.

Dévaványa plain A1, A4i, B1i, B1iv, B2 029

Admin region Békés, Jász-Nagykun-Szolnok
Coordinates 47°10'N 21°09'E
Altitude 86–94 m **Area** 85,000 ha

Site description

An area of reclaimed marshland on the Great Plain, bordered by the River Berettyó in the east, the River Körös in the south, the Hortobágy-Berettyó Canal in the west, and the Ó-Berettyó Canal in the north. Secondary saline grasslands and arable lands exist on poor soil. Human activities include cattle- and sheep-grazing, haymaking, goose-farming and arable farming.

Habitats Grassland (30%; steppe/dry calcareous grassland), Artificial landscape (70%; arable land)
Land-use Agriculture (90%), Nature conservation/research (20%)

Birds

Species	Season	Year	Pop min	Pop max	Acc	Criteria
Nycticorax nycticorax Night Heron	B	1996	80	80	—	B2
Falco vespertinus Red-footed Falcon	B	1996	30	40	A	B2
Otis tarda Great Bustard	R	1996	436	—	A	A1, B2
Limosa limosa Black-tailed Godwit	B	1996	30	50	A	B2
Limosa limosa Black-tailed Godwit	P	1996	13,000	13,000	A	A4i, B1i

An important area for farmland birds, associated with extensive forms of land-use, including the most significant local population of *Otis tarda* in Hungary. Daily maximum totals of between 1,500 and 2,000 *Grus grus* have been recorded migrating over the Hortobágy-Berettyó Canal. Species of global conservation concern that do not meet IBA criteria: *Haliaeetus albicilla* (4–12 wintering birds).

Protection status

National Partial **International** None
12,144 ha of IBA covered by National Park (Körös-Maros, 42,636 ha).

Conservation issues

Threats Afforestation (B), Agricultural intensification/expansion (B), Burning of vegetation (C), Disturbance to birds (B), Drainage (B), Industrialization/urbanization (C), Natural events (B)

The conservation authority runs a farming business within the National Park, and is therefore able to minimize the threat posed to *Otis tarda* by mechanical cultivation. Drainage and drought threaten wetlands, whilst farmland habitats could be lost if agriculture were to become economically viable. Two artificially flooded meadows provide habitat for waders, and there is a management plan dating from the late 1970s.

Biharugra fish-ponds A1, A4i, B1i, B2 030

Admin region Békés
Coordinates 46°58'N 21°34'E
Altitude 90–95 m **Area** 16,000 ha

Site description

A low-lying flood-plain situated between Körösnagyharsány and Mezőgyán, which is not subject to the effects of natural flooding. A large system of fish-ponds are managed for conservation, and are surrounded by extensive saline grasslands and arable fields. Human activities include sheep- and cattle-grazing, reed-harvesting ('Other' land-use, below), hunting, angling and birdwatching.

Habitats Forest and woodland (5%; alluvial/very wet forest), Grassland (30%; steppe/dry calcareous grassland; humid grassland), Wetland (40%; standing fresh water; water-fringe vegetation), Artificial landscape (25%; highly improved reseeded grassland; arable land; forestry plantation)
Land-use Agriculture (50%), Fisheries/aquaculture (40%), Forestry (10%), Military (5%), Nature conservation/research (40%), Other (5%), Water management (5%)

Birds

Species	Season	Year	Pop min	Pop max	Acc	Criteria
Phalacrocorax carbo Cormorant	P	1996	1,000	2,000	A	B1i
Ixobrychus minutus Little Bittern	B	1994	74	—	A	B2
Platalea leucorodia Spoonbill	B	1996	20	25	—	B2
Platalea leucorodia Spoonbill	P	1994	250	250	A	A4i, B1i
Anser albifrons White-fronted Goose	P	1996	5,000	10,000	A	B1i
Anser anser Greylag Goose	P	1996	500	2,000	—	B1i
Anas platyrhynchos Mallard	P	1994	—	50,000	—	B1i
Aythya nyroca Ferruginous Duck	B	1994	30	—	—	A1, B2
Haliaeetus albicilla White-tailed Eagle	W	1994	15	20	A	A1
Limosa limosa Black-tailed Godwit	P	1996	6,000	6,000	A	A4i, B1i
Chlidonias hybridus Whiskered Tern	B	1996	11	100	—	B2

An important area for both breeding and passage waterbirds. Species of global conservation concern that do not meet IBA criteria: *Aythya nyroca* (50–150 birds on passage), *Otis tarda* (5–10 breeding individuals).

Protection status

National Partial **International** Partial
7,899 ha of IBA covered by National Park (Körös-Maros, 42,636 ha). 2,791 ha of IBA covered by Ramsar Site (Biharugra fish-ponds, 2,791 ha).

Conservation issues

Threats Abandonment/reduction of land management (B), Afforestation (C), Agricultural intensification/expansion (B), Aquaculture/fisheries (A), Burning of vegetation (A), Consequences of animal/plant introductions (C), Disturbance to birds (B), Drainage (C), Recreation/tourism (C)

Threats include inappropriate timing of the flooding and draining of fish-ponds, uncontrolled burning and the introduction of grass carp *Ctenopharyngodon*. Artificial nest-sites are provided, and wildfowl shooting is prohibited. A management plan exists for the area.

Bihar plain A1, B2 031

Admin region Hajdú-Bihar
Coordinates 47°06′N 21°26′E
Altitude 80–100 m **Area** 48,000 ha

Site description

A poorly-drained flood-plain situated between alluvial fans, and overlain by meadow and saline soils. The site is made up of four subsites: Nagy-Kerek-tiszta at Püspökladány, Gatály at Szerep, Ásvány at Bihardancsháza, and the largest site, which lies between Váncsod and Csökmő. The area is dominated by saline grasslands, which form a mosaic with extensive arable lands. Human activities include cattle- and sheep-grazing, goose and arable farming.

Habitats Grassland (50%; steppe/dry calcareous grassland), Artificial landscape (50%; highly improved reseeded grassland; arable land; forestry plantation)
Land-use Agriculture (95%)

Birds

Species	Season	Year	Pop min	Pop max	Acc	Criteria
Falco vespertinus Red-footed Falcon	B	1996	40	—	C	B2
[1] ***Otis tarda*** Great Bustard	R	1996	140	140	A	A1, B2
Limosa limosa Black-tailed Godwit	B	1996	40	—	C	B2

1. Stable 1993–1996.

An important area for birds of lowland dry grassland. The site is part of a wider network of grassland fragments and includes areas that hold important local populations of *Otis tarda*. Species of global conservation concern that do not meet IBA criteria: *Haliaeetus albicilla* (8 wintering birds), *Aquila heliaca* (1–5 birds on passage).

Protection status

National Partial **International** None
711 ha of IBA covered by Nature Conservation Area (Bihari legelő, 711 ha). 17,095 ha of IBA covered by Landscape Protected Area (Bihari-sík, 17,095 ha).

Conservation issues

Threats Abandonment/reduction of land management (C), Agricultural intensification/expansion (B), Disturbance to birds (B)

Due to poor soil conditions the majority of the area is regarded as less favoured for agriculture purposes. Protection measures for *Otis tarda* are in place through MME, although extensive land-reclamation works, including the installation of irrigation systems, have destroyed areas of *O. tarda* habitat.

Hortobágy A1, A4i, A4ii, B1i, B1iii, B2 032

Admin region Hajdú-Bihar, Jász-Nagykun-Szolnok
Coordinates 47°37′N 21°04′E
Altitude 86–98 m **Area** 136,300 ha

Site description

Situated on the Hungarian Great Plain east of the River Tisza, Hortobágy is the largest puszta landscape in Hungary; an open plain dominated by saline grasslands and extensive arable lands. Depressions contain saline marshlands, with seasonal wetlands appearing in spring and autumn. Fish-ponds cover about 5,000 ha. Human activities include arable farming, cattle- and sheep-grazing, goose-farming, haymaking, angling, birdwatching, hunting and reed-harvesting ('Other' land-use, below). This area includes four sites that were treated as separate IBAs in the previous international IBA inventory (Grimmett and Jones 1989): 'Hortobágy' (former site HU016), 'Pusztakócsi mocsarak' (former site HU018), 'Ágota-puszta' (former site HU020) and 'Nagyszék' (former site HU043).

Habitats Grassland (60%; steppe/dry calcareous grassland; mesophile grassland), Wetland (10%; standing fresh water; water-fringe vegetation), Artificial landscape (30%; arable land; forestry plantation; ruderal land)
Land-use Agriculture (90%), Fisheries/aquaculture (5%), Military (10%), Nature conservation/research (80%), Other (5%), Tourism/recreation (10%)

Birds

Species	Season	Year	Pop min	Pop max	Acc	Criteria
Phalacrocorax pygmeus Pygmy Cormorant	B	1996	15	15	A	A1
Botaurus stellaris Bittern	R	1996	80	100	B	B2
Ixobrychus minutus Little Bittern	B	1996	80	—	C	B2
Nycticorax nycticorax Night Heron	B	1996	300	—	B	B2
Ardeola ralloides Squacco Heron	B	1996	10	15	B	B2
Ardea purpurea Purple Heron	B	1996	200	300	C	B2
Ciconia ciconia White Stork	B	1996	200	200	A	B2
Platalea leucorodia Spoonbill	B	1996	200	250	A	A4i, B1i, B2
Anser fabalis Bean Goose	P	1996	5,000	6,000	B	A4i, B1i
[1] ***Anser albifrons*** White-fronted Goose	P	1996	10,000	200,000	A	A4i, B1i
Anser erythropus Lesser White-fronted Goose	P	1996	70	450	A	A1, A4i, B1i
Anser anser Greylag Goose	B	1996	400	450	B	B1i
[2] ***Anser anser*** Greylag Goose	P	1996	2,000	2,000	A	B1i
[3] ***Anas crecca*** Teal	P	1996	30,000	30,000	—	A4i, B1i
[3] ***Anas clypeata*** Shoveler	P	1996	5,000	5,000	A	A4i, B1i
Aythya nyroca Ferruginous Duck	B	1996	50	70	C	A1, B2
Bucephala clangula Goldeneye	P	1996	700	800	—	B1i
Haliaeetus albicilla White-tailed Eagle	W	1996	50	70	A	A1
Aquila heliaca Imperial Eagle	P	1993	6	—	B	A1
[4] ***Falco vespertinus*** Red-footed Falcon	B	1996	100	500	A	A4ii, B1iii, B2
Falco cherrug Saker	R	1996	10	12	A	B2
Porzana pusilla Baillon's Crake	B	1996	100	—	—	B2
[5] ***Grus grus*** Crane	P	1996	50,000	60,000	A	A4i, B1i
Otis tarda Great Bustard	R	1996	170	170	A	A1, B2
[6] ***Philomachus pugnax*** Ruff	P	1996	150,000	200,000	B	A4i, B1i
Limosa limosa Black-tailed Godwit	P	1996	10,000	30,000	A	A4i, B1i
[7] ***Numenius tenuirostris*** Slender-billed Curlew	P	1996	—	—	A	A1
Numenius arquata Curlew	P	1996	6,000	10,000	A	A4i, B1i
Tringa erythropus Spotted Redshank	P	1996	1,000	1,000	—	A4i, B1i
Chlidonias hybridus Whiskered Tern	B	1996	300	360	A	A4i, B1i, B2
Chlidonias niger Black Tern	B	1996	100	150	A	B2
Acrocephalus paludicola Aquatic Warbler	B	1996	520	520	—	A1, B2
Lanius minor Lesser Grey Shrike	B	1996	200	200	A	B2

1. Up to 200,000 in spring, up to 10,000 in autumn.
2. Average of annual peaks 1992–1996.
3. 5 year average.
4. Large decrease (formerly 800 pairs).
5. Autumn passage (large increase); 500–1,000 oversummer.
6. Spring passage only.
7. 1–2 records in most years.

The most important site in Hungary for steppic birds and waterbirds.

Protection status

National Partial **International** Partial
62,358 ha of IBA covered by National Park (Hortobágy, 68,506 ha). 16,109 ha of IBA covered by Ramsar Site (Hortobágy, 19,473 ha). 52,000 ha of IBA covered by Biosphere Reserve (Hortobágy National Park, 52,000 ha).

Conservation issues

Threats Abandonment/reduction of land management (B), Aquaculture/fisheries (B), Burning of vegetation (B), Disturbance to birds (B), Drainage (B), Recreation/tourism (B)

Undergrazing and increasing disturbance are the main threats, whilst wildfowl are threatened by both hunting and reed-burning. Management plans exist, and artificial flooding is carried out for *Anser erythropus* and *Numenius tenuirostris*. Grasslands and arable fields are managed for *Acrocephalus paludicola* and *Otis tarda*. Artificial nest-sites are provided, and wildfowl shooting is banned within the protected areas.

Kisköre reservoir A4i, B1i, B2 033

Admin region Borsod-Abaúj-Zemplén, Heves, Jász-Nagykun-Szolnok
Coordinates 47°30′N 20°30′E
Altitude 89–89 m **Area** 12,700 ha

Site description

The reservoir, 35 km long and 6 km wide, was established in the 1970s on the River Tisza between Kisköre and Tiszadorogma. There is a large expanse of open water, in addition to numerous islands, extensive reedbeds (*Phragmites* and *Typha*) and floating vegetation. Human activities include angling, boating and hunting.

Habitats Forest and woodland (20%; alluvial/very wet forest), Scrub (5%; scrub), Grassland (5%; humid grassland), Wetland (70%; standing fresh water; river/stream; water-fringe vegetation)
Land-use Fisheries/aquaculture (50%), Nature conservation/research (20%), Tourism/recreation (60%), Urban/industrial/transport (5%), Water management (100%)

Birds

Species	Season	Year	Pop min	Pop max	Acc	Criteria
Nycticorax nycticorax Night Heron	B	1996	80	100	—	B2
Egretta alba Great White Egret	R	1995	150	—	—	A4i, B1i
Egretta alba Great White Egret	B	1995	40	50	—	A4i, B1i
[1] ***Anser albifrons*** White-fronted Goose	W	1995	1,008	1,008	—	B1i
[2] ***Anser anser*** Greylag Goose	P	1995	382	382	A	B1i

1. Average of annual peaks 1986–1994.
2. Average of annual peaks 1986–1995.

An important site for both breeding and passage waterbirds. Species of global conservation concern that do not meet IBA criteria: *Haliaeetus albicilla* (5–10 wintering birds).

Protection status

National Partial **International** Partial
6,148 ha of IBA covered by National Park (Hortobágy, 68,506 ha). 7,012 ha of IBA covered by Ramsar Site (Hortobágy, 19,473 ha).

Conservation issues

Threats Aquaculture/fisheries (C), Disturbance to birds (A), Intensified forest management (C), Recreation/tourism (B)

Wildfowl hunting is banned on 2,500 ha of the bird sanctuary. In the rest of the site, hunting is allowed according to general rules.

Upper Tisza — B2 — 034

Admin region Borsod-Abaúj-Zemplén, Szabolcs-Szatmár-Bereg
Coordinates 48°15′N 21°45′E
Altitude 95–110 m **Area** 10,000 ha

Site description

An inundated flood-plain bordering the River Tisza between Tuzsér and Rakamaz. Oxbow lakes and marshes exist alongside large blocks of softwood and hardwood flood-plain forests. Human activities include grazing, mowing and arable farming.

Habitats Forest and woodland (15%; alluvial/very wet forest), Grassland (35%; humid grassland), Wetland (15%; standing fresh water; river/stream), Artificial landscape (35%; arable land; forestry plantation; other urban/industrial areas)
Land-use Agriculture (40%), Fisheries/aquaculture (5%), Forestry (35%), Nature conservation/research (5%), Tourism/recreation (15%), Water management (10%)

Birds

Species	Season	Year	Pop min	Pop max	Acc	Criteria
Ciconia nigra Black Stork	B	1996	6	10	B	B2
Aquila pomarina Lesser Spotted Eagle	B	1996	5	10	B	B2
Alcedo atthis Kingfisher	B	1996	40	50	A	B2
Riparia riparia Sand Martin	B	1996	10,000	14,000	A	B2

The most important breeding site for *Riparia riparia* in central Europe.

Protection status

National Partial **International** None
1,564 ha of IBA covered by Nature Conservation Area (Tiszatelek-Tiszaberceli ártér, 1,564 ha).

Conservation issues

Threats Afforestation (B), Consequences of animal/plant introductions (B), Firewood-collection (A), Industrialization/urbanization (C), Intensified forest management (A), Natural events (B), Recreation/tourism (B)

The main threats are intensified forest management and the construction of second-home developments. An integrated monitoring project is being carried out on *Riparia riparia*.

Szatmár-Bereg plains — A1, B2 — 035

Admin region Szabolcs-Szatmár-Bereg
Coordinates 48°00′N 22°40′E
Altitude 110–219 m **Area** 117,300 ha

Site description

A low-lying flood-plain of the Tisza, Szamos and Túr rivers, situated between Vásárosnamény and the border with Ukraine.

Habitats Forest and woodland (20%; broadleaved deciduous forest; alluvial/very wet forest), Grassland (35%; humid grassland; mesophile grassland), Wetland (10%), Artificial landscape (35%; highly improved reseeded grassland; arable land; perennial crops/orchards/groves; forestry plantation)
Land-use Agriculture (50%), Forestry (30%), Nature conservation/research (10%), Tourism/recreation (10%), Urban/industrial/transport (5%), Water management (5%)

Birds

Species	Season	Year	Pop min	Pop max	Acc	Criteria
Ciconia ciconia White Stork	B	1993	182	200	—	B2
Falco vespertinus Red-footed Falcon	B	1993	40	50	—	B2
Crex crex Corncrake	B	1997	130	150	A	A1

An important site for breeding waterbirds, raptors and forest birds. Breeding species of global conservation concern that do not meet IBA criteria: *Aythya nyroca* (10–15 pairs).

Protection status

National Partial **International** None
22,246 ha of IBA covered by Landscape Protected Area (Szatmar-Beregi, 22,246 ha).

Conservation issues

Threats Abandonment/reduction of land management (A), Afforestation (B), Agricultural intensification/expansion (A), Burning of vegetation (B), Consequences of animal/plant introductions (B), Disturbance to birds (B), Drainage (A), Filling-in of wetlands (C), Firewood collection (B), Industrialization/urbanization (C), Infrastructure (A), Intensified forest management (B), Recreation/tourism (C)

The main threats to the grasslands are abandonment of management practices, invasion by *Amorpha fruticosa*, and conversion to arable land or forest. Approximately one third of *Crex crex* habitat within the IBA is threatened by the construction of Motorway 3.

Heves plains — A1, A4ii, B1iii, B2 — 036

Admin region Heves, Jász-Nagykun-Szolnok
Coordinates 47°37′N 20°27′E
Altitude 88–102 m **Area** 49,000 ha

Site description

A well-drained flood-plain overlain by clay soils, situated between Füzesabony and Jászkisér to the north of the River Tisza. The IBA includes extensive areas of arable land and saline grassland, with some forestry plantations. Human activities include sheep- and cattle-grazing.

Habitats Grassland (30%; steppe/dry calcareous grassland; humid grassland), Artificial landscape (70%; arable land; forestry plantation)
Land-use Agriculture (90%), Hunting (5%), Nature conservation/research (5%)

Birds

Species	Season	Year	Pop min	Pop max	Acc	Criteria
Aquila heliaca Imperial Eagle	R	1995	2	3	A	A1
Falco vespertinus Red-footed Falcon	B	1993	40	—	B	B2
Falco vespertinus Red-footed Falcon	P	1996	800	1,100	—	A4ii, B1iii
Falco cherrug Saker	R	1993	5	11	—	B2
Otis tarda Great Bustard	R	1993	64	64	A	A1

An important area for steppic birds.

Protection status

National Partial **International** None

8,199 ha of IBA covered by Landscape Protected Area (Hevesi füves puszták, 8,199 ha).

Conservation issues

Threats Abandonment/reduction of land management (B), Afforestation (B), Agricultural intensification/expansion (A), Disturbance to birds (B), Drainage (B), Firewood collection (B), Infrastructure (B)

The main threat to the site is agricultural intensification, with the introduction of irrigation and mechanized cultivation. Artificial nest-sites are provided for both raptors and *Coracias garrulus*, and protection measures for *Otis tarda* are in place through MME. Wetland restoration work has been undertaken, and a management plan exists for the area.

Borsodi-Mezőség — A1, A4i, B1i, B2 — 037

Admin region Borsod-Abaúj-Zemplén
Coordinates 47°45'N 20°48'E
Altitude 86–93 m **Area** 28,900 ha

Site description

An area of flood-plain underlain by loess soils, adjacent to the natural levees of the River Tisza between Mezőcsát and Mezőnagymihály. The main habitats are saline grasslands and arable fields, with some small areas of forest. Human activities include sheep- and cattle-grazing, arable farming and hunting.

Habitats Forest and woodland (5%; alluvial/very wet forest; wooded steppe), Grassland (60%; steppe/dry calcareous grassland), Wetland (5%; water-fringe vegetation), Artificial landscape (30%; highly improved reseeded grassland; arable land; forestry plantation)
Land-use Agriculture (90%), Forestry (5%), Water management (5%)

Birds

Species	Season	Year	Pop min	Pop max	Acc	Criteria
Platalea leucorodia Spoonbill	P	1996	150	200	A	B1i
Falco vespertinus Red-footed Falcon	B	1996	90	100	A	B2
Falco cherrug Saker	R	1996	5	6	A	B2
Perdix perdix Partridge	R	1996	200	300	C	B2
Grus grus Crane	P	1996	1,000	2,000	—	A4i, B1i
Otis tarda Great Bustard	R	1996	35	45	A	A1
Coracias garrulus Roller	B	1996	60	65	—	B2
Lanius minor Lesser Grey Shrike	B	1996	50	120	C	B2

An important area for steppic birds. Species of global conservation concern that do not meet IBA criteria: *Aythya nyroca* (5–10 breeding pairs), *Haliaeetus albicilla* (wintering), *Aquila heliaca* (breeding).

Protection status

National Partial **International** None

17,932 ha of IBA covered by Landscape Protected Area (Borsodi Mezőségi, 17,932 ha).

Conservation issues

Threats Abandonment/reduction of land management (A), Drainage (A), Dredging/canalization (B), Groundwater abstraction (B), Other (B)

The main threats are those of mechanized agriculture and the persecution of birds by hunters ('Other' threat, above). Wetland restoration work is in progress, and artificial nesting-sites are provided for both raptors and *Coracias garrulus*. Measures for the protection of *Otis tarda* are carried out by MME, and a management plan exists for the area.

Kesznyéten — B2 — 038

Admin region Borsod-Abaúj-Zemplén, Szabolcs-Szatmár-Bereg
Coordinates 48°01'N 21°03'E
Altitude 90–95 m **Area** 7,000 ha

Site description

An area of flood-plain bordering the River Tisza, located near to the town of Tiszaújváros. The IBA includes areas of wet forest and grassland, and oxbow lakes. Human activities include arable farming, grazing, mowing, reed-harvesting ('Other' land-use, below) and angling. This area includes two sites that were treated as separate IBAs in the previous international IBA inventory (Grimmett and Jones 1989): 'Tiszadobiártér' (former site HU034) and 'Tiszaluc-Kesznyéten puszta' (former site HU037).

Habitats Forest and woodland (10%; alluvial/very wet forest), Grassland (40%; steppe/dry calcareous grassland; humid grassland), Wetland (15%; standing fresh water; water-fringe vegetation), Artificial landscape (35%; arable land; ruderal land)
Land-use Agriculture (85%), Forestry (10%), Nature conservation/research (5%), Other (5%)

Birds

Species	Season	Year	Pop min	Pop max	Acc	Criteria
Nycticorax nycticorax Night Heron	B	1996	5	400	A	B2
Ardeola ralloides Squacco Heron	B	1996	—	15	B	B2

An important breeding site for herons (Ardeidae) and other waterbirds. Species of global conservation concern that do not meet IBA criteria: *Aythya nyroca* (3–5 breeding pairs), *Haliaeetus albicilla* (wintering), *Crex crex* (breeding).

Protection status

National Partial **International** None

6,083 ha of IBA covered by Landscape Protected Area (Kesznyéteni, 6,083 ha).

Conservation issues

Threats Agricultural intensification/expansion (A), Burning of vegetation (A), Construction/impact of dyke/dam/barrage (B), Deforestation (commercial) (B), Drainage (A), Firewood collection (C)

The Tiszatáj Land Trust has acquired the heronry and important areas of grassland, and a management plan for the LPA was prepared in 1997. An area of good-quality *Ardeola ralloides* habitat, owned by MME, was accidentally burned in 1996.

Bodrog flood-plain — A1, A4i, B1i, B2 — 039

Admin region Borsod-Abaúj-Zemplén, Szabolcs-Szatmár-Bereg
Coordinates 48°15'N 21°20'E
Altitude 99–105 m **Area** 10,000 ha

Site description

A permanently flooded area of flood-plain, c.30 km long and c.2–3 km wide, adjacent to the Bodrog and Tisza rivers. The high groundwater level results from the effects of the dam at Tiszalök. The area includes natural and artificial oxbow lakes, marshes, wet grasslands, flood-plain forests and several non-native poplar *Populus* plantations. Human activities include grazing, mowing, arable farming, angling and hunting.

Habitats Forest and woodland (5%; alluvial/very wet forest), Grassland (70%; humid grassland), Wetland (10%; standing fresh water; river/stream; water-fringe vegetation), Artificial landscape (15%; arable land; forestry plantation)
Land-use Agriculture (65%), Fisheries/aquaculture (5%), Forestry (10%), Tourism/recreation (20%), Water management (5%)

Birds

Species	Season	Year	Pop min	Pop max	Acc	Criteria
Nycticorax nycticorax Night Heron	B	1993	200	250	A	B2
Ciconia nigra Black Stork	P	1993	200	300	B	A4i, B1i
Platalea leucorodia Spoonbill	B	1993	25	30	A	B2
Aythya nyroca Ferruginous Duck	B	1993	15	—	—	B2
[1] ***Crex crex*** Corncrake	B	1993	100	—	C	A1
Grus grus Crane	P	1993	500	1,000	—	B1i

1. 256 calling males in 1997.

An important site for wetland birds and raptors.

Protection status

National Partial **International** Partial

1,123 ha of IBA covered by Nature Conservation Area (Long-erdő, 1,123 ha). 3,782 ha of IBA covered by Landscape Protected Area

(Tokaj-Bodrogzug, 4,242 ha). 3,782 ha of IBA covered by Ramsar Site (Bodrogzug, 3,782 ha).

Conservation issues

Threats Abandonment/reduction of land management (A), Afforestation (A), Agricultural intensification/expansion (B), Burning of vegetation (B), Disturbance to birds (B), Intensified forest management (C), Natural events (B), Recreation/tourism (C)

The main threats are the abandonment of land-management practices and the early cutting of hay, both of which have a detrimental effect on *Crex crex* numbers. Afforestation is another major threat, as grassland management is now considered to be uneconomic. Burning can destroy the nesting-sites of herons unless properly controlled. The enlargement of the existing LPA is in process.

Mátra mountains — A1, B2 — 040

Admin region Heves, Nógrád
Coordinates 47°49′N 20°00′E
Altitude 200–1,014 m **Area** 58,100 ha

Site description

A range of narrow mountain ridges situated between the Zagyva river and Tarna creek, including the Kékes summit, the highest point in Hungary. The vegetation is dominated by oak *Quercus* forest on acidic soils, with submontane *Quercus*, *Carpinus*, *Fagus* forests at higher altitudes. Large tracts of land are planted with spruce *Picea*, and the forested area as a whole is intensively managed, whilst orchards and grasslands are mostly abandoned. A popular area for recreation.

Habitats Forest and woodland (90%; broadleaved deciduous forest), Scrub (5%; scrub), Artificial landscape (5%; perennial crops/orchards/groves; forestry plantation; other urban/industrial areas)
Land-use Forestry (90%), Nature conservation/research (50%), Tourism/recreation (30%), Urban/industrial/transport (5%)

Birds

Species	Season	Year	Pop min	Pop max	Acc	Criteria
[1] ***Aquila heliaca*** Imperial Eagle	B	1996	4	6	A	A1
Falco cherrug Saker	R	1996	—	4	A	B2

1. 5 year average.

An important area for raptors and other forest species.

Protection status

National Partial **International** None
186 ha of IBA covered by Nature Conservation Area (Gyöngyösi Sárhegy, 186 ha). 11,863 ha of IBA covered by Landscape Protected Area (Mátrai, 11,863 ha).

Conservation issues

Threats Abandonment/reduction of land management (B), Forest grazing (A), Infrastructure (B), Intensified forest management (A)

High levels of grazing pressure from game species such as red deer *Cervus elaphus* are restricting forest regeneration.

Bükk mountains — A1, B2 — 041

Admin region Borsod-Abaúj-Zemplén, Heves
Coordinates 48°10′N 20°30′E
Altitude 200–958 m **Area** 61,000 ha

Site description

A mountain range between Eger and Miskolc, with deep valleys and a wide plateau. Deciduous oak *Quercus* forests exist between 200 m and 600 m, with submontane forests of *Quercus*, *Carpinus* and *Fagus* above 700 m. Oak forest, scrub and rocky grassland on the southern slopes experience a more continental micro-climate. Human activities include intensive forest management and hunting.

Habitats Forest and woodland (90%; broadleaved deciduous forest; mixed forest), Scrub (5%; scrub), Artificial landscape (5%; forestry plantation)
Land-use Forestry (90%), Military (5%), Nature conservation/research (90%), Tourism/recreation (20%), Urban/industrial/transport (5%)

Birds

Species	Season	Year	Pop min	Pop max	Acc	Criteria
Circaetus gallicus Short-toed Eagle	B	1996	5	5	A	B2
Aquila pomarina Lesser Spotted Eagle	B	1996	2	8	A	B2
Aquila heliaca Imperial Eagle	R	1996	5	5	A	A1, B2
Falco cherrug Saker	R	1996	2	4	A	B2

An important area for raptors and forest birds.

Protection status

National Partial **International** None
41,833 ha of IBA covered by National Park (Bükki, 41,833 ha).

Conservation issues

Threats Intensified forest management (A), Selective logging/cutting (C)

A long-term research and conservation project on raptors has been carried out, including the safeguarding of nests and provision of artificial nest-sites. The enlargement of Bükk National Park is underway, and a management plan exists for the area.

Zemplén mountains — A1, B2 — 042

Admin region Borsod-Abaúj-Zemplén
Coordinates 48°30′N 21°25′E
Altitude 163–895 m **Area** 102,000 ha

Site description

Situated near to the town of Sárospatak, and surrounded by the flood-plains of the Bodrog and Hernád rivers. *Quercus*, *Carpinus* and *Fagus* dominate the submontane forests, with montane *Fagus* forests existing on the northern slopes. Old, traditional orchards, vineyards and pastures occur in valleys and on slopes. Human activities include haymaking, grazing, hunting and mining.

Habitats Forest and woodland (80%; broadleaved deciduous forest), Scrub (5%; scrub), Grassland (5%; steppe/dry calcareous grassland; humid grassland), Artificial landscape (10%; arable land; perennial crops/orchards/groves; forestry plantation)
Land-use Agriculture (5%), Forestry (90%), Nature conservation/research (10%), Tourism/recreation (20%)

Birds

Species	Season	Year	Pop min	Pop max	Acc	Criteria
[1] ***Ciconia nigra*** Black Stork	B	1995	7	13	A	B2
[2] ***Ciconia ciconia*** White Stork	B	1995	125	148	A	B2
[1] ***Circaetus gallicus*** Short-toed Eagle	B	1995	5	15	A	B2
[1] ***Aquila pomarina*** Lesser Spotted Eagle	B	1995	7	16	A	B2
[1] ***Aquila heliaca*** Imperial Eagle	B	1995	9	9	A	A1, B2

1. 1985–1995.
2. 1994 and 1995 census data.

The most important site for raptors in Hungary, with large numbers of woodpeckers also present: *Picus canus*, *Dendrocopos medius* and *D. syriacus*.

Protection status

National Partial **International** None
2 ha of IBA covered by Landscape Protected Area (Bodrogszegi Várhegy, 2 ha). 26,496 ha of IBA covered by Landscape Protected Area (Zempléni, 26,496 ha).

Conservation issues

Threats Abandonment/reduction of land management (B), Afforestation (B), Construction/impact of dyke/dam/barrage (B), Disturbance to birds (A), Extraction industry (B), Firewood collection (C), Industrialization/urbanization (C), Infrastructure (B), Intensified forest management (A), Recreation/tourism (B)

The abandonment of pasture management and afforestation with spruce *Picea* has reduced the feeding area available to raptors. The damming of the Aranyos valley and a quarry at Pálháza have been proposed, and gold-prospecting in Telkibánya district and unregulated tourism are additional problems. The existing protected area is being enlarged. MME Zemplén Chapter protects raptor nest-sites, and the Directorate of Bükk National Park is re-creating mountain meadows.

Birds

Species	Season	Year	Pop min	Pop max	Acc	Criteria
[1] ***Aquila heliaca*** Imperial Eagle	B	1995	1	3	A	A1, B2
Crex crex Corncrake	B	1995	20	50	A	A1, B2

1. 1990–1995.

An important area for raptors and forest birds, with *Crex crex* breeding in the valleys.

Protection status

National Partial **International** Partial

19,763 ha of IBA covered by National Park (Aggteleki, 19,763 ha). 19,247 ha of IBA covered by Biosphere Reserve (Aggtelek Biosphere Reserve, 19,247 ha). The IBA also includes the Caves of Aggtelek and Slovak Karst World Heritage Site.

Conservation issues

Threats Abandonment/reduction of land management (B), Agricultural intensification/expansion (B), Burning of vegetation (C), Disturbance to birds (C), Firewood collection (A), Infrastructure (C), Intensified forest management (A), Recreation/tourism (B), Selective logging/cutting (C)

The Aggtelek National Park Authority undertook a basic ecological survey between 1991 and 1994. By 1996 management planning was in process, along with monitoring of *Crex crex* and *Lanius* numbers.

Aggtelek karst and Bódva valley A1, B2 043

Admin region Borsod-Abaúj-Zemplén
Coordinates 48°29′N 20°33′E
Altitude 155–604 m **Area** 28,600 ha

Site description

The site is situated on the Hungarian part of the Gömör-Torna karst plateau, adjacent to the Slovakian border. The main rock formation is Triassic limestone, which is overlain by forest soils. These soils support extensive areas of native deciduous forest, rocky grasslands and wet meadows, whilst the valleys are extensively farmed.

Habitats Forest and woodland (70%; broadleaved deciduous forest; wooded steppe), Scrub (5%; scrub), Grassland (15%; steppe/dry calcareous grassland; humid grassland), Artificial landscape (10%; arable land; forestry plantation)

Land-use Agriculture (5%), Forestry (80%), Nature conservation/research (100%), Tourism/recreation (70%)

REFERENCES

BANKOVICS, A. (1985) A Balaton átvonuló és áttelelő vízimadarai. [Migratory and wintering birds of Lake Balaton.] *Aquila* 92: 55–64. (In Hungarian, with English summary.)

CSIHAR, L. (1993) [Status report for Rétszilas fish-ponds 1993.] Unpublished. (In Hungarian.)

CSIHAR, L. (1993) [Status report for Sárviz Valley 1993]. Unpublished. (In Hungarian.)

CSIMA, P. (1994) *A Zempléni Tájvédelmi Körzet Regionális Rendezési Terv*. Budapest: Kertészeti és Élelmiszeripari Egyetem.

FARAGÓ, S. (1996) [Database of geese in Hungary 1984–1994: a long-term monitoring.] *Hungarian Waterfowl Publications* 2: 3–168. (In Hungarian.)

FARAGÓ, S. (1996) [Waterbird stocks on the Danube (reach Gönyű–Szob) river km 1791–1708: investigation conducted over a 10-year period (1982–1992).] *Hungarian Waterfowl Publications* 1: 1–461. (In Hungarian.)

FARAGÓ, S. AND NAGY, F. (1996) [*Game and nature conservation projects in the area possessed by the Lajta-hanság Joint Stock Company*.] Mosonmagyaróvár, Hungary: Lajta-Hanság Joint Stock Company. (In Hungarian.)

FINTHA, I. (1993) [Autumn crane migration in Hungary with special reference to recent records.] *Aquila* 100: 137–150. (In Hungarian, with English summary.)

FÜRI, A. AND URBÁN, S. (1995) [*Natural values of Tápió Region*.] Szolnok, Hungary: Jászkun Természetvédelmi Szervezet. (In Hungarian.)

GRIMMETT, R. F. A. AND JONES, T. A. (1989) *Important Bird Areas in Europe*. Cambridge, UK: International Council for Bird Preservation (Techn. Publ. 9).

HADARICS, T. AND PELLINGER A. (1993) [List of bird species recorded in the Hungarian part of Fertő 1970–1992.] *Szélkiáltó* 6: 3–16. (In Hungarian.)

HAVRANEK, L. (1996) A Balaton déli partján előforduló vízimadarak 1995. [Waterfowl on the southern shore of Lake Balaton in 1995.] *Anser* 2: 32–37. (In Hungarian.)

HAVRANEK, L. (1996) Ludak (*Anser*) előfordulása a Balaton déli partján. [Geese (*Anser*) on southern shore of Lake Balaton.] *Anser* 2: 52–59. (In Hungarian.)

HAVRANEK, L., SZABÓ, B. AND SZÁSZ, S. (1995) Balatoni madárvonulások az 1991–1994 években. [Bird migration in the vicinity of Lake Balaton during 1991–1994.] *Anser* 1: 12–28. (In Hungarian.)

HEATH, M. F. AND BORGGREVE, C. (2000) *BirdLife International/EBCC European Bird Database 1998*. Cambridge, UK: BirdLife International.

KÁRPÁTI, L. (1983) [Ecological research on the avifauna of Lake Fertő.] *EFE Tud. Közl.* 1982/2. (In Hungarian.)

KEVE, A. (1968) Aythyae és Merginidae-fajok előfordulásai és vonulási ciklusuk a Balatonon. [Occurrence and migration phenology of *Aythya* and *Mergus* species on Lake Balaton.] *Aquila* 75: 21–44. (In Hungarian, with English summary.)

KEVE, A. (1973) A Balaton úszórécéi. [Dabbling ducks (*Anas* sp.) of Lake Balaton.] *Aquila* 76–77: 117–139. (In Hungarian, with English summary.)

KEVE, A. (1983) A bukórécék jelentősége a Balaton életében. [The importance of diving ducks in the ecology of Lake Balaton.] *Aquila* 90: 105–112. (In Hungarian, with English summary.)

KOVÁCS, G. (1993) [Study of the population and habitats of the Great Bustard (*Otis tarda*) in the region of Hortobágy between 1975 and 1992.] *Aquila* 100: 151–159. (In Hungarian, with English summary.)

KOVÁCS, G. (1994) [Population increase and expansion of the Aquatic Warbler (*Acrocephalus paludicola*) in Hortobágy between 1977 and 1994.] *Aquila* 101: 133–139. (In Hungarian.)

KOVÁCS, G. (1995) [Birds of Hortobágy Fish-pond 1974–1994.] In I. Major, ed. *Alföldi mozaik*. Budapest: Természet Búvár Alapítvány Kiadó. A KTM Természetvédelmi Hivatalának Tanulmánykötetei 2. (In Hungarian.)

MUSICZ, L. (1990) [Wild goose movements on Lake Öreg at Tata 1984–1989.] *Aquila* 96–97: 30–35. (In Hungarian.)

MUSICZ, L. (1993) [Conservation effects of rural development in the Tata Basin.] Budapest, Hungary: ELTE, Faculty of Human Ecology (MSc. Thesis). (In Hungarian.)

MUSICZ, L. (1996) [Strategy for restoration of water quality of Lake Öreg. Proposal for Municipality of Tata.] Unpublished. (In Hungarian.)

NOVOTNY, ZS. (1996) A Látrányi Természetvédelmi Terület. [The Látrány Nature Conservation Area.] *Anser* 2: 49–51. (In Hungarian.)

PEST KÖRNYÉKI MADARÁSZ KÖR (1996) [Ornithofauna of Szentendre Island.] Pp.109–153 in O. Merkl, ed. [Closing and supplementary notes to zoological survey in the area of the planned Duna-Ipoly National Park.] Report to Ministry of Environment and Regional Policy. Unpublished. (In Hungarian.)

RADETCZKY, J. (1982) [*Birds and landscapes in Fejér county*.] Székesfehérvár. (In Hungarian.)

SÁNDOR, A., ED. (1983) [*Bükk National Park*.] Budapest: Mezőgazdasági Kiadó. (In Hungarian.)

SELMECZI KOVÁCS Á. (1995) Az Ipoly völgyének madárfaunája az 1993–1994 között végzett kutatások tükrében. [Birds of Ipoly Valley on the basis of research carried out during 1993–1994.] Pp.72–108 in O. Merkl, ed. Zoológiai vizsgálatok a tervezett Duna-Ipoly Nemzeti Park térségében 1994 során. [Zoological research in the planned Duna-Ipoly National Park in 1994.] Report to Ministry of Environment and Regional Policy. Unpublished. (In Hungarian.)

SZABÓ, B. AND POLYÁK, F. (1991) Mesterséges halastavak természetvédelmi problémái Somogyban. [Nature conservation conflicts on fish-ponds in Somogy county.] Kaposvár, Hungary: Pannon University (MSc. Thesis). (In Hungarian.)

SZÉLL, A. (1990) [Migration of Crane (*Grus grus* L., 1758) at Szeged-Fehértó in autumn 1986.] *Aquila* 96–97: 91–96. (In Hungarian, with English summary.)

SZÉP, T. (1991) [Numbers and distribution of the Hungarian Sand Martin (*Riparia riparia* L., 1758) population breeding along the Hungarian reaches of the River Tisza.] *Aquila* 98: 123–124. (In Hungarian.)

SZÉP, T. (1995) Relationship between West African rainfall and the survival of Central European Sand Martin (*Riparia riparia*) population. *Ibis* 137: 162–168.

SZÉP, T. (1995) Survival rates of Hungarian Sand Martins and their relationship with Sahel rainfall. *J. Applied Statistics* 22: 891–904.

SZÉP, T. (1996) [Report from the pilot ornithological field research carried out in the frame of the Hungarian National Biodiversity Monitoring Programme in Tiszabercel–Gávavencsellő 1996.] Report to Hungarian National Biodiversity Programme, European Commission PHARE Programme. Unpublished. (In Hungarian.)

TÓTH, K., ED. (1979) [*National Park in the Kiskunság*.] Budapest: Natura. (In Hungarian.)

TUCKER, G. M. AND HEATH, M. F. (1994) *Birds in Europe: their conservation status*. Cambridge, UK: BirdLife International (BirdLife Conservation Series 3).

VINCZE, B. (1996) Jégmadár (*Alcedo atthis*) előfordulása és költése az Irmapusztai halastavakon. [Occurrence and breeding of kingfisher (*Alcedo atthis*) on Irmapuszta fish-ponds.] *Anser* 2: 47–48. (In Hungarian.)

ITALY

Fabio Casale, Umberto Gallo-Orsi and Vincenzo Rizzi

Gargano National Park (IBA 129), a mountainous promontory along the Adriatic coast important for breeding raptors and some open-country species.
(PHOTO: ALBERTO NARDI/NHPA)

GENERAL INTRODUCTION

Italy covers a land area of 301,302 km² (including the large islands of Sicily and Sardinia), and in 1991 had a population of 56.7 million, resulting in an average density of c.188 persons per km² (ISTAT 1991). Plains cover 23% of the country and are mainly concentrated in the north (Po valley), along the coasts, and in the Puglia region, while mountains and hilly areas cover 35% and 41% of the land respectively.

The climate varies considerably with latitude. In the south it is warm temperate, with almost no rain in summer, but the north is cool temperate, often experiencing snow and freezing temperatures in winter, with rainfall more evenly distributed through the year. The 5,000-km-long coastline comprises a mixture of sandy beaches with sand-dunes, brackish lagoons, and marshes, and sections of cliffs and offshore rocky islands and islets. The three main estuaries are those of the Po, the Tiber and the Arno rivers.

Out of 262,028 km² of agricultural land, 90,123 km² are arable and 48,776 km² are permanent pasture. The average agricultural holding covers 7.5 ha, and the number and average area of holdings have decreased by 7.5% and 5% respectively following land abandonment in marginal areas in recent years (ISTAT 1991). In the lowlands, agriculture is very intensive and devoted mainly to arable monoculture (maize, wheat and rice being the three major crops), while in the hills and mountains traditional, and less intensive agriculture is still practised although land abandonment is spreading.

A total of 192 Important Bird Areas (IBAs) are listed in the present inventory (Table 1, Map 1), covering a total area of 46,270 km², equivalent to c.15% of the national land area. This compares with 140 IBAs identified in Italy in the previous pan-European IBA inventory (Grimmett and Jones 1989; LIPU 1992), covering some 35,100 km². Some of these original IBAs have been grouped for the current inventory: former site IT081 has been merged with IT082 to form current site 119; former sites IT086 and IT087 now form site 133; IT089 and IT100 are now site 141; and IT104 and IT106 are now site 149. Of the 17 subsites listed in 1989 for the IBA 'River Po from Dora Baltea mouth to Scrivia mouth' (former site IT005), four have been excluded from the current inventory (they do not meet the revised IBA criteria) while the remaining 13 have been grouped into four discrete IBAs ('new' sites 017, 020, 025 and 027). The name or boundaries of some

Table 1. Summary of Important Bird Areas in Italy. — 192 IBAs covering 46,270 km²

IBA code	1989 code	International name	National name	Administrative region	Area (ha)	Criteria (see p. 11)
001	—	Mont Blanc (Italian side)	Monte Bianco (lato italiano)	Valle d'Aosta	11,300	A3, C6
002	—	Monte Rosa (Valle d'Aosta side)	Monte Rosa (versante valdostano)	Valle d'Aosta	8,200	C6
003	IT002	Anzasca and Antrona valleys	Valle Anzasca e Val Antrona	Piemonte	42,600	A3, C6
004	IT001	Antigorio and Formazza valleys	Val Antigorio e Val Formazza	Piemonte	50,000	A3, B2, C6
005	IT003	Val Grande	Val Grande	Piemonte	29,000	C7
006	IT014	Mount Tabor	Monte Tabor	Lombardia	28,000	C7
007	IT012	Pian di Spagna and Lake Mezzola	Pian di Spagna e Lago di Mezzola	Lombardia	3,400	C7
008	IT004	Gran Paradiso	Gran Paradiso	Piemonte, Valle d'Aosta	110,000	A3, B2, C6

Table 1 ... continued. Summary of Important Bird Areas in Italy. 192 IBAs covering 46,270 km²

IBA code	1989 code	International name	National name	Administrative region	Area (ha)	Criteria (see p. 11)
009	—	Lake Maggiore	Lago Maggiore	Piemonte, Lombardia	21,600	C6
010	IT017	Campo dei Fiori, Valganna and Cuvia valley	Campo dei Fiori, Valganna e Val Cuvia	Lombardia	24,500	C7
011	IT016	Grigne	Grigne	Lombardia	14,038	B2, C6
012	IT013	Orobie	Orobie	Lombardia	107,000	A3, B2, C6
013	IT015	Camonica valley	Val Camonica	Lombardia	15,000	B2, C6
014	IT018	Brabbia peatland and Lake Varese	Palude Brabbia e Lago di Varese	Lombardia	1,960	A1, B2, B3, C1, C6
015	—	Lake Garda	Lago di Garda	Lombardia, Trentino-Alto Adige, Veneto	38,000	A4i, B1i, C3
016	—	Lake Viverone	Lago di Viverone	Piemonte	930	B3
017	IT005	Novara heronries	Garzaie del Novarese	Piemonte	50	A4i, B1i, B2, C2, C6
018	—	Ticino river	Fiume Ticino	Piemonte, Lombardia	25,000	A4i, B1i, B2, C2, C3, C6
019	IT020	Iseo peatlands	Torbiere d'Iseo	Lombardia	325	B3, C6
020	IT005	River Sesia heronries	Garzaie del Sesia	Piemonte	500	A1, A4i, B1i, B2, C1, C2, C6
021	IT006	Lake Candia	Lago di Candia	Piemonte	295	B3
022	IT023	Heronries of Pavia province	Garzaie del Pavese	Lombardia	1,173	A4i, A4iii, B1i, B2, C2, C4, C6
023	IT022	Adda Sud Park heronries	Garzaie del Parco Adda Sud	Lombardia	1,000	A4i, B1i, B2, C2, C6
024	IT007	Susa and Chisone valleys	Valle di Susa e Val Chisone	Piemonte	60,946	A3, B2, C6
025	IT005	Vercelli rice-fields	Risaie del Vercellese	Piemonte	15,000	A4i, B1i, B2, C2, C6
026	—	Trebbia river (lower course)	Fiume Trebbia (basso corso)	Emilia Romagna	30	C6
027	IT005	River Po: from Dora Baltea mouth to Scrivia mouth	Fiume Po: da Dora Baltea a Scrivia	Piemonte	14,000	B2, B3, C2, C6
028	IT024	River Po: from Ticino mouth to Adda mouth	Fiume Po: da Ticino a Adda	Lombardia, Emilia Romagna	25,000	A4i, B1i, B2, C2, C6
029	—	Marengo heronry	Garzaia di Marengo	Piemonte	30	A4i, B1i, B2, C2, C6
030	—	Stirone stream	Torrente Stirone	Emilia Romagna	2,400	B2
031	IT042	Taro river	Fiume Taro	Emilia Romagna	6,000	B2, C2, C6
032	IT043	Torrile LIPU reserve	Oasi LIPU Torrile	Emilia Romagna	200	C6
033	—	Scrivia stream	Torrente Scrivia	Piemonte	3,000	B2, C6
034	IT008	Crava–Morozzo lakes	Laghetti di Crava-Morozzo	Piemonte	300	C7
035	IT010	Maritime Alps	Alpi Marittime	Piemonte	95,400	A3, B1iv, B2, C5, C6
036	IT009	Mount Beigua	Monte Beigua	Liguria	10,000	B1iv, C5, C6
037	—	Finalese	Finalese	Liguria	3,750	C6
038	IT011	Ligurian Alps	Alpi Liguri	Liguria	15,000	B2, C6
039	—	Alpi Apuane	Alpi Apuane	Toscana	16,800	B2, C6
040	IT049	Apennines from Passo del Cerreto to Monte Caligi	Appennino dal Passo del Cerreto a Monte Caligi	Toscana, Emilia Romagna	45,000	B2, B3, C6
041	IT027	Stelvio National Park	Parco Nazionale dello Stelvio	Lombardia, Trentino-Alto Adige	134,621	A3, B2, C6
042	—	Dolomites	Dolomiti	Trentino-Alto Adige, Veneto	48,507	A3, C6
043	IT036	Carniche Alps	Alpi Carniche	Veneto, Friuli-Venezia Giulia	30,000	B2, C6
044	IT037	Tarvisio forest	Foresta di Tarvisio	Friuli-Venezia Giulia	23,300	B2, B3, C6
045	IT028	Adamello–Brenta	Adamello–Brenta	Lombardia, Trentino-Alto Adige	112,800	A3, B2, C6
046	IT029	Lagorai mountains	Catena dei Lagorai	Trentino-Alto Adige	40,000	B2, C6
047	IT038	Carniche pre-Alps	Prealpi Carniche	Veneto, Friuli-Venezia Giulia	63,000	A1, B2, C1, C6
048	—	Middle course of Tagliamento river	Medio corso del Tagliamento	Friuli-Venezia Giulia	3,579	C6
049	—	Cividalese and upper Torre valley	Cividalese e Alta Val Torre	Friuli-Venezia Giulia	28,400	A1, B3, C1, C2, C6
050	IT031	Belluno Dolomites	Dolomiti Bellunesi	Veneto	31,512	A1, B2, C1, C6
051	IT032	Lago di Busche and Vincheto di Cellarda	Lago di Busche e Vincheto di Cellarda	Veneto	125	C7
052	—	Giulie pre-Alps	Prealpi Giulie	Friuli-Venezia Giulia	39,400	C6
053	—	Magredi of Pordenone	Magredi di Pordenone	Friuli-Venezia Giulia	6,814	B2, C6
054	—	Mount Grappa	Monte Grappa	Veneto	40,000	A1, B1iv, B2, C1, C5, C6
055	—	Piave river	Medio corso del Fiume Piave	Veneto	10,000	B1iv, B2, C5, C6
056	—	Assa valley	Val d'Assa	Veneto	2,000	A1, C1, C6
057	IT030	Lessini mountains	Monti Lessini	Trentino-Alto Adige, Veneto	39,000	C6
058	IT019	Alto Garda Bresciano	Alto Garda Bresciano	Lombardia	38,269	C6
059	—	Middle course of River Brenta	Medio corso del Fiume Brenta	Veneto	1,100	B2, C6
060	—	Quinto heronry	Garzaia di Quinto	Veneto	50	B2, C2, C6
061	IT033	Caorle lagoon	Laguna di Caorle	Veneto	2,200	A4i, B1i, B2, B3, C2, C6
062	IT039	Grado and Marano lagoon	Laguna di Grado e Marano	Friuli-Venezia Giulia	21,000	A4i, A4iii, B1i, B2, C2, C3, C4, C6
063	IT040	Isonzo mouth, Cona island and Panzano Gulf	Foci dell'Isonzo, Isola della Cona e Golfo di Panzano	Friuli-Venezia Giulia	5,000	B1i, B2, C2, C3, C6
064	IT034	Venice lagoon	Laguna di Venezia	Veneto	50,000	A1, A4i, A4iii, B1i, B2, B3, C1, C2, C3, C4, C6
065	IT021	River Mincio and Bosco Fontana	Fiume Mincio e Bosco Fontana	Lombardia	8,000	A4i, B1i, B2, C2, C3, C6
066	IT041	Carso	Carso	Friuli-Venezia Giulia	18,950	B2, C6
067	IT026	Boscone island	Isola Boscone	Lombardia	200	A4i, B1i, B2, C2, C6

Table 1... continued. Summary of Important Bird Areas in Italy. 192 IBAs covering 46,270 km²

IBA code	1989 code	International name	National name	Administrative region	Area (ha)	Criteria (see p. 11)
068	IT025	River Po: from the mouth of Adda river to Ostiglia	Fiume Po: da Adda a Ostiglia	Lombardia, Veneto, Emilia Romagna	30,000	A4i, B1i, B2, B3, C2, C6
069	—	Codigoro heronry	Garzaia di Codigoro	Emilia Romagna	60	A4i, B1i, B2, C2, C6
070	IT035	Po delta	Delta del Po	Veneto	17,000	A4i, A4iii, B1i, B2, C2, C3, C4, C6
071	—	Valle Bertuzzi and Goro lagoon	Valle Bertuzzi e Sacca di Goro	Emilia Romagna	4,000	B1i, B2, B3, C2, C3, C6
072	IT044	Valli di Comacchio and Bonifica del Mezzano	Valli di Comacchio e Bonifica del Mezzano	Emilia Romagna	28,570	A1, A4i, A4iii, B1i, B2, C1, C2, C3, C4, C6
073	IT045	Valli di Argenta	Valli di Argenta	Emilia Romagna	2,000	A1, B1i, B2, C1, C2, C3, C6
074	IT046	Punte Alberete and Valle della Canna, Pineta San Vitale and Pialassa della Baiona	Punte Alberete, Valle della Canna, Pineta San Vitale e Pialassa della Baiona	Emilia Romagna	3,000	A1, A4i, B1i, B2, C1, C2, C3, C6
075	IT047	Ortazzo and Ortazzino	Ortazzo e Ortazzino	Emilia Romagna	900	B1i, B2, C2, C6
076	IT048	Cervia saltpans	Salina di Cervia	Emilia Romagna	1,000	A4i, B1i, B2, B3, C2, C3, C6
077	IT050	Lake Massaciuccoli	Lago di Massaciuccoli	Toscana	1,700	A1, A4i, B1i, B2, B3, C1, C2, C6
078	—	Foreste Casentinesi	Foreste Casentinesi	Toscana, Emilia Romagna	36,300	B2, B3, C6
079	—	Calvana mountains	Monti della Calvana	Toscana	3,900	C6
080	IT052	Fucecchio marsh	Palude di Fucecchio	Toscana	2,500	B2, B3, C2, C6
081	—	Arezzo heathlands	Brughiere Aretine	Toscana	4,394	B2, B3, C6
082	IT051	Migliarino–San Rossore	Migliarino–San Rossore	Toscana	7,000	A1, B2, C1
083	—	Pools of Florence plain	Stagni della Piana Fiorentina	Toscana	1,000	A1, B2, C1, C2, C6
084	IT053	Ponte Buriano–Lago della Penna	Ponte Buriano–Lago della Penna	Toscana	200	C6
085	—	Mount Conero	Monte Conero	Marche	1,140	B1iv, C5, C6
086	—	Furlo gorge and Mount Catria	Gola del Furlo e Monte Catria	Marche	36,000	B3, C6
087	—	Sentina	Sentina	Marche	180	C6
088	IT055	River Cecina	Fiume Cecina	Toscana	2,400	B3
089	IT054	Bolgheri	Bolgheri	Toscana	562	C6
090	—	Crete Senesi	Crete Senesi	Toscana	50,000	B2, C6
091	IT062	Lake Trasimeno	Lago Trasimeno	Umbria	13,200	A4iii, B1i, C3, C4, C6
092	IT057	Farma valley	Valle del Farma	Toscana	10,000	B2
093	IT058	Lakes Montepulciano and Chiusi	Laghi di Montepulciano e Chiusi	Toscana	900	B1i, B2, C3, C6
094	—	Colfiorito	Colfiorito	Umbria	500	B2, C6
095	IT063	Sibillini mountains	Monti Sibillini	Umbria, Marche	71,374	A3, B2, C6
096	IT056	Tuscan archipelago	Arcipelago Toscano	Toscana	28,000	A1, A3, A4i, A4ii, B1i, B1ii, B2, B3, C1, C2, C3, C6
097	IT059	Diaccia Botrona	Diaccia Botrona	Toscana	1,350	A1, B1i, C1, C2, C3, C6
098	IT060	Uccellina mountains, Trappola marshes and Ombrone mouth	Monti dell'Uccellina, Stagni della Trappola e Bocca d'Ombrone	Toscana	5,670	A1, B1i, C1, C3
099	—	Lake Bolsena	Lago di Bolsena	Lazio	16,000	B2, C6
100	IT065	Lake Alviano	Lago di Alviano	Umbria, Lazio	900	C6
101	IT064	Laga mountains	Monti della Laga	Lazio, Abruzzo, Marche	60,000	A1, B2, C1, C6
102	IT066	Selva del Lamone	Selva del Lamone	Lazio	2,200	B2, B3, C6
103	—	Argentario	Argentario	Toscana	5,200	A3
104	IT061	Orbetello lagoon and Burano lake	Laguna di Orbetello e Lago di Burano	Toscana	3,680	A1, A4i, A4iii, B1i, B2, C1, C2, C4, C6
105	IT068	Rieti lakes	Laghi Reatini	Lazio	3,278	C6
106	IT067	Rieti mountains	Monti Reatini	Lazio	22,980	B2, C6
107	IT077	Gran Sasso d'Italia	Gran Sasso d'Italia	Abruzzo	75,000	A3, B2, C6
108	IT069	Vico lake	Lago di Vico	Lazio	4,000	C6
109	IT078	Cicolano	Cicolano	Lazio, Abruzzo	49,485	B2, B3, C6
110	IT070	Lake Bracciano and surrounding areas	Lago di Bracciano e aree circostanti	Lazio	31,300	B2, B3, C6
111	IT072	Tolfa hills	Monti della Tolfa	Lazio	75,000	B2, B3, C6
112	—	Tarquinia saltpans	Saline di Tarquinia	Lazio	170	C2
113	IT071	Lucretili mountains	Monti Lucretili	Lazio	21,000	C6
114	IT079	Sirente, Velino and Duchessa mountains	Sirente, Velino e Montagne della Duchessa	Lazio, Abruzzo	50,000	A1, B2, C1, C6
115	IT080	Maiella, Pizzi and Frentani mountains	Maiella, Monti Pizzi e Monti Frentani	Abruzzo	177,500	B2, B3, C2, C6
116	—	Coast from Palo to Santa Marinella	Costa da Palo a Santa Marinella	Lazio	6,300	B1ii, C2
117	—	Litorale Romano	Litorale Romano	Lazio	19,529	B1i, B2, C3, C6
118	IT073	Ernici and Simbruini mountains	Monti Ernici e Simbruini	Lazio, Abruzzo	70,000	B2, C6
119	IT081, IT082	Abruzzo National Park	Parco Nazionale d'Abruzzo	Lazio, Abruzzo, Molise	110,000	B2, B3, C2, C6
120	IT074	Lepini mountains	Monti Lepini	Lazio	49,600	B2, C6
121	IT075	Circeo National Park	Parco Nazionale del Circeo	Lazio	8,350	A1, B2, C1, C2, C6
122	IT076	Ponziane islands	Isole Ponziane	Lazio	1,130	B2, C6
123	—	Ausoni and Aurunci mountains	Monti Ausoni e Aurunci	Lazio	137,200	B2, C6
124	IT084	Matese	Matese	Campania, Molise	41,718	B2, B3, C6
125	IT083	River Biferno	Fiume Biferno	Molise	30,000	B2, C6
126	—	Daunia mountains	Monti della Daunia	Puglia	50,000	B2, C6
127	IT092	Tremiti islands	Isole Tremiti	Puglia	2,116	A4ii, B1ii, B2, C3, C6

Table 1... continued. Summary of Important Bird Areas in Italy. 192 IBAs covering 46,270 km²

IBA code	1989 code	International name	National name	Administrative region	Area (ha)	Criteria (see p. 11)
128	IT093	Lakes Lesina and Varano	Laghi di Lesina e Varano	Puglia	20,353	A1, A4i, A4iii, B1i, C1, C2, C3, C4
129	IT094	Gargano promontory	Promontorio del Gargano	Puglia	240,000	A1, B2, B3, C1, C2, C6
130	IT095	Manfredonia Gulf wetlands	Zone umide del Golfo di Manfredonia	Puglia	8,000	A1, A4i, A4iii, B1i, B2, B3, C1, C2, C3, C4, C6
131	IT085	Capri island	Isola di Capri	Campania	1,036	C6
132	—	Serre Persano	Serre Persano	Campania	300	C6
133	IT086, IT087	Picentini mountains	Monti Picentini	Campania	64,000	C6
134	IT088	Alburni mountains	Monti Alburni	Campania	16,000	B2, C6
135	IT096	Murge	Murge	Puglia	90,000	A1, A3, A4ii, B1iii, B2, C1, C2, C6
136	IT090	Mount Cervati	Monte Cervati	Campania	44,000	B2, C6
137	IT099	Dolomiti di Pietrapertosa	Dolomiti di Pietrapertosa	Basilicata	2,500	B2, C6
138	—	Manferrara forest	Bosco della Manferrara	Basilicata	500	B2, C6
139	IT097	Gravine	Gravine	Puglia, Basilicata	30,000	A1, A4ii, B1iii, B2, C1, C2, C6
140	IT091	Coast between Marina di Camerota and Policastro Bussentino	Costa tra Marina di Camerota e Policastro Bussentino	Campania	5,700	B2, C6
141	IT089, IT100	Lagonegrese and gorges of River Calore	Lagonegrese e Gole del Fiume Calore	Campania, Basilicata	50,000	B2, C6
142	IT103	Mounts Orsomarso and Verbicaro	Monte Orsomarso e Monte Verbicaro	Calabria	48,000	C6
143	IT101	Pollino	Pollino	Basilicata, Calabria	115,814	B2, C6
144	IT102	Ferro valley	Valle del Ferro	Calabria	8,500	C6
145	—	Sant'Andrea island	Isola di Sant'Andrea	Puglia	52	A1, B2, C1, C6
146	IT098	Le Cesine	Le Cesine	Puglia	620	C7
147	—	Cape Otranto	Capo d'Otranto	Puglia	3,100	B1iv, C1, C5
148	IT105	Sila Grande	Sila Grande	Calabria	28,450	C6
149	IT104, IT106	Marchesato and River Neto	Marchesato e Fiume Neto	Calabria	44,125	B1iii, B2, C2, C6
150	—	Costa Viola	Costa Viola	Calabria	19,750	A1, B1iii, B1iv, C1, C5
151	IT107	Aspromonte	Aspromonte	Calabria	35,650	B1iv, C5, C6
152	IT127	Eolie islands	Isole Eolie	Sicilia	11,610	A4ii, B1iii, B2, C2, C6
153	IT128	Peloritani mountains	Monti Peloritani	Sicilia	17,100	A4iv, B1iv, C5
154	IT129	Nebrodi	Nebrodi	Sicilia	85,587	B2, C6
155	IT132	Mount Pecoraro and Pizzo Cirina	Monte Pecoraro e Pizzo Cirina	Sicilia	12,000	B2, C7
156	IT133	Mount Cofano, Cape San Vito and Mount Sparagio	Monte Cofano, Capo San Vito e Monte Sparagio	Sicilia	18,000	A3, B2, C6
157	IT134	Egadi islands	Isole Egadi	Sicilia	3,700	B1ii, B2, C2, C6
158	IT135	Stagnone di Marsala and Trapani saltpans	Stagnone di Marsala e Saline di Trapani	Sicilia	3,700	A1, A4i, A4iii, B1i, B2, B3, C1, C2, C3, C4, C6
159	IT131	Rocca Busambra	Rocca Busambra	Sicilia	100,000	A1, B2, C1, C6
160	—	Sicani mountains	Monti Sicani	Sicilia	18,000	A1, A4ii, B1iii, B2, C1, C2, C6
161	—	Castelluzzo	Castelluzzo	Sicilia	7,500	A1, B2, C1, C2
162	IT136	Cape Feto	Capo Feto	Sicilia	150	C7
163	IT137	Simeto mouth and Biviere di Lentini	Foce del Simeto e Biviere di Lentini	Sicilia	1,500	A1, A4i, A4iii, B1i, B2, C1, C2, C3, C4, C6
164	IT130	Madonie	Madonie	Sicilia	39,941	A1, C1, C6
165	IT138	Anapo gorges	Gole dell'Anapo	Sicilia	3,712	A3, B2, B3, C6
166	—	Biviere and Plain of Gela	Biviere e Piana di Gela	Sicilia	500	A1, B1i, C1, C2, C6
167	IT139	Pantani di Capo Passero	Pantani di Capo Passero	Sicilia	3,000	A4i, A4iii, B1i, B2, C2, C3, C4, C6
168	IT140	Pantelleria and Isole Pelagie	Pantelleria e Isole Pelagie	Sicilia	2,445	A4ii, B1ii, B1iii, B2, B3, C2, C6
169	—	Coast from Coghinas river mouth to Cape Testa	Costa da foce Coghinas a Capo Testa	Sardegna	300	A1, B1i, C1, C2
170	IT108	Maddalena archipelago	Arcipelago della Maddalena	Sardegna	800	A1, B1i, B2, B3, C1, C2, C6
171	IT109	Asinara island and Stintino peninsula cliffs	Isola Asinara e Peninsola di Stintino	Sardegna	4,900	A1, A4i, B1i, B2, B3, C1, C2, C6
172	IT111	Pilo and Casaraccio pools	Stagni di Pilo e di Casaraccio	Sardegna	500	B2
173	IT112	Campo d'Ozieri	Campo d'Ozieri	Sardegna	20,000	B2, B3, C6
174	IT113	Tavolara island	Isola Tavolara	Sardegna	3,000	A1, A4ii, B1i, B1ii, B2, B3, C1, C2, C3, C6
175	IT110	Cape Caccia	Capo Caccia	Sardegna	35,000	A1, B1i, B1ii, B2, B3, C1, C2, C6
176	IT115	Coast between Bosa and Alghero	Costa da Bosa a Alghero	Sardegna	20,000	B2, B3, C6
177	IT114	Altopiano di Campeda	Altopiano di Campeda	Sardegna	19,500	A1, B2, C1, C2, C6
178	IT116	Central Campidano	Campidano centrale	Sardegna	20,000	B2, C6
179	IT119	Altopiano di Abbasanta and Lake Omodeo	Altopiano di Abbasanta e Lago Omodeo	Sardegna	20,000	A1, B2, C1, C2, C6
180	IT117	Cuglieri coast	Costa di Cuglieri	Sardegna	2,800	B1i, B2, C2, C6
181	IT118	Orosei Gulf and Gennargentu mountains	Golfo di Orosei e Monti del Gennargentu	Sardegna	59,102	A1, A3, A4ii, B1i, B1iii, B2, B3, C1, C2, C6
182	IT120	Oristano wetlands	Stagni di Oristano	Sardegna	7,780	A1, A4i, A4iii, B1i, B2, C1, C2, C3, C4, C6

Table 1 ... continued. Summary of Important Bird Areas in Italy. 192 IBAs covering 46,270 km²

IBA code	1989 code	International name	National name	Administrative region	Area (ha)	Criteria (see p. 11)
183	IT121	Mount Ferru of Tertenia	Monte Ferru di Tertenia	Sardegna	1,000	A1, B2, B3, C1, C6
184	—	Cape San Marco	Capo San Marco	Sardegna	200	A1, B2, B3, C1, C2, C6
185	IT122	Flumendosa and Colostrai pools	Stagni del Flumendosa e di Colostrai	Sardegna	3,400	A1, B2, C1
186	IT123	Mount Sette Fratelli and Sarrabus	Monte dei Sette Fratelli e Sarrabus	Sardegna	50,000	B2, B3, C6
187	—	Coast between Cape Boi and Cape Ferrato	Costa tra Capo Boi e Capo Ferrato	Sardegna	200	A1, C1
188	IT124	Cagliari wetlands	Stagni di Cagliari	Sardegna	5,600	A1, A4i, A4iii, B1i, B2, B3, C1, C2, C3, C4, C6
189	IT126	Mount Arcosu	Monte Arcosu	Sardegna	4,500	B2, C6
190		Palmas Gulf wetlands	Stagni del Golfo di Palmas	Sardegna	4,000	A4i, B1i, B2, B3, C2, C3, C6
191	IT125	San Pietro and Sant'Antioco islands	Isole di San Pietro e Sant'Antioco	Sardegna	6,000	A1, A3, A4ii, B1ii, B1iii, B2, B3, C1, C2, C6
192	—	Coast between Cape Teulada and Pula	Costa tra Capo Teulada e Pula	Sardegna	500	A1, A4i, B1i, B2, B3, C1, C2, C6

Map 1. Location, area and criteria category of Important Bird Areas in Italy.

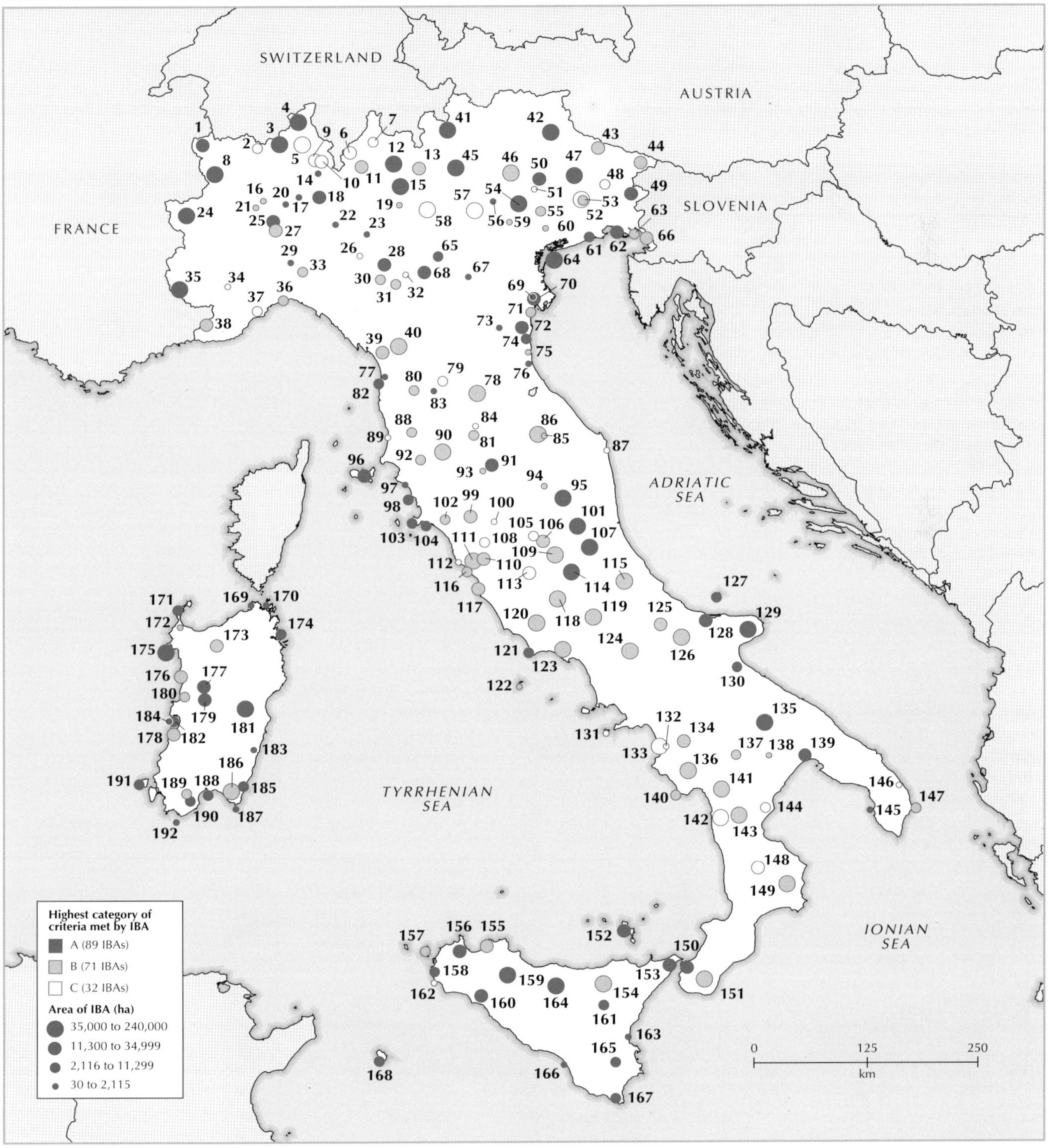

other IBAs have also been changed since 1989, to facilitate identification or protection according to administrative or ecological boundaries.

The IBAs are evenly spread across the country although they occur in a wide variety of sizes. As a general rule, where the IBAs are more spaced apart, they tend to be larger, e.g. in southern Italy or the Alps (see Map 1, Table 1).

ORNITHOLOGICAL IMPORTANCE

A total of 137 species of European conservation concern (SPECs) breed regularly in the country, of which 80 have an unfavourable conservation status in Europe (Tucker and Heath 1994), the latter including four species of global conservation concern (*Aythya nyroca*, *Falco naumanni*, *Crex crex* and *Larus audouinii*). Forty SPECs occur regularly in Italy during migration or in winter, including three globally threatened species (*Aquila clanga*, *Numenius tenuirostris* and *Acrocephalus paludicola*).

A total of 51 sites are IBAs due to their importance for such species of global conservation concern (Table 2). The following sites qualify for more than one such species: Simeto mouth and Biviere di Lentini (site 163) for *Aythya nyroca*, *Numenius tenuirostris* and *Acrocephalus paludicola*; Orbetello lagoon and Burano lake (104), Lakes Lesina and Varano (128) and Manfredonia Gulf wetlands (130) for *A. nyroca* and *N. tenuirostris*; Oristano wetlands (182) for *A. nyroca* and *Falco naumanni*; and Costa Viola (150) for *Aquila clanga* and *Falco naumanni*.

The IBA network is particularly important for congregatory species, especially breeding waterbirds and seabirds. In total, 60 IBAs support 50 congregatory bird species in important numbers, i.e. meeting A4/B1 criteria (Table 3). Of these: 39 IBAs support important numbers of 20 breeding species; 24 IBAs support important numbers of 26 wintering species; and nine IBAs support

Table 2. Important Bird Areas in Italy that are important for species of global conservation concern (meeting criterion A1).

Species	IBA code
Aythya nyroca Ferruginous Duck	014, 064, 072, 073, 074, 083, 101, 104, 128, 130, 158, 163, 166, 182
Aquila clanga Greater Spotted Eagle	150
Falco naumanni Lesser Kestrel	114, 135, 139, 150, 159, 160, 161, 164, 182
Crex crex Corncrake	047, 049, 050, 054, 056
Tetrax tetrax Little Bustard	129, 177, 179
Numenius tenuirostris Slender-billed Curlew	020, 077, 082, 097, 098, 104, 121, 128, 130, 163
Larus audouinii Audouin's Gull	096, 145, 169, 170, 171, 174, 175, 181, 183, 184, 185, 187, 188, 191, 192
Acrocephalus paludicola Aquatic Warbler	163

Table 3. Important Bird Areas in Italy that support important numbers of one or more congregatory species (i.e. meeting criteria A4 and/or B1). IBAs meeting both criteria A4 and B1 for the species are shown in **bold**. IBAs meeting only criterion B1 for the species concerned, and not A4, are shown in normal type. For key to 'Season', see p. 7.

Species	Season	IBA code
Tachybaptus ruficollis Little Grebe	W	**074**
Podiceps nigricollis Black-necked Grebe	W	**015**, **064**, **070**, **188**
Calonectris diomedea Cory's Shearwater	R	**168**, 191
	B	096, 175
	N	116
Puffinus yelkouan Yelkouan Shearwater	R	**174**
	B	**096**, **127**
Hydrobates pelagicus Storm Petrel	R	157, 168
	B	175
Phalacrocorax carbo Cormorant	W	018, 062, 064, 065, 070, 071, 072, 091, 093, 117, 128, 163, **182**, 190
Phalacrocorax aristotelis Shag	R	169, 170, 171, 174, 175, 180, 181, 192
Nycticorax nycticorax Night Heron	B	**017**, **018**, **020**, **022**, **023**, **025**, **028**, **029**, **067**, **068**, **069**, **072**
Ardeola ralloides Squacco Heron	B	025
Egretta garzetta Little Egret	B	**017**, **018**, **020**, **022**, 023, **025**, **061**, **064**, 067, 069, 070, **074**, **188**
	W	**064**, 182
	N	**076**
Egretta alba Great White Egret	W	**064**, **065**, **104**
Plegadis falcinellus Glossy Ibis	P	**163**, **167**
Platalea leucorodia Spoonbill	P	**158**, **163**, **167**
Phoenicopterus ruber Greater Flamingo	R	**188**
	W	104, **182**, **188**, **190**
	N	**188**
Anser anser Greylag Goose	W	062, 072, 098
Tadorna tadorna Shelduck	W	**130**
Anas penelope Wigeon	W	**062**, 063, 064, 070, 097, 130
Anas strepera Gadwall	W	**062**, 073, **130**
Anas crecca Teal	W	064
Anas querquedula Garganey	P	**158**
Anas clypeata Shoveler	W	**130**, **188**
Aythya ferina Pochard	W	**072**, **182**
Aythya nyroca Ferruginous Duck	W	**128**, 130
	P	163, 166, 182
Aythya fuligula Tufted Duck	W	128
Clangula hyemalis Long-tailed Duck	W	128
Bucephala clangula Goldeneye	W	062, 063, 064, 072
Mergus serrator Red-breasted Merganser	W	062, 128
Falco naumanni Lesser Kestrel	B	**135**, **139**, **160**
Falco vespertinus Red-footed Falcon	P	149, 150
Falco eleonorae Eleonora's Falcon	B	**152**, 168, **181**, **191**
Fulica atra Coot	W	**062**, 064, **128**
Himantopus himantopus Black-winged Stilt	B	**025**, 072, 076
Recurvirostra avosetta Avocet	R	**188**
	B	076, **130**
	W	076, **130**, 182
	P	**130**
Glareola pratincola Collared Pratincole	B	072, 182
Charadrius alexandrinus Kentish Plover	W	130, 182, 188, **190**
	P	130, **158**, **167**
Calidris minuta Little Stint	P	**167**
Calidris ferruginea Curlew Sandpiper	P	**167**
Calidris alpina Dunlin	W	**064**
Tringa erythropus Spotted Redshank	P	**062**
Larus melanocephalus Mediterranean Gull	W	064
Larus ridibundus Black-headed Gull	W	074
	N	076
Larus genei Slender-billed Gull	R	**188**
	W	130, 188, 190
	P	**167**
Larus audouinii Audouin's Gull	R	**171**, **192**
Larus cachinnans Yellow-legged Gull	R	**062**, **096**
	W	**070**
	N	**076**
Gelochelidon nilotica Gull-billed Tern	R	**188**
	B	072, **182**
Sterna caspia Caspian Tern	P	**167**
Sterna sandvicensis Sandwich Tern	B	072
Sterna hirundo Common Tern	B	072
Sterna albifrons Little Tern	B	**070**, **072**, 075, **076**, **130**, 182
	N	**064**
Chlidonias niger Black Tern	P	163, **167**
	N	**064**, **077**

Table 4. Species of European conservation concern and species listed on Annex I of the EC Birds Directive with significant breeding populations at IBAs in Italy (meeting any IBA criteria).

Species[1]	Minimum national breeding population (pairs)[2]	Proportion (%) of national population breeding at all IBAs in Italy
Calonectris diomedea Cory's Shearwater	15,000	78
Puffinus yelkouan Yelkouan Shearwater	11,000	59
Hydrobates pelagicus Storm Petrel	1,500	100
Phalacrocorax aristotelis Shag	1,600	76
Phalacrocorax pygmeus Pygmy Cormorant	7	100
Botaurus stellaris Bittern	20	100[3]
Ixobrychus minutus Little Bittern	1,000	46
Nycticorax nycticorax Night Heron	12,000	82
Ardeola ralloides Squacco Heron	250	100[3]
Egretta garzetta Little Egret	25,500	25
Egretta alba Great White Egret	30	73
Ardea purpurea Purple Heron	700	100[3]
Plegadis falcinellus Glossy Ibis	1	100
Platalea leucorodia Spoonbill	2	100[3]
Phoenicopterus ruber Greater Flamingo	1,600	63
Aythya nyroca Ferruginous Duck	25	100[3]
Pernis apivorus Honey Buzzard	500	28
Milvus migrans Black Kite	700	29
Milvus milvus Red Kite	130	79
Neophron percnopterus Egyptian Vulture	10	100[3]
Gyps fulvus Griffon Vulture	35	86
Circaetus gallicus Short-toed Eagle	200	33
Circus aeruginosus Marsh Harrier	50	100[3]
Circus pygargus Montagu's Harrier	200	27
Aquila chrysaetos Golden Eagle	300	47
Hieraaetus fasciatus Bonelli's Eagle	15	73
Falco naumanni Lesser Kestrel	500	100[3]
Falco tinnunculus Kestrel	5,000	9
Falco eleonorae Eleonora's Falcon	400	85
Falco biarmicus Lanner	150	40
Falco peregrinus Peregrine	430	76
Bonasa bonasia Hazel Grouse	5,000	8
Tetrao tetrix Black Grouse	10,000	11
Tetrao urogallus Capercaillie	2,000	9
Alectoris graeca Rock Partridge	10,000	25
Alectoris barbara Barbary Partridge	3,000	20
Coturnix coturnix Quail	5,000	6
Porzana porzana Spotted Crake	50	8
Crex crex Corncrake	350	46
Porphyrio porphyrio Purple Gallinule	300	100[3]
Tetrax tetrax Little Bustard	350 individuals	82
Himantopus himantopus Black-winged Stilt	900	100[3]
Recurvirostra avosetta Avocet	1,200	87
Burhinus oedicnemus Stone Curlew	200	95
Glareola pratincola Collared Pratincole	81	100[3]
Charadrius alexandrinus Kentish Plover	1,800	34
Larus melanocephalus Mediterranean Gull	223	100[3]
Larus audouinii Audouin's Gull	550	100[3]
Gelochelidon nilotica Gull-billed Tern	200	100[3]
Sterna sandvicensis Sandwich Tern	138	100[3]
Sterna hirundo Common Tern	4,000	68
Sterna albifrons Little Tern	3,000	97
Chlidonias hybridus Whiskered Tern	180	36
Chlidonias niger Black Tern	100	100
Streptopelia turtur Turtle Dove	50,000	3
Otus scops Scops Owl	4,000	5
Bubo bubo Eagle Owl	200	28
Glaucidium passerinum Pygmy Owl	300	10
Athene noctua Little Owl	10,000	4
Strix aluco Tawny Owl	5,000	24
Aegolius funereus Tengmalm's Owl	1,000	6
Caprimulgus europaeus Nightjar	5,000	13
Alcedo atthis Kingfisher	4,000	6
Merops apiaster Bee-eater	2,000	37
Coracias garrulus Roller	300	18
Picus canus Grey-headed Woodpecker	500	21
Picus viridis Green Woodpecker	5,000	10
Dryocopus martius Black Woodpecker	1,500	12
Dendrocopos medius Middle Spotted Woodpecker	300	15
Dendrocopos leucotos White-backed Woodpecker	200	100[3]
Picoides tridactylus Three-toed Woodpecker	50	76
Melanocorypha calandra Calandra Lark	5,000	15
Calandrella brachydactyla Short-toed Lark	15,000	4
Galerida cristata Crested Lark	200,000	1
Lullula arborea Woodlark	20,000	4
Anthus campestris Tawny Pipit	15,000	4
Erithacus rubecula Robin	1,000,000	1
Saxicola torquata Stonechat	200,000	1
Oenanthe hispanica Black-eared Wheatear	1,000	17
Monticola saxatilis Rock Thrush	5,000	4
Monticola solitarius Blue Rock Thrush	10,000	12
Turdus torquatus Ring Ouzel	10,000	1
Turdus philomelos Song Thrush	100,000	1
Locustella luscinioides Savi's Warbler	1,000	13
Acrocephalus melanopogon Moustached Warbler	1,000	68
Acrocephalus palustris Marsh Warbler	10,000	8
Acrocephalus scirpaceus Reed Warbler	30,000	5
Sylvia sarda Marmora's Warbler	5,000	13
Sylvia undata Dartford Warbler	10,000	14
Sylvia cantillans Subalpine Warbler	30,000	9
Sylvia melanocephala Sardinian Warbler	300,000	2
Sylvia hortensis Orphean Warbler	1,000	3
Sylvia nisoria Barred Warbler	1,000	6
Regulus ignicapillus Firecrest	100,000	2
Ficedula albicollis Collared Flycatcher	1,000	3
Parus cristatus Crested Tit	20,000	1
Parus caeruleus Blue Tit	100,000	5
Certhia brachydactyla Short-toed Treecreeper	100,000	2
Lanius collurio Red-backed Shrike	30,000	6
Lanius minor Lesser Grey Shrike	1,000	21
Lanius senator Woodchat Shrike	5,000	9
Pyrrhocorax pyrrhocorax Chough	500	100[3]
Corvus monedula Jackdaw	50,000	2
Sturnus unicolor Spotless Starling	50,000	2
Serinus citrinella Citril Finch	5,000	2
Emberiza cia Rock Bunting	30,000	3
Emberiza hortulana Ortolan Bunting	4,000	7
Emberiza melanocephala Black-headed Bunting	2,000	4

1. Only those species of European conservation concern (see Box 1, p. 12) that meet IBA criteria in Italy are listed, together with those species listed on Annex I of the EC Birds Directive that fulfil criterion C6 in IBAs in Italy.
2. Data are taken from the BirdLife/EBCC European Bird Database 1998 (Heath and Borggreve 2000).
3. The percentage of the national population in IBAs exceeds 100%. Usually this is because the national population estimate has not been updated recently whilst the IBA population estimate has been recently updated with new data as a result of comprehensive surveys of IBAs themselves. Also, the individual site count for a species may be the maximum or average over recent years, and summing these may record more birds than are present nationally in any single year.

important numbers of 13 species staging on migration. Three IBAs (all coastal wetlands) each host ten or more congregatory species in important numbers: Venice lagoon (site 064) has 12 species, Valli di Comacchio/Bonifica del Mezzano (072) has 11 species and Oristano wetlands (182) has 10 species.

For 21 species that are SPECs and/or are listed on Annex I of the EC Birds Directive, the IBA network supports the entire national breeding population, and there are 39 such species with more than half of their national population breeding within IBAs (Table 4).

Figure 1. Habitats at Important Bird Areas in Italy (see Appendix 3 for definitions of habitats).

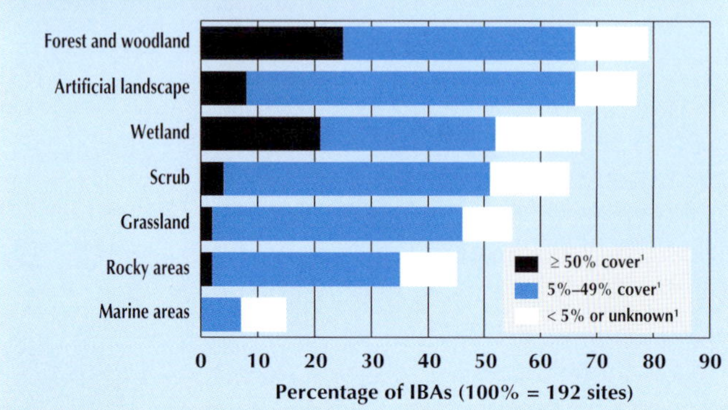

1. percentage of area of individual IBA covered by habitat

Figure 2. Land-uses at Important Bird Areas in Italy (see Appendix 3 for definitions of land-uses).

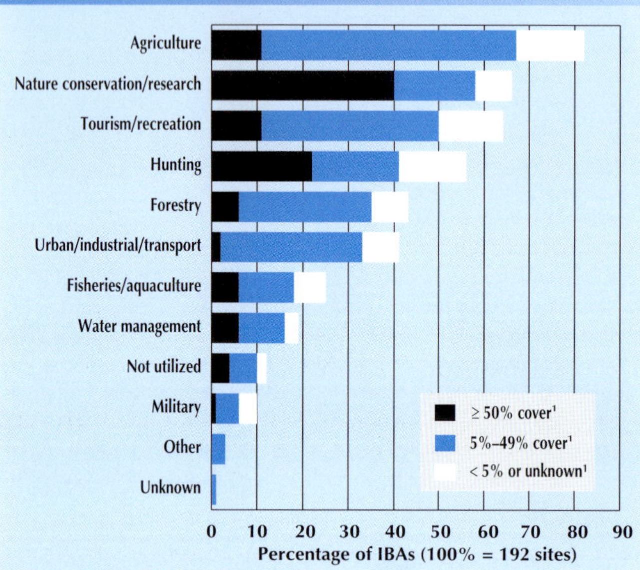

1. percentage of area of individual IBA covered by land-use

HABITATS

Mainland Italy may be divided into four major geographical and vegetational zones: the Alps, in the north of the peninsula; the Po valley in the north-west; the Apennine mountains, which includes most of mainland Italy; and the coastal zone.

The Alpine area is relatively unspoiled, with deciduous, mixed and coniferous forests (as semi-natural stands and as plantations), alpine pastures above the timberline (grazed by cattle and sheep), and snowfields and glaciers on the higher peaks. The Po valley is the largest lowland area in Italy, mainly covered by intensive agriculture, notably cereal and rice fields, and stock-farming, but also including areas of deciduous forest, lakes and rivers. Large areas of the Apennines are still covered by semi-natural deciduous forest, predominately oak *Quercus*, beech *Fagus* and sweet chestnut *Castanea*, although large areas have been cleared for agriculture. The coastal areas of Italy are dominated by typical Mediterranean scrub, usually referred to as 'macchia' or, when less dense, as 'gariga'; scrub of evergreen oak *Quercus ilex* is the characteristic climax vegetation-type.

According to figures from the National Forest Inventory (1985, quoted by Ministero dell'Ambiente [1997]) woodland covers 8.7 million ha (c.29% of the national land surface), being concentrated in mountain areas, with only 5% on the plains. Most woodland is mixed broadleaved coppice dominated by *Quercus* species (a minority by *Castanea* or *Fagus*), while forests are mainly coniferous, dominated by spruce *Picea*, larch *Larix* and fir *Abies*. Twenty-five percent of the IBAs in Italy are mainly forest, on an individual basis (Figure 1), and this is due to the importance of this habitat for raptors, gamebirds, owls and woodpeckers.

Official statistics concerning agricultural land cover (ISTAT 1991) do not match perfectly the habitat classification used in this work, but a general comparison between land cover at the national level and in the IBAs can be made anyway. 'Artificial landscapes' in Italy include 90,123 km² of arable fields and 29,598 km² of orchards and vineyards, but the so-called 'unproductive areas' covering 65,210 km² include not only roads and developed/urban areas but also some fallow land and rocky areas. 'Permanent grassland' accounts for 48,776 km² nationally, but the actual area may be larger since some poor (rocky or arid) grasslands are classified under other categories. Lowland grassland is mainly represented by dry calcareous pastures in Sardinia, Puglia, Sicily and Latium, which are important for SPECs such as *Neophron percnopterus*, *Falco naumanni*, *Falco biarmicus*, *Tetrax tetrax*, *Melanocorypha calandra*, *Calandrella brachydactyla* and *Merops apiaster*. In hilly and mountainous areas, grasslands contain key species such as *Alectoris graeca*, *Alectoris barbara* in the drier parts of Italy, *Crex crex* in hayfields at 500–1,200 m elevation, and *Pyrrhocorax graculus* on calcareous mountains.

All of the more important Italian wetlands have been included in the IBA list and a high proportion of the waterbirds which breed or congregate in them are represented. Wetlands are mainly coastal (estuaries, deltas, lagoons, intertidal mudflats, salt-pans, beaches and dunes) but some rivers in northern Italy, e.g. the Po (sites 027, 067 and 068), support important numbers of herons, and inland lakes and peatlands are important for wintering duck, e.g. Lake Garda (015), or breeding *Botaurus stellaris*, e.g. Colfiorito (094).

Scrub and rocky areas are often found together with broadleaved evergreen woodland in the Mediterranean zone, and are important for *Sylvia* warblers, *Lanius senator*, *Lanius collurio*, *Caprimulgus europaeus* and many other birds. 'Rocky areas' includes many sea cliffs and islets, on which *Larus audouinii* and *Falco eleonorae* nest.

IMPACTS ON IBAs – LAND-USE AND THREATS

Agriculture is the most frequent land-use at IBAs in Italy, although it is predominant at only c.10% of IBAs (Figure 2). About 40% of IBAs receive legal protection over more than half of their individual area, and so are devoted to nature conservation, but hunting is allowed to some extent in more than 50% of all sites. Recreation and tourism occur in the majority of IBAs, but this use is predominant at less than 15% of all IBAs. Despite the fact that most IBAs are forested or wooded to some extent (Figure 1), forestry is the main land-use in only a few IBAs, reflecting the fact that most Italian woods are often poorly managed (Figure 2). Fisheries and aquaculture occur at 25% of IBAs, often together with water management. Urban, industrial and transport are nearly always minor land-uses, although they occur at more than 40% of the sites.

The intensifying use of natural resources is the main threat to birds at IBAs in Italy (Figure 3). The high incidence of 'recreation/tourism' and 'disturbance to birds' as threats is often due to the lack of a suitable infrastructure that would allow the general public to enjoy natural areas without interrupting birds' essential activities such as breeding or foraging. Other sources of disturbance are hunting (excluding gamebirds) and forestry activities when performed near the breeding sites of some raptors. There is evidence that hunting is reducing the abundance and species-richness of breeding waterbirds at some sites, and cases have been reported of nest abandonment by *Hieraaetus fasciatus* due to tree-felling or the creation of forestry roads in previously unspoiled forest areas. 'Unsustainable exploitation' refers mainly to over-hunting and is the second most important threat to birds at IBAs in Italy, not only to those species that are legally hunted but also many other species, due to the lack of controls and proper hunting management procedures.

Agricultural intensification and changes in crops or management procedures (e.g. towards soya, sunflowers, 'dry' rice cultivation) are affecting bird populations, but the magnitude of the impact is not yet clear, due to a lack of appropriate studies and monitoring. Agri-environmental schemes do not seem to have had any significant positive effects on bird populations, while land abandonment is a serious threat in some areas, because the previous management tends to have been traditional, low-impact agriculture,

Figure 3. Threats at Important Bird Areas in Italy (see Appendix 3 for definitions of threat types and impact categories).

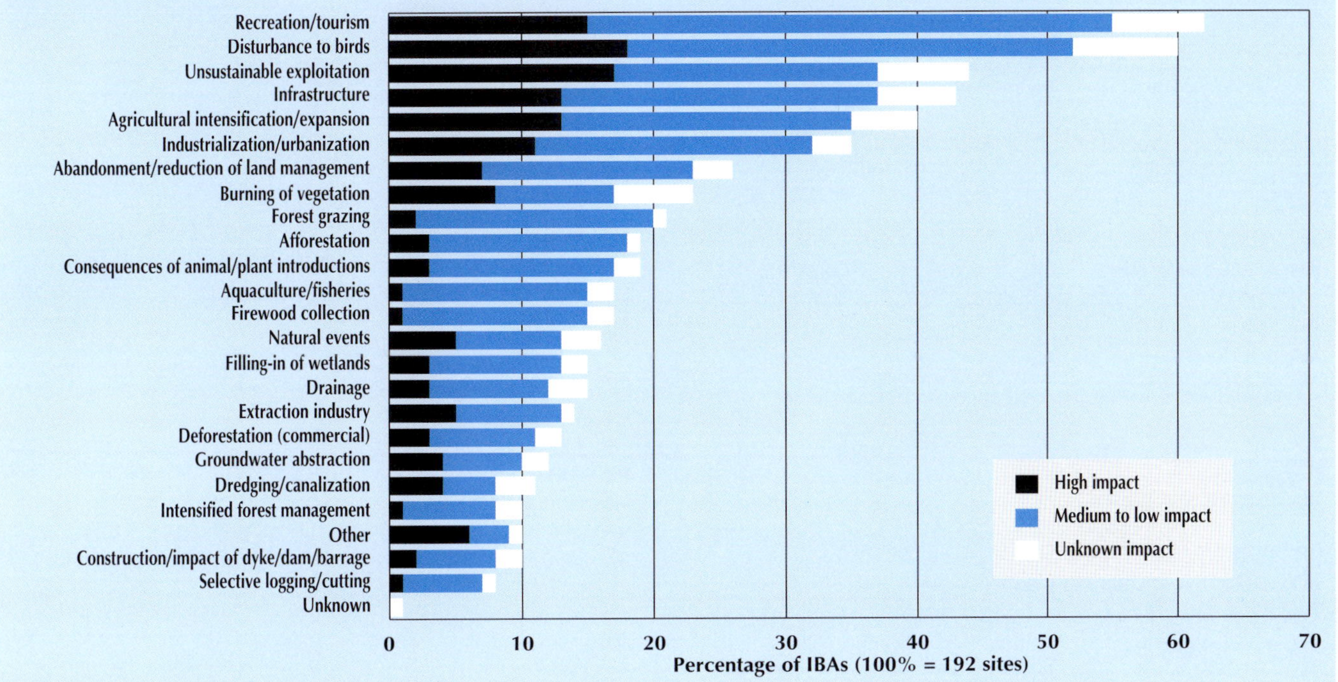

which is often beneficial for birds. Afforestation of areas that are marginal for agriculture (by deliberate tree-planting and natural plant succession following land abandonment) is having an important impact on birds in some areas, reducing the richness and abundance of open-country species. Poor or absent management of forests and woodlands often increases the risk of fire, which every year burns hundreds of thousands of woods in Italy.

Agricultural, industrial and tourist development and infrastructure, often aided by EU funding, have in the past destroyed large areas of semi-natural habitat throughout the country, and in some areas this is still occurring. Large areas of wetland have been drained for land-claim: while there were 7,640 km² of wetland in Italy in 1865, there are now only 1,800 km², and drainage remains a threat to some wetland IBAs (Figure 3).

PROTECTION STATUS

National protection

The national law on protected areas (no. 394/92) identifies several kinds of protected areas: (1) National Parks (17 covering an area of 1,231,351 ha), (2) National Nature Reserves (147 covering 40,020 ha), (3) National Marine Reserves (seven covering 28,290 ha), (4) Regional Parks and Regional Reserves (218 covering 683,964 ha). Other protected areas (referred to as 'Wildlife Reserves' in this chapter) are established according to the national hunting law (no. 157/92) which endorses the EC Birds Directive. The total surface extent of areas protected for nature conservation in Italy is 1,981,287 ha, representing 6.6% of the national land surface; this figure is not the direct sum of the previously listed areas since some types of protected area overlap and other areas are also protected according to other legal tools.

The national protection status of IBAs is far from satisfactory (Table 5). Only 30% overlap to a high extent with the national protected-area system, while 29% have no protection at all (Figure 4).

International protection

Italy is a contracting party to all of the international conventions that are relevant to nature conservation at particular sites (Box 1). The protection of IBAs by these instruments is very low, with only 9% of IBAs receiving a high level of coverage individually and 57% being totally unprotected at the international level (Figure 5).

The Italian government has designated 45 sites as 'wetlands of international importance' (Ramsar Sites), with a total area of about 56,500 ha, of which 44 are IBAs. The current IBA review has identified a total of 39 wetland IBAs that meet Ramsar criteria, of which 27 currently overlap with one or more (up to five) existing Ramsar Sites. At four such IBAs, the Ramsar Site needs to be enlarged to cover the whole of the wetland, as agreed in the Montrose Declaration of the Contracting Parties which calls for the protection

Figure 4. The national protection status of Important Bird Areas in Italy.

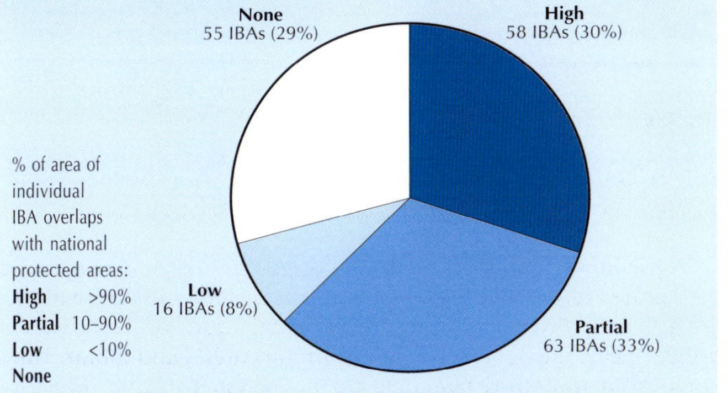

Total area of overlap between IBA network in Italy and national protected-area system (see Table 5 for categories) = 19,334–21,109 km² (42–46% of total IBA area).

Figure 5. The international protection status of Important Bird Areas in Italy.

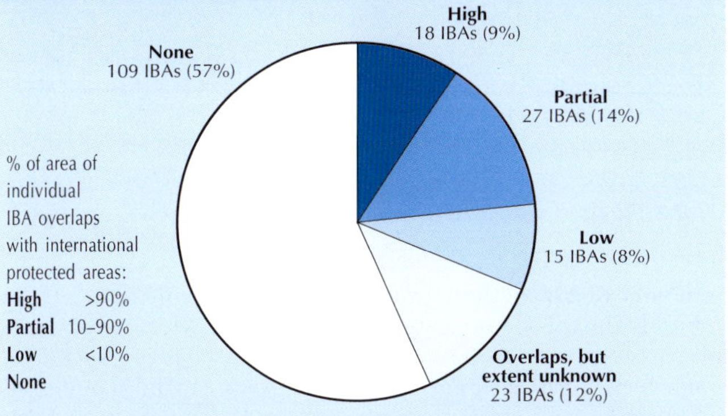

Total area of overlap between IBA network in Italy and international protected-area system (see Table 5 for categories) = 4,314 km² (9% of total IBA area). Sites of unknown status overlap with one or more of the 99 SPAs declared by the Italian government in late 1998.

Table 5. Protection status of Important Bird Areas in Italy. A tick (✔) indicates that an IBA overlaps with a protected area (to any extent). For the meaning of '+' sign against SPAs, see 'Analytical methods'.

IBA code	International name	National: National Park	National: National Nature Reserve[1]	National: National Marine Reserve[2]	National: Regional Park/Reserve[3]	National: Wildlife Reserve	National: Private Reserve	National: Nature Monument	International: Ramsar Site	International: Special Protection Area	International: Biogenetic Reserve
001	Mont Blanc (Italian side)										
002	Monte Rosa (Valle d'Aosta side)										
003	Anzasca and Antrona valleys										
004	Antigorio and Formazza valleys				✔						
005	Val Grande	✔	✔							✔	
006	Mount Tabor										
007	Pian di Spagna and Lake Mezzola				✔				✔	✔	
008	Gran Paradiso	✔			✔					✔	
009	Lake Maggiore										
010	Campo dei Fiori, Valganna and Cuvia valley				✔						
011	Grigne				✔						
012	Orobie				✔						
013	Camonica valley										
014	Brabbia peatland and Lake Varese				✔				✔	✔	
015	Lake Garda										
016	Lake Viverone					✔					
017	Novara heronries										
018	Ticino river				✔						
019	Iseo peatlands				✔				✔	✔	
020	River Sesia heronries				✔						
021	Lake Candia					✔					
022	Heronries of Pavia province				✔						
023	Adda Sud Park heronries				✔						
024	Susa and Chisone valleys				✔						
025	Vercelli rice-fields				✔						
026	Trebbia river (lower course)										
027	River Po: from Dora Baltea mouth to Scrivia mouth				✔						
028	River Po: from Ticino mouth to Adda mouth				✔						
029	Marengo heronry				✔						
030	Stirone stream				✔						
031	Taro river				✔					✔	
032	Torrile LIPU reserve						✔				
033	Scrivia stream					✔					
034	Crava–Morozzo lakes				✔						
035	Maritime Alps				✔						
036	Mount Beigua				✔						
037	Finalese				✔						
038	Ligurian Alps										
039	Alpi Apuane				✔	✔				✔	
040	Apennines from Passo del Cerreto to Monte Caligi		✔		✔					✔	✔
041	Stelvio National Park	✔								✔	
042	Dolomites				✔						
043	Carniche Alps					✔					
044	Tarvisio forest		✔								
045	Adamello–Brenta				✔				✔	✔	
046	Lagorai mountains				✔					+	
047	Carniche pre-Alps									+	
048	Middle course of Tagliamento river				✔						
049	Cividalese and upper Torre valley										
050	Belluno Dolomites	✔								✔	
	Subtotal of IBAs	4	3	0	29	5	1	0	4	13	1
051	Lago di Busche and Vincheto di Cellarda		✔						✔	✔	
052	Giulie pre-Alps				✔						
053	Magredi of Pordenone										
054	Mount Grappa										
055	Piave river										
056	Assa valley										
057	Lessini mountains				✔					+	
058	Alto Garda Bresciano				✔						
059	Middle course of River Brenta										
060	Quinto heronry				✔						
061	Caorle lagoon										
062	Grado and Marano lagoon				✔				✔	✔	
063	Isonzo mouth, Cona island and Panzano Gulf				✔					+	
064	Venice lagoon		✔			✔			✔	✔	
065	River Mincio and Bosco Fontana		✔		✔				✔	✔	
066	Carso				✔						
067	Boscone island				✔				✔	+	
068	River Po: from the mouth of Adda river to Ostiglia				✔				✔	✔	
069	Codigoro heronry					✔				✔	
070	Po delta		✔		✔		✔				
071	Valle Bertuzzi and Goro lagoon		✔		✔				✔	✔	
072	Valli di Comacchio and Bonifica del Mezzano		✔		✔				✔	✔	
073	Valli di Argenta				✔	✔			✔	✔	
074	Punte Alberete and Valle della Canna, Pineta San Vitale and Pialassa della Baiona		✔		✔				✔	✔	
075	Ortazzo and Ortazzino		✔		✔	✔			✔	✔	
076	Cervia saltpans		✔						✔	✔	
077	Lake Massaciuccoli				✔					✔	
078	Foreste Casentinesi	✔									
079	Calvana mountains										
080	Fucecchio marsh				✔						
081	Arezzo heathlands										
082	Migliarino–San Rossore				✔					+	
083	Pools of Florence plain										
084	Ponte Buriano–Lago della Penna					✔					
085	Mount Conero				✔						
086	Furlo gorge and Mount Catria										
087	Sentina										
088	River Cecina					✔					
089	Bolgheri					✔			✔	✔	
090	Crete Senesi										
091	Lake Trasimeno				✔						
092	Farma valley				✔						
093	Lakes Montepulciano and Chiusi					✔					
094	Colfiorito				✔				✔	✔	
095	Sibillini mountains	✔								✔	
096	Tuscan archipelago	✔								✔	
097	Diaccia Botrona				✔				✔	✔	
098	Uccellina mountains, Trappola marshes and Ombrone mouth				✔					+	
099	Lake Bolsena										
	Subtotal of IBAs	7	12	0	54	13	2	0	19	36	1

Table 5 ... continued. Protection status of Important Bird Areas in Italy. A tick (✔) indicates that an IBA overlaps with a protected area (to any extent). For the meaning of '+' sign against SPAs, see 'Analytical methods'.

IBA code	International name	National Park	National Nature Reserve[1]	National Marine Reserve[2]	Regional Park/Reserve[3]	Wildlife Reserve	Private Reserve	Nature Monument	Ramsar Site	Special Protection Area	Biogenetic Reserve
		National							International		
100	Lake Alviano					✔				✔	
101	Laga mountains	✔	✔		✔					✔	
102	Selva del Lamone				✔						
103	Argentario										
104	Orbetello lagoon and Burano lake		✔						✔	✔	
105	Rieti lakes				✔						
106	Rieti mountains									✔	
107	Gran Sasso d'Italia	✔			✔						
108	Vico lake				✔						
109	Cicolano					✔					
110	Lake Bracciano and surrounding areas										
111	Tolfa hills				✔					✔	
112	Tarquinia saltpans		✔							✔	
113	Lucretili mountains				✔					✔	
114	Sirente, Velino and Duchessa mountains		✔		✔					✔	
115	Maiella, Pizzi and Frentani mountains	✔	✔		✔		✔			✔	
116	Coast from Palo to Santa Marinella		✔		✔	✔				✔	
117	Litorale Romano		✔				✔			+	
118	Ernici and Simbruini mountains				✔					✔	
119	Abruzzo National Park	✔					✔			✔	
120	Lepini mountains									✔	
121	Circeo National Park	✔							✔	✔	
122	Ponziane islands	✔	✔	✔						✔	
123	Ausoni and Aurunci mountains							✔			
124	Matese				✔		✔				
125	River Biferno					✔					
126	Daunia mountains										
127	Tremiti islands	✔		✔							
128	Lakes Lesina and Varano	✔	✔							✔	
129	Gargano promontory	✔	✔							✔	✔
130	Manfredonia Gulf wetlands	✔	✔						✔	✔	
131	Capri island		✔								
132	Serre Persano						✔				
133	Picentini mountains				✔						
134	Alburni mountains	✔									
135	Murge									+	
136	Mount Cervati	✔									
137	Dolomiti di Pietrapertosa				✔					+	
138	Manferrara forest										
139	Gravine					✔				+	
140	Coast between Marina di Camerota and Policastro Bussentino	✔									
141	Lagonegrese and gorges of River Calore	✔								✔	
142	Mounts Orsomarso and Verbicaro	✔	✔							✔	
143	Pollino	✔	✔							✔	
144	Ferro valley										
145	Sant'Andrea island									+	
146	Le Cesine		✔						✔	✔	
147	Cape Otranto										
148	Sila Grande	✔								✔	
149	Marchesato and River Neto					✔					
	Subtotal of IBAs	24	27	2	68	18	7	1	23	64	2
150	Costa Viola										
151	Aspromonte	✔								✔	
152	Eolie islands				✔					+	
153	Peloritani mountains									+	
154	Nebrodi				✔					+	
155	Mount Pecoraro and Pizzo Cirina									+	
156	Mount Cofano, Cape San Vito and Mount Sparagio				✔					+	
157	Egadi islands			✔						+	
158	Stagnone di Marsala and Trapani saltpans				✔					+	
159	Rocca Busambra				✔					+	
160	Sicani mountains				✔						
161	Castelluzzo									+	
162	Cape Feto									✔	
163	Simeto mouth and Biviere di Lentini				✔					+	
164	Madonie				✔					+	
165	Anapo gorges				✔						
166	Biviere and Plain of Gela				✔				✔	+	
167	Pantani di Capo Passero				✔				✔	✔	
168	Pantelleria and Isole Pelagie									+	
169	Coast from Coghinas river mouth to Cape Testa										
170	Maddalena archipelago	✔									
171	Asinara island and Stintino peninsula cliffs	✔									
172	Pilo and Casaraccio pools					✔					
173	Campo d'Ozieri										
174	Tavolara island			✔							
175	Cape Caccia					✔					
176	Coast between Bosa and Alghero										
177	Altopiano di Campeda										
178	Central Campidano										
179	Altopiano di Abbasanta and Lake Omodeo										
180	Cuglieri coast										
181	Orosei Gulf and Gennargentu mountains	✔									
182	Oristano wetlands			✔					✔	✔	
183	Mount Ferru of Tertenia										
184	Cape San Marco										
185	Flumendosa and Colostrai pools										
186	Mount Sette Fratelli and Sarrabus										
187	Coast between Cape Boi and Cape Ferrato										
188	Cagliari wetlands								✔	✔	
189	Mount Arcosu						✔			✔	
190	Palmas Gulf wetlands										
191	San Pietro and Sant'Antioco islands					✔					
192	Coast between Cape Teulada and Pula										
	Total number of IBAs	**28**	**27**	**5**	**79**	**22**	**8**	**1**	**27**	**83**	**2**

1. Includes Nature Reserve, Integral Nature Reserve and State Nature Reserve.
2. Includes Marine Reserve and Marine Nature Reserve.
3. Includes Nature Park, Regional Park, Regional Nature Park, Regional Urban Park and Regional Nature Reserve.

Box 1. International legislation and initiatives that are relevant to site conservation in Italy (see Appendix 1 for a general description of these agreements).

Global	
Biodiversity Convention	✔
Ramsar Convention	✔
Bonn Convention	✔
World Heritage Convention	✔
MAB Programme	✔
Pan-European	
Bern Convention	✔
Regional	
EC Birds Directive	✔
EC Habitats Directive	✔
Barcelona Convention	✔

✔ Convention ratified/initiative supported
(✔) Convention signed

of the ecological units rather than sub-units of wetlands. These sites are Venice lagoon (site 064), Po delta (070), Grado and Marano lagoon (062) and Orbetello lagoon (104).

Of the 20 IBAs which meet Ramsar criteria but which have not yet been declared as Ramsar Sites, seven are heronries (017, 020, 022, 023, 025, 029, 069) while the others are inland lakes (015, 091, 093), coastal lagoons (061, 128, 158, 190), estuaries (063, 098, 163), rivers (018, 028) or coasts (117).

The designation of IBAs as Special Protection Areas (SPAs) under the EC Birds Directive has been very poor until recently, with only 108 SPAs (503,657 ha) having been declared by November 1998, of which only 58 were IBAs. Following an infringement procedure by the European Commission, the Italian government is now declaring a new batch of SPAs. During November and December 1998, 99 new SPAs were listed, raising the total SPA area to 898,028 ha. This information came too late to be fully incorporated into this chapter, but see 'Analytical methods'.

CONSERVATION

- IBAs and protected areas represent a potentially good network for monitoring many bird species and this is actually happening for some species at some regional or national protected areas (see 'Analytical methods'). Monitoring and research mainly cover specific issues such as acid rain in alpine forests, excessive erosion at coastal pine woods and sand-dunes, fires in Mediterranean evergreen woodland, nutrient pollution in lagoons and lakes, and river pollution.
- Projects funded by the EU's LIFE Regulation and managed by non-governmental organizations (often in collaboration with protected-area authorities) have allowed the writing and implementation of management plans at some IBAs: Brabbia peatland (site 014), Mount Beigua (036), Po delta (070), Orbetello lagoon (104), and those IBAs in the central Apennine protected areas and on the Sardinian steppes. Management plans are also being implemented at some National Parks (i.e. Gran Paradiso and Abruzzo), and by the Regional Park Service in some regions (i.e. Piedmont, Lombardy, Emilia-Romagna, Tuscany). Conservation activities in national reserves are implemented by the Corpo Forestale dello Stato (National Forest Guards), but many national protected areas still do not have clear management plans. Surveillance and law enforcement at protected areas is often, especially in the newly established National Parks, inadequate. Some of the Ramsar Sites do not have a management plan, and/or the responsibility for management and surveillance is not clearly allocated.

ANALYTICAL METHODS

- Data used for IBA identification included site-specific research carried out by the compilers of IBA data-sheets for individual sites.
- About 75% of the bird data displayed in the site accounts date from the period 1994–1996, and only about 25% of the bird data are of 'poor' or 'unknown' accuracy, indicating that the data-set is substantially up-to-date and of good quality. Counts of migrating raptors that are quoted in the site-accounts tend to be seasonal totals, while counts of waterbirds tend to be daily maxima. Where the quoted figures are averages over a number of years, the time-period is usually indicated as a footnote to the figure.
- Wintering waterbird data come mainly from the Midwinter Waterbird Count coordinated by INFS (National Wildlife Institute). Data on breeding herons were provided by the Annual Breeding Heron Record Scheme by the University of Pavia (courtesy of Prof. M. Fasola). Data on *Falco biarmicus*, *Falco eleonorae*, *Milvus milvus*, *Hieraaetus fasciatus*, *Neophron percnopterus* and *Falco naumanni* came from a specific LIFE-funded project that was managed by LIPU. National, regional and local bird atlas databases provided useful data for many areas as well.
- Breeding population sizes are poorly known for most birds in Italy (see Table 4). Reliable data are available for only a limited selection of species, including colonial waterbirds (herons, gulls and terns), some rare or localized raptors (e.g. *Falco biarmicus*, *Falco eleonorae*), some newly established species (e.g. *Platalea leucorodia*, *Phoenicopterus ruber*), some nationally endangered species (*Neophron percnopterus*, *Crex crex*, *Hieraaetus fasciatus*), and gamebirds (Tetraonidae). National population figures for most passerines are only rough estimates. However, data collection as part of this IBA review allowed the revision of national population figures for some bird species.
- Data on habitats, land-uses, threats and protection status are the most up-to-date available. Some difficulties were encountered in identifying the provincial protected areas referred to as Wildlife Reserves since they may change quite often.
- The unavailability of maps for the 99 new SPAs (declared in November/December 1998) while compiling this chapter did not allow them to be fully incorporated into the individual site-accounts, but Table 5 identifies the IBAs involved (with a '+' sign), and the existence of the new SPAs has also been incorporated into Figure 5.

GLOSSARY

LIFE The European Union's all-encompassing fund for nature conservation.
LIPU Lega Italiana Protezione Uccelli (the BirdLife Partner in Italy).
SPA Special Protection Area (designated under Article 4 of the EC Birds Directive).
NUTS Nomenclature des Unités Territoriales Statistiques (the administrative regions of the European Union are called NUTS regions, and are designated by the EC Statistical Office.

ACKNOWLEDGEMENTS

LIPU acknowledges the Ministry of Agriculture, (Comitato Tecnico Faunistico Venatorio Nazionale) for partially funding the IBA programme in Italy. We gratefully acknowledge the work of the contributors to the IBA review in Italy, listed below. M. Accinelli, C. Allegri, A. Ambrogio, M. Azzolini, N. Baccetti, A. Badami, L. Baghino, R. Balestrini, A. Bernardoni, M. Bernoni, G. Bianchi, R. Bionda, M. Biondi, G. Boano, M. Bocca, P. Bonvicini, A. Bozzetti, P. Brichetti, M. Brunelli, F. Bulgarini, D. Burdisso, E. Calvario, G. Campo, C. Carere, F. Casale, L. Casini, F. Castiglione, R. Cataudella, B. Cauli, V. Cavallaro, G. Cherubini, A. Ciaccio, A. Cirelli, L. Concezzi, P. Cortone, M. Costa, V. Cripezzi, P. De Franceschi, M. Della Toffola, F. Diana, E. Faraglia, U. Faralli, M. Fasola, L. Favilli, L. Felcher, G. Floris, S. Follini, M. Fraissinet, G. Framarin, U. Gallo-Orsi, A. Gariboldi, V. Giacoia, M. Gioiosa, E. Giudice, A. Golfré Andreasi, Gruppo NISORIA, M. Grussu, W. Guenzani, G. Guerrieri, M. Gustin, L. Ianniello, C.Iapichino, Istituto Nazionale Fauna Selvatica, F. La Carbonara, G. La Gioia, G. Lacarbonara, LIPU Abruzzo, P. Lodeserto, F. Lugarini, G. Manganelli, Martignago, C. Meo, A. Meschini, F. Mezzavilla, M. Milone, A. Minganti, T. Mingozzi, F. Mogavero, G. Monterosso, G. Motta, C. Movalli, M. Nicolino, G. Nuovo, PAI, G. Palumbo, M. Pandolfi, Parco Nazionale Circeo, R. Parodi, A. Pepponi, F. Perco, M. Pesente, M. Pesente, F. Pezzo, S. Piciocchi, F. Presutti, L. Puglisi, A. Putignani, E. Quarenghi, G. Rannisi, M. Ravasini, F. Riga, V. Rizzi, F. Roscelli, S. Russo, M. Sanna, R. Santopaolo, P. Santucci, S. Sarrocco, N. Scala, F. Scarton, G. Silveri, F. Simmi, A. Sorace, P. Spada, P. Sposimo, M. Stangalini, F. Stiglitz, G. Tallone, G. Tellini, R. Toffoli, A. Urban, P. Utmar, F. Velatta, P. Ventrella, M. Visentin, S. Zanghellini, S. Zerunian, L. Ziotti, T. Zorzenon.

SITE ACCOUNTS

Mont Blanc (Italian side) — A3, C6 — 001

Admin region Valle d'Aosta
Coordinates 45°50'N 6°51'E
Altitude 1,400–4,810 m **Area** 11,300 ha

Site description

The Italian side of Mont Blanc, the highest mountain of the Alps, from Col de la Seigne up to Col Ferret. The site includes typical alpine habitats, especially glaciers (50% of the area) and rocky areas.

Habitats Scrub (10%; scrub), Grassland (20%; alpine/subalpine/boreal grassland), Rocky areas (70%; scree/boulders; inland cliff)
Land-use Agriculture (10%), Not utilized (50%), Tourism/recreation (50%)

Birds

Species	Season	Year	Pop min	Pop max	Acc	Criteria
[1] ***Lagopus mutus*** Ptarmigan	R	1997	Frequent		—	C6
[2] ***Alectoris graeca*** Rock Partridge	R	1997	Frequent		—	C6
Prunella collaris Alpine Accentor	R	1997	Common		—	A3
Tichodroma muraria Wallcreeper	R	1997	Common		—	A3
Pyrrhocorax graculus Alpine Chough	R	1997	Frequent		—	A3
Montifringilla nivalis Snowfinch	R	1997	Common		—	A3
Serinus citrinella Citril Finch	R	1997	Frequent		—	A3

1. *L. m. helveticus.*
2. *A. g. saxatilis.*

An important breeding site for five of the 10 species of the Eurasian high-montane biome, and for two gamebirds.

Protection status

National None **International** None

Conservation issues

Threats Abandonment/reduction of land management (B), Recreation/tourism (A)

The main threats are posed by tourism and land abandonment. There are thousands of visitors each year, putting strong pressure on such a fragile environment. The abandonment of grazing means that many grassland areas are reverting to scrub. The area is a proposed International Parc between Italy and France, and is a proposed Special Area for Conservation.

Monte Rosa (Valle d'Aosta side) — C6 — 002

Admin region Valle d'Aosta
Coordinates 45°54'N 7°47'E
Altitude 2,100–4,531 m **Area** 8,200 ha

Site description

A large, high-altitude mountain area (the second highest in the Alps), dominated by rocky areas, glaciers and alpine grasslands. The main land-uses are tourism and cattle-grazing, in summer.

Habitats Scrub (10%; scrub), Grassland (20%; alpine/subalpine/boreal grassland), Rocky areas (70%; scree/boulders; inland cliff)
Land-use Agriculture (10%), Not utilized (60%), Tourism/recreation (40%)

Birds

Species	Season	Year	Pop min	Pop max	Acc	Criteria
[1] ***Lagopus mutus*** Ptarmigan	R	1994	Frequent		—	C6

1. *L. m. helveticus.*

Protection status

National None **International** None

Conservation issues

Threats Recreation/tourism (C)

The building of new ski-resorts poses a threat. The area is a proposed Special Area for Conservation.

Anzasca and Antrona valleys — A3, C6 — 003

Admin region Piemonte
Coordinates 46°00'N 8°06'E
Altitude 247–4,637 m **Area** 42,600 ha

Site description

A mountainous area in the south-west Alps, on the east side of the Monte Rosa complex, covered mainly by broadleaved and coniferous woodlands and by alpine grasslands. The main land-uses are cattle-rearing, cultivation, forestry and tourism.

Habitats Forest and woodland (60%; broadleaved deciduous forest; native coniferous forest; mixed forest; treeline ecotone), Scrub (9%; scrub), Grassland (15%; dry siliceous grassland; alpine/subalpine/boreal grassland), Wetland (1%; standing fresh water; river/stream), Rocky areas (5%; scree/boulders; inland cliff), Artificial landscape (10%; highly improved reseeded grassland; perennial crops/orchards/groves; forestry plantation; urban parks/gardens; other urban/industrial areas)
Land-use Agriculture (7%), Forestry (10%), Hunting (10%), Not utilized, Tourism/recreation (1%), Urban/industrial/transport (5%)

Birds

Species	Season	Year	Pop min	Pop max	Acc	Criteria
Bonasa bonasia Hazel Grouse	R	1997	Uncommon		—	C6
Dryocopus martius Black Woodpecker	R	1997	Frequent		—	C6
Prunella collaris Alpine Accentor	R	1996	30	—	B	A3
Tichodroma muraria Wallcreeper	R	1996	10	—	—	A3
Pyrrhocorax graculus Alpine Chough	R	1998	20	—	C	A3
Montifringilla nivalis Snowfinch	R	1996	20	—	C	A3
Serinus citrinella Citril Finch	R	1988	—	—	—	A3

Five of the 10 species of the Eurasian high-montane biome breed in important numbers, as do grouse (Tetraonidae).

Protection status

National None **International** None

Conservation issues

Threats Abandonment/reduction of land management (A), Construction/impact of dyke/dam/barrage (B), Disturbance to birds (B), Recreation/tourism (B), Unsustainable exploitation (U)

The abandonment of land management is a major threat to grouse.

Antigorio and Formazza valleys — A3, B2, C6 — 004

Admin region Piemonte
Coordinates 46°16'N 8°18'E
Altitude 337–3,552 m **Area** 50,000 ha

Site description

A mountainous area in the south-western Alps, which includes the valleys of Antigorio and Formazza (the main ones), Devero, Carrasca and Isorno. Broadleaved and coniferous woodlands are extensive, with relatively unspoilt alpine grasslands and rocky areas. The main land-uses are cattle-rearing, forestry and tourism.

Habitats Forest and woodland (45%; broadleaved deciduous forest; native coniferous forest; mixed forest; treeline ecotone), Scrub (6%; scrub), Grassland (22%; alpine/subalpine/boreal grassland), Wetland (1%; standing fresh water; river/stream), Rocky areas (15%; scree/boulders; inland cliff), Artificial landscape (10%; highly improved reseeded grassland; perennial crops/orchards/groves; forestry plantation; urban parks/gardens; other urban/industrial areas)
Land-use Agriculture (7%), Forestry (10%), Hunting (50%), Nature conservation/research (15%), Not utilized, Tourism/recreation (5%), Urban/industrial/transport (5%), Water management (1%)

Birds

Five of the 10 species of the Eurasian high-montane biome breed in important numbers. The area is also important for *Aquila chrysaetos* and grouse (Tetraonidae).

Species	Season	Year	Pop min	Pop max	Acc	Criteria
Aquila chrysaetos Golden Eagle	R	1996	5	6	A	B2, C6
Falco peregrinus Peregrine	R	1996	3	5	B	C6
Bonasa bonasia Hazel Grouse	R	1997	10	—	B	C6
Lagopus mutus Ptarmigan	R	1998	Common		—	C6
[1] ***Tetrao tetrix*** Black Grouse	R	—	Common		—	B2
Bubo bubo Eagle Owl	R	1998	4	—	B	C6
Glaucidium passerinum Pygmy Owl	R	1998	5	—	B	C6
Dryocopus martius Black Woodpecker	R	1997	10	—	B	C6
Prunella collaris Alpine Accentor	R	1996	Common		—	A3
Tichodroma muraria Wallcreeper	R	1996	Uncommon		—	A3
Pyrrhocorax graculus Alpine Chough	R	1996	Frequent		—	A3
Montifringilla nivalis Snowfinch	R	1996	Uncommon		—	A3
Serinus citrinella Citril Finch	R	1996	Uncommon		—	A3

1. Numbers not yet estimated.

Protection status

National Partial **International** None

11,000 ha of IBA covered by Regional Nature Park (Alpe Veglia e Devero, 11,000 ha).

Conservation issues

Threats Abandonment/reduction of land management (A), Construction/impact of dyke/dam/barrage (B), Disturbance to birds (B), Extraction industry (A), Firewood collection (C), Natural events (U), Recreation/tourism (A), Unsustainable exploitation (U)

The main threats are abandonment of land management, extraction industry (quarries), and tourism, the last one especially linked to disturbance of grouse. There is research on the population density of *Tetrao tetrix* in the Regional Nature Park.

Val Grande

C7 **005**

Admin region Piemonte
Coordinates 46°03′N 8°26′E
Altitude 196–2,301 m **Area** 29,000 ha

Site description

A mountainous Alpine area, on the west side of Lake Maggiore, including the following valleys: Val Grande, Val Pogallo, Val Cannobina and the south side of Val Vigezzo. The area is almost uninhabited, and comprises one of the largest wildernesses in the Alps. It is mainly covered by broadleaved woodlands (beech *Fagus*), with coniferous woodlands, alpine grasslands, streams and rocky areas also present. The only current land-use is nature conservation.

Habitats Forest and woodland (60%; broadleaved deciduous forest; native coniferous forest; mixed forest; treeline ecotone), Scrub (3%; scrub), Grassland (20%; alpine/subalpine/boreal grassland), Wetland (2%; river/stream), Rocky areas (5%; inland cliff), Artificial landscape (10%)
Land-use Agriculture (2%), Nature conservation/research (80%), Not utilized (18%)

Birds

The area is rich in breeding species of forests and mountains, including raptors, gamebirds, owls and woodpeckers.

Protection status

National Partial **International** Partial

12,210 ha of IBA covered by National Park (Val Grande, 12,210 ha). 2,410 ha of IBA covered by State Nature Reserve (Monte Mottac, 2,410 ha). 973 ha of IBA covered by State Nature Reserve (Val Grande, 973 ha). 973 ha of IBA covered by Integral Nature Reserve (Val Grande, 973 ha). 11,856 ha of IBA covered by Special Protection Area (Riserva Naturale Monte Mottac e Val Grande, 11,856 ha).

Conservation issues

Threats Abandonment/reduction of land management (C), Burning of vegetation (B)

There are no major threats. There was a LIFE-funded project (1994–1996) to prepare a management plan for the National Park.

Mount Tabor

C7 **006**

Admin region Lombardia
Coordinates 46°04′N 9°09′E
Altitude 500–2,245 m **Area** 28,000 ha

Site description

A mountainous area in the south-central Alps, on the west side of Lake Como, and covered mainly by broadleaved woodlands. The main land-uses are hunting and forestry.

Habitats Forest and woodland (60%; broadleaved deciduous forest; mixed forest), Scrub (10%), Grassland (15%; alpine/subalpine/boreal grassland), Wetland (5%; standing fresh water; river/stream), Rocky areas (5%; inland cliff), Artificial landscape (5%; arable land)
Land-use Agriculture (5%), Forestry (20%), Hunting (20%), Tourism/recreation (5%), Unknown (45%), Water management (5%)

Birds

The area is notable for its high diversity of breeding raptors and for its woodland passerines.

Protection status

National None **International** None

Conservation issues

No major threats are present. It is a candidate Special Protection Area.

Pian di Spagna and Lake Mezzola

C7 **007**

Admin region Lombardia
Coordinates 46°13′N 9°27′E
Altitude 198–698 m **Area** 3,400 ha

Site description

A medium-sized freshwater lake (Lago di Mezzola) in a mountain valley in the foothills of the south-central Alps, north of Lago di Como and connected with it by the Mera river. The lake and its associated wetlands are surrounded by cultivated areas. Pian di Spagna is an alluvial flood-plain, lying between the two lakes. The site comprises mainly open water, reedbeds and intensively cultivated land. The main land-uses are nature conservation, agriculture, and fisheries.

Habitats Forest and woodland (5%; broadleaved deciduous forest; mixed forest), Scrub (5%; scrub), Grassland (2%; dry siliceous grassland; alpine/subalpine/boreal grassland), Wetland (55%; shingle/stony beach; standing fresh water; river/stream; water-fringe vegetation), Rocky areas (2%; scree/boulders; inland cliff), Artificial landscape (31%; highly improved reseeded grassland; arable land; perennial crops/orchards/groves; urban parks/gardens; other urban/industrial areas; ruderal land)
Land-use Agriculture (30%), Fisheries/aquaculture (30%), Hunting (10%), Nature conservation/research (50%), Tourism/recreation (2%)

Birds

A notable area for migrating and wintering waterbirds. Species of global conservation concern that do not meet IBA criteria: *Aythya nyroca* (1, winter).

Protection status

National Partial **International** Partial

1,740 ha of IBA covered by Regional Nature Reserve (Lago di Mezzola-Pian di Spagna, 1,740 ha). 1,740 ha of IBA covered by Special Protection Area (Lago di Mezzola e Pian di Spagna, 1,740 ha). 1,740 ha of IBA covered by Ramsar Site (Pian di Spagna–Lago di Mezzola, 1,740 ha).

Conservation issues

Threats Agricultural intensification/expansion (A), Burning of vegetation (A), Disturbance to birds (A), Drainage (A), Dredging/canalization (A), Industrialization/urbanization (A), Recreation/tourism (B)

Important threats are posed by the large reedbed fires that happen every year, the lowering of the water-level, and water pollution from domestic, agricultural and industrial sources. The protection of the reserve is not implemented. Field studies are under way to compile a management plan for the reserve. The site is a bird-ringing field station.

Gran Paradiso

A3, B2, C6 **008**

Admin region Piemonte, Valle d'Aosta
Coordinates 45°33′N 7°16′E
Altitude 450–4,061 m **Area** 110,000 ha

Site description

A mountain area in the western Alps of Italy, between the rivers Dora Baltea and Orco. The site includes the area of Monte Emilius. A wide range of montane habitats are present, relatively intact and little disturbed by man—from deciduous forests to glaciers. The main land-use is nature conservation. The most managed habitats are the forests.

Habitats Forest and woodland (22%; broadleaved deciduous forest; native coniferous forest; mixed forest; treeline ecotone), Scrub (10%; scrub; heathland), Grassland (8%; dry siliceous grassland; alpine/subalpine/boreal grassland), Rocky areas (55%; scree/boulders; inland cliff), Artificial landscape (5%; other urban/industrial areas)
Land-use Forestry (10%), Hunting (20%), Nature conservation/research (80%), Tourism/recreation (30%), Water management (1%)

Birds

Species	Season	Year	Pop min	Pop max	Acc	Criteria
Pernis apivorus Honey Buzzard	B	1996	10	20	C	C6
Aquila chrysaetos Golden Eagle	R	1996	11	13	A	B2, C6
[1] ***Lagopus mutus*** Ptarmigan	R	1996	200	—	B	C6
Tetrao tetrix Black Grouse	R	1996	150	200	B	B2, C6
[2] ***Alectoris graeca*** Rock Partridge	R	1996	150	200	C	B2, C6
Aegolius funereus Tengmalm's Owl	R	1996	10	—	C	C6
Dryocopus martius Black Woodpecker	R	1996	12	15	C	C6
Prunella collaris Alpine Accentor	R	1996	500	1,000	C	A3
Tichodroma muraria Wallcreeper	R	1996	100	—	C	A3
Pyrrhocorax graculus Alpine Chough	R	1996	300	—	C	A3
Pyrrhocorax pyrrhocorax Chough	R	1996	50	—	C	C6
Montifringilla nivalis Snowfinch	R	1996	200	—	C	A3
Serinus citrinella Citril Finch	R	1996	30	—	C	A3

1. *L. m. helveticus.*
2. *A. g. saxatilis.*

This is one of the best areas in the Alps for birds of the Eurasian high-montane biome; five species occur in good numbers, out of the 10 possible species. Also important for *Aquila chrysaetos*. *Gypaetus barbatus* is present, having been reintroduced at the nearby protected areas of Vanoise National Park (France; where it bred for the first time in 1997), Mercantour Natural Park (France) and Argentera Natural Park (Italy).

Protection status

National Partial **International** Partial
70,200 ha of IBA covered by National Park (Gran Paradiso, 70,200 ha). 3,521 ha of IBA covered by Regional Nature Park (Mont Avic, 3,521 ha). 70,000 ha of IBA covered by Special Protection Area (Parco Nazionale del Gran Paradiso, 70,000 ha).

Conservation issues

Threats Deforestation (commercial) (B), Industrialization/urbanization (C), Infrastructure (B), Recreation/tourism (B)

In some areas, forests are periodically cleared and cut, while new forestry roads are continually being made. Research focuses mainly on ibex *Capra ibex* (numbering about 4,000, the most important population in the world) and chamois *Rupicapra rupicapra*; the only studied bird species is *Aquila chrysaetos*.

Lake Maggiore

C6 **009**

Admin region Piemonte, Lombardia
Coordinates 45°55′N 8°40′E
Altitude 198–283 m **Area** 21,600 ha

Site description

A large freshwater lake of glacial origin in a pre-Alpine area of northern Italy, bordered by broadleaved deciduous forest, very wet and alluvial forest, reedbeds and rocky cliffs. The main land-uses are tourism and recreation.

Habitats Wetland (100%; shingle/stony beach; standing fresh water; river/stream; water-fringe vegetation)
Land-use Hunting, Nature conservation/research (5%), Tourism/recreation (100%), Water management (100%)

Birds

Species	Season	Year	Pop min	Pop max	Acc	Criteria
Gavia arctica Black-throated Diver	W	1998	2	5	A	C6
Milvus migrans Black Kite	P	1997	100	—	B	C6
Pandion haliaetus Osprey	P	1997	10	—	B	C6

An important site for wetland raptors on passage and for wintering waterbirds. The reedbeds support good numbers of breeding waterbirds.

Protection status

National None **International** None

Conservation issues

Threats Recreation/tourism (B), Unsustainable exploitation (A)

The main threat is hunting—if this was forbidden, the site would hold many more wintering waterbirds.

Campo dei Fiori, Valganna and Cuvia valley

C7 **010**

Admin region Lombardia
Coordinates 45°55′N 8°46′E
Altitude 350–1,226 m **Area** 24,500 ha

Site description

A subalpine, mountainous area north of the town of Varese, covered mainly by broadleaved woodlands, with some streams and small lakes. The main land-uses are recreation and forestry.

Habitats Forest and woodland (80%; broadleaved deciduous forest; treeline ecotone), Grassland (5%; alpine/subalpine/boreal grassland), Wetland (1%; standing fresh water; river/stream; water-fringe vegetation), Rocky areas (1%; inland cliff), Artificial landscape (13%; highly improved reseeded grassland; arable land; forestry plantation; other urban/industrial areas)
Land-use Agriculture (3%), Forestry (10%), Hunting (20%), Nature conservation/research (22%), Tourism/recreation (90%), Urban/industrial/transport (10%)

Birds

The area is important mainly for species of woodland passerine and for some breeding raptors.

Protection status

National Partial **International** None
5,400 ha of IBA covered by Regional Nature Park (Campo dei Fiori, 5,400 ha).

Conservation issues

Threats Abandonment/reduction of land management (B), Industrialization/urbanization (C), Infrastructure (B)

The main threats are urbanization, including the building of new infrastructure, and the abandonment of land management. The site is a candidate Special Protection Area.

Grigne

B2, C6 **011**

Admin region Lombardia
Coordinates 45°57′N 9°22′E
Altitude 199–2,409 m **Area** 14,038 ha

Site description

Habitats Forest and woodland (45%; broadleaved deciduous forest; mixed forest), Scrub (5%; scrub), Grassland (35%; steppe/dry calcareous grassland; alpine/subalpine/boreal grassland), Wetland (5%; river/stream), Rocky areas (10%; inland cliff; caves)
Land-use Agriculture (10%), Nature conservation/research (100%), Tourism/recreation (5%), Urban/industrial/transport (5%)

The Grigne is a mountain complex located in the south-central Alps, on the east side of Lake Como, north of the town of Lecco. It includes broadleaved forests, alpine grasslands and rocky areas. The main land-uses are cattle-rearing and nature conservation.

Birds

Species	Season	Year	Pop min	Pop max	Acc	Criteria
Milvus migrans Black Kite	B	1995	20	—	A	B2, C6
Alectoris graeca Rock Partridge	R	1995	50	100	B	B2, C6

An important site for *Milvus migrans* and mountain bird species.

Protection status

National High **International** None
14,038 ha of IBA covered by Nature Park (Grigne, 14,038 ha).

Conservation issues

Threats Abandonment/reduction of land management (A), Consequences of animal/plant introductions (B), Disturbance to birds (A), Infrastructure (B), Recreation/tourism (B), Unsustainable exploitation (A)

The most important threats are the reduction in traditional agricultural activities and the lack of management for conservation.

Orobie — A3, B2, C6 — 012

Admin region Lombardia
Coordinates 46°03′N 9°50′E
Altitude 658–3,050 m **Area** 107,000 ha

Site description

A subalpine, mountainous area in the south-central Alps. It includes the southern part of low Valtellina and the high parts of Val Brembana, Val Seriana and Val di Scalve. Extensive broadleaved and coniferous woodlands, alpine grasslands and rocky areas predominate. The main land-uses are linked to nature conservation and to agriculture and forestry in mountain habitats.

Habitats Forest and woodland (50%; broadleaved deciduous forest; native coniferous forest; mixed forest; treeline ecotone), Scrub (15%), Grassland (20%; alpine/subalpine/boreal grassland), Wetland (5%; river/stream), Rocky areas (5%; inland cliff), Artificial landscape (5%; arable land; forestry plantation)
Land-use Agriculture (5%), Forestry (20%), Nature conservation/research (100%), Tourism/recreation (10%)

Birds

Species	Season	Year	Pop min	Pop max	Acc	Criteria
Aquila chrysaetos Golden Eagle	R	1996	6	—	A	B2, C6
Prunella collaris Alpine Accentor	R	—	Common		—	A3
Tichodroma muraria Wallcreeper	R	—	Frequent		—	A3
Pyrrhocorax graculus Alpine Chough	R	—	Common		—	A3
Montifringilla nivalis Snowfinch	R	—	Frequent		—	A3
Serinus citrinella Citril Finch	R	1996	Common		—	A3

The area is important for five of the 10 species of the Eurasian high-montane biome, and for *Aquila chrysaetos*.

Protection status

National High **International** None
63,000 ha of IBA covered by Regional Nature Park (Orobie Bergamasche, 63,000 ha). 44,000 ha of IBA covered by Regional Nature Park (Orobie Valtellinesi, 44,000 ha).

Conservation issues

No major threats are present.

Camonica valley — B2, C6 — 013

Admin region Lombardia
Coordinates 46°03′N 10°13′E
Altitude 328–2,743 m **Area** 15,000 ha

Site description

A mountainous area of the south-central Alps, including the north-west corner of the Camonica valley, south of the town of Edolo. Main habitats are broadleaved and coniferous woodlands. The main land-uses are hunting, agriculture and forestry.

Habitats Forest and woodland (50%; broadleaved deciduous forest; native coniferous forest; mixed forest; treeline ecotone), Scrub (10%; scrub), Grassland (20%; alpine/subalpine/boreal grassland), Wetland (standing fresh water; river/stream), Rocky areas (5%; inland cliff), Artificial landscape (15%; arable land)
Land-use Agriculture (10%), Forestry (30%), Hunting (30%), Not utilized (15%), Tourism/recreation (15%)

Birds

Species	Season	Year	Pop min	Pop max	Acc	Criteria
Alectoris graeca Rock Partridge	R	—	—	—	—	B2, C6

Protection status

National None **International** None

Conservation issues

Threats Disturbance to birds (A), Unsustainable exploitation (A)

The main threats are poaching and hunting, including illegal trapping of passerines. The area has one of the highest concentrations of poachers and hunters in Italy.

Brabbia peatland and Lake Varese — A1, B2, B3, C1, C6 — 014

Admin region Lombardia
Coordinates 45°47′N 8°45′E
Altitude 238–241 m **Area** 1,960 ha

Site description

Palude Brabbia is a fen peatland located at the foot of the pre-Alps, west of the town of Varese. The area includes open water, marsh vegetation (mainly *Phragmites* and *Carex*), humid meadows and wet woodlands, and it is fed by the Brabbia Canal which links it to two lakes. One of them is Lago di Varese, a medium-sized freshwater lake surrounded by a fringe of reedbed. The main land-uses are nature conservation in Palude Brabbia, and tourism and hunting in Lago di Varese.

Habitats Forest and woodland (10%; broadleaved deciduous forest; alluvial/very wet forest), Scrub (2%; scrub), Grassland (2%; humid grassland; mesophile grassland), Wetland (83%; standing fresh water; river/stream; water-fringe vegetation; fen/transition mire/spring), Artificial landscape (3%; highly improved reseeded grassland; arable land; perennial crops/orchards/groves; forestry plantation)
Land-use Agriculture (3%), Hunting (52%), Nature conservation/research (48%), Tourism/recreation (30%), Urban/industrial/transport (5%), Water management (50%)

Birds

Species	Season	Year	Pop min	Pop max	Acc	Criteria
Ixobrychus minutus Little Bittern	B	1996	30	40	A	B2, C6
Ardea purpurea Purple Heron	B	1996	15	17	A	C6
[1] ***Aythya nyroca*** Ferruginous Duck	B	1996	4	6	A	A1, C1, C6
Locustella luscinioides Savi's Warbler	B	1996	10	15	A	B3

1. Since 1992 (also a record of 1 in winter).

An important site for breeding *Aythya nyroca*, herons (including a mixed heronry of *Nycticorax nycticorax* and *Ardea cinerea*) and reedbed passerines, and for foraging raptors. Numbers are based on regular counts during 1987–1996, apart from breeding herons which started to colonize the area in 1993.

Protection status

National Partial **International** High
128 ha of IBA covered by Regional Nature Reserve (Lago di Biandronno, 128 ha). 459 ha of IBA covered by Regional Nature Reserve (Palude Brabbia, 459 ha). 1,889 ha of IBA covered by Special Protection Area (Valli del Mincio, Paludi di Ostiglia, Torbiere d'Iseo, 1,889 ha). 459 ha of IBA covered by Ramsar Site (Palude Brabbia, 459 ha).

Conservation issues

Threats Burning of vegetation (A), Consequences of animal/plant introductions (C), Disturbance to birds (A), Drainage (C), Filling-in of wetlands (B), Groundwater abstraction (B), Industrialization/urbanization (B), Infrastructure (C), Recreation/tourism (C)

At present the main threats are reedbed-burning, disturbance to birds, and water pollution from domestic and industrial sources. Palude Brabbia is a LIPU reserve and is covered by a management plan.

Lake Garda — A4i, B1i, C3 — 015

Admin region Lombardia, Trentino-Alto Adige, Veneto
Coordinates 45°50′N 10°00′E
Altitude 65–65 m **Area** 38,000 ha

Site description

A large freshwater lake of glacial origin, in a pre-Alpine area of northern Italy.

Habitats Wetland (100%; shingle/stony beach; standing fresh water; river/stream; water-fringe vegetation)
Land-use Tourism/recreation (100%)

Birds

Species	Season	Year	Pop min	Pop max	Acc	Criteria
Podiceps nigricollis Black-necked Grebe	W	1995	1,468	2,087	A	A4i, B1i, C3

An important site for wintering waterbirds (16,800 in January 1994), and especially for *Podiceps nigricollis*.

Protection status

National None **International** None

Conservation issues

Threats Industrialization/urbanization (B), Recreation/tourism (C)

There are no major threats. Water pollution comes from domestic sewage, especially in summer when the area receives many tourists. Regulation of boating on the lake is necessary.

Lake Viverone — B3 — 016

Admin region Piemonte
Coordinates 45°23′N 8°03′E
Altitude 220–235 m **Area** 930 ha

Site description

A freshwater lake of glacial origin on the western Padana plain, 15 km south of the town of Biella. The site includes large reedbeds and a wet woodland (*Alnus*). Tourism is the main human activity.

Habitats Forest and woodland (10%; alluvial/very wet forest), Wetland (80%; standing fresh water; water-fringe vegetation), Artificial landscape (10%)
Land-use Agriculture (5%), Fisheries/aquaculture (20%), Nature conservation/research (5%), Tourism/recreation (70%), Urban/industrial/transport (5%)

Birds

Species	Season	Year	Pop min	Pop max	Acc	Criteria
Locustella luscinioides Savi's Warbler	B	1995	Frequent		—	B3

An important site for migrating and wintering waterbirds (up to 15,000 in winter) and for breeding reedbed passerines. Species of global conservation concern that do not meet IBA criteria: *Aythya nyroca* (max. 5 on passage).

Protection status

National Low **International** None
49 ha of IBA covered by Wildlife Reserve (Lago di Viverone, 49 ha).

Conservation issues

Threats Agricultural intensification/expansion (B), Burning of vegetation (B), Disturbance to birds (B), Recreation/tourism (C)

The main threats are agricultural intensification in the surrounding land, disturbance of birds by motor boats, and the regular (and apparently deliberate) burning of reedbeds.

Novara heronries — A4i, B1i, B2, C2, C6 — 017

Admin region Piemonte
Coordinates 45°30′N 8°33′E
Altitude 80–185 m **Area** 50 ha

Site description

The heronries are located in small woodlands surrounded by rice-fields, on the west Padana plain, north-west of the town of Novara. The site was listed in the previous pan-European IBA inventory (Grimmett and Jones 1989) as subsite 4 of the IBA 'Heronries in the catchment of the upper Po and Sesia Rivers' (former site IT005).

Habitats Forest and woodland (80%; broadleaved deciduous forest; alluvial/very wet forest), Grassland (5%), Artificial landscape (15%; arable land)
Land-use Agriculture (15%), Hunting (50%), Not utilized (35%)

Birds

Species	Season	Year	Pop min	Pop max	Acc	Criteria
Nycticorax nycticorax Night Heron	B	1995	—	1,219	A	A4i, B1i, B2, C2, C6
Egretta garzetta Little Egret	B	1995	—	507	A	A4i, B1i, C2, C6

Two mixed heronries: Garzaia di Morghengo–San Bernardino, and Garzaia di Casaleggio.

Protection status

National None **International** None

Conservation issues

Threats Agricultural intensification/expansion (B), Disturbance to birds (A), Firewood collection (B), Unsustainable exploitation (C)

The main threat is disturbance to birds by activity at a game reserve at Garzaia di Morghengo–San Bernardino and by gravel extraction close to the same heronry. This is causing breeding pairs to move to Garzaia di Casaleggio. The Garzaia di Morghengo is included in a game reserve.

Ticino river — A4i, B1i, B2, C2, C3, C6 — 018

Admin region Piemonte, Lombardia
Coordinates 45°33′N 8°51′E
Altitude 60–350 m **Area** 25,000 ha

Site description

A lowland wooded area on the west Padana plain, with small freshwater wetlands and cultivated areas, along the valley of the Ticino river, a tributary of the River Po. The main land-uses are agriculture, nature conservation and recreational activities.

Habitats Forest and woodland (75%; broadleaved deciduous forest; native coniferous forest; mixed forest; alluvial/very wet forest), Scrub (3%; scrub; heathland), Wetland (5%; standing fresh water; river/stream; water-fringe vegetation), Artificial landscape (17%; highly improved reseeded grassland; arable land; forestry plantation; urban parks/gardens)
Land-use Agriculture (15%), Fisheries/aquaculture (3%), Forestry (20%), Hunting (30%), Military (1%), Nature conservation/research (100%), Tourism/recreation (30%), Urban/industrial/transport (5%), Water management (5%)

Birds

Species	Season	Year	Pop min	Pop max	Acc	Criteria
Phalacrocorax carbo Cormorant	W	1996	1,000	2,000	A	B1i, C3
Ixobrychus minutus Little Bittern	B	1996	10	—	C	C2
Nycticorax nycticorax Night Heron	B	1995	—	1,995	A	A4i, B1i, B2, C2, C6
Ardeola ralloides Squacco Heron	B	1996	20	—	B	C2

Species ... continued	Season	Year	Pop min	Pop max	Acc	Criteria
Egretta garzetta Little Egret	B	1995	—	889	A	A4i, B1i, C2, C6
Ardea purpurea Purple Heron	B	1996	20	—	C	C6
Caprimulgus europaeus Nightjar	B	1996	100	—	C	C6
Alcedo atthis Kingfisher	R	1996	50	—	C	C6

There are three main heronries: Cascina Portalupa, La Zelata and San Massimo, all of which were listed in the previous pan-European IBA inventory (Grimmett and Jones 1989) as subsites 3–5 of the former site IT023, 'Heronries close to Pavia and the catchment of the Fiume Ticino'. Counts for *Nycticorax nycticorax* and *Egretta garzetta* refer to the period 1990–1995.

Protection status

National High **International** None

Part of IBA covered by Regional Nature Park (Valle del Ticino (Lombardia), 90,640 ha). 6,136 ha of IBA covered by Regional Nature Park (Valle del Ticino (Piemonte), 6,136 ha).

Conservation issues

Threats Agricultural intensification/expansion (B), Consequences of animal/plant introductions (C), Deforestation (commercial) (C), Disturbance to birds (B), Firewood collection (C), Groundwater abstraction (U), Industrialization/urbanization (B), Natural events (C), Recreation/tourism (C), Selective logging/cutting (C), Unsustainable exploitation (C)

The main threats are from agricultural intensification, industrialization and urbanization. All of the site is covered by a management plan. At Parco del Ticino, a project to reintroduce Eurasian otter *Lutra lutra*, roe deer *Capreolus capreolus* and European pond turtle *Emys orbicularis* (globally near-threatened) is ongoing, and there are censuses of wintering ducks and *Phalacrocorax carbo*, and of breeding herons.

Iseo peatlands — B3, C6 — 019

Admin region Lombardia
Coordinates 45°38′N 10°02′E
Altitude 186–230 m **Area** 325 ha

Site description

Freshwater pools formed by peat-cutting in the pre-Alps, south of Lago d'Iseo. The pools are fringed by reedbeds, wet meadows and woodlands. Most of the peat was extracted after the Second World War. The main land-uses are nature conservation and recreation activities.

Habitats Forest and woodland (2%; broadleaved deciduous forest), Scrub (10%; scrub), Grassland (3%; humid grassland), Wetland (75%; standing fresh water; water-fringe vegetation; fen/transition mire/spring), Artificial landscape (10%)
Land-use Agriculture (5%), Nature conservation/research (100%), Tourism/recreation (5%)

Birds

Species	Season	Year	Pop min	Pop max	Acc	Criteria
Circus aeruginosus Marsh Harrier	B	1996	2	2	A	C6
Porzana porzana Spotted Crake	B	1996	2	2	A	C6
Locustella luscinioides Savi's Warbler	B	1996	8	10	A	B3

A breeding and wintering site for waterbirds and reedbed passerines.

Protection status

National High **International** High

325 ha of IBA covered by Regional Nature Reserve (Torbiere d'Iseo, 325 ha). 325 ha of IBA covered by Special Protection Area (Valli del Mincio, Paludi di Ostiglia, Torbiere d'Iseo, 1,889 ha). 324 ha of IBA covered by Ramsar Site (Torbiere d'Iseo, 324 ha).

Conservation issues

The main problems are posed by poor control over recreational activities, and over water management in the surrounding area ('Other' threat), as well as water pollution from domestic and agricultural sources. Some or all of the site is covered by a management plan. Research on breeding bird populations is carried out regularly.

Threats Agricultural intensification/expansion (C), Burning of vegetation (C), Consequences of animal/plant introductions (C), Disturbance to birds (B), Filling-in of wetlands (C), Other (A), Recreation/tourism (A)

River Sesia heronries — A1, A4i, B1i, B2, C1, C2, C6 — 020

Admin region Piemonte
Coordinates 45°24′N 8°23′E
Altitude 143–162 m **Area** 500 ha

Site description

The site lies along the Sesia river and its tributaries, on the west Padana plain, north-west of the town of Vercelli. The main land-use is nature conservation. The site was listed in the previous pan-European IBA inventory (Grimmett and Jones 1989) as subsites 1–3 of the IBA 'Heronries in the catchment of the upper Po and Sesia Rivers' (former site IT005).

Habitats Forest and woodland (70%; broadleaved deciduous forest; alluvial/very wet forest), Scrub (5%; scrub), Wetland (10%; standing fresh water; river/stream; water-fringe vegetation), Artificial landscape (15%; arable land)
Land-use Agriculture (15%), Nature conservation/research (100%), Tourism/recreation (10%)

Birds

Species	Season	Year	Pop min	Pop max	Acc	Criteria
Nycticorax nycticorax Night Heron	B	1995	—	2,979	A	A4i, B1i, B2, C2, C6
Egretta garzetta Little Egret	B	1995	—	1,570	A	A4i, B1i, C2, C6
[1] ***Numenius tenuirostris*** Slender-billed Curlew	W	1995	1	1	A	A1, C1

1. One record, in February 1986.

The site includes four large mixed heronries: Garzaia di Carisio, Garzaia di Vettignè, Garzaia di Villarboit (Torrente Druma), and Garzaia di Oldenico/Parco Naturale Lame del Sesia.

Protection status

National High **International** None

91 ha of IBA covered by Regional Nature Reserve (Garzaia di Carisio, 91 ha). 10 ha of IBA covered by Regional Nature Reserve (Garzaia di Villarboit, 10 ha). 52 ha of IBA covered by Regional Nature Reserve (Isolone di Oldenico, 52 ha). 500 ha of IBA covered by Regional Nature Park (Lame del Sesia, 845 ha).

Conservation issues

Threats Agricultural intensification/expansion (B)

The main threat is the change from 'wet' to 'dry' cultivation in the rice-fields around the sites.

Lake Candia — B3 — 021

Admin region Piemonte
Coordinates 45°18′N 7°58′E
Altitude 280–280 m **Area** 295 ha

Site description

A freshwater, nutrient-rich lake of modest dimensions, located on the western Padana plain, 10 km south of the town of Ivrea. It is bordered by reedbeds and wet woodlands, and surrounded by meadows and maize-fields. The main human activities are nature conservation and tourism.

Habitats Forest and woodland (alluvial/very wet forest), Wetland (standing fresh water; water-fringe vegetation)
Land-use Agriculture, Nature conservation/research (100%), Tourism/recreation

Birds

Species	Season	Year	Pop min	Pop max	Acc	Criteria
Acrocephalus palustris Marsh Warbler	B	1995	Common		—	B3

The area is important for reedbed-breeding passerines.

Protection status

National High **International** None

295 ha of IBA covered by Wildlife Reserve (Parco Provinciale Lago di Candia, 295 ha).

Conservation issues

Threats Abandonment/reduction of land management (C), Disturbance to birds (B), Drainage (C), Filling-in of wetlands (C)

Possible threats are the destruction of wetland habitats by agricultural activities, and the disturbance to birds due to tourism.

Heronries of Pavia province

A4i, A4iii, B1i, B2, C2, C4, C6 **022**

Admin region Lombardia
Coordinates 45°18′N 9°10′E
Altitude 63–107 m **Area** 1,173 ha

Site description

A complex of heronries on the western Padana plain, isolated from each other and situated mainly inside small broadleaved woodlands surrounded by cultivated land, especially rice-fields. The main land-uses are nature conservation and agriculture. 'Water management' is linked to rice-fields.

Habitats Forest and woodland (70%; broadleaved deciduous forest; alluvial/very wet forest), Grassland (3%; humid grassland), Wetland (7%; standing fresh water; river/stream; water-fringe vegetation), Artificial landscape (20%; highly improved reseeded grassland; arable land; ruderal land)
Land-use Agriculture (10%), Fisheries/aquaculture (5%), Forestry (30%), Hunting (20%), Nature conservation/research (80%), Tourism/recreation (10%), Water management (10%)

Birds

Species	Season	Year	Pop min	Pop max	Acc	Criteria
[1] ***Nycticorax nycticorax*** Night Heron	B	1995	—	5,443	A	A4i, B1i, B2, C2, C6
[1] ***Ardeola ralloides*** Squacco Heron	B	1995	—	91	A	B2, C2
[1] ***Egretta garzetta*** Little Egret	B	1995	—	4,811	A	A4i, B1i, C2, C6
[1] ***Ardea purpurea*** Purple Heron	B	1995	—	61	A	C2, C6

1. 1990–1995.

There are 12 main heronries: Abbazia Acqualunga, Bosco Basso, Carola, Cascina Notizia, Cascina Villarasca, Celpenchio, Porta Chiossa, Rinalda, Roggia Torbida, Sant'Alessandro, Verminesca, and Villa Biscossi. The site holds internationally important numbers of breeding *Nycticorax nycticorax* (second-largest breeding colony in Europe) and *Egretta garzetta*. The average total number of breeding herons exceeds 20,000 birds.

Protection status

National High **International** None

42 ha of IBA covered by Regional Nature Reserve (Garzaia del Bosco Basso, 42 ha). 32 ha of IBA covered by Regional Nature Reserve (Garzaia della Carola, 32 ha). 76 ha of IBA covered by Regional Nature Reserve (Garzaia della Cascina Isola, 76 ha). 75 ha of IBA covered by Regional Nature Reserve (Garzaia della Cascina Notizia, 75 ha). 53 ha of IBA covered by Regional Nature Reserve (Garzaia della Cascina Villarasca, 53 ha). 37 ha of IBA covered by Regional Nature Reserve (Garzaia della Rinalda, 37 ha). 13 ha of IBA covered by Regional Nature Reserve (Garzaia della Roggia Torbida, 13 ha). 173 ha of IBA covered by Regional Nature Reserve (Garzaia della Verminesca, 173 ha). 122 ha of IBA covered by Regional Nature Reserve (Garzaia di Acqualunga, 122 ha). 140 ha of IBA covered by Regional Nature Reserve (Garzaia di Celpenchio, 140 ha). 83 ha of IBA covered by Regional Nature Reserve (Garzaia di Porta Chiossa, 83 ha). 256 ha of IBA covered by Regional Nature Reserve (Garzaia di Sant'Alessandro, 256 ha). 72 ha of IBA covered by Regional Nature Reserve (Garzaia di Villa Biscossi, 72 ha).

Conservation issues

Threats Agricultural intensification/expansion (A), Deforestation (commercial) (A), Disturbance to birds (C)

Threats are posed by agricultural intensification and by clear-cutting of trees (mainly illegal, by local people) for commercial reasons or to increase the amount of cultivable land. Most of the heronries are protected as Regional Nature Reserves, but some heronries are located inside game reserves, where the habitat management is sometimes not favourable for the nesting herons. Censuses of breeding pairs are regularly carried out by the Università di Pavia on behalf of Regione Lombardia. Every heronry has a management plan, prepared by Regione Lombardia.

Adda Sud Park heronries

A4i, B1i, B2, C2, C6 **023**

Admin region Lombardia
Coordinates 45°16′N 9°39′E
Altitude 60–100 m **Area** 1,000 ha

Site description

Riparian habitats of the Adda river, a tributary of the River Po. Habitats are dominated by shallow standing water, aquatic vegetation and wet woodlands, and the site is surrounded by cultivated land and poplar *Populus* plantations. The main land-uses are nature conservation and agriculture.

Habitats Forest and woodland (30%; broadleaved deciduous forest; alluvial/very wet forest), Wetland (40%; standing fresh water; river/stream; water-fringe vegetation), Artificial landscape (30%; arable land)
Land-use Agriculture (30%), Nature conservation/research (100%)

Birds

Species	Season	Year	Pop min	Pop max	Acc	Criteria
[1] ***Nycticorax nycticorax*** Night Heron	B	1995	70	564	A	A4i, B1i, B2, C2
[1] ***Egretta garzetta*** Little Egret	B	1995	57	299	A	B1i, C2, C6
[1] ***Ardea purpurea*** Purple Heron	B	1995	6	59	A	C2, C6

1. 1990–1995.

There are two large mixed heronries, called Garzaia di Cascina Pioppo and Garzaia di Zerbaglia. The latter was listed in the previous pan-European IBA inventory (Grimmett and Jones 1989) as 'La Zerbaglia' (former site IT022).

Protection status

National High **International** None

1,000 ha of IBA covered by Regional Nature Park (Adda Sud, 23,600 ha).

Conservation issues

Threats Agricultural intensification/expansion (C), Disturbance to birds (C)

No major threats are present.

Susa and Chisone valleys

A3, B2, C6 **024**

Admin region Piemonte
Coordinates 45°03′N 7°00′E
Altitude 500–3,538 m **Area** 60,946 ha

Site description

A mountainous area in the south-west Alps, west of the town of Torino and east of the French border. There is a high diversity of habitats, with extensive coniferous woodlands, calcareous rocky areas and alpine grasslands.

Habitats Forest and woodland (broadleaved deciduous forest; native coniferous forest; treeline ecotone), Grassland (alpine/subalpine/boreal grassland), Wetland (river/stream), Rocky areas (inland cliff), Artificial landscape
Land-use Agriculture, Forestry, Nature conservation/research, Not utilized, Tourism/recreation, Urban/industrial/transport

Birds

An important breeding area for *Aquila chrysaetos* and other raptors, and for five out of the 10 species of the Eurasian high-montane biome.

Species	Season	Year	Pop min	Pop max	Acc	Criteria
Pernis apivorus Honey Buzzard	P	1992	600	—	B	C6
Circaetus gallicus Short-toed Eagle	B	1995	4	5	B	B2, C6
Aquila chrysaetos Golden Eagle	R	1995	8	10	A	B2, C6
Bubo bubo Eagle Owl	R	1995	4	7	A	B2, C6
Aegolius funereus Tengmalm's Owl	R	1995	Common		—	C6
Dryocopus martius Black Woodpecker	R	1995	6	—	C	C6
Lullula arborea Woodlark	B	1995	Frequent		—	C6
Prunella collaris Alpine Accentor	R	1995	Frequent		—	A3
Tichodroma muraria Wallcreeper	R	1995	Frequent		—	A3
Pyrrhocorax graculus Alpine Chough	R	1995	Frequent		—	A3
Pyrrhocorax pyrrhocorax Chough	R	1995	35	70	B	B2, C6
Montifringilla nivalis Snowfinch	R	1995	Frequent		—	A3
Serinus citrinella Citril Finch	R	1995	Frequent		—	A3

Protection status

National Partial **International** None

23 ha of IBA covered by Regional Nature Reserve (Orrido e Stazione di Leccio di Chianocco, 23 ha). 2,030 ha of IBA covered by Regional Nature Park (Gran Bosco di Salbertrand, 2,030 ha). 10,927 ha of IBA covered by Regional Nature Park (Orsiera-Rocciavre, 10,927 ha). 3,265 ha of IBA covered by Regional Nature Park (Val Troncea, 3,265 ha).

Conservation issues

No major threats are present. There is research on *Petronia petronia*.

Vercelli rice-fields — A4i, B1i, B2, C2, C6 — 025

Admin region Piemonte
Coordinates 45°12'N 8°16'E
Altitude 136–182 m **Area** 15,000 ha

Site description

A lowland agricultural area on the western Padana plain, south-west of the town of Vercelli, covered mainly by rice-fields. The area includes a broadleaved woodland called Bosco della Partecipanza and a wetland called Palude di San Genuario. Part of the site was listed in the previous pan-European IBA inventory (Grimmett and Jones 1989) as subsite 6 of the IBA 'Heronries in the catchment of the upper Po and Sesia Rivers' (former site IT005).

Habitats Forest and woodland (5%), Wetland (5%), Artificial landscape (90%; arable land)
Land-use Agriculture (75%), Fisheries/aquaculture (5%), Forestry (5%), Hunting (25%), Nature conservation/research (10%), Urban/industrial/transport (5%)

Birds

Species	Season	Year	Pop min	Pop max	Acc	Criteria
Botaurus stellaris Bittern	R	1995	5	5	A	C6
Ixobrychus minutus Little Bittern	B	1995	8	—	B	C6
Nycticorax nycticorax Night Heron	B	1995	1,764	2,952	A	A4i, B1i, B2, C2, C6
Ardeola ralloides Squacco Heron	B	1995	24	164	A	B1i, B2, C2, C6
Egretta garzetta Little Egret	B	1995	468	1,372	A	A4i, B1i, C2
Himantopus himantopus Black-winged Stilt	B	1995	300	330	A	A4i, B1i, C2, C6
Philomachus pugnax Ruff	P	1995	3,500	5,000	B	C6
Chlidonias niger Black Tern	B	1995	100	150	C	C2, C6

A large mixed heronry (Garzaia di Montarolo) is located in Bosco della Partecipanza. The rice-fields are important for breeding and migrating waders.

Protection status

National Low **International** None

1,068 ha of IBA covered by Regional Nature Park (Bosco delle Sorti della Partecipanza di Trino, 1,068 ha).

Conservation issues

Threats Agricultural intensification/expansion (A), Aquaculture/fisheries (C), Disturbance to birds (C), Filling-in of wetlands (C), Intensified forest management (C)

The change from 'wet' to 'dry' cultivation of rice is a very significant threat.

Trebbia river (lower course) — C6 — 026

Admin region Emilia Romagna
Coordinates 45°02'N 9°37'E
Altitude 60–90 m **Area** 30 ha

Site description

The lower course of a tributary of the River Po. The site is dominated by riparian and dry habitats, mainly wet woodlands (*Populus* and *Salix*), ponds, pseudo-steppe, gravel-pits, and sandy areas. The main land-use is gravel extraction.

Habitats Forest and woodland (alluvial/very wet forest), Scrub (scrub), Grassland (steppe/dry calcareous grassland), Wetland (shingle/stony beach; standing fresh water; river/stream; water-fringe vegetation), Artificial landscape (arable land)
Land-use Agriculture, Military, Not utilized, Urban/industrial/transport

Birds

Species	Season	Year	Pop min	Pop max	Acc	Criteria
Burhinus oedicnemus Stone Curlew	B	1996	10	20	B	C6

A notable breeding site for birds of dry areas (especially *Burhinus oedicnemus*, *Calandrella brachydactylata* and *Anthus campestris*) and for terns (*Sterna hirundo* and *S. albifrons*).

Protection status

National None **International** None

Conservation issues

Threats Disturbance to birds, Dredging/canalization, Extraction industry (A), Filling-in of wetlands

The lack of protection allows commercial gravel extraction (the main threat), disturbance to birds by uncontrolled four-wheel drive vehicles and motorbikes and, occasionally, by sheep-grazing, and illegal dumping. LIPU has proposed the area as a Regional Nature Reserve, and is carrying out an awareness/education campaign by means of a video shown in schools. Some or all of the site is covered by a management plan.

River Po: from Dora Baltea mouth to Scrivia mouth — B2, B3, C2, C6 — 027

Admin region Piemonte
Coordinates 45°07'N 8°20'E
Altitude 90–150 m **Area** 14,000 ha

Site description

A stretch of the River Po on the west Padana plain, including riparian woodlands, freshwater wetlands, poplar *Populus* plantations and other cultivation.

Habitats Forest and woodland (15%; broadleaved deciduous forest; alluvial/very wet forest), Scrub (5%; scrub), Wetland (60%; standing fresh water; river/stream; water-fringe vegetation), Artificial landscape (20%; highly improved reseeded grassland; arable land; perennial crops/orchards/groves; forestry plantation; ruderal land)
Land-use Agriculture (15%), Hunting (50%), Nature conservation/research (50%), Tourism/recreation (25%), Urban/industrial/transport (5%), Water management

Birds

Species	Season	Year	Pop min	Pop max	Acc	Criteria
Nycticorax nycticorax Night Heron	B	1995	—	329	A	B2, C2, C6
[1] ***Ardea purpurea*** Purple Heron	B	1995	5	57	A	C2, C6
Sterna hirundo Common Tern	B	1995	100	220	B	C6
Sterna albifrons Little Tern	B	1995	60	120	B	C6
Acrocephalus palustris Marsh Warbler	B	1995	100	200	C	B3
Emberiza hortulana Ortolan Bunting	B	1995	10	40	C	C6

1. Largest colony (Valenza) destroyed by flood November 1994.

Four heronries define this area: Garzaia della Tenuta Baraccone, Garzaia di Camino (or di Ghiaia Grande), Garzaia di Valmacca (or G. di C. Mezzano) and Garzaia di Valenza. The first and the last heronries were listed in the previous pan-European IBA inventory (Grimmett and Jones 1989) as subsites 5 and 13 respectively of the IBA 'Heronries in the catchment of the upper Po and Sesia Rivers' (former site IT005).

Protection status

National High **International** None
14,000 ha of IBA covered by Regional Nature Park (Po e Orba, 24,854 ha).

Conservation issues

Threats Afforestation (C), Agricultural intensification/expansion (B), Consequences of animal/plant introductions (C), Construction/impact of dyke/dam/barrage (B), Extraction industry (U), Recreation/tourism (C)

Some or all of the site is covered by a management plan.

River Po: from Ticino mouth to Adda mouth — A4i, B1i, B2, C2, C6 — 028

Admin region Lombardia, Emilia Romagna
Coordinates 45°00′N 10°00′E
Altitude 40–50 m **Area** 25,000 ha

Site description

A stretch of the River Po, including large islands of shingle and sand, together with the surrounding wooded and cultivated areas. Main land-uses are agriculture and recreation. The area includes a nuclear power station (close to the village of Caorso), not working at present.

Habitats Forest and woodland (10%; broadleaved deciduous forest; alluvial/very wet forest), Scrub (5%; scrub), Wetland (50%; standing fresh water; river/stream; water-fringe vegetation), Artificial landscape (35%; highly improved reseeded grassland; arable land; forestry plantation)
Land-use Agriculture (25%), Fisheries/aquaculture (50%), Hunting (50%), Nature conservation/research (5%), Tourism/recreation (50%), Urban/industrial/transport (10%), Water management (50%)

Birds

Species	Season	Year	Pop min	Pop max	Acc	Criteria
Ixobrychus minutus Little Bittern	B	1996	10	20	B	C6
Nycticorax nycticorax Night Heron	B	1995	412	681	A	A4i, B1i, B2, C2, C6
Circus aeruginosus Marsh Harrier	R	1996	1	3	A	C6
Alcedo atthis Kingfisher	R	1996	15	20	C	C6

There are two important heronries: Monticchie (or Somaglia) and Isola de Pinedo.

Protection status

National Low **International** None
200 ha of IBA covered by Regional Nature Reserve (Monticchie, 200 ha).

Conservation issues

Threats Afforestation (A), Agricultural intensification/expansion (A), Consequences of animal/plant introductions (C), Construction/impact of dyke/dam/barrage (B), Drainage (A), Filling-in of wetlands (A), Industrialization/urbanization (A)

The main threats are posed by: agricultural intensification, with consequent drainage; filling-in of wetlands and creation of poplar *Populus* plantations; water pollution by agricultural activities and by untreated urban sewage from the town of Milano. Isola de Pinedo is a proposed LIPU reserve. Monticchie is a WWF reserve.

Marengo heronry — A4i, B1i, B2, C2, C6 — 029

Admin region Piemonte
Coordinates 44°49′N 8°39′E
Altitude 100–120 m **Area** 30 ha

Site description

An area of woodland along the Orba river, on the west Padana plain, 10 km south of the town of Alessandria. The main land-use is nature conservation.

Habitats Forest and woodland
Land-use Agriculture, Forestry, Nature conservation/research

Birds

A large, mixed heronry.

Species	Season	Year	Pop min	Pop max	Acc	Criteria
Nycticorax nycticorax Night Heron	B	1995	—	968	A	A4i, B1i, B2, C2, C6

Protection status

National High **International** None
30 ha of IBA covered by Nature Reserve (Torrente Orba, 246 ha).

Conservation issues

No major threats are present.

Stirone stream — B2 — 030

Admin region Emilia Romagna
Coordinates 44°50′N 9°59′E
Altitude 76–474 m **Area** 2,400 ha

Site description

The middle course of an Apennine stream, on the Padana plain. The main land-uses are agriculture and nature conservation.

Habitats Forest and woodland (20%; broadleaved deciduous forest), Scrub (5%; scrub; heathland), Wetland (30%; standing fresh water; river/stream; water-fringe vegetation), Artificial landscape (65%; highly improved reseeded grassland; arable land; perennial crops/orchards/groves; other urban/industrial areas)
Land-use Agriculture (60%), Nature conservation/research (100%), Tourism/recreation (10%), Urban/industrial/transport (10%)

Birds

Species	Season	Year	Pop min	Pop max	Acc	Criteria
Merops apiaster Bee-eater	B	1995	70	100	A	B2

An important breeding site for *Merops apiaster* which also supports good numbers of terns and *Riparia riparia*.

Protection status

National Partial **International** None
1,769 ha of IBA covered by Regional Nature Park (Fiume Stirone, 1,769 ha).

Conservation issues

Threats Agricultural intensification/expansion (B), Firewood collection (C), Infrastructure (B), Natural events (B), Recreation/tourism (B)

There are no high-impact threats.

Taro river — B2, C2, C6 — 031

Admin region Emilia Romagna
Coordinates 44°49′N 10°13′E
Altitude 27–172 m **Area** 6,000 ha

Site description

The middle course of the Taro river, on the southern Padana plain, a few kilometres west of the town of Parma. The alluvial plain is rich in gravel areas and sandbanks. The area is surrounded by cultivation. The main land-uses are nature conservation, agriculture, tourism, hunting and gravel extraction.

Habitats Forest and woodland (8%; broadleaved deciduous forest), Scrub (12%; scrub; heathland), Grassland (6%; steppe/dry calcareous grassland), Wetland (28%; standing fresh water; river/stream; water-fringe vegetation), Artificial landscape (46%; highly improved reseeded grassland; arable land; forestry plantation; other urban/industrial areas; ruderal land)
Land-use Agriculture (35%), Forestry (15%), Hunting (60%), Nature conservation/research (30%), Tourism/recreation (20%), Urban/industrial/transport (20%)

Birds

Species	Season	Year	Pop min	Pop max	Acc	Criteria
[1] ***Nycticorax nycticorax*** Night Heron	B	1995	—	700	A	B2, C2, C6
Burhinus oedicnemus Stone Curlew	B	1995	20	30	A	C6
Sterna hirundo Common Tern	B	1995	150	200	A	C6
Sterna albifrons Little Tern	B	1995	35	40	A	B2, C2

Species ... continued	Season	Year	Pop min	Pop max	Acc	Criteria
Caprimulgus europaeus Nightjar	B	1995	30	50	A	B2, C6
Calandrella brachydactyla Short-toed Lark	B	1995	40	60	A	C6

1. Ozzano Taro and Gramignazzo heronries.

Important numbers of *Burhinus oedicnemus* (also many *Merops apiaster* and *Riparia riparia*) breed on the alluvial plain.

■ Protection status

National Partial **International** Partial

2,500 ha of IBA covered by Regional Nature Park (Taro, 2,500 ha). 3,144 ha of IBA contains Special Protection Area (Fiume Taro da Fornovo di Taro all'Autostrada del Sole, 3,144 ha).

■ Conservation issues

Threats Consequences of animal/plant introductions (B), Disturbance to birds (B), Dredging/canalization (A), Extraction industry (A), Groundwater abstraction (A), Industrialization/urbanization (A), Recreation/tourism (B)

The main threat is of wetland destruction by groundwater extraction, dredging and canalization, and gravel extraction. The increase in industrial activities around the IBA is another important threat. Some or all of the site is covered by a management plan. The University of Parma is carrying out a long-term study of *Riparia riparia*.

Torrile LIPU reserve

C6 **032**

Admin region Emilia Romagna
Coordinates 44°56'N 10°20'E
Altitude 29–35 m **Area** 200 ha

■ Site description

An artificially created wetland, in a heavily cultivated, lowland area of the Padana plain, 15 km north of the city of Parma. Habitats include wet grasslands, arable land, streams and rivers.

Habitats Forest and woodland (5%; broadleaved deciduous forest), Scrub (10%; scrub), Wetland (20%; standing fresh water; river/stream; water-fringe vegetation), Artificial landscape (65%; highly improved reseeded grassland; arable land; other urban/industrial areas)
Land-use Agriculture (45%), Hunting (15%), Nature conservation/research (15%), Tourism/recreation (25%), Urban/industrial/transport (25%)

■ Birds

Species	Season	Year	Pop min	Pop max	Acc	Criteria
Himantopus himantopus Black-winged Stilt	B	1996	80	120	A	C6

A notable site for migrating waterbirds and for breeding *Himantopus himantopus* and terns *Sterna*.

■ Protection status

National Partial **International** None

35 ha of IBA covered by Private Reserve (Torrile, 35 ha).

■ Conservation issues

Threats Consequences of animal/plant introductions (A), Disturbance to birds (B), Groundwater abstraction (A), Industrialization/urbanization (B), Infrastructure (B)

The main threats are the building of a power-station in the surroundings, the abundance of coypu *Myocastor coypu* (an introduced, non-native aquatic rodent), groundwater extraction by the large sugar factory next to the reserve, and hunting in the surroundings. The reserve is owned by LIPU, and there are projects on *Ciconia ciconia* (captive breeding), *Tyto alba* and *Himantopus himantopus*.

Scrivia stream

B2, C6 **033**

Admin region Piemonte
Coordinates 44°45'N 8°51'E
Altitude 120–250 m **Area** 3,000 ha

■ Site description

A stretch of the Scrivia stream, between the villages of Arquata and Tortona. Habitats include large areas of shingle, dry grasslands, and riverine woodlands (poplar *Populus*, oak *Quercus*). The main land-uses are nature conservation and research, and gravel extraction.

Habitats Forest and woodland (10%; broadleaved deciduous forest), Scrub (5%; scrub), Grassland (5%), Wetland (20%; standing fresh water; river/stream), Rocky areas (40%; scree/boulders), Artificial landscape (20%; highly improved reseeded grassland; arable land; ruderal land)
Land-use Agriculture (20%), Fisheries/aquaculture (5%), Hunting (30%), Nature conservation/research (30%), Urban/industrial/transport (15%)

■ Birds

Species	Season	Year	Pop min	Pop max	Acc	Criteria
Caprimulgus europaeus Nightjar	B	1996	10	20	B	C6
Merops apiaster Bee-eater	B	1996	80	160	A	B2
Calandrella brachydactyla Short-toed Lark	B	1996	10	80	A	C6

The area is important for the presence of breeding *Merops apiaster* and there is a small mixed heronry. Species of global conservation concern that do not meet IBA criteria: *Aquila clanga* (min. 1, winter).

■ Protection status

National Partial **International** None

900 ha of IBA covered by Wildlife Reserve (Torrente Scrivia, 900 ha).

■ Conservation issues

Threats Agricultural intensification/expansion (C), Aquaculture/fisheries (C), Disturbance to birds (C), Extraction industry (A), Industrialization/urbanization (B), Recreation/tourism (C)

The main threat is from gravel-extraction activities. Bird-ringing has been carried out regularly since 1990 by Museo di Storia Naturale di Carmagnola.

Crava–Morozzo lakes

C7 **034**

Admin region Piemonte
Coordinates 44°26'N 7°50'E
Altitude 384–420 m **Area** 300 ha

■ Site description

An area of broadleaved woodland and wetland in north-west Italy, 10 km east of the town of Cuneo. Habitats include wet grasslands, freshwater ponds and cultivated areas. The main human activities are nature conservation, tourism and agriculture.

Habitats Forest and woodland (30%; broadleaved deciduous forest), Grassland (10%; humid grassland), Wetland (40%; standing fresh water; river/stream; water-fringe vegetation), Artificial landscape (20%)
Land-use Agriculture (20%), Nature conservation/research (100%), Tourism/recreation (50%)

■ Birds

Notable species include woodland-breeding passerines and migrating waterbirds.

■ Protection status

National High **International** None

300 ha of IBA covered by Regional Nature Park (Alta Valle Pesio, 3,955 ha).

■ Conservation issues

No major threats are present. The site is a candidate Special Protection Area and has been a LIPU reserve since the 1970s.

Maritime Alps

A3, B1iv, B2, C5, C6 **035**

Admin region Piemonte
Coordinates 44°18'N 7°10'E
Altitude 650–3,841 m **Area** 95,400 ha

■ Site description

A large tract of the south-west Italian Alps, from Punta Marguaréis to Monviso. The area is divided into three sites: (1) Alta Valle Pesio e

Tanaro (Nature Park), (2) Alpi Marittime (Nature Park) including Media e Alta Valle Stura and Alta Valle Gesso, and (3) Alta Valle Varaita. The main habitats are coniferous and broadleaved woodlands, glaciers and rocky areas, and alpine grasslands. Cattle-breeding and tourism are the main economic activities for local people.

Habitats Forest and woodland (45%; broadleaved deciduous forest; native coniferous forest), Scrub (15%), Grassland (19%; alpine/subalpine/boreal grassland), Wetland (1%), Rocky areas (10%), Artificial landscape (10%)
Land-use Agriculture (10%), Forestry (10%), Hunting (10%), Nature conservation/research (30%), Tourism/recreation (10%), Urban/industrial/transport (2%)

Birds

Species	Season	Year	Pop min	Pop max	Acc	Criteria
Circaetus gallicus Short-toed Eagle	B	1995	5	9	B	B2, C6
Circaetus gallicus Short-toed Eagle	P	1994	33	62	A	C6
Aquila chrysaetos Golden Eagle	R	1995	16	20	A	B2, C6
Falco peregrinus Peregrine	R	1995	7	10	A	C6
Lagopus mutus Ptarmigan	R	1995	100	—	B	C6
Tetrao tetrix Black Grouse	R	1995	230	410	B	B2, C6
Alectoris graeca Rock Partridge	B	1995	350	550	B	B2, C6
Bubo bubo Eagle Owl	B	1995	4	7	A	B2, C6
Aegolius funereus Tengmalm's Owl	R	1995	13	18	B	C6
Dryocopus martius Black Woodpecker	R	1995	20	—	C	C6
Lullula arborea Woodlark	R	1995	Frequent	—		C6
Prunella collaris Alpine Accentor	R	1995	Common	—		A3
Tichodroma muraria Wallcreeper	R	1995	Frequent	—		A3
Pyrrhocorax graculus Alpine Chough	R	1995	Abundant	—		A3
Pyrrhocorax pyrrhocorax Chough	R	1995	55	85	B	B2, C6
Montifringilla nivalis Snowfinch	R	1995	Frequent	—		A3
Serinus citrinella Citril Finch	R	1995	Frequent	—		A3
Emberiza hortulana Ortolan Bunting	B	1995	50	70	B	C6

The site supports good breeding numbers of some raptors and gamebirds, as well as five of the ten species of the Eurasian high-montane biome. The locality 'Media e Alta Valle Stura' is an important 'bottleneck' for migrating raptors (2,900–3,200 per autumn), mainly *Pernis apivorus* (2,000–3,000). Species of global conservation concern that do not meet IBA criteria: *Aquila clanga* (max. 1, passage).

Protection status

National Partial **International** None
1,061 ha of IBA covered by Regional Nature Reserve (Bosco e Laghi di Palanfre, 1,061 ha). 25,643 ha of IBA covered by Regional Nature Park (Alpi Marittime, 25,643 ha). 3,955 ha of IBA covered by Regional Nature Park (Alta Valle Pesio, 3,955 ha).

Conservation issues

Threats Abandonment/reduction of land management (B), Afforestation (C), Disturbance to birds (C), Firewood collection (C), Infrastructure (A), Recreation/tourism (C), Selective logging/cutting (C), Unsustainable exploitation (C)

Major threats are posed by a project to build a highway to France, and by the large and ongoing loss of traditionally cultivated areas and of alpine grasslands. Projects in the Parco Naturale delle Alpi Marittime include the ongoing reintroduction of *Gypaetus barbatus* (and in the nearby Parc National du Mercantour in France), management of ungulates and fish, and studies on migratory birds, on breeding raptors, on *Nucifraga caryocatactes* and on *Aegolius funereus*.

Mount Beigua — 036

B1iv, C5, C6

Admin region Liguria
Coordinates 44°25'N 8°40'E
Altitude 200–1,287 m **Area** 10,000 ha

Site description

Habitats Forest and woodland (25%; broadleaved deciduous forest; treeline ecotone), Scrub (20%; sclerophyllous scrub/garrigue/maquis), Grassland (15%; alpine/subalpine/boreal grassland), Rocky areas (10%; inland cliff), Artificial landscape (30%; highly improved reseeded grassland; arable land; forestry plantation; other urban/industrial areas)
Land-use Agriculture (17%), Forestry (10%), Nature conservation/research (70%), Tourism/recreation (30%)

A mountainous area in the northern Apennines, 10 km north-east of the town of Savona. The main habitats are broadleaved and coniferous forests, cultivated and ruderal lands, streams and, in the highest areas, alpine grasslands and rocky cliffs. The main land-uses are nature conservation, forest management, sheep-rearing, and tourism.

Birds

Species	Season	Year	Pop min	Pop max	Acc	Criteria
Caprimulgus europaeus Nightjar	B	1995	20	—	C	C6
Sylvia undata Dartford Warbler	R	1996	100	—	A	C6

An important 'bottleneck' site for birds migrating along the French coast towards the Padana plain in spring, especially for raptors (more than 3,000 per season, especially *Pernis apivorus*, *Circaetus gallicus* and *Circus aeruginosus*), pigeons/doves, swifts and passerines (swallows, etc.).

Protection status

National Partial **International** None
7,000 ha of IBA covered by Regional Nature Park (Monte Beigua, 17,500 ha).

Conservation issues

Threats Abandonment/reduction of land management (B), Afforestation (C), Burning of vegetation (A), Consequences of animal/plant introductions (C), Forest grazing (B), Recreation/tourism (C)

The main threat to the area is from fires caused by man, e.g. by shepherds to create 'fresh' grazing. Attempts are being made to prevent and stop woodland fires. Some or all of the site is covered by a management plan. A LIFE project has been funded for the period 1999–2000. There are ornithological studies on migratory species, especially raptors.

Finalese — 037

C6

Admin region Liguria
Coordinates 44°15'N 8°20'E
Altitude 0–300 m **Area** 3,750 ha

Site description

A coastal area, rich in small valleys and chalk cliffs (inland as well as coastal). Vegetation cover is mainly maquis and coniferous woodland. The main land-uses are agriculture and tourism.

Habitats Forest and woodland (27%; broadleaved deciduous forest; native coniferous forest; broadleaved evergreen forest), Scrub (45%; sclerophyllous scrub/garrigue/maquis), Rocky areas (8%; sea cliff/rocky shore; inland cliff), Artificial landscape (20%; highly improved reseeded grassland; perennial crops/orchards/groves; ruderal land)
Land-use Agriculture (15%), Nature conservation/research (100%), Tourism/recreation (15%)

Birds

Species	Season	Year	Pop min	Pop max	Acc	Criteria
Circaetus gallicus Short-toed Eagle	B	1995	1	3	A	C6
Falco peregrinus Peregrine	R	1995	3	3	A	C6
Bubo bubo Eagle Owl	R	1995	3	3	A	C6

A notable breeding site for raptors.

Protection status

National High **International** None
3,750 ha of IBA covered by Regional Nature Park (Finale, 3,750 ha).

Conservation issues

Threats Burning of vegetation (A), Disturbance to birds (B), Infrastructure (C), Recreation/tourism (C)

The main threats are the burning of vegetation, and disturbance to breeding raptors by free-climbers (very common in the area). A regional park was declared in the area but is not yet operative. LIPU has proposed that free-climbing in the area should be regulated.

Ligurian Alps

B2, C6 **038**

Admin region Liguria
Coordinates 44°00'N 7°40'E
Altitude 670–2,200 m **Area** 15,000 ha

Site description

A mountain area in the south-west corner of the Italian Alps, from Monte Alto to Monte Tanarello, at the border with France. The area includes both Mediterranean and alpine habitats, from broadleaved evergreen woodlands to alpine grasslands. Tourism and cattle-breeding are the most important human activities.

Habitats Forest and woodland (50%), Scrub (10%), Grassland (30%), Rocky areas (10%)
Land-use Agriculture (10%), Forestry (10%), Hunting (5%), Not utilized (60%), Tourism/recreation (15%)

Birds

Species	Season	Year	Pop min	Pop max	Acc	Criteria
Pernis apivorus Honey Buzzard	B	1986	Abundant		—	C6
Circaetus gallicus Short-toed Eagle	B	1996	5	—	A	B2, C6
Aquila chrysaetos Golden Eagle	R	1986	Frequent		—	C6
Tetrao tetrix Black Grouse	R	1986	Frequent		—	C6
Alectoris graeca Rock Partridge	R	1996	100	—	C	B2, C6
Bubo bubo Eagle Owl	R	1986	5	—	C	C6
Dryocopus martius Black Woodpecker	R	1986	5	—	C	C6
Lullula arborea Woodlark	R	1986	Frequent		—	C6
Pyrrhocorax pyrrhocorax Chough	R	1986	Common		—	C6

There are important breeding numbers of several raptors, gamebirds and species of forest or open country.

Protection status

National None **International** None

Conservation issues

No major threats are present. A Regional Park is proposed.

Alpi Apuane

B2, C6 **039**

Admin region Toscana
Coordinates 44°06'N 10°14'E
Altitude 800–1,947 m **Area** 16,800 ha

Site description

A mountain complex in Tuscany, east of the towns of Massa and Carrara. The area is particularly rich in rocky areas and alpine grasslands. The main land-uses are grazing, nature conservation, and marble extraction.

Habitats Forest and woodland (5%; broadleaved deciduous forest), Scrub (10%; scrub), Grassland (75%; steppe/dry calcareous grassland; dry siliceous grassland; alpine/subalpine/boreal grassland), Rocky areas (5%; inland cliff), Artificial landscape (5%)
Land-use Agriculture (65%), Nature conservation/research (100%), Urban/industrial/transport (5%)

Birds

Species	Season	Year	Pop min	Pop max	Acc	Criteria
Aquila chrysaetos Golden Eagle	R	1995	1	3	A	C6
Anthus campestris Tawny Pipit	B	1996	30	60	B	C6
Monticola saxatilis Rock Thrush	B	1996	40	60	B	B2
Pyrrhocorax pyrrhocorax Chough	R	1996	15	45	B	C6

Important breeding site for montane passerines.

Protection status

National High **International** High
IBA contains Wildlife Reserve (Campocatino, area not known). 16,800 ha of IBA covered by Regional Nature Park (Alpi Apuane, 60,000 ha). 16,800 ha of IBA covered by Special Protection Area (Praterie primarie e secondarie delle Apuane, 17,414 ha).

Conservation issues

Threats Abandonment/reduction of land management (B), Afforestation (C), Extraction industry (A), Infrastructure (C), Natural events (C), Recreation/tourism (C)

The main threat is from marble quarrying. Some or all of the site is covered by a management plan.

Apennines from Passo del Cerreto to Monte Caligi

B2, B3, C6 **040**

Admin region Toscana, Emilia Romagna
Coordinates 44°12'N 10°27'E
Altitude 450–2,050 m **Area** 45,000 ha

Site description

A large, mountainous area in the northern Apennines, including broadleaved (mainly beech *Fagus*) and coniferous woodlands, alpine grasslands and rocky areas, crossed by rivers and streams. The main land-uses are nature conservation, forestry, cattle-rearing and hunting.

Habitats Forest and woodland (60%; broadleaved deciduous forest; native coniferous forest), Scrub (7%), Grassland (20%; alpine/subalpine/boreal grassland), Wetland (1%; river/stream), Rocky areas (2%), Artificial landscape (10%)
Land-use Agriculture (5%), Forestry (20%), Hunting (60%), Nature conservation/research (50%), Urban/industrial/transport (5%)

Birds

Species	Season	Year	Pop min	Pop max	Acc	Criteria
Pernis apivorus Honey Buzzard	B	1996	2	10	C	C6
Aquila chrysaetos Golden Eagle	R	1995	2	3	A	C6
Strix aluco Tawny Owl	R	1995	500	—	C	B3
Caprimulgus europaeus Nightjar	B	1993	50	—	C	B2
Picus viridis Green Woodpecker	R	1995	100	—	C	B2
Monticola saxatilis Rock Thrush	B	1995	40	60	B	B2
Parus caeruleus Blue Tit	R	1995	1,000	—	C	B3

Important for breeding species of forest and mountain.

Protection status

National Partial **International** Low
98 ha of IBA covered by State Nature Reserve (Campolino, 98 ha). 167 ha of IBA covered by State Nature Reserve (Lamarossa, 167 ha). 192 ha of IBA covered by State Nature Reserve (Orrido di Botri, 192 ha). 135 ha of IBA covered by State Nature Reserve (Pania di Corfino, 135 ha). 590 ha of IBA covered by State Nature Reserve (Pian degli Ontani, 590 ha). 217 ha of IBA covered by Nature Reserve (Orecchiella, 217 ha). Part of IBA covered by Regional Nature Park (Alto Appennino Modenese, 14,844 ha). Part of IBA covered by Regional Nature Park (Alto Appennino Reggiano, 16,981 ha). 584 ha of IBA covered by Biogenetic Reserve (Abetone, 584 ha). 584 ha of IBA covered by Special Protection Area (Abetone, 584 ha). 98 ha of IBA covered by Special Protection Area (Campolino, 98 ha). 192 ha of IBA covered by Special Protection Area (Orrido di Botri, 192 ha). 135 ha of IBA covered by Special Protection Area (Pania di Corfino, 135 ha). 590 ha of IBA covered by Special Protection Area (Pian degli Ontani, 590 ha).

Conservation issues

Threats Abandonment/reduction of land management (A), Afforestation (B), Infrastructure (C), Recreation/tourism (C), Unsustainable exploitation (C)

The most important threat is the abandonment of traditional cultivation and cattle-rearing activities, with consequent loss of pastures and grasslands. A National Park is proposed in the area.

Stelvio National Park

A3, B2, C6 **041**

Admin region Lombardia, Trentino-Alto Adige
Coordinates 46°30'N 10°32'E
Altitude 639–3,905 m **Area** 134,621 ha

Site description

A large area of the central-southern Alps, with all the typical mountain habitats, and where coniferous woodlands, alpine grasslands and rocky areas predominate. The main land-use is nature conservation (National Park). The main human activities are tourism, nature conservation, forestry and cattle-rearing.

Habitats Forest and woodland (20%; broadleaved deciduous forest; native coniferous forest; mixed forest), Scrub (10%; scrub), Grassland (30%; alpine/subalpine/boreal grassland), Wetland (river/stream; fen/transition mire/spring), Rocky areas (35%; inland cliff), Artificial landscape (5%; highly improved reseeded grassland; forestry plantation)
Land-use Agriculture (20%), Forestry (20%), Nature conservation/research (100%), Tourism/recreation (50%), Water management (5%)

Birds

Species	Season	Year	Pop min	Pop max	Acc	Criteria
Aquila chrysaetos Golden Eagle	R	1996	20	22	A	B2, C6
Falco peregrinus Peregrine	R	1989	10	—	B	C6
Bonasa bonasia Hazel Grouse	R	1989	170	—	B	C6
Tetrao tetrix Black Grouse	R	1989	300	—	B	B2, C6
Tetrao urogallus Capercaillie	R	1989	72	—	A	C6
Alectoris graeca Rock Partridge	B	1989	280	—	B	B2, C6
Picus canus Grey-headed Woodpecker	R	1996	50	—	C	C6
Picoides tridactylus Three-toed Woodpecker	R	1996	10	—	C	C6
Prunella collaris Alpine Accentor	R	1996	100	—	C	A3
Tichodroma muraria Wallcreeper	R	1996	20	—	C	A3
Pyrrhocorax graculus Alpine Chough	R	1996	100	—	C	A3
Montifringilla nivalis Snowfinch	R	1996	100	—	C	A3
Serinus citrinella Citril Finch	R	1996	50	—	C	A3

The area is important for gamebirds, raptors and five out of the 10 species of the Eurasian high-montane biome.

Protection status

National High **International** Partial
134,621 ha of IBA covered by National Park (Stelvio, 148,271 ha). 62,824 ha of IBA covered by Special Protection Area (Parco Nazionale dello Stelvio (the part in Lombardia), 62,824 ha).

Conservation issues

Threats Abandonment/reduction of land management (B), Construction/impact of dyke/dam/barrage (B), Disturbance to birds (C), Infrastructure (B), Intensified forest management (U), Other (A), Recreation/tourism (C), Unsustainable exploitation (B)

The most important threats are posed by the abandonment of mountain pastures, the building of tourist resorts, intensified forest management, and the poor management of the National Park, due to the lack of funds and a management plan, and its division into three parts managed by three different public bodies ('Other' threat).

Dolomites

A3, C6 **042**

Admin region Trentino-Alto Adige, Veneto
Coordinates 46°40'N 12°05'E
Altitude 1,268–3,225 m **Area** 48,507 ha

Site description

Various groups of mountains in the Alps, with extensive scree and boulders, cliffs, alpine grasslands, scrub and coniferous woodlands.

Habitats Forest and woodland (25%; native coniferous forest; treeline ecotone), Scrub (5%; scrub), Grassland (25%; alpine/subalpine/boreal grassland), Rocky areas (40%; scree/boulders; inland cliff)
Land-use Agriculture (15%), Forestry (25%), Hunting (77%), Nature conservation/research (100%), Tourism/recreation (25%)

Birds

Species	Season	Year	Pop min	Pop max	Acc	Criteria
Tetrao tetrix Black Grouse	R	1995	50	—	B	C6
Tetrao urogallus Capercaillie	R	1995	15	—	B	C6
Picus canus Grey-headed Woodpecker	R	1995	Frequent		—	C6
Picoides tridactylus Three-toed Woodpecker	R	1995	5	—	B	C6
Prunella collaris Alpine Accentor	R	1995	15	—	B	A3
Tichodroma muraria Wallcreeper	R	1995	10	—	B	A3
Pyrrhocorax graculus Alpine Chough	R	1995	100	—	B	A3
Montifringilla nivalis Snowfinch	R	1995	20	—	B	A3
Serinus citrinella Citril Finch	U	1995	Rare		—	A3

An important breeding site for at least four of the 10 species of the Eurasian high-montane biome, and for grouse and woodpeckers.

Protection status

National High **International** None
11,192 ha of IBA covered by Regional Nature Park (Dolomiti d'Ampezzo, 11,192 ha). 11,635 ha of IBA covered by Regional Nature Park (Dolomiti di Sesto, 11,635 ha). 25,680 ha of IBA covered by Regional Nature Park (Fanes Sennes Braies, 25,680 ha).

Conservation issues

Threats Deforestation (commercial) (C), Disturbance to birds (B), Infrastructure (C), Intensified forest management (C), Recreation/tourism (B), Selective logging/cutting (C), Unsustainable exploitation (A)

The main threats are from tourism and hunting of grouse. The latter practice is unsustainable, especially for *Lagopus mutus helveticus*.

Carniche Alps

B2, C6 **043**

Admin region Veneto, Friuli-Venezia Giulia
Coordinates 46°35'N 12°50'E
Altitude 553–2,780 m **Area** 30,000 ha

Site description

A mountain area in the south-east Alps, bordering Austria to the north. The site includes extensive coniferous woodlands of spruce *Picea abies*, alpine grasslands, streams and rocky areas. Forestry and hunting are the main land-uses.

Habitats Forest and woodland (50%; broadleaved deciduous forest; native coniferous forest; mixed forest), Grassland (40%; steppe/dry calcareous grassland; dry siliceous grassland; alpine/subalpine/boreal grassland), Wetland (river/stream), Rocky areas (10%; scree/boulders; inland cliff; caves)
Land-use Agriculture (6%), Forestry (50%), Hunting (75%), Nature conservation/research (15%), Not utilized (15%), Tourism/recreation (30%)

Birds

Species	Season	Year	Pop min	Pop max	Acc	Criteria
Aquila chrysaetos Golden Eagle	B	1995	3	—	B	C6
Lagopus mutus Ptarmigan	R	1995	20	—	B	C6
Tetrao tetrix Black Grouse	R	1995	20	—	C	C6
Tetrao urogallus Capercaillie	R	1995	10	—	C	C6
Glaucidium passerinum Pygmy Owl	R	1995	10	—	C	C6
Aegolius funereus Tengmalm's Owl	R	1995	20	—	C	C6
Picus canus Grey-headed Woodpecker	R	1996	5	—	C	B2, C6
Picoides tridactylus Three-toed Woodpecker	R	1995	3	—	C	C6

An important breeding site for forest/mountain species, especially gamebirds, owls, raptors and woodpeckers.

Protection status

National Low **International** None
2,000 ha of IBA covered by Wildlife Reserve (Bordaglia, 2,000 ha).

Conservation issues

Threats Abandonment/reduction of land management (B), Afforestation (C), Disturbance to birds (B), Firewood collection (C), Forest grazing (C), Intensified forest management (B), Recreation/tourism (B), Selective logging/cutting (U), Unsustainable exploitation (B)

There are no major threats. There is research on *Tetrao tetrix* by Osservatorio Faunistico di Udine.

Tarvisio forest

B2, B3, C6 **044**

Admin region Friuli-Venezia Giulia
Coordinates 46°30'N 13°30'E
Altitude 580–2,753 m **Area** 23,300 ha

Site description

One of the largest forests in the Alps, relatively intact and undisturbed. The area is comprised mainly of native coniferous woodland of spruce *Picea abies* and mixed woodland, with some alpine pastures and grasslands. Cattle-grazing and forestry are the main human activities.

Habitats Forest and woodland (62%; broadleaved deciduous forest; native coniferous forest; mixed forest; treeline ecotone), Scrub (3%), Grassland (15%; steppe/dry calcareous grassland; dry siliceous grassland; alpine/subalpine/boreal grassland), Wetland (5%; river/stream; fen/transition mire/spring), Rocky areas (15%; inland cliff; caves)
Land-use Agriculture (10%), Forestry (65%), Nature conservation/research (100%), Not utilized (15%), Tourism/recreation (10%)

■ Birds

Species	Season	Year	Pop min	Pop max	Acc	Criteria
[1] ***Gyps fulvus*** Griffon Vulture	N	1995	20	30	B	C6
Aquila chrysaetos Golden Eagle	R	1995	3	4	A	C6
Bonasa bonasia Hazel Grouse	R	1995	120	200	A	C6
Lagopus mutus Ptarmigan	R	1995	40	50	A	C6
Tetrao tetrix Black Grouse	R	1995	65	70	A	C6
Tetrao urogallus Capercaillie	R	1995	60	70	—	C6
Picus canus Grey-headed Woodpecker	R	1995	20	25	A	B2, C6
Dryocopus martius Black Woodpecker	R	1995	20	30	A	C6
Picoides tridactylus Three-toed Woodpecker	R	1995	20	40	B	C6
Turdus torquatus Ring Ouzel	B	1995	Common		—	B3
Parus cristatus Crested Tit	R	1995	Common		—	B3

1. Oversummering birds from Austrian Alps (and Croatia?).

An important site for forest/mountain species of grouse, raptor and woodpecker. Counts refer mostly to 1995. Breeding species of global conservation concern that does not meet IBA criteria: *Crex crex* (2–4 pairs).

■ Protection status

National Low **International** None
21 ha of IBA covered by State Nature Reserve (Cucco, 21 ha). 378 ha of IBA covered by State Nature Reserve (Rio Bianco, 378 ha).

■ Conservation issues

Threats Abandonment/reduction of land management (A), Afforestation (C), Firewood collection (C), Forest grazing (C), Industrialization/urbanization (B), Infrastructure (A), Natural events (A), Other (C), Recreation/tourism (B), Selective logging/cutting (C), Unsustainable exploitation (B)

Industrial/urban development and infrastructure, and hunting, are threats in the zone around the IBA, but there are no major threats within the site. The whole area is protected and managed by the National Forestry Department. Some or all of the site is covered by a management plan. Research has been carried out since the early 1980s on birds (mainly gamebirds), mammals (ungulates, *Ursus arctos*), reptiles and invertebrates.

Adamello–Brenta A3, B2, C6 045

Admin region Lombardia, Trentino-Alto Adige
Coordinates 46°08′N 10°46′E
Altitude 280–3,556 m **Area** 112,800 ha

■ Site description

A large, mountainous area in the central-southern Alps. Half of the area is covered by woodlands, mainly spruce *Picea abies*, while higher areas are dominated by alpine grasslands, bushes (*Pinus*, *Rhododendron*, *Alnus*, *Juniperus*), rocky areas and glaciers. Cattle-rearing, forestry and tourism are the dominant land-uses.

Habitats Forest and woodland (50%; broadleaved deciduous forest; native coniferous forest), Scrub (10%; scrub), Grassland (25%; alpine/subalpine/boreal grassland), Rocky areas (10%; inland cliff), Artificial landscape (5%)
Land-use Agriculture (20%), Forestry (50%), Hunting (90%), Nature conservation/research (55%), Tourism/recreation (20%)

■ Birds

Species	Season	Year	Pop min	Pop max	Acc	Criteria
Aquila chrysaetos Golden Eagle	R	1996	8	11	B	B2, C6
Bonasa bonasia Hazel Grouse	R	1996	50	—	B	C6
[1] ***Lagopus mutus*** Ptarmigan	R	1996	50	—	B	C6
Tetrao tetrix Black Grouse	R	1996	100	—	B	B2, C6
Alectoris graeca Rock Partridge	R	1996	100	—	B	B2
Glaucidium passerinum Pygmy Owl	R	1996	10	—	C	C6
Aegolius funereus Tengmalm's Owl	R	1996	15	—	C	C6
Dryocopus martius Black Woodpecker	R	1995	40	—	B	C6
Prunella collaris Alpine Accentor	R	1996	300	—	B	A3
Tichodroma muraria Wallcreeper	R	1996	20	—	C	A3
Pyrrhocorax graculus Alpine Chough	R	1996	200	—	B	A3
Montifringilla nivalis Snowfinch	R	1996	100	—	C	A3
Serinus citrinella Citril Finch	R	1996	Uncommon		—	A3

1. *L. m. helveticus*.

The area is important for gamebirds, raptors and five out of the 10 species of the Eurasian high-montane biome.

■ Protection status

National High **International** Low
48,000 ha of IBA covered by Regional Nature Park (Adamello, 48,000 ha). 61,800 ha of IBA covered by Regional Nature Park (Adamello Brenta, 61,864 ha). 37 ha of IBA covered by Special Protection Area (Lago di Tovel, 37 ha). 37 ha of IBA covered by Ramsar Site (Lago di Tovel, 37 ha).

■ Conservation issues

Threats Abandonment/reduction of land management (C), Infrastructure (B), Recreation/tourism (C)

The building of new infrastructure, especially for tourism and forest exploitation (roads), is the main threat. A project to reintroduce *Gypaetus barbatus* is ongoing, as are studies on *Tetrao urogallus*.

Lagorai mountains B2, C6 046

Admin region Trentino-Alto Adige
Coordinates 46°11′N 11°36′E
Altitude 1,000–2,754 m **Area** 40,000 ha

■ Site description

A mountain chain in the central-southern Alps. One of the largest areas of wilderness in the Alps, with almost no human settlement. The alpine habitats are relatively intact and undisturbed by man, ranging from coniferous woodlands (*Picea*, *Larix*, *Pinus*) to grasslands, rocky areas, and many small lakes. Cattle-rearing and forestry are important human activities.

Habitats Forest and woodland (broadleaved deciduous forest; native coniferous forest; mixed forest; treeline ecotone), Scrub (scrub), Grassland (alpine/subalpine/boreal grassland), Wetland (river/stream; fen/transition mire/spring), Rocky areas (inland cliff)
Land-use Agriculture, Forestry, Hunting, Tourism/recreation

■ Birds

Species	Season	Year	Pop min	Pop max	Acc	Criteria
Alectoris graeca Rock Partridge	R	—	—	—	—	B2
Picus canus Grey-headed Woodpecker	R	1996	5	—	C	B2, C6

■ Protection status

National Partial **International** Unknown
10,000 ha of IBA covered by Regional Nature Park (Paneveggio-Pale di San Martino, 19,097 ha). Overlaps with SPA.

■ Conservation issues

Threats Abandonment/reduction of land management (C), Infrastructure (C), Recreation/tourism (C)

No major threats are present.

Carniche pre-Alps A1, B2, C1, C6 047

Admin region Veneto, Friuli-Venezia Giulia
Coordinates 46°16′N 12°33′E
Altitude 313–2,479 m **Area** 63,000 ha

■ Site description

A mountainous area of the pre-Alps in north-east Italy, characterized by extensive pastures, streams, rivers, and coniferous and broadleaved woodlands. Agriculture is the main land-use.

Habitats Forest and woodland (broadleaved deciduous forest; native coniferous forest; mixed forest), Scrub (scrub), Grassland (alpine/subalpine/boreal grassland; humid grassland), Rocky areas (inland cliff)
Land-use Agriculture, Forestry, Hunting, Urban/industrial/transport

Birds

Species	Season	Year	Pop min	Pop max	Acc	Criteria
Crex crex Corncrake	B	1995	21	—	A	A1, C1, C6
Picus canus Grey-headed Woodpecker	R	1988	5	—	C	B2, C6

An important breeding area for *Crex crex*.

Protection status

National None **International** Unknown
Overlaps with SPA.

Conservation issues

Threats Abandonment/reduction of land management (C), Disturbance to birds (U), Industrialization/urbanization (C), Infrastructure (U), Intensified forest management (U), Natural events (C), Recreation/tourism (U), Unsustainable exploitation (C)

The main threat is the abandonment of traditional land-use on grasslands.

Middle course of Tagliamento river

C6 **048**

Admin region Friuli-Venezia Giulia
Coordinates 46°13′N 13°01′E
Altitude 158–1,478 m **Area** 3,579 ha

Site description

The middle course of Tagliamento river, in north-eastern Italy, is a pre-Alpine area with various dry and wet habitats (steppe grasslands, boulders, broadleaved and wet woodlands, springs, rocky cliffs). The main land-uses are hunting and recreational activities.

Habitats Forest and woodland (52%; broadleaved deciduous forest; native coniferous forest; alluvial/very wet forest), Grassland (30%; steppe/dry calcareous grassland; humid grassland), Wetland (13%; shingle/stony beach; standing fresh water; river/stream; water-fringe vegetation; fen/transition mire/spring), Rocky areas (5%; scree/boulders; inland cliff)
Land-use Hunting (50%), Tourism/recreation (19%)

Birds

Species	Season	Year	Pop min	Pop max	Acc	Criteria
Milvus migrans Black Kite	B	1996	3	5	B	C6
[1] ***Bonasa bonasia*** Hazel Grouse	R	1996	40	40	A	C6
Burhinus oedicnemus Stone Curlew	B	1996	2	10	B	C6
Bubo bubo Eagle Owl	R	1996	2	4	A	C6
Anthus campestris Tawny Pipit	B	1996	10	50	B	C6
Lanius minor Lesser Grey Shrike	B	1996	10	30	B	C6
Emberiza hortulana Ortolan Bunting	B	1996	Uncommon		—	C6

1. Individuals.

An important breeding site for birds of dry or open habitats, and a foraging area for large raptors. Breeding species of global conservation concern that do not meet IBA criteria: *Crex crex* (2–4 pairs).

Protection status

National Partial **International** None
487 ha of IBA covered by Regional Nature Reserve (Lago di Cornino, 487 ha).

Conservation issues

Threats Burning of vegetation (C), Infrastructure (C), Unsustainable exploitation (C)

No major threats are present. The area is a proposed Regional Park. A feeding station for large raptors, especially *Gyps fulvus*, is maintained in the area.

Cividalese and upper Torre valley

A1, B3, C1, C2, C6 **049**

Admin region Friuli-Venezia Giulia
Coordinates 46°10′N 13°26′E
Altitude 180–1,641 m **Area** 28,400 ha

Site description

An area of hills and mountains in north-eastern Italy, with large areas of agriculture and broadleaved woodland. Agriculture and forestry are the main land-uses.

Habitats Forest and woodland (broadleaved deciduous forest), Scrub (scrub), Grassland (alpine/subalpine/boreal grassland; humid grassland), Wetland (river/stream), Artificial landscape (highly improved reseeded grassland; perennial crops/orchards/groves; forestry plantation; ruderal land)
Land-use Agriculture, Forestry, Hunting, Urban/industrial/transport

Birds

Species	Season	Year	Pop min	Pop max	Acc	Criteria
Crex crex Corncrake	B	1995	67	—	A	A1, C1, C2, C6
Picus canus Grey-headed Woodpecker	R	1990	10	—	C	C6
Turdus pilaris Fieldfare	W	1990	1,000	—	C	B3
Turdus iliacus Redwing	W	1990	1,000	—	C	B3

An important breeding area for *Crex crex*. The only breeding site in Italy for *Strix uralensis*.

Protection status

National None **International** None

Conservation issues

Threats Abandonment/reduction of land management (A), Disturbance to birds (U), Natural events (C), Other (C), Recreation/tourism (U)

Abandonment of land management is the main threat to *Crex crex*.

Belluno Dolomites

A1, B2, C1, C6 **050**

Admin region Veneto
Coordinates 46°10′N 12°02′E
Altitude 300–2,600 m **Area** 31,512 ha

Site description

A large mountain area in the south-eastern Alps, rich in pastures, rocky cliffs, gorges, narrow valleys and coniferous woodlands. The main land-use is nature conservation.

Habitats Forest and woodland (50%), Grassland (15%), Wetland (5%), Rocky areas (15%), Artificial landscape (15%)
Land-use Agriculture (10%), Forestry (30%), Nature conservation/research (100%), Tourism/recreation (20%), Water management (5%)

Birds

Species	Season	Year	Pop min	Pop max	Acc	Criteria
Aquila chrysaetos Golden Eagle	R	1995	3	5	A	B2, C6
Falco peregrinus Peregrine	R	1995	3	5	A	C6
[1] ***Lagopus mutus*** Ptarmigan	R	1995	20	40	B	C6
Tetrao tetrix Black Grouse	R	1996	20	40	A	C6
Tetrao urogallus Capercaillie	R	—	10	20	—	C6
Crex crex Corncrake	B	1995	10	20	B	A1, C1, C6
Picus canus Grey-headed Woodpecker	R	1995	10	20	B	B2, C6
Dryocopus martius Black Woodpecker	R	1995	30	50	B	C6

1. *L. m. helveticus.*

Protection status

National High **International** Partial
31,000 ha of IBA covered by National Park (Dolomiti Bellunesi, 31,000 ha). 17,455 ha of IBA covered by Special Protection Area (Riserve Naturali delle Dolomiti Bellunesi, 17,455 ha).

Conservation issues

Threats Recreation/tourism (C)

There are no major threats.

Lago di Busche and Vincheto di Cellarda

C7 **051**

Admin region Veneto
Coordinates 46°03′N 11°59′E
Altitude 230–240 m **Area** 125 ha

Site description
A wetland complex of the Piave river, located 15 km south-west of the town of Belluno. Habitats include an artificial lake (Lago di Busche), a wet woodland (Vincheto di Cellarda) of willow *Salix* and alder *Alnus*, a peatland and a wet grassland. The main land-uses are nature conservation, recreation and water management.

Habitats Forest and woodland (10%; alluvial/very wet forest), Scrub (scrub), Wetland (80%; standing fresh water; river/stream; water-fringe vegetation; fen/transition mire/ spring)
Land-use Nature conservation/research (80%), Tourism/recreation (30%), Water management (70%)

Birds
An important stop-over site for waterbirds migrating along the valley of Piave river (crossing the Alps).

Protection status
National Partial **International** Partial
92 ha of IBA covered by State Nature Reserve (Vincheto di Cellarda, 92 ha). 92 ha of IBA covered by Nature Reserve (Vincheto di Cellarda, 92 ha). 99 ha of IBA covered by Special Protection Area (Vincheto di Cellarda, 99 ha). 99 ha of IBA covered by Ramsar Site (Vincheto di Cellarda, 99 ha).

Conservation issues
Threats Disturbance to birds (C)

No major threats are present. Wintering waterbirds are monitored regularly.

Giulie pre-Alps

C6 **052**

Admin region Friuli-Venezia Giulia
Coordinates 46°02′N 12°43′E
Altitude 360–2,587 m **Area** 39,400 ha

Site description
A pre-Alpine area in north-eastern Italy, with mountains and hills dominated by broadleaved woodland (beech *Fagus*) and scrub (*Pinus mugo*), with some alpine grasslands. The current area incorporates part of a site that was listed in the previous pan-European IBA inventory (Grimmett and Jones 1989), 'Area between Gemona, Valle Resia, Monte Canin and Foresta del Tarvisio' (part of former site IT037).

Habitats Forest and woodland (60%; broadleaved deciduous forest; native coniferous forest; mixed forest; treeline ecotone), Scrub (10%; scrub), Grassland (10%; steppe/dry calcareous grassland; dry siliceous grassland; alpine/subalpine/boreal grassland), Wetland (5%; river/stream; water-fringe vegetation), Rocky areas (5%; scree/boulders; inland cliff), Artificial landscape (10%; highly improved reseeded grassland; arable land; perennial crops/orchards/groves; other urban/industrial areas)
Land-use Agriculture (10%), Forestry (50%), Hunting (25%), Nature conservation/ research (22%), Tourism/recreation (20%), Urban/industrial/transport (5%)

Birds

Species	Season	Year	Pop min	Pop max	Acc	Criteria
Lagopus mutus Ptarmigan	R	1990	Uncommon		—	C6
Alectoris graeca Rock Partridge	R	1990	Uncommon		—	C6
Crex crex Corncrake	B	1995	16	16	A	C6

The site is particularly important for *Crex crex*, which breeds in the grasslands, and for gamebirds.

Protection status
National Partial **International** None
9,402 ha of IBA covered by Regional Nature Park (Prealpi Giulie, 9,402 ha).

Conservation issues
Threats Abandonment/reduction of land management (A), Infrastructure (C), Intensified forest management (C), Recreation/tourism (C), Unsustainable exploitation (A)

The main threats are the abandonment of grassland management, which allows shrubs and trees to invade, and illegal hunting of gamebirds.

Magredi of Pordenone

B2, C6 **053**

Admin region Friuli-Venezia Giulia
Coordinates 46°01′N 12°44′E
Altitude 69–273 m **Area** 6,814 ha

Site description
A lowland, almost uninhabited area in north-east Italy, north-east of the town of Pordenone. It includes the localities Magredi di S. Foca, Magredi del Cellina and Magredi di Vivaro. It is a very dry and gravelly site, dominated by pseudo-steppe vegetation, and used mainly for military purposes.

Habitats Forest and woodland, Grassland (steppe/dry calcareous grassland), Wetland (shingle/stony beach; river/stream; water-fringe vegetation), Artificial landscape
Land-use Agriculture, Hunting, Military, Tourism/recreation

Birds

Species	Season	Year	Pop min	Pop max	Acc	Criteria
Circus pygargus Montagu's Harrier	B	1996	3	5	A	C6
Burhinus oedicnemus Stone Curlew	B	1996	5	15	A	C6
Caprimulgus europaeus Nightjar	B	1997	50	—	B	B2, C6
Calandrella brachydactyla Short-toed Lark	B	1996	10	20	B	C6
Anthus campestris Tawny Pipit	B	1997	150	—	B	B2, C6
Lanius collurio Red-backed Shrike	B	1996	20	50	A	C6
Emberiza hortulana Ortolan Bunting	B	1996	50	100	A	C6

An area of notable importance for breeding birds of open country, and for migrating and wintering raptors.

Protection status
National None **International** None

Conservation issues
Threats Agricultural intensification/expansion (A), Disturbance to birds (U), Dredging/ canalization (U), Extraction industry (U), Industrialization/urbanization (U), Infrastructure (U), Recreation/tourism (U)

Agricultural intensification is the main threat.

Mount Grappa

A1, B1iv, B2, C1, C5, C6 **054**

Admin region Veneto
Coordinates 45°53′N 11°48′E
Altitude 200–1,775 m **Area** 40,000 ha

Site description
A large pre-Alpine area, part of the Prealpi Venete, north of the town of Bassano del Grappa. Large areas of broadleaved woodland and pasture dominate the site.

Habitats Forest and woodland (45%), Scrub (5%), Grassland (30%), Wetland (5%), Rocky areas (10%), Artificial landscape (5%)
Land-use Agriculture (5%), Forestry (5%), Hunting (80%), Tourism/recreation (10%), Urban/industrial/transport (10%)

Birds

Species	Season	Year	Pop min	Pop max	Acc	Criteria
Bonasa bonasia Hazel Grouse	R	1995	5	20	B	C6
Tetrao tetrix Black Grouse	R	1995	30	50	A	C6
Tetrao urogallus Capercaillie	R	1995	6	10	B	C6
Alectoris graeca Rock Partridge	R	1995	50	75	A	C6
Crex crex Corncrake	B	1995	22	—	A	A1, C1, C6

Species ... continued	Season	Year	Pop min	Pop max	Acc	Criteria
Bubo bubo Eagle Owl	R	1995	3	5	A	B2, C6
Glaucidium passerinum Pygmy Owl	R	1995	3	10	B	C6
Caprimulgus europaeus Nightjar	B	1995	50	—	B	B2
Dryocopus martius Black Woodpecker	R	1995	15	25	B	C6

The site is important for breeding *Crex crex*, and is also a 'bottleneck' site for migrating raptors in spring, especially *Pernis apivorus* (3,000–3,500 per season).

Protection status

National None **International** None

Conservation issues

Threats Abandonment/reduction of land management (A), Afforestation (C), Disturbance to birds (A), Infrastructure (B), Recreation/tourism (B), Unsustainable exploitation (A)

The abandonment of pastures, with consequent natural reforestation, and the early cutting of grass are the main threats for *Crex crex*.

Piave river — B1iv, B2, C5, C6 — 055

Admin region Veneto
Coordinates 45°50'N 12°08'E
Altitude 10–496 m **Area** 10,000 ha

Site description

The middle course of the Piave river, north of the town of Treviso, including the surrounding wooded hills (Colli Maser). Dominant riparian habitats are wet grasslands and woodlands (*Alnus*, *Salix*, *Populus*), alluvial flats of gravel, sand and mud, and also steppic areas.

Habitats Forest and woodland (30%; alluvial/very wet forest), Scrub (10%), Grassland (10%; steppe/dry calcareous grassland; humid grassland), Wetland (40%; river/stream), Artificial landscape (10%)
Land-use Agriculture (10%), Fisheries/aquaculture (20%), Forestry (5%), Hunting (95%), Tourism/recreation (20%), Urban/industrial/transport (5%)

Birds

Species	Season	Year	Pop min	Pop max	Acc	Criteria
Caprimulgus europaeus Nightjar	B	1995	70	100	A	B2, C6

There is an important flyway of migrating raptors along Colli Maser, mainly *Pernis apivorus* (3,000–5,000 per season). Breeding species of global conservation concern that do not meet IBA criteria: *Crex crex* (6 pairs). A small heronry is present.

Protection status

National None **International** None

Conservation issues

Threats Agricultural intensification/expansion (B), Aquaculture/fisheries (B), Consequences of animal/plant introductions (C), Construction/impact of dyke/dam/barrage (B), Disturbance to birds (A), Extraction industry (A), Unsustainable exploitation (A)

The most important threat is the excavation of new gravel-pits. The area is a proposed Regional Park.

Assa valley — A1, C1, C6 — 056

Admin region Veneto
Coordinates 45°51'N 11°24'E
Altitude 300–1,100 m **Area** 2,000 ha

Site description

An Alpine valley that is surrounded by a high plain, on the west side of the Altopiano di Asiago. The area is covered mainly by cultivated grassland.

Habitats Forest and woodland (45%; native coniferous forest; mixed forest), Grassland (40%; mesophile grassland), Rocky areas (15%; inland cliff), Artificial landscape (arable land)
Land-use Agriculture (40%), Forestry (20%), Hunting (40%), Urban/industrial/transport (5%)

Birds

Species	Season	Year	Pop min	Pop max	Acc	Criteria
Crex crex Corncrake	B	1996	15	30	A	A1, C1, C6
Sylvia nisoria Barred Warbler	B	1996	5	10	B	C6

Important site for breeding *Crex crex*.

Protection status

National None **International** None

Conservation issues

Threats Abandonment/reduction of land management (A), Disturbance to birds (B)

The main threat for *Crex crex* is the abandonment of grassland management. There is a yearly census of *Crex crex* by a local ornithological group (NISORIA).

Lessini mountains — C6 — 057

Admin region Trentino-Alto Adige, Veneto
Coordinates 45°44'N 11°10'E
Altitude 658–2,259 m **Area** 39,000 ha

Site description

A mountain area located north of the town of Verona. Habitats include broadleaved, coniferous and mixed forests, rocky areas and alpine grasslands. Cattle-rearing and forestry are the main economic activities.

Habitats Forest and woodland (45%; broadleaved deciduous forest; native coniferous forest; mixed forest; treeline ecotone), Scrub (15%; scrub), Grassland (10%; alpine/subalpine/boreal grassland), Wetland (1%; river/stream), Rocky areas (4%; inland cliff), Artificial landscape (25%; arable land; forestry plantation; urban parks/gardens; other urban/industrial areas)
Land-use Agriculture (20%), Forestry (50%), Hunting (80%), Tourism/recreation (60%), Urban/industrial/transport (5%)

Birds

Species	Season	Year	Pop min	Pop max	Acc	Criteria
Aquila chrysaetos Golden Eagle	R	1996	2	4	A	C6
Falco peregrinus Peregrine	R	1996	4	6	A	C6

An important site for two raptors.

Protection status

National Partial **International** Unknown
10,368 ha of IBA covered by Regional Nature Park (Lessinia, 10,368 ha). Overlaps with SPA.

Conservation issues

Threats Afforestation (B), Infrastructure (C), Unsustainable exploitation (B)

Hunting is an important threat.

Alto Garda Bresciano — C6 — 058

Admin region Lombardia
Coordinates 45°39'N 10°28'E
Altitude 65–1,977 m **Area** 38,269 ha

Site description

A pre-Alpine mountain area on the west side of Lago di Garda, north of the town of Salò. The site is rich in broadleaved woodlands, grasslands, gorges and rocky cliffs. It is almost uninhabited and mainly used for nature conservation.

Habitats Forest and woodland (50%; broadleaved deciduous forest; native coniferous forest; mixed forest; broadleaved evergreen forest; treeline ecotone), Scrub (5%), Grassland (20%; alpine/subalpine/boreal grassland), Wetland (5%; standing fresh water; river/stream), Rocky areas (10%; inland cliff), Artificial landscape (10%; arable land; perennial crops/orchards/groves; forestry plantation; ruderal land)
Land-use Agriculture (10%), Forestry (20%), Nature conservation/research (100%), Tourism/recreation (10%)

Birds

Species	Season	Year	Pop min	Pop max	Acc	Criteria
Circaetus gallicus Short-toed Eagle	B	1996	2	3	A	C6
Alectoris graeca Rock Partridge	R	1996	50	—	A	C6

A notable breeding site for raptors and gamebirds.

Protection status

National High **International** None

38,269 ha of IBA covered by Regional Nature Park (Alto Garda Bresciano, 38,269 ha).

Conservation issues

Threats Burning of vegetation (C), Deforestation (commercial) (C), Firewood collection (C), Recreation/tourism (C), Selective logging/cutting (C)

The main threat is the possibility that the park will be abolished or that its status will be downgraded so as to allow hunting again, a topic that is presently (1997) under discussion. Some or all of the site is covered by a management plan.

Middle course of River Brenta

B2, C6 **059**

Admin region Veneto
Coordinates 45°40′N 11°42′E
Altitude 35–80 m **Area** 1,100 ha

Site description

The middle course of the River Brenta, from the bridge at Nove to the bridge at Carturo, about 15 km north-east of the town of Vicenza. The main habitats are large stretches of shingle, pseudo-steppes, shrubs and wet woodlands. Most of the area is not utilized by man.

Habitats Forest and woodland (5%; broadleaved deciduous forest), Scrub (40%; scrub), Grassland (10%; steppe/dry calcareous grassland), Wetland (45%; shingle/stony beach; river/stream; water-fringe vegetation)
Land-use Fisheries/aquaculture (15%), Nature conservation/research (10%), Not utilized (75%)

Birds

Species	Season	Year	Pop min	Pop max	Acc	Criteria
Alcedo atthis Kingfisher	R	1995	20	50	C	B2, C6
Sylvia nisoria Barred Warbler	B	1995	50	100	B	C6
Lanius collurio Red-backed Shrike	B	1995	100	200	B	C6

Notable numbers of several passerines breed in the area. Species of global conservation concern that do not meet IBA criteria: *Aquila clanga* (min. 1, winter).

Protection status

National None **International** None

Conservation issues

Threats Dredging/canalization (B), Extraction industry (C), Forest grazing (C), Industrialization/urbanization (C), Recreation/tourism (B), Unsustainable exploitation (B)

The main threats are the proposed dredging of the riverbed and the proposed re-opening of hunting in the area, which is currently a no-hunting zone (so as to allow game species to reproduce). The area is a proposed Regional Park. Some or all of the site is covered by a management plan. Bird-ringing takes place.

Quinto heronry

B2, C2, C6 **060**

Admin region Veneto
Coordinates 45°40′N 12°15′E
Altitude 60–60 m **Area** 50 ha

Site description

The site lies on the Sile river, close to the village of Quinto, 10 km west of the town of Treviso, on the eastern Padana plain. Main habitats are freshwater wetland and wet woodland. Main land-uses are nature conservation and fish-farming.

Habitats Forest and woodland (20%; alluvial/very wet forest), Wetland (80%; standing fresh water; river/stream; water-fringe vegetation)
Land-use Fisheries/aquaculture (5%), Nature conservation/research (100%), Tourism/recreation (10%)

Birds

Species	Season	Year	Pop min	Pop max	Acc	Criteria
Nycticorax nycticorax Night Heron	B	1995	200	300	A	B2, C2, C6

An important colony of *Nycticorax nycticorax*.

Protection status

National High **International** None

50 ha of IBA covered by Regional Nature Park (Fiume Sile, 3,097 ha).

Conservation issues

Threats Aquaculture/fisheries (B), Drainage (C), Recreation/tourism (C)

Fish-farming is the main threat: herons that try to feed at the artificial ponds are sometimes trapped by the nets that cover the ponds.

Caorle lagoon

A4i, B1i, B2, B3, C2, C6 **061**

Admin region Veneto
Coordinates 45°39′N 12°55′E
Altitude 0–1 m **Area** 2,200 ha

Site description

A large, brackish lagoon complex on the west coast of the northern Adriatic, between the larger Venice lagoon and the lagoon of Grado and Marano. The lagoon is used for fish-farming and for hunting. The surroundings are mainly large, cultivated fields on claimed land. Main land-uses are fishing (especially fish-farming), hunting and agriculture.

Habitats Wetland (75%; tidal river/enclosed tidal water; mudflat/sandflat; saltmarsh; sand-dunes/sand beach; coastal lagoon; standing brackish and salt water), Artificial landscape (25%; highly improved reseeded grassland; urban parks/gardens; other urban/industrial areas)
Land-use Agriculture (10%), Fisheries/aquaculture (60%), Hunting (70%), Tourism/recreation (20%)

Birds

Species	Season	Year	Pop min	Pop max	Acc	Criteria
Nycticorax nycticorax Night Heron	B	1996	200	250	B	B2, C2, C6
Ardeola ralloides Squacco Heron	B	1996	10	30	B	C2, C6
Egretta garzetta Little Egret	B	1995	—	350	B	A4i, B1i, C2, C6
Ardea purpurea Purple Heron	B	1996	10	10	B	C6
Acrocephalus palustris Marsh Warbler	B	—	100	—	C	B3

An important site for breeding herons and wintering waterbirds. The average number of wintering waterbirds during 1993–1996 was 18,800. Wintering species of global conservation concern that do not meet IBA criteria: *Aythya nyroca* (max. 50, but that was in 1975), *Aquila clanga* (1).

Protection status

National None **International** None

Conservation issues

Threats Agricultural intensification/expansion (B), Aquaculture/fisheries (C), Disturbance to birds (C), Industrialization/urbanization (B), Recreation/tourism (B), Unsustainable exploitation (B)

Important threats are agricultural intensification, hunting and consequent disturbance to birds, and projects for urban, industrial and touristic development, such as the proposal for the construction of a marina close to the mouth of Tagliamento river.

Grado and Marano lagoon 062

A4i, A4iii, B1i, B2, C2, C3, C4, C6

Admin region Friuli-Venezia Giulia
Coordinates 45°45′N 13°19′E
Altitude 0–5 m **Area** 21,000 ha

Site description

A large, shallow tidal lagoon on the north Adriatic coast, fed by six rivers, 30 km in length and about 5 km in width, with six main connections with the open sea and a number of small islands and sandbars along the coast. The area is rich in tidal mudflats, sand-dunes and large reedbeds. The deeper parts of the lagoon are covered by eel-grass *Zostera* and sea lettuce *Ulva*; the mudflats have cord-grass *Spartina* and *Ruppia*; permanently exposed banks have *Salicornia* and the dunes are vegetated by grasses (*Agropyron*, *Ammophila*), etc. There are fringing stands of *Scirpus*, *Phragmites* and *Salix* around the freshwater channels and rivers which enter the lagoon. The site also includes the final stretch (and mouth of) the Stella river, a lowland woodland, some agricultural areas, and the sea up to 5 m depth. Fishing, fish-farming, hunting, tourism and agriculture are the main uses of the area.

Habitats Forest and woodland (3%; broadleaved deciduous forest; alluvial/very wet forest), Wetland (75%; tidal river/enclosed tidal water; mudflat/sandflat; saltmarsh; sand-dunes/sand beach; coastal lagoon; standing brackish and salt water; river/stream; water-fringe vegetation), Marine areas (14%; sea inlet/coastal features), Artificial landscape (8%; arable land; other urban/industrial areas)

Land-use Agriculture (2%), Fisheries/aquaculture (80%), Hunting (70%), Nature conservation/research (15%), Tourism/recreation (5%), Urban/industrial/transport (3%)

Birds

Species	Season	Year	Pop min	Pop max	Acc	Criteria
Phalacrocorax carbo Cormorant	W	1996	1,000	—	A	B1i, C3
Ixobrychus minutus Little Bittern	B	1996	10	20	A	C6
[1] **Nycticorax nycticorax** Night Heron	B	1994	100	100	A	C6
Egretta garzetta Little Egret	B	—	—	—	—	C6
Ardea purpurea Purple Heron	B	1996	80	80	A	C2, C6
Anser anser Greylag Goose	W	1994	133	221	A	B1i, C3
Anas penelope Wigeon	W	1995	18,202	18,202	A	A4i, B1i, C3
Anas strepera Gadwall	W	1996	3,188	3,188	A	A4i, B1i, C3
[2] **Bucephala clangula** Goldeneye	W	1996	1,276	2,603	A	B1i, C3
[3] **Mergus serrator** Red-breasted Merganser	W	1994	245	882	A	B1i, C3
Circus aeruginosus Marsh Harrier	B	1996	30	30	A	C6
Fulica atra Coot	W	1996	11,730	64,795	A	A4i, B1i, C3
Charadrius alexandrinus Kentish Plover	B	1995	25	25	A	B2
Numenius arquata Curlew	W	1996	450	450	A	B2
Tringa erythropus Spotted Redshank	P	1993	1,000	—	B	A4i, B1i, C3
Larus cachinnans Yellow-legged Gull	R	1996	4,500	4,500	A	A4i, B1i, C3
Sterna albifrons Little Tern	B	1996	200	200	A	B2, C2, C6
Alcedo atthis Kingfisher	R	1996	10	15	B	C6

1. Isola del Belli heronry; 1994 data.
2. 4,050 in winter 1982.
3. 882 in 1981.

A very important site for waterbirds in winter (more than 100,000) as well as when breeding and on passage, with internationally important numbers of a wide variety of species. Species of global conservation concern that do not meet IBA criteria: *Aythya nyroca* (max. 1, winter). A total of 280 species recorded, of which about 100 breed. Winter waterbird counts are from January, from the period 1992–1994 unless otherwise stated.

Protection status

National Low **International** Low

1,450 ha of IBA covered by Regional Nature Reserve (Foci dello Stella, 1,450 ha). 128 ha of IBA covered by Regional Nature Reserve (Valle Canal Novo, 128 ha). 340 ha of IBA covered by Regional Nature Reserve (Valle Cavanata, 340 ha). 1,400 ha of IBA covered by Special Protection Area (Marano Lagunare e Foce dello Stella, 1,400 ha). 243 ha of IBA covered by Special Protection Area (Valle Cavanata, 243 ha). 243 ha of IBA covered by Ramsar Site (Valle Cavanata, 243 ha).

Conservation issues

Threats Aquaculture/fisheries (B), Disturbance to birds (A), Drainage (B), Dredging/canalization (B), Filling-in of wetlands (B), Industrialization/urbanization (C), Infrastructure (C), Natural events (C), Recreation/tourism (B), Unsustainable exploitation (B)

The main threats are: drainage; increased sedimentation of wetlands by mud from the dredging of channels; disturbance by tourists of breeding birds along sandbars; intensification of fish-farming; the extension of industrial and touristic infrastructure; illegal hunting; and motor-boating. Some or all of the site is covered by a management plan. There has been a regular census of wintering waterbirds and breeding gulls and herons since the 1970s. Since 1987 the Osservatori Faunistici di Gorizia e Udine has monitored breeding species regularly. *Cygnus olor* and *Anser anser* have been introduced as breeding species.

Isonzo mouth, Cona island and Panzano Gulf 063

B1i, B2, C2, C3, C6

Admin region Friuli-Venezia Giulia
Coordinates 45°45′N 13°34′E
Altitude 0–5 m **Area** 5,000 ha

Site description

The Isonzo river meets the north Adriatic Sea some 30 km west of Trieste. There are saltmarshes, reedbeds, mudflats and sandbars, riparian woods, wet grasslands and agricultural areas. Panzano Gulf is an open sea bay. The main human activities are nature conservation, tourism, hunting, fishing, mussel-collecting and agriculture.

Habitats Forest and woodland (5%; broadleaved deciduous forest), Wetland (55%; tidal river/enclosed tidal water; mudflat/sandflat; saltmarsh; sand-dunes/sand beach; river/stream; water-fringe vegetation), Marine areas (10%; sea inlet/coastal features), Rocky areas (1%; sea cliff/rocky shore), Artificial landscape (29%; arable land)

Land-use Agriculture (30%), Fisheries/aquaculture (40%), Hunting (45%), Nature conservation/research (50%), Tourism/recreation (5%)

Birds

Species	Season	Year	Pop min	Pop max	Acc	Criteria
Ixobrychus minutus Little Bittern	B	1996	5	10	A	C6
Anas penelope Wigeon	W	1996	5,000	—	B	B1i, C3
[1] **Bucephala clangula** Goldeneye	W	1994	120	400	A	B1i, C3
Numenius arquata Curlew	W	1996	260	260	A	B2
Sterna albifrons Little Tern	B	1996	50	50	A	B2, C2, C6

1. Data 1992–1994.

Panzano Gulf is important for wintering waterbirds, with internationally important numbers of *Anas penelope* and *Bucephala clangula*. Species of global conservation concern that do not meet IBA criteria: *Aythya nyroca* (max. 4, winter). More than 280 species have been recorded, of which 70 breed.

Protection status

National Partial **International** Unknown

2,400 ha of IBA covered by Regional Nature Reserve (Foci Isonzo, 2,400 ha). Overlaps with SPA.

Conservation issues

Threats Agricultural intensification/expansion (B), Aquaculture/fisheries (B), Consequences of animal/plant introductions (U), Disturbance to birds (A), Recreation/tourism (A), Unsustainable exploitation (B)

An important threat is the disturbance of birds by visitors and mussel-collectors along the seashore. Some or all of the site is covered by a management plan. The area is a proposed Ramsar Site and is already partially managed as a Wildlife Sanctuary (Isola della Cona), with ongoing projects to restore wetland habitat and to introduce *Anser anser* as a breeding species. Since 1988 the Osservatorio Faunistico della Provincia di Gorizia has been managing a bird-ringing station in the area.

Venice lagoon 064

A1, A4i, A4iii, B1i, B2, B3, C1, C2, C3, C4, C6

Admin region Veneto
Coordinates 45°22′N 12°28′E
Altitude 0–1 m **Area** 50,000 ha

Site description

An extensive brackish coastal lagoon along the eastern coast of the northern Adriatic Sea, between the mouths of the Brenta and Piave rivers. Three large, open water-bodies form the actual lagoon, surrounded by a chain of smaller, closed lagoons (called 'valli') of lower salinity. The bottoms of the large water-bodies are covered by eel-grass *Zostera* and *Ruppia*. There are tidal mudflats with sea-lettuce *Ulva* and *Enteromorpha*, and cord-grass *Spartina* is locally common. The surrounding areas are largely cultivated intensively (mainly maize). Main land-uses are fish-farming, aquaculture and hunting inside the 'valli', and tourism.

Habitats Forest and woodland, Wetland (93%; tidal river/enclosed tidal water; mudflat/sandflat; saltmarsh; coastal lagoon; standing brackish and salt water), Artificial landscape (7%; highly improved reseeded grassland; arable land; urban parks/gardens; other urban/industrial areas)
Land-use Agriculture (5%), Fisheries/aquaculture (30%), Hunting (20%), Nature conservation/research (5%), Tourism/recreation (5%), Urban/industrial/transport (10%), Water management (30%)

Birds

Species	Season	Year	Pop min	Pop max	Acc	Criteria
Podiceps nigricollis Black-necked Grebe	W	1994	1,833	3,552	A	A4i, B1i, C3
Phalacrocorax carbo Cormorant	W	1994	1,006	1,197	—	B1i, C3
Ixobrychus minutus Little Bittern	B	1996	50	—	B	B2, C6
Nycticorax nycticorax Night Heron	B	1995	300	300	A	B2, C2, C6
Ardeola ralloides Squacco Heron	B	1995	60	60	A	B2, C2, C6
Egretta garzetta Little Egret	B	1996	1,008	1,008	A	A4i, B1i, C2, C6
Egretta garzetta Little Egret	W	1994	612	1,522	A	A4i, B1i, C2
Egretta alba Great White Egret	W	1995	362	362	A	A4i, B1i, C2
Ardea purpurea Purple Heron	B	1995	215	355	A	B2, C2, C6
Anas penelope Wigeon	W	1995	7,269	7,269	A	B1i, C3
Anas crecca Teal	W	1995	10,256	10,259	A	B1i, C3
[1] ***Aythya nyroca*** Ferruginous Duck	W	1994	3	33	A	A1, C1
Bucephala clangula Goldeneye	W	1994	96	295	A	B1i, C3
Circus pygargus Montagu's Harrier	B	1996	10	—	B	B3, C6
Fulica atra Coot	W	1994	28,880	32,957	A	B1i, C3
Himantopus himantopus Black-winged Stilt	B	—	150	—	A	C6
Recurvirostra avosetta Avocet	B	1996	60	60	A	C6
Recurvirostra avosetta Avocet	W	1995	115	115	A	B2
Charadrius alexandrinus Kentish Plover	B	1992	145	145	A	B2
Calidris alpina Dunlin	W	1995	22,310	22,310	A	A4i, B1i, B2, C3
Numenius arquata Curlew	W	1996	459	459	A	B2
[2] ***Larus melanocephalus*** Mediterranean Gull	W	1995	163	4,053	A	B1i, C2
Sterna sandvicensis Sandwich Tern	R	1996	200	—	A	C2, C6
Sterna hirundo Common Tern	B	1996	800	—	A	C6
Sterna albifrons Little Tern	B	1993	128	128	A	B2, C2
[3] ***Sterna albifrons*** Little Tern	N	1992	5,000	7,000	A	A4i, B1i, C2
[3] ***Chlidonias niger*** Black Tern	N	1992	6,000	10,000	A	A4i, B1i, C2
Acrocephalus palustris Marsh Warbler	B	1996	100	—	C	B3

1. 3 in 1994, 33 in 1985.
2. 1993–1995; a few pairs breed.
3. Roosting July–September.

A very important site for more than 100,000 wintering waterbirds (*Fulica atra* and *Calidris alpina* are particularly common), for breeding herons, and for post-breeding terns which collect to roost here in large numbers from all over the eastern Adriatic Sea (especially *Sterna albifrons* and *Chlidonias niger*, which form their largest concentrations in Europe here). The maximum total number of wintering waterbirds in January was 74,642 in 1993 and 100,048 in 1994.

Protection status

National Partial **International** Low

IBA contains State Nature Reserve (Valle Millecampi, area not known). 245 ha of IBA covered by Wildlife Reserve (Barene di S. Giuliano, 245 ha). 233 ha of IBA covered by Wildlife Reserve (Boschi di Cà Savio-Punta Sabbioni, 233 ha). 1,450 ha of IBA covered by Wildlife Reserve (Casse di Colmata, 1,450 ha). 40 ha of IBA covered by Wildlife Reserve (Cà Roman, 40 ha). 115 ha of IBA covered by Wildlife Reserve (Dune degli Alberoni, 115 ha). 14,500 ha of IBA covered by Wildlife Reserve (Isolotto petrolchimico e Laguna Viva, 14,500 ha). 45 ha of IBA covered by Wildlife Reserve (Laghetti Decal, 45 ha). 90 ha of IBA covered by Wildlife Reserve (Le Vignole-La Certosa, 90 ha). 158 ha of IBA covered by Wildlife Reserve (Pineta di Cà Ballarin, 158 ha). 137 ha of IBA covered by Wildlife Reserve (Seca del Bacan, 137 ha). 52 ha of IBA covered by Wildlife Reserve (Vallesina-Canale Casson, 52 ha). 500 ha of IBA covered by Special Protection Area (Valle Averto, 500 ha). 200 ha of IBA covered by Ramsar Site (Valle Averto, 200 ha).

Conservation issues

Threats Agricultural intensification/expansion (B), Aquaculture/fisheries (B), Disturbance to birds (A), Industrialization/urbanization (B), Natural events (A), Recreation/tourism (B), Unsustainable exploitation (A)

The main threats are posed by natural erosion, by increasing industrial activities, by increasing water pollution from agriculture and from the industrial area of Porto Marghera, and by disturbance to birds caused mainly by tourism and hunting. Only a small part of the lagoon (Valle Averto) is a Special Protection Area. Numerous small wildlife reserves were created in the 1990s by the provincial authority (Provincia di Venezia), following the national hunting law of 1991 that obliges a certain percentage of the provincial territory to be closed to hunting. Some or all of the site is covered by a management plan.

River Mincio and Bosco Fontana

A4i, B1i, B2, C2, C3, C6 **065**

Admin region Lombardia
Coordinates 45°11'N 10°45'E
Altitude 18–32 m **Area** 8,000 ha

Site description

A river and its surroundings on the Padana plain, bordering the town of Mantova. The site is rich in freshwater wetland habitat, large reedbeds and areas of open water (Valli del Mincio) close to the river, while agricultural land and broadleaved forests (Bosco Fontana) predominate in the surroundings. Agriculture is the main human activity.

Habitats Forest and woodland (10%; broadleaved deciduous forest; alluvial/very wet forest), Scrub (10%; scrub), Wetland (50%; standing fresh water; river/stream; water-fringe vegetation; fen/transition mire/spring), Artificial landscape (30%; highly improved reseeded grassland; arable land; forestry plantation; urban parks/gardens; other urban/industrial areas)
Land-use Agriculture (30%), Fisheries/aquaculture (10%), Forestry (3%), Nature conservation/research (60%), Tourism/recreation (50%), Urban/industrial/transport (10%), Water management (30%)

Birds

Species	Season	Year	Pop min	Pop max	Acc	Criteria
Phalacrocorax carbo Cormorant	W	1995	1,310	1,310	A	B1i, C3
[1] ***Nycticorax nycticorax*** Night Heron	B	1995	266	843	A	B2, C2
[1] ***Ardeola ralloides*** Squacco Heron	B	1995	7	25	A	C2, C6
Egretta alba Great White Egret	W	1995	124	124	A	A4i, B1i
Ardea purpurea Purple Heron	B	1995	114	125	A	B2, C2, C6
[2] ***Milvus migrans*** Black Kite	B	1996	—	—	—	C6

1. Breeding at Garolda, Valdaro and Lago Superiore heronries.
2. Reduced since 1980s (see text).

An important site for breeding, migrating and wintering waterbirds. There is a large colony of tree-nesting herons (Garolda) and a colony of *Chlidonias niger* (Valli del Mincio). There used to be a large number of breeding *Milvus migrans* (min. 30 pairs), mostly at Bosco Fontana, until the 1980s. Counts for heronries refer to the 1990–1995 period.

Protection status

National High **International** Partial

233 ha of IBA covered by State Nature Reserve (Bosco Fontana, 233 ha). 1,081 ha of IBA covered by Regional Nature Reserve (Valli del Mincio, 1,082 ha). 8,000 ha of IBA covered by Regional Nature Park (Mincio, 13,708 ha). 1,889 ha of IBA covered by Special Protection Area (Valli del Mincio, Paludi di Ostiglia, Torbiere d'Iseo, 1,889 ha). 1,081 ha of IBA covered by Ramsar Site (Valli del Mincio, 1,081 ha).

Conservation issues

Threats Agricultural intensification/expansion (B), Consequences of animal/plant introductions (C), Disturbance to birds (B), Industrialization/urbanization (B), Infrastructure (B), Natural events (C), Recreation/tourism (C), Unsustainable exploitation (B)

The main threats to the site come from unsustainable use of water resources for agricultural purposes, and from water pollution by agricultural sources (also much more from domestic and industrial sources in the past).

Carso

B2, C6 **066**

Admin region Friuli-Venezia Giulia
Coordinates 45°44'N 13°46'E
Altitude 0–670 m **Area** 18,950 ha

Site description

An area of karst between the northern Adriatic coast and the Slovenian border, north of the town of Trieste. It is rich in forests, with inland and coastal rocky cliffs, and a small lake.

Habitats Forest and woodland (60%; broadleaved deciduous forest), Grassland (5%; steppe/dry calcareous grassland), Rocky areas (5%; sea cliff/rocky shore; inland cliff), Artificial landscape (30%; perennial crops/orchards/groves; forestry plantation)
Land-use Agriculture (15%), Forestry (20%), Hunting (80%), Military (5%), Nature conservation/research (5%), Tourism/recreation (80%), Urban/industrial/transport (10%)

Birds

Species	Season	Year	Pop min	Pop max	Acc	Criteria
Bubo bubo Eagle Owl	R	1996	3	3	A	B2, C6
[1] ***Emberiza cia*** Rock Bunting	R	1996	400	—	B	B2
Emberiza hortulana Ortolan Bunting	B	1996	10	—	C	C6

1. Possibly the most important breeding site in Italy.

The area supports both Mediterranean and alpine species, as well as numerous species of raptor.

Protection status

National Low **International** None
760 ha of IBA covered by Regional Nature Reserve (Doberdò e Pietrarossa, 760 ha). 100 ha of IBA covered by Regional Nature Reserve (Falesie di Duino, 100 ha). 300 ha of IBA covered by Regional Nature Reserve (Monte Lanaro, 300 ha). 166 ha of IBA covered by Regional Nature Reserve (Monte Orsario, 166 ha). 500 ha of IBA covered by Regional Nature Park (Val Rosandra, 500 ha).

Conservation issues

Threats Deforestation (commercial) (C), Disturbance to birds (C), Industrialization/urbanization (B), Infrastructure (B), Intensified forest management (C), Recreation/tourism (B), Selective logging/cutting (C)

Boscone island

A4i, B1i, B2, C2, C6 **067**

Admin region Lombardia
Coordinates 45°02'N 11°15'E
Altitude 25–25 m **Area** 200 ha

Site description

An alluvial island on the south bank of the River Po, connected to the shore by a narrow isthmus. It is covered mainly by shrubby *Salix* woodland, and bordered by reedbeds. The main land-use is nature conservation. The surrounding land is agricultural, mainly poplar *Populus* plantations.

Habitats Forest and woodland (50%; alluvial/very wet forest), Wetland (50%; standing fresh water; river/stream)
Land-use Nature conservation/research (100%)

Birds

Species	Season	Year	Pop min	Pop max	Acc	Criteria
[1] ***Nycticorax nycticorax*** Night Heron	B	1995	564	682	A	A4i, B1i, B2, C2, C6
[2] ***Ardeola ralloides*** Squacco Heron	B	1988	5	14	A	C2
[1] ***Egretta garzetta*** Little Egret	B	1995	5	328	A	B1i, C2, C6

1. 1990–1995.
2. Censused 1981, 1985, 1986, 1988.

The island holds an important mixed heronry of *Nycticorax nycticorax*, *Egretta garzetta* and *Ardeola ralloides*.

Protection status

National High **International** High
200 ha of IBA covered by Regional Nature Reserve (Isola Boscone, 200 ha). 200 ha of IBA covered by Ramsar Site (Isola Boscone, 201 ha). Overlaps with SPA.

Conservation issues

Threats Natural events (A)

The main threat is posed by the large floods of the River Po that periodically destroy or modify parts of the island. The area is managed by LIPU for Regione Lombardia. Some or all of the site is covered by a management plan. Conservation activities include excavation works for wetland restoration, education and birdwatching.

River Po: from the mouth of Adda river to Ostiglia

A4i, B1i, B2, B3, C2, C6 **068**

Admin region Lombardia, Veneto, Emilia Romagna
Coordinates 45°00'N 10°36'E
Altitude 16–41 m **Area** 30,000 ha

Site description

A stretch of the River Po, including large islands of shingle and sand, riparian woodlands, poplar *Populus* plantations and flat, cultivated areas. The area includes a lowland freshwater wetland (169 ha) located a few kilometres north of the River Po, called Palude di Ostiglia. This marshland of reed *Phragmites* and sedge *Carex* is separated from the surrounding intensively cultivated countryside by a 2 m high dyke. Main land-uses are intensive agriculture (especially maize), forestry (poplar), and recreational activities. In Palude di Ostiglia the main land-use is the sustainable exploitation of sedge for producing traditional handcrafts.

Habitats Forest and woodland (10%; broadleaved deciduous forest; alluvial/very wet forest), Wetland (50%; standing fresh water; river/stream; water-fringe vegetation), Artificial landscape (40%; arable land; forestry plantation)
Land-use Agriculture (25%), Fisheries/aquaculture (30%), Forestry (5%), Hunting (20%), Nature conservation/research (5%), Other (5%), Tourism/recreation (30%), Urban/industrial/transport (5%), Water management (30%)

Birds

Species	Season	Year	Pop min	Pop max	Acc	Criteria
Ixobrychus minutus Little Bittern	B	1994	5	20	B	C6
Nycticorax nycticorax Night Heron	B	1994	—	2,045	A	A4i, B1i, B2, C2, C6
Circus aeruginosus Marsh Harrier	R	1994	3	3	B	C6
Circus pygargus Montagu's Harrier	B	1994	1	5	B	C6
Sterna albifrons Little Tern	B	1994	—	80	C	B2, C2, C6
Athene noctua Little Owl	R	1996	50	100	B	B2
Alcedo atthis Kingfisher	R	1996	20	50	B	B2, C6
Acrocephalus palustris Marsh Warbler	B	1996	150	300	B	B3

This site is important for three large heronries (Pomponesco, Gramignazzo and Bellaguarda), and especially for *Nycticorax nycticorax*. The area also supports notable numbers of a wide variety of other wetland species. Species of global conservation concern that do not meet IBA criteria: *Aythya nyroca* (max. 1, passage), *Acrocephalus paludicola* (max. 1, passage).

Protection status

National Partial **International** Low
23 ha of IBA covered by Regional Nature Reserve (Garzaia di Pomponesco, 23 ha). 123 ha of IBA covered by Regional Nature Reserve (Paludi di Ostiglia, 123 ha). IBA also covered by four Regional Nature Reserves of unknown area (Bosco Ronchetti, Lanca di Gerole, Lancone di Gussola, Spiaggioni di Spinadesco). 1,889 ha of IBA covered by Special Protection Area (Valli del Mincio, Paludi di Ostiglia, Torbiere d'Iseo, 1,889 ha). 123 ha of IBA covered by Ramsar Site (Palude di Ostiglia, 123 ha).

Conservation issues

Threats Agricultural intensification/expansion (B), Disturbance to birds (B), Drainage (A), Dredging/canalization (A), Filling-in of wetlands (A), Firewood collection (C), Forest grazing (C), Infrastructure (A), Natural events (A), Recreation/tourism (B), Selective logging/cutting (C)

The main threats are posed by the drainage, dredging, canalization and filling-in of wetlands. Building of civil infrastructure and periodic large floods also have a large negative impact on the area. Palude di Ostiglia is a LIPU Reserve, and the area has been proposed as a Natural Park. Istituto Nazionale Fauna Selvatica carries out bird-ringing regularly at Ostiglia in spring and autumn.

Codigoro heronry

A4i, B1i, B2, C2, C6 **069**

Admin region Emilia Romagna
Coordinates 44°56'N 12°14'E
Altitude 1–1 m **Area** 60 ha

Site description

The site is located on the east Padana plain, close to the village of Codigoro, inland of the Po delta. The pools of a sugar factory are the main land-use of the area, together with nature conservation.

Habitats Forest and woodland (broadleaved deciduous forest), Wetland (standing fresh water; water-fringe vegetation)
Land-use Nature conservation/research (67%), Urban/industrial/transport (67%)

Birds

Species	Season	Year	Pop min	Pop max	Acc	Criteria
Nycticorax nycticorax Night Heron	B	1995	—	785	A	A4i, B1i, B2, C2, C6
Ardeola ralloides Squacco Heron	B	1995	—	75	A	B2, C2, C6
Egretta garzetta Little Egret	B	1995	—	863	A	A4i, B1i, C2, C6

A large mixed heronry. The presence of the colony is linked to the sugar-factory pools.

Protection status

National Partial **International** High
31 ha of IBA covered by Wildlife Reserve (Bacini dello Zuccherificio di Codigoro, 31 ha). 60 ha of IBA covered by Special Protection Area (Garzaia dello zuccherificio di Codigoro, 60 ha).

Conservation issues

No major threats are present.

Po delta

A4i, A4iii, B1i, B2, C2, C3, C4, C6 **070**

Admin region Veneto
Coordinates 44°56'N 12°16'E
Altitude 0–10 m **Area** 17,000 ha

Site description

A large Mediterranean delta on the north-east coast of the Adriatic Sea, between the Adige river-mouth and the Sacca di Goro. The delta is fed by six major branches of the River Po. Its main habitats are closed brackish lagoons ('valli'), open coastal lagoons ('sacche'), sand beaches and sandbanks, rivers and riverine vegetation, and farmland. Vegetation ranges from wet woodlands to reedbeds and plants of sand-dunes and saline soils. The site is surrounded by intensively cultivated arable land. Fish-farming, agriculture and hunting are the main human activities.

Habitats Forest and woodland (2%; alluvial/very wet forest), Wetland (85%; tidal river/enclosed tidal water; mudflat/sandflat; saltmarsh; sand-dunes/sand beach; coastal lagoon; standing fresh water; standing brackish and salt water; river/stream; water-fringe vegetation), Marine areas (8%; open sea; sea inlet/coastal features), Artificial landscape (5%; highly improved reseeded grassland; arable land)
Land-use Agriculture (3%), Fisheries/aquaculture (70%), Hunting (70%), Tourism/recreation (20%), Urban/industrial/transport (1%), Water management

Birds

Species	Season	Year	Pop min	Pop max	Acc	Criteria
Podiceps nigricollis Black-necked Grebe	W	1994	2,933	2,933	A	A4i, B1i, C3
Phalacrocorax carbo Cormorant	W	1994	2,496	2,496	—	B1i, C3
[1] ***Nycticorax nycticorax*** Night Heron	B	1994	385	385	A	B2, C2, C6
[2] ***Ardeola ralloides*** Squacco Heron	B	1994	54	54	A	B2, C2, C6
[1] ***Egretta garzetta*** Little Egret	B	1994	913	913	A	B1i, C2, C6
[3] ***Ardea purpurea*** Purple Heron	B	1994	151	151	A	B2, C2, C6
Anas penelope Wigeon	W	1994	12,839	12,839	A	B1i, C3
[4] ***Himantopus himantopus*** Black-winged Stilt	B	1994	64	79	A	C6
Recurvirostra avosetta Avocet	W	1994	112	112	A	B2
[5] ***Charadrius alexandrinus*** Kentish Plover	B	1994	65	140	A	B2
Calidris alpina Dunlin	W	1994	1,012	1,012	A	B2
Larus cachinnans Yellow-legged Gull	W	1994	5,921	5,921	A	A4i, B1i, C3
Sterna albifrons Little Tern	B	1994	1,249	1,249	A	A4i, B1i, B2, C2, C6
Phalacrocorax carbo Cormorant	W	1995	1,250	1,250	A	B1i, C3

1. Maistra, Bastimento, Pila and Dindona heronries.
2. Maistra and Dindona heronries.
3. Maistra, Dindona, Tramontana and Dritta heronries.
4. 1987–1994.
5. 1991–1994.

A very important site for wintering waterbirds (maximum count is 48,931 for all species) and for colony-breeding waterbirds (5,500 pairs of herons, 5,000 pairs of gulls, 5,200 pairs of terns). In particular, this is the most important breeding site in the western Palearctic for *Sterna albifrons*. Only 1994 counts are available for wintering waterbirds (January) and for breeding herons. Species of global conservation concern that do not meet IBA criteria: *Aythya nyroca* (rare breeder and winterer).

Protection status

National Partial **International** None
457 ha of IBA covered by State Nature Reserve (Bocche di Po, 457 ha). 14 ha of IBA covered by State Nature Reserve (Bonello Cà Zuliani, 14 ha). IBA also contains three State Nature Reserves of unknown area (Bonello Bacucco, Bonello Cornerona, Bonello Giozzette). 35 ha of IBA covered by Private Reserve (Biotopo Bonello, 35 ha). 32 ha of IBA covered by Private Reserve (Cà Mello, 32 ha). Part of IBA covered by Regional Nature Park (Delta del Po, area unknown).

Conservation issues

Threats Agricultural intensification/expansion (A), Aquaculture/fisheries (B), Consequences of animal/plant introductions (C), Construction/impact of dyke/dam/barrage (B), Disturbance to birds (A), Drainage (B), Dredging/canalization (B), Filling-in of wetlands (B), Groundwater abstraction (B), Industrialization/urbanization (A), Natural events (A), Recreation/tourism (B), Unsustainable exploitation (A)

Important threats are: the discharge into the River Po of urban and agricultural (especially pig-farming) sewage, causing nutrient pollution; the erosion of coastal sandbars (excessive water extraction is causing subsidence); illegal hunting; and disturbance of birds by hunting and tourism (the former at the 'valli', the latter along the coast). Some or all of the site is covered by a management plan.

Valle Bertuzzi and Goro lagoon

B1i, B2, B3, C2, C3, C6 **071**

Admin region Emilia Romagna
Coordinates 44°47'N 12°15'E
Altitude 1–1 m **Area** 4,000 ha

Site description

This area lies on the eastern Adriatic coast, and comprises Valle Bertuzzi (a large brackish lagoon) and Sacca di Goro (an open lagoon), as well as some smaller sites: Lago delle Nazioni, Taglio della Falce, Scannone di Volano and Valli Cannaviè e Porticino. The main activities are fishing and fish-farming.

Habitats Forest and woodland (1%; broadleaved deciduous forest), Scrub (2%; scrub), Wetland (97%; tidal river/enclosed tidal water; mudflat/sandflat; saltmarsh; sand-dunes/sand beach; coastal lagoon; standing brackish and salt water; river/stream; water-fringe vegetation)
Land-use Fisheries/aquaculture (80%), Hunting (80%), Nature conservation/research (80%)

Birds

Species	Season	Year	Pop min	Pop max	Acc	Criteria
[1] ***Egretta garzetta*** Little Egret	B	1994	126	220	A	C2, C6
[2] ***Ardea purpurea*** Purple Heron	B	1995	238	355	A	B2, C2, C6
Circus pygargus Montagu's Harrier	B	1994	5	5	A	B3, C6
[1] ***Himantopus himantopus*** Black-winged Stilt	B	1994	40	70	B	C6
Numenius arquata Curlew	W	1994	124	124	A	B2
Larus melanocephalus Mediterranean Gull	B	1994	641	641	A	C2, C6
[3] ***Sterna hirundo*** Common Tern	B	1994	663	1,213	A	B1i, C2, C6

Species ... continued	Season	Year	Pop min	Pop max	Acc	Criteria
[4] ***Sterna albifrons*** Little Tern	B	1994	153	153	A	B2, C2, C6
Acrocephalus scirpaceus Reed Warbler	B	1996	100	300	A	B3

1. 1993–1994.
2. 1994–1995.
3. 1990–1994.
4. 1994 only.

An important breeding and wintering site for waterbirds, notably for colony-breeding gulls and terns, for breeding *Ardea purpurea* (the largest colony in Italy and one of the largest in the western Paleartic), and for roosting *Phalacrocorax carbo* in winter (Valle Bertuzzi is the largest such roost on the Adriatic coast). Winter data refer to January 1994 count only. Wintering waders occur on the sandbanks and beaches of Sacca di Goro.

Protection status

National Partial **International** Partial

221 ha of IBA covered by State Nature Reserve (Po di Volano, 221 ha). Part of IBA covered by Regional Nature Park (Delta del Po, 59,118 ha). Part of IBA covered by Special Protection Area (Valle Gorino, Bertuzzi, Comacchio, Ortazzo, Baiona, 20,400 ha). 3,100 ha of IBA covered by Ramsar Site (Valle Bertuzzi, 3,100 ha).

Conservation issues

Threats Aquaculture/fisheries (C), Disturbance to birds (A), Natural events (A)

Important threats to the waterbird colonies are: erosion and subsidence of the islets; predation by rats *Rattus*; predation, disturbance and competition for breeding sites by *Larus cachinnans*; disturbance by aquaculture workers. Some or all of the site is covered by a management plan. There have been studies on the colony-breeding waterbirds (since the 1970s), on the impact of *Phalacrocorax carbo* on fish-farming, and on plants and invertebrates.

Valli di Comacchio and Bonifica del Mezzano

A1, A4i, A4iii, B1i, B2, C1, C2, C3, C4, C6

072

Admin region Emilia Romagna
Coordinates 44°37'N 12°10'E
Altitude 1–1 m **Area** 28,570 ha

Site description

A complex of wetlands and cultivated land on the eastern Padana plain, being the remnants of a large complex of lagoons, most of which were drained from 1850 onwards. The main sites are Valli di Comacchio, Sacca di Bellocchio, Bonifica del Mezzano and three heronries (Garzaia di Lepri, Garzaia di Bando, Garzaia di Ostellato). The shallow, brackish lagoon of Valli di Comacchio is the largest in Italy and one of the largest in the Mediterranean basin. Sacca di Bellocchio is a complex of saltmarshes (*Salicornia*, *Suaeda*) located in between the Valli and the Adriatic Sea. Bonifica del Mezzano is an agricultural area, formerly a lagoon (linked to the Valli di Comacchio) which was drained in the 1960s. The most important human activities are fish-farming, hunting and tourism in Valli di Comacchio, agriculture in Bonifica del Mezzano and nature conservation in Sacca di Bellocchio.

Habitats Forest and woodland (5%; alluvial/very wet forest), Wetland (50%; saltmarsh; sand-dunes/sand beach; coastal lagoon; standing fresh water; standing brackish and salt water; river/stream; water-fringe vegetation), Artificial landscape (45%; arable land; urban parks/gardens)

Land-use Agriculture (40%), Fisheries/aquaculture (40%), Hunting (50%), Military (5%), Nature conservation/research (50%), Tourism/recreation (20%), Urban/industrial/transport (5%), Water management (10%)

Birds

Species	Season	Year	Pop min	Pop max	Acc	Criteria
Phalacrocorax carbo Cormorant	W	1995	1,157	1,157	A	B1i, C3
Nycticorax nycticorax Night Heron	B	1995	455	730	A	A4i, B1i, B2, C2, C6
Ardeola ralloides Squacco Heron	B	1995	—	35	A	C2, C6
Ardea purpurea Purple Heron	B	1995	100	300	A	B2, C2, C6
Anser anser Greylag Goose	W	1994	—	230	A	B1i, C3
Aythya ferina Pochard	W	1976	16,000	16,000	A	A4i, B1i, C3
Aythya nyroca Ferruginous Duck	W	1976	95	95	A	A1, C1
Bucephala clangula Goldeneye	W	1995	686	686	A	B1i, C3
Himantopus himantopus Black-winged Stilt	B	1994	—	140	A	B1i, C2, C6
Recurvirostra avosetta Avocet	B	1994	—	149	A	C6
Glareola pratincola Collared Pratincole	B	1996	—	30	B	B1i, C2, C6
Charadrius alexandrinus Kentish Plover	B	1994	36	96	A	B2
Larus melanocephalus Mediterranean Gull	R	1994	—	1,550	A	C2, C6
Gelochelidon nilotica Gull-billed Tern	B	1993	—	145	A	B1i, B2, C2, C6
Sterna sandvicensis Sandwich Tern	B	1994	—	569	A	B1i, C2, C6
Sterna hirundo Common Tern	B	1994	—	1,010	A	B1i, C2, C6
Sterna albifrons Little Tern	B	1994	178	1,750	—	A4i, B1i, B2, C2, C6
Acrocephalus melanopogon Moustached Warbler	R	1996	10	20	A	C6
Lanius collurio Red-backed Shrike	B	1996	50	100	B	C6
Emberiza hortulana Ortolan Bunting	B	1996	50	100	B	C6

A very important site for breeding, migrating and wintering waterbirds, especially for *Sterna albifrons*. Valli di Comacchio is important for breeding terns, gulls and waders and for passage and wintering ducks and *Fulica atra*. Bonifica del Mezzano is important as a feeding area for wintering geese *Anser*, for the presence of heronries, and for its potential value for wetland-restoration projects.

Protection status

National Partial **International** Partial

48 ha of IBA covered by State Nature Reserve (Destra foce fiume Reno, 48 ha). 45 ha of IBA covered by State Nature Reserve (Foce fiume Reno, 45 ha). 163 ha of IBA covered by Nature Reserve (Sacca di Bellocchio (I), 163 ha). 83 ha of IBA covered by Nature Reserve (Sacca di Bellocchio (II), 83 ha). 38 ha of IBA covered by Nature Reserve (Sacca di Bellocchio (III), 38 ha). 13,500 ha of IBA covered by Regional Nature Park (Delta del Po, 59,118 ha). Part of IBA covered by Special Protection Area (Valle Gorino, Bertuzzi, Comacchio, Ortazzo, Baiona, 20,400 ha). 223 ha of IBA covered by Ramsar Site (Sacca di Bellocchio, 223 ha). 13,500 ha of IBA covered by Ramsar Site (Valli residue del comprensorio di Comacchio, 13,500 ha).

Conservation issues

Threats Agricultural intensification/expansion (C), Aquaculture/fisheries (B), Disturbance to birds (B), Groundwater abstraction (B), Natural events (B), Unsustainable exploitation (B)

Major threats are: the lowering of the water-table and consequent erosion of the small islets used by breeding waterbirds; disturbance of birds by hunters, birdwatchers and photographers; intensive fish-farming involving the creation of artificial pools in place of the traditional 'valli', especially along the southern border of the Valli di Comacchio, using EU funds; pollution of water with nutrients from cultivated land and especially from a large 26,000-pig farm. Waterbirds are censused in midwinter and during the breeding season (colonial species). Botanical and entomological studies have been carried out by Università di Ferrara. LIPU is carrying out a wetland-restoration project at Volta Scirocco, along the southern border of the Valli di Comacchio, with support from the EU (LIFE programme).

Valli di Argenta

A1, B1i, B2, C1, C2, C3, C6

073

Admin region Emilia Romagna
Coordinates 44°34'N 11°50'E
Altitude 4–15 m **Area** 2,000 ha

Site description

A complex of freshwater wetlands on the eastern Padana plain. Habitats include open water, reedbed, riverine woodland, wet grassland and agricultural areas crossed by three rivers. The area can be divided into the following subsites: Valle Santa, a wetland with reedbeds, *Typha* beds and wet grassland; Val Campotto, a wetland half-covered by water-lilies *Nymphaea*; Bassarone, an open water-body. The surroundings are mainly cultivated with maize, wheat, and poplar *Populus* plantations. The main use of the site is for nature conservation and research—it is the main bird-ringing station in Italy.

Habitats Forest and woodland (7%; alluvial/very wet forest), Grassland (10%; humid grassland), Wetland (75%; standing fresh water; river/stream; water-fringe vegetation), Artificial landscape (8%; highly improved reseeded grassland; perennial crops/orchards/groves)
Land-use Agriculture (3%), Forestry (5%), Nature conservation/research (100%), Water management (80%)

Birds

Species	Season	Year	Pop min	Pop max	Acc	Criteria
Botaurus stellaris Bittern	B	1996	2	3	B	C6
Ixobrychus minutus Little Bittern	B	1996	50	—	A	B2, C6
[1] ***Nycticorax nycticorax*** Night Heron	B	1995	10	277	A	B2, C2
Ardeola ralloides Squacco Heron	B	1995	—	70	A	C2, C6
[2] ***Ardea purpurea*** Purple Heron	B	1995	9	70	A	C2
Anas strepera Gadwall	W	1994	1,022	1,022	A	B1i, C3
Aythya nyroca Ferruginous Duck	W	1989	12	12	A	A1, C1
[3] ***Chlidonias hybridus*** Whiskered Tern	B	1996	15	20	B	B2, C6

1. 600–900 pairs in 1980s.
2. 80–100 pairs in 1980s.
3. 150–250 pairs in 1980s.

There is a large heronry of *Nycticorax nycticorax*, *Egretta garzetta* and *Ardeola ralloides*, as well as the largest Italian colony of *Phalacrocorax carbo*. Winter waterbird data refer to January 1994 counts only. Species of global conservation concern that do not meet IBA criteria: *Aythya nyroca* (2–3 pairs breed), *Acrocephalus paludicola* (rare on passage).

Protection status

National High **International** High
IBA contains Wildlife Reserve (Valli Argenta e Marmorta, area not known). 2,000 ha of IBA covered by Regional Nature Park (Delta del Po, 59,118 ha). 1,830 ha of IBA covered by Special Protection Area (Valle Santa e Val Campotto, 1,830 ha). 1,363 ha of IBA covered by Ramsar Site (Valle Campotto e Bassarone, 1,363 ha). 261 ha of IBA covered by Ramsar Site (Valle Santa, 261 ha).

Conservation issues

Threats Agricultural intensification/expansion (A), Consequences of animal/plant introductions (A)

There are two main threats. Coypu *Myocastor coypus* (an introduced, non-native aquatic rodent) have caused breeding *Chlidonias hybridus* to decline by largely destroying the cover of water-lilies *Nymphaea*. Nitrates and phosphates from agricultural fields are polluting the wetlands with excess nutrients. Some or all of the site is covered by a management plan. Coypu numbers are controlled by the local management body (Consorzio di Bonifica Renana). There is regular monitoring of herons and *Phalacrocorax carbo* by the Università di Bologna. Species listed in the EC Habitats and Birds Directives are censused by Regione Emilia Romagna.

Punte Alberete and Valle della Canna, Pineta San Vitale and Pialassa della Baiona
074 A1, A4i, B1i, B2, C1, C2, C3, C6

Admin region Emilia Romagna
Coordinates 44°30′N 12°15′E
Altitude 1–10 m **Area** 3,000 ha

Site description

An area of mixed wetland and wooded habitats, a few kilometres from the northern Adriatic coast, north of the town of Ravenna. Punte Alberete (190 ha) is a flooded forest; Pialassa della Baiona (1,200 ha) is a brackish lagoon; Valle della Canna (260 ha) is a large reedbed; Pineta di San Vitale is a pinewood *Pinus* (not native, but planted in Roman times). Punte Alberete and Valle della Canna are used mainly for nature conservation, Pialassa della Baiona and Pineta di San Vitale mainly for recreation and hunting.

Habitats Forest and woodland (55%; native coniferous forest; alluvial/very wet forest), Wetland (35%; saltmarsh; standing brackish and salt water; river/stream; water-fringe vegetation), Artificial landscape (10%; arable land)
Land-use Agriculture (10%), Fisheries/aquaculture (40%), Hunting (85%), Nature conservation/research (15%), Tourism/recreation (40%)

Birds

Species	Season	Year	Pop min	Pop max	Acc	Criteria
[1] ***Tachybaptus ruficollis*** Little Grebe	W	1996	994	1,000	A	A4i, B1i, C3
[2,3] ***Phalacrocorax pygmeus*** Pygmy Cormorant	B	1995	7	7	A	C6
Botaurus stellaris Bittern	R	1995	—	3	B	C6
Ixobrychus minutus Little Bittern	B	1996	18	20	A	C6
[2,4] ***Nycticorax nycticorax*** Night Heron	B	1994	267	361	A	B2, C2, C6
[2,4] ***Ardeola ralloides*** Squacco Heron	B	1994	25	57	A	B2, C2, C6
Egretta garzetta Little Egret	B	1994	690	977	A	A4i, B1i, C2, C6
Egretta alba Great White Egret	B	1995	22	22	A	C6
[5] ***Ardea purpurea*** Purple Heron	B	1994	81	82	A	C2, C6
[2] ***Plegadis falcinellus*** Glossy Ibis	B	1996	1	8	A	C6
Platalea leucorodia Spoonbill	B	1991	5	5	A	C6
[2,6] ***Aythya nyroca*** Ferruginous Duck	B	1996	—	20	B	A1, C1, C6
Himantopus himantopus Black-winged Stilt	B	1995	30	40	A	C6
Recurvirostra avosetta Avocet	B	1995	40	50	A	C6
[7] ***Larus ridibundus*** Black-headed Gull	W	1996	12,350	12,500	A	B1i, C3
Sterna hirundo Common Tern	B	1996	250	300	B	C6
Sterna albifrons Little Tern	B	1995	10	20	B	C2
Chlidonias hybridus Whiskered Tern	B	1997	50	—	A	B2, C6
Acrocephalus melanopogon Moustached Warbler	R	1996	50	70	A	C6

1. Pialassa della Baiona.
2. Punte Alberete.
3. 1995 only.
4. 1992–1994.
5. Valle della Canna (80), Pialassa (1–2).
6. Valle Mandriole.
7. 1996 only.

This is an important breeding site for a wide variety of waterbirds, notably herons and terns, and is the best breeding site in Italy for *Aythya nyroca* and *Acrocephalus melanopogon*. Winter waterbird counts refer to winters of 1993, 1994 and 1996.

Protection status

National High **International** High
709 ha of IBA covered by Nature Reserve (Pineta di Ravenna (San Vitale), 709 ha). 3,000 ha of IBA covered by Regional Nature Park (Delta del Po, 59,118 ha). 400 ha of IBA covered by Special Protection Area (Punte Alberete e Valle della Canna, 400 ha). Part of IBA covered by Special Protection Area (Valle Gorino, Bertuzzi, Comacchio, Ortazzo, Baiona, 20,400 ha). 1,630 ha of IBA covered by Ramsar Site (Piallassa della Baiona e Risega, 1,630 ha). 480 ha of IBA covered by Ramsar Site (Punte Alberete, 480 ha).

Conservation issues

Threats Aquaculture/fisheries (B), Consequences of animal/plant introductions (B), Disturbance to birds (A), Drainage (B), Filling-in of wetlands (B), Groundwater abstraction (A), Industrialization/urbanization (A), Other (U), Unsustainable exploitation (A)

Coypu *Myocastor coypus*, a non-native aquatic rodent, is destroying the floating vegetation. 'Other' threats (above) are (1) the unsustainable collection of mushrooms and other products from the pinewood and collection of mussels in the Piallassa della Baiona, and (2) the destruction of terns' nests built on the roofs of hunting huts in Piallassa della Baiona. The nearby industrial area of Ravenna causes water pollution. Some or all of the site is covered by a management plan. Bird studies are carried out by WWF-Italia, Università di Ferrara and Università di Pavia, while breeding gulls and terns are censused by LIPU and Provincia di Ravenna. Research is also carried out by Regione Emilia Romagna on those species and habitats that are listed in the EC Habitats and Birds Directives.

Ortazzo and Ortazzino
075 B1i, B2, C2, C6

Admin region Emilia Romagna
Coordinates 44°22′N 12°19′E
Altitude 0–3 m **Area** 900 ha

Site description

A complex of brackish wetlands, close to the mouths of the rivers Bevano and Ghiaia, along the northern Adriatic coast, 10 km south-east of the town of Ravenna. The site includes the mouths of the rivers and the related coastal area including a stretch of open sea, the natural dune system, small lagoons, a pinewood *Pinus*, maquis, wet grasslands, mudflats and cultivated land. The main land-use is nature conservation.

Habitats Forest and woodland (10%; mixed forest), Scrub (10%; sclerophyllous scrub/garrigue/maquis), Grassland (10%; humid grassland), Wetland (50%; tidal river/enclosed tidal water; mudflat/sandflat; sand-dunes/sand beach; coastal lagoon; standing brackish and salt water; river/stream; water-fringe vegetation), Marine areas (10%; sea inlet/coastal features), Artificial landscape (10%; highly improved reseeded grassland; other urban/industrial areas; ruderal land)

Land-use Agriculture (8%), Nature conservation/research (100%), Tourism/recreation (10%), Urban/industrial/transport (5%)

Birds

Species	Season	Year	Pop min	Pop max	Acc	Criteria
Recurvirostra avosetta Avocet	B	1996	35	40	A	C6
Charadrius alexandrinus Kentish Plover	B	1996	35	45	A	B2
[1] ***Gelochelidon nilotica*** Gull-billed Tern	B	1996	24	24	A	B2, C2, C6
Sterna albifrons Little Tern	B	1996	200	250	A	B1i, B2, C2, C6

1. First breeding in 1996.

An important breeding site for several waterbirds, and also notable for its migrating waders and wintering ducks and *Fulica atra*.

Protection status

National High **International** High

6 ha of IBA covered by State Nature Reserve (Foce Bevano,6 ha). Part of IBA covered by State Nature Reserve (Pineta di Ravenna, 709 ha). 700 ha of IBA covered by Wildlife Reserve (Ortazzo-Ortazzino, 700 ha). 900 ha of IBA covered by Regional Nature Park (Delta del Po, 59,118 ha). 900 ha of IBA covered by Special Protection Area (Valle Gorino, Bertuzzi, Comacchio, Ortazzo, Baiona, 20,400 ha). 440 ha of IBA covered by Ramsar Site (Ortazzo e Ortazzino, 440 ha).

Conservation issues

Threats Disturbance to birds (B), Industrialization/urbanization (B), Other (A), Recreation/tourism (A), Unsustainable exploitation (B)

Birds are disturbed by tourists, especially along the beach where *Haematopus ostralegus*, *Sterna albifrons* and *Charadrius alexandrinus* might otherwise breed. Rapid changes in water-level, due to lack of water management, lead to the loss of hundreds of nests of terns and waders ('Other' threat). Illegal huts are being built on the beach. Research is carried out by Regione Emilia Romagna on those species and habitats that are listed in the EC Habitats and Birds Directives.

Cervia saltpans A4i, B1i, B2, B3, C2, C3, C6 076

Admin region Emilia Romagna
Coordinates 44°15′N 12°21′E
Altitude 1–1 m **Area** 1,000 ha

Site description

A former lagoon on the north-eastern Adriatic coast near the town of Cervia, now isolated from the sea as an active saltpan. The area is bounded by two canals and is surrounded by agricultural land.

Habitats Forest and woodland, Wetland (80%; saltmarsh; standing brackish and salt water; water-fringe vegetation), Artificial landscape (20%; highly improved reseeded grassland; arable land)

Land-use Agriculture (10%), Nature conservation/research (80%), Tourism/recreation, Urban/industrial/transport (80%)

Birds

Species	Season	Year	Pop min	Pop max	Acc	Criteria
Egretta garzetta Little Egret	N	1990	87	1,856	A	A4i, B1i, C2
Himantopus himantopus Black-winged Stilt	B	1994	95	150	A	B1i, C2, C6
Recurvirostra avosetta Avocet	B	1994	55	240	A	B1i, B3, C2
Recurvirostra avosetta Avocet	W	1994	181	700	A	B1i, B2, C2
Charadrius alexandrinus Kentish Plover	B	1994	45	70	A	B2
Calidris alpina Dunlin	W	1994	2,070	6,211	A	B2
Philomachus pugnax Ruff	P	1991	—	5,257	A	C6
Larus melanocephalus Mediterranean Gull	N	1990	—	699	A	C6
Larus ridibundus Black-headed Gull	N	1990	—	5,650	A	B1i, C3
Larus cachinnans Yellow-legged Gull	N	1990	—	16,089	A	A4i, B1i, C3
Sterna albifrons Little Tern	B	1994	200	370	—	A4i, B1i, B2, C2, C6

An important site for breeding and wintering waders, and for non-breeding *Egretta garzetta*, *Recurvirostra avosetta* and gulls *Larus*. A total of 10,704 waterbirds were recorded in January 1994. Breeding data were collected during 1990–1994, wintering data in winter 1990/1991 and in January 1992, 1993 and 1994, while counts of passage birds refer to spring 1991.

Protection status

National Partial **International** Partial

785 ha of IBA covered by Nature Reserve (Salina di Cervia, 785 ha). 765 ha of IBA covered by Special Protection Area (Riserva Naturale Salina di Cervia, 765 ha). 785 ha of IBA covered by Ramsar Site (Saline di Cervia, 785 ha).

Conservation issues

Threats Disturbance to birds (A), Natural events (B), Other (A)

The management of water-levels during salt-extraction activities affects the breeding success of waders. All of the site is covered by a management plan. Waterbirds are censused in midwinter and while breeding.

Lake Massaciuccoli A1, A4i, B1i, B2, B3, C1, C2, C6 077

Admin region Toscana
Coordinates 43°51′N 10°20′E
Altitude 1–5 m **Area** 1,700 ha

Site description

A shallow lake, close to the town of Viareggio, 4 km from the Tyrrhenian coast. The lake is surrounded by a belt of reeds *Phragmites* of variable width and there is a large freshwater marshland (Padule) adjacent to the lake on the north side, consisting of stands of sedge *Cladium* with scattered pools and canals. The surrounding areas are intensively cultivated. Main land-uses are agriculture, nature conservation and tourism.

Habitats Wetland (95%; standing fresh water; water-fringe vegetation), Artificial landscape (5%; arable land)

Land-use Agriculture (20%), Fisheries/aquaculture (40%), Nature conservation/research (100%), Tourism/recreation (15%)

Birds

Species	Season	Year	Pop min	Pop max	Acc	Criteria
Botaurus stellaris Bittern	R	1996	20	22	A	C2, C6
Ixobrychus minutus Little Bittern	B	1996	10	50	C	B2, C6
Ardea purpurea Purple Heron	B	1996	85	—	A	C2, C6
Circus aeruginosus Marsh Harrier	B	1991	8	9	A	C6
Numenius tenuirostris Slender-billed Curlew	P	—	—	—	—	A1, C1
Chlidonias niger Black Tern	N	1996	2,000	5,000	B	A4i, B1i, C2
Locustella luscinioides Savi's Warbler	B	1996	50	100	A	B3
Acrocephalus melanopogon Moustached Warbler	R	1996	500	1,200	B	C2, C6
Acrocephalus scirpaceus Reed Warbler	B	1996	500	—	B	B3

An important breeding site for reedbed species, notably *Botaurus stellaris* (the main breeding site in Italy), and a key staging site for the globally threatened *Numenius tenuirostris*. Formerly much more important for wintering waterbirds, especially *Fulica atra* (tens of thousands); decline due especially to hunting and a reduction in the extent of submerged vegetation. Species of global conservation concern that do not meet IBA criteria: *Aythya nyroca* (max. 4 birds in winter).

Protection status

National High **International** High

1,700 ha of IBA covered by Regional Nature Park (Migliarino–San Rossore–Massaciuccoli, 22,000 ha). 1,700 ha of IBA covered by Special Protection Area (Massaciuccoli, 2,000 ha).

Conservation issues

Threats Abandonment/reduction of land management (C), Agricultural intensification/expansion (A), Aquaculture/fisheries (C), Disturbance to birds (C), Filling-in of wetlands (B), Industrialization/urbanization (B), Infrastructure (B)

The main threat is the intensification of agriculture in the surrounding area, which is leading to water pollution, lowering of the water-table, and shrinkage of wetlands. Some or all of the site is covered by a management plan. There is a LIPU reserve on the east side of the lake.

Foreste Casentinesi — B2, B3, C6 — 078

Admin region Toscana, Emilia Romagna
Coordinates 43°52′N 11°42′E
Altitude 500–1,658 m **Area** 36,300 ha

Site description

A large tract of broadleaved and coniferous forest in the northern Apennines, dominated by beech *Fagus* and fir *Abies*. The main land-use is forestry.

Habitats Forest and woodland (88%; broadleaved deciduous forest; mixed forest), Artificial landscape (12%; highly improved reseeded grassland; arable land)
Land-use Agriculture (10%), Forestry (80%), Nature conservation/research (10%)

Birds

Species	Season	Year	Pop min	Pop max	Acc	Criteria
Strix aluco Tawny Owl	R	1993	300	—	C	B3
Caprimulgus europaeus Nightjar	B	1993	50	100	A	B2, C6
Picus viridis Green Woodpecker	R	1993	50	100	A	B2
Lullula arborea Woodlark	R	1993	100	200	A	B2, C6
Erithacus rubecula Robin	R	1993	3,000	—	C	B3
Turdus philomelos Song Thrush	B	1993	1,000	—	C	B3
Sylvia cantillans Subalpine Warbler	B	1993	200	400	A	B3
Regulus ignicapillus Firecrest	R	1993	2,000	4,000	A	B3
Parus caeruleus Blue Tit	R	1993	2,000	—	C	B3
Emberiza cia Rock Bunting	R	1993	150	300	A	B2

An important breeding site for woodland species.

Protection status

National Partial **International** None
30,000 ha of IBA covered by National Park (Foreste Casentinesi, 30,000 ha). 69 ha of IBA covered by State Nature Reserve (Scodella, 69 ha). 1,168 ha of IBA covered by State Nature Reserve (Camaldoli, 1,168 ha). 1,631 ha of IBA covered by State Nature Reserve (Badia Prataglia, 1,631 ha). 764 ha of IBA covered by State Nature Reserve (Sasso Fratino, 764 ha). 1,191 ha of IBA covered by State Nature Reserve (Campigna, 1,191 ha).

Conservation issues

Threats Abandonment/reduction of land management (A), Infrastructure (B), Recreation/tourism (B)

The main threat is from the abandonment of pastures. Some or all of the site is covered by a management plan.

Calvana mountains — C6 — 079

Admin region Toscana
Coordinates 43°55′N 11°10′E
Altitude 79–916 m **Area** 3,900 ha

Site description

A hilly area of Tuscany, 15 km north-west of Florence. Habitats are dominated by pastures, heathland, ruderal land, grassland and woodland. Vegetation cover is highly influenced by fire and grazing. The area is mainly used for stock-grazing.

Habitats Forest and woodland (44%), Scrub (heathland), Grassland (22%), Artificial landscape (34%)
Land-use Agriculture (50%), Forestry (10%), Not utilized (40%)

Birds

Species	Season	Year	Pop min	Pop max	Acc	Criteria
Lullula arborea Woodlark	R	1996	50	100	B	C6
Anthus campestris Tawny Pipit	B	1996	40	80	B	C6
Lanius collurio Red-backed Shrike	B	1996	60	120	B	C6

A notable site for passerine species.

Protection status

National None **International** None

Conservation issues

Threats Abandonment/reduction of land management (C), Agricultural intensification/expansion (U), Afforestation (B), Burning of vegetation (B), Extraction industry (C), Recreation/tourism (C)

Afforestation, uncontrolled fires and overgrazing are the main threats.

Fucecchio marsh — B2, B3, C2, C6 — 080

Admin region Toscana
Coordinates 43°46′N 10°47′E
Altitude 13–33 m **Area** 2,500 ha

Site description

A freshwater wetland in Tuscany, a few kilometres south of the town of Montecatini. There are reedbeds, open waters, a wet woodland (Bosco di Chiusi) and surrounding cultivated areas.

Habitats Forest and woodland (15%; broadleaved deciduous forest; mixed forest), Grassland (5%; humid grassland), Wetland (40%; standing fresh water; water-fringe vegetation), Artificial landscape (40%; arable land; forestry plantation)
Land-use Agriculture (35%), Hunting (10%), Nature conservation/research (10%), Water management (10%)

Birds

Species	Season	Year	Pop min	Pop max	Acc	Criteria
Ixobrychus minutus Little Bittern	B	1994	10	30	A	C6
Nycticorax nycticorax Night Heron	B	1995	230	270	A	B2, C2, C6
Ardeola ralloides Squacco Heron	B	1996	4	10	A	C6
Egretta garzetta Little Egret	B	1995	100	150	A	C6
Locustella luscinioides Savi's Warbler	B	1996	20	50	A	B3
Acrocephalus melanopogon Moustached Warbler	R	1996	100	—	A	C6

An important site for breeding herons and reedbed passerines.

Protection status

National Partial **International** None
300 ha of IBA covered by Regional Nature Reserve (Padule di Fucecchio e Lago di Sibolla, 560 ha).

Conservation issues

Threats Afforestation (C), Agricultural intensification/expansion (B), Consequences of animal/plant introductions (B), Drainage (B), Filling-in of wetlands (A), Infrastructure (C), Recreation/tourism (C), Unsustainable exploitation (A)

The main threats are hunting and water management.

Arezzo heathlands — B2, B3, C6 — 081

Admin region Toscana
Coordinates 43°25′N 11°45′E
Altitude 400–1,300 m **Area** 4,394 ha

Site description

A hilly area of Tuscany, mainly covered by heathland. The area can be divided into three subsites: Alpe di Poti, Monte Ginezzo, Pian di Scò. Main land-uses are forestry, hunting and heath-cutting ('Other').

Habitats Forest and woodland (50%), Scrub (35%; heathland), Grassland (5%), Rocky areas (3%), Artificial landscape (7%)
Land-use Agriculture, Forestry (35%), Hunting (100%), Other (15%)

Birds

Species	Season	Year	Pop min	Pop max	Acc	Criteria
Circus pygargus Montagu's Harrier	B	1997	6	11	A	B3, C6
Sylvia undata Dartford Warbler	R	1995	350	800	A	B2, C6
Sylvia cantillans Subalpine Warbler	B	1995	250	550	C	B3

An important site for breeding raptors and passerines, especially *Sylvia* warblers.

Protection status
National None **International** None

Conservation issues

Threats Abandonment/reduction of land management (A), Afforestation (A), Burning of vegetation (A), Consequences of animal/plant introductions (B), Unsustainable exploitation (A)

The main threats are burning of vegetation, afforestation, abandonment of land management, and hunting.

Migliarino–San Rossore A1, B2, C1 082

Admin region Toscana
Coordinates 43°46′N 10°16′E
Altitude 0–5 m **Area** 7,000 ha

Site description
A wetland complex on the Tyrrhenian coast, east of the city of Pisa. The site comprises open (San Rossore) and wooded (San Rossore and Migliarino) coastal marshes, the estuaries of two large rivers (Arno and Serchio) and a tract of very shallow sea offshore. There are dunes between the marshes and the sea and extensive arable/pasture areas at San Rossore. The main land-use is nature conservation.

Habitats Forest and woodland (60%), Wetland (12%; sand-dunes/sand beach; river/stream; water-fringe vegetation), Artificial landscape (28%; arable land)
Land-use Forestry (25%), Nature conservation/research (70%)

Birds

Species	Season	Year	Pop min	Pop max	Acc	Criteria
Lymnocryptes minimus Jack Snipe	W	1995	—	5	A	B2
Numenius tenuirostris Slender-billed Curlew	P	—	—	—	—	A1, C1

An important site for some migrating and wintering waders. Species of global conservation concern that do not meet IBA criteria: *Aythya nyroca* (max. 2 in winter, also occurs on passage).

Protection status
National High **International** Unknown
7,000 ha of IBA covered by Regional Nature Park (Migliarino–San Rossore–Massaciuccoli, 22,000 ha). Overlaps with SPA.

Conservation issues

Threats Afforestation (C), Consequences of animal/plant introductions (B), Natural events (A), Other (A)

The main threats are coastal erosion and river pollution ('Other' threat, above). Some or all of the site is covered by a management plan.

Pools of Florence plain A1, B2, C1, C2, C6 083

Admin region Toscana
Coordinates 43°48′N 11°04′E
Altitude 30–42 m **Area** 1,000 ha

Site description
A complex of small freshwater pools, wetlands, abandoned pits and cultivated land on a plain west of Florence. The pools are used mainly as game reserves. Agriculture is the most important human activity away from the pools.

Habitats Wetland (40%; standing fresh water; water-fringe vegetation), Artificial landscape (60%; highly improved reseeded grassland; arable land; other urban/industrial areas; ruderal land)
Land-use Agriculture (55%), Hunting (30%), Urban/industrial/transport (5%)

Birds

Species	Season	Year	Pop min	Pop max	Acc	Criteria
Nycticorax nycticorax Night Heron	B	1996	200	300	A	B2, C2, C6
Aythya nyroca Ferruginous Duck	P	1996	—	20	C	A1, C1

Important especially for the colony of *Nycticorax nycticorax*.

Protection status
National None **International** None

Conservation issues

Threats Agricultural intensification/expansion (B), Burning of vegetation (C), Construction/impact of dyke/dam/barrage (B), Disturbance to birds (B), Extraction industry (C), Industrialization/urbanization (A), Infrastructure (A), Recreation/tourism (C)

The main threats are industrialization, urbanization, hunting and the development of infrastructure. One of the pools is a proposed LIPU reserve.

Ponte Buriano–Lago della Penna C6 084

Admin region Toscana
Coordinates 43°31′N 11°45′E
Altitude 208–208 m **Area** 200 ha

Site description
An artificial lake and associated freshwater wetland, partly covered by reedbeds and sedges, located about 10 km north-west of the town of Arezzo. The surrounding areas are mainly cultivated. Main land-use is water management for producing electricity.

Habitats Forest and woodland (20%; broadleaved deciduous forest; alluvial/very wet forest), Scrub (5%; scrub), Grassland (5%), Wetland (70%; standing fresh water; water-fringe vegetation)
Land-use Fisheries/aquaculture (50%), Nature conservation/research (100%), Tourism/recreation (10%), Water management (70%)

Birds

Species	Season	Year	Pop min	Pop max	Acc	Criteria
Ardeola ralloides Squacco Heron	P	1995	5	20	A	C6
Aythya nyroca Ferruginous Duck	P	1995	1	10	A	C6

Species of global conservation concern that do not meet IBA criteria: *Aythya nyroca* (1, winter).

Protection status
National High **International** None
200 ha of IBA covered by Wildlife Reserve (Ponte Buriano–Lago della Penna, 200 ha).

Conservation issues

Threats Burning of vegetation (A), Consequences of animal/plant introductions (B), Construction/impact of dyke/dam/barrage (A), Disturbance to birds (B), Drainage (B), Recreation/tourism (B)

The main threats are posed by water-level regulation, burning of reedbeds, wetland reclamation, and disturbance of birds by fishermen.

Mount Conero B1iv, C5, C6 085

Admin region Marche
Coordinates 43°32′N 12°45′E
Altitude 151–572 m **Area** 1,140 ha

Site description
A mountain promontory along the west-central Adriatic coast, about 10 km south-east of the town of Ancona. The main habitats are broadleaved evergreen woodland and maquis, with many rocky cliffs and offshore islets. Main land-uses are nature conservation, forestry and tourism.

Habitats Forest and woodland (70%; broadleaved evergreen forest), Scrub (10%; sclerophyllous scrub/garrigue/maquis), Grassland (20%), Rocky areas (sea cliff/rocky shore; rock stacks/islets)
Land-use Agriculture (20%), Forestry (50%), Military (5%), Nature conservation/research (100%), Tourism/recreation (100%), Urban/industrial/transport (15%)

Birds

Species	Season	Year	Pop min	Pop max	Acc	Criteria
Emberiza hortulana Ortolan Bunting	B	1996	15	—	B	C6

An important 'bottleneck' site for migrating raptors, with 4,000–5,000 regularly passing each spring, mainly *Pernis apivorus*, *Milvus migrans*, *Circaetus gallicus* and *Circus aeruginosus*.

■ Protection status

National High **International** None
1,140 ha of IBA covered by Regional Nature Park (Conero, 5,820 ha).

■ Conservation issues

Threats Afforestation (C), Burning of vegetation (B), Deforestation (commercial) (C), Disturbance to birds (B), Filling-in of wetlands (C), Firewood collection (B), Recreation/tourism (B)

The main threats are tourism and the burning of vegetation, although none have a high impact.

Furlo gorge and Mount Catria B3, C6 086

Admin region Marche
Coordinates 43°33'N 12°40'E
Altitude 200–1,701 m **Area** 36,000 ha

■ Site description

An area of hills and mountains, south of the town of Urbino, mainly covered by broadleaved woodland. Gola del Furlo is a gorge. Main land-uses are forestry, hunting and tourism.

Habitats Forest and woodland (70%; broadleaved deciduous forest; treeline ecotone), Scrub (10%; scrub; heathland), Grassland (10%; alpine/subalpine/boreal grassland), Rocky areas (5%; scree/boulders; inland cliff), Artificial landscape (5%)
Land-use Agriculture (10%), Forestry (35%), Hunting, Nature conservation/research, Tourism/recreation, Urban/industrial/transport (5%)

■ Birds

Species	Season	Year	Pop min	Pop max	Acc	Criteria
Aquila chrysaetos Golden Eagle	R	1996	2	3	A	C6
Falco peregrinus Peregrine	R	1996	2	4	A	C6
Alectoris graeca Rock Partridge	R	1996	15	25	A	C6
Anthus campestris Tawny Pipit	B	1996	Common		—	C6
Pyrrhocorax pyrrhocorax Chough	R	1996	20	30	B	C6
Corvus monedula Jackdaw	R	—	300	500	B	B3

The site is important for breeding raptors.

■ Protection status

National None **International** None

■ Conservation issues

Threats Abandonment/reduction of land management (A), Afforestation (B), Consequences of animal/plant introductions (B), Deforestation (commercial) (B), Disturbance to birds (C), Extraction industry (B), Firewood collection (C), Recreation/tourism (C)

The main threat is the reduction in extent of open grassland, due to the abandonment of pastures which leads to afforestation with non-native conifers as well as natural forest regeneration.

Sentina C6 087

Admin region Marche
Coordinates 43°29'N 13°40'E
Altitude 0–5 m **Area** 180 ha

■ Site description

A coastal area of sand-dunes, grasslands and small brackish pools. Most of the site is now cultivated but, after heavy rains or influxes of seawater, the fields are flooded and temporary pools are created. The area had been largely drained by the mid-1980s.

Habitats Grassland (10%; humid grassland), Wetland (20%; saltmarsh; sand-dunes/sand beach; standing brackish and salt water; river/stream; water-fringe vegetation), Marine areas (10%; open sea), Artificial landscape (60%; arable land; ruderal land)
Land-use Agriculture (60%), Nature conservation/research (10%), Tourism/recreation (10%)

■ Birds

Species	Season	Year	Pop min	Pop max	Acc	Criteria
Calandrella brachydactyla Short-toed Lark	B	1998	3	3	A	C6

For many migrating waterbirds this is the only stop-over site along the Adriatic coast between the Gargano promontory and the wetland complexes of the Emilia Romagna region.

■ Protection status

National None **International** None

■ Conservation issues

Threats Abandonment/reduction of land management (A), Disturbance to birds (A), Industrialization/urbanization (A), Recreation/tourism (B)

There are many important threats: disturbance of birds due to the frequent presence of people (even for illegal activities); projects to develop tourist resorts in the area; pools dry up in the absence of the water management that was formerly done by hunters' associations. The area is a wildlife reserve, designated by the provincial authority, and thus hunting is now forbidden. The site is a proposed 'Important Flora Area' at the regional level. There have been scattered bird counts by LIPU, and in 1998 LIPU started a bird-ringing scheme at the site.

River Cecina B3 088

Admin region Toscana
Coordinates 43°20'N 10°52'E
Altitude 50–450 m **Area** 2,400 ha

■ Site description

The site lies about 10 km south of the town of Volterra, in Tuscany, and comprises a large broadleaved evergreen woodland, called Berignone, and a stretch of the River Cecina. Habitats include rocky cliffs, rivers, streams, grasslands and cultivated areas. Main land-uses are forestry, agriculture and hunting.

Habitats Forest and woodland (60%; broadleaved deciduous forest; broadleaved evergreen forest), Scrub (5%; sclerophyllous scrub/garrigue/maquis), Wetland (15%; river/stream; water-fringe vegetation), Rocky areas (inland cliff), Artificial landscape (20%; highly improved reseeded grassland)
Land-use Agriculture (20%), Forestry (50%), Hunting (40%)

■ Birds

Species	Season	Year	Pop min	Pop max	Acc	Criteria
Sylvia cantillans Subalpine Warbler	B	1995	200	400	A	B3

The site is also notable for its breeding raptors and woodland passerines.

■ Protection status

National Partial **International** None
IBA contains Wildlife Reserve (Berignone, area not known).

■ Conservation issues

Threats Abandonment/reduction of land management (B), Agricultural intensification/expansion (A), Construction/impact of dyke/dam/barrage (B), Infrastructure (B), Other (A)

The main threats are linked to agricultural intensification, abandonment of traditional pastures, and unsustainable exploitation of woodland resources ('Other' threat). The site is a proposed Regional Nature Reserve.

Bolgheri C6 089

Admin region Toscana
Coordinates 43°14'N 10°33'E
Altitude 0–12 m **Area** 562 ha

■ Site description

A wetland on the Tyrrhenian coast, about 10 km south of the village of Cecina. Habitats include sand beaches and dunes, maquis, wet woodlands and grasslands, and open water. The surroundings include arable land. Its main use is for nature conservation.

Habitats Forest and woodland (30%; alluvial/very wet forest), Grassland (20%), Wetland (30%; sand-dunes/sand beach; standing fresh water), Artificial landscape (20%; arable land)
Land-use Agriculture (20%), Nature conservation/research (100%), Water management (30%)

Birds

Species	Season	Year	Pop min	Pop max	Acc	Criteria
Pluvialis apricaria Golden Plover	W	1994	70	137	A	C6

Site is also notable for breeding and wintering waterbirds and migrating waders. Species of global conservation concern that do not meet IBA criteria: *Aythya nyroca* (4, winter).

Protection status

National Partial **International** High
500 ha of IBA covered by Wildlife Reserve (Bolgheri, 500 ha). 518 ha of IBA covered by Special Protection Area (Bolgheri, 518 ha). 562 ha of IBA covered by Ramsar Site (Palude di Bolgheri, 562 ha).

Conservation issues

Threats Agricultural intensification/expansion (A), Filling-in of wetlands (B), Natural events (B)

The main threat is the unsustainable use of water resources for agriculture.

Crete Senesi — B2, C6 — 090

Admin region Toscana
Coordinates 43°12′N 11°22′E
Altitude 130–400 m **Area** 50,000 ha

Site description

A hilly area of Tuscany, south-east of the town of Siena. There are large tracts of cultivated land, small woodlands, gorges and highly eroded slopes. Agriculture is the main land-use.

Habitats Forest and woodland (15%), Scrub (15%), Grassland (5%), Rocky areas (inland cliff), Artificial landscape (65%)
Land-use Agriculture (85%), Hunting (70%), Urban/industrial/transport (10%)

Birds

Species	Season	Year	Pop min	Pop max	Acc	Criteria
Falco biarmicus Lanner	R	1995	1	3	A	B2, C6
Burhinus oedicnemus Stone Curlew	B	1995	5	10	B	C6

Protection status

National None **International** None

Conservation issues

Threats Abandonment/reduction of land management (B), Agricultural intensification/expansion (A), Disturbance to birds (C), Firewood collection (C), Forest grazing (B)

The main threat is agricultural intensification, which destroys the eroded land-forms that are important nest-sites for breeding raptors.

Lake Trasimeno — A4iii, B1i, C3, C4, C6 — 091

Admin region Umbria
Coordinates 43°09′N 12°05′E
Altitude 258–262 m **Area** 13,200 ha

Site description

A freshwater eutrophic lake, about 10 km west of the town of Perugia. It is the largest lake in central Italy (12,000 ha) and the fourth-largest in the country. There is a belt of reeds *Phragmites*, of variable width, all around the lakeshore, and submerged beds of waterweed (*Potamogeton*, *Ceratophyllum*) offshore. Wet woodlands (*Salix*, *Alnus*, *Ulmus*) and grasslands are also present. Tourism is the main activity; pig-farming and intensive agriculture dominate the surrounding area.

Habitats Wetland (95%; standing fresh water; water-fringe vegetation), Artificial landscape (5%; highly improved reseeded grassland; other urban/industrial areas)
Land-use Agriculture (5%), Fisheries/aquaculture (80%), Nature conservation/research (100%), Tourism/recreation (5%), Urban/industrial/transport (5%)

Birds

Species	Season	Year	Pop min	Pop max	Acc	Criteria
Phalacrocorax carbo Cormorant	W	1996	1,459	2,400	A	B1i, C3
Nycticorax nycticorax Night Heron	B	1996	100	100	A	C6
Ardeola ralloides Squacco Heron	B	1996	13	—	A	C6
Egretta garzetta Little Egret	B	1996	100	100	A	C6

An area of international importance for wintering *Phalacrocorax carbo* and other waterbirds collectively (more than 20,000 occur on a regular basis, mainly *Fulica atra*) and for the presence of a mixed heronry. Winter count data refer to January 1993, 1994 and 1996. The number of wintering birds has greatly increased (especially *Fulica atra*) following the declaration of the Regional Park in 1995 and the lowering of water-levels.

Protection status

National High **International** None
13,200 ha of IBA covered by Regional Nature Park (Trasimeno, 13,200 ha).

Conservation issues

Threats Agricultural intensification/expansion (A), Burning of vegetation (B), Consequences of animal/plant introductions (B), Filling-in of wetlands (A), Natural events (B), Recreation/tourism (A)

The main threats are: the pollution of the wetland by nutrients and pesticides from the surrounding agricultural areas; the reduction in reedbed extent due to erosion; the creation of new beaches; the creation of new arable land; fires in the reedbeds; coypu *Myocastor coypus*, a non-native aquatic rodent, has dramatically reduced the extent of floating water-lilies *Nymphaea*; extraction of water for agriculture. Some or all of the site is covered by a management plan. Research is carried out on hydrobiology (Università di Perugia), botany (Università di Perugia, Università di Camerino) and ornithology (Provincia di Perugia).

Farma valley — B2 — 092

Admin region Toscana
Coordinates 43°04′N 11°04′E
Altitude 167–525 m **Area** 10,000 ha

Site description

A large, hilly, wooded valley in Tuscany, with highly intact beech woodlands (*Fagus*) and crossed by the Farma river. Main land-use is forestry.

Habitats Forest and woodland (64%; broadleaved deciduous forest), Scrub (8%), Wetland (3%; river/stream), Rocky areas (1%; inland cliff), Artificial landscape (22%)
Land-use Forestry (50%)

Birds

Species	Season	Year	Pop min	Pop max	Acc	Criteria
Picus viridis Green Woodpecker	R	1996	80	—	B	B2

Also a notable breeding site for raptors and woodland passerines.

Protection status

National Partial **International** None
1,509 ha of IBA covered by Regional Nature Reserve (Valle del Farma, 1,509 ha).

Conservation issues

Threats Infrastructure (C), Recreation/tourism (C)

Threats are associated with the development of tourism. Some or all of the site is covered by a management plan.

Lakes Montepulciano and Chiusi — B1i, B2, C3, C6 — 093

Admin region Toscana
Coordinates 43°04′N 11°58′E
Altitude 248–253 m **Area** 900 ha

Site description
Two freshwater lakes, 15 km east of the village of Montepulciano. The lakes are surrounded by extensive reedbeds, cultivated land and small broadleaved woodlands, and there are also some small artificial ponds called Vasche di Dolciano. Lake water-levels fluctuate greatly during the year.

Habitats Forest and woodland (5%; alluvial/very wet forest), Wetland (80%; standing fresh water; water-fringe vegetation), Artificial landscape (10%; arable land; perennial crops/orchards/groves)
Land-use Agriculture (10%), Hunting (60%), Nature conservation/research (40%), Tourism/recreation (20%), Water management (70%)

Birds

Species	Season	Year	Pop min	Pop max	Acc	Criteria
Phalacrocorax carbo Cormorant	W	1996	600	1,200	A	B1i, C3
Ixobrychus minutus Little Bittern	B	1996	30	50	A	B2, C6
Nycticorax nycticorax Night Heron	B	1995	120	150	A	C6
Ardeola ralloides Squacco Heron	B	1995	6	10	A	C6
Acrocephalus melanopogon Moustached Warbler	R	1996	20	—	B	C6

An important area for breeding *Ixobrychus minutus*, wintering *Phalacrocorax carbo*, and for a mixed heronry at Lago di Chiusi. Herons and cormorants roost at Vasche di Dolciano. Species of global conservation concern that do not meet IBA criteria: *Aythya nyroca* (1–2 pairs breed, max. 4 in winter).

Protection status
National Partial **International** None
300 ha of IBA covered by Wildlife Reserve (Lago di Montepulciano, 300 ha).

Conservation issues
Threats Abandonment/reduction of land management (B), Agricultural intensification/expansion (B), Aquaculture/fisheries (B), Burning of vegetation (B), Disturbance to birds (A), Drainage (B), Other (A), Recreation/tourism (C), Unsustainable exploitation (A)

The main threats are: the lack of management of the fluctuating water-levels ('Other' threat, above); water pollution from surrounding cultivated areas; disturbance of birds by hunting in the surroundings; burning of vegetation. Some or all of the site is covered by a management plan. There is a LIPU reserve covering part of Lake Montepucliano. A LIFE-supported project at Lake Montepulciano for conservation of *Aythya nyroca* and *Botaurus stellaris*.

Colfiorito — B2, C6 — 094

Admin region Umbria
Coordinates 43°01′N 12°52′E
Altitude 752–988 m **Area** 500 ha

Site description
A freshwater wetland (157 ha) in the central Apennines, close to the village of Colfiorito. The wetland is fringed by a reedbed and the site includes the surrounding hilly areas (Piani di Ricciano, Piani di Arvello, Piano di Annifo).

Habitats Forest and woodland (30%; broadleaved deciduous forest), Grassland (15%), Wetland (30%; standing fresh water; water-fringe vegetation), Rocky areas (5%), Artificial landscape (20%; arable land; perennial crops/orchards/groves; forestry plantation)
Land-use Agriculture (10%), Nature conservation/research (30%), Tourism/recreation (5%)

Birds

Species	Season	Year	Pop min	Pop max	Acc	Criteria
Botaurus stellaris Bittern	R	1996	7	9	B	C6
Ixobrychus minutus Little Bittern	B	1996	7	—	C	C6
Circus pygargus Montagu's Harrier	B	1996	2	2	A	C6
Falco biarmicus Lanner	R	1996	1	2	A	B2, C6

An important breeding site for reedbed species, especially *Botaurus stellaris*.

Protection status
National Partial **International** Partial
157 ha of IBA covered by Regional Nature Park (Colfiorito, 157 ha). 157 ha of IBA covered by Special Protection Area (Palude di Colfiorito, 157 ha). 157 ha of IBA covered by Ramsar Site (Palude di Colfiorito, 157 ha).

Conservation issues
Threats Agricultural intensification/expansion (A), Disturbance to birds (B), Drainage (C), Infrastructure (A), Unsustainable exploitation (A)

The main threats are agricultural intensification (leading to wetland drainage, and water pollution with nutrients and pesticides), hunting around the wetland, and a road-building project which will cross the area. Some or all of the site is covered by a management plan. Ornithological research takes place.

Sibillini mountains — A3, B2, C6 — 095

Admin region Umbria, Marche
Coordinates 42°56′N 13°10′E
Altitude 604–2,448 m **Area** 71,374 ha

Site description
A mountain area in the central Apennines, with large tracts of broadleaved woodland and especially extensive alpine grasslands. Main land-uses are nature conservation, tourism and forestry.

Habitats Forest and woodland (30%; broadleaved deciduous forest; treeline ecotone), Scrub (5%; scrub; heathland), Grassland (45%; alpine/subalpine/boreal grassland), Wetland (5%; river/stream), Rocky areas (10%; inland cliff), Artificial landscape (5%)
Land-use Agriculture (5%), Forestry (15%), Nature conservation/research (100%), Tourism/recreation (60%), Urban/industrial/transport (5%)

Birds

Species	Season	Year	Pop min	Pop max	Acc	Criteria
Falco biarmicus Lanner	R	—	—	—	—	B2, C6
Alectoris graeca Rock Partridge	R	1995	10	100	C	B2, C6
Prunella collaris Alpine Accentor	R	—	Frequent		—	A3
Tichodroma muraria Wallcreeper	R	—	Uncommon		—	A3
Pyrrhocorax graculus Alpine Chough	R	—	Frequent		—	A3
Pyrrhocorax pyrrhocorax Chough	R	1996	10	30	B	C6
Montifringilla nivalis Snowfinch	R	—	Frequent		—	A3

A breeding site for four of the ten species of the Eurasian high-montane biome, as well as raptors and owls.

Protection status
National High **International** Partial
71,374 ha of IBA covered by National Park (Sibillini, 71,374 ha). 17,810 ha of IBA covered by Special Protection Area (Monti Sibillini (versante umbro), 17,810 ha).

Conservation issues
Threats Construction/impact of dyke/dam/barrage (B), Disturbance to birds (B), Groundwater abstraction (A), Infrastructure (A), Intensified forest management (B), Recreation/tourism (A)

Recreation, tourism and the building of new infrastructure (e.g. roads) are the main threats to the area.

Tuscan archipelago — A1, A3, A4i, A4ii, B1i, B1ii, B2, B3, C1, C2, C3, C6 — 096

Admin region Toscana
Coordinates 42°48′N 10°20′E
Altitude 0–1,000 m **Area** 28,000 ha

Site description
A complex of seven small islands between Tuscany and Corse (Montecristo, Giannutri, Gorgona, Capraia, Elba, Giglio e Pianosa). They are mainly covered by maquis, garrigue, sclerophyllous scrub, pine woodlands and have many rocky cliffs. Some have evergreen

woodlands, grasslands and agricultural land. The main land-uses are tourism, fishing, boating and agriculture.

Habitats Forest and woodland (15%; native coniferous forest), Scrub (45%; sclerophyllous scrub/garrigue/maquis), Grassland (5%), Wetland (1%), Rocky areas (7%; sea cliff/rocky shore), Artificial landscape (27%; arable land; other urban/ industrial areas)
Land-use Agriculture (15%), Fisheries/aquaculture (5%), Nature conservation/research (100%), Tourism/recreation (5%)

■ Birds

Species	Season	Year	Pop min	Pop max	Acc	Criteria
Calonectris diomedea Cory's Shearwater	B	1994	100	800	B	B1ii, B2, C2, C6
Puffinus yelkouan Yelkouan Shearwater	B	1994	200	1,000	B	A4ii, B1ii, C3
Phalacrocorax aristotelis Shag	R	1994	20	25	A	C6
Falco peregrinus Peregrine	R	1994	15	25	A	B2, C6
Larus audouinii Audouin's Gull	R	1994	50	85	A	A1, C1, C6
Larus cachinnans Yellow-legged Gull	R	1994	—	10,000	B	A4i, B1i, C3
Oenanthe hispanica Black-eared Wheatear	B	1994	10	30	B	A3
Monticola solitarius Blue Rock Thrush	R	1994	100	200	B	B2
Sylvia sarda Marmora's Warbler	R	1994	80	150	B	A3, B3, C6
Sylvia undata Dartford Warbler	R	1994	500	—	B	B2, C6
Sylvia conspicillata Spectacled Warbler	B	1994	10	100	B	A3
Sylvia cantillans Subalpine Warbler	B	1994	800	—	B	A3, B3
Sylvia melanocephala Sardinian Warbler	R	1994	1,000	—	B	A3

The islands support five of the 21 species of the Mediterranean biome, as well as breeding *Calonectris diomedea*, *Puffinus yelkouan* and *Larus audouinii*, among other seabirds.

■ Protection status

National High **International** Low
28,000 ha of IBA covered by National Park (Arcipelago Toscano, 67,500 ha). 500 ha of IBA covered by Special Protection Area (Isola di Capraia, 500 ha).

■ Conservation issues

Threats Abandonment/reduction of land management (B), Afforestation (C), Burning of vegetation (B), Disturbance to birds (B), Extraction industry (C), Industrialization/ urbanization (C), Infrastructure (C), Recreation/tourism (B), Unsustainable exploitation (C)

The main threats are: the disturbance of breeding seabirds by tourists, especially using motorboats; the development of tourist resorts and associated infrastructure; burning of vegetation. Mediterranean monk seal *Monachus monachus* (Critically Endangered) was recorded here until the 1980s.

Diaccia Botrona — A1, B1i, C1, C2, C3, C6 — 097

Admin region Toscana
Coordinates 42°47'N 10°55'E
Altitude 1–1 m **Area** 1,350 ha

■ Site description

A large wetland on the Tyrrhenian coast, close to the village of Castglione della Pescaia, that was formerly freshwater but which is now becoming brackish due to seawater inflow. The main habitats are large reedbeds, open water, and a small coniferous wood. On the southern and eastern borders are present large areas of cultivated fields on reclaimed land. The main land-use is nature conservation.

Habitats Forest and woodland (5%; native coniferous forest), Grassland (10%; humid grassland), Wetland (85%; standing fresh water; standing brackish and salt water; water-fringe vegetation)
Land-use Fisheries/aquaculture (3%), Nature conservation/research (100%)

■ Birds

Species	Season	Year	Pop min	Pop max	Acc	Criteria
[1] ***Botaurus stellaris*** Bittern	R	1996	3	18	A	C2, C6
Egretta garzetta Little Egret	B	1992	138	—	A	C6
Anas penelope Wigeon	W	1994	4,000	9,000	A	B1i, C3
[2] ***Numenius tenuirostris*** Slender-billed Curlew	N	1995	1	1	A	A1, C1

1. 3 in 1996, 18 in 1991.
2. Two records: 1 in April 1978, 2 in January 1984.

An important site for wintering and migrating waterbirds (more than 15,000), for breeding *Botaurus stellaris*, and for *Numenius tenuirostris* in winter and on passage. There is a heronry in the wood. Species of global conservation concern that do not meet IBA criteria: *Aythya nyroca* (max. 2, winter).

■ Protection status

National High **International** High
1,350 ha of IBA covered by Regional Nature Reserve (Diaccia Botrona, 1,350 ha). Part of IBA covered by Special Protection Area (Diaccia Botrona, 1,200 ha). 1,350 ha of IBA covered by Ramsar Site (Palude della Diaccia Botrona, 2,500 ha).

■ Conservation issues

Threats Aquaculture/fisheries (A), Disturbance to birds (A), Drainage (C), Groundwater abstraction (B)

The main threat is from aquaculture, which is causing saltwater intrusion along the southern border of the site. Some or all of the site is covered by a management plan.

Uccellina mountains, Trappola marshes and Ombrone mouth — A1, B1i, C1, C3 — 098

Admin region Toscana
Coordinates 42°38'N 11°00'E
Altitude 0–417 m **Area** 5,670 ha

■ Site description

The site comprises wooded hills, mainly covered by *Pinus* and maquis (Monti dell'Uccellina), as well as a large brackish marsh (Stagni della Trappola) with very little vegetation, mainly *Salicornia*, and a river mouth (Bocca d'Ombrone) on the Tyrrhenian coast of Tuscany, about 10 km south of the town of Grosseto. Habitats also include rocky sea cliffs and sand beaches. The main land-use is nature conservation.

Habitats Forest and woodland (40%; native coniferous forest), Scrub (24%; sclerophyllous scrub/garrigue/maquis), Grassland (3%), Wetland (15%; saltmarsh; river/ stream), Marine areas, Rocky areas (sea cliff/rocky shore), Artificial landscape (12%)
Land-use Agriculture (5%), Nature conservation/research (100%), Tourism/recreation (30%)

■ Birds

Species	Season	Year	Pop min	Pop max	Acc	Criteria
Anser anser Greylag Goose	W	1995	422	422	A	B1i, C3
Numenius tenuirostris Slender-billed Curlew	N	—	—	—	—	A1, C1

The site is important for wintering waterbirds (especially *Anser anser*) and is a key non-breeding site for *Numenius tenuirostris*.

■ Protection status

National High **International** Unknown
5,670 ha of IBA covered by Regional Nature Park (Maremma, 9,800 ha). Overlaps with SPA.

■ Conservation issues

Threats Afforestation (C), Agricultural intensification/expansion (B), Disturbance to birds (B), Natural events (A), Other (B), Recreation/tourism (C)

Important threats are: coastal erosion; the disturbance of breeding and wintering birds by tourists who walk among the sand-dunes; the poor water-quality of the Ombrone river, which affects Stagni della Trappola ('Other' threat); and overgrazing, which threatens *Burhinus oedicnemus*. Some or all of the site is covered by a management plan.

Lake Bolsena — B2, C6 — 099

Admin region Lazio
Coordinates 42°35'N 11°55'E
Altitude 305–550 m **Area** 16,000 ha

■ Site description

A freshwater lake of volcanic origin, 15 km north of the town of Viterbo, with large areas of reedbeds. The surroundings include large oak *Quercus* woodlands. The main land-uses are fishing, hunting and tourism.

Habitats Forest and woodland (25%; broadleaved deciduous forest; broadleaved evergreen forest), Wetland (75%; standing fresh water; river/stream; water-fringe vegetation)
Land-use Fisheries/aquaculture (75%), Hunting (100%), Tourism/recreation (80%), Urban/industrial/transport (20%)

Birds

Species	Season	Year	Pop min	Pop max	Acc	Criteria
Ixobrychus minutus Little Bittern	B	1995	15	20	A	C6
Milvus migrans Black Kite	B	1995	10	15	A	B2, C6
Otus scops Scops Owl	B	1995	30	40	A	B2
Alcedo atthis Kingfisher	R	1995	20	25	A	C6

The reedbeds are important for breeding *Ixobrychus minutus*. The woodlands are important for raptors.

Protection status

National None **International** None

Conservation issues

Threats Agricultural intensification/expansion (B), Aquaculture/fisheries (C), Consequences of animal/plant introductions (C), Disturbance to birds (B), Industrialization/urbanization (B), Infrastructure (B), Recreation/tourism (B)

Urbanization and water pollution are the main threats. Urbanization refers to new buildings along the lake shore and in the woodland. Water is polluted by pesticides and nutrients from the surrounding cultivated land. Disturbance to birds is also an important threat, caused by hunters (at the lake and in the woodlands) and by boats during the tourism season.

Lake Alviano — C6 — 100

Admin region Umbria, Lazio
Coordinates 42°34′N 12°15′E
Altitude 80–90 m **Area** 900 ha

Site description

A man-made lake in central Italy, created by the construction of a hydroelectric dam on the Tevere river. Habitats include open water, a large reedbed (450 ha), wet woodland of poplar *Populus*, alder *Alnus* and willow *Salix* (150 ha), and freshwater marshes. The surroundings are agricultural. The main land-use is nature conservation.

Habitats Forest and woodland (15%; alluvial/very wet forest), Grassland (10%; humid grassland), Wetland (55%; standing fresh water; river/stream; water-fringe vegetation), Artificial landscape (20%; highly improved reseeded grassland; perennial crops/orchards/groves)
Land-use Agriculture (20%), Fisheries/aquaculture (30%), Nature conservation/research (80%), Tourism/recreation, Urban/industrial/transport (5%)

Birds

Species	Season	Year	Pop min	Pop max	Acc	Criteria
Milvus migrans Black Kite	B	1995	4	8	A	C6
Alcedo atthis Kingfisher	R	1995	2	10	A	C6

An important stop-over point for some wetland species, mainly those on passage through central Italy.

Protection status

National High **International** High
900 ha of IBA covered by Wildlife Reserve (Lago di Alviano, 900 ha). 900 ha of IBA covered by Special Protection Area (Valle del Tevere: Laghi Corbara-Alviano [contains Lago di Alviano SPA], 7,200 ha).

Conservation issues

Threats Agricultural intensification/expansion (C), Aquaculture/fisheries (C), Industrialization/urbanization (C), Other (B)

The main threat is the low water-quality of the Tevere river, which carries high levels of nutrients and heavy metals. Some or all of the site is covered by a management plan. The area is managed by WWF-Italia. There are ongoing projects on the reintroduction of *Aythya nyroca* (WWF-Italia and University of Tuscia), studies of parasites (University of Pisa) and of aquatic insects (University of Perugia).

Laga mountains — A1, B2, C1, C6 — 101

Admin region Lazio, Abruzzo, Marche
Coordinates 42°40′N 13°26′E
Altitude 600–2,458 m **Area** 60,000 ha

Site description

A mountain area in the central Apennines, between Monti Sibillini (in the north) and Gran Sasso d'Italia (in the south). The main habitats are extensive broadleaved forests (oak *Quercus*, beech *Fagus*), alpine grasslands, cultivated land, and a freshwater lake (Lago di Campotosto). The main land-use is nature conservation and tourism.

Habitats Forest and woodland (70%; broadleaved deciduous forest; treeline ecotone), Scrub (9%), Grassland (14%; alpine/subalpine/boreal grassland), Wetland (6%; standing fresh water; river/stream), Rocky areas (1%; inland cliff)
Land-use Agriculture (10%), Nature conservation/research (100%), Tourism/recreation (70%)

Birds

Species	Season	Year	Pop min	Pop max	Acc	Criteria
[1] ***Aythya nyroca*** Ferruginous Duck	W	1995	10	10	A	A1, C1
Falco tinnunculus Kestrel	R	1996	50	—	B	B2
Alectoris graeca Rock Partridge	R	1996	100	—	B	B2, C6
Bubo bubo Eagle Owl	R	1996	1	2	B	C6
Picus viridis Green Woodpecker	R	1996	80	—	B	B2
Anthus campestris Tawny Pipit	B	1995	Common		—	C6
Lanius collurio Red-backed Shrike	B	1996	100	—	B	C6
Pyrrhocorax pyrrhocorax Chough	R	1996	60	—	B	B2
Emberiza hortulana Ortolan Bunting	B	1996	80	—	B	C6

1. 10 in 1987 and 1995.

The lake holds nationally important numbers of wintering waterbirds.

Protection status

National High **International** Low
1,600 ha of IBA covered by State Nature Reserve (Lago di Campotosto, 1,600 ha). 800 ha of IBA covered by Regional Nature Reserve (Gole del Salinello, 800 ha). 60,000 ha of IBA covered by National Park (Gran Sasso–Monti della Laga, 145,000 ha). IBA contains Special Protection Area (Monti della Laga, 4,841 ha).

Conservation issues

Threats Abandonment/reduction of land management (C), Disturbance to birds (C), Infrastructure (B), Recreation/tourism (B)

The main threats are activities linked to tourism and the building of new infrastructure (e.g. roads). Regular bird monitoring is carried out at Lago di Campotosto.

Selva del Lamone — B2, B3, C6 — 102

Admin region Lazio
Coordinates 42°30′N 11°34′E
Altitude 200–460 m **Area** 2,200 ha

Site description

A large old broadleaved woodland of oak *Quercus*, crossed by streams and surrounded by cultivated land. The area is located in the upper part of River Fiora, in central Italy, west of Lake Bolsena. Main land-uses are nature conservation and forestry.

Habitats Forest and woodland (80%; broadleaved deciduous forest), Scrub (10%), Artificial landscape (10%; arable land)
Land-use Agriculture (10%), Forestry (20%), Nature conservation/research (80%)

Birds

Species	Season	Year	Pop min	Pop max	Acc	Criteria
Streptopelia turtur Turtle Dove	B	1996	500	—	A	B2
Lullula arborea Woodlark	R	1996	200	—	A	B2, C6
Parus caeruleus Blue Tit	R	1996	1,000	—	B	B3

An important breeding area for woodland species.

Protection status

National Partial **International** None

1,600 ha of IBA covered by Regional Nature Reserve (Selva del Lamone, 1,600 ha).

Conservation issues

Threats Agricultural intensification/expansion (C), Deforestation (commercial) (C), Forest grazing (C)

Argentario — A3 — 103

Admin region Toscana
Coordinates 42°24′N 11°07′E
Altitude 0–450 m **Area** 5,200 ha

Site description

A calcareous promontory along the Tyrrhenian coast. The main habitats are maquis, garrigue, broadleaved woodland and rocky cliffs. Land-uses are mainly tourism and forestry.

Habitats Forest and woodland (75%; broadleaved evergreen forest), Scrub (20%; sclerophyllous scrub/garrigue/maquis), Rocky areas (5%; sea cliff/rocky shore)
Land-use Forestry (5%), Hunting (70%), Tourism/recreation (50%)

Birds

Species	Season	Year	Pop min	Pop max	Acc	Criteria
Oenanthe hispanica Black-eared Wheatear	B	1990	10	—	C	A3
Sylvia sarda Marmora's Warbler	R	1990	20	—	C	A3
Sylvia conspicillata Spectacled Warbler	R	1990	10	—	C	A3
Sylvia cantillans Subalpine Warbler	B	1990	500	—	C	A3
Sylvia melanocephala Sardinian Warbler	R	1990	1,000	—	C	A3
Emberiza melanocephala Black-headed Bunting	B	1990	5	—	A	A3

An important site for six of the 21 species of the Mediterranean biome. Breeding species of global conservation concern that do not meet IBA criteria: *Falco naumanni* (rare).

Protection status

National None **International** None

Conservation issues

Threats Afforestation (B), Burning of vegetation (A), Industrialization/urbanization (A), Infrastructure (A), Recreation/tourism (A), Unsustainable exploitation (A)

The main threats are the fires that happen every year, the development of tourist resorts and associated infrastructure and urbanization, and illegal hunting.

Orbetello lagoon and Burano lake — A1, A4i, A4iii, B1i, B2, C1, C2, C4, C6 — 104

Admin region Toscana
Coordinates 42°24′N 11°19′E
Altitude 0–5 m **Area** 3,680 ha

Site description

Two wetlands on the Tyrrhenian coast of southern Tuscany. Orbetello lagoon is large and brackish, divided in two by a strip of land and separated from the sea by two sandbars (Tombolo di Feniglia, Tombolo della Giannella) and by the promontory of Mount Argentario. Burano is a brackish lake, parallel to the coast, 7 km south of the lagoon. The main land-uses are nature conservation and aquaculture (the latter in Orbetello lagoon).

Habitats Scrub (2%; sclerophyllous scrub/garrigue/maquis), Wetland (81%; saltmarsh; sand-dunes/sand beach; coastal lagoon; standing brackish and salt water), Artificial landscape (17%; highly improved reseeded grassland; forestry plantation; other urban/industrial areas)
Land-use Fisheries/aquaculture (50%), Hunting (60%), Nature conservation/research (20%), Urban/industrial/transport (5%), Water management (60%)

Birds

Species	Season	Year	Pop min	Pop max	Acc	Criteria
Egretta garzetta Little Egret	B	1995	216	216	A	C6
Egretta alba Great White Egret	W	1995	203	203	A	A4i, B1i, C2
[1] ***Phoenicopterus ruber*** Greater Flamingo	W	1994	616	773	A	B1i, C2
[2] ***Aythya nyroca*** Ferruginous Duck	W	1995	4	19	A	A1, C1
Recurvirostra avosetta Avocet	W	1995	40	87	A	B2
[3] ***Numenius tenuirostris*** Slender-billed Curlew	P	1995	3	3	A	A1, C1
Sterna sandvicensis Sandwich Tern	W	1995	57	96	A	C2

1. 26 pairs bred in 1994.
2. 4 in 1995, 19 in 1986.
3. One record, at Orbetello in April 1980.

An important site for wintering waterbirds (more than 20,000 on a regular basis, e.g. 21,691 in January 1994), notably for *Phoenicopterus ruber*, and a key site for *Numenius tenuirostris*.

Protection status

National Partial **International** Partial

474 ha of IBA covered by State Nature Reserve (Duna Feniglia, 474 ha). 360 ha of IBA covered by State Nature Reserve (Lago di Burano, 360 ha). 887 ha of IBA covered by State Nature Reserve (Laguna di ponente di Orbetello, 887 ha). 410 ha of IBA covered by Nature Reserve (Lago di Burano, 410 ha). 950 ha of IBA covered by Nature Reserve (Laguna di Ponente di Orbetello, 950 ha). 474 ha of IBA covered by Special Protection Area (Foresta di Protezione Duna Feniglia, 474 ha). 410 ha of IBA covered by Special Protection Area (Lago di Burano, 410 ha). 887 ha of IBA covered by Special Protection Area (Laguna di Orbetello (parte), 887 ha). 410 ha of IBA covered by Ramsar Site (Lago di Burano, 410 ha). 887 ha of IBA covered by Ramsar Site (Laguna di Orbetello, 887 ha).

Conservation issues

Threats Abandonment/reduction of land management (C), Afforestation (C), Aquaculture/fisheries (B), Disturbance to birds (C), Dredging/canalization (B), Industrialization/urbanization (B), Natural events (B), Other (B), Recreation/tourism (B)

The main threats are the further development of fish-farming, urbanization and tourism. 'Other' threat is water pollution.

Rieti lakes — C6 — 105

Admin region Lazio
Coordinates 42°29′N 12°50′E
Altitude 370–400 m **Area** 3,278 ha

Site description

Two freshwater lakes in central Italy, 5 km north of the town of Rieti. The lakes are fringed by beds of *Phragmites*, *Scirpus* and *Typha*. The site has many small rivers, streams, canals and ditches, while the area around the lakes is mainly agricultural. The main land-use is agriculture.

Habitats Wetland (30%; standing fresh water; river/stream; water-fringe vegetation; fen/transition mire/spring), Artificial landscape (70%; highly improved reseeded grassland; arable land)
Land-use Agriculture (70%), Fisheries/aquaculture (10%), Nature conservation/research (100%), Water management (30%)

Birds

Species	Season	Year	Pop min	Pop max	Acc	Criteria
Ixobrychus minutus Little Bittern	B	1995	8	12	A	C6

A notable site for wintering waterbirds, including *Aythya nyroca*.

Protection status

National High **International** None

3,278 ha of IBA covered by Regional Nature Reserve (Laghi Lungo e Ripasottile, 3,278 ha).

Conservation issues

Threats Agricultural intensification/expansion (A), Aquaculture/fisheries (B), Dredging/canalization (B)

Agricultural intensification is the main threat. Some or all of the site is covered by a management plan. The avifauna is regularly monitored for the local and regional authorities.

Rieti mountains

B2, C6 **106**

Admin region Lazio
Coordinates 42°27′N 12°59′E
Altitude 500–2,216 m **Area** 22,980 ha

Site description

A large area of broadleaved forest in the central Apennines, north-east of the town of Rieti, dominated by oak *Quercus* and beech *Fagus*. Shrubs and alpine grasslands dominate in the upper parts of the area. The main land-uses are hunting and nature conservation.

Habitats Forest and woodland (80%; broadleaved deciduous forest), Scrub (9%; scrub), Grassland (9%; alpine/subalpine/boreal grassland), Wetland (1%; river/stream), Rocky areas (1%; inland cliff)
Land-use Agriculture (10%), Hunting (50%), Nature conservation/research (50%), Tourism/recreation (40%)

Birds

Species	Season	Year	Pop min	Pop max	Acc	Criteria
Aquila chrysaetos Golden Eagle	R	1995	2	2	A	C6
Alectoris graeca Rock Partridge	R	1995	10	30	C	C6
Pyrrhocorax pyrrhocorax Chough	R	1995	30	60	B	B2, C6

Important breeding site for montane species.

Protection status

National None **International** Low
IBA contains Special Protection Area (Monti Reatini, area not known).

Conservation issues

Threats Deforestation (commercial) (C), Disturbance to birds (B), Firewood collection (B), Forest grazing (B), Groundwater abstraction (B), Intensified forest management (C), Recreation/tourism (A), Unsustainable exploitation (B)

Winter-sport resorts are the main threat to the area.

Gran Sasso d'Italia

A3, B2, C6 **107**

Admin region Abruzzo
Coordinates 42°28′N 13°39′E
Altitude 450–2,914 m **Area** 75,000 ha

Site description

The highest area in the Apennines. Montane habitats include alpine meadows, rocky areas, glaciers and extensive beech woodlands (*Fagus*). The site includes Lago di Penne, a small artificial lake. The main land-uses are nature conservation and tourism.

Habitats Forest and woodland (50%; broadleaved deciduous forest; treeline ecotone), Scrub (7%; scrub), Grassland (25%; alpine/subalpine/boreal grassland), Rocky areas (18%; scree/boulders; inland cliff)
Land-use Agriculture (5%), Nature conservation/research (100%), Other (20%), Tourism/recreation (80%)

Birds

Species	Season	Year	Pop min	Pop max	Acc	Criteria
Nycticorax nycticorax Night Heron	B	1995	30	40	A	C6
Aquila chrysaetos Golden Eagle	R	1996	2	2	A	C6
Falco biarmicus Lanner	R	1996	1	2	A	B2
Falco peregrinus Peregrine	R	1996	10	15	A	B2, C6
Alectoris graeca Rock Partridge	B	—	—	—	—	C6
Anthus campestris Tawny Pipit	B	1996	30	—	B	C6
Prunella collaris Alpine Accentor	R	1996	Common		—	A3
Tichodroma muraria Wallcreeper	R	1996	Frequent		—	A3
Lanius collurio Red-backed Shrike	B	1996	30	—	B	C6
Pyrrhocorax graculus Alpine Chough	R	1996	Common		—	A3
Montifringilla nivalis Snowfinch	R	1996	Common		—	A3

An important breeding site for four of the 10 species of the Eurasian high-montane biome, and for *Alectoris graeca* as well as various raptors, passerines, and *Nycticorax nycticorax* at Lago di Penne.

Protection status

National High **International** None
75,000 ha of IBA covered by National Park (Gran Sasso-Monti della Laga, 145,000 ha). 150 ha of IBA covered by Regional Nature Reserve (Lago di Penne, 150 ha).

Conservation issues

Threats Abandonment/reduction of land management (B), Disturbance to birds (B), Infrastructure (A), Recreation/tourism (A)

The main threat is the building of new tourist (ski) resorts and related infrastructure. Chamois *Rupicapra rupicapra* are being reintroduced to the area. Università dell'Aquila is researching the invertebrates and high-altitude flora, and the avifauna is monitored. There is a WWF reserve at Lago di Penne.

Vico lake

C6 **108**

Admin region Lazio
Coordinates 42°20′N 12°10′E
Altitude 510–965 m **Area** 4,000 ha

Site description

A freshwater lake in central Italy, in a volcanic crater about 10 km south-east of the town of Viterbo. The borders of the lake are fringed with *Phragmites* and *Juncus*. The surrounding area is mainly broadleaved forest (dominated by oak *Quercus*, beech *Fagus*, hornbeam *Carpinus*, chestnut *Castanea*, maple *Acer*) and agricultural land. Main land-use is agriculture.

Habitats Forest and woodland (30%; broadleaved deciduous forest), Scrub (5%; scrub), Grassland (5%; humid grassland), Wetland (35%; standing fresh water; water-fringe vegetation), Artificial landscape (25%; highly improved reseeded grassland; arable land; perennial crops/orchards/groves)
Land-use Agriculture (25%), Fisheries/aquaculture (5%), Forestry (20%), Military (5%), Nature conservation/research (70%), Tourism/recreation (20%), Urban/industrial/transport (5%)

Birds

Species	Season	Year	Pop min	Pop max	Acc	Criteria
Aythya nyroca Ferruginous Duck	W	1995	4	4	A	C6

The lake is notable for wintering duck, the surroundings for breeding owls, woodpeckers and passerines.

Protection status

National Partial **International** None
3,200 ha of IBA covered by Regional Nature Reserve (Lago di Vico, 3,200 ha).

Conservation issues

Threats Agricultural intensification/expansion (B), Aquaculture/fisheries (C), Deforestation (commercial) (C), Firewood collection (C), Forest grazing (C), Groundwater abstraction (U), Recreation/tourism (U)

Agricultural intensification is the major threat to the site.

Cicolano

B2, B3, C6 **109**

Admin region Lazio, Abruzzo
Coordinates 42°20′N 13°11′E
Altitude 483–1,820 m **Area** 49,485 ha

Site description

A site in the central Apennines. Gole di Antrodoco–Monte Giano and Monte Cabbia constitute a predominantly mountainous area, while Cicolano–Val Malito and Pian di Rascino are mainly cultivated uplands. The principal habitats are broadleaved woodlands, rocky areas, scrub, grasslands and cultivated land (mainly wheat). Dominant land-uses are agriculture, forestry and hunting.

Habitats Forest and woodland (30%; broadleaved deciduous forest; mixed forest), Scrub (20%; scrub), Grassland (20%; steppe/dry calcareous grassland; dry siliceous grassland; humid grassland), Wetland (5%; standing fresh water), Rocky areas (5%; inland cliff; caves), Artificial landscape (20%; arable land)

Land-use Agriculture (20%), Forestry (10%), Hunting (25%), Nature conservation/research (5%), Unknown (40%)

Birds

Species	Season	Year	Pop min	Pop max	Acc	Criteria
Circus pygargus Montagu's Harrier	B	1990	10	—	B	B3, C6
Aquila chrysaetos Golden Eagle	R	1996	2	2	A	C6
Alectoris graeca Rock Partridge	R	1995	15	—	B	C6
Coturnix coturnix Quail	B	1995	45	80	A	B2
Bubo bubo Eagle Owl	R	1995	6	7	B	C6
Pyrrhocorax pyrrhocorax Chough	R	1990	20	—	B	C6

An important breeding site for birds of prey and lowland gamebirds.

Protection status

National Partial **International** None
12,338 ha of IBA covered by Wildlife Reserve (Terminillo, 12,338 ha).

Conservation issues

Threats Abandonment/reduction of land management (B), Afforestation (B), Agricultural intensification/expansion (C), Consequences of animal/plant introductions (C), Deforestation (commercial) (A), Disturbance to birds (A), Extraction industry (B), Firewood collection (A), Forest grazing (A), Industrialization/urbanization (A), Recreation/tourism (A)

The main threats are from tourism, urbanization, and unsustainable exploitation of forest resources.

Lake Bracciano and surrounding areas — B2, B3, C6 — 110

Admin region Lazio
Coordinates 42°10′N 12°12′E
Altitude 164–612 m **Area** 31,300 ha

Site description

The area includes two freshwater lakes of volcanic origin (Bracciano and Martignano), located 20 km north-west of Rome, surrounded by extensive tracts of broadleaved woodland (oak *Quercus*, chestnut *Castanea*) and of arable land. The main land-uses are forestry, hunting and agriculture.

Habitats Forest and woodland (28%; broadleaved deciduous forest), Scrub (5%; sclerophyllous scrub/garrigue/maquis), Grassland (5%; steppe/dry calcareous grassland), Wetland (20%; standing fresh water; water-fringe vegetation), Artificial landscape (42%; highly improved reseeded grassland; arable land; perennial crops/orchards/groves)
Land-use Agriculture (42%), Fisheries/aquaculture (20%), Forestry (25%), Hunting (75%), Military (5%), Water management (18%)

Birds

Species	Season	Year	Pop min	Pop max	Acc	Criteria
Ixobrychus minutus Little Bittern	B	1995	5	10	B	C6
Milvus migrans Black Kite	B	1995	8	10	B	C6
Coturnix coturnix Quail	B	1995	100	150	B	B2
Saxicola torquata Stonechat	R	1996	1,000	2,000	—	B2
Sylvia cantillans Subalpine Warbler	B	1996	500	1,000	B	B3
Parus caeruleus Blue Tit	R	1996	800	2,000	B	B3
Certhia brachydactyla Short-toed Treecreeper	R	1996	1,000	1,500	B	B3

Fulica atra and grebes also winter in some numbers on the lakes. Species of global conservation concern that do not meet IBA criteria: *Aythya nyroca* (max. 1, winter).

Protection status

National None **International** None

Conservation issues

Tourism, infrastructure development and urbanization are important threats.

Threats Agricultural intensification/expansion (C), Aquaculture/fisheries (C), Deforestation (commercial) (B), Disturbance to birds (C), Filling-in of wetlands (A), Forest grazing (B), Industrialization/urbanization (B), Intensified forest management (B), Recreation/tourism (B), Selective logging/cutting (B)

Tolfa hills — B2, B3, C6 — 111

Admin region Lazio
Coordinates 42°08′N 12°04′E
Altitude 10–650 m **Area** 75,000 ha

Site description

A hilly area along the Tyrrhenian coast, east of the town of Civitavecchia. Habitats include maquis, broadleaved woodlands of beech *Fagus*, and extensive wooded pastures. The main land-use is cattle- and horse-rearing.

Habitats Forest and woodland (40%; broadleaved deciduous forest; mixed forest; broadleaved evergreen forest), Scrub (15%; sclerophyllous scrub/garrigue/maquis), Grassland (15%; steppe/dry calcareous grassland), Wetland (5%; river/stream), Rocky areas (5%; inland cliff), Artificial landscape (20%; highly improved reseeded grassland; arable land; ruderal land)
Land-use Agriculture (20%), Forestry (30%), Hunting (60%), Military (5%), Nature conservation/research (100%), Urban/industrial/transport (8%)

Birds

Species	Season	Year	Pop min	Pop max	Acc	Criteria
Pernis apivorus Honey Buzzard	B	1995	25	35	A	C6
Milvus migrans Black Kite	B	1995	20	30	A	B2, C6
Milvus milvus Red Kite	R	1995	4	5	A	C6
Circaetus gallicus Short-toed Eagle	B	1995	10	15	A	B2, C6
Circus pygargus Montagu's Harrier	B	1995	5	10	A	B3, C6
Falco tinnunculus Kestrel	R	1995	100	120	A	B2
Burhinus oedicnemus Stone Curlew	B	1995	2	5	A	C6
Athene noctua Little Owl	R	1995	100	120	A	B2
Merops apiaster Bee-eater	B	1995	200	300	A	B2
Coracias garrulus Roller	B	1995	30	40	A	B2
Melanocorypha calandra Calandra Lark	R	1995	45	55	A	C6
Calandrella brachydactyla Short-toed Lark	B	1995	30	40	A	C6
Lanius collurio Red-backed Shrike	B	1995	50	—	C	C6

An important breeding site for many SPECs, especially raptors (the best site in Italy for *Circaetus gallicus*), owls, larks and shrikes. In the 1970s three pairs of *Neophron percnopterus* also bred. Species of global conservation concern that do not meet IBA criteria: *Falco naumanni* (occurs on passage).

Protection status

National Low **International** Partial
900 ha of IBA covered by Regional Nature Reserve (Monterano, 900 ha). 1,450 ha of IBA covered by Regional Urban Park (Marturanum, 1,450 ha). 8,373 ha of IBA covered by Special Protection Area (Comprensorio meridionale dei Monti della Tolfa, 8,373 ha).

Conservation issues

Threats Deforestation (commercial) (B), Disturbance to birds (C), Firewood collection (C), Forest grazing (C), Industrialization/urbanization (A), Infrastructure (A), Unsustainable exploitation (B)

The main threats are house-building and road-construction in the area. The site is a proposed Regional Park. A project to study and protect *Milvus milvus* is run by LIPU, and individuals have been reintroduced from Spain.

Tarquinia saltpans — C2 — 112

Admin region Lazio
Coordinates 42°12′N 11°45′E
Altitude 0–0 m **Area** 170 ha

Site description

Small saltpans on the Tyrrhenian coast, close to the village of Tarquinia. The main vegetation is *Salicornia* bushes. The main land-use is salt extraction.

Habitats Scrub (5%; sclerophyllous scrub/garrigue/maquis), Wetland (95%; sand-dunes/sand beach; standing brackish and salt water)
Land-use Nature conservation/research (100%), Urban/industrial/transport (100%)

Birds

An important site for wintering waterbirds.

Species	Season	Year	Pop min	Pop max	Acc	Criteria
Larus melanocephalus Mediterranean Gull	W	1995	300	1,000	A	C2

Protection status

National High **International** High

170 ha of IBA covered by State Nature Reserve (Salina di Tarquinia, 170 ha). 170 ha of IBA covered by Special Protection Area (Riserva Naturale Saline di Tarquinia, 170 ha).

Conservation issues

Threats Aquaculture/fisheries (B), Disturbance to birds (A), Natural events (C)

The main threats are intensive aquaculture projects, disturbance to birds by unauthorized entry of cars into the site, and dune erosion. Regular ornithological surveys are carried out, of wintering waterbirds and of migrating waders especially.

Lucretili mountains

C6 **113**

Admin region Lazio
Coordinates 42°06'N 12°52'E
Altitude 300–1,368 m **Area** 21,000 ha

Site description

A mountain area in central Italy, north of the town of Tivoli, mainly covered by broadleaved forest. There are rocky cliffs, alpine grasslands, streams and cultivated areas. The main land-use is agriculture, including stock-rearing.

Habitats Forest and woodland (45%; broadleaved deciduous forest; broadleaved evergreen forest; treeline ecotone), Scrub (14%; scrub; sclerophyllous scrub/garrigue/maquis), Grassland (10%; alpine/subalpine/boreal grassland; mesophile grassland), Wetland (6%; standing fresh water; river/stream), Rocky areas (5%; inland cliff; caves), Artificial landscape (20%; highly improved reseeded grassland; arable land; perennial crops/orchards/groves; other urban/industrial areas; ruderal land)

Land-use Agriculture (30%), Hunting (25%), Nature conservation/research (85%)

Birds

Species	Season	Year	Pop min	Pop max	Acc	Criteria
Falco biarmicus Lanner	R	—	—	—	—	C6
Alectoris graeca Rock Partridge	R	—	—	—	—	C6

The site is also notable for breeding raptors and passerines (especially warblers), based on a qualitative ornithological survey in the 1991 breeding season.

Protection status

National Partial **International** Partial

18,000 ha of IBA covered by Regional Nature Park (Monti Lucretili, 18,000 ha). 8,750 ha of IBA covered by Special Protection Area (Monti Lucretili, 8,750 ha).

Conservation issues

Threats Abandonment/reduction of land management (B), Disturbance to birds (C), Firewood collection (B), Forest grazing (C), Groundwater abstraction (C)

The main threat is the abandonment of traditional land management. Some or all of the site is covered by a management plan.

Sirente, Velino and Duchessa mountains

A1, B2, C1, C6 **114**

Admin region Lazio, Abruzzo
Coordinates 42°11'N 13°27'E
Altitude 500–2,486 m **Area** 50,000 ha

Site description

A mountain area in the central Apennines, including Monte Velino, Monte Sirente and Montagne della Duchessa. The main habitats are extensive broadleaved forests (*Fagus*), alpine grasslands and rocky areas. The main land-uses are nature conservation, cultivation, stock-rearing, and tourism.

Habitats Forest and woodland (40%; broadleaved deciduous forest), Scrub (10%), Grassland (25%; alpine/subalpine/boreal grassland), Rocky areas (15%; scree/boulders; inland cliff), Artificial landscape (10%)

Land-use Agriculture (35%), Forestry (7%), Hunting (4%), Nature conservation/research (95%), Tourism/recreation (100%)

Birds

Species	Season	Year	Pop min	Pop max	Acc	Criteria
Aquila chrysaetos Golden Eagle	R	1996	4	4	A	B2, C6
Falco naumanni Lesser Kestrel	P	1996	—	130	A	A1, C1
Falco biarmicus Lanner	R	1996	2	3	A	B2, C6
Falco peregrinus Peregrine	R	1996	6	10	A	C6
Alectoris graeca Rock Partridge	R	1996	200	250	A	B2, C6
Bubo bubo Eagle Owl	R	1996	1	2	C	C6
[1] ***Pyrrhocorax pyrrhocorax*** Chough	R	1996	30	—	B	C6

1. Up to 250–400 in the area.

An important site for birds of prey and mountain species.

Protection status

National High **International** Low

3,550 ha of IBA covered by State Nature Reserve (Monte Velino, 3,550 ha). 3,900 ha of IBA covered by Regional Nature Reserve (Montagne della Duchessa, 3,900 ha). 50,000 ha of IBA covered by Regional Nature Park (Sirente Velino, 50,000 ha). 3,550 ha of IBA covered by Special Protection Area (Monte Velino, 3,550 ha).

Conservation issues

Threats Disturbance to birds (B), Forest grazing (C), Intensified forest management (B), Recreation/tourism (A), Unsustainable exploitation (B)

The main threat is uncontrolled touristic development, although illegal hunting is also important at Montagne della Duchessa. Some or all of the site is covered by a management plan. Projects by the National Forestry Department (CFS), that manages the area, include: introduction of *Gyps fulvus*; reintroduction of *Corvus corax*; studies on raptor populations; reintroduction of red deer *Cervus elaphus*.

Maiella, Pizzi and Frentani mountains

B2, B3, C2, C6 **115**

Admin region Abruzzo
Coordinates 42°10'N 14°12'E
Altitude 40–2,795 m **Area** 177,500 ha

Site description

A large, mountainous area in the central Apennines, including Maiella, Monti Pizzi, Bacino Aventino e medio Sangro, Monti Frentani, Alto Vastese and Lago di Serranella. The main habitats are broadleaved and mixed woodlands (dominated by beech *Fagus* and oak *Quercus*), very large boulders (especially in Maiella), gorges, scrub, grasslands, and cultivated areas, as well as rivers, streams and a small lake. The main land-uses are cultivation, nature conservation, cattle-rearing and tourism. Part of this area was listed as a separate IBA in the previous pan-European IBA inventory (Grimmett and Jones 1989), 'Mountains and hills between Capracotta and Rosello' (former site IT081).

Habitats Forest and woodland (50%; broadleaved deciduous forest; native coniferous forest; mixed forest; broadleaved evergreen forest; treeline ecotone), Scrub (10%; scrub; heathland), Grassland (15%; steppe/dry calcareous grassland), Wetland (5%; standing fresh water; river/stream; water-fringe vegetation), Rocky areas (10%; scree/boulders; inland cliff), Artificial landscape (10%; highly improved reseeded grassland; forestry plantation; other urban/industrial areas; ruderal land)

Land-use Agriculture (30%), Forestry (15%), Hunting (10%), Nature conservation/research (50%), Tourism/recreation (40%), Urban/industrial/transport (5%)

Birds

Species	Season	Year	Pop min	Pop max	Acc	Criteria
Pernis apivorus Honey Buzzard	B	1996	9	14	B	C6
Milvus migrans Black Kite	B	1996	10	15	A	C6
Milvus milvus Red Kite	R	1996	45	55	A	C6

Species ... continued	Season	Year	Pop min	Pop max	Acc	Criteria
Aquila chrysaetos Golden Eagle	R	1996	2	4	A	C6
Falco biarmicus Lanner	R	1996	10	15	A	B2, C2, C6
Falco peregrinus Peregrine	B	1996	20	28	A	B2, C6
Alectoris graeca Rock Partridge	R	1996	100	—	C	B2, C6
Dendrocopos medius Middle Spotted Woodpecker	R	1996	10	—	C	B3
Anthus campestris Tawny Pipit	B	1996	Common		—	C6
Pyrrhocorax pyrrhocorax Chough	R	1996	380	506	A	B2, C2, C6

The site is important for breeding raptors and some species of mountain or forest.

■ Protection status

National Partial **International** Low

80,708 ha of IBA covered by National Park (Maiella-Morrone, 80,708 ha). 4,202 ha of IBA covered by State Nature Reserve (Fara S. Martino–Palombaro, 4,202 ha). 1,563 ha of IBA covered by State Nature Reserve (Feudo Ugni, 1,563 ha). 1,407 ha of IBA covered by State Nature Reserve (Lama Bianca di S. Eufemia a Maiella, 1,407 ha). 1,452 ha of IBA covered by State Nature Reserve (Monte Rotondo, 1,452 ha). 1,920 ha of IBA covered by State Nature Reserve (Valle dell'Orfento, 1,920 ha). 130 ha of IBA covered by Regional Nature Reserve (Bosco di S. Antonio, 130 ha). 1,680 ha of IBA covered by Regional Nature Reserve (Maiella Orientale, 1,680 ha). 472 ha of IBA covered by Regional Nature Reserve (Valle del Foro, 472 ha). 378 ha of IBA covered by Regional Nature Reserve (Valle dell'Orta, 378 ha). 485 ha of IBA covered by Nature Reserve (Quarto S. Chiara, 485 ha). 200 ha of IBA covered by Private Reserve (Abetina di Rosello, 200 ha). 9,973 ha of IBA covered by Special Protection Area (Maiella, 9,973 ha). 1,452 ha of IBA covered by Special Protection Area (Monte Rotondo, 1,452 ha [enlargement ongoing]).

■ Conservation issues

Threats Abandonment/reduction of land management (C), Agricultural intensification/expansion (B), Consequences of animal/plant introductions (C), Deforestation (commercial) (B), Disturbance to birds (B), Dredging/canalization (C), Extraction industry (C), Filling-in of wetlands (C), Firewood collection (B), Forest grazing (C), Groundwater abstraction (C), Industrialization/urbanization (B), Recreation/tourism (A), Selective logging/cutting (C), Unsustainable exploitation (C)

The main threats are posed by the development of touristic resorts, especially for winter sports in the Maiella area. No major threats are present in the Monti Pizzi e Monti Frentani areas. Some or all of the site is covered by a management plan. Projects of the Maiella National Park include: reintroduction of red deer *Cervus elaphus*, chamois *Rupicapra rupicapra* and roe deer *Capreolus capreolus*; studies on *Alectoris graeca*, *Accipiter gentilis*, *Bubo bubo* and *Pyrrhocorax pyrrhocorax*; conservation of wolf *Canis lupus*. Lago di Serranella and Abetina di Rosello are WWF reserves.

Coast from Palo to Santa Marinella — B1ii, C2 — 116

Admin region Lazio
Coordinates 42°00'N 12°00'E
Altitude 0–10 m **Area** 6,300 ha

■ Site description

A coastal area in Lazio, between Palo and Santa Marinella. It includes open sea, wetlands, sand beaches, grasslands, woodlands, urban and cultivated areas. The main land-uses are tourism and agriculture.

Habitats Forest and woodland (15%; broadleaved deciduous forest; broadleaved evergreen forest), Scrub (25%; scrub; sclerophyllous scrub/garrigue/maquis), Grassland (10%; steppe/dry calcareous grassland; mesophile grassland), Wetland (20%; mudflat/sandflat; saltmarsh; sand-dunes/sand beach; coastal lagoon; standing fresh water; standing brackish and salt water; water-fringe vegetation), Marine areas (10%; sea inlet/coastal features), Artificial landscape (10%; arable land; perennial crops/orchards/groves; urban parks/gardens)
Land-use Agriculture (30%), Fisheries/aquaculture (5%), Hunting (30%), Military (5%), Nature conservation/research (10%), Tourism/recreation, Urban/industrial/transport

■ Birds

A notable foraging area for shearwaters and gulls. Species of global conservation concern that do not meet IBA criteria: *Larus audouinii* (1–30 on passage).

Species	Season	Year	Pop min	Pop max	Acc	Criteria
Calonectris diomedea Cory's Shearwater	N	1995	—	1,000	B	B1ii, C2

■ Protection status

National Low **International** Low

250 ha of IBA covered by Regional Nature Reserve (Macchiatonda, 250 ha). 120 ha of IBA covered by Nature Reserve (Bosco di Palo, 120 ha). 37 ha of IBA covered by Wildlife Reserve (Torre Flavia, 37 ha). 96 ha of IBA covered by Special Protection Area (Macchiatonda, 96 ha). 33 ha of IBA covered by Special Protection Area (Torre Flavia, 33 ha).

■ Conservation issues

Threats Agricultural intensification/expansion (B), Disturbance to birds (A), Drainage (B), Dredging/canalization (A), Filling-in of wetlands (C), Groundwater abstraction (A), Industrialization/urbanization (A), Infrastructure (A), Recreation/tourism (A)

The main threats are industrialization, urbanization, building of infrastructure, dredging and canalization, disturbance to birds during the breeding season, and touristic development. Three WWF-Italia reserves are present. Some or all of the site is covered by a management plan. There are bird-ringing schemes for passerines and waders.

Litorale Romano — B1i, B2, C3, C6 — 117

Admin region Lazio
Coordinates 41°51'N 12°12'E
Altitude 0–80 m **Area** 19,529 ha

■ Site description

An area on the Tyrrhenian coast, south-west of Rome, including open sea, sand-dunes and beaches, and a mosaic of maquis, woodlands, wetlands, urban and cultivated areas. The main land-uses are agriculture, tourism and nature conservation.

Habitats Forest and woodland (50%; broadleaved deciduous forest; native coniferous forest; broadleaved evergreen forest), Scrub (15%; sclerophyllous scrub/garrigue/maquis), Grassland (steppe/dry calcareous grassland; dry siliceous grassland; mesophile grassland), Wetland (5%; sand-dunes/sand beach; standing fresh water; standing brackish and salt water; river/stream; water-fringe vegetation), Marine areas (5%; open sea), Artificial landscape (25%; arable land; perennial crops/orchards/groves; forestry plantation; urban parks/gardens; ruderal land)
Land-use Agriculture (35%), Hunting (5%), Nature conservation/research (25%), Tourism/recreation (10%), Urban/industrial/transport (20%)

■ Birds

Species	Season	Year	Pop min	Pop max	Acc	Criteria
Phalacrocorax carbo Cormorant	W	1995	1,000	1,500	A	B1i, C3
Ixobrychus minutus Little Bittern	B	1995	10	18	A	C6
Milvus migrans Black Kite	B	1995	25	30	A	B2, C6
Merops apiaster Bee-eater	B	1995	200	230	A	B2
Coracias garrulus Roller	B	1995	5	6	A	B2

An important breeding site for open-country birds such as *Milvus migrans* and *Merops apiaster*. Species of global conservation concern that do not meet IBA criteria: *Aythya nyroca* (max. 2, winter).

■ Protection status

National Partial **International** Unknown

IBA contains State Nature Reserve (Litorale Romano, area not known). 5,989 ha of IBA covered by Private Reserve (Castelporziano e Capracotta, 5,989 ha). 280 ha of IBA covered by Private Reserve (Macchiagrande di Focene, 280 ha). Overlaps with SPA.

■ Conservation issues

Threats Abandonment/reduction of land management (U), Agricultural intensification/expansion (A), Consequences of animal/plant introductions (U), Disturbance to birds (B), Drainage (C), Dredging/canalization (A), Extraction industry (B), Filling-in of wetlands (C), Firewood collection (C), Forest grazing (C), Groundwater abstraction (B), Industrialization/urbanization (A), Infrastructure (A), Natural events (U), Recreation/tourism (A), Unsustainable exploitation (B)

Although human impact is already quite strong in the area, the main threats are from further urbanization and industrialization, agricultural intensification, and the building of tourist resorts and associated infrastructure. There is a bird-ringing station at Macchiagrande di Focene. There is a project to protect the nests of *Charadrius alexandrinus* and *Charadrius dubius*.

Ernici and Simbruini mountains B2, C6 118

Admin region Lazio, Abruzzo
Coordinates 41°53′N 13°19′E
Altitude 400–2,156 m **Area** 70,000 ha

Site description

A mountainous area of limestone karst in the central Apennines, covered mainly by broadleaved forests (especially beech *Fagus*) and extensive rocky pastures. The site is rich in rocky cliffs and caves. The main land-use is the rearing of cattle and horses.

Habitats Forest and woodland (70%; broadleaved deciduous forest; treeline ecotone), Scrub (5%; scrub; heathland), Grassland (5%; steppe/dry calcareous grassland; alpine/subalpine/boreal grassland), Rocky areas (15%; scree/boulders; inland cliff; caves), Artificial landscape (5%; perennial crops/orchards/groves; forestry plantation)
Land-use Agriculture (10%), Forestry (10%), Nature conservation/research (100%), Tourism/recreation (20%)

Birds

Species	Season	Year	Pop min	Pop max	Acc	Criteria
Aquila chrysaetos Golden Eagle	R	—	3	3	A	C6
Falco biarmicus Lanner	R	1996	—	2	A	B2, C6
Falco peregrinus Peregrine	R	1996	8	—	B	C6
Lullula arborea Woodlark	B	1995	100	—	A	C6
Anthus campestris Tawny Pipit	B	1995	200	—	A	B2, C6
Lanius collurio Red-backed Shrike	B	1995	100	—	A	C6

An important site for some raptor and passerine species.

Protection status

National Partial **International** Partial
1,025 ha of IBA covered by Regional Nature Reserve (Zompo lo Schioppo, 1,025 ha). 29,000 ha of IBA covered by Regional Nature Park (Monti Simbruini, 29,000 ha). 51,514 ha of IBA covered by Special Protection Area (Monti Simbruini ed Ernici, 51,514 ha).

Conservation issues

Threats Abandonment/reduction of land management (B), Afforestation (C), Consequences of animal/plant introductions (A), Deforestation (commercial) (C), Disturbance to birds (B), Firewood collection (B), Forest grazing (A), Industrialization/urbanization (B), Infrastructure (B), Intensified forest management (B), Other (A), Recreation/tourism (B), Unsustainable exploitation (B)

Threats include hunting and off-road driving ('Other' threat). Some or all of the site is covered by a management plan.

Abruzzo National Park B2, B3, C2, C6 119

Admin region Lazio, Abruzzo, Molise
Coordinates 41°45′N 13°51′E
Altitude 500–2,249 m **Area** 110,000 ha

Site description

Habitats Forest and woodland (60%; broadleaved deciduous forest; treeline ecotone), Scrub (10%; scrub), Grassland (10%; alpine/subalpine/boreal grassland), Wetland (1%; standing fresh water; river/stream), Rocky areas (10%; scree/boulders; inland cliff), Artificial landscape (9%; arable land)
Land-use Agriculture (25%), Forestry (5%), Hunting (2%), Nature conservation/research (98%), Other (20%), Tourism/recreation (90%)

A large, mountainous area in the central Apennines, including the National Park and the adjacent Gole del Sagittario. The main habitats are broadleaved and mixed woodlands, scrub, alpine grasslands, streams, rivers and small lakes, cliffs and gorges. The main land-uses are nature conservation, cultivation, stock-farming and tourism.

Birds

The site is important for breeding raptors, woodpeckers and other species of mountain or forest.

Species	Season	Year	Pop min	Pop max	Acc	Criteria
Aquila chrysaetos Golden Eagle	R	1996	4	4	A	B2, C6
Falco biarmicus Lanner	R	1996	1	—	A	B2
Falco peregrinus Peregrine	R	1996	7	10	A	C6
Alectoris graeca Rock Partridge	R	1996	—	—	—	B2
Dendrocopos medius Middle Spotted Woodpecker	R	1996	5	—	C	B3
Dendrocopos leucotos White-backed Woodpecker	R	1996	230	280	A	C2, C6

Protection status

National Partial **International** Partial
43,950 ha of IBA covered by National Park (Abruzzo, 43,950 ha). 450 ha of IBA covered by Private Reserve (Gole del Sagittario, 450 ha). 40,000 ha of IBA covered by Special Protection Area (Parco Nazionale d'Abruzzo, 40,000 ha).

Conservation issues

Threats Disturbance to birds (B), Infrastructure (C), Recreation/tourism (B)

The main threat is touristic development with its associated activities and infrastructure. Some or all of the site is covered by a management plan. The Park is carrying out research on raptors, brown bear *Ursus arctos* and wolf *Canis lupus*.

Lepini mountains B2, C6 120

Admin region Lazio
Coordinates 41°36′N 13°05′E
Altitude 80–1,536 m **Area** 49,600 ha

Site description

A mountain complex in central Italy, isolated from the Apennines, located between the towns of Latina and Frosinone. The main habitats are broadleaved deciduous woodland and agricultural land. The main land-uses are cultivation, stock-rearing, hunting and forestry.

Habitats Forest and woodland (70%; broadleaved deciduous forest; broadleaved evergreen forest), Scrub (10%; sclerophyllous scrub/garrigue/maquis), Grassland (10%; steppe/dry calcareous grassland), Rocky areas (5%), Artificial landscape (5%)
Land-use Agriculture (10%), Forestry (20%), Hunting (30%), Not utilized (50%), Tourism/recreation (5%)

Birds

Species	Season	Year	Pop min	Pop max	Acc	Criteria
Falco peregrinus Peregrine	R	1989	20	20	A	B2, C6
Lanius collurio Red-backed Shrike	B	1989	50	—	C	C6

Protection status

National None **International** Partial
8,514 ha of IBA covered by Special Protection Area (Monti Lepini centrali, 8,514 ha).

Conservation issues

Threats Deforestation (commercial) (B), Infrastructure (A), Unsustainable exploitation (B)

Road-building and hunting are the main threats. The area is a proposed Regional Park.

Circeo National Park A1, B2, C1, C2, C6 121

Admin region Lazio
Coordinates 41°16′N 13°02′E
Altitude 0–540 m **Area** 8,350 ha

Site description

An area on the Tyrrhenian coast, between Roma and Napoli, including four brackish lakes (Laghi Pontini) and surrounding marshes, rocky cliffs, sand-dunes and beaches, broadleaved and seasonally flooded woodlands, and maquis. Isola di Zannone, a small island that is part of the Circeo National Park, is included in the 'Ponziane islands'

IBA (site 122). The main land-uses are nature conservation, cattle-raising, tourism, cultivation and fish-farming.

Habitats Forest and woodland (40%; broadleaved deciduous forest; alluvial/very wet forest; broadleaved evergreen forest), Grassland (10%; humid grassland; mesophile grassland), Wetland (30%; mudflat/sandflat; saltmarsh; coastal lagoon; standing fresh water; water-fringe vegetation), Rocky areas (5%; sea cliff/rocky shore), Artificial landscape (15%; highly improved reseeded grassland; arable land; other urban/industrial areas; ruderal land)
Land-use Agriculture (10%), Fisheries/aquaculture (25%), Military (5%), Nature conservation/research (100%), Tourism/recreation (40%), Urban/industrial/transport (5%)

Birds

Species	Season	Year	Pop min	Pop max	Acc	Criteria
Ixobrychus minutus Little Bittern	B	1986	3	—	—	C6
[1] ***Numenius tenuirostris*** Slender-billed Curlew	W	1995	1	4	A	A1, C1
Sterna sandvicensis Sandwich Tern	W	1995	13	73	A	C2
Merops apiaster Bee-eater	B	1986	20	300	B	B2
Monticola solitarius Blue Rock Thrush	R	—	100	—	C	B2

1. Two records: 4 from December 1980 to January 1981; 1 in February 1989.

The site is important for wintering and migrating waterbirds, most notably *Numenius tenuirostris*.

Protection status

National High **International** High
8,350 ha of IBA covered by National Park (Circeo, 8,622 ha). 8,350 ha of IBA covered by Special Protection Area (Parco Nazionale del Circeo, 8,400 ha). 94 ha of IBA covered by Ramsar Site (Lago dei Monaci, 94 ha). 229 ha of IBA covered by Ramsar Site (Lago di Caprolace, 229 ha). 395 ha of IBA covered by Ramsar Site (Lago di Fogliano, 395 ha). 1,474 ha of IBA covered by Ramsar Site (Lago di Sabaudia, 1,474 ha).

Conservation issues

Threats Agricultural intensification/expansion (A), Groundwater abstraction (A), Industrialization/urbanization (B), Other (B), Recreation/tourism (B), Unsustainable exploitation (B)

The main threats are water pollution ('Other' threat), especially of Lago Monaci, tourism, illegal house construction, agricultural intensification, groundwater abstraction, and illegal hunting. Some or all of the site is covered by a management plan. There are various research projects by the National Park in collaboration with universities and Ente Nazionale Energie Alternative.

Ponziane islands — B2, C6 — 122

Admin region Lazio
Coordinates 40°54′N 12°59′E
Altitude 0–280 m **Area** 1,130 ha

Site description

An archipelago of small islands, islets and rock stacks in the Tyrrhenian Sea, close to the Lazio coast. It includes Isola di Ponza, Isola Palmarola, Isola Ventotene, Isola Zannone, Isola Santo Stefano, and Isola di Gavi. Maquis and garrigue are the dominant habitats. The main land-uses are fishing, tourism, hunting, stock-farming.

Habitats Forest and woodland (broadleaved evergreen forest), Scrub (sclerophyllous scrub/garrigue/maquis), Wetland (shingle/stony beach), Marine areas, Rocky areas (sea cliff/rocky shore; rock stacks/islets), Artificial landscape (highly improved reseeded grassland; perennial crops/orchards/groves; urban parks/gardens; ruderal land)
Land-use Agriculture, Fisheries/aquaculture, Hunting, Nature conservation/research, Tourism/recreation

Birds

Species	Season	Year	Pop min	Pop max	Acc	Criteria
Calonectris diomedea Cory's Shearwater	R	1986	—	—	—	C6
Phalacrocorax aristotelis Shag	R	1986	—	—	—	C6
[1,2] ***Circus pygargus*** Montagu's Harrier	P	1997	—	200	C	C6
Falco peregrinus Peregrine	R	1996	15	20	A	B2, C6
Monticola solitarius Blue Rock Thrush	R	1996	50	100	C	B2

1. 200 is maximum in a day.
2. Non-systematic counts.

The site is important for cliff-breeding species and for migrants, especially raptors and passerines. Species of global conservation concern that do not meet IBA criteria: *Acrocephalus paludicola* (1–5 on passage).

Protection status

National High **International** High
110 ha of IBA covered by National Park (Circeo, 8,622 ha). IBA contains State Nature Reserve (Isole di Ponza e Palmarola, area not known). IBA contains State Nature Reserve (Isole di Ventotene e S. Stefano, area not known). IBA contains Marine Nature Reserve (Ventotene e S. Stefano, area not known). 1,130 ha of IBA covered by Special Protection Area (Isole di Ponza, Palmarola, Zannone, Ventotene e S. Stefano, 17,069 ha).

Conservation issues

Threats Consequences of animal/plant introductions (A), Recreation/tourism (C), Unsustainable exploitation (A)

The main threats are hunting and illegal bird-trapping during the spring migration. European mouflon *Ovis orientalis musimon* has been introduced on Zannone and is a globally threatened subspecies (Vulnerable), but its overgrazing is affecting the vegetation. The Instituto Nazionale Fauna Selvatica carries out bird-ringing on Ventotene.

Ausoni and Aurunci mountains — B2, C6 — 123

Admin region Lazio
Coordinates 41°22′N 13°30′E
Altitude 0–1,533 m **Area** 137,200 ha

Site description

A mountain complex bordering the Tyrrhenian Sea, north of the towns of Terracina and Gaeta. Habitats include maquis, broadleaved evergreen woodland, garrigue, pastures and rocky cliffs. The main human activities are agriculture, forestry and hunting.

Habitats Forest and woodland (25%; broadleaved evergreen forest), Scrub (35%; sclerophyllous scrub/garrigue/maquis), Grassland (5%), Rocky areas (5%; inland cliff), Artificial landscape (30%)
Land-use Agriculture (25%), Forestry (10%), Hunting (50%), Tourism/recreation (10%), Urban/industrial/transport (10%)

Birds

Species	Season	Year	Pop min	Pop max	Acc	Criteria
Falco peregrinus Peregrine	R	1995	12	—	A	B2, C6
Monticola solitarius Blue Rock Thrush	R	1995	300	400	A	B2
Lanius minor Lesser Grey Shrike	B	1995	15	25	A	B2
Lanius senator Woodchat Shrike	B	1996	300	400	A	B2

Protection status

National Low **International** None
1,500 ha of IBA covered by Nature Monument (Sonnino, 1,500 ha).

Conservation issues

Threats Agricultural intensification/expansion (A), Burning of vegetation (A), Deforestation (commercial) (B), Disturbance to birds (B), Extraction industry (C), Firewood collection (C), Forest grazing (B), Selective logging/cutting (B), Unsustainable exploitation (A)

The main threats are the strong hunting pressure; fires that are illegally started every year to burn scrub and create new pasture; and overgrazing.

Matese — B2, B3, C6 — 124

Admin region Campania, Molise
Coordinates 41°27′N 14°25′E
Altitude 1,011–2,050 m **Area** 41,718 ha

Site description

A mountain area in the central Apennines, south-east of the town of Isernia. Habitats include extensive broadleaved woodlands (oak

Quercus, beech *Fagus*), rocky areas, agricultural land, and a karstic lake, Lago del Matese (500 ha). The main land-uses are forestry and agriculture.

Habitats Forest and woodland (45%; broadleaved deciduous forest; native coniferous forest; mixed forest; broadleaved evergreen forest; treeline ecotone), Scrub (7%; scrub; sclerophyllous scrub/garrigue/maquis), Grassland (8%; steppe/dry calcareous grassland; humid grassland; mesophile grassland), Wetland (15%; standing fresh water; river/stream), Rocky areas (10%; inland cliff), Artificial landscape (15%; perennial crops/orchards/groves; urban parks/gardens; other urban/industrial areas; ruderal land)
Land-use Agriculture (20%), Forestry (45%), Nature conservation/research (63%), Urban/industrial/transport (10%), Water management (15%)

Birds

Species	Season	Year	Pop min	Pop max	Acc	Criteria
Pernis apivorus Honey Buzzard	B	1989	3	—	—	C6
Milvus milvus Red Kite	R	1993	3	6	A	B3, C6
Circaetus gallicus Short-toed Eagle	B	1990	2	—	B	C6
Aquila chrysaetos Golden Eagle	R	1995	2	—	B	C6
Falco biarmicus Lanner	R	1996	1	2	B	B2, C6
Bubo bubo Eagle Owl	R	1993	1	2	B	C6

An important site for breeding birds of prey.

Protection status

National Partial **International** None
25,000 ha of IBA covered by Regional Park (Matese, 25,000 ha). 1,650 ha of IBA covered by Private Reserve (Monte Mutria, 1,650 ha).

Conservation issues

Threats Abandonment/reduction of land management (C), Disturbance to birds (B), Extraction industry (C), Firewood collection (C), Forest grazing (C), Industrialization/urbanization (B), Infrastructure (A), Recreation/tourism (B)

The site is threatened mainly by the building of new infrastructure (roads and a dam) and new tourist resorts, especially for winter sports. There have been studies on shrikes *Lanius*.

River Biferno — B2, C6 — 125

Admin region Molise
Coordinates 41°46′N 14°47′E
Altitude 50–889 m **Area** 30,000 ha

Site description

A hilly area in central Italy, 15 km north-east of the town of Campobasso. Habitats include a stretch of river, a man-made lake, streams, maquis, broadleaved woodlands, and large tracts of agricultural land. The main land-uses are agriculture, hunting and forestry.

Habitats Forest and woodland (25%; broadleaved deciduous forest), Scrub (5%; sclerophyllous scrub/garrigue/maquis), Wetland (10%; standing fresh water; river/stream; water-fringe vegetation), Artificial landscape (60%)
Land-use Agriculture (50%), Forestry (20%), Hunting (30%), Nature conservation/research (6%), Water management (5%)

Birds

Species	Season	Year	Pop min	Pop max	Acc	Criteria
Pernis apivorus Honey Buzzard	B	1995	3	—	B	C6
Milvus migrans Black Kite	B	1995	18	22	B	B2, C6
Milvus milvus Red Kite	R	1995	3	4	A	C6
Coracias garrulus Roller	B	1995	2	—	B	C6

An important site for breeding raptors. Species of global conservation concern that do not meet IBA criteria: *Crex crex* (rare on passage).

Protection status

National Low **International** None
140 ha of IBA covered by Wildlife Reserve (Casacalenda, 140 ha). 1,700 ha of IBA covered by Wildlife Reserve (Lago di Guardialfiera, 1,700 ha).

Conservation issues

The main threat is the building of infrastructure, especially new roads. Oasi Casacalenda (Wildlife Reserve) is managed by LIPU.

Threats Afforestation (C), Agricultural intensification/expansion (B), Aquaculture/fisheries (C), Consequences of animal/plant introductions (B), Infrastructure (A), Intensified forest management (A), Natural events (B), Selective logging/cutting (C)

Daunia mountains — B2, C6 — 126

Admin region Puglia
Coordinates 41°40′N 15°06′E
Altitude 200–1,150 m **Area** 50,000 ha

Site description

A pre-Apennine mountain area in southern Italy, mainly covered by remnants of broadleaved woodland and extensive cultivated areas. The site also includes an artificial lake (Lago di Occhito). The main land-uses are agriculture and forestry.

Habitats Forest and woodland (30%), Scrub (15%; scrub; sclerophyllous scrub/garrigue/maquis), Grassland (20%; steppe/dry calcareous grassland; humid grassland; mesophile grassland), Artificial landscape (35%)
Land-use Agriculture (20%), Forestry (30%), Hunting (90%), Tourism/recreation (2%), Urban/industrial/transport (5%), Water management (10%)

Birds

Species	Season	Year	Pop min	Pop max	Acc	Criteria
Falco biarmicus Lanner	R	—	1	2	A	B2, C6
Coracias garrulus Roller	B	—	3	6	B	B2

Important breeding area for raptors. Breeding species of global conservation concern that do not meet IBA criteria: *Falco naumanni*.

Protection status

National None **International** None

Conservation issues

Threats Abandonment/reduction of land management (U), Afforestation (A), Agricultural intensification/expansion (A), Burning of vegetation (A), Construction/impact of dyke/dam/barrage (A), Disturbance to birds (A), Drainage (B), Dredging/canalization (C), Filling-in of wetlands (C), Firewood collection (U), Forest grazing (B), Groundwater abstraction (U), Industrialization/urbanization (A), Infrastructure (U), Intensified forest management (U), Natural events (U), Recreation/tourism (B), Unsustainable exploitation (A)

Hunting, fires and urbanization are the main threats. Within the area the provincial authorities periodically forbid hunting at some sites, but these are not included on the national list of protected areas (e.g. the wildlife refuge of Lago di Occhito, of 1,200 ha, in Molise).

Tremiti islands — A4ii, B1ii, B2, C3, C6 — 127

Admin region Puglia
Coordinates 42°07′N 15°30′E
Altitude 0–70 m **Area** 2,116 ha

Site description

A small archipelago in the southern Adriatic Sea. The main habitats are sea cliffs, islets, rock stacks, maquis and native pinewoods *Pinus*. The main land-uses are tourism, agriculture, fishing and hunting.

Habitats Forest and woodland (native coniferous forest), Scrub (sclerophyllous scrub/garrigue/maquis), Grassland (steppe/dry calcareous grassland), Wetland (shingle/stony beach), Marine areas (open sea; sea inlet/coastal features), Rocky areas (sea cliff/rocky shore; rock stacks/islets; caves), Artificial landscape (highly improved reseeded grassland; arable land; urban parks/gardens)
Land-use Agriculture, Fisheries/aquaculture, Hunting, Tourism/recreation

Birds

Species	Season	Year	Pop min	Pop max	Acc	Criteria
Calonectris diomedea Cory's Shearwater	B	1988	100	150	A	B2, C6
Puffinus yelkouan Yelkouan Shearwater	B	1988	250	350	A	A4ii, B1ii, C3

An important breeding site for *Calonectris diomedea* and *Puffinus yelkouan*.

Protection status

National High **International** None

2,116 ha of IBA covered by National Park (Gargano, 133,000 ha [boundaries are under discussion]). 2,116 ha of IBA covered by Marine Reserve (Isole Tremiti, 2,116 ha).

Conservation issues

Threats Agricultural intensification/expansion (B), Aquaculture/fisheries (C), Burning of vegetation (B), Disturbance to birds (A), Industrialization/urbanization (B), Recreation/tourism (A), Unsustainable exploitation (A)

The main threats are tourism and consequent disturbance of breeding birds, burning of vegetation, agricultural intensification, and illegal hunting.

Lakes Lesina and Varano A1, A4i, A4iii, B1i, C1, C2, C3, C4 128

Admin region Puglia
Coordinates 41°52'N 15°24'E
Altitude 0–5 m **Area** 20,353 ha

Site description

Two large lakes on the southern Adriatic coast, north of the Gargano promontory. They are both connected to the sea and are very close to each other. Lesina is relatively narrow, shallow and parallel to the coast; large reedbeds fringe it. Varano is deeper, the banks are mostly artificial and there virtually no reedbeds. The strip of sea in front of the lakes and beach is also important for birds. The main uses of the area are fisheries, hunting, nature conservation and agriculture, the last one to the east of Varano and to the south-west of Lesina.

Habitats Forest and woodland (7%; native coniferous forest; broadleaved evergreen forest), Scrub (sclerophyllous scrub/garrigue/maquis), Grassland (2%; steppe/dry calcareous grassland; dry siliceous grassland), Wetland (69%; mudflat/sandflat; saltmarsh; sand-dunes/sand beach; coastal lagoon; standing brackish and salt water; water-fringe vegetation), Marine areas (15%), Artificial landscape (7%; highly improved reseeded grassland; arable land; perennial crops/orchards/groves; forestry plantation)
Land-use Agriculture (30%), Fisheries/aquaculture (50%), Hunting (70%), Nature conservation/research (80%), Tourism/recreation (25%), Urban/industrial/transport (2%)

Birds

Species	Season	Year	Pop min	Pop max	Acc	Criteria
Phalacrocorax carbo Cormorant	W	1994	4,033	4,033	A	B1i, C3
[1] ***Aythya nyroca*** Ferruginous Duck	W	1985	1,000	1,000	A	A1,A4i,B1i,C1,C2
[2] ***Aythya fuligula*** Tufted Duck	W	1983	6,000	6,000	A	B1i, C3
Clangula hyemalis Long-tailed Duck	W	1995	206	206	A	B1i, C3
Mergus serrator Red-breasted Merganser	W	1995	554	554	A	B1i, C3
Fulica atra Coot	W	1995	27,718	27,718	A	A4i, B1i, C3
[3] ***Numenius tenuirostris*** Slender-billed Curlew	W	1995	2	2	A	A1, C1

1. Winter 1985 only (Lake Lesina).
2. Winter 1983 only.
3. One record, at Lake Lesina in January 1989.

A very important site for wintering waterbirds (more than 20,000 occur regularly, mainly *Fulica atra*), for example a total of 27,424 in January 1994.

Protection status

National Partial **International** Low

14,000 ha of IBA covered by National Park (Gargano, 133,000 ha [boundaries are under discussion]). 930 ha of IBA covered by Nature Reserve (Lago di Lesina, 930 ha). 930 ha of IBA covered by Special Protection Area (Lago di Lesina, 930 ha).

Conservation issues

Threats Afforestation (C), Agricultural intensification/expansion (B), Aquaculture/fisheries (A), Construction/impact of dyke/dam/barrage (C), Disturbance to birds (A), Drainage (A), Dredging/canalization (A), Filling-in of wetlands (B), Forest grazing (C), Groundwater abstraction (B), Recreation/tourism (A), Unsustainable exploitation (A)

The main threats are water pollution, illegal hunting and agricultural intensification (drainage of wetlands, overgrazing and agricultural expansion on steppes and dry grasslands). A management plan for the area will be produced by LIPU for the Gargano National Park as part of a programme to reintroduce *Oxyura leucocephala*.

Gargano promontory A1, B2, B3, C1, C2, C6 129

Admin region Puglia
Coordinates 41°50'N 16°01'E
Altitude 0–1,055 m **Area** 240,000 ha

Site description

A mountainous promontory along the south Adriatic coast. The main habitats are broadleaved and coniferous woodlands and agricultural land, with rocky sea cliffs. The main land-uses are forestry, agriculture and tourism.

Habitats Forest and woodland (60%; broadleaved deciduous forest; native coniferous forest; broadleaved evergreen forest), Scrub (10%; sclerophyllous scrub/garrigue/maquis), Grassland (10%; steppe/dry calcareous grassland), Wetland (standing fresh water; river/stream), Rocky areas (5%; sea cliff/rocky shore; rock stacks/islets), Artificial landscape (15%; highly improved reseeded grassland; arable land; perennial crops/orchards/groves)
Land-use Agriculture (15%), Forestry (40%), Military (3%), Nature conservation/research (100%), Tourism/recreation (25%), Urban/industrial/transport (5%)

Birds

Species	Season	Year	Pop min	Pop max	Acc	Criteria
Pernis apivorus Honey Buzzard	B	1990	7	19	B	C6
Neophron percnopterus Egyptian Vulture	B	1996	2	2	A	B2, C6
Circaetus gallicus Short-toed Eagle	B	1990	5	9	B	B2, C6
Falco biarmicus Lanner	R	1990	11	18	B	B2, C2, C6
Falco peregrinus Peregrine	R	1990	5	10	C	C6
[1] ***Tetrax tetrax*** Little Bustard	R	1996	7	60	B	A1, C1, C6
Burhinus oedicnemus Stone Curlew	B	1990	80	170	B	C6
Coracias garrulus Roller	B	1990	5	10	C	C6
Dendrocopos medius Middle Spotted Woodpecker	R	1990	30	—	C	B3, C6
Monticola solitarius Blue Rock Thrush	R	1996	200	—	B	B2
Lanius minor Lesser Grey Shrike	B	1990	100	—	A	C2, C6

1. 100 in 1980s; 30–60 in 1990; 7–15 in 1996.

An important breeding site for raptors and for some open-country species. Breeding species of global conservation concern that do not meet IBA criteria: *Falco naumanni* (1–5 pairs).

Protection status

National Partial **International** Low

120,000 ha of IBA covered by National Park (Gargano, 133,000 ha [boundaries are under discussion]). 48 ha of IBA covered by State Nature Reserve (Falascone, 48 ha). 124 ha of IBA covered by State Nature Reserve (Monte Barone, 124 ha). 56 ha of IBA covered by Integral Nature Reserve (Sfilzi, 56 ha). 399 ha of IBA covered by Biogenetic Reserve (Foresta Umbra, 399 ha). 48 ha of IBA covered by Special Protection Area (Falascone, 48 ha). 399 ha of IBA covered by Special Protection Area (Foresta Umbra, 399 ha). 124 ha of IBA covered by Special Protection Area (Monte Barone, 124 ha). 56 ha of IBA covered by Special Protection Area (Sfilzi, 56 ha).

Conservation issues

Threats Afforestation (B), Agricultural intensification/expansion (A), Burning of vegetation (U), Consequences of animal/plant introductions (B), Construction/impact of dyke/dam/barrage (U), Deforestation (commercial) (U), Disturbance to birds (A), Dredging/canalization (U), Filling-in of wetlands (B), Firewood collection (B), Forest grazing (B), Industrialization/urbanization (B), Infrastructure (A), Recreation/tourism (A), Unsustainable exploitation (A)

Manfredonia Gulf wetlands A1, A4i, A4iii, B1i, B2, B3, C1, C2, C3, C4, C6 130

Admin region Puglia
Coordinates 41°25'N 15°55'E
Altitude 0–5 m **Area** 8,000 ha

Site description

A complex of brackish wetlands along the southern Adriatic coast, between the town of Manfredonia and the village of Margherita di

Savoia. Carapelle–San Floriano is a complex of small ponds scattered across a drained agricultural area, usually managed as game reserves or fish-farms. Saline di Margherita di Savoia is a large active saltpan. Alma Dannata is a lagoon used for fish-farming. Paludi del Candelaro e Cervaro is the remains of what was formerly the largest wetland, dominated by reedbeds (Daunia Risi) and by *Salicornia* bushes (Frattarolo). The whole complex is surrounded by cultivated land.

Habitats Wetland (40%; saltmarsh; coastal lagoon; standing fresh water; standing brackish and salt water; river/stream; water-fringe vegetation), Artificial landscape (60%; arable land)
Land-use Agriculture (60%), Hunting (70%), Nature conservation/research (57%), Tourism/recreation (5%), Urban/industrial/transport (52%)

Birds

Species	Season	Year	Pop min	Pop max	Acc	Criteria
Tadorna tadorna Shelduck	W	1994	3,873	11,287	A	A4i, B1i, C3
Anas penelope Wigeon	W	1994	10,394	15,060	A	B1i, C3
[1] ***Anas strepera*** Gadwall	W	1978	—	1,500	A	A4i, B1i, C3
[1] ***Anas clypeata*** Shoveler	W	1978	—	5,000	B	A4i, B1i, C3
[1] ***Aythya nyroca*** Ferruginous Duck	W	1978	—	250	A	A1, B1i, C1, C2
Recurvirostra avosetta Avocet	B	1993	304	600	A	A4i, B1i, B3, C2
Recurvirostra avosetta Avocet	P	1993	700	10,601	A	A4i, B1i, C2
Recurvirostra avosetta Avocet	W	1995	4,770	4,770	A	A4i, B1i, B2, C2
Charadrius alexandrinus Kentish Plover	B	1993	60	80	A	B2
Charadrius alexandrinus Kentish Plover	P	1993	160	273	A	B1i, C3
Charadrius alexandrinus Kentish Plover	W	1995	706	706	A	B1i, C3
[2] ***Numenius tenuirostris*** Slender-billed Curlew	N	1996	1	20	A	A1, C1, C2
Numenius arquata Curlew	W	1994	263	263	A	B2
Larus genei Slender-billed Gull	B	1993	40	204	A	C2, C6
Larus genei Slender-billed Gull	W	1995	110	230	A	B1i, C2
Gelochelidon nilotica Gull-billed Tern	B	1993	2	24	A	B2, C2, C6
Sterna albifrons Little Tern	B	1993	180	514	A	A4i, B1i, B2, C2, C6

1. 1978 only.
2. Five records, August–May (including wintering birds), 1988–1995.

A very important site for many wintering and migrating waterbird species and for some breeding waders and terns. This is a key site for *Numenius tenuirostris* on passage and in winter. The total number of waterbirds in January 1993 and 1994 was 26,775 and 33,152.

Protection status

National Partial **International** Partial
750 ha of IBA covered by National Park (Gargano, 133,000 ha [boundaries are under discussion]). 257 ha of IBA covered by State Nature Reserve (Palude di Frattarolo, 257 ha). 3,871 ha of IBA covered by State Nature Reserve (Salina di Margherita di Savoia, 3,871 ha). 257 ha of IBA covered by Special Protection Area (Palude di Frattarolo, 257 ha). 3,871 ha of IBA covered by Special Protection Area (Salina di Margherita di Savoia, 3,871 ha). 3,871 ha of IBA covered by Ramsar Site (Saline di Margherita di Savoia, 3,871 ha).

Conservation issues

Threats Agricultural intensification/expansion (A), Drainage (U), Unsustainable exploitation (U)

The main threats are illegal hunting, agricultural intensification, drainage and filling-in of wetlands.

Capri island

C6 **131**

Admin region Campania
Coordinates 40°33′N 14°15′E
Altitude 0–589 m **Area** 1,036 ha

Site description

A small island in the Tyrrhenian Sea, covered mainly by maquis and rich in rocky cliffs. Tourism is the most important human activity.

Habitats Forest and woodland (20%; broadleaved evergreen forest), Scrub (20%; sclerophyllous scrub/garrigue/maquis), Marine areas, Rocky areas (10%; sea cliff/rocky shore), Artificial landscape (50%; perennial crops/orchards/groves; urban parks/gardens; other urban/industrial areas)
Land-use Agriculture (20%), Not utilized (30%), Tourism/recreation (20%), Urban/industrial/transport (30%)

Birds

Species	Season	Year	Pop min	Pop max	Acc	Criteria
Anthus campestris Tawny Pipit	P	1994	20	—	C	C6
Ficedula albicollis Collared Flycatcher	P	1994	30	—	C	C6

Important stop-over site for migratory birds, especially passerines.

Protection status

National Partial **International** None
IBA contains State Nature Reserve (Isola di Capri, area not known).

Conservation issues

Threats Burning of vegetation (A), Disturbance to birds (A), Industrialization/urbanization (B), Infrastructure (B), Recreation/tourism (A), Unsustainable exploitation (A)

The main threats are the illegal, large-scale trapping and shooting of birds, and the burning of vegetation.

Serre Persano

C6 **132**

Admin region Campania
Coordinates 40°30′N 15°05′E
Altitude 49–85 m **Area** 300 ha

Site description

An alluvial plain formed by the Sele river, between Monti Alburni and Monti Picentini. A dam on the river has created an artificial lake which is fringed by wet riparian woodland. Other habitats include maquis, broadleaved woodland, grassland and arable fields. The main land-use is nature conservation.

Habitats Forest and woodland (20%; broadleaved deciduous forest; alluvial/very wet forest), Grassland (20%; humid grassland), Wetland (50%; standing fresh water; river/stream; water-fringe vegetation), Artificial landscape (10%; arable land)
Land-use Agriculture (10%), Nature conservation/research (100%), Tourism/recreation (50%), Water management (50%)

Birds

Species	Season	Year	Pop min	Pop max	Acc	Criteria
Milvus migrans Black Kite	B	1995	2	4	A	C6
Grus grus Crane	P	1995	—	500	A	C6
Alcedo atthis Kingfisher	R	1995	Common		—	C6

A notable site for breeding and staging waterbirds. Breeding species of global conservation concern that do not meet IBA criteria: *Aythya nyroca* (6–8 pairs).

Protection status

National High **International** None
300 ha of IBA covered by Private Reserve (Serre Persano, 300 ha).

Conservation issues

Threats Recreation/tourism (C)

There are no major threats. The site is managed as a reserve by WWF-Italia, and is covered by a management plan. There are ongoing projects to reintroduce *Aythya nyroca*, to conserve Eurasian otter *Lutra lutra*, and to ring migrating passerines.

Picentini mountains

C6 **133**

Admin region Campania
Coordinates 40°30′N 15°00′E
Altitude 280–1,809 m **Area** 64,000 ha

Site description

A mountain area in southern Italy, north-east of the town of Salerno. There are extensive broadleaved woodlands (especially beech *Fagus*), as well as streams, rivers, grasslands, cultivated areas, and rocky areas of karst. The main land-uses are forestry, cultivation and stock-farming. This IBA includes 'Monte Terminio and Monte Polveracchio', which was listed as a separate IBA (former site IT087) in the previous pan-European IBA inventory (Grimmett and Jones 1989).

Habitats Forest and woodland (65%; broadleaved deciduous forest; broadleaved evergreen forest), Scrub (5%; sclerophyllous scrub/garrigue/maquis), Grassland (5%; mesophile grassland), Wetland (5%; river/stream), Artificial landscape (20%; perennial crops/orchards/groves; ruderal land)
Land-use Agriculture (15%), Forestry (30%), Tourism/recreation (5%), Water management (5%)

Birds

Species	Season	Year	Pop min	Pop max	Acc	Criteria
Milvus migrans Black Kite	B	1993	5	8	A	C6
Milvus milvus Red Kite	R	1993	3	4	A	C6
Bubo bubo Eagle Owl	R	1993	2	—	B	C6
[1] ***Dryocopus martius*** Black Woodpecker	R	1993	—	—	—	C6

1. Very few breeding sites in NUTS region.

The site is notable for breeding birds of prey and woodpeckers.

Protection status

National High **International** None
64,000 ha of IBA covered by Regional Nature Park (Monti Picentini, 64,000 ha).

Conservation issues

Threats Deforestation (commercial) (A), Extraction industry (B), Firewood collection (C), Forest grazing (B), Groundwater abstraction (B), Industrialization/urbanization (B), Infrastructure (A), Intensified forest management (B), Recreation/tourism (B), Selective logging/cutting (B)

The site is threatened especially by deforestation and by the building of new roads. Restocking of *Alectoris graeca* occurs.

Alburni mountains B2, C6 134

Admin region Campania
Coordinates 40°35′N 15°19′E
Altitude 204–1,742 m **Area** 16,000 ha

Site description

A calcareous mountain area in southern Italy, crossed by the rivers Sele and Sanagro. There are broadleaved woodlands, agricultural land, grasslands, and maquis. The main land-uses are agriculture, forestry and tourism.

Habitats Forest and woodland (35%; broadleaved deciduous forest; native coniferous forest; mixed forest; broadleaved evergreen forest; treeline ecotone), Scrub (20%; sclerophyllous scrub/garrigue/maquis), Grassland (15%; mesophile grassland), Wetland (5%; river/stream), Rocky areas (5%; inland cliff), Artificial landscape (10%; highly improved reseeded grassland; arable land; urban parks/gardens; other urban/industrial areas; ruderal land)
Land-use Agriculture (20%), Forestry (15%), Not utilized (42%), Tourism/recreation (15%), Urban/industrial/transport (8%)

Birds

Species	Season	Year	Pop min	Pop max	Acc	Criteria
Falco biarmicus Lanner	R	1989	1	—	C	C6
[1] ***Anthus campestris*** Tawny Pipit	B	1993	Uncommon		—	C6
Lanius senator Woodchat Shrike	B	1997	50	—	C	B2

1. Very few breeding sites in NUTS region.

Notable for several open-country species, and rich in woodland species.

Protection status

National High **International** None
16,000 ha of IBA covered by National Park (Cilento e Vallo di Diano, 227,052 ha).

Conservation issues

Threats Agricultural intensification/expansion (B), Deforestation (commercial) (A), Disturbance to birds (B), Extraction industry (B), Firewood collection (C), Forest grazing (C), Groundwater abstraction (C), Industrialization/urbanization (A), Infrastructure (A), Recreation/tourism (A)

The site is threatened by industrial and urban development and by the building of new infrastructure. The National Park is not yet fully operative.

Murge A1, A3, A4ii, B1iii, B2, C1, C2, C6 135

Admin region Puglia
Coordinates 40°53′N 16°24′E
Altitude 340–679 m **Area** 90,000 ha

Site description

A very large area of steppe in southern Italy, including the village of Altamura, with areas of maquis and agricultural land.

Habitats Forest and woodland (12%; broadleaved deciduous forest; native coniferous forest; wooded steppe), Scrub (5%; scrub; sclerophyllous scrub/garrigue/maquis), Grassland (55%; steppe/dry calcareous grassland), Wetland (1%; standing fresh water; river/stream), Rocky areas (4%; inland cliff; caves), Artificial landscape (23%; highly improved reseeded grassland; arable land; perennial crops/orchards/groves; forestry plantation; urban parks/gardens; other urban/industrial areas; ruderal land)
Land-use Agriculture (75%), Forestry (5%), Hunting (75%), Tourism/recreation (5%), Urban/industrial/transport (5%)

Birds

Species	Season	Year	Pop min	Pop max	Acc	Criteria
Circaetus gallicus Short-toed Eagle	B	1996	2	5	C	C6
Falco naumanni Lesser Kestrel	B	1997	1,500	1,500	A	A1, A4ii, B1iii, B2, C1, C2, C6
Falco biarmicus Lanner	R	1994	3	—	A	B2, C6
Burhinus oedicnemus Stone Curlew	B	1996	—	40	B	C6
Melanocorypha calandra Calandra Lark	R	1996	500	—	C	C6
Calandrella brachydactyla Short-toed Lark	B	1996	300	—	—	B2, C6
Galerida cristata Crested Lark	R	1996	1,000	3,000	C	B2
Oenanthe hispanica Black-eared Wheatear	B	1996	100	—	C	A3
Sylvia conspicillata Spectacled Warbler	B	1996	200	—	C	A3
Sylvia cantillans Subalpine Warbler	B	1996	100	—	—	A3
Sylvia melanocephala Sardinian Warbler	R	1996	1,500	—	C	A3
Lanius minor Lesser Grey Shrike	B	1996	50	—	C	B2, C2, C6
Lanius senator Woodchat Shrike	B	1996	50	—	C	B2
Emberiza melanocephala Black-headed Bunting	B	1996	50	—	C	A3

An important area for breeding raptors (especially *Falco naumanni*) and for larks and other steppic species, as well as for five of the 21 species of the Mediterranean biome. Species of global conservation concern that do not meet IBA criteria: *Tetrax tetrax* (rare resident).

Protection status

National None **International** Unknown
Overlaps with SPA.

Conservation issues

Threats Abandonment/reduction of land management (U), Afforestation (U), Agricultural intensification/expansion (A), Consequences of animal/plant introductions (U), Construction/impact of dyke/dam/barrage (A), Dredging/canalization (C), Extraction industry (B), Forest grazing (B), Industrialization/urbanization (A), Infrastructure (A), Recreation/tourism (C), Unsustainable exploitation (A)

Part of the site has been proposed as a National Park. There are censuses and ongoing research on breeding *Falco naumanni*; research on raptors by LIPU was funded by the LIFE programme of the EU.

Mount Cervati B2, C6 136

Admin region Campania
Coordinates 40°17′N 15°26′E
Altitude 192–1,899 m **Area** 44,000 ha

Site description

A karstic mountain area in the southern Apennines. Mount Cervati is the highest mountain in the area. The site is mainly covered by broadleaved woodland, scrub and cultivated land, and is rich in deep gorges and karstic land-forms. The main land-uses are traditional forestry, cultivation and cattle-rearing.

Habitats Forest and woodland (25%; broadleaved deciduous forest; native coniferous forest; mixed forest; treeline ecotone), Scrub (25%; scrub; sclerophyllous scrub/garrigue/maquis), Grassland (5%; steppe/dry calcareous grassland), Wetland (5%; river/stream), Rocky areas (5%; inland cliff), Artificial landscape (35%; highly improved reseeded grassland; arable land; perennial crops/orchards/groves; forestry plantation; ruderal land)
Land-use Agriculture (15%), Forestry (30%), Nature conservation/research (50%), Urban/industrial/transport (5%)

Birds

Species	Season	Year	Pop min	Pop max	Acc	Criteria
Milvus milvus Red Kite	R	1993	4	7	A	C6
Circaetus gallicus Short-toed Eagle	B	1995	3	3	A	C6
Falco biarmicus Lanner	R	1993	Uncommon		—	C6
Falco peregrinus Peregrine	R	1995	9	9	A	C6
Alectoris graeca Rock Partridge	R	1995	72	—	A	C6
Otus scops Scops Owl	B	1997	100	—	C	B2
Bubo bubo Eagle Owl	R	1993	1	2	B	C6
Athene noctua Little Owl	R	1997	100	—	C	B2
Pyrrhocorax pyrrhocorax Chough	R	1995	60	60	A	B2, C6

The area is rich in breeding birds of prey, with some mountain species.

Protection status

National High **International** None
44,000 ha of IBA covered by National Park (Cilento e Vallo di Diano, 227,052 ha).

Conservation issues

Threats Abandonment/reduction of land management (B), Afforestation (C), Agricultural intensification/expansion (C), Consequences of animal/plant introductions (C), Deforestation (commercial) (A), Disturbance to birds (C), Firewood collection (B), Forest grazing (A), Groundwater abstraction (A), Infrastructure (B), Intensified forest management (C), Natural events (B), Selective logging/cutting (A)

The main threat is unsustainable exploitation of forest resources. Some or all of the site is covered by a management plan. The National Park is not yet fully operative.

Dolomiti di Pietrapertosa — B2, C6 — 137

Admin region Basilicata
Coordinates 40°30′N 16°03′E
Altitude 500–1,350 m **Area** 2,500 ha

Site description

A wooded, mountainous area in southern Italy, very rich in rocky cliffs and gorges.

Habitats Forest and woodland (75%; broadleaved deciduous forest; native coniferous forest), Rocky areas (15%; inland cliff; caves), Artificial landscape (10%; highly improved reseeded grassland; arable land)
Land-use Agriculture (9%), Hunting (20%), Nature conservation/research (71%)

Birds

Species	Season	Year	Pop min	Pop max	Acc	Criteria
Milvus milvus Red Kite	R	1995	3	5	A	C6
Neophron percnopterus Egyptian Vulture	B	1995	1	1	A	B2, C6
Falco biarmicus Lanner	R	1993	1	3	A	B2, C6
Dryocopus martius Black Woodpecker	R	1996	Frequent		—	C6

An important site for breeding raptors.

Protection status

National High **International** Unknown
2,500 ha of IBA covered by Regional Nature Park (Gallipoli–Cognato–Piccole Dolomiti Lucane, 2,500 ha). Overlaps with SPA.

Conservation issues

Threats Disturbance to birds (C), Firewood collection (C), Forest grazing (C), Unsustainable exploitation (C)

There are no major threats. Some or all of the site is covered by a management plan.

Manferrara forest — B2, C6 — 138

Admin region Basilicata
Coordinates 40°32′N 16°30′E
Altitude 390–475 m **Area** 500 ha

Site description

A wooded area (mainly oak *Quercus*) surrounded by cultivated land, in southern Italy, close to the village of Pomarico. The main land-uses are forestry and agriculture.

Habitats Forest and woodland (60%; broadleaved deciduous forest; mixed forest; broadleaved evergreen forest), Scrub (10%; sclerophyllous scrub/garrigue/maquis), Artificial landscape (30%; arable land; forestry plantation; ruderal land)
Land-use Agriculture (20%), Forestry (10%), Hunting (10%), Nature conservation/research (90%), Tourism/recreation (20%), Urban/industrial/transport (10%)

Birds

Species	Season	Year	Pop min	Pop max	Acc	Criteria
Milvus migrans Black Kite	R	1994	10	12	A	B2
Milvus milvus Red Kite	R	1996	3	3	A	C6

An important site for kites *Milvus*.

Protection status

National None **International** None

Conservation issues

Threats Disturbance to birds (C), Extraction industry (B), Forest grazing (B), Industrialization/urbanization (B), Infrastructure (C), Recreation/tourism (C), Unsustainable exploitation (B)

Urbanization is the main threat to the area. The area is a proposed LIPU reserve. Some or all of the site is covered by a management plan.

Gravine — A1, A4ii, B1iii, B2, C1, C2, C6 — 139

Admin region Puglia, Basilicata
Coordinates 40°35′N 17°00′E
Altitude 10–450 m **Area** 30,000 ha

Site description

An area of steppe and maquis in southern Italy, north-west of the town of Taranto. The site includes the historical town of Matera, and is incised by streams, which have created canyons and gorges. Most of the land is agricultural.

Habitats Forest and woodland (10%; broadleaved deciduous forest; broadleaved evergreen forest), Scrub (5%; sclerophyllous scrub/garrigue/maquis), Grassland (15%; steppe/dry calcareous grassland), Rocky areas (5%; scree/boulders; inland cliff; caves), Artificial landscape (65%; highly improved reseeded grassland; arable land; perennial crops/orchards/groves; forestry plantation; other urban/industrial areas; ruderal land)
Land-use Agriculture (55%), Urban/industrial/transport (10%)

Birds

Species	Season	Year	Pop min	Pop max	Acc	Criteria
Falco naumanni Lesser Kestrel	B	1997	600	600	A	A1, A4ii, B1iii, B2, C1, C2, C6
Falco biarmicus Lanner	R	1996	2	2	A	B2, C6

Important numbers of *Falco naumanni* breed in Matera town.

Protection status

National Partial **International** Unknown
1,115 ha of IBA covered by Wildlife Reserve (Corno della Strega, 1,115 ha). 200 ha of IBA covered by Wildlife Reserve (Gravina di Castellaneta, 200 ha). 800 ha of IBA covered by Wildlife Reserve (Gravina di Laterza, 800 ha). 2,380 ha of IBA covered by Wildlife Reserve (Santa Trinità, 2,380 ha). Overlaps with SPA.

Conservation issues

Threats are posed by fires, agricultural intensification and illegal hunting. The area is a proposed Regional Park.

Threats Abandonment/reduction of land management (B), Agricultural intensification/expansion (A), Burning of vegetation (A), Disturbance to birds (A), Forest grazing (C), Industrialization/urbanization (A), Unsustainable exploitation (B)

Coast between Marina di Camerota and Policastro Bussentino

B2, C6 **140**

Admin region Campania
Coordinates 40°00′N 15°18′E
Altitude 0–1,224 m **Area** 5,700 ha

Site description

A stretch of rocky coast along the southern Tyrrhenian Sea. Maquis and rocky cliffs are the main habitats. The main land-uses are agriculture and tourism.

Habitats Forest and woodland (30%; broadleaved deciduous forest; native coniferous forest; mixed forest; broadleaved evergreen forest), Scrub (20%; sclerophyllous scrub/garrigue/maquis), Grassland (10%), Wetland (10%), Marine areas (10%), Rocky areas (10%; sea cliff/rocky shore; caves), Artificial landscape (10%; highly improved reseeded grassland; arable land; perennial crops/orchards/groves; forestry plantation; urban parks/gardens; other urban/industrial areas; ruderal land)
Land-use Agriculture (20%), Not utilized (50%), Tourism/recreation (20%), Urban/industrial/transport (10%)

Birds

Species	Season	Year	Pop min	Pop max	Acc	Criteria
Falco tinnunculus Kestrel	R	1997	50	—	A	B2
Falco eleonorae Eleonora's Falcon	B	1997	5	—	C	C6
Falco peregrinus Peregrine	R	1997	10	—	C	C6
Monticola solitarius Blue Rock Thrush	R	1997	100	200	C	B2
Sylvia undata Dartford Warbler	B	1997	50	—	B	C6
Lanius senator Woodchat Shrike	B	1997	50	—	C	B2

Protection status

National High **International** None
5,700 ha of IBA covered by National Park (Cilento e Vallo di Diano, 227,052 ha).

Conservation issues

Threats Abandonment/reduction of land management (C), Burning of vegetation (A), Disturbance to birds (B), Extraction industry (B), Forest grazing (B), Industrialization/urbanization (A), Infrastructure (A), Recreation/tourism (A), Unsustainable exploitation (B)

The main threats are tourism, urban development, building of new infrastructure, and burning of vegetation. The National Park is not yet fully operative.

Lagonegrese and gorges of River Calore

B2, C6 **141**

Admin region Campania, Basilicata
Coordinates 40°08′N 15°55′E
Altitude 600–2,005 m **Area** 50,000 ha

Site description

A mountain complex in southern Italy, including the sites Monte Sirino, Monte Raparo, Monte dell'Agresto, Monte Caldarosa, Monte Volturino, and Gole del fiume Calore. The main habitats are broadleaved and mixed woodlands, cliffs and agricultural land, crossed by rivers and streams. Land-uses include forestry, cultivation and cattle-rearing. 'Gole del Fiume Calore' was listed as a separate IBA (former site IT089) in the previous pan-European IBA inventory (Grimmett and Jones 1989).

Habitats Forest and woodland (65%; broadleaved deciduous forest; native coniferous forest; mixed forest), Wetland (5%; standing fresh water; river/stream), Rocky areas (15%; scree/boulders; inland cliff; caves), Artificial landscape (15%; perennial crops/orchards/groves; forestry plantation; ruderal land)
Land-use Agriculture (10%), Hunting (10%), Nature conservation/research (100%), Urban/industrial/transport (5%)

Birds

Species	Season	Year	Pop min	Pop max	Acc	Criteria
Milvus migrans Black Kite	B	1995	10	—	B	C6
Falco biarmicus Lanner	R	—	1	2	A	B2, C6
Dryocopus martius Black Woodpecker	R	1995	Uncommon		—	C6
Lullula arborea Woodlark	R	1995	300	—	B	B2, C6
Sylvia hortensis Orphean Warbler	B	1996	10	—	B	B2
Pyrrhocorax pyrrhocorax Chough	R	1996	5	—	B	C6

Protection status

National High **International** Low
50,000 ha of IBA covered by National Park (Val d'Agri–Lagonegrese, area not known). 1,650 ha of IBA covered by Special Protection Area (Gole del Fiume Calore, 1,650 ha).

Conservation issues

Threats Extraction industry (B), Firewood collection (C), Forest grazing (C), Industrialization/urbanization (B), Infrastructure (B), Unsustainable exploitation (A)

The main threat is intense illegal hunting.

Mounts Orsomarso and Verbicaro

C6 **142**

Admin region Calabria
Coordinates 39°50′N 15°57′E
Altitude 50–1,987 m **Area** 48,000 ha

Site description

A large, mountainous area (both coastal and inland) in the southern Apennines, west of Castrovillari, practically uninhabited and mainly covered by broadleaved woodland and maquis. The main land-use is nature conservation.

Habitats Forest and woodland (65%; broadleaved deciduous forest; native coniferous forest; mixed forest; broadleaved evergreen forest), Scrub (30%; sclerophyllous scrub/garrigue/maquis), Rocky areas (5%; inland cliff)
Land-use Nature conservation/research (100%), Tourism/recreation (5%)

Birds

Species	Season	Year	Pop min	Pop max	Acc	Criteria
Alectoris graeca Rock Partridge	R	1996	Uncommon		—	C6
Lullula arborea Woodlark	R	1996	Frequent		—	C6
Ficedula albicollis Collared Flycatcher	B	1996	3	—	B	C6

The site is important for breeding raptors of forest, and possibly holds the southernmost breeding *Dryocopus martius* in Italy.

Protection status

National High **International** Partial
48,000 ha of IBA covered by National Park (Pollino, 196,437 ha). 3,980 ha of IBA covered by State Nature Reserve (Valle del Fiume Argentino, 3,980 ha). 5,533 ha of IBA covered by State Nature Reserve (Valle del Fiume Lao, 5,533 ha). 3,980 ha of IBA covered by Special Protection Area (Valle del Fiume Argentino, 3,980 ha). 5,533 ha of IBA covered by Special Protection Area (Valle del Fiume Lao, 5,533 ha).

Conservation issues

Threats Burning of vegetation (B), Infrastructure (C), Unsustainable exploitation (A)

The main threat is illegal hunting and fires. Fires are started by man, mainly by shepherds. There is a project for the conservation of roe deer *Capreolus capreolus*.

Pollino

B2, C6 **143**

Admin region Basilicata, Calabria
Coordinates 39°53′N 16°12′E
Altitude 190–2,267 m **Area** 115,814 ha

Site description

The largest mountain complex in the southern Apennines, with a high diversity of habitats including maquis, broadleaved, mixed and coniferous woodlands, dry and alpine grasslands, pastures, and cultivated land. The presence and percentage cover of habitats (listed in the header) refer only to the part of the site within Calabria region.

Habitats Forest and woodland (40%; broadleaved deciduous forest; native coniferous forest; mixed forest; broadleaved evergreen forest; treeline ecotone), Scrub (scrub; sclerophyllous scrub/garrigue/maquis), Grassland (25%; steppe/dry calcareous grassland; alpine/subalpine/boreal grassland), Wetland (river/stream), Rocky areas (10%; inland cliff), Artificial landscape (25%; highly improved reseeded grassland; perennial crops/orchards/groves; ruderal land)
Land-use Agriculture (50%), Hunting (20%), Nature conservation/research (80%), Tourism/recreation, Urban/industrial/transport (10%)

Birds

Species	Season	Year	Pop min	Pop max	Acc	Criteria
Milvus migrans Black Kite	B	1996	10	20	A	B2, C6
Milvus milvus Red Kite	R	1996	10	20	B	C6
Neophron percnopterus Egyptian Vulture	B	1996	2	2	B	B2, C6
Aquila chrysaetos Golden Eagle	R	1996	1	2	A	C6
Falco peregrinus Peregrine	R	1996	2	4	A	C6
Alectoris graeca Rock Partridge	R	1996	Uncommon	—		C6
Bubo bubo Eagle Owl	R	1996	1	2	C	C6
Dryocopus martius Black Woodpecker	R	1996	3	—	B	C6
Calandrella brachydactyla Short-toed Lark	B	1996	10	15	A	C6
Lullula arborea Woodlark	R	1996	Abundant	—		C6
Anthus campestris Tawny Pipit	B	1996	Common	—		C6
Pyrrhocorax pyrrhocorax Chough	R	1996	10	—	B	C6

An important breeding area for raptors and species of forest, mountain and open country. Ornithological data only refer to the part of Pollino that is in Calabria region.

Protection status

National High **International** Low
112,014 ha of IBA covered by National Park (Pollino, 196,437 ha). 1,600 ha of IBA covered by State Nature Reserve (Gole del Raganello, 1,600 ha). 1,600 ha of IBA covered by Special Protection Area (Gole del Raganello, 1,600 ha).

Conservation issues

Threats Afforestation (B), Extraction industry (A), Forest grazing (B), Industrialization/urbanization (A), Infrastructure (A), Recreation/tourism (B), Unsustainable exploitation (B)

The main threats are urban development and the building of new infrastructure.

Ferro valley — C6 — 144

Admin region Calabria
Coordinates 39°59′N 16°32′E
Altitude 10–666 m **Area** 8,500 ha

Site description

A small valley facing the Ionian Sea, north of the village of Amendolara, rich in maquis, garrigue, grasslands and agricultural areas. The main land-use is agriculture.

Habitats Scrub (sclerophyllous scrub/garrigue/maquis), Grassland, Rocky areas (scree/boulders; inland cliff), Artificial landscape (arable land; perennial crops/orchards/groves)
Land-use Agriculture, Hunting, Urban/industrial/transport

Birds

Species	Season	Year	Pop min	Pop max	Acc	Criteria
Falco biarmicus Lanner	R	1997	1	—	B	C6

An important breeding site for raptors, and stop-over site for many migrants.

Protection status

National None **International** None

Conservation issues

The main threats are illegal hunting, and burning of vegetation.

Threats Burning of vegetation (B), Industrialization/urbanization (C), Infrastructure (C), Unsustainable exploitation (B)

Sant'Andrea island — A1, B2, C1, C6 — 145

Admin region Puglia
Coordinates 40°03′N 17°59′E
Altitude 0–6 m **Area** 52 ha

Site description

A small island not far off the north-west coast of Puglia, close to the town of Gallipoli, and mainly covered by *Salicornia* bushes. At present the island is not used by man.

Habitats Grassland (90%; steppe/dry calcareous grassland), Wetland (10%; coastal lagoon)
Land-use Not utilized (100%)

Birds

Species	Season	Year	Pop min	Pop max	Acc	Criteria
Larus audouinii Audouin's Gull	R	1996	20	30	A	A1, B2, C1, C6

The island holds a colony of *Larus audouinii*.

Protection status

National None **International** Unknown
Overlaps with SPA.

Conservation issues

Threats Disturbance to birds (A), Recreation/tourism (A)

The main threat is the disturbance of breeding birds by tourists. There is a proposal to declare it a nature reserve.

Le Cesine — C7 — 146

Admin region Puglia
Coordinates 40°21′N 18°20′E
Altitude 0–5 m **Area** 620 ha

Site description

An area on the Adriatic coast in southern Italy, about 10 km east of the town of Lecce. It is a mosaic of beaches, sand-dunes, maquis, freshwater and brackish pools, olive-groves and arable land. The main land-uses are nature conservation and agriculture.

Habitats Scrub (5%; sclerophyllous scrub/garrigue/maquis), Wetland (40%; sand-dunes/sand beach; standing fresh water; standing brackish and salt water; water-fringe vegetation), Artificial landscape (55%; highly improved reseeded grassland; arable land; perennial crops/orchards/groves; forestry plantation)
Land-use Agriculture (40%), Nature conservation/research (55%), Tourism/recreation (5%)

Birds

A notable site for migrating and wintering waterbirds. Species of global conservation concern that do not meet IBA criteria: *Aythya nyroca* (max. 3, winter).

Protection status

National Partial **International** High
348 ha of IBA covered by State Nature Reserve (La Cesine, 348 ha). 620 ha of IBA covered by Special Protection Area (Le Cesine, 620 ha). 620 ha of IBA covered by Ramsar Site (Le Cesine, 620 ha).

Conservation issues

Threats Agricultural intensification/expansion (C), Disturbance to birds (B), Filling-in of wetlands (C), Infrastructure (C), Natural events (A)

Dune erosion is the main threat. The site is a WWF reserve. Some or all of the site is covered by a management plan. Birds are monitored and ringed. There are wetland-restoration projects, financed by regional and EU funds.

Cape Otranto — B1iv, C1, C5 — 147

Admin region Puglia
Coordinates 40°08′N 18°35′E
Altitude 0–128 m **Area** 3,100 ha

■ Site description
A rocky area on the coast of south-east Puglia, south of the town of Otranto. Habitats are dominated by pseudo-steppes, garrigue, pastures and cultivated land. Cultivation and stock-rearing are the main land-uses.

Habitats Scrub (30%; sclerophyllous scrub/garrigue/maquis), Grassland (30%; steppe/dry calcareous grassland), Rocky areas (5%; sea cliff/rocky shore), Artificial landscape (35%; highly improved reseeded grassland; arable land; perennial crops/orchards/groves; forestry plantation; ruderal land)
Land-use Agriculture (65%), Hunting, Tourism/recreation (10%)

■ Birds

Species	Season	Year	Pop min	Pop max	Acc	Criteria
Falco naumanni Lesser Kestrel	P	1996	50	—	B	C1

An important 'bottleneck' site for raptors migrating along the Adriatic coast in spring. It is considered that more than 3,000 raptors pass each spring, although there is only incomplete data from one year (1989), when a total of about 1,500 birds were counted, mainly harriers *Circus* and *Pernis apivorus*.

■ Protection status
National None **International** None

■ Conservation issues
Threats Abandonment/reduction of land management (C), Agricultural intensification/expansion (B), Burning of vegetation (A), Industrialization/urbanization (B), Infrastructure (B), Recreation/tourism (B), Unsustainable exploitation (B)

The main threats are fires, illegal hunting, overgrazing and the building of tourist resorts.

Sila Grande — C6 — 148

Admin region Calabria
Coordinates 39°21′N 16°32′E
Altitude 1,127–1,928 m **Area** 28,450 ha

■ Site description
A mountain area in southern Italy, dominated by extensive woodlands (mainly pine *Pinus* and beech *Fagus*), traditionally cultivated land, pastures, grasslands and artificial lakes. The main land-uses are forestry, agriculture, hunting and tourism.

Habitats Forest and woodland (60%; broadleaved deciduous forest; native coniferous forest; mixed forest; treeline ecotone), Grassland (10%; dry siliceous grassland; mesophile grassland), Wetland (5%; standing fresh water; river/stream), Artificial landscape (25%; highly improved reseeded grassland; arable land; perennial crops/orchards/groves; ruderal land)
Land-use Agriculture (25%), Forestry (60%), Hunting (60%), Nature conservation/research (25%), Tourism/recreation (50%), Urban/industrial/transport (5%), Water management (5%)

■ Birds

Species	Season	Year	Pop min	Pop max	Acc	Criteria
Dryocopus martius Black Woodpecker	R	1996	Frequent		—	C6
Lullula arborea Woodlark	R	1996	Abundant		—	C6
Anthus campestris Tawny Pipit	B	1996	Common		—	C6

A notable breeding site for several woodland species.

■ Protection status
National Partial **International** Partial
7,084 ha of IBA covered by National Park (Calabria, 15,894 ha). 7,084 ha of IBA covered by Special Protection Area (Parco Nazionale della Calabria, 15,892 ha).

■ Conservation issues
Threats Agricultural intensification/expansion (A), Forest grazing (B), Recreation/tourism (C)

The most important threat is agricultural intensification.

Marchesato and River Neto — B1iii, B2, C2, C6 — 149

Admin region Calabria
Coordinates 39°13′N 16°57′E
Altitude 0–562 m **Area** 44,125 ha

■ Site description
A low, dry area of hills in southern Italy, facing the Ionian Sea, north-west of the town of Crotone and crossed by the Neto river. The site is rich in rocky areas and small villages. The main land-use is agriculture. The site incorporates an IBA that was listed separately in the previous pan-European IBA inventory (Grimmett and Jones 1989), 'Foce del Fiume Neto' (former site IT106).

Habitats Forest and woodland (10%; mixed forest; broadleaved evergreen forest), Scrub (20%; sclerophyllous scrub/garrigue/maquis), Grassland (20%; steppe/dry calcareous grassland), Wetland (5%; sand-dunes/sand beach; river/stream), Marine areas, Rocky areas (5%; inland cliff; caves), Artificial landscape (40%; highly improved reseeded grassland; arable land; perennial crops/orchards/groves; other urban/industrial areas; ruderal land)
Land-use Agriculture (50%), Forestry (10%), Hunting (50%), Nature conservation/research (3%), Urban/industrial/transport (5%)

■ Birds

Species	Season	Year	Pop min	Pop max	Acc	Criteria
Milvus migrans Black Kite	B	1996	1	3	B	C6
Milvus milvus Red Kite	B	1996	3	8	A	C6
Neophron percnopterus Egyptian Vulture	B	1996	3	4	A	B2, C6
Circaetus gallicus Short-toed Eagle	B	1996	2	3	A	C6
Falco vespertinus Red-footed Falcon	P	1996	1,000	—	C	B1iii
Falco biarmicus Lanner	B	1996	3	3	B	B2, C6
Falco peregrinus Peregrine	B	1996	1	5	B	C6
Grus grus Crane	P	1996	1	500	A	C6
Larus melanocephalus Mediterranean Gull	W	1995	300	300	A	C2
Coracias garrulus Roller	B	1996	5	10	B	B2, C6
Anthus campestris Tawny Pipit	B	1996	Common		—	C6

The area inland is important for breeding and migrating raptors while the coastal lagoons support high numbers of non-breeding waterbirds. Species of global conservation concern that do not meet IBA criteria: *Falco naumanni* (1–10 on passage).

■ Protection status
National Low **International** None
1,500 ha of IBA covered by Wildlife Reserve (Foce del fiume Neto, 1,500 ha).

■ Conservation issues
Threats Abandonment/reduction of land management (A), Afforestation (C), Agricultural intensification/expansion (C), Disturbance to birds (A), Drainage (A), Dredging/canalization (A), Extraction industry (A), Industrialization/urbanization (B), Infrastructure (B), Unsustainable exploitation (A)

The most important threats are illegal hunting, and disturbance to birds.

Costa Viola — A1, B1iii, B1iv, C1, C5 — 150

Admin region Calabria
Coordinates 38°13′N 15°45′E
Altitude 0–1,304 m **Area** 19,750 ha

■ Site description
Rocky coast and hills along the Calabrian coast of the Tyrrhenian Sea, on the east side of the Messina Strait.

Habitats Forest and woodland (10%; broadleaved deciduous forest; broadleaved evergreen forest), Scrub (30%; scrub; sclerophyllous scrub/garrigue/maquis), Wetland (5%; sand-dunes/sand beach; shingle/stony beach), Marine areas, Artificial landscape (50%; perennial crops/orchards/groves; urban parks/gardens; ruderal land)
Land-use Agriculture (30%), Hunting (100%), Not utilized (20%), Tourism/recreation (10%), Urban/industrial/transport (20%)

Birds

Species	Season	Year	Pop min	Pop max	Acc	Criteria
Aquila clanga Greater Spotted Eagle	P	1990	—	2	B	A1, C1
Falco naumanni Lesser Kestrel	P	1990	—	30	B	A1, C1
Falco vespertinus Red-footed Falcon	P	1990	28	672	B	B1iii

An important 'bottleneck' site for soaring birds on passage, with 10,000–20,000 being counted in total each spring. The most common species include *Pernis apivorus* (3,138–14,960 per season), *Circus aeruginosus* and *Falco vespertinus*. Counts have been made in 1987, 1990, 1993 and 1994, and the maximum total count (all species combined) for one season is 18,160 in spring 1993.

Protection status

National None **International** None

Conservation issues

Threats Disturbance to birds (A), Industrialization/urbanization (A), Infrastructure (A), Natural events (A), Recreation/tourism (C), Unsustainable exploitation (A)

There is massive and intense illegal hunting, which is the main threat to the migrating raptors. Counting of migrating raptors/storks and anti-poaching activities have been carried out by LIPU every year since the mid-1980s.

Aspromonte — B1iv, C5, C6 — 151

Admin region Calabria
Coordinates 38°07'N 16°00'E
Altitude 106–1,955 m **Area** 35,650 ha

Site description

A mountain area, in the southern Apennines, which includes the central area of the Aspromonte complex and two valleys which reach down to the coast. Extensive broadleaved and coniferous woodlands are the dominant habitats. The main land-uses are forestry and cattle-rearing.

Habitats Forest and woodland (75%; broadleaved deciduous forest; native coniferous forest; mixed forest), Scrub (15%; sclerophyllous scrub/garrigue/maquis), Grassland (5%; river/stream; inland cliff), Artificial landscape (5%)
Land-use Agriculture, Forestry, Nature conservation/research

Birds

Species	Season	Year	Pop min	Pop max	Acc	Criteria
[1] ***Dryocopus martius*** Black Woodpecker	R	1987	2	—	—	C6
Ficedula albicollis Collared Flycatcher	B	1987	20	50	—	C6

1. Breeding not proven.

An important migratory 'bottleneck' site for raptors. A few thousand are thought to pass regularly every spring, mostly *Pernis apivorus*, with many *Milvus migrans* and *Circus aeruginosus* as well, although there have been no systematic counts.

Protection status

National Partial **International** Partial
Part of IBA covered by National Park (Aspromonte, 50,000 ha).

Conservation issues

Threats Construction/impact of dyke/dam/barrage (A), Forest grazing (B), Groundwater abstraction (A), Recreation/tourism (C), Unsustainable exploitation (B)

Illegal killing of raptors during migration is a major threat.

Eolie islands — A4ii, B1iii, B2, C2, C6 — 152

Admin region Sicilia
Coordinates 38°30'N 14°50'E
Altitude 0–962 m **Area** 11,610 ha

Site description

An archipelago of small islands and islets, of volcanic origin, in the Tyrrhenian Sea, north of Sicily. The islands are Alicudi, Filicudi, Salina, Lipari, Vulcano, Panarea, Basiluzzo, Lisca Bianca, and Stromboli. The main habitats are maquis, rocky areas and garrigue. The principal land-uses are agriculture and tourism.

Habitats Forest and woodland (broadleaved deciduous forest), Scrub (sclerophyllous scrub/garrigue/maquis), Grassland, Marine areas, Rocky areas (sea cliff/rocky shore; rock stacks/islets), Artificial landscape (highly improved reseeded grassland; perennial crops/orchards/groves)
Land-use Agriculture, Fisheries/aquaculture, Hunting, Tourism/recreation, Urban/industrial/transport

Birds

Species	Season	Year	Pop min	Pop max	Acc	Criteria
Calonectris diomedea Cory's Shearwater	R	1996	100	—	—	C6
Falco eleonorae Eleonora's Falcon	B	1994	108	118	A	A4ii,B1iii,B2,C2,C6
Falco peregrinus Peregrine	R	1996	8	—	—	C6

An important area for breeding *Calonectris diomedea* and falcons *Falco*.

Protection status

National Partial **International** Unknown
370 ha of IBA covered by Regional Nature Reserve (Isola di Alicudi, 370 ha). 635 ha of IBA covered by Regional Nature Reserve (Isola di Filicudi, 635 ha). 1,585 ha of IBA covered by Regional Nature Reserve (Isola di Lipari, 1,585 ha). 282 ha of IBA covered by Regional Nature Reserve (Isola di Panarea e scogli viciniori, 282 ha). 1,053 ha of IBA covered by Regional Nature Reserve (Isola di Stromboli e Strombolicchio, 1,053 ha). 1,362 ha of IBA covered by Regional Nature Reserve (Isola di Vulcano, 1,362 ha). 1,521 ha of IBA covered by Regional Nature Reserve (Le Montagne delle Felci e dei Porri, 1,521 ha). Overlaps with SPA.

Conservation issues

Threats Disturbance to birds (B), Recreation/tourism (A)

Touristic development poses the main threat.

Peloritani mountains — A4iv, B1iv, C5 — 153

Admin region Sicilia
Coordinates 38°13'N 15°31'E
Altitude 10–1,286 m **Area** 17,100 ha

Site description

A mountain area in north-east Sicily, west of the town of Messina, dominated by maquis, garrigue, broadleaved evergreen woodlands and agricultural land. The main land-uses are cultivation and stock-rearing.

Habitats Forest and woodland (40%; broadleaved deciduous forest; broadleaved evergreen forest), Scrub (35%; sclerophyllous scrub/garrigue/maquis), Wetland (river/stream), Rocky areas (5%), Artificial landscape (20%; highly improved reseeded grassland; arable land; perennial crops/orchards/groves)
Land-use Agriculture (80%), Forestry (20%)

Birds

A very important 'bottleneck' site for migrating raptors, where more than 20,000 pass regularly in spring, especially *Pernis apivorus* (mean minimum of 6,032 pass per spring).

Protection status

National None **International** Unknown
Overlaps with SPA.

Conservation issues

Threats Afforestation (B), Burning of vegetation (A), Disturbance to birds (C), Infrastructure (A)

The main threats are fires and illegal hunting. In the past there was intense hunting of migrating raptors, which has now largely decreased. A major threat is the building of forestry roads. There are anti-poaching activities.

Nebrodi

B2, C6 **154**

Admin region Sicilia
Coordinates 37°56′N 14°42′E
Altitude 10–1,847 m **Area** 85,587 ha

Site description

A large mountain area in northern Sicily, south of the village of S. Agata di Militello. The main habitat is broadleaved woodland (beech *Fagus*, oak *Quercus*); the area is rich in small freshwater ponds. It includes an area that was listed in the previous pan-European IBA inventory (Grimmett and Jones 1989) as 'Rocche del Castro and surrounding valley' (former site IT129).

Habitats Forest and woodland (60%; broadleaved deciduous forest; broadleaved evergreen forest), Grassland (25%), Wetland (standing fresh water; river/stream; water-fringe vegetation), Rocky areas (15%), Artificial landscape (highly improved reseeded grassland; arable land; ruderal land)
Land-use Agriculture (30%), Forestry (60%), Nature conservation/research (100%)

Birds

Species	Season	Year	Pop min	Pop max	Acc	Criteria
Milvus milvus Red Kite	R	1989	4	4	—	C6
Falco biarmicus Lanner	R	1996	2	—	—	B2
[1] ***Alectoris graeca*** Rock Partridge	R	1996	200	—	—	C6
Monticola solitarius Blue Rock Thrush	R	1996	50	—	—	B2

1. *A. g. whitakeri.*

Protection status

National High **International** Unknown
4,396 ha of IBA covered by Regional Nature Reserve (Monte Soro, 4,396 ha). 85,578 ha of IBA covered by Regional Nature Park (Nebrodi, 85,578 ha). Overlaps with SPA.

Conservation issues

Threats Agricultural intensification/expansion (B), Burning of vegetation (B), Forest grazing (C), Infrastructure (C)

Inappropriate pastoral practices such as overgrazing and burning of vegetation are the main threats. Research on the mammals, reptiles and amphibians is ongoing, coordinated by the Regional Park. A project to reintroduce *Gyps fulvus*, coordinated by LIPU and the Regional Park, is on the way.

Mount Pecoraro and Pizzo Cirina

B2, C7 **155**

Admin region Sicilia
Coordinates 38°09′N 13°07′E
Altitude 47–964 m **Area** 12,000 ha

Site description

A coastal mountain area in northern Sicily, west of the city of Palermo. The main habitats are garrigue, maquis and broadleaved evergreen woodland.

Habitats Forest and woodland (5%; broadleaved deciduous forest; broadleaved evergreen forest), Scrub (90%; sclerophyllous scrub/garrigue/maquis), Rocky areas (5%)
Land-use Agriculture (50%), Not utilized (50%)

Birds

Species	Season	Year	Pop min	Pop max	Acc	Criteria
Monticola solitarius Blue Rock Thrush	R	—	—	—	—	B2

Protection status

National None **International** Unknown
Overlaps with SPA.

Conservation issues

Threats Burning of vegetation (A), Unsustainable exploitation (A)

Mount Cofano, Cape San Vito and Mount Sparagio

A3, B2, C6 **156**

Admin region Sicilia
Coordinates 38°03′N 12°47′E
Altitude 0–1,110 m **Area** 18,000 ha

Site description

A mountain peninsula along the northern coast of Sicily, west of the town of Castellamare del Golfo, and including the following localities: Monte Sparagio, Monte Cofano, Monte Inici, Capo Puntazza, Lo Zingaro, Monte Speziale, and S. Vito lo Capo. The main habitats are rocky sea cliffs, stony and sand beaches, maquis, garrigue and cultivated land. The main land-uses are cultivation, stock-rearing and tourism.

Habitats Forest and woodland (5%), Scrub (65%; scrub; sclerophyllous scrub/garrigue/maquis), Grassland (5%; steppe/dry calcareous grassland), Rocky areas (15%; sea cliff/rocky shore; caves), Artificial landscape (10%; forestry plantation; ruderal land)
Land-use Agriculture (20%), Forestry (5%), Hunting (25%), Nature conservation/research (23%), Tourism/recreation (20%)

Birds

Species	Season	Year	Pop min	Pop max	Acc	Criteria
Hieraaetus fasciatus Bonnelli's Eagle	R	1997	4	4	A	B2, C6
Falco biarmicus Lanner	R	1996	2	2	A	B2
Falco peregrinus Peregrine	R	1996	11	11	A	B2, C6
[1] ***Alectoris graeca*** Rock Partridge	R	1996	40	40	—	A3, C6
Sylvia conspicillata Spectacled Warbler	R	1996	Common		—	A3
Sylvia cantillans Subalpine Warbler	B	1996	Abundant		—	A3
Sylvia melanocephala Sardinian Warbler	R	1996	Abundant		—	A3
Sturnus unicolor Spotless Starling	R	1996	Common		—	A3

1. *A. g. whitakeri.*

An important breeding area for raptors and for five of the 20 species of the Mediterranean biome.

Protection status

National Partial **International** Unknown
537 ha of IBA covered by Regional Nature Reserve (Monte Cofano, 537 ha). 1,600 ha of IBA covered by Regional Nature Reserve (Zingaro, 1,600 ha). Overlaps with SPA.

Conservation issues

Threats Afforestation (A), Burning of vegetation (A), Disturbance to birds (C), Forest grazing (B), Intensified forest management (A), Recreation/tourism (B)

The main threats are the burning of vegetation (in 1996 practically all of 'Lo Zingaro' reserve was burned), and afforestation projects.

Egadi islands

B1ii, B2, C2, C6 **157**

Admin region Sicilia
Coordinates 37°59′N 12°20′E
Altitude 0–686 m **Area** 3,700 ha

Site description

A group of islands (Favignana, Levanzo and Marettimo) and islets to the west of Sicily. The main habitats are maquis, garrigue and rocky cliffs. The main land-uses are non-intensive agriculture, hunting, tourism and fishing.

Habitats Scrub (sclerophyllous scrub/garrigue/maquis), Grassland (steppe/dry calcareous grassland), Marine areas, Rocky areas (sea cliff/rocky shore; rock stacks/islets), Artificial landscape (perennial crops/orchards/groves; urban parks/gardens)
Land-use Agriculture (35%), Fisheries/aquaculture, Hunting, Nature conservation/research, Tourism/recreation

Birds

Species	Season	Year	Pop min	Pop max	Acc	Criteria
Calonectris diomedea Cory's Shearwater	R	1992	100	—	—	C6
Hydrobates pelagicus Storm Petrel	R	1992	1,000	—	B	B1ii, B2, C2, C6
Hieraaetus fasciatus Bonnelli's Eagle	R	1992	1	1	A	B2

Important breeding site for seabirds, especially *Hydrobates pelagicus*.

Protection status

National Low **International** None

Part of IBA covered by Marine Reserve (Isole Egadi, 35 ha).

Conservation issues

Threats Afforestation (B), Burning of vegetation (B), Consequences of animal/plant introductions (A), Disturbance to birds (A), Recreation/tourism (A), Unsustainable exploitation (A)

The main threat is illegal hunting, especially of migrating birds.

Stagnone di Marsala and Trapani saltpans — A1, A4i, A4iii, B1i, B2, B3, C1, C2, C3, C4, C6 — 158

Admin region Sicilia
Coordinates 37°47′N 12°30′E
Altitude 0–7 m **Area** 3,700 ha

Site description

A complex of brackish wetlands (Stagnone di Marsala), saltpans (Saline di Trapani) and islets (Isole dello Stagnone di Marsala) along the west coast of Sicily, south of the town of Trapani. The Stagnone di Marsala is one of the last remaining lagoons in Sicily. It measures 10 km (north–south) by 3 km. It has an average depth of 1 m (maximum is only 2.5 m). The principal vegetation is bushes of *Salicornia* and *Salsola*, whilst the surrounding areas are largely agricultural. The Saline di Trapani are north of the lagoon and comprise a complex of saltpans at the mouth of the Baiata river. The main land-uses are salt extraction and agriculture.

Habitats Grassland (10%; humid grassland), Wetland (80%; mudflat/sandflat; saltmarsh; standing brackish and salt water; river/stream), Artificial landscape (10%; highly improved reseeded grassland)
Land-use Agriculture (10%), Fisheries/aquaculture (30%), Urban/industrial/transport (30%), Nature conservation/research (84%)

Birds

Species	Season	Year	Pop min	Pop max	Acc	Criteria
Platalea leucorodia Spoonbill	P	1989	—	200	C	A4i, B1i, C2
[1] ***Anas querquedula*** Garganey	P	1989	—	20,000	C	A4i, B1i, C3
[2] ***Aythya nyroca*** Ferruginous Duck	W	1992	10	10	A	A1, C1
Himantopus himantopus Black-winged Stilt	B	1995	—	75	A	C6
Recurvirostra avosetta Avocet	B	1996	—	200	A	B3, C6
Charadrius alexandrinus Kentish Plover	B	1995	50	120	C	B2
Charadrius alexandrinus Kentish Plover	P	1989	—	3,000	C	A4i, B1i, C3
Sterna albifrons Little Tern	B	1996	6	100	A	B2, C2, C6

1. Spring total.
2. Single count.

An important site for wintering and migrating waterbirds and for breeding waders and terns. Over 21,000 waders have been recorded passing through the area in spring.

Protection status

National Partial **International** Unknown

2,012 ha of IBA covered by Regional Nature Reserve (Isole dello Stagnone di Marsala, 2,012 ha). 1,100 ha of IBA covered by Regional Nature Reserve (Saline di Trapani e Paceco, 1,100 ha). Overlaps with SPA.

Conservation issues

Threats Aquaculture/fisheries (B), Consequences of animal/plant introductions (B), Disturbance to birds (A), Drainage (B), Dredging/canalization (A), Filling-in of wetlands (C), Industrialization/urbanization (C), Unsustainable exploitation (A)

The main threats are from illegal hunting and from the water-level management associated with salt-extraction and fishing. Some or all of the site is covered by a management plan.

Rocca Busambra — A1, B2, C1, C6 — 159

Admin region Sicilia
Coordinates 37°53′N 13°27′E
Altitude 225–1,613 m **Area** 100,000 ha

Site description

An area of hills south of the city of Palermo, in north-west Sicily. The area lies between the following localities, Misilmeri, Vicari, Corleone, Roccamena and Monte Gradara, and comprises cultivated land, with patches of maquis, woodland and garrigue.

Habitats Forest and woodland (broadleaved deciduous forest; broadleaved evergreen forest), Scrub (sclerophyllous scrub/garrigue/maquis), Wetland (standing fresh water; river/stream; water-fringe vegetation), Artificial landscape (highly improved reseeded grassland; arable land; perennial crops/orchards/groves; urban parks/gardens; ruderal land)
Land-use Agriculture, Forestry, Hunting, Water management

Birds

Species	Season	Year	Pop min	Pop max	Acc	Criteria
[1] ***Falco naumanni*** Lesser Kestrel	B	1994	47	—	A	A1, B2, C1, C6
Falco biarmicus Lanner	R	1994	2	4	A	B2, C6
Pyrrhocorax pyrrhocorax Chough	R	1989	50	—	—	C6

1. Six small colonies.

An important site for breeding raptors, including *Falco naumanni*.

Protection status

National Partial **International** Unknown

IBA contains Regional Nature Reserve (Bagni di Cefalà Diana e Chiara Stella, area not known). 7,397 ha of IBA covered by Regional Nature Reserve (Bosco della Ficuzza, Rocca Busambra, ecc., 7,397 ha). IBA contains Regional Nature Reserve (Serre della Pizzuta, area not known). Overlaps with SPA.

Conservation issues

Threats Unknown

Sicani mountains — A1, A4ii, B1iii, B2, C1, C2, C6 — 160

Admin region Sicilia
Coordinates 37°37′N 13°09′E
Altitude 250–1,180 m **Area** 18,000 ha

Site description

A hilly area in southern Sicily, north of the coastal village of Sciacca. The main habitats are rocky cliffs, pastures and arable land. The main land-use is agriculture.

Habitats Forest and woodland (10%; broadleaved deciduous forest), Scrub (5%; sclerophyllous scrub/garrigue/maquis), Grassland (25%; steppe/dry calcareous grassland), Wetland (5%; standing fresh water; river/stream), Rocky areas (5%; inland cliff), Artificial landscape (50%; highly improved reseeded grassland; arable land; perennial crops/orchards/groves; forestry plantation)
Land-use Agriculture (25%), Forestry, Hunting

Birds

Species	Season	Year	Pop min	Pop max	Acc	Criteria
Neophron percnopterus Egyptian Vulture	B	1996	3	7	A	B2, C6
Falco naumanni Lesser Kestrel	B	1996	50	200	B	A1, A4ii, B1iii, B2, C1, C2, C6
Falco biarmicus Lanner	R	1996	5	7	B	B2, C6
Falco peregrinus Peregrine	R	1996	10	12	A	B2, C6

The most important breeding site for *Falco naumanni* in Sicily, and also important for other breeding raptors, especially *Falco peregrinus* and *Falco biarmicus*.

Protection status

National Partial **International** Unknown
2,553 ha of IBA covered by Regional Nature Reserve (Monte Genuardo e Santa Maria del Bosco, 2,553 ha). Overlaps with SPA.

Conservation issues

Threats Afforestation (A), Disturbance to birds (A), Extraction industry (A), Forest grazing (C), Industrialization/urbanization (B), Infrastructure (B), Unsustainable exploitation (A), Other (A)

Important threats are: afforestation by the Regional Forestry Department (Corpo Forestale Regionale) of natural prairies where raptors, especially *Falco naumanni*, feed; disturbance to breeding raptors (and cases of illegal hunting and poisoning); the presence and activity of numerous quarries; heavy and uncontrolled hunting pressure.

Castelluzzo A1, B2, C1, C2 161

Admin region Sicilia
Coordinates 37°42′N 14°44′E
Altitude 260–742 m **Area** 7,500 ha

Site description

A hilly, inland area in eastern Sicily, covered mainly by pastures and arable land.

Habitats Forest and woodland (broadleaved deciduous forest; broadleaved evergreen forest), Scrub (sclerophyllous scrub/garrigue/maquis), Grassland, Rocky areas (inland cliff), Artificial landscape (arable land; other urban/industrial areas)
Land-use Agriculture (90%), Hunting (50%), Urban/industrial/transport (5%)

Birds

Species	Season	Year	Pop min	Pop max	Acc	Criteria
Falco naumanni Lesser Kestrel	B	1996	20	32	A	A1, B2, C1, C2

An important breeding area for raptors, especially *Falco naumanni*.

Protection status

National None **International** Unknown
Overlaps with SPA.

Conservation issues

Threats Agricultural intensification/expansion (B), Unsustainable exploitation (A)

Cape Feto C7 162

Admin region Sicilia
Coordinates 37°40′N 12°31′E
Altitude 0–5 m **Area** 150 ha

Site description

Wetlands with sand-dunes and small marshes, along the south-west coast of Sicily, 5 km west of the town of Mazara del Vallo. The main land-use is agriculture.

Habitats Wetland (70%; sand-dunes/sand beach; standing brackish and salt water), Artificial landscape (30%; perennial crops/orchards/groves)
Land-use Agriculture (30%), Nature conservation/research (100%), Tourism/recreation (5%)

Birds

A notable stop-over site for migrating waterbirds.

Protection status

National None **International** High
150 ha of IBA covered by Special Protection Area (Capo Feto, 150 ha).

Conservation issues

Threats Agricultural intensification/expansion (A), Drainage (C), Filling-in of wetlands (A), Industrialization/urbanization (B), Recreation/tourism (C)

The main threats are filling-in of wetlands and agricultural intensification.

Simeto mouth and Biviere di Lentini A1, A4i, A4iii, B1i, B2, C1, C2, C3, C4, C6 163

Admin region Sicilia
Coordinates 37°24′N 15°06′E
Altitude 0–100 m **Area** 1,500 ha

Site description

A complex of wetlands along the south-east coast of Sicily, around the town of Catania, and including the following sites: Biviere di Lentini, a freshwater reservoir; Foce del Simeto, a river-mouth; Traversa di Ponte Barca, a stretch of the Simeto river. There are large reedbeds at Biviere di Lentini. The main land-uses are water-resource management at Biviere di Lentini (for domestic water-supply), and nature conservation at Foce del Simeto. The surrounding areas are largely agricultural.

Habitats Wetland (100%; saltmarsh; sand-dunes/sand beach; coastal lagoon; standing fresh water; river/stream; water-fringe vegetation)
Land-use Hunting (7%), Nature conservation/research (26%), Water management (67%)

Birds

Species	Season	Year	Pop min	Pop max	Acc	Criteria
Phalacrocorax carbo Cormorant	W	1996	500	4,000	A	B1i, C3
Ixobrychus minutus Little Bittern	B	1998	80	100	B	B2, C2, C6
Nycticorax nycticorax Night Heron	B	1997	50	50	B	C6
Ardeola ralloides Squacco Heron	B	1992	10	15	A	C2, C6
Ardea purpurea Purple Heron	B	1997	15	20	A	B2, C6
Plegadis falcinellus Glossy Ibis	P	1997	200	300	A	A4i, B1i, C2
Platalea leucorodia Spoonbill	P	1997	100	350	A	A4i, B1i, C2
Aythya nyroca Ferruginous Duck	B	1997	15	20	A	A1, C1
Aythya nyroca Ferruginous Duck	W	1995	67	67	A	A1, C1
Aythya nyroca Ferruginous Duck	P	1998	280	280	A	A1, B1i, C1, C2
Charadrius alexandrinus Kentish Plover	B	1997	13	18	A	B2
[1] ***Numenius tenuirostris*** Slender-billed Curlew	P	1997	1	1	A	A1, C1
Sterna sandvicensis Sandwich Tern	P	1997	100	100	A	C2
Sterna albifrons Little Tern	B	1997	50	50	A	B2, C2, C6
Chlidonias niger Black Tern	P	1997	700	700	A	B1i, C2
Acrocephalus paludicola Aquatic Warbler	P	1997	2	3	A	A1, C1

1. One record, at Biviere di Lentini in March 1996.

A very important site for waterbirds while breeding (especially *Aythya nyroca* and herons), migrating (e.g. *Plegadis falcinellus*, *Platalea leucorodia*) and wintering. The site holds more than 20,000 waterbirds on a regular basis, with *Larus ridibundus* especially common in winter.

Protection status

National Partial **International** Unknown
400 ha of IBA covered by Regional Nature Reserve (Oasi del Simeto, 1,859 ha). Overlaps with SPA.

Conservation issues

Threats Aquaculture/fisheries (B), Disturbance to birds (B), Infrastructure (B), Other (A), Recreation/tourism (A), Unsustainable exploitation (A)

Water management at Biviere di Lentini is the main threat ('Other' threat). The aqueduct for filling the reservoir has never been built, but its completion remains a possibility, and would destroy the wetland complex at this site. Some or all of the site is covered by a management plan.

Madonie A1, C1, C6 164

Admin region Sicilia
Coordinates 37°50′N 14°00′E
Altitude 10–1,979 m **Area** 39,941 ha

Site description

A coastal mountain complex in the north of Sicily, south of the town of Cefalù. There is a very varied mosaic of habitats, including broadleaved woodlands, pastures and rocky areas.

Habitats Forest and woodland (broadleaved deciduous forest; native coniferous forest; broadleaved evergreen forest), Scrub (scrub; sclerophyllous scrub/garrigue/maquis), Grassland, Wetland (river/stream), Rocky areas (inland cliff; caves), Artificial landscape
Land-use Agriculture, Forestry, Nature conservation/research (100%)

Birds

Species	Season	Year	Pop min	Pop max	Acc	Criteria
Milvus milvus Red Kite	R	1989	4	6	—	C6
Neophron percnopterus Egyptian Vulture	B	1989	3	3	—	C6
Aquila chrysaetos Golden Eagle	R	1989	3	3	—	C6
Hieraaetus fasciatus Bonelli's Eagle	R	1989	2	2	—	C6
Falco naumanni Lesser Kestrel	B	1994	23	—	A	A1, C1
Falco peregrinus Peregrine	R	1989	8	8	—	C6
[1] ***Alectoris graeca*** Rock Partridge	R	1996	200	—	—	C6

1. *A. g. whitakeri.*

An important breeding site for raptors.

Protection status

National High **International** Unknown

2,949 ha of IBA covered by Regional Nature Reserve (Faggeta Madonia, 2,949 ha). 2,010 ha of IBA covered by Regional Nature Reserve (Monte Quacella, 2,010 ha). 39,941 ha of IBA covered by Regional Nature Park (Madonie, 39,941 ha). Overlaps with SPA.

Conservation issues

Threats Agricultural intensification/expansion (B), Burning of vegetation (B), Forest grazing (C), Infrastructure (C)

There is a project to reintroduce *Gyps fulvus*, coordinated by LIPU and the Regional Nature Park.

Anapo gorges — A3, B2, B3, C6 — 165

Admin region Sicilia
Coordinates 37°07′N 14°59′E
Altitude 425–697 m **Area** 3,712 ha

Site description

Rocky gorges created by the Anapo river between the villages of Palazzolo Acreide and Sortino, in eastern Sicily. Habitats include maquis, grasslands and cultivated land.

Habitats Forest and woodland (10%), Scrub (50%; sclerophyllous scrub/garrigue/maquis), Rocky areas (5%; inland cliff), Artificial landscape (35%)
Land-use Agriculture (85%), Tourism/recreation

Birds

Species	Season	Year	Pop min	Pop max	Acc	Criteria
Falco biarmicus Lanner	R	1996	2	—	—	B2
[1] ***Alectoris graeca*** Rock Partridge	R	1996	50	—	C	A3, C6
Monticola solitarius Blue Rock Thrush	B	1996	50	—	C	B2
Sylvia conspicillata Spectacled Warbler	R	1996	Common		—	A3
Sylvia cantillans Subalpine Warbler	B	1996	Abundant		—	A3
Sylvia melanocephala Sardinian Warbler	R	1996	Abundant		—	A3
Sturnus unicolor Spotless Starling	R	—	—	—	—	A3, B3

1. *A. g. whitakeri.*

An important breeding site for raptors and for five of the 20 species of the Mediterranean biome.

Protection status

National High **International** None

3,712 ha of IBA covered by Regional Nature Reserve (Pantalica, Valle dell'Anapo e Torrente Cava Grande, 3,712 ha).

Conservation issues

Threats Afforestation (B), Recreation/tourism (B)

The main threat is the destruction of garrigue for reforestation projects.

Biviere and Plain of Gela — A1, B1i, C1, C2, C6 — 166

Admin region Sicilia
Coordinates 37°01′N 14°20′E
Altitude 0–5 m **Area** 500 ha

Site description

A complex of coastal brackish lakes, in the south of Sicily, separated from the sea by sand-dunes. There are large reedbeds and beds of submerged waterweed (*Potamogeton*, *Ceratophyllum*). The main land-uses are grazing, water management and nature conservation.

Habitats Scrub (30%; sclerophyllous scrub/garrigue/maquis), Wetland (40%; sand-dunes/sand beach; standing brackish and salt water; river/stream; water-fringe vegetation), Artificial landscape (30%; highly improved reseeded grassland; perennial crops/orchards/groves)
Land-use Agriculture (50%), Nature conservation/research (50%), Water management (50%)

Birds

Species	Season	Year	Pop min	Pop max	Acc	Criteria
Aythya nyroca Ferruginous Duck	P	1986	—	1,500	A	A1, B1i, C1, C2
[1] ***Glareola pratincola*** Collared Pratincole	B	1997	150	150	A	C6

1. Piana del Signore.

The site is important for migrating waterbirds, especially *Aythya nyroca* (also breeds, min. 3 pairs), herons, *Plegadis falcinellus*, *Platalea leucorodia*, and waders.

Protection status

National Partial **International** Partial

IBA contains Regional Nature Reserve (Biviere di Gela, area not known). 256 ha of IBA covered by Ramsar Site (Biviere di Gela, 256 ha). Overlaps with SPA.

Conservation issues

Threats Agricultural intensification/expansion (C), Filling-in of wetlands (B), Industrialization/urbanization (B), Infrastructure (B), Other (A)

The main threats are the infilling of wetlands, the unsuitable water management at Biviere di Gela ('Other' threat), and industrialization and infrastructure development at Piana del Signore.

Pantani di Capo Passero — A4i, A4iii, B1i, B2, C2, C3, C4, C6 — 167

Admin region Sicilia
Coordinates 36°44′N 15°03′E
Altitude 0–1 m **Area** 3,000 ha

Site description

A complex of brackish coastal wetlands, in the south-east corner of Sicily, around the village of Pachino. It includes two groups of lakes: Vendicari (Pantano Piccolo, Pantano Grande, Pantano Roveto), south of the town of Noto, and Pachino (Pantano Morghello, Salina di Marzameni, Pantano Pontenio, Pantano Baronello, Pantano Auruca, Pantano Cannone, Pantano Cuba, Pantano Longarini, Pantano Bruno, Pantano Arezzi).

Habitats Scrub (sclerophyllous scrub/garrigue/maquis), Wetland (50%; saltmarsh; sand-dunes/sand beach; coastal lagoon; standing brackish and salt water; water-fringe vegetation), Artificial landscape (50%)
Land-use Agriculture (50%), Nature conservation/research (100%), Other (5%), Tourism/recreation (10%)

Birds

Species	Season	Year	Pop min	Pop max	Acc	Criteria
Plegadis falcinellus Glossy Ibis	P	1990	250	—	—	A4i, B1i, C2
Platalea leucorodia Spoonbill	P	1990	250	—	—	A4i, B1i, C2
Charadrius alexandrinus Kentish Plover	B	1990	100	—	—	B2
Charadrius alexandrinus Kentish Plover	P	1990	—	1,000	—	A4i, B1i, C3
Calidris minuta Little Stint	P	1990	—	6,000	—	A4i, B1i, C3
Calidris ferruginea Curlew Sandpiper	P	1990	—	4,500	—	A4i, B1i, C3
Philomachus pugnax Ruff	P	1990	—	8,000	—	C6
Tringa glareola Wood Sandpiper	P	1990	—	3,000	—	C6
Larus genei Slender-billed Gull	P	1990	700	1,400	—	A4i, B1i, C2
Sterna caspia Caspian Tern	P	1990	—	150	—	A4i, B1i, C2
Sterna albifrons Little Tern	B	1990	10	25	—	B2, C2, C6
Chlidonias niger Black Tern	P	1990	2,500	—	—	A4i, B1i, C2

A very important site for migrating waterbirds (regularly more than 20,000 present in total), including large numbers of waders. Species of global conservation concern that do not meet IBA criteria: *Aythya nyroca* (max. 10, winter), *Larus audouinii* (max. 25 non-breeding birds).

Protection status

National High **International** Partial

1,512 ha of IBA covered by Regional Nature Reserve (Oasi Faunistica di Vendicari, 1,512 ha). 1,385 ha of IBA covered by Regional Nature Reserve (Pantani Sicilia S. O., 1,385 ha). 1,450 ha of IBA covered by Special Protection Area (Pantani di Vendicari, 1,450 ha). 1,450 ha of IBA covered by Ramsar Site (Vendicari, 1,450 ha).

Conservation issues

Threats Agricultural intensification/expansion (B), Disturbance to birds (B), Industrialization/urbanization (A)

Pantelleria and Isole Pelagie — 168

A4ii, B1ii, B1iii, B2, B3, C2, C6

Admin region Sicilia
Coordinates 36°45′N 12°00′E
Altitude 0–835 m **Area** 2,445 ha

Site description

An archipelago of small islands between Sicily and Tunisia. The main habitats are rocky islets and cliffs, maquis, garrigue and cultivated land.

Habitats Scrub (80%; sclerophyllous scrub/garrigue/maquis), Marine areas (sea cliff/rocky shore; rock stacks/islets), Artificial landscape (20%; highly improved reseeded grassland; perennial crops/orchards/groves)
Land-use Agriculture (15%), Hunting (40%), Nature conservation/research, Tourism/recreation (100%)

Birds

Species	Season	Year	Pop min	Pop max	Acc	Criteria
[1] ***Calonectris diomedea*** Cory's Shearwater	R	1989	10,000	—	B	A4ii, B1ii, B2, C2, C6
Hydrobates pelagicus Storm Petrel	R	1990	100	—	B	B1ii, B2, C2, C6
[2] ***Phalacrocorax aristotelis*** Shag	R	1992	30	—	B	C2, C6
Falco eleonorae Eleonora's Falcon	B	1994	35	40	A	B1iii, B2, C2, C6
Sylvia sarda Marmora's Warbler	R	—	—	—	—	B3

1. Isola di Linosa.
2. Isola di Lampedusa.

The most important breeding site in Italy for *Calonectris diomedea*.

Protection status

National None **International** Unknown
Overlaps with SPA.

Conservation issues

Threats Disturbance to birds (A), Industrialization/urbanization (A), Recreation/tourism (A)

Coast from Coghinas river mouth to Cape Testa — 169

A1, B1i, C1, C2

Admin region Sardegna
Coordinates 41°14′N 9°10′E
Altitude 0–130 m **Area** 300 ha

Site description

A coastal area in the north-west corner of Sardinia, between the mouth of the Coghinas river and Cape Testa. Habitats include areas of maquis and sandy and rocky coast.

Habitats Scrub (10%; sclerophyllous scrub/garrigue/maquis), Wetland (10%; sand-dunes/sand beach), Marine areas (20%; sea inlet/coastal features), Rocky areas (60%; sea cliff/rocky shore)
Land-use Fisheries/aquaculture, Forestry, Hunting, Tourism/recreation

Birds

Species	Season	Year	Pop min	Pop max	Acc	Criteria
Phalacrocorax aristotelis Shag	R	1995	80	—	A	B1i, C2
Larus audouinii Audouin's Gull	R	1995	25	—	A	A1, C1

Important site for breeding *Larus audouinii*.

Protection status

National None **International** None

Conservation issues

Threats Disturbance to birds (B), Infrastructure (B), Recreation/tourism (A)

Maddalena archipelago — 170

A1, B1i, B2, B3, C1, C2, C6

Admin region Sardegna
Coordinates 41°16′N 9°22′E
Altitude 0–212 m **Area** 800 ha

Site description

An archipelago of 15 islands and islets along the north-eastern coast of Sardinia. They are mainly covered by maquis, with some coniferous plantations, and there are many rocky cliffs. Tourism is the most important land-use.

Habitats Scrub (40%; sclerophyllous scrub/garrigue/maquis), Marine areas (40%), Rocky areas (10%; sea cliff/rocky shore; rock stacks/islets), Artificial landscape (10%; perennial crops/orchards/groves; other urban/industrial areas)
Land-use Hunting, Military, Nature conservation/research, Tourism/recreation

Birds

Species	Season	Year	Pop min	Pop max	Acc	Criteria
Calonectris diomedea Cory's Shearwater	B	1995	—	100	B	C6
Phalacrocorax aristotelis Shag	R	1995	—	350	A	B1i, B3, C2, C6
Larus audouinii Audouin's Gull	R	1995	20	40	A	A1, B2, C1
Sterna hirundo Common Tern	B	1995	20	30	A	C6

An important breeding site for seabirds, including *Larus audouinii*, shearwaters and *Hydrobates pelagicus*.

Protection status

National High **International** None
800 ha of IBA covered by National Park (Maddalena, area not known).

Conservation issues

Threats Disturbance to birds (B), Recreation/tourism (B)

Tourism is the most important threat.

Asinara island and Stintino peninsula cliffs — 171

A1, A4i, B1i, B2, B3, C1, C2, C6

Admin region Sardegna
Coordinates 41°01′N 8°12′E
Altitude 0–408 m **Area** 4,900 ha

Site description

Isola Asinara is a small island off the north-western corner of Sardinia. Its shore is partly sandy and partly rocky. Maquis and broadleaved evergreen woodland are the main habitats. The site also includes the western rocky coast of Penisola di Stintino, a small peninsula close to Isola Asinara. Some small islets are also present.

Habitats Forest and woodland (broadleaved evergreen forest), Scrub (sclerophyllous scrub/garrigue/maquis), Grassland (steppe/dry calcareous grassland; humid grassland), Wetland (sand-dunes/sand beach), Marine areas, Rocky areas (sea cliff/rocky shore; rock stacks/islets)
Land-use Agriculture, Fisheries/aquaculture, Military (95%), Nature conservation/research (5%)

Birds

Species	Season	Year	Pop min	Pop max	Acc	Criteria
Phalacrocorax aristotelis Shag	R	1996	50	220	B	B1i, B3, C2
Alectoris barbara Barbary Partridge	R	1997	30	—	C	B2
Larus audouinii Audouin's Gull	R	1996	145	203	A	A1, A4i, B1i, B2, C1, C2, C6
Sylvia sarda Marmora's Warbler	R	—	—	—	—	B3

An important breeding site for seabirds.

Protection status

National High **International** None
4,900 ha of IBA covered by National Park (Asinara, area not known).

Conservation issues

Threats Agricultural intensification/expansion (U), Aquaculture/fisheries (U), Burning of vegetation (U)

Threats include fires, overgrazing, and overfishing. The island is a proposed Regional Nature Reserve.

Pilo and Casaraccio pools

B2 **172**

Admin region Sardegna
Coordinates 40°51′N 8°17′E
Altitude 0–1 m **Area** 500 ha

Site description

Two pools to the west of the village of Porto Torres, at the north-western corner of Sardinia. Pilo pool is separated from the sea by sand-dunes and is fringed by a belt of reeds *Phragmites*. Casaraccio pool is the remnant of a large wetland. The site also includes a small salina (Saline di Stintino). The main activity is fishing.

Habitats Wetland (100%; mudflat/sandflat; saltmarsh; sand-dunes/sand beach; standing fresh water; standing brackish and salt water; water-fringe vegetation)
Land-use Fisheries/aquaculture (100%)

Birds

Species	Season	Year	Pop min	Pop max	Acc	Criteria
Porphyrio porphyrio Purple Gallinule	R	1997	5	—	B	B2

A wintering site for waterbirds (including *Phoenicopterus ruber*). Species of global conservation concern that do not meet IBA criteria: *Aythya nyroca* (max. 8, winter).

Protection status

National Partial **International** None
383 ha of IBA covered by Wildlife Reserve (Stagno di Pilo, 383 ha).

Conservation issues

Threats Disturbance to birds (B), Recreation/tourism (A)

The main threat is the building of tourist resorts in the area of Casaraccio.

Campo d'Ozieri

B2, B3, C6 **173**

Admin region Sardegna
Coordinates 40°45′N 8°50′E
Altitude — **Area** 20,000 ha

Site description

A large, steppic area in central Sardinia, including the villages of Ozieri and Oschiri. Some of the land is used for agriculture and cattle-rearing.

Habitats Forest and woodland (10%; wooded steppe), Scrub (20%; sclerophyllous scrub/garrigue/maquis), Grassland (50%; steppe/dry calcareous grassland), Artificial landscape (20%; arable land; ruderal land)
Land-use Agriculture (30%), Forestry, Hunting (80%), Not utilized (20%), Water management (5%)

Birds

Species	Season	Year	Pop min	Pop max	Acc	Criteria
Alectoris barbara Barbary Partridge	R	1996	Abundant		—	B2
[1] ***Tetrax tetrax*** Little Bustard	R	1997	80	100	A	C6
Sturnus unicolor Spotless Starling	R	1997	500	—	C	B3

1. Individuals.

Important breeding site for several open-country species.

Protection status

National None **International** None

Conservation issues

Threats Agricultural intensification/expansion (A), Unsustainable exploitation (B)

The main threat is agricultural intensification.

Tavolara island

A1, A4ii, B1i, B1ii, B2, B3, C1, C2, C3, C6 **174**

Admin region Sardegna
Coordinates 41°00′N 9°39′E
Altitude 0–565 m **Area** 3,000 ha

Site description

A small archipelago of three islands—Tavolara (the largest), Molara and Molarotto—off the north-east coast of Sardinia. Vegetation is mainly maquis. Tourism is the main land-use.

Habitats Forest and woodland (broadleaved evergreen forest), Scrub (sclerophyllous scrub/garrigue/maquis), Marine areas, Rocky areas (sea cliff/rocky shore; rock stacks/islets)
Land-use Military, Nature conservation/research, Tourism/recreation

Birds

Species	Season	Year	Pop min	Pop max	Acc	Criteria
Puffinus yelkouan Yelkouan Shearwater	R	1995	6,000	9,000	B	A4ii, B1ii, B3, C3
Phalacrocorax aristotelis Shag	R	1995	200	300	A	B1i, B3, C2, C6
Larus audouinii Audouin's Gull	R	1995	100	—	B	A1, B2, C1, C6

An important breeding site for seabirds.

Protection status

National Partial **International** None
IBA contains Marine Nature Reserve (Tavolara–Punta Coda Cavallo, area not known).

Conservation issues

Threats Burning of vegetation (U), Disturbance to birds (U), Recreation/tourism (U), Unsustainable exploitation (U)

Tourism is the main threat, along with fires, and disturbance to birds by military activities. The site is a proposed Regional Nature Reserve.

Cape Caccia

A1, B1i, B1ii, B2, B3, C1, C2, C6 **175**

Admin region Sardegna
Coordinates 40°34′N 8°11′E
Altitude 0–361 m **Area** 35,000 ha

Site description

A rocky area along the west coast of Sardinia, covered by maquis. The main land-use is tourism. It includes Capo Caccia, Punta Cristallo and Isola Foradada.

Habitats Forest and woodland (10%; wooded steppe), Scrub (40%; sclerophyllous scrub/garrigue/maquis), Wetland (5%; sand-dunes/sand beach; shingle/stony beach), Marine areas (10%), Rocky areas (20%; sea cliff/rocky shore; rock stacks/islets; caves), Artificial landscape (15%; arable land; perennial crops/orchards/groves)
Land-use Agriculture (5%), Fisheries/aquaculture, Hunting (30%), Military (5%), Nature conservation/research (30%), Tourism/recreation (40%)

Birds

Species	Season	Year	Pop min	Pop max	Acc	Criteria
Calonectris diomedea Cory's Shearwater	B	1995	800	1,000	B	B1ii, B2, C2, C6
Hydrobates pelagicus Storm Petrel	B	1994	400	500	A	B1ii, B2, C2, C6
Phalacrocorax aristotelis Shag	R	1994	400	550	B	B1i, B3, C2, C6
Alectoris barbara Barbary Partridge	R	1997	30	—	C	B2
Larus audouinii Audouin's Gull	R	1996	20	25	A	A1, B2, C1
Sylvia sarda Marmora's Warbler	B	1997	50	—	C	B3

Protection status

National Low **International** None
1,701 ha of IBA covered by Wildlife Reserve (Tramariglio, 1,701 ha).

Conservation issues

Threats Consequences of animal/plant introductions (B), Disturbance to birds (A), Intensified forest management (B), Recreation/tourism (A)

The main threat is posed by tourism.

Coast between Bosa and Alghero — B2, B3, C6 — 176

Admin region Sardegna
Coordinates 40°21′N 8°27′E
Altitude 0–802 m **Area** 20,000 ha

Site description

A rocky, coastal area of western Sardinia, between the towns of Alghero and Bosa. The main land-uses are stock-rearing, forestry and hunting.

Habitats Forest and woodland (10%; mixed forest), Scrub (50%; sclerophyllous scrub/garrigue/maquis), Marine areas (10%), Rocky areas (30%; sea cliff/rocky shore)
Land-use Agriculture (60%), Forestry (30%), Hunting (50%)

Birds

Species	Season	Year	Pop min	Pop max	Acc	Criteria
[1] ***Gyps fulvus*** Griffon Vulture	R	1994	30	30	A	C6
Falco naumanni Lesser Kestrel	B	1995	7	11	B	C6
Falco peregrinus Peregrine	R	1994	10	—	—	B2, C6
Sylvia sarda Marmora's Warbler	R	1997	50	—	C	B3

1. Colony is periodically 'restocked' by LIPU.

The site is important for breeding raptors, especially for the largest Italian colony of *Gyps fulvus*. Breeding species of global conservation concern that do not meet IBA criteria: *Larus audouinii* (frequent).

Protection status

National None **International** None

Conservation issues

Threats Burning of vegetation (B), Infrastructure (A), Recreation/tourism (B), Unsustainable exploitation (B)

The main threats are a new power line which is planned in front of the *Gyps fulvus* colony, touristic development projects along the coast, and fires. LIPU runs a project to conserve and restock the population of *Gyps fulvus*. The area is a proposed Regional Nature Reserve.

Altopiano di Campeda — A1, B2, C1, C2, C6 — 177

Admin region Sardegna
Coordinates 40°18′N 8°47′E
Altitude 425–845 m **Area** 19,500 ha

Site description

An area of grassland, with patches of cork oak *Quercus suber* woodland, located north of the village of Macomer. The area is rich in small, temporary pools, and there are also rocky areas and rivers. The main land-use is agriculture.

Habitats Forest and woodland (10%; wooded steppe), Scrub (20%; sclerophyllous scrub/garrigue/maquis), Grassland (40%; steppe/dry calcareous grassland), Artificial landscape (30%; arable land; perennial crops/orchards/groves)
Land-use Agriculture (40%), Forestry, Hunting (70%), Not utilized (20%)

Birds

Species	Season	Year	Pop min	Pop max	Acc	Criteria
Milvus milvus Red Kite	R	1996	2	6	B	C6
Alectoris barbara Barbary Partridge	R	1995	100	—	C	B2, C2, C6
Tetrax tetrax Little Bustard	R	1995	80	100	B	A1, B2, C1, C6
Burhinus oedicnemus Stone Curlew	R	1996	20	30	B	C6
Melanocorypha calandra Calandra Lark	R	1996	100	—	B	C6

An important breeding site for *Tetrax tetrax* and other open-country species. Breeding species of global conservation concern that do not meet IBA criteria: *Falco naumanni*.

Protection status

National None **International** None

Conservation issues

Threats Agricultural intensification/expansion (B), Disturbance to birds (B), Drainage (B), Infrastructure (A), Unsustainable exploitation (B)

The main threats are the drainage of the small wetlands, the building of new power lines, and the lack of legal site-protection.

Central Campidano — B2, C6 — 178

Admin region Sardegna
Coordinates 39°40′N 8°38′E
Altitude 0–274 m **Area** 20,000 ha

Site description

The central area of the largest lowland in Sardinia, including the villages of Sanluri, Sardana and Uras. The site is dominated by steppe-grassland on basalt bedrock, crossed by numerous streams. The main land-uses are agriculture and hunting.

Habitats Grassland (steppe/dry calcareous grassland), Wetland (river/stream), Artificial landscape (highly improved reseeded grassland; arable land; perennial crops/orchards/groves; forestry plantation; urban parks/gardens; other urban/industrial areas)
Land-use Agriculture, Hunting, Urban/industrial/transport, Water management

Birds

Species	Season	Year	Pop min	Pop max	Acc	Criteria
Alectoris barbara Barbary Partridge	R	1996	Common		—	B2
[1] ***Tetrax tetrax*** Little Bustard	R	1996	20	40	—	B2, C6
Burhinus oedicnemus Stone Curlew	R	1996	20	40	—	C6
Calandrella brachydactyla Short-toed Lark	B	1996	150	—	C	B2, C6

1. Males.

An important breeding site for *Tetrax tetrax*.

Protection status

National None **International** None

Conservation issues

Threats Agricultural intensification/expansion (U), Burning of vegetation (U), Consequences of animal/plant introductions (U), Disturbance to birds (U), Dredging/canalization (U), Industrialization/urbanization (U), Infrastructure (U), Unsustainable exploitation (U)

The main threats are fires started for stubble-burning, water pollution, building of new roads, and urbanization.

Altopiano di Abbasanta and Lake Omodeo — A1, B2, C1, C2, C6 — 179

Admin region Sardegna
Coordinates 40°10′N 8°50′E
Altitude 101–364 m **Area** 20,000 ha

Site description

A steppe area in central Sardinia, between the villages of Abbasanta and Macomer, crossed by numerous streams and rich in small, temporary wetlands. The site also includes the upper valley of the Tirso river and Lago Omodeo, an artificial lake that is temporarily dry. The main land-use is grazing.

Habitats Forest and woodland (broadleaved evergreen forest; wooded steppe), Scrub (sclerophyllous scrub/garrigue/maquis), Grassland (steppe/dry calcareous grassland), Wetland (standing fresh water; river/stream; water-fringe vegetation), Artificial landscape (highly improved reseeded grassland; perennial crops/orchards/groves; other urban/industrial areas; ruderal land)
Land-use Agriculture, Forestry, Hunting, Military, Urban/industrial/transport

Birds

An important site for breeding *Tetrax tetrax* and for wintering *Burhinus oedicnemus*.

Species	Season	Year	Pop min	Pop max	Acc	Criteria
Alectoris barbara Barbary Partridge	R	1994	100	—	C	B2, C2, C6
Tetrax tetrax Little Bustard	R	1994	100	—	C	A1, B2, C1, C6
Melanocorypha calandra Calandra Lark	R	1994	100	—	—	C6

Protection status

National None **International** None

Conservation issues

Threats Abandonment/reduction of land management (U), Agricultural intensification/expansion (U), Burning of vegetation (U), Deforestation (commercial) (U), Firewood collection (U), Infrastructure (U), Unsustainable exploitation (U)

The main threats are the burning of vegetation by farmers and shepherds, illegal hunting, and the opening of new roads.

Cuglieri coast

B1i, B2, C2, C6 **180**

Admin region Sardegna
Coordinates 40°11′N 8°25′E
Altitude 0–131 m **Area** 2,800 ha

Site description

A rocky coastal area, rich in sea cliffs and islets. The principal vegetation is maquis. The main land-use is agriculture.

Habitats Scrub (70%; sclerophyllous scrub/garrigue/maquis), Wetland (river/stream), Rocky areas (30%; sea cliff/rocky shore; rock stacks/islets)
Land-use Agriculture, Hunting, Tourism/recreation

Birds

Species	Season	Year	Pop min	Pop max	Acc	Criteria
Phalacrocorax aristotelis Shag	R	1995	50	—	—	B1i, C2
Falco peregrinus Peregrine	R	1995	6	8	—	C6
Alectoris barbara Barbary Partridge	R	1997	30	50	C	B2

An important breeding site for several species. Species of global conservation concern that do not meet IBA criteria: *Larus audouinii* (resident).

Protection status

National None **International** None

Conservation issues

Threats Agricultural intensification/expansion (U), Burning of vegetation (U), Recreation/tourism (U), Unsustainable exploitation (U)

The main threats are fires and the building of new tourist resorts along the beaches. The site is a proposed Regional Nature Reserve.

Orosei Gulf and Gennargentu mountains

A1, A3, A4ii, B1i, B1iii, B2, B3, C1, C2, C6 **181**

Admin region Sardegna
Coordinates 40°10′N 9°24′E
Altitude 0–1,829 m **Area** 59,102 ha

Site description

A large area of east-central Sardinia that includes coastal and mountain habitats. Vegetation is dominated by maquis along the coast (Orosei Gulf) and by broadleaved evergreen woodlands, pastures, gorges and rocky cliffs in the mountains inland (Gennargentu mountains). The main land-use is stock-rearing.

Habitats Forest and woodland (broadleaved evergreen forest), Scrub (sclerophyllous scrub/garrigue/maquis), Grassland (steppe/dry calcareous grassland), Wetland (sand-dunes/sand beach; river/stream), Rocky areas (sea cliff/rocky shore; rock stacks/islets; inland cliff; caves), Artificial landscape (perennial crops/orchards/groves; forestry plantation; other urban/industrial areas; ruderal land)
Land-use Agriculture, Forestry, Hunting, Nature conservation/research, Tourism/recreation, Urban/industrial/transport

Birds

Species	Season	Year	Pop min	Pop max	Acc	Criteria
Phalacrocorax aristotelis Shag	R	1995	200	300	—	B1i, B3, C2, C6
Accipiter gentilis Goshawk	R	1995	10	—	B	C6
Aquila chrysaetos Golden Eagle	R	1995	—	10	C	C6
Hieraaetus fasciatus Bonelli's Eagle	R	1995	—	—	A	B2, C6
Falco eleonorae Eleonora's Falcon	B	1995	140	—	A	A3, A4ii, B1iii, B2, C2, C6
Falco peregrinus Peregrine	R	1995	20	—	C	B2
Alectoris barbara Barbary Partridge	R	1997	30	—	C	B2
Larus audouinii Audouin's Gull	R	1995	20	40	C	A1, C1
Sylvia sarda Marmora's Warbler	R	1997	50	—	C	A3, B3
Sylvia cantillans Subalpine Warbler	B	1995	—	—	—	A3
Sylvia melanocephala Sardinian Warbler	R	1995	—	—	—	A3
Sturnus unicolor Spotless Starling	R	1995	—	—	—	A3

An important breeding site for raptors and for five of the 21 species of the Mediterranean biome.

Protection status

National High **International** None
59,102 ha of IBA covered by National Park (Gennargentu, Golfo di Orosei, area not known).

Conservation issues

Threats Abandonment/reduction of land management (U), Afforestation (U), Agricultural intensification/expansion (U), Burning of vegetation (U), Deforestation (commercial) (U), Disturbance to birds (U), Forest grazing (U), Infrastructure (U), Intensified forest management (U), Recreation/tourism (U)

Tourism along the coast, fires, new roads, and deforestation are the main threats.

Oristano wetlands

A1, A4i, A4iii, B1i, B2, C1, C2, C3, C4, C6 **182**

Admin region Sardegna
Coordinates 39°52′N 8°29′E
Altitude 0–53 m **Area** 7,780 ha

Site description

A complex of wetlands along the west-central coast of Sardinia, surrounding the town of Oristano, and divided into the following main subsites: Mistras, Cabras, Mar'e Pauli, Santa Giusta, S'Ena Arrubia, Pauli Maiori, Pauli Menta, San Giovanni, Marceddì, Corru s'Ittiri, Putzu Idu, Is Benas, Sa Salina Manna and Sale Porcus. These areas include freshwater and brackish lagoons, ponds, large reedbeds (especially at Santa Giusta) and saltpans, as well as sand-dunes, rocky coasts (between Cuccuru Mannu and Su Tingiosu and at Capo Mannu), and adjacent agricultural land. The main activities are fisheries and aquaculture.

Habitats Wetland (80%; saltmarsh; sand-dunes/sand beach; shingle/stony beach; coastal lagoon; standing fresh water; standing brackish and salt water; water-fringe vegetation), Rocky areas (2%; sea cliff/rocky shore; rock stacks/islets), Artificial landscape (18%; arable land; other urban/industrial areas; ruderal land)
Land-use Agriculture, Fisheries/aquaculture, Hunting, Nature conservation/research, Tourism/recreation, Urban/industrial/transport

Birds

Species	Season	Year	Pop min	Pop max	Acc	Criteria
Phalacrocorax carbo Cormorant	W	1995	8,000	8,000	A	A4i, B1i, C3
Egretta garzetta Little Egret	W	1994	743	907	A	B1i, C2
Ardea purpurea Purple Heron	B	1995	20	—	—	C6
Phoenicopterus ruber Greater Flamingo	W	1994	1,835	2,548	A	A4i, B1i, C2
Aythya ferina Pochard	W	1995	13,472	13,472	A	A4i, B1i, C3
Aythya nyroca Ferruginous Duck	P	1995	—	210	A	A1, B1i, C1, C2
Falco naumanni Lesser Kestrel	B	1995	—	10	—	A1, C1, C2
Alectoris barbara Barbary Partridge	R	1995	Common		—	B2
Porphyrio porphyrio Purple Gallinule	R	1995	195	260	A	B2, C2, C6
Recurvirostra avosetta Avocet	W	1995	—	800	A	B1i, B2, C2
Glareola pratincola Collared Pratincole	B	1995	40	—	A	B1i, C2, C6
Charadrius alexandrinus Kentish Plover	W	1995	442	442	A	B1i, C3

Species ... continued	Season	Year	Pop min	Pop max	Acc	Criteria
Larus genei Slender-billed Gull	R	1995	—	500	—	C2, C6
Larus genei Slender-billed Gull	W	1995	55	146	A	C2
Gelochelidon nilotica Gull-billed Tern	B	1995	100	—	B	A4i, B1i, B2, C2, C6
Sterna sandvicensis Sandwich Tern	W	1995	6	140	A	C2
Sterna albifrons Little Tern	B	1995	—	150	—	B1i, B2, C2, C6

A very important site for wintering waterbirds (up to 50,000 or more), especially wildfowl and *Phoenicopterus ruber*, as well as for migrating *Aythya nyroca* and for several breeding waterbirds. The total number of waterbirds in January 1993 was 35,775, with 58,367 in January 1994. Species of global conservation concern that do not meet IBA criteria: *Aythya nyroca* (1–3 pairs breed, max. 13 birds in winter), *Tetrax tetrax* (resident), *Larus audouinii* (non-breeding visitor).

Protection status

National Low **International** High

IBA contains Marine Nature Reserve (Penisola del Sinis–Isola di Mal di Ventre, area not known). 3,575 ha of IBA covered by Special Protection Area (Stagno di Cabras, 3,575 ha). 2,610 ha of IBA covered by Special Protection Area (Stagno di Corru s'Ittiri, 2,610 ha). 680 ha of IBA covered by Special Protection Area (Stagno di Mistras, 680 ha). 287 ha of IBA covered by Special Protection Area (Stagno di Pauli Maiori, 287 ha). 300 ha of IBA covered by Special Protection Area (Stagno di S'Ena Arrubia, 300 ha). 330 ha of IBA covered by Special Protection Area (Stagno di Sale Porcus, 330 ha). 2,610 ha of IBA covered by Ramsar Site (Stagno di Corru S'Ittiri, Stagni di San Giovanni e, 2,610 ha). 3,575 ha of IBA covered by Ramsar Site (Stagno di C bras, 3,575 ha). 680 ha of IBA covered by Ramsar Site (Stagno di Mistras, 680 ha). 287 ha of IBA covered by Ramsar Site (Stagno di Pauli Maiori, 287 ha). 300 ha of IBA covered by Ramsar Site (Stagno di S'Ena Arrubia, 300 ha). 324 ha of IBA covered by Ramsar Site (Stagno di Sale Porcus, 324 ha).

Conservation issues

Threats Agricultural intensification/expansion (U), Aquaculture/fisheries (B), Burning of vegetation (U), Construction/impact of dyke/dam/barrage (U), Disturbance to birds (U), Drainage (U), Industrialization/urbanization (U), Infrastructure (U), Recreation/tourism (B), Unsustainable exploitation (U)

The main threats are the burning of vegetation, illegal hunting, and water pollution due to untreated urban and industrial sewage. Overfishing and agricultural intensification in the surrounding areas are important potential threats to the site. Shooting of *Phalacrocorax carbo* was authorized in 1994, 1995 and 1996 by the regional authority and the Instituto Nazionale Fauna Selvatica, as the birds were thought to be reducing fish populations in the lagoons. A Regional Nature Park has been proposed for the area of Sinis, and a Regional Nature Reserve for Corru s'Ittiri. Sale Porcus is managed by LIPU.

Mount Ferru of Tertenia A1, B2, B3, C1, C6 183

Admin region Sardegna
Coordinates 39°45′N 9°37′E
Altitude 0–875 m **Area** 1,000 ha

Site description

A coastal area in south-east Sardinia, with sea cliffs, maquis and dry grasslands.

Habitats Scrub (scrub; sclerophyllous scrub/garrigue/maquis), Marine areas, Rocky areas (sea cliff/rocky shore)
Land-use Agriculture, Hunting

Birds

Species	Season	Year	Pop min	Pop max	Acc	Criteria
[1] ***Accipiter gentilis*** Goshawk	R	1995	Uncommon		—	C6
Alectoris barbara Barbary Partridge	R	1995	30	—	C	B2
Larus audouinii Audouin's Gull	N	1995	20	40	B	A1, C1
Sylvia sarda Marmora's Warbler	R	1995	50	—	C	B3

1. *A. g. arrigonii*.

A site for non-breeding *Larus audouinii*.

Protection status

National None **International** None

Conservation issues

Threats Disturbance to birds (U), Firewood collection (U)

General threats to site are burning of vegetation, overgrazing and development.

Cape San Marco A1, B2, B3, C1, C2, C6 184

Admin region Sardegna
Coordinates 39°51′N 8°25′E
Altitude 0–56 m **Area** 200 ha

Site description

A rocky coastal area in west-central Sardinia, which includes the small island Isola di Mal di Ventre, the islet Scoglio del Catalano, and the capes Turr'e Seu and Capo San Marco. The main habitats are rocky cliffs and shores, maquis and garrigue. The main land-use is tourism.

Habitats Scrub (sclerophyllous scrub/garrigue/maquis), Grassland (steppe/dry calcareous grassland), Marine areas (open sea; sea inlet/coastal features), Rocky areas (sea cliff/rocky shore; rock stacks/islets)
Land-use Fisheries/aquaculture, Hunting, Military, Nature conservation/research, Tourism/recreation

Birds

Species	Season	Year	Pop min	Pop max	Acc	Criteria
Phalacrocorax aristotelis Shag	R	1996	30	40	A	C2
Alectoris barbara Barbary Partridge	R	1996	Common		—	B2
Larus audouinii Audouin's Gull	R	1996	100	—	B	A1, B2, C1, C6
Sylvia sarda Marmora's Warbler	R	1996	Common		—	B3
Sylvia undata Dartford Warbler	R	1996	Common		—	B2

Important breeding site for seabirds, including *Larus audouinii*. Breeding species of global conservation concern that do not meet IBA criteria: *Falco naumanni* (4 pairs).

Protection status

National Partial **International** None

111 ha of IBA covered by Private Reserve (Seu, 111 ha).

Conservation issues

Threats Disturbance to birds (U), Recreation/tourism (U), Unsustainable exploitation (U)

The main threat is tourism. The area is a proposed Regional Park, 'Parco Regionale del Sinis-Montiferru'.

Flumendosa and Colostrai pools A1, B2, C1 185

Admin region Sardegna
Coordinates 39°21′N 9°35′E
Altitude 0–104 m **Area** 3,400 ha

Site description

A complex of brackish wetlands on the south-east Sardinian coast, close to the village of Muravera. There are sand-dunes and beaches.

Habitats Wetland (saltmarsh; sand-dunes/sand beach; coastal lagoon; river/stream), Marine areas
Land-use Agriculture, Fisheries/aquaculture, Hunting, Tourism/recreation, Urban/industrial/transport

Birds

Species	Season	Year	Pop min	Pop max	Acc	Criteria
Porphyrio porphyrio Purple Gallinule	R	1995	—	6	—	B2
Larus audouinii Audouin's Gull	R	1995	15	25	—	A1, C1

The site is important for breeding *Larus audouinii*. Breeding species of global conservation concern that do not meet IBA criteria: *Falco naumanni* (min. 2 pairs).

Protection status

National None **International** None

Conservation issues

Threats Agricultural intensification/expansion (U), Aquaculture/fisheries (U), Disturbance to birds (U), Drainage (U), Dredging/canalization (U), Industrialization/urbanization (U), Infrastructure (U), Recreation/tourism (U), Unsustainable exploitation (U)

The main threats are wetland drainage, water pollution by untreated domestic sewage, and the building of new infrastructure (new roads to tourist resorts). The area has been proposed as a Regional Nature Reserve (Stagno di Colostrai).

Mount Sette Fratelli and Sarrabus — B2, B3, C6 — 186

Admin region Sardegna
Coordinates 39°17′N 9°27′E
Altitude 0–1,067 m **Area** 50,000 ha

Site description

A mountainous area in the south-east corner of Sardinia, north-east of the town of Cagliari. The main habitats are broadleaved evergreen forest, maquis, grasslands, streams, rivers and cultivated land.

Habitats Forest and woodland (broadleaved evergreen forest), Scrub (sclerophyllous scrub/garrigue/maquis), Grassland, Wetland (river/stream), Artificial landscape (highly improved reseeded grassland; arable land)
Land-use Agriculture, Forestry, Hunting, Tourism/recreation

Birds

Species	Season	Year	Pop min	Pop max	Acc	Criteria
[1] ***Accipiter gentilis*** Goshawk	R	1997	2	4	A	C6
Aquila chrysaetos Golden Eagle	R	1997	3	5	A	C6
Alectoris barbara Barbary Partridge	R	1997	30	—	C	B2
Sylvia sarda Marmora's Warbler	R	1997	50	—	C	B3

1. *A. g. arrigonii.*

An important site for breeding raptors and species of rocky shrubland.

Protection status

National None **International** None

Conservation issues

Threats Burning of vegetation (U), Agricultural intensification/expansion (U)

Coast between Cape Boi and Cape Ferrato — A1, C1 — 187

Admin region Sardegna
Coordinates 39°06′N 9°33′E
Altitude 0–270 m **Area** 200 ha

Site description

A rocky coast in the south-east corner of Sardinia, between Capo Boi and Capo Ferrato. Habitats include maquis, some islets, and small temporary wetlands on the coast. The main land-uses are agriculture and tourism.

Habitats Scrub (60%; sclerophyllous scrub/garrigue/maquis), Grassland (steppe/dry calcareous grassland), Wetland (10%; mudflat/sandflat; saltmarsh; sand-dunes/sand beach; shingle/stony beach; standing brackish and salt water), Rocky areas (20%; sea cliff/rocky shore; rock stacks/islets), Artificial landscape (10%)
Land-use Agriculture, Hunting, Tourism/recreation, Urban/industrial/transport

Birds

Species	Season	Year	Pop min	Pop max	Acc	Criteria
Larus audouinii Audouin's Gull	R	1996	50	100	A	A1, C1

An important site for breeding *Larus audouinii*, also notable for migrants and holding 5,000–7,000 waterbirds in winter.

Protection status

National None **International** None

Conservation issues

Threats Disturbance to birds (U), Drainage (U), Filling-in of wetlands (U), Industrialization/urbanization (U), Infrastructure (U), Recreation/tourism (U), Unsustainable exploitation (U)

Cagliari wetlands — A1, A4i, A4iii, B1i, B2, B3, C1, C2, C3, C4, C6 — 188

Admin region Sardegna
Coordinates 39°11′N 9°09′E
Altitude 0–14 m **Area** 5,600 ha

Site description

A complex of brackish coastal wetlands and saltpans around Cagliari in south Sardinia. It includes the following sites: Stagno di Cagliari, Saline di Quartu, Saline di Macchiareddu, Laguna di Santa Gilla, and Stagno di Molentargius. The main activities are salt-extraction and fishing.

Habitats Wetland (100%; mudflat/sandflat; saltmarsh; sand-dunes/sand beach; standing brackish and salt water; water-fringe vegetation)
Land-use Fisheries/aquaculture, Nature conservation/research, Urban/industrial/transport, Water management

Birds

Species	Season	Year	Pop min	Pop max	Acc	Criteria
Podiceps nigricollis Black-necked Grebe	W	1995	238	3,400	A	A4i, B1i, C3
Egretta garzetta Little Egret	B	1995	—	780	A	A4i, B1i, C2, C6
Phoenicopterus ruber Greater Flamingo	R	1995	1,000	1,400	A	A4i, B1i, B2, C2
Phoenicopterus ruber Greater Flamingo	W	1995	2,500	5,043	A	A4i, B1i, C2
Phoenicopterus ruber Greater Flamingo	N	1995	—	11,500	A	A4i, B1i, C2
Anas clypeata Shoveler	W	1995	2,494	8,995	A	A4i, B1i, C3
Alectoris barbara Barbary Partridge	R	1995	—	—	—	B2
Porphyrio porphyrio Purple Gallinule	R	1995	100	140	A	B2, C2, C6
Recurvirostra avosetta Avocet	R	1995	500	—	B	A4i,B1i,B3,C2,C6
Glareola pratincola Collared Pratincole	B	1995	7	10	A	C6
Charadrius alexandrinus Kentish Plover	W	1995	902	902	A	B1i, C3
Larus genei Slender-billed Gull	R	1995	3,000	—	B	A4i, B1i, C2, C6
Larus genei Slender-billed Gull	W	1995	642	642	A	B1i, C2
[1] ***Larus audouinii*** Audouin's Gull	W	1995	24	24	A	A1, C1
Gelochelidon nilotica Gull-billed Tern	R	1995	—	150	B	A4i,B1i,B2,C2,C6

1. Also non-breeding visitor.

An important site for a wide variety of wintering waterbirds (30,443 on January 1993; 42,216 on January 1994) and for breeding herons, *Phoenicopterus ruber*, gulls and terns. Species of global conservation concern that do not meet IBA criteria: *Aythya nyroca* (max. 12, winter), *Larus audouinii* (non-breeding visitor).

Protection status

National None **International** Partial

3,466 ha of IBA covered by Special Protection Area (Stagno di Cagliari, 3,466 ha). 1,400 ha of IBA covered by Special Protection Area (Stagno di Molentargius, 1,400 ha). 3,466 ha of IBA covered by Ramsar Site (Stagno di Cagliari, 3,466 ha). 1,401 ha of IBA covered by Ramsar Site (Stagno di Molentargius, 1,401 ha).

Conservation issues

Threats Aquaculture/fisheries (U), Disturbance to birds (U), Drainage (U), Dredging/canalization (U), Filling-in of wetlands (U), Industrialization/urbanization (U), Infrastructure (U), Unsustainable exploitation (U)

The main threats are: water pollution by urban sewage and by industrial and harbour discharges; illegal hunting; building of illegal tourist resorts and associated infrastructure; drainage of wetlands. The site is a proposed Regional Nature Reserve.

Mount Arcosu B2, C6 189

Admin region Sardegna
Coordinates 39°09′N 8°52′E
Altitude 100–948 m **Area** 4,500 ha

Site description

A mountain area in southern Sardinia, 15 km south-west of the town of Cagliari. The largest evergreen broadleaved woodland of the region.

Habitats Forest and woodland (broadleaved evergreen forest), Scrub, Wetland (river/stream)
Land-use Nature conservation/research, Tourism/recreation

Birds

Species	Season	Year	Pop min	Pop max	Acc	Criteria
[1] ***Accipiter gentilis*** Goshawk	R	1995	2	3	—	C6
Hieraaetus fasciatus Bonnelli's Eagle	R	1995	1	2	A	B2, C6

1. *A. g. arrigonii.*

An important breeding site for raptors.

Protection status

National Partial **International** Partial
3,205 ha of IBA covered by Private Reserve (Monte Arcosu, 3,205 ha). 3,205 ha of IBA covered by Special Protection Area (Monte Arcosu, 3,205 ha).

Conservation issues

Threats Burning of vegetation (U), Disturbance to birds (U), Natural events (U), Recreation/tourism (U), Unsustainable exploitation (U)

The main threats are illegal hunting and the burning of vegetation. The area is a proposed Regional Nature Park. The area is owned and managed by WWF. Some or all of the site is covered by a management plan.

Palmas Gulf wetlands A4i, B1i, B2, B3, C2, C3, C6 190

Admin region Sardegna
Coordinates 39°05′N 8°56′E
Altitude 0–4 m **Area** 4,000 ha

Site description

A complex of coastal pools, lagoons and saltpans in the south-west corner of Sardinia. The site includes Stagno di Santa Caterina, Stagno di Porto Botte, Stagno Baiocco, Saline di Sant'Antioco, Stagno di Mulargia, and Laguna di Sant'Antioco. The main activities are fishing and salt extraction.

Habitats Wetland (mudflat/sandflat; saltmarsh; sand-dunes/sand beach; coastal lagoon; standing brackish and salt water), Marine areas, Artificial landscape)
Land-use Fisheries/aquaculture (62%), Hunting, Urban/industrial/transport (42%)

Birds

Species	Season	Year	Pop min	Pop max	Acc	Criteria
Phalacrocorax carbo Cormorant	W	1996	—	1,000	A	B1i, C3
Phoenicopterus ruber Greater Flamingo	W	1996	—	2,000	B	A4i, B1i, C2
Porphyrio porphyrio Purple Gallinule	R	1996	—	25	A	B2, C6
Recurvirostra avosetta Avocet	R	1996	—	200	A	B3, C6
Charadrius alexandrinus Kentish Plover	W	1996	800	1,000	B	A4i, B1i, C3
Larus genei Slender-billed Gull	W	1995	280	560	A	B1i, C2

An important wintering and breeding site for waders. Species of global conservation concern that do not meet IBA criteria: *Tetrax tetrax* (resident).

Protection status

National None **International** None

Conservation issues

Threats Agricultural intensification/expansion (C), Aquaculture/fisheries (B), Disturbance to birds (C), Industrialization/urbanization (B), Infrastructure (C), Recreation/tourism (B), Unsustainable exploitation (B)

The main threat is from the building of intensive aquaculture plants, and of a thermal spa and other tourist resorts and associated infrastructure. One of the wetlands, Stagno di Is Prunis, is polluted by industrial discharges.

San Pietro and Sant'Antioco islands A1, A3, A4ii, B1ii, B1iii, B2, B3, C1, C2, C6 191

Admin region Sardegna
Coordinates 39°07′N 8°12′E
Altitude 0–271 m **Area** 6,000 ha

Site description

Small islands, of volcanic origin, off the south-west corner of Sardinia. The site also includes the islets Isola del Toro and Isola della Vacca. Rocky cliffs, maquis and saltpans are the main habitats.

Habitats Forest and woodland, Scrub (sclerophyllous scrub/garrigue/maquis), Wetland (saltmarsh; sand-dunes/sand beach; standing brackish and salt water; river/stream), Marine areas, Rocky areas (sea cliff/rocky shore; rock stacks/islets), Artificial landscape (arable land; perennial crops/orchards/groves; other urban/industrial areas; ruderal land)
Land-use Agriculture, Fisheries/aquaculture, Hunting, Military, Nature conservation/research, Tourism/recreation, Urban/industrial/transport

Birds

Species	Season	Year	Pop min	Pop max	Acc	Criteria
Calonectris diomedea Cory's Shearwater	R	1995	500	700	—	B1ii, B2, C2, C6
Hydrobates pelagicus Storm Petrel	R	1995	2	10	A	B2, C6
Falco eleonorae Eleonora's Falcon	B	1995	—	185	A	A3, A4ii, B1iii, B2, C2, C6
Alectoris barbara Barbary Partridge	R	1997	50	—	C	B2
Larus audouinii Audouin's Gull	R	1995	36	56	B	A1, C1
Sylvia sarda Marmora's Warbler	R	1997	100	—	C	A3, B3, C6
Sylvia conspicillata Spectacled Warbler	R	1986	—	—	—	A3
Sylvia cantillans Subalpine Warbler	B	—	—	—	—	A3
Sylvia melanocephala Sardinian Warbler	R	1986	—	—	—	A3

An important breeding site for seabirds and for five of the 21 species of the Mediterranean biome.

Protection status

National Low **International** None
124 ha of IBA covered by Wildlife Reserve (Carloforte, 124 ha).

Conservation issues

Threats Abandonment/reduction of land management (U), Agricultural intensification/expansion (U), Burning of vegetation (U), Disturbance to birds (U), Infrastructure (U), Natural events (U), Recreation/tourism (U), Unsustainable exploitation (U)

The main threats are tourism and burning of vegetation. There is a LIPU reserve (Carloforte) at Isola San Pietro to protect the colony of *Falco eleonorae*. The area is a proposed Regional Nature Reserve, 'Isola di S. Pietro e Isole Piana, dei ratti, del Corno'.

Coast between Cape Teulada and Pula A1, A4i, B1i, B2, B3, C1, C2, C6 192

Admin region Sardegna
Coordinates 38°50′N 8°49′E
Altitude 0–230 m **Area** 500 ha

Site description

A rocky coastal area in the south-west corner of Sardinia, between Pula and Capo Teulada, including many small rocky islets and some small wetlands.

Habitats Scrub (sclerophyllous scrub/garrigue/maquis), Wetland (mudflat/sandflat; saltmarsh; sand-dunes/sand beach; shingle/stony beach; standing brackish and salt water; river/stream), Marine areas (open sea; sea inlet/coastal features), Rocky areas (sea cliff/rocky shore; rock stacks/islets), Artificial landscape (perennial crops/orchards/groves; forestry plantation)
Land-use Agriculture, Fisheries/aquaculture, Hunting, Military, Tourism/recreation, Urban/industrial/transport

Birds

Species	Season	Year	Pop min	Pop max	Acc	Criteria
Phalacrocorax aristotelis Shag	R	1995	50	200	A	B1i, B3, C2
[1] ***Larus audouinii*** Audouin's Gull	R	1995	100	150	—	A1, A4i, B1i, B2, C1, C2, C6

1. Up to 112 in winter.

Species of global conservation concern that do not meet IBA criteria: *Aythya nyroca* (rare on passage).

Protection status

National None **International** None

Conservation issues

Threats Construction/impact of dyke/dam/barrage (U), Disturbance to birds (U), Drainage (U), Filling-in of wetlands (U), Recreation/tourism (U)

As a military area, a large part of the site receives some informal protection.

REFERENCES

Agostini, M. (year unknown) La foresta di Tarvisio. Pp. 31–35 in *L'altra Italia*. Bologna, Italy: Caldeini.

Aleffi, M. (year unknown) Natura e ambienti della provincia di Taranto. Ricerche Umanesimo della Pietra Martina Franca (TA). Unpublished report.

Aleo, M., Basciano, G., Famà, M. L., Giambalvo, D., Ingianni, A. and Santulli, A. (year unknown) *Isola dello Stagnone di Marsala R.N.O.* Italy: Provincia Regionale di Trapani. 20 pp.

Alessandria, G., Della Toffola, M. and Carpegna, F. (1992) Nuovo sito di riproduzione di Tarabuso, *Botaurus stellaris*, e di Airone rosso, *Ardea purpurea*, in Piemonte (Aves). *Riv. Piem. St. Nat.* 13: 97–102.

Allavena, S. (1980) Aspetti dell'ornitofauna dei laghi Lungo e Ripasottile. Amm. Pov. di Rieti: 13. Unpublished report.

Amato, S., Semenzato, M., Borgoni, N., Richard, J. and Talota, G. (1995) Status attuale della popolazione di Ardeidi nidificanti nella laguna di Venezia. *Riv. Ital. Ornit.* 63: 200–205.

Anoè, N., Carpenè, B. and Zanaboni, A. (1988) *Flora e vegetazione del fiume Sile*. Treviso, Italy: Lega Italiana per la Protezione degli Uccelli (LIPU). 30 pp.

Arcà, G. (1993) Environment and protection of the Fiora river valley and of the Selva del Lamone. Pp. 251–256 in M. Olmi and M. Zapparoli, eds. *The environment of the Tuscia laziale*. Italy: Università della Tuscia Viterbo.

Arcà, G., Calvario, E., Fraticelli, F. and Petretti, F. (1995) Lista Rossa degli uccelli nidificanti nel Lazio. Seconda edizione. *Alula* 2: 201–205.

Arcamone, E. (1988) [Waterfowl wintering in Tuscany 1984–1988]. *Quad. Mus. Stor. Nat. Livorno* 10 (Suppl.). (In Italian.)

Arcamone, E., Baccetti, N. and Serra, L. (1994) Wintering waders in the coastal wetlands of Tuscany. *Riv. Ital. Ornit.* 64: 3–13.

Baccetti, N. (1988) Lo svernamento del Cormorano in Italia. *Suppl. Ric. Biol. Selvaggina* 15.

Baglieri, S. and Iapichino, C. (1974) The little tern (*Sterna albifrons*) breeds in Sicily. *Riv. Ital. Ornit.* 44: 293–295.

Baragliu, G. A. (1993) Vegetational aspects of the Fiora river valley and of the Selva del Lamone. Pp. 263–269 in M. Olmi and M. Zapparoli, eds. *The environment of the Tuscia laziale*. Italy: Università della Tuscia Viterbo.

Barbieri, F., Bogliani, G., Fasola, M., Pazzuconi, A. and Prigioni, C. (1977) L'Atlante ornitologico della provincia di Pavia. *Atti del VII Simp. Naz. Cons. Natura, Bari*: 87–99.

Basso, R. (1983) Observations of nesting birds in the lower part of Salento (Lecce) in 1982. *Gli Uccelli d'Italia* 3: 155–167.

Bello, B. (1989) Rilevamento geomorfologico dei dintorni di Laterza. Umanesimo della Pietra verde numero unico 41–47.

Bernoni, M. (1984) Il metodo del mappaggio in una zona umida del Lazio: le vasche di Maccarese. *Riv. Ital. Ornit.* 54: 235–245.

Biddan Luca (1995) L'avifauna acquatica svernante sul Lago di Candia. *Riv. Piem. St. Nat.* 16: 233–251.

Biondi, M. and Pietrelli, L. (1987) Birdwatching sul Litorale Romano. Comune di Roma-LIPU. Delegazione di Ostia Lido. Unpublished report.

Biondi, M., Guerrieri, G. and Pietrelli, L. (1989) La nicchia di foraggiamento in sei specie di Passeiformi presenti in inverno sul Litorale Romano. *Sitta* 3: 45–54.

Biondi, M., Guerrieri, G. and Pietrelli, L. (1990) Ciclo annuale della comunità ornitica di una zona umida artificiale dell'Italia centrale. *Avocetta* 14: 11–26.

Biondi, M., Guerrieri, G. and Pietrelli, L. (1991) Stress antropogenici ed evoluzione di una comunità ornitica di una zona umida artificiale dell'Italia centrale. *Avocetta* 17: 23–27.

Biondi, M., Pietrelli, L., Guerrieri, G. and Martucci, O. (1992a) Selezione di habitat e riproduzione del Corriere piccolo, *Charadrius dubius*, lungo la costa laziale. *Avocetta* 16: 41–43.

Biondi, M., Pietrelli, L., Guerrieri, G. and Martucci, O. (1992b) Distribuzione e riproduzione del Gruccione, *Merops apiaster*, nella fascia costiera laziale. *Avocetta* 16: 44–46.

Biondi, M., Corbi, F. Guerrieri, G., Gustin, M., Meschini, A. and Pietrelli, L. (1993) I limicoli svernanti nella fascia costiera della Lazio. *Riv. Ital. Ornit.* 63: 89–91.

Biondi, M., Pietrelli, L. and Guerrieri, G. (1994a) Presenze invernali e status degli Ardeidae lungo la costa del Lazio (1986–1992). *Riv. Ital. Ornit.* 63: 137–144.

Biondi, M., Pietrlli, L. and Guerrieri, G. (1994b) Avvistamenti di particolare interesse sul Litorale Romano. *Picus* 18: 133–140.

Biondi, M. and Pietrelli, L. (1995) I limicoli svernanti lungo la costa laziale: conteggi 1993–1995. *Avocetta* 19: 80.

Boano, A., Brunelli, M., Bulgarini, F., Montemaggiori, A., Sarrocco, S. and Visentin, M. (1995) Breeding atlas of birds in Latium. *Alula* (Spec. vol.) 1–2. 224 pp. (In Italian, with English summary.)

Bordignon, L. (1993) L'avifauna acquatica della 'Fontana gigante' di Tricerro. *Riv. Piem. St. Nat.* 14: 231–241.

Brichetti, P., ed. (1982–1988) Atlante degli uccelli nidificanti sulle Alpi italiane, I–IV. *Riv. Ital. Ornit.* 52: 3–50; 53: 101–144; 56: 3–39; 58: 3–39.

Brichetti, P. (1987) *Atlante degli uccelli delle Alpi italiane*. Brescia, Italy: Ramperto.

Brichetti, P. and Massa, B. (1984) Check-list of Italian birds. *Riv. Ital. Ornit.* 54(1/2): 3–37.

Brichetti, P. and Cambi, D. (1985) *Atlante degli uccelli nidificanti in Provincia di Brescia (Lombardia) 1980–1984*. Brescia, Italy: Museo Civico di Scienze Naturali di Brescia (Monografie di Natura Bresciana no. 8). 142 pp.

Brichetti, P. and Cambi, D. (1990) *Atlante degli uccelli svernanti in Provincia di Brescia (Lombardia). Inverni dal 1984–1985 al 1987–1988*. Brescia, Italy: Museo Civico di Scienze Naturali di Brescia (Monografie di Natura Bresciana no. 14). 112 pp.

Brichetti, P. and Fasola, M. (1990) *Atlante degli uccelli nidificanti in Lombardia 1983–1987*. Brescia, Italy: Ramperto. 242 pp.

Brichetti, P. and Massa, B. (1993a) [Check-list of Italian birds updated to 1992: non-passeriformes.] *Rivista Ital. di Birdwatching* 1(2): 61–73. (In Italian.)

Brichetti, P. and Massa, B. (1993b) [Check-list of Italian birds updated to 1992: passeriformes.] *Rivista Ital. di Birdwatching* 1(3): 20–26. (In Italian.)

Brunelli, M. and Sarrocco, S. (in press) Analisi del popolamento dell'avifauna a 10 anni dall'istituzione della Riserva Naturale Parziale dei laghi Lungo e Ripasattile (anni 1985–1995). Convegno per il decennale della Riserva (Cittaducale, RI). Unpublished report.

Bulgarini, F., Petretti, F. and Visentin, M. (1991) Elenco delle aree italiane di interesse naturalistico. WWF-Italia. Unpublished report.

Busnaro, G. and Lasen, C. (1994) *Incontri con il Grappa 'Il paesaggio vitale'. Centro incontri con la natura 'Don P. Chivacci'*. Crespano del Grappa, Italy: Moro.

Calchetti, L., Cianchi, F. and Giannella, C. (1987) [The avifauna of Orbetello lagoon.] *Picus* 13: 81–126. (In Italian.)

Caldonazzi, M. and Fasola, M. (1989) La comunità di uccelli nidificanti nell'altopiano del Monte Calisio (Trentino). *Studi Trentini Scienze Naturali, Acta Biologica* 65: 177–202.

Calò, C. M. (1981) I monti della Duchessa Una scheda dell'avifauna. *L'upupa, LIPU Lazio* 9(4): 2.

Calò, C. M. (1982) Note sulla fauna di maggior interesse nei Monti della Duchessa (cat. di M. Velino). WWF-Roma. Unpublished report.

Calò, C. M. (1983) La situazione ambientale nei Monti della Duchessa. *Natura e Montagna (Bologna)* 1: 39–54.

Calò, C. M. (1983–1986) Dati personali (schede rilevamento Progetto Atlante Uccelli Nidificanti). Roma.

Carpegna, F., Alessandria, G. and Della Toffola, M. (1987) La confluenza tra i fiumi Dora Baltea e Po: contributo alla conoscenza dell'avifauna. *Riv. Piem. St. Nat.* 8: 187–213.

Cassol, M. (1987) L'avifauna del piave nella val Belluna. *Dolomiti* 10(2): 33–39.

Cassol, M., De Faveri, A. and Nadalet, G. (1987) Accertata nidificazione di moretta *Aythya fuligula*, in provincia di Belluno. *Riv. Ital. Ornit.* 59: 112–113.

Castaldi, A. and Guerrieri, G. (1995) Distribuzione altitudinale del genere *Lanius* nel Lazio. *Avocetta* 19: 136.

Cataldini, G. E. and Scarpina, L. (1994) Nesting of *Larus audouinii* in S. Andrea Island, Gallipoli, Puglia. *Riv. Ital. Ornit.* 217–219.

CHERUBINI, G. AND PANZARIN, F. (1993) Il fratino nidificante lungo il litorale della provincia di Venezia. Pp. 111–112 in F. E. Mezzavilla and E. Stival, eds. *Atti del I Convegno Faunis. Veneti C.O.V.O. Montebelluna TV.*

CHIARETTA, F. (1994) Tarvisio. La foresta d'Europa. *ALP* 109: 45–59.

CORBETTA, F. AND LORENZONI, G. G. (1976) La vegetazione degli stagni del Golfo di Oristano (Sardegna). *Suppl. Ric. Biol. Selvaggina* 7: 759–783.

CORSETTI, L. (1989) *Atlante ornitologico dei Monti Lepini 1982–1988.* Italy: Ypothékai Bollettino del consorzio Biblioteche dei monti Lepini. Anno V. 222 pp.

CORSI, F. AND GIOVACCHINI, P., ED. (1995) *Atlante degli uccelli svernanti in provincia di Grosseto. Inverni 1988/89–1993/94.* Italy: Caletra. 251 pp.

CORTONE, P. AND LIBERATORI, F. (1989) Updated status of the Egyptian vulture (*Neophron percnopterus*) on the Italian mainland. *Riv. Ital. Ornit.* 50(1–2): 49 -59.

CUCCO, M., LEVI, L., MAFFEI, G. AND PULCHER, C. (1996) Atlante degli uccelli di Piemonte e Valle d'Aosta in inverno (1986–1992). *Mus. Reg. Sci. Nat.* 19.

DE ANGELIS, G. (1988) *Monti Lucretili.* Terza edizione. Rome: Assessorato Ambiente Provincia di Roma. (In Italian.)

DE FRANCESCHI, P. (1986) Il tetraonidi della foresta di Tarvisio. Verona, Italy: Cierre. 141 pp.

DE FRANCESCHI, P. (1991) *Atlante degli uccelli nidificanti in Provincia di Verona (Veneto) 1983–1987.* Verona, Italy: Museo Civico di Storia Naturale di Verona (Memorie no. 9). 154 pp.

DE FRANCESCHI, P. AND BOTTAZZO, M. (1988) Caratteristiche vegetazionali delle arene di canto del Gallo Cedrone (*Tetrao urogallus*, L.) nel Tarvisiano (Alpi orientali, Friuli-Venezia-Giulia, italia). Pp. 169–181 in M. Spagnesi and S. Toso, eds. *Atti del I Convegno Nazionale Biologia Ogi Selvaggina, Bologna* (*Suppl. Ric. Biol. Selvaggina* 14).

DE FRANCESCHI, P. AND BOTTAZZO, M. (1991) Capercaillie *Tetrao urogallus* and forest management in the Tarvisio Forest (Easter Alps, Italy) in 1982–1988. Pp. 192–196 in *Proceed. Vth Int. Symp. on Grouse, Eleverum, Norway* (*Ornis Scandinavia* 22).

DE FRANCESCHI, P. F. AND BOTTAZZO, M. (1992) Caratteristiche vegetazionali di alcuni biotopi di riproduzione di Pernice bianca (*Lagopus mutus* Montin) nel tarvisiano e in Carnia (Alpi orientali, Friuli-Venezia-Giulia). Pp. 165–180 in M. Spagnesi and S. Toso, eds. *Atti del II Convegno Nazionale Biologia Ogi Selvaggina* (*Suppl. Ric. Biol. Selvaggina* 19).

DI CARLO, E. A. (1956) Sulla nidificazione di alcuni uccelli non comuni nell'Appennino Centrale umbro-laziale-abruzzese. *Riv. Ital. Ornit.* 26: 177–188.

DI CARLO, E. A. (1958) Risultati di ricerche ornitologiche sulle montagne d'Abruzzo. *Riv. Ital. Ornit.* 28: 145–217.

DI CARLO, E. A. AND CASTIGLIA, G. (1981) Risultati di ricerche ornitologiche effettuate nell'area dei laghi Velini (piano Reatino, Rieti, Lazio). *Gli Uccelli d'Italia* 6: 127–170.

FRAMARIN, F. (1996) Gli uccelli del Gran Paradiso. Unpublished manuscript.

FARALLI, U. (1991) [Confirmed breeding of Shoveler *Anas clypeata* and Ferruginous Duck *Aythya nyroca* in the lake of Montepulciano (Tuscany).] *Riv. Ital. Ornit.* 61: 61–62. (In Italian.)

FARALLI, U. AND LAMBERTINI M. (1991) Effects of hunting on the bird communities of Montepulciano lake. *Suppl. Ric. Biol. Selvaggina* 19: 113–124.

FASOLA, M. (1986) Distribuzione e popolazione deo Laridi e Sternidi nidificanti in Italia. *Suppl. Ric. Biol. Selvaggina* 11.

FASOLA, M., ALFIERI, R. AND ZANDONELLA NECCA, D. (1992) Strategia per la conservazione delle colonie di Ardeinae e modello per la gestione di riserve specifiche riserve naturali. *Ric. Biol. Selvaggina* 90.

FENAROLI, L. (1938) [The vegetation and the flora of the Grigne.] *L'Alpe Firenze* 25: 3–17. (In Italian.)

FORNASARI, L., BOTTONI, L., MASSA, R., FASOLA, M., BRICHETTI, P. AND VIGORITA, V. (1992) *Atlante degli uccelli svernandi in Lombardia.* Milan, Italy: Regione Lombardia e Università degli Studi di Milano. 378 pp.

FOSCHI, U. F. AND GELLINI, S. (1987) *Atlante degli uccelli nidificanti in Provincia di Forlì 1982–1986.* Italy: Maggioli.

FRAISSINET, M. (1985, 1986) Atlante ornitologico degli uccelli e svernanti in Provincia di Napoli. II-III Parte. *Gli Uccelli d'Italia* 10: 119–127, 11: 51–56.

FRAISSINET, M. AND CAPUTO, E. (1984) Atlante ornitologico degli uccelli nidificanti e svernanti in Provincia di Napoli. I Parte. *Gli Uccelli d'Italia* 9: 57–75, 135–150.

FRAISSINET, M. AND KALBY, M. (1989) *Atlante degli uccelli nidificanti in Campania 1983–1987.* Naples, Italy: ASOIM Regione Campania (Monografia no. 1).

FRAISSINET, M. AND MILONE, M. (1992) *Migration and ringing birds in Campania.* Naples, Italy: ASOIM Regione Campania (Monografia no. 2).

FRAISSINET, M., GROTTA, M. AND PICIOCCHI, S. (1994) *The Red List of birds in Campania.* Naples, Italy: ASOIM Regione Campania (Monografia no. 3).

FRUGIS, S. AND SCHENK, H. (1981) Red List of Italian birds. *Avocetta* 5: 133–141.

G.A.R.O.L. (1995) Progetto Atlante Invernale del Litorale Laziale (PAILL): dati riassuntivi. *Avocetta* 19: 143.

G.E.P.S.O. (1982–1995) Resoconti ornitologici per la regione Piemonte e Valle d'Aosta. *Riv. Piem. St. Nat.*

GEROUDET, P. (1972) Contribution a l'ornitologia dei Gran Paradis. *Nos Oiseaux* 31: 269–296.

GIANNELLA, C. AND RABACCHI, R. (1992) *Atlante degli uccelli nidificanti in Provincia di Modena 1982–1990. Relazione sullo stato dell'ambiente in Prov. di Modena vol. 3.* Modena, Italy: Prov. di Modena and S.O.M. 196 pp.

GIORDA, M. (1990) Il parco regionale della laguna di Corado e Marano. Quaderni dell'istituto di geografia della facoltà di economia e commercio dell'università di Trieste, no. 13. Laurea thesis.

GIUSTI, F. (1993) [*The natural history of southern Tuscany*.] Milan, Italy: Amilcare Pilli. (In Italian.)

GRIMMETT, R. F. A. AND JONES, T. A. (1989) *Important Bird Areas in Europe.* Cambridge, UK: International Council for Bird Preservation (Techn. Publ. 9). 888 pp.

GRUPPO VICENTINO STUDIO ORNITOLOGICI [NISORIA] (1994) *Atlante degli uccelli nidificanti in Provincia di Vicenza.* Vicenza, Italy: Gilberto Padovan. 206 pp.

GRUSSO, M. (1995a) Status, popolazione e distribuzione degli Uccelli nidificanti in Sardegna al 1995. *Gli Uccelli d'Italia* 20(2) and 21(1).

GRUSSO, M. (1995b) Birdwatching a Cagliari. *Rivista Ital. di Birdwatching* 9(3): 6–17.

GUENZANI, W. AND SAPORETTI, F. (1988) *Atlante degli uccelli nidificanti in Provincia di Varese (Lombardia) 1983–1987.* Vareze, Italy: Lativa. 176 pp.

GUSTIN, M. (1991) Aspects of spring raptor migration 1989 at Capo Otranto (Lecce). Pp. 457–460 in S.R.O.P.U., eds. *Atti del V Convegno Italiano di Ornitologia* (*Suppl. Ric. Biol. Selvaggina* 17).

HEATH, M. F. AND BORGGREVE, C. (2000) *BirdLife International/EBCC European Bird Database 1998.* Cambridge, UK: BirdLife International.

ISTAT (1991) *Censimento Nazionale ISTAT 1990.* Rome: Poligrafico dello Stato.

LAMBERTINI, M. (1987) [The avifauna of Montepulciano Lake. 1. Annual cycle of the community.] *Avocetta* 11: 17–35. (In Italian.)

LARDELLI, R. (1988) Atlante degli uccelli nidificanti nel Mendrisiotto 1981–1985. *Memorie Soc. Ticinese Sc. Nat.* 2: 1–222.

LARDELLI, R. (1992) *Atlante degli uccelli del Ticino in inverno.* Graversano, Italy: Ficedula. 302 pp.

LAURENTI, S. AND DI CARLO, E. A. (1994) Alviano, oasis for the birds. P.551 in *Atti del VI Convegno Italiano di Ornitologia, Torino, Tedesco 1991* (*Mus. Reg. Sci. Nat.*).

LAURENTI, S., CONCEZZI, L. AND TAGLIONI, T. (1995) Gadwall, *Anas strepera*, breeds in Alviano Lake. *Gli Uccelli d'Italia* 20: 46.

LIBERARORI, F., PENTERIANI, V. AND PINCHERA, F. (1991) Census of the population of the Marsh Harrier, *Circus aeruginosus*, in Tuscany. *Avocetta* 15: 51–54.

LIPU (Lega Italiana Protezione Uccelli) (1992) *Aree di importanza Europea per gli uccelli selvatica in Italia.* Parma, Italy: Fava Cassinelli.

LO VALVO, M., MASSA, B. AND SARÀ, M., EDS. (1994) Uccelli e paesaggio in Sicilia alle soglie del terzo Millennio. *Naturalista Siciliano* (Suppl.) 17 (1993): 7–238.

LO VALVO, M. AND SARÀ, M. (1994) Breeding of Black-winged stilt (*Himantopus himantopus*) in western Sicily. *Naturalista Siciliano* (Suppl.) 17 (1993).

LUPOLI, A., PALUMBO, G. AND RIZZI, V. (1995) Avifauna nel territorio del comune di Pomarico. Pp. 653–656 in *Atti del VII Convegno di Ornitologia, Università di Urbino, 1993* (*Suppl. Ric. Biol. Selvaggina* 17).

MASSA, B., ED. (1985) Atlante degli uccelli nidificanti in Sicilia 1979–1983. Atlas faunae Siciliae (Aves). *Naturalista Siciliano* 9 (Spec. issue).

MASSI, A., MARSILI, L., FOCARDI, S., LEONZIO, C. AND CORTONE, P. (1990) Trace metals and chlorinated hydrocarbons in tissues of Egyptian vulture. *Riv. Ital. Ornit.* 60(1/2): 29–36.

MESCHINI, E. AND FRUGIS, S. (1993) *Atlante degli uccelli nidificanti in Italia.* (*Suppl. Ric. Biol. Selvaggina* 20). 346 pp.

MEZZALIRA, G. (1983) Gli uccelli delle riserve naturali delle dolomiti di Feltre e Belluno. (dattiloscritto) Italy: ASFA Belluno.

MEZZAVILLA, F. (1984) *Uccelli del fiume Sile.* Treviso, Italy: Lega Italiana per la Protezione degli Uccelli (LIPU).

MEZZAVILLA, F. (1989) *Atlante degli uccelli nidificanti nelle provincie di Treviso e Belluno (Veneto) 1983–1988.* Italy: Museo Civico di Storia e Scienze Naturale di Montebelluna. 116 pp.

MEZZAVILLA, F., BATTISTELLA, V. AND COSTA, R. (1992) Censimento degli ardeidi in provincia di Treviso negli anni 1985–1990. *Soc. Ven. Sc. Nat.* 17.

MINGOZZI, T., BOANO, G., PULCHER, C. *ET AL.* (1988) *Atlante degli uccelli nidificanti in Piemonte e Val d'Aosta 1980–1984.* Torino, Italy: Museo Regionale di Scienze Naturali di Torino (Monografia VIII). 514 pp.

MINISTERO DELL'AMBIENTE (1997) *Relazione sullo stato dell'Ambiente*. Rome: Franco Angeli.

MONCINI, E. (1994) La popolazione del topino (*Riparia riparia*) del parco regionale del Taro. Rapporto 1993 Consorzio del parco del Taro. Unpublished report.

MONCINI, E. AND POLLONARA, E. (1995) Quale futuro per il topino *Riparia riparia* nella provincia di Parma? *Avocetta* 19(1): 44.

MORIMANDO, F., PEZZO, F., DRAGHIS, S. AND FRATALOCCHI, G. (1994) First breeding of lanner falcon *Falco biarmicus* in Siena country and notes on the local historical distribution. *Avocetta* 18: 157–159.

MUSI, F., ED. (1991) *Inventario Faunistico Regionale Permanente. Primi risultati relativi al periodo riproduttivo 1986–1990*. Udine, Italy: Regione Autonoma Friuli-Venezia-Giulia. 232 pp.

MUSI, F., PERCO, F. AND UTMAR, P. (1992) Loss restoration and management of wetlands in F.U.G.N.E., Italy. *IWRB Spec. Publ.* 20: 257–262.

NATALI, M. (1993) *The fishes of lake Trasimeno*. Italy: Perugia Provincia di Perugia, Assessorato Programmazione e Gestione Faunistica.

PANDOLFI, M. AND GIACCHINI, P. (1995) *Avifauna della Provincia di Pesaro e Urbino*. Italy: Amministrazione Provinciale di Pesaro e Urbino. 270 pp.

PARODI, R. (1987) *Atlante degli uccelli nidificanti in Provincia di Pordenone (Friuli-Venezia-Giulia) 1981–1986*. Italy: Museo Civico di Storia Naturale di Pordenone (Quaderno no. 1). 118 pp.

PECORINI, G., RAGE, J.-C. AND THALER, L. (1973) La formation continentale de Capo Mannu, sa faune de vèrtèbres pliocènes et la question du Messinien en Sardaigne. *Rend. Sem. Fac. Sci. Univ. Cagliari* 43 (Suppl.): 305–319.

PEDROTTI, F. AND PETTIROSSI, L. (1969) Survey map of the vegetation of the wetland of Colfiorito. *Ostalp. Din. Pflanzensoz arbeistsgem. Camerino*: 365–388.

PENATI, F. (1988) La foresta Incantata. *OASIS* 4/5: 56–75.

PERCO, F., MUSI, F. AND PARODI, R. (1983) L'oasi avifaunistica di Marano lagunare. Fondo mondiale per la natura delegazioni del Friuli-Venezia-Giulia, Udine. Unpublished report.

PERCO, F. AND UTMAR, P. (1989a) L'avifauna delle provincie di Trieste e Gorizia fino all'Isonzo. *Biogeographia* 13 (1987): 801–843.

PERCO, F. AND UTMAR, P. (1989b) Il censimento degli acquatici svernanti nelle principali zone umide del Friuli-Venezia-Giulia. Fino al 1987. *Fauna, bollettino degli osservatori faunistici* 1: 4–31.

PERCO, F. AND UTMAR, P. (1993) L'avifauna della valle Cavanata. *Fauna, bollettino degli osservatori faunistici* 3: 7–38.

PERCO, F. AND UTMAR, P. (in press) Gli acquatici svernanti nelle zone umide del Friuli-Venezia-Giulia, 1988–1996. *Fauna, bollettino degli osservatori faunistici* 4.

PETRETTI, F. (1976) Studio ornitologico sul territorio di Maccarese. Pp. 535–577 in *Scritti in memoria di A. Toschi* (*Suppl. Ric. Biol. Selvaggina* 7).

PETRETTI, F. (1979) Osservazione su una popolazione di Cannareccione (*Acrocephalus arundinaceus*). *Avocetta* 3: 29–46.

PETRETTI, F., ED. (1987) *I rapaci nel Lazio. Status e conservazione*. Italy: Regione Lazio (Quaderno Lazionatura no. 6). 86 pp.

PIETRELLI, L., BIONDI, M. AND GUERRIERI, G. (1994) Contrazione stagionale della distribuzione del Martin pescatore, *Alcedo atthis*, sul Litorale Romano. Pp. 505–506 in *Atti del VI Convegno Italiano di Ornitologia, Torino 1991* (*Mus. Reg. Sci. Nat.*).

PIETRELLI, L. AND BIONDI, M. (1995) Svernamento del Fratino *Charadrius alexandrinus* lungo la costa laziale. *Avocetta* 19: 94.

PIOVESAN, A. AND MEZZAVILLA, F. (1992) *Sile, il fiume e il sud parco*. Italy: Celio libri. 175 pp.

PUGLISI, L., FONTANELLI, A. AND BALDACCINI, N. E. (1995a) The avifauna of Diaccia Botrona marsh (Italy): present status and recent modifications. *Ric. Biol. Selvaggina* 95: 1–50.

PUGLISI, L., FONTANELLI, A., PERFETTI, A. AND TAVERNI, M. (1995b) The population of the Bittern (*Botaurus stellaris*) in the Diaccia Botrona marsh (central Italy): fours years of census (1991–1994). *Avocetta* 19: 182–188.

PULCHER, C. (1983) Osservazioni durante un ciclo annuale sull'avifauna della Garzaia di Valenza. *Riv. Piem. St. Nat.* 4: 111–125.

PULCHER, C. (1987) La variabilità della comunità ornitica svernante in un ambiente padano. *Riv. Piem. St. Nat.* 8: 161–169.

RETEUNA, D. (1994) La migrazione degli Accipitriformes, Falconiformes e Ciconiiformes attraverso la Valle di Laura. *Riv. Piem. St. Nat.* 15: 127–153.

RIZZI, V. AND CRIPEZZI, V. (1994) Dati preliminari sull'attuale distribuzione della Gallina prataiola *Tetrax tetrax* e dell'Occhione *Burhinus oedicnemus* in provincia di Foggia. Pp. 501–502 in *Atti del VI Convegno Italiano di Ornitologia, Torino, 1991*.

RIZZI, V. AND CRIPEZZI, V. AND PALUMBO, G. (1994) Dati preliminari sulla densità dell'Occhione *Burhinus oedicnemus* in alcuni agroecosistemi marginali della Puglia, Basilicata e Calabria. Pp. 97–98 in *Atti del Convegno Nazionale L'Avifauna degli ecosistemi di origine antropica: zone umide artificiali, coltivi, aree urbane. Organizzato ASOIM, Università di Napoli 26–27 novembre 1994* (*Monografia no. 5 dell'Associazione Studi Ornitologici Italia Meridionale*).

ROMA, S. AND ROSSETTI, M. (1990) Gli uccelli della provincia di Frosinone. *Gli Uccelli d'Italia* 15(1–4).

RUSSO, S. (1993) The birds of the salt-pits of Trapani. Palermo, Italy: Università degli Studi di Palermo. Unpublished graduation thesis.

SANTONE, P. (1994–1995) *Uccelli d'Abruzzo*, vol. 2. Italy: Regione Abruzzo. 420 pp.

SANTONE, P. AND DI CARLO, E. A. (1994) Check-list degli uccelli d'Abruzzo e Molise. *Gli Uccelli d'Italia* 19: 25–38.

SARROCCO, S. (1981–1982) Il valore ornitologico della Piana di Rieti e dei suoi ambienti umidi. Università degli studi di Roma 'La Sapienza'. Laurea thesis.

SARROCCO, S. (1986) Alcuni dati sulla biologia riproduttiva dello Svasso maggiore in due bacini dell'Italia centrale, Laghi Reatini (Rieti). *Riv. Ital. Ornit.* 56: 197–202.

SCARTON, F., VALLE, R. AND BORELLA, S. (1995) Il Fraticello nidificante nella laguna di Venezia: anni 1989–1993. *Lav. Soc. Ven. Sc. Nat.* 20: 81–88.

SCEBBA, S. (1993) *The birds of Campania*. Naples, Italy: Esselibri Napoli.

SCEBBA, S., MOSCHETTI, G., CORTONE, P. AND GIORGIO, A. (1992–1993) Check-list degli uccelli della Calabria aggiornata a gennaio 1993. *Sitta* 6: 33–45.

SIGISMONDI, A. (1986) L'Ultimo nido del Capovaccaio. Umanesimo della Pietra verde numero unico 31–41.

SIGISMONDI, A. (1987) Il Gufo reale Signore della Notte. Umanesimo della Pietra verde numero unico 37–43.

SIGISMONDI, A. AND TEDESCO, N. (1988) Avifauna nidificante nelle gravine e ambienti frequentati. Umanesimo della Pietra verde numero unico 33–40.

SIGISMONDI, A. AND TEDESCO, N. (1990) *Natura in Puglia*. Italy: Mario Adda.

SIGISMONDI, A., CASIZZI, G., CILLO, N., LATERZA, M., RIZZI, V. AND VENTURA, T. (1995) Distribuzione e consistenza delle popolazioni di Accipitriformi e Falconiformi nidificanti in Puglia e Basilicata. Pp. 707–711 in *Atti del VII Convegno di Ornitologia, Università di Urbino, 1993* (*Suppl. Ric. Biol. Selvaggina* 21).

SILVANO, F. (1974) Elenco degli uccelli della Valle Scrivia. *Riv. Ital. Ornit.* 44: 165–192.

SILVANO, F. (1985) Aggiornamento all'elenco degli uccelli della Valle Scrivia, Alessandria (Piemonte). *Riv. Ital. Ornit.* 55: 99–102.

SIMONETTI, G. (1991) L'isola della Cona, analisi naturalistica. Enel direzione per la relazione con enti ed istituzioni, gruppo ambiente. Unpublished.

SPANÒ, S., TRUFFI, G. ET AL. (1989) *Atlante degli uccelli nidificanti in Liguria 1981–1986*. Genova, Italy: Regione Liguria (Cataloghi dei beni naturali no. 1). 210 pp.

SPOSIMO, P. (1996) Breeding birds and environmental quality evaluation in a suburban area. Pp. 50–55 in *L'avifauna degli ecosistemi di origine antropica*. Naples, Italy: Electa.

SPOSIMO, P. (1998) Breeding bird communities of the Monti della Calvana (Firenze). *Quad. Mus. Storn. Nat. Livorno* 9: 105–129.

TEDESCO, N. (1991) Un grande parco naturale per l'arco jonico delle gravine. Umanesimo della Pietra verde numero unico 11–19.

TELLINI, G., ARCAMONE, E., BACCETTI, N., MESCHINI, E. AND SPOSIMO, P. (in prep.) [Atlas of birds breeding and wintering in Tuscany.] *Quad. Mus. Storn. Nat. Livorno* (Monografia no. 1). (In Italian.)

TRALONCO, S. ET AL. (1995) Dati preliminari sul Gruccione *Merops apiaster* nel parco regionale dello Stirone. *Avocetta* 19(1): 161.

TUCKER, G. M. AND HEATH, M. F. (1994) *Birds in Europe: their conservation status*. Cambridge, UK: BirdLife International (BirdLife Conservation Series no. 3). 600 pp.

UTMAR, P. (1993) La nidificazione del Falco di palude (*Circus aeruginosus*) nel Friuli-Venezia-Giulia. *Fauna, bollettino degli osservatori faunistici* 3: 77–90.

UTMAR, P. AND PARODI, R. (1990) Primi dati sull'avifauna dell'alta val Torre (Italia nord-orientale-prealpi Giulie). *Gortania–Atti delmus. fr. st. nat. Udine 11 (1989)*: 297–240.

VALLE, R., SCARTON, F., BORELLA, S. AND TILOCA, G. (1994) Nidificazione di avocetta nella laguna di Venezia. *Lav. Soc. Ven. Sc. Nat.* 19: 99–102.

VALLE, R., PIACENTINI, D., SCARTON, F., SERRA, L., GRUSSU, M., CORSO, A. AND UTMAR, P. (1995) Status e distribuzione della Volpoca *Tadorna tadorna* nidificante in Italia. *Avocetta* 19: 163.

VALLE, R., SCARTON, F., GRUSSU, M., TINARELLI, R., UTMAR, P. AND BORELLA, S. (1995) Primo censimento di Pettegola (*Tringa totanus*) nidificante in Italia. In M. Pandolfi and U. F. Foschi, eds. *Atti del VII Convegno di Ornitologia, Università di Urbino, 1993* (*Suppl. Ric. Biol. Selvaggina* 22).

VELATTA, F. (1990) The riparian bird community of lake Trasimeno. *Riv. Idrobiol.* 29(2): 617–631.

VELATTA, F. (1995) Wintering waterfowl at lake Trasimeno. *Avocetta* 19: 29.

VIOLANTE, F., SCARATTI, A. AND TEDESCO, N. (1993) *Guida Naturalistica delle Gravine*. Brindisi, Italy: Schena Editore—Fasano Brindisi.

FORMER YUGOSLAV REPUBLIC OF MACEDONIA

Lammergeier *Gypaetus barbatus*. (PHOTO: NIGEL J. DENNIS/NHPA)

OVERVIEW

The Former Yugoslav Republic of Macedonia occupies 25,715 km^2 with a population of 2,066,000 people. New data on Important Bird Areas (IBAs) are not available and therefore the 10 IBAs included here (Table 1, Map 1) are identical to those in the previous pan-European inventory (Grimmett and Jones 1989). The IBAs occupy 1,907 km^2, or 7% of the country's surface area.

Four globally threatened or near-threatened species (Collar *et al.* 1994) occur. *Aquila heliaca* (003 possible breeder, 004) and *Falco naumanni* (007 and 008) meet the A1 criterion as breeding species. *Phalacrocorax pygmeus* (005, 006, 010) and *Pelecanus crispus* (006) occur as non-breeding birds or in winter.

The majority of IBAs are mountainous areas or river gorges. They hold important populations of species with an unfavourable conservation status in Europe (meeting the B2/B3 criteria), particularly breeding raptors, including *Gyps fulvus, Neophron percnopterus* and *Aquila chrysaetos*. In 1989 hunting, poisoning and trapping were recorded as common threats across the mountain IBAs. Three IBAs (005, 006, 010) are lakes of particular importance for wintering waterbirds, with one (site 005) holding more than 20,000 waterbirds in winter.

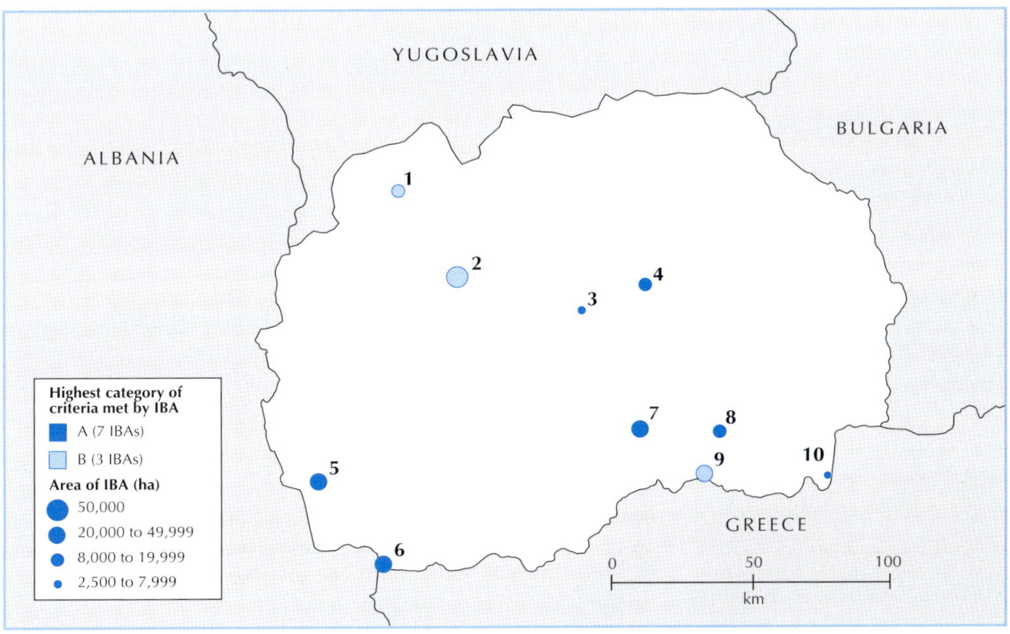

Map 1. Location, area and criteria category of Important Bird Areas in the Former Yugoslav Republic of Macedonia.

Table 1. Summary of Important Bird Areas in the Former Yugoslav Republic of Macedonia. 10 IBAs covering 1,907 km²

IBA code	1989 code	International name	National name	Area (ha)	Criteria (see p. 11)
001	YU048	Sara mountains [parts of]	Sar planina [parts of]	12,000	B2
002	YU049	Korab mountain and Radika gorge	Planina Korab i klisura Radike	50,000	B2
003	YU050	Babuna gorge, Topolka gorge, and Crn Kamen	Klisura reke Babune i Topolke i Crn Kamen	2,500	A1
004	YU051	River Bregalnica	Reka Bregalnica	10,000	A1, B2
005	YU055	Lake Ohrid	Ohridsko jezero	25,100	A1, A4i, A4iii, B1i
006	YU056	Lake Prespa	Prespansko jezero	40,000	A1
007	YU052	River Crna gorge	Klisura Crna reka	40,000	A1, B2
008	YU053	Demir kapija gorge	Demir kapija	8,000	A1, B2
009	YU054	Kozuf i Bosava mountains (Kozuf mountain and Bosava region)	Kozuf i Bosava mountains (Kozuf mountain and Bosava region)	20,000	B2
010	YU057	Lake Dojran	Dojransko jezero	4,200	A1

SITE ACCOUNTS

Sara mountains [parts of] — B2 001

Admin region Macedonia Kosovo
Coordinates 42°00′N 21°00′E
Altitude — **Area** 12,000 ha

Site description

High mountainous region (includes the peaks of Titov Vrh, Popova Šapka, and Kobilica), with river gorges, cliffs and rocky hillsides, alpine pastures, and woodland (coniferous and deciduous). Land-uses include cattle-grazing and forestry. The area extends into Kosovo (the given area of 12,000 ha includes the part in Kosovo).

Habitats Forest and woodland (broadleaved deciduous forest; native coniferous forest), Grassland (alpine/subalpine/boreal grassland), Wetland (river/stream), Rocky areas (scree/boulders; inland cliff)
Land-use Agriculture, Forestry

Birds

Species	Season	Year	Pop min	Pop max	Acc	Criteria
Aquila chrysaetos Golden Eagle	B	1989	3	4	—	B2
Bubo bubo Eagle Owl	B	1989	3	5	—	B2

Other breeding species include *Pernis apivorus* (1–2 pairs), *Falco peregrinus* (1 pair), *Bonasa bonasia*, *Alectoris graeca*, *Dryocopus martius*, *Saxicola rubetra*, *Lanius collurio* and *Pyrrhocorax pyrrhocorax* (50–100 pairs), and possibly include *Tetrao urogallus*, *Strix uralensis*, *Aegolius funereus*, *Dendrocopos leucotos*, *Picoides tridactylus* and *Pyrrhocorax graculus*. *Gypaetus barbatus* (1) and *Gyps fulvus* (15–20) also occur.

Protection status

National None **International** None

Conservation issues

Threats Infrastructure (U), Intensified forest management (U), Unsustainable exploitation (U)

There is uncontrolled hunting in the Macedonian part, and forestry practices and road building are adversely affecting the area. The area is a proposed Regional Park.

Korab mountain and Radika gorge — B2 002

Admin region —
Coordinates 41°45′N 21°15′E
Altitude — **Area** 50,000 ha

Site description

Cliffs, rocky gorges, mountains, woodland (deciduous and coniferous), and alpine pastures. Land-uses include cattle-grazing and forestry.

Habitats Forest and woodland (broadleaved deciduous forest; native coniferous forest), Grassland (alpine/subalpine/boreal grassland), Rocky areas (scree/boulders; inland cliff)
Land-use Agriculture, Forestry

Birds

Species	Season	Year	Pop min	Pop max	Acc	Criteria
Gyps fulvus Griffon Vulture	B	1989	—	10	—	B2
Aquila chrysaetos Golden Eagle	B	1989	6	7	—	B2
Bubo bubo Eagle Owl	B	1989	—	10	—	B2

Other breeding species include *Circaetus gallicus* (1 pair), *Aquila clanga*/*A. pomarina* (1 pair), *Hieraaetus pennatus* (possibly 1–2 pairs), *Falco biarmicus* (1 pair), *F. peregrinus* (2–3 pairs), *Bonasa bonasia*, *Caprimulgus europaeus*, *Alcedo atthis*, *Dryocopus martius*, and possibly *Dendrocopos leucotos*, *Dendrocopos medius* and *Picoides tridactylus*.

Protection status

National Partial **International** None
IBA covered by National Park (Mavrovo).

Conservation issues

Threats Unsustainable exploitation (U)

Some illegal hunting and trapping of wildlife takes place.

Babuna gorge, Topolka gorge, and Crn Kamen — A1 003

Admin region —
Coordinates 41°40′N 21°45′E
Altitude — **Area** 2,500 ha

Site description

The gorges of the Topolka, Vardar and Babuna rivers with cliffs, riverine vegetation, and scrub-covered hillsides. Land-uses include hunting and cattle-grazing.

Habitats Scrub (scrub), Wetland (river/stream), Rocky areas (inland cliff)
Land-use Agriculture, Hunting

Birds

Species	Season	Year	Pop min	Pop max	Acc	Criteria
[1] ***Aquila heliaca*** Imperial Eagle	B	1989	—	—	—	A1

1. Possibly breeds.

Other breeding species include *Ciconia nigra* (1 pair), *Neophron percnopterus* (2–3 pairs), *Circaetus gallicus* (1 pair), *Buteo rufinus* (2 pairs), *Aquila chrysaetos* (1 pair), *Falco naumanni* (a few pairs), *F. biarmicus* (1 pair), *F. peregrinus* (1 pair), *Alectoris graeca*, *Bubo bubo* (1–2 pairs), *Caprimulgus europaeus*, *Alcedo atthis* and *Coracias garrulus*. *Gypaetus barbatus* and *Gyps fulvus* also occur.

Protection status

National None **International** None

Conservation issues

Threats Other (U), Unsustainable exploitation (U)

There is possibly illegal poisoning of wildlife ('Other' threat) and the eggs and young of birds of prey are regularly stolen by foreigners. The area is a proposed Ornithological Reserve.

River Bregalnica — A1, B2 — 004

Admin region —
Coordinates 41°45'N 22°00'E
Altitude — **Area** 10,000 ha

Site description

A river valley with damp woodland, steppe habitat, bare eroded land, bushes and trees. There is a small area of cultivation and cattle-grazing in winter. The area is used for hunting.

Habitats Forest and woodland (alluvial/very wet forest), Scrub (scrub), Grassland (steppe/dry calcareous grassland)
Land-use Agriculture, Hunting

Birds

Species	Season	Year	Pop min	Pop max	Acc	Criteria
Aquila heliaca Imperial Eagle	B	1989	2	3	—	A1
Bubo bubo Eagle Owl	B	1989	2	3	—	B2

Other breeding species include *Milvus migrans* (1–2 pairs), *Neophron percnopterus* (2–3 pairs), *Circaetus gallicus* (1–2 pairs), *Circus pygargus*, *Aquila heliaca* (2–3 pairs), *A. chrysaetos* (1 pair), *Falco peregrinus* (1 pair), *Caprimulgus europaeus*, *Alcedo atthis* and *Coracias garrulus*; *Buteo rufinus*, *Falco cherrug* and *Tetrax tetrax* possibly breed. *Gyps fulvus* (18 birds) also occurs.

Protection status

National None **International** None

Conservation issues

Threats Other (U)

There is possibly illegal poisoning of wildlife ('Other' threat).

Lake Ohrid — A1, A4i, A4iii, B1i — 005

Admin region —
Coordinates 41°07'N 20°45'E
Altitude 750–750 m **Area** 25,100 ha

Site description

A large inland freshwater lake in a mountain basin, part of which is in Albania (see Albanian site 002).

Habitats Wetland (standing fresh water)
Land-use Unknown

Birds

Species	Season	Year	Pop min	Pop max	Acc	Criteria
Podiceps nigricollis Black-necked Grebe	W	1988	—	2,635	—	A4i, B1i
Phalacrocorax pygmeus Pygmy Cormorant	W	1988	—	600	—	A1
Netta rufina Red-crested Pochard	W	1988	—	1,150	—	A4i, B1i

No data are available for breeding season. The International Waterfowl Census counts (in the Macedonian sector) in winter 1987 and 1988 recorded (1988 data unless stated) *Tachybaptus ruficollis* (1,090), *Phalacrocorax carbo sinensis* (70), *P. aristotelis* (10), *Cygnus olor* (100 in 1987), *Aythya ferina* (7,050), *A. fuligula* (2,200) and *Fulica atra* (32,100).

Protection status

National Partial **International** High
IBA covered by Ornithological Reserve. 25,100 ha of IBA covered by World Heritage Site (Ohrid Region; 25,100 ha).

Conservation issues

Threats Unknown

With an age of 2–4 million years, this is one of the most ancient lakes in the world, harbouring many endemic species among its aquatic fauna.

Lake Prespa — A1 — 006

Admin region —
Coordinates 40°49'N 21°00'E
Altitude — **Area** 40,000 ha

Site description

A very large inland freshwater lake, surrounded by mountains and extending into Albania (Albanian site 003) and Greece (Limni Megali Prespa, Greek site 047). The area given is the entire area of the lake.

Habitats Wetland (standing fresh water)
Land-use Unknown

Birds

Species	Season	Year	Pop min	Pop max	Acc	Criteria
Phalacrocorax pygmeus Pygmy Cormorant	W	1988	—	80	—	A1
Phalacrocorax pygmeus Pygmy Cormorant	N	1989	—	200	—	A1
Pelecanus crispus Dalmatian Pelican	N	1989	—	58	—	A1

No data are available for breeding birds. The International Waterfowl Census counts (in the Macedonian sector) in winters 1987 and 1988 recorded (1988 data unless stated) *Phalacrocorax carbo sinensis* (40), *Netta rufina* (80; 150 in 1987), *Aythya fuligula* (8,000) and *Fulica atra* (5,950). Additional reports of non-breeding birds include *Podiceps cristatus* (1,300), *Pelecanus crispus* (58), *Anser anser* (200) and *Mergus merganser* (90).

Protection status

National None **International** None

Conservation issues

Threats Unknown

River Crna gorge — A1, B2 — 007

Admin region —
Coordinates 41°19'N 22°00'E
Altitude — **Area** 40,000 ha

Site description

A gorge with cliffs, scrub and woodland (including *Fagus* and *Quercus*), bare and eroded land, an artificial lake without vegetation, and grazing pasture. Land-uses include hunting, fishing, and cattle-grazing.

Habitats Forest and woodland (broadleaved deciduous forest), Scrub (scrub), Grassland, Wetland (standing fresh water; river/stream), Rocky areas (inland cliff)
Land-use Agriculture, Fisheries/aquaculture, Hunting

Birds

Species	Season	Year	Pop min	Pop max	Acc	Criteria
Neophron percnopterus Egyptian Vulture	B	1989	8	15	—	B2
Gyps fulvus Griffon Vulture	B	1989	9	12	—	B2
Circaetus gallicus Short-toed Eagle	B	1989	6	7	—	B2
Aquila chrysaetos Golden Eagle	B	1989	4	7	—	B2
Falco naumanni Lesser Kestrel	B	1989	15	20	—	A1
Bubo bubo Eagle Owl	B	1989	3	5	—	B2

Breeding species include *Ciconia nigra* (2–3 pairs), *Buteo rufinus* (1 pair), *Aquila chrysaetos* (4–7 pairs), *Hieraaetus pennatus* (1 pair), *H. fasciatus* (1 pair), *Falco biarmicus* (1–2 pairs), *F. peregrinus* (2–3 pairs), *Bonasa bonasia*, *Alectoris graeca* and *Coracias garrulus*. *Gypaetus barbatus* (1), *Aegypius monachus* (2) and *Aquila heliaca* (2–3) also occur.

Protection status

National Low **International** None
IBA partly covered by small Hunting Reserve.

Conservation issues

Threats Industrialization/urbanization (U), Other (U), Unsustainable exploitation (U)

Problems at the site include illegal trapping and probably poisoning of wildlife ('Other' threat), and house-building.

Demir kapija gorge

A1, B2 **008**

Admin region —
Coordinates 41°19′N 22°19′E
Altitude — **Area** 8,000 ha

Site description

The site includes the Vardar gorge, Celevecka river and Krastavec ridge with cliffs and rocky hillsides, scrub, and woodland; there are also small areas of marsh, damp woodland, and some steppe habitat in the Vardar valley. Mainly inaccessible, some of the area is used as agricultural land, while hunting and cattle-grazing take place.

Habitats Forest and woodland, Scrub (scrub), Grassland (steppe/dry calcareous grassland), Rocky areas
Land-use Agriculture, Hunting

Birds

Species	Season	Year	Pop min	Pop max	Acc	Criteria
Neophron percnopterus Egyptian Vulture	B	1989	3	4	—	B2
Gyps fulvus Griffon Vulture	B	1989	—	18	—	B2
Aquila heliaca Imperial Eagle	B	1989	1	2	—	A1
Falco naumanni Lesser Kestrel	B	1989	5	10	—	A1

Breeding species include *Ciconia nigra* (1–2 pairs), *Pernis apivorus* (1–2 pairs), *Milvus migrans* (2–3 pairs), *Circaetus gallicus* (2 pairs), *Accipiter brevipes* (1 pair), *Buteo rufinus* (1 pair), *Aquila chrysaetos* (2 pairs), *Hieraaetus pennatus* (1 pair), *Falco biarmicus* (1–2 pairs), *F. peregrinus* (1 pair), *Alectoris graeca*, *Bubo bubo* (1–2 pairs), *Coracias garrulus* (3–4 pairs) and *Cercotrichas galactotes*. *Gypaetus barbatus*, *Aegypius monachus* and *Falco cherrug* also occur.

Protection status

National None **International** None

Conservation issues

Threats Infrastructure (U), Other (U), Selective logging/cutting (U), Unsustainable exploitation (U)

Hunting, theft of raptor eggs and young by foreigners, road and quarry building, and cutting of trees and bushes are problems at the site. Illegal poisoning and trapping of wildlife probably also occurs. The area is a proposed Ornithological Reserve.

Kozuf i Bosava mountains (Kozuf mountain and Bosava region)

B2 **009**

Admin region —
Coordinates 41°10′N 22°15′E
Altitude — **Area** 20,000 ha

Site description

A mountainous area with alpine pastures, cliffs, rocky hillsides, bare and eroded land, coniferous woodland (*Pinus nigra*, *P. sylvestris*, *Abies alba*), and *Fagus* woodland. Land-uses include forestry and cattle-grazing.

Habitats Forest and woodland (broadleaved deciduous forest; native coniferous forest), Grassland (alpine/subalpine/boreal grassland), Rocky areas (inland cliff)
Land-use Agriculture, Forestry

Birds

Species	Season	Year	Pop min	Pop max	Acc	Criteria
Neophron percnopterus Egyptian Vulture	B	1989	—	4	—	B2
Bubo bubo Eagle Owl	B	1989	3	5	—	B2

Breeding species include *Ciconia nigra* (2–3 pairs), *Pernis apivorus* (1–3 pairs), *Gypaetus barbatus* (1 pair), *Aquila pomarina* (1 pair), *A. chrysaetos* (3 pairs), *Hieraaetus fasciatus* (1 pair), *Falco peregrinus* (1 pair), *Bonasa bonasia*, *Alectoris graeca*, and possibly *Aegolius funereus*, *Coracias garrulus*, *Dendrocopos leucotos* and *Picoides tridactylus*. Small flocks of *Falco eleonorae* also occur during the summer.

Protection status

National None **International** None

Conservation issues

Threats Intensified forest management (U), Other (U), Unsustainable exploitation (U)

Problems at the site include hunting, illegal trapping and poisoning, and forestry.

Lake Dojran

A1 **010**

Admin region —
Coordinates 41°12′N 22°45′E
Altitude — **Area** 4,200 ha

Site description

Shallow eutrophic lake, with extensive reedbeds extending into Greece (site 023) There is an intensive traditional fishing industry.

Habitats Wetland (standing fresh water; water-fringe vegetation)
Land-use Fisheries/aquaculture

Birds

Species	Season	Year	Pop min	Pop max	Acc	Criteria
Phalacrocorax pygmeus Pygmy Cormorant	W	1988	—	100	—	A1

Formerly important for breeding waterfowl including *Pelecanus onocrotalus*, *Egretta alba*, *Ardea purpurea*, *Cygnus olor* and *Oxyura leucocephala*; however, no data on breeding species are available for recent years. The International Waterfowl Census counts in winter 1987 and 1988 recorded (1988 data unless stated) *Gavia arctica* (10), *Podiceps nigricollis* (40), *Phalacrocorax carbo sinensis* (10), *Aythya ferina* (1,150), and *Fulica atra* (6,350).

Protection status

National Partial **International** None
IBA protected by Ornithological Reserve.

Conservation issues

Threats Unknown

REFERENCES

COLLAR, N. J., CROSBY, M. J. AND STATTERSFIELD, A. J. (1994) *Birds to watch 2: the world list of threatened birds*. Cambridge, UK: BirdLife International (Conservation Series 4).

GRIMMETT, R. F. A. AND JONES T. A. (1989) *Important Bird Areas in Europe*. Cambridge, UK: International Council for Bird Preservation (Tech. Publ. 9).

MALTA

IAN MIFSUD

Ta'Cenc cliffs (IBA 001). (PHOTO: JOHN BORG)

GENERAL INTRODUCTION

Five Important Bird Areas (IBAs) have been identified in the Maltese archipelago, which is situated in the centre of the Mediterranean Sea at approximately 36°N 14°E. The principal islands are Malta, Gozo and Comino. The IBAs cover 302 ha (Table 1), which is 0.9% of the total terrestrial area of the islands (320 km²). The islands support a human population of 360,000 inhabitants, making Malta the most densely populated country in Europe and in the Mediterranean basin.

Human activity has had a widespread impact on the countryside. The rural landscape is dominated by terraced agricultural land along hill-sides giving way to lush valley bottoms. Many valley floors have been dammed to retain water for irrigation. Habitats are mostly semi-natural, dominated by garrigue. A few pockets of maquis and remnants of native woodland, characterized by evergreen oak *Quercus* and sandarac tree *Tetraclinis*, still remain.

There is at least one IBA on each of the three main islands. Four of the IBAs are coastal, with one woodland site lying inland. The previous pan-European IBA inventory (Grimmett and Jones 1989) identified five sites; four are the same as in the current inventory and one previous site was not confirmed as internationally

Map 1. Location, area and criteria category of Important Bird Areas in Malta.

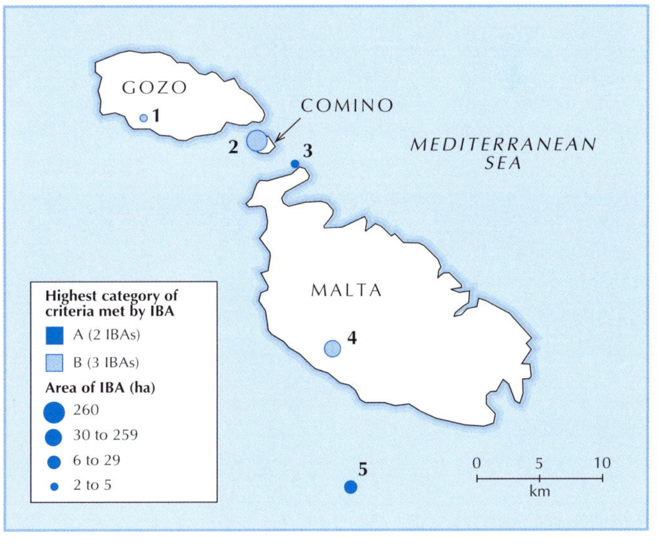

Table 1. Summary of Important Bird Areas in Malta. — 5 IBAs covering 3 km²

IBA code	1989 code	International name	National name	Administrative region	Area (ha)	Criteria (see p. 11)
001	MT001	Ta'Cenc cliffs	Rdum Ta'Cenc	Gozo	2	B1ii, B2
002	—	Comino island	Gzejjer ta'Kemmuna	Malta	260	B3
003	MT005	Rdum Tal-Madonna	Rdum Tal-Madonna, L-Ahrax	Malta	4	A4ii, B1ii, B2, B3
004	MT003	Buskett and Wied il-Luq	Buskett and Wied il-Luq	Malta	30	B1iv
005	MT004	Filfla islet	Filfla islet	Malta	6	A4ii, B1ii, B2

Site identified in the previous inventory of IBAs in Europe (Grimmett and Jones 1989) but no longer considered to be an IBA
MT002 Ghadira Pool.

Table 2. Important Bird Areas in Malta that support important numbers of one or more congregatory species (i.e. meeting criteria A4 and/or B1). IBAs meeting both criteria A4 and B1 for the species are shown in **bold**. IBAs meeting only criterion B1 for the species concerned, and not A4, are shown in normal type. For key to 'Season', see p. 7.

Species	Season	IBA code
Calonectris diomedea Cory's Shearwater	B	001
Puffinus yelkouan Yelkouan Shearwater	B	**003**
Hydrobates pelagicus Storm Petrel	B	**005**

Table 3. Species of European conservation concern with significant breeding populations at Important Bird Areas in Malta (meeting any IBA criteria).

Species [1]	Minimum national breeding population (pairs) [2]	Proportion (%) of national population breeding at all IBAs in Malta
Calonectris diomedea Cory's Shearwater	8,000	11
Puffinus yelkouan Yelkouan Shearwater	500	100[3]
Hydrobates pelagicus Storm Petrel	3,000	100[3]
Monticola solitarius Blue Rock Thrush	200	7

1. Only those species of European conservation concern (see Box 1, p. 12) that meet IBA criteria in Malta are listed.
2. Data are taken from the BirdLife/EBCC European Bird Database 1998 (Heath and Borggreve 2000).
3. The percentage of the national population in IBAs exceeds 100%. Usually this is because the national population estimate has not been updated recently whilst the IBA population estimate has been recently updated with new data as a result of comprehensive surveys of IBAs themselves. Also, the individual site count for a species may be the maximum or average over recent years, and summing these may record more birds than are present nationally in any single year.

important during this review (see Table 1). One new site has been identified (Comino island, 002).

ORNITHOLOGICAL IMPORTANCE

A total of 15 species of European conservation concern (SPECs) breed regularly in the Maltese archipelago (Tucker and Heath 1994). None are globally threatened or endemic, but eight have an unfavourable conservation status in Europe (Tucker and Heath 1994). Most of the IBAs are important because of their numbers of breeding seabirds, in particular *Calonectris diomedea*, *Puffinus yelkouan* and *Hydrobates pelagicus*. IBAs hold the entire national breeding population of the latter two species (Tables 2 and 3). The coniferous woodland at Buskett (004) is an important roosting site for migrating raptors, especially *Pernis apivorus*, *Falco subbuteo*, *Falco eleonorae* and *Circus aeruginosus*. The national bird, *Monticola solitarius*, is particularly common at two IBAs (Ta'Cenc cliffs, 001; Rdum Tal-Madonna, 003).

HABITATS

Most of the sites are coastal (Map 1, Figure 1), located along sea cliffs on the main islands (Ta'Cenc cliffs, 001; Rdum tal-Madonna, 003; Comino island, 002) or on an offshore islet (Filfla islet, 005). Buskett and Wied il-Luq (004) is the only site that is predominantly non-marine, with its woodland, maquis and cultivation. There are no wetland IBAs, but two marshy areas, at Ghadira and Is-Simar, are of great national importance for resting waders and passerines during migration, and for breeding *Ixobrychus minutus* and *Charadrius dubius*.

IMPACTS ON IBAs – LAND-USE AND THREATS

The main land-uses at IBAs are agriculture and tourism/recreation. Human disturbance of wildlife is intense over most of the island. Illegal shooting is a serious problem at Buskett and Wied il-Luq (004) as well as in other protected areas and several other IBAs, and recreation/tourism also cause problems at most of the sites (Figure 2).

PROTECTION STATUS

Nature conservation is the responsibility of the governmental Environment Protection Department. The Environment Protection Act of 1991 is the main framework of law which deals with bird protection, including site protection. There are two types of national protected area: Bird Sanctuary and Nature Reserve. Table 4 shows the protection status of the IBAs. One of the IBAs is unprotected (Rdum Tal-Madonna, 003) while two are Bird Sanctuaries and three are Nature Reserves. Overall, 298 ha (99%) of the total IBA area is covered by these protected areas.

There are over 20 countryside areas where hunting and trapping of birds are currently forbidden. Ghadira and Is-Simar are managed as Nature Reserves and offer the only protection on the island for breeding and migrating waterbirds, as well as being used for environmental education.

Malta has ratified various international environmental agreements with relevance to site protection (Box 1). Filfla islet (005) has been designated as a Mediterranean Specially Protected

Figure 1. Habitats at Important Bird Areas in Malta (see Appendix 3 for definitions of habitats).

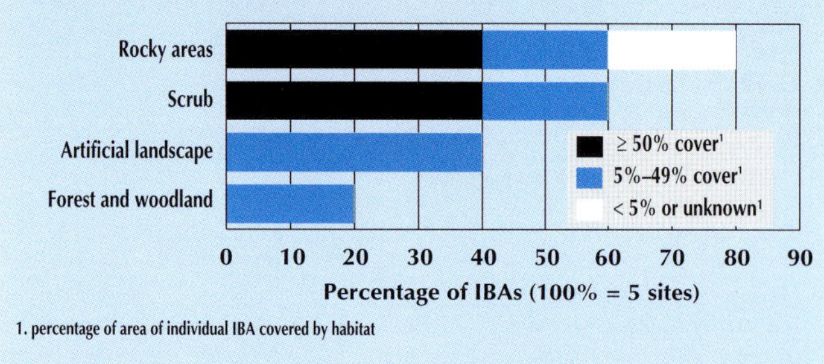

Table 4. Protection status of Important Bird Areas in Malta. A tick (✔) indicates that an IBA overlaps with a protected area (to any extent).

IBA code	International name	National: Bird Sanctuary	National: Nature Reserve	International: Mediterranean Specially Protected Area
001	Ta'Cenc cliffs		✔	
002	Comino island	✔		
003	Rdum Tal-Madonna			
004	Buskett and Wied il-Luq	✔	✔	
005	Filfla islet		✔	✔
Total number of IBAs		**2**	**3**	**1**

Figure 2. Threats at Important Bird Areas in Malta (see Appendix 3 for definitions of threat types and impact categories).

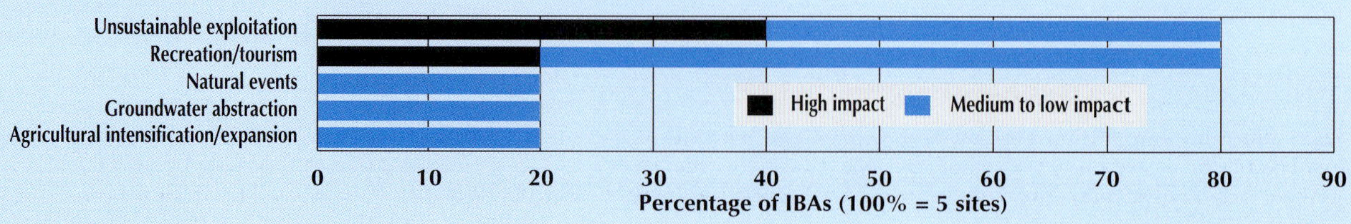

Box 1. International legislation and initiatives that are relevant to site conservation in Malta (see Appendix 1 for a general description of these agreements).

Global	
Biodiversity Convention	(✔)
Ramsar Convention	✔
Bonn Convention	
World Heritage Convention	✔
MAB Programme	✔
Pan-European	
Bern Convention	✔
Regional	
Barcelona Convention	✔

✔ Convention ratified/initiative supported
(✔) Convention signed

Area under the relevant Protocol of the Barcelona Convention, and is a strictly protected area, with boat-landing prohibited.

CONSERVATION

- None of the IBAs are managed in terms of nature conservation, although a project for habitat restoration in a small valley on Comino island (002) has been put forward by BirdLife Malta. Buskett and Wied il-Luq (004) fall under the jurisdiction of the Agriculture Department, which manages the area for agricultural and recreational purposes. Some consultation between the Agriculture Department and the Environment Protection Department takes place occasionally.
- Monitoring of seabird populations at coastal and islet IBAs is carried out within the framework of the Research Working Group of BirdLife Malta. Visits to Filfla islet (005) require prior authorization from the Environment Protection Department and such visits are routinely carried out in collaboration with the Nature Protection Division of the Environment Protection Department. Monitoring of raptor migration is also carried out by volunteers from BirdLife Malta, and is concentrated in September, coincident with the peak autumn migration. A systematic list of recent bird observations at IBAs and other sites is regularly published in the scientific journal of BirdLife Malta, *Il-Merrill*.
- Regulations published in 1993 under the Environment Protection Act (1991) have afforded increased protection to various bird species over the previous regulations which dated back to 1980. A partial close season in spring was also introduced. This regulation was superceded by a new law in 1997 which deregulated some areas where hunting and trapping were previously forbidden, and removed protection from some of the protected species.

ANALYTICAL METHODS

- The seabird population sizes quoted in the site-accounts are estimates carried out over the four-year period 1995–1998. Due to the facts that pelagic birds are nocturnal and that many of the breeding sites are on inaccessible sea cliffs, estimates are given as a range (minimum–maximum).
- Apart from Filfla (005), which is a small isolated islet, data on the habitats and land-uses at the IBAs are estimates based on actual visits to the sites.

ACKNOWLEDGEMENTS

The author would like to thank Mr Alex Casha for helping in collating the data, Mr John Borg for the updated information on seabird populations, and Mr Joe Sultana for his comments on the final draft of the overview.

SITE ACCOUNTS

Ta'Cenc cliffs — B1ii, B2 — 001

Admin region Gozo
Coordinates 36°02'N 14°14'E
Altitude 0–135 m **Area** 2 ha

Site description

Sheer sea cliffs (135 m high), on the island of Gozo, with profuse vegetation on some ledges.

Habitats Scrub (100%; sclerophyllous scrub/garrigue/maquis), Rocky areas (sea cliff/rocky shore)
Land-use Not utilized (100%)

Birds

Species	Season	Year	Pop min	Pop max	Acc	Criteria
Calonectris diomedea Cory's Shearwater	B	1994	800	1,000	A	B1ii, B2
Monticola solitarius Blue Rock Thrush	R	1989	10	10	A	B2

The cliffs support the largest breeding colony of *Calonectris diomedea* in the Maltese islands, and are a stronghold for *Monticola solitarius*. One of the last known breeding pairs of *Falco peregrinus* in Malta nested here, but was shot in the early 1980s, and a pair of *Tyto alba* used to breed here in the 1970s. The rocky cliff-top area is one of the best breeding areas in Malta for *Calandrella brachydactyla*, *Sylvia conspicillata* and *Miliaria calandra*.

Protection status

National High **International** None
2 ha of IBA covered by Nature Reserve (Ta'Cenc Cliffs, 2 ha).

Conservation issues

Threats Recreation/tourism (A), Unsustainable exploitation (A)

Illegal shooting and trapping of birds is a problem. The site is threatened by proposed tourist development. The cliffs are of interest for their undisturbed garrigue vegetation and associated animal communities.

Comino island — B3 — 002

Admin region Malta
Coordinates 36°01'N 14°20'E
Altitude 0–75 m **Area** 260 ha

Site description

A small island lying mid-way between mainland Malta and Gozo. The karst landscape supports sclerophyllous scrub. An afforestation programme has been carried out on a limited scale with *Pinus*. There is a sand-dune system at Santa Maria bay with typical native vegetation, including trees of *Tamarix* and *Vitex*.

Habitats Scrub (65%; sclerophyllous scrub/garrigue/maquis), Rocky areas (10%; sea cliff/rocky shore), Artificial landscape (25%; arable land)
Land-use Agriculture (5%), Not utilized (70%), Tourism/recreation (5%)

Birds

Species	Season	Year	Pop min	Pop max	Acc	Criteria
Puffinus yelkouan Yelkouan Shearwater	B	—	50	80	—	B3

An important site for breeding *Puffinus yelkouan*. Many trans-Saharan passerine migrants seek refuge on the island, especially during bad weather conditions in spring, and birds of prey occur regularly on migration, including *Circus aeruginosus*, *Milvus migrans* and *Falco vespertinus*. *Falco peregrinus* used to breed, and a pair of *Asio flammeus* nested in 1983.

Protection status

National High **International** None
260 ha of IBA covered by Bird Sanctuary (Comino Island, 260 ha).

Conservation issues

Threats Agricultural intensification/expansion (C), Recreation/tourism (C), Unsustainable exploitation (C)

There is a management plan for the site. BirdLife Malta participates in an ongoing research programme coordinated by the Wildlife Institute of Bologna, the main aim of which is to study the spring migration of passerines across the western Mediterranean. BirdLife Malta also manages part of a valley where it is intended to set up a bird observatory.

Rdum Tal-Madonna

A4ii, B1ii, B2, B3 **003**

Admin region Malta
Coordinates 36°00'N 14°22'E
Altitude 0–25 m **Area** 4 ha

Site description

A promontory in the L-Ahrax area, where there is a 200 m stretch of sea cliffs (25 m high) at the base of which there is a lot of rock debris.

Habitats Rocky areas (100%; sea cliff/rocky shore)
Land-use Not utilized (100%)

Birds

Species	Season	Year	Pop min	Pop max	Acc	Criteria
Puffinus yelkouan Yelkouan Shearwater	B	1994	500	500	—	A4ii, B1ii, B3
Monticola solitarius Blue Rock Thrush	R	—	4	4	—	B2

The cliffs support one of the largest colonies of *Puffinus yelkouan* in the species's range. Notable numbers of *Calandrella brachydactyla* and *Sylvia conspicillata* (locally decreasing) breed along the cliff-tops.

Protection status

National None **International** None

Conservation issues

Threats Recreation/tourism (C), Unsustainable exploitation (B)

Shooting and trapping of birds takes place along the cliff-tops. The colony of *Puffinus yelkouan* is monitored through bird-ringing activities.

Buskett and Wied il-Luq

B1iv **004**

Admin region Malta
Coordinates 35°52'N 14°24'E
Altitude 130–200 m **Area** 30 ha

Site description

Buskett is the most extensive area of coniferous woodland on the island, through which runs a valley of deciduous woodland (Wied il-Luq).

Habitats Forest and woodland (30%; broadleaved deciduous forest; native coniferous forest), Scrub (30%; sclerophyllous scrub/garrigue/maquis), Artificial landscape (40%; perennial crops/orchards/groves)
Land-use Agriculture (40%), Tourism/recreation (40%)

Birds

This area is one of the major known passage sites for birds of prey in the Maltese archipelago. Over 3,000 raptors pass regularly per season, with the site being especially important for roosting *Pernis apivorus* and *Circus aeruginosus* in autumn, and *Falco tinnunculus*, *F. subbuteo* and *F. eleonorae* also occur in notable numbers. The extensive undergrowth provides feeding sites for wintering passerines as well as many trans-Saharan autumn migrants.

Protection status

National High **International** None
30 ha of IBA covered by Nature Reserve (Buskett Gardens, 30 ha). 30 ha of IBA covered by Bird Sanctuary (Buskett Gardens, 30 ha).

Conservation issues

Threats Groundwater abstraction (C), Recreation/tourism (C), Unsustainable exploitation (A)

Illegal bird-shooting is a serious problem. There is a management plan for the site.

Filfla islet

A4ii, B1ii, B2 **005**

Admin region Malta
Coordinates 35°46'N 14°25'E
Altitude 0–60 m **Area** 6 ha

Site description

An offshore islet of calcareous rock, mainly upper coralline limestone, which is bounded by cliffs and screes. The top of the islet is a plateau which is not easily accessible from below. The screes of loose boulders and rock debris surrounding the base of the cliff result mainly from past bombing practice by the military (up to 1971). Shore erosion due to wind and wave action is a predominant feature.

Habitats Rocky areas (100%; rock stacks/islets)
Land-use Not utilized (100%)

Birds

Species	Season	Year	Pop min	Pop max	Acc	Criteria
Calonectris diomedea Cory's Shearwater	B	1994	100	200	A	B2
Hydrobates pelagicus Storm Petrel	B	1994	5,000	8,000	A	A4ii, B1ii, B2

The screes have proved to be ideal breeding sites for *Hydrobates pelagicus*, and the islet supports one of the largest colonies in the Mediterranean. Other breeding seabirds include *Calonectris diomedea* and *Larus cachinnans*; the latter is the main predator on *Hydrobates pelagicus* and several tens of pairs breed below the cliffs.

Protection status

National High **International** High
6 ha of IBA covered by Nature Reserve (Filfla Islet, 6 ha). 2 ha of IBA covered by Mediterranean Specially Protected Area (Filfla Islet, 2 ha).

Conservation issues

Threats Natural events (B)

Numbers of *Hydrobates pelagicus* are probably decreasing after storms washed away some of the scree. Scientists visit the islet two or three times a year during the breeding season to ring and census birds. The Nature Reserve was established in 1988, but no management plan exists. Permission has to be obtained in order to land on the island and to carry out any activities; permits are only granted for scientific and educational purposes. Legal Notice to Mariners No. 16 of 1987 prohibits the berthing or navigation of any craft within an area of one nautical mile radius off Filfla as well as swimming, underwater activities and any activity connected with fishing and trawling. This notice was superceded by Government Notice No. 173 of 1990 which permits fishing from surface vessels within one nautical mile off Filfla.

REFERENCES

Anon. (1990) *Centro – an international journal of environment studies in the Mediterranean* 1(5). (University of Malta.)

Borg, J. (1989) The storm petrel *Hydrobates pelagicus* in the Maltese Islands: a review. *Naturalista sicil.* S. IV, XIII (1-2): 45–52.

Borg, J., Cachia Zammit, R. and Sultana, J. (in press) The status and distribution of Procellariiformes in Malta.

Grimmett, R. F. A. and Jones, T. A. (1989) *Important Bird Areas in Europe*. Cambridge, UK: International Council for Bird Protection (Tech. Publ. 9).

Heath, M. F. and Borggreve, C. (2000) *BirdLife International/EBCC European Bird Database 1998*. Cambridge, UK: BirdLife International.

Schembri, P. J., Lanfranco, E., Farrugia, P., Schembri, S. and Sultana, J. (1987) *Localities with conservation value in the Maltese Islands*. Floriana, Malta: Ministry of Education.

Schembri, P. J. and Sultana, J. (1989) *Red Data Book for the Maltese Islands*. Valetta: Department of Information.

Sultana, J. (1993) *Important seabird sites in the Mediterranean*. Valletta: Malta Ornithological Society.

Sultana, J. and Gauci, C. (1982) *A new guide to the birds of Malta*. Valletta: The Ornithological Society.

Tucker, G. M. and Heath, M. F. (1994) *Birds in Europe: their conservation status*. Cambridge, UK: BirdLife International (BirdLife Conservation Series no. 3).

MOLDOVA

A. MUNTEANU AND N. ZUBCOV

The Prut valley, near Brinza (IBA 002). (PHOTO: PAUL GORIUP)

OVERVIEW

Moldova is a small land-locked country of 33,700 km^2 and a population of 4,356,000 people. It borders Ukraine to the north, south and east and Romania to the west. Twelve Important Bird Areas (IBAs) are identified, covering 509 km^2. Eleven of these are new sites, with a single site having been included in the previous pan-European inventory (Grimmett and Jones 1989), in the then USSR (Table 1, Map 1). IBAs occupy 1.5% of the country's surface area.

No bird population minima and maxima data are available for sites in Moldova. Therefore, criteria have been applied at a site level, based on species lists for individual sites. As a result, no species tables accompany the site accounts. Too little data are available to include several proposed sites as IBAs in this inventory. Further research is necessary to establish whether they meet criteria. These include: Bazinul Congaz, Curchi–Seliste–Ivancia, Pădure Ilânceşti, Căprieni–Scoreni, Caracuşeni–Rosoşeni, Rudi–Arioneşti, Saharna, and Furceni–Trebujeni.

Ten globally threatened or near-threatened bird species (Collar *et al.* 1994) occur: *Phalacrocorax pygmeus* (breeding, non-breeding), *Pelecanus crispus* (non-breeding), *Anser erythropus* (passage), *Branta ruficollis* (passage), *Aythya nyroca* (breeding), *Aquila heliaca* (non-breeding), *Aquila clanga* (breeding, passage), *Haliaeetus albicilla* (passage), *Crex crex* (breeding, passage), and *Gallinago media* (passage). Of particular note are sites 010 and 011 which hold eight and nine such species respectively. At least 103 bird species of European conservation concern (SPECs) breed regularly in Moldova (Tucker and Heath 1994).

Of the 12 IBAs selected, nine include extensive wetland habitat and eight are thought to be important as stop-over sites for waterbirds (meeting the Bli criterion at the site level). Broadleaved forests are also widespread in IBAs, often in association with river flood-plains and meadows. Such habitat mosaics support breeding SPECs with an unfavourable conservation status in Europe, such as *Aquila pomarina, Hieraaetus pennatus* and *Ciconia nigra*, for which the B2 and B3 criteria have been applied at the site level.

Map 1. Location, area and criteria category of Important Bird Areas in Modova.

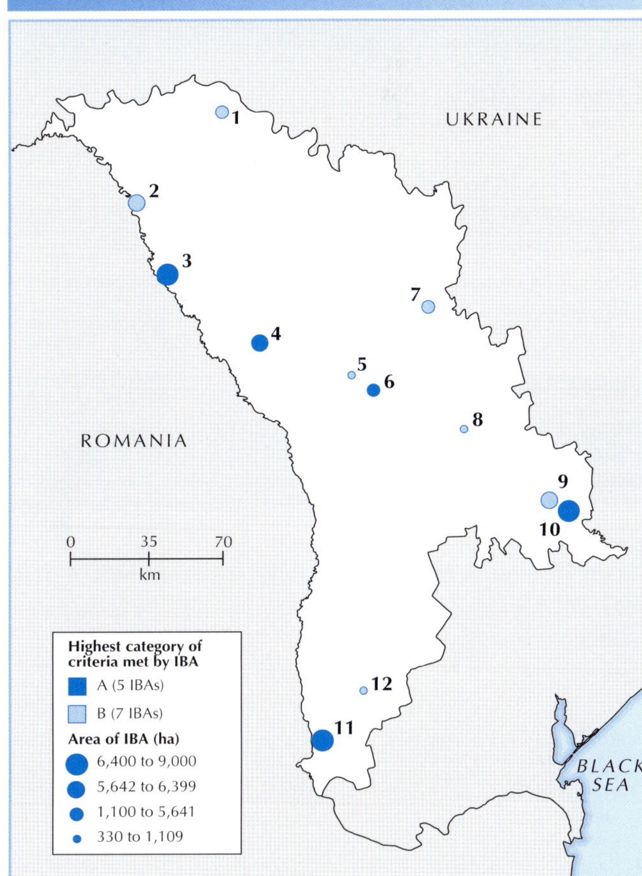

Table 1. Summary of Important Bird Areas in Moldova. 12 IBAs covering 509 km²

IBA code	1989 code	International name	National name	Administrative region	Area (ha)	Criteria (see p. 11)
001	—	Otaci-Goloşnita (bazinul Dubăsari)	Otaci-Goloşnita (bazinul Dubăsari)	Edineţi, Soroca	1,100	B2
002	—	Bazinul Costeşti-Stînca	Bazinul Costeşti-Stînca	Bălţi, Edineţi	5,900	B1i
003	—	Balatina	Balatina	Bălţi	9,000	A1, B2, B3
004	—	Plaiul Fagului	Plaiul Fagului	Ungheni	5,642	A1, B2, B3
005	—	Bazinul Ghidighici	Bazinul Ghidighici	Chişinău	900	B1i
006	SU031	Codrii	Codrii	Chişinău	5,177	A1, B2, B3
007	—	Golf Goeni	Golf Goeni	Dubăsari	1,500	B1i
008	—	Lacul Salaş	Lacul Salaş	Chişinău	330	B1i
009	—	Copanca-Talmaz	Copanca-Talmaz	Tighina	6,000	B1i
010	—	Bazinul Cuciurgan	Bazinul Cuciurgan	Dubăsari	6,400	A1, B1i, B2, B3
011	—	Manta flood-plain-Beleu	Manta-Beleu	Cahul	8,364	A1, B1i, B2, B3
012	—	Bazinul Taraclia	Bazinul Taraclia	Cahul	550	B1i, B2

SITE ACCOUNTS

Otaci–Goloşnita (bazinul Dubăsari) — B2 — 001

Admin region Edineţi, Soroca
Coordinates 48°20′N 27°46′E
Altitude — **Area** 1,100 ha

Site description
The site covers part of the Dubăsari reservoir. Fringing habitats include cliffs, coppices, meadows, patches of steppe and transitional areas. The site is of scientific interest and has social and economic importance.

Habitats Wetland (shingle/stony beach; standing fresh water)
Land-use Unknown

Birds
The B2 criterion has been assigned at the site level as SPECs are present, but in unknown numbers. Stretches of shingle and sand provide nesting habitats for sandpipers (Scolopacidae) and gulls (Laridae). Coppices are used by breeding *Grus grus*. Other breeding species include *Hieraaetus pennatus* and *Picus viridis*.

Protection status
National None **International** None

Conservation issues
Threats Unknown

The construction of Verhnednestvovstoe reservoir has led to changes in the hydrological regime within the IBA, with consequent implications for fauna.

Bazinul Costeşti–Stînca — B1i — 002

Admin region Bălţi, Edineţi
Coordinates 47°58′N 27°12′E
Altitude — **Area** 5,900 ha

Site description
The site comprises a reservoir in the Prut valley that reaches 20 m in depth. Its waters are very important for fishing and irrigation.

Habitats Wetland (standing fresh water)
Land-use Fisheries/aquaculture, Water management

Birds
The B1i criterion has been assigned at the site level as the IBA is known to be an important stop-over site for passage waterbirds, but numbers are unknown.

Protection status
National None **International** None

Conservation issues
Threats Unknown

Balatina — A1, B2, B3 — 003

Admin region Bălţi
Coordinates 47°40′N 27°23′E
Altitude — **Area** 9,000 ha

Site description
An extensive woodland (including *Quercus*, *Acer*, *Carpinus* and *Fraxinus*) with areas of marsh. Other vegetation includes *Corylus*, *Crataegus*, *Staphylea*, *Cornus* and *Vitis*. When the Prut river floods favourable conditions are created for waterbirds and waders.

Habitats Forest and woodland (82%; broadleaved deciduous forest), Scrub (2%), Wetland (30%; river/stream; water-fringe vegetation), Artificial landscape (32%; forestry plantation)
Land-use Nature conservation/research

Birds
A1/B2/B3 criteria have been assigned at the site level as many globally threatened species and SPECs are present, but in unknown numbers. Six globally threatened species occur: *Phalacrocorax pygmeus* (non-breeding), *Anser erythropus* (passage), *Aythya nyroca* (breeding), *Aquila clanga* (breeding), *Crex crex* (breeding) and *Gallinago media* (passage). Breeding visitors and residents also include *Egretta alba*, *Pernis apivorus*, *Aquila pomarina* and *Hieraaetus pennatus*. Passage visitors include *Ciconia nigra*, *Plegadis falcinellus*, *Cygnus cygnus*, *Haliaeetus albicilla*, *Aquila chrysaetos*, *Pandion haliaetus* and *Falco cherrug*.

Protection status
National Partial **International** None
6,032 ha of IBA covered by Nature Reserve (Padurea Domneasca).

Conservation issues
Threats Afforestation (B), Agricultural intensification/expansion (B), Deforestation (commercial) (C), Disturbance to birds (B), Drainage (C), Dredging/canalization (C), Filling-in of wetlands (C), Forest grazing (C), Selective logging/cutting (C), Unsustainable exploitation (C)

The area is well preserved.

Plaiul Fagului — A1, B2, B3 — 004

Admin region Ungheni
Coordinates 47°22′N 27°57′E
Altitude — **Area** 5,642 ha

Site description
An extensive area of broadleaved forest coinciding with the Plaiul Fagului Reserve in the northwestern part of Codri forest. According to

the 1985 forest inventory, the area of natural forest in the reserve is 4,639 ha; comprising *Quercus* (36% of IBA area), *Fraxinus* (21%), *Carpinus* (19%) and *Fagus* forest (5%). Forestry plantations, covering 12% of the IBA area, are predominantly of *Quercus*. The Byc river and numerous brooks have their sources in the IBA, the landscape of which is prone to soil erosion and landslides.

Habitats Forest and woodland (83%; broadleaved deciduous forest), Grassland (8%; mesophile grassland), Wetland (2%; standing fresh water; river/stream), Artificial landscape (12%; arable land; forestry plantation)
Land-use Forestry (95%), Hunting, Nature conservation/research (80%)

■ Birds

A1/B2/B3 criteria have been assigned at the site level as many globally threatened species and SPECs are present, but in unknown numbers. Four globally threatened species occur: *Aquila clanga* (passage), *Aquila heliaca* (non-breeding), *Crex crex* (breeding) and *Gallinago media* (passage). Eighty-five species breed at the site, including *Pernis apivorus*, *Aquila pomarina* and *Hieraaetus pennatus*.

■ Protection status

National High **International** None
5,642 ha of IBA covered by Reserve (Plaiul Fagului, 5,642 ha).

■ Conservation issues

Threats Afforestation (B), Disturbance to birds (C), Firewood collection (C), Recreation/tourism (C), Unsustainable exploitation (C)

The Plaiul Fagului Reserve was established in 1991 to protect its unique *Fagus* vegetation.

Bazinul Ghidighici

B1i **005**

Admin region Chişinău
Coordinates 47°13′N 28°31′E
Altitude — **Area** 900 ha

■ Site description

A reservoir situated in the Bac valley, in the vicinity of the town of Ghidighici and the city of Chişinău. Fringing habitats comprise meadows and arable land; reedbeds (*Phragmites*) occur upstream. The beach and recreational facilities are within easy reach of the inhabitants of Chisinau.

Habitats Grassland (humid grassland), Wetland (standing fresh water; water-fringe vegetation), Artificial landscape (arable land)
Land-use Tourism/recreation

■ Birds

The B1i criterion has been assigned at the site level as the IBA is known to be an important stop-over site for passage waterbirds, but numbers are unknown.

■ Protection status

National None **International** None

■ Conservation issues

Threats Unknown

Codrii

A1, B2, B3 **006**

Admin region Chişinău
Coordinates 47°09′N 28°39′E
Altitude 130–382 m **Area** 5,177 ha

■ Site description

Habitats Forest and woodland (90%; broadleaved deciduous forest), Wetland (2%; river/stream), Artificial landscape (8%; forestry plantation)
Land-use Nature conservation/research, Unknown

The site is situated in the Central Moldovan Heights, straddling the watershed of the Botna and Byc rivers. The main tree species are *Quercus* (49% of forest area) and *Fagus–Fraxinus*, *Tilia*, *Acer* and *Carpinus* are also present. The main shrub species are *Cornus*, *Viburnum* and *Euonymus*. Meadow communities occupy small areas.

■ Birds

A1/B2/B3 criteria have been assigned at the site level as many globally threatened species and SPECs are present, but in unknown numbers. Two globally threatened species occur: *Aquila clanga* (breeding) and *Crex crex* (breeding). Eighteen species with an unfavourable conservation status in Europe occur, out of a total of 150 species.

■ Protection status

National Partial **International** None
IBA overlaps with Nature Reserve (Codrii, 12,300 ha).

■ Conservation issues

Threats Deforestation (commercial) (C), Disturbance to birds (B), Firewood collection (C), Forest grazing (C), Groundwater abstraction (C), Recreation/tourism (C), Selective logging/cutting (C), Unsustainable exploitation (C)

The Nature Reserve is divided into three functional zones: strict protection, buffer and transition. Research is undertaken by Reserve staff and the Academy of Sciences, and covers: development of a scientific basis for protection, conservation and restoration of threatened species; assessment of ecosystem productivity; determination of changes taking place within ecosystems under anthropogenic stress; and sustainable exploitation of plant and animal resources.

Golf Goeni

B1i **007**

Admin region Dubăsari
Coordinates 47°29′N 29°01′E
Altitude — **Area** 1,500 ha

■ Site description

The site includes part of the Dubăsari reservoir, adjacent to the village of Goeni, and the Iagorlac Nature Reserve and surrounding area. The site is predominantly wetland with some trees, grasses and reedbeds (*Phragmites*), and is of high importance for fishing.

Habitats Wetland (standing fresh water; water-fringe vegetation)
Land-use Fisheries/aquaculture

■ Birds

The B1i criterion has been assigned at the site level as the IBA is known to be an important breeding and stop-over site for waterbirds; numbers are unknown.

■ Protection status

National Partial **International** None
IBA overlaps with Nature Reserve (Iagorlac).

■ Conservation issues

Threats Unknown

Lacul Salaş

B1i **008**

Admin region Chişinău
Coordinates 46°58′N 29°12′E
Altitude — **Area** 330 ha

■ Site description

A water-body in the Bac river valley, located near to its confluence with the Dniester river. The site also includes fish-ponds, and is of social and economic value as well as being of wildlife conservation importance.

Habitats Wetland (standing fresh water)
Land-use Unknown

■ Birds

The B1i criterion has been assigned at the site level as the IBA is known to be an important stop-over site for passage waterbirds; numbers are unknown.

■ Protection status
National None **International** None

■ Conservation issues

Threats Unknown

Copanca–Talmaz
B1i **009**

Admin region Tighina
Coordinates 46°39′N 29°42′E
Altitude — **Area** 6,000 ha

■ Site description
A complex of wetland habitats in the flood-plain of the lower Dniester river. Landscape elements include flood-plain forest, meadows, former riverbeds and rich aquatic and near-shore vegetation. The site is of social and economic value, as well as being important for wildlife conservation.

Habitats Forest and woodland (alluvial/very wet forest), Grassland, Wetland (river/stream; water-fringe vegetation)
Land-use Unknown

■ Birds
The B1i criterion has been assigned at the site level as the IBA is known to be important for waterbirds, but numbers are unknown.

■ Protection status
National None **International** None

■ Conservation issues

Threats Unknown

Bazinul Cuciurgan
A1, B1i, B2, B3 **010**

Admin region Dubăsari
Coordinates 46°36′N 29°49′E
Altitude 3–23 m **Area** 6,400 ha

■ Site description
A wetland situated in the Turunchuk, near to the mouth of the Kuchurgan river, extending from the town of Pervomayskoe to the village of Nezavertaylovka (60 km from Odessa). The site, part of which adjoins the border with Ukraine, covers the Kuchurgan reservoir, flood-plain lakes, rivers and streams—some of the latter are overgrown with aquatic vegetation. Water-levels are affected by a hydroelectric power-station; there are eight dams (total length 7 km) around the reservoir. The wetland is also used for recreation, hunting and scientific research activities.

Habitats Forest and woodland (23%; alluvial/very wet forest; scrub), Grassland (humid grassland), Wetland (71%; standing fresh water; river/stream; water-fringe vegetation), Artificial landscape (3%; arable land; perennial crops/orchards/groves; forestry plantation; other urban/industrial areas), Unknown (3%)
Land-use Agriculture (3%), Fisheries/aquaculture (65%), Forestry (23%), Not utilized (9%)

■ Birds
A1/B2/B3 criteria have been assigned at the site level as many globally threatened species and SPECs are present, but in unknown numbers. B1i has also been assigned, as the IBA is known to be an important stop-over site for passage wildfowl and waders. Eight globally threatened species occur: *Phalacrocorax pygmeus* (non-breeding), *Anser erythropus* (passage), *Branta ruficollis* (passage), *Aythya nyroca* (breeding), *Haliaeetus albicilla* (passage), *Aquila clanga* (passage), *Crex crex* (breeding) and *Gallinago media* (passage).

■ Protection status
National None **International** None

■ Conservation issues
Threats include the expansion of activities at the power-station, leading to possible increases in water temperature and subsequent nutrient enrichment of the water-body. Poaching is also a threat. The Institute of Zoology at the Moldovan Academy of Sciences has undertaken ornithological, ichthyological, hydrobiological and hydrochemical investigations at the site. At the beginning of the 1990s plans were made to create a seasonal reserve for waterbirds and waders. Unfortunately, subsequent changes in the national economy and policy halted the project.

Threats Abandonment/reduction of land management (U), Agricultural intensification/expansion (B), Aquaculture/fisheries (B), Burning of vegetation (C), Consequences of animal/plant introductions (U), Construction/impact of dyke/dam/barrage (C), Deforestation (commercial) (C), Disturbance to birds (A), Drainage (U), Forest grazing (C), Industrialization/urbanization (B), Recreation/tourism (B), Unsustainable exploitation (C)

Manta flood-plain–Beleu
A1, B1i, B2, B3 **011**

Admin region Cahul
Coordinates 45°42′N 28°08′E
Altitude 4–11 m **Area** 8,364 ha

■ Site description
A wetland in the Prut valley, situated 150 km south-west of Chişinău and extending from the town of Cahul to the village of Slobozia Mare. The site includes several flood-plain lakes fed by freshwater springs, natural wet meadows and patches of shrub-dominated woodland. Fish-farming is a major land-use with over 2,000 ha of ponds supporting intensive carp *Cyprinus* production. Mineral springs in the area are used recreationally.

Habitats Forest and woodland (15%; alluvial/very wet forest), Grassland (14%; humid grassland), Wetland (60%; standing fresh water; river/stream; water-fringe vegetation), Artificial landscape (9%; arable land; perennial crops/orchards/groves; other urban/industrial areas), Unknown (2%)
Land-use Agriculture (8%), Fisheries/aquaculture (24%), Forestry (15%), Nature conservation/research (26%), Not utilized (27%), Tourism/recreation

■ Birds
A1/B2/B3 criteria have been assigned at the site level as many globally threatened species and SPECs are present, but detailed population estimates are unknown. B1i has also been assigned, as the IBA is known to be an important stop-over site for passage wildfowl and waders. Nine globally threatened species occur: *Phalacrocorax pygmeus* (breeding), *Pelecanus crispus* (non-breeding), *Anser erythropus* (passage), *Branta ruficollis* (passage), *Aythya nyroca* (resident), *Haliaeetus albicilla* (passage), *Aquila clanga* (passage), *Crex crex* (passage) and *Gallinago media* (passage). Approximately 70 species breed, and c.50 species of waterbirds stop over on spring and autumn migration.

■ Protection status
National Partial **International** None
IBA includes the reservation 'Prutul de Jos' (1,691 ha) and surrounding buffer-zone (1.5 km wide).

■ Conservation issues

Threats Agricultural intensification/expansion (B), Aquaculture/fisheries (A), Burning of vegetation (C), Disturbance to birds (A), Drainage (B), Extraction industry (A), Filling-in of wetlands (B), Forest grazing (B), Groundwater abstraction (B), Recreation/tourism (C), Unsustainable exploitation (B)

Various laboratories and institutes within the Moldovan Academy of Sciences undertake faunal inventories, research into the migration and ecology of taxa, and studies of water quality and hydrological, hydrochemical and hydrobiological regimes.

Bazinul Taraclia
B1i, B2 **012**

Admin region Cahul
Coordinates 45°54′N 28°30′E
Altitude — **Area** 550 ha

■ Site description
The reservoir is situated in the Ialpug valley, near the town of the same name in the south of Moldova. Aquatic and near-shore vegetation is present.

Habitats Wetland (standing fresh water; water-fringe vegetation)
Land-use Fisheries/aquaculture

■ Birds

The B1i criterion has been assigned at the site level, as the IBA is known to be an important stop-over site for passage waterbirds. Passage species include *Phalacrocorax pygmeus*, *Pelecanus onocrotalus*, *Pelecanus crispus*, *Cygnus olor*, *Haematopus ostralegus*, *Himantopus himantopus* and *Recurvirostra avosetta*. B2 has also been assigned at the site level as SPECs are present but in unknown numbers.

■ Protection status

National None **International** None

■ Conservation issues

Threats Unknown

REFERENCES

ANON. (1973) [*Cuciurgan lagoon—cooling reservoirs of hydropower stations*.] Chişinău: Stiinta. (In Russian.).

ANON. (1978) [*Geographical atlas of Moldavian SSR*.] Moscow: Glavnoe upravlenie geodezii i cartografii pri sovete ministrov SSSR. (In Russian.)

ANON. (1981) [*Fishes, reptiles, amphibians*.] Chişinău: Stiinta. (In Russian.)

ANON. (1984) [*The nature of the Reserve "Codrii"*.] Chişinău: Stiinta. (In Russian.)

ANON. (1988) [*Processes of biological production in cooling reservoirs of hydropower stations*.] Chişinău: Stiinta. (In Russian.)

ANON. (1990) [*Ecosystem of the lower Dniester river under high anthropogenic pressure*.] Chişinău: Stiinta. (In Russian.)

ANON. (1991) [*Territorial scheme for environmental protection and utilisation of natural resources of Republic of Moldova up to 2010*.] Chişinău: Department of Ecology. (In Romanian.)

ANON. (1995) [Protection, rehabilitation and sustainable exploitation of wildlife biodiversity.] In [*Proceedings of the 3rd international conference of Zoologists of Moldova*.] Chişinău: Institutul de Zoologie. (In Romanian.)

AVERIN, I. V., GANEA, I. M., ZUBCOV, N. I. *ET AL.* (1981) [*Birds*.] Chişinău: Stiinta. (In Russian.)

CAPCELEA, A. (1995) [*Republic of Moldova on the way to sustainable development: achievements and problems*.] Chişinău: Ministry of Economy. (In Romanian.)

COLLAR, N. J., CROSBY, M. J. AND STATTERSFIELD, A. J. (1994) *Birds to watch 2: the world list of threatened birds*. Cambridge, UK: BirdLife International (Conservation Series 4).

DOLGHII, V. N. (1993) [*Fish fauna of the Dniester and Prut river basins: current status, genesis, ecology and biological basis for fishery needs*.] Chişinău: Stiinta. (In Russian.)

GANEA, I. M., ZUBCOV, N. I. AND COTEATI, M. I. (1991) [*Radiolocation ornithology*.] Chişinău: Stiinta. (In Russian.)

GRIMMETT, R. F. A. AND JONES, T. A. (1989) *Important Bird Areas in Europe*. Cambridge, UK: International Council for Bird Preservation (Techn. Publ. 9).

HEIDEMAN, T. S., OSTAPENCO, B. F., NICOLAEVA, L. P. *ET AL.* (1964) [*Forest types and forest associations of Moldavian SSR*.] Chişinău: Cartea Moldoveneasca. (In Russian.)

POSTOLACHE, G. (1995) [*Vegetation of the Republic of Moldova*.] Chişinău: Stiinta. (In Romanian.)

SMIRNOVA-GARAEVA, N. V. (1980) [*The Dniester river vegetation and its commercial importance*.] Chişinău: Cartea Moldoveneasca. (In Russian.)

TUCKER, G. M. AND HEATH, M. F. (1994) *Birds in Europe: their conservation status*. Cambridge, UK: BirdLife International (BirdLife Conservation Series no. 3).

USPENSKII, G. A., AVERIN, I. V., GANEA, I. M. *ET AL.* (1979) [*Mammals*.] Chişinău: Stiinta. (In Russian.)

PORTUGAL

Rui Rufino and Renato Neves

Little Bustard *Tetrax tetrax* at Campo Maior (IBA 019). (Photo: Márcia Pinto)

GENERAL INTRODUCTION

The Portuguese territory comprises a continental area located in the south-west corner of mainland Europe, and two archipelagos, the Azores and Madeira, both situated in the Atlantic Ocean. Portugal itself extends over 89,000 km², and is delimited by the Atlantic to the west and south and by Spain to the east and north. This chapter is concerned with the continental territory; the Atlantic islands are dealt with in separate chapters (Azores p. 463, Madeira p. 473).

Mainland Portugal has 34 Important Bird Areas (IBAs) covering 9,421 km² or 10.6% of the continental area (Table 1). These are unevenly distributed in the country, with a larger proportion of sites in the south and east (Map 1).

Due to its geographic location, Portugal's climate is subject to strong Atlantic and Mediterranean influences. A great variety of landscapes therefore occur, ranging from the wet north-west to the dry south-east, and from the mild south-west to the continental extremes of the east and north-east. Most of the IBAs fall within those areas experiencing Mediterranean conditions, and many are situated along the Spanish border. These are areas of low population, where the landscape, although shaped by man, is not intensively managed. In comparison the large coastal wetlands have been affected by intense human activity.

A reassessment of the 36 IBAs identified in 1989 (Grimmett and Jones 1989) was undertaken for the current review. Six sites were removed from the 1989 list, and five new sites were identified (Table 1). The sites removed were found to be less important for birds than was previously thought. The new sites allow for better coverage of the distribution of certain species, as habitat types that were poorly covered by the former list, such as *Quercus suber* montados, have now been included. The boundaries of several sites were reviewed. Four IBAs identified in 1989 were merged to form two new IBAs:

Table 1. Summary of Important Bird Areas in Portugal. 34 IBAs covering 9,421 km²

IBA code	1989 code	International name	National name	Administrative region	Area (ha)	Criteria (see p. 11)
001	PT001	Minho and Coura estuaries	Estuários do Minho e Coura	Norte	3,393	C6
002	PT002	Gerês mountains	Parque Nacional da Peneda-Gerês	Norte	62,922	B2, C6
003	PT003, PT004	Montesinho and Nogueira mountains	Serras de Montesinho e Nogueira	Norte	108,089	B2, B3, C6
004	PT006	Sabôr and Maçãs	Sabôr e Maçãs	Norte	50,748	B2, C6
005	PT006	Upper River Douro and Águeda valleys	Douro Internacional e Vale do Águeda	Norte, Centro	50,744	B2, C2, C6
006	—	Côa valley	Vale do Côa	Centro	23,727	B2, C6
007	PT010	Ria de Aveiro	Ria de Aveiro	Centro	52,720	A4i, A4iii, B1i, B2, C2, C3, C4, C6
008	—	Malcata mountains	Serra da Malcata	Centro	16,361	C6
009	PT036	Paul do Taipal	Paul do Taipal	Centro	233	C6
010	PT011	Paul de Arzila	Paul de Arzila	Centro	482	C6
011	PT012	Paul de Madriz	Paul de Madriz	Centro	89	C6
012	PT035	Toulões and Idanha-a-Nova	Toulões e Idanha-a-Nova	Centro	10,920	A1, B2, C1, C6

Table 1 ... continued. Summary of Important Bird Areas in Portugal. 34 IBAs covering 9,421 km²

IBA code	1989 code	International name	National name	Administrative region	Area (ha)	Criteria (see p. 11)
013	PT013	Upper River Tejo	Tejo Internacional	Centro	26,129	B2, C6
014	PT016	Berlenga and Farilhões islands	Berlenga e Farilhões	Lisboa e Vale do Tejo	9,560	A4ii, A4iii, B1ii, B2, C2, C4, C6
015	PT014	Paul do Boquilobo	Paul do Boquilobo	Lisboa e Vale do Tejo	433	A4i, B1i, C2, C3, C6
016	—	Cabeção	Cabeção	Alentejo	20,227	A4iii, B2, C4, C6
017	PT031	Alter do Chão	Alter do Chão	Alentejo	7,959	A1, A4iii, B2, C1, C4, C6
018	PT029	Monforte plains	Planície de Monforte	Alentejo	9,426	A1, A4iii, B2, B3, C1, C4, C6
019	—	Campo Maior	Campo Maior	Alentejo	9,355	A1, B1i, B2, C1, C2, C6
020	PT028	Vila Fernando	Vila Fernando	Alentejo	8,166	A1, B2, C1, C6
021	PT017	Tejo estuary	Estuário do Tejo	Lisboa e Vale do Tejo	45,071	A1, A4i, A4iii, B1i, B2, B3, C1, C2, C3, C4, C6
022	—	Coast of the Setúbal peninsula	Costa da Península de Setúbal	Lisboa e Vale do Tejo	22,930	C6
023	PT018	Sado estuary	Estuário do Sado	Alentejo	24,633	A4i, A4iii, B1i, B2, C2, C3, C4, C6
024	PT019	Murta dam	Açude da Murta	Alentejo	498	A4i, B1i, C2, C3, C6
025	PT033	Évora plains	Planície de Évora	Alentejo	33,693	A1, B2, C1, C6
026	PT034	Alvito	Alvito	Alentejo	2,606	A1, B2, C1, C6
027	PT030	Mourão, Moura e Barrancos	Mourão, Moura e Barrancos	Alentejo	82,667	A1, B1i, B2, B3, C1, C2, C6
028	PT021, PT022	Santo André and Sancha lagoons	Lagoas de Santo André e Sancha	Alentejo Litoral	2,638	A4i, B1i, C3, C6
029	PT027	Castro Verde plains	Castro Verde	Alentejo	79,066	A1, B2, B3, C1, C2, C6
030	PT020	River Guadiana	Rio Guadiana	Alentejo	76,578	A1, B1iii, B2, B3, C1, C2, C6
031	PT023	South-west coast of Portugal	Costa Sudoeste	Alentejo Litoral, Algarve	74,563	B1iv, B2, C6
032	PT024	Leixão da Gaivota	Leixão da Gaivota	Algarve	1	C6
033	PT025	Ria Formosa (Faro lagoon)	Ria Formosa	Algarve	23,296	A4i, A4iii, B1i, B2, C2, C4, C6
034	PT026	Castro Marim	Castro Marim	Algarve	2,147	A4i, B1i, B2, C2, C6

Sites identified in the previous inventory of IBAs in Europe (Grimmett and Jones 1989) but no longer considered to be IBAs
PT005 Alvão mountain; PT007 Montemuro mountain; PT008 Arada and Freita mountains; PT009 Estrela mountain; PT015 Candeeiros mountain; PT032 Moura–Safara

Map 1. Location, area and criteria category of Important Bird Areas in Portugal.

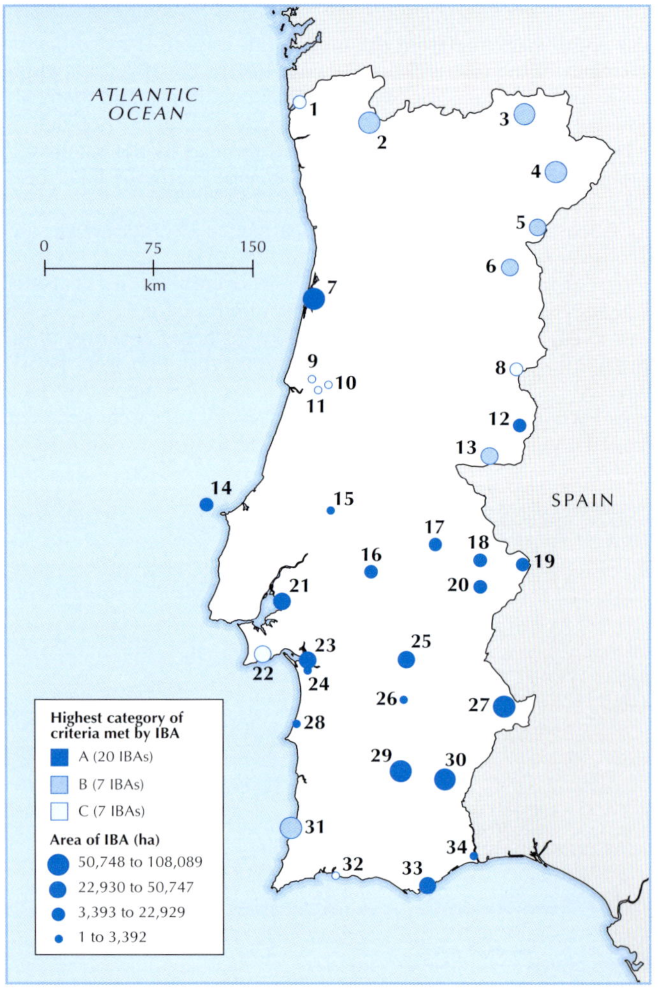

003 and 028 (Table 1). One IBA identified in 1989 (former site PT006) was redrawn and divided into two different sites: 004 and 005.

ORNITHOLOGICAL IMPORTANCE

Table 1 details the criteria met by each IBA, and Map 1 provides a summary. Two globally threatened species, *Falco naumanni* and *Otis tarda*, and one near-threatened species, *Tetrax tetrax*, breed in mainland Portugal (Table 2), with over 90% of the national population of *Otis tarda* occurring within IBAs (Table 4). Coastal wetlands (for example, 007, 021, 023, 033 and 034) and inland wetlands (for example, 015 and 024) hold internationally important numbers of both breeding and wintering waterbirds (Table 3).

One hundred and fifteen species of European conservation concern (SPECs) breed regularly in Portugal (Tucker and Heath 1994) and 46 SPECs occur outside of the breeding season. Portugal holds the only known breeding population of *Oceanodroma castro* near mainland Europe, located within the Berlenga and Farilhões islands (014).

Lowland plains hold important numbers of breeding steppic species (for example, the SPECs *Tetrax tetrax*, *Burhinus oedicnemus*, *Pterocles orientalis* and *Melanocorypha calandra*) and wintering *Grus grus*, and inland cliffs support important numbers of breeding raptors. The national breeding populations of large and colonially nesting raptors (for example, *Aquila chrysaetos*, *Hieraaetus fasciatus*, *Gyps fulvus*, *Neophron percnopterus* and *Falco naumanni*) are largely included within the IBA network (Table 4). The following sites are particularly important in this context: 002, 005, 013 and 030.

Table 2. Important Bird Areas in Portugal that are important for species of global conservation concern (meeting criterion A1).

Species	IBA code
Falco naumanni Lesser Kestrel	019, 029, 030
Tetrax tetrax Little Bustard	017, 018, 019, 020, 021, 025, 026, 027, 029
Otis tarda Great Bustard	012, 017, 018, 019, 020, 025, 026, 027, 029

Table 3. Important Bird Areas in Portugal that support important numbers of one or more congregatory species (i.e. meeting criteria A4 and/or B1). IBAs meeting both criteria A4 and B1 for the species are shown in **bold**. IBAs meeting only criterion B1 for the species concerned, and not A4, are shown in normal type. For key to 'Season', see p. 7.

Species	Season	IBA code
Oceanodroma castro Madeiran Petrel	B	**014**
Phalacrocorax carbo Cormorant	W	023
Bubulcus ibis Cattle Egret	R	**015**
	B	**024**
Egretta garzetta Little Egret	B	**015, 024**
Platalea leucorodia Spoonbill	B	015
	W	**033**
	P	**034**
Phoenicopterus ruber Greater Flamingo	W	**021, 034**
Anser anser Greylag Goose	W	021
Anas penelope Wigeon	W	**033**
Anas crecca Teal	W	021
Anas clypeata Shoveler	W	021
Netta rufina Red-crested Pochard	W	**028**
Melanitta nigra Common Scoter	W	007
Falco naumanni Lesser Kestrel	B	030
Grus grus Crane	W	019, 027
Himantopus himantopus Black-winged Stilt	R	**033**
	B	**021, 023, 034**
Recurvirostra avosetta Avocet	W	**007, 021, 023**
Glareola pratincola Collared Pratincole	B	**021**
Charadrius hiaticula Ringed Plover	W	**033**
Charadrius alexandrinus Kentish Plover	R	**033**
	W	**033**
Pluvialis squatarola Grey Plover	W	007, **021**, **033**
Calidris alpina Dunlin	W	021, 023, **033**
Limosa limosa Black-tailed Godwit	W	**021**, 023
Limosa lapponica Bar-tailed Godwit	W	**033**
Tringa totanus Redshank	W	**021**, 033
Arenaria interpres Turnstone	W	**033**
	P	**033**
Sterna albifrons Little Tern	B	**033**

Overall, the majority of gregarious species such as shearwaters and petrels, colonially nesting herons, raptors and waders, and wintering ducks and waders are well covered by the IBA network (Table 3). The network does not adequately cover dispersed species such as *Ciconia ciconia*, forest raptors (for example, *Elanus caeruleus* and *Circaetus gallicus*) and many passerines, for which conservation of the wider environment is required.

HABITATS

The Portuguese landscape has been shaped by intense human activity over a long period of time. A variety of farmland types are found in the different regions, from the small intensively farmed fields of the north-west to the vast extensively farmed cereal fields of the south. Forests are a significant feature of the landscape, and are dominated by *Pinus* and *Eucalyptus* plantations and managed stands of *Quercus suber* and *Q. ilex*, known as montados.

Figure 1. Occurrence of habitats at Important Bird Areas in Portugal (see Appendix 3 for definitions of habitats).

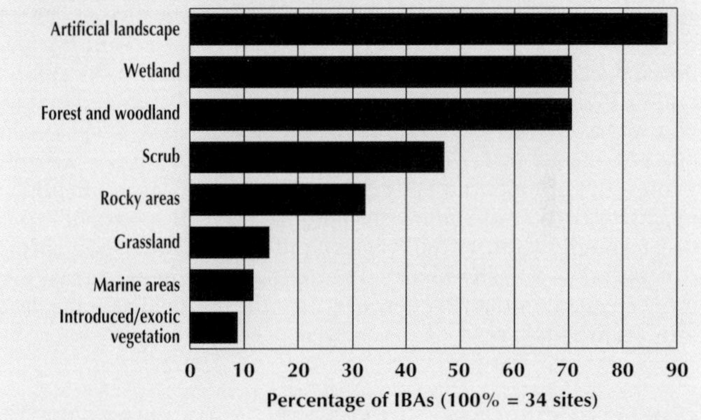

Table 4. Species of European conservation concern and species listed on Annex I of the EC Birds Directive with significant breeding populations at IBAs in Portugal (meeting any IBA criteria).

Species [1]	Minimum national breeding population (pairs) [2]	Proportion (%) of national population breeding at all IBAs in Portugal
Calonectris diomedea Cory's Shearwater	180	100
Oceanodroma castro Madeiran Petrel	200	100
Botaurus stellaris Bittern	5	100
Ixobrychus minutus Little Bittern	500	50
Nycticorax nycticorax Night Heron	200	15
Ardeola ralloides Squacco Heron	10	10
Egretta garzetta Little Egret	1,500	94
Ardea purpurea Purple Heron	100	100[3]
Ciconia nigra Black Stork	30	100[3]
Ciconia ciconia White Stork	1,000	50
Platalea leucorodia Spoonbill	1	100[3]
Elanus caeruleus Black-winged Kite	100	36
Milvus migrans Black Kite	400	8
Neophron percnopterus Egyptian Vulture	30	100[3]
Gyps fulvus Griffon Vulture	100	100[3]
Aegypius monachus Cinereous Vulture	1	100
Circaetus gallicus Short-toed Eagle	100	34
Circus aeruginosus Marsh Harrier	25	100[3]
Circus cyaneus Hen Harrier	1	100[3]
Circus pygargus Montagu's Harrier	900	22
Aquila chrysaetos Golden Eagle	20	100[3]
Hieraaetus pennatus Booted Eagle	100	72
Hieraaetus fasciatus Bonnelli's Eagle	83	30
Falco naumanni Lesser Kestrel	150	63
Falco peregrinus Peregrine	25	100[3]
Porzana pusilla Baillon's Crake	10	100
Porphyrio porphyrio Purple Gallinule	5	100[3]

Species [1]	Minimum national breeding population (pairs) [2]	Proportion (%) of national population breeding at all IBAs in Portugal
Tetrax tetrax Little Bustard	10,000	14
Otis tarda Great Bustard	500 individuals	100[3]
Himantopus himantopus Black-winged Stilt	1,000	100[3]
Burhinus oedicnemus Stone Curlew	1,000	27
Glareola pratincola Collared Pratincole	250	54
Charadrius alexandrinus Kentish Plover	1,000	80
Sterna albifrons Little Tern	150	100[3]
Chlidonias hybridus Whiskered Tern	20	100[3]
Pterocles orientalis Black-bellied Sandgrouse	100	31
Bubo bubo Eagle Owl	100	72
Coracias garrulus Roller	100	42
Melanocorypha calandra Calandra Lark	10,000	2
Calandrella brachydactyla Short-toed Lark	100,000	–[4]
Anthus campestris Tawny Pipit	1,000	–[4]
Oenanthe leucura Black Wheatear	100	–[4]
Lanius collurio Red-backed Shrike	100	100[3]
Pyrrhocorax pyrrhocorax Chough	100	60
Emberiza hortulana Ortolan Bunting	1,000	–[4]

1. Only those species of European conservation concern (see Box 1, p. 12) that meet IBA criteria in Portugal are listed, together with those species listed on Annex I of the EC Birds Directive that fulfil criterion C6 in IBAs in Portugal.
2. Data are taken from the BirdLife/EBCC European Bird Database 1998 (Heath and Borggreve 2000).
3. The percentage of the national population in IBAs exceeds 100%. Usually this is because the national population estimate has not been updated recently whilst the IBA population estimate has been recently updated with new data as a result of comprehensive surveys of IBAs themselves. Also, the individual site count for a species may be the maximum or average over recent years, and summing these may record more birds than are present nationally in any single year.
4. No population data available for Portuguese IBAs.

Figure 2. Occurrence of land-uses at Important Bird Areas in Portugal (see Appendix 3 for definitions of land-uses).

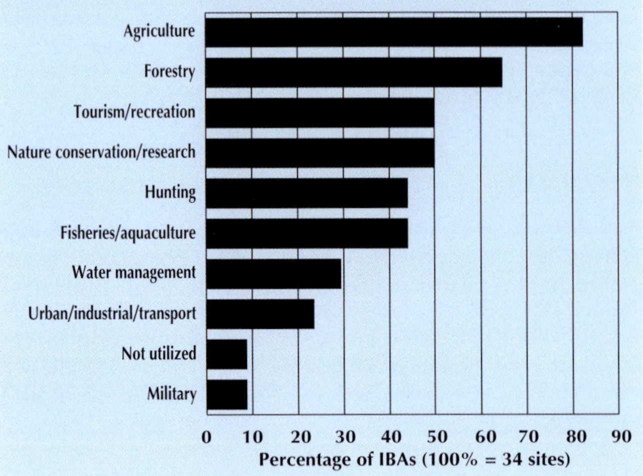

Whilst habitats within Portuguese IBAs reflect the dominance of farmland and forest (Figure 1), wetlands and scrub are also strongly represented (71% and 47% of IBAs respectively). This is because many Portuguese wetlands hold nationally and internationally important numbers of waterbirds and because, in remote areas of the country, farming has been abandoned and the fallows have given way to scrub. A number of habitat-types occur more frequently within IBAs as compared to the country as a whole, as they support important bird populations. Rocky areas are present in 32% of IBAs, and are the stronghold of several species of raptor (for example, vultures and eagles) and other cliff-nesting species (for example, *Ciconia nigra*). The southern farmland areas, also more frequent in the IBA network than the national average, hold important numbers of steppic species.

IMPACTS ON IBAs – LAND-USE AND THREATS

As already mentioned, the vast majority of Portugal's land area is covered by either farmland or forest, or a combination of both as in the montados. Land-use in IBAs reflects this fact (Figure 2). In coastal areas tourism/recreation and fisheries/aquaculture are particularly important (50% and 44% of IBAs respectively). As the majority of Portuguese protected areas are included within the IBA network, nature conservation/research is also a major land-use (50% of IBAs).

Agricultural intensification is the most significant threat to many bird species, and affects 50% of IBAs (Figure 3). It is a particular problem for species dependent on non-intensive farming systems (for example *Otis tarda* and *Pterocles orientalis*), and for farmland species in areas that have undergone intensification and are now seeing further measures being taken to increase productivity.

The construction of large reservoirs threatens 38% of IBAs, particularly those in the north-east of the country. Some of these projects will lead to the flooding of a significant proportion of the breeding habitat of several cliff-nesting raptors, leaving these species with few alternative breeding sites in the country.

Other threats of note are the effects of dredging and canalization on estuarine sediment balances and erosion events (21% of IBAs); the replacement of native forest with *Eucalyptus* plantations (24% of IBAs); and the abandonment of land (26% of IBAs) which, at least in the first years of succession, can result in monospecific scrub stands and a consequent reduction in bird-species diversity and density.

Over the last 20 years Portugal has undergone a process of modernization, which is still ongoing. Further agricultural intensification is therefore expected, as are new infrastructure projects, such as roads, reservoirs and urban developments.

PROTECTION STATUS

Table 5 and Figures 4 and 5 summarize the national and international protection status of Portuguese IBAs.

■ National protection

Most of the land in Portuguese protected areas is privately owned. A small proportion either belongs to the managing authority, Instituto da Conservação da Natureza (ICN), or is in the public domain (river, estuarine and coastal waters). There are three main site-protection designations in Portugal:

1. National Park
These cover habitats and ecosystems of national importance, which are essentially unaffected by human activities and are representative of characteristic natural regions. This designation allows for the implementation of management actions for the maintenance of the integrity of habitats and ecosystems.

2. Nature Reserve
These are designed for the protection of fauna and flora and their habitats. This classification allows for the implementation of

Figure 3. Threats at Important Bird Areas in Portugal (see Appendix 3 for definitions of threat types and impact categories).

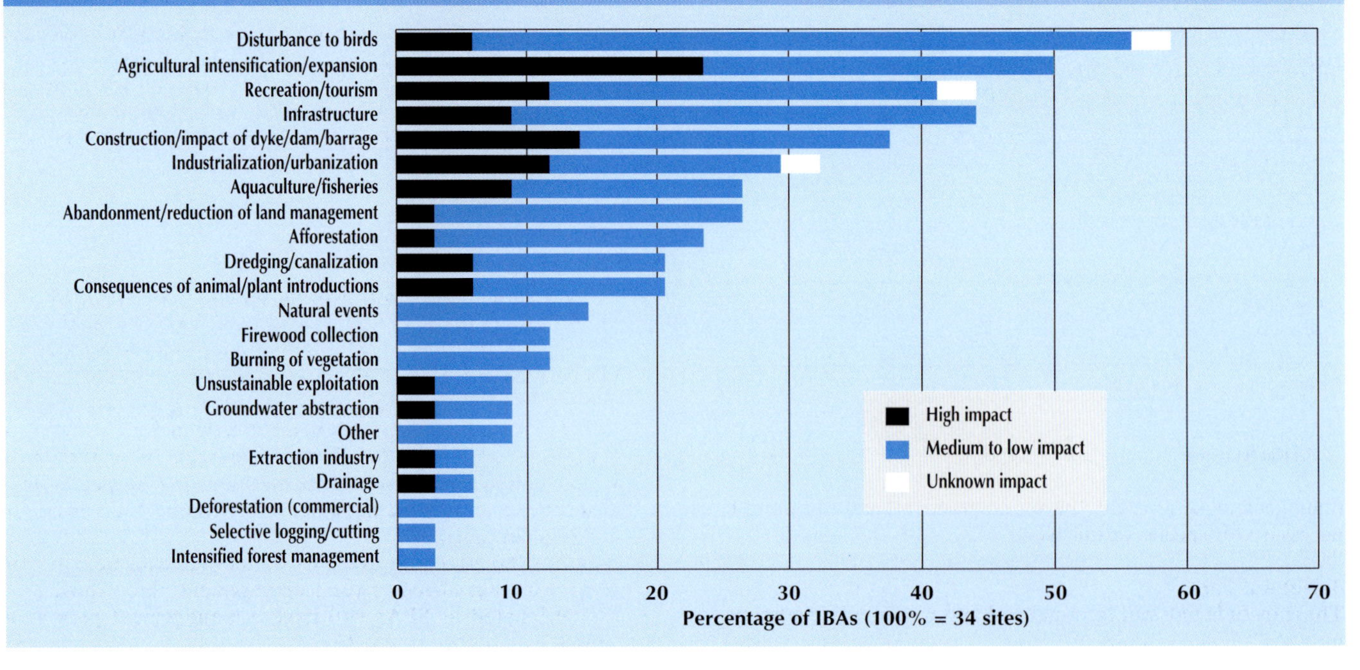

Figure 4. The national protection status of Important Bird Areas in Portugal.

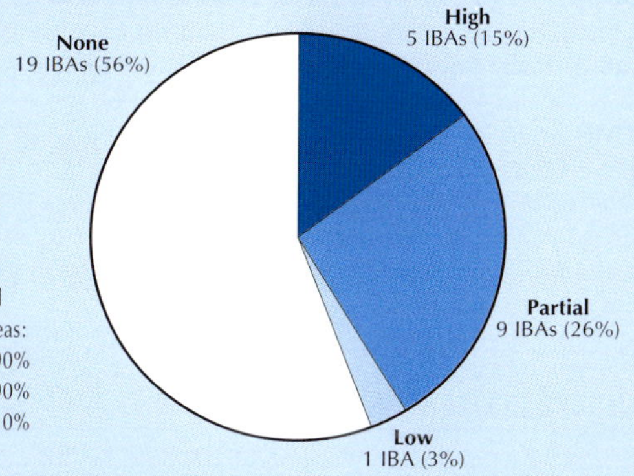

Total area of overlap between IBA network in Portugal and national protected-area system (see Table 5 for categories) = 4,036 km² (43% of total IBA area).

Figure 5. The international protection status of Important Bird Areas in Portugal.

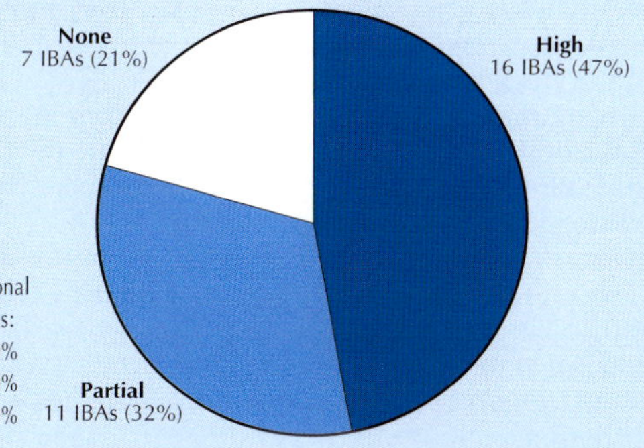

Total area of overlap between IBA network in Portugal and international protected-area system (see Table 5 for categories) = 3,751–3,780 km² (40% of total IBA area).

Table 5. Protection status of Important Bird Areas in Portugal. A tick (✔) indicates that an IBA overlaps with a protected area (to any extent).

IBA code	International name	National: National Park	National: Nature Reserve	National: Natural Park	International: Ramsar Site	International: Biosphere Reserve	International: Special Protection Area
001	Minho and Coura estuaries						✔
002	Gerês mountains	✔					✔
003	Montesinho and Nogueira mountains			✔			✔
004	Sabôr and Maçãs						✔
005	Upper River Douro and Águeda valleys			✔			✔
006	Côa valley						✔
007	Ria de Aveiro		✔				✔
008	Malcata mountains		✔				✔
009	Paul do Taipal						✔
010	Paul de Arzila		✔		✔		✔
011	Paul de Madriz				✔		✔
012	Toulões and Idanha-a-Nova						
013	Upper River Tejo						✔
014	Berlenga and Farilhões islands		✔				✔
015	Paul do Boquilobo		✔		✔	✔	✔
016	Cabeção						
017	Alter do Chão						
018	Monforte plains						
019	Campo Maior						✔
020	Vila Fernando						
021	Tejo estuary		✔		✔		✔
022	Coast of the Setúbal peninsula			✔	✔		✔
023	Sado estuary		✔		✔		✔
024	Murta dam				✔		✔
025	Évora plains						
026	Alvito						
027	Mourão, Moura e Barrancos						✔
028	Santo André and Sancha lagoons						✔
029	Castro Verde plains						✔
030	River Guadiana			✔			✔
031	South-west coast of Portugal			✔			✔
032	Leixão da Gaivota						✔
033	Ria Formosa (Faro lagoon)			✔	✔		✔
034	Castro Marim		✔		✔		✔
Total number of IBAs		1	8	6	9	1	27

management actions for the maintenance of habitat conditions necessary for species to survive.

3. Natural Park
These include natural, semi-natural and man-made landscapes of national importance, and provide examples of sustainable use. This designation allows for the implementation of management actions for the maintenance and enhancement of landscape features.

Box 1. International legislation and initiatives that are relevant to site conservation in Portugal (see Appendix 1 for a general description of these agreements).

Global	
Biodiversity Convention	✔
Ramsar Convention	✔
Bonn Convention	✔
World Heritage Convention	✔
MAB Programme	✔
Pan-European	
Bern Convention	✔
Regional	
EC Birds Directive	✔
EC Habitats Directive	✔

✔ Convention ratified/initiative supported
(✔) Convention signed

Fifteen out of the 34 IBAs overlap, either wholly or partly, with one or more national protected areas that confer legal protection (Table 5, Figure 4). Most of the sites that remain outside of the national protected-area network are located in the southern lowlands, and hold important populations of steppic species.

All of the IBAs face threats, regardless of their protection status. This is because the majority of protected land is privately owned, making management and implementation of restrictions difficult. Several state authorities have jurisdiction over protected areas, and decision-making power is therefore contested.

■ International protection

Box 1 shows the relevant international conventions and directives ratified by Portugal. A total of 27 IBAs have been wholly or partly designated as Special Protection Areas (SPAs) under the EC Birds Directive. Nine IBAs overlap with Ramsar Sites, and Paul do Boquilobo (015) was declared a Biosphere Reserve in 1981 (Table 5).

CONSERVATION

- A good knowledge of the distribution and numbers of SPECs aided the delimitation of the Portuguese IBAs. During the last 20 years a great deal of information concerning these species has been collected by ICN (or its predecessors: SEA-CEMPA and SNPRCN-CEMPA). Regular monitoring schemes were set up for wintering waterbirds in cooperation with the International Waterfowl and Wetlands Research Bureau (now Wetlands International). Special surveys of more threatened species were also conducted, in order to gather information on distributions and population sizes.
- Some of the sites that are part of the national network of protected areas already have a management plan. Those that are designated as SPAs will have a management plan as a consequence of the designation.

ANALYTICAL METHODS

- The majority of the wildfowl, wader and seaduck counts are accurate and are averages of data collected over the period 1993–1997. Wildfowl and wader counts began on a regular basis in 1977, whilst seaduck counts were initiated in 1991.
- Information on migrating raptors comes from irregular counts organized by ICN and the Portuguese Society for the Study of Birds (SPEA), which have produced a large quantity of data on migration at Cape São Vicente.
- Breeding data for colonially nesting herons, storks, waders, cliff-nesting raptors and seabirds have come mostly from specially designed surveys. Data for forest raptors are from surveys conducted at the IBAs.
- For some of the remaining species, namely passerines with restricted distributions, data were collected either for the first national breeding bird atlas, regional atlases or during impromptu surveys.
- Data on threats and land-uses were derived from information available in 1997.

GLOSSARY

ICN Instituto da Conservação da Natureza.

maquis dense, mostly evergreen shrub community 1–3 m high, characteristic of the Mediterranean region.

montado open, wooded pastoral habitat, dominated by evergreen oak *Quercus*, similar to dehesa in Spain. Growth of scrub is controlled and pasture is maintained by extensive rotational cultivation and/or grazing.

SEA-CEMPA Serviço de Estudos do Ambiente—Centro de Estudos de Migrações e Protecção de Aves.

SNPRCN-CEMPA Serviço Nacional de Parques, Reservas e Conservação da Natureza—Centre de Estudos de Migrações e Protecção de Aves.

SPA Special Protection Area (designated under Article 4 of the EC Birds Directive).

ACKNOWLEDGEMENTS

The present inventory was prepared by a group of people working at ICN, including Luis Costa, Vitor Encarnação, João Carlos Farinha, José Pedro Granadeiro, Renato Neves, Márcia Pinto, Miguel Pimenta, Gonçalo Rosa, Rui Rufino, Luis Silva and António Teixeira, as well as P. X. Catry, P. C. Dias, Nuno Onofre and Luis Palma. Data were compiled by L. Costa, R. Neves and R. Rufino. The compilation work were funded by the Royal Society for the Protection of Birds (BirdLife International Partner in the United Kingdom).

SITE ACCOUNTS

Minho and Coura estuaries — C6 — 001

Admin region Norte
Coordinates 41°55′N 8°47′W
Altitude 0–50 m **Area** 3,393 ha

Site description

The medium-sized estuary of a major river (Minho) and its tributary (Coura), including non-estuarine areas upstream, in the north-west corner of Portugal (the Minho demarcates the border with Spain here). Habitats include saltmarshes, sand- and mudflats, extensive reedbeds and alluvial forest.

Habitats Forest and woodland (broadleaved deciduous forest; mixed forest; alluvial/very wet forest), Scrub (scrub), Grassland (humid grassland), Wetland (tidal river/enclosed tidal water; mudflat/sandflat; saltmarsh; sand-dunes/sand beach; standing fresh water; river/stream; water-fringe vegetation), Marine areas (sea inlet/coastal features), Artificial landscape (arable land)
Land-use Agriculture, Fisheries/aquaculture, Forestry, Hunting, Tourism/recreation, Urban/industrial/transport

Birds

Species	Season	Year	Pop min	Pop max	Acc	Criteria
Circus aeruginosus Marsh Harrier	R	1996	4	10	A	C6

The IBA is internationally important for one species of raptor, and is also nationally notable for its wintering waterfowl and for some breeding passerines of very wet forest or reedbeds, e.g. *Locustella luscinioides*.

Protection status

National None **International** High
3,393 ha of IBA covered by Special Protection Area (Estuários do Minho e Coura, 3,393 ha).

Conservation issues

Threats Agricultural intensification/expansion (C), Aquaculture/fisheries (B), Burning of vegetation (B), Construction/impact of dyke/dam/barrage (B), Drainage (B), Dredging/canalization (B), Firewood collection (B), Industrialization/urbanization (B), Infrastructure (B), Recreation/tourism (C)

Excessive dredging is changing the sediment distribution in the estuary and thus possibly reducing the feeding densities of wintering waders and ducks. Over-fishing is certainly affecting piscivorous birds at the estuary. The construction of a large dam upstream from the site will reduce the fine sediment input to the site, which will reduce the extent of the marginal vegetation and the productivity of intertidal areas.

Gerês mountains — B2, C6 — 002

Admin region Norte
Coordinates 41°52′N 8°08′W
Altitude 500–1,545 m **Area** 62,922 ha

Site description

A very large site in the north-west of Portugal, covering several mountain ranges (Soajo, Peneda, Amarela and Gerês). Each range has a different exposure to the Atlantic climatic influence and therefore there is a wide variety of habitat-types.

Habitats Forest and woodland (broadleaved deciduous forest; mixed forest; broadleaved evergreen forest; treeline ecotone), Scrub (scrub; heathland), Grassland (alpine/subalpine/boreal grassland; humid grassland), Wetland (standing fresh water; river/stream; raised bog), Rocky areas (scree/boulders; inland cliff), Artificial landscape (arable land; forestry plantation)
Land-use Agriculture, Fisheries/aquaculture, Forestry, Nature conservation/research, Tourism/recreation

Birds

Species	Season	Year	Pop min	Pop max	Acc	Criteria
Circus cyaneus Hen Harrier	B	—	3	4	A	B2, C6
Circus pygargus Montagu's Harrier	B	—	20	25	A	C6
Aquila chrysaetos Golden Eagle	B	—	3	4	A	B2, C6
Falco peregrinus Peregrine	B	—	10	15	A	B2, C6
Bubo bubo Eagle Owl	B	—	10	—	B	B2, C6
Lanius collurio Red-backed Shrike	B	—	600	800	A	B2, C6
Emberiza hortulana Ortolan Bunting	B	—	Frequent		—	C6

The area supports important numbers of breeding raptors, and is the main Portuguese stronghold for *Lanius collurio* and *Emberiza hortulana*. Several birds are at their south-western limit in Europe, including *L. collurio* and *Saxicola rubetra*. The avifauna is rich, with more than 120 breeding species.

Protection status

National Partial **International** High
52,883 ha of IBA covered by National Park (Peneda–Gerês, 69,869 ha). 62,922 ha of IBA covered by Special Protection Area (Serra do Gerês, 62,922 ha).

Conservation issues

Threats Abandonment/reduction of land management (B), Burning of vegetation (C), Consequences of animal/plant introductions (C), Infrastructure (B), Recreation/tourism (B)

An introduced species of *Acacia* tree is invading native vegetation and is considered a plague. There is a management plan for the National Park.

Montesinho and Nogueira mountains B2, B3, C6 003

Admin region Norte
Coordinates 41°50′N 6°55′W
Altitude 450–1,481 m **Area** 108,089 ha

Site description

A series of mountain ranges (Coroa, Montesinho and Nogueira), located in the north-east corner of Portugal, with a continental-type climate. There is a vast forest of deciduous oak *Quercus pyrenaica*, together with heathland and other shrublands. The site is an amalgamation of two sites featured in the previous pan-European IBA inventory (Grimmett and Jones 1989): Montesinho Mountain (former site PT003) and Nogueira Mountain (former site PT004).

Habitats Forest and woodland (broadleaved deciduous forest; mixed forest; treeline ecotone), Scrub (scrub; heathland; sclerophyllous scrub/garrigue/maquis), Grassland (alpine/subalpine/boreal grassland; humid grassland), Wetland (standing fresh water; river/stream), Rocky areas (scree/boulders; inland cliff), Artificial landscape (arable land; forestry plantation)

Land-use Agriculture, Forestry, Hunting, Nature conservation/research, Tourism/recreation

Birds

Species	Season	Year	Pop min	Pop max	Acc	Criteria
Circaetus gallicus Short-toed Eagle	B	1994	11	13	A	B2, C6
Circus cyaneus Hen Harrier	B	1994	5	8	A	B2, C6
Circus pygargus Montagu's Harrier	B	1994	40	60	A	B3, C6
Aquila chrysaetos Golden Eagle	R	1994	3	4	A	B2, C6
Anthus campestris Tawny Pipit	B	1994	Common		—	C6
Emberiza hortulana Ortolan Bunting	B	1994	Common		—	C6

These mountains support important numbers of breeding raptors. The avifauna is rich, with more than 125 breeding species.

Protection status

National Partial **International** Partial
56,844 ha of IBA covered by Natural Park (Montesinho, 74,618 ha). 56,844 ha of IBA covered by Special Protection Area (Serra de Montesinho, 56,844 ha).

Conservation issues

Threats Abandonment/reduction of land management (B), Burning of vegetation (C), Construction/impact of dyke/dam/barrage (C), Firewood collection (B), Infrastructure (C), Recreation/tourism (C)

The area is relatively unaffected by human activities, although the abandonment of agriculture is reducing landscape diversity, through regeneration of forest-cover and loss of the patchwork of open areas. Firewood-collection in the deciduous forests, allied to lack of management, is preventing the development of forest stands of high ecological quality.

Sabôr and Maçãs B2, C6 004

Admin region Norte
Coordinates 41°28′N 6°40′W
Altitude 200–840 m **Area** 50,748 ha

Site description

Steep valleys with high cliffs in the watershed of the Sabôr river (a tributary of the Douro), in north-east Portugal. There is sclerophyllous scrub and farmland on the intervening plateaus.

Habitats Scrub (52%; sclerophyllous scrub/garrigue/maquis), Wetland (8%; standing fresh water; water-fringe vegetation), Rocky areas (10%; inland cliff), Artificial landscape (30%; arable land; perennial crops/orchards/groves)

Land-use Agriculture, Hunting, Water management

Birds

Species	Season	Year	Pop min	Pop max	Acc	Criteria
[1] ***Ciconia nigra*** Black Stork	B	1996	1	5	B	B2, C6
Ciconia ciconia White Stork	B	1996	1	5	A	C6
Milvus migrans Black Kite	B	1996	1	5	A	C6
Neophron percnopterus Egyptian Vulture	B	1996	11	20	A	B2, C6
Aquila chrysaetos Golden Eagle	R	1996	11	20	A	B2, C6
Hieraaetus fasciatus Bonnelli's Eagle	R	1996	6	10	A	B2, C6
Bubo bubo Eagle Owl	R	1996	11	20	B	B2, C6
Oenanthe leucura Black Wheatear	R	1996	Uncommon		—	C6

1. 2–5% of national breeding population.

The site holds important numbers of cliff-nesting bird species.

Protection status

National None **International** Partial
IBA overlaps with Special Protection Area

Conservation issues

Threats Abandonment/reduction of land management (B), Construction/impact of dyke/dam/barrage (A), Extraction industry (B), Infrastructure (B), Recreation/tourism (C)

There are plans to build several large dams in the IBA, which will flood most of the area, and the construction of major roads is also planned. The riverbed is exploited for shingle.

Upper River Douro and Águeda valleys B2, C2, C6 005

Admin region Norte, Centro
Coordinates 41°04′N 6°46′W
Altitude 150–771 m **Area** 50,744 ha

Site description

The site comprises two valleys, located in the north-east of Portugal and marking the border with Spain. There is a vast area of high cliffs.

Habitats Forest and woodland (broadleaved deciduous forest; broadleaved evergreen forest), Scrub (scrub; sclerophyllous scrub/garrigue/maquis), Wetland (standing fresh water; river/stream), Rocky areas (inland cliff), Artificial landscape (arable land; perennial crops/orchards/groves)

Land-use Agriculture, Fisheries/aquaculture, Hunting, Nature conservation/research, Tourism/recreation, Water management

Birds

Species	Season	Year	Pop min	Pop max	Acc	Criteria
Ciconia nigra Black Stork	B	—	11	25	A	B2, C6
Neophron percnopterus Egyptian Vulture	B	—	51	75	A	B2, C6
Gyps fulvus Griffon Vulture	R	1996	320	325	A	B2, C6
Aquila chrysaetos Golden Eagle	R	—	11	15	A	B2, C6
Hieraaetus fasciatus Bonnelli's Eagle	R	—	10	20	A	B2, C2, C6
Falco peregrinus Peregrine	R	—	20	25	A	B2, C6
Bubo bubo Eagle Owl	R	—	11	50	B	B2, C6
Oenanthe leucura Black Wheatear	R	—	Frequent		—	C6
Pyrrhocorax pyrrhocorax Chough	R	—	50	100	B	B2, C6

The cliffs provide exceptionally good breeding habitat for raptors and other soaring birds. Breeding species of global conservation concern that do not meet IBA criteria: *Falco naumanni* (uncommon, probably extinct).

Protection status

National High **International** Partial
50,744 ha of IBA covered by Natural Park (Parque Natural do Douro Internacional, 85,146 ha). IBA overlaps with Special Protection Area.

Conservation issues

Threats Abandonment/reduction of land management (B), Disturbance to birds (B), Infrastructure (C), Recreation/tourism (C)

The area is almost undisturbed by human activities, although important cliff-nesting species are sometimes disturbed by boating and fishing in the reservoirs along the cliffs. The ongoing abandonment of cultivation and sheep-raising will reduce the food resources currently available for the large raptors.

Côa valley — B2, C6 — 006

Admin region Centro
Coordinates 40°53'N 7°04'W
Altitude 200–976 m **Area** 23,727 ha

Site description

A steep valley with cliffs, olive and almond plantations, and sclerophyllous scrub, located in north-east Portugal and including a part of the Côa valley (a tributary of the Douro).

Habitats Scrub (58%; sclerophyllous scrub/garrigue/maquis), Wetland (3%; river/stream), Rocky areas (9%; inland cliff), Artificial landscape (30%; arable land; perennial crops/orchards/groves)
Land-use Agriculture, Hunting, Tourism/recreation, Water management

Birds

Species	Season	Year	Pop min	Pop max	Acc	Criteria
[1] ***Ciconia nigra*** Black Stork	B	1996	1	5	A	B2, C6
Ciconia ciconia White Stork	B	1996	1	5	A	C6
Neophron percnopterus Egyptian Vulture	B	1996	6	10	A	B2, C6
Gyps fulvus Griffon Vulture	R	1996	1	5	A	C6
Aquila chrysaetos Golden Eagle	R	1996	1	5	A	B2, C6
Hieraaetus fasciatus Bonnelli's Eagle	R	1996	6	10	A	B2, C6
Bubo bubo Eagle Owl	R	1996	11	20	B	B2, C6
Oenanthe leucura Black Wheatear	R	1996	Uncommon		—	C6

1. 2–5% of national breeding population.

The site holds important numbers of cliff-nesting bird species.

Protection status

National None **International** Partial
IBA overlaps with Special Protection Area.

Conservation issues

Threats Abandonment/reduction of land management (B), Construction/impact of dyke/dam/barrage (B), Infrastructure (B)

There are plans to build several large dams in the IBA, which will flood part of the area.

Ria de Aveiro — A4i, A4iii, B1i, B2, C2, C3, C4, C6 — 007

Admin region Centro
Coordinates 40°42'N 8°40'W
Altitude 0–79 m **Area** 52,720 ha

Site description

This site, located in the west and north-west of Portugal, includes a large estuarine area, salinas, freshwater marshes and alluvial forest associated with the main rivers, and shallower coastal waters offshore (to a depth of 20 m).

Habitats Forest and woodland (4%; native coniferous forest; alluvial/very wet forest), Grassland (10%; humid grassland), Wetland (44%; tidal river/enclosed tidal water; mudflat/sandflat; saltmarsh; coastal lagoon; water-fringe vegetation), Marine areas (29%; open sea), Artificial landscape (13%; arable land)
Land-use Agriculture, Fisheries/aquaculture, Forestry, Hunting, Military, Nature conservation/research, Tourism/recreation, Urban/industrial/transport

Birds

Species	Season	Year	Pop min	Pop max	Acc	Criteria
Ixobrychus minutus Little Bittern	B	—	Frequent		—	C6
Ardea purpurea Purple Heron	B	—	20	25	A	C6
Platalea leucorodia Spoonbill	P	—	Uncommon		—	C2
Melanitta nigra Common Scoter	W	1996	3,000	25,000	A	B1i, C3
Circus aeruginosus Marsh Harrier	R	1996	15	30	A	C6
Himantopus himantopus Black-winged Stilt	B	1996	50	100	A	C6
Recurvirostra avosetta Avocet	W	1996	800	1,800	A	A4i, B1i, B2, C2
Pluvialis squatarola Grey Plover	W	1996	800	1,400	A	B1i, C3
Calidris alpina Dunlin	W	1996	10,000	12,000	A	B2
Sterna albifrons Little Tern	B	1996	10	15	A	B2, C6

The site holds 20,000 or more wintering waterbirds on a regular basis, with important congregations of sea-duck offshore (*Melanitta nigra*) and waders on the intertidal flats (notably *Calidris alpina*). The freshwater marshes and estuarine areas support important breeding numbers of several waterbirds.

Protection status

National Low **International** High
681 ha of IBA covered by Nature Reserve (Dunas de S. Jacinto, 681 ha). 52,720 ha of IBA covered by Special Protection Area (Ria de Aveiro, 52,720 ha).

Conservation issues

Threats Agricultural intensification/expansion (A), Aquaculture/fisheries (A), Consequences of animal/plant introductions (C), Construction/impact of dyke/dam/barrage (A), Dredging/canalization (A), Industrialization/urbanization (B), Infrastructure (B)

The site is subject to several important impacts. The constant dredging of the channel that connects the whole area to the sea has increased tidal amplitude inside the lagoon system as well as water-flow. These two effects combine to increase erosion of sediments within the lagoon. The drainage and conversion to agriculture of inner parts of the lagoon are reducing the extent of alluvial forest and thus reducing the landscape diversity. Salinas are being destroyed by erosion or transformed into fish-farms. An invasive species of *Acacia* tree is strongly competing with native vegetation.

Malcata mountains — C6 — 008

Admin region Centro
Coordinates 40°15'N 7°02'W
Altitude 425–1,078 m **Area** 16,361 ha

Site description

A medium-sized mountain range with rolling hills covered with scrub and conifer plantations. There is a sharp difference in vegetation-types between the northern and southern slopes, with the former having a colder and more continental microclimate, while the latter is more Mediterranean in character.

Habitats Forest and woodland (broadleaved deciduous forest; broadleaved evergreen forest), Scrub (sclerophyllous scrub/garrigue/maquis), Wetland (standing fresh water; river/stream; water-fringe vegetation), Rocky areas (inland cliff), Artificial landscape (arable land; perennial crops/orchards/groves; forestry plantation)
Land-use Agriculture, Fisheries/aquaculture, Forestry, Hunting, Nature conservation/research, Tourism/recreation, Water management

Birds

Species	Season	Year	Pop min	Pop max	Acc	Criteria
Aegypius monachus Cinereous Vulture	R	1996	1	1	A	C6
Circaetus gallicus Short-toed Eagle	B	1996	2	3	A	C6
Circus pygargus Montagu's Harrier	B	1996	4	6	A	C6
Hieraaetus pennatus Booted Eagle	B	1996	3	4	A	C6
Bubo bubo Eagle Owl	R	1996	—	2	B	C6
Coracias garrulus Roller	B	1996	1	2	B	C6

An important area for breeding raptors, being one of only two breeding sites known for *Aegypius monachus* in Portugal in recent years.

Protection status

National High **International** High
15,523 ha of IBA covered by Nature Reserve (Serra da Malcata, 16,348 ha). 16,361 ha of IBA covered by Special Protection Area (Serra da Malcata, 16,361 ha).

Conservation issues

Habitat loss due to afforestation with non-native trees, namely *Pseudotsuga* and some *Pinus* species, is causing problems. Birds are disturbed by hunting.

Threats Afforestation (B), Burning of vegetation (B), Construction/impact of dyke/dam/barrage (C), Deforestation (commercial) (B), Disturbance to birds (C), Firewood collection (C), Intensified forest management (B), Recreation/tourism (C)

Paul do Taipal

C6 **009**

Admin region Centro
Coordinates 40°12′N 8°41′W
Altitude 13–116 m **Area** 233 ha

Site description

A small, densely vegetated freshwater marsh located in the lower Mondego valley, west of the town of Coimbra in central Portugal. An important part of a network of small wetlands in the Mondego valley, which includes two other IBAs: Paul de Arzila (010) and Paul de Madriz (011).

Habitats Wetland (standing fresh water; water-fringe vegetation)
Land-use Not utilized

Birds

Species	Season	Year	Pop min	Pop max	Acc	Criteria
Ixobrychus minutus Little Bittern	B	—	Abundant		—	C6
Egretta garzetta Little Egret	R	—	100	120	A	C6
Ardea purpurea Purple Heron	B	—	5	6	A	C6
Ciconia ciconia White Stork	B	—	6	10	A	C6
Milvus migrans Black Kite	B	—	Common		—	C6
Circus aeruginosus Marsh Harrier	R	—	2	5	A	C6
Himantopus himantopus Black-winged Stilt	B	—	Common		—	C6

The site supports a wide variety of breeding waterbirds, and is nationally important for wintering duck, as well as holding large numbers of migrating passerines during the autumn migration.

Protection status

National None **International** Partial
IBA overlaps with Special Protection Area.

Conservation issues

Threats Disturbance to birds (U)

Paul de Arzila

C6 **010**

Admin region Centro
Coordinates 40°10′N 8°33′W
Altitude 5–10 m **Area** 482 ha

Site description

A small freshwater marsh with an extensive reedbed, located in central-west Portugal, very close to the town of Coimbra. The marsh is bordered by pine woodland, and is situated in a relatively densely populated region. An important part of a network of small wetlands in the Mondego valley, which includes two other IBAs: Paul do Taipal (009) and Paul de Madriz (011).

Habitats Wetland (standing fresh water; water-fringe vegetation), Artificial landscape (forestry plantation)
Land-use Forestry, Nature conservation/research

Birds

Species	Season	Year	Pop min	Pop max	Acc	Criteria
Ixobrychus minutus Little Bittern	B	—	Common		—	C6
Ardea purpurea Purple Heron	B	—	—	—	—	C6
[1] ***Milvus migrans*** Black Kite	B	—	Common		—	C6

1. Probably the top site in Portugal.

The area supports important breeding numbers of two herons and a raptor, as well as large numbers of migratory passerines during autumn migration.

Protection status

National Partial **International** High
315 ha of IBA covered by Nature Reserve (Paul de Arzila, 580 ha). 482 ha of IBA covered by Special Protection Area (Paul de Arzila, 482 ha). IBA also overlaps with Ramsar Site (Paul de Arzila, 585 ha).

Conservation issues

Threats Disturbance to birds (C), Infrastructure (B)

There are plans to build a new road on the edge of the site.

Paul de Madriz

C6 **011**

Admin region Centro
Coordinates 40°08′N 8°38′W
Altitude 5–10 m **Area** 89 ha

Site description

A small freshwater marsh bordered by pine woodland, located in central Portugal, very close to the town of Coimbra, on a tributary of the Mondego river. An important part of a network of small wetlands in the Mondego valley, which includes two other IBAs: Paul do Taipal (009) and Paul de Arzila (010).

Habitats Wetland (100%; standing fresh water; water-fringe vegetation)
Land-use Not utilized

Birds

Species	Season	Year	Pop min	Pop max	Acc	Criteria
Ixobrychus minutus Little Bittern	B	—	Common		—	C6
[1] ***Ardea purpurea*** Purple Heron	B	—	—	—	—	C6
[1] ***Circus aeruginosus*** Marsh Harrier	B	—	—	—	—	C6

1. Possibly breeds.

The reedbeds are important for three breeding waterbirds, and support large numbers of migratory passerines during autumn migration. Wintering duck numbers regularly exceed 2,000 birds.

Protection status

National None **International** High
89 ha of IBA covered by Special Protection Area (Paul de Madriz, 89 ha). IBA also overlaps with Ramsar Site (Paul de Madriz, 226 ha).

Conservation issues

No serious threats are known at the site, which is presently undisturbed by human activities.

Toulões and Idanha-a-Nova

A1, B2, C1, C6 **012**

Admin region Centro
Coordinates 39°54′N 7°01′W
Altitude 200–407 m **Area** 10,920 ha

Site description

An area of steppic habitats and open woodland of evergreen oak *Quercus*, in central-eastern Portugal, close to Castelo-Branco.

Habitats Forest and woodland (27%; broadleaved evergreen forest), Scrub (25%; sclerophyllous scrub/garrigue/maquis), Wetland (3%; standing fresh water; water-fringe vegetation), Artificial landscape (45%; arable land; perennial crops/orchards/groves)
Land-use Agriculture, Forestry

Birds

Species	Season	Year	Pop min	Pop max	Acc	Criteria
Milvus migrans Black Kite	B	—	—	—	—	C6
Circaetus gallicus Short-toed Eagle	B	1996	1	5	B	C6
Circus pygargus Montagu's Harrier	B	—	—	—	—	C6
Hieraaetus pennatus Booted Eagle	B	1996	1	5	B	C6
[1] ***Otis tarda*** Great Bustard	B	1996	11	11	A	A1, C1, C6
Burhinus oedicnemus Stone Curlew	B	1996	20	50	B	B2, C6
Pterocles orientalis Black-bellied Sandgrouse	R	1996	Frequent		—	C6
Coracias garrulus Roller	B	1996	1	5	B	C6
Melanocorypha calandra Calandra Lark	B	1996	Common		—	C6

1. Large decrease.

The area supports a wide diversity of bird species of steppic habitats.

■ **Protection status**

National None **International** None

■ **Conservation issues**

Threats Agricultural intensification/expansion (B)

The intensification of agriculture (including irrigation) will reduce the extent of existing steppic habitats.

Upper River Tejo B2, C6 013

Admin region Centro
Coordinates 39°40′N 7°15′W
Altitude 150–316 m **Area** 26,129 ha

■ **Site description**

Steep valleys with cliffs and Mediterranean scrub, located at the eastern border of central Portugal. The riverbed is much wider than it was 25 years ago, due to the construction of a large dam, and now allows the small-boat traffic of professional fishermen and recreational purposes.

Habitats Forest and woodland (10%; broadleaved evergreen forest), Scrub (55%; sclerophyllous scrub/garrigue/maquis), Wetland (15%; standing fresh water; water-fringe vegetation), Rocky areas (5%; scree/boulders; inland cliff), Artificial landscape (15%; arable land; perennial crops/orchards/groves; forestry plantation)
Land-use Agriculture, Fisheries/aquaculture, Forestry, Hunting, Nature conservation/research, Tourism/recreation, Water management

■ **Birds**

Species	Season	Year	Pop min	Pop max	Acc	Criteria
Ciconia nigra Black Stork	B	1995	15	20	A	B2, C6
Milvus migrans Black Kite	B	1996	6	10	B	B2
Neophron percnopterus Egyptian Vulture	B	1996	15	17	A	B2, C6
Gyps fulvus Griffon Vulture	R	1996	70	75	A	B2, C6
Circaetus gallicus Short-toed Eagle	B	1996	1	5	B	C6
Hieraaetus pennatus Booted Eagle	B	1996	11	50	B	B2, C6
Hieraaetus fasciatus Bonnelli's Eagle	R	1996	1	5	B	B2, C6
Bubo bubo Eagle Owl	R	1996	6	10	B	C6
Oenanthe leucura Black Wheatear	R	—	Uncommon		—	C6

Important numbers of cliff-nesting species breed here, especially large raptors.

■ **Protection status**

National None **International** Partial
IBA overlaps with Special Protection Area.

■ **Conservation issues**

Threats Afforestation (B), Aquaculture/fisheries (C), Disturbance to birds (C), Infrastructure (C), Recreation/tourism (C), Other (C)

Important cliff-nesting species are disturbed by boating and fishing activity in the reservoirs along the cliffs. In recent years several forest plantations (*Eucalyptus*) and special hunting areas have been established along the edge of the site, which have affected birds. In particular, there have been attempts in the hunting areas to control predators, including large raptors ('Other' threat).

Berlenga and Farilhões islands A4ii, A4iii, B1ii, B2, C2, C4, C6 014

Admin region Lisboa e Vale do Tejo
Coordinates 39°25′N 9°31′W
Altitude 0–90 m **Area** 9,560 ha

■ **Site description**

Habitats Wetland (shingle/stony beach), Marine areas (open sea; sea inlet/coastal features), Rocky areas (sea cliff/rocky shore; rock stacks/islets), Introduced/exotic vegetation
Land-use Fisheries/aquaculture, Military, Nature conservation/research, Tourism/recreation

A small archipelago of rocky islets, the only sizeable such group off the Portuguese mainland, located c.8 km off the west coast. The largest of the islets covers only 2.5 km². Tourists and anglers visit the area in spring and summer.

■ **Birds**

Species	Season	Year	Pop min	Pop max	Acc	Criteria
Calonectris diomedea Cory's Shearwater	B	1996	180	220	A	C6
Oceanodroma castro Madeiran Petrel	B	1995	200	400	A	A4ii,B1ii,B2,C2,C6

This is the only known breeding colony of *Oceanodroma castro* near the European mainland. The site holds 10,000 or more pairs of seabirds on a regular basis. The number of breeding *Larus cachinnans* has increased dramatically in recent years and this colony is now controlled by culling. Berlenga marks the southern limit in Europe for breeding *Uria aalge*, with 45 individuals (declined from 6,000 pairs in 1939), and the northern limit for breeding *Calonectris diomedea*. Significant proportion (≥1%) of national population breeding at site: *Falco peregrinus* (2 pairs).

■ **Protection status**

National High **International** High
9,560 ha of IBA covered by Nature Reserve (Berlenga, 9,560 ha). 9,560 ha of IBA covered by Special Protection Area (Ilhas Berlengas, 9,560 ha).

■ **Conservation issues**

Threats Consequences of animal/plant introductions (B), Disturbance to birds (C), Recreation/tourism (A)

Tourists and anglers cause a certain amount of disturbance to breeding birds. The introduction of the non-native and highly invasive creeper *Carpobrotus* has affected the natural vegetation cover.

Paul do Boquilobo A4i, B1i, C2, C3, C6 015

Admin region Lisboa e Vale do Tejo
Coordinates 39°23′N 8°32′W
Altitude 10–15 m **Area** 433 ha

■ **Site description**

Freshwater marshes with alluvial forest and other water-fringe vegetation, located in the centre of the country, c.90 km north-east of Lisbon. The site is drained every year so that the outer areas can be cultivated but the core area is left flooded in most years. This IBA is located very close to the expanding town of Golegã.

Habitats Forest and woodland (alluvial/very wet forest), Wetland (standing fresh water; river/stream; water-fringe vegetation), Artificial landscape (arable land), Introduced/exotic vegetation
Land-use Agriculture, Nature conservation/research, Water management

■ **Birds**

Species	Season	Year	Pop min	Pop max	Acc	Criteria
Ixobrychus minutus Little Bittern	B	1996	Common		—	C6
Nycticorax nycticorax Night Heron	B	1996	30	100	A	C6
Ardeola ralloides Squacco Heron	B	1995	Rare		—	C6
Bubulcus ibis Cattle Egret	R	1995	1,000	2,000	A	A4i, B1i, C3
Egretta garzetta Little Egret	B	1996	200	400	B	A4i, B1i, C2, C6
Ardea purpurea Purple Heron	B	1995	6	12	A	C6
Platalea leucorodia Spoonbill	B	1995	20	30	A	B1i, C2, C6
Milvus migrans Black Kite	B	1996	Common		—	C6
Chlidonias hybridus Whiskered Tern	B	1996	40	120	A	C6

One of the largest heronries in Iberia, also holding important numbers of breeding *Platalea leucorodia*, *Milvus migrans* and *Chlidonias hybridus*, and nationally important numbers of wintering duck.

■ **Protection status**

National Partial **International** High
369 ha of IBA covered by Nature Reserve (Paul do Boquilobo, 549 ha). 433 ha of IBA covered by Special Protection Area (Paúl do Boquilobo, 433 ha). 395 ha of IBA covered by Biosphere Reserve (Paul do Boquilobo, 395 ha). IBA also overlaps with Ramsar Site (Paul do Boquilobo, 529 ha).

Conservation issues

Threats Agricultural intensification/expansion (B), Consequences of animal/plant introductions (A), Drainage (A), Industrialization/urbanization (B), Natural events (B)

The presence of the introduced and highly invasive water-hyacinth *Eichhornia crassipes* causes a certain degree of ecological disruption. The site is also subject to pressure from farmers who want to intensify farming practices and increase the extent of cultivation (e.g. by further drainage). Urban expansion (Golegã) also causes problems.

Cabeção — A4iii, B2, C4, C6 — 016

Admin region Alentejo
Coordinates 39°00'N 8°13'W
Altitude 100–200 m **Area** 20,227 ha

Site description

An extensive, evergreen woodland of cork oak *Quercus suber*, in central Portugal, inland and south of the River Tejo. The woodland is managed and there is some cultivation.

Habitats Forest and woodland (75%; broadleaved evergreen forest), Artificial landscape (25%; arable land; perennial crops/orchards/groves; forestry plantation)
Land-use Agriculture (20%), Forestry (80%), Hunting, Nature conservation/research, Tourism/recreation, Urban/industrial/transport, Water management

Birds

Species	Season	Year	Pop min	Pop max	Acc	Criteria
Elanus caeruleus Black-winged Kite	R	1996	4	7	A	B2, C6
Milvus migrans Black Kite	B	1996	2	4	A	C6
Circaetus gallicus Short-toed Eagle	B	1996	10	13	A	B2, C6
Hieraaetus pennatus Booted Eagle	B	1996	36	40	A	B2, C6

The site supports important numbers of breeding birds of prey. Significant proportion (≥1%) of national population breeding at site: *Circus pygargus* (2–4 pairs).

Protection status

National None **International** None

Conservation issues

Threats Afforestation (C), Agricultural intensification/expansion (B), Disturbance to birds (C), Natural events (C), Other (B), Selective logging/cutting (C)

'Other' threats (see above) are forest degradation, tree-disease and soil degradation.

Alter do Chão — A1, A4iii, B2, C1, C4, C6 — 017

Admin region Alentejo
Coordinates 39°10'N 7°42'W
Altitude 200–300 m **Area** 7,959 ha

Site description

Steppic agricultural land of wheat-fields and dry pasture, with scattered evergreen oak *Quercus* woodland.

Habitats Forest and woodland (10%; broadleaved evergreen forest), Artificial landscape (90%; arable land; perennial crops/orchards/groves)
Land-use Agriculture (90%), Forestry (10%)

Birds

Species	Season	Year	Pop min	Pop max	Acc	Criteria
Circus pygargus Montagu's Harrier	B	1996	Common		—	C6
Tetrax tetrax Little Bustard	R	1996	100	150	A	A1, B2, C1
Otis tarda Great Bustard	R	1996	25	25	A	A1, C1

Important for three species of steppic habitats.

Protection status

National None **International** None

Conservation issues

Threats Agricultural intensification/expansion (A), Consequences of animal/plant introductions (B), Disturbance to birds (B)

Birds are disturbed by increasing road traffic and by aircraft used in aerial spraying for agriculture.

Monforte plains — A1, A4iii, B2, B3, C1, C4, C6 — 018

Admin region Alentejo
Coordinates 39°04'N 7°21'W
Altitude 250–350 m **Area** 9,426 ha

Site description

Open country with cultivation and grassland, alternating with open woodland of evergreen oak *Quercus ilex* and olive-groves. Fallow fields are grazed by sheep.

Habitats Forest and woodland (10%; broadleaved evergreen forest), Artificial landscape (90%; arable land; perennial crops/orchards/groves)
Land-use Agriculture (90%), Forestry (10%)

Birds

Species	Season	Year	Pop min	Pop max	Acc	Criteria
Elanus caeruleus Black-winged Kite	R	1996	6	6	A	B2, C6
Circus pygargus Montagu's Harrier	B	1996	Common		—	B3
Tetrax tetrax Little Bustard	R	1996	80	100	A	A1, B2, C1
Otis tarda Great Bustard	R	1996	15	15	A	A1, C1
Burhinus oedicnemus Stone Curlew	R	1996	Common		—	B2

Important for breeding birds of steppic habitats.

Protection status

National None **International** None

Conservation issues

Threats Afforestation (A), Agricultural intensification/expansion (A), Construction/impact of dyke/dam/barrage (B), Disturbance to birds (B)

Nesting birds are disturbed by increasing road-traffic. Agricultural practices are undergoing moderate intensification, which will lead to a greater human influence on the site.

Campo Maior — A1, B1i, B2, C1, C2, C6 — 019

Admin region Alentejo
Coordinates 39°02'N 7°01'W
Altitude 190–341 m **Area** 9,355 ha

Site description

Open plains used for cultivation of cereals (mostly wheat) in rotation with melon, sunflower and tomato. Pastoral woodland and fallow areas are grazed by sheep and cattle. Low level of human activity.

Habitats Forest and woodland (30%; broadleaved evergreen forest), Artificial landscape (70%; arable land)
Land-use Agriculture (70%), Forestry (30%)

Birds

Species	Season	Year	Pop min	Pop max	Acc	Criteria
[1] ***Circus pygargus*** Montagu's Harrier	B	1996	15	20	A	C6
Falco naumanni Lesser Kestrel	B	1996	14	18	A	A1, C1, C6
Grus grus Crane	W	1996	800	1,000	A	B1i, C2
Tetrax tetrax Little Bustard	R	1996	200	250	A	A1, B2, C1, C6
Otis tarda Great Bustard	R	1996	90	100	A	A1, B2, C1, C6
Burhinus oedicnemus Stone Curlew	B	1996	20	30	B	B2, C6

1. Large decrease.

An important area for the conservation of *Otis tarda* in Portugal, as well as one of the most important wintering sites for *Grus grus*.

Protection status

National None **International** Partial
IBA overlaps with Special Protection Area.

Conservation issues

Threats Afforestation (B), Agricultural intensification/expansion (A), Construction/impact of dyke/dam/barrage (A), Disturbance to birds (B), Dredging/canalization (A), Infrastructure (B)

The non-intensive cultivation of cereals maintains favourable habitat for steppic bird species, but the advent of irrigated crops in the IBA and the construction of a large dam nearby, to provide water for further irrigation, are strong indicators that further agricultural intensification will take place in the future. Birds are disturbed by the increasing road-traffic and by irrigation structures operating in the vicinity.

Vila Fernando A1, B2, C1, C6 020

Admin region Alentejo
Coordinates 38°54′N 7°21′W
Altitude 300–470 m **Area** 8,166 ha

Site description

Steppic agricultural land of wheat-fields and dry pasture, with scattered evergreen oak *Quercus* woodland. Land-uses include cultivation and livestock-grazing. There is scattered human habitation. In the previous international IBA inventory (Grimmett and Jones 1989), this site was called Elvas Plains (former site PT028).

Habitats Forest and woodland (20%; broadleaved evergreen forest), Artificial landscape (80%; arable land)
Land-use Agriculture (80%), Forestry (20%)

Birds

Species	Season	Year	Pop min	Pop max	Acc	Criteria
Elanus caeruleus Black-winged Kite	R	1994	6	6	A	B2, C6
Milvus migrans Black Kite	B	1994	4	4	A	C6
Tetrax tetrax Little Bustard	R	1996	80	100	A	A1, B2, C1
[1] ***Otis tarda*** Great Bustard	R	1996	40	45	A	A1, B2, C1, C6
Burhinus oedicnemus Stone Curlew	B	1994	15	20	A	B2, C6
[2] ***Pterocles orientalis*** Black-bellied Sandgrouse	R	1996	Common		—	C6
Coracias garrulus Roller	B	1996	Common		—	C6
Melanocorypha calandra Calandra Lark	B	1996	Abundant		—	C6

1. Large decrease.
2. Breeds regularly (numbers unknown).

Important for breeding birds of steppic habitats. Significant proportion (≥1%) of national population breeding at site: *Circus pygargus* (8–10 pairs).

Protection status

National None **International** None

Conservation issues

Threats Agricultural intensification/expansion (A), Construction/impact of dyke/dam/barrage (B), Disturbance to birds (A)

Nesting birds are disturbed by increasing road-traffic in the region.

Tejo estuary A1, A4i, A4iii, B1i, B2, B3, C1, C2, C3, C4, C6 021

Admin region Lisboa e Vale do Tejo
Coordinates 38°49′N 8°55′W
Altitude 0–43 m **Area** 45,071 ha

Site description

The largest estuary in the country, located in central Portugal. The site includes the upper and central parts of the River Tejo estuary and the valley of a small tributary (Ria Enguias). Large expanses of mudflats, saltmarshes and rice-fields predominate. Lisbon is located at the mouth of the River Tejo, and more than two million people live around this IBA.

Habitats Forest and woodland (15%; broadleaved evergreen forest), Wetland (65%; tidal river/enclosed tidal water; mudflat/sandflat; saltmarsh; coastal lagoon; water-fringe vegetation), Artificial landscape (20%; arable land; forestry plantation)
Land-use Agriculture, Fisheries/aquaculture, Forestry, Hunting, Nature conservation/research, Urban/industrial/transport, Water management

Birds

Species	Season	Year	Pop min	Pop max	Acc	Criteria
Ixobrychus minutus Little Bittern	B	—	Common		—	C6
Ardea purpurea Purple Heron	B	—	150	200	A	C2, C6
Ciconia ciconia White Stork	R	1996	50	100	B	B2, C6
Phoenicopterus ruber Greater Flamingo	W	1995	1,000	3,000	A	A4i, B1i, C2
[1] ***Anser anser*** Greylag Goose	W	1996	1,200	3,400	A	B1i, C3
Anas crecca Teal	W	1996	5,000	11,000	A	B1i, C3
Anas clypeata Shoveler	W	1996	1,500	4,500	A	B1i, C3
Elanus caeruleus Black-winged Kite	R	1996	6	10	B	B2, C6
Circus aeruginosus Marsh Harrier	R	1996	30	50	B	C6
Circus pygargus Montagu's Harrier	B	1996	20	40	B	B3, C6
Hieraaetus pennatus Booted Eagle	B	1996	10	20	B	B2, C6
Porzana pusilla Baillon's Crake	B	—	Rare		—	C6
[2] ***Tetrax tetrax*** Little Bustard	W	1992	400	800	A	A1, C1
Himantopus himantopus Black-winged Stilt	B	1996	500	750	A	A4i, B1i, C2, C6
[3] ***Recurvirostra avosetta*** Avocet	W	1996	4,000	6,000	A	A4i, B1i, B2, C2
Glareola pratincola Collared Pratincole	B	—	100	250	A	A4i, B1i, B2, C2, C6
Pluvialis squatarola Grey Plover	W	1996	3,000	5,000	A	A4i, B1i, C3
[3] ***Calidris alpina*** Dunlin	W	1996	10,000	20,000	A	B1i, B2, C3
Limosa limosa Black-tailed Godwit	W	1997	30,000	80,000	A	A4i, B1i, C3
Tringa totanus Redshank	W	1996	2,000	3,000	A	A4i, B1i, C3
Sterna albifrons Little Tern	B	1996	20	20	A	B2, C6
Melanocorypha calandra Calandra Lark	R	—	Common		—	C6
Calandrella brachydactyla Short-toed Lark	B	—	Abundant		—	B2, C6

1. Large increase.
2. Breeds (c.5 males).
3. Large decrease.

The site holds 20,000 or more wintering waterbirds on a regular basis, and is the most important locality in Portugal for wintering flamingos, ducks, waders and gulls. The area also holds important numbers of a wide variety of breeding waterbirds. The reedbeds in the upper part of the estuary support large numbers of migrating passerines during autumn.

Protection status

National Partial **International** High
14,192 ha of IBA covered by Nature Reserve (Estuário do Tejo, 14,192 ha). 45,071 ha of IBA covered by Special Protection Area (Estuário do Tejo, 45,071 ha). 14,563 ha of IBA covered by Ramsar Site (Estuário do Tejo, 14,563 ha).

Conservation issues

Threats Agricultural intensification/expansion (B), Aquaculture/fisheries (B), Consequences of animal/plant introductions (A), Disturbance to birds (B), Dredging/canalization (C), Industrialization/urbanization (A), Infrastructure (A)

Disturbance to birds can be moderately intense locally, due to the high human population density, and due to the use of aircraft to seed and spray chemicals onto rice-fields during the breeding season. Several construction projects, including a very large bridge across the Tejo, are underway or are proposed for the site or along its edges. The introduced and highly invasive water-hyacinth *Eichhornia crassipes* is affecting the natural vegetation cover in the network of channels and is reducing the area of open water. There is a management plan for the Nature Reserve.

Coast of the Setúbal peninsula C6 022

Admin region Lisboa e Vale do Tejo
Coordinates 38°29′N 9°04′W
Altitude 0–501 m **Area** 22,930 ha

Site description

The site, between Setúbal and Lisbon on the west coast, incorporates a wide variety of habitats and landscapes, from Mediterranean maquis to limestone cliffs, and from pine woodland to reedbeds (*Phragmites*), as well as a coastal lagoon.

Habitats Forest and woodland (17%; broadleaved deciduous forest; native coniferous forest), Scrub (17%; sclerophyllous scrub/garrigue/maquis), Wetland (4%; sand-dunes/sand beach; coastal lagoon; water-fringe vegetation), Marine areas (15%; open sea), Rocky areas (10%; sea cliff/rocky shore), Artificial landscape (37%; arable land; perennial crops/orchards/groves; other urban/industrial areas)
Land-use Agriculture, Fisheries/aquaculture, Forestry, Hunting, Nature conservation/research, Tourism/recreation, Urban/industrial/transport

Birds

Species	Season	Year	Pop min	Pop max	Acc	Criteria
Ixobrychus minutus Little Bittern	B	1996	1	5	B	C6
Hieraaetus fasciatus Bonnelli's Eagle	R	1996	1	1	A	C6
Falco peregrinus Peregrine	R	1996	2	5	B	C6
Sterna sandvicensis Sandwich Tern	P	1996	300	1,000	B	C6
Bubo bubo Eagle Owl	R	1996	6	10	B	C6

The site is important throughout the year for some waterbirds and raptors. The reedbeds that border the inner parts of the coastal lagoon support large numbers of migratory passerines during the autumn migration.

Protection status

National Partial **International** Partial
10,821 ha of IBA covered by Natural Park (Arrábida, 11,054 ha). IBA also overlaps with Ramsar Site. IBA overlaps with Special Protection Area.

Conservation issues

Threats Extraction industry (A), Industrialization/urbanization (A), Infrastructure (A), Recreation/tourism (A), Unsustainable exploitation (B)

There is a management plan for the Natural Park.

Sado estuary — A4i, A4iii, B1i, B2, C2, C3, C4, C6 — 023

Admin region Alentejo
Coordinates 38°27'N 8°43'W
Altitude 0–30 m **Area** 24,633 ha

Site description

A large estuary with extensive mudflats and saltmarshes, located in central Portugal, c.50 km south of Lisbon. Setúbal is the main town in its vicinity. The inner part of the estuary is bordered by rice-fields and reedbeds, whereas the north-western limits are densely populated and the south-western parts are sheltered from the open sea by a sandbar.

Habitats Forest and woodland (mixed forest; broadleaved evergreen forest), Scrub (sclerophyllous scrub/garrigue/maquis), Wetland (tidal river/enclosed tidal water; mudflat/sandflat; saltmarsh; sand-dunes/sand beach; coastal lagoon; standing fresh water; water-fringe vegetation), Artificial landscape (arable land; forestry plantation; other urban/industrial areas)
Land-use Agriculture, Fisheries/aquaculture, Forestry, Hunting, Nature conservation/research, Tourism/recreation, Urban/industrial/transport

Birds

Several waterbird species (mainly waders) occur in winter in large, internationally significant numbers, and the site holds 20,000 or more wintering waterbirds on a regular basis. There are also important numbers of breeding waterbirds, notably *Himantopus himantopus* and *Sterna albifrons*. The rice-fields are an important feeding ground for the egrets and herons from the nearby breeding colony at Murta dam (024). More than 500,000 *Columba palumbus* roost within the IBA in winter.

Species	Season	Year	Pop min	Pop max	Acc	Criteria
[1] ***Phalacrocorax carbo*** Cormorant	W	1996	1,000	2,200	A	B1i, C3
Ixobrychus minutus Little Bittern	B	1996	Common		—	C6
[2] ***Egretta garzetta*** Little Egret	N	1996	Common		—	C6
Ardea purpurea Purple Heron	B	—	15	15	—	C6
Ciconia ciconia White Stork	R	1996	50	70	A	C6
[3] ***Circus aeruginosus*** Marsh Harrier	R	1996	5	6	B	C6
[4] ***Himantopus himantopus*** Black-winged Stilt	B	1996	200	800	A	A4i, B1i, C2, C6
[5] ***Recurvirostra avosetta*** Avocet	W	1996	2,000	4,000	A	A4i, B1i, B2, C2
[6] ***Glareola pratincola*** Collared Pratincole	B	1996	5	20	B	C6
Calidris alpina Dunlin	W	1996	12,000	16,000	A	B1i, B2, C3
Limosa limosa Black-tailed Godwit	W	1996	2,000	11,000	A	B1i, C3
Sterna albifrons Little Tern	B	1996	50	100	A	B2, C6

1. Large increase; *P. c. sinensis* predominates.
2. c.1,000 wintering.
3. 20–30 birds wintering.
4. 150–350 birds wintering.
5. 1–2 pairs breed irregularly.
6. Breeding seems to be irregular.

Protection status

National Partial **International** High
20,011 ha of IBA covered by Nature Reserve (Estuário do Sado, 23,971 ha). 24,633 ha of IBA covered by Special Protection Area (Estuário do Sado, 24,633 ha). IBA also overlaps with Ramsar Site (Estuário do Sado, 25,588 ha).

Conservation issues

Threats Agricultural intensification/expansion (C), Aquaculture/fisheries (A), Dredging/canalization (B), Groundwater abstraction (C), Industrialization/urbanization (B), Infrastructure (C), Natural events (C), Recreation/tourism (C)

Over 30% of the available area of saltpans has been transformed into fish-farms in the course of the last ten years. There are plans for large-scale dredging at the mouth of the estuary that may affect water-flow and sediment deposition inside the estuary, thus indirectly affecting the bird community.

Murta dam — A4i, B1i, C2, C3, C6 — 024

Admin region Alentejo
Coordinates 38°23'N 8°43'W
Altitude 20–35 m **Area** 498 ha

Site description

A small reservoir and marshes, very close to the west coast of Portugal, at the southern edge of the Sado estuary (023). The wetland is surrounded by inland sand-dunes which are planted with stone pine *Pinus pinea*.

Habitats Scrub (scrub; sclerophyllous scrub/garrigue/maquis), Wetland (standing fresh water; water-fringe vegetation), Artificial landscape (forestry plantation)
Land-use Forestry, Water management (100%)

Birds

Species	Season	Year	Pop min	Pop max	Acc	Criteria
[1] ***Botaurus stellaris*** Bittern	B	—	Rare		—	C6
Bubulcus ibis Cattle Egret	B	—	1,000	2,000	A	A4i, B1i, C3
Egretta garzetta Little Egret	B	—	600	1,000	A	A4i, B1i, C2, C6
Ardea purpurea Purple Heron	B	—	5	6	A	C6

1. The only confirmed breeding site in Portugal in recent years.

An important heronry. Significant proportion (≥1%) of national population breeding at site: *Hieraaetus pennatus* (1 pair). The site also holds nationally important numbers of wintering duck.

Protection status

National None **International** High
498 ha of IBA covered by Special Protection Area (Açude da Murta, 498 ha). IBA also overlaps with Ramsar Site (Estuário do Sado, 25,588 ha).

Conservation issues

Threats Disturbance to birds (C), Unsustainable exploitation (C), Other (B)

There is illegal bird-hunting in autumn and winter, and uncontrolled visitors to the area can cause significant disturbance to breeding herons and egrets. In some years the water of the reservoir, which is used for irrigating the nearby rice-fields, reaches very low levels during the breeding season, allowing predators to reach the nesting birds ('Other' threat, above).

Évora plains
A1, B2, C1, C6 **025**

Admin region Alentejo
Coordinates 38°27'N 7°57'W
Altitude 175–300 m **Area** 33,693 ha

Site description
Open agricultural land used for cereals (mainly wheat) and other crops in rotation such as melon, sunflower and beans. There is sparse evergreen woodland of oak *Quercus ilex* and cork oak *Q. suber*. Fallow land is used for grazing sheep and cattle.

Habitats Forest and woodland (30%; broadleaved evergreen forest), Artificial landscape (70%; arable land)
Land-use Agriculture (70%), Forestry (30%)

Birds

Species	Season	Year	Pop min	Pop max	Acc	Criteria
Elanus caeruleus Black-winged Kite	R	—	Common		—	C6
Falco naumanni Lesser Kestrel	B	1996	Common		—	C6
Grus grus Crane	W	1996	100	200	A	C6
Tetrax tetrax Little Bustard	R	1996	200	250	A	A1, B2, C1, C6
Otis tarda Great Bustard	R	1996	25	30	A	A1, B2, C1, C6
Melanocorypha calandra Calandra Lark	R	1996	Common		—	C6

The site supports good numbers of breeding birds of steppic habitats, and is important for wintering *Grus grus*.

Protection status
National None **International** None

Conservation issues
Threats Agricultural intensification/expansion (A), Construction/impact of dyke/dam/barrage (B), Disturbance to birds (B)

Birds are disturbed by the increasing road-traffic. Current habitats, favourable for steppic birds, are maintained by the traditional farming system based on non-intensive cultivation of cereals, but the weakness of the rural economy might lead farmers to change land-uses.

Alvito
A1, B2, C1, C6 **026**

Admin region Alentejo
Coordinates 38°12'N 7°58'W
Altitude 100–179 m **Area** 2,606 ha

Site description
A steppic agricultural area of wheat-fields and dry pasture, with scattered evergreen oak *Quercus* woodland. In the previous pan-European IBA inventory (Grimmett and Jones 1989) this site was called Vidigueira–Beja (former site PT034).

Habitats Forest and woodland (20%; broadleaved evergreen forest), Artificial landscape (80%; arable land; perennial crops/orchards/groves)
Land-use Agriculture (80%), Forestry (20%)

Birds

Species	Season	Year	Pop min	Pop max	Acc	Criteria
Tetrax tetrax Little Bustard	R	1996	150	200	B	A1, B2, C1, C6
Otis tarda Great Bustard	R	1996	15	15	A	A1, C1

An important area for two bustard species. Significant proportion (≥1%) of national population breeding at site: *Ciconia ciconia* (14–16 pairs).

Protection status
National None **International** None

Conservation issues
Threats Agricultural intensification/expansion (A), Disturbance to birds (B)

Birds are disturbed by the increasing road-traffic.

Mourão, Moura e Barrancos
A1, B1i, B2, B3, C1, C2, C6 **027**

Admin region Alentejo
Coordinates 38°09'N 7°11'W
Altitude 150–535 m **Area** 82,667 ha

Site description
The area is a mixture of non-intensive cereal cultivation, semi-permanent pasture, and evergreen woodland of oak *Quercus ilex* and *Q. suber*. Rivers are bordered locally by vegetation and cliffs. The area is grazed by sheep and cattle, and there are small olive-groves and vineyards. Housing density is very low.

Habitats Forest and woodland (35%; broadleaved evergreen forest), Grassland (18%; steppe/dry calcareous grassland), Artificial landscape (47%; arable land; perennial crops/orchards/groves)
Land-use Agriculture (65%), Forestry (35%)

Birds

Species	Season	Year	Pop min	Pop max	Acc	Criteria
Ciconia nigra Black Stork	B	1996	5	8	A	B2, C6
Ciconia ciconia White Stork	B	1996	60	100	A	B2, C6
Elanus caeruleus Black-winged Kite	R	1996	6	10	B	B2, C6
Milvus migrans Black Kite	B	1996	8	15	B	B2
Circaetus gallicus Short-toed Eagle	B	1996	8	15	A	B2, C6
Circus pygargus Montagu's Harrier	B	1996	20	40	B	B3, C6
Aquila adalberti Spanish Imperial Eagle	N	1996	2	4	—	C6
Aquila chrysaetos Golden Eagle	R	1996	3	3	A	B2, C6
Hieraaetus pennatus Booted Eagle	B	1996	10	15	A	B2, C6
Grus grus Crane	W	1996	500	1,500	A	B1i, C2, C6
Tetrax tetrax Little Bustard	R	1996	250	500	A	A1, B2, C1, C6
Otis tarda Great Bustard	R	1996	80	100	A	A1, B2, C1, C6
Burhinus oedicnemus Stone Curlew	R	1996	100	250	A	B2, C6
Glareola pratincola Collared Pratincole	B	1996	6	10	B	B2, C6
Pterocles orientalis Black-bellied Sandgrouse	R	1996	11	50	B	C6
Bubo bubo Eagle Owl	R	1996	6	10	B	B2, C6

An important breeding area for many species of steppic habitats, especially raptors, as well as the most important wintering locality for *Grus grus* in Portugal.

Protection status
National None **International** Partial
IBA overlaps with Special Protection Area.

Conservation issues
Threats Abandonment/reduction of land management (C), Afforestation (B), Agricultural intensification/expansion (B), Construction/impact of dyke/dam/barrage (A), Deforestation (commercial) (B), Disturbance to birds (C), Dredging/canalization (C), Industrialization/urbanization (A), Infrastructure (A), Recreation/tourism (C)

Birds are disturbed by increasing road traffic in the region.

Santo André and Sancha lagoons
A4i, B1i, C3, C6 **028**

Admin region Alentejo Litoral
Coordinates 38°03'N 8°48'W
Altitude 0–47 m **Area** 2,638 ha

Site description
A complex of two coastal lagoons and a stretch of well-consolidated sand-dunes, on the south-west coast of Portugal, just north of the town of Sines. Both lagoons have extensive reedbeds. This site is an amalgamation of two sites that featured in the previous pan-European IBA inventory (Grimmett and Jones 1989): Santo André Lagoon (former site PT021) and Sancha Lagoon (former site PT022).

Habitats Scrub (sclerophyllous scrub/garrigue/maquis), Wetland (sand-dunes/sand beach; coastal lagoon; river/stream; water-fringe vegetation), Artificial landscape (forestry plantation)
Land-use Agriculture, Fisheries/aquaculture, Hunting, Tourism/recreation, Urban/industrial/transport, Water management

Birds
Important numbers of breeding and wintering waterbirds occur, and the site also harbours large numbers of migrating passerines during autumn migration.

Species	Season	Year	Pop min	Pop max	Acc	Criteria
Ixobrychus minutus Little Bittern	B	1996	Frequent		—	C6
Ardea purpurea Purple Heron	B	1994	9	9	A	C6
[1] ***Netta rufina*** Red-crested Pochard	W	—	—	1,000	—	A4i, B1i, C3
Elanus caeruleus Black-winged Kite	R	—	Frequent		—	C6
Circus aeruginosus Marsh Harrier	R	—	Common		—	C6
Porzana pusilla Baillon's Crake	B	—	Uncommon		—	C6
Himantopus himantopus Black-winged Stilt	R	—	10	20	B	C6
Sterna albifrons Little Tern	B	—	Common		—	C6

1. 10–20 pairs breed.

Protection status

National None **International** High

409 ha of IBA covered by Special Protection Area (Lagoa da Sancha, 409 ha). 2,165 ha of IBA covered by Special Protection Area (Lagoa da Sto. André, 2,165 ha).

Conservation issues

Threats Aquaculture/fisheries (B), Consequences of animal/plant introductions (C), Disturbance to birds (B), Industrialization/urbanization (B), Infrastructure (B)

Hunting, grazing, and fishing with an excessive number of nets are the main causes of disturbance to birds. Tourist development along the northern border is causing habitat loss.

Castro Verde plains — A1, B2, B3, C1, C2, C6 — 029

Admin region Alentejo
Coordinates 37°45'N 8°00'W
Altitude 160–289 m **Area** 79,066 ha

Site description

Open plains used for non-intensive cultivation of cereals (mainly wheat), with sparse woodland of evergreen oak *Quercus ilex*. Due to the poor soils, cereals are grown on crop rotations of long duration. Fallow land is used for grazing sheep and cattle. There is scattered human habitation.

Habitats Forest and woodland (22%; broadleaved evergreen forest), Artificial landscape (78%; arable land)
Land-use Agriculture (78%), Forestry (22%)

Birds

Species	Season	Year	Pop min	Pop max	Acc	Criteria
Ciconia ciconia White Stork	B	1996	120	180	B	B2, C6
Elanus caeruleus Black-winged Kite	R	1996	8	10	A	C6
Milvus migrans Black Kite	B	1996	6	10	B	B2, C6
Circus pygargus Montagu's Harrier	B	1996	50	80	B	B3, C6
Falco naumanni Lesser Kestrel	B	1996	30	40	B	A1, B2, C1, C6
Grus grus Crane	W	1996	80	140	A	C6
Tetrax tetrax Little Bustard	R	1996	300	360	A	A1, B2, C1, C6
Otis tarda Great Bustard	R	1996	300	320	A	A1, B2, C1, C2, C6
Burhinus oedicnemus Stone Curlew	B	1996	100	150	B	B2, C6
Pterocles orientalis Black-bellied Sandgrouse	B	1996	20	40	B	C6
Coracias garrulus Roller	B	1996	40	60	A	C6
Melanocorypha calandra Calandra Lark	B	1994	200	400	B	C6

The area is of great importance for breeding birds of steppic habitats, especially *Otis tarda* (the most important site in Portugal).

Protection status

National None **International** Partial

IBA overlaps with Special Protection Area.

Conservation issues

Threats Abandonment/reduction of land management (B), Afforestation (B), Agricultural intensification/expansion (B), Disturbance to birds (C)

Nesting birds are disturbed by the increasing road-traffic in the region and by the aircraft used in aerial spraying for agriculture. The weakness of the rural economy may lead farmers to change land-uses, which will affect the present high natural value of this landscape. The continuance of low-intensity land-uses is vital for the conservation of the steppe avifauna.

River Guadiana — A1, B1iii, B2, B3, C1, C2, C6 — 030

Admin region Alentejo
Coordinates 37°42'N 7°39'W
Altitude 170–310 m **Area** 76,578 ha

Site description

This recently designated Natural Park includes a stretch of the lower Guadiana valley and the surrounding areas, close to the south-east border of Portugal. The valley is steep, with cliffs and sclerophyllous scrub. In the remaining areas there are mainly steppic habitats, non-intensive cereal fields and scrub.

Habitats Forest and woodland (12%; broadleaved evergreen forest), Scrub (50%; sclerophyllous scrub/garrigue/maquis), Wetland (4%; river/stream; water-fringe vegetation), Artificial landscape (34%; arable land; perennial crops/orchards/groves)
Land-use Agriculture, Fisheries/aquaculture, Hunting, Nature conservation/research, Tourism/recreation

Birds

Species	Season	Year	Pop min	Pop max	Acc	Criteria
Ardea purpurea Purple Heron	B	—	—	—	—	C6
Ciconia nigra Black Stork	B	1996	2	5	A	B2, C6
Ciconia ciconia White Stork	R	1996	51	100	A	B2, C6
Milvus migrans Black Kite	B	1996	1	5	B	C6
Neophron percnopterus Egyptian Vulture	B	1996	1	5	A	B2, C6
[1] ***Aegypius monachus*** Cinereous Vulture	N	1996	Frequent		—	C6
Circus pygargus Montagu's Harrier	B	1996	11	50	B	B3, C6
Hieraaetus fasciatus Bonnelli's Eagle	R	1996	1	5	A	B2, C6
Falco naumanni Lesser Kestrel	B	1996	51	100	A	A1, B1iii, B2, C1, C2, C6
Grus grus Crane	W	1996	101	250	A	C6
[2] ***Pterocles orientalis*** Black-bellied Sandgrouse	R	1996	Uncommon		—	C6
Bubo bubo Eagle Owl	R	1996	11	50	B	B2, C6

1. Regular visitor; a potential new breeding site.
2. Small numbers thought to breed regularly.

An important area for breeding raptors, especially cliff-nesting species, but with good numbers of some birds of steppic habitats and of wetlands as well. Significant proportion (≥1%) of national population breeding at site: *Circaetus gallicus* (1–5 pairs).

Protection status

National High **International** Partial

69,733 ha of IBA covered by Natural Park (Guadiana, 69,733 ha). IBA overlaps with Special Protection Area.

Conservation issues

Threats Abandonment/reduction of land management (B), Construction/impact of dyke/dam/barrage (A), Firewood collection (B)

Ongoing abandonment of agriculture will lead to an increase in the extent of scrub and a reduction in extent of steppic habitats, reducing densities of open-country bird species. There are plans to build a dam downriver, which will partially flood the area, and another large reservoir is planned upstream, which will affect water-flow through the site.

South-west coast of Portugal — B1iv, B2, C6 — 031

Admin region Alentejo Litoral, Algarve
Coordinates 37°24'N 8°51'W
Altitude 0–156 m **Area** 74,563 ha

Site description

A long stretch of rocky coast, from Sines to Sagres, including a large proportion of the south-west coast of Portugal, as well as the extreme south-west point of Europe at Cape São Vicente. There are small beaches, river-mouths, and farmland and scrub on inland plateaus.

Habitats Forest and woodland (mixed forest; broadleaved evergreen forest), Scrub (sclerophyllous scrub/garrigue/maquis), Wetland (tidal river/enclosed tidal water; saltmarsh; sand-dunes/sand beach; shingle/stony beach; standing fresh water), Rocky areas (sea cliff/rocky shore; rock stacks/islets; caves), Artificial landscape (arable land; forestry plantation)
Land-use Agriculture, Fisheries/aquaculture, Hunting, Nature conservation/research, Tourism/recreation

Birds

Species	Season	Year	Pop min	Pop max	Acc	Criteria
Ciconia ciconia White Stork	R	1996	20	40	B	C6
[1] ***Circaetus gallicus*** Short-toed Eagle	B	1996	Frequent		—	C6
Falco peregrinus Peregrine	R	1996	Frequent		—	C6
Pyrrhocorax pyrrhocorax Chough	R	1996	10	20	B	B2, C6

1. Common during migration.

Important numbers of cliff-nesting species breed, and Cape São Vicente concentrates significant numbers of migrating raptors, particularly *Gyps fulvus* and *Hieraaetus pennatus*. Species of global conservation concern that do not meet IBA criteria: *Falco naumanni* (probably extinct), *Tetrax tetrax* (resident, frequent). The area is used as an emergency stop-over site by migrating passerines.

Protection status

National High **International** High

67,768 ha of IBA covered by Natural Park (Costa Vicentina e Sudoeste Alentejano, 89,617 ha). 74,563 ha of IBA covered by Special Protection Area (Costa Sudoeste, 74,563 ha).

Conservation issues

Threats Abandonment/reduction of land management (A), Afforestation (B), Agricultural intensification/expansion (B), Aquaculture/fisheries (C), Disturbance to birds (B), Natural events (B), Recreation/tourism (A)

Due to ongoing land-abandonment, the extent of scrub is increasing at the expense of non-intensive farmland. Coastal erosion is a problem and the development of tourist infrastructure poses strong pressures. Cliff-nesting birds are disturbed by professional and amateur anglers. A few endemic plant species occur at this site.

Leixão da Gaivota — C6 — 032

Admin region Algarve
Coordinates 37°06′N 8°30′W
Altitude 0–30 m **Area** 1 ha

Site description

A small rocky islet with a flat top covered by scrub, very close to Portimão on the south coast of Portugal.

Habitats Rocky areas (100%; rock stacks/islets)
Land-use Not utilized

Birds

Species	Season	Year	Pop min	Pop max	Acc	Criteria
Egretta garzetta Little Egret	B	—	5	150	—	C6

The most important heronry on offshore rocky islets in Portugal.

Protection status

National None **International** High

1 ha of IBA covered by Special Protection Area (Leixão da Gaivota, 1 ha).

Conservation issues

Threats Industrialization/urbanization (U), Recreation/tourism (U)

The heronry is inaccessible and therefore not prone to disturbance. The egrets do however feed in a nearby estuary that is subject to pressure from port development and tourism activities.

Ria Formosa (Faro lagoon) — A4i, A4iii, B1i, B2, C2, C3, C4, C6 — 033

Admin region Algarve
Coordinates 37°02′N 7°48′W
Altitude 0–25 m **Area** 23,296 ha

Site description

An extensive lagoon system with large areas of sandflats, mudflats and saltmarshes, located on the south coast of Portugal. The lagoons are protected from the open sea by a long, thin, discontinuous belt of sand-dunes. Several thousand people live around the IBA, and numbers increase significantly during the summer months due to the influx of tourists.

Habitats Forest and woodland (native coniferous forest), Scrub (sclerophyllous scrub/garrigue/maquis), Wetland (tidal river/enclosed tidal water; mudflat/sandflat; saltmarsh; sand-dunes/sand beach; coastal lagoon; standing fresh water; standing brackish and salt water; river/stream; water-fringe vegetation), Artificial landscape (arable land; perennial crops/orchards/groves; forestry plantation; ruderal land), Introduced/exotic vegetation
Land-use Agriculture, Fisheries/aquaculture, Forestry, Military, Nature conservation/research, Tourism/recreation, Urban/industrial/transport

Birds

Species	Season	Year	Pop min	Pop max	Acc	Criteria
Ixobrychus minutus Little Bittern	B	1996	Common		—	C6
Egretta garzetta Little Egret	R	1996	500	700	A	C6
Ciconia ciconia White Stork	R	1996	10	15	A	C6
[1] ***Platalea leucorodia*** Spoonbill	W	1996	150	350	A	A4i, B1i, C2
Anas penelope Wigeon	W	1996	1,500	11,500	A	A4i, B1i, C3
Porphyrio porphyrio Purple Gallinule	R	1996	10	15	A	C6
[1,2] ***Himantopus himantopus*** Black-winged Stilt	R	1996	350	400	A	A4i, B1i, C2, C6
[3] ***Recurvirostra avosetta*** Avocet	W	1996	650	750	A	B2
Glareola pratincola Collared Pratincole	B	1996	25	45	A	B2, C6
Charadrius hiaticula Ringed Plover	W	1996	2,000	4,500	A	A4i, B1i, C3
Charadrius alexandrinus Kentish Plover	R	1996	800	1,200	A	A4i, B1i, B2, C3
Charadrius alexandrinus Kentish Plover	W	—	2,000	4,000	—	A4i, B1i, C3
Pluvialis squatarola Grey Plover	W	1996	1,700	3,000	A	A4i, B1i, C3
Calidris alpina Dunlin	W	1996	12,000	22,500	A	A4i, B1i, B2, C3
Limosa lapponica Bar-tailed Godwit	W	1996	1,900	4,200	A	A4i, B1i, B2, C2
Tringa totanus Redshank	W	1996	1,000	2,000	A	B1i, C3
Arenaria interpres Turnstone	W	—	500	1,000	—	A4i, B1i, C3
Arenaria interpres Turnstone	P	1996	700	1,200	A	A4i, B1i, C3
Sterna albifrons Little Tern	B	1996	200	300	A	A4i, B1i, B2, C2, C6

1. Large increase.
2. 600–750 wintering.
3. 40–50 pairs breed regularly.

The site is important for numerous waterbirds, especially for beach-nesting *Charadrius alexandrinus* and *Sterna albifrons* and for wintering waders and duck, holding 20,000 or more wintering waterbirds on a regular basis. The reedbeds of the site are important for several nesting waterbird species, as well as for large numbers of migrating passerines during both autumn and spring migration.

Protection status

National Partial **International** High

14,373 ha of IBA covered by Natural Park (Ria Formosa, 14,373 ha). 23,296 ha of IBA covered by Special Protection Area (Ria Formosa, 23,296 ha). 16,000 ha of IBA covered by Ramsar Site (Ria Formosa, 16,000 ha).

Conservation issues

Threats Agricultural intensification/expansion (A), Aquaculture/fisheries (A), Disturbance to birds (A), Dredging/canalization (B), Groundwater abstraction (A), Industrialization/urbanization (A), Natural events (B), Recreation/tourism (A), Unsustainable exploitation (A)

Tourists cause disturbance to birds, which can be moderately high locally, especially for beach-nesting species. The intensification of farming in the vicinity is increasing the flow into the lagoon of pollutants (herbicides, pesticides and nutrients), and is also using excessive amounts of groundwater, thus disturbing the water balance at the site. The bivalves and polychaete worms of the intertidal flats are very intensively exploited by man, thus reducing the food supply for some waterbird species. Natural erosion is an additional threat. There is a management plan for the Natural Park.

Castro Marim — A4i, B1i, B2, C2, C6 — 034

Admin region Algarve
Coordinates 37°13′N 7°26′W
Altitude 0–50 m **Area** 2,147 ha

Site description

The Portuguese side of the Guadiana estuary, at the country's south-east corner. There is an extensive area of saltpans and of claimed land (used for cattle and sheep-grazing) that was formerly saltmarsh.

Habitats Wetland (tidal river/enclosed tidal water; mudflat/sandflat; saltmarsh; coastal lagoon; standing brackish and salt water), Artificial landscape (arable land; perennial crops/orchards/groves; ruderal land)
Land-use Agriculture, Fisheries/aquaculture, Nature conservation/research, Tourism/recreation

Birds

Important numbers of breeding, wintering and migrating waterbirds use this site, as do some steppic species such as *Tetrax tetrax*. Species of global conservation concern that do not meet IBA criteria: *Falco naumanni* (extinct; 10–20 pairs bred formerly).

Species	Season	Year	Pop min	Pop max	Acc	Criteria
[1] ***Ciconia ciconia*** White Stork	R	1996	100	150	B	B2, C6
Platalea leucorodia Spoonbill	P	1996	200	500	B	A4i, B1i, C2
[2] ***Phoenicopterus ruber*** Greater Flamingo	W	1996	500	1,000	A	A4i, B1i, C2
Tetrax tetrax Little Bustard	B	1996	Frequent		—	C6
Himantopus himantopus Black-winged Stilt	B	1996	300	400	B	A4i, B1i, C2, C6
[3] ***Recurvirostra avosetta*** Avocet	W	1996	400	600	A	B2
Glareola pratincola Collared Pratincole	B	1996	Frequent		—	C6
Sterna albifrons Little Tern	B	1996	50	75	B	B2, C6
Melanocorypha calandra Calandra Lark	B	—	Common		—	C6

1. Large increase.
2. Similar numbers present during migratory periods.
3. 50–100 pairs breed.

Protection status

National Partial **International** High
1,876 ha of IBA covered by Nature Reserve (Sapal de Castro Marim, 2,154 ha). 2,147 ha of IBA covered by Special Protection Area (Sapais de Castro Marim, 2,147 ha). IBA also overlaps with Ramsar Site (Sapais de Castro Marim, 2,235 ha).

Conservation issues

Threats Aquaculture/fisheries (B), Construction/impact of dyke/dam/barrage (C), Groundwater abstraction (B), Industrialization/urbanization (C), Recreation/tourism (C)

REFERENCES

Anastácio, P. and Amaro, P. (1992) Caracterização das populações de ardeídeos do Paul da Quinta do Taipal. *AIRO* 3: 13–17.

Araújo, A., Rosa, G., Monteiro, A. and Poças, M. (in press) The situation of the Black Stork (*Ciconia nigra*) in Portugal. *II International Conference on the Black Stork. Trujillo, Spain, March 1996.*

Brito, P. (1996) Aspectos sobre a selecção do habitat e biologia de reprodução do Alcaravão *Burhinus oedicnemus* numa região do Alto-Alentejo. Lisbon: Universidade de Lisboa.

Cardoso, A. N. (1996) *Importância do Paul do Boquilobo como habitat de alimentação e relevância do Lagostim-do-rio (*Procambarus clarkii*) como recurso alimentar da colónia de garças.* Rep. R. N. Paul do Boquilobo/ICN.

Costa, H., Santos, B. Leitão, D. and Catry, P. (1993) Marsh Harrier *Circus aeruginosus* populations in Tejo and Sado estuary. *AIRO* 4(1): 1–6.

Costa, L. T. and Guedes, R. S. (1994) *Contagens de anatídeos invernantes em Portugal–1992/93.* Lisbon: ICN.

Costa, L. T. and Guedes, R. S. (1996) *Contagens de anatídeos invernantes em Portugal. Invernos de 1993/94 a 1995/96.* Lisbon: ICN (Estudos de Biologia e conservação da Natureza no. 20).

Costa, L. T. and Rufino, R. (1994) Mid-winter waterfowl counts in Portugal—January 1994. *AIRO* 5(1): 8–16.

Costa, L. T. and Rufino, R. (1996) Mid-winter waterfowl counts in Portugal—January 1995. *AIRO* 7(1): 36–43.

Encarnação, V. (1996) *Censos de aves aquáticas no Parque Natural da Ria Formosa. Nov. 1991 to Dez. 1994.* Faro, Portugal: PNRF/ICN (Estudos Técnicos no. 2).

Farinha, J. C. and Trindade, A. (1994) *Contribuição para o inventário e caracterização de zonas húmidas em Portugal Continental.* Lisbon: ICN/MedWet.

Fráguas, B. (1996) Distribuição e situação actual da população de Águia de Bonelli *Hieraaetus fasciatus* no noroeste de Portugal. *Actas do I Congresso de Ornitologia da Sociedade Portuguesa para o Estudo das Aves. Vila Nova de Cerveira, Lisboa.* P. 102.

Franco, C. (1995) Estudo das comunidades de aves na bacia hidrográfica do rio Guadiana/Região de Mértola. Lisbon: Universidade de Lisboa, Fac. Ciências.

Granadeiro, J. P. (1990) The breeding biology of Cory's Shearwater *Calonectris diomedea borealis* on Berlenga Island, Portugal. *Seabird* 13: 30–39.

Granadeiro, J. P., Morais, L., Silva, M. C., Grade, N., Lecoq, M. and Santos, C. (in press) Breeding Madeiran Storm Petrels *Oceanodroma castro* on the Farilhões Islands, Portugal. *4th Pan-Mediterranean Symposium of Medmaravis, Hammamet, Tunisia, 11–16 April 1995.*

Grimmett, R. F. A. and Jones T. A. (1989) *Important Bird Areas in Europe.* Cambridge, UK: International Council for Bird Preservation (Tech. Publ. 9).

Heath, M. F. and Borggreve, C. (2000) *BirdLife International/EBCC European Bird Database 1998.* Cambridge, UK: BirdLife International.

Luís, A. (1989) Limícolas na Ria de Aveiro. Aveiro, Portugal: Universidade de Aveiro.

Monteiro, A., Carvalho, A. and Fráguas, B. (1996) Demografia do Abutre do Egipto (*Neophron percnopterus*) em Portugal. *Actas do I Congresso de Ornitologia da Sociedade Portuguesa para o Estudo das Aves. Vila Nova de Cerveira, Lisboa.* Pp. 90–91.

Monteiro, A., Berliner, A., Carvalho, A. and Araújo, A. (1996) Situação populacional e distribuição do Grifo (*Gyps fulvus*) em Portugal no ano de 1996. *Actas do I Congresso de Ornitologia da Sociedade Portuguesa para o Estudo das Aves. Vila Nova de Cerveira, Lisboa.* Pp. 92–93.

Monteiro, A., Fráguas, B., Pombal, E., Berliner, A. and Jambas, J. (in press) Demography and conservation of five cliff-nesting raptors in the northeast of Portugal. *Proc. of II International Conference on Raptors. Urbino, Italy, October 1996.*

Neto, J. (1996) Biologia e ecologia reprodutiva do Corvo-marinho-de-crista *Phalacrocorax aristotelis* na Reserva Natural da Berlenga. *Actas do I Congresso de Ornitologia da Sociedade Portuguesa para o Estudo das Aves. Vila Nova de Cerveira, Lisboa.* Pp. 9–10.

Oliveira, L. F. (1991) Estudo preliminar da avifauna de uma zona da Serra da Arrábida. Lisbon: SNPRCN.

Palma, L., Pais, M. and Fráguas, B. (in press) Situation and distribution of Bonelli's Eagle (*Hieraetus fasciatus*) in Portugal. *Proc. of II International Conference on Raptors. Urbino, Italy, October 1996.*

Patacho, D. (in prep.) *Atlas das aves da Serra da Nogueira.* Bragança, Portugal: Escola Superior Agrária da Bragança.

Pimenta, M. and Santarém, M. L. (1996) *Atlas of the birds of the Peneda-Gerês National Park.* Braga, Portugal: ICN/PNPG.

Pinto, M. and Almeida, J. (1989) A survey-program for the Common Crane *Grus grus* wintering in Portugal. *Proc. of the Palearctic Crane Workshop, Estonia.*

Pinto, M. and Hellmich, J. (1996) Grosstrappenzahlungen an der Spanisch-Portugiesischen Grenzen im winter 1994 und frujahr 1995. *Naturschutz und landschaftspflege in Brandenburg* 1/2.

Pinto, M. (1991) Des problèmes pour l'avifaune de steppe dûs à des changements d'utilization du sol au Portugal. *Proc. implementation of agri-environmental policies in the EC, Brussels.*

Reino, L. (1994) Atlas das aves nidificantes do Parque Natural de Montesinho. Scholar thesis, Escola Superior Agrária de Bragança, Bragança, Portugal.

Rocha, P. (1995) O Peneireiro-de-dorso-liso *Falco naumanni* na região de Mértola-Castro Verde. Biologia e ecologia de uma ave de presa colonial. Lisbon: Universidade de Lisboa.

Rosa, G. (1997) Recenseamento de Cegonha-preta *Ciconia nigra* no sul de Portugal. Lisbon: CEMPA.

Rufino, R. and Costa, L. T. (1993) Mid-winter waterfowl counts in Portugal—January 1994. *AIRO* 4(2): 51–55.

Rufino, R., Neves, R. and Pina, J. P. (1996) Recenseamentos aéreos na costa portuguesa. *Actas do I Congresso de Ornitologia da Sociedade Portuguesa para o Estudo das Aves. Vila Nova de Cerveira, Lisboa.* P. 103.

Silva, L. (1998) *Atlas das aves da Reserva Natural da Serra da Malcata.* Lisbon: ICN.

Tucker, G. M. and Heath, M. F. (1994) *Birds in Europe: their conservation status.* Cambridge, UK: BirdLife International (BirdLife Conservation Series no. 3).

THE AZORES

Luís R. Monteiro

Roseate Tern *Sterna dougallii* colony at the Alagoa complex on the coast of Flores (IBA 001). (Photo: L. R. Monteiro/ImagDOP)

GENERAL INTRODUCTION

The Azores archipelago (36–39°N, 25–31°W) lies c.1,450 km west of mainland Portugal, and has a land surface area of 2,333 km². It comprises nine volcanic islands, forming three groups (western, central and eastern) along a tectonic zone running west-north-west–east-south-east for c.600 km. The c.700 km long coastline is rocky with boulder shores, and the cliffs frequently reach 500 m in height. Within 1 km of the main islands lie 26 small islets (0.1–10 ha) and stacks.

The Azores is an integral part of Portugal but has its own autonomous government. The islands were first colonized by people over a period of several decades at the end of the 15th century. The human population has now stabilized at around 240,000 inhabitants.

The Azores has 15 Important Bird Areas (IBAs) with a total area of 12,805 ha, equivalent to 5.5% of the land area of the archipelago (Table 1, Map 1). These IBAs are evenly distributed across the main islands, the total by island varying between one and four. Seven IBAs were identified in 1989 (Grimmett and Jones

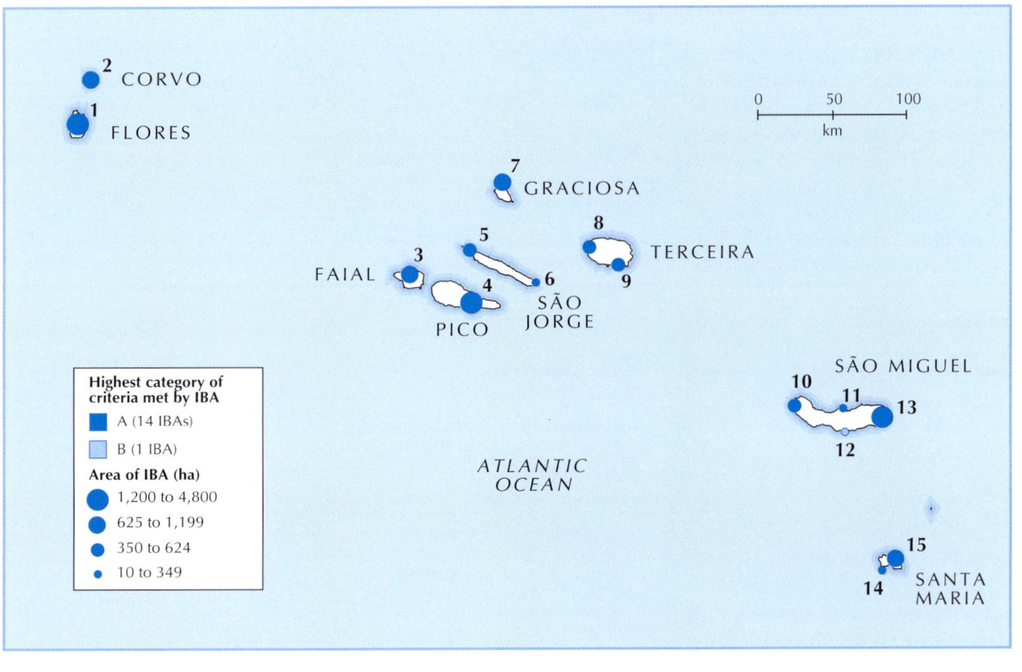

Map 1. Location, area and criteria category of Important Bird Areas in the Azores.

Table 1. Summary of Important Bird Areas in the Azores. 15 IBAs covering 128 km²

IBA code	1989 code	International name	National name	Administrative region	Area (ha)	Criteria (see p.11)
001	QQ001	Coast of Flores	Costa das Flores	Santa Cruz das Flores, Lages das Flores	1,200	A2, A4i, A4ii, B1i, B1ii, B2, B3, C2, C6
002	QQ002	Coast of Corvo	Costa e Caldeirão do Corvo	Vila Nova do Corvo	690	A4i, A4ii, B1i, B1ii, B2, C2, C6
003	—	Coast of Faial	Costa do Faial	Horta	940	A2, A4i, A4ii, B1i, B1ii, B2, B3, C2, C6
004	—	Coast of Pico	Costa do Pico	São Roque do Pico, Lajes do Pico	1,500	A2, A4i, A4ii, B1i, B1ii, B2, B3, C2, C6
005	—	Rosais	Rosais	Velas	500	A4i, A4ii, B1i, B1ii, B2, C2, C6
006	—	Topo	Topo	Calheta	340	A4i, A4ii, B1i, B1ii, B2, C2, C6
007	QQ003, QQ004	Coast and islets of Graciosa	Costa e Ilhéus da Graciosa	Santa Cruz da Graciosa	680	A4i, A4ii, B1i, B1ii, B2, C2, C6
008	—	West coast of Terceira	Costa Oeste da Terceira	Angra do Heroísmo, Praia da Vitória	600	A4ii, B1ii, B2, C2
009	—	South-east coast of Terceira	Costa Sudeste da Terceira	Angra do Heroísmo	350	A4i, A4ii, B1i, B1ii, B2, C2, C6
010	—	West coast of São Miguel	Costa Oeste de São Miguel	Ponta Delgada	360	A4ii, B1ii, B2, C2
011	—	Cintrão	Cintrão	Ribeira Grande	200	A4ii, B1ii, B2, C2
012	—	Vila Franca islet	Ilhéu de Vila Franca do Campo	Vila Franca do Campo	10	B1ii, C2
013	QQ007	East of São Miguel	Nordeste de São Miguel	Nordeste e Povoação	4,800	A2, A4ii, B1ii, B2, B3, C2, C6
014	QQ005	Vila islet	Ilhéu da Vila	Vila do Porto	10	A4i, A4ii, B1i, B1ii, B2, C2, C6
015	QQ006	North and east coast of Santa Maria	Costa Norte e Leste de Santa Maria	Vila do Porto	625	A4i, A4ii, B1i, B1ii, B2, C2, C6

1989), with a total area of 2,844 ha. Former sites QQ003 and QQ004 have been merged into present site 007. Recent seabird surveys have resulted in the identification of a further nine IBAs and the overall IBA area has been enlarged by 9,961 ha.

ORNITHOLOGICAL IMPORTANCE

Table 1 presents the criteria on which each IBA is qualifying. 'East of São Miguel' (IBA 013) extends to a vast mountainous sector that includes the entire restricted breeding range of *Pyrrhula murina*, endemic to the island and a proposed species of global conservation concern (BirdLife International, in prep.). Overall, the following sites are especially important due to the high number of species of European conservation concern (SPECs) and size of populations they hold: Coast of Flores (001), Coast of Corvo (002), Coast and islets of Graciosa (007) and Vila islet (014).

All 15 IBAs hold important breeding congregations of waterbirds and seabirds (qualifying under criteria A4, B1 and B2). Of particular significance is that the Azores holds the largest breeding concentration of *Calonectris diomedea* in the world, the largest European breeding population of *Sterna dougallii*, and two temporally distinct (hot- and cool-season) and genetically divergent breeding populations of *Oceanodroma castro* (Monteiro and Furness 1998). This importance is reflected in the identification of 14 IBAs which hold at least 1% of the global (or in the case of *Sterna dougallii* 1% of the European) breeding population of one or more of these three species (criterion A4, Table 2). Some of these 14 IBAs also hold at least 1% of distinct breeding populations of *Bulweria bulwerii* and *Puffinus assimilis baroli*. The remaining IBA (Vila Franca islet, 012) qualifies under criterion B1ii because it holds more than 1% of the European population of breeding *Calonectris diomedea*.

Seventeen SPECs breed in the Azores. Nine of these have an unfavourable conservation status, of which six are seabirds (Tucker and Heath 1994). Several of these species are covered well by the IBA network (Table 3).

The Azorean subspecies of *Columba palumbus azorica*, presumed to be declining and listed on Annex I of the EC Birds Directive, and other distinct forest bird subspecies are not well covered by the IBA network. This is due to a lack of information on their status and distribution.

HABITATS

The vegetation cover of the Azores has been considerably altered since humans first colonized the archipelago. Clearance of the native broadleaved evergreen forest (dominated by *Laurus azorica* and *Juniperus brevifolia*) took place at a fast rate. Later introductions of invasive alien plant species (for example, *Pittosporum undulatum*, *Hedychium gardneranum*, *Arundo donax*, and *Clethra arborea*) caused further major changes in the vegetation (Sjögren 1973). Native forest now covers less than 5% of the islands' surface area, and more than half of the vascular plant species have been introduced.

Table 2. Important Bird Areas in the Azores that support important numbers of one or more congregatory species (i.e. meeting criteria A4 and/or B1). IBAs meeting both criteria A4 and B1 for the species are shown in **bold**. IBAs meeting only criterion B1 for the species concerned, and not A4, are shown in normal type. For key to 'Season', see p. 7.

Species	Season	IBA code
Calonectris diomedea Cory's Shearwater	B	**001**, **002**, **003**, **004**, **005**, **006**, **007**, **008**, **009**, **010**, **011**, 012, **013**, 014, **015**
Puffinus assimilis Little Shearwater	B	001, 002, 003, 004, 005, 006, 007, 010, 011, 014, 015
Oceanodroma castro Madeiran Petrel	B	**007**, **014**, 015
Sterna dougallii Roseate Tern	B	**001**, **002**, **003**, **004**, **005**, **006**, **007**, **009**, **014**, **015**

Table 3. Species of European conservation concern and species listed on Annex I of the EC Birds Directive with significant breeding populations at IBAs in the Azores (meeting any IBA criteria).

Species [1]	Minimum national breeding population (pairs) [2]	Proportion (%) of national population breeding at all IBAs in the Azores
Bulweria bulwerii Bulwer's Petrel	500	10
Calonectris diomedea Cory's Shearwater	50,000	100[3]
Puffinus assimilis Little Shearwater	500	100[3]
Oceanodroma castro Madeiran Petrel	1,000	90
Sterna dougallii Roseate Tern	1,000	100[3]
Sterna hirundo Common Tern	4,000	91
Columba palumbus Woodpigeon	250	–[4]
Serinus canaria Canary	30,000	8
Pyrrhula murina Azores Bullfinch	100	100

1. Only those species of European conservation concern (see Box 1, see p. 12) that meet IBA criteria in the Azores are listed, together with those species listed on Annex I of the EC Birds Directive that fulfil criterion C6 in IBAs in the Azores.
2. Data are taken from the BirdLife/EBCC European Bird Database 1998 (Heath and Borggreve 2000).
3. The percentage of the national population in IBAs exceeds 100%. Usually this is because the national population estimate has not been updated recently whilst the IBA population estimate has been recently updated with new data as a result of comprehensive surveys of IBAs themselves. Also, the individual site count for a species may be the maximum or average over recent years, and summing these may record more birds than are present nationally in any single year.
4. No population data available for Azorean IBAs.

Figure 1. Habitats at Important Bird Areas in the Azores (see Appendix 3 for definitions of habitats).

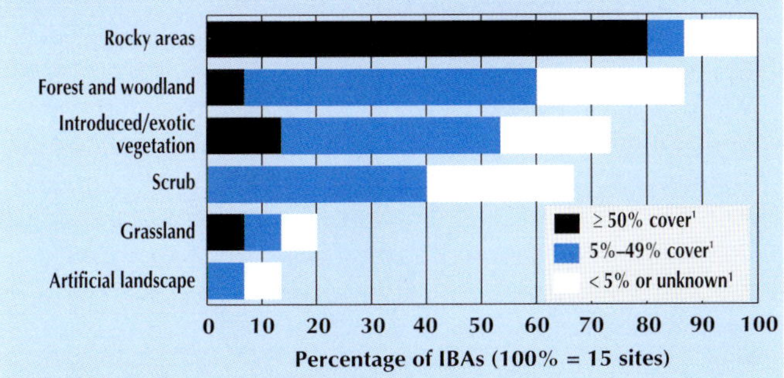

1. percentage of area of individual IBA covered by habitat

Figure 2. Land-uses at Important Bird Areas in the Azores (see Appendix 3 for definitions of land-uses).

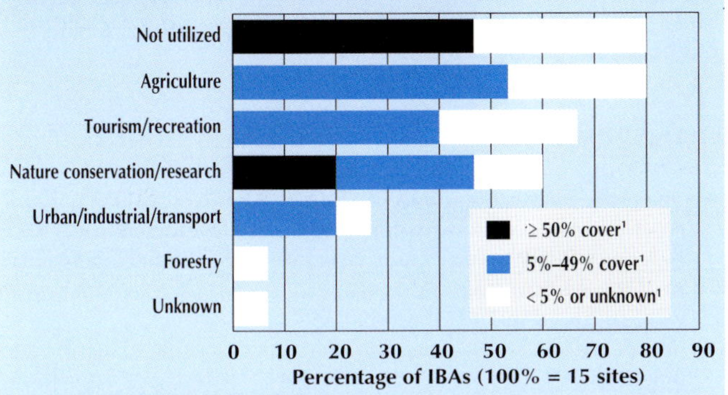

1. percentage of area of individual IBA covered by land-use

All 15 Azorean IBAs contain some rocky habitats (for example, sea cliffs, rocky shores, rock stacks and islets), and in 80% of sites, these cover more than 50% of the IBA area (Figure 1). Forest, introduced vegetation and scrub each occur in over 60% of IBAs, and generally constitute less than 50% of the IBA area (Figure 1). An exception is 'East of São Miguel' (013), which includes the extensive forest area of Pico da Vara, habitat of *Pyrrhula murina*. Under-representation of forest and other terrestrial habitats within the IBA network is due to a lack of information on the status and distribution of characteristic bird species. Coastal heathland comprises *Erica scoparia azorica*, with *Myrica faya* and exotics such as *Arundo donax* and *Pittosporum undulatum*, and is present at IBAs 002, 003, 004, 005, 007, 008, 009, 010, 012 and 013.

IMPACTS ON IBAs – LAND-USE AND THREATS

Forty-seven percent of Azorean IBAs are not utilized over more than 50% of their area (Figure 2). Agricultural activity takes place in 80% of IBAs, and virtually all the coastal fringe land and most of the accessible islets are used as pasture for cattle. Tourism and recreation is present at 67% of IBAs, with coastal sites under intense and increasing pressure from activities including fishing, boating, scuba-diving, crab and limpet collecting, and picnicking. Nature conservation and research activities take place at 60% of IBAs, but coverage is less than 50% at most of these. Forestry is restricted to 'East of São Miguel' (013), and consists mainly of plantations of introduced *Cryptomeria japonica*.

The majority of the IBAs are privately owned, including coastal sectors and islets, except for the forests of Vila Franca islet (012) and 'East of São Miguel' (013), which are state owned. The local government is purchasing some islets and important coastal habitats.

Animal and plant introductions pose a widespread threat, affecting over 90% of IBAs, and having a high impact in 60% of IBAs (Figure 3). Colonization resulted in the introduction of bird predators (for example, black rat *Rattus rattus*, brown rat *Rattus norvegicus*, weasel *Mustela nivalis*, ferret *Mustela furo*, and cat *Felis catus*), which were widely spread by the 16th century (Fructuoso 1561). Introduced carnivores are a major threat to seabirds, and those burrow-nesting species that are at colonies for long periods are the most vulnerable. As a result of predation, only residual colonies of the smaller shearwaters and petrels (Procellariidae) now remain on the main Azorean islands (Monteiro *et al.* 1996).

Major changes in coastal and cliff vegetation have taken place following invasion by densely rooted exotics, such as *Arundo donax*. This is thought to have resulted in losses of suitable burrowing ground for breeding shearwaters and petrels (Monteiro *et al.* 1996).

Recreation/tourism and disturbance to birds are the second most common threats, each affecting 53% of Azorean IBAs, and having a high impact in 33% of IBAs (Figure 3). Tern colonies along the Coast of Flores (001), Coast and islets of Graciosa (007), Coast of Faial (003) and South-east coast of Terceira (009) are most vulnerable to these threats, which have led to colony desertion and low productivity. Recreation is a serious problem at Vila Franca islet (012), where losses of *Calonectris diomedea* nest-sites due to trampling by tourists have been observed during the summer. Human exploitation of adults and young of the same species for fishing bait and food occurs in the most accessible areas of some IBAs (Monteiro *et al.* 1996).

The pressure on the littoral zone has increased in recent years, in the form of dumping of solid wastes, fisheries, recreation, infrastructure (for example, harbours and roads) and tourist developments (Santos *et al.* 1995).

PROTECTION STATUS

Table 4 and Figures 4 and 5 summarize the national and international protection status of all Azorean IBAs.

■ National protection

Two independent networks of protected areas coexist in the Azores, and are administered by the Regional Government through either the Environment Department or the Forestry Department. Those protected areas designated by the Environment Department are Nature Reserves (Strict or Partial), Natural Parks and Protected Landscapes. The Forestry Department is responsible for designating Strict Forest Nature Reserves, Partial Forest Nature Reserves and Recreational Forest Nature Reserves. Protected areas are not necessarily owned by the state and some are privately owned.

Only five of the 15 IBAs are afforded some legal protection through the national protected-areas network; 91% of the total

Figure 3. Threats at Important Bird Areas in the Azores (see Appendix 3 for definitions of threat types and impact categories).

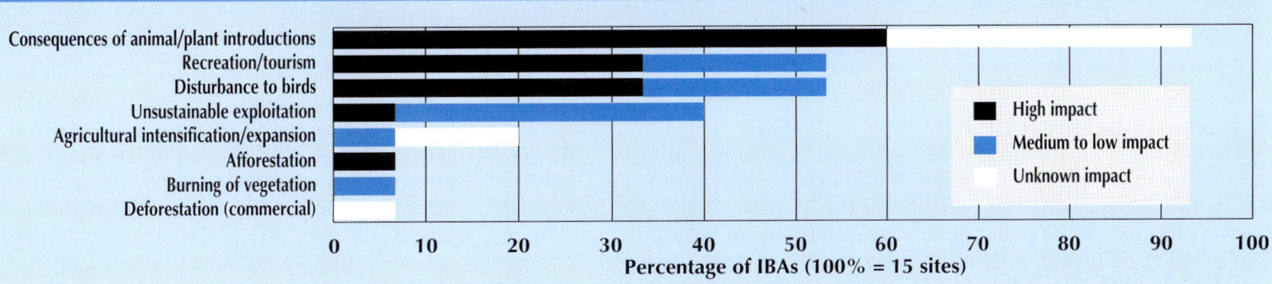

Figure 4. The national protection status of Important Bird Areas in the Azores.

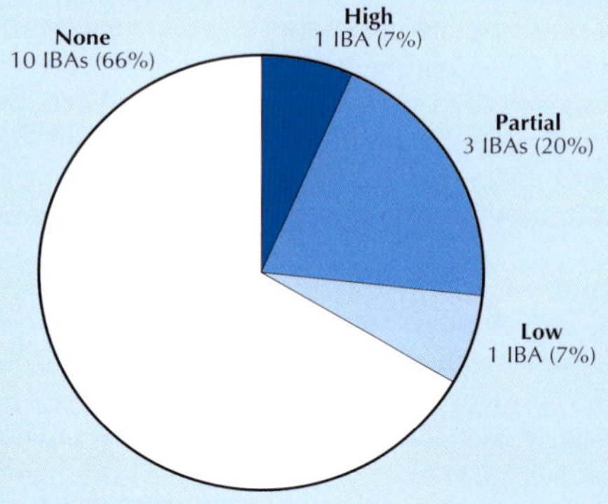

% of area of individual IBA overlaps with national protected areas:
High >90%
Partial 10–90%
Low <10%
None

Total area of overlap between IBA network in the Azores and national protected-area system (see Table 4 for categories) = 11–12 km² (9% of total IBA area).

Figure 5. The international protection status of Important Bird Areas in the Azores.

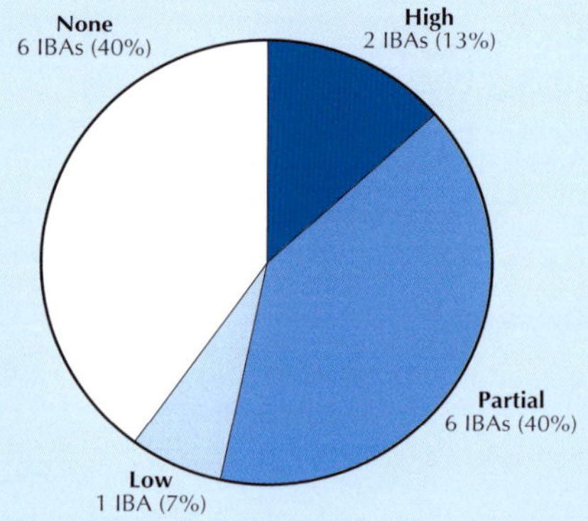

% of area of individual IBA overlaps with international protected areas:
High >90%
Partial 10–90%
Low <10%
None

Total area of overlap between IBA network in the Azores and international protected-area system (see Table 4 for categories) = 34–36 km² (27–28% of total IBA area).

Table 4. Protection status of Important Bird Areas in the Azores. A tick (✔) indicates that an IBA overlaps with a protected area (to any extent).

		National		International
IBA code	International name	Natural Park	Nature Reserve	Special Protection Area
001	Coast of Flores			✔
002	Coast of Corvo			✔
003	Coast of Faial		✔	✔
004	Coast of Pico			✔
005	Rosais			
006	Topo		✔	✔
007	Coast and islets of Graciosa			✔
008	West coast of Terceira			
009	South-east coast of Terceira			✔
010	West coast of São Miguel			
011	Cintrão			
012	Vila Franca islet	✔		
013	East of São Miguel		✔	✔
014	Vila islet			✔
015	North and east coast of Santa Maria	✔		
Total number of IBAs		2	3	9

Box 1. International legislation and initiatives that are relevant to site conservation in the Azores (see Appendix 1 for a general description of these agreements).

Global	
Biodiversity Convention	✔
Ramsar Convention	✔
Bonn Convention	✔
World Heritage Convention	✔
MAB Programme	✔
Pan-European	
Bern Convention	✔
Regional	
EC Birds Directive	✔
EC Habitats Directive	✔

✔ Convention ratified/initiative supported
(✔) Convention signed

IBA area is not legally protected at the national level (Table 4, Figure 4).

International protection

As part of the Portuguese State, the Azores is covered by various pieces of international legislation (Box 1). Fifteen Special Protection Areas (SPAs) have been designated under the EC Birds Directive, covering an area of 11,825 ha. These overlap partially or fully with nine of the 15 IBAs (Table 4, Figure 5), but cover only 28.3% of the total IBA area. Proper enforcement of protection measures within SPAs will not be achieved until national regulatory legislation is published in the near future.

CONSERVATION

- Scientific research and monitoring of Azorean birds began only recently. Annual monitoring of tern colonies started in 1989 (del Nevo *et al.* 1993) and extensive surveys of other seabird species in 1995. Monitoring and ecological studies of *Pyrrhula murina* were initiated in 1991 (Ramos 1993).
- Wardening at *Sterna dougallii* colonies vulnerable to disturbance began in 1993.
- A management plan exists for Praia islet (within 007), which has been subject to a restoration programme since 1995.
- A habitat restoration programme was also initiated in 1995 in East of São Miguel (013), which will benefit *Pyrrhula murina*.

ANALYTICAL METHODS

- Bird counts are given in pairs, except data for *Larus cachinnans atlantis* which are given as individuals.
- *Bulweria bulwerii*, *Calonectris diomedea*, *Larus cachinnans*, *Sterna dougallii*, *Sterna hirundo* and *Pyrrhula murina* counts are of good quality; most counts of *Puffinus puffinus*, *Puffinus assimilis* and *Oceanodroma castro* are less accurate.
- The method used for calculating *Calonectris diomedea* numbers is outlined in Monteiro *et al.* (1999).
- All bird data are from 1989 onwards.
- Threat and land-use estimates are based on information from the 1990s.

GLOSSARY

ACE a now superseded EU financial instrument for funding biotope projects.
EU European Union.
LIFE the EU's all-encompassing fund for nature conservation.
SPA Special Protection Area (designated under Article 4 of the EC Birds Directive).

ACKNOWLEDGEMENTS

The information contained in this work was collected between 1989 and 1995 through research and conservation projects. These were funded by the European Commission (for example, Programme LIFE contract B4-3200/95/351), Junta Nacional de Investigação Científica (Lisbon, research contract STRDB/C/MAR/228/92) and the Royal Society for the Protection of Birds (BirdLife International Partner in the United Kingdom). All those who gave valuable assistance in the field are acknowledged: J. A. Ramos, E. Sola, R. W. Furness, P. Monteiro, R. Feio, D. R. Thompson, S. Bearhop, N. Ratcliffe, J. Pereira, G. Hilton, L. Wilson, L. Hewitson, A. Tavares, M. Laranjo, M. P. Groz, V. Neves and K. Thompson.

SITE ACCOUNTS

Coast of Flores — 001

A2, A4i, A4ii, B1i, B1ii, B2, B3, C2, C6

Admin region Santa Cruz das Flores, Lages das Flores
Coordinates 39°27'N 31°12'W
Altitude 0–500 m **Area** 1,200 ha

Site description

The narrow, rocky coastal strip of the island of Flores, reaching from the shoreline to the cliff-tops (average width 250 m). There are numerous islets and stacks. Human activities include fishing, yachting, and tourism.

Habitats Forest and woodland (55%; broadleaved evergreen forest), Rocky areas (100%; sea cliff/rocky shore; rock stacks/islets), Introduced/exotic vegetation (40%)
Land-use Agriculture (20%), Nature conservation/research (30%), Not utilized (50%), Tourism/recreation (20%)

Birds

Species	Season	Year	Pop min	Pop max	Acc	Criteria
Calonectris diomedea Cory's Shearwater	B	1996	13,883	16,080	A	A4ii, B1ii, B2, C2, C6
Puffinus assimilis Little Shearwater	B	1996	150	300	C	B1ii, B2, C2, C6
Oceanodroma castro Madeiran Petrel	B	1997	20	50	C	B2, C6
Sterna dougallii Roseate Tern	B	1996	500	500	A	A4i, B1i, B2, C2, C6
Sterna hirundo Common Tern	B	1996	750	750	A	C6
Serinus canaria Canary	R	—	300	—	—	A2, B3

Important mainly for breeding seabirds, including the largest concentration of breeding *Sterna dougallii* in the eastern Atlantic (concentrated mainly between Santa Cruz and Ponta Ruiva).

Protection status

National None **International** Partial
130 ha of IBA covered by Special Protection Area (Costa Nordeste, 130 ha). 230 ha of IBA covered by Special Protection Area (Costa Sul e Sudeste, 230 ha).

Conservation issues

Threats Consequences of animal/plant introductions (A), Disturbance to birds (A), Recreation/tourism (A)

The tern colonies are affected by human disturbance and vandalism. Introduced ground predators (rats, mustelids, feral cats and dogs) possibly limit the breeding of most seabirds, especially of smaller shearwaters/petrels. Invasion by densely rooted exotic plants, such as cane grass *Arundo donax*, results in the loss of suitable burrowing ground for shearwaters/petrels. There is a management plan for Costa Nordeste SPA. The wardening of *Sterna dougallii* colonies at Baixa do Moinho and at Alagoa was initiated in 1993. Numbers of terns have been monitored since 1989 and other seabirds since 1995, with EU funding (ACE and LIFE projects).

Coast of Corvo — 002

A4i, A4ii, B1i, B1ii, B2, C2, C6

Admin region Vila Nova do Corvo
Coordinates 39°42'N 31°06'W
Altitude 0–715 m **Area** 690 ha

Site description

The coastal strip (average width 250 m) with islets and sea cliffs, extending inland to the cliff-tops and to the slopes of the Caldeirão basin.

Habitats Forest and woodland (20%; broadleaved evergreen forest), Scrub (20%; heathland), Rocky areas (100%; sea cliff/rocky shore; rock stacks/islets; scree/boulders), Introduced/exotic vegetation (10%)
Land-use Agriculture (20%), Not utilized (80%)

Birds

Important mainly for breeding seabirds, including the largest congregations of breeding *Calonectris diomedea* and *Puffinus puffinus* in the Azores.

Species	Season	Year	Pop min	Pop max	Acc	Criteria
Calonectris diomedea Cory's Shearwater	B	1996	20,350	25,611	A	A4ii, B1ii, B2, C2, C6
Puffinus assimilis Little Shearwater	B	1996	100	200	C	B1ii, B2, C2, C6
Sterna dougallii Roseate Tern	B	1996	30	30	B	A4i, B1i, B2, C2

Protection status

National None **International** High
642 ha of IBA covered by Special Protection Area (Costa a Caldeirão, 642 ha).

Conservation issues

Threats Consequences of animal/plant introductions (A)

Introduced ground predators (rats, mustelids, feral cats and dogs) possibly limit the breeding of most seabirds, especially of smaller shearwaters/petrels. Invasion by densely rooted exotic plants, such as cane grass *Arundo donax*, results in the loss of suitable burrowing ground for shearwaters/petrels. The numbers of terns have been monitored since 1989, and other seabirds since 1995, with EU funding (ACE and LIFE projects).

Coast of Faial — 003

A2, A4i, A4ii, B1i, B1ii, B2, B3, C2, C6

Admin region Horta
Coordinates 38°36'N 28°43'W
Altitude 0–300 m **Area** 940 ha

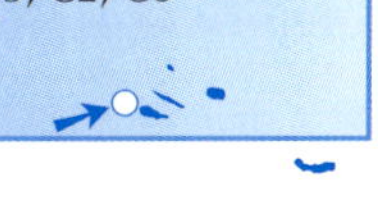

Site description

The IBA extends along the coastline from Castelo Branco harbour to Ribeirinha harbour, encompassing a 250 m-wide band from the shoreline to the cliff-tops.

Habitats Forest and woodland (30%; broadleaved evergreen forest), Scrub (15%; heathland), Rocky areas (100%; sea cliff/rocky shore; scree/boulders), Introduced/exotic vegetation (60%)
Land-use Agriculture (15%), Nature conservation/research (10%), Not utilized (70%), Tourism/recreation (20%)

Birds

Species	Season	Year	Pop min	Pop max	Acc	Criteria
Calonectris diomedea Cory's Shearwater	B	1996	11,375	14,528	A	A4ii, B1ii, B2, C2, C6
Puffinus assimilis Little Shearwater	B	1998	20	50	C	B1ii, B2, C2
Sterna dougallii Roseate Tern	B	1996	100	100	A	A4i, B1i, B2, C2, C6
Sterna hirundo Common Tern	B	1996	420	420	A	C6
Serinus canaria Canary	R	1996	300	—	—	A2, B3

Important mainly for breeding seabirds.

Protection status

National Partial **International** Partial
100 ha of IBA covered by Nature Reserve (Vulcão dos Capelinhos, 100 ha). 100 ha of IBA covered by Special Protection Area (Caldeira e Capelinhos, 2,067 ha).

Conservation issues

Threats Consequences of animal/plant introductions (A), Disturbance to birds (A), Recreation/tourism (A), Unsustainable exploitation (C)

Introduced ground predators (rats, mustelids, feral cats and dogs) possibly limit the breeding of most seabirds, especially of smaller shearwaters/petrels. Invasion by densely rooted exotic plants, such as cane grass *Arundo donax*, results in the loss of suitable burrowing ground for shearwaters/petrels. Unauthorized recreational activities on Capelinhos volcano cause disturbance to breeding terns. Numbers of terns have been monitored since 1989 and other seabirds since 1995, with EU funding (ACE and LIFE projects).

Coast of Pico — 004

A2, A4i, A4ii, B1i, B1ii, B2, B3, C2, C6

Admin region São Roque do Pico, Lajes do Pico
Coordinates 38°26′N 28°15′W
Altitude 0–150 m **Area** 1,500 ha

Site description

The IBA extends along the coast from Cabrito to Lajes, encompassing a 250 m-wide band from the shoreline to the cliff-tops.

Habitats Forest and woodland (20%; broadleaved evergreen forest), Scrub (10%; heathland), Rocky areas (100%; sea cliff/rocky shore; rock stacks/islets; scree/boulders), Introduced/exotic vegetation (50%)
Land-use Agriculture (30%), Not utilized (60%), Urban/industrial/transport (10%)

Birds

Species	Season	Year	Pop min	Pop max	Acc	Criteria
Calonectris diomedea Cory's Shearwater	B	1995	15,491	18,544	A	A4ii,B1ii,B2,C2,C6
Puffinus assimilis Little Shearwater	B	1996	40	120	C	B1ii, B2, C2, C6
Sterna dougallii Roseate Tern	B	1996	20	20	A	A4i, B1i, B2, C2
Serinus canaria Canary	R	—	300	—	—	A2, B3

Important mainly for breeding seabirds.

Protection status

National None **International** Partial
23 ha of IBA covered by Special Protection Area (Furnas de Santo António, 23 ha). 29 ha of IBA covered by Special Protection Area (Lajes do Pico, 29 ha). 150 ha of IBA covered by Special Protection Area (Ponta da Ilha, 342 ha).

Conservation issues

Threats Consequences of animal/plant introductions (A), Disturbance to birds (A), Recreation/tourism (A), Unsustainable exploitation (C)

Recreation at Furnas de Santo António threatens the single colony of *Sterna dougallii* in the IBA. Introduced ground predators (rats, mustelids, feral cats and dogs) possibly limit the breeding of most seabirds, especially of smaller shearwaters/petrels. Invasion by densely rooted exotic plants, such as cane grass *Arundo donax*, results in the loss of suitable burrowing ground for shearwaters/petrels. Numbers of terns have been monitored since 1989, and other seabirds since 1995, with EU funding (ACE and LIFE projects).

Rosais — 005

A4i, A4ii, B1i, B1ii, B2, C2, C6

Admin region Velas
Coordinates 38°44′N 28°16′W
Altitude 0–400 m **Area** 500 ha

Site description

The IBA extends along the coast from Morro de Lemos to Fajã João Dias, encompassing a 250 m-wide band from the shoreline to the cliff-tops.

Habitats Forest and woodland (20%; broadleaved evergreen forest), Scrub (10%; heathland), Rocky areas (100%; sea cliff/rocky shore; rock stacks/islets; scree/boulders)
Land-use Agriculture (10%), Not utilized (90%)

Birds

Species	Season	Year	Pop min	Pop max	Acc	Criteria
Calonectris diomedea Cory's Shearwater	B	1996	23,108	27,314	A	A4ii,B1ii,B2,C2,C6
Puffinus assimilis Little Shearwater	B	1996	25	50	C	B1ii, B2, C2
Sterna dougallii Roseate Tern	B	1992	50	50	A	A4i, B1i, B2, C2

Important mainly for breeding seabirds, including the second largest concentration of *Calonectris diomedea* in the Azores.

Protection status

National None **International** None

Conservation issues

Threats Consequences of animal/plant introductions (A)

Introduced ground predators (rats, mustelids, feral cats and dogs) possibly limit the breeding of most seabirds, especially of smaller shearwaters/petrels. Invasion by densely rooted exotic plants, such as cane grass *Arundo donax*, results in loss of suitable burrowing ground for shearwaters/petrels. Numbers of terns have been monitored since 1989, and other seabirds since 1995, with EU funding (ACE and LIFE projects).

Topo — 006

A4i, A4ii, B1i, B1ii, B2, C2, C6

Admin region Calheta
Coordinates 38°33′N 27°46′W
Altitude 0–350 m **Area** 340 ha

Site description

The IBA extends along the coast from Fajã dos Cúberes (Cabeço da Lagoa) to Ponta dos Monteiros (including Topo islet), encompassing a 250 m-wide band from the shoreline to the cliff-tops.

Habitats Forest and woodland (30%; broadleaved evergreen forest), Rocky areas (100%; sea cliff/rocky shore; rock stacks/islets; scree/boulders), Introduced/exotic vegetation (40%)
Land-use Agriculture (20%), Nature conservation/research (80%)

Birds

Species	Season	Year	Pop min	Pop max	Acc	Criteria
Calonectris diomedea Cory's Shearwater	B	1996	2,351	2,847	A	A4ii, B1ii, B2, C2
Puffinus assimilis Little Shearwater	B	1996	25	60	C	B1ii, B2, C2
Oceanodroma castro Madeiran Petrel	B	1996	5	—	C	C6
Sterna dougallii Roseate Tern	B	1996	22	22	A	A4i, B1i, B2, C2

Important mainly for breeding seabirds.

Protection status

National Low **International** Partial
20 ha of IBA covered by Nature Reserve (Topo, 20 ha). 170 ha of IBA covered by Special Protection Area (Ilhéu do Topo e Costa Adjacente, 346 ha).

Conservation issues

Threats Agricultural intensification/expansion (C), Consequences of animal/plant introductions (A), Disturbance to birds (C)

Cows and sheep are present on Topo islet and should be removed to avoid the trampling of burrows and tern nests. Within the coastal sector of the main island, introduced ground predators (rats, mustelids, feral cats and dogs) possibly limit the breeding of most seabirds, especially of small shearwaters/petrels. Invasion by densely rooted exotic plants, such as cane grass *Arundo donax*, has resulted in the loss of suitable burrowing ground for shearwaters/petrels. Numbers of terns have been monitored since 1989, and of other seabirds since 1995, with EU funding (ACE and LIFE projects).

Coast and islets of Graciosa — 007

A4i, A4ii, B1i, B1ii, B2, C2, C6

Admin region Santa Cruz da Graciosa
Coordinates 39°07′N 28°01′W
Altitude 0–203 m **Area** 680 ha

Site description

Habitats Forest and woodland (10%; broadleaved evergreen forest), Scrub (15%; heathland), Rocky areas (100%; sea cliff/rocky shore; rock stacks/islets; scree/boulders), Introduced/exotic vegetation (30%)
Land-use Agriculture (20%), Nature conservation/research (6%), Not utilized (55%), Tourism/recreation (13%), Urban/industrial/transport (10%)

The coastal strip (average width 200 m) of Graciosa, extending inland to the cliff-tops, and including the islets of Praia and Baixo (listed as sites 003 and 004 in the previous international IBA inventory: Grimmett and Jones 1989), and of Gaivota and Ponta da Barca.

Birds

Species	Season	Year	Pop min	Pop max	Acc	Criteria
Calonectris diomedea Cory's Shearwater	B	1996	11,537	15,088	A	A4ii, B1ii, B2, C2, C6
Puffinus assimilis Little Shearwater	B	1996	100	100	B	B1ii, B2, C2, C6
Oceanodroma castro Madeiran Petrel	B	1996	700	700	B	A4ii, B1ii, B2, C2, C6
Sterna dougallii Roseate Tern	B	1996	100	100	A	A4i, B1i, B2, C2, C6

Important mainly for breeding seabirds, including the largest concentration of *Oceanodroma castro* in the Azores (there are two temporally segregated breeding populations, possibly two sibling species). The first and only known breeding locality of *Phaethon aethereus* in Europe. The only locality in the Azores where the globally threatened *Pterodroma feae* has been recorded.

Protection status

National None **International** Low

12 ha of IBA covered by Special Protection Area (Ilhéu da Praia, 12 ha). 30 ha of IBA covered by Special Protection Area (Ilhéu de Baixo, 30 ha).

Conservation issues

Threats Consequences of animal/plant introductions (A), Disturbance to birds (A), Recreation/tourism (A), Unsustainable exploitation (C)

Praia islet is under excessive pressure from the recreational activities of the local human population. Adult and young *Calonectris diomedea* are regularly harvested by fishermen. Introduced ground predators (rats, mustelids, feral cats and dogs) possibly limit the breeding of most seabirds, especially of small shearwaters/petrels. Invasion by densely rooted exotic plants, such as cane grass *Arundo donax*, results in the loss of suitable burrowing ground for shearwaters/petrels. Introduced rabbits *Oryctolagus cuniculus* on Praia islet also have a detrimental effect on the burrows of shearwaters/petrels and accelerate soil-erosion due to overgrazing. Numbers of terns have been monitored since 1989 and other seabirds since 1995, with EU funding (ACE and LIFE projects). There is a management plan for Praia islet, where wardening and habitat restoration are in progress (LIFE project). The area forms part of an ongoing study of seabird ecology.

West coast of Terceira — A4ii, B1ii, B2, C2 — 008

Admin region Angra do Heroísmo, Praia da Vitória
Coordinates 38°45'N 27°22'W
Altitude 0–155 m **Area** 600 ha

Site description

The IBA extends along the coast from Porto das Cinco to Ponta da Selvagem, comprising a band (average width 200 m) from the shoreline to the cliff-tops.

Habitats Forest and woodland (broadleaved evergreen forest), Scrub (heathland), Grassland), Rocky areas (sea cliff/rocky shore; rock stacks/islets; scree/boulders), Introduced/exotic vegetation)
Land-use Agriculture, Not utilized, Tourism/recreation

Birds

Species	Season	Year	Pop min	Pop max	Acc	Criteria
Calonectris diomedea Cory's Shearwater	B	1996	3,752	5,276	A	A4ii, B1ii, B2, C2

Important mainly for breeding *Calonectris diomedea*.

Protection status

National None **International** None

Conservation issues

Threats Agricultural intensification/expansion (U), Consequences of animal/plant introductions (U)

Introduced ground predators (rats, mustelids, feral cats and dogs) possibly limit the breeding of most seabirds, especially of smaller shearwaters/petrels. Invasion by densely rooted exotic plants, such as cane grass *Arundo donax*, results in the loss of suitable burrowing ground for shearwaters/petrels. Agricultural intensification and expansion may result in a loss of nesting habitat. Numbers of terns have been monitored since 1989, and other seabirds since 1995, with EU funding (ACE and LIFE projects).

South-east coast of Terceira — A4i, A4ii, B1i, B1ii, B2, C2, C6 — 009

Admin region Angra do Heroísmo
Coordinates 38°39'N 27°09'W
Altitude 0–150 m **Area** 350 ha

Site description

The IBA extends along the coast from Monte Brasil to Porto Novo, including the islets off Ponta das Contendas and Cabras, and encompasses a 200 m-wide band from the shoreline to the cliff-tops.

Habitats Forest and woodland (broadleaved evergreen forest), Scrub (heathland), Rocky areas (100%; sea cliff/rocky shore; rock stacks/islets), Introduced/exotic vegetation)
Land-use Agriculture, Nature conservation/research, Not utilized, Tourism/recreation, Urban/industrial/transport

Birds

Species	Season	Year	Pop min	Pop max	Acc	Criteria
Calonectris diomedea Cory's Shearwater	B	1996	2,244	2,700	A	A4ii, B1ii, B2, C2
Sterna dougallii Roseate Tern	B	1996	70	70	A	A4i, B1i, B2, C2
Sterna hirundo Common Tern	B	1996	305	305	A	C6

Important mainly for breeding seabirds.

Protection status

National None **International** Partial

28 ha of IBA covered by Special Protection Area (Ilhéus das Cabras, 28 ha). 93 ha of IBA covered by Special Protection Area (Ponta das Contendas, 93 ha).

Conservation issues

Threats Agricultural intensification/expansion (U), Consequences of animal/plant introductions (U), Disturbance to birds (A), Recreation/tourism (A)

Vandalism and recreation cause disturbance to nesting birds, notably on the islets off Ponta das Contendas. Over 100 sheep were introduced in 1995 to the Cabras islets. There is a management plan for Ponta das Contendas SPA. Numbers of terns have been monitored since 1989, and other seabirds since 1995, with EU funding (ACE and LIFE projects). The islets off Ponta das Contendas have been wardened during the tern breeding season since 1994.

West coast of São Miguel — A4ii, B1ii, B2, C2 — 010

Admin region Ponta Delgada
Coordinates 37°51'N 25°50'W
Altitude 0–150 m **Area** 360 ha

Site description

The IBA extends along the coast from Feteiras to Ponta Bretanha, including Mosteiros islets, and encompasses a band (average width 200 m) from the shoreline to the cliff-tops.

Habitats Forest and woodland (broadleaved evergreen forest), Scrub (heathland), Rocky areas (100%; sea cliff/rocky shore; rock stacks/islets)
Land-use Agriculture, Not utilized, Tourism/recreation

Birds

Species	Season	Year	Pop min	Pop max	Acc	Criteria
Calonectris diomedea Cory's Shearwater	B	1996	8,844	12,130	A	A4ii, B1ii, B2, C2
Puffinus assimilis Little Shearwater	B	1996	30	60	C	B1ii, B2, C2

Important mainly for breeding seabirds.

Protection status

National None **International** None

Conservation issues

Threats Consequences of animal/plant introductions (U)

Introduced ground predators (rats, mustelids, feral cats and dogs) possibly limit the breeding of most seabirds, especially of smaller shearwaters/petrels. Invasion by densely rooted exotic plants, such as cane grass *Arundo donax*, results in the loss of suitable burrowing ground for shearwaters/petrels. Numbers of terns have been monitored since 1989, and other seabirds since 1995, with EU funding (ACE and LIFE projects).

Cintrão — A4ii, B1ii, B2, C2 — 011

Admin region Ribeira Grande
Coordinates 37°50′N 25°28′W
Altitude 0–150 m **Area** 200 ha

Site description

The IBA extends along the coast from Calhau do Cabo to Ponta Formosa, comprising a 200 m-wide band from the shoreline to the cliff-tops.

Habitats Rocky areas (100%; sea cliff/rocky shore; rock stacks/islets)
Land-use Agriculture, Not utilized, Unknown

Birds

Species	Season	Year	Pop min	Pop max	Acc	Criteria
Calonectris diomedea Cory's Shearwater	B	1996	3,177	4,467	A	A4ii, B1ii, B2, C2
Puffinus assimilis Little Shearwater	B	1996	20	40	C	B1ii, B2, C2

Important for breeding seabirds.

Protection status

National None **International** None

Conservation issues

Threats Consequences of animal/plant introductions (U)

Introduced ground predators (rats, mustelids, feral cats and dogs) possibly limit the breeding of most seabirds, especially of smaller shearwaters/petrels. Invasion by densely rooted exotic plants, such as cane grass *Arundo donax*, results in the loss of suitable burrowing ground for shearwaters/petrels. Numbers of terns have been monitored since 1989, and other seabirds since 1995, with EU funding (ACE and LIFE projects).

Vila Franca islet — B1ii, C2 — 012

Admin region Vila Franca do Campo
Coordinates 37°42′N 25°27′W
Altitude 0–61 m **Area** 10 ha

Site description

The islet is an extinct volcanic crater with about half of the area taken up by a shallow lagoon opening to the sea. There are steep slopes and cliffs of sandstone, with varied vegetation of evergreen exotic trees, native heathland, cane grass *Arundo donax*, tussocks of dune grass, and abandoned fields. The area is uninhabited but receives up to 600 visitors per day in summer.

Habitats Forest and woodland (20%; broadleaved evergreen forest), Scrub (20%; heathland), Grassland (20%; steppe/dry calcareous grassland), Rocky areas (30%; sea cliff/rocky shore; rock stacks/islets), Introduced/exotic vegetation (40%)
Land-use Nature conservation/research (100%), Tourism/recreation (30%)

Birds

Important for breeding *Calonectris diomedea*. The presence of *Oceanodroma castro* was detected in October 1996, but breeding has not yet been confirmed.

Species	Season	Year	Pop min	Pop max	Acc	Criteria
Calonectris diomedea Cory's Shearwater	B	1989	300	—	B	B1ii, C2

Protection status

National High **International** None
10 ha of IBA covered by Natural Park (Ilhéu de Vila Franca do Campo, 10 ha).

Conservation issues

Threats Consequences of animal/plant introductions (U), Recreation/tourism (B), Unsustainable exploitation (A)

There is intense recreational pressure during summer. Harvesting of *Calonectris diomedea* was intense in the recent past. A management plan is being implemented, the goals of which include complete eradication of rats *Rattus*, control of invasive alien plants, and prohibition of tourist access to the most sensitive areas.

East of São Miguel — A2, A4ii, B1ii, B2, B3, C2, C6 — 013

Admin region Nordeste e Povoação
Coordinates 37°47′N 25°10′W
Altitude 0–1,100 m **Area** 4,800 ha

Site description

The coastal area from Ponta do Arnel to Faial da Terra, extending inland to include the peaks of Pico da Vara (1,100 m) and Pico Verde (931 m) as well as the valley of Ribeira do Guilherme. One of the few remaining areas of laurel forest in the Azores. In the previous international IBA inventory (Grimmett and Jones 1989), this IBA (former site QQ007) encompassed only the inland part.

Habitats Forest and woodland (broadleaved evergreen forest), Scrub (heathland), Rocky areas (sea cliff/rocky shore; scree/boulders), Artificial landscape (forestry plantation), Introduced/exotic vegetation
Land-use Forestry, Nature conservation/research, Not utilized, Tourism/recreation

Birds

Species	Season	Year	Pop min	Pop max	Acc	Criteria
Calonectris diomedea Cory's Shearwater	B	1996	8,832	12,849	A	A4ii, B1ii, B2, C2
Puffinus assimilis Little Shearwater	B	1996	5	20	C	B2
[1] ***Columba palumbus*** Woodpigeon	R	1989	—	—	—	C2
Serinus canaria Canary	R	—	300	—	—	A2, B3
Pyrrhula murina Azores Bullfinch	R	1993	100	120	A	A2, C6

1. *C. p. azorica*

Important for breeding seabirds and native landbirds, including the only known population of *Pyrrhula murina* in the world.

Protection status

National Partial **International** Partial
6.5 ha of IBA covered by Nature Reserve (Graminhais, 6.5 ha). 875 ha of IBA covered by Nature Reserve (Pico da Vara, 875 ha). 1,982 ha of IBA covered by Special Protection Area (Pico da Vara/Ribeira do Guilherme, 1,982 ha).

Conservation issues

Threats Afforestation (A), Consequences of animal/plant introductions (A), Deforestation (commercial) (U)

The inland part of this site is severely encroached upon by commercial forestry and has been invaded by exotic plants. In the coastal sector, introduced ground predators (rats, mustelids, feral cats and dogs) possibly limit the breeding of most seabirds, especially of smaller shearwaters/petrels. Invasion by densely rooted exotic plants, such as cane grass *Arundo donax*, has resulted in the loss of suitable burrowing ground for shearwaters/petrels. There is a management plan for the inland part of the site. Restoration of the habitat of *Pyrrhula murina* is in progress, with EU funding (LIFE project).

Vila islet — A4i, A4ii, B1i, B1ii, B2, C2, C6 — 014

Admin region Vila do Porto
Coordinates 36°55′N 25°10′W
Altitude 0–60 m **Area** 10 ha

Site description

A rocky islet of basalt, with steep slopes and cliffs. On top of the islet and on steep slopes the rock is overlaid with soil, which supports annual plants. The island is uninhabited, but is visited regularly by fishermen and occasionally by tourists. In the previous international IBA inventory (Grimmett and Jones 1989) this site also included the adjacent coastal sector but it is now limited to the islet only, as the adjacent coastal sector does not support sufficient numbers of birds for qualification under the revised IBA criteria. The site was used for grazing livestock until 1993.

Habitats Grassland (50%), Rocky areas (100%; rock stacks/islets)
Land-use Nature conservation/research (100%), Tourism/recreation (10%)

Birds

Species	Season	Year	Pop min	Pop max	Acc	Criteria
Bulweria bulwerii Bulwer's Petrel	B	1996	50	50	A	B2, C6
Calonectris diomedea Cory's Shearwater	B	1996	400	400	A	B1ii, C2
Puffinus assimilis Little Shearwater	B	1996	50	50	B	B1ii, B2, C2
Oceanodroma castro Madeiran Petrel	B	1996	150	250	B	A4ii,B1ii,B2,C2,C6
Sterna dougallii Roseate Tern	B	1996	180	180	A	A4i,B1i,B2,C2,C6
Sterna hirundo Common Tern	B	1996	600	600	A	C6

Important for breeding seabirds, including the only known breeding site for *Bulweria bulwerii* in the Azores, the largest colony of *Sterna hirundo* in the Azores, and the first and only known breeding locality for *Sterna fuscata* in Europe.

Protection status

National None **International** High
10 ha of IBA covered by Special Protection Area (Ilhéu da Vila e Costa Adjacente, 48 ha).

Conservation issues

Threats Disturbance to birds (B), Recreation/tourism (B), Unsustainable exploitation (B)

Goats were eradicated in 1993. Chicks and adults of *Calonectris diomedea* are regularly harvested by fishermen. Recreation disturbs nesting terns. Numbers of terns have been monitored since 1989 and other seabirds since 1995, with EU funding (ACE and LIFE projects). The area forms part of an ongoing study of seabird ecology.

North and east coast of Santa Maria — A4i, A4ii, B1i, B1ii, B2, C2, C6 — 015

Admin region Vila do Porto
Coordinates 36°59′N 25°04′W
Altitude 0–300 m **Area** 625 ha

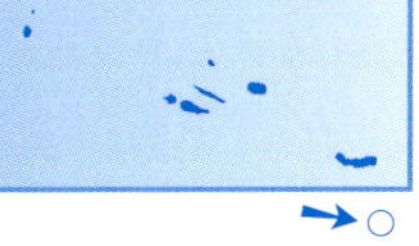

Site description

The stretch of coast extending from Ponta dos Frades to Malbusca, and from the shoreline to the cliff-tops (average width 250 m).

Habitats Forest and woodland (20%; broadleaved evergreen forest), Rocky areas (80%; sea cliff/rocky shore; rock stacks/islets), Artificial landscape (20%; perennial crops/orchards/groves), Introduced/exotic vegetation (20%)
Land-use Agriculture (30%), Nature conservation/research (25%), Not utilized (60%), Tourism/recreation (10%), Urban/industrial/transport (5%)

Birds

Species	Season	Year	Pop min	Pop max	Acc	Criteria
Calonectris diomedea Cory's Shearwater	B	1996	9,563	13,524	A	A4ii, B1ii, B2, C2
Puffinus assimilis Little Shearwater	B	1996	100	250	C	B1ii, B2, C2, C6
Oceanodroma castro Madeiran Petrel	B	1996	20	100	C	B1ii, B2, C2, C6
Sterna dougallii Roseate Tern	B	1996	100	100	A	A4i,B1i,B2,C2,C6
Sterna hirundo Common Tern	B	1996	310	310	A	C6

Important mainly for breeding seabirds.

Protection status

National Partial **International** None
50 ha of IBA covered by Natural Park (Baía de Maia, 50 ha). 113 ha of IBA covered by Natural Park (Baía de São Lourenço, 113 ha).

Conservation issues

Threats Burning of vegetation (C), Consequences of animal/plant introductions (A), Disturbance to birds (C), Recreation/tourism (C), Unsustainable exploitation (C)

Intense harvesting of adult and young *Calonectris diomedea* is a common practice on São Lourenço and Malbusca. Vegetation control by burning takes place during the breeding season and may affect burrow-nesting seabirds. Vandalism and recreation cause disturbance to the most accessible tern colonies. Introduced ground predators (rats, mustelids, feral cats and dogs) possibly limit the breeding of most seabirds, especially of smaller shearwaters/petrels. Invasion by densely rooted exotic plants, such as cane grass *Arundo donax*, results in the loss of suitable burrowing ground for shearwaters/petrels. Numbers of terns have been monitored since 1989, and other seabirds since 1995, with EU funding (ACE and LIFE projects).

REFERENCES

Anon. (1991) *Zonas de protecção especial*. Horta: Direcção Regional do Ambiente (Açores).

Bibby, C. J. and Charlton, T. D. (1991) Observations of the São Miguel bullfinch. *Açoreana* 7: 297–304.

del Nevo, A. J., Dunn, E. K., Medeiros, F. M., Le Grand, G., Akers, P., Avery, M. I. and Monteiro, L. R. (1990) Status, distribution and conservation of garajau-rosada (*Sterna dougalli*) and garajou-comum (*Sterna hirundo*) in the Açores. Preliminary report by the Royal Society for the Protection of Birds and the Universidade dos Açores. Unpublished.

del Nevo, A. J., Dunn, E. K., Medeiros, F. M., Le Grand, G., Akers, P., Avery, M. I. and Monteiro, L. R. (1993) The status of Roseate terns (*Sterna dougalli*) and Common terns (*Sterna hirundo*) in the Azores. *Seabird* 15: 30–37.

Fructuoso, G. (1561) *Saudades da terra*. 2nd edition published in 6 volumes from 1978 to 1983, ed. J. B. O. Rodrigues. Ponta Delgada: Instituto Cultural de Ponta Delgada.

Furness, R. W. and Monteiro, L. R. (1995) Red-billed tropicbird *Phaethon aethereus* in the Azores: first breeding record for Europe. *Bull. British Orn. Club* 115: 6–8.

Grimmett, R. F. A. and Jones, T. A. (1989) *Important Bird Areas in Europe*. Cambridge, UK: International Council for Bird Preservation (Techn. Publ. 9).

Hamer, K. C., Baber, I., Furness, *et al.* (1989) Ecology and conservation of Azorean seabirds. Report of the 1989 Glasgow University Expedition to the Azores Islands. Applied Ornithology Unit, University of Glasgow, Glasgow, UK Unpublished report.

Heath, M. F. and Borggreve, C. (2000) *BirdLife International/EBCC European Bird Database 1998*. Cambridge, UK: BirdLife International.

Monteiro, L. R. and Furness, R. W. (1995) Fea's Petrel *Pterodroma feae* in the Azores. *Bull. Brit. Orn. Club* 115: 9–14.

Monteiro, L. R. and Furness, R. W. (1998) Speciation through temporal segregation of Madeiran storm petrel (*Oceanodroma castro*) populations in the Azores? *Phil. Trans. Roy. Soc. Lond. B* 353: 945–953

Monteiro, L. R., Ramos, J. A. and Furness, R. W. (1996) Past and present status and conservation of the seabirds breeding in the Azores archipelago. *Biol. Conserv.* 78: 319–328.

Monteiro, L. R., Ramos, J. A., Sola, E., Furness, R. W., Monteiro, P., Feio, R., Thompson, D. R., Bearhop, S., Ratcliffe, N., Pereira, J., Hilton, G., Neves, V. C., Wilson, L., Hewitson, L., Tavares, A., Thompson, K., Laranjo M. and Groz, M. P. (1998) Conservation of seabird communities and habitats in the Azores archipelago, Final technical report to the European Commission (Programme LIFE, contract B4-3200/95/351). University of the Azores, Horta. Unpublished report.

Monteiro, L. R., Feio, R. S., Pereira, J. C., Wilson, L., Hewitson, L., Ramos, J. A., Monteiro, P. R., Tavares, A., Neves, V. C. and Furness, R. W. (1999) Using raft counts to census Cory's Shearwater *Calonectris diomedea*: method development and application in the Azores archipelago. Unpublished manuscript.

Ramos, J. A. (1993) The status and ecology of the Priolo or Azores bullfinch, *Pyrrhula murina*. Ph.D. thesis, University of Oxford, UK.

Ramos, J. R., Monteiro, L. R., Sola, E. and Moniz, Z. (1997) Characteristics and competition for nest cavities in burrowing Procellariiformes. *Condor* 99: 634–641

Santos, R. S., Hawkins, S. J., Monteiro, L. R., Alves, M., Isidro, E. J. and Martins, H. R. (1995) Marine research, resources and conservation in the Azores. *Aquatic Conserv.: Marine and Freshwater Ecosystems* 5: 311–354

Sjögren, E. (1973) Recent changes in the vascular flora and vegetation of the Azores islands. *Mem. Soc. Broteriana* XXII: 1–515

Tucker, G. M. and Heath, M. F. (1994) *Birds in Europe: their conservation status*. BirdLife International: Cambridge, UK (BirdLife Conservation Series no. 3).

MADEIRA

FRANCIS ZINO, MANUEL JOSÉ BISCOITO AND PAULO OLIVEIRA

Madeiran Petrel *Oceanodroma castro*. (PHOTO: BERNIE ZONFRILLO/IMAGDOP)

GENERAL INTRODUCTION

The Madeiran archipelago is situated in the North-east Atlantic Ocean and comprises the island of Madeira, on the northern limit with an area of 736 km², the island of Porto Santo and its surrounding islets, and the Desertas, consisting of three islands, Ilheu Chão, Deserta Grande and Bugio. Some 160 nautical miles to the south lies the sub-archipelago of the Selvagens, made up of Selvagem Grande, Selvagem Pequena and Ilheu de Fora.

There are 11 Important Bird Areas (IBAs) within the archipelago, covering an area of 133 km²; six on the island of Madeira, one on Porto Santo, two on the Desertas and two on the Selvagem islands (Table 1, Map 1). This compares to 13 sites identified in the previous pan-European inventory (Grimmett and Jones 1989). Changes have been made to the Fajã da Nogueira and Ribeiro Frio IBA (previously ZZ006) which has been joined with the Areeiro region (site 005), with the exception of Ribeiro Frio which is no longer considered part of an IBA. Also, all the rugged offshore islets of Porto Santo (previously ZZ007 and ZZ008) have been consolidated into one site (007).

ORNITHOLOGICAL IMPORTANCE

Two hundred and ninety-five species or subspecies of birds have been recorded from the archipelago, of which 42 are breeders (Zino *et al.* 1995).

Table 1. Summary of Important Bird Areas in Madeira. 11 IBAs covering 133 km²

IBA code	1989 code	International/national name	Administrative region	Area (ha)	Criteria (see p. 11)
001	ZZ001	Vale da Ribeira da Janela, Fanal and Ribeira Funda	Madeira	3,969	A1, A2, B2, C1, C2, C6
002	ZZ002	Rabaças and Folhadal and Tis amarelos	Madeira	2,082	A1, A2, B2, C1, C2, C6
003	ZZ003	Montado dos Pessegueiros, Cabeceira da Ribeira do Seixal and Lombo Barbinhas	Madeira	1,357	A1, A2, B2, C1, C2, C6
004	ZZ004	Caldeirão Verde, Moquinhas, Pico Casado	Madeira	1,773	A1, A2, B2, C1, C2, C6
005	ZZ005, ZZ006p	Montado de Areeiro, Montado do Sabugal, Montado da Fajã da Nogueira, Montado dos Piquetes, Montado do Cidrão	Madeira	1,900	A1, A2, B2, B3, C1, C2, C6
006	ZZ007	Ponta da São Lourenço	Madeira	340	B2, C6
007	ZZ008, ZZ009	Ilheu de Cima, Ilheu de Baixo, Ilheu de Ferro	Porto Santo	174	B2, C6
008	ZZ010	Bugio 'Deserta Islands'	Madeira	333	A1, A2, A4ii, B1ii, B2, B3, C1, C2, C6
009	ZZ011	Ilheu Chão and Deserta Grande	Madeira	1,089	A2, A4ii, B1ii, B2, B3, C2, C6
010	ZZ012	Selvagem Grande	Selvagens	245	A2, A4ii, B1ii, B2, B3, C2, C6
011	ZZ013	Selvagem Pequena and Ilheu de Fora	Selvagens	20	A4ii, B1ii, B2, C2, C6

Site identified in the previous inventory of IBAs in Europe (Grimmett and Jones 1989) but no longer considered to be an IBA

ZZ006p Fajã da Nogueira and Ribeiro Frio.

p = part of IBA

Map 1. Location, area and criteria category of Important Bird Areas in Madeira.

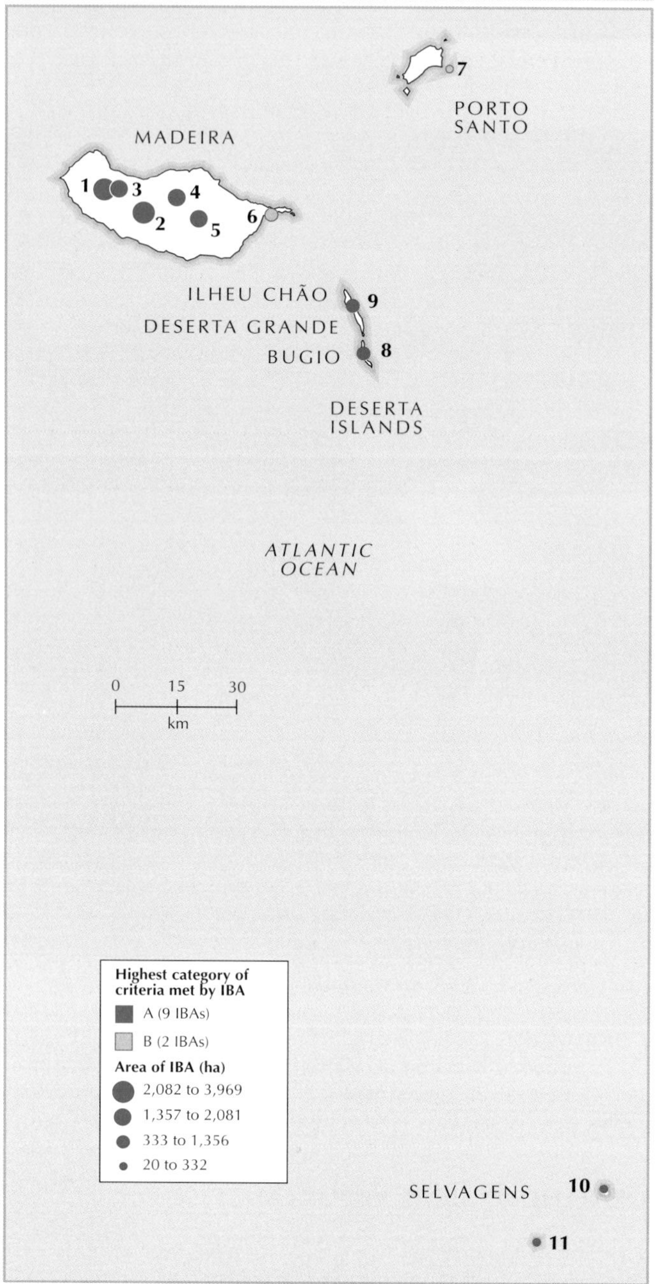

Table 2. Important Bird Areas in Madeira that are important for species of global conservation concern (meeting criterion A1).

Species	IBA code
Pterodroma feae Fea's Petrel	008
Pterodroma madeira Zino's Petrel	005
Columba trocaz Long-toed Pigeon	001, 002, 003, 004, 005

Table 3. Important Bird Areas in Madeira that support important numbers of one or more congregatory species (i.e. meeting criteria A4 and/or B1). IBAs meeting both criteria A4 and B1 for the species are shown in **bold**. IBAs meeting only criterion B1 for the species concerned, and not A4, are shown in normal type. For key to 'Season', see p. 7.

Species	Season	IBA code
Bulweria bulwerii Bulwer's Petrel	B	**008, 009, 010**
Calonectris diomedea Cory's Shearwater	B	**008, 009, 010**
Puffinus assimilis Little Shearwater	B	008, 009, **010**
Pelagodroma marina White-faced Petrel	B	**010, 011**
Oceanodroma castro Madeiran Petrel	B	**008, 009, 010**

Table 4. Species of European conservation concern and species listed on Annex I of the EC Birds Directive with significant breeding populations at IBAs in Madeira (meeting any IBA criteria).

Species [1]	Minimum national breeding population (pairs) [2]	Proportion (%) of national population breeding at all IBAs in Madeira
Pterodroma feae Fea's Petrel	150	100
Pterodroma madeira Zino's Petrel	20	100
Bulweria bulwerii Bulwer's Petrel	6,000	100[3]
Calonectris diomedea Cory's Shearwater	16,500	40
Puffinus assimilis Little Shearwater	1,800	100[3]
Pelagodroma marina White-faced Petrel	16,000	100[3]
Oceanodroma castro Madeiran Petrel	2,000	100[3]
Sterna dougallii Roseate Tern	5	100[3]
Sterna hirundo Common Tern	100	65
Columba trocaz Long-toed Pigeon	3,500 individuals	99
Apus unicolor Plain Swift	1,000	5
Anthus berthelotii Berthelot's Pipit	1,000	26
Serinus canaria Canary	3,000	10

1. Only those species of European conservation concern (see Box 1, p. 12) that meet IBA criteria in Madeira are listed, together with those species listed on Annex I of the EC Birds Directive that fulfil criterion C6 at IBAs in Madeira.
2. Data are taken from the BirdLife/EBCC European Bird Database 1998 (Heath and Borggreve 2000).
3. The percentage of the national population in IBAs exceeds 100%. Usually this is because the national population estimate has not been updated recently whilst the IBA population estimate has been recently updated with new data as a result of comprehensive surveys of IBAs themselves. Also, the individual site count for a species may be the maximum or average over recent years, and summing these may record more birds than are present nationally in any single year.

Two birds are endemic to the archipelago, *Pterodroma madeira* and *Columba trocaz*. Both these species, as well as *Pterodroma feae*, are considered globally threatened. Six IBAs support significant populations of one or more of these species (Table 2). *Pterodroma madeira* is Europe's most endangered seabird and is known to nest only in the Montado region (IBA 005), where the Freira Conservation Project estimates only 20–30 nesting pairs. There are various IBAs in which large numbers of *Columba trocaz* are to be found (001–005). The island of Bugio, in the Deserta Islands (008), holds the only known European breeding colony of *Pterodroma feae*, with a population of 150–200 breeding pairs (Zino and Biscoito 1994). If the Bugio population of *Pterodroma feae* is considered an endemic subspecies, or even species, in the future then this colony will become even more important.

Several IBAs are very important for large concentrations of breeding petrels (Procellariidae) (Table 3). The Selvagem Islands, hold two very important IBAs (010 and 011), supporting massive breeding colonies of Procellariidae. Since the islands were designated as a Strict Nature Reserve in 1971, the breeding numbers have increased. There is also a single breeding record of *Sterna fuscata*, and there are regular records of this species visiting since then.

The island of Bugio (008) supports internationally important breeding colonies of five species of Procellariidae. The remaining two Deserta islands (Deserta Grande and Ilheu Chão) make up IBA 009. In 1996, Deserta Grande was cleared of all rabbits and goats and since that time the flora has made a dramatic recovery. It is hoped that these conditions will aid the increase of populations of Procellariidae, which are already present in internationally significant numbers. This site also has significant numbers of *Anthus berthelotii madeirensis* and *Serinus canaria* and an in-depth study of these two species is currently under way on Ilheu Chão.

As well as supporting *Columba trocaz*, several forest areas are also of considerable importance to a number of Madeiran subspecies, notably the birds of prey and passerines living within the forest canopy such *Fringilla coelebs maderensis* and *Regulus ignicapillus madeirensis*.

Ponta de São Lourenço (006) is of considerable importance to the vagrants that visit Madeira, but it also supports four species of breeding Procellaridae, a small colony of *Sterna dougallii* and several important populations of breeding landbirds such as *Anthus berthelotii madeirensis*, *Petronia petronia*, and *Sylvia conspicillata orbitalis*.

The islets around the island of Porto Santo (007) are deserted and the lighthouses on two of the islands have been automated. As Strict Nature Reserves, the islands offer ideal conditions for

Figure 1. Occurrence of habitats at Important Bird Areas in Madeira (see Appendix 3 for definitions of habitats).

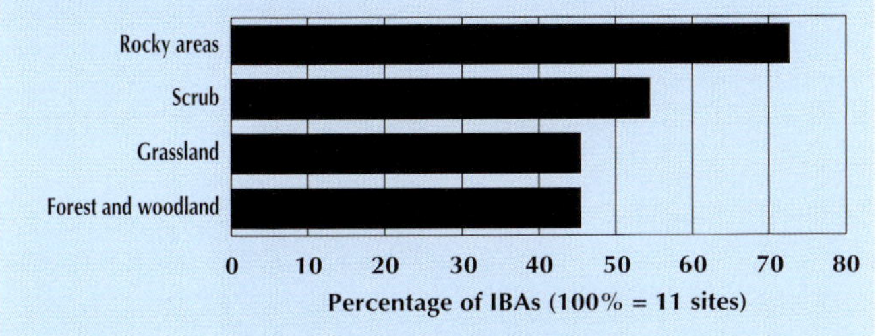

Figure 2. Occurrence of land-uses at Important Bird Areas in Madeira (see Appendix 3 for definitions of land-uses).

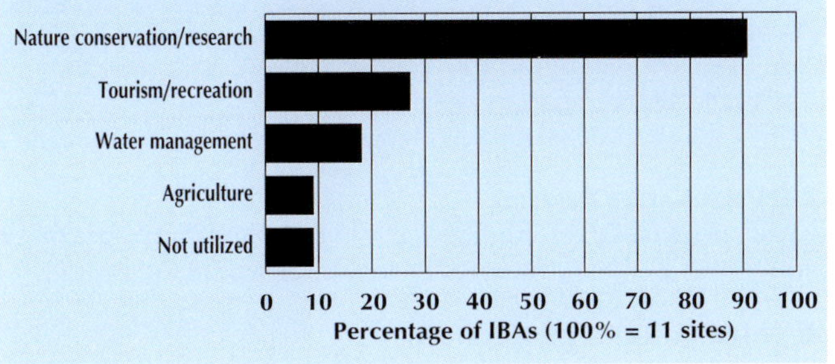

breeding Procellariidae; *Bulweria bulwerii* and *Oceanodroma castro* are especially abundant.

Many species have significant populations within IBAs (Table 4) and thus are very dependent on such areas being adequately managed and conserved.

HABITATS

The 11 IBAs within the Madeiran archipelago are spread over a number of islands and thus cover a broad range of habitats. Many IBAs contain several different major habitat-types; rocky areas, scrub and forest/woodland are the most common, and grassland is predominant in four IBAs (Figure 1).

Laurisilva (laurel forest) dominates four of the six IBAs on the island of Madeira (001, 002, 003 and 004), being especially important for *Columba trocaz*. Laurisilva was once common in southern Europe, in particular on the northern edge of the Mediterranean and the Atlantic islands, but due to climatic changes in the last glaciation, this forest disappeared from Europe and is now only to be found on the Atlantic islands, with the best examples on Madeira. Significantly, the laurisilva of Madeira is being considered for inclusion in the World Heritage List.

Some of the best high-altitude alpine scrub and grassland of Madeira are to be found in IBA 005, the only known breeding ground for *Pterodroma madeira*, with some 33 endemic plants from a total of 77 plants identified in the area. The habitat on the breeding ledges of *Pterodroma madeira* supports an incredible 30 endemic species from a total of 40 plant species recorded on the ledges, thus further illustrating the importance of keeping these areas free of herbivores.

Different types of scrub habitat are also present in many IBAs. Sites 001, 002 and 003 have coastal cliffs extending from sea-level to c.300 m, and are covered in a typical *Aeonion-Lythanthion* community. Ponta de São Lourenço (006) is low-lying with sea cliffs and flatter ground which is covered in scrub and grassland. The islets off Porto Santo (007) have a characteristic xerophytic flora, composed mainly of scrub and grassland. In the past Ilheu da Cal was mined for lime, which has resulted in many caves being dug into the side of the cliffs. These abandoned caves are much used as nest-sites by *Calonectris diomedea borealis*.

The Desertas (008, 009) are in many respects similar to the islets off Porto Santo. However, due to the large size and high altitude of the islands they offer a greater variety of habitats, ranging from vertical cliffs to a lush valley on the northern half of Deserta Grande. Ilheu Chão is flat on top, as the name indicates, and holds the densest vegetation, with the jasmine bush (*Jasminum odoratissimum*) being the most prominent plant. In contrast, the top of Bugio is almost a knife-edge, except for two small plateaus, which are covered in grass, and present the preferred breeding ground of *Pterodroma feae*.

The two IBAs on the Selvagens are largely rocky. Selvagem Grande (010), is a semi-desertic volcanic plateau surrounded by sea cliffs. The presence of goats in the past and the continuing presence of rabbits have destroyed much of the natural vegetation, which is now dominated by the introduced *Nicotiana glauca* and *Mesembryanthemum* spp. This island is very important for *Calonectris diomedea borealis*, *Bulweria bulwerii*, *Puffinus assimilis*, *Oceanodroma castro* and *Pelagodroma marina*. Selvagem Pequena and Ilheu de Fora (011), are low-lying islets covered in virgin vegetation, containing various endemic species, a unique situation in the North Atlantic. The most common breeding bird is *Pelagodroma marina*.

IMPACTS ON IBAs – LAND-USE AND THREATS

The inclusion of all IBAs within the Parque Natural da Madeira means that nature conservation and research is a major land-use in all Madeiran IBAs (Figure 2). Other forms of land-use are prohibited in many of these areas and, where permitted, are controlled or monitored by law and therefore do not constitute major threats.

Three areas are used for tourism and recreation. In recent years an increasing number of people have taken an interest in nature and open spaces and this has resulted in an increase of human pressure on fragile habitats. The absence of regulations to control this influx may turn this into a major threat in the foreseeable future.

Herbivores (goats, sheep and rabbits) constitute a major threat to all areas, particularly in IBAs 005, 007 and 010, and in areas above 1,400 m; there will always be the need for control measures to exclude herbivores from the IBAs (Figure 3). Cats and rats also present a major threat to the avifauna on the island of Madeira.

PROTECTION STATUS

■ National protection

Nature conservation policy-making and law enforcement are the exclusive responsibility of the Parliament and Government of the Autonomous Region of Madeira. All IBAs are subject to regional legislation on nature conservation and are under the jurisdiction of the Parque Natural da Madeira, the agency of the Regional Government responsible for the implementation of the Government's policy on the conservation of nature. Ten of the 11 IBAs fall within the Parque Natural da Madeira (and Porto Santo

Figure 3. Occurrence of threats at Important Bird Areas in Madeira (see Appendix 3 for definitions of threat types and impact categories).

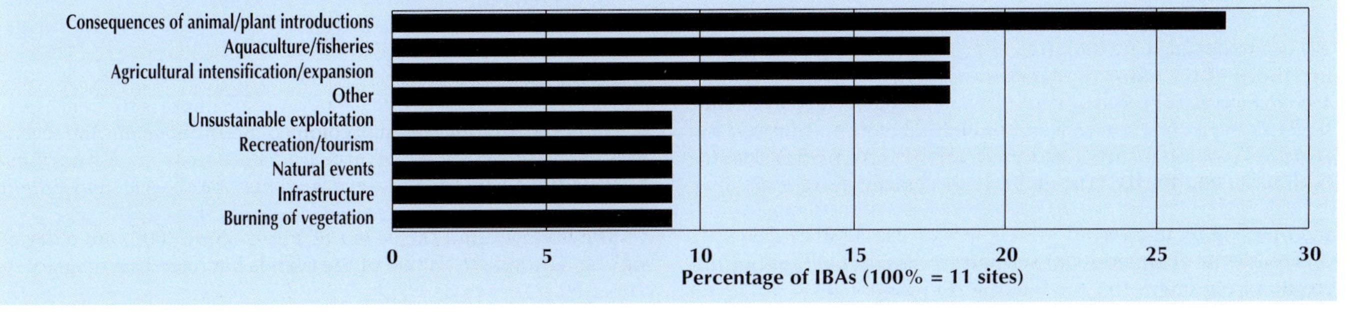

Figure 4. The national protection status of Important Bird Areas in Madeira.

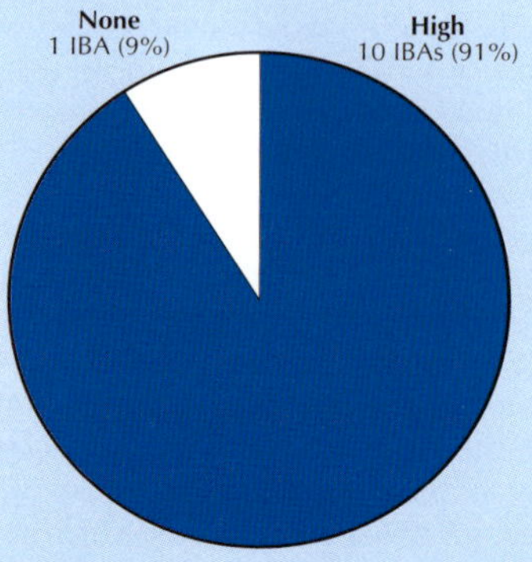

% of area of individual IBA overlaps with national protected areas:
High >90%
Partial 10–90%
Low <10%
None

Not possible to calculate total area of overlap between IBA network in Madeira and national protected-area system (see Table 5 for categories) due to incomplete overlap-area data.

Figure 5. The international protection status of Important Bird Areas in Madeira.

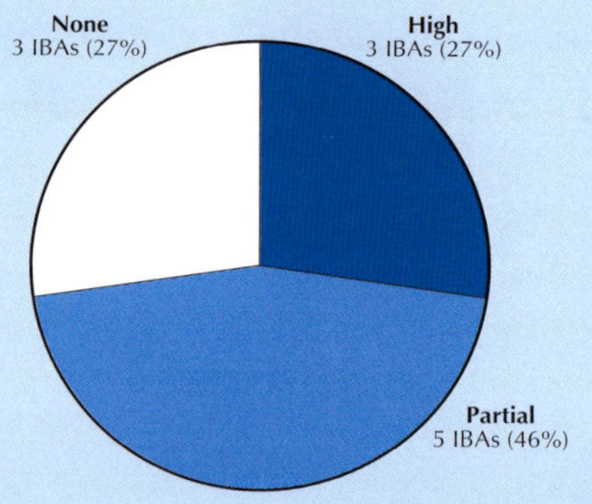

% of area of individual IBA overlaps with international protected areas:
High >90%
Partial 10–90%
Low <10%
None

Not possible to calculate total area of overlap between IBA network in Madeira and international protected-area system (see Table 5 for categories) due to incomplete overlap-area data.

Table 5. Protection status of Important Bird Areas in Madeira. A tick (✔) indicates that an IBA overlaps with a protected area (to any extent).

		National			International	
IBA code	International name	Strict Nature Reserve	Partial Nature Reserve	Natural Park	Biogenetic Reserve	Special Protection Area
001	Vale da Ribeira da Janela, Fanal and Ribeira Funda			✔	✔	
002	Rabaças and Folhadal and Tis amarelos	✔	✔	✔		
003	Montado dos Pessegueiros, Cabeceira da Ribeira do Seixal and Lombo Barbinhas	✔	✔	✔		
004	Caldeirão Verde, Moquinhas, Pico Casado	✔	✔	✔	✔	
005	Montado de Areeiro, Montado do Sabugal, Montado da Fajã da Nogueira, Montado dos Piquetes, Montado do Cidrão	✔		✔	✔	
006	Ponta da São Lourenço	✔	✔	✔		
007	Ilheu de Cima, Ilheu de Baixo, Ilheu de Ferro					
008	Bugio 'Deserta Islands'	✔		✔	✔	
009	Ilheu Chão and Deserta Grande	✔		✔	✔	
010	Selvagem Grande	✔		✔		✔
011	Selvagem Pequena and Ilheu de Fora	✔		✔		✔
Total number of IBAs		9	4	10	5	2

Box 1. International legislation and initiatives that are relevant to site conservation in Madeira (see Appendix 1 for a general description of these agreements).

Global	
Biodiversity Convention	✔
Ramsar Convention	✔
Bonn Convention	✔
World Heritage Convention	✔
MAB Programme	
Pan-European	
Bern Convention	✔
Regional	
EC Birds Directive	✔
EC Habitats Directive	✔

✔ Convention ratified/initiative supported
(✔) Convention signed

007 will be included in 1999) (Figure 4), a Natural Park which covers two thirds of the island of Madeira and virtually all the laurisilva. Two main types of protected area provide stricter protection within the Park; Strict Nature Reserves and Partial Nature Reserves; five IBAs are covered by Strict Nature Reserves and a further four are both Strict and Partial Nature Reserves (Table 5, Figure 4).

1. Strict Nature Reserve
Areas of total protection. Access only allowed for study/scientific purposes; cattle-grazing and hunting are prohibited.

2. Partial Nature Reserve
Areas of special protection. Access is restricted; cattle-grazing is prohibited; hunting, tree-cutting and construction are restricted.

■ International protection

The archipelago of Madeira is an integral part of Portugal but has its own autonomous government. Box 1 shows the legal instruments and initiatives that apply to Portugal, and also to Madeira.

Two sites are covered by Special Protection Areas under the EC Birds Directive (010, 011) and five by the Council of Europe's network of Biogenetic Reserves (Table 5, Figure 5)

CONSERVATION

- The Freira Conservation Project, initiated in 1986 when all attempted breeding of *Pterodroma madeira* failed, has been an important conservation project over the last 13 years. It involves the Museu Municipal do Fuchal and the Parque Natural da Madeira. The project continues to this day, wardening and trying to control the rats and cats in the area. Whilst rat and cat control over such a vast area can only be partially effective, the program has been a success and breeding continues.
- All rabbits and goats were removed from Deserta Grande (009) in 1996, facilitated by a LIFE project, and the flora is fast recovering, making the island more attractive for nesting marine birds.
- In-depth studies have been conducted on several species in the Madeiran archipelago, including on Selvagem Grande where the *Calonectris diomedea borealis* colony has been under continuous study since 1969; an ongoing study using satellite tags is trying to obtain more accurate information on the species's movements. In-depth studies of *Bulweria bulwerii*, *Puffinus assimilis*, *Oceanodroma castro* and *Pelagodroma marina* have all been carried out on the island.

ANALYTICAL METHODS

- Population estimates, threats and land-use data are based mainly on good quality data from 1995 to 1999.

GLOSSARY

ICBP International Council for Bird Preservation.
laurisilva laurel forest.
LIFE an EU funding line (Regulation 1973/92/EEC) which supports EU environmental legislation, including the Birds and Habitats Directives.
PNM Parque Natural da Madeira.
SAC Special Area for Conservation (proposed under the EC Habitats Directive).

SITE ACCOUNTS

Vale da Ribeira da Janela, Fanal and Ribeira Funda
A1, A2, B2, C1, C2, C6 **001**

Admin region Madeira
Coordinates 32°47′N 17°09′W
Altitude 0–1,200 m **Area** 3,969 ha

Site description
The largest river valley in Madeira, with extensive areas of laurisilva covering the steep valley sides. There are many large trees in the valley bottom, along with scattered human habitation. Human activities include grazing, fishing and hunting.

Habitats Forest and woodland (90%; broadleaved evergreen forest), Scrub (10%; heathland)
Land-use Nature conservation/research (95%)

Birds

Species	Season	Year	Pop min	Pop max	Acc	Criteria
[1] ***Bulweria bulwerii*** Bulwer's Petrel	B	—	—	—	—	B2
[1] ***Calonectris diomedea*** Cory's Shearwater	B	—	—	—	—	B2
Columba trocaz Long-toed Pigeon	R	1996	400	800	A	A1,A2,B2,C1,C2,C6

1. Numbers unknown.

An important area for *Columba trocaz*; the forest also holds breeding *Regulus ignicapillus madeirensis* and *Fringilla coelebs maderensis*. On the sea cliffs *Calonectris diomedea borealis, Bulweria bulwerii* and *Puffinus puffinus* breed. *Anthus berthelotii madeirensis* breeds at Fanal.

Protection status
National High **International** Partial
3,969 ha of IBA covered by Natural Park (Parque Natural da Madeira). IBA covered by Biogenetic Reserve.

Conservation issues
There are no serious threats at present as agriculture is declining throughout the island. Goats are a minor threat and there is the possibility of road construction which could become a problem and should be strictly controlled. The control of illegal hunting of *Columba trocaz* and special attention to the introduction of sheep, goats and cows is required. A management plan exists for the area. The IBA is currently in the process of being designated as Strict and Partial Nature Reserves, and is a proposed SAC.

Rabaças and Folhadal and Tis amarelos
A1, A2, B2, C1, C2, C6 **002**

Admin region Madeira
Coordinates 32°44′N 17°03′W
Altitude 0–1,640 m **Area** 2,082 ha

Site description
The site comprises mainly laurisilva. Human activities include large-scale goat-herding ('Agriculture' land-use).

Habitats Forest and woodland (broadleaved evergreen forest), Scrub
Land-use Agriculture, Nature conservation/research, Tourism/recreation, Water management

Birds

Species	Season	Year	Pop min	Pop max	Acc	Criteria
Columba trocaz Long-toed Pigeon	R	1996	100	200	A	A1,A2,B2,C1,C2,C6

This is an important area for *Columba trocaz* (present at very high density). The site also holds breeding *Bulweria bulweria, Puffinus puffinus, Anthus berthelotii, Apus unicolor, Fringilla coelebs maderensis* and *Regulus ignicapillus madeirensis*.

Protection status
National High **International** Partial
2,082 ha IBA covered by Natural Park (Parque Natural da Madeira). 2,082 ha of IBA covered by Strict and Partial Nature Reserves. IBA covered by Biogenetic Reserve.

Conservation issues
Threats Agricultural intensification/expansion (U)

Goat-herding poses a serious threat to native vegetation, with over-browsing restricting new tree growth ('Agricultural intensification/ expansion' threat). This situation is gradually being brought under control by the government. There is a management plan and the site is a proposed SAC.

Montado dos Pessegueiros, Cabeceira da Ribeira do Seixal and Lombo Barbinhas
A1, A2, B2, C1, C2, C6 **003**

Admin region Madeira
Coordinates 32°47′N 17°07′W
Altitude 0–1,500 m **Area** 1,357 ha

Site description
The best-preserved area of laurisilva on Madeira, being almost free of sheep and goats.

Habitats Forest and woodland (broadleaved evergreen forest), Rocky areas (sea cliff/ rocky shore), Scrub
Land-use Nature conservation/research (100%)

Birds

Species	Season	Year	Pop min	Pop max	Acc	Criteria
Bulweria bulwerii Bulwer's Petrel	B	—	—	—	—	B2
Columba trocaz Long-toed Pigeon	R	1996	150	400	A	A1,A2,B2,C1,C2,C6

This is an important area for *Columba trocaz. Fringilla coelebs maderensis, Regulus ignicapillus madeirensis* and *Scolopax rusticola* also breed within the forest. *Calonectris diomedea borealis* and *Bulweria bulwerii* breed at sea level with *Puffinus puffinus* a little higher up. Raptors are also present in the area.

Protection status
National High **International** None
1,357 ha of IBA covered by Natural Park (Parque Natural da Madeira). 1,357 ha of IBA covered by Strict Nature Reserve (Montado dos Peesgueiros and Lombo Barbinhas) and Partial Nature Reserve (Cabeceira da Ribeira do Seixal).

Conservation issues
No serious threats are known to be present. Nearly all sheep and goats have been removed from the area and it is vital that pressure in this direction is continued, to completely free the site of herbivores. A management plan exists and the area is a proposed SAC.

Caldeirão Verde, Moquinhas, Pico Casado
A1, A2, B2, C1, C2, C6 **004**

Admin region Madeira
Coordinates 32°46′N 16°58′W
Altitude 400–1,700 m **Area** 1,773 ha

Site description
An area of steep-sided valleys with dense laurisilva, high-altitude grasslands and large expanses of exposed cliffs. Human activities include goat-herding and fishing. Tis Amarelos was included within this site in the previous European IBA inventory (1989 site ZZ004) but is now included in site 002.

Habitats Forest and woodland (broadleaved evergreen forest), Grassland (mesophile grassland), Rocky areas (inland cliff)
Land-use Nature conservation/research (100%)

Birds

Species	Season	Year	Pop min	Pop max	Acc	Criteria
Columba trocaz Long-toed Pigeon	R	1996	400	800	A	A1,A2,B2,C1,C2,C6

This is probably the most important area in Madeira for *Columba trocaz*. *Regulus ignicapillus madeirensis*, *Fringilla coelebs maderensis* and *Scolopax rusticola* also breed in the forest. Four species of breeding raptor are present and *Puffinus puffinus* breeds on the inland cliffs.

Protection status

National High **International** Partial

1,773 ha of IBA covered by Natural Park (Parque Natural da Madeira). 1,773 ha of IBA covered by Strict Nature Reserve (Caldeirão Verde and Pico Casado) and Partial Nature Reserve (Moquinhas). IBA covered by Biogenetic Reserve.

Conservation issues

Threats Agricultural intensification/expansion (U), Infrastructure (U)

Some parts of the area, despite being strictly protected, are heavily overgrazed by goats and sheep ('Agricultural intensification/ expansion' threat). The problem is slowly but surely being brought under control by the government. There is also a threat of water authorities building roads in the area to facilitate their work. A management plan exists and the site is a proposed SAC.

Montado de Areeiro, Montado do Sabugal, Montado da Fajã da Nogueira, Montado dos Piquetes, Montado do Cidrão

A1, A2, B2, B3, C1, C2, C6 **005**

Admin region Madeira
Coordinates 32°40'N 16°56'W
Altitude 900–1,810 m **Area** 1,900 ha

Site description

A precipitous mountain area with inland cliffs and ridges. Habitats include alpine scrub and grassland, below which is transitional forest. Laurisilva occurs in the lower parts of the area. The site receives a large number of walkers in the spring and summer months (400+ walkers per day) and the area is a water catchment for the town of Funchal. Fajã da Nogueira, included in the previous pan-European inventory (Grimmett and Jones 1989) as site 006, is also included within this IBA.

Habitats Scrub (scrub), Grassland (alpine/subalpine/boreal grassland), Rocky areas (inland cliff), Forest and woodland (broadleaved evergreen forest)
Land-use Nature conservation/research (100%), Tourism/recreation (5%), Water management (100%)

Birds

Species	Season	Year	Pop min	Pop max	Acc	Criteria
Pterodroma madeira Zino's Petrel	B	1996	10	30	B	A1,B2,C1,C2,C6
Columba trocaz Long-toed Pigeon	R	1996	100	300	A	A1,A2,B2,C1,C2,C6
Apus unicolor Plain Swift	B	—	50	100	C	A2, B3
Anthus berthelotii Berthelot's Pipit	R	—	15	30	C	A2, B3

The area supports the only known breeding population of *Pterodroma madeira* in the world. *Puffinus puffinus*, all four raptors that breed in Madeira and *Regulus ignicapillus* all also breed here.

Protection status

National High **International** Partial

1,900 ha of IBA covered by Natural Park (Parque Natural da Madeira). 1,900 ha of IBA covered by Strict Nature Reserve. IBA covered by Biogenetic Reserve.

Conservation issues

Threats Burning of vegetation (A), Consequences of animal/plant introductions (A), Recreation/tourism (C)

Thought to be extinct, *Pterodroma madeira* was rediscovered in 1969, but the population soon fell prey to collectors. By 1986 all attempted breeding on the known ledges failed, due to rat predation. As a result the Freira Conservation Project was started in 1986, involving the Museu Municipal do Fuchal and the Parque Natural da Madeira, and financed by Zeneca Agrochemical through ICBP (the body preceding BirdLife International). The project continues to this day, wardening and trying to control the rats and cats in the area. Whilst rat and cat control over such a vast area can only be partially effective, the program has been a success and breeding continues. Overgrazing has been a severe problem in the past, but over recent years the Madeiran government has been making a concerted effort to acquire these Montados and remove all sheep and goats. The recovery of the flora is vital to the survival of *Pterodroma madeira*, which only breeds on vegetation-covered ledges, free of sheep and goats. The burning of vegetation to produce pasture for sheep and goats is also a threat. There is a management plan and the site is a proposed SAC.

Ponta da São Lourenço

B2, C6 **006**

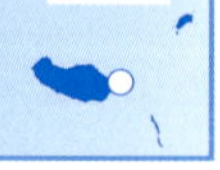

Admin region Madeira
Coordinates 32°44'N 16°42'W
Altitude 0–163 m **Area** 340 ha

Site description

A rocky peninsula, sea cliffs and associated islands that together comprise the eastern tip of Madeira. Covered mainly by scrub and grassland, with scattered human habitation. Human activities include hunting and tourism.

Habitats Scrub (scrub), Grassland (steppe/dry calcareous grassland), Rocky areas (sea cliff/rocky shore)
Land-use Nature conservation/research, Tourism/recreation

Birds

Species	Season	Year	Pop min	Pop max	Acc	Criteria
[1] ***Bulweria bulwerii*** Bulwer's Petrel	B	1996	—	—	—	B2
[1] ***Calonectris diomedea*** Cory's Shearwater	B	1996	—	—	—	B2
[1] ***Oceanodroma castro*** Madeiran Petrel	B	1996	—	—	—	B2
Sterna hirundo Common Tern	B	1996	10	30	B	C6

1. Numbers unknown.

Puffinus assimilis also breeds and the first record in Madeira of breeding *Sterna dougallii* is from this area. On land, *Anthus berthelotii* (common), *Petronia petronia*, *Sylvia conspicillata* and *Serinus canarius* (abundant) are of particular note. This is also an important stop-over site for migrating passerines.

Protection status

National High **International** None

340 ha of IBA covered by Natural Park (Parque Natural da Madeira). 340 ha of IBA covered by Partial Nature Reserve (Ponta de São Lourenço) and Strict Nature Reserve (Ilheu dos Desembarcadouros).

Conservation issues

The main threat was the taking of seabirds by fishermen. Since the purchase of the area by the Regional Government of Madeira in 1996 and the management by the Parque Natural da Madeira (PNM) this problem has almost disappeared and it is hoped that the seabird populations will make a full recovery. Grazing and the extraction of sand are now totally prohibited. The site is a proposed SAC.

Ilheu de Cima, Ilheu de Baixo, Ilheu de Ferro

B2, C6 **007**

Admin region Porto Santo
Coordinates 33°03'N 16°16'W
Altitude 0–0 m **Area** 174 ha

Site description

Three rocky islands covered with scrub and Macaronesian coastal flora, located off the coast of Porto Santo. Two of the islands have lighthouses and the third was once used for mining lime. Two IBAs identified in the previous European IBA inventory (Grimmett and Jones 1989) (former sites 008 and 009) have been merged to form this one site.

Habitats Scrub (sclerophyllous scrub/garrigue/maquis)
Land-use Not utilized

Birds

Species	Season	Year	Pop min	Pop max	Acc	Criteria
[1] ***Bulweria bulwerii*** Bulwer's Petrel	B	1996	Abundant		—	B2, C6
[1] ***Calonectris diomedea*** Cory's Shearwater	B	1996	Common		—	B2, C6
[1] ***Oceanodroma castro*** Madeiran Petrel	B	1996	—	—	—	B2

1. Numbers unknown.

Four species of shearwater and petrel (Procellariidae) nest on these islands but as yet no in-depth study has been carried out. *Sterna dougallii, S. hirundo* and *Larus cachinnans* also breed. On land, *Anthus berthelotii madeirensis* (common) and *Apus unicolor* are of note.

Protection status

National None **International** None

Conservation issues

Threats Consequences of animal/plant introductions (U), Unsustainable exploitation (U), Other (U)

Nesting seabirds are subject to human predation. The introduction of European rabbits *Oryctolagus cuniculus* has resulted in considerable damage to the vegetation and subsequent erosion. The lighthouses are a danger to migratory birds and also to the juveniles of the resident seabirds ('Other' threat). The island is in the process of being designated as a Strict Nature Reserve and is a proposed SAC, and there are also plans to include it within the Parque Natural da Madeira.

Bugio 'Deserta Islands' A1, A2, A4ii, B1ii, B2, B3, C1, C2, C6 008

Admin region Madeira
Coordinates 32°26′N 16°29′W
Altitude 0–380 m **Area** 333 ha

Site description

An elongated island of basalt rock with steep and unstable cliffs. The island is made up of two main areas, each with a small grassland plateau at the top.

Habitats Grassland (steppe/dry calcareous grassland), Rocky areas (sea cliff/rocky shore)
Land-use Nature conservation/research (100%)

Birds

Species	Season	Year	Pop min	Pop max	Acc	Criteria
Pterodroma feae Fea's Petrel	B	1995	70	200	B	A1, B2, C1, C2, C6
Bulweria bulwerii Bulwer's Petrel	B	1995	2,000	3,500	B	A4ii, B1ii, B2, C2
Calonectris diomedea Cory's Shearwater	B	1995	500	2,000	C	A4ii, B1ii, B2, C2, C6
Puffinus assimilis Little Shearwater	B	—	100	300	—	B1ii, B2, C2, C6
Oceanodroma castro Madeiran Petrel	B	—	500	2,000	C	A4ii, B1ii, B2, C2, C6
Anthus berthelotii Berthelot's Pipit	R	—	15	30	B	A2, B3
Serinus canaria Canary	R	—	200	400	B	A2, B3

This is a very important site for five breeding shearwaters and petrels (Procellariidae) found in Madeira. This is the only European breeding ground of the Macaronesian endemic *Pterodroma feae*, which nests mainly on the grassy southern plateau. Other breeding birds include terns *Sterna* and *Larus cachinnans*.

Protection status

National High **International** Partial
333 ha of IBA covered by Natural Park (Parque Natural da Madeira). 333 ha of IBA covered by Strict Nature Reserve. IBA covered by Biogenetic Reserve.

Conservation issues

Threats Consequences of animal/plant introductions (A)

Overgrazing by introduced goats and rabbits causes severe problems, damaging vegetation and resulting in erosion. Occasionally *Calonectris diomedea* may still be taken by fishermen. The area is a Total Land Reserve, and landing is controlled by means of a special permit. A management plan exists and the area is a proposed SAC.

Ilheu Chão and Deserta Grande A2, A4ii, B1ii, B2, B3, C2, C6 009

Admin region Desertas
Coordinates 32°32′N 16°31′W
Altitude 0–408 m **Area** 1,089 ha

Site description

Two islands of volcanic origin which were once joined and are now separated by a small channel. The islands are wild and deserted with high sea cliffs. Ilheu Chão is flat on top and holds a wide variety of flora. Large parts of Deserta Grande have suffered desertification, although a valley in the north which was once cultivated now supports grassland with Macaronesian maritime flora.

Habitats Unknown
Land-use Nature conservation/research (100%)

Birds

Species	Season	Year	Pop min	Pop max	Acc	Criteria
Bulweria bulwerii Bulwer's Petrel	B	1996	2,000	5,000	A	A4ii, B1ii, B2, C2
Calonectris diomedea Cory's Shearwater	B	1996	1,500	3,000	A	A4ii, B1ii, B2, C2, C6
Puffinus assimilis Little Shearwater	B	1996	250	500	B	B1ii, B2, C2, C6
Oceanodroma castro Madeiran Petrel	B	1996	750	1,500	B	A4ii, B1ii, B2, C2, C6
Sterna hirundo Common Tern	B	1996	25	50	A	C6
Anthus berthelotii Berthelot's Pipit	R	1996	200	400	B	A2, B3
Serinus canaria Canary	R	1996	300	600	B	A2, B3

The area is important for four species of breeding shearwater and petrel (Procellariidae). Ilheu Chão has one of the largest breeding colonies of *Larus cachinnans* in Madeira. Other breeding species include *Sterna dougallii* (possibly breeding) and *Apus unicolor*.

Protection status

National High **International** High
1,089 ha of IBA covered by Natural Park (Parque Natural da Madeira). 1,089 ha of IBA covered by Strict Nature Reserve. IBA covered by Biogenetic Reserve.

Conservation issues

Threats Natural events (U)

Ilheu Chão once had a population of hares *Lepus capensis*, but these died out some 20 years ago and since then the flora has made a dramatic recovery. All rabbits and goats were removed from Deserta Grande in 1996, facilitated by a LIFE project, and the flora is fast recovering, making the island more attractive for nesting marine birds. There is also continuing erosion from wind and water ('Natural events' above). In the past, the four breeding Procellariidae suffered from predation by fishermen. Since the area came under the control of the PNM, and the islands are permanently wardened, all predation has stopped and the colony is making a good recovery, with numbers having more than doubled since the reserve was established. However, there is constant pressure on politicians to open the marine reserves to fishermen, which should be resisted. The site is a proposed SAC.

Selvagem Grande A2, A4ii, B1ii, B2, B3, C2, C6 010

Admin region Selvagens
Coordinates 30°09′N 15°52′W
Altitude 0–163 m **Area** 245 ha

Site description

A rocky island of volcanic origin, characterized by sparse vegetation and a total lack of surface fresh water. The top of the island is flat and has sand deposits from the raised sea floor.

Habitats Rocky areas (sea cliff/rocky shore)
Land-use Nature conservation/research (100%)

Birds

Species	Season	Year	Pop min	Pop max	Acc	Criteria
Bulweria bulwerii Bulwer's Petrel	B	1996	4,000	6,000	B	A4ii, B1ii, B2, C2, C6
Calonectris diomedea Cory's Shearwater	B	1996	4,000	18,000	A	A4ii, B1ii, B2, C2, C6
Puffinus assimilis Little Shearwater	B	1996	1,500	3,000	A	A4ii, B1ii, B2, C2, C6
Pelagodroma marina White-faced Petrel	B	1996	9,000	15,000	B	A4ii, B1ii, B2, C2, C6
Oceanodroma castro Madeiran Petrel	B	1996	1,500	2,000	B	A4ii, B1ii, B2, C2, C6
Anthus berthelotii Berthelot's Pipit	R	1996	40	60	B	A2, B3

One of the most important shearwater and petrel (Procellariidae) colonies in the North Atlantic.

Protection status

National High **International** High
245 ha of IBA covered by Natural Park (Parque Natural da Madeira). 245 ha covered by Strict Nature Reserve. 245 ha of IBA covered by Special Protection Area (Ilhas Selvagens, 30,400 ha).

Conservation issues

Threats Aquaculture/fisheries (C)

The *Calonectris diomedea borealis* colony is making a slow recovery. At one time 20,000 chicks were culled each year and in 1976 all birds on the island were killed. There are now some 14,000 breeding pairs, probably only 25% of past numbers but the population is making a good recovery. The only threat is from fishermen, who have, through their dumping of fish and offal, produced an increasing population of *Larus cachinnans*, which predates on other marine birds. The *Calonectris diomedea borealis* colony has been under continuous study since 1969; an ongoing study using satellite tags is trying to obtain more accurate information on the species' movements. In-depth studies of *Bulweria bulwerii*, *Puffinus assimilis*, *Oceanodroma castro* and *Pelagodroma marina* have all been carried out on the island. Attempts have been made in the past to build a landing strip for light and military aircraft. The permanent wardening of the island and the building of a lighthouse, in addition to visiting scientists, have caused a large increase in human population and the resultant accumulation of rubbish; great care needs to be taken in the disposal of this rubbish. The site is a proposed SAC.

Selvagem Pequena and Ilheu de Fora — A4ii, B1ii, B2, C2, C6 — 011

Admin region Selvagens
Coordinates 30°02′N 16°02′W
Altitude 0–49 m **Area** 20 ha

Site description

Islands of volcanic origin, with rock stacks and inland sand-dunes. The islands are surrounded by reefs, which act to double the island's area when exposed at low tide.

Habitats Rocky areas (rock stacks/islets; inland sand-dunes)
Land-use Nature conservation/research (100%)

Birds

Species	Season	Year	Pop min	Pop max	Acc	Criteria
Bulweria bulwerii Bulwer's Petrel	B	1996	30	60	A	B2, C6
Calonectris diomedea Cory's Shearwater	B	1996	160	220	A	B2, C6
Pelagodroma marina White-faced Petrel	B	1996	10,000	30,000	A	A4ii, B1ii, B2, C2, C6
Oceanodroma castro Madeiran Petrel	B	1996	30	60	A	B2, C6
Sterna dougallii Roseate Tern	B	1996	—	10	A	C6
Sterna hirundo Common Tern	B	1996	30	60	A	C6

This is an important site for breeding shearwaters and petrels (Procellariidae) and terns *Sterna*, including a large population of *Pelagodroma marina*. Breeding *Sterna fuscata* have been recorded. There are no land birds.

Protection status

National High **International** High
20 ha of IBA covered by Natural Park (Parque Natural da Madeira). 20 ha of IBA covered by Strict Nature Reserve. 20 ha of IBA covered by Special Protection Area (Ilhas Selvagens, 30,400 ha).

Conservation issues

Threats Aquaculture/fisheries (U), Other (U)

The islands' vegetation is in a natural state, having never suffered any plant or animal introductions. Every effort should be made to ensure that this situation is maintained. The islands are permanently wardened from April to November and this has controlled the illegal taking of seabirds. The remaining threats are from fishermen and oil spills ('Other' threat). The site is a proposed SAC.

REFERENCES

Grimmett, R. F. A. and Jones, T. A. (1989) *Important Bird Areas in Europe*. Cambridge, UK: International Council for Bird Preservation (Tech. Publ. 9).

Heath, M. F. and Borggreve, C. (2000) *BirdLife International/EBCC European Bird Database 1998*. Cambridge, UK: BirdLife International.

Oliveira, P. (1999) *Gestão e Conservação das aves do Archipelago da Madeira*. Madeira; Parque Natural da Madeira.

Zino, F., Biscoito, M. J. and Zino, P. A. (1995) Birds of the achipelago of Madeira and the Selvagens: new records and checklist. *Bol. Mus. Mun. Funchal* 47:63–100.

Zino, F. and Biscoito, M. (1994) Breeding seabirds in the Madeira archipelago. Pp. 172–185 in D. N. Nettleship, J. Burger and M. Gochfeld, eds. *Seabirds on islands: threats, case studies and action plans*. Cambridge, UK: BirdLife International (Conservation Series 1).

ROMANIA

Dan Munteanu

Dalmatian Pelican *Pelecanus crispus*. (PHOTO: DOMINIQUE ROBERT)

GENERAL INTRODUCTION

Romania is situated in south-east Europe and has an area of 237,500 km^2. It borders Ukraine to the north, Moldova to the north-east, Bulgaria to the south, Yugoslavia to the south-west and Hungary to the west. The human population in 1993 was 22,789,000. Romania is divided into eight historical provinces—Maramureş, Crişana, Transylvania, Banat, Moldavia, Dobrogea (Dobruja), Oltenia and Muntenia (the two latter were once known as Walachia)—and 40 counties.

Romania's topography is dominated by the Carpathian Mountains, and can be divided into three parts: the Eastern, Southern and Western Carpathians. The Eastern Carpathians extend from the northern frontier to the Prahova river valley, reaching their maximum height in the Rodna mountains. The Southern Carpathians lie between the Prahova river valley to the east and the Timiş and Cerna river valleys to the west. The Western Carpathians form a discontinuous range between the Danube and Someş river valleys.

Lying on the northern edge of the Dobruja region, the Danube delta covers c.5,054 km^2, of which 4,536 km^2 are in Romania. It is the largest European wetland after the Volga delta, and includes one of the most extensive reedbeds (*Phragmites*) in the world. Sand-dunes, large sandy beaches and several fresh or salt-water lakes are found along the Black Sea coast.

Forty-four Important Bird Areas (IBAs) have been identified in Romania (Table 1), covering a total area of 6,557 km^2, or 3% of the country's land area. The sites are not uniformly distributed (Map 1)—most are wetlands found along the Danube and other main river corridors. The previous international IBA inventory identified

Table 1. Summary of Important Bird Areas in Romania. 44 IBAs covering 6,557 km^2

IBA code	1989 code	International name	National name	Administrative region	Area (ha)	Criteria (see p. 11)
001	RO001, RO001.1–RO001.6	Danube delta and Razelm-Sinoe complex	Delta Dunării şi complexul lagunar Razelm-Sinoe	Tulcea	442,000	A1, A4i, A4ii, A4iii, B1i, B1iii, B2, B3
002	—	Parcheş-Somova wetland	Ghiolurile Parcheş-Somova	Tulcea	3,100	A1, A4i, B1i, B2, B3
003	—	Lake Beibugeac	Lacul Beibugeac	Tulcea	180	A1, B1i, B3
004	RO002	Niculiţel and Babadag forest	Pădurea Babadag-Niculiţel	Tulcea	10,000	A1, B2
005	—	Black Sea coast-Chituc	Litoralul Mării Negre-Chituc	Constanţa	5,000	A1, A4i, A4iii, B1i, B2, B3
006	RO003	Lake Taşaul	Lacul Taşaul	Constanţa	1,830	B1i
007	RO004	Lake Siutghiol	Lacul Siutghiol	Constanţa	2,000	A1, A4i, B1i
008	RO005	Lake Techirghiol	Lacul Techirghiol	Constanţa	1,170	A1, A4i, A4iii, B1i, B2, B3
009	RO006	Hagieni forest	Pădurea Hagieni	Constanţa	600	B2, B3
010	—	Vădeni-Maţa-Cârja-Rădeanu wetlands	Bălţile Vădeni-Maţa-Cârja-Rădeanu	Galaţi, Vaslui	380	A1
011	—	Odobeşti hill	Măgura Odobeştilor	Vrancea	12,000	B2
012	—	Bârnova-Repedea woodland	Pădurile Bârnova-Repedea	Iaşi	15,000	B2
013	RO011	Bicaz gorge and Lake Roşu	Cheile Bicazului şi Lacu Roşu	Neamţ, Harghita	2,200	B2

Table 1 ... continued. Summary of Important Bird Areas in Romania. 44 IBAs covering 6,557 km²

IBA code	1989 code	International name	National name	Administrative region	Area (ha)	Criteria (see p.11)
014	—	Vlădeni fish-ponds	Eleşteele Vlădeni	Iaşi	1,200	A1, B1i, B2
015	—	Stânca-Ştefăneşti reservoir	Acumularea Stânca-Ştefăneşti	Botoşani	7,000	B1i, B2
016	—	Slătioara forest	Codrul Secular Slătioara	Suceava	397	B2
017	RO013	Pietrosul Rodnei mountain	Muntele Pietrosul Rodnei	Maramureş	3,300	B2
018	—	Dobron forest	Pădurea Dobron	Mureş	220	B2, B3
019	—	Zau de Câmpie-Tăureni fish-ponds	Eleşteele Zau de Câmpie-Tăureni	Mureş	520	B2
020	—	Fizeş valley	Valea Fizeşului	Cluj	450	B2
021	—	Vlădeasa mountain	Masivul Vlădeasa	Cluj, Bihor	14,000	B2
022	RO016	Întregalde and Râmeţi gorges	Cheile Întregalde şi Cheile Râmeţilor	Alba	4,000	B2, B3
023	—	Rotbav fish-ponds	Eleşteele Rotbav	Braşov	180	B2
024	RO021	Retezat National Park	Parcul National Retezat	Hunedoara	54,400	B2, B3
025	RO018	Cefa fish-ponds and Rădvani wood	Eleşteele Cefa-Pădurea Rădvani	Bihor	1,000	A1, A4i, A4iii, B1i, B2
026	—	Lunca wood	Pădurea Lunca	Arad	174	B2
027	—	Socodor wood	Pădurea Socodor	Arad	500	B2
028	RO019	Mureş river flood-plain: Arad-Pecica-Cenad	Lunca Mureşului: Arad-Pecica-Cenad	Arad	12,000	B1i, B2, B3
029	RO020	Satchinez marsh	Mlaştina Satchinez	Timiş	236	A1, B2
030	—	Murani lake and Pişchia forest	Lacul Murani şi Pădurea Pişchia	Timiş	1,500	A1, B2
031	RO022	Domogled mountain	Muntele Domogled	Caraş-Severin	2,382	A1, B2
032	—	Porţile de Fier reservoir	Lacul de acumulare Porţile de Fier	Caraş-Severin, Mehedinţi	32,000	A1, A4i, A4iii, B1i
033	—	Mehedinţi fish-ponds-Izvoarele	Eleşteele Mehedinţi-Izvoarele	Mehedinţi	210	A1
034	—	Mehedinţi fish-ponds-Gruia	Eleşteele Mehedinţi-Gruia	Mehedinţi	200	B2
035	—	Mehedinţi fish-ponds-Gârla Mare	Eleşteele Mehedinţi-Gârla Mare	Mehedinţi	140	A1, B2
036	—	Lake Dunăreni-Bistreţ	Lacul Dunăreni-Bistreţ	Dolj	1,936	A1, A4i, B1i, B2
037	—	Strejeşti reservoir	Acumularea Strejeşti	Valcea	2,000	A1, B1i
038	—	Lake Comana	Lacul Comana	Gorj	800	A1, B2, B3
039	—	Ciocăneşti fish-farm	Ferma piscicolă Ciocăneşti	Călăraşi	300	A1, A4i, B1i, B2
040	—	Lake Strachina	Lacul Strachina	Ialomiţa	1,050	A1, B2
041	—	Lake Fundata	Lacul Fundata	Ialomiţa	510	B1i
042	—	Lake Tătaru	Lacul Tătaru	Brăila	120	A1, B2
043	—	The Little Island of Brăila	Insula Mică a Brăilei	Brăila	14,862	A1, A4i, B1i, B2
044	RO009	Balta Albă, Amara and Jirlău lakes	Balta Albă, Lacul Amara şi Lacul Jirlău	Brăila, Buzău	2,680	A1, A4i, A4iii, B1i, B2

Sites identified in the previous inventory of IBAs in Europe (Grimmett and Jones 1989) but no longer considered to be IBAs
RO007 Canaraua Fetii; RO008 Lacul Dunăreni; RO010 Eleşteele Sînpaul (Sînpaul Ponds); RO012 Lacul Bicaz (Izvoru Muntelui); RO014 Fărăgău-Glodeni Fish-ponds; RO015 Cheile Turzii (Turda Gorge); RO017 Iezerele Cindrelului (Cindrel Lakes); RO023 Zona Mostiştea (Mostiştea wetlands); RO024 Lacul Snagov.

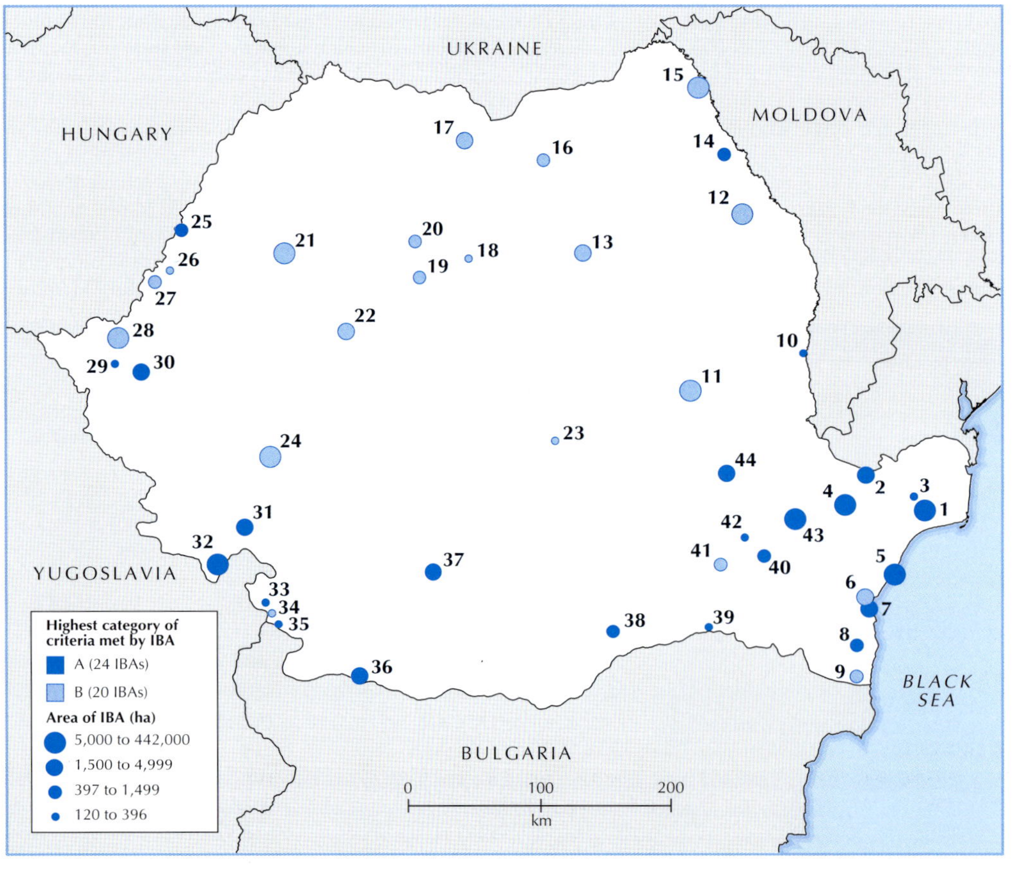

Map 1. Location, area and criteria category of Important Bird Areas in Romania.

24 IBAs (Grimmett and Jones 1989). Of these, nine have been dropped (Table 1), either because of (adverse) changes to the habitats present at the site, or because they do not meet the revised IBA criteria. Twenty-nine new sites have been identified between 1994 and 1997.

A further six areas are subject to monitoring in order to determine whether they meet IBA criteria: Herghelia Marsh (Dobruja), Lake Bugeac (Dobruja), Lake Brateş (Moldavia), Lake Galbeni (Moldavia), Lower valley of the Tur river (Transylvania) and Reservoirs of the Argeş river (Muntenia).

ORNITHOLOGICAL IMPORTANCE

There are 144 species of European conservation concern (SPECs) breeding regularly in Romania. Of these, seven are globally threatened (*Pelecanus crispus*, *Aythya nyroca*, *Aquila clanga*, *A. heliaca*, *Falco naumanni*, *Crex crex* and *Otis tarda*) and 75 have an unfavourable conservation status in Europe (Tucker and Heath 1994). Twenty-one SPECs regularly occur on passage or overwinter, and include the globally threatened *Pelecanus crispus*, *Branta ruficollis* and *Aythya nyroca*, and the near-threatened *Phalacrocorax pygmeus*.

Table 2. Important Bird Areas in Romania that are important for species of global conservation concern (meeting criterion A1).

Species	IBA code
Phalacrocorax pygmeus Pygmy Cormorant	001, 002, 005, 007, 008, 032, 035, 036, 037, 039, 042, 043
Pelecanus crispus Dalmatian Pelican	001, 036, 039, 044
Anser erythropus Lesser White-fronted Goose	014
Branta ruficollis Red-breasted Goose	001, 002, 003, 008
Aythya nyroca Ferruginous Duck	001, 010, 025, 029, 030, 032, 033, 038, 040, 043, 044
Oxyura leucocephala White-headed Duck	001, 008
Circus macrourus Pallid Harrier	001
Aquila clanga Greater Spotted Eagle	001
Aquila heliaca Imperial Eagle	001, 004, 031
Crex crex Corncrake	001, 038, 043
Glareola nordmanni Black-winged Pratincole	001

Twenty-four IBAs hold internationally important numbers of at least one species of global conservation concern and therefore meet the A1 criterion (Table 2). Two endemic subspecies occur in Romania—*Tetrao urogallus rudolfi* and *Parus montanus transylvanicus*—but no restricted-range species.

As a result of its location towards the east of the European continent, Romania supports many breeding species either at the western or northern limits of their ranges. Most are waterbirds (for example, *Phalacrocorax pygmeus*, *Bubulcus ibis*, *Plegadis falcinellus*, *Tadorna ferruginea* and *Chlidonias leucopterus*) or species of dry grassland/steppe (for example, *Circus macrourus*, *Accipiter brevipes*, *Apus melba*, *Hirundo daurica*, *Cercotrichas galactotes*, *Oenanthe isabellina*, *O. pleschanka*, *O. hispanica*, *Acrocephalus agricola*, *Parus lugubris*, *Emberiza cirlus* and *E. melanocephala*). Romania is also on the migration route of many northern breeders—especially ducks (for example, *Anas penelope*, *A. acuta* and *A. clypeata*) and waders (for example, *Calidris* spp., *Philomachus pugnax*, *Gallinago* spp., *Limosa limosa*, *Tringa erythropus*, *T. nebularia*, *T. ochropus* and *T. glareola*), and provides wintering quarters for some species (*Cygnus cygnus*, *Anser albifrons*, *Branta ruficollis*, *Anas crecca*, *Aythya fuligula*, *Bucephala clangula*, *Mergus* spp., *Circus cyaneus*, *Buteo lagopus*, *Larus fuscus*, *Turdus pilaris*, *Lanius excubitor*, *Fringilla montifringilla* and *Carduelis flammea*).

The Danube delta (IBA 001) supports a variety of species, with 320 having been recorded, some in very large numbers (for example, *Phalacrocorax carbo*, *Pelecanus onocrotalus*, *Ixobrychus minutus*, *Nycticorax nycticorax*, *Egretta garzetta*, *Anser anser*, *Anas platyrhynchos*, *Aythya ferina*, *Aythya nyroca*, *Falco subbuteo*, *Porzana* spp., *Vanellus vanellus*, *Larus ridibundus*, *Chlidonias* spp., *Locustella* spp., *Acrocephalus* spp. and *Panurus biarmicus*). Along with other wetlands in Moldavia and Dobruja, the delta acts as a

Table 3. Important Bird Areas in Romania that support important numbers of one or more congregatory species (i.e. meeting criteria A4 and/or B1). IBAs meeting both criteria A4 and B1 for the species are shown in **bold**. IBAs meeting only criterion B1 for the species concerned, and not A4, are shown in normal type. For key to 'Season', see p. 7.

Species	Season	IBA code
Podiceps grisegena Red-necked Grebe	B	**001**
Podiceps nigricollis Black-necked Grebe	B	**001**, **008**
	W	**005**
Phalacrocorax carbo Cormorant	B	**001**, **002**
	N	036
Phalacrocorax pygmeus	B	**001**, **002**, **043**
	W	**008**
	N	**036**
Pelecanus onocrotalus White Pelican	B	**001**
Pelecanus crispus Dalmatian Pelican	B	**001**
	P	**044**
	N	**036**, **039**
Nycticorax nycticorax Night Heron	B	**001**
Ardeola ralloides Squacco Heron	B	**001**, 036
Egretta alba Great White Egret	B	**001**
	N	**025**
Ardea cinerea Grey Heron	B	**001**
Ciconia nigra Black Stork	P	**001**
Platalea leucorodia Spoonbill	B	**001**, 043, 044
	P	014, 025
Cygnus olor Mute Swan	B	**001**
Cygnus cygnus Whooper Swan	W	**001**, 003, 005
Anser albifrons White-fronted Goose	W	**001**, 002, 003, **008**, 014, 028, 037
	P	025, 036, 039
Anser anser Greylag Goose	B	043
	W	003
	P	015, 025
Branta ruficollis Red-breasted Goose	W	**001**, **002**, **008**
Anas penelope Wigeon	P	001
Anas strepera Gadwall	B	**001**
Anas crecca Teal	P	001
Anas platyrhynchos Mallard	B	**001**, 044
Anas clypeata Shoveler	P	**044**
Netta rufina Red-crested Pochard	B	**001**
Aythya ferina Pochard	B	**001**
	W	**005**, **032**
	P	006
Aythya nyroca Ferruginous Duck	B	**001**
	P	**032**, **044**
Aythya fuligula Tufted Duck	W	001, **005**
Aythya marila Scaup	W	**005**
Bucephala clangula Goldeneye	W	001, 005
Mergus albellus Smew	W	**001**, **005**
Mergus merganser Goosander	W	001
Oxyura leucocephala White-headed Duck	W	**008**
Falco vespertinus Red-footed Falcon	B	**001**
Gallinula chloropus Moorhen	B	**001**
Fulica atra Coot	B	**001**
Grus grus Crane	P	001, 025
Recurvirostra avosetta Avocet	B	001
Larus minutus Little Gull	P	**007**, **008**
Larus cachinnans Yellow-legged Gull	R	006
Sterna hirundo Common Tern	B	**001**
Sterna albifrons Little Tern	B	**001**
Chlidonias hybridus Whiskered Tern	B	**001**, 002
	P	041
Chlidonias niger Black Tern	B	**001**
Chlidonias leucopterus White-winged Black Tern	B	**001**

stop-over site for migrants passing north–south along the western coast of the Black Sea. Several of these wetlands have been identified as IBAs (001, 002, 005, 006, 007, 008, 010, 014, and 015), forming a linear network of great importance to wildfowl and waders. A total of eleven IBAs support more than 1% of the biogeographic populations of those congregatory species listed in Table 3, and therefore meet the A4i criterion.

Key migration routes also run in a north-west–south-east diagonal over the middle of Transylvania (IBAs 019, 020 and 023) and eastern Muntenia (IBAs 039, 040, 041, 042 and 043), and in a north-north-east–south-south-west direction over Crişana and Banat in western Romania (IBAs 025, 028, 029 and 030)—a secondary route links to the Danube valley (IBAs 032, 033, 034, 035 and 036).

Thirty-seven IBAs support important breeding populations of SPECs (meeting criteria B2 and B3)—the entire national breeding populations of seven waterbirds and one raptor are found within IBAs (Table 4).

Species with a dispersed distribution (for example, *Circaetus gallicus*, *Aquila heliaca*, *A. chrysaetos*, *Falco tinnunculus* and *F. vespertinus*) or those with restricted local distributions (for example, *Tetrao tetrix*, *Otis tarda*, *Monticola saxatilis*, *Emberiza cia* and *E. hortulana*) are not adequately covered by the IBA inventory.

Table 4. Species of European conservation concern with significant breeding populations at Important Bird Areas in Romania (meeting any IBA criteria).

Species [1]	Minimum national breeding population (pairs) [2]	Proportion (%) of national population breeding at all IBAs in Romania
Phalacrocorax pygmeus Pygmy Cormorant	6,000	100[3]
Pelecanus onocrotalus White Pelican	3,000	100
Pelecanus crispus Dalmatian Pelican	200	100
Botaurus stellaris Bittern	500	100[3]
Ixobrychus minutus Little Bittern	10,000	53
Nycticorax nycticorax Night Heron	5,000	100[3]
Ardeola ralloides Squacco Heron	3,500	95
Ardea purpurea Purple Heron	800	68
Ciconia nigra Black Stork	40	8
Ciconia ciconia White Stork	4,000	2
Plegadis falcinellus Glossy Ibis	3,600	1
Platalea leucorodia Spoonbill	650	41
Anas strepera Gadwall	4,500	89
Netta rufina Red-crested Pochard	1,000	30
Aythya ferina Pochard	20,000	100[3]
Aythya nyroca Ferruginous Duck	8,000	40
Circaetus gallicus Short-toed Eagle	30	20
Aquila pomarina Lesser Spotted Eagle	100	21
Aquila heliaca Imperial Eagle	10	40
Falco tinnunculus Kestrel	3,000	12
Falco vespertinus Red-footed Falcon	500	100[3]
Porzana porzana Spotted Crake	5,000	82
Porzana parva Little Crake	3,000	35
Crex crex Corncrake	20,000	1
Recurvirostra avosetta Avocet	200	100[3]
Burhinus oedicnemus Stone Curlew	300	60
Glareola pratincola Collared Pratincole	200	24
Sterna albifrons Little Tern	400	75
Chlidonias hybridus Whiskered Tern	12,000	89
Chlidonias niger Black Tern	7,000	86
Otus scops Scops Owl	1,000	1
Bubo bubo Eagle Owl	200	25
Strix aluco Tawny Owl	3,000	2
Caprimulgus europaeus Nightjar	2,000	7
Alcedo atthis Kingfisher	2,000	4
Merops apiaster Bee-eater	10,000	2
Coracias garrulus Roller	3,000	2
Picus canus Grey-headed Woodpecker	20,000	1
Dendrocopos medius Middle Spotted Woodpecker	3,000	8
Picoides tridactylus Three-toed Woodpecker	2,000	7
Lullula arborea Woodlark	30,000	<1
Riparia riparia Sand Martin	30,000	2
Parus lugubris Sombre Tit	1,000	2
Parus cristatus Crested Tit	100,000	1
Lanius collurio Red-backed Shrike	600,000	<1
Lanius minor Lesser Grey Shrike	30,000	<1
Emberiza hortulana Ortolan Bunting	10,000	11

1. Only those species of European conservation concern (see Box 1, p. 12) that meet IBA criteria in Romania are listed.
2. Data are taken from the BirdLife/EBCC European Bird Database 1998 (Heath and Borggreve 2000).
3. The percentage of the national population in IBAs exceeds 100%. Usually this is because the national population estimate has not been updated recently whilst the IBA population estimate has been recently updated with new data as a result of comprehensive surveys of IBAs themselves. Also, the individual site count for a species may be the maximum or average over recent years, and summing these may record more birds than are present nationally in any single year.

HABITATS

Between the Black Sea coast and the Carpathian peaks (2,543 m above sea-level) there are a wide range of vegetation-types and plant associations. Forests cover about 27% of the land area—*Quercus* predominates up to c.800 m, with *Fagus* between 800 and 1,400 m, *Picea* between 1,400 and 1,800 m and dwarf *Pinus* above the treeline. At the highest altitudes, subalpine and alpine grasslands occur. Cultivation has, to a large degree, replaced natural vegetation in the foothills and plain regions. Approximately 45% of Romania's land area is under agriculture, including arable land, pasture and hay-meadows, and perennial crops (orchards and vineyards). Plantations comprise mainly *Picea*, *Pinus*, *Populus* and *Robinia*.

Wetland habitats, in the form of natural wetlands, fish-ponds and reservoirs, are present at 30 IBAs (68%), with cover exceeding 50% at 26 of these (Figure 1). On the lower Danube alone, wetlands total approximately half a million hectares. Key breeding species in wetland habitats include *Pelecanus onocrotalus*, *Pelecanus crispus*,

Figure 1. Habitats at Important Bird Areas in Romania (see Appendix 3 for definitions of habitats).

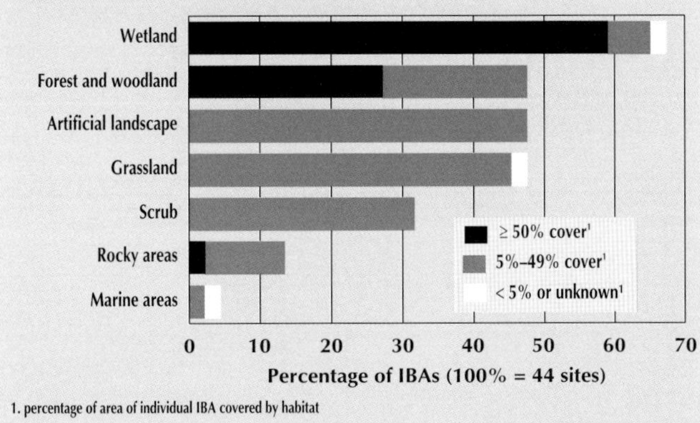

1. percentage of area of individual IBA covered by habitat

Figure 2. Land-uses at Important Bird Areas in Romania (see Appendix 3 for definitions of land-uses).

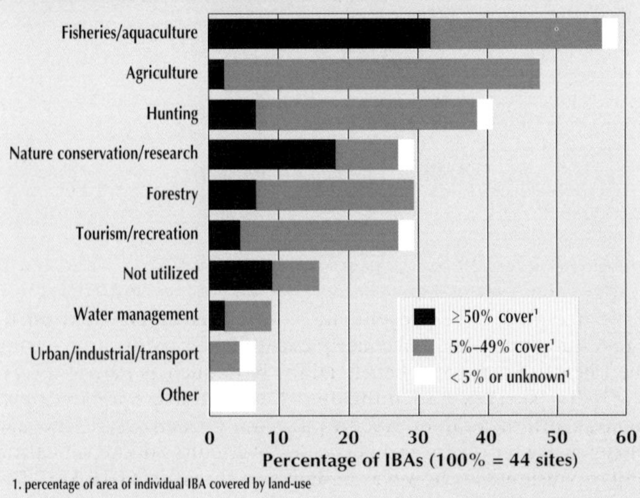

1. percentage of area of individual IBA covered by land-use

Figure 3. Threats at Important Bird Areas in Romania (see Appendix 3 for definitions of threat types and impact categories).

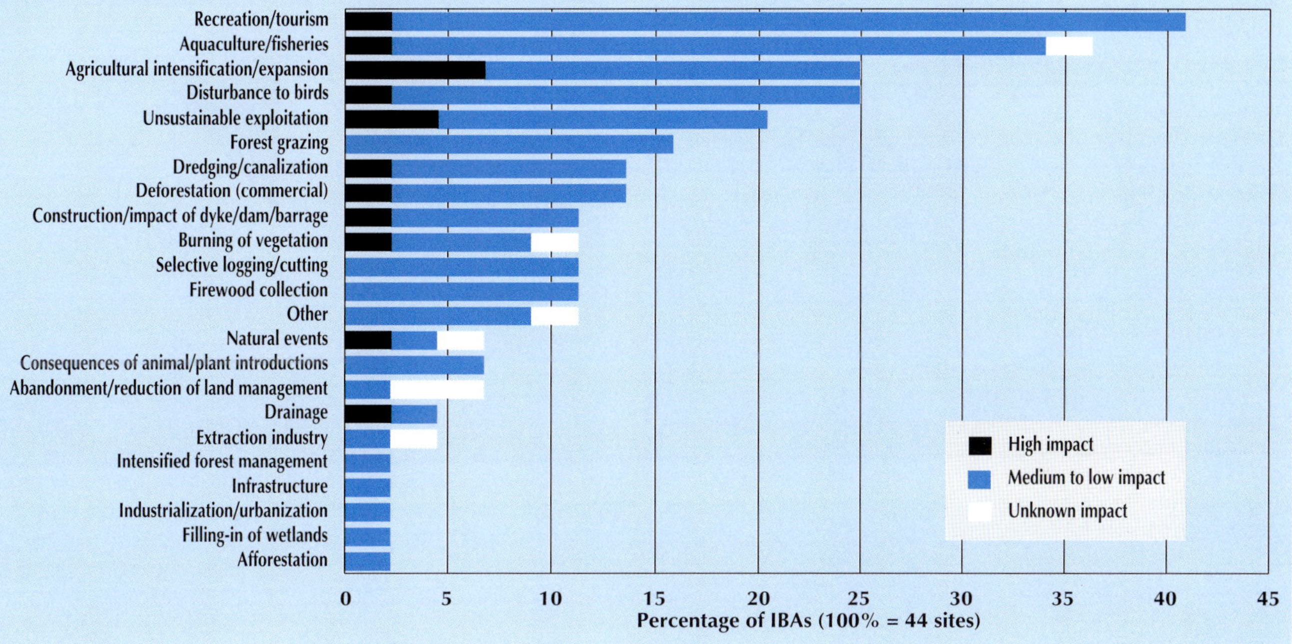

Phalacrocorax pygmeus, *Plegadis falcinellus*, *Nycticorax nycticorax*, *Ardeola ralloides*, *Ardea purpurea*, *Aythya nyroca*, *Recurvirostra avosetta*, *Himantopus himantopus*, *Glareola pratincola*, *Chlidonias hybridus*, *C. niger* and *C. leucopterus*.

There are fewer IBAs covering forest and/or mountainous habitats (Figure 1). Those that have been identified are nevertheless representative of the extensive forests found on the Carpathians (IBAs 013, 016, 017, 021, 022, 024 and 031). These coniferous forests support boreal species (for example, *Tetrao urogallus*, *Glaucidium passerinum*, *Strix uralensis*, *Aegolius funereus*, *Picoides tridactylus* and *Nucifraga caryocatactes*); the alpine grasslands at higher altitudes support tundra species (for example, *Charadrius morinellus*).

Artificial landscapes occur at 21 IBAs (48%), and take the form of agricultural land (including perennial crops), secondary pasture (used for grazing or hay production), tree plantations and human settlements.

IMPACTS ON IBAs – LAND-USE AND THREATS

The dominant land-use at Romanian IBAs is fisheries/aquaculture (59% of IBAs) (Figure 2), reflecting the predominance of wetland habitats referred to in the previous section. Several types of threat are associated with this land-use—over-fishing causes serious reductions in fish populations, and the burning and cutting of reedbeds reduces the amount of breeding habitat available to waterbirds. The destruction of the nests of piscivorous species (for example, *Phalacrocorax* spp., *Ardea cinerea* and gulls (Laridae)) by fishermen is also encountered (Figure 3).

Agriculture is widespread, occuring at 21 IBAs (48%). During the post-war period the Romanian state took action to extend the land area under agriculture, mainly by draining wetlands and turning natural grasslands over to cultivation. As a result the area of natural wetland and grassland has decreased, although losses of the former have been partly compensated for by the creation of many large reservoirs and fish-ponds. These changes have had a direct impact on the diversity and abundance of birds—waterbirds and grassland species (for example, *Tetrax tetrax* and *Otis tarda*) have declined. The present move towards diversification of agricultural practices, as a consequence of land privatization during the 1990s, is likely to be beneficial for birds such as *Perdix perdix*, warblers, shrikes and buntings. Less intensive agricultural production should be accompanied by declines in the use of expensive fertilizers and pesticides. Agricultural intensification and expansion does however remain as a threat to 11 IBAs (25%) (Figure 3).

Hunting takes place at 18 IBAs (41%) (Figure 2), and the pressure on ducks and geese (Anatidae) can be intense, especially at the beginning of the hunting season or during cold spells. The hunting season is long—from 15 August to 15 March for ducks, for example. It should be noted that the number of Romanian hunters is relatively small (about 65,000 people—0.3% of the total population), and that the fines and punishment handed down to poachers can be very severe. There is some illegal trade in songbirds (for example, *Carduelis chloris*, *C. carduelis* and *C. spinus*) caught in the wild.

Tourism/recreational activities are reasonably well developed, especially in the vicinity of large towns, in the mountains and on the coast. The overall impact of such activities is relatively low (Figure 3), although in recent years the increasing popularity of angling has resulted in greater levels of disturbance to waterbirds.

The most threatened IBAs are wetlands, especially those that are the subject of intensive exploitation or those that suffer pollution. Despite being protected, the Danube delta (IBA 001) receives pollutants in the form of chemicals transported down river for great distances, as well as from industrial discharges from plants in Tulcea. Many dykes and canals built in the past are still causing perturbations to the hydrological regime in the delta region.

PROTECTION STATUS

Table 5 and Figures 4 and 5 summarize the national and international protection status of all Romanian IBAs.

■ National protection

The protected-area system is under the control of the Ministry of Waters, Forests and Environment Protection, which cooperates with the Commission of Natural Monuments (Romanian Academy) and with local authorities.

1. National Parks

Comprise natural geomorphological units that support extensive ecosystems subject to minimal human pressure. Public access and any other human interference are banned in limited scientific areas (core zones)—only scientific research is permitted.

2. Nature Reserves

A network exists, comprising various types: botanical, zoological, ornithological, mixed, palaeontological, geological or speleological. All are state-owned. In addition to the reserves at the national level there are local reserves that are created by local authorities.

Figure 4. The national protection status of Important Bird Areas in Romania.

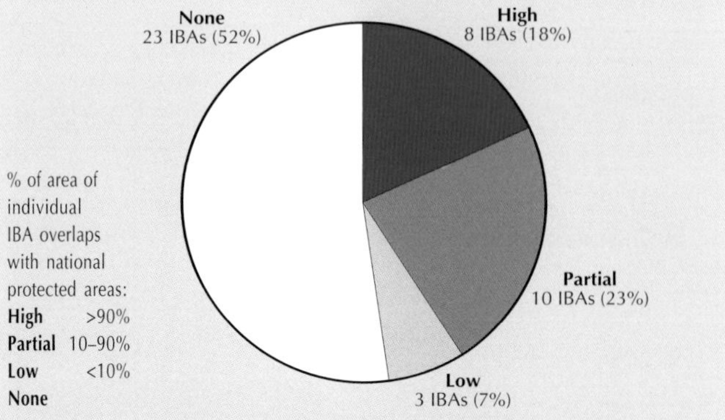

Total area of overlap between IBA network in Romania and national protected-area system (see Table 5 for categories) = 4,653–4,665 km² (71% of total IBA area).

Figure 5. The international protection status of Important Bird Areas in Romania.

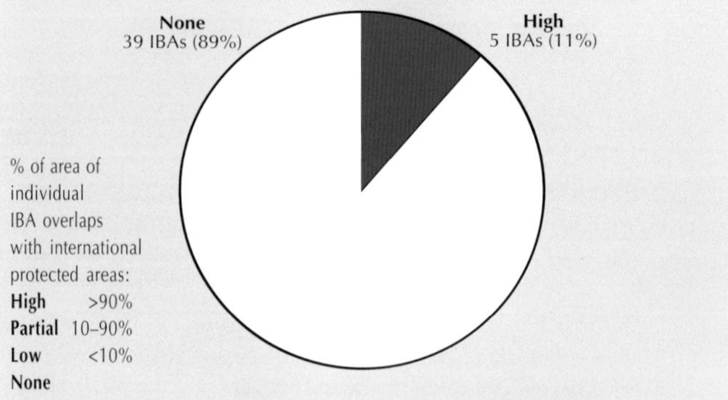

Total area of overlap between IBA network in Romania and international protected-area system (see Table 5 for categories) = 5,026 km² (77% of total IBA area).

Table 5. Protection status of Important Bird Areas in Romania. A tick (✔) indicates that an IBA overlaps with a protected area (to any extent).

		National				International		
IBA code	International name	National Park	Nature Reserve	Forestry area	Bird Sanctuary	Ramsar Site	World Heritage Site	Biosphere Reserve
001	Danube delta and Razelm–Sinoe complex					✔	✔	✔
002	Parcheş–Somova wetland							✔
003	Lake Beibugeac							
004	Niculiţel and Babadag forest		✔					
005	Black Sea coast–Chituc							✔
006	Lake Taşaul							
007	Lake Siutghiol							
008	Lake Techirghiol							
009	Hagieni forest		✔					
010	Vădeni–Maţa–Cârja–Rădeanu wetlands							
011	Odobeşti hill							
012	Bârnova–Repedea woodland							
013	Bicaz gorge and Lake Roşu		✔					
014	Vlădeni fish-ponds							
015	Stânca–Ştefăneşti reservoir							
016	Slătioara forest		✔					
017	Pietrosul Rodnei mountain		✔					✔
018	Dobron forest			✔				
019	Zau de Câmpie–Tăureni fish-ponds							
020	Fizeş valley							
021	Vlădeasa mountain							
022	Întregalde and Râmeţi gorges		✔					
	Subtotal of IBAs	0	6	1	0	1	1	4
023	Rotbav fish-ponds							
024	Retezat National Park	✔						✔
025	Cefa fish-ponds and Rădvani wood				✔			
026	Lunca wood			✔				
027	Socodor wood			✔				
028	Mureş river flood-plain: Arad–Pecica–Cenad		✔					
029	Satchinez marsh		✔					
030	Murani lake and Pişchia forest		✔					
031	Domogled mountain		✔					
032	Porţile de Fier reservoir							
033	Mehedinţi fish-ponds–Izvoarele							
034	Mehedinţi fish-ponds–Gruia							
035	Mehedinţi fish-ponds–Gârla Mare							
036	Lake Dunăreni–Bistreţ							
037	Strejeşti reservoir							
038	Lake Comana		✔					
039	Ciocăneşti fish-farm							
040	Lake Strachina							
041	Lake Fundata							
042	Lake Tătaru							
043	The Little Island of Brăila		✔					
044	Balta Albă, Amara and Jirlău lakes		✔					
	Total number of IBAs	**1**	**13**	**3**	**1**	**1**	**1**	**5**

3. Bird Sanctuaries
Small areas that support large numbers of breeding, wintering or passage birds.

4. Forestry areas
Local forestry authorities protect large areas of woodland from exploitation.

Of great importance for bird conservation is the Law on Hunting and Game Protection (1996), which relates to 200 bird species (including raptors and passerines) and takes into consideration almost all stipulations of the Bern Convention. As there is no effective bird-protection law, this is the only piece of national legislation to provide protection for wild birds.

The Environment Protection Law was adopted in 1995, and it is expected that a more definitive law will be adopted concerning protected areas. Additional tools for environmental protection include the 1996 Forestry Code and the Law of Waters (also 1996).

■ International protection

Box 1 lists those international conventions and initiatives ratified by Romania. Full implementation of Romania's obligations under these has not, however, been achieved.

The Danube delta is designated as a World Heritage Site and Biosphere Reserve, and is the only Ramsar Site in Romania. It is managed by the Authority of the Danube Delta Reserve. The

Box 1. International legislation and initiatives that are relevant to site conservation in Romania (see Appendix 1 for a general description of these agreements).

Global	
Biodiversity Convention	✔
Ramsar Convention	✔
Bonn Convention	✔
World Heritage Convention	✔
MAB Programme	✔
Pan-European	
Bern Convention	✔

✔ Convention ratified/initiative supported
(✔) Convention signed

responsible authorities for other international reserves are the reserve administration (if any), the forestry service or local administrative bodies. Non-governmental organizations (NGOs) are involved in the protection of several national and local reserves.

CONSERVATION

- The institutional framework for environmental protection at the national level is inadequate—too few institutions exist and there is a lack of human, financial and capital resources. Consequently, the majority of 'less important' reserves have been abandoned by the authorities and are now threatened with alteration or destruction.
- Once the law on protected areas is passed, the Romanian Ornithological Society (ROS) will try to obtain Nature Reserve designation for as many IBAs as possible.
- The Danube Delta Biosphere Reserve (IBAs 001, 002 and 005) has a management plan issued under the auspices of the Ministry of Waters, Forests and Environment Protection. Other sites have management plans performed by their owners or administrators (for example, fish-farms, the local forestry service or reserve administrations).
- Between 1995 and 1999 the Danube Delta Biosphere Reserve was subject to a conservation management programme financed by the Global Environment Facility. Public awareness (education and training) and bird monitoring were the primary aims. BirdLife International supported this programme both directly and through the work of ROS. ROS now has a local office in Tulcea and works with local authorities to promote BirdLife policies in the Danube delta.
- The Ministry of Waters, Forest and Environment Protection, together with other institutions (including ROS), have published a 'National Strategy for biodiversity conservation' (Anon. 1996), but there is a long way to go before implementation is complete.
- The biological research institutes, museums of natural sciences and university biology departments undertake research on biodiversity.
- NGOs are taking more of a role in the process of studying and conserving nature.

ANALYTICAL METHODS

- The midwinter wildfowl count (Wetlands International programme) began in 1988 and takes place annually. Since 1991 it has been included in the activities of ROS, and is run in cooperation with the Danube Delta Biosphere Reserve Administration, several institutions and a large number of volunteers. BirdLife International and ROS are also involved in monitoring numbers of breeding birds.
- Data concerning Romanian IBAs were updated during the period 1992–1999. Bird counts were undertaken by individual researchers and, at the national level, by ROS. Nevertheless the level of bird monitoring currently undertaken is insufficient and data are therefore incomplete for some sites. This is due mainly to the small numbers of ornithologists in large parts of the country.
- Species that do not necessarily meet IBA criteria—due, for example, to dispersed distributions—were also monitored, including *Ciconia ciconia*, diurnal raptors, *Tetrao tetrix*, *Tetrao urogallus*, *Perdix perdix*, *Charadrius morinellus*, *Riparia riparia* and *Hirundo rustica*. Special attention was paid to those species whose main distributions lie to the north (in most cases) (for example, *Tetrao tetrix* and *Turdus pilaris*) or south of Romania (for example, *Apus melba*, *Passer hispaniolensis*, *Sturnus roseus*, *Oenanthe isabellina*, *Hirundo rupestris* and *H. daurica*).
- Wintering and passage bird figures are minimum/maximum daily counts.
- Bird count averages have usually been calculated over a number of years. The year given in the species tables is the last year of the count period.
- Numbers of breeding birds are given as pairs.
- The following sites of national importance were proposed as IBAs but did not meet the current criteria: Branişte Reserve, Nucet fish-farm, Mândra fish-ponds, Jijila lake, Bucecea reservoir, Vermeş marsh, Goşmanu mountain, Livada fish-farm, Drocea mountain, Bucegi mountains, Lakes Ianca and Plopu, Pârâul de Câmpie valley, and Lake Sărături–Murighiol.
- It is possible that in the southern part of the country, especially in the Danube valley, areas exist that might meet IBA criteria. Their identification will be one of the key tasks in the future activities of ROS, with an immediate goal being the publication of a national IBA inventory.

GLOSSARY

ROS Romanian Ornithological Society (Societatea Ornitologică Română, the BirdLife International Partner in Romania).

ACKNOWLEDGEMENTS

The IBA programme started in Romania in 1987 following discussions between Richard Grimmett (International Council for Bird Preservation) and Dan Munteanu. After the foundation of the Romanian Ornithological Society in 1990, the IBA programme became a priority in ROS activities. The IBA Officer between 1996 and 1998 was Sorin Dîrjan, who was helped in his work by the following people who provided valuable information: Adrian Andrei, Libus András, Gabriel Bănică, Hodor Călin, Marcel Ciobanu, József Béres, Carmen Gache, Mirela Gheorghe, Iordache Ion, Dan Traian Ionescu, Ladislau Kalabér, Andrei Kiss, János Botond Kiss, Ştefan Kohl, Mihai Marinov, Dan Munteanu, Nicolae Onea, Péter László Papp, Aurel Papadopol, Petre Teodor, Angela Petrescu, Raluca Răşinariu, Rozalia Poliă, Attila D. Sándor, Imre Simó, József Szabó, József Szabó jun., Nicolae Toniuc and Peter Weber. The programme was financed between 1992 and 1996 by BirdLife International and then by the Royal Society for the Protection of Birds (BirdLife International Partner in the UK). The national coordinator thanks all financial supporters and all volunteers who have given their best efforts (and continue to do so) to the development of the Romanian IBA programme.

SITE ACCOUNTS

Danube delta and Razelm–Sinoe complex

A1, A4i, A4ii, A4iii, B1i, B1iii, B2, B3 **001**

Admin region Tulcea
Coordinates 44°56′N 29°12′E
Altitude 0–12 m **Area** 442,000 ha

Site description

The most extensive wetland in Europe after the Volga delta, comprising three arms of the Danube (Chilia, Sulina and Sf. Gheorghe) interspersed with marine and river sandbanks, fields, marshes, lakes and canals. Reedbeds (*Phragmites*) cover large areas (totalling 160,000 ha), and include a type of floating reed habitat known as 'plauri'. Two forests, Letea and Caraorman, cover 22,700 ha, and *Salix* is widespread. Also present are sand-dunes and beaches, wet and dry grassland, salt vegetation (in the south), agricultural lands (some abandoned), fish-ponds and human settlements. The Razim–Sinoe lagoon complex is situated to the south of the delta. Formally brackish, the lagoons have become fresh following the closure of the link to the sea. Reed-harvesting takes place ('Other' land-use).

Habitats Forest and woodland (5%; alluvial/very wet forest), Wetland (70%; mudflat/sandflat; saltmarsh; coastal lagoon; standing fresh water; river/stream; water-fringe vegetation), Marine areas (5%), Artificial landscape (18%; arable land; other urban/industrial areas)
Land-use Agriculture (12%), Fisheries/aquaculture (57%), Forestry (8%), Hunting (10%), Not utilized (23%), Other

Birds

Species	Season	Year	Pop min	Pop max	Acc	Criteria
Podiceps grisegena Red-necked Grebe	B	1996	2,000	—	—	A4i, B1i
Podiceps nigricollis Black-necked Grebe	B	1996	2,000	—	—	A4i, B1i
Phalacrocorax carbo Cormorant	B	1998	13,500	—	A	A4i, B1i
Phalacrocorax pygmeus Pygmy Cormorant	B	1998	6,400	—	A	A1, A4i, B1i, B2
Pelecanus onocrotalus White Pelican	B	1998	3,000	—	A	A4i, B1i, B2
Pelecanus crispus Dalmatian Pelican	B	1998	200	—	A	A1, A4i, B1i, B2
Botaurus stellaris Bittern	B	1996	1,500	—	C	B2
Ixobrychus minutus Little Bittern	B	1996	5,000	—	B	B2
Nycticorax nycticorax Night Heron	B	1998	4,400	—	A	A4i, B1i, B2
Ardeola ralloides Squacco Heron	B	1998	3,200	—	A	A4i, B1i, B2
Egretta alba Great White Egret	B	1998	360	—	A	A4i, B1i
Ardea cinerea Grey Heron	B	1996	1,600	—	B	A4i, B1i
Ardea purpurea Purple Heron	B	1998	500	—	A	B2
Ciconia nigra Black Stork	P	1996	300	—	C	A4i, B1i
Platalea leucorodia Spoonbill	B	1996	216	—	B	A4i, B1i, B2
Cygnus olor Mute Swan	B	1996	1,200	—	B	A4i, B1i
Cygnus cygnus Whooper Swan	W	1996	1,755	—	B	A4i, B1i, B3
Anser albifrons White-fronted Goose	W	1996	100,000	300,000	A	A4i, B1i
Branta ruficollis Red-breasted Goose	W	1996	20,000	55,000	A	A1, A4i, B1i
Anas penelope Wigeon	P	1996	6,000	—	A	B1i
Anas strepera Gadwall	B	1996	4,000	—	B	A4i, B1i, B2
Anas crecca Teal	P	1996	10,000	—	B	B1i
Anas platyrhynchos Mallard	B	1998	30,000	80,000	A	A4i, B1i
Netta rufina Red-crested Pochard	B	1996	300	—	B	A4i, B1i
Aythya ferina Pochard	B	1996	55,000	—	B	A4i, B1i, B3
Aythya nyroca Ferruginous Duck	B	1996	3,000	—	B	A1, A4i, B1i, B2
Aythya fuligula Tufted Duck	W	1994	10,000	—	B	B1i
Bucephala clangula Goldeneye	W	1994	3,000	—	B	B1i
Mergus albellus Smew	W	1994	1,500	—	B	A4i, B1i
Mergus merganser Goosander	W	1994	200	—	B	B1i
Oxyura leucocephala White-headed Duck	W	1994	10	—	C	A1
Circus macrourus Pallid Harrier	P	1996	100	—	B	A1
Aquila clanga Greater Spotted Eagle	P	1996	20	—	B	A1
Aquila heliaca Imperial Eagle	P	1996	80	—	B	A1
Falco tinnunculus Kestrel	B	1996	300	—	B	B2
Falco vespertinus Red-footed Falcon	B	1996	600	—	B	A4ii, B1iii, B2
Porzana porzana Spotted Crake	B	1996	4,000	—	B	B3
Porzana parva Little Crake	B	1996	1,000	—	B	B3
Crex crex Corncrake	B	1996	60	—	B	A1, B2
Gallinula chloropus Moorhen	B	1996	10,000	—	B	A4i, B1i
Fulica atra Coot	B	1996	50,000	100,000	B	A4i, B1i
Grus grus Crane	P	1996	1,000	—	C	B1i
Recurvirostra avosetta Avocet	B	1996	150	—	B	B1i
Burhinus oedicnemus Stone Curlew	B	1996	180	—	B	B2
Glareola nordmanni Black-winged Pratincole	U	1996	10	—	C	A1
Sterna hirundo Common Tern	B	1996	15,000	—	B	A4i, B1i
Sterna albifrons Little Tern	B	1996	300	—	B	A4i, B1i, B2
Chlidonias hybridus Whiskered Tern	B	1996	10,000	—	B	A4i, B1i, B2
Chlidonias niger Black Tern	B	1996	6,000	40,000	A	A4i, B1i, B2
Chlidonias leucopterus White-winged Black Tern	B	1996	2,000	—	B	A4i, B1i

This is one of Europe's most important sites for breeding, passage and wintering waterbirds, particularly wintering Anatidae. It is also an important site for breeding and passage raptors and passerines. The site regularly holds more than 20,000 waterbirds breeding, on passage and in winter. Species of global conservation concern that do not meet IBA criteria: *Anser erythropus* (wintering), *Haliaeetus albicilla* (breeding), *Falco naumanni* (passage), *Gallinago media* (passage) and *Acrocephalus paludicola* (passage).

Protection status

National High **International** High

IBA overlaps with 16 strictly protected (or core) areas (totalling 50,600 ha). 442,000 ha of IBA covered by Biosphere Reserve (Danube Delta, 580,000 ha). IBA overlaps with Ramsar Site (Danube Delta). IBA overlaps with World Heritage Site (Danube Delta).

Conservation issues

Threats Abandonment/reduction of land management (U), Agricultural intensification/expansion (B), Aquaculture/fisheries (C), Burning of vegetation (C), Consequences of animal/plant introductions (C), Disturbance to birds (B), Drainage (B), Dredging/canalization (B), Filling-in of wetlands (B), Firewood collection (B), Other (B), Recreation/tourism (B), Unsustainable exploitation (B)

The main threats are from drainage, canalization and dredging, wetland infilling and pollution ('Other' threat). Additional threats include overfishing, disturbance from boat traffic, reedbed fires, grazing and other agricultural practices. Hunting is permitted in several areas, including buffer zones. The Danube Delta Biosphere Reserve Authority (DDBRA) is responsible for conservation and protection projects, and the Danube Delta Institute undertakes research and management work. The DDBRA management plan covers biodiversity conservation, bird monitoring and public awareness. The recent restoration of 3,680 ha of former agricultural land has been successfully completed.

Parcheş–Somova wetland A1, A4i, B1i, B2, B3 002

Admin region Tulcea
Coordinates 45°12'N 28°38'E
Altitude 2–3 m **Area** 3,100 ha

Site description

An area of lakes and reedbeds (*Phragmites*) bordered by the Danube and situated upstream from Tulcea. Reedbeds cover approximately 65% of the site; a few *Salix* trees and *Populus* plantations occur along the Danube. Human activities include fishing and reed-harvesting by local people ('Other' land-use).

Habitats Wetland (100%; standing fresh water; water-fringe vegetation)
Land-use Fisheries/aquaculture (10%), Nature conservation/research (90%), Other

Birds

Species	Season	Year	Pop min	Pop max	Acc	Criteria
Phalacrocorax carbo Cormorant	B	1993	1,400	—	—	A4i, B1i
Phalacrocorax pygmeus Pygmy Cormorant	B	1999	80	—	A	A1, A4i, B1i, B2
Cygnus cygnus Whooper Swan	W	1993	60	120	—	B3
Anser albifrons White-fronted Goose	W	1996	2,000	—	—	B1i
Branta ruficollis Red-breasted Goose	W	1996	1,000	—	—	A1, A4i, B1i
Chlidonias hybridus Whiskered Tern	B	1999	300	—	—	B1i, B2

The site is important for breeding and passage waterbirds, with a bird fauna similar to that of the Danube delta. Species of global conservation concern that do not meet IBA criteria: *Pelecanus crispus* (20 passage birds) and *Aythya nyroca* (18 breeding pairs).

Protection status

National High **International** High

3,100 ha of IBA covered by Biosphere Reserve (Danube Delta, 580,000 ha).

Conservation issues

Threats Construction/impact of dyke/dam/barrage (B)

The dyke constructed along the Danube hinders water inflow to the IBA. A management plan exists for the site.

Lake Beibugeac A1, B1i, B3 003

Admin region Tulcea
Coordinates 45°02'N 29°06'E
Altitude 8–8 m **Area** 180 ha

Site description

A shallow brackish lake with a maximum depth of 1.5 m. The water-level and lake area varies considerably depending upon the amount of rainfall. Along the western shore there is a narrow fringe of reed *Phragmites*. Natural grassland, agricultural land and human settlements surround the IBA.

Habitats Wetland (98%; standing brackish and salt water; water-fringe vegetation)
Land-use Not utilized (98%)

Birds

Species	Season	Year	Pop min	Pop max	Acc	Criteria
Cygnus cygnus Whooper Swan	W	1997	—	280	A	B1i, B3

Species ... continued	Season	Year	Pop min	Pop max	Acc	Criteria
Anser albifrons White-fronted Goose	W	1997	—	2,000	B	B1i
Anser anser Greylag Goose	W	1997	—	200	A	B1i
Branta ruficollis Red-breasted Goose	W	1997	—	600	A	A1

This is an important wintering site for Anatidae and is also a breeding site for waders including *Himantopus himantopus* and *Recurvirostra avosetta*. Passage species of global conservation concern that do not meet IBA criteria: *Pelecanus crispus* and *Aythya nyroca*.

Protection status

National None **International** None

Conservation issues

Threats Disturbance to birds (C)

There is limited human disturbance of birds; grazing animals may disturb nests around the edge of the lake. Managed by ROS.

Niculiţel and Babadag forest — A1, B2 — 004

Admin region Tulcea
Coordinates 45°00′N 28°25′E
Altitude 20–300 m **Area** 10,000 ha

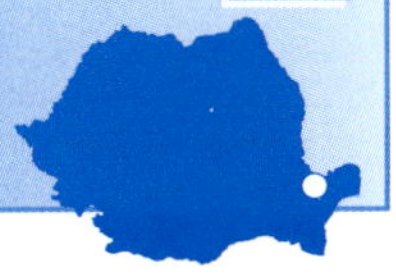

Site description

An area of forest found on the Niculiţel and Babadag hills. The forest is dominated by *Quercus* but includes *Tilia*, *Ulmus*, *Populus*, and *Acer*, with forest composition primarily dependent upon the slope aspect. The valleys contain grasslands and areas under agriculture.

Habitats Forest and woodland (95%; broadleaved deciduous forest), Scrub (5%)
Land-use Forestry (40%), Hunting (30%), Tourism/recreation (30%)

Birds

Species	Season	Year	Pop min	Pop max	Acc	Criteria
Aquila heliaca Imperial Eagle	P	1992	10	—	B	A1
Bubo bubo Eagle Owl	B	1992	16	20	A	B2
Caprimulgus europaeus Nightjar	B	1992	100	200	A	B2
Coracias garrulus Roller	B	1992	60	100	A	B2
Emberiza hortulana Ortolan Bunting	B	1992	1,000	2,000	A	B2

The site is particularly important for breeding raptors and for woodland species in general. Raptors also occur in large numbers on passage. Species of global conservation concern that do not meet IBA criteria: *Falco naumanni* (2–4 breeding pairs).

Protection status

National Low **International** None
11 ha of IBA covered by Nature Reserve (Valea Fagilor, 11 ha).

Conservation issues

Threats Deforestation (commercial) (C), Forest grazing (B), Intensified forest management (B)

The main threats are from overgrazing in the forest, which prevents regeneration, and intensive forest management and exploitation. A project is underway to establish a Nature Park in the Macin mountains, which will include part of the IBA.

Black Sea coast–Chituc — A1, A4i, A4iii, B1i, B2, B3 — 005

Admin region Constanţa
Coordinates 44°30′N 28°52′E
Altitude 0–0 m **Area** 5,000 ha

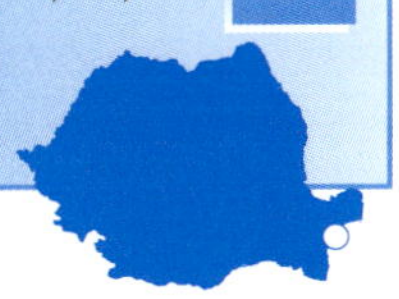

Site description

The site includes the narrow sand beach of Chituc and associated coastal waters; there is very little tidal fluctuation. There are only two temporary fishing settlements along a 22 km stretch of coastline.

Habitats Wetland (sand-dunes/sand beach), Marine areas (sea inlet/coastal features)
Land-use Fisheries/aquaculture (30%), Nature conservation/research (100%)

Birds

Species	Season	Year	Pop min	Pop max	Acc	Criteria
Podiceps nigricollis Black-necked Grebe	W	1998	200	1,200	A	A4i, B1i
Phalacrocorax pygmeus Pygmy Cormorant	W	1998	10	30	A	A1
Cygnus cygnus Whooper Swan	W	1998	30	230	A	B1i, B3
Aythya ferina Pochard	W	1998	2,000	40,000	A	A4i, B1i
Aythya fuligula Tufted Duck	W	1998	5,000	80,000	A	A4i, B1i
Aythya marila Scaup	W	1998	400	4,000	A	A4i, B1i, B2
Bucephala clangula Goldeneye	W	1998	300	1,200	A	B1i
Mergus albellus Smew	W	1998	200	1,100	A	A4i, B1i

An important area for wintering wildfowl, holding 20,000 or more on a regular basis. Wintering species of global conservation concern that do not meet IBA criteria: *Aythya nyroca*, *Oxyura leucocephala* and *Haliaeetus albicilla*.

Protection status

National High **International** High
IBA overlaps with Biosphere Reserve (Danube Delta).

Conservation issues

Threats Other (B)

The main threat is pollution, which is reducing the food resource available to birds ('Other' threat, above). The Danube Delta Research and Management Institute study the reserve as a whole; ROS study the bird fauna. A management plan exists for the site.

Lake Taşaul — B1i — 006

Admin region Constanţa
Coordinates 44°21′N 28°34′E
Altitude 1–10 m **Area** 1,830 ha

Site description

A coastal lake at the end of the Casimcea valley with a steep limestone shoreline, except on the eastern side where it is separated from the sea by a sandbank. The lake reaches a depth of 5.6 m and has two islands. Sparse reedbeds (*Phragmites*) are found in the north-west of the lake.

Habitats Wetland (88%; sand-dunes/sand beach; coastal lagoon; water-fringe vegetation), Artificial landscape (8%)
Land-use Agriculture (8%), Fisheries/aquaculture (80%), Urban/industrial/transport (8%)

Birds

Species	Season	Year	Pop min	Pop max	Acc	Criteria
Aythya ferina Pochard	P	1996	50	10,000	B	B1i
Larus cachinnans Yellow-legged Gull	R	1996	20	1,000	A	B1i

This is an important site for passage and wintering waterbirds.

Protection status

National None **International** None

Conservation issues

Threats Agricultural intensification/expansion (B), Aquaculture/fisheries (B), Disturbance to birds (B), Industrialization/urbanization (B)

The main threats are from intensive agricultural practices in the lake surrounds, and disturbance, particularly along the eastern shore, from sheep-grazing, fishing, a factory and quarry. Air pollution from the Navodari oil refinery is an occasional problem.

Lake Siutghiol — A1, A4i, B1i — 007

Admin region Constanţa
Coordinates 44°16′N 28°36′E
Altitude 0–10 m **Area** 2,000 ha

Site description

The site is a former marine inlet now separated from the sea by a narrow sand-spit. The lake is predominantly freshwater with a maximum depth of 18 m. The western shore is formed by a steep limestone promontory, and narrow reedbeds (*Phragmites*) occur. In winter up to 90% of the lake area may freeze.

Habitats Wetland (94%; sand-dunes/sand beach; coastal lagoon; water-fringe vegetation)
Land-use Fisheries/aquaculture (60%), Tourism/recreation (50%), Urban/industrial/transport (20%)

Birds

Species	Season	Year	Pop min	Pop max	Acc	Criteria
Phalacrocorax pygmeus Pygmy Cormorant	P	1996	10	100	B	A1
Larus minutus Little Gull	P	1996	20	2,000	A	A4i, B1i

This is an important site for passage waterbirds, particularly Anatidae and Laridae; the surrounds are a stop-over site for migrating passerines. Species of global conservation concern that do not meet IBA criteria: *Aythya nyroca* (2–40 passage birds).

Protection status

National None **International** None

Conservation issues

Threats Aquaculture/fisheries (C), Recreation/tourism (A)

The tourist towns of Constanţa, Mamaia and Ovidiu surround the lake. Disturbance of birds during the summer is high.

Lake Techirghiol A1, A4i, A4iii, B1i, B2, B3 008

Admin region Constanţa
Coordinates 44°01′N 28°28′E
Altitude 0–6 m **Area** 1,170 ha

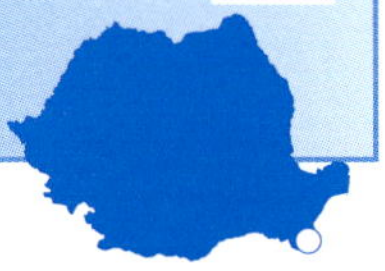

Site description

This predominantly saltwater lake lies 18 km from Constanţa; it is separated from the Black Sea by a 200 m sand-spit. The high salt concentrations in the south-eastern part of the lake prevent it from freezing during winter. Narrow reedbeds (*Phragmites*) are found in the western part of the lake. Surrounding slopes support fragmented steppic vegetation and arable land.

Habitats Wetland (93%; sand-dunes/sand beach; coastal lagoon; water-fringe vegetation), Artificial landscape (5%; arable land)
Land-use Agriculture (5%), Hunting (30%), Tourism/recreation (10%)

Birds

Species	Season	Year	Pop min	Pop max	Acc	Criteria
Podiceps nigricollis Black-necked Grebe	B	1996	100	2,000	A	A4i, B1i
[1] ***Phalacrocorax pygmeus*** Pygmy Cormorant	W	1995	25	800	A	A1, A4i, B1i
Cygnus cygnus Whooper Swan	W	1996	2	40	A	B3
Anser albifrons White-fronted Goose	W	1996	100	20,000	A	A4i, B1i
Branta ruficollis Red-breasted Goose	W	1996	—	7,000	A	A1, A4i, B1i
Aythya ferina Pochard	R	1996	20	3,000	A	B3
Oxyura leucocephala White-headed Duck	W	1998	—	800	A	A1, A4i, B1i
Larus minutus Little Gull	P	1996	100	5,200	A	A4i, B1i
Riparia riparia Sand Martin	B	1996	200	1,000	C	B2

1. Large increase over last 20 years.

This is a very important site for wintering and passage waterbirds—sometimes holding more than 20,000 wintering individuals. Species of global conservation concern that do not meet IBA criteria: *Anser erythropus* (2–10 wintering birds) and *Circus macrourus* (passage).

Protection status

National None **International** None

Conservation issues

Threats Agricultural intensification/expansion (A), Infrastructure (C), Natural events (A), Recreation/tourism (B), Unsustainable exploitation (A)

The main threat is posed by decreasing salt-levels in the lake due to the irrigation of surrounding agricultural land and inflow from freshwater springs. In the western third of the lake two dams have been built to prevent the invasion of freshwater. Ducks and geese are hunted in winter.

Hagieni forest B2, B3 009

Admin region Constanţa
Coordinates 43°48′N 28°27′E
Altitude 60–80 m **Area** 600 ha

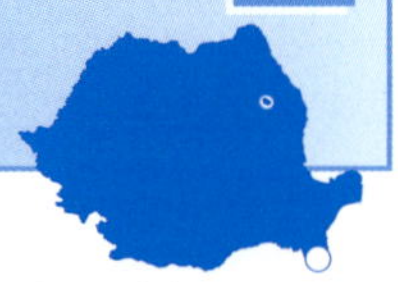

Site description

An isolated area of forest in a steppe zone, dominated by *Quercus*. The forest contains many small clearings with wetlands and patches of grassland.

Habitats Forest and woodland (65%; broadleaved deciduous forest), Scrub (10%), Artificial landscape (15%; forestry plantation)
Land-use Forestry (20%), Hunting (10%), Nature conservation/research (70%)

Birds

Species	Season	Year	Pop min	Pop max	Acc	Criteria
Circaetus gallicus Short-toed Eagle	B	1992	2	4	B	B2
Caprimulgus europaeus Nightjar	B	1995	20	30	B	B2
Dendrocopos medius Middle Spotted Woodpecker	R	1995	10	30	A	B3
[1] ***Lanius collurio*** Red-backed Shrike	B	1990	100	250	B	B2
[1] ***Lanius minor*** Lesser Grey Shrike	B	1990	24	50	B	B2

1. Threshold not met but population size significant at European level.

This is an important site for breeding forest species and migrating passerines. Species of global conservation concern that do not meet IBA criteria: *Aythya nyroca* (4–8 passage birds).

Protection status

National Partial **International** None
393 ha of IBA covered by Nature Reserve (Hagieni, 393 ha).

Conservation issues

Threats Afforestation (C), Consequences of animal/plant introductions (C)

The most important threat is from afforestation with *Pinus*—fewer species have been recorded in plantations compared with natural forest.

Vădeni–Maţa–Cârja–Rădeanu wetlands A1 010

Admin region Galaţi, Vaslui
Coordinates 46°04′N 28°07′E
Altitude 40–50 m **Area** 380 ha

Site description

An area of wetland with numerous pools and fish-ponds close to the River Prut. Aquatic vegetation is well developed and includes reedbeds (*Phragmites*); there are also areas of *Salix* and *Populus*.

Habitats Wetland (95%; standing fresh water; water-fringe vegetation)
Land-use Fisheries/aquaculture (40%), Not utilized (60%)

Birds

Species	Season	Year	Pop min	Pop max	Acc	Criteria
Aythya nyroca Ferruginous Duck	B	1996	19	40	B	A1

This is an important breeding, wintering and stop-over site for waterbirds. Seventy-four species have been recorded. Species of global conservation concern that do not meet IBA criteria: *Phalacrocorax pygmeus* (2–20 non-breeding birds).

Protection status

National None **International** None

Conservation issues

Threats Aquaculture/fisheries (C), Burning of vegetation (C)

The main threats are disturbance from fish-farming activities and reedbed burning. Access to the site is partly restricted due to its position on the state boundary—a situation beneficial to nature conservation interests.

Odobeşti hill — B2 — 011

Admin region Vrancea
Coordinates 45°50′N 26°55′E
Altitude 240–999 m **Area** 12,000 ha

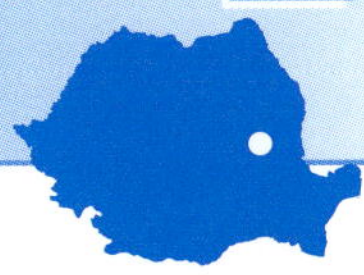

Site description

A hilly area covered with a mosaic of deciduous woodland, scrub, grassland, orchards, vineyards and arable land. Human activities include arable agriculture and livestock-grazing.

Habitats Forest and woodland (40%; broadleaved deciduous forest), Scrub (20%), Grassland (30%), Artificial landscape (10%; arable land; perennial crops/orchards/groves)
Land-use Agriculture (6%), Forestry (20%), Hunting (30%)

Birds

Species	Season	Year	Pop min	Pop max	Acc	Criteria
Ciconia nigra Black Stork	B	1998	1	2	—	B2

The site is important as a staging post for migratory species. Species of global conservation concern that do not meet IBA criteria: *Aquila clanga* (passage) and *Crex crex* (breeding). A total of 143 species have been recorded.

Protection status

National None **International** None

Conservation issues

Threats Agricultural intensification/expansion (C), Deforestation (commercial) (C), Firewood collection (C)

Agricultural expansion and deforestation are low-level threats.

Bârnova–Repedea woodland — B2 — 012

Admin region Iaşi
Coordinates 47°03′N 27°30′E
Altitude 120–407 m **Area** 15,000 ha

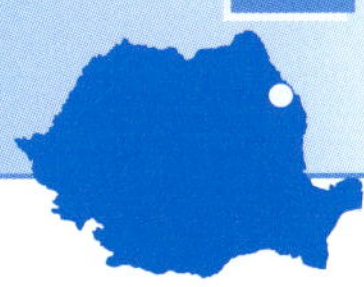

Site description

An area of low hills on the Central Moldavian Plateau. The area is predominantly covered by deciduous forests with *Fagus* dominant at higher altitudes and *Quercus* on lower slopes. There are also patches of grassland, agricultural fields, and a few orchards and vineyards.

Habitats Forest and woodland (80%; broadleaved deciduous forest), Scrub (6%), Artificial landscape (10%; arable land; perennial crops/orchards/groves)
Land-use Agriculture (10%), Forestry (50%), Not utilized (40%), Tourism/recreation

Birds

Species	Season	Year	Pop min	Pop max	Acc	Criteria
Bubo bubo Eagle Owl	R	1993	8	—	C	B2

This is an important site for migratory species, particularly raptors such as *Pernis apivorus*, *Milvus milvus* and *Aquila pomarina*. Species of global conservation concern that do not meet IBA criteria: *Circus macrourus* (passage). A total of 210 species have been recorded.

Protection status

National None **International** None

Conservation issues

Threats Deforestation (commercial) (B), Firewood collection (B), Forest grazing (C), Recreation/tourism (B)

The main threats include tree-cutting for firewood, local deforestation and pressure from tourism on weekends. A management plan exists for the area.

Bicaz gorge and Lake Roşu — B2 — 013

Admin region Neamţ, Harghita
Coordinates 46°48′N 25°51′E
Altitude 700–1,100 m **Area** 2,200 ha

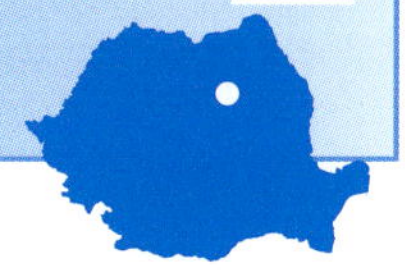

Site description

The 8 km long Bicaz gorge lies in the eastern Carpathians. The dominant vegetation is *Abies* forest with traditionally grazed alpine grassland also present. Lake Roşu is naturally dammed by landslide debris and occupies 12.6 ha. The area is crossed by a major highway; human activities include rock-climbing.

Habitats Forest and woodland (83%; broadleaved deciduous forest; native coniferous forest), Grassland (5%; alpine/subalpine/boreal grassland), Rocky areas (7%; scree/boulders)
Land-use Forestry (79%), Nature conservation/research (10%), Tourism/recreation (19%)

Birds

Species	Season	Year	Pop min	Pop max	Acc	Criteria
Bubo bubo Eagle Owl	R	1994	8	12	A	B2

This is an important site for species of upland conifer forests and open alpine habitats. Species of global conservation concern that do not meet IBA criteria: *Falco naumanni* (breeding).

Protection status

National High **International** None
2,128 ha of IBA covered by Nature Reserve (Cheile Bicazului–Lacu Roşu, 2,128 ha).

Conservation issues

Threats Extraction industry (C), Recreation/tourism (B), Selective logging/cutting (C)

The main threat is from the development of tourism and its associated infrastructure. Stone extraction is a threat in peripheral areas. A management plan exists for the site.

Vlădeni fish-ponds — A1, B1i, B2 — 014

Admin region Iaşi
Coordinates 47°28′N 27°20′E
Altitude 42–60 m **Area** 1,200 ha

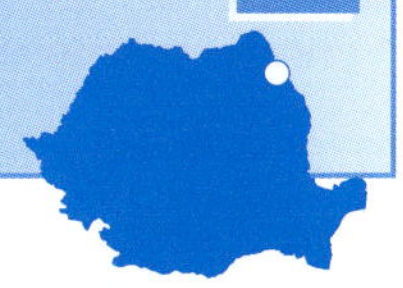

Site description

The site includes a series of fish-ponds associated with the rivers Jijia and Miletin and a reservoir used for irrigation. The northern and eastern shores of the reservoir are reinforced with concrete; the southern shore supports cereal cultivation. The western shore is marshy, with c.450 ha of reedbed (*Phragmites*). The Jijia plain is subject to flooding; grassland and agricultural land comprise the surrounds. The site also includes the Borsa and Catachi forests.

Habitats Grassland (10%), Wetland (87%; standing fresh water; water-fringe vegetation)
Land-use Fisheries/aquaculture (70%), Hunting (10%), Water management (20%)

Birds

Species	Season	Year	Pop min	Pop max	Acc	Criteria
Botaurus stellaris Bittern	B	1996	10	20	A	B2
Nycticorax nycticorax Night Heron	B	1996	40	40	A	B2
Ardeola ralloides Squacco Heron	B	1996	40	50	A	B2
Platalea leucorodia Spoonbill	P	1996	50	92	A	B1i
Anser albifrons White-fronted Goose	W	1994	2,000	2,200	B	B1i
Anser erythropus Lesser White-fronted Goose	P	1996	30	46	A	A1

The site is particularly important for breeding waterbirds, wintering and passage Anatidae; waders occur in large numbers on passage. Species of global conservation concern that do not meet IBA criteria: *Phalacrocorax pygmeus* (passage), *Aythya nyroca* (8 breeding pairs), *Haliaeetus albicilla* (passage), *Aquila heliaca* (passage), *Crex crex* (8–10 breeding pairs) and *Gallinago media* (passage). A total of 190 species have been recorded.

Protection status

National None **International** None

Conservation issues

Threats Agricultural intensification/expansion (C), Aquaculture/fisheries (C), Disturbance to birds (B), Other (U), Recreation/tourism (C)

The main threats are destruction of nests of piscivorous birds ('Other' threat), hunting, and degradation of wet meadows caused by grazing.

Stânca–Ştefăneşti reservoir

B1i, B2 **015**

Admin region Botoşani
Coordinates 47°57′N 27°10′E
Altitude 61–92 m Area 7,000 ha

Site description

A large reservoir situated in the valley of the River Prut, on the border with Moldova. One side of the river has slopes dominated by arable crops and small patches of grassland. The old river valley contains riverine forest dominated by *Salix*, *Fagus*, *Quercus* and *Ulmus*. There is also a *Populus* plantation.

Habitats Forest and woodland (23%), Scrub (12%), Wetland (38%; standing fresh water), Artificial landscape (27%; arable land; forestry plantation)
Land-use Agriculture (20%), Fisheries/aquaculture (34%), Tourism/recreation (10%), Water management (34%)

Birds

Species	Season	Year	Pop min	Pop max	Acc	Criteria
Ciconia ciconia White Stork	B	1996	7	50	B	B2
Anser anser Greylag Goose	P	1996	50	1,200	B	B1i

This is an important stop-over site for passage waterbirds due to its location on a migration route for birds moving to the Danube delta and the Black Sea. 124 species have been recorded in the IBA. Passage species of global conservation concern that do not meet IBA criteria: *Aythya nyroca*, *Aquila heliaca* and *Crex crex*.

Protection status

National None **International** None

Conservation issues

Threats Agricultural intensification/expansion (B), Aquaculture/fisheries (A), Dredging/canalization (C)

No serious threats are known at the site, which is to be proposed as a protected area. Hunting is not permitted due to the IBA's proximity to an international border. The Biology Faculty from Iasi is researching the flora and fauna of the Prut catchment.

Slătioara forest

B2 **016**

Admin region Suceava
Coordinates 47°27′N 25°27′E
Altitude 800–1,320 m Area 397 ha

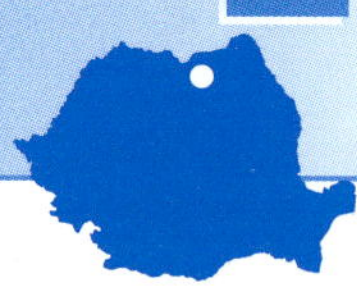

Site description

Situated on the eastern side of the Rarău mountains in the East Carpathians, this is thought to be one of the oldest natural conifer forests in Europe (some trees are up to 350 years old).

Habitats Forest and woodland (82%; native coniferous forest), Scrub (8%), Rocky areas (2%)
Land-use Nature conservation/research (100%)

Birds

Species	Season	Year	Pop min	Pop max	Acc	Criteria
Aquila pomarina Lesser Spotted Eagle	B	1996	2	3	A	B2
Bubo bubo Eagle Owl	R	1996	4	6	B	B2

This is an important site for breeding raptors, particularly owls. 128 species have been recorded. Species of global conservation concern that do not meet IBA criteria: *Crex crex* (non-breeding).

Protection status

National High **International** None
397 ha of IBA covered by Nature Reserve (Slătioara, 854 ha).

Conservation issues

No serious threats are known at the site.

Pietrosul Rodnei mountain

B2 **017**

Admin region Maramureş
Coordinates 47°35′N 24°37′E
Altitude 700–2,303 m Area 3,300 ha

Site description

Pietrosul mountain lies in the northern part of the Rodna mountains. Four vegetation belts occur: *Fagus* forest up to 1,000 m; *Abies* forest up to 1,700 m; dwarf *Pinus* shrubs at the treeline; and alpine grasslands and rocky areas above that.

Habitats Forest and woodland (60%; broadleaved deciduous forest; native coniferous forest; treeline ecotone), Grassland (25%; alpine/subalpine/boreal grassland), Rocky areas (14%)
Land-use Nature conservation/research (50%), Tourism/recreation (50%)

Birds

Species	Season	Year	Pop min	Pop max	Acc	Criteria
Bubo bubo Eagle Owl	R	1996	4	6	C	B2
Picoides tridactylus Three-toed Woodpecker	R	1995	10	20	A	B2

This is an important site for species characteristic of upland deciduous forests, montane conifer forests and open, rocky alpine habitats.

Protection status

National High **International** High
3,068 ha of IBA covered by Nature Reserve (Pietrosul Mare, 3,068 ha).
3,068 ha of IBA covered by Biosphere Reserve (Pietrosul Mare Nature Reserve, 3,068 ha).

Conservation issues

Threats Disturbance to birds (C), Recreation/tourism (B), Selective logging/cutting (B)

The main threats are from tourism, selective logging, livestock-grazing and disturbance as a result of chamois *Rupicapra rupicapra* hunting. A management plan exists for the site. The Biology Institute from Cluj, sponsored by the Environment Ministry, has carried out research in the Nature Reserve.

Dobron forest

B2, B3 **018**

Admin region Mureş
Coordinates 46°46′N 24°40′E
Altitude 390–440 m Area 220 ha

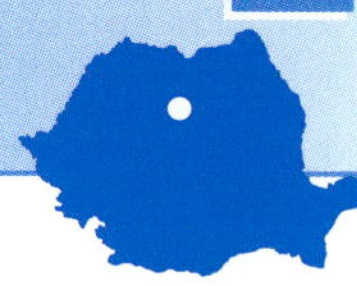

Site description

A forest predominantly composed of *Quercus*, *Carpinus*, *Populus*, *Tilia* and *Acer*. Most trees are 30–80 years old. There is a well-developed shrub layer. Grasslands, agricultural fields and orchards surround the forest.

Habitats Forest and woodland (75%; broadleaved deciduous forest), Scrub (5%), Artificial landscape (16%; arable land; perennial crops/orchards/groves)
Land-use Forestry (30%), Hunting (50%), Tourism/recreation (20%)

Birds

Species	Season	Year	Pop min	Pop max	Acc	Criteria
Bubo bubo Eagle Owl	R	1992	5	8	A	B2
Strix aluco Tawny Owl	R	1992	40	80	A	B3

This is an important site for breeding and passage raptors. Species of global conservation concern that do not meet IBA criteria: *Aquila heliaca* (passage) and *Crex crex* (4–6 breeding pairs).

Protection status

National Partial **International** None

Protected by the local forestry authority (functional category no.1).

Conservation issues

Threats Recreation/tourism (B)

The main threat is from recreational activities although the occasional felling of old trees, hunting and grazing also occur in the forest.

Zau de Câmpie–Tăureni fish-ponds

B2 **019**

Admin region Mureş
Coordinates 46°38′N 24°09′E
Altitude 282–300 m Area 520 ha

Site description

The Zau de Campie–Taureni are situated on the Transylvanian Plain along a tributary of the Mureş river. There are three large lakes and several smaller ponds with about 50 ha of reedbed (*Phragmites*). There is also a *Pinus nigra* plantation which supports a mixed heron colony. The main use of the ponds is fish production, with some angling.

Habitats Wetland (91%; standing fresh water; water-fringe vegetation), Artificial landscape (9%; forestry plantation)
Land-use Agriculture (5%), Fisheries/aquaculture (91%), Hunting (10%)

Birds

Species	Season	Year	Pop min	Pop max	Acc	Criteria
Nycticorax nycticorax Night Heron	B	1996	130	320	A	B2
Falco tinnunculus Kestrel	B	1996	20	42	A	B2
Chlidonias niger Black Tern	B	1996	14	48	A	B2
Otus scops Scops Owl	B	1996	7	10	A	B2
[1] ***Lanius minor*** Lesser Grey Shrike	B	1996	10	45	A	B2

1. Threshold not met but population size significant at European level.

This is an important site for passage waterbirds and breeding herons. Breeding species of global conservation concern that do not meet IBA criteria: *Aythya nyroca* (1–4 pairs) and *Crex crex* (3–4 pairs).

Protection status

National None **International** None

Conservation issues

Threats Agricultural intensification/expansion (C), Aquaculture/fisheries (B), Disturbance to birds (B), Other (B), Recreation/tourism (B)

The main threats are from recreation (particularly disturbance caused by angling), disturbance as a result of maintenance work around the ponds and reed-cutting during the autumn ('Other' threat, above), and the destruction of heron nests. A management plan exists for the site.

Fizeş valley

B2 **020**

Admin region Cluj
Coordinates 46°53′N 24°06′E
Altitude 280–300 m Area 450 ha

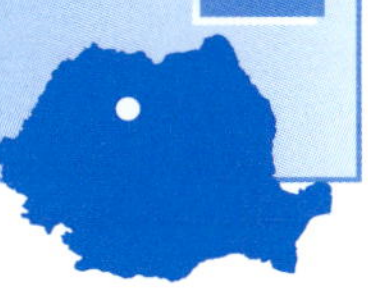

Site description

The site comprises nine large fish-ponds in the Fizeş valley, created in areas of old marsh. Some ponds support reedbeds (*Phragmites*). The surrounding hills have grassland, arable fields and small woodlands. Human activities include angling and traditional grazing.

Habitats Scrub (5%), Grassland (5%), Wetland (80%; standing fresh water; water-fringe vegetation), Artificial landscape (10%)
Land-use Fisheries/aquaculture (80%), Hunting (20%)

Birds

Species	Season	Year	Pop min	Pop max	Acc	Criteria
Botaurus stellaris Bittern	B	1995	5	—	B	B2

This is an important site for breeding and passage waterbirds. Wildfowl use the ponds for roosting. There are few wintering birds as the lakes usually freeze over. Species of global conservation concern that do not meet IBA criteria: *Branta ruficollis* (4 wintering birds), *Aythya nyroca* (10–36 passage birds).

Protection status

National None **International** None

Conservation issues

Threats Aquaculture/fisheries (C), Recreation/tourism (C)

Vlădeasa mountain

B2 **021**

Admin region Cluj, Bihor
Coordinates 46°47′N 22°45′E
Altitude 400–1,836 m Area 14,000 ha

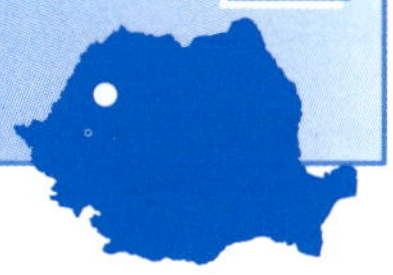

Site description

The site is situated in the northern part of the Apuseni mountains. Forest predominates, with deciduous forest composed mainly of *Quercus* and *Carpinus*, mixed forest of *Fagus* and *Abies*, and coniferous forest of *Abies* and *Juniperus*. Grasslands and small agricultural fields are also present, and a large reservoir is situated on the western boundary. The area is well populated and is a popular tourist destination.

Habitats Forest and woodland (70%; broadleaved deciduous forest; native coniferous forest; mixed forest), Scrub (6%), Grassland (10%), Rocky areas (10%)
Land-use Agriculture (10%), Forestry (15%), Tourism/recreation (5%)

Birds

Species	Season	Year	Pop min	Pop max	Acc	Criteria
[1] ***Picus canus*** Grey-headed Woodpecker	R	1996	50	100	B	B2
Picoides tridactylus Three-toed Woodpecker	R	1996	20	80	A	B2

1. Threshold not met but population size significant at European level.

This site is important for species of montane *Abies* and *Fagus* forest, and as a staging site for migrants crossing the Carpathians. Species of global conservation concern that do not meet IBA criteria: *Crex crex* (5–8 breeding pairs).

Protection status

National None **International** None

Conservation issues

Threats Agricultural intensification/expansion (B), Deforestation (commercial) (B), Recreation/tourism (C)

The main threats are from deforestation and pressure from tourism—there are many access roads and marked trails.

Întregalde and Râmeţi gorges

B2, B3 **022**

Admin region Alba
Coordinates 46°15′N 23°24′E
Altitude 550–900 m Area 4,000 ha

Site description

Two limestone gorges in the Trascău mountains. The steep slopes are partially covered by herbs, bushes and *Pinus* trees. *Quercus*, *Fagus* and *Carpinus* dominated woodlands, grasslands and hay-meadows also occur. The surrounds are used for agriculture and cattle-grazing, with tourism in the gorges.

Habitats Forest and woodland (31%; broadleaved deciduous forest; native coniferous forest), Grassland (35%; steppe/dry calcareous grassland), Rocky areas (14%; scree/boulders; inland cliff), Artificial landscape (20%)
Land-use Agriculture (65%), Fisheries/aquaculture (25%), Forestry (20%), Hunting, Nature conservation/research

■ Birds

Species	Season	Year	Pop min	Pop max	Acc	Criteria
Circaetus gallicus Short-toed Eagle	B	1996	2	4	B	B2
Aquila pomarina Lesser Spotted Eagle	B	1996	6	15	B	B2
[1] ***Picus canus*** Grey-headed Woodpecker	R	1996	25	50	B	B2
Lullula arborea Woodlark	B	1996	100	400	B	B2
Parus lugubris Sombre Tit	R	1996	20	30	B	B3
[1] ***Lanius collurio*** Red-backed Shrike	B	1996	300	500	B	B2

1. Threshold not met but population size significant at European level.

This is an important area for breeding raptors, forest species and birds associated with rocky habitats.

■ Protection status

National Partial **International** None

355 ha of IBA covered by Nature Reserve (Întregalde, 355 ha). 200 ha of IBA covered by Nature Reserve (Râmeţi, 200 ha).

■ Conservation issues

Threats Recreation/tourism (C), Unsustainable exploitation (C)

The main threat is from increasing and currently uncontrolled tourism. Unregulated climbing threatens breeding species such as *Aquila chrysaetos*. Poaching of game species occurs.

Rotbav fish-ponds B2 023

Admin region Braşov
Coordinates 45°30′N 25°33′E
Altitude 486–487 m **Area** 180 ha

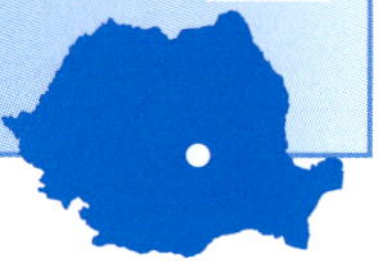

■ Site description

A fish-farm situated on the flood-plain of the River Olt. There are five large ponds and 9–10 smaller ones. The ponds have an extensive covering of reed *Phragmites*, bulrush *Scirpus* and other aquatic plants. Cultivated fields and patches of *Salix* scrub surround the ponds.

Habitats Wetland (90%; standing fresh water; water-fringe vegetation), Artificial landscape (10%; arable land)
Land-use Fisheries/aquaculture (100%)

■ Birds

Species	Season	Year	Pop min	Pop max	Acc	Criteria
Ardea purpurea Purple Heron	B	1998	10	20	B	B2

The site is important for breeding and passage waterbirds, particularly herons, ducks and waders. Species of global conservation concern that do not meet IBA criteria: *Phalacrocorax pygmeus* (non-breeding) and *Aythya nyroca* (passage).

■ Protection status

National None **International** None

■ Conservation issues

Threats Aquaculture/fisheries (B), Burning of vegetation (U), Dredging/canalization (B)

The main threats are fluctuations in water-level and the cutting and burning of reed.

Retezat National Park B2, B3 024

Admin region Hunedoara
Coordinates 45°22′N 22°40′E
Altitude 700–2,482 m **Area** 54,400 ha

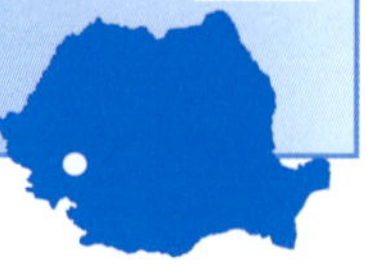

■ Site description

A mountainous area in the south-eastern Carpathians supporting a diversity of habitat-types. *Fagus* and *Abies* forests predominate with dwarf *Pinus* scrub at the treeline and alpine pastures above that.

Habitats Forest and woodland (78%; broadleaved deciduous forest; native coniferous forest; treeline ecotone), Grassland (9%; alpine/subalpine/boreal grassland), Rocky areas (10%)
Land-use Agriculture (10%), Nature conservation/research (95%), Tourism/recreation (5%)

■ Birds

Species	Season	Year	Pop min	Pop max	Acc	Criteria
Aquila pomarina Lesser Spotted Eagle	B	1984	6	—	C	B2
Parus cristatus Crested Tit	R	1985	1,000	4,000	B	B3

This is an important site for species associated with open alpine habitats and extensive montane *Fagus* and coniferous forests.

■ Protection status

National High **International** High

54,400 ha of IBA covered by National Park (Retezat, 54,400 ha). 54,400 ha of IBA covered by Biosphere Reserve (Retezat National Park, 54,400 ha).

■ Conservation issues

Threats Forest grazing (C), Recreation/tourism (C)

No serious threats are known at the site. The World Bank (GEF) began sponsoring a programme of research and management in 1999; a management plan exists for the site.

Cefa fish-ponds and Rădvani wood A1, A4i, A4iii, B1i, B2 025

Admin region Bihor
Coordinates 46°55′N 21°41′E
Altitude 96–108 m **Area** 1,000 ha

■ Site description

The site consists of a fish-pond complex in an area of old marsh, fed by the Crisul Negru river and surrounded by the Corhana canal. Reedbeds (*Phragmites*) grow along the dykes and edges of the fish-ponds. The site also includes agricultural fields, grasslands and hay meadows and to the south the Rădvani forest which is dominated by *Quercus* and *Ulmus*.

Habitats Forest and woodland (20%; broadleaved deciduous forest), Scrub (5%), Grassland (5%; humid grassland), Wetland (60%; standing fresh water; river/stream; water-fringe vegetation), Artificial landscape (10%; arable land)
Land-use Agriculture (10%), Fisheries/aquaculture (70%), Hunting (10%), Nature conservation/research (5%), Tourism/recreation (5%)

■ Birds

Species	Season	Year	Pop min	Pop max	Acc	Criteria
Nycticorax nycticorax Night Heron	B	1996	30	160	A	B2
Egretta alba Great White Egret	N	1996	—	174	A	A4i, B1i
Platalea leucorodia Spoonbill	P	1996	—	53	B	B1i
Anser albifrons White-fronted Goose	P	1996	—	7,000	B	B1i
Anser anser Greylag Goose	P	1996	—	230	—	B1i
[1] ***Aythya nyroca*** Ferruginous Duck	B	1996	2	20	A	A1
Grus grus Crane	P	1996	9	325	A	B1i
[1] ***Chlidonias hybridus*** Whiskered Tern	B	1996	40	150	A	B2
Chlidonias niger Black Tern	B	1996	2	37	B	B2

1. Also occurs on passage.

This is a very important site for breeding and passage waterbirds. The site holds 20,000 or more waterbirds on passage on a regular basis (includes 7,000 *Anser albifrons*, 2,500 *Anas crecca*, 1,160 *Anas platyrhynchos*, 2,000 *Calidris alpina*). Rădvani wood holds a heronry. Species of global conservation concern that do not meet IBA criteria: *Phalacrocorax pygmeus* (passage), *Branta ruficollis* (wintering) and *Haliaeetus albicilla* (wintering).

■ Protection status

National Low **International** None

3 ha of IBA covered by Bird Sanctuary (Rădvani Wood Mixed Heron Colony, 3 ha).

■ Conservation issues

Threats Agricultural intensification/expansion (C), Aquaculture/fisheries (B), Construction/impact of dyke/dam/barrage (C), Disturbance to birds (B), Dredging/canalization (B), Recreation/tourism (C), Selective logging/cutting (B), Unsustainable exploitation (B)

Phalacrocorax carbo and game species (geese and ducks) are hunted. A management plan exists for the site, which is a proposed Ramsar Site.

Lunca wood

B2 **026**

Admin region Arad
Coordinates 46°38′N 21°35′E
Altitude 95–95 m **Area** 174 ha

Site description

A *Quercus–Fraxinus* woodland in an agricultural landscape, approximately 1 km north of the village of Misca. Wet meadows are present along the River Teuz.

Habitats Forest and woodland (98%; broadleaved deciduous forest)
Land-use Forestry (10%)

Birds

Species	Season	Year	Pop min	Pop max	Acc	Criteria
Nycticorax nycticorax Night Heron	B	1996	200	250	A	B2

The site holds an important breeding colony of *Nycticorax nycticorax*, *Egretta garzetta* and *Ardea cinerea*, which covers c.4 ha.

Protection status

National Partial **International** None
Protected by the local forestry authority (functional category no.1).

Conservation issues

Threats Forest grazing (C)

No serious threats are known at the site. Local people are allowed to cut wood from dead trees for their own purposes but there is no commercial forestry. A management plan exists for the site.

Socodor wood

B2 **027**

Admin region Arad
Coordinates 46°33′N 21°26′E
Altitude 90–90 m **Area** 500 ha

Site description

A wood on the flood-plain of the Crisul Alb river, dominated by *Quercus*. In places *Robinia* plantations have replaced native species. *Salix* scrub grows along the banks of the river. Hunting is of mammals only.

Habitats Forest and woodland (85%; alluvial/very wet forest; broadleaved deciduous forest), Scrub (10%), Grassland (5%)
Land-use Forestry (10%), Hunting (80%)

Birds

Species	Season	Year	Pop min	Pop max	Acc	Criteria
Nycticorax nycticorax Night Heron	B	1996	120	180	A	B2

The site is important for a stable, mixed breeding colony of *Nycticorax nycticorax*, *Egretta garzetta* and *Ardea cinerea*.

Protection status

National Partial **International** None
Protected by the local forestry authority (functional category no.1).

Conservation issues

Threats Firewood collection (C), Forest grazing (C)

No serious threats are known at the site, for which a management plan exists.

Mureş river flood-plain: Arad–Pecica–Cenad

B1i, B2, B3 **028**

Admin region Arad
Coordinates 46°09′N 21°05′E
Altitude 90–95 m **Area** 12,000 ha

Site description

A mixture of meadows and almost unbroken woodland situated along the Mureş river between Arad and the Hungarian border. In wetter areas *Salix* and *Populus* dominate with *Quercus*, *Ulmus* and *Fraxinus* on drier sites. Only 14% of the remaining woods are natural, most having been replaced by plantations. The woods are interspersed with clearings, pastures and permanent small marshes and lakes. Flooding occurs each spring.

Habitats Forest and woodland (52%; broadleaved deciduous forest; alluvial/very wet forest), Grassland (7%; humid grassland), Wetland (5%; river/stream), Artificial landscape (30%; arable land)
Land-use Agriculture (30%), Forestry (52%), Not utilized (18%)

Birds

Species	Season	Year	Pop min	Pop max	Acc	Criteria
Botaurus stellaris Bittern	B	1994	4	10	B	B2
Anser albifrons White-fronted Goose	W	1994	2,000	3,000	A	B1i
Aquila pomarina Lesser Spotted Eagle	B	1999	3	6	A	B2
Falco vespertinus Red-footed Falcon	B	1999	5	10	A	B2
Strix aluco Tawny Owl	B	1994	20	40	B	B3
Caprimulgus europaeus Nightjar	B	1994	20	30	B	B2
Alcedo atthis Kingfisher	B	1996	20	30	B	B2
Merops apiaster Bee-eater	B	1994	100	120	B	B2
[1] ***Picus canus*** Grey-headed Woodpecker	R	1994	40	60	B	B2
Dendrocopos medius Middle Spotted Woodpecker	R	1998	200	300	B	B3
Riparia riparia Sand Martin	B	1999	500	1,000	A	B2
[1] ***Lanius collurio*** Red-backed Shrike	B	1999	200	300	A	B2

1. Threshold not met but population size significant at European level.

This is an important site for breeding, passage and wintering waterbirds and raptors. Species of global conservation concern that do not meet IBA criteria: *Aythya nyroca* (10 breeding pairs), *Haliaeetus albicilla* (3–4 wintering birds) and *Crex crex* (breeding).

Protection status

National Low **International** None
25 ha of IBA covered by Nature Reserve (Bezdin Lake, 25 ha). 17 ha of IBA covered by Nature Reserve (Prundul Mare Island), 17 ha).

Conservation issues

Threats Deforestation (commercial) (A), Dredging/canalization (B), Unsustainable exploitation (B)

The main threats are from deforestation, dyke construction and hunting. The area has been proposed as a transboundary reserve.

Satchinez marsh

A1, B2 **029**

Admin region Timiş
Coordinates 45°58′N 21°04′E
Altitude 90–92 m **Area** 236 ha

Site description

A remnant wetland associated with the Ier river. Extensive areas of reedbed (*Phragmites*) are interspersed with open water; wet meadows occur in the surrounds. *Salix* trees and bushes provide breeding sites for herons.

Habitats Wetland (93%; standing fresh water; water-fringe vegetation)
Land-use Nature conservation/research (100%)

Birds

Species	Season	Year	Pop min	Pop max	Acc	Criteria
Nycticorax nycticorax Night Heron	B	1996	10	120	A	B2
Ardea purpurea Purple Heron	B	1996	2	25	A	B2
Aythya nyroca Ferruginous Duck	B	1996	40	50	A	A1

An important site for breeding and passage waterbirds, particularly herons and ducks. There are 69–75 breeding species and a further 79 species have been recorded on passage. Species of global conservation concern that do not meet IBA criteria: *Crex crex* (5–15 breeding pairs).

Protection status

National High **International** None
236 ha of IBA covered by Nature Reserve (Satchinez Bird Reserve, 236 ha).

Conservation issues

Threats Agricultural intensification/expansion (A), Burning of vegetation (B), Construction/impact of dyke/dam/barrage (A), Drainage (A)

As a result of agricultural intensification, dyke construction and drainage around the IBA, the wetland is drying out as water-levels drop and organic matter builds up. The wetland is also being grazed and is subject to fires. A warden is responsible for the site, for which a management plan exists. Between 1996 and 1997 research on the avifauna was carried out, particularly on a mixed colony of herons.

Murani lake and Pişchia forest — A1, B2 — 030

Admin region Timiş
Coordinates 45°55′N 21°20′E
Altitude 50–135 m **Area** 1,500 ha

Site description

The site includes a lake, deciduous forest and both wet and dry grassland. At the end of the lake is a 1–2 ha area of reed *Phragmites* and *Salix* scrub. The shores of the lake are dominated by grassland and *Prunus spinosa* scrub.

Habitats Forest and woodland (40%; broadleaved deciduous forest), Scrub (5%), Grassland (40%; steppe/dry calcareous grassland; humid grassland), Wetland (15%; standing fresh water; water-fringe vegetation)
Land-use Agriculture (10%), Fisheries/aquaculture (10%), Forestry (30%), Hunting (50%)

Birds

Species	Season	Year	Pop min	Pop max	Acc	Criteria
Botaurus stellaris Bittern	R	1997	4	6	A	B2
Ciconia nigra Black Stork	B	1997	2	3	A	B2
Aythya nyroca Ferruginous Duck	B	1997	6	41	A	A1
Aquila pomarina Lesser Spotted Eagle	B	1997	2	3	B	B2

This is an important passage site for waterbirds and raptors. Species of global conservation concern that do not meet IBA criteria: *Phalacrocorax pygmeus* (10–11 passage birds), *Anser erythropus* (passage), *Haliaeetus albicilla* (2–11 wintering birds), *Aquila heliaca* (passage) and *Crex crex* (passage). A total of 189 species have been recorded.

Protection status

200 ha of IBA covered by Nature/Ornithological Reserve (Mlastinile Murani, 200 ha).

Conservation issues

Threats Abandonment/reduction of land management (B), Dredging/canalization (A), Recreation/tourism (B), Selective logging/cutting (B), Unsustainable exploitation (A)

The main threats are posed by high hunting pressure, angling, wood cutting, and drying of the surrounding marsh as a result of dredging and canalization. A proposal has existed for many years for the whole area to be declared a reserve.

Domogled mountain — A1, B2 — 031

Admin region Caraş-Severin
Coordinates 44°52′N 22°26′E
Altitude 190–1,105 m **Area** 2,382 ha

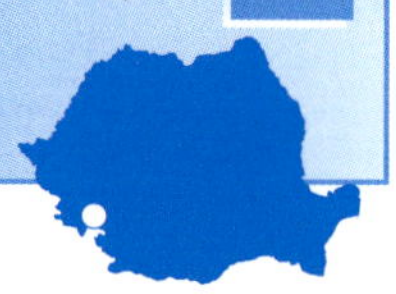

Site description

An area of limestone valleys with vegetation reflecting both central European and Mediterranean influences. Woodland comprises conifers such as *Pinus nigra*, as well as areas dominated by deciduous *Fagus*, *Fraxinus* and *Carpinus*. Steep limestone slopes, clearings and grasslands are also present.

Habitats Forest and woodland (15%; broadleaved deciduous forest; native coniferous forest), Scrub (10%), Rocky areas (70%), Artificial landscape (5%)
Land-use Nature conservation/research (90%), Tourism/recreation (10%)

Birds

Species	Season	Year	Pop min	Pop max	Acc	Criteria
Circaetus gallicus Short-toed Eagle	B	1996	2	3	B	B2
Aquila pomarina Lesser Spotted Eagle	B	1992	2	2	B	B2
Aquila heliaca Imperial Eagle	R	1995	2	4	B	A1, B2

An important site for breeding raptors and other species associated with upland rocky habitats and forest. The area is also used by migrants.

Protection status

National Partial **International** None
IBA overlaps with Nature Reserve (Domogled Mountain, 2,382 ha).

Conservation issues

Threats Abandonment/reduction of land management (U), Firewood collection (C), Forest grazing (C), Recreation/tourism (C), Selective logging/cutting (C)

The main threat is from tourism and recreational activities. Plans exist for the creation of Cerna Gorge–Domogled National Park—a designation that would be beneficial for the purposes of nature conservation.

Porţile de Fier reservoir — A1, A4i, A4iii, B1i — 032

Admin region Caraş-Severin, Mehedinţi
Coordinates 44°34′N 22°12′E
Altitude 60–70 m **Area** 32,000 ha

Site description

A large reservoir created on the Danube, on the Romanian/Yugoslavian border (adjacent to Đerdap gorge, Yugoslavian site 034). The shores are steep and predominantly rocky; deciduous scrub and small areas of grassland dominate the slopes. Reedbeds (*Phragmites*) are present in the upper part of the reservoir.

Habitats Wetland (98%; standing fresh water)
Land-use Fisheries/aquaculture (10%), Hunting (5%), Nature conservation/research (5%)

Birds

Species	Season	Year	Pop min	Pop max	Acc	Criteria
Phalacrocorax pygmeus Pygmy Cormorant	P	1996	12	36	B	A1
Aythya ferina Pochard	W	1996	26	26,500	A	A4i, B1i
Aythya nyroca Ferruginous Duck	P	1996	2	850	A	A1, A4i, B1i

The IBA lies on an important flyway between the Pannonian plain and the Balkan peninsula, and is an important staging post for migrating waterbirds. The site holds 20,000 or more wintering waterbirds on a regular basis.

Protection status

National None **International** None

Conservation issues

Threats Other (C)

Ship traffic poses a threat ('Other' threat). Fishing and hunting activities have minimal impact on birds due to the large area of the reservoir. A management plan exists for the area.

Mehedinţi fish-ponds–Izvoarele — A1 — 033

Admin region Mehedinţi
Coordinates 44°21′N 22°40′E
Altitude 55–55 m **Area** 210 ha

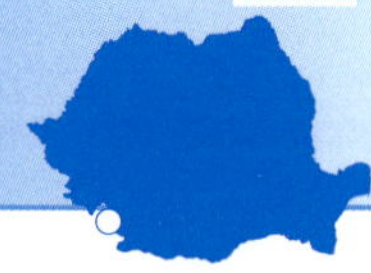

Site description

A natural wetland which has partially been converted into over 20 fish-ponds. Aquatic vegetation, including reedbeds (*Phragmites*), is abundant and covers approximately 60% of the site. Surrounding the ponds are sand-hills, uncultivated areas, a *Robinia* plantation and grassland.

Habitats Scrub (5%), Grassland (5%), Wetland (90%; standing fresh water; water-fringe vegetation)
Land-use Agriculture (5%), Fisheries/aquaculture (70%)

Birds

Species	Season	Year	Pop min	Pop max	Acc	Criteria
Aythya nyroca Ferruginous Duck	B	1996	20	—	B	A1

This is an important site for breeding wetland species. Species of global conservation concern that do not meet IBA criteria: *Phalacrocorax pygmeus* (passage).

Protection status
National None **International** None

Conservation issues

Threats Burning of vegetation (A), Natural events (U)

The main threat to birds is posed by reedbed burning. In recent years the reed has begun to encroach on areas of open water and muddy shore, reducing the overall area of these habitats. A management plan exists for the site.

Mehedinţi fish-ponds–Gruia B2 034

Admin region Mehedinţi
Coordinates 44°16′N 22°41′E
Altitude 55–55 m **Area** 200 ha

Site description
The fish-farm consists of one large and six small fish-ponds with shallow water, situated in a previously natural wetland. Aquatic vegetation is generally poorly-developed although in the largest pond there are abundant submerged plants.

Habitats Grassland (10%), Wetland (80%; standing fresh water), Artificial landscape (10%)
Land-use Fisheries/aquaculture (80%), Hunting (20%)

Birds

Species	Season	Year	Pop min	Pop max	Acc	Criteria
Chlidonias hybridus Whiskered Tern	B	1996	60	—	B	B2

The site is of importance chiefly for passage waterbirds. Species of global conservation concern that do not meet IBA criteria: *Phalacrocorax pygmeus* (passage).

Protection status
National None **International** None

Conservation issues

Threats Aquaculture/fisheries (B), Disturbance to birds (B)

The main threats are from human disturbance and fish-farming activities.

Mehedinţi fish-ponds–Gârla Mare A1, B2 035

Admin region Mehedinţi
Coordinates 44°12′N 22°48′E
Altitude 65–66 m **Area** 140 ha

Site description
The IBA is situated on a low-lying terrace of the Danube river. The site includes natural wetland in the east, a shallow lake with reedbeds (*Phragmites*), and small pools for fish breeding. There are *Salix* trees and *Populus* plantations in the south.

Habitats Grassland (10%), Wetland (70%; standing fresh water; water-fringe vegetation), Artificial landscape (20%; forestry plantation)
Land-use Agriculture (10%), Fisheries/aquaculture (70%), Hunting (10%), Water management (10%)

Birds

Species	Season	Year	Pop min	Pop max	Acc	Criteria
Phalacrocorax pygmeus Pygmy Cormorant	P	1996	60	—	B	A1
Chlidonias hybridus Whiskered Tern	B	1996	60	—	B	B2

This is an important site for passage and breeding waterbirds. Passage species of global conservation concern that do not meet IBA criteria: *Aythya nyroca* and *Haliaeetus albicilla*.

Protection status
National None **International** None

Conservation issues

Threats Aquaculture/fisheries (C), Disturbance to birds (B)

It is important to conserve the areas of natural wetland and reedbed (*Phragmites*) that remain.

Lake Dunăreni–Bistreţ A1, A4i, B1i, B2 036

Admin region Dolj
Coordinates 43°51′N 23°36′E
Altitude 8–8 m **Area** 1,936 ha

Site description
A natural lake situated near to the Danube river, and supplied by the Desnatui and Baboia rivers. There is limited aquatic vegetation; mudflats occur in neighbouring marshy areas. The lake is managed for pisciculture.

Habitats Grassland (15%), Wetland (85%; standing fresh water; water-fringe vegetation)
Land-use Agriculture (15%), Fisheries/aquaculture (85%)

Birds

Species	Season	Year	Pop min	Pop max	Acc	Criteria
Phalacrocorax carbo Cormorant	N	1997	100	1,500	A	B1i
Phalacrocorax pygmeus Pygmy Cormorant	N	1997	200	1,000	A	A1, A4i, B1i
Pelecanus crispus Dalmatian Pelican	N	1997	10	25	A	A1, A4i, B1i
Botaurus stellaris Bittern	B	1997	10	20	B	B2
Ardeola ralloides Squacco Heron	B	1997	11	50	A	B1i, B2
Ardea purpurea Purple Heron	B	1997	2	16	A	B2
Anser albifrons White-fronted Goose	P	1997	500	2,000	A	B1i

This is an important breeding and stop-over site for large numbers of waterbirds; also utilized by non-breeding waterbirds as a feeding area. Species of global conservation concern that do not meet IBA criteria: *Branta ruficollis* (2–20 wintering birds), *Aythya nyroca* (10–15 passage birds) and *Haliaeetus albicilla* (1–2 passage birds).

Protection status
National None **International** None

Conservation issues

Threats Agricultural intensification/expansion (A), Unsustainable exploitation (B)

Hunting is encouraged, particularly of piscivorous birds.

Strejeşti reservoir A1, B1i 037

Admin region Valcea
Coordinates 44°35′N 24°19′E
Altitude 140–140 m **Area** 2,000 ha

Site description
One of a series of reservoirs created by damming the Olt river. Small islands and aquatic vegetation are present due to a build-up of silt.

Habitats Grassland (5%), Wetland (90%; standing fresh water), Artificial landscape (5%)
Land-use Fisheries/aquaculture, Urban/industrial/transport, Water management (100%)

Birds
This is an important site for wintering waterbirds. Species of global

conservation concern that do not meet IBA criteria: *Aythya nyroca* (2 breeding pairs).

Species	Season	Year	Pop min	Pop max	Acc	Criteria
Phalacrocorax pygmeus Pygmy Cormorant	P	1996	50	60	A	A1
Anser albifrons White-fronted Goose	W	1996	1,000	—	B	B1i

■ Protection status

National None **International** None

■ Conservation issues

Threats Aquaculture/fisheries (U), Construction/impact of dyke/dam/barrage (B), Extraction industry (U)

Sand extraction poses a major threat to *Riparia riparia* colonies. Over-fishing of the waters surrounding the islands causes disturbance to breeding birds. A management plan exists for the site.

Lake Comana — A1, B2, B3 — 038

Admin region Gorj
Coordinates 44°10′N 26°06′E
Altitude 55–60 m **Area** 800 ha

■ Site description

The site includes the largest natural lake on the Romanian plain, associated with the Neajlov river. Approximately 75% of its surface is covered by reedbeds (*Phragmites*). The shallow waters increase in depth during periods of flood; intense silting is taking place. An area of forest lies to the south of the lake.

Habitats Forest and woodland (10%), Wetland (90%; standing fresh water; water-fringe vegetation)
Land-use Fisheries/aquaculture (35%), Hunting (35%), Tourism/recreation (30%)

■ Birds

Species	Season	Year	Pop min	Pop max	Acc	Criteria
Botaurus stellaris Bittern	B	1993	4	6	A	B2
Ixobrychus minutus Little Bittern	B	1993	120	160	A	B2
Nycticorax nycticorax Night Heron	B	1993	100	200	A	B2
Ardeola ralloides Squacco Heron	B	1993	20	30	A	B2
Ardea purpurea Purple Heron	B	1996	10	12	B	B2
Aythya nyroca Ferruginous Duck	B	1993	20	30	A	A1
Porzana porzana Spotted Crake	B	1993	60	80	B	B3
Porzana parva Little Crake	B	1993	40	60	B	B3
Crex crex Corncrake	B	1993	16	20	A	A1, B2
Alcedo atthis Kingfisher	B	1993	16	20	B	B2

This is an important site for breeding and passage waterbirds. 212 species have been recorded from the lake and forest. Passage species of global conservation concern that do not meet IBA criteria: *Phalacrocorax pygmeus*, *Pelecanus crispus* and *Gallinago media*.

■ Protection status

National Partial **International** None
80 ha of IBA covered by Nature Reserve (Grădinari Wood, 245 ha).

■ Conservation issues

Threats Aquaculture/fisheries (B), Natural events (B), Recreation/tourism (B)

The main threats are from fishing and recreational activities. The reed is encroaching on areas of open water, reducing the overall area of this habitat. A management plan exists for the site.

Ciocăneşti fish-farm — A1, A4i, B1i, B2 — 039

Admin region Călăraşi
Coordinates 44°11′N 27°02′E
Altitude 10–14 m **Area** 300 ha

■ Site description

A fish-farm that comprises c.12 ponds surrounded by dykes and canals, situated in the Danube valley. Narrow reedbeds (*Phragmites*) fringe the ponds; the total area of open water is 220 ha. Pasture, arable fields and *Populus* plantations form the surrounds.

Habitats Grassland (5%), Wetland (93%; standing fresh water; water-fringe vegetation)
Land-use Fisheries/aquaculture (93%)

■ Birds

Species	Season	Year	Pop min	Pop max	Acc	Criteria
Phalacrocorax pygmeus Pygmy Cormorant	N	1997	10	40	A	A1
Pelecanus crispus Dalmatian Pelican	N	1993	10	30	A	A1, A4i, B1i
Nycticorax nycticorax Night Heron	B	1993	40	—	B	B2
Anser albifrons White-fronted Goose	P	1993	—	1,000	B	B1i

This is an important feeding area for waterbirds breeding at Sreberna reserve (Bulgaria), in riverine forest along the Danube and at other wetland sites. It is also an important stop-over site during spring and autumn migration. Passage species of global conservation concern that do not meet IBA criteria: *Branta ruficollis*, *Aythya nyroca*, *Haliaeetus albicilla* and *Gallinago media*.

■ Protection status

National None **International** None

■ Conservation issues

Threats Disturbance to birds (A)

Piscivorous birds, particularly *Phalacrocorax carbo*, are driven away with gunshots and are sometimes killed. In late autumn the water is drained from the majority of the ponds until early spring—it is not a suitable site for wintering waterbirds. A management plan exists for the site.

Lake Strachina — A1, B2 — 040

Admin region Ialomiţa
Coordinates 44°40′N 27°36′E
Altitude 18–22 m **Area** 1,050 ha

■ Site description

A large brackish lake on the Romanian plain, fed by fresh water from the Ialomita river. There are small areas of aquatic vegetation, primarily along the western shore. Agricultural lands and fish-ponds form the surrounds.

Habitats Wetland (95%; standing brackish and salt water), Artificial landscape (5%)
Land-use Agriculture (5%), Not utilized (95%)

■ Birds

Species	Season	Year	Pop min	Pop max	Acc	Criteria
Aythya nyroca Ferruginous Duck	B	1993	20	—	A	A1
Glareola pratincola Collared Pratincole	B	1993	40	—	B	B2

This is an important breeding site for waterbirds, particularly ducks, and also passage and wintering waterbirds.

■ Protection status

National None **International** None

■ Conservation issues

No serious threats are known at the site.

Lake Fundata — B1i — 041

Admin region Ialomiţa
Coordinates 44°37′N 27°10′E
Altitude 35–40 m **Area** 510 ha

■ Site description

Lake Fundata lies in the Reviga valley. The southern part of the lake is brackish whereas the northern part is fresh and holds most of the aquatic vegetation; few fish species are present. Water depth reaches 5 m and the shores are generally steep. The lake is surrounded by arable land with a village to the east.

Habitats Wetland (90%; standing fresh water; standing brackish and salt water; water-fringe vegetation), Artificial landscape (10%; arable land)
Land-use Agriculture (10%), Fisheries/aquaculture (40%)

■ Birds

Species	Season	Year	Pop min	Pop max	Acc	Criteria
Chlidonias hybridus Whiskered Tern	P	1993	800	—	B	B1i

This is an important site for roosting, wintering and passage waterbirds. Species of global conservation concern that do not meet IBA criteria: *Phalacrocorax pygmeus* (passage), *Branta ruficollis* (wintering) and *Aythya nyroca* (non-breeding).

■ Protection status

National None **International** None

■ Conservation issues

No serious threats are known at the site.

Lake Tătaru — A1, B2 — 042

Admin region Brăila
Coordinates 44°48′N 27°25′E
Altitude 16–20 m **Area** 120 ha

■ Site description

This brackish lake reaches a maximum depth of 12 m. The shores are a mixture of mud and sand covered with halophytic vegetation.

Habitats Grassland (humid grassland), Wetland (80%; standing brackish and salt water), Artificial landscape (20%)
Land-use Agriculture (20%), Fisheries/aquaculture (10%)

■ Birds

Species	Season	Year	Pop min	Pop max	Acc	Criteria
Phalacrocorax pygmeus Pygmy Cormorant	N	—	—	180	—	A1
Glareola pratincola Collared Pratincole	B	1995	8	12	B	B2

This is an important site for breeding and passage wildfowl, waders and gulls (Laridae). Ducks use the site for moulting. Species of global conservation concern that do not meet IBA criteria: *Aythya nyroca* (8 breeding pairs).

■ Protection status

National None **International** None

■ Conservation issues

No serious threats are known at the lake, which has no economic or touristic value.

The Little Island of Brăila — A1, A4i, B1i, B2 — 043

Admin region Brăila
Coordinates 44°55′N 27°55′E
Altitude 2–8 m **Area** 14,862 ha

■ Site description

A wetland complex of many lakes linked by canals and marshes, situated in the Danube valley. There are open grassy areas with lines of *Salix* trees and also areas of riverine forest. The water-level varies according to the Danube's level and spring floods are common. The island also has agricultural crops, meadows and pasture. There are only a few small human settlements.

Habitats Forest and woodland (30%; alluvial/very wet forest), Grassland (10%; humid grassland), Wetland (60%; standing fresh water; water-fringe vegetation)
Land-use Agriculture (10%), Fisheries/aquaculture (50%), Hunting (10%), Nature conservation/research (30%)

■ Birds

This is an important area for breeding, passage and wintering waterbirds and breeding wetland passerines. Species of global conservation concern that do not meet IBA criteria: *Branta ruficollis* (wintering).

Species	Season	Year	Pop min	Pop max	Acc	Criteria
Phalacrocorax pygmeus Pygmy Cormorant	B	1993	100	—	A	A1, A4i, B1i, B2
Nycticorax nycticorax Night Heron	B	1993	40	50	B	B2
Ardeola ralloides Squacco Heron	B	1993	40	45	B	B2
Plegadis falcinellus Glossy Ibis	B	1993	52	60	A	B2
Platalea leucorodia Spoonbill	B	1993	28	30	A	B1i, B2
Anser anser Greylag Goose	B	1993	80	100	A	B1i
Aythya nyroca Ferruginous Duck	B	1993	46	50	A	A1
Crex crex Corncrake	B	1997	10	20	—	A1, B2
Chlidonias hybridus Whiskered Tern	B	1993	150	180	B	B2
Alcedo atthis Kingfisher	R	1993	18	30	C	B2

■ Protection status

National Partial **International** None
5,336 ha of IBA covered by Nature Reserve (Little Island of Brăila, 5,336 ha).

■ Conservation issues

Threats Aquaculture/fisheries (C), Consequences of animal/plant introductions (C), Deforestation (commercial) (C), Forest grazing (C), Unsustainable exploitation (C)

The main threats are from deforestation, overgrazing and hunting. A local NGO undertakes conservation work.

Balta Albă, Amara and Jirlău lakes — A1, A4i, A4iii, B1i, B2 — 044

Admin region Brăila, Buzău
Coordinates 45°15′N 27°15′E
Altitude 40–40 m **Area** 2,680 ha

■ Site description

The site has three lakes associated with the River Buzău. Balta Albă is a salt lake (max. depth 4 m) and does not freeze in winter. Amara is brackish and supports areas of reedbed (*Phragmites*). Jirlău is fresh and has extensive reedbeds. The lakes are surrounded by agricultural fields and both wet and dry meadows. Reed is harvested at Jirlău ('Other' land-use).

Habitats Grassland (10%), Wetland (90%; standing brackish and salt water; water-fringe vegetation)
Land-use Agriculture (10%), Fisheries/aquaculture (30%), Not utilized (60%), Other

■ Birds

Species	Season	Year	Pop min	Pop max	Acc	Criteria
Pelecanus crispus Dalmatian Pelican	P	1995	—	40	B	A1, A4i, B1i
Platalea leucorodia Spoonbill	B	1995	25	40	A	B1i, B2
Anas platyrhynchos Mallard	B	1995	40	8,000	A	B1i
Anas clypeata Shoveler	P	1995	—	5,000	B	A4i, B1i
Aythya nyroca Ferruginous Duck	P	1995	—	1,000	B	A1, A4i, B1i

This is an important site for breeding and passage waterbirds, particularly grebes, ducks and herons. The site holds 20,000 or more passage waterbirds on a regular basis (including 8,000 *Anas crecca*, 5,000 *Anas clypeata*, and 8,000 *Aythya ferina*). Species of global conservation concern that do not meet IBA criteria: *Crex crex* (breeding).

■ Protection status

National Partial **International** None
400 ha of IBA covered by Nature Reserve (Vişani Bird Sanctuary (Lake Jirlău), 400 ha). 600 ha of IBA covered by Nature Reserve (Lake Balta Albă, 600 ha). 900 ha of IBA covered by Nature Reserve (Lake Amara, 900 ha).

■ Conservation issues

Threats Aquaculture/fisheries (C), Construction/impact of dyke/dam/barrage (B), Disturbance to birds (B), Recreation/tourism (B), Unsustainable exploitation (B)

The main threats are from angling, tourism and hunting. A management plan exists for the site.

REFERENCES

ANDONE, G., ALMĂŞAN, H., RADU, D., ANDONE, L., CHIRIAC, E. AND SCĂRLĂTESCU, G. (1969) Cercetări asupra faunei ihtiofage din Delta Dunării. Pp. 133–183 in *Studii şi cercetări ICSPS vol. 27, caiet 1 (Vânătoare)*. Bucharest.

ANON. (1960) *Monografia geografică a R.P.R. I Geografia fizică*. Bucharest: Ed. Academiei R.P.R.

ANON. (1995) Management objectives for biodiversity conservation and sustainable development in the Danube Delta Biosphere Reserve, Romania (draft). Tulcea, Romania: Danube Delta Biosphere Reserve Authority.

ANON. (1996) National Strategy for biodiversity conservation. Bucharest: Ministry of Waters, Forest and Environment Protection.

BADEA, L. ET AL. (1981) *Valea Cernei*. Bucharest: Ed. Academiei R.S.R.

BALLON, E., KISS, J. B., CONDAK, M. AND MIHĂLCIOIU, S. (1996) Citeva observaţii privind iernarea în Delta Dunării în 1995–1996. Pp. 141–153 in *Ecosistemul Dunărea şi conservarea biodiversităţii europene*. Brăila, Romania: Ed. Sapiens.

BĂNICĂ, G. (1996) Contribuţii la studiul avifaunei lacului Techirghiol. Pp. 376–379 in *Lucrările celei de—a treia Conferinţe Naţionale pentru Protecţia Mediului prin Metode şi Mijloace Bilogice şi Biotehnice*. Braşov, Romania: Ed. Universittea "Transilvania" din Braşov.

BREHME, S., MÜLLER, T. AND REDLICH, J. (1992) *Bird observations in the Danube Delta and in the Dobrogea (Romania)*. Zeist, Netherlands: Foundation International Working Group Wader and Waterfowl Research (WIWO) (Rep. No. 43).

BUJOREANU, G. AND POPESCU, P. C. (1996) Rezervaţia naturală Domogled. *Ocrot. nat. med. înconj.* 10(1): 5–30.

CĂTUNEANU, I. AND MIHALCIUC, M. (1987) Contribuţii la cunoaşterea ornitofaunei Măgura. Pp. 267–285 in *Odobeşti Vrancea Studii şi Comunicări* 5–6. Focşani, Romania.

COLDEA, G. (1990) *Munţii Rodnei, studiu geobotanic*. Bucharest: Ed. Academiei Române.

DOMBROWSKI, R. (1946) *Păsările României*. Bucharest.

DRAGOMIR, N. AND STARAŞ, M. (1992) Dinamica unor efective de păsări de importanţa faunistica din Delta Dunării şi impactul biologic asupra producţiei piscicole în perioda 1945–1989. *Ocrot. nat. med. înconj.* 36(2): 97–104.

FILIPAŞCU, A. (1969) Fauna ornitologică a lacurilor de la Geaca după 100 ani. *Ocrotirea naturii* 12(1): 21–41.

GÂŞTESCU, P. (1971) *Lacurile din România*. Bucharest: Ed. Academiei Române.

GRIMMETT, R. F. A. AND JONES T. A. (1989) *Important Bird Areas in Europe*. Cambridge, UK: International Council for Bird Preservation (Tech. Publ. 9).

HEATH, M. F. AND BORGGREVE, C. (2000) *BirdLife International/EBCC European Bird Database 1998*. Cambridge, UK: BirdLife International.

VAN IMPE, J. (1969) Concentration enorme de *Podiceps nigricollis* Brehm. en Dobroudja—Romania. *Alauda* 37(1): 77–78.

VAN IMPE, J. (1970) Migration d'automne de laro-limicoles en Dobrudja maritime du sud (Roumanie). *Gerfaut* 60: 3–4.

IONESCU, D. (1996) Colonia de pescăruşi de la Rotbav. *Bul. S.O.R.*: 4.

IORDACHE, I. AND GACHE, C. (1995) Aspecte avifaunistice din "Codrul secular Slătioara". *Anal. Bucovinei*: 2.

IORDACHE, I. AND GACHE, C. (1993) Observations sur la dispersion zonale de l'avifaune du basin de la riviére Prut. *Anal. Şt. Univ. Iaşi* 38–39: 163–166.

KALABÉR, L. (1980) Rapitoarele din regiunea Reghin. *Vân. si Pesc. Sportiv.* 2: 2.

KALABÉR, L. (1982) La situation des oiseaux rapaces diurnes et le probléme de leur protection en Roumanie. Pp. 557–568 in *Studii şi Comunicări SSB Reghin*.

KISS, A. (1999) *Avifauna din zonele umede ale Banatului*. Timişoara, Romania: Editura Mirton.

KISS, J. B. (1997) *Cartea Deltei. Aves*. Odorhei, Romania.

KLEMM, W. AND KOHL, S. (1988) *Die Ornis Siebenbürgens*. Köln, Germany: Böhlau Verlag.

LINŢIA, D. (1954/1955) *Păsările din R.P.R.*, Vol. I–II. Bucharest: Ed. Academiei R.P.R.

MARINOV, M. (1995) Tendinţe recente in evoluţia avifaunei în Rezervaţia Biosferei Delta Dunării. *An. şt. Inst. Delta Dunării* 4(1): 119–122.

MARINOV, M. AND HULEA, D. (1995) Dinamica coloniilor mixte de cormorani si stârci din Delta Dunării, între 1959–1995. *An. şt. Inst. Delta Dunării* 5: 211–226.

MARINOV, M., HULEA, D. AND CONDAC, M. (1996) Dinamica avifaunei din R.B.D.D. in corelatie cu principalii factori de mediu. *An. şt. Inst. Delta Dunării* 6: 25–28.

MOHAN, G., ARDELEAN, A. AND GEORGESCU, M. (1993) Pp. 46–48 in *Rezervaţii si monumente naturale din Romania*. Bucharest: Editura Scaiul.

MUNTEANU, D. (1960) La situation actuélle de l'avifaune dans le Delta du Danube. *Nos Oiseaux* 269: 209–223.

MUNTEANU, D. (1974) Analiza zoogeografică a avifaunei României. *Nymphaea* 2: 27–69.

MUNTEANU, D. (1980) Abundenţa si biomasa populaţiilor de păsări din Masivul Vlădeasa. *Studii cercet. biol., s. biol. anim.* 32(2): 185–190.

MUNTEANU, D. (1982) Avifauna Masivului Vlădeasa. *Nymphaea* 10: 269–281.

MUNTEANU, D. (1983) Caracterizarea ornitologică a văii Cernei. *An. Banatului, Şt. nat.* 1: 217–225.

MUNTEANU, D. (1985) Zoogeographical connections of the Romanian avifauna. *Rev. Rom. Biol., Geophys., Geogr., Geographie* 29: 27–32.

MUNTEANU, D. (1985) Răspândirea altitudinală a avifaunei pe valea Zlătuii (Parcul Naţional Retezat). *Ocrot. nat. med. înconj.* 29(1): 32–36.

MUNTEANU, D. (1986) Studii asupra populaţiilor de păsări (Aves) din Munţii Retezat (I). *Studii cercet. biol., s. biol. anim.* 38(2): 87–90 .

MUNTEANU, D. (1986) Studii asupra populaţiilor de păsări (Aves) din Munţii Retezat (II). *Studii cercet. biol., s. biol. anim.* 39(2): 171–174.

MUNTEANU, D. (1998) Romania. Pp. 201–217 in *Managing Nature in the CEEC: strategy for development and supporting nature conservation management in Central and Eastern European countries, Vol. 2, Country Reports*. Wimereax, France: Eurosite.

MUNTEANU, D. (1998) *Status of birds in Romania*. Cluj, Romania: Romanian Ornithological Society.

MUNTEANU, D. AND MĂTIEŞ, M. (1969) Aperçu sur l'avifaune des étanges de Geaca. *Lucr. Staţ. biol. geol., geogr. Stejarul* 2: 185–200.

MUNTEANU, D., PAPADOPOL, A. AND WEBER, P., EDS. (1999) *Atlasul provizoriu al păsărilor din România*. Second edition. Cluj, Romania: Romanian Ornithological Society.

MUNTEANU, D. AND TONIUC, N. (1992) The present and the future state of the Danube Delta. Pp. 43–46 in *Managing Mediterranean wetlands and their birds. Proc. IWRB International Symp., Grado, Italy, 1991*. Slimbridge, UK: International Waterfowl and Wetland Research Bureau (Spec. Publ. No. 20).

PAPADOPOL, A. (1963) Cercetări asupra păsărilor de pe litoralul Mării Negre şi a lacurilor litorale—Dobrogea. Pp. 159–181 in *Comunicări de Zoologie, Vol. II*. Bucharest: Ed. Acadedmiei R.P.R.

PAPADOPOL, A. (1971) Recherches ornithologiques dans la zone de lacs salés du Nord-Est de la Plaine Roumaine (I). *Travaux du Mus. d'Hist. Nat. Gr. Antipa Bucureşti* 11: 363–392.

PAPADOPOL, A. (1972) Recherches ornithologiques dans la zone de lacs salés du Nord-Est de la Plaine Roumaine (II).—Les bassins inférieurs des riviéres Buzău, Călmăţui et Ialomoiţa. *Travaux du Mus. d'Hist. Nat. Gr. Antipa Bucureşti* 12: 320–341.

PAPADOPOL, A. (1973) Recherches ornithologiques dans le Nord-Est de la Plaine Roumaine (III). Considération aviphénologiques, sur l'importance biologique et économique des oiseaux dans le sector situé entre les riviéres Buzău et Ialomiţa. *Travaux du Mus. d'Hist. Nat. Gr. Antipa Bucureşti* 13: 346–368.

PAPADOPOL, A. (1974) Recherches ornithologiques dans le Nord-Est de la Plaine Roumaine (IV). La zone des étangs situés dans la partie centrale de la Plaine du Bărăgan, entre les cours inférieurs des riviéres Călmăţui et Ialomiţa. *Travaux du Mus. d'Hist. Nat. Gr. Antipa Bucureşti* 14: 360–383.

PAPADOPOL, A. (1974) Quelques considerations sur l'importance des eaux stagnantes et courantes du Nord-Est de la Plaine Roumanie, pour la dynamique des oiseaux (V). *Travaux du Mus. d'Hist. Nat. Gr. Antipa Bucureşti* 15: 332–344.

PAPADOPOL, A. (1976) Contributions a la connaissance de la faune du Nord-Est de la Plaine Roumaine, entre Siret, Le Danube et la Ialomiţa. *Travaux du Mus. d'Hist. Nat. Gr. Antipa Bucureşti* 17: 185–193.

PAPADOPOL, A. (1989) L'avifaune de la zone des cours inférieures des rivieres Argeş, Dâmboviţa, Neajlov et leurs affluents (Roumanie); dynamique saisonniére et aspecte écologiques. *Travaux du Mus. d'Hist. Nat. Gr. Antipa Bucureşti* 30: 317–334.

PAPADOPOL, A. (1989) L'avifaune de la zone des cours inférieurs des riviéres Argeş, Dîmboviţa, Neajlov et leurs affluents (Roumanie); dynamique saisonniére et aspects écologiques. *Travaux du Mus. d'Hist. Nat. Gr. Antipa Bucureşti* 30: 317–334.

PAPADOPOL, A. AND PETRESCU, A. (1995) Lacustrine and forest complex Comana (Romania), a very important area in the avifauna dynamics. *Travaux du Mus. d'Hist. Nat. Gr. Antipa Bucureşti* 35: 609–626.

PAPADOPOL, A. AND CODREANU, A. (1996) Données de syntyhése sur l'avufaune existant entre les zones inférieures des riviéres Siret et Prut et les lacs du Nord du Bărăgan et de la Dobrogea (Roumanie). *Travaux du Mus. d'Hist. Nat. Gr. Antipa Bucureşti* 36: 349–360.

PAPADOPOL, A. AND TĂLPEANU, M. (1979) Considération phénologiques et écologiques sur les oiseaux du départment d'Ilfov (Roumanie). *Travaux du Mus. d'Hist. Nat. Gr. Antipa Bucureşti* 20: 423–440.

PASPALEVA, M., KISS, J. B. AND TĂLPEANU, M. (1984) Sur la dynamique de quelques espéces d'oiseaux dominants dans le Delta du Danube. *Travaux du Mus. d'Hist. Nat. Gr. Antipa Bucureşti* 25: 313–329.

PASPALEVA, M., KISS, J. B. AND TĂLPEANU, M. (1985) Oiseaux coloniaux dans le Delta du Danube. *Travaux du Mus. d'Hist. Nat. Gr. Antipa Bucureşti* 27: 289–304.

PETRESCU, A. AND BÉRES, I (1997) Ornithological remarks in the basins of Iza and Săpânta and the Biosphere Reserve Pietrosul Rodnei, Maramureş (Romania), in 1995–1996. *Travaux du Mus. d'Hist. Nat. Gr. Antipa Bucureşti* 39: 379–400.

POLIŞ, R. (1973) Colonii de Ardeidae din nord—vestul României. *Nymphaea*: 41–69.

POLIŞ, R. (1976) Observatii asupra avifaunei de la Cefa (judeţul Bihor). *Nymphaea*: 195–225 .

POP, E. AND SĂLĂGEAN, N. (1969) *Monumente ale naturii din România*. Bucharest: Ed. Meridiane.

POPOVICI, I., ED. (1983) Parcul Naţional Retezat. Studii Ecologice. Braşov, Romania: Ed. West Side Computers.

PUŞCARIU, V. (1967) La situation de la protection des zones humides en Roumanie. Compte rendu. Pp. 104–109 in *Deuxiéme reunion europeenne sur la conservation de la sauvagine. Noordwijk aan Zee, Pays Bas, 9–14 1966*. Netherlands: Salverda.

RADU, D. (1979) *Păsările din Delta Dunării*. Bucharest: Ed. Academiei R.S.R.

SEGHEDIN, T. AND RĂDUCU A. (1969) *Rezervaţiile naturale din ţinuturile Sucevei*. Suceava: Cons. jud. ptr. ocrot. nat.

TĂLPEANU, M. AND PASPALEVA, M. (1973) *Aripi asupra Deltei*. Bucharest: Edit. Ştiinţifică.

TONIUC, N. AND OLTEAN, M. (1992) List of protected areas in Romania (1932–1991). *Ocrot. nat. med. înconj.*

TUCKER, G. M. AND HEATH, M. F. (1994) *Birds in Europe: their conservation status*. Cambridge, UK: BirdLife International (BirdLife Conservation Series no. 3).

TUSCHERER, K. AND FÖRSTER, D. (1965) Ornitologische Beobachtungen in der Umgebung von Constanţa. *Der Falke* 12(7): 236–241.

■ SLOVENIA

SLAVKO POLAK

Lake Cerknica (IBA 008), situated on the largest polje (karst field) in Slovenia. (PHOTO: FRANCI NOVAK)

GENERAL INTRODUCTION

Despite its small extent, covering only 20,256 km^2, Slovenia has a relatively diverse landscape. It lies at the geographic centre of Europe, where four biogeographic regions meet. The mountainous Alpine region (covering 30% of the surface area of Slovenia) is restricted to the north of the country, with a number of peaks reaching 2,000 m altitude. To the south lie forests of the Dinaric region (30%), which are managed according to nature-friendly practices. The lowlands of eastern Slovenia belong to the sub-Pannonian region (30%), where there are some intact riverine forests. In the very south-west corner of Slovenia, the 46 km of coast along the Adriatic Sea, backed by shrubland, belong to the sub-Mediterranean region (10%).

This inventory lists 14 Important Bird Areas (IBAs) in Slovenia, covering approximately 3,413 km^2, about 17% of the land area (Table 1, Map 1). This is an increase on the six sites that were identified in the previous pan-European IBA inventory (Grimmett and Jones 1989), following the splitting of one site (Severovzhodna Slovenija, 'site YU004' in 1989) into three, and the identification of six new sites, four of them due mainly to their importance for *Crex crex* (Table 1).

The IBAs are well distributed across the regions and cover all of the main and most important habitats.

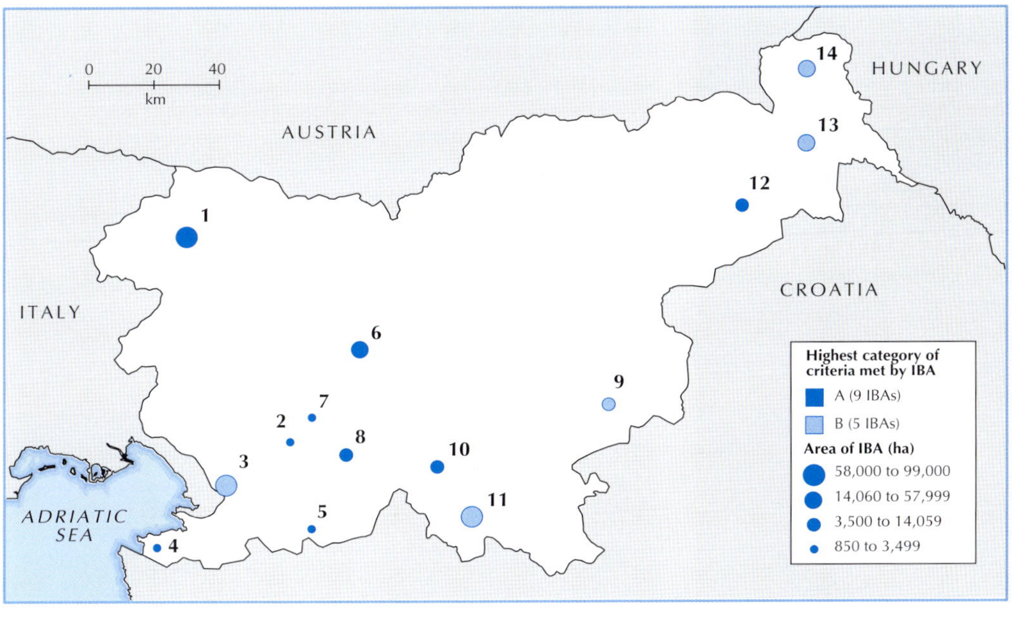

Map 1. Location, area and criteria category of Important Bird Areas in Slovenia.

Table 1. Summary of Important Bird Areas in Slovenia. 14 IBAs covering 3,413 km²

IBA code	1989 code	International name	National name	Administrative region	Area (ha)	Criteria (see p. 11)
001	YU005	Triglav National Park	Triglavski narodni park	Primorska, Gorenjska	84,805	A3, B2, B3
002	—	Nanosčica river basin	Porečje Nanosčice	Notranjska	1,425	A1
003	—	Karst	Kras	Primorska	58,000	B2, B3
004	YU003	Sečovlje saltpans	Sečoveljske soline	Primorska	850	A4i, A4ii, B1i, B1ii
005	—	Reka valley	Dolina Reke	Primorska	1,700	A1
006	YU006	Ljubljansko moor	Ljubljansko barje	Notranjska	14,560	A1, B2, B3
007	—	Planina polje	Planinsko polje	Notranjska	1,600	A1
008	YU001	Lake Cerknica	Cerkniško jezero	Notranjska	3,500	A1
009	YU002	Krakovski forest	Krakovski gozd	Dolenjska	4,000	B3
010	YU004	Ribnica valley	Ribniška dolina	Dolenjska	3,980	A1, B3
011	—	Kočevsko-Kolpa	Kočevsko-Kolpa	Dolenjska	99,000	B2, B3
012	YU004	River Drava from Maribor to Zavrč	Drava ed Maribor od Sredisče	Čtajerska	8,300	A4i, B1i, B2, B3
013	YU004	River Mura from Veržej to Gibina	Mura	Prekmurje, Pomurje	14,060	B2, B3
014	—	Goričko	Goričko	Prekmurje	45,500	B2

ORNITHOLOGICAL IMPORTANCE

A total of 103 species of European conservation concern (SPECs) breed regularly in Slovenia (Tucker and Heath 1994). Regarding SPECs that are also of global conservation concern (SPEC1), *Crex crex* is the only such species which breeds in sufficient numbers at sites as to satisfy the A1 criterion (Table 2). The minimum total population in Slovenia is estimated at 500 calling males, of which the majority breed in IBAs, as follows: 200 pairs at Ljubljansko moor (site 006), 80 pairs at Lake Cerknica (008), 20 pairs at Reka valley (005), 20 pairs at Planina polje (007), 20 pairs at Ribnica valley (010), and 19 pairs at Nanoščica river basin (002). *Aythya nyroca* also breeds at some IBAs, such as River Drava (012) and Lake Cerknica (008), but not in sufficient numbers to meet criteria. The globally threatened *Falco naumanni*, which until recently used to breed at Ljubljansko moor (006), is now extinct in Slovenia. The near-threatened *Haliaeetus albicilla* also breeds in Kočevsko-Kolpa (011) and probably sporadically in Lake Cerknica (008) and River Drava (012).

Two wetlands, River Drava (012) and Sečovlje saltpans (004), are particularly important for their congregations of waterbirds (Table 3). Triglav National Park (001) holds more than 50% of the national population of several species restricted to the Eurasian high-montane biome (criterion A3), such as *Prunella collaris*, *Tichodroma muraria*, *Pyrrhocorax graculus* and *Montifringilla nivalis*. In total, 52 of the 103 SPECs in Slovenia have an unfavourable conservation status in Europe, and the remaining 51 have a favourable conservation status but are concentrated in Europe. Many of these species are well represented in the larger Slovene IBAs.

Alcedo atthis meets the B2 criterion at three lowland IBAs (006, 012 and 013) with relatively intact riverbanks (Table 4). *Alectoris graeca* and *Monticola saxatilis* meet the B2 criterion in alpine rocky pastures in Triglav National Park (001) and in low karst at Karst (003). *Falco tinnunculus*, *Otus scops*, *Jynx torquilla*, *Picus canus*, *Hirundo rustica*, *Saxicola torquata* and *Lanius collurio* are all still common in non-intensively managed agricultural land, grasslands,

Table 2. Important Bird Areas in Slovenia that are important for species of global conservation concern (meeting criterion A1).

Species	IBA code
Crex crex Corncrake	002, 005, 006, 007, 008, 010

Table 3. Important Bird Areas in Slovenia that support important numbers of one or more congregatory species (i.e. meeting criteria A4 and/or B1). IBAs meeting both criteria A4 and B1 for the species are shown in **bold**. IBAs meeting only criterion B1 for the species concerned, and not A4, are shown in normal type. For key to 'Season', see p. 7.

Species	Season	IBA code
Tachybaptus ruficollis Little Grebe	R	**012**
Puffinus yelkouan Yelkouan Shearwater	P	**004**
Phalacrocorax carbo Cormorant	W	012
Egretta alba Great White Egret	W	**012**
Anser fabalis Bean Goose	W	**012**
Anser albifrons White-fronted Goose	W	012
Anser anser Greylag Goose	P	012
Anas platyrhynchos Mallard	W	012
Aythya fuligula Tufted Duck	W	012
Bucephala clangula Goldeneye	W	012
Mergus merganser Goosander	W	012
Larus melanocephalus Mediterranean Gull	P	**004**
Larus cachinnans Yellow-legged Gull	P	**004**
Chlidonias niger Black Tern	P	**012**

Table 4. Species of European conservation concern with significant breeding populations at Important Bird Areas in Slovenia (meeting any IBA criteria).

Species [1]	Minimum national breeding population (pairs) [2]	Proportion (%) of national population breeding at all IBAs in Slovenia
Pernis apivorus Honey Buzzard	600	20
Falco tinnunculus Kestrel	1,500	10
Alectoris graeca Rock Partridge	200	15
Crex crex Corncrake	500	73
Otus scops Scops Owl	500	100[3]
Alcedo atthis Kingfisher	400	21
Jynx torquilla Wryneck	2,000	36
Picus canus Grey-headed Woodpecker	1,000	39
Dendrocopos medius Middle Spotted Woodpecker	200	100[3]
Hirundo rustica Swallow	200,000	3
Saxicola torquata Stonechat	8,000	18
Monticola saxatilis Rock Thrush	200	25
Turdus torquatus Ring Ouzel	2,000	53
Turdus merula Blackbird	200,000	7
Turdus philomelos Song Thrush	100,000	11
Turdus viscivorus Mistle Thrush	20,000	21
Sylvia atricapilla Blackcap	300,000	20
Regulus regulus Goldcrest	100,000	44
Regulus ignicapillus Firecrest	50,000	62
Ficedula albicollis Collared Flycatcher	2,000	65
Parus cristatus Crested Tit	30,000	24
Certhia brachydactyla Short-toed Treecreeper	30,000	12
Lanius collurio Red-backed Shrike	20,000	35
Fringilla coelebs Chaffinch	500,000	17
Carduelis chloris Greenfinch	50,000	13

1. Only those species of European conservation concern (see Box 1, p. 12) that meet IBA criteria in Slovenia are listed.
2. Data are taken from the BirdLife/EBCC European Bird Database 1998 (Heath and Borggreve 2000).
3. The percentage of the national population in IBAs exceeds 100%. Usually this is because the national population estimate has not been updated recently whilst the IBA population estimate has been recently updated with new data as a result of comprehensive surveys of IBAs themselves. Also, the individual site count for a species may be the maximum or average over recent years, and summing these may record more birds than are present nationally in any single year.

Figure 1. Habitats at Important Bird Areas in Slovenia (seeAppendix 3 for definitions of habitats).

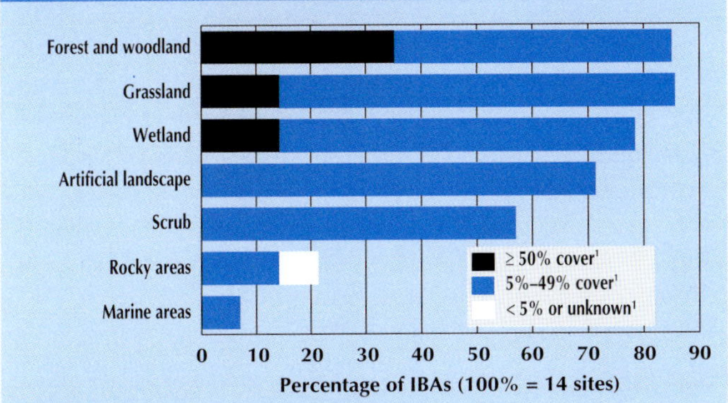

1. percentage of area of individual IBA covered by habitat

Figure 2. Land-uses at Important Bird Areas in Slovenia (see Appendix 3 for definitions of land-uses).

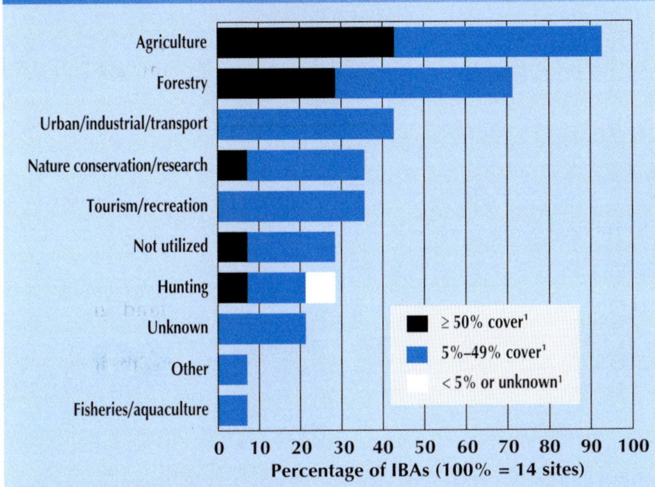

1. percentage of area of individual IBA covered by land-use

pastures and orchards, and meet the B2 criterion at some of the larger IBAs. A further 15 species, mainly forest-related (montane, temperate lowland, and riverine), meet the B3 criterion and are well represented in nearly all of the larger IBAs, with several species having more than 40% of the national population within IBAs (Table 4).

Three of the SPECs, *Scolopax rusticola*, *Numenius arquata* and *Turdus pilaris* occur in important numbers during migration and in winter, despite breeding in only small numbers.

HABITATS

Individual Slovenian IBAs generally comprise a mixture of habitats. About 53% of Slovenia is covered by forest, and the landscape is forest-dominated over nearly 80% of the country, making it one of the most forested countries in Europe. It is therefore not surprising that 86% of IBAs contain forest and woodland, and that forests (of various types) predominate in 36% of IBAs (Figure 1). Broadleaved deciduous and mixed temperate forests predominate in Karst (003) and Kočevsko-Kolpa (011), montane forests in Triglav National Park (001), and riverine forests at Krakovo forest (009) and at River Mura (013).

About 40% of Slovenia is agricultural land, of which 12% is arable land and the remainder natural meadow, pasture and natural grassland. Woodlands, grasslands and artificial landscapes usually form a complex mosaic in the Slovenian landscape. About 85% of IBAs contain grassland but in only 14% of them does grassland predominate (Figure 1). More than 70% of IBAs contain artificial habitats but mainly as a small proportion of each site. Similarly, 57% of IBAs contain shrubland but this is normally present in relatively small patches at each site, admixed with other habitats. The six IBAs with good numbers of breeding *Crex crex* are dominated by wet meadows and wetlands. Wetlands with lakes, such as River Drava (012), Lake Cerknica (008) and Sečovlje saltpans (004), are important for congregations of waterbirds. The last-mentioned IBA contains significant marine habitat. Rocky areas occur in 21% of the IBAs (Figure 1) but are only well-represented in the montane Triglav National Park (001) and at Karst (003).

IMPACTS ON IBAs – LAND-USE AND THREATS

Agriculture and forestry are the main land-uses in Slovenian IBAs (Figure 2). Agriculture is practised at nearly all sites, and at nearly half of them (43%) it is the predominant land-use. Slovenia is well forested, thus forestry is practised at 72% of IBAs and is the main land-use at 29% of all IBAs. In general, forest-harvesting does not present a threat to IBAs, since nature-friendly techniques are used and clear-cutting is prohibited. However, Figure 3 shows that intensive afforestation of traditionally used agricultural land is a problem in some IBAs, e.g. Karst (site 003) and Kočevsko-Kolpa (011). This is connected to the abandonment of agricultural management which affects 57% of IBAs. On the other hand, agricultural intensification and expansion presents the main threat to birds in many lowland sites, affecting 79% of all IBAs (Figure 3).

Wet meadows are the most threatened habitat for birds in Slovenia, being seriously affected by drainage, dredging and canalization. Such water-management, together with the construction of dykes and dams and attempts to agriculturally improve arable land, all have an influence on groundwater-levels and thus indirectly on wetlands. For riverine forests (River Drava,

Figure 3. Threats at Important Bird Areas in Slovenia (see Appendix 3 for definitions of threat types and impact categories).

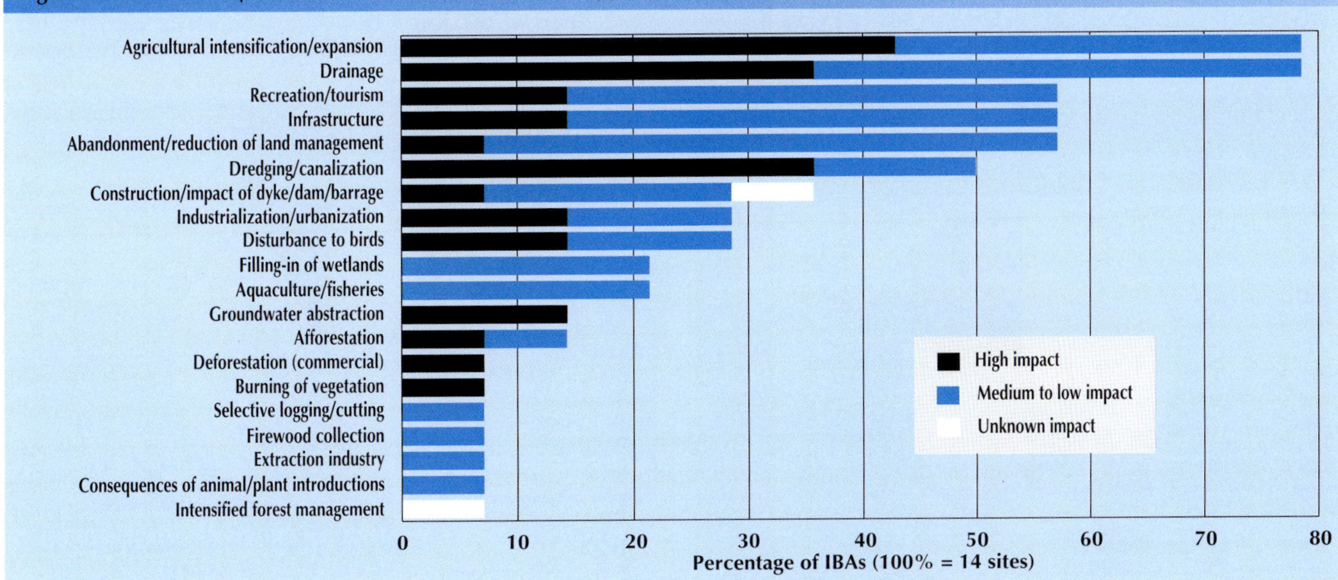

012; River Mura, 013) and Krakovo forest (009), such activities are one of the main threats.

Other types of land-use such as urban, industrial and transport developments, tourism and recreation, hunting and fisheries are of less wide occurrence and are usually not predominant but do pose threats, with recreation/tourism and infrastructure development each threatening over 50% of the sites (Figure 3). In Triglav National Park (001) and Sečovlje saltpans (004) visitors cause serious disturbance to birds. Urbanization and industrialization is also a serious threat, affecting 28% of IBAs (Figure 3), mainly in the lowlands. In about 36% of IBAs, mainly in protected areas, nature conservation is a land-use too (Figure 2)

PROTECTION STATUS

Table 5 and Figures 4 and 5 summarize the national and international protection status and coverage at the 14 Slovenian IBAs.

National protection

Categories of national protected areas in Slovenia (MOP–URSVN 1996) are as follows (IUCN protected-area category in brackets).

1. **Nature Reserve** (I/IV)
2. **National Park** (II/V)
3. **Natural Monument** (III)
4. **Regional Park** (V/III)
5. **Landscape Park** (V)

Slovenia has relatively few protected areas. They cover about 8% of the territory, mainly due to the large extent of Triglav National Park. Additionally a national network of strict protected forest reserves has been established, covering 1% of all forests (60,000 ha). Currently about 70% of forests and more than 80% of arable land are privately owned and therefore ownership of protected forests is also predominately private.

Figure 4 shows the national protection status of the IBAs. Six IBAs have no national protection by law at all. Three of the IBAs include smaller Nature Reserves: primeval forest remnants are strictly protected in Krakovo forest (009) and Kočevsko-Kolpa (011), but these Nature Reserves are small and do not offer protection to the rest of the IBA. Similarly, a small Natural Monument occurs within Ljubljansko moor (006) and River Drava (012). The only fully protected IBA in Slovenia is Triglav National Park (001) which is the only National Park in the country.

At the moment there is only one Regional Park in Slovenia, Škocjanske jame, and this is fully contained within the IBA of Karst (003). Karst (003), Sečovlje saltpans (004), Planina polje (007) and River Drava (012) are all partly protected by Landscape Parks, which lie fully within these larger IBAs.

International protection

Slovenia is a party or signatory to most international legislation that is relevant to site protection in Europe (Box 1). Figure 5 shows the international protection status of the IBAs. The only internationally protected area in the country is Sečovlje saltpans (004), which is a designated Ramsar Site.

Table 5. Protection status of Important Bird Areas in Slovenia. A tick (✔) indicates that an IBA overlaps with a protected area (to any extent).

IBA code	International name	National: Nature Reserve	National Park	Natural Monument	Regional Park	Landscape Park	International: Ramsar Site
001	Triglav National Park		✔				
002	Nanosčica river basin						
003	Karst				✔	✔	
004	Sečovlje saltpans					✔	✔
005	Reka valley						
006	Ljubljansko moor	✔		✔			
007	Planina polje					✔	
008	Lake Cerknica						
009	Krakovski forest	✔					
010	Ribnica valley						
011	Kočevsko-Kolpa	✔					
012[1]	River Drava from Maribor to Zavrč			✔		✔	
013	River Mura from Veržej to Gibina						
014	Goričko						
Total number of IBAs		3	1	2	1	4	1

1. IBA 012 also has protected areas on unknown status

Box 1. International legislation and initiatives that are relevant to site conservation in Slovenia (see Appendix 1 for a general description of these agreements).

Global	
Biodiversity Convention	✔
Ramsar Convention	✔
Bonn Convention	✔
World Heritage Convention	✔
MAB Programme	(✔)
Pan-European	
Bern Convention	(✔)
Regional	
Barcelona Convention	✔

✔ Convention ratified/initiative supported
(✔) Convention signed

New proposals

Slovenia is in transition towards a new system of protected areas. Proposals and plans for new protected areas are the responsibility of the Nature Protection Authority under the Ministry of Environment and Physical Planning. New proposals will affect the majority of Slovenian IBAs. The identification of several large Regional Parks is under way, at Snezniška, Kočevsko-Kolpa, Pohorje, Kraški, Karavanško–Kamniško Savinjski, and Kozjanski.

Two Ramsar Sites have recently been proposed: Ljubljansko Karst River Basin (which includes IBAs Lake Cerknica (008), Ljubljansko moor (006) and Planina Polje (007)); and Skocjan Caves (part of the Karst IBA (003)). Several areas have been discussed and proposed as Biosphere Reserves, e.g. the Drava–

Figure 4. The national protection status of Important Bird Areas in Slovenia.

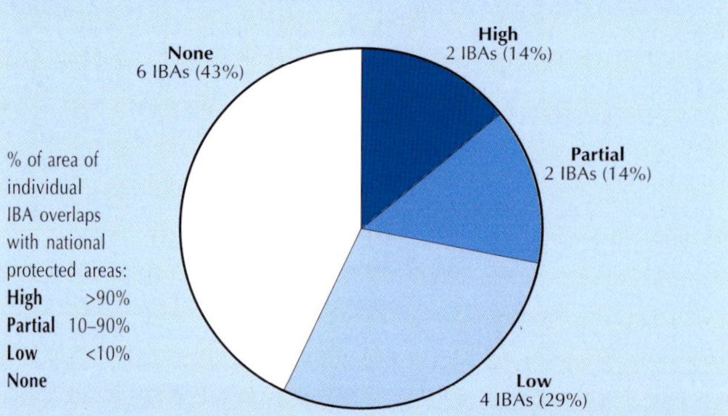

Total area of overlap between IBA network in Slovenia and national protected-area system (see Table 5 for categories) = 894–900 km² (26% of total IBA area).

Figure 5. The international protection status of Important Bird Areas in Slovenia.

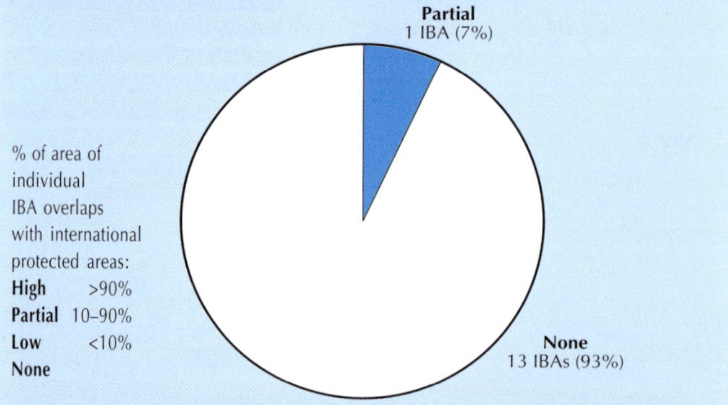

Total area of overlap between IBA network in Slovenia and international protected-area system (see Table 5 for categories) = 7 km² (<1% of total IBA area).

Mura area, which would include the IBAs of River Drava (012) and River Mura (013), as well as an area within the huge forests of southern Slovenia, which would encompass at least five IBAs.

CONSERVATION

Triglav National Park (001) is the only IBA in Slovenia covered entirely by a management plan, due to its overlap with a protected area which already has such a plan. Management plans for other protected areas are currently in preparation.

Scientific research and monitoring of bird numbers at Slovenian IBAs have mostly been carried out by volunteer members of DOPPS (Društvo za opazovanje in proucevanje ptic Slovenije–BirdLife Slovenia), the BirdLife Partner Designate in Slovenia. Such bird data, especially on waterbirds, have been collected since 1979. Waterbird counts and annual monitoring counts of migratory and wintering birds have been made especially at River Drava (012), Lake Cerknica (008) and Sečovlje saltpans (004). DOPPS also initiated atlas projects to map the breeding and wintering birds of Triglav National Park (001), Ljubljansko barje (006) and River Drava (012).

Surveys of bird numbers at other IBAs have been carried out especially for this IBA inventory. Over the last 15 years DOPPS has made many attempts to protect important sites for birds, by collecting data on bird numbers and by working to increase public awareness and by lobbying decision-makers. Euronatur has supported DOPPS in some of its conservation projects, such as at Lake Cerknica (008) and River Drava (012), as well as in holding an international conference on the proposed Drava–Mura Biosphere Reserve. DOPPS has also had good cooperation with the governmental Nature Protection Authority and the Ministry for Environment and Physical Planning.

ANALYTICAL METHODS

- Waterbird numbers at IBAs have been counted annually, but not systematically, for the past 15 years. The counts are published in *Acrocephalus*, the magazine of DOPPS.
- Figures for passage birds at sites are based on the maximum daily count in any one year, averaged over a number of years. For some species, because of a lack of precise counts, numbers at a site were estimated and are presented as a range. Such estimates are based on transect counts in the field and are calculated according to the habitat-type.
- Bird numbers in Triglav National Park (001) are based on data from the local atlas of breeding and wintering birds that was made by DOPPS. In other IBAs, figures for bird numbers were deliberately collected for this IBA inventory during field surveys carried out between 1992 and 1996. Some additional counts were made in 1997. *Crex crex* and owls were mostly counted at night.
- The majority of sites meet more than one criteria: four sites meet two, two meet three, and two sites meets four criteria (Table 1).
- Data on habitats, land-uses and threats at IBAs are, in general, considered to have been accurately assigned. However, there were many problems in assigning categories of habitat, land-use and threat. A characteristic of land-ownership in Slovenia is that private properties are extremely small and fragmented, thus estimates of the percentage cover of habitats and land-uses, and of the impact of threats, can be difficult. The compilers of the IBA data (relevant local IBA coordinators) therefore made estimates based on common sense.
- The data on site protection are based on the *Inventory of the most important natural heritage of Slovenia* (Anon. 1988, 1991) published by the Nature Protection Authority. DOPPS is preparing to publish a national inventory of IBAs.

GLOSSARY

barje mire with remnants of peatbog.
DOPPS Društvo za opazovanje in proucevanje ptic Slovenije–BirdLife Slovenia (Bird Watching and Bird Study Association of Slovenia), the BirdLife Partner Designate in Slovenia.
polje periodically flooded depression (closed basin) in karst area.
SEN/Euronatur European Natural Heritage Fund (Stiftung Europäisches Naturerbe).

ACKNOWLEDGEMENTS

IBA data-sheets were compiled by the DOPPS expert group of local IBA coordinators, comprising Peter Trontelj (006, 003), Davorin Tome (006), Andrej Sovinc (006), Primož Kmecl (008), Miro Perušek (010, 011), Tomaž Jančar (001), Borut Rubinič (004), Borut Štumberger (012), Franc Bračko (013), Damjan Denac (014) and Andrej Hudoklin (009). Borut Mozetič and Nataša Adlešič helped with inputting data into the computer database.

The data on birds held in the database were originally recorded by numerous ornithologists (members of DOPPS).

Funds for the national IBA project were partly provided by the Slovene Ministry of Environment and Physical Planning.

SITE ACCOUNTS

Triglav National Park — A3, B2, B3 — 001

Admin region Primorska, Gorenjska
Coordinates 46°20′N 13°46′E
Altitude 190–2,864 m **Area** 84,805 ha

Site description

Habitats Forest and woodland (65%; broadleaved deciduous forest; native coniferous forest; mixed forest; treeline ecotone), Grassland (25%; alpine/subalpine/boreal grassland; mesophile grassland), Rocky areas (10%; inland cliff)
Land-use Agriculture (10%), Forestry (30%), Hunting (50%), Nature conservation/research (90%), Tourism/recreation (30%)

The only National Park in Slovenia, Triglav National Park (TNP) includes almost all of the Julian Alps, and is named after the highest Slovenian mountain, Triglav, which is located inside the park. The core of the park consists of all the high mountain crests, peaks and glacier valleys. The area also includes two forested plateaus, Pokljuka and Mežakla, as well as Bohinj lake, the largest lake (c.300 ha) in Slovenia, and several smaller high-mountain glacier lakes. Two of the longest rivers in Slovenia, Soča and Sava, rise in the park. There are 24 settlements and hamlets in the area, with a resident population of 2,200. Triglav National Park is also important for its cultural heritage, especially the folk architecture of alpine villages and the traditional highland pastures. About 63% of the park is state property, but this is expected to be reduced to 10–20% after denationalization claims have been concluded, when private ownership will predominate, and c.20% will return to the hands of the farming community. The park is managed by 'Zavod TNP' which developed from the game-hunting reservation 'Zavod za gojitev divjadi Triglav'. Commercial hunting is still an important source of income for the park.

Birds

Species	Season	Year	Pop min	Pop max	Acc	Criteria
Pernis apivorus Honey Buzzard	B	1995	11	30	B	B3
Falco tinnunculus Kestrel	R	1995	31	100	B	B2
Alectoris graeca Rock Partridge	R	1995	30	100	B	B2
Picus canus Grey-headed Woodpecker	R	1995	10	30	B	B2
Prunella collaris Alpine Accentor	B	1995	300	1,000	A	A3
Monticola saxatilis Rock Thrush	B	1995	30	100	B	B2
Turdus torquatus Ring Ouzel	B	1995	1,000	3,000	A	B3
Turdus merula Blackbird	R	1995	3,000	10,000	B	B3
Turdus pilaris Fieldfare	W	1995	1,000	3,000	B	B3
Turdus philomelos Song Thrush	B	1995	3,000	10,000	C	B3
Turdus viscivorus Mistle Thrush	B	1995	1,000	3,000	C	B3
Sylvia atricapilla Blackcap	B	1995	10,000	30,000	B	B3
Regulus ignicapillus Firecrest	B	1995	1,000	3,000	B	B3
Parus cristatus Crested Tit	R	1995	3,000	10,000	B	B3
Tichodroma muraria Wallcreeper	B	1995	30	100	B	A3
Pyrrhocorax graculus Alpine Chough	R	1995	100	300	B	A3

Species ... continued	Season	Year	Pop min	Pop max	Acc	Criteria
Montifringilla nivalis Snowfinch	R	1995	100	300	B	A3
Fringilla coelebs Chaffinch	R	1995	10,000	30,000	B	B3
[1] ***Serinus citrinella*** Citril Finch	U	1995	0	3	A	A3
Carduelis chloris Greenfinch	R	1995	300	1,000	B	B3

1. Breeding not confirmed during the last 10 years.

The site supports internationally important numbers of 15 SPECs of woodland and mountain, as well as 4–5 of the 10 species in Europe that are restricted to the Eurasian high-montane biome (when breeding). Breeding species of global conservation concern that do not meet IBA criteria: *Crex crex* (4–10 pairs, based on diurnal surveys only). Significant proportion (≥1%) of national population breeding at site: *Lanius collurio* (100–300 pairs). During 1991–1995 the ornithological atlas of TNP recorded 115 species, of which 81 were confirmed breeders, 22 probable and 11 possible, with one non-breeding summer visitor, and 85 species in winter.

Protection status

National High **International** None

84,805 ha of IBA covered by National Park (Triglavski narodni park, 84,805 ha).

Conservation issues

Threats Abandonment/reduction of land management (C), Disturbance to birds (C), Intensified forest management (U), Recreation/tourism (C)

Increasing tourism and outdoor activities such as skiing, rafting and mountain climbing cause serious disturbance and other threats to birds and other wildlife. The National Park was designated in 1981 but the area has been protected since 1924. There is a management plan for the site. Recent research has included an ornithological atlas of TNP (initiated by DOPPS), an ongoing project to estimate the population sizes of birds in TNP, and a study on *Aquila chrysaetos*.

Nanoščica river basin — A1 — 002

Admin region Notranjska
Coordinates 45°47′N 14°10′E
Altitude 530–568 m **Area** 1,425 ha

Site description

The Nanoščica river basin is situated in the lower Pivka valley near Postojna at the edge of a large depression. Habitats are relatively intact and undisturbed by man. The river has numerous meanders with well-preserved meadows, and small areas of reed and sedge. Pugled hill, in the middle of the site, supports relict remnants of neutral-soil beech *Fagus* forest. Because of the wet ground, extensive farming is possible only on the driest border of the site, and human settlement is concentrated here.

Habitats Forest and woodland (10%; mixed forest; alluvial/very wet forest), Scrub (10%; scrub), Grassland (30%; humid grassland; mesophile grassland), Wetland (45%; river/stream; water-fringe vegetation), Artificial landscape (5%; arable land)
Land-use Agriculture (80%), Forestry (5%), Hunting (5%), Unknown (10%)

Birds

Species	Season	Year	Pop min	Pop max	Acc	Criteria
Crex crex Corncrake	B	1995	19	25	A	A1

An important site for breeding *Crex crex*. A total of 63 bird species have been recorded, among them 45 breeders and a further seven possible breeders. The site is also nationally important for birds on migration.

Protection status

National None **International** None

Conservation issues

Threats Agricultural intensification/expansion (A), Drainage (A), Dredging/canalization (B), Recreation/tourism (C)

Large cattle-farms sometimes pollute nearby waters with liquid manure. Agricultural improvement and drainage of wet land should be controlled. Intensive ornithological and entomological research is planned.

Karst — B2, B3 — 003

Admin region Primorska
Coordinates 45°40′N 13°55′E
Altitude 150–1,028 m **Area** 58,000 ha

Site description

The north-western ridge of the Dinaric mountains. A flat to hilly limestone area, after which the term 'karst' was named, with a rich diversity of karst features such as caves, sink-holes, dolinas and canyons. Rocky and sparsely vegetated until the beginning of the 20th century, the area is now partly afforested (*Pinus nigra*). Land-use has been abandoned over vast areas which are now undergoing natural succession to shrubland and forest. The region is now a mosaic of dry and rocky meadows and pastures, shrublands in various stages of succession, and woods. There is a strong Mediterranean component to the flora and fauna.

Habitats Forest and woodland (50%; native coniferous forest), Scrub (30%; scrub), Grassland (10%; steppe/dry calcareous grassland), Rocky areas (5%; inland cliff; caves), Artificial landscape (30%; arable land; forestry plantation; other urban/industrial areas)
Land-use Agriculture (10%), Forestry (40%), Not utilized (40%), Urban/industrial/transport (10%)

Birds

Species	Season	Year	Pop min	Pop max	Acc	Criteria
Otus scops Scops Owl	B	1996	300	600	B	B2
Jynx torquilla Wryneck	B	1997	200	300	—	B2
Monticola saxatilis Rock Thrush	B	1996	20	50	B	B2
Turdus merula Blackbird	R	1996	2,500	—	B	B3
Sylvia atricapilla Blackcap	B	1996	3,500	—	B	B3
Regulus regulus Goldcrest	R	1996	1,000	—	B	B3
Parus cristatus Crested Tit	R	1996	300	—	B	B3
Certhia brachydactyla Short-toed Treecreeper	R	1996	300	—	B	B3
Lanius collurio Red-backed Shrike	B	1996	900	—	A	B2
Fringilla coelebs Chaffinch	R	1996	5,000	—	B	B3
Emberiza cia Rock Bunting	R	1996	800	1,000	A	B2

Several species of European conservation concern (SPECs), typical of woodland, meadow and pasture, breed in important numbers, and numerous other SPECs breed in less significant numbers. Significant proportion (≥1%) of national population breeding at site: *Carduelis chloris* (min. 500 pairs). Nationally, Karst is also notable as a wintering site for many species because of its favourable climate (low snow-cover, etc.).

Protection status

National Low **International** None

413 ha of IBA covered by Regional Park (Skocjanske Jame, 413 ha). 93 ha of IBA covered by Landscape Park (Vremscica, 93 ha).

Conservation issues

Threats Abandonment/reduction of land management (A), Afforestation (A), Agricultural intensification/expansion (C), Infrastructure (C)

Intensive afforestation and agricultural abandonment of the rocky steppe-grasslands and open agricultural landscapes pose a high threat to these habitats and their associated birds. The long-term conservation of these areas will only be possible if traditional forms of land-use are maintained or reintroduced. For the landscape to support maximum biodiversity, a traditional mosaic should be maintained of the different habitats such as small woods, grasslands, pastures, vineyards and cultivated fields. A management plan exists for the site.

Sečovlje saltpans — A4i, A4ii, B1i, B1ii — 004

Admin region Primorska
Coordinates 45°30′N 13°39′E
Altitude 0–5 m **Area** 850 ha

Site description

On the Slovenia–Croatia border, this is one of the largest areas of salinas on the Adriatic Sea. The major part lies in Slovenia. Desert-like areas are frequently flooded by seawater or brackish water. Salt-tolerant invertebrates are common, and up to 16 species of salt-tolerant plants are present. Most of the salinas (Fontanigge) were abandoned after the Second World War, but a smaller part (Lera), covering some 5–10% of the site, is still active in producing salt.

Habitats Wetland (70%; standing brackish and salt water), Marine areas (30%; open sea)
Land-use Fisheries/aquaculture (15%), Nature conservation/research (5%), Not utilized (65%), Tourism/recreation (15%)

Birds

Species	Season	Year	Pop min	Pop max	Acc	Criteria
Puffinus yelkouan Yelkouan Shearwater	P	1995	10	1,000	B	A4ii, B1ii
[1] ***Larus melanocephalus*** Mediterranean Gull	P	1995	3,200	36,000	A	A4i, B1i
[2] ***Larus cachinnans*** Yellow-legged Gull	P	1995	1,000	20,000	A	A4i, B1i

1. Seasonal total.
2. Daily count.

Important numbers of migratory and post-breeding *Larus melanocephalus* and *Larus cachinnans* congregate at the saltpans, and the adjacent sea is a migration route for significantly large numbers of *Puffinus yelkouan*. Species of global conservation concern that do not meet IBA criteria: *Phalacrocorax pygmeus* (3–23 birds at the pans in winter; numbers have been increasing), *Acrocephalus paludicola* (rare on passage, max. 1). The salinas are also notable nationally for their concentrations of waders on passage, for some breeding waterbirds and for wintering wildfowl, gulls and herons.

Protection status

National High **International** Partial
835 ha of IBA covered by Landscape Park (Sečovljske soline, 850 ha). 650 ha of IBA covered by Ramsar Site (Sečovljske soline [Sečovlje saltpans], 650 ha).

Conservation issues

Threats Abandonment/reduction of land management (C), Aquaculture/fisheries (C), Construction/impact of dyke/dam/barrage (B), Disturbance to birds (B), Drainage (B), Infrastructure (A), Recreation/tourism (B)

The only important problems are the disturbance to birds caused by visitors and the increasing pressures for infrastructure development (for aquaculture, tourism, sport). The area is the only Ramsar Site in Slovenia (designated in 1993). Waterbird hunting has been prohibited. The site is well researched and frequently visited by ornithologists, but permanent monitoring of birds has not yet been established.

Reka valley

A1 **005**

Admin region Primorska
Coordinates 45°33'N 14°15'E
Altitude 400–490 m **Area** 1,700 ha

Site description

The Reka river basin lies between karstic limestone (Snežnik plateau) and fluvial terrain. The river has created a flat plain around the biggest town in the region, Ilirska Bistrica, and industry and dense infrastructure are developing here, although wet grasslands, meadows and small marshes remain. There is extensive farming and orchard cultivation. Land subsidence is occurring where there are abandoned coal mines.

Habitats Forest and woodland (5%; mixed forest), Scrub (5%; scrub), Grassland (50%; humid grassland; mesophile grassland), Wetland (20%; river/stream; water-fringe vegetation; fen/transition mire/spring), Artificial landscape (20%; arable land; other urban/industrial areas)
Land-use Agriculture (70%), Urban/industrial/transport (30%)

Birds

Species	Season	Year	Pop min	Pop max	Acc	Criteria
Crex crex Corncrake	B	1995	20	30	A	A1

An important site for breeding *Crex crex*. Nationally rare and threatened bird species occur in the site in small numbers. Significant proportion (≥1%) of national population breeding at site: *Jynx torquilla* (10–20 pairs), *Picus canus* (5–10 pairs).

Protection status

National None **International** None

Conservation issues

Threats Agricultural intensification/expansion (A), Drainage (A), Dredging/canalization (B), Filling-in of wetlands (B), Industrialization/urbanization (A), Infrastructure (B)

The most difficult problem is the heavy pollution of the Reka river. However, the water condition is improving since some clean-up measures were taken. Some reedbeds and marshes were destroyed in the past, but restoration is possible. Intensive ornithological and entomological research is planned.

Ljubljansko Moor

A1, B2, B3 **006**

Admin region Notranjska
Coordinates 46°02'N 14°26'E
Altitude 286–390 m **Area** 14,560 ha

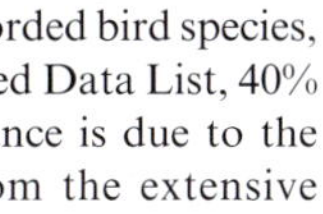

Site description

Ljubljansko barje is a large 150 km^2 depression, traversed by the Ljubljanica river. Once a vast mire, today it is mainly cultivated (maize). The area is a traditional cultural landscape with extensive wet grasslands, flood-plain woodlands, thickets and small, wooded hills. Groundwater-levels are high, some areas are flooded regularly, and there are dense networks of ditches, channels and brooks. Settlements are generally concentrated along the driest edges of the site, with the Slovenian capital Ljubljana lying on the northern border.

Habitats Forest and woodland (10%; mixed forest; alluvial/very wet forest), Scrub (10%; scrub), Grassland (30%; humid grassland), Wetland (10%; river/stream; raised bog), Artificial landscape (40%; arable land; other urban/industrial areas)
Land-use Agriculture (70%), Forestry (10%), Nature conservation/research (5%), Tourism/recreation (5%), Unknown (5%), Urban/industrial/transport (5%)

Birds

Species	Season	Year	Pop min	Pop max	Acc	Criteria
Falco tinnunculus Kestrel	B	1995	20	40	A	B2
[1] ***Crex crex*** Corncrake	B	1995	200	250	A	A1
Alcedo atthis Kingfisher	R	1995	5	10	A	B2
Jynx torquilla Wryneck	B	1995	70	100	A	B2
Hirundo rustica Swallow	B	1995	1,500	3,000	A	B2
Saxicola torquata Stonechat	B	1995	300	600	A	B2
Sylvia atricapilla Blackcap	B	1995	3,000	5,000	A	B3
Lanius collurio Red-backed Shrike	B	1995	500	—	B	B2
Carduelis chloris Greenfinch	R	1995	500	1,000	A	B3

[1] The largest population in the country.

The site is the most important breeding place in Slovenia for the globally threatened *Crex crex* and is one of the most important sites for seven other SPECs. Species of global conservation concern that do not meet IBA criteria: *Falco naumanni* (max. 8 pairs, the last Slovenian colony and nearing national extinction). During migration, large flocks of waterbirds and raptors rest in the area. Out of 210 recorded bird species, 107 have bred (including 53 species on the national Red Data List, 40% of the total). This high species-richness and abundance is due to the great variety of vegetation-types which result from the extensive management of the landscape.

Protection status

National Low **International** None
3 ha of IBA covered by Natural Monument (Draga pri Igu, 3 ha). 3 ha of IBA covered by Nature Reserve (Koslerjeva gošča, 3 ha).

Conservation issues

Threats Abandonment/reduction of land management (C), Agricultural intensification/expansion (A), Aquaculture/fisheries (B), Drainage (B), Dredging/canalization (A), Industrialization/urbanization (B), Recreation/tourism (C)

Agricultural improvement and conversion of crops to maize monoculture has caused a number of negative effects. There is a no management plan for the site. Recent research at the site has included a breeding bird atlas project (almost completed) initiated by DOPPS, census counts and mapping of birds, and a detailed analysis of bird communities.

Planina polje

A1 **007**

Admin region Notranjska
Coordinates 45°51′N 14°15′E
Altitude 440–453 m **Area** 1,600 ha

Site description

Planinsko polje is part of the karst basin of the Ljubljanica river, and displays typical karst features such as caves (more than 100 in the vicinity) and sink-holes. Higher hills surround the IBA, from where water collects and sinks underground. Floods periodically cover about 16 km² of the area. Besides the numerous small streams, the Unica river with its beautiful meanders is the main water-body at the site. Arable land and human settlements occur at the outermost, higher edges of the site. Land-use is mainly mowing of grass and hay; some farmers use fertilizer to increase grass production. This has been a classic area for karst research since the 17th century.

Habitats Forest and woodland (5%; broadleaved deciduous forest), Scrub (5%; scrub), Grassland (70%; steppe/dry calcareous grassland; humid grassland; mesophile grassland), Wetland (15%; river/stream; water-fringe vegetation), Rocky areas (caves), Artificial landscape (5%; highly improved reseeded grassland)
Land-use Agriculture (90%), Nature conservation/research (5%), Tourism/recreation (5%)

Birds

Species	Season	Year	Pop min	Pop max	Acc	Criteria
Crex crex Corncrake	B	1994	20	35	A	A1

An important site for breeding *Crex crex*. Other wet meadow and grassland bird species breed at the site in nationally important numbers, including many SPECs. Significant proportion (≥1%) of national population breeding at site: *Alcedo atthis* (2–5 pairs). The site is known to support notable numbers of birds on migration as well.

Protection status

National Partial **International** None
960 ha of IBA covered by Landscape Park (Krajinski park Planinsko polje, 960 ha).

Conservation issues

Threats Agricultural intensification/expansion (A), Drainage (A), Dredging/canalization (A), Recreation/tourism (B)

The main threat to birds is the drainage of wet depressions and protected meadows. During 1970–1979 the national Natural Protection Service organized field research concerning the proposal for Notranjski Krajinski Park (Protected Landscape). There is a management plan for the Landscape Park. Bird data is incomplete, and intensive ornithological and entomological research is planned.

Lake Cerknica

A1 **008**

Admin region Notranjska
Coordinates 45°45′N 14°23′E
Altitude 548–623 m **Area** 3,500 ha

Site description

Lake Cerknica is situated on the largest polje (karst field) in Slovenia. A temporary eutrophic lake which is usually totally dry in summer and late winter, but with an area of 2,600 ha when full in spring and autumn. There are huge expanses of reed, sedge and wet grasslands. On the eastern border of the lake, there are remnants of a raised bog. The habitats have been little affected by man. There are a few settlements along the border of the IBA, especially to the north and east, the largest being Cerknica town with 3,000 inhabitants. The lake is historically famous for its underground water system, first described in 1689.

Habitats Grassland (30%; humid grassland; mesophile grassland), Wetland (70%; standing fresh water; river/stream; raised bog; water-fringe vegetation; fen/transition mire/spring)
Land-use Agriculture (40%), Not utilized (40%), Tourism/recreation (20%)

Birds

Species	Season	Year	Pop min	Pop max	Acc	Criteria
[1] ***Crex crex*** Corncrake	B	1996	80	100	A	A1

[1] Average counts from 1992–1996 period.

Important numbers of the globally threatened *Crex crex* breed. Other species of global conservation concern that do not meet IBA criteria: *Aythya nyroca* (1–5 pairs, the most important breeding site in Slovenia; 10–25 birds on passage), *Haliaeetus albicilla* (2–5 birds feed in the area but do not breed). A total of 230 bird species have been recorded so far, of which 94 species have bred. Besides its international importance, the site has great value for nationally rare breeding species and nationally important numbers of waterbirds in winter and on passage.

Protection status

National None **International** None

Conservation issues

Threats Agricultural intensification/expansion (B), Aquaculture/fisheries (C), Burning of vegetation (A), Construction/impact of dyke/dam/barrage (B), Disturbance to birds (A), Drainage (A), Dredging/canalization (A), Filling-in of wetlands (C), Infrastructure (B), Recreation/tourism (A)

Wind-surfing, boating and angling cause disturbance to birds. To improve conservation of the lake, legal protection should be enforced, hunting on the lake should be completely banned, angling should be allowed only in certain areas, and planned tourism developments should be nature-friendly. Abandonment of extensive farming would lead to loss of wet grasslands and may be a problem in future, thus involvement of (and support for) local people is necessary. DOPPS founded a local group in 1995 to promote conservation in the area, and DOPPS has gathered and publicized information and proposals that justify the inclusion of Lake Cerknica in a proposed Regional Park, with support from SEN (Euronatur). Field surveys are ongoing, including annual nocturnal counting of *Crex crex* over the whole area.

Krakovski forest

B3 **009**

Admin region Dolenjska
Coordinates 45°53′N 15°24′E
Altitude 150–159 m **Area** 4,000 ha

Site description

Krakovo forest is one of the largest intact examples of lowland alluvial forest to remain in Slovenia. Totally uninhabited and almost untouched by man, it lies in the central Krško valley, on the left bank of the Krka river, between Radulja and Velikovaški potok stream. The Ljubljana–Zagreb highway runs along the northern border of the site. The Krka river draws off upwelling groundwater from the IBA, which is permanently waterlogged, and there are many small streams along the edge of the forest, some of which are regulated. Broadleaved deciduous trees (especially *Quercus*) dominate the forest. The mean age of trees is 150 years, and the largest *Salix alba* tree and one of the largest *Quercus robur* trees in the country occur here. There are also well-preserved marshes, meadows and wet grasslands. Human settlements and intensively farmed land occur outside the IBA.

Habitats Forest and woodland (56%; alluvial/very wet forest), Grassland (25%; humid grassland), Wetland (6%; river/stream), Artificial landscape (12%; arable land; other urban/industrial areas)
Land-use Agriculture (30%), Forestry (52%), Hunting (9%), Nature conservation/research (1%), Other (5%), Water management (3%)

Birds

Species	Season	Year	Pop min	Pop max	Acc	Criteria
Dendrocopos medius Middle Spotted Woodpecker	R	1990	20	30	B	B3
Ficedula albicollis Collared Flycatcher	B	1990	50	100	B	B3

Two specialists of deciduous broadleaved woodland breed in important numbers. A total of 44 certain or probable breeding species have been recorded in the forest and in the immediate surroundings, including numerous nationally important species such as *Aquila pomarina* (the only known breeding locality in Slovenia).

Protection status

National Low **International** None

41 ha of IBA covered by Nature Reserve (Krakovski gozd, 41 ha).

Conservation issues

Threats Agricultural intensification/expansion (C), Construction/impact of dyke/dam/barrage (U), Drainage (A), Dredging/canalization (A), Groundwater abstraction (A), Infrastructure (B)

Groundwater-levels have fallen over the whole IBA by 1 m, due to drainage in the surrounding Krško karst field, and this is the main reason behind the death of old oak *Quercus* trees. Other threats to water quality and quantity are the intensive use of pesticides on the Krško karst field, a planned chain of hydroelectric power-stations along the Sava river, and the extension of a road in the direction of Zamesko–Malence. Ornithologically, Krakovski forest is not well known.

Ribnica valley — A1, B3 — 010

Admin region Dolenjska
Coordinates 45°43'N 14°44'E
Altitude 500–600 m **Area** 3,980 ha

Site description

The IBA lies in the middle of a large karst depression in the western and southern part of the Ribnica valley, and approximately 50% of the area is wet meadows and grasslands near the streams Bistrica, Sajevec, Ribnica and Obrh. Streams in the area are partly regulated. Besides nearly ten smaller villages, the only town is Ribnica. Some fields have been abandoned and are in the initial stages of succession to scrub and forest.

Habitats Forest and woodland (10%; mixed forest), Grassland (35%; humid grassland; mesophile grassland), Wetland (5%; river/stream), Artificial landscape (45%; highly improved reseeded grassland; arable land; urban parks/gardens; ruderal land)
Land-use Agriculture (80%), Forestry (10%), Not utilized (10%)

Birds

Species	Season	Year	Pop min	Pop max	Acc	Criteria
Pernis apivorus Honey Buzzard	B	1996	6	10	B	B3
Crex crex Corncrake	B	1996	20	30	A	A1

An important site for breeding *Crex crex* and one woodland species. Significant proportion (≥1%) of national population breeding at site: *Jynx torquilla* (30–40 pairs), *Picus canus* (10–15 pairs). The site is of national importance for migrating birds—numerous ducks, gulls and birds of prey rest on the water surface at Prigorica.

Protection status

National None **International** None

Conservation issues

Threats Abandonment/reduction of land management (B), Agricultural intensification/expansion (A), Drainage (B)

The intensification of grass mowing (silage production) is the main threat to *Crex crex*. The fauna and flora of the site have not been well investigated. Intensive field research is planned.

Kočevsko-Kolpa — B2, B3 — 011

Admin region Dolenjska
Coordinates 45°35'N 14°52'E
Altitude 150–1,298 m **Area** 99,000 ha

Site description

Kočevsko-Kolpa is a huge area of limestone karst on the high Dinaric plateau in southern Slovenia. The coniferous, deciduous and mixed forests, which connect with Snežnik plateau and Gorski Kotar in Croatia, represent one of the largest and least populated areas of dense forest in Europe. Wooded land (including plantations) covers 95% of the area, and includes remnants of primeval forest, while pastures and grasslands cover 5%. The whole area was suddenly and almost entirely deserted by its inhabitants in 1941, and for political reasons many human settlements in the area remained abandoned after the Second World War. Formerly open areas (grasslands, orchards and villages) are thus reverting naturally to scrub and forest. The Kolpa river canyon on the international border has large cliffs on the southern slopes.

Habitats Forest and woodland (80%; broadleaved deciduous forest; mixed forest), Scrub (5%; scrub), Artificial landscape (15%; highly improved reseeded grassland; forestry plantation)
Land-use Agriculture (5%), Forestry (90%), Hunting (3%), Military (2%)

Birds

Species	Season	Year	Pop min	Pop max	Acc	Criteria
Pernis apivorus Honey Buzzard	B	1992	15	25	B	B3
Falco tinnunculus Kestrel	R	1992	10	15	B	B2
Otus scops Scops Owl	B	1992	10	20	A	B2
Jynx torquilla Wryneck	B	1992	200	250	B	B2
Picus canus Grey-headed Woodpecker	R	1992	100	115	B	B2
Turdus torquatus Ring Ouzel	B	1992	50	100	A	B3
Turdus merula Blackbird	R	1992	5,000	7,000	A	B3
Turdus pilaris Fieldfare	W	1992	2,000	5,000	A	B3
Turdus philomelos Song Thrush	R	1992	4,000	6,000	A	B3
Turdus viscivorus Mistle Thrush	R	1992	3,000	5,000	A	B3
Sylvia atricapilla Blackcap	B	1992	20,000	30,000	A	B3
Regulus regulus Goldcrest	R	1992	40,000	60,000	A	B3
Regulus ignicapillus Firecrest	R	1992	30,000	50,000	A	B3
Ficedula albicollis Collared Flycatcher	B	1992	50	80	A	B3
Parus cristatus Crested Tit	R	1992	4,000	5,000	B	B3
Certhia brachydactyla Short-toed Treecreeper	R	1992	1,500	2,000	A	B3
Lanius collurio Red-backed Shrike	B	1992	600	800	A	B2
Fringilla coelebs Chaffinch	R	1992	60,000	80,000	A	B3
Carduelis chloris Greenfinch	R	1992	1,500	1,800	B	B3

The site supports internationally important numbers of several SPECs of woodland/forest (even some boreal-type species). Breeding species of global conservation concern that do not meet IBA criteria: *Haliaeetus albicilla* (1 resident pair, the only known nest in Slovenia).

Protection status

National Low **International** None

218 ha of IBA covered by six Nature Reserves (Kočevski pragozdovi ['virgin forest remnants'], 218 ha).

Conservation issues

Threats Afforestation (B), Recreation/tourism (B), Selective logging/cutting (B)

Ornithologically, the Kočevsko forests are not well known. Only one ornithologist has made quantitative surveys. The well preserved remnants of primeval forest are of great scientific importance.

River Drava from Maribor to Zavrč — A4i, B1i, B2, B3 — 012

Admin region Štajerska
Coordinates 46°25'N 15°55'E
Altitude 172–291 m **Area** 8,300 ha

Site description

Habitats Forest and woodland (9%; alluvial/very wet forest), Grassland (5%), Wetland (14%; standing fresh water; river/stream), Artificial landscape (46%; arable land)
Land-use Agriculture (52%), Forestry (95%), Hunting, Nature conservation/research (18%), Urban/industrial/transport (5%)

The River Drava is the largest river to flow through Slovenia. Due to numerous development schemes in the past, the river and its flood-plains are now one of the most ecologically degraded waterways in the country. The IBA is located between Maribor and Središče ob Dravi on the Croatian border. Along this section the river is exploited in full, with large reservoirs and associated channels dividing up the entirety of the riverine countryside. Between the reservoirs (Ptujsko jezero, Ormoško jezero) there are still some large flood-plains which are periodically inundated. Flooded woodlands with meadows and wet grasslands are also present.

Birds

Species	Season	Year	Pop min	Pop max	Acc	Criteria
Tachybaptus ruficollis Little Grebe	R	1995	700	1,000	A	A4i, B1i
Phalacrocorax carbo Cormorant	W	1995	800	1,400	A	B1i
Egretta alba Great White Egret	W	1995	80	120	A	A4i, B1i
[1] ***Anser fabalis*** Bean Goose	W	1995	1,500	4,000	A	A4i, B1i
Anser albifrons White-fronted Goose	W	1995	100	1,800	A	B1i
Anser anser Greylag Goose	P	1995	—	700	A	B1i
Anas platyrhynchos Mallard	W	1995	15,000	30,000	A	B1i
Aythya fuligula Tufted Duck	W	1995	4,000	6,500	A	B1i
Bucephala clangula Goldeneye	W	1995	1,500	3,500	A	B1i
Mergus merganser Goosander	W	1995	80	140	A	B1i
[2] ***Pernis apivorus*** Honey Buzzard	B	1995	28	35	A	B3
Falco tinnunculus Kestrel	B	1995	35	40	A	B2
Chlidonias niger Black Tern	P	1995	5,500	10,000	A	A4i, B1i
[3] ***Alcedo atthis*** Kingfisher	R	1995	20	30	A	B2
Jynx torquilla Wryneck	B	1995	90	110	A	B2
Picus canus Grey-headed Woodpecker	R	1995	35	40	A	B2
Dendrocopos medius Middle Spotted Woodpecker	R	1995	12	16	A	B3
Saxicola torquata Stonechat	B	1995	320	380	A	B2
Ficedula albicollis Collared Flycatcher	B	1995	300	320	A	B3
Carduelis chloris Greenfinch	R	1995	550	650	A	B3

1. Level of hunting influences numbers.
2. >500 on spring migration.
3. >120 on migration.

The site supports important numbers of wintering waterbirds (at the reservoirs) and of breeding birds of flooded forest, meadows, and gravel-banks, and is also an important bird-migration route. Species of global conservation concern that do not meet IBA criteria: *Phalacrocorax pygmeus* (max. 37 on passage), *Aythya nyroca* (15–60 on passage). Significant proportion (≥1%) of national population breeding at site: *Sylvia atricapilla* (3,000–3,400 pairs), *Lanius collurio* (420–460 pairs). A total of 266 species have been recorded so far, of which 89 are regular or occasional breeders. *Nycticorax nycticorax*, *Burhinus oedicnemus* and *Sterna albifrons* were lost as breeders at the end of the 1970s after the construction of the last hydroelectric power-station on the river.

Protection status

National Partial **International** None

2,595 ha of IBA covered by Landscape Parks (Stara Drava, 2,150 ha; Šturmovci, 280 ha), Natural Monument (Ormoško jezero, 140 ha) and smaller protected areas (25 ha).

Conservation issues

Threats Abandonment/reduction of land management (B), Agricultural intensification/expansion (A), Consequences of animal/plant introductions (C), Construction/impact of dyke/dam/barrage (A), Deforestation (commercial) (A), Disturbance to birds (A), Drainage (B), Extraction industry (B), Filling-in of wetlands (C), Groundwater abstraction (A), Industrialization/urbanization (A), Infrastructure (A), Recreation/tourism (A)

The IBA is particularly threatened by intensive water management and the intensification of agriculture. The growth of tourist activities such as wind-surfing and hunting also causes serious disturbance to birds. DOPPS and Euronatur cooperated to protect the Drava–Mura area (see also 'River Mura', IBA 013) between 1992 and 1994. A management plan for the area was prepared, publications were produced to increase public awareness of the area and its importance, a conference on the Drava–Mura Biosphere Reserve was held, and a local ornithological atlas project was initiated.

River Mura from Veržej to Gibina

B2, B3 **013**

Admin region Prekmurje, Pomurje
Coordinates 46°35′N 16°10′E
Altitude 159–234 m **Area** 14,060 ha

Site description

The 80-km stretch of the River Mura which flows through Slovenia. Seasonal flooding creates a characteristic landscape with numerous natural features, from gravel riverbanks to dense marshlands to old meanders along the lower stretch of the river, as well as huge alluvial forests. Črni log (1,200 ha) is the largest *Alnus* forest in central Europe, and is dense and uninhabited. Because of the danger of flooding, dykes have been built along the riverbanks. The Mura river meadows, wet grasslands, flood-plain and forests remain relatively intact, but cultivated fields and grasslands are intensively managed.

Habitats Forest and woodland (70%; alluvial/very wet forest), Scrub (20%), Grassland (20%; humid grassland), Wetland (30%; standing fresh water; river/stream; water-fringe vegetation)
Land-use Agriculture (25%), Forestry (70%), Urban/industrial/transport (5%)

Birds

Species	Season	Year	Pop min	Pop max	Acc	Criteria
Pernis apivorus Honey Buzzard	B	1995	60	100	B	B3
Falco tinnunculus Kestrel	B	1995	60	80	B	B2
Alcedo atthis Kingfisher	R	1994	60	100	A	B2
Picus canus Grey-headed Woodpecker	R	1994	150	200	A	B2
Dendrocopos medius Middle Spotted Woodpecker	R	1995	300	400	B	B3
Saxicola torquata Stonechat	B	1995	600	1,000	B	B2
Turdus merula Blackbird	R	1995	3,000	5,000	B	B3
Turdus pilaris Fieldfare	W	1994	3,000	6,000	A	B3
Turdus philomelos Song Thrush	B	1995	4,000	6,000	B	B3
Turdus viscivorus Mistle Thrush	R	1995	200	400	B	B3
Sylvia atricapilla Blackcap	B	1995	20,000	30,000	A	B3
Ficedula albicollis Collared Flycatcher	B	1995	900	1,700	A	B3
Certhia brachydactyla Short-toed Treecreeper	R	1995	1,500	2,000	A	B3
Lanius collurio Red-backed Shrike	B	1995	2,500	5,000	A	B2
Fringilla coelebs Chaffinch	R	1995	10,000	12,000	A	B3
Carduelis chloris Greenfinch	R	1995	3,000	4,000	A	B3

The site supports important breeding numbers of numerous species of European conservation concern that are typical of flooded forest, meadows, and riverbanks. The population densities of many species are high due to the relatively natural state of the alluvial forest habitats. Significant proportion (≥1%) of national population breeding at site: *Jynx torquilla* (30–50 pairs). The site also holds nationally important numbers of breeding *Ciconia ciconia* and riverine forest passerines.

Protection status

National None **International** None

Conservation issues

Threats Abandonment/reduction of land management (C), Agricultural intensification/expansion (C), Construction/impact of dyke/dam/barrage (C), Drainage (C), Dredging/canalization (A), Firewood collection (C), Infrastructure (C)

In 1985 the regional Nature Conservation Service prepared a proposal for some Landscape Parks in the area, but at the same time hydroelectric power-stations were planned, and the latter continue to be the main threat to the site. Remaining wet grasslands are threatened with drainage. Gravel extraction should be limited and some habitat restoration projects could be done. A planned highway between Slovenia and Hungary will probably cross the area. In 1992 DOPPS and Euronatur initiated the Drava–Mura nature conservation project (see also 'River Drava', IBA 012). Conservation issues were publicized, and at an international conference this IBA was proposed as a Biosphere Reserve.

Goričko

B2 **014**

Admin region Prekmurje
Coordinates 46°47′N 16°10′E
Altitude 174–416 m **Area** 45,500 ha

Site description

Goričko is in the north-easternmost part of Slovenia, on the border with Austria and Hungary. It is a mainly hilly region of Eocene sandstone, traversed by numerous streams, and with lower and flatter land in the eastern part where riverine vegetation and wet meadows are present. In summer the area is very dry. The soil in the east is not fertile, so farming is non-intensive, but is much more fertile in the west, where orchards, vineyards, dry grasslands and small cultivated fields are the main land-uses. Small settlements are common. The area is economically less developed and therefore habitats and wildlife remain in good condition.

Habitats Forest and woodland (15%; native coniferous forest; mixed forest), Scrub (10%; scrub), Grassland (30%; mesophile grassland), Wetland (15%; standing fresh water; river/stream; fen/transition mire/spring), Artificial landscape (30%; arable land; perennial crops/orchards/groves)
Land-use Agriculture (40%), Forestry (30%), Unknown (10%), Urban/industrial/transport (20%)

■ Birds

Species	Season	Year	Pop min	Pop max	Acc	Criteria
[1] ***Otus scops*** Scops Owl	B	1997	220	240	A	B2
Jynx torquilla Wryneck	B	1997	80	100	C	B2
Picus canus Grey-headed Woodpecker	R	1997	80	100	C	B2
Hirundo rustica Swallow	B	1997	1,500	2,000	C	B2
Saxicola torquata Stonechat	B	1997	150	200	C	B2
Lanius collurio Red-backed Shrike	B	1997	1,800	2,000	C	B2

1. One of the largest populations in central Europe.

The traditionally managed habitats support important numbers of several SPECs, as well as many more in less significant numbers, e.g. *Coracias garrulus*, *Lanius minor*, *Phoenicurus phoenicurus*, some of which may meet IBA criteria once quantitative surveys have been made.

■ Protection status

National None **International** None

■ Conservation issues

Threats Abandonment/reduction of land management (B), Agricultural intensification/expansion (B), Drainage (B), Industrialization/urbanization (C), Infrastructure (B)

The main threat is the abandonment of traditional, non-intensive agricultural land-use, which may lead to a decline in numbers of some birds, as would any intensification of farming (e.g. use of pesticides, drainage, and renovation of old orchards). The fauna and flora have not been surveyed in detail, and intensive field surveys are needed to provide quantitative and qualitative bird population data. The non-intensive farming seems of great importance in conserving old varieties of fruit trees and nationally threatened arable weeds.

REFERENCES

Anko, B. (1978) [About the new forest reserves in Slovenia.] *Varstvo narave* 11: 57–63. (In Slovenian.)

Bračko, F., Sovinc, A., Štumberger, B., Jančar, T., and Vogrin, M. (1994) [Red List of endangered breeding birds in Slovenia.] *Acrocephalus* 15(67): 166–180. (In Slovenian.)

Bračko, F. (1997) [Ornithological Atlas of the Drava River between Maribor and Ptuj for the 1989–1992 period]. *Acrocephalus* 18(82): 57–97. (In Slovenian.)

Economic Commission for Europe (1997) *Environmental Performance Reviews—Slovenia*. Series No. 2. New York and Geneva: United Nations.

Čelik, T. (1994) [Butterflies (Lep. Papilionidea in Hesperidae) as Bioindicator Group of Natural Conservation value of Planinsko polje.] University of Ljubljana, BTF (graduation thesis). (In Slovenian.)

Čelik, T. and Rebeušek, F. (1996) [*Atlas of threatened butterflies of Slovenia.*] Ljubljana: SED. (In Slovenian.)

Fabjan, I., ed. (1985) [*The Triglav National Park, field guide.*] First edition. Bled, Slovakia: Triglavski narodni park. (In Slovenian.)

Geister, I. (1995) [*Ornithological atlas of Slovenia—Breeding birds.*] Ljubljana: DZS. (In Slovenian.)

Gregori, J. (1975) [The birds of oak virgin forest and the immediate surroundings in Krakovian Forest.] *Acrocephalus* 13(52): 66–75. (In Slovenian.)

Gregori, J. (1976) Okvirni ekološki in favnistični pregled ptičev Sečoveljskih solin in bližnje okolice. *Varstvo narave* 9: 81–102.

Gregori, J. (1978) [A contribution to knowledge of birds of Cerknica Lake and its vicinity]. *Acta carsologica* 8: 301–329. (In Slovenian.)

Gregori, J. (1987) [Ecological and faunistic overview of birds of north-west Slovenia.] *Larus* 29/30: 33–81. (In Slovenian.)

Grimmett, R. F. A. and Jones T. A. (1989) *Important Bird Areas in Europe*. Cambridge, UK: International Council for Bird Preservation (Tech. Publ. 9).

Heath, M. F. and Borggreve, C. (2000) *BirdLife International/EBCC European Bird Database 1998*. Cambridge, UK: BirdLife International.

Jančar, T.(1995) [Monitoring of aquatic birds at Sečoveljske soline.] *Acrocephalus* 71: 108–112. (In Slovenian.)

Jančar, T. and Kmecl, P. (1996) [A review of current ornithological research in Triglav National Park (Julian Alps, northwestern Slovenia)]. *Avocetta* 20: 40–45.

Kmecl, P. and Rižner, K. (1993) [Survey of the waterfowl and birds of prey of Lake Cerknica with emphasis on their passage and wintering.] *Acrocephalus* 14(56/57): 4–31. (In Slovenian.)

Mlinšek, D. (1996) Silviculture in Slovenia is based on 'biodiversity'. *International Biodiversity Seminar, June 30–July 4, Gozd Martuljek, Slovenia*. Ljubljana.

MOP—URSVN (1997) *Convention on Biological Diversity—National Report of the Republic of Slovenia.* Ljubljana: MOP—URSVN.

Mršić, N. (1997) [*Biotic diversity in Slovenia. Slovenia—the 'hot spot' of Europe.*] Ljubljana: MOP—URSVN. (In Slovenian.)

Perušek, M. (1988) Vpliv nekaterih ekoloških dejavnikov na razporeditev in gostoto ptic v gozdu. Univerza v Ljubljani, BTF. (Graduation thesis.)

Perušek, M. (1992) [Birds of Rajhenavski Rog and Pečka, the remains of Slovene virgin forests.] *Acrocephalus* 12(49): 124–136. (In Slovenian.)

Polak, S. (1993) [Breeders of Lake Cerknica and its vicinity.] *Acrocephalus* 14(56/57): 32–62. (In Slovenian.)

Schneider-Jacoby, M. (1995) *Drau und Mur—Leben durch Flußdynamik*. Germany: European Natural Heritage Fund (Euronatur)/Jürgen Resch Verlag.

Schneider-Jacoby, M. and Thilcke, G. (1992) [*Projeckt Cerknica lake Radolfzell.*] Germany: European Natural Heritage Fund (Euronatur). (In Slovenian and German.)

Sket, B. (1997) [The cave fauna in the triangle Cerknica—Postojna—Planina; its conservational importance.] *Varstvo narave* 12: 45–59. (In Slovenian.)

Skobrne, P. and Petrlin, S. (1988) *Inventar najpomembnejše naravne dediščine Slovenije*. 1 del. Ljubljana: ZRSVNKD.

Skobrne, P. and Petrlin, S. (1991) *Inventar najpomembnejše naravne dediščine Slovenije*. 2 del. Ljubljana: ZRSVNKD.

Smerdu, R. (1979) [The Planinsko polje as a part of the most significant heritage of Slovenia.] *Varstvo narave* 12: 65–73. (In Slovenian.)

Sovinc, A., Tome, D. and Trontelj, P. (1993) [Ornithological atlas of Ljubljansko barje—preliminary report.] *Acrocephalus* 14(60): 145–151. (In Slovenian.)

Sovinc, A. (1994) [*The atlas of wintering birds in Slovenia (ZOAS)*]. First edition. Ljubljana: Tehniška založba Slovenije. (In Slovenian.)

Škornik, I. (1987) Pomembno ornitološko območje Evrope—Sečoveljske soline. Koper *Falco* 2: 3–14.

Škornik, I. s sod. (1990) Favnistični pregled ptic slovenske obale. Ljubljana, *Varstvo narave* 16: 49–99.

Škornik, I., Makovec, T. and Lipej, L. (1995) Sečoveljske soline—ornitološka ocena slovenskih priobalnih mokrišč. Koper. *Annales* 7: 89–94. (In Slovenian, with Italian summary.)

Štrumbelj, C. (1992) [Large beasts in Kočevski Rog.] *Dolenjski zbornik*: 204–212. (In Slovenian.)

Štumberger, B. (1992) Avifauna der Slowenischen Drau und deren Schutz. Internationale Drau Konferenz. *Vortrag anläßlich der Mur–Drau–Donau Konferenz*, Kaposvar. SEN (EURONATUR). (In German.)

Štumberger, B. (1995) [The Drava river between Maribor and Središče ob Dravi—the sphere of conflict between nature conservation and development policy]. *Acrocephalus* 16(68/69/70): 3–43. (In Slovenian.)

Trontelj, P. (1993) [About the conservational concept of Lake Cerknica with emphasis on protection of birds.] *Acrocephalus* 14(56/57): 63–80. (In Slovenian.)

Trontelj, P. (1994) [The importance of birds as ecological indicators on the Ljubljansko Barje (Slovenia).] *Scopolia* 32: 1–61. (In Slovenian.)

Trontelj, P. (1995) [Census of the Corn Crake *Crex crex* in Slovenia in 1992–1993.] *Acrocephalus* 16(73): 174–180.

Tucker, G. M. and Heath, M. F. (1994) *Birds in Europe: their conservation status*. Cambridge, UK: BirdLife International (BirdLife Conservation Series no. 3).

SPAIN
(INCLUDING THE CANARY ISLANDS)

Carlota Viada

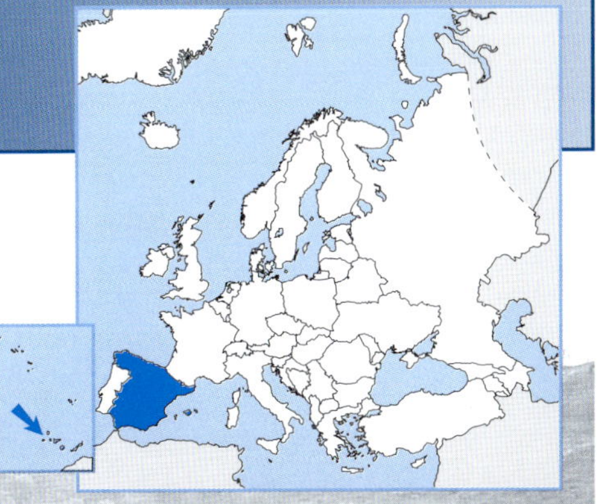

Spanish dehesas. (PHOTO: FERNANDO BARRIO, SEO/BIRDLIFE)

GENERAL INTRODUCTION

Spain is one of the largest states in Europe. Most of the country (97.5%, 492,463 km^2) lies on the Iberian peninsula, at the southern tip of Europe. In addition, Spanish territory includes the Balearic Islands in the Mediterranean Sea (5,014 km^2), the Canary Islands in the Atlantic Ocean (7,447 km^2), and the two small territories of Ceuta and Melilla on the North African coast.

The Iberian peninsula comprises several distinct regions, which gives rise to a great variety of habitats across the country. The north of the peninsula includes the Pyrenees and Cantabric mountain ranges together with the wide northern plateau, the Iberian mountain range and the valley of the River Ebro. The central region is dominated by the Central mountain range which lies to the north of the southern plateau. The southern third of the peninsula is occupied by the Sierra Morena mountain range, the Guadalquivir valley and the Betic mountain ranges. The Mediterranean coast is dominated by sand beaches, and includes two large deltas, those of the Rivers Llobregat and Ebro. It is separated from inland regions by a coastal mountain range.

The Balearic Islands (four main islands and a number of small islands and islets) are surrounded by sea cliffs and beaches; only Mallorca, the largest island, has a significant mountain range. The territory of Ceuta is on the African side of the Strait of Gibraltar and Melilla (Chafarinas Islands) on the Mediterranean coast of Morocco. The Canary Islands (seven main islands) are mountainous (the Pico del Teide, 3,718 m, is the highest peak in Spain) with deep ravines, cliffs and precipices.

Spain has 391 Important Bird Areas (IBAs), covering 157,689 km^2 (32% of Spanish territory). Of these, the Balearics hold 20 IBAs (1,312 km^2), the Canaries 65 (1,334 km^2), Ceuta a single IBA covering all its territory (1,900 ha) and Melilla (Chafarinas Islands) one IBA (50 ha) (Table 1, Maps 1 and 2).

In the previous pan-European inventory (Grimmett and Jones 1989), 352 IBAs were identified, covering 95,000 km^2. In 1992, SEO/BirdLife reviewed the inventory for the Iberian peninsula and the Balearics, increasing the number of IBAs to 386 and the area covered to 134,289 km^2. A second national inventory was published in 1998, with the number of sites being identical to the present inventory (Viada 1998). The area covered by the present inventory and by the second national inventory has increased significantly as a result of the identification of new IBAs, an increase in the area of some existing IBAs and also improved accuracy of measurement.

Ninety-six percent of IBAs from the previous pan-European inventory (Grimmett and Jones 1989) are confirmed in the present inventory. Fifteen (4%) IBAs, identified for both the previous European and national inventories, are not confirmed. Sixty-seven new IBAs are identified, mainly as a result of increased effort and improved knowledge of sites and their birds (Table 1).

ORNITHOLOGICAL IMPORTANCE

There are 175 species of European conservation concern (SPECs 1–4) breeding regularly in Spain (Tucker and Heath 1994). Of these, eleven are globally threatened (*Marmaronetta angustirostris, Aythya nyroca, Oxyura leucocephala, Aquila adalberti, Falco naumanni, Crex crex, Otis tarda, Larus audouinii, Columba bollii, C. junoniae, Fringilla teydea*), two are near-threatened (*Tetrax tetrax, Saxicola dacotiae*) and a further 162 breeding species are SPECs 2–4 (Tucker and Heath 1994). Two non-breeding globally threatened species occur: *Numenius tenuirostris* (wintering) and *Acrocephalus paludicola* (passage).

One hundred and forty-six sites were selected as IBAs on the basis of internationally important breeding, wintering or passage populations for thirteen globally threatened or near-threatened species, meeting criterion A1 (Table 2). Of this group of species, Spain holds the entire European population or a major proportion of the European breeding populations of *Aegypius monachus, Aquila adalberti, Falco naumanni, Otis tarda, Tetrax tetrax, Larus audouinii, Columba bollii, C. junoniae, Saxicola dacotiae* and *Fringilla teydea*.

Map 1. Location, area and criteria category of Important Bird Areas in Spain.

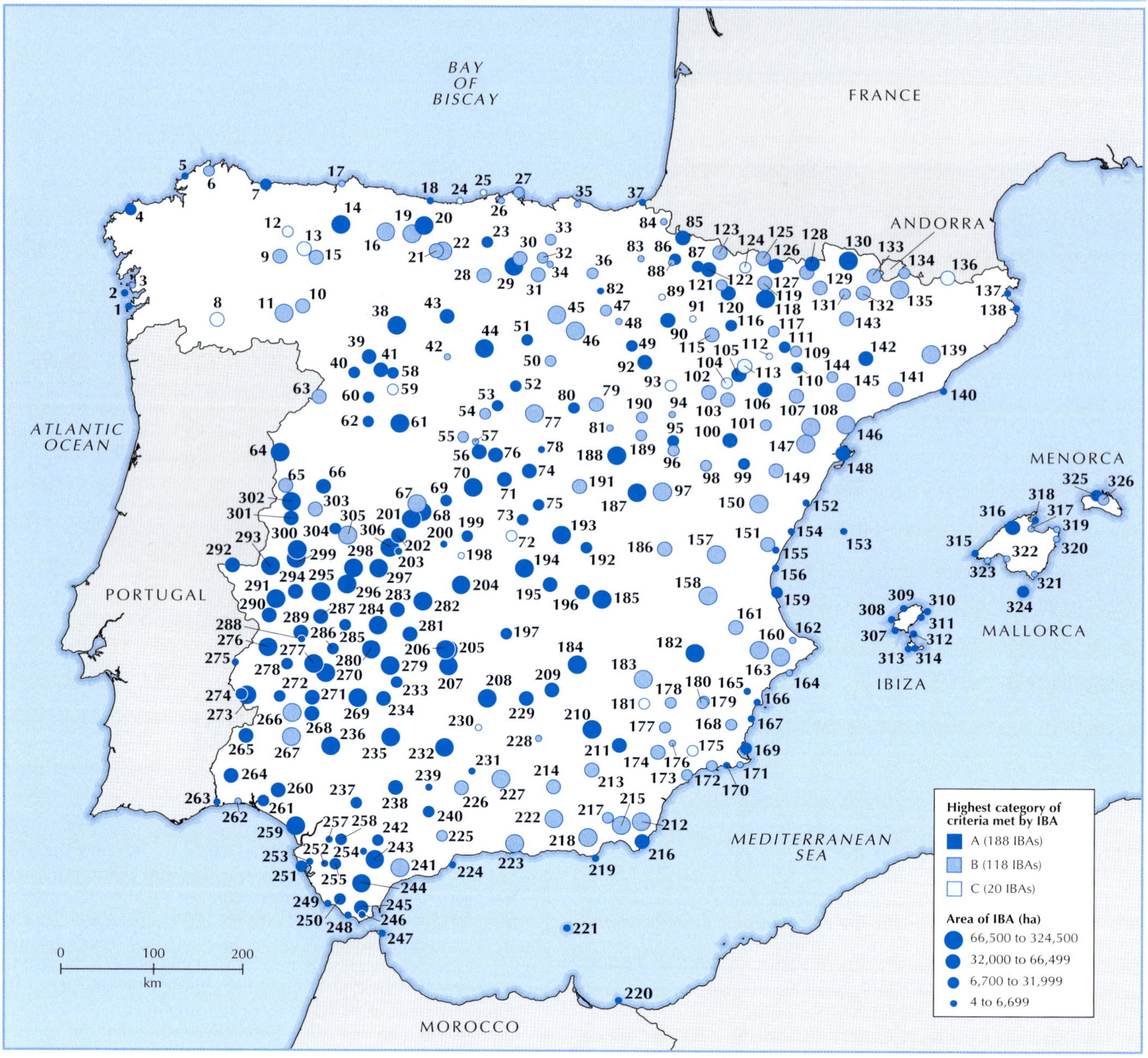

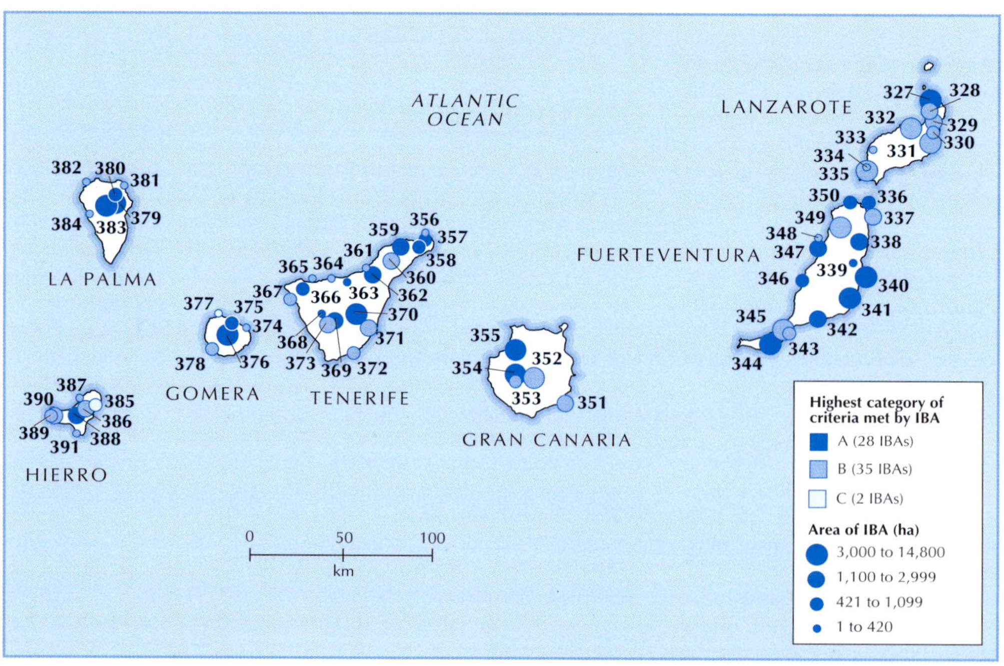

Map 2. Location, area and criteria category of Important Bird Areas in the Canary Islands.

Table 1. Summary of Important Bird Areas in Spain. 391 IBAs covering 157,689 km²

IBA code	National code	1989 code	International name	National name	Administrative region	Area (ha)	Criteria (see p. 11)
001	001	ES001	Cíes islands	Islas Cíes	Galicia	433	A4i, B1i, B3, C2, C3, C6
002	002	ES002	Ons islands	Islas Ons	Galicia	500	A4i, B1i, B3, C3
003	003	ES003	Ría de Arosa (O Grove) estuary	Ría de Arosa (O Grove)	Galicia	2,561	B1i, B2, C2
004	004	ES005, ES006	Costa de la Muerte (North coast)	Costa de la Muerte (Norte)	Galicia	9,650	A1, A4i, B1i, B2, C1, C2, C3, C6
005	005	—	Ferrolterra-Valdoviño coast	Costa de Ferrolterra-Valdoviño	Galicia	4,400	A4i, B1i, C3, C6
006	006	ES007	Cape Candelaria–Ortigueira estuary–Cape Estaca de Bares	Punta Candelaria–Ría de Ortigueira–Estaca de Bares	Galicia	6,700	B2, C7
007	007	ES008	Eo river estuary–Barayo beach–Ría de Foz	Ría del Eo–Playa de Barayo–Ría de Foz	Galicia, Asturias	7,500	A4i, B1i, B2, C3, C6
008	008	ES010	La Limia	La Limia	Galicia	32,800	C6
009	009	ES011	Ancares mountains	Sierra de los Ancares	Galicia, Castilla y León	50,600	B2, C6
010	010	ES027	Montes Aquilanos range	Montes Aquilanos	Galicia, Castilla y León	55,000	B2, B3, C6
011	011	ES026	La Cabrera mountain range	Sierra de La Cabrera	Galicia, Castilla y León	73,700	B2, C6
012	012	ES012	Muniellos	Muniellos	Asturias	24,100	C6
013	013	ES013	Degaña–Hermo	Degaña–Hermo	Asturias	32,000	C6
014	014	ES014	Babia–Somiedo	Babia–Somiedo	Asturias, Castilla y León	166,860	A3, B1iii, B2, B3, C2, C6
015	015	ES015	Gistreo and Coto mountain ranges	Sierras de Gistreo y Coto	Castilla y León	59,000	B2, B3, C6
016	016	ES016, ES017, ES018	Central ranges of the Cantabric mountains	Sierras Centrales de la Cordillera Cantábrica	Asturias, Castilla y León	117,500	B2, C6
017	017	ES009	Cape Busto–Luanco	Cabo Busto–Luanco	Asturias	6,000	B1i, C3, C6
018	018	ES297	Ribadesella–Tina Mayor	Ribadesella–Tina Mayor	Asturias	3,600	A4i, B1i, B2, C3, C6
019	019	ES019	Riaño	Riaño	Castilla y León	70,000	B2, B3, C2, C6
020	020	ES020	Picos de Europa mountain range	Picos de Europa	Asturias, Cantabria, Castilla y León	132,000	A3, B1iii, B2, B3, C2, C6
021	021	ES021	Fuentes Carrionas	Fuentes Carrionas	Castilla y León	66,000	B2, C6
022	022	ES022	Peña Labra and Cordel mountain ranges	Sierras de Peña Labra y del Cordel	Cantabria, Castilla y León	110,000	B2, B3, C6
023	023	ES023	River Ebro dam	Embalse del Ebro	Cantabria, Castilla y León	11,600	A4i, B1i, B2, C3
024	024	—	Tina Menor–Sarnosa island–Oyambre	Tina Menor–Isla Sarnosa–Oyambre	Cantabria	1,100	C6
025	025	—	Portios islets–Conejera island	Islotes de Portios–Isla Conejera	Cantabria	10	C6
026	026	—	Santander bay–Mouro islet	Bahía de Santander–Isla de Mouro	Cantabria	3,300	B1i, C2, C6
027	027	ES024	Santoña marshes	Marismas de Santoña	Cantabria	6,907	B1i, B2, C2, C3
028	028	ES028	Humada–Peña Amaya	Humada–Peña Amaya	Castilla y León	41,000	B1iii, B2, C2
029	029	ES029, ES030	Upper Ebro and Rudrón canyons	Hoces del Alto Ebro y Rudrón	Castilla y León	74,500	A4ii, B1iii, B2, C2, C6
030	030	ES032	Oña and Tesla mountain ranges	Sierras de Oña y de la Tesla	Castilla y León	37,800	B2, C2, C6
031	031	ES035	Obarenes mountains	Montes Obarenes	Castilla y León, La Rioja	35,000	B2, C2, C6
032	032	—	Valdegovía-Arcena mountain range	Valdegovía–Sierra de Arcena	Castilla y León, País Vasco	20,000	B1iii, B2, C2, C6
033	033	ES033	La Losa–Orduña (Salvada mountain range)	La Losa–Orduña (Sierra Salvada)	Castilla y León, País Vasco	20,000	B1iii, B2, C2, C6
034	034	ES034	Sobrón canyon	Hoz de Sobrón	Castilla y León, País Vasco	4,800	B1iii, B2, C2, C6
035	035	ES025	Guernica estuary–Cape Ogoño	Ría de Guernica–Cabo de Ogoño	País Vasco	4,000	B1i, C2, C6
036	036	—	Izki mountains	Montes de Izki	País Vasco	9,250	B3, C2, C6
037	037	—	Chingudi–Bidasoa estuary	Chingudi–Estuario del Bidasoa	País Vasco	525	A1, A4i, A4ii, B1i, B1ii, B2, C1, C2, C3
038	038	ES036, ES037	Tierra de Campos steppes	Tierra de Campos	Castilla y León	268,000	A1, A4i, A4ii, B1i, B1iii, B2, B3, C1, C2, C3, C6
039	039	ES038	Villafáfila	Villafáfila	Castilla y León	32,682	A1, A4i, A4ii, B1i, B1iii, B2, C1, C2, C3, C6
040	040	ES301	Esla dam	Embalse del Esla	Castilla y León	22,300	A1, C1
041	041	ES039	Belver de los Montes–Gallegos del Pan	Belver de los Montes–Gallegos del Pan	Castilla y León	40,300	A1, B2, C1, C2, C6
042	042	—	River Pisuerga at Dueñas	Río Pisuerga en Dueñas	Castilla y León	2,300	B2, C6
043	043	—	Carrión–Frómista	Carrión–Frómista	Castilla y León	58,500	A1, B2, C1, C2, C6
044	044	ES059	Cerrato high moor	Páramos del Cerrato	Castilla y León	108,000	A1, B2, C1, C2, C6
045	045	ES139	La Demanda mountains	Sierra de la Demanda	Castilla y León, La Rioja	133,000	B2, C6
046	046	ES140	Urbión and Cebollera mountain ranges	Sierras de Urbión y Cebollera	Castilla y León, La Rioja	142,000	B2, B3, C6
047	047	ES141	Gorges of Iregua, Leza and Jubera	Hoces del Iregua, el Leza y el Jubera	La Rioja	23,400	B1iii, B2, C2, C6
048	048	ES142	Arnedillo-Isasa crag	Arnedillo–Peña Isasa	La Rioja	5,530	B1iii, B2, C2, C6
049	049	ES143	Alcarama mountain and Alhama river	Sierra de Alcarama y Río Alhama	Castilla y León, La Rioja	10,500	A4ii, B1iii, B2, C2, C6
050	050	ES057	River Lobos canyon	Cañón del Río Lobos	Castilla y León	9,580	B1iii, B2, C2
051	051	ES058	Arlanza valley–Peñas de Cervera	Valle del Arlanza–Peñas de Cervera	Castilla y León	26,000	A4ii, B1iii, B2, C2, C6
052	052	ES054	Montejo de la Vega–Riaza canyon	Montejo de la Vega–Hoces del Riaza	Castilla y León	9,400	A4ii, B1iii, B2, C2, C6
053	053	ES053	Duratón canyon	Cañón del Duratón	Castilla y León	8,800	A4ii, B1iii, B2, C2, C6
054	054	—	River Cega–Tierra de Pinares–Cantalejo	Río Cega–Tierra de Pinares–Cantalejo	Castilla y León	25,500	B2, B3, C6
055	055	ES051	River Moros	Río Moros	Castilla y León	11,500	B2, B3, C6
056	056	ES062	Northern slope of Guadarrama mountain range	Umbría de Guadarrama	Castilla y León	54,500	A1, A4ii, B1iii, B2, C1, C2, C6

Table 1 ... continued. Summary of Important Bird Areas in Spain. 391 IBAs covering 157,689 km²

IBA code	National code	1989 code	International name	National name	Administrative region	Area (ha)	Criteria (see p. 11)
057	057	ES052	Segovia	Segovia	Castilla y León	2,200	B1i, B2, C2
058	058	ES040	Tordesillas–Mota del Marqués	Tordesillas–Mota del Marqués	Castilla y León	20,000	A1, C1, C6
059	059	ES041	Castronuño–Zamora	Castronuño–Zamora	Castilla y León	19,500	C6
060	060	ES045	Fuentelapeña–Jambrina	Fuentelapeña–Jambrina	Castilla y León	25,000	A1, B2, C1, C2, C6
061	061	ES042, ES043	Tierra de Campiñas steppes	Tierra de Campiñas	Castilla y León	189,600	A1, A4ii, B1i, B1iii, B2, B3, C1, C2, C6
062	062	—	Topas	Topas	Castilla y León	28,000	A1, B2, C1, C2
063	063	ES046	Arribes del Duero–Fermoselle	Arribes del Duero–Fermoselle	Castilla y León	53,000	B1i, B1iii, B2, C2, C6
064	064	ES047	Río Huebra–Arribes del Duero	Río Huebra–Arribes del Duero	Castilla y León	170,000	A4ii, B1i, B1iii, B2, C2, C6
065	065	ES231	Campo de Argañán	Campo de Argañán	Castilla y León	47,000	B1i, B2, C2, C6
066	066	ES049	Peña de Francia	Peña de Francia	Castilla y León, Extremadura	65,500	A1, A4ii, B1i, B1iii, B2, C1, C2, C6
067	067	ES050	Gredos and Candelario mountain ranges	Sierras de Gredos y Candelario	Castilla y León, Extremadura	173,600	B1i, B2, B3, C2, C6
068	068	—	Tiétar valley	Valle del Tiétar	Castilla y León	86,000	A1, B1i, B2, C1, C2, C6
069	069	ES050	Iruelas valley	Valle de Iruelas	Castilla y León	8,440	A1, A4ii, B1iii, B2, C1, C2, C6
070	070	ES061	El Escorial–San Martín de Valdeiglesias	El Escorial–San Martín de Valdeiglesias	Castilla y León, Comunidad de Madrid	198,400	A1, B1i, B2, C1, C2, C6
071	071	ES063	El Pardo–Viñuelas	El Pardo–Viñuelas	Comunidad de Madrid	34,000	A1, A3, B1i, B2, C1, C2, C6
072	072	—	Redbeeds and copses of Aranjuez	Carrizales y Sotos de Aranjuez	Comunidad de Madrid	18,000	C6
073	073	ES065	Cortados del Jarama	Cortados del Jarama	Comunidad de Madrid	25,000	A1, A4i, B1i, B2, C1, C2, C6
074	074	ES064	Talamanca–Camarma	Talamanca–Camarma	Comunidad de Madrid, Castilla-La Mancha	52,000	A1, B2, B3, C1, C2, C6
075	075	—	Alcarria de Alcalá steppes	Alcarria de Alcalá	Comunidad de Madrid	8,000	A1, C1, C6
076	076	ES062	Alto Lozoya–La Pedriza	Alto Lozoya–La Pedriza	Comunidad de Madrid	55,000	A1, B1iii, B2, C1, C2, C6
077	077	ES066	Ayllón mountain range	Sierra de Ayllón	Castilla y León, Comunidad de Madrid, Castilla-La Mancha	187,000	B2, C6
078	078	ES067	Puebla de Beleña lagoons	Puebla de Beleña	Castilla-La Mancha	3,100	A4i, B1i, B1iv, C2, C5
079	079	ES055	Tiermes–Caracena	Tiermes–Caracena	Castilla y León	45,300	B1iii, B2, C2, C6
080	080	ES056	Altos de Barahona high moor	Altos de Barahona	Castilla y León	29,000	A1, B2, C1, C2, C6
081	081	ES070	Layna high moors	Páramos de Layna	Castilla y León, Castilla-La Mancha	2,100	B2, C2, C6
082	082	ES144	Las Cañas lake	Laguna de las Cañas	Navarra	101	A4i, B1i, B2, C2, C6
083	083	ES148	Echauri crags	Peñas de Echauri	Navarra	1,400	B2, C6
084	084	ES151	Gorramendi mountain and Itxusi crags	Monte Gorramendi y Peñas de Itxusi	Navarra	6,300	B2, C6
085	085	ES152	Roncesvalles–Irati–Abodi mountain range	Roncesvalles–Irati–Sierra de Abodi	Navarra	39,000	A4i, A4iv, B1i, B1iv, B2, C2, C5, C6
086	086	ES149	Areta, Archuba and Zariquieta mountain ranges	Montes de Areta, Archuba y Zariquieta	Navarra	19,500	A4ii, B1iii, B2, C2, C6
087	087	ES163	Leyre, Illón and San Miguel mountain ranges	Sierras de Leyre, Illón y San Miguel	Navarra	23,500	A4ii, B1iii, B2, C2, C6
088	088	ES147	Izaga crag	Peña Izaga	Navarra	1,746	B1iii, B2, C2, C6
089	089	ES145	Pitillas and Dos Reinos lagoons	Lagunas de Pitillas y Dos Reinos	Navarra, Aragón	260	C6
090	090	ES145	Bardenas Reales	Bardenas Reales	Navarra, Aragón	55,900	A3, B1iii, B2, C2, C6
091	091	—	Cinco Villas ponds	Carrizales y estancas de las Cinco Villas	Aragón	860	C6
092	092	ES188	Moncayo mountain range	Sierra del Moncayo	Aragón, Castilla y León	47,600	A4ii, B1iii, B2, C2, C6
093	093	ES190	River Jalón canyon	Hoces del Jalón	Aragón	23,500	C6
094	094	—	Langa–Mara–Belmonte de Calatayud plains	Llanos de Langa–Mara–Belmonte de Catalayud	Aragón	6,100	B2, C2, C6
095	095	ES192	Gallocanta lake	Laguna de Gallocanta	Aragón	30,280	A1, A4i, A4iv, B1i, B1iv, B2, C1, C2, C3, C5, C6
096	096	ES193	Blancas–Torralba de los Sisones	Blancas–Torralba de los Sisones	Aragón	7,850	B2, C2, C6
097	097	ES198	Montes Universales–Albarracín mountain ranges	Montes Universales–Sierra de Albarracín	Aragón, Castilla-La Mancha	80,300	B2, C6
098	098	ES194	Campo Visiedo	Campo Visiedo	Aragón	18,000	B2, C2, C6
099	099	ES197	River Guadalope	Río Guadalope	Aragón	31,800	A4ii, B1iii, B2, C2, C6
100	100	ES195, ES196	River Martín canyons and Arcos mountain range	Cañones del Río Martín y Sierra de Arcos	Aragón	38,800	A4ii, B1iii, B2, C2, C6
101	101	ES187	Alcañiz salt-lakes	Saladas de Alcañiz	Aragón	7,360	B1i, C2, C6
102	102	ES185	Lower course of the River Huerva	Bajo Huerva	Aragón	37,000	B2, C6
103	103	ES186	Belchite–Mediana	Belchite–Mediana	Aragón	53,000	B2, C2, C6
104	104	ES182	River Ebro oxbow lakes	Galachos del Río Ebro	Aragón	7,400	C6
105	105	ES177	Monegrillo–Pina steppe area–Pina	Estepas de Monegrillo–Pina	Aragón	43,100	A1, B2, C1, C2, C6
106	106	ES178	Los Monegros (South)	Los Monegros (Sur)	Aragón	49,000	A1, A4ii, B1iii, B2, C1, C2, C6
107	107	ES183	Valdurrios–Serreta Negra and Los Rincones mountain ranges	Sierras de Valdurrios–Serreta Negra y Los Rincones	Aragón	49,500	B2, C6
108	108	—	Lower course of the River Matarraña–Ribarroja	Bajo Matarraña–Ribarroja	Aragón, Cataluña	73,500	B1i, B2, C2, C6

Table 1 ... continued. Summary of Important Bird Areas in Spain. 391 IBAs covering 157,689 km²

IBA code	National code	1989 code	International name	National name	Administrative region	Area (ha)	Criteria (see p. 11)
109	109	—	Rivers Cinca and Alcanadre riverine forest	Sotos de los ríos Cinca y Alcanadre	Aragón, Cataluña	9,450	B1i, C2, C6
110	110	ES179	Ballobar–Candasnos	Ballobar–Candasnos	Aragón	26,000	A1, B2, C1, C2, C6
111	111	—	Cinca river rice-fields and steppe area	Arrozales y estepas del Cinca medio	Aragón	14,400	A4i, B1i, B2, C2, C6
112	112	—	Sariñena lake and La Estación pond	Laguna de Sariñena y Balsa de la Estación	Aragón	610	C6
113	113	ES175	Alcubierre and Sigena mountain ranges	Sierras de Alcubierre y Sigena	Aragón	64,500	C6
114	114	—	Confidential	Confidencial	Aragón	4,600	C6
115	115	ES174	Zuera mountains	Montes de Zuera	Aragón	35,000	B2, C6
116	116	ES173	Tormos reservoir (La Sotonera)	Embalse de Tormos (La Sotonera)	Aragón	7,500	A4i, A4iv, B1i, B1iv, C2, C5
117	117	ES176	Bajo Alcanadre–Serreta de Tramaced	Bajo Alcanadre–Serreta de Tramaced	Aragón	17,000	B1iii, B2, C2
118	118	ES166	Guara mountain range	Sierra de Guara	Aragón	95,000	A4ii, B1iii, B2, B3, C2, C6
119	119	ES167	Oturia–Canciás	Oturia–Canciás	Aragón	33,520	B2, C6
120	120	ES165	Santo Domingo–Riglos–Gratal	Santo Domingo–Riglos–Gratal	Aragón	45,000	A4ii, B1iii, B2, B3, C2, C6
121	121	ES164	San Juan de la Peña–Oroel crag	San Juan de la Peña–Peña Oroel	Aragón	22,400	B2, B3, C6
122	122	ES150	Dos Ríos and Orba mountain ranges	Sierras de los Dos Ríos y de Orba	Aragón	34,700	A4ii, B1iii, B2, C2, C6
123	123	ES153	Belagua–Ansó–Hecho	Belagua–Ansó–Hecho	Navarra, Aragón	44,500	B1iii, B2, C2, C6
124	124	ES154	Collarada–Telera	Collarada–Telera	Aragón	21,000	C6
125	125	ES155	Panticosa–Vignemale–Tendeñera	Panticosa–Viñamala–Tendeñera	Aragón	32,000	B2, C6
126	126	ES156	Ordesa–Bielsa	Ordesa–Bielsa	Aragón	43,000	A3, B2, C6
127	127	ES157	Gistaín–Cotiella	Gistaín–Cotiella	Aragón	42,000	B2, C6
128	128	ES158	Posets–Maladeta	Posets–Maladeta	Aragón	44,700	A3, B2, C6
129	129	ES168	Turbón–Espés–Sis	Turbón–Espés–Sis	Aragón	43,000	B2, B3, C2, C6
130	130	ES159	San Mauricio–Bohí–Beret	San Mauricio–Bohí–Beret	Cataluña	120,000	A3, B2, C2, C6
131	131	ES169	Sant Gervás mountains	Sierra de Sant Gervàs	Cataluña	24,500	B2, C2, C6
132	132	ES170	Boumort mountains	Sierra del Boumort	Cataluña	54,000	B1iii, B2, C2, C6
133	133	ES160	Monteixo–L'Orri–Tornafort	Monteixo–L'Orri–Tornafort	Cataluña	65,000	B2, C6
134	134	ES161	Cerdaña	Cerdaña	Cataluña	22,000	B2, C6
135	135	ES171	Cadí mountains	Sierra del Cadí	Cataluña	69,600	B1iv, B2, C5, C6
136	136	ES162	Fresser–Setcases	Fresser–Setcases	Cataluña	38,500	C6
137	137	ES199	Ampurdán marshes	Aiguamolls del Ampurdán	Cataluña	5,454	A4i, B1i, B2, C2, C3, C6
138	138	ES200	Medas islands	Islas Medas	Cataluña	21	A4i, B1i, C3, C6
139	139	—	Mountains of Barcelona	Sierras Prelitorales de Barcelona	Cataluña	124,000	B1i, B2, B3, C2, C3, C6
140	140	—	Llobregat delta	Delta del Llobregat	Cataluña	2,200	A1, A4i, A4ii, B1i, B1ii, B2, C1, C2, C3, C6
141	141	—	Montagut and Montmell mountains	Sierras de Montagut y Montmell	Cataluña	44,300	B2, C6
142	142	—	Lerida steppes	Secanos de Lérida	Cataluña	62,500	A1, B1iii, B2, C1, C2, C6
143	143	ES172	Montsech and Montgai mountains	Sierras del Montsech y Montgai	Aragón, Cataluña	49,000	B2, C2, C6
144	144	ES181	Cogul–Alfés steppes	Cogul–Alfés	Cataluña	18,000	B2, C2, C6
145	145	ES201	Montsant and Prades mountains	Sierras del Montsant y de Prades	Cataluña	80,000	B2, C2, C6
146	146	—	Cardó, Tivissa and Llabería mountains	Sierras de Cardó, Tivissa y Llaberia	Cataluña	68,500	B2, C2, C6
147	147	ES203	Pass of Beceite–Turmell mountain	Puertos de Beceite–Monte Turmell	Aragón, Cataluña, País Valenciano	107,000	B2, C6
148	148	ES202	Ebro delta	Delta del Ebro	Cataluña	32,000	A1, A4i, B1i, B2, B3, C1, C2, C3, C6
149	149	ES204	Morella pass	Puertos de Morella	País Valenciano	43,000	B2, C6
150	150	ES205	Peñagolosa	Peñagolosa	Aragón, País Valenciano	83,000	B2, B3, C6
151	151	ES206	Espadán mountain range	Sierra de Espadán	País Valenciano	59,400	B2, C6
152	152	ES211	Prat de Cabanes–Torreblanca	Prat de Cabanes–Torreblanca	País Valenciano	860	A1, A4i, B1i, B2, C1, C2, C6
153	153	ES210	Columbretes islands	Islas Columbretes	País Valenciano	2,500	A1, A4i, B1i, B2, C1, C2, C6
154	154	ES212	River Mijares mouth	Desembocadura del Río Mijares	País Valenciano	375	A4i, B1i, C2, C3, C6
155	155	ES213	Almenara marshes	Lagunas y Marjal de Almenara	País Valenciano	3,000	A4i, B1i, B2, C2, C3, C6
156	156	—	El Moro marshes	Marjal de El Moro	País Valenciano	350	A4i, B1i, B2, C2, C3, C6
157	157	ES207	Turia canyon and Los Serranos	Hoces del Turia y Los Serranos	Castilla-La Mancha, País Valenciano	135,000	B2, B3, C2, C6
158	158	ES208	Rivers Cabriel and Júcar canyons	Hoces del Cabriel y del Júcar	Castilla-La Mancha, País Valenciano	163,000	B2, C2, C6
159	159	ES214	Albufera de Valencia marshes	Albufera de Valencia	País Valenciano	21,120	A1, A4i, B1i, B2, C1, C2, C3, C6
160	160	—	La Safor and North Alicante mountain ranges	Sierras de La Safor y Norte de Alicante	País Valenciano	102,000	B2, B3, C6
161	161	—	Enguera mountain range–La Canal de Navarrés	Sierrra de Enguera–La Canal de Navarrés	País Valenciano	52,500	B2, C6
162	162	—	Pego–Oliva marshes	Marjales de Pego–Oliva	País Valenciano	1,290	B1i, B2, C2, C6
163	163	ES209	Marina mountain ranges	Sierras de la Marina	País Valenciano	102,300	B2, C2, C6
164	164	—	Alicante islets	Islotes de Alicante	País Valenciano	40	B1ii, B2, C2, C6
165	165	ES215	El Hondo wetland	El Hondo	País Valenciano	2,387	A1, A4i, B1i, B2, C1, C2, C3, C6

Table 1 ... continued. Summary of Important Bird Areas in Spain. 391 IBAs covering 157,689 km²

IBA code	National code	1989 code	International name	National name	Administrative region	Area (ha)	Criteria (see p.11)
166	166	ES216	Santa Pola saltpans	Salinas de Santa Pola	País Valenciano	2,496	A1, A4i, B1i, B2, B3, C1, C2, C3, C6
167	167	ES217	Mata and Torrevieja lagoons	Lagunas de la Mata y Torrevieja	País Valenciano	3,755	A1, A4i, B1i, B2, C1, C2, C3, C6
168	168	—	El Valle, Altaona and Escalona mountains	Monte El Valle y Sierras de Altaona y Escalona	Murcia, País Valenciano	24,000	B1iii, C2, C6
169	169	ES218	Mar Menor coastal lagoon	Mar Menor	Murcia	14,933	A1, A4i, B1i, B2, B3, C1, C2, C3, C6
170	170	ES219	Littoral islets of Murcia and Almería	Islotes litorales de Murcia y Almeria	Andalucía, Murcia	300	A1, A4i, B1i, B2, C1, C2, C6
171	171	—	Fausilla mountains	Sierra de la Fausilla	Murcia	900	B2, C6
172	172	—	Muela mountain range and Cape Tiñoso	Sierra de la Muela y Cabo Tiñoso	Murcia	11,700	B2, C6
173	173	—	La Almenara mountain range–Cape Cope	Sierra de La Almenara–Cabo Cope	Murcia	25,850	B2, C6
174	174	—	Torrecilla and Gigante mountain ranges	Sierra de la Torrecilla y del Gigante	Murcia	42,000	B2, C6
175	175	ES222	Guadalentín saltmarshes	Saladares del Guadalentín	Murcia	6,800	C6
176	176	—	Las Cabras plain	Llanos de las Cabras	Murcia	1,300	B2, C6
177	177	—	Burete, Cambrón and Espuña mountain ranges	Sierras de Burete, del Cambrón y de Espuña	Murcia	47,000	B2, C6
178	178	ES221	Quipar reservoir, Molino mountain range and Cagitán plains	Embalse de Quipar, Sierra del Molino y Llanos del Cagitán	Murcia	22,000	B2, C6
179	179	—	Ricote and Nevela mountain ranges	Sierras de Ricote y Nevela	Murcia	9,500	B2, C6
180	180	—	Pila mountain range	Sierra de la Pila	Murcia	8,900	B2, C6
181	181	ES220	Moratalla mountain range	Sierra de Moratalla	Murcia	21,000	C6
182	182	ES077	Pétrola–Almansa–Yecla	Pétrola–Almansa–Yecla	Murcia, Castilla-La Mancha	75,800	A1, B1i, B2, C1, C2, C6
183	183	ES078, ES079	River Mundo and River Segura canyons	Hoces del Río Mundo y del Río Segura	Castilla-La Mancha	78,000	B2, C6
184	184	ES080	Campo de Montiel	Campo de Montiel	Castilla-La Mancha	153,500	A1, B2, C1, C2, C6
185	185	ES081	San Clemente–Villarrobledo	San Clemente–Villarrobledo	Castilla-La Mancha	103,000	A1, B2, C1, C2, C6
186	186	ES076	Canyons on the middle section of the Cabriel river	Hoces del Cabriel Medio	Castilla-La Mancha	52,300	B2, C6
187	187	ES072	Cuenca mountain ranges	Serranía de Cuenca	Castilla-La Mancha	102,000	A4ii, B1iii, B2, C2, C6
188	188	ES069	Upper sections of the Rivers Tajo and Tajuña	Alto Tajo y Tajuña	Castilla-La Mancha	157,000	A4ii, B1iii, B2, C2, C6
189	189	ES071	Embid–Molina high moors	Paramera de Embid–Molina	Castilla-La Mancha	13,500	B2, C2, C6
190	190	ES191	River Mesa canyons	Cortados del Río Mesa	Aragón, Castilla-La Mancha	12,500	B1iii, B2, C2, C6
191	191	ES068	Entrepeñas and Buendía reservoirs	Embalses de Entrepeñas y Buendía	Castilla-La Mancha	38,500	B1i, B2, C2
192	192	ES083	El Hito	El Hito	Castilla-La Mancha	24,000	A4i, B1i, B3, C2, C6
193	193	ES083	Tarancón–Ocaña–Corral de Almaguer	Tarancón–Ocaña–Corral de Almaguer	Castilla-La Mancha	129,100	A1, B2, C1, C2, C6
194	194	ES084, ES085	Tembleque–La Guardia plains	Llanos de Tembleque–La Guardia	Castilla-La Mancha	128,000	A1, A4i, B1i, B2, C1, C2, C3, C6
195	195	ES086	Alcázar de San Juan–Quero endorreic lagoons	Complejo lagunar de Alcázar de San Juan–Quero	Castilla-La Mancha	58,500	A1, A4i, B1i, B1iii, B2, C1, C2, C3, C6
196	196	ES082	Pedro Muñoz–Manjavacas endorreic lagoons	Complejo lagunar de Pedro Muñoz–Manjavacas	Castilla-La Mancha	41,500	A1, A4i, B1i, B2, C1, C2, C3, C6
197	197	ES087	Tablas de Daimiel marshes; 'Vicario' and 'Gasset' reservoirs and Malagón lakes	Tablas de Daimiel, Embalses del Vicario y Gasset y Navas de Malagón	Castilla-La Mancha	31,500	A4i, B1i, B2, C2, C3, C6
198	198	ES092	Castrejón reservoir	Embalse de Castrejón	Castilla-La Mancha	3,800	C6
199	199	ES093	Torrijos	Torrijos	Castilla-La Mancha	28,600	A1, B2, C1, C2, C6
200	200	—	Malpica island (River Tagus)	Isla de Malpica (Río Tajo)	Castilla-La Mancha	4	A4i, B1i, C3, C6
201	201	ES096	Rosarito and Navalcán reservoirs–La Iglesuela (Tiétar valley)	Embalses de Rosarito y Navalcán–La Iglesuela (Valle del Tiétar)	Castilla-La Mancha	70,000	A4i, B1i, B2, C2, C3, C6
202	202	ES095	Oropesa plains	Llanos de Oropesa	Castilla-La Mancha	45,500	A1, B2, C1, C6
203	203	ES094	Azután reservoir	Embalse de Azután	Castilla-La Mancha	933	A4i, B1i, B2, C3, C6
204	204	ES090, ES091	Toledo mountains–Cabañeros	Montes de Toledo-Cabañeros	Castilla-La Mancha	324,500	A1, A3, A4ii, B1i, B1iii, B2, C1, C2, C6
205	205	ES089	Canalizos mountain range (Saceruela)	Sierra de los Canalizos (Saceruela)	Castilla-La Mancha	100,000	A1, A4ii, B1iii, B2, C1, C2, C6
206	206	—	Campo de Calatrava	Campo de Calatrava	Castilla-La Mancha	102,000	A1, A4ii, B1iii, B2, C1, C2, C6
207	207	ES088	Alcudia valley and mountain range	Valle y Sierra de Alcudia	Andalucía, Castilla-La Mancha	226,300	A1, B1i, B1iii, B2, C1, C2, C6
208	208	ES236	Madrona and Andújar mountain ranges	Sierra Madrona–Sierra de Andújar	Andalucía, Castilla-La Mancha	212,300	A1, B1i, B1iii, B2, C1, C2, C6
209	209	ES237	Aldeaquemada–Dañador	Aldeaquemada–Dañador	Andalucía, Castilla-La Mancha	46,000	A1, B2, C1, C6
210	210	ES238	Cazorla and Segura mountain ranges	Sierras de Cazorla y Segura	Andalucía, Castilla-La Mancha	250,000	A4ii, B1iii, B2, C2, C6
211	211	ES223	Puebla de Don Fadrique–Las Cañadas	Puebla de Don Fadrique–Las Cañadas	Andalucía	52,140	A1, B2, C1, C6
212	212	—	María mountain range–Los Vélez	Sierra de María–Los Vélez	Andalucía	82,880	B2, C6
213	213	ES224	Hoya de Baza	Hoya de Baza	Andalucía	38,600	B2, C2, C6
214	214	ES225	Hoya de Guadix	Hoya de Guadix	Andalucía	47,000	B2, C2, C6

Table 1 ... continued. Summary of Important Bird Areas in Spain. 391 IBAs covering 157,689 km²

IBA code	National code	1989 code	International name	National name	Administrative region	Area (ha)	Criteria (see p.11)
215	215	ES228	Alhamilla mountain range–Campo de Níjar plain	Sierra Alhamilla–Campo de Níjar	Andalucía	80,000	B2, C2, C6
216	216	ES226, ES227	Mountain range and saltpans at Cabo de Gata	Sierra y salinas de Cabo de Gata	Andalucía	46,100	A1, A3, A4i, B1i, B2, B3, C1, C2, C6
217	217	ES229	Tabernas desert	Desierto de Tabernas	Andalucía	25,000	B2, C6
218	218	—	Gádor mountain range	Sierra de Gádor	Andalucía	57,210	B2, C2, C6
219	219	ES230	Wetlands of western Almería	Humedales del Poniente Almeriense	Andalucía	3,000	A1, A4i, B1i, B2, C1, C2, C3, C6
220	220	ES270	Chafarinas Islands	Islas Chafarinas	Melilla	50	A1, A4i, B1i, B1ii, B2, C1, C2, C3, C6
221	221	—	Alborán island	Isla de Alborán	Andalucía	7	A1, A4i, B1i, B2, C1, C2, C6
222	222	ES269	Sierra Nevada mountain range	Sierra Nevada	Andalucía	245,000	B2, C2, C6
223	223	ES268	Prelittoral mountain ranges of Granada	Sierras prelitorales de Granada	Andalucía	125,000	B2, C2, C6
224	224	—	Mouth of the River Guadalhorce	Desembocadura del Río Guadalhorce	Andalucía	125	A1, A4i, B1i, C1, C2, C6
225	225	ES266	Antequera mountain range–El Chorro	Sierra de Antequera–El Chorro	Andalucía	23,500	B2, C2, C6
226	226	ES242	Mountain ranges of Córdoba	Sierras Subbéticas de Córdoba	Andalucía	63,000	B2, C6
227	227	ES241	Mountain ranges south of Jaén	Sierras al sur de Jaén	Andalucía	200,000	B2, C2, C6
228	228	ES239	Middle section of River Guadalquivir reservoirs	Embalses del Tramo medio del Guadalquivir	Andalucía	663	B1iii, B2, C2, C6
229	229	—	Jaén countryside	Campiña de Jaén	Andalucía	35,500	A1, B1iii, B2, C1, C2, C6
230	230	ES240	Marmolejo reservoir–La Ropera	Embalse de Marmolejo–La Ropera	Andalucía	675	C6
231	231	—	Conde, Chinche and Honda lakes	Lagunas de Conde, Chinche y Honda	Andalucía	420	A1, B1i, B2, C1, C2, C6
232	232	—	Córdoba countryside	Campiña alta de Córdoba	Andalucía	116,000	A1, A4ii, B1iii, B2, B3, C1, C2, C6
233	233	ES235	Hinojosa del Duque–El Viso	Hinojosa del Duque–El Viso	Andalucía	30,000	A1, B1i, B2, C1, C2
234	234	ES234	Los Blázquez–La Granjuela–Fuenteovejuna	Los Blázquez–La Granjuela–Fuenteovejuna	Andalucía	32,000	A4i, B1i, C2, C6
235	235	ES233	Sierra Morena de Córdoba	Sierra Morena de Córdoba	Andalucía	119,715	A1, A3, A4ii, B1i, B1iii, B2, C1, C2, C6
236	236	ES232	Sierra Morena de Sevilla	Sierra Morena de Sevilla	Andalucía	164,840	A1, B1i, B2, C1, C2, C6
237	237	ES244	Carmona countryside	Campiña de Carmona	Andalucía	18,200	A1, A4ii, B1iii, B2, C1, C2, C6
238	238	—	Écija–Osuna plain	Llanura cerealista de Écija–Osuna	Andalucía	62,200	A1, A4ii, B1iii, B2, C1, C2, C6
239	239	ES243	Wetlands at south Córdoba	Zonas húmedas del sur de Córdoba	Andalucía	3,054	A1, B1i, B2, C1, C2, C6
240	240	ES267	Fuente de Piedra, Gosque and Campillos lakes	Lagunas de Fuente de Piedra, Gosque y Campillos	Andalucía	10,600	A1, A4i, B1i, B2, C1, C2, C3, C6
241	241	ES264, ES265	Mountain ranges of Ronda, Bermeja and Crestellina	Serranía de Ronda, Sierras Bermeja y Crestellina	Andalucía	105,000	B2, C6
242	242	ES263	Líjar mountain range–Zaframagon crag	Sierra de Líjar–Peñón de Zaframagón	Andalucía	6,850	A4ii, B1iii, B2, C2, C6
243	243	ES262	Ubrique and Grazalema mountain ranges	Sierras de Ubrique y Grazalema	Andalucía	90,500	A4ii, B1iii, B2, C2, C6
244	244	ES261	Cabras, Aljibe and Montecoche mountain range	Sierras de las Cabras, del Aljibe y de Montecoche	Andalucía	133,500	A1, A4iv, B1iii, B1iv, B2, C1, C2, C5, C6
245	245	ES260	Bujeo, Ojén, del Niño and Blanquilla mountain ranges	Sierras del Bujeo, Ojén, del Niño y Blanquilla	Andalucía	42,100	A4ii, A4iv, B1iii, B1iv, B2, C2, C5, C6
246	246	ES259	Tarifa	Tarifa	Andalucía	5,000	A1, A4i, A4ii, A4iv, B1i, B1iii, B1iv, B2, C1, C2, C5
247	247	—	Ceuta	Ceuta	Ceuta	1,900	A1, A4i, A4ii, A4iv, B1i, B1iii, B1iv, C1, C2, C3, C5
248	248	ES258	De la Plata mountain range	Sierra de la Plata	Andalucía	3,000	A4iv, B1iv, C5
249	249	ES256	Tajo de Barbate coastal cliffs	Tajo de Barbate	Andalucía	2,017	A4i, B1i, C3
250	250	ES257	La Janda	La Janda	Andalucía	20,000	A4i, A4iv, B1i, B1iv, B2, C2, C3, C5
251	251	ES251	Cádiz Bay	Bahía de Cádiz	Andalucía	12,500	A4i, B1i, B2, B3, C2, C3, C6
252	252	ES252, ES253	Medina and Puerto Real lagoons	Laguna de Medina y de Puerto Real	Andalucía	4,900	A1, A4i, B1i, B2, C1, C2, C3, C6
253	253	ES250	Terry lagoons	Lagunas de Terry	Andalucía	350	A1, B1i, B2, C1, C2, C6
254	254	ES254	Bornos reservoir	Embalse de Bornos	Andalucía	2,500	A4i, B1i, C2, C3, C6
255	255	ES255	Medina-Sidonia	Medina-Sidonia	Andalucía	22,000	A1, A4i, B1i, C1, C3, C6
256	256	—	Confidential	Confidencial	Andalucía	2,000	A1, C1
257	257	—	Los Tollos lake	Laguna de Los Tollos	Andalucía	100	A1, A4i, B1i, C1, C2
258	258	ES249	Lebrija, Las Cabezas and Espera lagoons	Lagunas de Lebrija, Las Cabezas y Espera	Andalucía	7,600	A1, A4i, B1i, B2, C1, C2, C6
259	259	ES247	Guadalquivir marshes	Marismas del Guadalquivir	Andalucía	230,000	A1, A4i, A4iv, B1i, B1iv, B2, B3, C1, C2, C3, C5, C6
260	260	—	Condado–Campiña	Condado–Campiña	Andalucía	56,500	A1, A4ii, B1iii, B2, B3, C1, C2, C6
261	261	ES246	Odiel and Tinto marshes and Huelva coastal lagoons	Marismas del Tinto y el Odiel y Lagunas costeras de Huelva	Andalucía	14,900	A4i, B1i, B2, C2, C3, C6
262	262	—	River Piedras marshes and Rompido sandbank	Marismas del Río Piedras y Flecha del Rompido	Andalucía	3,120	B1i, B2, C2
263	263	ES245	Isla Cristina and Ayamonte marshes and Prado lagoon	Marismas de Isla Cristina y Ayamonte y Laguna del Prado Hondo	Andalucía	3,390	A4i, B1i, B2, B3, C2, C3, C6

Table 1 ... continued. Summary of Important Bird Areas in Spain. 391 IBAs covering 157,689 km²

IBA code	National code	1989 code	International name	National name	Administrative region	Area (ha)	Criteria (see p. 11)
264	264	—	West Andévalo	Andévalo Occidental	Andalucía	45,800	A1, B1i, B1iii, B2, C1, C2, C6
265	265	ES231	Sierra Pelada mountain range	Sierra de Pelada	Andalucía	48,000	A1, A4ii, B1iii, B2, C1, C2, C6
266	266	ES231	Las Contiendas mountain range	Sierra de las Contiendas	Andalucía	35,000	B1i, B1iii, C2
267	267	ES231	East mountain range of Huelva	Sierra oriental de Huelva	Andalucía	71,000	B1i, B2, C2, C6
268	268	ES131	Fuente de Cantos–Montemolín	Fuente de Cantos–Montemolín	Extremadura	49,000	A1, B1iii, B2, C1, C2, C6
269	269	ES134, ES135	Azuaga–Llerena–Peraleda de Zaucejo	Azuaga–Llerena–Peraleda de Zaucejo	Extremadura	147,000	A1, A4i, A4ii, B1i, B1iii, B2, B3, C1, C2, C6
270	270	ES133	Central Badajoz mountain ranges	Sierras Centrales de Badajoz	Extremadura	90,000	A1, B1i, B2, C1, C2, C6
271	271	ES132	Bienvenida–Usagre–Ribera del Fresno	Bienvenida–Usagre–Ribera del Fresno	Extremadura	6,700	A1, A4i, B1i, B2, C1, C2, C6
272	272	ES130	Valuengo reservoir	Embalse de Valuengo	Extremadura	11,200	A4i, B1i, C2, C3
273	273	ES129	Jerez de los Caballeros dehesas	Dehesas de Jerez de los Caballeros	Extremadura	160,000	A1, A4i, B1i, B1iii, B2, C1, C2, C6
274	274	ES128	Villanueva del Fresno	Villanueva del Fresno	Extremadura	7,800	A1, B1i, B2, C1, C2
275	275	ES127	Valongo	Valongo	Extremadura	2,800	A4i, B1i, C3, C6
276	276	ES125, ES126	Olivenza–La Albuera	Olivenza–La Albuera	Extremadura	79,000	A1, B1i, B2, C1, C2, C6
277	277	—	Alange	Alange	Extremadura	66,500	A1, B1i, B2, C1, C2
278	278	—	Villalba de los Barros	Villalba de los Barros	Extremadura	12,500	A1, B2, C1, C2
279	279	ES136	Puerto de Mejoral–Almorchón–Cabeza del Buey	Puerto de Mejoral–Almorchón–Cabeza del Buey	Extremadura	82,500	A4i, B1i, B2, C2, C6
280	280	ES137	La Serena	La Serena	Extremadura	109,800	A1, A4i, B1i, B2, B3, C1, C2, C3, C6
281	281	ES138	Siruela–Agudo	Siruela–Agudo	Extremadura	66,000	A4i, B1i, C2
282	282	ES107	Cijara reservoir	Embalse de Cíjara	Extremadura, Castilla-La Mancha	73,500	A1, B1i, B2, C1, C2, C6
283	283	ES108	Puerto Peña reservoir–Valdecaballeros	Embalse de Puerto Peña–Valdecaballeros	Extremadura	32,000	A1, B1i, B2, B3, C1, C2, C6
284	284	ES109, ES110	Pela mountain range–Orellana reservoir–Zorita	Sierra de Pela–Embalse de Orellana–Zorita	Extremadura	140,000	A1, A4i, A4ii, B1i, B1iii, B2, B3, C1, C2, C3, C6
285	285	—	Don Benito–Guareña	Don Benito–Guareña	Extremadura	31,000	A1, B1i, B2, C1, C2
286	286	—	Valdehornillos–Santa Amalia	Valdehornillos–Santa Amalia	Extremadura	23,000	A4i, B1i, B2, C2, C3, C6
287	287	ES120	Montánchez mountain range–Cornalvo reservoir	Sierra de Montánchez–Embalse de Cornalvo	Extremadura	54,500	A1, A4i, B1i, C1, C2
288	288	ES122	Mérida–Montijo reservoir	Mérida–Embalse de Montijo	Extremadura	4,100	A1, A4i, B1i, B2, C1, C3, C6
289	289	ES121	Lácara–Morante	Lácara–Morante	Extremadura	56,900	A1, A4i, B1i, B1iii, B2, C1, C2, C6
290	290	ES124	Botoa–Villar del Rey	Botoa–Villar del Rey	Extremadura	47,800	A1, B1i, B2, C1, C2
291	291	ES118	San Pedro mountain range	Sierra de San Pedro	Extremadura	295,000	A1, A4i, A4ii, B1i, B1iii, B2, C1, C2, C3, C6
292	292	ES116	Cedillo reservoir	Embalse del Cedillo	Extremadura	53,000	A1, B1i, B1iii, B2, C1, C2, C6
293	293	ES115	Brozas–Membrío	Brozas–Membrío	Extremadura	94,000	A1, A4i, B1i, B2, C1, C2, C6
294	294	ES114	Malpartida de Cáceres–Arroyo de la Luz	Malpartida de Cáceres–Arroyo de la Luz	Extremadura	43,900	A1, A4i, B1i, B2, C1, C2, C3, C6
295	295	ES118	Plain between Cáceres and Trujillo–Aldea del Cano	Llanos entre Cáceres y Trujillo–Aldea del Cano	Extremadura	103,000	A1, A4i, A4ii, B1i, B1iii, B2, B3, C1, C2, C6
296	296	ES111	Trujillo–Torrecillas de la Tiesa	Trujillo–Torrecillas de la Tiesa	Extremadura	108,500	A1, A4i, A4ii, B1i, B1iii, B2, B3, C1, C2, C3, C6
297	297	ES106	Las Villuercas mountain range	Sierras de las Villuercas	Extremadura	216,230	A1, B1i, B1iii, B2, C1, C2, C6
298	298	ES103	Monfragüe	Monfragüe	Extremadura	155,000	A1, A3, A4ii, B1i, B1iii, B2, C1, C2, C6
299	299	ES101, ES102	Alcántara reservoir–Cuatro Lugares	Embalse de Alcántara–Cuatro Lugares	Extremadura	116,000	A1, A4i, B1i, B1iii, B2, C1, C2, C6
300	300	ES100	Coria mountain ranges	Sierras de Coria	Extremadura	71,500	A1, B1i, B2, C1, C2, C6
301	301	ES099	Borbollón reservoir	Embalse del Borbollón	Extremadura	48,400	A1, A4i, B1i, C1, C2, C3
302	302	ES097	East Gata mountain range	Solana de la Sierra de Gata	Extremadura	86,600	A1, B1i, B1iii, B2, C1, C2, C6
303	303	ES098	Gabriel y Galán reservoir	Embalse de Gabriel y Galán	Extremadura	55,000	B1i, C2
304	304	—	Plasencia and San Bernabé mountain range	Plasencia y Sierra de San Bernabé	Extremadura	18,000	A1, B1i, B2, C1, C2
305	305	—	Lower Tiétar river and La Vera valley	Bajo Tiétar y rampa de La Vera	Extremadura	73,000	B1i, B2, C2, C6
306	306	ES104, ES105	Campo Arañuelo–Valdecañas reservoir	Campo Arañuelo–Embalse de Valdecañas	Extremadura, Castilla-La Mancha	81,950	A4i, B1i, B2, C2, C3, C6
307	307	ES271	Islets Vedrá and Vedranell	Islotes de Vedrá y Vedranell	Baleares	100	A4ii, B1ii, B1iii, B2, B3, C2, C6
308	308	—	Isle of Conejera and islets of Bledes and Espartar (Ibiza)	Isla Conejera e islotes de Bledes y Espartar (Ibiza)	Baleares	147	A1, A4i, A4ii, B1i, B1ii, B2, B3, C1, C2, C6
309	309	ES272	Cape Nonó–Isle of Murada	Cabo Nonó–Isla Murada	Baleares	1,000	A4ii, B1ii, B1iii, B2, B3, C2, C6
310	310	ES273	Tagomago island	Isla de Tagomago	Baleares	90	A4ii, B1ii, B1iii, B2, B3, C2, C6
311	311	—	Islets of Santa Eulalia, Redona and Es Canar (Ibiza)	Islotes de Santa Eulalia, Redona y Es Canar (Ibiza)	Baleares	6	A1, B1ii, C1, C2, C6
312	312	ES274, ES275	Saltpans of Ibiza and Formentera and Freus isles	Salinas de Ibiza y Formentera e islas de los Freus	Baleares	2,600	A1, A4i, A4ii, B1i, B1ii, B2, B3, C1, C2, C3, C6
313	313	—	Cape Barbaria	Cabo de Barbaria	Baleares	100	A4ii, B1i, B1ii, B3, C2, C6

Table 1 ... continued. Summary of Important Bird Areas in Spain. 391 IBAs covering 157,689 km²

IBA code	National code	1989 code	International name	National name	Administrative region	Area (ha)	Criteria (see p. 11)
314	314	ES276	La Mola of Formentera	La Mola de Formentera	Baleares	1,800	A4i, A4ii, B1i, B1ii, B3, C2, C3, C6
315	315	ES277	Dragonera island–La Trapa	Isla Dragonera–La Trapa	Baleares	2,660	A1, A4i, A4ii, B1i, B1ii, B1iii, B2, B3, C1, C2, C6
316	316	ES278	Tramuntana mountains	Sierra de Tramuntana (Norte)	Baleares	54,953	A1, A4ii, B1i, B1iii, B2, B3, C1, C2, C6
317	317	ES279	Cape Pinar	Cabo Pinar	Baleares	2,500	B3, C6
318	318	ES280	Albufera de Mallorca and Albufereta de Pollença marshes	Albufera de Mallorca y Albufereta de Pollença	Baleares	2,800	A1, B1i, B2, B3, C1, C2, C3, C6
319	319	ES281	Cape Freu–Cape Farrutx	Cabo del Freu–Cabo Farrutx	Baleares	500	B1i, C2, C6
320	320	ES282	Cape Vermell	Cabo Vermell	Baleares	150	B1i, C2
321	321	ES283	Salobrar de Campos lagoons–Sa Vall	Lagunas de Salobrar de Campos–Sa Vall	Baleares	2,397	B2, B3, C6
322	322	ES284	Coastal cliffs between Cape Enderrocat and Cala Pí	Acantilados entre Cap Enderrocat y Cala Pí	Baleares	900	B1i, B2, B3, C2, C6
323	323	ES285	Coastal cliffs between isles of Malgrats and Cala Figuera	Acantilados costeros entre las Islas Malgrats y Cala Figuera	Baleares	440	B1i, B1ii, B3, C2, C6
324	324	ES286	Archipelago of Cabrera	Archipiélago de Cabrera	Baleares	10,000	A1, A3, A4i, B1i, B1ii, B2, B3, C1, C2, C3, C6
325	325	ES287	North and east coasts of Minorca and Island of Aire	Costa Norte y Este de Menorca e Isla del Aire	Baleares	17,900	A1, A4ii, B1i, B1ii, B2, B3, C1, C2, C6
326	326	ES288	Ravines and pinewoods in central Minorca	Barrancos y pinares centrales de Menorca	Baleares	30,200	B1i, B2, C2, C6
327	327	CI001	Lanzarote islets	Los islotes de Lanzarote	Las Palmas	4,000	A4ii, B1ii, B1iii, B2, B3, C2, C6
328	328	CI002	Famara crag	Riscos de Famara	Las Palmas	1,350	B2, C2, C6
329	329	CI004	Haría–Tabayesco	Haría–Tabayesco	Las Palmas	2,350	B2, C6
330	330	CI010	Tenegüime gully	Barranco de Tenegüime	Las Palmas	421	B2, C6
331	331	CI007	Plains of La Corona–Las Honduras	Llanos de La Corona–Las Honduras	Las Palmas	3,470	B2, B3, C2, C6
332	332	CI006	Famara sandy plain	Jable de Famara	Las Palmas	5,700	B2, B3, C2, C6
333	333	CI008	El Mojón	El Mojón	Las Palmas	60	B2, C6
334	334	CI003	Janubio saltpans	Salinas de Janubio	Las Palmas	169	B2, C6
335	335	CI009	Plains of Mareta–Hoya de la Yegua	Llanos de Mareta–Hoya de la Yegua	Las Palmas	3,600	B2, B3, C2, C6
336	336	CI023	Island of Lobos	Isla de Lobos	Las Palmas	468	A4ii, B1ii, B2, C2, C6
337	337	CI024	Sandy plain of Corralejo	Jable de Corralejo	Las Palmas	2,668	B2, B3, C2, C6
338	338	CI015	Morro Tabaiba–Morro de los Rincones-Vallebrón mountains	Morro Tabaiba–Morro de los Rincones-Vallebrón	Las Palmas	1,150	A1, A2, B2, C1, C2, C6
339	339	CI026	Río Cabras gully	Barranco de Río Cabras	Las Palmas	100	A1, B2, C1
340	340	CI020	Cuchillete de Buenavista–gully of La Torre–Los Alares	Cuchillete de Buenavista–Barranco de La Torre–Los Alares	Las Palmas	7,500	A1, A2, B2, B3, C1, C2, C6
341	341	CI022	Pozo Negro mountain – Vigán	Macizo de Pozo Negro – Vigán	Las Palmas	8,000	A1, B2, C1, C2, C6
342	342	CI019	Tarajalejo mountain range	Macizo de Tarajalejo	Las Palmas	2,200	A1, B2, C1, C2, C6
343	343	CI025	Sotavento beach	Playa de Sotavento	Las Palmas	450	B2, C7
344	344	CI021	Peninsula of Jandía	Península de Jandía	Las Palmas	12,400	A1, B2, B3, C1, C6
345	345	CI012	Isthmus of Jandía sandy plain	Jable del Istmo de Jandía	Las Palmas	3,900	B2, B3, C2, C6
346	346	CI016	Gully of Ajuí–Betancuria	Barranco de Ajuí–Betancuria	Las Palmas	700	A1, A2, B2, B3, C1, C6
347	347	CI017	Los Molinos gully–Plain of La Laguna	Barranco de Los Molinos–Llano de La Laguna	Las Palmas	1,700	A1, B2, B3, C1, C6
348	348	CI018	Coast of Esquinzo–Puertito de Los Molinos	Costa de Esquinzo–Puertito de Los Molinos	Las Palmas	250	B2, C6
349	349	CI013	Lajares sandy plain–Cotillo–Ezquinzo	Jable de Lajares–Cotillo–Ezquinzo	Las Palmas	6,800	B2, B3, C2, C6
350	350	CI014	Coast between Corralejo and Tostón	Costa de Corralejo a Tostón	Las Palmas	530	A1, B2, C1, C6
351	351	CI041	Coast between Arinaga and Castillo del Romeral	Costa de Arinaga–Castillo del Romeral	Las Palmas	1,100	B2, C7
352	352	CI046	Tirajana pine woodland	Pinar de Tirajana	Las Palmas	3,600	B2, C6
353	353	CI045	Tauro pine woodland	Pinar de Tauro	Las Palmas	800	B2, C6
354	354	CI044	Pajonales, Ojeda, Inagua and La Data pine woodlands	Pinares de Pajonales, Ojeda, Inagua y la Data	Las Palmas	3,700	A1, A2, B2, B3, C1, C2, C6
355	355	CI043	Tamadaba pine woodland	Pinar de Tamadaba	Las Palmas	3,400	A2, B2, C6
356	356	CI029	Anaga rocky islets	Roques de Anaga	Santa Cruz de Tenerife	10	B1ii, B2, C2, C6
357	357	CI034	San Andrés, Pijaral and Anaga mountains	Montes de San Andrés, Pijaral y Anaga	Santa Cruz de Tenerife	860	A1, B2, C1
358	358	CI032	Las Vueltas, Aguas Negras and Quebrades mountains	Monte de Las Vueltas, Aguas Negras y Quebrades	Santa Cruz de Tenerife	450	A1, B2, B3, C1
359	359	CI035	Las Mercedes, Mina y Yedra, Aguirre, La Goleta, and Pedro Alvarez mountains	Montes de Las Mercedes, Mina y Yedra, Aguirre, La Goleta y Pedro Alvarez	Santa Cruz de Tenerife	2,000	A1, B2, B3, C1, C6
360	360	CI040	Los Rodeos–La Esperanza	Los Rodeos–La Esperanza	Santa Cruz de Tenerife	1,200	B2
361	361	CI031	La Playa islet	Roque de la Playa	Santa Cruz de Tenerife	1	B2, C6
362	362	CI037	Santa Úrsula and La Victoria laurel forest	Monte Verde de Santa Úrsula y La Victoria	Santa Cruz de Tenerife	1,800	A1, B2, C1, C6
363	363	CI033	Tigaiga slope	Ladera de Tigaiga	Santa Cruz de Tenerife	370	A1, B2, C1, C6

Table 1 ... continued. Summary of Important Bird Areas in Spain. 391 IBAs covering 157,689 km²

IBA code	National code	1989 code	International name	National name	Administrative region	Area (ha)	Criteria (see p. 11)
364	364	—	Santo Domingo coastal cliffs	Acantilados de Santo Domingo	Santa Cruz de Tenerife	7	B2, C6
365	365	CI030	Garachico rocky islet	Roque de Garachico	Santa Cruz de Tenerife	5	B2, C6
366	366	CI036	Mountain of Agua, gully of Los Cochinos and gully of Cuevas Negras	Monte del Agua, Barranco de los Cochinos y Barranco de Cuevas Negras	Santa Cruz de Tenerife	800	A1, A2, B2, B3, C1, C6
367	367	CI028	Los Gigantes coastal cliff	Acantilado de Los Gigantes	Santa Cruz de Tenerife	1,000	B2, B3, C6
368	368	CI027	Tágara gully	Barranco de Tágara	Santa Cruz de Tenerife	400	A1, B2, C1, C6
369	369	CI039	Vilaflor pine woodland	Pinar de Vilaflor	Santa Cruz de Tenerife	1,750	A1, A2, B2, B3, C1, C6
370	370	CI038	Arico pine woodland	Pinar de Arico	Santa Cruz de Tenerife	3,000	A1, A2, B2, C1, C6
371	371	—	Centinela mountain and La Esquina plain	Montaña Centinela y Llano de la Esquina	Santa Cruz de Tenerife	1,600	B2, C6
372	372	—	El Médano coast	El Médano	Santa Cruz de Tenerife	1,000	B2, C6
373	373	—	Rasca lava-fields–Guaza mountain–Las Mesas plain	Malpaís de Rasca–Montaña de Guaza–Llano de las Mesas	Santa Cruz de Tenerife	1,400	B2, C2
374	374	CI047	Majona coast	Costa de Majona	Santa Cruz de Tenerife	170	B2, C2
375	375	CI050	Step rocks of Hermigua and Agulo	Riscos de Hermigua y Agulo	Santa Cruz de Tenerife	450	A1, B2, C1, C6
376	376	CI051	Garajonay National Park	Parque Nacional de Garajonay	Santa Cruz de Tenerife	3,984	A1, A2, B2, B3, C1, C2, C6
377	377	CI049	Vallehermoso coast	Costa de Vallehermoso	Santa Cruz de Tenerife	59	C6
378	378	CI048	South-west coast of La Gomera	Costa meridional de La Gomera	Santa Cruz de Tenerife	600	B1ii, B2, C2, C6
379	379	CI052	La Palma laurel forest	Monte Verde de La Palma	Santa Cruz de Tenerife	14,800	A1, A2, B2, B3, C1, C2, C6
380	380	CI056	El Canal y Los Tiles laurel forest	El Canal y Los Tiles	Santa Cruz de Tenerife	511	A1, A2, B2, B3, C1, C6
381	381	CI054	Negro rocky islet	Roque Negro	Santa Cruz de Tenerife	1	B2, C6
382	382	CI053	Garafía rocky islets	Roques de Garafía	Santa Cruz de Tenerife	4	B2, C2, C6
383	383	CI055	La Caldera de Taburiente National Park	Parque Nacional de La Caldera de Taburiente	Santa Cruz de Tenerife	4,690	A1, B2, B3, C1, C6
384	384	—	El Roque coastal cliffs	El Roque	Santa Cruz de Tenerife	40	B2, C6
385	385	CI058	Ventejís mountain	Macizo de Ventejís	Santa Cruz de Tenerife	870	C6
386	386	CI063	Nizdafe plains	Llanos de Nizdafe	Santa Cruz de Tenerife	2,000	B3
387	387	CI062	Salmor rocky islets	Roques de Salmor	Santa Cruz de Tenerife	5	B1ii, B2, C2, C6
388	388	CI059	Frontera laurel forest	Monte Verde de Frontera	Santa Cruz de Tenerife	2,300	A1, B2, B3, C1, C6
389	389	CI061	Western coast of El Hierro	Costa occidental de El Hierro	Santa Cruz de Tenerife	600	B1ii, B2, C2, C6
390	390	CI057	La Dehesa grasslands	La Dehesa	Santa Cruz de Tenerife	2,000	B2, C7
391	391	CI060	Naos Bay–Hoya de Tacorón	Bahía de Naos–Hoya de Tacorón	Santa Cruz de Tenerife	170	B2, C6

Sites identified in the previous inventory of IBAs in Europe (Grimmett and Jones 1989) but no longer considered to be IBAs
ES004 Cabo Finisterre–Costa de Touriñán; ES031 Páramo de Masa; ES048 Ledesma–Río Tormes; ES060 Río Carrión en Husillos; ES073 Valeria; ES074 Carboneras de Guadazaón; ES075 Moya; ES146 Gallipienzo–Sierra de la Peña; ES184 Urrea de Jalón; ES189 Río Manubles; CI005 Litoral de Arrecife; CI011 Costa Punta Lima–Playa de Matagorda; CI042 Roque Partido; CI064 Name and location confidential.

Several IBAs hold more than one globally threatened or near-threatened species, including the Guadalquivir marshes (IBA 259; *Marmaronetta angustirostris, Oxyura leucocephala, Aquila adalberti* and wintering *Larus audouinii*), the San Pedro mountain range (IBA 291; *Aegypius monachus, Aquila adalberti, Otis tarda*, and *Tetrax tetrax*) and the Alcántara reservoir–Cuatro Lugares (IBA 299; *Aquila adalberti, Falco naumanni, Otis tarda* and *Tetrax tetrax*).

Two IBAs (004, 037) have been identified on the migratory route of the globally threatened *Acrocephalus paludicola* along the Cantabric and Atlantic coasts, however, site coverage for this species is lacking in this inventory as a result of a paucity of data. Similarly a lack of data and small numbers of birds have prevented the selection of sites for *Numenius tenuirostris* (wintering) and *Crex crex* (breeding).

The Canary Islands are included in one of the three Endemic Bird Areas (EBAs) identified in Europe (Stattersfield *et al.* 1998): the Madeira and Canary Islands EBA, selected for its 10 restricted-range species. Nine of these species are present in the Canaries, with five species being restricted to these islands alone. Eleven IBAs have been identified for restricted-range species, meeting criterion A2, with sites holding 2–6 of the relevant species. Tenerife has the largest number of restricted-range species and Fuerteventura one single island endemic, *Saxicola dacotiae*. The central and western

Table 2. Important Bird Areas in Spain that are important for species of global conservation concern (meeting criterion A1).

Species	IBA code
Marmaronetta angustirostris Marbled Duck	165, 166, 252, 257, 258, 259
Oxyura leucocephala White-headed Duck	165, 194, 195, 196, 219, 231, 239, 240, 253, 257, 258, 259, 318
Aegypius monachus Cinereous Vulture	056, 066, 069, 070, 076, 204, 205, 207, 208, 235, 236, 264, 265, 282, 289, 291, 292, 297, 298, 302, 316
Aquila adalberti Spanish Imperial Eagle	056, 068, 070, 071, 204, 205, 207, 208, 209, 235, 236, 252, 255, 256, 258, 259, 264, 270, 283, 291, 297, 298, 299, 300
Falco naumanni Lesser Kestrel	038, 039, 041, 061, 073, 074, 106, 110, 185, 195, 202, 206, 207, 211, 229, 232, 233, 237, 238, 244, 246, 247, 260, 268, 269, 273, 284, 288, 291, 293, 295, 296, 299, 304
Tetrax tetrax Little Bustard	038, 044, 061, 074, 075, 105, 106, 142, 182, 184, 185, 193, 194, 199, 202, 206, 232, 268, 269, 271, 274, 276, 280, 284, 285, 287, 290, 293, 294, 295, 296, 299
Otis tarda Great Bustard	038, 039, 040, 041, 043, 044, 058, 060, 061, 062, 074, 075, 080, 095, 106, 182, 184, 185, 193, 194, 199, 202, 206, 238, 269, 271, 274, 276, 277, 278, 280, 284, 285, 290, 291, 293, 294, 295, 296, 299, 301
Larus audouinii Audouin's Gull	140, 148, 152, 153, 159, 167, 169, 170, 216, 219, 220, 221, 224, 246, 247, 259, 308, 311, 312, 315, 324, 325
Columba bollii Dark-tailed Laurel Pigeon	357, 358, 359, 362, 363, 366, 375, 376, 379, 380, 388
Columba junoniae White-tailed Laurel Pigeon	357, 358, 359, 362, 363, 366, 375, 376, 379, 380, 383
Saxicola dacotiae Fuerteventura Chat	338, 339, 340, 341, 342, 344, 346, 347, 350
Acrocephalus paludicola Aquatic Warbler	004, 037
Fringilla teydea Blue Chaffinch	354, 368, 369, 370

Table 3. Important Bird Areas in Spain that support important numbers of one or more congregatory species (i.e. meeting criteria A4 and/or B1). IBAs meeting both criteria A4 and B1 for the species are shown in **bold**. IBAs meeting only criterion B1 for the species concerned, and not A4, are shown in normal type. For key to 'Season', see p. 7.

Species	Season	IBA code
Gavia immer Great Northern Diver	P	**004**
Podiceps nigricollis Black-necked Grebe	W	**167**, **169**, **261**, **312**
Bulweria bulwerii Bulwer's Petrel	B	327, 356
Calonectris diomedea Cory's Shearwater	B	220, 324, 325, **327**, **336**, 389
Puffinus yelkouan Yelkouan Shearwater	R	307
	B	308, 310, **312**, **313**, **314**, **315**, 323, 324, **325**
	W	**140**
	P	**037**
Puffinus assimilis Little Shearwater	B	327, 336, 378
Hydrobates pelagicus Storm Petrel	B	164, **308**, 309, 311, 312, 324, 327, 336, 378, 387
Oceanodroma castro Madeiran Petrel	B	327, 356
Phalacrocorax carbo Cormorant	W	001, 148, 159, 201, 251, **306**
Phalacrocorax aristotelis Shag	R	**001**, 153, 313, 316, 322, 323, 324, 325, 326
	B	314, 319, 320
Nycticorax nycticorax Night Heron	B	**082**
Ardeola ralloides Squacco Heron	B	**148**, **159**
Bubulcus ibis Cattle Egret	R	**138**, **148**, **159**, **200**, **249**, **250**, **254**, **255**, **263**, **288**, **306**
	B	**203**, **272**, **275**, **280**, **284**, **291**, **294**, **296**, **301**
	W	**148**, 155, **159**, **247**, **288**
	P	**148**, **247**
	N	**137**, 154
Egretta garzetta Little Egret	R	**148**, **159**, **259**, 261, **263**
	W	**148**, **159**, **251**
	P	**148**
Ciconia nigra Black Stork	R	208, 306
	B	063, 064, 065, 066, 067, 068, 070, 201, 204, 207, 235, 236, 264, 267, 273, 279, 282, 283, 284, 287, 291, 292, 293, 296, 297, 298, 299, 300, 302, 303, 305
	W	259
	P	037, 068, 070, 071, 085, 108, 204, 224, **246**, **247**, 266, 269, 270, 272, **273**, 274, 280, 281, 282, 283, 284, 287, 291, 293, 294, **295**, 296, 297, 299, 301, 304, 306
	N	201
Ciconia ciconia White Stork	R	263
	B	284, 296, 299
	P	**246**, **247**
	N	057
Platalea leucorodia Spoonbill	R	**259**, **261**, **263**
	P	003, 026, 027, 035, **037**
	N	262
Phoenicopterus ruber Greater Flamingo	R	**148**, **259**
	B	**165**, **240**
	W	**148**, **165**, **166**, **240**, **259**
	P	**148**
	N	**216**, **219**, **251**, **252**, **261**
Anser anser Greylag Goose	W	038, **039**, **259**
Tadorna tadorna Shelduck	W	**259**
Anas penelope Wigeon	W	027, 148, **259**
Anas strepera Gadwall	W	**095**, **148**, **197**, **259**
Anas crecca Teal	W	**148**, **259**
Anas platyrhynchos Mallard	R	039, 148
	W	038, 137, 148, 197
Anas acuta Pintail	W	007, **259**
Anas clypeata Shoveler	W	**148**, **159**, **165**, 196, **259**
Marmaronetta angustirostris Marbled Duck	R	**165**, **166**, **259**
	W	**165**, **257**, 259
	N	**252**, **258**
Netta rufina Red-crested Pochard	R	**148**, 159, **166**, **195**
	B	156, **197**
	W	**095**, **148**, **155**, **159**, **165**, 166, 194, **195**, 196, 197, **259**
	N	**023**
Aythya ferina Pochard	W	**095**, **148**
Oxyura leucocephala White-headed Duck	R	165, 195, 196, **219**, 231, 258, 259, 318
	B	194, 239, 240
	W	**165**, 194, 196, 239, **257**, **259**
	N	**219**, 252, 253, 258
Milvus migrans Black Kite	N	228
Milvus milvus Red Kite	W	**038**, **061**, **295**
Neophron percnopterus Egyptian Vulture	R	063
	B	014, **064**, 090, 297, 299
	P	**246**, 247
	N	029, 090, 122
Gyps fulvus Griffon Vulture	R	020, 032, 034, 047, 048, **049**, 050, **051**, **052**, **053**, **064**, 079, **086**, **087**, 088, **092**, **099**, **100**, **118**, **120**, **122**, 123, 132, **187**, **188**, 190, **210**, 235, **242**, **243**, 244, **245**, 297, **298**
	B	028, **029**, 033
	W	**052**
	P	**246**, **247**
	N	117, 264, 265, 266
Aegypius monachus Cinereous Vulture	R	**056**, **066**, **069**, 076, **204**, **205**, 207, 208, **235**, **265**, **291**, 292, 297, **298**, 302, 316
	N	264, 289
Hieraaetus pennatus Booted Eagle	W	168
Hieraaetus fasciatus Bonnelli's Eagle	N	142, 168
Falco naumanni Lesser Kestrel	B	038, **039**, 061, **106**, 195, **206**, 207, 229, **232**, **237**, **238**, **260**, 268, **269**, 273, **284**, **295**, **296**, 299
	P	246
Falco eleonorae Eleonora's Falcon	B	**307**, **309**, **310**, **315**, **316**, **327**
Fulica atra Coot	W	**095**, 148, 252
Grus grus Crane	W	061, 068, **095**, **201**, 233, **234**, **240**, **269**, **271**, 274, 276, 277, **279**, **280**, **281**, 283, **284**, 285, **286**, **287**, 289, 290, **293**, **295**, 296, **299**, 301, 303, **306**
	P	039, 071, **078**, **085**, **095**, 101, 109, **111**, **116**, 191, **192**
Himantopus himantopus Black-winged Stilt	R	**154**, **259**, 286
	B	**039**, **111**, **148**, 155, **156**, **159**, 162, **165**, 167, 182, **192**, 195, 196, 197, 219, **251**, 284, 312
	W	251
	P	**073**
Recurvirostra avosetta Avocet	R	**148**, **166**, **259**
	B	**251**
	W	**148**, 165, **166**, **251**
	P	**261**
Glareola pratincola Collared Pratincole	R	**259**, **286**
	B	**148**, **152**, 195, **250**, **251**, **263**, **284**, **289**
	P	**159**
Charadrius hiaticula Ringed Plover	W	**263**
Charadrius alexandrinus Kentish Plover	R	**148**, **166**, 167, **259**, 312, 318
	B	251, **263**
	W	148, **251**
	P	**148**
	N	219
Charadrius morinellus Dotterel	W	**038**
Calidris alba Sanderling	P	**004**, **005**
Calidris minuta Little Stint	W	**263**
	P	**148**
Calidris ferruginea Curlew Sandpiper	P	**148**
Calidris alpina Dunlin	W	**148**
	P	**148**
Limosa limosa Black-tailed Godwit	W	**148**, **259**, 284
	P	**148**, **286**
Numenius arquata Curlew	W	**037**
Tringa totanus Redshank	R	**259**
	W	251, **263**
	P	**148**
Larus melanocephalus Mediterranean Gull	W	**140**, **148**

Figure 1. Occurrence of habitats at Important Bird Areas in Spain (see Appendix 3 for definitions of habitats).

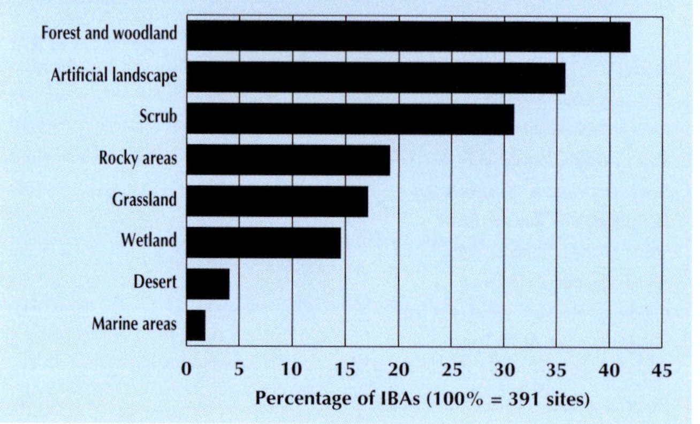

Extensive areas of artificial landscape occur in 36% of IBAs (Figure 1). Non-irrigated arable cultivation, which occurs in more than 22% of IBAs, is of great importance for the conservation of *Otis tarda, Tetrax tetrax* and *Falco naumanni*. Rice crops, linked to coastal wetlands, for example in the Ebro delta (148), Ampurdán marshes (137) and Valencia marshes (159), hold important numbers of breeding and wintering waterfowl.

Scrub is a significant habitat-type in almost 31% of the IBAs (Figure 1). Associated with the Mediterranean region, it includes characteristic formations such as garrigue and maquis. It is of particular importance for *Burhinus oedicnemus* and species of Sylviidae. In upland areas, scrub is important for breeding *Perdix perdix* and foraging raptors, and on the Canary Islands is vital for *Saxicola dacotiae*.

Extensive rocky areas occur in almost 20% of IBAs (Figure 1). Of these 47% contain inland cliffs, used by cliff-nesting raptors, such as *Gypaetus barbatus, Gyps fulvus, Aquila chrysaetos, Hieraaetus fasciatus*, as well as *Pyrrhocorax pyrrhocorax, P. graculus* and *Tichodroma muraria*. A further 26% include coastal cliffs, holding seabird and coastal raptor colonies and 27% are rocky islets, important for *Calonectris diomedea, Puffinus yelkouan mauretanicus, Larus audouinii*, and raptors such as *Pandion haliaetus* and *Falco eleonorae*.

Grasslands are a major habitat-type in 17% of IBAs (Figure 1). Alpine and subalpine grasslands are important for *Perdix perdix*. At lower altitudes, grasslands hold breeding *Circus* spp. and some steppe species such as *Chersophilus duponti*, and are used by foraging raptors.

Despite occurring as an extensive habitat in under 15% of IBAs (Figure 1), wetlands are the most important areas for birds in Spain, in particular the Guadalquivir marshes (IBA 259) and the Ebro delta (148). Although most major wetlands are highly threatened, they are of major international significance for breeding, passage and wintering waterbirds. Most wetlands are natural or semi-natural, but also include some man-made reservoirs used by wildfowl, post-breeding congregations of *Ciconia nigra* and roosting *Grus grus*.

Located mainly in the south-eastern part of the Iberian peninsula and the Canary Islands, desert and semi-desert habitats are extensive in less than 5% of IBAs (Figure 1). They are characterized by low scrub and are highly affected by erosion. Ornithologically such areas are important for *Burhinus oedicnemus, Chlamydotis undulata, Bucanetes githagineus* and other steppic species.

IMPACTS ON IBAs – LAND-USE AND THREATS

Agriculture, both arable and livestock production, is the most widespread land-use in IBAs, occurring in over 60% (Figure 2). Agricultural expansion and intensification is also a high-impact threat to nearly 14% of IBAs and a low/medium/unknown-impact threat in a further 22% (Figure 3). Extensive livestock-farming influences the ecology of a wide range of habitats (pseudosteppes, dehesas, mountain grasslands and scrub) with animals moved around mosaics of habitats to exploit favourable conditions. Sheep-grazing is widespread but cattle are limited to mountainous areas, the northern seaboard and some areas of the west and south-west. European Union livestock headage payments have caused overgrazing problems in steppe areas of high value, destroying vegetation, increasing erosion and disturbing breeding birds. Arable production has become more intensive, particularly through the irrigation of crops, resulting in serious losses of steppe habitats and traditional non-irrigated crops. Huge irrigation projects threaten the most important areas for *Otis tarda, Tetrax tetrax* and *Falco naumanni*. Traditional cultivation is increasingly dependent on EU subsidies.

Infrastructure development is a high-impact threat in 14% of IBAs and a medium/low/unknown-impact threat in a further 27% (Figure 3). Most of these threats have arisen since Spain joined the European Union in 1986. Since then, transport and water-related infrastructure (reservoirs, hydro-power, irrigation) improvements have been made but much of the investment has detrimentally affected natural areas. Environmental Impact Assessments have failed to adequately examine alternative projects. Many water-related developments are planned that will affect IBAs, some of them without a real justification and with a potentially high impact on sites and their species. Hydro-power plants are proliferating on rivers in mountainous areas with few remaining unregulated. New wind-energy plants are being constructed in mountainous areas where birds migrate, breed or feed. Although impact-mitigation measures are now applied to all newly erected powerlines, collisions continue to affect birds.

Infrastructural developments are often connected to recreation and tourism activities, the most common threat to IBAs (Figure 3). Impacts tend to be concentrated in coastal areas, particularly along the Mediterranean coast and on the Balearic and Canary Islands, where urbanization has severely affected many habitats that were previously of interest and has destroyed others, including wetlands. Connected developments such as major roads are often funded by the EU. Away from coastal areas recreational activities such as climbing, parasailing, and hiking cause disturbance, particularly to breeding raptors in mountainous areas.

Hunting is widespread throughout Spain, and is a threat in over 40% of IBAs (Figures 2 and 3), As well as shooting, methods include non-selective poisoning, the use of lime, and trapping. In wetlands, lead poisoning of waterbirds, as a result of ingesting spent shot, is a problem. Nest robbing, of chicks and eggs, is still an activity that affects some raptor species, especially *Falco peregrinus*.

Forestry occurs in over 20% of IBAs and intensified forest management and afforestation are threats in 8% and 12% of IBAs respectively (Figures 2 and 3). In the past there have been large losses of natural forest to plantations of exotic species. The rate of change has declined but natural forests continue to be exploited causing loss of habitat and disturbance to breeding species. In naturally treeless areas (e.g. steppic habitats), EU-promoted afforestation schemes are causing the loss of habitat for species

Figure 2. Occurrence of land-uses at Important Bird Areas in Spain (see Appendix 3 for definitions of land-uses).

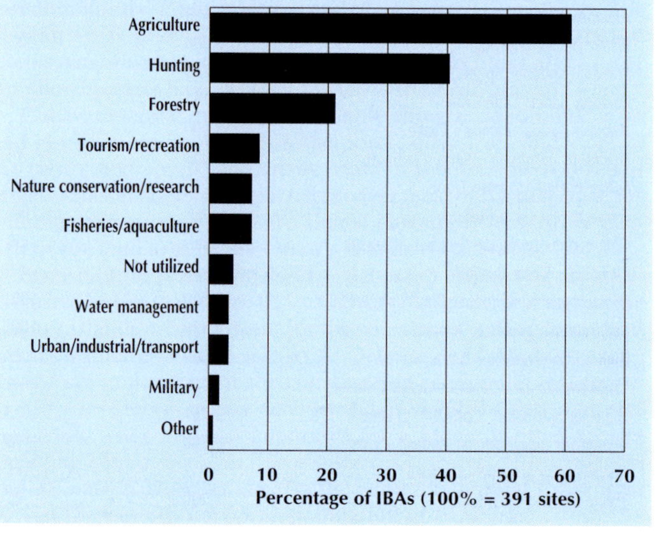

Figure 3. Threats at Important Bird Areas in Spain (see Appendix 3 for definitions of threat types and impact categories).

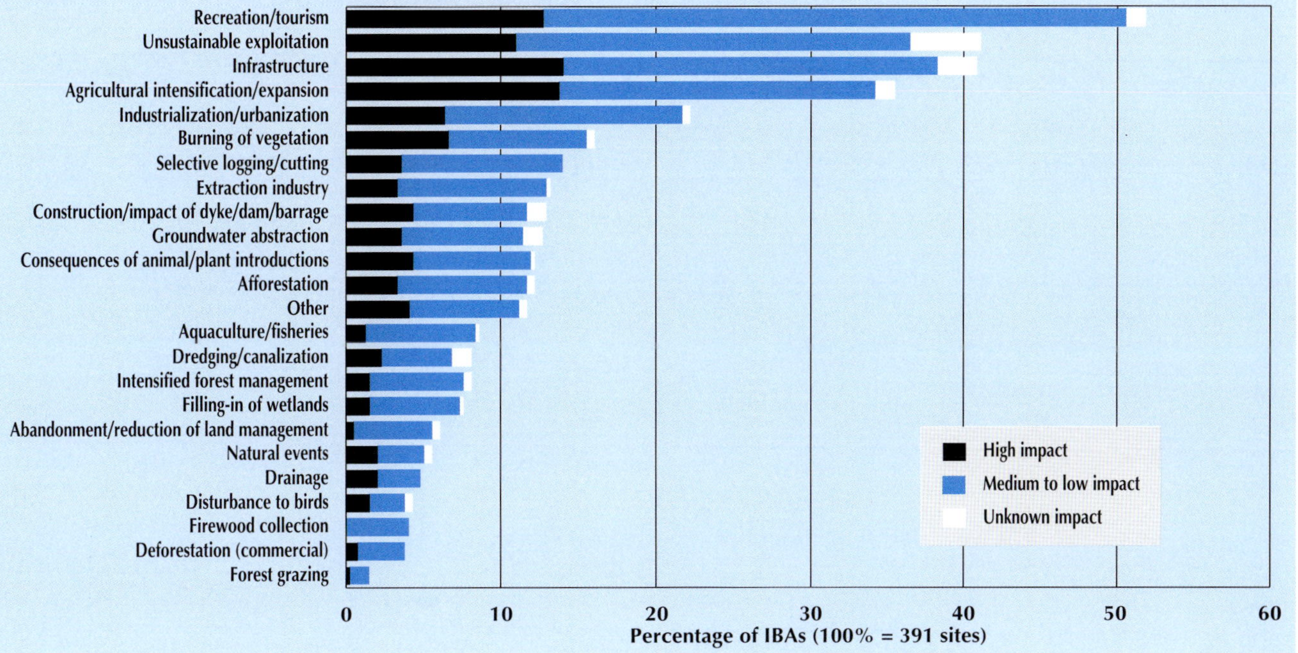

such as *Chersophilus duponti*. In the Canary Islands, wood-cutting for use in vegetable gardens is one of the greatest threats to *Columba bollii* and *C. junoniae*.

Some forested areas are also threatened by fire, with the highest incidence of fire in the Mediterranean area. Forests in mountainous areas, important for *Hieraaetus pennatus* and other raptors, are particularly threatened. Other habitats in which anthropogenic fire is a threat to breeding birds include water-fringe vegetation and alpine grasslands.

The effects of fires are compounded by the effects of drought, the main natural event affecting birds in Spain. A severe drought between 1992–1996 resulted in the drying-out of many small wetlands. Erosion in semi-desert areas appears to be increasing, often magnified by human activities.

The proliferation of introduced rats on islands and islets where seabirds breed is an important threat and a number of rat-eradication campaigns are underway. Feral cats and dogs are also a problem, particularly in the laurel forest in the Canary Islands, where they predate *Columba bollii* and *C. junoniae*. Hybridization with introduced *Oxyura jamaicensis* threatens *O. lecucocephala*. A large increase in the Mediterranean population of *Larus cachinnans* is a threat, particularly to *L. audouinii*. In the Balearic Islands and in Cataluña culling campaigns to control *L. cachinnans* are carried out every year.

PROTECTION STATUS

National protection

Forty-seven percent of IBAs in Spain are, at least in part, covered by national protection areas (Figure 4). However, in only 16% of IBAs is 90% or more of the area of the IBA protected. Overlaps with different national protection designations are listed in Table 5.

The protected-area designations listed below and in Table 5 were established under National Law 4/1989 (which concerned the conservation of protected sites and wild fauna and flora and included five protection designations) and by several regional laws. Since 1987, responsibility for the designation of sites has been with regional governments, which have increased the number of designation types to 21. The most important of these relating to bird conservation are considered in this inventory. The exception to the regional approach are National Parks which are designated by the National Parliament and administered by a mixed body, comprising representatives of regional and national government. In the last 10 years, the number of sites receiving protection has increased spectacularly. Currently in Spain there are 12 National Parks, 111 Parks (natural, regional), 137 Reserves (natural, integral, partial) and 242 other protection titles (Europarc España 1998). National protected-area designations considered in this inventory, and their aims, are listed as follows.

Figure 4. The national protection status of Important Bird Areas in Spain.

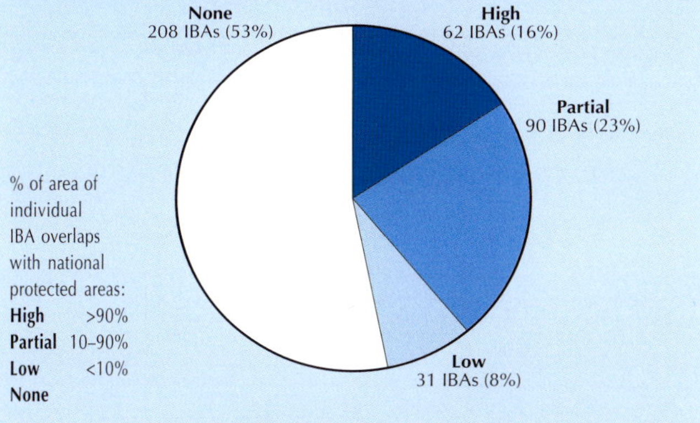

Total area of overlap between IBA network in Spain and national protected-area system (see Table 5 for categories) = 29,141–34,266 km² (18–22% of total IBA area).

Figure 5. The international protection status of Important Bird Areas in Spain.

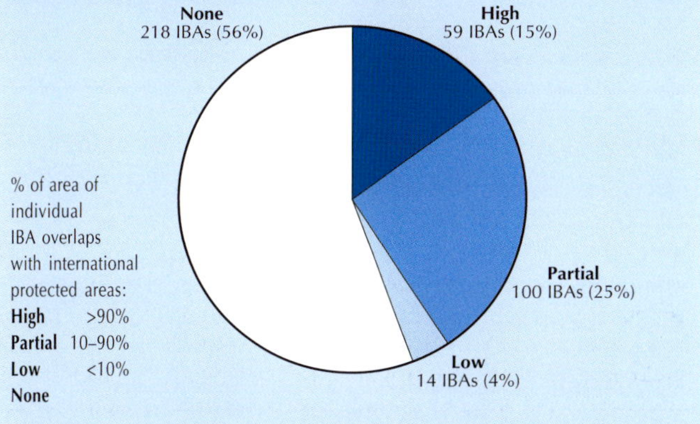

Total area of overlap between IBA network in Spain and international protected-area system (see Table 5 for categories) = 30,544–31,871 km² (19–20% of total IBA area).

Table 5. Protection status of Important Bird Areas in Spain.
A tick (✓) indicates that an IBA overlaps with a protected area (to any extent).

		National																International			
IBA code	International name	Integral Nature Reserve	Special Nature Reserve	National Biological Reserve	National Park	Natural Reserve	Natural Park	Regional Park	Rural Park	Site of Scientific Interest	Natural Landscape	National Game Reserve	Protected Landscape	Natural Area	Natural Monument	Natural Site of National Interest	Wild Fauna Reserve	Ramsar Site	World Heritage Site	Biosphere Reserve#	Special Protection Area
001	Cíes islands						✓														✓
002	Ons islands																				
003	Ría de Arosa (O Grove) estuary																	✓			✓
004	Costa de la Muerte (North coast)															✓					
005	Ferrolterra–Valdoviño coast																	✓			
006	Cape Candelaria–Ortigueira estuary–Cape Estaca de Bares															✓		✓			✓
007	Eo river estuary–Barayo beach–Ría de Foz											✓						✓			✓
008	La Limia																				
009	Ancares mountains											✓									
010	Montes Aquilanos range														✓				✓		
011	La Cabrera mountain range						✓								✓						
012	Muniellos			✓																	✓
013	Degaña–Hermo											✓									✓
014	Babia–Somiedo						✓					✓									✓
015	Gistreo and Coto mountain ranges																				
016	Central ranges of the Cantabric mountains						✓					✓									
017	Cape Busto–Luanco											✓									
018	Ribadesella–Tina Mayor																				
019	Riaño							✓				✓									
020	Picos de Europa mountain range				✓			✓				✓									✓
021	Fuentes Carrionas											✓									
022	Peña Labra and Cordel mountain ranges						✓					✓									
023	River Ebro dam																				
024	Tina Menor–Sarnosa island–Oyambre						✓														
025	Portios islets–Conejera island																				
026	Santander bay–Mouro islet																				
027	Santoña marshes					✓												✓			✓
028	Humada–Peña Amaya																				
029	Upper Ebro and Rudrón canyons																				
030	Oña and Tesla mountain ranges																				
031	Obarenes mountains																				✓
032	Valdegovía–Arcena mountain range						✓														
033	La Losa–Orduña (Salvada mountain range)																				
034	Sobrón canyon																				
035	Guernica estuary–Cape Ogoño																	✓		✓	✓
036	Izki mountains																				
037	Chingudi–Bidasoa estuary																				
038	Tierra de Campos steppes																				
039	Villafáfila											✓						✓			✓
040	Esla dam																				
041	Belver de los Montes–Gallegos del Pan																				
042	River Pisuerga at Dueñas																				
043	Carrión–Frómista																				
044	Cerrato high moor																				
045	La Demanda mountains											✓									✓
046	Urbión and Cebollera mountain ranges											✓									✓
047	Gorges of Iregua, Leza and Jubera																				✓
048	Arnedillo–Isasa crag																				✓
049	Alcarama mountain and Alhama river																				✓
050	River Lobos canyon						✓														✓
051	Arlanza valley–Peñas de Cervera																				
052	Montejo de la Vega–Riaza canyon																				✓
053	Duratón canyon						✓														✓
054	River Cega–Tierra de Pinares–Cantalejo																				
055	River Moros																				
056	Northern slope of Guadarrama mountain range															✓					✓
057	Segovia																				
Subtotal of IBAs		0	0	1	1	1	9	2	0	0	0	13	0	0	2	3	0	7	1	1	21

Table 5 ... continued. Protection status of Important Bird Areas in Spain.
A tick (✔) indicates that an IBA overlaps with a protected area (to any extent).

		National																International			
IBA code	International name	Integral Nature Reserve	Special Nature Reserve	National Biological Reserve	National Park	Natural Reserve	Natural Park	Regional Park	Rural Park	Site of Scientific Interest	Natural Landscape	National Game Reserve	Protected Landscape	Natural Area	Natural Monument	Natural Site of National Interest	Wild Fauna Reserve	Ramsar Site	World Heritage Site	Biosphere Reserve#	Special Protection Area
058	Tordesillas–Mota del Marqués																				
059	Castronuño–Zamora																				✔
060	Fuentelapeña–Jambrina																				
061	Tierra de Campiñas steppes																				
062	Topas																				
063	Arribes del Duero–Fermoselle																				✔
064	Río Huebra–Arribes del Duero																				✔
065	Campo de Argañán																				
066	Peña de Francia											✔									✔
067	Gredos and Candelario mountain ranges							✔				✔									
068	Tiétar valley																				
069	Iruelas valley					✔															✔
070	El Escorial–San Martín de Valdeiglesias																				✔
071	El Pardo–Viñuelas						✔													✔	✔
072	Redbeeds and copses of Aranjuez					✔															✔
073	Cortados del Jarama							✔													✔
074	Talamanca–Camarma																				✔
075	Alcarria de Alcalá steppes																				
076	Alto Lozoya–La Pedriza						✔													✔	✔
077	Ayllón mountain range						✔					✔				✔					
078	Puebla de Beleña lagoons																				
079	Tiermes–Caracena																				
080	Altos de Barahona high moor																				
081	Layna high moors																				
082	Las Cañas lake					✔															✔
083	Echauri crags																✔				
084	Gorramendi mountain and Itxusi crags					✔											✔				✔
085	Roncesvalles–Irati–Abodi mountain range	✔				✔											✔				✔
086	Areta, Archuba and Zariquieta mountain ranges					✔															✔
087	Leyre, Illón and San Miguel mountain ranges					✔											✔				✔
088	Izaga crag																✔				✔
089	Pitillas and Dos Reinos lagoons					✔								✔							✔
090	Bardenas Reales					✔															
091	Cinco Villas ponds																				
092	Moncayo mountain range						✔														
093	River Jalón canyon																				
094	Langa–Mara–Belmonte de Calatayud plains																				
095	Gallocanta lake																✔	✔			✔
096	Blancas–Torralba de los Sisones																				
097	Montes Universales–Albarracín mountain range											✔									
098	Campo Visiedo																				
099	River Guadalope																				
100	River Martín canyons and Arcos mountain range																				
101	Alcañiz salt-lakes																				
102	Lower course of the River Huerva																				
103	Belchite–Mediana																				✔
104	River Ebro oxbow lakes					✔															✔
105	Monegrillo–Pina steppe area–Pina																				
106	Los Monegros (South)																				
107	Valdurrios–Serreta Negra and Los Rincones mountain ranges																				
108	Lower course of the River Matarraña–Ribarroja																				
109	Rivers Cinca and Alcanadre riverine forest																				
110	Ballobar–Candasnos																				
111	Cinca river rice-fields and steppe area																				
112	Sariñena lake and La Estación pond																✔				
113	Alcubierre and Sigena mountain ranges																				
114	Confidential																				
Subtotal of IBAs		1	0	1	1	11	13	4	0	0	0	17	0	1	2	4	7	8	1	3	42

Table 5 ... continued. Protection status of Important Bird Areas in Spain.
A tick (✔) indicates that an IBA overlaps with a protected area (to any extent).

		National																International			
IBA code	International name	Integral Nature Reserve	Special Nature Reserve	National Biological Reserve	National Park	Natural Reserve	Natural Park	Regional Park	Rural Park	Site of Scientific Interest	Natural Landscape	National Game Reserve	Protected Landscape	Natural Area	Natural Monument	Natural Site of National Interest	Wild Fauna Reserve	Ramsar Site	World Heritage Site	Biosphere Reserve#	Special Protection Area
115	Zuera mountains																				
116	Tormos reservoir (La Sotonera)																				
117	Bajo Alcanadre–Serreta de Tramaced																				
118	Guara mountain range						✔														✔
119	Oturia–Canciás																				
120	Santo Domingo–Riglos–Gratal																				
121	San Juan de la Peña–Oroel crag															✔					
122	Dos Ríos and Orba mountain ranges																				
123	Belagua–Ansó–Hecho	✔				✔						✔									✔
124	Collarada–Telera																				
125	Panticosa–Vignemale–Tendeñera				✔							✔									
126	Ordesa–Bielsa				✔							✔								✔	✔
127	Gistaín–Cotiella						✔					✔									✔
128	Posets–Maladeta						✔					✔									✔
129	Turbón–Espés–Sis																				
130	San Mauricio–Bohí–Beret				✔	✔															✔
131	Sant Gervás mountains																				
132	Boumort mountains					✔						✔									
133	Monteixo–L'Orri–Tornafort																				
134	Cerdaña											✔									
135	Cadí mountains						✔					✔									✔
136	Fresser–Setcases											✔									
137	Ampurdán marshes	✔					✔											✔			✔
138	Medas islands																				
139	Mountains of Barcelona						✔														
140	Llobregat delta					✔															✔
141	Montagut and Montmell mountains																				
142	Lerida steppes																				
143	Montsech and Montgai mountains					✔															
144	Cogul–Alfés steppes					✔															✔
145	Montsant and Prades mountains																				
146	Cardó, Tivissa and Llabería mountains																				
147	Pass of Beceite–Turmell mountain											✔									
148	Ebro delta					✔	✔											✔			✔
149	Morella pass																				
150	Peñagolosa																				
151	Espadán mountain range																				
152	Prat de Cabanes–Torreblanca						✔											✔			✔
153	Columbretes islands					✔															✔
154	River Mijares mouth																				
155	Almenara marshes																				
156	El Moro marshes																				✔
157	Turia canyon and Los Serranos																				
158	Rivers Cabriel and Júcar canyons					✔						✔									✔
159	Albufera de Valencia marshes						✔											✔			✔
160	La Safor and North Alicante mountain ranges						✔														
161	Enguera mountain range–La Canal de Navarrés																				
162	Pego–Oliva marshes						✔														✔
163	Marina mountain ranges																				
164	Alicante islets					✔															✔
165	El Hondo wetland						✔											✔			✔
166	Santa Pola saltpans						✔											✔			✔
167	Mata and Torrevieja lagoons						✔											✔			✔
168	El Valle, Altaona and Escalona mountains							✔													
169	Mar Menor coastal lagoon							✔					✔					✔			
170	Littoral islets of Murcia and Almería																				
171	Fausilla mountains																				
172	Muela mountain range and Cape Tiñoso																				
Subtotal of IBAs		3	0	1	4	21	27	6	0	0	0	28	1	1	2	5	7	16	1	4	63

Table 5 ... continued. Protection status of Important Bird Areas in Spain.
A tick (✔) indicates that an IBA overlaps with a protected area (to any extent).

		National																International			
IBA code	International name	Integral Nature Reserve	Special Nature Reserve	National Biological Reserve	National Park	Natural Reserve	Natural Park	Regional Park	Rural Park	Site of Scientific Interest	Natural Landscape	National Game Reserve	Protected Landscape	Natural Area	Natural Monument	Natural Site of National Interest	Wild Fauna Reserve	Ramsar Site	World Heritage Site	Biosphere Reserve#	Special Protection Area
173	La Almenara mountain range–Cape Cope							✔					✔								
174	Torrecilla and Gigante mountain ranges																				
175	Guadalentín saltmarshes																				
176	Las Cabras plain																				
177	Burete, Cambrón and Espuña mountain ranges							✔				✔									
178	Quipar reservoir, Molino mountain range and Cagitán plains																				
179	Ricote and Nevela mountain ranges																				
180	Pila mountain range							✔													
181	Moratalla mountain range																				
182	Pétrola–Almansa–Yecla																				✔
183	River Mundo and River Segura canyons																				
184	Campo de Montiel																				
185	San Clemente–Villarrobledo																				
186	Canyons on the middle section of the Cabriel river																				
187	Cuenca mountain ranges											✔									✔
188	Upper sections of the Rivers Tajo and Tajuña																				✔
189	Embid–Molina high moors																				
190	River Mesa canyons																				
191	Entrepeñas and Buendía reservoirs																				✔
192	El Hito																				✔
193	Tarancón–Ocaña–Corral de Almaguer																				✔
194	Tembleque–La Guardia plains																				✔
195	Alcázar de San Juan–Quero endorreic lagoons																	✔			✔
196	Pedro Muñoz–Manjavacas endorreic lagoons																	✔			✔
197	Tablas de Daimiel marshes; 'Vicario' and 'Gasset' reservoirs and Malagón lakes	✔			✔													✔		✔	✔
198	Castrejón reservoir																				✔
199	Torrijos																				
200	Malpica island (River Tagus)																				✔
201	Rosarito and Navalcán reservoirs–La Iglesuela (Tiétar valley)																				✔
202	Oropesa plains																				✔
203	Azután reservoir																✔				✔
204	Toledo mountains–Cabañeros				✔																✔
205	Canalizos mountain range (Saceruela)																				✔
206	Campo de Calatrava																				✔
207	Alcudia valley and mountain range																				✔
208	Madrona and Andújar mountain ranges						✔														✔
209	Aldeaquemada–Dañador										✔										✔
210	Cazorla and Segura mountain ranges						✔					✔								✔	✔
211	Puebla de Don Fadrique–Las Cañadas																				
212	María mountain range–Los Vélez						✔														
213	Hoya de Baza																				
214	Hoya de Guadix																				
215	Alhamilla mountain range–Campo de Níjar plain										✔										✔
216	Mountain range and saltpans at Cabo de Gata						✔											✔			✔
217	Tabernas desert										✔										✔
218	Gádor mountain range																				
219	Wetlands of western Almería					✔					✔							✔			✔
220	Chafarinas Islands											✔									✔
221	Alborán island																				
222	Sierra Nevada mountain range						✔					✔								✔	
223	Prelittoral mountain ranges of Granada										✔										
224	Mouth of the River Guadalhorce										✔										
225	Antequera mountain range–El Chorro										✔										✔
226	Mountain ranges of Córdoba						✔														
227	Mountain ranges south of Jaén						✔														
228	Middle section of River Guadalquivir reservoirs										✔										
Subtotal of IBAs		4	0	1	6	22	34	9	0	0	8	33	2	1	2	5	8	21	1	7	91

Table 5 ... continued. Protection status of Important Bird Areas in Spain.
A tick (✔) indicates that an IBA overlaps with a protected area (to any extent).

		National																International			
IBA code	International name	Integral Nature Reserve	Special Nature Reserve	National Biological Reserve	National Park	Natural Reserve	Natural Park	Regional Park	Rural Park	Site of Scientific Interest	Natural Landscape	National Game Reserve	Protected Landscape	Natural Area	Natural Monument	Natural Site of National Interest	Wild Fauna Reserve	Ramsar Site	World Heritage Site	Biosphere Reserve#	Special Protection Area
229	Jaén countryside																				
230	Marmolejo reservoir–La Ropera																				
231	Conde, Chinche and Honda lakes					✔															✔
232	Córdoba countryside																				
233	Hinojosa del Duque–El Viso																				
234	Los Blázquez–La Granjuela–Fuenteovejuna																				
235	Sierra Morena de Córdoba						✔														✔
236	Sierra Morena de Sevilla						✔														✔
237	Carmona countryside																				
238	Écija–Osuna plain					✔															
239	Wetlands at south Córdoba					✔					✔							✔			✔
240	Fuente de Piedra, Gosque and Campillos lakes					✔												✔			✔
241	Mountain ranges of Ronda, Bermeja and Crestellina						✔				✔	✔									
242	Líjar mountain range–Zaframagon crag					✔															
243	Ubrique and Grazalema mountain ranges						✔													✔	✔
244	Cabras, Aljibe and Montecoche mountain range						✔														✔
245	Bujeo, Ojén, del Niño and Blanquilla mountain ranges						✔														✔
246	Tarifa										✔										
247	Ceuta																				
248	De la Plata mountain range																				
249	Tajo de Barbate coastal cliffs						✔														
250	La Janda																				
251	Cádiz Bay					✔	✔				✔										✔
252	Medina and Puerto Real lagoons					✔															✔
253	Terry lagoons					✔												✔			✔
254	Bornos reservoir										✔										
255	Medina-Sidonia																				
256	Confidential																				
257	Los Tollos lake																				
258	Lebrija, Las Cabezas and Espera lagoons					✔															✔
259	Guadalquivir marshes				✔		✔				✔							✔	✔	✔	✔
260	Condado–Campiña																				
261	Odiel and Tinto marshes and Huelva coastal lagoons					✔					✔							✔		✔	✔
262	River Piedras marshes and Rompido sandbank										✔										
263	Isla Cristina and Ayamonte marshes and Prado lagoon										✔										
264	West Andévalo																				
265	Sierra Pelada mountain range										✔										✔
266	Las Contiendas mountain range						✔														✔
267	East mountain range of Huelva						✔														✔
268	Fuente de Cantos–Montemolín																				
269	Azuaga–Llerena–Peraleda de Zaucejo																				
270	Central Badajoz mountain ranges																				✔
271	Bienvenida–Usagre–Ribera del Fresno																				
272	Valuengo reservoir																				
273	Jerez de los Caballeros dehesas																				
274	Villanueva del Fresno																				
275	Valongo																				
276	Olivenza–La Albuera																				
277	Alange																				
278	Villalba de los Barros																				
279	Puerto de Mejoral–Almorchón–Cabeza del Buey																				
280	La Serena																				
281	Siruela–Agudo																				
282	Cijara reservoir											✔									
283	Puerto Peña reservoir–Valdecaballeros																				
Subtotal of IBAs		4	0	1	7	32	45	9	0	0	18	35	2	1	2	5	8	26	2	10	109

Table 5 ... continued. Protection status of Important Bird Areas in Spain.
A tick (✔) indicates that an IBA overlaps with a protected area (to any extent).

		National																International			
IBA code	International name	Integral Nature Reserve	Special Nature Reserve	National Biological Reserve	National Park	Natural Reserve	Natural Park	Regional Park	Rural Park	Site of Scientific Interest	Natural Landscape	National Game Reserve	Protected Landscape	Natural Area	Natural Monument	Natural Site of National Interest	Wild Fauna Reserve	Ramsar Site	World Heritage Site	Biosphere Reserve#	Special Protection Area
284	Pela mountain range–Orellana reservoir–Zorita																	✔			✔
285	Don Benito–Guareña																				
286	Valdehornillos–Santa Amalia																				
287	Montánchez mountain range–Cornalvo reservoir						✔														✔
288	Mérida–Montijo reservoir																				
289	Lácara–Morante																				
290	Botoa–Villar del Rey																				
291	San Pedro mountain range																				✔
292	Cedillo reservoir																				
293	Brozas–Membrío																				
294	Malpartida de Cáceres–Arroyo de la Luz																				
295	Plain between Cáceres and Trujillo–Aldea del Cano																				✔
296	Trujillo–Torrecillas de la Tiesa																				
297	Las Villuercas mountain range																				
298	Monfragüe						✔														✔
299	Alcántara reservoir–Cuatro Lugares																				
300	Coria mountain ranges																				
301	Borbollón reservoir																				
302	East Gata mountain range																				
303	Gabriel y Galán reservoir																				
304	Plasencia and San Bernabé mountain range																				
305	Lower Tiétar river and La Vera valley																				
306	Campo Arañuelo–Valdecañas reservoir																				✔
307	Islets Vedrá and Vedranell																				✔
308	Isle of Conejera and islets of Bledes and Espartar (Ibiza)																				
309	Cape Nonó–Isle of Murada																				
310	Tagomago island																				✔
311	Islets of Santa Eulalia, Redona and Es Canar (Ibiza)																				
312	Salt-plans of Ibiza and Formentera and Freus isles					✔												✔			✔
313	Cape Barbaria																				
314	La Mola of Formentera																				
315	Dragonera island–La Trapa						✔														✔
316	Tramuntana mountains																				✔
317	Cape Pinar																				✔
318	Albufera de Mallorca and Albufereta de Pollença marshes						✔											✔			✔
319	Cape Freu–Cape Farrutx																				✔
320	Cape Vermell																				✔
321	Salobrar de Campos lagoons–Sa Vall																				✔
322	Coastal cliffs between Cape Enderrocat and Cala Pí																				✔
323	Coastal cliffs between isles of Malgrats and Cala Figuera																				✔
324	Archipelago of Cabrera				✔																✔
325	North and east coasts of Minorca and Island of Aire						✔													✔	✔
326	Ravines and pinewoods in central Minorca																				
327	Lanzarote islets	✔					✔														✔
328	Famara crag						✔														✔
329	Haría–Tabayesco						✔														
330	Tenegüime gully												✔								
331	Plains of La Corona–Las Honduras																				
332	Famara sandy plain						✔														
333	El Mojón				✔																✔
334	Janubio saltpans									✔											✔
335	Plains of Mareta–Hoya de la Yegua																				
336	Island of Lobos						✔														✔
337	Sandy plain of Corralejo						✔														✔
338	Morro Tabaiba–Morro de los Rincones-Vallebrón mountains																				
Subtotal of IBAs		5	0	1	9	33	56	9	0	1	18	35	3	1	2	5	8	29	2	11	135

Table 5 ... continued. Protection status of Important Bird Areas in Spain.
A tick (✔) indicates that an IBA overlaps with a protected area (to any extent).

		National																International			
IBA code	International name	Integral Nature Reserve	Special Nature Reserve	National Biological Reserve	National Park	Natural Reserve	Natural Park	Regional Park	Rural Park	Site of Scientific Interest	Natural Landscape	National Game Reserve	Protected Landscape	Natural Area	Natural Monument	Natural Site of National Interest	Wild Fauna Reserve	Ramsar Site	World Heritage Site	Biosphere Reserve#	Special Protection Area
339	Río Cabras gully																				
340	Cuchillete de Buenavista–gully of La Torre–Los Alares																				
341	Pozo Negro mountain – Vigán														✔						✔
342	Tarajalejo mountain range																				
343	Sotavento beach						✔														✔
344	Peninsula of Jandía						✔			✔											✔
345	Isthmus of Jandía sandy plain						✔														✔
346	Gully of Ajuí–Betancuria								✔												✔
347	Los Molinos gully–Plain of La Laguna								✔												✔
348	Coast of Esquinzo–Puertito de Los Molinos								✔												✔
349	Lajares Reserve–Cotillo–Ezquinzo																				✔
350	Coast between Corralejo and Tostón																				
351	Coast between Arinaga and Castillo del Romeral									✔											✔
352	Tirajana pine woodland						✔														✔
353	Tauro pine woodland														✔						✔
354	Pajonales, Ojeda, Inagua and La Data pine woodlands	✔																			✔
355	Tamadaba pine woodland						✔		✔												✔
356	Anaga rocky islets	✔							✔												✔
357	San Andrés, Pijaral and Anaga	✔							✔												✔
358	Las Vueltas, Aguas Negras and Quebradas mountains								✔												✔
359	Las Mercedes, Mina y Yedra, Aguirre, La Goleta, and Pedro Alvarez mountains								✔												✔
360	Los Rodeos–La Esperanza																				
361	La Playa islet																				
362	Santa Úrsula and La Victoria laurel forest	✔	✔										✔								
363	Tigaiga slope												✔								✔
364	Santo Domingo coastal cliffs																				
365	Garachico rocky islet														✔						
366	Mountain of Agua, gully of Los Cochinos and gully of Cuevas Negras								✔												✔
367	Los Gigantes coastal cliff								✔												✔
368	Tágara gully						✔														✔
369	Vilaflor pine woodland						✔														✔
370	Arico pine woodland						✔														✔
371	Centinela mountain and La Esquina plain														✔						
372	El Médano coast		✔																		
373	Rasca lava-fields–Guaza mountain–Las Mesas plain		✔												✔						
374	Majona coast		✔				✔														
375	Step rocks of Hermigua and Agulo																				
376	Garajonay National Park				✔														✔		✔
377	Vallehermoso coast														✔						✔
378	South-west coast of La Gomera								✔	✔			✔								✔
379	La Palma laurel forest	✔	✔				✔			✔			✔								✔
380	El Canal y Los Tiles laurel forest						✔													✔	✔
381	Negro rocky islet																				
382	Garafía rocky islets																				
383	La Caldera de Taburiente National Park				✔																✔
384	El Roque coastal cliffs																				
385	Ventejís mountain												✔								✔
386	Nizdafe plains																				
387	Salmor rocky islets	✔	✔																		✔
388	Frontera laurel forest	✔	✔						✔												✔
389	Western coast of El Hierro								✔												✔
390	La Dehesa grasslands								✔												✔
391	Naos Bay–Hoya de Tacorón								✔												✔
Total number of IBAs		12	7	1	11	33	67	9	15	5	18	35	8	1	8	5	8	29	3	12	170

1. Integral Nature Reserve
Preservation of ecosystems. Human presence is highly restricted and is only allowed for scientific research and awareness.

2. Special Natural Reserve
For preservation of outstanding habitats, particular species, geological formations or natural ecological processes of special interest. Human presence only allowed for scientific, educational and, exceptionally, recreational or traditional activities.

3. National Biological Reserve
For strict maintenance of the natural biological balance and the conservation of the integrity of existing faunal and floral communities and their genetic resources. Exploitation of all resources is prohibited. The only permitted activities are for management and improvement of the habitat.

4. National Park
Large areas with a high conservation status, not significantly modified by human activities, with scientific, landscape and educational value. The aim of the designation is to protect areas from any intervention that could alter their physiognomy, integrity or the evolution of natural processes. Exploitation of natural resources and alteration of the landscape are not allowed, except if the activity is compatible with protection. Also prohibited are hunting, fishing, disturbance to animals and the extraction industry.

5. Natural Reserve
Protection of ecosystems, biological or geological communities or their elements because of their rarity, fragility, importance or singularity. Exploitation of natural resources is limited to those compatible with the conservation of the site. Collection of biological material is prohibited.

6. Natural Park/Regional Park
Areas of notable natural value and outstanding biological quality, where human presence and activities compatible with dynamic natural processes are permitted.

7. Rural Park
Conservation of natural value and promotion of harmonic development of local populations and their lives.

8. Site of Scientific Interest
Isolated sites with natural elements of scientific interest with specimens or populations of endangered animals or plants that need specific conservation measures.

9. Natural Landscape (Andalucía region only)
Protection of outstanding fauna, flora, geological formations, special beauty or other natural elements. Traditional activities are allowed, under regulation.

10. National Game Reserve
Protection, conservation and promotion of game richness, that should be exploited in a wise way in order to allow a maximum economic, recreational and social profit.

Box 1. International legislation and initiatives that are relevant to site conservation in Spain (see Appendix 1 for a general description of these agreements).

Global	
Biodiversity Convention	✔
Ramsar Convention	✔
Bonn Convention	✔
World Heritage Convention	✔
MAB Programme	✔
Pan-European	
Bern Convention	✔
Regional	
EC Birds Directive	✔
EC Habitats Directive	✔
Barcelona Convention	✔

✔ Convention ratified/initiative supported
(✔) Convention signed

11. Protected Landscape
Protection of sites because of their aesthetic and cultural value.

12. Natural Area
Protection of landscape or ecological values. Human activities are allowed, under regulation.

13. Natural Site of National Interest/Natural Monument
Sites or elements of nature, of small size/dimension, of great singularity, rarity or beauty, that are subject to special protection (mainly geological forms, fossil beds).

14. Wild Fauna Reserve
For the protection of fauna; hunting is prohibited.

International protection

Forty-four percent of IBAs in Spain have at least part of their area under international protection (Figure 5, Table 5). In only 15% of IBAs is more than 90% of the site protected. Spain has ratified the most important international conventions related to site-based nature conservation (Box 1).

By April 1998 Spain had designated 170 Special Protection Areas (under the EC Birds Directive) that overlapped with or were contained by IBAs, representing 3,344,197 ha (21% of the total IBA area in Spain). Since data for this inventory were analysed this has increased to 175 SPAs covering 3,388,003 ha of the IBA network. However, the current SPA network is insufficient to ensure the conservation of species listed on Annex I of the Birds Directive. It is particularly important to improve the SPA coverage of some steppe species, as well as for *Ciconia nigra, Circus aeruginosus, Hieraaetus fasciatus, Tetrao urogallus cantabricus, Perdix perdix, Grus grus* and *Dryocopus martius*.

Seventy-seven wetlands in Spain meet Ramsar criteria for birds; 38 of them are included on the Ramsar List. Twenty-nine IBAs overlap in part or totally with one or more of these Ramsar Sites.

Spain's 13 Biosphere Reserves (MAB-UNESCO) have all been designated within or encompass IBAs. Under the World Heritage Convention, 28 World Heritage Sites have been designated in Spain, with three of them related to wild landscapes and overlapping with IBAs (Table 5).

CONSERVATION

- The status of wildlife conservation varies across regions in Spain. In general, in the last 10 years, monitoring and conservation actions have increased as a result of the development of regional conservation policies.
- The National Government promotes task forces for some species and for conservation issues. These include representatives of all relevant regions and also conservation NGOs. Task forces act as coordinating groups for monitoring and conservation activities. They exist for *Gypaetus barbatus, Aquila adalberti, Marmaronetta angustirostris, Oxyura leucocephala, Larus audouinii*, and also for issues relating to ecotoxicology, wetlands and electric powerlines.
- An increasing number of conservation projects funded by the EU LIFE-Nature programme have improved the knowledge and conservation status of Spanish populations of threatened species including *Aquila adalberti, Gypaetus barbatus, Aegypius monachus, Puffinus yelkouan, Falco naumanni, Otis tarda, Hieraetus fasciatus, Larus audouinii, Botaurus stellaris* and *Marmaronetta angustirostris*. Some species recovery plans exist at a regional level, for example for *Gypaetus barbatus* in Aragón, Cataluña and Navarra; *Ciconia nigra* in Castilla y León; *Hieraaetus fasciatus* in La Rioja; *Haematopus ostralegus* in Asturias.
- The Environmental Impact Assessment procedure allows NGOs to examine information on development projects before their final approval and to submit recommendations to reduce environmental impacts.
- One of the greatest problems for site protection is the lack of management and finances to ensure that protection is effective. From a total of 490 protected sites in Spain only 124 (25%) have management plans (Europarc España 1998). This means that

there are statutory protected sites that are highly threatened by different human activities.

ANALYTICAL METHODS

- More than 75% of species data in the inventory have been collected since 1995. The quality of data is variable for different groups of birds. For IBAs that overlap with protected sites, data are accurate. Accurate data exist for protected and rare species monitored by regional governments, for example large raptors and wildfowl. Thirty-eight percent of bird population data are of high accuracy, 44% of medium accuracy and 18% are low accuracy. Most passerines, common raptors and owls are not well covered by the present inventory due to a lack of data.
- Recent data are scarce for bird numbers in some migratory bottleneck areas, particularly in the Strait of Gibraltar. Data are also lacking for migratory species in some of the main wetlands in Spain. The Guadalquivir marshes (IBA 259), the most important IBA in Spain, is monitored by different administrations (the National Park by the National Government technical staff, the Natural Park by Regional Government, and also the Doñana Biological Station). There is no coordination between these bodies so it has been almost impossible to obtain total numbers for certain migratory species.
- Population estimates do not exist for most species in the Canary Islands. IBA selection has been based on the areas of suitable habitat for species in the Canaries.
- Data on congregatory species have been presented in different ways. Minimum wintering numbers of *Grus grus* in Extremadura are the average of annual censuses between 1990 and 1996, and maximum numbers are the highest number of individuals counted in one year of the same period. For wildfowl, these can be minimum and maximum estimates for one year or minimum and maximum figures over other periods; in the case of the Guadalquivir marshes, minimum wildfowl numbers are the average of the period 1978–1989, and maximum values represent the highest figures for the same period. Data for all breeding birds are given in pairs except *Otis tarda, Chlamydotis undulata, Tetrax tetrax* and *Tetrao urogallus*.
- The criteria A4ii (*Milvus milvus* only), B1iii and C2 have been applied to wintering or non-breeding *Milvus migrans, Milvus milvus, Hieraaetus fasciatus* and *H. pennatus*, where numbers are particularly high and therefore the species has been deemed sufficiently congregatory.
- Spain holds the largest European populations of the globally threatened *Falco naumanni* and *Otis tarda* and the near-threatened *Tetrax tetrax*. As these three species are relatively abundant in some areas of Spain, the thresholds for meeting the A1 criterion were raised in order to identify the most important sites in the country. The threshold for *Falco naumanni* was raised from 10 to 50 pairs at a site, for *Otis tarda* 50 individuals instead of 30 individuals and *Tetrax tetrax* 200 birds instead of 60 birds.
- Marine areas have also not been well covered, due to lack of data.
- Quantitative data on the percentage cover of habitats and land-uses is lacking for many sites. Analysis in this chapter has been limited to those habitats and land-uses judged to be the most extensive by IBA compilers. These are marked with '*' in the site accounts. A more complete list of the land-uses and habitats is presented in the site accounts.
- Site boundaries were delimited on official topographic maps (scales 1:200,000 and 1:100,000; and 1:50,000 for the Canary Islands). Areas of IBAs have been measured using a digital planimeter, so accuracy is high. The digitised IBA boundaries are also available through a project carried out by SEO/BirdLife and the Universidad Autónoma de Madrid (contracted by the Ministry of Environment).
- NUTS-level regions used for the application of C criteria correspond to the autonomous regions in Spain (17 in total).

GLOSSARY

dehesa wooded pastoral habitats on the Iberian peninsula dominated by oak *Quercus*. Scrub growth is controlled by extensive rotational cultivation or grazing.
endorreic of, or pertaining to, a closed drainage basin, i.e. closed from the sea.
garrigue a community of low, scattered, often spiny and aromatic shrubs, characteristic of the Mediterranean region.
LIFE an EU funding line (Regulation 1973/92/EEC) which supports EU environmental legislation, including the Birds and Habitats Directives.
maquis dense, mostly evergreen shrub community 1–3 m high, characteristic of the Mediterranean region.
NUTS Nomenclature des Unités Territoriales Statistiques – the administrative regions of the European Union are called NUTS regions, and are designated by the EC Statistical Office.
SEO/BirdLife Sociedad Española de Ornitología, BirdLife International partner in Spain.
SPA Special Protection Area (designated under Article 4 of the EC Birds Directive).

ACKNOWLEDGEMENTS

More than 300 ornithologists have collaborated in collecting information. SEO/BirdLife would like to particularly acknowledge their help, without which the inventory would not be accurate. Unfortunately, they cannot be listed here as a result of space restrictions but a full list can be found in Viada (1998).

Special thanks to the volunteers who collaborated in the review: Dolores Hedo, Manuel Hernández, Ignacio Jaime Arranz, David López, Juan Antonio Lorenzo, Paloma Martín and José María Pérez; all SEO/BirdLife staff, and particularly the Regional Officers Juan Carlos Cirera (Aragón), Mario Jiménez (Valencia), Cristina González (Canary Islands) and Francisco Romero (Andalucía).

SITE ACCOUNTS

Cíes islands — A4i, B1i, B3, C2, C3, C6 — 001

Admin region Galicia
Coordinates 42°15'N 8°53'W
Altitude 0–197 m **Area** 433 ha

Site description

Two islands in the Ría de Vigo estuary with extensive areas of cliff. The vegetation is mainly scrub and *Pinus* and *Eucalyptus* plantations. The main human activities are tourism and recreation.

Habitats Forest and woodland (10%; broadleaved deciduous forest), Scrub (40%; scrub), Wetland (5%; sand-dunes/sand beach), Rocky areas (20%; sea cliff/rocky shore; rock stacks/islets), Artificial landscape (25%; forestry plantation)
Land-use Fisheries/aquaculture, Nature conservation/research (100%), Tourism/recreation

Birds

This is an important site for breeding seabirds with one of the largest colonies of *Larus cachinnans* in the world and of *Phalacrocorax aristotelis* in Europe.

Species	Season	Year	Pop min	Pop max	Acc	Criteria
[1] ***Phalacrocorax carbo*** Cormorant	W	1996	500	1,500	B	B1i, C3
[1] ***Phalacrocorax aristotelis*** Shag	R	1996	900	1,100	A	A4i, B1i, B3, C2
[1] ***Larus cachinnans*** Yellow-legged Gull	R	1996	19,500	19,500	—	A4i, B1i, C3
Uria aalge Guillemot	R	1995	0	1	B	C6

1. Large increase.

Protection status

National High **International** High
433 ha of IBA covered by Natural Park (Islas Cíes, 433 ha). 433 ha of IBA covered by Special Protection Area (Islas Cíes, 433 ha).

Conservation issues

Threats Aquaculture/fisheries (C), Burning of vegetation (B), Recreation/tourism (B)

The main threats are from tourism and fire. A management plan exists.

Ons islands A4i, B1i, B3, C3 002

Admin region Galicia
Coordinates 42°23′N 8°57′W
Altitude 0–119 m **Area** 500 ha

Site description

Two small islands in the Ría de Pontevedra estuary, covered by scrub and grasslands. The main human activity is tourism.

Habitats Scrub (*; scrub), Marine areas, Rocky areas (*; sea cliff/rocky shore; rock stacks/islets)
Land-use Fisheries/aquaculture, Not utilized (*), Tourism/recreation

Birds

Species	Season	Year	Pop min	Pop max	Acc	Criteria
Phalacrocorax aristotelis Shag	R	1992	400	600	B	B3
Larus cachinnans Yellow-legged Gull	R	1992	2,000	—	C	A4i, B1i, C3

This is an important site for breeding seabirds.

Protection status

National None **International** None

Conservation issues

Threats Burning of vegetation (B), Recreation/tourism (B)

The main threats are from tourism and fires. The regional government is in charge of management and restricts tourist visits to protect seabirds.

Ría de Arosa (O Grove) estuary B1i, B2, C2 003

Admin region Galicia
Coordinates 42°28′N 8°51′W
Altitude 0–50 m **Area** 2,561 ha

Site description

A small harbour to the south of the Ría de Arosa, with large intertidal mudflats and sandbanks. Human activities include fishing, shellfish collecting and tourism.

Habitats Wetland (50%; tidal river/enclosed tidal water; mudflat/sandflat; sand-dunes/sand beach; coastal lagoon), Marine areas (40%; sea inlet/coastal features), Artificial landscape (10%; arable land; forestry plantation)
Land-use Agriculture (5%), Fisheries/aquaculture (30%), Not utilized (60%), Tourism/recreation (5%)

Birds

Species	Season	Year	Pop min	Pop max	Acc	Criteria
Platalea leucorodia Spoonbill	P	1994	—	34	A	B1i, C2
Calidris alpina Dunlin	W	1994	4,267	—	A	B2

This is an important wintering and passage site for waterbirds.

Protection status

National None **International** High
2,420 ha of IBA covered by Ramsar Site (Complejo intermareal Umia–Grove, 2,420 ha). 2,420 ha of IBA covered by Special Protection Area (Complejo intermareal Umia–O Grove, La Lanzada, Pun, 2,420 ha).

Conservation issues

Threats Industrialization/urbanization (B), Recreation/tourism (B)

Threats include an industrial waste dump and residential land-use. Tourism and outdoor sports cause disturbance to shorebirds.

Costa de la Muerte (North coast) A1, A4i, B1i, B2, C1, C2, C3, C6 004

Admin region Galicia
Coordinates 43°14′N 8°56′W
Altitude 0–291 m **Area** 9,650 ha

Site description

A coastal site including some marine areas. It includes several small islands, cliffs, beaches, sand-dunes, littoral lagoons, estuaries and marshes. The main human activity is fishing.

Habitats Wetland (5%), Marine areas (80%; open sea; sea inlet/coastal features), Artificial landscape (15%; highly improved reseeded grassland; arable land; perennial crops/orchards/groves)
Land-use Agriculture (25%), Fisheries/aquaculture (10%), Hunting, Tourism/recreation, Urban/industrial/transport

Birds

Species	Season	Year	Pop min	Pop max	Acc	Criteria
Gavia immer Great Northern Diver	P	1995	50	—	C	A4i, B1i, C2
Hydrobates pelagicus Storm Petrel	B	1995	25	100	C	C6
Himantopus himantopus Black-winged Stilt	R	1996	5	10	A	C6
Calidris alba Sanderling	P	1995	1,200	—	B	A4i, B1i, C3
Limosa lapponica Bar-tailed Godwit	W	1993	110	110	A	B2
[1] ***Larus cachinnans*** Yellow-legged Gull	R	1992	13,500	13,500	A	A4i, B1i, C3
[2] ***Uria aalge*** Guillemot	R	1995	10	10	A	C6
Acrocephalus paludicola Aquatic Warbler	P	1996	30	—	C	A1, C1
[2] ***Pyrrhocorax pyrrhocorax*** Chough	R	1995	80	—	B	C6

1. Large increase.
2. Large decrease.

This is an important site for breeding and passage seabirds and waders (Charadriiformes).

Protection status

National Low **International** None
7 ha of IBA covered by Natural Site of National Interest (Cabo Vilán, 7 ha).

Conservation issues

Threats Aquaculture/fisheries (B), Burning of vegetation (A), Recreation/tourism (B), Unsustainable exploitation (B)

The main threats are from fisheries, fires, tourism and illegal hunting.

Ferrolterra–Valdoviño coast A4i, B1i, C3, C6 005

Admin region Galicia
Coordinates 43°36′N 8°12′W
Altitude 0–260 m **Area** 4,400 ha

Site description

An area of cliffs and sand beaches with coastal lagoons and a number of small offshore islands. The main human activities are forestry, fisheries and tourism.

Habitats Scrub (50%), Grassland, Wetland (10%; sand-dunes/sand beach; coastal lagoon), Marine areas (open sea; sea inlet/coastal features), Rocky areas (sea cliff/rocky shore; rock stacks/islets), Artificial landscape (25%)
Land-use Agriculture (10%), Fisheries/aquaculture, Forestry (15%), Military, Tourism/recreation (20%)

Birds

Species	Season	Year	Pop min	Pop max	Acc	Criteria
Hydrobates pelagicus Storm Petrel	B	1995	25	115	C	C6
Calidris alba Sanderling	P	1995	1,200	—	B	A4i, B1i, C3

This is an important site for breeding seabirds and passage waders.

Protection status

National Low **International** None
255 ha of IBA covered by Ramsar Site (Laguna y Arenal de Valdoviño, 255 ha).

Conservation issues

Threats Afforestation (C), Deforestation (commercial) (B), Recreation/tourism (B)

Disturbance from visitors and deforestation are the most important threats.

Cape Candelaria–Ortigueira estuary–Cape Estaca de Bares

B2, C7 **006**

Admin region Galicia
Coordinates 43°33'N 7°04'W
Altitude 0–613 m **Area** 6,700 ha

Site description

An estuary between Cape Ortegal and the Estaca de Bares. It includes marine areas, wide intertidal mudflats, a belt of halophytic vegetation, and some islets.

Habitats Scrub (25%), Wetland (10%; tidal river/enclosed tidal water; mudflat/sandflat; sand-dunes/sand beach), Marine areas (5%; open sea; sea inlet/coastal features), Rocky areas (10%; sea cliff/rocky shore; rock stacks/islets), Artificial landscape (50%; highly improved reseeded grassland; arable land; perennial crops/orchards/groves; forestry plantation)
Land-use Agriculture (*), Fisheries/aquaculture, Forestry (*), Hunting, Tourism/recreation

Birds

Species	Season	Year	Pop min	Pop max	Acc	Criteria
Limosa lapponica Bar-tailed Godwit	W	1995	125	125	A	B2
Numenius arquata Curlew	W	1995	536	536	A	B2

This is an important site for breeding seabirds and passage and wintering waders.

Protection status

National Low **International** Partial
1 ha of IBA covered by Natural Site of National Interest (Estaca de Bares, 1 ha). 2,920 ha of IBA covered by Ramsar Site (Rias de Ortigueira y Ladrido, 2,920 ha). 2,940 ha of IBA covered by Special Protection Area (Ría de Ortigueira y Ladrido, 2,940 ha).

Conservation issues

Threats Aquaculture/fisheries (B), Extraction industry (B), Infrastructure (B), Intensified forest management (B), Recreation/tourism (B), Unsustainable exploitation (B)

The main threats are from intensive shellfish collecting, illegal hunting, the construction of windmills, building, and the cutting of marsh vegetation.

Eo river estuary–Barayo beach–Ría de Foz

A4i, B1i, B2, C3, C6 **007**

Admin region Galicia, Asturias
Coordinates 43°40'N 7°04'W
Altitude 0–50 m **Area** 7,500 ha

Site description

This site includes intertidal mudflats and sandbanks, saltmarsh, cliffs, beaches and a marine section.

Habitats Wetland (*; tidal river/enclosed tidal water; mudflat/sandflat; river/stream; water-fringe vegetation), Marine areas (*; open sea; sea inlet/coastal features), Artificial landscape (highly improved reseeded grassland; arable land; forestry plantation; other urban/industrial areas)
Land-use Agriculture, Fisheries/aquaculture (*), Forestry, Urban/industrial/transport

Birds

Species	Season	Year	Pop min	Pop max	Acc	Criteria
[1] ***Anas acuta*** Pintail	W	1996	250	770	B	B1i, C3
Porzana pusilla Baillon's Crake	B	1996	30	—	B	B2, C6
Larus cachinnans Yellow-legged Gull	R	1996	1,500	—	A	A4i, B1i, C3
Larus cachinnans Yellow-legged Gull	W	1996	5,400	—	A	A4i, B1i, C3

1. Large decrease.

This is an important area for wintering and passage wildfowl and waders and breeding seabirds. It is also one of the most important sites for waterbirds on the Cantabric coast.

Protection status

National Partial **International** Partial
1,800 ha of IBA covered by National Game Reserve (Ría del Eo, 2,023 ha). 1,740 ha of IBA covered by Ramsar Site (Ría del Eo, 1,740 ha). 1,740 ha of IBA covered by Special Protection Area (Ría del Eo, 1,740 ha).

Conservation issues

Threats Abandonment/reduction of land management (B), Aquaculture/fisheries (B), Construction/impact of dyke/dam/barrage (A), Recreation/tourism (B)

Threats include intensive fishing and shellfish collection, the abandonment of traditional land-use, and recreation and tourism. A recently built dyke has probably caused a decline in bird numbers.

La Limia

C6 **008**

Admin region Galicia
Coordinates 42°10'N 7°40'W
Altitude 610–905 m **Area** 32,800 ha

Site description

The site encompasses a former lagoon that has now been drained. Relict humid grasslands and patches of *Quercus robur* woodland remain, surrounded by arable agriculture.

Habitats Forest and woodland (15%; broadleaved deciduous forest), Scrub (20%; scrub; heathland), Grassland (5%; humid grassland), Artificial landscape (60%; highly improved reseeded grassland; arable land)
Land-use Agriculture (70%), Hunting, Not utilized (10%)

Birds

Species	Season	Year	Pop min	Pop max	Acc	Criteria
[1] ***Ciconia ciconia*** White Stork	B	1992	13	13	A	C6
Circus pygargus Montagu's Harrier	B	1995	15	—	C	C6
[2] ***Tetrax tetrax*** Little Bustard	N	1996	60	—	C	C6
[2] ***Burhinus oedicnemus*** Stone Curlew	R	1989	15	50	C	C6

1. Large increase.
2. Large decrease.

The site is important for breeding species associated with a mosaic of arable land, grassland and woodland.

Protection status

National None **International** None

Conservation issues

Threats Agricultural intensification/expansion (A), Drainage (B), Infrastructure (C), Selective logging/cutting (C), Unsustainable exploitation (C)

The main threat is from continued intensification of arable agriculture.

Ancares mountains

B2, C6 **009**

Admin region Galicia, Castilla y León
Coordinates 42°50'N 6°50'W
Altitude 400–1,910 m **Area** 50,600 ha

Site description

A rugged mountainous area with many valleys covered by deciduous woodland, scrub and grasslands. The main human activities are cattle-farming and hunting.

Habitats Forest and woodland (*; broadleaved deciduous forest), Scrub (*; heathland), Artificial landscape (highly improved reseeded grassland; arable land; forestry plantation)
Land-use Agriculture, Forestry (*), Hunting (*)

Birds

Species	Season	Year	Pop min	Pop max	Acc	Criteria
Pernis apivorus Honey Buzzard	B	1996	10	—	A	C6
[1,2] ***Tetrao urogallus*** Capercaillie	R	1996	5	—	B	C6
[3] ***Perdix perdix*** Partridge	R	1996	50	80	B	C6
Monticola saxatilis Rock Thrush	B	1996	50	—	A	B2

1. Breeding males.
2. Large decrease.
3. *P. p. hispaniensis*.

This is an important area for breeding forest species.

Protection status

National High **International** None

38,300 ha of IBA covered by National Game Reserve (Ancares Leoneses, 38,300 ha). 7,975 ha of IBA covered by National Game Reserve (Ancares de Lugo, 7,975 ha).

Conservation issues

Threats Burning of vegetation (B), Construction/impact of dyke/dam/barrage (A), Deforestation (commercial) (B), Infrastructure (B), Recreation/tourism (A)

The main threats are from increasing tourism development, hydro-power projects, fires and potential infrastructure developments. A Game Management Plan is under preparation.

Montes Aquilanos range

B2, B3, C6 **010**

Admin region Galicia, Castilla y León
Coordinates 42°20'N 6°30'W
Altitude 750–2,188 m **Area** 55,000 ha

Site description

A mountainous area comprising the Montes Aquilanos and the Sierra del Teleno ranges. The vegetation is a mixture of deciduous and mixed forest, scrub, *Pinus* plantations and alpine grasslands. Human activities include hunting and forestry.

Habitats Forest and woodland (*; broadleaved deciduous forest; mixed forest; broadleaved evergreen forest), Scrub (*; scrub; heathland), Grassland (dry siliceous grassland; alpine/subalpine/boreal grassland), Wetland (river/stream), Rocky areas (scree/boulders; inland cliff), Artificial landscape (forestry plantation; ruderal land)
Land-use Agriculture (5%), Forestry (20%), Hunting (90%), Military (10%), Tourism/recreation (5%), Water management (5%)

Birds

Species	Season	Year	Pop min	Pop max	Acc	Criteria
Pernis apivorus Honey Buzzard	B	1996	25	—	B	B3, C6
Circaetus gallicus Short-toed Eagle	B	1996	15	20	B	B2
Circus pygargus Montagu's Harrier	B	1996	10	—	B	C6
Aquila chrysaetos Golden Eagle	R	1996	7	11	A	C6
Falco peregrinus Peregrine	R	1996	14	25	B	B2, C6
[1] ***Perdix perdix*** Partridge	R	1990	50	55	B	C6
Monticola saxatilis Rock Thrush	B	1996	50	60	B	B2
Pyrrhocorax pyrrhocorax Chough	R	1996	100	120	B	B2, C6

1. Large decrease.

This is an important area for cliff-nesting raptors and forest birds. Species of global conservation concern that do not meet IBA criteria: *Falco naumanni* (100 non-breeding individuals).

Protection status

National Low **International** Partial

1,115 ha of IBA covered by Natural Monument (Las Médulas, 1,115 ha). IBA overlaps with World Heritage Site (Las Médulas).

Conservation issues

Threats Abandonment/reduction of land management (B), Afforestation (A), Burning of vegetation (A), Construction/impact of dyke/dam/barrage (A), Infrastructure (C), Recreation/tourism (B)

The main threats are from frequent fires, development of skiing infrastructure, hydro-power schemes and the afforestation of grasslands.

La Cabrera mountain range

B2, C6 **011**

Admin region Galicia, Castilla y León
Coordinates 42°15'N 6°45'W
Altitude 900–2,125 m **Area** 73,700 ha

Site description

A mountainous area in the western Cantabrian Range, with extensive areas of scrub, patches of *Quercus* woodland, plantations of *Pinus sylvestris*, and alpine grasslands grazed by sheep and cows. The site includes 13 lakes.

Habitats Forest and woodland (broadleaved deciduous forest; alluvial/very wet forest; broadleaved evergreen forest), Scrub (*; scrub; heathland), Grassland (*; dry siliceous grassland; alpine/subalpine/boreal grassland; humid grassland), Wetland (standing fresh water; raised bog), Rocky areas (scree/boulders), Artificial landscape (arable land; forestry plantation)
Land-use Agriculture (*), Forestry, Hunting, Not utilized, Urban/industrial/transport

Birds

Species	Season	Year	Pop min	Pop max	Acc	Criteria
Circaetus gallicus Short-toed Eagle	B	1990	10	20	B	B2
Circus pygargus Montagu's Harrier	B	1990	5	10	B	C6
[1] ***Perdix perdix*** Partridge	R	1990	100	200	B	C6
Monticola saxatilis Rock Thrush	B	1990	50	70	B	B2
Pyrrhocorax pyrrhocorax Chough	R	1990	50	80	B	C6

1. Large decrease.

This is a very important area for forest and alpine grassland species, particularly *Perdix perdix* and raptors. Species of global conservation concern that do not meet IBA criteria: *Falco naumanni* (non-breeding).

Protection status

National Partial **International** None

22,365 ha of IBA covered by Natural Park (Lago de Sanabria y alrededores, 22,365 ha). 1,066 ha of IBA covered by Natural Monument (Lago de Truchillas, 1,066 ha). 731 ha of IBA covered by Natural Monument (Lago de La Baña, 731 ha).

Conservation issues

Threats Afforestation (A), Burning of vegetation (A), Construction/impact of dyke/dam/barrage (B), Extraction industry (A), Infrastructure (B), Recreation/tourism (B)

The most important threat is the enlargement of open-cast slate quarries. Other threats include fires, hydro-power schemes and unplanned tourism. The lakes are protected at a regional level.

Muniellos

C6 **012**

Admin region Asturias
Coordinates 43°05'N 6°44'W
Altitude 600–1,685 m **Area** 24,100 ha

Site description

An area of the Cantabrian mountains covered by deciduous woodlands dominated by *Quercus* and *Fagus* forest. The IBA includes the Muniellos forest (3,000 ha), probably the best preserved *Quercus robur* forest in Spain. The main human activities are hunting, mining and nature conservation.

Habitats Forest and woodland (70%; broadleaved deciduous forest; alluvial/very wet forest), Scrub (25%; scrub; heathland), Rocky areas (5%; scree/boulders)
Land-use Hunting, Nature conservation/research (*), Not utilized, Urban/industrial/transport

Birds

Species	Season	Year	Pop min	Pop max	Acc	Criteria
[1,2] ***Tetrao urogallus*** Capercaillie	R	1992	25	30	B	C6
[3] ***Perdix perdix*** Partridge	R	1996	20	—	B	C6
Dryocopus martius Black Woodpecker	R	1991	7	—	B	C6

1. Males.
2. Large decrease.
3. *P. p. hispaniensis*.

The IBA is important for its good populations of forest species.

Protection status

National Partial **International** Partial

5,542 ha of IBA covered by National Biological Reserve (Bosque de Muniellos, 5,542 ha). 2,975 ha of IBA covered by Special Protection Area (Bosque de Muniellos, 2,975 ha).

Conservation issues

Threats Burning of vegetation (B), Extraction industry (A), Industrialization/urbanization (B), Infrastructure (B), Recreation/tourism (C), Unsustainable exploitation (B)

There are a number of threats to the site including coal and gold mining, illegal hunting, acid rain from industrial pollution, forest fires and road building.

Degaña–Hermo C6 013

Admin region Asturias
Coordinates 42°55′N 6°30′W
Altitude 600–2,007 m **Area** 32,000 ha

Site description

A mountainous area in the Cordillera Cantábrica with extensive deciduous woods dominated by *Fagus sylvatica*. The area is used for game hunting, coal mining and cattle-grazing.

Habitats Forest and woodland (55%; broadleaved deciduous forest), Scrub (35%; scrub; heathland), Rocky areas (10%; scree/boulders)
Land-use Agriculture, Hunting (*), Urban/industrial/transport

Birds

Species	Season	Year	Pop min	Pop max	Acc	Criteria
[1] ***Tetrao urogallus*** Capercaillie	R	1994	15	25	B	C6
Perdix perdix Partridge	R	1994	30	—	A	C6
Dryocopus martius Black Woodpecker	R	1992	5	—	B	C6

1. Large decrease.

This is an important site for breeding species associated with upland forests and open alpine habitats.

Protection status

National Partial **International** Partial

6,500 ha of IBA covered by National Game Reserve (Degaña, 8,657 ha). 8,500 ha of IBA covered by Special Protection Area (Degaña–Hermo, 11,659 ha).

Conservation issues

Threats Burning of vegetation (A), Extraction industry (A), Firewood collection (B), Infrastructure (B), Recreation/tourism (C)

Coal mining is a serious threat to the site. A project to enlarge the ski-station at Leitariegos and construct new roads poses an additional threat. Illegal hunting and forest fires also occur.

Babia–Somiedo A3, B1iii, B2, B3, C2, C6 014

Admin region Asturias, Castilla y León
Coordinates 43°10′N 6°00′W
Altitude 800–2,417 m **Area** 166,860 ha

Site description

A large mountainous area in the Cordillera Cantábrica. The site is characterized by deciduous woodland, scrub, grassland and narrow valleys with many small villages and some cultivation. The area is used for cattle-grazing, hunting, forestry and tourism and some mining.

Habitats Forest and woodland (15%; broadleaved deciduous forest; alluvial/very wet forest), Scrub (20%; scrub; heathland), Grassland (40%; alpine/subalpine/boreal grassland; humid grassland), Wetland (5%; river/stream), Rocky areas (30%; scree/boulders; inland cliff)
Land-use Agriculture (60%), Forestry, Hunting (90%), Not utilized (30%), Tourism/recreation, Urban/industrial/transport, Water management (6%)

Birds

Species	Season	Year	Pop min	Pop max	Acc	Criteria
Pernis apivorus Honey Buzzard	B	1996	20	—	C	B3, C6
Neophron percnopterus Egyptian Vulture	B	1996	30	40	B	B1iii, B2, C2, C6
Circaetus gallicus Short-toed Eagle	B	1996	19	24	B	B2, C6
Circus cyaneus Hen Harrier	R	1996	12	16	B	B2, C6
Aquila chrysaetos Golden Eagle	R	1996	6	7	B	C6
Falco peregrinus Peregrine	R	1996	12	16	B	B2, C6
[1,2] ***Tetrao urogallus*** Capercaillie	R	1996	30	40	B	C6
Perdix perdix Partridge	R	1996	110	—	A	C6
Dryocopus martius Black Woodpecker	R	1996	5	—	C	C6
Dendrocopos medius Middle Spotted Woodpecker	R	1996	1	4	B	C6
Prunella collaris Alpine Accentor	R	1996	Common		—	A3
Monticola saxatilis Rock Thrush	B	1996	35	50	C	B2
Tichodroma muraria Wallcreeper	R	1996	Common		—	A3
Pyrrhocorax graculus Alpine Chough	R	1996	Common		—	A3
Montifringilla nivalis Snowfinch	R	1996	Common		—	A3
Serinus citrinella Citril Finch	R	1996	Common		—	A3

1. Males.
2. Large decrease.

Breeding birds include five of the 10 species in Europe that are restricted to the Eurasian high-montane biome (when breeding).

Protection status

National Partial **International** Partial

86,000 ha of IBA covered by National Game Reserve (Somiedo, 87,900 ha). 29,122 ha of IBA covered by Natural Park (Somiedo, 29,122 ha). 29,122 ha of IBA covered by Special Protection Area (Somiedo, 29,122 ha).

Conservation issues

Threats Abandonment/reduction of land management (B), Extraction industry (C), Firewood collection (C), Infrastructure (A), Recreation/tourism (C), Selective logging/cutting (B), Unsustainable exploitation (A)

The most important threats are over-exploitation of game, mining and increased infrastructural development for tourism.

Gistreo and Coto mountain ranges B2, B3, C6 015

Admin region Castilla y León
Coordinates 42°50′N 6°20′W
Altitude 700–2,111 m **Area** 59,000 ha

Site description

A mountainous area in the southern part of the Cordillera Cantábrica, with slopes covered by scrub, grassland, and deciduous woodland. Human activities include cattle-grazing, game hunting and forestry.

Habitats Forest and woodland (*; broadleaved deciduous forest; mixed forest; alluvial/very wet forest; broadleaved evergreen forest; treeline ecotone), Scrub (*; scrub; heathland; sclerophyllous scrub/garrigue/maquis), Grassland (dry siliceous grassland; alpine/subalpine/boreal grassland; humid grassland), Wetland (river/stream; raised bog), Rocky areas (scree/boulders; inland cliff), Artificial landscape (arable land; forestry plantation)
Land-use Agriculture, Forestry (*), Hunting, Urban/industrial/transport

Birds

Species	Season	Year	Pop min	Pop max	Acc	Criteria
Pernis apivorus Honey Buzzard	B	1996	10	20	B	B3, C6
Circaetus gallicus Short-toed Eagle	B	1996	15	20	B	B2
Circus cyaneus Hen Harrier	R	1996	10	15	A	C6
[1] ***Tetrao urogallus*** Capercaillie	R	1992	25	30	A	C6
Perdix perdix Partridge	R	1996	120	150	A	C6
Monticola saxatilis Rock Thrush	B	1996	50	60	B	B2

1. Males.

This is a very important site for breeding alpine and forest species, particularly *Perdix perdix hispaniensis* and forest raptors.

Protection status

National None **International** None

Conservation issues

Threats Afforestation (B), Burning of vegetation (A), Construction/impact of dyke/dam/barrage (C), Extraction industry (A), Unsustainable exploitation (B)

The main threats are from uncontrolled fires, open-cast mining, hydroelectric schemes, illegal hunting, and exploitation of the forest resource.

Central ranges of the Cantabric mountains
B2, C6 **016**

Admin region Asturias, Castilla y León
Coordinates 43°06′N 5°22′W
Altitude 200–2,189 m **Area** 117,500 ha

Site description
A mountainous area in the central Cordillera Cantábrica, between Asturias and León. The area is crossed by deep valleys, with grasslands and deciduous woodland, and scrub and rocky areas at higher altitudes. The main human activities are forestry, game hunting and tourism.

Habitats Forest and woodland (*; broadleaved deciduous forest; alluvial/very wet forest), Scrub (*; scrub; heathland), Grassland (*; alpine/subalpine/boreal grassland; humid grassland), Wetland (river/stream; raised bog), Rocky areas (scree/boulders; inland cliff), Artificial landscape (arable land)
Land-use Agriculture (85%), Forestry, Hunting (90%), Tourism/recreation

Birds

Species	Season	Year	Pop min	Pop max	Acc	Criteria
Neophron percnopterus Egyptian Vulture	B	1996	10	18	B	B2, C6
Aquila chrysaetos Golden Eagle	R	1996	13	16	A	B2, C6
[1] ***Tetrao urogallus*** Capercaillie	R	1996	17	27	B	C6
Perdix perdix Partridge	R	1996	65	—	B	C6
Dryocopus martius Black Woodpecker	R	1995	6	8	B	C6
Dendrocopos medius Middle Spotted Woodpecker	R	1996	6	7	A	C6

1. Males.

This is an important area for breeding raptors, particularly *Neophron percnopterus* and *Aquila chrysaetos*, and for breeding forest species.

Protection status
National Partial **International** None
29,834 ha of IBA covered by National Game Reserve (Caso, 29,834 ha). 1,500 ha of IBA covered by National Game Reserve (Piloña, 10,492 ha). 20,082 ha of IBA covered by National Game Reserve (Ponga, 20,082 ha). 30,000 ha of IBA covered by Natural Park (Reres, 45,000 ha). 150 ha of IBA covered by National Game Reserve (Rioseco, 150 ha).

Conservation issues
Threats Afforestation (B), Burning of vegetation (A), Construction/impact of dyke/dam/barrage (B), Extraction industry (B), Infrastructure (A), Recreation/tourism (B), Selective logging/cutting (B), Unsustainable exploitation (A)

Enlargement of the ski infrastructure, tourism and dams are the main threats. Woodland fires and grassland fires, used to provide grazing, are also problems.

Cape Busto–Luanco
B1i, C3, C6 **017**

Admin region Asturias
Coordinates 43°35′N 6°00′W
Altitude 0–100 m **Area** 6,000 ha

Site description
A rocky stretch of coast with cliffs and scattered islets close to the shore, and estuaries with mudflats.

Habitats Wetland (tidal river/enclosed tidal water; mudflat/sandflat), Marine areas (sea inlet/coastal features), Rocky areas (*; sea cliff/rocky shore)
Land-use Fisheries/aquaculture (*), Forestry, Not utilized (*)

Birds

Species	Season	Year	Pop min	Pop max	Acc	Criteria
Hydrobates pelagicus Storm Petrel	B	1996	150	—	B	C6
Falco peregrinus Peregrine	R	1993	8	9	A	C6
Larus cachinnans Yellow-legged Gull	W	1996	4,200	—	B	B1i, C3

This is an important passage and wintering area for waders and breeding site for seabirds.

Protection status
National Low **International** None
543 ha of IBA covered by National Game Reserve (Cabo Busto, 543 ha).

Conservation issues
Threats Aquaculture/fisheries (B), Deforestation (commercial) (C)

The main threats are from fishing and shellfish collection. Pollution may also be a problem.

Ribadesella–Tina Mayor
A4i, B1i, B2, C3, C6 **018**

Admin region Asturias
Coordinates 43°25′N 4°45′W
Altitude 0–207 m **Area** 3,600 ha

Site description
The site includes a 45-km stretch of coastal cliffs and several small islets. The vegetation is mainly grassland with areas of deciduous woodland. The main human activities are fishing and forestry.

Habitats Forest and woodland (*; broadleaved deciduous forest), Scrub, Grassland, Wetland (tidal river/enclosed tidal water; mudflat/sandflat; sand-dunes/sand beach), Rocky areas (*; sea cliff/rocky shore; rock stacks/islets), Artificial landscape (*; forestry plantation)
Land-use Fisheries/aquaculture, Forestry (*), Not utilized (*)

Birds

Species	Season	Year	Pop min	Pop max	Acc	Criteria
Hydrobates pelagicus Storm Petrel	B	1995	700	—	B	B2, C6
Larus cachinnans Yellow-legged Gull	R	1996	1,800	—	—	A4i, B1i, C3

This is an important area for breeding, wintering and passage seabirds.

Protection status
National None **International** None

Conservation issues
Threats Aquaculture/fisheries (U), Industrialization/urbanization (B), Infrastructure (B), Recreation/tourism (B)

The most important threats are industrial and tourism-related development. The Ribadesella estuary is a Game Refuge (200 ha).

Riaño
B2, B3, C2, C6 **019**

Admin region Castilla y León
Coordinates 43°05′N 5°00′W
Altitude 960–2,190 m **Area** 70,000 ha

Site description
A mountainous area in the Cordillera Cantábrica with steep slopes covered by scrub and grasslands, deciduous woodland and limestone cliffs at higher altitudes. The summer pastures are grazed by sheep and cattle. Other activities include game hunting and forestry.

Habitats Forest and woodland (30%; broadleaved deciduous forest; native coniferous forest; mixed forest; alluvial/very wet forest; treeline ecotone), Scrub (40%; scrub; heathland; sclerophyllous scrub/garrigue/maquis), Grassland (30%; alpine/subalpine/boreal grassland; humid grassland), Wetland (raised bog), Rocky areas (inland cliff), Artificial landscape (forestry plantation)
Land-use Agriculture (80%), Forestry (30%), Hunting (90%), Nature conservation/research (10%), Water management (10%)

Birds

Species	Season	Year	Pop min	Pop max	Acc	Criteria
Pernis apivorus Honey Buzzard	B	1996	15	—	B	C6
[1] ***Tetrao urogallus*** Capercaillie	R	1981	153	153	A	C6
[2] ***Perdix perdix*** Partridge	R	1993	50	100	B	C6
Dryocopus martius Black Woodpecker	R	1993	15	—	B	C6
Dendrocopos medius Middle Spotted Woodpecker	R	1993	10	15	B	B3, C6
Monticola saxatilis Rock Thrush	B	1993	40	—	B	B2

Species ... continued	Season	Year	Pop min	Pop max	Acc	Criteria
Pyrrhocorax pyrrhocorax Chough	R	1993	160	—	B	B2, C2

1. Males in Riaño National Game Reserve.
2. Large increase.

This is an important site for species associated with forests.

Protection status

National High **International** None

70,000 ha of IBA covered by Regional Park (Spain) (Picos de Europa, 120,760 ha). 23,000 ha of IBA covered by National Game Reserve (Mampodre, 29,238 ha). 47,000 ha of IBA covered by National Game Reserve (Riaño, 71,214 ha).

Conservation issues

Threats Agricultural intensification/expansion (C), Construction/impact of dyke/dam/barrage (B), Deforestation (commercial) (C), Infrastructure (B), Recreation/tourism (C)

The main threats include hydroelectric schemes, the development of the ski infrastructure, deforestation to create new pastures, overgrazing and the pressure from tourism.

Picos de Europa mountain range

A3, B1iii, B2, B3, C2, C6 **020**

Admin region Asturias, Cantabria, Castilla y León
Coordinates 43°10'N 4°50'W
Altitude 350–2,617 m **Area** 132,000 ha

Site description

A rugged limestone karst landscape, with expanses of rocks, cliffs, two deep gorges and two small mountain lakes. Grasslands predominate, but in the lower areas there are extensive deciduous woods. Human activities include tourism, livestock-farming and game hunting.

Habitats Forest and woodland (30%; broadleaved deciduous forest; alluvial/very wet forest), Scrub (20%; scrub; heathland), Grassland (10%; alpine/subalpine/boreal grassland; humid grassland), Rocky areas (25%; scree/boulders; inland cliff), Artificial landscape (10%; highly improved reseeded grassland)
Land-use Agriculture (45%), Forestry (5%), Hunting (20%), Nature conservation/research (50%), Tourism/recreation (10%)

Birds

Species	Season	Year	Pop min	Pop max	Acc	Criteria
Gyps fulvus Griffon Vulture	R	1996	100	—	A	B1iii, B2, C2, C6
Aquila chrysaetos Golden Eagle	R	1996	7	—	A	C6
[1,2] ***Tetrao urogallus*** Capercaillie	R	1995	5	—	A	C6
Perdix perdix Partridge	R	1996	60	—	A	C6
Dryocopus martius Black Woodpecker	R	1987	5	—	B	C6
Dendrocopos medius Middle Spotted Woodpecker	R	1996	15	20	A	B3, C6
Prunella collaris Alpine Accentor	R	1996	Common	—		A3
Tichodroma muraria Wallcreeper	R	1996	Common	—		A3
Pyrrhocorax graculus Alpine Chough	R	1996	Abundant	—		A3
Montifringilla nivalis Snowfinch	R	1996	Common	—		A3

1. Males.
2. Large decrease.

Breeding birds include four of the 10 species in Europe that are restricted to the Eurasian high-montane biome (when breeding).

Protection status

National Partial **International** Partial

64,660 ha of IBA covered by National Park (Picos de Europa, 64,660 ha). 7,630 ha of IBA covered by National Game Reserve (Picos de Europa, 7,630 ha). 20,088 ha of IBA covered by Regional Park (Picos de Europa, 120,760 ha). 16,925 ha of IBA covered by Special Protection Area (Montaña de Covadonga, 16,925 ha).

Conservation issues

Threats Burning of vegetation (B), Infrastructure (A), Unsustainable exploitation (B)

A funicular railway is being built in the core area of the National Park and other infrastructure is being developed for tourism. Grassland burning, illegal hunting and the unsustainable exploitation of game species are also threats.

Fuentes Carrionas

B2, C6 **021**

Admin region Castilla y León
Coordinates 42°55'N 4°40'W
Altitude 1,000–2,520 m **Area** 66,000 ha

Site description

A mountainous area in the Cordillera Cantábrica with limestone cliffs, *Quercus* and *Fagus* woodland, and alpine grasslands grazed by cattle. The site also includes five protected lakes.

Habitats Forest and woodland (broadleaved deciduous forest; mixed forest), Scrub (*; scrub; heathland), Grassland (alpine/subalpine/boreal grassland; humid grassland), Rocky areas (scree/boulders; inland cliff), Artificial landscape (highly improved reseeded grassland; forestry plantation)
Land-use Agriculture (60%), Forestry (20%), Hunting (80%), Not utilized (20%), Water management (5%)

Birds

Species	Season	Year	Pop min	Pop max	Acc	Criteria
Circaetus gallicus Short-toed Eagle	B	1992	20	—	B	B2, C6
Perdix perdix Partridge	R	1995	110	120	A	C6
Dendrocopos medius Middle Spotted Woodpecker	R	1995	5	10	B	C6
Monticola saxatilis Rock Thrush	B	1995	140	—	C	B2

This is an important site for cliff-nesting and forest species.

Protection status

National Partial **International** None

47,755 ha of IBA covered by National Game Reserve (Fuentes Carrionas, 47,755 ha).

Conservation issues

Threats Afforestation (B), Burning of vegetation (B), Extraction industry (C), Recreation/tourism (B)

The main threats are from tourism, fires and quarrying.

Peña Labra and Cordel mountain ranges

B2, B3, C6 **022**

Admin region Cantabria, Castilla y León
Coordinates 43°00'N 4°20'W
Altitude 600–2,175 m **Area** 110,000 ha

Site description

A mountainous site in the Cantabrian range with extensive areas of scrub and grassland, and some deciduous forest. Human activities include tourism and recreation, hunting and cattle-grazing.

Habitats Forest and woodland (broadleaved deciduous forest; alluvial/very wet forest), Scrub (*; scrub; heathland), Grassland (*; alpine/subalpine/boreal grassland; humid grassland), Wetland (river/stream)
Land-use Agriculture (*), Hunting (*), Tourism/recreation (*)

Birds

Species	Season	Year	Pop min	Pop max	Acc	Criteria
Pernis apivorus Honey Buzzard	B	1992	35	—	B	B3, C6
Circaetus gallicus Short-toed Eagle	B	1992	20	—	B	B2, C6
[1] ***Tetrao urogallus*** Capercaillie	R	1991	5	6	A	C6
[2] ***Perdix perdix*** Partridge	R	1996	—	100	B	C6
Dryocopus martius Black Woodpecker	R	1992	14	—	B	C6
Dendrocopos medius Middle Spotted Woodpecker	R	1992	45	—	B	B3, C6
Pyrrhocorax pyrrhocorax Chough	R	1992	100	—	B	B2

1. Breeding males.
2. *P. p. hispaniensis*.

This is an important site for breeding forest raptors.

Protection status

National Partial **International** None

70,000 ha of IBA covered by National Game Reserve (Saja, 180,186 ha). 23,932 ha of IBA covered by Natural Park (Saja–Besaya, 23,932 ha).

Conservation issues

Threats Agricultural intensification/expansion (C), Recreation/tourism (C)

The main threat is from the development of tourism infrastructure, particularly for skiing.

River Ebro dam A4i, B1i, B2, C3 023

Admin region Cantabria, Castilla y León
Coordinates 43°00′N 3°57′W
Altitude 900–920 m **Area** 11,600 ha

Site description

A wide, generally shallow reservoir with surrounding wet grassland and bogs. Human activities include livestock-farming and arable agriculture.

Habitats Scrub (10%; heathland), Grassland (10%; humid grassland), Wetland (80%; standing fresh water; raised bog)
Land-use Agriculture (25%), Fisheries/aquaculture (70%), Tourism/recreation (15%), Urban/industrial/transport (10%), Water management (80%)

Birds

Species	Season	Year	Pop min	Pop max	Acc	Criteria
Anas strepera Gadwall	R	1995	100	120	A	B2
Netta rufina Red-crested Pochard	N	1994	3,000	5,000	A	A4i, B1i, C3

This is a very important site for resident and passage waterbirds. Species of global conservation concern that do not meet IBA criteria: *Tetrax tetrax* (38–63 non-breeding individuals).

Protection status

National None **International** None

Conservation issues

Threats Agricultural intensification/expansion (A), Infrastructure (B), Other (B), Recreation/tourism (C)

Threats come from the transfer of water from the River Ebro to the River Besaya which is reducing the water-level of the reservoir ('Other' threats, above), and a project to create a golf course. Hunting is forbidden.

Tina Menor–Sarnosa island–Oyambre C6 024

Admin region Cantabria
Coordinates 43°25′N 4°20′W
Altitude 0–25 m **Area** 1,100 ha

Site description

A section of coastline on the Cantabric Sea, with cliffs, estuaries and sand-dune systems. Human activities include agriculture and fisheries.

Habitats Wetland (65%; tidal river/enclosed tidal water; mudflat/sandflat; sand-dunes/sand beach; water-fringe vegetation), Rocky areas (30%; sea cliff/rocky shore)
Land-use Agriculture (80%), Fisheries/aquaculture (10%), Not utilized

Birds

Species	Season	Year	Pop min	Pop max	Acc	Criteria
Falco peregrinus Peregrine	R	1992	5	5	B	C6

This is a nationally important site for breeding seabirds and passage and wintering ducks and waders.

Protection status

National High **International** None
5,758 ha of IBA covered by Natural Park (Oyambre, 5,758 ha).

Conservation issues

Threats Aquaculture/fisheries (A), Industrialization/urbanization (B), Recreation/tourism (B)

The main threats are from disturbance from fishing and tourists.

Portios islets–Conejera island C6 025

Admin region Cantabria
Coordinates 43°30′N 4°00′W
Altitude 0–50 m **Area** 10 ha

Site description

The site includes a group of small islands close to the north coast of Spain and mainland cliffs. These are dominated by scrub.

Habitats Marine areas (*; sea inlet/coastal features), Rocky areas (*; sea cliff/rocky shore; rock stacks/islets)
Land-use Fisheries/aquaculture (80%), Tourism/recreation (15%)

Birds

Species	Season	Year	Pop min	Pop max	Acc	Criteria
Hydrobates pelagicus Storm Petrel	B	1995	80	—	B	C6

The site is important for breeding seabirds.

Protection status

National None **International** None

Conservation issues

Threats Aquaculture/fisheries (A), Industrialization/urbanization (B), Recreation/tourism (B)

Urban development threatens the coastal cliffs.

Santander bay–Mouro islet B1i, C2, C6 026

Admin region Cantabria
Coordinates 43°25′N 3°45′W
Altitude 0–50 m **Area** 3,300 ha

Site description

A coastal bay with sandbanks and a small island. The site includes a port. Human activities include fishing and tourism.

Habitats Grassland (5%; humid grassland), Wetland (85%; tidal river/enclosed tidal water; mudflat/sandflat; sand-dunes/sand beach), Artificial landscape (5%; arable land; other urban/industrial areas)
Land-use Fisheries/aquaculture (50%), Tourism/recreation (25%), Urban/industrial/transport (75%)

Birds

Species	Season	Year	Pop min	Pop max	Acc	Criteria
Hydrobates pelagicus Storm Petrel	B	1995	40	—	C	C6
Platalea leucorodia Spoonbill	P	1995	30	—	B	B1i, C2

This is an important site for migrant waterbirds.

Protection status

National None **International** None

Conservation issues

Threats Aquaculture/fisheries (B), Dredging/canalization (U), Filling-in of wetlands (B), Industrialization/urbanization (A), Infrastructure (A), Recreation/tourism (A)

Main threats are from tourism and urban and industrial development.

Santoña marshes B1i, B2, C2, C3 027

Admin region Cantabria
Coordinates 43°30′N 3°30′W
Altitude 0–378 m **Area** 6,907 ha

Site description

A coastal marsh with sand-dunes and extensive intertidal mudflats. Human activities include fishing and shellfish collecting.

Habitats Wetland (60%; tidal river/enclosed tidal water; mudflat/sandflat; standing brackish and salt water), Artificial landscape (50%; highly improved reseeded grassland; other urban/industrial areas)

Land-use Agriculture (60%), Fisheries/aquaculture (10%), Forestry (40%), Urban/industrial/transport

Birds

Species	Season	Year	Pop min	Pop max	Acc	Criteria
Platalea leucorodia Spoonbill	P	1995	52	110	A	B1i, C2
Anas penelope Wigeon	W	1995	2,391	5,828	A	B1i, C3
Numenius arquata Curlew	W	1995	426	1,450	A	B2

The most important wintering and passage site for waterbirds on the north coast of the Iberian peninsula.

Protection status

National Partial **International** High

3,866 ha of IBA covered by Natural Reserve (Marismas de Santoña y Noja, 3,866 ha). 6,907 ha of IBA covered by Ramsar Site (Marismas de Santoña, 6,907 ha). 6,907 ha of IBA covered by Special Protection Area (Marismas de Santoña, 6,907 ha).

Conservation issues

Threats Aquaculture/fisheries (A), Construction/impact of dyke/dam/barrage (A), Filling-in of wetlands (A), Industrialization/urbanization (A), Infrastructure (A), Recreation/tourism (A), Unsustainable exploitation (A)

There are numerous current and potential threats to this site. These include proposals for industrial development with infilling, a road project, aquaculture projects, urban and industrial dumping and intense hunting and tourism. A management plan was approved in 1997.

Humada–Peña Amaya B1iii, B2, C2 028

Admin region Castilla y León
Coordinates 42°40′N 4°00′W
Altitude 900–1,362 m **Area** 41,000 ha

Site description

A mountainous area with many cliffs, slopes dominated by maquis and *Pinus* plantations. Human activities include arable agriculture, livestock-farming and hunting.

Habitats Forest and woodland (25%; broadleaved deciduous forest; broadleaved evergreen forest), Scrub (25%; scrub; heathland), Grassland (30%; steppe/dry calcareous grassland; humid grassland), Rocky areas (20%; scree/boulders; inland cliff), Artificial landscape (15%; arable land; forestry plantation)
Land-use Agriculture (40%), Forestry (10%), Hunting (70%), Not utilized (15%)

Birds

Species	Season	Year	Pop min	Pop max	Acc	Criteria
Gyps fulvus Griffon Vulture	R	1989	100	120	B	B1iii, B2, C2

This is an important area for cliff-nesting birds, especially raptors.

Protection status

National None **International** None

Conservation issues

Threats Afforestation (B), Drainage (A), Dredging/canalization (A), Extraction industry (A), Infrastructure (B), Recreation/tourism (B)

The main threats are from the construction of new tracks and roads, afforestation with *Pinus* plantations, disturbance from tourism and the development of mining and wind-farms.

Upper Ebro and Rudrón canyons A4ii, B1iii, B2, C2, C6 029

Admin region Castilla y León
Coordinates 42°45′N 3°35′W
Altitude 700–1,260 m **Area** 74,500 ha

Site description

The area comprises the canyons of the Rivers Ebro and Rudrón with high limestone cliffs. The vegetation is diverse and includes deciduous forest, evergreen forest, and alluvial forest at the base of the cliffs and low open scrub elsewhere. The main human activities are livestock-farming, arable agriculture and hunting.

Habitats Forest and woodland (20%; broadleaved deciduous forest; alluvial/very wet forest; broadleaved evergreen forest), Scrub (15%; scrub; heathland), Grassland (10%; steppe/dry calcareous grassland), Wetland (5%; river/stream), Rocky areas (25%; inland cliff), Artificial landscape (25%; arable land; perennial crops/orchards/groves; ruderal land)
Land-use Agriculture (50%), Forestry (10%), Hunting (60%), Not utilized (25%)

Birds

Species	Season	Year	Pop min	Pop max	Acc	Criteria
Neophron percnopterus Egyptian Vulture	B	1995	23	26	A	B2, C6
[1] ***Neophron percnopterus*** Egyptian Vulture	N	1995	90	—	A	B1iii, C2
Gyps fulvus Griffon Vulture	R	1995	200	220	A	A4ii, B1iii, B2, C2, C6
Circus cyaneus Hen Harrier	R	1995	10	15	A	C6
[2] ***Hieraaetus fasciatus*** Bonnelli's Eagle	R	1996	3	4	A	C6
Falco peregrinus Peregrine	B	1995	11	12	A	C6
Bubo bubo Eagle Owl	R	1995	12	—	A	C6

1. Post-breeding juveniles
2. Large decrease.

This is an important site for cliff-nesting raptors, particularly *Gyps fulvus* and *Hieraaetus fasciatus*.

Protection status

National None **International** None

Conservation issues

Threats Afforestation (B), Infrastructure (B), Recreation/tourism (A)

Major threats include the establishment of wind-farms near a *Gyps fulvus* colony and increasing disturbance from tourism and related infrastructural developments. There are plans to construct a road and a viaduct.

Oña and Tesla mountain ranges B2, C2, C6 030

Admin region Castilla y León
Coordinates 42°50′N 3°30′W
Altitude 750–1,332 m **Area** 37,800 ha

Site description

A mountain range in the north of Burgos province with numerous cliffs and two small gorges. The slopes are covered by scrub, *Pinus* plantations and small *Quercus* and *Fagus* woodlands. The area is used for livestock grazing, forestry and hunting.

Habitats Forest and woodland (20%; broadleaved deciduous forest; mixed forest; alluvial/very wet forest), Scrub (20%; scrub; heathland), Wetland (5%; river/stream), Rocky areas (65%; inland cliff), Artificial landscape (35%; arable land; perennial crops/orchards/groves; forestry plantation)
Land-use Agriculture (35%), Forestry (55%), Hunting (60%), Not utilized (5%)

Birds

Species	Season	Year	Pop min	Pop max	Acc	Criteria
Neophron percnopterus Egyptian Vulture	B	1995	15	—	A	B2
Circus cyaneus Hen Harrier	R	1995	15	—	A	B2, C6
[1] ***Hieraaetus fasciatus*** Bonnelli's Eagle	R	1996	2	3	A	C6
Pyrrhocorax pyrrhocorax Chough	R	1995	300	325	A	B2, C2, C6

1. Large decrease.

This is an important site for cliff-nesting birds.

Protection status

National None **International** None

Conservation issues

Threats Afforestation (B), Deforestation (commercial) (B), Extraction industry (C), Infrastructure (A), Other (B), Unsustainable exploitation (U)

The main threats include the enlargement of a road in one of the gorges, the growing use of poison baits, afforestation and game hunting.

Obarenes mountains B2, C2, C6 031

Admin region Castilla y León, La Rioja
Coordinates 42°40′N 3°15′W
Altitude 600–1,434 m **Area** 35,000 ha

Site description

A low mountain range to the west of Miranda del Ebro, which includes many cliffs and a small gorge. The main vegetation-types are scrub, grassland and *Quercus* and *Fagus* woodland. The area is used for livestock-farming, arable agriculture and hunting.

Habitats Forest and woodland (30%; broadleaved deciduous forest; alluvial/very wet forest; broadleaved evergreen forest), Scrub (35%; scrub; sclerophyllous scrub/garrigue/maquis), Rocky areas (10%; inland cliff), Artificial landscape (35%; arable land; perennial crops/orchards/groves)
Land-use Agriculture (35%), Forestry (7%), Hunting (65%), Not utilized (14%), Other (30%)

Birds

Species	Season	Year	Pop min	Pop max	Acc	Criteria
Neophron percnopterus Egyptian Vulture	B	1995	15	—	B	B2, C6
Aquila chrysaetos Golden Eagle	R	1995	9	10	B	C6
Hieraaetus fasciatus Bonnelli's Eagle	R	1995	3	4	B	C6
Falco peregrinus Peregrine	R	1995	9	—	B	C6
Bubo bubo Eagle Owl	R	1995	15	—	B	C6
Pyrrhocorax pyrrhocorax Chough	R	1995	150	200	B	B2, C2, C6

This is an important site for cliff-nesting raptors.

Protection status

National None **International** Partial
4,000 ha of IBA covered by Special Protection Area (Obarenes–Sierra de Cantabria, 5,350 ha).

Conservation issues

Threats Afforestation (B), Extraction industry (B)

Main threats are from the expansion of mining and from afforestation.

Valdegovía–Arcena mountain range B1iii, B2, C2, C6 032

Admin region Castilla y León, País Vasco
Coordinates 42°50′N 3°11′W
Altitude 480–1,333 m **Area** 20,000 ha

Site description

A mountainous area with abundant cliffs. There are extensive areas of *Pinus* forest in the valleys together with grassland and arable crops. The main human activities are forestry and hunting.

Habitats Forest and woodland (47%; broadleaved deciduous forest; native coniferous forest; mixed forest; alluvial/very wet forest; broadleaved evergreen forest), Scrub (48%; scrub; heathland; sclerophyllous scrub/garrigue/maquis), Grassland (49%; steppe/dry calcareous grassland; humid grassland), Wetland (2%; standing fresh water; river/stream; raised bog), Rocky areas (2%; inland cliff; caves), Artificial landscape (16%; highly improved reseeded grassland; arable land)
Land-use Agriculture (65%), Forestry (50%), Hunting (100%), Nature conservation/research (22%), Tourism/recreation (10%)

Birds

Species	Season	Year	Pop min	Pop max	Acc	Criteria
Neophron percnopterus Egyptian Vulture	B	1996	7	7	A	C6
Gyps fulvus Griffon Vulture	R	1996	95	100	A	B1iii, B2, C2, C6
Circaetus gallicus Short-toed Eagle	B	1995	5	6	A	C6
Falco peregrinus Peregrine	R	1996	4	4	A	C6
Pyrrhocorax pyrrhocorax Chough	R	1995	80	100	A	C6

This is an important area for cliff-nesting raptors.

Protection status

National Partial **International** None
3,496 ha of IBA covered by Natural Park (Valderejo, 3,496 ha).

Conservation issues

Threats Agricultural intensification/expansion (C), Intensified forest management (C), Recreation/tourism (C), Selective logging/cutting (C)

No serious threats but some disturbance is caused by forestry activities, hunting and visitors. A management plan exists for the Natural Park.

La Losa–Orduña (Salvada mountain range) B1iii, B2, C2, C6 033

Admin region Castilla y León, País Vasco
Coordinates 43°01′N 3°04′W
Altitude 260–1,187 m **Area** 20,000 ha

Site description

A mountain range with cliffs, large expanses of scrub and grassland, a few *Fagus* and *Quercus* woodlands and *Pinus* plantations. The area is used for livestock-farming and hunting.

Habitats Forest and woodland (40%; broadleaved deciduous forest; native coniferous forest; mixed forest), Scrub (20%; heathland), Grassland (20%; alpine/subalpine/boreal grassland; humid grassland), Rocky areas (5%; inland cliff; caves), Artificial landscape (15%; highly improved reseeded grassland; forestry plantation; ruderal land)
Land-use Agriculture (50%), Forestry (25%), Hunting (80%), Tourism/recreation (20%)

Birds

Species	Season	Year	Pop min	Pop max	Acc	Criteria
Neophron percnopterus Egyptian Vulture	B	1992	5	7	B	C6
Gyps fulvus Griffon Vulture	R	1997	85	110	B	B1iii, B2, C2, C6
Falco peregrinus Peregrine	R	1996	6	—	B	C6
Pyrrhocorax pyrrhocorax Chough	R	1997	400	550	B	B2, C2, C6

This is an important site for cliff-nesting birds.

Protection status

National None **International** None

Conservation issues

Threats Firewood collection (B), Forest grazing (C), Infrastructure (B), Intensified forest management (B), Recreation/tourism (C), Selective logging/cutting (B)

The main threats are from intensive forest management, overgrazing in the *Fagus* forest and hunting. There is a project to erect 40 wind-farms in the site. The pressure from recreational activities is increasing.

Sobrón canyon B1iii, B2, C2, C6 034

Admin region Castilla y León, País Vasco
Coordinates 42°46′N 3°05′W
Altitude 480–1,173 m **Area** 4,800 ha

Site description

The area includes a gorge of the River Ebro, with cliffs and slopes covered by *Fagus* and *Quercus* woodland, scrub and *Pinus* plantations. The main human activities are arable agriculture, livestock-farming and hunting.

Habitats Forest and woodland (65%; broadleaved deciduous forest; native coniferous forest; alluvial/very wet forest; broadleaved evergreen forest), Scrub (10%; heathland; sclerophyllous scrub/garrigue/maquis), Wetland (5%; river/stream), Rocky areas (25%; scree/boulders; inland cliff; caves), Artificial landscape (15%; arable land; forestry plantation)
Land-use Agriculture (30%), Forestry (20%), Hunting (80%), Not utilized (25%), Tourism/recreation (10%), Water management (10%)

Birds

This is an important area for resident raptors.

Species	Season	Year	Pop min	Pop max	Acc	Criteria
Gyps fulvus Griffon Vulture	R	1992	119	119	A	B1iii, B2, C2, C6

Protection status

National None **International** None

Conservation issues

Threats Deforestation (commercial) (B), Infrastructure (B), Recreation/tourism (B)

The major threats are from intensive forestry operations and a high hunting pressure. Threats are also posed by the erection of powerlines.

Guernica estuary–Cape Ogoño B1i, C2, C6 035

Admin region País Vasco
Coordinates 43°22′N 2°41′W
Altitude 0–50 m **Area** 4,000 ha

Site description

A deep estuary with some remnant brackish marshes, extensive mudflats and sandbanks. There is a small rocky islet at the mouth of the estuary and a large coastal cliff in the west of the area.

Habitats Wetland (90%; tidal river/enclosed tidal water), Marine areas (10%; sea inlet/coastal features)
Land-use Agriculture (10%), Fisheries/aquaculture (25%), Nature conservation/research (90%)

Birds

Species	Season	Year	Pop min	Pop max	Acc	Criteria
Hydrobates pelagicus Storm Petrel	B	1996	200	—	C	C6
Platalea leucorodia Spoonbill	P	1996	—	120	B	B1i, C2
Falco peregrinus Peregrine	R	1996	7	8	A	C6

This is an important site for passage and wintering waders and *Platalea leucorodia* and resident seabirds.

Protection status

National None **International** Partial
945 ha of IBA covered by Ramsar Site (Ría de Mundaka–Guernika, 945 ha). 1,020 ha of IBA covered by Biosphere Reserve (Urdaibai, 23,000 ha). 1,020 ha of IBA covered by Special Protection Area (Urdaibai, 23,000 ha).

Conservation issues

Threats Aquaculture/fisheries (A), Consequences of animal/plant introductions (B), Drainage (C), Dredging/canalization (A), Filling-in of wetlands (B), Recreation/tourism (A)

The main threats include disturbance from tourists and intense shellfish gathering and trawling. The bottom of the estuary has been damaged by dredging. There is a management plan for the Biosphere Reserve. Hunting is forbidden.

Izki mountains B3, C2, C6 036

Admin region País Vasco
Coordinates 42°40′N 2°30′W
Altitude 700–1,057 m **Area** 9,250 ha

Site description

Mountains with streams, bogs and a large *Quercus pyrenaica* woodland. Cattle are grazed in the woodland which contains many small patches of grassland. The main human activities are game hunting and forestry.

Habitats Forest and woodland (80%; broadleaved deciduous forest; alluvial/very wet forest; broadleaved evergreen forest), Scrub (40%; scrub), Grassland (15%; dry siliceous grassland), Wetland (river/stream; raised bog), Rocky areas (scree/boulders; inland cliff), Artificial landscape (5%; perennial crops/orchards/groves)
Land-use Agriculture (5%), Forestry (100%), Hunting (100%), Tourism/recreation (5%)

Birds

Species	Season	Year	Pop min	Pop max	Acc	Criteria
Neophron percnopterus Egyptian Vulture	B	1995	2	4	A	C6
Circaetus gallicus Short-toed Eagle	B	1994	3	5	B	C6
Dendrocopos medius Middle Spotted Woodpecker	R	1995	200	300	A	B3, C2, C6

This is an important area for woodland species. The site holds probably the best population of *Dendrocopos medius* in Spain.

Protection status

National None **International** None

Conservation issues

Threats Extraction industry (C), Firewood collection (C), Infrastructure (C), Intensified forest management (C), Selective logging/cutting (C)

The main threats are from intensive forest management, including the construction of new tracks which allow visitor access to the site.

Chingudi–Bidasoa estuary A1, A4i, A4ii, B1i, B1ii, B2, C1, C2, C3 037

Admin region País Vasco
Coordinates 43°22′N 1°47′W
Altitude 0–50 m **Area** 525 ha

Site description

A highly developed estuary on the Atlantic border between Spain and France. The site includes sandy beaches, cliffs, large areas of intertidal mud, grasslands, reedbeds (*Phragmites*) and some riverine woodland. Main human activities are agriculture, fisheries, tourism and industry.

Habitats Forest and woodland (10%; alluvial/very wet forest), Grassland (5%; humid grassland), Wetland (90%; tidal river/enclosed tidal water; mudflat/sandflat; sand-dunes/sand beach; river/stream; water-fringe vegetation), Marine areas (20%; sea inlet/coastal features), Rocky areas (5%; sea cliff/rocky shore), Artificial landscape (25%; highly improved reseeded grassland; arable land; other urban/industrial areas)
Land-use Agriculture (20%), Fisheries/aquaculture (5%), Tourism/recreation (20%), Urban/industrial/transport (30%)

Birds

Species	Season	Year	Pop min	Pop max	Acc	Criteria
Puffinus yelkouan Yelkouan Shearwater	P	1996	300	2,000	B	A4ii, B1ii, C2
Ciconia nigra Black Stork	P	1996	1	11	B	B1i, C2
Platalea leucorodia Spoonbill	P	1996	69	255	A	A4i, B1i, C2
Numenius arquata Curlew	W	1997	32	5,000	A	A4i, B1i, B2, C3
Larus cachinnans Yellow-legged Gull	P	1996	900	5,000	B	A4i, B1i, C3
Sterna sandvicensis Sandwich Tern	P	1987	1,100	1,500	B	B1i, C2
Sterna hirundo Common Tern	P	1996	200	2,000	B	B1i, C2
Acrocephalus paludicola Aquatic Warbler	P	1996	9	40	B	A1, C1

A very important area for passage and wintering seabirds and waders.

Protection status

National None **International** None

Conservation issues

Threats Dredging/canalization (A), Filling-in of wetlands (B), Industrialization/urbanization (B), Infrastructure (A), Recreation/tourism (B), Unsustainable exploitation (B)

The main threats are from dredging, industrial, transport and urban development, tourism and illegal hunting. A management and recovery plan exists but has not been implemented.

Tierra de Campos steppes A1, A4i, A4ii, B1i, B1iii, B2, B3, C1, C2, C3, C6 038

Admin region Castilla y León
Coordinates 42°09′N 5°12′W
Altitude 750–910 m **Area** 268,000 ha

Site description

Large plains dominated by arable cultivation with some grasslands. The site includes La Nava lagoon.

Habitats Forest and woodland (alluvial/very wet forest; broadleaved evergreen forest), Grassland (5%; humid grassland), Wetland (standing fresh water; river/stream; water-fringe vegetation), Artificial landscape (90%; arable land; forestry plantation; ruderal land)
Land-use Agriculture (100%), Hunting (100%)

Birds

Species	Season	Year	Pop min	Pop max	Acc	Criteria
Nycticorax nycticorax Night Heron	B	1995	30	30	A	C6
[1] ***Ardea purpurea*** Purple Heron	R	1996	30	31	A	C6
[1] ***Anser anser*** Greylag Goose	W	1995	1,500	2,000	A	B1i, C3
Anas platyrhynchos Mallard	W	1996	6,000	10,000	A	B1i, C3
Milvus migrans Black Kite	B	1995	50	80	B	C6
Milvus milvus Red Kite	W	1994	600	1,400	A	A4ii, B1iii, C2
Circus aeruginosus Marsh Harrier	B	1995	35	45	C	C6
Circus pygargus Montagu's Harrier	B	1996	80	120	C	B3, C6
Falco naumanni Lesser Kestrel	B	1996	100	—	B	A1, B1iii, B2, C1, C2, C6
[1] ***Tetrax tetrax*** Little Bustard	R	1996	120	150	—	C6
Tetrax tetrax Little Bustard	W	1995	250	300	B	A1, C1
[2] ***Otis tarda*** Great Bustard	R	1997	2,000	2,500	A	A1, B2, C1, C2, C6
Himantopus himantopus Black-winged Stilt	B	1996	77	90	A	C6
Burhinus oedicnemus Stone Curlew	R	1996	80	100	B	C6
Charadrius morinellus Dotterel	W	1992	1,000	—	B	A4i, B1i, C2
Pterocles orientalis Black-bellied Sandgrouse	R	1996	100	200	B	B2, C6
Melanocorypha calandra Calandra Lark	R	1996	4,500	11,000	B	B2

1. Large increase.
2. Individuals.

The site holds steppic species at high densities.

Protection status

National None **International** None

Conservation issues

Threats Abandonment/reduction of land management (C), Afforestation (C), Agricultural intensification/expansion (A), Infrastructure (C)

Available habitat for steppic species is being reduced by changes of land-use and agricultural intensification, particularly following irrigation. A zonal plan exists for the site.

Villafáfila — A1, A4i, A4ii, B1i, B1iii, B2, C1, C2, C3, C6 — 039

Admin region Castilla y León
Coordinates 41°50'N 5°35'W
Altitude 676–733 m **Area** 32,682 ha

Site description

An large, semi-arid, treeless plain in the north-east of Zamora province. Mainly non-irrigated arable cultivation with patches of grassland between the cultivated fields and surrounding seasonal saline lagoons.

Habitats Wetland (10%; standing brackish and salt water), Artificial landscape (90%; arable land)
Land-use Agriculture (95%), Hunting (80%), Nature conservation/research (100%)

Birds

Species	Season	Year	Pop min	Pop max	Acc	Criteria
[1] ***Ciconia ciconia*** White Stork	B	1996	136	—	A	C6
[1] ***Anser anser*** Greylag Goose	W	1995	—	34,183	A	A4i, B1i, C3
Anas platyrhynchos Mallard	R	1995	5,867	—	A	B1i, C3
Circus aeruginosus Marsh Harrier	R	1996	8	—	A	C6
Circus pygargus Montagu's Harrier	B	1993	80	—	A	C6
Falco naumanni Lesser Kestrel	B	1996	374	—	A	A1, A4ii, B1iii, B2, C1, C2, C6
Grus grus Crane	P	1995	811	—	A	B1i, C2
Otis tarda Great Bustard	R	1996	1,026	2,198	A	A1, B2, C1, C2, C6
Himantopus himantopus Black-winged Stilt	B	1996	457	—	A	A4i, B1i, C2, C6
Recurvirostra avosetta Avocet	B	1995	41	—	A	C6
Pterocles orientalis Black-bellied Sandgrouse	B	1996	40	—	B	C6
Pterocles alchata Pin-tailed Sandgrouse	R	1987	50	—	B	C6

1. Large increase.

This is a very important area for breeding steppic species. The lagoons are important for breeding and passage waterbirds. Species of global conservation concern that do not meet IBA criteria: *Tetrax tetrax* (40 breeding individuals).

Protection status

National High **International** High
32,682 ha of IBA covered by National Game Reserve (Lagunas de Villafafila, 32,682 ha). 2,854 ha of IBA covered by Ramsar Site (Lagunas de Villafáfila, 2,854 ha). 32,682 ha of IBA covered by Special Protection Area (Villafáfila, 32,682 ha).

Conservation issues

Threats Agricultural intensification/expansion (A), Groundwater abstraction (C), Infrastructure (A), Recreation/tourism (B)

The main threats are from a new irrigation project.

Esla dam — A1, C1 — 040

Admin region Castilla y León
Coordinates 41°40'N 5°47'W
Altitude 686–750 m **Area** 22,300 ha

Site description

A long, narrow reservoir on the River Esla surrounded by cereal crops.

Habitats Forest and woodland (10%; mixed forest), Scrub (10%; scrub), Wetland (standing fresh water), Artificial landscape (80%; arable land)
Land-use Agriculture (90%), Water management

Birds

Species	Season	Year	Pop min	Pop max	Acc	Criteria
Otis tarda Great Bustard	R	1995	44	—	A	A1, C1

An important site for wintering wildfowl and resident steppic species. It is the only site for wintering *Anser fabalis* in the Iberian peninsula.

Protection status

National None **International** None

Conservation issues

Threats Agricultural intensification/expansion (B), Firewood collection (C), Recreation/tourism (A)

The main threats are from disturbance from visitors and irrigation plans.

Belver de los Montes–Gallegos del Pan — A1, B2, C1, C2, C6 — 041

Admin region Castilla y León
Coordinates 41°42'N 5°25'W
Altitude 670–818 m **Area** 40,300 ha

Site description

An area of plains to the west of Villafáfila, mainly covered by non-irrigated cereal crops with small areas of scrub and grassland.

Habitats Forest and woodland (5%; broadleaved evergreen forest), Grassland (10%; steppe/dry calcareous grassland), Wetland (river/stream), Artificial landscape (85%; arable land; ruderal land)
Land-use Agriculture (95%), Hunting (50%)

Birds

Species	Season	Year	Pop min	Pop max	Acc	Criteria
Nycticorax nycticorax Night Heron	B	1994	10	20	B	C6

Species ... continued	Season	Year	Pop min	Pop max	Acc	Criteria
Circus aeruginosus Marsh Harrier	R	1994	3	4	A	C6
Falco naumanni Lesser Kestrel	B	1996	50	60	B	A1, B2, C1, C6
Otis tarda Great Bustard	R	1996	200	250	A	A1, B2, C1, C2, C6

This is an important site for breeding steppic species.

Protection status

National None **International** None

Conservation issues

Threats Afforestation (B), Agricultural intensification/expansion (A), Drainage (B), Groundwater abstraction (C), Recreation/tourism (C), Unsustainable exploitation (A)

The main threats are from irrigation, afforestation of agricultural land and a high hunting pressure.

River Pisuerga at Dueñas — B2, C6 — 042

Admin region Castilla y León
Coordinates 41°50′N 4°31′W
Altitude 700–710 m **Area** 2,300 ha

Site description

An area of flood-plain with alluvial woodland, cultivation and some industrial sites.

Habitats Forest and woodland (15%; alluvial/very wet forest), Wetland (10%; river/stream), Artificial landscape (75%; arable land; other urban/industrial areas)
Land-use Agriculture (75%), Hunting (40%), Urban/industrial/transport (5%), Water management (5%)

Birds

Species	Season	Year	Pop min	Pop max	Acc	Criteria
Nycticorax nycticorax Night Heron	R	1995	211	211	A	B2, C6

This is an important area for species characteristic of alluvial woodland and for breeding herons.

Protection status

National None **International** None

Conservation issues

Threats Agricultural intensification/expansion (B), Filling-in of wetlands (B), Industrialization/urbanization (B), Infrastructure (A)

The main threats are from development of an existing railway, agricultural intensification and industrial activities.

Carrión–Frómista — A1, B2, C1, C2, C6 — 043

Admin region Castilla y León
Coordinates 42°15′N 4°31′W
Altitude 750–859 m **Area** 58,500 ha

Site description

Large plains dominated by non-irrigated cereal cultivation with some dry grasslands. Riverine forest occurs along the banks of the River Carrión.

Habitats Forest and woodland (alluvial/very wet forest), Scrub (scrub), Grassland (dry siliceous grassland), Artificial landscape (*; arable land)
Land-use Agriculture (*), Hunting (*)

Birds

Species	Season	Year	Pop min	Pop max	Acc	Criteria
Circus pygargus Montagu's Harrier	B	1996	80	—	B	C6
Otis tarda Great Bustard	R	1996	300	400	B	A1, B2, C1, C2, C6
Burhinus oedicnemus Stone Curlew	R	1996	50	—	C	C6

This is an important site for steppic species. Species of global conservation concern that do not meet IBA criteria: *Falco naumanni* (6 resident pairs) and *Tetrax tetrax* (resident).

Protection status

National None **International** None

Conservation issues

Threats Agricultural intensification/expansion (A)

The main threats are from future plans for irrigating crops and other forms of agricultural intensification.

Cerrato high moor — A1, B2, C1, C2, C6 — 044

Admin region Castilla y León
Coordinates 41°55′N 4°00′W
Altitude 800–931 m **Area** 108,000 ha

Site description

A plateau, mainly used for cereal cultivation, with small areas of scrub and maquis. The main human activities are livestock-farming, arable cultivation and hunting.

Habitats Forest and woodland (30%; broadleaved deciduous forest; broadleaved evergreen forest), Artificial landscape (70%; arable land)
Land-use Agriculture (70%), Forestry (30%), Hunting (100%)

Birds

Species	Season	Year	Pop min	Pop max	Acc	Criteria
Hieraaetus pennatus Booted Eagle	B	1996	30	—	B	B2
Falco peregrinus Peregrine	R	1996	12	—	B	C6
Tetrax tetrax Little Bustard	R	1996	200	—	B	A1, C1, C6
Otis tarda Great Bustard	R	1996	120	150	A	A1, B2, C1, C2
Burhinus oedicnemus Stone Curlew	R	1992	100	—	B	C6

The site is important for steppic species.

Protection status

National None **International** None

Conservation issues

Threats Agricultural intensification/expansion (B), Firewood collection (C), Groundwater abstraction (C)

The main threat to the site is from agricultural intensification.

La Demanda mountains — B2, C6 — 045

Admin region Castilla y León, La Rioja
Coordinates 42°15′N 3°00′W
Altitude 800–2,271 m **Area** 133,000 ha

Site description

A mountainous area which comprises a large part of the Sistema Ibérico Norte. Much of area has been deforested and scrub and alpine grasslands predominate although in some areas there are extensive *Quercus* and *Fagus* woodlands. The area of *Pinus* plantations is increasing. Human activities include livestock-farming, game hunting and tourism and recreation.

Habitats Forest and woodland (35%; broadleaved deciduous forest; mixed forest; alluvial/very wet forest; treeline ecotone), Scrub (15%; heathland), Grassland (25%; alpine/subalpine/boreal grassland; humid grassland), Wetland (5%; river/stream), Rocky areas (5%; scree/boulders), Artificial landscape (35%; arable land; forestry plantation)
Land-use Agriculture (40%), Fisheries/aquaculture (5%), Forestry (60%), Hunting, Water management (5%)

Birds

Species	Season	Year	Pop min	Pop max	Acc	Criteria
Pernis apivorus Honey Buzzard	B	1995	15	—	B	C6
Circaetus gallicus Short-toed Eagle	B	1995	20	—	B	B2, C6
Circus cyaneus Hen Harrier	B	1995	20	—	B	B2, C6
Aquila chrysaetos Golden Eagle	R	1996	9	9	B	C6
Hieraaetus pennatus Booted Eagle	B	1995	19	—	B	B2, C6
[1] ***Perdix perdix*** Partridge	R	1989	35	41	A	C6

1. *P. p. hispaniensis.*

This is an important area for species associated with upland forest and alpine grasslands.

Protection status

National Partial **International** Partial

15,000 ha of IBA covered by National Game Reserve (Cameros, 92,918 ha). 35,000 ha of IBA covered by National Game Reserve (Sierra de la Demanda, 73,819 ha). 58,000 ha of IBA covered by Special Protection Area (Sierra de la Demanda, 58,000 ha).

Conservation issues

Threats Afforestation (A), Construction/impact of dyke/dam/barrage (A), Recreation/tourism (A), Selective logging/cutting (A), Unsustainable exploitation (B)

The site is threatened by poor forestry practice, particularly the creation of *Pinus* plantations and the indiscriminate logging of the *Fagus* forest. Hunting pressure is high. A dam project is planned within the IBA.

Urbión and Cebollera mountain ranges — B2, B3, C6 — 046

Admin region Castilla y León, La Rioja
Coordinates 42°05′N 2°45′W
Altitude 900–2,228 m **Area** 142,000 ha

Site description

The site encompasses the central mountains of the Sistema Ibérico Norte with large *Pinus*, *Quercus* and *Fagus* woodlands, and extensive areas of grassland and scrub. The main human activities are forestry, livestock-farming, and recreational activities.

Habitats Forest and woodland (30%; broadleaved deciduous forest; native coniferous forest; mixed forest; broadleaved evergreen forest), Scrub (30%; scrub; heathland), Grassland (20%; steppe/dry calcareous grassland), Artificial landscape (20%; forestry plantation)
Land-use Agriculture (40%), Forestry (50%), Hunting (100%)

Birds

Species	Season	Year	Pop min	Pop max	Acc	Criteria
Pernis apivorus Honey Buzzard	B	1996	17	—	B	B3, C6
Circaetus gallicus Short-toed Eagle	B	1996	15	—	B	C6
Circus cyaneus Hen Harrier	R	1996	16	—	B	B2, C6
Aquila chrysaetos Golden Eagle	R	1996	5	—	A	C6
Hieraaetus pennatus Booted Eagle	B	1996	20	—	B	B2, C6
Perdix perdix Partridge	R	1996	100	—	—	C6

This is an important area for species associated with upland forest and grassland, particularly raptors.

Protection status

National High **International** Partial

82,000 ha of IBA covered by National Game Reserve (Cameros, 92,918 ha). 80,000 ha of IBA covered by National Game Reserve (Urbion, 100,023 ha). 81,250 ha of IBA covered by Special Protection Area (Sierras de Urbión, Cebollera y Cameros, 81,250 ha).

Conservation issues

Threats Afforestation (A)

Afforestation is the main threat to the area. Management plans for the Sierra de Cebollera and Alto Najerilla exist.

Gorges of Iregua, Leza and Jubera — B1iii, B2, C2, C6 — 047

Admin region La Rioja
Coordinates 42°17′N 2°20′W
Altitude 600–1,328 m **Area** 23,400 ha

Site description

An area of mountains and the valley of the River Ebro with many cliffs and three small gorges. There are extensive areas of scrub as well as grassland and broadleaved deciduous woodland.

Habitats Forest and woodland (40%; broadleaved deciduous forest), Scrub, Grassland, Rocky areas (60%; inland cliff)
Land-use Agriculture (25%), Not utilized (60%)

Birds

Species	Season	Year	Pop min	Pop max	Acc	Criteria
Gyps fulvus Griffon Vulture	R	1994	155	155	A	B1iii, B2, C2, C6
Aquila chrysaetos Golden Eagle	R	1994	6	6	A	C6
Falco peregrinus Peregrine	R	1994	12	12	A	C6
Bubo bubo Eagle Owl	R	1994	10	10	A	C6

The site is important for cliff-nesting raptors.

Protection status

National None **International** Partial

8,000 ha of IBA covered by Special Protection Area (Peñas de Iregua, Leza y Jubera, 8,000 ha).

Conservation issues

Threats Extraction industry (C), Recreation/tourism (B)

The main threats are from disturbance to birds by visitors and from the extraction industry. There is a management plan for Camero Viejo.

Arnedillo–Isasa crag — B1iii, B2, C2, C6 — 048

Admin region La Rioja
Coordinates 42°10′N 2°10′W
Altitude 650–1,471 m **Area** 5,530 ha

Site description

An area of limestone cliffs along a spur of the Northern Iberian mountain range with slopes covered by scrub and grassland. The area is grazed by sheep.

Habitats Scrub (scrub), Grassland (steppe/dry calcareous grassland), Rocky areas (inland cliff), Artificial landscape (*; perennial crops/orchards/groves; ruderal land)
Land-use Agriculture, Forestry (*), Hunting (*), Not utilized

Birds

Species	Season	Year	Pop min	Pop max	Acc	Criteria
[1] ***Gyps fulvus*** Griffon Vulture	R	1994	145	145	A	B1iii, B2, C2, C6
Bubo bubo Eagle Owl	R	1994	6	6	A	C6

1. Large increase.

The site is important for cliff-nesting raptors.

Protection status

National None **International** Partial

2,400 ha of IBA covered by Special Protection Area (Peñas de Arnedillo, Peña Isasa y Turruncún, 2,400 ha).

Conservation issues

Threats Recreation/tourism (B)

Tourism and recreational activities cause some disturbance.

Alcarama mountain and Alhama river — A4ii, B1iii, B2, C2, C6 — 049

Admin region Castilla y León, La Rioja
Coordinates 41°55′N 2°00′W
Altitude 550–1,187 m **Area** 10,500 ha

Site description

A mountainous area in the Northern Iberian mountain range, including the Alhama river valley. The site comprises limestone cliffs, deforested slopes with open scrub, dry grassland and occasional patches of *Quercus* woodland. There are also small areas of cereal cultivation, olive-groves and orchards. The area is used for cattle-grazing.

Habitats Forest and woodland (alluvial/very wet forest; broadleaved evergreen forest), Scrub (*; sclerophyllous scrub/garrigue/maquis), Grassland (*; steppe/dry calcareous grassland), Rocky areas (inland cliff), Artificial landscape (arable land; perennial crops/orchards/groves; ruderal land)
Land-use Agriculture, Not utilized (*)

Birds

Species	Season	Year	Pop min	Pop max	Acc	Criteria
Neophron percnopterus Egyptian Vulture	B	1994	7	7	A	C6
Gyps fulvus Griffon Vulture	R	1994	189	189	A	A4ii,B1iii,B2,C2,C6
[1] ***Hieraaetus fasciatus*** Bonnelli's Eagle	R	1994	2	2	A	C6
Falco peregrinus Peregrine	R	1994	5	5	A	C6
Bubo bubo Eagle Owl	R	1994	9	9	A	C6

1. Large decrease.

The site is important for cliff-nesting raptors, with the largest *Gyps fulvus* colony in La Rioja.

Protection status

National None **International** Partial
8,600 ha of IBA covered by Special Protection Area (Sierra de Alcarama y Río Alhama, 8,600 ha).

Conservation issues

Threats Construction/impact of dyke/dam/barrage (C)

The only major threat is a plan to enlarge a dam.

River Lobos canyon — B1iii, B2, C2 — 050

Admin region Castilla y León
Coordinates 41°47′N 3°06′W
Altitude 950–1,200 m **Area** 9,580 ha

Site description

A limestone canyon with slopes that are covered by *Pinus* forest, maquis, scrub and grassland. The area is used for forestry, tourism and sheep-grazing.

Habitats Forest and woodland (alluvial/very wet forest; broadleaved evergreen forest), Scrub (*; scrub; sclerophyllous scrub/garrigue/maquis), Grassland (humid grassland), Wetland (river/stream), Rocky areas (*; inland cliff), Artificial landscape (forestry plantation)
Land-use Agriculture (*), Forestry, Nature conservation/research (*), Not utilized

Birds

Species	Season	Year	Pop min	Pop max	Acc	Criteria
[1] ***Gyps fulvus*** Griffon Vulture	R	1996	92	92	A	B1iii, B2, C2

1. Large increase.

A very important area for cliff-nesting raptors, particularly *Gyps fulvus*.

Protection status

National High **International** High
9,580 ha of IBA covered by Natural Park (Spain) (Cañón del Río Lobos, 9,580 ha). 9,580 ha of IBA covered by Special Protection Area (Cañón del Río Lobos, 9,580 ha).

Conservation issues

Threats Recreation/tourism (B)

The main threats are from the high pressure from tourism and sand extraction from the river.

Arlanza valley–Peñas de Cervera — A4ii, B1iii, B2, C2, C6 — 051

Admin region Castilla y León
Coordinates 42°00′N 3°25′W
Altitude 860–1,455 m **Area** 26,000 ha

Site description

A group of limestone cliffs carved by the Rivers Arlanza and Mataviejas. The slopes and plateaus are covered by *Juniperus* and *Quercus* woodland, scrub and arable agriculture in the valleys.

Habitats Forest and woodland (30%; broadleaved deciduous forest; alluvial/very wet forest; broadleaved evergreen forest), Scrub (20%; scrub; sclerophyllous scrub/garrigue/maquis), Rocky areas (20%; scree/boulders; inland cliff; caves), Artificial landscape (35%; arable land; forestry plantation; other urban/industrial areas; ruderal land)
Land-use Agriculture (35%), Forestry (25%), Hunting (65%)

Birds

Species	Season	Year	Pop min	Pop max	Acc	Criteria
Neophron percnopterus Egyptian Vulture	B	1992	17	17	A	B2
Gyps fulvus Griffon Vulture	R	1992	219	219	A	A4ii,B1iii,B2,C2,C6

This is an important site for cliff-nesting raptors.

Protection status

National None **International** None

Conservation issues

Threats Afforestation (A), Dredging/canalization (A), Recreation/tourism (B), Unsustainable exploitation (B)

The main threats are from afforestation with *Eucalyptus*, canalization of the rivers, and hunting.

Montejo de la Vega–Riaza canyon — A4ii, B1iii, B2, C2, C6 — 052

Admin region Castilla y León
Coordinates 41°32′N 3°35′W
Altitude 864–1,166 m **Area** 9,400 ha

Site description

A limestone canyon with *Populus* plantations, small woods of *Quercus* and *Juniperus*, open scrub, grassland and cultivation.

Habitats Forest and woodland (*; broadleaved evergreen forest), Scrub (sclerophyllous scrub/garrigue/maquis), Grassland (dry siliceous grassland), Rocky areas (*; inland cliff), Artificial landscape (*; arable land; ruderal land)
Land-use Agriculture, Nature conservation/research (*), Water management

Birds

Species	Season	Year	Pop min	Pop max	Acc	Criteria
Neophron percnopterus Egyptian Vulture	B	1993	19	19	A	B2, C6
[1] ***Gyps fulvus*** Griffon Vulture	R	1995	336	336	A	A4ii,B1iii,B2,C2,C6
[1] ***Gyps fulvus*** Griffon Vulture	W	1995	740	790	A	A4ii, B1iii, C2
Chersophilus duponti Dupont's Lark	R	1992	150	150	B	B2, C2, C6

1. Large increase.

This is an important area for cliff-nesting raptors.

Protection status

National None **International** Partial
2,100 ha of IBA covered by Special Protection Area (Montejo de la Vega, 2,100 ha).

Conservation issues

Threats Infrastructure (B), Recreation/tourism (B)

The main threats are from overhead powerlines and high visitor numbers.

Duratón canyon — A4ii, B1iii, B2, C2, C6 — 053

Admin region Castilla y León
Coordinates 41°20′N 3°50′W
Altitude 850–1,076 m **Area** 8,800 ha

Site description

A limestone canyon with *Populus* plantations and *Salix* and *Ulmus* woodland along the river. The valley sides are covered by scrub and dry grasslands. The main human activities are hunting, forestry, agriculture and sheep-grazing.

Habitats Forest and woodland (30%; native coniferous forest; alluvial/very wet forest), Scrub (scrub), Grassland (30%; steppe/dry calcareous grassland), Rocky areas (10%; inland cliff), Artificial landscape (30%; arable land; ruderal land)
Land-use Agriculture (15%), Forestry (25%), Hunting (60%), Nature conservation/research (90%), Water management (5%)

■ Birds

Species	Season	Year	Pop min	Pop max	Acc	Criteria
[1] ***Gyps fulvus*** Griffon Vulture	R	1996	364	—	A	A4ii,B1iii,B2,C2,C6
Chersophilus duponti Dupont's Lark	R	1995	280	—	B	B2, C2, C6
Pyrrhocorax pyrrhocorax Chough	R	1994	400	—	B	B2, C2, C6

1. Large increase.

An important site for cliff-nesting raptors and *Chersophilus duponti*.

■ Protection status

National Partial **International** Partial
5,037 ha of IBA covered by Natural Park (Hoces del Río Duratón, 5,037 ha). 5,037 ha of IBA covered by Special Protection Area (Hoces del Río Duratón, 5,037 ha).

■ Conservation issues

Threats Abandonment/reduction of land management (B), Recreation/tourism (C)

The main threat, particularly to *Chersophilus duponti*, is the abandonment of traditional agricultural practice.

River Cega–Tierra de Pinares–Cantalejo

B2, B3, C6 **054**

Admin region Castilla y León
Coordinates 41°15′N 4°00′W
Altitude 790–1,068 m **Area** 25,500 ha

■ Site description

Plains in the province of Segovia with large *Pinus pinaster* woods and areas of humid grassland used for cattle farming.

Habitats Forest and woodland (*; native coniferous forest; alluvial/very wet forest), Scrub (scrub), Wetland (standing fresh water; river/stream)
Land-use Agriculture, Forestry (*), Hunting (*)

■ Birds

Species	Season	Year	Pop min	Pop max	Acc	Criteria
Ciconia ciconia White Stork	B	1996	120	—	B	C6
Milvus migrans Black Kite	B	1996	75	120	B	B2, C6
Milvus milvus Red Kite	R	1996	40	50	B	B3, C6

An important site for breeding forest raptors and *Ciconia ciconia*.

■ Protection status

National None **International** None

■ Conservation issues

Threats Afforestation (A), Agricultural intensification/expansion (B), Construction/impact of dyke/dam/barrage (C), Deforestation (commercial) (A), Dredging/canalization (B), Filling-in of wetlands (A), Groundwater abstraction (B), Infrastructure (A), Recreation/tourism (A), Selective logging/cutting (A), Unsustainable exploitation (B)

The main threats are from forestry, and construction of a dam, railway and camping area.

River Moros

B2, B3, C6 **055**

Admin region Castilla y León
Coordinates 41°01′N 4°18′W
Altitude 850–964 m **Area** 11,500 ha

■ Site description

The site includes conifer forests, riverine forest along the River Moros and some grassland and arable cultivation. The main human activities are arable cultivation, livestock-farming and hunting.

Habitats Forest and woodland (alluvial/very wet forest), Artificial landscape (*; arable land; forestry plantation)
Land-use Agriculture (*), Forestry (*), Hunting (*), Not utilized, Urban/industrial/transport

■ Birds

Species	Season	Year	Pop min	Pop max	Acc	Criteria
[1] ***Ciconia ciconia*** White Stork	R	1996	250	—	A	B2, C6
Milvus milvus Red Kite	R	1996	30	—	B	B3, C6

1. Large increase.

This is an important site for breeding raptors and breeding *Ciconia ciconia*. Species of global conservation concern that do not meet IBA criteria: *Aegypius monachus* (non-breeding).

■ Protection status

National None **International** None

■ Conservation issues

Threats Agricultural intensification/expansion (B), Extraction industry (A), Selective logging/cutting (B)

The main threats are increasing sand extraction, increasing disturbance from traffic and the clearance of forest.

Northern slope of Guadarrama mountain range

A1, A4ii, B1iii, B2, C1, C2, C6 **056**

Admin region Castilla y León
Coordinates 40°52′N 4°05′W
Altitude 880–2,428 m **Area** 54,500 ha

■ Site description

Northern slope of the Guadarrama mountain range in the Sistema Central with *Quercus pyrenaica* and *Pinus sylvestris* woodland and *Q. rotundifolia* below 1,000 m. There are also large areas of degraded scrub and above 1,800 m *Cytisus* and *Juniperus* scrub. The main human activities are cattle-farming, game hunting and skiing.

Habitats Forest and woodland (*; broadleaved deciduous forest; native coniferous forest; broadleaved evergreen forest), Scrub (*; scrub), Grassland (*; dry siliceous grassland; humid grassland), Wetland (river/stream), Artificial landscape (arable land)
Land-use Agriculture, Forestry (*), Hunting, Tourism/recreation

■ Birds

Species	Season	Year	Pop min	Pop max	Acc	Criteria
Milvus milvus Red Kite	B	1996	20	—	C	C6
Aegypius monachus Cinereous Vulture	R	1996	50	55	A	A1, A4ii, B1iii, B2, C1, C2, C6
Aquila adalberti Spanish Imperial Eagle	R	1996	4	4	A	A1, B2, C1, C6

This is a very important area for breeding raptors. Species of global conservation concern that do not meet IBA criteria: *Falco naumanni* (4–5 breeding pairs.

■ Protection status

National Partial **International** Partial
587 ha of IBA covered by Natural Site of National Interest (Pinar de la Acebeda, 587 ha). 4,792 ha of IBA covered by Special Protection Area (El Espinar, 4,792 ha). 10,427 ha of IBA covered by Special Protection Area (Pinar de Valsaín, 10,427 ha).

■ Conservation issues

Threats Industrialization/urbanization (B), Infrastructure (A), Recreation/tourism (A), Selective logging/cutting (B)

The main threats are from recreational activities, building developments and infrastructural developments.

Segovia

B1i, B2, C2 **057**

Admin region Castilla y León
Coordinates 40°58′N 4°08′W
Altitude 900–1,009 m **Area** 2,200 ha

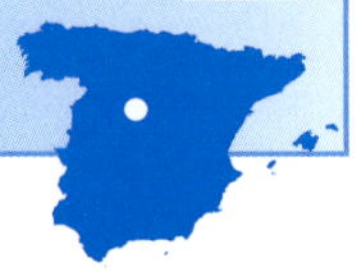

■ Site description

The site comprises the city of Segovia and its surroundings with non-irrigated cereal crops, *Pinus* copses and riverine vegetation.

Habitats Forest and woodland (*; alluvial/very wet forest; broadleaved evergreen forest), Grassland (steppe/dry calcareous grassland), Wetland (river/stream), Artificial landscape (*; arable land; urban parks/gardens; other urban/industrial areas)
Land-use Agriculture (*), Hunting, Urban/industrial/transport (*)

■ Birds

Species	Season	Year	Pop min	Pop max	Acc	Criteria
[1] ***Ciconia ciconia*** White Stork	N	1996	1,500	2,000	B	B1i, C2
Pyrrhocorax pyrrhocorax Chough	R	1996	100	—	B	B2

1. Feeding at town rubbish-tip.

An important site for *Ciconia ciconia* and *Pyrrhocorax pyrrhocorax*.

■ Protection status

National None **International** None

■ Conservation issues

Threats Agricultural intensification/expansion (A), Infrastructure (B)

The main threats are from road construction, overgrazing and intensive agriculture.

Tordesillas–Mota del Marqués

A1, C1, C6 **058**

Admin region Castilla y León
Coordinates 41°40′N 5°15′W
Altitude 550–750 m **Area** 20,000 ha

■ Site description

An area of plains dominated by cereal production, interspersed with small *Pinus* woodlands.

Habitats Forest and woodland (15%; native coniferous forest; alluvial/very wet forest), Grassland (5%; humid grassland), Wetland (20%; standing fresh water; river/stream; water-fringe vegetation), Artificial landscape (95%; arable land; forestry plantation; ruderal land)
Land-use Agriculture (90%), Hunting (60%), Urban/industrial/transport (25%), Water management (5%)

■ Birds

Species	Season	Year	Pop min	Pop max	Acc	Criteria
Tetrax tetrax Little Bustard	B	1996	60	100	B	C6
Otis tarda Great Bustard	B	1996	100	—	B	A1, C1
Himantopus himantopus Black-winged Stilt	B	1996	60	—	B	C6
Pterocles orientalis Black-bellied Sandgrouse	R	1996	30	—	C	C6
[1] ***Pterocles alchata*** Pin-tailed Sandgrouse	R	1996	20	—	B	C6

1. Large decrease.

This is an important area for breeding steppic species and raptors associated with *Pinus* woodland. Species of global conservation concern that do not meet IBA criteria: *Falco naumanni* (breeding).

■ Protection status

National None **International** None

■ Conservation issues

Threats Agricultural intensification/expansion (A), Groundwater abstraction (B), Infrastructure (B), Unsustainable exploitation (B)

The main threats are from an increased use of irrigation, new road construction and a high hunting pressure.

Castronuño–Zamora

C6 **059**

Admin region Castilla y León
Coordinates 41°30′N 5°15′W
Altitude 625–750 m **Area** 19,500 ha

■ Site description

The River Duero between Pollos and Zamora. The site includes alluvial woodland dominated by *Salix*, *Populus* and *Ulmus* and an area of marshland. The main human activity is arable cultivation.

Habitats Forest and woodland (50%; alluvial/very wet forest; broadleaved evergreen forest), Wetland (15%; river/stream; water-fringe vegetation), Artificial landscape (60%; arable land; forestry plantation)
Land-use Agriculture (50%), Forestry (5%), Hunting (60%), Urban/industrial/transport (10%), Water management (10%)

■ Birds

Species	Season	Year	Pop min	Pop max	Acc	Criteria
Nycticorax nycticorax Night Heron	B	1996	61	61	A	C6
[1] ***Egretta garzetta*** Little Egret	B	1996	22	22	A	C6
Ardea purpurea Purple Heron	B	1996	3	3	A	C6
[2] ***Ciconia ciconia*** White Stork	B	1996	110	125	A	C6
Milvus migrans Black Kite	B	1995	50	75	B	C6
Circus aeruginosus Marsh Harrier	R	1996	7	11	A	C6

1. Large decrease.
2. Large increase.

This is an important site for breeding herons (Ardeidae) and *Ciconia ciconia*. Species of global conservation concern that do not meet IBA criteria: *Falco naumanni* (4–6 breeding pairs).

■ Protection status

National None **International** Partial
4,671 ha of IBA covered by Special Protection Area (Riberas de Castronuño, 4,671 ha).

■ Conservation issues

Threats Agricultural intensification/expansion (B), Consequences of animal/plant introductions (B), Deforestation (commercial) (B), Extraction industry (B), Filling-in of wetlands (B), Industrialization/urbanization (B), Infrastructure (B), Recreation/tourism (B), Selective logging/cutting (B)

The greatest threats are from the cutting of riverine forest, urbanization, quarrying and the introduction of exotic fish species.

Fuentelapeña–Jambrina

A1, B2, C1, C2, C6 **060**

Admin region Castilla y León
Coordinates 41°25′N 5°35′W
Altitude 680–864 m **Area** 25,000 ha

■ Site description

A plain to the south-east of Zamora, dominated by non-irrigated cereal fields with some sheep pasture.

Habitats Grassland (7%; steppe/dry calcareous grassland), Artificial landscape (80%; arable land)
Land-use Agriculture (80%), Hunting

■ Birds

Species	Season	Year	Pop min	Pop max	Acc	Criteria
Otis tarda Great Bustard	B	1996	150	250	A	A1, B2, C1, C2, C6

This is an important area for steppic species.

■ Protection status

National None **International** None

■ Conservation issues

Threats Agricultural intensification/expansion (B), Unsustainable exploitation (A)

The main threats are plans to introduce irrigation schemes and also illegal hunting.

Tierra de Campiñas steppes — A1, A4ii, B1i, B1iii, B2, B3, C1, C2, C6 — 061

Admin region Castilla y León
Coordinates 41°09′N 5°09′W
Altitude 690–933 m **Area** 189,600 ha

Site description

An area of plains to the south-west of Valladolid, used for non-irrigated cereal cultivation and sheep-grazing. There are also small *Pinus* and *Quercus* copses and a few lakes.

Habitats Forest and woodland (5%; native coniferous forest; broadleaved evergreen forest), Artificial landscape (90%; arable land; ruderal land)
Land-use Agriculture (100%), Forestry (5%), Hunting (70%)

Birds

Species	Season	Year	Pop min	Pop max	Acc	Criteria
Milvus milvus Red Kite	W	1993	1,500	—	B	A4ii, B1iii, C2
[1] ***Circus pygargus*** Montagu's Harrier	B	1994	200	—	B	B3, C6
Falco naumanni Lesser Kestrel	B	1995	100	—	B	A1, B1iii, B2, C1, C2, C6
Grus grus Crane	W	1995	500	700	A	B1i, C2
[1] ***Tetrax tetrax*** Little Bustard	R	1992	1,500	—	B	A1, B2, C1, C6
[1] ***Tetrax tetrax*** Little Bustard	W	1992	1,000	—	B	A1, C1
Otis tarda Great Bustard	R	1992	2,300	2,500	A	A1, B2, C1, C2, C6
[1] ***Burhinus oedicnemus*** Stone Curlew	R	1992	100	—	B	C6
Pterocles orientalis Black-bellied Sandgrouse	R	1992	200	—	B	B2, C6
Pterocles alchata Pin-tailed Sandgrouse	R	1992	50	—	B	C6

1. Large decrease.

This is a very important area for species of steppic habitats and as a feeding area for raptors. Species of global conservation concern that do not meet IBA criteria: *Aegypius monachus* (non-breeding) and *Aquila adalberti* (non-breeding).

Protection status

National None **International** None

Conservation issues

Threats Afforestation (B), Agricultural intensification/expansion (A), Firewood collection (B), Groundwater abstraction (A), Infrastructure (A), Selective logging/cutting (B), Unsustainable exploitation (B)

Powerlines are a threat to *Otis tarda*, as is hunting. Other threats are from irrigation, agricultural intensification and afforestation. A zonal plan exists for the area.

Topas — A1, B2, C1, C2 — 062

Admin region Castilla y León
Coordinates 41°10′N 5°35′W
Altitude 800–926 m **Area** 28,000 ha

Site description

An area of plains, north of Salamanca, dominated by non-irrigated cereal crops and patches of *Quercus* woodland.

Habitats Forest and woodland (15%; broadleaved evergreen forest), Grassland (6%; steppe/dry calcareous grassland), Wetland (8%; river/stream), Artificial landscape (70%; arable land; perennial crops/orchards/groves)
Land-use Agriculture (65%), Forestry (10%), Hunting (30%), Urban/industrial/transport (10%)

Birds

Species	Season	Year	Pop min	Pop max	Acc	Criteria
Otis tarda Great Bustard	R	1997	73	150	B	A1, B2, C1, C2

This is an area important for steppic species.

Protection status

National None **International** None

Conservation issues

Threats Agricultural intensification/expansion (A), Firewood collection (C), Infrastructure (A), Unsustainable exploitation (B)

The main threats are from road construction, the impact of combine-harvesters on breeding *Circus pygargus*, illegal hunting and powerlines.

Arribes del Duero–Fermoselle — B1i, B1iii, B2, C2, C6 — 063

Admin region Castilla y León
Coordinates 41°25′N 6°15′W
Altitude 500–823 m **Area** 53,000 ha

Site description

An area of granitic cliffs along the River Duero. The slopes covered in *Quercus* woodland, sclerophyllous scrub and grassland. The main human activities are arable cultivation and livestock-farming.

Habitats Forest and woodland (30%; native coniferous forest; broadleaved evergreen forest), Scrub (10%; scrub), Grassland (10%; dry siliceous grassland), Rocky areas (10%; inland cliff), Artificial landscape (25%; arable land)
Land-use Agriculture (55%), Urban/industrial/transport

Birds

Species	Season	Year	Pop min	Pop max	Acc	Criteria
Ciconia nigra Black Stork	B	1996	8	8	A	B1i, C2, C6
Milvus milvus Red Kite	R	1996	11	16	A	C6
Neophron percnopterus Egyptian Vulture	R	1996	58	58	A	B1iii, B2, C2, C6
Aquila chrysaetos Golden Eagle	R	1996	11	12	A	B2, C6
[1] ***Hieraaetus fasciatus*** Bonnelli's Eagle	R	1996	7	7	A	B2, C6
Falco peregrinus Peregrine	R	1996	10	10	A	C6
Bubo bubo Eagle Owl	R	1992	15	—	A	C6

1. Large decrease.

This is an important site for cliff-nesting raptors and breeding *Ciconia nigra*.

Protection status

National None **International** Partial
16,000 ha of IBA covered by Special Protection Area (Arribes del Duero, 86,000 ha).

Conservation issues

Threats Abandonment/reduction of land management (B), Infrastructure (A), Recreation/tourism (A), Unsustainable exploitation (A)

The main threats are from the abandonment of traditional land-use and disturbance along the river from recreational activities.

Río Huebra–Arribes del Duero — A4ii, B1i, B1iii, B2, C2, C6 — 064

Admin region Castilla y León
Coordinates 40°50′N 6°45′W
Altitude 400–837 m **Area** 170,000 ha

Site description

A area of granitic cliffs around the River Duero and its tributaries. The site is dominated by *Quercus*, *Cistus* and *Juniperus* scrub. The main human activities are arable agriculture and livestock-farming.

Habitats Forest and woodland (30%; alluvial/very wet forest; broadleaved evergreen forest), Scrub (10%; scrub), Grassland (10%; dry siliceous grassland), Rocky areas (10%; inland cliff), Artificial landscape (25%; arable land)
Land-use Agriculture (45%), Hunting

Birds

Species	Season	Year	Pop min	Pop max	Acc	Criteria
Ciconia nigra Black Stork	B	1996	8	10	A	B1i, C2, C6
Neophron percnopterus Egyptian Vulture	B	1996	110	—	B	A4ii, B1iii, B2, C2, C6

Species ... continued	Season	Year	Pop min	Pop max	Acc	Criteria
Gyps fulvus Griffon Vulture	R	1996	250	300	B	A4ii,B1iii,B2,C2,C6
Aquila chrysaetos Golden Eagle	R	1996	14	—	A	B2, C6
Hieraaetus fasciatus Bonnelli's Eagle	R	1996	10	12	A	B2, C2, C6
Falco peregrinus Peregrine	R	1996	10	12	A	C6
Pyrrhocorax pyrrhocorax Chough	R	1996	100	160	C	B2, C2

This is an important site for cliff-nesting raptors and breeding *Ciconia nigra*.

Protection status

National None **International** Partial

75,000 ha of IBA covered by Special Protection Area (Arribes del Duero, 86,000 ha).

Conservation issues

Threats Construction of dyke/dam/barrage (A), Extraction industry (B), Intensified forest management (A), Selective logging/cutting (A)

The main threats are from forestry, quarrying and the abandonment of traditional land-uses.

Campo de Argañán — B1i, B2, C2, C6 — 065

Admin region Castilla y León
Coordinates 40°30′N 6°40′W
Altitude 650–915 m **Area** 47,000 ha

Site description

An area of plains to the south-west of Salamanca with dehesas, arable crops and grasslands. The site is crossed by the River Azaba with alluvial forest. The main human activities are arable cultivation and livestock-farming.

Habitats Forest and woodland (*; native coniferous forest; alluvial/very wet forest; broadleaved evergreen forest), Scrub (sclerophyllous scrub/garrigue/maquis), Grassland (dry siliceous grassland), Artificial landscape (arable land)
Land-use Agriculture (*), Forestry

Birds

Species	Season	Year	Pop min	Pop max	Acc	Criteria
Ciconia nigra Black Stork	B	1996	8	12	A	B1i, C2, C6
Circaetus gallicus Short-toed Eagle	B	1996	10	20	B	B2
Hieraaetus pennatus Booted Eagle	B	1996	10	20	B	B2

This is an important site for breeding raptors and *Ciconia nigra*. Species of global conservation concern that do not meet IBA criteria: *Tetrax tetrax* (5 resident pairs).

Protection status

National None **International** None

Conservation issues

Threats Construction/impact of dyke/dam/barrage (B), Extraction industry (A), Infrastructure (C)

The most important threats are plans to build a wind-farm and a hydroelectric power-station, and gravel extraction.

Peña de Francia — A1, A4ii, B1i, B1iii, B2, C1, C2, C6 — 066

Admin region Castilla y León, Extremadura
Coordinates 40°30′N 6°10′W
Altitude 650–1,730 m **Area** 65,500 ha

Site description

A mountainous area with slopes covered by *Quercus* and *Castanea* woodland, *Cytisus* scrub and a few *Pinus* plantations. The main human activities are forestry, game hunting and tourism.

Habitats Forest and woodland (broadleaved deciduous forest; mixed forest; broadleaved evergreen forest), Scrub (25%; scrub), Wetland (river/stream), Rocky areas (3%; scree/boulders; inland cliff), Artificial landscape (70%; forestry plantation)
Land-use Agriculture, Forestry (70%), Hunting

Birds

Species	Season	Year	Pop min	Pop max	Acc	Criteria
Ciconia nigra Black Stork	B	1996	6	7	A	B1i, C2, C6
Aegypius monachus Cinereous Vulture	R	1996	45	50	B	A1, A4ii, B1iii, B2, C1, C2, C6
Falco peregrinus Peregrine	R	1996	10	—	B	C6

This is an important site for breeding forest raptors. Species of global conservation concern that do not meet IBA criteria: *Aquila adalberti* (resident).

Protection status

National Partial **International** Partial

20,976 ha of IBA covered by National Game Reserve (Las Batuecas, 20,976 ha). 2,200 ha of IBA covered by Special Protection Area (Arca–Buitrera, 2,200 ha). 21,500 ha of IBA covered by Special Protection Area (Las Batuecas, 21,500 ha).

Conservation issues

Threats Afforestation (A), Recreation/tourism (B), Selective logging/cutting (A)

The main threats are from afforestation and human disturbance.

Gredos and Candelario mountain ranges — B1i, B2, B3, C2, C6 — 067

Admin region Castilla y León, Extremadura
Coordinates 40°20′N 4°55′W
Altitude 400–2,592 m **Area** 173,600 ha

Site description

A mountain range in the Sistema Central. The dominant vegetation-types are *Quercus* woodland, *Pinus* plantations, *Genista* and *Cytisus* scrub, and alpine grassland. Human activities include forestry, cattle-grazing in summer pastures, hunting and tourism.

Habitats Forest and woodland (40%; broadleaved deciduous forest; native coniferous forest; mixed forest; alluvial/very wet forest; broadleaved evergreen forest), Scrub (40%; scrub), Grassland (15%; alpine/subalpine/boreal grassland; humid grassland), Wetland (standing fresh water; river/stream; raised bog), Rocky areas (scree/boulders), Artificial landscape (5%; highly improved reseeded grassland; arable land; perennial crops/orchards/groves; forestry plantation)
Land-use Forestry (*), Hunting (*), Water management

Birds

Species	Season	Year	Pop min	Pop max	Acc	Criteria
Ciconia nigra Black Stork	B	1996	5	—	B	B1i, C2
Milvus migrans Black Kite	B	1996	100	—	B	B2, C6
[1] ***Milvus milvus*** Red Kite	R	1996	50	—	B	B3, C6
[1] ***Aquila adalberti*** Spanish Imperial Eagle	R	1996	1	1	B	B2
Aquila chrysaetos Golden Eagle	R	1996	25	30	B	B2, C6
Hieraaetus pennatus Booted Eagle	B	1996	65	—	B	B2
Bubo bubo Eagle Owl	R	1996	25	—	B	B2, C6

1. Large decrease.

This is an important area for forest- and cliff-nesting raptors. Species of global conservation concern that do not meet IBA criteria: *Aegypius monachus* (3 resident pairs).

Protection status

National Partial **International** None

87,500 ha of IBA covered by Regional Park (Sierra de Gredos, 87,717 ha). 22,815 ha of IBA covered by National Game Reserve (Sierra de Gredos, 22,815 ha).

Conservation issues

Threats Afforestation (C), Intensified forest management (A), Other (U), Recreation/tourism (A), Selective logging/cutting (A), Unsustainable exploitation (A)

Building, an increase in the number of roads, inappropriate forestry management, hunting, and scrub burning are threats to the site. The use of poison baits is deleteriously affecting raptors.

Tiétar valley A1, B1i, B2, C1, C2, C6 068

Admin region Castilla y León
Coordinates 40°15'N 4°51'W
Altitude 355–2,008 m **Area** 86,000 ha

Site description

A large valley, crossed by River Tiétar, with dehesas, *Pinus* woodland and grassland. The northern part of the IBA is mountainous with *Quercus* woodland, *Pinus* plantations, scrub and alpine grassland. The main human activities are forestry, cattle-grazing, hunting and tourism.

Habitats Forest and woodland (50%; broadleaved deciduous forest; native coniferous forest; broadleaved evergreen forest), Scrub (15%; sclerophyllous scrub/garrigue/maquis), Artificial landscape (30%; arable land; perennial crops/orchards/groves; forestry plantation)
Land-use Agriculture (30%), Forestry (45%), Hunting

Birds

Species	Season	Year	Pop min	Pop max	Acc	Criteria
Ciconia nigra Black Stork	B	1995	10	—	A	B1i, C2, C6
Ciconia nigra Black Stork	P	1995	12	22	A	B1i, C2
Ciconia ciconia White Stork	B	1995	100	—	B	C6
Elanus caeruleus Black-winged Kite	R	1995	10	—	B	B2
Circaetus gallicus Short-toed Eagle	B	1995	20	—	B	B2, C6
Aquila adalberti Spanish Imperial Eagle	R	1995	3	3	A	A1, B2, C1, C6
Grus grus Crane	W	1995	1,500	—	B	B1i, C2

This is an important site for breeding raptors and *Ciconia nigra*. Species of global conservation concern that do not meet IBA criteria: *Aegypius monachus* (non-breeding).

Protection status

National None **International** None

Conservation issues

Threats Agricultural intensification/expansion (B), Burning of vegetation (B), Construction/impact of dyke/dam/barrage (B), Industrialization/urbanization (B), Infrastructure (A), Intensified forest management (A)

The main threats are from urban development, infrastructure development, inappropriate forest management, agricultural intensification, forest fires and overgrazing.

Iruelas valley A1, A4ii, B1iii, B2, C1, C2, C6 069

Admin region Castilla y León
Coordinates 40°22'N 4°32'W
Altitude 850–1,959 m **Area** 8,440 ha

Site description

A mountain valley with numerous gorges and streams. The dominant vegetation is scrub and *Pinus* forest with small areas of *Quercus* and *Castanea* woodland and riverine forest. The main human activities are forestry, hunting and livestock-farming.

Habitats Forest and woodland (50%; broadleaved deciduous forest; native coniferous forest; mixed forest; alluvial/very wet forest; broadleaved evergreen forest), Scrub (40%; scrub), Grassland (10%; alpine/subalpine/boreal grassland), Rocky areas (scree/boulders), Artificial landscape (forestry plantation)
Land-use Agriculture (10%), Forestry (80%), Hunting, Tourism/recreation

Birds

Species	Season	Year	Pop min	Pop max	Acc	Criteria
Aegypius monachus Cinereous Vulture	R	1996	90	90	A	A1, A4ii, B1iii, B2, C1, C2, C6

An important area for resident raptors. Species of global conservation concern that do not meet IBA criteria: *Aquila adalberti* (resident).

Protection status

National High **International** High
8,440 ha of IBA covered by Natural Reserve (Valle de Iruelas, 8,440 ha). 8,440 ha of IBA covered by Special Protection Area (Valle de Iruelas, 8,440 ha).

Conservation issues

Threats Burning of vegetation (A), Firewood collection (B), Intensified forest management (B), Recreation/tourism (B)

The main threats are from intensive forestry, forest fires and disturbance from parachuting.

El Escorial–San Martín de Valdeiglesias A1, B1i, B2, C1, C2, C6 070

Admin region Castilla y León, Comunidad de Madrid
Coordinates 40°30'N 4°10'W
Altitude 650–1,754 m **Area** 198,400 ha

Site description

The site includes the Sistema Central between Puerto de Guadarrama and Gredos. Vegetation includes dehesa woodland, grassland and scrub, small areas of *Quercus* and *Pinus* woodland and arable cultivation.

Habitats Forest and woodland (*; broadleaved deciduous forest; alluvial/very wet forest; broadleaved evergreen forest; treeline ecotone), Scrub (*; sclerophyllous scrub/garrigue/maquis), Grassland (dry siliceous grassland), Wetland (standing fresh water; river/stream), Rocky areas (scree/boulders), Artificial landscape (arable land; perennial crops/orchards/groves; ruderal land)
Land-use Agriculture (*), Forestry (*), Hunting (*), Tourism/recreation (*), Urban/industrial/transport, Water management

Birds

Species	Season	Year	Pop min	Pop max	Acc	Criteria
Ciconia nigra Black Stork	B	1996	5	6	A	B1i, C2, C6
Ciconia nigra Black Stork	P	1996	33	33	A	B1i, C2
[1] ***Ciconia ciconia*** White Stork	B	1992	93	98	B	C6
Gyps fulvus Griffon Vulture	R	1996	28	28	A	C6
[2] ***Aegypius monachus*** Cinereous Vulture	R	1996	9	9	A	A1, B2, C1, C6
Aquila adalberti Spanish Imperial Eagle	R	1996	22	22	A	A1, B2, C1, C2, C6
Aquila chrysaetos Golden Eagle	R	1996	6	6	A	C6
Hieraaetus pennatus Booted Eagle	B	1992	15	15	B	C6
Hieraaetus fasciatus Bonnelli's Eagle	R	1992	1	1	A	C6

1. Large increase.
2. Large decrease.

This is an important area for breeding and passage raptors and storks.

Protection status

National None **International** Partial
61,600 ha of IBA covered by Special Protection Area (Encinares de los ríos Alberche y Cofio, 61,600 ha).

Conservation issues

Threats Extraction industry (A), Industrialization/urbanization (A), Infrastructure (A), Recreation/tourism (A), Selective logging/cutting (C), Unsustainable exploitation (B)

The site is close to Madrid and is therefore highly threatened by urban and infrastructure developments and tourism. Open-cast mining is a particular threat to *Aquila adalberti*.

El Pardo–Viñuelas A1, A3, B1i, B2, C1, C2, C6 071

Admin region Comunidad de Madrid
Coordinates 40°35'N 3°45'W
Altitude 500–746 m **Area** 34,000 ha

Site description

The site covers the foothills of the Sistema Central, immediately to the north of Madrid. It is an area of dehesa and grassland.

Habitats Forest and woodland (60%; broadleaved evergreen forest), Scrub (15%; sclerophyllous scrub/garrigue/maquis), Grassland (20%; dry siliceous grassland), Wetland (10%; river/stream), Artificial landscape (35%; forestry plantation; ruderal land)
Land-use Military (5%), Tourism/recreation (5%), Urban/industrial/transport (5%), Water management (10%)

Birds

Species	Season	Year	Pop min	Pop max	Acc	Criteria
Ciconia nigra Black Stork	P	1996	10	20	A	B1i, C2
Aegypius monachus Cinereous Vulture	R	1996	3	—	B	C6
Aquila adalberti Spanish Imperial Eagle	R	1996	9	10	B	A1, B2, C1, C6
Grus grus Crane	P	1995	500	1,000	B	B1i, C2
Bubo bubo Eagle Owl	R	1996	15	20	B	C6
Caprimulgus ruficollis Red-necked Nightjar	B	1996	50	—	B	A3
Oenanthe hispanica Black-eared Wheatear	B	1996	10	—	B	A3
Oenanthe leucura Black Wheatear	R	1995	3	—	C	A3
[1] ***Sylvia conspicillata*** Spectacled Warbler	B	1996	Common		—	A3
Sylvia cantillans Subalpine Warbler	B	1996	25	—	B	A3
Sturnus unicolor Spotless Starling	R	1996	200	—	B	A3

1. Large decrease.

Breeding birds include six of the 20 species in Europe that are restricted to the Mediterranean biome (when breeding).

Protection status

National Partial **International** Partial

15,000 ha of IBA covered by Natural Park (Cuenca Alta del Rio Manzanares, 46,323 ha). 15,000 ha of IBA covered by Biosphere Reserve (Cuenca Alta del Rio Manzanares, 101,300 ha). 14,774 ha of IBA covered by Special Protection Area (Monte de El Pardo, 14,774 ha). 4,532 ha of IBA covered by Special Protection Area (Soto de Viñuelas, 4,532 ha).

Conservation issues

Threats Abandonment/reduction of land management (C), Consequences of animal/plant introductions (U), Firewood collection (B), Industrialization/urbanization (B), Infrastructure (A), Recreation/tourism (C), Selective logging/cutting (C)

The main threat is the expansion of nearby conurbations and associated infrastructure such as the construction of new roads. The river is threatened by industrial pollution. Introduced mink *Mustela vison* may also be a threat.

Redbeeds and copses of Aranjuez C6 072

Admin region Comunidad de Madrid
Coordinates 40°00′N 3°40′W
Altitude 450–700 m **Area** 18,000 ha

Site description

A section of the Tagus river with well conserved riverine forests and several reedbeds (*Phragmites*).

Habitats Forest and woodland (alluvial/very wet forest), Scrub (8%; sclerophyllous scrub/garrigue/maquis), Grassland (40%; steppe/dry calcareous grassland), Wetland (standing fresh water; river/stream), Artificial landscape (50%; arable land; forestry plantation)
Land-use Agriculture (*)

Birds

Species	Season	Year	Pop min	Pop max	Acc	Criteria
Nycticorax nycticorax Night Heron	B	1995	17	17	A	C6
Circus aeruginosus Marsh Harrier	R	1994	7	7	A	C6
Otis tarda Great Bustard	R	1994	20	24	B	C6
Himantopus himantopus Black-winged Stilt	R	1992	110	115	B	C6
Pyrrhocorax pyrrhocorax Chough	R	1994	50	60	B	C6

This is an important site for breeding and wintering waterbirds and wetland raptors. Species of global conservation concern that do not meet IBA criteria: *Aythya nyroca* (wintering) and *Tetrax tetrax* (resident).

Protection status

National Low **International** Partial

56 ha of IBA covered by Natural Reserve (Carrizal de Villamejor, 56 ha). 635 ha of IBA covered by Natural Reserve (El Regajal–Mar de Ontígola, 635 ha). 15,520 ha of IBA covered by Special Protection Area (Carrizales y Sotos de Aranjuez, 15,520 ha).

Conservation issues

Threats Agricultural intensification/expansion (A), Drainage (B), Extraction industry (A), Filling-in of wetlands (A), Industrialization/urbanization (B), Selective logging/cutting (B)

The main threats are the logging of riverine forest, industrial and urban development and overgrazing. Management plans exist for both Natural Reserves.

Cortados del Jarama A1, A4i, B1i, B2, C1, C2, C6 073

Admin region Comunidad de Madrid
Coordinates 40°10′N 3°31′W
Altitude 560–695 m **Area** 25,000 ha

Site description

The site includes sections of the Rivers Jarama and Manzanares to the south-east of Madrid. Patches of riverine forest remain and the site includes a number of small lakes, the result of gravel extraction. The main human activities are arable cultivation and livestock-farming.

Habitats Forest and woodland (alluvial/very wet forest; broadleaved evergreen forest; treeline ecotone), Scrub (scrub; sclerophyllous scrub/garrigue/maquis), Grassland (steppe/dry calcareous grassland), Wetland (*; standing fresh water; water-fringe vegetation), Rocky areas (*; inland cliff), Artificial landscape (*; highly improved reseeded grassland; arable land; perennial crops/orchards/groves; forestry plantation; other urban/industrial areas; ruderal land)
Land-use Agriculture (*), Hunting (*), Military, Nature conservation/research, Tourism/recreation, Urban/industrial/transport (*), Water management

Birds

Species	Season	Year	Pop min	Pop max	Acc	Criteria
Ixobrychus minutus Little Bittern	B	1996	5	10	B	C6
Nycticorax nycticorax Night Heron	B	1994	60	—	A	C6
Ardea purpurea Purple Heron	B	1994	5	—	B	C6
Ciconia ciconia White Stork	B	1994	164	164	A	C6
Milvus migrans Black Kite	B	1996	50	50	A	C6
Circus aeruginosus Marsh Harrier	B	1996	3	3	A	C6
[1] ***Falco naumanni*** Lesser Kestrel	B	1995	38	47	A	A1, B2, C1, C6
Falco peregrinus Peregrine	R	1996	10	10	A	C6
[2] ***Porphyrio porphyrio*** Purple Gallinule	R	1996	5	10	B	C6
Otis tarda Great Bustard	R	1996	25	35	C	C6
Himantopus himantopus Black-winged Stilt	B	1996	76	85	A	C6
Himantopus himantopus Black-winged Stilt	P	1996	1,500	—	B	A4i, B1i, C2
Bubo bubo Eagle Owl	R	1996	8	10	B	C6
Merops apiaster Bee-eater	B	1996	230	—	B	B2
Pyrrhocorax pyrrhocorax Chough	R	1996	250	300	A	B2, C2, C6

1. Large decrease.
2. Large increase.

This is an important site for species associated with riverine habitats and also for migrant passerines. Species of global conservation concern that do not meet IBA criteria: *Tetrax tetrax* (resident).

Protection status

National High **International** High

25,000 ha of IBA covered by Regional Park (Sureste, 31,550 ha). 25,000 ha of IBA covered by Special Protection Area (Cortados y cantiles de los ríos Manzanares y Jaram, 29,205 ha).

Conservation issues

Threats Abandonment/reduction of land management (C), Agricultural intensification/expansion (B), Extraction industry (A), Filling-in of wetlands (A), Industrialization/urbanization (A), Infrastructure (A), Recreation/tourism (B), Selective logging/cutting (C)

The main threats are from urban and transport developments, sand and gravel extraction from the rivers, river canalization, refuse dumping in the lakes, the erection of powerlines and uncontrolled hunting. A management plan exists for the Regional Park of Sureste but has not been applied.

Talamanca–Camarma A1, B2, B3, C1, C2, C6 074

Admin region Comunidad de Madrid, Castilla-La Mancha
Coordinates 40°40′N 3°25′W
Altitude 750–833 m **Area** 52,000 ha

Site description

An area of plains and low hills with extensive cereal cultivation and small patches of scrub.

Habitats Forest and woodland (5%; alluvial/very wet forest; broadleaved evergreen forest), Scrub (10%; sclerophyllous scrub/garrigue/maquis), Wetland (standing fresh water), Artificial landscape (85%; arable land)
Land-use Agriculture (80%), Hunting (100%), Urban/industrial/transport (5%)

Birds

Species	Season	Year	Pop min	Pop max	Acc	Criteria
[1] ***Ciconia ciconia*** White Stork	B	1996	44	—	A	C6
Circus aeruginosus Marsh Harrier	B	1996	6	7	A	C6
Circus cyaneus Hen Harrier	B	1996	21	—	A	B2, C6
Circus pygargus Montagu's Harrier	B	1996	118	—	A	B3, C6
Falco naumanni Lesser Kestrel	B	1995	74	78	A	A1, B2, C1, C6
Tetrax tetrax Little Bustard	B	1990	300	—	B	A1, C1, C6
Otis tarda Great Bustard	R	1996	484	484	A	A1, B2, C1, C2, C6
Burhinus oedicnemus Stone Curlew	R	1992	80	—	C	C6

1. Large increase.

An important site for steppic species and for foraging and dispersing raptors. Species of global conservation concern that do not meet IBA criteria: *Aegypius monachus* (non-breeding) and *Aquila adalberti* (non-breeding).

Protection status

National None **International** Partial
33,200 ha of IBA covered by Special Protection Area (Estepas cerealistas de los ríos Jarama y Henares, 33,520 ha).

Conservation issues

Threats Agricultural intensification/expansion (A), Industrialization/urbanization (A), Infrastructure (A), Unsustainable exploitation (B)

Threats include illegal house construction and urban infrastructure developments. A management plan for the SPA, produced by SEO, has yet to be applied by the regional government.

Alcarria de Alcalá steppes A1, C1, C6 075

Admin region Comunidad de Madrid
Coordinates 40°19′N 3°18′W
Altitude 700–800 m **Area** 8,000 ha

Site description

An area of steppe to the east of Madrid with extensive arable cultivation and also olive-groves, vineyards and patches of scrub.

Habitats Scrub (5%; sclerophyllous scrub/garrigue/maquis), Artificial landscape (95%; arable land; perennial crops/orchards/groves; ruderal land)
Land-use Agriculture (80%), Hunting (100%), Urban/industrial/transport (20%)

Birds

Species	Season	Year	Pop min	Pop max	Acc	Criteria
Tetrax tetrax Little Bustard	R	1997	160	300	A	A1, C1, C6
Otis tarda Great Bustard	R	1997	80	120	A	A1, C1, C6

This is an important area for steppic species.

Protection status

National None **International** None

Conservation issues

Threats Extraction industry (B), Industrialization/urbanization (A), Infrastructure (A), Recreation/tourism (A), Unsustainable exploitation (A)

The main threats include illegal house construction, industrial expansion, a planned airport and disturbance from off-road driving.

Alto Lozoya–La Pedriza A1, B1iii, B2, C1, C2, C6 076

Admin region Comunidad de Madrid
Coordinates 40°50′N 3°52′W
Altitude 850–2,428 m **Area** 55,000 ha

Site description

This site includes the southern slopes of the Guadarrama mountain range. The main vegetation-types are *Quercus* and *Pinus* woodland, degraded scrub and above 1,800 m alpine grassland.

Habitats Forest and woodland (68%; broadleaved deciduous forest; native coniferous forest), Scrub (10%; scrub; sclerophyllous scrub/garrigue/maquis), Grassland (15%; alpine/subalpine/boreal grassland; humid grassland), Wetland (10%; standing fresh water), Rocky areas (15%; scree/boulders), Artificial landscape (30%; forestry plantation)
Land-use Agriculture (50%), Forestry (80%), Hunting (5%), Nature conservation/research (55%), Water management (8%)

Birds

Species	Season	Year	Pop min	Pop max	Acc	Criteria
Ciconia ciconia White Stork	B	1992	35	38	A	C6
Gyps fulvus Griffon Vulture	R	1995	32	35	A	C6
Aegypius monachus Cinereous Vulture	R	1996	35	40	B	A1, B1iii, B2, C1, C2, C6
Aquila chrysaetos Golden Eagle	R	1996	5	6	A	C6
Falco peregrinus Peregrine	R	1996	5	7	A	C6

An important area for forest raptors. Species of global conservation concern that do not meet IBA criteria: *Aquila adalberti* (resident).

Protection status

National Partial **International** Partial
8,000 ha of IBA covered by Natural Park (Cuenca Alta del Rio Manzanares, 46,323 ha). 14,745 ha of IBA covered by Natural Park (Cumbre, Circo y Lagunas de Peñalara, 768 ha). 8,000 ha of IBA covered by Biosphere Reserve (Cuenca Alta del Rio Manzanares, 101,300 ha). 5,500 ha of IBA covered by Special Protection Area (Alto Lozoya, 5,500 ha).

Conservation issues

Threats Industrialization/urbanization (A), Infrastructure (A), Recreation/tourism (A), Selective logging/cutting (B)

The main threat is from urban development. Other threats are from the high tourist pressure and the use of the site for recreation. A management plan exists for part of the site.

Ayllón mountain range B2, C6 077

Admin region Castilla y León, Comunidad de Madrid, Castilla-La Mancha
Coordinates 41°15′N 3°20′W
Altitude 1,000–2,262 m **Area** 187,000 ha

Site description

A massif at the eastern end of the Sistema Central mountains. The dominant vegetation is sclerophyllous scrub with small *Quercus* and *Fagus* woods and extensive *Pinus* plantations. Hunting and forestry are the most important human activities.

Habitats Forest and woodland (10%; broadleaved deciduous forest), Scrub (15%; scrub), Grassland (10%; dry siliceous grassland; humid grassland), Rocky areas (inland cliff), Artificial landscape (50%; arable land; forestry plantation)
Land-use Agriculture, Forestry (60%), Hunting (100%)

Birds

Species	Season	Year	Pop min	Pop max	Acc	Criteria
Aquila chrysaetos Golden Eagle	R	1995	16	19	A	B2, C6
Hieraaetus fasciatus Bonnelli's Eagle	R	1995	3	3	A	C6
Falco peregrinus Peregrine	R	1995	16	17	A	B2, C6

This is an important site for breeding raptors.

Protection status

National Partial **International** None

1,641 ha of IBA covered by Natural Park (Hayedo de Tejera Negra, 1,641 ha). 68,106 ha of IBA covered by National Game Reserve (Sonsaz, 68,106 ha). 87 ha of IBA covered by Natural Site of National Interest (Riofrío de Riaza, 87 ha). 250 ha of IBA covered by Natural Site of National Interest (Hayedo de Montejo de la Sierra, 250 ha).

Conservation issues

Threats Construction/impact of dyke/dam/barrage (B), Dredging/canalization (C), Infrastructure (B), Intensified forest management (B), Recreation/tourism (C)

The main threats are from forestry activities, particularly the construction of new tracks and roads, increasing tourism and dams on rivers.

Puebla de Beleña lagoons — A4i, B1i, B1iv, C2, C5 — 078

Admin region Castilla-La Mancha
Coordinates 40°53′N 3°15′W
Altitude 950–1,010 m **Area** 3,100 ha

Site description

Two small, shallow lagoons surrounded by meadows and cultivated land.

Habitats Grassland, Wetland (standing fresh water), Artificial landscape (*; arable land)
Land-use Agriculture (100%)

Birds

Species	Season	Year	Pop min	Pop max	Acc	Criteria
Grus grus Crane	P	1997	500	3,000	A	A4i, B1i, C2

The site is a migratory bottleneck, where 3,000 *Grus grus* regularly pass. Birds are moving to wintering sites in dehesas in Extremadura.

Protection status

National None **International** None

Conservation issues

Threats Groundwater abstraction (U), Natural events (A), Unsustainable exploitation (B)

Over-exploitation of aquifers and illegal hunting are important threats. There are also natural drought periods during which the lagoons dry out. Bird numbers fluctuate depending upon water-levels in the lagoons.

Tiermes–Caracena — B1iii, B2, C2, C6 — 079

Admin region Castilla y León
Coordinates 41°20′N 2°30′W
Altitude 987–1,361 m **Area** 45,300 ha

Site description

A rugged landscape with many small gorges, small *Quercus* and *Juniperus* woods, scrub and arable cultivation. The main human activities are sheep-grazing and forestry.

Habitats Forest and woodland (*; broadleaved deciduous forest; broadleaved evergreen forest), Scrub (*; scrub), Grassland (steppe/dry calcareous grassland), Rocky areas (inland cliff), Artificial landscape (*; arable land)
Land-use Agriculture (*), Forestry (*)

Birds

Species	Season	Year	Pop min	Pop max	Acc	Criteria
Neophron percnopterus Egyptian Vulture	B	1994	12	13	A	B2
Gyps fulvus Griffon Vulture	R	1995	148	148	A	B1iii, B2, C2, C6
Bubo bubo Eagle Owl	R	1995	8	11	A	C6

This is an important breeding area for cliff-nesting raptors, particularly *Gyps fulvus* and *Neophron percnopterus*.

Protection status

National None **International** None

Conservation issues

Threats Recreation/tourism (C)

No serious threats known at the site.

Altos de Barahona high moor — A1, B2, C1, C2, C6 — 080

Admin region Castilla y León
Coordinates 41°18′N 2°48′W
Altitude 1,100–1,313 m **Area** 29,000 ha

Site description

A deforested limestone plain in southern Soria, with small cliffs along rivers and streams. The site is dominated by non-irrigated cereal crops with some low scrub and dry grassland.

Habitats Forest and woodland (10%; broadleaved evergreen forest), Scrub (scrub), Grassland (60%; steppe/dry calcareous grassland), Artificial landscape (30%; arable land)
Land-use Agriculture (100%), Hunting (90%)

Birds

Species	Season	Year	Pop min	Pop max	Acc	Criteria
Tetrax tetrax Little Bustard	R	1995	60	—	A	C6
Otis tarda Great Bustard	N	1996	46	50	A	A1, C1
Burhinus oedicnemus Stone Curlew	R	1996	50	50	A	C6
Pterocles orientalis Black-bellied Sandgrouse	R	1992	150	150	A	B2, C6
Chersophilus duponti Dupont's Lark	R	1992	2,200	2,200	A	B2, C2, C6

This is an important area for steppic species, including the most important population of *Chersophilus duponti* in Spain.

Protection status

National None **International** None

Conservation issues

Threats Abandonment/reduction of land management (A), Afforestation (B), Unsustainable exploitation (B)

The main threats are from changes in management of agricultural land, afforestation and hunting.

Layna high moors — B2, C2, C6 — 081

Admin region Castilla y León, Castilla-La Mancha
Coordinates 41°05′N 2°20′W
Altitude 1,150–1,211 m **Area** 2,100 ha

Site description

An upland area of scrub and grassland. The main human activity is livestock-farming.

Habitats Scrub (scrub), Grassland (90%; steppe/dry calcareous grassland), Artificial landscape (ruderal land)
Land-use Agriculture (90%), Not utilized

Birds

Species	Season	Year	Pop min	Pop max	Acc	Criteria
Pterocles orientalis Black-bellied Sandgrouse	R	1992	20	20	A	C6
Chersophilus duponti Dupont's Lark	R	1992	500	500	A	B2, C2, C6

The site is one of the most important for *Chersophilus duponti* in the Iberian peninsula.

Protection status

National None **International** None

Conservation issues

Threats Afforestation (B), Agricultural intensification/expansion (B), Unsustainable exploitation (B)

The main threats are from afforestation, agricultural intensification and illegal hunting.

Las Cañas lake — A4i, B1i, B2, C2, C6 — 082

Admin region Navarra
Coordinates 42°29′N 2°24′W
Altitude 370–370 m **Area** 101 ha

Site description

An irrigation pond on the River Ebro close to Logroño, with surrounding marshy vegetation (*Typha, Phragmites*) and small patches of *Tamarix* scrub. The surrounding areas are used for vine and cereal cultivation and there is an industrial zone.

Habitats Wetland (100%; standing brackish and salt water; water-fringe vegetation)
Land-use Nature conservation/research (100%), Water management (100%)

Birds

Species	Season	Year	Pop min	Pop max	Acc	Criteria
Nycticorax nycticorax Night Heron	B	1994	470	470	A	A4i, B1i, B2, C2, C6
Egretta garzetta Little Egret	R	1994	4	11	A	C6
Ardea purpurea Purple Heron	B	1994	52	52	A	C2, C6
Circus aeruginosus Marsh Harrier	B	1994	5	5	A	C6

The site is particularly important for breeding herons.

Protection status

National High **International** High
101 ha of IBA covered by Natural Reserve (Embalse de El Salobre o de Las Cañas, 101 ha). 101 ha of IBA covered by Special Protection Area (Embalse de Las Cañas, 101 ha).

Conservation issues

Threats Agricultural intensification/expansion (C), Aquaculture/fisheries (C), Recreation/tourism (B)

The main threats include a high human presence, fires and eutrophication. A management plan was officially approved in 1991.

Echauri crags — B2, C6 — 083

Admin region Navarra
Coordinates 42°48′N 1°50′W
Altitude 400–1,138 m **Area** 1,400 ha

Site description

An area of limestone cliffs, scrub, and *Quercus* woodland. Human activities include livestock-farming, hunting and recreation.

Habitats Forest and woodland (40%; broadleaved deciduous forest; broadleaved evergreen forest), Scrub (60%; scrub; sclerophyllous scrub/garrigue/maquis), Rocky areas (10%; scree/boulders; inland cliff)
Land-use Agriculture (100%), Hunting (50%), Tourism/recreation (20%)

Birds

Species	Season	Year	Pop min	Pop max	Acc	Criteria
[1] ***Gyps fulvus*** Griffon Vulture	R	1994	72	82	A	B2
Hieraaetus fasciatus Bonnelli's Eagle	R	1996	1	1	A	C6

1. Large increase.

An important site for cliff-nesting raptors, particularly *Gyps fulvus*.

Protection status

National Low **International** None
76 ha of IBA covered by Wild Fauna Reserve (Roquedos de Echauri, 76 ha).

Conservation issues

Threats Recreation/tourism (A)

The most serious threat is disturbance from climbers. A management plan exists for the reserve area.

Gorramendi mountain and Itxusi crags — B2, C6 — 084

Admin region Navarra
Coordinates 43°10′N 1°30′W
Altitude 150–1,097 m **Area** 6,300 ha

Site description

A mountainous area along the border with France. The site contains scrub, wet grassland and *Fagus* forest. The main human activities are livestock-farming, hunting and forestry.

Habitats Forest and woodland (30%; broadleaved deciduous forest; mixed forest), Scrub (30%; heathland), Grassland (30%; humid grassland), Artificial landscape (10%; highly improved reseeded grassland)
Land-use Agriculture (60%), Forestry (35%), Hunting (95%)

Birds

Species	Season	Year	Pop min	Pop max	Acc	Criteria
Gyps fulvus Griffon Vulture	R	1994	77	82	A	B2, C6

An important site for breeding raptors, particularly for *Gyps fulvus*.

Protection status

National Low **International** High
266 ha of IBA covered by Wild Fauna Reserve (Iparla, 266 ha). 107 ha of IBA covered by Natural Reserve (Irubetakaskoa, 107 ha). 115 ha of IBA covered by Natural Reserve (Itxusi, 115 ha). 5,909 ha of IBA covered by Special Protection Area (Aritzakun–Urrizate–Gorramendi, 5,909 ha).

Conservation issues

Threats Infrastructure (B), Intensified forest management (B), Recreation/tourism (A), Unsustainable exploitation (B)

The main threats are from intensive forest management, disturbance from visitors and hunting. There is a recovery plan for *Gypaetus barbatus*, and there are management plans for the reserves.

Roncesvalles–Irati–Abodi mountain range — A4iv, B1i, B1iv, B2, C2, C5, C6 — 085

Admin region Navarra
Coordinates 43°00′N 1°15′W
Altitude 530–2,021 m **Area** 39,000 ha

Site description

A mountainous area along the border with France. The site has extensive *Fagus* and *Abies* forests, including the largest *Fagus* forest in Spain. The main human activities are forestry, livestock-farming and hunting.

Habitats Forest and woodland (60%; broadleaved deciduous forest; mixed forest), Scrub (20%; heathland), Grassland (20%; alpine/subalpine/boreal grassland), Artificial landscape (20%; highly improved reseeded grassland; arable land; forestry plantation)
Land-use Agriculture (80%), Forestry (60%), Hunting (95%)

Birds

Species	Season	Year	Pop min	Pop max	Acc	Criteria
Ciconia nigra Black Stork	P	1995	70	125	B	B1i, C2
Gypaetus barbatus Lammergeier	R	1996	1	1	A	B2
Dryocopus martius Black Woodpecker	R	1994	80	100	B	C6
Dendrocopos leucotos White-backed Woodpecker	R	1995	50	60	B	C2, C6

The site is very important for forest species. In particular it holds 90% of the Iberian population of *Dendrocopos leucotos* as well as breeding *Dryocopus martius*. The site is a major migratory bottleneck, where more than 20,000 cranes and more than 3,000 raptors regularly pass in autumn, including *Pernis apivorus* (mean 5,235 per season), *Milvus milvus* (mean 2,072 per season) and *Grus grus* (maximum 20,100).

Protection status

National Low **International** Partial
232 ha of IBA covered by Wild Fauna Reserve (Arrollandieta, 232 ha). 64 ha of IBA covered by Integral Nature Reserve (Lizardoya, 64 ha).

119 ha of IBA covered by Natural Reserve (Mendilaz, 119 ha). 56 ha of IBA covered by Natural Reserve (Tristuibartea, 56 ha). 18,684 ha of IBA covered by Special Protection Area (Selva de Irati–Roncesvalles, 18,684 ha).

■ Conservation issues

Threats Infrastructure (B), Intensified forest management (A), Recreation/tourism (B), Selective logging/cutting (A), Unsustainable exploitation (B)

The most important threats relate to the deleterious impact of intensive forestry and logging. There have been several studies on the effects of forestry activities on species of woodpecker. There are management plans for all protected sites.

Areta, Archuba and Zariquieta mountain ranges — A4ii, B1iii, B2, C2, C6 — 086

Admin region Navarra
Coordinates 42°47'N 1°22'W
Altitude 500–1,475 m **Area** 19,500 ha

■ Site description

Two mountainous areas to the east of Pamplona, with limestone cliffs, *Pinus* woodland, grassland and scrub. The main human activities are livestock-farming, hunting and forestry.

Habitats Forest and woodland (40%; broadleaved deciduous forest; native coniferous forest; mixed forest; broadleaved evergreen forest), Scrub (40%; scrub; heathland), Artificial landscape (15%; arable land)
Land-use Agriculture (90%), Forestry (40%), Hunting (80%)

■ Birds

Species	Season	Year	Pop min	Pop max	Acc	Criteria
Gypaetus barbatus Lammergeier	R	1996	2	2	A	B2, C6
Neophron percnopterus Egyptian Vulture	B	1991	15	15	A	B2, C6
[1] ***Gyps fulvus*** Griffon Vulture	R	1994	182	182	A	A4ii,B1iii,B2,C2,C6
[2] ***Tetrao urogallus*** Capercaillie	R	1991	2	3	B	C6
Dryocopus martius Black Woodpecker	R	1992	10	15	B	C6

1. Large increase.
2. Males.

This is an important site for cliff-nesting raptors and forest birds.

■ Protection status

National Low **International** Partial
72 ha of IBA covered by Natural Reserve (Foz de Gaztelu, 72 ha). 266 ha of IBA covered by Natural Reserve (Foz de Iñarbe, 266 ha). 40 ha of IBA covered by Natural Reserve (Poche de Chinchurrenea, 40 ha). 3,804 ha of IBA covered by Special Protection Area (Sierra de Archuba y Sierra de Zariquieta, 3,804 ha). 1,109 ha of IBA covered by Special Protection Area (Arabarko, 1,109 ha). 11,994 ha of IBA covered by Special Protection Area (Montes de Areta, 11,994 ha).

■ Conservation issues

Threats Construction/impact of dyke/dam/barrage (A), Infrastructure (A), Selective logging/cutting (B), Unsustainable exploitation (C)

Forestry operations and the creation of new forest tracks are threats to the site. The construction of the Itoiz reservoir will affect the area. The IBA also includes nine Wild Fauna Reserves covering 1,770 ha, declared by the regional government in 1995. Management plans exist for the protected sites.

Leyre, Illón and San Miguel mountain ranges — A4ii, B1iii, B2, C2, C6 — 087

Admin region Navarra
Coordinates 42°42'N 1°04'W
Altitude 400–1,420 m **Area** 23,500 ha

■ Site description

An area of limestone cliffs and several large gorges. The vegetation includes alpine grassland, scrub, and *Quercus*, *Fagus* and *Pinus* forests. The main human activities are forestry, hunting and livestock-grazing.

Habitats Forest and woodland (80%; broadleaved deciduous forest; native coniferous forest; mixed forest; broadleaved evergreen forest), Scrub (20%; scrub; heathland; sclerophyllous scrub/garrigue/maquis), Grassland (alpine/subalpine/boreal grassland), Rocky areas (inland cliff), Artificial landscape (arable land)
Land-use Agriculture (90%), Forestry (80%), Hunting (90%)

■ Birds

Species	Season	Year	Pop min	Pop max	Acc	Criteria
Gypaetus barbatus Lammergeier	R	1996	4	4	A	B2, C6
Neophron percnopterus Egyptian Vulture	B	1991	25	25	A	B2, C6
[1] ***Gyps fulvus*** Griffon Vulture	R	1994	668	668	A	A4ii,B1iii,B2,C2,C6
Aquila chrysaetos Golden Eagle	R	1991	7	7	A	C6
Falco peregrinus Peregrine	R	1991	9	9	A	C6
Dryocopus martius Black Woodpecker	R	1992	6	10	B	C6

1. Large increase.

This is an important site for cliff-nesting birds and a wintering site for *Perdix perdix*. The site holds the largest *Gyps fulvus* colonies in Spain.

■ Protection status

National Low **International** Partial
393 ha of IBA covered by Natural Reserve (Acantilados de la Piedra y San Adrián, 393 ha). 1,164 ha of IBA covered by Natural Reserve (Foz de Arbayun, 1,164 ha). 153 ha of IBA covered by Natural Reserve (Foz de Benasa, 153 ha). 115 ha of IBA covered by Natural Reserve (Foz de Burgui, 115 ha). 40 ha of IBA covered by Natural Reserve (Foz de Lumbier, 40 ha). 160 ha of IBA covered by Wild Fauna Reserve (Mendibeltza, 160 ha). 2,987 ha of IBA covered by Special Protection Area (Sierra de San Miguel, 2,987 ha). 9,020 ha of IBA covered by Special Protection Area (Foz de Arbayun–Sierra de Leyre, 9,020 ha). 4,619 ha of IBA covered by Special Protection Area (Foz de Burgui–Sierra de Illón, 4,619 ha).

■ Conservation issues

Threats Infrastructure (B), Intensified forest management (B), Recreation/tourism (B), Selective logging/cutting (B), Unsustainable exploitation (B)

The most important threats are from intensive forestry and logging, hunting and disturbance from visitors. A management plan exists for the reserves.

Izaga crag — B1iii, B2, C2, C6 — 088

Admin region Navarra
Coordinates 42°45'N 1°25'W
Altitude 650–1,361 m **Area** 1,746 ha

■ Site description

A small mountain range to the south-east of Pamplona, with limestone cliffs, *Quercus* and *Fagus*, scrub and *Pinus* plantations. The area is grazed by sheep.

Habitats Forest and woodland (20%; broadleaved deciduous forest), Scrub (50%; scrub), Grassland (20%), Rocky areas (10%; inland cliff), Artificial landscape (40%; arable land; forestry plantation)
Land-use Agriculture (100%), Forestry (40%), Hunting (100%)

■ Birds

Species	Season	Year	Pop min	Pop max	Acc	Criteria
Gypaetus barbatus Lammergeier	R	1996	1	1	A	B2
[1] ***Gyps fulvus*** Griffon Vulture	R	1994	163	163	A	B1iii, B2, C2, C6

1. Large increase.

This is an important site for cliff-nesting raptors, particularly *Gyps fulvus*. *Gypaetus barbatus* has recently started breeding.

■ Protection status

National Partial **International** High
314 ha of IBA covered by Wild Fauna Reserve (Peña Izaga, 314 ha). 1,746 ha of IBA covered by Special Protection Area (Peña Izaga, 1,746 ha).

Conservation issues

Threats Afforestation (B), Agricultural intensification/expansion (B), Burning of vegetation (B), Intensified forest management (B), Unsustainable exploitation (B)

Intensive forest management, afforestation with plantations, fires, overgrazing and hunting are major threats in the IBA. A management plan has existed since 1995.

Pitillas and Dos Reinos lagoons C6 089

Admin region Navarra, Aragón
Coordinates 42°24'N 1°34'W
Altitude 350–370 m **Area** 260 ha

Site description

Small endorreic lagoons, now used for irrigation, with water-fringe vegetation.

Habitats Wetland (100%; standing fresh water; standing brackish and salt water; water-fringe vegetation)
Land-use Water management (100%)

Birds

Species	Season	Year	Pop min	Pop max	Acc	Criteria
Botaurus stellaris Bittern	R	1996	1	2	A	C6
Ardea purpurea Purple Heron	B	1996	18	22	A	C6
Circus aeruginosus Marsh Harrier	R	1992	35	40	A	C6

This is an important site for breeding herons and wetland raptors and passage site for *Grus grus* and wildfowl.

Protection status

National High **International** High
21 ha of IBA covered by Natural Area (Laguna de Dos Reinos, 21 ha). 216 ha of IBA covered by Natural Reserve (Laguna de Pitillas, 216 ha). 21 ha of IBA covered by Special Protection Area (Laguna de Dos Reinos, 21 ha). 216 ha of IBA covered by Special Protection Area (Laguna de Pitillas, 216 ha).

Conservation issues

Threats Agricultural intensification/expansion (B), Aquaculture/fisheries (C)

The main threats are from sedimentation and nutrient pollution, both a result of intensive agriculture around the lagoons. A management plan exists for both lagoons.

Bardenas Reales A3, B1iii, B2, C2, C6 090

Admin region Navarra, Aragón
Coordinates 42°10'N 1°30'W
Altitude 290–646 m **Area** 55,900 ha

Site description

An area of steppe along the River Ebro. The site has a mixture of non-irrigated cereal cultivation, dry grassland, scrub and small *Pinus* woods.

Habitats Scrub (40%; sclerophyllous scrub/garrigue/maquis), Grassland (30%; dry siliceous grassland), Rocky areas (10%; inland cliff), Artificial landscape (50%; arable land)
Land-use Agriculture (100%), Hunting (95%), Military (5%)

Birds

Species	Season	Year	Pop min	Pop max	Acc	Criteria
Neophron percnopterus Egyptian Vulture	B	1995	50	55	A	B1iii, B2, C2, C6
[1] ***Neophron percnopterus*** Egyptian Vulture	N	1994	200	200	A	B1iii, C2
Circus pygargus Montagu's Harrier	B	1995	6	10	B	C6
Aquila chrysaetos Golden Eagle	R	1995	6	7	A	C6
Tetrax tetrax Little Bustard	R	1995	10	20	A	C6
Pterocles orientalis Black-bellied Sandgrouse	R	1990	40	60	B	C6
Pterocles alchata Pin-tailed Sandgrouse	R	1990	30	50	B	C6
Bubo bubo Eagle Owl	R	1990	15	20	A	C6
Caprimulgus ruficollis Red-necked Nightjar	B	1995	Frequent		—	A3
Chersophilus duponti Dupont's Lark	R	1994	400	—	A	B2, C2, C6
Calandrella rufescens Lesser Short-toed Lark	R	1994	1,000	—	B	B2
Oenanthe hispanica Black-eared Wheatear	B	1995	Common		—	A3
Oenanthe leucura Black Wheatear	R	1995	100	—	C	A3, B2, C6
Sylvia conspicillata Spectacled Warbler	B	1995	Common		—	A3
Pyrrhocorax pyrrhocorax Chough	R	1995	500	—	B	B2, C2, C6

1. Juveniles.

Breeding birds include four of the 20 species in Europe that are restricted to the Mediterranean biome (when breeding). Species of global conservation concern that do not meet IBA criteria: *Otis tarda* (non-breeding).

Protection status

National Low **International** None
1,926 ha of IBA covered by Natural Reserve (Caidas de la Negra, 1,926 ha). 460 ha of IBA covered by Natural Reserve (El Bu, 460 ha). 500 ha of IBA covered by Natural Reserve (Vedado de Egüaras, 500 ha).

Conservation issues

Threats Afforestation (C), Agricultural intensification/expansion (A), Other (B), Recreation/tourism (A)

There is a high level of disturbance from visitors, severe overgrazing and an increasing use of irrigation of arable crops.

Cinco Villas ponds C6 091

Admin region Aragón
Coordinates 42°10'N 1°10'W
Altitude 250–650 m **Area** 860 ha

Site description

Small ponds used for irrigation, with emergent vegetation (*Phragmites, Typha, Scirpus*), surrounded by irrigated crops.

Habitats Wetland (100%; standing fresh water; water-fringe vegetation)
Land-use Hunting (10%), Water management (80%)

Birds

Species	Season	Year	Pop min	Pop max	Acc	Criteria
Botaurus stellaris Bittern	R	1996	2	—	B	C6
Ardea purpurea Purple Heron	B	1996	15	20	B	C6
Circus aeruginosus Marsh Harrier	R	1996	15	20	B	C6

This is an important site for breeding waterbirds and wintering ducks.

Protection status

National None **International** None

Conservation issues

Threats Agricultural intensification/expansion (A), Burning of vegetation (B), Dredging/canalization (B), Recreation/tourism (B), Unsustainable exploitation (B)

The main threats are from agricultural intensification, dredging, disturbance and hunting.

Moncayo mountain range A4ii, B1iii, B2, C2, C6 092

Admin region Aragón, Castilla y León
Coordinates 41°45'N 1°50'W
Altitude 600–2,316 m **Area** 47,600 ha

Site description

An isolated massif that has been subject to deforestation. Some *Quercus, Fagus* and *Pinus* forests remain on the northern slopes. The main human activities are livestock-farming and hunting.

Habitats Forest and woodland (40%; broadleaved deciduous forest; mixed forest; broadleaved evergreen forest), Scrub (5%; sclerophyllous scrub/garrigue/maquis), Grassland (15%; steppe/dry calcareous grassland; alpine/subalpine/boreal grassland), Rocky areas (5%; inland cliff), Artificial landscape (30%; forestry plantation; ruderal land)
Land-use Agriculture (80%), Forestry (30%), Hunting (95%)

Birds

Species	Season	Year	Pop min	Pop max	Acc	Criteria
[1] ***Gyps fulvus*** Griffon Vulture	R	1994	269	275	A	A4ii,B1iii,B2,C2,C6

1. Large increase.

This is an important site for cliff-nesting species.

Protection status

National Partial **International** None
9,848 ha of IBA covered by Natural Park (Dehesa del Moncayo, 9,848 ha).

Conservation issues

Threats Burning of vegetation (B), Construction/impact of dyke/dam/barrage (B), Selective logging/cutting (B)

The main threats are from logging, a dam project and forest fires. A management plan exists for the Natural Park.

River Jalón canyon — C6 — 093

Admin region Aragón
Coordinates 41°30′N 1°30′W
Altitude 500–1,433 m **Area** 23,500 ha

Site description

A sparsely populated mountainous site with scrub, garrigue, grasslands and *Pinus* forest. The area is used for sheep- and goat-grazing and forestry

Habitats Forest and woodland (*; native coniferous forest), Scrub (*; scrub; sclerophyllous scrub/garrigue/maquis), Rocky areas (*; scree/boulders; inland cliff)
Land-use Agriculture (*), Forestry (*)

Birds

Species	Season	Year	Pop min	Pop max	Acc	Criteria
Aquila chrysaetos Golden Eagle	R	1995	5	8	B	C6

This is an important site for cliff-nesting raptors.

Protection status

National None **International** None

Conservation issues

Threats Construction/impact of dyke/dam/barrage (B), Infrastructure (B)

Plans to construct a dam and a railway are the main threats to the site.

Langa–Mara–Belmonte de Calatayud plains — B2, C2, C6 — 094

Admin region Aragón
Coordinates 41°12′N 1°30′W
Altitude 780–950 m **Area** 6,100 ha

Site description

A large area of scrub and arable cultivation.

Habitats Scrub (scrub), Grassland (steppe/dry calcareous grassland), Artificial landscape (*; arable land; ruderal land)
Land-use Agriculture (*), Hunting

Birds

Species	Season	Year	Pop min	Pop max	Acc	Criteria
Chersophilus duponti Dupont's Lark	R	1989	100	200	C	B2, C2, C6

This is an important site for species associated with scrub and open habitats, particularly *Chersophilus duponti*.

Protection status

National None **International** None

Conservation issues

Threats Agricultural intensification/expansion (A)

The main threat is conversion of the remaining natural vegetation to arable cultivation.

Gallocanta lake — A1, A4i, A4iv, B1i, B1iv, B2, C1, C2, C3, C5, C6 — 095

Admin region Aragón
Coordinates 40°56′N 1°30′W
Altitude 1,000–1,057 m **Area** 30,280 ha

Site description

A large brackish lake with marsh vegetation in some places and many islets. The surrounding area is dominated by cereal fields.

Habitats Grassland (15%; humid grassland; mesophile grassland), Wetland (15%; standing brackish and salt water; water-fringe vegetation), Artificial landscape (70%; arable land; ruderal land)
Land-use Agriculture (*), Hunting, Tourism/recreation

Birds

Species	Season	Year	Pop min	Pop max	Acc	Criteria
Anas strepera Gadwall	W	1992	1,017	3,800	A	A4i, B1i, C3
Netta rufina Red-crested Pochard	W	1992	8,318	35,600	A	A4i, B1i, C3
Aythya ferina Pochard	W	1992	22,392	79,200	A	A4i, B1i, C3
Circus pygargus Montagu's Harrier	B	1995	8	—	B	C6
Fulica atra Coot	W	1992	12,241	40,900	—	A4i, B1i, C3
Grus grus Crane	W	1994	4,098	7,748	A	A4i, B1i, C2
Grus grus Crane	P	1996	26,975	54,114	A	A4i, B1i, C2
Tetrax tetrax Little Bustard	R	1995	—	20	B	C6
Otis tarda Great Bustard	R	1995	52	—	B	A1, C1
Himantopus himantopus Black-winged Stilt	B	1992	100	—	B	C6
Recurvirostra avosetta Avocet	B	1992	70	—	B	C6
Gelochelidon nilotica Gull-billed Tern	B	1992	41	—	B	B1i, B2, C2, C6
Chlidonias hybridus Whiskered Tern	B	1992	—	—	—	C6

The most important passage site for *Grus grus* in Europe, it is a major migratory bottleneck, where more than 20,000 cranes regularly pass in spring and autumn. It is also a very important site for wintering wildfowl and for breeding steppic birds.

Protection status

National Partial **International** Partial
6,720 ha of IBA covered by Wild Fauna Reserve (Laguna de Gallocanta, 6,720 ha). 6,720 ha of IBA covered by Ramsar Site (Laguna de Gallocanta, 6,720 ha). 6,720 ha of IBA covered by Special Protection Area (Gallocanta, 6,720 ha).

Conservation issues

Threats Agricultural intensification/expansion (B), Groundwater abstraction (A), Recreation/tourism (A)

The main threats are from over-abstraction of water for agriculture and increasing disturbance from visitors, particularly of *Grus grus*.

Blancas–Torralba de los Sisones — B2, C2, C6 — 096

Admin region Aragón
Coordinates 40°50′N 1°30′W
Altitude 1,000–1,171 m **Area** 7,850 ha

Site description

An upland area to the south-east of Gallocanta lake, covered by scrub, small areas of cereal cultivation and small patches of *Pinus* woodland.

Habitats Forest and woodland (5%; native coniferous forest), Scrub (sclerophyllous scrub/garrigue/maquis), Grassland (55%; steppe/dry calcareous grassland), Artificial landscape (40%; arable land)
Land-use Agriculture (80%), Hunting (80%), Not utilized

■ Birds

Species	Season	Year	Pop min	Pop max	Acc	Criteria
Burhinus oedicnemus Stone Curlew	R	1996	220	—	B	B2, C6
Pterocles orientalis Black-bellied Sandgrouse	R	1990	100	—	B	B2, C6
Chersophilus duponti Dupont's Lark	R	1990	500	—	A	B2, C2, C6

This is an important area for steppic birds, particularly *Chersophilus duponti*. Species of global conservation concern that do not meet IBA criteria: *Tetrax tetrax* (resident).

■ Protection status
National None **International** None

■ Conservation issues

Threats Abandonment/reduction of land management (B), Agricultural intensification/expansion (B), Unsustainable exploitation (B)

The abandonment of traditional agriculture, irrigation schemes and illegal hunting threaten the area.

Montes Universales–Albarracín mountain range   B2, C6   097

Admin region Aragón, Castilla-La Mancha
Coordinates 40°25′N 1°40′W
Altitude 1,100–1,900 m **Area** 80,300 ha

■ Site description
A mountainous area with large *Pinus* woodlands. There are also areas of maquis, scrub, and sheep-grazed dry grassland.

Habitats Forest and woodland (*; broadleaved deciduous forest; native coniferous forest; broadleaved evergreen forest), Scrub (sclerophyllous scrub/garrigue/maquis), Grassland (steppe/dry calcareous grassland), Rocky areas (inland cliff), Artificial landscape (ruderal land)
Land-use Agriculture (*), Forestry (*)

■ Birds

Species	Season	Year	Pop min	Pop max	Acc	Criteria
Aquila chrysaetos Golden Eagle	R	1995	8	10	B	C6
Hieraaetus pennatus Booted Eagle	B	1996	20	—	C	B2, C6

This is an important site for breeding raptors

■ Protection status
National Partial **International** None
59,260 ha of IBA covered by National Game Reserve (Montes Universales, 59,260 ha).

■ Conservation issues

Threats Recreation/tourism (C), Selective logging/cutting (C)

Some threat is posed by tourism and logging.

Campo Visiedo   B2, C2, C6   098

Admin region Aragón
Coordinates 40°40′N 1°05′W
Altitude 1,100–1,404 m **Area** 18,000 ha

■ Site description
An upland area dominated by scrub and cereal cultivation.

Habitats Forest and woodland (5%; native coniferous forest), Scrub (40%; scrub), Artificial landscape (55%; arable land)
Land-use Agriculture (60%)

■ Birds

Species	Season	Year	Pop min	Pop max	Acc	Criteria
Tetrax tetrax Little Bustard	R	1996	50	—	C	C6
Pterocles orientalis Black-bellied Sandgrouse	R	1990	100	—	B	B2, C6
Chersophilus duponti Dupont's Lark	R	1990	400	—	A	B2, C2, C6

This is an important area for steppic birds, particularly *Chersophilus duponti*. Species of global conservation concern that do not meet IBA criteria: *Otis tarda* (resident).

■ Protection status
National None **International** None

■ Conservation issues

Threats Agricultural intensification/expansion (B), Construction/impact of dyke/dam/barrage (U)

The main threats are from a dam project and from overgrazing. Within the site there is a private ornithological reserve managed by SEO (800 ha).

River Guadalope   A4ii, B1iii, B2, C2, C6   099

Admin region Aragón
Coordinates 40°40′N 0°35′W
Altitude 600–1,467 m **Area** 31,800 ha

■ Site description
The site includes a gorge which runs along a 40-km stretch of the river. There are limestone cliffs with *Pinus* woodland, scrub and grassland areas. The main human activities include sheep-farming and hunting.

Habitats Forest and woodland (30%; native coniferous forest; alluvial/very wet forest), Scrub (40%; scrub), Grassland (10%; humid grassland), Rocky areas (20%; inland cliff)
Land-use Agriculture (*), Hunting

■ Birds

Species	Season	Year	Pop min	Pop max	Acc	Criteria
[1] ***Gyps fulvus*** Griffon Vulture	R	1996	250	300	B	A4ii,B1iii,B2,C2,C6
Oenanthe leucura Black Wheatear	R	1996	50	—	B	B2
Pyrrhocorax pyrrhocorax Chough	R	1996	100	—	B	B2

1. Large increase.

This is an important site for cliff-nesting raptors.

■ Protection status
National None **International** None

■ Conservation issues

Threats Burning of vegetation (U), Construction/impact of dyke/dam/barrage (B), Infrastructure (B), Other (A)

The main threats include the construction of forest tracks, fires and the illegal use of poisoned baits.

River Martín canyons and Arcos mountain range   A4ii, B1iii, B2, C2, C6   100

Admin region Aragón
Coordinates 40°55′N 0°45′W
Altitude 400–1,237 m **Area** 38,800 ha

■ Site description
Limestone cliffs along the River Martín in the northern foothills of the Sistema Ibérico. The vegetation is composed of Mediterranean scrub, maquis, xerophytic grasslands and a few *Pinus* woodlands.

Habitats Forest and woodland (native coniferous forest), Scrub (*; sclerophyllous scrub/garrigue/maquis), Rocky areas (*; inland cliff)
Land-use Agriculture (*), Hunting, Not utilized

Birds

Species	Season	Year	Pop min	Pop max	Acc	Criteria
Neophron percnopterus Egyptian Vulture	B	1996	15	—	A	B2, C6
[1] ***Gyps fulvus*** Griffon Vulture	R	1996	200	300	B	A4ii,B1iii,B2,C2,C6
Aquila chrysaetos Golden Eagle	R	1996	10	—	B	C6
[2] ***Hieraaetus fasciatus*** Bonnelli's Eagle	R	1996	3	4	A	C6
Falco peregrinus Peregrine	R	1996	10	—	B	C6
Oenanthe leucura Black Wheatear	R	1996	40	—	B	B2
Pyrrhocorax pyrrhocorax Chough	R	1996	300	—	B	B2, C2

1. Large increase.
2. Large decrease.

This is an important site for cliff-nesting raptors.

Protection status

National None **International** None

Conservation issues

Threats Industrialization/urbanization (C), Infrastructure (A), Other (A), Recreation/tourism (C)

Threats include the illegal use of poisoned baits, birds colliding with powerlines and rubbish-dumping in the river.

Alcañiz salt-lakes — B1i, C2, C6 — 101

Admin region Aragón
Coordinates 41°03′N 0°16′W
Altitude 360–400 m **Area** 7,360 ha

Site description

The site contains a number of small, seasonal salt-lakes and associated halophytic vegetation. The surrounding areas are dominated by cereal cultivation.

Habitats Wetland (*; standing brackish and salt water), Artificial landscape (*; arable land)
Land-use Agriculture (*), Hunting (*)

Birds

Species	Season	Year	Pop min	Pop max	Acc	Criteria
Grus grus Crane	P	1996	1,000	—	B	B1i, C2
Tetrax tetrax Little Bustard	R	1996	16	20	A	C6

This is an important site for steppic birds and *Grus grus* on passage. Species of global conservation concern that do not meet IBA criteria: *Falco naumanni* (10–15 breeding pairs).

Protection status

National None **International** None

Conservation issues

Threats Agricultural intensification/expansion (A), Unsustainable exploitation (B)

The main threats are irrigation schemes and illegal hunting.

Lower course of the River Huerva — B2, C6 — 102

Admin region Aragón
Coordinates 41°25′N 1°00′W
Altitude 400–747 m **Area** 37,000 ha

Site description

An area south of Zaragoza with plains and low hills. The area is dominated by arable cultivation interspersed with steppe vegetation and *Pinus* woodland. Human activities include livestock-farming, arable cultivation and hunting.

Habitats Forest and woodland (20%; native coniferous forest; broadleaved evergreen forest), Grassland (20%; steppe/dry calcareous grassland), Wetland (10%; river/stream), Rocky areas (25%; scree/boulders; inland cliff), Artificial landscape (45%; arable land; forestry plantation)
Land-use Agriculture (*), Forestry, Hunting (*), Tourism/recreation

Birds

Species	Season	Year	Pop min	Pop max	Acc	Criteria
Bubo bubo Eagle Owl	R	1996	30	35	A	B2, C6
Oenanthe leucura Black Wheatear	R	1996	70	—	B	B2

This is an important site for wintering raptors, particularly juvenile *Aquila chrysaetos*.

Protection status

National None **International** None

Conservation issues

Threats Agricultural intensification/expansion (C), Construction/impact of dyke/dam/barrage (C), Infrastructure (A), Other (A)

Threats include the use of poisons and the construction of wind-farms.

Belchite–Mediana — B2, C2, C6 — 103

Admin region Aragón
Coordinates 41°20′N 0°45′W
Altitude 200–704 m **Area** 53,000 ha

Site description

One of the best preserved steppe areas in the River Ebro basin. The vegetation is a mixture of steppe-grassland and arable agriculture.

Habitats Grassland (35%; steppe/dry calcareous grassland), Artificial landscape (60%; arable land)
Land-use Agriculture (80%), Hunting (90%)

Birds

Species	Season	Year	Pop min	Pop max	Acc	Criteria
Tetrax tetrax Little Bustard	R	1995	60	80	B	C6
Burhinus oedicnemus Stone Curlew	R	1995	150	150	A	C6
Pterocles orientalis Black-bellied Sandgrouse	R	1995	800	1,000	B	B2, C2, C6
Pterocles alchata Pin-tailed Sandgrouse	R	1995	700	800	B	B2, C2, C6
Chersophilus duponti Dupont's Lark	R	1995	800	—	B	B2, C2, C6
Calandrella rufescens Lesser Short-toed Lark	R	1995	1,400	2,000	B	B2

This is an important site for breeding steppic species. Species of global conservation concern that do not meet IBA criteria: *Otis tarda* (15 non-breeding individuals).

Protection status

National None **International** Low
961 ha of IBA covered by Special Protection Area (La Lomaza de Belchite, 961 ha).

Conservation issues

Threats Agricultural intensification/expansion (B), Extraction industry (B), Groundwater abstraction (U), Natural events (U), Unsustainable exploitation (U)

Threats include agricultural intensification, particularly through irrigation, ploughing of steppe vegetation, sand extraction and, in some areas, natural erosion. Illegal hunting is also a problem.

River Ebro oxbow lakes — C6 — 104

Admin region Aragón
Coordinates 41°30′N 0°45′W
Altitude 160–206 m **Area** 7,400 ha

Site description

A 25-km stretch of the River Ebro downstream from Zaragoza, including a number of abandoned meanders. Some areas of alluvial woodland, ponds and small areas of marsh vegetation remain in an otherwise intensively cultivated area.

Habitats Forest and woodland (30%; alluvial/very wet forest), Wetland (25%; shingle/stony beach; standing fresh water; river/stream; water-fringe vegetation), Artificial landscape (60%; arable land; forestry plantation)

Land-use Agriculture (50%), Forestry (30%), Hunting (85%), Nature conservation/research (15%)

Birds

Species	Season	Year	Pop min	Pop max	Acc	Criteria
Nycticorax nycticorax Night Heron	B	1996	90	100	B	C6
Egretta garzetta Little Egret	B	1996	10	12	A	C6
Ardea purpurea Purple Heron	B	1996	10	12	A	C6
Circus aeruginosus Marsh Harrier	R	1996	3	4	A	C6

This is an important site for breeding herons and migrating passerines.

Protection status

National Partial **International** Partial

777 ha of IBA covered by Natural Reserve (Galachos de la Alfranca de Pastriz, la Cartuja y El, 777 ha). 777 ha of IBA covered by Special Protection Area (Galachos de la Alfranca de Pastriz,la Cartuja y El, 777 ha).

Conservation issues

Threats Afforestation (B), Agricultural intensification/expansion (A), Construction/impact of dyke/dam/barrage (B), Deforestation (commercial) (B), Extraction industry (B), Filling-in of wetlands (B), Industrialization/urbanization (A), Infrastructure (A), Unsustainable exploitation (B)

Major threats include the destruction of riverine forest, afforestation with *Populus* plantations, sand and gravel extraction, hunting, and industrial and urban pollution.

Monegrillo–Pina steppe area–Pina

A1, B2, C1, C2, C6 **105**

Admin region Aragón
Coordinates 41°35′N 0°35′W
Altitude 300–448 m **Area** 43,100 ha

Site description

A steppe area to the west of Los Monegros. The area includes cereal crops, patches of steppe-grassland and areas of former saltpans.

Habitats Grassland (30%; steppe/dry calcareous grassland), Rocky areas (inland cliff), Artificial landscape (70%; arable land; ruderal land)
Land-use Agriculture (70%), Hunting (100%)

Birds

Species	Season	Year	Pop min	Pop max	Acc	Criteria
Falco naumanni Lesser Kestrel	B	1997	40	—	A	B2, C6
Tetrax tetrax Little Bustard	W	1990	145	346	B	A1, C1, C6
Otis tarda Great Bustard	R	1997	10	20	B	C6
Burhinus oedicnemus Stone Curlew	R	1996	—	450	B	B2, C2, C6
Pterocles alchata Pin-tailed Sandgrouse	R	1996	150	230	B	B2, C2, C6
Chersophilus duponti Dupont's Lark	R	1993	225	—	B	B2, C2, C6
Calandrella brachydactyla Short-toed Lark	R	1993	15,000	—	B	B2, C6
Calandrella rufescens Lesser Short-toed Lark	R	1996	4,500	—	B	B2

This is an important area for steppic species. Species of global conservation concern that do not meet IBA criteria: *Tetrax tetrax* (65–125 breeding individuals).

Protection status

National None **International** None

Conservation issues

Threats Agricultural intensification/expansion (A), Infrastructure (B), Unsustainable exploitation (B)

The main threats are from railway construction, agricultural intensification, particularly irrigation, ploughing of natural vegetation and overgrazing, and a high hunting pressure.

Los Monegros (South)

A1, A4ii, B1iii, B2, C1, C2, C6 **106**

Admin region Aragón
Coordinates 41°25′N 0°15′W
Altitude 350–417 m **Area** 49,000 ha

Site description

An area of extensive endorreic plains with some seasonal brackish lagoons and steppe vegetation. The major land-use is non-irrigated cereal cultivation, livestock-farming and hunting.

Habitats Grassland (10%; steppe/dry calcareous grassland), Wetland (5%; standing brackish and salt water), Artificial landscape (85%; arable land; ruderal land)
Land-use Agriculture (100%), Forestry (5%), Hunting (100%)

Birds

Species	Season	Year	Pop min	Pop max	Acc	Criteria
Falco naumanni Lesser Kestrel	B	1996	230	230	A	A1, A4ii, B1iii, B2, C1, C2, C6
Tetrax tetrax Little Bustard	R	1996	50	100	A	A1, C1, C6
Otis tarda Great Bustard	R	1996	75	80	A	A1, C1, C6
Pterocles alchata Pin-tailed Sandgrouse	R	1996	Common		—	C6

This is an important area for steppic species, wildfowl and passage waders.

Protection status

National None **International** None

Conservation issues

Threats Agricultural intensification/expansion (A), Construction/impact of dyke/dam/barrage (B), Dredging/canalization (B), Filling-in of wetlands (B), Infrastructure (B), Other (B)

The site is threatened by huge irrigation schemes, the use of poison baits, and overgrazing.

Valdurrios–Serreta Negra and Los Rincones mountain ranges

B2, C6 **107**

Admin region Aragón
Coordinates 41°20′N 0°10′E
Altitude 200–409 m **Area** 49,500 ha

Site description

The site includes a group of low hills crossed by the River Ebro and a reservoir. The vegetation is a mixture of *Pinus* woodland, scrub and grassland. Alluvial woodland occurs around parts of the reservoir. The main human activities are livestock-farming and fishing.

Habitats Forest and woodland (*; native coniferous forest), Scrub (*; sclerophyllous scrub/garrigue/maquis), Grassland (steppe/dry calcareous grassland), Wetland (standing fresh water; water-fringe vegetation), Artificial landscape (*; arable land; ruderal land)
Land-use Agriculture (*), Fisheries/aquaculture, Hunting (*), Tourism/recreation, Urban/industrial/transport, Water management

Birds

Species	Season	Year	Pop min	Pop max	Acc	Criteria
Neophron percnopterus Egyptian Vulture	B	1995	11	16	B	B2, C6

This is an important site for raptors and species characteristic of Mediterranean habitats.

Protection status

National None **International** None

Conservation issues

Threats Agricultural intensification/expansion (B), Burning of vegetation (B), Recreation/tourism (B), Selective logging/cutting (B), Unsustainable exploitation (B)

The main threats are from irrigation schemes, hydro-power schemes, fires, illegal hunting, logging and recreational activities.

Lower course of the River Matarraña–Ribarroja

B1i, B2, C2, C6 **108**

Admin region Aragón, Cataluña
Coordinates 41°00′N 0°20′E
Altitude 100–500 m **Area** 73,500 ha

Site description

A mountainous area dominated by *Pinus* forests, *Quercus* woodland and sclerophyllous scrub. Three rivers with associated riverine vegetation cross the area. The main human activities are forestry and hunting.

Habitats Forest and woodland (*; native coniferous forest; alluvial/very wet forest), Scrub (*; sclerophyllous scrub/garrigue/maquis), Wetland (river/stream), Rocky areas (*; inland cliff), Artificial landscape (arable land; perennial crops/orchards/groves; forestry plantation)
Land-use Agriculture (30%), Forestry (70%), Hunting (100%)

Birds

Species	Season	Year	Pop min	Pop max	Acc	Criteria
Ciconia nigra Black Stork	P	1996	20	30	B	B1i, C2
Milvus migrans Black Kite	B	1997	40	50	B	C6
Circaetus gallicus Short-toed Eagle	B	1997	20	30	B	B2, C6
[1] ***Hieraaetus fasciatus*** Bonnelli's Eagle	R	1997	5	10	B	B2, C6
Bubo bubo Eagle Owl	R	1997	25	30	C	B2, C6

1. Large decrease.

This is an important site for breeding raptors.

Protection status

National None **International** None

Conservation issues

Threats Agricultural intensification/expansion (B), Groundwater abstraction (A), Unsustainable exploitation (A)

The main threats are from the high hunting pressure and the expansion of irrigated arable cultivation.

Rivers Cinca and Alcanadre riverine forest

B1i, C2, C6 **109**

Admin region Aragón, Cataluña
Coordinates 41°47′N 0°12′E
Altitude 78–400 m **Area** 9,450 ha

Site description

A long stretch of riverine forest along the Rivers Cinca and Alcanadre, which also have areas of shingle, cliffs and emergent vegetation. The main human activities are hunting and forestry.

Habitats Forest and woodland (40%; alluvial/very wet forest), Grassland (10%; humid grassland), Wetland (55%; shingle/stony beach; river/stream; water-fringe vegetation)
Land-use Forestry (25%), Hunting (80%)

Birds

Species	Season	Year	Pop min	Pop max	Acc	Criteria
Botaurus stellaris Bittern	B	1996	1	2	B	C6
Nycticorax nycticorax Night Heron	B	1990	70	85	A	C6
Egretta garzetta Little Egret	B	1990	40	45	A	C6
[1] ***Ardea purpurea*** Purple Heron	B	1995	21	21	A	C6
Circus aeruginosus Marsh Harrier	B	1995	7	10	B	C6
Grus grus Crane	P	1995	—	1,000	C	B1i, C2
Alcedo atthis Kingfisher	R	1995	20	30	B	C6

1. Large decrease.

This is an important area for riverine species including the best *Nycticorax nycticorax* colony in Spain.

Protection status

National None **International** None

Conservation issues

Threats Afforestation (A), Dredging/canalization (B), Extraction industry (B), Industrialization/urbanization (C)

The main threats are from afforestation with *Populus* plantations and regulation of the river channel.

Ballobar–Candasnos

A1, B2, C1, C2, C6 **110**

Admin region Aragón
Coordinates 41°37′N 0°12′E
Altitude 300–400 m **Area** 26,000 ha

Site description

An area of steppe vegetation, amongst large areas of cereal cultivation, in the semi-arid region of Los Monegros.

Habitats Grassland (steppe/dry calcareous grassland), Artificial landscape (100%; arable land; ruderal land)
Land-use Agriculture (100%), Hunting (100%)

Birds

Species	Season	Year	Pop min	Pop max	Acc	Criteria
Falco naumanni Lesser Kestrel	B	1997	72	—	A	A1, B2, C1, C6
Pterocles orientalis Black-bellied Sandgrouse	R	1995	—	100	B	B2, C6
Pterocles alchata Pin-tailed Sandgrouse	R	1995	900	1,000	B	B2, C2, C6

This is an important area for steppic species, particularly breeding *Pterocles alchata*. Species of global conservation concern that do not meet IBA criteria: *Tetrax tetrax* (11–40 resident individuals).

Protection status

National None **International** None

Conservation issues

Threats Agricultural intensification/expansion (B), Infrastructure (B), Other (B)

The main threats are from irrigation plans to allow agricultural intensification, the construction of a railway and the use of poison baits.

Cinca river rice-fields and steppe area

A4i, B1i, B2, C2, C6 **111**

Admin region Aragón
Coordinates 41°50′N 0°03′E
Altitude 200–300 m **Area** 14,400 ha

Site description

An area of irrigated rice and cereal cultivation crossed by three gullies with reedbeds (*Phragmites*). A small area of natural steppe vegetation remains.

Habitats Wetland (10%; river/stream; water-fringe vegetation), Artificial landscape (90%; arable land)
Land-use Agriculture (80%), Hunting (100%)

Birds

Species	Season	Year	Pop min	Pop max	Acc	Criteria
Circus aeruginosus Marsh Harrier	R	1996	15	15	A	C6
Circus pygargus Montagu's Harrier	B	1996	10	15	A	C6
Falco naumanni Lesser Kestrel	B	1996	36	36	A	C6
Grus grus Crane	P	1996	300	2,000	B	A4i, B1i, C2
Himantopus himantopus Black-winged Stilt	B	1996	250	350	A	A4i, B1i, C2, C6
Merops apiaster Bee-eater	B	1996	300	—	B	B2

This is an important site for breeding waders, raptors and passage *Grus grus*.

Protection status

National None **International** None

Conservation issues

Threats Agricultural intensification/expansion (A), Deforestation (commercial) (C), Dredging/canalization (A)

The main threat is agricultural intensification.

Sariñena lake and La Estación pond — C6 — 112

Admin region Aragón
Coordinates 41°45′N 0°10′W
Altitude 280–300 m **Area** 610 ha

Site description

A former salty, endorreic lake that now is a freshwater wetland with reedbeds (*Phragmites*).

Habitats Grassland (humid grassland), Wetland (100%; mudflat/sandflat; standing fresh water; water-fringe vegetation)
Land-use Water management (100%)

Birds

Species	Season	Year	Pop min	Pop max	Acc	Criteria
Ardea purpurea Purple Heron	B	1996	20	25	B	C6
Circus aeruginosus Marsh Harrier	B	1996	2	6	B	C6

This is also a nationally important wintering site for waterbirds.

Protection status

National High **International** None
604 ha of IBA covered by Wild Fauna Reserve (Laguna de Sariñena, 604 ha).

Conservation issues

Threats Agricultural intensification/expansion (B)

The main threat is pesticide pollution from agricultural areas around the wetland.

Alcubierre and Sigena mountain ranges — C6 — 113

Admin region Aragón
Coordinates 41°40′N 0°30′W
Altitude 400–822 m **Area** 64,500 ha

Site description

An eroded range of hills dominated by sclerophyllous scrub and dry grassland, with *Juniperus* and *Pinus* forest. The main human activity is livestock-farming.

Habitats Forest and woodland (*; native coniferous forest), Scrub (*; sclerophyllous scrub/garrigue/maquis), Grassland (*), Rocky areas (inland cliff), Artificial landscape (arable land; ruderal land)
Land-use Agriculture (*), Forestry, Hunting (*)

Birds

Species	Season	Year	Pop min	Pop max	Acc	Criteria
Milvus migrans Black Kite	B	1996	20	—	C	C6
Circaetus gallicus Short-toed Eagle	B	1996	10	15	C	C6
Aquila chrysaetos Golden Eagle	R	1996	8	10	C	C6
Hieraaetus pennatus Booted Eagle	B	1996	10	15	C	C6

This is an important site for breeding raptors.

Protection status

National None **International** None

Conservation issues

Threats Burning of vegetation (U), Infrastructure (U), Unsustainable exploitation (U)

The main threats are from forest fires, localized hunting pressure and infrastructural developments.

Confidential — C6 — 114

Admin region Aragón
Coordinates Co-ordinates withheld
Altitude 280–459 m **Area** 4,600 ha

Site description

Steppic area with extensive cereal crops, fallow and ruderal land.

Habitats Artificial landscape (100%; arable land; ruderal land)
Land-use Agriculture (100%), Hunting (100%)

Birds

Species	Season	Year	Pop min	Pop max	Acc	Criteria
Falco naumanni Lesser Kestrel	B	1997	21	25	A	C6

This is an important site for steppic species. Species of global conservation concern that do not meet IBA criteria: *Tetrax tetrax* (30 resident individuals).

Protection status

National None **International** None

Conservation issues

Threats Other (B)

The lack of suitable rural buildings and the degradation of those that exist are the most important threat to *Falco naumanni*.

Zuera mountains — B2, C6 — 115

Admin region Aragón
Coordinates 42°00′N 0°55′W
Altitude 450–769 m **Area** 35,000 ha

Site description

A range of hills to the north-west of Zaragoza. They are covered with *Pinus* woodland, sclerophyllous scrub and arable crops. Human activities include hunting, agriculture and military manoeuvres.

Habitats Forest and woodland (20%; native coniferous forest), Scrub (30%; scrub; sclerophyllous scrub/garrigue/maquis), Grassland (20%; dry siliceous grassland), Artificial landscape (20%; arable land)
Land-use Agriculture (40%), Forestry (20%), Hunting (100%), Military (40%)

Birds

Species	Season	Year	Pop min	Pop max	Acc	Criteria
Milvus migrans Black Kite	B	1995	20	—	C	C6
Hieraaetus pennatus Booted Eagle	B	1995	15	20	A	B2, C6
Oenanthe leucura Black Wheatear	R	1996	40	—	B	B2

This is also a nationally important site for resident raptors.

Protection status

National None **International** None

Conservation issues

Threats Afforestation (U), Burning of vegetation (A), Industrialization/urbanization (C)

The greatest threat is from fires.

Tormos reservoir (La Sotonera) — A4i, A4iv, B1i, B1iv, C2, C5 — 116

Admin region Aragón
Coordinates 42°05′N 0°39′W
Altitude 450–533 m **Area** 7,500 ha

Site description

A reservoir with a strongly fluctuating water-level. There is marsh vegetation along some of the banks and the reservoir is surrounded by cereal fields.

Habitats Grassland (humid grassland), Wetland (*; standing fresh water), Artificial landscape (*; arable land)
Land-use Agriculture (*), Hunting, Tourism/recreation, Water management

Birds

Species	Season	Year	Pop min	Pop max	Acc	Criteria
[1] ***Grus grus*** Crane	P	1995	15,000	20,000	B	A4i, B1i, C2

1. Large increase.

An important wintering site for wildfowl. The site is a major migratory bottleneck, where more than 20,000 cranes regularly pass in spring.

Protection status

National None **International** None

Conservation issues

Threats Agricultural intensification/expansion (B), Dredging/canalization (C), Filling-in of wetlands (U), Recreation/tourism (B)

Threats include high hunting pressure and disturbance from tourists.

Bajo Alcanadre–Serreta de Tramaced

B1iii, B2, C2 **117**

Admin region Aragón
Coordinates 42°00'N 0°05'W
Altitude 400–500 m **Area** 17,000 ha

Site description

A large semi-arid plain with eroded limestone plateaus. There are areas of non-irrigated cereal crops, low sclerophyllous scrub, and sheep-grazed pastures. Apiculture ('Other' land-use, below) is widely practised.

Habitats Forest and woodland (alluvial/very wet forest; broadleaved evergreen forest), Scrub (sclerophyllous scrub/garrigue/maquis), Rocky areas (*; inland cliff), Artificial landscape (*; arable land; perennial crops/orchards/groves)
Land-use Agriculture (*), Hunting (*), Other (*)

Birds

Species	Season	Year	Pop min	Pop max	Acc	Criteria
[1] ***Gyps fulvus*** Griffon Vulture	N	1995	300	—	B	B1iii, C2
Merops apiaster Bee-eater	B	1996	300	—	B	B2

1. Large increase.

This is an important site for raptors and species characteristic of Mediterranean habitats, particularly *Merops apiaster*.

Protection status

National None **International** None

Conservation issues

Threats Agricultural intensification/expansion (B)

The main threat is from agricultural intensification. There are conflicts between apiculture and *Merops apiaster*.

Guara mountain range

A4ii, B1iii, B2, B3, C2, C6 **118**

Admin region Aragón
Coordinates 42°20'N 0°10'W
Altitude 600–2,077 m **Area** 95,000 ha

Site description

A massif at the southern edge of the Pyrenees with large limestone cliffs and deep gorges. There are extensive areas of scrub and grassland. Main human activities are livestock-farming, forestry, hunting, and recreation.

Habitats Forest and woodland (native coniferous forest; alluvial/very wet forest; broadleaved evergreen forest), Scrub (*; sclerophyllous scrub/garrigue/maquis), Grassland (alpine/subalpine/boreal grassland), Rocky areas (*; scree/boulders; inland cliff), Artificial landscape (arable land; perennial crops/orchards/groves; forestry plantation)
Land-use Agriculture (*), Forestry, Hunting (*), Military, Tourism/recreation

Birds

Species	Season	Year	Pop min	Pop max	Acc	Criteria
Milvus milvus Red Kite	R	1996	30	—	C	B3, C6
Gypaetus barbatus Lammergeier	R	1996	7	7	A	B2, C6
Neophron percnopterus Egyptian Vulture	B	1995	16	—	C	B2, C6
[1] ***Gyps fulvus*** Griffon Vulture	R	1996	500	—	B	A4ii,B1iii,B2,C2,C6
Circaetus gallicus Short-toed Eagle	B	1993	14	18	B	B2, C6
Aquila chrysaetos Golden Eagle	R	1995	9	—	B	C6
Hieraaetus fasciatus Bonnelli's Eagle	R	1996	3	4	A	C6
Falco peregrinus Peregrine	R	1996	15	20	B	B2, C6
Monticola saxatilis Rock Thrush	B	1996	50	—	C	B2
Pyrrhocorax pyrrhocorax Chough	R	1996	200	—	C	B2, C2

1. Large increase.

This is a very important area for cliff-nesting and forest species.

Protection status

National Partial **International** Partial
81,350 ha of IBA covered by Natural Park (Sierra y Cañones de Guara, 81,350 ha). 81,350 ha of IBA covered by Special Protection Area (Sierra y Cañones de Guara, 81,350 ha).

Conservation issues

Threats Abandonment/reduction of land management (B), Burning of vegetation (A), Infrastructure (U), Recreation/tourism (A), Unsustainable exploitation (U)

Threats include uncontrolled outdoor sports and recreation, hunting and fires. A reduction in the area of arable cultivation has resulted in a drop in the rabbit *Oryctolagus cuniculus* population which has affected raptor populations. A management plan was approved in 1997.

Oturia–Canciás

B2, C6 **119**

Admin region Aragón
Coordinates 42°30'N 0°10'W
Altitude 780–1,928 m **Area** 33,520 ha

Site description

A mountainous area between Sabiñánigo and Boltaña with cliffs, *Quercus* and *Pinus* forest and scrub. Livestock-farming and game hunting are the most important human activities.

Habitats Forest and woodland (*; broadleaved deciduous forest; broadleaved evergreen forest), Scrub (sclerophyllous scrub/garrigue/maquis), Grassland (alpine/subalpine/boreal grassland), Rocky areas (inland cliff), Artificial landscape (forestry plantation)
Land-use Agriculture, Forestry, Hunting (*), Military, Not utilized (*)

Birds

Species	Season	Year	Pop min	Pop max	Acc	Criteria
Gypaetus barbatus Lammergeier	R	1996	3	3	A	B2, C6

This is a nationally important site for alpine species.

Protection status

National None **International** None

Conservation issues

Threats Abandonment/reduction of land management (C), Infrastructure (A), Recreation/tourism (A)

The most important threats are the construction of new roads and powerlines and disturbance from recreational activities, particularly of *Gypaetus barbatus*.

Santo Domingo–Riglos–Gratal

A4ii, B1iii, B2, B3, C2, C6 **120**

Admin region Aragón
Coordinates 42°25'N 0°40'W
Altitude 800–1,595 m **Area** 45,000 ha

Site description

A mountainous area on the southern edge of the Pyrenees. The vegetation includes maquis, *Pinus* plantations, and extensive scrub and grassland areas. The main human activities are forestry, livestock-farming and recreation.

Habitats Forest and woodland (*; broadleaved deciduous forest; native coniferous forest; broadleaved evergreen forest), Scrub (*; sclerophyllous scrub/garrigue/maquis), Grassland (steppe/dry calcareous grassland), Rocky areas (*; scree/boulders; inland cliff), Artificial landscape (arable land; perennial crops/orchards/groves; forestry plantation)
Land-use Agriculture (*), Forestry, Hunting, Tourism/recreation

Birds

Species	Season	Year	Pop min	Pop max	Acc	Criteria
Milvus milvus Red Kite	R	1995	20	30	B	B3, C6
Gypaetus barbatus Lammergeier	R	1996	3	3	A	B2, C6
[1] ***Gyps fulvus*** Griffon Vulture	R	1995	250	—	B	A4ii,B1iii,B2,C2,C6
Pyrrhocorax pyrrhocorax Chough	R	1996	100	—	B	B2

1. Large increase.

This is an important area for raptors.

Protection status

National None **International** None

Conservation issues

Threats Construction/impact of dyke/dam/barrage (B), Intensified forest management (B), Recreation/tourism (B)

Threats are from intensive forest management, particularly the construction of new tracks, disturbance caused by climbers, particularly of nesting raptors, and a plan to build a dam at Biscarrués.

San Juan de la Peña–Oroel crag — B2, B3, C6 — 121

Admin region Aragón
Coordinates 42°30′N 0°45′W
Altitude 800–1,769 m **Area** 22,400 ha

Site description

The site comprises cliffs, large *Pinus* plantations and areas of scrub and maquis. The area is used for tourism, forestry and game hunting.

Habitats Forest and woodland (*; broadleaved deciduous forest; native coniferous forest; alluvial/very wet forest), Scrub (sclerophyllous scrub/garrigue/maquis), Grassland (steppe/dry calcareous grassland), Rocky areas (scree/boulders; inland cliff)
Land-use Agriculture, Forestry (*), Hunting (*), Military, Tourism/recreation

Birds

Species	Season	Year	Pop min	Pop max	Acc	Criteria
Milvus milvus Red Kite	R	1994	25	30	A	B3, C6
Gypaetus barbatus Lammergeier	R	1996	2	2	A	B2
Dryocopus martius Black Woodpecker	R	1995	10	—	C	C6
Pyrrhocorax pyrrhocorax Chough	R	1995	125	—	B	B2

This is an important site for cliff-nesting and forest raptors.

Protection status

National Low **International** None
245 ha of IBA covered by Natural Site of National Interest (San Juan de la Peña, 245 ha).

Conservation issues

Threats Infrastructure (C), Recreation/tourism (A)

The main problems are from a very high tourist pressure and plans to enlarge an existing airfield.

Dos Ríos and Orba mountain ranges — A4ii, B1iii, B2, C2, C6 — 122

Admin region Aragón
Coordinates 42°40′N 0°55′W
Altitude 600–1,477 m **Area** 34,700 ha

Site description

The site includes the Sierra de los Dos Ríos mountain range, crossed by two gorges, with limestone cliffs, dense maquis and *Pinus* woodland. The main human activities are forestry and sheep-grazing.

Habitats Forest and woodland (*; broadleaved deciduous forest; native coniferous forest), Scrub (*; scrub; sclerophyllous scrub/garrigue/maquis), Grassland (*), Rocky areas (inland cliff)
Land-use Agriculture, Forestry (*), Hunting (*), Not utilized

Birds

Species	Season	Year	Pop min	Pop max	Acc	Criteria
Gypaetus barbatus Lammergeier	R	1996	3	3	A	B2, C6
Neophron percnopterus Egyptian Vulture	B	1996	8	11	B	C6
Neophron percnopterus Egyptian Vulture	N	1994	22	100	A	B1iii, C2
Gyps fulvus Griffon Vulture	R	1996	200	—	B	A4ii,B1iii,B2,C2,C6
Pyrrhocorax pyrrhocorax Chough	R	1996	200	—	A	B2, C2

This is an important site for cliff-nesting raptors and an important roosting site for *Neophron percnopterus*.

Protection status

National None **International** None

Conservation issues

Threats Aquaculture/fisheries (C), Selective logging/cutting (B), Unsustainable exploitation (B)

The main threats are from logging and hunting.

Belagua–Ansó–Hecho — B1iii, B2, C2, C6 — 123

Admin region Navarra, Aragón
Coordinates 42°50′N 0°45′W
Altitude 728–2,670 m **Area** 44,500 ha

Site description

A Pyrenean high-mountain zone with *Fagus*, *Abies* and *Pinus* forests. The main human activities are forestry, hunting, livestock-grazing and tourism and recreation.

Habitats Forest and woodland (40%; broadleaved deciduous forest; native coniferous forest; mixed forest), Scrub (20%; scrub; heathland), Grassland (40%; alpine/subalpine/boreal grassland)
Land-use Agriculture (80%), Forestry (40%), Hunting (90%), Not utilized (10%)

Birds

Species	Season	Year	Pop min	Pop max	Acc	Criteria
Gypaetus barbatus Lammergeier	R	1996	6	6	A	B2, C6
Neophron percnopterus Egyptian Vulture	B	1995	10	—	B	C6
[1] ***Gyps fulvus*** Griffon Vulture	R	1995	150	150	B	B1iii, B2, C2, C6
Aquila chrysaetos Golden Eagle	R	1995	6	6	B	C6
Falco peregrinus Peregrine	R	1995	6	10	B	C6
Lagopus mutus Ptarmigan	R	1996	47	50	A	C6
[2] ***Tetrao urogallus*** Capercaillie	R	1992	16	20	A	C6
[3] ***Perdix perdix*** Partridge	R	1992	40	50	B	C6
Aegolius funereus Tengmalm's Owl	R	1995	2	5	B	C6
Dryocopus martius Black Woodpecker	R	1992	30	40	B	C6
Dendrocopos leucotos White-backed Woodpecker	R	1995	10	—	B	C6
Pyrrhocorax pyrrhocorax Chough	R	1995	200	—	B	B2, C2, C6

1. Large increase.
2. Males.
3. *P. p. hispaniensis.*

This is an important site for cliff-nesting raptors and forest birds.

Protection status

National Partial **International** Partial
175 ha of IBA covered by Integral Nature Reserve (Aztaparreta, 175 ha). 322 ha of IBA covered by Integral Nature Reserve (Ukerdi, 322 ha). 2,353 ha of IBA covered by Natural Reserve (Larra, 2,353 ha). 28,765 ha of IBA covered by National Game Reserve (Los Valles, 28,765 ha). 28,757 ha of IBA covered by Special Protection Area (Reserva Nacional de Caza de Los Valles, 28,757 ha). 5,913 ha of IBA covered by Special Protection Area (Sierra de Arrigorrieta–Peña de

Ezkaurre, 5,913 ha). 3,767 ha of IBA covered by Special Protection Area (Larra–Aztaparreta, 3,767 ha).

Conservation issues

Threats Infrastructure (B), Intensified forest management (B), Recreation/tourism (B), Selective logging/cutting (B), Unsustainable exploitation (C)

The main threat is from intensive forestry operations. Hunting and visitor pressure are also threats.

Collarada–Telera C6 124

Admin region Aragón
Coordinates 42°40'N 0°25'W
Altitude 1,500–2,886 m **Area** 21,000 ha

Site description

A high-mountain zone dominated by alpine grasslands. The area is used for summer livestock-grazing and for skiing and hiking.

Habitats Forest and woodland (15%; broadleaved deciduous forest; native coniferous forest; mixed forest), Grassland (75%; alpine/subalpine/boreal grassland), Rocky areas (10%; inland cliff)
Land-use Agriculture (70%), Forestry (10%), Hunting (100%), Tourism/recreation (5%)

Birds

Species	Season	Year	Pop min	Pop max	Acc	Criteria
Lagopus mutus Ptarmigan	R	1995	10	—	B	C6

This is an important site for alpine birds.

Protection status

National None **International** None

Conservation issues

Threats Construction/impact of dyke/dam/barrage (B), Infrastructure (A), Recreation/tourism (A)

The main threats are from development of the skiing infrastructure, disturbance from tourists and the development of hydroelectric plants.

Panticosa–Vignemale–Tendeñera B2, C6 125

Admin region Aragón
Coordinates 42°45'N 0°10'W
Altitude 1,500–3,303 m **Area** 32,000 ha

Site description

An alpine area in the Pyrenees with numerous small glacial lakes and conifer forest. The area is used for livestock-grazing in summer, game hunting, hiking, and skiing.

Habitats Forest and woodland (15%; native coniferous forest), Grassland (70%; alpine/subalpine/boreal grassland), Wetland (5%; standing fresh water), Rocky areas (10%; inland cliff)
Land-use Agriculture (70%), Forestry (5%), Hunting (100%), Tourism/recreation (5%)

Birds

Species	Season	Year	Pop min	Pop max	Acc	Criteria
Gypaetus barbatus Lammergeier	R	1996	1	1	A	B2
Lagopus mutus Ptarmigan	R	1995	10	—	B	C6

This is an important site for alpine birds.

Protection status

National High **International** None

3,500 ha of IBA covered by National Park (Ordesa y Monte Perdido, 35,287 ha). 30,000 ha of IBA covered by National Game Reserve (Viñamala, 49,230 ha).

Conservation issues

Threats Recreation/tourism (B)

The main threat is from the development of the tourism and ski infrastructures.

Ordesa–Bielsa A3, B2, C6 126

Admin region Aragón
Coordinates 42°40'N 0°00'
Altitude 1,200–3,355 m **Area** 43,000 ha

Site description

The site encompasses a spectacular area of the central Pyrenees with many peaks over 3,000 m. There are large areas of bare rock and canyons and also *Pinus* and *Fagus* forests. Human activities include tourism, forestry, game hunting, and cattle-grazing.

Habitats Forest and woodland (broadleaved deciduous forest; native coniferous forest; broadleaved evergreen forest), Grassland (*; alpine/subalpine/boreal grassland), Rocky areas (*; scree/boulders; inland cliff)
Land-use Agriculture, Forestry, Hunting, Nature conservation/research, Tourism/recreation (*)

Birds

Species	Season	Year	Pop min	Pop max	Acc	Criteria
Gypaetus barbatus Lammergeier	R	1996	6	6	A	B2, C6
Lagopus mutus Ptarmigan	R	1996	20	—	B	C6
[1] ***Tetrao urogallus*** Capercaillie	R	1996	20	—	B	C6
Perdix perdix Partridge	R	1996	50	—	B	C6
Dryocopus martius Black Woodpecker	R	1996	30	—	B	C6
Prunella collaris Alpine Accentor	R	1996	Abundant	—		A3
Tichodroma muraria Wallcreeper	R	1996	Abundant	—		A3
Pyrrhocorax graculus Alpine Chough	R	1996	Abundant	—		A3
Pyrrhocorax pyrrhocorax Chough	R	1996	100	—	B	B2
Montifringilla nivalis Snowfinch	R	1996	Abundant	—		A3
Serinus citrinella Citril Finch	R	1996	Abundant	—		A3

1. Males.

Breeding birds include five of the 10 species in Europe that are restricted to the Eurasian high-montane biome (when breeding).

Protection status

National Partial **International** Partial

31,800 ha of IBA covered by National Park (Ordesa y Monte Perdido, 35,287 ha). 13,000 ha of IBA covered by National Game Reserve (Los Circos, 27,119 ha). 15,000 ha of IBA covered by National Game Reserve (Viñamala, 49,230 ha). 20,000 ha of IBA covered by Biosphere Reserve (Ordesa–Viñamala, 51,396 ha). 15,608 ha of IBA covered by Special Protection Area (Ordesa y Monte Perdido, 15,608 ha).

Conservation issues

Threats Infrastructure (C), Recreation/tourism (B)

The main threat is disturbance from skiers and the development of the ski infrastructure.

Gistaín–Cotiella B2, C6 127

Admin region Aragón
Coordinates 42°35'N 0°25'E
Altitude 1,000–3,375 m **Area** 42,000 ha

Site description

This site includes the Gistaín valley and the northern and southern massifs of the central Pyrenees. There are spectacular cliffs in the southern section and also conifer forests. The main human activities are forestry and livestock-farming.

Habitats Forest and woodland (native coniferous forest), Grassland (*; humid grassland), Rocky areas (*; inland cliff)
Land-use Agriculture (*), Forestry (*), Urban/industrial/transport

Birds

Species	Season	Year	Pop min	Pop max	Acc	Criteria
Gypaetus barbatus Lammergeier	R	1996	3	3	A	B2, C6
Lagopus mutus Ptarmigan	R	1996	20	—	B	C6
[1] ***Tetrao urogallus*** Capercaillie	R	1996	20	—	B	C6
Perdix perdix Partridge	R	1996	50	—	B	C6
Dryocopus martius Black Woodpecker	R	1996	20	—	B	C6

1. Males.

This is a very important site for alpine and forest birds.

Protection status

National Partial **International** Partial

10,000 ha of IBA covered by National Game Reserve (Los Circos, 27,119 ha). 8,300 ha of IBA covered by Natural Park (Posets–Maladeta, 33,267 ha). 8,300 ha of IBA covered by Special Protection Area (Parque Posets–Maladeta, 33,267 ha).

Conservation issues

Threats Infrastructure (B)

The greatest threat is from infrastructural developments, particularly roads and powerlines.

Posets–Maladeta — A3, B2, C6 — 128

Admin region Aragón
Coordinates 42°40′N 0°30′E
Altitude 1,100–3,404 m **Area** 44,700 ha

Site description

A high-altitude region of the Pyrenees with large areas over 3,000 m and numerous small mountain lakes and rocky areas. Lower down, there are *Pinus*, *Abies* and *Fagus* forests. The main human activities are forestry, livestock-farming and recreation.

Habitats Forest and woodland (20%; broadleaved deciduous forest; native coniferous forest), Grassland (35%; alpine/subalpine/boreal grassland), Rocky areas (30%; scree/boulders; inland cliff), Artificial landscape (15%; highly improved reseeded grassland)
Land-use Agriculture (40%), Forestry, Hunting (80%), Tourism/recreation (15%)

Birds

Species	Season	Year	Pop min	Pop max	Acc	Criteria
Gypaetus barbatus Lammergeier	R	1996	2	2	A	B2
Lagopus mutus Ptarmigan	R	1996	40	—	B	C6
[1] ***Tetrao urogallus*** Capercaillie	R	1996	50	—	B	C6
Perdix perdix Partridge	R	1996	70	—	B	C6
Aegolius funereus Tengmalm's Owl	R	1996	7	10	C	C6
Dryocopus martius Black Woodpecker	R	1996	25	—	B	C6
Prunella collaris Alpine Accentor	R	1996	Common		—	A3
Tichodroma muraria Wallcreeper	R	1996	Common		—	A3
Pyrrhocorax graculus Alpine Chough	R	1995	Abundant		—	A3
Pyrrhocorax pyrrhocorax Chough	R	1995	100	—	B	B2
Serinus citrinella Citril Finch	R	1996	Abundant		—	A3

1. Males.

Breeding birds include four of the 10 species in Europe that are restricted to the Eurasian high-montane biome (when breeding).

Protection status

National Partial **International** Partial

23,913 ha of IBA covered by National Game Reserve (Benasque, 23,913 ha). 25,200 ha of IBA covered by Natural Park (Posets–Maladeta, 33,267 ha). 25,200 ha of IBA covered by Special Protection Area (Parque Posets–Maladeta, 33,267 ha.

Conservation issues

Threats Abandonment/reduction of land management (A), Infrastructure (B), Recreation/tourism (B)

The main threats are from abandonment of traditional alpine grazing and disturbance from the large number of tourists visiting the site.

Turbón–Espés–Sis — B2, B3, C2, C6 — 129

Admin region Aragón
Coordinates 42°25′N 0°35′E
Altitude 800–2,492 m **Area** 43,000 ha

Site description

A mountainous area with cliffs and gorges, extensive *Pinus* forests, scrub and grasslands.

Habitats Forest and woodland (native coniferous forest), Scrub (*; scrub; sclerophyllous scrub/garrigue/maquis), Grassland (alpine/subalpine/boreal grassland), Rocky areas (*; inland cliff)
Land-use Agriculture (*), Forestry

Birds

Species	Season	Year	Pop min	Pop max	Acc	Criteria
Gypaetus barbatus Lammergeier	R	1996	4	4	A	B2, C6
Tetrao urogallus Capercaillie	R	1995	15	—	C	C6
Perdix perdix Partridge	R	1995	20	—	C	C6
Dryocopus martius Black Woodpecker	R	1995	10	—	C	C6
Dendrocopos medius Middle Spotted Woodpecker	R	1995	10	—	C	B3, C6
Pyrrhocorax pyrrhocorax Chough	R	1995	200	—	C	B2, C2

An important site for forest species and for resident *Gypaetus barbatus*.

Protection status

National None **International** None

Conservation issues

No serious threats known at the site.

San Mauricio–Bohí–Beret — A3, B2, C2, C6 — 130

Admin region Cataluña
Coordinates 42°40′N 1°00′E
Altitude 650–3,032 m **Area** 120,000 ha

Site description

A high-altitude mountain area in the Pyrenees with cliffs, numerous small mountain lakes, extensive alpine grasslands and *Pinus* and *Abies* forest. The area is used for forestry, stock-grazing in summer and tourism and recreation.

Habitats Forest and woodland (40%; broadleaved deciduous forest; native coniferous forest; mixed forest; alluvial/very wet forest), Scrub (15%; scrub; heathland), Grassland (25%; alpine/subalpine/boreal grassland; humid grassland), Wetland (15%; standing fresh water; river/stream; raised bog), Rocky areas (10%; scree/boulders)
Land-use Agriculture (25%), Forestry (30%), Hunting (30%), Nature conservation/research (20%), Tourism/recreation (25%), Water management (10%)

Birds

Species	Season	Year	Pop min	Pop max	Acc	Criteria
Milvus migrans Black Kite	B	1995	B	—	—	C6
Gypaetus barbatus Lammergeier	R	1996	5	5	A	B2, C6
Circaetus gallicus Short-toed Eagle	B	1995	10	—	C	C6
Aquila chrysaetos Golden Eagle	R	1996	18	—	B	B2, C6
[1] ***Lagopus mutus*** Ptarmigan	R	1996	70	—	B	C6
[2] ***Tetrao urogallus*** Capercaillie	R	1996	215	215	B	C6
[3] ***Perdix perdix*** Partridge	R	1996	150	—	B	C6
Aegolius funereus Tengmalm's Owl	R	1996	15	—	B	C6
Dryocopus martius Black Woodpecker	R	1995	70	—	B	C6
Prunella collaris Alpine Accentor	R	1995	400	—	C	A3
Monticola saxatilis Rock Thrush	B	1995	50	—	C	B2
Tichodroma muraria Wallcreeper	R	1996	5	—	C	A3
Pyrrhocorax graculus Alpine Chough	R	1996	400	—	B	A3
Pyrrhocorax pyrrhocorax Chough	R	1996	300	—	B	B2, C2, C6
Serinus citrinella Citril Finch	R	1995	Common		—	A3

1. *L. m. pyrenaicus*.
2. Males.
3. *P. p. hispaniensis*.

Breeding birds include four of the 10 species in Europe that are restricted to the Eurasian high-montane biome (when breeding).

Protection status

National Low **International** Low

10,230 ha of IBA covered by National Park (Aiguestortes i Estany de Sant Maurici, 10,230 ha). 346 ha of IBA covered by Natural Reserve (Alt Aneu, 346 ha). 450 ha of IBA covered by Natural Reserve (Noguera Pallaresa–Bonaigua, 450 ha). 10,230 ha of IBA covered by Special Protection Area (Aiguestortes i Estany de Sant Maurici, 10,230 ha).

Conservation issues

Threats Afforestation (A), Construction/impact of dyke/dam/barrage (A), Firewood collection (B), Infrastructure (A), Recreation/tourism (B), Selective logging/cutting (B), Unsustainable exploitation (B)

The main threats are from high tourist pressure, particularly damaging activities such as off-road driving. Hydro-power schemes, illegal hunting and logging are also threats.

Sant Gervás mountains — B2, C2, C6 — 131

Admin region Cataluña
Coordinates 42°20′N 0°55′E
Altitude 800–1,839 m **Area** 24,500 ha

Site description

A mountain range dominated by scrub and grassland. The area is used for grazing sheep

Habitats Forest and woodland, Scrub (*), Grassland (*), Rocky areas (*)
Land-use Agriculture (*), Hunting

Birds

Species	Season	Year	Pop min	Pop max	Acc	Criteria
Gypaetus barbatus Lammergeier	R	1996	3	3	A	B2, C6
Neophron percnopterus Egyptian Vulture	B	1996	5	5	A	C6
[1] ***Gyps fulvus*** Griffon Vulture	R	1996	73	73	A	C6
Pyrrhocorax pyrrhocorax Chough	R	1996	300	—	B	B2, C2, C6

1. Large increase.

This is an important site for breeding raptors.

Protection status

National None **International** None

Conservation issues

Threats Abandonment/reduction of land management (C), Recreation/tourism (B)

The main threats are uncontrolled tourism and the abandonment of traditional agriculture.

Boumort mountains — B1iii, B2, C2, C6 — 132

Admin region Cataluña
Coordinates 42°20′N 1°10′E
Altitude 550–2,076 m **Area** 54,000 ha

Site description

A rugged area of limestone rock and a narrow gorge. The vegetation consists of *Pinus* forest, scrub and alpine grasslands. The main human activities are game hunting, forestry, apiculture ('Other' land-use, below) and some livestock-farming.

Habitats Forest and woodland (65%; broadleaved deciduous forest; native coniferous forest; alluvial/very wet forest; treeline ecotone), Grassland (10%; steppe/dry calcareous grassland; alpine/subalpine/boreal grassland), Rocky areas (10%; scree/boulders), Artificial landscape (forestry plantation)
Land-use Agriculture (30%), Forestry (55%), Hunting (40%), Other (10%)

Birds

Species	Season	Year	Pop min	Pop max	Acc	Criteria
[1] ***Gypaetus barbatus*** Lammergeier	R	1996	3	3	A	B2, C6
Neophron percnopterus Egyptian Vulture	B	1996	9	9	A	C6
[1] ***Gyps fulvus*** Griffon Vulture	R	1996	100	—	B	B1iii, B2, C2, C6
Aquila chrysaetos Golden Eagle	R	1996	9	9	A	C6
[2,3] ***Tetrao urogallus*** Capercaillie	R	1996	45	45	A	C6
Dryocopus martius Black Woodpecker	R	1996	15	20	B	C6
Pyrrhocorax pyrrhocorax Chough	R	1996	400	—	B	B2, C2, C6

1. Large increase.
2. *T. u. aquitanicus*.
3. Males.

This is an important site for cliff-nesting raptors and forest birds.

Protection status

National Partial **International** None
13,097 ha of National Game Reserve (Serra de Boumort, 13,097 ha). 55 ha of IBA covered by Natural Reserve (Noguera Pallaresa–Collegats, 55 ha).

Conservation issues

There is a high level of disturbance from sporting and recreational activities, particularly off-road driving. Intensive forest management, fires, and hunting are also serious threats.

Threats Abandonment/reduction of land management (B), Afforestation (B), Agricultural intensification/expansion (C), Burning of vegetation (C), Firewood collection (B), Recreation/tourism (B), Selective logging/cutting (A), Unsustainable exploitation (A)

Monteixo–L'Orri–Tornafort — B2, C6 — 133

Admin region Cataluña
Coordinates 42°30′N 1°20′E
Altitude 700–3,144 m **Area** 65,000 ha

Site description

An extensive area in the Pyrenees with large *Pinus* forests and more limited areas of *Betula* woodland. There are also extensive areas of scrub, pasture, and alpine grassland. The main human activities are forestry, stock-grazing, hiking, and skiing.

Habitats Forest and woodland (45%; native coniferous forest; mixed forest), Scrub (30%; scrub), Grassland (25%; alpine/subalpine/boreal grassland; mesophile grassland)
Land-use Agriculture (*), Forestry (*), Tourism/recreation (*)

Birds

Species	Season	Year	Pop min	Pop max	Acc	Criteria
Gypaetus barbatus Lammergeier	R	1996	2	2	A	B2, C6
Circaetus gallicus Short-toed Eagle	B	1994	8	15	B	C6
Aquila chrysaetos Golden Eagle	R	1994	11	11	B	C6
[1] ***Lagopus mutus*** Ptarmigan	R	1996	50	—	B	C6
[2,3] ***Tetrao urogallus*** Capercaillie	R	1996	175	175	A	C6
[4] ***Perdix perdix*** Partridge	R	1996	150	—	B	C6
Aegolius funereus Tengmalm's Owl	R	1993	10	—	B	C6
Dryocopus martius Black Woodpecker	R	1994	40	—	B	C6
Monticola saxatilis Rock Thrush	B	1996	35	—	B	B2

1. *L. m. pyrenaicus*.
2. *T. u. aquitanicus*.
3. Males.
4. *P. p. hispaniensis*.

This is an important area for upland forest species.

Protection status

National None **International** None

Conservation issues

Threats Infrastructure (B), Recreation/tourism (B), Selective logging/cutting (A)

The main threat is from development of the skiing and visitor related infrastructure. In some areas intensive forestry is also a threat.

Cerdaña — B2, C6 — 134

Admin region Cataluña
Coordinates 42°30′N 1°45′E
Altitude 1,450–2,898 m **Area** 22,000 ha

Site description

A mountainous area on the border with Andorra, with large *Pinus* woodlands and alpine pastures. The area is used for livestock-farming, game hunting, hiking, and skiing.

Habitats Forest and woodland (*), Scrub, Grassland (*), Rocky areas (*)
Land-use Agriculture (*), Forestry, Hunting (*), Tourism/recreation (*)

Birds

Species	Season	Year	Pop min	Pop max	Acc	Criteria
[1] ***Lagopus mutus*** Ptarmigan	R	1996	60	90	C	C6
[2,3,4] ***Tetrao urogallus*** Capercaillie	R	1996	17	20	A	C6
[5] ***Perdix perdix*** Partridge	R	1996	—	100	C	C6
Charadrius morinellus Dotterel	B	1996	2	10	C	C6
Pyrrhocorax pyrrhocorax Chough	R	1996	100	—	C	B2

1. *L. m. pyrenaicus*.
2. *T. u. aquitanicus*.
3. Males.
4. Large decrease.
5. *P. p. hispaniensis*.

This is an important area for alpine species.

Protection status

National Partial **International** None

19,437 ha of IBA covered by National Game Reserve (Cerdana, 19,437 ha).

Conservation issues

Threats Infrastructure (A), Intensified forest management (B), Recreation/tourism (B)

The main threat is from skiing inside the forest and from continuing development of the skiing related infrastructure. Intensive forestry is also a threat.

Cadí mountains — B1iv, B2, C5, C6 — 135

Admin region Cataluña
Coordinates 42°20′N 1°40′E
Altitude 500–2,562 m **Area** 69,600 ha

Site description

A mountain range with large limestone and extensive *Pinus*- and *Abies*-dominated forests. The main human activities are game hunting, forestry, livestock-farming and recreation.

Habitats Forest and woodland (75%; broadleaved deciduous forest; native coniferous forest; mixed forest; alluvial/very wet forest; treeline ecotone), Scrub (12%; scrub), Grassland (19%; alpine/subalpine/boreal grassland; mesophile grassland), Rocky areas (7%; scree/boulders; caves), Artificial landscape (5%; arable land)
Land-use Agriculture (3%), Forestry (80%), Tourism/recreation (5%)

Birds

Species	Season	Year	Pop min	Pop max	Acc	Criteria
Gypaetus barbatus Lammergeier	R	1996	1	2	A	B2, C6
[1,2,3] ***Tetrao urogallus*** Capercaillie	R	1996	33	35	A	C6
Aegolius funereus Tengmalm's Owl	R	1996	4	6	B	C6
Dryocopus martius Black Woodpecker	R	1996	21	25	A	C6
Pyrrhocorax pyrrhocorax Chough	R	1996	100	—	B	B2

1. *T. u. aquitanicus*.
2. Males.
3. Large decrease.

This is a nationally important area for alpine species. The site is a migratory bottleneck, where 3,000 or more raptors regularly pass, notably *Pernis apivorus* (3,500–5,000 per season).

Protection status

National Partial **International** Partial

50,361 ha of IBA covered by National Game Reserve (Cadi, 50,361 ha). 41,342 ha of IBA covered by Natural Park (Cadi Moixero, 41,342 ha). 41,342 ha of IBA covered by Special Protection Area (Cadí–Moixerò, 41,342 ha).

Conservation issues

Threats Extraction industry (C), Intensified forest management (A), Recreation/tourism (A), Selective logging/cutting (B)

The main threats are from intensive forestry and disturbance from tourists. There is a forestry management plan and a management plan for the Natural Park.

Fresser–Setcases — C6 — 136

Admin region Cataluña
Coordinates 42°25′N 2°20′E
Altitude 1,200–2,913 m **Area** 38,500 ha

Site description

A mountainous area in the eastern Pyrenees, with large conifer forests, scrub, grasslands and rocky areas. The area is used for game hunting, forestry, livestock-farming, and hiking.

Habitats Forest and woodland (*), Scrub (*), Grassland (*), Rocky areas
Land-use Agriculture (*), Forestry, Hunting (*), Tourism/recreation (*)

Birds

Species	Season	Year	Pop min	Pop max	Acc	Criteria
Circus cyaneus Hen Harrier	B	1996	2	6	B	C6
[1] ***Lagopus mutus*** Ptarmigan	R	1996	20	30	C	C6
[2] ***Perdix perdix*** Partridge	R	1996	50	—	B	C6
Charadrius morinellus Dotterel	B	1996	3	5	B	C6
Aegolius funereus Tengmalm's Owl	R	1996	4	8	B	C6

1. *L. m. pyrenaicus*.
2. *P. p. hispaniensis*.

This is an important site for forest and alpine species.

Protection status

National Partial **International** None

20,200 ha of IBA covered by National Game Reserve (Fresser–Setcases, 20,200 ha).

Conservation issues

Threats Industrialization/urbanization (B), Infrastructure (A), Recreation/tourism (B)

The main threats are from infrastructural developments such as roads and powerlines, urban development and disturbance from visitors. Monitoring of *Tetrao urogallus* and *Perdix perdix* is conducted by the regional government.

Ampurdán marshes — A4i, B1i, B2, C2, C3, C6 — 137

Admin region Cataluña
Coordinates 42°12′N 3°07′E
Altitude 0–45 m **Area** 5,454 ha

Site description

A coastal wetland complex with saltmarshes, coastal lagoons and freshwater ponds, humid grasslands and riverine forest. The area is used for sheep-grazing and rice cultivation.

Habitats Forest and woodland (30%; alluvial/very wet forest), Grassland (35%; humid grassland; mesophile grassland), Wetland (55%; mudflat/sandflat; sand-dunes/sand beach; coastal lagoon; standing fresh water; standing brackish and salt water; river/stream; water-fringe vegetation), Artificial landscape (70%; highly improved reseeded grassland; arable land)
Land-use Agriculture (60%), Forestry, Nature conservation/research (90%), Tourism/recreation (20%)

Birds

Species	Season	Year	Pop min	Pop max	Acc	Criteria
Botaurus stellaris Bittern	R	1996	5	6	A	C6
Ixobrychus minutus Little Bittern	B	1993	35	40	A	C6
[1] ***Bubulcus ibis*** Cattle Egret	N	1994	5,800	—	A	A4i, B1i, C3
[1] ***Ardea purpurea*** Purple Heron	B	1996	49	49	A	C2, C6
[1] ***Anas platyrhynchos*** Mallard	W	1994	18,000	18,000	A	B1i, C3
[1] ***Circus aeruginosus*** Marsh Harrier	R	1995	14	14	A	C6
[1] ***Porphyrio porphyrio*** Purple Gallinule	R	1996	35	—	B	B2, C2, C6
Himantopus himantopus Black-winged Stilt	B	1993	109	117	A	C6
[1] ***Larus cachinnans*** Yellow-legged Gull	W	1997	5,000	—	A	A4i, B1i, C3
Chlidonias hybridus Whiskered Tern	P	1996	250	—	A	B1i, C2
Acrocephalus melanopogon Moustached Warbler	R	1997	Common		—	C6
Lanius minor Lesser Grey Shrike	B	1997	12	20	A	C6

1. Large increase.

This is an important site for breeding, wintering and passage herons and wildfowl.

Protection status

National High **International** High

4,784 ha of IBA covered by Natural Park (Aiguamolls de l'Emporda, 4,784 ha). 867 ha of IBA covered by Integral Nature Reserve (Aiguamolls de l'Emporda, 867 ha). 4,803 ha of IBA covered by Ramsar Site (Aiguamolls de l'Empordà, 4,803 ha). 867 ha of IBA covered by Special Protection Area (Aiguamolls de l'Empordà, 867 ha).

Conservation issues

Threats Agricultural intensification/expansion (A), Filling-in of wetlands (B), Groundwater abstraction (A), Industrialization/urbanization (B), Infrastructure (B), Other (B)

Threats include degradation of the site by urbanization and tourism activities, pollution from the use of pesticides in the surrounding crops, and electric powerlines. Water management in the Integral Reserve is difficult because the land is privately-owned ('Other' threats, above). *Ciconia ciconia* and *Porphyrio porphyrio* have been reintroduced.

Medas islands — A4i, B1i, C3, C6 — 138

Admin region Cataluña
Coordinates 42°02'N 3°13'E
Altitude 0–75 m **Area** 21 ha

Site description

A small, scrub-covered island and six more islets 1 km off Cape Estartit, on the Costa Brava. The islands are uninhabited. The main activity is tourism.

Habitats Scrub (sclerophyllous scrub/garrigue/maquis), Marine areas (sea inlet/coastal features), Rocky areas (90%; sea cliff/rocky shore; rock stacks/islets)
Land-use Tourism/recreation (100%)

Birds

Species	Season	Year	Pop min	Pop max	Acc	Criteria
Phalacrocorax aristotelis Shag	R	1996	8	8	A	C6
Nycticorax nycticorax Night Heron	B	1996	120	120	A	C6
[1] ***Bubulcus ibis*** Cattle Egret	R	1996	1,081	1,081	A	A4i, B1i, C3
Egretta garzetta Little Egret	R	1996	49	49	A	C6
Larus cachinnans Yellow-legged Gull	R	1996	13,666	13,666	A	A4i, B1i, C3

1. Large increase.

This is an important site for breeding seabirds and for a heron colony. The site has one of the largest *Larus cachinnans* colonies in the Mediterranean

Protection status

National None **International** None

Conservation issues

Threats Recreation/tourism (B)

The main threat to the islands is from increasing tourism. The surroundings of the island are protected as a Marine Reserve (418 ha).

Mountains of Barcelona — B1i, B2, B3, C2, C3, C6 — 139

Admin region Cataluña
Coordinates 41°39'N 2°01'E
Altitude 200–1,236 m **Area** 124,000 ha

Site description

Calcareous mountains covered by *Pinus* forests and above 600 m by *Quercus* forest. The adjacent plain contains cereal crops, maquis and alluvial forest.

Habitats Forest and woodland (*; broadleaved deciduous forest; native coniferous forest; mixed forest; broadleaved evergreen forest), Scrub (*; sclerophyllous scrub/garrigue/maquis), Grassland (steppe/dry calcareous grassland), Rocky areas (scree/boulders; inland cliff; caves), Artificial landscape (*; arable land; perennial crops/orchards/groves; other urban/industrial areas; ruderal land)
Land-use Agriculture (30%), Forestry (50%), Hunting, Tourism/recreation, Urban/industrial/transport

Birds

Species	Season	Year	Pop min	Pop max	Acc	Criteria
[1] ***Circaetus gallicus*** Short-toed Eagle	B	1996	16	17	A	B2, C6
[2] ***Hieraaetus fasciatus*** Bonnelli's Eagle	R	1996	3	—	A	C6
[1] ***Falco peregrinus*** Peregrine	R	1996	22	—	A	B2, C6
[1] ***Larus cachinnans*** Yellow-legged Gull	N	1996	2,000	4,000	A	B1i, C3
Columba oenas Stock Dove	B	1996	500	700	B	B3
Columba palumbus Woodpigeon	R	1996	10,000	40,000	B	B3
Streptopelia turtur Turtle Dove	B	1996	6,000	8,000	B	B2
Tyto alba Barn Owl	R	1996	500	—	B	B2
Bubo bubo Eagle Owl	R	1996	40	60	A	B2, C6
Athene noctua Little Owl	R	1996	600	800	A	B2
Strix aluco Tawny Owl	R	1996	700	1,500	A	B3
Caprimulgus europaeus Nightjar	B	1996	4,000	6,000	A	B2, C2
Merops apiaster Bee-eater	B	1996	800	1,000	B	B2
Picus viridis Green Woodpecker	R	1996	700	1,000	B	B2
Hirundo rustica Swallow	B	1996	4,000	10,000	B	B2
Monticola solitarius Blue Rock Thrush	R	1996	300	400	B	B2

1. Large increase.
2. Large decrease.

This is an important area for species characteristic of the Mediterranean region, and cliff-nesting species.

Protection status

National Partial **International** None
3,630 ha of IBA covered by Natural Park (Muntanya de Montserrat, 3,630 ha). 9,638 ha of IBA covered by Natural Park (Sant Llorenc de Munt i L'Obac, 9,638 ha).

Conservation issues

Threats Abandonment/reduction of land management (B), Afforestation (A), Burning of vegetation (B), Industrialization/urbanization (A), Infrastructure (A), Intensified forest management (B), Recreation/tourism (A), Selective logging/cutting (B), Unsustainable exploitation (A)

Threats in the mountains include land abandonment, fires, hunting, the erection of powerlines, and uncontrolled recreational activities which cause particular disturbance to raptors. In the plains, increasing urban development is a major threat.

Llobregat delta — A1, A4i, A4ii, B1i, B1ii, B2, C1, C2, C3, C6 — 140

Admin region Cataluña
Coordinates 41°16'N 2°08'E
Altitude 0–10 m **Area** 2,200 ha

Site description

The site includes several small wetlands and lagoons close to Barcelona. Also included are sand-dunes covered by *Pinus* woodland, an area of shoreline and areas of cultivation.

Habitats Forest and woodland (5%; native coniferous forest; alluvial/very wet forest), Wetland (30%; sand-dunes/sand beach; coastal lagoon; standing brackish and salt water; river/stream; water-fringe vegetation), Marine areas (open sea; sea inlet/coastal features), Artificial landscape (65%; arable land; other urban/industrial areas; ruderal land)
Land-use Agriculture (40%), Forestry (10%), Nature conservation/research (10%), Not utilized (10%), Tourism/recreation (10%), Urban/industrial/transport (15%)

Birds

Species	Season	Year	Pop min	Pop max	Acc	Criteria
[1] ***Puffinus yelkouan*** Yelkouan Shearwater	W	1996	2,000	7,080	A	A4ii, B1ii, C2
Ixobrychus minutus Little Bittern	B	1996	22	25	A	C6
Circus aeruginosus Marsh Harrier	R	1995	1	2	A	C6
Porzana parva Little Crake	B	1995	1	—	C	C6
Porzana pusilla Baillon's Crake	B	1996	5	—	C	C6
Himantopus himantopus Black-winged Stilt	B	1996	42	45	A	C6
Charadrius alexandrinus Kentish Plover	R	1996	68	101	A	B2
Lymnocryptes minimus Jack Snipe	W	1996	20	—	—	B2
Larus melanocephalus Mediterranean Gull	W	1996	—	10,343	A	A4i, B1i, C2
Larus minutus Little Gull	P	1996	1,000	3,000	B	A4i, B1i, C3
Larus audouinii Audouin's Gull	P	1996	200	400	A	A1, A4i, B1i, C1, C2
[2] ***Larus cachinnans*** Yellow-legged Gull	W	1996	—	4,451	A	A4i, B1i, C3
Sterna sandvicensis Sandwich Tern	W	1996	107	223	A	C2
Chlidonias niger Black Tern	P	1996	500	2,000	B	A4i, B1i, C2
Acrocephalus melanopogon Moustached Warbler	R	—	—	—	—	C6

1. *P. y. mauretanicus.*
2. Large increase.

This is an important area for breeding, wintering and passage seabirds and other waterbirds. Species of global conservation concern that do not meet IBA criteria: *Aythya nyroca* (2–5 wintering individuals) and *Larus audouinii* (8 wintering individuals).

Protection status

National Partial **International** Partial

110 ha of IBA covered by Natural Reserve (El Remolar–Filipines, 110 ha). 177 ha of IBA covered by Natural Reserve (La Ricarda–Ca l'Arana, 177 ha). 573 ha of IBA covered by Special Protection Area (Delta del Llobregat, 573 ha).

Conservation issues

Threats Agricultural intensification/expansion (B), Consequences of animal/plant introductions (C), Groundwater abstraction (C), Industrialization/urbanization (B), Infrastructure (A), Recreation/tourism (C)

The main threat to the site is from a large development plan for the delta which would mean extension of the airport, enlargement of the harbour and many associated infrastructural developments. Feral cats and water pollution from agricultural chemicals are further threats.

Montagut and Montmell mountains B2, C6 141

Admin region Cataluña
Coordinates 41°20'N 1°30'E
Altitude 400–963 m **Area** 44,300 ha

Site description

A mountain range dominated by *Pinus* and *Quercus* woodland and Mediterranean scrub.

Habitats Forest and woodland (*; broadleaved deciduous forest; native coniferous forest), Scrub (*; sclerophyllous scrub/garrigue/maquis), Rocky areas (inland cliff)
Land-use Agriculture, Forestry, Hunting (*), Not utilized

Birds

Species	Season	Year	Pop min	Pop max	Acc	Criteria
Hieraaetus fasciatus Bonnelli's Eagle	R	1996	5	6	A	B2, C6
Bubo bubo Eagle Owl	R	1996	30	—	C	B2, C6

An important site for species associated with Mediterranean habitats.

Protection status

National None **International** None

Conservation issues

Threats Burning of vegetation (B), Industrialization/urbanization (C), Infrastructure (B)

The site is threatened by spreading urban development, forest fires and the erection of powerlines.

Lerida steppes A1, B1iii, B2, C1, C2, C6 142

Admin region Cataluña
Coordinates 41°40'N 1°08'E
Altitude 200–829 m **Area** 62,500 ha

Site description

A steppe area in the Ebro valley dominated by cereal crops and locally olive- and almond-groves. In the higher areas *Quercus* and *Pinus* woodlands occur.

Habitats Forest and woodland (10%; broadleaved deciduous forest; native coniferous forest; mixed forest; alluvial/very wet forest; broadleaved evergreen forest), Scrub (5%; sclerophyllous scrub/garrigue/maquis), Grassland (5%; steppe/dry calcareous grassland), Wetland (1%; river/stream; water-fringe vegetation), Artificial landscape (85%; arable land; perennial crops/orchards/groves; other urban/industrial areas)
Land-use Agriculture (90%), Forestry (10%), Hunting (100%)

Birds

Species	Season	Year	Pop min	Pop max	Acc	Criteria
Circaetus gallicus Short-toed Eagle	B	1995	10	15	C	C6
Circus pygargus Montagu's Harrier	B	1996	15	20	A	C6
[1] ***Hieraaetus fasciatus*** Bonnelli's Eagle	N	1994	20	70	A	B1iii, C2
[1] ***Alectoris rufa*** Red-legged Partridge	R	1995	10,000	15,000	C	B2
[2] ***Tetrax tetrax*** Little Bustard	R	1995	600	1,000	A	A1, B2, C1, C6
Burhinus oedicnemus Stone Curlew	B	1995	600	700	A	B2, C2, C6
Pterocles orientalis Black-bellied Sandgrouse	R	1995	10	—	A	C6
Merops apiaster Bee-eater	B	1995	850	—	B	B2
Coracias garrulus Roller	B	1995	40	50	A	B2, C6
Melanocorypha calandra Calandra Lark	R	1995	5,200	7,900	A	B2

1. Large decrease.
2. Territorial males.

This is an important area for steppic species and dispersing juvenile raptors. Species of global conservation concern that do not meet IBA criteria: *Falco naumanni* (7–15 breeding pairs).

Protection status

National None **International** None

Conservation issues

Threats Agricultural intensification/expansion (A), Consequences of animal/plant introductions (C), Deforestation (commercial) (C), Drainage (C), Extraction industry (C), Groundwater abstraction (B), Industrialization/urbanization (B), Infrastructure (B), Selective logging/cutting (A), Shifting agriculture (C), Unsustainable exploitation (A)

The main threats are from hunting, plans to irrigate the cereal crops and forestry operations.

Montsech and Montgai mountains B2, C2, C6 143

Admin region Aragón, Cataluña
Coordinates 42°05'N 0°55'E
Altitude 400–1,678 m **Area** 49,000 ha

Site description

A mountainous area of limestone cliffs and gorges. Sclerophyllous scrub predominates, with some scattered *Quercus* woodland, arable cultivation and olive-groves.

Habitats Forest and woodland (broadleaved deciduous forest), Scrub (*; sclerophyllous scrub/garrigue/maquis), Wetland (river/stream), Rocky areas (*; scree/boulders; inland cliff), Artificial landscape (arable land; perennial crops/orchards/groves)
Land-use Agriculture (*), Hunting, Tourism/recreation, Urban/industrial/transport (*)

Birds

Species	Season	Year	Pop min	Pop max	Acc	Criteria
Gypaetus barbatus Lammergeier	R	1996	2	2	A	B2, C6
Neophron percnopterus Egyptian Vulture	B	1996	5	—	B	C6
Aquila chrysaetos Golden Eagle	R	1996	7	—	A	C6
Falco peregrinus Peregrine	R	1966	10	—	B	C6
Pyrrhocorax pyrrhocorax Chough	R	1996	200	—	B	B2, C2, C6

An important area for breeding raptors.

Protection status

National Low **International** None

1 ha of IBA covered by Natural Reserve (Noguera Ribagorcana–Mont–Remei, 1 ha).

Conservation issues

Threats Abandonment/reduction of land management (U), Recreation/tourism (B)

The main threats are from the abandonment of traditional agricultural practices and disturbance from tourists and associated outdoor sports.

Cogul–Alfés steppes B2, C2, C6 144

Admin region Cataluña
Coordinates 41°30'N 0°40'E
Altitude 100–352 m **Area** 18,000 ha

Site description

A steppe area in Cataluña comprising undulating plains with non-irrigated cereal cultivation, interspersed with garrigue-like low scrub and almond trees.

Habitats Scrub (*; sclerophyllous scrub/garrigue/maquis), Grassland (*; steppe/dry calcareous grassland), Wetland (river/stream), Artificial landscape (*; arable land; perennial crops/orchards/groves)
Land-use Agriculture (*), Hunting, Military, Tourism/recreation, Water management

Birds

Species	Season	Year	Pop min	Pop max	Acc	Criteria
Ardea purpurea Purple Heron	R	1996	35	40	A	C6
Circus aeruginosus Marsh Harrier	R	1996	4	4	A	C6
[1] ***Tetrax tetrax*** Little Bustard	R	1995	115	185	A	C6
Burhinus oedicnemus Stone Curlew	R	1995	275	335	A	B2, C6
Pterocles alchata Pin-tailed Sandgrouse	R	1995	80	110	A	B2, C2, C6
Merops apiaster Bee-eater	B	1996	200	—	C	B2
Coracias garrulus Roller	B	1995	45	—	A	B2, C6
[2] ***Chersophilus duponti*** Dupont's Lark	R	1995	44	53	A	C6
Melanocorypha calandra Calandra Lark	R	1995	4,500	6,300	A	B2
Lanius minor Lesser Grey Shrike	B	1995	10	15	A	C6

1. Males.
2. Large decrease.

This is an important area for steppic species, particularly *Chersophilus duponti*.

Protection status

National Low **International** Low
1,140 ha of IBA covered by Natural Reserve (Mas de Melons, 1,140 ha). 1,140 ha of IBA covered by Special Protection Area (Mas de Melons, 1,140 ha).

Conservation issues

Threats Agricultural intensification/expansion (B), Extraction industry (C), Infrastructure (B), Recreation/tourism (C)

The area is seriously threatened by irrigation schemes to permit agricultural intensification. A private airport, that holds a population of *Chersophilus duponti*, has plans to expand. There is a zonal plan approved by the European Commission but not implemented by the Catalonian regional government.

Montsant and Prades mountains — B2, C2, C6 — 145

Admin region Cataluña
Coordinates 41°20′N 0°50′E
Altitude 400–1,201 m **Area** 80,000 ha

Site description

An area of low coastal mountains with many cliffs. The site has extensive *Pinus* forests and large areas of scrub. Agricultural areas are mainly used for olive-groves, vineyards and orchards. The main human activities are hunting, forestry, and agriculture.

Habitats Forest and woodland (40%; native coniferous forest), Scrub (30%; sclerophyllous scrub/garrigue/maquis), Rocky areas (20%; scree/boulders; inland cliff), Artificial landscape (10%; perennial crops/orchards/groves)
Land-use Agriculture (10%), Forestry (30%), Hunting (75%), Not utilized

Birds

Species	Season	Year	Pop min	Pop max	Acc	Criteria
Hieraaetus fasciatus Bonnelli's Eagle	R	1996	14	—	B	B2, C2, C6
Falco peregrinus Peregrine	R	1995	17	21	B	B2, C6
Bubo bubo Eagle Owl	R	1996	50	—	C	B2, C6

This is an important area for resident raptors.

Protection status

National None **International** None

Conservation issues

Threats Abandonment/reduction of land management (B), Construction/impact of dyke/dam/barrage (U), Infrastructure (B), Intensified forest management (U), Other (B), Recreation/tourism (A), Unsustainable exploitation (B)

The main threats are from abandonment of traditional land management, erection of powerlines, disturbance from recreational activities, particularly off-road driving, and illegal hunting and poisoning.

Cardó, Tivissa and Llabería mountains — B2, C2, C6 — 146

Admin region Cataluña
Coordinates 41°00′N 0°48′E
Altitude 100–938 m **Area** 68,500 ha

Site description

An area of mountains dominated by *Pinus* forests and Mediterranean scrub. The site includes an area of plain dominated by olive-groves.

Habitats Forest and woodland (*; native coniferous forest), Scrub (*; sclerophyllous scrub/garrigue/maquis), Rocky areas (inland cliff), Artificial landscape (*; perennial crops/orchards/groves)
Land-use Agriculture (*), Not utilized, Urban/industrial/transport

Birds

Species	Season	Year	Pop min	Pop max	Acc	Criteria
Hieraaetus fasciatus Bonnelli's Eagle	R	1996	18	19	A	B2, C2, C6
Falco peregrinus Peregrine	R	1996	14	15	A	C6
Bubo bubo Eagle Owl	R	1996	80	100	C	B2, C2, C6

This is an important site for breeding raptors and species characteristic of Mediterranean habitats.

Protection status

National None **International** None

Conservation issues

Threats Burning of vegetation (B), Infrastructure (B), Natural events (B)

The main threats are from both natural and man-induced fires and the erection of powerlines.

Pass of Beceite–Turmell mountain — B2, C6 — 147

Admin region Aragón, Cataluña, País Valenciano
Coordinates 40°50′N 0°15′E
Altitude 150–1,447 m **Area** 107,000 ha

Site description

A mountainous area with many cliffs and gullies. Vegetation includes extensive *Pinus* woodlands, areas of maquis and large areas of scrub. The area is used for forestry, livestock-grazing, and hunting.

Habitats Forest and woodland (50%; broadleaved deciduous forest; native coniferous forest), Scrub (20%; sclerophyllous scrub/garrigue/maquis), Rocky areas (30%; inland cliff)
Land-use Agriculture (10%), Forestry (50%), Hunting (30%), Not utilized (10%)

Birds

Species	Season	Year	Pop min	Pop max	Acc	Criteria
Gyps fulvus Griffon Vulture	R	1995	59	—	A	C6
Circaetus gallicus Short-toed Eagle	B	1996	10	—	B	C6
Aquila chrysaetos Golden Eagle	R	1996	10	—	B	C6
Hieraaetus fasciatus Bonnelli's Eagle	R	1996	7	—	B	B2, C6
Falco peregrinus Peregrine	R	1995	25	30	B	B2, C6
Bubo bubo Eagle Owl	R	1996	25	—	C	B2, C6

This is an important site for breeding raptors.

Protection status

National Partial **International** None
30,418 ha of IBA covered by National Game Reserve (Puertos de Beceite, 30,418 ha).

Conservation issues

Threats Abandonment/reduction of land management (U), Burning of vegetation (B), Construction/impact of dyke/dam/barrage (B), Infrastructure (B), Other (B)

Threats include a dam project on the River Matarraña, fires and the use of poisoned baits.

Ebro delta A1, A4i, B1i, B2, B3, C1, C2, C3, C6 148

Admin region Cataluña
Coordinates 40°43′N 0°44′E
Altitude 0–185 m **Area** 32,000 ha

Site description

A large delta that protrudes into the Mediterranean Sea forming a complex of shallow brackish lagoons, saltmarshes, salt-lakes and sand beaches with dunes. The site includes approximately 1,500 ha of rice cultivation. Human activities include hunting, fishing, reed-harvesting, arable cultivation and livestock-farming.

Habitats Forest and woodland (alluvial/very wet forest), Wetland (30%; tidal river/enclosed tidal water; saltmarsh; sand-dunes/sand beach; coastal lagoon; water-fringe vegetation), Artificial landscape (75%; arable land; other urban/industrial areas)
Land-use Agriculture (70%), Fisheries/aquaculture, Hunting (65%), Nature conservation/research (25%), Tourism/recreation, Urban/industrial/transport

Birds

Species	Season	Year	Pop min	Pop max	Acc	Criteria
[1,2] ***Phalacrocorax carbo*** Cormorant	W	1996	1,600	2,500	A	B1i, C3
Botaurus stellaris Bittern	R	1996	2	4	A	C6
Ixobrychus minutus Little Bittern	B	1996	700	—	A	B2, C2, C6
[2] ***Nycticorax nycticorax*** Night Heron	R	1996	160	160	A	B2, C6
Ardeola ralloides Squacco Heron	B	1996	450	450	A	A4i, B1i, B2, C2, C6
[2] ***Bubulcus ibis*** Cattle Egret	R	1996	4,300	4,300	A	A4i, B1i, C3
Bubulcus ibis Cattle Egret	W	1996	2,000	3,000	A	A4i, B1i, C3
Bubulcus ibis Cattle Egret	P	1996	7,000	14,000	A	A4i, B1i, C3
Egretta garzetta Little Egret	R	1996	1,000	1,000	A	A4i, B1i, C2, C6
Egretta garzetta Little Egret	W	1996	500	1,200	A	A4i, B1i, C2
Egretta garzetta Little Egret	P	1996	2,000	6,000	A	A4i, B1i, C2
[2] ***Ardea purpurea*** Purple Heron	B	1995	500	500	A	B2, C2, C6
Phoenicopterus ruber Greater Flamingo	R	1996	500	1,500	A	A4i, B1i, B2, C2, C6
[2] ***Phoenicopterus ruber*** Greater Flamingo	W	1996	1,500	5,000	A	A4i, B1i, C2
Phoenicopterus ruber Greater Flamingo	P	1996	6,000	7,000	A	A4i, B1i, C2
Anas penelope Wigeon	W	1996	3,000	13,700	A	B1i, C3
Anas strepera Gadwall	R	1996	200	300	A	B2
Anas strepera Gadwall	W	1996	530	4,100	A	A4i, B1i, C3
[3] ***Anas crecca*** Teal	W	1996	1,800	18,500	A	A4i, B1i, C3
Anas platyrhynchos Mallard	R	1996	9,000	9,000	A	B1i, C3
Anas platyrhynchos Mallard	W	1996	17,200	42,800	A	B1i, C3
Anas clypeata Shoveler	W	1996	5,000	20,600	A	A4i, B1i, C3
[2] ***Netta rufina*** Red-crested Pochard	R	1996	1,500	2,000	A	A4i, B1i, B2, C3
Netta rufina Red-crested Pochard	W	1996	1,000	6,100	A	A4i, B1i, C3
Aythya ferina Pochard	W	1996	1,000	15,000	A	A4i, B1i, C3
Porzana pusilla Baillon's Crake	B	1990	50	—	C	B2, C2, C6
Porphyrio porphyrio Purple Gallinule	R	1996	10	—	A	C6
[2] ***Fulica atra*** Coot	W	1996	10,000	25,000	A	B1i, C3
Himantopus himantopus Black-winged Stilt	B	1996	—	5,000	A	A4i, B1i, C2, C6
Recurvirostra avosetta Avocet	R	1996	400	550	A	A4i, B1i, B3, C2, C6
Recurvirostra avosetta Avocet	W	1996	700	1,500	A	A4i, B1i, B2, C2
Glareola pratincola Collared Pratincole	B	1996	39	99	A	A4i, B1i, B2, C2, C6
Charadrius alexandrinus Kentish Plover	R	1996	1,500	1,700	A	A4i, B1i, B2, C3
Charadrius alexandrinus Kentish Plover	W	1996	200	800	A	B1i, C3
Charadrius alexandrinus Kentish Plover	P	1996	Common		—	A4i, B1i, C3
[2] ***Calidris minuta*** Little Stint	P	1996	Abundant		—	A4i, B1i, C3
Calidris ferruginea Curlew Sandpiper	P	1996	Abundant		—	A4i, B1i, C3
[2] ***Calidris alpina*** Dunlin	W	1996	10,000	21,000	A	A4i, B1i, B2, C3
Calidris alpina Dunlin	P	1996	Abundant		—	A4i, B1i, C3
Limosa limosa Black-tailed Godwit	W	1996	2,000	4,600	A	A4i, B1i, C3
Limosa limosa Black-tailed Godwit	P	1996	—	10,000	A	A4i, B1i, C3
Limosa lapponica Bar-tailed Godwit	W	1996	50	500	A	B2
Tringa totanus Redshank	P	1996	—	—	B	A4i, B1i, C3
Larus melanocephalus Mediterranean Gull	W	1996	8,000	16,000	A	A4i, B1i, C2
Larus genei Slender-billed Gull	R	1996	600	600	A	A4i, B1i, C2, C3, C6
[2] ***Larus audouinii*** Audouin's Gull	R	1996	11,300	11,300	A	A1, A4i, B1i, B2, C1, C2, C6
Larus audouinii Audouin's Gull	W	1996	Abundant		—	A1, A4i, B1i, C1, C2
Larus audouinii Audouin's Gull	P	1996	Common		—	A1, A4i, B1i, C1, C2
[2] ***Gelochelidon nilotica*** Gull-billed Tern	B	1996	300	300	A	A4i, B1i, B2, C2, C6
[2] ***Sterna sandvicensis*** Sandwich Tern	B	1996	1,900	1,900	A	A4i, B1i, B2, C2, C6
Sterna hirundo Common Tern	B	1996	5,000	5,500	A	A4i, B1i, C2, C6
Sterna hirundo Common Tern	P	1996	Abundant		—	B1i, C2
Sterna albifrons Little Tern	B	1996	450	650	A	A4i, B1i, B2, C2, C6
Chlidonias hybridus Whiskered Tern	B	1996	1,500	1,500	A	A4i, B1i, B2, C2, C6
Acrocephalus melanopogon Moustached Warbler	R	1996	—	150	A	C2, C6

1. *Ph. c. sinensis.*
2. Large increase.
3. Large decrease.

One of the most important sites in the Mediterranean for breeding, passage and wintering waterbirds—27,000 pairs of breeding waterbird occur in the summer and up to 180,000 waterbirds winter at the site. The site holds more than 60% of the world population of *Larus audouinii*.

Protection status

National Partial **International** Partial
7,736 ha of IBA covered by Natural Park (Delta del Ebro, 7,736 ha). 2,700 ha of IBA covered by Natural Reserve (La Banya, 2,700 ha). 7,736 ha of IBA covered by Ramsar Site (Delta del Ebro, 7,736 ha). 7,736 ha of IBA covered by Special Protection Area (Delta del Ebro, 7,736 ha).

Conservation issues

Threats Agricultural intensification/expansion (B), Aquaculture/fisheries (B), Drainage (B), Dredging/canalization (C), Extraction industry (C), Filling-in of wetlands (C), Industrialization/urbanization (A), Other (B), Recreation/tourism (B), Unsustainable exploitation (B)

The delta is poorly protected outside the Natural Park and suffers from a large number of competing interests for using the land. Threats include: increasing urban development; disturbance of waterbird colonies by tourists; a high hunting pressure and lead poisoning of birds from spent shot ('Other' threats, above); pollution from chemicals used in rice production; and a decline in sedimentation rates as a result of dams along the River Ebro. Bird monitoring is carried out by the regional government.

Morella pass B2, C6 149

Admin region País Valenciano
Coordinates 40°35′N 0°10′W
Altitude 800–1,318 m **Area** 43,000 ha

Site description

A low mountain range with *Pinus halepensis* forest, small *Quercus* woodlands and large expanses of Mediterranean scrub. The area is used for forestry, hunting and sheep-grazing. The most important threat are forest fires, also new roads and high hunting pressure.

Habitats Forest and woodland (50%; broadleaved deciduous forest; native coniferous forest; mixed forest; broadleaved evergreen forest), Scrub (30%; sclerophyllous scrub/garrigue/maquis), Grassland (20%; steppe/dry calcareous grassland), Rocky areas (5%; inland cliff), Artificial landscape (5%; highly improved reseeded grassland)
Land-use Agriculture (10%), Forestry (45%), Hunting (80%)

Birds

Species	Season	Year	Pop min	Pop max	Acc	Criteria
Neophron percnopterus Egyptian Vulture	B	1997	2	2	A	C6
[1] ***Gyps fulvus*** Griffon Vulture	R	1997	85	—	A	B2, C6
Hieraaetus fasciatus Bonnelli's Eagle	R	1997	—	5	B	B2
Oenanthe leucura Black Wheatear	R	1997	40	—	C	B2
Pyrrhocorax pyrrhocorax Chough	R	1997	100	—	B	B2

1. Large increase.

This is an important site for breeding raptors.

Protection status
National None **International** None

Conservation issues

Threats Burning of vegetation (A), Infrastructure (B), Unsustainable exploitation (B)

The greatest threat is from forest fires. Other threats include new road construction and high hunting pressure.

Peñagolosa B2, B3, C6 150

Admin region Aragón, País Valenciano
Coordinates 40°15′N 0°25′W
Altitude 420–1,814 m **Area** 83,000 ha

Site description
A group of limestone mountains crossed by deep valleys and with many cliffs. Predominantly covered by *Pinus* forests with some areas dominated by *Quercus*. There are also extensive areas of Mediterranean scrub. The main human activities are forestry and livestock-farming.

Habitats Forest and woodland (75%; broadleaved deciduous forest; native coniferous forest; mixed forest; broadleaved evergreen forest), Scrub (30%; scrub; sclerophyllous scrub/garrigue/maquis), Grassland (10%; steppe/dry calcareous grassland), Rocky areas (10%; scree/boulders; inland cliff), Artificial landscape (15%; arable land; perennial crops/orchards/groves)
Land-use Agriculture (15%), Forestry (55%), Hunting (60%)

Birds

Species	Season	Year	Pop min	Pop max	Acc	Criteria
Aquila chrysaetos Golden Eagle	R	1996	4	6	B	C6
Columba oenas Stock Dove	R	1996	360	—	C	B3
Oenanthe leucura Black Wheatear	R	1996	40	—	C	B2
Monticola saxatilis Rock Thrush	B	1996	35	—	C	B2
Pyrrhocorax pyrrhocorax Chough	R	1996	100	—	C	B2

This is an important site for breeding raptors.

Protection status
National None **International** None

Conservation issues

Threats Construction/impact of dyke/dam/barrage (C), Deforestation (commercial) (B), Dredging/canalization (C), Extraction industry (B), Infrastructure (B), Recreation/tourism (B)

The most important threats are forestry operations and infrastructural developments for visitors.

Espadán mountain range B2, C6 151

Admin region País Valenciano
Coordinates 39°50′N 0°20′W
Altitude 400–1,106 m **Area** 59,400 ha

Site description
A mountainous area covered by *Pinus* forest, Mediterranean scrub and olive- and almond-groves. The main human activities are forestry, livestock-farming and hunting.

Habitats Forest and woodland (60%; native coniferous forest; mixed forest; broadleaved evergreen forest), Scrub (30%; scrub; sclerophyllous scrub/garrigue/maquis), Grassland (20%; steppe/dry calcareous grassland; dry siliceous grassland), Rocky areas (10%; inland cliff), Artificial landscape (15%; perennial crops/orchards/groves)
Land-use Agriculture (15%), Forestry (12%), Hunting (80%), Urban/industrial/transport (5%)

Birds

Species	Season	Year	Pop min	Pop max	Acc	Criteria
Hieraaetus fasciatus Bonnelli's Eagle	R	1997	10	—	A	B2, C6
Falco peregrinus Peregrine	R	1997	6	—	B	C6
Bubo bubo Eagle Owl	R	1997	15	—	B	C6
Oenanthe leucura Black Wheatear	R	1997	—	40	B	B2
Pyrrhocorax pyrrhocorax Chough	R	1997	100	—	B	B2

This is an important site for breeding raptors.

Protection status
National None **International** None

Conservation issues

Threats Burning of vegetation (A), Extraction industry (B), Groundwater abstraction (C), Unsustainable exploitation (B)

The main threats are from fires and hunting.

Prat de Cabanes–Torreblanca A1, A4i, B1i, B2, C1, C2, C6 152

Admin region País Valenciano
Coordinates 40°14′N 0°12′E
Altitude 0–10 m **Area** 860 ha

Site description
A coastal marshland, flooded in winter, with emergent vegetation (*Phragmites, Typha, Cladium*), meadows with halophytic vegetation and areas of shingle and sand-dunes.

Habitats Wetland (90%; shingle/stony beach; standing fresh water; water-fringe vegetation), Artificial landscape (10%; perennial crops/orchards/groves)
Land-use Agriculture (10%), Hunting (90%), Urban/industrial/transport

Birds

Species	Season	Year	Pop min	Pop max	Acc	Criteria
Ixobrychus minutus Little Bittern	B	1996	20	26	B	C6
Ardea purpurea Purple Heron	B	1996	6	6	A	C6
Circus pygargus Montagu's Harrier	B	1996	32	34	A	C6
Glareola pratincola Collared Pratincole	B	1996	63	90	A	A4i, B1i, B2, C2, C6
Larus audouinii Audouin's Gull	P	1997	30	120	A	A1, C1
Acrocephalus melanopogon Moustached Warbler	R	1997	400	—	B	C2, C6

This is an important site for breeding herons and *Glareola pratincola*.

Protection status
National High **International** High
812 ha of IBA covered by Natural Park (Prat de Cabanes–Torreblanca, 812 ha). 812 ha of IBA covered by Ramsar Site (Prat de Cabanes–Torreblanca, 812 ha). 860 ha of IBA covered by Special Protection Area (Prat de Cabanes–Torreblanca, 860 ha).

Conservation issues

Threats Burning of vegetation (C), Disturbance to birds (C), Extraction industry (B), Groundwater abstraction (B), Infrastructure (B), Recreation/tourism (B), Unsustainable exploitation (B)

Main threats are from disturbance from cattle-grazing, peat extraction, high hunting pressure, fire, tourism, and new harbour development.

Columbretes islands A1, A4i, B1i, B2, C1, C2, C6 153

Admin region País Valenciano
Coordinates 39°55′N 0°40′E
Altitude 0–67 m **Area** 2,500 ha

Site description
An archipelago 70 km from the Castellón coast. The islands (18 ha) have steep cliffs and are predominantly vegetated with scrub.

Habitats Scrub, Rocky areas (rock stacks/islets), Marine areas (open sea)
Land-use Nature conservation/research (100%), Urban/industrial/transport (5%)

Birds
This is an important site for breeding seabirds and *Falco eleonorae* and passage migrants.

Species	Season	Year	Pop min	Pop max	Acc	Criteria
Calonectris diomedea Cory's Shearwater	B	1995	113	113	A	C6
Phalacrocorax aristotelis Shag	R	1995	10	15	C	B1i, C2, C6
Falco eleonorae Eleonora's Falcon	B	1995	31	31	A	B2, C6
Larus audouinii Audouin's Gull	R	1995	625	625	A	A1, A4i, B1i, B2, C1, C2, C6

Protection status

National Low **International** High

18 ha of IBA covered by Natural Reserve (Islas Columbretes, 18 ha). 2,500 ha of IBA covered by Special Protection Area (Islas Columbretes, 2,500 ha).

Conservation issues

Threats Recreation/tourism (B)

The main threat is uncontrolled tourism.

River Mijares mouth A4i, B1i, C2, C3, C6 154

Admin region País Valenciano
Coordinates 39°57′N 0°01′W
Altitude 0–40 m **Area** 375 ha

Site description

A river mouth with a stony riverbed and small shallow lagoons, with considerable expanses of *Typha*.

Habitats Grassland (50%; mesophile grassland), Wetland (70%; standing fresh water; river/stream; water-fringe vegetation), Artificial landscape (10%; arable land; perennial crops/orchards/groves)
Land-use Agriculture (5%), Not utilized, Water management (80%)

Birds

Species	Season	Year	Pop min	Pop max	Acc	Criteria
Bubulcus ibis Cattle Egret	N	1995	1,936	—	A	B1i, C3
[1] ***Himantopus himantopus*** Black-winged Stilt	R	1993	395	395	A	A4i, B1i, C2, C6

1. Large increase.

This is an important site for breeding waders, wildfowl and herons. Species of global conservation concern that do not meet IBA criteria: *Larus audouinii* (54 passage individuals).

Protection status

National None **International** None

Conservation issues

Threats Agricultural intensification/expansion (B), Drainage (B), Dredging/canalization (A), Extraction industry (B), Filling-in of wetlands (B), Industrialization/urbanization (A), Unsustainable exploitation (A)

The main threats are pollution from urban areas and from spraying against mosquitos. Drainage, canalization and agricultural intensification are also threats.

Almenara marshes A4i, B1i, B2, C2, C3, C6 155

Admin region País Valenciano
Coordinates 39°46′N 0°14′W
Altitude 0–10 m **Area** 3,000 ha

Site description

Formerly an area of marshland with only a few lagoons remaining. The main human activities are arable cultivation and hunting.

Habitats Grassland (15%; humid grassland), Wetland (30%; sand-dunes/sand beach; shingle/stony beach; coastal lagoon; water-fringe vegetation), Artificial landscape (60%; arable land; perennial crops/orchards/groves)
Land-use Agriculture (60%), Hunting (80%), Urban/industrial/transport (10%)

Birds

Species	Season	Year	Pop min	Pop max	Acc	Criteria
Bubulcus ibis Cattle Egret	W	1997	599	2,026	A	B1i, C3
Netta rufina Red-crested Pochard	W	1997	26	2,600	A	A4i, B1i, C3
Himantopus himantopus Black-winged Stilt	B	1996	156	156	A	B1i, C2
Glareola pratincola Collared Pratincole	B	1996	12	12	A	C6
Charadrius alexandrinus Kentish Plover	R	1996	59	59	A	B2
Chlidonias hybridus Whiskered Tern	B	1996	35	35	A	C6

This is an important site for breeding waders, breeding and wintering herons and wintering ducks. Species of global conservation concern that do not meet IBA criteria: *Aythya nyroca* (five wintering individuals).

Protection status

National None **International** None

Conservation issues

Threats Agricultural intensification/expansion (A), Drainage (A), Extraction industry (C), Filling-in of wetlands (B), Groundwater abstraction (A), Industrialization/urbanization (A), Recreation/tourism (B), Unsustainable exploitation (B)

Threats include intensive hunting in winter, urban development and rubbish dumping, over-exploitation of water and agricultural intensification.

El Moro marshes A4i, B1i, B2, C2, C3, C6 156

Admin region País Valenciano
Coordinates 39°35′N 0°15′W
Altitude 0–10 m **Area** 350 ha

Site description

A small wetland with abundant water-fringe vegetation. Formerly a coastal lagoon, The site now contains salt and freshwater marshes, rice-fields and reedbeds (*Phragmites*).

Habitats Wetland (90%; shingle/stony beach; standing fresh water; standing brackish and salt water; water-fringe vegetation), Artificial landscape (10%; arable land)
Land-use Agriculture (10%), Hunting (90%)

Birds

Species	Season	Year	Pop min	Pop max	Acc	Criteria
Ixobrychus minutus Little Bittern	B	1996	30	—	B	C6
Marmaronetta angustirostris Marbled Duck	B	1996	2	2	A	B2, C6
Netta rufina Red-crested Pochard	B	1996	119	119	A	B1i, B2, C3
Porphyrio porphyrio Purple Gallinule	R	1996	18	18	B	C6
Himantopus himantopus Black-winged Stilt	B	1996	86	300	A	A4i, B1i, C2, C6
Glareola pratincola Collared Pratincole	B	1996	13	15	A	C6
Charadrius alexandrinus Kentish Plover	B	1996	56	56	A	B2
Chlidonias hybridus Whiskered Tern	B	1996	108	108	A	B1i, B2, C2, C6

An important area for breeding herons, ducks and waders.

Protection status

National None **International** High

350 ha of IBA covered by Special Protection Area (Marjal del Moro, 350 ha).

Conservation issues

Threats Agricultural intensification/expansion (B), Filling-in of wetlands (C), Groundwater abstraction (A), Industrialization/urbanization (A), Recreation/tourism (B), Unsustainable exploitation (B)

Industrial development, drainage, dumping and the use of pesticides are the most important threat to the marsh. Illegal hunting and recreational activities also cause disturbance. Ten percent of the site is a controlled hunting area.

Turia canyon and Los Serranos B2, B3, C2, C6 157

Admin region Castilla-La Mancha, País Valenciano
Coordinates 39°45'N 1°00'W
Altitude 400–1,551 m **Area** 135,000 ha

Site description

A mountainous area with large *Pinus* forests and areas of scrub. The main human activities are forestry, hunting and arable cultivation.

Habitats Forest and woodland (50%; native coniferous forest; alluvial/very wet forest; broadleaved evergreen forest), Scrub (30%; scrub; sclerophyllous scrub/garrigue/maquis), Wetland (25%; river/stream), Rocky areas (15%; inland cliff), Artificial landscape (30%; arable land; perennial crops/orchards/groves; forestry plantation)
Land-use Agriculture (15%), Forestry (65%), Hunting (90%), Tourism/recreation, Water management

Birds

Species	Season	Year	Pop min	Pop max	Acc	Criteria
Circaetus gallicus Short-toed Eagle	B	1997	20	30	B	B2, C6
Hieraaetus fasciatus Bonnelli's Eagle	R	1997	10	10	A	B2, C6
Columba oenas Stock Dove	R	1997	400	—	C	B3
Bubo bubo Eagle Owl	R	1997	40	50	B	B2, C6
Merops apiaster Bee-eater	B	1996	300	500	B	B2
Oenanthe leucura Black Wheatear	R	1996	40	—	B	B2
Pyrrhocorax pyrrhocorax Chough	R	1996	150	200	B	B2, C2, C6

An important site for raptors characteristic of Mediterranean habitats.

Protection status

National None **International** None

Conservation issues

Threats Extraction industry (C), Firewood collection (B), Recreation/tourism (C), Selective logging/cutting (B)

The main threats are from forestry activities and visitor pressure.

Rivers Cabriel and Júcar canyons B2, C2, C6 158

Admin region Castilla-La Mancha, País Valenciano
Coordinates 39°20'N 1°08'W
Altitude 200–1,108 m **Area** 163,000 ha

Site description

A mountainous area crossed by the valleys of the Rivers Cabriel and Júcar, with *Pinus* forest and scrub.

Habitats Forest and woodland (70%; native coniferous forest; broadleaved evergreen forest), Scrub (30%; sclerophyllous scrub/garrigue/maquis), Grassland (20%; steppe/dry calcareous grassland), Rocky areas (5%; inland cliff), Artificial landscape (25%; perennial crops/orchards/groves)
Land-use Agriculture (25%), Forestry (50%), Hunting (80%)

Birds

Species	Season	Year	Pop min	Pop max	Acc	Criteria
Circaetus gallicus Short-toed Eagle	B	1997	20	—	C	B2, C6
Aquila chrysaetos Golden Eagle	R	1997	12	—	C	B2, C6
Hieraaetus fasciatus Bonnelli's Eagle	R	1997	11	—	B	B2, C2, C6
Falco peregrinus Peregrine	R	1997	20	—	C	B2, C6
Bubo bubo Eagle Owl	R	1997	25	—	C	B2, C6
Merops apiaster Bee-eater	B	1997	300	—	B	B2
Pyrrhocorax pyrrhocorax Chough	R	1997	100	—	C	B2

This is an important site for cliff-nesting raptors and species characteristic of Mediterranean woodland and scrub.

Protection status

National Partial **International** Low
30,000 ha of IBA covered by National Game Reserve (Muela de Cortes, 36,009 ha). 1,662 ha of IBA covered by Natural Reserve (Hoces del Cabriel en Cuenca, 1,662 ha). 16,089 ha of IBA covered by Special Protection Area (Hoces del Cabriel, Guadazaōn y Ojos de Moya, 56,136 ha).

Conservation issues

Threats Burning of vegetation (A), Infrastructure (B), Intensified forest management (B), Recreation/tourism (U), Unsustainable exploitation (U)

The main threats are from intensive forest management and fires.

Albufera de Valencia marshes A1, A4i, B1i, B2, C1, C2, C3, C6 159

Admin region País Valenciano
Coordinates 39°20'N 0°15'W
Altitude 0–60 m **Area** 21,120 ha

Site description

A freshwater coastal lagoon with abundant emergent vegetation and several islands, and a coastal *Pinus* woodland. The area also includes 14,000 ha of rice cultivation.

Habitats Forest and woodland, Wetland (22%; sand-dunes/sand beach; coastal lagoon; water-fringe vegetation), Artificial landscape (73%; arable land)
Land-use Agriculture (72%), Fisheries/aquaculture (10%), Hunting (60%), Nature conservation/research (25%)

Birds

Species	Season	Year	Pop min	Pop max	Acc	Criteria
[1,2] ***Phalacrocorax carbo*** Cormorant	W	1997	2,382	3,000	A	B1i, C3
Ixobrychus minutus Little Bittern	B	1996	100	100	C	B2, C2, C6
Nycticorax nycticorax Night Heron	B	1995	320	320	A	B2, C2, C6
Ardeola ralloides Squacco Heron	B	1995	47	280	A	A4i, B1i, B2, C2, C6
Bubulcus ibis Cattle Egret	R	1995	3,800	3,800	A	A4i, B1i, C3
Bubulcus ibis Cattle Egret	W	1997	632	3,460	A	A4i, B1i, C3
Egretta garzetta Little Egret	R	1995	1,750	1,750	A	A4i, B1i, C2, C6
Egretta garzetta Little Egret	W	1997	493	2,605	A	A4i, B1i, C2
Ardea purpurea Purple Heron	B	1995	60	60	A	B2, C2, C6
Anas clypeata Shoveler	W	1997	380	28,732	A	A4i, B1i, C3
Marmaronetta angustirostris Marbled Duck	B	1996	3	4	A	B2, C6
Netta rufina Red-crested Pochard	R	1996	104	114	A	B1i, B2, C3
Netta rufina Red-crested Pochard	W	1997	380	11,100	A	A4i, B1i, C3
Himantopus himantopus Black-winged Stilt	B	1996	130	342	A	A4i, B1i, C2, C6
Glareola pratincola Collared Pratincole	P	1996	100	200	B	A4i, B1i, C2
Charadrius alexandrinus Kentish Plover	B	1996	81	135	A	B2
Larus ridibundus Black-headed Gull	W	1997	10,000	58,287	A	B1i, C3
Larus audouinii Audouin's Gull	P	1997	100	400	A	A1, A4i, B1i, C1, C2
Gelochelidon nilotica Gull-billed Tern	B	1996	163	163	A	A4i, B1i, B2, C2, C6
Sterna sandvicensis Sandwich Tern	B	1996	471	471	A	B1i, B2, C2, C6
[2] ***Sterna hirundo*** Common Tern	B	1996	79	1,650	A	B1i, C2, C6
[2] ***Sterna albifrons*** Little Tern	B	1996	12	206	A	B1i, B2, C2, C6
Chlidonias hybridus Whiskered Tern	P	1997	250	—	B	B1i, C2

1. *Ph. c. sinensis.*
2. Large increase.

This is a very important site for breeding and wintering waterbirds.

Protection status

National High **International** High
21,120 ha of IBA covered by Natural Park (Albufera de Valencia, 21,120 ha). 21,000 ha of IBA covered by Ramsar Site (Albufera de Valencia, 21,000 ha). 21,000 ha of IBA covered by Special Protection Area (Albufera de Valencia, 21,000 ha).

Conservation issues

Threats Agricultural intensification/expansion (B), Dredging/canalization (B), Filling-in of wetlands (C), Industrialization/urbanization (A), Infrastructure (C), Recreation/tourism (B), Unsustainable exploitation (B)

Pollution from agricultural, urban and industrial sources is a serious problem. Other threats include illegal drainage of rice cultivation, high levels of hunting, urban development and tourism development.

La Safor and North Alicante mountain ranges — B2, B3, C6 — 160

Admin region País Valenciano
Coordinates 38°45′N 0°31′W
Altitude 230–1,390 m **Area** 102,000 ha

Site description

An area of mountains between the provinces of Valencia and Alicante with well conserved Mediterranean forest. Human activities include forestry, hunting and agriculture.

Habitats Forest and woodland (35%; native coniferous forest; broadleaved evergreen forest), Scrub (50%; sclerophyllous scrub/garrigue/maquis), Grassland (30%; steppe/dry calcareous grassland), Rocky areas (5%; scree/boulders; inland cliff), Artificial landscape (20%; perennial crops/orchards/groves)
Land-use Agriculture (20%), Forestry (30%), Hunting (75%)

Birds

Species	Season	Year	Pop min	Pop max	Acc	Criteria
Aquila chrysaetos Golden Eagle	R	1995	5	—	A	C6
[1] ***Hieraaetus fasciatus*** Bonnelli's Eagle	R	1995	8	9	A	B2, C6
Falco peregrinus Peregrine	R	1995	10	—	A	C6
Columba oenas Stock Dove	R	1995	360	—	C	B3
Bubo bubo Eagle Owl	R	1995	12	—	B	C6
Merops apiaster Bee-eater	B	1995	230	—	C	B2
Pyrrhocorax pyrrhocorax Chough	R	1995	100	—	C	B2

1. Large decrease.

This is an important site for species typical of Mediterranean habitats and cliff-nesting raptors.

Protection status

National None **International** Low
2,450 ha of IBA covered by Natural Park (Carrascal de la Font Roja, 2,450 ha).

Conservation issues

Threats Abandonment/reduction of land management (B), Burning of vegetation (A), Intensified forest management (B), Unsustainable exploitation (U)

Forestry exploitation and forest fires are the most important threats. A management plan for the Natural Park has existed since 1993.

Enguera mountain range–La Canal de Navarrés — B2, C6 — 161

Admin region País Valenciano
Coordinates 39°00′N 0°48′W
Altitude 300–1,126 m **Area** 52,500 ha

Site description

A mountainous area dominated by *Pinus* forest and scrub, degraded by forest fires.

Habitats Forest and woodland (50%; native coniferous forest), Scrub (50%; sclerophyllous scrub/garrigue/maquis), Grassland (30%; steppe/dry calcareous grassland), Rocky areas (5%; inland cliff), Artificial landscape (10%; perennial crops/orchards/groves)
Land-use Agriculture (10%), Forestry (50%), Hunting (80%)

Birds

Species	Season	Year	Pop min	Pop max	Acc	Criteria
Circaetus gallicus Short-toed Eagle	B	1997	10	—	B	C6
[1] ***Hieraaetus fasciatus*** Bonnelli's Eagle	R	1997	4	5	A	B2, C6
Merops apiaster Bee-eater	B	1997	230	—	B	B2

1. Large decrease.

This is an important area for breeding raptors.

Protection status

National None **International** None

Conservation issues

Threats Burning of vegetation (A), Intensified forest management (B), Unsustainable exploitation (U)

The most important threats are forest fires and illegal hunting.

Pego–Oliva marshes — B1i, B2, C2, C6 — 162

Admin region País Valenciano
Coordinates 38°50′N 0°05′W
Altitude 0–87 m **Area** 1,290 ha

Site description

An area of reedbed (*Phragmites*) and humid grassland that is prone to flooding. Hunting and agriculture are the main human activities.

Habitats Scrub (10%; sclerophyllous scrub/garrigue/maquis), Grassland (15%; humid grassland), Wetland (55%; standing fresh water; river/stream; water-fringe vegetation), Artificial landscape (20%; perennial crops/orchards/groves)
Land-use Agriculture (20%), Hunting (100%)

Birds

Species	Season	Year	Pop min	Pop max	Acc	Criteria
Ixobrychus minutus Little Bittern	R	1995	30	50	B	B2, C6
Ardea purpurea Purple Heron	R	1996	12	15	B	C6
Marmaronetta angustirostris Marbled Duck	B	1996	3	3	B	B2, C6
[1] ***Himantopus himantopus*** Black-winged Stilt	B	1996	13	166	A	B1i, C2, C6

1. Large decrease.

This is an important site for breeding herons and waders and wintering and passage waterbirds.

Protection status

National High **International** High
1,290 ha of IBA covered by Natural Park (Marjal de Pego–Oliva, 1,290 ha). 1,290 ha of IBA covered by Special Protection Area (Marjal de Pego–Oliva, 1,290 ha).

Conservation issues

Threats Agricultural intensification/expansion (C), Drainage (A), Dredging/canalization (B), Groundwater abstraction (B), Unsustainable exploitation (A)

The main threats are from water management and abstraction by local farmers and the high hunting pressure.

Marina mountain ranges — B2, C2, C6 — 163

Admin region País Valenciano
Coordinates 38°40′N 0°15′W
Altitude 0–1,558 m **Area** 102,300 ha

Site description

A mountainous area covered by *Pinus* forests, Mediterranean scrub and fruit cultivation.

Habitats Forest and woodland (25%; native coniferous forest; broadleaved evergreen forest), Scrub (50%; scrub; sclerophyllous scrub/garrigue/maquis), Grassland (20%; steppe/dry calcareous grassland), Rocky areas (10%; sea cliff/rocky shore; inland cliff), Artificial landscape (20%; perennial crops/orchards/groves)
Land-use Agriculture (20%), Forestry (20%), Hunting (75%)

Birds

Species	Season	Year	Pop min	Pop max	Acc	Criteria
Aquila chrysaetos Golden Eagle	R	1996	4	5	A	C6
[1] ***Hieraaetus fasciatus*** Bonnelli's Eagle	R	1997	4	5	A	B2, C6
Falco peregrinus Peregrine	R	1996	12	—	B	C6
Merops apiaster Bee-eater	B	1996	300	—	B	B2
Oenanthe leucura Black Wheatear	R	1996	300	—	C	B2, C2
Monticola solitarius Blue Rock Thrush	R	1996	200	—	B	B2
Pyrrhocorax pyrrhocorax Chough	R	1996	400	—	B	B2, C2, C6

1. Large decrease.

This is an important site for raptors and species characteristic of Mediterranean habitats.

Protection status

National None **International** None

Conservation issues

Threats Burning of vegetation (B), Infrastructure (B), Unsustainable exploitation (B)

The main threats are from forest fires and a high hunting pressure.

Alicante islets — B1ii, B2, C2, C6 — 164

Admin region País Valenciano
Coordinates 38°30′N 0°09′E
Altitude 0–73 m **Area** 40 ha

Site description

Three small rocky islets near the coast of Alicante with cliffs and some areas of scrub.

Habitats Scrub (50%; scrub), Rocky areas (100%; sea cliff/rocky shore; rock stacks/islets)
Land-use Fisheries/aquaculture

Birds

Species	Season	Year	Pop min	Pop max	Acc	Criteria
Hydrobates pelagicus Storm Petrel	B	1993	350	460	B	B1ii, B2, C2, C6

This is an important site for breeding seabirds and migrant passerines.

Protection status

National High **International** Partial
40 ha of IBA covered by Natural Reserve (Isla de Tabarca, 50 ha). 7 ha of IBA covered by Special Protection Area (Islote de Benidorm y Peñas de Arabi, 7 ha).

Conservation issues

Threats Recreation/tourism (A)

The most important threat is disturbance from fishermen and tourists.

El Hondo wetland — A1, A4i, B1i, B2, C1, C2, C3, C6 — 165

Admin region País Valenciano
Coordinates 38°20′N 0°42′W
Altitude 0–10 m **Area** 2,387 ha

Site description

Two reservoirs with reedbeds (*Phragmites*) and surrounded by small brackish and freshwater lakes. The main human activities are hunting and fishing.

Habitats Wetland (95%; standing fresh water; standing brackish and salt water; water-fringe vegetation), Artificial landscape (5%; arable land)
Land-use Agriculture (5%), Fisheries/aquaculture (60%), Hunting (80%), Nature conservation/research (10%), Water management (60%)

Birds

Species	Season	Year	Pop min	Pop max	Acc	Criteria
Ixobrychus minutus Little Bittern	B	1995	35	106	A	B2, C2, C6
Nycticorax nycticorax Night Heron	B	1993	40	40	A	C6
Ardeola ralloides Squacco Heron	B	1993	25	25	A	C2, C6
Ardea purpurea Purple Heron	B	1995	22	26	A	C6
Phoenicopterus ruber Greater Flamingo	B	1997	300	300	—	A4i, B1i, B2, C2, C6
Phoenicopterus ruber Greater Flamingo	W	1997	50	2,150	A	A4i, B1i, C2
[1] ***Anas clypeata*** Shoveler	W	1997	620	10,040	A	A4i, B1i, C3
[1] ***Marmaronetta angustirostris*** Marbled Duck	R	1996	35	47	A	A1, A4i, B1i, B2, C1, C2, C6
[1] ***Marmaronetta angustirostris*** Marbled Duck	W	1996	—	486	A	A1, A4i, B1i, C1, C2
Netta rufina Red-crested Pochard	R	1995	21	78	A	B2
Netta rufina Red-crested Pochard	W	1997	55	915	A	A4i, B1i, C3

Species ... continued	Season	Year	Pop min	Pop max	Acc	Criteria
[1] ***Oxyura leucocephala*** White-headed Duck	R	1996	10	15	A	A1, B1i, B2, C1, C2, C6
[1] ***Oxyura leucocephala*** White-headed Duck	W	1997	97	155	A	A1, A4i, B1i, C1, C2
[1] ***Himantopus himantopus*** Black-winged Stilt	B	1995	301	331	A	A4i, B1i, C2, C6
Recurvirostra avosetta Avocet	R	1997	40	53	A	C6
Recurvirostra avosetta Avocet	W	1997	10	712	A	B1i, B2, C2
Charadrius alexandrinus Kentish Plover	R	1995	108	136	A	B2
[2] ***Sterna albifrons*** Little Tern	B	1995	4	75	A	B2, C2, C6
Chlidonias hybridus Whiskered Tern	B	1995	219	219	A	B1i, B2, C2, C6

1. Large increase.
2. Large decrease.

This is a very important wetland for breeding and wintering waterbirds.

Protection status

National High **International** High
2,387 ha of IBA covered by Natural Park (El Hondo, 2,387 ha). 2,387 ha of IBA covered by Ramsar Site (El Hondo, 2,387 ha). 2,387 ha of IBA covered by Special Protection Area (El Hondo, 2,387 ha).

Conservation issues

Threats Agricultural intensification/expansion (B), Consequences of animal/plant introductions (A), Drainage (B), Filling-in of wetlands (B), Natural events (A), Other (A), Recreation/tourism (C), Unsustainable exploitation (A)

Threats include a channel in which juvenile ducks become trapped every year, lead poisoning from spent shot ('Other' threats, above) and a recent severe drought. *Oxyura jamaicensis* has been hybridizing with *O. leucocephala* since 1992, although all hybrids have been killed. Conservation NGOs have successfully campaigned for a hunting ban although there are still some problems.

Santa Pola saltpans — A1, A4i, B1i, B2, B3, C1, C2, C3, C6 — 166

Admin region País Valenciano
Coordinates 38°13′N 0°35′W
Altitude 0–10 m **Area** 2,496 ha

Site description

An area of saltpans surrounded by artificial freshwater and brackish ponds with dense plant cover.

Habitats Wetland (100%; saltmarsh; sand-dunes/sand beach; standing fresh water; standing brackish and salt water; water-fringe vegetation)
Land-use Hunting (90%), Nature conservation/research (5%), Urban/industrial/transport (50%)

Birds

Species	Season	Year	Pop min	Pop max	Acc	Criteria
Nycticorax nycticorax Night Heron	B	1995	21	21	A	C6
Ardeola ralloides Squacco Heron	B	1995	4	4	A	B2
Ardea purpurea Purple Heron	B	1995	26	26	A	C6
Phoenicopterus ruber Greater Flamingo	W	1997	614	1,922	A	A4i, B1i, C2
Marmaronetta angustirostris Marbled Duck	R	1996	6	10	A	A1, A4i, B1i, B2, C1, C2, C6
Netta rufina Red-crested Pochard	R	1995	321	321	A	A4i, B1i, B2, C3
Netta rufina Red-crested Pochard	W	1997	32	421	A	B1i, C3
Recurvirostra avosetta Avocet	R	1995	593	593	A	A4i, B1i, B3, C2, C6
Recurvirostra avosetta Avocet	W	1997	875	1,595	A	A4i, B1i, B2, C2
Charadrius alexandrinus Kentish Plover	R	1995	328	—	—	A4i, B1i, B2, C3
Sterna hirundo Common Tern	B	1995	174	174	A	C6
Sterna albifrons Little Tern	B	1995	148	—	A	B1i, B2, C2, C6
Chlidonias hybridus Whiskered Tern	B	1995	80	90	A	B1i, B2, C2, C6

This is an important site for breeding and wintering waterbirds.

Protection status

National High **International** High
2,496 ha of IBA covered by Natural Park (Salinas de Santa Pola, 2,496 ha). 2,496 ha of IBA covered by Ramsar Site (Salinas de Santa Pola, 2,496 ha). 2,496 ha of IBA covered by Special Protection Area (Salinas de Santa Pola, 2,496 ha).

Conservation issues

Threats Aquaculture/fisheries (C), Industrialization/urbanization (C), Infrastructure (C), Recreation/tourism (B), Unsustainable exploitation (A)

Threats come from the high hunting pressure, a central access road and urbanization along the beach zone. A management plan for the Natural Park has existed since 1994.

Mata and Torrevieja lagoons — A1, A4i, B1i, B2, C1, C2, C3, C6 — 167

Admin region País Valenciano
Coordinates 38°03'N 0°40'W
Altitude 0–10 m **Area** 3,755 ha

Site description

Two coastal lagoons which have been transformed for salt extraction. There is a narrow belt of reed (*Phragmites*) and halophytic vegetation.

Habitats Scrub (10%; scrub), Wetland (90%; standing brackish and salt water; water-fringe vegetation)
Land-use Not utilized, Urban/industrial/transport (40%), Water management (20%)

Birds

Species	Season	Year	Pop min	Pop max	Acc	Criteria
Podiceps nigricollis Black-necked Grebe	W	1997	844	2,100	A	A4i, B1i, C3
Himantopus himantopus Black-winged Stilt	B	1995	40	140	A	B1i, C2
Recurvirostra avosetta Avocet	B	1995	78	83	A	C6
Charadrius alexandrinus Kentish Plover	R	1995	210	222	A	B1i, B2, C3
[1] ***Larus audouinii*** Audouin's Gull	W	1997	482	1,900	A	A1, A4i, B1i, C1, C2
Sterna albifrons Little Tern	B	1995	211	223	A	B1i, B2, C2, C6
Chlidonias hybridus Whiskered Tern	B	1991	57	78	A	B2, C6

1. Large increase.

The site is important for breeding, wintering and passage waterbirds. The site acts as an important refuge when there is hunting in the El Hondo (215) and Santa Pola (216) IBAs.

Protection status

National High **International** High
3,755 ha of IBA covered by Natural Park (Lagunas de la Mata y Torrevieja, 3,755 ha). 3,700 ha of IBA covered by Ramsar Site (Lagunas de la Mata y Torrevieja, 3,700 ha). 2,100 ha of IBA covered by Special Protection Area (Salinas de la Mata y Torrevieja, 2,100 ha).

Conservation issues

Threats Industrialization/urbanization (C), Infrastructure (C), Recreation/tourism (B)

Threats include building construction, water pollution and disturbance from the saltworks and from tourists. A management plan for the Natural Park has existed since 1995.

El Valle, Altaona and Escalona mountains — B1iii, C2, C6 — 168

Admin region Murcia, País Valenciano
Coordinates 38°00'N 0°55'W
Altitude 200–647 m **Area** 24,000 ha

Site description

The site includes both mountains and plains south-east of the city of Murcia. The vegetation is mainly *Pinus* forest, scrub and steppe habitats.

Habitats Forest and woodland (native coniferous forest), Scrub (*; sclerophyllous scrub/garrigue/maquis), Grassland (steppe/dry calcareous grassland), Desert (*; semi-desert), Rocky areas (inland cliff), Artificial landscape (*; arable land)
Land-use Agriculture, Hunting (*), Tourism/recreation

Birds

This is an important area for dispersing juvenile raptors and also for steppic species. Species of global conservation concern that do not meet IBA criteria: *Aegypius monachus* (passage), *Aquila clanga* (passage) and *Falco naumanni* (14 wintering individuals).

Species	Season	Year	Pop min	Pop max	Acc	Criteria
[1] ***Hieraaetus pennatus*** Booted Eagle	R	1996	6	8	B	C6
[1,2] ***Hieraaetus pennatus*** Booted Eagle	W	1996	25	100	B	B1iii, C2
[1,2] ***Hieraaetus fasciatus*** Bonnelli's Eagle	N	1996	15	50	A	B1iii, C2
Falco naumanni Lesser Kestrel	B	1996	6	6	A	C6
Burhinus oedicnemus Stone Curlew	R	1996	30	50	C	C6
Bubo bubo Eagle Owl	R	1996	9	10	A	C6
Coracias garrulus Roller	B	1996	5	10	C	C6

1. Large increase.
2. Juveniles.

Protection status

National Low **International** None
2,000 ha of IBA covered by Regional Park (El Valle, 10,500 ha).

Conservation issues

Threats Agricultural intensification/expansion (B), Industrialization/urbanization (B), Infrastructure (A), Other (A)

The main threats are from plans for irrigation schemes, urbanization, the use of poison ('Other' threats, above) and powerlines.

Mar Menor coastal lagoon — A1, A4i, B1i, B2, B3, C1, C2, C3, C6 — 169

Admin region Murcia
Coordinates 37°45'N 0°45'W
Altitude 0–5 m **Area** 14,933 ha

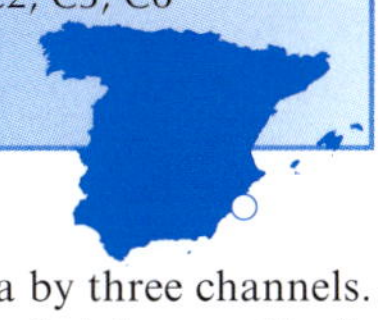

Site description

A large saline coastal lagoon connected to the sea by three channels. There are several areas of saltpans surrounded by reedbeds (*Phragmites*) and saltmarshes.

Habitats Wetland (100%; mudflat/sandflat; saltmarsh; sand-dunes/sand beach; coastal lagoon; water-fringe vegetation)
Land-use Fisheries/aquaculture (90%), Military (10%), Tourism/recreation (95%)

Birds

Species	Season	Year	Pop min	Pop max	Acc	Criteria
[1] ***Podiceps nigricollis*** Black-necked Grebe	W	1990	263	2,000	A	A4i, B1i, C3
Marmaronetta angustirostris Marbled Duck	B	1995	0	3	C	B2, C6
[1] ***Himantopus himantopus*** Black-winged Stilt	B	1994	68	68	A	C6
[1] ***Recurvirostra avosetta*** Avocet	B	1995	126	126	A	B3, C6
Charadrius alexandrinus Kentish Plover	B	1994	155	155	A	B2
[2] ***Larus audouinii*** Audouin's Gull	W	1996	96	96	A	A1, C1
Gelochelidon nilotica Gull-billed Tern	B	1995	25	25	A	B2, C2
Sterna albifrons Little Tern	B	1994	138	138	A	B1i, B2, C2, C6

1. Large increase.
2. Large decrease.

This is an important site for breeding and wintering ducks, waders and seabirds. Species of global conservation concern that do not meet IBA criteria: *Marmaronetta angustirostris* (wintering).

Protection status

National Partial **International** Partial
856 ha of IBA covered by Regional Park (Salinas y arenales de San Pedro del Pinatar, 856 ha). 50 ha of IBA covered by Regional Park (Calblanque, Monte de las Cenizas y Peña del Águila, 2,453 ha). 900 ha of IBA covered by Protected Landscape (Espacios abiertos e islas del Mar Menor, 1,154 ha). 14,933 ha of IBA covered by Ramsar Site (Mar Menor, 14,933 ha).

Conservation issues

Threats Agricultural intensification/expansion (B), Aquaculture/fisheries (C), Drainage (B), Dredging/canalization (A), Filling-in of wetlands (C), Groundwater abstraction (B), Industrialization/urbanization (A), Recreation/tourism (B)

The most important threats are urbanization, particularly along the main sandbar, in-filling, dredging and canalization, road building and disturbance caused by tourists. A management plan is being prepared and a LIFE project has examined conservation of the area.

Littoral islets of Murcia and Almería — A1, A4i, B1i, B2, C1, C2, C6 — 170

Admin region Andalucía, Murcia
Coordinates 37°35'N 1°00'W
Altitude 0–100 m **Area** 300 ha

Site description
This site includes six small, rocky, sparsely vegetated islets, close to the coast of Murcia and Almería.

Habitats Rocky areas (100%; rock stacks/islets)
Land-use Military, Nature conservation/research

Birds

Species	Season	Year	Pop min	Pop max	Acc	Criteria
[1] ***Calonectris diomedea*** Cory's Shearwater	B	1996	100	100	A	C6
[2] ***Hydrobates pelagicus*** Storm Petrel	B	1996	29	—	A	C6
Larus audouinii Audouin's Gull	R	1996	450	450	A	A1, A4i, B1i, B2, C1, C2, C6

1. Large increase.
2. Large decrease.

This is an important site for breeding seabirds.

Protection status
National None **International** None

Conservation issues

Threats Disturbance to birds (A), Other (A)

The main threat is from disturbance from tourists and from military manoeuvres. There is competition between *Larus cachinnans* and *L. audouinii* ('Other' threats, above).

Fausilla mountains — B2, C6 — 171

Admin region Murcia
Coordinates 37°35'N 0°50'W
Altitude 0–368 m **Area** 900 ha

Site description
An area of coastal hills with dry scrub and rocky areas.

Habitats Desert (*; semi-desert), Rocky areas (*; sea cliff/rocky shore; inland cliff), Artificial landscape (other urban/industrial areas)
Land-use Hunting, Urban/industrial/transport

Birds

Species	Season	Year	Pop min	Pop max	Acc	Criteria
Bucanetes githagineus Trumpeter Finch	R	1996	15	—	C	B2, C6

This is an important site for species associated with Mediterranean scrub.

Protection status
National None **International** None

Conservation issues

Threats Industrialization/urbanization (A)

The expansion of an industrial area within the site is the main threat.

Muela mountain range and Cape Tiñoso — B2, C6 — 172

Admin region Murcia
Coordinates 37°35'N 1°11'W
Altitude 0–629 m **Area** 11,700 ha

Site description
An area of semi-arid coastal mountains dominated by dry scrub. There are also some small *Pinus* woodlands, non-irrigated crops, ruderal areas and crops under plastic.

Habitats Forest and woodland (native coniferous forest), Scrub (*; scrub; sclerophyllous scrub/garrigue/maquis), Grassland (steppe/dry calcareous grassland), Rocky areas (sea cliff/rocky shore; inland cliff), Artificial landscape (*; arable land; perennial crops/orchards/groves; other urban/industrial areas; ruderal land)
Land-use Agriculture (*), Hunting (*), Tourism/recreation, Urban/industrial/transport

Birds

Species	Season	Year	Pop min	Pop max	Acc	Criteria
Hieraaetus fasciatus Bonnelli's Eagle	R	1996	2	2	A	C6
Falco peregrinus Peregrine	R	1996	13	13	A	C6
Bucanetes githagineus Trumpeter Finch	R	1996	5	—	B	B2, C6

This is an important area for species associated with Mediterranean scrub.

Protection status
National None **International** None

Conservation issues

Threats Agricultural intensification/expansion (B), Extraction industry (C), Industrialization/urbanization (B), Other (A), Recreation/tourism (B)

The main threats are from urbanization, particularly of the cape area, new quarries, disturbance from recreational activities and the use of poison ('Other' threats, above).

La Almenara mountain range–Cape Cope — B2, C6 — 173

Admin region Murcia
Coordinates 37°30'N 1°30'W
Altitude 0–888 m **Area** 25,850 ha

Site description
Low mountains including a coastal section. The main habitats are scrub, orchards, *Pinus halepensis* woodland and arable crops. Human activities include hunting, forestry and agriculture.

Habitats Forest and woodland (*; native coniferous forest), Scrub (scrub; sclerophyllous scrub/garrigue/maquis), Grassland (*; steppe/dry calcareous grassland), Desert (semi-desert), Rocky areas (*; sea cliff/rocky shore; inland cliff), Artificial landscape (arable land; perennial crops/orchards/groves; ruderal land)
Land-use Agriculture, Forestry (*), Hunting (*)

Birds

Species	Season	Year	Pop min	Pop max	Acc	Criteria
Hieraaetus fasciatus Bonnelli's Eagle	R	1996	4	4	A	C6
Bubo bubo Eagle Owl	R	1991	10	10	B	C6
Pyrrhocorax pyrrhocorax Chough	R	1992	100	—	C	B2, C6
Bucanetes githagineus Trumpeter Finch	R	1996	Common	—		B2, C6

This is an important area for raptors and steppic species. The site holds a new population of *Bucanetes githagineus*.

Protection status
National Low **International** None
250 ha of IBA covered by Regional Park (Cabo Cope–Puntas Calnegre, 2,936 ha). 1,960 ha of IBA covered by Protected Landscape (Sierra de las Moreras, 1,960 ha).

Conservation issues

Threats Agricultural intensification/expansion (B), Groundwater abstraction (B), Infrastructure (C), Unsustainable exploitation (C)

The main threats are from the abandonment of traditional crops and an increase use of crops grown under plastic, increasing use of game fences and the creation of new tracks.

Torrecilla and Gigante mountain ranges
B2, C6 **174**

Admin region Murcia
Coordinates 37°45′N 1°51′W
Altitude 400–1,494 m **Area** 42,000 ha

Site description
Mountains, in the west of the Murcia region, with *Pinus, Quercus* and *Juniperus* woodland, steppic habitats and non-irrigated crops. Agriculture, forestry, mining and hunting are the most important human activities.

Habitats Forest and woodland (*; native coniferous forest), Scrub (*; scrub; sclerophyllous scrub/garrigue/maquis), Grassland (steppe/dry calcareous grassland), Rocky areas (inland cliff), Artificial landscape (perennial crops/orchards/groves; forestry plantation; ruderal land)
Land-use Agriculture (*), Forestry, Hunting (*)

Birds

Species	Season	Year	Pop min	Pop max	Acc	Criteria
Circaetus gallicus Short-toed Eagle	B	1991	11	11	A	C6
Aquila chrysaetos Golden Eagle	R	1996	7	8	A	C6
Hieraaetus pennatus Booted Eagle	B	1991	8	8	A	C6
Falco peregrinus Peregrine	R	1991	8	8	A	C6
Bubo bubo Eagle Owl	R	1995	15	—	B	C6
Pyrrhocorax pyrrhocorax Chough	R	1992	100	—	C	B2, C6

This is an important area for cliff-nesting birds. Species of global conservation concern that do not meet IBA criteria: *Falco naumanni* (breeding).

Protection status
National None **International** None

Conservation issues

Threats Agricultural intensification/expansion (B), Extraction industry (B), Groundwater abstraction (C), Recreation/tourism (B)

The main threats are from the expansion of crops grown under plastic, disturbance from sporting activities, the expansion of mining, and potential groundwater extraction for crop irrigation.

Guadalentín saltmarshes
C6 **175**

Admin region Murcia
Coordinates 37°45′N 1°25′W
Altitude 100–193 m **Area** 6,800 ha

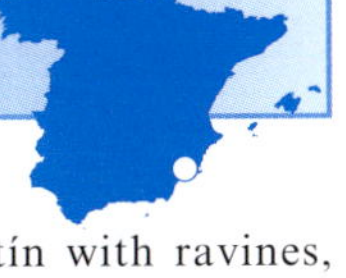

Site description
Wide plains both sides of the River Guadalentín with ravines, saltmarshes and small expanses of unploughed saline ground. There are also orchards, areas of cereal cultivation and almond-groves.

Habitats Scrub (10%; sclerophyllous scrub/garrigue/maquis), Grassland (20%; steppe/dry calcareous grassland; humid grassland), Wetland (5%; saltmarsh; water-fringe vegetation), Artificial landscape (65%; arable land; perennial crops/orchards/groves; forestry plantation; other urban/industrial areas; ruderal land)
Land-use Agriculture (60%), Hunting (80%), Military (30%), Urban/industrial/transport (10%), Water management

Birds

Species	Season	Year	Pop min	Pop max	Acc	Criteria
Tetrax tetrax Little Bustard	B	1995	6	58	A	C6
Himantopus himantopus Black-winged Stilt	B	1995	52	52	A	C6
Burhinus oedicnemus Stone Curlew	R	1995	37	37	A	C6
Pterocles orientalis Black-bellied Sandgrouse	R	1995	23	30	A	C6
Coracias garrulus Roller	B	1994	15	25	B	C6

This is an important area for breeding steppe birds, although numbers are declining. Species of global conservation concern that do not meet IBA criteria: *Marmaronetta angustirostris* (2–7 passage individuals) and *Falco naumanni* (1–5 passage individuals).

Protection status
National None **International** None

Conservation issues

Threats Agricultural intensification/expansion (A), Filling-in of wetlands (B), Groundwater abstraction (A), Industrialization/urbanization (A), Infrastructure (A)

The most important threats are from crop irrigation, agricultural expansion and industrial developments.

Las Cabras plain
B2, C6 **176**

Admin region Murcia
Coordinates 37°50′N 1°40′W
Altitude 500–700 m **Area** 1,300 ha

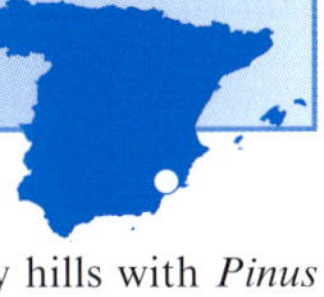

Site description
A plain dominated by dry scrub and surrounded by hills with *Pinus* forests. The main human activity is sheep-farming.

Habitats Scrub (*; scrub; sclerophyllous scrub/garrigue/maquis), Grassland (*; steppe/dry calcareous grassland)
Land-use Agriculture (*), Hunting, Not utilized

Birds

Species	Season	Year	Pop min	Pop max	Acc	Criteria
Chersophilus duponti Dupont's Lark	R	1996	118	130	A	B2, C6

This is an important site for species associated with open habitats, particularly *Chersophilus duponti.*

Protection status
National None **International** None

Conservation issues

Threats Afforestation (U), Agricultural intensification/expansion (B)

The main threats are from the ploughing of steppe habitat and the exploitation of the forested areas.

Burete, Cambrón and Espuña mountain ranges
B2, C6 **177**

Admin region Murcia
Coordinates 38°00′N 1°45′W
Altitude 500–1,585 m **Area** 47,000 ha

Site description
A mountainous area with cliffs and a large area of *Pinus* forest. At lower altitudes dry scrub dominates. The main human activities are forestry, hunting and mining.

Habitats Forest and woodland (*; native coniferous forest), Scrub (*; sclerophyllous scrub/garrigue/maquis), Grassland (steppe/dry calcareous grassland), Rocky areas (inland cliff), Artificial landscape (*; forestry plantation)
Land-use Forestry (*), Hunting, Urban/industrial/transport

Birds

Species	Season	Year	Pop min	Pop max	Acc	Criteria
Circaetus gallicus Short-toed Eagle	R	1996	20	—	A	B2, C6
Aquila chrysaetos Golden Eagle	R	1996	6	8	A	C6
Hieraaetus pennatus Booted Eagle	R	1996	30	40	B	B2, C6

Species ... continued	Season	Year	Pop min	Pop max	Acc	Criteria
Falco peregrinus Peregrine	R	1996	10	15	A	C6
Bubo bubo Eagle Owl	R	1996	20	20	A	C6
Pyrrhocorax pyrrhocorax Chough	R	1996	100	—	B	B2, C6

This is a very important site for forest- and cliff-nesting raptors.

Protection status

National Partial **International** None

13,855 ha of IBA covered by National Game Reserve (Sierra Espuña, 13,855 ha). 10,786 ha of IBA covered by Regional Park (Sierra Espuña, 10,786 ha).

Conservation issues

Threats Agricultural intensification/expansion (C), Extraction industry (B), Intensified forest management (B)

The main threats are from mining and forestry operations. Ploughing, in the lower parts of the site threatens *Chersophilus duponti*.

Quipar reservoir, Molino mountain range and Cagitán plains — B2, C6 — 178

Admin region Murcia
Coordinates 38°15′N 1°40′W
Altitude 400–826 m **Area** 22,000 ha

Site description

An area of low semi-arid hills with arable crops, some scrub and reedbeds (*Phragmites*) around a reservoir.

Habitats Scrub (*; scrub), Wetland, Rocky areas (inland cliff), Artificial landscape (*; arable land; perennial crops/orchards/groves)
Land-use Agriculture (*), Forestry, Hunting (*), Water management

Birds

Species	Season	Year	Pop min	Pop max	Acc	Criteria
Circus pygargus Montagu's Harrier	B	1996	6	—	C	C6
Burhinus oedicnemus Stone Curlew	B	1996	220	—	C	B2, C6

This is an important site for species characteristic of open and scrub-dominated habitats. Species of global conservation concern that do not meet IBA criteria: *Falco naumanni* (breeding) and *Tetrax tetrax* (breeding).

Protection status

National None **International** None

Conservation issues

Threats Abandonment/reduction of land management (B)

The abandonment of arable cultivation threatens typical steppe species.

Ricote and Nevela mountain ranges — B2, C6 — 179

Admin region Murcia
Coordinates 38°15′N 1°15′W
Altitude 100–1,121 m **Area** 9,500 ha

Site description

A group of low mountains dominated by *Pinus* woodland and maquis and with many cliffs and ravines. The main human activities are hunting and forestry.

Habitats Forest and woodland (*; native coniferous forest), Scrub (*; sclerophyllous scrub/garrigue/maquis), Rocky areas (inland cliff)
Land-use Forestry (*), Hunting (*)

Birds

This is an important area for cliff-nesting raptors.

Species	Season	Year	Pop min	Pop max	Acc	Criteria
Falco peregrinus Peregrine	R	1996	18	18	A	B2, C6

Protection status

National None **International** None

Conservation issues

Threats Unsustainable exploitation (A)

The main threat is from high hunting pressure, and poaching of chicks.

Pila mountain range — B2, C6 — 180

Admin region Murcia
Coordinates 38°14′N 1°16′W
Altitude 300–1,023 m **Area** 8,900 ha

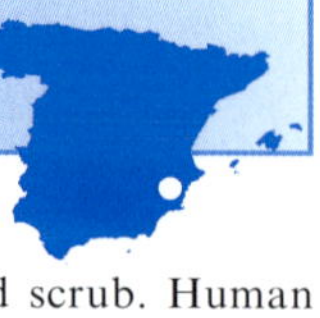

Site description

A mountainous area covered by *Pinus* forest and scrub. Human activities include forestry, mining and hunting.

Habitats Forest and woodland (*; native coniferous forest), Scrub (*; sclerophyllous scrub/garrigue/maquis), Rocky areas (inland cliff), Artificial landscape (*; forestry plantation)
Land-use Forestry (*), Hunting, Urban/industrial/transport

Birds

Species	Season	Year	Pop min	Pop max	Acc	Criteria
Aquila chrysaetos Golden Eagle	R	1996	5	5	A	C6
Falco peregrinus Peregrine	R	1996	9	9	A	C6
Pyrrhocorax pyrrhocorax Chough	R	1996	100	—	A	B2, C6

This is an important site for cliff-nesting birds.

Protection status

National Partial **International** None

7,858 ha of IBA covered by Regional Park (Sierra de la Pila, 7,858 ha).

Conservation issues

Threats Extraction industry (B), Intensified forest management (B), Other (A), Unsustainable exploitation (A)

The main threats are from mining, forestry, hunting and the use of poison ('Other' threats, above).

Moratalla mountain range — C6 — 181

Admin region Murcia
Coordinates 38°15′N 2°00′W
Altitude 500–1,475 m **Area** 21,000 ha

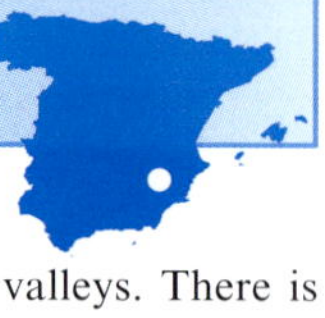

Site description

A group of limestone mountains crossed by deep valleys. There is cultivation in the valleys and mixed forest, *Quercus* forest and scrub in the mountains. The main human activity is livestock-farming.

Habitats Forest and woodland (*; native coniferous forest; broadleaved evergreen forest), Scrub (*; sclerophyllous scrub/garrigue/maquis), Rocky areas (inland cliff), Artificial landscape (*; arable land)
Land-use Agriculture (*)

Birds

Species	Season	Year	Pop min	Pop max	Acc	Criteria
Circaetus gallicus Short-toed Eagle	B	1994	7	7	A	C6
Aquila chrysaetos Golden Eagle	R	1996	5	5	A	C6
Falco peregrinus Peregrine	R	1996	10	10	A	C6

This is an important area for species characteristic of Mediterranean forest and scrub.

■ Protection status

National None **International** None

■ Conservation issues

Threats Burning of vegetation (A), Unsustainable exploitation (A)

The main threat is from forest fires. Persecution of *Hieraaetus fasciatus* has led to its extinction in the site.

Pétrola–Almansa–Yecla A1, B1i, B2, C1, C2, C6 182

Admin region Murcia, Castilla-La Mancha
Coordinates 38°45'N 1°20'W
Altitude 800–1,000 m **Area** 75,800 ha

■ Site description

An undulating plain dominated by arable agriculture with scrub, grassland and small areas of *Pinus* woodland and endorreic lagoons.

Habitats Scrub (scrub), Grassland (steppe/dry calcareous grassland), Wetland (standing fresh water), Artificial landscape (*; arable land; perennial crops/orchards/groves; ruderal land)
Land-use Agriculture (*), Hunting (*)

■ Birds

Species	Season	Year	Pop min	Pop max	Acc	Criteria
[1] ***Tetrax tetrax*** Little Bustard	R	1993	1,800	—	B	A1, B2, C1, C6
Otis tarda Great Bustard	R	1994	176	176	A	A1, B2, C1, C2, C6
Himantopus himantopus Black-winged Stilt	B	1992	188	—	A	B1i, C2, C6
Recurvirostra avosetta Avocet	B	1992	62	—	A	C6
Pterocles orientalis Black-bellied Sandgrouse	R	1996	150	—	B	B2, C6

1. Males.

This is an important site for steppic birds and breeding waders.

■ Protection status

National None **International** Partial
12,707 ha of IBA covered by Special Protection Area (Área Esteparia del Este de Albalete, 14,257 ha).

■ Conservation issues

Threats Agricultural intensification/expansion (B), Infrastructure (C)

The main threats are from agricultural intensification, particularly through irrigation, and the construction of a new railway.

River Mundo and River Segura canyons B2, C6 183

Admin region Castilla-La Mancha
Coordinates 38°30'N 2°00'W
Altitude 500–1,700 m **Area** 78,000 ha

■ Site description

Two river valleys with well conserved riverine forest and also *Pinus* forest with some areas of scrub. The main human activities are livestock-farming, forestry, tourism and hunting.

Habitats Forest and woodland (25%; broadleaved deciduous forest; native coniferous forest; mixed forest; alluvial/very wet forest; broadleaved evergreen forest), Scrub (10%; sclerophyllous scrub/garrigue/maquis), Grassland (10%; steppe/dry calcareous grassland), Wetland (10%; standing fresh water), Rocky areas (25%; scree/boulders; inland cliff), Artificial landscape (25%; arable land; perennial crops/orchards/groves; forestry plantation)
Land-use Agriculture (35%), Forestry (20%), Hunting (50%)

■ Birds

Species	Season	Year	Pop min	Pop max	Acc	Criteria
Hieraaetus fasciatus Bonnelli's Eagle	R	1995	7	8	A	B2, C6
Pyrrhocorax pyrrhocorax Chough	R	1995	150	—	C	B2, C6

The site is important for forest- and cliff-nesting raptors.

■ Protection status

National None **International** None

■ Conservation issues

Threats Burning of vegetation (A), Infrastructure (B), Other (B), Recreation/tourism (B), Unsustainable exploitation (B)

The main threats are from increasing tourism, the development of the transport infrastructure, illegal hunting and poisoning, and forest fires.

Campo de Montiel A1, B2, C1, C2, C6 184

Admin region Castilla-La Mancha
Coordinates 38°40'N 2°50'W
Altitude 880–1,042 m **Area** 153,500 ha

■ Site description

An undulating plain with arable cultivation and patches of scrub and grassland.

Habitats Forest and woodland (broadleaved evergreen forest), Scrub (scrub), Grassland, Artificial landscape (*; arable land; perennial crops/orchards/groves)
Land-use Agriculture (*)

■ Birds

Species	Season	Year	Pop min	Pop max	Acc	Criteria
[1] ***Tetrax tetrax*** Little Bustard	R	1994	3,000	—	A	A1, B2, C1, C2, C6
Otis tarda Great Bustard	R	1996	23	230	A	A1, B2, C1, C2, C6
Pterocles orientalis Black-bellied Sandgrouse	R	1994	100	—	B	B2, C6
Pterocles alchata Pin-tailed Sandgrouse	R	1994	1,000	—	A	B2, C2, C6

1. Males.

This is an important site for steppic birds.

■ Protection status

National None **International** None

■ Conservation issues

Threats Agricultural intensification/expansion (U)

The main threat is from agricultural intensification, particularly irrigation of crops.

San Clemente–Villarrobledo A1, B2, C1, C2, C6 185

Admin region Castilla-La Mancha
Coordinates 39°20'N 2°30'W
Altitude 700–852 m **Area** 103,000 ha

■ Site description

An area of cultivated plains, with a few *Quercus* woodlands.

Habitats Forest and woodland (broadleaved evergreen forest), Grassland (steppe/dry calcareous grassland), Artificial landscape (*; arable land; perennial crops/orchards/groves; forestry plantation)
Land-use Agriculture (*), Forestry

■ Birds

Species	Season	Year	Pop min	Pop max	Acc	Criteria
Falco naumanni Lesser Kestrel	B	1995	95	95	A	A1, B2, C1, C6
Tetrax tetrax Little Bustard	R	1994	2,500	—	A	A1, B2, C1, C2, C6
Otis tarda Great Bustard	R	1994	37	120	A	A1, C1
Pterocles alchata Pin-tailed Sandgrouse	R	1994	150	—	B	B2, C2, C6

The site is important for species associated with steppic habitats.

■ Protection status

National None **International** None

Conservation issues

Threats Agricultural intensification/expansion (U), Dredging/canalization (U), Infrastructure (U)

The main threats are from the irrigation of crops, road building and the canalization of rivers.

Canyons on the middle section of the Cabriel river — B2, C6 — 186

Admin region Castilla-La Mancha
Coordinates 39°50'N 1°40'W
Altitude 800–1,120 m **Area** 52,300 ha

Site description

Vegetation comprises *Pinus*, *Quercus* and *Juniperus* woodland, scrub and grassland.

Habitats Forest and woodland (*; native coniferous forest; broadleaved evergreen forest), Scrub (*; sclerophyllous scrub/garrigue/maquis), Grassland (steppe/dry calcareous grassland), Rocky areas (*; scree/boulders; inland cliff)
Land-use Agriculture, Forestry (*), Hunting (*)

Birds

Species	Season	Year	Pop min	Pop max	Acc	Criteria
Hieraaetus fasciatus Bonnelli's Eagle	R	1995	5	7	A	B2, C6
Falco peregrinus Peregrine	R	1995	10	16	B	B2, C6

This is an important site for cliff-nesting raptors.

Protection status

National None **International** None

Conservation issues

Threats Recreation/tourism (B), Unsustainable exploitation (U)

The main threat is from increasing recreational use of the river.

Cuenca mountain ranges — A4ii, B1iii, B2, C2, C6 — 187

Admin region Castilla-La Mancha
Coordinates 40°25'N 2°00'W
Altitude 1,200–1,600 m **Area** 102,000 ha

Site description

Flat-topped mountains to the north of Cuenca. The landscape is dominated by cliffs and gorges and the vegetation cover is almost completely *Pinus* forest. The main human activities are forestry, livestock-farming and hunting.

Habitats Forest and woodland (70%; broadleaved deciduous forest; native coniferous forest), Scrub (25%; scrub), Rocky areas (scree/boulders; inland cliff), Artificial landscape (*; forestry plantation; ruderal land)
Land-use Agriculture, Forestry (10%), Hunting (*)

Birds

Species	Season	Year	Pop min	Pop max	Acc	Criteria
Neophron percnopterus Egyptian Vulture	B	1995	13	13	A	B2, C6
[1] ***Gyps fulvus*** Griffon Vulture	R	1995	224	224	A	A4ii,B1iii,B2,C2,C6
Aquila chrysaetos Golden Eagle	R	1995	12	12	A	B2, C6
Falco peregrinus Peregrine	R	1995	17	17	A	B2, C6
Pyrrhocorax pyrrhocorax Chough	R	1995	148	148	A	B2, C6

1. Large increase.

This is an important site for cliff-nesting raptors.

Protection status

National Partial **International** Partial
25,724 ha of IBA covered by National Game Reserve (Serrania de Cuenca, 25,724 ha). 70,590 ha of IBA covered by Special Protection Area, 128,345 ha).

Conservation issues

No serious threats known at the site.

Upper sections of the Rivers Tajo and Tajuña — A4ii, B1iii, B2, C2, C6 — 188

Admin region Castilla-La Mancha
Coordinates 40°48'N 2°15'W
Altitude 800–1,744 m **Area** 157,000 ha

Site description

A limestone canyon along the upper reaches of the Rivers Tajo and Tajuña. Vegetation includes *Pinus* forest, garrigue and grassland areas.

Habitats Forest and woodland (*; native coniferous forest; alluvial/very wet forest), Scrub (scrub), Grassland, Wetland (river/stream), Rocky areas (*; inland cliff), Artificial landscape (forestry plantation)
Land-use Agriculture, Forestry, Tourism/recreation

Birds

Species	Season	Year	Pop min	Pop max	Acc	Criteria
Neophron percnopterus Egyptian Vulture	B	1995	27	27	A	B2, C6
[1] ***Gyps fulvus*** Griffon Vulture	R	1995	525	525	A	A4ii,B1iii,B2,C2,C6
Aquila chrysaetos Golden Eagle	R	1995	21	—	A	B2, C6
Falco peregrinus Peregrine	R	1995	35	35	A	B2, C6
Pyrrhocorax pyrrhocorax Chough	R	1995	288	293	A	B2, C2, C6

1. Large increase.

The site is important for cliff-nesting raptors.

Protection status

National None **International** Partial
87,269 ha of IBA covered by Special Protection Area (Alto Tajo, 125,988 ha). 8,900 ha of IBA covered by Special Protection Area (Parameras de Maranchón, 44,819 ha).

Conservation issues

Threats Recreation/tourism (B)

Threats are from more intensive forest management and disturbance from increasing recreational activities, particularly angling and canoeing.

Embid–Molina high moors — B2, C2, C6 — 189

Admin region Castilla-La Mancha
Coordinates 41°00'N 1°55'W
Altitude 1,000–1,203 m **Area** 13,500 ha

Site description

A treeless, flat, upland area with non-irrigated cereal crops and patches of grassland grazed by sheep.

Habitats Scrub (scrub), Grassland (*; steppe/dry calcareous grassland), Artificial landscape (*; arable land; ruderal land)
Land-use Agriculture (*)

Birds

Species	Season	Year	Pop min	Pop max	Acc	Criteria
Chersophilus duponti Dupont's Lark	R	1992	400	—	C	B2, C2, C6

This is an important site for *Chersophilus duponti*. Species of global conservation concern that do not meet IBA criteria: *Tetrax tetrax* (resident).

Protection status

National None **International** None

Conservation issues

Threats Afforestation (B)

Afforestation is a threat.

River Mesa canyons B1iii, B2, C2, C6 190

Admin region Aragón, Castilla-La Mancha
Coordinates 41°11'N 1°54'W
Altitude 800–1,151 m **Area** 12,500 ha

Site description

A steep-sided valley with many limestone cliffs. The vegetation consists primarily of maquis and sheep-grazed grassland.

Habitats Scrub (*; sclerophyllous scrub/garrigue/maquis), Grassland (*), Rocky areas (*; inland cliff), Artificial landscape (ruderal land)
Land-use Agriculture (*)

Birds

Species	Season	Year	Pop min	Pop max	Acc	Criteria
[1] ***Gyps fulvus*** Griffon Vulture	R	1995	125	125	A	B1iii, B2, C2, C6

1. Large increase.

An important site for cliff-nesting raptors, particularly *Gyps fulvus*.

Protection status

National None **International** None

Conservation issues

No serious threats known at the site.

Entrepeñas and Buendía reservoirs B1i, B2, C2 191

Admin region Castilla-La Mancha
Coordinates 40°30'N 2°45'W
Altitude 800–1,189 m **Area** 38,500 ha

Site description

The site includes two reservoirs with numerous limestone cliffs, maquis, *Pinus* woodland and sclerophyllous scrub. The main human activity is livestock-farming.

Habitats Forest and woodland (25%; broadleaved evergreen forest), Scrub (20%; scrub), Wetland (10%; standing fresh water), Rocky areas (inland cliff), Artificial landscape (40%; arable land; perennial crops/orchards/groves)
Land-use Agriculture (40%), Water management

Birds

Species	Season	Year	Pop min	Pop max	Acc	Criteria
Hieraaetus fasciatus Bonnelli's Eagle	R	1995	5	5	A	B2
Grus grus Crane	P	1997	—	1,000	B	B1i, C2

This is an important site for breeding *Hieraaetus fasciatus* and for *Grus grus* on passage.

Protection status

National None **International** Partial
23,000 ha of IBA covered by Special Protection Area (Sierra de Altomira, 29,018 ha).

Conservation issues

Threats Recreation/tourism (C)

Visitors to the reservoir dam cause some disturbance.

El Hito A4i, B1i, B3, C2, C6 192

Admin region Castilla-La Mancha
Coordinates 39°52'N 2°41'W
Altitude 835–835 m **Area** 24,000 ha

Site description

A seasonal lagoon surrounded by cereal crops and patches of grassland.

Habitats Wetland (standing fresh water), Artificial landscape (100%; arable land)
Land-use Agriculture (100%), Hunting

Birds

Species	Season	Year	Pop min	Pop max	Acc	Criteria
Grus grus Crane	P	1994	—	9,950	A	A4i, B1i, C2
Himantopus himantopus Black-winged Stilt	B	1997	272	286	A	A4i, B1i, C2, C6
Recurvirostra avosetta Avocet	B	1997	109	149	A	B3, C6

This is an important site for passage *Grus grus*, breeding waders and winter waterbirds.

Protection status

National None **International** Partial
915 ha of IBA covered by Special Protection Area (El Hito, 915 ha).

Conservation issues

Threats Construction/impact of dyke/dam/barrage (U), Disturbance to birds (A), Dredging/canalization (B), Infrastructure (U)

The main threat is disturbance of *Grus grus* by farmers.

Tarancón–Ocaña–Corral de Almaguer A1, B2, C1, C2, C6 193

Admin region Castilla-La Mancha
Coordinates 40°00'N 3°00'W
Altitude 740–762 m **Area** 129,100 ha

Site description

A mainly cultivated plain with patches of grassland and *Quercus* woodland.

Habitats Wetland (standing fresh water; river/stream), Artificial landscape (95%; arable land; perennial crops/orchards/groves)
Land-use Agriculture (95%), Hunting (60%)

Birds

Species	Season	Year	Pop min	Pop max	Acc	Criteria
Circus aeruginosus Marsh Harrier	R	1990	4	4	A	C6
Tetrax tetrax Little Bustard	R	1994	600	—	A	A1, B2, C1
Otis tarda Great Bustard	R	1994	419	579	A	A1, B2, C1, C2, C6

This is an important site for breeding steppic species. Species of global conservation concern that do not meet IBA criteria: *Falco naumanni* (39 breeding pairs).

Protection status

National None **International** Partial
40,046 ha of IBA covered by Special Protection Area (Área Esteparia de la Mancha Norte, 97,973 ha).

Conservation issues

Threats Afforestation (B), Agricultural intensification/expansion (A), Dredging/canalization (B), Filling-in of wetlands (C), Groundwater abstraction (A), Infrastructure (B)

A number of problems threaten this site including irrigation of crops, illegal hunting, river canalization, afforestation and over-exploitation of groundwater.

Tembleque–La Guardia plains A1, A4i, B1i, B2, C1, C2, C3, C6 194

Admin region Castilla-La Mancha
Coordinates 39°40'N 3°30'W
Altitude 700–877 m **Area** 128,000 ha

Site description

A cultivated area with seasonal saline lagoons.

Habitats Scrub, Wetland (standing fresh water; standing brackish and salt water; river/stream), Artificial landscape (95%; arable land; perennial crops/orchards/groves)
Land-use Agriculture (95%)

Birds

Species	Season	Year	Pop min	Pop max	Acc	Criteria
Netta rufina Red-crested Pochard	W	1997	274	274	A	B1i, C3
Oxyura leucocephala White-headed Duck	B	1995	16	29	A	A1, B1i, B2, C1, C2, C6
Oxyura leucocephala White-headed Duck	W	1997	9	9	A	B1i, C2
[1] ***Tetrax tetrax*** Little Bustard	R	1995	1,800	—	A	A1, B2, C1, C6
Otis tarda Great Bustard	R	1995	707	1,205	A	A1, B2, C1, C2, C6
Gelochelidon nilotica Gull-billed Tern	B	1995	293	308	A	A4i, B1i, B2, C2, C6
Pterocles orientalis Black-bellied Sandgrouse	R	1995	40	—	B	C6
Pterocles alchata Pin-tailed Sandgrouse	R	1995	600	—	B	B2, C2, C6

1. Males.

This is an important area for breeding steppic species.

Protection status

National None **International** Partial

48,900 ha of IBA covered by Special Protection Area (Área Esteparia de la Mancha Norte, 97,973 ha). 1,156 ha of IBA covered by Special Protection Area (Humedales de la Mancha, 11,520 ha).

Conservation issues

Threats Unsustainable exploitation (B)

The main threat is from illegal hunting.

Alcázar de San Juan–Quero endorreic lagoons

A1, A4i, B1i, B1iii, B2, C1, C2, C3, C6 **195**

Admin region Castilla-La Mancha
Coordinates 39°30′N 3°10′W
Altitude 700–750 m **Area** 58,500 ha

Site description

An area of arable cultivation with several natural and artificial lagoons.

Habitats Wetland (*; standing fresh water; standing brackish and salt water; river/stream), Artificial landscape (85%; arable land; perennial crops/orchards/groves)
Land-use Agriculture (85%), Hunting

Birds

Species	Season	Year	Pop min	Pop max	Acc	Criteria
Ardea purpurea Purple Heron	B	1996	20	30	A	C6
Anas strepera Gadwall	B	1997	45	45	A	B2
Netta rufina Red-crested Pochard	R	1996	338	338	A	A4i, B1i, B2, C3
Netta rufina Red-crested Pochard	W	1997	4,638	4,638	A	A4i, B1i, C3
Oxyura leucocephala White-headed Duck	R	1996	20	20	A	A1, B1i, B2, C1, C2, C6
Circus aeruginosus Marsh Harrier	R	1996	52	52	A	C6
Falco naumanni Lesser Kestrel	B	1996	100	—	B	A1, B1iii, B2, C1, C2, C6
Himantopus himantopus Black-winged Stilt	B	1996	201	210	A	B1i, C2, C6
Recurvirostra avosetta Avocet	B	1996	65	65	A	C6
Glareola pratincola Collared Pratincole	B	1996	47	47	A	B1i, B2, C2, C6
Gelochelidon nilotica Gull-billed Tern	B	1996	182	182	A	A4i, B1i, B2, C2, C6

The site is important for steppic species and wildfowl. Species of global conservation concern that do not meet IBA criteria: *Tetrax tetrax* (22 resident pairs) and *Otis tarda* (14–27 resident pairs).

Protection status

National None **International** Partial

250 ha of IBA covered by Ramsar Site (Lagunas de Alcázar de San Juan, 250 ha). 8,800 ha of IBA covered by Special Protection Area (Humedales de la Mancha, 11,520 ha).

Conservation issues

Threats Construction/impact of dyke/dam/barrage (U), Dredging/canalization (U), Groundwater abstraction (U), Industrialization/urbanization (B), Natural events (B)

Threats include over-exploitation of water resources, urban development, river canalization, and droughts which lead to the lagoons drying out.

Pedro Muñoz–Manjavacas endorreic lagoons

A1, A4i, B1i, B2, C1, C2, C3, C6 **196**

Admin region Castilla-La Mancha
Coordinates 39°25′N 2°45′W
Altitude 680–720 m **Area** 41,500 ha

Site description

A group of lakes surrounded by arable land.

Habitats Wetland (10%; standing fresh water; standing brackish and salt water; river/stream), Artificial landscape (90%; arable land; perennial crops/orchards/groves; ruderal land)
Land-use Agriculture (100%)

Birds

Species	Season	Year	Pop min	Pop max	Acc	Criteria
Anas clypeata Shoveler	W	1997	4,229	—	A	B1i, C3
Netta rufina Red-crested Pochard	W	1997	253	253	A	B1i, C3
[1] ***Oxyura leucocephala*** White-headed Duck	R	1995	17	17	A	A1, B1i, B2, C1, C2, C6
Oxyura leucocephala White-headed Duck	W	1997	8	32	A	A1, B1i, C1, C2
Circus aeruginosus Marsh Harrier	B	1992	5	—	C	C6
Himantopus himantopus Black-winged Stilt	B	1996	134	134	A	B1i, C2, C6
Recurvirostra avosetta Avocet	B	1996	62	62	A	C6
Gelochelidon nilotica Gull-billed Tern	B	1996	190	260	A	A4i, B1i, B2, C2, C6

1. Large increase.

This is an important site for breeding and wintering ducks.

Protection status

National None **International** Low

231 ha of IBA covered by Ramsar Site (Laguna de Manjavacas, 231 ha). 34 ha of IBA covered by Ramsar Site (Laguna de la Vega (o del Pueblo), 34 ha). 1,565 ha of IBA covered by Special Protection Area (Humedales de la Mancha, 11,520 ha).

Conservation issues

Threats Dredging/canalization (U), Groundwater abstraction (U), Natural events (B)

The main threats are from droughts, abstraction of groundwater and river canalization.

Tablas de Daimiel marshes; 'Vicario' and 'Gasset' reservoirs and Malagón lakes

A4i, B1i, B2, C2, C3, C6 **197**

Admin region Castilla-La Mancha
Coordinates 39°00′N 3°45′W
Altitude 600–800 m **Area** 31,500 ha

Site description

A group of wetlands to the north of the Ciudad Real, comprising the flood-plain of the River Guadiana, two reservoirs, and two small endorreic lagoons. Arable cultivation surrounds the wetlands.

Habitats Forest and woodland (broadleaved evergreen forest), Wetland (*; standing fresh water; river/stream; water-fringe vegetation), Artificial landscape (70%; arable land; perennial crops/orchards/groves; forestry plantation)
Land-use Agriculture (70%), Nature conservation/research (10%), Water management (*)

Birds

This is an important site for breeding herons and breeding and wintering ducks. Species of global conservation concern that do not meet IBA criteria: *Aythya nyroca* (breeding) and *Oxyura leucocephala* (wintering).

Species	Season	Year	Pop min	Pop max	Acc	Criteria
Ixobrychus minutus Little Bittern	B	1996	19	19	A	C6
Nycticorax nycticorax Night Heron	B	1996	30	30	A	C6
Ardeola ralloides Squacco Heron	B	1996	4	4	A	C6
Ardea purpurea Purple Heron	B	1996	88	88	A	B2, C2, C6
Anas strepera Gadwall	B	1996	63	63	A	B2
Anas strepera Gadwall	W	1997	712	2,010	A	A4i, B1i, C3
Anas platyrhynchos Mallard	W	1997	1,061	21,275	A	B1i, C3
Netta rufina Red-crested Pochard	B	1996	237	1,150	A	A4i, B1i, B2, C3
Netta rufina Red-crested Pochard	W	1997	276	5,365	A	A4i, B1i, C3

Species ... continued	Season	Year	Pop min	Pop max	Acc	Criteria
Oxyura leucocephala White-headed Duck	B	1996	2	2	A	B2
Circus aeruginosus Marsh Harrier	B	1992	5	—	C	C6
Himantopus himantopus Black-winged Stilt	B	1997	110	210	A	B1i, C2, C6

Protection status

National Low **International** Partial

1,928 ha of IBA covered by National Park (Tablas de Daimiel, 1,928 ha). 318 ha of IBA covered by Integral Nature Reserve (Las Tablas de Daimiel, 318 ha). 25,000 ha of IBA covered by Biosphere Reserve (La Mancha Húmeda, 25,000 ha). 1,928 ha of IBA covered by Ramsar Site (Las Tablas de Daimiel, 1,928 ha). 1,928 ha of IBA covered by Special Protection Area (Tablas de Daimiel, 1,928 ha).

Conservation issues

Threats Agricultural intensification/expansion (A), Construction/impact of dyke/dam/barrage (B), Drainage (A), Dredging/canalization (A), Filling-in of wetlands (A), Groundwater abstraction (A), Infrastructure (A), Natural events (A), Recreation/tourism (U)

Over-exploitation of water for agriculture has caused the flood-plain area to dry out. Further threats come from intensification of agriculture, canalization of rivers and droughts.

Castrejón reservoir — C6 — 198

Admin region Castilla-La Mancha
Coordinates 39°48′N 4°20′W
Altitude 400–569 m **Area** 3,800 ha

Site description

An irrigation reservoir with steep banks on the River Tajo. There are areas of marshland and alluvial woodland.

Habitats Forest and woodland (*; alluvial/very wet forest; broadleaved evergreen forest), Grassland, Wetland (*; standing fresh water; river/stream; water-fringe vegetation), Artificial landscape (arable land; perennial crops/orchards/groves)
Land-use Agriculture, Tourism/recreation, Water management (*)

Birds

Species	Season	Year	Pop min	Pop max	Acc	Criteria
Ixobrychus minutus Little Bittern	B	1996	4	4	A	C6
[1] ***Nycticorax nycticorax*** Night Heron	B	1995	18	18	A	C6
Ardea purpurea Purple Heron	B	1995	13	13	A	C6
Circus aeruginosus Marsh Harrier	B	1997	13	15	B	C6
Porphyrio porphyrio Purple Gallinule	R	1995	20	20	A	C6

1. Large decrease.

This is an important site for breeding and wintering waterbirds, particularly wintering *Phalacrocorax carbo*.

Protection status

National None **International** Partial

1,200 ha of IBA covered by Special Protection Area (Río Tajo en Castrejon, Islas de Malpica del Tajo y Azután, 1,960 ha).

Conservation issues

Threats Agricultural intensification/expansion (B), Industrialization/urbanization (B), Recreation/tourism (A), Unsustainable exploitation (C)

There are high levels of disturbance as a result of recreational activities. Hunting is also a threat. Some parts of the reservoir are subject to sedimentation.

Torrijos — A1, B2, C1, C2, C6 — 199

Admin region Castilla-La Mancha
Coordinates 40°00′N 4°15′W
Altitude 450–579 m **Area** 28,600 ha

Site description

Plains to the north-west of Toledo dominated by non-irrigated arable cultivation.

Habitats Artificial landscape (100%; arable land; perennial crops/orchards/groves; ruderal land)
Land-use Agriculture (100%), Hunting

Birds

Species	Season	Year	Pop min	Pop max	Acc	Criteria
Tetrax tetrax Little Bustard	R	1994	1,200	—	A	A1, B2, C1, C6
Otis tarda Great Bustard	R	1994	139	201	A	A1, B2, C1, C2, C6

This is an important site for breeding steppic species.

Protection status

National None **International** None

Conservation issues

Threats Agricultural intensification/expansion (A), Construction/impact of dyke/dam/barrage (A), Infrastructure (C)

Agricultural intensification, particularly through the use of irrigation is a major threat. Two new reservoirs and irrigation channels are under construction.

Malpica island (River Tagus) — A4i, B1i, C3, C6 — 200

Admin region Castilla-La Mancha
Coordinates 39°55′N 4°34′W
Altitude 390–390 m **Area** 4 ha

Site description

A small island in the River Tajo dominated by *Tamarix* and *Populus* riverine forest.

Habitats Forest and woodland (alluvial/very wet forest), Wetland (*; river/stream; water-fringe vegetation)
Land-use Not utilized (100%)

Birds

Species	Season	Year	Pop min	Pop max	Acc	Criteria
Nycticorax nycticorax Night Heron	R	1995	43	43	A	C6
Bubulcus ibis Cattle Egret	R	1995	1,416	1,416	A	A4i, B1i, C3
Egretta garzetta Little Egret	R	1995	39	39	A	C6

This is a very important site for breeding herons.

Protection status

National None **International** High

4 ha of IBA covered by Special Protection Area (Río Tajo en Castrejon, Islasde Malpica del Tajo y Azután, 1,960 ha).

Conservation issues

Threats Natural events (B), Recreation/tourism (B)

The main threats are from disturbance from tourists and from seasonal fluctuations in water-levels.

Rosarito and Navalcán reservoirs–La Iglesuela (Tiétar valley) — A4i, B1i, B2, C2, C3, C6 — 201

Admin region Castilla-La Mancha
Coordinates 40°10′N 5°00′W
Altitude 300–764 m **Area** 70,000 ha

Site description

The site includes a river and two reservoirs surrounded by large expanses of dehesa and scrub. There is some arable cultivation and sheep-grazed grassland.

Habitats Forest and woodland (70%; broadleaved evergreen forest), Scrub (sclerophyllous scrub/garrigue/maquis), Grassland, Wetland (*; standing fresh water), Artificial landscape (arable land; perennial crops/orchards/groves)

Land-use Agriculture (75%), Hunting (*), Water management (*)

Birds

Species	Season	Year	Pop min	Pop max	Acc	Criteria
[1] ***Phalacrocorax carbo*** Cormorant	W	1993	400	1,500	B	B1i, C3
Ciconia nigra Black Stork	B	1996	2	4	A	B1i, C2, C6
Ciconia nigra Black Stork	N	1996	41	100	B	B1i, C2
[2] ***Aquila adalberti*** Spanish Imperial Eagle	R	1996	1	1	A	B2, C6
Grus grus Crane	W	1996	1,000	5,244	A	A4i, B1i, C2

1. *Ph. c. sinensis.*
2. Large decrease.

This is an important site for resident raptors and wintering waterbirds, particularly *Grus grus*.

Protection status

National None **International** Partial

36,620 ha of IBA covered by Special Protection Area (Valle del Tiétar–Embalses de Rosarito y Navalcán, 53,167 ha).

Conservation issues

Threats Construction/impact of dyke/dam/barrage (B), Disturbance to birds (A), Other (A)

The site is threatened by a massive dam project. Disturbance affects *Grus grus* around the Rosarito reservoir. The use of poisoned baits is a threat to *Aquila adalberti*.

Oropesa plains A1, B2, C1, C6 202

Admin region Castilla-La Mancha
Coordinates 40°00′N 5°10′W
Altitude 350–400 m **Area** 45,500 ha

Site description

An area of plains in the Tajo valley dominated by grassland and cereal cultivation with areas of dehesa.

Habitats Forest and woodland (45%; broadleaved evergreen forest), Scrub, Grassland (*), Artificial landscape (50%; arable land)
Land-use Agriculture (*), Hunting (*)

Birds

Species	Season	Year	Pop min	Pop max	Acc	Criteria
Ciconia ciconia White Stork	B	1994	40	—	C	C6
Circus pygargus Montagu's Harrier	B	1992	20	—	C	C6
Falco naumanni Lesser Kestrel	B	1992	60	—	A	A1, B2, C1, C6
[1,2] ***Tetrax tetrax*** Little Bustard	R	1994	1,200	—	A	A1, B2, C1, C6
Otis tarda Great Bustard	R	1994	24	50	A	A1, C1

1. Males.
2. Large decrease.

This is an important site for steppic birds.

Protection status

National None **International** Partial

9,780 ha of IBA covered by Special Protection Area (Llanuras de Oropesa, Lagartera y Calera y Chozas, 14,948 ha).

Conservation issues

Threats Agricultural intensification/expansion (A)

The main threat is from agricultural intensification, particularly in association with irrigation.

Azután reservoir A4i, B1i, B2, C3, C6 203

Admin region Castilla-La Mancha
Coordinates 39°50′N 5°10′W
Altitude 354–360 m **Area** 933 ha

Site description

A reservoir with steep banks and some partially flooded islets with riverine vegetation.

Habitats Wetland (100%; standing fresh water; water-fringe vegetation)
Land-use Water management (100%)

Birds

Species	Season	Year	Pop min	Pop max	Acc	Criteria
Nycticorax nycticorax Night Heron	B	1995	217	—	A	B2, C6
[1] ***Bubulcus ibis*** Cattle Egret	B	1995	3,026	—	—	A4i, B1i, C3
Egretta garzetta Little Egret	R	1995	229	229	A	C6
Ardea purpurea Purple Heron	B	1995	23	—	A	C6

1. Large increase.

This is an important site for breeding herons.

Protection status

National High **International** Partial

933 ha of IBA covered by Wild Fauna Reserve (Embalse de Azután, 1,250 ha). 620 ha of IBA covered by Special Protection Area (Río Tajo en Castrejón, Islas de Malpica del Tajo y Azután, 1,960 ha).

Conservation issues

Threats Agricultural intensification/expansion (B), Burning of vegetation (A), Filling-in of wetlands (B)

The main threats are from ploughing of land on the islands for crops, dumping of material in the wetland, fires and illegal hunting.

Toledo mountains–Cabañeros A1, A3, A4ii, B1i, B1iii, B2, C1, C2, C6 204

Admin region Castilla-La Mancha
Coordinates 39°30′N 4°20′W
Altitude 600–1,447 m **Area** 324,500 ha

Site description

In the mountains the vegetation comprises large expanses of sclerophyllous scrub, areas of dense maquis, and small *Quercus* woodlands. In the plains there are extensive dehesas and grasslands.

Habitats Forest and woodland (*; mixed forest; alluvial/very wet forest; broadleaved evergreen forest), Scrub (*; sclerophyllous scrub/garrigue/maquis), Grassland, Wetland (river/stream), Rocky areas (scree/boulders; inland cliff)
Land-use Agriculture (*), Hunting (*), Nature conservation/research (*)

Birds

Species	Season	Year	Pop min	Pop max	Acc	Criteria
Ciconia nigra Black Stork	B	1995	4	4	A	B1i, C2, C6
Ciconia nigra Black Stork	P	1994	100	100	B	B1i, C2
Aegypius monachus Cinereous Vulture	R	1995	126	126	A	A1, A4ii, B1iii, B2, C1, C2, C6
Aquila adalberti Spanish Imperial Eagle	R	1995	10	12	A	A1, B2, C1, C2, C6
Aquila chrysaetos Golden Eagle	R	1992	20	20	B	B2, C6
Bubo bubo Eagle Owl	R	1995	35	—	B	B2, C6
Oenanthe hispanica Black-eared Wheatear	B	1996	Abundant	—		A3
Sylvia conspicillata Spectacled Warbler	R	1996	Common	—		A3
Sylvia cantillans Subalpine Warbler	B	1996	Common	—		A3
Sylvia melanocephala Sardinian Warbler	R	1996	Abundant	—		A3

Breeding birds include four of the 20 species in Europe that are restricted to the Mediterranean biome (when breeding).

Protection status

National Partial **International** Partial

40,000 ha of IBA covered by National Park (Cabañeros, 40,000 ha). 117,191 ha of IBA covered by Special Protection Area (Cabañeros, 117,191 ha).

Conservation issues

Threats Other (A)

The main threats are from road and track improvements, the use of poison baits and the use of metal fences, all as part of hunting activities ('Other' threats).

Canalizos mountain range (Saceruela) A1, A4ii, B1iii, B2, C1, C2, C6 205

Admin region Castilla-La Mancha
Coordinates 38°50'N 4°30'W
Altitude 550–884 m **Area** 100,000 ha

Site description
An area of dehesa, scrub and *Quercus* woodland. The main human activity is livestock-farming.

Habitats Forest and woodland (*; broadleaved evergreen forest), Scrub (*; sclerophyllous scrub/garrigue/maquis), Grassland, Wetland (river/stream), Rocky areas (inland cliff)
Land-use Agriculture (100%)

Birds

Species	Season	Year	Pop min	Pop max	Acc	Criteria
Ciconia ciconia White Stork	B	1994	25	25	A	C6
Aegypius monachus Cinerous Vulture	R	1992	60	60	B	A1, A4ii, B1iii, B2, C1, C2, C6
Aquila adalberti Spanish Imperial Eagle	R	1995	3	3	A	A1, B2, C1, C6

The site is important for breeding raptors.

Protection status
National None **International** Partial
23,545 ha of IBA covered by Special Protection Area (Sierra de los Canalizos, 24,135 ha).

Conservation issues

Threats Afforestation (B)

The main threat is from afforestation.

Campo de Calatrava A1, A4ii, B1iii, B2, C1, C2, C6 206

Admin region Castilla-La Mancha
Coordinates 38°50'N 4°33'W
Altitude 638–996 m **Area** 102,000 ha

Site description
An undulating area with scrub surrounded by intensive cereal cultivation, olive-groves and vineyards.

Habitats Scrub (40%), Artificial landscape (60%; arable land; perennial crops/orchards/groves; other urban/industrial areas; ruderal land)
Land-use Agriculture (*), Hunting

Birds

Species	Season	Year	Pop min	Pop max	Acc	Criteria
Falco naumanni Lesser Kestrel	B	1995	269	300	A	A1, A4ii, B1iii, B2, C1, C2, C6
[1] ***Tetrax tetrax*** Little Bustard	R	1996	300	—	A	A1, C1
Otis tarda Great Bustard	R	1996	100	—	A	A1, C1

1. Males.

This is an important site for steppic species.

Protection status
National None **International** Low
6,545 ha of IBA covered by Special Protection Area (Campo de Calatrava, 6,545 ha).

Conservation issues

Threats Other (B)

The restoration of old buildings, therefore destroying nesting sites for *Falco naumanni*, is the most important threat ('Other' threats, above).

Alcudia valley and mountain range A1, B1i, B1iii, B2, C1, C2, C6 207

Admin region Andalucía, Castilla-La Mancha
Coordinates 38°40'N 4°30'W
Altitude 400–1,323 m **Area** 226,300 ha

Site description
The site consists of a wide valley dominated by pastures, some dehesas and a mountainous area with cliffs, scrub and maquis. Human activities include agriculture, livestock-grazing, lead mining and game hunting.

Habitats Forest and woodland (*; alluvial/very wet forest; broadleaved evergreen forest), Scrub (*; scrub; heathland; sclerophyllous scrub/garrigue/maquis), Grassland (*; dry siliceous grassland), Wetland (river/stream; water-fringe vegetation), Rocky areas (scree/boulders; inland cliff), Artificial landscape (arable land; perennial crops/orchards/groves; forestry plantation; ruderal land)
Land-use Agriculture (*), Forestry, Hunting (*), Urban/industrial/transport

Birds

Species	Season	Year	Pop min	Pop max	Acc	Criteria
Ciconia nigra Black Stork	B	1996	8	—	C	B1i, C2, C6
Ciconia ciconia White Stork	B	1996	80	—	B	C6
Elanus caeruleus Black-winged Kite	R	1992	5	—	C	C6
Aegypius monachus Cinereous Vulture	R	1996	20	—	A	A1, B1iii, B2, C1, C2, C6
Aquila adalberti Spanish Imperial Eagle	R	1996	6	—	A	A1, B2, C1, C6
Aquila chrysaetos Golden Eagle	R	1996	10	13	A	B2, C6
Hieraaetus fasciatus Bonnelli's Eagle	R	1996	10	13	A	B2, C2, C6
Falco naumanni Lesser Kestrel	B	1996	100	—	B	A1, B1iii, B2, C1, C2, C6

This is an important site for upland birds, particularly raptors, and steppic species.

Protection status
National None **International** Partial
68,528 ha of IBA covered by Special Protection Area (Umbría del Valle de Alcudia, 97,528 ha). 5,122 ha of IBA covered by Special Protection Area (Sierras de Almadén–Chillõn–Guadalmez, 5,122 ha).

Conservation issues

Threats Agricultural intensification/expansion (A)

The main threats are from agricultural intensification and overgrazing.

Madrona and Andújar mountain ranges A1, B1i, B1iii, B2, C1, C2, C6 208

Admin region Andalucía, Castilla-La Mancha
Coordinates 38°20'N 4°00'W
Altitude 350–1,300 m **Area** 212,300 ha

Site description
A mountainous area in the Sierra Morena. The vegetation comprises Mediterranean scrub, dehesas, *Pinus* plantations and small areas of cultivation. The main human activities are livestock-farming, game hunting and forestry.

Habitats Forest and woodland (*; alluvial/very wet forest; broadleaved evergreen forest), Scrub (*; sclerophyllous scrub/garrigue/maquis), Grassland (dry siliceous grassland), Wetland (standing fresh water; river/stream), Rocky areas (scree/boulders; inland cliff), Artificial landscape (*; arable land; forestry plantation; ruderal land)
Land-use Agriculture (*), Forestry, Hunting (*), Tourism/recreation

Birds

Species	Season	Year	Pop min	Pop max	Acc	Criteria
Ciconia nigra Black Stork	R	1992	7	9	B	B1i, C2, C6
Aegypius monachus Cinereous Vulture	R	1996	25	35	A	A1, B1iii, B2, C1, C2, C6
Aquila adalberti Spanish Imperial Eagle	R	1996	8	8	A	A1, B2, C1, C6

Species ... continued	Season	Year	Pop min	Pop max	Acc	Criteria
Hieraaetus pennatus Booted Eagle	B	1992	20	—	C	B2
Hieraaetus fasciatus Bonnelli's Eagle	R	1992	10	—	C	B2, C6

This is an important area for breeding raptors

Protection status

National Partial **International** Partial

1,500 ha of IBA covered by Natural Park (Despeñaperros, 6,000 ha). 65,000 ha of IBA covered by Natural Park (Sierra de Andujar, 68,000 ha). 41,212 ha of IBA covered by Natural Park (Sierra de Cardeña y Montoro, 41,212 ha). 27,700 ha of IBA covered by Special Protection Area (Sierra Morena, 97,528 ha).

Conservation issues

Threats Afforestation (A), Construction/impact of dyke/dam/barrage (B), Firewood collection (B), Infrastructure (B), Recreation/tourism (C), Unsustainable exploitation (A)

Threats include forestry, particularly the use of plantations and dam and road projects. Tourism is a potential threat. A management plan has been officially approved for the Sierra de Andujar Natural Park. *Aquila adalberti* is monitored.

Aldeaquemada–Dañador A1, B2, C1, C6 209

Admin region Andalucía, Castilla-La Mancha
Coordinates 38°25′N 3°10′W
Altitude 600–1,001 m **Area** 46,000 ha

Site description

A mountainous area with *Quercus* woodland, dehesas, and extensive areas of scrub. The main human activities are game hunting, livestock-grazing, apiculture and cork harvesting ('Other' land-use, below).

Habitats Forest and woodland (*; mixed forest; alluvial/very wet forest; broadleaved evergreen forest), Scrub (*; scrub; sclerophyllous scrub/garrigue/maquis), Wetland (river/stream), Rocky areas (inland cliff), Artificial landscape (highly improved reseeded grassland; arable land; perennial crops/orchards/groves; forestry plantation)
Land-use Agriculture (*), Forestry (*), Hunting (*), Other

Birds

Species	Season	Year	Pop min	Pop max	Acc	Criteria
Aquila adalberti Spanish Imperial Eagle	R	1995	4	5	B	A1, B2, C1, C6

This is an important site for breeding raptors, particularly *Aquila adalberti*. Species of global conservation concern that do not meet IBA criteria: *Aegypius monachus* (2 resident pairs).

Protection status

National Low **International** None

534 ha of IBA covered by Natural Landscape (Cascada de Cimbarra, 534 ha). 4,045 ha of IBA covered by Special Protection Area (Río Guadalén, 4,045 ha).

Conservation issues

Threats Afforestation (B), Agricultural intensification/expansion (B), Other (B), Recreation/tourism (B), Selective logging/cutting (A), Unsustainable exploitation (B)

The main threats are from the use of pesticides, hunting, including the use of poison, and forestry plantations.

Cazorla and Segura mountain ranges A4ii, B1iii, B2, C2, C6 210

Admin region Andalucía, Castilla-La Mancha
Coordinates 38°00′N 2°40′W
Altitude 450–2,383 m **Area** 250,000 ha

Site description

Mountains with cliffs, extensive *Pinus* forests, small *Quercus* woodlands, grasslands and extensive areas of scrub. The main human activities are game hunting, forestry, sheep-grazing, apiculture ('Other' land-use, below) and tourism.

Habitats Forest and woodland (65%; broadleaved deciduous forest; native coniferous forest; mixed forest; alluvial/very wet forest; broadleaved evergreen forest), Scrub (35%; scrub; sclerophyllous scrub/garrigue/maquis), Grassland (humid grassland; mesophile grassland), Wetland (standing fresh water; river/stream), Rocky areas (scree/boulders; inland cliff; caves), Artificial landscape (arable land; perennial crops/orchards/groves; forestry plantation)
Land-use Agriculture (50%), Forestry (40%), Hunting (50%), Nature conservation/research (80%), Other, Tourism/recreation, Urban/industrial/transport, Water management

Birds

Species	Season	Year	Pop min	Pop max	Acc	Criteria
Neophron percnopterus Egyptian Vulture	B	1996	15	20	A	B2, C6
[1] ***Gyps fulvus*** Griffon Vulture	R	1996	190	200	A	A4ii, B1iii, B2, C2, C6
Aquila chrysaetos Golden Eagle	R	1996	15	20	A	B2, C6
Hieraaetus pennatus Booted Eagle	B	1996	50	—	B	B2, C6
Falco peregrinus Peregrine	R	1996	40	—	B	B2, C6
Bubo bubo Eagle Owl	R	1996	25	—	B	B2, C6
Pyrrhocorax pyrrhocorax Chough	R	1996	200	—	B	B2, C2, C6

1. Large increase.

This is an important site for breeding raptors.

Protection status

National High **International** Partial

76,000 ha of IBA covered by National Game Reserve (Sierras de Cazorla y Segura, 76,000 ha). 214,300 ha of IBA covered by Natural Park (Sierras de Cazorla, Segura y las Villas, 214,300 ha). 12,000 ha of IBA covered by Natural Park (Sierra de Castril, 12,265 ha). 214,300 ha of IBA covered by Biosphere Reserve (Sierras de Cazorla, Segura y las Villas, 214,300 ha). 214,300 ha of IBA covered by Special Protection Area (Sierras de Cazorla, Segura y las Villas, 214,300 ha).

Conservation issues

Threats Agricultural intensification/expansion (B), Groundwater abstraction (C), Infrastructure (B), Recreation/tourism (B)

The main threats are from the large numbers of tourists visiting the area and infrastructure developments in the Guadalquivir valley. The site has a management plan. Conservation activities have included the reintroduction of *Gypaetus barbatus*. This is one of the largest protected areas in Europe.

Puebla de Don Fadrique–Las Cañadas A1, B2, C1, C6 211

Admin region Andalucía
Coordinates 37°50′N 2°20′W
Altitude 970–1,467 m **Area** 52,140 ha

Site description

A high plateau with extensive cereal cultivation and areas of matweed *Stipa tenacissima* and sclerophyllous scrub. There are some small *Quercus* and *Pinus* woodlands. The main human activities are arable cultivation and livestock-farming.

Habitats Forest and woodland (native coniferous forest), Scrub (sclerophyllous scrub/garrigue/maquis), Grassland (steppe/dry calcareous grassland), Artificial landscape (*; arable land)
Land-use Agriculture (100%)

Birds

Species	Season	Year	Pop min	Pop max	Acc	Criteria
Falco naumanni Lesser Kestrel	B	1996	70	80	A	A1, B2, C1
Burhinus oedicnemus Stone Curlew	R	1996	220	—	C	B2, C6
Pterocles orientalis Black-bellied Sandgrouse	R	1996	50	—	C	C6

This is an important area for steppe birds. Species of global conservation concern that do not meet IBA criteria: *Tetrax tetrax* (resident).

Protection status

National None **International** None

Conservation issues

Threats Agricultural intensification/expansion (B)

The most important threat is agricultural intensification, particularly the use of plastic covered crops in the foraging areas of *Falco naumanni*, and also overgrazing.

María mountain range–Los Vélez — B2, C6 — 212

Admin region Andalucía
Coordinates 37°02′N 2°05′W
Altitude 700–2,045 m **Area** 82,880 ha

Site description

Mountains dominated by *Pinus* and *Quercus* forests and *Pinus* plantations, with *Juniperus* scrub at higher altitudes. The main human activities are agriculture and hunting.

Habitats Forest and woodland (*; broadleaved deciduous forest; native coniferous forest; broadleaved evergreen forest), Scrub (*; scrub; sclerophyllous scrub/garrigue/maquis)
Land-use Agriculture, Forestry (*), Hunting

Birds

Species	Season	Year	Pop min	Pop max	Acc	Criteria
Circaetus gallicus Short-toed Eagle	B	1993	21	22	A	B2, C6
Aquila chrysaetos Golden Eagle	R	1993	10	12	A	B2
Hieraaetus pennatus Booted Eagle	R	1993	50	50	A	B2, C6
Falco peregrinus Peregrine	R	1993	13	17	A	B2, C6
Bubo bubo Eagle Owl	R	1993	15	15	A	C6

This is an important site for breeding raptors.

Protection status

National Partial **International** None
18,962 ha of IBA covered by Natural Park (Sierra de Maria–Los Vélez, 18,962 ha).

Conservation issues

Threats Afforestation (C), Forest grazing (B), Infrastructure (B)

The main threats are from overgrazing and infrastructure developments. A management plan exists for the Natural Park.

Hoya de Baza — B2, C2, C6 — 213

Admin region Andalucía
Coordinates 37°35′N 2°41′W
Altitude 700–1,443 m **Area** 38,600 ha

Site description

A semi-arid area mostly covered by cereal cultivation, interspersed with scrub. The main human activities are arable cultivation and hunting.

Habitats Desert (semi-desert), Rocky areas (inland cliff), Artificial landscape (*; arable land; ruderal land)
Land-use Agriculture (60%), Hunting (80%), Nature conservation/research

Birds

Species	Season	Year	Pop min	Pop max	Acc	Criteria
Burhinus oedicnemus Stone Curlew	R	1996	150	—	C	C6
Coracias garrulus Roller	B	1996	40	—	C	B2
Chersophilus duponti Dupont's Lark	R	1990	20	20	B	C6
Melanocorypha calandra Calandra Lark	R	1996	Abundant	—		C6
Calandrella rufescens Lesser Short-toed Lark	R	1996	Abundant	—		B2
Oenanthe leucura Black Wheatear	R	1996	200	—	C	B2, C2

This is an important area for steppic species and passerines characteristic of semi-arid agricultural sites, particularly *Calandrella rufescens*. Species of global conservation concern that do not meet IBA criteria: *Tetrax tetrax* (resident).

Protection status

National None **International** None

Conservation issues

Threats Agricultural intensification/expansion (B), Unsustainable exploitation (U)

The main threat is from agricultural intensification and expansion and illegal hunting.

Hoya de Guadix — B2, C2, C6 — 214

Admin region Andalucía
Coordinates 37°25′N 3°10′W
Altitude 650–1,100 m **Area** 47,000 ha

Site description

A semi-arid area with non-irrigated cereal crops, open *Quercus* woodland and extensive areas of low scrub and dry grasslands.

Habitats Desert (*; semi-desert), Wetland (river/stream), Rocky areas (inland cliff; caves), Artificial landscape (arable land)
Land-use Agriculture (40%), Hunting (*)

Birds

Species	Season	Year	Pop min	Pop max	Acc	Criteria
Burhinus oedicnemus Stone Curlew	R	1995	150	—	C	C6
Pterocles orientalis Black-bellied Sandgrouse	R	1995	150	—	B	B2, C6
Coracias garrulus Roller	B	1995	40	—	C	B2
Melanocorypha calandra Calandra Lark	R	1995	Abundant	—		B2, C6
Calandrella brachydactyla Short-toed Lark	B	1995	Common	—		B2
Oenanthe leucura Black Wheatear	R	1995	300	—	B	B2, C2
Pyrrhocorax pyrrhocorax Chough	R	1995	100	—	B	B2

An important area for steppic species and passerines characteristic of semi-arid agricultural sites. Species of global conservation concern that do not meet IBA criteria: *Tetrax tetrax* (100 resident individuals).

Protection status

National None **International** None

Conservation issues

Threats Agricultural intensification/expansion (B), Construction/impact of dyke/dam/barrage (C), Natural events (B), Other (B), Unsustainable exploitation (A)

The main threats are from illegal hunting and poisoning, overgrazing, drought and erosion

Alhamilla mountain range–Campo de Níjar plain — B2, C2, C6 — 215

Admin region Andalucía
Coordinates 37°00′N 2°20′W
Altitude 0–1,387 m **Area** 80,000 ha

Site description

A mountainous area with deep gullies and plains. The vegetation comprises small *Quercus* woodlands in the highest zones; otherwise scrub and bare, stony ground. Some areas have been afforested with *Pinus* plantations.

Habitats Forest and woodland (broadleaved evergreen forest), Scrub (*; scrub), Desert (*; semi-desert), Rocky areas (inland cliff; caves), Artificial landscape (forestry plantation; other urban/industrial areas)
Land-use Agriculture (*), Forestry, Hunting (*), Military

Birds

Species	Season	Year	Pop min	Pop max	Acc	Criteria
[1] ***Hieraaetus fasciatus*** Bonnelli's Eagle	R	1996	7	7	A	B2
Pterocles orientalis Black-bellied Sandgrouse	R	1992	30	—	B	C6
Chersophilus duponti Dupont's Lark	R	1996	50	50	A	C6
Oenanthe leucura Black Wheatear	R	1996	50	—	C	B2
Bucanetes githagineus Trumpeter Finch	R	1996	100	—	C	B2, C2, C6

1. Large decrease.

This is an important site for steppic birds, particularly *Bucanetes githagineus*, and for resident raptors in the mountains. Species of global conservation concern that do not meet IBA criteria: *Tetrax tetrax* (breeding).

Protection status

National Partial **International** Partial

2,375 ha of IBA covered by Natural Landscape (Karst de Yesos de Sorbas, 2,375 ha). 8,500 ha of IBA covered by Natural Landscape (Sierra Alhamilla, 8,500 ha). 8,500 ha of IBA covered by Special Protection Area (Sierra Alhamilla, 8,500 ha).

Conservation issues

Threats Agricultural intensification/expansion (A), Deforestation (commercial) (B), Disturbance to birds (A), Dredging/canalization (B), Unsustainable exploitation (A)

The main threats are from agricultural intensification, illegal hunting and disturbance from military manoeuvres.

Mountain range and saltpans at Cabo de Gata — A1, A3, A4i, B1i, B2, B3, C1, C2, C6 — 216

Admin region Andalucía
Coordinates 36°50′N 2°05′W
Altitude 0–478 m **Area** 46,100 ha

Site description

The site includes coastal mountains with dry vegetation and an area of saltpans, that supports steppe vegetation. The main human activities are arable and livestock-farming, salt production and tourism.

Habitats Scrub (*; scrub; sclerophyllous scrub/garrigue/maquis), Grassland (*; steppe/dry calcareous grassland), Desert (*; semi-desert), Wetland (sand-dunes/sand beach; standing brackish and salt water), Rocky areas (*; sea cliff/rocky shore; rock stacks/islets; scree/boulders; inland cliff)
Land-use Agriculture, Fisheries/aquaculture, Tourism/recreation, Urban/industrial/transport

Birds

Species	Season	Year	Pop min	Pop max	Acc	Criteria
Phoenicopterus ruber Greater Flamingo	N	1996	982	—	A	A4i, B1i, C2
Hieraaetus fasciatus Bonnelli's Eagle	R	1996	5	6	A	B2
Falco peregrinus Peregrine	R	1993	17	20	A	B2, C6
Recurvirostra avosetta Avocet	R	1996	193	193	A	B3, C6
Larus audouinii Audouin's Gull	N	1996	350	—	A	A1, A4i, B1i, C1, C2
Sterna albifrons Little Tern	B	1996	35	35	A	B2, C2, C6
Pterocles orientalis Black-bellied Sandgrouse	R	1996	Abundant	—		C6
Bubo bubo Eagle Owl	R	1993	20	20	A	C6
Caprimulgus ruficollis Red-necked Nightjar	B	1996	Common	—		A3
Chersophilus duponti Dupont's Lark	R	1996	130	—	B	B2, C6
Oenanthe hispanica Black-eared Wheatear	B	1996	Abundant	—		A3
Oenanthe leucura Black Wheatear	R	1996	Abundant	—		A3
Sylvia conspicillata Spectacled Warbler	R	1996	Abundant	—		A3
Bucanetes githagineus Trumpeter Finch	R	1996	50	—	C	B2, C2, C6

Breeding birds include four of the 20 species in Europe that are restricted to the Mediterranean biome (when breeding).

Protection status

National Partial **International** Partial

38,000 ha of IBA covered by Natural Park (Cabo de Gata–Nijar, 38,000 ha). 300 ha of IBA covered by Ramsar Site (Salinas del Cabo de Gata, 300 ha). 30,000 ha of IBA covered by Special Protection Area (Cabo de Gata–Níjar, 30,000 ha).

Conservation issues

Threats Agricultural intensification/expansion (B), Extraction industry (C), Groundwater abstraction (B), Industrialization/urbanization (B), Recreation/tourism (B)

Threats include sand extraction, building and tourism development. There are also problems caused by overgrazing and agricultural intensification. The regional government has management plans for the protected areas.

Tabernas desert — B2, C6 — 217

Admin region Andalucía
Coordinates 37°05′N 2°30′W
Altitude 260–952 m **Area** 25,000 ha

Site description

An arid depression between the Sierra de los Filabres and the Sierra Alhamilla, with seasonal rivers and cliffs. There is sparse scrub and steppic vegetation and much bare ground. The main human activities are tourism, hunting and film making.

Habitats Forest and woodland, Scrub (sclerophyllous scrub/garrigue/maquis), Grassland (steppe/dry calcareous grassland), Desert (*; semi-desert), Artificial landscape (arable land; perennial crops/orchards/groves)
Land-use Hunting, Tourism/recreation, Urban/industrial/transport

Birds

Species	Season	Year	Pop min	Pop max	Acc	Criteria
Burhinus oedicnemus Stone Curlew	R	1996	Common	—		C6
[1] ***Bucanetes githagineus*** Trumpeter Finch	R	1996	20	—	C	B2, C6

1. Large decrease.

This is an important area for steppic birds.

Protection status

National Partial **International** Partial

11,625 ha of IBA covered by Natural Landscape (Desierto de Tabernas, 11,625 ha). 11,625 ha of IBA covered by Special Protection Area (Desierto de Tabernas, 11,625 ha).

Conservation issues

Threats Infrastructure (C), Natural events (A), Other (B), Recreation/tourism (U)

Threats are from a new road, the use of poison baits, and erosion.

Gádor mountain range — B2, C2, C6 — 218

Admin region Andalucía
Coordinates 36°53′N 2°45′W
Altitude 300–2,236 m **Area** 57,210 ha

Site description

A mountainous area in the Sistema Bético. It is mostly covered by scrub and relict *Quercus* woods. Some areas have been afforested with *Pinus*. Human activities include the grazing of cattle and the collection of aromatic plants ('Other' land-use, below).

Habitats Forest and woodland (native coniferous forest; mixed forest; alluvial/very wet forest; broadleaved evergreen forest), Scrub (*; scrub; sclerophyllous scrub/garrigue/maquis), Grassland (steppe/dry calcareous grassland), Wetland (river/stream), Rocky areas (scree/boulders; inland cliff), Artificial landscape (arable land; perennial crops/orchards/groves; forestry plantation)
Land-use Agriculture (*), Forestry, Hunting, Other

Birds

Species	Season	Year	Pop min	Pop max	Acc	Criteria
Hieraaetus fasciatus Bonnelli's Eagle	R	1995	8	12	B	B2, C2, C6

This is an important site for breeding raptors.

Protection status

National None **International** None

Conservation issues

Threats Agricultural intensification/expansion (A), Infrastructure (U), Natural events (C)

The main threats are from the intensification of cattle-farming, erosion and desertification, and the erection of wind-farms.

Wetlands of western Almería — A1, A4i, B1i, B2, C1, C2, C3, C6 — 219

Admin region Andalucía
Coordinates 36°40'N 2°40'W
Altitude 0–50 m **Area** 3,000 ha

Site description

A group of five wetlands south-west of the city of Almería. The site includes a coastal strip 15 km long, to the south of Roquetas del Mar, with a sand beach, sand-dunes and large seasonal lagoons with sparse halophytic vegetation and reedbeds (*Phragmites*). Saltpans occupy 450 ha.

Habitats Scrub (scrub), Wetland (100%; saltmarsh; sand-dunes/sand beach; coastal lagoon; standing brackish and salt water; water-fringe vegetation)
Land-use Tourism/recreation, Agriculture

Birds

Species	Season	Year	Pop min	Pop max	Acc	Criteria
Ixobrychus minutus Little Bittern	B	1995	6	—	A	C6
[1] ***Phoenicopterus ruber*** Greater Flamingo	N	1995	1,200	—	A	A4i, B1i, C2
Marmaronetta angustirostris Marbled Duck	R	1995	1	1	A	B2
Oxyura leucocephala White-headed Duck	R	1996	61	61	A	A1, A4i, B1i, B2, C1, C2, C6
Oxyura leucocephala White-headed Duck	N	1995	561	—	A	A1, A4i, B1i, C1, C2
Himantopus himantopus Black-winged Stilt	B	1995	208	208	A	B1i, C2, C6
Charadrius alexandrinus Kentish Plover	R	1995	126	126	A	B2
Charadrius alexandrinus Kentish Plover	N	1995	800	—	B	B1i, C3
Larus audouinii Audouin's Gull	N	1995	2,000	—	A	A1, A4i, B1i, C1, C2
Sterna albifrons Little Tern	B	1995	36	36	A	B2, C2, C6

1. Large decrease.

This is an important site for breeding, passage and wintering waterbirds. Species of global conservation concern that do not meet IBA criteria: *Marmaronetta angustirostris* (9 wintering individuals).

Protection status

National High **International** High
65 ha of IBA covered by Natural Reserve (Albufera de Adra, 65 ha). 785 ha of IBA covered by Natural Reserve (Punta Entinas–Sabinar, 785 ha). 1,960 ha of IBA covered by Natural Landscape (Punta Entinas–Sabinar, 1,960 ha). 65 ha of IBA covered by Ramsar Site (Albuferas de Adra, 65 ha). 2,745 ha of IBA covered by Special Protection Area (Punta Entinas–Sabinar, 2,745 ha).

Conservation issues

Threats Agricultural intensification/expansion (A), Consequences of animal/plant introductions (A), Drainage (A), Filling-in of wetlands (B), Groundwater abstraction (A), Industrialization/urbanization (B), Recreation/tourism (B)

The major threats include huge numbers of tourists, over-exploitation of groundwater, drainage, infilling of wetland areas, urban development and agricultural intensification. *Oxyura jamaicensis* is regularly recorded and threatens *Oxyura leucocephala*.

Chafarinas Islands — A1, A4i, B1i, B1ii, B2, C1, C2, C3, C6 — 220

Admin region Melilla
Coordinates 35°11'N 2°26'W
Altitude 0–137 m **Area** 50 ha

Site description

Three small islands close to the north-east coast of Morocco, dominated by scrub. The central island is occupied by a small army detachment, the other two are uninhabited.

Habitats Scrub (sclerophyllous scrub/garrigue/maquis), Rocky areas (100%; sea cliff/rocky shore; rock stacks/islets)
Land-use Military (30%), Nature conservation/research (100%)

Birds

Species	Season	Year	Pop min	Pop max	Acc	Criteria
Calonectris diomedea Cory's Shearwater	B	1996	1,000	—	C	B1ii, B2, C2, C6
Larus audouinii Audouin's Gull	R	1997	2,698	2,698	A	A1, A4i, B1i, B2, C1, C2, C6
Larus cachinnans Yellow-legged Gull	R	1997	3,500	3,500	A	A4i, B1i, C3

This is an important site for breeding seabirds, including the world's second largest *Larus audouinii* colony.

Protection status

National High **International** High
50 ha of IBA covered by National Game Reserve (Islas Chafarinas, 50 ha). 50 ha of IBA covered by Special Protection Area (Islas Chafarinas, 50 ha).

Conservation issues

Threats Other (A)

Larus cachinnans predate the eggs and chicks of *L. audouinii* ('Other' threats, above). Conservation research is carried out by institutes and universities.

Alborán island — A1, A4i, B1i, B2, C1, C2, C6 — 221

Admin region Andalucía
Coordinates 35°57'N 3°02'W
Altitude 0–19 m **Area** 7 ha

Site description

A small volcanic island with cliffs and two beaches.

Habitats Scrub (scrub), Rocky areas (100%; sea cliff/rocky shore; rock stacks/islets)
Land-use Fisheries/aquaculture (30%), Military (70%)

Birds

Species	Season	Year	Pop min	Pop max	Acc	Criteria
[1] ***Larus audouinii*** Audouin's Gull	R	1996	160	160	A	A1, A4i, B1i, B2, C1, C2, C6

1. Large increase.

An important area for breeding seabirds and migrating passerines.

Protection status

National None **International** None

Conservation issues

Threats Aquaculture/fisheries (B), Disturbance to birds (A), Infrastructure (A)

The main threat is disturbance, particularly from military activities. The island is surrounded by a Marine Reserve (429 ha).

Sierra Nevada mountain range — B2, C2, C6 — 222

Admin region Andalucía
Coordinates 37°05'N 3°10'W
Altitude 350–3,481 m **Area** 245,000 ha

Site description

A mountain range dominated by scrub and grassland. There are also *Pinus* plantations, some native *Pinus* forest and broadleaved deciduous forest. The main human activities are livestock-farming, game hunting and recreation.

Habitats Forest and woodland (*; broadleaved deciduous forest; native coniferous forest; mixed forest; alluvial/very wet forest; broadleaved evergreen forest), Scrub (*; scrub), Grassland (steppe/dry calcareous grassland; dry siliceous grassland), Wetland (standing fresh water; river/stream), Rocky areas (*; scree/boulders; inland cliff), Artificial landscape (arable land; perennial crops/orchards/groves; forestry plantation)
Land-use Agriculture (*), Forestry, Hunting (*), Nature conservation/research (*), Tourism/recreation

Birds

Species	Season	Year	Pop min	Pop max	Acc	Criteria
Aquila chrysaetos Golden Eagle	R	1995	15	20	A	B2, C6
Hieraaetus pennatus Booted Eagle	B	1995	20	25	C	B2
Hieraaetus fasciatus Bonnelli's Eagle	R	1995	7	10	A	B2
Oenanthe leucura Black Wheatear	R	1995	300	—	C	B2, C2
Monticola solitarius Blue Rock Thrush	R	1995	150	—	C	B2
Pyrrhocorax pyrrhocorax Chough	R	1995	150	—	B	B2, C6

This is an important area for breeding raptors and species characteristic of Mediterranean habitats.

Protection status

National Partial **International** Partial
35,430 ha of IBA covered by National Game Reserve (Sierra Nevada, 35,430 ha). 171,646 ha of IBA covered by Natural Park (Sierra Nevada, 171,646 ha). 171,646 ha of IBA covered by Biosphere Reserve (Sierra Nevada, 171,646 ha).

Conservation issues

Threats Agricultural intensification/expansion (B), Industrialization/urbanization (B), Infrastructure (B), Recreation/tourism (B), Unsustainable exploitation (B)

The main threats are from overgrazing and development associated with tourism. There has been a management plan since 1994.

Prelittoral mountain ranges of Granada — B2, C2, C6 — 223

Admin region Andalucía
Coordinates 36°50′N 3°40′W
Altitude 800–2,065 m **Area** 125,000 ha

Site description

A mountain range along the Mediterranean coast of the provinces of Granada and Malaga. The vegetation comprises sclerophyllous scrub, dry grassland, *Pinus* plantations and patches of *Quercus* woodland. The main human activities are hunting and livestock-grazing.

Habitats Forest and woodland (broadleaved evergreen forest), Scrub (*; scrub; sclerophyllous scrub/garrigue/maquis), Grassland (steppe/dry calcareous grassland; dry siliceous grassland), Rocky areas (*; scree/boulders; inland cliff), Artificial landscape (forestry plantation)
Land-use Agriculture (*), Hunting

Birds

Species	Season	Year	Pop min	Pop max	Acc	Criteria
Circaetus gallicus Short-toed Eagle	B	1996	10	15	B	C6
Aquila chrysaetos Golden Eagle	R	1996	10	15	A	B2
Hieraaetus pennatus Booted Eagle	B	1996	20	—	B	B2
Hieraaetus fasciatus Bonnelli's Eagle	R	1996	11	14	A	B2, C2, C6
Falco peregrinus Peregrine	R	1996	12	16	B	B2, C6
Bubo bubo Eagle Owl	R	1996	25	—	B	B2, C6
Oenanthe leucura Black Wheatear	R	1996	300	—	B	B2, C2
Monticola solitarius Blue Rock Thrush	R	1996	200	—	B	B2
Pyrrhocorax pyrrhocorax Chough	R	1996	100	—	B	B2

This is an important area for breeding raptors, particularly *Hieraaetus fasciatus*. Species of global conservation concern that do not meet IBA criteria: *Aquila adalberti* (resident).

Protection status

National Low **International** None
395 ha of IBA covered by Natural Landscape (Acantilados de Maro–Cerro Gordo, 395 ha).

Conservation issues

Threats Recreation/tourism (B)

The main threat is from disturbance from visitors, particularly of *Hieraaetus fasciatus*.

Mouth of the River Guadalhorce — A1, A4i, B1i, C1, C2, C6 — 224

Admin region Andalucía
Coordinates 36°37′N 4°27′W
Altitude 0–5 m **Area** 125 ha

Site description

An alluvial plain with artificial salt and freshwater lagoons, a sand beach, reedbeds (*Phragmites*) and *Tamarix* scrub. Inland the site is mainly covered by irrigated crops and riverine forest.

Habitats Forest and woodland (5%; alluvial/very wet forest), Scrub (5%; scrub), Grassland (10%; humid grassland; mesophile grassland), Wetland (25%; sand-dunes/sand beach; coastal lagoon; standing fresh water; standing brackish and salt water; river/stream; water-fringe vegetation), Artificial landscape (55%; arable land; forestry plantation)
Land-use Agriculture (55%)

Birds

Species	Season	Year	Pop min	Pop max	Acc	Criteria
Ixobrychus minutus Little Bittern	B	1996	6	6	A	C6
Ciconia nigra Black Stork	P	1997	20	—	B	B1i, C2
Larus audouinii Audouin's Gull	P	1996	227	412	A	A1, A4i, B1i, C1, C2

This is an important passage site for seabirds and other waterbirds.

Protection status

National Partial **International** None
67 ha of IBA covered by Natural Landscape (Desembocadura del Guadalhorce, 67 ha).

Conservation issues

Threats Dredging/canalization (B), Industrialization/urbanization (C), Recreation/tourism (B), Unsustainable exploitation (A)

Canalization of the river, alteration of the river course, disturbance from visitors and illegal hunting are all threats.

Antequera mountain range–El Chorro — B2, C2, C6 — 225

Admin region Andalucía
Coordinates 36°55′N 4°35′W
Altitude 275–1,637 m **Area** 23,500 ha

Site description

A low mountain range to the north of Málaga. The IBA has been largely deforested, with some patches of *Quercus* and *Pinus* woodland. The slopes are heavily cultivated with perennial crops. Remaining areas are dominated by scrub.

Habitats Forest and woodland (broadleaved deciduous forest; native coniferous forest; broadleaved evergreen forest), Scrub (*; scrub), Rocky areas (*; scree/boulders; inland cliff), Artificial landscape (*)
Land-use Agriculture (*), Hunting, Tourism/recreation

Birds

Species	Season	Year	Pop min	Pop max	Acc	Criteria
[1] ***Hieraaetus fasciatus*** Bonnelli's Eagle	R	1996	6	8	B	B2
Pyrrhocorax pyrrhocorax Chough	R	1996	300	—	B	B2, C2, C6

1. Large decrease.

This is an important area for cliff-nesting raptors.

Protection status

National Partial **International** Low
2,016 ha of IBA covered by Natural Landscape (Desfiladero de los Gaitanes, 2,016 ha). 1,171 ha of IBA covered by Natural Landscape (Torcal de Antequera, 1,171 ha). 1,710 ha of IBA covered by Special Protection Area (Torcal de Antequera, 1,710 ha).

Conservation issues

The main threats are from the high hunting pressure and the use of poison ('Other' threats, below), powerlines and tourism development.

Threats Afforestation (C), Agricultural intensification/expansion (C), Infrastructure (A), Other (U), Recreation/tourism (B), Unsustainable exploitation (U)

Mountain ranges of Córdoba — B2, C6 — 226

Admin region Andalucía
Coordinates 37°25′N 4°20′W
Altitude 400–1,570 m **Area** 63,000 ha

Site description

A mountainous area with many cliffs. The slopes are covered with Mediterranean scrub, some *Quercus* woodland, arable crops and olive-groves.

Habitats Forest and woodland (20%; alluvial/very wet forest; broadleaved evergreen forest), Scrub (30%; scrub; sclerophyllous scrub/garrigue/maquis), Grassland (10%; steppe/dry calcareous grassland; alpine/subalpine/boreal grassland; humid grassland), Wetland (river/stream), Rocky areas (inland cliff), Artificial landscape (40%; highly improved reseeded grassland; arable land; perennial crops/orchards/groves; forestry plantation)
Land-use Agriculture (40%), Forestry (15%), Hunting (95%)

Birds

Species	Season	Year	Pop min	Pop max	Acc	Criteria
[1] ***Hieraaetus fasciatus*** Bonnelli's Eagle	R	1996	10	10	A	B2, C6
Pyrrhocorax pyrrhocorax Chough	R	1996	150	—	B	B2

1. Large increase.

This is an important area for resident raptors.

Protection status

National Partial **International** None
31,568 ha of IBA covered by Natural Park (Sierras Subbeticas de Córdoba, 31,568 ha).

Conservation issues

Threats Agricultural intensification/expansion (A), Dredging/canalization (B), Extraction industry (B), Groundwater abstraction (C), Infrastructure (B), Other (B), Recreation/tourism (B), Selective logging/cutting (B), Unsustainable exploitation (A)

The main threats are from the enlargement of olive-groves, overgrazing, the high hunting pressure and use of poison, uncontrolled tourism, and disturbance from climbers. A management plan exists.

Mountain ranges south of Jaén — B2, C2, C6 — 227

Admin region Andalucía
Coordinates 37°30′N 3°50′W
Altitude 600–1,872 m **Area** 200,000 ha

Site description

A mountainous area dominated by maquis, *Quercus* forest and olive-groves. The main human activities are livestock-farming and hunting.

Habitats Forest and woodland (alluvial/very wet forest; broadleaved evergreen forest), Scrub (*; sclerophyllous scrub/garrigue/maquis), Wetland (river/stream), Rocky areas (scree/boulders; inland cliff), Artificial landscape (*; arable land; perennial crops/orchards/groves; forestry plantation)
Land-use Agriculture (*), Forestry, Hunting (*)

Birds

Species	Season	Year	Pop min	Pop max	Acc	Criteria
Hieraaetus fasciatus Bonnelli's Eagle	R	1996	25	25	A	B2, C2, C6
Oenanthe leucura Black Wheatear	R	1996	40	—	B	B2
Pyrrhocorax pyrrhocorax Chough	R	1996	100	—	B	B2

This is an important area for breeding raptors, particularly *Hieraaetus fasciatus*. Species of global conservation concern that do not meet IBA criteria: *Falco naumanni* (18–21 breeding pairs).

Protection status

National Low **International** None
710 ha of IBA covered by Natural Park (Sierra Mágina, 19,900 ha).

Conservation issues

Threats Afforestation (B), Agricultural intensification/expansion (B), Intensified forest management (B), Recreation/tourism (A), Unsustainable exploitation (B)

There are many threats to the site including disturbance from visitors, intensive forest management, and overgrazing.

Middle section of River Guadalquivir reservoirs — B1iii, B2, C2, C6 — 228

Admin region Andalucía
Coordinates 37°55′N 3°21′W
Altitude 320–361 m **Area** 663 ha

Site description

A group of three shallow reservoirs with abundant emergent vegetation and reedbeds (*Phragmites*). An area of riverine forest remains. The surrounds are dominated by arable cultivation and olive-groves.

Habitats Forest and woodland (*; alluvial/very wet forest), Wetland (*; standing fresh water; water-fringe vegetation)
Land-use Urban/industrial/transport (*), Water management

Birds

Species	Season	Year	Pop min	Pop max	Acc	Criteria
Milvus migrans Black Kite	N	1996	—	2,000	B	B1iii, C2
Circus aeruginosus Marsh Harrier	R	1996	3	3	A	C6
Porphyrio porphyrio Purple Gallinule	R	1996	30	50	B	B2, C2, C6
[1] ***Recurvirostra avosetta*** Avocet	W	1996	—	300	A	B2

1. Large increase.

This is an important site for breeding, passage and wintering waterbirds, particularly breeding *Porphyrio porphyrio*, and also for non-breeding *Milvus migrans*. Species of global conservation concern that do not meet IBA criteria: *Oxyura leucocephala* (3–5 passage individuals).

Protection status

National High **International** None
663 ha of IBA covered by Natural Landscape (Alto Guadalquivir, 663 ha).

Conservation issues

Threats Agricultural intensification/expansion (B), Construction/impact of dyke/dam/barrage (A), Natural events (B), Unsustainable exploitation (C)

The main threats are from sedimentation and the possible construction of a larger dam which would destroy the area.

Jaén countryside — A1, B1iii, B2, C1, C2, C6 — 229

Admin region Andalucía
Coordinates 38°20′N 3°30′W
Altitude 300–550 m **Area** 35,500 ha

Site description

An area of cereal crops and olive-groves.

Habitats Grassland (15%; dry siliceous grassland), Wetland (10%; river/stream), Artificial landscape (75%; arable land; perennial crops/orchards/groves)
Land-use Agriculture (75%), Hunting (100%)

Birds

Species	Season	Year	Pop min	Pop max	Acc	Criteria
[1] ***Circus pygargus*** Montagu's Harrier	B	1996	60	—	A	C6
Falco naumanni Lesser Kestrel	B	1996	100	150	A	A1, B1iii, B2, C1, C2, C6
Melanocorypha calandra Calandra Lark	B	1996	Abundant	—		C6

1. Large decrease.

This is an important area for steppic species. Species of global conservation concern that do not meet IBA criteria: *Tetrax tetrax* (resident) and *Otis tarda* (breeding).

Protection status

National None **International** None

Conservation issues

Threats Agricultural intensification/expansion (A), Groundwater abstraction (B), Infrastructure (A), Natural events (A)

The main threats are from the conversion of cereal crops to olive-groves, which led to the extinction of *Otis tarda* from the site, infrastructure development and drought.

Marmolejo reservoir–La Ropera C6 230

Admin region Andalucía
Coordinates 38°02'N 4°07'W
Altitude 190–210 m **Area** 675 ha

Site description

A reservoir with abundant emergent vegetation, including reedbeds (*Phragmites*). The surrounding area is used for arable cultivation and cattle-farming.

Habitats Forest and woodland (10%; alluvial/very wet forest), Wetland (80%; standing fresh water; river/stream; water-fringe vegetation), Artificial landscape (15%; arable land; perennial crops/orchards/groves)
Land-use Agriculture (20%), Hunting (40%), Water management (60%)

Birds

Species	Season	Year	Pop min	Pop max	Acc	Criteria
Ixobrychus minutus Little Bittern	B	1996	15	20	B	C6
Porphyrio porphyrio Purple Gallinule	R	1996	10	15	B	C6

This is an important site for breeding *Porphyrio porphyrio*.

Protection status

National None **International** None

Conservation issues

Threats Agricultural intensification/expansion (B), Drainage (A), Natural events (A)

Agricultural intensification and sedimentation are the main threats.

Conde, Chinche and Honda lakes A1, B1i, B2, C1, C2, C6 231

Admin region Andalucía
Coordinates 37°35'N 4°12'W
Altitude 450–450 m **Area** 420 ha

Site description

Three small brackish lakes along the River Guadajoz. They are surrounded by a belt of *Tamarix* scrub and reedbed (*Phragmites*).

Habitats Wetland (100%; standing brackish and salt water; water-fringe vegetation)
Land-use Agriculture

Birds

Species	Season	Year	Pop min	Pop max	Acc	Criteria
Oxyura leucocephala White-headed Duck	R	1996	5	7	A	A1, B1i, B2, C1, C2, C6

This is an important site for breeding waterbirds, particularly *Oxyura leucocephala*.

Protection status

National Partial **International** Partial
65 ha of IBA covered by Natural Reserve (Laguna Honda, 65 ha). 7 ha of IBA covered by Natural Reserve (Laguna del Chinche, 7 ha). 11 ha of IBA covered by Natural Reserve (Laguna el Conde, 11 ha). 89 ha of IBA covered by Special Protection Area (Lagunas del Sur de Côrdoba, 1,107 ha).

Conservation issues

Threats Agricultural intensification/expansion (A), Disturbance to birds (B), Drainage (A)

The lakes are threatened by sedimentation as a result of ploughing of surrounding land. Disturbance is caused by hunters and visitors.

Córdoba countryside A1, A4ii, B1iii, B2, B3, C1, C2, C6 232

Admin region Andalucía
Coordinates 37°50'N 4°33'W
Altitude 180–572 m **Area** 116,000 ha

Site description

An area of cereal crops and olive-groves in the Gaudajoz river valley. The site includes lakes and also some urban areas.

Habitats Grassland (steppe/dry calcareous grassland), Wetland (river/stream; water-fringe vegetation), Artificial landscape (90%; highly improved reseeded grassland; arable land; perennial crops/orchards/groves; other urban/industrial areas; ruderal land)
Land-use Agriculture (98%), Hunting (90%), Urban/industrial/transport

Birds

Species	Season	Year	Pop min	Pop max	Acc	Criteria
Nycticorax nycticorax Night Heron	B	1994	41	41	A	C6
[1] ***Circus pygargus*** Montagu's Harrier	B	1996	50	100	B	B3, C6
Falco naumanni Lesser Kestrel	B	1996	240	260	A	A1, A4ii, B1iii, B2, C1, C2, C6
[1] ***Tetrax tetrax*** Little Bustard	R	1996	500	—	B	A1, B2, C1, C6
[1] ***Otis tarda*** Great Bustard	R	1996	30	—	B	C6
Burhinus oedicnemus Stone Curlew	R	1996	220	—	B	B2, C6
Merops apiaster Bee-eater	B	1996	500	—	B	B2
Coracias garrulus Roller	B	1996	100	—	B	B2, C2

1. Large decrease.

This is an important site for steppic species and riverine forest species.

Protection status

National None **International** None

Conservation issues

Threats Agricultural intensification/expansion (A), Recreation/tourism (B), Unsustainable exploitation (B)

The main threats are from high levels of pesticide use, the enlargement of olive-groves, plans to irrigate crops, sedimentation of lakes and ploughing of natural vegetation.

Hinojosa del Duque–El Viso A1, B1i, B2, C1, C2 233

Admin region Andalucía
Coordinates 38°30'N 5°10'W
Altitude 500–630 m **Area** 30,000 ha

Site description

An undulating area with dehesas, olive-groves, grasslands and cereal fields. The area is used for sheep-grazing and hunting.

Habitats Forest and woodland (*; broadleaved evergreen forest), Grassland (*), Artificial landscape (*; arable land; perennial crops/orchards/groves)
Land-use Agriculture (*), Hunting (*)

Birds

Species	Season	Year	Pop min	Pop max	Acc	Criteria
Falco naumanni Lesser Kestrel	B	1992	60	—	C	A1, B2, C1
Grus grus Crane	W	1992	—	1,200	B	B1i, C2

An important wintering site for *Grus grus*. Species of global conservation concern that do not meet IBA criteria: *Tetrax tetrax* (resident).

Protection status

National None **International** None

Conservation issues

No serious threats known at the site.

Los Blázquez–La Granjuela–Fuenteovejuna A4i, B1i, C2, C6 234

Admin region Andalucía
Coordinates 38°20′N 5°20′W
Altitude 550–781 m **Area** 32,000 ha

■ Site description

An undulating area covered by grasslands, dehesas, olive-groves and cereal cultivation.

Habitats Forest and woodland (*; broadleaved evergreen forest), Grassland (*), Artificial landscape (*; arable land; perennial crops/orchards/groves)
Land-use Agriculture (*), Hunting (*)

■ Birds

Species	Season	Year	Pop min	Pop max	Acc	Criteria
Grus grus Crane	W	1992	3,200	—	B	A4i, B1i, C2
Otis tarda Great Bustard	R	1992	20	—	B	C6

This is an important wintering site for *Grus grus*.

■ Protection status

National None **International** None

■ Conservation issues

Threats Selective logging/cutting (A), Unsustainable exploitation (B)

The main threats are from hunting and the logging of *Quercus* trees from the dehesas.

Sierra Morena de Córdoba A1, A3, A4ii, B1i, B1iii, B2, C1, C2, C6 235

Admin region Andalucía
Coordinates 37°56′N 5°14′W
Altitude 190–722 m **Area** 119,715 ha

■ Site description

A range of hills with Mediterranean scrub, dehesas and a few olive-groves. The main human activities are livestock-grazing and game hunting.

Habitats Forest and woodland (70%; alluvial/very wet forest; broadleaved evergreen forest), Scrub (30%; sclerophyllous scrub/garrigue/maquis), Grassland (dry siliceous grassland), Wetland (river/stream), Rocky areas (inland cliff), Artificial landscape (forestry plantation; ruderal land)
Land-use Agriculture, Forestry, Hunting (90%), Tourism/recreation

■ Birds

Species	Season	Year	Pop min	Pop max	Acc	Criteria
Ciconia nigra Black Stork	B	1996	2	4	B	B1i, C2
Gyps fulvus Griffon Vulture	R	1996	125	—	B	B1iii, B2, C2, C6
Aegypius monachus Cinereous Vulture	R	1996	55	—	B	A1, A4ii, B1iii, B2, C1, C2, C6
Aquila adalberti Spanish Imperial Eagle	R	1997	3	4	A	A1, B2, C1, C6
Aquila chrysaetos Golden Eagle	R	1996	8	10	B	C6
Hieraaetus fasciatus Bonnelli's Eagle	R	1996	6	7	B	B2
Caprimulgus ruficollis Red-necked Nightjar	B	1996	Abundant	—		A3
Oenanthe hispanica Black-eared Wheatear	B	1996	Frequent	—		A3
Oenanthe leucura Black Wheatear	R	1996	Frequent	—		A3
Sylvia conspicillata Spectacled Warbler	R	1996	Abundant	—		A3
Sylvia cantillans Subalpine Warbler	B	1996	Abundant	—		A3
Sylvia melanocephala Sardinian Warbler	R	1996	Abundant	—		A3
Sturnus unicolor Spotless Starling	R	1996	Abundant	—		A3

Breeding birds include seven of the 20 species in Europe that are restricted to the Mediterranean biome (when breeding).

■ Protection status

National Partial **International** Partial
67,202 ha of IBA covered by Natural Park (Sierra de Hornachuelos, 67,202 ha). 67,202 ha of IBA covered by Special Protection Area (Sierra de Hornachuelos, 67,202 ha).

■ Conservation issues

Threats Afforestation (B), Construction/impact of dyke/dam/barrage (A), Forest grazing (A), Infrastructure (B), Selective logging/cutting (B)

The main threats are from two planned dam projects, overgrazing in forest areas, and plantations of non-native trees.

Sierra Morena de Sevilla A1, B1i, B2, C1, C2, C6 236

Admin region Andalucía
Coordinates 37°50′N 6°00′W
Altitude 260–968 m **Area** 164,840 ha

■ Site description

A rugged mountainous area to the north of Sevilla, with steep-sided valleys. Vegetation comprises Mediterranean scrub, grassland, *Quercus* woodland and dehesas. The main human activities are livestock-farming and hunting.

Habitats Forest and woodland (*; broadleaved deciduous forest; mixed forest; alluvial/very wet forest; broadleaved evergreen forest), Scrub (*; sclerophyllous scrub/garrigue/maquis), Grassland (dry siliceous grassland), Rocky areas (scree/boulders; inland cliff), Artificial landscape (arable land; forestry plantation)
Land-use Agriculture (*), Forestry, Hunting (*)

■ Birds

Species	Season	Year	Pop min	Pop max	Acc	Criteria
Ciconia nigra Black Stork	B	1996	7	8	A	B1i, C2, C6
Aegypius monachus Cinereous Vulture	R	1996	5	—	B	A1, C1
Circaetus gallicus Short-toed Eagle	B	1992	12	—	B	C6
Aquila adalberti Spanish Imperial Eagle	R	1994	4	4	A	A1, B2, C1, C6
Aquila chrysaetos Golden Eagle	R	1996	10	20	B	B2, C6
Hieraaetus fasciatus Bonnelli's Eagle	R	1996	5	—	B	B2

This is an important site for raptors and passerines characteristic of Mediterranean habitats. Species of global conservation concern that do not meet IBA criteria: *Falco naumanni* (20 breeding pairs).

■ Protection status

National High **International** High
164,840 ha of IBA covered by Natural Park (Sierra Norte de Sevilla, 164,840 ha). 164,840 ha of IBA covered by Special Protection Area (Sierra Norte de Sevilla, 164,840 ha).

■ Conservation issues

Threats Afforestation (B), Construction/impact of dyke/dam/barrage (A), Recreation/tourism (A), Selective logging/cutting (B)

The main threats are from tourism and a large dam project. A management plan was officially approved in 1994.

Carmona countryside A1, A4ii, B1iii, B2, C1, C2, C6 237

Admin region Andalucía
Coordinates 37°15′N 5°40′W
Altitude 75–254 m **Area** 18,200 ha

■ Site description

An area of the Guadalquivir valley with extensive cereal cultivation, sunflower fields and olive-groves.

Habitats Artificial landscape (100%; arable land; perennial crops/orchards/groves; other urban/industrial areas)
Land-use Agriculture (100%), Hunting (60%), Urban/industrial/transport

■ Birds

An important area for steppic species. Species of global conservation concern that do not meet IBA criteria: *Tetrax tetrax* (resident).

Species	Season	Year	Pop min	Pop max	Acc	Criteria
Falco naumanni Lesser Kestrel	B	1996	409	—	B	A1, A4ii, B1iii, B2, C1, C2, C6
Otis tarda Great Bustard	R	1996	—	22	B	C6

Protection status

National None **International** None

Conservation issues

Threats Agricultural intensification/expansion (A), Drainage (A), Dredging/canalization (B), Groundwater abstraction (A), Natural events (B)

The main threats are the intensification of agriculture, the over-exploitation of water resources, and drought.

Écija–Osuna plain A1, A4ii, B1iii, B2, C1, C2, C6 238

Admin region Andalucía
Coordinates 37°25'N 5°10'W
Altitude 50–307 m **Area** 62,200 ha

Site description

A plain, dominated by cereal cultivation, in the Guadalquivir valley. There are patches of dehesa, olive-groves and a complex of saline lakes.

Habitats Forest and woodland (*; alluvial/very wet forest; broadleaved evergreen forest), Wetland (standing fresh water; standing brackish and salt water; water-fringe vegetation), Artificial landscape (*; arable land; perennial crops/orchards/groves)
Land-use Agriculture (75%), Hunting (70%)

Birds

Species	Season	Year	Pop min	Pop max	Acc	Criteria
Falco naumanni Lesser Kestrel	B	1996	306	306	B	A1, A4ii, B1iii, B2, C1, C2, C6
Otis tarda Great Bustard	R	1996	—	50	B	A1, C1, C6

This is an important area for steppic species. Species of global conservation concern that do not meet IBA criteria: *Oxyura leucocephala* (non-breeding) and *Tetrax tetrax* (20 resident pairs).

Protection status

National Low **International** None
66 ha of IBA covered by Natural Reserve (Complejo Endorreico de la Lantejuela, 66 ha).

Conservation issues

Threats Agricultural intensification/expansion (A), Groundwater abstraction (A), Natural events (B)

The main threats are agricultural intensification, over-exploitation of water resources and drought.

Wetlands at south Córdoba A1, B1i, B2, C1, C2, C6 239

Admin region Andalucía
Coordinates 37°25'N 4°45'W
Altitude 190–430 m **Area** 3,054 ha

Site description

A complex of brackish lagoons and two small reservoirs, generally with wide belts of emergent vegetation. They are surrounded by olive-groves and cereal cultivation.

Habitats Wetland (80%; standing brackish and salt water; water-fringe vegetation), Artificial landscape (20%; arable land; perennial crops/orchards/groves)
Land-use Agriculture (*), Nature conservation/research (*)

Birds

Species	Season	Year	Pop min	Pop max	Acc	Criteria
Oxyura leucocephala White-headed Duck	B	1996	7	45	A	A1, B1i, B2, C1, C2, C6

Species ... continued	Season	Year	Pop min	Pop max	Acc	Criteria
Oxyura leucocephala White-headed Duck	W	1997	40	100	A	A1, B1i, C1, C2
Porphyrio porphyrio Purple Gallinule	R	1992	10	—	B	C6

This is an important area for breeding and wintering waterbirds, particularly *Oxyura leucocephala*.

Protection status

National Partial **International** Partial
1,460 ha of IBA covered by Natural Landscape (Embalse de Cordobilla, 1,460 ha). 512 ha of IBA covered by Natural Landscape (Embalse de Malpasillo, 512 ha). 13 ha of IBA covered by Natural Reserve (Laguna Amarga, 13 ha). 21 ha of IBA covered by Natural Reserve (Laguna de Tiscar, 21 ha). 66 ha of IBA covered by Natural Reserve (Laguna de Zóñar, 66 ha). 5 ha of IBA covered by Natural Reserve (Laguna de los Jarales, 5 ha). 7 ha of IBA covered by Natural Reserve (Laguna del Rincón, 7 ha). 86 ha of IBA covered by Ramsar Site (Lagunas del sur de Córdoba (Zóñar, Rincón y Amarga, 86 ha). 1,972 ha of IBA covered by Ramsar Site (Embalses de Cordobillo y Melpasillo, 1.972 ha). 1,018 ha of IBA covered by Special Protection Area (Lagunas del Sur de Córdoba, 1,107 ha).

Conservation issues

Threats Agricultural intensification/expansion (A), Consequences of animal/plant introductions (A), Groundwater abstraction (U)

The main threats are the presence of *Oxyura jamaicensis* and over-abstraction of water for agriculture. Monitoring of *O. leucocephala* and control of *O. jamaicensis* is undertaken by the regional government.

Fuente de Piedra, Gosque and Campillos lakes A1, A4i, B1i, B2, C1, C2, C3, C6 240

Admin region Andalucía
Coordinates 37°10'N 4°45'W
Altitude 400–550 m **Area** 10,600 ha

Site description

A wetland complex comprising a group of shallow, seasonal, brackish lakes, surrounded by cultivation.

Habitats Wetland (15%; standing brackish and salt water; water-fringe vegetation), Artificial landscape (85%; arable land; perennial crops/orchards/groves)
Land-use Agriculture (85%), Hunting (85%), Nature conservation/research (15%)

Birds

Species	Season	Year	Pop min	Pop max	Acc	Criteria
Phoenicopterus ruber Greater Flamingo	B	1996	2,000	16,500	A	A4i, B1i, B2, C2, C6
Phoenicopterus ruber Greater Flamingo	W	1994	3,549	—	B	A4i, B1i, C2
Oxyura leucocephala White-headed Duck	B	1996	2	5	B	A1, B1i, B2, C1, C2, C6
Grus grus Crane	W	1996	500	2,140	A	A4i, B1i, C2
Charadrius alexandrinus Kentish Plover	R	1996	100	—	B	B2
Larus fuscus Lesser Black-backed Gull	W	1994	4,629	16,000	B	A4i, B1i, C3
Gelochelidon nilotica Gull-billed Tern	B	1992	400	676	A	A4i, B1i, B2, C2, C6

This is an important site for breeding and wintering waterbirds. It is the only regular breeding area for *Phoenicopterus ruber* in Spain. Species of global conservation concern that do not meet IBA criteria: *Tetrax tetrax* (83 wintering individuals).

Protection status

National Partial **International** Partial
1,554 ha of IBA covered by Natural Reserve (Laguna de Fuentepiedra, 1,554 ha). 80 ha of IBA covered by Natural Reserve (Laguna de Campillos, 80 ha). 23 ha of IBA covered by Natural Reserve (Laguna de la Ratosa, 23 ha). 28 ha of IBA covered by Natural Reserve (Laguna del Gosque, 28 ha). 1,554 ha of IBA covered by Ramsar Site (Laguna de Fuente de Piedra, 1,554 ha). 1,364 ha of IBA covered by Special Protection Area (Laguna de Fuentepiedra, 1,364 ha).

Conservation issues

Threats Agricultural intensification/expansion (B), Burning of vegetation (B), Drainage (B), Filling-in of wetlands (B), Groundwater abstraction (B)

Threats include the expansion of olive cultivation, sedimentation in the lakes, burning of the reedbeds, drainage and over-extraction of groundwater and infilling with rubbish. A management plan is being produced.

Mountain ranges of Ronda, Bermeja and Crestellina — B2, C6 — 241

Admin region Andalucía
Coordinates 36°35′N 5°06′W
Altitude 335–1,919 m **Area** 105,000 ha

Site description

A limestone massif with large areas of exposed rock. The vegetation comprises alpine pastures, scrub, maquis and *Quercus*, *Pinus* and *Abies* forest. The main human activities are livestock-farming, hunting and tourism.

Habitats Forest and woodland (*; native coniferous forest; broadleaved evergreen forest), Scrub (*; scrub; sclerophyllous scrub/garrigue/maquis), Grassland (steppe/dry calcareous grassland), Rocky areas (*; scree/boulders; inland cliff), Artificial landscape (arable land)
Land-use Agriculture (*), Hunting, Tourism/recreation

Birds

Species	Season	Year	Pop min	Pop max	Acc	Criteria
Hieraaetus pennatus Booted Eagle	B	1995	20	—	C	B2, C6
Hieraaetus fasciatus Bonnelli's Eagle	R	1995	6	6	B	B2

This is an important area for breeding raptors.

Protection status

National Partial **International** None
1,000 ha of IBA covered by National Game Reserve (Serrania de Ronda, 21,982 ha). 18,530 ha of IBA covered by Natural Park (Sierra de las Nieves, 18,530 ha). 5,316 ha of IBA covered by Natural Landscape (Los Reales de Sierra Bermeja, 1,236 ha). 478 ha of IBA covered by Natural Landscape (Sierra Crestellina, 478 ha).

Conservation issues

Threats Burning of vegetation (B), Forest grazing (B), Groundwater abstraction (B)

The main threats are from groundwater extraction, overgrazing preventing regeneration in forested areas, and forest fires.

Líjar mountain range–Zaframagon crag — A4ii, B1iii, B2, C2, C6 — 242

Admin region Andalucía
Coordinates 36°52′N 5°23′W
Altitude 400–1,051 m **Area** 6,850 ha

Site description

A mountainous area with limestone cliffs. The vegetation comprises scrub, grassland and *Quercus* woodland.

Habitats Forest and woodland (*; alluvial/very wet forest; broadleaved evergreen forest), Scrub (*; scrub), Grassland (*; steppe/dry calcareous grassland), Wetland (river/stream; water-fringe vegetation), Rocky areas (inland cliff), Artificial landscape (highly improved reseeded grassland; arable land)
Land-use Agriculture (*), Hunting, Tourism/recreation

Birds

Species	Season	Year	Pop min	Pop max	Acc	Criteria
1 ***Gyps fulvus*** Griffon Vulture	R	1995	230	—	A	A4ii,B1iii,B2,C2,C6

1. Large increase.

This is an important site for breeding raptors, particularly *Gyps fulvus*.

Protection status

National Low **International** None
311 ha of IBA covered by Natural Reserve (Peñón de Zaframagón, 311 ha).

Conservation issues

Threats Recreation/tourism (A), Unsustainable exploitation (B)

The main threats are from disturbance from visitors, particularly of breeding *Gyps fulvus*, and high hunting pressure. A management plan is being prepared.

Ubrique and Grazalema mountain ranges — A4ii, B1iii, B2, C2, C6 — 243

Admin region Andalucía
Coordinates 36°40′N 5°25′W
Altitude 180–1,654 m **Area** 90,500 ha

Site description

A mountain massif with many high limestone cliffs. The vegetation comprises scrub, grassland and *Quercus* and *Abies* forests. The main human activities are livestock-farming, hunting and tourism.

Habitats Forest and woodland (35%; native coniferous forest; alluvial/very wet forest; broadleaved evergreen forest), Scrub (10%; scrub; sclerophyllous scrub/garrigue/maquis), Grassland (10%), Wetland (5%; standing fresh water; river/stream), Rocky areas (20%; scree/boulders; inland cliff), Artificial landscape (20%; arable land; perennial crops/orchards/groves)
Land-use Agriculture (90%), Forestry, Hunting (60%)

Birds

Species	Season	Year	Pop min	Pop max	Acc	Criteria
Neophron percnopterus Egyptian Vulture	B	1995	13	—	A	B2, C6
1 ***Gyps fulvus*** Griffon Vulture	R	1996	320	320	A	A4ii,B1iii,B2,C2,C6
Hieraaetus fasciatus Bonnelli's Eagle	R	1996	12	—	A	B2, C2, C6

1. Large increase.

This is an important area for cliff-nesting raptors, particularly *Gyps fulvus* and *Hieraaetus fasciatus*. Species of global conservation concern that do not meet IBA criteria: *Aquila adalberti* (non-breeding) and *Falco naumanni* (7 breeding pairs).

Protection status

National Partial **International** Partial
51,695 ha of IBA covered by Natural Park (Sierra de Grazalema, 51,695 ha). 51,695 ha of IBA covered by Biosphere Reserve (Sierra de Grazalema, 51,695 ha). 47,210 ha of IBA covered by Special Protection Area (Sierra de Grazalema, 47,210 ha).

Conservation issues

Threats Burning of vegetation (B), Construction/impact of dyke/dam/barrage (C), Other (A), Recreation/tourism (B)

The main threats are disturbance from tourists and recreational activities, and the use of poisoned baits ('Other' threats, above). A management plan is being prepared by the regional government.

Cabras, Aljibe and Montecoche mountain range — A1, A4iv, B1iii, B1iv, B2, C1, C2, C5, C6 — 244

Admin region Andalucía
Coordinates 36°25′N 5°35′W
Altitude 100–1,091 m **Area** 133,500 ha

Site description

A range of hills to the east of Alcalá de los Gazules. The site includes some large *Quercus* woodlands and areas of scrub. The main human activities are livestock-farming, hunting and cork harvesting ('Other' land-use, below).

Habitats Forest and woodland (60%; alluvial/very wet forest; broadleaved evergreen forest), Scrub (20%; sclerophyllous scrub/garrigue/maquis), Grassland (20%; dry siliceous grassland), Rocky areas (scree/boulders; inland cliff), Artificial landscape (highly improved reseeded grassland; arable land; perennial crops/orchards/groves)
Land-use Agriculture (35%), Hunting (90%), Other (60%)

Birds

Species	Season	Year	Pop min	Pop max	Acc	Criteria
Ciconia ciconia White Stork	B	1996	52	52	A	C6
Neophron percnopterus Egyptian Vulture	B	1995	10	14	A	B2, C6
Gyps fulvus Griffon Vulture	R	1995	94	94	A	B1iii, B2, C2
Hieraaetus pennatus Booted Eagle	B	1996	34	—	B	B2, C6
Hieraaetus fasciatus Bonnelli's Eagle	R	1996	13	13	A	B2, C2, C6
[1] ***Falco naumanni*** Lesser Kestrel	B	1994	62	69	A	A1, B2, C1
Falco peregrinus Peregrine	R	1996	22	25	A	B2, C6
Apus caffer White-rumped Swift	B	1996	Common		—	C6

1. Large increase.

This is an important site for breeding raptors and passage migrants. It is also a major migratory bottleneck, where 20,000 storks and raptors regularly pass. Species of global conservation concern that do not meet IBA criteria: *Aquila adalberti* (non-breeding).

Protection status

National High **International** High
127,925 ha of IBA covered by Natural Park (Los Alcornocales, 170,025 ha). 127,925 ha of IBA covered by Special Protection Area (Los Alcornocales, 170,025 ha).

Conservation issues

Threats Burning of vegetation (B), Construction/impact of dyke/dam/barrage (B), Groundwater abstraction (B), Infrastructure (B)

The main threats are a large dam project on the Hozgarganta river, and fires.

Bujeo, Ojén, del Niño and Blanquilla mountain ranges

A4ii, A4iv, B1iii, B1iv, B2, C2, C5, C6 **245**

Admin region Andalucía
Coordinates 36°10′N 5°35′W
Altitude 100–786 m **Area** 42,100 ha

Site description

A chain of limestone hills covered by *Quercus* woodland and scrub. The main human activities are livestock-farming and cork harvesting ('Other' land-use, below).

Habitats Forest and woodland (*; broadleaved deciduous forest; alluvial/very wet forest; broadleaved evergreen forest), Scrub (scrub; sclerophyllous scrub/garrigue/maquis), Rocky areas (inland cliff), Artificial landscape (arable land)
Land-use Agriculture (*), Hunting, Military, Other

Birds

Species	Season	Year	Pop min	Pop max	Acc	Criteria
Neophron percnopterus Egyptian Vulture	B	1996	13	—	A	B2, C6
Gyps fulvus Griffon Vulture	R	1996	277	277	A	A4ii,B1iii,B2,C2,C6
Hieraaetus fasciatus Bonnelli's Eagle	R	1996	8	8	A	B2
Bubo bubo Eagle Owl	R	1996	20	—	C	C6
Apus caffer White-rumped Swift	B	1996	5	—	C	C6

The site is a major migratory bottleneck, where 20,000 storks and raptors regularly pass.

Protection status

National High **International** High
42,100 ha of IBA covered by Natural Park (Los Alcornocales, 170,025 ha). 42,100 ha of IBA covered by Special Protection Area (Los Alcornocales, 170,025 ha).

Conservation issues

Threats Deforestation (commercial) (A), Firewood collection (B), Infrastructure (A), Intensified forest management (C), Selective logging/cutting (B), Unsustainable exploitation (B)

The main threat is from infrastructural developments, including a new road and windmills for wind-power. There are additional problems from illegal building and illegal hunting. There is a management plan for the Natural Park.

Tarifa

A1, A4i, A4ii, A4iv, B1i, B1iii, B1iv, B2, C1, C2, C5 **246**

Admin region Andalucía
Coordinates 36°05′N 5°35′W
Altitude 0–178 m **Area** 5,000 ha

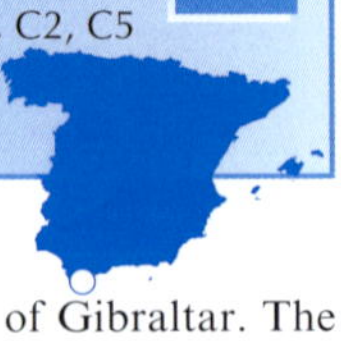

Site description

A coastal area around Tarifa bordering the Strait of Gibraltar. The site includes plains and low hills covered by Mediterranean scrub and grassland. The main human activities are arable cultivation, livestock-farming, tourism, and military use.

Habitats Scrub (*; sclerophyllous scrub/garrigue/maquis), Wetland (mudflat/sandflat; sand-dunes/sand beach; river/stream; water-fringe vegetation), Artificial landscape (arable land; ruderal land)
Land-use Agriculture (*), Military, Tourism/recreation

Birds

Species	Season	Year	Pop min	Pop max	Acc	Criteria
Ciconia nigra Black Stork	P	1995	2,000	—	C	A4i, B1i, C2
Ciconia ciconia White Stork	P	1995	40,000	120,000	C	A4i, B1i, C2
Neophron percnopterus Egyptian Vulture	P	1980	1,300	3,800	C	A4ii, B1iii, C2
Gyps fulvus Griffon Vulture	P	1980	700	1,000	C	A4ii, B1iii, C2
Falco naumanni Lesser Kestrel	P	1980	—	500	C	A1, B1iii, C1, C2
Charadrius alexandrinus Kentish Plover	R	1995	80	—	A	B2
Larus audouinii Audouin's Gull	P	1995	500	—	C	A1, A4i, B1i, C1, C2

The site is a major migratory bottleneck, where more than 20,000 storks and raptors regularly pass, including (figures are seasonal totals) *Pernis apivorus* (8,000–11,500 birds), *Milvus migrans* (45,000–60,000), *Circaetus gallicus* (4,000), *Buteo buteo* (3,000) and *Hieraaetus pennatus* (4,000).

Protection status

National Low **International** None
226 ha of IBA covered by Natural Landscape (Los Lances, 226 ha).

Conservation issues

Threats Groundwater abstraction (B), Industrialization/urbanization (B), Infrastructure (A), Recreation/tourism (B), Unsustainable exploitation (B)

The main threats are from new roads, illegal hunting, tourism development and the erection of windmills.

Ceuta

A1, A4i, A4ii, A4iv, B1i, B1iii, B1iv, C1, C2, C3, C5 **247**

Admin region Ceuta
Coordinates 35°54′N 5°19′W
Altitude 0–360 m **Area** 1,900 ha

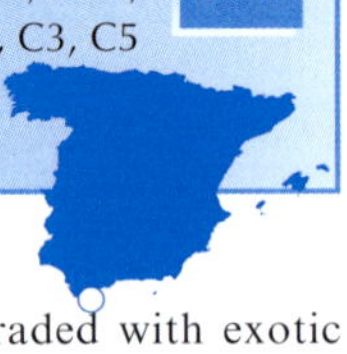

Site description

The habitat of Ceuta peninsula is partially degraded with exotic *Eucalyptus* and *Pinus* plantations. The city of Ceuta lies within the site.

Habitats Forest and woodland (40%; broadleaved evergreen forest), Scrub (20%; sclerophyllous scrub/garrigue/maquis), Wetland (shingle/stony beach; river/stream), Rocky areas (sea cliff/rocky shore), Artificial landscape (perennial crops/orchards/groves; forestry plantation; other urban/industrial areas)
Land-use Forestry (40%), Military (10%), Tourism/recreation, Urban/industrial/transport (35%)

Birds

The site is a major migratory bottleneck, where more than 20,000 storks and raptors regularly pass in spring and autumn, including (seasonal totals in brackets) *Pernis apivorus* (10,000–48,000 birds), *Milvus migrans* (10,000–20,000) and *Circaetus gallicus* (250–3,300).

Species	Season	Year	Pop min	Pop max	Acc	Criteria
Bubulcus ibis Cattle Egret	W	1995	2,500	7,000	B	A4i, B1i, C3
Bubulcus ibis Cattle Egret	P	1995	15,700	—	B	A4i, B1i, C3
Ciconia nigra Black Stork	P	1995	86	198	B	A4i, B1i, C2
Ciconia ciconia White Stork	P	1995	—	9,000	B	A4i, B1i, C2
Neophron percnopterus Egyptian Vulture	P	1995	100	200	B	B1iii, C2
Gyps fulvus Griffon Vulture	P	1995	420	1,000	B	A4ii, B1iii, C2
Falco naumanni Lesser Kestrel	P	1995	30	80	B	A1, C1
Larus audouinii Audouin's Gull	P	1995	400	—	B	A1, A4i, B1i, C1, C2
Merops apiaster Bee-eater	P	1995	7,000	28,000	B	B1iii, C3

Protection status
National None **International** None

Conservation issues
Threats Unsustainable exploitation (A)

The main threat is the intense spring hunting of migratory birds.

De la Plata mountain range — A4iv, B1iv, C5 — 248
Admin region Andalucía
Coordinates 36°05′N 5°45′W
Altitude 20–459 m **Area** 3,000 ha

Site description
An area of hills with Mediterranean scrub, patches of *Quercus* woodland and *Eucalyptus* plantations. The site includes a military base and a residential area.

Habitats Forest and woodland (broadleaved evergreen forest), Scrub (sclerophyllous scrub/garrigue/maquis), Rocky areas (scree/boulders; inland cliff), Artificial landscape (90%; arable land; forestry plantation)
Land-use Agriculture, Forestry (60%), Military (100%), Urban/industrial/transport

Birds
This is a very important site for passage migrants. The site is a major migratory bottleneck, where more than 20,000 storks and raptors regularly pass.

Protection status
National None **International** None

Conservation issues
Threats Burning of vegetation (B), Infrastructure (A), Selective logging/cutting (B)

The main threats are from forestry operations, fires, and the construction of wind-farms.

Tajo de Barbate coastal cliffs — A4i, B1i, C3 — 249
Admin region Andalucía
Coordinates 36°12′N 6°00′W
Altitude 0–169 m **Area** 2,017 ha

Site description
The site includes coastal cliffs and an inland plateau dominated by a large *Pinus* forest, scrub and dry grassland.

Habitats Forest and woodland (*; native coniferous forest), Scrub (scrub), Grassland (dry siliceous grassland), Rocky areas (*; sea cliff/rocky shore)
Land-use Agriculture

Birds

Species	Season	Year	Pop min	Pop max	Acc	Criteria
[1] ***Bubulcus ibis*** Cattle Egret	R	1996	400	700	A	A4i, B1i, C3

1. Large decrease.

An important site for breeding herons, particularly *Bubulcus ibis*.

Protection status
National High **International** None
2,017 ha of IBA covered by Natural Park (Acantilado y Pinar de Barbate, 2,017 ha).

Conservation issues
Threats Burning of vegetation (A), Industrialization/urbanization (C)

Fire is the main threat.

La Janda — A4i, A4iv, B1i, B1iv, B2, C2, C3, C5 — 250
Admin region Andalucía
Coordinates 36°15′N 5°51′W
Altitude 20–112 m **Area** 20,000 ha

Site description
The site includes a large expanse of land devoted to agriculture and cattle ranching, with large drainage channels that retain some aquatic vegetation. Many areas flood in winter.

Habitats Grassland (humid grassland), Wetland (standing fresh water; river/stream), Artificial landscape (*; arable land)
Land-use Agriculture (*), Urban/industrial/transport

Birds

Species	Season	Year	Pop min	Pop max	Acc	Criteria
Bubulcus ibis Cattle Egret	R	1994	2,000	3,000	A	A4i, B1i, C3
Glareola pratincola Collared Pratincole	B	1996	—	100	B	A4i, B1i, B2, C2

This is an important site for migrant raptors and storks and foraging area for dispersing juvenile raptors. It is also a major migratory bottleneck, where about 85,000 *Ciconia ciconia* pass regularly in spring. Species of global conservation concern that do not meet IBA criteria: *Aquila adalberti* (non-breeding), *Tetrax tetrax* (35 resident pairs) and *Otis tarda* (4 resident pairs).

Protection status
National None **International** None

Conservation issues
Threats Agricultural intensification/expansion (A), Dredging/canalization (A), Groundwater abstraction (A), Infrastructure (A)

Continued drainage is the major threat.

Cádiz bay — A4i, B1i, B2, B3, C2, C3, C6 — 251
Admin region Andalucía
Coordinates 36°35′N 6°20′W
Altitude 0–20 m **Area** 12,500 ha

Site description
A shallow bay with a wide intertidal zone and a large saltpan complex. There are rivers, small lagoons, artificial channels and halophytic scrub. The main human activities are aquaculture, shellfish collection, research and salt production.

Habitats Wetland (100%; tidal river/enclosed tidal water; mudflat/sandflat; saltmarsh; coastal lagoon; standing brackish and salt water; water-fringe vegetation)
Land-use Fisheries/aquaculture (80%), Nature conservation/research (100%), Urban/industrial/transport

Birds

Species	Season	Year	Pop min	Pop max	Acc	Criteria
[1,2] ***Phalacrocorax carbo*** Cormorant	W	1995	—	1,188	A	B1i, C3
Egretta garzetta Little Egret	W	1995	—	1,094	A	A4i, B1i, C2
Phoenicopterus ruber Greater Flamingo	N	1992	—	1,200	B	A4i, B1i, C2
Fulica cristata Crested Coot	B	1987	—	—	—	B2, C2, C6
Himantopus himantopus Black-winged Stilt	B	1992	2,000	—	C	A4i, B1i, C2, C6

Species ... continued	Season	Year	Pop min	Pop max	Acc	Criteria
Himantopus himantopus Black-winged Stilt	W	1991	436	436	A	B1i, C2
Recurvirostra avosetta Avocet	B	1992	1,000	—	C	A4i, B1i, B3, C2, C6
Recurvirostra avosetta Avocet	W	1991	—	1,399	A	A4i, B1i, B2, C2
Glareola pratincola Collared Pratincole	B	1992	140	—	C	A4i, B1i, B2, C2, C6
Charadrius alexandrinus Kentish Plover	B	1992	Abundant	—		B1i, B2, C3
[3] ***Charadrius alexandrinus*** Kentish Plover	W	1991	—	3,171	A	A4i, B1i, C3
Calidris alpina Dunlin	W	1991	—	12,176	A	B2
Limosa lapponica Bar-tailed Godwit	W	1991	—	373	A	B2
[3] ***Numenius arquata*** Curlew	W	1991	—	465	A	B2
[3] ***Tringa totanus*** Redshank	W	1991	—	1,630	A	B1i, C3
[2] ***Larus fuscus*** Lesser Black-backed Gull	W	1991	—	5,543	A	B1i, C3
[2] ***Larus cachinnans*** Yellow-legged Gull	W	1991	—	3,610	A	B1i, C3
Sterna albifrons Little Tern	B	1995	30	—	C	C2

1. *Ph. c. sinensis*.
2. Large increase.
3. Large decrease.

This is a very important area for breeding, passage and wintering waterbirds, particularly waders.

■ Protection status

National Partial **International** Low

10,000 ha of IBA covered by Natural Park (Bahia de Cadiz, 10,000 ha). 49 ha of IBA covered by Natural Reserve (Complejo endorreico de Chiclana, 49 ha). 525 ha of IBA covered by Natural Landscape (Isla de Trocadero, 525 ha). 170 ha of IBA covered by Natural Landscape (Marismas de Sancti Petri, 170 ha). 49 ha of IBA covered by Special Protection Area (Lagunas de Chiclana: Laguna de Montellano y de Jel, 49 ha).

■ Conservation issues

Threats Aquaculture/fisheries (A), Consequences of animal/plant introductions (B), Drainage (C), Dredging/canalization (U), Filling-in of wetlands (A), Industrialization/urbanization (A), Other (B), Recreation/tourism (B), Unsustainable exploitation (A)

There are a number of serious threats to the site. Aquaculture projects have caused the dredging of lagoons and saltpans and the destruction of islands used by breeding species. Shellfish collection impacts on the birds. The expansion of the *Larus cachinnans* colony ('Other' threats, above) and the presence of feral cats and dogs are also important threats. Other threats include intensive poaching, tourism developments and urban and industrial pollution.

Medina and Puerto Real lagoons

A1, A4i, B1i, B2, C1, C2, C3, C6 **252**

Admin region Andalucía
Coordinates 36°37′N 6°03′W
Altitude 20–117 m **Area** 4,900 ha

■ Site description

A complex of lagoons with fluctuating water-levels. Some become dry in summer. There are several areas of dense *Tamarix, Juncus, Typha* and *Phragmites*-dominated vegetation. The marshland is surrounded by scrub, grassland and cereal cultivation.

Habitats Scrub (5%; scrub), Grassland (25%; dry siliceous grassland), Wetland (15%; standing brackish and salt water; water-fringe vegetation), Artificial landscape (55%; arable land)
Land-use Agriculture (55%), Not utilized (10%), Tourism/recreation (5%)

■ Birds

An important site for wintering and non-breeding wildfowl and also for dispersing *Aquila adalberti*.

Species	Season	Year	Pop min	Pop max	Acc	Criteria
Phoenicopterus ruber Greater Flamingo	N	1992	—	800	B	A4i, B1i, C2
Marmaronetta angustirostris Marbled Duck	N	1990	—	110	A	A1, A4i, B1i, C1, C2
Oxyura leucocephala White-headed Duck	N	1997	104	—	A	B1i, C2
Aquila adalberti Spanish Imperial Eagle	N	1994	7	—	A	A1, C1

Species ... continued	Season	Year	Pop min	Pop max	Acc	Criteria
Porphyrio porphyrio Purple Gallinule	R	1992	10	—	B	C6
Fulica atra Coot	W	1992	25,000	—	B	B1i, C3
Fulica cristata Crested Coot	R	1996	3	—	B	B2, C2, C6

■ Protection status

National Low **International** Low

104 ha of IBA covered by Natural Reserve (Complejo endorreico de Puerto Real, 104 ha). 121 ha of IBA covered by Natural Reserve (Laguna de Medina, 121 ha). 4 ha of IBA covered by Natural Reserve (Lagunas de las Canteras y El Tejón, 4 ha). 121 ha of IBA covered by Special Protection Area (Laguna de Medina, 121 ha). 300 ha of IBA covered by Special Protection Area (Lagunas de Puerto Real: Taraje, Comisario y San Anto, 300 ha).

■ Conservation issues

Threats Agricultural intensification/expansion (A), Consequences of animal/plant introductions (A), Groundwater abstraction (C), Infrastructure (C), Unsustainable exploitation (C)

The main threat is sedimentation as a result of agricultural intensification. The presence of *Oxyura jamaicensis* is a threat to *O. leucocephala* population.

Terry lagoons

A1, B1i, B2, C1, C2, C6 **253**

Admin region Andalucía
Coordinates 36°38′N 6°14′W
Altitude 30–56 m **Area** 350 ha

■ Site description

Three small lagoons to the north of Puerto de Santa María. They are surrounded by irrigated crops and grassland.

Habitats Wetland (90%; standing fresh water; water-fringe vegetation)
Land-use Nature conservation/research (100%)

■ Birds

Species	Season	Year	Pop min	Pop max	Acc	Criteria
Oxyura leucocephala White-headed Duck	N	1997	54	—	A	A1, B1i, C1, C2
[1] ***Fulica cristata*** Crested Coot	R	1992	16	—	B	B2, C2, C6

1. Individuals.

This is an important site for breeding and non-breeding waterbirds. Species of global conservation concern that do not meet IBA criteria: *Marmaronetta angustirostris* (non-breeding)

■ Protection status

National Partial **International** Partial

291 ha of IBA covered by Natural Reserve (Lagunas Salada, Juncosa y Chica, 291 ha). 37 ha of IBA covered by Ramsar Site (Laguna Salada, 37 ha). 63 ha of IBA covered by Special Protection Area (Lagunas de Terry: Salada, Juncosa y Chica, 63 ha).

■ Conservation issues

Threats Agricultural intensification/expansion (B), Burning of vegetation (B), Infrastructure (B), Unsustainable exploitation (B)

The main threats are from fires in the surrounding vegetation, collisions with powerlines, illegal hunting and increased sedimentation as a result of agricultural intensification. A management plan exists.

Bornos reservoir

A4i, B1i, C2, C3, C6 **254**

Admin region Andalucía
Coordinates 36°45′N 5°34′W
Altitude 100–150 m **Area** 2,500 ha

■ Site description

A reservoir on the River Guadalete surrounded by maquis, grasslands and arable cultivation. At one end of the reservoir there is a large expanse of *Tamarix*.

Habitats Forest and woodland (alluvial/very wet forest), Wetland (100%; standing fresh water; water-fringe vegetation)
Land-use Water management (100%)

Birds

Species	Season	Year	Pop min	Pop max	Acc	Criteria
Nycticorax nycticorax Night Heron	B	1996	40	50	B	C6
Bubulcus ibis Cattle Egret	R	1996	2,000	2,000	B	A4i, B1i, C3
Egretta garzetta Little Egret	R	1996	150	150	B	C6
Ardea purpurea Purple Heron	R	1996	50	50	B	C2, C6

This is an important site for breeding herons.

Protection status

National Partial **International** None
630 ha of IBA covered by Natural Landscape (Cola del Embalse de Bornos, 630 ha).

Conservation issues

Threats Agricultural intensification/expansion (A)

The main threat is from agricultural intensification which is leading to increased sedimentation and pesticide pollution of the water.

Medina–Sidonia A1, A4i, B1i, C1, C3, C6 255

Admin region Andalucía
Coordinates 36°37′N 5°55′W
Altitude 75–750 m **Area** 22,000 ha

Site description

An area dominated by sheep- and cattle-grazed dry grasslands and wheat cultivation. There are a few streams and ponds surrounded by *Tamarisk* and *Salix* scrub.

Habitats Grassland (*; dry siliceous grassland), Artificial landscape (*; arable land)
Land-use Agriculture (100%)

Birds

Species	Season	Year	Pop min	Pop max	Acc	Criteria
Bubulcus ibis Cattle Egret	R	1996	3,700	3,700	B	A4i, B1i, C3
Egretta garzetta Little Egret	R	1996	90	100	A	C6
Aquila adalberti Spanish Imperial Eagle	N	1994	5	—	A	A1, C1

This is an important site for breeding herons and dispersing raptors, particularly juvenile *Aquila adalberti.*

Protection status

National None **International** None

Conservation issues

Threats Agricultural intensification/expansion (U)

The main threat is from agricultural intensification and expansion.

Confidential A1, C1 256

Admin region Andalucía
Coordinates Coordinates withheld
Altitude 150–165 m **Area** 2,000 ha

Site description

An area of dehesa used for livestock-farming and hunting.

Habitats Forest and woodland (100%; broadleaved evergreen forest), Grassland
Land-use Agriculture (100%), Hunting

Birds

This is an important site for dispersing juvenile raptors, particularly *Aquila adalberti.*

Species	Season	Year	Pop min	Pop max	Acc	Criteria
Aquila adalberti Spanish Imperial Eagle	N	1994	3	6	A	A1, C1

Protection status

National None **International** None

Conservation issues

Threats Unsustainable exploitation (U)

No serious threats are known at the site.

Los Tollos lake A1, A4i, B1i, C1, C2 257

Admin region Andalucía
Coordinates 36°52′N 6°00′W
Altitude 60–60 m **Area** 100 ha

Site description

An isolated endorreic lagoon with a fluctuating water-level. It is surrounded by *Tamarix* copses. Part of the lagoon is a gravel-pit.

Habitats Wetland (100%; standing fresh water; water-fringe vegetation)
Land-use Agriculture (*), Urban/industrial/transport (*)

Birds

Species	Season	Year	Pop min	Pop max	Acc	Criteria
Marmaronetta angustirostris Marbled Duck	W	1997	30	50	A	A1, A4i, B1i, C1, C2
Oxyura leucocephala White-headed Duck	W	1997	10	444	A	A1, A4i, B1i, C1, C2
Fulica cristata Crested Coot	N	1991	2	2	A	C2

This is an important site for wintering ducks, particularly *Marmaronetta angustirostris* and *Oxyura leucocephala.*

Protection status

National None **International** None

Conservation issues

Threats Agricultural intensification/expansion (A), Extraction industry (A)

The most important threats are from gravel extraction and ploughing of fields around the lake. Hunting is forbidden.

Lebrija, Las Cabezas and Espera lagoons A1, A4i, B1i, B2, C1, C2, C6 258

Admin region Andalucía
Coordinates 36°52′N 5°51′W
Altitude 75–410 m **Area** 7,600 ha

Site description

A complex of with dense emergent vegetation. The surrounding area is dominated by arable agriculture and dehesas.

Habitats Scrub (sclerophyllous scrub/garrigue/maquis), Grassland (dry siliceous grassland), Wetland (standing fresh water; water-fringe vegetation), Artificial landscape (*; arable land)
Land-use Agriculture (*)

Birds

Species	Season	Year	Pop min	Pop max	Acc	Criteria
Marmaronetta angustirostris Marbled Duck	N	1996	75	—	B	A1, A4i, B1i, C1, C2
Oxyura leucocephala White-headed Duck	R	1996	10	—	B	A1, B1i, B2, C1, C2, C6
Oxyura leucocephala White-headed Duck	N	1997	48	—	A	A1, B1i, C1, C2
Aquila adalberti Spanish Imperial Eagle	N	1994	5	10	A	A1, C1
Porphyrio porphyrio Purple Gallinule	R	1992	10	—	B	C6
Fulica cristata Crested Coot	B	1991	9	—	B	B2, C2, C6

This is an important area for breeding and wintering ducks and for dispersing juvenile *Aquila adalberti*, *Aquila chrysaetos* and *Hieraaetus fasciatus*. Species of global conservation concern that do not meet IBA criteria: *Aythya nyroca* (wintering).

Protection status

National Low **International** Low

59 ha of IBA covered by Natural Reserve (Complejo endorreico de Espera, 59 ha). 23 ha of IBA covered by Natural Reserve (Complejo endorreico de Lebrija-Las Cabezas, 23 ha). 438 ha of IBA covered by Special Protection Area (Lagunas de Espera, 438 ha).

Conservation issues

Threats Agricultural intensification/expansion (B), Disturbance to birds (B), Natural events (B), Recreation/tourism (B), Unsustainable exploitation (B)

The main threats are from the intensification and expansion of agriculture, disturbance from visitors and gamekeepers, drought and hunting.

Guadalquivir marshes — 259

A1, A4i, A4iv, B1i, B1iv, B2, B3, C1, C2, C3, C5, C6

Admin region Andalucía

Coordinates 37°00'N 6°25'W

Altitude 0–100 m **Area** 230,000 ha

Site description

One of the largest wetlands in Europe, at the mouth of the River Guadalquivir. In the north and east natural vegetation has been replaced with rice cultivation, irrigated cultivation, aquaculture, and saltpans, although there are still expanses of halophytic scrub. Marshes, Mediterranean scrub, woodland and sand-dunes occur to the south. The marshes are flooded only seasonally with some permanent rivers and lagoons. The main human activities include arable agriculture, cattle-grazing, hunting, fishing, fish-farming, research, apiculture ('Other' land-use, below) and tourism.

Habitats Forest and woodland (20%; native coniferous forest; alluvial/very wet forest; broadleaved evergreen forest), Scrub (20%; sclerophyllous scrub/garrigue/maquis), Grassland (humid grassland), Wetland (60%; saltmarsh; sand-dunes/sand beach; coastal lagoon; standing fresh water; standing brackish and salt water; river/stream; water-fringe vegetation), Artificial landscape (arable land; forestry plantation)

Land-use Agriculture (*), Fisheries/aquaculture, Forestry (*), Hunting (*), Military, Nature conservation/research (*), Other, Tourism/recreation, Urban/industrial/transport, Water management

Birds

Species	Season	Year	Pop min	Pop max	Acc	Criteria
[1] ***Nycticorax nycticorax*** Night Heron	R	1996	50	—	B	C6
[1] ***Ardeola ralloides*** Squacco Heron	R	1996	40	—	B	B2, C2, C6
[1] ***Egretta garzetta*** Little Egret	R	1996	300	400	B	A4i, B1i, C2, C6
Ardea purpurea Purple Heron	R	1996	300	—	B	B2, C2, C6
Ciconia nigra Black Stork	W	1996	30	—	—	B1i, C2
Ciconia ciconia White Stork	R	1996	250	—	B	B2, C6
[1] ***Plegadis falcinellus*** Glossy Ibis	R	1996	15	—	A	C6
[1] ***Platalea leucorodia*** Spoonbill	R	1996	1,100	—	A	A4i, B1i, B2, C2, C6
Phoenicopterus ruber Greater Flamingo	R	1992	2,000	5,000	B	A4i, B1i, B2, C2, C6
Phoenicopterus ruber Greater Flamingo	W	1996	5,000	—	B	A4i, B1i, C2
Anser anser Greylag Goose	W	1995	45,000	100,000	B	A4i, B1i, C3
Tadorna tadorna Shelduck	W	1989	2,300	5,000	B	A4i, B1i, C3
Anas penelope Wigeon	W	1992	80,000	100,000	C	A4i, B1i, C3
Anas strepera Gadwall	W	1992	3,000	8,000	C	A4i, B1i, C3
Anas crecca Teal	W	1989	70,000	160,000	C	A4i, B1i, C3
Anas acuta Pintail	W	1989	13,500	17,000	C	A4i, B1i, C3
Anas clypeata Shoveler	W	1989	40,000	80,000	C	A4i, B1i, C3
Marmaronetta angustirostris Marbled Duck	R	1996	5	10	A	A1, A4i, B1i, B2, C1, C2, C6
Marmaronetta angustirostris Marbled Duck	W	1996	25	—	A	A1, B1i, C1, C2
Netta rufina Red-crested Pochard	B	1982	70	—	B	B2
Netta rufina Red-crested Pochard	W	1982	2,000	5,000	C	A4i, B1i, C3
Oxyura leucocephala White-headed Duck	R	1996	10	—	A	A1, B1i, B2, C1, C2, C6
Oxyura leucocephala White-headed Duck	W	1996	100	400	A	A1, A4i, B1i, C1, C2
Milvus migrans Black Kite	R	1996	600	—	B	B2, C6
Circus aeruginosus Marsh Harrier	R	1996	10	—	C	C6
Aquila adalberti Spanish Imperial Eagle	R	1996	15	—	A	A1, B2, C1, C2, C6
Porphyrio porphyrio Purple Gallinule	R	1996	400	—	C	B2, C2, C6
Fulica cristata Crested Coot	R	1996	—	—	—	B2, C2, C6
Himantopus himantopus Black-winged Stilt	R	1996	8,000	—	A	A4i, B1i, C2, C6
Recurvirostra avosetta Avocet	R	1996	8,000	—	B	A4i, B1i, B3, C2, C6
Glareola pratincola Collared Pratincole	R	1990	2,000	4,000	B	A4i, B1i, B2, C2, C6
Charadrius alexandrinus Kentish Plover	R	1990	550	—	B	A4i, B1i, B2, C3
Limosa limosa Black-tailed Godwit	W	1990	—	20,000	C	A4i, B1i, C3
Tringa totanus Redshank	R	1996	1,000	—	B	A4i, B1i, C3
Larus genei Slender-billed Gull	R	1996	100	—	B	C2, C6
Larus audouinii Audouin's Gull	W	1996	200	—	B	A1, C1
Gelochelidon nilotica Gull-billed Tern	B	1996	1,000	—	B	A4i, B1i, B2, C2, C6
Sterna albifrons Little Tern	B	1996	25	—	B	C2
Chlidonias hybridus Whiskered Tern	B	1996	3,000	—	B	A4i, B1i, B2, C2, C6
Chlidonias niger Black Tern	B	1996	7	—	B	C6

1. Large increase.

The most important wetland in Spain for breeding, passage and wintering waterbirds and passerines. Over 360 species have been recorded. Wintering waterbird numbers reach 400,000 individuals climbing to over 6 million birds during migration periods. The site is a major migratory bottleneck, where more than 20,000 storks and raptors regularly pass. Species of global conservation concern that do not meet IBA criteria: *Aythya nyroca* (wintering).

Protection status

National Partial **International** Partial

50,720 ha of IBA covered by National Park (Doñana, 50,720 ha). 54,250 ha of IBA covered by Natural Park (Doñana, 54,250 ha). 1,336 ha of IBA covered by Natural Landscape (Brazo del Este, 1,336 ha). 77,260 ha of IBA covered by Biosphere Reserve (Doñana, 77,260 ha). 77,260 ha of IBA covered by Ramsar Site (Doñana, 77,260 ha). 50,720 ha of IBA covered by Special Protection Area (Doñana, 50,720 ha). IBA overlaps with World Heritage Site (Doñana National Park).

Conservation issues

Threats Afforestation (C), Agricultural intensification/expansion (A), Aquaculture/fisheries (B), Consequences of animal/plant introductions (B), Groundwater abstraction (A), Industrialization/urbanization (B), Infrastructure (A), Natural events (A), Recreation/tourism (B), Unsustainable exploitation (B)

There are numerous threats and problems. Perhaps the most important of these relate to the expansion and intensification of agriculture, particularly the uncontrolled use of pesticides and over-exploitation of groundwater. Other threats include a high poaching pressure, intensive crayfish fishing, an increase in fish-farming, industrial pollution with heavy metals which caused a major ecological disaster in 1998, urban development, uncontrolled tourism, and hunting and road construction. There is a research station, a water management plan and a sustainable development plan.

Condado–Campiña — 260

A1, A4ii, B1iii, B2, B3, C1, C2, C6

Admin region Andalucía

Coordinates 37°22'N 6°39'W

Altitude 10–90 m **Area** 56,500 ha

Site description

An area of non-irrigated cereal crops and olive-groves north-east of Huelva.

Habitats Artificial landscape (100%; arable land; perennial crops/orchards/groves)

Land-use Agriculture, Hunting

Birds

Species	Season	Year	Pop min	Pop max	Acc	Criteria
Circus pygargus Montagu's Harrier	B	1996	101	101	A	B3, C6
Falco naumanni Lesser Kestrel	B	1996	450	—	B	A1, A4ii, B1iii, B2, C1, C2, C6

Species ... continued	Season	Year	Pop min	Pop max	Acc	Criteria
Tetrax tetrax Little Bustard	R	1996	60	—	B	C6
Otis tarda Great Bustard	R	1996	10	—	B	C6

This is an important site for species associated with open habitats.

Protection status

National None **International** None

Conservation issues

Threats Agricultural intensification/expansion (B)

The main threat is from irrigation of arable crops.

Odiel and Tinto marshes and Huelva coastal lagoons

A4i, B1i, B2, C2, C3, C6 **261**

Admin region Andalucía
Coordinates 37°15′N 6°50′W
Altitude 0–40 m **Area** 14,900 ha

Site description

A large area of tidal marshes at the mouths of the Rivers Tinto and Odiel. There are extensive areas of mud and halophytic scrub and some saltpans. The main human activities are fishing, shellfish collecting and tourism.

Habitats Scrub (scrub), Grassland (dry siliceous grassland), Wetland (*; tidal river/enclosed tidal water; mudflat/sandflat; saltmarsh; sand-dunes/sand beach; coastal lagoon; standing fresh water; river/stream; water-fringe vegetation), Artificial landscape (arable land; other urban/industrial areas)
Land-use Agriculture, Fisheries/aquaculture (10%), Tourism/recreation, Urban/industrial/transport

Birds

Species	Season	Year	Pop min	Pop max	Acc	Criteria
Podiceps nigricollis Black-necked Grebe	W	1996	150	2,000	B	A4i, B1i, C3
Ixobrychus minutus Little Bittern	B	1996	50	—	C	B2, C2, C6
Egretta garzetta Little Egret	R	1996	120	300	B	B1i, C2, C6
Ardea purpurea Purple Heron	B	1996	6	20	B	C6
Platalea leucorodia Spoonbill	R	1996	400	600	B	A4i, B1i, B2, C2, C6
Phoenicopterus ruber Greater Flamingo	N	1996	150	2,000	B	A4i, B1i, C2
Recurvirostra avosetta Avocet	P	1996	1,550	—	C	A4i, B1i, C2
Charadrius alexandrinus Kentish Plover	R	1996	150	—	B	B2
Sterna albifrons Little Tern	B	1992	300	—	C	A4i, B1i, B2, C2, C6

This is an important wetland for breeding, wintering and passage waterbirds with the second most important *Platalea leucorodia* breeding colony in Spain.

Protection status

National Partial **International** Partial
162 ha of IBA covered by Natural Landscape (Enebrales de Punta Umbría, 162 ha). 480 ha of IBA covered by Natural Landscape (Estero de Domingo Rubio, 480 ha). 693 ha of IBA covered by Natural Landscape (Lagunas de Palos y las Madres, 693 ha). 7,185 ha of IBA covered by Natural Landscape (Marismas del Odiel, 7,185 ha). 480 ha of IBA covered by Natural Reserve (Isla de Enmedio, 480 ha). 16 ha of IBA covered by Natural Reserve (Laguna del Portil, 16 ha). 7,185 ha of IBA covered by Biosphere Reserve (Marismas del Odiel, 7,185 ha). 7,185 ha of IBA covered by Ramsar Site (Marismas del Odiel, 7,185 ha). 7,185 ha of IBA covered by Special Protection Area (Marismas del Odiel, 7,185 ha).

Conservation issues

Threats Dredging/canalization (B), Filling-in of wetlands (B), Industrialization/urbanization (A), Infrastructure (B), Recreation/tourism (A)

The main threats are from industrial pollution, infrastructural development and uncontrolled tourism. A management plan exists for part of the site.

River Piedras marshes and Rompido sandbank

B1i, B2, C2 **262**

Admin region Andalucía
Coordinates 37°14′N 7°09′W
Altitude 0–20 m **Area** 3,120 ha

Site description

The site includes a 12-km-long sandbank, productive sandflats and an estuarine area. The main human activities include fisheries and tourism.

Habitats Forest and woodland (10%; native coniferous forest), Scrub (10%; scrub), Wetland (100%; tidal river/enclosed tidal water; mudflat/sandflat; sand-dunes/sand beach; water-fringe vegetation), Marine areas (50%; sea inlet/coastal features)
Land-use Fisheries/aquaculture (60%), Tourism/recreation (20%)

Birds

Species	Season	Year	Pop min	Pop max	Acc	Criteria
Platalea leucorodia Spoonbill	N	1994	34	34	B	B1i, C2
Charadrius alexandrinus Kentish Plover	R	1994	50	100	B	B2
Calidris alpina Dunlin	W	1994	1,500	—	B	B2
Sterna albifrons Little Tern	B	1994	20	30	B	C2

This is an important area for wintering waders.

Protection status

National Partial **International** None
2,530 ha of IBA covered by Natural Landscape (Marismas del Rio Piedras y Flecha del Rompido, 2,530 ha).

Conservation issues

Threats Agricultural intensification/expansion (B), Aquaculture/fisheries (C), Industrialization/urbanization (C), Recreation/tourism (B)

The main threats are from disturbance from tourists and shellfish collectors and deteriorating water-quality as a result of agricultural pollution.

Isla Cristina and Ayamonte marshes and Prado lagoon

A4i, B1i, B2, B3, C2, C3, C6 **263**

Admin region Andalucía
Coordinates 37°13′N 7°25′W
Altitude 0–10 m **Area** 3,390 ha

Site description

Coastal marshes at the mouth of the River Guadiana with sand beaches, halophytic scrub, and numerous abandoned saltpans. Human activities include scrub collection, fishing and tourism.

Habitats Wetland (100%; tidal river/enclosed tidal water; mudflat/sandflat; saltmarsh)
Land-use Fisheries/aquaculture (15%), Tourism/recreation (80%), Urban/industrial/transport (20%)

Birds

Species	Season	Year	Pop min	Pop max	Acc	Criteria
Bubulcus ibis Cattle Egret	R	1996	1,000	1,000	B	A4i, B1i, C3
Egretta garzetta Little Egret	R	1996	500	500	B	A4i, B1i, C2, C6
Ciconia ciconia White Stork	R	1996	500	500	B	B1i, B2, C2, C6
Platalea leucorodia Spoonbill	R	1996	100	100	B	A4i, B1i, B2, C2, C6
Recurvirostra avosetta Avocet	B	1996	200	200	B	B3, C6
Burhinus oedicnemus Stone Curlew	R	1996	400	400	B	B2, C2, C6
Glareola pratincola Collared Pratincole	B	1996	100	100	B	A4i, B1i, B2, C2
Charadrius hiaticula Ringed Plover	W	1996	5,000	5,000	B	A4i, B1i, C3
Charadrius alexandrinus Kentish Plover	B	1996	500	500	B	A4i, B1i, B2, C3
Calidris minuta Little Stint	W	1996	5,000	5,000	—	A4i, B1i, C3
Calidris alpina Dunlin	W	1996	5,000	5,000	B	B2
Limosa lapponica Bar-tailed Godwit	W	1996	500	500	B	B2
Numenius arquata Curlew	W	1996	500	500	B	B2
Tringa totanus Redshank	W	1996	3,000	3,000	B	A4i, B1i, C3
Larus fuscus Lesser Black-backed Gull	W	1996	5,000	5,000	B	B1i, C3
Larus cachinnans Yellow-legged Gull	W	1996	5,000	5,000	B	A4i, B1i, C3

Species ... continued	Season	Year	Pop min	Pop max	Acc	Criteria
Sterna caspia Caspian Tern	R	1996	50	50	B	A4i, B1i, B2, C2, C6
Sterna sandvicensis Sandwich Tern	W	1996	500	500	B	C2
[1] **Sterna albifrons** Little Tern	B	1996	150	150	B	B1i, B2, C2, C6
Chlidonias hybridus Whiskered Tern	P	1996	300	300	B	B1i, C2

1. Large decrease.

A very important area for breeding, passage and wintering waterbirds.

Protection status

National Partial **International** None

2,145 ha of IBA covered by Natural Landscape (Marismas de Isla Cristina–Ayamonte, 2,145 ha).

Conservation issues

Threats Aquaculture/fisheries (B), Construction/impact of dyke/dam/barrage (U), Filling-in of wetlands (B), Industrialization/urbanization (B), Recreation/tourism (B)

The main threats are from uncontrolled building, irrigation schemes and aquaculture projects. Levels of disturbance from tourists are high.

West Andévalo — A1, B1i, B1iii, B2, C1, C2, C6 — 264

Admin region Andalucía
Coordinates 37°30′N 7°15′W
Altitude 50–397 m **Area** 45,800 ha

Site description

An area of dehesas and *Quercus* woodland with scrub and fallow land.

Habitats Forest and woodland (*; broadleaved evergreen forest), Scrub (scrub), Artificial landscape (*; arable land; forestry plantation)
Land-use Agriculture (*), Forestry, Hunting (*)

Birds

Species	Season	Year	Pop min	Pop max	Acc	Criteria
Ciconia nigra Black Stork	B	1996	4	5	B	B1i, C2, C6
Gyps fulvus Griffon Vulture	N	1996	300	—	B	B1iii, C2
Aegypius monachus Cinereous Vulture	N	1996	50	—	B	A1, B1iii, C1, C2
Aquila adalberti Spanish Imperial Eagle	N	1996	4	6	B	A1, C1
Alectoris rufa Red-legged Partridge	R	1996	20,000	—	B	B2
Tetrax tetrax Little Bustard	R	1996	120	—	B	C6
[1] **Otis tarda** Great Bustard	R	1996	40	—	B	C6

1. Large decrease.

This is an important area for steppic birds and feeding and roosting raptors and dispersing juveniles.

Protection status

National None **International** None

Conservation issues

Threats Afforestation (A), Agricultural intensification/expansion (A), Construction/impact of dyke/dam/barrage (A), Infrastructure (A), Other (A)

Major threats are from agricultural intensification, dam projects, the use of poison ('Other' threats, above), and the erection of powerlines.

Sierra Pelada mountain range — A1, A4ii, B1iii, B2, C1, C2, C6 — 265

Admin region Andalucía
Coordinates 37°55′N 7°05′W
Altitude 250–657 m **Area** 48,000 ha

Site description

An area of hills with slopes covered by scrub, *Quercus* woodland and *Eucalyptus* and *Pinus* plantations. Dehesas occupy the northern section of the site. The main human activities are forestry and hunting.

Habitats Forest and woodland (broadleaved deciduous forest; broadleaved evergreen forest), Scrub (scrub), Artificial landscape (*; forestry plantation)
Land-use Agriculture, Forestry (*), Hunting

Birds

Species	Season	Year	Pop min	Pop max	Acc	Criteria
Gyps fulvus Griffon Vulture	N	1996	300	—	B	B1iii, C2
[1] **Aegypius monachus** Cinereous Vulture	R	1996	61	80	A	A1, A4ii, B1iii, B2, C1, C2, C6
Circaetus gallicus Short-toed Eagle	B	1996	20	—	B	B2, C6

1. Large decrease.

This is an important site for breeding raptors, particularly for *Aegypius monachus*. Species of global conservation concern that do not meet IBA criteria: *Aquila adalberti* (non-breeding).

Protection status

National Partial **International** Partial

718 ha of IBA covered by Natural Landscape (Picos de Aroche, 718 ha). 12,980 ha of IBA covered by Natural Landscape (Sierra Pelada y Rivera del Aserador, 12,980 ha). 12,980 ha of IBA covered by Special Protection Area (Sierra Pelada y Rivera del Aserrador, 12,980 ha).

Conservation issues

Threats Afforestation (A), Deforestation (commercial) (A), Intensified forest management (A)

The main threat is from intensive forest management, which particularly threatens *Aegypius monachus*.

Las Contiendas mountain range — B1i, B1iii, C2 — 266

Admin region Andalucía
Coordinates 38°00′N 6°50′W
Altitude 400–712 m **Area** 35,000 ha

Site description

A mountainous area, north-west of Huelva, with deep gullies and river canyons. Vegetation includes dehesas, *Pinus* and *Eucalyptus* plantations, scrub and riverine vegetation around the Rivers Chanza and Múrtigas.

Habitats Forest and woodland (*; broadleaved evergreen forest), Scrub, Grassland, Artificial landscape (forestry plantation)
Land-use Agriculture, Forestry (*), Hunting

Birds

Species	Season	Year	Pop min	Pop max	Acc	Criteria
Ciconia nigra Black Stork	P	1997	30	—	B	B1i, C2
Gyps fulvus Griffon Vulture	N	1997	200	—	B	B1iii, C2

This is an important site for breeding and dispersing raptors and *Ciconia nigra*. Species of global conservation concern that do not meet IBA criteria: *Aegypius monachus* (resident) and *Aquila adalberti* (4 non-breeding individuals).

Protection status

National Partial **International** Partial

30,000 ha of IBA covered by Natural Park (Sierra de Aracena y Picos de Aroche, 184,000 ha). 30,000 ha of IBA covered by Special Protection Area (Sierra de Aracena y Picos de Aroche, 184,000 ha).

Conservation issues

Threats Intensified forest management (U), Recreation/tourism (U), Unsustainable exploitation (U)

Intensive forestry and the associated network of tracks is the most important threat. Illegal hunting and increasing visitor pressure are also problems.

East mountain range of Huelva — B1i, B2, C2, C6 — 267

Admin region Andalucía
Coordinates 37°55′N 6°30′W
Altitude 400–871 m **Area** 71,000 ha

Site description

A mountainous area north-east of Huelva. Dehesas dominate the area.

Riverine vegetation occurs along the main streams. The main human activity is livestock-farming.

Habitats Forest and woodland (*; broadleaved evergreen forest), Scrub, Grassland
Land-use Agriculture (*), Hunting

Birds

Species	Season	Year	Pop min	Pop max	Acc	Criteria
Ciconia nigra Black Stork	B	1997	4	—	B	B1i, C2
Circaetus gallicus Short-toed Eagle	B	1997	20	—	B	B2, C6
Hieraaetus pennatus Booted Eagle	B	1997	25	—	B	B2, C6

This is an important area for breeding and dispersing juvenile raptors and *Ciconia nigra*. It is also an important passage and wintering site for passerines. Species of global conservation concern that do not meet IBA criteria: *Aquila adalberti* (non-breeding) and *Falco naumanni* (2–4 breeding pairs).

Protection status

National High **International** High
71,000 ha of IBA covered by Natural Park (Sierra de Aracena y Picos de Aroche, 184,000 ha). 71,000 ha of IBA covered by Special Protection Area (Sierra de Aracena y Picos de Aroche, 184,000 ha).

Conservation issues

Threats Afforestation (B), Extraction industry (U), Recreation/tourism (C), Unsustainable exploitation (U)

The main threats are from high hunting pressure, mining, and uncontrolled tourism.

Fuente de Cantos–Montemolín

A1, B1iii, B2, C1, C2, C6 **268**

Admin region Extremadura
Coordinates 38°10′N 6°15′W
Altitude 500–731 m **Area** 49,000 ha

Site description

An area of non-irrigated cereal cultivation and grasslands, surrounded by dehesas.

Habitats Forest and woodland (*; broadleaved evergreen forest), Scrub, Grassland, Wetland (standing fresh water; river/stream), Artificial landscape (*; arable land)
Land-use Agriculture (*)

Birds

Species	Season	Year	Pop min	Pop max	Acc	Criteria
Falco naumanni Lesser Kestrel	B	1996	100	—	B	A1, B1iii, B2, C1, C2
Tetrax tetrax Little Bustard	R	1996	400	—	B	A1, C1
Burhinus oedicnemus Stone Curlew	R	1996	220	—	B	B2, C6

This is an important site for steppe birds. Species of global conservation concern that do not meet IBA criteria: *Otis tarda* (30 resident pairs).

Protection status

National None **International** None

Conservation issues

No serious threats known at this site.

Azuaga–Llerena–Peraleda de Zaucejo

A1, A4i, A4ii, B1i, B1iii, B2, B3, C1, C2, C6 **269**

Admin region Extremadura
Coordinates 38°20′N 5°40′W
Altitude 400–848 m **Area** 147,000 ha

Site description

An area of dry, cultivated plains with grasslands, small patches of *Quercus* woodland and dehesas. The main human activities are cereal and livestock-farming, hunting, and some mining.

Habitats Forest and woodland (broadleaved evergreen forest), Scrub, Grassland, Wetland (standing fresh water; river/stream), Artificial landscape (*; arable land; perennial crops/orchards/groves; forestry plantation)
Land-use Agriculture (*), Hunting

Birds

Species	Season	Year	Pop min	Pop max	Acc	Criteria
Ciconia nigra Black Stork	P	1995	15	20	B	B1i, C2
Circus pygargus Montagu's Harrier	B	1996	100	—	B	B3, C6
Falco naumanni Lesser Kestrel	B	1996	200	—	C	A1, A4ii, B1iii, B2, C1, C2, C6
Grus grus Crane	W	1992	6,935	12,939	A	A4i, B1i, C2
Tetrax tetrax Little Bustard	R	1996	3,000	—	B	A1, B2, C1, C2, C6
Otis tarda Great Bustard	R	1996	1,500	—	B	A1, B2, C1, C2, C6

This is an important area for wintering *Grus grus* and for breeding steppic species.

Protection status

National None **International** None

Conservation issues

Threats Disturbance to birds (B), Other (B), Unsustainable exploitation (B)

Localized poaching of nests, poisoning, and illegal hunting occurs. There is occasional disturbance of breeding birds by forestry work.

Central Badajoz mountain ranges

A1, B1i, B2, C1, C2, C6 **270**

Admin region Extremadura
Coordinates 38°35′N 6°05′W
Altitude 400–943 m **Area** 90,000 ha

Site description

A mountainous area in the centre of Badajoz province with high cliffs, extensive areas of broadleaved evergreen forest, sclerophyllous scrub and dehesas in the valleys. The area is used for livestock-grazing.

Habitats Forest and woodland (*; broadleaved evergreen forest), Scrub (sclerophyllous scrub/garrigue/maquis), Grassland, Rocky areas (inland cliff), Artificial landscape (arable land; perennial crops/orchards/groves; forestry plantation; ruderal land)
Land-use Agriculture (*), Forestry (*), Hunting

Birds

Species	Season	Year	Pop min	Pop max	Acc	Criteria
Ciconia nigra Black Stork	P	1995	—	15	B	B1i, C2
Circaetus gallicus Short-toed Eagle	B	1996	25	—	B	B2, C6
Aquila adalberti Spanish Imperial Eagle	R	1995	2	2	A	A1, B2, C1
Aquila adalberti Spanish Imperial Eagle	N	1996	8	—	B	A1, C1
Aquila chrysaetos Golden Eagle	R	1995	9	9	A	C6
Hieraaetus pennatus Booted Eagle	B	1996	25	—	B	B2
Hieraaetus fasciatus Bonnelli's Eagle	R	1995	11	11	A	B2, C2, C6

The site is important for raptors. Species of global conservation concern that do not meet IBA criteria: *Aegypius monachus* (25 non-breeding individuals).

Protection status

National None **International** Low
6,000 ha of IBA covered by Special Protection Area (Sierra Grande de Hornachos, 6,000 ha).

Conservation issues

Threats Disturbance to birds (U)

Forestry and agricultural operations cause disturbance to breeding birds.

Bienvenida–Usagre–Ribera del Fresno — A1, A4i, B1i, B2, C1, C2, C6 — 271

Admin region Extremadura
Coordinates 38°20′N 6°15′W
Altitude 350–655 m **Area** 6,700 ha

Site description
A plain to the east of Zafra with non-irrigated cereal cultivation, grasslands, and dehesas. There are also some vineyards and olive-groves.

Habitats Forest and woodland (broadleaved evergreen forest), Grassland, Wetland (standing fresh water; river/stream), Artificial landscape (65%; arable land; perennial crops/orchards/groves)
Land-use Agriculture (65%)

Birds

Species	Season	Year	Pop min	Pop max	Acc	Criteria
Circus pygargus Montagu's Harrier	B	1996	100	—	B	C6
Grus grus Crane	W	1995	1,200	3,000	B	A4i, B1i, C2
Tetrax tetrax Little Bustard	R	1996	500	—	B	A1, B2, C1
Otis tarda Great Bustard	R	1996	—	600	B	A1, B2, C1, C2
Burhinus oedicnemus Stone Curlew	R	1996	220	—	B	B2, C6
Pterocles orientalis Black-bellied Sandgrouse	R	1996	200	—	B	B2, C6

An important area for wintering *Grus grus* and breeding steppic birds.

Protection status
National None **International** None

Conservation issues

Threats Infrastructure (U)

The main threat is from powerlines.

Valuengo reservoir — A4i, B1i, C2, C3 — 272

Admin region Extremadura
Coordinates 38°20′N 6°40′W
Altitude 350–545 m **Area** 11,200 ha

Site description
The reservoir is surrounded by dehesas, grasslands and irrigated crops. Human activities include livestock-farming, arable cultivation and hunting.

Habitats Forest and woodland (broadleaved evergreen forest), Grassland (*), Wetland (standing fresh water; river/stream), Artificial landscape (*; arable land)
Land-use Agriculture (*), Hunting, Water management

Birds

Species	Season	Year	Pop min	Pop max	Acc	Criteria
Bubulcus ibis Cattle Egret	B	1994	1,500	1,500	A	A4i, B1i, C3
Ciconia nigra Black Stork	P	1996	115	—	B	B1i, C2

An important breeding site for herons and passage *Ciconia nigra.*

Protection status
National None **International** None

Conservation issues

Threats Agricultural intensification/expansion (B), Construction/impact of dyke/dam/barrage (A)

A planned new reservoir would destroy feeding areas of *Ciconia nigra.*

Jerez de los Caballeros dehesas — A1, A4i, B1i, B1iii, B2, C1, C2, C6 — 273

Admin region Extremadura
Coordinates 38°15′N 6°50′W
Altitude 200–776 m **Area** 160,000 ha

Site description
The site contains large areas of *Quercus*-dominated dehesas, which form mature woodland in some areas. Human activities include widespread livestock-farming and cork harvesting ('Other' land-use, below).

Habitats Forest and woodland (65%; broadleaved evergreen forest), Scrub, Grassland, Wetland (river/stream), Artificial landscape (arable land; perennial crops/orchards/groves; ruderal land)
Land-use Agriculture (*), Other

Birds

Species	Season	Year	Pop min	Pop max	Acc	Criteria
Ciconia nigra Black Stork	B	1995	40	40	A	B1i, B2, C2, C6
Ciconia nigra Black Stork	P	1996	215	—	B	A4i, B1i, C2
Milvus migrans Black Kite	B	1996	200	—	B	B2, C6
Circaetus gallicus Short-toed Eagle	B	1996	25	—	C	B2, C6
Hieraaetus pennatus Booted Eagle	B	1996	50	—	C	B2, C6
Falco naumanni Lesser Kestrel	B	1996	100	—	B	A1, B1iii, B2, C1, C2

This is an important site for breeding *Ciconia nigra* and raptors.

Protection status
National None **International** None

Conservation issues

Threats Disturbance to birds (B)

There is some disturbance of breeding species, especially *Ciconia nigra*, by livestock and farmers.

Villanueva del Fresno — A1, B1i, B2, C1, C2 — 274

Admin region Extremadura
Coordinates 38°20′N 7°10′W
Altitude 150–237 m **Area** 7,800 ha

Site description
An area to the west of Badajoz with open dehesas alternating with grasslands and cereal crops.

Habitats Forest and woodland (75%; broadleaved evergreen forest), Grassland, Artificial landscape (25%; arable land)
Land-use Agriculture (50%), Forestry (70%), Hunting (100%)

Birds

Species	Season	Year	Pop min	Pop max	Acc	Criteria
Ciconia nigra Black Stork	P	1991	8	12	B	B1i, C2
Grus grus Crane	W	1995	1,650	1,720	A	B1i, C2
Tetrax tetrax Little Bustard	R	1992	90	500	B	A1, B2, C1
Otis tarda Great Bustard	R	1995	320	320	A	A1, B2, C1, C2

This is an important area for steppic species and wintering *Grus grus.*

Protection status
National None **International** None

Conservation issues

Threats Construction/impact of dyke/dam/barrage (A), Forest grazing (B), Natural events (A)

The most important threat to the site is a plan to create the Alqueva reservoir in Portugal. Rain following a recent drought has caused widespread erosion. The area is generally overgrazed.

Valongo — A4i, B1i, C3, C6 — 275

Admin region Extremadura
Coordinates 38°40′N 7°15′W
Altitude 200–207 m **Area** 2,800 ha

Site description
The site includes a section of the River Guadiana with patches of alluvial woodland and surrounding dehesas and cereal crops. The main activity is livestock-farming.

Habitats Forest and woodland (*; alluvial/very wet forest; broadleaved evergreen forest), Grassland, Wetland (river/stream), Artificial landscape (arable land)
Land-use Agriculture (*)

Birds

Species	Season	Year	Pop min	Pop max	Acc	Criteria
Nycticorax nycticorax Night Heron	B	1994	7	7	A	C6
Bubulcus ibis Cattle Egret	B	1994	5,000	5,000	A	A4i, B1i, C3

This is an important area for species associated with riverine habitats, particularly breeding herons. The IBA holds one of Spain's most important colonies of *Bubulcus ibis*.

Protection status

National None **International** None

Conservation issues

Threats Construction/impact of dyke/dam/barrage (A)

The planned Alqueva dam in Portugal would probably flood the area.

Olivenza–La Albuera A1, B1i, B2, C1, C2, C6 276

Admin region Extremadura
Coordinates 38°50′N 6°50′W
Altitude 275–335 m **Area** 79,000 ha

Site description

Plains to the south-east of Badajoz with grasslands, dehesas and some non-irrigated cereal cultivation. There is an endorreic lagoon complex.

Habitats Forest and woodland (10%; broadleaved evergreen forest), Grassland (10%), Wetland (standing fresh water; river/stream), Artificial landscape (80%; arable land; other urban/industrial areas)
Land-use Agriculture (80%), Hunting (100%)

Birds

Species	Season	Year	Pop min	Pop max	Acc	Criteria
Ciconia ciconia White Stork	B	1996	200	—	B	B2
Elanus caeruleus Black-winged Kite	R	1996	25	—	B	B2, C2, C6
Grus grus Crane	W	1996	1,000	—	B	B1i, C2
Tetrax tetrax Little Bustard	R	1996	3,000	—	B	A1, B2, C1, C2, C6
Otis tarda Great Bustard	R	1996	1,500	—	B	A1, B2, C1, C2, C6

This is an important area for breeding steppic species, resident and wintering raptors and wintering *Grus grus*.

Protection status

National None **International** None

Conservation issues

Threats Agricultural intensification/expansion (A), Infrastructure (U)

New irrigation schemes and a new powerline are the main threats.

Alange A1, B1i, B2, C1, C2 277

Admin region Extremadura
Coordinates 38°40′N 6°15′W
Altitude 275–677 m **Area** 66,500 ha

Site description

An area of plains, in the centre of the province of Badajoz, dominated by cereal cultivation and dehesas.

Habitats Forest and woodland (*; broadleaved evergreen forest), Scrub, Grassland, Wetland (standing fresh water; river/stream), Rocky areas (inland cliff), Artificial landscape (*; arable land; perennial crops/orchards/groves)
Land-use Agriculture (100%), Water management

Birds

Species	Season	Year	Pop min	Pop max	Acc	Criteria
Hieraaetus fasciatus Bonnelli's Eagle	R	1995	6	6	A	B2
Grus grus Crane	W	1995	647	1,852	A	B1i, C2
Otis tarda Great Bustard	R	1996	36	100	B	A1, C1

An important area for raptors, steppic species, and wintering *Grus grus*.

Protection status

National None **International** None

Conservation issues

No serious threats known at the site.

Villalba de los Barros A1, B2, C1, C2 278

Admin region Extremadura
Coordinates 38°40′N 6°35′W
Altitude 350–425 m **Area** 12,500 ha

Site description

An area of plains, south of the city of Badajoz, dominated by non-irrigated cereal cultivation.

Habitats Forest and woodland (15%; broadleaved evergreen forest), Artificial landscape (85%; arable land)
Land-use Agriculture (100%)

Birds

Species	Season	Year	Pop min	Pop max	Acc	Criteria
Otis tarda Great Bustard	R	1996	200	200	B	A1, B2, C1, C2

This is an important area for steppic species. Species of global conservation concern that do not meet IBA criteria: *Tetrax tetrax* (150 resident pairs).

Protection status

National None **International** None

Conservation issues

Threats Agricultural intensification/expansion (U)

No serious threats known at the site.

Puerto de Mejoral–Almorchón–Cabeza del Buey A4i, B1i, B2, C2, C6 279

Admin region Extremadura
Coordinates 38°40′N 5°15′W
Altitude 375–938 m **Area** 82,500 ha

Site description

An area of mountains and a river valley. The mountain slopes are dominated by sclerophyllous scrub. The river valley has large dehesas, and some grassland, arable cultivation and olive-groves. The main human activities are livestock-grazing and hunting.

Habitats Forest and woodland (*; broadleaved evergreen forest), Scrub (*), Grassland, Wetland (standing fresh water; river/stream), Rocky areas (scree/boulders; inland cliff), Artificial landscape (*; arable land; perennial crops/orchards/groves; forestry plantation)
Land-use Agriculture (*), Hunting (*)

Birds

Species	Season	Year	Pop min	Pop max	Acc	Criteria
Ciconia nigra Black Stork	B	1997	9	9	A	B1i, C2
Aquila chrysaetos Golden Eagle	R	1997	9	9	A	C6
Hieraaetus fasciatus Bonnelli's Eagle	R	1997	8	8	A	B2, C6
Grus grus Crane	W	1995	2,361	6,000	B	A4i, B1i, C2

This is an important wintering area for *Grus grus* and breeding site for raptors.

Protection status

National None **International** None

Conservation issues

Threats Industrialization/urbanization (B)

The main threat is from industrial activity in Peñalgordo, which causes high levels of disturbance to breeding raptors.

La Serena A1, A4i, B1i, B2, B3, C1, C2, C3, C6 280

Admin region Extremadura
Coordinates 38°50'N 5°30'W
Altitude 300–744 m **Area** 109,800 ha

Site description

A large area of dry grassland with occasional areas of cereal cultivation. The grasslands are used for sheep grazing. There are three large dams on the site's rivers.

Habitats Scrub (5%), Grassland (50%; dry siliceous grassland)
Land-use Agriculture (100%), Hunting (100%)

Birds

Species	Season	Year	Pop min	Pop max	Acc	Criteria
Bubulcus ibis Cattle Egret	B	1994	1,140	—	A	A4i, B1i, C3
Ciconia nigra Black Stork	P	1992	140	—	A	B1i, C2
Circus pygargus Montagu's Harrier	B	1992	200	—	C	B3, C6
Grus grus Crane	W	1995	2,548	—	A	A4i, B1i, C2
[1] ***Tetrax tetrax*** Little Bustard	R	1996	2,500	—	B	A1, B2, C1, C2, C6
Otis tarda Great Bustard	R	1996	800	—	A	A1, B2, C1, C2, C6
[1] ***Burhinus oedicnemus*** Stone Curlew	R	1996	500	—	B	B2, C2, C6
Glareola pratincola Collared Pratincole	B	1992	25	—	C	C6
[1] ***Pterocles orientalis*** Black-bellied Sandgrouse	R	1996	167	167	B	B2, C6
[1] ***Pterocles alchata*** Pin-tailed Sandgrouse	R	1996	1,600	—	B	B2, C2, C6
Melanocorypha calandra Calandra Lark	R	1992	11,000	—	B	B2, C6

1. Large decrease.

This is an area of exceptional importance for steppic species, although their populations have been declining since the 1980s.

Protection status

National None **International** None

Conservation issues

Threats Agricultural intensification/expansion (A), Construction/impact of dyke/dam/barrage (C), Infrastructure (B), Natural events (U)

Enlargement of the area of irrigated crops, the building of another dam and several other infrastructure projects (oil pipeline, gas pipeline, electric powerlines) are the most important threats. Crop spraying is a further threat. The area is prone to drought.

Siruela–Agudo A4i, B1i, C2 281

Admin region Extremadura
Coordinates 39°00'N 5°00'W
Altitude 350–940 m **Area** 66,000 ha

Site description

Situated to the north of the Sierra de Siruela, the site is predominantly covered by dehesas, with grasslands and some cereal cultivation. Human activities include cattle-grazing.

Habitats Forest and woodland (*; broadleaved evergreen forest), Scrub, Grassland (*), Wetland (standing fresh water; river/stream), Artificial landscape (arable land; perennial crops/orchards/groves; forestry plantation)
Land-use Agriculture (*)

Birds

Species	Season	Year	Pop min	Pop max	Acc	Criteria
Ciconia nigra Black Stork	P	1996	—	100	B	B1i, C2
Grus grus Crane	W	1995	1,000	2,000	B	A4i, B1i, C2

This is an important area for wintering *Grus grus* and breeding raptors.

Protection status

National None **International** None

Conservation issues

Threats Other (C)

Occasional robbing of raptors' nests is reported.

Cijara reservoir A1, B1i, B2, C1, C2, C6 282

Admin region Extremadura, Castilla-La Mancha
Coordinates 39°20'N 4°50'W
Altitude 350–877 m **Area** 73,500 ha

Site description

A reservoir on the River Guadiana surrounded by areas of sclerophyllous scrub, *Quercus* woodland, *Pinus* plantations, dehesas, pasture and non-irrigated arable cultivation.

Habitats Forest and woodland (*; broadleaved evergreen forest), Scrub (scrub), Grassland (dry siliceous grassland), Wetland (standing fresh water), Rocky areas (scree/boulders), Artificial landscape (*; arable land; forestry plantation)
Land-use Agriculture, Forestry (*), Hunting (*)

Birds

Species	Season	Year	Pop min	Pop max	Acc	Criteria
Ciconia nigra Black Stork	B	1995	4	4	A	B1i, C2
Ciconia nigra Black Stork	P	1992	14	14	A	B1i, C2
Aegypius monachus Cinereous Vulture	R	1995	7	7	A	A1, C1
Aquila chrysaetos Golden Eagle	R	1995	10	12	A	B2, C6
Hieraaetus fasciatus Bonnelli's Eagle	R	1995	5	5	A	B2

This is an important site for breeding raptors and *Ciconia nigra*.

Protection status

National Partial **International** None
24,999 ha of IBA covered by National Game Reserve (Cijara, 24,999 ha).

Conservation issues

Threats Disturbance to birds (B)

The main threat is from disturbance by hunters, particularly to breeding *Aegypius monachus*.

Puerto Peña reservoir–Valdecaballeros A1, B1i, B2, B3, C1, C2, C6 283

Admin region Extremadura
Coordinates 39°15'N 5°10'W
Altitude 350–850 m **Area** 32,000 ha

Site description

A reservoir on the River Guadiana surrounded by *Quercus* woodland, sclerophyllous scrub and *Pinus* plantations. In the mountainous part of the site there are dehesas and areas of pasture. The main human activities are livestock-farming, hunting and forestry.

Habitats Forest and woodland (*; broadleaved deciduous forest), Scrub (*; sclerophyllous scrub/garrigue/maquis), Grassland (dry siliceous grassland), Wetland (standing fresh water; river/stream), Rocky areas (inland cliff), Artificial landscape (*; arable land; forestry plantation)
Land-use Agriculture (80%), Forestry (10%), Hunting (90%), Water management

Birds

Species	Season	Year	Pop min	Pop max	Acc	Criteria
Ciconia nigra Black Stork	B	1995	6	7	A	B1i, C2
Ciconia nigra Black Stork	P	1992	25	40	B	B1i, C2
Milvus milvus Red Kite	R	1996	30	—	A	B3
Gyps fulvus Griffon Vulture	R	1995	38	38	A	C6
Aquila adalberti Spanish Imperial Eagle	R	1995	2	2	A	A1, B2, C1
Hieraaetus pennatus Booted Eagle	B	1996	25	—	B	B2
Grus grus Crane	W	1995	800	1,400	A	B1i, C2

This is an important site for breeding raptors and wintering *Grus grus*.

Protection status

National None **International** None

Conservation issues

Threats Infrastructure (B), Intensified forest management (B), Recreation/tourism (B)

Threats include inadequate forest management, disturbance to breeding *Ciconia nigra* from anglers and boats, new powerlines and new road building.

Pela mountain range–Orellana reservoir–Zorita A1, A4i, A4ii, B1i, B1iii, B2, B3, C1, C2, C3, C6 284

Admin region Extremadura
Coordinates 39°05′N 5°25′W
Altitude 340–751 m **Area** 140,000 ha

Site description

An isolated mountain range with sclerophyllous scrub and plains with extensive dehesas, pastures, olive-groves, and arable cultivation. The site includes a reservoir.

Habitats Forest and woodland (10%; broadleaved evergreen forest), Grassland (15%; steppe/dry calcareous grassland), Wetland (10%; standing fresh water; river/stream), Artificial landscape (65%; arable land; forestry plantation)
Land-use Agriculture (80%), Hunting (90%), Water management (10%)

Birds

Species	Season	Year	Pop min	Pop max	Acc	Criteria
Bubulcus ibis Cattle Egret	B	1996	5,000	6,000	B	A4i, B1i, C3
Egretta garzetta Little Egret	B	1996	50	60	A	C6
Ciconia nigra Black Stork	B	1995	6	6	A	B1i, C2
Ciconia nigra Black Stork	P	1995	47	52	A	B1i, C2
Ciconia ciconia White Stork	B	1996	342	362	A	B1i, B2, C2, C6
Elanus caeruleus Black-winged Kite	R	1996	12	14	A	B2, C6
Circus pygargus Montagu's Harrier	B	1996	105	120	A	B3, C6
Hieraaetus fasciatus Bonnelli's Eagle	R	1995	7	7	A	B2, C6
Falco naumanni Lesser Kestrel	B	1996	310	340	A	A1, A4ii, B1iii, B2, C1, C2, C6
Grus grus Crane	W	1996	13,500	24,000	A	A4i, B1i, C2
Tetrax tetrax Little Bustard	R	1992	200	—	B	A1, C1
Otis tarda Great Bustard	R	1995	448	448	A	A1, B2, C1, C2
Himantopus himantopus Black-winged Stilt	B	1992	140	—	B	B1i, C2, C6
Glareola pratincola Collared Pratincole	B	1996	60	120	B	A4i, B1i, B2, C2, C6
Limosa limosa Black-tailed Godwit	W	1996	2,000	4,000	B	B1i, C3
Larus fuscus Lesser Black-backed Gull	W	1996	5,000	6,000	B	A4i, B1i, C3
Gelochelidon nilotica Gull-billed Tern	B	1995	400	450	B	A4i, B1i, B2, C2, C6
Coracias garrulus Roller	B	1996	40	—	B	B2

An important site for resident raptors and particularly for wintering *Grus grus*, which has 16 roost sites within the IBA. The site is also important for steppic species and waterbirds around the reservoir.

Protection status

National None **International** Partial
5,500 ha of IBA covered by Ramsar Site (Embalse de Orellana, 5,500 ha). 25,000 ha of IBA covered by Special Protection Area (Embalse de Orellana y Sierra de Pela, 25,000 ha).

Conservation issues

Threats Agricultural intensification/expansion (B), Construction/impact of dyke/dam/barrage (B), Infrastructure (B)

The main threats are from irrigation of arable cultivation and a new road.

Don Benito–Guareña A1, B1i, B2, C1, C2 285

Admin region Extremadura
Coordinates 38°50′N 6°00′W
Altitude 200–425 m **Area** 31,000 ha

Site description

An area of plains south of the River Guadiana, in central Badajoz. The site is dominated by arable agriculture and dehesas.

Habitats Forest and woodland (*; broadleaved evergreen forest), Scrub, Grassland, Artificial landscape (*; arable land; perennial crops/orchards/groves)
Land-use Agriculture (*)

Birds

Species	Season	Year	Pop min	Pop max	Acc	Criteria
Grus grus Crane	W	1995	620	1,302	B	B1i, C2
Tetrax tetrax Little Bustard	R	1996	800	—	B	A1, B2, C1
Otis tarda Great Bustard	R	1996	50	60	B	A1, C1

This site is important for wintering *Grus grus*

Protection status

National None **International** None

Conservation issues

Threats Agricultural intensification/expansion (A)

The main threat is from the expansion of irrigation schemes.

Valdehornillos–Santa Amalia A4i, B1i, B2, C2, C3, C6 286

Admin region Extremadura
Coordinates 39°05′N 5°51′W
Altitude 250–323 m **Area** 23,000 ha

Site description

A large area of rice-fields in the north of Badajoz province.

Habitats Forest and woodland (broadleaved evergreen forest), Grassland, Artificial landscape (*; arable land)
Land-use Agriculture (90%)

Birds

Species	Season	Year	Pop min	Pop max	Acc	Criteria
Grus grus Crane	W	1992	1,433	3,000	A	A4i, B1i, C2
Himantopus himantopus Black-winged Stilt	R	1996	150	150	B	B1i, C2, C6
Glareola pratincola Collared Pratincole	R	1996	110	110	B	A4i, B1i, B2, C2, C6
Limosa limosa Black-tailed Godwit	P	1996	6,000	12,000	B	A4i, B1i, C3

This is an important site for passage waders and wintering *Grus grus.*

Protection status

National None **International** None

Conservation issues

Species abundance fluctuates depending upon the timing of the flooding of areas under rice cultivation.

Montánchez mountain range–Cornalvo reservoir A1, A4i, B1i, C1, C2 287

Admin region Extremadura
Coordinates 39°10′N 6°10′W
Altitude 250–698 m **Area** 54,500 ha

Site description

The site contains extensive areas of dehesa areas and also grasslands and scrub. A small reservoir lies within the site. The main human activities are livestock-farming, hunting and cork harvesting ('Other' land-use).

Habitats Forest and woodland (*; alluvial/very wet forest; broadleaved evergreen forest), Grassland, Wetland (standing fresh water; river/stream), Artificial landscape (perennial crops/orchards/groves)
Land-use Agriculture (*), Hunting, Other

Birds

Species	Season	Year	Pop min	Pop max	Acc	Criteria
Ciconia nigra Black Stork	B	1995	5	—	B	B1i, C2
Ciconia nigra Black Stork	P	1995	—	12	B	B1i, C2
Grus grus Crane	W	1995	471	3,000	B	A4i, B1i, C2
Tetrax tetrax Little Bustard	R	1996	350	—	B	A1, C1

An important site for raptors, steppic birds and wintering *Grus grus.*

Protection status

National Partial **International** Partial

10,570 ha of IBA covered by Natural Park (Cornalvo, 10,570 ha). 10,570 ha of IBA covered by Special Protection Area (Embalse de Cornalvo y Sierra Bermeja, 10,570 ha).

Conservation issues

Threats Infrastructure (B)

The main threats are from infrastructural development projects including new roads, a gas pipeline, golf course and a reservoir.

Mérida–Montijo reservoir

A1, A4i, B1i, B2, C1, C3, C6 **288**

Admin region Extremadura
Coordinates 38°55'N 6°25'W
Altitude 200–250 m **Area** 4,100 ha

Site description

The site comprises the town of Mérida, its surrounding area and nearby reservoirs. There is a mosaic of habitats with arable cultivation, olive-groves, dehesas, *Quercus* woodland, and alluvial vegetation. The area is used for grazing and hunting.

Habitats Forest and woodland (broadleaved evergreen forest), Wetland (standing fresh water; river/stream), Artificial landscape (*; arable land; perennial crops/orchards/groves)
Land-use Agriculture (*), Hunting

Birds

Species	Season	Year	Pop min	Pop max	Acc	Criteria
Nycticorax nycticorax Night Heron	B	1992	18	18	A	C6
Bubulcus ibis Cattle Egret	R	1994	1,825	1,825	A	A4i, B1i, C3, C6
Bubulcus ibis Cattle Egret	W	1994	2,500	—	A	A4i, B1i, C3
Egretta garzetta Little Egret	B	1992	59	59	A	C6
Falco naumanni Lesser Kestrel	B	1996	50	—	B	A1, B2, C1

This is an important site for breeding and wintering herons.

Protection status

National None **International** None

Conservation issues

Threats Agricultural intensification/expansion (U), Industrialization/urbanization (U), Infrastructure (U)

Urbanization and infrastructural developments are the main threats.

Lácara–Morante

A1, A4i, B1i, B1iii, B2, C1, C2, C6 **289**

Admin region Extremadura
Coordinates 39°00'N 6°25'W
Altitude 200–317 m **Area** 56,900 ha

Site description

Plains to the north of Mérida, mainly covered by dehesas interspersed with dry grasslands and cereal crops. The main human activity is livestock-farming.

Habitats Forest and woodland (*; alluvial/very wet forest; broadleaved evergreen forest), Grassland, Wetland (standing fresh water; river/stream), Artificial landscape (*; arable land; perennial crops/orchards/groves)
Land-use Agriculture (75%), Hunting, Water management

Birds

Species	Season	Year	Pop min	Pop max	Acc	Criteria
Elanus caeruleus Black-winged Kite	R	1996	25	—	C	B2, C2, C6
Aegypius monachus Cinereous Vulture	N	1997	23	30	B	A1, B1iii, C1, C2
Grus grus Crane	W	1995	788	1,883	A	B1i, C2
[1] ***Glareola pratincola*** Collared Pratincole	B	1997	70	—	B	A4i, B1i, B2, C2, C6
Sterna albifrons Little Tern	B	1992	31	31	A	C2

1. Large increase.

This is an important site for breeding raptors and wintering *Grus grus*. Species of global conservation concern that do not meet IBA criteria: *Falco naumanni* (6–20 breeding pairs) and *Otis tarda* (10 resident pairs).

Protection status

National None **International** None

Conservation issues

Threats Agricultural intensification/expansion (B), Burning of vegetation (B), Industrialization/urbanization (C), Recreation/tourism (A), Unsustainable exploitation (A)

The most important threat to the site is an expansion in the area of irrigated arable land. Other threats include illegal hunting and trapping, fires, and the destruction of riverine vegetation.

Botoa–Villar del Rey

A1, B1i, B2, C1, C2 **290**

Admin region Extremadura
Coordinates 39°10'N 6°50'W
Altitude 175–443 m **Area** 47,800 ha

Site description

Plains to the north of Badajoz with cereal cultivation, grasslands and dehesas. The area is grazed by sheep.

Habitats Forest and woodland (*; broadleaved evergreen forest), Scrub, Grassland (*), Wetland (river/stream), Artificial landscape (*; arable land; other urban/industrial areas)
Land-use Agriculture (75%)

Birds

Species	Season	Year	Pop min	Pop max	Acc	Criteria
Grus grus Crane	W	1995	339	792	A	B1i, C2
Tetrax tetrax Little Bustard	R	1996	500	—	B	A1, B2, C1
Otis tarda Great Bustard	R	1995	332	332	A	A1, B2, C1, C2

This is an important site for steppic species.

Protection status

National None **International** None

Conservation issues

Threats Agricultural intensification/expansion (B)

The main threat is from increases in the area of irrigated crops.

San Pedro mountain range

A1, A4i, A4ii, B1i, B1iii, B2, C1, C2, C3, C6 **291**

Admin region Extremadura
Coordinates 39°20'N 6°45'W
Altitude 270–703 m **Area** 295,000 ha

Site description

A large mountainous area on the border with Portugal. The site contains well-preserved Mediterranean vegetation, with *Quercus* woodland, dehesas, grassland, scrub and maquis. The main human activities are livestock-farming, game hunting and cork harvesting ('Other' land-use).

Habitats Forest and woodland (broadleaved evergreen forest), Scrub, Wetland (standing brackish and salt water; river/stream), Rocky areas (inland cliff), Artificial landscape (perennial crops/orchards/groves; forestry plantation)
Land-use Agriculture, Hunting, Other

Birds

Species	Season	Year	Pop min	Pop max	Acc	Criteria
Bubulcus ibis Cattle Egret	B	1994	1,100	1,100	A	A4i, B1i, C3
Ciconia nigra Black Stork	B	1995	9	9	A	B1i, C2
Ciconia nigra Black Stork	P	1996	100	—	C	B1i, C2
Milvus migrans Black Kite	B	1996	100	—	C	B2, C6
Neophron percnopterus Egyptian Vulture	B	1995	21	21	A	B2, C6
Gyps fulvus Griffon Vulture	R	1995	40	40	A	C6

Species ... continued	Season	Year	Pop min	Pop max	Acc	Criteria
[1] ***Aegypius monachus*** Cinereous Vulture	R	1995	168	168	A	A1, A4ii, B1iii, B2, C1, C2, C6
Aquila adalberti Spanish Imperial Eagle	R	1995	20	20	A	A1, B2, C1, C2, C6
Hieraaetus pennatus Booted Eagle	B	1995	50	—	C	B2, C6
Hieraaetus fasciatus Bonnelli's Eagle	R	1995	10	10	A	B2, C6
Falco naumanni Lesser Kestrel	B	1996	50	—	B	A1, B2, C1
Otis tarda Great Bustard	R	1996	150	—	C	A1, B2, C1, C2
Bubo bubo Eagle Owl	R	1995	25	—	C	B2, C6

1. Large increase.

An important site for breeding storks, raptors and steppic species.

Protection status

National None **International** Partial

83,217 ha of IBA covered by Special Protection Area (Sierra de San Pedro, 83,217 ha).

Conservation issues

Threats Infrastructure (B), Recreation/tourism (B), Unsustainable exploitation (B)

The main threats are hunting, disturbance by climbers and photographers, the construction of a gas pipeline, and hunting pressure.

Cedillo reservoir — A1, B1i, B1iii, B2, C1, C2, C6 — 292

Admin region Extremadura
Coordinates 39°40′N 7°20′W
Altitude 220–372 m **Area** 53,000 ha

Site description

A reservoir on the River Tajo with steep banks and occasional cliffs. The surrounding vegetation comprises Mediterranean scrub, *Quercus* woodland, dehesas, and some olive-groves. The main human activities are livestock-farming and hunting.

Habitats Forest and woodland (*; broadleaved evergreen forest), Scrub, Grassland, Wetland (standing fresh water; river/stream), Artificial landscape (highly improved reseeded grassland)
Land-use Agriculture (*), Hunting, Water management

Birds

Species	Season	Year	Pop min	Pop max	Acc	Criteria
Ciconia nigra Black Stork	B	1995	11	14	A	B1i, C2
Aegypius monachus Cinereous Vulture	R	1995	15	15	A	A1, B1iii, B2, C1, C2, C6
Hieraaetus fasciatus Bonnelli's Eagle	R	1995	6	6	A	B2

This is an important area for cliff-nesting species, particularly raptors and storks, and steppic species.

Protection status

National None **International** None

Conservation issues

Threats Recreation/tourism (B)

There is some disturbance to breeding species by anglers and boats.

Brozas–Membrío — A1, A4i, B1i, B2, C1, C2, C6 — 293

Admin region Extremadura
Coordinates 39°40′N 6°50′W
Altitude 300–521 m **Area** 94,000 ha

Site description

A wide plain between the River Tajo and the Sierra de San Pedro with many endorreic ponds. The vegetation includes extensive grasslands with occasional cereal fields and large dehesas.

Habitats Forest and woodland (*; broadleaved evergreen forest), Scrub, Grassland (*), Wetland (river/stream), Artificial landscape (*; arable land)
Land-use Agriculture (*), Hunting (*)

Birds

Species	Season	Year	Pop min	Pop max	Acc	Criteria
Ciconia nigra Black Stork	B	1995	5	5	A	B1i, C2
Ciconia nigra Black Stork	P	1996	35	—	C	B1i, C2
Ciconia ciconia White Stork	B	1994	200	—	B	B2
Falco naumanni Lesser Kestrel	B	1992	50	—	C	A1, B2, C1
Grus grus Crane	W	1996	2,295	2,741	A	A4i, B1i, C2
Tetrax tetrax Little Bustard	R	1996	1,000	—	C	A1, B2, C1
Otis tarda Great Bustard	R	1996	500	800	B	A1, B2, C1, C2, C6
Burhinus oedicnemus Stone Curlew	R	1996	500	—	B	B2, C2, C6
Pterocles orientalis Black-bellied Sandgrouse	R	1996	500	—	B	B2, C2, C6
Pterocles alchata Pin-tailed Sandgrouse	R	1996	500	—	B	B2, C2, C6

This is a very important area for steppic species.

Protection status

National None **International** None

Conservation issues

Threats Recreation/tourism (C)

The main threat is from recreational activities.

Malpartida de Cáceres–Arroyo de la Luz — A1, A4i, B1i, B2, C1, C2, C3, C6 — 294

Admin region Extremadura
Coordinates 39°25′N 6°30′W
Altitude 350–515 m **Area** 43,900 ha

Site description

A wide, undulating plain with granite hillocks and numerous ponds. There is a mosaic of dehesas, grasslands and arable cultivation, with some patches of *Pinus* woodland. The area is extensively grazed by sheep.

Habitats Forest and woodland (*; broadleaved evergreen forest), Scrub, Grassland (*), Wetland (standing fresh water; river/stream), Artificial landscape (arable land; ruderal land)
Land-use Agriculture (100%)

Birds

Species	Season	Year	Pop min	Pop max	Acc	Criteria
Bubulcus ibis Cattle Egret	B	1992	1,083	—	A	A4i, B1i, C3
Egretta garzetta Little Egret	B	1992	40	—	C	C6
Ciconia nigra Black Stork	P	1992	31	—	B	B1i, C2
Ciconia ciconia White Stork	B	1992	200	—	B	B2
Hieraaetus pennatus Booted Eagle	B	1996	20	—	C	B2
Tetrax tetrax Little Bustard	R	1996	200	—	C	A1, C1
Otis tarda Great Bustard	R	1996	25	70	B	A1, C1

This is an important area for species associated with steppe habitats. Species of global conservation concern that do not meet IBA criteria: *Falco naumanni* (20 breeding pairs).

Protection status

National None **International** None

Conservation issues

Threats Agricultural intensification/expansion (A), Industrialization/urbanization (B), Infrastructure (A)

Agricultural intensification, particularly using irrigation, urban development and road-building are the main threats.

Plain between Cáceres and Trujillo–Aldea del Cano — A1, A4i, A4ii, B1i, B1iii, B2, B3, C1, C2, C6 — 295

Admin region Extremadura
Coordinates 39°25′N 6°10′W
Altitude 270–644 m **Area** 103,000 ha

Site description

The site includes an area of undulating plains dissected by steep-sided rivers. The main vegetation-types are dry grassland, garrigue, arable cultivation and patches of dehesa. The main human activities include livestock-farming, arable agriculture and game hunting.

Habitats Forest and woodland (broadleaved evergreen forest), Scrub (*), Grassland (*), Wetland (standing fresh water; river/stream), Artificial landscape (*; arable land)
Land-use Agriculture (*), Hunting

Birds

Species	Season	Year	Pop min	Pop max	Acc	Criteria
Ciconia nigra Black Stork	P	1996	250	—	B	A4i, B1i, C2
Ciconia ciconia White Stork	B	1992	220	—	C	B2, C6
Milvus milvus Red Kite	W	1996	500	—	B	A4ii, B1iii, C2
Circus pygargus Montagu's Harrier	B	1996	140	—	B	B3, C6
Hieraaetus pennatus Booted Eagle	B	1996	25	—	C	B2
Falco naumanni Lesser Kestrel	B	1996	450	—	B	A1, A4ii, B1iii, B2, C1, C2, C6
Grus grus Crane	W	1996	900	2,200	B	A4i, B1i, C2
Tetrax tetrax Little Bustard	R	1996	3,000	—	B	A1, B2, C1, C2, C6
Otis tarda Great Bustard	R	1996	1,300	—	B	A1, B2, C1, C2, C6
Burhinus oedicnemus Stone Curlew	R	1996	220	—	B	B2, C6
Chlidonias hybridus Whiskered Tern	B	1996	60	—	B	B2, C6
Pterocles orientalis Black-bellied Sandgrouse	R	1996	100	—	C	B2, C6
Pterocles alchata Pin-tailed Sandgrouse	R	1996	100	—	C	B2, C2, C6
Coracias garrulus Roller	B	1996	150	—	C	B2, C2, C6

This is an important site for steppic birds, also passage *Ciconia nigra* and wintering *Grus grus*. Species of global conservation concern that do not meet IBA criteria: *Aquila adalberti* (resident).

Protection status

National None **International** Partial
48,000 ha of IBA covered by Special Protection Area (Llanos de Cáceres y Sierra de Fuentes, 48,000 ha).

Conservation issues

Threats Agricultural intensification/expansion (A), Industrialization/urbanization (A), Unsustainable exploitation (B)

The main threats are from agricultural intensification and urbanization. A management plan exists.

Trujillo–Torrecillas de la Tiesa 296

A1, A4i, A4ii, B1i, B1iii, B2, B3, C1, C2, C3, C6

Admin region Extremadura
Coordinates 39°30′N 5°50′W
Altitude 350–560 m **Area** 108,500 ha

Site description

A wide, undulating plain around Trujillo dominated by dehesas and dry grassland, with some non-irrigated arable land.

Habitats Forest and woodland (*; broadleaved evergreen forest), Scrub (scrub), Grassland (*; dry siliceous grassland), Wetland (river/stream), Artificial landscape (arable land)
Land-use Agriculture (80%), Hunting

Birds

Species	Season	Year	Pop min	Pop max	Acc	Criteria
Bubulcus ibis Cattle Egret	B	1994	8,600	8,600	A	A4i, B1i, C3
Ciconia nigra Black Stork	B	1995	15	15	A	B1i, B2, C2, C6
Ciconia nigra Black Stork	P	1994	52	—	B	B1i, C2
Ciconia ciconia White Stork	B	1996	350	—	B	B1i, B2, C2, C6
Elanus caeruleus Black-winged Kite	R	1994	10	—	C	B2, C6
Neophron percnopterus Egyptian Vulture	B	1995	10	10	A	C6
Circus pygargus Montagu's Harrier	B	1996	100	—	C	B3, C6
Hieraaetus pennatus Booted Eagle	B	1996	20	—	C	B2
Falco naumanni Lesser Kestrel	B	1996	250	—	B	A1, A4ii, B1iii, B2, C1, C2, C6
Grus grus Crane	W	1992	682	1,080	A	B1i, C2
Tetrax tetrax Little Bustard	B	1996	1,000	—	B	A1, B2, C1
Otis tarda Great Bustard	B	1996	300	—	B	A1, B2, C1, C2
Pterocles orientalis Black-bellied Sandgrouse	R	1996	100	—	C	B2, C6

An important site for breeding steppic birds, breeding and passage *Ciconia nigra* and wintering *Grus grus*. Species of global conservation concern that do not meet IBA criteria: *Aquila adalberti* (resident).

Protection status

National None **International** None

Conservation issues

Threats Disturbance to birds (B), Infrastructure (A)

The main threats are from new infrastructure projects, disturbance and a new reservoir.

Las Villuercas mountain range 297

A1, B1i, B1iii, B2, C1, C2, C6

Admin region Extremadura
Coordinates 39°40′N 5°25′W
Altitude 350–1,443 m **Area** 216,230 ha

Site description

A mountainous area with dense oak *Quercus* forest, *Castanea sativa* woodland, *Pinus* plantations and extensive areas of scrub. Around villages there are arable fields, olive-groves, grasslands and dehesas.

Habitats Forest and woodland (*; broadleaved evergreen forest), Scrub (*; sclerophyllous scrub/garrigue/maquis), Wetland (river/stream), Rocky areas (inland cliff), Artificial landscape (arable land; forestry plantation)
Land-use Agriculture (*), Hunting (*)

Birds

Species	Season	Year	Pop min	Pop max	Acc	Criteria
Ciconia nigra Black Stork	B	1995	20	20	A	B1i, B2, C2, C6
Ciconia nigra Black Stork	P	1992	30	—	B	B1i, C2
Neophron percnopterus Egyptian Vulture	B	1995	38	38	A	B1iii, B2, C2, C6
[1] ***Gyps fulvus*** Griffon Vulture	R	1995	146	146	A	B1iii, B2, C2, C6
[1] ***Aegypius monachus*** Cinereous Vulture	R	1995	15	15	A	A1, B1iii, B2, C1, C2, C6
[1] ***Aquila adalberti*** Spanish Imperial Eagle	R	1995	6	6	A	A1, B2, C1, C6
Aquila chrysaetos Golden Eagle	R	1995	19	19	A	B2, C6
Hieraaetus pennatus Booted Eagle	B	1992	20	—	B	B2
[1] ***Hieraaetus fasciatus*** Bonnelli's Eagle	R	1995	21	21	A	B2, C2, C6

1. Large increase.

This is an important site for breeding raptors and *Ciconia nigra*.

Protection status

National None **International** None

Conservation issues

Threats Construction/impact of dyke/dam/barrage (A), Disturbance to birds (B), Industrialization/urbanization (B), Infrastructure (A), Unsustainable exploitation (B)

The main threats are the construction of new dams and roads, urban development, and hunting pressure. *Aquila adalberti* is vulnerable to disturbance by livestock.

Monfragüe 298

A1, A3, A4ii, B1i, B1iii, B2, C1, C2, C6

Admin region Extremadura
Coordinates 39°40′N 5°45′W
Altitude 320–839 m **Area** 155,000 ha

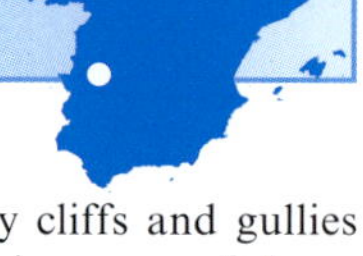

Site description

The site encompasses a range of hills with many cliffs and gullies and the Rivers Tajo and Tiétar. There are extensive areas of dense sclerophyllous scrub, small *Quercus*-dominated woodlands and areas of dehesa, as well as some *Eucalyptus* plantations.

Habitats Forest and woodland (*; mixed forest; alluvial/very wet forest; broadleaved evergreen forest), Scrub (*; sclerophyllous scrub/garrigue/maquis), Grassland (dry siliceous grassland), Wetland (standing fresh water; river/stream), Artificial landscape (arable land; forestry plantation)
Land-use Agriculture (*), Forestry, Hunting, Other, Tourism/recreation

Birds

Species	Season	Year	Pop min	Pop max	Acc	Criteria
Ardea purpurea Purple Heron	B	1994	60	60	A	B2, C2, C6
Ciconia nigra Black Stork	B	1995	31	—	A	B1i, B2, C2, C6
Elanus caeruleus Black-winged Kite	R	1992	10	—	C	B2, C6
Neophron percnopterus Egyptian Vulture	B	1995	22	—	A	B2, C6
Gyps fulvus Griffon Vulture	R	1995	377	—	A	A4ii, B1iii, B2, C2, C6
[1] ***Aegypius monachus*** Cinereous Vulture	R	1995	214	214	A	A1, A4ii, B1iii, B2, C1, C2, C6
Circaetus gallicus Short-toed Eagle	B	1992	15	—	C	C6
Aquila adalberti Spanish Imperial Eagle	R	1995	10	11	A	A1, B2, C1, C2, C6
Aquila chrysaetos Golden Eagle	R	1995	7	8	A	C6
Hieraaetus fasciatus Bonnelli's Eagle	R	1995	7	7	A	B2, C6
Bubo bubo Eagle Owl	R	1992	20	—	C	C6
Caprimulgus ruficollis Red-necked Nightjar	B	1996	Frequent		—	A3
Oenanthe hispanica Black-eared Wheatear	B	1996	Common		—	A3
Oenanthe leucura Black Wheatear	R	1996	Frequent		—	A3
Sylvia conspicillata Spectacled Warbler	R	1996	Abundant		—	A3
Sylvia cantillans Subalpine Warbler	B	1996	Abundant		—	A3
Sylvia melanocephala Sardinian Warbler	R	1996	Abundant		—	A3
Sturnus unicolor Spotless Starling	R	1996	Abundant		—	A3

1. Large increase.

This is a very important area for breeding raptors, particularly *Aegypius monachus* and also for *Ciconia nigra*. Breeding birds include seven of the 20 species in Europe that are restricted to the Mediterranean biome (when breeding).

Protection status

National Partial **International** Partial

17,852 ha of IBA covered by Natural Park (Monfragüe, 17,852 ha). 17,852 ha of IBA covered by Special Protection Area (Monfragüe, 17,852 ha).

Conservation issues

Threats Infrastructure (U), Recreation/tourism (C)

The main threats to the site are disturbance from visitors and potential infrastructural developments, especially road building and river regulation. A management plan is being produced. *Eucalyptus* plantations are being removed.

Alcántara reservoir–Cuatro Lugares — A1, A4i, B1i, B1iii, B2, C1, C2, C6 — 299

Admin region Extremadura
Coordinates 39°45'N 6°30'W
Altitude 250–815 m **Area** 116,000 ha

Site description

The site includes a huge reservoir and surrounding dry grasslands, sclerophyllous scrub, dehesa, olive-groves and arable land. There are some old *Pinus* plantations. Human activities include sheep-grazing, forestry, hunting and recreation.

Habitats Forest and woodland (*; broadleaved evergreen forest), Scrub (*; scrub), Grassland (humid grassland), Wetland (standing fresh water; river/stream), Artificial landscape (arable land; forestry plantation)
Land-use Agriculture (*), Forestry, Tourism/recreation

Birds

Species	Season	Year	Pop min	Pop max	Acc	Criteria
Ciconia nigra Black Stork	B	1995	18	—	A	B1i, B2, C2, C6
Ciconia nigra Black Stork	P	1992	17	—	B	B1i, C2
[1] ***Ciconia ciconia*** White Stork	B	1996	400	—	B	B1i, B2, C2, C6
Milvus migrans Black Kite	B	1996	100	—	B	B2, C6
Neophron percnopterus Egyptian Vulture	B	1995	32	—	A	B1iii, B2, C2, C6
Aquila adalberti Spanish Imperial Eagle	R	1995	2	3	A	A1, B2, C1, C6
Hieraaetus pennatus Booted Eagle	B	1996	50	—	B	B2, C6

Species ... continued	Season	Year	Pop min	Pop max	Acc	Criteria
Hieraaetus fasciatus Bonnelli's Eagle	R	1995	6	7	A	B2, C6
Falco naumanni Lesser Kestrel	B	1996	150	—	B	A1, B1iii, B2, C1, C2, C6
Grus grus Crane	W	1995	1,558	3,000	B	A4i, B1i, C2
Tetrax tetrax Little Bustard	R	1996	2,000	—	B	A1, B2, C1, C6
Otis tarda Great Bustard	R	1996	150	3,000	B	A1, B2, C1, C2, C6

1. Large increase.

This is an important site for breeding storks, raptors and steppic species and wintering *Grus grus*.

Protection status

National None **International** None

Conservation issues

Threats Agricultural intensification/expansion (B), Infrastructure (U), Recreation/tourism (B)

The main threats are from irrigation of arable land, disturbance from recreational use of the reservoir, and plans for a new road.

Coria mountain ranges — A1, B1i, B2, C1, C2, C6 — 300

Admin region Extremadura
Coordinates 39°50'N 6°30'W
Altitude 300–625 m **Area** 71,500 ha

Site description

A plain to the south-east of Coria crossed by a small range of hills. Most of the area is dehesa with areas of arable land, scrub, grassland, *Quercus* woodland and *Pinus* and *Eucalyptus* plantations. The main human activity is livestock-farming.

Habitats Forest and woodland (50%; broadleaved evergreen forest), Scrub (30%), Grassland, Wetland (river/stream), Rocky areas (inland cliff), Artificial landscape (arable land; forestry plantation)
Land-use Agriculture (*), Hunting (*)

Birds

Species	Season	Year	Pop min	Pop max	Acc	Criteria
Ciconia nigra Black Stork	B	1995	4	7	A	B1i, C2
Gyps fulvus Griffon Vulture	R	1995	68	—	A	C6
Aquila adalberti Spanish Imperial Eagle	R	1995	2	2	A	A1, B2, C1
Hieraaetus pennatus Booted Eagle	B	1996	25	—	B	B2

This is an important site for forest- and cliff-nesting raptors.

Protection status

National None **International** None

Conservation issues

Threats Afforestation (B), Agricultural intensification/expansion (A), Infrastructure (B)

The main threats are afforestation, irrigation of crops, the erection of powerlines, and water-management schemes.

Borbollón reservoir — A1, A4i, B1i, C1, C2, C3 — 301

Admin region Extremadura
Coordinates 40°10'N 6°35'W
Altitude 300–510 m **Area** 48,400 ha

Site description

This site comprises a reservoir surrounded by dehesa, non-irrigated arable land and grassland. There is a wooded island in the reservoir. The main human activities are livestock-grazing and hunting.

Habitats Forest and woodland (*; broadleaved evergreen forest), Scrub (scrub), Grassland (*), Wetland (standing fresh water), Artificial landscape (highly improved reseeded grassland; arable land)
Land-use Agriculture (*), Hunting (*)

Birds

Species	Season	Year	Pop min	Pop max	Acc	Criteria
Bubulcus ibis Cattle Egret	B	1994	750	750	A	A4i, B1i, C3
Ciconia nigra Black Stork	P	1992	40	50	B	B1i, C2
Grus grus Crane	W	1995	853	1,620	B	B1i, C2
Otis tarda Great Bustard	W	1995	40	50	A	A1, C1

The site is important for post-breeding *Ciconia nigra*, breeding herons and wintering *Grus grus*.

Protection status

National None **International** None

Conservation issues

Threats Agricultural intensification/expansion (A), Disturbance to birds (B)

The major threat to the site is from agricultural intensification, in particular an increase in the area of irrigated crops. An irrigation scheme threatens a wooded island in the reservoir used by breeding birds. Disturbance is also a threat.

East Gata mountain range — A1, B1i, B1iii, B2, C1, C2, C6 — 302

Admin region Extremadura
Coordinates 40°20′N 6°35′W
Altitude 350–1,412 m **Area** 86,600 ha

Site description

A mountain range in northern Extremadura. The vegetation consists of *Pinus pinaster* plantations, scrub and grassland. There are also some olive-groves and orchards. The main human activities are livestock-farming and forestry.

Habitats Forest and woodland (broadleaved deciduous forest; broadleaved evergreen forest), Scrub (60%; sclerophyllous scrub/garrigue/maquis), Grassland (mesophile grassland), Rocky areas (inland cliff), Artificial landscape (25%; perennial crops/orchards/groves; forestry plantation)
Land-use Agriculture (*), Forestry

Birds

Species	Season	Year	Pop min	Pop max	Acc	Criteria
Ciconia nigra Black Stork	B	1995	6	—	A	B1i, C2
[1] ***Aegypius monachus*** Cinereous Vulture	R	1995	43	—	A	A1, B1iii, B2, C1, C2, C6
Hieraaetus pennatus Booted Eagle	B	1996	25	—	A	B2

1. Large increase.

An important site for resident raptors. Species of global conservation concern that do not meet IBA criteria: *Aquila adalberti* (resident).

Protection status

National None **International** None

Conservation issues

Threats Afforestation (B), Burning of vegetation (B), Infrastructure (B)

Afforestation with *Pinus* plantations, fires, and new road construction are the main threats to the site.

Gabriel y Galán reservoir — B1i, C2 — 303

Admin region Extremadura
Coordinates 40°16′N 6°16′W
Altitude 300–900 m **Area** 55,000 ha

Site description

A reservoir on the River Alagón surrounded by dehesas, plantations of *Pinus pinaster*, grassland and arable cultivation.

Habitats Forest and woodland (60%; broadleaved evergreen forest), Grassland (10%; dry siliceous grassland), Wetland (20%; standing fresh water), Artificial landscape (10%; arable land; forestry plantation)
Land-use Agriculture (60%), Hunting, Water management

Birds

Species	Season	Year	Pop min	Pop max	Acc	Criteria
Ciconia nigra Black Stork	B	1995	8	8	A	B1i, C2
Grus grus Crane	W	1995	787	1,200	A	B1i, C2

This is an important site for breeding *Ciconia nigra and* wintering *Grus grus*. Species of global conservation concern that do not meet IBA criteria: *Aegypius monachus* (2 resident pairs).

Protection status

National None **International** None

Conservation issues

Threats Afforestation (B), Agricultural intensification/expansion (A), Infrastructure (B)

Afforestation with *Eucalyptus*, increasing use of irrigation in arable crops and plans for a new road are the major threats to the site.

Plasencia and San Bernabé mountain range — A1, B1i, B2, C1, C2 — 304

Admin region Extremadura
Coordinates 40°04′N 6°00′W
Altitude 350–1,250 m **Area** 18,000 ha

Site description

A small mountain range with *Quercus* woodland and dehesas at lower altitudes. The site includes the city of Plasencia and the Plasencia reservoir.

Habitats Forest and woodland (*; broadleaved deciduous forest; broadleaved evergreen forest), Scrub, Wetland (standing fresh water; river/stream), Rocky areas (inland cliff), Artificial landscape (other urban/industrial areas)
Land-use Agriculture (*), Hunting

Birds

Species	Season	Year	Pop min	Pop max	Acc	Criteria
Ciconia nigra Black Stork	P	1997	10	—	B	B1i, C2
Falco naumanni Lesser Kestrel	B	1997	70	70	A	A1, B2, C1

This is an important area for *Falco naumanni* and passage *Ciconia nigra*. Species of global conservation concern that do not meet IBA criteria: *Aegypius monachus* (non-breeding).

Protection status

National None **International** None

Conservation issues

Threats Industrialization/urbanization (U), Other (U)

The main threats are from the restoration of old buildings used as nesting sites by *Falco naumanni* ('Other' threats, above), and increasing urban development.

Lower Tiétar river and La Vera valley — B1i, B2, C2, C6 — 305

Admin region Extremadura
Coordinates 40°05′N 5°35′W
Altitude 200–1,175 m **Area** 73,000 ha

Site description

A fertile valley with riverine forest, *Pinus* forest and dehesas.

Habitats Forest and woodland (40%; broadleaved deciduous forest; native coniferous forest; mixed forest; alluvial/very wet forest; broadleaved evergreen forest), Scrub (15%; sclerophyllous scrub/garrigue/maquis), Grassland (10%; dry siliceous grassland), Wetland (10%; river/stream), Artificial landscape (50%; arable land; perennial crops/orchards/groves)
Land-use Agriculture (90%), Forestry (40%), Hunting (40%)

Birds

This is an important site for breeding forest raptors and *Ciconia nigra*.

Species of global conservation concern that do not meet IBA criteria: *Falco naumanni* (20 breeding pairs).

Species	Season	Year	Pop min	Pop max	Acc	Criteria
Ciconia nigra Black Stork	B	1995	12	—	B	B1i, C2
Milvus migrans Black Kite	B	1995	150	—	B	B2, C6

Protection status

National None **International** None

Conservation issues

Threats Agricultural intensification/expansion (B), Construction/impact of dyke/dam/barrage (B), Intensified forest management (B)

Agricultural intensification, particularly pesticide pollution and illegal use of water resources, together with a project to build a new dam are the main threats.

Campo Arañuelo–Valdecañas reservoir

A4i, B1i, B2, C2, C3, C6 **306**

Admin region Extremadura, Castilla-La Mancha
Coordinates 39°53′N 5°16′W
Altitude 275–403 m **Area** 81,950 ha

Site description

A wide plain to the south of the River Tiétar with many creeks and small ponds. The reservoir is surrounded by extensive dehesas and areas of grassland and arable cultivation.

Habitats Forest and woodland (*; broadleaved evergreen forest), Grassland (dry siliceous grassland), Wetland (standing fresh water), Artificial landscape (arable land)
Land-use Agriculture (*), Hunting (*)

Birds

Species	Season	Year	Pop min	Pop max	Acc	Criteria
[1] ***Phalacrocorax carbo*** Cormorant	W	1996	4,500	—	B	A4i, B1i, C3
Bubulcus ibis Cattle Egret	R	1992	1,400	1,400	A	A4i, B1i, C3
Ciconia nigra Black Stork	R	1995	4	4	A	B1i, C2, C6
Ciconia nigra Black Stork	P	1992	15	—	A	B1i, C2
Ciconia ciconia White Stork	B	1992	200	—	A	B2, C6
Elanus caeruleus Black-winged Kite	R	1996	10	—	A	B2, C6
Grus grus Crane	W	1995	2,200	5,300	A	A4i, B1i, C2
Gelochelidon nilotica Gull-billed Tern	B	1992	100	—	A	A4i, B1i, B2, C2, C6

1. *Ph. c. sinensis.*

This is an important site for breeding herons and raptors and wintering *Grus grus*. Species of global conservation concern that do not meet IBA criteria: *Aquila adalberti* (resident).

Protection status

National None **International** Low

5,167 ha of IBA covered by Special Protection Area (Llanvras de Oropesa, Lagartera y Calera y Chozas, 14,948 ha). 732 ha of IBA covered by Special Protection Area (Valle del Tiétar y Embalses de Rosarito y Navalcán, 53,167 ha).

Conservation issues

Threats Agricultural intensification/expansion (A), Industrialization/urbanization (B), Infrastructure (B)

The main threat is agricultural intensification as a result of more extensive use of irrigation, plans for a new road, powerlines and river regulation. There is a nuclear power-station at Almaraz.

Islets Vedrá and Vedranell

A4ii, B1ii, B1iii, B2, B3, C2, C6 **307**

Admin region Baleares
Coordinates 38°52′N 1°14′E
Altitude 0–381 m **Area** 100 ha

Site description

Two rocky, limestone islets close to the south-western coast of Ibiza.

Habitats Rocky areas (100%; sea cliff/rocky shore; rock stacks/islets)
Land-use Agriculture (100%)

Birds

Species	Season	Year	Pop min	Pop max	Acc	Criteria
Puffinus yelkouan Yelkouan Shearwater	B	1991	26	75	B	B1ii, B3, C2
Falco eleonorae Eleonora's Falcon	B	1993	76	—	A	A4ii, B1iii, B2, C2, C6

This is an important site for breeding seabirds and coastal raptors. *Puffinus yelkouan* probably no longer breeds. *Larus audouinii* is an occasional breeder. Species of global conservation concern that do not meet IBA criteria: *Larus audouinii* (25 resident pairs).

Protection status

National None **International** High

100 ha of IBA covered by Special Protection Area (Islas de Vedrá y Vedranell, 100 ha).

Conservation issues

Threats Consequences of animal/plant introductions (B), Other (B), Recreation/tourism (A)

The main threats are from disturbance from tourism in the summer, predation by *Larus cachinnans* ('Other' threats, above) and problems caused by introduced mammals, particularly rabbits and goats. A proposal exists to declare the site a Natural Reserve. A recovery plan for *Puffinus yelkouan* has been carried out by the Balearic Government. *L. cachinnans* is culled annually. The IBA overlaps with a Natural Area of Special Interest (Cap Llentrisca–Sa Talaia de Sant Josep).

Isle of Conejera and islets of Bledes and Espartar (Ibiza)

A1, A4i, A4ii, B1i, B1ii, B2, B3, C1, C2, C6 **308**

Admin region Baleares
Coordinates 38°59′N 1°12′E
Altitude 0–74 m **Area** 147 ha

Site description

A group of islets to the west of Ibiza. The largest island, Conejera, has cliffs and is dominated by *Juniperus* scrub. There are no permanent inhabitants but there are many visitors during the summer. The rest of the islets are rocky with sparse scrub cover.

Habitats Rocky areas (100%; sea cliff/rocky shore; rock stacks/islets; caves)
Land-use Hunting (10%), Military (10%), Not utilized (90%)

Birds

Species	Season	Year	Pop min	Pop max	Acc	Criteria
Calonectris diomedea Cory's Shearwater	B	1991	100	300	B	B2, C6
Puffinus yelkouan Yelkouan Shearwater	B	1991	50	100	B	B1ii, B3, C2
Hydrobates pelagicus Storm Petrel	B	1991	2,100	4,000	B	A4ii, B1ii, B2, C2, C6
Larus audouinii Audouin's Gull	B	1996	200	250	A	A1, A4i, B1i, B2, C1, C2, C6

This is an important site for breeding seabirds.

Protection status

National None **International** None

Conservation issues

Threats Consequences of animal/plant introductions (B), Disturbance to birds (A), Recreation/tourism (A), Unsustainable exploitation (A)

The main threats are from disturbance of *Larus audouinii* by rabbit hunters, military activities and tourists, hunting, and the collection of *Puffinus yelkouan* chicks for food. *Larus cachinnans* colonies are culled annually. The IBA overlaps with a Natural Area of Special Interest (Isoltes de Ibiza y Formentera).

Cape Nonó– Isle of Murada — A4ii, B1ii, B1iii, B2, B3, C2, C6 — 309

Admin region Baleares
Coordinates 39°05′N 1°22′E
Altitude 0–368 m **Area** 1,000 ha

Site description

Limestone cliffs and small islets along Ibiza's north-western coast. The dominant vegetation if *Pinus halepensis* woodland. The main human activities are hunting and tourism.

Habitats Forest and woodland (native coniferous forest), Scrub, Rocky areas (*; sea cliff/rocky shore; rock stacks/islets)
Land-use Tourism/recreation

Birds

Species	Season	Year	Pop min	Pop max	Acc	Criteria
Hydrobates pelagicus Storm Petrel	B	1991	26	75	C	B1ii, C2, C6
Phalacrocorax aristotelis Shag	R	1991	13	45	C	C2
Falco eleonorae Eleonora's Falcon	B	1994	50	—	A	A4ii,B1iii,B2,C2,C6
Falco peregrinus Peregrine	R	1992	6	—	A	C6
Sylvia sarda Marmora's Warbler	R	1996	150	—	C	B3

This is an important site for breeding seabirds and *Falco eleonorae*.

Protection status

National None **International** None

Conservation issues

Threats Aquaculture/fisheries (C), Industrialization/urbanization (A), Recreation/tourism (B)

The main threats are from tourism and spreading urban development, whilst marine species are affected by fishing practices. There is a proposal to declare the area a nature reserve. *Larus cachinnans* is culled annually. The IBA overlaps with a Natural Area of Special Interest (Els Amunts).

Tagomago island — A4ii, B1ii, B1iii, B2, B3, C2, C6 — 310

Admin region Baleares
Coordinates 39°02′N 1°40′E
Altitude 0–114 m **Area** 90 ha

Site description

A steep limestone islet close to Ibiza's north-east coast, dominated by maquis and coastal vegetation.

Habitats Scrub (*; sclerophyllous scrub/garrigue/maquis), Rocky areas (*; sea cliff/rocky shore)
Land-use Not utilized (100%)

Birds

Species	Season	Year	Pop min	Pop max	Acc	Criteria
Puffinus yelkouan Yelkouan Shearwater	B	1991	76	200	C	B1ii, B3, C2, C6
Falco eleonorae Eleonora's Falcon	B	1994	52	—	A	A4ii,B1iii,B2,C2,C6

This is an important site for breeding seabirds and *Falco eleonorae*.

Protection status

National None **International** High
90 ha of IBA covered by Special Protection Area (Isla de Tagomago, 90 ha).

Conservation issues

Threats Consequences of animal/plant introductions (B), Recreation/tourism (A), Unsustainable exploitation (A)

The main threats are disturbance from tourism, hunting and rat predation of breeding birds. *Larus cachinnans* is culled annually. The IBA overlaps with a Natural Area of Special Interest (Isla de Tagomago).

Islets of Santa Eulalia, Redona and Es Canar (Ibiza) — A1, B1ii, C1, C2, C6 — 311

Admin region Baleares
Coordinates 38°59′N 1°35′E
Altitude 0–31 m **Area** 6 ha

Site description

Very small rocky islets off the south-east coast of Ibiza.

Habitats Rocky areas (100%; rock stacks/islets; caves)
Land-use Not utilized (100%)

Birds

Species	Season	Year	Pop min	Pop max	Acc	Criteria
Calonectris diomedea Cory's Shearwater	B	1991	50	100	A	C6
Hydrobates pelagicus Storm Petrel	B	1991	50	100	C	B1ii, C2, C6
Larus audouinii Audouin's Gull	B	1996	57	65	A	A1, C1

This is an important site for breeding seabirds.

Protection status

National None **International** None

Conservation issues

Threats Consequences of animal/plant introductions (A), Recreation/tourism (A)

The main threats are from high levels of disturbance from tourists during the summer and predation of breeding birds by rats. There has been an annual cull of *Larus cachinnans* since 1993. IBA overlaps with a Natural Area of Special Interest (Islotes de Ibiza y Formentera).

Saltpans of Ibiza and Formentera and Freus isles — A1, A4i, A4ii, B1i, B1ii, B2, B3, C1, C2, C3, C6 — 312

Admin region Baleares
Coordinates 38°49′N 1°28′E
Altitude 0–159 m **Area** 2,600 ha

Site description

The site includes working and abandoned saltpans, two lakes and several small islets. There are small areas of scrub.

Habitats Forest and woodland (35%; native coniferous forest), Scrub (5%; sclerophyllous scrub/garrigue/maquis), Wetland (45%; mudflat/sandflat; sand-dunes/sand beach; shingle/stony beach; standing brackish and salt water; water-fringe vegetation), Marine areas (30%; open sea; sea inlet/coastal features), Rocky areas (10%; sea cliff/rocky shore; rock stacks/islets), Artificial landscape (30%; arable land; other urban/industrial areas)
Land-use Agriculture, Hunting (50%), Tourism/recreation (15%), Urban/industrial/transport (40%)

Birds

Species	Season	Year	Pop min	Pop max	Acc	Criteria
Podiceps nigricollis Black-necked Grebe	W	1995	765	—	A	A4i, B1i, C3
Puffinus yelkouan Yelkouan Shearwater	B	1991	102	275	B	A4ii,B1ii,B3,C2,C6
Hydrobates pelagicus Storm Petrel	B	1991	152	400	B	B1ii, B2, C2, C6
[1] ***Himantopus himantopus*** Black-winged Stilt	B	1996	79	135	A	B1i, C2, C6
Charadrius alexandrinus Kentish Plover	R	1996	273	—	A	B1i, B2, C3
Larus audouinii Audouin's Gull	R	1996	268	—	B	A1, A4i, B1i, B2, C1, C2, C6
Sylvia sarda Marmora's Warbler	R	1996	150	—	C	B3

1. Large decrease.

This is an important site for breeding seabirds and waterbirds.

Protection status

National High **International** Partial
1,500 ha of IBA covered by Natural Reserve (Ses Salines de Ibiza, Formentera e Islas des Freus, 1,500 ha). 1,640 ha of IBA covered by Ramsar Site (Salinas de Ibiza y Formentera, 1,640 ha). 1,179 ha of IBA covered by Special Protection Area (Islas de los Freus–Salinas de Ibiza y Formentera, 1,179 ha).

Conservation issues

Threats Aquaculture/fisheries (C), Industrialization/urbanization (B), Infrastructure (C), Recreation/tourism (B), Unsustainable exploitation (A)

The main threats are from disturbance by tourists and the poaching of *Calonectris diomedea* and *Puffinus yelkouan* chicks. *Larus cachinnans* is culled annually. The IBA overlaps with a Natural Area of Special Interest (Ses Salines).

Cape Barbaria A4ii, B1i, B1ii, B3, C2, C6 313

Admin region Baleares
Coordinates 38°40'N 1°23'E
Altitude 0–90 m **Area** 100 ha

Site description

Coastal cliffs with higher areas dominated by scrub and small patches of *Pinus halepensis* and *Juniperus phoenicea* forest. The southern area of the cape has practically no vegetation. The main human activities are agriculture and hunting.

Habitats Scrub (40%; sclerophyllous scrub/garrigue/maquis), Rocky areas (60%; sea cliff/rocky shore; caves)
Land-use Agriculture (10%), Hunting (40%), Not utilized (60%)

Birds

Species	Season	Year	Pop min	Pop max	Acc	Criteria
Puffinus yelkouan Yelkouan Shearwater	B	1991	500	1,500	C	A4ii, B1ii, B3, C2, C6
Phalacrocorax aristotelis Shag	R	1991	30	50	A	B1i, C2
Sylvia sarda Marmora's Warbler	R	1996	140	—	C	B3

An important site for breeding seabirds, particularly *Puffinus yelkouan*.

Protection status

National None **International** None

Conservation issues

Threats Consequences of animal/plant introductions (B), Recreation/tourism (B), Unsustainable exploitation (A)

Main threats are from very high hunting pressure, collection of *Puffinus yelkouan* chicks, illegal house building, disturbance from visitors, and predation from introduced cats. *Larus cachinnans* colonies are culled. IBA overlaps with Natural Area of Special Interest (Cap de Barbaria).

La Mola of Formentera A4i, A4ii, B1i, B1ii, B3, C2, C3, C6 314

Admin region Baleares
Coordinates 38°40'N 1°28'E
Altitude 0–200 m **Area** 1,800 ha

Site description

A peninsula in eastern Formentera with high coastal cliffs. The main human activities are arable agriculture and hunting.

Habitats Forest and woodland (20%; native coniferous forest), Scrub (15%; sclerophyllous scrub/garrigue/maquis), Rocky areas (25%; sea cliff/rocky shore; caves), Artificial landscape (40%; arable land; perennial crops/orchards/groves; other urban/industrial areas)
Land-use Agriculture

Birds

Species	Season	Year	Pop min	Pop max	Acc	Criteria
Puffinus yelkouan Yelkouan Shearwater	B	1991	853	1,400	C	A4ii, B1ii, B3, C2, C6
Phalacrocorax aristotelis Shag	R	1995	50	60	A	B1i, C2, C6
Larus cachinnans Yellow-legged Gull	R	1995	2,100	2,200	A	A4i, B1i, C3
Sylvia sarda Marmora's Warbler	R	1996	150	—	C	B3

An important site for breeding seabirds, particularly *Puffinus yelkouan*, and cliff-nesting raptors. Species of global conservation concern that do not meet IBA criteria: *Larus audouinii* (10 resident pairs).

Protection status

National None **International** None

Conservation issues

Threats Consequences of animal/plant introductions (C), Industrialization/urbanization (B), Recreation/tourism (C), Unsustainable exploitation (A)

The main threats are from widespread legal and illegal hunting, collection of *Puffinus yelkouan* chicks, illegal housing, predation from feral cats, and disturbance from tourism. The IBA overlaps with Natural Area of Special Interest (Cala Compte–Cala Bassa).

Dragonera island–La Trapa A1, A4i, A4ii, B1i, B1ii, B1iii, B2, B3, C1, C2, C6 315

Admin region Baleares
Coordinates 39°35'N 2°21'E
Altitude 0–480 m **Area** 2,660 ha

Site description

An island close to Mallorca's western tip, with very high limestone cliffs, dominated by Mediterranean scrub.

Habitats Forest and woodland (30%; native coniferous forest), Scrub (40%; sclerophyllous scrub/garrigue/maquis), Marine areas (20%; sea inlet/coastal features), Rocky areas (10%; sea cliff/rocky shore)
Land-use Fisheries/aquaculture

Birds

Species	Season	Year	Pop min	Pop max	Acc	Criteria
Calonectris diomedea Cory's Shearwater	B	1993	85	180	B	B2, C6
Puffinus yelkouan Yelkouan Shearwater	B	1996	150	250	B	A4ii, B1ii, B3, C2, C6
Phalacrocorax aristotelis Shag	R	1991	11	25	B	B3
Pandion haliaetus Osprey	R	1996	1	1	A	C6
Falco eleonorae Eleonora's Falcon	B	1994	188	—	A	A4ii, B1iii, B2, C2, C6
Larus audouinii Audouin's Gull	R	1994	218	218	A	A1, A4i, B1i, B2, C1, C2, C6

This is an important site for breeding seabirds, *Falco eleonorae* and migrant passerines.

Protection status

National Partial **International** Partial
290 ha of IBA covered by Natural Park (Isla Dragonera, 290 ha). 975 ha of IBA covered by Special Protection Area (Isla Dragonera–La Trapa, 975 ha).

Conservation issues

Threats Recreation/tourism (A)

The main threat is from increased tourism, disturbance to *Larus audouinii* and occasional poaching of *F. eleonorae*. *Larus cachinnans* is culled annually.

Tramuntana mountains A1, A4ii, B1i, B1iii, B2, B3, C1, C2, C6 316

Admin region Baleares
Coordinates 39°49'N 2°52'E
Altitude 0–1,436 m **Area** 54,953 ha

Site description

The site lies on the north-west coast of Mallorca, with high cliffs, islets, and inland mountains. *Quercus* and *Pinus* woodlands and maquis dominate. The site is extensively grazed by sheep and goats.

Habitats Forest and woodland (*; native coniferous forest; mixed forest; broadleaved evergreen forest), Scrub (*; sclerophyllous scrub/garrigue/maquis), Rocky areas (*; sea cliff/rocky shore; scree/boulders), Artificial landscape (perennial crops/orchards/groves)
Land-use Agriculture (20%), Hunting (50%), Not utilized (30%)

Birds

Species	Season	Year	Pop min	Pop max	Acc	Criteria
Phalacrocorax aristotelis Shag	R	1991	72	210	B	B1i, B3, C2, C6
Milvus milvus Red Kite	R	1993	20	30	A	B3, C6
Aegypius monachus Cinereous Vulture	R	1996	8	12	A	A1, B1iii, B2, C1, C2, C6

Species ... continued	Season	Year	Pop min	Pop max	Acc	Criteria
Hieraaetus pennatus Booted Eagle	R	1993	20	25	B	B2, C6
Pandion haliaetus Osprey	R	1993	7	7	A	C6
Falco eleonorae Eleonora's Falcon	B	1994	196	—	A	A4ii,B1iii,B2,C2,C6
Falco peregrinus Peregrine	B	1992	25	30	B	B2, C6
Sylvia sarda Marmora's Warbler	R	1996	150	—	C	B3

This is an important site for seabirds, raptors and passerines. Species of global conservation concern that do not meet IBA criteria: *Larus audouinii* (20–30 resident pairs).

Protection status

National None **International** Partial

48,000 ha of IBA covered by Special Protection Area (Sierra de Tramuntana de Mallorca, 48,000 ha).

Conservation issues

Threats Burning of vegetation (B), Forest grazing (C), Infrastructure (B), Other (A), Recreation/tourism (B), Unsustainable exploitation (B)

Threats include disturbance from visitors, the use of poison baits, particularly a problem for *Aegypius monachus*, and poaching, particularly of *Falco peregrinus* and *F. eleonorae* ('Other' threats, above). Uncontrolled fires and overgrazing are also hazards. Colonies of *Larus cachinnans* are culled. IBA overlaps with Natural Area of Special Interest (Areas Naturales de la Serra de Tramuntana).

Cape Pinar — B3, C6 — 317

Admin region Baleares
Coordinates 39°52'N 3°10'E
Altitude 0–444 m **Area** 2,500 ha

Site description

A small rocky peninsula in northern Mallorca with limestone cliffs and a small range of hills. The dominant vegetation is Mediterranean scrub and *Pinus* woodland. Main human activities are military and tourism.

Habitats Forest and woodland (50%; native coniferous forest), Scrub (30%; sclerophyllous scrub/garrigue/maquis), Rocky areas (20%; sea cliff/rocky shore)
Land-use Military (10%), Tourism/recreation (20%)

Birds

Species	Season	Year	Pop min	Pop max	Acc	Criteria
Pandion haliaetus Osprey	R	1994	1	1	A	C6
Sylvia sarda Marmora's Warbler	R	1996	140	—	C	B3

This is an important site for breeding coastal raptors, seabirds and species characteristic of Mediterranean scrub.

Protection status

National None **International** High

2,500 ha of IBA covered by Special Protection Area (Cap des Pinar, 2,500 ha).

Conservation issues

Threats Burning of vegetation (C), Industrialization/urbanization (C)

The main threat is from uncontrolled fires. The IBA overlaps with Natural Area of Special Interest (La Victoria).

Albufera de Mallorca and Albufereta de Pollença marshes — A1, B1i, B2, B3, C1, C2, C3, C6 — 318

Admin region Baleares
Coordinates 39°47'N 3°06'E
Altitude 0–6 m **Area** 2,800 ha

Site description

Two wetlands in northern Mallorca. The Albufera has a large coastal lagoon connected to the sea but also areas of standing freshwater. There are extensive areas of marsh vegetation dominated by *Phragmites*, *Scirpus* and *Cladium*. There are also abandoned saltpans and sand-dunes with Mediterranean scrub and *Pinus* forest. The Albufereta has open water and also areas of reedbed (*Phragmites*) and *Tamarix* scrub.

Habitats Forest and woodland (10%; native coniferous forest), Wetland (80%; mudflat/sandflat; saltmarsh; sand-dunes/sand beach; coastal lagoon; standing fresh water; river/stream; water-fringe vegetation), Artificial landscape (10%; arable land)
Land-use Agriculture (20%), Fisheries/aquaculture (10%), Hunting (10%), Nature conservation/research (90%)

Birds

Species	Season	Year	Pop min	Pop max	Acc	Criteria
[1] ***Botaurus stellaris*** Bittern	R	1994	1	10	B	C6
Ixobrychus minutus Little Bittern	R	1992	11	50	B	C6
Nycticorax nycticorax Night Heron	R	1994	11	11	A	C6
Egretta garzetta Little Egret	R	1992	1	10	B	C6
Ardea purpurea Purple Heron	R	1994	35	55	B	C2, C6
Oxyura leucocephala White-headed Duck	R	1994	3	8	A	A1, B1i, B2, C1, C2, C6
Circus aeruginosus Marsh Harrier	R	1994	8	9	A	C6
Porzana porzana Spotted Crake	B	1992	51	250	B	B3, C2, C6
Porphyrio porphyrio Purple Gallinule	R	1993	8	10	B	C6
Himantopus himantopus Black-winged Stilt	B	1994	87	110	A	C6
Charadrius alexandrinus Kentish Plover	R	1992	51	250	B	B1i, B2, C3
Acrocephalus melanopogon Moustached Warbler	R	1992	1,000	—	B	C2, C6

1. Large increase.

This is a very important site for breeding, passage and wintering waterbirds, raptors and passerines—190 bird species have been recorded. Species of global conservation concern that do not meet IBA criteria: *Aythya nyroca* (wintering).

Protection status

National Partial **International** High

1,900 ha of IBA covered by Natural Park (S'Albufera de Mallorca, 1,900 ha). 1,700 ha of IBA covered by Ramsar Site (S'Albufera de Mallorca, 1,700 ha). 2,584 ha of IBA covered by Special Protection Area (S'Albufera de Mallorca, 2,584 ha).

Conservation issues

Threats Industrialization/urbanization (B), Unsustainable exploitation (A)

The main threats are from hunting in the Albufereta and urban expansion. The IBA overlaps with Natural Area of Special Interest (S'Albufereta).

Cape Freu–Cape Farrutx — B1i, C2, C6 — 319

Admin region Baleares
Coordinates 39°45'N 3°26'E
Altitude 0–432 m **Area** 500 ha

Site description

A rocky section of coast at the north-eastern tip of Mallorca with high limestone cliffs and Mediterranean scrub.

Habitats Rocky areas (100%; sea cliff/rocky shore)
Land-use Tourism/recreation (10%)

Birds

Species	Season	Year	Pop min	Pop max	Acc	Criteria
[1] ***Phalacrocorax aristotelis*** Shag	R	1991	37	100	B	B1i, C2, C6

1. Large decrease.

This is an important site for breeding seabirds. Species of global conservation concern that do not meet IBA criteria: *Larus audouinii* (20–25 breeding pairs).

Protection status

National None **International** Partial

80 ha of IBA covered by Special Protection Area (Cap des Freu–Cap Farruch, 80 ha).

Conservation issues

Threats Recreation/tourism (A)

Disturbance from anglers has caused the abandonment of a *Larus audouinii* colony. *L. cachinnans* is culled annually. The IBA overlaps with Natural Area of Special Interest (Montañas de Arta).

Cape Vermell — B1i, C2 — 320

Admin region Baleares
Coordinates 39°39′N 3°25′E
Altitude 0–252 m **Area** 150 ha

Site description

A small section of coast with limestone cliffs in eastern Mallorca.

Habitats Forest and woodland (50%; native coniferous forest), Scrub (30%; sclerophyllous scrub/garrigue/maquis), Rocky areas (20%; sea cliff/rocky shore)
Land-use Not utilized (90%), Tourism/recreation (10%)

Birds

Species	Season	Year	Pop min	Pop max	Acc	Criteria
Phalacrocorax aristotelis Shag	R	1991	26	75	B	B1i, C2

This is an important site for breeding seabirds.

Protection status

National None **International** Partial
80 ha of IBA covered by Special Protection Area (Cap Vermell, 80 ha).

Conservation issues

No serious threats are known at the site. *Larus cachinnans* is culled annually. The IBA overlaps with Natural Area of Special Interest (Cap Vermell).

Salobrar de Campos lagoons–Sa Vall — B2, B3, C6 — 321

Admin region Baleares
Coordinates 39°19′N 3°05′E
Altitude 0–21 m **Area** 2,397 ha

Site description

A plain, at Mallorca's southern tip, occupied by two small natural lagoons and some larger saltpans surrounded by Mediterranean scrub. There are also sand-dunes and some small islets. The main human activities are agriculture, hunting and salt production.

Habitats Forest and woodland (30%; native coniferous forest; alluvial/very wet forest), Scrub (20%; sclerophyllous scrub/garrigue/maquis), Wetland (20%; saltmarsh; sand-dunes/sand beach; standing brackish and salt water), Rocky areas (rock stacks/islets), Artificial landscape (30%; arable land; perennial crops/orchards/groves)
Land-use Agriculture (30%), Hunting (30%), Not utilized (20%), Tourism/recreation (10%), Urban/industrial/transport (20%)

Birds

Species	Season	Year	Pop min	Pop max	Acc	Criteria
Himantopus himantopus Black-winged Stilt	B	1991	96	—	A	C6
Charadrius alexandrinus Kentish Plover	B	1995	85	105	B	B2
Sylvia sarda Marmora's Warbler	R	1996	140	—	C	B3

This is an important wetland for breeding and wintering waterbirds.

Protection status

National None **International** Partial
900 ha of IBA covered by Special Protection Area (Lagunas del Salobrar de Campos-Playa d'es Trenc, 900 ha).

Conservation issues

Threats Industrialization/urbanization (B), Recreation/tourism (A), Unsustainable exploitation (B)

The main threats are from illegal hunting, tourism, particularly off-road driving, and new urban development. The IBA overlaps with Natural Area of Special Interest (Es Trenc–Salobrar de Campos).

Coastal cliffs between Cape Enderrocat and Cala Pí — B1i, B2, B3, C2, C6 — 322

Admin region Baleares
Coordinates 39°30′N 2°45′E
Altitude 0–102 m **Area** 900 ha

Site description

Limestone cliffs and a scrub-covered plain along the southern coast of Mallorca.

Habitats Forest and woodland (20%; native coniferous forest), Scrub (50%; sclerophyllous scrub/garrigue/maquis), Grassland (20%; steppe/dry calcareous grassland), Rocky areas (5%; sea cliff/rocky shore), Artificial landscape (20%; arable land; perennial crops/orchards/groves; other urban/industrial areas; ruderal land)
Land-use Agriculture (20%), Hunting (80%), Military (5%)

Birds

Species	Season	Year	Pop min	Pop max	Acc	Criteria
[1] ***Phalacrocorax aristotelis*** Shag	R	1991	74	200	B	B1i, B3, C2, C6
Burhinus oedicnemus Stone Curlew	R	1993	220	—	C	B2
Sylvia sarda Marmora's Warbler	R	1993	140	—	—	B3

1. Large decrease.

This is an important site for breeding seabirds, cliff-nesting species and species characteristic of coastal scrub.

Protection status

National None **International** Partial
150 ha of IBA covered by Special Protection Area (Acantilados de Cap Blanc y Cap Enderrocat, 150 ha).

Conservation issues

Threats Industrialization/urbanization (A), Infrastructure (B), Recreation/tourism (B), Unsustainable exploitation (A)

The most important threats to the site are from the development of tourism-related infrastructure, urban development, disturbance from visitors, and a high hunting pressure. The IBA overlaps with two Natural Areas of National Interest (Cap Enderrocat and Marina de Llucmajor).

Coastal cliffs between isles of Malgrats and Cala Figuera — B1i, B1ii, B3, C2, C6 — 323

Admin region Baleares
Coordinates 39°30′N 2°30′E
Altitude 0–64 m **Area** 440 ha

Site description

Limestone cliffs and several rocky islets in south-eastern Mallorca.

Habitats Forest and woodland (40%; native coniferous forest), Scrub (20%; sclerophyllous scrub/garrigue/maquis), Rocky areas (50%; sea cliff/rocky shore; rock stacks/islets)
Land-use Military (80%), Tourism/recreation (10%)

Birds

Species	Season	Year	Pop min	Pop max	Acc	Criteria
Puffinus yelkouan Yelkouan Shearwater	B	1993	80	100	B	B1ii, B3, C2, C6
Hydrobates pelagicus Storm Petrel	B	1992	30	50	—	C2, C6
Phalacrocorax aristotelis Shag	R	1994	20	50	B	B1i, C2

An important area for breeding seabirds. Species of global conservation concern that do not meet IBA criteria: *Larus audouinii* (resident).

Protection status

National None **International** Partial
100 ha of IBA covered by Special Protection Area (Islas Malgrats, Isla del Toro y Cap de Cala Figuera, 100 ha).

Conservation issues

Threats Consequences of animal/plant introductions (A), Recreation/tourism (A)

Disturbance from anglers has caused the abandonment of a *Larus audouinii* colony. Rats are a major threat to the breeding seabirds. *L. cachinnans* is culled annually. The IBA overlaps with Natural Area of Special Interest (Cap de Cala Figuera).

Archipelago of Cabrera — 324

A1, A3, A4i, B1i, B1ii, B2, B3, C1, C2, C3, C6

Admin region Baleares
Coordinates 39°09′N 2°55′E
Altitude 0–172 m **Area** 10,000 ha

Site description

A small archipelago, to the south-east of Mallorca, formed by two main islands and many rocky islets. Mediterranean scrub and small *Pinus* woodlands occur on the largest islands. The main human activities are military and fishing.

Habitats Forest and woodland (10%; native coniferous forest), Scrub (70%; sclerophyllous scrub/garrigue/maquis), Marine areas (20%; open sea; sea inlet/coastal features), Rocky areas (10%; sea cliff/rocky shore; rock stacks/islets; inland cliff; caves), Artificial landscape (1%; arable land)
Land-use Fisheries/aquaculture (100%), Military (100%), Nature conservation/research (100%), Tourism/recreation (100%)

Birds

Species	Season	Year	Pop min	Pop max	Acc	Criteria
Calonectris diomedea Cory's Shearwater	B	1996	400	500	A	B1ii, B2, C2, C6
Puffinus yelkouan Yelkouan Shearwater	B	1996	60	120	B	B1ii, B3, C2, C6
Hydrobates pelagicus Storm Petrel	B	1996	300	500	B	B1ii, B2, C2, C6
[1] ***Phalacrocorax aristotelis*** Shag	R	1995	85	85	A	B1i, C2, C6
Pandion haliaetus Osprey	R	1996	2	2	A	C6
Falco eleonorae Eleonora's Falcon	B	1996	25	35	A	A3, B2, C6
Falco peregrinus Peregrine	R	1996	8	10	A	C6
Larus audouinii Audouin's Gull	R	1995	381	381	A	A1, A4i, B1i, B2, C1, C2, C6
Larus cachinnans Yellow-legged Gull	R	1996	1,300	1,700	B	A4i, B1i, C3
Sylvia sarda Marmora's Warbler	R	1995	315	361	A	A3, B3
Sylvia cantillans Subalpine Warbler	B	1995	40	40	A	A3
Sylvia melanocephala Sardinian Warbler	R	1995	358	424	A	A3

1. Large decrease.

Breeding birds include four of the 20 species in Europe that are restricted to the Mediterranean biome (when breeding).

Protection status

National High **International** High
10,000 ha of IBA covered by National Park (Archipielago de Cabrera, 10,000 ha). 10,000 ha of IBA covered by Special Protection Area (Archipiélago de Cabrera, 10,000 ha).

Conservation issues

Threats Agricultural intensification/expansion (C), Aquaculture/fisheries (C), Consequences of animal/plant introductions (A), Recreation/tourism (C)

The main threats are from predation of seabirds by introduced rats, disturbance from illegal fishing, and overgrazing by sheep on the main island. *Larus cachinnans* are culled annually.

North and east coasts of Minorca and Island of Aire — 325

A1, A4ii, B1i, B1ii, B2, B3, C1, C2, C6

Admin region Baleares
Coordinates 40°03′N 4°00′E
Altitude 0–205 m **Area** 17,900 ha

Site description

The site includes limestone, the Es Grao marshes and a coastal lagoon. The vegetation is mainly garrigue and maquis with *Quercus ilex* and *Pinus halepensis* forest.

Habitats Forest and woodland (50%; native coniferous forest; mixed forest), Scrub (30%; sclerophyllous scrub/garrigue/maquis), Wetland (10%; coastal lagoon), Rocky areas (10%; sea cliff/rocky shore; rock stacks/islets)
Land-use Agriculture (50%), Forestry (50%), Hunting (70%), Tourism/recreation (20%), Urban/industrial/transport (5%)

Birds

Species	Season	Year	Pop min	Pop max	Acc	Criteria
Calonectris diomedea Cory's Shearwater	B	1995	500	1,500	B	B1ii, B2, C2, C6
Puffinus yelkouan Yelkouan Shearwater	B	1995	300	—	C	A4ii, B1ii, B3, C2, C6
Phalacrocorax aristotelis Shag	R	1995	200	250	B	B1i, B3, C2, C6
[1] ***Milvus milvus*** Red Kite	R	1996	8	8	A	C6
Neophron percnopterus Egyptian Vulture	R	1995	25	50	B	B2, C2, C6
Hieraaetus pennatus Booted Eagle	R	1995	10	—	—	C6
Pandion haliaetus Osprey	R	1995	6	6	A	C6
Falco peregrinus Peregrine	R	1995	10	20	B	B2, C6
Larus audouinii Audouin's Gull	R	1995	50	100	B	A1, C1, C6

1. Large decrease.

This is an important area for seabirds and coastal raptors.

Protection status

National Partial **International** High
1,947 ha of IBA covered by Natural Park (Albufera des Grau, 1,947 ha). 17,900 ha of IBA covered by Biosphere Reserve (Isla de Menorca Biosphere Reserve, 70,000 ha). 900 ha of IBA covered by Special Protection Area (Costa Norte de Menorca, 900 ha).

Conservation issues

Threats Extraction industry (C), Infrastructure (A), Other (A), Recreation/tourism (C), Selective logging/cutting (C), Unsustainable exploitation (B)

Threats include mass tourism, building, poisoning, particularly of *Milvus milvus* ('Other' threats, above), and the erection of new powerlines. A reduction in the rabbit *Oryctolagus cunniculus* population has reduced prey availability for raptors. Nine *Larus cachinnans* colonies are culled annually. The IBA overlaps with six Natural Areas of Special Interest.

Ravines and pinewoods in central Minorca — 326

B1i, B2, C2, C6

Admin region Baleares
Coordinates 40°00′N 4°05′E
Altitude 0–357 m **Area** 30,200 ha

Site description

The IBA includes limestone cliffs along ravines, *Pinus halepensis* forest, garrigue and some cultivation.

Habitats Forest and woodland (20%; native coniferous forest; broadleaved evergreen forest), Scrub (20%; sclerophyllous scrub/garrigue/maquis), Wetland (3%; sand-dunes/sand beach; standing brackish and salt water; water-fringe vegetation), Marine areas (5%; sea inlet/coastal features), Rocky areas (15%; sea cliff/rocky shore; inland cliff), Artificial landscape (35%; highly improved reseeded grassland; arable land; other urban/industrial areas)
Land-use Agriculture (35%), Forestry (20%), Hunting (80%)

Birds

Species	Season	Year	Pop min	Pop max	Acc	Criteria
Phalacrocorax aristotelis Shag	R	1991	74	74	A	B1i, C2, C6
[1] ***Milvus milvus*** Red Kite	R	1996	2	3	B	C6
Neophron percnopterus Egyptian Vulture	R	1992	13	15	B	B2, C6
Hieraaetus pennatus Booted Eagle	R	1996	20	35	B	B2, C6
Falco peregrinus Peregrine	R	1992	10	12	A	C6

1. Large decrease.

This is an important area for breeding raptors. Species of global conservation concern that do not meet IBA criteria: *Falco naumanni* and *Larus audouinii* (non-breeding).

Protection status

National None **International** None

Conservation issues

Threats Firewood collection (B), Industrialization/urbanization (B), Infrastructure (A), Other (A), Unsustainable exploitation (A)

The main threats are from poison baits ('Other' threats, above) and new powerlines. A reduction in the rabbit *Oryctolagus cuniculus* population has reduced prey availability for raptors ('Other' threats, above). The IBA overlaps with three Natural Areas of Special Interest.

Lanzarote islets — A4ii, B1ii, B1iii, B2, B3, C2, C6 — 327

Admin region Las Palmas
Coordinates 29°14′N 13°30′W
Altitude 0–266 m **Area** 4,000 ha

Site description

A group of five rocky islands that support drought-resistant vegetation: La Graciosa (coordinates as above), Montaña Clara (29°17′N, 13°32′W), Roque del Este (29°18′N, 13°31′W), Roque del Oeste (29°16′N, 13°20′W) and Alegranza (29°23′N, 13°30′W). Human activities are mainly related to conservation and research, and fishing.

Habitats Desert (semi-desert), Wetland (sand-dunes/sand beach; shingle/stony beach), Rocky areas (*; rock stacks/islets; inland sand-dunes)
Land-use Fisheries/aquaculture, Nature conservation/research (100%)

Birds

Species	Season	Year	Pop min	Pop max	Acc	Criteria
Bulweria bulwerii Bulwer's Petrel	B	1987	185	210	B	B1ii, B2, C2, C6
Calonectris diomedea Cory's Shearwater	B	1987	8,575	11,075	B	A4ii, B1ii, B2, C2, C6
Puffinus assimilis Little Shearwater	B	1987	50	60	B	B1ii, B2, C2, C6
Pelagodroma marina White-faced Petrel	B	1987	10	15	B	C6
Hydrobates pelagicus Storm Petrel	B	1987	340	540	B	B1ii, B2, C2, C6
Oceanodroma castro Madeiran Petrel	B	1987	70	100	B	B1ii, B2, C2, C6
Neophron percnopterus Egyptian Vulture	B	1988	3	4	B	B2, C6
Pandion haliaetus Osprey	R	1995	2	3	B	C6
Falco eleonorae Eleonora's Falcon	B	1988	66	100	B	A4ii, B1iii, B2, C2, C6
Alectoris barbara Barbary Partridge	R	1996	6	—	C	B2
Anthus berthelotii Berthelot's Pipit	R	1996	Abundant		—	B3

The site holds some of the highest numbers of breeding seabirds and raptors in the Canaries.

Protection status

National High **International** Partial
4,000 ha of IBA covered by Natural Park (Archipiélago Chinijo, 9,112 ha). 165 ha of IBA covered by Integral Nature Reserve (Los Islotes, 165 ha). 1,722 ha of IBA covered by Special Protection Area (Islotes del Norte de Lanzarote y Riscos de Famara, 1,722 ha).

Conservation issues

Threats Consequences of animal/plant introductions (C), Recreation/tourism (B), Unsustainable exploitation (A)

Feral cats *Felis catus*, introduced rabbits *Oryctolagus cuniculus* and illegal hunting pose the main threats to breeding birds. Some 2,000–3,000 *Calonectris diomedea* chicks are poached every year, which impacts heavily on the local population. The area is a game refuge. La Laguna University undertakes research work, and monitoring of bird numbers is ongoing.

Famara crag — B2, C2, C6 — 328

Admin region Las Palmas
Coordinates 29°11′N 13°30′W
Altitude 0–600 m **Area** 1,350 ha

Site description

High sea cliffs on Lanzarote with accumulations of landslide and erosion debris at the base of many sectors. The cliffs support drought-resistant vegetation. Human activities include fishing and tourism, and there is a designated military zone.

Habitats Desert (semi-desert), Wetland (saltmarsh; sand-dunes/sand beach), Rocky areas (*; sea cliff/rocky shore; inland cliff)
Land-use Fisheries/aquaculture, Military, Tourism/recreation

Birds

Species	Season	Year	Pop min	Pop max	Acc	Criteria
Calonectris diomedea Cory's Shearwater	B	1994	300	—	C	B2, C2
Oceanodroma castro Madeiran Petrel	B	1996	3	—	C	B2
Neophron percnopterus Egyptian Vulture	B	1994	1	2	B	B2, C6
Pandion haliaetus Osprey	R	1992	2	2	A	C6
Falco eleonorae Eleonora's Falcon	B	1987	10	—	C	B2, C6
Alectoris barbara Barbary Partridge	R	1996	Abundant		—	B2, C6

The site is important for breeding seabirds and raptors.

Protection status

National High **International** High
1,350 ha of IBA covered by Natural Park (Archipiélago Chinijo, 9,112 ha). 1,350 ha of IBA covered by Special Protection Area (Islotes del Norte de Lanzarote y Riscos de Famara, 1,722 ha).

Conservation issues

Threats Consequences of animal/plant introductions (C), Industrialization/urbanization (B), Infrastructure (B), Recreation/tourism (A), Unsustainable exploitation (B)

The main threats are posed by urbanization, electric powerlines, disturbance caused to birds such as *Falco pelegrinoides* by recreation and tourism, and illegal hunting. Introduced rats *Rattus* and feral cats *Felis catus* also pose problems for breeding birds. The area is a game refuge, and only rabbits *Oryctolagus cuniculus* can be hunted. The local government ran several marine bird conservation campaigns between 1992 and 1996.

Haría–Tabayesco — B2, C6 — 329

Admin region Las Palmas
Coordinates 29°08′N 13°29′W
Altitude 100–581 m **Area** 2,350 ha

Site description

An area of scrub, steppe-grassland and agricultural land situated in the north of Lanzarote, and interspersed with several urban developments.

Habitats Scrub (*), Grassland (*; steppe/dry calcareous grassland), Rocky areas (inland cliff), Artificial landscape (*; arable land; forestry plantation; ruderal land)
Land-use Agriculture (*), Hunting (*), Tourism/recreation (*)

Birds

Species	Season	Year	Pop min	Pop max	Acc	Criteria
Falco tinnunculus Kestrel	R	1995	Abundant		—	B2
[1] ***Falco eleonorae*** Eleonora's Falcon	B	1995	—	—	—	C6
Alectoris barbara Barbary Partridge	R	1995	10	—	C	B2, C6

1. Probably breeds.

The site is important for breeding steppic species and seabirds, and for migrant passerines. The area also provides feeding opportunities for raptors.

Protection status

National Low **International** None
40 ha of IBA covered by Natural Park (Archipiélago Chinijo, 9,112 ha).

Conservation issues

Threats Extraction industry (C), Industrialization/urbanization (B), Unsustainable exploitation (B)

Gravel extraction, urban expansion and illegal hunting pose threats to the site.

Tenegüime gully — B2, C6 — 330

Admin region Las Palmas
Coordinates 29°05′N 13°29′W
Altitude 100–400 m **Area** 421 ha

Site description

A gully with scrub, drought-resistant grassland, exposed bedrock and inland cliffs situated in the north-east of Lanzarote. Human activities include goat-grazing.

Habitats Scrub (*), Grassland (*; steppe/dry calcareous grassland), Rocky areas (inland cliff)
Land-use Agriculture (*), Hunting (*)

Birds

Species	Season	Year	Pop min	Pop max	Acc	Criteria
Alectoris barbara Barbary Partridge	R	1995	10	—	C	B2, C6

The site is also important for breeding seabirds and raptors.

Protection status

National High **International** None
421 ha of IBA covered by Protected Landscape (Tenegüime, 421 ha).

Conservation issues

Threats Recreation/tourism (A), Unsustainable exploitation (A)

Threats are posed by the presence of people taking part in open-air sports (climbing has the highest impact), and the poaching of *Calonectris diomedea* chicks (more than 70% of chicks are captured every year).

Plains of La Corona–Las Honduras — B2, B3, C2, C6 — 331

Admin region Las Palmas
Coordinates 29°02′N 13°30′W
Altitude 0–235 m **Area** 3,470 ha

Site description

A steppic area in the north-east of Lanzarote with scrub, drought-resistant grassland and saltpans. Human activities include cattle-grazing, and there is a designated military zone.

Habitats Scrub (*), Grassland (*; steppe/dry calcareous grassland), Wetland (standing brackish and salt water), Rocky areas (inland cliff), Artificial landscape (arable land; ruderal land)
Land-use Agriculture (*), Military, Tourism/recreation (*)

Birds

Species	Season	Year	Pop min	Pop max	Acc	Criteria
Alectoris barbara Barbary Partridge	R	1995	10	—	C	B2, C6
Chlamydotis undulata Houbara Bustard	R	1995	72	82	B	B2, C2, C6
Burhinus oedicnemus Stone Curlew	R	1995	5	—	C	B2, C6
Cursorius cursor Cream-coloured Courser	R	1995	Common		—	B2, C6
Anthus berthelotii Berthelot's Pipit	R	1995	Common		—	B3

The site is one of the best sites in Lanzarote for *Chlamydotis undulata* and other steppic species. The saltpans provide habitat for migrating and wintering waders and gulls (Laridae).

Protection status

National None **International** None

Conservation issues

Threats Infrastructure (C), Recreation/tourism (B), Unsustainable exploitation (A)

Threats include electric powerlines, jeep-safaris that cause disturbance to breeding birds, illegal hunting and poaching of *Chlamydotis undulata* chicks. The area is a game refuge.

Famara sandy plain — B2, B3, C2, C6 — 332

Admin region Las Palmas
Coordinates 29°06′N 13°36′W
Altitude 0–339 m **Area** 5,700 ha

Site description

A sandy landscape comprising scrub, drought-resistant grassland and rocky habitats located in the north of Lanzarote. Goat-grazing is the main human activity.

Habitats Scrub (*), Grassland (*; steppe/dry calcareous grassland), Rocky areas (*; inland sand-dunes), Artificial landscape (arable land)
Land-use Agriculture (*), Tourism/recreation (*)

Birds

Species	Season	Year	Pop min	Pop max	Acc	Criteria
Falco tinnunculus Kestrel	R	1988	Abundant		—	B2
Alectoris barbara Barbary Partridge	R	1995	6	—	B	B2
Chlamydotis undulata Houbara Bustard	R	1995	28	44	B	B2, C2, C6
Burhinus oedicnemus Stone Curlew	R	1995	Common		—	B2, C6
Cursorius cursor Cream-coloured Courser	R	1995	Common		—	B2, C6
Calandrella rufescens Lesser Short-toed Lark	R	1995	Common		—	B2
Anthus berthelotii Berthelot's Pipit	R	1995	Common		—	B3

The site is important for *Chlamydotis undulata* and other steppic species.

Protection status

National Partial **International** None
1,900 ha of IBA covered by Natural Park (Archipiélago Chinijo, 9,112 ha).

Conservation issues

Threats Agricultural intensification/expansion (A), Extraction industry (B), Infrastructure (C), Recreation/tourism (A)

Goat-grazing, gravel extraction, electric powerlines and disturbance to birds caused by safaris are the most important threats to the site. The area is a game refuge, and the local government has undertaken five censuses of *Chlamydotis undulata* since 1978.

El Mojón — B2, C6 — 333

Admin region Las Palmas
Coordinates 29°00′N 13°48′W
Altitude 0–64 m **Area** 60 ha

Site description

Sea cliffs and lava-fields ('malpaís') in the west of Lanzarote that support semi-desertic vegetation. There is no relevant human activity, other than some fishing.

Habitats Desert (*; semi-desert), Rocky areas (*; sea cliff/rocky shore; rock stacks/islets; caves)
Land-use Fisheries/aquaculture

Birds

Species	Season	Year	Pop min	Pop max	Acc	Criteria
Bulweria bulwerii Bulwer's Petrel	B	1994	10	15	B	B2
Calonectris diomedea Cory's Shearwater	B	1994	650	750	B	B2, C6
Oceanodroma castro Madeiran Petrel	B	1994	Frequent		—	B2
Alectoris barbara Barbary Partridge	R	1995	Common		—	B2

The site is important for breeding seabirds and some steppic species.

Protection status

National Partial **International** High

35 ha of IBA covered by National Park (Timanfaya, 5,107 ha). 25 ha of IBA covered by Special Protection Area (La Geria, 15,189 ha). 35 ha of IBA covered by Special Protection Area (Parque Nacional de Timanfaya, 5,107 ha).

Conservation issues

Threats Consequences of animal/plant introductions (A), Recreation/tourism (B), Unsustainable exploitation (A)

Introduced rats *Rattus* and feral cats *Felis catus* impact on numbers of *Calonectris diomedea*. Tourists camping on the beaches disturb seabirds, and illegal hunting is also a problem.

Janubio saltpans — B2, C6 — 334

Admin region Las Palmas
Coordinates 28°55'N 13°50'W
Altitude 0–40 m **Area** 169 ha

Site description

The site comprises saltpans, lagoons, sand and shingle beaches, and a small area of steppe, on Lanzarote. Human activities include exploitation of the saltpans.

Habitats Wetland (*; saltmarsh; shingle/stony beach; standing brackish and salt water)
Land-use Fisheries/aquaculture, Tourism/recreation, Urban/industrial/transport (*)

Birds

Species	Season	Year	Pop min	Pop max	Acc	Criteria
Himantopus himantopus Black-winged Stilt	B	1995	5	10	B	C6
Cursorius cursor Cream-coloured Courser	R	1995	2	—	C	B2
Charadrius alexandrinus Kentish Plover	R	1995	34	34	A	B2
Calandrella rufescens Lesser Short-toed Lark	R	1995	Common		—	B2

This site is important for breeding and wintering waterbirds. It is also one of the best sites in the Canary Islands for migrant waders, both in terms of the variety of species present and the numbers of individuals.

Protection status

National High **International** High
169 ha of IBA covered by Site of Scientific Interest (Janubio, 169 ha). 157 ha of IBA covered by Special Protection Area (Salinas de Janubio, 157 ha).

Conservation issues

Threats Consequences of animal/plant introductions (B), Industrialization/urbanization (C), Other (B), Recreation/tourism (B)

Threats include feral cats *Felis catus* that prey upon birds at night, the dumping of car oil into the lagoon ('Other' threat), and disturbance to birds caused by uncontrolled recreation and tourism. The area is a game refuge, and the local government undertook a census of wintering birds during 1993 and 1994.

Plains of Mareta–Hoya de la Yegua — B2, B3, C2, C6 — 335

Admin region Las Palmas
Coordinates 28°54'N 13°50'W
Altitude 0–100 m **Area** 3,600 ha

Site description

An area of scrub, steppe-grassland and sea cliffs on the south-west tip of Lanzarote. Human activities include livestock-farming and residential use.

Habitats Scrub, Grassland (*; steppe/dry calcareous grassland), Desert (*; semi-desert), Rocky areas (*; sea cliff/rocky shore)
Land-use Agriculture (*), Hunting (*), Military, Tourism/recreation (*)

Birds

Species	Season	Year	Pop min	Pop max	Acc	Criteria
Bulweria bulwerii Bulwer's Petrel	B	1987	10	—	C	B2
[1] ***Chlamydotis undulata*** Houbara Bustard	R	1995	20	20	A	B2, C2, C6
Cursorius cursor Cream-coloured Courser	R	1995	Common		—	B2
Pterocles orientalis Black-bellied Sandgrouse	R	1995	Frequent		—	C6
Anthus berthelotii Berthelot's Pipit	R	1995	Common		—	B3
Bucanetes githagineus Trumpeter Finch	R	1995	Common		—	B2, C6

1. Small decrease over last 20 years.

This is an important area for *Chlamydotis undulata* and other steppic birds.

Protection status

National None **International** None

Conservation issues

Threats Extraction industry (A), Industrialization/urbanization (B), Infrastructure (A), Recreation/tourism (B), Unsustainable exploitation (U)

Threats include soil extraction, habitat fragmentation due to new roads and powerlines, and disturbance to birds caused by jeep-safaris. The area is a game refuge. The local government has undertaken four censuses of *Chlamydotis undulata* since 1979, and monitoring of the impacts of electric powerlines on birds takes place.

Island of Lobos — A4ii, B1ii, B2, C2, C6 — 336

Admin region Las Palmas
Coordinates 28°45'N 13°49'W
Altitude 0–127 m **Area** 468 ha

Site description

Island situated off the north coast of Fuerteventura, with lava fields and some sandy areas. Human activities include fishing and tourism.

Habitats Scrub, Desert (*; semi-desert), Wetland (*; sand-dunes/sand beach), Marine areas (*; sea inlet/coastal features), Rocky areas (*; rock stacks/islets; inland sand-dunes)
Land-use Fisheries/aquaculture (*), Tourism/recreation (*)

Birds

Species	Season	Year	Pop min	Pop max	Acc	Criteria
Calonectris diomedea Cory's Shearwater	B	1987	1,000	2,000	A	A4ii, B1ii, B2, C2, C6
Puffinus assimilis Little Shearwater	B	1987	20	30	A	B1ii, B2, C2, C6
Hydrobates pelagicus Storm Petrel	B	1987	20	30	A	B1ii, B2, C2, C6
Oceanodroma castro Madeiran Petrel	B	1987	5	—	C	B2, C6
Charadrius alexandrinus Kentish Plover	R	1995	16	22	A	B2

The site is very important for seabirds, and also for raptors and some steppic species. The coastline provides feeding opportunities for migrating herons (Ardeidae), waders and gulls (Laridae).

Protection status

National High **International** High
468 ha of IBA covered by Natural Park (Isla de Lobos, 468 ha). 468 ha of IBA covered by Special Protection Area (Dunas de Corralejo e Isla de Lobos, 2,482 ha).

Conservation issues

Threats Consequences of animal/plant introductions (A), Recreation/tourism (B), Unsustainable exploitation (B)

Threats include feral cats *Felis catus* preying upon breeding seabirds, disturbance to birds and habitat degradation caused by the c.100 people that visit the island every day, and poaching of *Calonectris diomedea*. The area is a game reserve. In 1992 the local government undertook a cat cull, and a census of *Chlamydotis undulata* followed in 1995.

Sandy plain of Corralejo — B2, B3, C2, C6 — 337

Admin region Las Palmas
Coordinates 28°41′N 13°48′W
Altitude 0–312 m **Area** 2,668 ha

Site description
Inland sand-dunes and sand beaches on the north-east coast of Fuerteventura. Human activities include sport boating.

Habitats Scrub, Desert (*; semi-desert), Wetland (*; sand-dunes/sand beach; coastal lagoon), Rocky areas (*; inland sand-dunes)
Land-use Agriculture, Fisheries/aquaculture (*), Tourism/recreation (*)

Birds

Species	Season	Year	Pop min	Pop max	Acc	Criteria
Chlamydotis undulata Houbara Bustard	R	1995	20	24	A	B2, C2, C6
Burhinus oedicnemus Stone Curlew	R	1995	Common		—	B2, C6
Cursorius cursor Cream-coloured Courser	R	1995	10	15	B	B2, C2, C6
Anthus berthelotii Berthelot's Pipit	R	1995	Common		—	B3

The site is important for breeding steppic species and wintering and migrating waders. The area also supports breeding *Calonectris diomedea* and *Charadrius alexandrinus*.

Protection status
National High **International** Partial
2,668 ha of IBA covered by Natural Park (Corralejo, 2,668 ha). 600 ha of IBA covered by Special Protection Area (Dunas de Corralejo e Isla de Lobos, 2,482 ha).

Conservation issues

Threats Consequences of animal/plant introductions (C), Industrialization/urbanization (B), Infrastructure (A), Recreation/tourism (A)

Threats include bird collisions with electric powerlines, and disturbance caused to both breeding and wintering birds by tourists. A road follows the coastline, hampering the natural landward movement of coastal sand-dunes. The area is a game reserve, and the local government has undertaken five censuses of *Chlamydotis undulata* since 1979.

Morro Tabaiba–Morro de los Rincones-Vallebrón mountains — A1, A2, B2, C1, C2, C6 — 338

Admin region Las Palmas
Coordinates 28°34′N 13°52′W
Altitude 250–527 m **Area** 1,150 ha

Site description
Situated in a mountainous area in the north of Fuerteventura, Vallebrón is a large valley with steep slopes and terracing in places. Scrub and drought-resistant grassland are interspersed with arable land. Human activities include goat-grazing.

Habitats Scrub (*), Grassland (*), Rocky areas (inland cliff), Artificial landscape (arable land)
Land-use Agriculture (*)

Birds

Species	Season	Year	Pop min	Pop max	Acc	Criteria
Anthus berthelotii Berthelot's Pipit	R	1995	Common		—	A2
Saxicola dacotiae Fuerteventura Chat	R	1994	20	—	C	A1, A2, B2, C1, C2, C6
Bucanetes githagineus Trumpeter Finch	R	1995	Common		—	B2, C6

One of the most important areas on Fuerteventura for *Saxicola dacotiae*.

Protection status
National None **International** None

Conservation issues

Threats Agricultural intensification/expansion (B), Infrastructure (C), Unsustainable exploitation (C)

Overgrazing, electric powerlines and illegal hunting are the most important threats to this area.

Río Cabras gully — A1, B2, C1 — 339

Admin region Las Palmas
Coordinates 28°28′N 13°54′W
Altitude 20–160 m **Area** 100 ha

Site description
Located in the east of Fuerteventura, this is a good example of the deep rocky gullies found in the eastern Canary Islands, with running water and areas of dense vegetation including a *Tamarix canariensis* copse. Human activities include goat-grazing.

Habitats Scrub, Wetland (river/stream; water-fringe vegetation), Rocky areas (inland cliff)
Land-use Agriculture (*), Hunting

Birds

Species	Season	Year	Pop min	Pop max	Acc	Criteria
Falco tinnunculus Kestrel	R	1988	Abundant		—	B2
Saxicola dacotiae Fuerteventura Chat	R	1995	Common		—	A1, B2, C1

The site is important for breeding *Saxicola dacotiae*, and also supports wintering and migrating waders and passerines.

Protection status
National None **International** None

Conservation issues

Threats Agricultural intensification/expansion (B), Industrialization/urbanization (C), Infrastructure (C), Recreation/tourism (C), Unsustainable exploitation (C)

High goat numbers adversely affect the vegetation. Domestic waste disposal results in an increase in the rat *Rattus* population, leading to increased predation of breeding birds. Electric powerlines, disturbance caused by people, and illegal hunting pose additional threats.

Cuchillete de Buenavista–gully of La Torre–Los Alares — A1, A2, B2, B3, C1, C2, C6 — 340

Admin region Las Palmas
Coordinates 28°24′N 13°50′W
Altitude 0–416 m **Area** 7,500 ha

Site description
The site comprises a mountainous area with rocky plains, and includes a gully with running water, and saltpans and sea cliffs at the coast. Scrub, drought-resistant grassland and crops all occur. Human activities include cattle-grazing.

Habitats Scrub, Grassland (*; steppe/dry calcareous grassland), Desert (semi-desert), Wetland (river/stream), Rocky areas (inland cliff), Artificial landscape (*; arable land; ruderal land)
Land-use Agriculture (*), Fisheries/aquaculture, Hunting (*), Tourism/recreation

Birds

Species	Season	Year	Pop min	Pop max	Acc	Criteria
Neophron percnopterus Egyptian Vulture	R	1988	Frequent		—	B2, C6
Falco tinnunculus Kestrel	R	1988	50	—	B	B2
Chlamydotis undulata Houbara Bustard	R	1995	124	124	A	B2, C2, C6
Burhinus oedicnemus Stone Curlew	R	1988	10	—	B	B2, C6
Cursorius cursor Cream-coloured Courser	R	1994	10	15	B	B2, C2, C6
Pterocles orientalis Black-bellied Sandgrouse	R	1994	40	—	B	B2, C6

Species ... continued	Season	Year	Pop min	Pop max	Acc	Criteria
Apus unicolor Plain Swift	R	1995	Frequent	—		A2, B3
Anthus berthelotii Berthelot's Pipit	R	1996	Abundant	—		A2, B3
Saxicola dacotiae Fuerteventura Chat	R	1995	Common	—		A1, A2, B2, C1, C6
Bucanetes githagineus Trumpeter Finch	R	1995	Common	—		B2, C6

An important area for raptors, El Cuchillete de Buenavista holds the largest number of breeding *Neophron percnopterus* in the Canaries. It is also the most important area on Fuerteventura for *Pterocles orientalis*. The coast and saltpans support wintering and migrating waders.

Protection status

National None **International** None

Conservation issues

Threats Infrastructure (B), Recreation/tourism (A), Unsustainable exploitation (C)

The main threat is posed by tourism (for example, jeep-safaris). Additional threats include bird collisions with electric powerlines, the possible construction of an aerodrome and illegal hunting. Five censuses of *Chlamydotis undulata* have been undertaken since 1979, and bird collisions with powerlines are monitored.

Pozo Negro mountain–Vigán A1, B2, C1, C2, C6 341

Admin region Las Palmas
Coordinates 28°18′N 13°55′W
Altitude 0–460 m **Area** 8,000 ha

Site description

A mountainous coastal area in the south-east of Fuerteventura, with dispersed scrub vegetation. Human activities include livestock-grazing.

Habitats Scrub, Desert (*; semi-desert), Rocky areas (sea cliff/rocky shore; inland cliff), Artificial landscape (arable land)
Land-use Agriculture (*), Fisheries/aquaculture, Tourism/recreation (*)

Birds

Species	Season	Year	Pop min	Pop max	Acc	Criteria
Calonectris diomedea Cory's Shearwater	B	1987	300	—	C	B2, C2
Neophron percnopterus Egyptian Vulture	R	1988	10	—	A	B2, C6
Cursorius cursor Cream-coloured Courser	R	1995	Common	—		B2
Saxicola dacotiae Fuerteventura Chat	R	1995	Common	—		A1, B2, C1, C6

The site is important for breeding raptors and for *Calonectris diomedea*.

Protection status

National Partial **International** Partial
5,800 ha of IBA covered by Natural Monument (Cuchillos de Vigan, 6,090 ha). 4,000 ha of IBA covered by Special Protection Area (Pozo Negro, 9,237 ha).

Conservation issues

Threats Agricultural intensification/expansion (B), Recreation/tourism (C), Unsustainable exploitation (B)

Threats include overgrazing, disturbance to birds by tourists, and illegal hunting.

Tarajalejo mountain range A1, B2, C1, C2, C6 342

Admin region Las Palmas
Coordinates 28°12′N 14°05′W
Altitude 0–464 m **Area** 2,200 ha

Site description

A mountainous area of inland cliffs and exposed bedrock, with sea cliffs at the coast. The landscape supports drought-resistant scrub and crops, and goat-farming and tourism are the main land-uses.

Habitats Scrub, Desert (semi-desert), Rocky areas (inland cliff), Artificial landscape (*; arable land)
Land-use Agriculture (*), Tourism/recreation

Birds

Species	Season	Year	Pop min	Pop max	Acc	Criteria
Calonectris diomedea Cory's Shearwater	B	1987	Abundant	—		B2
Saxicola dacotiae Fuerteventura Chat	R	1995	20	—	C	A1, B2, C1, C2, C6

The site holds an important population of *Calonectris diomedea*.

Protection status

National None **International** None

Conservation issues

Threats Consequences of animal/plant introductions (B), Industrialization/urbanization (C), Recreation/tourism (B)

The main threats are tourist activities, which can cause disturbance to birds and habitat degradation, and the effects of introduced rats *Rattus*, cats *Felis catus*, goats *Capra hircus*, and sheep.

Sotavento beach B2, C7 343

Admin region Las Palmas
Coordinates 28°08′N 14°14′W
Altitude 0–50 m **Area** 450 ha

Site description

Located in the south of Fuerteventura, the site comprises a stretch of sandy beach and a line of sandy, muddy offshore bars. Human activities include sand extraction.

Habitats Desert (semi-desert), Wetland (*; mudflat/sandflat; sand-dunes/sand beach; coastal lagoon)
Land-use Fisheries/aquaculture (*), Tourism/recreation (*), Urban/industrial/transport

Birds

Species	Season	Year	Pop min	Pop max	Acc	Criteria
Charadrius alexandrinus Kentish Plover	R	1995	60	—	B	B2

The site is very important for breeding *Charadrius alexandrinus*, and is one of the best sites in the Canaries for wintering and migrating waders, gulls and terns.

Protection status

National Partial **International** Partial
370 ha of IBA covered by Natural Park (Jandía, 14,318 ha). 370 ha of IBA covered by Special Protection Area (Jandía, 2,754 ha).

Conservation issues

Threats Extraction industry (A), Industrialization/urbanization (B), Recreation/tourism (A)

Residential and tourism developments pose a threat to the habitat and its birds.

Peninsula of Jandía A1, B2, B3, C1, C6 344

Admin region Las Palmas
Coordinates 28°05′N 14°20′W
Altitude 0–807 m **Area** 12,400 ha

Site description

A rugged semi-arid massif located on the southern tip of Fuerteventura, which includes the highest peak on the island. The plains support drought-resistant scrub vegetation and include sandy areas. Human activities include stock-farming, sport boating and residential usage.

Habitats Scrub, Grassland (*), Desert (*; semi-desert), Wetland (sand-dunes/sand beach), Rocky areas (*; sea cliff/rocky shore; inland cliff; inland sand-dunes), Artificial landscape (arable land)
Land-use Agriculture (*), Fisheries/aquaculture, Hunting, Tourism/recreation, Urban/industrial/transport

Birds

Species	Season	Year	Pop min	Pop max	Acc	Criteria
Calonectris diomedea Cory's Shearwater	B	1987	Common		—	B2
Falco tinnunculus Kestrel	R	1988	60	—	B	B2
Chlamydotis undulata Houbara Bustard	R	1994	8	—	C	B2
Cursorius cursor Cream-coloured Courser	R	1995	4	—	C	B2
Pterocles orientalis Black-bellied Sandgrouse	R	1995	Common		—	C6
Anthus berthelotii Berthelot's Pipit	R	1995	Common		—	B3
Saxicola dacotiae Fuerteventura Chat	R	1995	Abundant		—	A1, B2, C1, C6
Bucanetes githagineus Trumpeter Finch	R	1995	Common		—	B2, C6

One of the most important areas on Fuerteventura for *Saxicola dacotiae*. The IBA is also important for raptors and steppic birds.

Protection status

National Partial **International** Partial
10,300 ha of IBA covered by Natural Park (Jandía, 14,318 ha). 116 ha of IBA covered by Site of Scientific Interest (Playa del Matorral, 116 ha). 2,500 ha of site covered by Special Protection Area (Jandía, 2,754 ha).

Conservation issues

Threats Abandonment/reduction of land management (B), Consequences of animal/plant introductions (B), Extraction industry (C), Industrialization/urbanization (C), Infrastructure (C), Recreation/tourism (B), Unsustainable exploitation (U)

The abandonment of stock-farming reduces feeding opportunities for some birds (for example, *Corvus corax* and *Neophron percnopterus*), and allows vegetation to recover, resulting in a loss of habitat for steppic species (for example, *Chlamydotis undulata*). Additional threats are posed by new residential developments and electric powerlines, disturbance to birds caused by tourists and illegal hunting. Five censuses of *C. undulata* have been undertaken since 1979.

Isthmus of Jandía sandy plain B2, B3, C2, C6 345

Admin region Las Palmas
Coordinates 28°09′N 14°16′W
Altitude 0–300 m **Area** 3,900 ha

Site description

An extensive sandy plain in the south of Fuerteventura that supports salt-tolerant and drought-resistant scrub vegetation. Human activities include goat-grazing.

Habitats Desert (*; semi-desert), Wetland (*; sand-dunes/sand beach), Rocky areas (*; inland sand-dunes), Artificial landscape (other urban/industrial areas)
Land-use Agriculture (*), Tourism/recreation

Birds

Species	Season	Year	Pop min	Pop max	Acc	Criteria
Chlamydotis undulata Houbara Bustard	R	1995	24	24	A	B2, C2, C6
Burhinus oedicnemus Stone Curlew	R	1995	10	—	C	B2, C6
Cursorius cursor Cream-coloured Courser	R	1994	30	40	B	B2, C2, C6
Pterocles orientalis Black-bellied Sandgrouse	R	1995	Common		—	C6
Anthus berthelotii Berthelot's Pipit	R	1995	Abundant		—	B3

The site is important for breeding steppic species and is a stopover site for migrating herons, raptors and passerines.

Protection status

National Partial **International** Partial
1,800 ha of IBA covered by Natural Park (Jandía, 14,318 ha). 1,800 ha of IBA covered by Special Protection Area (Jandía, 2,754 ha).

Conservation issues

Threats Consequences of animal/plant introductions (C), Industrialization/urbanization (B), Infrastructure (A), Recreation/tourism (A), Unsustainable exploitation (B)

The uncontrolled development of wind-farms, electric powerlines and new roads poses a major threat to both habitats and fauna. Some of this development is for tourism, with jeep-safaris an additional threat. Feral cats are also a problem. The area is a game refuge, and the local government has undertaken six censuses of *Chlamydotis undulata* since 1979.

Gully of Ajuí–Betancuria A1, A2, B2, B3, C1, C6 346

Admin region Las Palmas
Coordinates 28°23′N 14°10′W
Altitude 0–292 m **Area** 700 ha

Site description

The gully of Ajuí is located in the west of Fuerteventura, and holds running water, scrub, grassland, agricultural land and small copses of *Tamarix*. Betancuria is a mountainous area, semi-desertic with sparse vegetation. Human activities include goat-grazing.

Habitats Scrub, Desert (semi-desert), Wetland (river/stream), Rocky areas (inland cliff), Artificial landscape (*; arable land; ruderal land)
Land-use Agriculture (*), Tourism/recreation (*), Water management

Birds

Species	Season	Year	Pop min	Pop max	Acc	Criteria
Neophron percnopterus Egyptian Vulture	R	1994	1	2	A	B2, C6
Alectoris barbara Barbary Partridge	R	1995	10	—	B	B2, C6
Burhinus oedicnemus Stone Curlew	R	1995	6	—	B	C6
Apus unicolor Plain Swift	R	1995	Common		—	A2, B3
Anthus berthelotii Berthelot's Pipit	R	1995	Common		—	A2
Saxicola dacotiae Fuerteventura Chat	R	1995	10	—	C	A1, A2, B2, C1, C6
Bucanetes githagineus Trumpeter Finch	R	1995	Common		—	B2, C6

The site is one of the most important sites on the island for breeding *Saxicola dacotiae*, raptors and steppic species, and is also important as a stop-over site for migrating passerines.

Protection status

National High **International** High
700 ha of IBA covered by Rural Park (Betancuria, 16,544 ha). 700 ha of IBA covered by Special Protection Area (Betancuria, 15,538 ha).

Conservation issues

Threats Agricultural intensification/expansion (B), Consequences of animal/plant introductions (C), Construction/impact of dyke/dam/barrage (C), Dredging/canalization (B), Infrastructure (C), Recreation/tourism (B), Unsustainable exploitation (B)

The construction of water infrastructure, dredging and canalization, electric powerlines, disturbance to birds by tourists, and illegal hunting are the main threats to the site. Overgrazing is also a problem, and introduced rats *Rattus*, feral cats *Felis catus*, and dogs pose a threat to breeding birds. A portion of the site is a game refuge.

Los Molinos gully– Plain of La Laguna A1, B2, B3, C1, C6 347

Admin region Las Palmas
Coordinates 28°32′N 14°05′W
Altitude 0–323 m **Area** 1,700 ha

Site description

A dammed valley (Los Molinos gully) and a stony plain (Plain of La Laguna) situated in the north-west of Fuerteventura. Dry calcareous grassland and drought-resistant scrub exist with crops and a small copse of *Tamarix canariensis*. Human activities include livestock-farming (dromedaries *Camelus dromedarius* and goats).

Habitats Grassland (*; steppe/dry calcareous grassland), Desert (*; semi-desert), Wetland (river/stream), Rocky areas (inland cliff), Artificial landscape (*; arable land)
Land-use Agriculture (*), Military, Tourism/recreation (*), Water management

Birds

Species	Season	Year	Pop min	Pop max	Acc	Criteria
Chlamydotis undulata Houbara Bustard	R	1995	6	6	A	B2
Cursorius cursor Cream-coloured Courser	R	1995	Frequent		—	B2
Apus unicolor Plain Swift	R	1995	—	—	—	B3
Calandrella rufescens Lesser Short-toed Lark	R	1995	—	—	—	B2
Anthus berthelotii Berthelot's Pipit	R	1996	Abundant		—	B3
Saxicola dacotiae Fuerteventura Chat	R	1995	10	—	C	A1, B2, C1, C6

The plain supports breeding steppic species (very important for *Chlamydotis undulata*); the valley is important for wintering and migrating aquatic birds (herons, waders and ducks).

Protection status

National Partial **International** Partial
915 ha of IBA covered by Rural Park (Betancuria, 16,544 ha). 915 ha of IBA covered by Special Protection Area (Betancuria, 15,538 ha).

Conservation issues

Threats Agricultural intensification/expansion (B), Recreation/tourism (B), Unsustainable exploitation (B)

The main threats are overgrazing, disturbance to birds caused by tourists and illegal hunting (especially on the reservoir).

Coast of Esquinzo–Puertito de Los Molinos

B2, C6 **348**

Admin region Las Palmas
Coordinates 28°35′N 14°05′W
Altitude 0–50 m **Area** 250 ha

Site description

Sea cliffs along the western coast of Fuerteventura.

Habitats Marine areas (sea inlet/coastal features), Rocky areas (100%; sea cliff/rocky shore; inland cliff)
Land-use Fisheries/aquaculture (*), Tourism/recreation (*)

Birds

Species	Season	Year	Pop min	Pop max	Acc	Criteria
Calonectris diomedea Cory's Shearwater	B	1987	Frequent		—	B2
Neophron percnopterus Egyptian Vulture	R	1995	2	2	B	B2, C6

The site also holds breeding *Apus unicolor* and *Corvus corax*.

Protection status

National Partial **International** Partial
30 ha of IBA covered by Rural Park (Betancuria, 16,544 ha). 30 ha of IBA covered by Special Protection Area (Betancuria, 15,538 ha).

Conservation issues

Threats Aquaculture/fisheries (B), Recreation/tourism (B), Unsustainable exploitation (C)

Disturbance caused by tourists and fishermen, and illegal hunting are the main threats to the site.

Lajares sandy plain–Cotillo–Ezquinzo

B2, B3, C2, C6 **349**

Admin region Las Palmas
Coordinates 28°38′N 13°58′W
Altitude 0–308 m **Area** 6,800 ha

Site description

An extensive semi-desertic rocky plain holding salt-tolerant and drought-resistant scrub, with sandy areas, hills, gullies and small patches of arable land. Human activities include cattle-grazing and military manoeuvres.

Habitats Scrub, Desert (*; semi-desert), Rocky areas (*; inland sand-dunes), Artificial landscape (arable land; ruderal land)
Land-use Agriculture (*), Military, Tourism/recreation

Birds

Species	Season	Year	Pop min	Pop max	Acc	Criteria
Alectoris barbara Barbary Partridge	R	1995	10	—	B	B2
Chlamydotis undulata Houbara Bustard	R	1995	44	46	B	B2, C2, C6
Burhinus oedicnemus Stone Curlew	R	1995	Common		—	C6
Cursorius cursor Cream-coloured Courser	R	1994	15	20	B	B2, C2, C6
Pterocles orientalis Black-bellied Sandgrouse	R	1995	Frequent		—	C6
Calandrella rufescens Lesser Short-toed Lark	R	1995	Abundant		—	B2
Anthus berthelotii Berthelot's Pipit	R	1995	Abundant		—	B3

One of the most important areas in Fuerteventura for *Chlamydotis undulata* and other steppic birds. Areas of arable land provide habitat for migrating and wintering passerines.

Protection status

National None **International** Partial
2,700 ha of IBA covered by Special Protection Area (Lajares, 2,700 ha).

Conservation issues

Threats Agricultural intensification/expansion (B), Extraction industry (C), Industrialization/urbanization (B), Infrastructure (C), Other (B), Recreation/tourism (B), Unsustainable exploitation (C)

Threats include overgrazing, urban development and infrastructure improvements related to tourism, and disturbance to birds caused by military activity ('Other' threat) and tourism (for example, jeep-safaris). The area is a game refuge. The local government has undertaken six censuses of *Chlamydotis undulata* since 1979, and is implementing a recovery plan for this species.

Coast between Corralejo and Tostón

A1, B2, C1, C6 **350**

Admin region Las Palmas
Coordinates 28°45′N 13°55′W
Altitude 0–40 m **Area** 530 ha

Site description

Sand-dunes, beaches and rock stacks located along the northern coast of Fuerteventura. Human activities include fishing and tourism.

Habitats Scrub (scrub), Desert (semi-desert), Wetland (*; sand-dunes/sand beach; shingle/stony beach)
Land-use Fisheries/aquaculture (*), Tourism/recreation (*)

Birds

Species	Season	Year	Pop min	Pop max	Acc	Criteria
Burhinus oedicnemus Stone Curlew	R	1996	5	—	B	C6
Charadrius alexandrinus Kentish Plover	R	1995	47	—	B	B2
Saxicola dacotiae Fuerteventura Chat	R	1996	Common		—	A1, B2, C1

This site is also important for wintering herons, waders and other waterbirds.

Protection status

National None **International** None

Conservation issues

Threats Aquaculture/fisheries (C), Industrialization/urbanization (B), Recreation/tourism (B)

Threats include degradation of the habitat through rubbish disposal, and disturbance caused to birds by tourism and marine sports.

Coast between Arinaga and Castillo del Romeral

B2, c7 **351**

Admin region Las Palmas
Coordinates 27°48′N 15°25′W
Altitude 0–20 m **Area** 1,100 ha

Site description

A stretch of coast comprising mudflats and sandflats, sand and shingle-beaches, rock stacks and brackish standing water. High levels of industrial activity occur adjacent to the northern tip of the site.

Habitats Grassland, Wetland (*; shingle/stony beach; coastal lagoon), Artificial landscape (ruderal land)
Land-use Fisheries/aquaculture (*), Tourism/recreation (*), Urban/industrial/transport (*)

Birds

Species	Season	Year	Pop min	Pop max	Acc	Criteria
Charadrius alexandrinus Kentish Plover	R	1995	20	20	A	B2

The most important area on Gran Canaria for wintering and migrating birds, especially herons and waders.

Protection status

National Low **International** Low

83 ha of IBA covered by Site of Scientific Interest (Juncalillo del Sur, 192 ha). 83 ha of IBA covered by Special Protection Area (Juncalillo del Sur, 179 ha).

Conservation issues

Threats Aquaculture/fisheries (B), Filling-in of wetlands (B), Industrialization/urbanization (A), Infrastructure (A), Unsustainable exploitation (C)

Tourist activity (for example, cars being driven on beaches and windsurfing), illegal hunting, fishing and industrial activity cause disturbance and destroy habitats. The building of a new commercial harbour in Arinaga will have a high impact on littoral dynamics. The section of the site that runs from Arinaga saltplant to Formas Bay is a game reserve.

Tirajana pine woodland

B2, C6 **352**

Admin region Las Palmas
Coordinates 27°55′N 15°35′W
Altitude 600–1,400 m **Area** 3,600 ha

Site description

Arid *Pinus canariensis* forest located in the south of Gran Canaria. Regeneration is very slow due to low humidity, and undergrowth is sparse. Human activities include stock-farming and forestry practices.

Habitats Forest and woodland (100%; native coniferous forest), Scrub (scrub)
Land-use Agriculture (*), Forestry (*), Hunting

Birds

Species	Season	Year	Pop min	Pop max	Acc	Criteria
Dendrocopos major Great Spotted Woodpecker	R	1996	Common		—	C6
[1] ***Fringilla teydea*** Blue Chaffinch	R	1996	—	—	—	B2

1. Extinct – to be reintroduced.

One of the priority areas to be improved as part of the *Fringilla teydea* conservation programme, which provides for the release of captive-bred individuals.

Protection status

National High **International** High

3,600 ha of IBA covered by Natural Park (Pilancones, 5,794 ha). 3,600 ha of IBA covered by Special Protection Area (Ayagaures y Pilancones, 10,166 ha).

Conservation issues

Threats Agricultural intensification/expansion (B), Burning of vegetation (A), Natural events (C)

Forest-fires (both natural and man-made) are a major threat. Overgrazing by feral goats *Capra hircus* restricts vegetation regeneration. A conservation programme for *Fringilla teydea polatzeki* has run since 1991 (with LIFE funding in 1995 and 1996).

Tauro pine woodland

B2, C6 **353**

Admin region Las Palmas
Coordinates 27°54′N 15°41′W
Altitude 700–1,200 m **Area** 800 ha

Site description

An area of regenerating arid *Pinus canariensis* forest located in the south-west of Gran Canaria. Human activities include livestock-farming and the collection of pine-needles for use in cultivation.

Habitats Forest and woodland (100%; native coniferous forest), Scrub (scrub)
Land-use Agriculture (*), Forestry (*)

Birds

Species	Season	Year	Pop min	Pop max	Acc	Criteria
Dendrocopos major Great Spotted Woodpecker	R	1995	Common		—	C6
Fringilla teydea Blue Chaffinch	R	1995	Rare		—	B2

This is one of the priority areas to be improved for *Fringilla teydea* under the conservation programme. The species no longer breeds in the site, and reintroductions will take place.

Protection status

National Partial **International** Partial

500 ha of IBA covered by Natural Monument (Tauro, 1,256 ha). 550 ha of IBA covered by Special Protection Area (Macizo de Tauro, 1,179 ha).

Conservation issues

Threats Burning of vegetation (A), Other (C)

Forest fires pose the major threat. A conservation programme for *Fringilla teydea polatzeki* has run since 1991 (with LIFE funding in 1995 and 1996).

Pajonales, Ojeda, Inagua and La Data pine woodlands

A1, A2, B2, B3, C1, C2, C6 **354**

Admin region Las Palmas
Coordinates 27°56′N 15°41′W
Altitude 850–1,500 m **Area** 3,700 ha

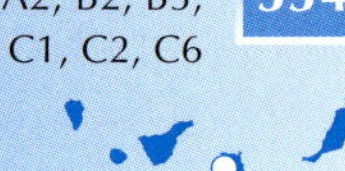

Site description

Arid *Pinus canariensis* forest located in the west of Gran Canaria. Human activities include stock-farming, the collection of pine-needles for use in cultivation, and conservation/education.

Habitats Forest and woodland (*; native coniferous forest), Scrub (scrub), Artificial landscape (forestry plantation)
Land-use Agriculture (*), Forestry (*), Nature conservation/research (*)

Birds

Species	Season	Year	Pop min	Pop max	Acc	Criteria
Dendrocopos major Great Spotted Woodpecker	R	1996	Abundant		—	C6
Anthus berthelotii Berthelot's Pipit	R	1995	—	—	—	A2
[1] ***Fringilla teydea*** Blue Chaffinch	R	1996	84	150	A	A1, A2, B2, C1, C2, C6
Serinus canaria Canary	R	1996	Frequent		—	A2, B3

1. Large decrease over last 20 years.

The site is the most important area for *Fringilla teydea polatzeki*.

Protection status

National High **International** High

3,700 ha of IBA covered by Integral Nature Reserve (Inagua, 3,920 ha). 3,700 ha of IBA covered by Special Protection Area (Ojeda, Inagua, Pajonales y La Data, 3,742 ha).

Conservation issues

Threats Burning of vegetation (A), Consequences of animal/plant introductions (A), Natural events (C), Other (B), Recreation/tourism (C)

Forest fires (both natural and man-made) are a major threat. Feral goats *Capra hircus* adversely affect the vegetation, and feral cats *Felis catus* prey upon *Fringilla teydea*. A Conservation Programme for *F. teydea polatzeki* has run since 1991 (with LIFE funding in 1995 and 1996). There is an interpretation centre and an educational programme for schools. The area is a game refuge.

Tamadaba pine woodland

A2, B2, C6 **355**

Admin region Las Palmas
Coordinates 28°03'N 15°41'W
Altitude 400–1,400 m **Area** 3,400 ha

Site description

Heterogeneous *Pinus canariensis* forest situated in the north-west of Gran Canaria. The most humid areas exhibit great floristic richness. Human activities include hunting and the collection of pine-needles for use in cultivation.

Habitats Forest and woodland (*; native coniferous forest), Scrub (heathland), Artificial landscape (forestry plantation)
Land-use Forestry (*), Hunting, Tourism/recreation

Birds

Species	Season	Year	Pop min	Pop max	Acc	Criteria
Apus unicolor Plain Swift	R	1996	Frequent		—	A2
Dendrocopos major Great Spotted Woodpecker	R	1996	Common		—	C6
Anthus berthelotii Berthelot's Pipit	R	1996	—	—	—	A2
[1] ***Fringilla teydea*** Blue Chaffinch	R	1996	1	3	B	A2, B2, C6
Serinus canaria Canary	R	1996	—	—	—	A2

1. Large decrease over last 20 years.

This is one of the priority areas to be improved for *Fringilla teydea* under the conservation programme (with LIFE funding).

Protection status

National High **International** High

810 ha of IBA covered by Rural Park (Roque Nublo, 26,307 ha). 2,600 ha of IBA covered by Natural Park (Tamadaba, 7,538 ha). 3,200 ha of IBA covered by Special Protection Area (Tamadaba, 8,010 ha).

Conservation issues

Threats Burning of vegetation (B), Other (B), Recreation/tourism (B)

Forest fires and massive and uncontrolled human presence threaten the IBA. The local government has run a conservation programme for *Fringilla teydea polatzeki* since 1991.

Anaga rocky islets

B1ii, B2, C2, C6 **356**

Admin region Santa Cruz de Tenerife
Coordinates 28°36'N 16°10'W
Altitude 0–180 m **Area** 10 ha

Site description

Two rocky islets (Roque de Dentro and Roque de Fuera) located off the north-eastern tip of Tenerife, which support drought-resistant and salt-tolerant vegetation. A copse of dragon's blood tree *Dracaena draco* exists on one of the islets. Human activities include gathering shellfish and boating.

Habitats Rocky areas (100%; rock stacks/islets)
Land-use Fisheries/aquaculture (*), Hunting, Tourism/recreation (*)

Birds

Species	Season	Year	Pop min	Pop max	Acc	Criteria
Bulweria bulwerii Bulwer's Petrel	B	1987	300	—	B	B1ii, B2, C2, C6
Puffinus assimilis Little Shearwater	B	1987	20	—	B	B2, C6
Oceanodroma castro Madeiran Petrel	B	1987	100	—	B	B1ii, B2, C2, C6

The site is very important for breeding seabirds, and holds the largest *Bulweria bulwerii* breeding colony in the Canary Islands.

Protection status

National High **International** High

10 ha of IBA covered by Rural Park (Anaga, 14,119 ha). 10 ha of IBA covered by Integral Nature Reserve (Roques de Anaga, 10 ha). 10 ha of IBA covered by Special Protection Area (Anaga, 14,119 ha).

Conservation issues

Threats Aquaculture/fisheries (B), Recreation/tourism (A), Unsustainable exploitation (U)

The main threat is disturbance caused to birds by fishermen, people gathering shellfish, tourism and illegal hunting.

San Andrés, Pijaral and Anaga mountains

A1, B2, C1 **357**

Admin region Santa Cruz de Tenerife
Coordinates 28°34'N 16°10'W
Altitude 550–900 m **Area** 860 ha

Site description

Laurel *Laurus* forest in the north-east of Tenerife, ranging from degraded and exploited areas to areas with a well-developed tree canopy (the forest in Pijaral is one of the best conserved in the Canary Islands). The cutting of branches for use in tomato/banana cultivation is an important human activity.

Habitats Forest and woodland (*; broadleaved evergreen forest), Scrub (scrub; heathland), Wetland (river/stream), Rocky areas (inland cliff), Artificial landscape (arable land; forestry plantation)
Land-use Forestry (*), Nature conservation/research (*), Tourism/recreation (*)

Birds

Species	Season	Year	Pop min	Pop max	Acc	Criteria
Columba bollii Dark-tailed Laurel Pigeon	R	1996	Abundant		—	A1, B2, C1
Columba junoniae White-tailed Laurel Pigeon	R	1996	Rare		—	A1, C1

The site is important for both endemic *Columba* species as well as for other forest passerines.

Protection status

National High **International** High

860 ha of IBA covered by Rural Park (Anaga, 14,119 ha). 250 ha of IBA covered by Integral Nature Reserve (Pijaral, 301 ha). 860 ha of IBA covered by Special Protection Area (Anaga, 14,119 ha).

Conservation issues

Threats Burning of vegetation (C), Consequences of animal/plant introductions (A), Groundwater abstraction (C), Recreation/tourism (B), Selective logging/cutting (B), Unsustainable exploitation (B)

The main threats are forest fires, which can have a very high impact on habitat and bird numbers, predation by rats *Rattus* of *Columba* eggs and chicks, and illegal hunting. Tourism can cause disturbance to breeding birds and degradation of the forest. Groundwater abstraction limits water availability, making birds vulnerable to hunters when they congregate at the few remaining drinking sites. The area is a game reserve, and a management plan exists for the site. Various studies have been undertaken on the biology, distribution, ecology and conservation of the two *Columba* species.

Las Vueltas, Aguas Negras and Quebradas mountains — A1, B2, B3, C1 — 358

Admin region Santa Cruz de Tenerife
Coordinates 28°32'N 16°12'W
Altitude 350–900 m **Area** 450 ha

Site description

Well-conserved laurel *Laurus* forest and inland cliffs in the north-east of Tenerife, with a well-developed tree canopy.

Habitats Forest and woodland (*; broadleaved evergreen forest), Scrub (scrub; heathland), Wetland (river/stream), Rocky areas (inland cliff), Artificial landscape (arable land; perennial crops/orchards/groves)
Land-use Forestry (*), Tourism/recreation

Birds

Species	Season	Year	Pop min	Pop max	Acc	Criteria
Columba bollii Dark-tailed Laurel Pigeon	R	1995	Abundant		—	A1, B2, C1
Columba junoniae White-tailed Laurel Pigeon	R	1995	Rare		—	A1, C1
Serinus canaria Canary	R	1995	Abundant		—	B3

The site is important for forest passerines, especially for *Columba bollii*.

Protection status

National High **International** High
450 ha of IBA covered by Rural Park (Anaga, 14,119 ha). 450 ha of IBA covered by Special Protection Area (Anaga, 14,119 ha).

Conservation issues

Threats Burning of vegetation (C), Consequences of animal/plant introductions (A), Groundwater abstraction (B), Recreation/tourism (B), Selective logging/cutting (A), Unsustainable exploitation (U)

Forest fires, the cutting of branches (for use in tomato/banana cultivation), and hunting have a great impact on *Columba* numbers and habitat. The main threat to these species, however, comes from rats *Rattus*, which prey mainly on *C. junoniae*. Disturbance from tourism, and predation by feral cats *Felis catus*, can also adversely affect numbers. Groundwater abstraction limits water availability, making birds vulnerable to hunters when they congregate at the few remaining drinking sites. Various studies have been undertaken on the biology, distribution, ecology and conservation of the two *Columba* species.

Las Mercedes, Mina y Yedra, Aguirre, La Goleta, and Pedro Alvarez mountains — A1, B2, B3, C1, C6 — 359

Admin region Santa Cruz de Tenerife
Coordinates 28°32'N 16°18'W
Altitude 400–1,000 m **Area** 2,000 ha

Site description

Laurel *Laurus* forest in the north-east of Tenerife, ranging from degraded and exploited areas to areas with a well-developed tree canopy. Human activities include the cutting of branches for use in tomato/banana cultivation.

Habitats Forest and woodland (*; broadleaved evergreen forest), Scrub (scrub; heathland), Wetland (river/stream), Rocky areas (inland cliff), Artificial landscape (arable land; forestry plantation)
Land-use Forestry (*), Tourism/recreation

Birds

Species	Season	Year	Pop min	Pop max	Acc	Criteria
Columba bollii Dark-tailed Laurel Pigeon	R	1996	Abundant		—	A1, B2, C1, C6
Columba junoniae White-tailed Laurel Pigeon	R	1996	Rare		—	A1, C1
Regulus teneriffae Tenerife Goldcrest	R	1996	Abundant		—	B3

The site is important for both endemic *Columba* species (numbers of *C. junoniae* are low) as well as for other forest passerines.

Protection status

National High **International** High
2,000 ha of IBA covered by Rural Park (Anaga, 14,119 ha). 2,000 ha of IBA covered by Special Protection Area (Anaga, 14,119 ha).

Conservation issues

Threats Burning of vegetation (C), Consequences of animal/plant introductions (A), Groundwater abstraction (B), Intensified forest management (B), Recreation/tourism (B), Selective logging/cutting (C), Unsustainable exploitation (B)

The main threats are forest fires, predation by rats *Rattus* of eggs, chicks and sometimes adults of both *Columba* species, and illegal hunting. Tourism and recreation cause disturbance to breeding birds and degradation of the forest. Groundwater abstraction limits water availability, making birds vulnerable to hunters when they congregate at the few remaining drinking sites. The area is a game reserve, and a management plan exists for the site. Various studies have been undertaken on the biology, distribution, ecology and conservation of the two *Columba* species.

Los Rodeos–La Esperanza — B2 — 360

Admin region Santa Cruz de Tenerife
Coordinates 28°28'N 16°21'W
Altitude 510–960 m **Area** 1,200 ha

Site description

An area of grassland, scrub and agricultural land located in north-east Tenerife, bordered by a residential area and the airport.

Habitats Scrub (*; scrub), Grassland (*; steppe/dry calcareous grassland), Artificial landscape (*; arable land; forestry plantation; other urban/industrial areas), Introduced/exotic vegetation
Land-use Agriculture (70%), Military (5%), Urban/industrial/transport (25%)

Birds

Species	Season	Year	Pop min	Pop max	Acc	Criteria
[1] ***Calandrella rufescens*** Lesser Short-toed Lark	R	1995	25	40	A	B2

1. Large decrease over last 20 years.

The only site for *Calandrella rufescens rufescens* in the Canaries and the best area on Tenerife for grassland species. The site is also a wintering and stop-over site for herons, waders and passerines and a feeding site for species that breed nearby (mainly raptors and passerines).

Protection status

National None **International** None

Conservation issues

Threats Industrialization/urbanization (A), Infrastructure (A), Unsustainable exploitation (B)

Plans exist for the construction of an industrial area and a road that will traverse the site. Additional threats are posed by illegal hunting, and bird strikes on nearby roads. Monitoring of *Calandrella rufescens* is undertaken by local ornithologists.

La Playa islet — B2, C6 — 361

Admin region Santa Cruz de Tenerife
Coordinates 28°26'N 16°29'W
Altitude 0–20 m **Area** 1 ha

Site description

A small rocky islet situated off the north coast of Tenerife, which has no infrastructure and supports salt-tolerant vegetation.

Habitats Rocky areas (100%; rock stacks/islets)
Land-use Fisheries/aquaculture (100%)

Birds

Species	Season	Year	Pop min	Pop max	Acc	Criteria
Bulweria bulwerii Bulwer's Petrel	B	1996	20		— A	B2, C6

The site is important for breeding *Bulweria bulwerii*. In 1996 20 chicks were observed, and more than 20 pairs probably breed.

Protection status

National None **International** None

Conservation issues

Threats Aquaculture/fisheries (C)

A few incidents have been recorded of disturbance to breeding birds caused by fishermen.

Santa Úrsula and La Victoria laurel forest — A1, B2, C1, C6 — 362

Admin region Santa Cruz de Tenerife
Coordinates 28°24′N 16°27′W
Altitude 700–1,400 m **Area** 1,800 ha

Site description

Laurel *Laurus* forest that has been heavily exploited and is now regenerating. *Pinus canariensis* forest occurs at higher altitudes, bordered by patches of terraced cultivation and *Castanea* coppice. Streams provide natural drinking sites. Human activities include the cutting of branches for use in tomato/banana cultivation.

Habitats Forest and woodland (*; broadleaved evergreen forest), Scrub (scrub; heathland), Wetland (river/stream), Artificial landscape (arable land; perennial crops/ orchards/groves; forestry plantation), Introduced/exotic vegetation
Land-use Agriculture (*), Forestry (*), Tourism/recreation

Birds

Species	Season	Year	Pop min	Pop max	Acc	Criteria
Accipiter nisus Sparrowhawk	R	1988	Frequent		—	C6
Columba bollii Dark-tailed Laurel Pigeon	R	1996	Common		—	A1, B2, C1, C6
Columba junoniae White-tailed Laurel Pigeon	R	1996	Rare		—	A1, B2, C1, C6

The site supports both endemic *Columba* species, as well as endemic subspecies of several forest passerines.

Protection status

National High **International** None
450 ha of IBA covered by Protected Landscape (La Resbala, 777 ha). 1,350 ha of IBA covered by Protected Landscape (Las Lagunetas, 3,800 ha). 400 ha of IBA covered by Special Natural Reserve (Las Palomas, 584 ha). 50 ha of IBA covered by Integral Nature Reserve (Pinoleris, 181 ha).

Conservation issues

Threats Consequences of animal/plant introductions (A), Selective logging/cutting (A), Unsustainable exploitation (A)

Introduced rats *Rattus* prey upon the eggs and chicks of both *Columba* species, with *C. junoniae* worst affected. Forestry practices destroy and degrade the *Laurus* forest. Illegal hunting of *C. bollii* and *C. junoniae* takes place. Various studies have been undertaken on the biology, distribution, ecology and conservation of the two *Columba* species.

Tigaiga slope — A1, B2, C1, C6 — 363

Admin region Santa Cruz de Tenerife
Coordinates 28°22′N 16°35′W
Altitude 725–1,450 m **Area** 370 ha

Site description

Habitats Forest and woodland (*; broadleaved evergreen forest), Scrub (scrub; heathland), Wetland (river/stream), Rocky areas (inland cliff), Artificial landscape (arable land; forestry plantation), Introduced/exotic vegetation
Land-use Forestry (*)

The greater part of the site (located in the north of Tenerife) is covered by a laurel *Laurus* forest growing on steep slopes and amongst cliffs. Introduced *Pinus radiata* is interspersed with *Laurus* at higher altitudes, and copses of *Castanea sativa* and exotic conifers at lower altitudes.

Birds

Species	Season	Year	Pop min	Pop max	Acc	Criteria
Columba bollii Dark-tailed Laurel Pigeon	R	1996	Abundant		—	A1, B2, C1, C6
Columba junoniae White-tailed Laurel Pigeon	R	1996	Frequent		—	A1, B2, C1, C6

The site is important for both endemic *Columba* species, as well as for other laurel forest birds.

Protection status

National High **International** High
370 ha of IBA covered by Protected Landscape (Los Campeches, Tigaiga y Ruiz, 691 ha). 370 ha of IBA covered by Special Protection Area (Tigaiga, 1,735 ha).

Conservation issues

Threats Burning of vegetation (A), Consequences of animal/plant introductions (A), Groundwater abstraction (C), Recreation/tourism (C), Selective logging/cutting (B), Unsustainable exploitation (B)

Major threats are posed by forest fires and the presence of rats *Rattus* and feral cats *Felis catus* that prey upon *Columba* chicks and eggs. Additional threats include the cutting of branches for use in tomato/ banana cultivation and illegal hunting. Groundwater abstraction limits water availability, making birds vulnerable to hunters when they congregate at the few remaining drinking sites. The area is a game reserve. Various studies have been undertaken on the biology, distribution, ecology and conservation of the two *Columba* species.

Santo Domingo coastal cliffs — B2, C6 — 364

Admin region Santa Cruz de Tenerife
Coordinates 28°23′N 16°40′W
Altitude 0–100 m **Area** 7 ha

Site description

Coastal cliffs with beaches and rock outcrops. The drought-resistant vegetation includes *Limonium* and *Kdeinia*.

Habitats Wetland (shingle/stony beach), Rocky areas (*; sea cliff/rocky shore)
Land-use Fisheries/aquaculture (100%)

Birds

Species	Season	Year	Pop min	Pop max	Acc	Criteria
Puffinus assimilis Little Shearwater	B	1996	20	—	A	B2, C6

The site holds an important *Puffinus assimilis* breeding colony that has only recently been discovered.

Protection status

National None **International** None

Conservation issues

Threats Consequences of animal/plant introductions (A), Industrialization/ urbanization (C), Unsustainable exploitation (A)

Major threats include the illegal poaching of *Calonectris diomedea* chicks, and predation of the eggs and chicks of seabirds by introduced rats *Rattus*. Several houses have been built on the cliff tops, resulting in concrete spilling over the cliff edge.

Garachico rocky islet — B2, C6 — 365

Admin region Santa Cruz de Tenerife
Coordinates 28°23′N 16°46′W
Altitude 0–77 m **Area** 5 ha

Site description

A small volcanic islet situated very close to the northern coast of

Tenerife, with no infrastructure. It supports salt-tolerant vegetation, and an endemic *Euphorbia* species that is typical of semi-desertic habitats. Minor land-uses include hunting.

Habitats Rocky areas (100%; rock stacks/islets)
Land-use Fisheries/aquaculture (100%), Tourism/recreation (100%)

Birds

Species	Season	Year	Pop min	Pop max	Acc	Criteria
Bulweria bulwerii Bulwer's Petrel	B	1990	40	50	A	B2, C6
Oceanodroma castro Madeiran Petrel	B	1995	Common		—	B2

The site is important for breeding seabirds, and as a roosting site for wintering herons.

Protection status

National High **International** None
5 ha of IBA covered by Natural Monument (Roque de Garachico, 5 ha).

Conservation issues

Threats Other (B), Recreation/tourism (B), Unsustainable exploitation (C)

Threats include disturbance caused by fishermen, boating and other recreational activities. A football ground on the main island is illuminated at night, causing *Bulweria bulwerii* and *Calonectris diomedea* to fall to the ground dazzled by the lights ('Other' threat).

Mountain of Agua, gully of Los Cochinos and gully of Cuevas Negras — A1, A2, B2, B3, C1, C6 — 366

Admin region Santa Cruz de Tenerife
Coordinates 28°20′N 16°49′W
Altitude 250–1,200 m **Area** 800 ha

Site description

One of the best-conserved laurel *Laurus* forests in Tenerife. Formerly cultivated areas have been invaded by scrub, and vertical cliffs support characteristic vegetation. Human activities include the cutting of branches for use in tomato/banana cultivation.

Habitats Forest and woodland (*; broadleaved evergreen forest), Scrub (scrub; heathland), Wetland (river/stream), Rocky areas (inland cliff), Artificial landscape (arable land; perennial crops/orchards/groves; forestry plantation)
Land-use Agriculture (*), Forestry (*), Tourism/recreation

Birds

Species	Season	Year	Pop min	Pop max	Acc	Criteria
Accipiter nisus Sparrowhawk	R	1995	5	7	A	C6
Alectoris barbara Barbary Partridge	R	1996	Common		—	B2
Columba bollii Dark-tailed Laurel Pigeon	R	1996	Abundant		—	A1, A2, B2, C1, C6
Columba junoniae White-tailed Laurel Pigeon	R	1996	Frequent		—	A1, A2, B2, C1, C6
Apus unicolor Plain Swift	R	1996	Abundant		—	A2, B3
Anthus berthelotii Berthelot's Pipit	R	1996	Common		—	A2
Regulus teneriffae Tenerife Goldcrest	R	1995	Abundant		—	A2, B3
Serinus canaria Canary	R	1996	Abundant		—	A2, B3

This site holds the most significant numbers of *Columba junoniae* on Tenerife, and the main *Puffinus puffinus* breeding colony on the island. It is also important for other endemic birds.

Protection status

National High **International** High
800 ha of IBA covered by Rural Park (Teno, 8,064 ha). 800 ha of IBA covered by Special Protection Area (Teno, 7,645 ha).

Conservation issues

Threats Burning of vegetation (B), Consequences of animal/plant introductions (B), Groundwater abstraction (B), Recreation/tourism (B), Selective logging/cutting (B), Unsustainable exploitation (B)

Introduced rats *Rattus* affect both *Columba* populations, especially *C. junoniae*. Forest fires (some man-made) cause habitat loss and affect breeding species. Additional threats include open-air sports, illegal hunting and the clearing of tracks and roads. The area is a game reserve. Various studies have been undertaken on the biology, distribution, ecology and conservation of the two *Columba* species.

Los Gigantes coastal cliff — B2, B3, C6 — 367

Admin region Santa Cruz de Tenerife
Coordinates 28°17′N 16°53′W
Altitude 0–600 m **Area** 1,000 ha

Site description

Sea cliffs and large gullies on the north-west side of Tenerife island. They support typical rocky vegetation that includes endemic *Limonium* species. Human activities include sport boating.

Habitats Scrub (scrub), Marine areas (sea inlet/coastal features), Rocky areas (*; sea cliff/rocky shore; inland cliff)
Land-use Fisheries/aquaculture, Hunting, Tourism/recreation (*)

Birds

Species	Season	Year	Pop min	Pop max	Acc	Criteria
Calonectris diomedea Cory's Shearwater	B	1996	400	—	B	B2, C6
Pandion haliaetus Osprey	R	1996	2	2	A	C6
Apus unicolor Plain Swift	R	1996	Abundant		—	B3

One of the most significant sites in Tenerife for cliff-nesting seabirds and raptors.

Protection status

National High **International** High
1,000 ha of IBA covered by Rural Park (Teno, 8,064 ha). 1,000 ha of IBA covered by Special Protection Area (Teno, 7,645 ha).

Conservation issues

Threats Aquaculture/fisheries (B), Consequences of animal/plant introductions (B), Infrastructure (C), Recreation/tourism (A), Unsustainable exploitation (B)

Rats *Rattus* prey on the eggs and chicks of breeding birds, which are also sensitive to disturbance caused by tourist activities (such as open-air sports), illegal hunting and fishing (for example, *Pandion haliaetus*). Several studies and censuses of seabirds and raptors have taken place, and the local government undertakes monitoring of breeding *P. haliaetus*.

Tágara gully — A1, B2, C1, C6 — 368

Admin region Santa Cruz de Tenerife
Coordinates 28°13′N 16°43′W
Altitude 1,300–1,800 m **Area** 400 ha

Site description

One of the most intact *Pinus canariensis* forests in Tenerife, located in the west of the island. The forest and scrub cover a number of gullies.

Habitats Forest and woodland (*; native coniferous forest), Scrub (scrub), Rocky areas (inland cliff)
Land-use Forestry (*), Hunting

Birds

Species	Season	Year	Pop min	Pop max	Acc	Criteria
Dendrocopos major Great Spotted Woodpecker	R	1996	—	—	—	C6
Fringilla teydea Blue Chaffinch	R	1996	Abundant		—	A1, B2, C1, C6

The site is important for *Dendrocopos major canariensis*, *Fringilla teydea* and other pine-forest species.

Protection status

National High **International** High

400 ha of IBA covered by Natural Park (Corona Forestal, 40,545 ha). 400 ha of IBA covered by Special Protection Area (Corona Forestal de Tenerife, 37,173 ha).

Conservation issues

Threats Burning of vegetation (A), Consequences of animal/plant introductions (B), Groundwater abstraction (B), Unsustainable exploitation (B)

Threats include forest fires, feral cats *Felis catus* and dogs that prey on some bird species, and groundwater abstraction. Water is privately owned and can therefore be sold. As a result abstraction limits water availability, making birds vulnerable to hunters when they congregate at the few remaining drinking sites. Diverse censuses and studies on bird populations have been undertaken (for example, *Fringilla teydea*, *Dendrocopos major* and raptors).

Vilaflor pine woodland — A1, A2, B2, B3, C1, C6 — 369

Admin region Santa Cruz de Tenerife
Coordinates 28°11′N 16°39′W
Altitude 1,450–2,400 m **Area** 1,750 ha

Site description

Pinus canariensis woodland in the south of Tenerife, that are of high ecological and floristic value. *Spartocystus* and *Adenocarpus* scrub occurs at higher altitudes.

Habitats Forest and woodland (*; native coniferous forest), Scrub (*; scrub), Rocky areas (inland cliff)
Land-use Forestry (*)

Birds

Species	Season	Year	Pop min	Pop max	Acc	Criteria
Dendrocopos major Great Spotted Woodpecker	R	1996	—	—	—	C6
Regulus teneriffae Tenerife Goldcrest	R	1996	Common		—	A2
Fringilla teydea Blue Chaffinch	R	1996	Abundant		—	A1, A2, B2, C1, C6
Serinus canaria Canary	R	1996	Abundant		—	A2, B3

The site is important for endemic forest passerines, especially *Fringilla teydea*.

Protection status

National High **International** High
1,750 ha of IBA covered by Natural Park (Corona Forestal, 40,545 ha). 1,750 ha of IBA covered by Special Protection Area (Corona Forestal de Tenerife, 37,173 ha).

Conservation issues

Threats Burning of vegetation (A), Consequences of animal/plant introductions (C), Groundwater abstraction (B), Selective logging/cutting (C), Unsustainable exploitation (B)

Forest fires (some of them man-made) are a major threat. Illegal hunting and activities related to forestry practices (cutting of branches for use in cultivation) adversely affect the habitat, and rats *Rattus* and feral cats *Felis catus* prey on birds. Various studies have been undertaken on the biology and distribution of raptors, *Fringilla teydea* and *Dendrocopos major*.

Arico pine woodland — A1, A2, B2, C1, C6 — 370

Admin region Santa Cruz de Tenerife
Coordinates 28°13′N 16°32′W
Altitude 1,400–2,200 m **Area** 3,000 ha

Site description

Pinus canariensis forest in the east of Tenerife, with *Spartocystus* and *Adenocarpus* scrub at higher altitudes.

Habitats Forest and woodland (*; native coniferous forest), Scrub (*; scrub), Rocky areas (inland cliff)
Land-use Agriculture, Forestry (*)

Birds

Species	Season	Year	Pop min	Pop max	Acc	Criteria
Alectoris barbara Barbary Partridge	R	1996	Common		—	B2
Dendrocopos major Great Spotted Woodpecker	R	1996	—	—	—	C6
Anthus berthelotii Berthelot's Pipit	R	1996	Abundant		—	A2
Regulus teneriffae Tenerife Goldcrest	R	1996	Common		—	A2
Fringilla teydea Blue Chaffinch	R	1996	Abundant		—	A1, A2, B2, C1, C6
Serinus canaria Canary	R	1996	Common		—	A2

The site is important for forest species, especially *Fringilla teydea*.

Protection status

National High **International** High
3,000 ha of IBA covered by Natural Park (Corona Forestal, 40,545 ha). 3,000 ha of IBA covered by Special Protection Area (Corona Forestal de Tenerife, 37,173 ha).

Conservation issues

Threats Burning of vegetation (A), Selective logging/cutting (B), Unsustainable exploitation (B)

Forest fires (mainly man-made) are a major threat. Other threats are posed by forestry practices that can cause disturbance and habitat degradation, and illegal hunting. Studies have been undertaken on the biology and distribution of raptors, *Fringilla teydea* and *Dendrocopos major*.

Centinela mountain and La Esquina plain — B2, C6 — 371

Admin region Santa Cruz de Tenerife
Coordinates 28°09′N 16°28′W
Altitude 80–480 m **Area** 1,600 ha

Site description

A rocky plain located in the south-east of Tenerife, with some gullies supporting drought-tolerant vegetation (*Euphorbia*).

Habitats Scrub, Desert (semi-desert), Rocky areas (inland cliff), Artificial landscape (arable land; ruderal land)
Land-use Agriculture (*)

Birds

Species	Season	Year	Pop min	Pop max	Acc	Criteria
Alectoris barbara Barbary Partridge	R	1994	Frequent		—	B2, C6

The site is important for steppic birds, supporting species typical of semi-arid conditions.

Protection status

National Low **International** None
132 ha of IBA covered by Natural Monument (Montaña Centinela, 132 ha).

Conservation issues

Threats Agricultural intensification/expansion (B), Groundwater abstraction (B), Infrastructure (B), Recreation/tourism (U), Unsustainable exploitation (B)

Threats include the presence of a main road and electric powerlines and illegal hunting. Disturbance is likely due to weekend recreational activities.

El Médano coast — B2, C6 — 372

Admin region Santa Cruz de Tenerife
Coordinates 28°02′N 16°33′W
Altitude 0–171 m **Area** 1,000 ha

Site description

The site encompasses a coastal area in the south of Tenerife that includes rock stacks, sand beaches and some coastal lagoons. Inland lie sandy, rocky plains with deep gullies and cliffs. One site supports dune-adapted vegetation, the only area of such habitat on the island.

Habitats Grassland (steppe/dry calcareous grassland), Desert (semi-desert), Wetland (*; sand-dunes/sand beach; coastal lagoon), Rocky areas (sea cliff/rocky shore; inland cliff), Artificial landscape (arable land; ruderal land)
Land-use Agriculture, Fisheries/aquaculture (*), Military, Tourism/recreation (*)

■ Birds

Species	Season	Year	Pop min	Pop max	Acc	Criteria
Alectoris barbara Barbary Partridge	R	1993	6	—	B	B2
Burhinus oedicnemus Stone Curlew	R	1993	5	6	A	C6

The plains are important for steppic species and the gullies support breeding raptors. The site is also important for wintering waders, especially *Calidris alba*, and is the last remaining breeding site of *Charadrius alexandrinus* in the Santa Cruz de Tenerife province.

■ Protection status

National Partial **International** None
166 ha of IBA covered by Special Natural Reserve (Montaña Roja, 166 ha).

■ Conservation issues

Threats Agricultural intensification/expansion (C), Aquaculture/fisheries (C), Consequences of animal/plant introductions (B), Extraction industry (B), Industrialization/urbanization (A), Infrastructure (A), Other (C), Recreation/tourism (A), Unsustainable exploitation (B)

Threats include high-impact recreational activities involving off-road vehicles, urbanization and waste disposal, bird collisions with vehicles, the effects on birds of feral cats *Felis catus* and dogs, illegal hunting and nest destruction, the gathering of shellfish, and intensive goat-grazing. Other threats (above) are posed by military manoeuvres and pollution. Several ecological studies and censuses have been carried out and a management plan is almost complete. A habitat regeneration programme for this area has been proposed.

Rasca lava fields–Guaza mountain–Las Mesas plain — B2, C2 — 373

Admin region Santa Cruz de Tenerife
Coordinates 28°10′N 16°41′W
Altitude 0–428 m **Area** 1,400 ha

■ Site description

A semi-arid landscape located in the south of Tenerife, comprising volcanic cones, lava fields, coastal habitats and arable plains. The vegetation is mainly salt-tolerant and drought-resistant.

Habitats Scrub, Grassland (steppe/dry calcareous grassland), Desert (*; semi-desert), Wetland (shingle/stony beach), Rocky areas (sea cliff/rocky shore; inland cliff), Artificial landscape (arable land)
Land-use Agriculture (*), Tourism/recreation

■ Birds

Species	Season	Year	Pop min	Pop max	Acc	Criteria
Calonectris diomedea Cory's Shearwater	B	1996	Common	—		B2, C2
Alectoris barbara Barbary Partridge	R	1996	15	—	C	B2

One of the most important areas left for steppic birds in Tenerife.

■ Protection status

National Partial **International** None
315 ha of IBA covered by Special Natural Reserve (Malpaís de Rasca, 315 ha). 726 ha of IBA covered by Natural Monument (Montaña de Guaza, 726 ha).

■ Conservation issues

Threats Agricultural intensification/expansion (B), Disturbance to birds (U), Extraction industry (B), Industrialization/urbanization (C), Infrastructure (C), Recreation/tourism (B), Unsustainable exploitation (U)

Various human activities are damaging and degrading habitats: agricultural intensification, mining, infrastructure development and recreational activities. Illegal hunting also poses a threat. A management plan exists for the Special Natural Reserve.

Majona coast — B2, C2 — 374

Admin region Santa Cruz de Tenerife
Coordinates 28°09′N 17°07′W
Altitude 0–611 m **Area** 170 ha

■ Site description

Sea cliffs and rock stacks along the north-east coast of La Gomera, which support a typical rocky vegetation. Stony and sandy beaches are also present, along with a rocky intertidal shelf.

Habitats Scrub (scrub), Wetland (sand-dunes/sand beach; shingle/stony beach), Rocky areas (*; sea cliff/rocky shore; rock stacks/islets)
Land-use Fisheries/aquaculture (*), Tourism/recreation (*)

■ Birds

Species	Season	Year	Pop min	Pop max	Acc	Criteria
Calonectris diomedea Cory's Shearwater	B	1987	300	—	B	B2, C2

The site is important for breeding *Calonectris diomedea*, and is also used by wintering and passage waders.

■ Protection status

National Partial **International** None
70 ha of IBA covered by Natural Park (Majona, 1,757 ha). 40 ha of IBA covered by Special Natural Reserve (Puntallana, 292 ha).

■ Conservation issues

Threats Other (B), Recreation/tourism (B)

Pilgrims visiting a hermitage, and anglers can cause disturbance and generate rubbish. Shanties proliferate around the hermitage ('Other' threat). Wintering censuses take place irregularly.

Step rocks of Hermigua and Agulo — A1, B2, C1, C6 — 375

Admin region Santa Cruz de Tenerife
Coordinates 28°10′N 17°12′W
Altitude 300–1,034 m **Area** 450 ha

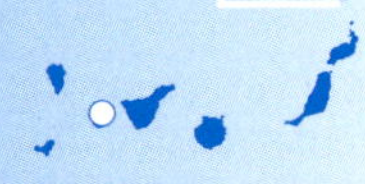

■ Site description

Steep-sided valley in the central part of La Gomera, with running water. Heath is present, in addition to exposed bedrock, inland cliffs and laurel *Laurus* forest. The site is adjacent to Garajonay National Park.

Habitats Forest and woodland (*; broadleaved evergreen forest), Scrub (*; heathland), Wetland (river/stream), Rocky areas (inland cliff), Artificial landscape (arable land)
Land-use Forestry (*)

■ Birds

Species	Season	Year	Pop min	Pop max	Acc	Criteria
Columba bollii Dark-tailed Laurel Pigeon	R	1985	Common	—		A1, B2, C1
Columba junoniae White-tailed Laurel Pigeon	R	1985	Common	—		A1, B2, C1, C6

One of the most important areas in La Gomera for forest passerines.

■ Protection status

National None **International** None

■ Conservation issues

Threats Consequences of animal/plant introductions (B), Selective logging/cutting (B), Unsustainable exploitation (A)

The cutting of branches for use in tomato cultivation has a large impact on the laurel forest. Illegal hunting and predation by introduced rats *Rattus* and feral cats *Felis catus* pose major threats to both *Columba* species. Various studies and conservation campaigns (SEO/BirdLife) have centred on the two *Columba* species.

Garajonay National Park — A1, A2, B2, B3, C1, C2, C6 — 376

Admin region Santa Cruz de Tenerife
Coordinates 28°07′N 17°13′W
Altitude 700–1,487 m **Area** 3,984 ha

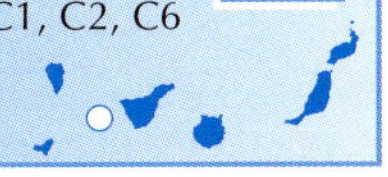

Site description

The largest continuous expanse of laurel *Laurus* forest in the Canary Islands, located in the central area of La Gomera. Other habitats include heath and scrub, and exotic coniferous woodland.

Habitats Forest and woodland (*; broadleaved evergreen forest), Scrub (*; scrub; heathland), Wetland (river/stream), Rocky areas (inland cliff), Artificial landscape (forestry plantation), Introduced/exotic vegetation
Land-use Nature conservation/research (*), Tourism/recreation (*)

Birds

Species	Season	Year	Pop min	Pop max	Acc	Criteria
Accipiter nisus Sparrowhawk	R	1993	7	10	A	C6
Alectoris barbara Barbary Partridge	R	1993	Frequent		—	B2
Columba bollii Dark-tailed Laurel Pigeon	R	1993	500	500	B	A1, A2, B2, C1, C2, C6
Columba junoniae White-tailed Laurel Pigeon	R	1993	Common		—	A1, A2, B2, C1, C6
[1] ***Apus unicolor*** Plain Swift	N	1993	Abundant		—	A2
Anthus berthelotii Berthelot's Pipit	R	1993	Common		—	A2
Regulus teneriffae Tenerife Goldcrest	R	1996	Abundant		—	A2, B3
Serinus canaria Canary	R	1993	Abundant		—	A2, B3

1. Probably breeds.

The site holds the highest numbers of *Columba bollii* in the Canary Islands, and is also important for other endemic birds.

Protection status

National High **International** High
3,984 ha of IBA covered by National Park (Garajonay, 3,984 ha). 3,984 ha of IBA covered by Special Protection Area (Garajonay, 3,984 ha). IBA overlaps with World Heritage Site (Garajonay National Park).

Conservation issues

Threats Burning of vegetation (B), Consequences of animal/plant introductions (B), Infrastructure (B), Recreation/tourism (B), Unsustainable exploitation (C)

Forest fires pose the most serious threat to the site. Additional threats include rat *Rattus* predation, and the illegal hunting of both *Columba* species, bird strikes on roads and with electric powerlines, and disturbance from visitors. A management plan exists for the site. Various censuses, studies and conservation campaigns have been carried out by the local government.

Vallehermoso coast — C6 — 377

Admin region Santa Cruz de Tenerife
Coordinates 28°13′N 17°16′W
Altitude 0–250 m **Area** 59 ha

Site description

Sea cliffs and rock stacks along the northern coast of La Gomera. The cliff tops support grassland, *Juniperus* bushes and pine *Pinus* plantations. Human activities include livestock-grazing.

Habitats Grassland, Rocky areas (100%; sea cliff/rocky shore), Artificial landscape (arable land; forestry plantation)
Land-use Agriculture (*), Fisheries/aquaculture, Tourism/recreation

Birds

Species	Season	Year	Pop min	Pop max	Acc	Criteria
Sterna hirundo Common Tern	B	1987	5	6	A	C6

The site is important due to the potential for recolonization by *Pandion haliaetus*, numbers of which are increasing on the island.

Protection status

National High **International** High
55 ha of IBA covered by Natural Monument (Los Organos, 154 ha). 55 ha of IBA covered by Special Protection Area (Los Organos, 106 ha).

Conservation issues

Threats Consequences of animal/plant introductions (C)

Introduced goats can cause disturbance, but no other important threats have been recorded.

South-west coast of La Gomera — B1ii, B2, C2, C6 — 378

Admin region Santa Cruz de Tenerife
Coordinates 28°03′N 17°18′W
Altitude 0–200 m **Area** 600 ha

Site description

Sea cliffs, rock stacks, small beaches and intertidal zones along the south-west coast of La Gomera (between Punta de la Calera and Playa de Santiago).

Habitats Wetland (shingle/stony beach), Marine areas (sea inlet/coastal features), Rocky areas (*; sea cliff/rocky shore; rock stacks/islets)
Land-use Fisheries/aquaculture (*), Tourism/recreation (*)

Birds

Species	Season	Year	Pop min	Pop max	Acc	Criteria
Bulweria bulwerii Bulwer's Petrel	B	1988	50	100	B	B2, C6
Calonectris diomedea Cory's Shearwater	B	1987	300	—	B	B2, C6
Puffinus assimilis Little Shearwater	B	1987	Frequent		—	B1ii, B2, C2, C6
Hydrobates pelagicus Storm Petrel	B	1993	50	—	B	B1ii, B2, C2, C6
Pandion haliaetus Osprey	R	1992	3	3	A	C6
Sterna hirundo Common Tern	B	1987	2	—	B	C6

One of the most important areas on La Gomera for breeding seabirds. The site holds the majority of the island's population of *Bulweria bulwerii* as well as one of the main *Puffinus assimilis* colonies in the Canary Islands.

Protection status

National Partial **International** Partial
250 ha of IBA covered by Site of Scientific Interest (Acantilados de Alajeró, 297 ha). 6 ha of IBA covered by Site of Scientific Interest (Charco del Cieno, 6 ha). 11 ha of IBA covered by Site of Scientific Interest (Charco del Conde, 11 ha). 50 ha of IBA covered by Protected Landscape (Orone, 1,788 ha). 140 ha of IBA covered by Rural Park (Valle Gran Rey, 1,192 ha). 250 ha of IBA covered by Special Protection Area (Acantilados de Alajeró, 306 ha).

Conservation issues

Threats Infrastructure (B), Recreation/tourism (C), Unsustainable exploitation (C)

Threats include the construction of a new airport and illegal hunting.

La Palma laurel forest — A1, A2, B2, B3, C1, C2, C6 — 379

Admin region Santa Cruz de Tenerife
Coordinates 28°44′N 17°49′W
Altitude 100–2,200 m **Area** 14,800 ha

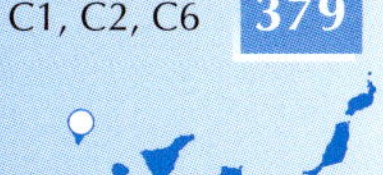

Site description

A large mountainous area in the north-east of La Palma with deep gullies, exposed bedrock, inland cliffs, laurel *Laurus* forest and agricultural land. There are few villages and roads in the area. Human activities include livestock-farming.

Habitats Forest and woodland (*; native coniferous forest; broadleaved evergreen forest), Scrub (scrub; heathland), Wetland (river/stream), Rocky areas (inland cliff), Artificial landscape (arable land), Introduced/exotic vegetation
Land-use Agriculture, Forestry (*), Hunting, Nature conservation/research, Tourism/recreation

Birds

Species	Season	Year	Pop min	Pop max	Acc	Criteria
Accipiter nisus Sparrowhawk	R	1988	Common		—	C6
Columba bollii Dark-tailed Laurel Pigeon	R	1985	125	150	B	A1, A2, B2, C1, C2, C6
Columba junoniae White-tailed Laurel Pigeon	R	1985	500	600	B	A1, A2, B2, C1, C2, C6
Apus unicolor Plain Swift	R	1996	—	—	—	A2
Anthus berthelotii Berthelot's Pipit	R	1996	Common		—	A2
Regulus teneriffae Tenerife Goldcrest	R	1996	—	—	—	A2, B3
Pyrrhocorax pyrrhocorax Chough	R	1996	—	—	—	B2, C6
Serinus canaria Canary	R	1996	—	—	—	A2

One of the most important areas in the Canary Islands for *Columba junoniae*. The site is also important for other endemic birds.

Protection status

National Partial **International** Partial
75 ha of IBA covered by Site of Scientific Interest (Barranco del Agua, 75 ha). 600 ha of IBA covered by Natural Park (Cumbre Vieja, 7,499 ha). 222 ha of IBA covered by Protected Landscape (El Tablao, 222 ha). 450 ha of IBA covered by Special Natural Reserve (Guelguén, 1,074 ha). 5,094 ha of IBA covered by Natural Park (Las Nieves, 5,094 ha). 984 ha of IBA covered by Integral Nature Reserve (Pinar de Garafía, 984 ha). 4,067 ha of IBA covered by Special Protection Area (Monte de los Sauces, Punta Llana y Pinar de Garafía, 4,067 ha).

Conservation issues

Threats Burning of vegetation (A), Consequences of animal/plant introductions (A), Groundwater abstraction (B), Intensified forest management (C), Selective logging/cutting (A), Unsustainable exploitation (A)

Forest fires (some man-made) pose a major threat. Although agricultural activity is on the decrease in forested areas, the cutting of branches for use in tomato/banana cultivation continues to have a large impact on the laurel forest. Illegal hunting and predation by introduced rats *Rattus* greatly affect both species of *Columba*. Parts of the site form a game reserve. Various censuses, studies and conservation campaigns have centred on the two *Columba* species.

El Canal y Los Tiles laurel forest — A1, A2, B2, B3, C1, C6 — 380

Admin region Santa Cruz de Tenerife
Coordinates 28°46′N 17°49′W
Altitude 250–1,800 m **Area** 511 ha

Site description

Several valleys that support the most valuable laurel *Laurus* forest on La Palma. Located in the north-east of the island, the site also contains mixed (*Laurus* and *Pinus canariensis*) and coniferous woodland, exposed bedrock and cliffs. In the past, forests were intensively exploited, water-ways were channelized, and livestock-grazing and hunting took place.

Habitats Forest and woodland (*; broadleaved evergreen forest), Scrub (*; heathland), Wetland (river/stream), Rocky areas (inland cliff; caves)
Land-use Nature conservation/research (100%), Tourism/recreation

Birds

Species	Season	Year	Pop min	Pop max	Acc	Criteria
Accipiter nisus Sparrowhawk	R	1988	Frequent		—	C6
Columba bollii Dark-tailed Laurel Pigeon	R	1985	Common		—	A1, A2, B2, C1, C6
Columba junoniae White-tailed Laurel Pigeon	R	1985	Abundant		—	A1, A2, B2, C1, C6
Regulus teneriffae Tenerife Goldcrest	R	1996	Abundant		—	A2, B3
Pyrrhocorax pyrrhocorax Chough	R	1996	—	—	—	B2, C6
Serinus canaria Canary	R	1996	Frequent		—	A2

One of the most important sites on the island for the endemic *Columba* species. The site is also important for other forest passerines.

Protection status

National Partial **International** High
250 ha of IBA covered by Natural Park (Las Nieves, 5,094 ha). 511 ha of IBA covered by Biosphere Reserve (El Canal y Los Tiles, 511 ha). 250 ha of IBA covered by Special Protection Area (Monte de los Sauces, Punta Llana y Pinar de Garafía, 4,067 ha).

Conservation issues

Threats Consequences of animal/plant introductions (B), Recreation/tourism (B), Unsustainable exploitation (B)

Predation by introduced rats *Rattus* adversely affects both *Columba* species. Recreational disturbance and illegal hunting pose additional threats. The area is a game reserve. Various studies and conservation campaigns have centred on the two *Columba* species.

Negro rocky islet — B2, C6 — 381

Admin region Santa Cruz de Tenerife
Coordinates 28°49′N 17°46′W
Altitude 0–100 m **Area** 1 ha

Site description

The site comprises a small offshore islet situated very close to the north-east coast of La Palma, and the main island cliffs. The islet supports salt-tolerant and drought-resistant vegetation.

Habitats Rocky areas (100%; sea cliff/rocky shore; rock stacks/islets)
Land-use Tourism/recreation (100%)

Birds

Species	Season	Year	Pop min	Pop max	Acc	Criteria
Bulweria bulwerii Bulwer's Petrel	B	1987	40	50	A	B2, C6

The islet supports an important *Bulweria bulwerii* breeding colony.

Protection status

National None **International** None

Conservation issues

Threats Recreation/tourism (C)

Anglers visit the islet but with little impact.

Garafía rocky islets — B2, C2, C6 — 382

Admin region Santa Cruz de Tenerife
Coordinates 28°50′N 17°58′W
Altitude 0–80 m **Area** 4 ha

Site description

A group of three small islets and rock stacks situated off the north-west coast of La Palma, that support salt-tolerant vegetation.

Habitats Marine areas (sea inlet/coastal features), Rocky areas (100%; rock stacks/islets)
Land-use Tourism/recreation (100%)

Birds

Species	Season	Year	Pop min	Pop max	Acc	Criteria
Bulweria bulwerii Bulwer's Petrel	B	1988	15	20	B	B2
Calonectris diomedea Cory's Shearwater	B	1987	300	—	B	B2, C2
Sterna hirundo Common Tern	B	1987	2	—	A	C6
Pyrrhocorax pyrrhocorax Chough	R	1996	Common		—	B2, C6

The site is important for breeding seabirds.

Protection status
National None **International** None

Conservation issues
Threats Aquaculture/fisheries (C)

Fishermen visit the islets to rest but with little impact.

La Caldera de Taburiente National Park — A1, B2, B3, C1, C6 — 383

Admin region Santa Cruz de Tenerife
Coordinates 28°43'N 17°52'W
Altitude 400–2,426 m **Area** 4,690 ha

Site description
A mountainous area situated in the middle of La Palma, with deep gullies supporting *Pinus canariensis* and mixed *Pinus* and *Laurus* woodland. High mountain vegetation, streams and springs are also found within the site.

Habitats Forest and woodland (*; native coniferous forest; broadleaved evergreen forest), Scrub (scrub), Wetland (river/stream), Rocky areas (inland cliff; caves)
Land-use Nature conservation/research (*), Tourism/recreation

Birds

Species	Season	Year	Pop min	Pop max	Acc	Criteria
Columba junoniae White-tailed Laurel Pigeon	R	1994	Frequent		—	A1, B2, C1
Pyrrhocorax pyrrhocorax Chough	R	1996	Common		—	B2, C6
Serinus canaria Canary	R	1996	Abundant		—	B3

The site is the most important area in the Canary Islands for breeding *Pyrrhocorax pyrrhocorax*, and is also important for forest passerines.

Protection status
National High **International** High
4,690 ha of IBA covered by National Park (La Caldera de Taburiente, 4,690 ha). 4,690 ha of IBA covered by Special Protection Area (La Caldera de Taburiente, 4,690 ha).

Conservation issues
Threats Burning of vegetation (B), Consequences of animal/plant introductions (B), Recreation/tourism (B), Unsustainable exploitation (C)

Introduced Barbary sheep *Ammotragus lervia* can cause damage to habitats, while other introduced species prey upon birds (e.g., rats *Rattus* and cats *Felis catus*). Forest fires and illegal hunting pose additional threats. The area is a game refuge, and there is a management plan.

El Roque coastal cliffs — B2, C6 — 384

Admin region Santa Cruz de Tenerife
Coordinates 28°41'N 17°57'W
Altitude 0–200 m **Area** 40 ha

Site description
A small stretch of coastal cliffs and a rock stack that supports drought-resistant vegetation, situated in the north-west of La Palma.

Habitats Wetland (shingle/stony beach), Marine areas (sea inlet/coastal features), Rocky areas (*; sea cliff/rocky shore; rock stacks/islets)
Land-use Not utilized (100%)

Birds

Species	Season	Year	Pop min	Pop max	Acc	Criteria
Calonectris diomedea Cory's Shearwater	B	1996	Frequent		—	B2
Alectoris barbara Barbary Partridge	R	1996	Frequent		—	B2
Sterna hirundo Common Tern	B	1996	6	—	A	C6

The site is also important as a possible breeding site for *Pyrrhocorax pyrrhocorax*.

Protection status
National None **International** None

Conservation issues
Human access is very difficult; no serious threats are known at the site.

Ventejís mountain — C6 — 385

Admin region Santa Cruz de Tenerife
Coordinates 27°47'N 17°55'W
Altitude 650–1,100 m **Area** 870 ha

Site description
A mountainous area surrounded by plains, situated in the north-east of El Hierro. Scrub, heath and grassland are interspersed with small woods of *Pinus canariensis* and introduced *P. radiata* and *Eucalyptus*. Human activities include stock-farming.

Habitats Forest and woodland (native coniferous forest; broadleaved evergreen forest), Scrub (*; scrub; heathland), Grassland (*; mesophile grassland), Rocky areas (inland cliff), Artificial landscape (arable land; forestry plantation; ruderal land)
Land-use Agriculture (*), Forestry

Birds

Species	Season	Year	Pop min	Pop max	Acc	Criteria
Fringilla coelebs Chaffinch	R	1996	—	—	—	C6

The site is important for endemic subspecies of various passerines and holds the most significant *Corvus corax* population in the Canaries.

Protection status
National Partial **International** Partial
750 ha of IBA covered by Protected Landscape (Ventijís, 1,143 ha). 750 ha of IBA covered by Special Protection Area (Garoe, 927 ha).

Conservation issues
Threats Abandonment/reduction of land management (B), Afforestation (C), Extraction industry (B), Selective logging/cutting (C), Unsustainable exploitation (B)

The abandonment of livestock-rearing affects some species, such as the carrion-feeding *Corvus corax*. Gravel-pits cause degradation of the habitat, and illegal hunting poses an additional threat.

Nizdafe plains — B3 — 386

Admin region Santa Cruz de Tenerife
Coordinates 27°46'N 17°58'W
Altitude 800–1100 m **Area** 2,000 ha

Site description
An area of mainly grassland, with some scrub and agricultural land, situated in the north-east of El Hierro. Human activities include stock-farming.

Habitats Scrub (*), Grassland (*; steppe/dry calcareous grassland), Rocky areas (inland cliff), Artificial landscape (*; arable land; ruderal land)
Land-use Agriculture (*)

Birds

Species	Season	Year	Pop min	Pop max	Acc	Criteria
Serinus canaria Canary	R	1996	Abundant		—	B3

The site is also important for *Burhinus oedicnemus*, endemic subspecies of several steppic passerines, and as a feeding area for raptors. It is one of the most important wintering and passage stop-over sites on the island.

Protection status
National None **International** None

Conservation issues

Threats Abandonment/reduction of land management (B), Infrastructure (B), Unsustainable exploitation (B)

Several main roads traverse the area, causing habitat degradation and disturbance to birds. Illegal hunting is also a problem.

Salmor rocky islets — B1ii, B2, C2, C6 — 387

Admin region Santa Cruz de Tenerife
Coordinates 27°49'N 18°00'W
Altitude 0–92 m **Area** 5 ha

Site description

A number of small rocky islets located to the north of El Hierro that support sparse salt-tolerant vegetation.

Habitats Rocky areas (100%; rock stacks/islets)
Land-use Fisheries/aquaculture (*)

Birds

Species	Season	Year	Pop min	Pop max	Acc	Criteria
Bulweria bulwerii Bulwer's Petrel	B	1987	25	50	B	B2, C6
Hydrobates pelagicus Storm Petrel	B	1987	235	350	B	B1ii, B2, C2, C6
Oceanodroma castro Madeiran Petrel	B	1987	Uncommon		—	B2, C6
Sterna dougallii Roseate Tern	B	1990	1	—	A	C6
Sterna hirundo Common Tern	B	1990	7	—	A	C6

The site is important for breeding seabirds, and is the second most important site for *Hydrobates pelagicus* in the Canary Islands.

Protection status

National High **International** High
4 ha of IBA covered by Integral Nature Reserve (Roques de Salmor, 4 ha). 5 ha of IBA covered by Special Natural Reserve (Tibataje, 601 ha). 5 ha of IBA covered by Special Protection Area (Gorreta y Salmor, 922 ha).

Conservation issues

Threats Aquaculture/fisheries (C)

Fishermen occasionally disturb breeding seabirds.

Frontera laurel forest — A1, B2, B3, C1, C6 — 388

Admin region Santa Cruz de Tenerife
Coordinates 27°44'N 18°01'W
Altitude 0–1,500 m **Area** 2,300 ha

Site description

A large area covering parts of northern and western El Hierro. Heath, scrub and laurel *Laurus* forest are interspersed with exposed bedrock and inland cliffs that support rocky vegetation.

Habitats Forest and woodland (*; broadleaved evergreen forest), Scrub (*; scrub; heathland), Rocky areas (sea cliff/rocky shore; inland cliff), Artificial landscape (perennial crops/orchards/groves)
Land-use Agriculture, Forestry (*), Tourism/recreation

Birds

Species	Season	Year	Pop min	Pop max	Acc	Criteria
Columba bollii Dark-tailed Laurel Pigeon	R	1993	Uncommon		—	A1, B2, C1
Fringilla coelebs Chaffinch	R	1996	—	—	—	C6
Serinus canaria Canary	R	1996	Abundant		—	B3

The site is the only area on El Hierro that supports *Columba bollii*. It is also important for endemic subspecies of several forest passerines.

Protection status

National Partial **International** High
1,400 ha of IBA covered by Rural Park (Frontera, 12,488 ha). 463 ha of IBA covered by Integral Nature Reserve (Mencáfite, 463 ha). 500 ha of IBA covered by Special Natural Reserve (Tibataje, 601 ha). 1,400 ha of IBA covered by Special Protection Area (El Hierro, 11,980 ha). 900 ha of IBA covered by Special Protection Area (Gorreta y Salmor, 922 ha).

Conservation issues

Threats Recreation/tourism (C), Selective logging/cutting (B), Unsustainable exploitation (B)

The main threats are posed by the cutting of branches for use in cultivation and illegal hunting. The local government has run conservation campaigns for *Calonectris diomedea* (1993–1996) and *Columba bollii*.

Western coast of El Hierro — B1ii, B2, C2, C6 — 389

Admin region Santa Cruz de Tenerife
Coordinates 27°44'N 18°09'W
Altitude 0–200 m **Area** 600 ha

Site description

West-facing cliffs and rocky islets that form part of the El Hierro coastline.

Habitats Scrub, Rocky areas (100%; sea cliff/rocky shore; rock stacks/islets)
Land-use Fisheries/aquaculture (*)

Birds

Species	Season	Year	Pop min	Pop max	Acc	Criteria
Calonectris diomedea Cory's Shearwater	B	1987	Abundant		—	B1ii, B2, C2, C6
Oceanodroma castro Madeiran Petrel	B	1987	Uncommon		—	B2, C6
Pandion haliaetus Osprey	R	1988	2	2	A	C6

A very important area for breeding seabirds and raptors. The *Calonectris diomedea* colony is among the most important on El Hierro.

Protection status

National High **International** High
600 ha of IBA covered by Rural Park (Frontera, 12,488 ha). 600 ha of IBA covered by Special Protection Area (El Hierro, 11,980 ha).

Conservation issues

Threats Infrastructure (B), Recreation/tourism (C), Unsustainable exploitation (B)

Threats include disturbance caused by tourists and illegal hunting. A track that traverses the IBA has recently been asphalted, making the area more easily accessible by tourists. The local government ran a conservation campaign for *Calonectris diomedea* from 1993–1996.

La Dehesa grasslands — B2, c7 — 390

Admin region Santa Cruz de Tenerife
Coordinates 27°44'N 18°08'W
Altitude 300–1,200 m **Area** 2,000 ha

Site description

An area of dry grassland and agricultural land in the north-west of El Hierro, with a small *Juniperus turbinata* copse. Human activities include cattle-grazing.

Habitats Forest and woodland, Scrub (scrub), Grassland (*; steppe/dry calcareous grassland), Artificial landscape (arable land), Introduced/exotic vegetation
Land-use Agriculture (*), Tourism/recreation

Birds

Species	Season	Year	Pop min	Pop max	Acc	Criteria
Falco tinnunculus Kestrel	R	1988	50	—	B	B2

The site is also important for steppic birds such as *Burhinus oedicnemus*, and is one of the two most important sites on the island for migrating passerines.

Protection status

National High **International** High

2,000 ha of IBA covered by Rural Park (Frontera, 12,488 ha). 2,000 ha of IBA covered by Special Protection Area (El Hierro, 11,980 ha).

Conservation issues

Threats Agricultural intensification/expansion (B), Consequences of animal/plant introductions (B), Recreation/tourism (C), Unsustainable exploitation (B)

Overgrazing reduces vegetation cover and can result in soil loss. The high density of feral cats *Felis catus* poses a threat to breeding birds, and illegal hunting is an additional problem.

Naos Bay–Hoya de Tacorón — B2, C6 — 391

Admin region Santa Cruz de Tenerife
Coordinates 27°39'N 18°01'W
Altitude 0–250 m **Area** 170 ha

Site description

A stretch of south-west facing coastline with sea cliffs and rock stacks. The cliffs support rocky vegetation. Human activities include fishing and sailing.

Habitats Rocky areas (100%; sea cliff/rocky shore; rock stacks/islets)
Land-use Fisheries/aquaculture (*)

Birds

Species	Season	Year	Pop min	Pop max	Acc	Criteria
Calonectris diomedea Cory's Shearwater	B	1987	Common		—	B2
Pandion haliaetus Osprey	R	1996	2	2	A	C6

The site is important for breeding seabirds and raptors.

Protection status

National Partial **International** Partial
30 ha of IBA covered by Rural Park (Frontera, 12,488 ha). 30 ha of IBA covered by Special Protection Area (El Hierro, 11,980 ha).

Conservation issues

Threats Unsustainable exploitation (B)

Illegal hunting is the main threat. The local government ran a conservation campaign for *Calonectris diomedea* from 1993–1996.

REFERENCES

AGUILAR, J. S. (1994) Atlas de aves marinas de Baleares 1992. Pp. 9–20 in *Trabajos en Aves Marinas. Documents Tècnics de Conservació*, 23. Govern Balear: Palma de Mallorca.

ALBERDI, M., IZQUIERDO, A., RICO, L. AND SÁNCHEZ, J. A. (1996) Distribución y problemática de las grandes águilas (*Hieraetus fasciatus* y *Aquila chrysaetos*) en el sector Suroccidental de la provincia de Valencia. Valencia, Spain: Conselleria de Agricultura y Medio Ambiente. Unpublished report.

ALONSO, J. A. AND ALONSO, J. C., EDS. (1990) *Distribución y demografía de la Grulla Común (*Grus grus*) en España.* Madrid: ICONA (Colección Técnica).

ALONSO, J. C. AND ALONSO, J. A. (1996) The great bustard in Spain: present status, recend trends and an evaluation of earlier censuses. *Biol. Conserv.* 77(1): 79–86.

ÁLVAREZ, J. *ET AL.* (1985) *Atlas de los vertebrados continentales de Álava, Vizcaya y Guipúzcoa (excepto Chiroptera).* Gobierno Vasco. Bilbao.

AMENGUAL, J. F. (1996) Las rapaces del Parque Nacional de Cabrera. Pp. 311–315 in J. Muntaner and J. Mayol eds. *Biología y conservación de las rapaces mediterráneas, 1994.* Madrid: Sociedad Española de Ornitología Monogr. 4).

ARAGÜÉS, A. (1992) Estudio de la Alondra de Dupont (*Chersophilus duponti*) en la Región Aragonesa. Universidad de Zaragoza (Doctoral Thesis).

ARAMBARRI, R. AND RODRÍGUEZ, A. (1995) Informe sobre los Pícidos del País Vasco. Gobierno Vasco. Unpublished report.

ARROYO, B. (1995) Censo de aves rupícolas en la provincia de Guadalajara. Consejería de Agricultura y Medio Ambiente, Junta de Comunidades de Castilla-La Mancha. Unpublished report.

ARROYO, B., FERREIRO, E. AND GARZA, V., EDS. (1990) *II Censo Nacional de Buitre Leonado (*Gyps fulvus*): población, distribución, demografía y conservación.* Madrid: ICONA (Colección Técnica).

BALLESTEROS, F., BENITO, J. L., GIL, J. AND GONZÁLEZ-QUIRÓS, P. (1991) Censo del Cormorán Moñudo (*Phalacrocorax aristotelis*) en Asturias. Consejería de Presidencia, Principado de Asturias. Unpublished report.

BALLESTEROS, F., BENITO, J. L., GIL, J. AND GONZÁLEZ-QUIRÓS, P. (1991) Situación y problemática del Halcón Peregrino (*Falco peregrinus*) en Asturias, 1991. Consejería de Presidencia, Principado de Asturias. Unpublished report.

BARONE, R. AND EMMERSON, K. W. (1995) Distribución, estatus y características del hábitat de la Terrera Marismeña *Calandrella rufescens* (Vieillot, 1820) en la isla de Tenerife, Canarias (Aves: Alaudidae). *Rev. Acad. Canar. Cienc.* VII (2, 3, 4): 47–61.

BERGERANDI, A. AND ARZOZ, M. J. (1991) Censo de aves acuáticas nidificantes en Navarra. Servicio de Medio Ambiente, Gobierno de Navarra. Unpublished report.

BERNIS, F. (1980) *La migración de las aves en el Estrecho de Gibraltar,* I. *Aves planeadoras.* Madrid: Cátedra de Zoología de Vertebrados, Universidad Complutense.

BLANCO, J. C. AND GONZÁLEZ, J. L., EDS. (1992) *Libro Rojo de los Vertebrados de España.* Madrid: ICONA (Colección Técnica).

BROS, V., MIRALLES, J. AND REAL, J. (1982) *La fauna del Vallès Occidental.* Sabadell, Spain: El Pot.

BROTONS, P. A. (1992) Inventariación de la avifauna en las zonas húmedas de la provincia de Cuenca, octubre 1990–junio 1991. Consejería de Agricultura y Medio Ambiente, Junta de Comunidades de Castilla-La Mancha. Unpublished report.

CAPELLÀ, L., JARA, J. L., MAYOL, J., MUNTANER, J. AND PONS, M. (1986) The 1986 census of the breeding population of Shags in the Balearic Islands. Pp. 505–508 in MEDMARAVIS and X. Monbailliu, eds. *Mediterranean Marine Avifauna: population studies and conservation.* Berlin: Springer.

CASTRO, H., NEVADO, J. C., PARACUELLOS, M. AND LÓPEZ-MARTOS, J. M. (1994) La Malvasía (*Oxyura leucocephala*) en la provincia de Almería. Evolución poblacional, nidificación y selección de hábitat. *Oxyura* 7(1): 119–133.

CEÑA, A. *ET AL.* (1994) Censo de las aves rupícolas en las ZEPAs de La Rioja. Dirección General de Montes y Conservación de la Naturaleza, Gobierno de La Rioja. Unpublished report.

CONSEJERÍA DE MEDIO AMBIENTE DE LA JUNTA DE ANDALUCÍA (1996) Censos periódicos de aves acuáticas en las lagunas de Fuente Piedra, Campillos, Ratosa. Periodo 1986–96. Consejería de Medio Ambiente, Junta de Andalucía. Unpublished report.

COSTA, M. AND CUERDA, S. (1993) Censo de Cernícalo Primilla (*Falco naumanni*) en la provincia de Cuenca. Consejería de Agricultura y Medio Ambiente, Junta de Comunidades de Castilla-La Mancha. Unpublished report.

DE LA CRUZ ALEMÁN, M. A., CÁMARA ORGAZ, F. AND DE LA CRUZ ALEMÁN, J. L. (1995) Censo de nidificación de ardeidas coloniales y otras aves ligadas a carrizales en Toledo occidental. Consejería de Agricultura y Medio Ambiente, Junta de Comunidades de Castilla-La Mancha. Unpublished report.

DÍAZ, M., ASENSIO, B. AND TELLERÍA, J. L. (1996) *Aves Ibéricas,* I. *No Paseriformes.* Madrid: Reyero.

DIES, B. (1995) El Calamón Común en la Comunidad Valenciana: situación actual. Consellería de Agricultura y Medio Ambiente, Generalitat Valenciana. Unpublished report.

DIES, J. I. AND DIES, B., COORDS. (1994, 1995 and 1997) *Anuario ornitológico de la comunidad Valenciana (año 1992, 1993, 1994)*, 5. Valencia, Spain: Estación Ornitológica L'Albufera (SEO).

DOLZ, J. C., GIMÉNEZ, M. AND HUERTAS, J. (1991) Status of some threatened Anatidae species in the Comunidad Valenciana, East Spain. *IWRB Threatened Waterfowl Res. Group Newsl.* 1: 7–8.

DONÁZAR, J. A., NAVESO, M. A., TELLA, J. E. AND CAMPIÓN, D. (1997) Extensive grazing and raptors in Spain. Pp. 117–149 in D. J. Pain and M. W. Pienkowski, eds. *Farming and birds in Europe: the Common Agricultural Policy and its implications for bird conservation.* London: Academic Press.

EGUÍA, S. AND SÁNCHEZ, J. A. (1993) Inventario de las aves de presa de los parques naturales almerienses. Agencia de Medio Ambiente. Almería. Unpublished report.

EMMERSON, K. (1985) Estudio de la biología y ecología de la Paloma Turqué (*Columba bollii*) y la Paloma Rabiche (*C. junoniae*) con vistas a su conservación, II. ICONA. Unpublished report.

EMMERSON, K. (1988) Estudio para la catalogación y valoración ecológica de las principales áreas del litoral canario como base para un futuro programa de conservación. Dirección General del Medio Ambiente y Conservación de la Naturaleza. Gobierno de Canarias. Unpublished report.

EMMERSON, K., BARONE, R., LORENZO, J.A. AND NARANJO, J. J. (1993) Censo y análisis de la comunidad ornítica del Parque Nacional de Garajonay (La Gomera). ICONA. Unpublished report.

EMMERSON, K., LORENZO, J. A., BARONE, R., TRUJILLO, D. AND DELGADO, G. (1991) Resultados del censo de las aves acuáticas nidificantes en Canarias. Tragsatec/Ornistudio S.L. Unpublished report.

ERVA, EDS. (1996) *3° Anuario das Aves de Galicia .1995*. Erva. Vigo.

EUROPARC ESPAÑA (1998) Actas del 3er seminario de Espacios Naturales Protegidos. Modelso de Planificación y Gestión en Espacios Naturales Protegidos. Sección del estado Español de la Federación de Parques Naturales y Nacionales de Europa. Casa Nova de l'Obac (Barcelona), 10–13 diciembre 1997.

FERNÁNDEZ, J. M. (1994) Censo y distribución del Búho Real (*Bubo bubo*) en Álava. *Estudios del Museo de Ciencias Naturales de Álava* 8: 231–235.

FERNÁNDEZ, M. AND HERNÁNDEZ, M. A. (1995) Revisión del inventario de aves rupícolas de la provincia de Cuenca. Consejería de Agricultura y Medio Ambiente, Junta de Comunidades de Castilla-La Mancha. Unpublished report.

FERNÁNDEZ, C. AND AZKONA, P. (1996) Distribución de la Perdiz Pardilla y la Perdiz Nival en Navarra. Servicio de Medio Ambiente, Diputación Foral de Navarra. Pamplona. Unpublished report.

FERNÁNDEZ, C. (1991) Censo de rapaces rupícolas en Navarra. Servicio de Medio Ambiente, Diputación Foral de Navarra. Pamplona. Unpublished report.

FERNÁNDEZ, C. (1994) Censo, evolución demográfica de las colonias y productividad del Buitre Leonado en Navarra. Servicio de Medio Ambiente, Diputación Foral de Navarra. Pamplona. Unpublished report.

FERRER, M. (1993) Juvenile dispersal behaviour and natal philopatry of a long-lived raptor, the Spanish Imperial Eagle *Aquila adalberti*. *Ibis* 135: 132–138.

FINLAYSON, C. (1992) *Birds of the Strait of Gibraltar*. London: Poyser.

GARCÍA, J. L. (1996) Estudio de la avifauna acuática de la Ría de Urdaibai: análisis crítico. Unpublished report.

GARCÍA, D., MARGALIDA, A., PARELLADA, X. AND CANUT, J. (1996) Evolución y parámetros reproductores del Quebrantahuesos en Cataluña (NE de España). *Alauda* 64(2): 229–238.

GARCIAS, P. (1991) Seguiment de la colònia de Avisadors (*Himantopus himantopus*) al Salobrar de Campos. *Anuari Ornitològic de les Balears*: 29–34.

GARCIAS, P. (1995) Situació del Tirolil.lo Camanegra (*Charadrius alexandrinus*) a Mallorca. *Anuari Ornitològic de les Balears*: 25–34.

GARONA, E. T. (1980) Estudio de las áreas de cría e invernada, bebederos y principales enclaves de aves esteparias: Sisón, Ganga y Ortega en el Valle del Ebro. Fasc.1. Departamento de Agricultura, Ganadería y Montes, Diputación General de Aragón.

GARRIDO, H. (1996) *Aves de las Marismas del Odiel y su entorno*. Madrid: Rueda.

GARZA, V. AND SUÁREZ, F. (1990) Distribución, población y selección de hábitat de la Alondra de Dupont en la Península Ibérica. *Ardeola* 37: 3–12.

GARZA, V. AND SUÁREZ, F. (1992) Estudio para la inclusión de zonas esteparias de la Comunidad de Castilla y León en la Red de Espacios Naturales. Informe de SEO/BirdLife. Consejería de Medio Ambiente y Ordenación del Territorio, Junta de Castilla y León. Unpublished report.

GÓMEZ, D. (1989) II censo Nacional de Buitre Leonado. Año 1989. Zona Oriental de la provincia de Huesca. Diputación General de Aragón. Unpublished report.

GONZÁLEZ, G., SANTIAGO, J. M. AND FERNÁNDEZ L., EDS. (1992) *El Águila Pescadora en España: censo, reproducción y conservación*. Madrid: ICONA (Colección Técnica).

GONZÁLEZ-QUIRÓS, P. AND BENITO, J. L. (1996) Seguimiento de la población de Buitre Leonado en Asturias. Consejería de Agricultura, Principado de Asturias. Unpublished report.

GONZÁLEZ-QUIRÓS, P. AND BENITO, J. L. (1996) Seguimiento de la población de Águila Real en Asturias. Consejería de Agricultura, Principado de Asturias. Unpublished report.

GONZÁLEZ-QUIRÓS, P. AND BENITO, J. L. (1996) Seguimiento de la población de Alimoche Común en Asturias. Consejería de Agricultura, Principado de Asturias. Unpublished report.

GOVERN BALEAR (1995) *Butlletí del Parc Natural de l'Albufera de Mallorca*, 2. Govern Balear: Palma de Mallorca.

GRIMMETT, R. F. A. AND JONES T. A. (1989) *Important Bird Areas in Europe*. Cambridge, UK: International Council for Bird Preservation (Tech. Publ. 9).

GRUPO DE ORNITOLOGIA BALEAR (1994) *Anuari Ornitològic de les Balears*. GOB. Palma de Mallorca.

GUTIÉRREZ, R. AND FIGUEROLA, J. (1995) Wintering distribution of the Balearic Shearwater (*Puffinus yelkouan mauretanicus*, Lowe 1921) of the Northeastern coast of Spain. *Ardeola* 42(2): 161–166.

GUTIÉRREZ, R. ET AL. (1995) *Els ocells del Delta del Llobregat*. Barcelona: Lynx.

GUZMÁN, J., HERNÁNDEZ, J. M., MURO, F. J., PALOMO, R. AND UCEDA, J. (1994) Estudio de la población de Cernícalo Primilla (*Falco naumanni*) en el Campo de Calatrava (Ciudad Real). Junta de Comunidades de Castilla-La Mancha. Unpublished report.

HEATH, M. F. AND BORGGREVE, C. (2000) *BirdLife International/EBCC European Bird Database 1998*. Cambridge, UK: BirdLife International.

HERNÁNDEZ, E., MARTÍN, A., DÍAZ, G., TRUJILLO, O. AND ASCANIO, M. (1985) Censo y datos sobre la biología del Halcón de Eleonor (*Falco eleonorae* Gené, 1839) en las Islas Canarias. Agosto-septiembre 1983. *Doñana Acta Vertebrata* 12(1): 63–73.

HERNÁNDEZ, E., MARTÍN, A., NOGALES, M., QUILIS, V., DELGADO, G. AND TRUJILLO, O. (1990) Distribution and status of Bulwer's Petrel (*Bulweria bulwerii* Jardine and Selby, 1828) in the Canary Islands. *Bol. Museu Mun. Funchal* 42(214): 5–16.

HERNÁNDEZ, E., NOGALES, M., QUILIS, V. AND DELGADO, G. (1990) Nesting of the Manx Shearwater (*Puffinus puffinus* Brünnich, 1764) on the Island of Tenerife (Canary Islands). *Bonn. zool. Beitr.* 41(1): 59–62.

HIRALDO, F. (1994) Colonias de nidificación en el Cernícalo Primilla (*Falco naumanni*) en Andalucía: estado actual de las mismas, problemas de conservación y normas para su manejo. Estación Biológica de Doñana. Sevilla. Unpublished report.

ILLANA, A. AND CALVO, M. (1994) El Alimoche (*Neophron percnopterus*) en Álava. Unpublished report.

ILLANA, A. ET AL. (1991) Estudio del status, biología y protección de las Falconiformes nidificantes en las zonas rupícolas de la provincia de Álava. Unpublished report.

INÉS, H., SEGOVIA, C. AND GALÁN, R. (1996) Inventario cartográfico de zonas e importancia para las aves esteparias en las provincias de Huelva y Sevilla. Andalus (Convenio FAADN-AMA). Unpublished report.

ITSAS ENERA ORNITOLOGI ELKARTEA (1993–1996) *Anuarios Ornitológicos de Gipuzkoa. Años 1991–1995*. Donostia.

JIMÉNEZ, J., DEL MORAL, A., MORILLO, C. AND SÁNCHEZ, M. J. (1992) *Las aves del Parque Nacional de las Tablas de Daimiel y otros humedales manchegos*. Barcelona: Lynx.

DE JUANA, E., ED. (1990) *Áreas Importantes para las Aves en España*. [*Important Bird Areas in Spain*]. Madrid: Sociedad Española de Ornitología (Monogr. 3).

DEL JUNCO, O. AND PAZ DE LA ROCHA, J. L. (1995) Censo de parejas reproductoras de Buitre Común, Alimoche, Águila Perdicera, Águila Real y Halcón Peregrino en la provincia de Cádiz. Unpublished report.

LARA, A. AND PICAZO, J. (1991) Informe sobre el censo primaveral de aves acuáticas en la provincia de Albacete. Primavera 1991. Delegación de Agricultura y Medio Ambiente. Albacete. Unpublished report.

LÓPEZ, M. AND MENOR, A. (1994) La Malvasía (*Oxyura leucocephala*) en Castilla-La Mancha: población y características. *Oxyura* 7(1): 135–146.

LORENZO, J.A. (1993) Descripción de la comunidad de aves limícolas de El Médano (Tenerife, Islas Canarias) durante un ciclo anual. *Ardeola* 40(1): 13–19.

LUCIO, A. J. ET AL. (1992) *La Perdiz Pardilla (*Perdix perdix*) en España*. Madrid: ICONA (Colección Técnica).

MANRIQUE, J. (1996) *Corología y ecogeografía de las aves nidificantes en la provincia de Almería (SE ibérico)*. Universidad de Granada (Doctoral Thesis). Unpublished report.

MARTÍ, R. (1992) Áreas Importantes para las Aves en España. Review of the inventory by the Sociedad Española de Ornitología/BirdLife for the the European Commission. Unpublished report.

MARTÍN, A. AND HERNÁNDEZ, E. (1985) Importante colonia de aves marinas en los Roques de Salmor (Isla de El Hierro, Canarias). *Ardeola* 32(1): 123–127.

MARTÍN, A. AND NOGALES, M. (1993) Ornithological importance of the island of Alegranza (Canary Islands). *Bol. Mus. Mun. Funchal Suppl.* 2: 167–179.

MARTÍN, A., HERNÁNDEZ, E., DELGADO, G. AND QUILIS, V. (1984) Nidificación del Paíño de Madeira *Oceanodroma castro* (Harcourt, 1851) en las islas Canarias. *Doñana Acta Vertebrata* 11(2): 337–341.

MARTÍN, A., HERNÁNDEZ, E., NOGALES, M., QUILIS, V., TRUJILLO, O. AND DELGADO, G. (1990) *El Libro Rojo de los Vertebrados Terrestres de Canarias*. Caja Canarias: Santa Cruz de Tenerife.

MARTÍN, A., HERNÁNDEZ, M. A., LORENZO, J.A. AND NOGALES, M. (1996) Estudio sobre la distribución y estado de las poblaciones de las palomas Turqué (*Columba bollii*) y Rabiche (*Columba junoniae*) en Tenerife, con especial referencia a la comarca Tacoronte-Santa Úrsula. Universidad de La Laguna. Viceconsejería de Medio Ambiente. Gobierno de Canarias. Unpublished report.

MARTÍN, A., LORENZO, J.A., HERNÁNDEZ, M. A., NOGALES, M., MEDINA, F. M., DELGADO, J. D., NARANJO, J. J., QUILIS, V., DELGADO, G. (1995) Censo de la Hubara Canaria (*Chlamydotis undulata fuertaventurae*) en las islas e islotes de Lanzarote y Fuerteventura (noviembre-diciembre de 1994). Departamento de Biología Animal (Zoología). Universidad de La Laguna. Unpublished report.

MARTÍN, A., NOGALES, M., QUILIS, V., DELGADO, G., HERNÁNDEZ, E., TRUJILLO, O. AND SANTANA, F. (1987) Distribución y Status de las Aves Marinas Nidificantes en el Archipiélago Canario con vistas a su conservación. Dirección General del Medio Ambiente y Conservación de la Naturaleza. Gobierno de Canarias. Unpublished report.

MARTÍNEZ, F. AND PACHECO, B. (1990) Censo de Parejas reproductoras de Aguilucho Lagunero (*Circus aeruginosus*) en Aragón, Cantabria, Castilla-León, Castilla-La Mancha, Madrid y La Rioja. ICONA. Unpublished report.

MARTÍNEZ, I. ET AL. (1995) Censo, distribución y protección de las rapaces forestales en el Territorio Histórico de Álava. Departamento de Agricultura y Pesca, Gobierno Vasco. Unpublished report.

MORENO, A. (1991) Distribución, ecología y características esenciales del hábitat del Pinzón Azul de Gran Canaria. Viceconsejería de Medio Ambiente. Gobierno de Canarias. Unpublished report.

MORENO, J. D., LAVEDÁN, J., RIVAS, J. L. AND GIL, J. A. (1995) Estatus, distribución y conservación del aguilucho cenizo en Aragón. *Alytes* 7:241–242.

MUNILLA, I. AND MOURIÑO, J., EDS. (1995) *Actas do II Congreso Galego de Ornitoloxía*. Santiago de Compostela, Spain: Universidade de Santiago de Compostela.

MUNTANER, J., FERRER, X. AND MARTÍNEZ-VILALTA, A. (1984) *Atlas dels ocells nidificants de Catalunya i Andorra*. Barcelona, Spain: Ketres.

NAVARRO MEDINA, J. D. AND ROBLEDANO, F. (COORDS.). (1995) *La Cerceta Pardilla* Marmaronetta angustirostris *en España*. Madrid: ICONA (Colección Técnica).

NOGALES, M., MARTÍN, A., QUILIS, V., DELGADO, G., HERNÁNDEZ, E. AND TRUJILLO, O. (1993) Estatus y distribución del Paíño Común (*Hydrobates pelagicus*) en las islas Canarias (15–24). *In* J. S. Aguilar, X. Monbailliu and A. M. Paterson, eds. *Estatus y conservación de aves marinas. Ecogeografía y plan de acción para el Mediterráneo*. Madrid: MEDMARAVIS/Sociedad Española de Ornitología.

DE PABLO, F. AND CATCHOT, S. (1992) El Corbmarí (*Phalacrocorax aristotelis desmarestii*) a Menorca: recompte de parelles reproductores i parametres reproductius. *Anuari Ornitològic de les Balears* 6: 13–16.

DE PABLO, F., CATCHOT, S. AND ORFILA, G. (1992) Informe sobre el cens de Falcó (*Falco peregrinus*) a l'illa de Menorca. Govern Balear. Unpublished report.

PALACIOS, A. (1994) Control de las colonias de ardeidas en Castilla y León. Junta de Castilla y León. Unpublished report.

PALACIOS, J. AND RODRÍGUEZ, M. (1996) La gestión en la reserva 'Las Lagunas de Villafáfila'. Pp. 315–325 in J. Fernández Gutiérrez. and J. Sanz-Zuasti, eds. *Conservación de las aves esteparias y su hábitat*. Valladolid, Spain: Junta de Castilla y León.

PARACUELLOS, M. AND NEVADO, J. C. (1995) Nidificación de Láridos en la provincia de Almería (SE ibérico). *Doñana Acta Vertebrata* 22(1–2): 102–106.

PARQUE NACIONAL DE LAS TABLAS DE DAIMIEL. (1995) Censo de anátidas y fochas invernantes y de parejas de somormujos, ardeidas, anátidas y fochas nidificantes en el Parque Nacional de las Tablas de Daimiel (1987–1995). ICONA. Unpublished report.

PELAYO, E. (1991) *Aves nidificantes de la comarca del Moncayo y Campo de Borja*. Zaragoza, Spain: Diputación Provincial de Zaragoza, Centro de Estudios Borjanos.

PÉREZ-HURTADO, A. AND HORTAS, F. (1992) Cádiz Bay, South-West Spain as a potential Ramsar site: Its importance for wintering waders. *Wader Study Group Bull.* 72: 34–38.

PÉREZ-HURTADO, A., HORTAS, F., RUIZ, J. AND SOLÍS, F. (1993) Importancia de la Bahía de Cádiz para las poblaciones de limícolas invernantes e influencia de las transformaciones humanas. *Ardeola* 40(2): 133–142.

PHILLIPS, B. , ED. (1986) *The Fuerteventura Stonechat project*. Cambridge, UK: International Council for Bird Preservation (Study Report 8).

PURROY, F. J. (1972) El Pico Dorsiblanco del Pirineo. *Ardeola* 20: 145–158.

REAL, J. AND RIBAS, J. (1985) Status, distribució i migració dels rapinyaires diurns (Falconiformes i Accipitriformes) al Vallès Occidental. *El Medi Natural del Vallès. Annals del CEEM* 1: 151–170.

REAL, J. (1991) L'àliga perdiguera *Hieraaetus fasciatus* a Catalunya: Status, ecologia tròfica, biologia reproductora i demografia. Universitat de Barcelona (Doctoral Thesis).

RICO, L. AND MARTÍN, C. (1995) Seguimiento de la población de Águila Perdicera *Hieraaetus fasciatus* en la provincia de Alicante, 1994–1995. Conselleria de Agricultura y Medio Ambiente, Generalitat Valenciana. Unpublished report.

ROBLEDANO, F. (1991) Invernada de patos marinos en el Mar Menor (Murcia). *Boletín GIAMI* 15: 4–5.

RODRÍGUEZ, J. (1995) *Aves rapaces del Parque Natural de Grazalema*. Gibraltar: Gibraltar Ornithological and Natural History Society and Wildlife Gibraltar.

RODRÍGUEZ, M. AND PALACIOS, J. (1991) *El Ánsar Campestre y el Ánsar Común en Castilla y León*. Valladolid, Spain: Dirección General del Medio Natural, Junta de Castilla y León.

ROMÁN, J. *ET AL.* (1996) *Atlas de las aves nidificantes de la provincia de Burgos*. Burgos, Spain: Caja Ahorros Círculo Católico de Obreros.

SALAVERRI, L. J. AND MUNILLA, I., EDS. (1995) *2º Anuario das Aves de Galicia*. Santiago de Compostela, Spain.

SAMPIETRO, F. J. AND PELAYO, E., EDS. (in press) *Aves de Aragón. Atlas de especies nidificantes*. Zaragoza, Spain: Diputación General de Aragón.

SAMPIETRO, F. J. AND PELAYO, E. (1994) Resultados de los censos invernales de aves acuáticas en la Comunidad Autónoma de Aragón (enero de 1991 y enero de 1992). *Rocín, Anuario Ornitológico de Aragón 1991–1992*: 76–89.

SAMPIETRO, F. J. AND PELAYO, E. (1995) Seguimiento de aves acuáticas y grullas en Aragón durante la temporada 1994/95. Dirección General de Conservación del Medio Natural, Diputación General de Aragón. Unpublished report.

SAMPIETRO, F. J. AND PELAYO, E. (1996) Informe sobre el núcleo de avutardas (*Otis tarda*) de Monegros. Ocupación espacial y uso del hábitat. Dirección General del Medio Natural, Departamento de Agricultura y Medio Ambiente, Diputación General de Aragón. Unpublished report.

SÁNCHEZ CODOÑER, A. (1995) Aves nidificantes de las Islas Columbretes. Informe 1995. Conselleria de Medio Ambiente, Generalitat Valenciana. Unpublished report.

SÁNCHEZ, A., GUTIÉRREZ, A., ÁLVAREZ, J. A., PLAZA, M. AND CALDERA, J. (1996) Situación de la Avutarda (*Otis tarda*) en Extremadura (1981–95). Pp. 279–282 in J. Fernández Gutiérrez and J. Sanz-Zuasti, eds. *Conservación de las aves esteparias y su hábitat*. Valladolid, Spain: Junta de Castilla y León.

SÁNCHEZ-LAFUENTE, A. M., REY, P., VALERA, F. AND MUÑOZ-COBO, J. (1992) Past and current distribution of the Purple Swamphen *Porphyrio porphyrio* L. in the Iberian Peninsula. *Biol. Conserv.* 61: 23–30.

SANZ, J., VELASDCO, T. AND SOTO, E. (1996) Censo (1995–96) de las ardeidas nidificantes en las zonas húmedas de Navarra. Unpublished report.

SECCIÓN DEL ESTADO ESPAÑOL DE LA FEDERACIÓN DE PARQUES NATURALES Y NACIONALES DE EUROPA (1994) *Inventario de Espacios Naturales Protegidos*. FPNE, Centro de Investigación de Espacios Naturales Protegidos Fernando González Bernáldez y Agencia de Medio Ambiente de la Comunidad de Madrid.

SEO/BIRDLIFE (1992) Áreas Importantes para las Aves en España. Revisión del inventario para la Comisión Europea. Unpublished report.

SEO/BIRDLIFE (1994) Censo de la población reproductora de Aguilucho Lagunero (*Circus aeruginosus*) en la Comunidad de Madrid. 1994. Agencia de Medio Ambiente, Comunidad de Madrid. Unpublished report.

SEO/BIRDLIFE (1994) V Censo Nacional de Cigüeña Blanca (*Ciconia ciconia*). ICONA. Unpublished report.

SEO/BIRDLIFE (1997) *Atlas de las aves de España (1975–1995)*. Barcelona: Lynx.

SEO-ESTACIÓ ORNITOLÒGICA DE L'ALBUFERA (1996–1997). Censos de aves acuáticas invernantes y nidificantes en la Comunidad Valenciana, 1996–1997. Conselleria de Agricultura y Medio Ambiente, Generalitat Valenciana. Unpublished report.

SILVA, P. , SOLANO, S., BALLESTEROS, F., BENITO, J. L., QUIRÓS, P. AND VÉLEZ, E. (1995) Cartografía de cantaderos de Urogallo. Consejería de Medio Ambiente y Urbanismo, Principado de Asturias. Unpublished report.

SOCIEDAD ALBACETENSE DE ORNITOLOGÍA (COORD. CAMPOS, B.) (1995) Población y selección de hábitat de dos especies de aves esteparias en el Campo de Montiel (Castilla-La Mancha): Sisón (*Tetrax tetrax*) y Alcaraván (*Burhinus oedicnemus*). Consejería de Agricultura y Medio Ambiente, Junta de Comunidades de Castilla-La Mancha. Unpublished report.

SOCIEDAD ORNITOLÓGICA LANIUS. (1994) Censo de Buitre Leonado (*Gyps fulvus*) en la Comunidad Autónoma del País Vasco. Unpublished report.

SOCIEDADE GALEGA DE HISTORIA NATURAL, ED. (1995) *Atlas de vertebrados de Galicia*, II. *Aves*. Consello da Cultura Galega. Santiago de Compostela.

TORRES-ESQUIVIAS, J. A., ALCALÁ-ZAMORA, A. AND MORENO, B. (1996) Resultados de los censos coordinados de Malvasía Común (*Oxyura leucocephala*) durante los años 1994 y 1995. *Oxyura* 8(1): 5–14.

TUCKER, G. M. AND HEATH, M. F. (1994) *Birds in Europe: their conservation status*. Cambridge, UK: BirdLife International (BirdLife Conservation Series no. 3).

VELASCO, T., OLLERO, J. AND GRIJALBO, J. (1987) Limícolas en la Mancha: estatus, fenología y estudio de un ciclo anual. Consejería de Agricultura y Medio Ambiente, Junta de Comunidades de Castilla-La Mancha. Unpublished report.

VELASCO, T. (1993) Inventariación de la avifauna en las zonas húmedas de la provincia de Cuenca, III. Censos primaveral y de nidificantes. Temporada 92/93. Consejería de Agricultura y Medio Ambiente, Junta de Comunidades de Castilla-La Mancha. Unpublished report.

VELASCO, T. (1995) Censo de las aves acuáticas nidificantes en las zonas húmedas de la provincia de Toledo. Año 1995. Consejería de Agricultura y Medio Ambiente, Junta de Comunidades de Castilla-La Mancha. Unpublished report.

VIADA, C. (1992–1994) Memorias del Plan de Conservación de las Rapaces de Baleares. Conselleria de Agricultura y Pesca, Govern Balear. Unpublished report.

VIADA, C., ED. (1998) *Areas Importantes para las Aves en España*. 2ª edición revisada y ampliada. Madrid: Sociedad Española de Ornitología (Monogr. 5).

TURKEY

GERNANT MAGNIN, GÜVEN EKEN AND MURAT YARAR

Sultansazlığı saltmarshes, Central Anatolia (IBA 052). (PHOTO: CUNEYT OGUZTÜZÜN)

GENERAL INTRODUCTION

Turkey covers an area of 779,452 km² and has 97 Important Bird Areas (IBAs) covering a total of 29,978 km² or 4% of the total land area (Map 1, Table 1). The country is divided into seven major geographical regions, each with its own distinct climate, topography and consequently habitats, flora and fauna: Marmara, Black Sea, Aegean, Mediterranean, Central Anatolia, Eastern Anatolia and South-eastern Anatolia.

Birdwatching and other research activities in Turkey are, by European standards, limited in extent. Large parts of the country, especially in the east and south-east, have hardly been visited by birdwatchers, due to access being limited for safety reasons over the last decade. In addition, many parts of the country are inaccessible during the wet season without proper equipment. Accordingly, the western half of Turkey, especially the coastal areas, is better known and IBA coverage is more or less complete (Map 1). Additional IBAs may lie undetected in the eastern part of the country.

The previous international IBA inventory (Grimmett and Jones 1989) identified 79 sites, of which 14 were excluded from the 1997 national inventory (Magnin and Yarar 1997) as they did not meet the revised IBA criteria (Table 1). Thirty-five new IBAs were identified by the 1997 inventory, with various splits and mergers also taking place (Table 1). The current international inventory follows the 1997 national inventory.

Table 1. Summary of Important Bird Areas in Turkey. 97 IBAs covering 29,978 km²

IBA code	National code	1989 code	International name	National name	Administrative region	Area (ha)	Criteria (see p. 11)
001	TR001	TR008	Meriç delta	Meriç Deltası	Edirne	7,000	A1, A4i, A4iii, B1i, B2
002	TR002	—	Iğneada forests	Iğneada Ormanları	Kırklareli	3,000	B1iv, B2
003	TR003	TR002	Büyükçekmece lake	Büyükçekmece Gölü	İstanbul	2,850	A1, A4i, A4iii, B1i
004	TR004	—	Küçükçekmece lake	Küçükçekmece Gölü	İstanbul	1,500	A1, A4i, A4iii, B1i
005	TR005	TR003	Bosphorus	Boğaziçi	İstanbul	55,000	A4ii, A4iv, B1ii, B1iv
006	TR006	—	Şile islands	Şile Adaları	İstanbul	5	B1i
007	TR007	TR004	Iznik lake	Iznik Gölü	Bursa	29,830	A1, B2
008	TR008	TR010	Uludağ	Uludağ	Bursa	20,000	B2
009	TR009	TR001	Uluabat lake	Uluabat Gölü	Bursa	13,500	A1, A4i, A4iii, B1i, B2
010	TR010	TR006	Kuş lake	Kuş Gölü	Balıkesir	16,000	A1, A4i, B1i, B2
011	TR011	TR005	Kocaçay delta	Kocaçay Deltası	Bursa	4,200	A1, A4i, A4iii, B1i, B2
012	TR012	—	Foça islands	Foça Adaları	İzmir	134	B1i
013	TR013	—	Murat Dağı	Murat Dağı	Kütahya, Uşak	52,500	B2
014	TR014	—	Demirköprü reservoir	Demirköprü Barajı	Manisa	5,060	A4i, B1i
015	TR015	TR041	Marmara lake	Marmara Gölü	Manisa	6,800	A1, A4i, A4iii, B1i, B2
016	TR016	TR036	Gediz delta	Gediz Deltası	İzmir	20,400	A1, A4i, A4iii, B1i, B2, B3

Table 1 ... continued. Summary of Important Bird Areas in Turkey. 97 IBAs covering 29,978 km²

IBA code	National code	1989 code	International name	National name	Administrative region	Area (ha)	Criteria (see p. 11)
017	TR017	—	Ildır Strait islands	Ildır Körfezi Adaları	İzmir	957	B1i
018	TR018	TR040	Küçük Menderes delta	Küçük Menderes Deltası	İzmir	1,500	A1
019	TR019	—	Akdağ	Akdağ	Afyon, Denizli	42,394	B2
020	TR020	TR038	Işıklı lake	Işıklı Gölü	Denizli	7,300	A1, A4i, A4iii, B1i, B2
021	TR021	TR034	Bafa lake	Bafa Gölü	Aydın, Muğla	12,281	A1, A4i, A4iii, B1i, B2
022	TR022	TR035	Büyük Menderes delta	Büyük Menderes Deltası	Aydin	9,800	A1, A4i, A4iii, B1i, B2
023	TR023	TR037	Güllük delta	Güllük Deltası	Muğla	1,400	A1, A4i, B1i
024	TR024	TR055	Köyceğiz lake	Köyceğiz Gölü	Muğla	8,000	A1, A4i, A4iii, B1i
025	TR025	TR043	Acıgöl lake	Acıgöl	Afyon, Denizli	21,000	A4i, B1i, B2, B3
026	TR026	TR049	Çorak lake	Çorak Gölü	Burdur	1,150	A1, A4i, A4iii, B1i
027	TR027	TR056	Salda lake	Salda Gölü	Burdur	4,370	A1, B1i
028	TR028	TR053	Karataş lake	Karataş Gölü	Burdur	1,190	A1, A4i, B1i
029	TR029	TR057	Yarışlı lake	Yarışlı Gölü	Burdur	1,400	A1, A4i, B1i
030	TR030	TR047	Burdur lake	Burdur Gölü	Burdur, Isparta	23,700	A1, A4i, A4iii, B1i, B2
031	TR031	TR051	Eğirdir lake	Eğirdir Gölü	Isparta	47,250	A1, A4i, A4iii, B1i
032	TR032	TR039	Karamık marshes	Karamık Sazlığı	Afyon	4,500	B2
033	TR033	TR019, TR022	Akşehir and Eber lakes	Akşehir and Eber Gölü	Afyon, Konya	53,600	A1, A4i, B1i, B2
034	TR034	TR021	Çavuşçu lake	Çavuşçu Gölü	Konya	1,200	A4i, A4iii, B1i, B2
035	TR035	—	Altıntaş plain	Altıntaş Ovası	Kütahya	19,200	A1, B2
036	TR036	—	Türkmenbaba Dağı	Türkmenbaba Dağı	Eskişehir, Kütahya	17,500	A1, B1iii, B2
037	TR037	—	Aliken	Aliken	Eskişehir	11,500	A1, B2
038	TR038	—	Balıkdamı	Balıkdamı	Eskişehir	1,470	A1, A4i, A4iii, B1i, B3
039	TR039	—	Uyuz lake	Uyuz Gölü	Konya	15	A1, B2
040	TR040	—	Çöl lake	Çöl Gölü	Ankara	4,700	A1, A4i, A4iii, B1i, B2
041	TR041	TR028	Mogan lake	Mogan Gölü	Ankara	1,500	A1, A4i, A4iii, B1i, B2
042	TR042	TR014	Kızılcahamam forests	Kızılcahamam Ormanları	Ankara	9,500	A1, B2
043	TR043	—	Kavaklı Dağı	Kavaklı Dağı	Ankara, Bolu	8,100	A1, B2
044	TR044	—	Inözü gorge	Inözü Vadisi	Ankara	50	B2
045	TR045	—	Sarıyar reservoir	Sarıyar Barajı	Ankara, Eskişehir	8,400	A4i, B1i, B2
046	TR046	—	Hamam Dağı	Hamam Dağı	Eskişehir	14,000	A1, B2
047	TR047	—	Sapanca lake	Sapanca Gölü	Kocaeli, Sakarya	4,700	A4i, A4iii, B1i
048	TR048	TR012	Ilgaz mountains	Ilgaz Dağları	Kastamonu, Çankırı	35,000	B2
049	TR049	TR079	Sarıkum lake	Sarıkum Gölü	Sinop	785	A1, A4iii
050	TR050	TR031	Tödürge lake	Tödürge Gölü	Sivas	750	B2
051	TR051	TR033	Palas lake	Palas Gölü	Kayseri	2,720	A4i, B1i, B2
052	TR052	TR030	Sultansazlığı	Sultansazlığı	Kayseri	39,000	A1, A4i, A4iii, B1i, B2, B3
053	TR053	TR029	Seyfe lake	Seyfe Gölü	Kirşehir	14,000	A1, A4i, A4iii, B1i, B2, B3
054	TR054	—	Hirfanlı reservoir	Hirfanlı Barajı	Aksaray, Ankara, Kirşehir	26,300	A1, A4i, A4iii, B1i
055	TR055	—	Beynam forest	Beynam Ormanı	Ankara	2,100	A1, B2
056	TR056	TR015	Kızılırmak delta	Kızılırmak Deltası	Samsun	16,110	A1, A4i, A4ii, A4iii, B1i, B1ii, B2
057	TR057	—	Yedikır reservoir	Yedikır Barajı	Amasya	593	A4i, B1i
058	TR058	TR018	Yeşilırmak delta	Yeşilırmak Deltası	Samsun	3,000	B1i, B1ii, B2
059	TR059	TR011	Akkuş island	Akkuş Adası	Ordu	2	B1i
060	TR060	TR013, TR016	North-east Turkey	Doğu Karadeniz Dağları	Artvin, Erzurum, Trabzon, Gümüşhane, Bayburt, Rize	1,230,000	A1, A3, A4iv, B1iii, B1iv, B2
061	TR061	—	Samsam lake	Samsam Gölü	Konya	830	B1i
062	TR062	—	Kozanlı Gökgöl	Kozanlı Gökgöl	Konya	650	A1, B2
063	TR063	TR027	Kulu lake	Kulu Gölü	Konya	860	A1, A4i, B1i, B2, B3
064	TR064	TR023	Ereğli marshes	Ereğli Sazlığı	Karaman, Konya	37,000	A1, A4i, B1i, B2, B3
065	TR065	TR025	Karapınar plain	Karapınar Ovası	Konya	15,200	A4i, B1i, B2
066	TR066	—	Eşmekaya marshes	Eşmekaya Sazlığı	Aksaray	11,250	A1, A4i, A4iii, B1i, B2, B3
067	TR067	TR046	Beyşehir lake	Beyşehir Gölü	Isparta, Konya	73,000	A4i, A4iii, B1i, B2
068	TR068	—	Suğla lake	Suğla Gölü	Konya	16,500	A4iii
069	TR069	TR024	Hotamış marshes	Hotamış Sazlığı	Konya	16,500	A1, A4i, B1i, B2
070	TR070	TR020	Bolluk lake	Bolluk Gölü	Konya	1,150	A4i, B1i, B2, B3
071	TR071	TR032	Tersakan lake	Tersakan Gölü	Konya	6,400	A4i, A4iii, B1i, B2, B3
072	TR072	TR032	Tuz lake	Tuz Gölü	Aksaray, Ankara, Konya	260,000	A1, A4i, B1i, B1iii, B2, B3
073	TR073	TR052	Göksu delta	Göksu Deltası	İçel	14,480	A1, A4i, A4iii, A4iv, B1i, B1iv, B2
074	TR074	TR048	Aydıncık islands	Aydıncık Adaları	İçel	2	A1, B2
075	TR075	TR045	Aladağlar	Aladağlar	Adana, Kayseri, Niğde	54,514	B2
076	TR076	TR050	Tuzla lake	Tuzla Gölü	Adana	2,800	A1, A4i, A4iii, B1i, B2
077	TR077	TR050	Akyatan lake	Akyatan Gölü	Adana	14,000	A1, A4i, A4iii, B1i, B2
078	TR078	TR050	Ağyatan lake	Ağyatan Gölü	Adana	2,200	A1, A4i, A4iii, B1i, B1iv, B2
079	TR079	TR050	Yumurtalık lagoons	Yumurtalık Lagünleri	Adana	16,430	A4i, A4iii, B1i, B2

Table 1 ... continued. Summary of Important Bird Areas in Turkey. 97 IBAs covering 29,978 km²

IBA code	National code	1989 code	International name	National name	Administrative region	Area (ha)	Criteria (see p. 11)
080	TR080	—	Gavur lake	Gavur Gölü	Kahramanmaraş	1,500	A1, A4iii
081	TR081	TR059	Nur mountains	Nur Dağları	Hatay, Osmaniye	21,500	A4iv, B1iv
082	TR082	—	Erzurum plain	Erzurum Ovası	Erzurum	3,300	A4i, B1i, B2
083	TR083	—	Sarısu plain	Sarısu Ovası	Ağrı	4,800	B2
084	TR084	—	Haçlı lake	Haçlı Gölü	Muş	2,500	A4i, B1i, B2
085	TR085	TR073	Bulanık plain	Bulanık Ovası	Muş	8,000	A1, A4i, B1i, B2
086	TR086	—	Ardahan forest	Ardahan Ormanı	Ardahan	2,500	B3
087	TR087	—	Aktaş lake	Aktaş Gölü	Ardahan	1,400	A1, A4i, B1i, B2
088	TR088	—	Çıldır lake	Çıldır Gölü	Ardahan	14,000	A4i, B1i
089	TR089	—	Kuyucuk lake	Kuyucuk Gölü	Kars	219	A4i, B1i, B2
090	TR090	—	Çalı lake	Çalı Gölü	Kars	25	A1, B2
091	TR091	TR066	Balık lake	Balık Gölü	Ağrı	3,400	B1i
092	TR092	TR075	Doğubeyazıt marshes	Doğubeyazıt Sazlığı	Ağrı	8,750	B3
093	TR093	TR074	Nemrut lake	Nemrut Gölü	Bitlis	4,500	B1i, B2
094	TR094	TR076	Sodalıgöl	Sodalı Göl	Bitlis	1,500	A1, A4i, B1i, B2
095	TR095	TR071	Erçek lake	Erçek Gölü	Van	9,520	A4i, B1i, B2, B3
096	TR096	TR065, TR067, TR069, TR070, TR072, TR077	Lake Van	Van Gölü	Bitlis, Van	390,000	A1, A4i, B1i, B2
097	TR097	TR078	Yüksekova	Yüksekova	Hakkari	24,900	B3

Sites identified in the previous inventory of IBAs in Europe (Grimmett and Jones 1989) but no longer considered to be IBAs
TR007 Saros Körfezi; TR009 Terkos Gölü; TR017 Yeniçağa Gölü; TR026 Krater Gölü and Meketuzlası; TR042 Samsun Dağı; TR044 Aksaz Adası; TR054 Kovada Gölü; TR058 Yılanlı Ada; TR060 Ceylanpınar; TR061 Fırat (Euphrates) at Birecik; TR062 Fırat (Euphrates) at Kargamış; TR063 Fırat (Euphrates) at Rumkale; TR064 Ağrı Ovası; TR068 Çaldıran Sazlığı.

Map 1. Location, area and criteria category of Important Bird Areas in Turkey.

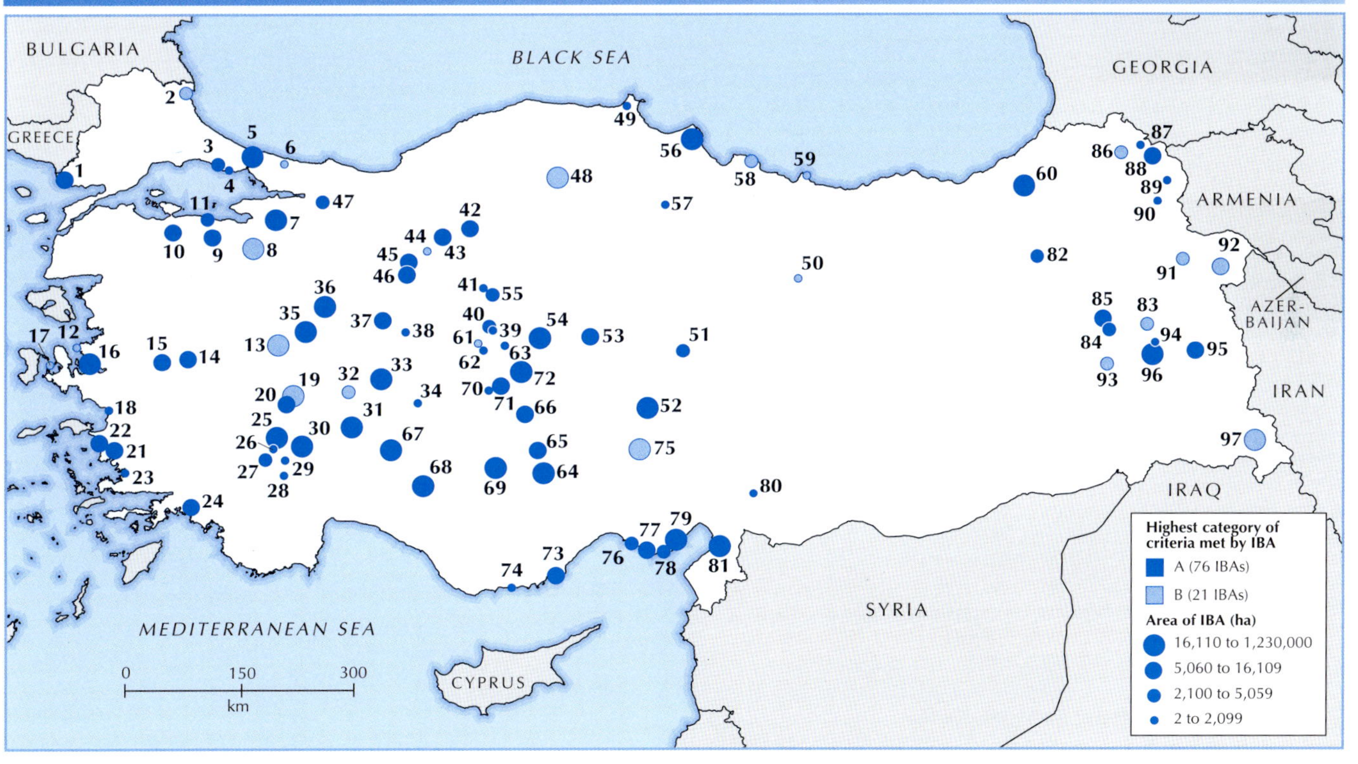

ORNITHOLOGICAL IMPORTANCE

A total of 164 species of European conservation concern (SPECs) regularly breed in Turkey, including nine species of global conservation concern: *Pelecanus crispus*, *Marmaronetta angustirostris*, *Aythya nyroca*, *Oxyura leucocephala*, *Aquila heliaca*, *Falco naumanni*, *Crex crex*, *Otis tarda* and *Larus audouinii*.

Fifty-six IBAs support significant numbers of one or more globally threatened or near-threatened species and therefore meet the A1 criterion (Table 2). Sixty-five sites regularly hold at least 1% of biogeographic or flyway populations of those waterbird species listed in Table 3, and thus qualify under criteria A4 and/or B1. Thirty-four sites meet the A4iii criterion (at least 20,000 waterbirds) (Table 1). *Pelecanus onocrotalus*, *P. crispus*, *Marmaronetta angustirostris* and *Oxyura leucocephala* breed in small numbers in wetlands across Turkey, with up to 70% of the known world population of the latter wintering at Burdur lake (030). *Phoenicopterus ruber* breeds at three Turkish wetlands, with the main colony at Tuz lake (072) holding up to 14,000 pairs.

Turkey is especially important in a European context for its steppic (and semi-steppic) avifauna (Table 4). *Otis tarda* breeds at several IBAs in Central and Eastern Anatolia and a new *Tetrax tetrax* breeding site has been discovered south of Tuz lake (072). *Charadrius leschenaultii* is a common breeder on salt-steppe in the vicinity of wetlands, whilst wetland-breeders such as *Grus grus* and *Circus pygargus* depend on surrounding steppe for feeding. Other

Table 2. Important Bird Areas in Turkey that are important for species of global conservation concern (meeting criterion A1).

Species	IBA code
Phalacrocorax pygmeus Pygmy Cormorant	001, 004, 007, 009, 010, 011, 015, 016, 018, 020, 021, 022, 023, 024, 031, 033, 052, 056, 064, 069, 073, 077, 080, 085
Pelecanus crispus Dalmatian Pelican	001, 009, 010, 015, 016, 021, 022, 064, 073, 087
Branta ruficollis Red-breasted Goose	003
Marmaronetta angustirostris Marbled Duck	052, 064, 069, 073, 076, 077, 096
Aythya nyroca Ferruginous Duck	009, 011, 015, 020, 026, 027, 041, 052, 056, 069, 073
Oxyura leucocephala White-headed Duck	010, 015, 026, 027, 028, 029, 030, 039, 040, 049, 052, 054, 056, 062, 063, 064, 069, 077, 078, 090, 094
Aegypius monachus Cinereous Vulture	036, 042, 043, 046, 060
Aquila clanga Greater Spotted Eagle	073
Aquila heliaca Imperial Eagle	055, 073
Falco naumanni Lesser Kestrel	016, 038, 040, 053, 062, 064, 066, 072
Tetrao mlokosiewiczi Caucasian Black Grouse	060
Tetrax tetrax Little Bustard	072
Otis tarda Great Bustard	035, 037, 072, 085, 096
Glareola nordmanni Black-winged Pratincole	085
Larus audouinii Audouin's Gull	074

Table 3. Important Bird Areas in Turkey that support important numbers of one or more congregatory species (i.e. meeting criteria A4 and/or B1). IBAs meeting both criteria A4 and B1 for the species are shown in **bold**. IBAs meeting only criterion B1 for the species concerned, and not A4, are shown in normal type. For key to 'Season', see p. 7.

Species	Season	IBA code
Tachybaptus ruficollis Little Grebe	W	**021**
Podiceps cristatus Great Crested Grebe	W	**004**, **021**
Podiceps nigricollis Black-necked Grebe	W	**021**
	P	**030**, **063**, **095**
Puffinus yelkouan Yelkouan Shearwater	W	**056**, 058
	P	**005**
Phalacrocorax carbo Cormorant	B	**010**
	W	**001**, **004**, 016, 021, 022, 024, 057
Phalacrocorax aristotelis Shag	B	006, 012, 017, 059
	W	012
Phalacrocorax pygmeus Pygmy Cormorant	B	**009**, **010**, **052**, **064**
	W	**001**, **009**, **016**, **022**, **052**
	P	**023**
Pelecanus onocrotalus White Pelican	P	**010**, **011**, **033**, **064**
Pelecanus crispus Dalmatian Pelican	B	**010**, **016**, **022**, **087**
	W	**001**, **010**, **015**, **016**, **021**, **022**, **073**
	N	**009**, **015**
	U	**064**
Ardeola ralloides Squacco Heron	B	**001**, 009, 010, **015**
Egretta garzetta Little Egret	B	**001**, **056**
Egretta alba Great White Egret	W	**001**, **015**, **022**, **024**, **056**, **077**, **078**
Ciconia nigra Black Stork	P	**073**
Ciconia ciconia White Stork	P	**038**, **045**, **066**, **073**, **078**
Plegadis falcinellus Glossy Ibis	B	**001**
	P	**033**, **056**, **073**
Platalea leucorodia Spoonbill	B	001, **009**, **010**, 033, **053**, **056**, 064, **070**, 084
	P	**052**, **070**
	N	**064**
Phoenicopterus ruber Greater Flamingo	B	**016**, **053**, **064**, **072**
	W	**016**, **022**, **077**
	P	**030**, **052**
	N	**053**, **071**
	U	**040**
Cygnus olor Mute Swan	W	**001**
Cygnus cygnus Whooper Swan	W	**001**
Anser albifrons White-fronted Goose	W	020, **033**, **040**, 053, 063, **065**, **066**, **071**, 072
Anser anser Greylag Goose	W	056, 073, 079
	P	052
	N	034
Tadorna ferruginea Ruddy Shelduck	B	**040**, **070**, **095**
	W	**014**, **015**, **016**, **025**, **028**, **029**, **030**, **038**, **040**, **045**, **052**, **053**, **057**, **064**, **065**
	P	**030**, **054**, **057**, **072**, **082**
	N	**022**, **052**, **063**, **084**, **087**, **089**
	U	**051**, **053**, **088**
Tadorna tadorna Shelduck	W	077
Anas penelope Wigeon	W	015, 022, 073, 077, 078, **079**
Anas strepera Gadwall	W	021
Anas crecca Teal	W	001, 014, 016, **026**, 040, **052**, **073**, 076
Anas clypeata Shoveler	W	**056**, 073
Marmaronetta angustirostris Marbled Duck	B	052, **064**, **069**, **073**, **076**, 077, 096
	P	**052**
	N	064
Netta rufina Red-crested Pochard	B	**052**, **064**
	W	**020**, 024, **030**, **031**, **034**, 041, **047**, **054**, **056**, 058, 067, 069
Aythya ferina Pochard	W	**001**, **003**, **009**, **015**, **020**, **021**, **026**, **030**, **031**, 047, 054, **056**, **067**, **077**
Aythya fuligula Tufted Duck	W	009, 030, 054
	U	031
Melanitta fusca Velvet Scoter	B	091, 093
	W	056, 058
	U	087
Oxyura leucocephala White-headed Duck	B	**064**, 069
	W	015, **026**, 027, **030**, 054, **056**, **063**, **077**, **078**
	P	**069**
	N	**063**, **064**, **094**
Aegypius monachus Cinereous Vulture	B	036, 060
Falco naumanni Lesser Kestrel	B	072
Fulica atra Coot	W	001, **009**, **011**, **020**, **021**, 022, 024, **030**, **031**, 047, **054**, **056**, **067**, 073, **077**, 078
	P	**030**
	U	**041**
Grus grus Crane	W	064
	P	052
Himantopus himantopus Black-winged Stilt	B	**001**, **033**, **040**, **052**, **063**, **064**, **069**, **094**
	U	077
Recurvirostra avosetta Avocet	B	052, **053**, 063, 072
	W	**015**, 022, **052**, 079
	P	061, **077**
	N	063
Glareola pratincola Collared Pratincole	B	**001**, **011**, 016, 021, **022**, 052, **056**, 064, **069**, **072**, **073**, **076**
	P	066, 078
Charadrius alexandrinus Kentish Plover	B	**016**, 022, 052, **072**, 073, **076**
	W	016, 076, 079
	P	**077**
	N	**025**
	U	063
Hoplopterus spinosus Spur-winged Plover	B	**016**, **069**, **073**, **076**
Calidris minuta Little Stint	W	**079**
	P	**077**
Larus melanocephalus Mediterranean Gull	P	**003**
Larus minutus Little Gull	P	**001**, **003**, **056**
Larus ridibundus Black-headed Gull	W	009, 016
Larus genei Slender-billed Gull	B	**070**
	W	**022**
	N	**071**
Larus cachinnans Yellow-legged Gull	W	**003**, 004, **016**
Larus armenicus Armenian Gull	B	**072**, **087**, **088**, **094**, **096**
Gelochelidon nilotica Gull-billed Tern	B	052, **053**, **063**, 066, **070**, **085**
Sterna caspia Caspian Tern	B	**016**
	N	022
Sterna albifrons Little Tern	B	001, 016, **022**, **053**, **076**
Chlidonias hybridus Whiskered Tern	B	**001**, **009**, **052**
Chlidonias niger Black Tern	P	**030**
Chlidonias leucopterus White-winged Black Tern	P	**056**, **082**, **096**

steppic species occurring in Turkey include *Buteo rufinus*, *Aquila heliaca*, *Falco naumanni*, *F. biarmicus* and *F. cherrug*. Site 060 holds significant numbers of two species whose distributions are largely or wholly confined to the Eurasian high-montane (alpine) biome, and therefore meets the A3 criterion.

Turkey lies on two major migratory flyways, with six IBAs qualifying under the A4iv/B1iv criteria due to the vast numbers of *Ciconia ciconia* and raptors that either pass or stop-over (Table 1).

Passerines are not adequately covered by the Turkish IBA inventory due to a lack of data regarding the distribution and abundance of many species.

Table 4. Species of European conservation concern with significant breeding populations at Important Bird Areas in Turkey (meeting any IBA criteria).

Species [1]	Minimum national breeding population (pairs) [2]	Proportion (%) of national population breeding at all IBAs in Turkey
Phalacrocorax aristotelis Shag	810	50
Phalacrocorax pygmeus Pygmy Cormorant	1,000	100[3]
Pelecanus onocrotalus White Pelican	250	69
Pelecanus crispus Dalmatian Pelican	150	93
Ixobrychus minutus Little Bittern	1,000	18
Nycticorax nycticorax Night Heron	1,000	88
Ardeola ralloides Squacco Heron	3,000	35
Ardea purpurea Purple Heron	2,000	35
Ciconia nigra Black Stork	500	20
Plegadis falcinellus Glossy Ibis	500	70
Platalea leucorodia Spoonbill	500	100[3]
Phoenicopterus ruber Greater Flamingo	14,000	100[3]
Tadorna ferruginea Ruddy Shelduck	4,000	65
Anas strepera Gadwall	500	52
Marmaronetta angustirostris Marbled Duck	150	82
Netta rufina Red-crested Pochard	1,000	100[3]
Aythya nyroca Ferruginous Duck	500	71
Oxyura leucocephala White-headed Duck	200	100[3]
Gypaetus barbatus Lammergeier	50	70
Neophron percnopterus Egyptian Vulture	1,000	3
Gyps fulvus Griffon Vulture	100	36
Aegypius monachus Cinereous Vulture	100	41
Circus pygargus Montagu's Harrier	200	34
Aquila heliaca Imperial Eagle	10	20
Aquila chrysaetos Golden Eagle	1,000	2
Hieraaetus pennatus Booted Eagle	100	18
Falco naumanni Lesser Kestrel	1,500	12
Falco biarmicus Lanner	10	20
Tetrao mlokosiewiczi Caucasian Black Grouse	1,000	13
Tetraogallus caspius Caspian Snowcock	200	1
Francolinus francolinus Black Francolin	300	48
Porphyrio porphyrio Purple Gallinule	350	86
Grus grus Crane	100	100[3]
Tetrax tetrax Little Bustard	100	20
Otis tarda Great Bustard	800 individuals	30
Recurvirostra avosetta Avocet	1,500	92
Burhinus oedicnemus Stone Curlew	1,000	16
Glareola pratincola Collared Pratincole	2,000	59
Charadrius alexandrinus Kentish Plover	6,000	48
Charadrius leschenaultii Greater Sand Plover	500	20
Hoplopterus spinosus Spur-winged Plover	1,050	52
Larus audouinii Audouin's Gull	20	100[3]
Gelochelidon nilotica Gull-billed Tern	2,200	81
Sterna caspia Caspian Tern	150	72
Sterna albifrons Little Tern	5,000	38
Chlidonias hybridus Whiskered Tern	1,000	100[3]

1. Only those species of European conservation concern (see Box 1, p. 12) that meet IBA criteria in Turkey are listed.
2. Data are taken from the BirdLife/EBCC European Bird Database 1998 (Heath and Borggreve 2000).
3. The percentage of the national population in IBAs exceeds 100%. Usually this is because the national population estimate has not been updated recently whilst the IBA population estimate has been recently updated with new data as a result of comprehensive surveys of IBAs themselves. Also, the individual site count for a species may be the maximum or average over recent years, and summing these may record more birds than are present nationally in any single year.

HABITATS

Turkey lies in several climatic zones, ranging from Mediterranean influences (hot and dry summers, mild and wet winters) to more continental conditions (Eastern Anatolia for example, where temperatures reach 40°C during summer and -30°C during winter). The varied climatic and topographical conditions mean that a great diversity of habitats are present.

About 12% of the country's land area is covered by large stretches of intact forest, mainly in the major mountain ranges in the north and south. Large parts of Anatolia are characterized by a mixture of arable and fallow land, rocky areas with scattered *Quercus* or *Juniperus* scrub and (normally overgrazed) grassland and steppe. Original salt-steppe is found around salt-lakes in Central Anatolia, including Tuz lake (072), whilst arid, montane and subalpine steppe prevail in Eastern Anatolia. Alpine habitats are confined to the higher mountain ranges in the eastern Black Sea, Taurus and Eastern Anatolia regions.

Although wetlands are well covered by Turkish IBAs (they comprise at least 50% of the area of 68 sites), coverage is less complete for forest, mountain, steppe and Mediterranean habitats (Figure 1). This is partly explained by the preferences of visiting birdwatchers and hence the availability of data, and means that passerines have generally been under-recorded. The importance for breeding birds of certain widespread artificial habitats (for example, arable land and improved grassland), as demonstrated in Tucker and Evans (1997), is also under-stated by the Turkish IBA list. In terms of setting conservation priorities, however, the high representation of wetlands mirrors the fact that this is the fastest disappearing habitat in Turkey (Magnin and Yarar 1997).

Figure 1. Habitats at Important Bird Areas in Turkey (see Appendix 3 for definitions of habitats).

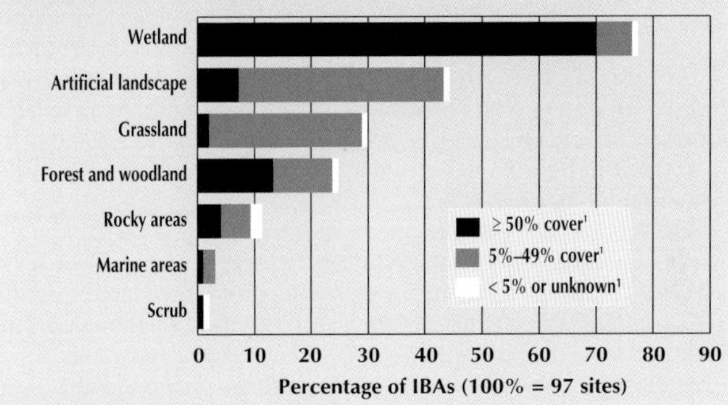

1. percentage of area of individual IBA covered by habitat

Figure 2. Land-uses at Important Bird Areas in Turkey (see Appendix 3 for definitions of land-uses).

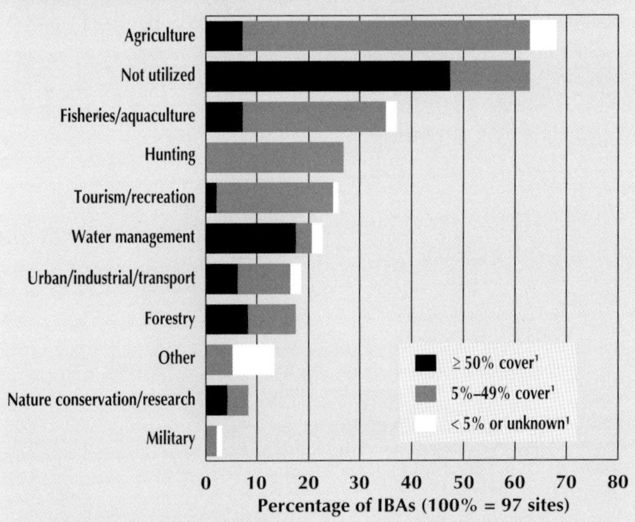

1. percentage of area of individual IBA covered by land-use

Figure 3. Threats at Important Bird Areas in Turkey (see Appendix 3 for definitions of threat types and impact categories).

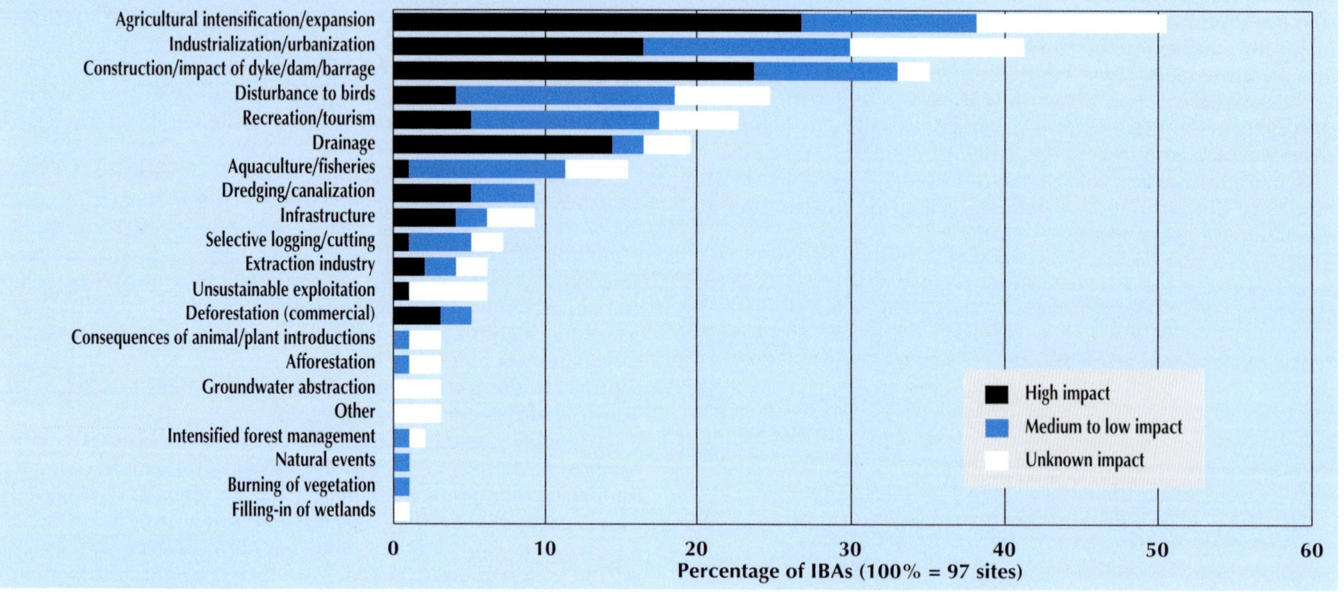

IMPACTS ON IBAs – LAND-USE AND THREATS

The main land-use type at Turkish IBAs is agriculture (Figure 2), reflecting the fact that around half of the country's land area is farmed. Agricultural activities range from highly specialized and labour-intensive crop production (for example, cotton) to non-intensive stock-grazing of alpine grasslands.

Agricultural intensification/expansion is a major threat to IBAs in Turkey, affecting 49 sites (51%) (Figure 3). Since the 1950s the land area under cultivation has more than doubled, with the expansion coming partly at the expense of available grazing land. This has led to an increase in grazing pressure and hence problems of overgrazing and erosion. Wetlands have been reclaimed or drained for conversion to arable land and many others have been adversely affected by water diversion schemes. At least 1,300,000 ha of wetland habitat have been lost this century, mainly since 1960 (Magnin and Yarar 1997).

The Konya Basin of Central Anatolia provides a good example of the consequences of large-scale irrigation projects. Two large wetlands totalling 30,000 ha have dried-up, and three large wetland IBAs totalling 106,000 ha (067, 068 and 069) have sustained serious damage. A further three IBAs (070, 071 and 072) now risk long-term damage as a result of pollution from untreated drainage water. To compensate for recent water shortages, the Turkish State Water Works plans to bring in water from the Göksu catchment by diverting the Göksu river, thus placing the Göksu delta (073) at risk.

Fisheries operate at many wetland IBAs (36 in total – Figure 2), although in recent years many have declined as a result of pollution, over-fishing and the introduction of non-native species (Figure 3).

The reduced economic importance of fisheries has provided the authorities with the opportunity to convert wetlands into sources of water for irrigation or industrial use, usually with negative consequences for birds. A number of lakes and reservoirs situated within IBAs are used to store drinking or irrigation water (classified under 'water management' and 'agriculture' respectively), and several are of importance to the wood-product industry (Figure 2).

Hunting takes place across Turkey, including at 26 IBAs (27%), the majority of which are wetlands. Reductions in the area of wetland as a result of agricultural intensification have meant that wetland birds are increasingly susceptible to both legal and illegal hunting (Figure 3).

Large-scale clear-felling and other unsympathetic forestry practices threaten those forest habitats that remain (Figure 3). The felling of trees used by raptors as nesting sites is a particular problem. Coastal IBAs and those situated near to major cities are under increasing pressure from tourism (Figure 3).

PROTECTION STATUS

Table 5 and Figures 4 and 5 summarize the national and international protection status of all Turkish IBAs.

National protection

1. National Park, Nature Reserve and Nature Park
Nature Reserves are strictly protected; limited recreational use is permitted in the other two. Responsible institution: Ministry of Forestry.

Figure 4. The national protection status of Important Bird Areas in Turkey.

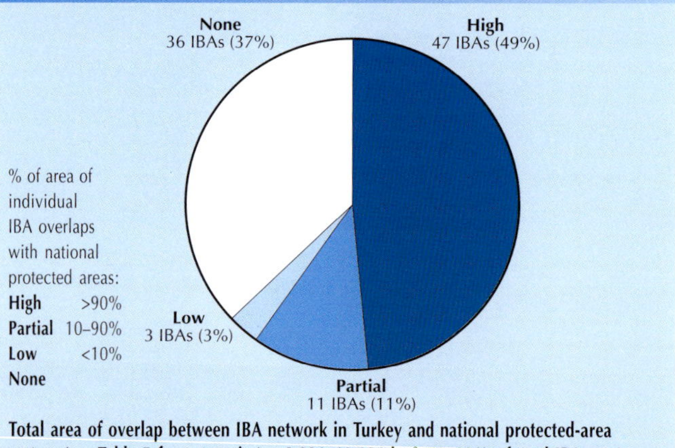

Total area of overlap between IBA network in Turkey and national protected-area system (see Table 5 for categories) = 8,996–10,026 km² (30–33% of total IBA area).

Figure 5. The international protection status of Important Bird Areas in Turkey.

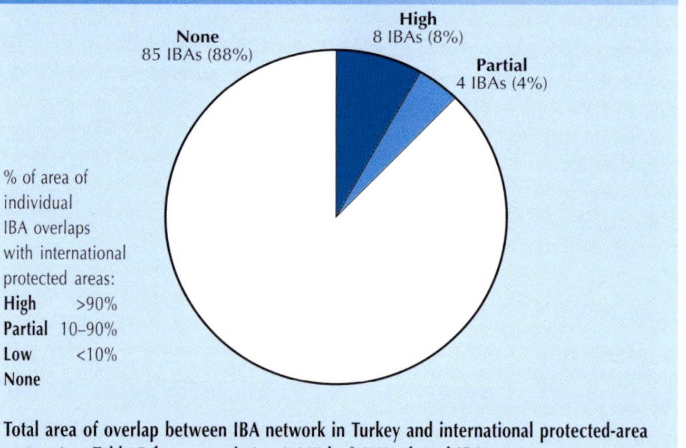

Total area of overlap between IBA network in Turkey and international protected-area system (see Table 5 for categories) = 1,447 km² (5% of total IBA area).

Table 5. Protection status of Important Bird Areas in Turkey. A tick (✓) indicates that an IBA overlaps with a protected area (to any extent).

IBA code	International name	National: National Park	Nature Reserve	Nature Park	Permanent Wildlife Reserve	Drinking Water Reservoir	Preservation Forest	SİT	International: Ramsar Site	Mediterranean Specially Protected Area
001	Meriç delta		✓					✓		
002	Iğneada forests		✓		✓			✓		
003	Büyükçekmece lake					✓				
004	Küçükçekmece lake									
005	Bosphorus			✓	✓		✓	✓		
006	Şile islands							✓		
007	Iznik lake							✓		
008	Uludağ	✓						✓		
009	Uluabat lake								✓	
010	Kus lake	✓			✓			✓	✓	
011	Kocaçay delta									
012	Foça islands							✓		✓
013	Murat Dağı									
014	Demirköprü reservoir									
015	Marmara lake									
016	Gediz delta				✓			✓	✓	
017	Ildır Strait islands				✓			✓		
018	Küçük Menderes delta				✓			✓		
019	Akdağ				✓					
020	Işıklı lake									
021	Bafa lake			✓				✓		
022	Büyük Menderes delta	✓						✓		
023	Güllük delta									
024	Köyceğiz lake									✓
025	Acıgöl lake				✓					
026	Çorak lake									
027	Salda lake							✓		
028	Karataş lake				✓					
029	Yarışlı lake									
030	Burdur lake				✓				✓	
031	Eğirdir lake					✓				
032	Karamık marshes							✓		
033	Akşehir and Eber lakes							✓		
034	Çavuşçu lake							✓		
035	Altıntaş plain				✓					
036	Türkmenbaba Dağı				✓			✓		
037	Aliken									
038	Balıkdamı				✓			✓		
039	Uyuz lake							✓		
040	Çöl lake									
041	Mogan lake									✓
042	Kızılcahamam forests	✓								
043	Kavaklı Dağı				✓					
044	Inözü gorge									
045	Sarıyar reservoir				✓					
046	Hamam Dağı									
047	Sapanca lake					✓		✓		
048	Ilgaz mountains	✓								
049	Sarıkum lake		✓					✓		
Subtotal of IBAs		5	3	2	15	3	1	22	4	3

IBA code	International name	National: National Park	Nature Reserve	Nature Park	Permanent Wildlife Reserve	Drinking Water Reservoir	Preservation Forest	SİT	International: Ramsar Site	Mediterranean Specially Protected Area
050	Tödürge lake									
051	Palas lake							✓		
052	Sultansazlığı		✓		✓			✓	✓	
053	Seyfe lake		✓					✓	✓	
054	Hirfanlı reservoir									
055	Beynam forest						✓	✓		
056	Kızılırmak delta				✓			✓	✓	
057	Yedikır reservoir				✓			✓		
058	Yeşilırmak delta				✓					
059	Akkuş island									
060	North-east Turkey	✓	✓	✓	✓					
061	Samsam lake							✓		
062	Kozanlı Gökgöl							✓		
063	Kulu lake							✓		
064	Ereğli marshes		✓					✓		
065	Karapınar plain							✓		
066	Eşmekaya marshes				✓			✓		
067	Beyşehir lake	✓				✓		✓		
068	Suğla lake									
069	Hotamış marshes							✓		
070	Bolluk lake							✓		
071	Tersakan lake							✓		
072	Tuz lake							✓		
073	Göksu delta				✓			✓	✓	✓
074	Aydıncık islands									
075	Aladağlar	✓			✓					
076	Tuzla lake				✓					
077	Akyatan lake				✓				✓	
078	Ağyatan lake									
079	Yumurtalık lagoons		✓					✓		
080	Gavur lake									
081	Nur mountains				✓					
082	Erzurum plain									
083	Sarısu plain									
084	Haçlı lake									
085	Bulanık plain									
086	Ardahan forest									
087	Aktaş lake									
088	Çıldır lake									
089	Kuyucuk lake				✓					
090	Çalı lake									
091	Balık lake					✓				
092	Doğubeyazıt marshes									
093	Nemrut lake									
094	Sodalıgöl									
095	Erçek lake									
096	Lake Van							✓		
097	Yüksekova									
Total number of IBAs		8	8	3	27	5	2	42	9	4

2. Permanent Wildlife Reserve
Often declared in wetlands to protect waterbirds. Responsible institution: Ministry of Forestry.

3. Drinking Water Reservoir
This designation is based on the Water Pollution Control Regulation and allows human activities to be controlled in the vicinity of reservoirs. Responsible institution: Ministry of the Environment.

4. Preservation Forest
Forests excluded from timber production. Responsible institution: Ministry of Forestry.

5. SİT
Three categories allow for varying degrees of development. Can be quickly granted to any site by regional committees, allowing for rapid legal action against developments. Responsible institution: Ministry of Culture.

In total 61 IBAs are fully or partially protected at the national level, leaving 37 unprotected (Table 5, Figure 4). Enforcement of these various national designations is, however, weak (Magnin and Yarar 1997). SİTs can easily be cancelled (for example, IBA 005), and the designation of National Parks, Nature Reserves, Nature Parks and Permanent Wildlife Reserves has often failed to improve matters *in situ* (for example, site 002 Iğneado forests and site 010 Kuş lake).

Box 1. International legislation and initiatives that are relevant to site conservation in Turkey (see Appendix 1 for a general description of these agreements).

Global	
Biodiversity Convention	✔
Ramsar Convention	✔
Bonn Convention	
World Heritage Convention	✔
MAB Programme	
Pan-European	
Bern Convention	✔
Regional	
Barcelona Convention	✔

✔ Convention ratified/initiative supported
(✔) Convention signed

International protection

Box 1 details the international conventions and initiatives that Turkey is party to. The Ministry of the Environment is responsible for administering Ramsar Sites and Specially Protected Areas (the latter having been designated under a Protocol of the Barcelona Convention).

CONSERVATION

- The Turkish government has issued a number of laws and decrees aimed specifically at protecting nature, notably the Coastal Law (1990), the Environmental Impact Assessment Regulation (1993) and the Prime Minister's Decree on the Conservation of Wetlands (1993). The EIA Regulation has been largely ineffective, and wetland destruction has continued unabated since the Prime Minister's Wetland Decree. In addition, a large number of old legal instruments contradict many conservation regulations.
- The government institutions responsible for nature conservation are hampered by a lack of trained staff and equipment, a lack of dialogue and uncertainties regarding responsibilities at an inter-ministerial level, a lack of scientific data, a lack of a long-term conservation strategy, and insufficient funds to implement even small components of existing conservation programmes.

ANALYTICAL METHODS

- This overview was produced using information from the recent Turkish national IBA inventory (Magnin and Yarar 1997). Data on birds and sites were gathered during extensive fieldwork undertaken over the period 1989–1996, and were used in conjunction with information from visiting birdwatchers and a large number of *in situ* conservation projects and ornithological studies. In addition, an extensive literature review was undertaken to collate all available information on Turkish IBAs.
- Passerines were under-recorded during the collection of data for this review, with observer effort being directed mainly towards wetland sites. No assessment has therefore been made of the importance of Turkish IBAs for such species. This has also meant that it has not been possible to apply properly the biome-restricted assemblage criterion (A3) in Turkey for the Mediterranean, Eurasian high-montane (alpine) and Eurasian steppe biomes. It is likely that as data become available, new and/or existing Turkish IBAs will have A3 assigned to them based on the presence of a significant component of the group of species whose breeding distributions are largely or wholly confined to one or other of these biomes.
- The numbers of breeding pairs given in the species' tables represent one-off counts rather than average numbers recorded over several years.
- The numbers of wintering and passage birds given in the species' tables are maximum daily counts.

GLOSSARY

DHKD Doğal Hayati Koruma Derneği (Society for the Protection of Nature, BirdLife International Partner in Turkey).
DSİ State Water Works.
İSKİ İstanbul Water and Sewerage Administration.
MedSPA Mediterranean Specially Protected Area (under the Barcelona Convention).
SİT an area protected under the Law on Protection of Cultural and Natural Assets (1983).
WIWO Stichting Werkgroep Internationaal wad-en Watervogelonderzoek. (Foundation Working Group International Wader and Waterfowl Research).

ACKNOWLEDGEMENTS

This country overview was produced with the help of many people who also assisted with the national inventory (Magnin and Yarar 1997). We would especially like to thank all DHKD staff, Gunesin Aydemir, Burak Gunes, and Geoff and Hilary Welch. We are grateful to the Garanti Bank, Turkey, for the financial support given to the Turkish IBA project over the last six years.

SITE ACCOUNTS

Meriç delta — A1, A4i, A4iii, B1i, B2 — 001

Admin region Edirne
Coordinates 40°47′N 26°14′E
Altitude 0–0 m **Area** 7,000 ha

Site description

The delta is situated on the Turkish–Greek border where the Meriç river flows into the Aegean Sea (adjacent to Greek site 006). The IBA comprises a complex of freshwater lakes, extensive reedbeds (*Phragmites*), lagoons and 3,400 ha of former flood-plain that has largely been converted to farmland. The lakes and marshes are surrounded by irrigated rice-fields. Sheep- and cattle-grazing is widespread.

Habitats Wetland (50%; tidal river/enclosed tidal water; coastal lagoon; standing fresh water; water-fringe vegetation), Artificial landscape (50%; arable land)
Land-use Agriculture (25%), Fisheries/aquaculture (20%), Hunting (5%), Military (25%), Water management (20%)

Birds

Species	Season	Year	Pop min	Pop max	Acc	Criteria
Phalacrocorax carbo Cormorant	W	1999	—	9,300	—	A4i, B1i
Phalacrocorax pygmeus Pygmy Cormorant	W	1998	—	20,000	—	A1, A4i, B1i
Pelecanus crispus Dalmatian Pelican	W	1998	—	610	—	A1, A4i, B1i
Ixobrychus minutus Little Bittern	B	—	50	50	—	B2
Nycticorax nycticorax Night Heron	B	—	200	200	—	B2
Ardeola ralloides Squacco Heron	B	—	300	300	—	A4i, B1i, B2
Egretta garzetta Little Egret	B	1994	377	470	—	A4i, B1i
Egretta alba Great White Egret	W	1993	260	260	—	A4i, B1i
Ardea purpurea Purple Heron	B	—	100	100	—	B2
Plegadis falcinellus Glossy Ibis	B	—	100	100	—	A4i, B1i, B2
Platalea leucorodia Spoonbill	B	—	40	40	—	B1i, B2
Cygnus olor Mute Swan	W	1999	—	8,900	—	A4i, B1i
Cygnus cygnus Whooper Swan	W	1998	—	1,200	—	A4i, B1i
Anas crecca Teal	W	1999	—	12,400	—	B1i
Aythya ferina Pochard	W	1996	15,528	15,528	—	A4i, B1i
Fulica atra Coot	W	1999	—	31,300	—	B1i
Himantopus himantopus Black-winged Stilt	B	1994	350	350	—	A4i, B1i
Glareola pratincola Collared Pratincole	B	1994	200	200	—	A4i, B1i, B2
Larus minutus Little Gull	P	—	0	700	—	A4i, B1i
Sterna albifrons Little Tern	B	—	200	200	—	B1i, B2
Chlidonias hybridus Whiskered Tern	B	1993	500	500	—	A4i, B1i, B2

An important site for both breeding and wintering waterbirds. Large numbers of *Pelecanus onocrotalus* roost on the lakes during migration.

Protection status

National High **International** None
2,369 ha of IBA covered by Nature Reserve (Meriç/Evros deltası, 2,369 ha). 7,000 ha of IBA covered by SİT (Meriç Deltası, 7,000 ha).

Conservation issues

Threats Agricultural intensification/expansion (B), Aquaculture/fisheries (C), Construction/impact of dyke/dam/barrage (B), Drainage (B), Industrialization/urbanization (A)

A large-scale DSİ irrigation and flood-control initiative (the Lower İpsala Project) includes provisions to prevent pollution and further loss of wetland habitats in the area. The Meriç river is heavily polluted by industrial waste, untreated sewage and agricultural run-off. Lake water is used for irrigation during periods of drought, leading to drops in water-levels, seawater incursion and mass fish-kills. In recent years the fishing industry in the lakes and lagoons has collapsed, with pollution, over-fishing, changes in water regime, and silting-up of the river mouth quoted as the main causal factors.

İğneada forests — B1iv, B2 — 002

Admin region Kırklareli
Coordinates 41°52′N 27°57′E
Altitude 0–0 m **Area** 3,000 ha

Site description

A complex of seasonally-flooded forests, swamps, freshwater lakes and sand-dunes on the Black Sea coast near the Turkish–Bulgarian border. Surface water accumulation behind the dunes feeds the largely intact flooded forests, which are below sea-level. The c.10 km long pristine dune and beach system is of high botanical importance. Human activities include cattle- and sheep-grazing, small-scale freshwater fisheries and reed-cutting ('Other' land-use).

Habitats Forest and woodland (95%; broadleaved deciduous forest; alluvial/very wet forest), Wetland (sand-dunes/sand beach; standing fresh water; water-fringe vegetation), Marine areas (5%; sea inlet/coastal features)
Land-use Agriculture (5%), Forestry (90%), Hunting (5%), Nature conservation/research (50%), Other

Birds

Species	Season	Year	Pop min	Pop max	Acc	Criteria
Ciconia nigra Black Stork	B	—	5	5	—	B2

The site is also a migratory bottleneck, where more than 8,000 *Ciconia ciconia* regularly pass in autumn. Although no comprehensive counts have been undertaken, available data suggest that the IBA is also a bottleneck for migrating raptors.

Protection status

National High **International** None
1,345 ha of IBA covered by Nature Reserve (İğneada, 1,345 ha). 3,000 ha of IBA covered by Permanent Wildlife Reserve (İğneada, 5,399 ha). 3,000 ha of IBA covered by SİT (İğneada, 3,000 ha).

Conservation issues

Threats Agricultural intensification/expansion (B), Construction/impact of dyke/dam/barrage (B), Deforestation (commercial) (B), Extraction industry (B), Infrastructure (A), Other (U), Recreation/tourism (C)

İSKİ plans to divert water from the nearby Istranca mountains by damming the five major streams feeding the flooded forests. Additional threats include the proposed Bulgaria–Turkey coastal highway, tourism development, deforestation to allow replanting with *Populus*, illegal sand extraction and reed-cutting regimes that reduce nesting site availability ('Other' threat).

Büyükçekmece lake — A1, A4i, A4iii, B1i — 003

Admin region İstanbul
Coordinates 41°03′N 28°34′E
Altitude 6–6 m **Area** 2,850 ha

Site description

A freshwater lake situated west of İstanbul where the Karasu stream reaches the Marmara Sea. The stream was dammed to create a drinking water reservoir; during dry years less than 25% of the lake area remains flooded. Mudflats, sandbanks and reedbeds (*Phragmites*) have formed where silt deposition occurs. South of the dam is a brackish lake with reed fringes and islands. The lake is largely surrounded by arable land and, increasingly, industrial and residential development. Motorways cut across both ends of the lake.

Habitats Wetland (100%; standing fresh water; water-fringe vegetation)
Land-use Water management (100%)

Birds

Species	Season	Year	Pop min	Pop max	Acc	Criteria
Branta ruficollis Red-breasted Goose	W	1998	—	80	—	A1
Aythya ferina Pochard	W	1996	14,036	14,036	—	A4i, B1i
Larus melanocephalus Mediterranean Gull	P	1995	10,000	10,000	—	A4i, B1i
Larus minutus Little Gull	P	1991	1,740	1,740	—	A4i, B1i
Larus cachinnans Yellow-legged Gull	W	1996	1,250	8,775	—	A4i, B1i

This site is important for wintering and passage waterbirds. It also supports a range of breeding waterbirds. During autumn the arable land surrounding the lake can hold thousands of migrating *Ciconia ciconia*, which roost here before crossing the Boğaziçi (site 005). Species of global conservation concern that do not meet IBA criteria: *Aythya nyroca* (thought to breed).

Protection status

National High **International** None
2,850 ha of IBA covered by Drinking Water Reservoir (Büyükçekmece Gölü, 2,850 ha).

Conservation issues

Threats Construction/impact of dyke/dam/barrage (C), Disturbance to birds (B), Industrialization/urbanization (A)

The ecological character of the lake was drastically altered following its conversion; large areas of reed and marsh along the northern and western shores were lost. Industrial and urban developments close to the lake may cause pollution. The nearby airport and motorways cause disturbance, and the area is subject to high hunting pressure. Commercial fishing is prohibited by İSKİ.

Küçükçekmece lake — A1, A4i, A4iii, B1i — 004

Admin region İstanbul
Coordinates 41°00′N 28°45′E
Altitude 0–0 m **Area** 1,500 ha

Site description

A natural brackish lagoon (20 m deep) situated west of İstanbul on the Sea of Marmara. Fringe vegetation is confined to an area of reedbed (*Phragmites*) in the north of the lagoon. Unplanned buildings surround the site; the only relatively undisturbed area lies within the grounds of the nuclear energy research centre and military camp on the east shore. A motorway passes to the north.

Habitats Wetland (100%; standing fresh water; water-fringe vegetation)
Land-use Fisheries/aquaculture (10%), Hunting (10%), Not utilized (80%)

Birds

Species	Season	Year	Pop min	Pop max	Acc	Criteria
Podiceps cristatus Great Crested Grebe	W	1993	5,750	5,750	—	A4i, B1i
Phalacrocorax carbo Cormorant	W	1993	10,200	10,200	—	A4i, B1i
[1] ***Phalacrocorax pygmeus*** Pygmy Cormorant	P	1995	126	160	—	A1
Larus cachinnans Yellow-legged Gull	W	1993	2,685	2,685	—	B1i

1. Also overwinters.

The site holds significant numbers of wintering waterbirds.

Protection status

National None **International** None

Conservation issues

Threats Construction/impact of dyke/dam/barrage (U), Industrialization/urbanization (A)

The construction of a drinking water reservoir on the Sazlı stream has reduced freshwater inflow—the lagoon may become more saline as a result of seawater intrusion. Urban and industrial pollution is a problem that will intensify with further development of the lake surrounds.

Bosphorus — A4ii, A4iv, B1ii, B1iv — 005

Admin region İstanbul
Coordinates 41°10′N 29°05′E
Altitude 0–409 m **Area** 55,000 ha

Site description

The Boğaziçi strait runs north–south for c.33 km, and is c.1.5 km wide. The site includes the forested areas east (Polonezköy Nature Park) and west (Belgrade forest) of the strait, which are known to be of importance to roosting migrants. It does not include the Çamlıca hills and the islands in the Sea of Marmara, which would qualify as an IBA in their own right due to the large numbers of migrants passing over, but which provide limited scope for bird conservation.

Habitats Forest and woodland (30%; mixed forest), Marine areas (10%; sea inlet/coastal features), Artificial landscape (60%; other urban/industrial areas)
Land-use Agriculture (5%), Forestry (15%), Military (5%), Tourism/recreation (20%), Urban/industrial/transport (60%)

Birds

Species	Season	Year	Pop min	Pop max	Acc	Criteria
[1] ***Puffinus yelkouan*** Yelkouan Shearwater	P	—	—	—	—	A4ii, B1ii

1. Thousands pass daily, throughout the year.

The site is a major migratory bottleneck, where more than 20,000 storks and raptors regularly pass in spring and autumn. Few counts covering the entire migration period have been undertaken. Total counts made in the mid-1960s and early 1970s and covering the entire autumn migration period are as follows: *Ciconia nigra* (8,318), *Ciconia ciconia* (338,353), *Pernis apivorus* (25,751), *Milvus migrans* (2,707), *Buteo buteo* (32,895) and *Aquila pomarina* (18,898). More recent autumn records include 11,703 *Aquila pomarina* at Çamlıca and 11,379 of the same species at Sarıyer, both on the same day in September 1990. *Ciconia* sp. often roost in flocks of thousands in fields outside the city. Disturbance by *Larus cachinnans* and possibly pollution are thought to be behind a drastic decline in numbers of breeding *Milvus migrans*.

Protection status

National Partial **International** None
3,006 ha of IBA covered by Nature Park (Polonezköy, 3,006 ha). 2,338 ha of IBA covered by Permanent Wildlife Reserve (Sarıyer, 2,338 ha). 5,296 ha of IBA covered by Preservation Forest (Belgrad Ormanı, 5,296 ha). IBA also overlaps with SİT (Boğaziçi).

Conservation issues

Threats Extraction industry (A), Industrialization/urbanization (A), Infrastructure (A), Recreation/tourism (A), Unsustainable exploitation (A)

The main threat is the rapid urban and industrial expansion of İstanbul. Mining and road building also result in the loss of natural habitats. The conservation of forested areas on both sides of the strait is crucial, in order to preserve raptor roosting sites. There are isolated instances of raptor persecution.

Şile islands — B1i — 006

Admin region İstanbul
Coordinates 41°11′N 29°36′E
Altitude 0–20 m **Area** 5 ha

Site description

A complex of four small islands at the entrance to Şile harbour, a holiday resort on the Black Sea coast, east of İstanbul. The islands are sparsely vegetated with low grass and scrub vegetation on the higher parts.

Habitats Marine areas (100%; rock stacks/islets)
Land-use Not utilized (100%)

Birds

Species	Season	Year	Pop min	Pop max	Acc	Criteria
Phalacrocorax aristotelis Shag	B	1995	175	175	—	B1i

Protection status

National High **International** None
5 ha of IBA covered by SİT (Şile Adaları, 5 ha).

Conservation issues

Threats Disturbance to birds (B)

Iznik lake — A1, B2 — 007

Admin region Bursa
Coordinates 40°26′N 29°32′E
Altitude 87–87 m **Area** 29,830 ha

Site description

A tectonic freshwater lake with a max. depth of 80 m. Small deltas supporting extensive reedbeds (*Phragmites*) have formed where streams enter the lake; a levee has been constructed to prevent flooding. Arable land, olive groves and *Populus* plantations surround the lake. Water is abstracted for industrial and irrigation purposes. The town of İznik attracts many tourists.

Habitats Wetland (95%; standing fresh water; water-fringe vegetation), Artificial landscape (5%; perennial crops/orchards/groves)
Land-use Agriculture (5%), Fisheries/aquaculture (10%), Tourism/recreation (10%), Urban/industrial/transport (25%), Water management (95%)

Birds

Species	Season	Year	Pop min	Pop max	Acc	Criteria
Phalacrocorax pygmeus Pygmy Cormorant	B	1995	30	30	—	A1, B2
Nycticorax nycticorax Night Heron	B	—	250	250	—	B2

The site is important for breeding waterbirds.

Protection status

National High **International** None
29,830 ha of IBA covered by SİT (Iznik Gölü, 29,830 ha).

Conservation issues

Threats Agricultural intensification/expansion (A), Construction/impact of dyke/dam/barrage (A), Dredging/canalization (B), Industrialization/urbanization (A)

From the small amount of data available, it appears that the lake was more important from an ornithological perspective prior to its conversion to a reservoir in 1963 (it is assumed that large areas of reed and marsh have been lost). Work is under way to substantially increase the land area under irrigation. Untreated industrial and urban waste is discharged to the lake—recent algal blooms and mass fish-kills indicate the scale of the problem.

Uludağ — B2 — 008

Admin region Bursa
Coordinates 40°04′N 29°13′E
Altitude 1,250–2,543 m **Area** 20,000 ha

Site description

Situated south of Bursa, this is the highest mountain in western Anatolia. *Pinus* predominates at lower altitudes; *Abies*/*Fagus* forest occurs between 1,500 and 2,000 m; alpine scrub and pasture prevail at higher altitudes. Limestone fields occupy the summit areas; dry montane cliff communities occur to the north of the summit ridge. Several small alpine lakes are found below these cliffs. The area is popular for winter sports.

Habitats Forest and woodland (70%; native coniferous forest; mixed forest), Grassland (10%; alpine/subalpine/boreal grassland), Rocky areas (20%; inland cliff)
Land-use Forestry (40%), Not utilized (10%), Tourism/recreation (50%)

Birds

Species	Season	Year	Pop min	Pop max	Acc	Criteria
Gypaetus barbatus Lammergeier	B	—	2	2	—	B2
Aquila chrysaetos Golden Eagle	B	—	2	2	—	B2

This is an important site for breeding raptors. *Sitta krueperi* and *Serinus pusillus* are at the western limits of their range on Uludağ; this is one of the few known Turkish sites for *Aegolius funereus*.

Protection status

National High **International** None
12,732 ha of IBA covered by National Park (Uludağ, 12,732 ha). 20,000 ha of IBA covered by SİT (Uludağ, 20,000 ha).

Conservation issues

Threats Industrialization/urbanization (U), Recreation/tourism (B)

A new ski-resort is planned, areas of forest have been cleared for ski-slopes, and water resources at higher elevations are scarce due to hotel demand. No study of the consequences of these developments has been carried out. Uludağ is affected by acid rain and industrial pollution. No grazing or forest management takes place within the National Park.

Uluabat lake A1, A4i, A4iii, B1i, B2 009

Admin region Bursa
Coordinates 40°10′N 28°35′E
Altitude 9–9 m **Area** 13,500 ha

Site description

A large, shallow (max. depth 6 m), nutrient-rich freshwater lake on the south side of the Sea of Marmara. The lake is fed principally by the Mustafakemalpaşa river and contains four islands. Silt deposition has formed an inland delta largely under agriculture; the western and southern shores of the lake support vast reedbeds (*Phragmites*). Fisheries are the main economic activity; lake water is used for irrigation.

Habitats Wetland (100%; standing fresh water; water-fringe vegetation)
Land-use Agriculture (10%), Fisheries/aquaculture (60%), Hunting (30%), Water management (90%)

Birds

Species	Season	Year	Pop min	Pop max	Acc	Criteria
Phalacrocorax pygmeus Pygmy Cormorant	B	1998	—	823	—	A1, A4i, B1i, B2
Phalacrocorax pygmeus Pygmy Cormorant	W	1996	170	1,078	—	A1, A4i, B1i
Pelecanus crispus Dalmatian Pelican	N	1994	136	136	—	A1, A4i, B1i
Ixobrychus minutus Little Bittern	B	1998	—	150	—	B2
Nycticorax nycticorax Night Heron	B	1998	—	105	—	B2
Ardeola ralloides Squacco Heron	B	1998	—	109	—	B1i, B2
Platalea leucorodia Spoonbill	B	1998	—	48	—	A4i, B1i, B2
Aythya ferina Pochard	W	1996	14,700	42,500	—	A4i, B1i
Aythya nyroca Ferruginous Duck	B	1998	—	32	—	A1, B2
Aythya fuligula Tufted Duck	W	1996	13,600	13,600	—	B1i
Fulica atra Coot	W	1996	45,840	321,550	—	A4i, B1i
Larus ridibundus Black-headed Gull	W	1996	6,200	6,200	—	B1i
Chlidonias hybridus Whiskered Tern	B	1998	—	660	—	A4i, B1i, B2

This site is important for both breeding and wintering waterbirds. Up to 429,500 waterbirds were present in winter 1996, representing the highest count at any Turkish site since 1970.

Protection status

National None **International** High
13,000 ha of IBA covered by Ramsar Site (Uluabat Gölü, 13,000 ha).

Conservation issues

Threats Agricultural intensification/expansion (B), Aquaculture/fisheries (A), Construction/impact of dyke/dam/barrage (B), Drainage (B), Extraction industry (A), Industrialization/urbanization (U)

Agricultural intensification threatens the remaining natural deltaic habitats. A further reduction in the lake's inflow will result from the planned construction of a reservoir on a major tributary of the Mustafakemalpaşa river, 20 km south-east of the lake. The possible effects on the lake's ecosystem are unknown. Large quantities of urban and industrial waste and agricultural run-off enter the lake, and further nutrient enrichment may pose a serious threat. Despite existing regulations, out-of-season and excessive fishing is widespread.

Kuş lake A1, A4i, B1i, B2 010

Admin region Balıkesir
Coordinates 40°11′N 27°58′E
Altitude 18–18 m **Area** 16,000 ha

Site description

A shallow nutrient-rich freshwater lake (average depth 3 m) fed by groundwater and four streams. Small deltas have formed where the latter enter the lake, comprising extensive marshes and tree-lined riverbanks. Narrow belts of reed *Phragmites* fringe much of the lake. Water is abstracted for factory use and for irrigation. Cattle- and sheep-grazing is common along the lake shores.

Habitats Forest and woodland (5%; alluvial/very wet forest), Wetland (95%; standing fresh water; water-fringe vegetation)
Land-use Agriculture (10%), Fisheries/aquaculture (10%), Urban/industrial/transport (80%), Water management (80%)

Birds

Species	Season	Year	Pop min	Pop max	Acc	Criteria
Phalacrocorax carbo Cormorant	B	1995	2,650	2,650	—	A4i, B1i
Phalacrocorax pygmeus Pygmy Cormorant	B	—	150	150	—	A1, A4i, B1i, B2
Phalacrocorax pygmeus Pygmy Cormorant	W	1996	58	147	—	A1
Pelecanus onocrotalus White Pelican	P	1991	1,500	1,500	—	A4i, B1i
Pelecanus crispus Dalmatian Pelican	B	—	35	35	—	A1, A4i, B1i, B2
Pelecanus crispus Dalmatian Pelican	W	1996	52	117	—	A1, A4i, B1i
Nycticorax nycticorax Night Heron	B	—	150	150	—	B2
Ardeola ralloides Squacco Heron	B	—	100	100	—	B1i, B2
Platalea leucorodia Spoonbill	B	1995	200	200	—	A4i, B1i, B2
Oxyura leucocephala White-headed Duck	W	1996	20	34	—	A1

This is an important site for breeding and wintering waterbirds; migrating *Pelecanus onocrotalus* often roost at the lake.

Protection status

National High **International** High
64 ha of IBA covered by National Park (Kuş Gölü, 64 ha). 16,000 ha of IBA covered by Permanent Wildlife Reserve (Kuş Gölü, 23,667 ha). 100 ha of IBA covered by SİT (Kuş Gölü, 100 ha). 16,000 ha of IBA covered by Ramsar Site (Kuş Gölü, 16,000 ha).

Conservation issues

Threats Agricultural intensification/expansion (A), Aquaculture/fisheries (C), Construction/impact of dyke/dam/barrage (A), Disturbance to birds (B), Drainage (A), Industrialization/urbanization (B)

Artificially high water-levels have resulted in the loss of feeding marshes. Nesting trees in the National Park have also begun to die as a result of prolonged inundation. Agricultural, industrial and household pollution enters the lake in large quantities. Due to over-fishing, disease, pollution and possibly the effects of the changing water regime, fish catches have declined drastically. Fish migration into the lake is now impossible following the construction of a regulator.

Kocaçay delta A1, A4i, A4iii, B1i, B2 011

Admin region Bursa
Coordinates 40°23′N 28°29′E
Altitude 0–0 m **Area** 4,200 ha

Site description

A complex of coastal lakes, marshes and forests on the southern shores of the Sea of Marmara, where the Kocaçay river enters the sea. The western half of the delta comprises two lagoons, reedbeds (*Phragmites*), extensive *Fraxinus*, *Alnus* and *Salix* alluvial forest, and a wide dune strip. The eastern half is more open with one large lagoon, arable land, orchards, dunes, reedbeds and *Salicornia*- and *Tamarix*-covered mudflats. Human activities include cattle-grazing.

Habitats Forest and woodland (40%; alluvial/very wet forest), Wetland (45%; sand-dunes/sand beach; coastal lagoon; water-fringe vegetation), Artificial landscape (15%; arable land)
Land-use Agriculture (15%), Fisheries/aquaculture (10%), Forestry (5%), Hunting (5%), Not utilized (60%), Tourism/recreation (5%)

Birds

Species	Season	Year	Pop min	Pop max	Acc	Criteria
Phalacrocorax pygmeus Pygmy Cormorant	P	1994	63	150	—	A1
Pelecanus onocrotalus White Pelican	P	—	800	800	—	A4i, B1i
Ciconia nigra Black Stork	B	—	10	10	—	B2
Aythya nyroca Ferruginous Duck	B	—	70	70	—	A1, B2
Fulica atra Coot	W	1996	42,610	42,610	—	A4i, B1i
Glareola pratincola Collared Pratincole	B	—	80	80	—	A4i, B1i, B2
Charadrius alexandrinus Kentish Plover	B	—	60	60	—	B2

This site is important for breeding, wintering and passage waterbirds. Several species of woodpecker also breed. Breeding species of global conservation concern that do not meet IBA criteria: *Haliaeetus albicilla* (one of the few confirmed Turkish sites).

Protection status

National None **International** None

Conservation issues

Threats Afforestation (B), Agricultural intensification/expansion (A), Industrialization/urbanization (A), Recreation/tourism (U)

Forests and other natural habitats are cleared for arable land and *Populus* plantations. Yeniköy is developing as a tourist centre and further expansion eastward should be halted. Sand-dunes are destroyed to fulfil the demand for sand. Housing complexes are encroaching from the east; the recent demolition of illegally built houses indicates this is now less of a threat. The effects of urban and industrial pollution on the delta's ecology are unknown. Effective management of the wetland is constrained by disputes over land ownership.

Foça islands B1i 012

Admin region İzmir
Coordinates 38°42′N 26°43′E
Altitude 0–80 m **Area** 134 ha

Site description

A group of six uninhabited islands off İzmir Bay: Orak, Hayırsız, Fener, İncir, Kartdere and Metalik. There is a long pebble beach on the south shore of Orak, with cliffs of up to 80 m found on Orak, Hayırsız and Kartdere. Herbaceous plants, garrigue and maquis cover the islands. Tourism is the major economic activity, and İncir in particular is heavily visited. There is no human activity on the other islands.

Habitats Scrub (85%; sclerophyllous scrub/garrigue/maquis), Rocky areas (15%; sea cliff/rocky shore)
Land-use Nature conservation/research (10%), Not utilized (80%), Tourism/recreation (10%)

Birds

Species	Season	Year	Pop min	Pop max	Acc	Criteria
Phalacrocorax aristotelis Shag	B	—	59	59	—	B1i
Phalacrocorax aristotelis Shag	W	1994	169	169	—	B1i

Protection status

National High **International** High
134 ha of IBA covered by SİT (Foça Adaları, 134 ha). 127 ha of IBA covered by Mediterranean Specially Protected Area (Foça, 2,750 ha).

Conservation issues

Threats Disturbance to birds (B)

Recreational activities such as angling and scuba diving are prohibited. The MedSPA designation halted several developments on the mainland and has secured the future of the islands. Hayırsız and Kartdere remain unprotected, however.

Murat Dağı B2 013

Admin region Kütahya, Uşak
Coordinates 38°56′N 29°43′E
Altitude 1,000–2,312 m **Area** 52,500 ha

Site description

An extensive mountain range comprising a high, barren, east–west running ridge. All slopes are extensively forested below 2,000 m with *Pinus* and *Quercus*. Large areas have been clear-felled and reforested—these are intensively managed and bisected by numerous tracks. Grazing pasture is found at higher altitudes; a holiday village is located around hot springs on the northern slopes.

Habitats Forest and woodland (90%; native coniferous forest), Grassland (10%; alpine/subalpine/boreal grassland), Artificial landscape (forestry plantation)
Land-use Forestry (80%)

Birds

Species	Season	Year	Pop min	Pop max	Acc	Criteria
Gypaetus barbatus Lammergeier	B	—	1	1	—	B2
Aegypius monachus Cinereous Vulture	B	—	2	2	—	B2
Aquila chrysaetos Golden Eagle	B	—	2	2	—	B2

This is an important site for breeding raptors.

Protection status

National None **International** None

Conservation issues

Threats Deforestation (commercial) (B)

A 1996 survey indicated that larger numbers of *Aegypius monachus* bred within the site before large-scale clear-felling and reforestation programmes were implemented. Forestry activities should take account of the distribution of breeding raptors (leaving an unmanaged buffer zone around each breeding valley), and old pine trees should not be felled.

Demirköprü reservoir A4i, B1i 014

Admin region Manisa
Coordinates 38°41′N 28°23′E
Altitude 244–244 m **Area** 5,060 ha

Site description

Situated on the Gediz river near Manisa and bordered by barren rocky hills. Used to generate electricity, control floodwaters, and to provide irrigation water for 99,000 ha.

Habitats Wetland (100%; standing fresh water)
Land-use Fisheries/aquaculture (10%), Urban/industrial/transport (100%), Water management (100%)

Birds

Species	Season	Year	Pop min	Pop max	Acc	Criteria
Tadorna ferruginea Ruddy Shelduck	W	1996	534	830	—	A4i, B1i
Anas crecca Teal	W	1992	10,410	10,410	—	B1i

This is an important site for wintering wildfowl.

Protection status

National None **International** None

Conservation issues

No serious threats are known at the site.

Marmara lake A1, A4i, A4iii, B1i, B2 015

Admin region Manisa
Coordinates 38°37′N 28°00′E
Altitude 79–79 m **Area** 6,800 ha

Site description

An irrigation reservoir averaging 3,400 ha in area (depth 3–4 m); reedbed (*Phragmites*) and marsh cover varies according to the water-level. Surrounded by hills to the north and north-east, with Gediz valley to the south and Akhisar valley to the north-west. Flood-plains are used for growing cotton, and cattle-grazing is widespread on the marshes.

Habitats Wetland (100%; mudflat/sandflat; standing fresh water; water-fringe vegetation)
Land-use Agriculture (10%), Fisheries/aquaculture (15%), Hunting (10%), Water management (100%)

Birds

Species	Season	Year	Pop min	Pop max	Acc	Criteria
Phalacrocorax pygmeus Pygmy Cormorant	W	1990	100	100	—	A1
Pelecanus crispus Dalmatian Pelican	W	1999	26	104	—	A1, A4i, B1i
Pelecanus crispus Dalmatian Pelican	N	1995	67	73	—	A1, A4i, B1i
Ixobrychus minutus Little Bittern	B	—	30	30	—	B2
Ardeola ralloides Squacco Heron	B	—	200	200	—	A4i, B1i, B2
Egretta alba Great White Egret	W	1996	194	428	—	A4i, B1i
Tadorna ferruginea Ruddy Shelduck	W	1996	231	290	—	A4i, B1i
Anas penelope Wigeon	W	1990	8,500	8,500	—	B1i
Aythya ferina Pochard	W	1996	18,690	24,000	—	A4i, B1i
Aythya nyroca Ferruginous Duck	B	—	25	25	—	A1, B2
Oxyura leucocephala White-headed Duck	W	1990	50	120	—	A1, B1i
Recurvirostra avosetta Avocet	W	1992	1,402	1,402	—	A4i, B1i, B2
Hoplopterus spinosus Spur-winged Plover	B	—	10	10	—	B2
Chlidonias hybridus Whiskered Tern	B	—	100	100	—	B2

This is an important area for breeding and wintering waterbirds. *Pelecanus crispus* is thought to breed.

Protection status

National None **International** None

Conservation issues

Threats Agricultural intensification/expansion (A), Aquaculture/fisheries (B), Construction/impact of dyke/dam/barrage (A), Industrialization/urbanization (A)

The lake virtually dried up in 1993 following a dry period and excessive abstraction. Further expansion of the irrigation network has led to a new channel being constructed to bring in more water from the Gediz river. Industrial and urban pollution entering the lake will increase as a result. DSİ also plans to construct a hydroelectric power-station on the existing channel connecting the Gediz river and the site. Local farmers temporarily convert ecologically valuable mudflats to arable land during periods of low water-level. Out-of-season fishing is common.

Gediz delta A1, A4i, A4iii, B1i, B2, B3 016

Admin region İzmir
Coordinates 38°30′N 26°55′E
Altitude 0–0 m **Area** 20,400 ha

Site description

An extensive coastal wetland situated 25 km north-west of İzmir. Shoals and islands separate four saline lagoons (totalling 3,450 ha) from the sea. The site also includes 5,000 ha of saltmarsh and 650 ha of freshwater marsh. Dry grassland and arable land occur north of the commercial saltpans. A fishing cooperative operates in the Çilazmak lagoon; the IBA is a popular weekend destination from İzmir.

Habitats Grassland (30%; steppe/dry calcareous grassland), Wetland (50%; saltmarsh; sand-dunes/sand beach; coastal lagoon), Artificial landscape (20%; arable land)
Land-use Agriculture (10%), Fisheries/aquaculture (30%), Not utilized (40%), Tourism/recreation, Urban/industrial/transport (20%)

Birds

Species	Season	Year	Pop min	Pop max	Acc	Criteria
Phalacrocorax carbo Cormorant	W	1999	—	1,368	—	B1i
Phalacrocorax pygmeus Pygmy Cormorant	W	1993	85	1,000	—	A1, A4i, B1i
Pelecanus crispus Dalmatian Pelican	B	—	35	50	—	A1, A4i, B1i, B2
Pelecanus crispus Dalmatian Pelican	W	1995	35	341	—	A1, A4i, B1i
Phoenicopterus ruber Greater Flamingo	B	—	1,450	1,450	—	A4i, B1i, B2
Phoenicopterus ruber Greater Flamingo	W	1996	5,000	21,300	—	A4i, B1i
Tadorna ferruginea Ruddy Shelduck	W	1996	250	450	—	A4i, B1i
Anas crecca Teal	W	1999	—	11,532	—	B1i
Falco naumanni Lesser Kestrel	B	—	25	25	—	A1
Recurvirostra avosetta Avocet	B	—	55	55	—	B3
Burhinus oedicnemus Stone Curlew	B	—	30	30	—	B2
Glareola pratincola Collared Pratincole	B	—	35	35	—	B1i, B2
Charadrius alexandrinus Kentish Plover	B	—	1,000	—	—	A4i, B1i, B2
Charadrius alexandrinus Kentish Plover	W	1993	532	809	—	B1i
Hoplopterus spinosus Spur-winged Plover	B	—	50	50	—	A4i, B1i, B2
Larus ridibundus Black-headed Gull	W	1999	—	30,990	—	B1i
Larus cachinnans Yellow-legged Gull	W	1999	—	40,000	—	A4i, B1i
Sterna caspia Caspian Tern	B	—	100	100	—	A4i, B1i, B2
Sterna albifrons Little Tern	B	—	205	205	—	B1i, B2

The IBA is important for breeding and wintering waterbirds. *Sterna sandvicensis* bred in 1996 (53 pairs)—the first Turkish breeding record.

Protection status

National Partial **International** Partial

8,000 ha of IBA covered by Permanent Wildlife Reserve (Gediz Deltası, 8,000 ha). 8,000 ha of IBA covered by SİT (Gediz Deltası, 8,000 ha). 15,000 ha of IBA covered by Ramsar Site (Gediz, 15,000 ha).

Conservation issues

Threats Agricultural intensification/expansion (A), Disturbance to birds (B), Dredging/canalization (B), Industrialization/urbanization (A), Recreation/tourism (B)

The Gediz river is heavily polluted by industrial and household waste. İzmir is encroaching upon the area. Lorries cause disturbance to the *Phoenicopterus ruber* colony; repairs to the saltpans may delay breeding. Plans exist for a 2,000 ha sewage treatment plant, a new seaport, and the dumping of 7,000,000 m^3 of polluted mud within the IBA boundary. Fish catches have declined, possibly due to increasing salinity in the lagoons as a result of reduced freshwater inflows.

Ildır Strait islands B1i 017

Admin region İzmir
Coordinates 38°27′N 26°21′E
Altitude 0–218 m **Area** 957 ha

Site description

Ildır Bay is situated between the Çeşme and Karaburun peninsulas, west of İzmir, and holds two groups of uninhabited islands. Of the four main islands situated at the mouth of the bay, Karaada is the largest (670 ha) and has steep cliffs on its western and eastern shores. All are covered in maquis. The second group is situated closer to the mainland and comprises seven small, sparsely vegetated rocky islands (totalling 127 ha). Human activity around the islands is minimal.

Habitats Scrub (sclerophyllous scrub/garrigue/maquis), Rocky areas (100%; rock stacks/islets)
Land-use Fisheries/aquaculture, Hunting (10%), Not utilized (80%), Tourism/recreation (10%)

Birds

Species	Season	Year	Pop min	Pop max	Acc	Criteria
Phalacrocorax aristotelis Shag	B	1996	84	84	—	B1i

Breeding species of global conservation concern that do not meet IBA criteria: *Falco naumanni* (5 pairs).

Protection status

National High **International** None

670 ha of IBA covered by Permanent Wildlife Reserve (Ildır Körfezi Adaları, 670 ha). 957 ha of IBA covered by SİT (Ildır Körfezi Adaları, 957 ha).

■ Conservation issues

Threats Aquaculture/fisheries (U), Consequences of animal/plant introductions (U)

Introduced rabbits *Oryctolagus cuniculus* occur on the islands nearest the mainland and are illegally hunted throughout the year, especially on Çifte Island where the largest number of *Phalacrocorax aristotelis* breed. This causes disturbance to the birds. Plans exist to build fish-farms in the bays of Karaada Island.

Küçük Menderes delta — A1 — 018

Admin region İzmir
Coordinates 37°59′N 27°18′E
Altitude 0–0 m **Area** 1,500 ha

■ Site description

A wetland complex comprising Çakal lake (a 74-ha, 4-m-deep freshwater lake), Gebekirse lake (a 75-ha, 5-m-deep brackish lake) and marshes on the alluvial plain north of the Küçük Menderes river. The former Cevaşir lake floods only in winter.

Habitats Wetland (90%; sand-dunes/sand beach; standing fresh water; water-fringe vegetation), Artificial landscape (10%; arable land)
Land-use Agriculture (5%), Fisheries/aquaculture (5%), Not utilized (90%)

■ Birds

Species	Season	Year	Pop min	Pop max	Acc	Criteria
Phalacrocorax pygmeus Pygmy Cormorant	P	1995	126	160	—	A1

■ Protection status

National High **International** None
1,000 ha of IBA covered by Permanent Wildlife Reserve (Küçük Menderes Deltası, 1,000 ha). 1,500 ha of IBA covered by SİT (Küçük Menderes Deltası, 1,500 ha).

■ Conservation issues

Threats Agricultural intensification/expansion (A), Construction/impact of dyke/dam/barrage (A), Drainage (A), Infrastructure (U), Recreation/tourism (A)

DSİ plans to construct four dams on the Küçük Menderes river and its tributaries for irrigation purposes. This will greatly reduce inflows to the wetland and may lead to the drying up of the Elaman and Akgöl marshes. Coastal tourism developments are encroaching from the north and south. The effects of the airstrip are unknown.

Akdağ — B2 — 019

Admin region Afyon, Denizli
Coordinates 38°20′N 30°00′E
Altitude 900–2,446 m **Area** 42,394 ha

■ Site description

A high mountain adjoining Işıklı lake (IBA 020). A gorge bisects it in a north–south direction, with high cliffs on either side and steep forested valleys to the north. The flatter ground above the valleys supports coniferous forest. Forestry activities take place in accessible areas.

Habitats Forest and woodland (60%), Rocky areas (40%)
Land-use Agriculture (10%), Forestry (50%), Not utilized (40%)

■ Birds

Species	Season	Year	Pop min	Pop max	Acc	Criteria
Gypaetus barbatus Lammergeier	B	—	1	1	—	B2
Gyps fulvus Griffon Vulture	B	—	3	3	—	B2
Aegypius monachus Cinereous Vulture	B	—	3	3	—	B2
Hieraaetus pennatus Booted Eagle	B	—	4	4	—	B2

Important for breeding raptors; *Pyrrhocorax pyrrhocorax* also breeds (50 pairs). Breeding species of global conservation concern that do not meet IBA criteria: *Haliaeetus albicilla* (1 pair).

■ Protection status

National High **International** None
42,394 ha of IBA covered by Permanent Wildlife Reserve (Akdağ, 42,394 ha).

■ Conservation issues

Threats Disturbance to birds (B), Selective logging/cutting (C)

Forestry activities should take account of the distribution of breeding raptors (leaving an unmanaged buffer zone around each breeding valley), and old pine trees should not be felled.

Işıklı lake — A1, A4i, A4iii, B1i, B2 — 020

Admin region Denizli
Coordinates 38°14′N 29°55′E
Altitude 821–821 m **Area** 7,300 ha

■ Site description

Situated at the head of the Büyük Menderes river south of Akdağ (IBA 019), this freshwater lake (max. depth 7 m) is fed by streams and springs. The lake stores irrigation water for the surrounding agricultural plains, and high water-levels mean that marsh vegetation is only present where the Kufi channel deposits silt into the lake. The IBA includes Gökgöl marshes (including a max. 300 ha lake) 3 km east of Işıklı Gölü.

Habitats Wetland (90%; standing fresh water; water-fringe vegetation), Artificial landscape (10%; arable land)
Land-use Fisheries/aquaculture (50%), Hunting (40%), Water management (90%)

■ Birds

Species	Season	Year	Pop min	Pop max	Acc	Criteria
Phalacrocorax pygmeus Pygmy Cormorant	W	1996	40	130	—	A1
Ardeola ralloides Squacco Heron	B	—	30	30	—	B2
Anser albifrons White-fronted Goose	W	1993	8,178	8,178	—	B1i
Netta rufina Red-crested Pochard	W	1993	931	1,060	—	A4i, B1i
Aythya ferina Pochard	W	1996	13,740	18,370	—	A4i, B1i
Aythya nyroca Ferruginous Duck	W	1992	97	97	—	A1
Fulica atra Coot	W	1996	25,500	190,690	—	A4i, B1i
Gelochelidon nilotica Gull-billed Tern	B	—	—	—	—	B2
Chlidonias hybridus Whiskered Tern	B	—	110	110	—	B2

This site is important for breeding waterbirds and large numbers of wintering wildfowl. Species of global conservation concern that do not meet IBA criteria: *Haliaeetus albicilla* (seen throughout the year).

■ Protection status

National None **International** None

■ Conservation issues

Threats Aquaculture/fisheries (C), Disturbance to birds (A), Other (U)

Poor water management has led to the lake being emptied on several occasions. High spring water-levels do not permit the development of suitable breeding habitat, and mudflats, important as feeding areas for birds during winter and spring, have been largely eliminated ('Other' threat). Fisheries are uncontrolled, with out-of-season fishing widespread. Hunting is a significant problem, with shooting taking place throughout the winter.

Bafa lake — A1, A4i, A4iii, B1i, B2 — 021

Admin region Aydın, Muğla
Coordinates 37°30′N 27°26′E
Altitude 5–5 m **Area** 12,281 ha

■ Site description

A slightly saline lake (max. 6,708 ha, max. depth 25 m) situated south-east of the Büyük Menderes delta (site 022) and enclosed by mountain ranges, except in the west where the land opens towards the Menderes plain. The lake is fed principally by river water, and contains four islands. Its western shores are shallow and support extensive marshes. Fisheries are the main economic activity, and the lake is a popular tourist destination.

Habitats Forest and woodland (20%; mixed forest), Wetland (75%; standing fresh water; water-fringe vegetation), Artificial landscape (5%; arable land; perennial crops/orchards/groves)
Land-use Agriculture (10%), Fisheries/aquaculture (20%), Forestry (20%), Not utilized (40%), Tourism/recreation (10%)

Birds

Species	Season	Year	Pop min	Pop max	Acc	Criteria
Tachybaptus ruficollis Little Grebe	W	1990	1,800	1,805	—	A4i, B1i
Podiceps cristatus Great Crested Grebe	W	1996	6,240	6,240	—	A4i, B1i
Podiceps nigricollis Black-necked Grebe	W	1992	1,136	1,136	—	A4i, B1i
Phalacrocorax carbo Cormorant	W	1993	1,172	2,000	—	B1i
Phalacrocorax pygmeus Pygmy Cormorant	W	1989	140	140	—	A1
Pelecanus crispus Dalmatian Pelican	W	1993	25	74	—	A1, A4i, B1i
Anas strepera Gadwall	W	1989	1,250	1,250	—	B1i
Aythya ferina Pochard	W	1996	27,085	27,085	—	A4i, B1i
Fulica atra Coot	W	1996	43,100	68,500	—	A4i, B1i
Glareola pratincola Collared Pratincole	B	—	50	50	—	B1i, B2
Hoplopterus spinosus Spur-winged Plover	B	—	30	30	—	B2

An important site for breeding waders and large numbers of wintering waterbirds. *Pelecanus crispus* occurs throughout the year, with birds coming mainly from site 022. Species of global conservation concern that do not meet IBA criteria: *Haliaeetus albicilla* (2 breeding pairs).

Protection status

National High **International** None
12,281 ha of IBA covered by Nature Park (Bafa Gölü, 12,281 ha). 12,281 ha of IBA covered by SİT (Bafa Gölü, 12,281 ha).

Conservation issues

Threats Agricultural intensification/expansion (B), Aquaculture/fisheries (B), Construction/impact of dyke/dam/barrage (A), Industrialization/urbanization (B), Unsustainable exploitation (U)

Reductions in inflow have occurred as a result of levee and dam construction and abstraction for irrigation. The water-level had to be artificially raised by DSİ after large areas dried up in 1990. Fish stocks have declined as a result of over-fishing and the lake/river connection being severed. A large housing complex is planned to the south of the lake. Illegal hunting is frequently observed. DHKD is involved in developing a management plan in cooperation with the Nature Park authorities, and runs a bird observatory.

Büyük Menderes delta A1, A4i, A4iii, B1i, B2 022

Admin region Aydın
Coordinates 37°34′N 27°12′E
Altitude 0–0 m **Area** 9,800 ha

Site description

Situated at the junction between the Büyük Menderes river and the Aegean, the delta comprises lagoons (separated from the sea by a narrow sandbar), extensive salt-steppe and mudflats. A north–south running motorway forms the eastern border of the IBA; west of this salt-steppe has been partially converted to arable land. To the north the delta is bordered by the hills and mountains of the Dilek peninsula. Human activities include cattle- and sheep-grazing.

Habitats Wetland (80%; mudflat/sandflat; sand-dunes/sand beach; coastal lagoon), Artificial landscape (20%; arable land)
Land-use Agriculture (20%), Fisheries/aquaculture (30%), Hunting (15%), Not utilized (35%)

Birds

The area is important for breeding and wintering waterbirds.

Species	Season	Year	Pop min	Pop max	Acc	Criteria
Phalacrocorax carbo Cormorant	W	1996	1,000	3,005	—	B1i
Phalacrocorax pygmeus Pygmy Cormorant	W	1989	130	350	—	A1, A4i, B1i
Pelecanus crispus Dalmatian Pelican	B	—	42	42	—	A1, A4i, B1i, B2
Pelecanus crispus Dalmatian Pelican	W	1996	39	434	—	A1, A4i, B1i
Egretta alba Great White Egret	W	1996	138	312	—	A4i, B1i
Phoenicopterus ruber Greater Flamingo	W	1996	6,290	8,650	—	A4i, B1i
Tadorna ferruginea Ruddy Shelduck	N	1992	200	200	—	A4i, B1i
Anas penelope Wigeon	W	1992	5,623	5,623	—	B1i
Fulica atra Coot	W	1999	—	32,535	—	B1i
Recurvirostra avosetta Avocet	W	1989	550	1,000	—	B1i, B2
Glareola pratincola Collared Pratincole	B	—	85	85	—	A4i, B1i, B2
Charadrius alexandrinus Kentish Plover	B	—	150	150	—	B1i, B2
Larus genei Slender-billed Gull	W	1989	1,284	1,284	—	A4i, B1i
Sterna caspia Caspian Tern	B	—	6	6	—	B2
Sterna caspia Caspian Tern	N	1988	97	97	—	B1i
Sterna albifrons Little Tern	B	—	300	300	—	A4i, B1i, B2

Protection status

National High **International** None
9,800 ha of IBA covered by National Park (Menderes Deltası, 27,675 ha). 9,800 ha of IBA covered by SİT (Menderes Deltası, 9,800 ha).

Conservation issues

Threats Agricultural intensification/expansion (A), Aquaculture/fisheries (B), Construction/impact of dyke/dam/barrage (A), Industrialization/urbanization (A), Recreation/tourism (B)

Irrigation projects, including 13 reservoirs, have been and are being developed on the Büyük Menderes river with severe consequences for large tracts of natural habitat. Industrial and untreated urban waste enters the lagoons via the river, and, along with water-level fluctuations and over-fishing, has caused fisheries to collapse. There is considerable pressure to build holiday homes. From 1991 to 1995 DHKD ran an education project both in this site and Bafa lake (site 021), and have assisted the National Park authorities in drawing up a management plan.

Güllük delta A1, A4i, B1i 023

Admin region Muğla
Coordinates 37°15′N 27°38′E
Altitude 0–0 m **Area** 1,400 ha

Site description

The delta of the Hamzabey and the Sarıçay streams, where they meet the Aegean Sea. The site consists of a 200-ha lagoon, two smaller lakes, reedbeds (*Phragmites*), *Salicornia* flats and *Tamarix* scrub. The lagoon is becoming less saline as a result of freshwater input. Agriculture now occupies most of the delta, and there is a basic fish-trap at the river mouth.

Habitats Wetland (70%; saltmarsh; coastal lagoon; standing fresh water; water-fringe vegetation), Artificial landscape (30%; arable land)
Land-use Agriculture (25%), Fisheries/aquaculture (30%), Not utilized (60%), Tourism/recreation (10%)

Birds

Species	Season	Year	Pop min	Pop max	Acc	Criteria
Phalacrocorax pygmeus Pygmy Cormorant	P	1996	65	221	—	A1, A4i, B1i

The site has lost most of its ornithological importance due to wetland drainage.

Protection status

National None **International** None

Conservation issues

Threats Drainage (A), Infrastructure (A), Recreation/tourism (B)

Threats include the effects of the numerous drainage channels that bisect the delta, the construction of a yacht harbour and airport, and large-scale tourist development.

Köyceğiz lake A1, A4i, A4iii, B1i 024

Admin region Muğla
Coordinates 36°54′N 28°38′E
Altitude 0–0 m **Area** 8,000 ha

Site description

A brackish lake (6,300 ha; max. depth 30 m) fed by springs and several streams. Extensive reedbeds (*Phragmites*) are found along the east, south-east and northern shores. The delta consists of a labyrinth of (former) flood-plains, reedbeds, lakes and sand beaches. Arable land and forested hills surround the IBA, which also includes a few hundred hectares of *Liquidambar* forest. Economic activities include greenhouse farming and cattle-grazing.

Habitats Forest and woodland (alluvial/very wet forest), Wetland (100%; sand-dunes/sand beach; coastal lagoon; standing fresh water; water-fringe vegetation)
Land-use Agriculture (10%), Fisheries/aquaculture (40%), Tourism/recreation (50%)

Birds

Species	Season	Year	Pop min	Pop max	Acc	Criteria
Phalacrocorax carbo Cormorant	W	1996	1,525	1,525	—	B1i
Phalacrocorax pygmeus Pygmy Cormorant	W	1996	40	40	—	A1
Egretta alba Great White Egret	W	1996	122	122	—	A4i, B1i
Netta rufina Red-crested Pochard	W	1996	528	528	—	B1i
Fulica atra Coot	W	1996	34,120	34,120	—	B1i

This is an important site for wintering waterbirds. Species of global conservation concern that do not meet IBA criteria: *Haliaeetus albicilla* (thought to breed in the surrounds).

Protection status

National None **International** High
8,000 ha of IBA covered by Mediterranean Specially Protected Area (Köyceğiz Dalyan, 38,500 ha).

Conservation issues

Threats Agricultural intensification/expansion (B), Industrialization/urbanization (A), Recreation/tourism (A)

The wetland is polluted by agricultural run-off and untreated urban waste, although sewage treatment plants for Dalyan and Köyceğiz are now being constructed. Heavy tourist-boat traffic on the canals between the lake and the sea causes disturbance to birds and wave-damage to reedbeds. The MedSPA authority has developed a management plan for the entire area but only partial implementation has been achieved.

Acıgöl lake A4i, B1i, B2, B3 025

Admin region Afyon, Denizli
Coordinates 37°49′N 29°48′E
Altitude 836–836 m **Area** 21,000 ha

Site description

A shallow, stream- and spring-fed saline lake (max. 16,500 ha) of tectonic origin. Freshwater marshes have developed around the springs; the lake supplies irrigation water and largely dries up during summer. The plain to the south-west comprises salt-steppe; the land to the north-east is intensively farmed. The site includes a complex of small freshwater lakes 10 km south-west of Acıgöl. Human activities include stock-grazing, salt extraction and reed-cutting ('Other' land-use).

Habitats Wetland (80%; mudflat/sandflat; saltmarsh; standing brackish and salt water), Artificial landscape (20%; arable land)
Land-use Agriculture (10%), Hunting (5%), Not utilized (65%), Other, Urban/industrial/transport (20%), Water management

Birds

Species	Season	Year	Pop min	Pop max	Acc	Criteria
[1] ***Phoenicopterus ruber*** Greater Flamingo	B	—	150	150	—	B2
Tadorna ferruginea Ruddy Shelduck	W	1994	240	240	—	A4i, B1i
Otis tarda Great Bustard	B	—	—	—	—	B2
Recurvirostra avosetta Avocet	B	—	50	50	—	B3
Charadrius alexandrinus Kentish Plover	B	—	50	50	—	B2
Charadrius alexandrinus Kentish Plover	N	1991	1,010	1,010	—	A4i, B1i
Hoplopterus spinosus Spur-winged Plover	B	—	—	—	—	B2
Gelochelidon nilotica Gull-billed Tern	B	—	—	—	—	B2

1. Breeding does not occur annually.

The site is important for breeding waterbirds, although numbers have fallen in recent years, possibly due to low water-levels. *Tadorna ferruginea* occurs throughout the year, and breeds in the surrounding mountains. Acıgöl was formerly important for breeding *Otis tarda*; it seems certain that a small population survives.

Protection status

National Low **International** None
1,100 ha of IBA covered by Permanent Wildlife Reserve (Çaltı Gölü, 1,100 ha).

Conservation issues

Threats Agricultural intensification/expansion (A), Construction/impact of dyke/dam/barrage (C), Drainage (U), Industrialization/urbanization (A)

DSİ plan to drain 13,620 ha of land between Acıgöl and the freshwater lakes, thereby destroying the largest intact salt-steppe ecosystem in the region. No environmental impact assessment has been carried out to date. Salt from the saltpans blows onto adjacent agricultural areas and reportedly damages crops and causes human respiratory problems. In 1996 work began on an industrial complex west of the area, which will use water from artesian wells. The possible effects of this scheme with regard to pollution and the lowering of groundwater levels are unknown.

Çorak lake A1, A4i, A4iii, B1i 026

Admin region Burdur
Coordinates 37°41′N 29°46′E
Altitude 1,050–1,050 m **Area** 1,150 ha

Site description

A small salt-lake of tectonic origin bordered by mountains to the north and agricultural lands to the south, east and west. It is fed by a number of small streams; reedbeds (*Phragmites*) have developed at the inflows. During the last decade the lake has been almost permanently dry, with a maximum of 5% flooded at the deepest points during March 1994.

Habitats Wetland (100%; mudflat/sandflat; standing brackish and salt water; river/stream)
Land-use Not utilized (100%)

Birds

Species	Season	Year	Pop min	Pop max	Acc	Criteria
Anas crecca Teal	W	1986	20,200	20,200	—	A4i, B1i
Aythya ferina Pochard	W	1969	40,000	40,000	—	A4i, B1i
Aythya nyroca Ferruginous Duck	W	1970	100	100	—	A1
Oxyura leucocephala White-headed Duck	W	1974	85	930	—	A1, A4i, B1i

The IBA has supported large numbers of wildfowl in the past. Recent counts have been low due to falling water-levels.

Protection status

National None **International** None

Conservation issues

Threats Construction/impact of dyke/dam/barrage (A)

During 1988 DSİ widened the outlet of the lake. It is unclear whether this, or the lack of rainfall in recent years, is responsible for the low water-levels.

Salda lake
A1, B1i **027**

Admin region Burdur
Coordinates 37°33'N 29°40'E
Altitude 1,139–1,139 m **Area** 4,370 ha

Site description
A slightly saline, stream- and spring-fed lake (max. depth 184 m) of tectonic origin surrounded by forested hills, rocky terrain and alluvial plains. There are several sandbanks and limited fringe vegetation. Hotels, restaurants and campsites are present on the southern and south-western shores.

Habitats Wetland (100%; standing brackish and salt water)
Land-use Not utilized (95%), Tourism/recreation (5%)

Birds

Species	Season	Year	Pop min	Pop max	Acc	Criteria
Aythya nyroca Ferruginous Duck	W	1990	400	400	—	A1
Oxyura leucocephala White-headed Duck	W	1993	40	128	—	A1, B1i

The site is important for wintering wildfowl.

Protection status
National High **International** None
4,370 ha of IBA covered by SİT (Salda Gölü, 4,370 ha).

Conservation issues
Threats Agricultural intensification/expansion (U), Construction/impact of dyke/dam/barrage (B), Recreation/tourism (B)

Inflow water is used to irrigate fields around the lake; a DSİ scheme will see the construction of a small reservoir on one of the inflow streams, allowing for the irrigation of a further 182 ha via a 3 km long channel. No assessment of the possible impacts has been made. Some hotels discharge untreated sewage into the lake, and more tourist facilities are planned. The SİT designation permits limited construction activity.

Karataş lake
A1, A4i, B1i **028**

Admin region Burdur
Coordinates 37°23'N 29°58'E
Altitude 1,053–1,053 m **Area** 1,190 ha

Site description
A shallow, stream-fed freshwater lake situated in the north-east Tefenni plain, surrounded by arable land. Fringing vegetation is restricted to the western shore.

Habitats Wetland (100%; standing fresh water; water-fringe vegetation)
Land-use Fisheries/aquaculture (10%), Hunting (5%), Water management (100%)

Birds

Species	Season	Year	Pop min	Pop max	Acc	Criteria
Tadorna ferruginea Ruddy Shelduck	W	1990	212	212	—	A4i, B1i
Oxyura leucocephala White-headed Duck	W	1995	47	82	—	A1

This is an important area for wintering wildfowl.

Protection status
National High **International** None
1,190 ha of IBA covered by Permanent Wildlife Reserve (Karataş Gölü, 4,720 ha).

Conservation issues
Threats Agricultural intensification/expansion (A), Construction/impact of dyke/dam/barrage (B)

Inflows to the lake were artificially increased in 1982 to enable the irrigation of surrounding land. The increase in water-level probably led to a significant loss of marginal vegetation. Increasing demand for irrigation water and drought has meant that the lake has been half-empty in recent years; fisheries have become less important as a result. The construction of a reservoir on an inflow to the lake will further reduce water-levels.

Yarışlı lake
A1, A4i, B1i **029**

Admin region Burdur
Coordinates 37°34'N 29°58'E
Altitude 915–915 m **Area** 1,400 ha

Site description
A shallow, spring-fed saline lake separated from Burdur lake (site 030) by a range of hills. The water-level fluctuates greatly and in summer the lake almost entirely dries up, leaving extensive saltmarshes and mudflats. To the north and west the lake is bordered by farmland, otherwise the shores are rocky and steep in places. Small vegetated areas occur around freshwater inputs.

Habitats Wetland (100%; mudflat/sandflat; saltmarsh; standing brackish and salt water; water-fringe vegetation)
Land-use Not utilized (100%)

Birds

Species	Season	Year	Pop min	Pop max	Acc	Criteria
Tadorna ferruginea Ruddy Shelduck	W	1996	206	1,736	—	A4i, B1i
Oxyura leucocephala White-headed Duck	W	1989	46	46	—	A1

This site is important for wintering wildfowl.

Protection status
National None **International** None

Conservation issues
No serious threats are known at the site.

Burdur lake
A1, A4i, A4iii, B1i, B2 **030**

Admin region Burdur, Isparta
Coordinates 37°44'N 30°11'E
Altitude 857–857 m **Area** 23,700 ha

Site description
A large, stream-fed saline lake (max. depth 110 m) of tectonic origin situated in a closed basin. Extensive alluvial flats lie to the north-east and south-west. The lake is flanked by forested hills and mountains; its near surrounds are under agriculture.

Habitats Wetland (100%; mudflat/sandflat; standing fresh water)
Land-use Not utilized (90%), Urban/industrial/transport (10%)

Birds

Species	Season	Year	Pop min	Pop max	Acc	Criteria
Podiceps nigricollis Black-necked Grebe	P	1995	5,163	25,280	—	A4i, B1i
Phoenicopterus ruber Greater Flamingo	P	—	—	6,500	—	A4i, B1i
Tadorna ferruginea Ruddy Shelduck	W	1996	235	550	—	A4i, B1i
Tadorna ferruginea Ruddy Shelduck	P	1995	—	405	—	A4i, B1i
Netta rufina Red-crested Pochard	W	1995	750	2,814	—	A4i, B1i
Aythya ferina Pochard	W	1995	11,160	57,555	—	A4i, B1i
Aythya fuligula Tufted Duck	W	1987	6,000	6,000	—	B1i
Oxyura leucocephala White-headed Duck	W	1996	342	10,927	—	A1, A4i, B1i
Fulica atra Coot	W	1995	34,067	138,925	—	A4i, B1i
Fulica atra Coot	P	—	—	300,000	—	A4i, B1i
Hoplopterus spinosus Spur-winged Plover	B	—	—	—	—	B2
Chlidonias niger Black Tern	P	—	—	2,105	—	A4i, B1i

The IBA is very important for wintering waterbirds, and is the single most important wintering site for *Oxyura leucocephala* (over two-thirds of the world wintering population), with large numbers occurring throughout the year except during the breeding season.

Protection status
National High **International** High

23,700 ha of IBA covered by Permanent Wildlife Reserve (Burdur Gölü, 38,125 ha). 23,700 ha of IBA covered by Ramsar Site (Burdur Gölü, 23,700 ha).

Conservation issues

Threats Agricultural intensification/expansion (U), Construction/impact of dyke/dam/barrage (B), Dredging/canalization (A), Extraction industry (U), Industrialization/urbanization (A), Infrastructure (A)

A long-term reduction in water-level has led to wetland loss and increasing salinity. Inflows have been reduced by reservoir construction, and average annual precipitation has fallen. The planned Bozçay–Karaçal dam/irrigation project will further reduce inflows and lead to alluvial flats south-west of the lake being converted to permanent arable land. Additional threats include disturbance and pollution from airport and industrial developments, pollution caused by mining waste, agricultural run-off and sewage. A hunting ban was introduced in 1993; educational/wardening activities take place.

Eğirdir lake — A1, A4i, A4iii, B1i — 031

Admin region Isparta
Coordinates 38°00'N 30°54'E
Altitude 918–918 m **Area** 47,250 ha

Site description

A large, nutrient-poor stream- and spring-fed lake of tectonic origin (max. depth 13 m); the northern part is shallower than the southern part. The northern and eastern shores support fringing reedbeds (*Phragmites*) and are bordered by arable land; elsewhere hills and mountains surround the lake. Lake water is used for irrigation and to provide drinking water for nearby towns.

Habitats Wetland (100%; standing fresh water; water-fringe vegetation)
Land-use Fisheries/aquaculture (50%), Water management (100%)

Birds

Species	Season	Year	Pop min	Pop max	Acc	Criteria
Phalacrocorax pygmeus Pygmy Cormorant	W	1990	81	81	—	A1
Netta rufina Red-crested Pochard	W	1996	600	1,503	—	A4i, B1i
Aythya ferina Pochard	W	1996	12,290	15,198	—	A4i, B1i
Aythya fuligula Tufted Duck	U	1994	6,500	12,665	—	B1i
Fulica atra Coot	W	1996	64,419	64,419	—	A4i, B1i

The IBA is important for wintering waterbirds.

Protection status

National High **International** None
47,250 ha of IBA covered by Drinking Water Reservoir (Eğirdir Gölü, 47,250 ha).

Conservation issues

Threats Agricultural intensification/expansion (U), Aquaculture/fisheries (B), Consequences of animal/plant introductions (B), Construction/impact of dyke/dam/barrage (A), Industrialization/urbanization (U)

The lake's water-level has fallen over the last 25 years, possibly as a result of human intervention and reduced precipitation. Further irrigation and hydroelectric schemes do not appear feasible. Former areas of marsh and steppe have been drained and reclaimed. Additional threats include pollution caused by agricultural run-off and industrial discharge (leading to fish-kills), and the effects of introduced pike-perch *Stizostedion lucioperca* on native fish stocks.

Karamık marshes — B2 — 032

Admin region Afyon
Coordinates 38°26'N 30°50'E
Altitude 1,002–1,002 m **Area** 4,500 ha

Site description

A freshwater marsh comprising extensive *Phragmites* and *Typha* beds interspersed with areas of open water. The only water-supply is run-off from the surrounding hills and a number of springs. Arable fields and pasture border the marsh in the east and west; extensive poplar plantations are found to the north and north-east. Cattle graze the marshes, and reed-cutting takes place ('Other' land-use).

Habitats Wetland (90%; standing fresh water; water-fringe vegetation; fen/transition mire/spring), Artificial landscape (10%; arable land)
Land-use Agriculture, Other

Birds

Species	Season	Year	Pop min	Pop max	Acc	Criteria
Ixobrychus minutus Little Bittern	B	—	25	25	—	B2
Aythya nyroca Ferruginous Duck	B	—	10	10	—	B2

This site is important for breeding waterbirds. Species of global conservation concern that do not meet IBA criteria: *Oxyura leucocephala* (thought to breed). The site was of much greater importance before water-levels were lowered.

Protection status

National High **International** None
4,100 ha of IBA covered by SİT (Karamık Gölü, 4,100 ha).

Conservation issues

Threats Agricultural intensification/expansion (A), Disturbance to birds (C), Industrialization/urbanization (U)

Irrigation water is extracted from the marshes, and as a result the wetland area has decreased drastically over recent decades. The SEKA state paper factory discharges waste water into Karamık Sazlığı, with the risk of nutrient enrichment. Local people are complaining about the decreasing quality and height of reed as a result of this pollution. Out-of-season fishing takes place on a large scale.

Akşehir and Eber lakes — A1, A4i, B1i, B2 — 033

Admin region Afyon, Konya
Coordinates 38°36'N 31°18'E
Altitude 966–967 m **Area** 53,600 ha

Site description

Two connected lakes in the Akarçay closed basin. Akşehir (max. 35,300 ha, max. depth 7 m) is a stream-fed tectonic lake, fresh in the west and south, becoming increasingly saline towards the north. Eber (max. 16,800 ha, max. depth 6 m) is situated 4 km to the north-west and is fed by the Akarçay stream. Large areas of reed *Phragmites* occur in both lakes, which were treated as separate IBAs in the previous international IBA inventory (Grimmett and Jones 1989): Akşehir Gölü (former site TR019) and Eber Gölü (former site TR022).

Habitats Wetland (100%; standing fresh water; water-fringe vegetation)
Land-use Agriculture (10%), Fisheries/aquaculture (50%), Hunting (5%), Not utilized (35%)

Birds

Species	Season	Year	Pop min	Pop max	Acc	Criteria
Phalacrocorax pygmeus Pygmy Cormorant	B	—	50	50	—	A1, B2
Pelecanus onocrotalus White Pelican	P	1988	2,000	2,000	—	A4i, B1i
Pelecanus crispus Dalmatian Pelican	B	—	1	—	—	B2
Ardeola ralloides Squacco Heron	B	—	70	70	—	B2
Ardea purpurea Purple Heron	B	—	50	50	—	B2
Plegadis falcinellus Glossy Ibis	B	—	50	50	—	B2
Plegadis falcinellus Glossy Ibis	P	1988	680	1,536	—	A4i, B1i
Platalea leucorodia Spoonbill	B	—	15	15	—	B1i, B2
Anser albifrons White-fronted Goose	W	1990	7,568	15,900	—	A4i, B1i
Aythya nyroca Ferruginous Duck	B	—	10	10	—	B2
Himantopus himantopus Black-winged Stilt	B	1995	307	311	—	A4i, B1i
Charadrius leschenaultii Greater Sand Plover	B	—	—	—	—	B2
Gelochelidon nilotica Gull-billed Tern	B	—	20	20	—	B2

The IBA is important for breeding waterbirds, with large mixed heronries on reed islands in Akşehir lake. It used to be of major importance for wintering wildfowl before controls were introduced to the water regime.

Protection status

National High **International** None
52,100 ha of IBA covered by SİT (Akşehir and Eber Gölü, 52,100 ha).

Conservation issues

Threats Agricultural intensification/expansion (U), Construction/impact of dyke/dam/barrage (A), Dredging/canalization (A), Groundwater abstraction (U), Industrialization/urbanization (U)

Two dams are under construction on the Akarçay stream, leading to a further reduction in water input to both lakes. Given the reduction in size and increasing salinity of Akşehir lake over the last two decades, there is little hope for its long-term future. Industrial, agricultural and untreated urban pollutants enter the lakes from a variety of sources, leading to severe nutrient loading in Eber lake. Local fisheries have declined as a result of the pollution and decreasing water-levels.

Çavuşçu lake — A4i, A4iii, B1i, B2 — 034

Admin region Konya
Coordinates 38°21′N 31°53′E
Altitude 1,026–1,026 m **Area** 1,200 ha

Site description

A freshwater lake (the remnants of a 5,000 ha wetland) fed by the Battal and Çebişçi streams, containing scattered reedbeds (mainly *Phragmites*). Water leaves through a controlled outlet in the south and from a pumping station on the western shore. The lake is surrounded by sheep-grazed arid hills, with arable land, *Salix* stands, orchards and cattle-grazed meadows immediately to the north and south.

Habitats Wetland (90%; mudflat/sandflat; standing fresh water), Artificial landscape (10%; arable land)
Land-use Agriculture (10%), Fisheries/aquaculture (25%), Water management (100%)

Birds

Species	Season	Year	Pop min	Pop max	Acc	Criteria
Anser anser Greylag Goose	N	1995	330	330	—	B1i
Netta rufina Red-crested Pochard	W	1993	1,000	1,000	—	A4i, B1i
Chlidonias hybridus Whiskered Tern	B	—	100	100	—	B2

The site is important for breeding terns (Sterninae) and large numbers of wintering wildfowl. There are indications that the lake was of greater importance prior to manipulation of its water regime.

Protection status

National High **International** None
1,200 ha of IBA covered by SİT (Çavuşçu Gölü, 1,200 ha).

Conservation issues

Threats Industrialization/urbanization (A)

The high water-level maintained within the embanked lake has resulted in an impoverished flora. Some new marsh habitat may form as a result of silt accumulation in the lake, provided such areas are protected against agricultural encroachment. During 1995 an investigation was begun into the planned construction of a coal-fired power-station in the lake environs.

Altıntaş plain — A1, B2 — 035

Admin region Kütahya
Coordinates 39°07′N 30°07′E
Altitude 970–1,698 m **Area** 19,200 ha

Site description

An extensive agricultural and upland area south of Kütahya, bordered to the south by the Kokar stream. Low-lying agricultural areas are in the main non-irrigated; nearly half of such areas lie fallow in each rotation. Where groundwater reaches the surface small marshy areas have developed (some have been drained in recent years). The northern parts of the IBA comprise low mountains that support extensive areas of *Juniperus* forest and scrub.

Habitats Forest and woodland (30%; wooded steppe), Artificial landscape (70%; arable land)
Land-use Agriculture (25%), Not utilized (75%)

Birds

Species	Season	Year	Pop min	Pop max	Acc	Criteria
[1] ***Otis tarda*** Great Bustard	B	—	20	20	—	A1, B2

1. Pairs.

The availability of fallow land, absence of agricultural intensification schemes and relatively low human disturbance have made this one of the last *Otis tarda* strongholds in western Turkey—that the birds appear to be resident and breeding has been confirmed every year since 1993.

Protection status

National High **International** None
19,200 ha of IBA covered by Permanent Wildlife Reserve (Altıntaş Ovası, 19,200 ha).

Conservation issues

Threats Agricultural intensification/expansion (A)

A DSİ irrigation project requires the damming of the Kokar and Murat streams and will lead to the irrigation of 17,763 ha, including the entire lower part of the IBA. Agricultural intensification and increasing human disturbance will result, undoubtedly affecting *Otis tarda*. Local hunters successfully campaigned for the area to be protected and now warden the site. Poaching has been virtually eliminated, and educating local farmers has proved successful. Restoration of marsh areas is being discussed.

Türkmenbaba Dağı — A1, B1iii, B2 — 036

Admin region Eskişehir, Kütahya
Coordinates 39°26′N 30°23′E
Altitude 1,200–1,826 m **Area** 17,500 ha

Site description

Situated at the northern end of the Yazılıkaya range, between the Kütahya, Eskişehir and Afyon plains. In contrast to its barren surroundings, the mountain is covered by extensive coniferous forests (including large old-growth stands) and meadows; deciduous forests prevail at lower altitudes. Cliffs are found in the summit area. Controlled grazing takes place, and forestry management is extensive, with small-scale selective felling.

Habitats Forest and woodland (100%; broadleaved deciduous forest; native coniferous forest), Rocky areas (inland cliff)
Land-use Agriculture (5%), Forestry (95%)

Birds

Species	Season	Year	Pop min	Pop max	Acc	Criteria
Ciconia nigra Black Stork	B	—	5	5	—	B2
Gypaetus barbatus Lammergeier	B	—	1	1	—	B2
Aegypius monachus Cinereous Vulture	B	1995	10	10	—	A1, B1iii, B2
Hieraaetus pennatus Booted Eagle	B	—	5	5	—	B2

The undisturbed and remote nature of the IBA accounts for its impressive diversity of raptors; it holds some of the highest numbers of *Aegypius monachus* known in Turkey.

Protection status

National Partial **International** None
5,000 ha of IBA covered by Permanent Wildlife Reserve (Türkmenbaba Dağı, 5,000 ha). IBA also overlaps with SİT (Türkmenbaba Dağı).

Conservation issues

Threats Disturbance to birds (B), Selective logging/cutting (C)

Forestry workers can cause disturbance to breeding raptors; forest tracks have been opened throughout the IBA. Forestry activities should take account of the distribution of breeding raptors (leaving an unmanaged buffer zone around each breeding valley), and old pine trees should not be felled.

Aliken

A1, B2 **037**

Admin region Eskişehir
Coordinates 39°19′N 31°17′E
Altitude 875–909 m **Area** 11,500 ha

Site description

Situated 30 km south-west of Sivrihisar, the southern border is formed by the Aliken stream. The higher parts of this plateau are barren and agriculture is non-intensive (with land being utilized in alternate years). In a typical year 50% of the area supports rain-fed wheat and barley fields, 40% is fallow and 10% uncultivated rocky steppe. Human disturbance is minimal. In contrast, the c.1 km wide strip of land adjoining the Aliken stream is more intensively farmed.

Habitats Grassland (10%; steppe/dry calcareous grassland), Artificial landscape (90%; arable land)
Land-use Agriculture (50%), Not utilized (50%)

Birds

Species	Season	Year	Pop min	Pop max	Acc	Criteria
[1] ***Otis tarda*** Great Bustard	B	1996	20	20	—	A1, B2

1. Pairs. Up to 45 individuals recorded.

Protection status

National None **International** None

Conservation issues

Threats Agricultural intensification/expansion (B), Drainage (U)

A large drainage channel was dug in 1948 and present construction works will result in the irrigation of 3,970 ha of land in the Aliken valley. 2,630 ha of marsh along the Aliken stream (partially outside the IBA) have already been claimed for agriculture; any remaining marsh will disappear following completion of the project.

Balıkdamı

A1, A4i, A4iii, B1i, B3 **038**

Admin region Eskişehir
Coordinates 39°12′N 31°39′E
Altitude 799–799 m **Area** 1,470 ha

Site description

Balıkdamı is one of the few remaining wetlands on the Sakarya river, elsewhere flood-plain and marsh habitats have been reclaimed for agricultural use. The eastern part of the IBA comprises 503 ha of reedbed (*Phragmites*), open water and mudflats. The western part consists of extensive areas of marsh and grassland. Human activities include cereal production and extensive stock-grazing.

Habitats Grassland (humid grassland), Wetland (80%; mudflat/sandflat; standing fresh water; water-fringe vegetation), Artificial landscape (20%; arable land)
Land-use Agriculture (25%), Hunting (25%), Not utilized (50%)

Birds

Species	Season	Year	Pop min	Pop max	Acc	Criteria
Ciconia ciconia White Stork	P	1993	17,915	17,915	—	A4i, B1i
Tadorna ferruginea Ruddy Shelduck	W	1996	302	302	—	A4i, B1i
Circus pygargus Montagu's Harrier	B	—	15	15	—	B3
Falco naumanni Lesser Kestrel	B	—	10	10	—	A1

The IBA is important for breeding raptors and herons (Ardeidae) and as a stop-over site for passage *Ciconia ciconia* and waterbirds. Species of global conservation concern that do not meet IBA criteria: *Phalacrocorax pygmeus* (thought to breed).

Protection status

National High **International** None
1,470 ha of IBA covered by Permanent Wildlife Reserve (Balıkdamı, 1,470 ha). 1,470 ha of IBA covered by SİT (Balıkdamı, 1,470 ha).

Conservation issues

Threats Agricultural intensification/expansion (A), Construction/impact of dyke/dam/barrage (A)

DSİ plans to irrigate the upper reaches of the Sakarya river, a project that will involve the reclamation of the entire western part of the IBA. A permanent lake of c.500 ha will also be created in the eastern part of the IBA, resulting in the loss of virtually all the ecologically important marsh and mudflat areas.

Uyuz lake

A1, B2 **039**

Admin region Konya
Coordinates 39°15′N 32°57′E
Altitude 1,185–1,185 m **Area** 15 ha

Site description

A small freshwater lake (max. depth 1.5 m) situated south of Çöl lake (IBA 040), fringed by reedbeds (*Phragmites*) and fed by groundwater. It is surrounded by arable land, which reaches the shoreline in places. Local people harvest reed ('Other' land-use).

Habitats Wetland (100%; standing fresh water; water-fringe vegetation)
Land-use Hunting (10%), Not utilized (85%), Other (5%)

Birds

Species	Season	Year	Pop min	Pop max	Acc	Criteria
Oxyura leucocephala White-headed Duck	B	1994	10	10	—	A1, B2

The area is important for breeding waterbirds. *Falco naumanni* breeds just outside the IBA.

Protection status

National High **International** None
15 ha of IBA covered by SİT (Uyuz Gölü, 15 ha).

Conservation issues

No serious threats are known at the site.

Çöl lake

A1, A4i, A4iii, B1i, B2 **040**

Admin region Ankara
Coordinates 39°18′N 32°54′E
Altitude 1,045–1,045 m **Area** 4,700 ha

Site description

A shallow saline lake (max. 1,500 ha) lying in a closed basin (Çalıkdüzü) 75 km south of Ankara. It is fed by small streams and largely dries up during summer. Wet grazing-marshes and extensive *Typha* beds are found at the southern and northern ends of the lake. Arable cultivation takes place in the higher, drier areas.

Habitats Wetland (70%; saltmarsh; standing brackish and salt water, water-fringe vegetation), Artificial landscape (30%; arable land)
Land-use Agriculture (25%), Not utilized (75%)

Birds

Species	Season	Year	Pop min	Pop max	Acc	Criteria
Phoenicopterus ruber Greater Flamingo	U	1986	5,500	5,500	—	A4i, B1i
Anser albifrons White-fronted Goose	W	1970	9,365	14,000	—	A4i, B1i
Tadorna ferruginea Ruddy Shelduck	B	1990	300	300	—	A4i, B1i, B2
Tadorna ferruginea Ruddy Shelduck	W	1970	1,296	6,847	—	A4i, B1i
Anas crecca Teal	W	1969	10,486	10,486	—	B1i
Oxyura leucocephala White-headed Duck	N	1991	27	27	—	A1
Falco naumanni Lesser Kestrel	B	1998	—	20	—	A1
Himantopus himantopus Black-winged Stilt	B	1986	1,000	1,000	—	A4i, B1i
Charadrius leschenaultii Greater Sand Plover	B	—	—	—	—	B2

The site supports significant numbers of breeding *Charadrius leschenaultii*. Large numbers of wildfowl winter, whilst *Phoenicopterus ruber*, *Oxyura leucocephala* and *Himantopus himantopus* occur principally during the post-breeding season.

Protection status

National None **International** None

Conservation issues
No serious threats are known at the site.

Mogan lake
A1, A4i, A4iii, B1i, B2 **041**

Admin region Ankara
Coordinates 39°46'N 32°48'E
Altitude 973–973 m **Area** 1,500 ha

Site description
A nutrient-rich, stream-fed freshwater lake (max. 759 ha, max. depth 2.5 m) 20 km south of Ankara. Reedbeds (*Phragmites*) fringe much of the lake shore, with marshland and wet meadows to the south. The town of Gölbaşı is situated on the north-east shore. Numerous holiday homes have been constructed overlooking the lake, which is also popular for rowing, sailing and angling.

Habitats Grassland (10%; humid grassland), Wetland (90%; standing fresh water; water-fringe vegetation)
Land-use Agriculture (5%), Not utilized (45%), Tourism/recreation (30%), Urban/industrial/transport (20%)

Birds

Species	Season	Year	Pop min	Pop max	Acc	Criteria
Ardeola ralloides Squacco Heron	B	—	30	30	—	B2
Netta rufina Red-crested Pochard	B	—	50	50	—	B2
Netta rufina Red-crested Pochard	W	1993	673	673	—	B1i
Aythya nyroca Ferruginous Duck	B	—	10	10	—	B2
Aythya nyroca Ferruginous Duck	W	1994	150	200	—	A1
Oxyura leucocephala White-headed Duck	B	—	2	2	—	B2
Fulica atra Coot	U	1994	25,000	70,100	—	A4i, B1i

Important site for breeding waterbirds, with large numbers of wildfowl occurring in late autumn and early spring (the lake freezes in winter).

Protection status
National None **International** High
1,500 ha of IBA covered by Mediterranean Specially Protected Area (Mogan Gölü, 24,500 ha).

Conservation issues
Threats Construction/impact of dyke/dam/barrage (B), Industrialization/urbanization (B), Recreation/tourism (A)

Formerly the lake size varied between 500 and 1,020 ha—as part of the Ankara flood-protection scheme a regulator was built, allowing a constant water-level to be maintained. The growth of Gölbaşı has resulted in increased disturbance and the release of industrial and untreated urban waste to the lake. Sewage-treatment plants and interceptors are now under construction.

Kızılcahamam forests
A1, B2 **042**

Admin region Ankara
Coordinates 40°28'N 32°33'E
Altitude 900–1,849 m **Area** 9,500 ha

Site description
A forested area situated at the northern edge of the barren Central Anatolian plateau, dominated by *Pinus nigra* with *P. sylvestris* at higher altitudes. Pasture and scrub (mainly *Quercus*) occur throughout. The forests are extensively managed except within the National Park, which attracts large numbers of visitors from Ankara.

Habitats Forest and woodland (90%; mixed forest), Grassland (10%; alpine/subalpine/boreal grassland)
Land-use Forestry (50%), Nature conservation/research (25%), Tourism/recreation (25%)

Birds

Species	Season	Year	Pop min	Pop max	Acc	Criteria
Ciconia nigra Black Stork	B	—	5	5	—	B2
Gypaetus barbatus Lammergeier	B	—	2	2	—	B2
Neophron percnopterus Egyptian Vulture	B	—	15	15	—	B2
Gyps fulvus Griffon Vulture	B	—	2	2	—	B2
Aegypius monachus Cinereous Vulture	B	—	6	6	—	A1, B2
Aquila chrysaetos Golden Eagle	B	1998	0	2	—	B2
Hieraaetus pennatus Booted Eagle	B	—	3	3	—	B2

This is an important site for a number of species of raptor.

Protection status
National Partial **International** None
1,050 ha of IBA covered by National Park (Kızılcahamam, 1,050 ha).

Conservation issues
Threats Disturbance to birds (U), Selective logging/cutting (U)

A dense network of roads and tracks means that disturbance from tourists and foresters is high. Forestry activities should take account of the distribution of breeding raptors and old pine trees should not be felled. A reduction in forestry activities within the National Park (hence maintaining the availability of tall nesting trees) is thought to be the main reason behind increasing numbers of *Aegypius monachus*.

Kavaklı Dağı
A1, B2 **043**

Admin region Ankara, Bolu
Coordinates 40°21'N 32°08'E
Altitude 1,000–1,983 m **Area** 8,100 ha

Site description
Kavaklı Dağı, located 100 km north-west of Ankara, constitutes the southern extremity of the vast Köroğlu mountain range. The lower parts consist of barren hills with *Quercus* scrub whilst the upper slopes are covered with conifer forests.

Habitats Forest and woodland (100%; broadleaved deciduous forest; native coniferous forest)
Land-use Forestry (25%), Not utilized (75%)

Birds

Species	Season	Year	Pop min	Pop max	Acc	Criteria
Gypaetus barbatus Lammergeier	B	1998	0	1	—	B2
Aegypius monachus Cinereous Vulture	B	—	5	5	—	A1, B2

Aegypius monachus breeds in a steep-sided south-facing valley.

Protection status
National High **International** None
8,100 ha of IBA covered by Permanent Wildlife Reserve (Kavaklı, 8,200 ha).

Conservation issues
Threats Selective logging/cutting (C)

Forestry activities should take account of the distribution of breeding raptors (leaving an unmanaged buffer zone around each breeding valley), and old pine trees should not be felled.

Inözü gorge
B2 **044**

Admin region Ankara
Coordinates 40°11'N 31°55'E
Altitude 675–750 m **Area** 50 ha

Site description
The site comprises rocky cliffs, c.100 m in height, that stretch for nearly 1 km along a narrow gorge north of Beypazarı. Orchards are found along the river.

Habitats Rocky areas (90%; inland cliff), Artificial landscape (10%; perennial crops/orchards/groves)
Land-use Agriculture (5%), Not utilized (95%)

Birds
The area is important for cliff-nesting species.

Species	Season	Year	Pop min	Pop max	Acc	Criteria
Ciconia nigra Black Stork	B	—	5	5	—	B2
Falco biarmicus Lanner	B	—	1	1	—	B2

Protection status

National None **International** None

Conservation issues

No serious threats are known at the site.

Sarıyar reservoir — A4i, B1i, B2 — 045

Admin region Ankara, Eskişehir
Coordinates 40°02′N 31°38′E
Altitude 475–475 m **Area** 8,400 ha

Site description

This IBA includes a reservoir (max. depth 79 m), temporary wetlands at its inflows and cliff-faces to the north. It lies between Hamam Dağı (site 046) to the south and the Kapıkırı mountains to the north. The reservoir is fed by three rivers and is used to produce hydroelectric power. *Populus* and *Salix* plantations border the seasonal wetlands.

Habitats Wetland (95%; mudflat/sandflat; standing fresh water), Rocky areas (5%; inland cliff)
Land-use Fisheries/aquaculture (50%), Urban/industrial/transport (100%), Water management (100%)

Birds

Species	Season	Year	Pop min	Pop max	Acc	Criteria
Nycticorax nycticorax Night Heron	B	—	120	120	—	B2
Ciconia nigra Black Stork	B	—	20	20	—	B2
Ciconia ciconia White Stork	P	1994	11,300	11,300	—	A4i, B1i
Tadorna ferruginea Ruddy Shelduck	W	1995	238	2,400	—	A4i, B1i
Neophron percnopterus Egyptian Vulture	B	—	10	10	—	B2
Falco biarmicus Lanner	B	—	1	1	—	B2

This site is important for tree- and cliff-nesting waterbirds and raptors. Large numbers of *Ciconia ciconia* roost here on migration. Species of global conservation concern that do not meet IBA criteria: *Haliaeetus albicilla* (frequently observed, thought to breed in the surrounds).

Protection status

National Low **International** None
425 ha of IBA covered by Permanent Wildlife Reserve (Nallıhan kuscenneti, 425 ha).

Conservation issues

Threats Disturbance to birds (B), Dredging/canalization (C), Industrialization/urbanization (U), Selective logging/cutting (U)

The felling of *Populus* stands threatens breeding species; a management plan could guarantee a rotation scheme whereby different sections of the stands are felled over successive years. The reservoir is heavily polluted, especially by untreated waste from Ankara (treatment facilities are being constructed). The effects of the pollution have not been studied.

Hamam Dağı — A1, B2 — 046

Admin region Eskişehir
Coordinates 39°52′N 31°36′E
Altitude 900–1,574 m **Area** 14,000 ha

Site description

The IBA lies at the easternmost extremity of the 130 km long Sündiken mountain range, which lies on an east–west axis between the Sakarya and Porsuk plains. Extensive forests of *Pinus nigra* are interspersed with *Quercus*/*Juniperus* scrub, a few scattered villages and areas of open grassland. The steeper northern slopes face Sarıyar reservoir (site 045); the gently sloping south-facing valleys meet the semi-arid, intensively cultivated Central Anatolian plateau.

Habitats Forest and woodland (90%; broadleaved deciduous forest; native coniferous forest), Grassland (10%; steppe/dry calcareous grassland)
Land-use Agriculture (10%), Forestry (50%)

Birds

Species	Season	Year	Pop min	Pop max	Acc	Criteria
Aegypius monachus Cinereous Vulture	B	1995	5	5	—	A1, B2
Hieraaetus pennatus Booted Eagle	B	—	1	1	—	B2

The IBA is important for breeding raptors.

Protection status

National None **International** None

Conservation issues

Threats Intensified forest management (U), Selective logging/cutting (B)

A newly constructed forest road appears to have forced two pairs of *Aegypius monachus* to nest higher up the valley. Forestry activities should take account of the distribution of breeding raptors (leaving an unmanaged buffer zone around each breeding valley), and old pine trees should not be felled.

Sapanca lake — A4i, A4iii, B1i — 047

Admin region Kocaeli, Sakarya
Coordinates 40°42′N 30°15′E
Altitude 31–31 m **Area** 4,700 ha

Site description

A freshwater lake (max. depth 61 m) formerly connected to the Sea of Marmara and fed by the Sakarya river. Arable land, fruit orchards and extensive *Populus* plantations surround the lake. Narrow reedbeds (*Phragmites*) fringe most of its shores; only at the eastern end does substantial marsh vegetation occur. Lake water is abstracted for industrial use.

Habitats Wetland (100%; standing fresh water; water-fringe vegetation)
Land-use Fisheries/aquaculture (25%), Urban/industrial/transport (100%), Water management (100%)

Birds

Species	Season	Year	Pop min	Pop max	Acc	Criteria
Netta rufina Red-crested Pochard	W	1995	1,002	1,002	—	A4i, B1i
Aythya ferina Pochard	W	1995	10,400	10,400	—	B1i
Fulica atra Coot	W	1996	30,150	30,700	—	B1i

The site is important for wintering wildfowl.

Protection status

National High **International** None
4,700 ha of IBA covered by Drinking Water Reservoir (Sapanca Gölü, 4,700 ha). 4,700 ha of IBA covered by SİT (Sapanca Gölü, 4,700 ha).

Conservation issues

Threats Agricultural intensification/expansion (U), Industrialization/urbanization (B), Recreation/tourism (B)

The lake is facing severe pollution problems from rapidly developing industries in its catchment, and from untreated sewage and agricultural run-off. Sapanca has become popular as a second-home resort; construction is taking place on the forested hillsides south of the lake.

Ilgaz mountains — B2 — 048

Admin region Kastamonu, Çankırı
Coordinates 41°06′N 33°52′E
Altitude 2,000–2,587 m **Area** 35,000 ha

Site description

Part of the Ilgaz mountain range that stretches 160 km on a south-west/north-east axis, forming the boundary between the wet Black Sea region and the dry Central Anatolian plateau. The IBA encompasses the highest and most intact parts of the range, including the two Ilgaz passes. *Pinus* and *Abies* forest covers large areas, with alpine grassland and rocky habitats at higher altitudes. A ski-resort is located within the National Park.

Habitats Forest and woodland (70%; native coniferous forest), Grassland (30%; alpine/subalpine/boreal grassland)
Land-use Agriculture (10%), Forestry (25%), Not utilized (35%), Tourism/recreation (25%), Urban/industrial/transport (5%)

Birds

Species	Season	Year	Pop min	Pop max	Acc	Criteria
Gypaetus barbatus Lammergeier	B	—	2	2	—	B2
Gyps fulvus Griffon Vulture	B	—	1	—	—	B2
Hieraaetus pennatus Booted Eagle	B	—	5	5	—	B2

This is an important site for breeding raptors. Other breeding species of interest include *Tadorna ferruginea*, *Dryocopus martius*, *Irania gutturalis*, *Sitta krueperi* and *Serinus pusillus*.

Protection status

National Low **International** None
1,090 ha of IBA covered by National Park (Ilgaz Dağı, 1,090 ha).

Conservation issues

Threats Deforestation (commercial) (A), Recreation/tourism (B), Selective logging/cutting (A)

Forests inside and outside the National Park are intensively managed.

Sarıkum lake A1, A4iii 049

Admin region Sinop
Coordinates 42°01′N 34°55′E
Altitude 0–0 m **Area** 785 ha

Site description

A complex of dune, lake and forest habitats situated in a broad valley. Large areas of a shallow, stream-fed brackish coastal lake are vegetated with *Juncus* and *Phragmites*. South of the lake is an extensive, seasonally flooded *Fraxinus* forest. *Quercus* and *Carpinus* forests surround the lake on drier ground; areas of dune have been forested with *Pinus*. Cattle graze the wetland.

Habitats Forest and woodland (50%; broadleaved deciduous forest), Wetland (50%; sand-dunes/sand beach; coastal lagoon; water-fringe vegetation)
Land-use Agriculture (5%), Nature conservation/research (90%)

Birds

Species	Season	Year	Pop min	Pop max	Acc	Criteria
Oxyura leucocephala White-headed Duck	W	1995	55	55	—	A1

The area is important for large numbers of wintering wildfowl.

Protection status

National High **International** None
785 ha of IBA covered by Nature Reserve (Sarıkum Gölü, 785 ha).
785 ha of IBA covered by SİT (Sarıkum Gölü, 785 ha).

Conservation issues

Threats Afforestation (U), Natural events (B), Recreation/tourism (U)

Large numbers of visitors cause considerable disturbance at weekends. The lake is reportedly silting up rapidly and the area of reedbed is expanding as a result. Afforestation of the dunes will lead to further loss of important dune vegetation. Fishing is banned according to Nature Reserve regulations.

Tödürge lake B2 050

Admin region Sivas
Coordinates 39°53′N 37°36′E
Altitude 1,295–1,295 m **Area** 750 ha

Site description

A slightly brackish lake (326 ha, average depth 4 m) situated in the upper Kızılırmak basin, fed by springs and the Acısu stream. Reed *Phragmites* fringes the lake and there is one small island in the east. The surrounds consist of arable/fallow land and pasture, except in the west where extensive wet grasslands are grazed by cattle.

Habitats Grassland (20%; humid grassland), Wetland (75%; standing fresh water; water-fringe vegetation), Artificial landscape (5%; arable land)
Land-use Agriculture (5%), Not utilized (90%), Tourism/recreation (5%)

Birds

Species	Season	Year	Pop min	Pop max	Acc	Criteria
Netta rufina Red-crested Pochard	B	—	40	40	—	B2

This is an important area for breeding and passage waterbirds. Other species of note include breeding *Motacilla citreola*.

Protection status

National None **International** None

Conservation issues

Threats Agricultural intensification/expansion (U)

An ongoing DSİ project will see the irrigation of 11,220 ha along the Kızılırmak river (including land around the lake).

Palas lake A4i, B1i, B2 051

Admin region Kayseri
Coordinates 39°02′N 35°49′E
Altitude 1,132–1,132 m **Area** 2,720 ha

Site description

A saline lake (max. 2,330 ha) north-east of Kayseri, separated from the Kızılırmak river by a low mountain range and fed mainly by surface run-off. Extensive areas of mudflat, saltmarsh and freshwater marsh occur along the eastern shore; vegetation along the southern and south-eastern shores is dominated by *Juncus*. The lake is surrounded by arable land, orchards and pasture, and largely dries up during summer.

Habitats Wetland (100%; mudflat/sandflat; saltmarsh; standing brackish and salt water; fen/transition mire/spring)
Land-use Agriculture (10%), Not utilized (90%)

Birds

Species	Season	Year	Pop min	Pop max	Acc	Criteria
[1] ***Tadorna ferruginea*** Ruddy Shelduck	U	1987	660	2,930	—	A4i, B1i
Charadrius leschenaultii Greater Sand Plover	B	—	—	—	—	B2

1. Occurs in autumn.

A WIWO study undertaken in spring 1988 revealed that the IBA is an important stop-over site for several wader species.

Protection status

National High **International** None
2,720 ha of IBA covered by SİT (Palas Gölü, 2,720 ha).

Conservation issues

Threats Agricultural intensification/expansion (C), Construction/impact of dyke/dam/barrage (A)

In 1995 DSİ started constructing the Bahçelik reservoir on the Zamantı river, 55 km south-east of the IBA. The lake's water-level may drop as a result of this scheme, leading to the permanent loss of valuable marsh areas.

Sultansazlığı A1, A4i, A4iii, B1i, B2, B3 052

Admin region Kayseri
Coordinates 38°20′N 35°16′E
Altitude 1,074–1,074 m **Area** 39,000 ha

Site description

The IBA lies in the Develi closed basin, surrounded by mountains. The wetland comprises a series of salt, brackish and freshwater lakes, and extensive marshes, and reaches 20,000 ha at its maximum extent. The salt-lakes are surrounded by *Salicornia* steppe; the southern marshes comprise vast reedbeds (*Phragmites*) interspersed with expanses of *Typha*, *Juncus* and *Carex*. Human activities include sheep- and cattle-grazing, and reed-cutting ('Other' land-use).

Habitats Grassland (40%; steppe/dry calcareous grassland), Wetland (55%; mudflat/sandflat; standing brackish and salt water; water-fringe vegetation), Artificial landscape (5%; arable land)
Land-use Agriculture (10%), Nature conservation/research (50%), Not utilized (20%), Other, Tourism/recreation (20%)

Birds

Species	Season	Year	Pop min	Pop max	Acc	Criteria
Phalacrocorax pygmeus Pygmy Cormorant	B	—	200	200	—	A1, A4i, B1i, B2
Phalacrocorax pygmeus Pygmy Cormorant	W	1996	50	850	—	A1, A4i, B1i
Ardeola ralloides Squacco Heron	B	—	70	70	—	B2
Plegadis falcinellus Glossy Ibis	B	—	75	75	—	B2
Platalea leucorodia Spoonbill	P	1988	138	138	—	A4i, B1i
[1] ***Phoenicopterus ruber*** Greater Flamingo	P	1994	6,500	59,150	—	A4i, B1i
Anser anser Greylag Goose	P	1994	250	250	—	B1i
Tadorna ferruginea Ruddy Shelduck	W	1990	267	539	—	A4i, B1i
[2] ***Tadorna ferruginea*** Ruddy Shelduck	N	1994	640	2,283	—	A4i, B1i
Anas strepera Gadwall	B	—	20	20	—	B2
Anas crecca Teal	W	1999	—	23,732	—	A4i, B1i
Marmaronetta angustirostris Marbled Duck	B	—	5	5	—	A1, B1i, B2
Marmaronetta angustirostris Marbled Duck	P	1998	—	121	—	A1, A4i, B1i
Netta rufina Red-crested Pochard	B	—	400	400	—	A4i, B1i, B2
Aythya nyroca Ferruginous Duck	B	—	20	20	—	A1, B2
Oxyura leucocephala White-headed Duck	B	—	20	20	—	A1, B2
Grus grus Crane	B	—	20	20	—	B2
Grus grus Crane	P	1994	1,200	1,200	—	B1i
Himantopus himantopus Black-winged Stilt	B	1994	390	390	—	A4i, B1i
Recurvirostra avosetta Avocet	B	—	150	150	—	B1i, B3
Recurvirostra avosetta Avocet	W	1994	2,115	2,115	—	A4i, B1i, B2
Glareola pratincola Collared Pratincole	B	—	50	50	—	B1i, B2
Charadrius alexandrinus Kentish Plover	B	—	100	100	—	B1i, B2
Charadrius leschenaultii Greater Sand Plover	B	—	—	—	—	B2
Hoplopterus spinosus Spur-winged Plover	B	—	20	20	—	B2
Gelochelidon nilotica Gull-billed Tern	B	—	50	50	—	B1i, B2
Sterna albifrons Little Tern	B	—	100	100	—	B2
Chlidonias hybridus Whiskered Tern	B	—	400	400	—	A4i, B1i, B2

1. Breeds irregularly.
2. Occurs in summer.

The site holds significant numbers of a range of breeding, wintering and passage waterbirds. Large numbers of *Phoenicopterus ruber*, *Grus grus* and *Recurvirostra avosetta* use the wetland in autumn. *Chettusia leucura* apparently bred in 1996.

Protection status

National High **International** Partial
17,200 ha of IBA covered by Nature Reserve (Sultansazlığı, 17,200 ha). 39,000 ha of IBA covered by Permanent Wildlife Reserve (Sultansazlığı, 45,000 ha). 17,200 ha of IBA covered by SİT (Sultansazlığı, 17,200 ha). 17,200 ha of IBA covered by Ramsar Site (Sultansazlığı, 17,200 ha).

Conservation issues

Threats Agricultural intensification/expansion (A), Drainage (A), Industrialization/urbanization (U), Other (U), Recreation/tourism (C), Unsustainable exploitation (U)

Agricultural intensification and associated water management in the basin poses a serious threat—reduced water inputs and a lack of rainfall led to the wetland drying up in 1990 and 1991. The wetland can only be preserved if further expansion of the irrigation scheme is cancelled. Industrial, agricultural and untreated urban waste enters through drainage channels. Tourists cause disturbance, and uncontrolled reed-cutting takes place ('Other' threat). Illegal hunting of falcons occurs on the surrounding steppe. The management plan covers developing ecotourism at the site.

Seyfe lake — A1, A4i, A4iii, B1i, B2, B3 — 053

Admin region Kirşehir
Coordinates 39°12'N 34°25'E
Altitude 1,110–1,110 m **Area** 14,000 ha

Site description

A shallow stream- and spring-fed brackish lake (max. 9,350 ha, average winter depth 1 m) situated at the southern end of the Seyfe basin. The shoreline is flat and muddy, and the lake area decreases during summer. Arable fields are present to the south and west, with steppe in the north and north-east. East of Kızıldağyeniyapan village are 1,500 ha of brackish–freshwater marsh. Twelve scrub- and grass-covered islands are present. Human activities include sheep- and cattle-grazing.

Habitats Grassland (10%; steppe/dry calcareous grassland), Wetland (80%; saltmarsh; standing brackish and salt water), Artificial landscape (10%; arable land)
Land-use Agriculture (10%), Hunting (5%), Not utilized (85%)

Birds

Species	Season	Year	Pop min	Pop max	Acc	Criteria
Pelecanus onocrotalus White Pelican	B	—	100	100	—	B2
Platalea leucorodia Spoonbill	B	—	50	50	—	A4i, B1i, B2
Phoenicopterus ruber Greater Flamingo	B	—	2,000	2,000	—	A4i, B1i, B2
Phoenicopterus ruber Greater Flamingo	N	1996	5,490	16,000	—	A4i, B1i
Anser albifrons White-fronted Goose	W	1989	7,200	7,200	—	B1i
Tadorna ferruginea Ruddy Shelduck	W	1996	610	978	—	A4i, B1i
Tadorna ferruginea Ruddy Shelduck	U	1993	352	500	—	A4i, B1i
Netta rufina Red-crested Pochard	B	—	15	15	—	B2
Falco naumanni Lesser Kestrel	U	1996	59	59	—	A1
Otis tarda Great Bustard	B	—	1	—	—	B2
Recurvirostra avosetta Avocet	B	—	500	500	—	A4i, B1i, B3
Hoplopterus spinosus Spur-winged Plover	B	—	10	10	—	B2
Gelochelidon nilotica Gull-billed Tern	B	—	500	500	—	A4i, B1i, B2
Sterna albifrons Little Tern	B	—	500	500	—	A4i, B1i, B2

This site is important for a range of breeding waterbirds and for large numbers of wintering wildfowl. Species of global conservation concern that do not meet IBA criteria: *Marmaronetta angustirostris* (may breed).

Protection status

National High **International** Partial
10,700 ha of IBA covered by Nature Reserve (Seyfe Gölü, 10,700 ha). 14,000 ha of IBA covered by SİT (Seyfe Gölü, 23,585 ha). 10,700 ha of IBA covered by Ramsar Site (Seyfe Gölü, 10,700 ha).

Conservation issues

Threats Agricultural intensification/expansion (U), Drainage (A), Dredging/canalization (A)

A DSİ project is underway to reduce the lake's area by diverting inflows. Land north and east of the lake will be reclaimed for agriculture, and most of the breeding islands and all of the brackish–freshwater marsh will disappear. Provisions will be made to divert water to the lake in times of drought, in theory creating a more stable lake area. Artificial breeding islands will be created. The ecological risks associated with the project are great, and a thorough environmental impact assessment has not been carried out.

Hirfanlı reservoir — A1, A4i, A4iii, B1i — 054

Admin region Aksaray, Ankara, Kirşehir
Coordinates 39°10'N 33°39'E
Altitude 851–851 m **Area** 26,300 ha

Site description

A large reservoir (max. depth 70 m) on the Kızılırmak river, north-east of Tuz lake (site 072). Up to 7,000 ha of mudflats appear during periods of low water-levels. The area is surrounded by arable land. The dam provides hydroelectric power and serves flood-control purposes.

Habitats Wetland (100%; mudflat/sandflat; standing fresh water)
Land-use Fisheries/aquaculture, Urban/industrial/transport, Water management

Birds

Species	Season	Year	Pop min	Pop max	Acc	Criteria
[1] ***Tadorna ferruginea*** Ruddy Shelduck	P	1996	2,070	2,070	—	A4i, B1i
Netta rufina Red-crested Pochard	W	1996	824	3,560	—	A4i, B1i
Aythya ferina Pochard	W	1996	13,430	13,430	—	B1i
Aythya fuligula Tufted Duck	W	1996	14,550	14,550	—	B1i
Oxyura leucocephala White-headed Duck	W	1996	19	122	—	A1, B1i
Fulica atra Coot	W	1996	68,350	68,350	—	A4i, B1i

1. Occurs in spring.

The site is very important for wintering waterbirds, with counts of up to 134,000 birds recorded in recent years. Terns (Sterninae) breed in small numbers on the islands.

Protection status

National None **International** None

Conservation issues

Threats Agricultural intensification/expansion (B), Disturbance to birds (B)

Proposals are being developed to irrigate 13,500 ha adjacent to the IBA with water from the reservoir. This may lead to agricultural run-off flowing back into the reservoir. Hunting causes disturbance.

Beynam forest A1, B2 055

Admin region Ankara
Coordinates 39°41′N 32°56′E
Altitude 1,200–1,521 m **Area** 2,100 ha

Site description

A small remnant of forest situated on the northern side of Kuyrukçu mountain, 30 km south of Ankara. The higher areas are cut by steep valleys and dry riverbeds. *Pinus* and *Quercus* cover 1,248 ha, the remainder is more open with low vegetation. The entire forest is fenced off to prevent grazing, and a small section is used for recreational purposes. No forestry activities are carried out although some reforestation has taken place on the south-western slopes.

Habitats Forest and woodland (100%; mixed forest)
Land-use Forestry (5%), Not utilized (85%), Tourism/recreation (10%)

Birds

Species	Season	Year	Pop min	Pop max	Acc	Criteria
Aquila heliaca Imperial Eagle	B	—	2	2	—	A1, B2

The area is important for breeding raptors.

Protection status

National High **International** None
2,100 ha of IBA covered by Preservation Forest (Beynam Ormanı, 2,100 ha). 2,100 ha of IBA covered by SİT (Beynam Ormanı, 2,100 ha).

Conservation issues

Threats Recreation/tourism (U)

The rapidly increasing population of Ankara could lead to intense recreational pressure in the future; the development of a tourist management plan would be beneficial.

Kızılırmak delta A1, A4i, A4ii, A4iii, B1i, B1ii, B2 056

Admin region Samsun
Coordinates 41°36′N 36°05′E
Altitude 0–0 m **Area** 16,110 ha

Site description

The largest area of intact wetland on the Turkish Black Sea coast, comprising open water, marsh vegetation (including vast reedbeds), sand-dunes, farmland and remnant woodland (including *Alnus/Fraxinus* seasonally-flooded forest). The lake edges support *Phragmites* and *Typha*; the dunes support maquis-like vegetation. Outside the IBA, the delta is used for grazing and intensive arable production. Reed-cutting is an important economic activity ('Other' land-use).

Habitats Forest and woodland (5%; alluvial/very wet forest), Wetland (90%; sand-dunes/sand beach; coastal lagoon; water-fringe vegetation), Artificial landscape (5%; arable land)
Land-use Agriculture (10%), Fisheries/aquaculture (75%), Hunting (25%), Other, Tourism/recreation (10%)

Birds

Species	Season	Year	Pop min	Pop max	Acc	Criteria
Puffinus yelkouan Yelkouan Shearwater	W	1995	1,000	1,000	—	A4ii, B1ii
Phalacrocorax pygmeus Pygmy Cormorant	W	1992	67	88	—	A1
Pelecanus crispus Dalmatian Pelican	B	—	6	6	—	B2
Egretta garzetta Little Egret	B	1992	460	3,200	—	A4i, B1i
Egretta alba Great White Egret	W	1989	210	210	—	A4i, B1i
Ardea purpurea Purple Heron	B	—	500	500	—	B2
Ciconia nigra Black Stork	B	—	50	50	—	B2
Plegadis falcinellus Glossy Ibis	P	1992	590	590	—	A4i, B1i
Platalea leucorodia Spoonbill	B	—	75	75	—	A4i, B1i, B2
Anser anser Greylag Goose	W	1993	322	322	—	B1i
Anas strepera Gadwall	B	—	200	200	—	B2
Anas clypeata Shoveler	W	1999	—	5,930	—	A4i, B1i
Netta rufina Red-crested Pochard	B	—	75	75	—	B2
Netta rufina Red-crested Pochard	W	1999	—	1,826	—	A4i, B1i
Aythya ferina Pochard	W	1996	14,952	14,952	—	A4i, B1i
Aythya nyroca Ferruginous Duck	B	—	150	150	—	A1, B2
Melanitta fusca Velvet Scoter	W	1996	75	97	—	B1i
Oxyura leucocephala White-headed Duck	W	1995	15	1,246	—	A1, A4i, B1i
Fulica atra Coot	W	1999	—	49,000	—	A4i, B1i
Grus grus Crane	B	—	50	50	—	B2
Burhinus oedicnemus Stone Curlew	B	—	50	50	—	B2
Glareola pratincola Collared Pratincole	B	—	100	100	—	A4i, B1i, B2
Larus minutus Little Gull	P	1993	3,000	41,000	—	A4i, B1i
Chlidonias leucopterus White-winged Black Tern	P	1987	3,000	3,000	—	A4i, B1i

The site is very important for breeding, wintering and passage waterbirds, with over 10,000 waders passing through the delta in spring. Thirty-five species of passerine breed in the forests and *Haliaeetus albicilla* has bred in the past.

Protection status

National High **International** Partial
4,000 ha of IBA covered by Permanent Wildlife Reserve (Kızılırmak Deltası, 4,000 ha). 11,000 ha of IBA covered by SİT (Kızılırmak Deltası, 11,000 ha). 11,000 ha of IBA covered by Ramsar Site (Kızılırmak Deltası, 11,000 ha).

Conservation issues

Threats Agricultural intensification/expansion (A), Aquaculture/fisheries (U), Burning of vegetation (C), Construction/impact of dyke/dam/barrage (A), Deforestation (commercial) (A), Disturbance to birds (C), Drainage (A), Dredging/canalization (A), Extraction industry (U), Industrialization/urbanization (U), Recreation/tourism (B), Unsustainable exploitation (U)

The Bafra irrigation project will claim a large part of the remaining wetland. Other major threats include pollution from agricultural run-off and untreated sewage, and the construction of holiday homes. Illegal hunting still takes place. Fish catches have declined in the eastern half of the delta, possibly as a result of over-fishing and pollution. Sand extraction, although illegal, is common. Reed-burning occurs throughout the year to improve grazing conditions. DHKD ran a conservation project from 1992 to 1997 and lobbied for suitable protection for the area. A 1996 management plan regulates all land-use in the delta, and places restrictions on the construction of holiday homes.

Yedikır reservoir A4i, B1i 057

Admin region Amasya
Coordinates 40°48′N 35°34′E
Altitude 517–517 m **Area** 593 ha

Site description

A small reservoir situated north-west of Amasya, fed by the Tersakan stream and used to irrigate 7,966 ha in the adjacent Suluova plain.

There is virtually no shoreline vegetation. DSİ operates a small fish-farm next to the dam and releases fish into the lake.

Habitats Wetland (100%; standing fresh water; water-fringe vegetation)
Land-use Fisheries/aquaculture (5%), Water management (100%)

■ Birds

Species	Season	Year	Pop min	Pop max	Acc	Criteria
Phalacrocorax carbo Cormorant	W	1999	—	2,000	—	B1i
Tadorna ferruginea Ruddy Shelduck	W	1995	205	610	—	A4i, B1i
Tadorna ferruginea Ruddy Shelduck	P	1995	744	1,099	—	A4i, B1i

This is an important site for wintering and passage waterbirds.

■ Protection status

National High **International** None
593 ha covered by Permanent Wildlife Reserve (Yedikır Barajı, 817 ha). 593 ha covered by SİT (Yedikır Barajı, 817 ha).

■ Conservation issues

No serious threats are known at the site.

Yeşilırmak delta B1i, B1ii, B2 058

Admin region Samsun
Coordinates 41°18′N 36°55′E
Altitude 0–0 m **Area** 3,000 ha

■ Site description

Part of the largest delta on the Turkish Black Sea coast, the majority of which is now under agriculture. The IBA comprises the largest remaining wetland—the Simenlik–Akgöl lake complex—a 1,900-ha area in the eastern part of the delta of which 200 ha is open water (max. depth 3 m), and the remainder is reed *Phragmites* and marsh vegetation. Also included are dunes and *Cladium* fen communities. Reed-cutting takes place ('Other' land-use).

Habitats Forest and woodland (5%; mixed forest), Wetland (65%; sand-dunes/sand beach; coastal lagoon; river/stream; water-fringe vegetation), Artificial landscape (30%; arable land)
Land-use Agriculture (10%), Fisheries/aquaculture (10%), Forestry (5%), Hunting (20%), Not utilized (45%), Other, Tourism/recreation (10%)

■ Birds

Species	Season	Year	Pop min	Pop max	Acc	Criteria
Puffinus yelkouan Yelkouan Shearwater	W	1995	500	500	—	B1ii
Ardeola ralloides Squacco Heron	B	—	30	30	—	B2
Netta rufina Red-crested Pochard	W	1995	520	520	—	B1i
Melanitta fusca Velvet Scoter	W	1996	57	870	—	B1i

The site is important for breeding waterbirds and congregations of wintering wildfowl. The majority of the latter are forced to spend the day at sea as a result of human disturbance, returning to feed in the delta at night. No information is available on the natural importance of the delta prior to large-scale drainage.

■ Protection status

National High **International** None
3,000 ha of IBA covered by Permanent Wildlife Reserve (Yeşilırmak Deltası, 16,043 ha).

■ Conservation issues

Threats Afforestation (U), Agricultural intensification/expansion (A), Disturbance to birds (A), Drainage (A), Infrastructure (C), Intensified forest management (B)

Ongoing drainage and irrigation projects threaten the remaining areas of natural wetland. *Pinus* forests designed to stabilize the dunes and *Salix*/*Alnus* plantations threaten to outcompete native vegetation. The lake complex is fed entirely by irrigation drains, leading to severe nutrient enrichment and the lakes becoming choked by reed. Fish catches are reported to have declined as a result of pollution. An airport is planned north-west of Çarsamba, 25 km from the delta. Although closed to hunters, the area is subject to heavy hunting pressure throughout the year.

Akkuş island B1i 059

Admin region Ordu
Coordinates 41°07′N 37°47′E
Altitude 0–20 m **Area** 2 ha

■ Site description

A small, sparsely-vegetated island situated in the Black Sea west of Ordu, less than 100 m from the shore. The IBA lies adjacent to a major freshwater outflow into the sea and has therefore become popular with anglers.

Habitats Rocky areas (100%; rock stacks/islets)
Land-use Not utilized (95%), Tourism/recreation (5%)

■ Birds

Species	Season	Year	Pop min	Pop max	Acc	Criteria
Phalacrocorax aristotelis Shag	B	—	90	90	—	B1i

The area is important for cliff-nesting species.

■ Protection status

National None **International** None

■ Conservation issues

Threats Disturbance to birds (C)

The anglers cause disturbance to breeding birds, a situation exacerbated by the installation of night-time lighting.

North-east Turkey A1, A3, A4iv, B1iii, B1iv, B2 060

Admin region Artvin, Erzurum, Trabzon, Gümüşhane, Bayburt, Rize
Coordinates 40°50′N 41°10′E
Altitude 0–3,932 m **Area** 1,230,000 ha

■ Site description

The IBA covers three major mountain ranges (Soğanlı, Kaçkar and Karçal). *Pinus* and *Abies* dominate the upper slopes, with *Quercus* scrub and open forest at lower altitudes. Above the treeline, extensive alpine grasslands and slopes covered with *Rhododendron* and *Juniperus* scrub prevail; summit areas are characterized by cliffs, screes and lakes. The thin coastal strip is densely populated, and the IBA as a whole is an increasingly popular tourist destination. This area includes two sites that were treated as separate IBAs in the previous international IBA inventory (Grimmett and Jones 1989): 'Kaçkar Dağları' (former IBA TR013) and 'North-east Turkey' (former IBA TR016).

Habitats Forest and woodland (60%; broadleaved deciduous forest; native coniferous forest), Grassland (30%; alpine/subalpine/boreal grassland), Rocky areas (inland cliff), Artificial landscape (10%; arable land)
Land-use Agriculture (10%), Forestry (50%), Not utilized (35%), Tourism/recreation (5%)

■ Birds

Species	Season	Year	Pop min	Pop max	Acc	Criteria
Gypaetus barbatus Lammergeier	B	—	20	20	—	B2
Gyps fulvus Griffon Vulture	B	—	20	20	—	B2
Aegypius monachus Cinereous Vulture	B	—	10	10	—	A1, B1iii, B2
Aquila chrysaetos Golden Eagle	B	—	10	10	—	B2
[1] ***Tetrao mlokosiewiczi*** Caucasian Black Grouse	B	1993	134	—	—	A1, A3, B2
Tetraogallus caspius Caspian Snowcock	B	—	—	—	—	A3, B2

1. Lekking males.

Important for breeding raptors and species of upland habitats (the only Turkish IBA to qualify as a Eurasian high montane (alpine) biome site); contains the entire Turkish range of *Tetrao mlokosiewiczi*. Includes a major migratory bottleneck, where large numbers of raptors regularly pass in autumn. A count of 380,220 birds was made in 1976, including 138,000 *Pernis apivorus*, 5,775 *Milvus migrans* and 205,000 *Buteo buteo*. Actual numbers may be higher as important new passage points have been discovered in recent years. Spring passage involves smaller numbers (max. 205,131 during an eight-week count in 1994), including 25,183 *Pernis apivorus*, 9,069 *Milvus migrans* and 136,327 *Buteo buteo*.

Protection status

National Partial **International** None

17,104 ha of IBA covered by National Park (Hatila Vadisi, 17,104 ha). 51,550 ha of IBA covered by National Park (Kaçkar Dağları, 51,550 ha). 1,625 ha of IBA covered by Nature Park (Uzungöl, 1,625 ha). 180 ha of IBA covered by Nature Reserve (Çamburnu, 180 ha). 8,700 ha of IBA covered by Permanent Wildlife Reserve (Çoruh Vadisi, 8,700 ha). 50,458 ha of IBA covered by Permanent Wildlife Reserve (Verçenik, 50,458 ha). 3,492 ha of IBA covered by Permanent Wildlife Reserve (Şavşat Balıklı, 3,492 ha). 4,143 ha of IBA covered by Permanent Wildlife Reserve (Kaçkar, 4,143 ha).

Conservation issues

Threats Construction/impact of dyke/dam/barrage (A), Deforestation (commercial) (A), Disturbance to birds (A), Extraction industry (B), Unsustainable exploitation (U)

A large dam is to be constructed on the Çoruh river near Artvin. Large areas of forest have been degraded by commercial deforestation; DHKD has recommended changes in management techniques that should lead to the preservation of remaining old-growth forest. Mining pollution is a major problem, and raptors following the eastern Black Sea migration route are subject to heavy persecution by local people. DHKD has carried out educational activities in the main trapping area, and ran a WWF-funded conservation project in the IBA from 1993 to 1997.

Samsam lake — B1i — 061

Admin region Konya
Coordinates 39°06′N 32°45′E
Altitude 980–980 m **Area** 830 ha

Site description

A shallow freshwater lake fed mainly by surface run-off, and surrounded by grassy plains and arable land. When the water-level drops, the entire lake is gradually colonized by temporary vegetation—in late summer it is practically dry. Human activities include cattle-grazing.

Habitats Wetland (100%; mudflat/sandflat; standing fresh water)
Land-use Agriculture (10%)

Birds

Species	Season	Year	Pop min	Pop max	Acc	Criteria
Recurvirostra avosetta Avocet	P	1995	616	616	—	B1i

The area is also important for wintering waterbirds. No information is available on the ornithological importance of the lake prior to drainage.

Protection status

National High **International** None

830 ha of IBA covered by SİT (Samsam Gölü, 830 ha).

Conservation issues

Threats Disturbance to birds (U), Drainage (A)

The IBA is situated on a route frequently used by low-flying military jets, which can cause considerable disturbance to birds. It is over 25 years since the lake was partly drained—the soils created are still largely saline and not utilized by farmers, and it is unclear whether they will ever be of sufficient quality to justify the drainage scheme. Restoration would be technically feasible.

Kozanlı Gökgöl — A1, B2 — 062

Admin region Konya
Coordinates 39°01′N 32°50′E
Altitude 925–925 m **Area** 650 ha

Site description

The site comprises a 50 ha freshwater lake (partly covered by reedbeds (*Phragmites*)) and grazing-marsh between the lake and Kozanlı village, 20 km west of Kulu. The lake is fed by groundwater and a drainage channel from Samsam lake (site 061), 8 km to the north-west. A low barrier at the lake's outlet maintains an artificially high water-level; without this it would probably dry up during the summer. Human activities include cereal production and cattle-grazing.

Habitats Wetland (90%; standing fresh water; water-fringe vegetation), Artificial landscape (10%; arable land)
Land-use Agriculture (30%), Not utilized (70%)

Birds

Species	Season	Year	Pop min	Pop max	Acc	Criteria
Oxyura leucocephala White-headed Duck	B	—	10	10	—	A1, B2
Falco naumanni Lesser Kestrel	B	—	10	10	—	A1

The site is also important for several species of breeding waterbird. Up to 2,000 *Phoenicopterus ruber* from the Tuz lake colony (site 072) regularly feed in the marsh.

Protection status

National High **International** None

650 ha of IBA covered by SİT (Kozanlı Gökgöl, 650 ha).

Conservation issues

Threats Agricultural intensification/expansion (B), Disturbance to birds (U)

Grazing is intensive and should be limited both in extent and the number of animals. The IBA is situated on a frequently used low-flying route for military jets based in Ankara; considerable disturbance is caused when these pass low over the wetland.

Kulu lake — A1, A4i, B1i, B2, B3 — 063

Admin region Konya
Coordinates 39°05′N 33°09′E
Altitude 950–950 m **Area** 860 ha

Site description

A shallow brackish lake situated north-west of Tuz lake (site 072), 5 km east of the town of Kulu. Fed mainly by the Kulu stream, the lake is largely devoid of fringe vegetation, except for a small freshwater area in the south. Most of the nine islands are covered with grass in spring and early summer. Non-irrigated arable land, barren steppe and cattle-grazed grass plains surround the lake.

Habitats Wetland (100%; mudflat/sandflat; saltmarsh; standing brackish and salt water)
Land-use Agriculture (10%), Not utilized (90%)

Birds

Species	Season	Year	Pop min	Pop max	Acc	Criteria
[1] ***Podiceps nigricollis*** Black-necked Grebe	P	1994	2,055	2,055	—	A4i, B1i
Anser albifrons White-fronted Goose	W	1992	9,718	9,718	—	B1i
Tadorna ferruginea Ruddy Shelduck	N	1994	268	567	—	A4i, B1i
Marmaronetta angustirostris Marbled Duck	B	—	3	3	—	B2
Netta rufina Red-crested Pochard	B	—	20	20	—	B2
Aythya nyroca Ferruginous Duck	B	—	10	10	—	B2
Oxyura leucocephala White-headed Duck	B	—	30	30	—	A1, B2
Oxyura leucocephala White-headed Duck	W	1996	56	600	—	A1, A4i, B1i
Oxyura leucocephala White-headed Duck	N	1993	85	319	—	A1, A4i, B1i
Himantopus himantopus Black-winged Stilt	B	1986	400	4,000	—	A4i, B1i
Recurvirostra avosetta Avocet	B	—	250	250	—	B1i, B3
Recurvirostra avosetta Avocet	N	1987	1,000	1,000	—	B1i
Charadrius alexandrinus Kentish Plover	U	1993	300	300	—	B1i
Charadrius leschenaultii Greater Sand Plover	B	—	—	—	—	B2
Gelochelidon nilotica Gull-billed Tern	B	—	450	450	—	A4i, B1i, B2

1. Also breeds.

The IBA is important for breeding waterbirds. In winter the lake normally freezes but occasionally holds significant numbers of wildfowl.

Protection status

National High **International** None
860 ha of IBA covered by SİT (Kulu Gölü, 860 ha).

Conservation issues

Threats Disturbance to birds (U), Filling-in of wetlands (U), Industrialization/urbanization (U), Recreation/tourism (B)

Untreated waste water from Kulu enters the lake and the town's rubbish is dumped in the Kulu stream. Plans exist to improve access to the lake and to construct a road and watch towers around it, in order to promote tourism. This would undoubtedly threaten gulls and terns (Laridae) breeding on the islands, and breeding ducks. Hunting pressure would also increase.

Ereğli marshes — A1, A4i, B1i, B2, B3 — 064

Admin region Karaman, Konya
Coordinates 37°32′N 33°45′E
Altitude 998–998 m **Area** 37,000 ha

Site description

A complex of shallow marshes, reedbeds (*Phragmites*), freshwater lakes and salt-steppe situated in the Ereğli plain and bordered by mountains to the north and south. The lakes cover 6,400 ha at their maximum extent and are fed by drainage channels. The area also includes surrounding agricultural land, parts of which flood during winter. Human activities include crop production, sheep-grazing and reed-cutting ('Other' land-use).

Habitats Grassland (20%; steppe/dry calcareous grassland), Wetland (60%; mudflat/sandflat; saltmarsh; standing fresh water; water-fringe vegetation), Artificial landscape (20%; arable land)
Land-use Agriculture, Military, Other

Birds

Species	Season	Year	Pop min	Pop max	Acc	Criteria
Phalacrocorax pygmeus Pygmy Cormorant	B	—	600	600	—	A1, A4i, B1i, B2
Pelecanus onocrotalus White Pelican	B	—	23	23	—	B2
Pelecanus onocrotalus White Pelican	P	1993	1,000	1,000	—	A4i, B1i
[1] ***Pelecanus crispus*** Dalmatian Pelican	B	—	1	—	—	B2
Pelecanus crispus Dalmatian Pelican	U	1991	38	62	—	A1, A4i, B1i
Ardeola ralloides Squacco Heron	B	—	30	30	—	B2
Plegadis falcinellus Glossy Ibis	B	—	50	50	—	B2
Platalea leucorodia Spoonbill	B	—	20	20	—	B1i, B2
Platalea leucorodia Spoonbill	N	1991	250	250	—	A4i, B1i
Phoenicopterus ruber Greater Flamingo	B	—	300	300	—	A4i, B1i, B2
Tadorna ferruginea Ruddy Shelduck	W	1996	460	3,016	—	A4i, B1i
Anas strepera Gadwall	B	—	20	20	—	B2
Marmaronetta angustirostris Marbled Duck	B	—	5	5	—	A1, A4i, B1i, B2
Marmaronetta angustirostris Marbled Duck	N	1993	10	40	—	A1, B1i
Netta rufina Red-crested Pochard	B	—	500	500	—	A4i, B1i, B2
Aythya nyroca Ferruginous Duck	B	—	10	10	—	B2
Oxyura leucocephala White-headed Duck	B	—	50	50	—	A1, A4i, B1i, B2
Oxyura leucocephala White-headed Duck	N	1996	80	508	—	A1, A4i, B1i
Falco naumanni Lesser Kestrel	B	—	20	20	—	A1
Grus grus Crane	W	1993	1,000	1,000	—	B1i
Himantopus himantopus Black-winged Stilt	B	1991	600	600	—	A4i, B1i
Recurvirostra avosetta Avocet	B	—	75	75	—	B3
Glareola pratincola Collared Pratincole	B	—	50	50	—	B1i, B2
Charadrius leschenaultii Greater Sand Plover	B	—	—	—	—	B2
Hoplopterus spinosus Spur-winged Plover	B	—	20	20	—	B2
Sterna albifrons Little Tern	B	—	60	60	—	B2

1. Breeds irregularly.

The IBA is very important for a range of breeding waterbirds, although numbers have fallen in recent years due to low water-levels. Winter counts that have produced up to 30,000 waterbirds in the past have also shown a large decline, again due to low water-levels and the effects of severe winters.

Protection status

National High **International** None
6,787 ha of IBA covered by Nature Reserve (Ereğli Sazlığı, 6,787 ha).
37,000 ha of IBA covered by SİT (Ereğli Sazlığı, 37,000 ha).

Conservation issues

Threats Agricultural intensification/expansion (A), Construction/impact of dyke/dam/barrage (U), Disturbance to birds (U), Drainage (A), Industrialization/urbanization (B)

As a result of reduced water inputs the lake has retreated drastically, and thousands of hectares of wetland have been converted to agriculture. Further irrigation is planned. Inputs are expected to cease entirely once two new reservoirs have been constructed. Agricultural, industrial and untreated urban pollutants enter the lake via drainage channels. Military usage of the Ereğli plain may increase.

Karapınar plain — A4i, B1i, B2 — 065

Admin region Konya
Coordinates 37°49′N 33°40′E
Altitude 960–960 m **Area** 15,200 ha

Site description

A largely barren plain containing a shallow lake (max. 6,000 ha) fed by run-off from the Karacadağ mountain range to the east. Waste water from the township of Karapınar enters the lake via a drain, around the mouth of which a 200 ha fresh/brackish *Phragmites* marsh has developed. The lake is surrounded by grasslands and mud plains, and is virtually inaccessible during wet periods. During summer it may dry up completely. Human activities include sheep-grazing.

Habitats Grassland (30%; steppe/dry calcareous grassland), Wetland (70%; mudflat/sandflat; saltmarsh; standing fresh water)
Land-use Agriculture (60%), Hunting (25%), Not utilized (15%)

Birds

Species	Season	Year	Pop min	Pop max	Acc	Criteria
Anser albifrons White-fronted Goose	W	1987	14,000	14,000	—	A4i, B1i
Tadorna ferruginea Ruddy Shelduck	W	1996	477	900	—	A4i, B1i
Charadrius leschenaultii Greater Sand Plover	B	1998	30	30	—	B2

The site is important for breeding waders and wintering wildfowl. 30–50 pairs of *Pelecanus onocrotalus* were found in 1985 but have apparently not bred since.

Protection status

National Partial **International** None
6,000 ha of IBA covered by SİT (Karapınar Ovası, 6,000 ha).

Conservation issues

No serious threats are known at the site.

Eşmekaya marshes — A1, A4i, A4iii, B1i, B2, B3 — 066

Admin region Aksaray
Coordinates 38°15′N 33°28′E
Altitude 945–945 m **Area** 11,250 ha

Site description

The area comprises a permanent freshwater lake (groundwater-fed), large expanses of marsh and rush pasture, two salt-lakes (Akgöl and Bezirci) and a number of small satellite wetlands east of the main lake. *Juncus* covers most of the dry, salty soils, especially in the south. *Typha* and *Cladium* are common to the north with *Phragmites* being the dominant species around the main lake.

Habitats Wetland (80%; standing fresh water; water-fringe vegetation), Artificial landscape (20%; arable land)
Land-use Agriculture (10%), Hunting (10%), Not utilized (60%), Water management (20%)

Birds

This is an important area for breeding raptors and several wetland species. Significant numbers of wildfowl occur in winter. Species of global conservation concern that do not meet IBA criteria: *Circus macrourus* (only known Turkish breeding locality).

Species	Season	Year	Pop min	Pop max	Acc	Criteria
Ciconia ciconia White Stork	P	1994	9,550	9,550	—	A4i, B1i
Anser albifrons White-fronted Goose	W	1993	19,400	19,400	—	A4i, B1i
Netta rufina Red-crested Pochard	B	—	30	30	—	B2
Oxyura leucocephala White-headed Duck	B	1998	—	2	—	B2
Circus pygargus Montagu's Harrier	B	—	15	15	—	B3
Falco naumanni Lesser Kestrel	B	—	12	12	—	A1
Glareola pratincola Collared Pratincole	P	1994	150	150	—	B1i
Hoplopterus spinosus Spur-winged Plover	B	—	10	10	—	B2
Gelochelidon nilotica Gull-billed Tern	B	1998	—	50	—	B1i, B2

Protection status

National High **International** None

4,500 ha of IBA covered by Permanent Wildlife Reserve (Eşmekaya Sazlığı, 4,500 ha). 11,250 ha of IBA covered by SİT (Eşmekaya Sazlığı, 11,250 ha).

Conservation issues

Threats Agricultural intensification/expansion (A), Construction/impact of dyke/dam/barrage (A)

Work has begun on a DSİ project to turn Eşmekaya lake into a reservoir for irrigation purposes; the lake's ecological character will be radically altered as a result.

Beyşehir lake — A4i, A4iii, B1i, B2 — 067

Admin region Isparta, Konya
Coordinates 37°45'N 31°30'E
Altitude 1,123–1,123 m **Area** 73,000 ha

Site description

The largest freshwater lake in Turkey (max. depth 10 m), 75 km west of Konya. Fluctuating water-levels give a surface area of between 60,000 and 73,000 ha; the lake is fed by mountain streams and several springs. To the north is the Kızıldağ *Cedrus* forest; to the east the land is under agriculture. Three of the 30 islands are inhabited, others are farmed, the majority are deserted. Narrow reedbeds (*Phragmites*) fringe the shores with extensive tracts in bays in the east and south-west. Human activities include reed-cutting ('Other' land-use).

Habitats Wetland (95%; standing fresh water; water-fringe vegetation), Artificial landscape (5%; arable land)
Land-use Fisheries/aquaculture (20%), Other (10%), Water management (100%)

Birds

Species	Season	Year	Pop min	Pop max	Acc	Criteria
Ixobrychus minutus Little Bittern	B	1998	—	30	—	B2
Netta rufina Red-crested Pochard	B	1998	—	15	—	B2
Netta rufina Red-crested Pochard	W	1996	603	603	—	B1i
Aythya ferina Pochard	W	1996	15,700	47,883	—	A4i, B1i
Fulica atra Coot	W	1996	130,200	163,378	—	A4i, B1i

The site is important for large numbers of wintering wildfowl. Breeding colonies of *Phalacrocorax carbo*, *Pelecanus crispus*, *Nycticorax nycticorax* and *Egretta garzetta* were formerly present on the islands.

Protection status

National High **International** None

73,000 ha of IBA covered by Drinking Water Reservoir (Beyşehir Gölü, 73,000 ha). 73,000 ha of IBA covered by National Park (Beyşehir, 88,750 ha). 73,000 ha of IBA covered by SİT (Beyşehir Gölü, 73,000 ha).

Conservation issues

Threats Agricultural intensification/expansion (U), Aquaculture/fisheries (U), Consequences of animal/plant introductions (U), Construction/impact of dyke/dam/barrage (A), Industrialization/urbanization (U), Recreation/tourism (U)

Water-levels have fallen considerably in recent years, mainly as a result of abstraction—DSİ intends to increase the amount of water abstracted for irrigation schemes. The lake is polluted by untreated industrial discharges, waste from nearby villages and agricultural run-off. Holiday-home construction and associated sand extraction have begun on the shores west of Beyşehir. Fish diversity declined following the 1978 introduction of pike-perch *Stizostedion lucioperca*, and out-of-season fishing is a problem.

Suğla lake — A4iii — 068

Admin region Konya
Coordinates 37°20'N 32°00'E
Altitude 1,095–1,095 m **Area** 16,500 ha

Site description

The present stream-fed lake extends to 2,500 ha in winter, during exceptionally wet periods this can increase to 8,000 ha. Large parts of the former lakebed have been converted to arable land; reedbeds (*Phragmites*) occur only in the northern part. Forested mountains border the lake to the west, whilst orchards and *Populus* stands are located to the north and east. Cattle graze throughout.

Habitats Wetland (30%; mudflat/sandflat; standing fresh water), Artificial landscape (70%; arable land)
Land-use Agriculture (100%)

Birds

24,852 waterbirds were counted in March 1996 after an exceptionally wet winter, indicating the potential of the site.

Protection status

National None **International** None

Conservation issues

Threats Agricultural intensification/expansion (A), Construction/impact of dyke/dam/barrage (A), Drainage (A)

The wetland has been seriously damaged as a result of historical water-manipulation schemes. A DSİ project to convert Suğla Gölü into a 5,150 ha reservoir began in 1996; surrounding land will be drained and converted for arable use.

Hotamış marshes — A1, A4i, B1i, B2 — 069

Admin region Konya
Coordinates 37°35'N 33°03'E
Altitude 999–999 m **Area** 16,500 ha

Site description

A large freshwater marsh situated at the southern edge of the Konya plain, fed by surface run-off, stream and spring water. Areas of reedbed (*Phragmites*) are interspersed with patches of open water. The IBA includes Süleymanhacı, a groundwater-fed saline lake, 6 km south of the main wetland. Extensive areas of steppe and arable land surround the IBA. Reed-cutting is the main economic activity ('Other' land-use).

Habitats Grassland (40%; steppe/dry calcareous grassland), Wetland (40%; standing fresh water; water-fringe vegetation), Artificial landscape (20%; arable land)
Land-use Agriculture (60%), Fisheries/aquaculture (10%), Hunting (15%), Other (15%)

Birds

Species	Season	Year	Pop min	Pop max	Acc	Criteria
Phalacrocorax pygmeus Pygmy Cormorant	B	—	25	25	—	A1, B2
Ixobrychus minutus Little Bittern	B	—	30	30	—	B2
Ardeola ralloides Squacco Heron	B	—	50	50	—	B2
Plegadis falcinellus Glossy Ibis	B	—	75	75	—	B2
Marmaronetta angustirostris Marbled Duck	B	—	20	20	—	A1, A4i, B1i, B2
Netta rufina Red-crested Pochard	B	—	30	30	—	B2
Netta rufina Red-crested Pochard	W	1992	600	600	—	B1i
Aythya nyroca Ferruginous Duck	U	1994	89	89	—	A1

Species ... continued	Season	Year	Pop min	Pop max	Acc	Criteria
Oxyura leucocephala White-headed Duck	B	—	40	40	—	A1, B1i, B2
Oxyura leucocephala White-headed Duck	P	1991	37	354	—	A1, A4i, B1i
Himantopus himantopus Black-winged Stilt	B	1985	950	950	—	A4i, B1i
Glareola pratincola Collared Pratincole	B	—	100	100	—	A4i, B1i, B2
Charadrius leschenaultii Greater Sand Plover	B	—	—	—	—	B2
Hoplopterus spinosus Spur-winged Plover	B	—	40	40	—	A4i, B1i, B2

This is an important site for breeding waterbirds, numbers of which have fallen considerably in recent years as a result of low water-levels. *Pelecanus crispus* is thought to have bred in small numbers until 1990 and *Chettusia leucura* bred until 1986.

Protection status

National Partial **International** None
The northern half of the IBA covered by SİT (Hotamış Sazlığı).

Conservation issues

Threats Agricultural intensification/expansion (A), Construction/impact of dyke/dam/barrage (A), Drainage (A)

The area of wet marsh has decreased in size from c.16,000 ha in 1985 to a few hundred hectares today, as a result of inputs being diverted for irrigation. Current plans involve conversion to a reservoir; embankments will be constructed and the water-level raised to 15 m above its present level. All remaining natural habitats will undoubtedly disappear as a result. DSİ has promised mitigation measures (details unavailable at the time of writing). No environmental impact assessment has been carried out.

Bolluk lake — A4i, B1i, B2, B3 — 070

Admin region Konya
Coordinates 38°32′N 32°56′E
Altitude 925–925 m **Area** 1,150 ha

Site description

A highly saline lake located west of Tuz lake (site 072) and surrounded by arid hills and dry steppe-grasslands. The lake is fed by surface flow with additional inputs from sulphurous springs in the north and a channel that brings drainage water from Konya city. Economic activities include salt extraction.

Habitats Wetland (100%; mudflat/sandflat; standing brackish and salt water)
Land-use Not utilized (80%), Urban/industrial/transport (20%)

Birds

Species	Season	Year	Pop min	Pop max	Acc	Criteria
Platalea leucorodia Spoonbill	B	1998	45	45	—	A4i, B1i, B2
Platalea leucorodia Spoonbill	P	1994	400	400	—	A4i, B1i
Tadorna ferruginea Ruddy Shelduck	B	1994	310	310	—	A4i, B1i, B2
Recurvirostra avosetta Avocet	B	—	55	55	—	B3
Charadrius leschenaultii Greater Sand Plover	B	—	—	—	—	B2
Larus genei Slender-billed Gull	B	1994	1,942	2,000	—	A4i, B1i
Gelochelidon nilotica Gull-billed Tern	B	—	400	400	—	A4i, B1i, B2

The site is important for a range of breeding waterbirds.

Protection status

National High **International** None
1,150 ha of IBA covered by SİT (Bolluk Gölü, 1,150 ha).

Conservation issues

Threats Industrialization/urbanization (B)

The nutrient-rich inflow from the Konya drain is causing the salt-tolerant vegetation at the southern end of the lake to be replaced by freshwater marsh vegetation. The lake's hydrology is managed to suit industrial needs, and this, along with possible pollution from industrial processes, may have a negative effect on the lake's ecology.

Tersakan lake — A4i, A4iii, B1i, B2, B3 — 071

Admin region Konya
Coordinates 38°35′N 33°06′E
Altitude 920–920 m **Area** 6,400 ha

Site description

A saline lake situated to the west of Tuz lake (site 072), fed by springs, surface run-off and drainage. Large areas of mudflat and small areas of freshwater marsh occur; otherwise the lake has virtually no fringe vegetation. Non-irrigated arable fields surround the lake, sheep graze the shores and mining takes place at the Alkim saltpans to the north.

Habitats Wetland (100%; mudflat/sandflat; saltmarsh; standing brackish and salt water)
Land-use Agriculture, Hunting (10%), Not utilized (80%), Urban/industrial/transport (25%)

Birds

Species	Season	Year	Pop min	Pop max	Acc	Criteria
[1] *Phoenicopterus ruber* Greater Flamingo	N	—	—	24,000	—	A4i, B1i
Anser albifrons White-fronted Goose	W	1993	14,700	14,700	—	A4i, B1i
Recurvirostra avosetta Avocet	B	1998	50	50	—	B3
Charadrius leschenaultii Greater Sand Plover	B	—	—	20	—	B2
[1] *Larus genei* Slender-billed Gull	N	—	—	1,800	—	A4i, B1i

1. Feeds at IBA during breeding season.

In winter, significant numbers of wildfowl occur.

Protection status

National High **International** None
6,400 ha of IBA covered by SİT (Tersakan Gölü, 6,400 ha).

Conservation issues

Threats Industrialization/urbanization (B)

It seems likely that the lake was previously less saline. However, no data exist on the ornithological importance of the lake before conversion to salt production.

Tuz lake — A1, A4i, B1i, B1iii, B2, B3 — 072

Admin region Aksaray, Ankara, Konya
Coordinates 38°45′N 33°23′E
Altitude 905–905 m **Area** 260,000 ha

Site description

Two major streams, groundwater, and surface water feed the second largest lake in Turkey (max. 190,000 ha, max. spring depth 1.5 m). Brackish marshes have formed where channels and streams enter the lake. Tuz lake is extremely saline and almost completely dries up during the summer. The lake is surrounded by arable fields, except in the south and south-west where extensive seasonally-flooded salt-steppe occurs. Economic activities include salt extraction.

Habitats Grassland (10%; steppe/dry calcareous grassland), Wetland (90%; mudflat/sandflat; standing brackish and salt water)
Land-use Agriculture (5%), Hunting (5%), Not utilized (60%), Urban/industrial/transport (30%)

Birds

Species	Season	Year	Pop min	Pop max	Acc	Criteria
Platalea leucorodia Spoonbill	B	1998	—	5	—	B2
Phoenicopterus ruber Greater Flamingo	B	1992	14,000	14,000	—	A4i, B1i, B2
Anser albifrons White-fronted Goose	W	1992	6,618	12,500	—	B1i
Tadorna ferruginea Ruddy Shelduck	P	1998	—	350	—	A4i, B1i
Circus pygargus Montagu's Harrier	B	1998	—	40	—	B3
Falco naumanni Lesser Kestrel	B	1998	100	100	—	A1, B1iii, B2
Grus grus Crane	B	1998	—	20	—	B2
Tetrax tetrax Little Bustard	B	1998	—	20	—	A1
Otis tarda Great Bustard	B	1998	—	20	—	A1, B2
Recurvirostra avosetta Avocet	B	1998	150	150	—	B1i, B3

Species ... continued	Season	Year	Pop min	Pop max	Acc	Criteria
Glareola pratincola Collared Pratincole	B	1998	—	200	—	A4i, B1i, B2
Charadrius alexandrinus Kentish Plover	B	1998	—	400	—	A4i, B1i, B2
Charadrius leschenaultii Greater Sand Plover	B	1998	40	40	—	B2
Larus armenicus Armenian Gull	B	1998	—	450	—	A4i, B1i
Sterna caspia Caspian Tern	B	1998	—	5	—	B2

The main Turkish breeding colony of *Phoenicopterus ruber* is present on a group of islands in the southern part of the lake. *Falco naumanni* is a common breeder in surrounding villages.

■ Protection status

National Partial **International** None
190,000 ha of IBA covered by SİT (Tuz Gölü, 190,000 ha).

■ Conservation issues

Threats Agricultural intensification/expansion (A), Dredging/canalization (A), Industrialization/urbanization (U)

Industrial, agricultural and untreated urban waste enters the lake via the main Konya channel. The effects of the pollution are made worse by reduced inputs of fresh water following the construction of two reservoirs on the main inflows. Large tracts of salt-steppe have been converted to arable land. Plans exist to triple the size of the saltpans, increasing output to three million tons per annum.

Göksu delta A1, A4i, A4iii, A4iv, B1i, B1iv, B2 073

Admin region İçel
Coordinates 36°18′N 33°58′E
Altitude 0–0 m **Area** 14,480 ha

■ Site description

The majority of the delta is used intensively by man and comprises agricultural land, settlements, wetlands, sand-dunes and beaches. Akgöl lake is a permanent nutrient-rich freshwater lake supporting rich marsh vegetation (including extensive *Potamogeton*, *Scirpus*, *Typha* and *Phragmites* beds). Saltmarsh surrounds the major lakes; the dunes are of particular conservation interest due to their intact and dynamic nature.

Habitats Wetland (50%; saltmarsh; sand-dunes/sand beach; coastal lagoon; standing fresh water; river/stream; water-fringe vegetation), Artificial landscape (50%; arable land)
Land-use Agriculture (50%), Fisheries/aquaculture (10%), Nature conservation/ research (10%), Not utilized (20%), Tourism/recreation (10%)

■ Birds

Species	Season	Year	Pop min	Pop max	Acc	Criteria
Phalacrocorax pygmeus Pygmy Cormorant	W	1996	46	180	—	A1
Phalacrocorax pygmeus Pygmy Cormorant	N	1994	30	70	—	A1
Pelecanus crispus Dalmatian Pelican	W	1992	28	56	—	A1, A4i, B1i
Ixobrychus minutus Little Bittern	B	—	25	25	—	B2
Nycticorax nycticorax Night Heron	B	—	150	150	—	B2
Ardeola ralloides Squacco Heron	B	—	70	70	—	B2
Ardea purpurea Purple Heron	B	—	50	50	—	B2
Ciconia nigra Black Stork	P	1993	268	268	—	A4i, B1i
Ciconia ciconia White Stork	P	1991	22,000	22,000	—	A4i, B1i
[1] ***Plegadis falcinellus*** Glossy Ibis	P	1991	2,500	2,500	—	A4i, B1i
Anser anser Greylag Goose	W	1995	680	1,450	—	B1i
Anas penelope Wigeon	W	1996	6,150	11,648	—	B1i
Anas crecca Teal	W	1992	14,952	14,952	—	A4i, B1i
Anas clypeata Shoveler	W	1992	4,325	4,325	—	B1i
Marmaronetta angustirostris Marbled Duck	B	1995	50	50	—	A1, A4i, B1i, B2
Aythya nyroca Ferruginous Duck	B	1995	30	30	—	A1, B2
Aquila clanga Greater Spotted Eagle	W	1995	6	20	—	A1
Aquila heliaca Imperial Eagle	W	1993	6	6	—	A1
Francolinus francolinus Black Francolin	B	—	50	50	—	B2
Porphyrio porphyrio Purple Gallinule	B	—	300	300	—	B2
Fulica atra Coot	W	1995	25,000	36,240	—	B1i
Burhinus oedicnemus Stone Curlew	B	—	60	60	—	B2
Glareola pratincola Collared Pratincole	B	—	300	300	—	A4i, B1i, B2
Charadrius alexandrinus Kentish Plover	B	—	200	200	—	B1i, B2
Hoplopterus spinosus Spur-winged Plover	B	—	300	300	—	A4i, B1i, B2
Sterna albifrons Little Tern	B	—	150	150	—	B2

1. Also breeds.

The site is of major importance for a range of breeding waterbirds, including the largest numbers of *Marmaronetta angustirostris* known in Turkey. It is also important for wintering raptors and waterbirds, and as a stop-over site for large numbers of *Pelecanus onocrotalus* and *Ciconia ciconia*. Species of global conservation concern that do not meet IBA criteria: *Haliaeetus albicilla* (wintering) and *Larus audouinii*. 332 bird species have been recorded, 90 are thought to breed.

■ Protection status

National High **International** High
4,350 ha of IBA covered by Permanent Wildlife Reserve (Göksu Deltası, 4,350 ha). 14,480 ha of IBA covered by SİT (Göksu Deltası, 14,480 ha). 8,650 ha of IBA covered by Ramsar Site (Göksu Deltası, 8,650 ha). 14,480 ha of IBA covered by Mediterranean Specially Protected Area (Göksu Deltası, 23,600 ha).

■ Conservation issues

Threats Agricultural intensification/expansion (A), Aquaculture/fisheries (C), Construction/impact of dyke/dam/barrage (A), Industrialization/urbanization (B), Infrastructure (U), Recreation/tourism (A)

Planned dam and reservoir construction on the Göksu river will stop sediment inputs to the delta. Large areas of dune and marsh have been illegally converted to agriculture. A new drainage and irrigation project will lead to further intensification, increased nutrient and agrochemical inputs, and changes in the hydrological regime. Industrial discharges may adversely affect fisheries; illegal trawling is common. From 1990 to 1998 DHKD and others ran a project aimed at developing an integrated management plan.

Aydıncık islands A1, B2 074

Admin region İçel
Coordinates 36°09′N 33°21′E
Altitude 0–30 m **Area** 2 ha

■ Site description

Two small islands near Aydıncık on the Mediterranean coast. The larger is up to 30 m high and has steep barren cliffs on the eastern and western sides, whilst the northern and southern sides are gently sloping and vegetated. The smaller island is up to 5 m high and is fully vegetated.

Habitats Rocky areas (100%; rock stacks/islets)
Land-use Agriculture (20%), Not utilized (80%)

■ Birds

Species	Season	Year	Pop min	Pop max	Acc	Criteria
Larus audouinii Audouin's Gull	B	1987	30	30	—	A1, B2

In 1996 only 6 pairs of *Larus audouinii* bred on the small island, possibly as a result of increasing numbers of breeding *Larus cachinnans*.

■ Protection status

National None **International** None

■ Conservation issues

Threats Agricultural intensification/expansion (U), Disturbance to birds (B)

Goats intensively graze the islands during the winter. Fishermen and shepherds land frequently, causing disturbance to birds. The collection of gull (Laridae) eggs was considered to be a major threat in the past but recent reports indicate that this practice has now stopped.

Aladağlar

B2 **075**

Admin region Adana, Kayseri, Niğde
Coordinates 37°50′N 35°09′E
Altitude 1,250–3,756 m **Area** 54,514 ha

Site description

An extensive mountainous area situated in the eastern Taurus range. The western part of the IBA is barren and rocky; the eastern part is covered with forests comprising *Pinus*, *Abies*, *Cedrus* and *Juniperus*. Alpine pasture and exposed stony fell occurs at higher altitudes.

Habitats Forest and woodland (30%; mixed forest), Grassland (40%; alpine/subalpine/boreal grassland), Rocky areas (30%; inland cliff)
Land-use Nature conservation/research (10%), Not utilized (80%), Tourism/recreation (10%)

Birds

Species	Season	Year	Pop min	Pop max	Acc	Criteria
Gypaetus barbatus Lammergeier	B	—	5	5	—	B2
Gyps fulvus Griffon Vulture	B	—	10	10	—	B2
Aquila chrysaetos Golden Eagle	B	—	4	4	—	B2
Tetraogallus caspius Caspian Snowcock	B	—	—	—	—	B2

The site is important for breeding raptors and other upland species.

Protection status

National High **International** None
54,514 ha of IBA covered by National Park (Aladağlar, 54,514 ha). Two Permanent Wildlife Reserves (totalling 66,854 ha) largely overlap with the IBA.

Conservation issues

No serious threats are known at the site.

Tuzla lake

A1, A4i, A4iii, B1i, B2 **076**

Admin region Adana
Coordinates 36°42′N 35°03′E
Altitude 0–0 m **Area** 2,800 ha

Site description

The westernmost lake in the Çukurova delta, situated east of the River Seyhan mouth. Brackish for most of the year, becoming fresher following winter rains. A narrow line of dunes separates the lake from the sea (excepting a small channel). A fish-trap operates at the sea opening. Also includes a complex of freshwater marshes, saltmarshes and ponds south-east of the lake. Large parts of the wetland dry completely during summer, when water is used for irrigation.

Habitats Wetland (80%; mudflat/sandflat; saltmarsh; sand-dunes/sand beach; standing brackish and salt water), Artificial landscape (20%; arable land)
Land-use Agriculture (10%), Fisheries/aquaculture (5%), Hunting (5%), Not utilized (50%), Tourism/recreation (10%), Water management (20%)

Birds

Species	Season	Year	Pop min	Pop max	Acc	Criteria
Anas crecca Teal	W	1994	12,600	13,000	—	B1i
Marmaronetta angustirostris Marbled Duck	B	—	30	50	—	A1, A4i, B1i, B2
Francolinus francolinus Black Francolin	B	—	5	5	—	B2
Burhinus oedicnemus Stone Curlew	B	—	20	20	—	B2
Glareola pratincola Collared Pratincole	B	—	—	150	—	A4i, B1i, B2
Charadrius alexandrinus Kentish Plover	B	—	1,000	1,000	—	A4i, B1i, B2
Charadrius alexandrinus Kentish Plover	W	1996	330	394	—	B1i
Hoplopterus spinosus Spur-winged Plover	B	—	50	50	—	A4i, B1i, B2
Sterna albifrons Little Tern	B	—	—	300	—	A4i, B1i, B2

The site is important for breeding waterbirds and for wintering wildfowl. A 1990 WIWO study revealed that large numbers of waders stop-over during spring passage, but remain for relatively short periods due to the poor feeding conditions.

Protection status

National High **International** None
2,800 ha of IBA covered by Permanent Wildlife Reserve (Tuzla Gölü, 5,769 ha).

Conservation issues

Threats Agricultural intensification/expansion (A), Aquaculture/fisheries (U), Construction/impact of dyke/dam/barrage (A), Groundwater abstraction (U), Industrialization/urbanization (B), Recreation/tourism (U)

Drainage water from the DSİ Lower Seyhan irrigation project will reduce salinity in the lake—no environmental impact assessment has been carried out. Dunes to the north have been illegally converted to arable land by local farmers—groundwater is pumped up for irrigation. Tourists create litter and erode dune systems. Human settlements around Tuzla are growing at a fast rate. The annual fish catch is to be increased. A 1996 management plan regulates all land-use in the delta, and places restrictions on the construction of holiday homes.

Akyatan lake

A1, A4i, A4iii, B1i, B2 **077**

Admin region Adana
Coordinates 36°37′N 35°16′E
Altitude 0–0 m **Area** 14,000 ha

Site description

A former oxbow of the River Seyhan, this is the largest lagoon in the Çukurova delta (max. 7,500 ha, max. depth 4 m). Its salinity fluctuates with wet season run-off and inputs of drainage water. A narrow belt of reed *Phragmites* and rush occurs along the shore; extensive mudflats appear during the summer. Sand-dunes separate the sea from the lake, which is otherwise surrounded by arable land. Also includes the seasonal Yemişli lake, which extends to 800 ha during wet winters.

Habitats Wetland (90%; mudflat/sandflat; saltmarsh; sand-dunes/sand beach; coastal lagoon), Artificial landscape (10%; arable land; forestry plantation)
Land-use Agriculture (10%), Fisheries/aquaculture (20%), Forestry (20%), Hunting (20%), Not utilized (30%)

Birds

Species	Season	Year	Pop min	Pop max	Acc	Criteria
Phalacrocorax pygmeus Pygmy Cormorant	U	1990	45	45	—	A1
Egretta alba Great White Egret	W	1992	416	416	—	A4i, B1i
Phoenicopterus ruber Greater Flamingo	W	1999	5,125	18,930	—	A4i, B1i
Tadorna tadorna Shelduck	W	1992	854	854	—	B1i
Anas penelope Wigeon	W	1993	10,073	13,900	—	B1i
Marmaronetta angustirostris Marbled Duck	B	—	5	5	—	A1, B1i, B2
Aythya ferina Pochard	W	1992	14,500	16,801	—	A4i, B1i
Oxyura leucocephala White-headed Duck	W	1993	230	978	—	A1, A4i, B1i
Francolinus francolinus Black Francolin	B	—	75	75	—	B2
Porphyrio porphyrio Purple Gallinule	B	—	—	—	—	B2
Fulica atra Coot	W	1996	28,100	46,000	—	A4i, B1i
Himantopus himantopus Black-winged Stilt	U	1993	300	300	—	B1i
Recurvirostra avosetta Avocet	P	1993	601	1,589	—	A4i, B1i, B2
Burhinus oedicnemus Stone Curlew	B	—	—	—	—	B2
Charadrius alexandrinus Kentish Plover	B	—	—	—	—	B2
Charadrius alexandrinus Kentish Plover	P	1990	400	3,918	—	A4i, B1i
Hoplopterus spinosus Spur-winged Plover	B	—	—	—	—	B2
Calidris minuta Little Stint	P	1990	4,835	5,165	—	A4i, B1i
Sterna albifrons Little Tern	B	—	100	100	—	B2

This is an important area for breeding waterbirds, large numbers of wintering waterbirds and as a stop-over site for passage waders. The dunes provide breeding and staging habitat for warblers. A 1990 WIWO study revealed that passage waders remain for relatively short periods due to the poor feeding conditions. Yemişli lake formerly held the only known Turkish breeding population of *Porphyrio porphyrio* outside of the Göksu delta (site 073).

■ Protection status

National Partial **International** High

11,244 ha of IBA covered by Permanent Wildlife Reserve (Akyatan Gölü, 11,244 ha). 14,000 ha of IBA covered by Ramsar Site (Akyatan Gölü, 14,700 ha).

■ Conservation issues

Threats Agricultural intensification/expansion (A), Aquaculture/fisheries (B), Disturbance to birds (B), Drainage (A), Industrialization/urbanization (B), Infrastructure (U)

Areas of wetland will be drained as a result of the DSİ Lower Seyhan Irrigation project. Agricultural run-off and industrial waste from factories south of Adana enters the lake via the YD3 canal. Small permanent settlements and tourism developments are being established throughout the area. A large airport is to be built, possibly on the former Yemişli lake. Fish catches declined over 25 years to 1995. Implementation of the Permanent Wildlife Reserve designation is insufficient and poaching is frequently reported. A 1996 management plan regulates all land-use in the delta, and places restrictions on the construction of holiday homes.

Ağyatan lake — A1, A4i, A4iii, B1i, B1iv, B2 — 078

Admin region Adana
Coordinates 36°36′N 35°31′E
Altitude 0–0 m **Area** 2,200 ha

■ Site description

A lagoon (max. 1,130 ha, max. depth 3 m) situated west of the River Ceyhan mouth, fed largely by groundwater, but augmented by river water during periods of flood. With the exception of a narrow strait it is separated from the sea by high sand-dunes. Extensive grasslands occur to the north. The water-level fluctuates to a lesser extent than in other lagoons in the Çukurova delta, resulting in smaller areas of mudflat and saltmarsh. The strait holds a traditional fish trap.

Habitats Grassland (10%; humid grassland), Wetland (90%; mudflat/sandflat; saltmarsh; sand-dunes/sand beach; coastal lagoon)
Land-use Agriculture (10%), Fisheries/aquaculture (10%), Not utilized (80%)

■ Birds

Species	Season	Year	Pop min	Pop max	Acc	Criteria
Egretta alba Great White Egret	W	1993	147	147	—	A4i, B1i
Ciconia ciconia White Stork	P	1990	12,439	12,439	—	A4i, B1i
Anas penelope Wigeon	W	1993	14,320	14,320	—	B1i
Oxyura leucocephala White-headed Duck	W	1993	191	191	—	A1, A4i, B1i
Francolinus francolinus Black Francolin	B	—	10	10	—	B2
Fulica atra Coot	W	1992	26,000	26,000	—	B1i
Glareola pratincola Collared Pratincole	P	1990	115	115	—	B1i
Charadrius alexandrinus Kentish Plover	B	—	—	—	—	B2
Hoplopterus spinosus Spur-winged Plover	B	—	—	—	—	B2
Sterna albifrons Little Tern	B	—	—	—	—	B2

Important for breeding and wintering waterbirds and as a migratory bottleneck site, where more 12,000 *Ciconia ciconia* regularly pass.

■ Protection status

National None **International** None

■ Conservation issues

Threats Agricultural intensification/expansion (B)

Land north of Ağyatan Gölü will be irrigated as part of the Lower Seyhan Irrigation Project, possibly resulting in the disappearance of the wet grasslands to the north of the lake. A 1996 management plan regulates all land-use in the delta, and places restrictions on the construction of holiday homes.

Yumurtalık lagoons — A4i, A4iii, B1i, B2 — 079

Admin region Adana
Coordinates 36°44′N 35°41′E
Altitude 0–0 m **Area** 16,430 ha

■ Site description

A complex of lagoons, salt and freshwater marshes, tidal mudflats, pastures, reedbeds (*Phragmites*), dunes and *Pinus* forest, situated between the River Ceyhan mouth and Yumurtalık bay. The area is connected to the sea via several large inlets. Saltmarshes have been converted to arable land in places, and cattle graze extensive meadows in the west. Fish traps operate at several locations.

Habitats Forest and woodland (5%; native coniferous forest), Wetland (85%; mudflat/sandflat; saltmarsh; sand-dunes/sand beach; coastal lagoon; water-fringe vegetation), Artificial landscape (10%; arable land)
Land-use Agriculture (20%), Fisheries/aquaculture (50%), Hunting (25%), Not utilized (25%), Tourism/recreation (10%)

■ Birds

Species	Season	Year	Pop min	Pop max	Acc	Criteria
Anser anser Greylag Goose	W	1990	301	301	—	B1i
Anas penelope Wigeon	W	1999	7,189	27,190	—	A4i, B1i
Francolinus francolinus Black Francolin	B	—	5	5	—	B2
Recurvirostra avosetta Avocet	W	1989	440	440	—	B1i, B2
Charadrius alexandrinus Kentish Plover	B	—	—	—	—	B2
Charadrius alexandrinus Kentish Plover	W	1992	356	805	—	B1i
Calidris minuta Little Stint	W	1996	2,200	2,200	—	A4i, B1i
Sterna albifrons Little Tern	B	—	—	—	—	B2

The area is important for breeding waders and terns *Sterna* and for large numbers of wintering waterbirds. A 1987 WIWO study revealed that large numbers of waders stop over during spring passage, but remain for relatively short periods due to the poor feeding conditions.

■ Protection status

National High **International** None

16,430 ha of IBA covered by Nature Reserve (Yumurtalık Lagünleri, 16,430 ha). 16,430 ha of IBA covered by SİT (Yumurtalık Lagünleri, 16,430 ha).

■ Conservation issues

Threats Agricultural intensification/expansion (B), Aquaculture/fisheries (B), Industrialization/urbanization (A)

Two major irrigation projects include provisions to prevent drainage water entering the lagoons, although no studies have been undertaken into the effects of reduced surface water inflow to the wetland. Large industrial areas are situated east of the area, and pollution is a threat. A new port is planned in the vicinity to accommodate oil tankers. Unsustainable fishing practices have caused a drastic decline in fish stocks. A 1996 management plan regulates all land-use in the delta, and places restrictions on the construction of holiday homes.

Gavur lake — A1, A4iii — 080

Admin region Kahramanmaraş
Coordinates 37°18′N 36°51′E
Altitude 490–490 m **Area** 1,500 ha

■ Site description

The site lies south of the town of Türkoğlu in the Sağlık plain, and comprises a complex of fields and small reedbeds (*Phragmites*) inundated during the winter but largely dry and cultivated during the summer. The area is fed by springs and streams. Rice-fields adjoin the lake to the north-east; a railway and motorway pass to the west.

Habitats Wetland (20%; standing fresh water; water-fringe vegetation), Artificial landscape (80%; arable land)
Land-use Agriculture (100%)

Birds

Species	Season	Year	Pop min	Pop max	Acc	Criteria
Phalacrocorax pygmeus Pygmy Cormorant	P	1996	36	121	—	A1

Prior to drainage the site held enormous numbers of passage waterbirds.

Protection status

National None **International** None

Conservation issues

Threats Drainage (A)

As part of its Ceyhan Menzelet scheme, DSİ plans to complete the land-claim and irrigation of the Sağlık plain, including this IBA. Gavur lake, although largely claimed, could be restored if excluded from the proposed scheme.

Nur mountains — A4iv, B1iv — 081

Admin region Hatay, Osmaniye
Coordinates 36°39′N 36°19′E
Altitude 1,000–2,262 m **Area** 21,500 ha

Site description

A high mountain ridge that forms the boundary between Mediterranean Turkey and the Syrian plateau. Wet areas at low altitude are clothed with dense deciduous forest; *Pinus* forest dominates the lower serpentine slopes. Extensive *Quercus* scrub occurs on the drier landward slopes, with dry grasslands and *Cedrus* at higher elevations. Human use of the area is limited to sheep-grazing and forestry. This area was treated in the previous international IBA inventory (Grimmett and Jones 1989) as Belen Geçidi (Belen pass) (former site TR059).

Habitats Forest and woodland (75%; mixed forest), Grassland (25%; alpine/subalpine/boreal grassland)
Land-use Agriculture, Forestry (50%), Not utilized (50%)

Birds

The majority of birds passing through site 005 (Bosphorus) during autumn and spring cross the Nur mountains, along with Turkish breeders. The only comprehensive autumn count was undertaken at the Belen pass over five weeks in 1976; few spring records are available. 6,203 *Pelecanus onocrotalus*, 3,303 *Ciconia nigra*, 82,887 *C. ciconia* and 26,756 raptors (mainly *Pernis apivorus*, *Accipiter brevipes* and *Aquila pomarina*) were counted. 9,950 *Ciconia ciconia* were counted over the Belen Pass on one day in spring 1992.

Protection status

National Partial **International** None
Southern part of IBA covered by Permanent Wildlife Reserve (Nur Dağları, 120,000 ha).

Conservation issues

No serious threats are known at the site.

Erzurum plain — A4i, B1i, B2 — 082

Admin region Erzurum
Coordinates 39°59′N 41°18′E
Altitude 1,750–1,750 m **Area** 3,300 ha

Site description

Habitats Grassland (10%; humid grassland), Wetland (90%; water-fringe vegetation; fen/transition mire/spring), Artificial landscape (10%; arable land)
Land-use Agriculture (20%), Not utilized (80%)

A complex of seasonally-flooded grasslands and marshes situated on the Karasu stream, a few kilometres north of the city of Erzurum. High mountains surround the plain on three sides. The Kösemehmet marshes in the east comprise extensive brackish clay grasslands with a few remaining hay-meadows. The Sitavuk and Çorak temporary marshes in the west merge during the wet season. Cattle graze the wetlands, with arable farming on the fertile Erzurum plain.

Birds

Species	Season	Year	Pop min	Pop max	Acc	Criteria
Tadorna ferruginea Ruddy Shelduck	P	1994	1,500	1,500	—	A4i, B1i
Grus grus Crane	B	—	15	15	—	B2
Chlidonias leucopterus White-winged Black Tern	P	1993	15,000	15,000	—	A4i, B1i

The site is important as a stop-over site for passage waterbirds.

Protection status

National None **International** None

Conservation issues

Threats Agricultural intensification/expansion (A), Construction/impact of dyke/dam/barrage (A), Drainage (U), Industrialization/urbanization (B), Infrastructure (C)

DSİ is improving drainage in the area, with the aim of creating more arable land. Erzurum is one of the fastest-growing cities in Eastern Anatolia, and increasing industrial activity and urbanization therefore pose a threat. A new highway is under construction, passing just south of the IBA.

Sarısu plain — B2 — 083

Admin region Ağrı
Coordinates 39°03′N 42°54′E
Altitude 1,850–1,850 m **Area** 4,800 ha

Site description

An extensive upland plain c.25 km north of Lake Van (site 096) and just north of Süphan mountain. It comprises wet meadows, upland grassland and a series of freshwater lakes and large reedbeds (*Phragmites*). Non-irrigated arable land and pasture surround the area, and cattle and water buffalo graze the wetland. Reed-cutting is an important economic activity ('Other' land-use).

Habitats Grassland (20%; humid grassland), Wetland (70%; standing fresh water; water-fringe vegetation), Artificial landscape (10%; arable land)
Land-use Agriculture (40%), Not utilized (60%), Other

Birds

Species	Season	Year	Pop min	Pop max	Acc	Criteria
Grus grus Crane	B	—	10	10	—	B2

Protection status

National None **International** None

Conservation issues

No serious threats are known at the site.

Haçlı lake — A4i, B1i, B2 — 084

Admin region Muş
Coordinates 39°01′N 42°18′E
Altitude 1,600–1,600 m **Area** 2,500 ha

Site description

A stream-fed freshwater lake (1,680 ha) situated south of the town of Bulanık. Extensive meadows (still hand-cut) adjoin the lake; marshes are found to the south and south-west. The lake is surrounded by hills, except in the south and east where it borders the Kazan and Tohmanik plains. Limited fishing activities take place; water is used for irrigation.

Habitats Grassland (10%; humid grassland), Wetland (90%; standing fresh water; water-fringe vegetation)
Land-use Agriculture (10%), Fisheries/aquaculture (5%), Water management (90%)

Birds

Important for breeding wildfowl. It is highly likely that the available information severely underestimates the importance of this IBA.

Species	Season	Year	Pop min	Pop max	Acc	Criteria
Platalea leucorodia Spoonbill	B	—	10	10	—	B1i, B2
Tadorna ferruginea Ruddy Shelduck	N	1980	650	650	—	A4i, B1i
Anas strepera Gadwall	B	—	10	10	—	B2

■ Protection status
National None **International** None

■ Conservation issues
No serious threats are known at the site. Although no ornithological data exist prior to the lake being converted for irrigation purposes, it seems likely that the artificially high water-levels have altered the character of the wetland system.

Bulanık plain — A1, A4i, B1i, B2 — 085
Admin region Muş
Coordinates 39°10′N 42°14′E
Altitude 1,460–1,460 m **Area** 8,000 ha

■ Site description
The valley and flood-plains of the Murat river, which here is fast-flowing with islands and gravel spits. One island complex lies to the north of the town of Bulanık, whilst another large vegetated island is situated to the west. To the north, the plain is bordered by hills; to the south of the river are large tracts of arable land and hay-meadows. Cattle graze on the flood-plain.

Habitats Grassland (5%; steppe/dry calcareous grassland), Wetland (65%; river/stream; water-fringe vegetation), Artificial landscape (30%; arable land; ruderal land)
Land-use Agriculture (20%), Not utilized (80%)

■ Birds

Species	Season	Year	Pop min	Pop max	Acc	Criteria
Phalacrocorax pygmeus Pygmy Cormorant	B	—	10	10	—	A1
Grus grus Crane	B	—	10	10	—	B2
Otis tarda Great Bustard	B	1986	30	30	—	A1, B2
Glareola nordmanni Black-winged Pratincole	P	1989	1,000	1,000	—	A1
Gelochelidon nilotica Gull-billed Tern	B	—	300	300	—	A4i, B1i, B2
Sterna caspia Caspian Tern	B	—	1	—	—	B2

The site is important for breeding waterbirds and raptors. *Otis tarda* occurs regularly in significant numbers and is thought to breed in the surrounding area. The IBA is the best-known Turkish locality for *Anthropoides virgo*—several pairs breed (max. 21 individuals).

■ Protection status
National None **International** None

■ Conservation issues
Threats Agricultural intensification/expansion (U), Construction/impact of dyke/dam/barrage (A)

DSİ plans to construct a large reservoir (Alpaslan 1) that will flood the western part of the IBA up to the point where the Hınıs and Murat rivers unite. Agriculture in the remainder of the area will be intensified.

Ardahan forest — B3 — 086
Admin region Ardahan
Coordinates 41°08′N 42°46′E
Altitude 1,930–2,291 m **Area** 2,500 ha

■ Site description
An upland area north-east of Ardahan that occupies the broad valley of the Kura stream. The area comprises extensive areas of high altitude *Pinus* forest and grazed montane steppe, together with hay-meadows (often cut by hand) and arable land. Putka lake is covered by extensive stands of emergent sedge *Carex*, with very little open water. Its margins are heavily grazed.

Habitats Forest and woodland (40%; native coniferous forest), Grassland (40%; alpine/subalpine/boreal grassland), Artificial landscape (20%; arable land)
Land-use Agriculture (10%), Not utilized (90%)

■ Birds

Species	Season	Year	Pop min	Pop max	Acc	Criteria
Circus pygargus Montagu's Harrier	B	—	10	10	—	B3

The site also supports small numbers of breeding waterbirds, and several species of raptor probably breed in the forests surrounding the marsh. Species of global conservation concern that do not meet IBA criteria: *Aythya nyroca* (breeding) and *Aegypius monachus* (thought to breed just outside the area).

■ Protection status
National None **International** None

■ Conservation issues
No serious threats are known at the site.

Aktaş lake — A1, A4i, B1i, B2 — 087
Admin region Ardahan
Coordinates 41°12′N 43°12′E
Altitude 1,798–1,798 m **Area** 1,400 ha

■ Site description
A shallow, stream-fed, soda-rich lake situated on the high plateau of the Turkish–Georgian border zone. About 1,400 of the lake's 2,700 ha lie within Turkey, and this area includes 12 uninhabited and sparsely vegetated islands. Arable fields and grazing land adjoin the lake; there are a few small reedbeds (*Phragmites*). Access is highly restricted for military reasons, and human disturbance is therefore minimal. Fishing is solely land-based.

Habitats Wetland (100%; standing fresh water)
Land-use Not utilized (100%)

■ Birds

Species	Season	Year	Pop min	Pop max	Acc	Criteria
Pelecanus onocrotalus White Pelican	B	—	50	50	—	B2
Pelecanus crispus Dalmatian Pelican	B	—	20	20	—	A1, A4i, B1i, B2
Tadorna ferruginea Ruddy Shelduck	N	1982	210	500	—	A4i, B1i
Melanitta fusca Velvet Scoter	U	1995	725	725	—	B1i
Larus armenicus Armenian Gull	B	1995	700	700	—	A4i, B1i

One of three Turkish *Pelecanus onocrotalus* breeding sites and one of seven *P. crispus* breeding sites. *Melanitta fusca* is also thought to breed.

■ Protection status
National None **International** None

■ Conservation issues
Threats Disturbance to birds (U)

Plans exist to open the Aktaş border post in order to improve trade and transport between the former USSR states and Turkey. This would increase traffic north of the lake.

Çıldır lake — A4i, B1i — 088
Admin region Ardahan
Coordinates 41°03′N 43°15′E
Altitude 1,962–1,962 m **Area** 14,000 ha

■ Site description
A freshwater lake (max. 12,350 ha, max. depth 22 m) situated in a volcanic depression between Kısır and Akbaba mountains. Extensive marshes and damp pastures border the lake to the north-west, otherwise it is surrounded by hill pasture and has very little fringe vegetation. Four islands are vegetated with low shrub. Human

activities include stock-grazing; water is used for hydroelectric power generation and irrigation.

Habitats Wetland (95%; standing fresh water; water-fringe vegetation), Artificial landscape (5%; arable land)
Land-use Agriculture (5%), Fisheries/aquaculture (5%), Urban/industrial/transport (95%), Water management (95%)

Birds

Species	Season	Year	Pop min	Pop max	Acc	Criteria
Tadorna ferruginea Ruddy Shelduck	U	1989	350	480	—	A4i, B1i
Larus armenicus Armenian Gull	B	1993	1,150	1,500	—	A4i, B1i

Pelecanus sp. formerly bred on the islands. There are recent records of *P. onocrotalus* and occasional records of *P. crispus*, but breeding has not been proven for either species. *Melanitta fusca* is also thought to breed.

Protection status

National None **International** None

Conservation issues

Threats Agricultural intensification/expansion (A), Disturbance to birds (A), Industrialization/urbanization (A)

DSİ plans to increase the lake volume by means of stream diversion. A second power-station will then be constructed, and the area of irrigated land will be increased. It is unclear what implications these works will have for the wetland system. Disturbance by visitors to the breeding islands may account for the disappearance of *Pelecanus* sp. which formerly bred. Further tourism development of the Akçakale area should be carefully planned.

Kuyucuk lake — A4i, B1i, B2 — 089

Admin region Kars
Coordinates 40°45′N 43°27′E
Altitude 1,627–1,627 m **Area** 219 ha

Site description

A stream- and spring-fed freshwater lake (max. depth 13 m) situated on the high Kars–Akyaka plateau, in the centre of an open, undulating landscape. Emergent vegetation is sparse, although *Phragmites* does occur in more sheltered bays. Human activities include cereal production and stock-grazing. The Kar–Akyaka road bisects the north end of the lake.

Habitats Wetland (100%; standing fresh water; water-fringe vegetation)
Land-use Nature conservation/research (100%), Urban/industrial/transport

Birds

Species	Season	Year	Pop min	Pop max	Acc	Criteria
[1] ***Tadorna ferruginea*** Ruddy Shelduck	N	—	—	6,000	—	A4i, B1i
Oxyura leucocephala White-headed Duck	B	—	2	2	—	B2

1. Post-breeding.

The area is important for breeding waterbirds.

Protection status

National High **International** None
219 ha of IBA covered by Permanent Wildlife Reserve (Kuyucuk Gölü, 219 ha).

Conservation issues

Threats Agricultural intensification/expansion (U), Construction/impact of dyke/dam/barrage (A)

The IBA is no longer an integral part of any major irrigation or drainage scheme, according to current DSİ plans. However, several thousand hectares of land in the vicinity of Kuyucuk lake are to be irrigated with water from the Kars reservoir, to be constructed on the Kars stream. Agricultural run-off is then likely to become a problem.

Çalı lake — A1, B2 — 090

Admin region Kars
Coordinates 40°31′N 43°16′E
Altitude 2,235–2,235 m **Area** 25 ha

Site description

A small wetland situated 20 km east of Kars, adjacent to the road between Kars and Digor. To the south of the road are 15 ha of open water, with a rich submerged flora (mainly *Potamogeton*) and a mosaic of emergent sedge-beds. The area to the north is largely seasonal. The lake lies on a vast elevated plateau, covered largely by extensively grazed steppe-grassland.

Habitats Wetland (100%; standing fresh water, fen/transition mire/spring)
Land-use Not utilized (100%)

Birds

Species	Season	Year	Pop min	Pop max	Acc	Criteria
Oxyura leucocephala White-headed Duck	B	—	10	10	—	A1, B2

The area is important for breeding waterbirds.

Protection status

National None **International** None

Conservation issues

Threats Groundwater abstraction (U)

Drinking water for Kars is pumped from a nearby well—the wetland seems to be unaffected.

Balık lake — B1i — 091

Admin region Ağrı
Coordinates 39°47′N 43°33′E
Altitude 2,241–2,241 m **Area** 3,400 ha

Site description

One of the highest lakes in Turkey, up to 70 m deep and fed by mountain streams, springs and groundwater. The lake is normally frozen for five months of the year, except for a small area in the north where a spring prevents freezing. The muddy shores have reed *Phragmites* fringes. The immediate surrounds comprise grazed meadows and arable land; the lake is a popular fishing ground for locals.

Habitats Wetland (100%; standing fresh water; water-fringe vegetation)
Land-use Water management (100%)

Birds

Species	Season	Year	Pop min	Pop max	Acc	Criteria
Melanitta fusca Velvet Scoter	B	—	—	—	—	B1i

Protection status

National High **International** None
3,400 ha of IBA covered by Drinking Water Reservoir (Balık Gölü, 3,400 ha).

Conservation issues

No serious threats are known at the site.

Doğubeyazıt marshes — B3 — 092

Admin region Ağrı
Coordinates 39°38′N 44°06′E
Altitude 1,520–1,520 m **Area** 8,750 ha

Site description

The IBA comprises two lakes separated by an extensive marsh with reedbeds (*Phragmites*) and wet meadows, situated at the foot of Büyük Ağrı mountain (Mount Ararat). Saz or Kurtkapan lake is fed mainly by surface run-off whilst Gölyüzü or Şeyhli lake is fed by Balık lake (site 091). The lakes are covered mainly by reed, with large parts drying up in summer. The plains surrounding the wetland are grazed, and reed-cutting takes place ('Other' land-use).

Habitats Grassland (30%; steppe/dry calcareous grassland), Wetland (70%; mudflat/sandflat; standing fresh water; water-fringe vegetation)
Land-use Agriculture (10%), Not utilized (80%), Other (10%)

Birds

Species	Season	Year	Pop min	Pop max	Acc	Criteria
Circus pygargus Montagu's Harrier	B	—	10	10	—	B3

The site is also important for breeding waterbirds. Breeding species of global conservation concern that do not meet IBA criteria: *Aythya nyroca*. According to the limited information available, the site was more important in the past.

Protection status

National None **International** None

Conservation issues

No serious threats are known at the site.

Nemrut lake — B1i, B2 — 093

Admin region Bitlis
Coordinates 38°37'N 42°14'E
Altitude 2,247–2,247 m **Area** 4,500 ha

Site description

A 10 km wide volcanic crater situated west of Lake Van (site 096). Three crater lakes are surrounded by low scrub; the cliffs of the crater walls reach 690 m above the crater floor. Nemrut lake (1,250 ha; 150 m deep) is largely unvegetated, is fed by springs and snowmelt, and freezes in winter. Ilıca lake (30 ha; 8 m deep) is fed by hot springs and supports reedbeds (*Phragmites*). Domestic stock graze the flatter parts of the crater.

Habitats Grassland (60%; alpine/subalpine/boreal grassland), Wetland (40%; standing fresh water)
Land-use Agriculture (5%), Not utilized (95%)

Birds

Species	Season	Year	Pop min	Pop max	Acc	Criteria
Melanitta fusca Velvet Scoter	B	1989	19	19	—	B1i
Aquila chrysaetos Golden Eagle	B	—	1	1	—	B2

Gyps fulvus also breeds.

Protection status

National None **International** None

Conservation issues

No serious threats are known at the site.

Sodalıgöl — A1, A4i, B1i, B2 — 094

Admin region Bitlis
Coordinates 38°49'N 42°59'E
Altitude 1,650–1,650 m **Area** 1,500 ha

Site description

A shallow, spring-fed saline lake separated from Lake Van (site 096) by an alluvial ridge. There is no outlet and up to 50% of the lake dries out during summer. Some marshland is present at the fringes where freshwater seepage occurs. Low hills and non-irrigated arable land surround the lake. Limited stock-grazing occurs along the shoreline.

Habitats Grassland (10%; steppe/dry calcareous grassland), Wetland (90%; mudflat/sandflat; standing brackish and salt water; water-fringe vegetation)
Land-use Agriculture (10%), Hunting (5%), Not utilized (85%)

Birds

Species	Season	Year	Pop min	Pop max	Acc	Criteria
Anas strepera Gadwall	B	—	10	10	—	B2
Netta rufina Red-crested Pochard	B	—	30	30	—	B2
Oxyura leucocephala White-headed Duck	B	—	30	30	—	A1, B2
Oxyura leucocephala White-headed Duck	N	1990	101	750	—	A1, A4i, B1i
Himantopus himantopus Black-winged Stilt	B	1995	440	440	—	A4i, B1i
Larus armenicus Armenian Gull	B	1995	310	310	—	A4i, B1i

The area is important for breeding waterbirds.

Protection status

National None **International** None

Conservation issues

No serious threats are known at the site.

Erçek lake — A4i, B1i, B2, B3 — 095

Admin region Van
Coordinates 38°40'N 43°35'E
Altitude 1,803–1,803 m **Area** 9,520 ha

Site description

A highly alkaline lake (max. depth 30 m) situated in a tectonic depression 20 km east of Lake Van (site 096). It has steep rocky shores on the northern and western sides; in the east and south the shores are shallow with mudflats. The Memedik river enters the lake in the east, forming a small delta with reedbeds (*Phragmites*) and wet meadows. Human activities include stock-grazing.

Habitats Wetland (100%; mudflat/sandflat; standing fresh water; water-fringe vegetation)
Land-use Agriculture (5%), Not utilized (95%)

Birds

Species	Season	Year	Pop min	Pop max	Acc	Criteria
Podiceps nigricollis Black-necked Grebe	P	1989	4,430	4,430	—	A4i, B1i
Tadorna ferruginea Ruddy Shelduck	B	1986	2,000	5,600	—	A4i, B1i, B2
Oxyura leucocephala White-headed Duck	B	—	2	2	—	B2
Recurvirostra avosetta Avocet	B	—	50	50	—	B3
Charadrius leschenaultii Greater Sand Plover	B	—	1	—	—	B2

This site is important for breeding and passage waterbirds.

Protection status

National None **International** None

Conservation issues

Threats Industrialization/urbanization (A)

The water-level has risen dramatically in recent years, inundating a village, 1,000 ha of arable land and meadows, and the Turkey–Iran railway. DSİ plans to construct a hydroelectric power-station west of the lake—the ecological impacts of this proposal are unknown.

Lake Van — A1, A4i, B1i, B2 — 096

Admin region Bitlis, Van
Coordinates 38°40'N 42°55'E
Altitude 1,650–1,650 m **Area** 390,000 ha

Site description

A soda lake of tectonic origin and the largest lake (371,700 ha, max. depth 451 m) in Turkey. It is fed by several streams and remains ice-free. Annual inflow is higher than evaporation, and the lake level is continuously rising as a result. The four islands are popular tourist destinations. Extensive meadows and arable land cover the flat ground between the lake and the surrounding mountains.

Human activities include intensive grazing. This area includes six sites that were treated as separate IBAs in the previous international IBA inventory (Grimmett and Jones 1989): 'Ahtamar Adası' (former site TR065), 'Bendimahi Deltası' (former site TR067), 'Çelebidağ Sazlığı' (former site TR069), 'Edremit Sazlığı' (former site TR070), 'Horkum Gölü' (former site TR072) and 'Van Sazlığı' (former site TR077).

Habitats Wetland (95%; standing brackish and salt water), Artificial landscape (5%; arable land)
Land-use Agriculture, Fisheries/aquaculture (5%), Not utilized (90%), Urban/industrial/transport (5%)

Birds

Species	Season	Year	Pop min	Pop max	Acc	Criteria
Marmaronetta angustirostris Marbled Duck	B	—	5	5	—	A1, B1i, B2
Otis tarda Great Bustard	B	—	—	—	—	A1, B2
Larus armenicus Armenian Gull	B	1995	3,285	3,285	—	A4i, B1i
Chlidonias leucopterus White-winged Black Tern	P	1989	3,450	3,450	—	A4i, B1i

The site is important for breeding waterbirds and for *Otis tarda* breeding on the peninsula east of Sodalıgöl (site 094), where up to 32 birds have been seen during the breeding season.

Protection status

National Partial **International** None
The islands are designated as a SİT.

Conservation issues

Threats Agricultural intensification/expansion (U), Dredging/canalization (B), Industrialization/urbanization (C), Recreation/tourism (B), Unsustainable exploitation (U)

Visitors to the islands cause disturbance to breeding birds, including the taking of gull (Laridae) eggs. Further tourist developments are planned. Untreated urban waste and agricultural run-off enters the lake.

Yüksekova B3 097

Admin region Hakkari
Coordinates 37°30′N 44°18′E
Altitude 1,925–1,925 m **Area** 24,900 ha

Site description

A large plain enclosed by high mountains and traversed by the Nehil stream. A large permanent marsh is located in the south-east; temporary marshes and grasslands are found along the river. Large areas are subject to spring flooding. Human activities include extensive grazing and reed-cutting ('Other' land-use).

Habitats Grassland (50%; humid grassland), Wetland (15%; water-fringe vegetation), Artificial landscape (35%; arable land)
Land-use Agriculture (80%), Other (20%)

Birds

Species	Season	Year	Pop min	Pop max	Acc	Criteria
Circus pygargus Montagu's Harrier	B	—	15	15	—	B3

Pre-1989 data indicate that *Ardea purpurea*, *Circus aeruginosus*, *Grus grus* and *Otis tarda* bred, and it is likely that they continue to do so.

Protection status

National None **International** None

Conservation issues

Threats Agricultural intensification/expansion (A), Construction/impact of dyke/dam/barrage (A), Drainage (A), Industrialization/urbanization (B)

In 1993 DSİ completed a flood-protection scheme by draining 2,095 ha—marshes and wet grassland in the western and central part of the plain were largely destroyed. An irrigation reservoir is under construction on one of the inflowing streams, possibly leading to the loss of several thousand hectares of natural habitat. The influx of new settlers into Yüksekova has accelerated in recent years, increasing pressure on the wetland.

REFERENCES

AKGT (1995) *Preliminary avifauna survey of the Bodrum peninsula*. A report for the Turkish National Committee on Coastal Zone Management. Ankara, Turkey: Ankara Bird Watching Group Report.

ANON. (1993) *Sultansazlığı Tabiatı Koruma Alanı Master Planı Raporu*. Ankara, Turkey: Milli Parklar ve Av-Yaban Hayatı Genel Müdürlüğü.

ANSTEY, S. (1989) *The status and conservation of the White-headed Duck* Oxyura leucocephala. Slimbridge, UK: International Waterfowl and Wetland Research Bureau (Spec. Publ. No. 10).

AYVAZ, Y. (1991) Çıldır Gölü Kuşları. *Doğa-Turkish J. Zool.* 15: 53–58.

BALİK, S. AND USTAOĞLU, M. R. (1989) Bafa Gölü'ndeki Ulubat Balığı'nın Biyoekolojik ve Ekonomik Yönden İncelenmesi. *Doğa-Turkish J. Zool.* 13(3): 141.

BARİŞ, Y. S., AKÇAKAYA, R. AND BILGIN, C. (1984) *Kızılcahamam. Birds of Turkey 3*. Heidelberg, Germany: Max Kasparek Verlag.

BARİŞ, Y. S. (1989) Turkey's bird habitats and ornithological importance. *Sandgrouse* 11: 42–52.

BERK, VAN DEN, V. M., VAN DORP, D., VAN HOORN, O. AND VOS, R. (1985) *Cranes and waterfowl counts of some Turkish wetlands*. Zeist, Netherlands: Foundation Working Group International Waterbird and Wetland Research (WIWO Report 10).

BERK, VAN DEN, V. M., CRONAU, J. P. AND VAN DER HAVE, T. M. (1993) *Waterbirds in the Van Province*. Zeist, Netherlands: Foundation Working Group International Waterbird and Wetland Research (WIWO Report 34).

BIJLSMA, R. G. (1987) *Bottleneck areas for migratory birds in the Mediterranean region*. Cambridge, UK: International Council for Bird Preservation (Study Report 18).

BIJLSMA, R. G. AND DE RODER, F. E. (1986) *Notes on the birds of some wetlands in Turkey*. Zeist, Netherlands: Foundation Working Group International Waterbird and Wetland Research (WIWO Report 12).

BRADSHAW, C. G. AND KIRWAN, G. (1994) The ornithological importance of the Akşehir Gölü wetlands, Central Plateau, Turkey. *Sandgrouse* 16(2): 118–139.

BRINKMANN, R., KÖHLER, B., HEINS, J. U. AND RÖSLER, S. (1991) *Menderes Delta—Zustand und Gefaehrdung eines ostmediterranen Flussdeltas*. Germany: Gesamthochschule Kassel.

DHKD (1991) *Akyatan Gölü*. Istanbul: Doğal Hayatı Koruma Derneği (Bird Section Report No. 4).

DHKD (1993) *Uluslararası Göksu Deltası Çevresel Kalkınma Semineri Bildiri Metinleri*. Istanbul: Doğal Hayatı Koruma Derneği.

DHKD (1993) *Towards the conservation and management of Istanbul's greenspace*. Istanbul: Doğal Hayatı Koruma Derneği.

DHKD (1995) *The old growth forests of north-east Turkey: a survey of the old growth Colchic Forests of Artvin, Rize and Trabzon provinces in north-eastern Turkey*. Istanbul: Doğal Hayatı Koruma Derneği.

DIJKSEN, L. J. AND KASPAREK, M. (1985) *Kızılırmak Deltası. Birds of Turkey 4*. Heidelberg, Germany: Max Kasparek Verlag.

DIJKSEN, L. J. AND KASPAREK, M (1988) *Acıgöl. Birds of Turkey 7*. Heidelberg, Germany: Max Kasparek Verlag.

DSİ (1995) *Dams and hydroelectric power plants in Turkey*. Ankara, Turkey: General Directorate of State Hydraulic Works.

EKEN, G. (1997) The breeding populations of some species of waterbirds at Gediz Delta, western Turkey. *Zool. Middle East* 14: 53–68.

EKEN, G. AND MAGNIN, G. (in press) A preliminary biodiversity atlas of the Konya Basin, central Turkey. Istanbul: Doğal Hayatı Koruma Derneği.

ERDEM, O. (1995) *Turkey's bird paradises*. Ankara, Turkey: Turkish Ministry of Environment.

ERTAN, A., KILIÇ, A. AND KASPAREK, M. (1989) *Türkiye'nin Önemli Kuş Alanları*. Istanbul: Doğal Hayatı Koruma Derneği.

ERTAN, K. T. (1996) *Kocaçay delta. Birds of Turkey 12*. Heidelberg, Germany: Max Kasparek Verlag.

FALDBORG, J. (1994) *Birds in Çamlıhemşin-Ardeşen area, north-east Turkey, April–May 1993*. Copenhagen: Dansk Ornithologisk Förening (DOF).

GÖRGÜN, E. O. (1994) Mogan Gölü Avifaunası. Ankara: Gazi Üniversitesi Fen Bilimleri Enstitüsü (MSc. thesis).

GREEN, A. J. (1996) *The summer ecology of the Marbled Teal* Marmaronetta angustirostris, *Ferruginous Duck* Aythya nyroca *and other ducks in the Göksu Delta, Turkey in 1995*. Spain: Donaña Biological Station.

GREEN, A. J., HILTON, G. M., HUGHES, B., FOX, A. D. AND YARAR, M. (1993) *The ecology and behaviour of the White-headed Duck* Oxyura leucocephala *at Burdur Gölü, Turkey*. Slimbridge, UK: The Wildfowl and Wetlands Trust.

GREEN, A. J. AND YARAR, M. (1996) Rapid decline of White-headed Ducks at Burdur Lake, Turkey. *Threatened Waterfowl Spec. Grp News* 9: 16–19.

GREEN, I. A. (1993) Caucasian Black Grouse and Caspian Snowcock in north-east Turkey. *OSME Bull.* 31: 14–18.

GRIMMETT, R. F. A. AND JONES, T. A. (1989) *Important Bird Areas in Europe*. Cambridge, UK: International Council for Bird Preservation (Techn. Publ. No. 9).

HAVE, VAN DER, T. M., VAN DEN BERK, V. M., CRONAU, J. P. AND LANGEVELD, M. J. (1988) *South Turkey Project: a survey of waders and waterfowl in the Çukurova deltas, spring 1987*. Zeist, Netherlands: Foundation Working Group International Waterbird and Wetland Research (WIWO Report 2).

HAVE, VAN DER, T. M., VAN DEN BERK, T. M., CRONAU, J. P. AND LANGEVELD, M. J. (1989) Importance of Çukurova deltas, southern Turkey, for migrating waders and other waterbirds in spring. *Sandgrouse* 11: 76–89.

HEATH, M. F. AND BORGGREVE, C. (2000) *BirdLife International/EBCC European Bird Database 1998*. Cambridge, UK: BirdLife International.

HUSBAND, C. AND KASPAREK, M. (1984) *Seyfe Gölü. Birds of Turkey 2*. Heidelberg, Germany: Max Kasparek Verlag.

HUSTINGS, F. AND VAN DIJK, K. (1994) *Bird census in the Kızılırmak delta, Turkey, spring 1992*. Zeist, Netherlands: Foundation Working Group International Waterbird and Wetland Research (WIWO Report 45).

İKİZ, R., KESİCİ, E. AND BECER, A. (1996) *Eğirdir Gölü*. Eğirdir, Turkey: Süleyman Demirel Üniversitesi Eğirdir Su Ürünleri Fakültesi.

JETZ, W. (1995) *Uludağ. Birds of Turkey 11*. Heidelberg, Germany: Max Kasparek Verlag.

KARADENIZ, N. (1995) Sultan Sazlığı Örneğinde Islak Alanların Çevre Koruma Açısından Önemi Üzerinde Bir Araştırma. Ankara, Turkey: Ankara Üniversitesi Fen Bilimleri Enstitüsü (PhD thesis).

KARATAŞ, A. (1993) *Küçük Menderes Deltası Kuşları*. Bornova, Turkey: Ege Üniversitesi Fen Fakültesi Biyoloji Bölümü Zooloji Anabilim Dalı.

KARAUZ KİRAÇ, S. (1993) *Çöl Gölü. Birds of Turkey 10*. Heidelberg, Germany: Max Kasparek Verlag.

KARAUZ KİRAÇ, S. AND KİRAÇ, C. (1995) A short breeding survey on the birds of Kulu Gölü, central Anatolia, Turkey in May 1995. *Sandgrouse* 18/2: 58–60.

KARAUZ KİRAÇ, S. AND KİRAÇ, C. (in press) Population and distribution of breeding seabird colonies on Foça Islands, Aegean Sea, Turkey. MEDMARAVIS Conference, 11–16 April 1995. Tunisia.

KASPAREK, M. AND VAN DER VEN, J. (1983) *Erçek Gölü. Birds of Turkey 1*. Heidelberg, Germany: Max Kasparek Verlag.

KASPAREK, M. (1985) *Die Sultansümpfe, Naturgeschichte eines Vogelparadieses in Anatolien*. Heidelberg, Germany: Max Kasparek Verlag.

KASPAREK, M. (1987) *Kulu Gölü. Birds of Turkey 5*. Heidelberg, Germany: Max Kasparek Verlag.

KASPAREK, M. (1988) *Bafa See, Natur und Geschichte in der Türkischen Ägäis*. Heidelberg, Germany: Max Kasparek Verlag.

KASPAREK, M. (1992) *Die Vögel der Türkei-Eine Übersicht*. Heidelberg, Germany: Max Kasparek Verlag.

KAZANCİ, N., PLASA, R. H., NEUBERT, E., İZBİRAK, A. (1992) On the limnology of Lake Köyceğiz (SW Anatolia). *Zool. Middle East* 6: 109–115.

KENCE, M. (1993) On the biology and management of a wetland, Lake Beyşehir. Pp. 391–402 in *The First International Conference on the Mediterranean Coastal Environment, MEDCOAST 93*. Antalya, Turkey.

KHAN, A. A. (1992) Activity budgets and population analyses of waterfowl and ecosystem management of Sultan Marshes. Ankara, Turkey: Middle East Technical University (PhD thesis).

KİLİÇ, A. AND KASPAREK, M. (1989) *Köyceğiz-Dalyan. Birds of Turkey 8*. Heidelberg, Germany: Max Kasparek Verlag.

KIRWAN, G. (1993) *Hotamış marshes. Birds of Turkey 9*. Heidelberg, Germany: Max Kasparek Verlag.

KIRWAN, G. (1994) The breeding status and distribution of the White-headed Duck *Oxyura leucocephala* on the Central Plateau, Turkey. *Sandgrouse* 16(2): 66–76.

KIVIT, H., NIJMEIJER, H. AND OVAA, A. (1994) *The eastern Mediterranean wader project. Wader and waterfowl migration in the Çukurova deltas, south Turkey, spring 1990*. Zeist, Netherlands: Foundation Working Group International Waterbird and Wetland Research (WIWO Report 48).

KIZIROĞLU, İ. (1989) *Türkiye Kuşları*. Ankara, Turkey: Orman Genel Müdürlüğü.

KIZIROĞLU, İ., TURAN, L. AND ERDOĞAN, A. (1992) Sultansazlığı'nın Eko-ornitolojisi ve Son Durumu. *Hacettepe Üniversitesi Eğitim Fakültesi Dergisi*. 7: 217–227.

KOK, M. AND ONGENAE, J. P. (1995) Raptor migration in the north-east Turkey, September 1990. *OSME Bull.* 34: 8–12.

KOZAN, A. T. (1985) Küçük Menderes Deltası ve Güneyinin Jeomorfolojisi. Istanbul: İstanbul Ünversitesi Coğrafya Fakültesi (MSc. thesis).

LANGEVELD, M. J. AND RUITERS, P. S. (1995) *Breeding bird census at Eber Gölü, Akşehir Gölü and Karamık Gölü 1994*. Istanbul: Doğal Hayatı Koruma Derneği (Bird and Wetland Section Report No.8).

MAGNIN, G. (1988) *Falconry and hunting in Turkey during 1987*. Cambridge, UK: International Council For Bird Preservation (Study Report 34).

MAGNIN, G. AND YARAR, M. (1994) Some notes on the breeding of Greater Flamingo *Phoenicopterus ruber* and White Pelican *Pelecanus onocrotalus* in Turkey. *OSME Bull.* 32: 28–31.

MAGNIN, G. AND YARAR, M. (1997) *Important Bird Areas in Turkey*. Istanbul: Doğal Hayatı Koruma Derneği.

MEININGER, P. L., SCHEPERS, F. J. AND STUART, J. J. (1989) *Bird observations Göksu delta, Turkey, spring 1989*. Istanbul: Doğal Hayatı Koruma Derneği.

MERIÇ, N. (1992) *Büyük Çekmece Baraj-Gölü Balıkları Üzerine Bir Ön Çalışma*. Istanbul: İstanbul Üniversitesi Biyoloji Bölümü.

MRLIK, V., BOBEK, M., POJER, F., SKOPEK, J. AND FORMÁNEK, J. (1995) Raptor Migration in north-east Turkey, autumn 1994. *OSME Bull.* 35: 41–46.

ONGAN, T., HOSANOĞLU, A., YÜCETAŞ, N. AND YAZİCİ, M. (1990) Gala Gölü ve Bunlarla İrtibatlı Lagün ve Kanalların Entegre Balıkçılık Durumu. Ankara, Turkey: TÜBİTAK (Türkiye Bilimsel ve Teknik Araştırma Kurumu). Unpublished research report.

ÖZER, A. AND KARAKOÇ, G. (1992) *Özel Çevre Koruma Bölgesi: Köyceğiz-Dalyan*. Ankara, Turkey: Özel Çevre Koruma Kurumu Başkanlığı (Özel Çevre Koruma Bölgesi Tanıtım Serisi 2).

ÖZER, A. AND SEVIM, F. (1994) *Özel Çevre Koruma Bölgesi: Göksu Deltası*. Ankara, Turkey: Özel Çevre Koruma Kurumu Başkanlığı (Özel Çevre Koruma Bölgesi Tanıtım Serisi 3).

ÖZER, A. AND SEVIM, F. (1994) *Özel Çevre Koruma Bölgesi: Mogan Gölü*. Ankara, Turkey: Özel Çevre Koruma Kurumu Başkanlığı (Özel Çevre Koruma Bölgesi Tanıtım Serisi 12).

ÖZESMI, U. (1989) Protection of Sultan marshes. *Sandgrouse* 11: 73–76.

PARR, S., COLLIN, P., SILK, S., WILBRAHAM, J., WILLIAMS, N. P. AND YARAR, M. (1995) A baseline survey of Lesser Kestrels (*Falco naumanni*) in central Turkey. *Biol. Conserv.* 72: 45–53.

ROBINSON, K., BUCKLEY, Y., HOLT, L. J., PULLEN, T., BOYLA, K. A. AND CAN, O. (1996) *A breeding survey of the White-headed Duck* Oxyura leucocephala *(Oxyurini) on the Central Plateau, Turkey*. Oxford, UK: Oxford University.

ROWECLIFFE, M., JOWITT, A., HUMPAGE, E., ATKINSON AND OĞURLU, İ. (1993) UEA Kaçkar Expedition 1993. Norwich, UK: University of East Anglia (preliminary report).

SCHEKKERMAN, H. AND VAN ROOMEN, M. W. J. (1993) *Migration of waterbirds through wetlands in central Anatolia, spring 1988*. Zeist, Netherlands: Foundation Working Group International Waterbird and Wetland Research (WIWO Report 32).

SİKİ, M. (1988) Çamaltı Tuzlası-Homa Dalyanı Kuş Türleri. *Doğa-Turkish J. Zool.* 12(3): 272–283.

TÇV (1993) *Wetlands of Turkey*. Ankara, Turkey: Türkiye Çevre Vakfı (Environment Foundation of Turkey).

TEMPLE LANG, J. (1989) International conservation treaties and Turkey's birdlife. *Sandgrouse* 11: 67–73.

TOĞULGA, M. (1992) Kuş Gölü'nün Deterjan Kirliliği Yönünden Araştırılması. *Turkish J. Zool.* 16: 249.

TUCKER, G. M. AND EVANS, M. I., EDS (1997) *Habitats for birds in Europe—a conservation strategy for the wider environment*. Cambridge, UK: BirdLife International.

TUCKER, G. M. AND HEATH, M. F. (1994) *Birds in Europe: their conservation status*. Cambridge, UK: BirdLife International (BirdLife Conservation Series no. 3).

TURAN, N. (1990) *Kuşlar*. Ankara, Turkey: Orman Genel Müdürlüğü.

TÜLÜMEN, E. (1994) *Enez Çevre Sempozyumu 1993*. Edirne, Turkey: Edirne Çevre Vakfı Yayınları (No. 1).

USLU, T. (1985) *Aydın'ın Batısında Küçük ve Büyük Menderes Nehirleri Arasında Kalan Bölge Vejetasyonunun Bitki Ekolojisi ve Sosyolojisi Yönünden Araştırılması*. Ankara, Turkey: Gazi Üniversitesi Fen Edebiyat Fakültesi Yayın (No. 8).

WELCH, G. AND WELCH, H. (1998) Breeding bird survey of Uluabat Gölü Ramsar Site. Unpublished report.

WINDEN, VAN , A., MOSTERT, K., RUITERS, P., SİKİ, M. AND DE WAARD, H. (1989) *Waders and waterfowl in spring 1988 at Eber Gölü*. Zeist, Netherlands: Foundation Working Group International Waterbird and Wetland Research (WIWO Report 28).

WITT, H. H. (1982) Ernährung und Brutverbreitung der Korallenmöwe (*Larus audouinii*) im Vergleich zur Mittelmeersilbermöwe (*Larus argentatus michahellis*). *Seevögel* (Special Issue): 87–91.

YARAR, M. (1995) Aktaş Gölü: a new pelican breeding site on the Turkish-Georgian border. *OSME Bull.* 35: 46–50.

YÜCEL, E. (1987) *Eber Gölü ve Yaban Ördekleri*. Eskişehir, Turkey: Anadolu Üniversitesi Fen Edebiyat Fakültesi Biyoloji Bölümü.

ZÜRCHER, E. J. (1993) *Turkey: a modern history*. London: I. B. Tauris & Co Ltd.

UKRAINE

ALEXANDER MIKITYUK

Steppe habitat, Bagerovo (IBA 102). (PHOTO: PAUL GORIUP)

GENERAL INTRODUCTION

Ukraine has a land area of 603,700 km², and is bounded to the west by Romania, Hungary, Slovakia and Poland, to the north by Belarus, to the east by Russia, and to the south by Moldova and the Black Sea and Sea of Azov. The human population is 51.6 million, giving an average density of 85 persons/km². The country has 102 Important Bird Areas (IBAs) covering 20,323 km², equivalent to 3.4% of the national land area (Table 1, Map 1). Ukraine lies between the latitudes of 53°N and 44°N, spanning three major vegetation zones, from forest in the north through forest-steppe to steppe in the south. Landscapes are highly diverse, with the Carpathian mountains in the west, the Crimean mountains in the south, extensive peatlands along the River Prip'yat flood-plain, forests in the north (mixed or pine *Pinus*-dominated), river-valley habitats in the forest-steppe and steppe zones in the centre of the country, brackish, fresh and saline wetlands along the coastal zone of the Black Sea and Sea of Azov, and a variety of steppe areas in the south and south-west. As a result of this landscape diversity, biological diversity is also high.

Twenty-seven IBAs were identified in Ukraine in the first pan-European IBA inventory (Grimmett and Jones 1989). The revision of the IBA criteria since 1989 has resulted in two of these 27 original sites (former sites SU046 and SU049) being merged (to form current site 067), three of the 27 (former sites SU052-1, SU054, SU055) being divided into several new IBAs, and two of the 27 no longer qualifying as IBAs (see Table 1). Organized surveys, literature searches and collaboration with other organizations during 1994–1998 have resulted in a further 76 IBAs being identified in Ukraine, the majority of which lie in the south of the country and in the river-basins of the Dnepr, Prip'yat, Desna, Dnestr and Severskiy Donec rivers (Table 1, Map 1).

Table 1. Summary of Important Bird Areas in Ukraine. 102 IBAs covering 20,323 km²

IBA code	1989 code	International name	National name	Administrative region	Area (ha)	Criteria (see p. 11)
001	SU033	Shatsk National Park	Shatskiye Ozera	Volin	32,850	A1, A4i, B1i, B2, B3
002	—	Zapadniy Bug river valley	Doluna r. Zakhidnuy Bug	Volin	1,450	A1
003	—	Orikhivski lake-system	Grupa Orikhivskukh ozer	Volin	3,000	B2
004	—	Turiya river valley	Zaplava reki Turiya	Volin	7,900	A1, B2, B3
005	—	River Prip'yat valley	Valley of r. Prip'yat	Volin	12,500	A1, A4i, B1i, B2, B3
006	—	Cheremske mire	Boloto Cheremske	Volin	3,007	B1i
007	—	Stokhod river valley	Zaplava richku Stokhid	Volin	17,800	A1, A4i, B1i, B2, B3
008	—	Sira Pogonya mire	Boloto Sira Pogonya	Rivne	12,718	A1, B2
009	SU034	Polesskiy Nature Reserve	Polesski	Zhitomyr	20,104	B2
010	—	Ushomirskoe Game Reserve	Okhotnichie khozyaistvo Ushomir	Zhitomyr	70,300	A1, B2
011	—	Novograd-Volunskoe Game Reserve	Okhotnichie khozyaistvo Novograd-Volunskoe	Zhitomyr	70,730	A1, B2
012	—	Korostushevskoe Game Reserve	Korostushevskoe okhotnichie khozyaistvo	Zhitomyr	44,130	A1, B1i, B2
013	—	Radomyshlskoe Game Reserve	Okhotnichie khozyaistvo Radomushlskoe	Zhitomyr	28,400	A1, B2

Table 1 ... continued. Summary of Important Bird Areas in Ukraine. 102 IBAs covering 20,323 km²

IBA code	1989 code	International name	National name	Administrative region	Area (ha)	Criteria (see p. 11)
014	—	Dneprovsko-Teterevskoe Game Reserve	Dneprovsko-Teterevskoe okhotnichie khozyaistvo	Kiev	30,627	A1, B2
015	—	Kiev reservoir	Verhovie Kievskogo vodokhranilishche	Kiev	102,400	A1, A4i, A4iii, B1i, B1ii, B2
016	—	Meadows near Procev village	Lug u sela Procev	Kiev	7,600	A1, B2
017	—	Rzhycshevskoe Game Reserve	Rzhycshevskoe okhotnichie khozyaistvo	Kiev	52,800	A1, B2
018	SU035	Kanev reservoir	Kanevskoye Vodokhranilishche	Kiev	200	B1i
019	—	Supoy marsh	Usovskiy zakaznik	Kiev	6,400	A1, B2
020	—	Forests in Dnepr and Desna valley lowlands	Forests in Dnepr and Desna valley lowlands	Chernigov	120,000	A1, B2, B3
021	—	Bondar marsh	Bondarevskoe boloto	Chernigov	6,400	A1, B2
022	—	Korotchenkov meadows	Korotchenkovo–Tshulatovckiy lug	Sumy	10,000	A1, A4i, B1i, B2, B3
023	—	Peschanskiy and Mogrichskiy forest	Pescanskoe i Mogrichskoe lesnichestva	Sumy	11,600	A1
024	—	Chaykovichi marsh	Boloto Chaykovichi	Lvov	8,300	A1, B2, B3
025	—	Ikva river valley	Doluna r. Ikva	Khmelnitskaya	83,500	A4i, B1i, B2
026	—	Vovk river valley near Derazhnya village	Doluna richku Vovk bilya sela Derazhnya	Khmelnitskaya	1,270	B2
027	—	Mikulinetskie fish-ponds	Mikulinetskie prudy	Vinnitsa	2,000	A1, B1i, B2, B3
028	SU036	Kanev Nature Reserve	Kanev	Cherkasu	2,027	A1, B1i
029	—	Uday river valley	Plavni reki Uday	Poltava	30,000	A4i, B1i, B2, B3
030	—	Krasnue marsh	Bolota Krasnue	Poltava	10,000	A1, B1i, B2
031	—	Sula bay, Kremenchug reservoir	Sula bay	Poltava	27,440	B1i, B2
032	—	Psel river valley near Komsomolsk	Zaplava r. Psel	Poltava	4,700	A1
033	—	Bodakovskoe marsh	Boloto Bodakivskoe	Poltava	1,500	A4i, B1i, B2
034	—	Meadows near Saru village in the Psel river valley	Lug u sela Saru v doline reki Psel	Poltava	1,200	A1, B2
035	—	Veluke marsh	Boloto Veluke	Poltava	1,000	B2
036	—	Dneprovsko-Orelskiy Nature Reserve	Dneprovsko-Orelskiy prirodniy zapovednik	Dnepropetrovskaya	3,766	A1, B2
037	—	Orel river valley	Dolina r. Orel	Dnepropetrovskaya	12,000	A1, A4i, B1i, B2, B3
038	—	Samarskiy forest	Samarskiy les	Dnepropetrovskaya	19,920	A1, A4i, B1i, B2, B3
039	—	Soleniy liman	Ozero Soleniy Liman	Dnepropetrovskaya	500	B1i
040	—	Debalcevskoe lake	Debalcevskiy liman	Dnepropetrovskaya	300	B1i, B2
041	—	Petropavlovskiy marsh	Petropavlovskiy liman	Dnepropetrovskaya	270	B1i
042	—	Bolshemikhailovskiy forest	Bolshemikhailovskiy les	Dnepropetrovskaya	1,079	B2
043	—	Samara river valley	Dolina r. Samara	Kharkov	1,700	A1
044	—	Mzha river valley	Dolina r. Mzha	Kharkov	7,200	A1, A4i, B1i, B2, B3
045	—	Liman lake-system	Sistema ozer Liman	Kharkov	4,730	A1, B1i, B2, B3
046	—	Pechenezhskoe reservoir	Pechenezhskoe vodokhranilishche	Kharkov	1,600	B2
047	—	Bakhtin river valley	Poyma r. Bakhtin	Kharkov	1,000	A1
048	—	Izyumskaya Luka forest	Izyumskaya Luka les	Kharkov	32,850	A1, B2
049	—	Kremenskoy forest	Kremenskoy Les	Lugansk	20,000	A1, B2, B3
050	—	Stanichno-Luganskoe fish-farm	Stanichno-Luganskiy rubkhoz	Lugansk	2,000	B2
051	—	Latorica river valley near Chop	Prichopskaya dolina r. Latorica	Zakarpatie	7,000	A1, B1i, B2
052	—	Carpathian Volkano mountains	Vulkanicheskiy khrebet	Zakarpatie	134,400	B2, B3
053	SU032	Carpathian Biosphere Reserve	Karpatski	Zakarpatie	12,507	B2, B3
054	—	Gorgany	Gorganu	Zakarpatie	75,200	B2, B3
055	—	Fish-ponds near Kukilniku and Bilshivtci villages	Ruborozplidni stavu	Ivano–Frankovsk	1,000	A4i, B1i, B2
056	—	Burshtyn reservoir	Burshtunskoe vodokhranilishche	Ivano–Frankovsk	1,600	B1i
057	—	Dnestr valley between Staruy Martuniv and Marinopil villages	Richka Dnister mizh selamu Staruy Martuniv ta Marunopil	Ivano–Frankovsk	3,000	B1i
058	—	Klivodin Protected Locality	Klivodinskiy ornitologicheskiy zakaznik	Chernivtci	50	A4i, B1i, B2
059	—	Derkul river valley	Dolina r. Derkul	Chernivtci	700	A1
060	—	Dgulin lake	Ozero Dgulin	Chernivtci	50	B2
061	—	Dranitckiy Protected Locality	Dranickiy ornitilogichniy zakaznik	Chernivtci	100	B2
062	—	Ingul river valley	Khristoforovskie plavni	Nikolaev	1,000	B1i, B2
063	—	Kinburn peninsula	Kinburskaya kosa	Nikolaev	4,000	A4i, B1i, B2
064	SU050	Lower Dnepr	Nizovya Dnepra	Kherson	26,000	A4i, A4iii, B1i, B2
065	SU052	Yagorlytskiy and Tendrovskiy Bays	Yagorlytski and Tendrovski Zaliv	Kherson	72,000	A1, A4i, A4iii, B1i, B2, B3
066	—	Kakhovsk reservoir	Ostrov Kozackiy, Kakhovsloe vodokhranilicshe	Kherson	1,000	A1, B1i, B2
067	SU046, SU049	Karkinitskiy and Dzharilgatchskiy Bays	Karkinitskiy i Dzharilgatchkiy zaliv	Kherson	87,000	A1, A4i, A4iii, B1i, B2
068	SU051	Askania-Nova Biosphere Reserve	Askania-Nova	Kherson	33,307	A1, A3, A4i, A4iii, A4iv, B1i, B1iv
069	SU044	Sivash Bay, Sea of Azov	Sivash Zaliv, Azovskoye More	Crimea, Kherson	245,000	A1, A3, A4i, A4iii, A4iv, B1i, B1iv, B2, B3
070	—	Utlyuk estuary	Utlyukskiy liman	Zaporozhie	75,200	A4i, A4iii, B1i
071	SU043	Molochniy liman	Molochny Liman	Zaporozhie	22,450	A4i, A4iii, B1i, B3
072	—	Molochnaya river valley	Dolina r. Molochnaya	Zaporozhie	3,800	A4i, B1i, B2, B3
073	SU042	Obitochnaya peninsula	Obitochnaya Kosa	Zaporozhie	2,000	A4i, B1i

Table 1 ... continued. Summary of Important Bird Areas in Ukraine. 102 IBAs covering 20,323 km²

IBA code	1989 code	International name	National name	Administrative region	Area (ha)	Criteria (see p. 11)
074	—	Elanchik river valley	Dolina reki Elanchik	Donetsk	400	B1i, B2
075	SU038	Ukrainian Steppe Nature Reserve	Ukrainskiy Stepnoy Zapovednik	Donetsk	1,134	A1, B2
076	SU040	Belosarayskaya peninsula	Belosarayskaya Kosa	Donetsk	2,000	A4i, B1i, B2
077	SU041	Berdyanskaya peninsula	Berdyanskaya Kosa	Zaporozhie	1,800	A4i, A4iii, B1i, B2
078	SU039	Krivaya peninsula	Krivaya Kosa	Donetsk	8,800	A4i, B1i, B2
079	—	Mokrye Yasli forestry plantation	Urochicshe Mokrue Yasli	Odessa	500	B2
080	—	Kagul lake	Ozero Kagul	Odessa	10,500	A4iii
081	—	Kugurluy and Kartal lakes	Ozera Kugurluy i Kartal	Odessa	19,200	A1, A4i, A4iii, B1i, B2
082	—	River Danube	Ruslo r. Dunay	Odessa	2,500	A4i, A4iii, B1i
083	—	Kitaj lakes	Ozero Kitaj	Odessa	5,000	A4i, A4iii, B1i
084	—	Zhebriyanovskie plavni	Zhebriyanovskie plavni	Odessa	6,800	A1, A4i, A4iii, B1i, B2, B3
085	—	Sasuk lake	Ozero Sasuk	Odessa	22,800	A4i, A4iii, B1i, B2
086	—	Jansheisk lakes	Jansheiskie ozera	Odessa	1,000	B1i
087	—	Shagany–Alibey–Burnas lake-system	Sistema ozer Shagany–Alibey–Burnas	Odessa	19,200	A1, A4i, A4iii, B1i, B2
088	—	Zmeinuy island	Ostrov Zmeinuy	Odessa	17	A1, A4i, B1i
089	—	Budakskiy estuary	Budakskiy liman	Odessa	2,700	A4i, B1i
090	—	Kuyalnik estuary	Kuyalnitskiy liman	Odessa	7,000	B1i, B2, B3
091	—	Dnestr delta	Dnestr delta	Odessa	27,600	A4i, A4iii, B1i, B2
092	SU053	Tiligul estuary	Tiligulski Liman	Odessa, Nikolaev	11,000	A4i, A4iii, B1i, B2
093	—	Cape Uret	Mys Uret	Crimea	9,600	B1i
094	—	Tarkhankut peninsula	Tarkhankutskiy poluostrov	Crimea	4,200	B1i, B2
095	—	Lake near Magazinka village	Lake near v. Magazinka	Crimea	800	B1i
096	—	Belogorie	Belogorie	Crimea	32,000	B2
097	SU048	Cape Martyan	Mys Martyan	Crimea	240	B1i
098	SU045	Crimean Nature Reserve	Krymski Zapovednik	Crimea	34,563	A1, B2, B3
099	SU056	Karadag Nature Reserve	Karadag Zapovednik	Crimea	2,855	B1i
100	—	Aktashskoe lake	Aktashko–Astaninskie plavni	Crimea	4,800	A4i, B1i, B2, B3
101	—	Uzunlarskoe lake	Uzunlarsky lake	Crimea	9,600	A1, B2
102	—	Bagerovo	Bagerovo	Crimea	20,510	A1, B2

Sites identified in the previous inventory of IBAs in Europe (Grimmett and Jones 1989) but no longer considered to be IBAs
SU037 Lugansk, Voroshilovgrad Oblast; SU047 Yalta Mountain Forest, Bolshaya Yalta, Krym

Map 1. Location, area and criteria category of Important Bird Areas in Ukraine.

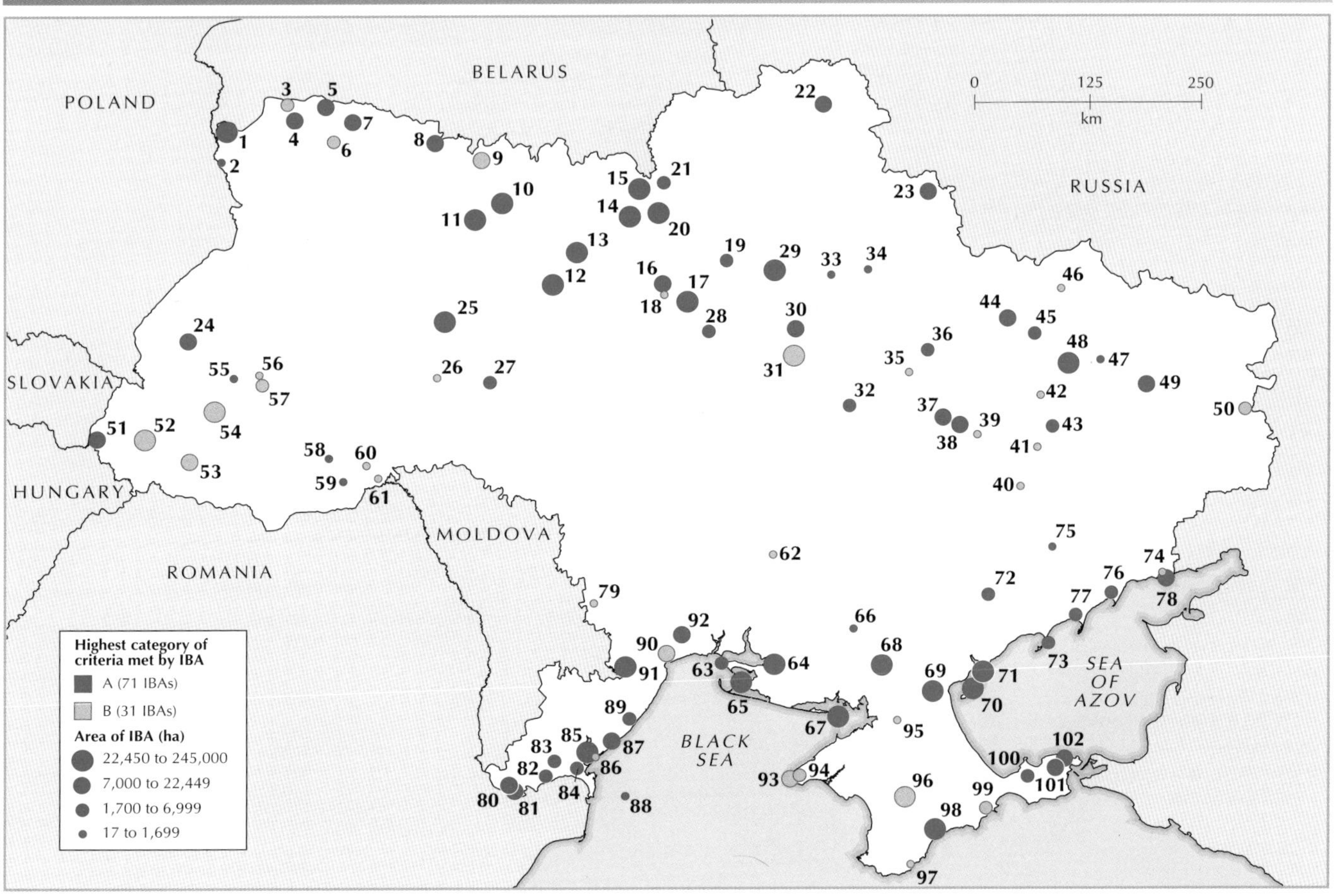

Table 2. Important Bird Areas in Ukraine that are important for species of global conservation concern (meeting criterion A1).

Species	IBA code
Phalacrocorax pygmeus Pygmy Cormorant	081, 084
Pelecanus crispus Dalmatian Pelican	084
Anser erythropus Lesser White-fronted Goose	069
Branta ruficollis Red-breasted Goose	066, 067, 068, 081, 087, 088
Aythya nyroca Ferruginous Duck	001, 051
Haliaeetus albicilla White-tailed Eagle	014, 015, 017, 020, 028
Aegypius monachus Cinereous Vulture	098
Aquila clanga Greater Spotted Eagle	020, 021
Aquila heliaca Imperial Eagle	038, 048, 049
Falco naumanni Lesser Kestrel	075
Crex crex Corncrake	001, 002, 004, 005, 007, 008, 010, 011, 012, 013, 016, 022, 023, 024, 027, 030, 032, 034, 036, 037, 038, 043, 044, 045, 047, 048, 049, 059
Tetrax tetrax Little Bustard	102
Otis tarda Great Bustard	069, 101, 102
Gallinago media Great Snipe	004, 005, 022, 044, 069
Numenius tenuirostris Slender-billed Curlew	065, 067
Acrocephalus paludicola Aquatic Warbler	001, 004, 005, 007, 019

Table 3. Important Bird Areas in Ukraine that support important numbers of one or more congregatory species (i.e. meeting criteria A4 and/or B1). IBAs meeting both criteria A4 and B1 for the species are shown in **bold**. IBAs meeting only criterion B1 for the species concerned, and not A4, are shown in normal type. For key to 'Season', see p. 7.

Species	Season	IBA code
Podiceps cristatus Great Crested Grebe	P	031, **071**
Podiceps grisegena Red-necked Grebe	B	055
Podiceps nigricollis Black-necked Grebe	B	**055**
	P	**071**
Phalacrocorax carbo Cormorant	B	**064**, **067**, **069**, **071**, 073, 077, **091**
	P	**064**, **065**, 067, 097
	N	063
Phalacrocorax aristotelis Shag	B	093, 094, 099
	N	094
Phalacrocorax pygmeus Pygmy Cormorant	B	**081**
Pelecanus onocrotalus White Pelican	P	**085**
Pelecanus crispus Dalmatian Pelican	P	**084**
Nycticorax nycticorax Night Heron	B	**058**, **091**
Ardeola ralloides Squacco Heron	B	**064**
Egretta garzetta Little Egret	B	**067**, **069**, **089**
	P	**065**
Egretta alba Great White Egret	B	**015**, **038**, **044**, **064**, **067**, **069**, **073**, **077**, **084**, **091**
	P	**067**, **092**
	N	**037**
Ardea cinerea Grey Heron	B	**064**
	P	**065**
Plegadis falcinellus Glossy Ibis	B	**064**, **069**, **091**, **092**
Platalea leucorodia Spoonbill	B	**081**, **084**
Cygnus olor Mute Swan	B	064, 081, 084
	W	005, **067**, 081, 084, 086, 087, 089, 092
	P	**064**, **065**, **070**, **071**, 084, **085**, 086
	N	069
Cygnus cygnus Whooper Swan	W	**067**
Anser fabalis Bean Goose	P	001, **033**, 051
Anser albifrons White-fronted Goose	W	**065**, **081**, **082**, **083**, 084, **087**, 092
	P	012, 025, 064, **065**, **068**, **069**, **070**, **071**, **085**, 092
Anser erythropus Lesser White-fronted Goose	P	**069**
Anser anser Greylag Goose	B	030, 031, 041, 062, 064, 084, 091
	W	065, **071**, 087
	P	**025**, 031, 039, 040, 045, **064**, 065, 066, **069**, 070
	N	001, 027, 031, 041, 085
Branta ruficollis Red-breasted Goose	W	**081**, **087**
Tadorna ferruginea Ruddy Shelduck	W	**068**
	N	100
Tadorna tadorna Shelduck	W	087, 090
	P	**069**
Anas penelope Wigeon	P	**069**
Anas crecca Teal	P	**091**
Anas platyrhynchos Mallard	W	**065**, 067, 071, 087
	P	015, 064, **065**
Anas acuta Pintail	P	**069**
Netta rufina Red-crested Pochard	B	**084**
Aythya ferina Pochard	W	**065**, **084**, **087**
	P	001, **015**, **065**, **085**
Aythya fuligula Tufted Duck	W	087
	P	015, 085
Aythya marila Scaup	W	**065**, **071**
	P	**065**, 070, **073**
Somateria mollissima Eider	R	065
	N	063
Bucephala clangula Goldeneye	W	018, 045, 056, 057
	P	051, 064
Mergus albellus Smew	W	**071**
	P	**085**
Mergus merganser Goosander	W	028, 056
Fulica atra Coot	W	**065**, 067
	P	**064**, **065**
Grus grus Crane	P	**001**, **005**, 006, 007, 064, 066, **068**, **069**, **088**
	N	**069**
Himantopus himantopus Black-winged Stilt	B	**069**, 072, 095
	P	**069**
Recurvirostra avosetta Avocet	B	065, **069**, 071
	P	**069**, **087**
Glareola pratincola Collared Pratincole	B	**069**, **072**, 085, 087, **100**
	P	**069**
Charadrius hiaticula Ringed Plover	P	**069**
Charadrius alexandrinus Kentish Plover	B	**069**
	P	**069**
Pluvialis squatarola Grey Plover	P	**069**
Calidris alba Sanderling	P	**069**
Calidris minuta Little Stint	P	**069**
Calidris alpina Dunlin	P	**069**
Limicola falcinellus Broad-billed Sandpiper	P	**069**
Philomachus pugnax Ruff	P	**069**
Limosa limosa Black-tailed Godwit	B	**005**
Limosa lapponica Bar-tailed Godwit	P	**069**
Tringa erythropus Spotted Redshank	P	**069**
Tringa totanus Redshank	B	**005**, 069
	P	**069**
Tringa stagnatilis Marsh Sandpiper	P	**069**
Arenaria interpres Turnstone	P	**069**
Larus ichthyaetus Great Black-headed Gull	B	**063**, **069**
Larus melanocephalus Mediterranean Gull	B	**069**, **100**
Larus minutus Little Gull	B	**022**, **065**, **067**
	P	**069**, **078**, **092**
Larus genei Slender-billed Gull	B	**065**, **069**, **100**
Larus cachinnans Yellow-legged Gull	B	**065**, **067**, **069**, **071**, 081, **087**, **100**
	W	**067**, **092**
	P	**070**, 085, **087**, 089, **092**
Gelochelidon nilotica Gull-billed Tern	B	**069**, **100**
	P	**100**
Sterna caspia Caspian Tern	B	**067**, **069**
Sterna sandvicensis Sandwich Tern	B	**065**, **067**, **069**, **076**, 077, **078**
Sterna hirundo Common Tern	B	**065**, 067, **069**, 074, 076, **078**, **092**
	P	078
Sterna albifrons Little Tern	B	**069**, **078**, 090
Chlidonias hybridus Whiskered Tern	B	022, 027, 055
	P	**069**
Chlidonias niger Black Tern	B	**005**, **029**
	P	**005**, **069**, **078**
Chlidonias leucopterus White-winged Black Tern	B	**005**, **007**, **029**
	P	**005**

Table 4. Species of European conservation concern with significant breeding populations at Important Bird Areas in Ukraine (meeting any IBA criteria).

Species [1]	Minimum national breeding population (pairs) [2]	Proportion (%) of national population breeding at all IBAs in Ukraine
Phalacrocorax aristotelis Shag	250	100[3]
Phalacrocorax pygmeus Pygmy Cormorant	10	100[3]
Botaurus stellaris Bittern	4,000	43
Ixobrychus minutus Little Bittern	5,000	29
Nycticorax nycticorax Night Heron	3,645	76
Ardeola ralloides Squacco Heron	200	8
Ardea purpurea Purple Heron	733	88
Ciconia nigra Black Stork	250	26
Plegadis falcinellus Glossy Ibis	5,500	15
Platalea leucorodia Spoonbill	300	100[3]
Tadorna ferruginea Ruddy Shelduck	125	31
Anas strepera Gadwall	700	10
Anas querquedula Garganey	26,000	10
Netta rufina Red-crested Pochard	40	100[3]
Aythya ferina Pochard	25,000	12
Aythya nyroca Ferruginous Duck	1,500	4
Milvus migrans Black Kite	650	5
Haliaeetus albicilla White-tailed Eagle	45	62
Aegypius monachus Cinereous Vulture	2	100[3]
Circaetus gallicus Short-toed Eagle	30	57
Circus macrourus Pallid Harrier	10	20
Circus pygargus Montagu's Harrier	200	33
Aquila pomarina Lesser Spotted Eagle	200	15
Aquila clanga Greater Spotted Eagle	40	30
Aquila heliaca Imperial Eagle	40	28
Hieraaetus pennatus Booted Eagle	450	2
Falco naumanni Lesser Kestrel	20	50
Falco tinnunculus Kestrel	2,800	6
Falco vespertinus Red-footed Falcon	400	5
Falco cherrug Saker	150	9
Coturnix coturnix Quail	10,000	7
Porzana porzana Spotted Crake	4,200	100[3]
Porzana parva Little Crake	3,500	100[3]
Crex crex Corncrake	25,000	9
Tetrax tetrax Little Bustard	8	63
Otis tarda Great Bustard	130	25
Recurvirostra avosetta Avocet	350	100[3]
Glareola pratincola Collared Pratincole	280	100[3]
Charadrius alexandrinus Kentish Plover	4,000	11
Gallinago media Great Snipe	400	28
Limosa limosa Black-tailed Godwit	4,500	27
Tringa totanus Redshank	6,000	61
Larus melanocephalus Mediterranean Gull	180,000	3
Larus minutus Little Gull	100	100[3]
Gelochelidon nilotica Gull-billed Tern	400	100[3]
Sterna caspia Caspian Tern	250	100[3]
Sterna sandvicensis Sandwich Tern	9,900	100[3]
Sterna albifrons Little Tern	1,200	100[3]
Chlidonias hybridus Whiskered Tern	8,000	8
Chlidonias niger Black Tern	3,500	100[3]
Streptopelia turtur Turtle Dove	20,000	27
Bubo bubo Eagle Owl	150	17
Strix aluco Tawny Owl	6,000	7
Asio flammeus Short-eared Owl	150	31
Caprimulgus europaeus Nightjar	12,000	9
Alcedo atthis Kingfisher	1,500	14
Merops apiaster Bee-eater	10,000	12
Coracias garrulus Roller	3,000	2
Jynx torquilla Wryneck	20,000	9
Picus canus Grey-headed Woodpecker	2,000	14
Dendrocopos medius Middle Spotted Woodpecker	1,000	15
Picoides tridactylus Three-toed Woodpecker	300	45
Lullula arborea Woodlark	4,500	32
Riparia riparia Sand Martin	140,000	12
Phoenicurus phoenicurus Redstart	9,000	1
Saxicola rubetra Whinchat	27,000	3
Turdus torquatus Ring Ouzel	16,000	4
Turdus philomelos Song Thrush	180,000	4
Locustella fluviatilis River Warbler	9,000	4
Locustella luscinioides Savi's Warbler	3,500	44
Acrocephalus paludicola Aquatic Warbler	2,500	76
Acrocephalus schoenobaenus Sedge Warbler	55,000	39
Sylvia nisoria Barred Warbler	2,700	28
Sylvia borin Garden Warbler	55,000	1
Phylloscopus sibilatrix Wood Warbler	150,000	<1
Muscicapa striata Spotted Flycatcher	95,000	3
Ficedula albicollis Collared Flycatcher	20,000	15
Parus caeruleus Blue Tit	85,000	4
Lanius minor Lesser Grey Shrike	3,000	9
Fringilla coelebs Chaffinch	2,500,000	2

1. Only those species of European conservation concern (see Box 1, p. 12) that meet IBA criteria in Ukraine are listed.
2. Data are taken from the BirdLife/EBCC European Bird Database 1998 (Heath and Borggreve 2000).
3. The percentage of the national population in IBAs exceeds 100%. Usually this is because the national population estimate has not been updated recently whilst the IBA population estimate has been recently updated with new data as a result of comprehensive surveys of IBAs themselves. Also, the individual site count for a species may be the maximum or average over recent years, and summing these may record more birds than are present nationally in any single year.

ORNITHOLOGICAL IMPORTANCE

There are 154 species of European conservation concern (SPECs) which regularly breed in Ukraine, including eight globally threatened species and seven near-threatened species (Tucker and Heath 1994). Of these latter 15 species of global conservation concern, all but *Glareola nordmanni* (near-threatened) are known to breed in important numbers at IBAs (Table 2). Out of the remaining 139 SPECs breeding in Ukraine that are not of global conservation concern, 86 have an unfavourable conservation status in Europe, however (Tucker and Heath 1994). A further four globally threatened species (*Pelecanus crispus*, *Anser erythropus*, *Branta ruficollis* and *Numenius tenuirostris*) do not breed in Ukraine but occur regularly during migration and/or in winter, with all using IBAs in important numbers (Table 2).

A total of 49 IBAs regularly support important numbers of globally threatened and near-threatened species at some stage in their annual life cycle, and thus qualify under the A1 criterion (Table 2). The majority of IBAs in Ukraine—62 out of 102 sites—support important numbers of congregatory waterbirds, whether in the breeding, wintering or migrating seasons, totalling 68 species, a very wide diversity (Table 3). Of these 61 sites, 37 host globally important congregations of waterbirds (qualifying under the A4i/A4iii criteria) and the other 24 support regionally important congregations (qualifying under B1i but not under A4i/A4iii). The wetlands are particularly important during the migration seasons, when 45 species occur in globally important congregations at 21 sites, mainly along the Azov–Black Sea coastline and along the Dnepr, Desna and Prip'yat valleys (Table 3, Map 1).

The greatest number of criteria that any one site meets is nine (one site only: Sivash Bay, site 069), while four sites (015, 065, 068, 084) meet more than five criteria (Table 1). Eleven sites out of the total of 102 have more than 15 qualifying species each, most notably Sivash Bay (site 069; 45 qualifying species), Samarskiy forest (038; 31 species), River Prip'yat valley (005; 21 species) Yagorlytskiy and Tendrovskiy Bays (065; 18 species), and Karkinitskiy and Dzharilgatchskiy Bays (067; 16 species). In reality these totals are probably higher still, since most of these sites have not been surveyed comprehensively.

Seventy-five of the 102 IBAs support important breeding numbers of SPECs, thus meeting criteria B2 and/or B3 (Table 4). A total of 80 SPECs qualify under these criteria and/or the A1 criterion (Table 4), and available evidence suggests that, for 18 of these 80 species, more than three-quarters of their national population breeds within the IBA network. All but one of these 18 species are waterbirds. On the other hand, for 41 of the 80 SPECs, less than 25% of the national population breeds within the IBA network (Table 4). Not surprisingly, most of these 41 species have highly dispersed distributions across forested, agricultural or steppe landscapes, but also included in this group are globally threatened species (*Aythya nyroca*, *Crex crex*, *Otis tarda*) and gregariously breeding waterbirds for which Ukraine holds globally important populations (*Larus melanocephalus*, *Chlidonias hybridus*).

HABITATS

Three main zones of vegetation cross Ukraine, approximately from north-east to south-west. The forest zone covers northern Ukraine and the Carpathian region, the forest–steppe zone covers central Ukraine, and the steppe zone covers southern Ukraine. Twenty-six IBAs are located in the forest zone, 33 IBAs in the forest–steppe zone, and 43 IBAs in the steppe zone. The predominant habitat at IBAs is wetland (Figure 1). At least 80 of the IBAs contain significant wetland habitat, of which 39 are predominantly wetland (by area covered). The majority of wetland sites are in the major river valleys, or along the Azov–Black Sea coastline, or occur as wetland–forest complexes. At least 57 IBAs contain some grassland habitat, usually as complexes with cultivated land or wetlands. At least 46 IBAs contain a significant amount of forest or woodland, and this habitat-type is predominant at 16 sites. Forty IBAs include some artificial landscape, mainly cultivated land, and at six sites this type of habitat covers more than 50% of the IBA area. Nine IBAs contain rocky habitats, mainly as coastal cliffs or mountain areas. Twenty IBAs include marine habitat, usually encompassing shallow sea bays with islands, with three sites being predominantly marine.

Peatlands and the various types of steppe habitat are not adequately covered by the current IBA network, in part because birds here tend to be highly dispersed, not concentrated in particular places, and thus particular IBAs are difficult to identify.

Figure 1. Habitats at Important Bird Areas in Ukraine (see Appendix 3 for definitions of habitats).

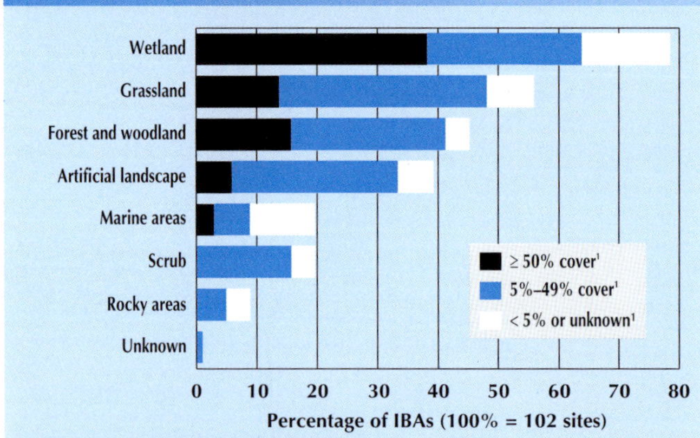

1. percentage of area of individual IBA covered by habitat

Figure 2. Land-uses at Important Bird Areas in Ukraine (seeAppendix 3 for definitions of land-uses).

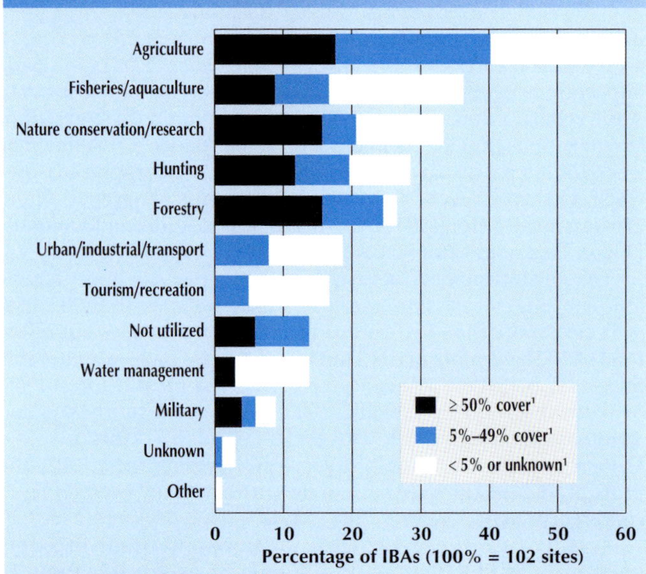

1. percentage of area of individual IBA covered by land-use

IMPACTS ON IBAs – LAND-USE AND THREATS

Figure 2 shows the different forms of land-use that occur in Ukrainian IBAs. At least 61 IBAs are used to some extent for agriculture, mainly for grazing and haymaking (but outside IBAs, arable cultivation is the predominant agricultural land-use). At least 27 of the IBAs support some forestry activity, and this is the predominant activity (by area) at 16 sites. At least 16 of the 102 IBAs are used predominantly for nature conservation. Fisheries/aquaculture is the predominant economic activity (by area) at only nine IBAs, but occurs to some extent at a minimum of 37 IBAs.

Most IBAs are state-owned. In recent years an increasing amount of state land has been transferred to private ownership, and most agricultural land is now privately owned. Wetlands, forests and woodland are still state-owned, however.

Figure 3 summarizes the key threats and impacts to IBAs. Disturbance to birds is the most frequently cited threat, affecting at least 62 sites to some extent. Legal hunting affects at least 29 IBAs to some extent (Figure 2) and poses a significant threat at some sites, while illegal hunting has an even greater impact, although this is difficult to estimate. Overall, hunting (termed 'unsustainable exploitation' in Figure 3) poses a threat, of variable intensity, at a minimum of 42 IBAs. The intensification of agriculture in recent decades has caused contamination of land and fresh water with fertilizers and chemicals, industrial activities are contaminating air, land and water in some regions of Ukraine, and urban/domestic waste-water is causing chemical and microbiological contamination of rivers and shallow marine waters. The direct impacts on birds are difficult to estimate but include higher mortality, shorter life-span, lowered reproductive potential, higher rates of illness, etc. Indirect impacts of pollution on birds occur through accelerated habitat succession (e.g. nutrient pollution accelerates the transition from grassland to woodland), the loss of suitable habitats, reductions in the productivity of prey species, etc.

The intensification of forestry activities, mainly commercial deforestation and selective logging, are causing a decrease in the national populations of some raptors. At the moment, the drainage of wetlands has largely halted, through lack of financial resources, but this is likely to pose the main threat to several IBAs as soon as the economic situation improves. The abandonment or reduction of management over large areas of agricultural land is having detrimental effects on numbers of open-country birds at seven IBAs, largely due to the overgrowth of meadows and fields by bushes, trees and tall grass, as a result of lowered grazing pressure and the cessation of haymaking and arable cultivation.

PROTECTION STATUS

Table 5 summarizes the national and international protection status of all Ukrainian IBAs.

■ National protection

Details of the principal categories of national protected area are given below.

1. Nature Reserve (Zapovednik)

This is the principal category of protected area in the countries of the former Soviet Union, including Ukraine. A Nature Reserve is an area of land withdrawn from economic use for the study and conservation of fauna and flora. Literally meaning 'forbidden/restricted area', Nature Reserves are closed to non-specialists, including tourists. Most Nature Reserves are administered by the

Figure 3. Threats at Important Bird Areas in Ukraine (see Appendix 3 for definitions of threat types and impact categories).

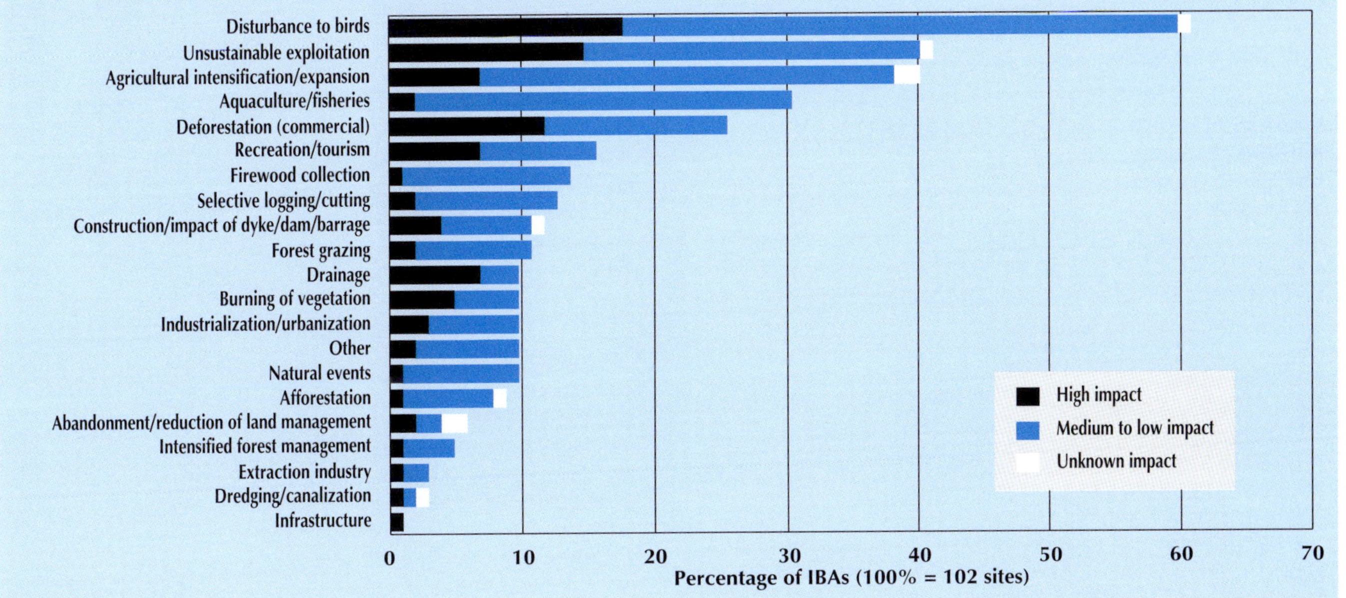

Ministry for Environmental Protection and Nuclear Safety or by the National Academy of Science.

2. Biosphere Reserve
Biosphere Reserves have core zones and buffer zones. Core zones have the same regime as Nature Reserves, while buffer zones are areas of semi-protected land where certain economic activities are allowed, provided that the reserve itself is unharmed. Biosphere Reserves are administered by the Ministry for Environmental Protection and Nuclear Safety or by the National Academy of Science.

3. National Park
National Parks have the same regime as Nature Reserves. Core zones, where certain economic activities are allowed, are surrounded by a large area where a greater variety of economic activities are possible, e.g. tourism, fishing, hunting, haymaking, cattle-grazing, etc. National Parks are administered by the Ministry for Environmental Protection and Nuclear Safety.

4. Regional Landscape Parks
Regional Landscape Parks have the same regime as National Parks but are administered by regional government. The first Regional Landscape Park on private land is currently being created and will be administered by the land-owner.

5. Protected Locality (Zakaznik)
Limited economic activities are possible within such an area. Protected Localities are administered by regional government.

All high-level protected areas (Nature Reserves, Biosphere Reserves and National Parks) are owned, administered and monitored by the state. Smaller areas, protected at the regional or district level, are state-owned but have no administrative staff or structure and are not monitored.

Nineteen IBAs are afforded some legal protection through their overlap with the national protected-area network, being partly or fully covered by Nature Reserves, Biosphere Reserves, National Parks, Regional Landscape Parks or Protected Localities (Table 5). It should be stressed that large areas of these 19 IBAs are not protected in real life because they do not have an effective law-enforcement system or management plan, and that the great majority of IBAs (83 out of 102) have no protection whatsoever under national law (Figure 4).

International protection

Ukraine is a signatory or contracting party to all of the major international conventions relevant to the protection of sites for nature conservation (Box 1), including the Agreement on the Conservation of African–Eurasian Migratory Waterbirds (AEWA). However, about three-quarters of the IBAs in Ukraine have no legal protection at the international level (Figure 5). No natural World Heritage Sites have been inscribed. Three Biosphere Reserves (owned, administered and monitored by the state) overlap with IBAs (Table 5). There are 22 designated Ramsar Sites in Ukraine, all of which qualify on the basis of their globally important congregations of waterbirds, and all of them overlap with, or are contained by, IBAs (Table 5). Given that at

Figure 4. The national protection status of Important Bird Areas in Ukraine.

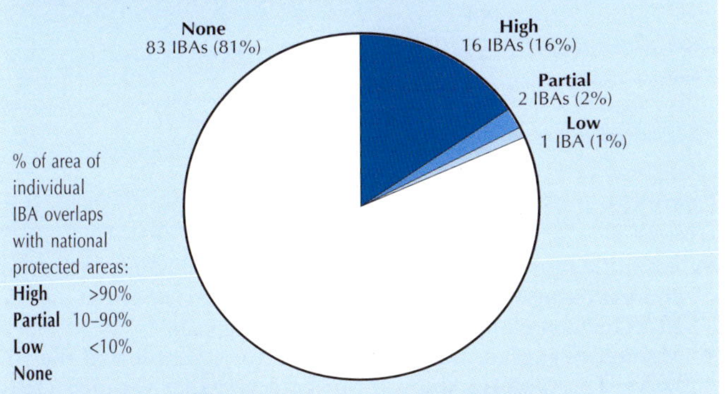

Total area of overlap between IBA network in Ukraine and national protected-area system (see Table 5 for categories) = 1,275 km² (6% of total IBA area).

Figure 5. The international protection status of Important Bird Areas in Ukraine.

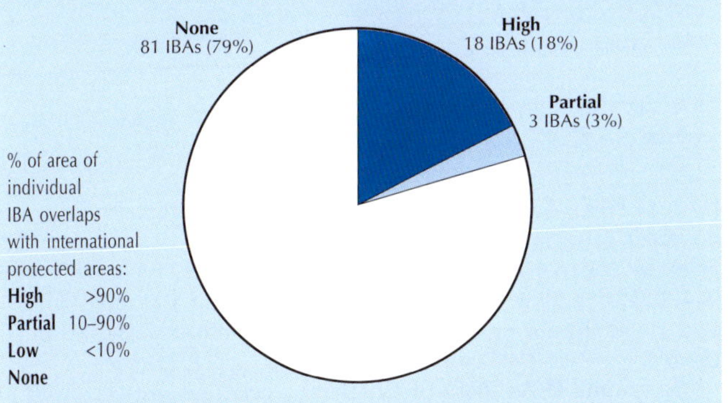

Total area of overlap between IBA network in Ukraine and international protected-area system (see Table 5 for categories) = 6,547 km² (32% of total IBA area).

Table 5. Protection status of Important Bird Areas in Ukraine.
A tick (✔) indicates that an IBA overlaps with a protected area (to any extent).

		National					International	
IBA code	International name	Nature Reserve	Biosphere Reserve	National Park	Regional Protected Area	Protected Locality	Ramsar Site	Biosphere Reserve
001	Shatsk National Park			✔			✔	
002	Zapadniy Bug river valley							
003	Orikhivski lake-system							
004	Turiya river valley							
005	River Prip'yat valley				✔		✔	
006	Cheremske mire							
007	Stokhod river valley						✔	
008	Sira Pogonya mire							
009	Polesskiy Nature Reserve	✔						
010	Ushomirskoe Game Reserve							
011	Novograd-Volunskoe Game Reserve							
012	Korostushevskoe Game Reserve							
013	Radomyshlskoe Game Reserve							
014	Dneprovsko-Teterevskoe Game Reserve							
015	Kiev reservoir							
016	Meadows near Procev village							
017	Rzhycshevskoe Game Reserve							
018	Kanev reservoir							
019	Supoy marsh					✔		
020	Forests in Dnepr and Desna valley lowlands							
021	Bondar marsh							
022	Korotchenkov meadows							
023	Peschanskiy and Mogrichskiy forests							
024	Chaykovichi marsh							
025	Ikva river valley							
026	Vovk river valley near Derazhnya village							
027	Mikulinetskie fish-ponds							
028	Kanev Nature Reserve	✔						
029	Uday river valley							
030	Krasnue marsh							
031	Sula bay, Kremenchug reservoir							
032	Psel river valley near Komsomolsk							
033	Bodakovskoe marsh							
034	Meadows near Saru village in the Psel river valley							
035	Veluke marsh							
036	Dneprovsko-Orelskiy Nature Reserve	✔						
037	Orel river valley							
038	Samarskiy forest							
039	Soleniy liman							
040	Debalcevskoe lake							
041	Petropavlovskiy marsh							
042	Bolshemikhailovskiy forest							
043	Samara river valley							
044	Mzha river valley							
045	Liman lake-system							
046	Pechenezhskoe reservoir							
047	Bakhtin river valley							
048	Izyumskaya Luka forest							
049	Kremenskoy forest							
050	Stanichno-Luganskoe fish-farm							
Subtotal of IBAs		3	0	1	1	1	3	0
051	Latorica river valley near Chop							
052	Carpathian Volkano mountains							
053	Carpathian Biosphere Reserve		✔					✔
054	Gorgany	✔						
055	Fish-ponds near Kukilniku and Bilshivtci villages							
056	Burshtyn reservoir							
057	Dnestr valley between Staruy Martuniv and Marinopil villages							
058	Klivodin Protected Locality					✔		
059	Derkul river valley							
060	Dgulin lake							
061	Dranitčkiy Protected Locality					✔		
062	Ingul river valley							
063	Kinburn peninsula				✔			
064	Lower Dnepr						✔	
065	Yagorlytskiy and Tendrovskiy Bays		✔				✔	✔
066	Kakhovsk reservoir							
067	Karkinitskiy and Dzharilgatchskiy Bays				✔		✔	
068	Askania-Nova Biosphere Reserve		✔					✔
069	Sivash Bay, Sea of Azov						✔	
070	Utlyuk estuary							
071	Molochniy liman						✔	
072	Molochnaya river valley							
073	Obitochnaya peninsula						✔	
074	Elanchik river valley							
075	Ukrainian Steppe Nature Reserve	✔						
076	Belosarayskaya peninsula						✔	
077	Berdyanskaya peninsula						✔	
078	Krivaya peninsula						✔	
079	Mokrye Yasli forestry plantation							
080	Kagul lake							
081	Kugurluy and Kartal lakes						✔	
082	River Danube							
083	Kitaj lakes							
084	Zhebriyanovskie plavni						✔	
085	Sasuk lake						✔	
086	Jansheisk lakes							
087	Shagany-Alibey-Burnas lake-system						✔	
088	Zmeinuy island							
089	Budakskiy estuary							
090	Kuyalnik estuary							
091	Dnestr delta						✔	
092	Tiligul estuary				✔		✔	
093	Cape Uret							
094	Tarkhankut peninsula							
095	Lake near Magazinka village							
096	Belogorie							
097	Cape Martyan	✔						
098	Crimean Nature Reserve	✔						
099	Karadag Nature Reserve	✔						
100	Aktashskoe lake							
101	Uzunlarskoe lake							
102	Bagerovo							
Total number of IBAs		**8**	**3**	**1**	**4**	**3**	**18**	**3**

least 17 IBAs do not overlap with designated Ramsar Sites but do meet criteria for designation as Ramsar Sites (under IBA criteria A4i/A4iii)—sites 015, 022, 025, 029, 033, 037, 038, 044, 055, 058, 063, 068, 072, 080, 088, 089, and 100 (see Tables 1 and 3)—it should be a priority to obtain Ramsar designation for them.

Sixty-nine IBAs (68% of all IBAs) receive no protection at all through national or international legislation (Table 5), and the majority of these are of global importance for waterbirds (meeting A1/A4i/A4iii criteria: see Tables 1–3).

CONSERVATION

- Scientific research into, and monitoring of, breeding, migrating and wintering waterbird populations are carried out at several IBAs located along the Azov–Black Sea coasts.
- Management plans have been written for only a few of the 102 IBAs. The plan for the soon-to-be-designated Danube Delta Biosphere Reserve (in site 084) is of good quality, while the plans for those other Nature Reserves and National Parks which overlap with IBAs are very simple and lack detail.

Box 1. International legislation and initiatives that are relevant to site conservation in Ukraine (see Appendix 1 for a general description of these agreements).		
	Global	
	Biodiversity Convention	✔
	Ramsar Convention	✔
	Bonn Convention	✔
	World Heritage Convention	✔
	MAB Programme	✔
	Pan-European	
	Bern Convention	(✔)
	✔ Convention ratified/initiative supported	
	(✔) Convention signed	

ANALYTICAL METHODS

- Most bird data quoted in this inventory were collected between 1994 and 1997.
- Most waterbird counts are of good quality (verification code A or B) and are based on transect-count or plot-count methods.
- Counts of staging birds on migration are, in most cases, maximum daily counts (rather than seasonal totals).
- The seriousness of threats and the percentage cover of land-uses at sites were assessed according to information available in 1997.
- Recent surveys of two globally threatened bird species with important populations in Ukraine (*Aquila heliaca* and *Acrocephalus paludicola*) have added sites to the IBA inventory.
- Maps (1:100,000) indicating IBA boundaries are available from the Ukrainian Union for Bird Conservation.

GLOSSARY

liman a water-body in a shallow depression in the steppe zone, normally but not necessarily saline (can be brackish or fresh).
plavni a large area of reedbeds with numerous narrow streams and small lakes; or, a stretch of open water in certain types of wetland (river flood-plain, river delta, estuary).
UTOP Ukrainian Union for Bird Conservation (Ukrainske Tovarystvo Okhorony Ptakhiv); the BirdLife Partner Designate in Ukraine.

ACKNOWLEDGEMENTS

The IBA expert group of the Ukrainian Union for Bird Conservation (UTOP) provided valuable input into the development of the IBA inventory for Ukraine, and comprised Sergey Zhila, Aleksey Lugovoy, Lyudvig Potish, Vladimir Cicyura, Andrey and Tat'yana Atemasovu, Igor Krivitskiy, Anna Nadtochiy, Anatoliy Poluda, Leonid Nazarenko, Nikolay Klestov, Vasiliy Kostyushin, Igor Gorban, Vladimir Pogranichnuy, Igor Shidlovskiy, Vladimir Chernikov, Petr Chegorko, Nikolay Prushinskiy, Igor Skilskiy, Tat'yana Ardamatskaya, Olga Yaremchenko, Vitaliy Grischenko, Maksim Gavrilyuk, Vladimir Buchko, Mikhail Khimin, Vitaliy Vetrov, Anatoliy Guziy, Anatoliy Gubkin, Vladimir Novak, Andrey Bokotey, Sergey Kostin, Mikhail Beskaravaynuy, Nikolay Knush, Alexander Voblenko, Alexander Malega, Petr Gorlov, Alexander Arkhipov, Vladimir Kinda, Yuriy Andryuschenko, Alexander Ponomarenko, Vitaliy Bulakhov, Nikolay Semenov, Vladimir Gulay, Victor Afanas'ev, Ivan Rusev, Oleg Potapov, Evgeniya Srebrodolskaya, Igor Martunyuk, Lena Dyadicheva, Rostislav Sorish, Igor Gerzhik, Alexander Macyura, Vladimir Popenko, Alexander Koshelev and Anatoliy Korzyukov.

UTOP also cooperated with the following organizations on data-gathering: Ministry of Environmental Protection and Nuclear Safety, Ukraine; Wetlands International–Ukraine; Azovo–Chernomorskaya Ornithological Station, Odessa State University; Institute of Zoology, National Academy of Science.

Much information on particular IBAs was taken from unpublished technical reports by the following authors: Afanas'ev (1996); Andryuschenko (1996); Ardamatskaya (1996); Atemasova and Atemasov (1996); Bashta (1996); Beskaravaynuy (1996); Bodnar (1997); Buchko (1996, 1997); Buchko and Pogranichnuy (1997); Bulakhov, Gubkin and Tarasenko (1996); Cherednyakov and Shaparenko (1996); Cicyura and Veselskiy (1996); Djamirzoev (1997); Dyadicheva and Macyura (1997); Gavris (1996); Gorban (1996, 1997); Gorban and Prushinskiy (1996); Gorlov (1996); Grischenko (1996, 1997); Gubkin (1997); Gubkin, Bulakhov and Tarasenko (1996); Gubkin, Bulakhov, Onuphriev, Chegorko and Tarasenko (1996); Gubkin, Bulakhov, Tarasenko, Obukhova and Chegorko (1997); Guziy (1996); Khimin (1995, 1996, 1997); Kinda (1996); Kinda and Grinchenko (1996); Klestov (1995); Knush (1997); Korzyukov (1997); Kostin and Beskaravaynuy (1996); Kostuyshin (1996, 1997); Krivitskiy (1997); Litvinenko (1996); Lugovoy (1997); Malega (1996, 1997); Mikityuk (1995, 1996); Novak (1997); Phedorenko (1995); Pisarev (1996); Poluda (1996); Prushunskiy (1997); Semenov (1997); Shaparenko (1996, 1997); Shidlovskiy (1997); Skilskiy (1996); Sorish (1997); Vergeles (1996); Vetrov (1996, 1997); Zhila (1996, 1997).

This review would have been almost impossible without substantial sponsorship from Vogelbescherming Nederland (BirdLife Partner in the Netherlands), the MATRA-Programme of the Netherlands Ministry of Foreign Affairs, the PIN-Programme of the Netherlands Ministry of Agriculture, Nature Conservation and Fisheries, and the Prince Bernhard Fonds, which provided special grants for setting up this work, purchasing equipment, coordination and fieldwork. We are grateful to BirdLife/Vogelbescherming Nederland and especially to Johanna Winkelman for all the practical help given during the project.

SITE ACCOUNTS

Shatsk National Park — A1, A4i, B1i, B2, B3 — 001

Admin region Volin
Coordinates 51°33'N 23°49'E
Altitude 162–189 m **Area** 32,850 ha

Site description

A lake-system with meadows, bordered by mixed forest dominated by pine *Pinus*.

Habitats Forest and woodland (40%; broadleaved deciduous forest; native coniferous forest; mixed forest; alluvial/very wet forest), Grassland (5%; humid grassland), Wetland (32%; sand-dunes/sand beach; standing fresh water; river/stream; water-fringe vegetation; fen/transition mire/spring), Artificial landscape (23%; arable land; other urban/industrial areas)
Land-use Agriculture (15%), Fisheries/aquaculture, Forestry (40%), Nature conservation/research (30%), Tourism/recreation (10%), Urban/industrial/transport (8%)

Birds

Species	Season	Year	Pop min	Pop max	Acc	Criteria
Botaurus stellaris Bittern	B	1994	80	—	B	B2
Anser fabalis Bean Goose	P	1989	2,000	3,000	—	B1i
Anser anser Greylag Goose	N	1993	350	—	B	B1i
Aythya ferina Pochard	B	1994	300	—	B	B3
Aythya ferina Pochard	P	1989	5,000	10,000	—	B1i
Aythya nyroca Ferruginous Duck	B	1994	20	—	B	A1
Circus pygargus Montagu's Harrier	B	1996	4	5	B	B3
Aquila pomarina Lesser Spotted Eagle	B	1996	3	4	B	B2
Porzana parva Little Crake	B	1989	50	70	B	B3
Crex crex Corncrake	B	1996	15	20	B	A1
Grus grus Crane	P	1989	2,000	3,000	—	A4i, B1i
Bubo bubo Eagle Owl	R	1989	2	3	B	B2
Asio flammeus Short-eared Owl	B	1989	2	4	B	B2
[1] ***Acrocephalus paludicola*** Aquatic Warbler	B	1994	40	60	B	A1, B2

1. Data for 10 km².

Large numbers of waterbirds stage here during migration (20,000–50,000).

Protection status

National High **International** High
32,515 ha of IBA covered by National Park (Shatsk, 32,515 ha). 32,850 ha of IBA covered by Ramsar Site (Shatsk Lakes, 32,850 ha).

Conservation issues

Threats Agricultural intensification/expansion (B), Aquaculture/fisheries (C), Deforestation (commercial) (A), Drainage (A), Firewood collection (B), Forest grazing (C), Industrialization/urbanization (B), Intensified forest management (C), Recreation/tourism (B), Selective logging/cutting (C), Unsustainable exploitation (C)

There are no serious threats.

Zapadniy Bug river valley A1 002

Admin region Volin
Coordinates 51°15'N 23°45'E
Altitude 162–164 m **Area** 1,450 ha

Site description

A stretch of the Zapadniy Bug valley, near the villages of Pidlissya, Rivne and Vushnivka. Habitats are dominated by grassland, with numerous lakes.

Habitats Grassland (83%; humid grassland), Wetland (7%; standing fresh water; river/stream; water-fringe vegetation), Artificial landscape (10%; highly improved reseeded grassland)
Land-use Agriculture (80%)

Birds

Species	Season	Year	Pop min	Pop max	Acc	Criteria
Crex crex Corncrake	B	1997	22	30	B	A1

Protection status

National None **International** None

Conservation issues

Threats Disturbance to birds (A)

Orikhivski lake-system B2 003

Admin region Volin
Coordinates 51°52'N 24°46'E
Altitude 148–162 m **Area** 3,000 ha

Site description

A group of natural lakes in the River Prip'yat valley, dominated by reedbeds *Phragmites* and meadows. Human activities include cattle-grazing, haymaking, and hunting.

Habitats Forest and woodland (10%; native coniferous forest; mixed forest), Scrub (5%; scrub), Grassland (5%; humid grassland), Wetland (55%; standing fresh water; water-fringe vegetation), Artificial landscape (25%; highly improved reseeded grassland; arable land)
Land-use Agriculture (75%), Fisheries/aquaculture (2%), Forestry (20%), Hunting, Not utilized (1%), Urban/industrial/transport (2%)

Birds

Species	Season	Year	Pop min	Pop max	Acc	Criteria
[1] ***Chlidonias niger*** Black Tern	B	1997	40	200	A	B2
[2] ***Bubo bubo*** Eagle Owl	R	1997	1	2	A	B2

1. Large decrease.
2. 1982–1997.

Breeding species of global conservation concern that do not meet IBA criteria: *Crex crex* (5–12 pairs).

Protection status

National None **International** None

Conservation issues

Threats Abandonment/reduction of land management (A), Afforestation (C), Agricultural intensification/expansion (B), Drainage (A), Firewood collection (C), Forest grazing (C), Other (C)

Haymaking and harvesting of sedge *Carex* ('Other') can cause problems.

Turiya river valley A1, B2, B3 004

Admin region Volin
Coordinates 51°43'N 24°54'E
Altitude 150–169 m **Area** 7,900 ha

Site description

A valley with marshes and meadows.

Habitats Forest and woodland (30%; mixed forest; alluvial/very wet forest), Grassland (54%; humid grassland), Wetland (16%; standing fresh water; river/stream; water-fringe vegetation; fen/transition mire/spring)
Land-use Agriculture (70%), Forestry (30%), Hunting (30%), Nature conservation/research (50%)

Birds

Species	Season	Year	Pop min	Pop max	Acc	Criteria
Circus pygargus Montagu's Harrier	B	1996	2	3	B	B3
Porzana porzana Spotted Crake	B	1996	500	—	B	B3
[1] ***Crex crex*** Corncrake	B	1996	32	—	B	A1
[1] ***Gallinago media*** Great Snipe	B	1996	40	50	—	A1
[1] ***Acrocephalus paludicola*** Aquatic Warbler	B	1996	150	250	—	A1, B2

1. Extrapolation.

Protection status

National None **International** None

Conservation issues

Threats Afforestation (C), Agricultural intensification/expansion (C), Aquaculture/fisheries (C), Deforestation (commercial) (B), Selective logging/cutting (C)

River Prip'yat valley A1, A4i, B1i, B2, B3 005

Admin region Volin
Coordinates 51°52'N 25°23'E
Altitude 145–152 m **Area** 12,500 ha

Site description

A river valley with meadows, lakes and forest tracts.

Habitats Forest and woodland (native coniferous forest; mixed forest; alluvial/very wet forest), Wetland (standing fresh water; river/stream; water-fringe vegetation; fen/transition mire/spring), Artificial landscape (arable land; forestry plantation)
Land-use Agriculture, Forestry, Nature conservation/research, Water management

Birds

Species	Season	Year	Pop min	Pop max	Acc	Criteria
[1] ***Botaurus stellaris*** Bittern	B	1996	400	600	B	B2
Ixobrychus minutus Little Bittern	B	1994	100	—	B	B2
Cygnus olor Mute Swan	W	1994	500	700	B	B1i
[1] ***Anas querquedula*** Garganey	B	1996	1,000	2,000	B	B2
[1] ***Circus pygargus*** Montagu's Harrier	B	1996	40	—	B	B3
[1] ***Porzana porzana*** Spotted Crake	B	1996	5,000	20,000	B	B3
[1] ***Porzana parva*** Little Crake	B	1996	5,000	10,000	B	B3
[1] ***Crex crex*** Corncrake	B	1996	700	1,400	B	A1, B2
Grus grus Crane	P	1994	600	2,000	B	A4i, B1i
[1] ***Gallinago media*** Great Snipe	B	1996	20	—	B	A1
Limosa limosa Black-tailed Godwit	B	1996	1,000	2,500	B	A4i, B1i, B2
Tringa totanus Redshank	B	1996	2,000	5,000	B	A4i, B1i, B2
Chlidonias hybridus Whiskered Tern	B	1996	60	100	A	B2
[1] ***Chlidonias niger*** Black Tern	B	1996	1,000	3,000	B	A4i, B1i, B2
Chlidonias niger Black Tern	P	1994	4,000	6,000	B	A4i, B1i
[1] ***Chlidonias leucopterus*** White-winged Black Tern	B	1996	20,000	30,000	B	A4i, B1i
Chlidonias leucopterus White-winged Black Tern	P	1996	4,000	6,000	B	A4i, B1i
Bubo bubo Eagle Owl	B	1994	4	8	B	B2
Locustella fluviatilis River Warbler	B	1996	100	—	—	B3
Locustella luscinioides Savi's Warbler	B	1996	1,000	2,000	B	B3
[1] ***Acrocephalus paludicola*** Aquatic Warbler	B	1996	1,200	2,200	B	A1, B2
Acrocephalus schoenobaenus Sedge Warbler	B	1996	20,000	30,000	B	B3
Sylvia nisoria Barred Warbler	B	1996	200	1,000	—	B3

1. Extrapolation.

Breeding waterbirds number 5,000–7,000 pairs in total, while up to 80,000 individuals congregate here on passage. The breeding bird data marked as 'Extrapolation' is taken from Tishechkin *et al.* (1996), with total population estimates based on extrapolations from counts of singing males or pairs per km of river valley or per km of swamp habitat. These estimates cover the Ratno–Zhveche part of the Upper Prip'yat, but there is also qualitative data available for other parts of the Prip'yat valley.

■ Protection status

National High **International** High

Part of IBA covered by regional protected areas (Khocynskiy, Cirskiy, Velikoglushanskiy and Vetlivskiy; exact area not known). 12,000 ha of IBA covered by Ramsar Site (Prypiat River Floodplains, 12,000 ha).

■ Conservation issues

Threats Construction/impact of dyke/dam/barrage (B), Deforestation (commercial) (B), Dredging/canalization (B)

Cheremske mire — B1i — 006

Admin region Volin
Coordinates 51°32'N 25°33'E
Altitude 159–164 m **Area** 3,007 ha

■ Site description

The site comprises forest with mire, near Lushnivka, Zamostya and Karasun villages.

Habitats Forest and woodland (67%; mixed forest; alluvial/very wet forest), Wetland (33%; standing fresh water; raised bog)
Land-use Forestry (67%), Hunting (33%), Nature conservation/research (67%)

■ Birds

Species	Season	Year	Pop min	Pop max	Acc	Criteria
Grus grus Crane	P	1996	150	400	B	B1i

■ Protection status

National None **International** None

■ Conservation issues

Threats Deforestation (commercial) (C), Disturbance to birds (C), Firewood collection (C), Forest grazing (C), Intensified forest management (C), Selective logging/cutting (C)

Stokhod river valley — A1, A4i, B1i, B2, B3 — 007

Admin region Volin
Coordinates 51°44'N 25°50'E
Altitude 140–171 m **Area** 17,800 ha

■ Site description

Lowland marsh and meadows in the Stokhod valley, near Sedlicshe and Stari Chervisha villages.

Habitats Forest and woodland (40%; mixed forest; alluvial/very wet forest), Grassland (6%; humid grassland), Wetland (54%; standing fresh water; river/stream; water-fringe vegetation; fen/transition mire/spring)
Land-use Agriculture (58%), Forestry (40%), Nature conservation/research (80%)

■ Birds

Species	Season	Year	Pop min	Pop max	Acc	Criteria
Botaurus stellaris Bittern	B	1996	30	50	B	B2
Ciconia nigra Black Stork	B	1996	5	8	B	B2
[1] ***Porzana porzana*** Spotted Crake	B	1996	500	—	B	B3
[1] ***Porzana parva*** Little Crake	B	1996	140	—	B	B3
[1] ***Crex crex*** Corncrake	B	1996	10	50	B	A1
Grus grus Crane	P	1996	200	600	B	B1i
Limosa limosa Black-tailed Godwit	B	1996	70	—	—	B2
Chlidonias niger Black Tern	B	1996	150	200	B	B2
Chlidonias leucopterus White-winged Black Tern	B	1996	500	1,200	—	A4i, B1i
Locustella fluviatilis River Warbler	B	1996	100	—	C	B3
Acrocephalus paludicola Aquatic Warbler	B	1996	10	30	—	A1, B2

1. Extrapolation.

■ Protection status

National None **International** Partial

10,000 ha of IBA covered by Ramsar Site (Stokhid River Flood-plains, 10,000 ha).

■ Conservation issues

Threats Agricultural intensification/expansion (C), Construction/impact of dyke/dam/barrage (C), Deforestation (commercial) (B), Drainage (C), Firewood collection (C), Forest grazing (C)

Sira Pogonya mire — A1, B2 — 008

Admin region Rivne
Coordinates 51°34'N 27°10'E
Altitude 142–163 m **Area** 12,718 ha

■ Site description

A mire surrounded by mixed forest, near the villages of Vezhitsa, Staroe Selo, Elno and Grabun.

Habitats Forest and woodland (55%; native coniferous forest; mixed forest; alluvial/very wet forest), Wetland (45%; raised bog)
Land-use Forestry (55%), Nature conservation/research

■ Birds

Species	Season	Year	Pop min	Pop max	Acc	Criteria
Circaetus gallicus Short-toed Eagle	B	1989	2	3	—	B2
Crex crex Corncrake	B	1990	25	30	B	A1

■ Protection status

National None **International** None

■ Conservation issues

Threats Afforestation (C), Deforestation (commercial) (C), Firewood collection (C), Forest grazing (B), Natural events (C), Unsustainable exploitation (C)

Flooding ('Natural events') is a threat.

Polesskiy Nature Reserve — B2 — 009

Admin region Zhitomyr
Coordinates 51°25'N 27°55'E
Altitude 147–198 m **Area** 20,104 ha

■ Site description

Forest with marshes, sand-dunes and small streams.

Habitats Forest and woodland (42%; broadleaved deciduous forest; native coniferous forest), Wetland (18%; standing fresh water; river/stream; raised bog; water-fringe vegetation; fen/transition mire/spring), Artificial landscape (40%; forestry plantation)
Land-use Nature conservation/research (100%)

■ Birds

Species	Season	Year	Pop min	Pop max	Acc	Criteria
Circaetus gallicus Short-toed Eagle	B	1995	1	2	A	B2

Forty-seven species that are listed in the Ukrainian Red Data Book have been recorded in the area, as have a total of 180 breeding species, including the globally threatened *Aquila clanga* (min. 1 pair) and boreal species such as *Glaucidium passerinum* and *Strix nebulosa*.

■ Protection status

National High **International** None

20,097 ha of IBA covered by Nature Reserve (Polesskiy, 20,097 ha).

■ Conservation issues

Threats Afforestation (B), Deforestation (commercial) (C), Drainage (C), Firewood collection (C), Forest grazing (C)

Ushomirskoe Game Reserve
A1, B2 **010**

Admin region Zhitomyr
Coordinates 51°00′N 28°15′E
Altitude 180–207 m **Area** 70,300 ha

Site description
Semi-natural mixed forest and agricultural habitats, with numerous small streams and marshes. Human activities include farming, cattle-grazing, haymaking and hunting.

Habitats Forest and woodland (40%; broadleaved deciduous forest; native coniferous forest; mixed forest), Scrub (9%), Grassland (20%; humid grassland), Wetland (20%; standing fresh water; river/stream; water-fringe vegetation), Artificial landscape (11%; urban parks/gardens)
Land-use Agriculture (30%), Forestry (60%), Hunting (100%)

Birds

Species	Season	Year	Pop min	Pop max	Acc	Criteria
Ciconia nigra Black Stork	B	1997	5	7	B	B2
Crex crex Corncrake	B	1997	80	100	B	A1

Protection status
National None **International** None

Conservation issues
Threats Deforestation (commercial) (A), Disturbance to birds (A), Firewood collection (B), Forest grazing (A)

Novograd–Volunskoe Game Reserve
A1, B2 **011**

Admin region Zhitomyr
Coordinates 50°50′N 27°50′E
Altitude 180–200 m **Area** 70,730 ha

Site description
Semi-natural forest and agricultural habitats, with numerous small rivers and streams (Sluch, Tnya, Smolka, Korchik, Perevrznya, etc.), situated near the villages of Pucshav, Sykhovolya, Gorodnitsa, Bronnitsa and Karpilovka. The forest is predominantly mixed. Human activities include cattle-grazing, haymaking, farming and hunting.

Habitats Forest and woodland (60%; broadleaved deciduous forest; native coniferous forest; mixed forest), Grassland (2%; humid grassland), Wetland (2%; river/stream; water-fringe vegetation), Artificial landscape (36%; arable land; other urban/industrial areas)
Land-use Agriculture (30%), Forestry (60%), Hunting (100%)

Birds

Species	Season	Year	Pop min	Pop max	Acc	Criteria
[1] ***Ciconia nigra*** Black Stork	B	1997	10	—	A	B2
Crex crex Corncrake	B	1997	60	80	B	A1

1. Up to 50–60 individuals by end of summer.

Protection status
National None **International** None

Conservation issues
Threats Deforestation (commercial) (A), Firewood collection (B), Forest grazing (A), Selective logging/cutting (A), Unsustainable exploitation (A)

Korostushevskoe Game Reserve
A1, B1i, B2 **012**

Admin region Zhitomyr
Coordinates 50°13′N 29°04′E
Altitude 180–200 m **Area** 44,130 ha

Site description
Semi-natural mixed forest and forestry plantations (the figure of 94% includes both forest-types) with small rivers and meadows along valley bottoms, near the villages of Gorodskoe, Kocherov, Krapivnya and Ivnica.

Habitats Forest and woodland (94%; broadleaved deciduous forest; native coniferous forest; mixed forest), Grassland (2%; humid grassland), Wetland (3%; standing fresh water; river/stream; water-fringe vegetation), Artificial landscape (1%; arable land; forestry plantation; other urban/industrial areas)
Land-use Agriculture (5%), Forestry (94%), Hunting (70%)

Birds

Species	Season	Year	Pop min	Pop max	Acc	Criteria
Anser albifrons White-fronted Goose	P	1997	2,000	10,000	B	B1i
Crex crex Corncrake	B	1997	30	80	B	A1
Alcedo atthis Kingfisher	B	1997	20	30	B	B2

Protection status
National None **International** None

Conservation issues
Threats Deforestation (commercial) (B), Selective logging/cutting (C)

Radomyshlskoe Game Reserve
A1, B2 **013**

Admin region Zhitomyr
Coordinates 50°32′N 29°26′E
Altitude 139–172 m **Area** 28,400 ha

Site description
An area of semi-natural mixed forest and forestry plantations with small rivers and meadows along valley bottoms, near the villages of Cudin, Krumok and Zabelocie.

Habitats Forest and woodland (89%; broadleaved deciduous forest; native coniferous forest; mixed forest), Grassland (1%; humid grassland), Wetland (1%; standing fresh water; river/stream; water-fringe vegetation), Artificial landscape (9%; arable land; forestry plantation; other urban/industrial areas)
Land-use Forestry (89%), Hunting (90%), Urban/industrial/transport (7%)

Birds

Species	Season	Year	Pop min	Pop max	Acc	Criteria
Crex crex Corncrake	B	1997	60	100	B	A1
Chlidonias niger Black Tern	B	1997	50	100	B	B2

Protection status
National None **International** None

Conservation issues
Threats Deforestation (commercial) (A), Selective logging/cutting (B)

Dneprovsko-Teterevskoe Game Reserve
A1, B2 **014**

Admin region Kiev
Coordinates 50°53′N 30°15′E
Altitude 104–159 m **Area** 30,627 ha

Site description
Semi-natural mixed forest with numerous marshes and meadows in the lower stretches of the Dnepr and Teterev valleys, near the villages of Sykholuchie, Pilyava, Bogdanu, Rutni, Rikhta and Avdeeva.

Habitats Forest and woodland (60%; broadleaved deciduous forest; native coniferous forest; mixed forest; alluvial/very wet forest), Wetland (35%; standing fresh water; river/stream; water-fringe vegetation; fen/transition mire/spring), Artificial landscape (5%; forestry plantation)
Land-use Hunting (100%)

Birds

Species	Season	Year	Pop min	Pop max	Acc	Criteria
Haliaeetus albicilla White-tailed Eagle	R	1996	8	12	A	A1, B2

Protection status

National None **International** None

Conservation issues

There may be serious threats but the current situation is unknown.

Kiev reservoir — A1, A4i, A4iii, B1i, B1ii, B2 — 015

Admin region Kiev
Coordinates 51°09′N 30°24′E
Altitude 112–156 m **Area** 102,400 ha

Site description

A reservoir on the River Dnepr. This site also includes parts of the River Prip'yat, with numerous lakes and mires.

Habitats Forest and woodland (10%; mixed forest), Scrub (5%), Grassland (15%; humid grassland), Wetland (50%; standing fresh water; river/stream; water-fringe vegetation; fen/transition mire/spring), Artificial landscape (20%; forestry plantation)
Land-use Fisheries/aquaculture (20%), Forestry (15%), Water management

Birds

Species	Season	Year	Pop min	Pop max	Acc	Criteria
Egretta alba Great White Egret	B	1992	50	150	A	A4i, B1i
Anas platyrhynchos Mallard	P	1992	5,000	30,000	A	B1i
Aythya ferina Pochard	P	1992	1,000	40,000	B	A4i, B1i
Aythya fuligula Tufted Duck	P	1992	1,000	10,000	B	B1i
Haliaeetus albicilla White-tailed Eagle	B	1992	2	6	A	A1, B2
Haliaeetus albicilla White-tailed Eagle	P	1995	26	—	C	A1
Sterna albifrons Little Tern	B	1992	10	150	A	B2
Chlidonias hybridus Whiskered Tern	B	1992	15	100	B	B2
Chlidonias niger Black Tern	B	1992	30	150	B	B2

Protection status

National None **International** None

Conservation issues

Threats Burning of vegetation (A), Deforestation (commercial) (C), Other (A)

The northern part of this site lies within the special zone (30-km-radius) centred on the Chernobyl atomic power-station, and thus is subject to very high levels of radioactive contamination ('Other' threat).

Meadows near Procev village — A1, B2 — 016

Admin region Kiev
Coordinates 50°14′N 30°46′E
Altitude 87–133 m **Area** 7,600 ha

Site description

Semi-natural meadows and small lakes in the Dnepr valley, dominated by meadow plants, sedge-beds (*Carex*) and reedbeds (*Phragmites*).

Habitats Grassland (85%; humid grassland), Wetland (15%; standing fresh water; water-fringe vegetation)
Land-use Agriculture (85%), Not utilized (10%), Urban/industrial/transport (5%)

Birds

Species	Season	Year	Pop min	Pop max	Acc	Criteria
Crex crex Corncrake	B	1997	200	300	B	A1, B2

Protection status

National None **International** None

Conservation issues

Threats Abandonment/reduction of land management (A), Construction/impact of dyke/dam/barrage (B), Industrialization/urbanization (A)

Rzhycshevskoe Game Reserve — A1, B2 — 017

Admin region Kiev
Coordinates 50°03′N 31°08′E
Altitude 91–120 m **Area** 52,800 ha

Site description

Semi-natural forest of pine *Pinus* on sand-dunes, with numerous marshes, in the Dnepr lowlands, near the villages of Staroe, Soshnikov, Kalnoe, Kovalin, Devichki, Golovurov, Erkovci and Mirnoe.

Habitats Forest and woodland (30%; broadleaved deciduous forest; native coniferous forest; mixed forest), Scrub (2%), Grassland (43%; humid grassland), Wetland (7%; standing fresh water; river/stream; water-fringe vegetation; fen/transition mire/spring), Artificial landscape (forestry plantation), Unknown (18%)
Land-use Agriculture, Fisheries/aquaculture, Hunting (100%), Military (50%)

Birds

Species	Season	Year	Pop min	Pop max	Acc	Criteria
Haliaeetus albicilla White-tailed Eagle	B	1996	3	5	A	A1, B2

Protection status

National None **International** None

Conservation issues

Threats Disturbance to birds (C)

Kanev reservoir — B1i — 018

Admin region Kiev
Coordinates 50°08′N 30°48′E
Altitude 91–92 m **Area** 200 ha

Site description

Kanev reservoir is situated on the outskirts of Kiev, near the town of Tripolie. Some areas of the reservoir remain unfrozen in winter due to warm-water discharge from a power-station.

Habitats Wetland (100%; standing fresh water)
Land-use Water management (100%)

Birds

Species	Season	Year	Pop min	Pop max	Acc	Criteria
Bucephala clangula Goldeneye	W	1995	300	—	—	B1i

Protection status

National None **International** None

Conservation issues

There may be serious threats but the current situation is unknown.

Supoy marsh — A1, B2 — 019

Admin region Kiev
Coordinates 50°27′N 31°46′E
Altitude 114–130 m **Area** 6,400 ha

Site description

A marsh in the lower reaches of the Supoy river valley, near Yagotin town.

Habitats Wetland (100%; standing fresh water; river/stream; water-fringe vegetation; fen/transition mire/spring)
Land-use Fisheries/aquaculture (5%), Not utilized (90%), Water management

Birds

Species	Season	Year	Pop min	Pop max	Acc	Criteria
Botaurus stellaris Bittern	B	1996	200	300	B	B2
Chlidonias niger Black Tern	B	1996	400	500	B	B2
Acrocephalus paludicola Aquatic Warbler	B	1996	500	1,000	B	A1, B2

Protection status

National Partial **International** None
209 ha of IBA covered by Protected Locality (Supoy Marsh, 209 ha).
2,000 ha of IBA covered by Protected Locality (Usovskiy, 2,000 ha).

Conservation issues

Threats Construction/impact of dyke/dam/barrage (C)

Forests in Dnepr and Desna valley lowlands

A1, B2, B3 **020**

Admin region Chernigov
Coordinates 50°55′N 30°42′E
Altitude 100–122 m **Area** 120,000 ha

Site description

Mixed forests with marshes and streams, between the valleys of the Dnepr and the Desna, near the villages of Khotyanovka and Tuzhar.

Habitats Forest and woodland (80%; mixed forest; alluvial/very wet forest), Wetland (10%; standing fresh water; river/stream; fen/transition mire/spring), Artificial landscape (10%; arable land; forestry plantation)
Land-use Agriculture (10%), Forestry (90%), Hunting (100%), Military (60%), Nature conservation/research (2%)

Birds

Species	Season	Year	Pop min	Pop max	Acc	Criteria
Ciconia nigra Black Stork	B	1996	5	8	B	B2
Haliaeetus albicilla White-tailed Eagle	B	1996	5	7	B	A1, B2
Haliaeetus albicilla White-tailed Eagle	W	1996	5	30	B	A1
Circaetus gallicus Short-toed Eagle	B	1996	10	15	B	B2
Circus pygargus Montagu's Harrier	B	1996	1	3	B	B3
[1] ***Aquila clanga*** Greater Spotted Eagle	B	1996	6	10	B	A1, B2
Hieraaetus pennatus Booted Eagle	B	1996	1	5	B	B2

1. Identification from *Aquila pomarina* needs confirmation.

Protection status

National None **International** None

Conservation issues

Threats Deforestation (commercial) (A), Disturbance to birds (A), Unsustainable exploitation (B)

Bondar marsh

A1, B2 **021**

Admin region Chernigov
Coordinates 51°13′N 30°48′E
Altitude 113–133 m **Area** 6,400 ha

Site description

A natural marshy area surrounded by mixed forest, near Tuzhar, on military land.

Habitats Forest and woodland (30%; mixed forest), Wetland (70%; standing fresh water; river/stream; water-fringe vegetation)
Land-use Military (100%)

Birds

Species	Season	Year	Pop min	Pop max	Acc	Criteria
Circaetus gallicus Short-toed Eagle	B	1995	1	2	A	B2
Aquila clanga Greater Spotted Eagle	B	1995	2	2	A	A1, B2

Protection status

National None **International** None

Conservation issues

Threats Deforestation (commercial) (B)

Korotchenkov meadows

A1, A4i, B1i, B2, B3 **022**

Admin region Sumy
Coordinates 51°56′N 33°23′E
Altitude 120–126 m **Area** 10,000 ha

Site description

Fen meadows in the lower valley of the River Desna, between Razletu and Trubchevsk and including Korotchenkovo-Chulatovskiy meadows near Korotchenkovo village.

Habitats Wetland (100%; river/stream; water-fringe vegetation; fen/transition mire/spring)
Land-use Agriculture

Birds

Species	Season	Year	Pop min	Pop max	Acc	Criteria
Circus macrourus Pallid Harrier	B	1996	1	2	A	B2
Crex crex Corncrake	B	1988	200	250	—	A1, B2
Gallinago media Great Snipe	B	1995	6	20	B	A1
Limosa limosa Black-tailed Godwit	B	1995	60	120	B	B2
Tringa totanus Redshank	B	1995	200	300	B	B2
Larus minutus Little Gull	B	1995	150	570	A	A4i, B1i, B2
Chlidonias hybridus Whiskered Tern	B	1995	40	300	A	B1i, B2
Locustella luscinioides Savi's Warbler	B	1996	80	120	B	B3

Protection status

National None **International** None

Conservation issues

Threats Disturbance to birds (B), Other (A)

Unstable water-levels pose a threat to nesting waterbirds ('Other' threat).

Peschanskiy and Mogrichskiy forests

A1 **023**

Admin region Sumy
Coordinates 51°02′N 35°00′E
Altitude 132–229 m **Area** 11,600 ha

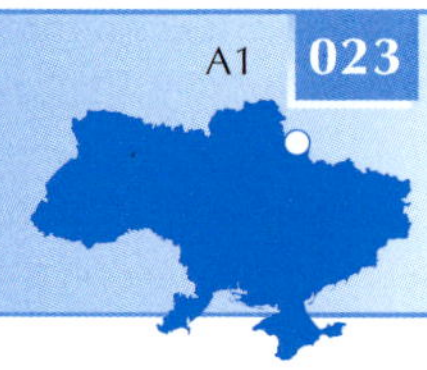

Site description

An area of semi-natural deciduous forest, with small rivers and associated meadows, near the villages of Radkovka, Ivolzhanskoe and Mogrica.

Habitats Forest and woodland (80%; broadleaved deciduous forest; mixed forest; alluvial/very wet forest), Scrub (2%; scrub), Grassland (4%; humid grassland; mesophile grassland), Wetland (2%; river/stream; water-fringe vegetation), Artificial landscape (12%; arable land; urban parks/gardens; other urban/industrial areas; ruderal land)
Land-use Agriculture (11%), Forestry (80%), Hunting (10%), Military (5%)

Birds

Species	Season	Year	Pop min	Pop max	Acc	Criteria
Crex crex Corncrake	B	1997	30	40	B	A1

Protection status

National None **International** None

Conservation issues

Threats Agricultural intensification/expansion (B), Deforestation (commercial) (A), Firewood collection (A), Selective logging/cutting (A), Unsustainable exploitation (A)

Chaykovichi marsh — A1, B2, B3 — 024

Admin region Lvov
Coordinates 49°30′N 23°31′E
Altitude 254–271 m **Area** 8,300 ha

Site description

An area in the Dnestr valley near Chaykovichi village. Drained 20 years ago, it is now predominantly wet grassland.

Habitats Forest and woodland (3%; alluvial/very wet forest), Scrub (7%; scrub), Grassland (90%; humid grassland)
Land-use Agriculture (90%)

Birds

Species	Season	Year	Pop min	Pop max	Acc	Criteria
Circus pygargus Montagu's Harrier	B	1997	7	10	B	B3
Aquila pomarina Lesser Spotted Eagle	B	1997	4	6	B	B2
Coturnix coturnix Quail	B	1996	330	400	B	B2
[1] ***Crex crex*** Corncrake	B	1996	300	400	B	A1, B2
Asio flammeus Short-eared Owl	B	1996	10	—	B	B2
Locustella fluviatilis River Warbler	B	1996	140	160	B	B3

1. 230–300 calling males in 45 km² in 1997.

Breeding species of global conservation concern that do not meet IBA criteria: *Gallinago media* (min. 3 pairs).

Protection status

National None **International** None

Conservation issues

Threats Afforestation (A), Burning of vegetation (C), Disturbance to birds (A), Drainage (A), Extraction industry (B)

Ikva river valley — A4i, B1i, B2 — 025

Admin region Khmelnitskaya
Coordinates 49°50′N 27°25′E
Altitude 250–350 m **Area** 83,500 ha

Site description

A stretch of the Ikva valley between the villages of Nova Sunyavka and Novoselucya, dominated by grassland and agricultural habitats with numerous fish-ponds.

Habitats Forest and woodland (1%; broadleaved deciduous forest), Grassland (7%; humid grassland), Wetland (20%; standing fresh water; river/stream; water-fringe vegetation), Artificial landscape (72%; arable land; forestry plantation; urban parks/gardens; other urban/industrial areas)
Land-use Agriculture (60%), Fisheries/aquaculture (20%), Urban/industrial/transport (20%)

Birds

Species	Season	Year	Pop min	Pop max	Acc	Criteria
[1] ***Anser albifrons*** White-fronted Goose	P	1997	4,000	6,000	B	B1i
[1] ***Anser anser*** Greylag Goose	P	1997	4,000	6,000	B	A4i, B1i
Anas strepera Gadwall	B	1997	5	15	B	B2
Chlidonias hybridus Whiskered Tern	B	1997	100	200	B	B2

1. Mixed flocks of *A. anser* and *A. albifrons* in spring.

Protection status

National None **International** None

Conservation issues

Threats Agricultural intensification/expansion (A)

Changing agricultural practices pose a threat.

Vovk river valley near Derazhnya village — B2 — 026

Admin region Khmelnitskaya
Coordinates 49°18′N 27°21′E
Altitude 281–300 m **Area** 1,270 ha

Site description

A stretch of the Vovk river valley, dominated by a mosaic of sedge-meadows *Carex*, reedbeds *Phragmites* and willow *Salix* thickets. Human activities include cattle-grazing, haymaking and hunting.

Habitats Wetland (100%; standing fresh water; river/stream; water-fringe vegetation)
Land-use Agriculture (30%), Hunting (100%)

Birds

Species	Season	Year	Pop min	Pop max	Acc	Criteria
Chlidonias niger Black Tern	B	1997	90	150	B	B2

Breeding species of global conservation concern that do not meet IBA criteria: *Acrocephalus paludicola* (min. 5 pairs).

Protection status

National None **International** None

Conservation issues

Threats Unsustainable exploitation (B)

Mikulinetskie fish-ponds — A1, B1i, B2, B3 — 027

Admin region Vinnitsa
Coordinates 49°16′N 28°09′E
Altitude 190–291 m **Area** 2,000 ha

Site description

Fish-ponds near Mikulinci village.

Habitats Wetland (100%; standing fresh water; river/stream; water-fringe vegetation)
Land-use Fisheries/aquaculture (100%), Hunting (100%)

Birds

Species	Season	Year	Pop min	Pop max	Acc	Criteria
Ixobrychus minutus Little Bittern	B	1996	40	60	—	B2
Anser anser Greylag Goose	N	1993	120	500	B	B1i
Aythya ferina Pochard	B	1996	250	300	—	B3
Crex crex Corncrake	B	1996	15	30	—	A1
Chlidonias hybridus Whiskered Tern	B	1996	250	300	—	B1i, B2
Chlidonias niger Black Tern	B	1996	150	170	—	B2

Protection status

National None **International** None

Conservation issues

There may be serious threats but the current situation is unknown.

Kanev Nature Reserve — A1, B1i — 028

Admin region Cherkasu
Coordinates 49°46′N 31°28′E
Altitude 91–244 m **Area** 2,027 ha

Site description

Semi-natural deciduous forest on the hills bordering the Dnepr valley near Kanev town. Some parts of the River Dnepr remain unfrozen in winter due to warm-water discharge from a power-station.

Habitats Forest and woodland (70%; broadleaved deciduous forest; alluvial/very wet forest), Wetland (30%; sand-dunes/sand beach; standing fresh water; river/stream; water-fringe vegetation; fen/transition mire/spring)
Land-use Nature conservation/research (100%)

Birds

Species	Season	Year	Pop min	Pop max	Acc	Criteria
[1] ***Mergus merganser*** Goosander	W	1997	50	500	A	B1i
Haliaeetus albicilla White-tailed Eagle	W	1996	4	16	A	A1
[1] ***Haliaeetus albicilla*** White-tailed Eagle	P	1997	5	20	B	A1

1. Data for several years.

Protection status

National High **International** None

2,027 ha of IBA covered by Nature Reserve (Kanev, 2,049 ha).

Conservation issues

Threats Afforestation (C), Deforestation (commercial) (C), Disturbance to birds (B), Forest grazing (C), Intensified forest management (C), Natural events (C), Unsustainable exploitation (C)

Uday river valley — A4i, B1i, B2, B3 — 029

Admin region Poltava
Coordinates 50°20′N 32°30′E
Altitude 99–134 m **Area** 30,000 ha

Site description

A natural wetland, comprising marshes and lakes in the lower reaches of the Uday valley, and dominated by reedbeds *Phragmites*.

Habitats Forest and woodland (native coniferous forest; mixed forest; alluvial/very wet forest), Wetland (standing fresh water; river/stream; water-fringe vegetation; fen/transition mire/spring)
Land-use Agriculture, Forestry, Hunting, Nature conservation/research

Birds

Species	Season	Year	Pop min	Pop max	Acc	Criteria
Botaurus stellaris Bittern	B	1996	300	500	B	B2
Ardea purpurea Purple Heron	B	1996	50	100	B	B2
[1] ***Aquila pomarina*** Lesser Spotted Eagle	B	1996	2	5	C	B2
Hieraaetus pennatus Booted Eagle	B	1996	1	5	B	B2
Porzana porzana Spotted Crake	B	1996	500	1,000	B	B3
Porzana parva Little Crake	B	1996	500	1,000	B	B3
Chlidonias hybridus Whiskered Tern	B	1996	50	100	B	B2
Chlidonias niger Black Tern	B	1996	1,000	2,000	B	A4i, B1i, B2
Chlidonias leucopterus White-winged Black Tern	B	1996	500	1,000	B	A4i, B1i

1. 1990–1996; identification from *Aquila clanga* needs confirmation.

Protection status

National None **International** None

Conservation issues

Threats Afforestation (C), Burning of vegetation (A), Construction/impact of dyke/dam/barrage (B), Deforestation (commercial) (B), Disturbance to birds (B), Firewood collection (C)

Krasnue marsh — A1, B1i, B2 — 030

Admin region Poltava
Coordinates 49°46′N 32°48′E
Altitude 105–107 m **Area** 10,000 ha

Site description

A system of marshes and lakes in the Sula valley, near the villages of Velikoseletskoe, Opishki, Plehov and Narizhzhya.

Habitats Forest and woodland (5%; alluvial/very wet forest), Wetland (95%; standing fresh water; river/stream; water-fringe vegetation; fen/transition mire/spring)
Land-use Agriculture (10%), Nature conservation/research (5%), Not utilized (85%)

Birds

Species	Season	Year	Pop min	Pop max	Acc	Criteria
Botaurus stellaris Bittern	B	1993	300	400	B	B2
Ixobrychus minutus Little Bittern	B	1993	700	—	B	B2
Anser anser Greylag Goose	B	1993	400	500	B	B1i
Crex crex Corncrake	B	1993	40	50	B	A1

Protection status

National None **International** None

Conservation issues

Threats Agricultural intensification/expansion (U), Burning of vegetation (B), Disturbance to birds (C), Extraction industry (C), Other (C), Unsustainable exploitation (A)

Hay-making and sedge-harvesting ('Other' threat) can cause problems.

Sula bay, Kremenchug reservoir — B1i, B2 — 031

Admin region Poltava
Coordinates 49°30′N 32°45′E
Altitude 86–109 m **Area** 27,440 ha

Site description

A large bay in Kremenchug reservoir, near the villages of Goroshino and Zhovnino, including adjacent lowlands of the Sula river valley.

Habitats Scrub (scrub), Grassland (humid grassland), Wetland (standing fresh water; river/stream; water-fringe vegetation)
Land-use Agriculture, Fisheries/aquaculture, Nature conservation/research, Water management

Birds

Species	Season	Year	Pop min	Pop max	Acc	Criteria
Podiceps cristatus Great Crested Grebe	P	1994	2,000	2,500	C	B1i
Nycticorax nycticorax Night Heron	B	1980	390	420	—	B2
Anser anser Greylag Goose	B	1991	300	—	B	B1i
Anser anser Greylag Goose	P	1991	2,000	—	C	B1i
Anser anser Greylag Goose	N	1991	1,200	1,500	B	B1i

Protection status

National None **International** None

Conservation issues

Threats Agricultural intensification/expansion (C), Recreation/tourism (C), Unsustainable exploitation (A)

Psel river valley near Komsomolsk — A1 — 032

Admin region Poltava
Coordinates 49°00′N 33°35′E
Altitude 50–90 m **Area** 4,700 ha

Site description

A stretch of the Psel river valley, near the River Dnepr, and dominated by grassland/meadow habitats with lakes and marshes.

Habitats Forest and woodland (12%; broadleaved deciduous forest; alluvial/very wet forest), Scrub (3%; scrub), Grassland (9%; humid grassland), Wetland (22%; standing fresh water; river/stream; water-fringe vegetation; fen/transition mire/spring), Artificial landscape (54%; highly improved reseeded grassland; arable land; forestry plantation; urban parks/gardens; other urban/industrial areas; ruderal land)
Land-use Agriculture (50%), Forestry (12%), Not utilized (10%), Urban/industrial/transport

Birds

Species	Season	Year	Pop min	Pop max	Acc	Criteria
[1] ***Crex crex*** Corncrake	B	1995	20	40	B	A1

1. Calling males (1993–1995).

Species of global conservation concern that do not meet IBA criteria: *Aythya nyroca* (uncommon on passage), *Haliaetus albicilla* (1–2 pairs breed, 2–3 birds winter), *Falco naumanni* (min. 1 pair breeds).

Protection status

National None **International** None

Conservation issues

Threats Agricultural intensification/expansion (A), Deforestation (commercial) (A), Disturbance to birds (A), Forest grazing (B), Recreation/tourism (A), Selective logging/cutting (C), Unsustainable exploitation (A)

Bodakovskoe marsh — A4i, B1i, B2 — 033

Admin region Poltava
Coordinates 50°17′N 33°24′E
Altitude 96–100 m **Area** 1,500 ha

Site description

Natural wetlands in the Sula river lowlands near the villages of Bodakva, Senha and Plaskivschina, dominated by beds of reed *Phragmites* and sedge *Carex*.

Habitats Wetland (100%; river/stream; water-fringe vegetation; fen/transition mire/spring)
Land-use Agriculture (10%), Not utilized (90%)

Birds

Species	Season	Year	Pop min	Pop max	Acc	Criteria
Botaurus stellaris Bittern	B	1996	50	—	B	B2
Ixobrychus minutus Little Bittern	B	1996	100	—	B	B2
Anser fabalis Bean Goose	P	1996	2,000	5,000	B	A4i, B1i

Protection status

National None **International** None

Conservation issues

Threats Agricultural intensification/expansion (C), Disturbance to birds (C), Other (C)

Hay-making and sedge-harvesting ('Other' threat) can cause problems.

Meadows near Saru village in the Psel river valley — A1, B2 — 034

Admin region Poltava
Coordinates 50°19′N 33°59′E
Altitude 96–124 m **Area** 1,200 ha

Site description

Natural lakes and associated meadows in the Psel river valley, dominated by meadow plants, reedbeds *Phragmites*, sedge-beds *Carex*, and willow *Salix*. Human activities include cattle-grazing, haymaking and hunting.

Habitats Grassland (90%; humid grassland), Wetland (10%; standing fresh water; river/stream; water-fringe vegetation)
Land-use Agriculture (90%), Hunting, Not utilized (10%)

Birds

Species	Season	Year	Pop min	Pop max	Acc	Criteria
Coturnix coturnix Quail	B	1997	100	200	C	B2
Crex crex Corncrake	B	1997	30	50	B	A1

Protection status

National None **International** None

Conservation issues

Threats Unsustainable exploitation (B)

Veluke marsh — B2 — 035

Admin region Poltava
Coordinates 49°18′N 34°32′E
Altitude 90–92 m **Area** 1,000 ha

Site description

A marsh surrounded by salt-meadows and forest, located in the Vorksla valley near the villages of Malaya Perecshepina, Velikoe Boloto and Mankivka.

Habitats Forest and woodland (5%; mixed forest), Grassland (10%; humid grassland), Wetland (85%; saltmarsh; water-fringe vegetation)
Land-use Agriculture (10%), Hunting (100%)

Birds

Species	Season	Year	Pop min	Pop max	Acc	Criteria
Ixobrychus minutus Little Bittern	B	1992	50	60	B	B2

Protection status

National None **International** None

Conservation issues

Threats Agricultural intensification/expansion (B), Burning of vegetation (B), Construction/impact of dyke/dam/barrage (A), Disturbance to birds (B), Other (B)

Hay-making and sedge-harvesting ('Other' threat) can cause problems.

Dneprovsko-Orelskiy Nature Reserve — A1, B2 — 036

Admin region Dnepropetrovskaya
Coordinates 49°30′N 34°50′E
Altitude 51–57 m **Area** 3,766 ha

Site description

A natural, permanently flooded lowland valley with lakes, sedge-beds *Carex* and reedbeds *Phragmites*, near the town of Dnepropetrovsk.

Habitats Forest and woodland (43%; alluvial/very wet forest), Grassland (4%; humid grassland), Wetland (25%; sand-dunes/sand beach; standing fresh water; river/stream; water-fringe vegetation; fen/transition mire/spring), Artificial landscape (28%; arable land; forestry plantation)
Land-use Nature conservation/research (100%)

Birds

Species	Season	Year	Pop min	Pop max	Acc	Criteria
Nycticorax nycticorax Night Heron	B	1996	70	—	A	B2
Crex crex Corncrake	B	1996	30	—	A	A1

Protection status

National High **International** None

3,766 ha of IBA covered by Nature Reserve (Dneprovsko-Orelskiy, 3,766 ha).

Conservation issues

Threats Natural events (C)

Orel river valley — A1, A4i, B1i, B2, B3 — 037

Admin region Dnepropetrovskaya
Coordinates 48°50′N 35°00′E
Altitude 79–148 m **Area** 12,000 ha

Site description

Habitats Forest and woodland (5%; broadleaved deciduous forest; alluvial/very wet forest), Grassland (30%; humid grassland), Wetland (61%; standing fresh water; river/stream; water-fringe vegetation; fen/transition mire/spring), Artificial landscape (4%; forestry plantation; other urban/industrial areas)
Land-use Agriculture (95%), Forestry (5%)

A stretch of the Orel river valley, near Gupalovka, Chernetchina and Krasnopolie villages. Habitats are dominated by grasslands, with numerous lakes and marshes.

Birds

Species	Season	Year	Pop min	Pop max	Acc	Criteria
Egretta alba Great White Egret	N	1996	160	180	B	A4i, B1i
Anas strepera Gadwall	B	1996	12	18	B	B2
Anas querquedula Garganey	B	1996	200	300	B	B2
Porzana porzana Spotted Crake	B	1996	140	160	B	B3
Crex crex Corncrake	B	1996	50	60	A	A1
Locustella luscinioides Savi's Warbler	B	1996	50	60	B	B3

Species of global conservation concern that do not meet IBA criteria: *Aythya nyroca* (1–2 birds, status unknown).

Protection status

National None **International** None

Conservation issues

Threats Agricultural intensification/expansion (A), Construction/impact of dyke/dam/barrage (A), Firewood collection (B), Unsustainable exploitation (A)

Samarskiy forest — A1, A4i, B1i, B2, B3 — 038

Admin region Dnepropetrovskaya
Coordinates 48°45′N 35°15′E
Altitude 77–121 m **Area** 19,920 ha

Site description

Mixed forest in the Samara valley lowlands.

Habitats Forest and woodland (89%; native coniferous forest; mixed forest; alluvial/very wet forest), Grassland (8%; humid grassland), Wetland (3%; standing fresh water; river/stream; water-fringe vegetation; fen/transition mire/spring)
Land-use Agriculture (8%), Forestry (90%), Military (40%), Nature conservation/research

Birds

Species	Season	Year	Pop min	Pop max	Acc	Criteria
[1,2] ***Ixobrychus minutus*** Little Bittern	B	1996	100	160	—	B2
[1] ***Egretta alba*** Great White Egret	B	1996	40	60	—	A4i, B1i
[1] ***Anas strepera*** Gadwall	B	1996	50	60	—	B2
[1,2] ***Anas querquedula*** Garganey	B	1996	600	1,100	—	B2
[1] ***Milvus migrans*** Black Kite	B	1996	12	13	A	B2
[1] ***Circus macrourus*** Pallid Harrier	B	1996	1	2	B	B2
[1] ***Circus pygargus*** Montagu's Harrier	B	1996	5	20	—	B3
[1] ***Aquila clanga*** Greater Spotted Eagle	B	1996	1	2	B	B2
Aquila heliaca Imperial Eagle	B	1996	5	—	A	A1, B2
[1] ***Falco tinnunculus*** Kestrel	B	1996	20	100	A	B2
[3] ***Falco cherrug*** Saker	B	1996	2	3	B	B2
[1,2] ***Coturnix coturnix*** Quail	B	1996	60	140	A	B2
[3] ***Crex crex*** Corncrake	B	1996	15	30	B	A1
[1] ***Chlidonias niger*** Black Tern	B	1996	120	200	B	B2
[1,2] ***Streptopelia turtur*** Turtle Dove	B	1996	5,300	14,500	A	B2
[1] ***Asio flammeus*** Short-eared Owl	B	1996	20	—	B	B2
[1,2] ***Caprimulgus europaeus*** Nightjar	B	1996	800	2,900	B	B2
[1] ***Alcedo atthis*** Kingfisher	B	1996	70	140	B	B2
[1,2] ***Merops apiaster*** Bee-eater	B	1996	1,000	3,000	B	B2
[1,2] ***Jynx torquilla*** Wryneck	B	1996	1,600	2,700	B	B2
[1] ***Dendrocopos medius*** Middle Spotted Woodpecker	B	1996	30	110	C	B3
[1,2] ***Lullula arborea*** Woodlark	B	1996	1,300	2,900	B	B2
Phoenicurus phoenicurus Redstart	B	1996	50	260	—	B2
[1,2] ***Turdus philomelos*** Song Thrush	B	1996	5,300	5,900	B	B3
[1,2] ***Sylvia nisoria*** Barred Warbler	B	1996	450	1,700	B	B3
[1,2] ***Sylvia borin*** Garden Warbler	B	1996	300	800	B	B3
[1,2] ***Phylloscopus sibilatrix*** Wood Warbler	B	1996	480	2,900	B	B3
[1,2] ***Muscicapa striata*** Spotted Flycatcher	B	1996	2,900	8,800	—	B2
[1,2] ***Ficedula albicollis*** Collared Flycatcher	B	1996	2,900	7,300	—	B3
[1,2] ***Parus caeruleus*** Blue Tit	B	1996	3,600	7,300	—	B3
[1,2] ***Fringilla coelebs*** Chaffinch	B	1996	32,000	44,000	—	B3

1. Individuals.
2. Extrapolation.
3. Pairs.

Protection status

National None **International** None

Conservation issues

Threats Deforestation (commercial) (B), Disturbance to birds (A), Selective logging/cutting (C)

Soleniy liman — B1i — 039

Admin region Dnepropetrovskaya
Coordinates 48°39′N 35°31′E
Altitude 62–62 m **Area** 500 ha

Site description

A lake in the Samara river valley, with a sandy shore. The vegetation is dominated by reedbeds *Phragmites* and meadows. Human activities include cattle-grazing and recreation.

Habitats Grassland (14%; humid grassland), Wetland (86%; sand-dunes/sand beach; standing brackish and salt water; water-fringe vegetation)
Land-use Agriculture (14%), Nature conservation/research (100%)

Birds

Species	Season	Year	Pop min	Pop max	Acc	Criteria
Anser anser Greylag Goose	P	1997	1,500	—	B	B1i

Breeding species of global conservation concern that do not meet IBA criteria: *Crex crex* (2–3 pairs).

Protection status

National None **International** None

Conservation issues

Threats Disturbance to birds (C)

Debalcevskoe lake — B1i, B2 — 040

Admin region Dnepropetrovskaya
Coordinates 48°07′N 36°06′E
Altitude 70–70 m **Area** 300 ha

Site description

A natural lowland lake in the Verkhnyaya Tersu river valley, dominated by reedbeds *Phragmites* and meadows. Human activities include cattle-grazing, fishing and hunting.

Habitats Scrub (4%; scrub), Grassland (30%; humid grassland), Wetland (56%; standing fresh water; water-fringe vegetation), Artificial landscape (2%; forestry plantation)
Land-use Agriculture, Fisheries/aquaculture (20%), Hunting (100%)

Birds

Species	Season	Year	Pop min	Pop max	Acc	Criteria
Ixobrychus minutus Little Bittern	B	1997	40	50	B	B2
Anser anser Greylag Goose	P	1997	1,000	1,500	B	B1i

Breeding species of global conservation concern that do not meet IBA criteria: *Crex crex* (5–7 pairs).

Protection status

National None **International** None

Conservation issues

Threats Aquaculture/fisheries (C), Unsustainable exploitation (B)

Petropavlovskiy marsh — B1i — 041

Admin region Dnepropetrovskaya
Coordinates 48°29′N 36°24′E
Altitude 73–160 m **Area** 270 ha

Site description

A natural marsh in the Samara valley lowlands.

Habitats Grassland (30%; humid grassland), Wetland (70%; standing fresh water; standing brackish/salt water; river/stream; water-fringe vegetation; fen/transition mire/spring)
Land-use Agriculture (40%), Nature conservation/research (100%)

Birds

Species	Season	Year	Pop min	Pop max	Acc	Criteria
Anser anser Greylag Goose	B	1996	200	250	B	B1i
[1] ***Anser anser*** Greylag Goose	N	1996	1,250	1,900	—	B1i

1. Non-breeders present in summer.

Protection status
National None **International** None

Conservation issues

Threats Agricultural intensification/expansion (B), Disturbance to birds (C), Natural events (C), Unsustainable exploitation (B)

Flooding ('Natural events') is a threat.

Bolshemikhailovskiy forest — B2 — 042
Admin region Dnepropetrovskaya
Coordinates 48°59′N 36°31′E
Altitude 79–165 m **Area** 1,079 ha

Site description
Mixed forest in river-valley lowlands, near the village of Veliko-Mikhailovka.

Habitats Forest and woodland (100%; broadleaved deciduous forest; native coniferous forest; mixed forest; alluvial/very wet forest)
Land-use Forestry (100%), Nature conservation/research (100%)

Birds

Species	Season	Year	Pop min	Pop max	Acc	Criteria
Hieraaetus pennatus Booted Eagle	B	1996	1	2	B	B2
Falco cherrug Saker	B	1996	1	—	B	B2

Protection status
National None **International** None

Conservation issues

Threats Deforestation (commercial) (C), Intensified forest management (C), Recreation/tourism (B), Selective logging/cutting (B)

Samara river valley — A1 — 043
Admin region Kharkov
Coordinates 48°40′N 36°39′E
Altitude 83–134 m **Area** 1,700 ha

Site description
A stretch of the Samara valley with numerous lakes and marshes, dominated by grassland habitats, and located near the villages of Samara, Varvarovka and Pavlovka.

Habitats Grassland (40%; humid grassland), Wetland (59%; standing fresh water; river/stream; water-fringe vegetation), Artificial landscape (1%; forestry plantation)
Land-use Agriculture (40%), Not utilized (60%)

Birds

Species	Season	Year	Pop min	Pop max	Acc	Criteria
Crex crex Corncrake	B	1997	15	20	B	A1

Protection status
National None **International** None

Conservation issues

Threats Agricultural intensification/expansion (B), Construction/impact of dyke/dam/barrage (A), Disturbance to birds (A), Unsustainable exploitation (A)

Mzha river valley — A1, A4i, B1i, B2, B3 — 044
Admin region Kharkov
Coordinates 49°45′N 36°06′E
Altitude 82–93 m **Area** 7,200 ha

Site description
A stretch of the Mzha river valley, near the villages of Chemuzhovka, Timchenki, Artyukhovka and Merefa. Habitats are dominated by grasslands, with numerous lakes and marshes.

Habitats Forest and woodland (10%; alluvial/very wet forest), Grassland (40%; humid grassland), Wetland (42%; standing fresh water; river/stream; water-fringe vegetation; fen/transition mire/spring), Artificial landscape (8%; arable land; forestry plantation; other urban/industrial areas)
Land-use Agriculture (90%), Forestry (10%)

Birds

Species	Season	Year	Pop min	Pop max	Acc	Criteria
Botaurus stellaris Bittern	B	1997	40	50	B	B2
Egretta alba Great White Egret	B	1997	42	54	A	A4i, B1i
Porzana porzana Spotted Crake	B	1996	70	80	B	B3
Crex crex Corncrake	B	1996	120	140	B	A1, B2
Gallinago media Great Snipe	B	1996	15	25	B	A1
Chlidonias niger Black Tern	B	1997	60	90	B	B2
Alcedo atthis Kingfisher	B	1996	10	15	B	B2

Protection status
National None **International** None

Conservation issues

Threats Agricultural intensification/expansion (B), Burning of vegetation (A), Construction/impact of dyke/dam/barrage (C), Deforestation (commercial) (B), Firewood collection (B), Other (B), Unsustainable exploitation (A)

Haymaking and harvesting of sedge *Carex* ('Other') are among the threats to the site.

Liman lake-system — A1, B1i, B2, B3 — 045
Admin region Kharkov
Coordinates 49°35′N 36°30′E
Altitude 90–112 m **Area** 4,730 ha

Site description
A system of lakes in the lower parts of the Severskiy Donec valley, near the village of Komsomolskoe.

Habitats Forest and woodland (5%; native coniferous forest; alluvial/very wet forest), Scrub (10%; scrub), Grassland (12%; humid grassland), Wetland (63%; standing fresh water; standing brackish and salt water; river/stream; water-fringe vegetation; fen/transition mire/spring), Artificial landscape (10%; other urban/industrial areas)
Land-use Agriculture (29%), Fisheries/aquaculture (19%), Hunting (22%), Not utilized (18%), Urban/industrial/transport (11%)

Birds

Species	Season	Year	Pop min	Pop max	Acc	Criteria
Ixobrychus minutus Little Bittern	B	1996	20	50	A	B2
Anser anser Greylag Goose	P	1996	100	450	B	B1i
Bucephala clangula Goldeneye	W	1996	150	415	A	B1i
Porzana porzana Spotted Crake	B	1996	70	180	B	B3
Crex crex Corncrake	B	1996	40	60	B	A1
Asio flammeus Short-eared Owl	B	1996	3	5	B	B2
Saxicola rubetra Whinchat	B	1996	200	280	B	B3
Locustella luscinioides Savi's Warbler	B	1996	70	110	B	B3

Species ... continued	Season	Year	Pop min	Pop max	Acc	Criteria
Acrocephalus schoenobaenus Sedge Warbler	B	1996	430	550	B	B3
Sylvia nisoria Barred Warbler	B	1996	80	100	B	B3

Protection status
National None **International** None

Conservation issues

Threats Agricultural intensification/expansion (B), Aquaculture/fisheries (C), Disturbance to birds (B), Drainage (B), Unsustainable exploitation (B)

Pechenezhskoe reservoir

B2 **046**

Admin region Kharkov
Coordinates 50°00′N 36°59′E
Altitude 103–193 m **Area** 1,600 ha

Site description
The site comprises a reservoir on the Severskiy Donec river with adjoining semi-natural broadleaved deciduous woodland. Human activities include forestry, fishing, cattle-grazing, hunting and recreation.

Habitats Forest and woodland (70%; broadleaved deciduous forest), Scrub (5%; scrub), Wetland (25%; standing fresh water; water-fringe vegetation)
Land-use Forestry (70%), Hunting (20%), Nature conservation/research (10%), Water management

Birds
Breeding species of global conservation concern that do not meet IBA criteria: *Haliaeetus albicilla* (min. 2 pairs).

Protection status
National None **International** None

Conservation issues

Threats Deforestation (commercial) (A), Disturbance to birds (A), Intensified forest management (A), Recreation/tourism (A), Unsustainable exploitation (A)

Bakhtin river valley

A1 **047**

Admin region Kharkov
Coordinates 49°16′N 37°28′E
Altitude 60–100 m **Area** 1,000 ha

Site description
A stretch of the Bakhtin river valley near the villages of Bakhtin, Komarovka, Nikolaevka and Chervonuy Oskol. Habitats are dominated by grassland, with numerous lakes, ponds and marshes.

Habitats Forest and woodland (2%; broadleaved deciduous forest), Grassland (70%; humid grassland), Wetland (20%; standing fresh water; river/stream; water-fringe vegetation; fen/transition mire/spring), Artificial landscape (8%; highly improved reseeded grassland; arable land; ruderal land)
Land-use Agriculture (60%), Fisheries/aquaculture (10%), Forestry (2%), Not utilized (28%)

Birds

Species	Season	Year	Pop min	Pop max	Acc	Criteria
Crex crex Corncrake	B	1997	35	45	A	A1

Protection status
National None **International** None

Conservation issues

Threats Agricultural intensification/expansion (A), Disturbance to birds (A)

Changing agricultural practices pose a threat.

Izyumskaya Luka forest

A1, B2 **048**

Admin region Kharkov
Coordinates 49°14′N 37°03′E
Altitude 58–103 m **Area** 32,850 ha

Site description
Mixed forest in the Severskiy Donec valley, near the villages of Norcovka, Vetrovka, Chenel and Protopopovka. Flood-meadows border the river.

Habitats Forest and woodland (95%; broadleaved deciduous forest; native coniferous forest; mixed forest; alluvial/very wet forest), Grassland (4%; humid grassland), Wetland (1%; standing fresh water; river/stream; water-fringe vegetation; fen/transition mire/spring)
Land-use Agriculture (5%), Forestry (95%)

Birds

Species	Season	Year	Pop min	Pop max	Acc	Criteria
Aquila heliaca Imperial Eagle	B	1996	2	4	B	A1, B2
Falco cherrug Saker	B	1996	1	—	B	B2
Crex crex Corncrake	B	1996	49	—	B	A1
Merops apiaster Bee-eater	B	1996	60	100	B	B2
Riparia riparia Sand Martin	B	1996	16,600	—	B	B2

Breeding species of global conservation concern that do not meet IBA criteria: *Haliaeetus albicilla* (1–2 pairs).

Protection status
National None **International** None

Conservation issues

Threats Abandonment/reduction of land management (U), Deforestation (commercial) (A), Disturbance to birds (C), Selective logging/cutting (B)

Kremenskoy forest

A1, B2, B3 **049**

Admin region Lugansk
Coordinates 48°59′N 38°08′E
Altitude 54–175 m **Area** 20,000 ha

Site description
Semi-natural forest and forestry plantation, dominated by wet forest in the Severskiy Donec river valley and by pine *Pinus* on sandy, drier soils.

Habitats Forest and woodland (20%; alluvial/very wet forest), Wetland (5%; standing fresh water; water-fringe vegetation), Artificial landscape (75%; forestry plantation)
Land-use Forestry (80%), Hunting, Tourism/recreation

Birds

Species	Season	Year	Pop min	Pop max	Acc	Criteria
Aquila heliaca Imperial Eagle	B	1997	4	—	A	A1, B2
Hieraaetus pennatus Booted Eagle	B	1997	5	6	A	B2
Falco vespertinus Red-footed Falcon	B	1997	10	25	B	B2
Crex crex Corncrake	B	1997	30	50	B	A1
Caprimulgus europaeus Nightjar	B	1997	200	300	B	B2
Alcedo atthis Kingfisher	B	1997	60	100	B	B2
Coracias garrulus Roller	B	1997	30	40	B	B2
Dendrocopos medius Middle Spotted Woodpecker	B	1997	50	—	B	B3

Breeding species of global conservation concern that do not meet IBA criteria: *Haliaeetus albicilla* (1–2 pairs).

Protection status
National None **International** None

Conservation issues

Threats Deforestation (commercial) (A), Drainage (A), Forest grazing (C), Selective logging/cutting (B), Unsustainable exploitation (B)

Stanichno-Luganskoe fish-farm B2 050

Admin region Lugansk
Coordinates 48°38'N 39°35'E
Altitude 42–56 m **Area** 2,000 ha

Site description
Fish-ponds in the Severskiy Donec valley.

Habitats Wetland (100%; standing fresh water; river/stream; water-fringe vegetation)
Land-use Fisheries/aquaculture (100%)

Birds

Species	Season	Year	Pop min	Pop max	Acc	Criteria
Nycticorax nycticorax Night Heron	B	1996	50	70	A	B2
Tadorna ferruginea Ruddy Shelduck	B	1996	1	2	A	B2

Protection status
National None **International** None

Conservation issues

Threats Aquaculture/fisheries (B), Disturbance to birds (B)

Latorica river valley near Chop A1, B1i, B2 051

Admin region Zakarpatie
Coordinates 48°28'N 22°18'E
Altitude 102–109 m **Area** 7,000 ha

Site description
A stretch of the Latorica river valley, near the villages of Velika Dobron and Geevci (near the town of Chop). Habitats are dominated by grasslands, which flood every year, and there are numerous lakes.

Habitats Forest and woodland (8%; alluvial/very wet forest), Grassland (30%; humid grassland), Wetland (22%; standing fresh water; river/stream; water-fringe vegetation; fen/transition mire/spring), Artificial landscape (40%; highly improved reseeded grassland; arable land; forestry plantation; urban parks/gardens; other urban/industrial areas)
Land-use Agriculture (70%), Forestry (8%), Nature conservation/research (10%), Not utilized (12%), Urban/industrial/transport (2%)

Birds

Species	Season	Year	Pop min	Pop max	Acc	Criteria
Anser fabalis Bean Goose	P	1997	1,000	3,000	B	B1i
Aythya nyroca Ferruginous Duck	B	1997	40	60	B	A1, B2
Bucephala clangula Goldeneye	P	1997	500	800	B	B1i
Coturnix coturnix Quail	B	1997	100	150	B	B2

An important site for foraging and roosting waterbirds during spring passage. Breeding species of global conservation concern that do not meet IBA criteria: *Crex crex* (4–10 pairs).

Protection status
National None **International** None

Conservation issues

Threats Disturbance to birds (A), Drainage (A), Firewood collection (B), Unsustainable exploitation (A)

Carpathian Volkano mountains B2, B3 052

Admin region Zakarpatie
Coordinates 48°30'N 23°00'E
Altitude 120–1,085 m **Area** 134,400 ha

Site description
The site comprises coniferous forest and subalpine meadows in the Carpathian mountains, in Perechin, Svalyava and Irshava districts.

Habitats Forest and woodland (80%; broadleaved deciduous forest; native coniferous forest; mixed forest; alluvial/very wet forest), Scrub (5%; scrub), Grassland (8%; alpine/subalpine/boreal grassland; humid grassland; mesophile grassland), Wetland (2%; river/stream; water-fringe vegetation), Rocky areas (inland cliff), Artificial landscape (5%; highly improved reseeded grassland; arable land; forestry plantation; other urban/industrial areas)
Land-use Agriculture (22%), Forestry (66%), Hunting (3%), Nature conservation/research (1%), Urban/industrial/transport (8%)

Birds

Species	Season	Year	Pop min	Pop max	Acc	Criteria
Ciconia nigra Black Stork	B	1997	10	15	B	B2
Aquila pomarina Lesser Spotted Eagle	B	1997	10	14	B	B2
Falco vespertinus Red-footed Falcon	B	1997	3	5	B	B2
Bubo bubo Eagle Owl	R	1997	8	10	B	B2
Strix aluco Tawny Owl	R	1997	60	80	B	B3
Alcedo atthis Kingfisher	B	1997	40	50	B	B2
Picus canus Grey-headed Woodpecker	R	1997	180	280	B	B2
Dendrocopos medius Middle Spotted Woodpecker	R	1997	50	70	B	B3

Protection status
National None **International** None

Conservation issues

Threats Deforestation (commercial) (A)

Carpathian Biosphere Reserve B2, B3 053

Admin region Zakarpatie
Coordinates 48°20'N 23°42'E
Altitude 200–1,400 m **Area** 12,507 ha

Site description
The site comprises two non-adjacent areas—Chernogorskiy (2,129 ha) and Ugolsko-Shirokoluszhanskiy (10,378 ha)—in the Carpathian Biosphere Reserve. Natural forest dominates, with streams and subalpine meadows.

Habitats Forest and woodland (native coniferous forest; mixed forest), Grassland (alpine/subalpine/boreal grassland), Wetland (standing fresh water; river/stream; water-fringe vegetation; fen/transition mire/spring)
Land-use Nature conservation/research (100%)

Birds

Species	Season	Year	Pop min	Pop max	Acc	Criteria
Picus canus Grey-headed Woodpecker	B	1996	90	—	B	B2
Dendrocopos medius Middle Spotted Woodpecker	B	1996	13	—	B	B3
Picoides tridactylus Three-toed Woodpecker	B	1996	85	—	B	B2
Turdus torquatus Ring Ouzel	B	1996	260	—	B	B3

Protection status
National High **International** High
12,507 ha of IBA covered by Biosphere Reserve (Carpathian, 63,245 ha).

Conservation issues
There are no serious threats.

Gorgany B2, B3 054

Admin region Zakarpatie
Coordinates 48°50'N 24°00'E
Altitude 700–1,788 m **Area** 75,200 ha

Site description
Mountain forest with numerous small rivers and streams, on Gorgany and Svidovec mountains (part of the Carpathian mountains), near the villages of Kalachava, Yasinya and Ust-Chorna.

Habitats Forest and woodland (70%; broadleaved deciduous forest; native coniferous forest; mixed forest; alluvial/very wet forest), Scrub (5%; scrub), Grassland (5%; alpine/subalpine/boreal grassland; humid grassland), Wetland (2%; river/stream), Rocky areas (8%; inland cliff), Artificial landscape (10%; highly improved reseeded grassland; arable land; forestry plantation; other urban/industrial areas)
Land-use Agriculture (4%), Forestry (88%), Hunting (85%), Nature conservation/research (5%), Not utilized (3%)

Birds

Species	Season	Year	Pop min	Pop max	Acc	Criteria
Aquila pomarina Lesser Spotted Eagle	B	1997	4	5	A	B2
Bubo bubo Eagle Owl	R	1997	8	10	B	B2
Strix aluco Tawny Owl	R	1997	100	150	B	B3
Picoides tridactylus Three-toed Woodpecker	R	1997	50	100	B	B2
Lullula arborea Woodlark	B	1997	100	150	B	B2
Turdus torquatus Ring Ouzel	B	1997	300	500	B	B3

Protection status

National Low **International** None
5,300 ha of IBA covered by Nature Reserve (Gorgany, 5,300 ha).

Conservation issues

Threats Afforestation (B), Deforestation (commercial) (A), Disturbance to birds (C)

Fish-ponds near Kukilniku and Bilshivtci villages

A4i, B1i, B2 **055**

Admin region Ivano–Frankovsk
Coordinates 49°11′N 24°16′E
Altitude 283–340 m **Area** 1,000 ha

Site description

Fish-ponds within the Dnestr and Gnila Lipa valleys.

Habitats Wetland (100%; standing fresh water; river/stream; water-fringe vegetation)
Land-use Fisheries/aquaculture (100%)

Birds

Species	Season	Year	Pop min	Pop max	Acc	Criteria
Podiceps grisegena Red-necked Grebe	B	1993	40	66	A	B1i
Podiceps nigricollis Black-necked Grebe	B	1993	120	450	A	A4i, B1i
Chlidonias hybridus Whiskered Tern	B	1993	10	300	A	B1i, B2

Protection status

National None **International** None

Conservation issues

Threats Aquaculture/fisheries (B), Disturbance to birds (C), Unsustainable exploitation (B)

Burshtyn reservoir

B1i **056**

Admin region Ivano–Frankovsk
Coordinates 49°14′N 24°39′E
Altitude 229–300 m **Area** 1,600 ha

Site description

A reservoir on the Lipa river, near the town of Burshtyn.

Habitats Forest and woodland (5%; broadleaved deciduous forest), Grassland (5%; humid grassland), Wetland (90%; standing fresh water; river/stream; water-fringe vegetation), Artificial landscape (arable land; other urban/industrial areas)
Land-use Water management (100%)

Birds

Species	Season	Year	Pop min	Pop max	Acc	Criteria
Bucephala clangula Goldeneye	W	1994	40	3,300	A	B1i
Mergus merganser Goosander	W	1994	15	250	A	B1i

Protection status

National None **International** None

Conservation issues

Threats Unsustainable exploitation (A)

Dnestr valley between Staruy Martuniv and Marinopil villages

B1i **057**

Admin region Ivano–Frankovsk
Coordinates 49°08′N 24°42′E
Altitude 214–341 m **Area** 3,000 ha

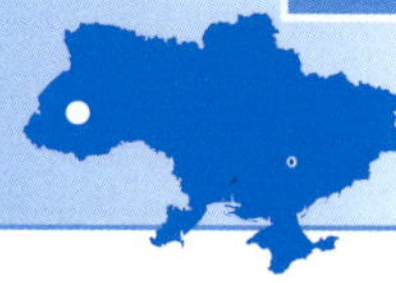

Site description

A river valley with lakes and islands, between the villages of Staruy Martuniv and Marinopil.

Habitats Forest and woodland, Wetland (100%; standing fresh water; river/stream; water-fringe vegetation)
Land-use Fisheries/aquaculture (70%), Tourism/recreation (30%)

Birds

Species	Season	Year	Pop min	Pop max	Acc	Criteria
Bucephala clangula Goldeneye	W	1996	10	1,500	A	B1i

Protection status

National None **International** None

Conservation issues

Threats Aquaculture/fisheries (A), Disturbance to birds (A), Extraction industry (A)

Klivodin Protected Locality

A4i, B1i, B2 **058**

Admin region Chernivtci
Coordinates 48°28′N 25°47′E
Altitude 89–90 m **Area** 50 ha

Site description

A rural landscape with small forestry plantations, near Klivodin village.

Habitats Artificial landscape (100%; arable land; forestry plantation)
Land-use Agriculture, Urban/industrial/transport

Birds

Species	Season	Year	Pop min	Pop max	Acc	Criteria
Nycticorax nycticorax Night Heron	B	1990	266	610	A	A4i, B1i, B2

Protection status

National High **International** None
50 ha of IBA covered by Protected Locality (Klivodin, 50 ha).

Conservation issues

Threats Firewood collection (B), Selective logging/cutting (B)

Derkul river valley

A1 **059**

Admin region Chernivtci
Coordinates 48°15′N 26°01′E
Altitude 157–227 m **Area** 700 ha

Site description

A stretch of the Derkul valley, near the villages of Ostrica and Moloduya, dominated by grassland with some scrub.

Habitats Scrub (35%; scrub), Grassland (35%; humid grassland), Wetland (5%; standing fresh water; river/stream), Artificial landscape (25%; arable land; forestry plantation; ruderal land)
Land-use Agriculture (20%), Not utilized (60%), Unknown (20%)

Birds

Species	Season	Year	Pop min	Pop max	Acc	Criteria
Crex crex Corncrake	B	1997	16	25	B	A1

Protection status

National None **International** None

Conservation issues

Threats Agricultural intensification/expansion (U), Disturbance to birds (A), Recreation/tourism (A), Other (B), Unsustainable exploitation (A)

Threats include haymaking and harvesting of sedge *Carex* ('Other').

Dgulin lake B2 060

Admin region Chernivtci
Coordinates 48°25′N 26°21′E
Altitude 210–210 m **Area** 50 ha

Site description

A natural freshwater lake near Cshirivci village.

Habitats Wetland (100%; standing fresh water; water-fringe vegetation)
Land-use Fisheries/aquaculture (100%)

Birds

Species	Season	Year	Pop min	Pop max	Acc	Criteria
Nycticorax nycticorax Night Heron	B	1991	12	272	A	B2

Protection status

National None **International** None

Conservation issues

Threats Aquaculture/fisheries (A), Disturbance to birds (B), Unsustainable exploitation (A)

Dranitckiy Protected Locality B2 061

Admin region Chernivtci
Coordinates 48°14′N 26°32′E
Altitude 161–162 m **Area** 100 ha

Site description

A lake within the Cherlena valley, near Dranitca village.

Habitats Wetland (100%; standing fresh water; river/stream; water-fringe vegetation)
Land-use Nature conservation/research (100%)

Birds

Species	Season	Year	Pop min	Pop max	Acc	Criteria
Nycticorax nycticorax Night Heron	B	1991	102	166	A	B2

Protection status

National High **International** None
100 ha of IBA covered by Protected Locality (Dranitckiy, 100 ha).

Conservation issues

Threats Aquaculture/fisheries (B), Disturbance to birds (B), Drainage (A), Unsustainable exploitation (A)

Ingul river valley B1i, B2 062

Admin region Nikolaev
Coordinates 47°35′N 32°22′E
Altitude 30–45 m **Area** 1,000 ha

Site description

A river with reedbeds *Phragmites*, near Khristoforovka village.

Habitats Grassland (30%; humid grassland), Wetland (70%; standing fresh water; river/stream; water-fringe vegetation)
Land-use Agriculture (40%), Fisheries/aquaculture

Birds

Species	Season	Year	Pop min	Pop max	Acc	Criteria
Nycticorax nycticorax Night Heron	R	1996	60	—	C	B2
[1] ***Anser anser*** Greylag Goose	B	1996	250	—	B	B1i

1. Individuals.

Protection status

National None **International** None

Conservation issues

Threats Other (B), Unsustainable exploitation (B)

Unstable river-levels pose a threat ('Other').

Kinburn peninsula A4i, B1i, B2 063

Admin region Nikolaev
Coordinates 46°32′N 31°34′E
Altitude 0–15 m **Area** 4,000 ha

Site description

A peninsula in the Black Sea, near the Dnepr delta. There are numerous lakes and lagoons (fresh, brackish and salt water), set in a mosaic of steppe, pseudo-steppe and small forestry plantations.

Habitats Grassland (80%; steppe/dry calcareous grassland), Wetland (15%; saltmarsh; sand-dunes/sand beach; coastal lagoon; standing fresh water; standing brackish and salt water; water-fringe vegetation), Artificial landscape (5%; forestry plantation; other urban/industrial areas)
Land-use Agriculture (15%), Hunting (20%), Nature conservation/research (100%), Tourism/recreation (10%)

Birds

Species	Season	Year	Pop min	Pop max	Acc	Criteria
Phalacrocorax carbo Cormorant	N	1996	2,380	3,480	A	B1i
[1] ***Tadorna ferruginea*** Ruddy Shelduck	B	1996	11	—	B	B2
Somateria mollissima Eider	N	1996	1,950	—	B	B1i
[2] ***Larus ichthyaetus*** Great Black-headed Gull	B	1996	2,090	2,290	B	A4i, B1i

1. 11 adults and 59 juveniles.
2. Individuals.

Protection status

National High **International** None
4,000 ha of IBA covered by Regional Landscape Park (Kinburn Peninsula, 17,890 ha).

Conservation issues

Threats Afforestation (U), Agricultural intensification/expansion (B), Disturbance to birds (A), Recreation/tourism (A)

Threats to the site include haymaking, overgrazing and afforestation.

Lower Dnepr A4i, A4iii, B1i, B2 064

Admin region Kherson
Coordinates 46°30′N 32°19′E
Altitude 2–7 m **Area** 26,000 ha

Site description

The delta of the River Dnepr, with numerous small islands of flood-plain forest and large reedbeds, near the town of Golaya Pristan.

Habitats Forest and woodland (5%; alluvial/very wet forest), Grassland (humid grassland), Wetland (standing fresh water; standing brackish and salt water; river/stream; water-fringe vegetation), Artificial landscape (forestry plantation)
Land-use Agriculture, Fisheries/aquaculture, Hunting, Nature conservation/research, Tourism/recreation, Urban/industrial/transport

Birds

Species	Season	Year	Pop min	Pop max	Acc	Criteria
Phalacrocorax carbo Cormorant	B	1992	3,000	4,500	B	A4i, B1i
Phalacrocorax carbo Cormorant	P	1992	10,000	—	B	A4i, B1i
Nycticorax nycticorax Night Heron	B	1992	450	—	A	B2
Ardeola ralloides Squacco Heron	B	1996	15	118	B	A4i, B1i, B2
Egretta alba Great White Egret	B	1992	350	700	A	A4i, B1i
Ardea cinerea Grey Heron	B	1992	1,800	—	B	A4i, B1i
Ardea purpurea Purple Heron	B	1996	60	—	B	B2
Plegadis falcinellus Glossy Ibis	B	1996	8	253	B	A4i, B1i, B2
Cygnus olor Mute Swan	B	1996	150	—	B	B1i
Cygnus olor Mute Swan	P	1996	8,000	—	B	A4i, B1i
Anser albifrons White-fronted Goose	P	1996	7,000	—	B	B1i
Anser anser Greylag Goose	B	1996	150	—	B	B1i
Anser anser Greylag Goose	P	1996	5,000	—	B	A4i, B1i
Anas platyrhynchos Mallard	P	1992	30,000	—	B	B1i
Bucephala clangula Goldeneye	P	1996	16	340	B	B1i
Fulica atra Coot	P	1992	90,000	—	B	A4i, B1i
Grus grus Crane	P	1996	120	700	B	B1i

The total number of breeding waterbirds is 6,000–8,000 pairs, while more than 140,000 waterbirds stage here during migration.

Protection status

National None **International** High

26,000 ha of IBA covered by Ramsar Site (Dnipro River delta, 26,000 ha).

Conservation issues

Threats Abandonment/reduction of land management (U), Agricultural intensification/expansion (B), Aquaculture/fisheries (C), Disturbance to birds (C), Recreation/tourism (C), Unsustainable exploitation (C)

Yagorlytskiy and Tendrovskiy Bays — A1, A4i, A4iii, B1i, B2, B3 — 065

Admin region Kherson
Coordinates 46°20′N 31°50′E
Altitude 0–5 m **Area** 72,000 ha

Site description

Yagorlytskiy (34,000 ha) and Tendrovskiy (38,000 ha) Bays, in the Black Sea, with sandy beaches.

Habitats Scrub, Grassland (humid grassland), Marine areas (sea inlet/coastal features)
Land-use Fisheries/aquaculture, Military, Nature conservation/research, Tourism/recreation

Birds

Species	Season	Year	Pop min	Pop max	Acc	Criteria
Phalacrocorax carbo Cormorant	P	1992	15,000	—	B	A4i, B1i
Egretta garzetta Little Egret	P	1992	76,000	—	B	A4i, B1i
Ardea cinerea Grey Heron	P	1992	5,000	—	B	A4i, B1i
Cygnus olor Mute Swan	P	1992	17,000	21,000	B	A4i, B1i
Anser albifrons White-fronted Goose	W	1992	15,000	40,000	B	A4i, B1i
Anser albifrons White-fronted Goose	P	1992	15,000	40,000	B	A4i, B1i
Anser anser Greylag Goose	W	1992	400	2,000	B	B1i
Anser anser Greylag Goose	P	1992	400	2,000	B	B1i
Anas platyrhynchos Mallard	W	1992	180,000	—	B	A4i, B1i
Anas platyrhynchos Mallard	P	1992	180,000	—	B	A4i, B1i
Aythya ferina Pochard	W	1992	85,000	100,000	B	A4i, B1i
Aythya ferina Pochard	P	1992	85,000	100,000	B	A4i, B1i
Aythya marila Scaup	W	1992	55,000	90,000	B	A4i, B1i
Aythya marila Scaup	P	1992	55,000	90,000	B	A4i, B1i
Somateria mollissima Eider	R	1992	700	—	A	B1i
Fulica atra Coot	W	1992	430,000	450,000	B	A4i, B1i
Fulica atra Coot	P	1992	430,000	450,000	B	A4i, B1i
Recurvirostra avosetta Avocet	B	1992	250	—	B	B1i, B3
[1] ***Numenius tenuirostris*** Slender-billed Curlew	P	1992	Rare		—	A1
Larus minutus Little Gull	B	1992	5,250	33,600	A	A4i, B1i, B2
Larus genei Slender-billed Gull	B	1992	6,800	37,450	B	A4i, B1i
Larus cachinnans Yellow-legged Gull	B	1992	4,000	—	B	A4i, B1i
Sterna sandvicensis Sandwich Tern	B	1992	8,300	28,800	B	A4i, B1i, B2
Sterna hirundo Common Tern	B	1992	15,000	15,000	B	A4i, B1i

1. Observed every year.

Bird data for the Danube is separate.

Protection status

National High **International** High

34,000 ha of IBA covered by Ramsar Site (Yagorlytska Bay, 34,000 ha). 38,000 ha of IBA covered by Ramsar Site (Tendrivska Bay, 38,000 ha). 72,000 ha of IBA covered by Biosphere Reserve (Black Sea, 87,348 ha).

Conservation issues

Threats Agricultural intensification/expansion (C), Disturbance to birds (C)

Kakhovsk reservoir — A1, B1i, B2 — 066

Admin region Kherson
Coordinates 46°50′N 33°30′E
Altitude 49–53 m **Area** 1,000 ha

Site description

A reservoir on the River Dnepr, with islands, near the town of Novaya Kakhovka.

Habitats Forest and woodland (45%; broadleaved deciduous forest; alluvial/very wet forest), Grassland (15%; humid grassland), Wetland (40%; standing fresh water; river/stream; water-fringe vegetation)
Land-use Agriculture (10%), Fisheries/aquaculture (40%), Hunting (40%), Water management

Birds

Species	Season	Year	Pop min	Pop max	Acc	Criteria
Anser anser Greylag Goose	P	1996	1,100	—	B	B1i
Branta ruficollis Red-breasted Goose	P	1996	8	60	B	A1
Haliaeetus albicilla White-tailed Eagle	B	1996	1	3	B	B2
Grus grus Crane	P	1996	150	400	B	B1i

Protection status

National None **International** None

Conservation issues

Threats Aquaculture/fisheries (C), Disturbance to birds (B), Natural events (C), Other (B), Unsustainable exploitation (B)

Unstable water levels ('Other') and storms ('Natural events') are among the threats to the site.

Karkinitskiy and Dzharilgatchskiy Bays — A1, A4i, A4iii, B1i, B2 — 067

Admin region Kherson
Coordinates 45°58′N 33°12′E
Altitude 0–9 m **Area** 87,000 ha

Site description

Large bays with sandy beaches, on the Black Sea coast near the towns of Skadovsk and Krasnoperekopsk. The site includes the Lebyazhi islands.

Habitats Grassland (25%; steppe/dry calcareous grassland), Marine areas (75%; sea inlet/coastal features)
Land-use Fisheries/aquaculture, Tourism/recreation

Birds

Species	Season	Year	Pop min	Pop max	Acc	Criteria
Phalacrocorax carbo Cormorant	B	1992	1,300	2,400	B	A4i, B1i
Phalacrocorax carbo Cormorant	P	1992	1,600	2,500	B	B1i
Egretta garzetta Little Egret	B	1992	480	900	B	A4i, B1i
Egretta alba Great White Egret	B	1992	200	700	B	A4i, B1i
Egretta alba Great White Egret	P	1992	800	—	B	A4i, B1i
Cygnus olor Mute Swan	W	1992	1,000	8,000	B	A4i, B1i
Cygnus cygnus Whooper Swan	W	1992	700	900	B	A4i, B1i
Branta ruficollis Red-breasted Goose	W	1994	520	—	A	A1
Anas platyrhynchos Mallard	W	1992	17,000	43,000	B	B1i
Fulica atra Coot	W	1992	32,500	37,500	B	B1i
Numenius tenuirostris Slender-billed Curlew	P	1995	Rare		—	A1

Species ... continued	Season	Year	Pop min	Pop max	Acc	Criteria
Larus minutus Little Gull	B	1992	6,800	17,000	B	A4i, B1i, B2
Larus cachinnans Yellow-legged Gull	B	1992	9,000	10,000	B	A4i, B1i
Larus cachinnans Yellow-legged Gull	W	1992	6,000	—	B	A4i, B1i
Gelochelidon nilotica Gull-billed Tern	B	1992	40	—	B	B2
Sterna caspia Caspian Tern	B	1992	200	570	B	A4i, B1i, B2
Sterna sandvicensis Sandwich Tern	B	1992	3,000	5,000	B	A4i, B1i, B2
Sterna hirundo Common Tern	B	1992	1,000	—	B	B1i
Sterna albifrons Little Tern	B	1992	150	—	B	B2

Protection status

National Partial **International** High

9,612 ha of IBA covered by unknown protected-area type (Lebyazhi Island, 9,612 ha). 87,000 ha of IBA covered by Ramsar Site (Karkinitska and Dzharylgatchska Bays, 87,000 ha).

Conservation issues

Threats Agricultural intensification/expansion (B), Aquaculture/fisheries (C), Disturbance to birds (C), Unsustainable exploitation (C)

Askania-Nova Biosphere Reserve — A1, A3, A4i, A4iii, A4iv, B1i, B1iv — 068

Admin region Kherson
Coordinates 46°27'N 33°52'E
Altitude 20–33 m **Area** 33,307 ha

Site description

Plots of natural steppe with man-made ponds, bordered by cultivated land.

Habitats Grassland (100%; steppe/dry calcareous grassland; humid grassland), Artificial landscape (urban parks/gardens)
Land-use Nature conservation/research (100%)

Birds

Species	Season	Year	Pop min	Pop max	Acc	Criteria
Anser albifrons White-fronted Goose	P	1996	500,000	—	B	A4i, B1i
Branta ruficollis Red-breasted Goose	P	1996	60	—	C	A1
Tadorna ferruginea Ruddy Shelduck	W	—	424	—	A	A4i, B1i
[1] ***Circus macrourus*** Pallid Harrier	B	1995	—	—	—	A3
[2] ***Grus grus*** Crane	P	1996	57,000	—	B	A4i, A4iv, B1i, B1iv
[1] ***Anthropoides virgo*** Demoiselle Crane	B	1995	—	—	—	A3
[1] ***Glareola nordmanni*** Black-winged Pratincole	U	1995	—	—	—	A3

1. No counts available.
2. Max. in one day is 42,000 at Chapliniy Pod.

An important site for birds of the steppe biome.

Protection status

National High **International** High

33,307 ha of IBA covered by Biosphere Reserve (Askania-Nova, 33,307 ha).

Conservation issues

There are no serious threats.

Sivash Bay, Sea of Azov — A1, A3, A4i, A4iii, A4iv, B1i, B1iv, B2, B3 — 069

Admin region Crimea, Kherson
Coordinates 46°10'N 34°35'E
Altitude 5–10 m **Area** 245,000 ha

Site description

An extensive system of lagoons with sandy islands and shores, surrounded by steppe, in the coastal zone of the Black Sea and Sea of Azov.

Habitats Grassland (steppe/dry calcareous grassland; humid grassland), Wetland (sand-dunes/sand beach; coastal lagoon; standing fresh water; standing brackish and salt water; water-fringe vegetation; fen/transition mire/spring)
Land-use Agriculture, Fisheries/aquaculture, Hunting, Tourism/recreation

Birds

Species	Season	Year	Pop min	Pop max	Acc	Criteria
Phalacrocorax carbo Cormorant	B	1992	8,270	—	B	A4i, B1i
Nycticorax nycticorax Night Heron	B	1992	150	200	B	B2
Egretta garzetta Little Egret	B	1992	400	500	B	A4i, B1i
Egretta alba Great White Egret	B	1992	400	500	B	A4i, B1i
Ardea purpurea Purple Heron	B	1992	120	200	B	B2
Plegadis falcinellus Glossy Ibis	B	1992	470	500	B	A4i, B1i, B2
Cygnus olor Mute Swan	N	1996	1,000	—	B	B1i
Anser albifrons White-fronted Goose	P	1992	200,000	—	B	A4i, B1i
Anser erythropus Lesser White-fronted Goose	P	1998	—	1,000	—	A1, A4i, B1i
Anser anser Greylag Goose	P	1996	5,000	—	B	A4i, B1i
Tadorna tadorna Shelduck	P	1992	12,000	—	B	A4i, B1i
Anas penelope Wigeon	P	1992	62,000	65,000	B	A4i, B1i
Anas acuta Pintail	P	1992	33,000	35,000	B	A4i, B1i
Grus grus Crane	P	1992	20,000	30,000	B	A4i, A4iv, B1i, B1iv
Grus grus Crane	N	1996	20,000	—	B	A4i, B1i
Anthropoides virgo Demoiselle Crane	B	1996	1,000	—	B	A3
[2] ***Otis tarda*** Great Bustard	B	1989	2	3	—	B2
[3] ***Otis tarda*** Great Bustard	P	1989	—	1,000	—	A1
Himantopus himantopus Black-winged Stilt	B	1992	600	800	B	A4i, B1i
Himantopus himantopus Black-winged Stilt	P	1992	1,000	2,500	B	A4i, B1i
Recurvirostra avosetta Avocet	B	1992	500	600	B	A4i, B1i, B3
Recurvirostra avosetta Avocet	P	1992	1,000	2,500	B	A4i, B1i
Glareola pratincola Collared Pratincole	B	1992	150	300	B	A4i, B1i, B2
Glareola pratincola Collared Pratincole	P	1992	300	600	B	A4i, B1i
Charadrius hiaticula Ringed Plover	P	1992	2,000	4,000	B	A4i, B1i
Charadrius alexandrinus Kentish Plover	B	1992	250	500	B	A4i, B1i, B2
Charadrius alexandrinus Kentish Plover	P	1992	500	1,000	B	A4i, B1i
Pluvialis squatarola Grey Plover	P	1992	3,500	5,000	B	A4i, B1i
Calidris alba Sanderling	P	1992	700	2,000	B	A4i, B1i
Calidris minuta Little Stint	P	1992	26,700	—	B	A4i, B1i
Calidris alpina Dunlin	P	1992	127,600	—	B	A4i, B1i
Limicola falcinellus Broad-billed Sandpiper	P	1992	6,000	8,000	B	A4i, B1i
Philomachus pugnax Ruff	P	1992	43,500	—	B	A4i, B1i
Gallinago media Great Snipe	P	1992	10	200	B	A1
Limosa lapponica Bar-tailed Godwit	P	1992	2,000	5,000	B	A4i, B1i
Tringa erythropus Spotted Redshank	P	1992	1,500	2,500	B	A4i, B1i
Tringa totanus Redshank	B	1992	400	500	B	B1i, B2
Tringa totanus Redshank	P	1992	11,800	—	B	A4i, B1i
Tringa stagnatilis Marsh Sandpiper	P	1992	1,000	3,000	B	A4i, B1i
Arenaria interpres Turnstone	P	1992	500	1,500	B	A4i, B1i
Larus ichthyaetus Great Black-headed Gull	B	1992	475	605	A	A3, A4i, B1i
Larus melanocephalus Mediterranean Gull	B	1992	2,450	—	B	A4i, B1i, B3
Larus minutus Little Gull	P	1992	3,000	10,000	B	A4i, B1i
Larus genei Slender-billed Gull	B	1992	1,850	2,150	A	A4i, B1i
Larus cachinnans Yellow-legged Gull	B	1992	3,800	—	B	A4i, B1i
Gelochelidon nilotica Gull-billed Tern	B	1992	315	—	A	A4i, B1i, B2
Sterna caspia Caspian Tern	B	1989	56	159	—	A4i, B1i, B2
Sterna sandvicensis Sandwich Tern	B	1992	4,700	—	B	A4i, B1i, B2
Sterna hirundo Common Tern	B	1992	2,900	3,080	A	A4i, B1i
Sterna albifrons Little Tern	B	1992	360	—	B	A4i, B1i, B2
Chlidonias hybridus Whiskered Tern	P	1992	250	1,000	B	A4i, B1i
Chlidonias niger Black Tern	P	1992	500	2,000	B	A4i, B1i

1. Autumn 1998.
2. 'Pairs'.
3. At Kuyuk-Tuk in spring.

An important site for waterbirds and birds of the steppe biome.

Protection status

National None **International** High

165,000 ha of IBA covered by Ramsar Site (Eastern Syvash, 165,000 ha). 80,000 ha of IBA covered by Ramsar Site (Central Syvash, 80,000 ha).

Conservation issues

Threats Agricultural intensification/expansion (B), Aquaculture/fisheries (B), Construction/impact of dyke/dam/barrage (C), Disturbance to birds (B), Recreation/tourism (B), Unsustainable exploitation (B)

Utlyuk estuary A4i, A4iii, B1i 070

Admin region Zaporozhie
Coordinates 46°10′N 35°10′E
Altitude 0–32 m **Area** 75,200 ha

Site description

An estuary on the coast of the Sea of Azov, with adjoining areas of steppe and arable land, near Genitchesk town.

Habitats Grassland (10%; steppe/dry calcareous grassland; humid grassland), Wetland (70%; sand-dunes/sand beach; standing brackish and salt water; water-fringe vegetation), Marine areas (10%; sea inlet/coastal features), Artificial landscape (10%; arable land)
Land-use Fisheries/aquaculture, Hunting, Unknown

Birds

Species	Season	Year	Pop min	Pop max	Acc	Criteria
Cygnus olor Mute Swan	P	1992	11,000	16,000	A	A4i, B1i
Anser albifrons White-fronted Goose	P	1992	19,000	22,000	A	A4i, B1i
Anser anser Greylag Goose	P	1992	200	3,000	B	B1i
Aythya marila Scaup	P	1992	2,000	3,000	B	B1i
Larus cachinnans Yellow-legged Gull	P	1992	6,000	7,000	B	A4i, B1i

The total number of breeding waterbirds is 600–800 pairs, while 30,000–70,000 waterbirds stage here during migration.

Protection status

National None **International** None

Conservation issues

Threats Disturbance to birds (B)

Molochniy liman A4i, A4iii, B1i, B3 071

Admin region Zaporozhie
Coordinates 46°19′N 35°19′E
Altitude 0–5 m **Area** 22,450 ha

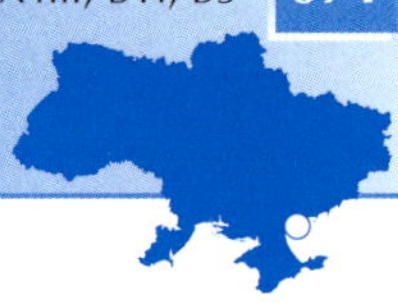

Site description

The estuary of the Molochnaya river, near Kirilovka village.

Habitats Grassland (steppe/dry calcareous grassland), Wetland (95%; tidal river/enclosed tidal water; sand-dunes/sand beach; standing fresh water; river/stream; water-fringe vegetation), Marine areas
Land-use Agriculture, Fisheries/aquaculture, Urban/industrial/transport

Birds

Species	Season	Year	Pop min	Pop max	Acc	Criteria
Podiceps cristatus Great Crested Grebe	P	1992	5,000	12,000	A	A4i, B1i
Podiceps nigricollis Black-necked Grebe	P	1992	5,000	10,000	A	A4i, B1i
Phalacrocorax carbo Cormorant	B	1992	1,000	3,000	B	A4i, B1i
Cygnus olor Mute Swan	P	1992	2,000	5,000	B	A4i, B1i
Anser albifrons White-fronted Goose	P	1992	25,000	40,000	B	A4i, B1i
Anser anser Greylag Goose	W	1992	10,000	12,000	B	A4i, B1i
Anas platyrhynchos Mallard	W	1992	25,000	30,000	B	B1i
Aythya marila Scaup	W	1992	80,000	100,000	B	A4i, B1i
Mergus albellus Smew	W	1992	5,000	12,000	B	A4i, B1i
Recurvirostra avosetta Avocet	B	1992	50	250	A	B1i, B3
Larus cachinnans Yellow-legged Gull	B	1992	4,000	5,500	B	A4i, B1i

The total number of breeding waterbirds is 12,000–15,000 pairs, while the total number of waterbirds staging on passage is 197,000–286,000 individuals, including up to 20,000 *Fulica atra*. The total staging is dependent on weather and water conditions.

Protection status

National None **International** High
22,400 ha of IBA covered by Ramsar Site (Molochnyi Liman, 22,400 ha).

Conservation issues

Threats Agricultural intensification/expansion (A), Aquaculture/fisheries (C), Disturbance to birds (B), Unsustainable exploitation (B)

Molochnaya river valley A4i, B1i, B2, B3 072

Admin region Zaporozhie
Coordinates 47°05′N 35°30′E
Altitude 25–105 m **Area** 3,800 ha

Site description

A stretch of the Molochnaya valley, near the villages of Svetlodolinskoe, Kamenskoe and Rubalovka, with numerous reedbeds *Phragmites*.

Habitats Grassland (65%; humid grassland), Wetland (23%; standing fresh water; river/stream; water-fringe vegetation), Artificial landscape (12%; arable land; forestry plantation; other urban/industrial areas)
Land-use Agriculture (80%), Hunting (10%), Not utilized (5%), Urban/industrial/transport (10%)

Birds

Species	Season	Year	Pop min	Pop max	Acc	Criteria
Botaurus stellaris Bittern	B	1997	140	160	B	B2
Ixobrychus minutus Little Bittern	B	1997	200	240	B	B2
Himantopus himantopus Black-winged Stilt	B	1997	120	200	B	B1i
Glareola pratincola Collared Pratincole	B	1997	60	80	B	A4i, B1i, B2
Tringa totanus Redshank	B	1997	275	315	A	B2
Saxicola rubetra Whinchat	B	1997	260	500	B	B3
Locustella luscinioides Savi's Warbler	B	1997	300	500	B	B3

Protection status

National None **International** None

Conservation issues

Threats Burning of vegetation (A), Disturbance to birds (B), Drainage (A), Dredging/canalization (A)

Obitochnaya peninsula A4i, B1i 073

Admin region Zaporozhie
Coordinates 46°34′N 36°19′E
Altitude 0–5 m **Area** 2,000 ha

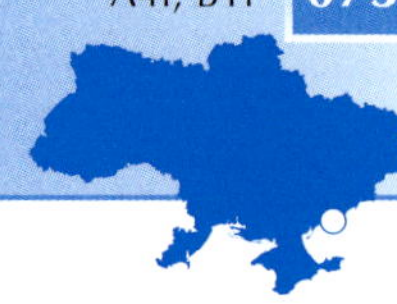

Site description

A bay and peninsula in the Sea of Azov.

Habitats Scrub, Wetland (sand-dunes/sand beach; coastal lagoon), Marine areas (sea inlet/coastal features)
Land-use Agriculture, Fisheries/aquaculture, Tourism/recreation, Urban/industrial/transport

Birds

Species	Season	Year	Pop min	Pop max	Acc	Criteria
Phalacrocorax carbo Cormorant	B	1992	1,330	—	B	B1i
Egretta alba Great White Egret	B	1992	120	—	B	A4i, B1i
Aythya marila Scaup	P	1992	4,000	6,000	B	A4i, B1i

Protection status

National None **International** High
2,000 ha of IBA covered by Ramsar Site (Obytochna Spit and Obytochna Bay, 2,000 ha).

Conservation issues

Threats Agricultural intensification/expansion (B), Aquaculture/fisheries (B), Disturbance to birds (B), Recreation/tourism (C)

Elanchik river valley B1i, B2 074

Admin region Donetsk
Coordinates 47°09′N 38°05′E
Altitude 50–55 m **Area** 400 ha

Site description

Valley lowlands near the town of Novoazovsk.

Habitats Grassland (humid grassland), Wetland (standing fresh water; standing brackish and salt water; river/stream; water-fringe vegetation; fen/transition mire/spring)
Land-use Agriculture

Birds

Species	Season	Year	Pop min	Pop max	Acc	Criteria
Charadrius alexandrinus Kentish Plover	B	1994	40	—	A	B2
Sterna hirundo Common Tern	B	1994	1,000	1,200	A	B1i

Protection status

National None **International** None

Conservation issues

Threats Agricultural intensification/expansion (B), Disturbance to birds (B)

Ukrainian Steppe Nature Reserve A1, B2 075

Admin region Donetsk
Coordinates 47°30′N 36°30′E
Altitude 53–163 m **Area** 1,134 ha

Site description

The IBA comprises the 'Krejdyana flora' parcel of the Ukrainian Steppe Nature Reserve. Natural steppe dominates, occurring as plots on the hills along a small river valley.

Habitats Forest and woodland (14%; broadleaved deciduous forest), Grassland (86%; steppe/dry calcareous grassland; humid grassland)
Land-use Nature conservation/research (100%)

Birds

Species	Season	Year	Pop min	Pop max	Acc	Criteria
[1] ***Falco naumanni*** Lesser Kestrel	B	1994	2	10	B	A1, B2

1. Data for Melovaya Flora parcel.

Protection status

National High **International** None
1,134 ha of IBA covered by Nature Reserve (Ukrainian Steppe, 2,764 ha).

Conservation issues

There may be serious threats but the current situation is unknown.

Belosarayskaya peninsula A4i, B1i, B2 076

Admin region Donetsk
Coordinates 47°00′N 37°18′E
Altitude 0–3 m **Area** 2,000 ha

Site description

A sandy peninsula in the Sea of Azov, near the village of Belosarayskaya Kosa, Yalta.

Habitats Wetland (saltmarsh; sand-dunes/sand beach; standing brackish and salt water; river/stream; water-fringe vegetation), Marine areas (sea inlet/coastal features)
Land-use Agriculture, Fisheries/aquaculture, Nature conservation/research, Tourism/recreation, Urban/industrial/transport

Birds

Species	Season	Year	Pop min	Pop max	Acc	Criteria
Ardea purpurea Purple Heron	B	1994	20	80	B	B2
Sterna sandvicensis Sandwich Tern	B	1994	500	1,000	B	A4i, B1i, B2
Sterna hirundo Common Tern	B	1994	600	2,000	B	B1i

Protection status

National None **International** High
2,000 ha of IBA covered by Ramsar Site (Bilosaraiska Bay and Bilosaraiska Spit, 2,000 ha).

Conservation issues

Threats Agricultural intensification/expansion (B), Aquaculture/fisheries (C), Disturbance to birds (A), Industrialization/urbanization (B), Recreation/tourism (B)

Berdyanskaya peninsula A4i, A4iii, B1i, B2 077

Admin region Zaporozhie
Coordinates 46°49′N 36°45′E
Altitude 0–6 m **Area** 1,800 ha

Site description

A sandy peninsula in the Sea of Azov, together with the estuary of the Berda river, near the town of Berdyansk.

Habitats Scrub (5%; scrub), Grassland (5%; steppe/dry calcareous grassland), Wetland (sand-dunes/sand beach; coastal lagoon; standing fresh water; river/stream; water-fringe vegetation), Marine areas (open sea; sea inlet/coastal features), Artificial landscape (forestry plantation; urban parks/gardens; other urban/industrial areas)
Land-use Agriculture, Fisheries/aquaculture, Tourism/recreation, Urban/industrial/transport

Birds

Species	Season	Year	Pop min	Pop max	Acc	Criteria
Phalacrocorax carbo Cormorant	B	1992	1,200	—	B	B1i
[1] ***Egretta alba*** Great White Egret	B	1992	70	—	A	A4i, B1i
[1] ***Sterna sandvicensis*** Sandwich Tern	B	1992	800	—	A	B1i, B2

1. Large decrease.

The total number of breeding waterbirds is 3,000–5,000 pairs, while the total number of waterbirds staging on passage is 30,000–40,000 individuals.

Protection status

National None **International** High
1,800 ha of IBA covered by Ramsar Site (Berda River Mouth and Berdianka Spit and Berdianska Bay, 1,800 ha).

Conservation issues

Threats Agricultural intensification/expansion (B), Aquaculture/fisheries (C), Industrialization/urbanization (B)

Krivaya peninsula A4i, B1i, B2 078

Admin region Donetsk
Coordinates 47°05′N 38°07′E
Altitude 0–2 m **Area** 8,800 ha

Site description

A sandy peninsula in the Sea of Azov, near Sedov town, adjacent to steppe and saline meadows.

Habitats Wetland (sand-dunes/sand beach; standing brackish and salt water; river/stream; water-fringe vegetation; fen/transition mire/spring), Marine areas (sea inlet/coastal features)
Land-use Fisheries/aquaculture, Nature conservation/research, Tourism/recreation, Urban/industrial/transport

Birds

Species	Season	Year	Pop min	Pop max	Acc	Criteria
Charadrius alexandrinus Kentish Plover	B	1994	18	80	B	B2
Larus minutus Little Gull	P	1992	5,000	7,000	B	A4i, B1i
Sterna sandvicensis Sandwich Tern	B	1994	3,000	20,000	B	A4i, B1i, B2
Sterna hirundo Common Tern	B	1994	4,800	12,000	B	A4i, B1i
Sterna hirundo Common Tern	P	1992	5,000	—	B	B1i
Sterna albifrons Little Tern	B	1994	440	—	B	A4i, B1i, B2
Chlidonias niger Black Tern	P	1992	2,500	—	B	A4i, B1i

Protection status

National None **International** Partial
1,400 ha of IBA covered by Ramsar Site (Kryva Bay and Kryva Spit, 1,400 ha).

Conservation issues

Threats Aquaculture/fisheries (C), Disturbance to birds (B), Industrialization/urbanization (B), Recreation/tourism (C), Unsustainable exploitation (C)

Mokrye Yasli forestry plantation — B2 — 079

Admin region Odessa
Coordinates 47°08′N 29°45′E
Altitude 150–174 m **Area** 500 ha

Site description

A forestry plantation on a hill, dominated by broadleaved deciduous trees. Land-uses within the IBA include cattle-grazing and hay-making.

Habitats Artificial landscape (100%; forestry plantation)
Land-use Agriculture, Forestry (100%)

Birds

Species	Season	Year	Pop min	Pop max	Acc	Criteria
Falco vespertinus Red-footed Falcon	B	1997	4	—	A	B2
Lanius minor Lesser Grey Shrike	B	1997	250	300	B	B2

Protection status

National None **International** None

Conservation issues

Threats Disturbance to birds (C)

Kagul lake — A4iii — 080

Admin region Odessa
Coordinates 45°13′N 28°26′E
Altitude 6–80 m **Area** 10,500 ha

Site description

A freshwater lake near the Danube delta.

Habitats Wetland (100%; standing fresh water; water-fringe vegetation)
Land-use Fisheries/aquaculture (100%)

Birds

The total number of migrating waterbirds that stage here exceeds 20,000 individuals, with a maximum during August–October. Species of global conservation concern that do not meet IBA criteria: *Phalacrocorax pygmeus* (min. 13 birds in winter).

Protection status

National None **International** None

Conservation issues

Threats Agricultural intensification/expansion (B), Aquaculture/fisheries (B), Natural events (B), Unsustainable exploitation (C)

Storms can pose a threat.

Kugurluy and Kartal lakes — A1, A4i, A4iii, B1i, B2 — 081

Admin region Odessa
Coordinates 45°17′N 28°39′E
Altitude 2–20 m **Area** 19,200 ha

Site description

Lakes surrounded by arable land, in the lower valley of the River Danube near the town of Izmail.

Habitats Wetland (100%; standing fresh water; water-fringe vegetation; fen/transition mire/spring)
Land-use Fisheries/aquaculture (100%), Water management

Birds

Species	Season	Year	Pop min	Pop max	Acc	Criteria
Phalacrocorax pygmeus Pygmy Cormorant	B	1992	150	—	A	A1, A4i, B1i, B2
[1] ***Phalacrocorax pygmeus*** Pygmy Cormorant	W	1995	55	—	A	A1
Platalea leucorodia Spoonbill	B	1992	300	—	A	A4i, B1i, B2
Cygnus olor Mute Swan	B	1992	180	—	A	B1i
[1] ***Cygnus olor*** Mute Swan	W	1995	540	—	—	B1i
[1] ***Anser albifrons*** White-fronted Goose	W	1995	50,000	—	B	A4i, B1i
[1] ***Branta ruficollis*** Red-breasted Goose	W	1995	1,000	—	A	A1, A4i, B1i
Larus cachinnans Yellow-legged Gull	B	1992	700	—	—	B1i

1. Data for two lakes (Yalpug and Kugurluj).

The total number of breeding waterbirds is 30,000 pairs, with more than 70,000 individuals staging here during migration.

Protection status

National None **International** Partial

6,500 ha of IBA covered by Ramsar Site (Kugurlui Lake, 6,500 ha). 500 ha of IBA covered by Ramsar Site (Kartal Lake, 500 ha).

Conservation issues

Threats Agricultural intensification/expansion (C), Aquaculture/fisheries (B), Disturbance to birds (C), Unsustainable exploitation (C)

River Danube — A4i, A4iii, B1i — 082

Admin region Odessa
Coordinates 45°23′N 29°07′E
Altitude 0–2 m **Area** 2,500 ha

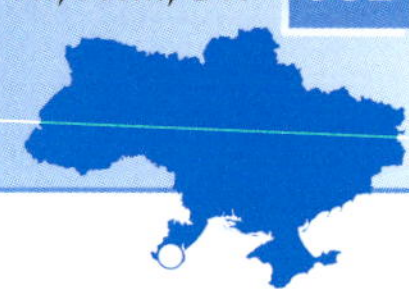

Site description

A stretch of the River Danube, near Kiliya town, that does not freeze in winter.

Habitats Forest and woodland (5%; alluvial/very wet forest), Scrub (5%), Grassland (humid grassland; mesophile grassland), Wetland (standing fresh water; river/stream; water-fringe vegetation; fen/transition mire/spring)
Land-use Agriculture, Fisheries/aquaculture, Nature conservation/research, Urban/industrial/transport, Water management

Birds

Species	Season	Year	Pop min	Pop max	Acc	Criteria
Anser albifrons White-fronted Goose	W	1992	20,000	—	B	A4i, B1i

The total number of waterbirds that stage here on migration is nearly 40,000.

Protection status

National None **International** None

Conservation issues

Threats Agricultural intensification/expansion (B), Aquaculture/fisheries (C), Burning of vegetation (B), Disturbance to birds (B), Unsustainable exploitation (U)

Kitaj lakes — A4i, A4iii, B1i — 083

Admin region Odessa
Coordinates 45°35′N 29°12′E
Altitude 16–29 m **Area** 5,000 ha

Site description

A group of lakes near the Danube delta.

Habitats Wetland (100%; standing fresh water; water-fringe vegetation)
Land-use Fisheries/aquaculture (100%)

Birds

Species	Season	Year	Pop min	Pop max	Acc	Criteria
Anser albifrons White-fronted Goose	W	1995	20,000	—	A	A4i, B1i

The total number of breeding waterbirds is nearly 2,500 pairs, and over 25,000 individuals stage here on passage, peaking during August–November.

Protection status

National None **International** None

Conservation issues

Threats Agricultural intensification/expansion (B), Aquaculture/fisheries (C), Disturbance to birds (C), Unsustainable exploitation (C)

Zhebriyanovskie plavni A1, A4i, A4iii, B1i, B2, B3 084

Admin region Odessa
Coordinates 45°31′N 29°31′E
Altitude 0–9 m **Area** 6,800 ha

Site description

A part of the Danube delta, near Vilkovo village.

Habitats Wetland (100%; standing fresh water; water-fringe vegetation)
Land-use Fisheries/aquaculture, Hunting, Water management

Birds

Species	Season	Year	Pop min	Pop max	Acc	Criteria
Phalacrocorax pygmeus Pygmy Cormorant	W	1995	20	—	A	A1
Pelecanus crispus Dalmatian Pelican	P	1992	1,500	—	B	A1, A4i, B1i
Egretta alba Great White Egret	B	1992	400	—	B	A4i, B1i
Platalea leucorodia Spoonbill	B	1992	350	—	B	A4i, B1i, B2
Cygnus olor Mute Swan	B	1992	350	—	B	B1i
Cygnus olor Mute Swan	W	1995	2,200	—	A	B1i
Cygnus olor Mute Swan	P	1992	2,000	—	B	B1i
Anser albifrons White-fronted Goose	W	1995	10,000	—	A	B1i
[1] ***Anser anser*** Greylag Goose	B	1993	—	600	B	B1i
Netta rufina Red-crested Pochard	B	1992	400	—	B	A4i, B1i
Aythya ferina Pochard	B	1992	1,800	—	B	B3
Aythya ferina Pochard	W	1995	13,500	—	A	A4i, B1i

1. Estimate covers two sites (Stencovsko-Zhebriyanovskie plavni and Kugurluy lake).

The total number of breeding waterbirds is 18,000 pairs, while more than 100,000 individuals stage here on migration.

Protection status

National None **International** High
6,800 ha of IBA covered by Ramsar Site (Kyliiske Mouth, 32,800 ha).

Conservation issues

Threats Agricultural intensification/expansion (B), Aquaculture/fisheries (C), Burning of vegetation (A), Construction/impact of dyke/dam/barrage (U)

The IBA is due to be included in the Danube Delta Biosphere Reserve, which is in the process of being designated.

Sasuk lake A4i, A4iii, B1i, B2 085

Admin region Odessa
Coordinates 45°40′N 29°39′E
Altitude 0–22 m **Area** 22,800 ha

Site description

A lake near the Danube delta.

Habitats Grassland (10%; steppe/dry calcareous grassland), Wetland (80%; coastal lagoon; standing brackish and salt water; water-fringe vegetation), Marine areas (10%; sea inlet/coastal features)
Land-use Fisheries/aquaculture (80%), Other, Urban/industrial/transport, Water management

Birds

Species	Season	Year	Pop min	Pop max	Acc	Criteria
Pelecanus onocrotalus White Pelican	P	1992	2,000	2,000	B	A4i, B1i
Cygnus olor Mute Swan	P	1992	6,000	6,000	B	A4i, B1i
Anser albifrons White-fronted Goose	P	1992	20,000	20,000	B	A4i, B1i
Anser anser Greylag Goose	N	1993	3,000	—	B	B1i
Aythya ferina Pochard	P	1992	18,000	18,000	B	A4i, B1i
Aythya fuligula Tufted Duck	P	1992	12,000	12,000	B	B1i
Mergus albellus Smew	P	1992	3,000	3,000	B	A4i, B1i
Glareola pratincola Collared Pratincole	B	1992	60	60	B	B1i, B2
Larus cachinnans Yellow-legged Gull	P	1992	4,000	4,000	B	B1i

The total number of breeding waterbirds is 10,000 pairs, while 100,000 individuals stage here on migration.

Protection status

National None **International** High
21,000 ha of IBA covered by Ramsar Site (Sasyk Lake, 21,000 ha).

Conservation issues

Threats Aquaculture/fisheries (B), Burning of vegetation (B), Disturbance to birds (U), Dredging/canalization (U)

Jansheisk lakes B1i 086

Admin region Odessa
Coordinates 45°38′N 29°47′E
Altitude 0–3 m **Area** 1,000 ha

Site description

Salt-lakes along the Black Sea coast.

Habitats Wetland (100%; sand-dunes/sand beach; standing brackish and salt water), Marine areas (sea inlet/coastal features)
Land-use Agriculture, Fisheries/aquaculture

Birds

Species	Season	Year	Pop min	Pop max	Acc	Criteria
Cygnus olor Mute Swan	W	1995	1,150	—	A	B1i
Cygnus olor Mute Swan	P	1995	1,150	—	A	B1i

Protection status

National None **International** None

Conservation issues

Threats Agricultural intensification/expansion (B), Aquaculture/fisheries (B), Disturbance to birds (B), Unsustainable exploitation (C)

Shagany–Alibey–Burnas lake-system A1, A4i, A4iii, B1i, B2 087

Admin region Odessa
Coordinates 45°47′N 30°00′E
Altitude 0–19 m **Area** 19,200 ha

Site description

A system of lakes in the lower reaches of the Danube valley, by the Black Sea coast near Zheltye Vodu village.

Habitats Scrub (8%; scrub), Wetland (82%; saltmarsh; sand-dunes/sand beach; standing brackish and salt water; river/stream; water-fringe vegetation), Marine areas (10%; sea inlet/coastal features)
Land-use Agriculture (10%), Fisheries/aquaculture (10%), Not utilized (60%), Tourism/recreation (20%)

Birds

Species	Season	Year	Pop min	Pop max	Acc	Criteria
Cygnus olor Mute Swan	W	1995	1,150	—	—	B1i
Anser albifrons White-fronted Goose	W	1992	13,000	—	B	A4i, B1i
Anser anser Greylag Goose	W	1992	700	—	B	B1i
Branta ruficollis Red-breasted Goose	W	1995	2,500	—	—	A1, A4i, B1i
Tadorna tadorna Shelduck	W	1995	820	—	—	B1i
Anas platyrhynchos Mallard	W	1992	25,000	—	B	B1i
Aythya ferina Pochard	W	1995	43,000	—	A	A4i, B1i
Aythya fuligula Tufted Duck	W	1995	14,000	—	B	B1i
Recurvirostra avosetta Avocet	P	1992	1,300	—	B	A4i, B1i
Glareola pratincola Collared Pratincole	B	1992	60	—	B	B1i, B2

Species ... continued	Season	Year	Pop min	Pop max	Acc	Criteria
Larus cachinnans Yellow-legged Gull	B	—	2,800	—	—	A4i, B1i
Larus cachinnans Yellow-legged Gull	P	1992	18,000	—	—	A4i, B1i

Protection status

National None **International** High

19,000 ha of IBA covered by Ramsar Site (Shagany-Alibei-Burnas Lakes System, 19,000 ha).

Conservation issues

Threats Aquaculture/fisheries (C), Construction/impact of dyke/dam/barrage (A), Disturbance to birds (B), Industrialization/urbanization (B), Natural events (C), Unsustainable exploitation (B)

Zmeinuy island — A1, A4i, B1i — 088

Admin region Odessa
Coordinates 45°15′N 30°12′E
Altitude 0–60 m **Area** 17 ha

Site description

A rocky island in the Black Sea, near the Danube delta, covered with grassland. It is a military training area.

Habitats Grassland (60%; steppe/dry calcareous grassland), Marine areas (open sea), Rocky areas (5%; rock stacks/islets), Artificial landscape (35%; other urban/industrial areas)
Land-use Military (100%)

Birds

Species	Season	Year	Pop min	Pop max	Acc	Criteria
Branta ruficollis Red-breasted Goose	P	1997	20	200	B	A1
[1] ***Grus grus*** Crane	P	1997	Frequent		—	A4i, B1i

1. Data from non-birdwatcher: thousands of birds gather here from time to time during storms.

Species of global conservation concern that do not meet IBA criteria: *Phalacrocorax pygmeus* (1–10 non-breeding birds).

Protection status

National None **International** None

Conservation issues

Threats Disturbance to birds (A), Infrastructure (A), Natural events (A), Unsustainable exploitation (A)

Storms can pose a threat.

Budakskiy estuary — A4i, B1i — 089

Admin region Odessa
Coordinates 46°00′N 30°15′E
Altitude 0–16 m **Area** 2,700 ha

Site description

An estuary, connected with the Dnestr delta, near Sergeevka and Kurortnoe villages.

Habitats Scrub (17%; scrub), Wetland (67%; sand-dunes/sand beach; coastal lagoon; standing brackish and salt water; river/stream; water-fringe vegetation), Marine areas (10%; sea inlet/coastal features), Artificial landscape (6%; other urban/industrial areas)
Land-use Agriculture (1%), Fisheries/aquaculture (2%), Military (2%), Tourism/recreation (20%), Urban/industrial/transport (10%)

Birds

Species	Season	Year	Pop min	Pop max	Acc	Criteria
[1] ***Egretta garzetta*** Little Egret	B	1996	350	400	B	A4i, B1i
Cygnus olor Mute Swan	W	1995	700	—	A	B1i
Larus cachinnans Yellow-legged Gull	P	1993	1,300	—	C	B1i

1. Data from early August.

Protection status

National None **International** None

Conservation issues

Threats Agricultural intensification/expansion (C), Aquaculture/fisheries (B), Disturbance to birds (B), Industrialization/urbanization (A), Natural events (C)

Storms ('Natural events') are a threat.

Kuyalnik estuary — B1i, B2, B3 — 090

Admin region Odessa
Coordinates 46°41′N 30°42′E
Altitude 30–85 m **Area** 7,000 ha

Site description

An estuary on the Black Sea coast near Odessa, with sandy beaches adjoined by steppe and arable land.

Habitats Grassland (10%; steppe/dry calcareous grassland; humid grassland), Wetland (90%; sand-dunes/sand beach; standing brackish and salt water; river/stream; water-fringe vegetation)
Land-use Tourism/recreation

Birds

Species	Season	Year	Pop min	Pop max	Acc	Criteria
Tadorna tadorna Shelduck	W	1992	2,200	—	B	B1i
Recurvirostra avosetta Avocet	B	1992	130	—	B	B3
Sterna albifrons Little Tern	B	1992	19	260	A	B1i, B2

Protection status

National None **International** None

Conservation issues

Threats Abandonment/reduction of land management (B), Disturbance to birds (C), Industrialization/urbanization (B), Recreation/tourism (B), Unsustainable exploitation (B)

Dnestr delta — A4i, A4iii, B1i, B2 — 091

Admin region Odessa
Coordinates 46°30′N 30°11′E
Altitude 15–35 m **Area** 27,600 ha

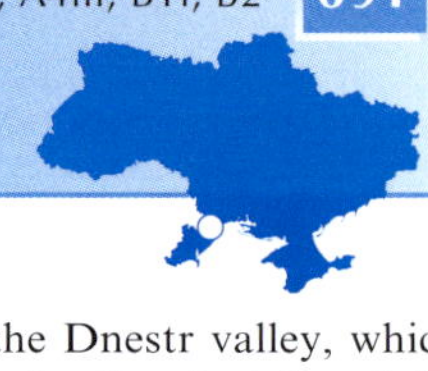

Site description

A natural wetland in the lower reaches of the Dnestr valley, which includes a man-made reservoir as well as the flood-plain of the Turunchuk river near Belyaevka town.

Habitats Forest and woodland (5%; alluvial/very wet forest), Scrub (5%; scrub), Grassland (10%; humid grassland), Wetland (80%; standing fresh water; river/stream; water-fringe vegetation)
Land-use Agriculture, Fisheries/aquaculture, Water management

Birds

Species	Season	Year	Pop min	Pop max	Acc	Criteria
Phalacrocorax carbo Cormorant	B	1993	2,000	2,500	B	A4i, B1i
Nycticorax nycticorax Night Heron	B	1993	1,000	2,500	B	A4i, B1i, B2
Egretta alba Great White Egret	B	1992	110	330	B	A4i, B1i
Ardea purpurea Purple Heron	B	1993	100	150	B	B2
Plegadis falcinellus Glossy Ibis	B	1993	120	1,500	B	A4i, B1i, B2
Anser anser Greylag Goose	B	1992	130	—	B	B1i
Anas crecca Teal	P	1992	25,000	—	C	A4i, B1i

Breeding waterbirds number more than 15,000 pairs in total.

Protection status

National None **International** High

20,000 ha of IBA covered by Ramsar Site (Northern Part of the Dniester Liman, 20,000 ha). 7,600 ha of IBA covered by Ramsar Site (Dniester-Turunchuk Crossrivers Area, 7,600 ha).

Conservation issues

Threats Agricultural intensification/expansion (B), Aquaculture/fisheries (C), Disturbance to birds (C), Unsustainable exploitation (C)

Tiligul estuary — A4i, A4iii, B1i, B2 — 092

Admin region Odessa, Nikolaev
Coordinates 46°49′N 31°00′E
Altitude 40–100 m **Area** 11,000 ha

Site description
The estuary/delta of the Tiligul river, on the Black Sea coast.

Habitats Scrub (scrub), Wetland (tidal river/enclosed tidal water; saltmarsh; coastal lagoon; standing brackish and salt water; river/stream; water-fringe vegetation)
Land-use Agriculture, Fisheries/aquaculture, Hunting, Nature conservation/research, Tourism/recreation

Birds

Species	Season	Year	Pop min	Pop max	Acc	Criteria
Egretta alba Great White Egret	P	1992	900	—	B	A4i, B1i
Plegadis falcinellus Glossy Ibis	B	1992	200	—	B	A4i, B1i, B2
Cygnus olor Mute Swan	W	1995	470	—	B	B1i
Anser albifrons White-fronted Goose	W	1992	8,000	—	B	B1i
Anser albifrons White-fronted Goose	P	1992	8,000	—	B	B1i
Tringa totanus Redshank	B	1992	400	—	B	B2
Larus minutus Little Gull	P	1992	7,000	—	B	A4i, B1i
Larus cachinnans Yellow-legged Gull	W	1992	5,000	—	B	A4i, B1i
Larus cachinnans Yellow-legged Gull	P	1992	5,000	—	B	A4i, B1i
Sterna hirundo Common Tern	B	1992	2,000	4,000	B	A4i, B1i

Protection status
National High **International** High
11,000 ha of IBA covered by Regional Landscape Park (Tiligul, 22,119 ha). 11,000 ha of IBA covered by Ramsar Site (Tyligulskyi Liman, 26,000 ha).

Conservation issues
Threats Agricultural intensification/expansion (B), Aquaculture/fisheries (C), Disturbance to birds (B), Industrialization/urbanization (B)

Cape Uret — B1i — 093

Admin region Crimea
Coordinates 45°23′N 32°30′E
Altitude 0–52 m **Area** 9,600 ha

Site description
The cape lies on the Black Sea coast near the town of Tshernomorsk.

Habitats Grassland (10%; steppe/dry calcareous grassland), Marine areas (80%; sea inlet/coastal features), Rocky areas (10%; sea cliff/rocky shore)
Land-use Unknown

Birds

Species	Season	Year	Pop min	Pop max	Acc	Criteria
Phalacrocorax aristotelis Shag	B	1992	350	500	A	B1i

Protection status
National None **International** None

Conservation issues
Threats Disturbance to birds (B)

Tarkhankut peninsula — B1i, B2 — 094

Admin region Crimea
Coordinates 45°25′N 32°38′E
Altitude 0–126 m **Area** 4,200 ha

Site description
Habitats Scrub (10%; scrub), Grassland (60%; steppe/dry calcareous grassland), Marine areas (10%; sea inlet/coastal features), Artificial landscape (20%; arable land; forestry plantation)
Land-use Agriculture (90%), Fisheries/aquaculture (3%)

Patches of natural steppe with numerous ravines, located within the Black Sea coastal zone and near the villages of Olenevka and Krasnoselskoe.

Birds

Species	Season	Year	Pop min	Pop max	Acc	Criteria
Phalacrocorax aristotelis Shag	N	1996	500	—	A	B1i
Phalacrocorax aristotelis Shag	B	1996	150	—	A	B1i
Falco naumanni Lesser Kestrel	B	1996	2	—	A	B2
Falco tinnunculus Kestrel	B	1996	120	—	B	B2
Falco cherrug Saker	B	1996	3	—	B	B2

Protection status
National None **International** None

Conservation issues
Threats Agricultural intensification/expansion (A), Disturbance to birds (A), Recreation/tourism (A)

Changes in agricultural practice pose a threat.

Lake near Magazinka village — B1i — 095

Admin region Crimea
Coordinates 45°55′N 34°04′E
Altitude 0–15 m **Area** 800 ha

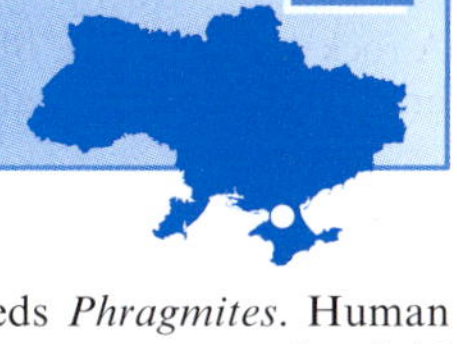

Site description
A natural steppe lake, dominated by reedbeds *Phragmites*. Human activities included cattle-grazing and water management for field irrigation.

Habitats Grassland (10%; steppe/dry calcareous grassland; humid grassland), Wetland (90%; standing brackish and salt water; water-fringe vegetation)
Land-use Agriculture (10%), Water management (90%)

Birds

Species	Season	Year	Pop min	Pop max	Acc	Criteria
Himantopus himantopus Black-winged Stilt	B	1997	300	350	B	B1i

Protection status
National None **International** None

Conservation issues
Threats Abandonment/reduction of land management (B), Aquaculture/fisheries (C), Natural events (C)

Storms ('Natural events') pose a threat.

Belogorie — B2 — 096

Admin region Crimea
Coordinates 45°09′N 34°06′E
Altitude 120–300 m **Area** 32,000 ha

Site description
A hilly landscape in the northern part of the Crimean mountains. Habitats are dominated by steppe and pseudo-steppe, with a low coverage of forestry plantations.

Habitats Forest and woodland (2%; broadleaved deciduous forest), Scrub (1%; scrub), Grassland (65%; steppe/dry calcareous grassland; dry siliceous grassland; humid grassland), Wetland (river/stream), Artificial landscape (30%; arable land; forestry plantation)
Land-use Agriculture (100%)

Birds

Species	Season	Year	Pop min	Pop max	Acc	Criteria
Falco cherrug Saker	B	1996	1	—	A	B2

Protection status
National None **International** None

Conservation issues

Threats Disturbance to birds (B)

Cape Martyan — B1i — 097

Admin region Crimea
Coordinates 44°30'N 34°09'E
Altitude 0–149 m **Area** 240 ha

Site description

The site lies on the Black Sea coast, near the town of Yalta, and includes Nikitskiy Botanic Garden.

Habitats Forest and woodland (40%; native coniferous forest; mixed forest), Marine areas (50%; sea inlet/coastal features), Rocky areas (5%; sea cliff/rocky shore), Artificial landscape (5%; urban parks/gardens; other urban/industrial areas)
Land-use Nature conservation/research, Tourism/recreation, Urban/industrial/transport

Birds

Species	Season	Year	Pop min	Pop max	Acc	Criteria
Phalacrocorax carbo Cormorant	P	1994	1,500	—	B	B1i

Protection status

National High **International** None
240 ha of IBA covered by Nature Reserve (Cape Martyan, 240 ha).

Conservation issues

Threats Disturbance to birds (A), Industrialization/urbanization (A), Recreation/tourism (A)

Crimean Nature Reserve — A1, B2, B3 — 098

Admin region Crimea
Coordinates 44°49'N 34°30'E
Altitude 282–1,253 m **Area** 34,563 ha

Site description

Mixed mountain forest with rocky areas.

Habitats Forest and woodland (95%; broadleaved deciduous forest; native coniferous forest; mixed forest), Rocky areas (5%)
Land-use Nature conservation/research (100%)

Birds

Species	Season	Year	Pop min	Pop max	Acc	Criteria
Aegypius monachus Cinereous Vulture	R	1982	5	8	—	A1, B2
Strix aluco Tawny Owl	R	1967	221	—	B	B3

Protection status

National High **International** None
34,563 ha of IBA covered by Nature Reserve (Crimean, 44,175 ha).

Conservation issues

Threats Disturbance to birds (C)

Karadag Nature Reserve — B1i — 099

Admin region Crimea
Coordinates 45°00'N 35°15'E
Altitude 0–576 m **Area** 2,855 ha

Site description

Habitats Forest and woodland (45%; mixed forest), Grassland (30%; mesophile grassland), Marine areas (sea inlet/coastal features), Rocky areas (sea cliff/rocky shore)
Land-use Nature conservation/research (100%)

Deciduous mountain forest in the coastal zone of the Black Sea.

Birds

Species	Season	Year	Pop min	Pop max	Acc	Criteria
[1] ***Phalacrocorax aristotelis*** Shag	B	1995	72	—	B	B1i

1. 94 pairs in 1988.

Protection status

National High **International** None
2,855 ha of IBA covered by Nature Reserve (Karadag, 2,855 ha).

Conservation issues

Threats Disturbance to birds (A), Recreation/tourism (A)

Aktashskoe lake — A4i, B1i, B2, B3 — 100

Admin region Crimea
Coordinates 45°17'N 35°52'E
Altitude 0–106 m **Area** 4,800 ha

Site description

This site, located near Lenino town, includes areas of steppe on the Kazantip peninsula together with wetland habitat along the coast of the Sea of Azov.

Habitats Grassland (17%; steppe/dry calcareous grassland), Wetland (74%; sand-dunes/sand beach; standing fresh water; standing brackish and salt water; water-fringe vegetation), Marine areas (6%; sea inlet/coastal features), Rocky areas (sea cliff/rocky shore), Artificial landscape (3%; arable land)
Land-use Agriculture, Fisheries/aquaculture

Birds

Species	Season	Year	Pop min	Pop max	Acc	Criteria
Tadorna ferruginea Ruddy Shelduck	B	1996	6	8	A	B2
Tadorna ferruginea Ruddy Shelduck	N	1994	160	160	B	B1i
Recurvirostra avosetta Avocet	B	1994	100	130	A	B3
Glareola pratincola Collared Pratincole	B	1994	60	70	A	A4i, B1i, B2
Larus melanocephalus Mediterranean Gull	B	1994	2,700	2,700	B	A4i, B1i, B3
Larus genei Slender-billed Gull	B	1994	3,200	3,200	A	A4i, B1i
Larus cachinnans Yellow-legged Gull	B	1994	2,200	—	A	A4i, B1i
Gelochelidon nilotica Gull-billed Tern	B	1994	360	360	A	A4i, B1i, B2
Gelochelidon nilotica Gull-billed Tern	P	1994	500	600	B	A4i, B1i

The total number of breeding waterbirds is 2,300–2,500 pairs, while 10,000–12,000 individuals stage here during migration (with maximum in October–November).

Protection status

National None **International** None

Conservation issues

Threats Agricultural intensification/expansion (B), Aquaculture/fisheries (C), Disturbance to birds (C), Unsustainable exploitation (B)

Uzunlarskoe lake — A1, B2 — 101

Admin region Crimea
Coordinates 45°20'N 36°15'E
Altitude 9–184 m **Area** 9,600 ha

Site description

A site on the Kerch peninsula of the Crimea, with salt-water and brackish lakes and areas of natural steppe on the hills. Uzunlarsky is the largest lake. There are rural plots of steppe and pseudo-steppe near the villages of Vulkanovka, Marievka and Marphovka. The area is a military zone.

Habitats Scrub (5%; scrub), Grassland (12%; steppe/dry calcareous grassland; humid grassland), Wetland (13%; standing brackish and salt water), Marine areas (sea inlet/coastal features), Rocky areas (sea cliff/rocky shore; inland sand-dunes), Artificial landscape (60%; arable land)
Land-use Agriculture (80%), Military, Nature conservation/research

Birds

Species	Season	Year	Pop min	Pop max	Acc	Criteria
[1] *Tadorna ferruginea* Ruddy Shelduck	B	1996	13	—	A	B2
Falco naumanni Lesser Kestrel	B	1996	4	—	B	B2
Falco cherrug Saker	B	1996	3	—	B	B2
[2] *Otis tarda* Great Bustard	R	1996	—	70	—	A1
Otis tarda Great Bustard	W	1994	500	—	B	A1

1. Breeding and non-breeding birds total 40 individuals.
2. Individuals (breeding and non-breeding).

Protection status

National None **International** None

Conservation issues

Threats Agricultural intensification/expansion (B), Disturbance to birds (B)

Military zonation provides *de facto* protection to some areas of steppe.

Bagerovo A1, B2 102

Admin region Crimea
Coordinates 45°30′N 36°20′E
Altitude 54–156 m **Area** 20,510 ha

Site description

An area of natural steppe (formerly included in a military zone) and agriculture (including patches of pseudo-steppe), on the Kerch peninsula near the town of Bagerovo.

Habitats Scrub (2%), Grassland (56%; steppe/dry calcareous grassland), Wetland (standing brackish and salt water; river/stream; fen/transition mire/spring), Marine areas (sea inlet/coastal features), Artificial landscape (33%; arable land)
Land-use Agriculture (69%), Military, Not utilized (7%)

Birds

Species	Season	Year	Pop min	Pop max	Acc	Criteria
Tadorna ferruginea Ruddy Shelduck	R	1996	8	12	B	B2
Falco cherrug Saker	B	1996	2	—	B	B2
[1] *Tetrax tetrax* Little Bustard	B	1996	5	7	B	A1
Otis tarda Great Bustard	B	1996	20	—	B	A1, B2
Otis tarda Great Bustard	N	1995	110	120	B	A1
Asio flammeus Short-eared Owl	B	1996	3	—	—	B2

1. 80–90 individuals in 1991.

Protection status

National None **International** None

Conservation issues

Threats Agricultural intensification/expansion (A)

Changes in agricultural practice pose a threat.

REFERENCES

AFANASIEF, V. T. (1992) Bird fauna of the river Desna valley. In *Vestnik zoologii.* Kiev.

AFANASIEF, V. T. (1995) Birds of Sumy region. Unpublished, available at UTOP office.

ANDRIENKO, T. L., POPOVICH, S. YU. AND SHELYAH-SOSONKO, YU. P. (1986) *Polesskiy State Zapovednik.* Kiev.

ANDRYUSCHENKO, YU. A., KINDA, V. V. AND STADNICHENKO, I. S. (1996) Proposal for the creation of a National Park on the northern part of Kerchenskiy peninsula. P. 163 in A. Mikityuk, ed. *'Important Bird Areas in Ukraine—IBA Programme' conference material, 7–9 April 1995, Nezhin.* Kiev: UTOP.

ANON. (1988) *Flora and fauna of the Zapovedniks of the USSR.* Moscow.

ARDAMATSKAYA, T. B. (1996) Changing of status and number of rare birds Dzharilgach island under human transformation of landscape. P. 5 in A. Mikityuk, ed. *'Important Bird Areas in Ukraine—IBA Programme' conference material, 7–9 April 1995, Nezhin.* Kiev: UTOP.

BESKARAVAINIY, M. M. (1995) Birds of 'Mys Martyan' zapovednik. In *Nature reserves in Ukraine*, vol. 1. Chernivtsi, Ukraine: Sklaviya.

BUCHKO, V. V. (1996) Burshtin reservoir and surroundings—a unique place of concentration of rare species of bird. Pp. 131–132 in A. Mikityuk, ed. *'Important Bird Areas in Ukraine—IBA Programme' conference material, 7–9 April 1995, Nezhin.* Kiev: UTOP.

CHERNICHKO, I. I. AND SIOCHIN, V. D., EDS. (1993) *Inventory and cadastral description of wetlands of the Azov—Black Sea coast of Ukraine.* Bulletin no. 1. Melitopol, Ukraine: Branta. 93 pp.

GAVRILENKO, V. S. (1997) Necessity for creating an ecological path for migrating birds in northern Black Sea coast. In *Ecosystem of wild nature.* Odessa, Ukraine: Nature Heritage Fund.

GAVRIS, G. G. (1995) Go West: Ausbreitung von Vogelarten in der Ukraine. *Der Falke* 42(2): 36–41.

GODOVANETS, B. I., LUGOVOY, A. E., GUZIY, A. I., SKILSKY, I. V. AND BUNDZYAK, P. V. (1995) Fauna, population and peculiarities of the breeding biology by birds of branch 'Narcissus Valley' of the Carpathian Nature Reserve and its environs. In *Nature reserves in Ukraine*, vol. 1. Chernivtsi, Ukraine: Sklaviya.

GORBAN, I. M. (1995) Dynamics of the Greylag Goose (*Anser anser*) population in the west of Ukraine. In *New investigations on Palearctic geese.* Zaporozhye, Ukraine: Dneprovskiy Metalurg.

GORLOV, P. I. (1996) Samara forest: problems of conservation of the most southern site of breeding *Grus grus*. P. 113 in A. Mikityuk, ed. *'Important Bird Areas in Ukraine—IBA Programme' conference material, 7–9 April 1995, Nezhin.* Kiev: UTOP.

GRIMMETT, R. F. A. AND JONES, T. A. (1989) *Important Bird Areas in Europe.* Cambridge, UK: International Council for Bird Preservation (Techn. Publ .9).

HEATH, M. F. AND BORGGREVE, C. (2000) *BirdLife International/EBCC European Bird Database 1998.* Cambridge, UK: BirdLife International.

KIJKO, A. O. (1996) Burshtin reservoir—wintering place of waterbirds. Pp. 145–148 in A. Mikityuk, ed. *'Important Bird Areas in Ukraine—IBA Programme' conference material, 7–9 April 1995, Nezhin.* Kiev: UTOP.

KLESTOV, N. L. (1995) The Sula bay of the Kremenchug reservoir as Greylag Goose (*Anser anser*) habitat. In *New investigations on Palearctic geese.* Zaporozhye, Ukraine: Dneprovskiy Metalurg.

KLESTOV, N. L., GAVRIS, G. G. AND ANDRIEVSKAYA, E. L. (1995) *Sula bay of Kremenchug water reservoir.* Kiev: UTOP (IBA technical publication).

KOSTIN, YU. V. (1983) *Birds of Crimea.* Moscow: Nauka. 239 pp.

KOSTYUSHIN, V. AND MIKHALEVICH, I. (1996) Results of midwinter bird counts. P. 302 in A. Mikityuk, ed. *'Important Bird Areas in Ukraine—IBA Programme' conference material, 7–9 April 1995, Nezhin.* Kiev: UTOP.

LOPAREV, S. V., MELNICHUK, V. A. AND RASUMOVSKY, B. I. (1995) The Greylag Goose (*Anser anser*) on fish-breeding ponds of the flood-lands of the river Zgar. In *New investigations on Palearctic geese.* Zaporozhye, Ukraine: Dneprovskiy Metalurg.

POLUDA, A. M. AND GAVRIS, G. G. (1996) Bondar's marsh is a place of concentration of rare species of birds. Pp. 173–176 in A. Mikityuk, ed. *'Important Bird Areas in Ukraine—IBA Programme' conference material, 7–9 April 1995, Nezhin.* Kiev: UTOP.

RUDENKO, A. (1996) Present status of gulls and terns nesting in the Black Sea Biosphere Reserve. *Colonial Waterbirds* 19 (Spec. Publ. 1): 41–45.

RUDENKO, A., YAREMCHENKO, O. AND RYBACHUK, K. (1996) Ornithocomplexes of the Black Sea Biosphere Reserve in 1994. In A. Mikityuk, ed. *'Important Bird Areas in Ukraine—IBA Programme' conference material, 7–9 April 1995, Nezhin.* Kiev: UTOP.

RUSEV, I. T. AND GERZHIK, I. P, EDS. (1996) Results of a midwinter bird count on the north-west Black Sea coast (11–26 January 1995). Pp. 290–296 in A. Mikityuk, ed. *'Important Bird Areas in Ukraine—IBA Programme' conference material, 7–9 April 1995, Nezhin.* Kiev: UTOP.

SAVCHUK, M. (1996) Wetlands of Berdyansk peninsula. In A. Mikityuk, ed. *'Important Bird Areas in Ukraine—IBA Programme' conference material, 7–9 April 1995, Nezhin.* Kiev: UTOP.

SCHOGOLEV, I. (1996) Fluctuations and trends in breeding populations of colonial waterbirds in the Dnestr Delta, Ukraine, Black Sea. *Colonial Waterbirds* 19 (Spec. Publ. 1): 91–97.

TISHECHKIN, A., FLADE, M., JEBRAM, J. AND LEINWEBER, T. (1996) Ornithological observations and population estimates. Upper Pripyat Expedition, 17 May to 25 May 1996. Unpublished report.

TUCKER, G. M. AND HEATH, M. F. (1994) *Birds in Europe: their conservation status.* BirdLife International: Cambridge, UK (BirdLife Conservation Series no. 3).

VAN DER WINDEN, J., VAN DER HAVE, T. M., STIKVOORT, E. AND ZWINSELMAN, B. (1993) Numbers of waterbirds in the Sivash, Ukraine, spring 1992. Pp. 7–32 in T. M. van der Have, S. van de Sant, Y. Verkuil and J. van der Winden, eds. *Waterbirds in the Sivash, Ukraine, spring 1992.* Zeist, the Netherlands: WIWO (WIWO-report 36).

VOVCHENKO, B. E. AND CHEREDNICHENKO, I. A. (1996) Number of *Cygnus olor* on the Utlyuk estuary in 1994. Pp. 16–17 in A. Mikityuk, ed. *'Important Bird Areas in Ukraine—IBA Programme' conference material, 7–9 April 1995, Nezhin.* Kiev: UTOP.

ZALEVSKIY, V. D. AND MOLODAN, G. N. (1996) Important bird areas in the Donetsk region. P. 176 in A. Mikityuk, ed. *'Important Bird Areas in Ukraine—IBA Programme' conference material, 7–9 April 1995, Nezhin.* Kiev: UTOP.

ZHMUD, M. E. (1995) Geese of the Ukrainian part of the Danube delta. In *New investigations on Palearctic geese.* Zaporozhye, Ukraine: Dneprovskiy Metalurg.

ZUBKO, V. (1996) Winter bird count data. In A. Mikityuk, ed. *'Important Bird Areas in Ukraine—IBA Programme' conference material, 7–9 April 1995, Nezhin.* Kiev: UTOP.

FEDERAL REPUBLIC OF YUGOSLAVIA

Slobodan Puzović and Bratislav Grubač

Long-legged Buzzard
Buteo rufinus.
(PHOTO: BRATISLAV GRUBAČ)

GENERAL INTRODUCTION

Yugoslavia lies mostly in the central Balkan peninsula, while its northern part spreads over the southern belt of the Pannonian plain. Yugoslavia covers an area of 102,173 km², of which Serbia comprises 88,361 km² and Montenegro comprises 13,812 km². The main geographic units in Serbia are the Pannonian region in the north, the Peripannonian region in the centre, and the Mountain-and-basin region in the south, while in Montenegro they are the Coastal and the Mountain-and-basin regions. More than 60% of Yugoslavia is made up of hilly to mountainous landscapes, above 500 m elevation. There are three main drainage basins in the country: the Black Sea basin, which comprises 88,919 km² (87% of the total territory of the country), and the Adriatic and Aegean basins. Montenegro has a c.200 km long coastline along the Adriatic Sea. Annual precipitation in Serbia is 550–1,000 mm, and in Montenegro is 700–5,000 mm.

This inventory identifies 40 Important Bird Areas (IBAs) in Yugoslavia (Table 1, Map 1). They have a total area of 8,617 km², equivalent to c.8% of the country's surface area. Within Serbia there are 35 IBAs with an area of 7,670 km², while in Montenegro there are five IBAs with an area of 947 km². Totals of IBAs by administrative region are: Vojvodina (16); Kosovo and Metohija (two); central Serbia (17); southern Montenegro (three); northern Montenegro (two). In general, Serbia is well-surveyed ornithologically and the identification of IBAs is comprehensive,

Table 1. Summary of Important Bird Areas in the Federal Republic of Yugoslavia. — 40 IBAs covering 8,617 km²

IBA code	National code	1989 code	International name	National name	Administrative region	Area (ha)	Criteria (see p. 11)
001	SER001	YU029	Gornje Podunavlje	Gornje Podunavlje	Serbia	30,000	A1, A4i, A4iii, B1i, B2, B3
002	SER002	YU032, YU033	Subotica lakes and sandy terrain	Suboticka jezera i pustare	Serbia	20,000	A1, B2, B3
003	SER003	—	Bečej fish-pond	Bečejski ribnjak	Serbia	4,000	A4i, B1i, B2, B3
004	SER004	—	Jegrička	Jegrička	Serbia	5,400	A1, A4i, B1i, B2, B3
005	SER005	YU037	Koviljski rit	Koviljski rit	Serbia	5,500	B2
006	SER006	YU031	Fruška Gora	Fruška Gora	Serbia	42,000	A1, B2
007	SER007	YU030	Bosutska forest	Bosutske sume	Serbia	17,500	A1, B1i, B2, B3
008	SER008	—	Zasavica	Zasavica	Serbia	5,200	B2, B3
009	SER009	YU041	Obedska bara	Obedska bara	Serbia	23,000	B1i, B2, B3
010	SER010	—	Danube loess bluffs	Dunavski lesni odsek	Serbia	2,500	A1, A4i, B1i, B2
011	SER011	YU034	Jazovo–Mokrin	Jazovo–Mokrin	Serbia	8,000	A1, B2
012	SER012	YU035	Slano kopovo	Slano kopovo	Serbia	2,660	B1iv, B2, B3
013	SER013	YU036	Carska bara	Carska bara	Serbia	9,300	A1, A4i, A4iii, B1i, B2, B3
014	SER014	—	Uzdin fish-pond	Uzdinski ribnjak	Serbia	5,500	A1, A4i, A4iii, B1i, B2, B3
015	SER015	YU040	Vršacki breg	Vršacki breg	Serbia	10,500	B2, B3
016	SER016	YU038	Deliblatska peščara	Deliblatska peščara	Serbia	38,000	B2, B3

Table 1 ... continued. Summary of Important Bird Areas in the Federal Republic of Yugoslavia. 40 IBAs covering 8,617 km²

IBA code	National code	1989 code	International name	National name	Administrative region	Area (ha)	Criteria (see p. 11)
017	SER017	—	Cer mountain	Cer planina	Serbia	10,200	B2, B3
018	SER018	—	Valjevske mountains	Valjevske planine	Serbia	11,000	B2
019	SER019	—	Tara mountain	Tara planina	Serbia	36,000	B2
020	SER020	—	Ovcar-Kablar gorge	Ovcarsko-Kablarska klisura	Serbia	5,400	A1
021	SER021	—	Uvac-Mileševka	Uvac-Mileševka	Serbia	30,000	A1, B2
022	SER022	—	Kopaonik	Kopaonik	Serbia	45,000	B2
023	SER023	—	Prokletije	Prokletije	Serbia	100,000	A3, B2, B3
024	SER024	YU048	Šara mountain	Šar planina	Serbia	80,000	A1, A3, B2
025	SER025	—	Pcinja	Pcinja	Serbia	16,000	B2, B3
026	SER026	—	Vlasina	Vlasina	Serbia	22,000	B2
027	SER027	—	Jerma	Jerma	Serbia	10,000	B3
028	SER028	—	Suva mountain	Suva planina	Serbia	20,000	B2
029	SER029	—	Stara mountain-Vidlic	Stara planina-Vidlic	Serbia	44,000	B2, B3
030	SER030	—	Sicevo gorge	Sicevacka klisura	Serbia	9,800	B2
031	SER032	YU047	Zlot gorge-Dubašnica	Zlotska klisura-Dubašnica	Serbia	10,000	B2
032	SER033	YU046	Resava gorge	Resavska klisura	Serbia	9,500	B2
033	SER034	YU039	Dubovac-Ram	Dubovac-Ram	Serbia	12,000	A1, A4i, A4iii, B1i, B2, B3
034	SER035	—	Đerdap gorge	Đerdapska klisura	Serbia	65,000	A1, A4i, A4iii, B1i, B2
035	SER036	—	Mala vrbica	Mala vrbica	Serbia	2,000	B2
036	YU042	—	Durmitor	Durmitor	Montenegro	33,000	B2, B3
037	—	—	Biogradska woods	Biogradska gora	Montenegro	20,000	B2, B3
038	YU043	—	Lake Skadar	Skadarsko jezero	Montenegro	40,000	A1, A3, A4i, A4iii, B1i, B2
039	YU044	—	Lake Sasko	Sasko jezero	Montenegro	350	B1i, B2, B3
040	YU045	—	Ulcinj saltpans	Ulcinjska solane	Montenegro	1,350	A1, A4i, B1i, B2

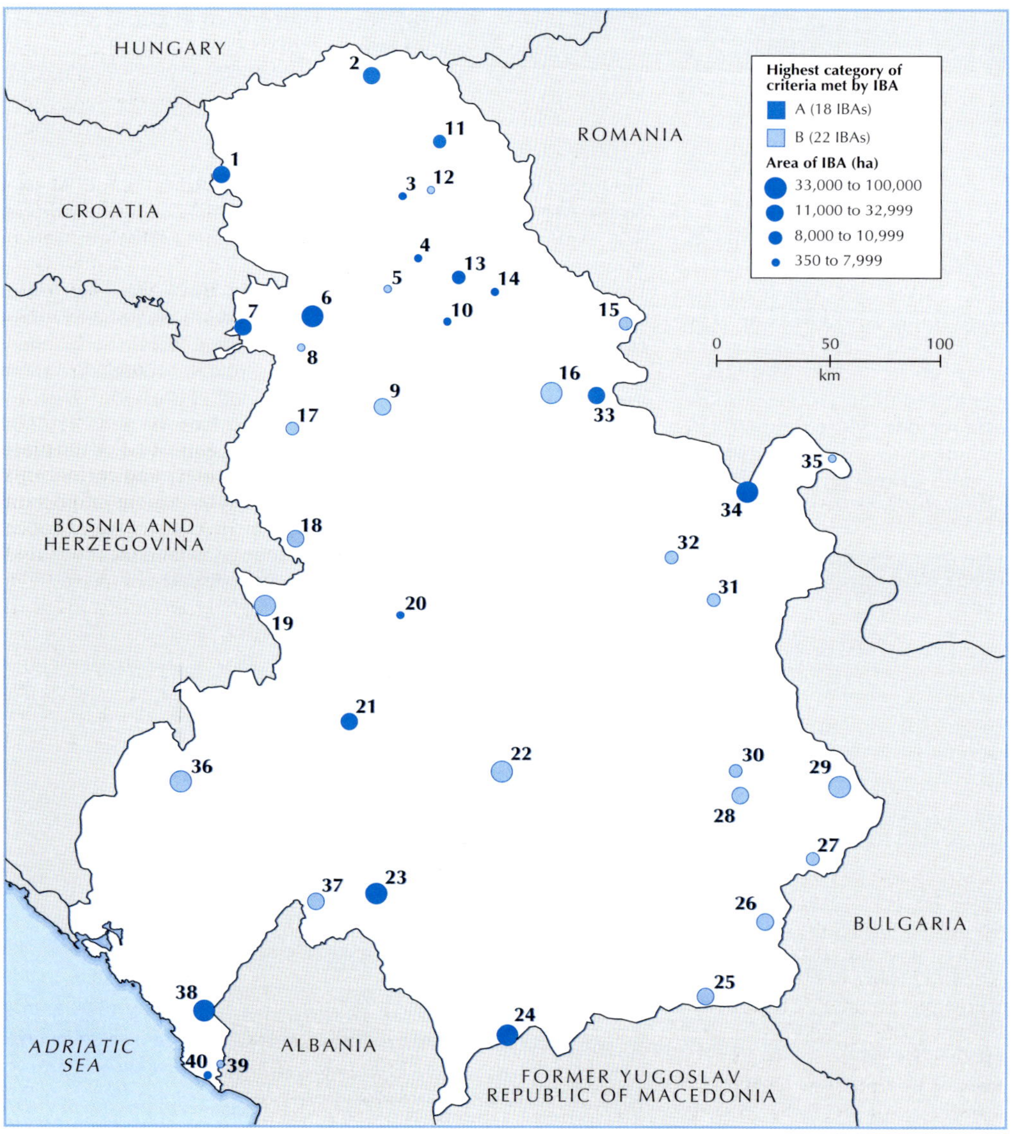

Map 1. Location, area and criteria category of Important Bird Areas in the Federal Republic of Yugoslavia.

Table 2. Important Bird Areas in the Federal Republic of Yugoslavia that are important for species of global conservation concern (meeting criterion A1).

Species	IBA code
Phalacrocorax pygmeus Pygmy Cormorant	010, 020, 033, 034, 038, 040
Pelecanus crispus Dalmatian Pelican	038
Aythya nyroca Ferruginous Duck	001, 002, 004, 013, 014, 033, 038
Haliaeetus albicilla White-tailed Eagle	001, 007
Aquila heliaca Imperial Eagle	006
Crex crex Corncrake	014, 021, 024
Otis tarda Great Bustard	011

Table 3. Important Bird Areas in the Federal Republic of Yugoslavia that support important numbers of one or more congregatory species (i.e. meeting criteria A4 and/or B1). IBAs meeting both criteria A4 and B1 for the species are shown in **bold**. IBAs meeting only criterion B1 for the species concerned, and not A4, are shown in normal type. For key to 'Season', see p. 7.

Species	Season	IBA code
Tachybaptus ruficollis Little Grebe	B	**013**
Podiceps cristatus Great Crested Grebe	W	**038**
Podiceps grisegena Red-necked Grebe	B	**004**
Podiceps nigricollis Black-necked Grebe	W	**038**
Phalacrocorax pygmeus Pygmy Cormorant	R	**038**
	B	**033**
	W	**010**, **014**, **033**, **034**, **038**
	P	**038**
Nycticorax nycticorax Night Heron	B	**003**
Ardeola ralloides Squacco Heron	B	**038**
Egretta alba Great White Egret	R	**003**, **013**, **014**
	W	**038**
Ciconia nigra Black Stork	P	**001**, 007, 009, **014**, **033**
Platalea leucorodia Spoonbill	B	039
	P	**001**, **013**
	N	**014**
Anas crecca Teal	W	**038**
Aythya ferina Pochard	W	**038**
	P	033
Aythya fuligula Tufted Duck	W	**038**
Bucephala clangula Goldeneye	P	033
Fulica atra Coot	W	**038**
Glareola pratincola Collared Pratincole	B	**040**
Limosa limosa Black-tailed Godwit	P	014
Larus ridibundus Black-headed Gull	W	**013**

while in Montenegro there are still areas which potentially fulfil the IBA criteria but which have not been confirmed as IBAs due to a scarcity of recent data.

The previous pan-European IBA inventory in 1989 (Grimmett and Jones 1989) identified 20 IBAs in Yugoslavia, with an area of 1,803 km², equivalent to c.2% of the country's surface area. Sixteen of the IBAs were in Serbia and four in Montenegro. All 20 of these sites have been confirmed as IBAs under the revised IBA criteria, and feature in this inventory (see Table 1), but mostly with new boundaries and with larger surface-areas.

ORNITHOLOGICAL IMPORTANCE

The total number of regularly occurring bird species in Yugoslavia is 333, of which 260 occur regularly during the breeding season (Vasić 1995). Among the latter group there are 140 species of European conservation concern (SPECs: Tucker and Heath 1994), which include five species of global conservation concern (*Aythya nyroca*, *Aquila heliaca*, *Falco naumanni*, *Otis tarda* and *Crex crex*) as well as a further 82 species that have an unfavourable conservation status in Europe (Tucker and Heath 1994). A total of 16 sites support internationally important numbers of globally threatened or near-threatened species, thus qualifying as IBAs under the A1 criterion (Table 2). Thirteen sites in Yugoslavia support important congregations of 18 species of migratory waterbird at some point during their annual life-cycle, with 10 of the 13 sites being globally important as they meet the A4i criterion (Table 3). Of special note for wintering waterbirds are Lake Skadar (site 038) in Montenegro where c.200,000 winter regularly (Vasić *et al.* 1992) and the River Danube near Dubovac–Ram in Serbia (site 033) where 80,000 winter on average (Paunović *et al.* 1994).

Three sites in Yugoslavia meet the A3 criterion (Table 1), being important for bird-species assemblages of particular biomes. One of the three sites holds six of the 21 species whose breeding distributions (in Europe) are largely or wholly confined to the Mediterranean biome, while the other two sites hold four of the 10 such species that are similarly restricted to the Eurasian high-montane biome. All but one of the IBAs (site 020) are considered to support important breeding numbers of SPECs, thus meeting B2 and/or B3 criteria, and 19 sites qualify as IBAs solely under these criteria (Table 1).

For many bird species that are rare and threatened in Yugoslavia, the majority of the national breeding population currently nests within the IBA network, notably *Phalacrocorax pygmeus* (100%), *Pelecanus crispus* (100%), *Ciconia nigra* (80%), *Plegadis falcinellus* (100%), *Platalea leucorodia* (100%), *Aythya nyroca* (50%), *Neophron percnopterus* (100%), *Gyps fulvus* (100%), *Haliaeetus albicilla* (75%), *Aquila heliaca* (80%), *Aquila pomarina* (50%), *Alectoris graeca* (70%), *Crex crex* (60%), *Otis tarda* (100%), *Limosa limosa* (90%), *Prunella collaris* (100%), *Eremophila alpestris balcanica* (80%), *Tichodroma muraria* (50%), *Montifringilla nivalis* (90%) and *Emberiza melanocephala* (60%).

HABITATS

Yugoslavia is an ecologically diverse country, and the biogeographical origins of its fauna and flora are unusually heterogeneous in a European context. Six main habitat-zones can be distinguished: high-montane rocky areas and pastures; coniferous forests; southern European and sub-Mediterranean forests (mainly deciduous); upland Mediterranean vegetation; steppe; and wooded steppe. In the north of Serbia, the Pannonian plain is a lowland landscape with large rivers (Danube, Sava, and Tisa), while to the south hilly/mountainous landscapes are intersected by river valleys (the mountain systems being the Dinaric, the Rhodope, the Šara–Pindus and the Carpathian). In Vojvodina there is a large area of sand-dunes—Deliblatska peščara—which is rare and unusual in inland Europe. In the south of Montenegro, Mediterranean and sub-Mediterranean habitats border the Adriatic coast, and one of the largest freshwater lakes in Europe, Lake Skadar, is also situated here.

At least 23 IBAs (58% of the total) have more than 5% of their extent covered by some form of wetland (Figure 1), and this extent is considered significant at 19 of the 23 sites, among them eight areas of fish-ponds. These 19 sites comprise almost all of the most important breeding colonies of herons (Ardeidae) and cormorants (Phalacrocoracidae) in Yugoslavia, as well as regions important for staging and wintering waterbirds. Four IBAs cover mainly open habitats with steppes, saltmarshes, saltpans, lowland meadows and pastures, which together support nationally threatened species such as *Otis tarda*, *Aquila heliaca*, *Falco cherrug*, *Charadrius alexandrinus*,

Figure 1. Habitats at Important Bird Areas in the Federal Republic of Yugoslavia (see Appendix 3 for definitions of habitats).

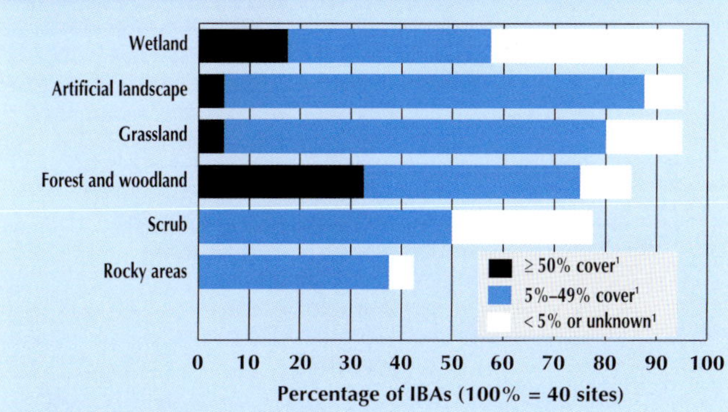

1. percentage of area of individual IBA covered by habitat

Figure 2. Land-uses at Important Bird Areas in the Federal Republic of Yugoslavia (see Appendix 3 for definitions of land-uses).

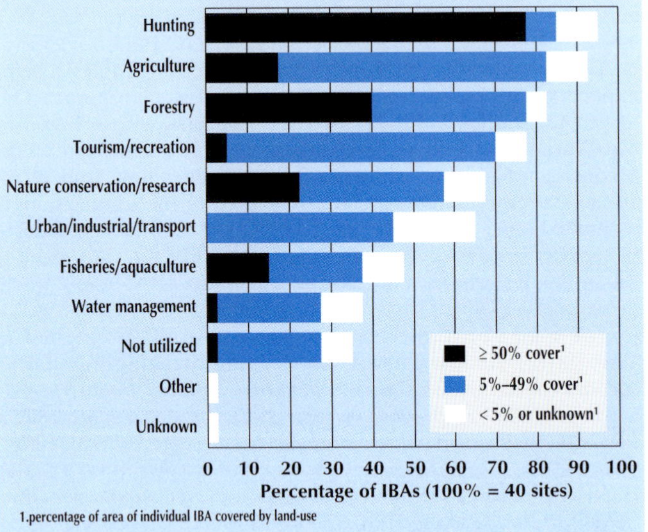

Recurvirostra avosetta, *Limosa limosa*, *Himantopus himantopus*, *Burhinus oedicnemus*, *Glareola pratincola*, *Sterna albifrons*, *Haematopus ostralegus* and *Tringa totanus*.

Forest covers 27% of Serbia (25,625 km^2), and in Montenegro 37% (5,100 km^2), with the proportion of conifers being only 10% in Serbia, and up to 30% in Montenegro. Seventeen IBAs (44%) contain large tracts of forest—deciduous forest in the lowlands and coniferous forest in high mountain regions—and forest is the predominant habitat within 13 of these 17 IBAs (Figure 1). Within seven IBAs there are protected gorges, mainly in Serbia, which are valuable refuges for a large number of raptors and for species typical of karst and rocky areas with scrub and low vegetation. Highland pastures and bare, alpine terrain, above the height of 1,700 m, are the major habitat within seven IBAs. All alpine terrain above 1,800 m in Serbia is included within IBAs, which thus hold most or all of the national breeding populations of such species as *Prunella collaris*, *Eremophila alpestris balcanica*, *Montifringilla nivalis*, *Pyrrhocorax graculus* and *Pyrrhocorax pyrrhocorax*.

Artificial landscape is present in varying quantities at most IBAs, but it is not usually predominant, and only 5% of IBAs have more than 50% cover of this habitat-type (Figure 1). Mostly it comprises plantations of poplar *Populus* or non-native conifers, arable land on plains, and perennial crops (orchards and groves) in hilly and mountainous regions.

IMPACTS ON IBAS—LAND-USE AND THREATS

Most IBAs are state-owned. Agriculture, hunting, forestry and tourism/recreation, being present at 75–90% of all IBAs, are the most widespread land-uses in Yugoslavian IBAs (Figure 2). Slightly less frequent uses are nature conservation/research and urban/industrial/transport uses, which are present at 55–70% of all IBAs. Agriculture is the dominant land-use in Vojvodina, while hunting and forestry are equally present in all regions where IBAs are located. Within agriculture, land is most commonly used for arable cultivation. Forest management is, in general, intensive and thus poses problems at some IBAs (Figure 3), being limited or forbidden only within protected areas such as National Parks and Nature Reserves. The spread of huge poplar plantations in the riparian zones of Vojvodina, currently covering about 500 km^2, is a great threat to riparian IBAs in this region (Figure 3). In general, within the IBAs, the area covered by forest is increasing gradually, but the average age of stands and the tree species-richness are both decreasing. Hunting is particularly widespread in Vojvodina, where there are numerous areas for hunting big game within IBAs. Hunting is a threat locally (Figure 3), especially when large birds of prey are hunted as trophies, but there are also cases where hunting clubs are caretakers of nature reserves and the rare birds within them (e.g. a reserve for *Otis tarda* in Banat).

Fisheries/aquaculture are present at 19 IBAs (Figure 2), including eight fish-pond areas where the persecution of waterbirds and burning of vegetation are very common. Disturbance of birds occurs at almost all IBAs (Figure 3). Disturbance has increased recently due to the economic crisis in Yugoslavia, which has led to increased collection of wood, medicinal herbs, mushrooms and forest fruits, as well as increased hunting, fishing and nomadic grazing of cattle. Drainage still poses a threat at some wetland IBAs in Vojvodina, and burning of vegetation is a problem in the mountains of Prokletije and Šara (Figure 3).

PROTECTION STATUS

At present 5% of Serbia is protected as National Parks and Nature Reserves, while in Montenegro the total protected area covers 10% of the region. A total of 29 IBAs overlap to some extent with areas protected by national law (Figure 4, Table 4). Eight of the nine

Figure 3. Threats at Important Bird Areas in the Federal Republic of Yugoslavia (see Appendix 3 for definitions of threat types and impact categories).

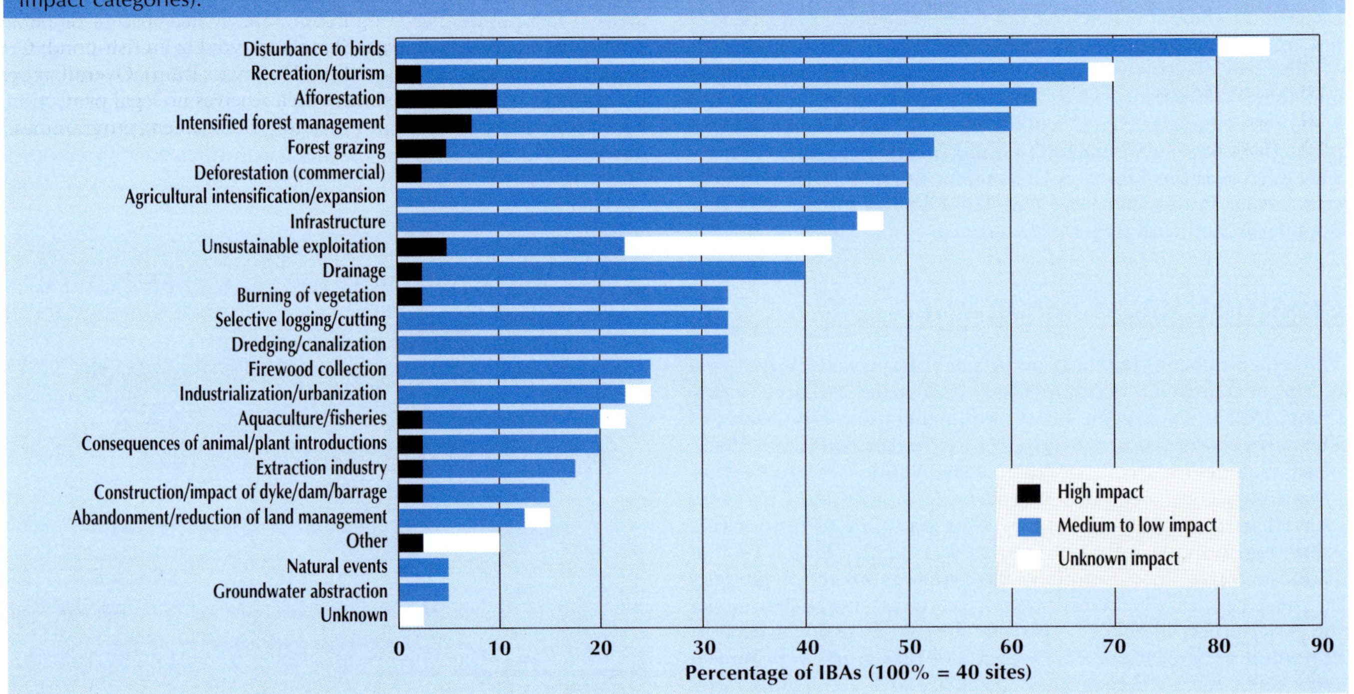

Figure 4. The national protection status of Important Bird Areas in the Federal Republic of Yugoslavia.

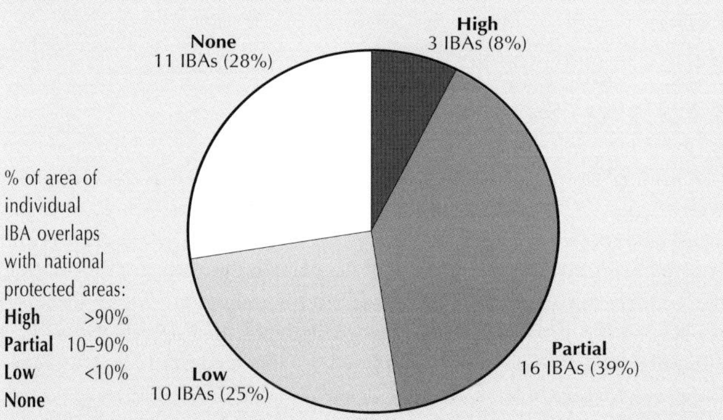

Total area of overlap between IBA network in the Federal Republic of Yugoslavia and national protected-area system (see Table 4 for categories) = 3,012–3,025 km² (35% of total IBA area).

Figure 5. The international protection status of Important Bird Areas in the Federal Republic of Yugoslavia.

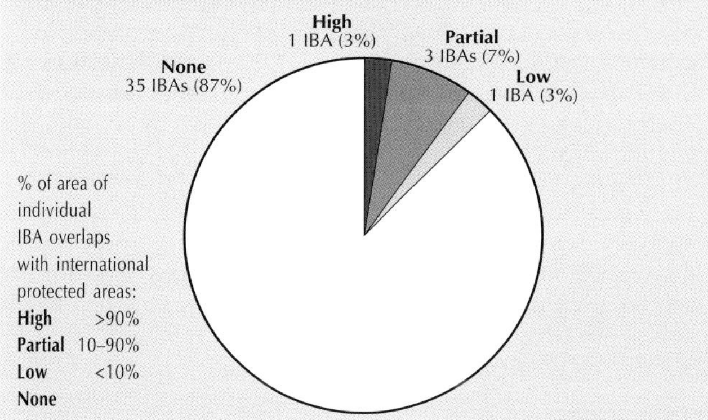

Total area of overlap between IBA network in the Federal Republic of Yugoslavia and international protected-area system (see Table 4 for categories) = 695 km² (8% of total IBA area).

Table 4. Protection status of Important Bird Areas in the Federal Republic of Yugoslavia. A tick (✔) indicates that an IBA overlaps with a protected area (to any extent).

IBA code	International name	National: National Park	National: Special/Strict Nature Reserve	National: Regional Nature Park	National: Other	International: Ramsar Site	International: World Heritage Site
001	Gornje Podunavlje			✔			
002	Subotica lakes and sandy terrain		✔	✔		✔	
003	Bečej fish-pond						
004	Jegrička		✔				
005	Koviljski rit		✔				
006	Fruška Gora	✔					
007	Bosutska forest		✔				
008	Zasavica		✔				
009	Obedska bara		✔			✔	
010	Danube loess bluffs						
011	Jazovo-Mokrin		✔				
012	Slano kopovo						
013	Carska bara		✔			✔	
014	Uzdin fish-pond						
015	Vršacki breg			✔			
016	Deliblatska peščara		✔				
017	Cer mountain						
018	Valjevske mountains		✔				
019	Tara mountain	✔					
020	Ovcar-Kablar gorge				✔		
021	Uvac-Mileševka		✔				
022	Kopaonik	✔					
023	Prokletije				✔		
024	Šara mountain	✔					
025	Pcinja				✔		
026	Vlasina						
027	Jerma				✔		
028	Suva mountain						
029	Stara mountain-Vidlic			✔			
030	Sicevo gorge			✔			
031	Zlot gorge-Dubašnica		✔		✔		
032	Resava gorge		✔				
033	Dubovac-Ram						
034	Đerdap gorge	✔					
035	Mala vrbica						
036	Durmitor	✔					✔
037	Biogradska woods	✔					
038	Lake Skadar	✔				✔	
039	Lake Sasko						
040	Ulcinj saltpans						
Total number of IBAs		8	13	5	5	4	1

Box 1. International legislation and initiatives that are relevant to site conservation in the Federal Republic of Yugoslavia (see Appendix 1 for a general description of these agreements).

Global	
Biodiversity Convention	(✔)
Ramsar Convention	✔
Bonn Convention	
World Heritage Convention	✔
MAB Programme	✔
Pan-European	
Bern Convention	
Regional	
Barcelona Convention	✔

✔ Convention ratified/initiative supported
(✔) Convention signed

National Parks in Yugoslavia are included within IBAs, as is the future Prokletije National Park which is likely to become legally protected soon, and 13 Strict/Special Nature Reserves lie within IBA boundaries too (Table 4). Yugoslavia is a contracting party to several international conventions and initiatives concerning the conservation of sites of nature conservation value (Box 1), and four IBAs fully enclose the four Ramsar Sites in the country (Table 4, Figure 5): Obedska Bara; Ludaško Lake; Stari Begej/Carska Bara; Skadarsko Jezero.

However, there are few instances where the boundaries of a protected area coincide with those of the IBA, the protected area usually being significantly smaller than the IBA. A total of 11 IBAs receive no protection at all through national or international legislation (Table 4), although some are in the process of being designated as protected areas (Slano kopovo; Uzdin fish-pond; Cer mountain; Vlasina; Suva mountain; Dubovac–Ram). Overall, more than 50% of the total IBA surface area receives no legal protection, which represents a serious problem for management programmes.

CONSERVATION

- Management plans exist or are in preparation for the 21 IBAs which overlap with a National Park or Nature Reserve, but the plans mostly do not embrace the total IBA surface area, only the part that is legally protected. In addition, the plans are of a standard whereby they cannot yet be implemented effectively, and therefore their improvement and advancement is one of the major priorities for the future. Since the existing plans only cover protected areas, a satisfactory management plan will also need to be drawn up for each IBA.

ANALYTICAL METHODS

- Within a country, the B2/B3 criteria are normally applied only to those species of European conservation concern (SPECs: Tucker and Heath 1994) which have 1% or more of their total European population breeding within the country concerned.

This is not possible for Yugoslavia, since the standardized European data-set of national population sizes of breeding birds during 1970–1990 (Tucker and Heath 1994) does not contain data for Yugoslavia. Thus B2/B3 criteria have been applied by different means in Yugoslavia. This involved examining, for each SPEC that was listed as breeding in one or more of the proposed Yugoslavian IBAs, its population size at the Yugoslavian site in question, and comparing this with the spread of similar values at all other sites in Europe where the B2 or B3 criterion had been met for this species. Where the population size at the Yugoslavian site was clearly of the same magnitude as the population sizes at most other B2-/B3-qualifying sites for the species elsewhere in Europe, then B2 or B3 was applied to the species for that site in Yugoslavia.

GLOSSARY

karst a limestone region with underground drainage and many cavities and passages caused by the dissolution of the rock.

loess a deposit of fine, light-coloured, wind-blown dust found especially in the basins of large rivers.

Pannonian region of, or pertaining to, a flat, lowland region centred on Hungary.

ACKNOWLEDGEMENTS

Work on the IBA project in Yugoslavia was accomplished at the Institute for the Protection of Nature of Serbia. Significant help with the compilation of general data and information on bird species at IBAs in Yugoslavia was given by the following collaborators: Ištvan Balog, Kanjo Bela, Beriša Dževad, Milan Dević, Boris Garovnikov (IBA consultant and former coordinator of Yugoslavian IBA project), Jožef Gergelj, Mihajlo Hadži-Pavlović, Ištvan Ham, Florijan Horvat, Ištvan Hulo, Milica Ivović, Dejan Krstojević, Jovan Lakatoš, Šandor Lukač, Saša Marinković, Zoran Milovanović, Milan Mirić, Aleksandar Nikolić, Borko Obućina, Aleksandar Pantelić, Milan Paunović, Zvonimir Popović, Javor Rašajski, Dragiša Savić, Oto Sekereš, Goran Sekulić, Mejdin Selmani, Dragan Simić, Nataša Simonov, Teodor Stefanović, Bogdan Šojić, Tibor Ternovac, Marko Tucakov, Antun Žuljević and Ondrej Vizi.

SITE ACCOUNTS

Gornje Podunavlje — A1, A4i, A4iii, B1i, B2, B3 — 001

Admin region Serbia
Coordinates 45°37′N 18°58′E
Altitude 82–95 m **Area** 30,000 ha

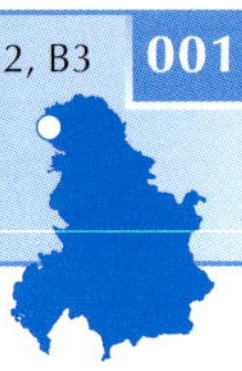

Site description

An inundated area along the eastern bank of the Danube, forming a transboundary IBA with sites across the border in Hungary ('Béda-Karapancsa'; Hungarian site 009) and Croatia ('Alluvial wetlands of the River Danube' and 'Kopacki rit'; Croatian sites 001 and 002 respectively). The site comprises extensive alluvial forest (oak *Quercus*, willow *Salix* and poplar *Populus*) interspersed with pools, rivers and streams, lowland forest of *Quercus* and ash *Fraxinus*, reedbeds and damp meadows (with sedge *Carex* and grass *Glyceria*) and large tracts of *Populus* plantation. Grazing-pasture, saltmarsh and arable fields occur in areas that remain free from flooding. The main economic activities are forestry and hunting.

Habitats Forest and woodland (35%; alluvial/very wet forest), Scrub (scrub), Grassland (5%), Wetland (15%; standing fresh water; river/stream; water-fringe vegetation; fen/transition mire/spring), Artificial landscape (55%; arable land; perennial crops/orchards/groves; forestry plantation)
Land-use Agriculture (7%), Fisheries/aquaculture, Forestry (70%), Hunting (75%), Nature conservation/research (30%), Not utilized, Tourism/recreation (5%), Urban/industrial/transport (5%)

Birds

Species	Season	Year	Pop min	Pop max	Acc	Criteria
Botaurus stellaris Bittern	B	1996	14	17	B	B2
Ixobrychus minutus Little Bittern	B	1996	80	100	B	B2
Nycticorax nycticorax Night Heron	B	1997	130	150	A	B2
Ardea purpurea Purple Heron	B	1997	60	80	A	B2
Ciconia nigra Black Stork	B	1997	35	45	A	B2
Ciconia nigra Black Stork	P	—	200	300	—	A4i, B1i
Ciconia ciconia White Stork	B	1997	30	40	A	B2
Platalea leucorodia Spoonbill	P	1993	200	300	—	A4i, B1i
[1] ***Aythya nyroca*** Ferruginous Duck	R	1997	30	50	B	A1, B2
Milvus migrans Black Kite	B	1997	15	18	A	B2
Haliaeetus albicilla White-tailed Eagle	R	1997	7	8	A	A1, B2
Ficedula albicollis Collared Flycatcher	B	1997	200	300	B	B3
Lanius collurio Red-backed Shrike	B	1997	200	250	B	B2

1. Mainly on Kolut fish-pond.

The site is important for breeding, wintering and passage waterbirds, holding 20,000 or more individuals on a regular basis during the winter.

Protection status

National Partial **International** None
10,000 ha of IBA covered by Regional Nature Park (Gornje Podunavlje, 10,000 ha).

Conservation issues

Threats Abandonment/reduction of land management (C), Afforestation (A), Agricultural intensification/expansion (C), Consequences of animal/plant introductions (A), Construction/impact of dyke/dam/barrage (B), Disturbance to birds (C), Drainage (A), Dredging/canalization (B), Firewood collection (C), Forest grazing (C), Intensified forest management (B), Recreation/tourism (C), Selective logging/cutting (B)

The main threats are losses of natural forest and meadows to afforestation, and drainage. Water is artificially transferred to marshes separated from the Danube by an embankment, with funding provided by domestic sources. The Institute for the Protection of Nature of Serbia has an ongoing research programme. A management plan exists for the site.

Subotica lakes and sandy terrain — A1, B2, B3 — 002

Admin region Serbia
Coordinates 46°04′N 19°49′E
Altitude 80–110 m **Area** 20,000 ha

Site description

Situated in the far north-east of Backa, in the vicinity of Subotica. Numerous wetlands are present, with extensive reedbeds and aquatic vegetation: Paličko jezero, Ludjaško jezero, Kapetanjski rit (now a fish-pond) and Velebit reservoir. The sandy ground in Selevenjske pustare and Mali pesak supports a complex of wooded steppe, steppe, and plant communities of sandy ground. Human activities include sport-fishing; recreational tourism has been developed at Paličko jezero and Ludjaško jezero. This site includes former sites YU032 and YU033 from the previous pan-European IBA inventory (Grimmett and Jones 1989).

Habitats Forest and woodland (5%; broadleaved deciduous forest; alluvial/very wet forest; wooded steppe), Scrub (5%; scrub), Grassland (30%; steppe/dry calcareous grassland; humid grassland; mesophile grassland), Wetland (20%; mudflat/sandflat; standing fresh water; river/stream; water-fringe vegetation), Artificial landscape (40%; arable land; perennial crops/orchards/groves; forestry plantation; urban parks/gardens)
Land-use Agriculture (60%), Fisheries/aquaculture (5%), Forestry (5%), Hunting (80%), Nature conservation/research (25%), Tourism/recreation (25%), Urban/industrial/transport (5%)

Birds

Species	Season	Year	Pop min	Pop max	Acc	Criteria
Ixobrychus minutus Little Bittern	B	1997	70	100	A	B2
Ardeola ralloides Squacco Heron	B	1997	35	50	A	B2
Ardea purpurea Purple Heron	B	1997	80	100	A	B2
[1] ***Aythya ferina*** Pochard	R	1997	100	140	B	B3
Aythya nyroca Ferruginous Duck	B	1997	15	25	A	A1, B2
Falco tinnunculus Kestrel	R	1997	80	100	A	B2
Falco vespertinus Red-footed Falcon	B	1997	20	25	A	B2

Species ... continued	Season	Year	Pop min	Pop max	Acc	Criteria
Merops apiaster Bee-eater	B	1997	60	70	B	B2
Riparia riparia Sand Martin	B	1997	1,000	1,500	B	B2
Acrocephalus scirpaceus Reed Warbler	B	1997	1,000	1,500	B	B3
[2] ***Lanius minor*** Lesser Grey Shrike	B	1997	30	40	A	B2

1. Main breeding site in Serbia. 3,000–5,000 individuals in winter.
2. Main site in Serbia.

This is an important site for breeding and passage waterbirds, and breeding species of open grassland/shrub habitats. Species of global conservation concern that do not meet IBA criteria: *Oxyura leucocephala* (last bred 1947; 1–2 on passage and in winter).

Protection status

National Low **International** Low

387 ha of IBA covered by Special Nature Reserve (Ludaško jezero, 387 ha). 677 ha of IBA covered by Special Nature Reserve (Selevenjske pustare, 677 ha). 712 ha of IBA covered by Regional Nature Park (Palič, 712 ha). 593 ha of IBA covered by Ramsar Site (Ludaško Lake, 593 ha).

Conservation issues

Threats Agricultural intensification/expansion (B), Disturbance to birds (B), Drainage (B), Dredging/canalization (B), Infrastructure (B), Other (U), Recreation/tourism (B), Unsustainable exploitation (B)

Threats include nutrient-enrichment and pollution ('Other' threat), erosion and disturbance. Members of the Bird Protection and Study Society of Vojvodina undertook intensive ornithological research between 1985 and 1997. Artificial nest-cavities are provided for *Otus scops*. A management plan exists for the site.

Bečej fish-pond

A4i, B1i, B2, B3 **003**

Admin region Serbia
Coordinates 45°35′N 20°03′E
Altitude 77–82 m **Area** 4,000 ha

Site description

Situated in an alluvial plain in Backa (Vojvodina), between Bečej and Backo Gradiste, and bordering the Tisa–Veliki canal. The fish-pond supports fringing vegetation, and is surrounded by pasture, arable land and a narrow strip of riverine forest (poplar *Populus*, willow *Salix* and ash *Fraxinus*). Human activities include crop production and cattle-rearing.

Habitats Forest and woodland (broadleaved deciduous forest; alluvial/very wet forest), Grassland (10%; humid grassland; mesophile grassland), Wetland (60%; standing fresh water; river/stream; water-fringe vegetation), Artificial landscape (30%; arable land; perennial crops/orchards/groves; urban parks/gardens)
Land-use Agriculture (30%), Fisheries/aquaculture (40%), Forestry (5%), Hunting (60%), Urban/industrial/transport (5%), Water management (5%)

Birds

Species	Season	Year	Pop min	Pop max	Acc	Criteria
[1] ***Nycticorax nycticorax*** Night Heron	B	1997	500	500	A	A4i, B1i, B2
[1] ***Ardeola ralloides*** Squacco Heron	B	1996	150	200	A	B2
[1,2] ***Egretta alba*** Great White Egret	R	1998	30	180	A	A4i, B1i
Ardea purpurea Purple Heron	B	1996	150	200	B	B2
[1] ***Platalea leucorodia*** Spoonbill	B	1996	70	80	A	B2
Recurvirostra avosetta Avocet	B	1994	30	30	A	B3

1. Main site in Serbia.
2. Fluctuating.

The site is important for breeding, wintering and passage waterbirds. Breeding species of global conservation concern that do not meet IBA criteria: *Phalacrocorax pygmeus* and *Aythya nyroca*.

Protection status

National None **International** None

Conservation issues

Threats Agricultural intensification/expansion (C), Aquaculture/fisheries (C), Disturbance to birds (B), Drainage (B), Dredging/canalization (B), Infrastructure (C), Unsustainable exploitation (U)

Threats include disturbance of birds, mowing of marsh vegetation and hunting. Members of the Bird Protection and Study Society of Vojvodina undertook intensive ornithological research between 1989 and 1997.

Jegrička

A1, A4i, B1i, B2, B3 **004**

Admin region Serbia
Coordinates 45°20′N 20°10′E
Altitude 76–86 m **Area** 5,400 ha

Site description

A narrow, 60-km stretch of river with fringing reed *Phragmites*; the lower reaches have been dammed to form a fish-pond. There are three major settlements in the vicinity of Jegrička, which is surrounded by arable land.

Habitats Scrub, Grassland, Wetland (80%; mudflat/sandflat; standing fresh water; river/stream; water-fringe vegetation), Artificial landscape (20%; arable land; perennial crops/orchards/groves; forestry plantation; other urban/industrial areas)
Land-use Agriculture (30%), Fisheries/aquaculture (60%), Hunting (70%), Nature conservation/research (10%), Tourism/recreation (5%), Water management (5%)

Birds

Species	Season	Year	Pop min	Pop max	Acc	Criteria
Podiceps grisegena Red-necked Grebe	B	1997	400	500	B	A4i, B1i
[1] ***Botaurus stellaris*** Bittern	R	1997	80	100	B	B2
Ixobrychus minutus Little Bittern	B	1997	200	400	B	B2
Nycticorax nycticorax Night Heron	B	1997	150	150	A	B2
Ardeola ralloides Squacco Heron	B	1997	60	70	A	B2
Ardea purpurea Purple Heron	B	1997	100	130	A	B2
Aythya nyroca Ferruginous Duck	B	1997	50	70	B	A1, B2
Porzana porzana Spotted Crake	B	1997	50	100	B	B3
[1] ***Porzana parva*** Little Crake	B	1997	300	400	B	B3
[1] ***Chlidonias hybridus*** Whiskered Tern	B	1997	400	400	A	B2
Merops apiaster Bee-eater	B	1997	50	70	A	B2
Lanius collurio Red-backed Shrike	B	1997	150	200	B	B2

1. Main site in Serbia.

An important site for breeding and passage waterbirds.

Protection status

National Low **International** None

20 ha of IBA covered by Strict Nature Reserve (Jegrička, 20 ha).

Conservation issues

Threats Afforestation (C), Agricultural intensification/expansion (B), Aquaculture/fisheries (U), Disturbance to birds (C), Drainage (B), Dredging/canalization (C), Infrastructure (C), Recreation/tourism (B), Unsustainable exploitation (U)

Threats include disturbance to birds, and the effects of fishing and hunting. One wet depression has been polluted by waste-water from a local cattle farm. Members of the Bird Protection and Study Society of Vojvodina undertook intensive ornithological research between 1989 and 1997.

Koviljski rit

B2 **005**

Admin region Serbia
Coordinates 45°12′N 20°00′E
Altitude 73–81 m **Area** 5,500 ha

Site description

A wetland bordering the River Danube in the vicinity of Novi Sad–Koviljski rit, near the settlements of Kovilj and Gardinovci in Back, and Petrovaradinski rit at the foot of Fruška Gora in Srem. A complex of naturally inundated forest (poplar *Populus*, willow *Salix* and ash *Fraxinus*), stagnant tributaries, canals, marshes, wet meadows, pasture and *Populus* plantations. Human activities include seasonal pig-grazing.

Habitats Forest and woodland (30%; broadleaved deciduous forest; alluvial/very wet forest), Grassland (5%; humid grassland), Wetland (30%; standing fresh water; river/stream; water-fringe vegetation), Artificial landscape (40%; arable land; perennial crops/orchards/groves; forestry plantation)

Land-use Agriculture (5%), Fisheries/aquaculture (10%), Forestry (70%), Hunting (80%), Nature conservation/research (80%), Tourism/recreation (5%)

Birds

Species	Season	Year	Pop min	Pop max	Acc	Criteria
Ciconia nigra Black Stork	B	1997	8	10	B	B2
Haliaeetus albicilla White-tailed Eagle	R	1997	2	2	A	B2

The site is important for breeding and passage waterbirds. Species of global conservation concern that do not meet IBA criteria: *Aythya nyroca* (breeding).

Protection status

National Partial **International** None

4,820 ha of IBA covered by Special Nature Reserve (Koviljsko-Petrovaradinski rit, 4,820 ha).

Conservation issues

Threats Afforestation (A), Aquaculture/fisheries (B), Construction/impact of dyke/dam/barrage (B), Disturbance to birds (B), Drainage (B), Dredging/canalization (B), Forest grazing (C), Intensified forest management (A), Selective logging/cutting (B), Unsustainable exploitation (U)

Threats include replacing natural woodland with plantations, afforestation of open areas, drainage, and uncontrolled hunting. A management plan exists for the site.

Fruška Gora — A1, B2 — 006

Admin region Serbia
Coordinates 45°04'N 19°34'E
Altitude 100–539 m **Area** 42,000 ha

Site description

A large, forest-covered massif situated in the north of Srem. Broadleaved forests (oak *Quercus*, lime *Tilia*, beech *Fagus*, hornbeam *Carpinus* and ash *Fraxinus*) are interspersed with steppe and wooded steppe, pasture, meadows, vineyards and arable land. Recreational tourism is well developed due to the proximity of Belgrade and Novi Sad.

Habitats Forest and woodland (70%; broadleaved deciduous forest), Scrub, Grassland (10%; steppe/dry calcareous grassland; mesophile grassland), Wetland (5%; river/stream), Artificial landscape (15%; arable land; perennial crops/orchards/groves; forestry plantation; urban parks/gardens; other urban/industrial areas; ruderal land)
Land-use Agriculture (10%), Forestry (75%), Hunting (60%), Nature conservation/research (80%), Tourism/recreation (90%), Urban/industrial/transport (5%), Water management

Birds

Species	Season	Year	Pop min	Pop max	Acc	Criteria
Aquila pomarina Lesser Spotted Eagle	B	1997	1	2	B	B2
[1] ***Aquila heliaca*** Imperial Eagle	B	1997	2	3	A	A1
Falco cherrug Saker	B	1997	2	3	B	B2
Otus scops Scops Owl	B	1997	20	40	C	B2
Picus viridis Green Woodpecker	R	1997	60	100	B	B2
Lullula arborea Woodlark	B	1997	60	80	C	B2
Lanius collurio Red-backed Shrike	B	1997	300	500	B	B2
Lanius minor Lesser Grey Shrike	B	1997	10	20	B	B2

1. Main site in Serbia.

Important site for forest-nesting species. Species of global conservation concern that do not meet IBA criteria: *Aquila clanga* (passage).

Protection status

National Partial **International** None

25,400 ha of IBA covered by National Park (Fruška Gora, 25,400 ha).

Conservation issues

Threats Afforestation (C), Agricultural intensification/expansion (C), Consequences of animal/plant introductions (C), Deforestation (commercial) (C), Disturbance to birds (B), Extraction industry (B), Industrialization/urbanization (C), Infrastructure (C), Intensified forest management (C), Recreation/tourism (B)

The main threats are disturbance from recreational activities and intensive forestry practices. A management plan exists for the site.

Bosutska forest — A1, B1i, B2, B3 — 007

Admin region Serbia
Coordinates 45°00'N 19°10'E
Altitude 79–85 m **Area** 17,500 ha

Site description

An area of lowland adjacent to the Croatian border in western Srem (Vojvodina), supporting vast forests of oak *Quercus*, with ash *Fraxinus*, poplar *Populus*, willow *Salix*, elm *Ulmus*, hornbeam *Carpinus* and maple *Acer*. Numerous depressions with swamp vegetation occur along with riverine forests along the Sava and Bosut rivers. Human activities primarily consist of intensive forestry and game hunting.

Habitats Forest and woodland (80%; broadleaved deciduous forest; alluvial/very wet forest), Grassland (mesophile grassland), Wetland (10%; mudflat/sandflat; standing fresh water; river/stream; water-fringe vegetation), Artificial landscape (10%; arable land; perennial crops/orchards/groves; forestry plantation; urban parks/gardens)
Land-use Agriculture (15%), Fisheries/aquaculture, Forestry (80%), Hunting (70%), Nature conservation/research, Urban/industrial/transport (5%)

Birds

Species	Season	Year	Pop min	Pop max	Acc	Criteria
Ciconia nigra Black Stork	B	1995	10	15	B	B2
Ciconia nigra Black Stork	P	—	100	200	—	B1i
Ciconia ciconia White Stork	B	1994	20	30	A	B2
Haliaeetus albicilla White-tailed Eagle	R	1994	4	5	A	A1
Strix aluco Tawny Owl	R	1993	B	—	—	B3

This is an important site for forest-nesting species.

Protection status

National Low **International** None

27 ha of IBA covered by Strict Nature Reserve (Majzecova Basta, 27 ha). 87 ha of IBA covered by Strict Nature Reserve (Radenovci, 87 ha). 11 ha of IBA covered by Strict Nature Reserve (Stara Vraticna, 11 ha). 39 ha of IBA covered by Strict Nature Reserve (Varos, 39 ha). 58 ha of IBA covered by Strict Nature Reserve (Vinicna, 58 ha).

Conservation issues

Threats Afforestation (B), Agricultural intensification/expansion (C), Disturbance to birds (U), Drainage (B), Dredging/canalization (B), Forest grazing (B), Infrastructure (C), Intensified forest management (A), Selective logging/cutting (B)

The main threats are disturbance, logging of old-growth forest and gradual drainage of wetland areas.

Zasavica — B2, B3 — 008

Admin region Serbia
Coordinates 44°56'N 19°31'E
Altitude 77–83 m **Area** 5,200 ha

Site description

A wetland situated along the River Sava in the plains of Macva, near to the town of Sremska Mitrovica. Fed by underground aquifers and with a depth of up to 2.5 m, it supports abundant marsh vegetation—mostly reed *Phragmites*. The site is surrounded by agricultural land, and remnants of forest remain (oak *Quercus*, ash *Fraxinus*, poplar *Populus* and willow *Salix*). Human activities include cattle-rearing; ecological tourism is a recent development.

Habitats Forest and woodland (30%; broadleaved deciduous forest; alluvial/very wet forest), Scrub, Grassland (10%; humid grassland; mesophile grassland), Wetland (30%; standing fresh water; river/stream; water-fringe vegetation), Artificial landscape (30%; arable land; perennial crops/orchards/groves; forestry plantation; urban parks/gardens)
Land-use Agriculture (60%), Fisheries/aquaculture (20%), Forestry (20%), Hunting (80%), Nature conservation/research (70%), Tourism/recreation (10%), Urban/industrial/transport

Birds

Species	Season	Year	Pop min	Pop max	Acc	Criteria
Nycticorax nycticorax Night Heron	B	1996	70	75	A	B2
Porzana porzana Spotted Crake	B	1996	20	50	C	B3

Important for breeding waterbirds. Wintering species of global conservation concern that do not meet IBA criteria: *Phalacrocorax pygmeus* and *Haliaeetus albicilla*.

■ Protection status

National Partial **International** None

670 ha of IBA covered by Special Nature Reserve (Zasavica, 670 ha).

■ Conservation issues

Threats Afforestation (B), Agricultural intensification/expansion (C), Deforestation (commercial) (B), Disturbance to birds (C), Drainage (B), Industrialization/urbanization (C), Recreation/tourism (B)

Artificial nest-sites are provided for *Ciconia ciconia*. A management plan exists for the site.

Obedska bara — B1i, B2, B3 — 009

Admin region Serbia
Coordinates 44°43′N 20°01′E
Altitude 71–82 m **Area** 23,000 ha

■ Site description

A vast area of wet forest stretching along the River Sava in southern Srem (Vojvodina). The site is a complex of forest (willow *Salix* and poplar *Populus* and oak *Quercus*), stagnant tributaries, marshes and wet meadows. Human activities include cattle-grazing.

Habitats Forest and woodland (60%; broadleaved deciduous forest; alluvial/very wet forest), Grassland, Wetland (15%; mudflat/sandflat; standing fresh water; river/stream; water-fringe vegetation), Artificial landscape (25%; arable land; perennial crops/orchards/groves; forestry plantation; urban parks/gardens)
Land-use Agriculture (30%), Forestry (70%), Hunting (80%), Nature conservation/research (50%), Tourism/recreation (10%), Urban/industrial/transport (5%), Water management (10%)

■ Birds

Species	Season	Year	Pop min	Pop max	Acc	Criteria
Ixobrychus minutus Little Bittern	B	1996	22	29	A	B2
Nycticorax nycticorax Night Heron	B	1996	195	195	A	B2
Ciconia nigra Black Stork	B	1996	17	20	A	B2
Ciconia nigra Black Stork	P	—	100	200	—	B1i
Ciconia ciconia White Stork	B	1996	30	32	A	B2
Haliaeetus albicilla White-tailed Eagle	R	1996	2	2	A	B2
Falco cherrug Saker	R	1996	1	1	A	B2
Porzana porzana Spotted Crake	B	1996	28	46	C	B3
Porzana parva Little Crake	B	1996	15	30	C	B3
Alcedo atthis Kingfisher	R	1994	14	22	B	B2
Picus viridis Green Woodpecker	R	1996	50	70	B	B2
Dendrocopos syriacus Syrian Woodpecker	R	1996	30	47	B	B3
Dendrocopos medius Middle Spotted Woodpecker	R	1996	140	220	B	B3
Locustella luscinioides Savi's Warbler	B	1996	70	90	B	B3
Sylvia atricapilla Blackcap	B	1996	7,500	8,600	B	B3
Ficedula albicollis Collared Flycatcher	B	1996	260	335	A	B3
Certhia brachydactyla Short-toed Treecreeper	R	1996	1,260	1,370	B	B3

The site is important for breeding waterbirds and forest-nesting species. Breeding species of global conservation concern that do not meet IBA criteria: *Aythya nyroca* (1–2 pairs) and *Aquila clanga*.

■ Protection status

National Partial **International** Partial

7,500 ha of IBA covered by Special Nature Reserve (Obedska Bara, 9,820 ha). 15,100 ha of IBA covered by Ramsar Site (Obedska Bara, 17,501 ha).

■ Conservation issues

Threats Afforestation (A), Agricultural intensification/expansion (B), Disturbance to birds (C), Drainage (B), Dredging/canalization (B), Forest grazing (A), Groundwater abstraction (B), Infrastructure (C), Intensified forest management (B), Other (U), Recreation/tourism (C), Selective logging/cutting (B)

The main threats are nutrient-enrichment ('Other' threat) and the afforestation of alluvial depressions and meadows with *Populus* plantations. Dams along the Vok canal, which supplies and drains Obedska bara, have raised the dry-season water-level by 60 cm and have significantly reduced a previous problem whereby vegetational succession was converting wetland and meadows to scrub. In addition, a work camp in 1997 cleared 10 ha of overgrown meadows. A management plan exists for the site.

Danube loess bluffs — A1, A4i, B1i, B2 — 010

Admin region Serbia
Coordinates 45°05′N 20°22′E
Altitude 75–131 m **Area** 2,500 ha

■ Site description

Steep loess bluffs up to 50 m in height, bordering the Danube between Banovci and Slankamen. Individual trees and abundant herbaceous vegetation survive along the upper edges of the bluffs; low-lying areas are undermined by the main river current at times of high flow. Human activities include extracting loess for use as building material.

Habitats Scrub (5%), Grassland (5%; steppe/dry calcareous grassland), Wetland (70%; standing fresh water; river/stream; water-fringe vegetation), Artificial landscape (20%; arable land; perennial crops/orchards/groves; forestry plantation; ruderal land)
Land-use Agriculture (10%), Fisheries/aquaculture (50%), Forestry (5%), Hunting (80%), Tourism/recreation (10%), Urban/industrial/transport, Water management (70%)

■ Birds

Species	Season	Year	Pop min	Pop max	Acc	Criteria
Phalacrocorax pygmeus Pygmy Cormorant	W	1996	100	200	—	A1, A4i, B1i
Falco tinnunculus Kestrel	R	1996	60	90	B	B2
Merops apiaster Bee-eater	B	1996	300	600	B	B2

Wintering species of global conservation concern that do not meet IBA criteria: *Branta ruficollis* and *Haliaeetus albicilla*.

■ Protection status

National None **International** None

■ Conservation issues

Threats Afforestation (B), Agricultural intensification/expansion (C), Aquaculture/fisheries (B), Disturbance to birds (B), Industrialization/urbanization (C), Infrastructure (B), Unsustainable exploitation (U)

The main threats are disturbance and hunting.

Jazovo–Mokrin — A1, B2 — 011

Admin region Serbia
Coordinates 45°49′N 20°15′E
Altitude 78–86 m **Area** 8,000 ha

■ Site description

The IBA covers an uninhabited area of loess terrace and alluvial plain in northern Banat, between Mokrin, Jazovo, Sajan and Crna Bara. Extensive steppic habitats are bordered by meadows and pasture, and contain a mosaic of relict Pannonian plant communities. Limited human activities include grazing and mowing of meadows.

Habitats Grassland (50%; steppe/dry calcareous grassland; humid grassland; mesophile grassland), Wetland (20%; mudflat/sandflat; saltmarsh; standing fresh water; river/stream; water-fringe vegetation), Artificial landscape (30%; arable land; perennial crops/orchards/groves; ruderal land)
Land-use Agriculture (60%), Fisheries/aquaculture (10%), Hunting (60%), Nature conservation/research (25%), Not utilized (5%)

■ Birds

The site is important for breeding species of open grassland. Species of global conservation concern that do not meet IBA criteria: *Aythya nyroca* (5–10 breeding pairs).

Species	Season	Year	Pop min	Pop max	Acc	Criteria
[1] ***Platalea leucorodia*** Spoonbill	B	1997	30	40	A	B2
Falco vespertinus Red-footed Falcon	B	1997	20	30	A	B2
[2] ***Otis tarda*** Great Bustard	R	1997	10	12	B	A1
Lanius minor Lesser Grey Shrike	B	1997	10	15	A	B2

1. Did not breed 1995–1997.
2. 36 individuals in winter 1996 (max. 45 individuals).

■ Protection status

National Partial **International** None

980 ha of IBA covered by Special Nature Reserve (*Otis tarda* pasture, 980 ha).

■ Conservation issues

Threats Agricultural intensification/expansion (B), Disturbance to birds (B), Drainage (B), Dredging/canalization (B), Forest grazing (B), Recreation/tourism (C)

The main threats are drainage of wet depressions and ploughing of steppe and saltmarsh habitats. Predator control and winter feeding of *Otis tarda* are undertaken. A management plan exists for the site.

Slano kopovo — B1iv, B2, B3 — 012

Admin region Serbia
Coordinates 45°37′N 20°13′E
Altitude 75–86 m **Area** 2,660 ha

■ Site description

An alluvial plain situated east of Novi Becej, containing seasonal bodies of salt water. One of the larger lakes, Slano kopovo, supports vegetation characterized by salt-tolerant succulents (for example, *Salicornia*, *Salsola* and *Suaeda*). It varies from 30 ha to 100 ha in size, depending on underground water-levels. Arable land, pasture and meadows surround the depressions. Human activities include cattle-grazing.

Habitats Grassland (10%; steppe/dry calcareous grassland; humid grassland), Wetland (30%; mudflat/sandflat; saltmarsh; standing brackish and salt water; water-fringe vegetation), Artificial landscape (60%; arable land; perennial crops/orchards/groves; ruderal land)
Land-use Agriculture (60%), Hunting (90%), Tourism/recreation (20%)

■ Birds

Species	Season	Year	Pop min	Pop max	Acc	Criteria
Recurvirostra avosetta Avocet	B	1997	15	20	A	B3
[1] ***Charadrius alexandrinus*** Kentish Plover	B	1997	10	20	B	B2

1. Main site in Serbia.

The site is important for breeding and passage waterbirds. A total of 5,000–10,000 *Grus grus* migrate over the area, with some stopping to rest and feed. Species of global conservation concern that do not meet IBA criteria: *Anser erythropus* (wintering), *Branta ruficollis* (passage), *Oxyura leucocephala* (breeding), *Aquila heliaca* (passage) and *Numenius tenuirostris* (passage).

■ Protection status

National None **International** None

■ Conservation issues

Threats Agricultural intensification/expansion (B), Disturbance to birds (B), Drainage (C), Dredging/canalization (C), Recreation/tourism (C)

The nearby Becej–Kikinda road is very busy. The area is proposed as a Special Nature Reserve (Slano Kopovo, 1,200 ha). A management plan for the site is in preparation.

Carska bara — A1, A4i, A4iii, B1i, B2, B3 — 013

Admin region Serbia
Coordinates 45°16′N 20°25′E
Altitude 75–85 m **Area** 9,300 ha

■ Site description

An inundation area between the Tisa and Begej rivers in the western part of Backa, south-west of the town of Zrenjanin. Habitats include pools, fish-ponds, meadows and willow/poplar *Salix*/*Populus* woodland. Includes Carska bara, Tiganjica, Perleska bara and Ečka fish-pond. Human activities include reed-cutting ('Other' land-use).

Habitats Forest and woodland (10%; broadleaved deciduous forest; alluvial/very wet forest), Grassland (5%; humid grassland; mesophile grassland), Wetland (65%; mudflat/sandflat; standing fresh water; river/stream; water-fringe vegetation), Artificial landscape (20%; arable land; perennial crops/orchards/groves; forestry plantation; urban parks/gardens)
Land-use Agriculture (15%), Fisheries/aquaculture (60%), Forestry (10%), Hunting (50%), Nature conservation/research (30%), Other, Tourism/recreation (5%), Urban/industrial/transport (5%), Water management

■ Birds

Species	Season	Year	Pop min	Pop max	Acc	Criteria
[1] ***Tachybaptus ruficollis*** Little Grebe	B	1996	400	600	B	A4i, B1i
Botaurus stellaris Bittern	R	1997	20	30	B	B2
Ixobrychus minutus Little Bittern	B	1997	70	100	B	B2
Nycticorax nycticorax Night Heron	B	1997	200	300	B	B2
Ardeola ralloides Squacco Heron	B	1997	90	100	A	B2
[2] ***Egretta alba*** Great White Egret	R	1997	70	100	A	A4i, B1i
Ardea purpurea Purple Heron	B	1997	150	200	A	B2
Platalea leucorodia Spoonbill	P	1997	200	400	—	A4i, B1i
[1] ***Anas strepera*** Gadwall	B	1996	30	40	B	B2
[1] ***Anas querquedula*** Garganey	B	1997	80	100	B	B2
[1] ***Aythya ferina*** Pochard	B	1997	200	300	B	B3
Aythya nyroca Ferruginous Duck	B	1997	30	50	B	A1, B2
Larus ridibundus Black-headed Gull	W	1996	50,000	80,000	—	A4i, B1i
Merops apiaster Bee-eater	B	1997	50	80	B	B2
Phoenicurus phoenicurus Redstart	B	1996	20	40	B	B2
Locustella luscinioides Savi's Warbler	B	1997	200	300	B	B3
Lanius minor Lesser Grey Shrike	B	1997	10	12	B	B2

1. Main site in Serbia.
2. Large increase.

An important site for breeding, wintering and passage waterbirds, regularly holding 20,000 or more individuals during winter and migration periods. Species of global conservation concern that do not meet IBA criteria: *Phalacrocorax pygmeus* (periodically breeding), *Anser erythropus* (wintering), *Haliaeetus albicilla* (resident), *Aquila clanga* (wintering), *Crex crex* (breeding) and *Numenius tenuirostris* (passage).

■ Protection status

National Partial **International** Partial

1,676 ha of IBA covered by Special Nature Reserve (Stari Begej–Carska Bara, 1,676 ha). 1,767 ha of IBA covered by Ramsar Site (Stari Begej/Carska Bara Special Nature Reserve, 1,767 ha).

■ Conservation issues

Threats Afforestation (B), Agricultural intensification/expansion (C), Construction/impact of dyke/dam/barrage (C), Disturbance to birds (C), Drainage (B), Dredging/canalization (B), Forest grazing (C), Industrialization/urbanization (B), Infrastructure (C), Intensified forest management (C), Other (U), Recreation/tourism (C)

Threats include disturbance from visitors and fishermen, and inappropriate water management ('Other' threat). A management plan exists for the site.

Uzdin fish-pond — A1, A4i, A4iii, B1i, B2, B3 — 014

Admin region Serbia
Coordinates 45°13′N 20°38′E
Altitude 75–82 m **Area** 5,500 ha

■ Site description

A fish-pond (460 ha in size) situated in saltmarsh pasture and marsh along the Tamis river. Numerous small islands and sandbars within the fish-pond are very important for nesting birds. The IBA also covers natural inundation forest (poplar *Populus*, willow *Salix* and ash *Fraxinus*), numerous stagnant tributaries, canals and arable fields.

Habitats Forest and woodland (5%; broadleaved deciduous forest; alluvial/very wet forest), Scrub, Grassland (10%; humid grassland; mesophile grassland), Wetland (65%; mudflat/sandflat; standing fresh water; river/stream; water-fringe vegetation), Artificial landscape (20%; arable land; perennial crops/orchards/groves; urban parks/gardens)
Land-use Agriculture (30%), Fisheries/aquaculture (70%), Forestry (5%), Hunting (80%), Urban/industrial/transport (5%), Water management (5%)

Birds

Species	Season	Year	Pop min	Pop max	Acc	Criteria
Phalacrocorax pygmeus Pygmy Cormorant	W	1996	800	1,000	—	A4i, B1i
[1] ***Botaurus stellaris*** Bittern	R	1997	20	25	B	B2
Nycticorax nycticorax Night Heron	B	1997	80	100	A	B2
[2] ***Egretta alba*** Great White Egret	R	1997	30	40	A	A4i, B1i
Ardea purpurea Purple Heron	B	1997	50	60	A	B2
[3] ***Ciconia nigra*** Black Stork	P	1997	—	200	—	A4i, B1i
Ciconia ciconia White Stork	B	1997	60	65	A	B2
[3] ***Platalea leucorodia*** Spoonbill	N	1997	300	400	—	A4i, B1i
Aythya ferina Pochard	R	1997	100	150	B	B3
[4] ***Aythya nyroca*** Ferruginous Duck	B	1997	100	150	B	A1, B2
Crex crex Corncrake	B	1997	10	20	B	A1, B2
Recurvirostra avosetta Avocet	B	1997	40	50	A	B3
[4] ***Limosa limosa*** Black-tailed Godwit	B	1997	40	60	B	B2
Limosa limosa Black-tailed Godwit	P	1997	1,000	3,000	—	B1i
[5] ***Chlidonias hybridus*** Whiskered Tern	B	1997	280	300	A	B2

1. 50–60 individuals in autumn/winter.
2. Large increase. 200–300 individuals in winter.
3. Autumn.
4. Main site in Serbia.
5. Fluctuating.

The site is important for breeding, wintering and passage waterbirds, holding 20,000 or more individuals on a regular basis during winter and migration periods. Species of global conservation concern that do not meet IBA criteria: *Haliaeetus albicilla* (resident).

Protection status

National None **International** None

Conservation issues

Threats Agricultural intensification/expansion (B), Disturbance to birds (C), Drainage (B), Dredging/canalization (B), Industrialization/urbanization (C), Unsustainable exploitation (U)

The main threats include agricultural intensification, disturbance and hunting. Members of the Bird Protection and Study Society of Vojvodina undertook intensive ornithological research between 1977 and 1997.

Vršacki breg — B2, B3 — 015

Admin region Serbia
Coordinates 45°07′N 21°25′E
Altitude 90–641 m **Area** 10,500 ha

Site description

A hilly area in the far south-east of Banat, on the boundary between the Pannonian lowlands and the Carpathian massif. Extensive plains comprising steppic habitats, agricultural land and vineyards surround forests of oak *Quercus*, beech *Fagus*, hornbeam *Carpinus*, lime *Tilia*, ash *Fraxinus* and maple *Acer*.

Habitats Forest and woodland (50%; broadleaved deciduous forest), Grassland (30%; steppe/dry calcareous grassland; mesophile grassland), Wetland, Artificial landscape (20%; arable land; perennial crops/orchards/groves; forestry plantation; urban parks/gardens; other urban/industrial areas)
Land-use Agriculture (50%), Forestry (50%), Hunting (80%), Nature conservation/research (20%), Tourism/recreation (5%), Urban/industrial/transport

Birds

Species	Season	Year	Pop min	Pop max	Acc	Criteria
Aquila pomarina Lesser Spotted Eagle	B	1996	2	2	A	B2
Picus canus Grey-headed Woodpecker	B	1996	20	40	B	B2
Picus viridis Green Woodpecker	R	1996	100	200	B	B2
Dendrocopos syriacus Syrian Woodpecker	R	1996	20	50	B	B3
Luscinia megarhynchos Nightingale	B	1996	300	600	C	B3

The site is important for forest-nesting species. Species of global conservation concern that do not meet IBA criteria: *Aquila heliaca* (non-breeding) and *Crex crex* (breeding).

Protection status

National Partial **International** None
4,180 ha of IBA covered by Regional Nature Park (Vrsacke Planine, 4,180 ha).

Conservation issues

Threats Agricultural intensification/expansion (C), Disturbance to birds (C), Dredging/canalization (C), Industrialization/urbanization (C), Infrastructure (C), Intensified forest management (B), Recreation/tourism (B), Selective logging/cutting (C)

Members of the Bird Protection and Study Society of Vojvodina undertook intensive ornithological research between 1977 and 1997. A management plan for the site is in preparation.

Deliblatska peščara — B2, B3 — 016

Admin region Serbia
Coordinates 44°49′N 21°00′E
Altitude 84–250 m **Area** 38,000 ha

Site description

Located in the south-east of Vojvodina, south Banat, the IBA comprises an elongated elliptical tract of sand, spreading from the Danube north towards Vladimirovac and Banatski Karlovac. It covers grazing-pasture, steppic habitats, woodland, scrub, conifer plantations, vineyards and agricultural land.

Habitats Forest and woodland (45%; broadleaved deciduous forest; wooded steppe; treeline ecotone; wooded desert/semi-desert), Scrub (30%; scrub; sclerophyllous scrub/garrigue/maquis), Grassland (20%; steppe/dry calcareous grassland; humid grassland; mesophile grassland), Wetland (river/stream; water-fringe vegetation), Artificial landscape (5%; arable land; perennial crops/orchards/groves; forestry plantation; urban parks/gardens; ruderal land)
Land-use Agriculture (20%), Forestry (80%), Hunting (90%), Nature conservation/research (80%), Tourism/recreation (5%), Urban/industrial/transport (5%)

Birds

Species	Season	Year	Pop min	Pop max	Acc	Criteria
[1] ***Caprimulgus europaeus*** Nightjar	B	1995	200	300	B	B2
Jynx torquilla Wryneck	B	1995	100	150	B	B2
Picus canus Grey-headed Woodpecker	B	1996	50	80	B	B2
Lullula arborea Woodlark	B	1996	400	700	B	B2
Alauda arvensis Skylark	B	1996	2,000	2,500	B	B2
[1] ***Luscinia megarhynchos*** Nightingale	B	1995	3,000	6,000	B	B3
Saxicola rubetra Whinchat	B	1996	150	250	B	B3
[1] ***Sylvia nisoria*** Barred Warbler	B	—	800	1,000	B	B3
[1] ***Lanius collurio*** Red-backed Shrike	B	1996	1,000	1,500	B	B2

1. Main site in Serbia.

An important site for breeding species of forest and scrub. Species of global conservation concern that do not meet IBA criteria: *Haliaeetus albicilla* (wintering), *Aquila heliaca* (breeding) and *Crex crex* (breeding).

Protection status

National Partial **International** None
29,640 ha of IBA covered by Strict Nature Reserve (Deliblatska Peščara, 29,640 ha).

Conservation issues

Threats Afforestation (A), Agricultural intensification/expansion (C), Burning of vegetation (C), Infrastructure (C), Intensified forest management (A), Recreation/tourism (C), Unsustainable exploitation (B)

The main threats are the afforestation of pasture and steppic habitats with pine *Pinus* and false acacia *Robinia*, overgrowth by scrub, and intensive hunting. Work to remove *Robinia* and *Crataegus* from grasslands has begun, and watering holes have been constructed. Plans exist to propose the area as a Biosphere Reserve. A management plan exists for the site.

Cer mountain — B2, B3 — 017

Admin region Serbia
Coordinates 44°36′N 19°30′E
Altitude 120–687 m **Area** 10,200 ha

Site description

A low mountain located in Macva at the edge of the Pannonian lowlands, between the towns of Loznica and Sabac. Almost completely covered by forest of oak *Quercus* and beech *Fagus*, with pasture and mown meadows. Numerous conifers have been planted in previously bare areas. Human activities include cattle-rearing.

Habitats Forest and woodland (80%; broadleaved deciduous forest), Scrub, Grassland (10%; dry siliceous grassland; humid grassland), Wetland (river/stream), Artificial landscape (10%; arable land; perennial crops/orchards/groves; forestry plantation; other urban/industrial areas)
Land-use Agriculture (15%), Forestry (80%), Hunting (60%), Not utilized (5%), Tourism/recreation (5%), Urban/industrial/transport (5%)

Birds

Species	Season	Year	Pop min	Pop max	Acc	Criteria
Dendrocopos syriacus Syrian Woodpecker	R	1997	40	70	B	B3
Lullula arborea Woodlark	B	1997	40	60	B	B2
Lanius collurio Red-backed Shrike	B	1997	150	250	B	B2

This is an important site for breeding species of forest and forest-edge habitats. Species of global conservation concern that do not meet IBA criteria: *Aquila heliaca* (breeding).

Protection status

National None **International** None

Conservation issues

Threats Afforestation (C), Consequences of animal/plant introductions (B), Deforestation (commercial) (C), Firewood collection (C), Intensified forest management (B)

The main threat is uncontrolled felling of old growth forest and subsequent replacement with younger plantations, as well as afforestation of the remaining bare terrain. Intensive ornithological research was undertaken between 1995 and 1997.

Valjevske mountains — B2 018

Admin region Serbia
Coordinates 44°09'N 19°34'E
Altitude 280–1,247 m **Area** 11,000 ha

Site description

Situated in the far west of Serbia along the Drina river, between the towns of Valjevo and Bejina Basta. The deciduous forests (predominantly beech *Fagus* and oak *Quercus*) are interspersed with pasture, meadows, limestone features, orchards and villages. Human activities include intensive agriculture and cattle-rearing.

Habitats Forest and woodland (60%; broadleaved deciduous forest; treeline ecotone), Scrub (5%; scrub), Grassland (10%; steppe/dry calcareous grassland; dry siliceous grassland; mesophile grassland), Wetland (river/stream), Rocky areas (10%; scree/boulders; inland cliff; caves), Artificial landscape (15%; arable land; perennial crops/orchards/groves; forestry plantation; ruderal land)
Land-use Agriculture (10%), Fisheries/aquaculture, Forestry (70%), Hunting (80%), Nature conservation/research (20%), Not utilized, Tourism/recreation (10%)

Birds

Species	Season	Year	Pop min	Pop max	Acc	Criteria
Aquila chrysaetos Golden Eagle	R	1996	2	3	A	B2
Otus scops Scops Owl	B	1996	20	30	B	B2

This is an important area for a variety of breeding raptors, including *Gyps fulvus* (together with Milesevka–Uvac, this is the only breeding site in Yugoslavia).

Protection status

National Low **International** None
596 ha of IBA covered by Special Nature Reserve (Klisura reke Tresnjice, 596 ha).

Conservation issues

Threats Afforestation (B), Burning of vegetation (C), Deforestation (commercial) (C), Disturbance to birds (C), Extraction industry (C), Forest grazing (C), Intensified forest management (C), Recreation/tourism (C), Unsustainable exploitation (U)

There are no immediate threats, other than the effects of disturbance and hunting. A feeding point for raptors was established in 1987 and is still operational. A management plan exists for part of the site.

Tara mountain — B2 019

Admin region Serbia
Coordinates 43°52'N 19°25'E
Altitude 230–1,673 m **Area** 36,000 ha

Site description

Located near to the Drina river in western Serbia, with coniferous, broadleaved and mixed forests (spruce *Picea*, fir *Abies*, pine *Pinus*, beech *Fagus* and oak *Quercus*) covering the majority of the land area. Human activities include cattle-grazing.

Habitats Forest and woodland (65%; broadleaved deciduous forest; native coniferous forest; mixed forest; treeline ecotone), Scrub (5%; scrub; sclerophyllous scrub/garrigue/maquis), Grassland (10%; steppe/dry calcareous grassland; mesophile grassland), Wetland (5%; river/stream; raised bog), Rocky areas (10%; scree/boulders; inland cliff; caves), Artificial landscape (5%; perennial crops/orchards/groves; forestry plantation; ruderal land)
Land-use Agriculture (10%), Forestry (70%), Hunting (90%), Nature conservation/research (70%), Tourism/recreation (30%), Water management (10%)

Birds

Species	Season	Year	Pop min	Pop max	Acc	Criteria
Aquila chrysaetos Golden Eagle	R	1997	3	4	A	B2
Falco peregrinus Peregrine	R	1997	4	4	A	B2
Otus scops Scops Owl	B	1997	20	30	B	B2
Picus canus Grey-headed Woodpecker	R	1993	40	60	C	B2
Picus viridis Green Woodpecker	R	1996	80	150	B	B2
Phoenicurus phoenicurus Redstart	B	1995	40	70	C	B2

This is an important for breeding raptors and forest-nesting species. Species of global conservation concern that do not meet IBA criteria: *Crex crex* (breeding).

Protection status

National Partial **International** None
19,200 ha of IBA covered by National Park (Tara, 19,200 ha).

Conservation issues

Threats Afforestation (C), Burning of vegetation (C), Construction/impact of dyke/dam/barrage (C), Deforestation (commercial) (C), Disturbance to birds (C), Extraction industry (C), Forest grazing (C), Infrastructure (C), Intensified forest management (B), Recreation/tourism (B)

The main threats are the effects of intensified forest management and recreation/tourism. A management plan exists for the National Park.

Ovcar–Kablar gorge — A1 020

Admin region Serbia
Coordinates 43°52'N 20°12'E
Altitude 245–985 m **Area** 5,400 ha

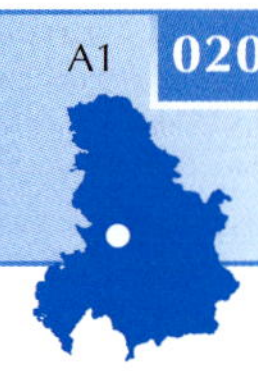

Site description

A gorge on the Zapadna Morava river, situated between the towns of Cacak and Pozega in western Serbia. Limestone cliffs and rocky terrain are interspersed with sparse tree cover and thickets comprising oak *Quercus*, ash *Fraxinus*, hornbeam *Carpinus*, elm *Ulmus* and lime *Tilia*. Areas of meadow and pasture, small reservoirs, houses and orchards are also present.

Habitats Forest and woodland (30%; broadleaved deciduous forest; treeline ecotone), Scrub (20%; scrub), Grassland (15%; steppe/dry calcareous grassland; mesophile grassland), Wetland (5%; standing fresh water; river/stream), Rocky areas (20%; scree/boulders; inland cliff), Artificial landscape (10%; arable land; perennial crops/orchards/groves; ruderal land)
Land-use Agriculture (5%), Fisheries/aquaculture (5%), Forestry (20%), Hunting (50%), Not utilized (30%), Tourism/recreation (5%), Urban/industrial/transport (5%), Water management (10%)

Birds

Species	Season	Year	Pop min	Pop max	Acc	Criteria
Phalacrocorax pygmeus Pygmy Cormorant	W	1993	50	70	—	A1

Also important for a variety of breeding raptors.

Protection status

National Low **International** None
IBA overlaps with protected area (unknown type).

Conservation issues

Threats Aquaculture/fisheries (C), Burning of vegetation (C), Deforestation (commercial) (B), Disturbance to birds (U), Extraction industry (C), Forest grazing (B), Industrialization/urbanization (B), Infrastructure (U), Recreation/tourism (C), Unsustainable exploitation (U)

The main threats are disturbance, hunting and infrastructure development.

Uvac–Mileševka A1, B2 021

Admin region Serbia
Coordinates 43°25′N 19°57′E
Altitude 500–1,625 m **Area** 30,000 ha

Site description

The limestone gorges of the Uvac and Mileševka rivers, situated between the Zlatar, Zlatibor and Jadovnik mountains and the Sjenicka plateau in western Serbia. The central part of the IBA is covered with dense coniferous or mixed forest (spruce *Picea*, fir *Abies*, pine *Pinus*, beech *Fagus* and oak *Quercus*); the periphery is mostly bare mountainous terrain. Extensive pasture and mown meadows exist towards the Sjenicka plateau. Human activities include cattle-rearing.

Habitats Forest and woodland (60%; broadleaved deciduous forest; native coniferous forest; mixed forest; treeline ecotone), Scrub (5%; scrub; sclerophyllous scrub/garrigue/maquis), Grassland (10%; steppe/dry calcareous grassland; humid grassland; mesophile grassland), Wetland (5%; standing fresh water; river/stream; fen/transition mire/spring), Rocky areas (10%; scree/boulders; inland cliff; caves), Artificial landscape (10%; perennial crops/orchards/groves; forestry plantation; urban parks/gardens; ruderal land)
Land-use Agriculture (10%), Fisheries/aquaculture (5%), Forestry (60%), Hunting (80%), Nature conservation/research (20%), Not utilized (10%), Tourism/recreation (10%), Urban/industrial/transport (5%), Water management (5%)

Birds

Species	Season	Year	Pop min	Pop max	Acc	Criteria
[1] ***Gyps fulvus*** Griffon Vulture	R	1997	36	36	A	B2
Circaetus gallicus Short-toed Eagle	B	1996	4	5	B	B2
Aquila pomarina Lesser Spotted Eagle	B	1996	2	2	A	B2
Aquila chrysaetos Golden Eagle	R	1996	5	6	A	B2
Falco peregrinus Peregrine	R	1996	4	5	A	B2
Alectoris graeca Rock Partridge	R	1996	50	80	B	B2
Crex crex Corncrake	B	1996	20	30	B	A1, B2
Otus scops Scops Owl	B	1996	30	40	B	B2
Bubo bubo Eagle Owl	R	1996	7	8	A	B2
Picus viridis Green Woodpecker	R	1996	100	150	B	B2
Lullula arborea Woodlark	B	1996	200	400	B	B2
Monticola saxatilis Rock Thrush	B	1996	15	20	B	B2
Emberiza cia Rock Bunting	R	1996	100	120	B	B2

1. 200–250 individuals in October 1997.

The site is important for a variety of breeding raptors. Species of global conservation concern that do not meet IBA criteria: *Phalacrocorax pygmeus* (wintering).

Protection status

National Low **International** None
1,200 ha of IBA covered by Special Nature Reserve (Klisura Reke Uvac, 1,200 ha).

Conservation issues

Threats Agricultural intensification/expansion (C), Aquaculture/fisheries (C), Deforestation (commercial) (C), Disturbance to birds (C), Intensified forest management (B), Recreation/tourism (C), Selective logging/cutting (B), Unsustainable exploitation (C)

Threats include disturbance, hunting and intensive forest exploitation. A feeding point for raptors (for example, *Gyps fulvus*, *Neophron percnopterus* and *Aquila chrysaetos*) was established in 1989 and has been very successful. A management plan exists for part of the site.

Kopaonik B2 022

Admin region Serbia
Coordinates 43°15′N 20°50′E
Altitude 650–2,016 m **Area** 45,000 ha

Site description

The largest mountain in central Serbia; a large coniferous forest (spruce *Picea*) in the Ravni Kopaonik predominates. A small area is covered with alpine vegetation.

Habitats Forest and woodland (50%; broadleaved deciduous forest; native coniferous forest; mixed forest; treeline ecotone), Scrub (5%; scrub), Grassland (30%; dry siliceous grassland; alpine/subalpine/boreal grassland; humid grassland; mesophile grassland), Wetland, Rocky areas (5%; scree/boulders; inland cliff), Artificial landscape (10%; perennial crops/orchards/groves; forestry plantation; ruderal land)
Land-use Agriculture (5%), Forestry (70%), Hunting (40%), Nature conservation/research (30%), Not utilized, Tourism/recreation (40%)

Birds

Species	Season	Year	Pop min	Pop max	Acc	Criteria
Alectoris graeca Rock Partridge	R	1995	50	70	B	B2
Otus scops Scops Owl	B	1996	20	30	C	B2
Lullula arborea Woodlark	R	1994	80	120	B	B2
Lanius collurio Red-backed Shrike	B	1996	400	600	B	B2

Species of global conservation concern that do not meet IBA criteria: *Crex crex* (5–10 breeding pairs).

Protection status

National Partial **International** None
11,800 ha of IBA covered by National Park (Kopaonik, 11,800 ha).

Conservation issues

Threats Abandonment/reduction of land management (C), Afforestation (C), Deforestation (commercial) (C), Disturbance to birds (B), Extraction industry (B), Forest grazing (C), Intensified forest management (B), Recreation/tourism (A), Selective logging/cutting (C)

Intensive ornithological research was undertaken between 1984 and 1997. A management plan exists for part of the site.

Prokletije A3, B2, B3 023

Admin region Serbia
Coordinates 42°43′N 20°10′E
Altitude 460–2,656 m **Area** 100,000 ha

Site description

Mainly a limestone massif comprising a number of peaks over 2,000 m, surrounded by the towns of Djakovica, Pec and Kosovska Mitrovica. Alpine vegetation occurs at high altitude; coniferous, mixed and broadleaved forests (spruce *Picea*, pine *Pinus*, maple *Acer*, beech *Fagus*, oak *Quercus* and chestnut *Castanea*) are present at lower levels.

Habitats Forest and woodland (50%; broadleaved deciduous forest; native coniferous forest; mixed forest; treeline ecotone), Scrub (10%; scrub; sclerophyllous scrub/garrigue/maquis), Grassland (25%; steppe/dry calcareous grassland; dry siliceous grassland; alpine/subalpine/boreal grassland; humid grassland; mesophile grassland), Wetland (5%; standing fresh water; river/stream; fen/transition mire/spring), Rocky areas (15%; scree/boulders; inland cliff), Artificial landscape (5%; perennial crops/orchards/groves; forestry plantation; ruderal land)
Land-use Agriculture (70%), Forestry (60%), Hunting (80%), Nature conservation/research (5%), Not utilized (20%)

Birds

The site is important for a variety of breeding raptors and other upland species, including four of the 10 species in Europe that are restricted (when breeding) to the Eurasian high-montane (alpine) biome. Species of global conservation concern that do not meet IBA criteria: *Crex crex* (breeding).

Species	Season	Year	Pop min	Pop max	Acc	Criteria
Pernis apivorus Honey Buzzard	B	1997	10	12	B	B3
Circaetus gallicus Short-toed Eagle	B	1997	4	6	B	B2
Aquila chrysaetos Golden Eagle	R	1997	6	7	A	B2
Falco peregrinus Peregrine	R	1997	6	7	A	B2
Alectoris graeca Rock Partridge	R	1997	150	200	B	B2
Otus scops Scops Owl	B	1997	50	80	B	B2
Bubo bubo Eagle Owl	R	1997	15	20	A	B2
Caprimulgus europaeus Nightjar	B	1997	100	150	C	B2
Lullula arborea Woodlark	R	1997	200	300	B	B2
Prunella collaris Alpine Accentor	R	1997	60	70	A	A3
Monticola saxatilis Rock Thrush	B	1997	30	40	B	B2
Turdus torquatus Ring Ouzel	R	1997	500	700	B	B3
Tichodroma muraria Wallcreeper	R	1997	10	12	A	A3
Lanius collurio Red-backed Shrike	B	1997	1,000	1,500	B	B2
Pyrrhocorax graculus Alpine Chough	R	1997	100	130	A	A3
Montifringilla nivalis Snowfinch	R	1997	10	15	A	A3
Emberiza cirlus Cirl Bunting	R	1997	1,000	2,000	B	B3
Emberiza cia Rock Bunting	R	1997	100	150	B	B2

Protection status

National Low **International** None

IBA overlaps with a few small protected areas (unknown type).

Conservation issues

Threats Burning of vegetation (A), Deforestation (commercial) (B), Disturbance to birds (C), Firewood collection (B), Forest grazing (A), Intensified forest management (B), Selective logging/cutting (B), Unsustainable exploitation (B)

Threats include disturbance, hunting, deforestation, burning and nomadic cattle-grazing. Proposed as a National Park (Prokletije, 100,000 ha).

Šara mountain — A1, A3, B2 — 024

Admin region Serbia
Coordinates 42°10′N 20°57′E
Altitude 450–2,651 m **Area** 80,000 ha

Site description

Situated in the far south of Serbia in Kosmet province, being a transboundary IBA that stretches across into Macedonia ('Sara mountains'; Macedonian site 001). The main mountain-ridge is surrounded by low-lying valleys and glacial lakes, with dry grassland slopes broken by numerous rivers/streams. Coniferous and broadleaved forest (pine *Pinus*, beech *Fagus* and oak *Quercus*) and alpine pastures are also present. Human activities include cattle-grazing.

Habitats Forest and woodland (40%; broadleaved deciduous forest; native coniferous forest; mixed forest; treeline ecotone), Scrub (5%; scrub), Grassland (40%; dry siliceous grassland; alpine/subalpine/boreal grassland; humid grassland; mesophile grassland), Wetland (5%; standing fresh water; river/stream; raised bog; water-fringe vegetation; fen/transition mire/spring), Rocky areas (5%; scree/boulders; inland cliff; caves), Artificial landscape (5%; forestry plantation; urban parks/gardens; ruderal land)
Land-use Agriculture (40%), Forestry (40%), Hunting (70%), Nature conservation/research (60%), Not utilized (20%), Tourism/recreation (10%)

Birds

Species	Season	Year	Pop min	Pop max	Acc	Criteria
Aquila chrysaetos Golden Eagle	R	1997	5	6	A	B2
Falco peregrinus Peregrine	R	1997	4	5	A	B2
[1] ***Alectoris graeca*** Rock Partridge	R	1996	150	250	B	B2
Crex crex Corncrake	B	1996	10	30	B	A1, B2
Bubo bubo Eagle Owl	R	1996	7	8	A	B2
Alauda arvensis Skylark	B	1996	2,000	3,000	B	B2
Prunella collaris Alpine Accentor	R	1997	50	70	A	A3
Tichodroma muraria Wallcreeper	R	1997	6	10	A	A3
Pyrrhocorax graculus Alpine Chough	R	1997	100	150	A	A3
[1] ***Pyrrhocorax pyrrhocorax*** Chough	R	1997	25	35	A	B2
Montifringilla nivalis Snowfinch	R	1996	15	25	A	A3

1. Main site in Serbia.

Important for breeding species of upland habitats, including four of the ten species in Europe that are restricted (when breeding) to the Eurasian high-montane (alpine) biome. Species of global conservation concern that do not meet IBA criteria: *Aegypius monachus* (non-breeding).

Protection status

National Partial **International** None

39,000 ha of IBA covered by National Park (Šar Planina, 39,000 ha).

Conservation issues

Threats Abandonment/reduction of land management (C), Afforestation (C), Burning of vegetation (C), Deforestation (commercial) (B), Disturbance to birds (C), Forest grazing (B), Intensified forest management (B), Natural events (B), Recreation/tourism (B), Selective logging/cutting (B)

A management plan exists for the part of the site.

Pcinja — B2, B3 — 025

Admin region Serbia
Coordinates 42°22′N 22°03′E
Altitude 410–1,345 m **Area** 16,000 ha

Site description

An extensive valley surrounded by the mountains of Kozjak, Rujen and Vlasinske in south-east Serbia, on the borders with Macedonia and Bulgaria. The diverse vegetation reflects sub-Mediterranean influences. Degraded oak *Quercus* forest occurs with open, dry grassland and areas of scrub. Human activities include cattle-rearing and some agriculture, as well as tourism at a low level.

Habitats Forest and woodland (35%; broadleaved deciduous forest; wooded steppe), Scrub (25%; scrub; sclerophyllous scrub/garrigue/maquis), Grassland (10%; steppe/dry calcareous grassland; dry siliceous grassland), Wetland (3%; river/stream), Rocky areas (15%; scree/boulders; inland cliff), Artificial landscape (10%; arable land; perennial crops/orchards/groves; forestry plantation; urban parks/gardens; ruderal land)
Land-use Agriculture (15%), Forestry (30%), Hunting, Not utilized (30%), Tourism/recreation (5%), Urban/industrial/transport (10%)

Birds

Species	Season	Year	Pop min	Pop max	Acc	Criteria
Picus viridis Green Woodpecker	R	1994	80	150	B	B2
Lullula arborea Woodlark	R	1995	200	500	B	B2
Luscinia megarhynchos Nightingale	B	1995	2,000	3,000	C	B3
[1] ***Parus lugubris*** Sombre Tit	R	1994	50	100	B	B3
Lanius collurio Red-backed Shrike	B	1994	400	600	B	B2
Lanius minor Lesser Grey Shrike	B	1994	10	20	B	B2
[1] ***Lanius senator*** Woodchat Shrike	B	1995	20	40	B	B2

1. Main site in Serbia.

This is an important site for a number of species associated with open, heterogeneous landscapes.

Protection status

National Low **International** None

120 ha of IBA covered by Landscape Nature Reserve (Prohor Pciniski, 120 ha).

Conservation issues

Threats Afforestation (B), Burning of vegetation (C), Consequences of animal/plant introductions (C), Deforestation (commercial) (B), Disturbance to birds (C), Drainage (C), Firewood collection (B), Forest grazing (B), Recreation/tourism (C), Unsustainable exploitation (U)

The main threat is the illegal hunting of birds. A management plan exists for the site.

Vlasina — B2 — 026

Admin region Serbia
Coordinates 42°41′N 22°22′E
Altitude 900–1,874 m **Area** 22,000 ha

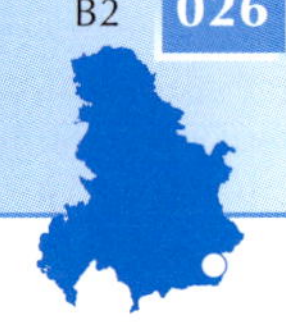

Site description

A plateau surrounded by the summits of Vardenik, Cemernik and Vrtop in south-east Serbia. The main part of the plateau has been turned into a large freshwater reservoir, with floating peat islands. The surroundings comprise coniferous and broadleaved (beech *Fagus*) forest and pasture.

Habitats Forest and woodland (25%; broadleaved deciduous forest; alluvial/very wet forest; treeline ecotone), Scrub (5%; scrub), Grassland (30%; dry siliceous grassland; alpine/subalpine/boreal grassland; humid grassland; mesophile grassland), Wetland (25%; standing fresh water; river/stream; raised bog; water-fringe vegetation; fen/transition mire/spring), Artificial landscape (15%; arable land; perennial crops/orchards/groves; forestry plantation; other urban/industrial areas; ruderal land)
Land-use Agriculture (70%), Fisheries/aquaculture (30%), Forestry (40%), Hunting (90%), Tourism/recreation (50%), Urban/industrial/transport (5%), Water management (30%)

Birds

Species	Season	Year	Pop min	Pop max	Acc	Criteria
Coturnix coturnix Quail	B	1992	500	1,000	B	B2
Lullula arborea Woodlark	B	1993	100	300	B	B2
Alauda arvensis Skylark	B	1993	3,000	4,000	B	B2
Lanius collurio Red-backed Shrike	B	1993	300	500	B	B2

An important site for breeding species of open, upland habitats.

Protection status

National None **International** None

Conservation issues

Threats Afforestation (B), Agricultural intensification/expansion (C), Aquaculture/fisheries (B), Construction/impact of dyke/dam/barrage (A), Industrialization/urbanization (C), Recreation/tourism (B)

Threats include rapid tourist development and infrastructure construction (for example, roads, hotels and cable railways).

Jerma — B3 — 027

Admin region Serbia
Coordinates 42°57′N 22°37′E
Altitude 620–1,442 m **Area** 10,000 ha

Site description

A river gorge situated in eastern Serbia, on the border with Bulgaria. Typical limestone karst features are present. Scarce wooded areas consist of oak *Quercus*, hornbeam *Carpinus* and beech *Fagus*, and there are extensive dry pastures on plateau areas. Human activities include crop production.

Habitats Forest and woodland (20%; broadleaved deciduous forest; treeline ecotone), Scrub (30%; scrub; sclerophyllous scrub/garrigue/maquis), Grassland (15%; steppe/dry calcareous grassland; mesophile grassland), Wetland, Rocky areas (25%; scree/boulders; inland cliff), Artificial landscape (10%; arable land; perennial crops/orchards/groves; urban parks/gardens; ruderal land)
Land-use Agriculture (10%), Forestry (20%), Hunting (60%), Nature conservation/research (70%), Tourism/recreation (20%), Urban/industrial/transport (5%)

Birds

Species	Season	Year	Pop min	Pop max	Acc	Criteria
Luscinia megarhynchos Nightingale	B	1996	C	—	—	B3

Protection status

National Low **International** None
IBA overlaps with protected area (unknown type).

Conservation issues

Threats Afforestation (C), Burning of vegetation (C), Deforestation (commercial) (B), Infrastructure (C), Intensified forest management (C), Recreation/tourism (C)

Proposed as a Landscape of Great Value (Jerma).

Suva mountain — B2 — 028

Admin region Serbia
Coordinates 43°12′N 22°12′E
Altitude 470–1,808 m **Area** 20,000 ha

Site description

A prominent limestone massif that adjoins the Juzna Morava river valley. Its southern side is covered with dense vegetation and pasture, whereas the northern slopes are barren. A range of habitats is present, ranging from sub-Mediterranean rocky terrain and scrub at lower levels, to subalpine pasture on the highest summits. Broadleaved forests (oak *Quercus* and beech *Fagus*) are significantly degraded. Human activities include husbandry and hay-cutting.

Habitats Forest and woodland (35%; broadleaved deciduous forest; treeline ecotone), Scrub (10%; scrub; sclerophyllous scrub/garrigue/maquis), Grassland (35%; steppe/dry calcareous grassland; alpine/subalpine/boreal grassland; mesophile grassland), Rocky areas (15%; scree/boulders; inland cliff; caves), Artificial landscape (5%; arable land; perennial crops/orchards/groves; forestry plantation; ruderal land)
Land-use Agriculture (5%), Forestry (40%), Hunting (100%), Nature conservation/research (5%), Tourism/recreation (10%)

Birds

Species	Season	Year	Pop min	Pop max	Acc	Criteria
Otus scops Scops Owl	B	1994	20	40	C	B2

A notable site (at the national level) for breeding species of upland habitats.

Protection status

National None **International** None

Conservation issues

Threats Afforestation (B), Burning of vegetation (C), Consequences of animal/plant introductions (C), Deforestation (commercial) (C), Disturbance to birds (C), Firewood collection (C), Forest grazing (C), Intensified forest management (C), Recreation/tourism (C), Unsustainable exploitation (C)

Stara mountain–Vidlic — B2, B3 — 029

Admin region Serbia
Coordinates 43°15′N 22°45′E
Altitude 400–2,169 m **Area** 44,000 ha

Site description

The largest massif in eastern Serbia, close to the Bulgarian border; this is a transboundary IBA, listed as 'Tchuprene' (Bulgarian site 002). Habitat-types present include sparse coniferous (spruce *Picea*, fir *Abies*) and broadleaved (beech *Fagus*) forest, sub-Mediterranean thickets, and alpine/steppic pastures. Human activities are limited.

Habitats Forest and woodland (30%; broadleaved deciduous forest; native coniferous forest; mixed forest), Scrub, Grassland (50%; steppe/dry calcareous grassland; dry siliceous grassland; alpine/subalpine/boreal grassland; humid grassland; mesophile grassland), Wetland (river/stream; raised bog), Rocky areas (8%), Artificial landscape (15%; arable land; perennial crops/orchards/groves; forestry plantation; ruderal land)
Land-use Agriculture (10%), Forestry (40%), Hunting (60%), Nature conservation/research (10%), Not utilized (40%), Tourism/recreation (5%)

Birds

Species	Season	Year	Pop min	Pop max	Acc	Criteria
Circaetus gallicus Short-toed Eagle	B	1997	2	3	A	B2
Aquila chrysaetos Golden Eagle	R	1997	2	3	A	B2
Falco peregrinus Peregrine	R	1997	3	4	A	B2
Alectoris graeca Rock Partridge	R	1997	50	70	C	B2
Bubo bubo Eagle Owl	R	1997	6	8	B	B2
[1] ***Alauda arvensis*** Skylark	B	1997	3,000	4,000	B	B2
Turdus torquatus Ring Ouzel	R	1997	40	60	C	B3

1. Main site in Serbia.

This is an important site for breeding raptors and other upland species.

Protection status

National Partial **International** None
IBA overlaps with Regional Nature Park (Stara planina, 142,000 ha).

Conservation issues

Threats Afforestation (C), Consequences of animal/plant introductions (C), Deforestation (commercial) (C), Disturbance to birds (C), Firewood collection (C), Forest grazing (B), Intensified forest management (B), Recreation/tourism (C)

A management plan exists for part of the site.

Sicevo gorge B2 030

Admin region Serbia
Coordinates 43°18′N 22°10′E
Altitude 230–1,334 m **Area** 9,800 ha

Site description

Situated in eastern Serbia between Nis and Bela Palanka, this limestone gorge contains the Nisava river and supports sub-Mediterranean vegetation. Vegetation is dominated by hornbeam *Carpinus*, *Syringa*, ash *Fraxinus* and oak *Quercus*. A heavily used road passes through the valley.

Habitats Forest and woodland (5%; broadleaved deciduous forest), Scrub (25%; scrub), Grassland (25%; steppe/dry calcareous grassland), Wetland, Rocky areas (25%; scree/boulders; inland cliff; caves), Artificial landscape (20%; arable land; perennial crops/orchards/groves; other urban/industrial areas; ruderal land)
Land-use Agriculture (15%), Forestry (5%), Not utilized (50%), Urban/industrial/transport (10%), Water management

Birds

Species	Season	Year	Pop min	Pop max	Acc	Criteria
Bubo bubo Eagle Owl	R	1997	5	6	A	B2

Protection status

National Low **International** None
860 ha of IBA covered by Regional Nature Park (860 ha).

Conservation issues

Threats Afforestation (C), Agricultural intensification/expansion (C), Burning of vegetation (B), Consequences of animal/plant introductions (C), Construction/impact of dyke/dam/barrage (C), Deforestation (commercial) (A), Disturbance to birds (B), Extraction industry (A), Firewood collection (B), Forest grazing (C), Industrialization/urbanization (C), Infrastructure (B), Intensified forest management (C), Natural events (C), Recreation/tourism (C), Selective logging/cutting (C)

Threats include the development of traffic infrastructure and quarrying. A management plan exists for part of the site.

Zlot gorge–Dubašnica B2 031

Admin region Serbia
Coordinates 44°00′N 22°00′E
Altitude 270–1,191 m **Area** 10,000 ha

Site description

Situated in eastern Serbia and surrounded by the Malinik and Dubašnica mountains. The cliffs in the limestone gorge are up to 500 m high; screes and diverse vegetation are also present. The Dubašnica plateau supports dry pasture and beech *Fagus* forest, interspersed with typical karst relief. Human activities include husbandry and hunting of large mammals.

Habitats Forest and woodland (50%; broadleaved deciduous forest; mixed forest), Scrub (20%; scrub; sclerophyllous scrub/garrigue/maquis), Grassland (10%; steppe/dry calcareous grassland; mesophile grassland), Wetland (river/stream; fen/transition mire/spring), Rocky areas (15%; scree/boulders; inland cliff; caves), Artificial landscape (5%; perennial crops/orchards/groves; forestry plantation; ruderal land)
Land-use Agriculture, Forestry (70%), Hunting (80%), Nature conservation/research (10%), Tourism/recreation (5%)

Birds

Species	Season	Year	Pop min	Pop max	Acc	Criteria
Lullula arborea Woodlark	B	1995	50	80	B	B2
[1] ***Phoenicurus phoenicurus*** Redstart	B	1993	50	80	B	B2

1. Main site in Serbia.

An important site for breeding species of forest and forest-edge habitats.

Protection status

National Partial **International** None
IBA overlaps with Strict Nature Reserve (Malinik). IBA overlaps with Speleological Monument (Lazareva Pecina).

Conservation issues

Threats Abandonment/reduction of land management (C), Afforestation (B), Burning of vegetation (C), Consequences of animal/plant introductions (C), Deforestation (commercial) (C), Firewood collection (C), Forest grazing (B), Groundwater abstraction (C), Intensified forest management (C)

Resava gorge B2 032

Admin region Serbia
Coordinates 44°10′N 21°45′E
Altitude 320–1,339 m **Area** 9,500 ha

Site description

A limestone gorge situated in the mountainous area of Juzni Kucaj in eastern Serbia, near Despotovac and Zagubica. Cliffs, rocky terrain, forest (beech *Fagus*, oak *Quercus*, hornbeam *Carpinus*) and pasture are present; a coniferous plantation was recently established. Human activities include cattle-grazing.

Habitats Forest and woodland (50%; broadleaved deciduous forest; treeline ecotone), Scrub (20%; scrub; sclerophyllous scrub/garrigue/maquis), Grassland (15%; steppe/dry calcareous grassland; mesophile grassland), Wetland (river/stream; fen/transition mire/spring), Rocky areas (10%; scree/boulders; inland cliff; caves), Artificial landscape (5%; perennial crops/orchards/groves; forestry plantation; ruderal land)
Land-use Agriculture (20%), Forestry (60%), Hunting (80%), Nature conservation/research (10%), Tourism/recreation (5%), Urban/industrial/transport

Birds

Species	Season	Year	Pop min	Pop max	Acc	Criteria
Otus scops Scops Owl	B	1993	20	30	C	B2

Protection status

National Partial **International** None
885 ha of IBA covered by Strict Nature Reserve (Klisura Gornje, 885 ha). 70 ha of IBA covered by Strict Nature Reserve (Vinatovaca, 70 ha).

Conservation issues

Threats Abandonment/reduction of land management (C), Afforestation (B), Burning of vegetation (C), Deforestation (commercial) (C), Disturbance to birds (C), Forest grazing (C), Intensified forest management (B), Recreation/tourism (U), Selective logging/cutting (B)

The main threats are intensified forest management and recreation/tourism.

Dubovac–Ram A1, A4i, A4iii, B1i, B2, B3 033

Admin region Serbia
Coordinates 44°49′N 21°16′E
Altitude 68–282 m **Area** 12,000 ha

Site description

A section of the Danube flood-plain, situated at the entrance to Đerdap gorge in the south-east Pannonian lowlands. Natural forests (willow *Salix*, poplar *Populus*, ash *Fraxinus*, oak *Quercus*), extensive *Populus* plantations, wet meadows and marsh, and dry grassland are present. There is heavy boat traffic on the Danube.

Habitats Forest and woodland (20%; broadleaved deciduous forest; alluvial/very wet forest; wooded steppe), Scrub (5%; scrub), Grassland (5%; steppe/dry calcareous grassland; humid grassland; mesophile grassland), Wetland (60%; tidal river/enclosed tidal water; sand-dunes/sand beach; standing fresh water; river/stream; water-fringe vegetation), Artificial landscape (10%; arable land; perennial crops/orchards/groves; forestry plantation; urban parks/gardens; ruderal land)
Land-use Agriculture (30%), Fisheries/aquaculture (60%), Forestry (30%), Hunting (80%), Tourism/recreation (5%), Urban/industrial/transport (5%), Water management (20%)

Birds

The site is important for breeding, wintering and passage waterbirds, regularly holding 60,000–100,000 individuals during winter and migration periods. This is also believed to be an important area for migrating storks—during autumn migration in 1985, 2,000 *Ciconia ciconia* and 100 *C. nigra* were counted over five days from part of the site—and also raptors, but further data on numbers are needed. Species of global conservation concern that do not meet IBA criteria: *Haliaeetus albicilla* (resident) and *Crex crex* (5–10 breeding pairs).

Species	Season	Year	Pop min	Pop max	Acc	Criteria
Phalacrocorax pygmeus Pygmy Cormorant	B	1997	150	250	B	A1, A4i, B1i, B2
Phalacrocorax pygmeus Pygmy Cormorant	W	—	500	1,000	—	A1, A4i, B1i
Ixobrychus minutus Little Bittern	B	1996	50	70	C	B2
Nycticorax nycticorax Night Heron	B	1997	150	200	A	B2
Ardeola ralloides Squacco Heron	B	1997	30	50	B	B2
Ciconia nigra Black Stork	P	1997	200	300	—	A4i, B1i
[1,2] ***Plegadis falcinellus*** Glossy Ibis	B	1997	10	15	A	B2
Aythya ferina Pochard	B	1996	30	50	B	B3
Aythya ferina Pochard	P	1996	10,000	20,000	—	B1i
Aythya nyroca Ferruginous Duck	B	1996	30	40	B	A1, B2
Bucephala clangula Goldeneye	P	1997	1,000	2,000	—	B1i
[1] ***Riparia riparia*** Sand Martin	B	1995	15,000	20,000	B	B2

1. Main site in Serbia.
2. Stable.

Protection status

National None **International** None

Conservation issues

Threats Afforestation (B), Agricultural intensification/expansion (C), Disturbance to birds (C), Drainage (C), Infrastructure (B), Intensified forest management (B), Unsustainable exploitation (B)

Part of the IBA is proposed as a Special Nature Reserve (Deliblatska Peščara, 40,000 ha) and Biosphere Reserve. A management plan exists for part of the site.

Đerdap gorge — A1, A4i, A4iii, B1i, B2 — 034

Admin region Serbia
Coordinates 44°27′N 22°10′E
Altitude 45–803 m **Area** 65,000 ha

Site description

Situated in north-east Serbia on the border with Romania, the site covers a section of the River Danube gorge and adjacent belt of land, about 6 km in width and some 100 km in length. This is a transboundary IBA shared with Romanian site 032 'Porţile de Fier reservoir'. Human activities include boat traffic on the Danube.

Habitats Forest and woodland (60%; broadleaved deciduous forest; treeline ecotone), Scrub (10%; scrub; sclerophyllous scrub/garrigue/maquis), Grassland (7%; steppe/dry calcareous grassland; mesophile grassland), Wetland (5%; standing fresh water; river/stream; fen/transition mire/spring), Rocky areas (5%; scree/boulders; inland cliff), Artificial landscape (5%; perennial crops/orchards/groves; forestry plantation; ruderal land)
Land-use Fisheries/aquaculture (5%), Forestry (60%), Hunting (5%), Nature conservation/research (95%), Not utilized (5%), Tourism/recreation (10%), Urban/industrial/transport (5%), Water management (5%)

Birds

Species	Season	Year	Pop min	Pop max	Acc	Criteria
Phalacrocorax pygmeus Pygmy Cormorant	W	1993	150	200	—	A1, A4i, B1i
[1] ***Aquila pomarina*** Lesser Spotted Eagle	B	1994	2	3	A	B2
[1] ***Hieraaetus pennatus*** Booted Eagle	B	1994	2	3	A	B2

1. Main site in Serbia.

The site is important for breeding raptors and wintering waterbirds, holding 20,000 or more individuals on a regular basis during winter (45,000 in January 1999). This is also believed to be an important area for migrating storks and raptors but further data on numbers are needed. Species of global conservation concern that do not meet IBA criteria: *Haliaeetus albicilla* (wintering), *Aquila heliaca* (passage) and *Crex crex* (3–7 breeding pairs).

Protection status

National High **International** None
63,500 ha of IBA covered by National Park (Đirdap, 63,500 ha).

Conservation issues

Threats Afforestation (C), Aquaculture/fisheries (B), Burning of vegetation (C), Deforestation (commercial) (C), Disturbance to birds (C), Extraction industry (C), Forest grazing (C), Infrastructure (C), Recreation/tourism (B)

A management plan exists for the site.

Mala vrbica — B2 — 035

Admin region Serbia
Coordinates 44°36′N 22°39′E
Altitude 40–90 m **Area** 2,000 ha

Site description

An area of steppe-like lowland in Negotinska Krajina, between the vast Daccian lowlands to the east and the Carpathian massif to the west. Habitat-types include a fish-pond, dry pasture, high loess bluffs and diverse shrub vegetation. Human activities include cattle-rearing.

Habitats Forest and woodland (5%; wooded steppe), Scrub (5%; scrub), Grassland (20%; steppe/dry calcareous grassland; mesophile grassland), Wetland (50%; standing fresh water; water-fringe vegetation), Artificial landscape (20%; arable land; perennial crops/orchards/groves; ruderal land)
Land-use Agriculture (20%), Fisheries/aquaculture (60%), Hunting (20%), Not utilized (5%)

Birds

Species	Season	Year	Pop min	Pop max	Acc	Criteria
Chlidonias hybridus Whiskered Tern	B	1992	80	100	B	B2
Merops apiaster Bee-eater	B	1995	50	60	A	B2

Species of global conservation concern that do not meet IBA criteria: *Phalacrocorax pygmeus* (wintering) and *Aythya nyroca* (breeding).

Protection status

National None **International** None

Conservation issues

Threats Disturbance to birds (B), Drainage (B), Dredging/canalization (C), Infrastructure (C)

Durmitor — B2, B3 — 036

Admin region Montenegro
Coordinates 43°07′N 19°01′E
Altitude 500–2,522 m **Area** 33,000 ha

Site description

A largely unspoiled mountainous area that includes several lakes and part of the Tara river gorge. The largest lake is Crno jezero. The nearest settlement, Zabljak, is a skiing centre.

Habitats Forest and woodland (broadleaved deciduous forest; native coniferous forest; mixed forest), Scrub (scrub), Grassland (steppe/dry calcareous grassland; alpine/subalpine/boreal grassland; mesophile grassland), Wetland (standing fresh water; river/stream), Rocky areas (scree/boulders; inland cliff), Artificial landscape
Land-use Agriculture, Forestry, Hunting, Nature conservation/research, Tourism/recreation, Urban/industrial/transport

Birds

The B2 criterion has been assigned at the site level as the following species of European conservation concern are present, but in unknown breeding numbers: *Pernis apivorus*, *Circaetus gallicus*, *Aquila chrysaetos*, *Falco tinnunculus*, *Falco peregrinus*, *Alectoris graeca*, *Perdix perdix*, *Coturnix coturnix*, *Bubo bubo*, *Otus scops*, *Caprimulgus europaeus*, *Picus canus*, *P. viridis*, *Lullula arborea*, *Alauda arvensis*, *Turdus torquatus*, *Saxicola rubetra*, *Monticola saxatilis*, *Phoenicurus phoenicurus*, *Lanius minor*, *Lanius collurio*, *Certhia brachydactyla*,

Parus lugubris, *Emberiza cirlus* and *Emberiza cia.* The B3 criterion has been assigned at the site level as the following species of European conservation concern are present, but in unknown breeding numbers: *Columba palumbus*, *Strix aluco*, *Dendrocopos syriacus*, *Prunella modularis*, *Erithacus rubecula*, *Luscinia megarhynchos*, *Turdus philomelos*, *Sylvia communis*, *Regulus regulus*, *Parus cristatus*, *Carduelis cannabina* and *Emberiza citrinella*. Species of global conservation concern that do not meet IBA criteria: *Crex crex* (small numbers breed regularly).

Protection status

National High **International** High

33,000 ha of IBA covered by National Park (Durmitor, 39,000 ha). 32,000 ha of IBA covered by World Heritage Site (Durmitor National Park, 32,000 ha).

Conservation issues

Threats Deforestation (commercial) (B), Disturbance to birds (B), Firewood collection (C), Forest grazing (B), Infrastructure (C), Intensified forest management (B), Recreation/tourism (B), Selective logging/cutting (C), Unsustainable exploitation (A)

Tourism/recreation has an impact which, though currently limited in its extent, could worsen if further development of skiing facilities takes place.

Biogradska woods — B2, B3 — 037

Admin region Montenegro
Coordinates 42°40′N 19°50′E
Altitude 800–2,116 m **Area** 20,000 ha

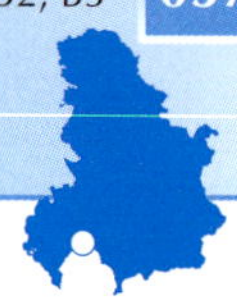

Site description

A largely mountainous area in the Bjelasica mountains, including several small lakes and large old-growth forests. The largest lake is Biogradsko jezero. The nearest settlements are Mojkovac and Kolasin, where there is a skiing centre.

Habitats Forest and woodland (broadleaved deciduous forest; native coniferous forest; mixed forest), Scrub, Grassland (steppe/dry calcareous grassland; alpine/subalpine/boreal grassland; mesophile grassland), Wetland (standing fresh water; river/stream), Rocky areas (scree/boulders; inland cliff), Artificial landscape
Land-use Agriculture, Forestry, Hunting, Nature conservation/research, Tourism/recreation, Urban/industrial/transport

Birds

The B2 criterion has been assigned at the site level as the following species of European conservation concern are present, but in unknown breeding numbers: *Circaetus gallicus*, *Aquila chrysaetos*, *Falco tinnunculus*, *Falco subbuteo*, *Falco peregrinus*, *Alectoris graeca*, *Perdix perdix*, *Coturnix coturnix*, *Bubo bubo*, *Otus scops*, *Caprimulgus europaeus*, *Picus canus*, *Picus viridis*, *Lullula arborea*, *Alauda arvensis*, *Turdus torquatus*, *Saxicola rubetra*, *Monticola saxatilis*, *Lanius collurio*, *Certhia brachydactyla*, *Parus lugubris*, *Emberiza cirlus* and *Emberiza cia*. The B3 criterion has also been assigned at the site level as the following species of European conservation concern are present, but in unknown breeding numbers: *Columba palumbus*, *Strix aluco*, *Dendrocopos syriacus*, *Prunella modularis*, *Erithacus rubecula*, *Luscinia megarhynchos*, *Turdus philomelos*, *Sylvia communis*, *Regulus regulus*, *Parus cristatus*, *Carduelis cannabina* and *Emberiza citrinella*.

Protection status

National Partial **International** None

5,400 ha of IBA covered by National Park (Biogradska Gora, 5,400 ha).

Conservation issues

Threats Abandonment/reduction of land management (U), Deforestation (commercial) (C), Disturbance to birds (B), Firewood collection (C), Forest grazing (B), Infrastructure (C), Intensified forest management (C), Recreation/tourism (B), Selective logging/cutting (C), Unsustainable exploitation (B)

Tourism and recreation impact on the area and could worsen if further development of skiing facilities takes place.

Lake Skadar — A1, A3, A4i, A4iii, B1i, B2 — 038

Admin region Montenegro
Coordinates 42°11′N 19°15′E
Altitude 5–100 m **Area** 40,000 ha

Site description

A large lake (also known as Lake Scutari) shared with Albania (Albanian site 001), situated near the town of Podgorica, with wetlands on the northern side; several rivers supply the lake.

Habitats Forest and woodland, Scrub, Grassland (humid grassland), Wetland (mudflat/sandflat; standing fresh water; river/stream; water-fringe vegetation), Artificial landscape
Land-use Agriculture, Fisheries/aquaculture, Hunting, Nature conservation/research, Tourism/recreation, Urban/industrial/transport, Water management

Birds

Species	Season	Year	Pop min	Pop max	Acc	Criteria
Podiceps cristatus Great Crested Grebe	W	1996	1,000	6,500	A	A4i, B1i
Podiceps nigricollis Black-necked Grebe	W	1996	4,000	22,000	A	A4i, B1i
Phalacrocorax pygmeus Pygmy Cormorant	R	—	1,500	2,000	B	A1, A4i, B1i, B2
Phalacrocorax pygmeus Pygmy Cormorant	W	1996	2,500	11,000	B	A4i, B1i
[1] ***Phalacrocorax pygmeus*** Pygmy Cormorant	P	—	8,000	10,000	—	A1, A4i, B1i
[2] ***Pelecanus crispus*** Dalmatian Pelican	R	1998	9	9	A	A1
Ardeola ralloides Squacco Heron	B	1998	300	400	B	A4i, B1i
Egretta alba Great White Egret	W	1996	100	600	B	A4i, B1i
Anas crecca Teal	W	1996	1,000	40,000	A	A4i, B1i
Aythya ferina Pochard	W	1996	16,000	55,000	A	A4i, B1i
Aythya nyroca Ferruginous Duck	B	1998	20	30	B	A1
Aythya fuligula Tufted Duck	W	1996	13,000	45,000	A	A4i, B1i
Alectoris graeca Rock Partridge	R	—	—	—	—	A3
Fulica atra Coot	W	1996	65,000	120,000	A	A4i, B1i
Oenanthe hispanica Black-eared Wheatear	B	—	—	—	—	A3
Hippolais olivetorum Olive-tree Warbler	B	—	—	—	—	A3
Sylvia cantillans Subalpine Warbler	B	—	—	—	—	A3
Sitta neumayer Rock Nuthatch	R	—	—	—	—	A3
Emberiza melanocephala Black-headed Bunting	B	—	—	—	—	A3

1. Autumn passage.
2. The only site in Yugoslavia.

This is a very important site for breeding, wintering and passage waterbirds, holding more than 20,000 individuals throughout the year. Numbers in winter range from 150,846 to 223,792, and in November numbers on passage regularly average 250,000. Breeding landbirds include six of the 21 species in Europe that are restricted (when breeding) to the Mediterranean biome.

Protection status

National High **International** Partial

40,000 ha of IBA covered by National Park (Skadarsko Jezero, 40,000 ha). 20,000 ha of IBA covered by Ramsar Site (Skadarsko Jezero, 20,000 ha).

Conservation issues

Threats Agricultural intensification/expansion (B), Aquaculture/fisheries (A), Disturbance to birds (B), Drainage (B), Other (A), Unsustainable exploitation (A), Recreation/tourism (B)

Large quantities of chemicals are thought to have entered the lake several years ago, with devastating consequences for fish and birds ('Other' threat). Many bird species are thought to have suffered large declines as a result. Additional threats include fishing and poaching.

Lake Sasko — B1i, B2, B3 — 039

Admin region Montenegro
Coordinates 41°58′N 19°22′E
Altitude — **Area** 350 ha

Site description

A small lake south of Lake Skadar (site 038) and part of the Bojana river catchment. It is situated about 10 km inland from the Adriatic coast, almost on the border with Albania, forming a transboundary IBA with 'Velipoja'—Albanian site 013.

Habitats Wetland (standing fresh water)
Land-use Unknown

Birds

Species	Season	Year	Pop min	Pop max	Acc	Criteria
Platalea leucorodia Spoonbill	B	—	25	25	—	B1i

A colony of breeding herons and cormorants existed at Lake Sasko formerly, but shifted in the late 1980s and early 1990s, probably to the Albanian side of the border, near the Bojana river. Birds from the colony still visit Lake Sasko regularly to feed. The B2 criterion has been assigned at the site level as the following species of European conservation concern are present during the breeding season, but in largely unknown numbers: *Phalacrocorax pygmeus*, *Botaurus stellaris*, *Ixobrychus minutus*, *Nycticorax nycticorax* (20 pairs), *Ardeola ralloides*, *Ardea purpurea*, *Plegadis falcinellus* (a few pairs), *Anas querquedula*, *Falco tinnunculus*, *Perdix perdix*, *Tringa totanus*, *Chlidonias niger*, *Chlidonias hybridus*, *Bubo bubo*, *Athene noctua*, *Otus scops*, *Picus viridis*, *Galerida cristata*, *Lanius senator* and *Emberiza hortulana*. The B3 criterion has been assigned at the site level as the following species of European conservation concern are present during the breeding season, but in unknown numbers: *Aythya ferina*, *Erithacus rubecula*, *Luscinia megarhynchos*, *Turdus merula*, *Locustella luscinioides*, *Sylvia melanocephala*, *Sylvia communis*, *Parus lugubris*, *Corvus monedula* and *Emberiza citrinella*.

Protection status

National None **International** None

Conservation issues

Threats Unknown

Ulcinj saltpans — A1, A4i, B1i, B2 — 040

Admin region Montenegro
Coordinates 41°55′N 19°18′E
Altitude 0–0 m **Area** 1,350 ha

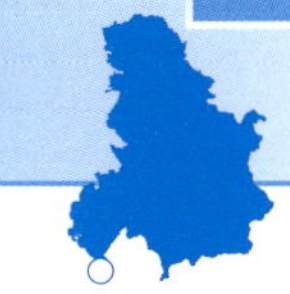

Site description

An area of salt-pans (with fresh water in winter and spring; saline in summer and autumn), largely without vegetation, bordered by agricultural land. Human activities include salt-extraction. The site lies close to Lake Sasko (site 039) and is a transboundary IBA with 'Velipoja' in Albania (site AL013).

Habitats Scrub, Wetland (standing brackish and salt water)
Land-use Hunting (80%), Urban/industrial/transport

Birds

Species	Season	Year	Pop min	Pop max	Acc	Criteria
[1] ***Phalacrocorax pygmeus*** Pygmy Cormorant	N	—	—	70	—	A1
[2] ***Glareola pratincola*** Collared Pratincole	B	1988	75	90	A	A4i, B1i, B2

1. Feeding birds from colonies elsewhere. Also occurs on passage.
2. Up to 150 pairs estimated.

The site is important for breeding and feeding waders, several of which are species of European conservation concern. In summer 1988, 15,000 waders (Charadrii) were counted.

Protection status

National None **International** None

Conservation issues

Threats Disturbance to birds (B), Industrialization/urbanization (U)

Due to the presence of the salt-extraction plant, hunting and human disturbance are controlled, but the industry itself causes some disturbance, and changes in water-level sometimes flood nests.

REFERENCES

BALOG, I. (1997) [Results of ringing herons at Bečej fishpond.] *Ciconia (Bull. Bird Prot. Study Soc. Voivodina)* 6: 17. (In Serbian, with English summary.)

BALOG, I. (1997) [Breeding of Pygmy Cormorant *Phalacrocorax pygmeus* on Bečej fishpond.] *Ciconia (Bull. Bird Prot. Study Soc. Voivodina)* 6: 40–43. (In Serbian, with English summary.)

BOZIĆ, J. (1995) [Bird fauna of Jablanica and Obnica (west Serbia).] *SNIRS Bull. Young Researchers of Serbia* 10: 37–49. (In Serbian.)

BUDAKOV, L. J. AND BRANKOVIĆ, D. (1996) [Proposal for the protection of the "Zasavica" as a Special Nature Reserve.] Belgrade: Institute for the Protection of Nature of Serbia. Unpublished report. (In Serbian.)

BUDAKOV, L. J. *ET AL.* (1994) [Proposal for the protection of the "Stari Begej—Carska bara" as a Special Nature Reserve.] Belgrade: Institute for the Protection of Nature of Serbia. Unpublished report. (In Serbian.)

BUDAKOV, L. J. *ET AL.* (1995) [Proposal for the protection of Palić lake as a Nature Park.] Belgrade: Institute for the Protection of Nature of Serbia. (In Serbian.)

BUTORAČ, B. AND HULO, I. (1992) [Phytocenological, floristic and ornithologic values of the "Selevenjska pustara" area as the base for the conservation.] *Protection Nature Bull., Inst. Protection of Nature of Serbia* 45: 65–76. (In Serbian, with English summary.)

BUTORAČ, B. *ET AL.* (1996) [Proposal for the protection of the "Koviljsko—Petrovaradinski rit" as a Special Nature Reserve.] Belgrade: Institute for the Protection of Nature of Serbia. (In Serbian.)

BUTORAČ, B. *ET AL.* (1996) [Proposal for the protection of the Selevenjske Pustare sand as a Special Nature Reserve.] Belgrade: Institute for the Protection of Nature of Serbia. (In Serbian.)

DEVIĆ, M. (1995) [Ornithofauna of the fishpond "Uzdin".] *Ciconia (Bull. Bird Prot. Study Soc. Voivodina)* 5: 32–44. (In Serbian, with English summary.)

DORDEVIĆ, Z. (1994) [Proposal for the protection of the "Klisura reke Tresnjice" as a Special Nature Reserve.] Belgrade: Institute for the Protection of Nature of Serbia. Unpublished report. (In Serbian.)

GAROVNIKOV, B. AND MARKOVIĆ, J. (1980) [Bird fauna of Kozjak—Koviljski rit.] Yugoslavia: The Institute for Protection of Nature of Voivodina. (In Serbian.)

GAROVNIKOV, B. AND POPOVIĆ, E. (1986) [Ornithofauna of the Carska bara marsh.] *Bull. Nature, Voivodina* 9/10/11: 39–54. (In Serbian, with English summary.)

GAROVNIKOV, B., GERGELJ, J., ZRNIĆ, D., LECIĆ, B. AND HULO, I. (1997) [Natural rarities on "Ludasko jezero" area.] *Catalogue J.P. "Palic-Ludas" Subotica* 1–29. (In Serbian and Hungarian.)

GERGELJ, J. AND SITE, T. (1989) [Red-footed falcon *Falco vespertinus* on Kapetanski rit and Jazovo.] *Ciconia (Bull. Bird Prot. Study Soc. Voivodina)* 1: 64–65. (In Serbian, with English summary.)

GERGELJ, J. AND ŠOTI, J. (1990) [Ornithofauna of the fishpond "Kapetanski rit".] *Ciconia (Bull. Bird Prot. Study Soc. Voivodina)* 2: 22–49. (In Serbian, with English summary.)

GRIMMETT, R. F. A. AND JONES T. A. (1989) *Important Bird Areas in Europe*. Cambridge, UK: International Council for Bird Preservation (Tech. Publ. 9).

GRUBAČ, B. (1988) The Golden Eagle (*Aquila chrysaetos*) in south-eastern Yugoslavia. *Larus (Zagreb)* 38/39: 95–135.

GRUBAČ, B. (1994) [Contributions on the status and breeding of the Long-legged Buzzard *Buteo rufinus* Cretzschmar 1827 in Serbia.] *Protection Nature Bull., Inst. Protection of Nature of Serbia* 46/47: 157–161. (In Serbian, with English summary.)

GRUBAČ, B. (1995) [Bird fauna of Zlot Gorge.] Belgrade: Institute for the Protection of Nature of Serbia. Unpublished report. (In Serbian.)

GRUBAČ, B. (1996) [Bird fauna of Đerdapska klisura Gorge.] Belgrade: Institute for the Protection of Nature of Serbia. Unpublished report. (In Serbian.)

GRUBAČ, B. (1996) [Bird fauna of Jerma Gorge.] Belgrade: Institute for the Protection of Nature of Serbia. Unpublished report. (In Serbian.)

GRUBAČ, B. (1996) [Bird fauna of Resava Gorge and Beljanica.] Belgrade: Institute for the Protection of Nature of Serbia. Unpublished report. (In Serbian.)

GRUBAČ, B. (1996) [Bird fauna of Sićevo Gorge.] Belgrade: Institute for the Protection of Nature of Serbia. Unpublished report. (In Serbian.)

GRUBAČ, B. (1996) [Bird fauna of Slono Kopovo.] Belgrade: Institute for the Protection of Nature of Serbia. Unpublished report. (In Serbian.)

GRUBAČ, B. (1996) [Situation of the Lanner Falcon *Falco biarmicus* in the Mediterranean and especially in the center of the Balkans.] Pp. 128–134 in *Biologia y conservación de las rapaces*. Madrid: Sociedad Española de Ornitologia. (In French.)

GRUBAČ, B. AND PUZOVIĆ, S. (1996) [Report of bird fauna of Prokletije Mountain.] Belgrade: Institute for the Protection of Nature of Serbia. Unpublished report. (In Serbian.)

HAM, I. (1986) [Evaluation of the Deliblato sand sections by ornithological values.] Pp. 87–104 in [*Proceedings of the Deliblato sands*.] (In Serbian, with English summary.)

HAM, I. (1989) [Survey of bird fauna of Deliblato sand—Voivodina.] P. 60 in [*III Symposium of fauna in Serbia Belgrade*.] (In Serbian.)

HAM, I. (1989) Changes in heron populations on wider area of the Carska bara. *Bull. Arh. Biol. Nauka* 41(3/4): 13–14.

HAM, I. (1989) Heron and Cormorant colonies on the Yugoslav sector of the Danube (588 km). *Bull. Arh. Biol. Nauka* 41(3/4): 17–18.

HAM, I. (1989) Influence of elevated water level of the Danube on colony size of Sand Martin *Riparia riparia*. *Bull. Arh. Biol. Nauka* 41(3/4): 17–18.

HAM, I. (1992) [Proposal for the protection of the Dubovac flooded area.] Belgrade: Institute for the Protection of Nature of Serbia. Unpublished report. (In Serbian.)

HAM, I., BUTORAČ, B. *ET AL.* (1993) [Proposal for the protection of Ludas lake as a Special Nature Reserve.] Belgrade: Institute for the Protection of Nature of Serbia. Unpublished report. (In Serbian.)

HAM, I., MIKUŠKA, J., SCHNEIDER, M. AND GEC, D. (1990) Recoveries and sightings of banded and wing-tagged White-tailed Eagle in Yugoslavia during 1985–1988. *Larus (Zagreb)* 41/42: 69–86.

HEATH, M. F. AND BORGGREVE, C. (2000) *BirdLife International/EBCC European Bird Database 1998*. Cambridge, UK: BirdLife International.

HULO, I. (1997) [Migration of birds from the orders of Gaviidae, Podicipedidae, and Anatidae on the Palić lake between 1981 and 1996.] *Ciconia (Bull. Bird Prot. Study Soc. Voivodina)* 6: 51–70. (In Serbian, with English summary.)

HULO, I. (1997) [Stone-curlew *Burhinus oedicnemus* on the Kapetanski rit.] *Ciconia (Bull. Bird Prot. Study Soc. Voivodina)* 6: 116. (In Serbian, with English summary.)

IRSAI, A. (1997) [Ornithological camp at the Kovilj—Petrovaradin meadow.] *Ciconia (Bull. Bird Prot. Study Soc. Voivodina)* 6: 11–14. (In Serbian, with English summary.)

KANJO, B. (1997) [Notes on ornithofauna of Ribarevo in Gornje Podunavlje area.] *Ciconia (Bull. Bird Prot. Study Soc. Voivodina)* 6: 96–99. (In Serbian, with English summary.)

LAKATOŠ, J. (1979) [*Bird fauna of Apatin*.] Apatin, Yugoslavia: NIIU Glas Komune. (In Serbian.)

LUKAČ, S. (1990) [Great White Egret *Egretta alba*, nester of Bečej fishpond.] *Ciconia (Bull. Bird Prot. Study Soc. Voivodina)* 2: 73. (In Serbian, with English summary.)

LUKAČ, S. AND LUKAČ, A. (1992) [Ornithofauna of the fishpond "Bečej".] *Ciconia (Bull. Bird Prot. Study Soc. Voivodina)* 4: 4–27. (In Serbian, with English summary.)

LUKAČ, S. AND TERNOVAC, T. (1990) [Notes on ornithofauna of Slono Kopovo in 1987–1989.] *Ciconia (Bull. Bird Prot. Study Soc. Voivodina)* 2: 50–63. (In Serbian, with English summary.)

LUKAČ, S. AND TERNOVAC, T. (1995) [Interesting nesting of Great Crested Grebe *Podiceps cristatus* and Coot *Fulica atra* on Jegrička.] *Ciconia (Bull. Bird Prot. Study Soc. Voivodina)* 5: 24. (In Serbian, with English summary.)

LUKAČ, S. AND TERNOVAC, T. (1997) [Breeding of Red-necked Grebe *Podiceps griseigena* on the Carska bara.] *Ciconia (Bull. Bird Prot. Study Soc. Voivodina)* 6: 111. (In Serbian, with English summary.)

LUKAČ, S. AND TERNOVAC, T. (1997) [Crane *Grus grus* migration on Slono Kopovo between 1993–1996.] *Ciconia (Bull. Bird Prot. Study Soc. Voivodina)* 6: 71–72. (In Serbian, with English summary.)

LUKAČ, S. AND TERNOVAC, T. (1997) [Heron colony on the Carska bara.] *Ciconia (Bull. Bird Prot. Study Soc. Voivodina)* 6: 115. (In Serbian, with English summary.)

MARINKOVIĆ, S. AND ORLANDIĆ, L. J. (1994) Status of the Griffon Vulture *Gyps fulvus* in Serbia. Pp. 163–171 in B.-U. Meyburg and R. D. Chancellor (eds) *Raptor conservation today*. Berlin: World Working Group for Birds of Prey.

MATVEJEV, S. AND PUNCER, I. (1989) *Map of biomes—Landscapes of Yugoslavia and their protection*. Belgrade: Natural History Museum (Special Issue No. 36).

MIKALACKI, V. (1997) [Birds of prey of Bečej and surroundings between 1994–1996.] *Ciconia (Bull. Bird Prot. Study Soc. Voivodina)* 7: 27–29. (In Serbian, with English summary.)

MILOVANOVIĆ, Z. (1996) [Bird fauna of National Park "Đerdapska klisura".] P. 55 in [*V Congress on the Ecology of Yugoslavia*.] (In Serbian.)

OBRADOVIĆ, R. (1991) [Saker falcon *Falco cherrug* in north-west Backa.] *Ciconia (Bull. Bird Prot. Study Soc. Voivodina)* 3: 48–49. (In Serbian, with English summary.)

OBRADOVIĆ, R. (1992) [Data on nesting of some bird species in the Gornje Podunavlje area.] *Ciconia (Bull. Bird Prot. Study Soc. Voivodina)* 4: 61–62. (In Serbian, with English summary.)

OBRADOVIĆ, R. (1994) [*Catalogue of birds of Backa*.] Apatin, Yugoslavia: EDG Nikola tesla. (In Serbian.)

OBRADOVIĆ, R. AND MIRIC, M. (1991) [Data on Honey Buzzard *Pernis apivorus* and Lesser Spotted Eagle *Aquila pomarina* in north-west Backa.] *Ciconia (Bull. Bird Prot. Study Soc. Voivodina)* 3: 50–51. (In Serbian, with English summary.)

PANJKOVIĆ, B. *ET AL.* (1997) [Natural values of the reserve on National Park "Kopaonik".] Belgrade: Institute for the Protection of Nature of Serbia. Unpublished report. (In Serbian.)

PAUNOVIĆ, M. AND STAJIĆ, D. (1988) [Bird fauna of Kopaonik in 1988.] *SNIRS Bull. Young Researchers of Serbia* 52–84. (In Serbian.)

PAUNOVIĆ, M., HAM, I. AND PUZOVIĆ, S. (1994) The wintering of waterfowl on the river Danube (Yugoslavia) 1988–1992. *Bios (Thessaloniki)* 2: 319-324.

PAUNOVIĆ, M., HAM, I. AND PUZOVIĆ, S. (1995) The wintering of waterfowl on the river Danube 1988–1992. *Bios (Thessaloniki)* 2: 319–324.

PAVKOV, G. AND VIDER-MILOSEVIĆ, V. (1993) [Proposal for the protection of the "Obedska bara" as a Special Nature Reserve.] Belgrade: Institute for the Protection of Nature of Serbia. Unpublished report. (In Serbian.)

PAVKOV, G. *ET AL.* (1996) [Proposal for the protection of the "Saniste velike droplje" as a Special Nature Reserve.] Belgrade: Institute for the Protection of Nature of Serbia. Unpublished report. (In Serbian.)

PELLE, I., ŠOTI, J., VASIC, V. AND HAM, I. (1977) [Survey of the ornithofauna of Slono Kopovo.] Pp. 36–37 in [*III Symposium of Biosystematics of Yugoslavia*.] (In Serbian.)

PURGER, J. AND MIKEŠ, M. (1984) [Survey of the recorded bird fauna on Fruška gora.] *Bull. Inst. Biol. PMF, Novi Sad* 51–70. (In Serbian.)

PUZOVIĆ, S. (1989) [First survey of bird fauna in Prijepolje area.] P. 92 in [*III Symposium for Fauna of Serbia, Belgrade*.] (In Serbian.)

PUZOVIĆ, S. (1995) [Historical changes of bird fauna in Obedska bara.] *SNIRS Bull. Young Researchers of Serbia* 10: 35–49. (In Serbian, with English summary.)

PUZOVIĆ, S. (1995) Balkan Shore Lark (*Eremophila alpestris balcanica*) Reichenow 1895, in Serbia and former Yugoslavia. *Bios (Thessaloniki)* 3: 72.

PUZOVIĆ, S. (in press) Diversity of breeding birds fauna in Serbia with special regards for National Parks. *FNNPE Conference, Tara National Park*.

PUZOVIĆ, S. AND HAM, I. (1991) [White-tailed Eagle *Haliaeetus albicilla* at the Obedska bara–Donja posavina area.] *Bull. Natl Hist. Mus.* B43/46: 163–188. (In Serbian, with English summary.)

PUZOVIĆ, S. AND PAUNOVIĆ, M. (1994) [Valorization of ornithofauna evaluation of Bor area (eastern Serbia).] Pp. 139–140 in [*Simposium "Nasa ekoloska istina", 2*.] (In Serbian, with English summary.)

PUZOVIĆ, S. AND PAUNOVIĆ, M. (1996) [Structure, number and distribution of the nesting ornithofauna on the southern slopes of Rtanj Mountain (eastern Serbia).] Pp. 223–228 in [*Simposium "Nasa ekoloska istina", 4*.] (In Serbian, with English summary.)

PUZOVIĆ, S. AND RAŠAJSKI, J. (in press) Black Stork (*Ciconia nigra*) in Serbia and former Yugoslavia: distribution and population. *II International Conference on the Black Stork, Extremadura, Spain*.

PUZOVIĆ, S. AND SEKULIĆ, G. (1995) [Bird fauna of Obedska bara 1990–1994.] *SNIRS Bull. Young Researchers of Serbia* 10: 72–90. (In Serbian, with English summary.)

PUZOVIĆ, S. AND SEKULIĆ, G. (1997) [Current diversity of breeding bird fauna in Obedska bara: population and distribution).] Belgrade: Institute for the Protection of Nature of Serbia. Unpublished report. (In Serbian.)

PUZOVIĆ, S., SEKULIĆ, V. AND PAVLOVIĆ, D. (1989) [Black stork *Ciconia nigra* at Obedska bara 1983–1987.] *Bull. Natl Hist. Mus.* B43/44: 161–174. (In Serbian, with English summary.)

RAŠAJSKI, J. (1986) [Contribution for breeding bird fauna of Voivodina.] *Larus (Zagreb)* 36/37: 225–230. (In Serbian, with English summary.)

RAŠAJSKI, J. AND PELLE, I. (1993) [*Birds of the Vršač mountains*.] Monographs of Matica Srpska, 108 pp. (In Serbian, with English summary.)

SEKULIĆ, G. (1995) [Enclouse to the knowledge of the ornithofauna of Stara planina.] *Ciconia (Bull. Bird Prot. Study Soc. Voivodina)* 5: 45–51. (In Serbian, with English summary.)

SEKULIĆ, G. AND CONGOR, K. (1995) [Bird fauna of central Stara planina in July 1994.] *SNIRS Bull. Young Researchers of Serbia* 10: 21–19. (In Serbian.)

SEKULIĆ, V. AND PERIĆ, V. *ET AL.* (1986) [Contribution to the ornithofauna of the Zlot Gorge area.] *Bull. Istrazivac* 1: 128–136. (In Serbian.)

ŠOTI, J. (1973) [Ecological characteristics of bird fauna in Koviljski rit. Notes on ornithofauna of Ribarevo in Gornje Podunavlje area.] *Bull. Prirodno—matematicki fakultet* 3: 109–127. (In Serbian, with Dutch summary.)

ŠOTI, J. (1985) [Ornithofauna of Fruška gora.] P. 112 in [*I Congress on Biosystematics of Yugoslavia*.] Yugoslavia: Popova Sapka. (In Serbian.)

STEFANOVIĆ, T. (1997) [Breeding of Imperial Eagle *Aquila heliaca* in the Fruška gora.] *Ciconia (Bull. Bird Prot. Study Soc. Voivodina)* 6: 23–24. (In Serbian, with English summary.)

STEFANOVIĆ, T. AND PUZOVIĆ, S. (1997) [Breeding of Honey Buzzard *Pernis apivorus* in Bestrement, near Apatin.] *Ciconia (Bull. Bird Prot. Study Soc. Voivodina)* 6: 25–26. (In Serbian, with English summary.)

TERNOVAC, T. (1991) [Data on the ornithofauna of Jegrička from 1989 to 1990.] *Ciconia (Bull. Bird Prot. Study Soc. Voivodina)* 3: 14–24. (In Serbian, with English summary.)

TERNOVAC, T. (1992) [Data of the nesting of some bird species from Jegrička.] *Ciconia (Bull. Bird Prot. Study Soc. Voivodina)* 4: 63–64. (In Serbian, with English summary.)

TERNOVAC, T. AND LUKAC, S. (1989) [Notes on the ornithofauna of Slono Kopovo in 1986.] *Ciconia (Bull. Bird Prot. Study Soc. Voivodina)* 1: 26–30. (In Serbian, with English summary.)

TUCKER, G. M. AND HEATH, M. F. (1994) *Birds in Europe: their conservation status*. Cambridge, UK: BirdLife International (BirdLife Conservation Series no. 3).

VASIĆ, V. (1979) [Survey of the bird fauna of the Tara Mountain area (west Serbia).] *Bull. Arh. Biol. Nauka* 29(1/2): 69–81. (In Serbian, with English summary.)

VASIĆ, V. (1980) [Survey of the avifauna of Lake Vlasina area.] *Biosistematika Belgrade* 6(1): 81–107. (In Serbian, with English summary.)

VASIĆ, V. (1995) Diverzitet ptica Jugoslavije sa pregledom vrsta od međunarodnog značaja. Pp. 471–510 in V. Stevanovič and V. Vasić, eds. *Biodiverzitet Jugoslavije sa pregledom vrsta od međunarodnog značaja*. Belgrade: Biološki fakultet i Ekolibri.

APPENDICES

Appendix 1. Descriptions of international agreements relevant to the protection and management of sites for nature conservation in Europe.

Biodiversity Convention: Convention on Biological Diversity (CBD)

The Convention on Biological Diversity has three objectives:

- the conservation of biological diversity,
- the sustainable use of its components, and
- the fair and equitable sharing of the benefits arising out of the use of genetic resources.

Covering almost every aspect of conservation and sustainable use, the CBD has become an important framework for conservation-related activities on the regional and national level.

For the conservation of biodiversity the primary approach is *in situ* conservation. Parties have to identify components of biodiversity important for its conservation and sustainable use (Article 7), *inter alia* ecosystems and habitats with large numbers of threatened species or required by migratory species (Annex I). Clearly, national lists of IBAs should become part of these inventories which should be included in national biodiversity strategies and action plans required from the parties by Article 6.

Article 8 outlines a framework for the *in situ* conservation activities by the parties. They are asked to establish a system of protected areas or areas where special measures need to be taken to conserve biodiversity (Article 8a). Thus, in cases where strong protection might not be appropriate they should find alternative ways to ensure the ecological integrity of the sites in question. This should be done in a rational way, using guidelines which need to be developed (Article 8b). As the lists of IBAs have been compiled following strict criteria on a scientific basis, harmonized with existing criteria under the EC Birds Directive and Ramsar Convention, they can serve well as part of a national system of protected and other areas with a focus on conservation.

Effective conservation of biological diversity needs plans and management strategies (Article 8c, f) for the rehabilitation and restoration of degraded ecosystems and the recovery of threatened species. IBAs in most cases need to be managed to enhance their ability to support important bird populations in the long-term. To mitigate negative effects on IBAs these strategies also need to take into account the adjacent areas and to promote their environmentally sound and sustainable development (Article 8e). The CBD also asks for the prevention of introduction and the control and eradication of alien species which threaten ecosystems, habitats and species (Article 8h) and urges parties to set up an effective system of environmental impact assessment. Impact assessment more generally is called for in Article 14.

Overall, the CBD clearly sets a framework for ensuring that IBAs obtain sufficient attention to ensure their long-term ecological integrity. The national biodiversity strategies and action plans which have recently been developed in almost all European countries should reflect the role of IBAs as scientifically-based elements of the national implementation of the provisions of the convention.

Pan-European Biological and Landscape Diversity Strategy

In October 1995, European environmental ministers endorsed the Pan-European Strategy for Biological and Landscape Diversity (PEBLDS), a framework for strengthening and building on existing initiatives and programmes, drawn up as a pan-European response to the Convention on Biological Diversity.

The Strategy is designed to strengthen and build on existing initiatives and programmes including the Convention on Biological Diversity, the Bern Convention, the Bonn Convention, the European Conservation Strategy (1990), the Dobris and Lucerne Ministerial Conferences (1991, 1993) and *inter alia* the United Nations Conference on Environment and Development (1992).

The Pan-European Strategy is defined for a 20 year period, and divided into four five-year Action Plans. Eleven Action Themes run through each Action Plan, of which one is to establish a Pan-European Ecological Network of protected areas (Council of Europe *et al.* 1996).

European Community Biodiversity Strategy

The European Community ratified the Convention on Biological Diversity in 1993 and in 1998, the European Commission adopted a communication on a European Community Biodiversity Strategy. The aims of this strategy are to anticipate, prevent and attack at source the causes of significant reductions in and losses of biodiversity.

The strategy defines general objectives in the four horizontal policy areas set out in the convention:

- Conservation and sustainable use of biodiversity
- Sharing benefits arising out of the use of genetic resources
- Research, identification, monitoring and exchange of information
- Education, awareness and training.

The strategy then focuses on the integration of biodiversity concerns into relevant sectoral policies, setting out sectoral policy objectives, which should be achieved through specific action plans or other measures for the following policy areas: conservation of natural resources, agriculture, fisheries, regional policies and spatial planning, forests, energy and transport, tourism and development and economic cooperation.

These objectives reflect a recognition that there is a major gap in existing Community conservation policies concerning the impact of relevant sectoral policy measures on biodiversity, both inside and outside of protected areas.

Ramsar Convention: Convention on Wetlands of International Importance especially as Waterfowl Habitat (Ramsar, 1971)

The Ramsar Convention is an inter-governmental treaty which provides the framework for international cooperation for the conservation and wise use of wetlands.

The main undertakings accepted by the contracting parties to the Ramsar Convention are to:

- designate suitable wetlands within (their) territory for inclusion in a List of Wetlands of International Importance (the Ramsar List) (Article 2.1)
- formulate and implement their planning so as to promote the conservation of the wetlands included in the List, and as far as possible, the wise use of wetlands in their territory (Article 3.1)
- promote the conservation of wetlands and waterfowl by establishing nature reserves on wetlands whether they are included in the List or not, and provide adequately for their wardening (Article 4.1).

Each contracting party must designate at least one site for inclusion in the List at the time it joins the Convention (Article 2.4). Specific criteria have been drawn up under the Convention to aid the identification of sites of international importance. One of the objectives of the IBA inventory is to facilitate the designation of wetlands of international importance under the Ramsar Convention. The criteria used to identify IBAs are consistent with the Ramsar criteria for waterbirds (see Box 4, 'Identifying IBAs' chapter). Therefore wetland IBAs meeting these criteria should be considered for addition to the Ramsar List.

Also of relevance is the Montreux Record, a register of selected wetlands on the Ramsar List where changes in ecological character have occurred, are occurring or are likely to occur as a result of technological developments, pollution or other human interference. This is a very important tool because it identifies specific sites for international conservation attention. Other site-conservation tools have been developed under the convention such as guidelines on management planning and on monitoring.

Appendix 1 ... continued. Descriptions of international agreements relevant to the protection and management of sites for nature conservation in Europe.

Bonn Convention: Convention on the conservation of migratory species of wild animals (CMS)

The fundamental objective of the Bonn Convention is to protect migratory species (not only birds but also migratory mammals, fish and invertebrates) in recognition of the fact that protection is needed throughout every part of their migratory ranges, and that this requires international cooperation and action. If a party to the Convention is a Range State of a migratory species listed in Appendix I or Appendix II, it accepts an obligation to provide strict protection for species in Appendix I and to endeavour to conclude Agreements with other Range States for the conservation and management of species in Appendix II. Appendix I species are in danger of extinction throughout all, or a major part, of their range, and Appendix II species are those which would benefit from international cooperation in their conservation and management.

For Appendix I species, Parties that are Range States are obliged to endeavour, amongst other things, to:

- conserve and, where feasible and appropriate, restore those habitats of the species which are of importance in removing the species from danger of extinction (Article III 4a); and
- with regard to the Agreements which Parties are encouraged to conclude for Appendix II species, each Agreement should, where appropriate and feasible, provide for the maintenance of a network of suitable habitats appropriately disposed in relation to the migration routes.

Agreement on the Conservation of African-Eurasian Migratory Waterbirds (AEWA)

AEWA is an Agreement under the Bonn Convention, which can be signed and ratified by a Range State, irrespective of whether the state has signed or ratified the Bonn Convention. AEWA aims to create the legal basis for a concerted conservation policy among the range states of all migratory waterbird species and populations which migrate in the African-Eurasian flyway, irrespective of their current conservation status. The Agreement provides a framework for conservation action, monitoring, research and management of several globally important bird-migration systems. As such, it has close links to the IBA Programme and the Ramsar Convention.

World Heritage Convention: Convention concerning the Protection of the World Cultural and Natural Heritage

The aim of the World Heritage Convention is the protection of natural and cultural areas of outstanding universal value. Such sites and monuments are considered to be of such exceptional value that their protection is the responsibility of all mankind, and thus international cooperation in order to contribute effectively to this protection is sought. Each site nominated by the Parties for inclusion in the World Heritage List is assessed by a World Heritage Committee, which in the case of natural sites, is advised by experts from IUCN.

The Convention imposes a legal duty on each party to do its utmost to protect designated sites. Each party is required to contribute to the World Heritage Fund which may be used to secure the protection of World Heritage Sites. A mechanism is provided for the transfer of resources from rich to poor countries for the safeguard of mankind's common heritage.

UNESCO's Man and the Biosphere Programme

One of the aims of the Man and the Biosphere Programme of UNESCO (United Nations Educational, Scientific and Cultural Organization) is to develop within the natural and social services a basis for the rational use and conservation of the resources of the biosphere.

MAB Project 8 (one of 14 international themes or projects in the Programme) is 'the conservation of natural areas and the genetic material they contain'. The objective is to create a worldwide network of reserves (Biosphere Reserves), with each reserve qualifying under one or more of the following categories:

- representative examples of natural biomes;
- unique communities or areas with unusual natural features of exceptional interest such as a population of a globally rare species;
- examples of harmonious landscapes resulting from traditional patterns of land-use;
- examples of modified or degraded ecosystems capable of being restored to more natural conditions.

Each Biosphere Reserve needs to be large enough to comprise an effective conservation unit and must have adequate long-term protection.

Bern Convention: Convention on the Conservation of European Wildlife and Natural Habitats

Contracting parties to the Bern Convention are required to maintain populations of wild flora and fauna and give particular emphasis to endangered and vulnerable species, including endangered and vulnerable migratory species. Specifically, each Contracting party undertakes to:

- take appropriate and necessary legislative and administrative measures to ensure the conservation of the habitats of the wild flora and fauna species especially those specified in Appendices I and II, and the conservation of endangered natural habitats (Article 4.1)
- give special attention to the protection of areas that are of importance for the migratory species specified in Appendix II and III and which are appropriately situated in relation to migration routes, as wintering, staging, feeding, breeding or moulting areas (Article 4.3)
- prohibit the deliberate damage to or destruction of breeding or resting sites of Appendix II species (Article 6.b).

Article 6.b is very specific and strict; however, the explanatory report of the Convention states that it should apply only to important breeding and resting sites. Implementation of Article 6.b necessitates the identification of important breeding and resting sites for species on Appendix II (Lyster 1985). Parties may make exceptions to their obligations under Articles 4 and 6, but only in very limited circumstances (such as interests of public health and safety) and use of the exception must never be detrimental to the survival of the population concerned (Article 9).

Appendix I is a list of plants. Appendix II is a long list of strictly protected faunal species which includes a high proportion of the European avifauna, with Appendix III covering almost all the bird species not included in Appendix II, with the exception of 11 species.

The Emerald Network

The Emerald network is an initiative, under the Bern Convention, to extend the Natura 2000 network of protected areas (see 'Habitats Directive', below) beyond the EU to cover member countries of the Council of Europe.

In June 1989 the Standing Committee of the Bern Convention held an extraordinary meeting exclusively devoted to habitat conservation within the Convention. One of the operative recommendations adopted at the meeting was Recommendation No. 16 (1989) 'on Areas of Special Conservation Interest' (ASCIs) within which the Standing Committee recommended Parties to take steps to designate Areas of Special Conservation Interest to ensure that necessary and appropriate conservation measures are taken for each area situated within their territory or under their responsibility where that area fits one or several of the following conditions:

a. It contributes substantially to the survival of threatened species, endemic species, or any species listed in Appendices I and II of the Convention;
b. It supports significant numbers of species in an area of high species diversity or supports important populations of one or more species;
c. It contains an important and/or representative sample of endangered habitat types;
d. It contains an outstanding example of a particular habitat-type or a mosaic of different habitat-types;

Appendix 1 ... continued. Descriptions of international agreements relevant to the protection and management of sites for nature conservation in Europe.

e. It represents an important area for one or more migratory species;
f. It otherwise contributes substantially to the achievement of the objectives of the Convention.

In 1996 Resolution No. 3 was adopted which resolved to 'set up a network (Emerald) which would include the Areas of Special Conservation Interest designated following its Recommendation No. 16'.

Member States of the European Union will satisfy the habitat protection requirements of the Bern Convention mostly through the designation of sites to the Natura 2000 network (see below). The Special Areas for Conservation and Special Protection Areas for birds in Natura 2000 automatically become Areas of Special Conservation Interest of the Emerald Network.

Most of the work to be done in building the Emerald Network will be concentrated in states which are not members of the European Union. In this way it is aimed to extend an homogenous network of areas over a wide area including the whole of Europe.

EC Birds Directive: Council Directive on the conservation of wild birds (79/409/EEC)

The Birds Directive, adopted in 1979, concerns the conservation and protection of all wild bird species in the European territory of the Member States of the European Union (currently 15 Member States, excluding Greenland). All Member States are committed to providing a sufficient variety of habitats and to preserve, maintain or restore an adequate proportion of these for all wild bird species in their territory. Special conservation measures shall be taken for the habitats of two categories of birds: species and subspecies listed in Annex I of the Directive and other migratory species not included in this Annex (Article 4.1–4.2). This requirement refers in particular to the classification of Special Protection Areas which are considered the 'most suitable' sites for the conservation of the species concerned. Each Member State has to designate such areas for all Annex I species and subspecies (currently 181 taxa in total, see Appendix 2b) occurring on its territory (Article 4.1). This concerns all species for which regularly occupied habitats can be identified so that the objective 'to ensure their survival and reproduction in their area of distribution' can be fulfilled. The number and the area of habitats to be designated as Special Protection Areas should be determined on the basis of the protection requirements and the conservation status of the species involved. Hence for an endangered, localized species a larger proportion of its habitats should be designated, than for a widespread, less threatened species.

With respect to regularly occurring migratory species not listed in Annex I, 'similar measures' have to be taken including the designation of Special Protection Areas as regards their breeding, moulting and wintering areas and staging posts along their migration routes (Article 4.2). Special attention shall be paid to the protection of wetlands and particularly to wetlands of international importance. The latter is considered a reference to the Ramsar Convention. In this IBA inventory the category of 'other migratory species' has been largely confined to waterbirds, seabirds and birds of prey during migration (criteria categories C3–C5) because it was considered appropriate to stay as close as possible to the criteria applied in the previous pan-European IBA inventory in 1989 (Grimmett and Jones 1989). However, other migratory species could be covered as well depending on their 'need of protection' within the territory of the Member States of the European Union.

The network of Special Protection Areas should form a coherent whole, which, together with the sites designated under the Habitats Directive (92/43/EEC), will form the ecological network Natura 2000 which is required to be completed in 2004. The protection regime applicable to Special Protection Areas, originally laid down in Article 4.4, has now been replaced by Article 6.2–6.4 of the Habitats Directive.

EC Habitats Directive: Council Directive on the conservation of natural habitats of wild fauna and flora (92/43/EEC)

The Habitats Directive aims, among other things, to create a coherent ecological network (Natura 2000) of Special Areas for Conservation (SACs), setting a minimum standard for biodiversity conservation in the EU. This network should be designed to maintain the distribution and abundance of threatened species and habitats, both terrestrial and marine. This network will contain all Special Protection Areas designated under the Birds Directive. Regarding habitats, each Member State is under an obligation to contribute to Natura 2000 in proportion to the representation within its terrritory of the natural habitat-types listed in Annex I and habitats of the species referred to in Annex II (animals other than birds). Member States are required to draw up lists of Sites of Community Interest (SCIs) following these criteria and from these the European Commission will establish a list of sites of Community importance. Each Member State shall then be required (by 2004) to designate the listed sites of Community importance as Special Areas for Conservation (SACs).

Member States are required to take necessary and appropriate conservation measures for SACs (Article 6). Necessary conservation measures include, if appropriate, management plans specifically designed for the sites or integrated into other development plans, and appropriate statutory, administrative or contractual measures. These measures should correspond to the ecological requirements of the natural habitat-types in Annex I and the species in Annex II present on the sites (note that this requirement does not apply to sites designated under the Birds Directive only). Member states are required to avoid significant disturbance and habitat deterioration in SACs with respect to the habitats (Annex I), habitats of the species, and the species (Annex II) for which the areas have been designated. They are also required to assess the implications for the SAC of any plan or project likely to have a significant effect thereon, and to ensure that the competent national authorities agree to the plan or project only after having ascertained that it will not adversely affect the integrity of the site and, if appropriate, after having obtained the opinion of the general public. Also they should take all compensatory measures necessary to ensure that the overall coherence of Natura 2000 is protected, if a plan or project must be carried out at or near a SAC for imperative reasons of overriding public interest, including those of a social or economic nature. Where the site concerned hosts a priority natural habitat-type and/or a priority species, the only reasons which may be considered are those relating to human health or public safety, to beneficial consequences of primary importance for the environment or, further to an opinion from the Commission, to other imperative reasons of overriding public interest.

LIFE-Nature
LIFE-Nature is an EU funding line (Regulation 1973/92/EEC, as modified by Regulation (EC) 1404/96 of 15 July 1996, under review to cover the period 2000–2003) which supports EU environmental legislation, including the Birds and Habitats Directives, and has supported work at many IBAs across Europe, within and outside the EU.

Source: European Commission (1998).

Appendix 1 ... continued. Descriptions of international agreements relevant to the protection and management of sites for nature conservation in Europe.

Barcelona Convention: Convention for the Protection of the Mediterranean Sea against Pollution

The objectives of the Barcelona Convention are to counteract the introduction of certain pollutants into the Mediterranean Sea area (including internal waters); and to conserve natural species and habitats. The 20 states adjoining the Mediterranean Sea, and the European Union, are contracting parties.

There is a Protocol to the Barcelona Convention concerning Mediterranean Specially Protected Areas. This specifies the action that Mediterranean States are bound to undertake to identify, establish and manage marine and coastal areas requiring special protection in the Mediterranean Sea Area.

Helsinki Convention: 1992 Helsinki Convention on the Protection of the Marine Environment of the Baltic Sea Area

The Helsinki Convention aims to counteract the introduction of certain pollutants into the Baltic Sea area (including internal waters), and to conserve natural species and habitats. The Convention brings the Baltic Sea (Czech Republic, Denmark, Estonia, Finland, Germany, Latvia, Lithuania, Norway, Poland, Russia, Slovakia, Sweden and Ukraine) together with the European Union under the Helsinki Commission (HELCOM).

HELCOM has recommended the establishment of a network of 65 Baltic Sea Protected Areas (BSPAs). Included within this are provisions for the identification of areas important for concentrations of feeding, wintering and staging seabirds (Hägerhäll and Skov 1998).

Parks for Life

Parks for Life is an initiative of the IUCN World Commission for Protected Areas, in association with the Federation of Nature and National Parks of Europe (FNNPE), World Wide Fund for Naure (WWF), World Conservation Monitoring Centre (WCMC) and BirdLife International (IUCN–CNPPA 1994). It sets out a vision for Europe's protected areas in terms of policies, recommendations and actions.

The main themes of the plan include the positioning of protected areas in a wider context, addressing priorities at all levels, and better planning and management for protected areas. There are chapters addressing the main impact of sectoral policies on protected areas, actions concerning some priority sub-regions and countries, and overviewing the legal framework and financial opportunities for better protection. The plan deals with securing public support for protected areas, and outlines the implementation process.

There are 30 priority projects listed in the plan, which need urgent attention. In 1995, a coordinator was appointed to oversee the implementation of these projects. The idea is that a wide range of organizations take up a leading or active participatory role in the projects, so as to ensure effective cooperation and broad support for these activities.

Program for the Conservation of Arctic Flora and Fauna

The Program for the Conservation of Arctic Flora and Fauna (CAFF) was established to address the special needs of Arctic species and their habitats in the rapidly developing Arctic region. It forms one of four programmes of The Arctic Environmental Protection Strategy (AEPS) which was adopted by the Arctic nations in 1991. The other programmes of the AEPS include the Arctic Monitoring and Assessment Program (AMAP) and the programmes for Emergency Prevention, Preparedness and Response (EPPR) and Protection of the Arctic Marine Environment (PAME).

CAFF's main goals, which are achieved in keeping with the concepts of sustainable development and utilization, are:

- to conserve Arctic flora and fauna, their diversity and their habitats;
- to protect the Arctic ecosystems from threats;
- to improve conservation management, laws, regulations and practices for the Arctic;
- to integrate Arctic interests into global conservation fora.

The majority of CAFF's Work Plan activities are directed at species and habitat conservation and at integrating indigenous peoples and their knowledge into CAFF. Some examples are: work on rare, vulnerable and endangered plants and animals of the Arctic; developing circumpolar conservation strategies for certain species (e.g. Scott 1998); work on Arctic vegetation; analysing and making recommendations on threats to Arctic species and habitat; an indigenous peoples, mapping project and development of an implementation strategy for the Convention on Biological Diversity in the Circumpolar Arctic (e.g. CAFF 1996a,b).

References

BirdLife International (1994) IBA Conservation in the European Union: an introduction to the mechanisms. Brussels, Belgium: 19-22 November Workshop proceedings. Cambridge, UK: BirdLife International (unpublished report).

CAFF [Conservation of Arctic Flora and Fauna] (1996a) *Proposed protected areas in the Circumpolar Arctic 1996.* Trondheim, Norway: Directorate for Nature Management (CAFF Habitat Conservation Report No. 2).

CAFF [Conservation of Arctic Flora and Fauna] (1996b) *Circumpolar Protected Areas Network (CPAN)—strategy and action plan.* Trondheim, Norway: Directorate for Nature Management (CAFF Habitat Conservation Report No. 6).

Council of Europe/UNEP/European Centre for Nature Conservation (1996) *The Pan-European Biological and Landscape Diversity Strategy, a vision for Europe's natural heritage.* Strasbourg, France: Council of Europe.

European Commission (1998) *Integrating biodiversity: A European Community strategy for action.* Luxembourg: Office for Official Publications of the European Communities.

Hägerhäll, B. and Skov, H. (1998). *Proposal for marine Baltic Sea Protected Areas (BSPAs).* Copenhagen: Ornis Consult (report to HELCOM Environment Committee).

Hecker, N. and Tomás Vives, P., eds. (1995) *The status of wetland inventories in the Mediterranean region.* Slimbridge, UK: MedWet/ International Waterfowl and Wetlands Research Bureau.

IUCN–CNPPA (1994) *Parks for life: action plan for protected areas in Europe.* Gland, Switzerland: International Union for Nature Conservation and Natural Resources (Commission on National Parks and Protected Areas).

Lyster, S. (1985) *International wildlife law.* Cambridge, UK: Grotius.

McCloskey, C. (1997) *P.E.B.L.D.S. explained.* Tilburg, Netherlands: IUCN European Programme.

Scott, D. A. (1998) *Global overview of the conservation of migratory Arctic breeding birds outside the Arctic.* Reykjavik: Conservation of Arctic Flora and Fauna (CAFF Technical Report No. 4).

Synge, H., ed. (1995) *Parks for life 97: Proceedings of the IUCN/WCPA European regional working session on protecting Europe's natural heritage.* Gland, Switzerland: IUCN—The World Conservation Union/German Federal Agency for Nature Conservation/Federal Ministry of the Environment, Nature Conservation and Nuclear Safety, Federal Republic of Germany.

Tucker, G. M. and Evans, M. I., eds. (1997) *Habitats for birds in Europe: a conservation strategy for the wider environment.* Cambridge, UK: BirdLife International (Conservation Series No. 6).

Appendix 2a. The 515 bird species occurring regularly in Europe, with the Important Bird Area criteria (at the global and European levels) that they can potentially fulfil and the numerical threshold(s) if applicable. Taxonomy and nomenclature follow Cramp *et al.* (1977–1994).

IBA criteria applicable at the European Union level (C) are shown in Appendix 2b.
For a full explanation of criteria, see the introductory chapter 'Identifying IBAs'.
Only criteria B1i, B1ii and Biii may be applied at the level of subspecies or population, all other criteria at the global (A) and European (B) levels apply to the full species.
All population thresholds are given in individuals, unless followed by 'p' indicating pairs.
To convert between individuals and pairs, a multiplying/dividing factor of 3 was used.

Species Subspecies or population applicable for criteria B1i or B1ii		Species of global conservation concern (A1)	Restricted-range species (A2)	Biome-restricted species (A3)	Congregations – global (1% biogeographic population of waterbird) (A4i)	(1% global population seabird or terrestrial species) (A4ii)	(20,000 waterbirds or 10,000 pairs seabirds) (A4iii)	(Bottleneck site of global importance) (A4iv)	Congregations – regional (1% flyway or distinct population of waterbird) (B1i)	(1% distinct population of seabird) (B1ii)	(1% distinct population of other species) (B1iii)	(Bottleneck site of regional importance) (B1iv)	Species of European conservation concern (categories 1–3) (B2)	Species of European conservation concern (category 4) (B3)
Gavia stellata	Red-throated Diver	—	—	—	750	—	W	—			—	—	★	—
NW Europe (win)									750	—				
Caspian/Black Sea/E Med (win)									?	—				
Gavia arctica	Black-throated Diver	—	—	—	1,200	—	W	—			—	—	★	—
W Siberia/Europe (bre)									✔	—				
Gavia immer	Great Northern Diver	—	—	—	50	—	W	—			—	—	—	—
Europe (win)									✔	—				
Gavia adamsii	White-billed Diver	—	—	Arctic	—[3]	—	W	—	—[3]	—	—	—	—	—
Tachybaptus ruficollis	Little Grebe	—	—	—	1,000	—	W	—			—	—	—	—
Western Palearctic									✔	—				
Podiceps cristatus	Great Crested Grebe	—	—	—	3,000	—	W	—			—	—	—	—
NW Europe (win)									1,500	—				
Black Sea/Med (win)									1,500	—				
Caspian Sea (win)									100	—				
Podiceps grisegena	Red-necked Grebe	—	—	—	400	—	W	—			—	—	—	—
NW Europe (win)									150	—				
Black Sea/Med (win)									100	—				
Caspian (win)									150	—				
Podiceps auritus	Slavonian Grebe	—	—	Boreal	300	—	W	—			—	—	—	—
NW Europe (long-billed form)									50	—				
NW Europe (short-billed form)									250	—				
Caspian									100	—				
Podiceps nigricollis	Black-necked Grebe	—	—	—	1,000	—	W	—			—	—	—	—
Western Palearctic									✔	—				
Fulmarus glacialis	Fulmar	—	—	—	—	45,000p	S	—	—	✔	—	—	—	—
Pterodroma feae	Fea's Petrel	5p	—	—	—	—[4]	S	—	—	—[4]	—	—	★	—
Pterodroma madeira	Zino's Petrel	5p	—	—	—	—[4]	S	—	—	—[4]	—	—	★	—
Bulweria bulwerii	Bulwer's Petrel	—	—	—	—	1,300p	S	—	—	150p	—	—	★	—
Calonectris diomedea	Cory's Shearwater	—	—	—	—	2,000p	S	—			—	—	★	—
C.d. diomedea									—	330p				
C.d. borealis									—	1,100p				
Puffinus gravis	Great Shearwater	—	—	—	—	No data	S	—	—	No data	—	—	—	—
Puffinus griseus	Sooty Shearwater	—	—	—	—	No data	S	—	—	No data	—	—	—	—
Puffinus puffinus	Manx Shearwater	—	—	—	—	2,700p	S	—	—	✔	—	—	★	—
Puffinus yelkouan	Yelkouan Shearwater	—	—	—	—	225p	S	—	—	✔	—	—	—	★
Puffinus assimilis	Little Shearwater	—	—	—	—	2,500p	S	—	—	30p	—	—	★	—
Pelagodroma marina	White-faced Petrel	—	—	—	—	10,000p	S	—			—	—	★	—
P.m. hypoleuca									—	160p				
Hydrobates pelagicus	Storm Petrel	—	—	—	—	2,800p	S	—			—	—	★	—
NW Europe									—	2,700p				
Med									—	70p				
Canary Islands									—	10p				
Oceanodroma leucorhoa	Leach's Petrel	—	—	—	—	70,000p	S	—	—	1,000p	—	—	★	—
Oceanodroma castro	Madeiran Petrel	—	—	—	—	150p	S	—			—	—	★	—
Atlantic									—	90p				
Sula bassana	Gannet	—	—	—	—	3,000p	S	—			—	—	★	—
E Atlantic									—	2,600p				
Phalacrocorax carbo	Cormorant	—	—	—	4,200	—	W	—			—	—	—	—
NW Europe									1,200	—				
N Europe/Central Europe									2,000	—				
Black Sea/Med									1,000	—				
Phalacrocorax aristotelis	Shag	—	—	—	2,700	—	W	—			—	—	—	★
N Europe									2,500	—				
Med									150	—				
Phalacrocorax pygmeus	Pygmy Cormorant	10p	—	—	210	—	W	—	✔	—	—	—	★	—
Pelecanus onocrotalus	White Pelican	—	—	—	700	—	W	—			—	—	★	—
Europe/W Asia (bre)									✔	—				

Appendix 2a ... continued. The 515 bird species occurring regularly in Europe, with the Important Bird Area criteria (at the global and European levels) that they can potentially fulfil and the numerical threshold(s) if applicable. Taxonomy and nomenclature follow Cramp *et al.* (1977–1994).

Species / Subspecies or population applicable for criteria B1i or B1ii		Species of global conservation concern (A1)	Restricted-range species (A2)	Biome-restricted species (A3)	Congregations – global (1% biogeographic population of waterbird) (A4i)	(1% global population seabird or terrestrial species) (A4ii)	(20,000 waterbirds or 10,000 pairs seabirds) (A4iii)	(Bottleneck site of global importance) (A4iv)	Congregations – regional (1% flyway or distinct population of waterbird) (B1i)	(1% distinct population of seabird) (B1ii)	(1% distinct population of other species) (B1iii)	(Bottleneck site of regional importance) (B1iv)	Species of European conservation concern (categories 1–3) (B2)	Species of European conservation concern (category 4) (B3)
Pelecanus crispus	Dalmatian Pelican	10p	—	—	25	—	W	—			—	—	★	—
Black Sea/Med (win)									✔	—				
Botaurus stellaris	Bittern	—	—	—	—[2]	—	W	—	—[2]	—	—	—	★	—
Ixobrychus minutus	Little Bittern	—	—	—	—[2]	—	W	—	—[2]	—	—	—	★	—
Nycticorax nycticorax	Night Heron	—	—	—	1,500	—	W	—			—	—	★	—
Europe/NW Africa (bre)									✔	—				
Ardeola ralloides	Squacco Heron	—	—	—	350	—	W	—			—	—	★	—
NW Africa/Med (bre)									100	—				
S & SW Asia/Black Sea (bre)									250	—				
Bubulcus ibis	Cattle Egret	—	—	—	2,100	—	W	—			—	—	—	—
SW Europe/NW Africa									2,000	—				
E Med/SW Asia									100	—				
Egretta garzetta	Little Egret	—	—	—	1,000	—	W	—			—	—	—	—
Black Sea/Med									✔	—				
Egretta alba	Great White Egret	—	—	—	120	—	W	—			—	—	—	—
Med/Black Sea (win)									✔	—				
Ardea cinerea	Grey Heron	—	—	—	4,500	—	W	—			—	—	—	—
N Europe (bre)									✔	—				
Ardea purpurea	Purple Heron	—	—	—	—[2]	—	W	—	—[2]	—	—	—	★	—
Ciconia nigra	Black Stork	—	—	—	210	—	W	✔			—	✔	★	—
SW Europe/W Africa									10	—				
Central/E Europe (bre)									200	—				
Ciconia ciconia	White Stork	—	—	—	4,850	—	W	✔			—	✔	★	—
Iberia/NW Africa (bre)									850	—				
Central/E Europe (bre)									4,000	—				
Plegadis falcinellus	Glossy Ibis	—	—	—	250	—	W	—			—	—	★	—
Black Sea/Med/W Africa									✔	—				
Platalea leucorodia	Spoonbill	—	—	—	130	—	W	—			—	—	★	—
E Atlantic									30	—				
Central SE Europe									100	—				
Phoenicopterus ruber	Greater Flamingo	—	—	—	800	—	W	—			—	—	★	—
W Med									✔	—				
Cygnus olor	Mute Swan	—	—	—	2,900	—	W	—			—	—	—	—
NW Mainland and Central Europe									2,100	—				
Britain									250	—				
Ireland									100	—				
Black Sea									450	—				
Cygnus columbianus	Bewick's Swan	—	—	Arctic	175	—	W	—			—	—	★W	—
W Siberia/NE & NW Europe									170	—				
N Siberia/Caspian									5	—				
Cygnus cygnus	Whooper Swan	—	—	—	730	—	W	—			—	—	—	★W
Iceland/UK/Ireland									160	—				
NW Mainland Europe									400	—				
N Europe/White Sea/Black Sea/E Med									170	—				
Anser fabalis	Bean Goose	—	—	—	3,800	—	W	—			—	—	—	—
W Siberia/NE & NW Europe									800	—				
W & Central Siberia/NE & SW Europe									3,000	—				
Anser brachyrhynchus	Pink-footed Goose	—	—	Arctic	2,600	—	W	—			—	—	—	★
E Greenland/Iceland/UK (bre)									2,250	—				
Svalbard/NW Europe (bre)									340	—				
Anser albifrons	White-fronted Goose	—	—	Arctic	12,800	—	W	—			—	—	—	—
NW Siberia/NE & NW Europe									6,000	—				
Central Europe/W Siberia									1,000	—				
W Siberia/Black Sea/Turkey									6,500	—				
Greenland/Ireland/UK (bre)									300	—				
Anser erythropus	Lesser White-fronted Goose	5p	—	Arctic	250	—	W	—			—	—	★	—
N Europe/W Siberia/Black Sea/Caspian									✔	—				

Appendix 2a ... continued. The 515 bird species occurring regularly in Europe, with the Important Bird Area criteria (at the global and European levels) that they can potentially fulfil and the numerical threshold(s) if applicable. Taxonomy and nomenclature follow Cramp *et al.* (1977–1994).

Species Subspecies or population applicable for criteria B1i or B1ii		Species of global conservation concern (A1)	Restricted-range species (A2)	Biome-restricted species (A3)	Congregations – global (1% biogeographic population of waterbird) (A4i)	(1% global population seabird or terrestrial species) (A4ii)	(20,000 waterbirds or 10,000 pairs seabirds) (A4iii)	(Bottleneck site of global importance) (A4iv)	Congregations – regional (1% flyway or distinct population of waterbird) (B1i)	(1% distinct population of seabird) (B1ii)	(1% distinct population of other species) (B1iii)	(Bottleneck site of regional importance) (B1iv)	Species of European conservation concern (categories 1–3) (B2)	Species of European conservation concern (category 4) (B3)
Anser anser	Greylag Goose	—	—	—	3,500	—	W	—			—	—	—	—
Iceland/UK/Ireland									1,000	—				
NW Scotland									50	—				
NW Europe/SW Europe									2,000	—				
Central Europe/N Africa									200	—				
Black Sea/Turkey									250	—				
Branta leucopsis	Barnacle Goose	—	—	Arctic	2,200	—	W	—			—	—	★W	★
E Greenland/Scotland/Ireland									320	—				
Svalbard/SW Scotland									120	—				
Russia/Germany/Netherlands									1,760	—				
Branta bernicla	Brent Goose	—	—	Arctic	3,250	—	W	—			—	—	★	—
W Siberia/W Europe									3,000	—				
Svalbard/Denmark/UK									50	—				
Canada/Greenland/Ireland									200	—				
Branta ruficollis	Red-breasted Goose	60	—	—	700	—	W	—			—	—	★	—
N. Siberia/Black Sea/Caspian									✔	—				
Tadorna ferruginea	Ruddy Shelduck	—	—	—	200	—	W	—			—	—	★	—
E Med/Black Sea/NE Africa									✔	—				
Tadorna tadorna	Shelduck	—	—	—	3,750	—	W	—			—	—	—	—
NW Europe									3,000	—				
Black Sea/Med									750	—				
Anas penelope	Wigeon	—	—	—	18,000	—	W	—			—	—	—	—
W Siberia & NW Europe/NE Europe									12,500	—				
W Siberia/NE Europe/Black Sea/Med									5,600	—				
Anas strepera	Gadwall	—	—	—	1,300	—	W	—			—	—	★	—
NW Europe									300	—				
NE Europe/Black Sea/Med									1,000	—				
Anas crecca	Teal	—	—	—	14,000	—	W	—			—	—	—	—
NW Europe									4,000	—				
W Siberia/NE Europe/Black Sea/Med									10,500	—				
Anas platyrhynchos	Mallard	—	—	—	83,000	—	W	—			—	—	—	—
NW Europe									50,000	—				
N Europe/W Med									10,000	—				
E Europe/Black Sea/E Med									22,500	—				
Greenland									150	—				
Anas acuta	Pintail	—	—	—	12,600	—	W	—			—	—	★	—
W Siberia/NE & E & S Europe/W Africa									12,000	—				
NW Europe									600	—				
Anas querquedula	Garganey	—	—	—	20,000	—	W	—			—	—	★	—
W Siberia/Europe/W Africa									✔	—				
Anas clypeata	Shoveler	—	—	—	4,900	—	W	—			—	—	—	—
NW/Central Europe									400	—				
W Siberia/NE & E & S Europe/W Africa									4,500	—				
Marmaronetta angustirostris	Marbled Duck	5p	—	—	40	—	W	—			—	—	★	—
W Med/W Africa									30	—				
E Med									10	—				
Netta rufina	Red-crested Pochard	—	—	—	750	—	W	—			—	—	★	—
SW/Central Europe & W Med									250	—				
Black Sea/E Med									500	—				
Aythya ferina	Pochard	—	—	—	13,500	—	W	—			—	—	—	★
Central & NE Europe/Black Sea/Med (win)									10,000	—				
NW Europe/NE Europe									3,500	—				
Aythya nyroca	Ferruginous Duck	20p	—	—	600	—	W	—			—	—	★	—
W Med/W Africa									100	—				
E Europe/E Med									500	—				
Aythya fuligula	Tufted Duck	—	—	—	16,000	—	W	—			—	—	—	—
Central Europe/Black Sea/Med									6,000	—				
NW Europe									10,000	—				

Appendix 2a ... continued. The 515 bird species occurring regularly in Europe, with the Important Bird Area criteria (at the global and European levels) that they can potentially fulfil and the numerical threshold(s) if applicable. Taxonomy and nomenclature follow Cramp *et al.* (1977–1994).

Species Subspecies or population applicable for criteria B1i or B1ii		Species of global conservation concern (A1)	Restricted-range species (A2)	Biome-restricted species (A3)	Congregations – global (1% biogeographic population of waterbird) (A4i)	(1% global population seabird or terrestrial species) (A4ii)	(20,000 waterbirds or 10,000 pairs seabirds) (A4iii)	(Bottleneck site of global importance) (A4iv)	Congregations – regional (1% flyway or distinct population of waterbird) (B1i)	(1% distinct population of seabird) (B1ii)	(1% distinct population of other species) (B1iii)	(Bottleneck site of regional importance) (B1iv)	Species of European conservation concern (categories 1–3) (B2)	Species of European conservation concern (category 4) (B3)
Aythya marila	Scaup	—	—	Arctic	4,600	—	W	—			—	—	★^{W}	—
N Europe/W Europe									3,100	—				
W Siberia/Black Sea/Caspian									1,500	—				
Somateria mollissima	Eider	—	—	—	18,830	—	W	—			—	—	—	—
Britain/Ireland									650	—				
Baltic/Denmark/Netherlands									13,500	—				
Norway/Russia									3,000	—				
White Sea									200	—				
Faroe Islands									60	—				
Shetland/Orkney Islands									120	—				
Svalbard/Franz Josef Land									400	—				
Greenland									300	—				
Iceland									600	—				
Somateria spectabilis	King Eider	—	—	Arctic	3,000	—	W	—			—	—	—	—
E Greenland/NE Europe/W Siberia									✔	—				
Polysticta stelleri	Steller's Eider	30	—	—	300	—	W	—			—	—	★	—
NE Europe/W Siberia									✔	—				
Histrionicus histrionicus	Harlequin Duck	—	—	—	90	—	W	—			—	—	★	—
Iceland									75	—				
Greenland									15	—				
Clangula hyemalis	Long-tailed Duck	—	—	Arctic	4,750	—	W	—			—	—	—	—
Iceland/Greenland									1,500	—				
W Siberia/N Europe									46,000	—				
Melanitta nigra	Common Scoter	—	—	Arctic	16,000	—	W	—			—	—	—	—
W Siberia/W & N Europe/NW Africa									✔	—				
Melanitta fusca	Velvet Scoter	—	—	—	10,000	—	W	—			—	—	★^{W}	—
Western Siberia/N Europe									✔	—				
Black Sea/Caspian									15	—				
Bucephala islandica	Barrow's Goldeneye	—	—	—	—[3]	—	W	—	—[3]	—	—	—	★	—
Bucephala clangula	Goldeneye	—	—	—	4,200	—	W	—			—	—	—	—
NW Europe/Central Europe									3,000	—				
W Siberia/NE Europe/Black Sea									200	—				
NE Europe/Adriatic									750	—				
W Siberia/Caspian									250	—				
Mergus albellus	Smew	—	—	Boreal	900	—	W	—			—	—	★	—
NW/Central Europe									250	—				
NE Europe/Black Sea/E Med									650	—				
Mergus serrator	Red-breasted Merganser	—	—	—	1,850	—	W	—			—	—	—	—
NW Europe/Central Europe									1,250	—				
NE Europe/Black Sea/Med									500	—				
E Greenland/Iceland/UK									200	—				
W Greenland									250	—				
Mergus merganser	Goosander	—	—	—	2,200	—	W	—			—	—	—	—
Iceland									10	—				
NW/Central Europe									2,000	—				
UK									65	—				
Central Europe (bre)									30	—				
Balkans									1	—				
NE Europe/Black Sea									100	—				
Oxyura leucocephala	White-headed Duck	5p	—	—	140	—	W	—			—	—	★	—
W Med									7	—				
E Med/SW Asia/Turkey									110	—				
Pernis apivorus	Honey Buzzard	—	—	—	—	—	—	✔	—	—	—	✔	—	★
Elanus caeruleus	Black-winged Kite	—	—	—	—	—	—	—[1]	—	—	—	—[1]	★	—
Milvus migrans	Black Kite	—	—	—	—	200p	—	✔	—	—	✔	✔	★	—
Milvus milvus	Red Kite	—	—	—	—	220p	—	✔	—	—	✔	✔	—	★
Haliaeetus albicilla	White-tailed Eagle	5p	—	—	—	—	—	✔	—	—	—	✔	★	—
Gypaetus barbatus	Lammergeier	—	—	—	—	—	—	—[1]	—	—	—	—[1]	★	—

Appendix 2a ... continued. The 515 bird species occurring regularly in Europe, with the Important Bird Area criteria (at the global and European levels) that they can potentially fulfil and the numerical threshold(s) if applicable. Taxonomy and nomenclature follow Cramp *et al.* (1977–1994).

Species	Subspecies or population applicable for criteria B1i or B1ii	Species of global conservation concern (A1)	Restricted-range species (A2)	Biome-restricted species (A3)	Congregations – global (1% biogeographic population of waterbird) (A4i)	(1% global population seabird or terrestrial species) (A4ii)	(20,000 waterbirds or 10,000 pairs seabirds) (A4iii)	(Bottleneck site of global importance) (A4iv)	Congregations – regional (1% flyway or distinct population of waterbird) (B1i)	(1% distinct population of seabird) (B1ii)	(1% distinct population of other species) (B1iii)	(Bottleneck site of regional importance) (B1iv)	Species of European conservation concern (categories 1–3) (B2)	Species of European conservation concern (category 4) (B3)
Neophron percnopterus	Egyptian Vulture	—	—	—	—	110p	—	✓	—	—	30p	✓	★	—
Gyps fulvus	Griffon Vulture	—	—	—	—	180p	—	✓	—	—	90p	✓	★	—
Aegypius monachus	Cinereous Vulture	5p	—	—	—	50p	—	—[1]	—	—	10p	—[1]	★	—
Circaetus gallicus	Short-toed Eagle	—	—	—	—	—	—	✓	—	—	—	✓	★	—
Circus aeruginosus	Marsh Harrier	—	—	—	—	—	—	✓	—	—	—	✓	—	—
Circus cyaneus	Hen Harrier	—	—	—	—	—	—	✓	—	—	—	✓	★	—
Circus macrourus	Pallid Harrier	10p	—	Steppe	—	—	—	✓	—	—	—	✓	★	—
Circus pygargus	Montagu's Harrier	—	—	—	—	—	—	✓	—	—	—	✓	—	★
Accipiter gentilis	Goshawk	—	—	—	—	—	—	—[1]	—	—	—	—[1]	—	—
Accipiter nisus	Sparrowhawk	—	—	—	—	—	—	✓	—	—	—	✓	—	—
Accipiter brevipes	Levant Sparrowhawk	—	—	Steppe	—	—	—	✓	—	—	—	✓	★	—
Buteo buteo	Buzzard	—	—	—	—	—	—	✓	—	—	—	✓	—	—
Buteo rufinus	Long-legged Buzzard	—	—	—	—	—	—	—[1]	—	—	—	—[1]	★	—
Buteo lagopus	Rough-legged Buzzard	—	—	Arctic	—	—	—	✓	—	—	—	✓	—	—
Aquila pomarina	Lesser Spotted Eagle	—	—	—	—	—	—	✓	—	—	—	✓	★	—
Aquila clanga	Greater Spotted Eagle	2p	—	—	—	—	—	✓	—	—	—	✓	★	—
Aquila nipalensis	Steppe Eagle	—	—	—	—	—	—	✓	—	—	—	✓	★	—
Aquila heliaca	Imperial Eagle	2p	—	Steppe	—	—	—	✓	—	—	—	✓	★	—
Aquila adalberti	Spanish Imperial Eagle	2p	—	—	—	—	—	—[1]	—	—	—	—[1]	★	—
Aquila chrysaetos	Golden Eagle	—	—	—	—	—	—	—[1]	—	—	—	—[1]	★	—
Hieraaetus pennatus	Booted Eagle	—	—	—	—	37p	—	✓	—	—	✓	✓	★	—
Hieraaetus fasciatus	Bonnelli's Eagle	—	—	—	—	10p	—	—[1]	—	—	✓	—[1]	★	—
Pandion haliaetus	Osprey	—	—	—	—	—	—	✓	—	—	—	✓	★	—
Falco naumanni	Lesser Kestrel	10p	—	—	—	200p	—	✓	—	—	100p	✓	★	—
Falco tinnunculus	Kestrel	—	—	—	—	—	—	✓	—	—	—	✓	★	—
Falco vespertinus	Red-footed Falcon	—	—	—	—	360p	—	✓	—	—	180p	✓	★	—
Falco columbarius	Merlin	—	—	—	—	—	—	✓	—	—	—	✓	—	—
Falco subbuteo	Hobby	—	—	—	—	—	—	✓	—	—	—	✓	—	—
Falco eleonorae	Eleonora's Falcon	—	—	Med	—	50p	—	✓	—	—	40p	✓	★	—
Falco biarmicus	Lanner	—	—	—	—	—	—	—[1]	—	—	—	—[1]	★	—
Falco cherrug	Saker	—	—	—	—	—	—	✓	—	—	—	✓	★	—
Falco rusticolus	Gyrfalcon	—	—	Arctic	—	—	—	—[1]	—	—	—	—[1]	★	—
Falco peregrinus	Peregrine	—	—	—	—	—	—	✓	—	—	—	✓	★	—
Falco pelegrinoides	Barbary Falcon	—	—	—	—	—	—	—[1]	—	—	—	—[1]	—	—
Bonasa bonasia	Hazel Grouse	—	—	—	—	—	—	—	—	—	—	—	—	—
Lagopus lagopus	Red/Willow Grouse	—	—	—	—	—	—	—	—	—	—	—	—	—
Lagopus mutus	Ptarmigan	—	—	—	—	—	—	—	—	—	—	—	—	—
Tetrao tetrix	Black Grouse	—	—	—	—	—	—	—	—	—	—	—	★	—
Tetrao mlokosiewiczi	Caucasian Black Grouse	20p	Caucasus	Alpine	—	—	—	—	—	—	—	—	★	—
Tetrao urogallus	Capercaillie	—	—	—	—	—	—	—	—	—	—	—	—	—
Tetraogallus caucasicus	Caucasian Snowcock	—	Caucasus	Alpine	—	—	—	—	—	—	—	—	—	★
Tetraogallus caspius	Caspian Snowcock	—	—	Alpine	—	—	—	—	—	—	—	—	★	—
Alectoris chukar	Chukar	—	—	—	—	—	—	—	—	—	—	—	★	—
Alectoris graeca	Rock Partridge	—	—	Med	—	—	—	—	—	—	—	—	★	—
Alectoris rufa	Red-legged Partridge	—	—	—	—	—	—	—	—	—	—	—	★	—
Alectoris barbara	Barbary Partridge	—	—	—	—	—	—	—	—	—	—	—	★	—
Ammoperdix griseogularis	See-see Partridge	—	—	—	—	—	—	—	—	—	—	—	—	—
Francolinus francolinus	Black Francolin	—	—	—	—	—	—	—	—	—	—	—	★	—
Perdix perdix	Partridge	—	—	—	—	—	—	—	—	—	—	—	★	—
Coturnix coturnix	Quail	—	—	—	—	—	—	—	—	—	—	—	★	—
Phasianus colchicus	Pheasant	—	—	—	—	—	—	—	—	—	—	—	—	—
Turnix sylvatica	Andalusian Hemipode	—	—	—	—	—	—	—	—	—	—	—	★	—
Rallus aquaticus	Water Rail	—	—	—	—	—	—	—	—	—	—	—	—	—
Porzana porzana	Spotted Crake	—	—	—	No data	—	W	—	60p	—	—	—	—	★
Porzana parva	Little Crake	—	—	—	55p	—	W	—	✓	—	—	—	—	★

Appendix 2a ... continued. The 515 bird species occurring regularly in Europe, with the Important Bird Area criteria (at the global and European levels) that they can potentially fulfil and the numerical threshold(s) if applicable. Taxonomy and nomenclature follow Cramp *et al.* (1977–1994).

Species Subspecies or population applicable for criteria B1i or B1ii		Species of global conservation concern (A1)	Restricted-range species (A2)	Biome-restricted species (A3)	Congregations – global (1% biogeographic population of waterbird) (A4i)	(1% global population seabird or terrestrial species) (A4ii)	(20,000 waterbirds or 10,000 pairs seabirds) (A4iii)	(Bottleneck site of global importance) (A4iv)	Congregations – regional (1% flyway or distinct population of waterbird) (B1i)	(1% distinct population of seabird) (B1ii)	(1% distinct population of other species) (B1iii)	(Bottleneck site of regional importance) (B1iv)	Species of European conservation concern (categories 1–3) (B2)	Species of European conservation concern (category 4) (B3)
Porzana pusilla	Baillon's Crake	—	—	—	No data	—	W	—	40p	—	—	—	★	—
Crex crex	Corncrake	20p	—	—	45p	—	W	—	✔	—	—	—	★	—
Gallinula chloropus	Moorhen	—	—	—	10,000	—	W	—			—	—	—	—
Europe/N Africa									✔	—				
Porphyrio porphyrio	Purple Gallinule	—	—	—	35p	—	W	—	✔	—	—	—	★	—
Fulica atra	Coot	—	—	—	40,000	—	W	—			—	—	—	—
NW Europe (win)									15,000	—				
Black Sea/Med (win)									25,000	—				
Fulica cristata	Crested Coot	—	—	—	10p	—	W	—	✔	—	—	—	★	—
Grus grus	Crane	—	—	—	1,550	—	W	✔			—	—	★	—
NW Europe (bre)									600	—				
NE/Central Europe (bre)									600	—				
Black Sea/E Med (win)									350	—				
Black Sea/Turkey									2	—				
Anthropoides virgo	Demoiselle Crane	—	—	Steppe	—[3]	—	W	✔	—[3]	—	—	—	—	—
Tetrax tetrax	Little Bustard	60	—	—	—	—	—	—	—	—	—	—	★	—
Chlamydotis undulata	Houbara Bustard	—	—	—	—	—	—	—	—	—	—	—	★	—
Otis tarda	Great Bustard	30	—	—	—	—	—	—	—	—	—	—	★	—
Haematopus ostralegus	Oystercatcher	—	—	—	11,000	—	W	—			—	—	—	—
Europe & NW Africa (win)									9,000	—				
Himantopus himantopus	Black-winged Stilt	—	—	—	700	—	W	—			—	—	—	—
W Med (bre)									400	—				
Black Sea/ E Med (bre)									300	—				
Recurvirostra avosetta	Avocet	—	—	—	1,100	—	W	—			—	—	★^{W}	★
W Europe/W Med (bre)									700	—				
Black Sea/ E Med (bre)									400	—				
Burhinus oedicnemus	Stone Curlew	—	—	—	398p	—	W	—	✔	—	—	—	★	—
Cursorius cursor	Cream-coloured Courser	—	—	—	10p	—	W	—	✔	—	—	—	★	—
Glareola pratincola	Collared Pratincole	—	—	—	200	—	W	—			—	—	★	—
W Med (bre)									100	—				
Black Sea/E Med (bre)									100	—				
Glareola nordmanni	Black-winged Pratincole	10p	—	Steppe	—[2]	—	W	—	—[2]	—	—	—	★	—
Charadrius dubius	Little Ringed Plover	—	—	—	3,200	—	W	—			—	—	—	—
Europe/W Africa									✔	—				
Charadrius hiaticula	Ringed Plover	—	—	—	2,400	—	W	—			—	—	—	—
Canada/W Europe									✔	—				
Charadrius alexandrinus	Kentish Plover	—	—	—	950	—	W	—			—	—	★	—
E Atlantic									700	—				
Black Sea/ E Med									250	—				
Charadrius leschenaultii	Greater Sand Plover	—	—	—	—[3]	—	W	—	—[3]	—	—	—	★	—
Charadrius asiaticus	Caspian Plover	—	—	—	—[3]	—	W	—	—[3]	—	—	—	★	—
Charadrius morinellus	Dotterel	—	—	—	1,000	—	W	—			—	—	—	—
Europe (bre)									✔	—				
Pluvialis apricaria	Golden Plover	—	—	—	10,000	—	W	—			—	—	—	★
NW Europe (bre)									✔	—				
Pluvialis squatarola	Grey Plover	—	—	Arctic	1,500	—	W	—			—	—	—	—
E Atlantic									✔	—				
Hoplopterus spinosus	Spur-winged Plover	—	—	—	100	—	W	—			—	—	★	—
Black Sea/Med									✔	—				
Hoplopterus indicus	Red-wattled Plover	—	—	—	—[3]	—	W	—	—[3]	—	—	—	—	—
Chettusia gregaria	Sociable Plover	10p	—	Steppe	—[3]	—	W	—	—[3]	—	—	—	★	—
Chettusia leucura	White-tailed Plover	—	—	—	—[3]	—	W	—	—[3]	—	—	—	—	—
Vanellus vanellus	Lapwing	—	—	—	70,000	—	W	—			—	—	—	—
Europe/W Africa									✔	—				
Calidris canutus	Knot	—	—	Arctic	8,500	—	W	—			—	—	★^{W}	—
NE Canada/Greenland/Iceland/NW Europe (win)									3,500	—				
W Africa/S Africa (win)									5,000	—				

Appendix 2a ... continued. The 515 bird species occurring regularly in Europe, with the Important Bird Area criteria (at the global and European levels) that they can potentially fulfil and the numerical threshold(s) if applicable. Taxonomy and nomenclature follow Cramp *et al.* (1977–1994).

Species Subspecies or population applicable for criteria B1i or B1ii		Species of global conservation concern (A1)	Restricted-range species (A2)	Biome-restricted species (A3)	Congregations – global (1% biogeographic population of waterbird) (A4i)	(1% global population seabird or terrestrial species) (A4ii)	(20,000 waterbirds or 10,000 pairs seabirds) (A4iii)	(Bottleneck site of global importance) (A4iv)	Congregations – regional (1% flyway or distinct population of waterbird) (B1i)	(1% distinct population of seabird) (B1ii)	(1% distinct population of other species) (B1iii)	(Bottleneck site of regional importance) (B1iv)	Species of European conservation concern (categories 1–3) (B2)	Species of European conservation concern (category 4) (B3)
Calidris alba	Sanderling	—	—	Arctic	1,200	—	W	—			—	—	—	—
E Atlantic/W & S Africa (win)									✔	—				
Calidris minuta	Little Stint	—	—	Arctic	2,100	—	W	—			—	—	—	—
Europe/Western Africa (win)									✔	—				
Calidris temminckii	Temminck's Stint	—	—	Arctic	3,000	—	W	—			—	—	—	—
Europe/Western Africa (win)									✔	—				
Calidris ferruginea	Curlew Sandpiper	—	—	—	4,500	—	W	—			—	—	—	—
W Africa/SW Europe (win)									✔	—				
Calidris maritima	Purple Sandpiper	—	—	Arctic	500	—	W	—			—	—	—	★
E Atlantic (win)									✔	—				
Calidris alpina	Dunlin	—	—	—	22,000	—	W	—			—	—	★W	—
N Siberia/Europe/W Africa (bre)									14,000	—				
Iceland/Greenland (bre)									8,000	—				
Baltic/UK/Ireland									200	—				
Greenland (bre)									150	—				
Limicola falcinellus	Broad-billed Sandpiper	—	—	—	250	—	W	—			—	—	★	—
N Europe/SW Asia/E Africa									✔	—				
Philomachus pugnax	Ruff	—	—	—	33,000	—	W	—			—	—	—	★
W Africa (win)									✔	—				
Lymnocryptes minimus	Jack Snipe	—	—	Boreal	660	—	W	—			—	—	★W	—
Europe (bre)									✔	—				
Gallinago gallinago	Snipe	—	—	—	54,000	—	W	—			—	—	—	—
Europe (bre)									47,000	—				
Iceland (bre)									7,500	—				
Gallinago media	Great Snipe	20p	—	—	300	—	W	—			—	—	★	—
Scandinavia (bre)									75	—				
NE Europe/W Siberia (bre)									250	—				
Gallinago stenura	Pintail Snipe	—	—	—	—[3]	—	W	—[3]	—	—	—	—	—	—
Scolopax rusticola	Woodcock	—	—	—	—[2]	—	W	—[2]	—	—	—	—	★W	—
Limosa limosa	Black-tailed Godwit	—	—	—	4,200	—	W	—			—	—	★	—
W Europe/W Africa									3,500	—				
Iceland (bre)									700	—				
Limosa lapponica	Bar-tailed Godwit	—	—	Arctic	1,000	—	W	—			—	—	★W	—
Western Palearctic (win)									✔	—				
Numenius phaeopus	Whimbrel	—	—	—	6,500	—	W	—			—	—	—	★
Europe/W Africa									✔	—				
Numenius tenuirostris	Slender-billed Curlew	✔	—	—	3	—	W	—	✔	—	—	—	★	—
Numenius arquata	Curlew	—	—	—	3,500	—	W	—			—	—	★W	—
Europe (bre)									✔	—				
Tringa erythropus	Spotted Redshank	—	—	Arctic	720	—	W	—			—	—	—	—
Europe/W Africa									✔	—				
Tringa totanus	Redshank	—	—	—	3,000	—	W	—			—	—	★	—
E Atlantic (win)									1,500	—				
Iceland/Faroe Islands (bre)									1,500	—				
Tringa stagnatilis	Marsh Sandpiper	—	—	—	300	—	W	—			—	—	—	—
Europe/W Africa									✔	—				
Tringa nebularia	Greenshank	—	—	Boreal	1,800	—	W	—			—	—	—	—
Europe/W Africa									✔	—				
Tringa ochropus	Green Sandpiper	—	—	—	10,000	—	W	—			—	—	—	—
Europe/W Africa									✔	—				
Tringa glareola	Wood Sandpiper	—	—	—	11,000	—	W	—			—	—	★	—
Europe (bre)									✔	—				
Xenus cinereus	Terek Sandpiper	—	—	—	440	—	W	—			—	—	—	—
SW Asia/Africa (win)									✔	—				
Actitis hypoleucos	Common Sandpiper	—	—	—	15,000	—	W	—			—	—	—	—
Europe (bre)									✔	—				
Arenaria interpres	Turnstone	—	—	—	700	—	W	—			—	—	—	—
Western Palearctic									✔	—				

Appendix 2a ... continued. The 515 bird species occurring regularly in Europe, with the Important Bird Area criteria (at the global and European levels) that they can potentially fulfil and the numerical threshold(s) if applicable. Taxonomy and nomenclature follow Cramp *et al.* (1977–1994).

Species Subspecies or population applicable for criteria B1i or B1ii		Species of global conservation concern (A1)	Restricted-range species (A2)	Biome-restricted species (A3)	Congregations – global (1% biogeographic population of waterbird) (A4i)	(1% global population seabird or terrestrial species) (A4ii)	(20,000 waterbirds or 10,000 pairs seabirds) (A4iii)	(Bottleneck site of global importance) (A4iv)	Congregations – regional (1% flyway or distinct population of waterbird) (B1i)	(1% distinct population of seabird) (B1ii)	(1% distinct population of other species) (B1iii)	(Bottleneck site of regional importance) (B1iv)	Species of European conservation concern (categories 1–3) (B2)	Species of European conservation concern (category 4) (B3)
Phalaropus lobatus	Red-necked Phalarope	—	—	Arctic	—[2]	—	W	—	—	—	—	—	—	—
Phalaropus fulicarius	Grey Phalarope	—	—	Arctic	—[2]	—	W	—	—	—	—	—	—	—
Stercorarius pomarinus	Pomarine Skua	—	—	Arctic	—	No data	S	—			—	—	—	—
Europe									—	No data				
Stercorarius parasiticus	Arctic Skua	—	—	—	—	3,000p	S	—	—	270p	—	—	—	—
Stercorarius longicaudus	Long-tailed Skua	—	—	Arctic	—	1,000p	S	—			—	—	—	—
Arctic/sub-Arctic/Scandinavia/Greenland									—	120p				
Stercorarius skua	Great Skua	—	—	—	—	140p	S	—	—	140p	—	—	—	★
Larus ichthyaetus	Great Black-headed Gull	—	—	Steppe	700	—	W	—			—	—	—	—
Black Sea/Caspian/SW Asia									✔	—				
Larus melanocephalus	Mediterranean Gull	—	—	—	5,500	—	W	—			—	—	—	★
W Europe/Med/NW Africa									✔	—				
Larus minutus	Little Gull	—	—	—	680	—	W	—			—	—	★	—
Central/E Europe (bre)									✔	—				
Larus sabini	Sabine's Gull	—	—	—	—[3]	—	W	—	—[3]	—	—	—	—	—
Larus ridibundus	Black-headed Gull	—	—	—	65,000	—	W	—			—	—	—	—
NW Europe (bre)									60,000	—				
Med (bre)									5,000	—				
Larus genei	Slender-billed Gull	—	—	—	1,200	—	W	—			—	—	—	—
Black Sea/Med (bre)									✔	—				
Larus audouinii	Audouin's Gull	20p	—	—	390	—	W	—			—	—	★	—
Europe									✔	—				
Larus canus	Common Gull	—	—	—	16,000	—	W	—			—	—	★	—
NW & Central Europe/Atlantic/Med									16,000	—				
NE Europe/W Siberia/Black Sea/Caspian									250	—				
Larus fuscus	Lesser Black-backed Gull	—	—	—	6,000	—	W	—			—	—	—	★
NE Europe/Black Sea/SW Asia/E Africa									2,000	—				
W Europe/Med/W Africa									4,000	—				
Larus argentatus	Herring Gull	—	—	—	27,000	—	W	—			—	—	—	—
NW Europe (bre)									14,000	—				
Iceland/W Europe (bre)									13,000	—				
Larus cachinnans	Yellow-legged Gull	—	—	—	4,500	—	W	—			—	—	—	—
Black Sea/Caspian/SW Asia									1,000	—				
Med (bre)									3,500	—				
Larus armenicus	Armenian Gull	—	—	—	300	—	W	—			—	—	—	—
Armenia/E Turkey/W Iran									✔	—				
Larus glaucoides	Iceland Gull	—	—	—	300	—	W	—			—	—	—	—
Greenland/N Atlantic									✔	—				
Larus hyperboreus	Glaucous Gull	—	—	Arctic	2,000	—	W	—			—	—	—	—
N Atlantic									✔	—				
Larus marinus	Great Black-backed Gull	—	—	—	4,800	—	W	—			—	—	—	★
NE Atlantic (bre)									✔	—				
Rhodostethia rosea	Ross's Gull	—	—	—	—[3]	—	W	—	—[3]	—	—	—	—	—
Rissa tridactyla	Kittiwake	—	—	—	84,000	—	W	—			—	—	—	—
E Atlantic (bre)									✔	—				
Pagophila eburnea	Ivory Gull	—	—	Arctic	50	—	W	—			—	—	★	—
Arctic									✔	—				
Gelochelidon nilotica	Gull-billed Tern	—	—	—	270	—	W	—			—	—	★	—
W Europe/W Africa									120	—				
Black Sea/E Med (bre)									150	—				
Sterna caspia	Caspian Tern	—	—	—	150	—	W	—			—	—	★	—
Europe (bre)									50	—				
Caspian (bre)									100	—				
Sterna sandvicensis	Sandwich Tern	—	—	—	2,800	—	W	—			—	—	★	—
W Europe/W Africa									1,500	—				
Black Sea/Med (bre)									1,300	—				
Sterna dougallii	Roseate Tern	—	—	—	50	—	W	—			—	—	★	—
Europe (bre)									✔	—				

Appendix 2a ... continued. The 515 bird species occurring regularly in Europe, with the Important Bird Area criteria (at the global and European levels) that they can potentially fulfil and the numerical threshold(s) if applicable. Taxonomy and nomenclature follow Cramp *et al.* (1977–1994).

Species Subspecies or population applicable for criteria B1i or B1ii		Species of global conservation concern (A1)	Restricted-range species (A2)	Biome-restricted species (A3)	Congregations – global (1% biogeographic population of waterbird) (A4i)	(1% global population seabird or terrestrial species) (A4ii)	(20,000 waterbirds or 10,000 pairs seabirds) (A4iii)	(Bottleneck site of global importance) (A4iv)	Congregations – regional (1% flyway or distinct population of waterbird) (B1i)	(1% distinct population of seabird) (B1ii)	(1% distinct population of other species) (B1iii)	(Bottleneck site of regional importance) (B1iv)	Species of European conservation concern (categories 1–3) (B2)	Species of European conservation concern (category 4) (B3)
Sterna hirundo	Common Tern	—	—	—	7,800	—	W	—			—	—	—	—
S Europe/W Europe (bre)									1,800	—				
N Europe/E Europe (bre)									6,000	—				
Sterna paradisaea	Arctic Tern	—	—	—	13,000	—	W	—			—	—	—	—
Europe									✔	—				
Sterna albifrons	Little Tern	—	—	—	900	—	W	—			—	—	★	—
E Atlantic (bre)									340	—				
Black Sea/E Med (bre)									600	—				
Chlidonias hybridus	Whiskered Tern	—	—	—	1,000	—	W	—			—	—	★	—
W Europe/W Med/NW Africa (bre)									250	—				
Black Sea/E Med (bre)									750	—				
Chlidonias niger	Black Tern	—	—	—	1,700	—	W	—			—	—	★	—
Europe/Asia (bre)									✔	—				
Chlidonias leucopterus	White-winged Black Tern	—	—	—	2,300	—	W	—			—	—	—	—
E Europe/W Asia/Africa									✔	—				
Uria aalge	Guillemot	—	—	—	—	80,000p	S	—			—	—	—	—
E Atlantic									—	20,000p				
Baltic									—	450p				
Uria lomvia	Brünnich's Guillemot	—	—	—	—	110,000p	S	—			—	—	—	—
Atlantic									—	70,000p				
Alca torda	Razorbill	—	—	—	—	5,400p	S	—			—	—	—	★
East Atlantic									—	250p				
Baltic									—	100p				
Iceland/Faroes/UK/Ireland									—	4,500p				
Cepphus grylle	Black Guillemot	—	—	—	—	3,000p	S	—			—	—	★	—
Faroes									—	40p				
Iceland									—	100p				
Baltic									—	150p				
Greenland									—	250p				
Russia/Svalbard									—	300p				
Norway									—	200p				
UK									—	190p				
Alle alle	Little Auk	—	—	—	—	110,000p	S	—			—	—	—	—
Greenland									—	100,000p				
Northern Europe									—	10,000p				
Fratercula arctica	Puffin	—	—	—	—	60,000p	S	—			—	—	★	—
Iceland/Norway/SW Greenland/N America									—	40,000p				
UK/Ireland/Faroe/France/Norway									—	8,500p				
Pterocles orientalis	Black-bellied Sandgrouse	—	—	—	—	—	—	—	—	—	—	—	★	—
Pterocles alchata	Pin-tailed Sandgrouse	—	—	—	—	—	—	—	—	—	—	—	★	—
Columba livia	Rock Dove	—	—	—	—	—	—	—	—	—	—	—	—	—
Columba oenas	Stock Dove	—	—	—	—	—	—	—	—	—	—	—	—	★
Columba palumbus	Woodpigeon	—	—	—	—	—	—	—	—	—	—	—	—	★
Columba trocaz	Long-toed Pigeon	10p	Madeira/Canaries	—	—	—	—	—	—	—	—	—	★	—
Columba bollii	Dark-tailed Laurel Pigeon	5p	Madeira/Canaries	—	—	—	—	—	—	—	—	—	★	—
Columba junoniae	White-tailed Laurel Pigeon	5p	Madeira/Canaries	—	—	—	—	—	—	—	—	—	★	—
Streptopelia decaocto	Collared Dove	—	—	—	—	—	—	—	—	—	—	—	—	—
Streptopelia turtur	Turtle Dove	—	—	—	—	—	—	—	—	—	—	—	★	—
Streptopelia orientalis	Oriental Turtle Dove	—	—	—	—	—	—	—	—	—	—	—	—	—
Streptopelia senegalensis	Laughing Dove	—	—	—	—	—	—	—	—	—	—	—	—	—
Clamator glandarius	Great Spotted Cuckoo	—	—	—	—	—	—	—	—	—	—	—	—	—
Cuculus canorus	Cuckoo	—	—	—	—	—	—	—	—	—	—	—	—	—
Cuculus saturatus	Oriental Cuckoo	—	—	—	—	—	—	—	—	—	—	—	—	—
Tyto alba	Barn Owl	—	—	—	—	—	—	—	—	—	—	—	★	—
Otus brucei	Striated Scops Owl	—	—	—	—	—	—	—	—	—	—	—	—	—
Otus scops	Scops Owl	—	—	—	—	—	—	—	—	—	—	—	★	—
Bubo bubo	Eagle Owl	—	—	—	—	—	—	—	—	—	—	—	★	—
Ketupa zeylonensis	Brown Fish Owl	—	—	—	—	—	—	—	—	—	—	—	—	—

Appendix 2a ... continued. The 515 bird species occurring regularly in Europe, with the Important Bird Area criteria (at the global and European levels) that they can potentially fulfil and the numerical threshold(s) if applicable. Taxonomy and nomenclature follow Cramp *et al.* (1977–1994).

Species Subspecies or population applicable for criteria B1i or B1ii		Species of global conservation concern (A1)	Restricted-range species (A2)	Biome-restricted species (A3)	Congregations – global (1% biogeographic population of waterbird) (A4i)	(1% global population seabird or terrestrial species) (A4ii)	(20,000 waterbirds or 10,000 pairs seabirds) (A4iii)	(Bottleneck site of global importance) (A4iv)	Congregations – regional (1% flyway or distinct population of waterbird) (B1i)	(1% distinct population of seabird) (B1ii)	(1% distinct population of other species) (B1iii)	(Bottleneck site of regional importance) (B1iv)	Species of European conservation concern (categories 1–3) (B2)	Species of European conservation concern (category 4) (B3)
Nyctea scandiaca	Snowy Owl	—	—	Arctic	—	—	—	—	—	—	—	—	★	—
Surnia ulula	Hawk Owl	—	—	Boreal	—	—	—	—	—	—	—	—	—	—
Glaucidium passerinum	Pygmy Owl	—	—	—	—	—	—	—	—	—	—	—	—	—
Athene noctua	Little Owl	—	—	—	—	—	—	—	—	—	—	—	★	—
Strix aluco	Tawny Owl	—	—	—	—	—	—	—	—	—	—	—	—	★
Strix uralensis	Ural Owl	—	—	—	—	—	—	—	—	—	—	—	—	—
Strix nebulosa	Great Grey Owl	—	—	Boreal	—	—	—	—	—	—	—	—	—	—
Asio otus	Long-eared Owl	—	—	—	—	—	—	—	—	—	—	—	—	—
Asio flammeus	Short-eared Owl	—	—	—	—	—	—	—	—	—	—	—	★	—
Aegolius funereus	Tengmalm's Owl	—	—	—	—	—	—	—	—	—	—	—	—	—
Caprimulgus europaeus	Nightjar	—	—	—	—	—	—	—	—	—	—	—	★	—
Caprimulgus ruficollis	Red-necked Nightjar	—	—	Med	—	—	—	—	—	—	—	—	—	—
Apus unicolor	Plain Swift	—	Madeira/Canaries	—	—	—	—	—	—	—	—	—	—	★
Apus apus	Swift	—	—	—	—	—	—	—	—	—	—	—	—	—
Apus pallidus	Pallid Swift	—	—	—	—	—	—	—	—	—	—	—	—	—
Apus melba	Alpine Swift	—	—	—	—	—	—	—	—	—	—	—	—	—
Apus caffer	White-rumped Swift	—	—	—	—	—	—	—	—	—	—	—	—	—
Apus affinis	Little Swift	—	—	—	—	—	—	—	—	—	—	—	—	—
Halcyon smyrnensis	White-breasted Kingfisher	—	—	—	—	—	—	—	—	—	—	—	—	—
Alcedo atthis	Kingfisher	—	—	—	—	—	—	—	—	—	—	—	★	—
Ceryle rudis	Pied Kingfisher	—	—	—	—	—	—	—	—	—	—	—	—	—
Merops superciliosus	Blue-cheeked Bee-eater	—	—	—	—	—	—	—	—	—	—	—	—	—
Merops apiaster	Bee-eater	—	—	—	—	40,000	—	—	—	—	13,000	—	★	—
Coracias garrulus	Roller	—	—	—	—	—	—	—	—	—	—	—	★	—
Upupa epops	Hoopoe	—	—	—	—	—	—	—	—	—	—	—	—	—
Jynx torquilla	Wryneck	—	—	—	—	—	—	—	—	—	—	—	★	—
Picus canus	Grey-headed Woodpecker	—	—	—	—	—	—	—	—	—	—	—	★	—
Picus viridis	Green Woodpecker	—	—	—	—	—	—	—	—	—	—	—	★	—
Dryocopus martius	Black Woodpecker	—	—	—	—	—	—	—	—	—	—	—	—	—
Dendrocopos major	Great Spotted Woodpecker	—	—	—	—	—	—	—	—	—	—	—	—	—
Dendrocopos syriacus	Syrian Woodpecker	—	—	—	—	—	—	—	—	—	—	—	—	★
Dendrocopos medius	Middle Spotted Woodpecker	—	—	—	—	—	—	—	—	—	—	—	—	★
Dendrocopos leucotos	White-backed Woodpecker	—	—	—	—	—	—	—	—	—	—	—	—	—
Dendrocopos minor	Lesser Spotted Woodpecker	—	—	—	—	—	—	—	—	—	—	—	—	—
Picoides tridactylus	Three-toed Woodpecker	—	—	—	—	—	—	—	—	—	—	—	★	—
Ammomanes deserti	Desert Lark	—	—	—	—	—	—	—	—	—	—	—	—	—
Chersophilus duponti	Dupont's Lark	—	—	—	—	—	—	—	—	—	—	—	★	—
Melanocorypha calandra	Calandra Lark	—	—	—	—	—	—	—	—	—	—	—	★	—
Melanocorypha bimaculata	Bimaculated Lark	—	—	—	—	—	—	—	—	—	—	—	—	—
Melanocorypha leucoptera	White-winged Lark	—	—	Steppe	—	—	—	—	—	—	—	—	—	★[W]
Melanocorypha yeltoniensis	Black Lark	—	—	Steppe	—	—	—	—	—	—	—	—	★	—
Calandrella brachydactyla	Short-toed Lark	—	—	—	—	—	—	—	—	—	—	—	★	—
Calandrella rufescens	Lesser Short-toed Lark	—	—	—	—	—	—	—	—	—	—	—	★	—
Galerida cristata	Crested Lark	—	—	—	—	—	—	—	—	—	—	—	★	—
Galerida theklae	Thekla Lark	—	—	—	—	—	—	—	—	—	—	—	★	—
Lullula arborea	Woodlark	—	—	—	—	—	—	—	—	—	—	—	★	—
Alauda arvensis	Skylark	—	—	—	—	—	—	—	—	—	—	—	★	—
Eremophila alpestris	Shore Lark	—	—	—	—	—	—	—	—	—	—	—	—	—
Riparia riparia	Sand Martin	—	—	—	—	120,000	—	—	—	—	60,000	—	★	—
Ptyonoprogne rupestris	Crag Martin	—	—	—	—	—	—	—	—	—	—	—	—	—
Hirundo rustica	Swallow	—	—	—	—	—	—	—	—	—	—	—	★	—
Hirundo daurica	Red-rumped Swallow	—	—	—	—	—	—	—	—	—	—	—	—	—
Delichon urbica	House Martin	—	—	—	—	—	—	—	—	—	—	—	—	—
Anthus campestris	Tawny Pipit	—	—	—	—	—	—	—	—	—	—	—	★	—
Anthus berthelotii	Berthelot's Pipit	—	Madeira/Canaries	—	—	—	—	—	—	—	—	—	—	★

Appendix 2a ... continued. The 515 bird species occurring regularly in Europe, with the Important Bird Area criteria (at the global and European levels) that they can potentially fulfil and the numerical threshold(s) if applicable. Taxonomy and nomenclature follow Cramp *et al.* (1977–1994).

Species Subspecies or population applicable for criteria B1i or B1ii		Species of global conservation concern (A1)	Restricted-range species (A2)	Biome-restricted species (A3)	Congregations – global (1% biogeographic population of waterbird) (A4i)	(1% global population seabird or terrestrial species) (A4ii)	(20,000 waterbirds or 10,000 pairs seabirds) (A4iii)	(Bottleneck site of global importance) (A4iv)	Congregations – regional (1% flyway or distinct population of waterbird) (B1i)	(1% distinct population of seabird) (B1ii)	(1% distinct population of other species) (B1iii)	(Bottleneck site of regional importance) (B1iv)	Species of European conservation concern (categories 1–3) (B2)	Species of European conservation concern (category 4) (B3)
Anthus hodgsoni	Olive-backed Pipit	—	—	—	—	—	—	—	—	—	—	—	—	—
Anthus trivialis	Tree Pipit	—	—	—	—	—	—	—	—	—	—	—	—	—
Anthus gustavi	Pechora Pipit	—	—	—	—	—	—	—	—	—	—	—	—	—
Anthus pratensis	Meadow Pipit	—	—	—	—	—	—	—	—	—	—	—	—	★
Anthus cervinus	Red-throated Pipit	—	—	Arctic	—	—	—	—	—	—	—	—	—	—
Anthus petrosus	Rock Pipit	—	—	—	—	—	—	—	—	—	—	—	—	—
Anthus spinoletta	Water Pipit	—	—	—	—	—	—	—	—	—	—	—	—	—
Motacilla flava	Yellow Wagtail	—	—	—	—	—	—	—	—	—	—	—	—	—
Motacilla citreola	Citrine Wagtail	—	—	—	—	—	—	—	—	—	—	—	—	—
Motacilla cinerea	Grey Wagtail	—	—	—	—	—	—	—	—	—	—	—	—	—
Motacilla alba	Pied Wagtail	—	—	—	—	—	—	—	—	—	—	—	—	—
Pycnonotus xanthopygos	Yellow-vented Bulbul	—	—	—	—	—	—	—	—	—	—	—	—	—
Bombycilla garrulus	Waxwing	—	—	Boreal	—	—	—	—	—	—	—	—	—	—
Cinclus cinclus	Dipper	—	—	—	—	—	—	—	—	—	—	—	—	—
Troglodytes troglodytes	Wren	—	—	—	—	—	—	—	—	—	—	—	—	—
Prunella modularis	Dunnock	—	—	—	—	—	—	—	—	—	—	—	—	★
Prunella montanella	Siberian Accentor	—	—	—	—	—	—	—	—	—	—	—	—	—
Prunella ocularis	Radde's Accentor	—	—	—	—	—	—	—	—	—	—	—	★	—
Prunella atrogularis	Black-throated Accentor	—	—	—	—	—	—	—	—	—	—	—	★	—
Prunella collaris	Alpine Accentor	—	—	Alpine	—	—	—	—	—	—	—	—	—	—
Cercotrichas galactotes	Rufous Bush Robin	—	—	—	—	—	—	—	—	—	—	—	—	—
Erithacus rubecula	Robin	—	—	—	—	—	—	—	—	—	—	—	—	★
Luscinia luscinia	Thrush Nightingale	—	—	—	—	—	—	—	—	—	—	—	—	★
Luscinia megarhynchos	Nightingale	—	—	—	—	—	—	—	—	—	—	—	—	★
Luscinia calliope	Siberian Rubythroat	—	—	—	—	—	—	—	—	—	—	—	—	—
Luscinia svecica	Bluethroat	—	—	—	—	—	—	—	—	—	—	—	—	—
Tarsiger cyanurus	Red-flanked Bluetail	—	—	—	—	—	—	—	—	—	—	—	—	—
Irania gutturalis	White-throated Robin	—	—	—	—	—	—	—	—	—	—	—	—	—
Phoenicurus ochruros	Black Redstart	—	—	—	—	—	—	—	—	—	—	—	—	—
Phoenicurus phoenicurus	Redstart	—	—	—	—	—	—	—	—	—	—	—	★	—
Phoenicurus erythrogaster	Güldenstädt's Redstart	—	—	—	—	—	—	—	—	—	—	—	★	—
Saxicola rubetra	Whinchat	—	—	—	—	—	—	—	—	—	—	—	—	★
Saxicola dacotiae	Fuerteventura Chat	5p	Madeira/Canaries	—	—	—	—	—	—	—	10p	—	★	—
Saxicola torquata	Stonechat	—	—	—	—	—	—	—	—	—	—	—	★	—
Oenanthe isabellina	Isabelline Wheatear	—	—	—	—	—	—	—	—	—	—	—	—	—
Oenanthe oenanthe	Wheatear	—	—	—	—	—	—	—	—	—	—	—	—	—
Oenanthe pleschanka	Pied Wheatear	—	—	—	—	—	—	—	—	—	—	—	—	—
Oenanthe cypriaca	Cyprus Pied Wheatear	—	Cyprus	Med	—	—	—	—	—	—	—	—	★	—
Oenanthe hispanica	Black-eared Wheatear	—	—	Med	—	—	—	—	—	—	—	—	★	—
Oenanthe deserti	Desert Wheatear	—	—	—	—	—	—	—	—	—	—	—	—	—
Oenanthe finschii	Finsch's Wheatear	—	—	—	—	—	—	—	—	—	—	—	—	—
Oenanthe xanthoprymna	Red-tailed Wheatear	—	—	—	—	—	—	—	—	—	—	—	—	—
Oenanthe leucura	Black Wheatear	—	—	Med	—	—	—	—	—	—	100p	—	★	—
Monticola saxatilis	Rock Thrush	—	—	—	—	—	—	—	—	—	—	—	★	—
Monticola solitarius	Blue Rock Thrush	—	—	—	—	—	—	—	—	—	—	—	★	—
Zoothera dauma	White's Thrush	—	—	—	—	—	—	—	—	—	—	—	—	—
Turdus torquatus	Ring Ouzel	—	—	—	—	—	—	—	—	—	—	—	—	★
Turdus merula	Blackbird	—	—	—	—	—	—	—	—	—	—	—	—	★
Turdus ruficollis	Black-throated Thrush	—	—	—	—	—	—	—	—	—	—	—	—	—
Turdus pilaris	Fieldfare	—	—	—	—	—	—	—	—	—	—	—	—	★W
Turdus philomelos	Song Thrush	—	—	—	—	—	—	—	—	—	—	—	—	★
Turdus iliacus	Redwing	—	—	—	—	—	—	—	—	—	—	—	—	★W
Turdus viscivorus	Mistle Thrush	—	—	—	—	—	—	—	—	—	—	—	—	★
Cettia cetti	Cetti's Warbler	—	—	—	—	—	—	—	—	—	—	—	—	—
Cisticola juncidis	Fan-tailed Warbler	—	—	—	—	—	—	—	—	—	—	—	—	—

Appendix 2a ... continued. The 515 bird species occurring regularly in Europe, with the Important Bird Area criteria (at the global and European levels) that they can potentially fulfil and the numerical threshold(s) if applicable. Taxonomy and nomenclature follow Cramp *et al.* (1977–1994).

Species Subspecies or population applicable for criteria B1i or B1ii		Species of global conservation concern (A1)	Restricted-range species (A2)	Biome-restricted species (A3)	Congregations – global (1% biogeographic population of waterbird) (A4i)	(1% global population seabird or terrestrial species) (A4ii)	(20,000 waterbirds or 10,000 pairs seabirds) (A4iii)	(Bottleneck site of global importance) (A4iv)	Congregations – regional (1% flyway or distinct population of waterbird) (B1i)	(1% distinct population of seabird) (B1ii)	(1% distinct population of other species) (B1iii)	(Bottleneck site of regional importance) (B1iv)	Species of European conservation concern (categories 1–3) (B2)	Species of European conservation concern (category 4) (B3)
Prinia gracilis	Graceful Warbler	—	—	—	—	—	—	—	—	—	—	—	—	—
Locustella lanceolata	Lanceolated Warbler	—	—	—	—	—	—	—	—	—	—	—	—	—
Locustella naevia	Grasshopper Warbler	—	—	—	—	—	—	—	—	—	—	—	—	★
Locustella fluviatilis	River Warbler	—	—	—	—	—	—	—	—	—	—	—	—	★
Locustella luscinioides	Savi's Warbler	—	—	—	—	—	—	—	—	—	—	—	—	★
Acrocephalus melanopogon	Moustached Warbler	—	—	—	—	—	—	—	—	—	—	—	—	—
Acrocephalus paludicola	Aquatic Warbler	10p	—	—	—	—	—	—	—	—	—	—	★	—
Acrocephalus schoenobaenus	Sedge Warbler	—	—	—	—	—	—	—	—	—	—	—	—	★
Acrocephalus agricola	Paddyfield Warbler	—	—	—	—	—	—	—	—	—	—	—	—	—
Acrocephalus dumetorum	Blyth's Reed Warbler	—	—	—	—	—	—	—	—	—	—	—	—	—
Acrocephalus palustris	Marsh Warbler	—	—	—	—	—	—	—	—	—	—	—	—	★
Acrocephalus scirpaceus	Reed Warbler	—	—	—	—	—	—	—	—	—	—	—	—	★
Acrocephalus arundinaceus	Great Reed Warbler	—	—	—	—	—	—	—	—	—	—	—	—	—
Hippolais pallida	Olivaceous Warbler	—	—	—	—	—	—	—	—	—	—	—	★	—
Hippolais caligata	Booted Warbler	—	—	—	—	—	—	—	—	—	—	—	—	—
Hippolais languida	Upcher's Warbler	—	—	—	—	—	—	—	—	—	—	—	—	—
Hippolais olivetorum	Olive-tree Warbler	—	—	Med	—	—	—	—	—	—	—	—	★	—
Hippolais icterina	Icterine Warbler	—	—	—	—	—	—	—	—	—	—	—	—	★
Hippolais polyglotta	Melodious Warbler	—	—	—	—	—	—	—	—	—	—	—	—	★
Sylvia sarda	Marmora's Warbler	—	—	Med	—	—	—	—	—	—	—	—	—	★
Sylvia undata	Dartford Warbler	—	—	—	—	—	—	—	—	—	—	—	★	—
Sylvia conspicillata	Spectacled Warbler	—	—	Med	—	—	—	—	—	—	—	—	—	—
Sylvia cantillans	Subalpine Warbler	—	—	Med	—	—	—	—	—	—	—	—	—	★
Sylvia mystacea	Mènètries's Warbler	—	—	—	—	—	—	—	—	—	—	—	—	—
Sylvia melanocephala	Sardinian Warbler	—	—	Med	—	—	—	—	—	—	—	—	—	★
Sylvia melanothorax	Cyprus Warbler	—	Cyprus	Med	—	—	—	—	—	—	—	—	★	—
Sylvia rueppelli	Rüppell's Warbler	—	—	Med	—	—	—	—	—	—	—	—	—	★
Sylvia nana	Desert Warbler	—	—	—	—	—	—	—	—	—	—	—	—	—
Sylvia hortensis	Orphean Warbler	—	—	—	—	—	—	—	—	—	—	—	★	—
Sylvia nisoria	Barred Warbler	—	—	—	—	—	—	—	—	—	—	—	—	★
Sylvia curruca	Lesser Whitethroat	—	—	—	—	—	—	—	—	—	—	—	—	—
Sylvia communis	Whitethroat	—	—	—	—	—	—	—	—	—	—	—	—	★
Sylvia borin	Garden Warbler	—	—	—	—	—	—	—	—	—	—	—	—	★
Sylvia atricapilla	Blackcap	—	—	—	—	—	—	—	—	—	—	—	—	★
Phylloscopus nitidus	Green Warbler	—	—	—	—	—	—	—	—	—	—	—	—	—
Phylloscopus trochiloides	Greenish Warbler	—	—	—	—	—	—	—	—	—	—	—	—	—
Phylloscopus borealis	Arctic Warbler	—	—	Boreal	—	—	—	—	—	—	—	—	—	—
Phylloscopus inornatus	Yellow-browed Warbler	—	—	—	—	—	—	—	—	—	—	—	—	—
Phylloscopus bonelli	Bonelli's Warbler	—	—	—	—	—	—	—	—	—	—	—	—	★
Phylloscopus sibilatrix	Wood Warbler	—	—	—	—	—	—	—	—	—	—	—	—	★
Phylloscopus lorenzii	Caucasian Chiffchaff	—	—	—	—	—	—	—	—	—	—	—	—	—
Phylloscopus collybita	Chiffchaff	—	—	—	—	—	—	—	—	—	—	—	—	—
Phylloscopus trochilus	Willow Warbler	—	—	—	—	—	—	—	—	—	—	—	—	—
Regulus regulus	Goldcrest	—	—	—	—	—	—	—	—	—	—	—	—	★
Regulus teneriffae	Tenerife Goldcrest	—	Madeira/Canaries	—	—	—	—	—	—	—	—	—	—	★
Regulus ignicapillus	Firecrest	—	—	—	—	—	—	—	—	—	—	—	—	★
Muscicapa striata	Spotted Flycatcher	—	—	—	—	—	—	—	—	—	—	—	★	—
Ficedula parva	Red-breasted Flycatcher	—	—	—	—	—	—	—	—	—	—	—	—	—
Ficedula semitorquata	Semi-collared Flycatcher	—	—	—	—	—	—	—	—	—	—	—	★	—
Ficedula albicollis	Collared Flycatcher	—	—	—	—	—	—	—	—	—	—	—	—	★
Ficedula hypoleuca	Pied Flycatcher	—	—	—	—	—	—	—	—	—	—	—	—	★
Panurus biarmicus	Bearded Tit	—	—	—	—	—	—	—	—	—	—	—	—	—
Aegithalos caudatus	Long-tailed Tit	—	—	—	—	—	—	—	—	—	—	—	—	—
Parus palustris	Marsh Tit	—	—	—	—	—	—	—	—	—	—	—	—	—
Parus lugubris	Sombre Tit	—	—	—	—	—	—	—	—	—	—	—	—	★

Appendix 2a ... continued. The 515 bird species occurring regularly in Europe, with the Important Bird Area criteria (at the global and European levels) that they can potentially fulfil and the numerical threshold(s) if applicable. Taxonomy and nomenclature follow Cramp *et al.* (1977–1994).

Species Subspecies or population applicable for criteria B1i or B1ii		Species of global conservation concern (A1)	Restricted-range species (A2)	Biome-restricted species (A3)	Congregations – global (1% biogeographic population of waterbird) (A4i)	(1% global population seabird or terrestrial species) (A4ii)	(20,000 waterbirds or 10,000 pairs seabirds) (A4iii)	(Bottleneck site of global importance) (A4iv)	Congregations – regional (1% flyway or distinct population of waterbird) (B1i)	(1% distinct population of seabird) (B1ii)	(1% distinct population of other species) (B1iii)	(Bottleneck site of regional importance) (B1iv)	Species of European conservation concern (categories 1–3) (B2)	Species of European conservation concern (category 4) (B3)
Parus montanus	Willow Tit	–	–	–	–	–	–	–	–	–	–	–	–	–
Parus cinctus	Siberian Tit	–	–	Boreal	–	–	–	–	–	–	–	–	–	–
Parus cristatus	Crested Tit	–	–	–	–	–	–	–	–	–	–	–	–	★
Parus ater	Coal Tit	–	–	–	–	–	–	–	–	–	–	–	–	–
Parus caeruleus	Blue Tit	–	–	–	–	–	–	–	–	–	–	–	–	★
Parus cyanus	Azure Tit	–	–	–	–	–	–	–	–	–	–	–	–	–
Parus major	Great Tit	–	–	–	–	–	–	–	–	–	–	–	–	–
Sitta krueperi	Krüper's Nuthatch	–	–	Med	–	–	–	–	–	–	–	–	–	★
Sitta whiteheadi	Corsican Nuthatch	10p	Corsica	Med	–	–	–	–	–	–	–	–	★	–
Sitta europaea	Nuthatch	–	–	–	–	–	–	–	–	–	–	–	–	–
Sitta tephronota	Great Rock Nuthatch	–	–	Alpine	–	–	–	–	–	–	–	–	–	–
Sitta neumayer	Rock Nuthatch	–	–	Med	–	–	–	–	–	–	–	–	–	★
Tichodroma muraria	Wallcreeper	–	–	Alpine	–	–	–	–	–	–	–	–	–	–
Certhia familiaris	Treecreeper	–	–	–	–	–	–	–	–	–	–	–	–	–
Certhia brachydactyla	Short-toed Treecreeper	–	–	–	–	–	–	–	–	–	–	–	–	★
Remiz pendulinus	Penduline Tit	–	–	–	–	–	–	–	–	–	–	–	–	–
Oriolus oriolus	Golden Oriole	–	–	–	–	–	–	–	–	–	–	–	–	–
Lanius collurio	Red-backed Shrike	–	–	–	–	–	–	–	–	–	–	–	★	–
Lanius minor	Lesser Grey Shrike	–	–	–	–	–	–	–	–	–	–	–	★	–
Lanius excubitor	Great Grey Shrike	–	–	–	–	–	–	–	–	–	–	–	★	–
Lanius senator	Woodchat Shrike	–	–	–	–	–	–	–	–	–	–	–	★	–
Lanius nubicus	Masked Shrike	–	–	Med	–	–	–	–	–	–	–	–	★	–
Garrulus glandarius	Jay	–	–	–	–	–	–	–	–	–	–	–	–	–
Perisoreus infaustus	Siberian Jay	–	–	Boreal	–	–	–	–	–	–	–	–	★	–
Cyanopica cyana	Azure-winged Magpie	–	–	–	–	–	–	–	–	–	–	–	–	–
Pica pica	Magpie	–	–	–	–	–	–	–	–	–	–	–	–	–
Nucifraga caryocatactes	Nutcracker	–	–	–	–	–	–	–	–	–	–	–	–	–
Pyrrhocorax graculus	Alpine Chough	–	–	Alpine	–	–	–	–	–	–	–	–	–	–
Pyrrhocorax pyrrhocorax	Chough	–	–	–	–	–	–	–	–	–	–	–	★	–
Corvus monedula	Jackdaw	–	–	–	–	–	–	–	–	–	–	–	–	★
Corvus frugilegus	Rook	–	–	–	–	–	–	–	–	–	–	–	–	–
Corvus corone	Carion Crow	–	–	–	–	–	–	–	–	–	–	–	–	–
Corvus corax	Raven	–	–	–	–	–	–	–	–	–	–	–	–	–
Sturnus vulgaris	Starling	–	–	–	–	–	–	–	–	–	–	–	–	–
Sturnus unicolor	Spotless Starling	–	–	Med	–	–	–	–	–	–	–	–	–	★
Sturnus roseus	Rose-coloured Starling	–	–	–	–	–	–	–	–	–	–	–	–	–
Passer domesticus	House Sparrow	–	–	–	–	–	–	–	–	–	–	–	–	–
Passer hispaniolensis	Spanish Sparrow	–	–	–	–	–	–	–	–	–	–	–	–	–
Passer moabiticus	Dead Sea Sparrow	–	–	–	–	–	–	–	–	–	–	–	–	–
Passer montanus	Tree Sparrow	–	–	–	–	–	–	–	–	–	–	–	–	–
Carpospiza brachydactyla	Pale Rock Sparrow	–	–	–	–	–	–	–	–	–	–	–	–	–
Petronia xanthocollis	Yellow-throated Sparrow	–	–	–	–	–	–	–	–	–	–	–	–	–
Petronia petronia	Rock Sparrow	–	–	–	–	–	–	–	–	–	–	–	–	–
Montifringilla nivalis	Snowfinch	–	–	Alpine	–	–	–	–	–	–	–	–	–	–
Fringilla coelebs	Chaffinch	–	–	–	–	–	–	–	–	–	–	–	–	★
Fringilla teydea	Blue Chaffinch	10p	Madeira/Canaries	–	–	–	–	–	–	–	–	–	★	–
Fringilla montifringilla	Brambling	–	–	Boreal	–	–	–	–	–	–	–	–	–	–
Serinus pusillus	Red-fronted Serin	–	–	–	–	–	–	–	–	–	–	–	–	–
Serinus serinus	Serin	–	–	–	–	–	–	–	–	–	–	–	–	★
Serinus canaria	Canary	–	Madeira/Canaries, Azores	–	–	–	–	–	–	–	–	–	–	★
Serinus citrinella	Citril Finch	–	–	Alpine	–	–	–	–	–	–	–	–	–	★
Carduelis chloris	Greenfinch	–	–	–	–	–	–	–	–	–	–	–	–	★
Carduelis carduelis	Goldfinch	–	–	–	–	–	–	–	–	–	–	–	–	–
Carduelis spinus	Siskin	–	–	–	–	–	–	–	–	–	–	–	–	★

Appendix 2a ... continued. The 515 bird species occurring regularly in Europe, with the Important Bird Area criteria (at the global and European levels) that they can potentially fulfil and the numerical threshold(s) if applicable. Taxonomy and nomenclature follow Cramp *et al.* (1977–1994).

Species Subspecies or population applicable for criteria B1i or B1ii		Species of global conservation concern (A1)	Restricted-range species (A2)	Biome-restricted species (A3)	Congregations – global (1% biogeographic population of waterbird) (A4i)	(1% global population seabird or terrestrial species) (A4ii)	(20,000 waterbirds or 10,000 pairs seabirds) (A4iii)	(Bottleneck site of global importance) (A4iv)	Congregations – regional (1% flyway or distinct population of waterbird) (B1i)	(1% distinct population of seabird) (B1ii)	(1% distinct population of other species) (B1iii)	(Bottleneck site of regional importance) (B1iv)	Species of European conservation concern (categories 1–3) (B2)	Species of European conservation concern (category 4) (B3)
Carduelis cannabina	Linnet	—	—	—	—	—	—	—	—	—	—	—	—	★
Carduelis flavirostris	Twite	—	—	—	—	—	—	—	—	—	—	—	—	—
Carduelis flammea	Redpoll	—	—	—	—	—	—	—	—	—	—	—	—	—
Carduelis hornemanni	Arctic Redpoll	—	—	Arctic	—	—	—	—	—	—	—	—	—	—
Loxia leucoptera	Two-barred Crossbill	—	—	Boreal	—	—	—	—	—	—	—	—	—	—
Loxia curvirostra	Crossbill	—	—	—	—	—	—	—	—	—	—	—	—	—
Loxia scotica	Scottish Crossbill	5p	Scotland	—	—	—	—	—	—	—	—	—	★	—
Loxia pytyopsittacus	Parrot Crossbill	—	—	Boreal	—	—	—	—	—	—	—	—	—	★
Rhodopechys sanguinea	Crimson-winged Finch	—	—	—	—	—	—	—	—	—	—	—	—	—
Rhodospiza obsoleta	Desert Finch	—	—	—	—	—	—	—	—	—	—	—	—	—
Bucanetes mongolicus	Mongolian Trumpeter Finch	—	—	—	—	—	—	—	—	—	—	—	—	—
Bucanetes githagineus	Trumpeter Finch	—	—	—	—	—	—	—	—	—	—	—	★	—
Carpodacus erythrinus	Scarlet Rosefinch	—	—	—	—	—	—	—	—	—	—	—	—	—
Carpodacus rubicilla	Great Rosefinch	—	—	Alpine	—	—	—	—	—	—	—	—	★	—
Pinicola enucleator	Pine Grosbeak	—	—	Boreal	—	—	—	—	—	—	—	—	—	—
Pyrrhula pyrrhula	Bullfinch	—	—	—	—	—	—	—	—	—	—	—	—	—
Pyrrhula murina	Azores Bullfinch	[not assessed]	Azores	—	—	—	—	—	—	—	—	—	[not assessed]	[not assessed]
Coccothraustes coccothraustes	Hawfinch	—	—	—	—	—	—	—	—	—	—	—	—	—
Calcarius lapponicus	Lapland Bunting	—	—	Arctic	—	—	—	—	—	—	—	—	—	—
Plectrophenax nivalis	Snow Bunting	—	—	Arctic	—	—	—	—	—	—	—	—	—	—
Emberiza citrinella	Yellowhammer	—	—	—	—	—	—	—	—	—	—	—	—	★
Emberiza cirlus	Cirl Bunting	—	—	—	—	—	—	—	—	—	—	—	—	★
Emberiza cia	Rock Bunting	—	—	—	—	—	—	—	—	—	—	—	★	—
Emberiza cineracea	Cinereous Bunting	5p	—	Med	—	—	—	—	—	—	—	—	★	—
Emberiza hortulana	Ortolan Bunting	—	—	—	—	—	—	—	—	—	—	—	★	—
Emberiza buchanani	Grey-necked Bunting	—	—	—	—	—	—	—	—	—	—	—	—	—
Emberiza caesia	Cretzschmar's Bunting	—	—	Med	—	—	—	—	—	—	—	—	—	★
Emberiza rustica	Rustic Bunting	—	—	Boreal	—	—	—	—	—	—	—	—	—	—
Emberiza pusilla	Little Bunting	—	—	—	—	—	—	—	—	—	—	—	—	—
Emberiza aureola	Yellow-breasted Bunting	—	—	—	—	—	—	—	—	—	—	—	—	—
Emberiza schoeniclus	Reed Bunting	—	—	—	—	—	—	—	—	—	—	—	—	—
Emberiza bruniceps	Red-headed Bunting	—	—	—	—	—	—	—	—	—	—	—	—	—
Emberiza melanocephala	Black-headed Bunting	—	—	Med	—	—	—	—	—	—	—	—	★	—
Miliaria calandra	Corn Bunting	—	—	—	—	—	—	—	—	—	—	—	—	★

✔ given for IBA criteria B1i indicates that the population threshold applied is the same as that given for A4i. For B1ii and B1iii the population threshold applied is the same as that given for A4ii.

★ given for IBA criteria B2/B3 indicates that the species is one of the 278 species of European conservation concern as listed in Tucker and Heath (1994). The numerical threshold for each species of European conservation concern varies by country – for full details on how to calculate thresholds see Box 2, 'Identifying IBAs' chapter.

For criterion A4iii 'S' indicates the species is classed as a 'seabird' and thus the collective threshold of 10,000 pairs of seabirds may apply, and 'W' indicates the species is considered a 'waterbird' following the definition of Rose and Scott (1997) (see p. 15, 'Identifying IBAs' chapter) and the collective threshold of 20,000 individuals may apply.

1. Non-migratory raptor in Europe
2. Non-congregatory in Europe
3. Marginal population in Europe
4. Small population

Superscript W – species status applies to wintering population

For definitions of regions listed for distinct populations of waterbirds under criterion B1i, see Rose and Scott (1997):

- Western Palearctic – As defined by Cramp *et al.* (1977–1994). For some species/populations, Iraq has been included.
- North-western Europe – Belgium, Denmark, Estonia, Finland, France, Germany, Iceland, Ireland, Latvia, Lithuania, Luxembourg, Netherlands, Norway, Poland, Russia (area around the Gulf of Finland and Kaliningrad), Sweden, United Kingdom.
- North-eastern Europe – Belarus, Russia (west of the Urals), Ukraine.
- Central Europe – Austria, Czech Republic, Hungary, Liechtenstein, Slovakia, Switzerland.
- Western Mediterranean – Algeria, France, Italy, Malta, Morocco, Portugal, San Marino, Spain, Tunisia.
- Eastern Mediterranean – Albania, Bosnia and Herzegovina, Croatia, Cyprus, Egypt, Greece, Israel, Lebanon, Libya, Slovenia, Syria, Former Yugoslav Republic of Macedonia, Turkey, Yugoslavia.
- Black Sea – Armenia, Bulgaria, Georgia, Moldova, Romania, Russia, Turkey, Ukraine.
- Caspian – Armenia, Azerbaijan, Iran, Kazakhstan, Russia, Turkmenistan, Uzbekistan.

Abbreviations given for distinct populations of waterbirds under criterion B1i:

bre	breeding
E	east
Med	Mediterranean
N	northern
NE	north-eastern
NW	north-western
S	southern
SE	south-eastern
SW	south-western
UK	United Kingdom
W	western
win	wintering

References

Cramp, S. *et al.* (1977–1994) *The birds of the western Palearctic.* Vols 1–9. Oxford, UK: Oxford University Press.

Rose, P. M. and Scott, D. A. (1997) *Waterfowl population estimates.* Second edition. Wageningen, Netherlands: Wetlands International.

Tucker, G. M. and Heath, M. F. (1994) *Birds in Europe: their conservation status.* Cambridge, UK: BirdLife International (BirdLife Conservation Series no. 3).

Appendix 2b. Bird species occurring regularly in the European Union that can fulfil Important Bird Area criteria (at the European Union level), and the numerical threshold(s) if applicable. Taxonomy and nomenclature follow Cramp *et al.* (1977–1994).

IBA criteria applicable at the global (A) and European (B) levels are shown in Appendix 2a.
For a full explanation of criteria see the introductory chapter 'Identifying IBAs'.
All population thresholds are given in individuals, unless followed by 'p' indicating pairs. To convert between individuals and pairs, a multiplying/dividing factor of 3 was used.
n/a = not applicable.

Annex I number	Speces		Species of global concervation concern (C1)	Concentrations of species threatened[2] at the EU level (C2)	Congregations of migratory species not threatened[2] at the EU level (C3)	Congregatory – large congregations 20,000 waterbirds or 10,000 pairs of seabirds (C4)	Congregatory – Bottleneck site for storks/raptors/cranes (C5)	'Top 5' sites in NUTS[1] region for species threatened[2] at EU level (C6)	Notes
1	***Gavia stellata***	Red-throated Diver	—	35p	—	W	—	★	
2	***Gavia arctica***	Black-throated Diver	—	✔	—	W	—	★	
3	***Gavia immer***	Great Northern Diver	—	160p	—	W	—	★	
	Tachybaptus ruficollis	Little Grebe	—	—	✔	W	—	—	
	Podiceps cristatus	Great Crested Grebe	—	—	✔	W	—	—	
	Podiceps grisegena	Red-necked Grebe	—	—	✔	W	—	—	
4	***Podiceps auritus***	Slavonian Grebe	—	60p	—	W	—	★	
	Podiceps nigricollis	Black-necked Grebe	—	—	1,000	W	—	—	
	Fulmarus glacialis	Fulmar	—	—	✔	S	—	—	
6	***Pterodroma feae***	Fea's Petrel	✔	n/a	—	S	—	★	
5	***Pterodroma madeira***	Zino's Petrel	✔	n/a	—	S	—	★	
7	***Bulweria bulwerii***	Bulwer's Petrel	—	✔	—	S	—	★	
8	***Calonectris diomedea***	Cory's Shearwater	—	✔	—	S	—	★	
	Puffinus puffinus	Manx Shearwater	—	—	✔	S	—	—	
9	***Puffinus yelkouan***	Yelkouan Shearwater	—	30p*	✔	S	—	★	**P. y. mauretanicus*
10	***Puffinus assimilis***	Little Shearwater	—	✔	—	S	—	★	
11	***Pelagodroma marina***	White-faced Petrel	—	✔	—	S	—	★	
12	***Hydrobates pelagicus***	Storm Petrel	—	1,600p*	—	S	—	★	*Italy, Greece 35p
13	***Oceanodroma leucorhoa***	Leach's Petrel	—	600p	—	S	—	★	
14	***Oceanodroma castro***	Madeiran Petrel	—	✔	—	S	—	★	
	Sula bassana	Gannet	—	—	✔	S	—	—	
	Phalacrocorax carbo sinensis	Cormorant	—	—	✔	W	—	—	
	Phalacrocorax aristotelis aristotelis	Shag	—	—	—	W	—	—	
15	***Phalacrocorax aristotelis desmarestii***	Shag	—	30p	—	W	—	★	
16	***Phalacrocorax pygmeus***	Pygmy Cormorant	✔	n/a	—	W	—	★	
17	***Pelecanus onocrotalus***	White Pelican	—	✔	—	W	—	★	
18	***Pelecanus crispus***	Dalmatian Pelican	✔	✔	—	W	—	★	
19	***Botaurus stellaris***	Bittern	—	15p	—	W	—	★	
20	***Ixobrychus minutus***	Little Bittern	—	55p	—	W	—	★	
21	***Nycticorax nycticorax***	Night Heron	—	270p	—	W	—	★	
22	***Ardeola ralloides***	Squacco Heron	—	15p	—	W	—	★	
	Bubulcus ibis	Cattle Egret	—	—	✔	W	—	—	
23	***Egretta garzetta***	Little Egret	—	✔	—	W	—	★	
24	***Egretta alba***	Great White Egret	—	✔	—	W	—	★	
25	***Ardea purpurea***	Purple Heron	—	50p	—	W	—	★	
26	***Ciconia nigra***	Black Stork	—	✔	—	W	✔	★	
27	***Ciconia ciconia***	White Stork	—	✔	—	W	✔	★	
28	***Plegadis falcinellus***	Glossy Ibis	—	n/a	—	W	—	★	
29	***Platalea leucorodia***	Spoonbill	—	✔	—	W	—	★	
30	***Phoenicopterus ruber***	Greater Flamingo	—	220p	—	W	—	★	
	Cygnus olor	Mute Swan	—	—	✔	W	—	—	Western European birds sedentary
31	***Cygnus columbianus***	Bewick's Swan	—	✔	—	W	—	★	
32	***Cygnus cygnus***	Whooper Swan	—	20p	—	W	—	★	
	Anser fabalis	Bean Goose	—	—	✔	W	—	—	
	Anser brachyrhynchus	Pink-footed Goose	—	—	✔	W	—	—	
33	***Anser albifrons flavirostris***	White-fronted Goose	—	✔	—	W	—	★	
	Anser albifrons albifrons	White-fronted Goose	—	—	✔	W	—	—	
34	***Anser erythropus***	Lesser White-fronted Goose	✔	✔	—	W	—	★	
	Anser anser	Greylag Goose	—	—	✔	W	—	—	
35	***Branta leucopsis***	Barnacle Goose	—	12p	—	W	—	★	
	Branta bernicla	Brent Goose	—	—	✔	W	—	—	
36	***Branta ruficollis***	Red-breasted Goose	✔	✔	—	W	—	★	
37	***Tadorna ferruginea***	Ruddy Shelduck	—	✔	—	W	—	★	
	Tadorna tadorna	Shelduck	—	—	✔	W	—	—	
	Anas penelope	Wigeon	—	—	✔	W	—	—	
	Anas strepera	Gadwall	—	—	✔	W	—	—	
	Anas crecca	Teal	—	—	✔	W	—	—	

Appendix 2b ... continued. Bird species occurring regularly in the European Union that can fulfil Important Bird Area criteria (at the European Union level), and the numerical threshold(s) if applicable. Taxonomy and nomenclature follow Cramp *et al.* (1977–1994).

Annex I number	Speces		Species of global concervation concern (C1)	Concentrations of species threatened² at the EU level (C2)	Congregations of migratory species not threatened² at the EU level (C3)	Congregatory – large congregations 20,000 waterbirds or 10,000 pairs of seabirds (C4)	Congregatory – Bottleneck site for storks/raptors/cranes (C5)	'Top 5' sites in NUTS¹ region for species threatened² at EU level (C6)	Notes
	Anas platyrhynchos	Mallard	—	—	✔	W	—	—	
	Anas acuta	Pintail	—	—	✔	W	—	—	
	Anas querquedula	Garganey	—	—	✔	W	—	—	
	Anas clypeata	Shoveler	—	—	✔	W	—	—	
38	***Marmaronetta angustirostris***	Marbled Duck	✔	✔	—	W	—	★	
	Netta rufina	Red-crested Pochard	—	—	✔	W	—	—	
	Aythya ferina	Pochard	—	—	✔	W	—	—	
39	***Aythya nyroca***	Ferruginous Duck	✔	✔	—	W	—	★	
	Aythya fuligula	Tufted Duck	—	—	✔	W	—	—	
	Aythya marila	Scaup	—	—	✔	W	—	—	
	Somateria mollissima	Eider	—	—	✔	W	—	—	
	Polysticta stelleri	Steller's Eider	✔	—	✔	W	—	—	
	Clangula hyemalis	Long-tailed Duck	—	—	✔	W	—	—	
	Melanitta nigra	Common Scoter	—	—	✔	W	—	—	
	Melanitta fusca	Velvet Scoter	—	—	✔	W	—	—	
	Bucephala clangula	Goldeneye	—	—	✔	W	—	—	
40	***Mergus albellus***	Smew	—	16p	—	W	—	★	
	Mergus serrator	Red-breasted Merganser	—	—	✔	W	—	—	
	Mergus merganser	Goosander	—	—	✔	W	—	—	
41	***Oxyura leucocephala***	White-headed Duck	✔	✔	—	W	—	★	
42	***Pernis apivorus***	Honey Buzzard	—	—	—	—	✔	★	
43	***Elanus caeruleus***	Black-winged Kite	—	✔	—	—	—[3]	★	
44	***Milvus migrans***	Black Kite	—	✔	—	—	✔	★	
45	***Milvus milvus***	Red Kite	—	✔	—	—	✔	★	
46	***Haliaeetus albicilla***	White-tailed Eagle	✔	—	—	—	✔	★	
47	***Gypaetus barbatus***	Lammergeier	—	—	—	—	—[3]	★	
48	***Neophron percnopterus***	Egyptian Vulture	—	✔	—	—	✔	★	
49	***Gyps fulvus***	Griffon Vulture	—	✔	—	—	✔	★	
50	***Aegypius monachus***	Cinereous Vulture	✔	✔	—	—	—[3]	★	
51	***Circaetus gallicus***	Short-toed Eagle	—	—	—	—	✔	★	
52	***Circus aeruginosus***	Marsh Harrier	—	—	—	—	✔	★	
53	***Circus cyaneus***	Hen Harrier	—	—	—	—	✔	★	
54	***Circus macrourus***	Pallid Harrier	—	—	—	—	✔	★	Rare in EU
55	***Circus pygargus***	Montagu's Harrier	—	—	—	—	✔	★	
56	***Accipiter gentilis arrigonii***	Goshawk	—	—	—	—	—[3]	★	
57	***Accipiter nisus granti***	Sparrowhawk	—	—	—	—	✔	★	
58	***Accipiter brevipes***	Levant Sparrowhawk	—	—	—	—	✔	★	
	Buteo buteo	Buzzard	—	—	—	—	✔	—	
59	***Buteo rufinus***	Long-legged Buzzard	—	—	—	—	—[3]	★	
	Buteo lagopus	Rough-legged Buzzard	—	—	—	—	✔	—	
60	***Aquila pomarina***	Lesser Spotted Eagle	—	—	—	—	✔	★	
	Aquila nipalensis	Steppe Eagle	—	—	—	—	✔	—	
61	***Aquila clanga***	Greater Spotted Eagle	—	—	—	—	✔	★	Rare in EU
62	***Aquila heliaca***	Imperial Eagle	✔	—	—	—	✔	★	
63	***Aquila adalberti***	Spanish Imperial Eagle	✔	—	—	—	—[3]	★	
64	***Aquila chrysaetos***	Golden Eagle	—	—	—	—	—[3]	★	
65	***Hieraaetus pennatus***	Booted Eagle	—	✔	—	—	✔	★	
66	***Hieraaetus fasciatus***	Bonnelli's Eagle	—	✔	—	—	—[3]	★	
67	***Pandion haliaetus***	Osprey	—	—	—	—	✔	★	
68	***Falco naumanni***	Lesser Kestrel	✔	✔	—	—	✔	★	
	Falco tinnunculus	Kestrel	—	—	—	—	✔	—	
	Falco vespertinus	Red-footed Falcon	—	—	✔	—	✔	—	
69	***Falco columbarius***	Merlin	—	—	—	—	✔	★	
	Falco subbuteo	Hobby	—	—	—	—	✔	—	
70	***Falco eleonorae***	Eleonora's Falcon	—	✔	—	—	✔	★	
71	***Falco biarmicus***	Lanner	—	—	—	—	—[3]	★	
72	***Falco rusticolus***	Gyrfalcon	—	—	—	—	—[3]	★	

Appendix 2b ... continued. Bird species occurring regularly in the European Union that can fulfil Important Bird Area criteria (at the European Union level), and the numerical threshold(s) if applicable. Taxonomy and nomenclature follow Cramp *et al.* (1977–1994).

Annex I number	Speces		Species of global concervation concern (C1)	Concentrations of species threatened[2] at the EU level (C2)	Congregations of migratory species not threatened[2] at the EU level (C3)	Congregatory – large congregations 20,000 waterbirds or 10,000 pairs of seabirds (C4)	Congregatory – Bottleneck site for storks/raptors/cranes (C5)	'Top 5' sites in NUTS[1] region for species threatened[2] at EU level (C6)	Notes
73	***Falco peregrinus***	Peregrine	—	—	—	—	✔	★	
74	***Bonasa bonasia***	Hazel Grouse	—	—	—	—	—	★	
76	***Lagopus mutus helveticus***	Ptarmigan	—	—	—	—	—	★	
75	***Lagopus mutus pyrenaicus***	Ptarmigan	—	—	—	—	—	★	
77	***Tetrao tetrix tetrix***	Black Grouse	—	—	—	—	—	★	Continental only
78	***Tetrao urogallus***	Capercaillie	—	—	—	—	—	★	
79	***Alectoris graeca saxatilis***	Rock Partridge	—	—	—	—	—	★	
80	***Alectoris graeca whitakeri***	Rock Partridge	—	—	—	—	—	★	
81	***Alectoris barbara***	Barbary Partridge	—	—	—	—	—	★	
82	***Perdix perdix italica***	Partridge	—	—	—	—	—	★	
83	***Perdix perdix hispaniensis***	Partridge	—	—	—	—	—	★	
90	***Turnix sylvatica***	Andalusian Hemipode	—	—	—	—	—	★	
84	***Porzana porzana***	Spotted Crake	—	100p	—	W	—	★	
85	***Porzana parva***	Little Crake	—	55p	—	W	—	★	
86	***Porzana pusilla***	Baillon's Crake	—	40p	—	W	—	★	
87	***Crex crex***	Corncrake	✔	45p	—	W	—	★	
88	***Porphyrio porphyrio***	Purple Gallinule	—	35p	—	W	—	★	
	Fulica atra	Coot	—	—	✔	W	—	—	
89	***Fulica cristata***	Crested Coot	—	✔	—	W	—	★	
91	***Grus grus***	Crane	—	✔	—	W	✔	★	
92	***Tetrax tetrax***	Little Bustard	✔	—	—	—	—	★	
93	***Chlamydotis undulata***	Houbara Bustard	—	—	—	—	—	★	
94	***Otis tarda***	Great Bustard	✔	—	—	—	—	★	
	Haematopus ostralegus	Oystercatcher	—	—	✔	W	—	—	
95	***Himantopus himantopus***	Black-winged Stilt	—	✔	—	W	—	★	
96	***Recurvirostra avosetta***	Avocet	—	✔	—	W	—	★	
97	***Burhinus oedicnemus***	Stone Curlew	—	✔	—	W	—	★	
98	***Cursorius cursor***	Cream-coloured Courser	—	✔	—	W	—	★	
99	***Glareola pratincola***	Collared Pratincole	—	✔	—	W	—	★	
	Charadrius dubius	Little Ringed Plover	—	—	✔	W	—	—	
	Charadrius hiaticula	Ringed Plover	—	—	✔	W	—	—	
	Charadrius alexandrinus	Kentish Plover	—	—	✔	W	—	—	
100	***Charadrius morinellus***	Dotterel	—	✔	—	W	—	★	
101	***Pluvialis apricaria***	Golden Plover	—	✔	—	W	—	★	
	Pluvialis squatarola	Grey Plover	—	—	✔	W	—	—	
102	***Hoplopterus spinosus***	Spur-winged Plover	—	✔	—	W	—	★	
	Vanellus vanellus	Lapwing	—	—	✔	W	—	—	
	Calidris canutus	Knot	—	—	✔	W	—	—	
	Calidris alba	Sanderling	—	—	✔	W	—	—	
	Calidris minuta	Little Stint	—	—	✔	W	—	—	
	Calidris temminckii	Temminck's Stint	—	—	✔	W	—	—	
	Calidris ferruginea	Curlew Sandpiper	—	—	✔	W	—	—	
	Calidris maritima	Purple Sandpiper	—	—	✔	W	—	—	
	Calidris alpina	Dunlin	—	—	✔	W	—	—	
	Limicola falcinellus	Broad-billed Sandpiper	—	—	✔	W	—	—	
103	***Philomachus pugnax***	Ruff	—	✔	—	W	—	★	
	Gallinago gallinago	Snipe	—	—	✔	W	—	—	
104	***Gallinago media***	Great Snipe	✔	15p	—	W	—	★	
	Limosa limosa	Black-tailed Godwit	—	—	✔	W	—	—	
105	***Limosa lapponica***	Bar-tailed Godwit	—	✔	—	W	—	★	
	Numenius phaeopus	Whimbrel	—	—	✔	W	—	—	
106	***Numenius tenuirostris***	Slender-billed Curlew	✔	—	—	W	—	★	
	Numenius arquata	Curlew	—	—	✔	W	—	—	
	Tringa erythropus	Spotted Redshank	—	—	✔	W	—	—	
	Tringa totanus	Redshank	—	—	✔	W	—	—	
	Tringa stagnatilis	Marsh Sandpiper	—	—	✔	W	—	—	
	Tringa nebularia	Greenshank	—	—	✔	W	—	—	

Appendix 2b ... continued. Bird species occurring regularly in the European Union that can fulfil Important Bird Area criteria (at the European Union level), and the numerical threshold(s) if applicable. Taxonomy and nomenclature follow Cramp *et al.* (1977–1994).

Annex I number	Speces		Species of global concervation concern (C1)	Concentrations of species threatened[2] at the EU level (C2)	Congregations of migratory species not threatened[2] at the EU level (C3)	Congregatory – large congregations 20,000 waterbirds or 10,000 pairs of seabirds (C4)	Congregatory – Bottleneck site for storks/raptors/cranes (C5)	'Top 5' sites in NUTS[1] region for species threatened[2] at EU level (C6)	Notes
	Tringa ochropus	Green Sandpiper	—	—	✔	W	—	—	
107	***Tringa glareola***	Wood Sandpiper	—	✔	—	W	—	★	
108	***Xenus cinereus***	Terek Sandpiper	—	✔	—	W	—	★	
	Actitis hypoleucos	Common Sandpiper	—	—	✔	W	—	—	
	Arenaria interpres	Turnstone	—	—	✔	W	—	—	
109	***Phalaropus lobatus***	Red-necked Phalarope	—	—	—	W	—	★	
	Stercorarius pomarinus	Pomarine Skua	—	—	—	S	—	—	
	Stercorarius parasiticus	Arctic Skua	—	—	✔	S	—	—	
	Stercorarius longicaudus	Long-tailed Skua	—	—	✔	S	—	—	
	Stercorarius skua	Great Skua	—	—	✔	S	—	—	
110	***Larus melanocephalus***	Mediterranean Gull	—	45p	—	W	—	★	
	Larus minutus	Little Gull	—	—	✔	W	—	—	
	Larus sabini	Sabine's Gull	—	—	—	W	—	—	
	Larus ridibundus	Black-headed Gull	—	—	✔	W	—	—	
111	***Larus genei***	Slender-billed Gull	—	20p	—	W	—	★	
112	***Larus audouinii***	Audouin's Gull	✔	✔	—	W	—	★	
	Larus canus	Common Gull	—	—	✔	W	—	—	
	Larus fuscus	Lesser Black-backed Gull	—	—	✔	W	—	—	
	Larus cachinnans	Yellow-legged Gull	—	—	✔	W	—	—	
	Larus glaucoides	Iceland Gull	—	—	✔	W	—	—	
	Larus hyperboreus	Glaucous Gull	—	—	✔	W	—	—	
	Larus marinus	Great Black-backed Gull	—	—	✔	W	—	—	
	Rissa tridactyla	Kittiwake	—	—	✔	W	—	—	
113	***Gelochelidon nilotica***	Gull-billed Tern	—	25p*	—	W	—	★	*Greece 50p (B1i)
114	***Sterna caspia***	Caspian Tern	—	✔	—	W	—	★	
115	***Sterna sandvicensis***	Sandwich Tern	—	✔*	—	W	—	★	*Greece 20p
116	***Sterna dougallii***	Roseate Tern	—	✔	—	W	—	★	
117	***Sterna hirundo***	Common Tern	—	✔*	—	W	—	★	*Sweden, Finland 650p
118	***Sterna paradisaea***	Arctic Tern	—	1,200p	—	W	—	★	
119	***Sterna albifrons***	Little Tern	—	✔*	—	W	—	★	*Greece 15p
120	***Chlidonias hybridus***	Whiskered Tern	—	✔*	—	W	—	★	*Greece 250p (B1i)
121	***Chlidonias niger***	Black Tern	—	40p	—	W	—	★	
	Chlidonias leucopterus	White-winged Black Tern	—	—	✔	W	—	—	
	Uria aalge	Guillemot	—	—	✔	S	—	—	
122	***Uria aalge ibericus***	Guillemot	—	n/a	—	S	—	★	
	Alca torda	Razorbill	—	—	✔	S	—	—	
	Fratercula arctica	Puffin	—	—	—	S	—	—	
123	***Pterocles orientalis***	Black-bellied Sandgrouse	—	—	—	—	—	★	
124	***Pterocles alchata***	Pin-tailed Sandgrouse	—	—	—	—	—	★	
125	***Columba palumbus azorica***	Woodpigeon	—	—	—	—	—	★	
126	***Columba trocaz***	Long-toed Pigeon	✔	—	—	—	—	★	
127	***Columba bollii***	Dark-tailed Laurel Pigeon	✔	—	—	—	—	★	
128	***Columba junoniae***	White-tailed Laurel Pigeon	✔	—	—	—	—	★	
129	***Bubo bubo***	Eagle Owl	—	—	—	—	—	★	
130	***Nyctea scandiaca***	Snowy Owl	—	—	—	—	—	★	
131	***Surnia ulula***	Hawk Owl	—	—	—	—	—	★	
132	***Glaucidium passerinum***	Pygmy Owl	—	—	—	—	—	★	
133	***Strix uralensis***	Ural Owl	—	—	—	—	—	★	
134	***Strix nebulosa***	Great Grey Owl	—	—	—	—	—	★	
135	***Asio flammeus***	Short-eared Owl	—	—	—	—	—	★	
136	***Aegolius funereus***	Tengmalm's Owl	—	—	—	—	—	★	
137	***Caprimulgus europaeus***	Nightjar	—	—	—	—	—	★	
138	***Apus caffer***	White-rumped Swift	—	—	—	—	—	★	
139	***Alcedo atthis***	Kingfisher	—	—	—	—	—	★	
	Merops apiaster	Bee-eater	—	—	✔	—	—	—	

Appendix 2b ... continued. Bird species occurring regularly in the European Union that can fulfil Important Bird Area criteria (at the European Union level), and the numerical threshold(s) if applicable. Taxonomy and nomenclature follow Cramp *et al.* (1977–1994).

Annex I number	Speces		Species of global concervation concern (C1)	Concentrations of species threatened[2] at the EU level (C2)	Congregations of migratory species not threatened[2] at the EU level (C3)	Congregatory – large congregations 20,000 waterbirds or 10,000 pairs of seabirds (C4)	Congregatory – Bottleneck site for storks/raptors/cranes (C5)	'Top 5' sites in NUTS[1] region for species threatened[2] at EU level (C6)	Notes
140	***Coracias garrulus***	Roller	—	—	—	—	—	★	
141	***Picus canus***	Grey-headed Woodpecker	—	—	—	—	—	★	
142	***Dryocopus martius***	Black Woodpecker	—	—	—	—	—	★	
143,144	***Dendrocopos major***	Great Spotted Woodpecker	—	—	—	—	—	★	*D.m. canariensis, D.m. thanneri*
145	***Dendrocopos syriacus***	Syrian Woodpecker	—	—	—	—	—	★	
146	***Dendrocopos medius***	Middle Spotted Woodpecker	—	—	—	—	—	★	
147	***Dendrocopos leucotos***	White-backed Woodpecker	—	—	—	—	—	★	
148	***Picoides tridactylus***	Three-toed Woodpecker	—	—	—	—	—	★	
149	***Chersophilus duponti***	Dupont's Lark	—	—	—	—	—	★	
150	***Melanocorypha calandra***	Calandra Lark	—	—	—	—	—	★	
151	***Calandrella brachydactyla***	Short-toed Lark	—	—	—	—	—	★	
152	***Galerida theklae***	Thekla Lark	—	—	—	—	—	★	
153	***Lullula arborea***	Woodlark	—	—	—	—	—	★	
154	***Anthus campestris***	Tawny Pipit	—	—	—	—	—	★	
155	***Troglodytes troglodytes fridanensis***	Wren	—	—	—	—	—	★	
156	***Luscinia svecica***	Bluethroat	—	—	—	—	—	★	
157	***Saxicola dacotiae***	Fuerteventura Chat	✔	—	—	—	—	★	
158	***Oenanthe leucura***	Black Wheatear	—	—	—	—	—	★	
159	***Acrocephalus melanopogon***	Moustached Warbler	—	—	—	—	—	★	
160	***Acrocephalus paludicola***	Aquatic Warbler	✔	—	—	—	—	★	
161	***Hippolais olivetorum***	Olive-tree Warbler	—	—	—	—	—	★	
162	***Sylvia sarda***	Marmora's Warbler	—	—	—	—	—	★	
163	***Sylvia undata***	Dartford Warbler	—	—	—	—	—	★	
164	***Sylvia rueppelli***	Rüppell's Warbler	—	—	—	—	—	★	
165	***Sylvia nisoria***	Barred Warbler	—	—	—	—	—	★	
166	***Ficedula parva***	Red-breasted Flycatcher	—	—	—	—	—	★	
167	***Ficedula semitorquata***	Semi-collared Flycatcher	—	—	—	—	—	★	
168	***Ficedula albicollis***	Collared Flycatcher	—	—	—	—	—	★	
169	***Sitta krueperi***	Krüper's Nuthatch	—	—	—	—	—	★	
170	***Sitta whiteheadi***	Corsican Nuthatch	✔	—	—	—	—	★	
171	***Lanius collurio***	Red-backed Shrike	—	—	—	—	—	★	
172	***Lanius minor***	Lesser Grey Shrike	—	—	—	—	—	★	
173	***Pyrrhocorax pyrrhocorax***	Chough	—	—	—	—	—	★	
174	***Fringilla coelebs ombriosa***	Chaffinch	—	—	—	—	—	★	
175	***Fringilla teydea***	Blue Chaffinch	✔	—	—	—	—	★	
176	***Loxia scotica***	Scottish Crossbill	✔	—	—	—	—	★	
177	***Bucanetes githagineus***	Trumpeter Finch	—	—	—	—	—	★	
178	***Pyrrhula murina***	Azores Bullfinch	[not assessed]	—	—	—	—	★	
179	***Emberiza cineracea***	Cinereous Bunting	✔	—	—	—	—	★	
180	***Emberiza hortulana***	Ortolan Bunting	—	—	—	—	—	★	
181	***Emberiza caesia***	Cretzschmar's Bunting	—	—	—	—	—	★	

1. NUTS — Nomenclature des Unités Territoriales Statistiques — the administrative regions of the European Union are called NUTS regions, and are designated by the EC Statistical Office.
2. 'Threatened' as listed on Annex I of the EC Birds Directive
3. Non-migratory raptor

✔ given for IBA criteria C1 indicates that the population threshold applied is the same as that given for A1 (see Appendix 2a).

✔ given for IBA criteria C2 indicates that the population threshold applied is the same as that given for B1i, B1ii or B1iii (see Appendix 2a).

C2 has also been applied to some more dispersed species on the basis that the site holds more than 1% of the EU population of the species. Thresholds are not given but are based on population data in the BirdLife International/European Bird Census Council European Bird Database, accessed in 1998.

✔ given for IBA criteria C3 indicates that the population threshold applied is the same as that given for B1i or B1ii (see Appendix 2a).

For criterion C4 'S' indicates the species is classed as a 'seabird' and thus the collective threshold of 10,000 pairs of seabirds may apply, and 'W' indicates the species is considered a 'waterbird' following the definition of Rose and Scott (1997) (see p. 15, 'Identifying IBAs' chapter) and the collective threshold of 20,000 individuals may apply.

✔ given for IBA criteria C5 indicates that the population threshold applied is the same as that given for A4iv (see Appendix 2a).

★ given for IBA criteria C6 indicates that the taxa is one of the 181 species or subspecies listed on Annex I of the EC Birds Directive.

Subspecies listed on Annex I of the EC Birds Directive (Commission Directive 97/49/EEC, OJL223 13.8.97)

Subspecies	**Breeding distribution**
Puffinus yelkouan mauritanicus	Balearic islands
Phalacrocorax aristotelis desmarestii	Mediterranean
Anser albifrons flavirostris	West Greenland
Accipiter gentilis arrigonii	Corsica and Sardinia
Accipiter nisus granti	Madeira and Canary Islands
Lagopus mutus pyrenaicus	Pyrenees
Lagopus mutus helveticus	Alps
Alectoris graeca saxatilis	Alps
Alectoris graeca whitakeri	Sicily
Perdix perdix italica	Italy
Perdix perdix hispaniensis	Pyrenees and northern Spain
Uria aalge ibericus	Atlantic coast of Spain and Portugal
Columba palumbus azorica	Azores
Dendrocopos major canariensis	Tenerife (Canary Islands)
Dendrocopos major thanneri	Gran Canaria (Canary Islands)
Troglodytes troglodytes fridanensis	Fair Isle
Fringilla coelebs ombriosa	Hierro (Canary Islands)

Appendix 3. The classification of habitats, land-uses and threats at Important Bird Areas in Europe.

HABITATS

This classification is based on *A classification of Palaearctic habitats* (Devillers and Devillers-Terschuren 1996), and is thus compatible with the habitat classification used by the CORINE Biotopes Programme (*Habitats of the European Community*, Devillers *et al.* 1991), which is the basis for EC legislation such as the Habitats Directive and also for other Europe-wide habitat-conservation initiatives.

The habitat-types in this classification are, in general, mutually exclusive. In some cases, the total percentage of all habitats at a site exceeds 100%, given that the distribution of habitats at individual IBAs is not always well known, and that estimates of habitat cover may be approximate and are sometimes derived independently of other habitats at the same site.

Forest and woodland

This category includes wooded grasslands and wooded scrublands (listed at primary level—'Forest and woodland'—only).

Broadleaved deciduous forest
Forests and woodlands of native deciduous trees, other than flood-plain or mire woods, in the boreal, nemoral (hemi-boreal; the transition zone between the boreal and temperate zones), continental steppe, warm-temperate humid and Mediterranean zones. Forests that are dominated by broadleaved deciduous trees, but include broadleaved evergreen trees, are included in this category.

Native coniferous forest
Forests and woodlands of native coniferous trees, other than flood-plain and mire woods, in the boreal, nemoral, continental steppe, warm-temperate humid and Mediterranean zones. Forests that are dominated by coniferous trees, but include broadleaved evergreen trees, are included in this category.

Mixed forest
Forests and woodlands of broadleaved deciduous and coniferous trees, in equal dominance, in the boreal, nemoral, continental steppe, warm-temperate humid and Mediterranean zones. Included here are plant communities of combined deciduous-tree and conifer dominance in which both constituents play a substantial part and that are related to primarily deciduous forests. Mixed coniferous and broadleaved evergreen forest is not included in this category, and should be listed under 'Native coniferous forest' or 'Broadleaved evergreen forest', depending on dominance.

Alluvial/very wet forest
Tree and shrub vegetation of riverine and lacustrine flood-plains, marshes, fens and bogs in the boreal, nemoral, continental steppe, warm-temperate humid, Mediterranean and subtropical desert zones.

Broadleaved evergreen forest
Forests and woodlands dominated by broadleaved evergreen trees (sclerophyllous, i.e. hard-leaved), or by palms. They are characteristic of the Mediterranean and warm-temperate humid zones, with a few representatives in the nemoral, continental steppe and transition to subtropical desert zones.

Wooded steppe
Formations of the transition zones between temperate/boreal forest and continental steppes, occurring in mid-latitudes south of, and inland from, the boreal and nemoral forest belts, in regions of reduced summer humidity, as well as in areas adjacent to, or under the influence of the Mediterranean and warm-temperate humid zones, represented by a macro-mosaic of steppe and connected, contiguous, disjunct or widely spaced woodland stands, the latter usually with a very developed grassy understorey, or by a scattering of trees within a steppe environment. The forest elements are often located on porous or slightly raised ground, valley sides or slopes, the grasslands occupying less well-drained soils and lower places.

Wooded tundra
Formations of the transition zone between boreal forest (taiga) and tundra, characterized by a scattering of stunted coniferous trees or deciduous shrubs within a tundra environment, or by a macro-mosaic of tundra with scattered islands of forest, or by open forest with scattered treeless tundra patches. These formations occur in a broad belt, up to several hundred kilometres wide, across the north of the northern continents and as a narrow belt in high-latitude mountains.

Treeline ecotone
Also called the timberline, krummholz zone, etc., this category comprises formations at the transition zone between montane forest and the alpine zone of mountains, where subalpine forests give way to alpine or boreal heaths and shrubs, and to alpine grasslands. They are characterized by a scattering of stunted, gnarled trees, punctuating an alpine shrub or grassland environment, by a macro-mosaic of alpine shrub and grass formations with scattered islands of forests, or by open or clear forest with an undergrowth composed of alpine elements such as ericaceous shrubs. The trees/shrubs occupy a narrow belt, varying in altitudinal location according to latitude, exposure and other climatic or edaphic conditions.

Wooded desert/semi-desert
Scattered trees/shrubs in the subtropical desert zones, resulting from the colonization by tropical woodland trees of shrubby or grassy semi-desert plant communities. Included in this category are desert savannas, pseudo-steppes and sand-dune open woodlands.

Scrub

Scrub
Shrub communities of boreal, nemoral or steppic affinities, such as high-montane and boreo-montane conifer scrubs, and subalpine and Mediterranean-mountain bush communities, as well as recolonization communities of deciduous forest and temperate conifer forest (including formations maintained artificially, e.g. coppicing). Xerophytic scrub is treated as 'Wooded desert/semi-desert'.

Heathland
Low vegetation dominated by dwarf-shrubs, where succession to forest is inhibited naturally by harsh climate or artificially by fire, grazing and turf-cutting. Such formations include coastal or oceanic heaths and moors (e.g. in Atlantic, Macaronesian or oro-Mediterranean zones), high-montane and boreo-montane heaths in alpine and Arctic zones, and continental tragacanthic (spiny dwarf-shrub) communities.

Sclerophyllous scrub/maquis/garrigue
Evergreen sclerophyllous (hard-leaved) or lauriphyllous shrub communities of Mediterranean or warm-temperate humid regions and their areas of influence, occurring mostly as recolonization and degradation stages of broadleaved evergreen forests (maintained through grazing and burning), but irradiating into deciduous forest, steppe and desert areas. This category includes garrigue (phrygana; batha): cushion-forming thermo-Mediterranean sclerophyllous formations, often thorny and summer-deciduous, lower and more open than other formations.

[Savanna/bushland]

Although this primary habitat-type was listed as an option in the habitat classification used in the IBA data-sheet and in the IBA database, it does not occur in Europe according to the habitat definition in Devillers and Devillers-Terschuren (1996), therefore the few registrations of this habitat-type that were encountered (on submitted data-sheets/database) were amended to a more suitable habitat at the data-checking stage.

Grassland

Steppe/dry calcareous grassland
Primary and secondary steppes, formations dominated by medium or tall perennial tuft-forming grasses or woody-stemmed herbs, with patchy

Appendix 3 ... continued. The classification of habitats, land-uses and threats at Important Bird Areas in Europe.

ground cover, together with their associated annual or ephemeral communities. Typical formations, with two seasons of dormancy (winter and summer) that are in a large part climactic, are characteristic of a steppe belt in continental mid-latitudes, with irradiations in the deciduous forest zones. Structurally similar formations, although perhaps largely man-made (secondary), and mostly with a single season of dormancy (summer), are widespread in the transition zones that form on the temperate side of the subtropical deserts. By extension, this category also includes dry grasslands, mostly grazing-maintained, and forest-fringe formations of the lowland, hill and montane altitudinal levels of deciduous and broadleaved evergreen temperate forest zones, on mostly calcareous soils, sands and decomposed rock surfaces.

In this publication, abandoned arable land in areas of primary or secondary steppe (i.e. regenerating steppe land) has been included under this category, while fallow arable land in such areas—so called 'pseudosteppe'—has been treated as 'Arable land'. Mountain steppe (e.g. in Armenia) and shrub-steppe (e.g. paramó in Spain) have also been included in this category.

Dry siliceous grassland
Short, mostly secondary (grazing-maintained), grasslands and annual or ephemeral communities colonizing acidic, well-drained soils of the lowland, hill and montane levels of temperate deciduous, conifer and broadleaved forest zones.

Alpine/subalpine/boreal grassland
Primary and secondary grass- or sedge-dominated formations of the alpine and subalpine levels of the boreal, nemoral, warm-temperate humid and steppic mountains, and of the higher levels of Mediterranean mountains.

Humid grassland
Unimproved or lightly improved wet meadows and tall herb communities of the boreal, nemoral, warm-temperate humid, continental steppe and Mediterranean zones. The water-table remains high for much of the year.

Mesophile grassland
Lowland and montane mesophile (i.e. not drought-adapted) permanent pastures and hay meadows of the boreal, nemoral, warm-temperate humid, continental steppe and Mediterranean zones.

Tundra
Low plant communities of grasses and shrubs developed over permafrost. Plant cover becomes increasingly moss-dominated, patchy and sparse towards the north, grading into polar desert in the most extreme (unglaciated) parts of the Arctic regions, this being characterized by mostly bare ground worked by solifluction (frost-heave) and colonized by algae and very species-poor, extremely open plant communities constituted by lichens, mosses, liverworts and a very restricted number of vascular plants. Tundra-like plant communities in boreal montane areas, away from the main permafrost areas (e.g. along the mountain spine of Norway and in the Ural mountains of Russia), are sometimes composed of the same plant species as tundra, but in this publication have generally been classified as 'boreal grassland' under the category 'Alpine/subalpine/boreal grassland'.

Machair
This is a distinctive, low sand-dune formation that is found nowhere else in the world but the north and west of Scotland and western Ireland. Machair is formed by a particular combination of physical factors, including climate and landform. Sand with a high shell content is blown onshore by the westerly winds that prevail in these areas, onto a low-lying coastal plain. Vegetation develops that is typical of calcareous to neutral sandy grassland. Traditionally, machair supports extensive grazing regimes and unique forms of cultivation that rely on low-intensity systems of rotational cropping.

Desert

Desert
Interior continental deserts of temperate areas lying beyond the reach of moisture-laden winds; rain-shadow deserts lying on the leeward side of high mountain ranges which intercept the rain-bearing winds; deserts situated within the Horse latitudes (subtropical belts of high atmospheric pressure); deserts of the west coast of continents, under the influence of cold ocean currents, almost without measurable rainfall but with high relative humidity in the form of fogs and mists.

Semi-desert
The transition zone between deserts (continental/subtropical/cool-coastal) and steppes or Mediterranean habitats or other formations, characterized by sparse xerophytic/halophytic shrub or dwarf-shrub communities, and/or xerophytic/ halophytic grasslands, etc.

Wetland

Tidal river/enclosed tidal water
A zone of broadening in rivers entering the oceans or their connected seas, and river channels, below the tidal limit. Included are all marine and marine-related pelagic and benthic plant communities, and all river-course and river-bed plant communities.

Mudflat/sandflat
Sands and muds of the coasts of the oceans, and their connected seas and associated lagoons, submerged for part of every tide or for part of the annual cycle, devoid of vascular plants, but usually coated by blue algae and diatoms. They are of particular importance as feeding grounds for wildfowl and waders.

Saltmarsh
Communities of flowering plants, for the most part halophytes (salt-tolerant), colonizing sites submerged by high tides at some stage of the annual tidal cycle of oceans and their connected seas. This category also includes similar halophyte communities colonizing inland, permanent or temporary, saline, hypersaline or brackish water-bodies and their periphery, including inland closed seas, lakes, pools, sebkhas, rivers, springs and seeps, as well as, by extension, azonal, strongly differentiated communities developing on habitually dry, alkali, chlorid or gypseous soils of the steppe or forest zones (e.g. salt-steppes, salt-scrubs, salt-forests). Zonal communities of the desert and semi-desert areas, composed, to varying degrees, of halophytes or gypsophytes, are listed under 'Desert' or 'Semi-desert'.

Sand-dunes/sand beach
Sand-covered shorelines of the oceans, their connected seas and associated coastal lagoons, fashioned by the action of wind or waves. Sand-bars and -banks on rivers should be classified under 'River/stream', and inland sand-dunes form their own category under 'Rocky areas'.

Shingle/stony beach
Beaches of the oceans, of their connected seas and their associated coastal lagoons, covered by pebbles, or sometimes boulders, usually formed by wave action.

Coastal lagoon
Saline or hypersaline waters of the vicinity of the oceans and their connected seas, often formed from sea inlets by silting and cut off from the sea by more or less effective obstacles such as sand- or mud-banks.

Standing fresh water
Lakes, ponds and pools of natural origin containing fresh (i.e. non-saline) water, as well as semi-natural aquatic plant communities occupying man-made freshwater bodies, including artificially created lakes, reservoirs and canals.

Standing brackish and salt water
Non-marine brackish, saline or hypersaline lakes, pools and ditches, their features, their associated pelagic and benthic communities, and their beds of macrophytic submerged or offshore vegetation. Included in this category are large inland 'seas', such as the Caspian Sea, although in this publication open areas of the Caspian have been (mistakenly but consistently) classified under 'Marine areas').

Appendix 3 ... continued. The classification of habitats, land-uses and threats at Important Bird Areas in Europe.

River/stream
All rivers and streams (running water), permanent or temporary, fresh or saline. In this publication, related bars and banks of mud, sand, gravel and rock have been included under this category.

Raised bog
Highly oligotrophic (nutrient-poor), strongly acidic plant communities composed mainly of sphagnum moss growing on (and forming) peat, and deriving moisture and nutrients only from rainfall (ombrotrophic). They form only in relatively cool climates with heavy rainfall. They are most widespread in the boreal zone and in the mountains and hills of the nemoral zone; they occur locally in the lowlands of the nemoral zone and rarely in the continental steppe zones. In this publication, aapa and palsa mires have been listed under this category.

Blanket bog
Plant communities similar to raised bogs, on flat or gentle sloping ground with poor surface drainage, in oceanic climates with heavy rainfall (north-west Europe only, including Norway).

Water-fringe vegetation
Reedbeds and large-sedge plant communities of the margins and floating rafts of lakes, rivers and brooks, and of fens and eutrophic (nutrient-rich) marshes.

Fen/transition mire/spring
Small-sedge and related plant communities of fens, transition mires and quaking bogs; vegetation of springs.

Marine areas

Open sea
Oceanic and continental-shelf waters of the world ocean and its connected seas, their associated open-water and bottom plant communities, and marine vascular vegetation beds; marine plant communities of the littoral zone and coastal lagoons.

Sea inlet/coastal features
Bays and narrow channels of the oceans and their connected seas, including sea lochs or loughs, fiords or fiards, rias and straits but excluding estuaries and lagoons.

Rocky areas

In this publication, some habitat-types have been listed under this primary category as there were no more suitable categories, including: 'eternal snow and ice' (high-mountain or high-latitude zones occupied by glaciers or by perennial snow); large, unvegetated expanses of outwash gravel (Iceland), mine-tailings (Russia), lava fields (Canary Islands/Spain).

Sea cliff/rocky shore
Rock exposures (including faces, ledges, caves and shore) adjacent to the oceans, their connected seas and associated coastal lagoons, or separated from them by a narrow shoreline.

In general, the listing of any habitat comprising less than 5% of the individual IBA area was discouraged during this inventory. Although this particular habitat-type usually comprised less than 5% of an IBA's area, it was often ornithologically important, therefore an effort was made to list this habitat whenever it was known to occur and/or be important, even if it comprised less than 5% of an IBA's area.

Rock stacks/islets
Permanently emerging, periodically uncovered, surface-breaking or near-surface raised features of the oceans, their connected seas and coastal waters, with their associated marine and terrestrial communities. Includes reefs, banks and shoals.

In general, the listing of any habitat comprising less than 5% of the individual IBA area was discouraged during this inventory. Although this particular habitat-type usually comprised less than 5% of an IBA's area, it is often ornithologically important, therefore an effort was made to list this habitat whenever it was known to occur and/or be important, even if it comprised less than 5% of an IBA's area.

Scree/boulders
Vegetated or sparsely vegetated and frequently unstable areas of stones, boulders or rubble on steep slopes, produced by erosion in mountainous terrain. They are characteristic of high mountains in all zones, extending to lower altitudes in the boreal and Mediterranean zones; a very few communities form in lowland areas elsewhere.

Inland cliff
Exposed rock of cliffs, rock-faces, limestone pavements, the plant communities that colonize their cracks, and their associated animal communities.

In general, the listing of any habitat comprising less than 5% of the individual IBA area was discouraged during this inventory. Although this particular habitat-type usually comprised less than 5% of an IBA's area, it was often ornithologically important, therefore an effort was made to list this habitat whenever it was known to occur and/or be important, even if it comprised less than 5% of an IBA's area.

Inland sand-dunes
Sand bodies of aeolian (wind-blown) origin, possessing constructional relief and separated from the coast and its dune cordons by non-dunal habitats, developed within the boreal, nemoral, continental steppe, warm-temperate humid, Mediterranean and subdesert steppe zones. They support a vegetation which differs markedly from coastal sand-dune communities.

Caves
Natural caves and cave systems.

In general, the listing of any habitat comprising less than 5% of the individual IBA area was discouraged during this inventory. Although this particular habitat-type usually comprised less than 5% of an IBA's area, it is sometimes ornithologically important (e.g. as a cliff-nesting site for raptors such as Lammergeier *Gypaetus barbatus*), therefore an effort was made to list this habitat whenever it was known to occur and/or be important, even if it comprised less than 5% of an IBA's area.

Artificial landscape

Highly improved reseeded grassland
Heavily fertilized or reseeded permanent grasslands, sometimes treated by selective herbicides, with very impoverished flora and fauna.

Arable land
Fields of cereals, beets, sunflowers, leguminous fodder, potatoes and other annually harvested plants. Faunal and floral quality and diversity depend on the intensity of agricultural use and on the presence of borders of natural vegetation between fields. This category can include tree-lines, hedges, and rural mosaics, i.e. ligneous formations of small size, arranged in a linear, reticulated or insular manner, closely interwoven with grassy or cultivated habitats. Also included are combinations of such elements and mixed agricultural formations, containing both ligneous and herbaceous layers. Very artificial, disturbed or heterogeneous systems, containing many planted or exogenous elements can be listed here, while more natural ensembles utilising many natural elements and covering substantial surfaces are better classified under 'Forest and woodland' or 'Scrub'.

Perennial crops/orchards/groves
Ligneous crops other than forestry plantations. Such areas, when non-intensively managed, may support a rich flora and fauna.

Forestry plantation
Areas of planted forest or woodland, or tree-based crops, of native or exotic tree species, deliberately planted to supply timber, here generally taken to mean areas planted less than about 100–150 years ago. Such areas, when old or non-intensively managed, may support a rich flora and fauna.

Urban parks/gardens
Large gardens. Usually varied formations, created for recreational use.

Appendix 3 ... continued. The classification of habitats, land-uses and threats at Important Bird Areas in Europe.

Other urban/industrial areas
Towns, villages, industrial sites. Areas used for human occupation and industrial activities.

Ruderal land
Fallow land, waste places. Fields set aside, abandoned or left to rest, roadsides and other interstitial spaces on disturbed ground. They are colonized by numerous pioneering, introduced or nitrophilous plants.

Introduced/exotic vegetation
This category includes all formations dominated by recently introduced or exotic plant species, apart from forestry plantations composed of such species, which are classified as 'Forestry plantation' under 'Artificial landscape'.

Unknown
This catch-all category was listed when the habitat covering part or all of an IBA's area was not known to the compiler of the data-sheet or database.

LAND-USES

The types of land-use in this classification are not mutually exclusive (e.g. forestry and recreation can use the same area of forest), therefore there is no requirement that the percentage covers of the different land-uses at an IBA should add up to 100%.

Agriculture
All land affected by cultivation, including perennial crops/groves/orchards, as well as all land affected by pastoral agriculture, including grazing lands and rangelands of livestock.

Fisheries/aquaculture
All land or water affected by commercial angling, fishing, aquaculture, mariculture, shellfish cultivation or harvesting, etc. Recreational angling was listed under 'Tourism/Recreation'.

Forestry
This term was taken to mean wooded land under active management for the extraction of timber and non-timber forest products, as well as for other non-extractive functions such as protection against erosion and avalanches, etc. and provision of amenity areas for recreation/tourism, etc.

Hunting
Although contributors were instructed to only list hunting as a land-use if the land was officially designated for hunting, this definition was not applied rigidly or consistently by contributors. This category tended to be listed for any land where hunting took place (i.e. the land was not necessarily designated officially for hunting *per se*).

Military
Any area of land or water used by the military, for any purpose.

Nature conservation/research
Contributors tended apply this category only to land or water where 'Nature conservation/research' was a primary land-use, i.e. only to those protected areas with a relatively high level of protection or management for nature conservation, tending to ignore 'protected landscapes' and other such 'broader', less exclusive designations.

Not utilized
The relevant area is not used. In this inventory, some contributors listed this category to indicate that none of the other listed uses occurred, not necessarily that the land was unused—this was corrected at the data-checking stage where possible.

Tourism/recreation
Although contributors were advised to only record this category if the land or water was allocated specifically for this use, e.g. on planning maps, in practice contributors usually listed this use for any areas of land or water that were used, directly or indirectly, for recreation, tourism or leisure activities, i.e. including infrastructural or built-up areas such as holiday villages and accommodation complexes.

Urban/industrial/transport
This category includes: general residential and built-up areas (not only in cities but also towns or villages); infrastructure other than for agriculture, forestry, fisheries, aquaculture, tourism/recreation or water management, i.e. mainly for transport and energy purposes, e.g. roads, bridges, railways, ports, airports, power-stations, pylon or pipeline networks, wind-farms, etc.; and non-built-up land used by commercial or industrial activities other than agriculture, forestry, fisheries, aquaculture, tourism/recreation or water management, i.e. such activities as extraction of oil/gas, mining of ores, peat, salt, gravel, etc.

Water management
This category includes the management of waterbodies such as rivers and lakes for purposes such as flood control, irrigation, storage (for drinking water, hydropower, cooling power-stations, etc.) and large-scale redistribution (to balance disturbed hydrology, facilitate engineering projects, etc.).

Unknown
This category was listed when the land-use, if any, over part or all of an IBA's area was not known to the compiler of the data-sheet or database.

Other
This catch-all category covers any human use of the IBA which cannot be classified under one of the other categories, and in this inventory included such activities/uses as: reed-cutting; firewood-gathering; traditional/artisanal salt-production; excavation of loess or marl in small quantities (i.e. not industrial/commercial); harvesting of animal/plant products such as cork, heather, berries, fungi, eggs/young/adults of birds; husbandry of wild species (e.g. Eider *Somateria mollissima* for eiderdown); ice-cutting.

THREATS

Threats can cause a reduction in bird populations at an IBA either directly, through causing increasing mortality at the site, or they can act indirectly by reducing the 'carrying capacity' of the site, by physically degrading or destroying habitats or by disturbing and interfering with birds' essential activities such as feeding, roosting, resting, moulting, nesting, and so on. Most of the categories of threat listed here are self-explanatory, or their listing at a site is accompanied by a specific justification/explanation in the text of the relevant IBA site-account in the inventory.

Threats tend to be posed mainly by human activities, and encompass not only the direct effects/impacts of the activity on the habitats and/or bird populations at the IBA but also those impacts/effects that are indirect or remote. For example, agricultural intensification/expansion quite far away from a wetland or heathland IBA, but upstream or upwind of it, can cause damaging nutrient pollution at the site, the nutrients being transported to the site by water flow or by atmospheric deposition.

Appendix 3 ... continued. The classification of habitats, land-uses and threats at Important Bird Areas in Europe.

Calculating the importance score of the threat

To calculate the seriousness of the threat, in terms of the magnitude of its actual or potential impact, the following method was used:

For habitat-related threats:
I. Effect of threat on the habitat
— destruction (scores 3)
— rapid deterioration (scores 2)
— slow deterioration (scores 1)

II. Spatial scale of the threat (in relation to the IBA)
— affects the IBA as a whole (3)
— affects a large part of the IBA but not critical sites for threatened species or a relatively small part of the IBA which is important for threatened species (2)
— affects a relatively small part of the IBA with no crucial site for threatened species (1)

III. Realization of threat
— threat already exists (3)
— threat is planned with realization expected in short term (2)
— threat is planned with realization expected in longer term (1)

For bird-related threats:
I. Expected/measured effect on threatened species
— majority of critical species are affected (3)
— some critical species are affected (2)
— only non-critical species are affected (1)

II. same as for habitat-related threats

III. same as for habitat-related threats

The combined level of the threat is calculated by summing up the values in I, II and III.

Score (threat) = score (I) + score (II) + score (III)

For ranking purposes, the threats can be classified into three groups:

1. **Level 'A' threats (high impact): scores 8 and 9**
2. **Level 'B' threats (medium impact): scores 6 and 7**
3. **Level 'C' threats (low impact): scores 3, 4 and 5**

Abandonment/reduction of land management
Includes: undergrazing; rural depopulation leading to reduced carrion for scavengers; abandonment of salt-pans. Abandonment of fish-ponds is treated under 'Aquaculture/fisheries'.

Afforestation
Afforestation with exotic tree species should be treated here rather than under 'Consequences of animal/plant introductions'.

Agricultural intensification/expansion
Includes: irrigation (including indirect impacts, e.g. draw-down of water leading to predators reaching breeding colonies on islands, or reservoir being kept artificially high in summer, etc.); high fertilizer input; excessive use of chemicals; changes in crop species or cultivation; loss of habitats; overgrazing; effects of pest control on non-target species; nutrient pollution of wetlands as a result of agricultural intensification.

Aquaculture/fisheries
The threat is not solely from intensification or expansion of aquaculture/fisheries—abandonment or reduction of traditional or non-intensive aquaculture can also affect some waterbirds (cf. agriculture). This category includes: the persecution (shooting, etc.) of waterbirds that can occur at some fisheries/aquaculture sites; reduction in fish-farming due to economic crisis. If recreational angling causes a threat, this is listed under 'Recreation/tourism'.

Burning of vegetation
Fires that are not caused by natural events, i.e. all man-made fires, whether purposeful (e.g. to clear vegetation for grazing) or accidental or malicious.

Consequences of animal/plant introductions
Afforestation with exotic tree species should be treated under 'Afforestation' rather than here.

Construction/impact of dyke/dam/barrage
Includes such phenomena as water-level fluctuations in, or downstream of, hydropower reservoirs, changing levels of reservoirs or storage basins, altered sedimentation patterns downstream, etc.

Deforestation (commercial)
Includes clear-cutting (selective felling is treated under 'Selective logging/cutting'), illegal as well as legal.

Disturbance to birds
This threat is usually a consequence of other human activities than deliberate scaring, and thus is often listed in combination with other threat categories (ultimate factors), e.g. recreation/tourism or intensified forest management. Contributors to this inventory have often listed both 'Disturbance to birds' and the responsible activity too, to highlight the ultimate source of the disturbance. Active persecution of birds is generally classed as 'Other' threat, but when disturbance effects are more important than mortality effects, e.g. farmers shooting at Cranes *Grus grus* or geese *Anser*/*Branta* just enough to scare them away, then such activity is better classed under 'Disturbance'.

Drainage

Dredging/canalization
Includes creation of canals for irrigation.

Appendix 3 ... continued. The classification of habitats, land-uses and threats at Important Bird Areas in Europe.

Extraction industry
Includes exploration as well as extraction activities/infrastructure and any resultant pollution, covering all kinds of extraction, e.g. not just mining but oil/gas, peat (commercial extraction only), etc., as well as pollution occurring during transport (accidental discharge/deliberate cleaning of bilges of oil tankers, etc.).

Filling-in of wetlands
Used for active filling-in only, e.g. land-fill using excess spoil or waste material. If the wetland is filling in due to increased sedimentation or other indirect processes, the threat should be classified under the cause of the increase in sedimentation.

Firewood collection

Forest grazing
Includes damaging grazing caused by over-population of wild deer (through winter-feeding by hunters, eradication of natural predators, etc.).

Groundwater abstraction

Industrialization/urbanization
Includes construction, chemical run-off or spillage, sewage effluent, wind-farms, etc. Ideally, only includes physical developments (or attendant phenomena/processes) that are not related to tourism/recreation, forestry, agriculture, extraction, aquaculture/fisheries, dyke/dam/barrage construction, since all of these can be classified separately. Thus, this category generally includes housing, offices, factories, transport and energy developments, and generally the threat is posed by the addition of infrastructure to an existing urban area or other centre, whereas isolated examples of such development are better classed under 'Infrastructure'. Examples of threatening processes or related phenomena, rather than physical developments, include the industrialization (intensification) of salina-management, the occurrence of acid rain downwind of industrial sources. Coastal land-claim, e.g. for expanding industrial or urban areas, is included here, but land-claim of inland wetlands is classed under 'Filling-in of wetlands'. Waste disposal (where source of waste is unspecified) is also classed here.

Infrastructure
Generally involves energy (power-lines) and transport (roads, railways). Includes roads, railways, overhead transmission lines, etc. This can be difficult to separate from 'Industrialization/urbanization', but tends to be listed where developments are more 'isolated', not involving the addition of infrastructure to some pre-existing 'centre' or focus (e.g. town, etc.). Again, ideally this category should include only those developments that are not related to tourism/recreation, forestry, extraction industry, agriculture, aquaculture/fisheries or construction of dyke/dam/barrage, since these can be categorized separately. Proposed wind-farm. Airport pollution.

Intensified forest management
Management generally refers to production of timber, not of non-timber forest products (e.g. mushrooms, honey). Excessive collection of the latter should be classified as 'Other' threat.

Natural events
Includes drought, erosion (at normal levels), storms, nest destruction by native predators, etc. It is considered possible for 'Natural events' to pose threats to birds and habitats, despite the adaptations of all organisms to the naturally dynamic environment over millions of years, because the environment in Europe has been changed drastically (generally, simplified in an ecological sense) by man, especially over the last 100 years or so. Landscapes and nature in Europe are now so heavily modified by man, and maintained in that state by his activities, that organisms cannot depend on normal 'ecosystem processes' to occur. For example, the alteration of sedimentation patterns in estuaries by upstream dam-construction puts offshore sandbanks and coastal spits (potentially valuable habitat, e.g. to nesting terns *Sterna*), etc. at more risk of destruction by natural storms, with less possibility of their natural regeneration elsewhere in the estuary. Following the eradication of top mammalian predators over much of Europe, there has been an increase in native nest-predators such as foxes, jackals and wild boars. Fires should be listed under this category only if they are stated to be natural—only 2–3% of boreal forest-fires, and less than 1% of Mediterranean forest-fires, are thought to be natural in origin (i.e. caused by lightning strike), all others being man-made according to European Commission reports, whether deliberate, accidental or malicious, and should be classed under 'Burning of vegetation'.

Recreation/tourism
As well as for direct impacts such as disturbance of birds, this category has been used to cover the impacts of 'hard' developments, e.g. building of hotels, holiday homes, etc.

Selective logging/cutting
Clear-cutting of forest/woodland should be treated under 'Deforestation'. This category includes the cutting/collection of branches as well as of whole trees, but applies only to trees and bushes, not to such vegetation as reeds, etc. (reed-cutting has been treated under 'Other' threat).

Shifting agriculture
This refers to swidden or slash-and-burn agriculture (more common in tropical forest countries than in Europe) or such practices as opportunistic ploughing and planting of desert/semi-desert (e.g. with barley/millet) after heavy rains, thus degrading or destroying the thin soil.

Unsustainable exploitation
'Exploitation' refers solely to birds, not to other natural resources such as reeds, forest fungi, etc., and thus includes unsustainable hunting (legal and illegal), egg-collection, husbandry of Eider *Somateria mollissima* for eiderdown, etc. Significant disturbance to birds caused by hunting should be classified under 'Disturbance to birds'.

Unknown
This category can only be used on its own, not with other threats—that is, it is not known what threats the site faces, if any.

Other
This catch-all category covers any threat to the IBA which cannot be classified under one of the other categories, and in this inventory included such threats as: various kinds of pollution for which the cause(s) cannot be (or were not) identified or which cannot be classed under other categories (e.g. fuel-oil pollution from unspecified kinds of boats; nutrient pollution or acid rain from unspecified sources; lead-shot pollution; radioactive contamination; water pollution by mosquito-control chemicals); deliberate persecution or incidental killing of birds (not exploitation *per se*), e.g. through poisoning of raptors, farmers shooting birds (except where disturbance effects are more important than mortality effects, when this threat should be classed under 'Disturbance to birds'); activities that threaten the site in an unexplained or unspecified way (military activity; heavy airport traffic; increased boat traffic; seaweed harvesting; hunting; hay-cutting; mussel-collecting; reed-cutting/harvesting); management of site is poor, inappropriate or lacking, especially water-management issues (water-level fluctuations; water transfer/abstraction of surface water); deer fences (effects unspecified); forest/soil degradation and tree disease (non-specific); excessive collection of non-timber forest products; transport of highly toxic chemicals through site (potential risk); outbreaks of botulism; algal blooms (cause unspecified).

References

DEVILLERS, P. AND DEVILLERS-TERSCHUREN, J. (1996) *A classification of Palaearctic habitats.* Strasbourg, France: Council of Europe.

DEVILLERS, P., DEVILLERS-TERSCHUREN, J. AND LEDANT, J.-P. (1991) Habitats of the European Community. In B. Wyatt *et al.*, eds. *CORINE Biotopes manual.* Luxembourg: Office of Official Publications of the European Communities.

Appendix 4. A comparison of Important Bird Area criteria categories applied in this publication with those used in the first pan-European IBA inventory (Grimmett and Jones 1989) and European Union study (Grimmett and Gammell 1989).

For a full explanation of the current criteria categories see the introductory chapter 'Identifying IBAs'.
For full lists of bird species and the criteria they may potentially fulfil see Appendices 2a and 2b.

	Species category to which IBA criteria applies	Criterion definition used in this publication (Heath and Evans 2000)	Criterion code (Heath and Evans 2000)	1989 criteria code(s) (Grimmett and Gammell 1989)	1989 criteria code(s) (Grimmett and Jones 1989)
GLOBAL	**Species of global conservation concern**	The site regularly holds significant numbers of a globally threatened species, or other species of global conservation concern.	A1	2 (part), 5 (part)	2.1
GLOBAL	**Restricted-range species**	The site is known or thought to hold a significant component of the restricted-range species whose breeding distributions define an Endemic Bird Area (EBA) or Secondary Area (SA).	A2	(part)[1]	4.1 (part), 4.2 (part)[1]
GLOBAL	**Biome-restricted species**	The site is known or thought to hold a significant assemblage of the species whose breeding distributions are largely or wholly confined to one biome.	A3	—[2]	—[2]
GLOBAL	**Congregations**	(i) The site is known or thought to hold, on a regular basis, ≥1% of a biogeographic population of a congregatory waterbird species.	A4i	1, 8	1.3 (waterbirds only)
GLOBAL		(ii) The site is known or thought to hold, on a regular basis, ≥1% of the global population of a congregatory seabird or terrestrial species.	A4ii	1, 8	1.1, 1.2 (waterbirds only)
GLOBAL		(iii) The site is known or thought to hold, on a regular basis, ≥20,000 waterbirds or ≥10,000 pairs of seabirds of one or more species.	A4iii	9 (waterbirds only)	1.3 (waterbirds only)
GLOBAL		(iv) The site is known or thought to be a 'bottleneck' site where at least 20,000 storks (Ciconiidae), raptors (Accipitriformes and Falconiformes) or cranes (Gruidae) regularly pass during spring or autumn migration.	A4iv	9[3]	1.4 (raptors only)
EUROPEAN	**Congregations**	(i) The site is known or thought to hold ≥1% of a flyway or other distinct population of a waterbird species.	B1i	1, 8	1.3 (waterbirds only)
EUROPEAN		(ii) The site is known or thought to hold ≥1% of a distinct population of a seabird species.	B1ii	1, 8	1.1, 1.2
EUROPEAN		(iii) The site is known or thought to hold ≥1% of a flyway or other distinct population of other congregatory species.	B1iii	1,8	1.1, 1.2
EUROPEAN		(iv) The site is a 'bottleneck' site where over 5,000 storks, or over 3,000 raptors or cranes regularly pass on spring or autumn migration.	B1iv	9[3]	1.4 (raptors 5,000)
EUROPEAN	**Species with an unfavourable conservation status in Europe**	The site is one of the 'n' most important in the country for a species with an unfavourable conservation status in Europe (SPEC 2, 3) and for which the site-protection approach is thought to be appropriate.	B2	0 (part), 3 (part)	3.1 (part), 3.2 (part), 3.3 (part), 4.1 (part), 4.2 (part)
EUROPEAN	**Species with a favourable conservation status in Europe**	The site is one of the 'n' most important in the country for a species with a favourable conservation status in Europe but concentrated in Europe (SPEC 4) and for which the site-protection approach is thought to be appropriate.	B3	0 (part), 3 (part)	3.1 (part), 3.2 (part), 3.3 (part), 4.1 (part), 4.2 (part)
EUROPEAN UNION	**Species of global conservation concern**	The site regularly holds significant numbers of a globally threatened species, or other species of global conservation concern.	C1	2 (part), 5 (part)	2.1
EUROPEAN UNION	**Concentrations of a species threatened at the EU level**	The site is known to regularly hold at least 1% of a flyway population or of the EU population of a species threatened at the EU level (listed on Annex I and referred to in Article 4.1 of the EC Birds Directive).	C2	1, 8[4]	1.1, 1.2, 1.3[4]
EUROPEAN UNION	**Congregations of migratory species not threatened at the EU level**	The site is known to regularly hold at least 1% of a flyway population of a migratory species not considered threatened at the EU level (as referred to in Article 4.2 of the EC Birds Directive) (not listed on Annex I).	C3	1, 8[4]	1.1, 1.2, 1.3[4]
EUROPEAN UNION	**Congregatory – large congregations**	The site is known to regularly hold at least 20,000 migratory waterbirds and/or 10,000 pairs of migratory seabirds of one or more species.	C4	9 (waterbirds only)	1.3 (waterbirds only)
EUROPEAN UNION	**Congregatory – bottleneck sites**	The site is a 'bottleneck' site where at least 5,000 storks (Ciconiidae) and/or at least 3,000 raptors (Accipitriformes and Falconiformes) and/or at least 3,000 cranes (Gruidae) regularly pass on spring or autumn migration.	C5	9[3]	1.4
EUROPEAN UNION	**Species threatened at the European Union level**	The site is one of the five most important in the European region (NUTS region) in question for a species or subspecies considered threatened in the European Union (i.e. listed on Annex I of the EC Birds Directive).	C6	0, 3 (part)[5]	3.1
EUROPEAN UNION	**Other ornithological criteria**	The site has been designated as a Special Protection Area (SPA) or selected as a candidate SPA based on ornithological criteria (similar to but not equal to C1–C6) in recognized use for identifying SPAs.	C7	0, 2 (part), 3, 4, 5 (part), 6, 7, 10, 11, 12	—

Footnotes
1. Some sites holding *Apus unicolor, Anthus berthelotii, Regulus teneriffae* and *Serinus canaria* are included under A2.
2. Three non-SPEC Annex I species are uniquely covered by this criterion: *Podiceps auritus, Surnia ulula* and *Strix nebulosa.*
3. 5,000 birds of prey, storks not covered.
4. 1% of EU populations (for a number of populations, these thresholds are lower, see Appendix 2b).
5. Concerning *Bulweria bulwerii, Puffinus assimilis, P. puffinus, P. yelkouan mauretanicus, Pelagodroma marina, Oceanodroma leucorhoa* and *O. castro*: sites holding less than the C2 threshold may still qualify under criterion C6.

Appendix 4 ... continued. A comparison of Important Bird Area criteria categories applied in this publication with those used in the first pan-European IBA inventory (Grimmett and Jones 1989) and European Union study (Grimmett and Gammell 1989).

CRITERIA APPLIED IN GRIMMETT AND GAMMELL (1989)

0. The site is amongst the five most important for such a species or subspecies in a NUTS region of the Community or is amongst the 100 most important in the Community for a vulnerable species or subspecies (Annex I).

Breeding sites

1. Sites supporting 1% or more of the breeding pairs of the biogeographical population. [The biogeographical population has been defined for the purposes of the European Community criteria, as a discrete population where this was identifiable or where it was not, the populations occurring in Europe and North-west Africa (Tunisia, Algeria and Morocco), but excluding for the most part, the USSR and Black Sea States.]
2. If Criterion 1 is not appropriate (because for example the biogeographical population is not clearly defined, is not known or the 1% criterion is too high to select important sites), criteria for the selection of breeding sites have been based on the specific characteristics of dispersion and habitat preference of the species.
3. If Criterion 2 also proved impossible to apply, all sites with proved breeding are selected (this Criterion is applied to six very poorly known seabird species only; *Bulweria bulwerii, Puffinus puffinus mauretanicus* [now treated as *Puffinus yelkouan mauretanicus*], *P. assimilis, Pelagodroma marina, Oceanodroma leucorhoa* and *O. castro*).
4. Sites of particular importance for marginal or isolated breeding population in the Community, with criteria based on specific characteristics of dispersion and habitat preference of the species.
5. All regular breeding sites of rare or endangered species or subspecies in the Community, or small and endangered distinct biogeographical populations: c.2,500 pairs or less (for some colonial species a level of five pairs is used to exclude irregular breeding sites.).
6. For widely dispersed species, breeding sites are selected on the basis of high densities or good numbers.
7. Regular breeding sites for significant numbers of three or more Annex I species.

Areas other than breeding areas

8. Sites having 1% (being at least 100 individuals) of the flyway or biogeographical population of one species.
9. Sites having (at least) 20,000 waterfowl, or 5,000 birds of prey on passage during a migration season.
10. Sites of particular importance for marginal or isolated populations in the Community, with criteria based on specific characteristics of dispersion and habitat preference of the species.
11. Sites which regularly hold five (for gregarious species, 25) individuals of rare and endangered species or subspecies in the Community, or small and endangered distinct biogeographical populations (less than 10,000 individuals in number).
12. Sites regularly holding numbers of three or more Annex I species.

CRITERIA APPLIED IN GRIMMETT AND JONES (1989)

Category 1: Sites for migratory species which congregate (either when breeding, or on passage, or in winter) in important numbers.

Criteria: 1. The site regularly holds 1% of a species's world population; or

2. the site regularly holds 1% of a species's European population (or EEC population for EEC Member States only); or

3[1]. the site regularly holds 1% of a species's biogeographical population; or

4. it is a 'bottleneck site' where over 5,000 storks (Ciconiidae) or over 3,000 raptors (Accipitridae) regularly pass on spring or autumn migration.

[1] Waterbird sites have also been selected where they regularly hold 20,000 or more waterbirds.

Category 2: Sites for globally threatened species.

Criterion: 1. The site regularly holds a significant number of the species.

Category 3: Sites for species and subspecies which are threatened throughout all or large parts of their range in Europe (but are not globally threatened).

Criteria: 1. The site is one of the five most important in the European region in question for the species or subspecies, or one of the ten most important in the European region in question for the species or subspecies (if the region is particularly large and is subdivided into comparatively small political units); or

2. the site is one of the 100 most important in Europe for the species or subspecies; or

3. the site is one of the 100 most important for the species or subspecies in the European Community.

Category 4: Sites for species which have relatively small total world ranges with important populations in Europe.

Criteria: 1. The site is one of the five most important in the European region in question for the species, or one of the ten most important in the European region in question for the species (if the region is particularly large and is subdivided into comparatively small political units); or

2. the site is one of the 100 most important in Europe for the species.

References

GRIMMETT R. F. A. AND GAMMELL, A. B. (1989) Inventory of Important Bird Areas in the European Community. (Unpublished report prepared for the Directorate-General for the Environment, Consumer Protection and Nuclear Safety of the European Community, Study contract B6610-54-88.) Cambridge, UK: International Council for Bird Preservation.

GRIMMETT, R. F. A. AND JONES T. A. (1989) *Important Bird Areas in Europe.* Cambridge, UK: International Council for Bird Preservation (Tech. Publ. 9).

Appendix 5. The regions of EU countries used in the application of Important Bird Area criteria for species threatened at the EU level (criterion C6).

Threatened species are those listed on Annex I of the EC Birds Directive (see Appendix 2b).

Country/territory	Biogeographic region [1]	NUTS level [2]	Administrative unit [2]	Number of regions [2]	Area (km^2) of largest region [2]
Austria	Alpine, Continental	1	Gruppen von Bundesländer	3	34,400
Belgium[3]	Atlantic, Continental	0	—	1	30,600
Denmark	Atlantic, Continental	1	—	1	43,100
Finland	Boreal	3	Maakunnat	19	98,900
France	Alpine, Atlantic, Continental, Mediterranean	2	Régions + DOM	22	83,900
Germany	Alpine, Atlantic, Continental	1	Länder	16	70,600
Greece[4]	Mediterranean	1	Groups of development regions	4	56,800
Italy	Alpine, Continental, Mediterranean	2	Regioni	20	25,700
Luxembourg	Continental	0	—	1	2,600
Netherlands	Atlantic	0	—	1	41,200
Portugal	Mediterranean	2	Comissaoes de coordenacao regiona	5	27,000
Portugal, Azores	Macaronesian	1	Regio autonomas	1	800
Portugal, Madeira	Macaronesian	1	Regio autonomas	1	2,400
Republic of Ireland[5]	Atlantic	0	—	1	68,900
Spain[6]	Alpine, Atlantic, Mediterranean	2	Comunidades autonomas	17	94,200
Spain, Canary Islands	Macaronesian	2	—	1	1,300
Sweden	Alpine, Boreal, Continental	3	Län	24	98,900
United Kingdom	Atlantic	1	Standard regions	11	77,100

1. Division of biogeographical regions as defined in the framework of the EC Habitats Directive (see Appendix 1).
2. NUTS regions and other data taken from Natura 2000 Standard Data Form.
   The Nomenclature of Territorial Units for Statistics (NUTS) was established by Eurostat, to provide a single uniform breakdown of territorial units for the production of regional statistics for the European Union. Although the NUTS has no legal value per se, it has been used since 1988 in the Community legislation. NUTS regions are based on human population density: regions are larger where population density is lower. Because many species of birds occur in remote areas the NUTS approach is not ideal for birds. Different NUTS levels have therefore been selected for the purpose of IBA identification, such that the size of the NUTS region used is roughly the same across the European Union (and preferably be within the size range 25,000–100,000 km^2). The Macaronesian islands belonging to Portugal and Spain have been considered as distinct regions.
3. In the previous pan-European IBA inventory (Grimmett and Jones 1989), NUTS level 1 was applied for Belgium, recognizing two distinct regions, Flanders (13,800 km^2) and Wallonne (16,800 km^2).
4. In the previous pan-European IBA inventory NUTS level 2 was applied for Greece, recognizing 13 distinct regions (average area 10,200 km^2, maximum area 19,100 km^2).
5. In the previous pan-European IBA inventory NUTS level 3 was applied for the Republic of Ireland, recognizing eight distinct regions (average area 7,700 km^2, maximum area 12,200 km^2).
6. Division according to the CORINE manual (CEC 1991).

References

CEC [COMMISSION OF THE EUROPEAN COMMUNITIES] (1991) *CORINE Biotopes Manual.* Luxembourg: Office for Official Publications of the European Communities.

GRIMMETT, R. F. A. AND JONES T. A. (1989) *Important Bird Areas in Europe.* Cambridge, UK: International Council for Bird Preservation (Tech. Publ. 9).

Appendix 6. The total number and total area of Important Bird Areas by country, and the reasons for importance of IBAs by country, indicated by the number of IBAs fulfilling each criterion.

			Criteria categories												
Country	Number of IBAs	Area of IBAs (ha)	Species of global conservation concern (A1)	Restricted-range species (A2)	Biome-restricted species (A3)	Congregations (global) (A4)	Congregations (regional) (B1)	Species of European conservation concern (B2/B3)	Species of global conservation concern (C1)	Concentrations of species threatened at the EU level (C2)	Congregations of migratory species not threatened at EU level (C3)	Congregatory – large congregations (C4)	Congregatory – Bottleneck sites (C5)	Species threatened at the EU level (C6)	Other ornithological criteria (C7)
Albania	15	90,309	9	—	—	7	8	6	—	—	—	—	—	—	—
Andorra	1	46,800	—	—	—	—	—	1	—	—	—	—	—	—	—
Armenia	5	182,076	4	—	1	3	3	2	—	—	—	—	—	—	—
Austria	55	1,238,756	6	—	—	3	2	32	6	1	2	1	—	48	3
Azerbaijan	52	616,141	30	—	9	15	16	28	—	—	—	—	—	—	—
Belarus	21	617,457	14	—	—	3	3	20	—	—	—	—	—	—	—
Belgium	48	648,298	1	—	—	9	15	10	1	8	—	1	1	37	4
Bosnia and Herzegovina	3	6,844	2	—	—	—	—	1	—	—	—	—	—	—	—
Bulgaria	50	700,185	29	—	4	23	26	40	—	—	—	—	—	—	—
Croatia	23	754,553	12	—	1	9	12	21	—	—	—	—	—	—	—
Cyprus	16	130,499	5	13	3	4	5	11	—	—	—	—	—	—	—
Czech Republic	16	627,853	7	—	—	3	5	15	—	—	—	—	—	—	—
Denmark	127	4,822,130 [1]	—	—	—	63	76	38	—	47	59	38	7	33	37
Faroe Islands	19	6,690	—	—	—	18	18	19	—	—	—	—	—	—	—
Greenland	55	2,541,570	—	—	3	32	45	18	—	—	—	—	—	—	—
Estonia	43	460,459	9	—	—	34	40	18	—	—	—	—	—	—	—
Finland	96	2,739,952	8	—	14	30	48	77	7	38	38	2	1	52	12
France	277	4,724,849	34	2	5	80	114	170	34	92	71	45	24	248	6
Georgia	11	1,546,275	6	7	8	2	2	5	—	—	—	—	—	—	—
Germany	285	3,506,487	24	—	1	87	116	131	24	50	109	67	11	76	77
Gibraltar	2	600	2	—	—	2	2	1	2	1	1	1	1	—	—
Greece	196	3,433,245	73	—	17	62	81	161	73	86	24	9	2	135	7
Hungary	43	1,466,244	25	—	—	26	27	34	—	—	—	—	—	—	—
Iceland	61	734,520	2	—	—	38	57	28	—	—	—	—	—	—	—
Italy	192	4,626,967	51	—	19	44	68	139	52	69	28	14	9	163	9
Latvia	58	1,086,499	19	—	—	16	41	27	—	—	—	—	—	—	—
Liechtenstein	2	1,040	—	—	—	—	—	2	—	—	—	—	—	—	—
Lithuania	35	317,652	8	—	—	5	6	32	—	—	—	—	—	—	—
Luxembourg	9	11,293	—	—	—	—	1	—	—	—	—	—	—	8	
FYR of Macedonia	10	190,700	7	—	—	1	1	6	—	—	—	—	—	—	—
Malta	5	302	—	—	—	2	5	4	—	—	—	—	—	—	—
Moldova	12	50,863	5	—	—	—	8	7	—	—	—	—	—	—	—
Netherlands	106	1,159,986	—	—	—	68	80	66	—	63	65	49	—	74	3
Norway	52	930,657	15	—	2	40	41	34	—	—	—	—	—	—	—
Svalbard	14	3,484,190	—	—	3	12	12	9	—	—	—	—	—	—	—
Poland	77	1,986,677	38	—	1	33	33	55	—	13	—	—	—	—	—
Portugal	34	942,070	11	—	—	9	13	24	11	—	7	5	—	34	—
Azores	15	12,805	—	4	—	14	15	14	—	15	—	—	—	11	—
Madeira	11	13,282	6	8	—	4	4	11	6	9	—	—	—	11	—
Republic of Ireland	140	421,660	4	—	—	29	76	83	4	47	54	18	—	37	44
Romania	44	655,727	24	—	—	11	18	37	—	—	—	—	—	—	—
Russia	218	17,451,921	152	2	25	102	126	144	—	—	—	—	—	—	—
Slovakia	32	1,216,737	23	—	1	6	6	29	—	—	—	—	—	—	—
Slovenia	14	341,280	6	—	1	2	2	9	—	—	—	—	—	—	—
Spain [2]	326	15,768,900	146	11	12	131	211	335	146	255	61	—	12	345	4
Sweden	63	1,267,809	9	—	5	30	44	45	9	39	35	3	3	40	1
Switzerland	31	547,390	—	—	14	10	13	21	—	—	—	—	—	—	—
Turkey	97	2,997,846	56	—	1	62	70	71	—	—	—	—	—	—	—
Ukraine	102	2,032,251	49	—	2	38	61	76	—	—	—	—	—	—	—
United Kingdom	295	3,146,767	21	—	—	132	168	193	21	85	145	89	—	187	18
Yugoslavia	40	861,660	16	—	3	10	14	39	—	—	—	—	—	—	—
Grand Total	**3,619**	**93,167,728**	**968**	**47**	**155**	**1,364**	**1,858**	**2,399**	**396**	**918**	**699**	**342**	**71**	**1,539**	**225**

An IBA may qualify under more than one criterion.
For definitions of criteria, see the introductory chapter 'Identifying IBAs'.
1. Total IBA area for Denmark includes several very large marine IBAs within the Exclusive Economic Zone.
2. Includes the Canary Islands.

Appendix 7. Transboundary Important Bird Areas.

The following text lists, for each country that has one or more transboundary sites, the number of such IBAs in the country, the name of the neighbouring country/countries with which it shares its transboundary site(s), followed by the codes for the native sites, each accompanied by the code(s) of the adjoining site(s) in the neighbouring country/countries.

For example,
"Ireland (3)
United Kingdom: IE004/UK278; IE027/UK282; IE122/UK274..."
indicates that, in Ireland, there are three transboundary sites: Irish site 004 adjoins UK site 278, Irish site 027 adjoins UK site 282, and Irish site 122 adjoins UK site 274.

This is a working list rather than a definitive set of sites, having been compiled at the end of the project, from the information immediately available to the editors (in terms of site descriptions, coordinates and site areas), without recourse to the original data-providers to confirm or check the exact relationships among all potential sites. It is often not clear whether the IBAs actually meet across the relevant border(s), or whether they are just very close but do not form an integral management unit. Thus the list provides an initial picture of the potential extent of transboundary IBAs within Europe, and will be refined with time.

In summary, there are approximately 207 transboundary IBAs in Europe (c.6% of the total number of IBAs in Europe). The protection and management of these areas will often require a higher degree of international cooperation than that already required for IBA conservation in general.

Albania (4)
Greece: AL003/GR047; AL004/GR047.
Macedonia: AL003/MK006.
Yugoslavia: AL001/YU038; AL013/YU039/YU040.

Andorra (1)
France: AD001/FR205.
Spain: AD001/ES133; AD001/ES134.

Austria (14)
Czech Republic: AT010/CZ012; AT026/CZ008; AT032/CZ005.
Germany: AT032/DE259; AT037/DE282.
Hungary: AT001/HU002; AT002/HU001; AT003/HU003; AT004/HU003.
Italy: AT039/IT042; AT046/IT044; AT048/IT044.
Lichtenstein: AT054/LI001.
Slovakia: AT002/SK003; AT010/SK001.
Slovenia: AT043/SI014; AT046/SI001; AT048/SI001.
Switzerland: AT052/CH025.

Azerbaijan (5)
Georgia: AZ003/GE011; AZ007/GE011.
Iran: AZ052/IR013/IR014.
Russia: AZ020/RU175; AZ023/RU173.

Belarus (2)
Poland: BY009/PL031.
Ukraine: BY018/UA008.

Belgium (4)
France: BE034/FR067.
Luxembourg: BE045/LU001.
Netherlands: BE006/NL033; BE013/NL105.

Bosnia-Herzegovina (2)
Croatia: BA001/HR023; BA003/HR008.

Bulgaria (2)
Romania: BG033/RO039.
Yugoslavia: BG002/YU029.

Croatia (5)
Bosnia-Herzegovina: HR008/BA003; HR023/BA001.
Hungary: HR001/HU009; HR011/HU007.
Yugoslavia: HR001/YU001; HR002/YU001.

Czech Republic (3)
Austria: CZ005/AT032; CZ008/AT026; CZ012/AT010.
Germany: CZ005/DE259.

Denmark (6)
Germany: DK064/DE004; DK120/[no site]; DK122/[no site]; DK123/[no site].
Norway: DK121/[no site].
Sweden: DK119/[no site]; DK121/[no site].

Estonia (3)
Latvia: EE017/LV014; EE031/LV036.
Russia: EE040/RU050.

France (17)
Andorra: FR205/AD001.
Belgium: FR067/BE034.
Germany: FR090/DE225; FR091/DE223.
Italy: FR169/IT008; FR261/IT035.
Spain: FR184/ES123; FR192/ES084; FR195/ES037; FR200/ES085; FR202/ES128; FR203/ES126; FR204/ES130; FR222/ES136; FR223/ES134.
Switzerland: FR095/CH001; FR170/CH007–009.

Georgia (3)
Azerbaijan: GE011/AZ003/AZ007.
Russia: GE001/RU163; GE002/RU163.

Germany (14)
Austria: DE259/AT032; DE282/AT037.
Czech Republic: DE259/CZ005.
Denmark: DE004/DK064; [no site]/DK120; [no site]/DK122; [no site]/DK123.
France: DE223/FR091; DE225/FR090.
Netherlands: DE058/NL017/NL018; DE063/NL017/NL018; DE080/NL084; DE154/NL104.
Poland: DE046/PL002; DE142/PL006.
Switzerland: DE227/CH015; DE231/CH014.

Greece (4) *Albania*: GR047/AL003/AL004.
Macedonia: GR023/MK010; GR038/MK009; GR047/MK006.
Turkey: GR006/TR001.

Hungary (10)
Austria: HU001/AT002; HU002/AT001; HU003/AT003/AT004.
Croatia: HU007/HR011; HU009/HR001.
Romania: HU030/RO025.
Slovakia: HU001/SK003; HU016/SK007; HU018/SK010; HU042/SK026; HU043/SK025.
Yugoslavia: HU009/YU001.

Ireland (3)
United Kingdom: IE004/UK278; IE027/UK282; IE122/UK274.

Italy (7)
Austria: IT042/AT039; IT044/AT046; IT044/AT048.
France: IT008/FR169; IT035/FR261.
Slovenia: IT044/SI001; IT066/SI005.
Switzerland: IT003/CH026; IT041/CH025.

Latvia (4)
Estonia: LV014/EE017; LV036/EE031.
Lithuania: LV005/LT001; LV043/LT016.

Liechtenstein (1)
Austria: LI001/AT054.

Appendix 7 ... continued. Transboundary Important Bird Areas.

Lithuania (4)
Latvia: LT001/LV005; LT016/LV043.
Russia: LT006/RU002; LT015/RU001.

Luxembourg (1)
Belgium: LU001/BE045.

Macedonia (4)
Albania: MK006/AL003.
Greece: MK006/GR047; MK009/GR038; MK010/GR023.
Yugoslavia: MK001/YU024.

Moldova (1)
Romania: MD002/RO015.

Netherlands (6)
Belgium: NL033/BE006; NL105/BE013.
Germany: NL017/DE058/DE063; NL018/DE058/DE063; NL084/DE080; NL104/DE154.

Norway (0)
Denmark: [no site]/DK121.

Poland (6)
Belarus: PL031/BY009.
Germany: PL002/DE046; PL006/DE142.
Slovakia: PL072/SK016; PL077/SK027.
Ukraine: PL067/UA002.

Portugal (5)
Spain: PT005/ES064; PT013/ES292; PT019/ES290; PT027/ES266/ES273; PT034/ES263.

Romania (5)
Bulgaria: RO039/BG033.
Hungary: RO025/HU030.
Moldova: RO015/MD002.
Ukraine: RO002/UA081.
Yugoslavia: RO032/YU034.

Russia (6)
Azerbaijan: RU173/AZ023; RU175/AZ020.
Estonia: RU050/EE040.
Georgia: RU163/GE001/GE002.
Lithuania: RU001/LT015; RU002/LT006.

Slovakia (9)
Austria: SK001/AT010; SK003/AT002.
Hungary: SK003/HU001; SK007/HU016; SK010/HU018; SK025/HU043; SK026/HU042.
Poland: SK016/PL072; SK027/PL077.
Ukraine: SK031/UA051

Slovenia (3)
Austria: SI001/AT046/AT048; SI014/AT043.
Italy: SI001/IT044; SI005/IT066.

Spain (16)
Andorra: ES133/AD001; ES134/AD001.
France: ES037/FR195; ES084/FR192; ES085/FR200; ES123/FR184; ES126/FR203; ES128/FR202; ES130/FR204; ES134/FR223; ES136/FR222.
Portugal: ES064/PT005; ES263/PT034; ES266/PT027; ES273/PT027; ES290/PT019; ES292/PT013.

Sweden (0)
Denmark: [no site]/DK119; [no site]/DK121.

Switzerland (8)
Austria: CH025/AT052.
France: CH001/FR095; CH007/FR170; CH008/FR170; CH009/FR170.
Germany: CH014/DE231; CH015/DE227.
Italy: CH025/IT041; CH026/IT003.

Turkey (1)
Greece: TR001/GR006.

Ukraine (4)
Belarus: UA008/BY018.
Poland: UA002/PL067.
Romania: UA081/RO002.
Slovakia: UA051/SK031.

United Kingdom (3)
Ireland: UK274/IE122; UK278/IE004; UK282/IE027.

Yugoslavia (7)
Albania: YU038/AL001; YU039/AL013; YU040/AL013.
Bulgaria: YU029/BG002.
Croatia: YU001/HR001/HR002.
Hungary: YU001/HU009.
Macedonia: YU024/MK001.
Romania: YU034/RO032.

Appendix 8. The number of Important Bird Areas in each country that overlap with Ramsar Sites, World Heritage Sites, Biosphere Reserves and Special Protection Areas.

For details of the international agreements under which these protected-area types have been designated, see Box 2 in the chapter 'Overview of results' and also Appendix 1. Details of the extent of overlap between these protected areas and individual IBAs, and the names of the protected areas concerned, are given in the relevant country chapters.

Country	Total number of IBAs in country	Number of IBAs overlapping (partly or wholly) with: Ramsar Site	World Heritage Site	Biosphere Reserve	Special Protection Area
Albania	15	1	—	—	—
Andorra	1	—	—	—	—
Armenia	5	2	—	—	—
Austria	55	8	—	—	38
Azerbaijan	52	1	—	—	—
Belarus	21	—	1	2	—
Belgium	48	6	—	—	36
Bosnia and Herzegovina	3	—	—	—	—
Bulgaria	50	5	1	5	—
Croatia	23	4	—	—	—
Cyprus	16	—	—	—	—
Czech Republic	16	7	—	5	—
Denmark	127	43	—	—	113
Faroe Islands	19	—	—	—	—
Greenland	55	10	—	8	—
Estonia	43	10	—	20	—
Finland	96	11	—	2	67
France	277	22	—	6	95
Georgia	11	1	—	—	—
Germany	285	29	—	12	149
Gibraltar	2	—	—	—	—
Greece	196	12	2	2	56
Hungary	43	17	1	4	—
Iceland	61	3	—	—	—
Italy	192	27	—	—	82
Latvia	58	3	—	6	—
Liechtenstein	2	1	—	—	—
Lithuania	35	5	—	—	—
Luxembourg	9	1	—	—	6
FYR of Macedonia	10	—	—	—	—
Malta	5	—	—	—	—
Moldova	12	—	—	—	—
Netherlands	106	62	—	1	104
Norway	52	3	—	—	—
Svalbard	14	3	—	1	—
Poland	77	7	1	4	—
Portugal	34	8	—	1	27
Azores	15	—	—	—	9
Madeira	11	—	—	—	2
Republic of Ireland	140	36	—	2	97
Romania	44	1	1	5	—
Russia	218	18	2	9	—
Slovakia	32	7	1	4	—
Slovenia	14	1	—	—	—
Spain [1]	391	29	3	12	170
Sweden	63	29	2	—	36
Switzerland	31	6	—	—	—
Turkey	97	9	—	—	—
Ukraine	102	18	—	3	—
United Kingdom	295	86	1	3	173
Yugoslavia	40	4	1	—	—
Total	**3,619**	**556**	**17**	**117**	**1,260**

1. Includes the Canary Islands.

INDEX TO SITES

A

Abodi 561
Abruzzo National Park 406
Absheron archipelago (north) 102
Absheron peninsula, other water bodies of 103
Acheloos estuary 302
Acheloos valley 299
Acheron delta 298
Acheron gorge 298
Acıgöl lake 666
Adamello 382
Adda mouth 377
Adda river 389
Adda Sud Park heronries 375
Adelphi forest 155
Aetoliko lagoon 302
Ag. Fokas islets 309
Ag. Vasiliou 285
Aggtelek karst 354
Agia Dinati mountains of Kefallinia 300
Agioi Theodori islet 326
Agios Efstratios island 314
Agios Mamas marsh 285
Agra artificial lake 287
Agrafa 305
Aguas Negras mountains 638
Agudo 616
Águeda valley 451
Aguirre 638
Agulo, step rocks of 642
Ağyatan lake 683
Aisimis forest 274
Akamas peninsula 154
Akarnanika mountains 302
Akdağ 664
Akhna Dam 152
Akkuş island 676
Akrotiri Cliffs 153
Akrotiri salt-lake 152
Akşehir lake 668
Akstafa-chai valley 94
Aktaş lake 685
Aktashskoe lake 722
Akyatan lake 682
Aladağlar 682
Alange 615
Alazani river valley 94
Albarracín mountain range 565
Alborán island 599
Albufera de Mallorca 626
Albufera de Valencia marshes 582
Albufereta de Pollença marshes 626
Alburni mountains 411
Alcañiz salt-lakes 566
Alcántara reservoir 621
Alcarama mountain 551
Alcarria de Alcalá steppes 559
Alcázar de San Juan 592
Alcubierre mountain range 569
Alcudia mountain range 595
Alcudia valley 595
Aldea del Cano 619
Aldeaquemada 596
Alfés steppes 577
Algeti 255
Alghero 423
Alhama river 551
Alhamilla mountain range 597
Aliakmon estuary 283
Aliakmon gorge 288
Alicante islets 584
Aliken 670
Aliki Psilis Ammou lagoon, Samos 318
Aljibe mountain range 605
Almansa 589
Almenara marshes 581
Almería, littoral islets of 586
Almería, wetlands of western 599
Almorchón 615
Almyrou delta 326
Alokteri 308
Alonissos, National Marine Park of 294
Alpi Apuane 380
Altaona mountains 585
Alter do Chão 455
Altıntaş plain 669
Alto Garda Bresciano 385
Alto Lozoya 559
Altopiano di Abbasanta 423
Altopiano di Campeda 423
Altos de Barahona high moor 560
Alty Agach area 100
Alvito 458
Alyki 323
Alyki Angelochoriou lagoon 284
Alyki Kitros lagoon 288
Alyki lagoon, Aegion 303
Amara lake 499
Amorgos island 320
Ampurdán marshes 575
Amvrakikos Gulf 299
Anafi island 321
Anaga mountains 637
Anaga rocky islets 637
Anapo gorges 420
Ancares mountains 540
Andújar mountain range 595
Anse de Fouras 198
Anse du Fiers d'Ars en Ré 199
Anse du Poulmic 172
Ansó 571
Antequera mountain range 600
Anthofito, area of 283
Antigorio valley 369
Antikhassia mountains 291
Antikithira island 313
Antimilos island 319
Antiparos island 319
Antipsara islands 316
Antrona valley 369
Anzasca valley 369
Apennines, from Passo del Cerrato to Monte Caligi 380
Apsalos gorge 287
Arad 495
Aranjuez, reedbeds and copses 558
Arcena mountain range 547
Archipel de Molène 171
Archipel des Glénan 172
Archipel des Sept-Iles 170
Archuba mountain range 562
Arcos mountain range 565
Ardahan forest 685
Ardas riparian forest 274
Areta mountain range 562
Arezzo heathlands 394
Argentario 401
Arico pine woodland 641
Arinaga and Castillo del Romeral, coast between 636
Arjuzanx 222
Arlanza valley 552
Armash fish-farm 85
Arnedillo 551
Aroania 303
Arribes del Duero 555
Arrière Côte de Dijon et de Beaune 207
Arroyo de la Luz 619
Artem bay 102
Artzan marshes 282
Asinara island 421
Askania-Nova Biosphere Reserve 715
Asprokremmos Dam 154
Aspromonte 416
Assa valley 385
Astara-chai valley 108
Astypalea island and nearby islets 322
Asvestades 274
Atanasovo lake 128
Athamanika mountains 298
Atokos 300
Aurunci mountains 407
Ausoni mountains 407
Axios estuary 283
Ayamonte marshes 611
Aydıncık islands 681
Ayllón mountain range 559
Azuaga 613
Azután reservoir 594

B

Babadag forest 489
Babia 542
Babuna gorge 432
Bafa lake 664
Bagerovo 723
Baie d'Authie 186
Baie de Bourgneuf 176
Baie de Carantec 173
Baie de Daoulas 172
Baie de Fontarabie 221
Baie de Goulven 175
Baie de l'Aiguillon 179
Baie de Morlaix 173
Baie de Quiberon 175
Baie de Saint-Brieuc 170
Baie de Somme 186
Baie de Vilaine 174
Baie des Veys 179
Baie du Mont Saint Michel 181
Baie d'Yves 198
Bajo Alcanadre 570
Bakhtin river valley 710
Balatina 440
Balık lake 686
Balıkdamı 670
Ballobar 568
Balta Albă lake 499
Banc d'Arguin 217
Barayo beach 540
Barbas mountains 303
Barda tugai forest 101
Bardaca 110
Bardenas Reales 563
Bârnova 491
Barois et Forêt de Clairvaux 189
Barrage d'Artix 220
Barrancos 458
Barthes de l'Adour 217
Basse Ardêche 213
Basse vallée de la Durance 238
Basse vallée de L'Indre 204
Basse vallée du Doubs: Dole Sud 197
Basse vallée du Var 240
Bassée et plaines adjacentes 183
Basses Corbières 227
Basses Vallées Angevines 177
Bassigny 190
Bassin d'Arcachon 217
Bassin du Drugeon: Pontarlier-Frasne 195
Bazinul Costeşti 440
Bazinul Cuciurgan 442
Bazinul Dubăsari 440
Bazinul Ghidighici 441
Bazinul Taraclia 442
Bazoncourt 192
Beauce centrale 202
Bec de Crigne 240
Beceite, pass of 578
Bečej fish-pond 731
Béda-Karapancsa 343
Belagua 571
Belchite 566
Belene islands complex 122
Beleu 442
Belluno Dolomites 383
Belmonte de Calatayud plains 564
Belogorie 721
Belosarayskaya peninsula 717
Belver de los Montes 549
Berdyanskaya peninsula 717
Beret 573
Berlenga island 454
Bermeja mountain range 605
Betancuria 634
Beynam forest 675
Beyşehir lake 679
Bicaz gorge 491
Bidasoa estuary 548
Bielsa 572
Bienvenida 614
Bihar plain 350

Biharugra fish-ponds 349
Bilshivtci village, fish ponds near 712
Biogradska woods 742
Bishop's Pool 152
Bistreţ 497
Biviere 420
Biviere di Lentini 419
Black Sea coast 489
Blancas 564
Blanquilla mountain range 606
Bledes, islet of 623
Bodakovskoe marsh 707
Bodrog flood-plain 352
Bódva valley 354
Bohí 573
Bois de Malvoisin 241
Bois de Palayson 241
Bois des Ayes 239
Bois du Chapitre 240
Bois du Défens 200
Bois du Roi 185
Bois du Rouet 241
Bolgheri 396
Bolluk lake 680
Bolshemikhailovskiy forest 709
Bondar marsh 704
Bonifica del Mezzano 391
Bonne Anse 199
Boracko jezero 110
Borbollón reservoir 621
Bornos reservoir 608
Borsodi-Mezőség 352
Börzsöny mountains 346
Bosa and Alghero, coast between 423
Bosava region 434
Bosco Fontana 388
Boscone island 389
Bosphorus 660
Bosutska forest 732
Botoa 618
Boucle de Moisson 182
Boucle de Pose et de Muids 182
Boumort mountains 574
Brabbia peatland 372
Brăila, the Little Island of 499
Brenne centrale 203
Brenne Sud 203
Brenne: Forêt de la Petite-Brenne 203
Brenne: Forêt de Lancosme 204
Brenta 382
Brozas 619
Budakskiy estuary 720
Buendía reservoir 591
Bugio 479
Bujeo mountain range 606
Bükk mountains 353
Bulanık plain 685
Burano lake 401
Burdur lake 667
Burete mountain range 587
Burgasko lake 128
Burshtyn reservoir 712
Buskett 438
Büyük Menderes delta 665
Büyükçekmece lake 659
Byala Reka 123

C

Cabañeros 594
Cabariot-Pons 197
Cabeção 455
Cabeceira da Ribeira do Seixal 477
Cabeza del Buey 615
Cabo de Gata, mountain range at 598
Cabo de Gata, saltpans at 598
Cabras mountain range 605
Cabrera, archipelago of 628
Cabriel river, canyons on the middle section of the 590
Cáceres and Trujillo, plain between 619
Cadí mountains 575
Cádiz bay 607
Cagitán plains 588
Cagliari wetlands 426
Cala Figuera 627
Cala Pí 627
Caldeirão Verde 477
Çalı lake 686
Calseraigne 236
Calvana mountains 394
Camargue 234
Camarma 559
Cambrón mountain range 587
Camonica valley 372
Campillos lake 604
Campiña 610
Campo Arañuelo 623
Campo de Argañán 556
Campo de Calatrava 595
Campo de Montiel 589
Campo de Níjar plain 597
Campo dei Fiori 371
Campo d'Ozieri 422
Campo Maior 455
Campo Visiedo 565
Canalizos mountain range 595
Canciás 570
Candasnos 568
Candelario mountain range 556
Cantabric mountains, central ranges of the 543
Cantalejo 553
Caorle lagoon 386
Cap Fagnet 182
Cap Frehel 174
Cap Gris-nez 187
Cap Sizun 172
Cape Aspro 153
Cape Barbaria 625
Cape Boi and Cape Ferrato, coast between 426
Cape Busto 543
Cape Caccia 422
Cape Candelaria 540
Cape Cope 586
Cape Enderrocat and Cala Pí, coastal cliffs between 627
Cape Estaca de Bares 540
Cape Farrutx 626
Cape Ferrato 426
Cape Feto 419
Cape Freu 626
Cape Greco 151
Cape Louros 323
Cape Martyan 722
Cape Nonó 624
Cape Ogoño 548
Cape Otranto 415
Cape Pinar 626
Cape San Marco 425
Cape San Vito 417
Cape Tenaro 311
Cape Testa 421
Cape Teulada and Pula, coast between 427
Cape Tiñoso 586
Cape Uret 721
Cape Vermell 627
Capri island 410
Caracena 560
Cardó mountains 578
Cârja 490
Carmona countryside 603
Carniche Alps 381
Carniche pre-Alps 382
Carpathian Biosphere Reserve 711
Carpathian Volkano mountains 711
Carrión 550
Carska bara 734
Carso 389
Casaraccio pool 422
Castelluzzo 419
Castillo del Romeral 636
Castrejón reservoir 593
Castro Marim 460
Castro Verde plains 459
Castronuño 554
Çavuşçu lake 669
Cazorla mountain range 596
Cebollera mountain range 551
Cedillo reservoir 619
Cefa fish-ponds 494
Cenad 495
Centinela mountain 641
Central Andros island 318
Central Badajoz mountain ranges 613
Central Balkan 120
Central Campidano 423
Central Rodopi mountains 279
Central Rodos 324
Cer mountain 735
Cerdaña 574
Cerrato high moor 550
Cervia saltpans 393
Ceuta 606
Cévennes, Parc National des 233
Chafarinas Islands 599
Chaîne des Alpilles 235
Chaîne des Côtes 237
Chalki island and nearby islets 324
Champ de Tir du Poteau 217
Chanoine 236
Chaykovichi marsh 705
Chengene Skele 131
Cheremske mire 701
Chinche lake 602
Chingudi 548
Chisone valley 375
Chituc 489
Christiana islets 321
Cicolano 402
Cíes islands 538
Çıldır lake 685
Cijara reservoir 616
Cika mountain 73
Cinca river rice-fields and steppe areas 568
Cinco Villas ponds 563
Cintrão 470
Ciocăneşti fish-farm 498
Circeo National Park 406
Cirque de Gavarnie 223
Cirque de Navacelles 228
Cividalese 383
Côa valley 452
Codigoro heronry 390
Codrii 441
Coghinas river mouth to Cape Testa, coast from 421
Cogul–Alfrés steppes 577
Col d'Aoueran 224
Col d'Artisagou 224
Col de l'Escrinet 213
Col de Lindux 221
Col de Lizarrieta 220
Çöl lake 670
Colfiorito 398
Collarada 572
Colostrai pools 425
Columbretes islands 580
Comino island 437
Compiègne forêts picardes 184
Cona island 387
Condado 610
Conde lake 602
Conejera island 545
Confluent des vallées de la Meuse etbde la Chiers 190
Congloue 236
Copanca 442
Çorak lake 666
Cordel mountain range 544
Córdoba countryside 602
Córdoba, mountain ranges of 601
Córdoba, wetlands at south 604
Cordon lagunaire de Sète à Agde 230
Coria mountain ranges 621
Cornalvo reservoir 617
Corral de Almaguer 591
Corralejo and Tostón, coast between 635
Corralejo, sandy plain of 632
Cortados del Jarama 558
Corvo, coast of 467
Costa de la Muerte (north coast) 539
Costa Viola 415
Côte de Tamarone à Centuri 244
Côte Médocaine: dunes boisées et dépressions humides 223
Cotiella 572
Cotillo 635
Coto mountain range 542
Coura estuary 450
Crau 234
Crava 378
Cres island, Special Ornithological Reserves on 143
Crestellina mountain range 605
Crete, north-eastern edge of 330
Crete Senesi 397
Crimean Nature Reserve 722
Crn Kamen 432
Cuatro Lugares 621
Cuchillete de Buenavista 632
Cuenca mountain ranges 590
Cuevas Negras, gully of 640
Cuglieri coast 424
Cuvia valley 371

D

Dadia, protected area of 275
Dadias 274
Dañador 596
Danube bend 345

Danube delta 487
Danube loess bluffs 733
Danube plain 347
Danube reach between Gönyű and Szob 345
Daunia mountains 408
De la Plata mountain range 607
Debalcevskoe lake 708
Degaña 542
del Niño mountain range 606
Deliblatska peščara 735
Delimidi mountains, Agrafa 305
Demir kapija gorge 434
Demirköprü reservoir 662
Derazhnya village 705
Đerdap gorge 741
Dereiou 274
Derkul river valley 712
Derventski Vuzvishenia 125
Deserta Grande 479
Deserta Islands 479
Détroit de Bonifaccio 241
Dévaványa plain 349
Dgulin lake 713
Dia island 329
Diaccia Botrona 399
Diapontia islands 299
Diapori fen 314
Dinara mountain 144
Dinnyés marshes 344
Dionisiades islands 329
Dipi marsh, Lesvos 315
Dirfi 308
Divari Pilou lagoon 310
Divichi liman 99
Dnepr and Desna valley lowlands, forests in 704
Dneprovsko-Orelskiy Nature Reserve 707
Dneprovsko-Teterevskoe Game Reserve 702
Dnestr delta 720
Dnestr valley between Staruy Martuniv and Marinop villages 712
Dobron forest 492
Doğubeyazıt marshes 686
Dolni Bogrov 118
Dolomites 381
Dolomiti di Pietrapertosa 412
Domaine d'Orx 221
Domogled mountain 496
Don Benito 617
Donji Miholjac fish-ponds 140
Dora Baltea mouth 376
Dos Reinos lagoons 563
Dos Ríos mountain range 571
Dragonera island 625
Dranitckiy Protected Locality 713
Drini delta 73
Dubašnica 740
Dubovac 740
Duchessa mountains 404
Dueñas 550
Durankulak lake 134
Duratón canyon 552
Durmitor 741
Dzharilgatchskiy Bay 714

E

East Gata mountain range 622
East Lakonia mountains 311
East Rodopi mountains 277
Eastern Caucasus 253
Eber lake 668
Ebro delta 579
Echauri crags 561
Echinades islands 300
Écija 604
Ecrins, Parc National des 241
Egadi islands 417
Eğirdir lake 668
El Canal y Los Tiles laurel forest 644
El Chorro 600
El Escorial 557
El Hierro, western coast of 646
El Hito 591
El Hondo wetland 584
El Médano coast 641
El Mojón 630
El Moro marshes 581
El Pardo 557
El Roque coastal cliffs 645
El Valle mountains 585
El Viso 602
Elanchik river valley 716
Elassona area 291
Embid–Molina high moors 590
Embouchure du Tech 231
Emine 131
Enguera mountain range 583
Entrepeñas reservoir 591
Eo river estuary 540
Eolie islands 416
Epanomi lagoon 284
Episkopi Cliffs 153
Erçek lake 687
Ereğli marshes 678
Erikoussa 299
Ernici mountains 406
Erzurum plain 684
Es Canar, islet of 624
Escalona mountain 585
Esla dam 549
Eşmekaya marshes 678
Espadán mountain range 580
Espartar, islet of 623
Espera lagoon 609
Espés 573
Espuña mountain range 587
Esquinzo, coast of 635
Estuaire de la Bidassoa 221
Estuaire de la Canche 186
Estuaire de la Charente 197
Estuaire de la Gironde: marais de la rive nord 201
Estuaire de la Gironde: marais du Blayais 222
Estuaire de la Gironde: marais du Nord Médoc 222
Estuaire de la Loire 176
Estuaire de l'Orne 180
Estuaire du Jaudy 171
Estuaire du Trieux 171
Estuaire et embouchure de la Seine 182
Estuaires picards: baies de Somme et d'Authie 186
Etang de Biguglia 243
Etang de Capestang 227
Étang de Combourg 200
Etang de Fos 238
Etang de Lapalme 226
Etang de Lavalduc 238
Etang de Lespignan 231
Etang de Leucate 226
Etang de Lindre 191
Etang de Pissevache 231
Etang de Pourra 238
Etang de Thau 230
Etang de Villeneuve-de-la-Raho 231
Etang d'Engrenier 238
Etang des Landes 208
Etang des Loges 203
Etang d'Estomac 238
Etang du Bagnas 230
Etang d'Urbino, et zones humides périphériques 243
Etang et Forêt de Villefermoy 183
Etangs d'Argonne 189
Etangs de Canet 231
Etangs de Citis 238
Etangs de la Woëvre: la Chaussée 191
Etangs de la Woëvre: lac de Madine 191
Etangs de Sologne: étang de L'Arche 205
Etangs de Sologne: St Viatre, Marcilly-en-Gault et forêt de Bruadan 205
Etangs de Vendres 231
Etangs et marais du bassin de la Somme 184
Etangs Montpellierains 228
Etangs Narbonnais 227
Eth Turoun des Aureys 219
Etier de Penerf 173
Evinos estuary 302
Évora plains 458
Evros delta 276
Evros prefecture, south forest complex of 275
Evrota delta 311
Ezquinzo 635

F

Fage-Cagnasses 232
Faial, coast of 467
Fakos peninsula 314
Falaise du Bessin 179
Falkonera islet 313
Famara crag 629
Famara sandy plain 630
Fanal 477
Farilhões islands 454
Farma valley 397
Faro lagoon 460
Farsala area 294
Fausilla mountains 586
Fermoselle 555
Ferro valley 414
Ferrolterra–Valdoviño coast 539
Filfla islet 438
Filiouri river valley 277
Finalese 379
Fizeş valley 493
Florence plain, pools of 395
Flores, coast of 467
Flumendosa pools 425
Foça islands 662
Folegandros island group 320
Folhadal 477
Foreste Casentinesi 394
Forêt de Bruadan 205
Forêt de Châtillon et environs 207
Forêt de Chaux 196
Forêt de Citeaux et environs 207
Forêt de Haguenau 193
Forêt de Jugny 207
Forêt de la Reine 193
Forêt de Moulière 200
Forêt de Saou 213
Forêt des Arbailles 218
Forêt des Coulmes 213
Forêt domaniale de la Harth 193
Forêt d'Irraty 218
Forêt d'Orion 223
Forêt d'Orléans: massifs d'Ingrannes et de Lorris 206
Forêt du Gavre 178
Forêts d'altitude du Jura: forêt du Risol et Mont d'Or 196
Forêts d'altitude du Jura: forêt du Massacre 196
Forêts d'altitude du Jura: forêt du Risoux 196
Forêts de Thiérache 188
Forêts domaniales de Corse 242
Forêts du Perche 181
Forêts picardes: Compiègne, Laigue, Ourscamp 184
Forêts picardes: massif de Retz 184
Forêts picardes: massif de Saint-Gobain 185
Forêts picardes: massif des Trois Forêts et bois du Rois 185
Formazza valley 369
Formentera 624
Fourmies 188
Fourni islands 317
Frencq, plateaux agricoles des environs de 187
Frentani mountains 404
Fresnes-en-Woëvre 192
Fresser 575
Freus isles 624
Frómista 550
Frontera laurel forest 646
Fruška Gora 732
Fteri mountains, Agrafa 305
Fucecchio marsh 394
Fuente de Cantos 613
Fuente de Piedra lake 604
Fuentelapeña 554
Fuenteovejuna 603
Fuentes Carrionas 544
Funiyata 119
Furlo gorge 396

G

Gabriel y Galán reservoir 622
Gádor mountain range 598
Galaxidi 307
Gallegos del Pan 549
Gallikos estuary 284
Gallocanta lake 564
Gamila 295
Garachico rocky islet 639
Garafía rocky islets 644
Garajonay National Park 643
Gargano promontory 409
Gârla Mare 497
Garonne: marais de Bordeaux 221
Garrigues de Lancon 237
Gasset reservoir 592
Gavdopoula island 327

Gavdos island 327
Gavur lake 683
Gediz delta 663
Gekchai Bozdag mountains 96
Gemenc 343
Gennargentu mountains 424
Georgioupolis beach 326
Gera gulf, Lesvos 315
Gerecse mountains 345
Gerês mountains 450
Geropotamos estuary 328
Gialova 310
Gibina 512
Gigante mountain range 587
Girkan forest 108
Gistaín 572
Gistreo mountain range 542
Giulie pre-Alps 384
Glynanyi island 104
Gobustan area 103
Göksu delta 681
Golf Goeni 441
Golfe de Galeria 242
Golfe de Porto 242
Golfe du Morbihan 173
Goloşnita 440
Gorgany 711
Gorges de la Dordogne 208
Gorges de la Dourbie et causses avoisinants 226
Gorges de la Frau et Belestat 225
Gorges de la Sioule 209
Gorges de la Truyère 211
Gorges de la Vis 228
Gorges d'Holzarte et d'Olhadubi 218
Gorges du Gardon 229
Gorges du Rieutord 232
Gorges du Tarn et de la Jonte 225
Goričko 512
Gormou valley 296
Gorni Tzibar, island near 119
Gornje Podunavlje 730
Goro lagoon 390
Gorramendi mountain 561
Gosque lake 604
Graciosa, coast and islets of 468
Grado lagoon 387
Gramvouses 325
Gran Paradiso 371
Gran Sasso d'Italia 402
Granada, prelittoral mountain ranges of 600
Grand et Petit Mez 203
Gratal 570
Gravine 412
Grazalema mountain range 605
Gredos mountain range 556
Grigne 371
Grudnjak fish-ponds 140
Gruia 497
Guadalentín saltmarshes 587
Guadalquivir marshes 610
Guadarrama mountain range, northern slope of 553
Guara mountain range 570
Guareña 617
Guaza mountain 642
Guernica estuary 548
Güllük delta 665
Gully of Ajuí 634
Gumista 252
Gusari area 99

H

Haçlı lake 684
Hagieni forest 490
Hamam Dağı 672
Hanság 340
Haramida marsh, Lesvos 315
Hardt, zones agricoles de la 194
Haría 629
Haut Giffre 216
Haut Val d'Allier 209
Haute Chaîne du Jura: défilé de l'écluse, Etournel et Mont Vuache 215
Haute Cize: Pic d'Errozate et Forêt d'Orion 223
Haute Soule: Forêt des Arbailles 218
Haute Soule: Forêt d'Irraty, Organbidexta et Pic des Escaliers 218
Haute Soule: massif de la Pierre-St-Martin 220
Haute Soule: Massif Forestier, Gorges d'Holzarte et d'Olhadubi 218
Hautes Corbières 227
Hautes garrigues du Montpellierais 230
Hautes Vallées d'Aspe 219
Hautes Vallées d'Ossau 219
Hauts Plateaux du Vercors 213
Havre de la Sienne 180
Hecho 571
Hermigua, step rocks of 642
Hermo 542
Heves plains 351
Hinojosa del Duque 602
Hirfanlı reservoir 674
Hirson et St-Michel 188
Honda lake 602
Hortobágy 350
Hotamış marshes 679
Hoya de Baza 597
Hoya de Guadix 597
Hoya de la Yegua 631
Hoya de Tacorón 647
Huelva coastal lagoons 611
Huelva, east mountain range of 612
Humada 546
Hursovska Reka 129
Hutovo blato 110

I

Ibisha island 119
Ibiza, saltpans of 624
Idanha-a-Nova 453
Iğneada forests 659
Ikaria island 317
Ikva river valley 705
Ildır Strait islands 663
Ile de Goulmedec 171
Ile de la Platière 214
Ile des Landes 181
Ile d'Oléron 198
Ile d'Ouessant 173
Iles Cerbicale 242
Iles Chausey 180
Iles de la Colombière 171
Iles de la Nellière 171
Iles des Haches 171
Iles d'Hyères 237
Iles du Haut-Rhône 214
Iles Finocchiarola 244
Iles Lavezzi 241
Iles Marseillaises: Maire, Jarron, Jarre, Riou, Calseraigne, Congloue et Pomègues 236
Iles Saint Marcouf 180
Iles Sanguinaires 244
Ilgaz mountains 672
Ilheu Chão 479
Ilheu de Baixo 478
Ilheu de Cima 478
Ilheu de Ferro 478
Ilheu de Fora 480
Ilisu area 93
Illón mountain range 562
Ilôts de Trévorc'h 171
Inagua pine woodland 636
Ingul river valley 713
Inner Somogy 342
Inözü gorge 671
Întregalde gorge 493
Ioannina city and neighbouring area 296
Ioanninon 296
Iori region 255
Ios 320
Irati 561
Iregua, Gorge of 551
Iruelas valley 557
Isasa crag 551
Iseo peatlands 374
Işıklı lake 664
Isla Cristina marshes 611
Island of Aire 628
Island of Lobos 631
Isle of Conejera 623
Isle of Murada 624
Isles of Malgrats and Cala Figueru, coastal cliffs between 627
Ismailly area 100
Ismarida 277
Isole Pelagie 421
Isonzo mouth 387
Isthmus of Jandía sandy plain 634
Itxusi crags 561
Izaga crag 562
Izki mountains 548
Iznik lake 660
Izvoarele 496
Izyumskaya Luka forest 710

J

Jaén countryside 601
Jaén, mountain ranges south of 601
Jambrina 554
Jandía, peninsula of 633
Jansheisk lakes 719
Janubio saltpans 631
Jarre 236
Jarron 236
Javakheti plateau 254
Jazovo 733
Jegricka 731
Jelas field 141
Jerez de los Caballeros dehesas 614
Jerma 739
Jirlău lake 499
Jubera, Gorge of 551

K

Kagul lake 718
Kaimaktchalan 287
Kakhovsk reservoir 714
Kalamaki gorge 291
Kalamas estuary 297
Kalamas gorge 297
Kalamos 300
Kalavrita 303
Kalimnos, north and east 322
Kalimok complex 126
Kallonis gulf 315
Kalogria lagoon 304
Kalohori lagoon 284
Kalolimnos islands and islets 322
Kamchia complex 132
Kamchijska mountain 131
Kandili mountain 307
Kanev Nature Reserve 705
Kanev reservoir 703
Karaburun peninsula 73
Karadag Nature Reserve 722
Karamık marshes 668
Karapınar plain 678
Karataş lake 667
Karavasta lagoon 72
Karayazi forest 94
Karkinitskiy Bay 714
Karpas peninsula 151
Karpathian Sea, islets of 325
Karst 508
Kartal lake 718
Karteri marsh 298
Kasos island 325
Katavia-Prassonisi 324
Kavaklı Dağı 671
Kavalli islets 330
Kazichene 118
Kerketion mountain 290
Kerkini 282
Kerkyra, lagoons of 300
Kesznyéten 352
Khosrov Reserve 85
Kiev reservoir 703
Killini 309
Kimi, vicinity of 308
Kinaros island 322
Kinburn peninsula 713
Kintrishi 252
Kis-balaton 341
Kisköre reservoir 350
Kitaj lakes 718
Kithira Island, north, east and south 313
Kizil Agach Bay 106
Kızılcahamam forests 671
Kızılırmak delta 675
Klidhes Islands 150
Klivodin Protected Locality 712
Klokos mountains 303
Kocaçay delta 661
Kočevsko-Kolpa 511
Kofinas 328
Kokkini Rachi mountains of Kefallinia 300
Kolkheti 252
Kompsatos valley 277
Komsomolsk 706
Koncanica fish-ponds and surrounding area 141
Konush reservoir 122
Kopacki Rit Special Zoological Reserve 139

Kopaonik 737
Korab mountain 432
Korchai area 95
Kornati National Park 143
Kornos peak 151
Koronia 285
Korostushevskoe Game Reserve 702
Korotchenkov meadows 704
Kotlenska mountain 126
Kotychi lagoon 304
Koufonisi island 330
Kourtaliotiko gorge 327
Koviljski rit 731
Köyceğiz lake 666
Kozanlı Gökgöl 677
Koziakas 290
Kozuf i Bosava mountains 434
Kozuf mountain 434
Krakovski forest 510
Krasnue marsh 706
Kremenchug reservoir 706
Kremenskoy forest 710
Krioneritis mountain 327
Krivaya peninsula 717
Krk island, Special Ornithological Reserve 143
Krka National Park 144
Krumovitza 120
Küçük Menderes delta 664
Küçükçekmece lake 659
Kugurluy lake 718
Kukilniku village, fish ponds near 712
Kulu lake 677
Kura delta 106
Kuş lake 661
Kusari area 99
Kuyalnik estuary 720
Kuyucuk lake 686
Kyrenia range, between Kornos and Pentadactylos peaks 151

L

La Albuera 615
La Almenara mountain range 586
La Cabrera mountain range 541
La Caldera de Taburiente National Park 645
La Canal de Navarrés 583
La Corona, plains of 630
La Data pine woodland 636
La Dehesa grasslands 646
La Demanda mountains 550
La Dombes 212
La Esperanza 638
La Esquina plain 641
La Estación pond 569
La Goleta 638
La Gomera, south-west coast of 643
La Granjuela 603
La Guardia plains 591
La Iglesuela 593
La Janda 607
La Laguna, plain of 634
La Limia 540
La Losa 547
La Mola of Formentera 625
La Palma laurel forest 643
La Pedriza 559
La Playa islet 638
La Ropera 602
La Safor mountain range 583
La Serena 616
La Sotonera 569
La Torre, gully of 632
La Trapa 625
La Vera valley 622
La Victoria laurel forest 639
Lac de Grand-Lieu 176
Lac de Rillé et forêts voisines 205
Lac du Der-Chantecoq et étangs latéraux 189
Lac et marais du Bourget 215
Lac Léman (partie française) 215
Lácara 618
Lachin area 97
Lacs de la Forêt d'Orient 188
Lacul Salaş 441
Laga mountains 400
Lago della Penna 395
Lago di Busche 384
Lagonegrese 413
Lagorai mountains 382
Lagouvardos islet 313
Laigue forêts picardes 184
Lajares sandy plain 635
Lake Alyki 314
Lake Aggel 101
Lake Akzibir 99
Lake Alviano 400
Lake Amvrakia 301
Lake Arpi 84
Lake Balaton 341
Lake Beibugeac 488
Lake Bolsena 399
Lake Boz-Koba 101
Lake Bracciano and surrounding area 403
Lake Butrinti 74
Lake Candia 374
Lake Cerknica 510
Lake Chimaditis 289
Lake Chiusi 398
Lake Comana 498
Lake Delvinaki 296
Lake Distos 308
Lake Doirani 282
Lake Dojran 434
Lake Dunăreni 497
Lake Fehér at Gátér 348
Lake Fehér at Kardoskút 349
Lake Fertő 341
Lake Fundata 498
Lake Garda 373
Lake Gey Gel 96
Lake Hadjikabul 104
Lake Kalodiki 298
Lake Karla, reservoirs of former 293
Lake Kastoria 289
Lake Kerkini 281
Lake Khortaro 314
Lake Kolon 347
Lake Kourna 326
Lake Krasnoye 103
Lake Langada 285
Lake Lesina 409
Lake Lysimachia 302
Lake Maggiore 371
Lake Mahmud-chala 105
Lake Massaciuccoli 393
Lake Megali Prespa 71, 289
Lake Mezzola 370
Lake Mikri Prespa 71, 289
Lake Mitrikou 277
Lake Montepulciano 398
Lake Ohrid 71, 433
Lake Omodeo 423
Lake Öreg at Tata 344
Lake Pamvotida 296
Lake Petron 288
Lake Pikrolimni 283
Lake Prespa 433
Lake Psalidi 323
Lake Roşu 491
Lake Sarysu 102
Lake Sasko 742
Lake Scadar 70
Lake Sevan 86
Lake Shkodra 70
Lake Siutghiol 489
Lake Skadar 742
Lake Strachina 498
Lake Stymphalia 309
Lake Taşaul 489
Lake Tătaru 499
Lake Techirghiol 490
Lake Trasimeno 397
Lake Trichonida 302
Lake Van 687
Lake Varano 409
Lake Varese 372
Lake Vegoritis 288
Lake Velence 344
Lake Vistonis 278
Lake Viverone 373
Lake Volvi 285
Lake Vransko 144
Lake Ych-chala 106
Lake Zazaris 289
Lakes of Thrace 278
Lalzi Bay 75
Lamia marshes 304
Langa–Mara–Belmonte de Calatayud plains 564
Lanzarote islets 629
Larnaca salt-lakes 151
Las Cabezas lagoon 609
Las Cabras plain 587
Las Cañadas 596
Las Cañas lake 561
Las Contiendas mountain range 612
Las Honduras 630
Las Mercedes 638
Las Mesas plain 642
Las Villuercas mountain range 620
Las Vueltas mountains 638
Latorica river valley near Chop 711
Layna high moors 560
Lazorema 306
Le Cesine 414
Le Pinail 200
Lebrija lagoon 609
Lefka mountains 326
Lefkimi 275
Leixão da Gaivota 460
Lepini mountains 406
Lerida steppes 577
Les Bauges 216
Les Couzes Nord 211
Les Couzes Sud 211
Lessini mountains 385
Lesvos, petrified forest 314
Lesvos, south-west penisular 314
Levitha island 322
Leyre mountain range 562
Leza, Gorge of 551
Liakhvi 253
Ligurian Alps 380
Líjar mountain range 605
Liman lake-system 709
Limnos island, rocky islets of 313
Lindos 324
Litorale Romano 405
Littoral Augeron 181
Ljubljansko Moor 509
Llabería mountains 578
Llerena 613
Llobregat delta 576
Lombo Barbinhas 477
Lomovete 125
L'Orri 574
Los Alares 632
Los Blázquez 603
Los Cochinos, gully of 640
Los Gigantes coastal cliff 640
Los island 104
Los Molinos gully 634
Los Monegros (south) 567
Los Rincones mountain range 567
Los Rodeos 638
Los Serranos 582
Los Tollos lake 609
Los Vélez 597
Loudias estuary 283
Lower Dnepr 713
Lower Tiétar river 622
Luanco 543
Lucretili mountains 404
Lunca wood 495
Lura 73

M

Maçãs 451
Maddalena archipelago 421
Madjarovo 121
Madonie 419
Madrona mountain range 595
Magazinka village, lake near 721
Magredi of Pordenone 384
Maiella 404
Maire 236
Majona coast 642
Makedonika Tempi 285
Mala vrbica 741
Maladeta 573
Malagón lakes 592
Malcata mountains 452
Maliakos gulf 305
Malko Sharkovo reservoir 125
Malpartida de Cáceres 619
Malpica island 593
Mandra–Poda complex 127
Manferrara forest 412
Manfredonia Gulf wetlands 409
Manjavacas endorreic lagoons 592
Manta flood-plain 442
Mar Menor coastal lagoon 585
Mara 564
Marais arrières littoraux picards 184
Marais breton 176
Marais de Balançon et de Villiers 187

Marais de Brière 175
Marais de Brouage 198
Marais de Fontenay-le-Vicomte 183
Marais de la Baie d'Audierne 172
Marais de la Souche 185
Marais de L'Ile Vieille 238
Marais de Manteyer 239
Marais de Mazerolles 178
Marais de Rochefort 198
Marais de Sacy 185
Marais d'Itteville 183
Marais du Cotentin 179
Marais entre Crau et Grand Rhône: Meyranne, Chanoine, Plan de Bourg et Salins du Caban 236
Marais et estuaire de la Seudre 201
Marais et forêt d'Olonne 177
Marais poitevin 179
Marais salants de Noirmoutier 177
Marano lagoon 387
Marash mountains 74
Marchesato 415
Marcilly-en-Gault 205
Marengo heronry 377
Margariti marsh 298
María mountain range 597
Maribor 511
Marina di Camerota and Policastro Bussentino, coast between 413
Marina mountain ranges 583
Maritime Alps 378
Marmara lake 663
Marmolejo reservoir 602
Mars sur Allier 206
Mars-la-Tour 192
Massif d'Aston 224
Massif de la Pierre-St-Martin 220
Massif de Retz 184
Massif de Saint-Gobain 185
Massif des Albères 228
Massif des Trois Forêts 185
Massif des Vosges: crêtes de Schneeberg au Donen 194
Massif des Vosges: Hautes Vosges 195
Massif du Canigou-Carança 229
Massif du Madres 232
Massif du Petit Lubéron 236
Massif Forestier 218
Massifs de Fontainebleau et zones humides adjacentes 183
Maţa 490
Mata lagoon 585
Matese 407
Mathraki 299
Mati of Tirnavos 292
Mátra mountains 353
Mavrovouni 308
Mavrovouni mountains, Larissa 293
Mechka fish-ponds 124
Medas islands 576
Mediana 566
Medina 609
Medina lagoon 608
Megalo Livari lagoon, Istieas 307
Megalou Emvolou 284
Mehedinţi fish-ponds 496, 497
Membrío 619
Mercantour, Parc National du 240
Meriç delta 658
Mérida 618
Meropis forest 296
Mesanagros 324
Meskheti 254
Mesolongi lagoon 302
Meteora 291
Meyranne 236
Migliarino 395
Mikres Kyklades 320
Mikro Livari lagoon, Istieas 307
Mikulinetskie fish-ponds 705
Mileševka 737
Mina y Yedra 638
Minervois 231
Minho estuary 450
Minorca, north and east coasts 628
Minorca, ravines and pinewoods of central 628
Modia plateau 326
Mogan lake 671
Moglenitsa gorge 287
Mogrichskiy forest 704
Mokrin 733
Mokrye Yasli forestry plantation 718
Molina high moors 590
Molino mountain range 588
Molochnaya river valley 716
Molochniy liman 716
Moncayo mountain range 563
Mondragon 220
Monegrillo 567
Monforte plains 455
Monfragüe 620
Mont Blanc (Italian side) 369
Mont Coronat 232
Mont Vallier 224
Montado da Fajã da Nogueira 478
Montado de Areeiro 478
Montado do Cidrão 478
Montado do Sabugal 478
Montado dos Pessegueiros 477
Montado dos Piquetes 478
Montagne de la Clape 226
Montagne de la Serre 210
Montagne de l'Espinouse 233
Montagne des Frêtes 216
Montagne du Caroux 233
Montagne Sainte-Victoire 235
Montagnes de Marcou 233
Montagut mountains 577
Montánchez mountain range 617
Monte Caligi 380
Monte Rosa (Valle d'Aosta side) 369
Montecoche mountain range 605
Monteixo 574
Montejo de la Vega 552
Montemolín 613
Montes Aquilanos range 541
Montes Universales 565
Montesinho mountains 451
Montgai mountains 577
Montijo reservoir 618
Montmell mountains 577
Monts et Plomb du Cantal 209
Montsant mountains 578
Montsech mountains 577
Moquinhas 477
Morante 618
Moratalla mountain range 588
Morella pass 579
Morozzo lakes 378
Morro de los Rincones-Vallebrón mountains 632
Morro Tabaiba 632
Moson plain 340
Mota del Marqués 554
Moudros gulf 314
Mount Ag. Dikaios 326
Mount Arcosu 427
Mount Artemisio 310
Mount Asterousia 328
Mount Athos 286
Mount Babadag 100
Mount Bazar-Duzu 98
Mount Beigua 379
Mount Beles 282
Mount Catria 396
Mount Cervati 411
Mount Cholomontas 285
Mount Cofano 417
Mount Conero 395
Mount Dalidag 97
Mount Dias 319
Mount Dikios 323
Mount Dikti 329
Mount Douskon 296
Mount Erimanthos 304
Mount Falakro 280
Mount Ferru of Tertenia 425
Mount Filiaton 297
Mount Foloi 305
Mount Ghiona 306
Mount Giamysh 96
Mount Gramos 290
Mount Grappa 384
Mount Gush-gaya 103
Mount Helmos 303
Mount Idi 328
Mount Ilandag 98
Mount Itamos 286
Mount Iti 306
Mount Jouchtas 328
Mount Kargabazar 103
Mount Kasidiaris 296
Mount Kato Olympos 292
Mount Kedros 327
Mount Kerkis 317
Mount Kohilas, Skyros island 309
Mount Koronos 319
Mount Kouskouras 281
Mount Koutroulis 326
Mount Kroussia 282
Mount Lepetimnos, Lesvos 315
Mount Lirkio 310
Mount Mavrovouni 282, 319
Mount Megali Rahi 297
Mount Mitsikeli 295
Mount Ochi and surrounding area 308
Mount Olympos 288
Mount Olympos, Lesvos 316
Mount Orliakas 290
Mount Ornon 329
Mount Orsomarso 413
Mount Ossa 292
Mount Othrys 305
Mount Pangaio 280
Mount Paramithias 297
Mount Parnassos, southern and eastern part of 307
Mount Parnitha 312
Mount Pecoraro 417
Mount Pharmakovouni 297
Mount Pilion 294
Mount Pinovo 286
Mount Sanghias 311
Mount Sette Fratelli 426
Mount Shahdag 98
Mount Sparagio 417
Mount Tabor 370
Mount Taigetos 310
Mount Thryptis 329
Mount Tsamanta 297
Mount Tsourgiakas 290
Mount Tzena 286
Mount Vardousia 306
Mount Verbicaro 413
Mount Voras 287
Mount Vourinos 290
Mount Ymittos 312
Mount Zakrou 330
Mount Zarkou 291
Mount Ziria 309
Mountain of Agua 640
Mountains of Barcelona 576
Moura 458
Mourão 458
Mouro islet 545
Moyenne Vallée de la Durance 234
Muela mountain range 586
Mugan steppe 105
Muniellos 541
Murani lake 496
Murat Dağı 662
Murcia, littoral islets of 586
Mureş river flood-plain 495
Murge 411
Murta dam 457
Mzha river valley 709

N

Nagyberek 342
Nanoščica river basin 508
Naos Bay 647
Narta lagoon 72
Nasicka Breznica fish-ponds 139
Navalcán reservoir 593
Nea Fokea marshes 286
Nebrodi 417
Negram mountains 98
Negro rocky islet 644
Nemrut lake 687
Neretva river, delta of the 144
Nestos gorge 279
Nestos valley 279
Nestou delta and coastal lagoons 279
Nevela mountain range 588
Niculiţel 489
Nisiros island and nearby islets 323
Nizdafe plains 645
Nogueira mountains 451
North Alicante mountain range 583
North Chios 316
North Dodekanisa, islets of 321
North Sporades 294

North-east Turkey 676
Northern Evros 274
Northern Karpathos 325
Northern Sifnos 319
Northern Syros 318
Novara heronries 373
Novogolovka-chala 106
Novograd–Volunskoe Game Reserve 702
Nur mountains 684

O

O Grove estuary 539
Obarenes mountains 547
Obedska bara 733
Obitochnaya peninsula 716
Ocaña 591
Odiel marshes 611
Odobeşti hill 491
Ojeda pine woodland 636
Ojén mountain range 606
Oliva marshes 583
Olivenza 615
Ombrone mouth 399
Oña mountain range 546
Ons islands 539
Orba mountain range 571
Orbetello lagoon 401
Ordesa 572
Ordubad area 97
Orduña 547
Orel river valley 707
Orellana reservoir 617
Orestiada 289
Organbidexta 218
Orikhivski lake-system 700
Oriokastro, area of 296
Oristano wetlands 424
Orobie 372
Oroel crag 571
Oropesa plain s 594
Orosei Gulf 424
Orsoya fish-ponds 119
Ortari mountains 308
Ortazzino 392
Ortazzo 392
Ortigueira estuary 540
Ostiglia 389
Osuna plain 604
Otaci 440
Othoni 299
Oturia 570
Ovcar–Kablar gorge 736
Ovtcharitza reservoir 124
Oyambre 545

P

Pacsmag fish-ponds 342
Pajonales pine woodland 636
Paklenica National Park 143
Palas lake 673
Palmas Gulf wetlands 427
Palo to Santa Marinella, coast from 405
Pantani di Capo Passero 420
Pantelleria 421
Panticosa 572
Panzano Gulf 387
Paphos plain 154
Paphos State Forest 155
Parcheş 488
Paros 319
Pass of Beceite 578
Passo del Cerreto 380
Patmos island and islets, north and south of 321
Patoku lagoon 74
Paul de Arzila 453
Paul de Madriz 453
Paul do Boquilobo 454
Paul do Taipal 453
Pavia province, heronries of 375
Pcinja 738
Pechenezhskoe reservoir 710
Pecica 495
Pedro Alvarez mountains 638
Pedro Muñoz 592
Pego–Olivo marshes 583
Pela mountain range 617
Peloritani mountains 416
Peña Amaya 546
Peña de Francia 556
Peña Labra mountain range 544
Peñagolosa 580
Peñas de Cervera 552
Pènes du Moulle de Jaut 219
Peraleda de Zaucejo 613
Peristeri mountain 296
Peschanskiy forest 704
Petit Mars 178
Petite Beauce 202
Petite Camargue fluvio-lacustre 232
Petite Camargue laguno-marine 233
Pétrola 589
Petropavlovskiy marsh 708
Phasouri reedbeds 153
Pialassa della Baiona 392
Pian di Spagna 370
Piave river 385
Pic d'Anie 219
Pic d'Errozate 233
Pic de l'Estibet 220
Pic des Escaliers 218
Picentini mountains 410
Pico Casado 477
Pico, coast of 468
Picos de Europa mountain range 544
Pietrosul Rodnei mountain 492
Pijaral mountains 637
Pila mountain range 588
Pilo pool 422
Pina 567
Pina steppe area 567
Pindos National Park 295
Pineta San Vitale 392
Pinios delta 292
Pirgulu 100
Pirineo de Andorra 78
Pirkuli 100
Pirsagat islands 104
Pişchia forest 496
Pitillas lagoon 563
Pit major d'Iguerande à Décize 210
Pizzi 404
Pizzo Cirina 417
Plain of Gela 420
Plaine calcaire du sud Vendée 178
Plaine d'Assais-les-Jumeaux 200
Plaine de Mirebeau 200
Plaine de Neuville-du-Poitou 200
Plaine de Niort (nord-ouest) 199
Plaine de Niort (sud-est) 199
Plaine de Saint-Jean-de-Sauves 201
Plaine de Saint-Jouin de Marne 200
Plaine de Villefagnan 201
Plaine des Maures 238
Plaine du Forez 214
Plains of Mareta 631
Plaiul Fagului 440
Plan de Bourg 236
Planèze de Saint-Flour 210
Planina polje 510
Plasencia and San Bernabé mountain range 622
Plateau Ardennais 188
Plateau de Bellefonds 200
Plateau de Delme 192
Plateau de l'Arbois 237
Plateau de Valensole 237
Plateau des Glières 216
Plateaux agricoles des environs de Frencq 187
Plateaux de Gentioux 208
Plateaux de Millevaches et de Gentioux 208
Plovdiv fish-ponds 122
Po delta 390
Podravska Slatina, alluvial forests 140
Pointe de Grave 217
Pokupsko depression 142
Polesskiy Nature Reserve 701
Policastro Bussentino 413
Pollino 413
Polyegos islands 319
Pombak mountain chain 85
Pomègues 236
Pomorie lake 129
Ponta da São Lourenço 478
Ponte Buriano 395
Pontikonisi 325
Ponziane islands 407
Porţile de Fier reservoir 496
Portios islets 545
Porto Lagos 278
Portugal, south-west coast of 459
Posets 573
Pozharevo island 127
Pozo Negro mountain 633
Prades mountains 578
Prado lagoon 611
Prasouda islet 313
Prassiano gorge 327
Prat de Cabanes 580
Préalpes de Grasse 241
Presqu'île de Crozon 174
Presqu'île de Scandola 242
Presqu'île Guérandaise, traicts et marais salants de la 175
Procev village, meadows near 703
Prokletije 737
Provadijsko 129
Prvic island 143
Psara island 316
Psel river valley, meadows near Saru village in the 707
Psel river valley near Komsomolsk 706
Pskhu 251
Puebla de Beleña lagoons 560
Puebla de Don Fadrique 596
Puertito de Los Molinos 635
Puerto de Mejoral 615
Puerto Peña reservoir 616
Puerto Real lagoon 608
Puig Carlit et environs 229
Pula 427
Punte Alberete 392
Puszta at Jászkarajenő 346
Puszta at Pitvaros 348
Pusztaszer Landscape Protection Area 348
Pyasachnik reservoir 120

Q

Quebradas mountains 638
Quero endorreic lagoons 592
Queue de Brenne: Etang des Loges, Grand et Petit Mez 203
Quinto heronry 386
Quipar reservoir 588

R

Rabaças 477
Rade de Brest 172
Rade de Lorient 174
Rădeanu wetlands 490
Radika gorge 432
Radinchevo 123
Radomyshlskoe Game Reserve 702
Rădvani wood 494
Ram 740
Râmeţi gorge 493
Rasca lava fields 642
Rayanovci meadows 117
Razelm 487
Razelm–Sinoe complex 487
Rdum Tal-Madonna 438
Redona, islet of 624
Region de Pressac 200
Reka ravine 306
Reka valley 509
Rentina gorge 285
Repedea woodland 491
Resava gorge 740
Réserve Naturelle du courant d'Huchet 222
Retezat National Park 494
Ría de Arosa estuary 539
Ria de Aveiro 452
Ría de Foz 540
Ria Formosa 460
Riaño 543
Riaza canyon 552
Ribadesella 543
Ribarroja 568
Ribeira Funda 477
Ribera del Fresno 614
Ribnica valley 511
Ricote mountain range 588
Ried de Colmar et Sélestat 193
Rieti lakes 401
Rieti mountains 402
Riglos 570
Río Cabras gully 632
Río Huebra 555
Riou 236
Ritsa 251
River Alcanadre riverine forest 568
River Axios 283

River Biferno 408
River Bregalnica 433
River Brenta, middle course of 386
River Cabriel canyon 582
River Calore, gorges of 413
River Cecina 396
River Cega 553
River Cinca riverine forest 568
River Crna gorge 433
River Danube 718
River Danube, alluvial wetlands of the 139
River Drava 139
River Drava, alluvial wetlands of the 142
River Drava from Maribor to Zavrč 511
River Ebro dam 545
River Ebro oxbow lakes 566
River Guadalhorce, mouth of the 600
River Guadalope 565
River Guadalquivir reservoirs, middle section of the 601
River Guadiana 459
River Huerva, lower course of the 566
River Ipoly, middle reach of 346
River Jalón canyon 564
River Júcar canyon 582
River Lobos canyon 552
River Martín canyons 565
River Matarraña, lower course of the 568
River Mesa canyons 591
River Mijares mouth 581
River Mincio 388
River Moros 553
River Mundo 589
River Mura from Veržej to Gibina 512
River Neto 415
River Piedras marshes 611
River Pisuerga at Dueñas 550
River Po: from Dora Baltea mouth to Scrivia mouth 376
River Po: from Ticino mouth to Adda mouth 377
River Po: from the mouth of Adda river to Ostiglia 389
River Prip'yat valley 700
River Sava, alluvial wetlands of the 141
River Segura canyons 589
River Sesia heronries 374
River Tagus 593
River Tajo, upper section of the 590
River Tajuña, upper section of the 590
Rocca Busambra 418
Rochers de Biarritz: le Bouccalot et la Roche Ronde 218
Rochers du Toulinguet 174
Rock of Gibraltar 259
Rodopos peninsula 325
Rompido sandbank 611
Roncesvalles–Irati–Abodi mountain range 561
Ronda mountain range 605
Ropotamo complex 130
Rosais 468
Rosarito reservoir 593
Rotbav fish-ponds 494
Royaksko Plato 129
Rozov Kladenetz reservoir 124
Rudrón canyon 546
Rzhycshevskoe Game Reserve 703

S

Sa Vall 627
Sabôr 451
Saceruela 595
Sado estuary 457
Saint-Agnant 198
Salda lake 667
Saligue du Gave de Pau 220
Salines de Fos 238
Salines de l'Etang de Berre 235
Salines de Rassuen 238
Salins de Pesquiers 237
Salins d'Hyères 237
Salins du Caban 236
Salmor rocky islets 646
Salobrar de Campos lagoons 627
Salvada mountain range 547
Samara river valley 709
Samarskiy forest 708
Samothraki island 276
Samsam lake 677
Samur delta 99
San Andrés mountains 637
San Bernabé mountain range 622
San Clemente 589
San Juan de la Peña 571
San Martín de Valdeiglesias 557
San Mauricio 573
San Miguel mountain range 562
San Pedro mountain range 618
San Pietro islands 427
San Rossore 395
Sancha lagoon 458
Sangachal Bay 104
Sant Gervás mountains 574
Santa Amalia 617
Santa Eulalia, islet of 624
Santa Maria, north and east coast of 471
Santa Marinella 405
Santa Pola saltpans 584
Santa Úrsula 639
Santander bay 545
Sant'Andrea island 414
Sant'Antioco islands 427
Santo André lagoon 458
Santo Domingo 570
Santo Domingo coastal cliffs 639
Santoña marshes 545
São Miguel, east of 470
São Miguel, west coast of 469
Sapanca lake 672
Šara mountain 738
Sara mountains [parts of] 432
Saria island 325
Sarıkum lake 673
Sariñena lake 569
Sarısu plain 684
Sarıyar reservoir 672
Sarnosa island 545
Sarrabus 426
Saru village in the Psel River valley, meadows near 707
Sárvíz valley 343
Sasuk lake 719
Satchinez marsh 495
Schinias marsh 312
Scrivia mouth 376
Scrivia stream 378
Sečovlje saltpans 508
Segovia 554
Segura mountain range 596
Selinous gorge 303
Selva del Lamone 400
Selvagem Grande 479
Selvagem Pequena 480
Sentina 396
Serifos island 318
Serre Persano 410
Serreta de Tramaced 570
Serreta Negra 567
Setcases 575
Setúbal peninsula, coast of the 456
Seyfe lake 674
Shabla lake complex 133
Shagany–Alibey–Burnas lake-system 719
Shah Cape 103
Shahbuz area 97
Shakhova Kosa 103
Shamkhor area 95
Shatsk National Park 699
Sheki upland 94
Shirvan reserve 105
Shorgel lakes 105
Sibillini mountains 398
Sicani mountains 418
Sicevo gorge 740
Sidonia 609
Sierra Morena de Córdoba 603
Sierra Morena de Sevilla 603
Sierra Nevada mountain range 599
Sierra Pelada mountain range 612
Sigena mountain range 569
Sikinos 320
Sila Grande 415
Şile islands 660
Simbruini mountains 406
Simeto mouth 419
Simi island and adjacent islets 324
Sinjsko field 144
Sinoe 487
Sira Pogonya mire 701
Sirente 404
Sirna island and nearby islets 323
Siruela 616
Sis 573
Sivash Bay, Sea of Azov 715
Skotini 308
Skyros, north-west 309
Slano kopovo 734
Slătioara forest 492
Smolikas mountains 295
Sobrón canyon 547
Socodor wood 495
Sodalıgöl 687
Soleniy liman 708
Sologne Bourbonnaise 210
Somiedo 542
Somova wetland 488
Sotavento beach 633
Soufli forest 275
Sounio area 312
Southern Mani 311
Southern Sakar 123
Sperchios delta 305
Sperchios valley 305
Spincourt, environs de 192
Srebarna lake 127
St-Sever-de-Saintonge 197
St Viatre 205
Stagnone di Marsala 418
Stânca 492
Stanichno-Luganskoe fish-farm 711
Stara mountain 739
Stavros tis Psokas Forest Station 155
Ştefăneşti reservoir 492
Stelvio National Park 380
Stenata 127
Stînca 440
Stintino peninsula cliffs 421
Stirone stream 377
Stokhod river valley 701
Strait of Gibraltar 259
Strandja 130
Strejeşti reservoir 497
Strofades islands 301
Strofilia forest 304
Strymon delta 281
Studen Kladenetz 121
Subotica lakes and sand terrain 730
Suğla lake 679
Suha Reka 133
Sula bay, Kremenchug reservoir 706
Sultansazlığı 673
Supoy marsh 703
Susa valley 375
Suva mountain 739
Szatmár-Bereg plains 351

T

Tabayesco 629
Tabernas desert 598
Tablas de Daimiel marshes 592
Ta'Cenc cliffs 437
Tágara gully 640
Tagliamento river, middle course of 383
Tagomago island 624
Tajo de Barbate coastal cliffs 607
Talamanca 559
Talmaz 442
Tamadaba pine woodland 637
Tápió valley 346
Tara mountain 736
Tarajalejo mountain range 633
Tarancón 591
Tarifa 606
Tarkhankut peninsula 721
Taro river 377
Tarquinia saltpans 403
Tarvisio forest 381
Tas de Pois 174
Tăureni fish-ponds 493
Tauro pine woodland 636
Tavolara island 422
Tchuprene 117
Tejo estuary 456
Telascica Nature Park 143
Telendos 322
Telera 572
Tembi ravine 292

Tembleque 591
Tendeñera 572
Tendrovskiy Bay 714
Tenegüime gully 630
Terceira, south-east coast of 469
Terceira, west coast of 469
Terry lagoons 608
Tersakan lake 680
Tesla mountain range 546
Thasos Island, north and north-east 280
Thessaly plain 293
Thethi 73
Ticino mouth 377
Ticino river 373
Tiermes 560
Tierra de Campiñas steppes 555
Tierra de Campos steppes 548
Tierra de Pinares 553
Tiétar valley 557, 593
Tigaiga slope 639
Tiganis peninsula 325
Tiligul estuary 721
Tilos island 323
Timfi mountains 295
Tina Mayor 543
Tina Menor 545
Tinos island, north-east 318
Tinto marshes 611
Tirajana pine woodland 636
Tis amarelos 477
Tissata 118
Tiszaalpár, flood plain at 347
Tivissa mountains 578
Tödürge lake 673
Tokmakia islets 315
Toledo mountains 594
Tolfa hills 403
Topas 555
Topo 468
Topolka gorge 432
Tordesillas 554
Tormos reservoir 569
Tornafort 574
Torralba de los Sisones 564
Torreblanca 580
Torrecilla mountain range 587
Torrecillas de la Tiesa 620
Torrevieja lagoon 585
Torrijos 593
Torrile LIPU reserve 378
Tostón 635
Toulões 453
Tramuntana mountains 625
Trapani saltpans 418
Trappola marshes 399
Trebbia river (lower course) 376
Trelon 188
Tremiti islands 408
Triglav National Park 507
Trujillo 619, 620
Turbón 573
Turia canyon 582
Turianchai 95
Turiya river valley 700
Turkey, north-east 676
Türkmenbaba Dağı 669
Turmell mountain 578
Turopolje 142
Tuscan archipelago 398
Tuz lake 680
Tuzla lake 682
Tyrnavos area 292
Tzerkovski reservoir 125
Tzoumerka 298

U

Ubrique mountain range 605
Uccellina mountains 399
Uday river valley 706
Ukrainian Steppe Nature Reserve 717
Ulcinj saltpans 743
Uluabat lake 661
Uludağ 660
Upper Ebro canyon 546
Upper River Douro valley 451
Upper River Tejo 454
Upper Tisza 351
Upper Torre valley 383
Urbión mountain range 551
Usagre 614
Ushomirskoe Game Reserve 702
Utlyuk estuary 716
Uvac 737
Uyuz lake 670
Uzdin fish-pond 734
Uzunlarskoe lake 722

V

Vădeni 490
Val d'Allier Bourbonnais 208
Val d'Allier: Saint-Yorre-Joze 212
Val de Chiers et environs de Spincourt 192
Val de Drôme: Les Ramières-Printegarde 213
Val de Petite Seille 192
Val de Saône 212
Val Grande 370
Valdecaballeros 616
Valdecañas reservoir 623
Valdegovía–Arcena mountain range 547
Valdehornillos 617
Valdoviño 539
Valdurrios–Serreta Negra mountain range 567
Vale da Ribeira da Janela 477
Valganna 371
Valia Kalda 295
Valjevske mountains 736
Valle Bertuzzi 390
Valle della Canna 392
Vallée d'Asco 242
Vallée d'Aspe, Rive Gauche de la 215
Vallée de la Charente et de la Seugne 197
Vallée de la Charente: amont d'Angoulême 198
Vallée de la Conie 202
Vallée de la Durance: de Tallard à Sisteron 239
Vallée de la Garonne: Boussens à Carbonne 224
Vallée de la Garonne: méandre de Grisolles 225
Vallée de la Garonne: Moissac 225
Vallée de la Garonne: Palayre et environs 224
Vallée de la Loire: confluence Loire-Vienne 204
Vallée de la Loire: de Nantes à Montsoreau 178
Vallée de la Loire: environs de Blois 202
Vallée de la Loire: environs de Montlouis-sur-Loir 204
Vallée de la Loire: gorges de la Loire 211
Vallée de la Loire: La Charité-sur-Loire 206
Vallée de la Loire: Orléanais 206
Vallée de la Loire: Pit majeur d'Iguerande à Décize 210
Vallée de la Meuse 191
Vallée de la Nive des Aldudes 221
Vallée de la Pique 223
Vallée de la Restonica 243
Vallée de la Saône de Corre à Broye 197
Vallée de la Scarpe 186
Vallée de la Seugne 197
Vallée de la Superbe 190
Vallée de l'Aisne 190
Vallée de l'Aube 190
Vallée de l'Escaut 186
Vallée de l'Oise de Thourotte à Vendeuil 185
Vallée de l'Yèvre 202
Vallée de Marigny 190
Vallée de Melles 224
Vallée du Fouzon 205
Vallée du Haut-Guil 239
Vallée du Lis 223
Vallée du Régino 243
Vallée du Rhin: Marckolsheim à Strasbourg 194
Vallée du Rhin: Strasbourg à Lauterbourg 195
Vallée du Rhin: Village-neuf à Bieschеim 194
Vallée du Verghello 244
Vallehermoso coast 643
Valli di Argenta 391
Valli di Comacchio 391
Valongo 614
Valtou mountains 301
Valuengo reservoir 614
Vanoise, Parc National de la 215
Vardim island 122
Varna–Beloslav lakes complex 132
Varvara reservoir 95
Vathia Lakka 306
Vedrá, islet 623
Vedranell, islet 623
Velino 404
Velipoja 74
Veluke marsh 707
Venetiko islet 317
Venice lagoon 387
Ventejís mountain 645
Veržej 512
Vercelli rice-fields 376
Vértes mountains 344
Vicario reservoir 592
Vico lake 402
Vidlic 739
Vigán 633
Vignemale 572
Vigy 192
Vila Fernando 456
Vila Franca islet 470
Vila islet 471
Vilaflor pine woodland 641
Vilajchai valley 107
Villafáfila 549
Villalba de los Barros 615
Villanueva del Fresno 614
Villar del Rey 618
Villarrobledo 589
Vincheto di Cellarda 384
Viñuelas 557
Vlădeasa mountain 493
Vlădeni fish-ponds 491
Vlasina 738
Vlora Bay 73
Vouraikos gorge 303
Vovk river valley 705
Vršacki breg 735
Vrisika area 274
Vuka, alluvial forests 140

W

West Andévalo 612
Western Milos 319
Wied il-Luq 438

X

Xirovouni 308

Y

Yagorlytskiy Bay 714
Yarışlı lake 667
Yashma island 102
Yatata 132
Yecla 589
Yedikır reservoir 675
Yeşilırmak delta 676
Yüksekova 688
Yumurtalık lagoons 683

Z

Zaframagon crag 605
Zagori, central part of 295
Zakataly 93
Zakynthos, west and north 300
Zámoly basin 344
Zamora 554
Zapadniy Bug river valley 700
Zariquieta mountain range 562
Zasavica 732
Zavrč 511
Zau de Câmpie 493
Zemplén mountains 353
Zhebriyanovskie plavni 719
Zimevitza meadows 118
Zlot gorge 740
Zmeinuy island 720
Zones agricoles de la Hardt 194
Zones rupestres du Tarasconnais 224
Zorita 617
Zuera mountains 569
Zuvand upland 107